# Processing Water, Wastewater, Residuals, and Excreta for Health and Environmental Protection

# Processing Water, Wastewater, Residuals, and Excreta for Health and Environmental Protection

## An Encyclopedic Dictionary

Nicolas G. Adrien

WILEY

A JOHN WILEY & SONS, INC., PUBLICATION

Published by John Wiley & Sons, Inc., Hoboken, New Jersey.
Published simultaneously in Canada.



***Library of Congress Cataloging-in-Publication Data:***

Adrien, Nicolas G.
Processing water, wastewater, residuals, and excreta for health and environmental protection : an encyclopedic dictionary / by Nicolas G. Adrien.
p. cm.
Includes index.
ISBN 978-0-470-26193-4 (cloth)
1. Water—Purification—Encyclopedias. I. Title.
TD208.A37 2008
628.1'62—dc22 2008011585

Printed in the United States of America.

10 9 8 7 6 5 4 3 2 1

*To my wife Vivianne*
*To my children Claude, Huguette, and Florence*
*To the memory of Mother Massoule*
*To the memory of Antoine Télémaque*

# Online Appendix

Readers can access an ftp site, at ftp://ftp.wiley.com/public/sci_tech_med/water, which contains additional information to complement this volume, including:

1. Additional terms
2. Bibliography
3. A list of illustrations

This book presents the definitions and illustrations of some basic terms used in the field of water, wastewater, residuals, and excreta processing. However, the vocabulary of these fields is more extensive. The downloadable list of additional terms contains approximately 12,700 additional terms that the reader is likely to find in the technical literature. These were not included in the printed book due to size limitations.

The downloadable bibliography file includes a full set of relevant literature that will be helpful to the reader. The References section in the printed book addresses those publications cited in the text.

# Acknowledgments

Sincere thanks are due to Marc P. Walch, P.E., who encouraged early efforts leading to my preceding work on this subject and reviewed constructively a publication proposal for the present manuscript; to the equipment manufacturers; Ms. Berinda Ross of The Water Environment Federation, who authorized the use of materials from their publications; and to personnel of John Wiley & Sons, who assisted in the publishing process, from proposal review to the actual production.

# About the Author

**Nicolas G. Adrien** is an independent consulting engineer in Florida. He has a civil engineering degree from the University of Haiti and two Master degrees, in environmental engineering and water resources, from Harvard University. He is a registered professional engineer, a certified environmental engineer, a Fellow of the American Society of Civil Engineers, and a Life Member of the American Water Works Association. He is listed in five *Who's Who* publications. For more than 40 years, he has provided engineering management and environmental planning services to well-known U.S. consulting firms, government agencies, and international organizations. He has authored or edited numerous technical reports. He recently published the predecessor volume of this book: *Computational Hydraulics and Hydrology—An Illustrated Dictionary,* published in 2004 by CRC Press LLC.

# A

**Å** Symbol of the unit of length angstrom, equal to one tenth of a millimicron or one ten millionth of a millimeter.

**A-7** *See* Duolite A-7.

**AADI** Acronym of adjusted acceptable daily intake.

**AA method** Abbreviation of atomic absorption spectrophotometry method.

**AAMI water** Water that meets the quality standards of the Association for the Advancement of Medical Instrumentation and used mainly in hemodialysis systems.

**AAS** Acronym of (1) atomic absorption spectroscopy and (2) atomic absorption spectrophotometry.

**AAS method** Abbreviation of atomic absorption spectrophotometry method.

**ABA-1000®** Proprietary aluminum oxide produced by Selecto, Inc., for the removal of phosphates.

**ABA-2000®** Proprietary aluminum oxide produced by Selecto, Inc., for the removal of lead and other heavy metals.

**ABA-8000®** Proprietary aluminum oxide produced by Selecto, Inc., for the removal of fluorides.

**abandoned mine** A mine where mining operations have occurred in the past and (a) the applicable reclamation bond or financial assurance has been released or forfeited, or (b) if no reclamation bond or other financial assurance has been posted, no mining operations have occurred for five years or more (EPA-40CFR434.11-r).

**abandoned water right** A water right that has not been used for a number of years, e.g., five to seven years; or a water right for which the owner states that it will not be used or takes action to prevent its beneficial use. *See also* forfeited water right.

**abandoned well** A well whose use has been permanently discontinued or that is in such disrepair that it cannot be used for its intended purpose. It is recommended to fill an abandoned well with cement or concrete to prevent groundwater pollution.

**abandonment** The definite cessation of the use of a water right, as may be implied for from a prolonged period of nonuse. *See also* abandoned water right, forfeiture.

**abatement** The reduction of the degree or intensity of pollution, or the elimination of pollution, by controls at the sources or by treatment of effluents before discharge.

**ABC** Acronym of Association of Boards of Certification.

**ABC Filter™** A cartridge filter, backwashable automatically, produced by the U.S. Filter Corporation of Rockford, Illinois.

**Abcor®** A proprietary ultrafiltration membrane produced by Koch Membrane Systems, Inc.

**ABF** Acronym of activated biofilter, activated biofiltration, or activated biological filtration.

**ABF-1** A system of wastewater treatment by activated biofiltration manufactured by Infilco Degremont, Inc.

**ABF-2** A traveling bridge produced by Aqua-Aerobic Systems, Inc. for the automatic backwashing of gravity sand filters.

**ABF process** Same as activated biofiltration.

**abiological** Characterized by an absence of biological activity; pertaining to processes or reactions that do not occur naturally; synthetic.

**abiota** Nonliving factors present in an ecosystem and affecting its characteristics.

**abiotic** Characterized by the absence of life or living organisms; not formed by biological processes.

**abiotic environment** An environment devoid of life or life-sustaining elements, for example, a sterilized dish.

**abiotic factor** A nonorganic characteristic that affects life in an ecosystem, for example, temperature, light, soil structure, sulfur dioxide emissions.

**ABJ** Products of Austgen-Biojet Waste Systems.

**ablation** Any process that removes snow, water, or ice from a glacier, snowfield, etc.; i.e., avalanche, calving, evaporation, melting, wind erosion. *See also* accumulation, firn line.

**ablation area** The portion of a glacier or snowfield where ablation exceeds accumulation.

**aboveground release** Any release to the surface of the land or to surface water. This includes releases from the aboveground portion of an underground storage tank (UST) system and aboveground releases associated with overfills and transfer of operations as the regulated substance moves to or from a UST system (EPA-40CFR280.12).

**aboveground storage facility** A tank or other container, the bottom of which is on a plane not more than 6 inches below the surrounding surface (EPA-40CFR113.3-a).

**aboveground tank** A tank situated in such a way that its entire surface area is completely above the plane of the surrounding surface and the entire surface area of the tank (including the tank bottom) can be visually inspected (EPA-40CFR260.10).

**A-B process** A two-stage activated sludge process that uses a high food-to-microorganism ratio and a short aeration time in the first stage (adsorption). The second stage provides biological oxidation of the effluent of the first stage. *See also* contact-stabilization and advanced primary treatment.

**ABR** Acronym of anaerobic baffled reactor.

**abrasion** Removal of soil from streambanks by water, ice, or debris. *See also* streambank erosion.

**abrasion number** A number that is used to define the resistance of granular activated carbon to damage during transportation or handling; it is expressed as the ratio of the final mean particle diameter of a sample to its initial mean particle diameter after contact with steel balls in a sieve-column vibrator.

**abrasives** Materials such as alumina, garnet, glass beads, and steel grit used in abrasive blasting to remove paints and other organic coatings from metallic and nonmetallic surfaces. Soft abrasives are recommended as a better pollution-prevention alternative than chemical-based strippers.

**ABS (1)** Acronym of acrylonitrite-butadiene-styrene.

**ABS (2)** Acronymn of alkyl-benzene-sulfonate.

**absolute filter rating** (1) A parameter used to define the sizes of particles that will be retained on a filter medium. *See* filter rating for more detail. (2) A filter rating that indicates that essentially all particles larger than the rating are retained on or within the filter.

**absolute humidity** A measure of the water vapor contained in the atmosphere or other volume of air, expressed in grams per cubic meter. *See also* relative humidity.

**absolute pressure** Total pressure, equal to gauge pressure plus atmospheric pressure.

**absolute purity water** Water with a specific resistance of 18.3 megohm-cm at a temperature of 25°C or 77°F.

**absolute rating** *See* absolute filter rating.

**absolute risk** The difference between the incidence of death or disease in the exposed populations and the incidence in the unexposed population.

**absolute temperature** A temperature expressed as a function of absolute zero, which is the temperature at which a substance does not possess any thermal energy, e.g., the molecules of an ideal gas are motionless. Absolute zero temperatures are, depending on the scale: 0° kelvin, 0° Rankine, –273.15°C, and –459.67°F.

**absolute viscosity ($\mu$)** A measure of the internal resistance of a fluid to tangential or shear stress, and thus to flow, equal to the ratio of the viscous shearing stress ($\tau$) to the velocity gradient ($\partial V/\partial s$).

Also called dynamic viscosity or coefficient of viscosity. Absolute viscosity decreases when temperature increases; e.g., for water from 1.8 centipoises (cp) at 0°C to 0.18 cp at 150°C. It is conveniently taken as 1.0 cp for water at room temperature (about 20°C).

**absolute zero** *See* absolute temperature.

**absorbance (*A*)** The capacity of a substance to absorb light or radiation, or the amount of light or radiation actually absorbed; defined as the common logarithm of the ratio of light intensity at the surface ($I_0$) to light intensity at a distance from the surface ($I$):

$$A = \log (I_0/I) \qquad \text{(A-01)}$$

Absorbance (in absorbance units/cm, i.e., a.u./cm) is measured with a spectrophotometer using a specified wavelength, typically 254 nm. It approaches zero ($A \rightarrow 0$) for a transparent solution and infinity ($A \rightarrow \infty$) for an opaque solution; average values vary from 0.03 a.u./cm for a reverse osmosis effluent to 0.42 a.u./cm for a primary effluent. Absorbance is also defined as the common logarithm of the reciprocal of transmittance ($T$):

$$A = \log (1/T) = -\log T \qquad \text{(A-02)}$$

The concepts of absorbance and transmittance are important for understanding colorimetric methods used in certain water and wastewater analyses. *See* Beer–Lambert law, UV absorbance.

**absorbate** The substance that is absorbed by a material (absorbent) in the absorption process. *See also* adsorbate.

**absorbed dose** (1) The amount of a chemical that enters the body of an exposed organism. (2) The amount of radiation that enters the body of an exposed organism, expressed in grays (Gy) or in rads, with 1 Gy = 100 rads.

**absorbed oxygen** *See* oxygen absorbed.

**absorbent** The material that absorbs a substance (the absorbate) in the absorption process. *See also* adsorbent.

**absorber** A material that can take in a substance as a sponge takes up water.

**absorption** Generally, the penetration or assimilation of atoms, ions, or molecules into the bulk mass of a substance or structure, without a chemical reaction; for example, the uptake of water or dissolved chemicals by a cell or an organism (as tree roots absorb dissolved nutrients in soil), the penetration of one substance into another substance, or the penetration of a substance through the skin. Another example of absorption, called scrubbing, is the use of mineral oil to absorb volatile organic chemicals or the use of a solution to absorb sulfur dioxide. Absorption is a common process in water and wastewater treatment, involving the transfer of compounds from the gas phase to the liquid phase. Absorption may also be chemical, as in a reaction between a compound and a solvent. *See also* sorption, adsorption, desorption, two-film theory.

**absorption analysis** A method used in the laboratory to measure the concentration of a substance in a solution by the variation in the signal from a source of radiation.

**absorption bed** *See* absorption field.

**absorption capacity** The amount of a substance (absorbate) that another substance (absorbent) can absorb, expressed, for example, in grams of absorbate per liter of absorbent or as a weight percentage. *See also* absorption test.

**absorption coefficient ($k^*$)** A proportionality constant ($k^*$) that relates the saturation concentration of a gas ($C^*$) in water to the partial pressure ($p$) of the gas in water, as indicated by Henry's law:

$$C^* = k^*p \qquad \text{(A-03)}$$

where $C^*$, $p$, and $k^*$ may be conveniently expressed in milliliters per liter, atmospheres, and milliliters per liter, respectively. The absorption coefficient is also defined as the volume of a gas that is dissolved at standard conditions of temperature and pressure (0°C and 760 mm of mercury). *See* Table A-01 and Henry's law constant (Fair et al., 1971).

**absorption factor** The fraction of a chemical that is absorbed by an organism when they make contact.

**absorption field** A subsurface area containing perforated pipes laid in gravel-lined looped or lateral trenches or in a bed of clean stones, through which treated wastewater may seep into the surrounding soil for further treatment and disposal. *See* soil absorption field for more detail.

**Table A-01.** Absorption coefficients of gases in water*

| | Absorption coefficient | | |
|---|---|---|---|
| Gas | 0°C | 20°C | 30°C |
| Ammonia, $NH_3$ | 1300 | 711 | — |
| Carbon dioxide, $CO_2$ | 1710 | 878 | 665 |
| Chlorine, $Cl_2$ | 4610 | 2260 | 1770 |
| Hydrogen, $H_2$ | 21.4 | 18.2 | 17.0 |
| Hydrogen sulfide, $H_2S$ | 4690 | 2670 | — |
| Methane, $CH_4$ | 55.6 | 33.1 | 27.6 |
| Oxygen, $O_2$ | 49.3 | 31.4 | 26.7 |

*mL of gas per liter of water, at 760 mm Hg.

**absorption hygrometer** An instrument that uses the shrinkage of an organic fiber or hair to determine the relative humidity of the atmosphere. Sometimes called chemical hygrometer.

**absorption loss** (1) The process or the quantity of water loss (volume per unit time) during priming and before stabilization of a reservoir or canal; thereafter, the loss is seepage. (2) The loss of water by seepage or infiltration during the initial irrigation of a field.

**absorption rate** *See* gas absorption rate.

**absorption system** A device used to receive and treat wastewater such as the effluent of a septic tank. *See* soil absorption system for more detail.

**absorption test** A test conducted on a porous material to determine its absorption capacity for water, by immersing the dry material in water and measuring its weight before and after.

**absorption tower** A structure used for gas absorption into a liquid, e.g., in the production of sulfuric acid ($H_2SO_4$) from sulfur dioxide ($SO_2$) and water ($H_2O$).

**absorption toxicokinetics** Refers to the bioavailability, i.e., the rate and extent of the test substance, and metabolism and excretion rates of the test substance after absorption (under the Toxic Substance Control Act, EPA-40CFR795.235-b).

**absorptive capacity** A measure of the ability of a site (e.g., soils or body of water) to accept waste without causing adverse environmental effects; the amount of waste that can be deposited at the site. *See* assimilative capacity.

**abstraction** The part of precipitation that does not contribute to runoff.

**abstractive use (of water)** An activity that removes water from further use, e.g., evaporation through a cooling tower.

**abutment** The part of a valley or a concrete gravity section against which a dam is contructed.

**abutment seepage** Seepage of water from a reservoir through seams or pores in the abutment.

**ABW®** A proprietary gravity sand filter with a traveling bridge manufactured by Infilco Degremont, Inc.

**abyssal zone** The bottom waters of a deep ocean.

**abyssopelagic habitat** The bottom of the pelagic zone, between the bathypelagic and benthopelagic zones.

***Acanthamoeba*** A genus of free-living soil and water amoebas that can cause severe infections (e.g., infection of the eye cornea through the use of soft or disposable contact lenses), chronic encephalitis in immunocompromised individuals, and even deaths in humans. Their cysts may be removed by filtration but are resistant to chlorine residual.

**ACC** Acronym of area control center.

**Accelapak® plant** A water treatment plant of Infilco Dgremont, Inc. that can be constructed in modular units.

**Accelator® clarifier** A premix-recirculation solids contact clarifier manufactured by Infilco Degremont, Inc. It provides for separate primary and secondary mixing.

**accelerated depreciation** A method of accounting for the depreciation of water, wastewater, and other facilities that allows a faster amortization than straight-line depreciation. *See also* double declining balance, sum of the year's digits, and units of production.

**accelerated erosion** Erosion increased by human activities beyond the geologic rate. *See also* geologic erosion, natural erosion.

**accelerated eutrophication** The condition of a water body that contains excessive levels of nutrients, as caused, e.g., by the discharge of wastewater treatment effluents. *See also* limiting nutrient, phosphorus fertilization.

**accelerated gravity separator** A large cylinder with a conical bottom, a closed top, and rotating flow that enters tangentially and creates a free vortex in the unit. It takes advantage of gravitational forces, centrifugal forces, and induced velocities to separate grit particles from wastewater and the lighter organic particles. *See also* hydrocyclone, swirl separator, Teacup™, vortex separator.

**accelerated gravity settling** The removal of grit and sand particles from wastewater by gravity settling in an acceleration flow field.

**accelerated sedimentation** The removal of grit and coarse solids from wastewater using such devices as accelerated gravity, swirl, and vortex separators.

**acceleration of gravity** $g = 32.2$ ft/sec$^2$ or 9.8 m/sec$^2$.

**Accelo-Biox®** A wastewater treatment plant manufactured by Infilco Degremont, Inc. It can be constructed in modular units.

**Accel-o-Fac™** A proprietary design of wastewater treatment plants by Lake Aid Systems.

**Accelo Hi-Cap** A filter underdrain system manufactured by Infilco Degremont, Inc.

**acceptable daily intake (ADI)** An estimate of the largest amount of chemical to which a person can be exposed on a daily basis without adverse effects even if continued exposure occurs over a lifetime (usually expressed in mg/kg/day). It is usually determined by dividing an experimentally established dose by a safety factor. Same as RfD or reference dose.

**acceptable risk** A level of risk determined by analysis and expected to correspond to minimal adverse effects, with due consideration of costs and benefits. For example, 1 in 10,000 is considered an acceptable risk for health and environmental effects in such activities as wastewater reuse and land application of sludge.

**acceptance criteria** *See* electroneutrality principle.

**acceptance limits criteria** Analytical limits established by the USEPA to assess the performance of testing laboratories that seek certification for specific contaminants under the Safe Drinking Water Act. The limits are usually set as a percentage range of a known concentration.

**access hole** *See* manhole.

**accessible equipment** In water and wastewater treatment works, accessible equipment can be inspected and cleaned without stoppage and with the use of only simple tools. Readily accessible equipment does not require any tools for cleaning and inspection.

**access port** *See* manhole.

**accidental release prevention program** A program of water supply, wastewater management, and other organizations that is designed to prevent and mitigate the accidental release of hazardous materials.

**accidental spill** The unplanned release of substances with notable environmental effects, due to natural or human causes.

**acclimated microorganism** A bacterium, fungus, etc. that survives by adjusting to changing environmental conditions such as temperature, oxygen levels, and disinfecting agents.

**acclimation** The physiological or behavioral adaptation of organisms or test animals to one or more environmental conditions such as temperature, hardness, pH, salinity, housing, and diet (EPA-40CFR797.1400/1600/1830/2050/21300); a phase in the growth curve of microorganisms when the initial or seed population adapts to the substrate and other environmental factors. Humans also acclimate, e.g., by increasing the number of red blood cells to increase oxygen-carrying capacity and compensate for lower oxygen levels at higher altitudes. Acclimation is an important factor in biological wastewater treatment as waste characteristics and other treatment parameters constantly change. It is also one of the factors that explain the lag phase in the growth of organisms. Same as acclimatization.

**acclimation period** The time necessary for a process or a system to reach design operation and performance. *See also* lag time.

**acclimatization** The physiological and behavioral adjustment of an organism to changes in its environment. *See* acclimation.

**account payable** (plural: **accounts payable)** A liability to a creditor, carried on the open account of a water or wastewater utility (or any other enterprise), resulting from the purchase of good or services from others.

**account receivable** (plural: **accounts receivable)** Money not yet received, carried on the open account of a water or wastewater utility (or any other enterprise), resulting from the sale of goods and services to others.

**accretion** A gradual increase in land area adjacent to a river due to deposition of waterborne sediments.

**accrual basis accounting** A method of accounting that records income and expenditures when they are earned or incurred instead of when they are actually received or paid. *See also* cash basis.

**accrued depreciation** The accumulated depreciation of assets such as water works and wastewater facilities; i.e., the difference between their original cost and their remaining monetary value.

**AccuCorr 3000 Digital Correlator** A leak detection device manufactured by Fluid Conservation Systems for use in in-house leak detection surveys.

**Accuguard™** A proprietary device manufactured by Leeds & Northrup for the automatic cleaning and calibration of pH electrodes.

**Accu-Mag** An electromagnetic flowmeter manufactured by Wallace & Tiernan.

**accumulation** The quantity of water in solid form added to a glacier or snowfield. *See also* ablation, firn line.

**accumulation area** The portion of a glacier or snowfield where accumulation exceeds ablation.

**accumulator** A balancing tank in a water supply to mitigate fluctuations in such chararcteristics as temperature, pressure, and flow rate.

**Accu-Pac®** A biological wastewater treatment system manufactured by Brentwood Industries, Inc. It uses a PVC cross-corrrugated medium.

**accuracy** The characteristic of a model that indicates to what extent it replicates the true or observed system, i.e., to what extent simulated values correspond to true or observed values. Accuracy (in general, the difference between a measurement or simulation and the true value) is different from precision, which is the ability to reproduce results. Systematic errors cause inaccuracy, whereas random errors cause imprecision; they can be evaluated and reduced through model validation and a sensitivity analysis. *See also* bias, level or limit of detection or quantification, practical quantification level or limit, method detection level or limit, minimum detection limit, precision, reliable detection or quantification level.

**Accura-flo®** A proprietary measuring flume manufactured by Hinde Engineering Co.

**Accu-Tab® chlorinator** A chlorination system designed by PPG Industries, Inc. of Pittsburgh, PA. It uses calcium hypochlorite tablets with 65% chlorine and patented erosion chlorinators.

**Accuvac** A reagent manufactured by the Hach Company for chemical analysis.

**acequia** A community-run irrigation canal, ditch, or channel in southwestern United States.

**acetaldehyde ($CH_3CHO$ or $C_2H_4O$)** A volatile, colorless liquid, soluble in water, with a pungent, apple-like flavor. It is an oxidation product of ethanol ($C_2H_5OH$) and can be oxidized to acetic acid ($CH_3COOH$). It may be formed during water disinfection, particularly with ozone. It is used industrially and also as a food additive. Also called ethanal.

**acetate** A salt or ester of acetic acid ($CH_3COOH$); a product of the fermentation of organic material, mostly biodegradable soluble COD.

**acetic acid ($CH_3COOH$ or $C_2H_4O_2$)** A clear or colorless corrosive liquid with a strong vinegar-like odor. Common name of ethanoic acid, of the carboxylic group, ionization constant K = 0.000,018; the acid in vinegar and a common industrial and laboratory solvent, produced by fermentation of ethanol; one of the three organic acids used to monitor the anaerobic digestion process. Also called ethylic acid, glacial acetic acid, methanecarboxylic acid, pyroligneous acid, vinegar acid.

**aceticlastic methanogen** *See* acetoclastic methanogen.

**acetochlor** An herbicide containing acetic acid or the acetyl group ($CH_3CO$—), registered by the USEPA's Pesticide Office.

**acetoclastic bacteria** Bacteria that use acetic acid ($CH_3COOH$), producing methane ($CH_4$) and carbon dioxide ($CO_2$), according to the following overall reaction:

$$CH_3COOH \rightarrow CH_4 + CO_2 \qquad \text{(A-03)}$$

Methane formation from acetic acid may also occur as follows:

$$CH_3COOH + 4\,H_2 \rightarrow 2\,CH_4 + 2\,H_2O \quad \text{(A-04)}$$

Also called acetophilic bacteria. *See also* acetogenic bacteria.

**acetoclastic methanogens** Microorganisms, mostly bacteria, that use acetic acid ($CH_3COOH$) to produce methane ($CH_4$) and carbon dioxide ($CO_2$). *See* acetoclastic bacteria. Most common methanogens active in wastewater treatment are the genera *Methanosarcina* and *Methanoaeta* (formerly *Methanothrix*). Also called aceticlastic methanogen.

**acetogen** A microorganism that produces acetic acid ($CH_3COOH$).

**acetogenesis** The formation of acetic acid ($CH_3COOH$) during anaerobic decomposition of organic matter.

**acetogenic bacteria** Bacteria that produce acetic acid ($CH_3COOH$).

**acetogenic microorganism** A microorganism that converts high-molecular-weight organic acids to acetic acid ($CH_3COOH$) plus hydrogen ($H_2$) in a free or bound form before the production of the end products (methane and carbon dioxide) of anaerobic decomposition.

**acetone ($C_3H_6O$ or $CH_3COCH_3$)** A clear or colorless volatile liquid, synthetic organic chemical of the group of ketones, miscible with water; also called dimethyl ketone or pyroacetic acid, it is a common household, industrial, and laboratory solvent, with formula as shown in Figure A-01. It is also used in the fabrication of plastics. Within the pH range of 6–7, it may react with the hypochlorite ion in water to form chloroform. Also called propanone.

**acetonitrile ($CH_3CN$ or $C_2H_3N$)** A colorless liquid with a pungent, vinegar-like odor; miscible with water. Its inhalation or ingestion may cause nausea, vomiting, gastrointestinal pain, etc. *See* Figure A-02.

**acetophilic bacteria** Bacteria that use acetic acid, producing methane and carbon dioxide. *See* acetoclastic bacteria for more detail.

**acetylene ($C_2H_2$ or H—C≡C—H)** A colorless, poisonous, gaseous, unsaturated hydrocarbon (olefin). It can be prepared from the reaction of calcium carbide ($CaC_2$) with water ($H_2O$):

$$CaC_2 + H_2O \leftrightarrow C_2H_2 + CaO \qquad \text{(A-05)}$$

Used in welding and in many chemicals (e.g., acetic acid); used also in denitrification assays to

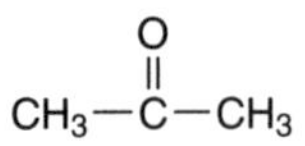


**Figure A-01.** Acetone.

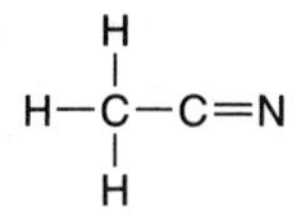


**Figure A-02.** Acetonitrile.

block the reduction of nitrous oxide ($N_2O$) to nitrogen gas ($N_2$). Also called ethyne.

**acetylene dichloride (ClCH═CHCl or $C_2H_2Cl_2$)** A volatile organic chemical made up of a mixture of two halogenated hydrocarbons: *cis*- and *trans*-1,2-dichloroethylenes; used as an industrial solvent and in the production of other organic compounds. It may affect the kidney and liver. Its components are regulated by the USEPA in drinking water. Also called 1,2-dichloroethylene.

**acetylene reduction assay** A method using acetylene ($C_2H_2$) instead of dinitrogen ($N_2$) as a substrate to determine the activity of nitrogenase. Acetylene is reduced to ethylene ($C_2H_4$) if there is nitrogenase activity. Gas chromatography can resolve both the substrate and the product.

**acetyl hydroxide** A quaternary equilibrium solution that contains acetic acid ($CH_3CO_2H$), hydrogen peroxide ($H_2O_2$), peracetic acid ($CH_3CO_3H$), and water ($H_2O$) that can be used as a chemical disinfectant. It corresponds to the following reaction:

$$CH_3CO_2H + H_2O_2 \leftrightarrow CH_3CO_3H + H_2O \quad (A\text{-}06)$$

Also called (commercially) ethaneperoxide acid, peroxyacetic acid, or peracetic acid.

***Achromatium*** A genus of strictly aerobic, chemoautotrophic, sulfur-oxidizing bacteria.

***Achromobacter*** A group of floc-forming, heterotrophic organisms found in activated sludge and active in biological denitrification and trickling filters; also removed on granular activated carbon.

**acid** A chemical substance that can donate a hydrogen ion ($H^+$) or proton; a substance that can accept an electron pair; or a substance that can react with a base to form a salt. Some (organic) acids are hydrocarbon derivatives in which a carboxyl group (CO · OH) replaces an oxygen atom; example: acetic acid ($CH_3$ · CO · OH). Wastewater discharges with high acid concentrations render receiving waters unsuitable for most uses and make them toxic to aquatic life. Acids also corrode pipes and treatment units, and interfere with treatment processes. The principal method of treatment of acid wastes is neutralization, for example, through an upflow lime bed. *See* strong acid and weak acid for examples of acids of interest in water chemistry. *See also* base, Bronsted concept, Lewis concept.

**acid aerosol** Airborne particles of 1–2 microns containing sulfates ($SO_4$), sulfuric acid ($H_2SO_4$), nitrates ($NO_3$), and nitric acid ($HNO_3$).

**acid alum** The combination of alum and sulfuric acid or another strong acid, used as a coagulant. It reacts with alkalinity to a greater extent than the nonacid fortified coagulants. Also called acidulated alum.

**acid attack** Corrosion caused by a high concentration of hydrogen ions, e.g.:

$$Fe + 2\,H^+ \rightarrow Fe^{2+} + H_2 \quad (A\text{-}07)$$

**acid–base indicator** A weak acid or a weak base used to titrate a solution. *See* indicator.

**acid–base reaction** A reaction that occurs when an acid and a base are added to water, resulting in the formation of a salt and water; one of the four major types of chemical reactions in aqueous solutions. *See also* precipitation, complexation, and redox reactions. The acid and base neutralize each other if added in equivalent amounts, e.g.:

$$H_2SO_4 + 2\,NaOH \rightarrow Na_2SO_4 + 2\,H_2O \quad (A\text{-}08)$$

Some acid–base reactions indicate the loss of a proton ($H^+$) from the acid species (HA) and the formation of the conjugate base ($A^-$):

$$HA \rightarrow H^+ + A^- \quad (A\text{-}09)$$

$$C_T = [HA] + [A^-] \quad (A\text{-}10)$$

$C_T$ is the total concentration of species A; square brackets denote molal concentrations.

**acid–chlorite process** A method used to generate chlorine dioxide ($ClO_2$) for water treatment, from the reaction of sodium chlorite ($NaClO_2$) with hydrochloric acid (HCl):

$$5\,NaClO_2 + 4\,HCl \rightarrow 4\,ClO_2 + 5\,NaCl + 2\,H_2O \quad (A\text{-}11)$$

**acid demand value (ADV)** A measure of the alkali content of sand as determined in sand reclamation testing. *See also* acid value.

**acid deposition** A complex chemical and atmospheric phenomenon that occurs when emissions of sulfur and nitrogen compounds and other substances are transformed by chemical processes in the atmosphere, often far from the original sources, and then deposited on earth in either a wet or dry form. The wet forms, popularly called "acid rain," can fall as rain, snow, or fog. The dry forms are acidic gases. The reactions involved in acid precipitation result in the formation of oxides of carbon, nitrogen, and sulfur; the absorption of gases into water; interaction of acids with ammonia; and the dissolution of aerosols into water. Acid depositions may compromise the ability of a water source to minimize the leaching and mobilization of such contaminants as aluminum, asbestos, cadmium, lead, mercury, and nitrates.

**acid dew point** (1) The temperature at which dilute sulfuric acid ($H_2SO_4$) appears as liquid droplets when a mixture of sulfur trioxide ($SO_3$) and water ($H_2O$) is cooled below the saturation temperature. To avoid corrosion of a stack, flue gases are released into the atmosphere above the acid dew point. (2) The temperature at which moisture condenses in an emission (from a stack or a flue) that contains sulfur oxides. *See also* water dew point.

**acid dissociation constant** The equilibrium constant for the dissociation reaction of an acid in water; it is a quantitative measure of the strength of an acid. *See* strong acid and acid strength.

**acid drainage** The drainage from mining operations, which becomes acidic in the presence of sulfur-bearing materials and air; it contains sulfuric acid ($H_2SO_4$). *See* acid mine drainage for detail.

**acid extractable concentration** The concentration of a chemical substance resulting from the treatment of a solution that contains it with a mineral acid. An acid extractable metal is obtained in a solution from the treatment of an unfiltered sample with a hot, dilute mineral acid. *See also* dissolved, suspended, and total metal concentrations.

**acid extractable metal** *See* acid extractable concentration.

**acid-fast** The property of organisms or substances that resist decolorization with acidified alcohol or with dilute mineral acids after staining with carbol fuchsin or another strong dye.

**acid-fast bacteria** Bacteria, mostly of the genus *Mycobacterium,* resistant to decolorizing by acidified alcohol after special staining (with hot carbol–fuchsin), and responsible for such diseases as leprosy and tuberculosis. Raw domestic wastewater contains approximately 100 acid-fast bacteria (as compared to 100,000,000 coliforms) per 100 ml. Because of its greater resistance to disinfectants, this group of bacteria is considered as an alternative indicator to the coliform group. *See*acid-fast stain.

**acid-fast stain** A special stain or technique used to identify mycobacteria, which do not stain with common dyes. This procedure uses a solution of ethanol (95%) and hydrochloric acid (3%) to wash the organisms, which eliminates the stain carbol fuchsin from all organisms except the acid-fast bacteria. Also called Ziehl–Neelsen stain. *See also* Gram stain.

**acid feed system** The apparatus used to add acids (hydrochloric [HCl], nitric [$HNO_3$], sulfuric [$H_2SO_4$]) for the neutralization of alkalinity. It may be of one of the following types: proportional feed or conastant rate, gravity or pressure feed, and concentrated feed or dilute feed.

**acid fermenter** A tank used in sidestream fermentation processes for the formation of volatile fatty acids, such as acetic and propionic acids. This fermentation may occur in a separate unit or in the same anaerobic basin that also provides for phosphorus release. Also called a sidestream fermenter or a sludge fermenter, the separate tank may include a compartment for storage of methane gas.

**acid fog** Airborne water drops of 3–30 microns composed of sulfuric acid ($H_2SO_4$) and/or nitric acid ($HNO_3$).

**acid formation** (1) The second stage of the anaerobic digestion process, after hydrolysis and before methane formation: saprophytic organisms convert the end products of hydrolysis to organic acids, mainly acetic, butyric, lactic, and propionic acids. If acid formation is substantial, it can lower pH and impede biological activity, unless there are sufficient methane bacteria to use the acids produced. (2) *See* distribution diagram.

**acid former** *See* acid-forming bacteria.

**acid-forming bacteria** Saprophytic bacteria that use complex organic substances to produce fatty acids of low molecular weight (e.g., acetic, butyric, propionic acids) during anaerobic digestion (in wastewater treatment, in bottom muds of reservoirs, or similar situations). *See also* methanogen.

**acid-forming material** Sulfide-containing material that forms sulfuric acid ($H_2SO_4$) when exposed to air, water, or weathering processes.

**acid gas** A gas stream of hydrogen sulfide ($H_2S$) and carbon dioxide ($CO_2$) that has been separated from sour natural gas by a sweetening unit (EPA-40CFR60.641).

**acid/gas digestion process** *See* acid/gas phased digestion.

**acid/gas phased digestion** One of a number of variations of the two-phased anaerobic digestion process, designed to optimize conditions for the separate groups of microorganisms involved by using two reactors in series. As shown in Figure A-03, acidogenesis takes place in the first reactor at a solids retention time (SRT) of 1–3 days and a pH of 6.0 or less, while the second reactor pro-

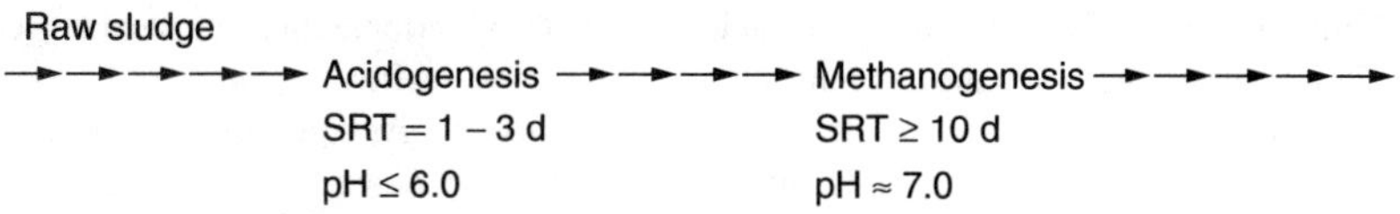


**Figure A-03.** Acid/gas phased digestion.

vides an SRT of 10 days or more and a neutral pH for methanogenesis, both under mesophilic or thermophilic conditions. *See also* staged mesophilic digestion, staged thermophilic digestion, temperature-phased digestion.

**acidic** The condition of water or soil that contains a sufficient amount of acid substances (or hydrogen ions $[H^+]$) to lower the pH below 7.0.

**acidic copper arsenite** *See* copper arsenite.

**acidic dye** A substance with a negative charge used to stain smears and identify cell membranes and other positively charged cell components.

**acidic rain** Same as acid rain.

**acidic solution** A solution that contains a sufficient amount of acid substances (or hydrogen ions $[H^+]$) to lower the pH below 7.0. The solution is neutral if pH = 7.0 or basic if pH > 7.0.

**acidic waste** Same as acid waste.

**acidification** (1) The addition of an acid (usually nitric or sulfuric) to a sample to lower the pH below 2.0. The purpose of acidification is to "fix" a sample so it will not change until it is analyzed. (2) A process used to recover iron and aluminum coagulants from water or wastewater treatment residuals by putting the metals back into solution. (3) The treatment of neutral (water-washed) alumina (alumina · HOH) with hydrochloric acid (HCl) or another acid to form protonated or acidic alumina (alumina · HCl); it is the first step in the adsorption–regeneration cycle of activated alumina. *See also* alumina regeneration, capacity restoration.

**acidified** Pertaining to a solution that has been converted to an acid or, more specifically, a solution whose pH has been lowered below 2.0 by acid addition for preservation purposes. *See also* acidized.

**acid ionization constant** Same as acid dissociation constant.

**acidity (Acy)** A measure of the capacity of an aqueous solution to react with hydroxyl ions ($OH^-$) or to neutralize strong bases, due to the presence of acids or hydrogen ions ($H^+$). It is usually expressed as mg/L of calcium carbonate ($CaCO_3$). Assuming that only species related to carbon dioxide ($CO_2$) are significant,

$$[Acy] = [H_2CO_3^*] + [HCO_3^-] + [H^+] \quad (A\text{-}12)$$

where the quantities between brackets denote concentrations in eq/L, and $[H_2CO_3^*]$ denotes the concentration of $H_2CO_3$ and $CO_2$, sometimes called free carbon dioxide. Other species that can contribute to acidity include HOCl, $HPO_4^{-2}$, and $H_2PO_4^-$. Note that these species, as well as the bicarbonate ion ($HCO_3^-$), can neutralize both bases and acids. *See also* alkalinity. Most natural waters are alkaline; acidity is a sign of pollution. It may be contributed by the activity of a hydrated metal ion such as hydrated aluminum:

$$Al(H_2O)_6^{3+} \leftrightarrow Al(H_2O)_5OH^{2+} + H^+ \quad (A\text{-}13)$$

Alkalies commonly used to neutralize acidity include lime, caustic soda, and sodium carbonate. *See also* acid value, H-acidity, mineral acidity.

**acidity analysis** An analysis, commonly by titration, to measure the concentration of carbon dioxide ($CO_2$) and other acids in an aqueous solution.

**acidity as $CaCO_3$** A conventional expression of the acidity of an aqueous solution, determined from the results of a titration that uses a base as titrant:

$$\text{Acidity as } CaCO_3 \text{ (mg/L)} = 50{,}000(\text{mL titrant}) \times (\text{normality of base})/(\text{mL of sample}) \quad (A\text{-}14)$$

**acidity constant** The equilibrium constant (K) corresponding to the dissociation of an acid ($HB^+$) in water, according to the reaction

$$HB^+ + H_2O = H_3O^+ + B \quad (A\text{-}15)$$

Depending on the conditions of dilution, the acidity constant is expressed in terms of activities (denoted by {.}), concentrations ([.]) or a combination of the two:

$$K = \{H^+\} \cdot \{B\}/\{HB^+\} \approx [H^+] \cdot [B]/[HB^+] \approx \{H^+\} \cdot [B]/[HB^+] \quad (A\text{-}16)$$

*See* Stumm and Morgan (1996) for more detail. *See also* conjugate acid–base pair, acid ionization constant, basicity constant, distribution diagram.

**acidity neutralization** The addition of an appropriate base to an aqueous solution to react with an acid. *See* acid-base reaction.

**acidity–pH relationship** The addition of an acid to an aqueous solution increases the concentration of hydrogen ions and thus reduces the pH of the solution but the resulting curve depends on the acid strength and the characteristics of the solution (e.g., pure water, presence of carbonates, etc.)

**acidity value** Same as acid value.

**acidized** Pertaining to a solution to which acid has been added; acidified.

**acidizing** Same as acid treatment or well acidization.

**acid mine drainage** Drainage of water from areas that have been mined for coal or other mineral ores. This water has a low pH because of its contact with sulfur-bearing material and because of biological oxidation of iron sulfide (pyrite, $FeS_2$) to sulfuric acid ($H_2SO_4$), a major water pollutant; ferrous sulfate (copperas, $FeSO_4$) or hydrated iron (III) oxide [$Fe(OH)_3$] is also formed:

$$2\,FeS_2 + 7\,O_2 + 2\,H_2O \rightarrow 2\,FeSO_4 + 2\,H_2SO_4 \quad \text{(A-17)}$$

$$4\,FeS_2 + 15\,O_2 + 14\,H_2O \rightarrow 4\,Fe(OH)_3 + 8\,H_2SO_4 \quad \text{(A-18)}$$

Another pollutant species of acid mine water is the hydrated iron (III) ion [$Fe(H_2O)_6^{3+}$]. Acid mine drainage is harmful to aquatic organisms; typically, it contains 100–6000 mg/L of sulfuric acid, 10–1500 mg/L of ferrous sulfate; 0–350 mg/L of aluminum sulfate, and 0–250 mg/L of manganese sulfate. A significant source of toxic metals, it also contains aluminum, arsenic, copper, lead, and zinc. It is sometimes used to clean raw coal. Wetlands treatment has been recommended for acid mine drainage at a rate of 200 square feet per gpm. Also called acid mine-drainage waste, acid mine-drainage water, acid mine waste, acid mine water. *See* strip mining, yellowboy.

**acid mine-drainage waste** *See* acid mine drainage.

**acid mine-drainage water** *See* acid mine drainage.

**acid mine waste** *See* acid mine drainage.

**acid mine water** *See* acid mine drainage.

**acid-neutralizing capacity (ANC)** A measure of the ability of water or soil to resist changes in pH; alkalinity. It measures the net deficiency of protons and can be determined from titration with a strong acid to a preselected equivalence point.

**acidogenesis** The second step or reaction in the anaerobic digestion process: the degradation of amino acids, sugars, and fatty acids results in the formation of soluble organic compounds and short-chain organic acids (acetate, butyrate, propionate); also called fermentation. The final products of fermentation are acetate, carbon dioxide, and hydrogen.

**acidophile** Any of a group of organisms that grow optimally at acidic pH values, or adjust to low pH conditions by exporting $H^+$ ions from their cells. They grow poorly or not at all under pH values higher than 7.0.

**acidophilic** Growing best under acidic (low pH) conditions.

**acid or ferruginous mine drainage** Mine drainage that, before any treatment, has either a pH of less than 6.0 or a total iron concentration equal to or greater than 10.0 mg/L (EPA-40CFR434.11-a).

**acid pickles** The industrial wastewater containing the spent liquor from metal-surface-cleaning operations and the rinse water used to wash the metal; acid pickles contain acids and other chemical compounds.

**acid pollution** Water pollution caused by wastewater from industries that produce acidic wastes, including the production of batteries, beer, chemicals, insecticides, and textiles, and mining and pickling operations.

**acid precipitation** Precipitation with a low pH, usually lower than 4.5. *See* acid deposition and acid rain for more detail.

**acid rain** Air pollution produced when acid chemicals are incorporated into rain, snow, fog, or mist. The "acid" in acid rain comes from sulfur and nitrogen oxides, products of burning coal and other fuels and from certain industrial processes. These oxides are related to two strong acids: sulfuric acid and nitric acid. (The Clean Air Act requires the reduction of emissions from power plants, which account for 80% of the sulfur dioxide emissions to the atmosphere.) When sulfur dioxide and nitrogen oxides are released from power plants and other sources, winds blow them far from their sources. If the acid chemicals in the air are blown into areas where the weather is wet, the acids can fall to earth in rain, snow, fog or mist. In areas where the weather is dry, the acid chemicals may become incorporated into dusts or smoke. Acid rain can damage the environment, human health, and property. *See also* acid deposition.

**acid retardation** The ability of strong-base anion exchange resins to adsorb whole molecules of strong acids (hydrochloric, nitric, sulfuric), in addition to their anion exchange capabilities. This property is used to recover acids from spent etchants and pickling liquors by elution with water. *See also* Recoflo™.

**acid sensitive** Characteristic of a substance that is easily affected by the addition of acidic material.

**acid shock** A temporary and significant reduction of pH levels in a water body caused by a sudden addition of acidic material such as acid runoff. Also called pH shock.

**acid spring** A spring that has a pH in the acid range, i.e., pH < 7.0.

**acid strength** A measure of the tendency of an acid to donate a proton, usually considered with respect to the substance accepting the proton. In aqueous solutions the relative strength of an acid is measured relative to water and is given by the equilibrium constant of the acid ionization reaction.

**acid treatment (of a well)** (1) The use of a solution of hydrochloric acid (HCl) and gelatin to remove incrustations from a well screen. The solution is placed above screen level and agitated intermittently for a few hours. Subsequent pumping of the well removes the acid and loosened incrustations. *See also* polyphosphate treatment, dry ice, chlorination. (2) A well stimulation procedure that uses chemicals (e.g., a solution of hydrochloric acid, HCl) to break down matter that clogs the rock pores in an injection zone, thereby improving the flow of injection fluids. Also called well acidization.

**acidulated alum** The combination of alum and sulfuric acid or another strong acid, used as a coagulant. It reacts with alkalinity to a greater extent than nonacid fortified coagulants. Also called acid alum.

**acid value** (1) A measure of the acidity of a waste, determined by titrating a 5-mL sample of the waste with an excess amount of 0.5 *N* sodium hydroxide (NaOH) and back-titrating with a 0.5 *N* hydrochloric acid (HCl) to the phenolphthalein end point. The acid value of a waste is used with the basicity factor of lime to determine the amount of alkaline agent required to neutralize the waste (Nemerow and Dasgupta, 1991). (2) A measure of the neutralizing capacity of a reagent; the weight of reagent equivalent in base neutralizing capacity to a unit weight of calcium oxide (CaO). It is determined by dividing the acidity of the reagent, expressed in mg/L as $CaCO_3$ required to neutralize 1.0 mg/L, by the weight of CaO expressed similarly, i.e., by 0.56. For example, the acid values of hydrochloric (HCl), nitric ($HNO_3$), and sulfuric ($H_2SO_4$) acids are respectively (a) 0.72/0.56 = 1.28, (b) 0.63/0.56 = 1.12, and 0.98/0.56 = 1.75. Also called acidity value. *See also* basicity factor (WEF & ASCE, 1991). Also called acidity value. *See also* acid demand value.

**acid-washed activated carbon** Activated carbon that has been cleaned from ashes using an acid solution.

**acid waste** A waste that contains acidic materials as evidenced by a low pH. In wastewater, such acidity may be caused by dilute industrial discharges of hydrochloric (HCl), nitric ($HNO_3$), and sulfuric ($H_2SO_4$) acids from manufacturing, apparel, and chemicals plants. Acid wastes are normally neutralized before processing in publicly owned treatment works.

**acifluorfen** A pesticide regulated by the USEPA, with MCLG = 0 and MCL = 0.002 mg/L.

**A/C index** *See Araphinidineae*-to-*Centrales* index.

**A/C pipe** Asbestos–cement pipe.

***Acinetobacter*** A genus of heterotrophic organisms active in biological denitrification and enhanced phosphorus uptake processes; called a bio-P microorganism; commonly removed on the surface of granular activated carbon. It accumulates phosphorus in excess of its growth requirements. It is also found in activated sludge flocs. Some species are opportunistic pathogens.

**ACM®** membrane A reverse osmosis membrane, with a thin film, manufactured by the TriSep Corporation.

**acoustic cavitation** The creation, growth, and sudden collapse of gas bubbles in response to ultrasonic compression. This phenomenon is being considered for the inactivation of pathogens in drinking water as an alternative (called ultrasonic disinfection, ultrasound inactivation, or sonication) to chemical oxidation and physical disinfection methods: when the bubbles collapse within fractions of a second and in the vicinity of solid particles, that creates elevated pressures, temperatures, and other conditions that can cause microbial inactivation. This method does not produce disinfection by-products, does not interfere with ultraviolet light transmittance, and does not cause pathogen enmeshment in flocs or fouling (Brown & Salveson, 2006).

**acquired immune deficiency syndrome (AIDS)** A disease caused by the human immunodeficiency virus (HIV) found in body fluids. There is no evidence that it can be transmitted through water or wastewater because of its fragility outside the human body. The virus may be stable in wastewater and distilled water for up to 12 hours; its survival is much lower than that of poliovirus, for example.

**acquired immunity** Immunity acquired through previous exposure to a pathogen or foreign substance, as opposed to natural immunity.

**acquired immunodeficiency syndrome (AIDS)** *See* acquired immune deficiency syndrome.

**acquisition** The act or process of gaining ownership and control of an organization such as a water or wastewater utility by another utility or a government agency; a utility so acquired.

**ACR** Acronym of acute-to-chronic ratio.

**acre-foot** The volume of water equivalent to a depth of one foot over an area of one acre, i.e., a volume of 43,560 cubic feet = 1,233,482 liters = about 325,851 gallons.

**acre-inch** The volume of water or other material equivalent to a depth of one inch over an area of one acre, i.e., one-twelfh of an acre-foot.

**acre-inch day** A measure of flow used mainly in irrigation and representing the volume of water equivalent to one inch over an area of one acre in a day, i.e., (1/12) × (43,560)/(24 × 60 × 60) or 0.042 cfs = 1.189 L/s.

**acridine ($C_{13}H_9N$)** A colorless crystalline solid obtained from coal tar, used in the production of dyes and drugs.

**acridine orange** A widely used stain for direct microscopic observation of bacteria. It associates with nucleic acids to stain bacteria green or orange, depending on whether they have RNA or high DNA. It is also used to differentiate between live and dead cells, with questionable accuracy. *See also* Petroff–Hauser counting chamber, electronic particle counter.

**acridine orange direct count** *See* acridine orange.

**acridine orange stain** *See* acridine orange.

**Acritet** Acrylonitrile.

**acrolein ($C_3H_4O$)** An aldehyde-based product used as a biocide and in organic chemicals. It is a yellow, flammable liquid that can be obtained from the decomposition of glycerol.

**Acro-Pac®** A package reverse osmosis plant for seawater conversion, manufactured by Aqua-Chem, Inc.

**acrylamide ($C_3H_5NO$, $H_2C{=}CHCONH_2$, or $CH_2CHCONH_2$)** A colorless, odorless, toxic, water-soluble, organic solid or flake-like crystalline substance used in the fabrication of polymers (including drinking water coagulants and filter aids), in wastewater treatment, in soil conditioning agents, and in industrial processes. It is on the EPA list of regulated synthetic organic chemicals; MCLG = 0. Also called acrylamide monomer, acrylic amide, propenamide, 2-propenamide.

**acrylamide monomer** Acrylamide.

**acrylic amide** Acrylamide.

**acrylic resin** A widely used, transparent synthetic resin derived from acrylic acid ($C_3H_4O_2$ or $CH_2CHCOOH$).

**acrylon** Acrylonitrile.

**acrylonitrile ($C_3H_3N$)** A clear, colorless, volatile liquid with a sweet irritating odor (threshold in air: 17 ppm); a synthetic organic chemical used in the manufacture of certain synthetic fibers; toxic to fish; regulated by the USEPA; MCLG = 0 and MCL = 0.003 mg/L. Also called acritet, acrylon, acrylonitrile monomer, carbacryl, cyanoethylene, fumigrain, Miller's fumigrain, nitrile, propenenitrile, 2-propenenitrile, ventox, vinyl cyanide.

**acrylonitrile-butadiene-styrene (ABS)** A black plastic used to manufacture pipes and other products.

**acrylonitrile monomer** Acrylonitrile.

**ACSC** Acronym of air cathode single chamber.

**ACS-Plus chemicals** Chemicals of high purity produced by The Hach Company for laboratory and other uses.

**act** Any one of various federal regulations, such as the (1) Ocupational Safety and Health Act (OSHA) of 1970, (2) Public Health Service Act as amended by the Safe Drinking Water Act (SDWA) of 1974, (3) National Environmental Policy Act (NEPA), (4) Clean Air Act (CAA), (5) Federal Water Pollution Control Act of 1972 as amended by Clean Water Act (CAA) of 1977 and the Water Quality Act of 1987, (6) Resource Conservation and Recovery Act (RCRA) of 1976, (7) Superfund Amendments and Reauthorization Act (SARA) of 1986, and (8) Toxic Substances Control Act (TSCA) of 1976.

**ACT-100®** tank A proprietary underground tank manufactured by the Steel Tank Association. It is made of a fiberglass-laminated steel.

**Actifil®** media Media manufactured by Aeration Engineering Resources Corp. for the packing of biological reactors.

**Actiflo® process** A proprietary drinking water treatment process developed by Krüger, Inc. It combines sand ballasting with inclined settling, following alum or ferric coagulation and flocculation. It requires a small footprint. *See also* Densadag®.

**Actiflo system** *See* Actiflo® process.

**actinide colloids** Colloidal particles generated by the polynuclear hydrolysis of actinide cations and their adsorption to the surface of natural colloids.

**actinides** Certain elements in the periodic table, including actinium, thorium, uranium, and others; some natural, others resulting from nuclear reactions, but all radioactive. Also called actinoids.

**actinograph** *See* actinometer.

**actinoid** (1) Actinide. (2) Raylike, radiate.

**actinometer** An instrument used to determine the intensity of the Sun's or other radiation, e.g., the intensity of ultraviolet light, based on a chemical reaction (*see* chemical actinometer) or on the heat absorbed by a blackened disk or chamber. An actinograph also records that measurement. *See also* pyrheliometer. (The prefix *actino* is from the Greek aktis, meaning ray, beam.)

**actinometry** The art or process of measuring the intensity of radiation with an actinometer.

**actinomyces** Filamentous, saprophytic, anaerobic bacteria of the genus *Actinomyces,* some of which are pathogenic to humans. They are also responsible for odor-causing metabolites in drunking water sources.

***Actinomycetales*** An order of filamentous, mostly Gram-positive bacteria found in soils, a few of which are pathogenic to humans. Some species produce the antibiotics actinomycin and streptomycin. They have some moldlike characteristics and include the genera *Actinomyces, Mycobacterium,* and *Nocardia. See also Actinomycetes.*

***Actinomycetes*** Any of numerous, generally filamentous or rod-shaped, prokaryotic, aerobic or anaerobic, and often pathogenic, microorganisms resembling both bacteria and fungi. They are found in soils and river muds, and are often detected follwing bluegreen algal blooms. They belong to the phylum *Chlamydobacteriae* or to the order *Actinomycetales.* Some species (*Actinomyces, Nocardia, Streptomyces*) produce substances (e.g., geosmin and 2-methylisoborneol or MIB) that cause objectionable earthy and musty odors in drinking water even at low concentrations. Storage in holding basins before treatment may reduce their concentration by a factor of 10 but that concentration may increase in slow sand filtration and granular activated carbon adsorption.

**action level** The concentration of a contaminant (e.g., lead or copper) in water specified in the Code of Federal Regulations 141.80(c), which determines, in some cases, the treatment requirements or corrective actions contained in subpart I of this part that a water system is required to complete. *See also* Lead and Copper Rule. Other regulatory agencies may also establish action levels. For example, the state of California has an action level of 18 μg/L for perchlorate in drinking water.

**activated algae process** A wastewater treatment process that removes phosphorus, nitrogen, and other nutrients by promoting the formation of algal cultures that incorporate these nutrients. The algae may also help stabilize organic matter by producing oxygen. For the removal of a specific nutrient (e.g., phosphorus) beyond normal metabolic activities, it may be necessary to add other nutrients (e.g., carbon and nitrogen) to the wastewater (McGhee, 1991). *See also* bacterial assimilation, luxury uptake.

**activated aliphatic substitution** *See* activated aromatic substitution.

**activated alumina** A charged, semicrystalline, highly porous, partially dehydrated form of aluminum oxide ($Al_2O_3$) used as an adsorbent, in chromatography, in pollution control, as a catalyst, or in ion exchange for fluoride removal in combination with a porous medium or for arsenic removal. It is a natural granular zeolite that has a selective preference not only for fluoride but also for arsenic, phosphate, selenium, and silica, depending on pH. A solution of up to 5% sodium hydroxide (NaOH) is used to regenerate activated alumina. The following equations (AWWA, 1999) represent the adsorption of fluoride ($F^-$) on alumina surface (≡Al) in an acid solution and fluoride desorption or alumina regeneration by hydroxide ($OH^-$):

$$\equiv Al{-}OH(s) + H^+ + F^- \rightarrow \equiv Al{-}F^-(s) + HOH \quad \text{(A-19)}$$

$$\equiv Al{-}F(s) + OH^- \rightarrow \equiv Al{-}OH(s) + F^- \quad \text{(A-20)}$$

**activated alumina adsorption** A water treatment process using packed beds of activated alumina to remove such contaminants as arsenic, fluoride, humic matter, selenium, and silica; the contaminant anions are exchanged for surface hydroxides on the activated alumina.

**activated aromatic (or aliphatic) substitution** An oxidation reaction involving the substitution of a halogen in an aromatic (or aliphatic) compound. *See also* ionic reaction, radical reaction.

**Activated Biofilm Method®** A fixed-film process developed by Ralph B. Carter Co. for wastewater treatment.

**activated biofilter (ABF)** A trickling filter, with redwood or plastic packing instead of rock, used in the activated biofiltration process; it receives the primary effluent, a portion of the filter underflow, and the return activated sludge from an aeration basin or a secondary clarifier. *See* note under activated biofiltration.

**activated biofiltration (ABF)** One of a few dual wastewater treatment processes that combine the trickling filtration and activated sludge methods. A trickling filter or biological tower of reduced size receives the primary effluent and is followed by a small aeration basin and a final clarifier with sludge recycled ahead of the filter. Also called activated biological filtration and activated biofiltration—activated sludge. *See also* combined filtration-aeration process. *Note:* Some references [e.g., Metcalf & Eddy (1991), Metcalf & Eddy (2003), and WEF & ASCE (1991)] indicate that the aeration basin is optional. However, without that unit, the plant is equivalent to a high-rate trickling filter and does not qualify as a dual process.

**activated biofiltration-activated sludge process** Same as activated biofiltration.

**activated biological filtration (ABF)** Same as activated biofiltration.

**activated carbon** A highly adsorbent form of carbon used to remove odors, tastes, and toxic substances from liquid or gaseous emissions. Activated carbon, which weighs on the average 12 pounds per cubic foot and is fed dry or as a slurry, is used to selectively remove certain trace and soluble materials from water or to remove dissolved organic matter (refractory organics) from wastewater. It is also used as a filter medium and as a coagulant aid in water treatment (*see* adsorbent-weighting agent). Adsorptive particles (powdered activated carbon or PAC) and granules (granulated activated carbon or GAC) are usually obtained by heating a carbonaceous material (bituminous coal, coconut shells, graphite, lignite, peat, pulp-mill char, wood) in the absence of air. Activation of the material at a temperature between 300°C and 900°C creates tiny fissures and pores. Sometimes called charcoal, activated charcoal, or simply carbon. *See also* blackout dosage, biophysical treatment, breakthrough, carbonization, chemical activation, carbon usage rate, char, column utilization, empty bed contact time, fixed bed, fluidized bed, granular activated carbon, iodine number, macropore, micropore, molasses number, phenol number (or value), packed bed, physical activation, powdered activated carbon, pyrolysis, reactivation, regeneration, uniformity coefficient.

**activated carbon adsorber** A device used to remove odors from air by adsorption on single beds or dual beds of activated carbon.

**activated carbon adsorption** *See* activated carbon treatment.

**activated carbon adsorption treatment** *See* activated carbon treatment.

**activated carbon block** A blend of fine activated carbon, water, and a suitable binder that is mixed, molded, and hardened or extruded to a cartridge filter. Specialized media may be added to the activated carbon. Activated carbon block filters are used for particular filtration, insoluble lead reduction, and the removal of *Giardia* and *Cryptosporidium*.

**activated carbon canister** A cartridge containing activated carbon to remove hazardous substances from a respirator.

**activated carbon contactor** A reactor containing a bed of granular activated carbon used for advanced wastewater treatment. Activated carbon contactors vary according to the type of flow (gravity or pressure, downflow, upflow, or countercurrent), the type of arrangement of individual columns (in series or in parallel), and the type of bed (expanded bed, fixed bed, or moving bed). *See* fixed-bed column, expanded-bed carbon contactor, empty-bed contact time, specific throughput, carbon usage rate, bed life.

**activated carbon dechlorination** The removal of combined and free residual chlorine by adsorption on gravity or pressure filter beds of granular activated carbon, according to the following reactions (Metcalf & Eddy, 2003):

$$C + 2\,Cl_2 + 2\,H_2O \rightarrow 4\,HCl + CO_2 \qquad \text{(A-21)}$$

$$C + 4\,NHCl_2 + 2\,H_2O \rightarrow 2\,N_2 + CO_2 + 8\,Cl^- + 8\,H^+ \qquad \text{(A-22)}$$

$$C + 2\,NH_2Cl + 2\,H_2O \rightarrow 2\,NH_4^+ + CO_2 + 2\,Cl^- \qquad \text{(A-23)}$$

**activated carbon process** *See* activated carbon treatment.

**activated carbon reactivation** *See* activated carbon regeneration.

**activated carbon regeneration** Reactivation or regeneration is the process of periodically removing adsorbed materials from spent activated carbon to restore its porous structure and adsorptive capacity, usually by a thermal or chemical means similar to the initial activation method. Thermal reactivation consists of (a) drying at a temperatures of up to 200°C, (b) vaporization and decomposition of volatile adsorbates between 200 and 500°C, (c) pyrolysis of nonvolatile adsorbates between 500 and 700°C, and (d) oxidation of the residue at a higher temperature. Chemical regeneration is accomplished by washing the carbon in organic solvents, mineral acid, or caustic soda. *See* rotary kiln and multiple-hearth furnace. A distinction is sometimes made between reactivation and regeneration: reactivation uses the same operations as the initial activation process (heating of the carbon to oxidize the absorbate and burning off the materials formed) while regeneration involves the chemical oxidation of the adsorbate, driving off the adsorbate by steam, and further regeneration by solvents and biological conversion.

**activated carbon treatment** A method of water or wastewater treatment that uses adsorption on granular activated carbon contactors or filters or on powdered activated carbon slurries for the removal of soluble substances. The process removes mostly organic compounds (e.g., MTBE from drinking water), but also some inorganics, through a combination of phenomena: adsorption of nonpolar molecules, filtration of larger particles, and deposition of colloids on the surface of the media. It is commonly used for the treatment of groundwater, industrial wastewater, and chemical spills. In wastewater treatment, activated carbon adsorption is used as a tertiary process following biological treatment or as an alternative to biological treatment, e.g., in the physical-chemical process. *See also* adsorbate, adsorbent, adsorbent dose, adsorption, adsorptive capacity, adsorption isotherms, bed life, breakthrough, carbon contactor, chemisorption, constant diffusivity model, desorption, empty-bed contact time (EBCT), exhaustion, fixed-bed column, high-pressure minicolumn, mass transfer zone, PACT®, proportional diffusivity model, rapid small-scale column test, regeneration cycle, specific throughput, specific volume, time to breakthrough, total organic carbon.

**activated carbon usage rate** One of six common parameters used to define the performance of granular activated carbon contactors (and sometimes powdered activated carbon units) in the treatment of water and wastewater. It represents the capacity of the activated cabon to treat a given liquid or to remove a specified contaminant and is often defined as the quantity (weight) of activated carbon used per unit volume treated. *See* carbon usage rate for more detail.

**activated charcoal** *See* activated carbon.

**activated silica ($SiO_2$)** A negatively charged colloid formed by treating a dilute solution of sodium silicate ($Na_2O \cdot (SiO_2)_x$] with sulfuric acid ($H_2SO_4$), aluminum sulfate [$Al_2(SO_4)_3$], carbon dioxide ($CO_2$), or chlorine ($Cl^-$). The solution contains polysilicates and other silicates that are effective coagulants. It is often used in water treatment as a coagulant or a coagulant aid after further dilution and aging for about 2 hours (it increases the rate of the chemical reactions and reduces the required dose of coagulant). Because silica is undesirable in cooling towers and boilers, it can be removed by adsorption on iron floc. Polyelectrolytes, easier to handle, have replaced activated silica in modern water treatment practice. Also called silica sol.

**activated sludge** A brown flocculent product that results when primary effluent or raw wastewater is mixed with bacteria-laden sludge and then agitated and aerated to promote biological treatment, speeding the breakdown of organic matter in the incoming waste. Activated sludge consists mostly of masses of microorganisms (biomass) and some inorganic solids; its density ranges from 1.01 to 1.10 grams per liter. These organisms are active in removing soluble organic matter from solution; hence the name activated sludge. Activated sludge in good condition has an inoffensive earthy odor but becomes septic rapidly with the odor of putrefaction. *See also* active biomass, completely mixed process, contact-stabilization, conventional activated sludge, endogenous respiration, extended aeration, food-to-microorganism ratio, high-purity oxygen, high-rate activated sludge, mean cell residence time, mixed liquor, mixed liquor volatile suspended solids, pin floc, sequencing batch reactor, short-term aeration, sludge age, sludge bulking, solids residence time, step aeration, tapered aeration.

**activated sludge age** The term commonly used is simply sludge age.

**activated sludge bacteria** A variety of bacteria found in activated sludge, particularly those growing in municipal wastewater, the most common belonging to the genera *Alcaligenes, Bacillus, Flavobacterium,* and *Pseudomonas.* Some are responsible for the poor settling characteristics of filamentous activated-sludge floc, e.g., *Sphaerotilus natans.*

**activated sludge bulking** A condition of activated sludge treatment plants in which sludge occupies a larger volume than normal, does not settle or concentrate well, and may carry over excessively in the effluent with a concomitant increase in BOD and suspended solids. *See* sludge bulking for more detail.

**activated sludge composition** Activated sludge is approximately 80% organic matter, generally represented by the formula $C_5H_7N\,O_2P_{0.2}$, and 20% inorganic substances that include calcium, iron, magnesium, potassium, sodium, sulfur, and trace elements such as cobalt, copper, molybdenum, and zinc. The formula $C_5H_7NO_2P_{0.074}$ has also been reported in the literature. *See also* activated sludge nutrients, biomass, and composition of organic matter.

**activated sludge effluent BOD ($BOD_e$)** The BOD concentration in the effluent of an activated sludge plant. For a system operating properly, it is approximately

$$BOD_e = sBOD + 0.6\ eTSS \qquad \text{(A-24)}$$

where sBOD is the effluent soluble BOD (= approximately 3.0 mg/L for municipal plants) and eTSS is the effluent total suspended solids (mg/L).

**activated sludge floc** The expression used by Arden and Lockett, the originators of the activated sludge process, to designate the mixture of wastewater and microorganisms contained in the recycled, settled sludge. Activated sludge flocs that settle well are large, nonbulking, and have a low sludge volume index. *See also* biological flocculation or bioflocculation, filamentous sludge, floc former, pinpoint floc.

**activated sludge foaming** A problem experienced in the operation of activated sludge plants and caused by the *Nocardia* and *Microthrix parvicella* groups of filamentous bacteria: their hydrophobic cells attach to and stabilize air bubbles to form the foam. Foaming control methods include spray chlorination of the foam, selector design to exclude the foaming organisms, and reducing the oil and grease content of the wastewater to be treated.

**activated sludge loading** The quantity of organic matter applied per unit volume of the aeration basin of an activated sludge plant, usually expressed in pounds of $BOD_5$ per 1000 cubic feet per day. This design and operating parameter varies from about 15 lb/day/1000 cu. ft. for extended-aeration plants to more than 120 lb/day/1000 cu. ft. for high-purity oxygen systems.

**Activated Sludge Model No. 1 (ASM1)** A mathematical model developed in a matrix format by a committee of the International Association of Water Pollution Research Control (IAWPRC) to simulate the activated sludge process. It includes 13 process variables and several formulas to determine the following kinetic or stoichiometric parameters or events: (1) aerobic growth of heterotrophs, (2) anoxic growth of heterotrophs, (3) aerobic growth of autotrophs, (4) death and lysis of heterotrophs, (5) death and lysis of autotrophs (6) ammonification of soluble organic nitrogen, (7) hydrolysis of particulate organics, (8) hydrolysis of particulate organic nitrogen. The ASM1 model is now called the IWA model because the IAWPRC has changed its name to International Water Association. *See also* lysis regrowth model.

**Activated Sludge Model No. 2 (ASM2)** A mathematical model similar to, but more complex than, the Activated Sludge Model No. 1. It includes formulas for the anaerobic, anoxic, and aerobic reactions of biological phosphorus removal.

**activated sludge nutrients** The approximate concentrations of nutrients that activated sludge microorganisms require for adequate growth. Otherwise, such undesirable phenomena as bulking may occur. They are mainly (in mg per mg of BOD): nitrogen (0.050), phosphorus (0.016), sulfur (0.004), sodium (0.004), potassium (0.003), calcium (0.004), magnesium (0.003), iron (0.001), and traces of cobalt, copper, molybdenum, and zinc.

**activated sludge operation problems** Common problems experienced in the operation of an activated sludge plant. *See* bulking sludge, rising sludge, foaming.

**activated sludge oxygen requirement** The quantity of oxygen required (Or, kg/day) for the degradation of organic matter in the activated sludge process (or any suspended growth process) as a function of the biodegradable matter (bCOD) removed and the COD of the waste sludge. It may be estimated from the following equation:

$$\begin{aligned} O_r &= \text{bCOD removed} - \text{COD of waste sludge} \\ &= Q(S_0 - S) - 1.42\, S_w \end{aligned} \quad \text{(A-25)}$$

where $Q$ is the hydraulic flow rate (m$^3$/day); $S_0$ and $S$ are, respectively, the influent and effluent substrate concentrations (mg/L BOD or biodegradable soluble COD); and $S_w$ is the wasted biomass (or sludge, kg/day).

**activated sludge parameters** A set of design and operation factors that determine the type and performance of the resulting variation of an activated sludge process. They include the aeration period or hydraulic detention time, solids retention time, food-to-microorganism ratio, BOD loading rate, mixed liquor total or volatile suspended solids, recycle ratio (recirculation ratio or return sludge rate), and the BOD removal efficiency. Other typical wastewater characterization parameters are alkalinity, various forms of nitrogen (total Kjehldal, $NH_4$, $NO_3$), and total phosphorus. The solids retention time, for example, can vary from an average of 10 days for conventional activated sludge to an average of 25 days for extended aeration.

**activated sludge process** An aerobic, suspended-growth, biological process that uses microorganisms in settled sludge to remove soluble organic matter from wastewater. *See* Figure A-04. It consists of an aeration basin followed by a secondary clarifier (a type III sedimentation tank). The mixed liquor formed by the wastewater and the sludge is aerated, agitated, and then allowed to settle; part of the settled sludge is wasted and part is returned to the aeration basin. It was reportedly developed in Manchester (England) in the early

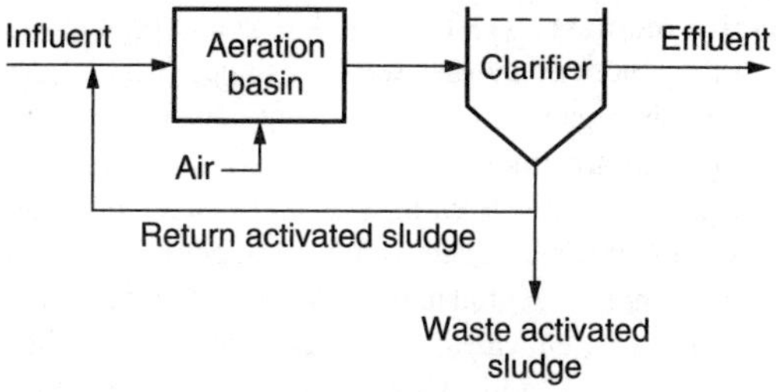


**Figure A-04.** Activated sludge process.

twentieth century (1914?) by Arden and Lockett on a fill-and-draw basis. However, it is also reported that Clark and Gage investigated the process in 1913 at the Lawrence Experiment Station in Massachusetts (Metcalf & Eddy, 2003). The organisms use the substrate (organic matter) partly for energy—converting it to stable end products, water and carbon dioxide—and partly for the synthesis of biomass. An important accompanying operation is the handling and disposal of the excess sludge. *See* activated sludge variations.

**activated sludge process control** Operation of an activated sludge plant is controlled to maintain adequate dissolved oxygen concentrations in the aeration basin and to regulate the quantities of return sludge and waste activated sludge through such parameters as oxygen uptake rate, solids retention time, and mixed liquor suspended solids.

**activated sludge process design** Activated sludge design considerations include raw waste characteristics, effluent limitations, reactor type (or process variation), kinetic relationships, solids retention time, food-to-microorgansim ratio, oxygen requirement, sludge production, nutrient requirement, and settling characteristics.

**activated sludge process modification** *See* activated sludge variations.

**activated sludge process safety factor** A factor used in the design and operation of activated sludge and other biological treatment processes to guard against system failure. The design solids retention time ($\theta_d$) is usually taken as a product of the safety factor (SF) and the minimum or critical SRT ($\theta_{min}$):

$$\theta_d = SF \cdot \theta_{min} \qquad \text{(A-26)}$$

In practice, SF varies between 2 and 20.

**activated sludge production** Activated sludge production (P, kg/day) depends on the substrate, the design and operating parameters of the system, and environmental conditions. It is usually estimated as a function of the observed yield factor for total solids ($Y$, mg TSS/mg $BOD_5$ removed), the rate of substrate removal ($R$, mg/L/day), and the volume ($V$, $m^3$) of the aeration basin:

$$P = -YRV \qquad \text{(A-27)}$$

(TSS = total suspended solids, $BOD_5$ = 5-day biochemical oxygen demand.)

**activated sludge recycle flows** Flows returned directly to the aeration tank or to preceding units from the mixed liquor, the clarifiers, digester supernatant, centrate or filtrate from dewatering units, backwash water from filtration units, and water from odor-control scrubbers.

**activated sludge simulation models** Computer programs that are developed to evaluate the activated sludge process, design an activated sludge system, or evaluate the performance or capacity of a given plant. *See* Activated Sludge Model No. 1 and Activated Sludge Model No. 2.

**activated sludge settleability** A characteristic of activated sludge that depends on plant operating parameters (e.g., F:M ratio and solids retention time) and affects plant performance. *See also* sludge bulking and pin floc.

**activated sludge system** The basic activated sludge system includes (a) one or more completely mixed or plugflow basins for the aeration of the mixed lquor suspended solids (MLSS), (b) equipment to disperse atmospheric air or pure oxygen to the aeration basins, (c) equipment for keeping the MLSS in suspension, (d) a clarifier for sedimentation of the MLSS, (e) an apparatus to collect settled solids and recyle them to the aeration basins, and (f) equipment for sludge wasting from the clarifier underflow or from the mixed liquor tank. *See* Figure A-04.

**activated sludge treatment parameters** *See* activated sludge parameters.

**activated sludge variations** Flowsheet variations of the activated sludge (AS) process include: complete mix AS, contact stabilization, conventional AS, deep-shaft AS, extended aeration, fixed-film AS, high-rate AS, modified aeration, plug flow AS, powdered activated-carbon AS, pure-oxygen AS, sequencing batch reactor, single-stage AS, step-feed aeration, and tapered aeration. The difference among most of these flowsheets is the solids retention time or the food-to-microorganism ratio.

**activated sludge wasting** The practice of wasting excess sludge to maintain a given solids retention time (SRT) in an activated sludge system. Under steady-state conditions, the flow rate of waste sludge ($Q_w$, $m^3$/day) is a function of the SRT ($\theta$, days), the volume of the aeration basin ($V$, $m^3$), the concentration of mixed liquor suspended solids ($X$, mg/L), the flow rate of the effluent ($Q_e$, $m^3$/day), the suspended solids concentrations in the return sludge ($X_r$, mg/L) and in the effluent ($X_e$, mg/L):

$$Q_w = (VX - \theta Q_e X_e)/\theta X_r \qquad \text{(A-28)}$$

**activated sludge with internal fixed packing** Any of a number of variations of the activated sludge process incorporating a fixed packing material in the aeration tank to increase performance. *See* e.g., Bio-2-Sludge®, BioMatrix®, and submerged rotating biological contactors.

**activated solids** Same as activated sludge (APHA, 1981).

**activating agent** A surface-active additive that reacts at the surface of particles to promote the adsorption of an another foaming agent (a collector) and their flotation. Also called an activator or activating reagent. *See also* collecting agent.

**activating reagent** Same as activating agent.

**activation** (1) The process of making a substance active or more reactive; for example, the generation of activated sludge and the production of activated carbon (by oxidation to develop the internal pore structure) or activated silica. *See also* chemical activation, physical activation. (2) The activated sludge process. (3) The testing of a chemical with and without enzymes that can produce a mutagenic metabolic product. *See also* Ames test.

**activation energy** The energy requirement for a process or reaction to start, a factor in the van't Hoff–Arrhenius equation. It is used to express the effect of temperature increases on a reaction rate. Values of activation energy for reactions in aqueous solutions vary from 1 to 30 kcal per mole. For example, the activation energy of the BOD reaction is about 7900 calories from 15 to 30°C; for the hydrolysis of carbon dioxide ($CO_2$), it is approximately 13 kcal/mole.

**activation polarization** A corrosion phenomenon whereby the activation energy of the reacting elements controls the rate of the reaction. *See also* activation energy, concentration polarization, corrosion, polarization, polarization curve.

**activation technique** *See* activation (3).

**activator** (1) A chemical added to a pesticide to increase its activity. (2) Same as activating agent.

**Activator® plant** A proprietary package wastewater treatment plant manufactured by Pollution Control, Inc.

**Activator III product** An oil recovery product of Sybron Chemicals, Inc.

**active biomass** (1) The live and viable organisms of the biomass in wastewater or sludge, which also contain organic solids that do not contribute to biological treatment activities. Volatile suspended solids (VSS) is often used as a practical and convenient measure of active biomass, but it also contains nonbiodegradable VSS from the influent, dead cells, enzymes, and other agents. Adenosine triphosphate and dehydrogenase measurements provide a more precise indication of biomass activity. *See also* biomass yield, cell debris, cell lysis, net biomass, net biomass yield, observed yield, true yield. (2) WEF & ASCE (1991) present the following equation to estimate the active mass applicable to denitrification rates:

$$M_{Xa} = M_b Y_s \theta/(1 + K_d \theta) \qquad \text{(A-29)}$$
$$= (1 - F_{us} - F_{up})\, M_T Y_s \theta/(1 + K_d \theta)$$

where $M_{Xa}$ = mass of active volatile solids, mg; $M_b$ = mass of influent biodegradable COD, mg; $Y_s$ = specific yield coefficient, COD in VSS/mg COD oxidized = 0.45; $\theta$ = solids retention time, days; $K_d$ = decay coefficient = 0.24/day; $F_{us}$ = nonbiodegradable soluble COD fraction of total COD = 0.05 – 0.08 mg/mg; $F_{up}$ = nonbiodegradable particulate COD fraction of total COD = 0.13 – 0.40 mg/mg; $M_T$ = mass of total influent COD, mg.

**active conservation storage** Storage of water for municipal or industrial water supply, irrigation, or hydroelectric power.

**active humoral immunity** Immunity acquired naturally through infection or artificially by inoculation.

**active immunity** Immunity of an organism resulting from its own production of antibodies. *See also* active humoral immunity, antibody-mediated immunity or humoral immunity, and passive immunity.

**active ingredient** In any pesticide product, the component that kills, or otherwise controls, target pests. Pesticides are regulated primarily on the basis of active ingredients.

**active intrusion** *See* saltwater intrusion.

**active life** The period of operation of a water, wastewater, solid waste, or other facility, beginning with start up and ending with decommission, abandonment, or closure. *See also* economic life, design life, and useful life.

**active mass** (1) The effective concentration of a substance in a solution. (2) *See* active biomass (2).

**active mine** An underground (uranium or other) mine that is being ventilated to allow workers to enter the mine for any purpose (EPA-40CFR61.21-a).

**active mining area** A place on or beneath land, used or disturbed in activities related to extraction, removal, or recovery of coal from its natural deposits or metal ore. This term excludes coal preparation plants and associated areas, post-mining areas, and any area of land on or in which grading has been completed to return the earth to desired contour and reclamation work has begun (EPA-40CFR434.11-b and 440.132-a).

**active nitrogen** Any of the reactive forms of nitrogen, collectively designated by $NO_x$, i.e., nitric oxide (NO) and nitrogen dioxide ($NO_2$), resulting from combustion, soil emissions, lightning, or the reaction of nitrous oxide ($N_2O$) with excited oxygen atoms.

**active portion** Areas of a water or wastewater facility where such operations as treatment, storage, and disposal are carried out.

**active power** The average instantaneous power ($P$, watts) over one period, a function of the number of phases ($N$), the root mean square voltage ($V$, volts), the root mean square current intensity ($I$, amperes), and the power factor ($F$):

$$P = N^{0.5} VIF \quad \text{(A-30)}$$

**active storage** (1) The portion of a pool, pond, or other detention/retention facility actually used for storage and subsequent release of stormwater, as opposed to dead storage or permanent pool. Extended detention dry ponds have only an active storage zone, whereas a wet pond has both an active storage zone and a permanent pool. The active storage zone is also called working volume, as opposed to permanent volume. (2) In water supply, active storage in a reservoir is required to meet water demand. In a multiple-purpose reservoir, active storage corresponds to all the intended uses such as drinking water supply, irrigation, flood control, hydroelectric power generation, and recreation. *See* reservoir storage and useful storage (1).

**active transport** An energy-expending mechanism by which a cell moves a chemical across the cell membrane from a point of lower concentration to a point of higher concentration, against the diffusion gradient.

**active water** Water that has corrosive characteristics. *See* aggressive water.

**activity** (1) The chemical activity ($\{A\}$) of an ion or species is the product of its molar concentration ($[C]$) by an activity coefficient ($\gamma$):

$$\{A\} = \gamma \cdot [C] \quad \text{(A-31)}$$

The concept of activity allows the differentiation between the effective concentration of a species in solution and the actual, measured concentration. (Such differences are due to solute–water interactions and specific ionic interactions.) In aqueous solutions, the activity of water is taken as unity. The activity coefficient, which depends on the ionic strength of a solution, approximates unity for dilute solutions, but may be less than 0.1 for seawater. In general, it is greater than 1.0 for nonelectrolytes and less than 1.0 for electrolytes. Activities and activity coefficients are used, instead of concentrations, in the expressions of chemical potential and equilibrium constant. *See* the Davies equation and the Guntelberg approximation for two estimates of the activity coefficient. *See also* fugacity, electrochemical activity. (2) The nuclear transformation of a radioactive substance within a given time period.

**activity coefficient** *See* activity.

**activity product** The product ($K_{ap}$) of the activities $\{A^+\}$ and $\{B^-\}$ of the two ionic species resulting from the dissociation of a molecule ($AB$):

$$AB \leftrightarrow A^+ + B^- \quad \text{(A-32)}$$
$$K_{ap} = \{A^+\}\{B^-\}$$

*See also* common ion effect, equilibrium constant, ion product, solubility product, solubility product constant

**activity ratio diagram** A plot versus pH of the ratios between the activities of the soluble and solid species in an aqueous solution. It indicates which solid species predominates as a stable phase for selected conditions. For example, the activity ratio diagram for Fe(II) in a $10^{-3}$ M carbonate system shows that below pH ≈ 10, the carbonate [$FeCO_3$(s)] predominates and controls the solubility, whereas above that pH it is the hydroxide [$Fe(OH)_2$(s)] that is more stable. *See also* solubility diagram.

**Activol™** A proprietary product of Probiotic Solutions used to emulsify grease in wastewater.

**ACT™ technologies** Proprietary aeration techniques designed by Aeration Industries, Inc.

**actual chlorine** *See* available chlorine.

**actual evaporation ($e_a$)** The volume of water that evaporates in a given time period, depending on measurements of cumulative precipitation ($P_c$), cumulative runoff ($R_c$), and cumulative soil moisture ($M_c$) over the period. It is different from pan evaporation. The following formula is often used, with all the quantities usually expressed as a length or depth, e.g., in inches or centimeters:

$$e_a = P_c - R_c + M_c \quad \text{(A-33)}$$

*See also* pan evaporation, soil moisture loss equation, and actual evapotranspiration.

**actual evapotranspiration** The evapotranspiration that results from actual field conditions (climate and soil moisture) as compared to the evapotranspiration potential.

**actual groundwater velocity** The average linear velocity ($V_a$) of groundwater as measured in the field; the ratio of the volume of groundwater ($v$) through a formation in unit time to the product of the cross-sectional area ($A$) by the effective porosity ($\theta$):

$$V_a = v/(At\theta) = Q/(A\theta) \quad \text{(A-34)}$$

where $t$ is the period of flow and $Q$ is the average discharge. Also called effective, field or, true groundwater velocity, and seepage velocity. *See also* hydraulic conductivity or coefficient of permeability.

**actual oxygen transfer rate** The rate of oxygen transfer ($A$, kg $O_2$/hr) of an aeration device, determined under field conditions for wastewater treatment. It affects the sizing of aeration equipment and depends on a number of factors such as wastewater characteristics, equipment characteristics, and temperature:

$$A = K(\beta C_w - C_L) \quad \text{(A-35)}$$

$$A = S[(\beta C_w - C_L)/C_{20}](1.024^{T-20})(\alpha)(F) \quad \text{(A-36)}$$

where $K$ = a transfer coefficient, per hour; $\beta$ = an oxygen saturation coefficient between 0.7 and 0.98, commonly 0.95 for wastewater; $C_w$ = average DO saturation concentration (mg/L) in clean water at temperature $T$°C and altitude H; $C_L$ = operating oxygen concentration, mg/L; $S$ = standard $O_2$ transfer rate in clean water at 20°C and 0 dissolved oxygen (DO), kg $O_2$/hr; $C_{20}$ = DO saturation concentration in clean water at 20°C and 1.0 atm = 9.2 mg/L; $T$ = operating temperature of the liquid, °C; $\alpha$ = oxygen transfer correction factor for the wastewater = 0.3 – 1.2; $F$ = fouling factor = 0.65 – 0.90. The difference ($\beta C_s - C_t$) represents the dissolved oxygen deficit. *See also* field oxygen transfer rate and standard oxygen transfer rate.

**actual residence time** A parameter ($T_a$, hr) used in the design and analysis of bulk media filters or biofilters; the product of the total volume of the filter bed ($V$, $m^3$) and the porosity of the bed ($\alpha$), divided by the volumetric flow rate ($Q$, $m^3$/hr):

$$T_a = V\alpha/Q \quad \text{(A-37)}$$

Also called true residence time. *See also* empty bed residence time, elimination capacity (rate).

**actual specific capacity** *See* specific capacity.

**actuator** A device placed inside the body of a valve or gate to automate it by acting on its throttling element. *See* electrical actuator, pneumatic actuator.

**Acumem® products** A line of reverse osmosis products of NWW Acumem, Inc.

**Acumer® polymer** A proprietary polymer produced by Rohm & Haas for use in water treatment installations.

**acute** Pertaining to a severe effect of a chemical or infectious agent over a short period of time; used to describe brief exposures and effects that appear promptly after exposure.

**Acutec** Equipment manufactured by Wallace & Tiernan for gas detection.

**acute delayed neurotoxicity** A prolonged, delayed-onset locomotor ataxia resulting from a single administration of the test substance, repeated once if necessary (EPA-40CFR798.6540).

**acute dermal $LD_{50}$** A statistically derived estimate of the single dermal dose of a substance that would cause 50% mortality to the test population under specified conditions (EPA-40CFR152.3-c).

**acute dermal toxicity** The adverse effects occurring within a short time of dermal application of a single dose of a substance or multiple doses given within 24 hours (EPA-40CFR798.1100-1).

**acute exposure** A single exposure, or multiple exposures within a short time, to a toxic substance that result in death or severe biological harm. Acute exposures are usually characterized as lasting no longer than a day, as compared to longer, continuing exposure over a period of time.

**acute gastroenteritis** A sporadic or epidemic inflammation of the stomach and intestines caused by a bacterial or viral infection characterized by one or more of the following symptoms: diarrhea, vomiting, dehydration, and septicemia.

**acute gastroenteritis of undetermined etiology** A waterborne illness whose causative agent has not been identified, usually for lack of laboratory analysis or insufficient epidemiologic data. More likely caused by a virus than a bacterial or protozoan agent.

**acute hazard or toxicity** *See* health hazard.

**acute hazardous industrial waste** A category of hazardous wastes that are dangerous to humans and animals at very low concentrations; usually invisible and difficult to detect, e.g., radioactive, biological, and asbestos wastes.

**acute hemorrhagic conjunctivitis** A disease caused by enteroviruses.

**acute lethal toxicity** The lethal effect produced on an organism, within a short period of time (days) of exposure to a chemical (EPA-40CFR797.1440-2).

**acutely toxic effects** Effects of a chemical substance that kills within a short period of time (usually 14 days) at least 50% of the exposed mammalian test animals following: (a) oral administration of a single dose of 25 mg or less per kg of body weight ($LD_{50}$), or (b) dermal administration of a single dose of 50 mg or less per kg of body weight ($LD_{50}$), or (c) administration of the substance for 8 hours or less by continuous inhalation at a steady concentration in air at 0.5 mg/L or less ($LC_{50}$) (EPA-40CFR721.3).

**acute-to-chronic ratio (ACR)** The ratio of the median lethal concentration to the no-observed-effect concentration; a parameter used in toxicity tests to estimate the concentration of toxicant that is safe for chronic or long-term exposure of a test organism. *See* toxic units, toxicity terms for detail.

**acute toxic exposure** *See* acute toxicity.

**acute toxicity** (1) The ability of a substance to cause poisonous effects resulting in a significant response soon after a single exposure or dose (e.g., death or severe biological harm within 48–96 hours). Also, any severe poisonous effect resulting from a single short-term exposure to a toxic substance. (2) A measure of toxicological responses that result from a single exposure to a substance or from multiple exposues within a short period of time (typically several days or less). Specific measures of acute toxicity used with the hazardous ranking system (HRS) include lethal dose-50 ($LD_{50}$) and lethal concentration-50 ($LC_{50}$), typically measured within a 24-hour to 96-hour period (EPA-40CFR300-AA). *See also* toxicity terms, chronic toxicity.

**acute violation** The failure of a public water supply system to meet a drinking water quality standard, which results in a public health risk, as specified by the USEPA.

**acute toxic unit** Same as toxic unit-acute. *See* toxic units.

**Adams, Julius W.** Designer of the first comprehensive sewerage system for Brooklyn, N.Y. in 1857, one of the significant achievements of the Great Sanitary Awakening.

**adaptation** (1) Changes in an organism structure or habits that help it adjust to its surroundings and make it fit for reproduction or survival. (2) The process by which a substance induces the synthesis of any degradative enzymes necessary to catalyze the transformation of that substance (EPA-40CFR796.3100-i).

**Addigest® plant** A package wastewater treatment plant manufactured by Smith & Loveless, and using the extended aeration process.

**addition compound** Same as adduct.

**additive** A chemical substance added to a process for a specific purpose such as coagulation or recarbonation.

**additive effect** Combined effect of two or more chemicals equal to the sum of their individual effects.

**address matching** A procedure used in planning studies, relating street addresses to such attributes as meter locations, census tracts, buildings, or emergency response units. *See also* water meter address matching.

**adduct** The product of a reaction between two or more independently stable compounds (e.g., a reactive chemical and a protein) by means of van der Waals forces or covalent bonds. Also called addition compound.

**adenosine diphosphate (ADP)** Chemical formula: $C_{10}H_{16}N_5O_9P_2$. *See* adenosine triphosphate.

**adenosine monophosphate (AMP)** Chemical formula: $C_{10}H_{16}N_5O_5P$. *See* adenosine triphosphate.

**adenosine triphosphate (ATP)** Chemical formula: $C_{10}H_{16}N_5O_{13}P_3$. A high-energy compound associated with energy transfer processes in all living cells and consisting of ribose sugar ($C_5H_{10}O_4$), nitrogen-containing adenine ($C_5H_6N_5$), and phosphate. The energy generated by biological activity is stored as ATP, which can be transformed during hydrolysis and redox reactions into lower-energy forms [adenosine diphosphate (ADP) and monophosphate (AMP)], thus releasing metabolism and growth energy. Aerobic and anaerobic microorganisms active in wastewater treatment contain about 1–2 mg ATP per gram of cell. The sum of these three forms is called total cellular adenylate. ATP activity measurements are used as an indicator of active biomass in biological processes. *See also* dehydrogenase, enzyme.

**adenoviridae** The family of organisms that include the adenoviruses.

**adenovirus** A nonenveloped, double-stranded DNA virus in a protein shell of about 70–110 nm in diameter, widespread in nature, infecting birds and mammals, found in large numbers in human feces, classified as an emerging clinical pathogen. Six human adenovirus species (A through F) cause a variety of diseases: acute febrile pharyngitis, acute hemorrhagic cystitis, acute respiratory disease, common cold, epidemic keratoconjunctivitis, gastroenteritis, hepatitis, meningoencephalitis, myocarditis, pertussis-like syndrome, pharyngoconjunctival fever, and pneumonia. There are documented outbreaks of conjunctivitis and pharyngitis linked to recreational waters and gastroenteritis through consumption of contaminated groundwater. Adenoviruses have an incubation period of 8–10 days and are effectively inactivated by drinking water chemical disinfectants such as chlorine, chlorine dioxide, and ozone, although some relative resistance to ordinary disinfecting agents may be occasionally noted. The USEPA requires an ultraviolet light dose of 186 $mJ/cm^2$, not including the validation factor, for 4-log inactivation of adenoviruses.

**adenovirus detection methods** *See* infectivity assay, molecular virus detection, integrated virus detection.

**adenylate** One of the three coenzymes ADP, AMP, and ATP. *See* adenosine triphosphate. The adenylate energy charge (AEC) is a weighted ratio of these three forms that is used to assess the physiological and nutritional status of microorganisms:

$$AEC = (ATP + 0.5\ ADP)/(ATP + ADP + AMP) \quad (A\text{-}38)$$

The higher the AEC, the more active the community.

**adenylate energy charge (AEC)** *See* adenylate.

**adfluvial** Characteristic of fish that migrate between lakes and rivers.

**adhesion** (1) Attachment of microorganisms or soluble substances to a solid surface during such processes as activated carbon adsorption and biofilm formation; it is the fourth and final step in the adsorption process, in which adsorbate and adsorbent form a bond. *See also* bulk solution transport, film diffusion, and internal (or pore) transport. (2) Attachment of suspended particles to the surface of a filtering medium, a major mechanism of removal of particulate matter within a granular filter. *See also* filtration mechanisms.

**adhesion water** Subsurface water that adheres to soil particles after drainage by gravity. Also called pellicular water or adhesive water, it is found between the soil and gravity subzones. It can be absorbed by roots and is subject to evapotranspiration. *See* subsurface water.

**adhesive water** Adhesion water.

**ADI** Acronym of acceptable daily intake.

**adiabatic** Pertaining to a thermodynamic process that occurs without any heat added to or withdrawn from the system concerned.

**adiabatic expansion** The increase in volume of an air mass without gain or loss of heat.

**adiabatic lapse rate** The constant rate at which temperature decreases with increasing altitude, e.g., about 1°C per 100 meters in dry air.

**adiabatic process** A process in which a component does not interact with surrounding components because of a temperature difference between them.

**ADI-BVF® reactor** A proprietary anaerobic lagoon of earthen and concrete construction, developed by ADI Systems, Inc. of Frederickton, NB (Canada), and consisting of a primary reaction zone, a secondary reaction zone, and a clarification zone. Wastewater enters at the bottom and exits at the top, just below the scum layer and the floating geomembrane cover. Sludge is recyled and wasted from the recycle line.

**adipate** A class of synthetic organic chemicals that are salts or esters of adipic acid ($C_6H_{10}O_4$), a white, crytalline, slightly soluble solid. Adipates have been found in treated wastewater effluent from meat processing plants. *See* di(2-ethylhexyl)adipate.

**adj $R_{Na}$** Abbreviation of adjusted sodium adsorption ratio.

**adjudication** A court proceeding to determine all rights to the use of water in a stream or aquifer.

**AdjustAir® diffuser** A proprietary adjustable coarse bubble diffuser manufactured by the FMC Corp.

**adjusted acceptable daily intake (AADI)** A parameter used in epidemiological/toxicological studies to establish limiting concentrations of contaminants in drinking water. Expressed in mg/L, it is equal to the No-observed-adverse-effect level (NOAEL, mg/kg/day) times the average weight of an adult (70 kg) divided by a dimensionless safety or uncertainty factor ($\varphi$) and by the assumed daily intake of water (2 liters per day):

$$AADI = (NOAEL)(70)/[(\varphi)(2)] \quad (A\text{-}39)$$
$$= 35(NOAEL)/(\varphi)$$

It may also be used to compute maximum contaminant levels (MCLs), recommended maximum contaminant levels (RMCLs), or maximum contaminant level goals (MCLGs), e.g.:

$$RMCL = AADI - \text{contributions from food and air} \quad (A\text{-}40)$$

**adjusted sodium adsorption ratio (adj $R_{Na}$)** A modification of the standard sodium adsorption ratio to account for changes in calcium solubility in soil water; the concentration of the calcium ion is adjusted in function of the ratio of bicarbonate to calcium ($HCO_3^-/Ca^{2+}$) and of the salinity of the applied water:

$$\text{adj } R_{Na} = [Na^+]/\{([Ca_x^{2+}] + [Mg^{2+}])/2\}^{0.5} \quad (A\text{-}41)$$

with all cation concentrations expressed in meq/L. Published tables are available for the adjusted calcium concentration $[Ca_x^{2+}]$; *see,* e.g., Metcalf & Eddy, 2003. *See also* sodium adsorption ratio.

**Adjust-O-Pitch® propellers** Adjustable mixing propellers manufactured by Walker Process Equipment Co.

**administrative order** A legal document issued by the USEPA to direct an individual, business, or other entity to take corrective action or refrain from an activity. It describes the violations and actions to be taken, and can be enforced in court. Such orders may be issued, for example, as a result of an administrative complaint whereby the respondent is ordered to pay a penalty for violations of a statute.

**administrative order on consent** A legal agreement signed by the USEPA and an individual, business, or other entity through which the violator agrees to pay for correction of violations, take the required corrective or cleanup actions, or refrain from an activity. It describes the actions to be taken, may be subject to a comment period, applies to civil actions, and can be enforced in court.

**admixture** A material, substance or product that is added to form a mixture; e.g. any substance other than cement, aggregate, or water added for the formation of concrete.

**Ad-Ox scrubber** An odor control product of Purafil, Inc.

**ADP** Acronym of adenosine diphosphate.

**Adpec filter** A horizontal vacuum filter of Komline-Sanderson.

**Adsep™ process** A proprietary process designed by the U.S. Filter Corp. for the separation of organic and inorganic compounds.

**Adsolv®** system Equipment manufactured by Ray-Solv, Inc. for the control of volatile organic compounds by activated carbon treatment.

**adsorbable organic halogen (AOX)** A surrogate measurement for the total amount of organic compounds that contain one or more halogen atoms in a sample of raw or treated water; often used in wastewater testing to indicate the overall level of bromine, chlorine, fluorine, and iodine. *See* halogen-substituted organic material, total organic halide analysis, and total organic halogen for more detail.

**adsorbate** The material (e.g., odor- and taste-producing ions and molecules) being removed from solution by the adsorption process.

**adsorbent** The material that is responsible for removing the undesirable substance in the adsorption process or exchanged for an unwanted ion in ion exchange.; also called presaturant ion. Common adsorbents in water and wastewater treatment include activated carbon, activated alumina, clays, resins, metal carbonates, hydroxides, and oxides.

**adsorbent dose** The quantity (or concentration) of adsorbent in gram per liter required to achieve a given pollutant removal; the reciprocal of the specific volume:

$$M/V = (C_0 - C_e)/q_e \qquad \text{(A-42)}$$

where $M$ is the mass of adsorbent (grams), $V$ the volume of liquid in the reactor (liters), $C_0$ the initial concentration of adsorbate (mg/L), $C_e$ the final equilibrium concentration of adsorbate after adsorption (mg/L), and $q_e$ the adsorbent phase concentration after equilibrium (mg adsorbate per gram adsorbent).

**adsorbent installation** A pollution control facility that uses the adsorption process.

**adsorbent resin** Any special resin that is used for the removal of selected organic compounds, e.g., styrene divinylbenzene (2 rings) and phenol-formaldehyde (3 rings), which do not have any functional groups beyond those that are included in a matrix.

**adsorbent-weighting agent** A coagulant aid used in the treatment of waters that yield light floc that is difficult to settle. Such waters may be high in color but low in mineral content. Weighting agents include activated carbon, bentonite clay, limestone, and powdered silica. Also called weighting agent.

**ADSORBIA™ GTO™** A water treatment system of Dow Water Solutions that incorporates arsenic removal media.

**adsorption** (1) Generally, the rapid and reversible attraction and retention of atoms, ions, molecules, or dissolved substances onto the surface of another substance; for example, the process by which chemicals are held on the surface of a mineral or soil particle. Adsorption involves the accumulation of an adsorbate at the interface between a liquid and a solid adsorbent or between a gas and a solid. (2) An advanced method of treating wastes in which activated carbon removes organic matter from wastewater. (3) The process of concentration of a gas or soluble susbstance on a surface. (4) A form of coprecipitation of secondary elements (e.g., heavy metals, radionuclides, and viruses) along with the main targets of chemical precipitation (e.g., calcium, magnesium, and organic contaminants) during water treatment. It involves the attachment of contaminants on the surface of precipitating particles. *See also* absorption, adhesion, adsorbent dose, adsorption constant, adsorption function, adsorptive capacity, adsorption zone, biological activated carbon, biological fluidized-bed reactor, breakthrough, breakthrough adsorption capacity, bulk solution transport, carbon adsorption, chemical adsorption, chemical-physical treatment, chemisorption, countercurrent adsorption, desorption, distribution coefficient, distribution ratio, electrostatic adsorption, exhaustion, film diffusion transport, fixed-bed adsorber, Frumkin equation, Gibbs equation, hydrodynamic boundary layer, hydrogen bonding, hydrophobic bonding, hydrophobic effect, inclusion, interaction coefficient, interfacial Gibbs free energy, interfacial tension, intraparticle diffusion, intraparticle transport, isotherm, linear isotherm, London–van der Waals dispersion interaction, macrotransport, mass transfer zone, microtransport, occlusion, operating line, phenol value, physical adsorption, polarity, pore diffusion, pore transport, pulse-bed adsorber, rate limiting step, sloughing, solid-solution formation, solute distribution parameter, solution force, sorbate, sorption, specific volume, surface film, surface tension, time to breakthrough, upflow expanded-bed mode, van der Waals attraction, Weber–Fechner law.

**adsorption affinity** The tendency of a cation to replace another cation on an ion exchange site. It varies with charge density, i.e., total charge and size of the cation. The following relation indicates the adsorption affinities of some cations in descending order (Maier et al., 2000):

$$Al^{3+} > Ca^{2+} = Mg^{2+} > K^{+} = NH_3^{+} > Na^{+} \quad \text{(A-43)}$$

*See also* cation exchange capacity and isomorphic substitution.

**adsorption bed exhaustion** The condition of a bed of activated carbon or other granular material when the concentration of a substance (e.g., a pollutant) has reached a high level, such as 95% of the influent concentration. Complete exhaustion is when effluent concentration = 100% influent concentration. *See also* breakthrough.

**adsorption capacity** Same as adsorptive capacity.

**adsorption clarifier** A water treatment device that provides flocculation while the coagulated liquid flows through a buoyant bed of granular plastic media, which also captures the flocculated solids.

**Adsorption Clarifier™** A proprietary flocculator/clarifier manufactured by Wheelabrator Engineered Systems, Inc.; it uses an upflow medium.

**adsorption constant** A coefficient that multiplies the equilibrium concentration of the adsorbate in the Frumkin equation or the Langmuir isotherm.

**adsorption-destabilization** One of the predominant mechanisms observed in the coagulation of particles using alum in water treatment; the particles lose their stability, stick together, and become settleable. The extent of this phenomenon depends on pH and alum dosage (C, mg/L), e.g., for $5.0 < \mathrm{pH} < 7.0$ and $0.5 < C < 40$. *See also* bridging (1), double-layer compression, restabilization, and sweep-floc coagulation.

**adsorption-elution** A procedure used in laboratory analysis for the separation and concentration of enteric viruses in water and wastewater samples. The procedure includes the use of chemicals to promote virus adsorption, pressure filtration through microporous cellulose nitrate or fiberglass, pumping the filtrate through a glycine-containing buffer or other eluent, and virus concentration by repeating these steps or by chemical precipitation and centrifugation.

**adsorption equilibria** Adsorption is a reversible process but equilibrium is reached rapidly. Adsorption isotherms (e.g., Freundlich's and Langmuir's) define equilibrium conditions between adsorbed and dissolved states.

**adsorption filtration** One of the mechanisms that contribute to the removal of particulate matter within a granular filter. *See* chemical adsorption (bonding and chemical interaction), physical adsorption, and surface biological adsorption.

**adsorption front** The leading edge of the adsorption zone in a fixed-bed adsorber. As it progresses through the column, pollutant removal occurs within the zone and adsosrptive capacity is consumed, leaving a saturated zone behind.

**adsorption function** An equation that is used to compute the adsorbent phase concentration and to develop adsorption isotherms, following an adsorption test. It expresses the relative amount adsorbed at equilibrium ($q_e$, mg of adsorbate per gram of adsorbent) as a function of the initial concentration of adsorbate ($C_0$, mg/L), the final equilibrium concentration in solution ($C_e$, mg/L), the volume of solution ($v$, L), and the mass of adsorbent ($m$, grams):

$$q_e = v(C_0 - C_e)/m \quad \text{(A-44)}$$

**adsorption heat** *See* heat of adsorption.

**adsorption isotherm** A graphical representation of the equilibrium relationship, based on experimental measurements, between adsorbate, adsorbent, and solution at a given temperature. For example, an activated carbon adsorption isotherm may be a diagram that shows the equilibrium distribution of a contaminant (or adsorbate) between the adsorbed or solid phase (or adsorbent) and the aqueous phase (or solution). However, other substances can affect this equilibrium by competing for the adsorption sites. Adsorption isotherms usually relate quantities of adsorbate per unit of adsorbent to the equilibrium concentration of adsorbate in solution. They do not indicate the rate of reaction. *See also* adsorption function, adsorption test, BET, and Freundlich, Langmuir, and linear isotherms.

**adsorption pore** Pore of an adsorbent or a pore that has adsorption capacity. *See* macropore, mesopore, micropore, transport pore.

**adsorption potential** A parameter ($\varepsilon$) used in the Dubinin–Raduskevich equation that correlates the isotherms of gas-phase volatile organic compounds (AWWA, 1999):

$$\varepsilon = RT \ln(P_s/P) \quad \text{(A-45)}$$

where $R$ is the universal gas constant, $T$ is temperature in °K, $P_s$ is the saturation vapor pressure of the volatile organic compound (VOC) at temperature $T$, and $P$ is the partial pressure of the VOC in the gas.

**adsorption process** A water or wastewater treatment method that removes substances (adsorbates) from solution through the mechanisms of bulk solution transport, film diffusion, pore transport, and attachment to sites on the outer surface or in the pores of an adsorbing material (adsorbent).

**adsorption rate** (1) Adsorption proceeds at a rate proportional to the square root of the initial solute concentration, inversely proportional to the square of particle size, and corresponding to a temperature quotient ($Q_{10}$) of about 1.3 or an activation energy of 4270 calories per mole. Actual uptake of a solute varies as the square root of time of exposure. The generalized purification equation is sometimes used to approximate adsorption rate (Fair et al., 1971):

$$C/C_0 = 1 - (1 + 3.6\, t)^{-0.25} \quad \text{(A-46)}$$

where $C$ = concentration of the substance being adsorbed at time $t$; $C_0$ = initial concentration; i.e., at $t = 0$; $t$ = time. *See also* adsorptive capacity. (2) Fick's first law of diffusion may also be used to formulate the rate of adsorption (Droste, 1997):

$$R = KA\,(C^* - C)/(V \cdot x) \quad \text{(A-47)}$$

where $R$ is the removal rate (mass/volume/time), $K$ the resistance to mass transfer, $A$ the adsorption area, $C^*$ the concentration when the liquid is in equilibrium with the adsorbent, $C$ the concentration in the liquid, $V$ the volume of the reactor, and $x$ the nominal distance.

**adsorption ratio ($k_d$)** The amount of test chemical adsorbed by a sediment or soil (i.e., the solid phase) divided by the amount of test chemical in the solution phase, which is in equilibrium with the solid phase, at a fixed solid/solution ratio (EPA-40CFR796.2750-vi).

**adsorption test** A test conducted to develop adsorption isotherms by exposing a given quantity of adsorbate and a given volume of solution to varying amounts of adsorbent and measuring the corresponding equilibrium concentrations. The results of the test are used in the adsorption function.

**adsorption water** Water that is held by molecular forces on the surface of solid particles.

**adsorption zone** (1) The zone in an adsorption or ion exchange column where most of the adsorption actually takes place; the concentration of adsorbate in the liquid decreases to a small value. As the operation progresses, the zone moves downward toward the breakthrough condition, preceded by an exhausted portion. *See* Figure A-05. It is practically defined as the zone where the concentration decreases from the initial concentration ($C_0$) to the breakthrough concentration ($C_b$). The depth ($D$) of the adsorption zone can be estimated from the cross-sectional area of the column and the volumes of liquid ($V_e$) and ($V_b$) processed at exhaustion and breakthrough, respectively (Droste, 1997):

$$D = (V_e - V_b)/A \quad \text{(A-48)}$$

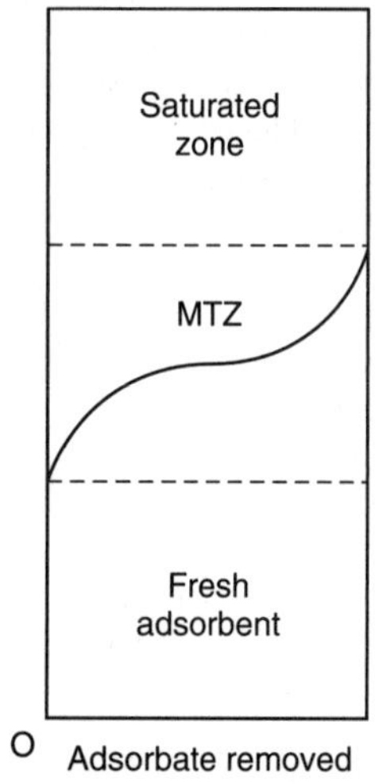


**Figure A-05.** Adsorption column (MTZ = mass transfer zone.

Actually, the depth of the zone depends also on other factors; the column approaches plugflow behavior as $D$ tends to 0. Another formula reported for the depth $D$ involves the height of the adsorption column ($Z$) instead of the cross-sectional area (Metcalf & Eddy, 2003):

$$D = 2\,Z\,(V_e - V_b)/(V_e + V_b) \qquad \text{(A-49)}$$

(2) A similar zone, in an aeration or air stripping column, where gases or liquids change from undersaturation to an equilibrium concentration. Also called mass transfer zone. *See also* adsorption front, breakthrough curve, carbon usage rate, critical depth, column utilization, time-to-bed exhaustion.

**adsorptive capacity** The ultimate amount of a solute that can be adsorbed by a unit weight of adsorbent; sometimes referred to as ultimate adsorptive capacity. For example, granular activated carbon can adsorb an approximate maximum (or saturation value) of 100 mg of phenol per gram. This can be determined from an adsorption isotherm such as Langmuir's equation. This theoretical carbon adsorption capacity differs from the breakthrough adsorption capacity. *See also* adsorption rate.

**ADV** Acronym of acid demand value.

**ad valorem tax** A tax based on the value of real or personal property. It is sometimes used to finance fully or partially improvements in water supply and wastewater disposal.

**advanced alkaline stabilization** Sludge stabilization using chemicals other than lime or the addition of other materials, e.g., cement kiln dust, fly ash, or pozzolanic materials. It includes also sludge pasteurization via an exothermic reaction of quicklime to achieve an adequate temperature.

**Advanced Fluidized Composting™** A proprietary process developed by the ERM Group combining biological and chemical sludge treatment.

**advanced oxidation process (AOP)** A process that uses such oxidants as ozone ($O_3$), hydrogen peroxide ($H_2O_2$), UV light, and hydroxyl ion ($OH^-$) to speed up or enhance oxidation–reduction reactions. These processes generate the hydroxyl radical, a strong oxidant, to achieve a higher level of oxidation and, thus, are sometimes called hydroxyl-radical-based processes. The following reactions illustrate the formation of hydroxyl free radical ($HO^{\cdot}$), the dot denoting the presence of an unpaired electron):

Ozone photolysis: $$O_3 + H_2O + UV \rightarrow O_2 + HO^{\cdot} + HO^{\cdot} \rightarrow O_2 + H_2O_2 \qquad \text{(A-50)}$$

Peroxide/ozone: $$H_2O_2 + 2\,O_3 \rightarrow HO^{\cdot} + HO^{\cdot} + 3\,O_2 \qquad \text{(A-51)}$$

Peroxide photolysis: $$H_2O_2 + UV \rightarrow H^{\cdot} + HO^{\cdot} \qquad \text{(A-52)}$$

AOPs such as UV photooxidation, UV/$H_2O_2$, and chemical oxidation with ozone and hydrogen peroxide are a common strategy considered for removing MTBE and synthetic organic chemicals (pesticides, xenobiotics, endocrine disruptors) from drinking water. AOPs are also used in the treatment of wastewater and toxic organic materials. *See also* PEROXONE.

**advanced pretreatment** A unit operation designed to remove excessive fouling materials from raw water so that the water can successfully undergo convential pretreatment and then membrane filtration. *See also* posttreatment.

**advanced primary treatment** A wastewater treatment process that enhances performance and reduces plant footprint by introducing a modification to the primary settling tank; for example, enhanced suspended solids and organic removal by chemical addition or filtration, or ballasted flocculation and the first stage of a dual-sludge process.

**advanced secondary treatment** Secondary wastewater treatment with additional or improved solids removal.

**advanced treatment** A level of water or wastewater treatment more stringent than usual or conventional. *See* advanced wastewater treatment and advanced water treatment plant.

**advanced waste treatment (AWT)** Same as advanced wastewater treatment.

**advanced wastewater treatment (AWT)** Any treatment that goes beyond the secondary or biological treatment stage, including the removal of nutrients such as phosphorus and nitrogen, a high percentage of (suspended, colloidal, and dissolved) solids, and refractory or other nonconventional pollutants. It uses such biological, chemical or physical processes as adsorption, advanced oxidation, air stripping, biological assimilation, biofiltration, breakpoint chlorination, chemical coagulation, chemical oxidation, chemical scrubbing, compost filtration, denitrification, distillation, electrodialysis, (diatomaceous earth, granular media, or membrane) filtration, ion exchange, luxury (phosphorus) uptake, microscreening, nitrification, phosphorus removal, reverse osmosis, stripping, ultrafiltration, and UV disinfection. Advanced wastewater treatment is usually recommended (a) to meet stringent discharge, reuse, or groundwater recharge requirements, (b) to limit the eutrophication of receiving waters, or (c) to remove harmful constituents for such industrial reuse as cooling or boiler makeup water. *See also* conventional treatment, Fred Hervey Water Reclamation Plant, primary treatment, secondary treatment, tertiary treatment, Water Factory 21.

**advanced water treatment plant** A plant that treats water to a level beyond that of the usual or conventional processes; e.g., disinfection for groundwater or coagulation–flocculation–sedimentation–filtration for surface water. Advanced water treatment processes include, for example, activated carbon adsorption and advanced oxidation.

**Advance® system** A proprietary chlorine gas feeder manufactured by Capital Controls Co.

**advance time** The time of travel of water along an irrigation furrow.

**advection** (1) The process of transfer of fluids (vapors or liquids) through a geologic formation in response to a pressure gradient that may be caused by changes in barometric pressure, water table levels, wind fluctuations, or infiltration. (2) Movement of dissolved or colloidal contaminants with water (longitudinal spreading) at the same velocity. *See* diffusion and dispersion.

**advection–dispersion equation** The partial differential formula that expresses the concentration of a substance or a microbe in terms of the flow and mixing conditions in a porous medium:

$$R_f\, \partial C/\partial t = \delta(\partial^2 C/\partial x^2) - V(\partial C/\partial x) \pm R_x \quad \text{(A-53)}$$

where $R_f$ = a retardation factor accounting for reversible interaction with the porous medium; $C$ = concentration of the substance or microbe in mg/l or g/m$^3$; $\delta$ = dispersion coefficient in m$^2$/sec; $t$ = time, sec; $x$ = distance in meters; $V$ = average velocity, m/s; and $R_x$ = microbial net decay, mass/time/volume. *See also* diffusion and dispersion.

**advection–dispersion model** *See* advection–dispersion equation.

**advective plume theory** One of three main hypotheses about the source of primary organic carbon and the process for its introduction into deep-sea hydrothermal vents: carbon production results from physicochemical factors such as heat energy; ionizing radiation; available supply of carbon dioxide ($CO_2$), hydrogen ($H_2$), ammonia ($NH_3$), methane ($CH_4$), nitrogen ($N_2$), etc.; advective flow for cooling; conversion of $CO_2$ to simple sugars; and production of biomass. *See also* chemoautotrophic theory and organic thermogenesis.

**advective transport** The primary transport mechanism of microorganisms in a flowing system; basically, water flow.

**Advent™ plant** A package water treatment plant manufactured by Infilco Degremont, Inc.

**adverse effect** (1) In risk analysis studies, an adverse effect refers to human health (death or disease), to the normal physiology of organisms, or, sometimes, to economic loss. (2) An apparent direct or indirect negative effect on the conservation and recovery of an ecosystem component listed as threatened or endangered.

**adverse environmental effect** Any significant and widespread adverse effect, which may reasonably be expected, to wildlife, aquatic life, or other natural resources, including adverse impacts on populations of endangered or threatened species or significant degradation of environmental quality over broad areas (Clean Air Act).

**advisory** A nonregulatory document that communicates risk information to those who may have to make risk management decisions.

**ADWF** Acronym of average dry weather flow.

**AEBR** Acronym of anaerobic expanded-bed reactor.

***Aedes*** A genus of mosquitoes that transmit a number of diseases, including dengue, filariasis, and yellow fever.

***Aedes aegypti*** The species of mosquitoes that transmit dengue, yellow fever, and other viral infections.

**Aeolian deposit** A deposit of soil by wind action from one area to another. Also called Aeolian soil, wind deposit.

**Aeolian soil** *See* Aeolian deposit.

**Aeralator unit** Equipment manufactured by General Filter Co. for the removal or reduction of iron and manganese from water.

**aerated channel** A channel equipped with aeration devices to serve as an aeration basin or to prevent deposition of solids during effluent transport.

**aerated contact bed** A biological treatment unit that consists of an aeration tank containing vertical plastic or asbestos-cement sheets on the surface of which microorganisms grow for wastewater stabilization. Diffusers at the bottom of the tank supply the necessary air, while settled wastewater flows through. Stones or other materials may be used to provide the contact surfaces. Filamentous bacteria sometimes grow over the surfaces of aerated contact beds and clog their spaces. *See* Figure A-06. Also called contact aerator or submerged-contact aerator. *See also* bacteria bed, contact bed, intermittent sand filter, nidus rack, percolating filter.

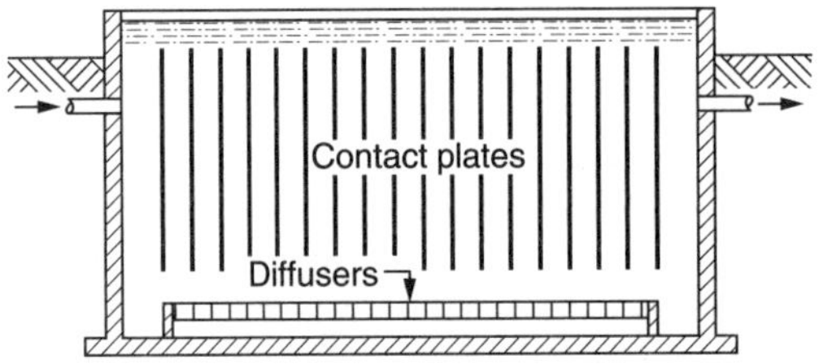


**Figure A-06.** Aerated contact bed.

**aerated facultative lagoon** *See* aerated facultative pond.

**aerated facultative pond** A shallow earthen basin used for wastewater treatment. The aeration is sufficient for aerobic treatment but not for mixing. At the bottom, it functions like a facultative lagoon. *See* facultative partially mixed lagoon for more detail.

**aerated grit chamber** A preliminary wastewater treatment unit that uses diffused air on one side to create a spiral flow pattern and a sufficient flow velocity throughout so that the inorganic solids are carried to the bottom of the aerated side and the organic matter is carried to the other side. *See also* detritus tank, horizontal-flow grit chamber, hydrocyclone, vortex grit removal.

**aerated impoundment** A surface impoundment equipped with aeration devices, commonly used for biological wastewater treatment or for solar evaporation.

**aerated lagoon** An artificial, shallow holding and/or treatment pond that speeds up the natural process of biological decomposition of organic waste by stimulating the growth and activity of bacteria that degrade organic wastes. To supplement the oxygen from algae and the atmosphere, the basin, 6–20 ft deep with 7–20 days of detention, is equipped with diffused-air or mechanical aerators that also provide partial or complete mixing. It is usually part of a system that includes facultative and maturation or polishing ponds. Aerated lagoons function like activated sludge systems. *See also* antierosion assembly, draft tube, and lagoons and ponds.

**aerated lagoon performance** The BOD reduction achieved in an aerated lagoon, assuming complete mixing, as a function of the aeration period ($t$, days) and a reaction rate parameter ($k$, per day) that depends on the temperature and biodegradability of the wastewater. The ratio ($r$) of effluent BOD to influent BOD is (Hammer & Hammer, 1996):

$$r = 1/(1 + k \cdot t) \qquad \text{(A-54)}$$

*See also* lagoons and ponds.

**aerated lagoon temperature** The water temperature ($T_w$) in an aerated lagoon based on the temperature of the influent wastewater ($T_i$), the ambient air temperature ($T_a$), the surface area (A) of the lagoon, the wastewater flow ($Q$), and a heat transfer coefficient ($\varphi$):

$$T_w = (A\varphi T_a + QT_i)/(A\varphi + Q) \qquad \text{(A-55)}$$

The heat transfer coefficient varies widely, for example, 0.06 – 2.6 m/d or 0.20 – 8.50 (ft/d). *See also* lagoons and ponds.

**aerated pond** A natural or artificial pond that treats wastewater using mechanical or diffused-air aeration. *See* aerated lagoon. *See also* lagoons and ponds.

**aerated skimming tank** A long, trough-shaped wastewater treatment structure with diffusers at the bottom to supply air to keep heavy solids from settling and to lift light solids.

**aerated static pile** A grid of aeration or exhaust piping supporting a mixture of material to be composted and a bulking agent, sometimes insulated by a layer of screened compost. The pile, 6 to 8 feet deep, is equipped with a blower for aeration. Air is supplied through flexible plastic drainage tubing or other type of pipe. Pugmills fed by mechanical bins or stationary devices provide the necessary mixing. If the unit is enclosed, it includes an apparatus to scrub process air and provide ventilation to the building. Also called static pile. *See also* agitated bed composting, windrow composting.

**aerated static pile sludge composting** One of three common methods of sludge composting, using an aerated static pile. The composting mixture is made up of dewatered sludge and wood chips as bulking agent. The process lasts approximately 60 days, half for composting and half for curing. The compost is sceened to recover 60–90% of the bulking agent. Sometimes called static pile or forced aeration composting. *See also* windrow composting and in-vessel composting, which includes a static-bed (not static-pile) variation.

**aeration** (1) The process of bringing air into contact with a liquid, usually by bubbling air through the liquid, spraying the liquid into the air, allowing the liquid to cascade down a waterfall, or by mechanical agitation. The process may be passive (as when waste is exposed to air), or active (as when a mixing or bubbling device introduces the air). The rate at which a gas transfers into solution can be described by Fick's first law. Aeration serves to strip dissolved gases from, and/or to oxygenate, groundwaters or other liquids. In water treatment, specifically, aeration is used for the removal or control of undesirable characteristics: algal growth, carbon dioxide, corrosiveness, hydrogen sulfide, iron and manganese, methane, taste and odor, and volatile organics. In wastewater treatment, aeration promotes biological degradation of organic matter. *See also* diffused-air aeration, extended aeration, mechanical aeration, modified aeration, step aeration, surface aeration, tapered aeration. (2) A technique used for removing volatile organic compounds in unsaturated soils; *see* soil-vapor extraction.

**aeration blower** *See* centrifugal blower, inlet guide vane-variable diffuser, and positive-displacement blower.

**aeration efficiency** The efficiency of oxygen transfer of a system or equipment as a rate ($r$) per unit of energy (Joule or hp-h) or a percentage ($E$) of oxygen supplied (Droste, 1997):

$$r = \text{kg } O_2 \text{ transf./J} \qquad \text{or} \qquad r = \text{lb } O_2 \text{ transf./hp-h} \tag{A-56}$$

$$E = 100 \text{ (mass of } O_2 \text{ utilized/mass of } O_2 \text{ supplied)} \tag{A-57}$$

**aeration kinetics** The kinetics of biological treatment in a process that uses aeration, e.g., activated sludge. *See* biological kinetics and kinetic constants.

**Aeration Panel™** A membrane diffuser manufactured by the Parkson Corp. and producing fine bubbles.

**aeration period** The theoretical time that water or wastewater is under aeration, without regard to recirculation. In the activated sludge process (or in an aerated lagoon), for example, it is the volume of the aeration basin (or the lagoon) divided by the flow rate. Also called detention time. *See also* hydraulic detention time and solids retention time.

**aeration tank** A chamber used to inject air into water, wastewater, or other liquid. In biological wastewater treatment, the volume of the aeration tank is usually determined by multiplying the selected detention time by the design flow rate. The tank can be rectangular, circular, or square and operate under completely mixed or plug-flow conditions. *See also* reactor.

**aeration zone** The area between the water table and the ground surface where soil pores are not fully saturated but contain some vadose water. Also called vadose zone, zone of aeration, unsaturated zone, or undersaturated zone. *See* subsurface water.

**aerator** A device used to introduce air or oxygen into liquids such as water or wastewater; commonly of four types: air diffusers, gravity aerators, mechanical aerators, and spray aerators. *See also* draft tube, packed-bed aerator, stripper (2).

**Aercor products** Aeration and packing products of the Aeration Engineering Resources Corp.

**Aer-Degritter** Aerated grit removal equipment manufactured by the FMC Corp.

**aereal loading rate** A conservative design parameter for facultative ponds, equal to the rate of BOD loading per unit surface area. It varies with the average winter air temperature, from a range of 11–22 kg/ha/day below 0°C to 45–90 kg/ha/day above 15°C (WEF & ASCE, 1991).

**AerFlare** An air diffuser manufactured by Walker Proccess Equipment Co.

**Aergrid™** A proprietary aeration system manufactured by Aeration Technologies, Inc.

**Aermax™** A fine-pore aeration diffuser manufactured by Aeration Technologies, Inc.

**Aero-Accelator®** A proprietary package wastewater treatment plant manufactured by Infilco Degremont, Inc. It uses the activated sludge process.

***Aerobacter aerogenes*** A group of short, rod-shaped, gas-producing, Gram-negative coliform bacteria of the genus *Aerobacter,* found in nature (e.g., in soil and on plants) but also in the intestinal tracts of humans and other animals. Also called *Enterobacter aerogenes.*

**aerobe** An aerobic organism.

**aerobic** A condition in which free (atmospheric) or dissolved oxygen is present; pertaining to life or processes that require, or are not destroyed by, the presence of oxygen. *See also* aerobic state, anaerobic, anoxic, and facultative.

**aerobically digested biosolids** A dark brown-to-black material with a faint odor and a large quantity of gas (a mixture of methane, carbon dioxide, and other gases) that has undergone anaerobic digestion following primary or secondary clarification. The material drains easily, leaving a cracked surface as the gases escape.

**aerobic–anaerobic interface** The area between aerobic and anaerobic activities in wastewater, sludge, or solid waste that undergo microbial decomposition; e.g., in a facultative lagoon or in composting.

**aerobic–anaerobic lagoon** A common wastewater stabilization pond that is deeper than an aerobic pond but not as deep as an anaerobic pond. It is designed so that aerobic conditions prevail in the upper layers due to photosynthesis, surface reaeration, and, sometimes, supplemental aeration, while the bottom layers (containing the settling solids) undergo anaerobic decomposition. An intermediate, facultative zone may separate the upper and lower layers. Also called facultative pond or facultative pond with aeration. Facultative ponds usually include three or more cells to reduce short-circuiting and have a depth of approximately 6 ft and detention times of 25–175 days. They remove a significant amount of organic matter (BOD), nitrogen, and pathogens, but, because it is difficult to control their operation and the release of algae in the effluent, their use is diminishing. *See also* amphi-aerobic or hetero-aerobic pond, lagoons and ponds, Wehner and Wilhelm's equation. The following equation (Cairncross and Feachem, 1992) is often used to determine the required surface area ($A$, $m^2$) of a facultative pond, based on the mean air temperature of the coldest month ($T$, °C), the wastewater flow ($Q$, $m^3$/day), and the influent BOD ($L_i$, mg/L):

$$A = 0.5\, QL_i/(T - 3) \qquad \text{(A-58)}$$

**aerobic–anaerobic pond** *See* aerobic–anaerobic lagoon, and lagoons and ponds.

**aerobic autotrophic respiration** The type of reaction carried out by *Nitrosomonas* and *Nitrobacter. See* nitrification.

**aerobic bacteria** Defined similarly to aerobic organism.

**aerobic composting** The decomposition of organic matter by microorganisms using oxygen, as on a forest floor or in a well-aerated compost pile. Aerobic composting proceeds at higher temperatures and faster than anaerobic composting; it minimizes the potential for nuisance odors. It is also effective in destroying pathogens in wastewater and sludge, even enteroviruses and the eggs of *Ascaris. See* sludge composting for detail.

**aerobic condition** A condition in which free (atmospheric) or dissolved oxygen is present. In an aerobic reactor, the oxidation–reduction potential is high and oxidized ions such as sulfate ($SO_4^{2-}$), nitrate ($NO_3^-$), and ferric iron ($Fe^{3+}$) predominate compared to reduced species such as $S^{2-}$, $NH_4^+$, and $Fe^{2+}$. *See also* anaerobic state, anoxic state.

**aerobic digester** A tank used for the stabilization (digestion) of waste sludge under aerobic conditions.

**aerobic digestion** The biochemical decomposition of organic matter into carbon dioxide and water by microorganisms in the presence of oxygen, usually in open-top tanks. A common step in the processing of sludge from wastewater treatment, it serves to reduce the volume of sludge ahead of other treatment processes, particularly in small installations, which require less skilled operators. The process, similar to completely mixed activated sludge treatment, involves hydrolysis of large molecules followed by endogenous decay; it is designed and operated at large solids-retention times, e.g., 40 days at 20°C or 60 days at 15°C, to meet the requirements of Class B sludge and a process to significantly reduce pathogens. *See also* anaerobic digestion, autothermal digestion, thermophilic digestion.

**aerobic digestion reactions** The various reactions that are considered to represent the biochemical changes in the aerobic digestion process, i.e., biomass ($C_5H_7NO_2$) destruction, formation of ammonia ($NH_3$) and carbon dioxide ($CO_2$), complete or partial nitrification and denitrification (Metcalf & Eddy, 2003 and WEF & ASCE, 1991):

$$C_5H_7NO_2 + 5\, O_2 \rightarrow 4\, CO_2 + H_2O + NH_4HCO_3 \qquad \text{(A-59)}$$

$$C_5H_7NO_2 + 5\, O_2 \rightarrow 5\, CO_2 + 2\, H_2O + NH_3 \qquad \text{(A-60)}$$

$$NH_4^+ + 2\, O_2 \rightarrow NO_3^- + H_2O + 2\, H^+ \qquad \text{(A-61)}$$

$$C_5H_7NO_2 + 7\, O_2 \rightarrow 5\, CO_2 + 3\, H_2O + HNO_3 \qquad \text{(A-62)}$$

$$C_5H_7NO_2 + 7\, O_2 \rightarrow 5\, CO_2 + 3\, H_2O + H^+ + NO_3^- \qquad \text{(A-63)}$$

$$C_5H_7NO_2 + 4\, NO_3^- + H_2O \rightarrow NH_4^+ + 5\, HCO_3^- + 2\, N_2 \qquad \text{(A-64)}$$

$$2\, C_5H_7NO_2 + 11.5\, O_2 \rightarrow 10\, CO_2 + 7\, H_2O + N_2 \qquad \text{(A-65)}$$

**aerobic fermentation** Decomposition of organic matter (e.g., sugar) under aerobic conditions with high cell yield and low or no production of alcohol. *See also* anaerobic fermentation.

**aerobic flow-through partially mixed lagoon** A shallow earthen basin, aerated by mechanical devices and used for wastewater treatment. The aeration is sufficient for aerobic treatment and for maintaining only a portion of the solids in suspension. A sedimentation unit is used to remove the effluent solids (Metcalf & Eddy, 2003). Also called flow-through aerated lagoon. *See also* aerated facultative lagoon, and lagoons and ponds.

**aerobic heterotrophic respiration** The type of respiration undergone by *Bacillus, Pseudomonas,* and many other organisms while decomposing organic matter:

$$C_6H_{12}O_6 + 6\, O_2 \rightarrow 6\, CO_2 + 6\, H_2O \qquad \text{(A-66)}$$

**aerobic lagoon** A shallow treatment lagoon (1–5 ft deep) whose entire contents are aerated naturally or mechanically and where bacteria and algae stabilize wastewater partially, producing sludge that must be processed like activated sludge. Aerobic lagoons may be high-rate ponds, low-rate ponds, or maturation (tertiary) ponds. Used for the treatment of soluble organic wastes and secondary effluents in warm sunny climates, they are designed based on organic loading and hydraulic residence time, often as complete-mix reactors. *See also* lagoons and ponds, Wehner and Wilhelm's equation.

**aerobic lagoon with solids recycle** A wastewater treatment system similar to the extended aeration variation of the activated sludge process, but in which the mixed liquor is aerated in an earthen basin and most of the solids are kept in suspension. *See also* nominal complete mix, and lagoons and ponds.

**aerobic microorganism** A microorganism that requires oxygen for growth. *See* aerobic organism.

**aerobic organism** An organism that requires air or free oxygen for respiration. Aerobic microorganisms, in activated sludge plants and trickling filters for example, use free dissolved oxygen in decomposing organic matter and for growth. *See also* anaerobe, facultative organism, strict or obligate aerobe.

**aerobic plate count** An estimate of the number of bacteria in a sample. *See* standard plate count.

**aerobic pond** Same as aerobic lagoon. *See also* lagoons and ponds.

**aerobic process** Biological wastewater treatment that occurs in the presence of oxygen, including suspended-growth, attached-growth, and combined (hybrid) processes. Microorganisms use the oxygen to oxidize organic matter, producing carbon dioxide, water, ammonia, and more microorganisms.

**aerobic respiration** The process of substrate utilization by microorganisms under aerobic conditions. The oxidation of organic matter, e.g., a carbohydrate like glucose ($C_6H_{12}O_6$), is a common bacterially mediated reaction in soil and water; the bacteria extract the energy needed for growth and reproduction, producing carbon dioxide ($CO_2$) and water ($H_2O$), as well as new cell mass ($C_5H_7NO_2$):

$$a\,C_6H_{12}O_6 + b\,NH_3 + c\,O_2 \rightarrow d\,C_5H_7NO_2 + e\,CO_2 + f\,H_2O \quad \text{(A-67)}$$

The overall, simplified biochemical reaction is:

$$C_6H_{12}O_6 + 6\,O_2 \rightarrow 6\,CO_2 + 6\,H_2O + \text{energy} \quad \text{(A-68)}$$

*See also* anaerobic respiration, fermentation, oxidation–reduction, photosynthesis.

**aerobic sludge digestion** *See* aerobic digestion.

**aerobic sporeformer** A member of a large group of bacteria of the genus *Bacillus* that form endospores to survive in adverse conditions; some species may cause food poisoning. They have been considered as an indicator of treatment efficiency in the control of *Cryptosporidium* and *Giardia* because they outlive most waterborne pathogens and are easy to detect.

**aerobic state** *See* aerobic condition.

**aerobic submerged fixed-film process** Any of a number of biological wastewater treatment methods that use a packed-bed reactor with diffusd aeration and no clarification. *See* the following variations for detail: Biocarbone®, Biofor®, Biolite®, biological aerated filter, Biostyr®, fluidized-bed bioreactor, Oxazur®.

**aerobic treatment** A process by which microbes decompose complex organic compounds in the presence of oxygen and use the liberated energy for reproduction and growth. Such processes include extended aeration, trickling filtration, and rotating biological contactors.

**aerobic treatment unit (ATU)** An onsite wastewater treatment unit similar to a septic tank, but with a mechanism to introduce air into the tank for aerobic instead of anaerobic treatment. Also called home aeration unit.

**aerobiosis** Life in an environment that requires oxygen; biological processes that require air or oxygen.

**Aeroburn** A wastewater treatment plant developed by the Walker Process Equipment Co.

**aerochlorination** A wastewater treatment process that uses compressed air and chlorine gas to remove grease (APHA et al., 1981).

**aeroculture** A method of growing plants without soils by suspending them above mists that constantly provide moisture and nutrients to the roots. Also called aeroponics.

**Aeroductor®** Aerated grit removal equipment manufactured by Lakeside Equipment Co.

**aerodynamic diameter** Term used to describe particles with common inertial properties to avoid the complications associated with the effects of particle size, shape, and physical density. It is the diameter of a sphere of unit density that behaves aerodynamically as the particle of the test substance. In contrast to optical, measured, or geometric properties, this term is used to compare particles of different sizes and densities and to predict where in the respiratory tract such particles may be deposited (EPA-40CFR798.1150-2 and 4350-2).

**aerodynamic force** The force exerted by a moving fluid on an immersed body, drag being its component parallel to the direction of flow.

**Aero-Filter** A rotary distributor manufactured by Lakeside Equipment Corp.

**Aer-O-Flo** Wastewater treatment equipment manufactured by Purestream, Inc.

**aerohydrous** Pertaining to minerals that contain air and water in pores or cavities.

**Aero-Max** A membrane diffuser manufactured by Aeration Research Company.

**aeromicrobiology** The branch of science that deals with the airborne transmission of microorganisms, particularly airborne pathogens. *See also* bioaerosol.

**Aero-Mod** Modular aeration and clarification units manufactured by Aero-Mod, Inc.

***Aeromonas*** A genus of Gram-negative, straight-rod or coccobacillus shaped, aerobic or facultatively anaerobic bacteria that include opportunistic pathogens found in uncontaminated water, wastewater, or sewage-contaminated water. The species *A. caviae, A. hydrophila,* and *A. sobria* are very similar and sometimes referred to as *A. hydrophila;* they cause diarrhea in humans, affecting mostly children and the elderly, sometimes with symptoms similar to those of cholera. Some *Aeromonas* spp. strains also produce cytotoxins. These bacteria can regrow in treated drinking water distribution pipes and storage tanks if the disinfectant residual is low and the water contains organic carbon. *See* regrowth and secondary disinfection.

***Aeromonas hydrophila*** *See Aeromonas*. This bacterial species is an indicator of nutrient-rich conditions in natural waters.

**aerophyte** A nonparasite plant that grows above ground on other plants; also called air plant or epiphyte.

**aeroponics** *See* aeroculture.

**Aeropure** An activated carbon unit developed by the American Norit Company for the filtration of vapors.

**AeroScrub** A flue gas scrubber manufactured by Aeropulse, Inc..

**aerosols** Small liquid or solid particles suspended in a gas, smoke, or fog, and falling at a much lower speed than the vertical component of air motion. They are larger than molecules and can be filtered from the air. An example is the colloidal-size droplets (from less than 1–50 microns in diameter) produced by land treatment of wastewater; if the latter is inadequately disinfected, the aerosols may pose a health risk. A buffer zone around spray irrigation sites may provide some protection against aerosols. *See also* chlorofluorocarbon.

**aerosolized bacteria** *See* aerosolized excreta bacteria.

**aerosolized excreted bacteria or viruses** Bacteria or viruses contained in airborne droplets from wastewater or excreta. Such droplets, generated when toilets are flushed or during wastewater treatment and disposal (e.g., activated sludge process and spray irrigation), may contain pathogens.

**Aero-Surf** Rotating biological contactor equipment driven by air and manufactured by Envirex, Inc.

**Aerotherm** An in-vessel composting system developed by Compost Systems Co.

**aerotolerant** Pertaining to anaerobic microorganisms that can grow in air, usually poorly.

**aerotolerant anaerobes** Microorganisms, particularly bacteria, that grow preferably in anaerobic conditions but are not appreciably affected when free oxygen is present. *See also* facultative, obligate, and true facultative anaerobes.

**AerResearch** Aeration diffuser equipment developed by Aeration Research Company.

**Aershear™** A proprietary coarse-bubble diffuser manufactured by Aeration Technologies, Inc.

**Aertec™** Proprietary air diffuser equipment manufactured by Aeration Technologies, Inc.

**Aertube™** Proprietary tube aerators manufactured by Aeration Technologies, Inc.

**AES** Acronym of atomic emission spectroscopy.

**aesthetic** *See* aesthetic water quality.

**aesthetic contaminant** A constituent that affects the taste, odor, color, and appearance of water, but does not necessarily have adverse health effects. USEPA's Secondary Drinking Water Regulations include suggested maximum contaminant levels for aesthetic contaminants. *See* aesthetic water quality.

**aesthetic issues** Aesthetic issues for drinking water include color (from algae, metals, and natural organic matter), hardness (from divalent metal ions), mineralization (from total dissolved solids, e.g., chlorides, sulfates), staining (mainly from iron, manganese, or copper compounds in solution), taste and odor (from chemicals, decaying vegetation, and metabolites of microorganisms), and turbidity (from colloidal and suspended matter).

**aesthetic quality** *See* aesthetic water quality.

**aesthetic water quality** Quality of water that relates to charatcteristics that can be perceived by our senses: odor, taste, color, turbidity. These characteristics do not necessary pose a risk to human health, but they may lead consumers to conclude that water is unhealthy and turn to more palatable and less safe sources.

**affected public** The population that lives or works near a source of pollutants.

**affinity constant** A measure of the strength of association between two compounds; the reciprocal of the dissociation constant. *See also* apparent dissociation constant, association constant, dissociation constant, equilibrium constant, ionization constant, Michaelis constant.

**affinity laws** *See* pump affinity laws.

**affordability** A criterion, under the 1996 amendments of the Safe Drinking Water Act, used by states or the USEPA to grant variances to small public water supply systems that cannot afford to comply with a primary drinking water standard. A recommended technology is presumed affordable if its costs do not cause median water bills to exceed an affordability threshold of 2.5% of the median household income. The EPA must specify an alternative technology when a variance is granted.

**affordability threshold** *See* affordability.

**afforestation** (1) The conversion of bare or cultivated land to forest. (2) Forest crop cultivation on land that has not recently grown trees.

**aflatoxins** A group of mycotoxins found in strains of the fungus *Aspergillis flavus.* Aflatoxin B1 (chemical formula: $C_{17}H_{12}O_6$) is the most potent of the group, but they all attack the liver and may be human carcinogens.

**AFM** Acronym of atomic force microscopy.

**African sleeping sickness** *See* Gambian sleeping sickness and Rhodesian sleeping sickness.

**African trypanosomiasis** *See* Gambian sleeping sickness and Rhodesian sleeping sickness. (Sleeping sickness is the common name of trypanosomiasis.)

**afterbay** A reservoir, pond, or pump station installed at the tail end of a hydroelectric power plant to regulate the flow below the plant.

**afterbay reservoir** *See* afterbay.

**AfterBlend** A chemical feed booster equipment manufactured by Stranco, Inc.

**afterburner** In incinerator technology, a burner located so that the combustion gases are made to pass through its flame in order to remove or reduce smoke- and odor-causing organic compounds. It may be attached to or be separated from the incinerator proper. A typical afterburner is a refractory-lined shell providing enough residence time at a sufficiently high temperature to destroy organic compounds in the off-gas stream.

**aftercondenser** A condenser located at the end of an evaporator system.

**aftergrowth** An increase in the number of microorganisms (mainly bacteria) in treated water or wastewater. *See* regrowth for more detail.

**afterprecipitation** The precipitation of chemical compounds after leaving the settling unit, e.g., the precipitation of calcium carbonate ($CaCO_3$) in filters or further downstream.

**Ag** Chemical symbol of silver.

**agar** A gelatinous extract from seaweed or red algae used in laboratories for the cultivation of bacteria and for solidifying certain culture media; e.g., the substance used in the MPN confirming test.

**agar plate** A glass plate containing agar or other nutrients for the cultivation of bacteria.

***Agasicles hygrophila*** *See* alligator weed.

**age** *See* water age, groundwater age.

**Agent Orange** A toxic organochlorine herbicide and defoliant used in the Vietnam conflict, containing 2,4,5-trichlorophenoxyacetic acid (2,4,5-T) and 2-4 dichlorophenoxyacetic acid (2,4-D) with trace amounts of dioxin.

**age tank** A tank used to store a chemical solution of known concentration for feed to a cheminal feeder. It usually stores sufficient chemical solution to properly treat water for at least one day. Also called a day tank.

**agglomeration** The collection, gathering or coalescence of suspended matter into larger flocs or particles that can be more easily separated from a solution by sedimentation or filtration. *See also* coagulation, flocculation.

**agglutination** The process of uniting by glue or some other substance, e.g., the clumping of microorganisms and red blood cells due to the application of an antibody.

**aggregate** (1) Coarse mineral material (e.g., sand, gravel, broken rock, pebbles) that is mixed with either a cementing agent to form concrete or plaster, or to tarry hydrocarbons to form asphalt. Sand is also called fine aggregate, whereas coarse aggregate particles measure from 6 to 18 mm. (2) A mass or cluster of soil particles, often having a characteristic shape.

**aggregate dead zone (ADZ) model** A dynamic water quality model that combines plug flow and complete mix flow in a stream or river reach, which is assumed to have sections where advection dominates (with plugflow transport) and sections where dispersion dominates (with dispersive mixing in the dead zones). Within a reach, all advective portions are assumed to be combined in one aggregate zone, and similarly for the dispersive sections.

**aggregate organic constituent** A mixture of individual but undistinguishable organic compounds. In wastewater analysis, tests to measure gross concentrations (e.g., > 1.0 mg/L) of aggregate organic matter include biochemical oxygen demand, chemical oxygen demand, theoretical oxygen demand, and total organic carbon. Gas chromatography and mass spectroscopy are used for trace organics.

**aggregate volume index (AVI)** The ratio of total aggregate volume to the volume of dry sludge solids as measured in sludge thickening tests. It is an indication of sludge dewaterability: low AVIs are desirable. *See also* sludge volume index and zone settling velocity.

**aggregation** (1) The combination of different characteristics into an average value. (2) The clumping together of solid particles, snow crystals, etc, following collision.

**aggressive** Pertaining to water that has corrosive characteristics. *See* aggressive water and aggressivity index.

**aggressive index (AI)** A substitute recommended by the USEPA in 1980 for the Langelier index to monitor corrosivity characteristics in asbestos-cement pipes:

$$AI = \text{pH} + \log [(A)(H)] \qquad \text{(A-69)}$$

where $A$ is of total alkalinity and $H$ is calcium hardness, both in mg/L as calcium hardness. Water is considered very corrosive, moderately corrosive, or nonagressive when $AI$ is < 10, between 10 and 12, or > 12 respectively. The EPA repealed this recommendation in 1994 because of the documented shortcomings of the $AI$. Also called aggressiveness index or aggressivity index (2). *See* corrosion index for a list of corrosion control parameters.

**aggressiveness index** Same as aggressive index.

**aggressive water** Water that tends to dissolve calcium carbonate ($CaCO_3$) and promote pipe corrosion by removing the protective $CaCO_3$ film. Water is corrosive naturally or through pH reduction during such treatment methods as coagulation or chlorine addition. The opposite is a water that deposits calcium carbonate. Stable water does not deposit or dissolve $CaCO_3$. *See also* chemical stabilization, Langelier index, and marble test.

**aggressivity index (AI)** (1) A performance indicator developed to help water utility managers compare the costs of infrastructure repairs against replacement. Based on historical network conditions and environmental factors, it allows the determination of the performance of a buried water main of a given age–length profile. (2) Same as aggressive index.

**Agidisc** A disc filter manufactured by Eimco Process Equipment Co.

**agitated bed composting** One of two variations of in-vessel composting: the material to be composted (e.g., a mixture of dewatered sludge and a bulking agent) is agitated periodically in an enclosed, long rectangular or circular container or vessel for aeration, temperature control, and mixing. Curing takes place in an outside static pile. Also called dynamic composting. *See* sludge composting for detail.

**agitator** A machine or device for mixing, agitating or aerating. *See* mechanical agitator.

**agrichemical** Same as agricultural chemicals.

**agricultural chemicals** Chemicals used to enhance agricultural production (nitrate and phosphate fertilizers, soil conditioners, animal feed additives, pharmaceuticals) or to control pests and weeds (fumigants, herbicides, insecticides, pesticides). Some of these chemicals cause water pollution by leaching through the soil while others can accumulate in animals and pass through the food chains; they are not used by organic farmers. Also called agrichemicals or agrochemicals.

**agricultural contaminants** *See* agricultural pollution.

**agricultural drainage** Runoff from farmed areas collected through surface channels or underground drains. *See* agricultural pollution.

**agricultural drought** A type of drought that results in insufficient moisture to agricultural crops, as compared to hydrological drought and meteorological drought.

**agricultural irrigation** The use of water or reuse of reclaimed water for agricultural production (e.g., for crop irrigation or in commercial nurseries). *See also* border-strip irrigation, spray irrigation, urban irrigation, land application, land disposal, water reuse applications.

**agricultural levee** A levee that protects agricultural lands (a lower requirement than for urban flood control).

**agricultural nonpoint pollutants** Sediment, dissolved solids, nutrients, bacteria, pathogenic organisms, and toxic materials generated by agricultural operations.

**agricultural pollution** Farming wastes, including runoff and leaching of insecticides, pesticides, and fertilizers; erosion and dust from plowing; improper disposal of animal manure and carcasses; crop residues and debris; nitrates, phosphates, and potassium from poultry farm wastes.

**agricultural reuse** The use of reclaimed water, wastewater, sludge, or night soil for agricultural purposes for growing cash crops, in commercial nurseries, etc. Agricultural reuse provides not only water, thus saving other sources for drinking purposes, but also the plant nutrients and soil amendments contained in the treated wastewater. This practice, however, does present a risk of groundwater contamination by nitrates and heavy metals. *See also* land application and land disposal.

**agricultural runoff** Runoff from agricultural areas, a source of nitrate in water sources; it contains an average of 0.05 mg/L of phosphorus.

**agricultural value of sludge** Wastewater sludge contains fertilizing elements essential for plant growth, mainly, nitrogen (N), phosphorus (P) or phosphoric acid ($P_2O_5$), and potassium (K) or potash ($K_2O$). For example, municipal primary sludge contains about 3% nitrogen and trickling filter humus about 3.5%. Phosphate content of sludge is about 2%. Sludge also acts as a soil conditioner.

**agricultural waste** Localized or diffuse, liquid and solid wastes resulting from agricultural activities, e.g., agricultural chemicals, irrigation return flows, crop residues, animal wastes. *See* agricultural pollution.

**agricultural water use** Water used for irrigation, livestock, and other farming needs.

**AGRI-PLUS 6-5-0™** A dried wastewater sludge (98% dry solids, 6% nitrogen, 5% phosphorus) produced at the W. B. Casey Pollution Control Plant of Clayton County, GA. It is shipped to Florida for use as a soil conditioner and fertilizer.

***Agrobacterium* genus** Heterotrophic organisms active in biological denitrification.

**agrochemicals** Same as agricultural chemicals.

**agro-ecosystem** Land used for crops, pasture, and livestock; the adjacent uncultivated land that supports other vegetation and wildlife; and the associated atmosphere, underlying soils, groundwater, and drainage networks.

**agronomic rate** The whole sludge application rate, on a dry weight basis, designed to (a) provide the amount of nitrogen needed by the food crop, feed crop, fiber crop, cover crop, or vegetation grown on the land, and (b) minimize the amount of nitrogen in the sewage sludge that passes below the root zone of the crop or vegetation grown on the land to the groundwater (EPA-40CFR503.11-b).

**A horizon** The light colored layer on a soil profile, just below the surface layer (or O horizon); it contains an accumulation of humified organic matter. It is a fertile soil layer from which salts and colloids are leached.

**AHS** Acronym of aquatic humic substance.

**AI or A.I** Acronym of (1) aggressivity index, (2) artificial intelligence.

**aid** A chemical used to improve the performance of a process. *See* coagulant or coagulation aid, filter aid, flocculant aid.

**AIDS** Acronym of acquired immune deficiency syndrome.

**air** A mixture of oxygen, nitrogen, carbon dioxide, water vapor, argon, and other gases that forms the Earth's atmosphere.

**air agitation** A method of backwashing gravity or pressure water filtration units by introducing compressed air to the backwash water to clean the media of impurities. Also called air backwash.

**Airamic®** An air diffuser manufactured by Ferro Corp.

**air and vacuum valve** A valve used in water and wastewater networks to let and maintain air into an empty pipe to counteract a vacuum. Also called a vacuum valve.

**air backwash** Same as air agitation.

**Airbeam™** A basin cover manufactured by Enviroquip, Inc.

**air binding** The situation in which air is released from water into soil interstices, filter beds, pipes, or pumps; or the effect of that release. It can affect the rate of infiltration into the soil, prevent the passage of water during the filtration process, resuspend solid particles, disrupt the filter's gravel base, lead to more rapid development of head loss, and cause the loss of filter media during the backwash process. In filter operation, air binding is associated with the development of negative heads, an increase in water temperature, or the release of oxygen by algal growth.

**airborne droplet** *See* aerosolized excreted bacteria.

**airborne human pathogen** Many of the human pathogens related to water and wastewater can be transmitted via the air, e.g., (a) the bacterial agents of Legionnaires disease (*Legionella* spp), typhoid fever (*Salmonella typhi*); (b) the fungi of aspergillosis (*Aspergillus fumigatus*); and (c) the viral agents of hemorrhagic fever (bunyavirus), hepatitis (hepatitis virus), yellow fever (flavivirus), dengue fever (flavivirus).

**airborne infection** Infections caused by airborne pathogens include respiratory infections (diphtheria, measles, meningitis, mumps) and pneumonic plagues.

**airborne particulates** Total suspended particulate matter found in the atmosphere as solid particles or liquid droplets. The chemical composition of particulates varies widely, depending on location and time of year. Airborne particulates include wind-blown dust, emissions from industrial processes, smoke from the burning of wood and coal, and motor vehicle or nonroad engine exhausts.

**airborne virus** *See* aerosolized excreted bacteria/virus.

**air-bound** Pertaining to a pipe or pump in which an air entrapment at a high point prevents water from flowing freely; also describes the condition of a filter that experiences air binding.

**air-bound filter** A water filter in which air fills the pore space or the volume of the underdrainage system, thus causing head losses to rise sharply and filter output to drop.

**Airbrush™** A rotor aerator manufactured by United Industries, Inc.

**air bump** A mechanism included in some diatomite filters to help dislodge accumulated materials during backwashing.

**air cathode single chamber (ACSC)** A technology currently in the experimental stage, intended to use bacteria to generate electricity while treating wastewater. It is a variation of the microbial fuel cell (MFC) in which bacteria oxidize organic matter and transport electrons from the cell surface to the anode. In contrast to other MFCs, the ACSC does not require the addition of dissolved oxygen and chemicals to the wastewater

**air chamber** A compact surge chamber or surge tower, with a closed upper end, installed on the discharge line of a pump to minimize flow and pressure variations. The liquid level in the chamber fluctuates to balance transient conditions in the line.

**air-chamber pump** A displacement pump using compressed air in lieu of pistons or plungers. Also called an air-displacement pump.

**air check** A device that retains air while allowing water to pass.

**Air Comb®** A coarse-bubble diffuser manufactured by Amwell, Inc.

**air compression requirement** The theoretical power requirement to compress a flow of air ($Q$, cfm) from atmospheric pressure (14.7 psia) to a pressure $p$ (psia) is (Fair et al., 1971):

$$P = 0.22\,Q\,[(p/14.7)^{0.283} - 1] \qquad \text{(A-70)}$$

**air conditioning condensate** A relatively innocuous flow that may be allowed into municipal stormwater systems, excluding water from cooling towers, heat exchangers, or other sources.

**air contaminant** Any solid, liquid, or gaseous matter, any odor, or any form of energy that can be released into the atmosphere from an emission source; any particulate matter, gas, or combination thereof, other than water vapor. *See also* air pollutant.

**air curtain** A method of containing oil spills. Air bubbling through a perforated pipe causes an upward water flow that slows the spread of oil. It can also be used to stop fish from entering polluted water.

**Aircushion** A flotation clarifier manufactured by Wilfley Weber, Inc.

**air diffuser** A device that transfers compressed air into water or another liquid through orifices or nozzles in pipes, diffuser plates, or tubes. Air diffusers of varied design are used in biological wastewater treatment processes such as activated sludge. They are porous plates, tubes, or nozzles installed at the bottom of an aeration tank or attached to pipe headers. Also called injection aerator. *See also* coarse bubble diffuser, diffused aeration, diffused-air aerator, fine bubble diffuser, membrane diffuser, nonporous diffuser, porous diffuser, rigid porous diffuser, static tube, valved orifice diffuser.

**air diffusion** The process of transferring air into water or another liquid using a mechanical device. It is used in wastewater treatment (e.g., activated sludge process), but also in receiving waters to raise their dissolved oxygen content and to prevent stratification. *See* air diffuser, floating compressor, jet aeration, U-tube aeration.

**air-displacement pump** *See* air-chamber pump.

**air dryer** A device that removes water vapor from air, e.g., in an air-fed ozonator.

**air drying** A sludge dewatering operation that removes moisture from (usually well-digested) sludges on sandbeds or other materials by letting it evaporate or drain to the drying bed. *See also* paved drying bed, freeze-assisted sand bed, sand drying bed, solar drying bed, vacuum-assisted drying bed, sludge lagoon, wedgewire bed. Air drying, which may include supernatant decanting, is listed by the USEPA as a process to significantly reduce pathogen when it lasts at least 3 months and the ambient temperature exceeds 0°C during at least 2 months. It increases solids content to about 40%. *See also* lagoon, mechanical dewatering.

**Aire-O2®** An aspirator aerator manufactured by Aeration Industries, Inc.

**air-fed ozonator** A device that generates ozone by using an electrical discharge through dry air instead of oxygen.

**air flotation** A treatment process that uses air and chemical additives to lift particles heavier than water to the surface, where they are removed by skimming. Air flotation is used in wastewater treatment to remove suspended particles and to concentrate biological or chemical sludge solids. It may also be used to lift oil suspensions or other particles that have a lower density than water. *See* dissolved-air flotation and dispersed-air flotation for more detail. *See also* natural flotation.

**air gap** An open vertical drop or empty space that separates a drinking water supply to be protected from another water system in a treatment plant or other location. The open gap protects the drinking water from contamination by backflow or back-siphonage. Air-gap protection is also used in wastewater renovation for reuse, e.g., between a filter bottom and the filter drain. *See also* backflow preventer and vacuum breaker.

**air-gap device** A device that uses an air gap to protect a drinking water supply.

**air-gap fitting** A device that creates an air gap.

**air-gap membrane distillation** Membrane distillation is a desalination method using a temperature-driven, hydrophobic membrane to separate water, in the form of condensed vapor, from contaminants. The air-gap configuration includes an air gap after the membrane and a cool surface for condensation.

**air-gap protection** *See* air gap.

**air-gap separation** Same as air gap.

**Air Grid** Air scour equipment manufactured by Roberts Filter Manufacturing Co. for sand filters.

**Air-Grit** Aerated grit removal equipment manufactured by Walker Process Equipment Co.

**air hole** An opening in the frozen surface of a water body.

**air injection** The introduction of compressed air into the soil to push water down from the unsaturated zone.

**AirLance™** A proprietary in-vessel composting technique developed by American Bio Tech, Inc.

**airlift (or air lift)** Same as airlift pump.

**airlift pump** A simple, low-efficiency device used mainly for lifting water (or wastewater) by forcing compressed air at the bottom of a well or sump through an eductor. Hydrostatic pressure forces up the resulting mixture into the outlet pipe. It may also consist of two pipes, one inside the other; the compressed air is forced into the inner pipe and the air–water mixture rises in the outer pipe. Airlift pumps are also used for lifting return activated sludge, waste-activated sludge, grit slurries, and in other low-head, low-volume wastewater treatment applications. Also simply called airlift.

**air/liquid ratio** A design and operation parameter of the stripping processes. *See* air-to-water ratio, theoretical air-to-liquid ratio.

**air lock** The situation in which an air pocket blocks or considerably impedes flow in a conduit, usually at a high point, in valve domes and fittings or in pump discharge lines.

**Air Mix** A process developed by Zimpro Environmental, Inc. to clean filter surfaces using a pulsing bed.

**air mixing** The use of air to create turbulence and mixing in a reactor or basin, e.g., a mixing tank or an aeration basin. *See* pneumatic mixing for detail.

**air mix-pulse mix** A feature of pulse bed filters that starts operating when the filter bed becomes clogged. It includes an air-mix probe to trigger the supply of low-pressure air and a pulse-mix probe to close the effluent valve and turn on the backwash pumps.

**Airmizer** An air diffuser manufactured by Enviro-Quip International Corp.

**Air-O-Lator®** A proprietary floating aerator manufactured by the Air-O-Lator Corp.

**air-operated diaphragm pump** A common type of positive-displacement pump used in handling sludge from wastewater treatment, particularly in sludge pumping from sedimentation tanks and gravity thickeners. It incorporates a spring-return diaphragm and an air pressure regulator.

**AiroPump** An airlift manufactured by Walker Process Equipment Co.

**AirOXAL®** A proprietary pure oxygen process developed by Liquid Air.

**air padding** Pumping dry air into a container to assist with the withdrawal of a liquid or to force a liquefied gas such as chlorine out of the container.

**air piping** The system of pipes, valves, meters, and other fittings used to carry compressed air from the blowers to the diffusers in wastewater treatment plants.

**air piping friction losses** Head losses due to friction in piping systems; it is usually determined through the Darcy–Weisbach equation. For steel pipes, the friction factor ($f$) may be calculated as follows (Metcalf & Eddy, 2003):

$$f = 0.029\, D^{0.027}/Q^{0.148} \qquad \text{(A-71)}$$

where $D$ is the pipe diameter in meters and $Q$ the airflow rate in $m^3$/min. *See* equivalent pipe and the Frische's formula for the applicable Darcy-Weisbach friction factor. *See also* minor losses.

**air plant** A nonparasitic plant that grows above ground on other plants; also called epiphyte or aerophyte.

**air pocket** An accumulation of air in piping or in a filtering medium, blocking or otherwise impeding the flow of water. The pocket may result from a pressure drop below atmospheric level that releases dissolved gases from the liquid, from leaky joints, or from a pump suction.

**air pollutant** Any substance in air that could, in high enough concentration, harm humans, other animals, vegetation, or material. Pollutants may include almost any natural or artificial composition of matter capable of being airborne. They may be in the form of solid particles, liquid droplets, gases, or in combination thereof. Generally, they fall into two main groups: (1) those emitted directly from identifiable sources and (2) those produced in the air by interaction between two or more primary pollutants, or by reaction with normal atmospheric constituents, with or without photoactivation. Exclusive of pollen, fog, and dust, which are of natural origin, about 100 contaminants have been identified and fall into the following categories: solids, sulfur compounds, volatile organic chemicals, nitrogen compounds, oxygen compounds, halogen compounds, radioactive compounds, and odors.

**air pollution** The presence of contaminant or pollutant substances in the air that do not disperse properly, that interfere with human health and welfare, or that produce other harmful environmental effects. Rainfall sometimes returns to land and water the substances causing air pollution. *See* acid rain.

**air-release valve** A device used to release trapped air and installed at air pocket locations (normally, high points in a pipeline or piping system). A common design consists of a ball floating at the top of a cylinder and sealing a small opening. The ball drops from the outlet to allow the release of trapped air. *See also* other pressure surge control devices, air-relief valve, air slam.

**air-relief valve** A valve installed at a high point in a long pipeline or other pressurized container to adjust up or down the internal pressure by admitting or releasing air automatically. Also called an air-vacuum valve. *See also* air-release valve.

**air requirement** In aerobic biological wastewater treatment, air is provided to satisfy the requirements of the influent wastewater (for BOD or COD oxidation) and of the biomass.

**Airsep** Aerated grit collection equipment manufactured by Aerators, Inc.

**air scour** (1) The use of air, blown upward at a rate of 3–5 cfm per square foot, to agitate filter media before or during backwashing and prevent the formation of mudballs; same as air wash. *See also* air scouring, scour (2), surface wash. (2) A technique for cleaning membrane equipment used in wastewater treatment: coarse bubble agitation beneath the membrane assemblies dislodges accumulated solids and carries them away. *See also* backpulsing, maintenance cleaning, recovery cleaninng, and empy-tank maintenance cleaning.

**air scouring** In water filtration, the practice of using forced air before or during backwashing for cleaning the media of impurities. Air scouring is used in underdrain systems consisting of pipe laterals with nozzles, whereas concurrent air-and-water scouring is for systems in precast concrete T-pees. *See* air scour (1).

**air scrubbing** (1) The use of an airstream as an auxiliary scour to improve backwashing in tertiary granular filtration of wastewaters. (2) A method of removing odorous compounds from airstreams in preliminary wastewater treatment, using adsorption on activated carbon, adsorption on soil mounds, biological towers, dissolution into liquid oxidants in packed towers, aerosol contact vessels, or ozonation. *See also* mist scrubber absorption, packed-bed scrubber, wet scrubber absorption. (3) A similar process using a liquid spray (e.g., water or caustic soda) to oxidize and eliminate undesirable gaseous contaminants.

**air scrubbing chemicals** Products used in liquid absorption systems (packed towers or aerosol contact vessels) for the oxidation of odorous compounds in wastewater treatment, including chlorine ($Cl_2$), chlorine dioxide ($ClO_2$), hydrogen peroxide ($H_2O_2$), potassium permanganate ($KMnO_4$), sodium hydroxide (NaOH), and sodium hypochlorite ($Na_2ClO_4$).

**Air Seal** A coarse-bubble diffuser manufactured by Jet, Inc.

**air slam** The phenomenon that occurs when all the air in the air pocket of a pipeline is released and the water column rejoins. It can be prevented by using a two- or three-stage air-release valve.

**air/solids ratio (A/S)** The ratio of the weight of air available for flotation to the weight of solids in the feed stream of a flotation thickener, an important parameter for the design and operation of such a unit, ranging from 0.005 to 0.06, with a common value of 0.02 (McGhee, 1991):

$$A/S = C_s(f \cdot P - 1)\, R/C_0 \qquad \text{(A-72)}$$

where $C_s$ = saturation concentration of air in wastewater at 1.0 atmosphere, from 37.2 mg/L at 0°C to 20.9 mg/L at 30°C; $f$ = fraction of saturation actually achieved (usually between 0.5 and

0.8); $P$ = pressure, in atm (usually between 3.5 and 5.0); $R$ = recycle rate or ratio of pressurized flow to wastewater flow; $C_0$ = concentration of solids in incoming sludge, mg/L. Two similar equations may be used to express $A/S$ in mL of air per mg of solids (Metcalf & Eddy, 2003), the first for a system with all flow pressurized and the second for a system with only the recycle pressurized:

$$A/S = 1.3\, C_s(f \cdot P - 1)/C_0 \qquad \text{(A-73)}$$

$$A/S = 1.3\, C_s(f \cdot P - 1)\, R/QC_0 \qquad \text{(A-74)}$$

where $C_s$ is expressed in mL/L, $R$ is the pressurized recycle in $m^3$/day, and $Q$ is the mixed-liquor flow also in $m^3$/day.

**air sparger well** The well used to inject air in an air sparging installation.

**air sparging** An innovative bioremediation technique used in combination with soil-vapor extraction (SVE) for removing volatile organic compounds from unsaturated soils. It consists of air injection below the water table of an aquifer to add oxygen to the saturated zone and force contaminants up via sparged air channels into the SVE wells. Also called in situ volatilization and in situ air stripping. *See also* biofiltration, bioventing

**air stripping** (1) A treatment process that, based on Henry's law, removes dissolved gases and volatile organic compounds (VOCs) from contaminated groundwater or surface water by forcing an airstream through the water and causing the compounds to evaporate. Air stripping increases the surface area of the liquid and, therefore, its exposure to air. *See also* packed tower aeration. For example, air stripping is used to remove MTBE and sulfides (but not sulfates) from drinking water. *See also* steam stripping. (2) An advanced treatment process for the removal of ammonia ($NH_3$) nitrogen, hydrogen sulfide ($H_2S$), and other gases from wastewater and digester supernatant. It uses stainless steel, aluminum, etc. for the construction of stripping towers that are packed with such materials as wood slats. *See* ammonia stripping for more detail.

**air stripping device** A device used to implement the air stripping process, e.g., diffused aeration or mechanical aeration devices, packed column, tray stripping column, and venturi air stripping column.

**air stripping effect** The result of mechanically aerating and mixing sludge, septage, or wastewater in a holding tank. While this practice improves treatability and prevents settling of organic solids, it also aggravates the odor problem. Enclosed holding tanks are, thus, recommended.

**air stripping of ammonia** *See* ammonia stripping.

**air stripping of VOCs** An aeration method of removing volatile organic compounds (VOCs) from groundwater; it uses a countercurrent packed tower in which water flows downward and air is forced up through the packing.

**air stripping operation** A desorption operation used to transfer one or more volatile components from a liquid mixture into a gas (air) with or without the application of heat to the liquid. Packed towers, spray towers, and bubble-cap, sieve, or valve-type plate towers are among the process configurations used for contacting the air and a liquid (EPA-40CFR264.1031).

**air surging** A method of well development that protects the screen and filter packing from clogging by removing fine sediment (clay, sand, silt) using compressed air.

**AirTainer™** A tank cover manufactured by NuTech Environmental Corp.

**air-to-solids ratio** *See* air/solids ratio.

**air-to-water ratio (A/W)** The ratio of the volume of air to the volume of water treated in an air-stripping unit. *See also* theoretical air-to-liquid ratio, minimum air-to-water ratio.

**Airvac®** Proprietary equipment manufactured by Airvac, Inc. for wastewater collection.

**air-vacuum valve** *See* air-relief valve.

**air vent** (1) An opening to allow air to flow into or out of pipes, tanks, etc. (2) A pipe that provides air to the outlet conduit of a dam to reduce turbulence and prevent negative pressures during water releases.

**air wash** (1) A supplementary high-pressure device that introduces air into the underdrainage system of a rapid granular filter to agitate and wash the medium. Air wash is used in addition to, and precedes, backwash. Its purpose is to fluidize the bed without expanding it and to scour the deposits on the media. *See also* air–water wash, surface wash, air scour. (2) The passing of an airstream through an ion exchange column or microporous membrane to remove accumulated impurities.

**air–water backwash** Same as air–water wash.

**air–water partition coefficient** The ratio of the partial pressure of a compound in air to its concentration in water at a given temperature under equilibrium conditions. Also called Henry's law constant, it may be calculated for a pressure of 1.0 atm or 760 mm Hg as (Montgomery, 1996):

$$K_H = PF_w/760\, S \qquad \text{(A-75)}$$

$$K_{H'} = K_H/RK = S_a/S_w \qquad \text{(A-76)}$$

where $K_H$ is the air-water partition coefficient in atm · m$^3$/mole, $P$ is vapor pressure in mm Hg, $F_w$ is the formula weight of the compound in g/mole, $S$ is the solubility of the compound in water in mg/L, $K_{H'}$ is the dimensionless Henry's law constant, $R = 8.20575 \times 10^{-5}$ is the universal gas law constant, $K$ is water temperature in degrees Kelvin, $S_a$ is the solute concentration in air in moles/L, and $S_w$ is the aqueous solute concentration in moles/L. The air–water partitioning coefficient is used in determining the volatilization rates of substances in groundwater and surface water; volatilization is very low for $K_H < 10^{-7}$ atm · m$^3$/mole, slow for $10^{-7} < K_H < 10^{-5}$ atm · m$^3$/mole, important for $10^{-5} < K_H < 10^{-3}$ atm · m$^3$/mole, and rapid for $K_H > 10^{-3}$ atm · m$^3$/mole. *See also* the table under Henry's law.

**air–water wash** The simultaneous application of air and water in backwashing granular, single-medium, unstratified filters to remove particles deposited on the media. This results in reduced washwater requirements compared to the use of water alone. However, water backwash for 2–3 minutes at subfluidization velocities is still necessary to remove remaining air bubbles. *See also* air wash and backwash.

**Akta Klor** A sodium chloride solution produced by Rio Linda Chemical Co..

**$Al_2Si_2O_5(OH)_4$** Chemical formula of kaolinite, a secondary silicate mineral.

**$Al_2(SO_4)_3$** Chemical formula of alum.

**$Al_4(Si_4O_{10})_2(OH)_4 \cdot xH_2O$** Chemical formula of montmorillonite, a clay mineral.

**ALA** Abbreviation of aminolevulinic acid.

**Alabama flocculator** A device designed to accomplish gentle mixing for flocculation. It consists of a series of chambers, each with an upward inlet, as shown in Figure A-07. The device was first used in Alabama but, having no mechanical components, it is appropriate for developing countries (Droste, 1997). Other hydraulic flocculators include paddle flocculators, baffled channels, pebble bed flocculators, and spiral flow tanks.

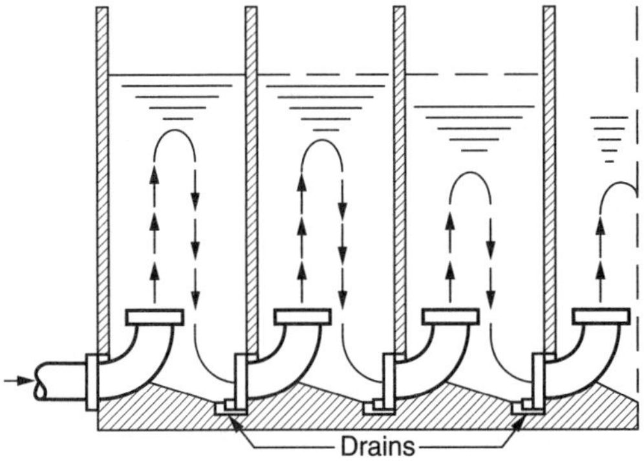


**Figure A-07.** Alabama flocculator (adapted from Droste, 1997).

**alachlor [2-chloro-2,6-diethyl-N-(methoxymethyl) acetanilide]** A widely used agricultural herbicide, marketed under the trade name Lasso, used mainly to control weeds in corn and soybean fields. Although slightly soluble in water, it has been detected in surface and ground water sources; the USEPA regulates it with an MCL = 0.002 mg/L. It can cause eye and skin irritations after an acute exposure. Chronic effects observed in the laboratory include hepatotoxicity, nephritis, hemolytic anemia, and degeneration of eye components.

**alachlor ethane sulfonic acid** A sulfonic acid ($SO_3H$) degradation product of alachlor.

**Alar** Trade name for daminozide, a pesticide that makes apples redder, firmer, and less likely to drop off trees before growers are ready to pick them. It is used to a lesser extent on peanuts, tart cherries, concord grapes, and other fruits.

**alarm contact** A switch that operates when some pre-set low, high, or abnormal condition exists.

**albite ($NaAlSi_3O_8$)** A light-colored silicate mineral found in igneous rocks; the sodium end member of the plagioclase feldspars. It dissolves in water to form montmorillonite [$Na_{1/3}Al_{7/3}Si_{11/3}O_310(OH)_2$]:

$$7\ NaAlSi_3O_8 + 6\ CO_2 + 26\ H_2O = 3\ Na_{1/3}Al_{7/3}Si_{11/3}O_{10}(OH)_2 + 6\ Na^+ + 6\ HCO_3^- + 10\ H_4SiO_4 \quad \text{(A-77)}$$

**albuminoid** A class of simple proteins (e.g., collagen, gelatin, keratin) or of substances that resemble true proteins, albumen, or albumin; they are insoluble in neutral solvents.

***Alcaligenes*** A genus of rod-shaped, aerobic or facultatively anaerobic bacteria found in the intestinal tract and in dairy products. They are floc-forming organisms, active in activated sludge, biological denitrification, and trickling filters; commonly removed on granular activated carbon.

***Alcaligenes faecalis*** A species of bacteria that has been used in in situ remediation of an aquifer contaminated by aromatic hydrocarbons (Freeman et al., 1998).

**$AlCl_3$** Chemical formula of aluminum trichloride.

**alcohol** A class of organic compounds formed from hydrocarbons by the substitution of one or more hydrogen atoms (H) by the hydroxyl group

(—OH), thus the general formula: ROH; e.g., ethanol ($CH_3CH_2OH$), methanol ($CH_3OH$), propanol ($CH_3CH_2CH_2OH$). Like water, the alcohol group is neither strongly acidic nor strongly basic.

**alcoholate** *See* alkoxide.

**aldehyde** An organic compound containing the carbonyl functional group (C═O) on the first or last carbon atom; e.g, formaldehyde ($CH_2O$ or HCHO), acetaldehyde ($CH_3CHO$), propionaldehyde ($CH_3CH_2CHO$), and bturaldehyde ($CH_3CH_2$ $CH_2CHO$). Also defined as a hydrocarbon derivative in which the oxygen atom replaces one or more hydrogen atoms; general formula: R · CHO. Some aldehydes are created as by-products of water disinfection by ozone and other agents. *See also* formaldehyde, ketone.

**aldicarb [$CH_3SC(CH_3)_2HC:NOCONHCH_3$]** Common name for 2-methyl-2-(methylthio) propionaldehyde O-(methylcarbamoyl) oxime, a crystalline insecticide or soil nematocide sold under the trade name Temek; a carbamate, very soluble in water, with a high acute toxicity. It is made from ethyl isocyanate.

**aldicarb sulfone** Same as aldoxycarb.

**aldicarb sulfoxide [$CH_3SOC(CH_3)_2HC:NOCONHCH_3$]** An intermediate oxidation by-product of aldicarb.

**aldoacid** An organic compound that contains an aldehyde group (HC═O) and a carboxylic acid group (COOH) in its structure. It may be formed during water disinfection by ozone and other chemical agents.

**aldoketone** An organic compound that contains an aldehyde group (HC═O) and a carbonyl group (C═O) in its structure. It may be formed during water disinfection by ozone and other chemical agents.

**aldoxycarb [$CH_3SO_2C(CH_3)_2HC:NOCONHCH_3$]** Common name of 2-methyl-2-(methylsulfonyl) propanal O{(methylamino)carbonyl} oxime, an insecticide or soil nematocide detected in drinking water. Also called aldicarb sulfone, it is an oxidation by-product of aldicarb.

**aldrin ($C_{12}H_8Cl_6$)** Common name for an insecticide that is mostly 1,2,3,4,10,10-hexachloro-1,4,4a,5,8,8a-hexahydro-exo-1,4-endo-5,8-dimeth-anonaphthalene; practically insoluble in water, has some of the same health effects as endrin. The USEPA does not currently (2005) regulate aldrin, but the USPHS recommended a limit of 17 ppb in 1968.

***Alexandrium*** A genus of cyanobacteria (blue-green algae) found in algal blooms and responsible for the production of toxins (saxitoxin) that can cause liver damage and gastrointestinal illness.

**alfalfa valve** A type of screw valve installed to regulate water flow at the end of a pipe.

**alga** Singular form of algae.

**algae** Simple rootless plants that grow in sunlit waters in proportion to the amount of available nutrients, from microscopic forms (that impart a green color to water), to visible branched forms appearing as attached green slime, to giant seaweeds. Algae belong in three large groups (brown, green, red) and are chiefly aquatic, eukaryotic one-celled or multicellular plants without true stems, roots, or leaves, that are typically autotrophic and photosynthetic. They contain chlorophyll and live floating or suspended in surface water, but are not usually found in groundwater. They may also be attached to structures, rocks, or other submerged surfaces. They are food for fish and small aquatic animals. Excess algal growths can impart tastes and odors to drinking water. Algae produce oxygen during sunlight hours and use oxygen during the night hours. They can appreciably affect pH and lower the dissolved oxygen in the water. Algae produce organic cell mass from inorganics. They are not significant in wastewater treatment, except in oxidation ponds, where they combine with saprophytic bacteria for the stabilization of organic matter. *See also* activated algae, blue-green algae, brown algae, diatom, green algae, plankton, red algae.

**algae bloom** *See* algal bloom.

**algaecide** Same as algicide.

**algae control** *See* algal control, algal inhibition, and algicide.

**algae harvesting or algal harvesting** The recovery of algal cells from tertiary wastewater treatment effluents. *See also* algal culture.

**AlgaeMonitor** A meter manufactured by Turner Designs to monitor algal concentrations online.

**algae production** Same as algal culture.

**Algae Sweep Automation** Equipment manufactured by Ford Hall Co., Inc. to sweep algae automatically from clarifiers.

**algal assay** A laboratory procedure that determines the limiting algal nutrient in water.

**algal–bacterial symbiosis** The relationship existing between algae and bacteria in aerobic stabilization ponds: algal photosynthesis produces oxygen that the bacteria use to degrade organic matter; in turn, the bacteria release nutrients and carbon dioxide for use by the algae. *See* Figure A-08.

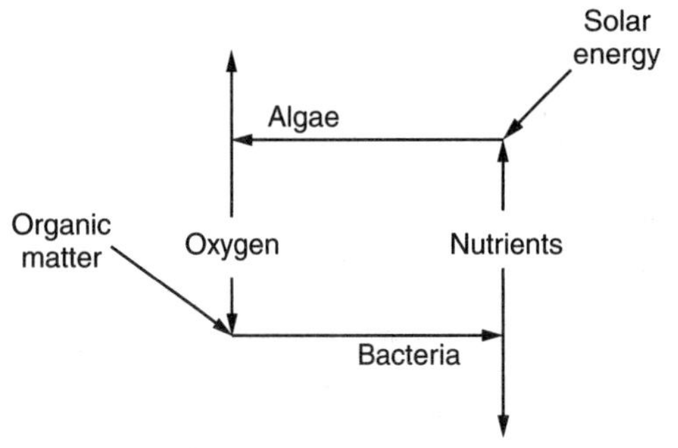


**Figure A-08.** Algal–bacterial symbiosis.

**algal bloom** Sudden spurts of algal growth (e.g. green and blue-green algae) in lakes and reservoirs, which can affect water quality adversely (e.g., oxygen depletion at night), cause fish kills, and indicate potentially hazardous changes in local water chemistry. In water treatment plants, algal masses decrease the hours of operation of filters and increase sludge production as well as the risk of taste and odor problems. Some blue-green algae (cyanobacteria) produce hepatotoxins and neurotoxins that water treatment does not easily remove. Only 0.005–0.05 mg/L of phosphorus, which may be available from natural sources, is necessary to support an algal bloom in surface waters.

**algal composition** One chemical formula proposed for algal cells is $C_{106}H_{263}N_{16}O_{110}P$, but algal biomolecules are sometimes represented by glucose ($C_6H_{12}O_6$) (Droste, 1997).

**algal control** The prevention or reduction of algal blooms by the application of chemicals: chlorine, copper sulfate (with or without the addition of sodium citrate), activated carbon, and lime. *See* blackout dosage.

**algal culture** The production of algae for direct harvesting in ponds enriched with wastewater, sludge, or nightsoil, as an alternative to fish production; one of three principal forms of aquaculture. *See also* algae harvesting, fish farming, and macrophyte production.

**algal growth potential (AGP)** The maximum algal biomass produced in a water sample under laboratory conditions, expressed in mg/L dry weight.

**algal growth rate** Algal productivity of a water body, measured as the mass of carbon used annually by algae per unit surface area of the water body, e.g., grams of carbon per square meter per year. It may be used to indicate the state of eutrophication of a lake. Also called primary productivity.

**algal harvesting** Same as algae harvesting.

**algal hepatotoxin** A product of some cyanobacteria (blue-green algae) that can cause livestock death and human illness and is not effectively removed by ordinary water treatment processes. *See* algal toxin, hepatotoxin.

**algal inhibition** The prevention or slowing down of algal growth. *See* algal control.

**algal mat** A layer of dead algae floating at the surface of a water body.

**algal neurotoxin** A product of some cyanobacteria (blue-green algae) that can cause livestock death and human illness and is not effectively removed by ordinary water treatment processes. *See* algal toxin, neurotoxin.

**algal protoplasm** The colloidal and liquid substance of the algal cell (cytoplasm and nucleus); commonly used formula: $C_{106}H_{263}N_{16}O_{110}P$ or $(CH_2O)_{106}(NH_3)_{16}(H_3PO_4)$. *See also* Redfield ratio.

**algal respiration** The use of dissolved oxygen and production of carbon dioxide by algae at night; the opposite of photosynthesis.

**algal toxin** A secondary metabolite produced by algae and some cyanobacteria that affects animals and aquatic biota. The most important algal toxins in the drinking water field are dermatotoxins, hepatotoxins (the most prevalent), and neurotoxins. Potential health effects include acute respiratory failure and other nerve damages, damages to organs, and malignant tumors.

**AlgaSORB®** A proprietary ion exchange medium produced by Bio-Recovery Systems, Inc. for the removal of heavy metals.

**algicide** A substance or chemical used specifically to kill or control algae, e.g., copper sulfate ($CuSO_4$), chlorine, cupric chloramines, and slurry of powdered activated carbon. Also called algaecide.

**algogenic organic matter** Natural organic matter produced by algae, a potential source of carbon for regrowth bacteria and a THM precursor.

**ALGOL** Acronym of algorithmic language.

**algorithm** A set of rules and procedures to solve a problem. In hydraulic modeling and other electronic data processing, computational algorithms perform the numerical procedures required to solve the mathematical representations of the system. A well-known algorithm used to solve nonlinear equations is the Newton–Raphson method, which approximates a function by truncating its Taylor series expansion after the first derivative term.

**algorithmic language (ALGOL)** A computer programming language used for mathematical problems: it expresses information in Boolean algebraic notations.

**$Al(H_2O)_6$** Chemical formula of the aquo aluminum ion.

**alimentary tract** The muscular tube that carries food and fluids from the throat to the stomach, the small intestine, the large intestine, and the anus.

**aliphatic** (1) Of or pertaining to a broad category of carbon compounds distinguished by a straight, or branched, open chain arrangement of the constituent carbon atoms. The carbon–carbon bonds may be either saturated or unsaturated. Alkanes, alkenes, and alkynes are aliphatic hydrocarbons. *See also* aromatic. (2) An aliphatic compound, mainly in the plural form.

**aliphatic aldehyde** An organic compound that contains the aldehyde (HC=O) group and an open chain of carbon atoms. *See* aldehyde.

**aliphatic compound** An organic, saturated or unsaturated, compound having straight or branched chains of carbon atoms, instead of a ring. Aliphatic compounds include three large groups (alkanes, alkenes, and alkynes), subdivided into acids, alcohols, aldehydes, alkyl halides, amides, amines, esters, ethers, hydrocarbons, and ketones. Examples: acetaldehyde ($CH_3 \cdot CHO$), acetic acid ($CH_3 \cdot CO \cdot OH$), ethane ($CH_3 \cdot CH_3$), ethanol ($CH_3 \cdot CH_2.OH$). *See also* aromatic hydrocarbon.

**aliphatic hydrocarbon** A hydrocarbon in which single covalent bonds consisting of two shared electrons join the carbon atoms; e.g., *n*-octane [$C_8H_{16}$ or $CH_3$—$(CH_2)_6$—$CH_3$]. *See* Figure A-09.

**aliphatic hydroxy acid** An organic acid with carbon atoms in open chains instead of rings; general formula: R—COOH, where R is a hydrogen atom or a carbon open chain. Also called aliphatic hydroxyl acid.

**aliphatic hydroxyl acid** Same as aliphatic hydroxy acid.

**aliphatic-type polyamide membrane** A thin, flat, synthetic organic membrane supported by a microporous substrate and used in reverse osmosis.

**aliquot** A representative portion (a known fraction) of a sample; often taken as an exact divisor. For example, 3 ml is an aliquot part of 15 ml. *See also* flow-weighted composite sample.

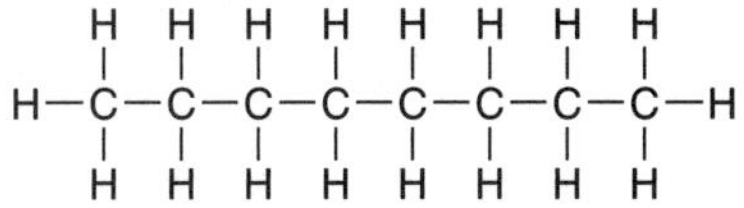

**Figure A-09.** Aliphatic hydrocarbon (*n*-octane).

**alizarin-visual test** A procedure used in the laboratory to determine fluoride concentrations. [Alizarin ($C1_4H_8O_4$) is a reddish-orange crystal or a brownish-yellow powder used in the synthesis of dyes.]

**alk** Abbreviation of alkali, alkaline, or alkalinity.

**[alk]** or **[Alk]** Abbreviation for the molar concentration of alkalinity.

**alkali** (1) Any of various soluble salts, principally of calcium, magnesium, potassium, and sodium, that have basic properties and can combine with acids to form neutral salts and may be used in chemical water treatment processes (e.g., for pH adjustment). (2) The hydroxide of an alkali metal, which forms salts by neutralizing acids and turns red litmus paper blue. (3) An alkali metal or a compound of an alkali metal, e.g., potassium or sodium carbonate. Sodium hydroxide (NaOH) and other alkalis may be detrimental to receiving streams and interfere with water treatment processes. (4) A soluble mineral salt found in soil, mostly in arid regions, and detrimental to plant growth.

**alkali accumulation** The accumulation of carbonates and hydroxides of alkali metals in the top layer of a soil due to evaporation of water that rises by capillarity.

**alkalic** Same as alkaline.

**alkali disease** A chronic livestock disease characterized by emaciation, loss of hair, deformation and shedding of hooves, loss of vitality and erosion of the joints of long bones; caused by an excessive intake of selenium (Se) at low levels and over a long duration. *See also* blind staggers.

**alkali–iodide–azide** A reagent used in the azide modification of the iodometric method for measuring dissolved oxygen.

**alkali metal** Any of a group of univalent, soft metals that readily react with water to form a basic solution (a hydroxide), including the elements cesium (Cs), francium (Fr), lithium (Li), potassium (K), rubidium (Rb), and sodium (Na). They form group IA or the first column of the periodic table and assume a +1 oxidation state in their compounds. Alkali metals react vigorously with water, producing caustic and corrosive hydroxides and explosive hydrogen gas, e.g., for sodium:

$$2\ Na + 2\ H_2O \rightarrow 2\ NaOH + H_2 \qquad (A\text{-}78)$$

*See also* alkali.

**alkaline** The condition of a solution or soil that contains a sufficient amount of alkali substances (or hydroxyl ions, $OH^-$) to raise the pH above 7.0. Also called basic, alkalic.

**alkaline agents** Chemicals used to neutralize acid wastes, including caustic soda or sodium hydroxide (NaOH), sodium carbonate ($Na_2CO_3$), magnesium oxide (MgO), high-calcium hydrated lime [Ca(OH)], high-calcium quicklime (CaO), high-calcium limestone, dolomitic hydrated lime, dolomitic limestone.

**alkaline chlorination** A process used to treat wastewater that contains cyanides (CN) by addition of chlorine gas ($Cl_2$) and, if necessary, enough lime [$Ca(OH)_2$] or sodium hydroxide (NaOH) to raise the pH to 11.0. The cyanide is oxidized as follows (Nemerow & Dasgupta, 1991):

$$2\ NaCN + 5\ Cl_2 + 12\ NaOH \rightarrow N_2 + 2\ Na_2CO_3 + 10\ NaCl + 6\ H_2O \quad \text{(A-79)}$$

**alkaline cleaning** A process that uses a solution (bath), usually detergent, to remove lard, oil, and other such compounds from a metal surface. Alkaline cleaning is usually followed by a water rinse. The rinse may consist of single or multiple stage rinsing (EPA-40 CFR471.02-c).

**alkaline cleaning bath** *See* alkaline cleaning.

**alkaline cleaning rinse** *See* alkaline cleaning.

**alkaline earth** The oxide of barium, calcium, magnesium, radium, or strontium.

**alkaline-earth metal** Any of a group of bivalent metals [including barium (Ba), beryllium (Be), calcium (Ca), magnesium (Mg), and strontium (Sr)] that are less soluble than the alkali metals but whose hydroxides are alkalis. They form group IIA or the second column of the periodic table; they assume an oxidation number of +2 in their compounds.

**alkaline soil** A soil that contains a sufficient amount of alkali substances (or hydroxyl ions, $OH^-$) to raise the pH above 7.0. Also called a basic soil. *See also* alkali soil.

**alkaline solution** A solution that contains a sufficient amount of alkali substances (or hydroxyl ions, $OH^-$) to raise the pH above 7.0. Also called a basic solution.

**alkaline stabilization** The addition of lime or another alkaline substance to render sludge unsuitable to microorganisms. *See* lime stabilization and advanced alkaline stabilization.

**alkaline waste** A waste that contains alkali concentrations; a waste with a high pH, e.g., from such industries as beet-sugar production, soap production, textiles, leather tanning, or rubber. Alkaline wastes with pH > 12.5 are corrosive and classified as hazardous. Various neutralization methods are available for the treatment of alkaline wastes. *See* alkaline waste fermentation, carbon dioxide treatment, submerged combustion, sulfuric acid treatment, waste boiler-flue gas treatment

**alkaline waste fermentation** A treatment process that consists in lowering the pH of an alkaline waste by fermenting the waste to produce carbon dioxide ($CO_2$) that lowers pH.

**alkaline water** Water that contains a sufficient amount of alkali substances (or hydroxyl ions, OH-) to raise the pH above 7.0. Also called basic water.

**alkalinity (Alk)** The capacity of water to neutralize acids. This capacity is caused by the water's content of carbonate ($CO_3^{2-}$), bicarbonate ($HCO_3^-$), hydroxide ($OH^-$), and occasionally borate, silicate, ammonia ($NH_3$), and phosphate ($PO_4$). When only the inorganic carbon species are significant,

$$\text{Alk} = (HCO_3^-) + (CO_3^{2-}) + (OH^-) \quad \text{(A-80)}$$

where all quantities are in meq/L. Total alkalinity (Talk) in drinking water is also described by (Metcalf & Eddy, 2003):

$$\text{Talk} = [HCO_3^-] + 2\ [CO_3^{2-}] + [OH^-] - [H^+] \quad \text{(A-81)}$$

where Talk is in meq/L and the quantities between brackets are in mg/L as $CaCO_3$. Other species that can contribute to acidity include HOCl, $HPO_4^{2-}$, and $H_2PO_4^-$. Note that these species, as well as the bicarbonate ion ($HCO_3^-$) can neutralize both acids and bases. In natural waters, alkalinity consists mainly of the bicarbonate ion [$HCO_3^-$]. In wastewater, alkalinity is caused mainly by the bicarbonates of calcium and magnesium. Alkalinity is expressed in milligrams per liter of equivalent calcium carbonate ($CaCO_3$). Alkalinity is not the same as pH because water does not have to be strongly basic (high pH) to have alkalinity. It is a measure of how much acid can be added to a liquid, or how many hydrogen ions can be added to it, without causing a great change in pH. Chemical sources of alkalinity include lime, sodium bicarbonate, sodium carbonate, sodium hydroxide, soaps, and salts of organic acids. *See also* acidity.

*Importance of alkalinity.* This parameter has no health significance. Some alkalinity (e.g., 20 mg/L as $CaCO_3$) is required in water coagulation. High alkalinity promotes corrosion of copper and lead, but low to moderate alkalinity reduces corrosion of most materials. Alkalinity is associated with carbonate hardness, affects the rate of ozone decomposition, increases during denitrification, and serves as an indicator of the stability of anaerobic digestion (*see* anaerobic alkalinity correction, usable alkalinity). Alkalinity affects biological nitrification.

**alkalinity analysis** Alkalinity of a sample is measured by titrating it with a 0.02 normal solution of sulfuric acid ($H_2SO_4$). Titration endpoints are determined by potentiometric titration or using colorimetric indicators (phenolphthalein, methyl orange, mixed bromocresol green–methyl red). Components of total alkalinity include hydroxide, carbonate, and bicarbonate.

**alkalinity as $CaCO_3$** An expression of the alkalinity of a sample:

$$\text{alk as } CaCO_3 = 50{,}000 \text{ (ml titrant)} \times \text{(acid normality)/(ml sample)} \quad \text{(A-82)}$$

**alkalinity correction** *See* anaerobic alkalinity correction.

**Alkalinity First™** The chemical sodium bicarbonate ($NaHCO_3$) as produced by Church & Dwight Co., Inc.

**alkalinity supplement** A basic chemical used to increase alkalinity of water or wastewater during coagulation, to adjust final pH, or reduce corrosivity of the product water. Lime is the most common alkalinity supplement, as quick lime (calcium oxide, CaO) or hydrated lime [calcium hydroxide, $Ca(OH)_2$]. Caustic soda (sodium hydroxide, NaOH) is also used; it is more expensive but easier to handle than lime.

**alkalinity test** An analytical procedure to determine the alkalinity of a sample. *See* alkalinity analysis.

**alkali pollution** Caustic condition in wastewater discharges from laundries, textiles, and other industries.

**alkali reactivity** The relative capacity of an aggregate to undergo an alkaline reaction. This applies to limestone aggregates used instead of granitic aggregates to increase the resistance of concrete to acid attack. Low-alkali cement is recommended for use with such aggregates.

**alkali soil** A soil that impedes plant growth because of high alkalinity ($pH \geq 8.5$) and/or a high sodium content (more than 15% of exchangeable sodium). *See also* alkaline soil, sodium absorption ratio.

**alkaloid** (1) A complex, basic, nitrogen-containing compound of vegetable origin, usually insoluble in water, soluble in alcohol; e.g., atropine, morphine, nicotine, and quinine. (2) Alkaline or resembling an alkali.

**alkanes** The homologous group of linear saturated aliphatic hydrocarbons having the basic building group (—$CH_2$—) and general formula $C_{(n)}H_{(2n+2)}$. Alkanes, which contain only single carbon–carbon bonds, can be straight chains, branched chains, or ring structures. Also referred to as paraffins. They are readily biodegradable and can be used by some microorganisms as sole source of carbon and energy. *See also* aliphatic hydrocarbon, alkenes, and alkynes.

**Alka-Pro®** Equipment manufactured by Davis Water & Waste Industries, Inc. for the control of biological wastewater treatment processes.

**alkenes** The group of unsaturated hydrocarbons having the general formula $C_{(n)}H_{(2n)}$ and characterized by high chemical reactivity. Alkenes contain at least one double carbon–carbon bond. Also referred to as olefins. They have the same biodegradation characteristics as alkanes. *See also* alkanes and alkynes.

**alkoxide** A compound derived from an alcohol by the substitution of a metal for the hydrogen atom in the hydroxyl group (OH), e.g., sodium methoxide ($NaOCH_3$) from methyl alcohol ($CH_3OH$). Also called alcoholate, metal alkoxide.

**alkylation** (1) A common mechanism of mutation involving the attachment of a small alkyl group (e.g., —$CH_3$ or —$C_2H_5$) to one of the nitrogenous bases in DNA. (2) The attachment of an alkyl group to a metal ion, which increases the availability of an element to organisms and may lead to increased toxicity. For example, methyl mercury is available as $CH_3HgCl$ in saline water and as $CH_3HgOH$ in fresh water. Also called methylation.

**alkyl benzene sulfonate (ABS)** A group of anionic surfactants, resistant to biological breakdown, previously used in synthetic detergents, now largely replaced by linear alkyl sulfonate or LAS, which is biodegradable; general formula: R—$C_6H_4SO_3$—, where R represents the alkyl portion, a branched chain of 11 to 15 carbon atoms. For example R = $C1_2H_{25}$ for dodecylbenzene. The World Health Organization sets a maximum acceptable limit of 0.5 mg/L and a maximum allowable limit of 1.0 mg/L of ABS in drinking water. *See* the methylene blue active substance and the cobalt thiocyanate active substance tests for the detection of surfactants in water and wastewater.

**alkyl compound** *See* alkylation.

**alkyl halide** An aliphatic compound resulting from the replacement of a hydrogen atom of an alkane by a bromine, chlorine, fluorine, or iodine atom.

**alkylmercury** A volatile, highly toxic organic compound of mercury, used as a fungicide and seed dressing. *See* methylmercury for detail.

**alkyl sulfonate** A class of compounds of the general formula R—SONa, where R is a hydrocarbon chain of 10 to 18 carbon atoms derived from tallow or coconut alcohol. Alkyl sulfonates are surface active agents used in synthetic detergents.

Some of them (e.g., alkyl benzene sulfonate) resist biodegradation and cause foaming in water supply sources and receiving waters.

**alkynes** The group of unsaturated hydrocarbons with a triple carbon–carbon bond having the general formula $C_{(n)}H_{(2n-2)}$. *See also* alkenes and alkanes.

**allelochemical** A substance released by a plant that is toxic to a nearby plant of the same or another species. This characteristic (allelopathy) may be used to control the growth of undesirable aquatic weeds or other undesirable species. Also called allelopathic chemical.

**allelopathic chemical** *See* allelochemical.

**allelopathy** *See* allelochemical.

**allergy** An abnormal reaction of the body to a previously encountered substance (allergen), from mild skin reactions to life-threatening reactions.

**alligator teeth** A short V-notch weir for the effluent of a sedimentation basin.

**alligator weed** An aquatic weed, originally from South America, *Althernanthera phioxeroides,* which can be controlled by the alligator flea beetle, *Agasicles hygrophila.*

**Allison screen** A rotating drum screen manufactured by KRC Hewitt, Inc.

**allochthonous** Not formed in the area where found; e.g., organisms introduced into soils or materials carried into a body of water from an outside source (humus, organic detritus, animals). Allochthonous organisms normally survive only for short periods. *See also* autochthonous.

**allogenic succession** Predictable changes in ecosystems based on floods, droughts, and other external factors.

**allotrope** Any of several forms of a substance; e.g., ozone ($O_3$) is an allotropic form of oxygen (O).

**allowable BOD loading** *See* allowable pollutional load.

**allowable infiltration** The infiltration rate that is judged reasonable for a new sewer pipe to be acceptable. The Ten States Standards recommend an allowable infiltration of 200 gpd/in · dia/mile. The acceptable infiltration rate before existing sewers are replaced or rehabilitated may be as high as 2000 gpd/in · dia/mile.

**allowable leakage** A parameter used to determine acceptability of a newly constructed water line. The American Water Works Association recommends an allowable leakage (L, in $m^3$/h or gal/h) less than (McGhee, 1991)

$$L = NDP^{0.5}/C \qquad \text{(A-83)}$$

where $N$ = number of joints in the length tested; $D$ = nominal pipe diameter (in mm or in); $P$ = average test pressure (in kPa or lb/in$^2$); and $C$ = constant depending on the system of units (32.6 for metric, 1850 for English units).

**allowable pollutional load** The maximum waste load that can be discharged to a receiving stream without violating some water quality standard. Parameters that affect the magnitude of an organic load, for example, include the deoxygenation and self-purification constants, as well as the initial and critical dissolved oxygen deficits. *See* oxygen sag analysis and the Streeter-Phelps formulation.

**alloy** A mixture of two or more metals, or a metal and a nonmetal. Some alloys are very stable and resistant to corrosion.

**alluvial** Relating to mud and/or sand deposited by flowing water. Alluvial deposits may occur after a heavy rainstorm.

**alluvial cone** *See* alluvial fan.

**alluvial fan** A fan-shaped deposit of coarse materials formed by a stream where its velocity abruptly decreases, e.g., at the mouth of a ravine, at the foot of a mountain, or as the stream flows onto a gentle plain. Also called alluvial cone.

**alluvial flat** A nearly level alluvial surface.

**alluvial land** An area of unconsolidated alluvium.

**alluvial-slope spring** A spring that issues from the boundary between two formations and at a point where the water table slope equals the surface gradient (APHA et al., 1981). Also called boundary spring or border spring.

**alluvial valley floors** Unconsolidated stream-laid deposits where available water is sufficient for such activities as subirrigation or flood irrigation; they do not include upland areas that are generally overlain by a thin veneer of colluvial deposits composed chiefly of debris from sheet erosion, deposits by unconcentrated runoff or slope wash, together with talus, other mass movement accumulation and windblown deposits.

**alluvion** (1) The gradual increase of land on a shore or a riverbank by alluvial deposits or by the recession of water. (2) Overflow of a shore or bank; flood.

**alluvium** Sediments deposited by erosional processes, usually by streams.

**$AlNH_4(SO_4)_2 \cdot 12H_2O$** Chemical formula of aluminum ammonium sulfate.

**$Al(OH)_3$** The chemical formula of aluminum hydroxide or the aluminum ore gibbsite (a constituent of bauxite and laterite).

**$Al(OH)_x(Cl)_y(SO_4)_z$** Chemical formula of polyaluminum chloride sulfate.

**$Al_2O_3$** Chemical formula of aluminum oxide or activated alumina.

**alpha (α)** aeration correction factor Same as alpha coefficient.

**alpha amino acid** *See* amino acid.

**alpha–beta brass** An alloy containing approximately 58 percent copper and 42 percent zinc. Also called Muntz metal.

**alpha (α) coefficient** A factor (α) used in the design of aeration equipment representing the ratio of oxygen transfer coefficients in water and wastewater under the same conditions. It adjusts the oxygen mass transfer coefficient for the effects of basin geometry, mixing intensity, and wastewater characteristics (mainly total dissolved solids); it also depends on the type of equipment considered, the air flow rate, location of the diffusers, operating parameters, and flow regime:

$$\alpha = (K_L\text{a wastewater})/(K_L\text{a tap water}) \quad \text{(A-84)}$$

(Aeration devices are rated using tap water.) Common α values for municipal wastewater range from 0.4 to 0.8 for diffused aeration and from 0.6 to 1.2 for mechanical aeration. *See* oxygen transfer rate. Higher α values are reported for some industrial wastes.

**alpha (α) decay** A radioactive process that emits an alpha particle (a twice-ionized helium nucleus) from an atom and decreases the atomic number by 2. It is common in nuclides with mass numbers greater than 209 and atomic numbers greater than 82. *See* alpha radiation.

**alpha (α) emission** Release of alpha particles. *See* alpha radiation.

**alpha factor** Same as alpha coefficient.

**α-FeOOH** Chemical formula of goethite.

**alpha-mesaprobic zone** An area of active decomposition of organic wastes that is partly aerobic, partly anaerobic in a slow-moving stream. *See also* saprobic classification.

**alphanumeric code (or variable)** A code (or variable), often found in computer programs, consisting of letters, numbers, punctuation marks, and mathematical and other symbols; for example, $S_{f2}$ for friction headloss at cross section 2.

**alpha particle ($^4_2\alpha$)** The nucleus of a helium atom (atomic number = 2, atomic mass = 4) or a positively charged particle consisting of two neutrons and two protons resulting from radioactive decay or nuclear fission; e.g., radon-222 decays to polonium-218 emitting helium-4. Despite their large mass and travel speeds as high as 10 million meters per second, alpha particles are less penetrating than beta and gamma particles. The USEPA has set an MCL of 15 pCi/L for gross alpha particle activity in drinking water. *See also* gross alpha particle.

**α-$PbO_2$** Formula of scrutinyite, one of two forms of lead dioxide.

**alpha (α) radiation** A stream of fast-moving, large, positively charged, easily absorbed alpha particles, emitted from the nucleus of a helium atom undergoing decay. It can be stopped by a sheet of paper and it does not penetrate the skin, but it can harm living organisms. Same as alpha ray.

**alpha (α)** ray *See* alpha radiation.

**$Al_2Si_2O_5(OH)_4$** Chemical formula of kaolinite.

***Alternanthera phioxeroides*** *See* alligator weed.

***Alternaria*** A waterborne fungus, pathogenic to humans. It has been identified in unchlorinated groundwater as well as surface water and service mains.

**alternate depths** The two depths in open-channel flow that correspond to a given specific energy: the subcritical depth and the supercritical depth. Same as alternate stages; used in the analysis of transitions in sewer lines.

**alternate stages** The upper alternate stage and lower alternate stage of flow at which a given discharge corresponds to a given energy head. Same as alternate depths.

**alternating aeration** A nitrogen removal process carried out in one basin (single-sludge activated sludge process or oxidation ditch) in which the (aerobic) nitrification and the (anoxic) denitrification steps occur alternately. This requires timing the aeration equipment for the two cycles, adequate basin volume for the two steps, and wastewater feeding arrangement for denitrification (WEF & ASCE, 1991). Also called intermittent aeration. *See also* Nitrox™, ORP knee, specific denitrification rate.

**alternating device** In a treatment plant or a pumping station, a device used to direct flow, manually or automatically, into different parallel units or through two or more pumps.

**alternative disinfectant** A chemical or physical agent other than chlorine used for the disinfection of water or wastewater; e.g., chloramines, chlorine dioxide, ozone, ultraviolet light. Some alternative disinfectants do not produce measurable disinfection by-products, such as the trihalomethanes and haloacetic acids.

**alternative energy** Energy derived from the sun, biomass, wave or geothermal power, or other renewable sources, as compared to traditional sources (e.g., fossil fuels).

**alternative filtration process** Any type of filtration other than the four processes defined in the USEPA's Surface Water Treatment Rule (SWTR): conventional filtration, diatomaceous earth filtration, direct filtration, and slow sand filtration.

When an alternative filtration process is recommended, the SWTR requires the determination of its performance in terms of the removal of *Giardia* cysts and viruses (AWWA, 1999).

**alternative indicator** An indicator of microbial quality of water other than the coliform group of bacteria. Alternative indicators considered include acid-fast bacteria, bacteriophages, *Clostridia,* endotoxins, and yeasts. *See also* heterotrophic plate count.

**alternative technology** (1) Generally, an approach that aims to use resources efficiently or to substitute resources in order to do minimum damage to the environment. This approach permits a large degree of personal user control over the technology. (2) In wastewater treatment, alternative technology includes proven processes and techniques that provide for the reclaiming and reuse of water, productively recycle wastewater constituents or otherwise eliminate the discharge of pollutants, or recover energy. Specifically, alternative technology includes land application of effluent and sludge, aquifer recharge, aquaculture, direct reuse (nonpotable), horticulture, revegetation of disturbed land, containment ponds, sludge composting and drying prior to land application, self-sustaining incineration, and methane recovery (EPA-40CFR35.2005-4).

**alternative to conventional treatment works for a small community** *See* alternative technology (2). Alternative treatment works for small communities also include individual onsite systems; small-diameter gravity, pressure, or vacuum sewers conveying treated or partially treated wastewater; and small-diameter gravity sewers carrying raw wastewater to cluster systems (EPA-40CFR35.2005-5).

**altitude-control valve** Same as altitude valve.

**altitude valve** A valve used to automatically shut off the flow into a container when the fluid in the container reaches a certain level; e.g., a valve operated based on the water level in an elevated storage tank. Also called an altitude-control valve.

**alum [$Al_2(SO_4)_3$]** (1) The common name of commercial-grade aluminum sulfate, a dusty, astringent, slightly hygroscopic coagulant used in water and wastewater treatment for the removal of solid particles, for phosphorus removal, and for sludge conditioning; densities vary between 600 and 1000 kg/m$^3$. A general formula for the commercial product is $Al_2(SO_4)_3 \cdot xH_2O$, with x = 14 or x = 18. It ionizes in water as follows:

$$Al_2(SO_4)_3 \cdot x\,H_2O \rightarrow 2\,Al^{3+} + 3\,SO_4^{2-} + x\,H_2O \qquad \text{(A-85)}$$

Alum dust irritates the eyes and mucous membranes. Dry alum is noncorrosive and slightly hygroscopic but the alum solution used in water treatment is corrosive. *See also* alumina sulfate, filter alum, waterworks aluminum. (2) Abbreviation of potassium aluminum sulfate [$K_2SO_4 \cdot Al_2(SO_4)_3 \cdot 24H_2O$], a crystalline solid used in medicine, industry, and technical processes; also called potash alum or potassium alum. (3) Abbreviation of aluminum.

**Alumadome** Self-supporting aluminum cover for circular tanks manufactured by Conservatek Industries, Inc.

**Alumavault** Self-supporting aluminum cover for rectangular tanks manufactured by Conservatek Industries, Inc.

**alum coagulation reactions** Hypothetical reactions that show the complex hydrolysis of aluminum in solution and lead to the precipitation of an alum floc, represented by $Al(OH)_3$. For example, the reactions with natural alkalinity [$Ca(HCO_3)_2$], lime [$Ca(OH)_2$], and soda ash ($Na_2CO_3$) are:

$$Al_2(SO_4)_3 \cdot 14\,H_2O + 3\,Ca(HCO_3)_2 \rightarrow 2\,Al(OH)_3 \downarrow + 3\,CaSO_4 + 14\,H_2O + 6\,CO_2 \qquad \text{(A-86)}$$

$$Al_2(SO_4)_3 \cdot 14\,H_2O + 3\,Ca(OH)_2 \rightarrow 2\,Al(OH)_3 \downarrow + 3\,CaSO_4 + 14\,H_2O \qquad \text{(A-87)}$$

$$Al_2(SO_4)_3 \cdot 14\,H_2O + 3\,Na_2CO_3 + 3\,H_2O \rightarrow 2\,Al(OH)_3 \downarrow + 3\,Na_2SO_4 + 14\,H_2O + 3\,CO_2 \qquad \text{(A-88)}$$

or

$$Al_2(SO_4)_3 \cdot 18\,H_2O + 6\,H_2O \rightarrow 2\,Al(OH)_3(s) + 6\,H^+ + 3\,SO_4^{2-} + 18\,H_2O \qquad \text{(A-89)}$$

$$Al_2(SO_4)_3 \cdot 18\,H_2O + 3\,Ca(HCO_3)_2 \rightarrow 2\,Al(OH)_3(s) + 3\,CaSO_4 + 6\,CO_2 + 18\,H_2O \qquad \text{(A-90)}$$

$$Al_2(SO_4)_3 \cdot 18\,H_2O + 3\,Ca(OH)_2 \rightarrow 2\,Al(OH)_3(s) + 3\,CaSO_4 + 18\,H_2O \qquad \text{(A-91)}$$

**Alumdum** A porous diffuser manufactured by Aeration Engineering Research Corp.

**alumina ($Al_2O_3$)** A form of aluminum oxide. *See also* activated alumina.

**alumina exhaustion** The adsorption of fluorides on activated alumina in an acid solution. *See* activated alumina.

**alumina regeneration** The desorption of fluoride by a hydroxide to restore the capacity of activated alumina. *See* activated alumina.

**alumina sulfate** Another name for aluminum sulfate.

**aluminium** British spelling of aluminum.

**alumino silicate (or aluminosilicate)** A natural or synthetic aluminum silicate that contains elements of alkali or alkaline-earth metals, as a feldspar, zeolite, or beryl; used as a cation exchange product in residential water softeners. *See also* gel zeolite.

**aluminum (Al)** An abundant, silver-white, ductile, malleable, lightweight but high strength:weight ratio, nonferrous metal with good corrosion resistance (but attackable by the alkali in concrete) as well as electrical and thermal conductivity. Atomic weight = 26.98. Atomic number = 13. Specific gravity = 2.70 at 20°C. Valence = 3. Melting point = 660°C; boiling point = 2,467°C. Aluminum may be used in a mixture with coal as a filter medium, without the addition of coagulants or coagulant aids, for the treatment of water of low alkalinity (McGhee, 1991). Aluminum may play a role in the aging process (e.g., neuropathological disorders such as Alzheimer's disease); in high oral doses it may cause irritation of the intestinal tract. A goal of 50 micrograms per liter of aluminum has been set for drinking water; however, the current median concentration of the metal is about 90 micrograms per liter in North American plants that use alum coagulation (Droste, 1997). Ion exchange, reverse osmosis, and electrodialysis are the most effective methods to remove this contaminant. Sometimes abbreviated as alum.

**aluminum ammonium sulfate [$AlNH_4(SO_4)_2 \cdot 12H_2O$]** A crystalline solid containing aluminum and ammonium; it is used as a coagulant in water treatment and in the manufacture of paper. Also called ammonium alum.

**aluminum floc** The gelatinous floc resulting from the coagulation of solid particles in water, using aluminum salts, particularly aluminum sulfate. The predominant form of the precipitate is a species of aluminum hydroxide [$Al(OH)_3$], which settles slowly.

**aluminum hydrate** An aluminum compound formed during water treatment using aluminum salts; it may cause turbidity in the treated water.

**aluminum hydrolysis product** Any of the chemical species formed when alum [$Al_2(SO_4)_3 \cdot xH_2O$] is added to water for coagulation (AWWA, 1999):

$$\text{Dissolution: } Al_2(SO_4)_3 \cdot 12\,H_2O \rightarrow 2\,Al(H_2O)_6^{+3} + 3\,SO_4^{-2} \quad \text{(A-92)}$$

$$\text{Hydrolysis: } Al(H_2O)_6^{+3} \rightarrow Al(H_2O)_5OH^{+2} + H^+ \quad \text{(A-93)}$$

$$Al(H_2O)_5OH^{+2} \rightarrow Al(H_2O)_4(OH)_2^{+1} + H^+ \quad \text{(A-94)}$$

$$Al(H_2O)_4(OH)_2^{+1} \rightarrow Al(H_2O)_3(OH)_3 + H^+ \quad \text{(A-95)}$$

$$Al(H_2O)_3(OH)_3 \rightarrow Al(H_2O)_2(OH)_4^- + H^+ \quad \text{(A-96)}$$

**aluminum hydroxide** An insoluble, inorganic chemical resulting from the use of aluminum sulfate as a coagulant in water treatment. The predominant form has the formula $Al(OH)_3$, corresponding to a molecular weight of 78.0 and an equivalent weight of 26.0. *See also* aluminum floc.

**aluminum oxide ($Al_2O_3$)** An abrasive compound used as construction material, as a grinding compound, or in coatings.

**aluminum silicate** A natural or synthesized substance, insoluble in water, containing oxides of aluminum ($Al_2O_3$) and silicon ($Si_2O_3$), used to manufacture glass, plastics, and other products. As pumicite, it is a water filtration medium.

**aluminum sulfate [$Al_2(SO_4)_3$]** An inorganic, grayish-white, crystalline, water-soluble solid, prepared by combining bauxite with sulfuric acid; used in water purification and in paper manufacturing. The commercial product used in water treatment contains approximately 17% water-soluble aluminum oxide ($Al_2O_3$). Molecular weight = 600. Equivalent weight = 100. *See* alum. Also called alumina sulfate, filter alum, waterworks aluminum.

**aluminum trichloride ($AlCl_3$)** A chemical that reacts with water to form aluminum hydroxide [$Al(OH)_3$] and hydrochloric acid (HCl). The latter is a corrosive and toxic product, which also reduces the pH of a solution:

$$AlCl_3 + 3\,H_2O \rightarrow Al(OH)_3 + 3\,HCl \quad \text{(A-97)}$$

**aluminum trioxide** Same as aluminum oxide; it contains three oxygen atoms.

**alum precipitation** Alum reacts with calcium and magnesium bicarbonate alkalinity to precipitate aluminum hydroxide. *See* alum coagulation for the reaction with calcium bicarbonate. Lime is sometimes added to supplement the alkalinity present. A similar reaction occurs with magnesium bicarbonate:

$$Al_2(SO_4)_3 \cdot 14\,H_2O + 3\,Mg(HCO_3)_2 \rightarrow 2\,Al(OH)_3 \downarrow + 3\,MgSO_4 + 14\,H_2O + 6\,CO_2 \quad \text{(A-98)}$$

**alum recovery** The recovery from sludge of the aluminum salt used in the coagulation process by addition of sulfuric acid:

$$2\,Al(OH)_3 + 3\,H_2SO_4 \rightarrow Al_2(SO_4)_3 + 6\,H_2O \quad \text{(A-99)}$$

**alum sludge** Amorphous solids produced by water or wastewater treatment that uses alum as coagulant; alum sludge consists of a floc with entrapped solid particles. Thickened alum sludge is usually dewatered mechanically and then processed for recovery of chemicals or discharged in drying beds.

**AlumStor** Modular storage and feed equipment manufactured by ModuTank, Inc.

**Alzheimer's disease** A form of dementia characterized by confusion, memory lapses, and progressive loss of mental ability; first described in 1907 by the German neurologist A. Alzheimer (1864–1915). *See* aluminum.

**Amberjet™** An ion exchange resin produced by Rohm & Haas Co.

**Amberpack™** An ion exchange resin manufactured by Rohm and Haas for use in countercurrent packed-bed systems; first intended for water treatment but also applicable to heavy metals recovery from wastewater.

**Ambersorb®** A product of Rohm & Haas Co. for the adsorption of volatile organic chemicals.

**ambient** Surrounding or encircling; environmental or surrounding conditions. Ambient temperature is the temperature of the surrounding medium, e.g., the temperature in a testing laboratory. Ambient water quality standards relate to designated criteria and uses of the water under consideration.

**ambient air** Any unconfined portion of the atmosphere: open air, surrounding air; often used in aeration and in ozone generation as a source of oxygen, which constitutes 21% of the volume of air. It also contains water vapor and particulates.

**ambient aquatic life advisory concentration (AALACS)** The EPA's advisory concentration limit for acute or chronic toxicity to aquatic organisms as established in the Clean Water Act. *See also* ambient water criterion.

**ambient monitoring** Monitoring conducted beyond the immediate influence of a discharge pipe or injection well, including sampling of sediments and living resources.

**ambient quality standard** *See* environmental quality standard.

**ambient temperature** Temperature of the surrounding air or water. *See* ambient.

**ambient water criterion** The concentration of a toxic pollutant in a navigable water that, based upon available data, will not result in adverse impact on important aquatic life or on consumers of such aquatic life, after exposure of that aquatic life for periods of time exceeding 96 hours and continuing at least through one reproductive cycle; and will not result in a significant risk of adverse health effects in a large human population based on available information such as mammalian laboratory toxicity data, epidemiological studies of human occupational exposures or human exposure data, or any other relevant data (EPA-40CFR129.2-g).

**ambient water criterion for aldrin/dieldrin in navigable waters** Concentration limit established by the USEPA for this toxic pollutant: 0.003 μg/L.

**ambient water criterion for benzidine in navigable waters** Concentration limit established by the USEPA for this toxic pollutant: 0.1 μg/L.

**ambient water criterion for DDT in navigable waters** Concentration limit established by the USEPA for this toxic pollutant: 0.001 μg/L.

**ambient water criterion for endrin in navigable waters** Concentration limit established by the USEPA for this toxic pollutant: 0.004 μg/L.

**ambient water criterion for PCBs in navigable waters** Concentration limit established by the USEPA for this toxic pollutant: 0.001 μg/L.

**ambient water criterion for toxaphene in navigable waters** Concentration limit established by the USEPA for this toxic pollutant: 0.005 μg/L.

**ambient water quality criteria** USEPA's maximum acute or chronic toxicity concentrations for the protection of aquatic life and its uses as established under the Clean Water Act.

**ambient water quality standards** *See* ambient.

**amblygonite** An important lithium ore, lithium aluminum fluorophosphates, $Li(AlF)PO_4$.

**AMBR® process** A simple and stable anaerobic treatment method that uses baffles to direct the flow of wastewater upward through sludge blanket compartments in series. The waste enters at the bottom of the first compartment and leaves at the top of the last compartment. It is similar to the anaerobic baffled reactor, except that (1) it incorporates mixing in each compartment and (2) it does not include recycling but the influent and effluent points periodically switch sides.

**Amcec** A line of products of Wheelabrator Clean Air Systems, Inc. for the control of volatile organic compounds.

**ameba** (plural: amebas, amoebae) A freshwater, marine, free-living or parasitic one-celled protozoan consisting of a granular nucleus surrounded by a jellylike cytoplasm. Amebas are sometimes found in trickling filter slime and on aeration basin walls. Some amebas cause amebic dysentery and other waterborne diseases. Granular filtration and stabilization ponds are effective against the cysts of pathogenic amebas, but these survive ordinary wastewater treatment and are found in the sludge. Also spelled amoeba. *See also Entamoeba histolytica.*

**amebae** Plural of ameba.

**amebal pathogen** A parasitic or free-living ameba that can cause infection or illness.

**amebiasis** An intestinal infection caused by waterborne or foodborne amebas. *See* amebic dysentery (its common name) for more detail.

**amebic abscess** One of the manifestations of waterborne amebiasis; *see* amebic dysentery.

**amebic cyst** Cysts of pathogenic amebae are relatively large and are removed by filtration and in stabilization ponds.

**amebic dysentery** A severely debilitating disease, common in tropical climates, caused by a waterborne or foodborne protozoan parasite, *Entamoeba histolytica.* The disease is asymptomatic in most infected persons, but it may cause diarrhea, bloody stools, sometimes fever, and even extraintestinal complications. The parasite invades the colon, forms ulcers, migrates to the liver and other organs, and develops an amebic abscess. This disease is transmitted via the fecal–oral route and is distributed worldwide. Between 1 and 10% of the U.S. population carries amebic cysts but the incidence of the disease is very low. *See also* bacillary dysentery.

**amendment** (1) A soil-conditioning substance added to improve soil qualities and promote plant growth. *See* soil amendment for detail. (2) An organic material, such as wood chips, manure, sawdust, straw, rice hulls, and recycled compost, added to sludge (or other feed substrate) in a composting operation to promote uniform air flow while reducing the bulk weight and the moisture content. *See also* bulking agent, organic amendment. (3) In bioremediation, an amendment is a substance added to soil or the waste being treated to enhance the process; e.g., bioaugmentation cultures, enzymes, fertilizers, hydrogen peroxide, moisture, nutrients, and surfactants.

**Amerfloc** A polyelectrolyte produced by Ashland chemical to enhance flocculation and settling.

**American-BFV®** A line of butterfly valves manufactured by Val-Matic.

**American degree** A unit of hardness, expressed in grains per gallon as $CaCO_3$, equivalent to approximately 17.2 mg/L as $CaCO_3$. *See also* British degree, French degree, German degree, Russian degree.

**American rule** The doctrine according to which a landowner is entitled to a reasonable amount of groundwater.

**American sieve series** *See* standard sieve series.

**Figure A-10.** American BFV®.

**American Standard Code for Information Exchange (ASCII)** A standard for describing computer-readable text, including alphabetic, numeric, and control characters, all coded in hexadecimal notation.

**American Standard fittings** The standardized types and dimensions of metal pipe fittings published by the American National Standards Institute.

**American trypanosomiasis** *See* its common name, Chagas' disease.

**Ameroid** A polyelectrolyte produced by Ashland chemical to enhance flocculation and settling.

**Ames, Bruce N.** American biochemist, born in 1928, who developed the Ames test in 1973.

**Ames Crosta** Sludge thickening equipment manufactured by Biwater Treatment Ltd.

**Ames DNA test** *See* Ames test.

**Ames test** A test used to infer potential carcinogenicity on the basis of mutagenic activity induced by an agent or to determine whether a carcinogenic chemical acts through a genotoxic mechanism. The test consists of inoculating special strains of *Salmonella typhimurium* bacteria into a histidine-free medium and observing whether these bacteria mutate to a form that synthesizes histidine. The Ames test has showed that, among disinfecting agents, the ranking for the production of mutagenic material is chlorine > chloramines > chlorine dioxide > ozone (Droste, 1997). Also called the *Salmonella* microsome assay. *See also* activation (3).

**A metal** *See* A-type metal.

**ametoecious parasite** A host-specific parasite; the opposite of a metoecious parasite. Also called monoxenic parasite.

**amide** A derivative of carboxylic acids, e.g., urea or diamide. Amides are hydrocarbon derivatives in which the amide group ($CO \cdot NH_2$) replaces a hydrogen atom.

**amide group** *See* amide.

**amine** An organic compound of nitrogen, derivative of ammonia ($NH_3$), in which carbon chains or rings of the alkyl groups replace hydrogen atoms; e.g., aminomethane (common name methylamine, $CH_3—NH_2$). General formula: $R_{3-n}NH_n$, with n = 0, 1, or 2. Amines are primary, secondary, or tertiary according to the number of hydrogen atoms substituted, respectively, 1, 2, or 3. *See also* alkanes.

**aminoacetic acid ($H_2NCH_2COOH$)** *See* glycine.

**amino acid** An organic compound that contains the carboxylic acid and amino groups (—COOH and —$NH_2$); it has both acidic and basic properties. Amino acids are linked together in huge molecular chains to form proteins. The 20 common natural amino acids (also called standard or alpha amino acids) in proteins have the general formula R—CH—COOH—$NH_2$, where R is a hydrogen atom (H) or a group containing carbon; e.g., glycine (H—CH—COOH—$NH_2$) and tryptophan (NH—$CH_2$—CH—COOH—$NH_2$). Microorganisms, including the bacteria active in wastewater treatment, require prefabricated amino acids for their growth. *See* Figure A-11.

**aminomethane** *See* amine. Also called methylamine.

**ammeter** An instrument that measures the intensity of electrical current (in amperes).

**ammonate** Same as ammoniate.

**ammonia ($NH_3$)** A colorless, pungent, suffocating, inorganic gaseous compound of hydrogen and nitrogen, very common in nature, highly soluble in water and wastewater as $NH_3$ and mostly as the protonated form (ammonium ion, $NH_4^+$) at low pH. It is used in laboratory reagents, in refrigeration, and in the production of commercial chemicals. Ammonia is an important parameter for water and wastewater treatment. (a) It is the principal form of nitrogen in untreated wastewater and exerts an oxygen demand of 4.6 mg/L per mg/L as N:

$$NH_4^+ + 2\,O_2 \rightarrow NO_3^- + 2\,H^+ + H_2O \quad \text{(A-100)}$$

(b) It may increase metal solubility and cause corrosion through the formation of complexes. (c) It can interfere with COD determinations in the presence of chlorides. (d) It is toxic to fish at high concentrations. (e) An average domestic wastewater contains 15–50 mg/L of $NH_3$—N. (f) Air stripping and ion exchange are effective methods of ammonia removal. (g) It forms chloramines with hypochlorous acid; *see* chloramination. (h) High concentrations of ammonia inhibit anaerobic metabolism. (i) When oxidized to nitrite and nitrate, ammonia in an effluent exerts a significant oxygen demand on receiving streams. (j) It increases chlorine demand. *See also* ammoniac, ammonia solution, ammonia water, ammonium chloride, ammonium hydroxide, ammonium sulfate, anhydrous ammonia, aqua ammonia, aqua ammoniae, aqueous ammonia, chloramination reaction sequence, Haber-Bosch process.

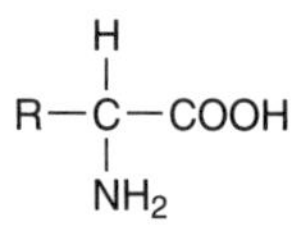


**Figure A-11.** Amino acid.

**ammonia air stripping** *See* ammonia stripping.

**ammonia alum** Same as ammonium alum.

**ammonia as nitrogen** *See* ammonia nitrogen.

**ammoniac or ammoniacum** A brownish-yellow gum resin, with an acrid taste, extracted from a plant and used in ceramics and in medicine. Also called gum ammoniac.

**ammonia N** *See* ammonia nitrogen.

**ammonia nitrogen (ammonia N, or $NH_3$-N)** The amount of elemental nitrogen present in the form of ammonium ion ($NH_4^+$) or ammonia gas ($NH_3$) depending on pH:

$$NH_4^+ \leftrightarrow NH_3 + H^+ \quad \text{(A-101)}$$

Also called ammonia as nitrogen. It is the value obtained by manual distillation (at pH 9.5) followed by the Nesslerization method. The concentration of ammonia nitrogen in a sample is equal to the concentration of nitrogen times 17/14.

**ammonia pollution** The discharge of ammonia-containing wastewater from such industries as gas and coke production, and chemical manufacture.

**ammonia solution** A solution of ammonia gas ($NH_3$) in water. *See* ammonium hydroxide.

**ammonia stripping** A wastewater treatment process that removes ammonia ($NH_3$), a volatile gas, by exposing the liquid to a large quantity of air in a tower and raising the pH to 10.8–11.5. The high pH is necessary to ensure that the ammonia is in the un-ionized form. Low temperature is a severe limitation on the performance of the process. *See* theoretical air-to-liquid ratio, approximate flooding, packing factor, steam stripping, stripping factor, and stripping tower design equations.

**ammoniate** A product obtained by stoichiometrically adding ammonia to another substance; e.g., the ammoniate of calcium chloride ($CaCl_2 \cdot 8$

$NH_3$) or copper sulfate ($CuSO_4 \cdot 4\ NH_3$). Also called ammonate.

**ammoniation** The process of treating or combining a substance with ammonia, for example, the reaction with chlorine to form chloramines.

**ammoniator** An orifice flowmeter or dosimeter, similar to a chlorinator, that is designed to apply ammonia or ammonium compounds to water or wastewater at a predetermined rate. This equipment allows the formation of chloramines and includes a protective apparatus with a ventilation outlet near the ceiling. *See also* chemical feeder.

**ammonia water** A solution of ammonia gas ($NH_3$) in water. *See* ammonium hydroxide.

**ammonification** (1) The bacterial decomposition of organic nitrogen to ammonia; the release of ammonia from dead and decaying cells. Also called ammonium mineralization. *See also* ammonium assimilation or immobilization. (2) The act of permeating or being permeated with ammonia, as in the formation of fertilizers.

**ammonifier** An organism (generally a bacterium) that converts organic nitrogen to ammonia in wastewater or solid waste.

**ammonifying bacteria** Bacteria that release ammonia from the protein of dead tissues.

**ammonium ($NH_4^+$)** A form of nitrogen that is usable by plants; a univalent ion ($NH_4^+$) or group that acts as a metal to form salts from reactions of ammonia with acids. Molecular weight = equivalent weight = 18. Electrical charge = 1+. It is formed by the reaction of ammonia ($NH_3$) with hydrogen (H).

**ammonium acetate [$NH_4(C_2H_3O_2)$]** A white, crystalline, water-soluble solid, used in dyes and preservatives.

**ammonium alum** A chemical product used as a coagulant in water treatment and containing aluminum. *See* aluminum ammonium sulfate.

**ammonium assimilation** The use by microbial cells of ammonium from nitrogen fixation to form proteins, cell components, and nucleic acids. Also called ammonium immob-ilization. *See also* assimilatory nitrate reduction and ammonification (Maier et al., 2000).

**ammonium bicarbonate ($NH_4HCO_3$)** A white, crystalline, water-soluble chemical, used in baking powder.

**ammonium bifluoride ($NH_4HF_2$)** A white, crystalline, water-soluble, poisonous solid used in cleaning and sterilizing equipment.

**ammonium carbamate ($CH_6N_2O_2$)** A white, volatile, crystalline, water-soluble powder, used as a fertilizer.

**ammonium chloride ($NH_4Cl$)** A white, crystalline, water-soluble powder, used in medicine and in industry. Also used in the production of chloramines for water disinfection and in water treatment as a dechlorinating agent. Also called sal ammoniac.

**ammonium cyanate ($CH_4N_2O$)** A white, crystalline solid that converts to urea when heated.

**ammonium fluosilicate [$(NH_4)_2SiF_6$]** A chemical used in water fluoridation. Molecular weight = 178.

**ammonium hydroxide ($NH_4OH$)** A strongly basic inorganic compound, a solution of ammonia gas ($NH_3$) in water ($H_2O$); sometimes used in the production of chloramines for water disinfection or as a chlorine reducing agent (antichlor). The commercial form, aqua ammonia, approximately 30% ammonia, is corrosive and hazardous. Also called ammonia solution, ammonia water, aqua ammonia, aqueous ammonia.

**ammonium immobilization** The use by microbial cells of the ammonium resulting from nitrogen fixation to form proteins, cell components, and nucleic acids. *See* ammonium assimilation (Maier et al., 2000).

**ammonium ion ($NH_4^+$)** A form of ammonia found in solution.

**ammonium lactate ($C_3H_9NO_3$)** A colorless or light-yellow, water-soluble liquid, used in electroplating.

**ammonium mineralization** The release of ammonia from dead and decaying cells. *See* ammonification for detail.

**ammonium molybdate** A reagent used in the laboratory procedure to determine the concentration of orthophosphates; it reacts with phosphate to produce a blue suspension.

**ammonium nitrate ($NH_4NO_3$)** A white, crystalline, water-soluble powder, produced from ammonia and nitric acid, used in fertilizers and explosives, or as a nitrogen source for nutrient-deficient industrial wastes. It is the main form of nitrate found in agricultural runoff.

**ammonium salt** The result of the reaction of ammonium hydroxide ($NH_4OH$) with an acid; e.g., ammonium sulfate from sulfuric acid:

$$H_2SO_4 + 2\ NH_4OH \rightarrow (NH_4)_2SO_4 + 2H_2O \quad \text{(A-102)}$$

**ammonium silicofluoride ($NH_4SiF_6$)** A chemical used to add approximately 1.0 mg/L of fluoride to drinking water after conventional treatment. *See also* hydrofluosilicic acid, sodium fluoride, sodium silicofluoride.

**ammonium stearate ($C_{18}H_{39}NO_2$)** A tan, waxlike solid, insoluble in water, used in cosmetics.

**ammonium sulfate [$(NH_4)_2SO_4$]** A colorless, crystalline, water-soluble solid, used mainly as a fertilizer, but sometimes in the production of chloramines for water disinfection. Molecular weight = 132. Equivalent weight = 66.

**ammonium thiocyanate ($CH_4N_2S$)** A colorless, crystalline, water-soluble solid, used as a fertilizer or in the textile industry.

**ammonium thiosulfate [$(NH_4)_2S_2O_3$]** A white, crystalline, water-soluble solid, used in cleaning substances for metal alloys.

**amoeba** Same as ameba.

**amoebae** Same as amebae.

**amoebal** Same as amebal.

**amoebiasis** Same as amebiasis.

**amoebic** Same as amebic.

**amorphous** Having no shape or form, noncrystalline.

**amperage** The strength of an electric current measured in amperes. The amount of electric current flow, similar to the flow of water in gallons per minute.

**ampere** The unit used to measure current strength; the current produced by electromotive force of one volt acting through a resistance of one ohm.

**amperometric** Based on the electric current that flows between two electrodes in a solution.

**amperometric method** An analytical method that uses an electrochemical reaction in a solution.

**amperometric titration** A method of measuring concentrations of certain substances in water (such as strong oxidizers and chlorine residuals) by titration with a reducing agent of known normality, using a constant electric current. *See also* iodometric method.

**amperometric titrator** A titration instrument that has an electrometric device to indicate when the reaction is complete; it is used in the laboratory analysis of chlorine, chloramines, and other oxidants.

**amphi-aerobic pond** A wastewater treatment pond in which organic matter is decomposed by microorganisms that function facultatively as aerobic or anaerobic, depending on diurnal or seasonal factors such as photosynthesis changes (Fair et al., 1971). Also called hetero-aerobic pond. *See also* facultative pond, and lagoons and ponds.

**amphibian** An animal that can live on land or in water.

**amphibiotic** Living in water during the larval stage and on land during the adult stage.

**amphibious** Living or capable of living in water or on land; able to operate both on land and in water.

**amphipathic molecule** A molecule that contains hydrophilic and hydrophobic groups.

**ampholyte polymer** A polymer that has both positive and negative sites. *See also* anionic, cationic, and nonionic polymers.

**ampholytic polymer** A long-chain, high-weight molecule bearing a large number of charged groups, with a neutral charge on the molecule, as opposed to cationic or anionic polymers that are positively or negatively charged. Figure A-12 is an example of the molecular structure of an ampholytic polymer (McGhee, 1991).

**amphoteric** Characteristic of a substance that can react in water as either a weak acid or a weak base, depending on prevailing pH; e.g., amino acids.

**amphoteric compound** A coordination compound that can exist in strong acids and strong bases. For example, aluminum hydroxide, [$Al(OH)_3$](s), can form aluminum ion, ($Al^{3+}$) in acidic solution or the aluminate ion [$Al(OH)_4^-$](aq), in a basic solution.

**amplifying host** An organism that serves as a host to multiple pathogens; beavers infected with *Giardia lamblia* from human excreta can return millions of cysts for every one ingested, with the potential for an outbreak of giardiasis.

**AMR** Acronym of automatic meter reading.

***Amstichthys nobilis*** A hardy fish species of the carp family that survives well in stabilization ponds used for aquaculture.

**AMU or amu** Acronym of atomic mass unit.

**AMW** Acronym of apparent molecular weight.

**AMWA** Acronym of Association of Metropolitan Water Agencies.

***Anabaena*** A genus of filamentous cyanobacteria or blue-green algae, found in polluted lakes and ponds and appearing to the naked eye as short grass clippings or strings of beads. They are responsible for undesirable fishy tastes and odors in drinking water sources. They also produce toxins (anatoxin a, hepatotoxin) that can cause liver damage and gastrointestinal illness. The formation of trihalomethanes has been observed in the laboratory from reactions of chlorine with *Anabaena* cells.

***Anabaena flos aquae*** A species of cyanobacteria (blue-green algae) commonly found in algal blooms and responsible for the production of hepatotoxins and neurotoxins.

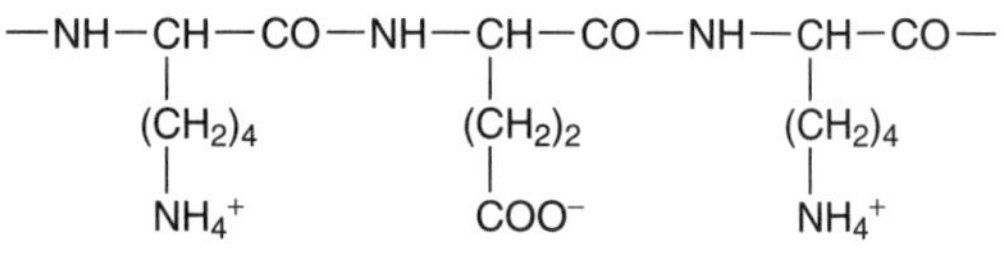


**Figure A-12.** Ampholytic polymer.

**anabolic reaction** A bacterial reaction leading to the synthesis of new cell constituents and metabolites; anabolism.

**anabolism** The fabrication of complex molecules from simpler ones; also called assimilation or synthesis. *See also* catabolism and metabolism.

**anabranch** A contraction of anastomotic branch. (1) An effluent branch that rejoins the main stream downstream, thus forming an island. (2) A stream divided by relatively large permanent islands, with branch channels wider and more distinctly separated than those of a braided stream. *See also* sedimentation terms.

***Anacystis*** A genus of cyanobacteria, similar to *Anabaena.*

**anadromous** Characteristic of fish, e.g., some salmon genera, that migrate from salt water to freshwater for spawning.

**anaerobe** An organism that can live in the absence of oxygen. *See* anaerobic organism.

**anaerobic** Pertaining to a life or process that occurs in, or is not destroyed by, the absence of oxygen; a condition in which "free" (atmospheric) or dissolved oxygen is not present. *See also* aerobic, anaerobic state, anoxic, and facultative.

**anaerobic alkalinity correction** In the anaerobic digestion process, an adjustment of the total alkalinity determined by titration to pH 4.3 ($Alk_{4.3}$) to obtain the total bicarbonate alkalinity to the same endpoint ($TBA_{4.3}$). It takes into consideration the contribution of volatile acids (VA) to alkalinity. The correction is equal to 85% of the concentration of volatile acids as $CaCO_3$ (Droste, 1997):

$$TBA_{4.3} = Alk_{4.3} - 0.85\ VA \qquad \text{(A-103)}$$

*See also* usable alkalinity.

**anaerobically digested biosolids** A dark brown to black material, from primary or secondary wastewater treatment, containing a significant amount of gas, with a relatively faint odor.

**anaerobic ammonium oxidation process** *See* Anammox process.

**anaerobic/anoxic/aerobic process** *See* A²O™ process.

**anaerobic bacteria** *See* anaerobic organism.

**anaerobic baffled reactor (ABR) process** A simple and stable anaerobic treatment method that uses baffles to direct the flow of wastewater upward through sludge blanket compartments in series. The waste enters at the bottom of the first compartment and leaves at the top of the last compartment, with recycling from the effluent to the influent lines. *See also* the AMBR® process and the upflow sludge blanket reactor (Metcalf & Eddy, 2003).

**anaerobic biological treatment** Waste treatment that uses anaerobic organisms to stabilize organic matter. *See* anaerobic treatment.

**anaerobic composting** Composting under anaerobic conditions, i.e., in the absence of air or oxygen. *See* anaerobic digestion, aerobic composting.

**anaerobic condition** A condition in which free (atmospheric) or dissolved oxygen is not available. Then, other substances can act as oxidants or electron acceptors. *See,* for example, sulfate reduction.

**anaerobic contact** A wastewater treatment process developed in the 1950s, using flocculation and digestion under anaerobic conditions. It includes the creation and maintenance of a biological floc, sludge return, disposal of excess sludge, and production of stable end products (e.g., methane and carbon dioxide). Also called anaerobic digestion or wastewater flocculation, it is used mainly for strong soluble organic wastes, e.g., from the meat- and poultry-packing industry. Its flowsheet is similar to that of the complete-mix activated sludge process; *see* Figure A-13.

**anaerobic covered-lagoon process** An anaerobic lagoon used for the treatment of strong industrial wastes, e.g., from the poultry- and meat-packing industries. It consists of a plug-flow or complete-mix reactor with a geomembrane cover and liner, including a gas-recovery system. Hydraulic detention and solids retention times are high, in the ranges of 30–50 and 50–100 days, respectively. *See also* the ADI-BVF® reactor (Metcalf & Eddy, 2003).

**anaerobic dechlorination** *See* anaerobic dehalogenation.

**anaerobic decomposition** Reduction of the net energy level and change in chemical composition of organic matter caused by microorganisms in an oxygen-free environment. The process requires an alternative electron acceptor: an organic compound in fermentation or an inorganic acceptor in anaerobic respiration. Also called anaerobic degradation. *See* anaerobic digestion.

**anaerobic degradation** Same as anaerobic decomposition.

**anaerobic dehalogenation** A method for the anaerobic degradation of toxic and recalcitrant halogenated organics, more particularly chlorinated compounds such as carbon tetrachloride, chlorohydrocarbons, PCBs, pentachlorophenol, tetrachloroethene, trichlorobenzene, and trichloroethene. In the anaerobically mediated reactions, the

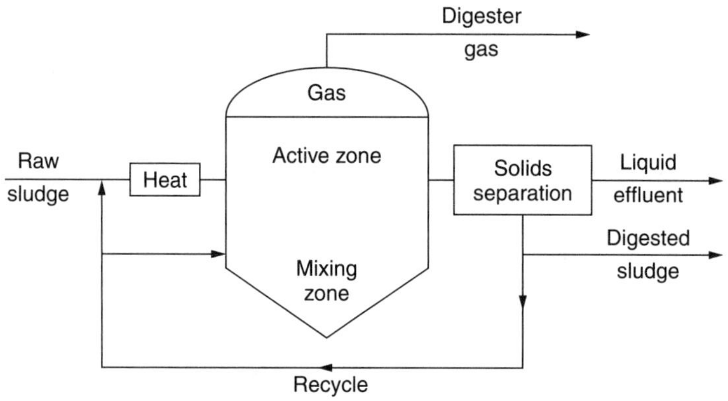


**Figure A-13.** Anaerobic contact (adapted from Water Environment Federation, 1991).

hydrogen produced by fermentation replaces chlorine in the compound in successive steps until complete dechlorination (Metcalf & Eddy, 2003), e.g., from tetrachloroethene ($CCl_2CCl_2$ or $C_2Cl_4$) to trichloroethene ($CHClCCl_2$ or $C_2HCl_3$) to dichloroethene ($CH_2CCl_2$ or $C_2H_2Cl_2$) to vinyl chloride ($CH_2CHCl$ or $C_2H_3Cl$) to ethene ($CH_2CH_2$ or $C_2H_4$).

**anaerobic digestion** The biochemical decomposition of organic matter in sewage sludge, organic mud, and other organic wastes into methane gas ($CH_4$, about 65% by volume), carbon dioxide ($CO_2$, about 30% of the gas volume), and other end products (organic acids, alcohols, etc.) by microorganisms in the absence of elemental oxygen. Anaerobic digestion is used mainly for the stabilization of sludge, but also for the reduction of pathogens. It is also effective in reducing the soluble BOD of some industrial wastes, e.g., from slaughterhouses, paper mills, and dairies. Also called fermentation, putrefaction, or anaerobic composting. *See also* anaerobic contact, anaerobic sludge digestion.

**anaerobic digestion gas** The gas mixture (digester gas, specific gravity = 0.86) released during the anaerobic decomposition of organic matter, more specifically during sludge digestion. Saturated with water vapor, it contains by volume approximately 65% of methane ($CH_4$), 30% of carbon dioxide ($CO_2$), and small amounts of nitrogen ($N_2$), hydrogen ($H_2$), and hydrogen sulfide ($H_2S$). The fuel value of sludge gas is usually used for heating, power production, etc.; because of its water vapor content and the gases other than methane, this fuel value is about 600 BTU/ft$^3$, i.e., 60% of that of natural gas (Metcalf & Eddy, 2003 and Fair et al., 1971).

**anaerobic downflow attached growth process** A method of treatment for strong wastewaters (e.g., brewery, cheese whey, citrus, molasses, and piggery slurry wastes) using an anaerobic downflow reactor filled with packing materials having a high void volume. A portion of the effluent may be recycled. COD removals vary from 40 to more than 95%, depending on the type of waste and the hydraulic detention time (Metcalf & Eddy, 2003).

**anaerobic expanded-bed reactor (AEBR) process** A method of treating municipal wastewater using an anaerobic upflow reactor containing a small packing such as silica sand of diameter 0.2–0.5 mm and operated at an expansion of 20%. *See* fluidized bed reactor.

**anaerobic fermentation** Decomposition of organic matter (e.g., sugar) under anaerobic conditions with low cell yield, production of alcohol, end products likely to be odorous, and, usually, an unstable effluent. It is used mainly for sludge stabilization. *See also* aerobic fermentation.

**anaerobic filter** An upward flow column filled with packing and operated under anaerobic conditions for the treatment of carbonaceous organic matter such as the warm, high-strength, biodegradable side streams of the thermal sludge-conditioning process. The anaerobic organisms are retained on the media, thus allowing high solids retention times (e.g., 100 days) with short hydraulic detention times.

**anaerobic fixed-film processes** *See* upflow anaerobic filter, downflow anaerobic filter, anaerobic fluidized-bed reactor, upflow anaerobic sludge blanket reactor.

**anaerobic flocculation** A wastewater treatment process that uses flocculation and digestion under anaerobic conditions. *See* anaerobic contact for more detail.

**anaerobic fluidized-bed process** A method of treating strong industrial wastewaters using an anaerobic reactor containing a packing material and operated at a hydraulic rate sufficient to fluidize the bed. It does not normally meet secondary treatment standards and may be part of a two-stage anaerobic–aerobic process. *See* fluidized bed reactor (McGhee, 1991).

**anaerobic fluidized-bed reactor (AFBR) process** A method of treating municipal wastewater using an anaerobic upflow reactor containing a small packing such as silica sand of diameter 0.2–0.5 mm (or diatomaceous earth, activated carbon, etc.). It is operated with effluent recycle, at an expansion of 100% and a fluid velocity of 20 meters per hour. The reactor has the expanded media at the bottom, below the clarifier zone. *See* Figure A-14 and fluidized bed reactor.

**anaerobic heterotrophic respiration** The type of reaction carried out by *Desulfovibrio* in the reduction of sulfur compounds in the presence of lactic acid ($CH_3CHOHCOOH$), with production of acetic acid ($CH_3COOH$) and hydrogen sulfide ($H_2S$):

$$2\ CH_3CHOHCOOH + SO_4^{2-} \rightarrow 2\ CH_3COOH + H_2S + 2\ HCO_3^- \quad \text{(A-104)}$$

**anaerobic lactobacilli** A genus of anaerobic microorganisms commonly found in sludge digesters. *See* bifidobacteria for detail.

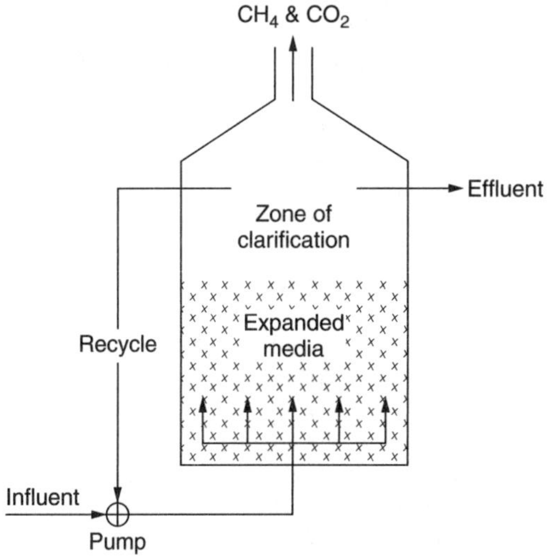


**Figure A-14.** Anaerobic fluidized-bed reactor.

**anaerobic lagoon** An earthen basin used for the treatment of municipal wastewater, industrial waste, or sludge under anaerobic conditions. Sometimes used as an anaerobic pretreatment pond for the processing of wastes of high organic strength (e.g., from meat processing), followed by facultative and maturation ponds. The process is similar to anaerobic digestion without mixing. Anaerobic ponds are deeper than other types of ponds (up to 30 ft), have long detention times (more than 50 days), do not require mechanical aeration devices, and generate only small quantities of sludge, but they produce hydrogen sulfide ($H_2S$) and other odoriferous substances unless their organic and sulfate loadings are low, e.g., below 400 $g/m^3/day$ of BOD and 100 mg/L of $SO_4^{2-}$ (Cairncross and Feachem, 1993). *See also* ADI-BVF® reactor, and lagoons and ponds.

**anaerobic migrating blanket reactor process** *See* AMBR® process.

**anaerobic organism** An organism, particularly a bacterium, that does not require air or free or molecular oxygen to live and grow. In the absence of dissolved oxygen, anaerobes can use the oxygen bound in carbon dioxide ($CO_2$), nitrates ($NO_3^-$), and sulfates ($SO_4^{2-}$), or other electron acceptors (such as oxidized metals) to oxidize organic matter. They then produce carbon dioxide, methane ($CH_4$), nitrogen, ammonia ($NH_3$), or hydrogen sulfide ($H_2S$). *See also* aerobe, facultative anaerobe, obligate anaerobe, strict anaerobe.

**anaerobic/oxic process** A proprietary biological treatment process developed by Kruger, Inc. (patented and marketed by Air Products and Chemicals, Inc.) for BOD removal and enhanced phosphorus uptake. *See* A/O® process for detail.

**anaerobic oxidation** The decomposition of organic matter under anaerobic conditions, proceeding in three steps: hydrolysis, fermentation or acid formation, and methanogenesis. *See also* interspecies hydrogen transfer, nuisance organism.

**anaerobic pond** *See* anaerobic lagoon, and lagoons and ponds.

**anaerobic pretreatment pond** An anaerobic pond used as a preliminary step in the treatment of high-strength waste and followed by facultative and aerobic ponds. *See also* anaerobic lagoon, and lagoons and ponds.

**anaerobic process** Any biological or chemical process carried out in the absence of air or oxygen, as compared to aerobic, anoxic, and facultative processes. Examples: anaerobic sludge digestion and benthal degradation in the absence of dissolved oxygen. Anaerobic processes are subdivided into attached growth, sludge blanket, suspended growth, and hybrid processes.

**anaerobic reaction** A reaction in which oxygen is not the electron acceptor.

**anaerobic respiration** Anaerobic degradation using any of a number of inorganic electron acceptors (as opposed to an organic compound in fermentation), a process that is less efficient than aerobic respiration. Inorganic electron acceptors include, in order of decreasing affinity: nitrate, manganese, iron, sulfate, carbonate, and certain heavy metals. *See also* anaerobiosis.

**Anaerobic Selector Process** A biological process developed by Davis Water & Waste Industries, Inc. for the removal of phosphorus and BOD.

**anaerobic sequencing batch reactor (ASBR)** An anaerobic reactor for wastewater treatment in a single vessel, in four steps: feed, react, settle, and decant, as in the aerobic SBR.

**anaerobic sludge blanket process** Any of the three current processes that use the sludge blanket concept for the anaerobic treatment of wastewater. *See* anaerobic baffled reactor, anaerobic migrating blanket reactor (AMBR®), and upflow sludge blanket reactor.

**anaerobic sludge digestion** The biochemical decomposition of organic matter in sewage sludge into methane gas ($CH_4$), carbon dioxide ($CO_2$), and other end products (organic acids, alcohols, etc.) by microorganisms in the absence of elemental oxygen: facultatively anaerobic organisms that convert organic matter to acids and alcohols, and anaerobic methane fermenters that use these intermediate products to methane and carbon dioxide. The process takes place in large covered tanks (digesters) over a period of weeks depending on temperature, reduces sludge volume by 30–50%, produces methane gas in excess of process requirements, and yields considerable pathogen reduction. It is widely used in treatment plants averaging wastewater flows larger than 5.0 mgd, but aerobic digestion is preferred for smaller plants (e.g., < 2.0 mgd). After digestion, the sludge is dried, burned, used as fertilizer, or disposed in a landfill. *See* the following process variations: anaerobic contact, high rate, low rate, and phase separation. *See also* acid formation, acid-forming bacteria, acid/gas phased digestion, acidogenesis, anaerobic digestion gas, cylindrical sludge digestion tank, digester loading, egg-shaped digester, fermentation, German digester, hydrolysis, hydrolytic bacteria, mesophilic anaerobic digestion, methane bacteria, methane formation, methane production, methanogenesis, separate sludge digestion, single-stage high-rate digestion, staged mesophilic digestion, staged thermophilic digestion, supernatant, temperature-phased digestion, thermophilic anaerobic digestion, two-phased anaerobic digestion, two-stage high-rate digestion, volatile solids destruction.

**anaerobic sporeformer** Any member of a group of anaerobic organisms that form endospores that are very resistant to disinfection and other adverse conditions (AWWA, 1999). This group includes the *Clostridium* genus, particularly the *C. perfringens* species, and the genus *Desulfotomaculum*. Also called sulfite-reducing *Clostridium*.

**anaerobic state** The condition of a water or wastewater reactor in which metabolic reactions have progressed to the point that reduced species predominate compared to oxidized species; the oxidation–reduction potential is then –300 mV > ORP > –400 mV. *See also* aerobic state, anoxic state.

**anaerobic stripper** An anaerobic reactor in which phosphorus is released to solution in enhanced phosphorus uptake processes; see, e.g., PhoStrip™. The stripped biomass returns to the aeration basin, whereas the supernatant is further processed for phosphorus precipitation. Also called stripper tank.

**anaerobic suspended growth process** A process used for the treatment of industrial wastewater, including such variations as the anaerobic contact, anaerobic sequencing batch reactor, and complete-mix suspended-growth anaerobic digester. The design of such units is similar to that of complete-mix activated sludge, the important parameters being solids retention time, solids concentration in the reactor, solids yield, decay coefficient, maximum specific growth rate, and half-velocity constant (Metcalf & Eddy, 2003).

**anaerobic toxicity assay** A test conducted on a waste sample to determine its toxicity relative to the production of methane ($CH_4$). It is the same as the biochemical methane production test, except that the serum bottle also contains acetate propionate as a source of readily degradable substrate (Droste, 1997).

**anaerobic treatment** Biological treatment of wastewater (such as high-strength agricultural or industrial wastes) or digestion of sludge by microorganisms in the absence of air or molecular oxygen. Compared to aerobic processes, anaerobic treatment produces less sludge for disposal, requires a lower energy input (which can be offset by its production of biogas), and requires fewer nutrients and a smaller reactor. Current major variations include conventional digester, upflow fixed-film reactor, digester with recycle, downflow fixed-film reactor, upflow anaerobic sludge blanket reactor, and fluidized-bed reactor. *See* anaerobic digestion.

**anaerobic treatment alkalinity formation** *See* anaerobic treatment ammonia production.

**anaerobic treatment ammonia production** The amount of ammonia ($NH_3$) produced under anaerobic conditions, as estimated from the Buswell–Boruff–Sykes relationship, knowing the composition ($C_vH_wO_xN_yS_z$) of the organic matter in the wastewater being treated (Metcalf & Eddy, 2003). Ammonia then reacts with water ($H_2O$) and carbon dioxide ($CO_2$) to form the ammonium ion ($NH_4^+$) and bicarbonate alkalinity ($HCO_3^-$):

$$NH_3 + H_2O + CO_2 \rightarrow NH_4^+ + HCO_3^- \quad \text{(A-105)}$$

**anaerobic treatment carbon dioxide production** *See* the Buswell–Boruff–Sykes relationship that is used to estimate the amount of carbon dioxide ($CO_2$) produced under anaerobic conditions, knowing the composition ($C_vH_wO_xN_yS_z$) of the organic matter in the wastewater being treated.

**anaerobic treatment gas production** *See* the Buswell–Boruff–Sykes relationship that is used to estimate the amount of ammonia ($NH_3$), carbon dioxide ($CO_2$), hydrogen sulfide ($H_2S$), and methane ($CH_4$) produced under anaerobic conditions, knowing the composition ($C_vH_wO_xN_yS_z$) of the organic matter in the wastewater being treated.

**anaerobic treatment hydrogen sulfide production** *See* the Buswell–Boruff–Sykes relationship that is used to estimate the amount of hydrogen sulfide ($H_2S$) produced under anaerobic conditions, knowing the composition ($C_vH_wO_xN_yS_z$) of the organic matter in the wastewater being treated.

**anaerobic treatment methane production** *See* the Buswell–Boruff–Sykes relationship that is used to estimate the amount of methane ($CH_4$) produced under anaerobic conditions, knowing the composition ($C_vH_wO_xN_yS_z$) of the organic matter in the wastewater being treated.

**anaerobic waste treatment** *See* anaerobic treatment.

**anaerobiosis** The prevalence of anaerobic conditions, i.e., a situation in which microorganisms degrade organic matter in the absence of air or molecular oxygen, or when the rate of microbial oxygen consumption exceeds the rate of oxygen diffusion or production, often creating undesirable odors; e.g., the condition of anaerobic lagoons treating wastewater with a high concentration of sulfate ions (Maier et al., 2000 and Hammer & Hammer, 1996). *See also* anaerobic respiration.

**anaerogenic** Pertaining to organisms that do not produce gas.

**Analite** A portable turbidimeter manufactured by Advanced Polymer Systems.

**anal–oral route of exposure** *See* fecal–oral route.

**analyzer** A device that conducts periodic or continuous measurement of some factor such as chlorine, fluoride or turbidity. Analyzers operate by any of several methods including photocells, conductivity, or complex instrumentation.

**analysis** The examination, as in a laboratory, of a sample of water, wastewater, soil, etc. to determine the type or concentration of its biological, chemical, or physical constituents.

**analyte** The element or substance for which an analysis is conducted.

**analytical acceptance criteria** *See* electroneutrality principle.

**analytical balance** *See* analytic balance.

**analytical detection limit** A measure of the precision of a method of analysis. *See* criterion of detection, critical detection level, instrumental detection level, level of quantification, lower level of detection, and method detection limit.

**analytical model; analytical solution** A model that classical methods, e.g., calculus or even elementary algebra, can solve. An example is the Laplace equation to represent groundwater flow. Analytical solutions are exact, as opposed to numerical solutions, but the analytical model itself is usually a crude approximation of the system it represents.

**analytical monitoring** Analysis of water and wastewater samples in the laboratory to determine the concentrations of various constituents.

**analytical solution** *See* analytical model.

**analytic balance** A precision device for weighing substances and having a sensitivity of 0.1 milligram; often used in water and wastewater laboratories.

**analytic grade** Property of a chemical product that is pure enough for use as a reagent or otherwise in laboratory analysis. *See also* pharmaceutical-grade water, reagent-grade water.

**analyzer** A device used to carry out measurements, e.g., a chlorine analyzer or a turbidity analyzer.

**Anammox bacteria** The anaerobic bacteria used in the Anammox process; order of *Planctomycetales.*

**Anammox process** A recent (mid-1990s), biological denitrification process applied to treat anaerobic digestion centrates with a high ammonia content. The process uses a novel group of slow-growing bacteria in a fluidized bed reactor to oxidize ammonia anaerobically (by nitrite), producing nitrogen gas and some nitrate (Metcalf & Eddy, 2003). (Anammox = anaerobic ammonia oxidation.)

**anaphylactic shock** The result of anaphylaxis.

**anaphylaxis** A severe, sometimes fatal, reacton due to previous exposure to a foreign protein or other substance.

**anastomotic branch** *See* anabranch.

**anatoxin** A toxin produced by the *Anabaena* genus of cyanobacteria (blue-green algae).

**ANC** Acronym of acid-neutralizing capacity.

**An-CAT®** Proprietary equipment manufactured by Norchem Industries for the control of polymer applications.

**anchor ice** Ice formed below the surface of a body of water, attached either to a submerged structure or to the bed; e.g., the ice that forms on screens, gates, and valves of water supply lake intakes. Its formation and clogging of the intake can be prevented by heating the intake's metallic surfaces or slightly raising the temperature of the water. Also called bottom ice, ground ice. *See also* frazil ice.

**ancillary data** Any information necessary for the interpretation of basic water quality data and for the formulation of courses of action., e.g., agricultural and domestic chemical applications, climatological data, geochemistry, geology, human health and ecological effects, land cover, land use, municipal and industrial waste discharges, population and demographics, soils, water use.

**ancillary equipment** Any devices such as piping, fittings, flanges, valves, and pumps used to distribute, meter, or control the flow of hazardous wastes or other regulated substances (EPA-40CFR260.10 and 280.12)

**Anco** A batch mixer manufactured by Enviropax, Inc.

***Ancylostoma*** The genus of hookworm.

***Ancylostoma braziliense*** An animal hookworm that causes incidental human infections unrelated to excreta disposal.

***Ancylostoma canicum*** An hookworm that causes incidental human infections unrelated to excreta disposal.

***Ancylostoma ceylanicum*** An that causes incidental human infections unrelated to excreta disposal.

***Ancylostoma duodenale*** A latent, persistent pathogen, about 10–12 mm long, excreted in human feces; transmissible in filarial form through drinking water, skin penetration, or from soil contact; one of two species that cause the hookworm disease. Each worm consumes a quarter of a milliliter of blood per day. Also called Old World hookworm. *See also Necator americanus.*

**ancylostomiasis** A water- or wastewater-related infection of the small intestine caused by one of two species of human hookworms: *Necator americanus* and *Ancylostoma duodenale;* frequently symptomless but sometimes characterized by anemia, gastrointestinal pain, edema, or other symptoms. *See* hookworm disease.

**Anderson–Domsch equation** A relationship proposed in 1978 to estimate biomass carbon (as a measure of microbial biomass) in soil:

$$Y = 40.04\,X + 0.37 \qquad \text{(A-106)}$$

where $Y$ is the biomass carbon, mg/100 g of dry soil, and $X$ is the respiration rate, ml $CO_2$/100 g dry sediment/hour (Maier et al., 2000).

**anecdotal data** Data based on descriptions of individual cases rather than on controlled studies.

**aneroid** Not using a fluid.

**angiostrongyliasis** An infection caused by the roundworm *Angiostrongylus cantonensis* or rat lungworm transmitted through land and aquatic snail vectors. Humans become accidental hosts by eating raw, infected plants, snails, fish, or crustaceans.

**angle gate valve** The combination of a gate valve and an elbow at one end. *See* angle valve.

**angle of repose** The greatest angle to the horizontal made by a bank of loose earth, gravel, or other unsupported granular material. Also called natural slope.

**anglesite** A common compound of lead, essentially lead sulfate, $PbSO_4$.

**angle valve** The combination of a valve and a 90° bend used primarily in household plumbing and for its low cost, e.g., in installing a water meter with its outlet at right angle to its inlet.

**angstrom, angstrom unit, Angstrom unit (Å)** A unit of length equal to 0.1 nanometer or 1/10,000,000,000 meter; the approximate size of an atom. It is used to express electromagnetic wavelengths. (After the Swedish physicist and astronomer Anders Jonas Ångström, 1814–1874.)

**anhydride** The product resulting from the elimination of water from another product. *See also* anhydrous product.

**anhydrite [$CaSO_4$(s)]** Anhydrous calcium sulfate, a usually whitish, sometimes colored, solid; a primary contributing mineral of calcium ion to freshwater sources.

**anhydrous** Characteristic of compounds that do not contain water.

**anhydrous ammonia ($NH_3$)** Liquefied ammonia gas that is almost pure (99–100%) and waterless; a hazardous substance, available in pressurized cylinders and tank cars, sometimes used in the production of chloramines for water disinfection or as a nitrogen source for nutrient-deficient industrial wastes. (The liquid vaporizes to a gas as it leaves the pressurized container.) Anhydrous ammonia reacts with water as a base to yield the hydroxyl ion ($OH^-$):

$$NH_3 + H_2O \rightarrow NH_4^+ + OH^- \qquad \text{(A-107)}$$

Anhydrous ammonia can cause irritation and damage to eyes, lungs, nose, and throat. Due to its basicity in water, it can cause scaling problems associated with the precipitation of calcium carbonate.

**anhydrous calcium sulfate** *See* anhydrite.

**anhydrous product** The theoretical product that would result if all water were removed from the actual product (EPA-40CFR417.11-b).

**aniline ($C_6H_7N$)** A colorless, oily liquid with a light ammonia-like odor; it becomes dark on exposure to light or air; degrades to catechol and eventually to carbon dioxide in surface water.

**animal bioassay** *See* carcinogenesis bioassay.

**animal feed** Any crop grown for consumption by animals, such as pasture crops, forage, and grain.

**animal feed production** An emerging sludge disposal method in which livestock are fed dried sludge solids, after appropriate treatment and mixing with straw, cottonseed hulls, and other feed components.

**Animalia kingdom** One of the five taxonomic categories of forms of life, including all animals, as compared to plants (Plantae kingdom), fungi (Fungi k.), protozoa and eukaryotic algae (Protista k.), and bacteria and blue-green algae (Monera k.).

**animal infectivity** The use of mice or other animals to detect disease that a certain biological agent can cause.

**animal manure** Excreta of domestic animals, usable as fertilizer after composting.

**animal starch** Same as glycogen.

**animal study** An investigation that uses animals as surrogates for humans with the expectation that the results are pertinent to humans.

**anion** A negatively charged ion in a solid, an electrolyte solution or other solvent, attracted to the anode (the positive electrode, where oxidation occurs) under the influence of a difference in electrical potential; sometimes represented by the symbol $X^{-c}$ where $c > 0$. Examples of anions in water and wastewater treatment include arsenate, chloride, chromate, fluoride, fulvate, humate, nitrate, selenate, sulfate.

**anion displacement** The substitution of one ion for another in a compound, analogous to ion exchange. For example, when acidified alumina reacts with hydrofluoric acid for defluoridation, the fluoride ion displaces the chloride ion:

$$\text{Alumina.HCl(s)} + HF \rightarrow \text{Alumina.HF(s)} + HCl \qquad \text{(A-108)}$$

**anion exchange** A water treatment process that removes undesirable anions from a liquid by exchanging them for anions on a resin or synthetic medium. When the exchanging anion is exhausted, the resin is backwashed and regenerated by passing it through a concentrated solution. An example is nitrate removal by a strongly basic resin using sodium chloride for regeneration:

$$RCl + NO_3^- \rightarrow RNO_3 + Cl^- \qquad \text{(A-109)}$$

**anion exchanger** A resin or similar material with a positively charged framework and pores containing negatively charged ions (anions), used in water treatment (e.g., softening, desalination, production of boiler-feed water). *See also* strong-base anion resin, weak-base anion resin, acid retardation.

**anion exchange treatment** The use of anion exchange to treat water or wastewater, a process that is expensive to operate because of the cost of waste brine disposal.

**anionic** Pertaining to anions; having a negative ionic charge; characteristic of polymers that have a net negative charge.

**anionic detergent** A synthetic compound of alkali or ammonium salts, e.g., sodium sulfates and sulfonates. They may also have such additives as phosphates, surfactants, and enzymes. They are widely used household detergents.

**anionic exchange** Same as anion exchange.

**anionic-exchange resin** A resin used to remove the carbonic and other acids that remain in the product water from ion exchange treatment: the resin exchanges its hydroxyl ion ($OH^-$) for the sulfate ($SO_4^{2-}$), carbonate ($CO_3^{2-}$), chloride ($Cl^-$), etc. of the acidic water (McGhee, 1991):

$$H_2CO_3 + A(OH)_2 \rightarrow 2\,H_2O + ACO_3 \qquad \text{(A-110)}$$

where $A(OH)_2$ and $ACO_3$ represent, respectively, the anion exchange bed (anionic-exchange resin) and the exhausted exchange bed. A strong base, like caustic soda (NaOH), is used for bed regeneration. Also called a hydroxyl-cycle resin.

**anionic flocculant** A polyelectrolyte with a net negative charge, used in water treatment. *See* anionic polymer.

**anionic metal** A metal such as arsenic or its compound arsenate ($AsO_4^{3-}$) that is attracted to positively charged surfaces, as opposed to a cationic metal such as lead or calcium that strongly interacts with negatively charged surfaces. The nature and intensity of the charge of a metal strongly affect its fate and bioavailability.

**anionic polyelectrolyte** *See* anionic polymer.

**anionic polymer** A synthetic polymer that has negatively charged groups of ions, i.e., a preponderance of negative sites; often used as a coagulant aid or filter aid to develop larger and tougher flocs by providing a bridge between particles, and for dewatering sludge. Figure A-15 is a representative molecular structure of an anionic polymer. *See also* ampholytic polymer, bridging effect, cationic polymer, nonionic polymer.

**anionic surfactant** A negatively charged ionic surface-active agent used in cleaning products. *See also* cationic surfactant, nonionic surfactant.

**anion membrane** A membrane used in electrodialysis to retain anions but not cations; it is impermeable to water and nonelectrolytes. Also called anion transfer membrane or cation-selective membrane. An example is a polystyrene–divinylbenzene membrane functionalized with a sulfonyl group ($—SO_3^-$).

**anion-selective membrane** A membrane used in electrodialysis to allow the passage of anions but not cations; it is impermeable to water and nonelectrolytes. Also called cation membrane or cation transfer membrane. An example is a polystyrene–divinylbenzene membrane functionalized with an ammonium group ($—NR_3^+$).

**anion transfer membrane** Same as anion membrane.

***Anisopus* fly** A genus of flies that are sometimes found around trickling filters; higher dosing rates on the filter contribute to reducing the fly population. *See also* Psychoda fly.

**anisotropic** The condition in which hydraulic or other properties of a medium (e.g., sedimentary deposits) are not equal when measured in all directions.

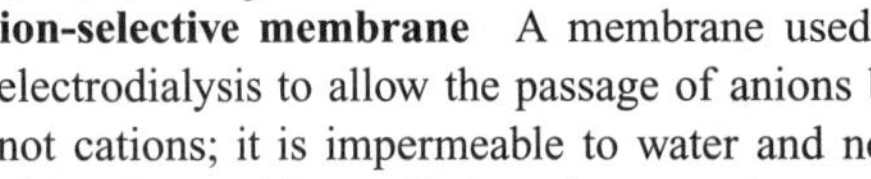


**Figure A-15.** Anionic polymers (molecular structure).

**anisotropic medium** A porous medium whose permeability or other hydraulic characteristics vary in different directions; a characteristic of sedimentary deposits.

**anisotropic membrane** A thin (less than 1.0 micron) reverse osmosis membrane that consists of a single polymer with nonuniform cross-sectional structure, such as cellulose acetate, supported by a thicker (up to 100 microns) substructure having larger pores. Also called asymmetric membrane. *See also* thin-film composite membrane.

**anisotropy** Characteristic of unequal physical properties along different axes; particularly the property of a porous medium whose permeability or other hydraulic characteristics vary in different directions. It is also a characteristic of light that originates from asymmetric crystals. *See also* isotropy.

**Anitron** A biological wastewater treatment process developed by Kruger, Inc., using fluidized beds.

**ANM™** A nanofiltration membrane manufactured by the TriSep Corp. for water softening.

**annealing** The process of forming double-stranded RNA from single-stranded RNA. *See also* hybridization.

**annual average daily flow** Same as annual average flow.

**annual average flow** The total volume of water consumption or wastewater discharge in a year, divided by the number of days in the year, usually expressed in million gallons per day (mgd), cubic meters per day ($m^3/d$), or liters per second (lps).

**annual evaporation** *See* annual precipitation.

**annual load factor** In services such as energy or water supply, the ratio of the maximum load over the average load for a period of one year.

**annual mass balance** An inventory of water input and output in a basin on an annual basis. *See* mass balance and water budget.

**annual pollutant loading rate** The maximum amount of a pollutant that can be applied to a unit area of land during a 365 day period under the Clean Water Act (EPA-40CFR503.1-c).

**annual precipitation and annual evaporation** The mean annual precipitation and mean annual lake evaporation, respectively, as established by the U.S. Department of Commerce, Environmental Science Services Administration, Environmental Data Services, or equivalent regional rainfall and evaporation data (EPA-40CFR440.132-b).

**annual risk** The risk of contracting one or more infections during a period of 365 days ($P_A$), a function of the probability ($P$) of infection from a single exposure. Assuming daily exposure to a constant concentration and a Poisson distribution of the contamination in water, for example (Maier et al., 2000):

$$P_A = 1 - (1 - P)^{365} \qquad \text{(A-111)}$$

*See* dose–response model for related distributions.

**annual whole sludge application rate** The maximum amount of sludge (dry weight basis) that can be applied to a unit area of land during a 365 day period (EPA-40CFR503.11-d).

**Annubar®** Proprietary flow monitoring equipment manufactured by Dieterich Standard.

**annular reactor** A small cylindrical vessel used to simulate the impact of biological treatment on microbial regrowth in a water supply distribution system.

**annular space** A ring-shaped space located between two concentric cylindrical objects, such as two pipes or the walls of a drilled hole and a casing. *See also* annulus.

**annulus** The space between two concentric circles on a plane; annular space.

**anode** The positive pole or electrode of an electrolytic system, such as a battery or voltaic cell. The anode attracts negatively charged particles or ions (anions), whereas electrons flow away from it. Dissolution of metals occurs at the anode as a result of an oxidation half-reaction. *See also* cathode.

**anodic** Pertaining to the anode or phenomena occurring at an anode, e.g., the site of an oxidation half-reaction.

**anodic current** The current that results from the oxidation of a metal in an electrochemical cell, corresponding to a loss of electrons and corrosion in the forward reaction:

$$\text{Metal} \leftrightarrow \text{Metal}^{z+} + ze^- \qquad \text{(A-112)}$$

**anodic inhibitor** A substance that migrates toward anodic sites and, by oxidation, forms a protective film against corrosion; e.g., the chromate ion.

**anodic protection** A method of protection against electrochemical corrosion, achieved through the use of an anode having a higher electrode potential than the metal to be protected. The introduction of an external voltage reduces the rate of corrosion. Other methods of corrosion control include cathodic protection (galvanic or impressed-current protection), chemical coating, inhibition, inert materials, and metallic coatings (McGhee, 1991).

**anodic stripping voltametry (ASV)** An electrochemical method used to determine the concentration of a metal in solution as a function of the current generated by oxidation at a specific voltage.

**anolyte** The portion of the electrolyte in the vicinity of the anode of an electrodialysis unit. *See also* catholyte.

**anomorphic zone** A zone of rock flowage in the subsurface.

***Anopheles*** A genus of mosquitoes many of which transmit the pathogens of such diseases as malaria and filariasis. They often breed in pools of sullage or rainwater on the ground, as well as in flood or irrigation water.

***Anopheles gambiae*** A major vector of malaria in Africa; it also transmits arboviral infections and Bancroftian filariasis. This mosquito may breed in lakes and irrigation water.

***Anopheles quadrimaculatus*** A malaria vector that breeds in static, shallow, shaded water. It can be controlled by water level fluctuation.

***Anopheles stephensi*** A common vector of malaria that breeds in water storage tanks on rooftops of urban areas of India.

**anopheline** Pertaining to the genus Anopheles and other mosquitoes of the anophelini subfamily.

**anorthite ($CaAl_2Si_2O_8$)** A white or grey silicate mineral found in igneous rocks; calcic plagioclase or calcium end member of the plagioclase feldspars. It dissolves in water to form kaolinite $[Al_2Si_2O_5(OH)_4]$:

$$CaAl_2Si_2O_8 + 2\,CO_2 + 3\,H_2O = Al_2Si_2O_5(OH)_4 + Ca^{2+} + 2\,HCO_3^- \qquad \text{(A-113)}$$

**anoxia** *See* anoxic condition.

**anoxic** Characterized by a severe depletion or an absence of free oxygen; sometimes used synonymously with anaerobic.

**anoxic/aerobic process** A wastewater treatment process that uses both aerobic and anoxic conditions, in a single reactor or in separate reactors; primarily used for denitrification, the rate of which is controlled by the utilization rate of substrate contained in the influent to the anoxic reactor. *See,* e.g., the Ludzack–Ettinger and Bardenpho™ processes.

**anoxic biological process** *See* anoxic process.

**anoxic condition** A condition in which free dissolved oxygen is absent or depleted but where such compounds as nitrates ($NO_3^-$) or sulfates ($SO_4^{2-}$) can act as terminal electron acceptors. Some reactors are purposely operated under anoxic conditions, e.g., to control activated sludge bulking or as part of a biological nutrient removal scheme (Droste, 1997 and WEF&ASCE, 1991). *See,* for example, the Ludzack–Ettinger, modified Ludzack–Ettinger, preanoxic denitrification, postanoxic denitrification, Bardenpho™, Wuhrmann, and Orbal Simpre™ processes. Two treatment trains designed for biological phosphorus removal include an anoxic reactor: the A²O™ and UCT processes. Anoxic conditions may exist in the upper layers of wetland soils. Also called anoxia. *See also* aerobic state, anaerobic state.

**anoxic process** Any process that uses anoxic reactions, e.g., the biological conversion of nitrate to nitrogen gas in the absence of oxygen. The anoxic reactor is normally one stage of a treatment process that includes also aerobic and/or anaerobic stages. *See* anoxic condition for a list of treatment trains that include an anoxic reactor (Metcalf & Eddy, 2003).

**anoxic reaction** A reaction mediated by microorganisms that use nitrite ($NO_2^-$) or nitrate ($NO_3^-$) as electron acceptors to oxidize organic matter, thus reducing these acceptors to nitrogen gas ($N_2$); also called denitrification reaction.

**anoxic state** *See* anoxic condition.

**ANOX-R** A process developed by Davis Water & Waste Industries, Inc. for the advanced treatment of industrial wastewater.

**anoxygenic** Unable to produce oxygen.

**antagonism** In toxicology, interference or inhibition of the effect of one chemical by the action of another. *See also* potentiation, synergism.

**antagonists** Biological agents that reduce the numbers of pathogens or restrict their activities through antibiosis, competition, or hyperparasitism.

**anthracite** A hard black coal containing a high percentage of fixed carbon (at least 86%) and a low percentage of volatile hydrocarbons (at most 14%); it burns slowly and almost without flame or smoke. It is used in single- or dual-media water filtration units. Crushed anthracite has an effective size of 0.8–2.0 mm and a uniformity coefficient of 1.3–1.8. Also called anthracite coal.

**anthracite coal** Same as anthracite.

**anthracite coal–sand filter** A dual-media filter consisting of a layer of crushed anthracite over a layer of sand, resulting in better performance than a single-medium sand filter.

**Anthrafilt®** Anthracite produced by Unifilt Corp. for use as a filter medium.

**anthrax bacillus** A type of bacteria pathogenic to humans and other bacteria; it is sometimes found in tannery wastes.

**anthropogenic** Caused or produced by humans.

**anthropogenic compounds** Man-made products, some of which are not biodegradable.

**antibiosis** An association between organisms that is injurious to one of them, for example, when an antagonist inhibits or lethally affects a pathogen by producing injurious substances such as acidic agents or antibiotics.

**antibiotic** A product, such as penicillin or streptomycin, derived from fungi (e.g., *Actinomycetes*) and other microorganisms, that can control or destroy bacteria and other organisms; used in the treatment of infectious diseases. Antibiotics come from live organisms, as compared to synthetic antimicrobials. When present in wastewater, antibiotics are considered as pharmaceutically active compounds as well as xenobiotics since they are not normally produced by the human body.

**antibiotic drug waste** A pharmaceutical waste that contains antibiotics; possible treatment includes activated sludge or evaporation and incineration.

**antibiotic resistance** Characteristic of a microorganism that neutralizes or resists the effect of an antibiotic.

**antibody** A protein molecule induced by the presence of an antigen (foreign substance) and produced as a primary immune defense against viruses, bacteria, and toxins. The relationship between antibody and antigen is useful in identifying pathogens. *See also* macrophage.

**antibody-mediated immunity** Immunity acquired through the activity of bone marrow cells that produce antibodies.

**antichlor** A chlorine-reducing compound; a substance, such as sodium thiosulfate ($NaS_2O_3 \cdot 5H_2O$) used for removing excess chlorine from paper pulp, textiles, etc., after bleaching. This and other reagents [sodium bisulfite ($NaHSO_3$) and sulfur dioxide ($SO_2$)] react with excess chlorine in water and wastewater to form inorganic salts. Ammonia ($NH_3$) is also an antichlor.

**anticholera O-group 1** A serum used to carry out tests to confirm suspected isolates of the cholera vibrio; this confirmation is necessary to diagnose the infection.

**anticorrosion treatment** Treatment of water to reduce or eliminate its corrosion potential. *See also* corrosion control.

**antidegradation clause** A clause of federal air quality and water quality requirements prohibiting deterioration where pollution levels are above the legal limit.

**antierosion assembly** A flat plate installed below the end of the intake of an aerator to prevent bottom erosion in a shallow aerated pond.

**antifoam agent** A surface active agent for the control of foaming, e.g., a substance added to detergents.

**antifoamant** Same as antifoam agent.

**antifoulant** A chemical additive used to prevent fouling and scaling. *See also* antiscalant.

**antigen** A foreign substance, including a protein, that can produce an immune reaction by stimulating the production of antibodies and combining with them.

**antigen capture** A procedure using antibodies to bind and recover organisms from a sample.

**antigenic determinant** The specific site on an antigen where an antibody can bind, the chemical structure of the site determining which antibody. Also called determinant.

**antiglobulin** An antibody produced by an immunized animal (e.g., a rabbit) and directed against another animal species' or another individual's antibodies.

**antimatter** *See* antiparticle.

**antimetabolite** A substance that resembles a metabolite but interferes with the growth of an organism by competing with or substituting for an essential nutrient.

**antimicrobial** A synthetic or naturally derived antibiotic, or similar substance that can control or destroy bacteria and other organisms. Such substances, used to promote the growth of poultry and livestock, are sometimes detected in wastewater and drinking water sources.

**antimony** Sb, from its Latin name, stibium. (1) A brittle, lustrous, bluish-white, trivalent or pentavalent metallic element (metalloid), found naturally free or combined as sulfide (stibnite), used in alloys (e.g., tin–antimony solder), glass, pesticides, medicine. Atomic weight = 121.75. Atomic number = 51. Specific gravity = 6.69. At sufficiently high concentrations, antimony may interfere with enzyme production and cause physiological effects in humans. It accumulates in the liver and in marine organisms. MCLG = MCL = 0.006 mg/L. (2) The total antimony present in the process wastewater stream exiting the wastewater treatment system, regulated by the USEPA (EPA-40CFR415.661-e).

**antineutron** A particle having no charge but with the magnetic moment opposite to that of a neutron.

**antinutrient** A substance that interferes with the utilization of a nutrient, e.g., oxalate and phytate that impede calcium absorption.

**antioxidant** A substance that prevents or impedes the oxidative deterioration of gasoline, plastics, etc., or an enzyme (e.g., vitamin E, ascorbic acid, beta carotene) that counteracts the damages of oxidation in animal tissues.

**antiparticle** A subatomic particle that is identical to other particles, but opposite in electrical charge or in magnetic moment. The collision of an antiparticle and its opposite converts them into photons of equivalent energy. *See also* antineutron, antiproton, photon, positron.

**antiproton** A proton with a negative charge.

**antiscalant** A chemical additive used to prevent the formation and buildup of scales, or to prevent the precipitation of a salt whose concentration exceeds solubility, e.g., to control fouling of reverse osmosis membranes. Typical antiscalants contain carboxylic acid or phosphate functional groups. Examples include polyacrylates and polyphosphonates. Sodium polyphosphonate ($Na_{n+2}P_nO_{3n+1}$) is used to reduce the precipitation of calcium carbonate ($CaCO_3$) in water. Also called scale inhibitor or scale prevention compound. *See also* antifoulant, dispersant.

**antiseptic** Compound that stops or inhibits the growth of microorganisms without killing them.

**anti-siphon valve** Same as siphon breaker.

**antiskid sand** A special aggregate designed to prevent the skidding of vehicles; the application of such materials on roads may lead to elevated concentrations of solids, phosphorus, lead, and zinc in water sources.

**anus-to-mouth pathway** *See* fecal–oral route.

**A/O®** or **A/O™** A proprietary biological treatment process developed by Kruger, Inc. (patented and marketed by Air Products and Chemicals, Inc.) for BOD removal and enhanced phosphorus uptake. It consists of an anaerobic basin, followed by an aerobic tank and a clarifier, with sludge recycle and wasting from the latter. *See* Figure A-16. Contrary to some similar nutrient removal processes, A/O does not provide nitrification. Also called anaerobic/oxic process and two-stage A/O™ process. *See also* Phostrip, Phoredox, A²O™, UCT process, and modified Bardenpho.

**AOC** Acronym of assimilable organic carbon.

**AOC test** *See* assimilable organic carbon test.

**AOP** Acronym of advanced oxidation process.

**A²O™ process** or **A²/O™ process** A three-stage (anaerobic-anoxic-aerobic) proprietary process developed for the biological removal of nitrogen and phosphorus from wastewater, including mixed liquor recycle from the aerobic reactor to the anoxic reactor and return activated sludge from the clarifier underflow to the anaerobic reactor (Metcalf & Eddy, 2003 and WEF & ASCE, 1991). The aerobic reactor is designed for nitrification; denitrification occurs in the anoxic reactor, which has a detention period of about 1.0 hr. (The anoxic reactor is deficient in dissolved oxygen but con-

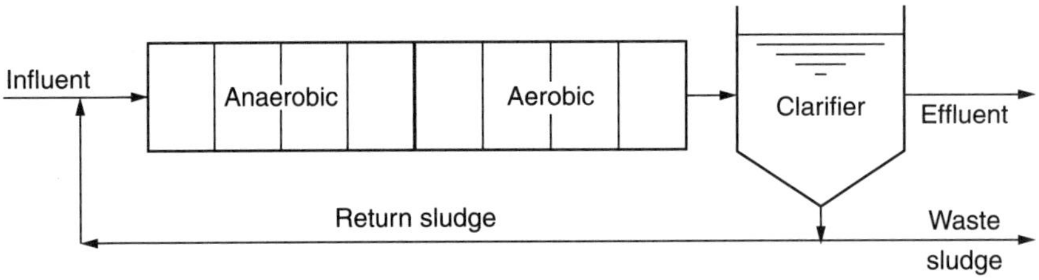


**Figure A-16.** A/O™ process.

tains chemically bound oxygen in nitrates and nitrites). This process can produce an effluent with less than 2.0 mg/L of total phosphorus and 8.0 mg/L of total nitrogen. *See* Figure A-17. *See also* the A/O (Phoredox), modified Bardenpho, UCT, modified UCT, VIP, Johannesburg, SBR, and PhoStrip processes.

**AOR** Acronym of area of review.

**AOTR** Acronym of actual oxygen transfer rate.

**AOX** Abbreviation of adsorbable organic halogen.

**apatite** An insoluble mineral that is a mixture of calcium compounds (phosphates, fluorides, carbonates, sulfates). With 3–7% fluoride, it is the principal source of this ion in water fluoridation. Also called calcium hydroxyphosphate, which is used as a specific ion-exchange medium for the removal of excess fluoride from drinking water. Apatite forms the enamel of teeth, is found in bones, and is the raw material in phosphate fertilizers. Suggested formulas: that of calcium fluorophosphates [$Ca_5F(PO_4)_3$], or more generally, $Ca_5(Cl, F, OH)(PO_4)_3$.

**APHA** Acronym of American Public Health Association.

***Aphanizomenon*** A genus of cyanobacteria similar to *Anabaena,* found in algal blooms and responsible for the production of toxins (saxitoxin, neosaxitoxin, hepatotoxin) that can cause liver damage and gastrointestinal illness.

***Aphanizomenon flos aquae*** One of the three most important species of cyanobacteria (blue-green algae) that produce neurotoxins.

**aphotic** Lightless, dark, as in the bottom layer of a lake or reservoir.

**aphotic zone** The deeper part of a sea, lake, or other water body, where light does not penetrate. The depth of the aphotic zone varies from one meter to about 200 meters depending on turbidity. *See* freshwater profile and marine profile for other related terms.

**API** Acronym of the American Petroleum Institute.

**API gravity** An index inversely proportional to the specific gravity of liquid hydrocarbons.

**API separator** A rectangular basin used for the separation of oil from water.

**apparel industry** Concerning the generation of wastewater, the apparel industry includes textiles, tanneries or leather goods, laundry, and dry cleaning. Some of the apparel industry wastes are characterized by high alkalinity, turbidity, organic content, total solids, and heavy metals.

**apparent alpha ($\alpha'$)** The product of the alpha factor ($\alpha$) in the formula of oxygen transfer rate by a

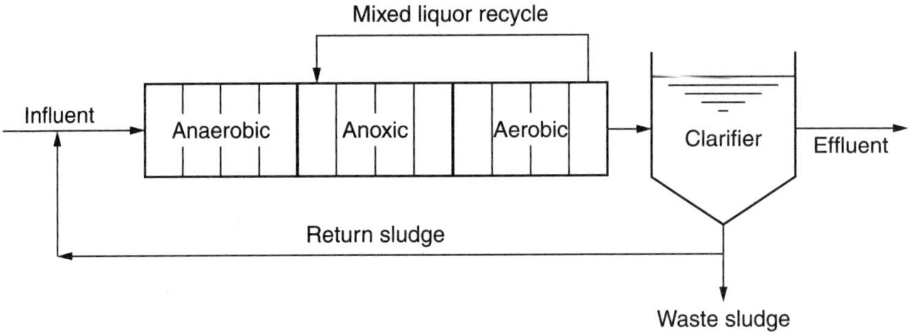


**Figure A-17.** A²/O™ process.

coefficient ($F$) to account for diffuser fouling and deterioration and for the effects of process water as opposed to clean water:

$$\alpha' = \alpha F \quad \text{(A-114)}$$

The coefficient $F$ varies from 0.5 to 1.0.

**apparent color** The color caused by the presence of suspended solids (turbidity) in water. It disappears when these solids are removed. Sometimes apparent color is defined as the total color, which includes both dissolved and suspended substances. Apparent color is sometimes used as a surrogate for total iron content because of the strong correlation between these two parameters; apparent color is easier to measure on-site. *See also* true color, red water.

**apparent density** The mass per unit volume of soil, including pores or the mass of unit volume of nonstratified dry activated carbon. *See also* apparent specific gravity and bed density—backwashed and drained.

**apparent dissociation constant** A dissociation constant related to the effects of substrate concentration on the rate of substrate transformation into product. Also called Michaelis constant. *See also* association constant, dissociation constant, equilibrium constant, ionization constant.

**apparent particle diameter (APD)** The opening of a screen mesh.

**apparent rejection** The inherent rejection rate (R*) of a membrane for a specific contaminant, taking into consideration flow and concentration variations on either side of the membrane at a specific location. *See* local rejection for detail.

**apparent solubility** The additional solubility (over the solubility in pure water) associated with humic colloids and other colloidal organic matter that absorb nonpolar organic chemicals.

**apparent specific gravity** The ratio of the mass of a volume of oven-dried soil to the mass of an equal volume of water under standard conditions. (This definition does not include the volume of intraparticle voids). It is a reproducible parameter, used in fluidization calculations instead of the bulk specific gravity. *See also* apparent density, bulk specific gravity.

**appendage** A subordinate part of a microbial cell serving for attachment or transport. *See* cilia, fimbriae, flagellae, pili.

**appendage phage** Bacteriophage that infect hosts through the sex pili or flagella. Also called F-specific phage. *See also* capsule phage, somatic phage.

**application factor** The ratio of the concentration of a substance that produces a given reaction to the concentration that causes death in 50% of the population in a given time period (the $LD_{50}$).

**approach velocity** The velocity of a liquid as it is entering a conduit, dam, Venturi tube, weir, or other hydraulic structure. For example, the approach velocity at water intake screens affects head losses through the screens and is typically 1–2 fps.

**appropriated water** The quantity of surface or ground water reserved from a source for a specific use.

**appropriate technology** Technology that fits the conditions of an application in terms of cost, performance, and requirements for operation and maintenance (e.g., fuel, operator skills, and other resources). Or, in the words of WHO and UNICEF, "technology that is scientifically sound and also acceptable to those who apply it and to those for whom it is used." In water supply and sanitation, e.g., many technologies common in developed countries are inappropriate for most developing areas of the world. *See* sustainable development, village-level operation and maintenance.

**appropriative rights** Water rights to, or ownership of, a water supply that is acquired for beneficial uses by following a specific legal procedure. It is currently the predominant body of law for the quantitative allocation of water in many western states of the United States.

**approximate flooding** The condition of a stripping tower in which the amounts of air and water applied are large enough to fill the pore spaces and cause water to start flooding within the tower (Metcalf & Eddy, 2003).

**appurtenances** Secondary appliances, instruments, or machinery attached to a main structure. Sewer appurtenances are auxiliary components other than pipes and conduits, such as manholes, inlets, inverted siphons, regulators, flap valves, and junctions. In general, appurtenances are not considered part of the main structure, but are necessary for its adequate operation.

**appurtenant water right** A water right that is incident to the ownership of land.

**apron** A covering in concrete, stone, timber, or other material to protect a hydraulic structure against erosion from flowing water.

**Aqua-4™** A plant manufactured by Smith & Loveless, Inc. for the treatment of surface water.

**AquaABF** A package filter manufactured by Aqua-Aerobic Systems, Inc., with automatic backwashing using a moveable, bridge-mounted device.

**aqua ammonia** A strongly basic inorganic compound, a solution of ammonia gas ($NH_3$) in water

($H_2O$); sometimes used in the production of chloramines for water disinfection. *See* more detail under ammonium hydroxide.

**aqua ammoniae** Same as aqua ammonia; Latin expression meaning water of ammonia.

**Aqua Bear** A foam of medium density produced by Girard Industries to clean pipelines. *See also* Aqua Criss Cross.

**Aquabelt®** A proprietary gravity-belt thickener manufactured by Ashbrook Corp.

**Aqua-Carb®** A proprietary activated carbon produced by Wheelabrator Clean Water Systems, Inc.

**Aquaclaire™** A proprietary wastewater treatment process developed by DAS International, Inc.

**Aqua Criss Cross** A coated foam of medium density produced by Girard Industries to clean pipelines. *See also* Aqua Bear.

**aquaculture** The managed production of aquatic animals and plants, e.g., fish, shellfish, water hyacinth, or seaweed, in natural or controlled marine or freshwater environments such as ponds, lagoons, or underwater bodies. Ducks and fish (e.g., carp and tilapia) thrive in ponds enriched with nightsoil and other similar wastes. Fish harvested from aquaculture are used as feed for pig or poultry farming as well as for human consumption. Aquaculture may be used to remove contaminants from wastewater, through the bacteria attached to the floating aquatic plants. Also called aquiculture, underwater agriculture. *See also* algal culture, fish farming, macrophyte production, pisciculture.

**aquaculture project** A defined managed water area that uses discharges of pollutants into that designated area for the maintenance or production of harvestable freshwater, estuarine, or marine plants, or animals (EPA-40CFR122.25-1).

**aquaculture reuse** The application of aquaculture to remove pollutants from wastewater, sludge, and nightsoil while growing aquatic plants and animals.

**AquaDDM** A direct-drive mixer manufactured by Aqua Aerobic Systems, Inc.

**Aquadene™** A line of proprietary products of Stiles-Kem Division of MetPro Corp., Waukegan, IL, for the control of scaling, corrosion, iron, and manganese.

**AquaDisk** A tertiary filter made of woven cloth by Aqua-Aerobic Systems, Inc.

**Aquafeed®** A line of products of B.F. Goodrich to prevent scale formation in reverse osmosis units.

**Aqua-Fer™** A treatment process developed by Smith & Loveless, Inc. for iron removal from groundwater.

**aquagenic organic matter** Natural organic matter from an aquatic environment. *See* aquatic fulvic acid, aquatic humic acid.

**AquaGrip® Connection System** A fitting manufactured by Mueller Co. of Decatur, IL for use in valve and hydrant installation without changing any component.

**Aqua Guard™ or Aquaguard® screen** A continuous self-cleaning bar screen manufactured by Parkson Corp.

**Aqua-Jet** A direct-drive aerator manufactured by Aqua-Aerobic Systems, Inc.

**Aqua-Lator®** A floating aerator manufactured by Aerators, Inc.

**Aqualenc** A product (aluminum chlorosulfate) of Rhone-Poulenc Basic Chemicals Co.

**AquaLift®** A screw pump manufactured by the Parkson Corp.

**Aquamag®** A proprietary product (magnesium hydroxide) of Premier Services Corp.

**Aquamite®** A process developed by Ionics, Inc. for water treatment using electrodialysis.

**Aqua Nuchar** A brand of activated carbon used in water and wastewater treatment.

**Aqua Pigs®** A polyurethane foam produced by Girard Industries to clean pipelines.

**AquaPipe®** A trenchless technology designed by the Sanexen Company for the structural rehabilitation of 6″–12″ water mains, including the reinstatement of service connections from within by robotic equipment, thus eliminating the need to excavate in front of every building.

**Aquaport®** Proprietary equipment manufactured by Ambient Technologies, Inc. for the desalination of seawater.

**aqua privy** A toilet (latrine) consisting of a squatting plate directly above a sealed settling chamber, a drop pipe, and a flushing chute; it usually receives only excreta and small volumes of flushing water, with retention times up to 60 days. The final effluent contains a relatively low concentration of pathogens and is disposed off through a soakaway, a small-bore sewerage system, or other means. *See also* pour-flush toilet, VIP latrine, water seal.

**Aquaray®** An ultraviolet disinfection system developed by Infilco Degremont, Inc.

**Aquaritrol®** Equipment manufactured by Wheelabrator Engineered Systems, Inc. for the programmable control of coagulants.

**Aquarius®** A modular water treatment plant manufactured by Wheelabrator Engineered Systems, Inc.

**AquaSBR** A sequencing batch reactor (SBR) manufactured by Aqua-Aerobic Systems, Inc.

**Aquascan** A device manufactured by Sentex Systems, Inc. for on-line monitoring of volatile organic compounds.

**Aqua-Scope** *See* Heath Aqua-Scope.

**Aqua-Scrub™** A proprietary treatment process developed by Wheelabrator Clean Water Systems, Inc. using adsorption on powdered activated carbon.

**Aqua-Sensor** Equipment manufactured by Culligan International Corp. to control the regeneration of water softeners.

**Aquashade®** A product of Applied Biochemists, Inc. for the control of aquatic plant growth.

**Aqua-Shear** A mixer manufactured by Flow Process Technology, Inc.

**AquaSorb®** Granular and powdered activated carbons produced by Jacobi Carbons, Inc. of Philadelphia, PA for the removal of organics, trihalomethanes, geosmin, MIB, and other taste and odor compounds.

**Aquasorb®** A treatment process developed by Hadley Industries using carbon adsorption treatment.

**Aquasource®** A membrane system developed by Infilco Degremont, Inc.

**Aquaspir®** A shaftless dewatering screw manufactured by Andritz-Ruthner, Inc.

**Aquastore®** A storage tank manufactured by A.O. Smith Harvestore Products, Inc.

**Aqua Swab** A soft polyurethane produced by Girard Industries to clean pipelines.

**Aquatair** A package plant manufactured by BCA Industrial Controls, Ltd. for biological wastewater treatment.

**aquatic** (1) Pertaining to plants or animals living in, growing in, associated with, or adapted to water. (2) An aquatic plant or animal.

**aquatic animal** An appropriately sensitive wholly aquatic animal that carries out respiration by means of a gill structure permitting gaseous exchange between the water and the circulatory system (EPA-40CFR116.3); or, simply, a member of the animal kingdom that lives or grows in water.

**aquatic chemistry** A branch of science dealing with the reactions and processes occurring in natural waters, drawing on other fields such as physical chemistry, biology, geology, thermodynamics, and hydrology. *See* water chemistry for more detail.

**aquatic community** An association of interacting populations of aquatic organisms in a given water body or habitat.

**aquatic earthworm** One of the grazing organisms supported by trickling filter films.

**aquatic ecosystem** The system of living and nonliving elements in a water system, with two major divisions—freshwater ecosystem and marine or saltwater ecosystem—and subdivisions such as surface and bottom zones. *See also* aquatic environment.

**aquatic environment and aquatic ecosystem** Waters (of the United States), including wetlands, that serve as habitat for interrelated and interacting communities and populations of plants and animals (EPA-40CFR230.3-c). Aquatic environments in general include the oceans, estuaries, harbors, river systems, lakes, wetlands, streams, springs, and aquifers.

**aquatic flora** Plant life associated with the aquatic ecosystem, including algae and higher plants (EPA-40CFR116.3).

**aquatic food chain** Animals in natural waters, from primary producers (algae) that use recycled nutrients and photosynthesis, to first-order consumers (nymphs, copepods, water fleas), second-order consumers (e.g., sunfishes), and third-order consumers such as the bass. Then come scavengers (like the crayfishes) and bacterial decomposers that help recycle nutrients. Eutrophication may cause the aquatic food chain to be unbalanced, resulting in floating masses of algae and absence of the higher-order consumers.

**aquatic fulvic acid** Aquatic fulvic acids and aquatic humic acids are aquagenic, nontoxic, complex organic substances resulting from decaying vegetation; they cause a brown color in water and constitute a major source of precursors of disinfection by-products. The fulvic acids have a lighter molecular weight but are more abundant than the humic acids. *See also* natural organic matter, fulvic acid, humic acid.

**aquatic growth** Plankton and any other floating or attached organisms in water; they can be controlled biologically or chemically. Aquatic weeds and higher aquatic plants can be removed by dredging, cutting, or similar methods, or destroyed by poisons.

**aquatic habitat** Same as aquatic ecosystem.

**aquatic humic acid** *See* aquatic fulvic acid.

**aquatic humic substance (AHS)** A humic substance that forms a true solution in water with colloidal properties. *See* aquatic fulvic acid, aquatic humic acid.

**aquatic life** Animals and plants that live in water.

**aquatic microbiology** The study of microorganisms and microbial communities in the water environment.

**aquatic plants** (1) Emergent, submerged, and floating plants found in wetlands, sometimes used in wastewater treatment, including bulrush, cattail, reeds, sedges, water hyacinth, and duckweed. *See also* constructed wetlands. (2) Plants such as algae and phytoplankton that live in water. (3) Water caltrop, chestnut, hyacinth, and similar plants that harbor metacercariae of such parasites as the liver fluke and the giant intestinal fluke. When the edible plants are consumed raw, the metacercariae excyst in the intestine.

**aquatic plant harvesting** *See* plant harvesting.

**Aquatreat™** A sequencing batch reactor (SBR) treatment system developed by EnviroSystems Supply, Inc.

**Aqua-Trim™** A tray air stripper manufactured by Delta Cooling Towers, Inc.

**Aqua UV™** A water disinfection process developed by Trojan Technologies, Inc., using ultraviolet light.

**Aquavap** An evaporator manufactured by Licon, Inc.

**Aqua-View** A measuring device manufactured by Particle Measuring Systems, Inc.

**Aquaward®** Equipment manufactured by Eltech International Corp. for water disinfection using tablets.

**Aquazur V®** A water or wastewater treatment system developed by Infilco Degremont, Inc., using rapid sand gravity filtration.

**aqueduct** A large, artificial channel for conveying water, usually a covered masonry conduit built in place at the hydraulic gradient and operated at atmospheric pressure. It may consist of several elements such as canals, pipelines, and tunnels. *See also* flume and stream for the differences among various watercourses.

**aqueous** (1) Pertaining to something made up of, similar to, or containing water; watery. (2) Designating a chemical constituent dissolved in water, with the symbol (aq), e.g., aqueous ammonia or $NH_3(aq)$.

**aqueous chemistry** A branch of science dealing with the reactions and processes occurring in natural waters, drawing on other fields such as physical chemistry, biology, geology, thermodynamics, and hydrology. *See* water chemistry for more detail.

**aqueous ammonia** Ammonia dissolved in water, $NH_3(aq)$; same as ammonium hydroxide; aqua ammonia.

**aqueous chlorine** Chlorine compounds dissolved in water for water or wastewater disinfection. Also called liquid chlorine, erroneously.

**aqueous chlorine species** The major chlorine compounds found in water, e.g., during chlorination, including molecular or gaseous chlorine ($Cl_2$), hypochlorous acid (HOCl), and hypochlorite ion ($OCl^-$) and some transient compounds: $H_2OCl^+$, $Cl^+$, and $Cl_3^-$. *See also* chloramines.

**aqueous solubility** The extent to which a compound will dissolve in water. The logarithm of solubility is generally inversely related to molecular weight.

**aqueous solution** The solution of a substance (solute) in water as the solvent.

**aqueous vapor** The gaseous form of water, subject to vapor pressure laws. Also called water vapor, moisture.

**aqueous waste** Wastewater, liquid waste, sludge.

**aquiclude** The low-permeability, upper or lower boundary of an aquifer or underground water course. It is a porous geologic formation, such as clay, which may contain water, but not of sufficient permeability for significant water transmission through wells and springs. *See also* subsurface water, aquifer, aquifuge, and aquitard.

**aquiculture** Same as aquaculture.

**aquifer** An underground geological formation, or group of formations, containing usable amounts of groundwater that can supply wells and springs. The water-bearing formation or structure usually consists of saturated sands, gravel, fractures, or cavernous and vesicular rock. *See also* subsurface water, aquiclude, aquifuge, and aquitard. Figure A-18 illustrates some of the terms listed below and related to aquifers and wells: artesian aquifer, artesian well, confined aquifer, flowing well, leaky aquifer, leaky confined aquifer, nonflowing well, perched aquifer, phreatic aquifer, piezometric surface, potentiometric surface, pumping test, pump well, recharge area, semiconfined aquifer, unconfined aquifer, water-table aquifer.

**aquifer assimilative capacity** The ability of an aquifer to dilute lower-quality water without significant impairment relative to the intended use.

**aquifer mining** Withdrawal of groundwater from an aquifer in excess of the rate of recharge of the aquifer.

**aquifer recharge area** The land area that contributes water to the aquifer.

**aquifer safe yield** The quantity of water that can be withdrawn from an aquifer year after year without depleting or otherwise impairing it. *See* safe yield.

**aquifer storage and recovery (ASR)** The process of artificially recharging an aquifer (usually in periods of surplus) and subsequently extracting the stored water (usually in periods of deficit or greater demand). ASR may also be used to manage groundwater quality. If the receiving aquifer is

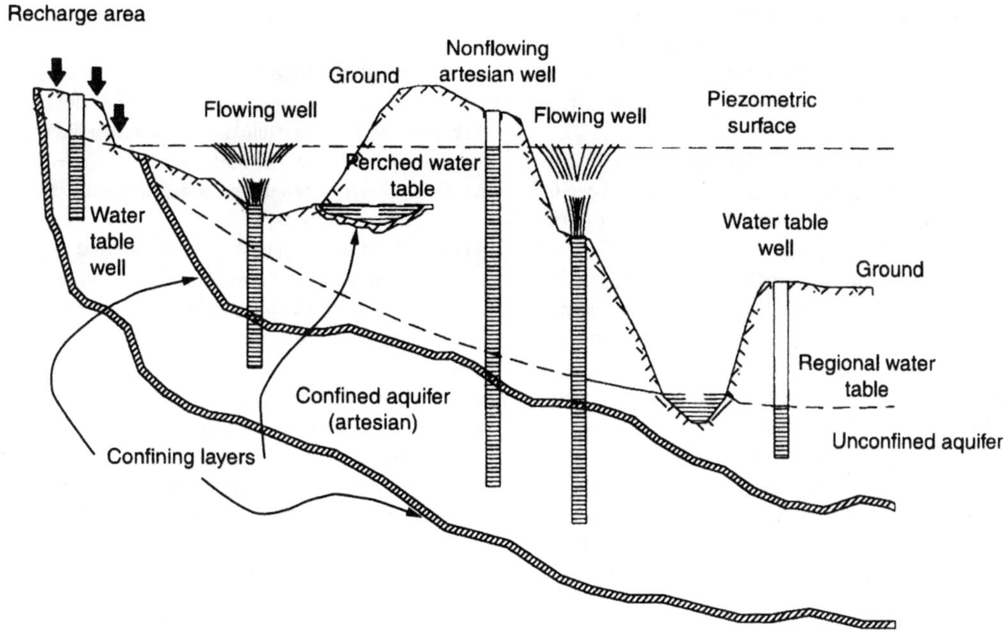


**Figure A-18.** Aquifer (courtesy CRC Press).

a drinking water source, the injected stream must meet drinking water standards, unless it is exempted by the Environmental Protection Agency. An ASR well, designed for that purpose, is sometimes used.

**aquifer storage and recovery (ASR) well** *See* aquifer storage and recovery.

**aquifer test** A technique used to determine the capacity of a well or borehole, or the hydraulic properties of an aquifer. The test is conducted by pumping a well at a constant rate over a period of time and recording the drop in the piezometric surface or water table in observation wells. These data are used to calculate the transmissivity, hydraulic conductivity, and storage coefficients. Also called pumping test or pumped-well technique. It is sometimes called an auger hole test, when it is performed in shallow auger holes. *See also* bail-down test, slug test.

**aquifuge** A geologic formation (e.g., unfissured granite) without interconnected fractures or interstices. It cannot absorb nor transmit water. *See also* aquiclude, aquifer, and aquitard.

**aquitard** A low-permeability geologic formation that may contain groundwater but is not capable of transmitting significant quantities of groundwater under normal hydraulic gradients. In some situations, aquitards may function as confining beds. However, an aquitard may transmit water vertically from one aquifer to another. *See also* aquiclude, aquifer, and aquifuge.

**aquo aluminum ion [$Al(H_2O)_6^{3+}$]** One of the products of aluminum hydrolysis during the coagulation process. It results from a strong bond between the aluminum ion and six water molecules.

**aquo aluminum ion deprotonation** The release of a hydrogen atom to an aqueous solution from an aquo aluminum ion [$Al(H_2O)_6^{3+}$] resulting from a weakened oxygen–hydrogen association of water molecules (AWWA, 1999):

$$Al(H_2O)_6^{3+} \rightarrow Al(OH)(H_2O)_5^{2+} + H^+ \quad \text{(A-115)}$$

**Aquox™** A proprietary potassium permanganate ($KMnO_4$) produced by Nalon Chemical.

**aragonite** A soluble, crystalline form of calcium carbonate ($CaCO_3$), slightly less stable than calcite. It is one of the primary minerals contributing calcium ions to natural waters.

***Araphinidineae*-to-*Centrales* (A/C) index** A biological indicator proposed to categorize lakes as oligotrophic ($0 < A/C < 1.0$), mesotrophic ($1.0 < A/C < 2.0$), or eutrophic ($A/C > 2.0$); it is more applicable to lakes of low alkalinity than others.

**arbitrary flow reactor** A reactor that has a flow regime between plug flow and complete mixing.

**arboviral disease** A disease caused by a virus and transmitted by blood-sucking arthropods. Arboviral diseases related to water include dengue, yellow fever, and encephalitic and hemorrhagic infections.

**arboviral infection** *See* arboviral disease.

**arbovirus** A virus carried by a blood-feeding arthropod (gnat, midge, mosquito, sandfly, tick) and transmitted by bite to a vertebrate. The word is an abbreviation of arthropod-borne virus. Arboviruses are responsible for such diseases as yellow fever and dengue. *See also* culicine mosquito.

**ARB®** Water Revenue System A water meter reading and information system designed by Neptune Technology Group, Inc. of Tallassee, AL.

**archaea** Microorganisms that have a simple cell structure, similar to bacteria and other prokaryotes, but with differences in cell wall and RNA composition. Some archaea are bacteria growing in adverse conditions or strict anaerobes responsible for the methanogenesis phase of anaerobic digestion. *See also* prokaryote.

**archaebacteria** Older term for archaea.

**Archimedean screw** Same as Archimedes screw.

**Archimedean screw pump** Same as Archimedes screw.

**Archimedes' principle** Buoyancy is the ability, tendency or power of a body to float or rise in a fluid. The upward pressure exerted by a fluid on an immersed body results from the buoyant force, i.e., the difference in the forces on the bottom and on the top of the body. Archimedes' principle states that the buoyant force is equal to the weight of the displaced fluid and acts vertically upward through the center of gravity of the displaced fluid. *See also* buoyancy.

**Archimedes screw** A low-lift, high-capacity, positive-displacement pump that raises water or wastewater by rotating a helical impeller in an inclined trough or cylinder. It can handle a wide flow range at a constant speed, prevents breakup of biological floc of return activated sludge, but requires considerable space. Also called Archimedean screw, Archimedean screw pump, Archimedes screw pump, screw pump, and water snail. *See also* enclosed screw pump, open screw pump, screw-feed pump.

**Archimedes screw pump** Same as Archimedes screw.

**architect/engineer procurement approach** A procurement process in which a consulting architect or engineer prepares plans and specifications for a project to be constructed by a contractor. *See also* turnkey procurement, privatization, combination procurement.

**architectural coating** A covering such as paint and roof tar that is used on building exteriors.

**Arc Screen™** A self-cleaning bar screen manufactured by Infilco Degremont, Inc.

**Arcticaer®** A submersible surface aerator manufactured by Aerators, Inc.

**Ardern, Edward** One of the two reported originators of the activated sludge process. The other is W. T. Lockett. They described their studies in a 1914 article entitled "Experiments on the Oxidation of Sewage without the Aid of Filters."

**area coefficient** *See* coefficient of area.

**area control center** An area in a water or wastewater treatment plant, such as a control room, in which an operator is responsible for the control of plant processes. The room is equipped with the appropriate instruments and devices such as distributed process controllers, operator consoles, alarms, computers, and printers.

**area filling** The disposal of water or wastewater treatment residuals in a mound, a layer, or a diked containment of a sanitary landfill (monofill). *See also* trench filling.

**area method** One of three common methods of landfilling. It uses a natural depression or one that has been excavated specifically for waste disposal. *See also* the ramp method and the trench method (WEF & ASCE, 1991).

**area of concern** A geographic area located within the great lakes, in which beneficial uses are impaired and which has been officially designated as such under the Great Lakes Water Quality Agreement (Clean Water Act).

**area of critical environmental concern** An area within public lands where special management attention is required (when such areas are developed or used or where no development is required) to protect and prevent irreparable damage to important historic, cultural, or scenic values, fish and wildlife resources or other natural systems or processes, or to protect life and safety from natural hazards.

**area of influence** The land area within the horizontal projection of the cone of depression, i.e., the land area where the water table is perceptibly lowered due to well pumping or other water withdrawal. For a wellfield or group of wells, it is called zone of influence. *See also* circle of influence.

**area of review** In the Underground Injection Control (UIC) program, the area surrounding an injection well that is reviewed during the permitting process to determine if flow between aquifers will be induced by the injection operation.

**area of review (AOR)** A circle of a quarter-mile radius around an injection well, used as a critical area for the determination of leakage.

**area-voids volume ratio ($A/V_v$)** A parameter used to characterize hydraulic properties of a bed or layer of granular materials; the reciprocal of the hydraulic radius. For a bed of porosity $f$ and consisting of equal perfect spheres of diameter $d$, the area-voids volume ratio is (Fair et al., 1971):

$$A/V_v = 6\,(1 - f)/(f \cdot d) \qquad \text{(A-116)}$$

**areawide agency** An agency designated under section 208 of the Water Pollution Control Act, which has responsibilities for water quality management planning within a specified area of a state (EPA-40CFR130.2-1).

**argentometric method** A common method of analysis for chlorides, using silver nitrate as the titration agent and chromate ion as the indicator.

**argentum** Latin name of silver (Ag).

**argyria** A bluish-gray discoloration of the skin due to colloidal silver in medication or long-term exposure to silver poisoning in industrial applications.

**ARI®** products A line of products of Wheelabrator Clean Air Systems, Inc. used for sulfur recovery and odor control.

**ARIES™ Managed Air Scour System** Modular or built-in equipment manufactured by Roberts Filter Group of Darby, PA for use in rehabilitation or new construction projects. The system provides cleaning of entire filter beds without modifying or replacing the underdrains.

**Arm & Hammer®** A proprietary sodium bicarbonate ($NaHCO_3$) produced by Church & Dwight Co., Inc.

**armoring** Formation of a layer of rock on a streambed, naturally or man made to prevent erosion.

**Arna®** Disinfection equipment manufactured by Arlat, Inc. using ultraviolet light.

**AroBIOS™ bioscrubber** A patented device (Duall Division of Met-Pro Corp., Owosso, Mich.) that combines biodegradation and absorption in one system to control odors at wastewater treatment plants and pumping stations. It consists of a water scrubber packed with microorganism-impregnated media. The acidic scrubber neutralizes ammonia ($NH_3$), while the bacteria digest hydrogen sulfide ($H_2S$) and other inorganics.

**aromatic** (1) Of or relating to organic compounds that resemble benzene in chemical behavior. These compounds are unsaturated and characterized by at least one six-carbon benzene ring. (2) Same as aromatic compound or aromatic hydrocarbon.

**aromatic compound** An organic, "closed-chain" substance that contains at least one benzene or equivalent ring. Many such compounds have an agreeable odor. Aromatic compounds include acids, alcohols, aldehydes, amines, hydrocarbons, ketones, nitro compounds, and phenols. *See also* aliphatic compound.

**aromatic content** The aromatic hydrocarbon content in volume percent as determined by the ASTM Standard Test Method for Hydrocarbon Types in Liquid Petroleum Products by Fluorescent Indicator Adsorption (EPA-40CFR80.2-z).

**aromatic hydrocarbon** A hydrocarbon, such as benzene or toluene, added to gasoline in order to increase octane. It has a six-carbon ring with alternating single and double bonds. Also called aromatic. Some aromatics are toxic.

**aromaticity** Characteristic of aromatic compounds, in particular their stability.

**aromatic polyamide** A synthetic organic material used in the manufacture of reverse osmosis membranes, exhibiting better mechanical and chemical properties than cellulose acetate.

**aromatic polyamide membrane** Asymmetric hollow fiber or thin-film composite flat sheets made of aromatic polyamide and used in water treatment by reverse osmosis or nanofiltration. *See* asymmetric membrane, thin-film composite membrane.

**aromatics** A group of hydrocarbon compounds, such as benzene, containing a closed ring structure. Same as aromatic hydrocarbons. *See also* aliphatic compound.

**aromatic sulfonate** A synthetic organic chemical used in the chemical industry (e.g., in making detergents), containing an aromatic hydrocarbon and the $SO_3^-$ radical.

**array** In pressure-driven membrane processes, an array consists of multiple interconnected stages in series and is sometimes controlled as a single unit. (A stage consists of parallel pressure vessels). Also called train. *See also* high-recovery array.

**Arrhenius constant** The dimensionless temperature coefficient $\theta$ in the Arrhenius equation. Also called Arrhenius parameter, temperature correction coefficient. *See* Table A-02 and also temperature correction factor.

**Arrhenius equation** A basic relationship of chemical reaction rate constants to temperature, formu-

**Table A-02.** Arrhenius constant

| Reported value | Application | Reference |
|---|---|---|
| $\theta = 0.962$ | Performance of plug-flow facultative ponds | Droste, 1997 |
| $\theta = 0.987$ | Performance of complete mix facultative ponds | Droste, 1997 |
| $1.00 < \theta < 1.04$ | CBOD removal in biological treatment systems | WEF & ASCE, 1991 |
| $1.0 < \theta < 1.8$ | Substrate removal rate in aerobic biological treatment | Droste, 1997 |
| $1.015 < \theta < 1.040$ | Typical range for diffused and mechanical aeration devices | Metcalf & Eddy, 2003 |
| $\theta = 1.02$ | Typical value for CBOD removal | WEF & ASCE, 1991 |
| $1.02 < \theta < 1.25$ | Range for biological treatment systems | Metcalf & Eddy, 2003 |
| $\theta = 1.024$ | Often used for biological wastewater treatment and stream reaeration and as an average value for diffused and mechanical aeration devices | Droste, 1997, Metcalf & Eddy, 2003 |
| $\theta = 1.026$ | Specific denitrification rate in anoxic tank design | Metcalf & Eddy, 2003 |
| $\theta = 1.035$ | Trickling filter design | Metcalf & Eddy, 2003, WEF & ASCE, 1991 |
| $\theta = 1.04$ | Typically for activated sludge; also used for endogenous decay | Droste, 1997, Metcalf & Eddy, 2003 |
| $\theta = 1.040 \pm 0.005$ | Rate of biological treatment (trickling filter) | Fair et al., 1971 |
| $\theta = 1.045$ | Rate of biological treatment in plastic media trickling filters | WEF & ASCE, 1991 |
| $\theta = 1.047$ | Stream studies | Metcalf & Eddy, 2003 |
| $\theta = 1.056$ | BOD exertion for 20°C < T < 30°C | Metcalf & Eddy, 1991, Metcalf & Eddy, 2003 |
| $1.06 < \theta < 1.12$ | CBOD removal in aerated lagoons | WEF & ASCE, 1991 |
| $\theta = 1.070 \pm 0.050$ | Rate of biological treatment (activated sludge) | Fair et al., 1971 |
| $\theta = 1.072$ | Commonly used for biological and chemical reactions | Hammer, 1996 |
| $\theta = 1.072$ | Reaction rate constant in the design of stabilization ponds | Fair et al., 1971 |
| $\theta = 1.085$ | Empirical equation for the required volume of a facultative pond | Droste, 1997 |
| $\theta = 1.097$ | Sludge digestion rate constant | Droste, 1997 |
| $\theta = 1.135$ | BOD exertion for 4°C < T < 20°C | Metcalf & Eddy, 1991, Metcalf & Eddy, 2003 |
| $\theta = 1.190$ | Rate constant for the removal of *E. coli* in ponds | Feachem, 1983 |

lated by Arrhenius in 1889, based on empirical observations:

$$\ln k = a - b/T \qquad \text{(A-117)}$$

or

$$k = k_0 e^{-E/RT} \qquad \text{(A-118)}$$

where $k$ is the reaction rate, $T$ is the temperature in degrees Kelvin (°K), $a = \ln k_0$ and $b = E/R$ are constants, $E$ is the activation energy, and $R$ is the universal gas constant.

The coefficient $k_0$ is called the frequency factor or preexponential factor. These equations were modified (see, e.g., the Boltzmann equation) and rearranged in the following usual form:

$$k_2 = k_1 \theta^{(T_2 - T_1)} \qquad \text{(A-119)}$$

where $k_1$ and $k_2$ are, respectively, the reaction rates at temperatures $T_1$ and $T_2$ (in °C), and $\theta$ is a dimensionless temperature coefficient (constant over small temperature ranges), also called Arrhenius constant, Arrhenius parameter, theta factor, or $\theta$ factor. This equation is used in many applications in the field of water quality and treatment. In biological wastewater treatment, the oxygen mass transfer coefficient ($k_T$) at temperature $T$ (in °C) is related to the coefficient ($k_20$) at 20°C:

$$k_T = k_{20}\, \theta^{(T-20)} \qquad \text{(A-120)}$$

The literature reports several values for the coefficient $\theta$; see Table A-02. The Arrhenius equation is sometimes called the Arrhenius relationship, the van't Hoff–Arrhenius equation, or the van't Hoff–Arrhenius relationship. *See also* the van't Hoff equation.

**Arrhenius parameters (*a, b, θ*)** The coefficients $a$ and $b$ in equation (A-117) and the coefficient $\theta$ in equation (A-120). *See* Arrhenius equation and Table A-02.

**Arrhenius relationship** Same as the Arrhenius equation.

**Arrhenius, Svante August** Swedish chemist and physicist, 1859–1927, won Nobel prize for chemistry in 1903.

**Arrhenius theory** The theory of ionization, i.e., electrolytes dissolved in water break into ions.

Electrolytes are acids, bases, and salts, whereas ions are atoms or groups of atoms, positively charged (cations), or negatively charged (anions) (DeZuane, 1997). *See also* ionization.

**Arro-Care®** A line of services provided by the U.S. Filter Corp. for the operation and maintenance of reverse osmosis units.

**Arro-Cleaning®** A line of services provided by the U.S. Filter Corp. for the operation and maintenance of membrane units.

**arroyo** A deep, water-carved channel or gully with steep banks in the southwest United States; dry most of the time, except after heavy rains. *See also* wadi, wash.

**Arrowhead®** A line of products of the U.S. Filter Corp. for use in water treatment operations.

**arsenate** A salt or ester of arsenic acid. In aqueous solutions, arsenate exists in four forms ($H_3AsO_4$, $H_2AsO_4^-$, $HAsO_4^{2-}$, and $AsO_4^{3-}$), and arsenite, in five forms ($H_4AsO_3^+$, $H_3AsO_3$, $H_2AsO_3^-$, $HAsO_3^{2-}$, and $AsO_3^{3-}$). *See* arsenic, arsenite.

**arsenic (As)** (1) A grayish-white, brittle, semimetallic element (or metalloid), with a metallic luster, vaporizing when heated, widely distributed in natural waters and in mine drainage, toxic to humans at very low concentrations; it may cause gastrointestinal and cardiac damages. Acute poisoning may result from the ingestion of 100 mg. At sufficiently high concentrations, arsenic may interfere with enzyme production and cause physiological effects in humans. The current maximum contaminant level for arsenic is 10 μg/L but some states (e.g., New Jersey) impose lower limits. The World Health Organization sets a guideline value of 0.05 mg/L for arsenic in drinking water. Atomic weight = 74.92. Atomic number = 33. Melting point = 817°C. Sublimation point = 613°C. Valence = +3, +4, and +5. Arsenic is found in nature as sulfides and as arsenides of heavy metals, oxides, arsenolites, and arsenates; see, e.g., arsenopyrite (or mispickel), orpiment, realgar. It originates in natural waters from arsenic-containing rocks and soils; generally found in trivalent form [As(III)] in groundwater and in oxidized, pentavalent form [As(V)] in surface water. Some formerly used pesticides contained lead arsenate [$Pb_3(AsO_4)_2$], sodium arsenite ($NaAsO_2$ or $Na_3AsO_3$), or Paris Green [copper arsenite, $Cu_3(AsO_3)_2$]. Its toxic methyl derivatives include dimethylarsine [$(CH_3)_2AsH$], dimethylarsinic acid [$(CH_3)_2AsO(OH)$], and methylarsinic acid [$CH_3AsO(OH)_2$]. Arsenic is used as an additive to lead and copper alloys. (2) The compound arsenic trioxide. (3) The total arsenic present in the process wastewater stream exiting the wastewater treatment system (EPA-40CFR415.671-c). Arsenic contamination affects primarily groundwater; its removal depends on its oxidation state, i.e., as arsenite [As(III)] or arsenate [As(V)]. It is found mainly as arsenious acid ($H_3AsO_3$), arsenite anion ($H_2AsO_3^{-1}$), arsenic acid ($H_3AsO_4$), monovalent arsenate anion ($H_2AsO_4^-$), and divalent arsenate ($HAsO_4^{-2}$). Available technologies for arsenic removal include (a) adsorption on granular media, ion exchange resins, or activated alumina, (b) coprecipitation with iron, and (c) ferric chloride coagulation followed by conventional or membrane filtration. *See also* Arsenic Rule, coagulation–microfiltration process, enhanced coagulation, iron CMF process.

**arsenic acid ($H_3AsO_4 \cdot ½\ H_2O$)** A white, crystalline, water-soluble powder. *See* arsenic.

**arsenical** Pertaining to arsenic; any pesticide that contains arsenic.

**arsenic butter** Another name of arsenic trichloride.

**arsenic disulfide ($As_4S_4$, $As_2S_2$, or AsS)** An orange-red powder, poisonous and insoluble in water, used in fireworks. Also called arsenic monosulfide.

**arsenic monosulfide** *See* arsenic disulfide.

**Arsenic Rule** A standard promulgated by the USEPA in January 2001 and requiring public drinking water systems to meet a maximum contaminant level (MCL) of 10 μg/L of arsenic by the end of January 2006. The previous MCL was 50 μg/L.

**arsenic trichloride ($AsCl_3$)** An oily, colorless or yellow, poisonous liquid, used in organic pesticides. Also called butter of arsenic.

**arsenic trioxide ($As_2O_3$)** A white, amorphous, tasteless, poisonous powder, slightly soluble in water, used in pigments, glass, and pesticides. Also called simply arsenic. *See also* arsenolite.

**arsenic trisulfide ($As_2S_3$)** A natural, yellow or red, crystalline compound of arsenic, used as a pigment and in fireworks. Also called orpiment.

**arsenious acid** *See* arsenic.

**arsenite** A salt or ester of arsenic acid. *See* arsenic, arsenate.

**arsenolite ($As_2O_3$)** A white incrustation of arsenic trioxide in arsenic ores.

**arsenopyrite (FeAsS)** The common white-to-gray mineral, iron arsenic sulfide, an arsenic ore found in hydrothermal deposits; also a gangue mineral of hypothermal deposits of gold and tin; formerly used in insecticides. Also called mispickel.

**arterial main** In a pipe network, one of the pipes that constitute its basic structure, carrying flow to and from storage tanks and usually laid in loops. Also called primary main. Secondary mains form smaller loops within primary loops, running from one primary line to another.

**artesian** (From Artois, a province of Northern France.) Capable of freely rising to the surface. Artesian discharge is the flow rate of an artesian well. Artesian pressure and artesian head refer to the pressure exerted by the aquifer against the overlying formation and to the elevation to which water will freely rise in an artesian well. In an artesian spring, water flows under pressure through openings in the formation above the aquifer.

**artesian aquifer** An aquifer in which groundwater is confined under pressure that is significantly greater than atmospheric pressure. It is a fully saturated formation of porous rock or soil, overlain by a confining layer. The potentiometric surface (or hydraulic head) of the water in a confined aquifer is at an elevation that is equal to or higher than the base of the overlying confining layer. Discharging wells in a confined aquifer lower the potentiometric surface that forms a cone of depression, but the saturated medium is not dewatered. Same as confined aquifer. *See* aquifer.

**artesian discharge** *See* artesian.

**artesian head** *See* artesian.

**artesian pressure** *See* artesian.

**artesian spring** *See* artesian.

**artesian well** A well that penetrates an artesian aquifer. It is a flowing or nonflowing well depending on whether the piezometric surface is above ground or not, i.e., whether or not it discharges above ground. *See* Figure A-18 under aquifer. The opposite is a phreatic or water-table well.

**artesian well water** Bottled water from a well drilled in a confined aquifer. Also called bottled artesian water, a labeling regulated by the U.S. Food and Drug Administration.

***Arthrobacter*** A group of heterotrophic organisms found in activated sludge flocs and active in biological denitrification. In conjunction with other microorganisms, they can enhance the biodegradation of petroleum hydrocarbons in soil.

**arthropod** An invertebrate belonging to the phylum Arthropoda; it has jointed limbs, a blood-containing body cavity, and a hard, segmented exoskeleton. Arthropods are abundant in water and certain species are disease vectors, e.g., bugs, flies, lice, mosquitoes, and ticks. *See also* arbovirus.

**Arthropoda** A broad group of animals that include insects, crustaceans, and spiders; the largest animal phylum, with more 700,000 species.

**arthropod-borne virus** *See* arbovirus.

**arthropod vector** *See* arthropod.

**articulated tug barge (ATB)** A surface vessel used to transport liquid sludge to an ocean disposal site; it consists of a barge pushed by a tug. *See also* barge in tow, barge transport, rubber barge in tow, and self-propelled sludge vessel.

**artificial destratification** An operation used to control nutrients and the effects of eutrophication in lakes and reservoirs. It consists of pumping cold water from the bottom to the surface, thereby adding dissolved oxygen to the hypolimnion, lowering the temperature of the epilimnion, and shifting the algal population from the blue-green algae (responsible for taste and odor problems) to the less troublesome green algae.

**artificial intelligence (AI** or **A.I.)** The capacity of a computer (or other machine) to perform such operations as learning, decision making, and updating existing databases with new information. *See also* expert system and intelligent system.

**artificial media sludge drying bed** *See* high-density polyurethane drying bed and wedge-wire drying bed.

**artificial rain** Water produced or applied artificially by such methods as cloud seeding and rainfall simulation. Artificial stimuli used include common salt (NaCl), dry ice [solid carbon dioxide, $CO_2(s)$] pellets, and silver iodide (AgI).

**artificial recharge** Intentional (artificial) replenishment of an aquifer by such techniques as injection wells, spreading basins, or induced infiltration of surface water. *See also* groundwater recharge.

**artificial sludge** Substances used in experiments as substitutes for return sludge in the activated sludge process, e.g., metal hydroxides and silica gels (APHA et al, 1981).

**artificial substrate** A device or material placed in water to provide living spaces for organisms or for their colonization.

**artificial wetland** A man-made wetland; also called constructed wetland. Sometimes used to provide tertiary wastewater treatment, with microorganisms transforming the pollutants into harmless by-products or essential nutrients for floating or submerged aquatic plants. *See also* wetland treatment.

**aryl hydrocarbon** An organic compound derived from an aromatic hydrocarbon by the removal of a hydrogen atom, such as phenyl ($C_6H_5$) from benzene ($C_6H_6$). Polynuclear aromatic hydrocarbons are condensed-ring aryl molecules.

**As** Chemical symbol of arsenic.

**As(III)** Arsenic in the form of arsenite.

**As(V)** Arsenic in the form of arsenate.

**A/S** Abbreviation of air-to-solids ratio.

**asbestos** A grayish mineral fiber that does not conduct heat or electricity; it can pollute air or water and cause cancer or asbestosis when inhaled. It is a variety of one of the following minerals: actinolite, amosite (cummingtonite-grunerite), anthophyllite, chrysotile (serpentine, the most common mineral mined, perhaps 90%), crocidolite (riebeckite), and tremolite. Asbestos consists of silica (40–60%) and oxides of such metals as iron and magnesium. The USEPA has banned or severely restricted the use of asbestos in manufacturing and construction.

**asbestos–cement pipe** A pipe made of a mixture of asbestos fibers and Portland cement; material used extensively in water supply because of its low cost, light weight, and easy installation. However, it is banned in some states. Low-pH waters may cause corrosion in asbestos–cement pipes. It is unlikely that asbestos–cement pipe used in water can cause gastrointestinal cancer.

**asbestosis** A chronic lung disease associated with exposure to or inhalation of asbestos fibers, which scar oxygen-absorbing tissues and reduce lung functions. The disease makes breathing progressively more difficult and can be fatal. It is unlikely that asbestos fibers in drinking water can cause the disease.

**ASBR** Acronym of anaerobic sequencing batch reactor.

**as-built drawings (plans)** Record drawings (plans) that are assumed to accurately represent existing works. They are usually prepared during or shortly after construction, by making appropriate revisions to the original construction drawings (plans).

**as $CaCO_3$** Abbreviation of as calcium carbonate.

**as calcium carbonate (as $CaCO_3$)** An expression of concentrations of chemical substances or parameters (acidity, alkalinity) found in water and wastewater in terms of their equivalent concentrations of calcium carbonate to facilitate comparisons. To convert a concentration ($C$, mg/L) of a substance to its concentration as calcium carbonate ($x$, mg/L), multiply $C$ by the molecular weight of calcium carbonate (100) and divide by the substance's molecular or atomic weight ($W$):

$$x = 100\,C/W \qquad \text{(A-121)}$$

*See* calcium carbonate equivalent. Noting that a milliequivalent (meq) of $CaCO_3$ corresponds to 50 mg (i.e., 100 mg/millimole divided by 2 meq/millimole), the conversions can be made from milliequivalent of a substance ($S$, meq/L) to concentration as calcium carbonate ($x$, mg/L):

$$x = 50\,S \qquad \text{(A-122)}$$

**ascariasis** A latent and persistent excreta-related infestation of *Ascaris lumbricoides* or other ascarids. *See* its common name, roundworm, for more detail.

**ascarid** A nematode of the family Ascaridae; roundworm.

***Ascaris*** The genus of roundworms.

***Ascaris lumbricoides*** A genus of nematodes or roundworms that cause ascariasis worldwide; sometimes called the stomach worm; characterized by a latency of 10 days, a persistence of one year, and a relatively low infective dose. For its mode of transmission, see *Ascaris lumbricoides* life cycle. *Ascaris* eggs are among the excreted pathogens most resistant to composting and other waste treatment methods.

***Ascaris lumbricoides* life cycle** The mode of transmission of the roundworm, from the feces of an infected host, through maturation in the soil, to ingestion and development in a new host.

***Ascaris suum*** The species of pig roundworms, which resemble the human roundworm, *Ascaris lumbricoides.*

**ASCE** Acronym of American Society of Civil Engineers.

**$AsCl_3$** Chemical formula of arsenic trichloride.

**ascorbic acid ($C_6H_8O_6$)** A white, crystalline, water-soluble substance, found naturally (e.g., in citrus fruits and vegetables) or produced synthetically; used as a dechlorinating agent. *See* quenching agent.

**aseptic** Free from the living germs of disease, fermentation or putrefaction; sterile.

**aseptic technique** Any common sense measure that minimizes the risk of contamination in a microbiological analysis.

**asexual reproduction** Reproduction that occurs without fertilization; includes budding, fission, spore formation, etc.

**ash** (1) The mineral content of a product remaining after complete combustion; used as a rough measure of the nonvolatile content of wastewater or sludge. (2) The nonvolatile inorganic constituents of activated carbon.

**Ashaire** A line of aerators manufactured by Aerators, Inc.

**Ashfix™** A process developed by Ashland Chemical, for the stabilization of heavy metals in sludge.

**Asiatic clam** A freshwater mollusk from Southeast Asia (*Corbicula fluminea*), now found in many waters of the United States; it causes clogging, taste, and odor in drinking water sources and long transmission lines; chlorine is an effective control agent. *See also* zebra mussel.

**Asiatic liver fluke disease** A helminthic infection of the bile ducts caused by the worm *Clonorchis sinensis,* found mainly in Southeast Asia and transmitted from the feces of infected persons or animals through aquatic snails and fish as intermediate hosts. *See* clonorchiasis for detail.

**ASM1 model** *See* Activated Sludge Model No. 1.

**ASM2 model** *See* Activated Sludge Model No. 2.

**$AsO_3$** Chemical formula of arsenic trioxide (arsenolite).

**Aspergillosis** An infection caused by the fungus species *Aspergillus fumigatus.*

***Aspergillus*** A waterborne fungus, pathogenic to humans. It has been identified in unchlorinated groundwater as well as surface water and service mains. Two of the species identified in drinking water sources are *Aspergillus fumigatus* and *Aspergillus niger.*

***Aspergillus fumigatus*** A common airborne fungus (a bioaerosol) that is found naturally in grass and leaves. It may develop in composting operations and cause human ear, lung, skin, and sinus infections in sensitive persons; it is a concern for the health of workers and residents around the composting site.

**asphalt** The dark-brown to black cementitious material (solid, semi-solid, or liquid in consistency) of which the main constituents are bitumens that occur naturally or as a residue of petroleum refining (40CFR52.741).

**asphalt blending** A technique used for the stabilization/solidification of solid and hazardous wastes by incorporating them into a hot or cold asphaltic material.

**asphyxia** The condition caused by a lack of oxygen and an excess of carbon dioxide, which may lead to unconsciousness or death.

**asphyxiant** A chemical that starves the cells of an individual from the oxygen needed to sustain metabolism.

**aspirating aerator** An aeration device that draws air and forms small bubbles in a liquid.

**aspirator** A hydraulic device that creates suction by forcing liquid through a restriction and increasing the velocity head; used as a vacuum pump, sump pump, in aeration, and in gas chlorinators.

**aspirator feeder** An aspirator used as a feeder.

**ASR** Acronym of aquifer storage and recovery.

**ASR well** *See* aquifer storage and recovery. *See also* plugging.

**AsS, or $As_2S_2$, or $As_4S_4$** Chemical formula of arsenic disulfide; realgar.

**$As_2S_3$** Chemical formula of arsenic trisulfide.

**assay** The qualitative or quantitative analysis of a chemical substance; a test for a particular chemical or effect.

**assessment district** A specific area of land within a state or local jurisdiction that is assessed a pro rata share of the cost of such improvements as water and wastewater facilities, thus securing the debt incurred by a direct lien against the benefiting properties.

**assessment district bond** A bond issued by an assessment district.

**asset management** A planning process and decision-making tool that maximizes the value of the assets of an enterprise or utility and ensures that adequate financial resources are available for their rehabilitation and replacement when necessary. It optimizes the mix of repair and replacement.

**Assets** The tangible and intangible resources of any entity, such as a water or wastewater agency; buildings, tools, equipment, furniture, materials, etc. used in the operation of a water or wastewater utility.

**assignment of water** The transfer of a water right application or permit from one person to another, often in conjunction with the sale of land.

**assimilable organic carbon (AOC)** The portion of organic carbon that microorganisms use very readily, or the biodegradable portion of organic matter, unlike dissolved organic carbon or total organic carbon. It consists of small compounds such as carboxylic and amino acids. Even at low concentrations (e.g., micrograms per liter), AOC of low molecular weight (acetate, amino acids, lactate, succinate) has the potential to promote microbial regrowth in water distribution systems, particularly after ozonation. Tap water contains between 0.1 and 9.0 % of its total organic carbon as AOC. In distribution systems, AOC correlates well with the density of heterotrophic bacteria. *See* biodegradable dissolved organic carbon, coliform growth response test.

**assimilable organic carbon test** A test conducted to determine the biodegradable matter content as a measure of the relative availability of food in a sample of water or wastewater. The test measures the cell growth of certain bacteria strains (P-17 or *Pseudomonas fluorescens,* and NOX or *Spirillum*) in the sample.

**assimilating nitrate reduction** The reduction of nitrate to ammonia for use in cell synthesis when ammonia nitrogen ($NH_4$-N) is not available. *See* assimilatory nitrate reduction. *See also* dissimilating nitrate reduction, biological denitrification.

**assimilation** (1) The natural process by which a body of water purifies itself of pollutants through microbial and other actions. (2) The fabrication of complex molecules from simpler ones; also called anabolism or synthesis. *See also* metabolism and catabolism. (3) The transformation of a nutrient into another form that can be digested by an organism. (4) The incorporation of food substances into a living organism.

**assimilation capacity** Same as assimilative capacity.

**assimilative capacity** (1) The ability of a water body to receive polluting materials (e.g., wastewater or toxic materials) without a degradation of water quality below a defined level; the ability of a stream to self-purify naturally. The desirable water quality, resulting from dilution, dispersion, and self-purification may relate to the protection of the aquatic biota and/or to human consumption. *See also* aquifer assimilative capacity. (2) The ability of the soils of an onsite disposal field to accept wastewater and let it percolate downward, flow laterally, or return to the hydrologic cycle by evapotranspiration. Darcy's law is commonly used to analyze the assimilative capacity of a site. *See also* shallow trench pump-in test, percolation test, soil profile examination, absorptive capacity.

**assimilatory nitrate reduction** The uptake of nitrate by plants and microorganisms, followed by the reduction of the nitrate to ammonium and its incorporation into the biomasss. This is an alternative to the process of ammonium assimilation and one of the two modes of biological nitrate removal from wastewater (see, e.g., biological denitrification and the Modified Ludzak–Ettinger process). Also called nitrate immobilization, dissimilating nitrate reduction. *See also* dissimilatory nitrate reduction (Metcalf & Eddy, 2003 and Maier et al., 2000).

**assimilatory sulfate reduction** The aerobic or anaerobic process used by plants and most microorganisms when they take up sulfur in the oxidized form (sulfate) and reduce it internally to sulfide, which they incorporate into an organic form (Maier et al, 2000). *See also* sulfur mineralization, sulfur respiration, dissimilatory sulfate reduction, sulfate reducing bacteria, sulfur-oxidizing bacteria, thiosulfate, *Thiobacillus thiooxidans, Thiobacillus ferrooxidans.*

**association colloid** One of three main classes of colloids, the other two being hydrophilic and hydrophobic colloids. The association colloid is an organized aggregate of hydrophilic and hydrophobic ions and molecules, like the clusters of stearate anions and their ionic heads formed in water by soaps, synthetic detergents, and dyestuffs. *See also* micelle.

**association constant** A measure of the strength of association between two compounds; the reciprocal of the dissociation constant. Also called affinity constant. *See also* apparent dissociation constant, dissociation constant, equilibrium constant, ionization constant, Michaelis constant.

**Association of Boards of Certification (ABC)** An international organization in Ames (Iowa) representing over 150 boards that certify the operators of waterworks and wastewater facilities.

***Asterionella*** A genus of filter-clogging algae sometimes found in drinking water sources; they may cause short direct filtration runs; diatoms frequently responsible for taste and odor problems in drinking water.

**Astrasand filter** A proprietary, upflow continuous backwash, denitrification filter manufactured by Paques bv of Balk, the Netherlands.

***Astroviridae*** A family of enteric viruses that comprises only the *Astrovirus* genus.

**astrovirus** An opportunistic, human enteric virus present in diarrheal stools and animal feces, measuring approximately 28 nm. It causes a mild gastrointestinal infection, with an incubation of 3 to 4 days, affecting mostly children, the elderly, and individuals with weakened immune systems (e.g., AIDS patients).

**Astrovirus genus** A group of human enteric viruses; *see* astrovirus.

**asymmetric membrane** A thin (less than 1.0 micron) reverse osmosis membrane that consists of a single polymer with nonuniform cross-sectional structure, such as cellulose acetate, supported by a thicker (up to 100 microns) substructure having larger pores; it is not reversible and can only desalinate efficiently in one direction. Also called anisotropic membrane. *See also* thin-film composite membrane.

**Asymptomatic** Characteristic of an infection or illness that does not show any evidence, clinical sign, or symptom in a host; an inapparent infection or subclinical disease. For example, some viruses may grow effectively without any noticeable symptoms and without killing the host cell, which continues to divide along with the virus, thereby causing an asymptomatic infection. *See also* inapparent infection, subclinical.

**asymptomatic carrier** A person or an animal that harbors a specific pathogen and serves as a potential source of infection but never develops the disease.

**asymptomatic infection** *See* asymptomatic.

**asymptote** A line that is considered to be the limit to a curve. As the curve approaches the asymptote, the distance separating the curve and the asymptote continues to decrease, but the curve never actually intersects the asymptote.

**ATAD** Acronym of autothermal thermophilic aerobic digestion.

**ATD™** A sludge processing apparatus manufactured by CBI Walker, Inc. using autothermal aerobic digestion.

**at-grade system** A soil absorption system with the bottom of its gravel envelope on the tilled soil surface; installed in conditions of fractured bedrock or high groundwater table.

**atlas** A sewer atlas is a bound set of maps of a sewer system, with such information as plans and profiles of the sewer lines, location of pumping stations, treatment plants, manholes and other appurtenances, sizes, elevations, right-of-way details, etc.

**atmophile element** A chemical element that has an affinity for the atmosphere, e.g., neon, helium. *See also* lithophile element.

**atmospheric corrosion** Corrosion due atmospheric exposure.

**atmospheric evaporator** One of two main categories of evaporators, the other being the vacuum evaporator. The atmospheric device effects evaporation by spraying the liquid over a packing medium and blowing air over the packing.

**atmospheric pressure** The pressure exerted by the weight of the atmosphere; it decreases with increasing elevation above sea level:

$$P_B/P_A = \exp[-gM(Z_B - Z_A)/RT] \quad \text{(A-123)}$$

where $P_A$ and $P_B$ are pressures at points A and B, respectively, $1.01325 \times 10^5$ N/m$^2$; $g = 9.81$ m/s$^2$ is the gravitational acceleration; $M$ = mole of air = 28.97 kg/kg.mole; $Z_A$ and $Z_B$ are elevations at points A and B, respectively, m; $R$ = 8314 kg · m$^2$/s$^2$ · kg-mole · °K = universal gas constant; and $T$ is temperature in °K. At mean sea level, it is one atmosphere (1 atm) = 14.7 psi under standard conditions. Also called standard pressure, standard atmospheric pressure. *See also* absolute pressure.

**at. no.** Abbreviation of atomic number.

**atom** (1) The smallest unit of a chemical element that has the chemical properties of the element; composed of (a) a nucleus containing positively charged protons and noncharged neutrons and (b) negatively charged electrons moving around the nucleus. Atoms combine to form molecules. (2) A hypothetical element that cannot be divided.

**Atomerator®** A pressure aerator manufactured by General Filter Co.

**atomic absorption spectrometric method (AA method or AAS method)** Same as atomic absorption spectrophotometric method.

**atomic absorption spectrophotometer** An instrument used to measure the concentration of trace metals and other elements in water and wastewater. It consists of three devices to convert the elements to free atoms in a flame, to isolate the light waves emitted, and to amplify the light waves. In the conventional or flame version, the cathode of the light source is of the same element being measured, so that the concentration of that element is equal to the reduction in light intensity after introduction of the sample into the flame. The flameless or graphite furnace technique uses carbon rods or graphite tubes; the concentration of the element of interest is proportional to the amount of light absorbed. The instrument requires only very small sample volumes and can detect very low concentrations; however, it measures total concentrations and does not distinguish between species of a metal. *See also* Beer–Lambert law, flame atomic absorption spectrophotometry, and graphite furnace atomic absorption spectrophotometry.

**atomic absorption spectrophotometric method (AA method or AAS method)** A laboratory analysis technique that uses an atomic absorption spectrophotometer to identify and measure the concentrations of toxic trace metals (and other common elements, e.g., sodium, calcium) in water and wastewater samples. It is based on detecting the quantity of light of a specific frequency or wavelength that the sample absorbs; each chemical element corresponds to a characteristic frequency. Also called atomic absorption spectrophotometry or atomic absorption spectroscopy.

**atomic absorption spectrophotometry** *See* atomic absorption spectrophotometric method.

**atomic absorption spectrophotometry (AAS) (flame)** Same as flame atomic absorption spectrophotometry.

**atomic absorption spectrophotometry (graphite furnace)** Same as graphite furnace atomic absorption spectrophotometry.

**atomic absorption spectroscopy** Same as atomic absorption spectrophotometric method.

**atomic emission spectroscopy** A laboratory analysis technique that uses the intensity of the radiation emitted by excited atoms to measure the concentrations of trace metals (and other common elements, e.g., sodium, calcium) in water and wastewater samples. *See* inductively coupled plasma atomic emission spectroscopy.

**atomic mass** The average mass of the atoms of an element, or the mass of an isotope of the element, expressed in atomic mass units. *See* periodic table.

**atomic mass unit (u, AMU or amu)** A mass equivalent to one-twelfth the atomic weight of the carbon-12 atom ($^{12}C$), used to express the masses of atoms, molecules, and subatomic particles; sometimes taken as AMU = $1.66054 \times 10^{-27}$ kg. Also called dalton. *See* proton.

**atomic number (at. no.)** The number of positive charges or protons in the nucleus of each atom of an element, which is also the number of electrons normally surrounding the nucleus; also called proton number.

**atomic weight (at. wt.)** The average weight of the atoms of an element, or the weight of an isotope of the element. It is approximately the number of protons and neutrons in the atom, but it reflects also the occurrence of isotopes of the element, which differ in the number of neutrons. *See* periodic table, gram atomic weight, equivalent (or combining) weight, and valence.

**atomization** The process of dividing a liquid into small droplets, as used, for example, in spray painting.

**atomized suspension** A sludge treatment and disposal technique consisting of the atomization of the waste liquor or slurry in a tower at high temperature, leading to quick evaporation and production of a solid residue. It includes the recovery of steam for reuse and by-product gases for further processing (Nemerow & Dasgupta, 1991).

**atomizer** A device used to spray a liquid in a fine mist.

**ATP** Acronym of adenosine triphosphate.

**ATP™** Aerobic thermophilic sludge treatment developed by CBI Walker, Inc.

**atrazine (2-chloro-4-ethylamino-6-isopropyl amino-1,3,5-triazine) ($C_5H_{16}N_5Cl$)** A white, crystalline, organic chemical used to control broadleaf and grassy weeds in fields of corn, soybeans and sorghum. It is found in surface waters receiving agricultural runoff. Atrazine's maximum contaminant level (MCL) has been set at 3.0 micrograms per liter. Activated carbon adsorption, advanced oxidation, and membrane filtration are the methods recommended for its removal from drinking water. *See also* s-triazine.

**atrazine-desethyl ($C_8H_{14}N_5Cl$)** A chemical compound resulting from the degradation of atrazine.

**attached growth** One of two types of microbial development used in biological wastewater treatment: microorganisms grow attached to natural or artificial media and use the contaminants of the liquid stream as substrate. *See* fixed growth for more detail.

**attached-growth biofilm** The biofilm formed on the support materials used in attached-growth processes. *See* biofilm.

**attached-growth denitrification** Any biological denitrification process using an attached-growth model, e.g., Biocarbone®, Biofor®, Biostyr®, fluidized-bed reactor.

**attached-growth process** A biological wastewater treatment process in which the microorganisms are attached to rock, slag, plastic, or other inert materials. *See* fixed-growth process for more detail.

**attachment** One of two principal mechanisms involved in rapid-rate filtration, the other being entrapment. In attachment, suspended particles attach to the surface of media, as the liquid passes through the media grains, and form larger flocs that accumulate in the filter. *See also* Brownian movement and interception.

**attachment probability** A dimensionless coefficient that indicates the probability that suspended particles in a moving fluid may attach to stationary particles or other fixed solid boundaries; determined as follows (Stumm & Morgan, 1996):

$$\alpha_d = \text{(rate at which particles attach to the collector)/(rate at which particles approach the collector)} \quad \text{(A-124)}$$

Also called sticking coefficient. *See also* collision efficiency.

**attack rate** The cumulative incidence of a disease during an outbreak or other special circumstances.

**attainability principle** A guideline used in establishing drinking water quality and other standards, which, to be meaningful, must be attainable, based on existing technology.

**attenuate and disperse landfill** The traditional type of landfill from which leachate produced seeps through soils into the underlying saturated zone for dilution.

**attenuated vaccine** Vaccine based on live infection material that has been treated in such a way that it creates an antibody response but does not cause the disease.

**attenuation** (1) The process by which a compound is reduced in concentration over time, through absorption, adsorption, degradation, dilution, and/or transformation; e.g., the reduction in the amount of contaminants in a plume as it migrates away from the source. (2) The loss of virulence in bacteria or other pathogens.

**attenuation coefficient** A parameter used in evaluating the germicidal effectiveness of ultraviolet light or other forms of radiant energy. The attenuation coefficient ($k_a$) is related to the Secchi depth ($d_s$) and radiation intensity (Droste, 1997):

$$1/k_a = 0.0222 + 0.261\, d_s \qquad \text{(A-125)}$$

$$p_e = 1 - \exp(-L \cdot k_a) \qquad \text{(A-126)}$$

where $p_e$ is the proportion of energy absorbed and $L$ is the length of the light beam. The attenuation coefficient of ultraviolet light can vary from 0.03/cm for filtered water to 0.20/cm for unfiltered water. Also called extinction coefficient.

**Atterberg limits** In soil mechanics, a set of three limits that, based on tests, indicate the water content (as a percentage of the dry weight) of soils when they pass from the liquid to the plastic state (liquid limit), or from the plastic to the solid state (plastic limit). No change in volume occurs below the shrinkage limit. The plasticity index is the difference between the liquid and plastic limits.

**Atterberg test** A test that measures the consistency of soils with respect to water content. When applied to water and wastewater sludge, only the liquid and plastic limits are of interest: the plastic limit corresponds to the transition from a semisolid sludge to a plastic sludge (e.g., from soft butter to stiff putty), whereas the liquid limit is the solids concentration of viscous sludge (e.g., from soft butter to a pea soup slurry). *See also* Büchner funnel test, capillary suction time test, compaction density, filterability constant, filterability index, shear strength, specific resistance, standard jar test, time-to-filter test.

**attributable risk** The difference between risk of exhibiting a certain adverse effect in the presence of a toxic substance and that risk in the absence of the substance; the proportion of a disease or other outcome in exposed individuals that can be attributed to the exposure of interest.

**attrition** The gradual lessening of capacity or effectiveness of a medium, due to friction, sacrificial properties, chemical attack, or contaminant saturation.

**ATU** Acronym of aerobic treatment unit.

**at. wt.** Abbreviation of atomic weight.

**A-type metal** Any of the lithophile metals, including aluminum (Al), antimony (Sb), chromium (Cr), cobalt (Co), manganese (Mn), nickel (Ni), titanium (Ti), and vanadium (V), whose mass transport to the oceans by streams exceeds their transport through the atmosphere. *See* B-type or atmophile metal and class A ion.

**A-type metal cation** Same as class A ion.

**audit** *See* model audit.

**aufwuchs** Periphyton.

**auger** (1) A simple tool for drilling or boring into unconsolidated earth materials, consisting of a spiral blade wound around a central stem or shaft that is commonly hollow (hollow-stem auger). Augers are commonly available in flights (sections) that are connected together to advance the depth of the borehole. (2) A device consisting of sharpened spiral blades attached to a hard metal central shaft for sampling hard or packed solid wastes or soil. *See also* bailer, composite liquid-waste sampler, dipper, thief, trier, weighted bottle.

**auger hole test** A test conducted in a shallow auger hole to determine an aquifer's hydraulic characteristics. *See also* aquifer test.

**Auger Monster®** Wastewater preliminary treatment equipment manufactured by JWC Environmental, combining screening and grinding operations.

**aurium** Latin name of gold (Au).

**autecology** The branch of ecology that studies the relationships between an individual organism or a species and its environment.

**AutoBelt** A rotary vacuum filter manufactured by Walker Process Equipment Co.

**AutoCAD** An automated program for computer-aided drafting (CAD), developed by Autodesk, Inc.; used to develop realistic, accurate two- and three-dimensional drawings.

**autocatalytic reaction** A reaction that accelerates with time, its rate depending on catalytic effects of the products; e.g., the decomposition of ozone (Droste, 1997):

$$d[O_3]/dt = -k_1\,[O_3] - k_2\,[O_3]^2 \qquad \text{(A-127)}$$

where $[O_3]$ is the concentration of ozone and $k_1$ and $k_2$ are rate constants.

**autochthonous** Pertaining to organisms, soils, or materials formed in the area where found. Autochthonous organisms develop slowly in soils and use nutrients that are present in low concentrations or are slowly released by the soil; also called K-strategists or oligotrophs. *See also* allochtonous, zymogenous.

**autoclave** An apparatus that sterilizes laboratory materials (culture media, dilution water, instruments, equipment, etc.) for bacteriological tests by exposure to pressurized steam. One recommended operation mode is 15 minutes at 15 psi, corresponding to 121.6°C.

**autoclaved** Sterilized in an autoclave.

**autoclaving** A sterilization procedure using an autoclave.

**Auto-Cleanse** A self-cleaning pumping station manufactured by ITT Flygt.

**autocombustible sludge** Sludge that does not require any auxiliary fuel to sustain the burning process, e.g., sludge that has undergone heat treatment ahead of incineration. *See also* thermal reduction.

**autocrine** The internal mechanism by which a cell regulates itself.

**Auto-Dox®** A weir manufactured by Purestream, Inc. for the control of wastewater aeration.

**autoecious parasite** A parasite that spends its entire life cycle on or in an individual host; the opposite of a heteroecious parasite.

**autoflotation** A natural process that removes algae from wastewater by gas supersaturation in stabilization ponds.

**autogenous combustion** An energy efficient sludge processing method that uses the heat of combustion of the wet material to vaporize the water without the addition of an auxiliary fuel. Also called autothermic combustion. The process operates at an autogenous temperature by maintaining an equilibrium between heat input and heat loss.

**autogenous temperature** *See* autogenous combustion.

**autoimmunity** The characteristic of an organism that makes an allergic or immune response toward its own tissues, cells, or components, sometimes resulting from exposure to toxic chemicals and causing chronic diseases.

**Auto-Jet®** A pressure leaf filter manufactured by U.S. Filter Corp.

**Autojust** An automatic gate controller manufactured by Aerators, Inc. for use in sludge collectors.

**automatic meter reading (AMR)** Reading and transmission of customer meters (via telephone lines, radio, or cables) to a remote location for billing.

**automatic regulator** A device used to regulate flow in combined or interceptor sewers: water level is controlled by float and gate travels. Also known as mechanical regulator or reverse taintor gate.

**automatic residual control** A procedure at water treatment plants that uses an apparatus to maintain a preset chlorine residual independent of flow variations. The chlorine feeder reacts to signals from the flow meter and from an analyzer below the point of application.

**automatic sampling** The collection of water or wastewater samples by a device without direct manual control, including preset sample volumes and sampling times. The device may be battery-operated for remote control. Automatic sampling requires less labor than manual sampling and is particularly effective in collecting composite samples, but automatically collected samples may not be appropriate for some analyses, e.g., oil and grease, volatile organic carbon, pH, temperature, chlorine, and fecal coliforms.

**automatic sluice gate** A device used to control water surface levels in combined sewers and capture stormflows completely or partially by providing in-system storage.

**automatic valve** A valve that operates according to preset conditions without human intervention.

**automatic water softener** A water softener that functions without human intervention, carrying out its various operations (backwashing, regeneration, bypassing, etc.) on a preset schedule.

**Automation** The use of electronic devices or other automatic means to operate or control such mechanisms or activities as design, process control, billing, or modeling.

**automobile manufacturing wastewater** Same as automotive wastewater.

**automotive wastewater** Wastewater generated at automobile assembly plants containing high concentrations of paints; zinc, lead, and other metals; and emulsified and free oils. Some of these pollutants as recovered and others are treated onsite before discharge to a sanitary sewer system.

**autooxidation** *See* autoxidation.

**Auto-Pulse™** A backpulse filter manufactured by U.S. Filter Corp.

**Auto-Rake®** A bar screen manufactured by Franklin Miller; it has a reciprocating rake.

**Auto-Retreat** Equipment manufactured by Infilco Degremont, Inc. for the automatic control of wastewater screening.

**autosampler** An instrument used for automatic handling of water or wastewater samples.

**AutoSDI™** An instrument manufactured by King Lee Technologies; it uses a portable computer to measure the silt density index or SDI.

**Auto-Shell™** A granular medium produced by the U.S. Filter Corp.

**Auto-Shok™** A vertical leaf filter manufactured by the U.S. Filter Corp.

**Auto-Skimmer™** A device manufactured by R.E. Wright Associates, Inc. for skimming hydrocarbons from water wells.

**AutoTherm™** Aerobic thermophilic digestion equipment manufactured by CBI Walker, Inc.

**autothermal digestion** A sludge stabilization process carried out in a covered and insulated reactor where the heat released by the metabolism of the organic matter maintains the required temperature. If the reactor operates in the thermophilic range, it can be designed for a detention time of 7 days or less.

**autothermal thermophilic aerobic digestion (ATAD)** A sludge stabilization system that converts soluble organics aerobically to lower-energy forms and produces sufficient heat to maintain a thermophilic temperature range in the reactor. It uses more oxygen than a conventional aerobic digester. It is a process to further reduce pathogens (PFRP) that produces Class A material; however, it may have objectionable odors and poor dewatering characteristics, and inhibit nitrification. A typical ATAD system includes a prethickening or blending tank, two ATAD reactors in series equipped with piping for gas release, and a biosolids storage tank. *See* Figure A-19. Also called autothermal thermophilic digestion. *See also* microaerobic condition.

**autothermal thermophilic digestion** *See* autothermal thermophilic aerobic digestion.

**autothermic combustion** Same as autogenous combustion.

**Autotravel®** Sludge collector of the traveling bridge type manufactured by Simon-Hartley, Ltd.

**autotroph** Any of a number of organisms that do not require a reduced organic (carbon) source of energy, including most plants, certain bacteria, many planktonic organisms, and certain protists; they can use carbon dioxide as a carbon source, inorganic materials as a source of nutrients, and photosynthesis or chemosynthesis as an energy mechanism. For example, iron bacteria generate energy by oxidizing ferrous iron to ferric iron; nitrifying bacteria are also autotrophs. Autotrophic organisms are the primary producers in any food chain, since they synthesize and store all the energy used by the other organisms. They generally have lower cell-mass yields than heterotrophs, as they spend more of their energy for synthesis. *See also* chemoautotrophs, chemoheterotrophs, chemotrophs, heterotrophs, phototrophs.

**autotroph–heterotroph relationship** Heterotrophs (including humans, from the top of the food chain) consume autotrophs, which produce and store food and energy.

**autotrophic** Designating or typical of organisms that derive nutrients from inorganic substances (carbon dioxide, nitrogen, nitrates, etc.) for the manufacture of cell mass.

**autotrophic bacteria** Bacteria that derive nutrients from inorganic substances and energy from sunlight or the oxidation of inorganics; e.g., cyanobacteria, iron bacteria, methanogens, nitrifiers, sulfur bacteria. They may be found in water distribution systems and in filters following iron or manganese precipitation. *See also* chemoautotrophs, heterotrophic bacteria, photoautotrophs.

**autotrophic biota** Autotrophic organisms.

**autotrophic organism** An organism that uses sunlight or chemical energy to convert inorganic matter into cells, e.g., algae and photosynthetic bacteria; *see* autotroph.

**Auto-Vac®** Rotary filter manufactured by Alar Engineering Corp.

**autoxidation** The spontaneous oxidation of a chemically unstable substance in the presence of air or other electron acceptors. In some biochemical reactions, autoxidation results in the conversion of cells to low-energy products and the release of additional energy. Also spelled autooxidation.

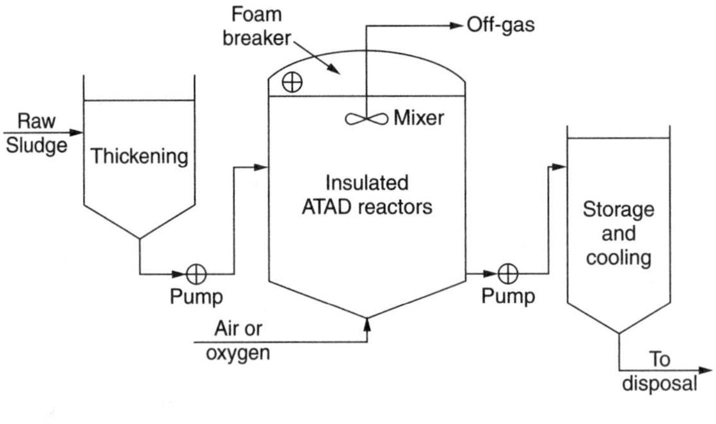


**Figure A-19.** Autothermal thermophilic aerobic digestion (ATAD).

**auxiliary scour** An effective method for cleaning granular filters by removing attached materials from the medium, using water or a mixture of air and water at high pressure to agitate the media and loosen the materials. *See* surface washing for more detail.

**auxiliary source** A source of water that has been approved as a backup supply and may be used when the primary source is inadequate.

**auxiliary tank valve** A valve that serves as a shut-off device in a chlorine tank when the main valve does not function properly.

**auxiliary water supply** A building's backup water supply, from an alternate purveyor or an auxiliary source, in case the public supply is not available. Auxiliary supplies sometimes pose the risk of contamination to the public supply by cross-connection.

**availability charge** A charge made by a water supply agency to a property owner for making the service available, whether the property uses or connects to the facilities or not.

**available chlorine** (1) A measure of the amount of chlorine and, thus, the oxidizing power, available in chlorinated lime, hypochlorite compounds, chloramines, and other chlorine-containing materials used for disinfection when compared with that of elemental (liquid or gaseous) chlorine. The literature is not always consistent in the computation of available chlorine.

(a) From Fair et al. (1971): available chlorine of a compound is expressed as elemental (liquid or gaseous) chlorine ($Cl_2$, molecular weight = 2 × 35.5 = 71) and, as such, depends on the moles of chlorine with the oxidizing capacity of one mole of the compound and on the purity of the commercial product. Example 1: dichloramine ($NHCl_2$, molecular weight = 86) has an available chlorine of 71/86 = 0.826 gram of chlorine per gram of dichloramine. Example 2: commercial calcium hypochlorite, which contains approximately 66.5% $Ca(OCl)_2$ (molecular weight = 143), has an available chlorine of 66.5% × 71/143 = 33.0%.

(b) From Metcalf & Eddy (2003): the percent of available chlorine [$(Cl_2)$available] of a compound X is the product of its chlorine equivalent (the electron change in the half reaction) and its percent actual chlorine [$(Cl_2)$actual, %]:

$$(Cl_2)\text{available} = (\text{Cl equivalent})\,[(Cl_2)_{\text{actual}}, \%] \quad \text{(A-128)}$$

$$(Cl_2)_{\text{actual}}, \% = 100 \times (\text{weight of } Cl_2 \text{ in X}) / (\text{molecular weight of X}) \quad \text{(A-129)}$$

Example 3: for hypochlorite [$Ca(OCl)_2$]:

$$(Cl_2)_{\text{actual}}, \% = 100 \text{ x } (2 \times 35.5/143) = 49.65\%$$

$$(Cl_2)_{\text{available}} = (2) \times (49.65\%) = 99.30\%$$

(c) From AWWA (1999), it is the electrochemical equivalent amount of chlorine ($Cl_2$), based on the number of electrons necessary to form inert chloride ($Cl^-$). Hence, one mole of hypochlorite ($OCl^-$) is equivalent to one mole of elemental chlorine and contains 71 g of available chlorine; one mole of calcium hypochlorite [$Ca(OCl)_2$] contains 142 g and 99.2% of available chlorine, whereas one mole of sodium hypochlorite (NaOCl) contains 71 g and 95.8% of available chlorine. *See also* chlorine equivalent, combined available chlorine, free available chlorine.

(2) A measure of the oxidizing capacity of nonchlorinous compounds, such as potassium permanganate ($KMnO_4$), in terms of available chlorine.

**available dilution** The ratio (at the point of disposal or at a point downstream) of the flow of receiving water to the flow of wastewater, treated effluent, or stormwater discharged. Also called dilution factor. *See also* dilution rate, dilution ratio.

**available expansion** The vertical distance from the sand surface to the underside of a trough in a sand filter. Also called freeboard.

**available head loss** The maximum head loss provided for in the design of a rapid-filtration plant.

**available heat** The amount of heat released by a fuel in a combustion chamber minus the dry flue gas and moisture losses. Available heat of a sludge depends on the characteristics of the sludge and those of the combustion chamber. Also called characteristic heating value, effective heating value, or net heating value. *See* sludge fuel value for detail.

**available nitrogen** The amount of nitrogen available for plant growth during the first year after land application of wastewater sludge depends on the mass fractions of the nitrogen species (nitrate, ammonia, organic) and the mineralization or volatilization factors. It can be estimated as follows (Droste, 1997):

$$N_1 = K[(NO_3^-) + V(NH_3) + M_1 N_o] \quad \text{(A-130)}$$

where $N_1$ = available nitrogen in pounds per ton dry solids; K = a conversion factor = 2000 lb/ton; $(NO_3^-)$ = mass fraction of nitrate-N in the sludge, lb of $NO_3^-$-N per lb of solids; V = volatilization factor, 0.5 for surface application, 1.0 for subsurface application; $(NH_3)$ = mass fraction of nitrate-N in the sludge, lb of $NH_3$-N per lb of solids; $M_1$ =

mineralized fraction of organic nitrogen in freshly applied sludge = 0.4 for raw sludge, 0.2 for digested aerobic sludge, and 0.1 for composted sludge; $N_o$ = mass fraction of organic nitrogen in the sludge at application, lb/lb. The available nitrogen ($N_n$) in any year $n$ is:

$$N_n = K(M_n)(N_{n-1})(1 - M_{n-1}) \qquad \text{(A-131)}$$

where $M_n$ (or $M_{n-1}$) is the mineralized fraction of organic nitrogen in year $n$ (or year $n - 1$) and $N_{n-1}$ is the available nitrogen in year $n - 1$. The available nitrogen is used to compute the nitrogen loading rate. *See* plant available nitrogen (PAN) for another formula.

**available oxygen** The quantity or concentration of dissolved oxygen available in a water body, treatment unit, etc.

**avalanche** A mass of snow (sometimes containing ice, rocks, and other objects) moving rapidly down a slope; a snowslide.

**average annual daily flow** The average water or wastewater flow rate over a 24-hour period, based on annual data.

**average annual hydraulic loading rate** For rapid wastewater infiltration systems, the average annual hydraulic loading rate ($L_w$, ft/yr) is the product of the minimum long-term infiltration rate ($I$, in/hr), an application factor ($F$, from 2% to 15% depending on the type of infiltration rate measurement), the number of operating days per year ($N$), and a unit conversion factor [(1 ft/12 in) × (24 h/1 d) = 2]:

$$L_w = 2\,IFN \qquad \text{(A-132)}$$

This variable is used to calculate the actual average application rate ($R$, ft/day), which also takes into account the number of days in the operating cycle ($C$, days) and the application period ($P$, days):

$$R = (L_w/365) \times (C/P) = 2\,IFNC/(365\,P) \qquad \text{(A-133)}$$

**average application rate** *See* average annual hydraulic loading rate.

**average daily demand** The annual water demand divided by the number of days in the year, corresponding to the average water demand in one day or 24 hours.

**average daily flow (ADF)** In a series of daily flows, it is their sum divided by the number of flows. *See* annual average flow.

**average-day demand** Same as average daily demand.

**average dry-weather flow (ADWF)** (1) In hydrology, the average streamflow in periods of dry weather, i.e., without the influence of runoff; it is mainly groundwater seepage. (2) In wastewater studies, it is the average flow of wastewater in a sanitary or combined sewer system during dry weather; *see* dry-weather flow for more detail.

**average-end-area method** A method used to calculate the volume ($V$) of water stored in a reservoir up to the nth contour (Fair et al., 1971):

$$V = 0.5\,H\left(A_0 + A_n + 2\sum_{i=1}^{n-1} A_i\right) \qquad \text{(A-134)}$$

where $V$ is contour interval and $A_0$, $A_1$ . . . $A_n$ are the successive contour areas.

**average flow rate** The average of instantaneous flow rates over a given period of time, e.g., a day or a year; same as annual average flow rate.

**average flow velocity** (or simply average velocity) **(V)** The ratio of discharge ($Q$) to cross-sectional area ($A$) at any point in a channel.

**average free available chlorine concentration** The average free chlorine concentration during a test of a water sample to evaluate disinfection byproduct formation, calculated as half the sum of the chlorine dose and the chlorine residual.

**average linear groundwater velocity** The ratio of the volume of groundwater through a formation in unit time to the product of the cross-sectional area by the effective porosity. *See* actual groundwater velocity for more detail.

**average monthly discharge limitation** The highest allowable average of daily discharges over a calendar month, calculated as the sum of all daily discharges measured during a calendar month divided by the number of daily discharges during that month (40CFR122.20).

**average retention time (ART) index** One of the terms used to characterize tracer response curves and describe the hydraulic performance of reactors; it is the ratio of the mean time to reach the centroid of the residence time distribution curve ($T_C$) to the theoretical hydraulic residence time ($\tau$):

$$ART = T_C/\tau \qquad \text{(A-135)}$$

*See also* residence time distribution (RTD) curve, time-concentration curve. The ART index indicates the degree of uniformity of flow distribution through the reactor; optimum performance corresponds to ART = 1.0.

**average UV irradiance** A measure of the intensity of radiant energy per unit area, which is used to determine UV doses in bench-scale experiments, after appropriate corrections. The average irradiance ($E$) is the product of the incident irradiance

($E_0$) by four factors: divergence ($d$), Petri ($p$), reflection ($r$), and water ($w$); thus (Batch et al., 2004):

$$E = dprwE_0 \quad \text{(A-136)}$$

**average velocity** *See* average flow velocity.

**average weekly discharge limitation** The highest allowable average of daily discharges over a calendar week, calculated as the sum of all daily discharges measured during a calendar week divided by the number of daily discharges measured during that week (40CFR122.2).

**average wet-weather flow** The average of daily flows during the wet-weather period, including dry-weather flow and infiltration/inflow.

**average year** A year in which hydrologic characteristics of a basin are approximately equal to their arithmetic mean over a long period (e.g., temperature, streamflow, precipitation). Sometimes called a normal year.

**AVGF®** Acronym of automatic valveless gravity filter, a product of the U.S. Filter Corp.

**AVI** (a) Acronym of aggregate volume index. (b) *See* inherent availability.

**AVIR™** A line of proprietary membrane products supplied by A/G Technology.

**AVO** *See* operating availability.

**Avogadro, Amedeo** Italian chemist and physicist, 1776–1856; stated Avogadro's law in a book published in 1811.

**Avogadro's constant** Same as Avogadro's number.

**Avogadro's hypothesis** Same as Avogadro's law.

**Avogadro's law** Equal volumes of different (ideal) gases, at the same temperature and pressure, contain equal numbers of molecules, specifically, $6.022 \times 10^{23}$ molecules per mole. (A mole is a mass, usually expressed in grams, equal to the molecular weight). It follows that the molal volume of an ideal gas at standard temperature (0°C) and pressure (760 mm Hg) is 22.412 liters. Avogadro's law is combined with other so-called gas laws to study the phenomena of gas absorption, desorption, precipitation, and release in water and wastewater treatment. *See also* Boyle's law, Charles' law, Dalton's law, Gay-Lussac's law, general gas law, Graham's law, Henry's law, Loschmidt's number.

**Avogadro's number** The number $N$ of molecules in a mole of gas or the number of atoms in a gram atom, with $N = 6.022 \times 10^{23}$, according to Avogadro's law. More generally, the number of atoms, molecules, or groups of ions that make up the smallest possible unit of an ionic compound, e.g., two sodium ($Na^+$) ions and one sulfide ($S^{2-}$) ion in sodium sulfide ($Na_2S$). Also called Avogadro's constant.

**A/W** Acronym of air-to-water ratio.

**awash** Nearly flush with or just above the water level so that waves break over the top.

**AWQC** Acronym of ambient water quality criteria.

**AWRA** Acronym of American Water Resources Association.

**AWT** Acronym of advanced waste treatment or advanced wastewater treatment.

**AWWA** Acronym of American Water Works Association.

**AWWA standard** "The minimum requirements for water supply products or procedures established by the American Water Works Association through a consensus process that involves representatives from different interest groups, including manufacturers, water utilities, consultants, governmental agencies, and others. AWWA standards are not legally binding and are not specifications, but they may be referenced in specifications of water utilities or others who are acquiring a particular type of product or procedure." (Symons et al., 2000)

**AWWF** Acronym of average wet-weather flow.

**axenic culture** An uncontaminated laboratory culture, except for a specific group or species of organisms.

**axial dispersion** The transport of material along an axis, brought about by velocity differences, turbulent eddies, and molecular diffusion.

**axial dispersion coefficient** A parameter that accounts for the effects of axial dispersion in the design of water and wastewater treatment facilities. For large Reynolds numbers, this coefficient ($D$, $m^2/s$) may be estimated as follows (Metcalf & Eddy, 2003):

$$D = 1.01\, \upsilon N_R^{0.875} \quad \text{(A-137)}$$

where $\upsilon$ = kinematic viscosity, $m^2/s$; $N_R$ = unitless Reynolds number = $4\, VR/\upsilon$; $V$ = flow velocity, m/s; $R$ = hydraulic radius, m.

**axial flow** (1) *See* axial-flow pump. (2) Movement of water up or down in a cartridge or loose-medium filter. Also called longitudinal flow.

**axial-flow mixer** A device that employs pumping action to disperse mixing energy, e.g., a pitched blade or high-efficiency hydrofoil impeller to direct discharge up or down. Baffles are sometimes used to reduce bulk rotation. *See* Figure A-20.

**axial-flow pump** A centrifugal pump that diverts liquids (clean water or treated effluent) in the axial direction of the pipeline in which its propeller-type impeller is installed. Also called propeller pump.

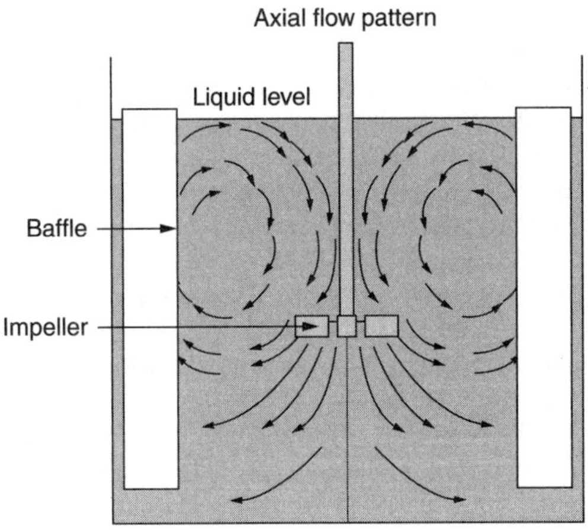


**Figure A-20.** Axial-flow mixer (courtesy Eimco Water Technologies).

**axial to impeller** The direction in which material being pumped flows around the impeller or flows parallel to the impeller shaft.

**axis of impeller** An imaginary line running along the center of the impeller shaft.

**azeotropic** Characteristic of a liquid mixture whose partial evaporation vapor has the same composition as the mixture.

**azide** A compound of an azido group, the univalent group —$N_3$; e.g., sodium azide, $NaN_3$.

**azide modification** A variation of the iodometric method for measuring dissolved oxygen concentrations. The test is conducted in a 300-ml BOD bottle, using a number of reagents, including alkali–iodide–azide reagent (Hammer & Hammer, 1996).

**azido group** *See* azide. Also called azido radical.

**azido radical** Same as azido group.

***Azolla*** A genus of small, free-floating aquatic weeds with short roots of about 12 mm, sometimes planted in constructed wetlands to provide habitat for organisms that metabolize organics in wastewater.

***Azotobacter*** A genus of aerobic, aquatic bacteria that can fix atmospheric nitrogen; by breaking down carbohydrates in soils, it contributes to increase in the nitrogen content of grasslands. *See* nitrogen fixation.

**Aztec** A monitoring device manufactured by Capital Controls Co.

**azurite [$Cu_3(CO_3)_2(OH)_2(s)$]** A blue mineral, hydrous copper carbonate; a copper ore.

# B

**$Ba(BrO_3)_2 \cdot H_2O$** Chemical formula of barium bromate.

**BAC** Acronym of (1) bacteriologically active carbon and (2) biological activated carbon.

**$Ba(C_{18}H_{35}O_2)_2$** Chemical formula of barium stearate.

**bacillary dysentery** A serious, acute waterborne bacterial infection caused by *Shigella dysenteriae* and transmitted through the fecal–oral route. Has a mortality rate as high as 25% in untreated cases. Also called shigellosis, it is distributed worldwide and is nonlatent, moderately persistent, with an infective dose between medium and high, and an incubation period from 3 to 7 days. *See also* amebic dysentery.

**bacilli** Plural of bacillus.

**bacillus** A rod-shaped or cylindrical, spore-forming bacterium of the genus *Bacillus;* one of the most common genera of bacteria found in activated sludge plants that treat domestic wastewater.

***Bacillus anthracis*** The anthrax spore, highly lethal (about 100 million lethal doses per gram), reportedly one of the most chemically and thermally resistant agents among pathogenic bacteria and viruses. It can be stored, remaining viable for decades. *See Bacillus subtilis.*

***Bacillus cereus*** A species of bacteria responsible for water-related infections through the consumption of contaminated shellfish. Common symptoms: diarrhea, vomiting.

***Bacillus* genus** A genus of enterobacteriaceae associated with gastroenteritis and active in biological denitrification, including aerobic sporeformers.

***Bacillus subtilis*** The predominant aerobic sporeforming bacteria in water, an opportunistic human pathogen; a nonlethal surrogate for the highly lethal *Bacillus anthracis.* It is used as a conservative indicator of the removal of *Cryptosporidium* oocysts in water treatment processes and in the validation of European ultraviolet treatment installations.

**backblowing** A reversal of flow under pressure to clean a conduit or well, e.g., backwashing of a filter.

**back clean/back return screen** A mechanically cleaned bar screen used to remove coarse materials from wastewater ahead of other treatment units. It consists of a chain-driven mechanism to move the rake teeth through the screen openings. The screen is raked from the back/downstream side, and the rakes return to the bottom from the back. *See* chain-driven screen.

**back-diffusion coefficient** A mass transfer coefficient used in membrane design to represent solute diffusion from the membrane surface back to the bulk in the feed stream. *See* film theory model.

**backflow** (1) A flow reversal due to a pressure differential; in particular, the flow of water or other liquid into a potable water distribution system from any unintended source. (2) A flow back into a container or device by gravity or siphonage. Also called backsiphonage. In drinking water systems, such devices or arrangements as check valves, vacuum breakers, double gates, or air gaps are used as backflow preventers. *See also* cross connection.

**backflow connection** A plumbing arrangement that makes a backflow possible. *See also* cross-connection and interconnection.

**backflow preventer** A device used to prevent cross-connection or backflow of nonpotable water into a potable water system; also any effective method or construction used to achieve the same objective. For example, an automatic valve installed to prevent backflow by closing when the flow reverses or when the pressure drops. Though less reliable, air gaps and vacuum breakers are sometimes used in residences. *See also* reduced-pressure backflow preventer.

**backflow-prevention device** *See* backflow preventer.

**backflushing** A reversal of flow direction in a water treatment unit (e.g., sand or membrane filter) for cleaning purposes. *See also* backwashing.

**background concentration** The general level of air or water pollution, ignoring local pollution sources. *See also* background conditions, background contamination, background level, and background pollution.

**background contamination** Air or water contamination from natural sources. Also, accidental contamination of dilution water, solvents or reagents, which affects the samples being analyzed in a laboratory. *See also* background conditions, background concentration, background level, and background pollution.

**background conditions** The biological, chemical, and physical conditions of a water body, upstream from the point or nonpoint source discharge under consideration. Background sampling location in an enforcement action will be upstream from the point of discharge, but not upstream from other inflows. If several discharges to any water body exist, and an enforcement action is being taken for possible violations to the standards, background sampling will be undertaken immediately upstream from any discharge (EPA-40CFR131.35.d-2).

**background correction (deuterium)** *See* deuterium background correction.

**background correction (Smith–Hieftje)** *See* Smith–Hieftje background correction

**background correction (Zeeman)** *See* Zeeman background correction

**background level** In air pollution control, the concentration of air pollutants in a definite area during a fixed period of time prior to the starting up or on the stoppage of a source of emission under control. In toxic substances monitoring, the average presence in the environment, originally referring to naturally occurring phenomena.

**background load** The level of pollution occurring naturally in a stream or a receiving water prior to effluent discharge, watershed development, or a similar event.

**background organic matter (BOM)** Natural organic matter exceeding the concentration of organic chemicals in a mixture; interferes with the removal of these chemicals.

**background pollution** Air or water pollution derived from natural sources, e.g., decaying organic matter, sediment, and dissolved minerals from uninhabited areas.

**background radiation** Natural radiation of the earth or the radiation extraneous to an experiment.

**back land** Land lying back from and not adjacent to a watercourse.

**backpressure** Pressure that can cause water to backflow into the water supply when a user's water system is at a higher pressure than the public system. A backpressure valve can be used to prevent backflows. *See also* backsiphonage.

**backpressure valve** A device that prevents backflows by limiting flow in a piping system to a single direction. Its hinged disc or flap opens in the direction of normal flow and closes to prevent flow reversal. Also called nonreturn valve or check valve.

**backpulse** A technique for cleaning membrane equipment used in water or wastewater treatment: solids accumulated on a membrane surface are removed by applying air and/or permeate water in the opposing flow direction of filtration through the membrane fiber; air may also be applied on the feed side of the membrane surface during reverse flow by water. Backpulse is usually automated, thus not requiring operator intervention. Also called backwash, backpulsing, or reverse filtration. Other membrane cleaning techniques include air scour, maintenance cleaning, recovery cleaning, and empty-tank maintenance cleaning.

**backpulsing** Same as backpulse. *See also* backblowing.

**backrush** The return of water toward the sea down the foreshore of a beach, following the uprush of a wave; sometimes called backwash.

**backset** An eddy or countercurrent.

**backsiphonage** A reverse flow condition created by a difference in water pressure that causes water to flow back into the distribution pipes of a drinking water supply from any source other than an intended one. Also called backflow, backpressure.

**back swamp** A marshy area at some distance and lower than the banks of a river.

**backup (or back up)** A system, part of a system, or unit used as a reserve, substitute, or alternate. Also called a standby. For example, in water and wastewater treatment plants, a diesel engine is often used as a backup system for an electric motor. In a pumping station, a backup pump is an extra pump to be used in case one of the units breaks down; it is usually specified as the largest pump of the station.

**backup pump** *See* backup.

**backup system** *See* backup.

**backwash** (1) A reversal of flow for cleaning or removing solids from a filter bed or screening medium; also called backwashing. Cleansing results from two actions: shear caused by the water flowing through the suspension and abrasion from interparticle contacts. The design of a backwash system includes consideration of medium characteristics, underdrainage arrangement, location of wash troughs, head losses, and any auxiliary scour or surface wash. *See* backwash stage. (2) Same as backpulse, a technique for cleaning membranes. (3) Same as backwash water. (4) Same as backrush.

**backwash bump, backwash bumping** *See* bump, bumping.

**backwash head loss** The sum ($H$) of the head losses in the expanded filter bed ($h_f$), the gravel layer ($h_g$) the underdrain ($h_u$), and the wash water pipe ($h_p$), all these terms being expressed in meters of water (McGhee, 1991):

$$H = h_f + h_g + h_u + h_p \quad \text{(B-01)}$$

$$h_f = L(1 - f)(\rho' - \rho) \quad \text{(B-02)}$$

$$h_g = 0.03\, L'V \quad \text{(B-03)}$$

$$h_u = (\tfrac{1}{2}\, g)(V/ar)^2 \quad \text{(B-04)}$$

where $a$ is an orifice coefficient, $f$ is the porosity of the bed, $g$ is the acceleration of gravity (m/s/s), $L$ is the depth of the unexpanded bed (m), $L'$ is the depth of the gravel layer (m), $r$ is the ratio of the orifice area to the filter bed area (usually between 0.002 and 0.007), $V$ is the backwash velocity (m/min), $\rho$ is the density of water, and $\rho'$ is the density of the medium.

**backwashing** Reversing the flow of water back through the filter media to remove the entrapped solids. The backwash water collects in wash troughs, which remove it from the filter box. Backwashing criteria include (a) a head loss limit across the filter, e.g., 8–10 ft of water, (b) a preset limit on the quality of the filtrate, and (c) a maximum time limit on the filter run, e.g., 3–4 days. *See also* air scour, backblowing, backpulsing, collapse pulsing, downflow contactor, jet action, mudball, porosity function, sand boil, surface wash, upflow wash.

**backwashing efficiency** A measure of the success of backwashing in restoring the original head loss and solids storage conditions of a filter.

**backwash launder** A channel that carries the backwash water. *See* launder.

**backwash rate** The flow rate used during filter backwashing, usually expressed as flow ($Q$) per unit area ($A$), which is equivalent to the water velocity ($V$). A recommended backwash rate is 1.3 times the minimum backwash velocity. Also called washwater rate. *See* backwash velocity.

**backwash stage** Any step in the backwash process: filter drawdown, initial media fluidization with or without auxiliary scour, actual backwashing, surface wash, air scour, partial fluidization, filter-to-waste. *See also* air scour, air wash, surface wash.

**backwash velocity** The velocity of water during backwash; also called backwash rate. It is usually between a minimum fluidization velocity and a terminal or washout velocity. The minimum fluidization velocity ($V_f$, in m/min) is the velocity that just begins to fluidize the bed but is insufficient to cleanse the filter of the suspended matter removed, whereas the washout or terminal velocity ($V_t$, in m/min) will cause the medium to wash out and thus damage the filter. They are related as follows:

$$V_f = V_t \times f^{4.5} \quad \text{(B-05)}$$

$V_t = 10\, D_{60}$ for sand (at 20°C and specific gravity 2.65) (B-06)

$V_t = 4.7\, D_{60}$ for anthracite (at 20°C and specific gravity 1.55) (B-07)

where $f$ is the porosity of the medium and $D_{60}$ is the 60% size in millimeters. *See also* backwash velocity equations, Carman–Kozeny equation, incipient fluidization, maximum abrasion velocity, minimum fluidization condition, Galileo number, minimum fluidization velocity, Richardson–Zaki equation.

**backwash velocity equations** A set of equations that can be used to estimate the required backwash velocity and depth of the expanded bed in filter backwashing:

$$L'/L = (1 - f)\ \Sigma[p/(1 - f')] \qquad \text{(B-08)}$$

$$L'/L = (1 - f)/(1 - f') \qquad \text{(B-09)}$$

$$f' = (V/V')^{0.22} \qquad \text{(B-10)}$$

where $f$ = bed porosity; $f'$ = porosity of expanded bed; $L$ = depth of filter bed or layer, m; $L'$ = depth of expanded bed, m; $p$ = fraction of particles within adjacent sieve sizes; $V$ = velocity of backwash water, m/s; and $V'$ = settling velocity of particles, m/s. Given the size distribution of a filter bed, a computation table can be set up to determine, successively, the summation term in the first equation, the settling velocity $V'$, the backwash velocity $V$, the ratio $L'/L$, and the expanded bed depth $L'$.

**backwash water** The water used to backwash filters; contains the particles that were retained by the media and removed by backwashing. Backwash water has a low solids concentration (in a range of 100–1000 mg/L) and represents 2–3% of the volume of water treated. Also called backwash or wash water. It may be (a) discharged to a recovery basin and recycled, (b) released to a sanitary sewer for treatment and disposal, or (c) discharged into sludge lagoons.

**backwater gate** A gate used to prevent the backflow into a combined sewer whose overflow outlet discharges below the high-water level of the receiving water. Also called tide gate.

**$BaCl_2$** Chemical formula of barium chloride

**BACM** Acronym of best available control measure.

**$BaCO_3$** Chemical formula of barium carbonate; witherite.

**$BaCrO_4$** Chemical formula of barium chromate or barium yellow.

**BACT** Acronym of best available control technology.

**bacteremia** Presence of viable bacteria in the blood.

**bacteria** Plural of bacterium. Colorless, unicellular, prokaryotic microorganisms that exist either as free-living organisms or as parasites and have a broad range of biochemical and often pathogenic properties. Among the most common and ubiquitous organisms on Earth, they can reproduce without sunlight and by binary fission, as often as every 20 minutes. Some bacterial species form spores. The bacterial cell contains the cytoplasm, a colloidal suspension of complex organic compounds. Ranging in size mostly from 0.5–1 μm in diameter by 1–2 μm long, bacteria can be grouped by form into five general categories: cocci (spherical), bacilli (rod-shaped), vibrio (curved rod-shaped), spirilla (spiral), and filamentous (thread-like). *Escherichia coli* is a commonly used indicator of bacteriological water quality. Bacteria participate actively in many biogeochemical cycles (e.g., the carbon, nitrogen, and sulfur cycles), can aid in pollution control by metabolizing organic matter in wastewater, sludge, oil spills, or other pollutants, and some species are used in food and industrial processes. However, bacteria in soil, water or air can also cause human, animal, and plant health problems. In water and wastewater treatment, bacteria may be considered as hydrophilic colloids carrying a negative charge; *see* bioflocculation. *See* the following types of bacteria: acidophile, activated sludge, aerobic, anaerobic, autotrophic, coliform, facultative, fecal, filamentous, Gram-negative, Gram-positive, halophile, heterotrophic, iron, methanogen, nitrifying, pathogenic, pigmented, saprophytic, sulfate-reducing, sulfur, xerophile. *See also* waterborne bacterial agent.

**bacteria bed** An early form of the trickling filter (or percolating filter) used for biological wastewater treatment. It consists of an aerobic bed of broken stone, gravel, or sand over which wastewater trickles down to an underdrainage system. *See also* aerated contact bed (or contact aerator or submerged contact aerator), contact bed, intermittent sand filter, nidus rack, percolating filter.

**bacteria cell components** Important components of a bacterial cell include cytoplasm, cytoplasmic inclusions, DNA, fimbriae, flagellae, pili, plasmid DNA, and ribosomes.

**bacteria enumeration and identification** *See* acridine orange stain, colony-forming unit, direct count, dot blot hybridization, electronic particle counter, enzymatic assay, fluorescence method, heterotrophic plate count, in situ hybridization, membrane-filter technique, most probable number, multiple-tube fermentation, nucleic acid probe, pour and spread plate count, presence–absence test, Thomas equation.

**bacterial adherence to hydrocarbon (BATH) test** A simple test used to determine the separation of bacterial cells between a water phase and an organic phase.

**bacterial aftergrowth** An increase in the number of bacteria in treated water or wastewater. *See* regrowth for more detail.

**bacterial analysis** Same as bacterial examination.

**bacterial assimilation** A wastewater treatment process that promotes nitrogen removal beyond normal metabolic activities by providing carbonaceous matter (e.g., methanol or glucose) to correspond to the nitrogen content of the waste, e.g., an approximate BOD:N ratio of 100:5. Bacterial assimilation may also be used for additional phosphorus removal. *See also* activated algae process, luxury uptake (McGhee, 1991).

**bacterial colony** A clump of bacterial cells, which usually do not exist in isolation. Colonies observed visually on a Petri plate consist each of about one million cells.

**bacterial count** An estimate of the number of bacteria in a sample, performed by filtering a measured volume of water through a cellulose acetate or glass filter and incubating the retained bacteria in a special medium. *See* standard plate count.

**bacterial denitrification** The anaerobic or anoxic oxidation of organic matter ($AH_2$) by bacteria using nitrate ($NO_3^-$) as a hydrogen acceptor and releasing nitrogen gas ($N_2$):

$$NO_3^- + AH_2 \xrightarrow[\text{denitrification}]{\text{anaerobic}} A + H_2O + N_2 \uparrow \quad \text{(unbalanced)} \quad \text{(B-11)}$$

*See also* anaerobic denitrification (Hammer & Hammer, 1996).

**bacterial density threshold** The minimum number of bacteria required before phage can multiply, e.g., 10,000 colony forming units for the bacteria species *Bacillus subtilis, Escherichia coli,* and *Staphylococcus aureus.*

**bacterial disease** A disease caused by bacteria. Important bacterial diseases related to water, wastewater, and excreta include bacillary dysentery, cholera, gastroenteritis, paratyphoid, and typhoid.

**bacterial enteritis** A water-related and excreta-related disease, also called diarrhea or gastroenteritis, caused by a number of bacteria (e.g., *Campylobacter jejuni, Escherichia coli, Salmonella spp, Yersinia enterocolitica*). It is transmitted through the fecal–oral route or by animals, has worldwide distribution, and affects particularly children.

**bacterial examination or bacterial analysis** An examination of water or wastewater to determine the type and number of bacteria present. *See also* bacteriological analysis, bacteriological count.

**bacterial gastroenteritis** A waterborne disease caused by bacteria. *See* gastroenteritis.

**bacterial growth** The development of bacteria in water distribution systems, a cause of taste and odor. In filters that remove iron and manganese, bacterial growth may reduce the metals and return them into solution. *See also* growth pattern and microbial growth.

**bacterial metabolism** The set of chemical reactions that bacteria carry out to survive and proliferate. *See also* adenosine triphosphate, aerobe, anaerobe, autotroph, chemoautotroph, chemoheterotroph, copiotroph, essential element, facultative anaerobe, heterotroph, macronutrient, micronutrient, nitrification, oligotroph, photoautotroph, photosynthesis, respiration, trace element, typical cell composition.

**bacterial mutagen** Any of a number of substances proven to be mutagenic in bacterial tests (e.g., Ames test). They include many drinking water disinfection by-products.

**bacterial mutagenesis** Mutagenic activity in bacteria; *see,* e.g., the Ames test.

**bacterial nutrients** Organic or inorganic materials used by bacteria for their nutritional needs.

**bacterial plate count** *See* bacterial count.

**bacterial pollution** Pollution of water sources by bacteria, particularly pathogens. *See* bacterial count.

**bacterial regrowth** An increase in the number of bacteria in treated water or wastewater. *See* regrowth for more detail.

**bacterial regrowth potential** The presence of conditions that promote the growth of bacteria in a drinking water distribution system, e.g., lack of an adequate disinfectant residual and presence of biodegradable organic matter.

**bacterial reproduction** *See* asexual reproduction, binary fission, budding, generation time, and growth patter.

**bacterial self-purification** The breakdown of degradable organic matter in receiving streams by saprophytic bacteria. Self-purification is slow and occurs over a great distance.

**bacterial slime** The gelatinous film of microbial growth covering the surface of a medium or spanning the interstices of a granular bed; e.g., the biofilm that is produced when wastewater is sprayed over fixed media, as in a trickling filter, or when water flows through a slow sand filter. *See* slime for detail.

**bacterial treatment** Wastewater treatment using bacteria to stabilize organic matter. A more precise term is biological treatment, because microorganisms other than bacteria participate also in treatment processes.

**bacterial virus** Same as bacteriophage.

**bacterial yield** In biological processes, organic matter is converted to new organisms and end products. Bacterial yield or biomass yield is the ratio of the amount of biomass produced to the amount of substrate consumed. *See* cell yield coefficient.

**bactericidal** Characteristic of substances that can kill bacteria.

**bactericide** Any substance or agent that can kill bacteria.

**bacteriological aftergrowth** An increase in the number of bacteria in treated water or wastewater. *See* regrowth for more detail.

**bacteriological analysis** Examination of water, wastewater, or food by microscopic or other methods to identify and quantify the presence of specific or general types of bacteria.

**bacteriological corrosion** Corrosion of pipes and containers caused by sulfate-reducing bacteria.

**bacteriological count** An examination of water or wastewater samples to determine the number of bacteria in a unit volume, using such techniques as the heterotrophic plate count, the most probable number, or the plate count.

**bacteriological tests** Tests conducted to determine the number and type of microorganisms in a sample of water or wastewater. *See* bacteria enumeration and identification.

**bacteriology** A branch of microbiology that studies bacteria and their applications in agriculture, medicine, etc.

**bacteriophage** (plural: bacteriophages or bacteriophage) Any of a group of viruses that infect and grow in specific bacteria, e.g., coliphages in *E. coli,* and are released when the host cell dies. They consist of nucleic acid (e.g., DNA) that forces the production mechanism of the bacterial cell to make the viral nucleic acid. Bacteriophages, such as the MS2 coliphages, are sometimes used as surrogates or pathogen indicators for enteric viruses, for fecal pollution, and for viral resistance to disinfectants. Also called bacterial virus, phage. *See also* appendage, F-specific, capsule, male-specific, and somatic phages.

**bacterioplankton** Suspended heterotrophic bacteria.

**bacteriostat** A substance that kills bacteria or otherwise retards or inhibits their growth.

**bacteriostatic** Characteristic of a bacteriostat, a substance or preparation that inhibits the further growth of bacteria without destroying those present.

**bacteriostatic water for injection** One of six grades of water as defined by the U.S. Pharmacopeia. *See* pharmaceutical-grade water.

**bacterium** Singular of bacteria.

**bacteroid** Rod-shaped or branched bacterium in the root nodule of nitrogen-fixing plants.

***Bacteroides*** A genus of rod-shaped, strict-anaerobic bacteria, found in the alimentary and genitourinary tracts of humans, considered as potential indicators of fecal pollution. They constitute the majority of the intestinal microorganisms, with a density larger than $10^{10}$ cells per gram of intestinal contents. Some species are pathogenic.

***Bacteroides fragilis*** A species of bacteria whose phage are considered an indicator of recent human fecal contamination of drinking water.

***Bacteroides* species phage** A virus specific to the *Bacteroides* species of anaerobic bacteria found in human feces and used as an indicator of pollution.

**Baden-Baden system** A composting system for the stabilization of a half-and-half homogenized mixture of wastewater sludge and municipal refuse. It consists of windrows placed over and aerated by perforated concrete pipes. The final compost is used in land reclamation (Sarnoff, 1971).

**bad water line** *See* fresh–salt water interface.

**BAF** Acronym of biological activated filter, biologically active filter, or biological aerated filter.

**baffle** A flat board, plate, deflector, grate, refractory wall, or other device in a stream, tank, basin, or the like used to regulate fluid flow, to obtain more uniform velocities, or to reduce short-circuiting or vortexing.

**baffle aerator** An apparatus used in wastewater aeration, consisting of baffles that provide turbulence and reduce short-circuiting.

**baffle chamber** In incinerator design, a chamber designed to promote the settling of fly ash and coarse particulate matter by changing the direction and/or reducing the velocity of the gases produced by the combustion of the refuse or sludge.

**baffled channel** A hydraulic structure used for mixing chemicals or flocculation in water treatment, with horizontal or vertical flow. *See* Figure B-01.

**baffled-channel system** An apparatus consisting of a channel equipped with baffles, used for flocculation and to reduce short-circuiting.

**baffed mixing basin** A basin equipped with baffles for mixing chemicals. *See* Figure B-02.

**Baffleflow** An oil removal tank with baffles manufactured by Walker Process Equipment Co.

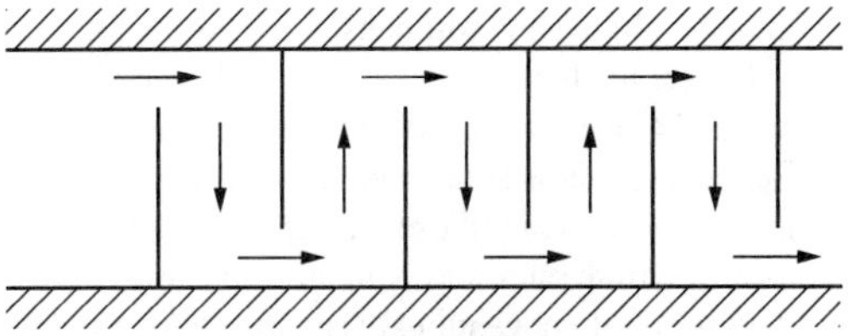

**Figure B-01.** Baffled channel.

**baffle pier** A pier used as an obstruction to dissipate energy of high-velocity waters.

**baffle plate** A plate used as a baffle.

**baffle wall** A wall used as a baffle to equalize flow distribution and prevent short-circuiting, for example, in a paddle flocculator. Baffle walls may be used in a reactor to approximate plug-flow conditions. Also called diffuser wall.

**baffling** Using baffles in channels and basins to improve flow distribution and reduce short-circuiting.

**bag filter** A device used in small plants or in point-of-use applications for the treatment of high-quality water. It consists of a pressure vessel containing a woven bag that retains impurities larger than its pores. It is not effective against viruses and is subject to frequent clogging by algae and high turbidity. When the filter run terminates, the bag is replaced by a new one. Disinfection should follow bag filtration. *See also* cartridge filter (AWWA, 1999).

**baghouse** A dust-collection chamber containing numerous permeable fabric filters through which the exhaust gases pass. Finer particles entrained in the exhaust gas stream are collected in the filters for subsequent treatment or disposal.

**bail-down test** A test conducted to determine the hydraulic properties of an aquifer by taking a known volume of water from a well using a bailer (Symons et al., 2000). *See also* aquifer test, auger hole test, slug test.

**bailer** (1) A 10- to-20-foot-long pipe with a valve at the lower end, used to remove slurry from the bottom or side of a well as it is being drilled. (2) A similar device used to remove water samples from small-diameter wells. *See also* auger, composite liquid-waste sampler, dipper, thief, trier, weighted bottle.

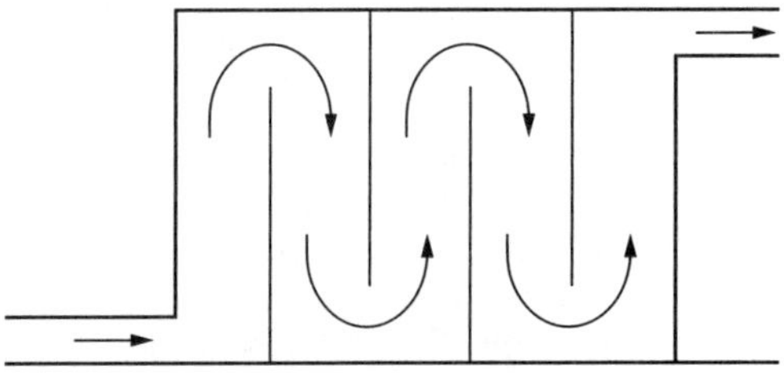

**Figure B-02.** Baffled mixing basin.

**bakery waste** Wastewater produced by bakeries and originating from pan greasing and washing and from floor washing. It contains high BOD, grease, sugars, flour, detergents, floor washings. Applicable wastewater treatment process: flotation, biological oxidation, acid treatment.

**Bakflo®** An oil skimmer installed on a barge, manufactured by Vikoma International.

**baking soda** A substance used in water treatment for pH adjustment and in many industrial applications. *See* sodium bicarbonate ($NaHCO_3$) for detail.

**balance** An instrument used to measure weights, consisting typically of a bar with a central fulcrum, with a scale or pan at both ends to hold the object to be weighed and a known weight.

**balanced equation** A chemical equation having the same number of atoms of each element in the right side (products) as in the left side (reactants); it is used to calculate the quantities of reactants and products, based on molecular relationships. *See also* stoichiometry.

**balanced gate** A gate used to operate a hydraulic structure automatically, releasing water based on the head behind the gate.

**balanced mechanical seal** A spring-loaded or rubber seal used to control leakage of water between the casing and the shaft of a pump. *See also* packing gland, packing materials.

**balanced valve** A valve that operates on the basis of water pressure differences on both sides of the closing mechanism.

**balancing basin** A basin used in a hydraulic system to render uniform the volume and composition of flow. It is often used in a wastewater or water supply system to stabilize influent characteristics ahead of treatment units or to facilitate water distribution to customers. Also called equalizing basin. A balancing reservoir (or equalizing basin) is defined similarly. Balancing storage refers to the volume of water that is transferred from the basin or reservoir back to the system.

**balancing reservoir** *See* balancing basin.

**balancing storage** *See* balancing basin.

**balantidial dysentery** *See* balantidiasis.

**balantidiasis** A relatively rare, water- and excreta-related infection, caused by the protozoal pathogen *Balantidium coli*. It is transmitted via the fecal–oral route or directly from pigs and humans; distributed worldwide. Symptoms include diarrhea (sometimes bloody), abdominal discomfort, colonic ulceration, and, less often, appendicitis, vaginitis, and cystitis. Sometimes called balantidial dysentery or simply diarrhea.

***Balantidium coli* or *Bal. coli*** A flattened oval protozoan whose main outside reservoirs for human infection are pigs and rats. Its ovoid or spherical cysts are found in the large intestine and passed in stools. *See* balantidiasis.

***Bal. coli*** Abbreviation of *Balantidium coli. See also B. coli.*

**balefill** (Combination of bale and landfill.) A land disposal site where solid waste material is compacted and conditioned in large bundles prior to disposal.

**baling** Compaction process for reducing the volume of waste to be stored or transported.

**ballast** (1) The flow of waters, from a ship, that is treated along with refinery wastewaters in the main treatment system (EPA-40CFR419.11-c). (2) An inert material that is added to water or wastewater to enhance flocculation and coagulation. Materials commonly used include microscopic sand particles (microsand), recycled settled sludge, bentonite, fly ash, powdered magnetite, Also called ballasting agent. *See* ballasted flocculation.

**ballasted floc** Floc produced by coagulating water containing a ballasting agent such as clay, bentonite, or powdered magnetite. The solid particles in the water attach to the ballast to form dense microfloc particles that settle more easily. *See* Figure B-03.

**ballasted flocculation** A variation of chemically assisted clarification processes, which uses a ballast to enhance the flocculation and sedimentation of particles. The process includes three separate steps: (a) addition of chemicals for coagulation, (b) maturation by the application of the ballast and formation of the flocculant, and (c) high-rate sedimentation in tube or plate settlers. Chemicals used include alum and iron salts for coagulation and polymers as flocculant aids. Often classified as an advanced primary wastewater treatment, ballasted flocculation is a small-footprint process that removes more than 90% of suspended solids at surface overflow rates approximately 70 times higher than those of conventional primary sedimentation. The process is used in such installations as Acheres (Paris, France, almost 500 mgd) and Lawrence, Kansas (40 mgd). Also called enhanced high-rate clarification. *See* Figure B-04. *See also* small-footprint technology, Actiflo system, Densadag system, Fluorapide system, Sirofloc.

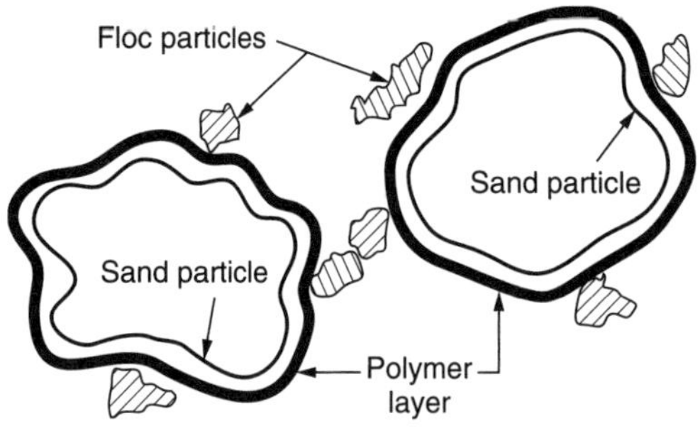


**Figure B-03.** Ballasted floc.

**ballasted flocculent settling** *See* ballasted flocculation.

**ballasted floc particle** *See* ballasted floc.

**ballasted particle** *See* ballasted floc.

**ballasted sedimentation** A treatment process combining grit removal, coagulation, flocculation, static and lamellar clarification, scum removal, and sludge thickening in one unit to remove suspended solids from wet-weather flows. It includes internal and external sludge recirculation. *See* Actiflo Process®, biological contact process.

**ballasting agent** Same as ballast (2).

**ballast water** Water used in a ship's hold for stabilization, often requiring treatment as an oily wastewater. Same as ballast (1).

**ball cock** A valve connected to a hollow floating ball for regulating the supply of water in a tank, cistern, toilet, etc. by rising and falling.

**ballistic separator** A machine that sorts organic from inorganic matter for composting.

**ball joint** A flexible and spherical pipe joint.

**balls** *See* rag balls.

**ball valve** A valve regulated by the position of a free-floating ball that moves in response to fluid or mechanical pressure; it allows a tight seal when closed and straight-through flow in the open position.

**Bancroftian filariasis** A mosquito-borne helminthic infection of the lymphatic ducts caused by the pathogen *Wuchereria bancrofti.* Distributed worldwide but mostly in the tropics, it is transmitted by *Aedes, Anopheles, Culex pipiens,* and other mosquitoes. The pathogens develop in the lymphatic system and release larvae in the blood, causing lymph obstruction, elephantiasis, and hydrocele. *See also* Malayan filariasis.

**band application** The spreading of chemicals over, or next to, each row of plants in a field.

**banded steel pipe** A steel pipe reinforced by bands around the shell, used for flows under very high heads.

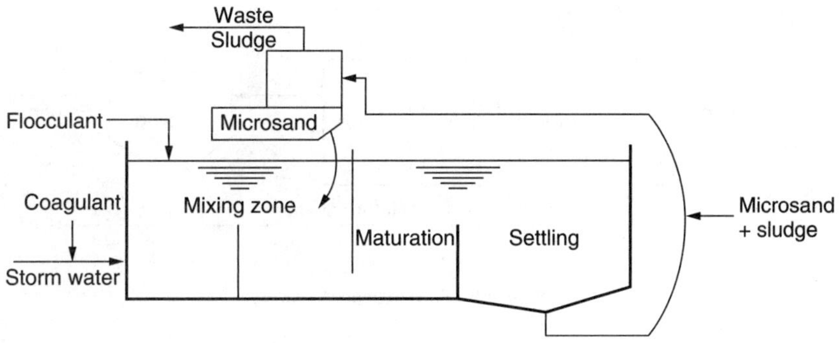


**Figure B-04.** Ballasted flocculation.

**bandscreen** An automatically cleaned screen for the removal of suspended or floating materials; it consists of a continuous band of wire mesh or similar screening medium. Also called belt screen. *See also* traveling water screen.

**bandspreading** The spreading of fertilizers in thick bands, 30 cm apart.

**Bangalore method** An anaerobic composting method whereby alternate layers of nightsoil and refuse are buried in trenches for several months; it is slow, does not yield high temperatures, and is not as efficient as aerobic composting in reducing pathogens.

**bank** (1) The continuous, sloping strip of land that borders a body of freshwater: river, stream, lake, or certain elevations on the floor of the sea. Bank protection by riprap or other means is sometimes provided against erosion. Bank storage is the volume of water stored in the banks of a body of water and released completely or partially when the water surface drops. (2) In pressure-driven membrane processes, a bank or stage consists of pressure vessels arranged in parallel. Interconnected stages constitute an array or train.

**bank filtration well** A well drilled near a lake or river to draw surface water that has been diluted and treated through the soil by filtration, sorption, and other processes. *See* riverbank filtration for detail.

**bank protection** *See* bank (1).

**bank storage** *See* bank (1).

***B. anthracis*** Same as *Bacillus anthracis.*

**BaO** Chemical formula of barium monoxide or barium protoxide.

**$BaO_2$** Chemical formula of barium dioxide or barium peroxide.

**$Ba(OH)_2$** Chemical formula of barium hydroxide, barium hydrate, or baryta.

**bar** (1) A centimeter-gram-second unit of pressure = 1,000,000 dynes per square centimeter = 14.504 psi = 100 $kN/m^2$. (2) A long ridge of sand, gravel, or other alluvial material near or slightly above the surface of the water at or near the mouth of a river or harbor entrance, often obstructing flow or navigation. *See also* braided stream, shoal.

**Bardenpho®** A proprietary biological wastewater treatment process of Eimco Process Equipment Co., developed in South Africa in the 1970s for the removal of nitrogen and phosphorus. It consists of four reactors in series: anoxic–aerobic–anoxic–aerobic and a secondary clarifier, with recycling from the first aerobic internally to the first anoxic and from the clarifier underflow to the influent line. The word Bardenpho is from the inventor's name (Barnard), denitrification, and phosphorus. Also called four-stage Bardenpho. *See also* biological nitrogen removal, modified Bardenpho, and Figure B-05.

**Bardenpho (4-stage) process** Same as Bardenpho®.

**Bardenpho (5-stage)** A modification of the Bardenpho® process with the addition of an anaerobic stage ahead of the first anoxic reactor; comparable to the $A^2O$ process.

**barges-in-tow** One of four main systems used to transport liquid sludge to an ocean disposal site; it consists of tugs pulling barges by hawsers, on a one-to-one basis, each barge divided into tank compartments. *See also* articulated tug barge, barge transport, rubber barges in tow, and self-propelled sludge vessel.

**barge transport** A means of transporting sludge from a treatment plant to an ocean disposal site. Sludge barging for ocean disposal is now prohibited by federal regulations and being phased out. *See also* articulated tug barge, barge-in-tow, rubber barges-in-tow, and self-propelled sludge vessel.

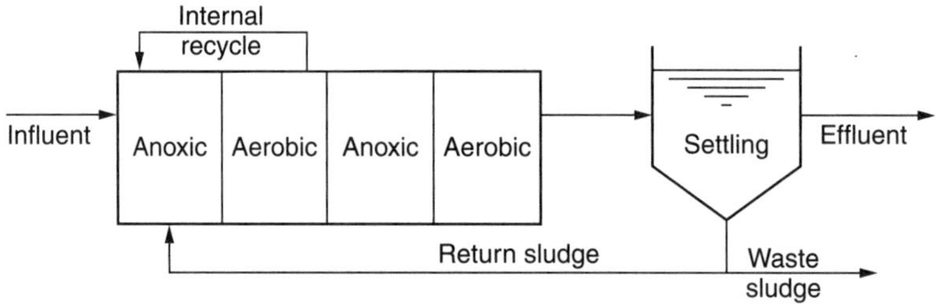


**Figure B-05.** Bardenpho® process.

**barite ($BaSO_4$)** The mineral barium sulfate, principal ore of barium; it is found as white, yellow, or colorless crystals. Also spelled barite and called heavy spar. *See also* baryta.

**barium (Ba)** A whitish, divalent, alkaline-earth, metallic element, found naturally as barite or as witherite and at low concentrations in surface waters and wastewater effluents, used in many industries. Atomic weight = 137.34. Atomic number = 56. Specific gravity = 3.5 at 20°C. Barium salts are toxic; at sufficiently high concentrations, barium may interfere with enzyme production and cause physiological effects in humans. Maximum contaminant level (MCL) set by the USEPA for drinking water is 2.0 mg/L.

**barium-140** The radioactive isotope of barium, mass number 140, half-life 12.8 days, used as tracer; sometimes found in water sources, originating from reactor and weapons fission.

**barium bromate [$Ba(BrO_3)_2 \cdot H_2O$]** Colorless, poisonous crystals, slightly soluble in water, used in the preparation of bromates.

**barium carbonate ($BaCO_3$)** A white, poisonous, soluble powder, used in rodenticides and paints.

**barium chloride ($BaCl_2 \cdot 2H_2O$)** A colorless, crystalline solid, poisonous and soluble in water, used in pigments, rodenticides, and pharmaceuticals.

**barium chromate ($BaCrO_4$)** A yellow, crystalline compound, as the pigment barium yellow.

**barium dioxide ($BaO_2$)** Same as barium peroxide.

**barium fouling** A condition in ion exchange deionization where an insoluble layer of barium sulfate ($BaSO_4$) coats the resin.

**barium hydrate [$Ba(OH)_2$]** Also called barium hydroxide. *See* baryta (2).

**barium hydroxide [$Ba(OH)_2$]** Also called barium hydrate. *See* baryta (2).

**barium monoxide (BaO)** Barium oxide, barium protoxide, calcined baryta. *See* baryta (1).

**barium oxide (BaO)** Barium monoxide, barium protoxide, calcined baryta. *See* baryta (1).

**barium peroxide ($BaO_2$)** A grayish-white, poisonous powder, insoluble in water, used in the production of hydrogen peroxide and in bleaching agents. Also called barium dioxide.

**barium protoxide (BaO)** Barium monoxide, barium oxide, calcined baryta. *See* baryta (1).

**barium stearate [$Ba(C_{18}H_{35}O_2)_2$]** A white, crystalline solid, insoluble in water, used in water proofing agents and lubricants.

**barium sulfate ($BaSO_4$)** A white, crystalline powder, insoluble in water, used in pigments, paints, and medicine (X-ray diagnosis).

**barium sulfide (BaS)** A gray or yellow-green, poisonous powder, used in pigments.

**barium thiosulfate ($BaS_2O_3 \cdot H_2O$)** A white, crystalline, poisonous solid, used in explosives, matches, and paints.

**barium titanate ($BaTiO_3$)** A crystalline compound used in capacitors and transducers.

**barium yellow ($BaCrO_4$)** *See* barium chromate.

**barminutor** A bar screen equipped with a high-speed rotating cutter moving up and down to shred and grind accumulated solids, which are then flushed with the wastewater. *See also* Barminutor®, comminutor, grinder, and shredder.

**Barminutor®** A bar screen-comminutor combination manufactured by Yeomans Chicago Corp.

**Barnes' formula** An empirical formula proposed in 1916 to calculate flow velocities (*V*, m/s) in sewers:

$$V = 107\,R^{0.7}S^{0.5} \tag{B-12}$$

where *R* is the hydraulic mean depth (m) and *S* is the sewer slope.

**barn wastewater** Wash waters from a barn or stable, containing animal wastes.

**barophiles** Microorganisms that live optimally at high pressures (more than 1000 bars) like those found in deep-sea trenches.

**barotolerant** Characteristic of an organism that tolerates high hydrostatic pressures, but grows better at normal pressures.

**bar rack** A device, consisting of parallel bars or rods, used to remove coarse debris from water source intakes or at the beginning of a treatment plant process; also used in wastewater treatment for the same purpose, e.g., at the intakes to wet wells. Materials removed on bar racks are usually disposed in a landfill. Also called coarse screen. *See* bar screen.

**barrage** An artificial obstruction in a watercourse to increase water level, divert water, facilitate irrigation, control peak flow for later release, etc. *See also* dam.

**Barrett equation** A commonly used equation, developed in 1960, to determine the required height ($H$, m) in cascade aeration devices (Metcalf & Eddy, 2003):

$$H = (R - 1)/[0.361\, AB(1 + 0.046\, T)] \quad \text{(B-13)}$$

$$R = (C_s - C_0)/(C_s - C) \quad \text{(B-14)}$$

where $R$ = deficit ratio, dimensionless; $A$ = a water quality parameter = 0.8 for wastewater treatment plant effluents; $B$ = a geometry parameter = 1.0 for a weir, 1.1 for steps, or 1.3 for a step weir; $T$ = water temperature, °C; $C_s$ = saturation concentration of dissolved oxygen at temperature $T$, mg/L; $C_0$ = concentration of dissolved oxygen in the influent liquid, mg/L; $C$ = concentration of dissolved oxygen in the effluent liquid, mg/L.

**barrier landscape water renovation system** A biological wastewater treatment process using percolation through a soil mound over a water barrier (Pankratz, 1996).

**Barry Rake®** A trash rake manufactured by Cross Machine, Inc.

**bar screen** A mechanically operated device used to remove large solids from intakes or waterways on bar racks. It consists of parallel steel bars welded to form a grid. In preliminary wastewater treatment, it is also called bar rack or rack, with clear bar openings of approximately one inch. *See also* screening.

**bar screen head loss** The head loss ($H$, ft) through a clean or partially clogged bar screen:

$$H = k(V^2 - v^2)/2g \quad \text{(B-15)}$$

where $k$ = friction coefficient ≈ 1.43; $V$ = velocity through screen, fps; $v$ = velocity upstream of screen, fps; $g$ = gravitational acceleration = 32.2 ft/sec$^2$. *See also* Kirschmer's formula (WEF & ASCE, 1991).

**baryta** One of two compounds of barium found under various names. (1) A white or yellowish, poisonous solid, barium oxide (BaO), used as a dehydrating agent or in glass; also called barium monoxide, barium protoxide, or calcined baryta. (2) The hydrated form, barium hydroxide [$Ba(OH)_2 \cdot 8H_2O$], used in sugar and oil refining; also called barium hydrate or caustic baryta.

**baryta water** A reagent consisting of an aqueous solution of barium hydroxide [$Ba(OH)_2$].

**barytes** Crystals of barium sulfate. *See* barite.

**BaS** Chemical formula of barium sulfide.

**$BaS_2O_3$** Chemical formula of barium thiosulfate.

**$BaSO_4$** Chemical formula of barium sulfate.

**basal medium** A medium that allows the growth of microorganisms that do not require the addition of special nutrients, e.g., nutrient broth.

**basal metabolism** The energy exchange of an animal at rest.

**basal respiration** One of the parameters used to determine the existence and distribution of microorganisms in a sample. It is measured as the biological production of carbon dioxide ($CO_2$) in a sample without nutrient addition. Basal respiration is also used as an indicator of soil condition (Maier et al., 2000).

**bascule gate** A device used to control water surface levels in combined sewers and capture stormflows completely or partially by providing in-system storage.

**base** A substance that can accept a proton, i.e., a proton ($H^+$) acceptor (Brønsted–Lowry theory) or a substance that can donate a lone pair of electrons (Lewis concept). Practically, a base is (a) a substance that ionizes or reacts with water to produce the hydroxyl ion ($OH^-$) and a cation, e.g., potassium hydroxide and ammonia:

$$KOH \rightarrow K^+ + OH^- \quad \text{(B-16)}$$

$$NH_3 + H_2O \rightarrow NH_4^+ + OH^- \quad \text{(B-17)}$$

(b) or a substance that yields a pH greater than 7.0 while dissolved in water; an alkaline substance; (c) or a substance that can react with an acid to form a salt, e.g., sodium hydroxide:

$$2\,NaOH + H_2SO_4 \rightarrow Na_2SO_4 + 2\,H_2O \quad \text{(B-18)}$$

(d) or any of certain nitrogenous organic compounds. *See also* alkali, strong base, and weak base.

**base-catalyzed hydrolysis** A chemical reaction between a substance and water, the hydroxyl ion ($OH^-$) serving as a catalyst, e.g., the conversion of trichloropropanone ($CCl_3COCH_3$) to chloroform ($CHCl_3$).

**base exchange** The exchange of positively charged ions (cations) between an aqueous solution and a mineral. *See* cation exchange for more detail. *See also* zeolite softening.

**base exchange process** A water treatment method that uses base exchange.

**base exchanger** Same as cation-exchange resin.

**base flow** (1) The sustained part of stream discharge, including groundwater runoff and delayed subsurface runoff, but not direct runoff from precipitation or melting snow. It is usually sustained by drainage from natural storage in groundwater, lakes, or swamps. Base flow is determined from a measured hydrograph at the basin outlet. Given that total runoff consists of base flow and direct runoff, for an approximate determination, base flow may be assumed constant throughout a storm and equal to the total flow before the storm. However, there are more accurate methods of hydrograph separation. Base flow recession occurs when the flow rate starts to decrease in a stream fed by groundwater, as can be observed on the corresponding base flow recession hydrograph. (2) Same as dry-weather flow (1). *See also* wastewater flow components, base runoff, delayed runoff, groundwater flow.

**base flow recession** *See* base flow.

**base flow recession hydrograph** *See* base flow.

**baseline** The condition of a system at a starting point, which can serve as a basis for measurement or for tracking future changes. Impact assessment studies usually include a baseline alternative, which is an extension of present conditions with minor improvements; it is different from the no-action alternative. Such studies usually include a baseline survey to describe existing environmental and socioeconomic conditions. Water supply and sewerage baseline data relate to historical water uses and wastewater flows. Baseline wastewater flow is the flow of wastewater in sanitary or combined sewer systems during dry weather.

**baseline alternative** *See* baseline.

**baseline data** *See* baseline.

**baseline process** A process that provides the minimum level of treatment that meets the established water quality goals in terms of particulates, dissolved solids, and other parameters. Baseline water treatment processes include clarification, softening, filtration, disinfection, oxidation, ion exchange, membrane filtration, activated carbon adsorption, and air stripping.

**baseline survey** *See* baseline.

**baseline test** A test conducted to establish a starting condition, e.g., for health statistics.

**baseline wastewater flow** *See* baseline. *See also* dry-weather flow (1) or base flow (2).

**baseline water treatment** (1) A combination of treatment processes to provide water that meets the established water quality goals. (2) With respect to disinfection goals, a water treatment method that provides some level of removal of viruses and *Giardia cysts*. *See* conventional water treatment, diatomaceous earth filtration, direct filtration, and slow sand filtration.

**base load** The minimum load or use over a given period of time.

**base map** A map containing the important elements common to all other maps or layers of a set; it shows basic data that usually remain constant.

**base metal** (1) Any metal (e.g., copper, lead, zinc, tin) other than a precious or noble metal. (2) The principal metal of an alloy. (3) A metal (such as iron) that reacts with dilute hydrochloric acid to form hydrogen. *See also* noble metal.

**base-neutralizing capacity** A measure of the ability of water or soil to resist changes in pH; acidity. It measures the net excess of protons over a reference level and can be determined from titration with a strong base to a preselected equivalence point.

**base saturation** The condition of a soil or other material whose cation exchange capacity is saturated with respect to calcium ($Ca^{2+}$), magnesium ($Mg^{2+}$), potassium ($K^+$), or sodium ($Na^+$).

**base strength** A measure of the tendency of a substance to accept a proton; a weak base has a weak tendency to accept protons, and conversely for a strong base.

**base temperature** An arbitrary reference temperature for determining liquid densities or adjusting the measured volume of a liquid quantity (EPA-40CFR60.431).

**base wastewater flow or baseline wastewater flow** *See* wastewater flow components, baseline, and dry-weather flow (1).

**basic** (1) Characteristic of a substance that reacts with an acid to form a salt. *See* base. (2) Characteristic of an igneous rock (e.g., feldspar) that contains about 50% silica, as compared to acidic, intermediate, and ultrabasic rocks.

**basic data** Recorded observations and measurements, with little processing, but sometimes including minimal computations. Basic hydrologic data include all forms of precipitation, stream flow, evaporation, transpiration, and water surface elevations.

**basic dye** A substance with a positive charge used to stain smears and identify nucleic acids and other negatively charged cell components.

**basic hydrologic data** *See* basic data.

**basic industry** An industry whose products are essential raw materials or inputs to the production of other industries instead of finished consumer goods, e.g., the chemical industry.

**basicity** (1) The state or condition of being a base. (2) The capacity of an acid to donate a hydrogen atom and react with a base. (3) In coagulation with prehydrolyzed metal salts, the degree to which the hydrogen ions produced by hydrolysis are preneutralized; expressed as a percentage of the weighted average of the ratio of hydroxide [OH] to metal hydrolysis products [M].

**basicity adjustment agent** A compound like crushed stone or lime added to sludge in a thermal destruction process (slagging furnace) to maintain the fluidity of the molten sludge.

**basicity constant** The equilibrium constant ($K_B$) corresponding to the dissociation of a base (B) in water, according to the reaction:

$$B + H_2O = OH^- + BH^+ \quad (B\text{-}19)$$

Depending on the conditions of dilution, the acidity constant is expressed in terms of activities (denoted by $\{\cdot\}$), concentrations ([·], or a combination of the two:

$$K_B = \{HB^+\} \cdot \{OH^-\}/\{B\} \approx [HB^+] \cdot [OH^-]/[B] \quad (B\text{-}20)$$

where the brackets denote activities ($\{\cdot\}$) or concentrations ([·]). *See* Stumm and Morgan (1996) for more detail. *See also* conjugate acid–base pair, basicity constant.

**basicity factor** (1) A measure of the alkali available for waste neutralization; a factor used to determine neutralization capabilities of alkaline reagents in treating acidic wastes. The basicity factor of lime (or any other neutralizing agent) is determined by titration with an excess of hydrochloric acid (HCl) and backtitrating with sodium hydroxide (NaOH) to the phenolphthalein end point. The basicity of an alkaline reagent is used with the acid value of an acidic waste to determine the amount of alkali required to neutralize the waste (Nemerow & Dasgupta, 1991). (2) The weight of an alkaline reagent equivalent to a unit weight of calcium oxide (CaO) in acid neutralizing capacity. It is determined by dividing the alkalinity of the reagent required to neutralize 1.0 mg/L, expressed as mg/L $CaCO_3$, by the alkalinity of CaO expressed similarly, i.e., by 0.56 (WEF & ASCE, 1991). For example the basicities of calcium carbonate ($CaCO_3$), calcium hydroxide [$Ca(OH)_2$], sodium hydroxide (NaOH), and sodium carbonate ($Na_2CO_3$) are, respectively: (a) 1.0/0.56 = 1.79, (b) 0.74/0.0.56 = 1.32, (c) 0.80/0.56 = 1.43, and (d) 1.06/0.56 = 1.89.

**basic pollution sources** The four basic groups of economic activities that cause pollution: materials extraction and transport, industrial production, product use, and product disposal (Nemerow and Dasgupta, 1991). *See also* polluting materials.

**basic polyaluminum chloride (BPACl)** A preformed solution of polyaluminum chloride, an inorganic polymer used in water treatment, alone or in combination with other coagulants.

**basic purification equation/function** Same as purification function.

**basic salt splitting** A typical first step in the reactions that occur during the treatment process of weak acid cation exchange (WACE), in which a basic salt such as calcium bicarbonate [$Ca(HCO_3)_2$] is split by a solid ion-exchange resin [$RCOOH_{(s)}$] while exchanging hydrogen ions for calcium and producing carbonic acid ($H_2CO_3$):

$$2\,RCOOH_{(s)} + Ca(HCO_3)_2 \rightarrow (RCOO^-)_2Ca^{2+}_{(s)} + 2\,H_2CO_3 \quad (B\text{-}21)$$

WACE is used to remove carbonate hardness. *See also* neutral salt splitting, strong acid cation exchange (AWWA, 1999).

**basic sciences** Fields of knowledge that contribute significantly to the understanding of water and wastewater treatment, e.g., biology, chemistry, hydraulics.

**basic solution** A solution whose pH exceeds 7.0 and that contains a large concentration of hydroxyl ions ($OH^-$).

**basic water** Water whose pH exceeds 7.0 and that contains a large concentration of hydroxyl ions ($OH^-$), e.g., as a result of the dissolution of a base.

**basic water requirement** The quantity of water required for fundamental human needs. One estimate by WHO is for a minimum freshwater of 50 liters per person per day for drinking (5 L), sanitation (20 L), food preparation (10 L), and bathing (15 L).

**basin** In general, a natural or artificial hollow place containing water. (1) A drainage basin or a river basin is the area drained by a river and its tributaries; a hollow or depression in the earth's surface wholly or partly surrounded by higher land. Also called catchment area, drainage area, or watershed. (2) A natural or artificial structure, on the surface or underground, that can hold water or other liquids for storage or treatment. Storage or sedimentation basins usually designate structures smaller than reservoirs but larger than tanks. *See also* lagoon, pond, pool.

**BASINS** Acronym of better assessment science integrating point and nonpoint sources.

**basket centrifuge** One of two centrifuge types used in sludge dewatering and thickening; a batch-type centrifuge in which sludge is introduced into a vertically mounted spinning basket and separation occurs as centrifugal force drives the solids to the wall of the basket. Because of operating difficulties and relative inefficiencies, the basket centrifuge has lost ground to the solid bowl conveyor centrifuge.

**BAT** Acronym of best available technology (achievable under the Clean Water Act).

**batch centrifuge** *See* basket centrifuge.

**batch composter** A composting toilet common in China and Vietnam; usually a double vault underneath a superstructure. It requires the addition of a carbon source such as garbage, vegetable leaves, and sawdust to adjust the carbon/nitrogen ratio. As the compost is applied to agricultural land, a minimum retention of three months is recommended. *See also* continuous composter (Feachem et al., 1983).

**batch composting toilet** Same as batch composter.

**batch culture** A defined medium to which a fixed amount of substrate is added to study the growth of an organism or a group of organisms. *See also* continuous culture.

**batch decant process** *See* sequencing batch reactor and ICEAS™ process.

**batch decant reactor** *See* sequencing batch reactor and ICEAS™ process.

**batch discharge** The discharge of a large volume of wastewater in a short period, usually from a wet manufacturing process, with concentrated contaminants and flow surges. To reduce the effects of a batch discharge, it can be retained in a holding pond and then allowed to flow uniformly over an extended period. Also called slug discharge.

**batch distillation operation** A noncontinuous distillation operation in which a discrete quantity or batch of liquid feed is charged into a distillation unit and distilled at one time. After the initial charging of the liquid feed, no liquid is added during the distillation operation (EPA-40CFR60.661). Also called differential distillation. *See also* continuous distillation.

**batch fill-and-draw thickening tank** *See* batch thickener.

**Batch-Master (1)** A basket centrifuge manufactured by Ketema, Inc.

**Batch Master (2)** A package wastewater treatment plant manufactured by Wastewater Treatment Systems, Inc.

**Batch-Miser** A horizontal plate filter manufactured by Ketema, Inc.

**batch mode** A type of plant or process operation that occurs in discrete steps instead of continuously. *See* batch process.

**Batch-O-Matic®** A basket centrifuge manufactured by Ketema, Inc.

**batch pan** A simple device in which a jacketed vessel supplies heat for batch evaporation. *See also* forced-circulation evaporator, natural-circulation evaporator, solar evaporator.

**batch pan evaporator** Same as batch pan.

**batch process** (1) A process that does not operate continuously, but rather in discrete steps or with a limited number of items; total processing time consists of fill, reaction, heating, cooling, and drain times. A batch concrete plant, e.g., operates with batchers and mixers for the production of concrete. Industries that use batch processes include pharmaceutical production, textile manufacturing, and pesticide production. (2) In a batch reactor, the vessel (usually stirred) is closed after the addition of desired quantities of reactants and catalysts and the contents are completely mixed. For example, the BOD test is carried out in a batch reactor. The materials balance for a batch reactor is usually represented by a first-order equation:

$$C/C_0 = e^{-kt} \quad \text{(B-22)}$$

where $C$ = constituent concentration at time $t = t$, mg/L; $C_0$ = constituent concentration initially, i.e., at time $t = 0$, mg/L; $k$ = first-order reaction rate constant, /time. (3) Water and wastewater can also receive a batch treatment: a fixed quantity of water or wastewater is processed in a tank, which is then emptied and refilled, if necessary; it is used sometimes in granular activated carbon adsorbers. *See also* continuous-flow system, sequencing batch reactor (SBR). (4) Fill-and-draw is sometimes used for a batch operation. The activated sludge process was first developed using fill-react-settle-draw cycles.

**BatchPro Designer** A computing procedure developed at the New Jersey Institute of Technology to simulate and evaluate chemical processes for pollution prevention. Based on a process simulator (BioPro Designer), it includes a manufacturing module, an end-of-pipe treatment module, and a graphical interface.

**batch reactor** *See* batch process.

**batch reactor model** An equation used to describe chemical and biological reactions that occur in wastewater treatment. Assuming a first-order reaction, it is similar to the BOD equation (Metcalf & Eddy, 2003):

$$C = C_0 e^{-kt} \quad \text{(B-23)}$$

where $C$ = effluent concentration, mg/L; $C_0$ = influent concentration, mg/L; $k$ = first-order rate constant, \time; and $t$ = time.

**batch thickener** A device used to thicken water or wastewater treatment residuals. It consists of a circular settling basin equipped with a scraping mechanism or a sludge hopper and is operated as a fill-and-draw tank.

**batch treatment** *See* batch process.

**BATEA** Acronym of best available technology economically available.

**bathing** According to USEPA, using water for personal hygiene in a home, school, etc., but excluding swimming or casual contacts with water in open bodies of water.

**bathroom retrofit kit** A kit used to repair bathroom devices or appliances to reduce water use and wastewater flow rates. It may include shower flow restrictors, toilet leak detector tablets, etc. *See also* flow-reduction devices and appliances.

**BATH test** *See* bacterial adherence to hydrocarbon test.

**bathtub curve** A curve shaped like a bathtub, characterizing the time-related reliability and deterioration of new process equipment. From start-up, the curve shows three periods: break-in period, during which the number of failures decreases; the normal operating period, during which performance is steady; and the wear-out period, during which the failures increase more or less rapidly.

**bathtub effect** The accumulation of leachate in a lined landfill that is not equipped for leachate collection and removal.

**bathypelagic zone** The third layer of the oceans, ranging in depth from approximately 200 m to 4000 m, between the mesopelagic zone and the benthic region. *See* freshwater profile and marine profile for other related terms.

**$BaTiO_3$** Chemical formula of barium titanate.

**BATNEEC** Acronym of best available techniques not entailing excessive costs.

**batter board** Any one of a number of horizontal boards that support strings outlining the foundations of a building. In the construction of sewers, batter boards are placed across trenches at 40-ft intervals to establish the centerline of the sewer.

**battery of wells** A series of wells connected to a single pump for water withdrawal. Also called a gang of wells.

**Bauer®** Screening equipment manufactured by Andritz-Ruthner, Inc.

**Baumé, A.** French chemist, 1728–1804.

**Baumé degree (°Bé)** A unit of measure of density or concentration of chemicals used in water treatment, e.g., aqua ammonia or ammonium chloride. *See* Baumé scale.

**Baumé scale** One of two scales for use with a hygrometer, calibrated in such a manner that the specific gravity or density of a given liquid may be easily computed. The origin, 0 Baumé degree (0° Bé) corresponds to a specific gravity of 1.0 (water at 4°C) in the heavy Baumé scale and to a solution of 10% sodium chloride (NaCl) in the light Baumé scale.

**bauxite** Principal aluminum ore, in the form of alumina monohydrate or trihydrate.

**Baylis, John R.** American sanitary engineer who developed in 1923 a fixed piping grid system for surface wash in water filtration plants. In the 1930s, he perfected the use of activated silica as a coagulant aid.

**bay salt** A relatively coarse salt derived from the evaporation of seawater in the sun.

**Bazalgette, John** Constructor of the main drainage system of London in 1850, one of the significant achievements of the Great Sanitary Awakening.

**BBDR modeling** Abbreviation of biologically based dose–response modeling.

**BCAA** Acronym of bromochloroacetic acid.

**BCAN** Acronym of bromochloroacetonitrile.

**B cell** A type of lymphocyte (white blood cell) that can synthesize specific antibodies in response to antigens. Also called B lymphocyte.

**BCF** Acronym of (1) bioconcentration factor and (2) brine concentration factor.

**bCOD** Abbreviation of biodegradable chemical oxygen demand.

***B. coli*** A member of the coliform group of bacteria; it is now called *E. coli*. *See also Bal. coli.*

**BCT** Acronym of best control technology (under the Clean Water Act), or best conventional pollutant control technology.

**BDAT** Acronym of best demonstrated available technology.

**BDCAA** Acronym of bromodichloroacetic acid.

**BDCM** Acronym of bromodichloromethane.

**BDOC** Acronym of biodegradable dissolved organic carbon.

**BDOM** Acronym of biodegradable dissolved organic matter.

**BDST** Acronym of bed depth service time.

**Be** Chemical symbol of beryllium.

**Bé** Abbreviation of Baumé degree.

**beach drift** The movement of marine and other sediments parallel to the contour of a beach, due to the action of waves and littoral currents. Also called littoral drift, longshore drift.

**bead count** A method of assessing the physical condition of the resin in a bed by determining the percent of whole, cracked, or broken beads in a wet sample of resin.

**Bead Mover™** A pump manufactured by IX Services Co. for the loading of resins and filter media.

**Bead Thief™** Equipment manufactured by IX Services Co. for the sampling of ion-exchange resins.

**beaker** A flat-bottomed cylindrical container with a pouring lip used for mixing chemicals in a laboratory.

**$Be_3Al_2Si_6O_{18}$** Chemical formula of beryllium aluminum silicate or beryl.

**bearer bond** A bond that is not registered in anyone's name; it is presumed owned by and payable to whoever possesses it. Such a bond has not been issued since the early 1980s. *See also* registered bond.

**beaver fever** Term commonly used in the western United States for giardiasis.

**becquerel (Bq)** The unit of radioactivity in the Système International. It is equal to one disintegration (one nuclear transformation) per second or approximately 27 pCi.

**bed** (1) The bottom of a water body, watercourse, etc., as in streambed, lake bed, river bed. (2) A layer of unconsolidated material or stratum; *see also* bedrock. (3) A quantity of material used in water or wastewater treatment, as in granular filtration or ion exchange.

**bed configuration** *See* bed form.

**bed density—backwashed and drained** The density of granular activated carbon that is stratified and free of water, typical of the carbon during actual operation, as compared to the apparent density of the product as shipped. The bed density is usually 10% lower than the apparent density (AWWA, 1999).

**bed depth** The depth of media in a filter, or the depth of materials in tanks, reactors, etc., e.g., the depth of ion-exchange resin.

**bed-depth-service-time (BDST) method** A practical method proposed for the design of continuous flow carbon adsorbers, using laboratory or pilot plant data to obtain breakthrough curves at different column lengths, minimum carbon exhaustion rates, and optimal contact times. Various design and operating parameters are determined as follows (Droste, 1997):

$$T_s = T_d - PAD/Q \tag{B-24}$$

$$R_e = D_p AD/T_d \tag{B-25}$$

$$EBCT = V_b/Q = D/(Q/A) \tag{B-26}$$

where $T_s$ = service time, days; $T_d$ = carbon regeneration time, days; $P$ = bed porosity; $A$ = area of carbon column, m$^2$; $D$ = depth of carbon column, m; $Q$ = influent flow rate, m$^3$/d; $R_e$ = carbon exhaustion rate, kg/day; $D_p$ = packed density of the carbon, g/cm$^3$; $EBCT$ = empty bed contact time, days; and $V_b$ = bed volume, m$^3$. The service time (*ST*) is determined from the carbon regeneration time (*RT*)

**bed expansion** The increase in bed volume during backwashing of a filter, usually expressed as a percentage of bed depth.

**bed form** A relief feature on a streambed, e.g., a dune. Also called bed configuration. *See also* sedimentation terms.

**bed irrigation** The application of wastewater through ditches across farmland for treatment and disposal. *See* land filtration for detail. *See also* flood irrigation.

**bed life** The period of time that a bed of activated carbon or other adsorbent is effective before it needs reactivation or replacement. It may be determined as the ratio of the calculated capacity of the activated carbon bed to the substrate loading rate.

**bed load, bed material load, and wash load** Sediment that moves on or near the streambed at a lower velocity than the water. The coarse sediment is called bed load or contact sediment. Wash load, also called suspended load or fine sediment load, is the suspended particles not found in the streambed. Bed material load comes from the bed, including bed load, but excluding wash load. *See also* suspended sediment, desilting basin.

**bed material** The constituent material of the bed of a stream, i.e., the geologic formations and deposits of the channel. It may be transported in contact with the bed or in suspension. *See also* sedimentation terms.

**bed material discharge** The part of total sediment discharge that consists of grain sizes found in the bed and corresponds to the transport capacity of the flow. *See also* sedimentation terms.

**bed material load** *See* bed load.

**bed material sampler** A tool for collecting samples from bed materials.

**bedpan washer** A shallow toilet pan used by persons confined to bed and installed in older buildings, with inlet below its water level and thus a potential source of contamination by cross-connection.

**bedrock** The consolidated rock lying below the superficial, loose material such as alluvium, rock fragments, or soil.

**bed volume (BV)** The volume of a bed of porous media or resin used in water treatment by adsorption, filtration, or a membrane process. It is also one of the parameters used in evaluating the performance of a treatment unit. *See* bed volumes to breakthrough, bed volumes treated, breakthrough, and empty bed contact time.

**bed volumes to 50% breakthrough ($BV_{50}$)** The volume of water treated in an activated carbon bed when the total organic carbon (TOC, mg/L) of the product water reaches 50% of the initial concentration ($TOC_0$, mg/L). It is used to evaluate the effects of empty bed contact time, carbon particle size, and source water characteristics on bed performance. It may be estimated as follows (Bond et al., 2004):

$$BV_{50} = 18{,}000/TOC_0 \qquad \text{(B-27)}$$

**bed volumes to breakthrough (BVB)** The volume of water treated in an activated carbon bed when it reaches breakthrough. *See also* carbon usage rate, bed volumes treated.

**bed volumes treated (BVT)** A parameter used in evaluating the performance of granular activated carbon units. BVT is the volume of water produced to a given fractional breakthrough divided by the volume of the bed of activated carbon. It is also the ratio of the service time to the empty bed contact time or EBCT.

**bed yield** A measure of the performance of a sand drying bed in terms of the weight of dry solids per unit surface area per year. *See* sand bed yield for detail.

**beef extract** A soluble paste of beef or beef blood extracts, or a powder of beef by-products. As a slightly alkaline, proteinaceous solution of about 1.5%, it is used in the elution and desorption of adsorbed viruses from filter surfaces. *See also* virus adsorption–elution.

**beef tapeworm** (1) The common name of *Taenia saginata,* a parasite that is distributed worldwide and is transmitted to persons who consume insufficiently cooked beef. (2) The excreted helminth infection caused by this parasite; taeniasis. The adult worm attaches to the small intestine. Occasional symptoms include irritation at the attachment site, abdominal pains, nausea, weight loss, increased appetite, headache, and intestinal obstruction. Cysticercosis is infection by the larval stage of the worm, which affects the muscles, the brain, and the heart. *See also Taenia saginata* life cycle, pork tapeworm.

**Beer's law** *See* Beer–Lambert law.

**Beer–Lambert law** The principle combining Beer's law and Lambert's law stating that the intensity of light is reduced by an absorbing medium (Lambert) and by the species in the medium (Beer). Thus, the transmittance ($T$) of a substance is expressed as follows:

$$T = I/I_0 = 10^{-kLC} \qquad \text{(B-28)}$$

The Beer–Lambert law is applied in the following equation to determine UV doses required for the inactivation of microorganisms (Metcalf & Eddy, 2003):

$$I = I_0\,(1 - e^{-kd})/(kd) \qquad \text{(B-29)}$$

where $I$ = light intensity at a distance from the surface (average UV intensity, $mW/cm^2$); $I_0$ = light intensity at the surface (average UV intensity at the surface of the sample, $mW/cm^2$); $k$ = an absorption coefficient (absorbance coefficient $k$ = 2.303 a.u./cm); $L$ = length of absorbing medium; $C$ = concentration of the absorbing species; $d$ = depth of sample, cm; and a.u. = absorbance units, /cm.

**beet-sugar waste** Industrial wastewater originating from screening and juicing operations, lime sludge draining, evaporator condensates, and sugar extraction. It is high in dissolved and suspended organic matter. The waste is amenable to reuse or treatment by coagulation and lagooning.

***Beggiatoa*** A group of strictly aerobic, chemoautotrophic, sulfur-oxidizing, whitish filamentous bacteria that form poorly settling suspensions; commonly associated with sludge bulking, which results from low dissolved oxygen levels, low ratios of organic matter to cell mass (low F/M ratios), and/or high sulfide levels. They may grow on contact aerators and clog the spaces between them. Beggiatoa may also develop a white/gray biofilm on rotating biological contactors that is difficult to slough off and may cause odors, process deterioration, and equipment failure. Also called filamentous sulfur bacteria, they derive energy by chemosynthesis:

$$H_2S + \tfrac{1}{2}\,O_2 \xrightarrow{\textit{Beggiatoa}} S + H_2O + \text{energy} \qquad \text{(B-30)}$$

$$S + 1\tfrac{1}{2}\,O_2 + H_2O \xrightarrow{\textit{Beggiatoa}} H_2SO_4 + \text{energy} \qquad \text{(B-31)}$$

*See also* filamentous sulfur bacteria.

**Belclene** An organic chemical produced by the FMC Corp. for scale control.

**Belcor** An organic chemical produced by the FMC Corp. for corrosion control.

**Belgard®** A chemical produced by the FMC Corp. for scale control in seawater evaporators.

**Belite®** A chemical produced by the FMC Corp. for foam control.

**Bellacide** An algicide produced by the FMC Corp.

**bell-and-spigot fitting** *See* bell-and-spigot joint.

**bell-and-spigot joint** An arrangement to join two or more clay pipes, directly or through a pipe fitting. Each pipe section has one end in a bell-like shape (with an enlarged diameter and polymeric rings) and the other in a spigot-like shape (with a polymeric sleeve). The spigot of one section fits into the bell of the next section. A joint compound is also added for water tightness. Clay pipes can also have plain ends and are then joined through polymeric sleeve casts and plastic corrugated rings. Some bell-and-spigot fittings include wyes, tees, reducers, and increasers.

**bell end** The inside threaded end of a pipe section that connects with the spigot end of another section. Also called inside threaded connection or female end.

**bell mouth** A flared or expanding, round entrance to a conduit or an orifice.

**BelloZon** Equipment manufactured by ProMinent Fluid Controls, Inc. for the generation of chlorine dioxide ($ClO_2$).

**Belspere** A chemical dispersant used in wastewater treatment, produced by the FMC Corp.

**belt conveyor** A conveying device that transports material from one location to another by means of an endless belt that is carried on a series of idlers and routed around a pulley at each end (EPA-40CFR60.671). *See also* bucket elevator.

**belt filter** Mechanical equipment used to dewater sludge. Same as a belt filter press.

**belt filter press** A sludge dewatering device consisting of two continuous, porous belts passing over rollers of decreasing diameters to squeeze water out of sludge. Dewatering occurs by gravity in a "free-drainage" zone, in a low-pressure zone, and in a high-pressure zone. The unit includes piping or boxes for sludge feeding, polymer addition, and mixing, as well as an apparatus for cake discharge and filtrate and washwater collection. Belt filter presses are very popular in wastewater treatment plants because they are economical and available in various sizes. They typically produce a sludge cake of about 20% solids. Also called belt filter or belt press. *See* sludge pressing for names commonly used and the following common features of a belt filter press: polymer conditioning, gravity drainage, low pressure, and high pressure.

**belt filter pressing** The use of a belt filter press, including chemical treatment, to dewater digested or raw sludge before it is trucked for landfill disposal or land application. Also called belt press dewatering, it comprises three basic operational stages: chemical conditioning, gravity drainage, and compaction of the thickened sludge. The filtrate and wash water is discharged to a sanitary sewer or recycled to the plant influent after appropriate treatment. *See also* gravity-belt thickening.

**belt filter press operating parameters** Parameters used in the operation of a belt filter press, mainly the hydraulic loading (e.g., gpm of sludge feed per meter of belt width) or solids loading (e.g., pounds of feed total dry solids per meter per hour) and the dosage of the chemicals (e.g., pounds of polymer per ton of total dry solids). The same parameters can also used for designing the unit.

**belt filter press performance parameters** Parameters used to evaluate the performance of a belt filter press, including solids recovery (SR, %), cake dryness (CD, % of dry solids by weight in the cake), washwater consumption and wastewater discharge, both in gpm per meter of belt:

$$SR = 100 \times (\text{total feed solids} - \text{suspended solids in wastewater})/\text{total feed solids} \quad \text{(B-32)}$$

Note that the wastewater discharge is the sum of the filtrate and the wash water.

**belt press** Mechanical equipment consisting of two fabric belts revolving over a series of rollers, used to dewater sludge. Same as a belt filter press.

**belt press dewatering** Using a belt filter press to dewater sludge; belt filter pressing.

**belt screen** An automatically cleaned screen for the removal of suspended or floating materials; it consists of a continuous band of wire mesh or similar screening medium. Also called bandscreen. *See also* traveling water screen.

**belt thickener** A mechanical processor where a horizontal filter belt thickens sludge before dewatering or disposal.

**benchmark concentration** The concentration that can cause a stormwater to impair water quality or affect human health or fish. It is not an effluent standard, but a target established by the USEPA for a facility to achieve through pollution prevention measures and to determine whether the discharge from the facility merits further monitoring.

**benchmarking** The process of (a) identifying and measuring operation and management indicators to assess a water or wastewater utility's strengths and areas for improvement and (b) comparing them to established performance standards. The American Water Works Association (AWWA, 2005) lists 22 such benchmarks: organizational best practices index, employee health and safety severity rate, training hours per employee, customer accounts per employee, debt ratio, system renewal/replacement rate, return on assets, customer service complaints, disruptions of service, residential cost of service, customer service cost per account, billing accuracy, drinking water compliance rate, distribution system water loss, water distribution system integrity, operation and maintenance cost ratio, planned maintenance ratio of water supply system, sewer overflow rate, wastewater collection system integrity, wastewater treatment effectiveness rate, wastewater operation and maintenance cost ratio, and planned maintenance ratio of sewer system.

**bench-scale development** The stage in a project development cycle in which required design data are collected, including potential environmental impacts. It precedes conceptual design, pilot-scale design, and preliminary engineering.

**bench-scale model** A hydraulic or other model used in a laboratory to represent a natural system at an appropriate scale. (In general, bench-scale testing is the testing of materials, methods, or processes on a small scale, such as on a laboratory worktable.) Examples are (a) the porous media model to study the movement of groundwater as in Darcy's experiment, and (b) a large tube and porous diffusers to study the activated sludge process. Bench-scale models try to replicate the natural system's characteristics: geometry, and other properties. With the advent of computers and the popularity of numerical models, bench-scale models are used mainly for teaching hydraulic engineering. *See also* porous-media model.

**bench-scale study** *See* bench-scale test.

**bench-scale test** (1) Laboratory testing of potential cleanup technologies. *See* treatability studies. (2) In general, a laboratory test or study conducted on a small scale to determine the feasibility of a technology in a specific application, usually following a concept definition and preliminary study stage; for example, a rapid small-scale column test for granular activated carbon. Also called bench test. *See also* pilot-scale test.

**bench-scale testing** *See* bench-scale model.

**bench test** *See* bench-scale test.

**beneficial organism** An organism that performs some beneficial action such as pollination, pest control, or predation of undesirable species.

**beneficial uses** The various ways in which water use or water reuse promotes economic and general wellbeing of the population, e.g., drinking water supply, fish and wildlife, water-contact recreation, wastewater disposal, transportation, and agricultural or industrial production. *See also* water reuse applications.

**beneficiation** A treatment process used for making ores more suitable for smelting by separating their useful fraction from a residue of tailings, which must be handled properly to avoid contamination. For example, iron beneficiation includes a variety of separation techniques: flotation, gravity separation, magnetic separation, and milling.

**benefit** A tangible or intangible advantage resulting from the implementation of a project, which may be as simple as the installation of a piece of equipment or as extensive as the construction and operation of an entire regional sewerage system.

**benefit–cost analysis** A technique used to compare the consequences of various alternatives of a project, various projects, or various courses of action in general. It is a quantitative evaluation of the overall benefits to society of a proposed action versus the costs that would be incurred, including not only monetary factors but also intangible benefits, which are often difficult to assess. The technique can be applied, e.g., to (a) the establishment of acceptable doses of toxic chemicals, (b) a decision among two or three sources for drinking water, (c) the choice between two pieces of equipment for sludge dewatering, (d) the construction of a multiple purpose reservoir, or (e) the selection among several local and regional wastewater management schemes. Benefit–cost analysis has been used to support certain health- or environment-related claims that are sometimes controversial, e.g., the nuclear industry's claim that zero radiation release is too expensive. Also called cost–benefit analysis. *See* financial indicators for related terms.

**benefit/cost ratio** For a given project, the benefit/cost ratio is the present value of benefits or cash inflows divided by the present value of costs or cash outflows over the planning period; the higher the ratio, the more attractive the project. Also called profitability index. *See* financial indicators for related terms.

**benthal demand** Same as benthal oxygen demand.

**benthal deposit** Same as benthic deposit.

**benthal oxygen demand** (From the Greek word *benthos,* meaning depth.) The oxygen demand exerted by the decomposition products of materials at the bottom of a body of water, resulting, for example, from the pollutional load that settles in the vicinity of the point of discharge. *See* the Streeter–Phelps formulation. Sediments deposited from wastewater in collection systems also exert an oxygen demand.

**benthic** (from the Greek word *benthos,* meaning depth) Pertaining to organisms that live, or materials that are, on or at the bottom of a body of water; benthal.

**benthic deposit** Accumulation on the bed of a water body of organic and inorganic materials from natural processes (erosion) and man-made discharges (e.g., wastewater effluent); benthal deposit. It is a diffuse source of contamination of water resources, e.g., soluble and suspended matter varying from a thin layer to a heavy sludge bank.

**benthic fauna** Organisms attached to or resting on the bottom or living in the bottom sediments of a water body. *See* benthos for detail.

**benthic habitat** *See* benthos.

**benthic region** The biogeographic region that includes the bottom of a lake, sea, or ocean, including the littoral zone of the shore. *See* freshwater profile and marine profile for other related terms.

**benthic zone** The bottom and associated sediments, below the profundal zone of a body of water. The large detrital biomass that it receives from the upper layers allows the benthic zone to be very active biologically, with a predominance of decomposers, detritovores, and anaerobic bacteria. *See* freshwater profile and marine profile for other related terms.

**benthon** The organisms living on or in the benthos.

**benthonic zone** *See* benthos.

**benthopelagic zone** The sea–sediment interface; same as benthos. *See* freshwater profile and marine profile for other related terms.

**benthos** (from the Greek word *benthos,* meaning depth) (1) Microbes and other organisms living on the bottom of a water body, e.g., worms, crustaceans, crabs, and clams. (2) More generally, the benthos is a biogeographic region that includes the bottom, the littoral, and supralittoral zones of a body of water; also called benthic zone, benthonic zone, or benthopelagic zone. It is also regarded as a transition between the water column and the mineral subsurface. The benthic habitat is much more fertile than the planktonic environment. *See* freshwater profile and marine profile for other related terms.

**BentoLiner™** A liner of clay composite made by SLT North America, Inc.

**Bentomat™** A geotextile–bentonite liner made by Colloid Environmental Technologies Co.

**bentonite** A colloidal clay formed by the decomposition of volcanic ash, largely made up of the mineral sodium montmorillonite, a hydrated aluminum silicate. Because of its ability to absorb water and expand to several times its normal volume, bentonite is commonly used to provide a tight seal around a well casing. It is also used (a) as a coagulant aid or weighting agent in water treatment (particularly water with high color, low turbidity, and low mineral content; *see* ballasted floc); (b) as a landfill liner; (c) as a major constituent of drilling mud; and (d) to reduce seepage in mines and channels . Also called bentonite clay or bentonitic clay. *See also* Fuller's earth. Named after Fort Benton, Indiana.

**bentonite clay** *See* bentonite.

**bentonite geotextile mat** A material used as a liner in wastewater treatment ponds to minimize percolation.

**bentonite liner** A compound of bentonite and select soils mixed at the site, compacted at the proper water content (at least 90% standard Proctor density), and applied uniformly on the bottom of a pond, landfill, etc. It allows water to pass through at a very slow rate, corresponding, e.g., to a permeability of $1 \times 10^{-8}$ cm/sec.

**bentonite soil** A typical material used as a liner in wastewater treatment ponds to minimize percolation.

**bentonitic clay** *See* bentonite.

**benzene ($C_6H_6$)** The parent compound of aromatic hydrocarbons, often characterized by its ring structure. A volatile, flammable, colorless liquid with a characteristic (aromatic or gasoline-like) odor and carcinogenic properties; used in explosives, insecticides, solvents, etc.; classified by the USEPA as priority pollutant No. 4, with an MCL = 0.005 mg/L in drinking water. Also called benzol, benzolene, carbon oil, coal naphta, coal tar naphta, mineral naphthalene, motor benzol, nitration benzene, phenyl hydride, pyrobenzol.

**benzene hexachloride ($C_6H_5Cl_6$)** *See* lindane.

**benzene ring** A hexagonal structure representative of benzene and other aromatic compounds, and showing six carbon atoms arranged in a ring, each with a hydrogen atom attached, sometimes with a circle inside to denote the stability of the bond among the participating carbon atoms. *See* Figure B-06.

**benzene–toluene–xylene (BTX)** Volatile organic compounds found in gasoline. Toluene's formula is $C_6H_5CH_3$, and xylene's is $C_6H_4(CH_3)_2$.

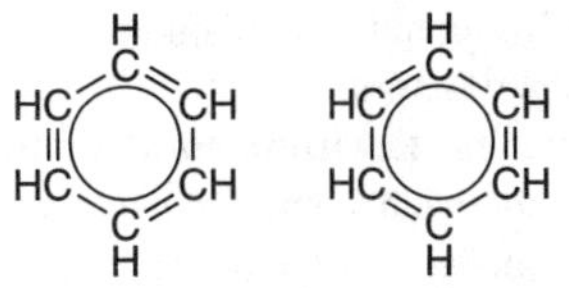

**Figure B-06.** Benzene rings.

**benzo[a]pyrene ($C_{20}H_{12}$)** A polynuclear aromatic hydrocarbon, containing five rings; a product of incomplete combustion present in coal tar and cigarette smoke; regulated by the USEPA, maximum contaminant level in drinking water MCL = 0.0002 mg/L (and MCLG = 0), but WHO recommends a limit of 0.00001 mg/L.

**benzoate of soda** Same as sodium benzoate.

**benzo[e]pyrene** Same molecular and formula weight, similar properties and uses as benzo-[a]pyrene.

**benzol** Another name of benzene.

**benzolene** Another name of benzene.

**Berkefield filter** A filtration device using a solid candle of diatomaceous earth for household water treatment.

**beryl ($Be_3Al_2Si_6O_{18}$)** Beryllium aluminum silicate, a green, blue, rose, white, or golden mineral, the principal ore of beryllium, but also a source of gems (emerald and aquamarine).

**berylliosis** Beryllium poisoning resulting from airborne concentrations as low as 20 micrograms per cubic meter for fewer than 50 days, characterized by the formation of granulomas, causing a lung inflammation, cough, fever, chill, chest pain, and shortness of breath. Beryllium dust also causes skin ailments.

**beryllium (Be)** A steel-gray, bivalent, light alkaline-earth metal, found naturally in ores of beryl, used in hard, corrosion-resistant copper and nickel alloys and in electrical parts. Only the chloride and nitrate are water-soluble. An airborne metal hazardous to human health when inhaled; *see* berylliosis. It is discharged by machine shops, ceramic and propellant plants, and foundries. Atomic weight = 9.0122. Atomic number = 4. Specific gravity = 1.85 at 20°C. Found in drinking water sources at low concentrations from the coal, nuclear power, and space industries; regulated as a hazardous pollutant by the USEPA: MCL = MCLG = 0.004 mg/L. At sufficiently high concentrations, beryllium may interfere with enzyme production and cause physiological human effects. Best available control technology: filtration and lime softening, activated alumina, ion exchange, and reverse osmosis.

**B.E.S.T.® amine extraction process** A proprietary solvent extraction process of Ionics RCC that treats hazardous soil, sludges, and sediments using cold aliphatic amines to separate the hazardous materials into their oil, water, and solid fractions. The flow chart features a cold extraction tank for sludges and sediments, an extracting dryer vessel for the soils, a decanter and water stripper for the extracted solution, and a solvent evaporator. All solvent is recycled, water discharged, oil recycled or destroyed, and solids backfilled or further treated before final disposal (Freeman, 1998). *See* inverse miscibility.

**best available control measure (BACM)** A term used to refer to the most effective measures (according to USEPA guidance) for controlling small or dispersed particulates from sources such as roadway dust, soot and ash from woodstoves and open burning of rush timber, grasslands, or trash.

**best available control technology (BACT)** Same as best available technology.

**best available demonstrated technology (BADT)** The effluent limitation technology required by the Clean Water Act for new source performance standards applicable to new industrial discharges.

**Best Available Techniques Not Entailing Excessive Costs (BATNEEC)** A British approach to exercise control over environmental emissions using currently available technology; required by the Environmental Protection Act as part of integrated pollution control.

**best available technology (BAT) or best available control technology (BACT)** The best technology, treatment techniques, or other means that the EPA Administrator finds, after examination for efficacy under field conditions and not solely under laboratory conditions, are available (taking cost into consideration) (EPA-40CFR141.2). Also, for any specific source, the necessary technology that would produce the greatest reduction of each pollutant regulated by the Clean Air Act, taking into account energy, environmental, economic, and other costs.

**best available technology economically achievable (BATEA)** A requirement of the USEPA to control toxins and nonconventional pollutants, with appropriate considerations of economic impacts.

**best conventional control technology (BCT)** The level of water pollution control technology required of existing dischargers of conventional pollutants.

**best conventional pollutant control technology (BCT)** A requirement of the USEPA to provide secondary treatment of wastewaters to control conventional pollutants, industrial pretreatment of toxic wastewaters, and effluent disinfection.

**best conventional technology** A technology-based requirement to control conventional pollutants in stormwater discharges.

**best demonstrated available technology (BDAT)** As identified by the USEPA, the most effective commercially available means of treating specific types of hazardous or other waste. The BDATs have statistically better performance than other technologies and may change with advances in treatment technologies.

**best management practice (BMP)** (1) A practice or combination of practices that are determined to be the most effective and practicable means of controlling point and nonpoint pollutants at levels compatible with environmental quality goals. They include technological, economic, and institutional considerations. (2) The schedules of activities, prohibitions of practices, maintenance procedures, and other management practices to prevent or reduce the pollution of waters of the United States from discharges of stormwater, wastewater effluents, dredged or fill materials, etc. BMPs include methods, measures, practices, or design and performance standards that facilitate compliance with such regulations as effluent limitations or prohibitions, and applicable water quality standards. BMPs also include treatment requirements, operating procedures, and practices to control plant site runoff, spillage or leaks, sludge or waste disposal, or drainage from raw material storage. (EPA-40CFR122.2 and 40CFR232.2). (3) BMPs are structural, nonstructural and managerial techniques that are recognized to be the most effective and practical means to control nonpoint source pollutants, yet are compatible with the productive use of the resource to which they are applied. (4) In urban stormwater management, BMPs include structural controls or devices for the treatment or storage of runoff to reduce flooding, remove pollutants, and provide other amenities. Stormwater quality ponds are considered among the most effective measures. On construction sites, for example, BMPs generally include measures to prevent erosion, trap pollutants before their discharge, and prevent construction material pollutants from mixing with stormwater. *See also* BMP Planner and stormwater pollution prevention plan.

**best practicable control technology (BPT)** A national goal of the Clean Water Act requiring industry to use the best practicable treatment methods, taking into account cost, plant and equipment age, among other factors.

**Best Practicable Environmental Option (BPEO)** A British approach to environmental quality requiring cross-media considerations, e.g., "appropriate measures to deal with any harmful discharges to water and for the treatment or disposal of other solid and liquid wastes to land. A BPEO should take into account the risk of transfer of pollutants from one medium to another" (Porteous, 1992).

**best practicable waste treatment technology** The cost-effective technology that can treat wastewater, combined sewer overflows, and nonexcessive infiltration and inflow in publicly owned or individual wastewater treatment works, to meet the applicable provisions of the Clean Water Act (EPA-40CFR35.2005-7).

**best usage** Use of fresh, tidal, or underground waters for such purposes as drinking, culinary, food processing, bathing, fishing, agricultural, industrial, cooling, etc., depending on the established classification.

**beta ($\beta$) aeration correction factor** Same as beta coefficient.

**beta ($\beta$) coefficient** A factor used in the design of aeration equipment, also called the oxygen saturation coefficient or the $\beta$ factor and representing the ratio of the dissolved oxygen saturation concentration in wastewater ($C_sWW$) to the saturation concentration in clean water ($C_sCW$) at the same temperature:

$$\beta = (C_sWW)/(C_sCW) \qquad \text{(B-33)}$$

It depends on the wastewater characteristics (e.g., organic matter and dissolved salts); in practice it varies between 0.70 and 0.98. *See* oxygen transfer rate.

**beta ($\beta$) decay** The emission of a positively charged beta particle from the nucleus of an atom, with an increase or decrease of 1 in the atomic number if the particle is negatively or positively charged, respectively; the loss of an electron.

**$\beta$-D-galactosidase** *See* galactosidase.

**$\beta$-D-galactoside** *See* galactoside.

**$\beta$-D-glucuronidase** *See* glucuronidase.

**$\beta$-D-glucuronide** *See* glucuronide.

**$\beta$ factor** *See* beta coefficient.

**$\beta$-GAL** *See* galactosidase.

**beta-mesaprobic zone** An area or reach of moderate decomposition of organic wastes in a slow-moving stream. *See also* saprobic classification.

**beta-minus decay** Radioactivity in which the nucleus of an atomic species changes or disintegrates by emitting a negative electron particle.

**beta (β) oxidation** The second step in the microbial oxidation of an alkane, following the formation of carboxylic acid:

$$CH_3CH_2CH_2CH_2CO_2H + 3\ O_2 \rightarrow CH_3CH_2CO_2H + 2\ CO_2 + 2\ H_2O \quad \text{(B-34)}$$

**beta (β) particle** An electron or positron emitted from an atomic nucleus during radioactive decay, with a single positive or negative charge. As an example, Radium-228 emits beta particles while decaying to Actinium-228. A thin sheet of metal can easily stop beta particles, which are more penetrating than alpha particles but produce less ionization per unit length. The USEPA sets a maximum contaminant level of 4 mrem/yr for beta particles and photon radioactivity in drinking water. *See also* gross beta particle.

**β-$PbO_2$** Formula of plattnerite, one of two forms of lead dioxide.

**beta-plus decay** Radioactivity in which the nucleus of an atomic species changes or disintegrates by emitting a positive electron antiparticle.

**beta (β)-Poisson distribution** A modified exponential model used to describe the probability of human infection by enteric microorganisms. Thus, the probability ($P$) of infection from a single exposure is:

$$P = 1 - (1 + N/\beta)^{-\alpha} \quad \text{(B-35)}$$

where $N$ is the number of organisms ingested in the exposure and $\alpha$ and $\beta$ are parameters characteristics of the dose–response curve. Also called modified exponential model. *See* dose–response model for related distributions.

**beta (β)-Poisson infection model** Same as beta-Poisson distribution.

**beta (β) radiation** The emission of beta particles, highly energetic negative or positive electrons (positrons).

**beta (β)-ray irradiation** A sludge processing technique that qualifies as a process to further reduce pathogens: beta rays from an accelerator irradiate the sludge solids at a minimum dosage of 1.0 megarad at approximately 20°C.

**BET isotherm** Abbreviation of Brunauer–Emmet–Teller isotherm.

**BET method** Abbreviation of Brunauer–Emmet–Teller method.

**BET surface area** A characteristic of an activated carbon provided by the manufacturer and determined by using the BET isotherm to analyze the adsorption of nitrogen gas molecules on the carbon surface. A smaller surface area is available to adsorbates having a larger molecule than nitrogen gas.

**better assessment science integrating point and nonpoint sources (BASINS)** A computer program that allows users (a) to organize geographic information for selected watersheds, (b) to model the impacts of pollutant loadings from point and nonpoint sources, and (c) to characterize the overall condition of the watersheds.

**BF-AS** Acronym of biofiltration-activated sludge process.

**BGM** Acronym of Buffalo green monkey.

**BGMK** Acronym of Buffalo green monkey kidney.

**Bhargava-Rajagopal equation** An equation proposed in 1989 for the performance of type I sedimentation tanks as a function of the uniformity coefficient of the particles ($U$), the 10-percentile ($P_{10}$, mm) and the design settling velocity ($V_0$, $m^3/m^2/sec$). The total removal ($R$, fraction by weight) of solid particles is such that (Droste, 1997):

$$1/R = U/(178 + 45\ U) + V_0\{1/\exp[(0.0032\ U + 2.04) \ln P_{10} + \exp(0.356\ U + 1.647)]\} \quad \text{(B-36)}$$

**BHC** Abbreviation of benzene hexachloride.

**BHN Probiotic** A chemical product made by Bio Huma Netics, Inc. for the oxidation of lagoon sludge.

**B horizon** A layer on a soil profile, beneath the O, A, and E horizons; basically a deposition of clay minerals, metallic oxides, and other substances from the E layer. *See* illuviation.

**bhp** Abbreviation of brake horsepower.

**bias** An inadequacy in experimental design that leads to results or conclusions not representative of the population under study. In monitoring or simulation, a tendency of an instrument or model to deviate constantly or systematically from the true value of a measured quantity. Causes bias errors. *See also* precision error.

**bias error** *See* bias.

**bias gain coefficient** A constant used in wastewater process control applications, equal to half the maximum value of a variable control parameter.

**bibenzene** Another name of biphenyl.

**Bibo** A pump manufactured by ITT Flygt and used for dewatering and drainage.

**bicarbonate** (1) The inorganic monovalent anion $HCO_3^-$ present in natural waters and, along with carbonate ($CO_3^{2-}$), resulting from the dissolution of carbonate rocks. Molecular weight = 61.0. Equivalent weight = 61.0. (2) A chemical compound containing the group $HCO_3$, e.g., calcium bicarbonate [$Ca(HCO_3)_2$], sodium bicarbonate ($NaHCO_3$), magnesium bicarbonate [$Mg(HCO_3)_2$]. Bicarbonates are the major component of alkalinity and provide an important buffer in natural waters; they also contribute to total dissolved solids in drinking water.

**bicarbonate alkalinity** The alkalinity (acid-neutralizing capacity) caused by bicarbonate ions ($HCO_3^-$). In ordinary drinking water, with a pH below 9.0, most of the alkalinity is caused by the bicarbonates of calcium and magnesium.

**bicarbonate endpoint** One of two carbon dioxide ($CO_2$) titration endpoints, corresponding to pH = 8.3, often used in the determination of acidity and alkalinity of water; also used for acidity by the wine industry. *See also* carbonic acid end point.

**bicarbonate hardness** Water hardness caused by the bicarbonates of divalent metals, mainly those of calcium [$Ca(HCO_3)_2$] and magnesium [$Mg(HCO_3)_2$]. *See* carbonate hardness.

**bicarbonate of soda** *See* sodium bicarbonate.

**Bi-Chem®** A proprietary bacterial culture produced by Sybron Chemicals, Inc. for use in wastewater treatment.

**bidentate ligand** An anion or molecule that forms a complex with a central metal atom and has two ligand atoms, i.e., two atoms responsible for its basic or nucleophilic nature. *See* multidentate ligand.

**biethylene** Another name of butadiene.

**BIF®** A product line of Leeds & Northrup.

**Bifad** A fill and draw activated sludge plant manufactured by Biwater Treatment, Ltd.

**bifidobacteria** Nonsporulating, strictly anaerobic, Gram-positive, V- or Y-shaped microorganisms, found in human intestines, and predominating in feces of mammals; they do not survive well in an aquatic environment. Previously known as anaerobic lactobacilli, they have been proposed as an indicator of fecal contamination for use in tropical waters because of their exclusively fecal origin.

***Bifidobacterium*** A genus of anaerobic microorganisms (bifidobacteria) commonly found in sludge digesters.

***Bifidobacterium adolescentis, Bifidobacterium longum*** The most common species of bifidobacteria.

**biflow filter** A granular media filter, developed in the former Soviet Union, characterized by water flow from both top and bottom to a collector located in the center of the filter bed. The flow is directed approximately 80% upward and 20% downward by restricting the expansion of the bed. *See* Figure B-07.

**Bilharz, Dr. Theodor** The German physician (1825–1862) who first identified one of the schistosome worms responsible for schistosomiasis or bilharziasis in Cairo in 1851.

**bilharzia** A waterborne disease of tropical and subtropical regions transmitted indirectly to humans by schistosomes. *See* schistosomiasis for more detail.

**bilharziasis** Same as bilharzia.

**bill frequency analysis** A periodic examination of water bills to determine consumption patterns by consumer categories or classes and to make necessary rate adjustments.

**bin** Any one of four risk levels defined by the USEPA to classify surface water sources with respect to *Cryptosporidium* monitoring results. Sources in bin 1 are required to maintain compliance with current Interim Enhanced Surface Water Treatment Rule, whereas sources in bins 2–4 must provide additional *Cryptosporidium* protection that may include riverbank filtration, slow sand filtration, membrane filtration, ozonation, chlorine dioxide inactivation, or ultraviolet irradiation.

**binary distillation** The separation by distillation of a substance from a solvent. For example, waste ink from newspaper printing contains also a solvent and water, which are first separated from the ink. Then the solvent is recovered from the mixture by binary distillation. *See also* flash distillation and batch distillation (Freeman, 1995).

**binary division** Same as binary fission.

**binary fission** A mode of bacterial and protozoan (asexual) reproduction whereby a cell divides, longitudinally or transversely, into two new independent cells approximately equal in size. *See also* budding, sexual reproduction, generation time.

**binary separation factor** One of two parameters that indicate quantitatively the preference of an

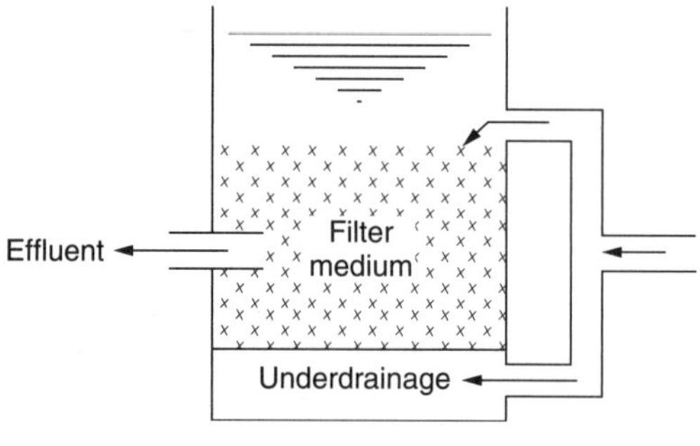


**Figure B-07.** Biflow filter.

ion exchange resin for specific ions. *See* separation factor for detail.

**Bingham plastic** A substance whose stress varies linearly with flow; characterized by yield stress and rigidity coefficient, two parameters that are also applied to wastewater sludge.

**Bingham plastic model** A model for non-Newtonian fluids that has been successfully applied to the laminar flow of wastewater sludges with known rheological properties. The model also predicts the transition point between laminar and turbulent flow conditions (WEF & ASCE, 1991).

**binomial distribution** The discrete probability distribution function that corresponds to Bernoulli's process: the probability that an event ($x$) of occurrence ($p$) will occur ($k$) times in ($n$) trials is:

$$f(x) = n!\, p^k(1-p)^{n-k}/k!(n-k)! \qquad \text{(B-37)}$$

The sign (!) indicates the factorial of the preceding number. The binomial distribution is used (with the Poisson distribution) in the most probable number (MPN) technique.

**binomial nomenclature** The scientific method of naming all organisms, except viruses, using two Latinized words usually written in italics or sometimes underlined. The first word, with a capital initial letter, denotes the genus and the second word denotes the species; e.g., *Escherichia coli. See also* biological classification conventions and biological species nomenclature for more detail.

**binomial system** Same as binomial nomenclature.

**Bio-2-Sludge® process** A proprietary, integrated fixed-film activated sludge method of wastewater treatment that uses a fixed PVC packing in the aeration tank. *See also* BioMatrix®, Captor®, Kaldnes®, Linpor®, moving bed biofilm reactor, Ringlace®, submerged rotating biological contactor.

**bioaccumulant** A substance that increases in concentration in living organisms as they take in contaminated air, water, or food because the substance is very slowly metabolized or excreted. *See also* biological magnification.

**bioaccumulation** The uptake and, at least temporary, storage of a substance (particularly a toxic substance) by an organism (especially fish, shellfish, and other aquatic organisms). The substance can be retained in its original form and/or as modified by enzymatic and nonenzymatic reactions in the body (EPA-40CFR798.7100). The process occurs by diffusion from aqueous solution and by ingestion from sediments, soil, food, etc. Bioaccumulation is a primary concern with trihalomethanes, xenobiotics, and other carcinogenic substances. Bioaccumulation is related to lipophilicity and the octanol-water partition coefficient is an indication for both. Also called biological accumulation, bioretention. *See also* bioaugmentation, bioconcentration, biological magnification, biomagnification, oil–water partition coefficient.

**bioaccumulative** Characteristic of a substance that living organisms can store, at least temporarily; i.e., the rate of intake exceeds the rate of excretion.

**Bio-Activation** A process developed by Amwell, Inc. for the treatment of wastewater by activated sludge combined with trickling filtration.

**bioaerosol** A solid and/or liquid airborne particle of 0.020–100 μm in diameter (such as dust or mist) containing microorganisms and/or microbial products; e.g., aerosolized coliforms from wastewater treatment plants, compost bioaerosols, endotoxins, or aerosolized *Legionella pneumophilia* in poorly ventilated buildings.

**bioassay** (1) Study of living organisms to assess qualitatively or quantitatively the effect or potency of a defined substance, factor, or condition using a change in biological activity, e.g., by comparing before-and-after exposure or other data. (2) A method used to determine the toxicity of specific chemical contaminants: a number of individuals of a sensitive species are placed in water containing specific concentrations of the contaminant for a specified period of time. Bioassays can be used to determine the toxic effects of certain wastewaters, e.g., in the mixing zone of an effluent discharge, by exposing in the laboratory selected aquatic organisms to the wastewaters. *See* biological assessment, biological monitoring, flow-through test, static toxicity test, test organisms.

**bioassay approach** The estimation of biomass in a water environment through the laboratory analysis of a sample by filtration, inoculation, incubation, and measurement of growth.

**bioassay toxicity test** Biological test conducted to determine the toxicity of a substance; bioassay.

**bioaugmentation** The introduction of nonindigenous or cultured microorganisms into the subsurface environment to enhance bioremediation of organic contaminants. Generally, the microorganisms are selected for their ability to degrade the organic compounds present at the remediation site. The culture can be either an isolated genus or a mix of more than one genus. Nutrients are usually blended with the aqueous solution containing the microbes to serve as a carrier and dispersant. The liquid is introduced into the subsurface under natural conditions (gravity) or injected under pressure. For example, microorganisms used in the degradation of mineral oils, aliphatic hydrocarbons, organics, and other oils include *Acinetobacter, Alcaligenes faecalis, Arthrobacter spp., Ochrobactrum anthropi, Phanerochaete chrysosporium, Pseudomonas aeruginosa, Pseudomonas cepacia, Pseudomonas putida, Rhodococcus,* and *Saccharomyces cerevisiae.* Bioaugmentation is also used in the remediation of metal-contaminated sites, the improvement of soil structure, the increased production of soils, and the control of plant pathogens. *See also* superbug.

**bioavailability** (1) The availability of a compound for biodegradation, influenced by the compound's location relative to microorganisms and its ability to dissolve in water. (2) The degree to which a nutrient (e.g., phosphorus, P) in manure, fertilizers, or other source is available for plant growth. (3) The degree to which a drug or other substance becomes available to the target organism or tissue after administration or exposure, depending on the substance's form, the route of exposure, and other factors.

**biobrick** A brick used in construction made with kiln-dried wastewater sludge substituted for sawdust or other organic substances; it contains 15-30% sludge by volume.

**Biocarb™** An activated carbon produced by Wheelabrator Clean Water Systems, Inc.

**Biocarbone®** process A wastewater treatment process of Biwater Treatment, Ltd., developed in France in the 1980s. It uses a fixed-bed downflow filter immersed in the liquid and packed with 3–5 mm fired clay material for the removal of carbonaceous BOD, nitrification, and denitrification. This is the same process as the biological aerated filter. *See* submerged attached growth process for related methods.

**biocenosis** A self-sufficient community of organisms living and interacting within a specific portion of a habitat. Also called biotic community, biocoenosis. *See also* biome.

**biochemical** Pertaining to chemical changes that result from biological activity, e.g., biochemical oxygen demand. Biochemical action is a result of the metabolism of living organisms. Biochemical oxidation is caused by biological activity and results in the chemical combination of oxygen with organic matter. The biochemical process or biological process consists of the metabolic activities of microorganisms converting complex organic matter into more stable substances; e.g., biological wastewater treatment processes, sludge digestion, and stream self-purification. *See* wastewater oxidation.

**biochemical action** *See* biochemical.

**biochemical fingerprinting** A procedure (e.g., glucose fermentation) that uses biochemical reactions to identify microorganisms.

**biochemical half-life** (1) The time required for the mass, concentration, or activity of a chemical or physical agent to be reduced by one-half through decay. *See* decay processes and half-life. (2) The time required for the elimination of one-half of a total dose from the body.

**biochemical methane potential** The potential of a waste to produce methane under anaerobic conditions. It is less than the theoretical yield of 5.6 $ft^3$/lb COD removed because of toxicity, refractory organics, and other reasons. *See also* methane yield, methane production, biochemical methane potential test, anaerobic toxicity assay.

**biochemical methane potential test** A procedure, similar to the BOD test, established in 1979 to determine the methane potential of a waste. It simulates anaerobic decomposition by inoculating a sample of the waste with an appropriate culture, required nutrients, and reagents in a serum bottle. Gas production and composition are monitored over an adequate period, typically 30 days. *See also* anaerobic toxicity assay.

**biochemical oxidation** *See* biochemical.

**biochemical oxygen demand (BOD)** A measure of the quantity of dissolved oxygen (DO) used in the oxidation of organic matter by microorganisms (mainly bacteria); the most commonly used parameter to define the strength or organic content of municipal wastewater. The greater the BOD, the greater the degree of pollution. Usually, the oxidation time and temperature are specified. For example, at 20°C or 68°F the five-day BOD or $BOD_5$ of an average municipal wastewater is approximately 240 milligrams per liter for a per-capita water consumption of 100 gallons per day. BOD is distinct from the chemical oxygen demand or COD, which measures the amount of oxygen used for the chemical oxidation of organic matter. $BOD_5$ can be calculated by multiplying the fats, proteins, and carbohydrates by the factors 0.890, 1.031, and 0.691 respectively. Organic acids (e.g., lactic acids) should be included as carbohydrates. The composition of input materials may be based on either direct analyses or generally accepted published values (EPA-40CFR405.11-b). The BOD process may be represented by two biological reactions that take place almost simultaneously:

$$\text{Organic matter} + \text{dissolved oxygen} + \text{bacteria} \rightarrow CO_2 + \text{bacterial cells} \quad \text{(B-38)}$$

$$\text{Bacterial cells} + \text{dissolved oxygen} + \text{protozoa} \rightarrow CO_2 + \text{protozoal cells} \quad \text{(B-39)}$$

The BOD (mg/L) of a wastewater sample, which is diluted in a bottle before testing, is the ratio of a difference (between the initial DO [$D_1$, mg/L] and the final DO [$D_2$, mg/L] of the diluted sample) to the dilution ratio ($P$):

$$\text{BOD} = (D_1 - D_2)/P \quad \text{(B-40)}$$

$$P = (\text{volume of sample, ml})/(\text{volume of dilution bottle, ml}) \quad \text{(B-41)}$$

Other measures of organic content are ThOD, TOC, and TOD. *See also* carbonaceous oxygen demand, first-stage oxygen demand, headspace BOD analysis, immediate chemical oxygen demand, nitrogenous oxygen demand, second-stage oxygen demand, standard oxygen demand, ultimate oxygen demand, Thomas method, Warburg apparatus.

**biochemical oxygen demand (BOD) loading** The BOD content of wastewater to a treatment unit or effluent to a receiving stream, usually expressed in pounds per day.

**biochemical process** *See* biochemical.

**biochemistry** The science that deals with the chemistry of living matter; the biochemical processes of plants and animals. The chemistry of living matter concerns mostly the study of biomolecules.

**Biocidal™** An apparatus manufactured by Scienco/FAST Systems for the injection of sodium hypochlorite.

**biocide** A substance capable of controlling or destroying living organisms (bacteria, fungi, etc.).

**Bio-Clarifier** A secondary clarifier manufactured by Envirex, Inc. for package rotating biological contactor plants.

**bioclastic** Pertaining to sediments formed from fragments of organic remains.

**Bioclean™** A chemical product of Argo Scientific for use as a biocide in reverse osmosis plants.

**Bioclere™** A package wastewater treatment plant manufactured by Ekoofinn Bioclere.

**biocoenosis** Same as biocenosis.

**biocoenotic model** One of three types of model developed to represent the biological, chemical, and physical processes of water quality. Biocoenotic models focus on the primary, secondary, and decomposition organisms that make up the food chain. They are complex and difficult to calibrate, and usually established only for selected groups of organisms. *See also* ecological models and kinetic models.

**biocolloids** Humic acids, some bacteria, and viruses, which can adsorb on the surfaces of inorganic particles. *See* hydrophilic colloids.

**bioconcentration** The passive or active accumulation of a chemical in tissues of an organism (such as a fish) to levels greater than in the surrounding medium (such as water) in which the organism lives, depending on metabolic energy and oil–water partition coefficient. *See also* bioaccumulation, bioaugmentation.

**bioconcentration factor (BCF)** The unitless measure of the tendency of a substance to accumulate in the tissue of an aquatic organism. BCF is determined by the extent of portioning of a substance, at equilibrium, between the tissue of an aquatic organism and water; it may be defined as the ratio of concentration of a substance in the organism divided by the concentration in water; higher BCF values reflect a tendency for substances to accumulate in the tissue of aquatic organisms. Any BCF value greater than 1.0 denotes a potential for bioaccumulation, but this potential is generally not significant for BCFs below 100. The EPA recommends a BCF limit of 3.75 for such THMs as BDCM and DBCM. BCFs are related to lipophilic compounds and to the octanol–water partition coefficient ($K_{OW}$):

$$\log \text{BCF} = a \log K_{OW} - b \quad \text{(B-42)}$$

where $a$ ($\leq 1$) and $b$ are constants.

**bioconcentration potential** The maximum concentration of a chemical resulting from absorption by an organism, taking into account its rate of metabolism and excretion.

**biocontactor** Contraction of biological contactor such as an activated sludge reactor, a trickling filter, lagoon, digester, or rotating biological contactor, in which microorganisms use organic matter for growth.

**biocontrol** *See* biological control.

**bioconversion** The conversion of biomass to usable energy, e.g., the production of heat from solid fuel, the production of liquid fuel (ethanol) by fermentation of plant materials, the production of methane by anaerobic digestion of wastewater sludge. According to the Department of Energy, bioconversion could produce 14% of the U.S. energy needs by 2020.

**biocriteria** Biological criteria.

**Biocube™** An aerobic filter manufactured by EG&G Biofiltration for the treatment of odors and volatile organics.

**biocycle** The cycle of living things, starting with birth, ending with death and restitution of elements to nature.

**Bio-D®** A line of nutrients made by Medina Products for use in bioremediation.

**biodegradability** The relative ease with which natural or man-made materials degrade as a result of biological metabolism. Virtually all petroleum hydrocarbons are biodegradable, but at a rate that depends on the type of hydrocarbon. In general, biodegradability increases with water solubility, which is inversely proportional to molecular weight. Biodegradability affects reaction rates and performance in biological treatment units. Also called biodegradation potential. *See* reaction-rate constant.

**biodegradable** Capable of decomposing (or breaking down into simpler substances) rapidly by biological action. *See* biodegradation.

**biodegradable COD (bCOD)** A measurable quantity used commonly to represent the concentration of organic compounds in municipal and industrial wastewaters; it consists of dissolved, colloidal, and particulate biodegradable components.

**biodegradable dissolved organic carbon (BDOC)** The portion of total organic carbon that can be degraded by microorganisms, a major source of bacterial food in water distribution systems (about 20% of dissolved organic carbon). It includes such large compounds as polysaccharides, humic acids, and fulvic acids. *See also* assimilable organic carbon.

**biodegradable material** A material that can be broken down into simple elements, usually by microorganisms; most organic wastes, like food.

**biodegradable organic carbon (BOC)** The portion of total organic carbon in water or wastewater that can be decomposed by microorganisms. Whereas biological filtration can reduce BOC, ozonation can increase it. BOC is sometimes used as a measure of assimilable organic carbon and determined as the difference between the initial and final dissolved organic concentrations (DOC) in a sample incubated in the dark until the DOC reaches a constant level:

$$\text{BOC} = \text{initial DOC} - \text{final DOC} \quad \text{(B-43)}$$

*See* biodegradable dissolved organic carbon, biodegradable organic matter, microbial regrowth.

**biodegradable organic matter (BOM)** The portion of total organic matter in water or wastewater that can be decomposed by microorganisms. Organic matter includes nitrogenous compounds in addition to organic carbon matter. BOM in wastewater is often represented by the formula $C_{10}H_{19}NO_3$, i.e., mainly a compound of carbon, hydrogen, oxygen, and nitrogen.

**biodegradable organics** Biodegradable organic materials, consisting mainly of carbohydrates, fats, and proteins, that exert an oxygen demand (usually measured as BOD or COD) for their stabilization. Very approximately, COD = 1.5 $BOD_5$.

**biodegradable plastics** Plastic materials built with/from microbial cells or based on such biodegradable materials as starch, cellulose, polyalcohols, and cellulose acetate. *See* biopol, polyhydroxybutyrate.

**biodegradable soluble COD (bsCOD)** The soluble portion of the biodegradable COD, used to represent the organic matter oxidized or used in cell synthesis.

**biodegradation** The decomposition of a substance into more elementary compounds or the alteration of its structure by the metabolic or enzymatic action of microorganisms such as bacteria in soils, water, or wastewater; under oxic or aerobic conditions (i.e., in the presence of molecular oxygen, $O_2$) or under anoxic conditions (i.e., in the absence of molecular oxygen). About 70% of the organic matter in wastewater is readily available for biodegradation. Some complex compounds (e.g., cellulose, hydrocarbons) require a relatively long time for microbial decomposition and are considered nonbiodegradable in biological wastewater treatment. *See also* biorefractory substance, bioremediation, biotransformation, detoxication, fermentation, hard detergent, metabolism, mineralization.

**biodegradation potential** Same as biodegradability.

**biodegradation reaction** A series of oxidation–reduction (redox) reactions in which microbial cells obtain energy for growth. Examples of biodegradation reactions (with an approximate order of their energy production) are: aerobic respiration (1.28), denitrification (1.22), nitrate reduction (0.83), and methanogenesis (0.22).

**BIOdek®** A synthetic media fabricated by Munters for use in fixed-film wastewater treatment.

**Bio-Denipho™ process** A proprietary wastewater treatment technique of Kruger, Inc. that combines the oxidation ditch and sequencing batch reactor processes for the removal of biochemical oxygen demand, suspended solids, total nitrogen, and total phosphorus. As shown in Figure B-08, it includes an anaerobic tank, two oxidation ditches in series equipped with rotor aerators, a clarifier, and sludge recycle/sludge wastage from the clarifier underflow. *See* phased isolation ditch for other related processes.

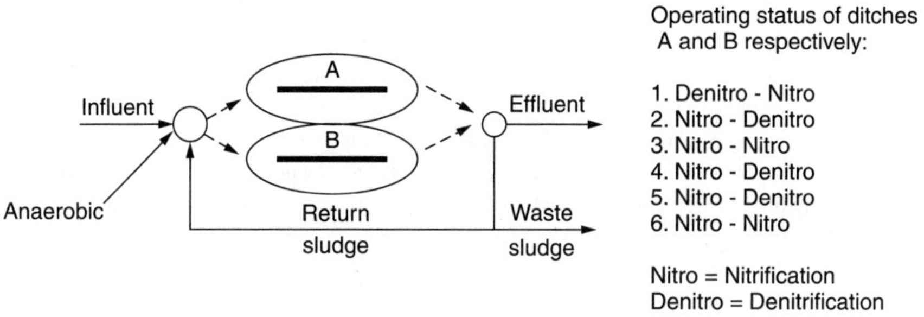


**Figure B-08.** Bio-Denipho™ process.

**Biodenit** A wastewater treatment process developed by Biwater Treatment, Ltd. to achieve biological denitrification in an immersed fixed-bed filter.

**biodenitrification** *See* biological denitrification.

**Bio-Denitro™ process** A proprietary wastewater treatment technique of Kruger, Inc. that combines the oxidation ditch and sequencing batch reactor processes for the removal of biochemical oxygen demand, suspended solids, and total nitrogen. As shown in Figure B-09, it includes two oxidation ditches in series equipped with rotor aerators, a clarifier, and sludge recycle/sludge wastage from the clarifier underflow. *See* phased isolation ditch for other related processes. *See also* biological nitrogen removal.

**biodisc** *See* biodisks.

**Bio-Disc system** A method of biological wastewater treatment developed in the United States and in Germany. *See* rotating biological contactor for detail.

**biodisks** Disks that are rotated in a tank for biological wastewater treatment. *See* rotating biological contactors for more detail.

**biodiversity** Refers to the variety and variability among living organisms and the ecological complexes in which they occur. Diversity can be defined as the number of different items and their relative frequencies. For biological diversity, these items are organized at many levels, ranging from complete ecosystems to the biochemical structures that are the molecular basis of heredity. Thus, the term encompasses different ecosystems, species, and genes.

**Biodiversity Convention** A multilateral environmental agreement for the conservation and sustainable use of global biological diversity.

**BioDoc®** A proprietary rotary distributor manufactured by WesTech Engineering, Inc. for use in trickling filters.

**biodosimetry** A test used to validate the performance of an ultraviolet (UV) system before it can be applied to disinfect drinking water. It involves adding nonpathogenic test microorganisms upstream of the UV reactor and measuring their inactivation rate as a function of flow, UV intensity, and water quality.

**Bio-Drum** A rotating drum manufactured by Ralph B. Carter Co. for the biological filtration of wastewater.

**bioenergetics** The study of energy transfers among living things (plants and animals). Bioenergetics is used to estimate cell yield through the

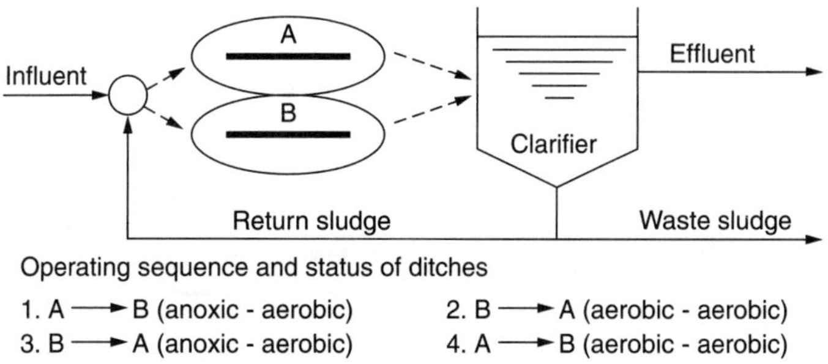


**Figure B-09.** Bio-Denitro™ process.

application of thermodynamic principles to biological reactions. *See also* endergonic reaction, exergonic reaction, Gibbs free energy.

**Bio-Energizer** An apparatus manufactured by Bio Huma Netics, Inc. for the oxidation of lagoon sludge.

**bioengineering** Biological science applied (a) to study the relation between workers and their environment or (b) to the design and application of instruments linked to life processes. *See also* biotechnology and ergonomics.

**bio-Fe** *See* biological iron and manganese removal.

**biofilm** An accumulation of microbial growth on a supporting surface; e.g., a layer of microorganisms at the interface between a solid substrate and the surrounding water. A biofilm may consist of one layer of cells or may be as thick as a 4-cm algal mat on the bottom of a reservoir. Characterized by bacterial extracellular polymers that sometimes create a slimy layer, biofilms are beneficial in trickling filters but detrimental in membrane filtration or in drinking water distribution systems where they may create problems of taste and odor, depletion of free chlorine, and corrosion. Chloramines are sometimes added to ozonated water to prevent microbial regrowth and biofilm formation. *See* attached-growth biofilm, biofilm mat, biofouling, glycocalyx, microbial biofilm, microbial mat, schmutzdecke, slime.

**biofilm mat** The layer of solids and biological growth that forms on the surface of slow sand filters, facilitating the removal of pathogens and suspended solids. This layer is periodically scraped off and removed or washed in place. *See also* schmutzdecke.

**biofilm process** A biological wastewater treatment process in which the microorganisms are attached to rock, slag, plastic, or other inert materials. *See* fixed-growth process for more detail.

**biofilm sloughing** The periodic shedding of the biofilm (and decomposed organic matter) in such fixed-growth processes as trickling filters and rotating biological contactors. Biofilm sloughing problems may occur when the organic loading exceeds the oxygen transfer capacity of the plant.

**biofilter** Same as biological filter. Materials removed by such a treatment unit are said to be biofiltered. *See* Figures B-10 and B-11. *See also* compost biofilter.

**biofilter-activated sludge** A wastewater treatment plant that uses the biofiltration-activated sludge process.

**biofiltered** *See* biofilter.

**biofilter performance** *See* destruction and removal efficiency.

**biofiltration** Same as biological filtration.

**biofiltration-activated sludge (BF-AS) process** One of a few dual wastewater treatment processes that combine the trickling filtration and activated sludge methods. A trickling filter or biological tower of reduced size receives the primary effluent and is followed by an aeration basin and a final clarifier. Some settled sludge and a portion of the filter underflow are recycled ahead of the filter. This process differs from the ABF only in the size of the aeration basin. The trickling filtration-solids contact process is also similar but includes a small unit for reaeration of recycle solids. *See also* combined filtration–aeration process.

**Bio*Fix®** A process developed by Wheelabrator Clean Water Systems, Inc. for the alkaline stabilization of sludge.

**bioflocculating agent** A microorganism that promotes the formation of floc in wastewater treat-

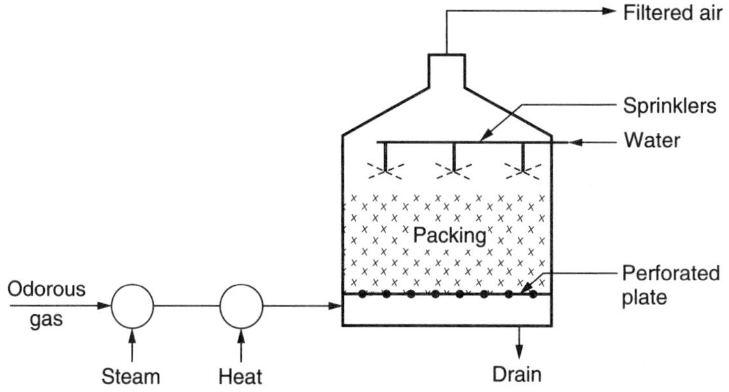


**Figure B-10.** Biofilter (enclosed). (Adapted from Metcalf & Eddy, 2003.)

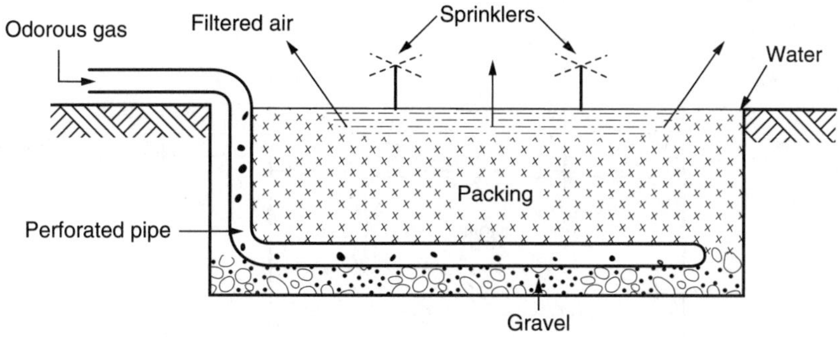


**Figure B-11.** Biofilter (open). (Adapted from Metcalf & Eddy, 2003.)

ment, e.g., return activated sludge in the trickling filter solids contact process. More exactly, polymers and other products excreted by the microorganisms or exposed at the cell surface are responsible for the microbial aggregation.

**bioflocculation** (1) The agglomeration of bacteria and other microorganisms in wastewater treatment works, as a result of surface charge phenomena, natural polyelectrolytes excreted by the organisms, bonding mechanisms of filamentous organisms, or added multivalent ions such as Fe(III) and Al(III). Bioflocculation produces larger particles and improves sedimentation, filtration, and sludge processing. *See* Figure B-12. (2) A term once proposed to replace the activated sludge process and avoid the connotation of waste product of the word "sludge." Another alternative would be "biological flocculation" (Fair et al., 1971).

**BioFlush** A trash rake manufactured by Beaudrey (E.) & Co.

**Biofor®** A wastewater treatment process developed by Infilco Degremont, Inc. using a biological bed of expanded clay (called Biolite®) immersed in the liquid and a proprietary air header (Oxazur®) for the removal of carbonaceous BOD, for nitrification and denitrification. *See* submerged attached growth process for related methods and Figure B-13.

**biofoul** Presence and growth of organic matter in a water system; biofouling.

**biofouling** The development of biological growth, e.g., a biofilm, which impairs the performance of membrane filters, other water treatment units, condensers, cooling towers, etc. Biofouling involves (a) the transport of soluble and particulate matter to a wetted surface, (b) the attachment of microorganisms to the surface and formation of a biofilm, (c) biochemical reactions and microbial growth on the surface, and (d) sloughing of portions of the growth by fluid shear stress. In reverse osmosis units, e.g., biofouling causes loss of flux, reduced solute rejection, increased headloss through the membranes, permeate contamination, degradation of membrane material, and reduced membrane life. Chlorination is often used to control biofouling. Also called biological fouling, microbiofouling, or, simply, fouling.

**biofuel** Renewable fuel, (e.g., alcohol, methanol, ethanol, wood), derived from biomass, corn, and other grains. *See also* biomass fuel.

**biogas** (1) Any combustible gas, usually containing methane as the main constituent, derived from the fermentation of organic matter. (2) The mixture of gases (approximately 65% methane and 35% carbon dioxide, with traces of hydrogen sulfide, ammonia, etc.) produced by the anaerobic decomposition of organic matter, e.g., the gas generated by mixing nightsoil with animal wastes. It derives its energy content (about 650 BTU/cubic foot) from methane. Biogas from anaerobic sludge digestion is normally used to heat the digesters or to meet some other energy needs of the wastewater treatment plant.

**biogas digester** A device used to produce biogas for domestic cooking and lighting, particularly in

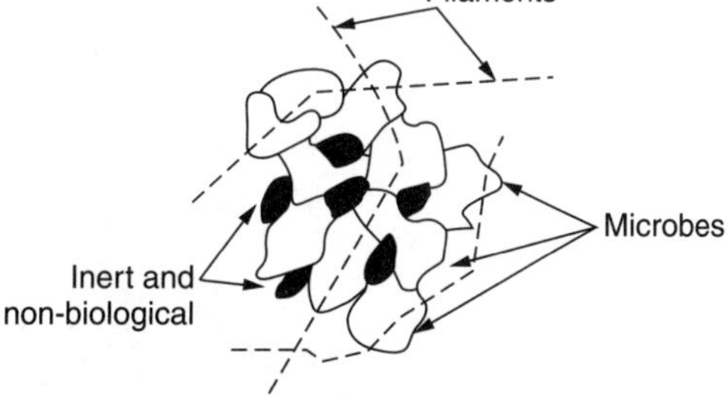


**Figure B-12.** Bioflocculation.

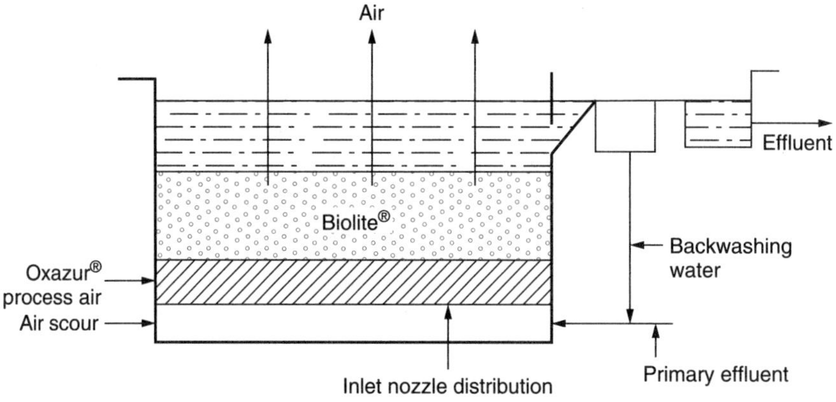


**Figure B-13.** Biofor®.

China, India, Korea, and Taiwan. It is basically an open tank for the anaerobic digestion of a mixture of diluted animal feces, human excreta, and vegetable refuse. The effluent slurry is used in agriculture or aquaculture. *See* Figure B-14.

**biogas potential** The potential of a waste to produce methane ($CH_4$) when treated under anaerobic conditions, as determined by the biochemical methane potential (BMP) test.

**biogas production** Production of biogas (methane) on a small scale by farmers or rural agencies.

**Bio Genesis™** A product of Bio Huma Netics, Inc. that uses microorganisms to reduce wastewater odors.

**biogenesis** The theory that living organisms arise only from other living organisms, or the actual production of living organisms from other living organisms. Also called biogeny.

**biogenic particle** Particle resulting from the activities of living organisms; also called biological debris.

**biogeny** Biogenesis.

**biogeochemical cycles** The circulation of matter (particularly plant and animal nutrients) between land, sea, and the atmosphere, involving living organisms and powered by solar energy. All major, minor, and trace elements cycle between biotic and abiotic forms in more or less definable ways. *See,* e.g., calcium cycle, carbon cycle, hydrologic cycle, nitrogen cycle, sulfur cycle. *See also* the Gaia hypothesis.

**biogeochemical cycling** The continuous circulation of chemicals in nature; *see,* for example, the carbon, nitrogen, and sulfur cycles.

**biogeochemistry** The study of microbially mediated chemical transformations in nature, e.g., nitrogen or sulfur cycling.

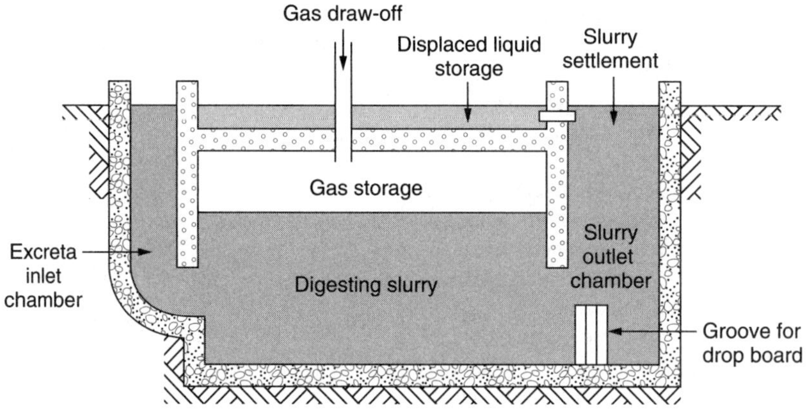


**Figure B-14.** Biogas digester (adapted from Feachem et al., 1983.)

**biohazard** The risk or potential risk to human health posed by the possible release of an infectious agent into the environment, especially a pathogen used or produced in biological research. *See also* hazardous material.

**bioindicator** Same as biological indicator.

**Bio Jet-7** A solution of seven strains of bacteria made by Jet, Inc.

**biokinetic coefficient** Any of a number of kinetic parameters used in model simulation of biological treatment processes; for example, biomass COD, cell debris, denitrifying fraction, endogenous decay, half-velocity, maximum specific growth rate or maximum specific substrate utilization rate, and yield.

**biokinetic constant** *See* biokinetic coefficient.

**Biolac®** A wastewater treatment process developed by the Parkson Corp. using the extended aeration variation of activated sludge for BOD removal and nitrification. It combines long solids retention time with submerged aeration.

**BioLift® reactor** A vessel manufactured by Eimco Process Equipment Co. of Salt Lake City, UT; it uses an airlift, an auxiliary mixer, and diffused aeration to treat sludge and soil slurry mixtures.

**Biolift™** An apparatus manufactured by Humboldt Decanter, Inc. of Norcross, GA for thickening of activated sludge.

**Bio*Lime™** An agricultural liming agent made by Wheelabrator Clean Water Systems, Inc.

**Biolite®** The proprietary packing used in the Biofor® process; an expanded clay material of 2–4 mm in diameter, with a density greater than 1.0.

**Biologic™** A nutrient supplement produced by SciCorp Systems, Inc. for use in wastewater treatment.

**biological accumulation** Same as bioaccumulation. *See also* biological magnification.

**biological activated filter (BAF)** An upflow or downflow granular filter used for biological wastewater treatment. Microorganisms attached to the media degrade the organic matter in the wastewater as it passes through the filter. *See also* biological aerated filter and biological filter.

**biological activated carbon** Same as biologically enhanced activated carbon.

**biological additive** A culture of bacteria, enzymes, or nutrients used to promote the decomposition of oil discharges or other wastes.

**biological aerated filter (BAF)** An upflow filtration unit where the submerged medium provides both a surface for biological activity and a means of solids separation. Ordinarily, a BAF follows primary settling and includes fine-bubble aeration introduced at the bottom of the medium. BAF is a small-footprint technology that allows loading rates 15 times higher than conventional processes (e.g., more than 250 pounds of $BOD_5$ per day per 1000 cubic feet). It has some nitrogen removal capability and does not produce filamentous or bulking sludge. There are also similar filtration units operating in a downflow mode; see Biocarbone® process.

**biological aeration** Any of the variations of the activated sludge and related processes that use an aeration basin to treat wastewater under aerobic conditions. *See* conventional activated sludge, step aeration, contact stabilization, high-purity oxygen, extended aeration, completely mixed aeration, aerobic lagoon.

**biological amplification** Same as biological magnification.

**biological analysis** The examination of a sample to determine the presence of macroscopic and microscopic organisms.

**biological assessment** A study to determine the effects of wastewater effluent disposal or other actions on aquatic life and other environmental aspects. Assessment factors include accumulation of toxins in fish and shellfish; species diversity, productivity and stability; and human health. *See also* bioassay, biological monitoring.

**biological assimilation** *See* bacterial assimilation.

**biological bed** A more appropriate term proposed for the trickling filter to acknowledge the microbial oxidation mechanism of the process instead of the straining action of a filter.

**biological characteristics (wastewater)** Wastewater constituents that have a biological significance because of the role they play in treatment processes or in the transmission of human diseases. *See* wastewater microorganisms (algae, archaea, bacteria, fungi, protozoa, rotifers, and viruses), pathogenic organism, indicator organism, bacteria enumeration and identification, virus enumeration and identification.

**biological–chemical process** A wastewater treatment method designed to remove organic matter and phosphorus. It consists of a primary clarifier, an aerobic basin, a secondary clarifier, a phosphate stripper, and a lime mixing tank. The final effluent exits from the secondary clarifier. The primary sludge is withdrawn for disposal, while the phosphorus-enriched sludge from the secondary clarifier is withdrawn (excess sludge) or routed to the phosphate stripper. The phosphorus-stripped sludge is recycled through the stripper and through the aeration basin, whereas the stripper's supernatant is recycled ahead of the primary clarifier through the lime mixing tank.

**biological classification** The description, identification, naming and classification of organisms, based on structural criteria and according to a hierarchy, e.g., species, genus, family, etc. *See also* environmental classification, phylogenetic classification, and taxonomic classification.

**biological classification conventions** A set of conventions generally adopted for the designation of the species of living organisms. The plant and animal kingdoms are divided into phyla (plural of phylum), subdivided into classes, orders, families, genera (plural of genus), and species. Also called binomial nomenclature, biological species nomenclature.

**biological colloid** *See* biocolloid.

**biological concentration** The mechanism by which organisms concentrate persistent substances (e.g., pesticides and heavy metals) that are present in dilute concentrations in seawater, freshwater, or wastewater. *See* biological accumulation.

**biological contactor** (1) Same as biocontactor. (2) *See* rotating biological contactor (RBC).

**biological contact process** A high-rate biological process that uses a small contact chamber and a short detention time to treat wet-weather flows by mixing them with mixed liquor or return activated sludge without chemical addition. Sedimentation occurs in a secondary clarifier. *See also* ballasted sedimentation.

**biological control** In pest control, the use of animals and other living organisms that eat or otherwise kill or outcompete pests. The introduction of sterile males in a population is also a biological control technique. Also called biocontrol. *See also* aquatic growth.

**biological conversion** *See* bioconversion.

**biological corrosion** Corrosion induced by microorganisms, e.g., corrosion of water pipes and storage tanks. *See also* concrete sewer corrosion.

**biological criteria** Numerical values or narrative expressions describing the reference biological integrity of aquatic communities that inhibit water of a given designated aquatic life use.

**biological cycle** *See* biocycle.

**biological debris** *See* biogenic particle.

**biological decay, decomposition,** or **degradation** The breakdown of complex, large organic molecules into small, simple molecules by microorganisms.

**biological degradability** *See* biodegradability.

**biological degradation** *See* biodegradation.

**biological denitrification** Conversion of nitrates ($NO_3^-$) and nitrites ($NO_2^-$) to nitrogen gas ($N_2$) in the absence of molecular oxygen ($O_2$) and the presence of an electron ($e^-$) donor such as carbon (C):

$$2\,NO_3^- \rightarrow N_2 + 3\,O_2 + 2\,e^- \qquad \text{(B-44)}$$

Biological denitrification is actually a two-step process: nitrification (aerobic) and denitrification (anoxic); the overall reduction of nitrates is a sequence of reactions that includes the formation of nitric oxide (NO) and nitrous oxide ($N_2O$):

$$NO_3^- \rightarrow NO_2^- \rightarrow NO \rightarrow N_2O \rightarrow N_2 \qquad \text{(B-45)}$$

Also called biological nitrogen removal. *See also* the following related terms: A²/O™ process, Anammox bacteria, Anammox process, anoxic/aerobic process, assimilating (assimilatory) nitrate reduction, attached growth denitrification, Bardenpho™ process, Biocarbone® process, Bio-Denipho™ process, Bio-Denitro™ process, Biofor® process, biological nitrogen removal reactions, Bionutre™ process, Biostyr® process, bump, deep-bed filter denitrification, denitrification filter, denitrifying bacteria, dissimilating (dissimilatory) nitrate reduction, downflow packed-bed denitrification filter, dual-sludge process, filter bumping, five-stage Bardenpho™ process, fluidized-bed denitrification, fluidized-bed reactor, intermittent aeration, Johannesburg process, low-DO oxidation ditch, Ludzack–Ettinger process, modified Bardenpho™ process, modified Ludzack–Ettinger process, modified UCT process, Orbal™ process, phased isolation ditch, Phoredox process, Phostrip™ process, PhoStrip™ II process, postanoxic denitrification, preanoxic denitrification, prestripper tank, simultaneous nitrification–denitrification, submerged rotating biological contactor, substrate denitrification, SymBio™ process, Tetra® filter, Trio-Denipho™ process, triple-sludge process, UCT process, VIP process, Wuhrmann process.

**biological dinitrogen fixation** The microbial conversion of nitrogen gas (i.e., dinitrogen, $N_2$) to ammonia by nitrogen-fixing organisms such as bacteria, cyanobacteria, and actinomycetes. *See also* diazotroph and nitrogen fixation.

**biological disk** Another name for rotating biological contactor.

**biological diversity** Same as biodiversity.

**biological engineering** *See* bioengineering.

**biological film** *See* biofilm.

**biological filter** (1) A bed of sand, stone, or other media through which water or wastewater flows.

The liquid develops a microbial film that contributes to the removal of fine particulate and dissolved solids. Various terms are used to designate similar devices: biofilter, biological activated filter, biological aerated filter, biologically active filter. *See also* biotower, biotrickling filter, compost biofilter, trickling filter. (2) A similar device (biofilter) used in air pollution or odor control.

**biological filtration** (1) A water treatment method that combines particle removal through granular filtration and removal of organic matter by microorganisms grown on the filter media, thereby reducing the concentration of precursors and the production of disinfection-by-products. It can be designed for nitrification and nitrate removal. The process can also use fluidized beds or granular activated carbon beds. Also called biofiltration. (2) Same as trickling filtration. (3) An air pollution control process in which a gas stream passes through a bed of biologically active media and attached microorganisms break down organic and inorganic compounds into carbon dioxide, water, inorganic acids, and microbial biomass. *See also* bioremediation, bioventing, air sparging.

**biological floc** In suspended-growth biological wastewater treatment, a settleable clump or mass of microorganisms held together by the extracellular biopolymers that they produce.

**biological flocculation** Same as bioflocculation.

**biological flotation** A method used to concentrate sludge solids by letting decomposition gases lift them to the surface; for example, the flotation of primary sludge in 5 days at 35°C and withdrawal of the supernatant as a by-product (Fair et al., 1971).

**biological fluidized-bed reactor** A vessel containing activated carbon supporting a biofilm for the treatment of industrial wastewater or contaminated groundwater; it combines microbial degradation and adsorption processes. *See* biologically activated carbon and PACT® system.

**biological fouling** The development of microbial growth on a surface, which may result in the clogging of filter media, filtration membranes, or well casings. Biological fouling may also affect the capacity of heat exchangers. *See* biofouling for detail.

**biological fuel** *See* biofuel.

**biological gas** *See* biogas.

**biological growth** (1) The activity and growth of any living organism. (2) A minor mechanism that contributes to the removal of particulate matter within a granular filter by reducing pore volume; also called surface biological adsorption. *See* filtration mechanisms.

**biological hazard** *See* biohazard.

**biological indicator** A species or organism that produces a specific response on exposure to a given substance or to certain conditions. Biological indicators are used to assess environmental quality or change. *See,* e.g., phytoplankton quotient and A/C index. Also called bioindicator.

**biological integrity** The condition of the aquatic community that inhabits unimpaired water bodies of a specified habitat as measured by community structure and function.

**biological iron** Iron that is involved in a redox reaction of a biological process.

**biological iron and manganese removal** A process that uses filters supporting the growth of iron- and manganese-oxidizing bacteria for the removal of these metals from water or wastewater (one filter for each metal, with appropriate pH and ORP conditions). Sometimes abbreviated as bio-Fe or bio-Mn removal.

**biological kinetics** The study of growth patterns, particularly growth rates, of organisms. *See* Monod equation.

**biologically activated carbon (BAC)** Same as biologically enhanced activated carbon.

**biologically active carbon (BAC)** Same as biologically enhanced activated carbon.

**biologically active filter** Same as biological filter.

**biologically active filtration** Same as biological filtration.

**biologically active floc** Activated sludge or other floc formed by microorganisms.

**biologically available** Pertaining to a substance that an organism can use directly or immediately.

**biologically degradable** *See* biodegradable.

**biologically enhanced activated carbon (BAC)** Granular activated carbon used as a filter medium and on which microorganisms grow for water or wastewater treatment. Such a unit removes contaminants through carbon adsorption and microbial degradation; *see,* e.g., biological fluidized-bed reactor. Note that granular or powdered activated carbon fines sometimes provide a transport mechanism for coliform bacteria to discharge into treated water or wastewater. Also called biological activated carbon, biologically activated carbon, biologically active carbon. *See also* steady-state removal.

**biologically pure water** *See* biopure water.

**biologically refractory** Difficult or slow to degrade biologically.

**biologically stable water** Treated water that does not support a significant microbial growth in the distribution system. *See* regrowth.

**biological magnification** The process whereby certain persistent substances such as pesticides, organochlorine compounds, or heavy metals move up the food chain, work their way into rivers or lakes, and are eaten by aquatic organisms such as fish, which in turn are eaten by large birds, animals, or humans. The substances become concentrated in tissues or internal organs as they move up the chain; e.g., an aquatic plant that contains a heavy metal concentration 100 times higher than water is consumed by a mollusk that concentrates the metal by a factor of 10, and so on through detritovores and carnivores. Also called biomagnification. *See* bioaccumulation.

**biological manipulation** *See* biomanipulation.

**biological marker** *See* biomarker.

**biological mass** *See* biomass.

**biological mat** *See* biomat.

**biological metal removal** Removal of heavy and other metals in biological wastewater treatment processes by adsorption or complexation with microorganisms.

**biological microconstituent** A microorganism. Pathogens of most concern in drinking water include bacteria (*Aeromonas, Cyanobacteria, Helicobacter, Legionella, Mycobacterium avum,* pathogenic *E. coli*), protozoa (*Cryptosporidium, Giardia,* Microsporidia, *Toxoplasma*), and viruses (Adenovirus, Calicivirus, Coxsackievirus, Echovirus).

**biological mineralization** *See* biomineralization.

**biological molecule** *See* biomolecule.

**biological monitor** *See* biomonitor.

**biological monitoring** (1) The determination of the effects on aquatic life, including accumulation of pollutants in tissue, in receiving waters due to the discharge of pollutants: (a) by techniques and procedures, including sampling of organisms representative of appropriate levels of the food chain, considering the volume and the physical, chemical, and biological characteristics of the effluent, and (b) at appropriate locations and frequencies (Clean Water Act). (2) Also defined as the use of living organisms to test the suitability of effluents for discharge into receiving waters and to test the quality of such waters downstream from the discharge. (3) Analysis of blood, urine, tissues, etc., to measure chemical exposure in humans. (4) An emerging technology that uses living organisms to detect environmental contamination. Also called biomonitoring. *See also* biomonitor, bioassay, biological assessment.

**biological nitrification** The conversion of ammonia nitrogen ($NH_4$-N) to nitrates ($NO_3^-$) by microorganisms.

**biological nitrification–denitrification** The conversion of ammonia nitrogen to nitrites, nitrates, and finally nitrogen gas by microorganisms, a process used to remove nitrogen from wastewater. *See* biological nitrogen removal.

**biological nitrification reactions** Nitrification involves the conversion of ammonia nitrogen ($NH_4$-N) to nitrite $NO_2^-$) and then nitrate ($NO_3^-$), which requires a certain amount of alkalinity ($HCO_3^-$) and results in the production of cell tissue ($C_5H_7O_2N$) as well as energy. The total oxidation and cell synthesis equations are respectively:

$$NH_4^+ + 2\,O_2 \rightarrow NO_3^- + 2\,H^+ + H_2O \qquad \text{(B-46)}$$

$$NH_4^+ + 4\,CO_2 + HCO_3^- + H_2O \rightarrow C_5H_7O_2N + 5\,O_2 \qquad \text{(B-47)}$$

**biological nitrogen fixation** The conversion by microorganisms of elemental nitrogen ($N_2$, gas) from the air to organic or available nitrogen; e.g., when cyanobacteria (or blue-green algae) abstract nitrogen from the atmosphere.

**biological nitrogen removal (BNR)** The removal of nitrogenous compounds from wastewater using microbial activity. *See* biological (or bacterial) assimilation and nitrification-denitrification.

**biological nitrogen removal reactions** The overall biochemical reactions occurring in biological denitrification, depending on (Metcalf & Eddy, 2003 and WEF & ASCE, 1991) (a) the source of electrons (carbon) such as carbon dioxide ($H_2CO_3$), soluble organic matter from the influent or from endogenous decay as represented by the formula $C_{10}H_{19}O_3N$, an exogenous source such as methanol ($CH_3OH$), or acetate ($CH_3COOH$); (b) nitrate ($NO_3^-$) and/or ammonia ($NH_4^+$) as the nitrogen source; and (c) whether or not cell synthesis ($C_5H_7O_2N$) is included:

$$5\,CH_3OH + 6\,NO_3^- \rightarrow 3\,N_2 + 5\,CO_2 + 7\,H_2O + 6\,OH^- \qquad \text{(B-48)}$$

$$1.08\,CH_3OH + NO_3^- + 0.24\,H_2CO_3 \rightarrow 0.47\,N_2 + 0.056\,C_5H_7O_2N + 1.68\,H_2O + HCO_3^- \qquad \text{(B-49)}$$

$$C_{10}H_{19}O_3N + 10\,NO_3^- \rightarrow 5\,N_2 + 10\,CO_2 + 3\,H_2O + NH_3 + 10\,OH^- \qquad \text{(B-50)}$$

$$5\,CH_3OH + 2\,NO_3^- + NH_4^+ + H_2CO_3 \rightarrow N_2 + C_5H_7O_2N + 9\,H_2O + HCO_3^- \qquad \text{(B-51)}$$

$$5\,CH_3COOH + 8\,NO_3^- \rightarrow 4\,N_2 + 10\,CO_2 + 6\,H_2O + 8\,OH^- \qquad \text{(B-52)}$$

$$0.345\ C_{10}H_{19}O_3N + NO_3^- + H^+ + 0.267\ HCO_3^- + 0.267\ NH_4^+ \rightarrow 0.5\ N_2 + 0.655\ CO_2 + 0.612\ C_5H_7O_2N + 2.3\ H_2O \quad \text{(B-53)}$$

**biological nomenclature** The scientific method of naming all organisms, except viruses, using two Latinized words usually written in italics or sometimes underlined. *See* biological species nomenclature for detail.

**biological nutrient removal (BNR)** Removal of nutrients, mainly nitrogen and phosphorus, in biological wastewater treatment processes. *See* biological nitrogen removal and biological phosphorus removal.

**biological odor control** Odor management in treatment plants and other installations using biological methods. In activated sludge reactors, odorous gases can be combined with process air to remove compounds responsible for odor. In trickling filters, odorous compounds can be removed by passing the gases through the media bed. *See* biofilters, biotrickling filters, and compost filters.

**biological oxidation** Decomposition of complex organic materials by microorganisms into a more stable or mineral form. It occurs in self-purification of water bodies, in activated sludge, in trickling filters and other biological wastewater treatment processes.

**biological oxygen demand (BOD)** An indirect measure of the concentration of biologically degradable material present in organic wastes. It usually reflects the amount of oxygen consumed in five days by biological processes breaking down organic waste. The more commonly used term is biochemical oxygen demand.

**biological phosphorus removal (BPR)** The removal of phosphorus compounds from wastewater using microbial activity; phosphorus accumulates in cell biomass and is subsequently removed during solids separation and wastage. Some bacteria can store phosphorus in excess of metabolic requirements. Many BPR flowsheets include a sequence of anaerobic, anoxic, and aerobic reactors. *See* biological phosphorus removal model, enhanced phosphorus. *See also* activated algae, A/O process, $A^2O$ process, biological (or bacterial) assimilation, biological–chemical process, luxury uptake, Phoredox process, phosphorus-accumulating organisms, PhoStrip™ process, UCT process.

**biological phosphorus removal (BPR) biosolids** Biosolids produced via a biological phosphorus removal process, which usually have greater total phosphorus and soluble phosphorus concentrations than non-BPR biosolids.

**biological phosphorus removal model** A parametric model that is based on extensive tests to predict the performance of enhanced phosphorus (P) uptake processes (WEF & ASCE, 1991):

$$R = S_0\{[(1 - F_1 - F_2)Y_h/(1 + \theta K_h)](\gamma + \theta F F_p K_h) + F_p F_2/F_3)\} \quad \text{(B-54)}$$

where $R$ = phosphorus removal by incorporation in the sludge, mg P/L; $S_0$ = total influent COD, mg/L; $F_1$ = fraction of influent nonbiodegradable soluble COD, mg/mg COD (e.g., 0.05); $F_2$ = fraction of influent nonbiodegradable particulate COD, mg/mg COD (e.g., 0.13); $Y_h$ = yield coefficient for heterotrophic organisms, mg VSS/mg COD (e.g., 0.45) (VSS = volatile suspended solids); $\theta$ = solids retention time, days; $K_h$ = decay coefficient for heterotrophic organisms, /day (e.g., 0.24); $\gamma$ = coefficient of excess phosphorus removal, mg P/mg VSS; $F$ = nonbiodegradable fraction of active mass, mg/mg VSS (e.g., 0.20); $F_p$ = phosphorus content of nonbiodegradable volatile solids, mg P/mg VSS (e.g., 0.015); and $F_3$ = ratio of COD to VSS = 1.48 mg COD/mg VSS.

**biological phosphorus storage** The accumulation by certain bacteria of phosphorus as polyphosphates in their cells in excess of their metabolic requirements.

**biological polymer** *See* biopolymer.

**biological population dynamics** *See* population dynamics.

**biological process** *See* biochemical.

**biological productivity** *See* bioproductivity.

**biological purification** The conversion of organic matter and other contaminants by living organisms into stable, mineral, or innocuous form.

**biological reactor** A vessel, tank, or other container in which biological reactions take place, e.g., the aeration basin of the activated sludge process or a stabilization pond; bioreactor.

**biological regeneration** *See* bioregeneration.

**biological regrowth** An increase in the number of bacteria and other organisms in treated water or wastewater. *See* regrowth for more detail.

**biological ripening period** The initial period of operation during which the microorganisms in a biological treatment process adjust to temperature and other environmental factors; it varies from a few weeks in the summer to a few months in the winter for trickling filters, and from a few days to a few weeks for activated sludge. *See also* filter ripening.

**biologicals** Vaccines, cultures, and other preparations made from living organisms and their products, intended for use in diagnosing, immunizing, or treating humans or animals, or in related research (EPA-40CFR259.10-b).

**biological score** A parameter designed to provide engineers and planners an idea about the type of biota in a stream with respect to pollution. In this parameter, the presence of any pollution tolerant species carries a weight of 1; that of any "facultative" species, 2; and any intolerant species, 3. The score of a medium varies from 0 (no life form can survive) to 6 (acceptable to all forms, e.g., an unpolluted stream). *See also* biotic index.

**biological scrubber** *See* bioscrubber.

**biological selector** A small contact tank included in a biological wastewater treatment plant to limit the growth of undesirable organisms or to promote the growth of selected organisms. *See* selector for detail.

**biological slime** The gelatinous film of microbial growth covering the surface of a medium or spanning the interstices of a granular bed; e.g., the biofilm that is produced when wastewater is sprayed over fixed media, as in a trickling filter, or when water flows through a slow sand filter. *See* slime for detail.

**biological sludge conditioning** The use of biological methods to improve wastewater sludge dewaterability, including aerobic digestion, anaerobic digestion, and composting.

**biological sludge solids** Solid particles in the sludge from biological wastewater treatment (e.g., activated sludge process or trickling filter) as compared to sludge from primary treatment. Their dry weight ($W_b$, pounds per day) is proportional to the applied BOD load in pounds per day:

$$W_b = K(\text{BOD}) \quad \text{(B-55)}$$

where $K$ is the fraction of applied BOD that appears as excess biological solids, a parameter that varies between 0.30 and 0.52 depending on the food-to-microorganism ratio (0.05–0.50).

**biological solids** *See* biosolids.

**biological sorption** *See* biosorption.

**biological sparging** *See* biosparging.

**biological species nomenclature** The scientific method of naming all organisms, except viruses, using two Latinized words, usually written in italics or sometimes underlined. The first word, with a capital initial letter, denotes the genus and the second word denotes the species; e.g., *Escherichia coli*. The second and subsequent times a genus is mentioned in a text, it is usually abbreviated; e.g., *E. coli*. Subspecies or varieties of a species are designated by the abbreviation "ssp" or by a third name. Serotyping allows further differentiation of these organisms; e.g., *E. coli* 0157:H7. *See also* biological classification conventions.

**biological stability** The condition of a treated water that does not support significant microbial growth in the distribution system. Ozonation may affect stability by making organic matter more available for microbial growth in parts of a distribution system with depleted disinfectant residual. *See* regrowth.

**biological stabilizer** *See* biostabilizer.

**biological statistics** *See* biostatistics.

**biological stimulant** *See* biostimulant.

**biological storage** *See* (a) biological phosphorus storage and (b) biostorage.

**biological stripping tower** Same as biotrickling filter.

**biological surfactant** *See* biosurfactant.

**biological survey** The collection, processing, and analysis of representative portions of a resident aquatic community to determine its structure and function.

**biological synthesis** The production of new cells by microorganisms during wastewater treatment or other biological processes. Also called biosynthesis.

**biological taxonomy** (1) The conventional classification of diseases according to the characteristics of the pathogens that cause them, as compared to an environmental classification. Also called generic classification. (2) The description, identification, naming and classification of organisms; also simply called taxonomy. For detail, see biological classification, biological classification convention.

**biological technology** *See* biotechnology.

**biological tower** A trickling filter that uses plastic packing, allowing taller devices, higher loading rates, and less land area. Also called biotrickling filter, or biotower. *See also* nitrification tower.

**biological toxin** *See* biotoxin.

**biological treatment** A treatment technology that uses bacteria and other microorganisms to consume organic waste. Biological waste treatment involves the combination of a series of mechanisms: the transfer of nutrients from the liquid waste to the biomass (film or floc), the oxidation of organic matter and synthesis of new cells, bioflocculation or the conversion of biomass into removable solids, and processing and disposal of the sludge produced.

**biological treatment processes** *See* the following types of treatment processes: (a) attached growth vs. suspended growth; (b) aerobic vs. anaerobic vs. anoxic vs. facultative; (c) lagoon processes; (d) combined attached growth/suspended growth; (e) combined aerobic/anaerobic/anoxic. Examples of specific processes include: bed irrigation, subsurface irrigation, intermittent sand filtration, trickling filtration, activated sludge, stabilization pond.

**biological treatment safety factor** A factor used in the design of biological treatment systems to take into account a number of uncertainties such as variability of wastewater flow, waste characteristics, and weather conditions. *See* safety factor for detail.

**biological treatment system** A wastewater treatment system that uses microorganisms in controlled natural processes to remove soluble and colloidal organic matter. The treatment system or treatment plant includes not only the biological reactors but also solids separation units (clarifiers), sludge processing, and other units.

**biological type** *See* biotype.

**biological unit process** A water or wastewater treatment method, such as biological oxidation in activated sludge or trickling filtration, in which biological activity brings about the removal or modification of constituents, as compared to physical unit operations and chemical unit processes. A wastewater treatment plant usually incorporates two or all three treatment methods.

**biological uptake** (1) The incorporation of nutrients such as nitrogen and phosphorus to meet normal metabolic requirements of microorganisms. *See* biological synthesis and luxury uptake. (2) The rate of dissolved oxygen utilization by microorganisms in wastewater treatment, a function of organic loading (e.g., food-to-microorganism ratio) and temperature. *See* oxygen uptake rate (OUR) for detail.

**biological vector** An agent involved in the transmission of an infection; it is actually infected by the pathogen. Contrary to a mechanical vector, the biological vector serves as a place for pathogens to develop and/or multiply. For example, the tsetse fly is a biological vector of African sleeping sickness.

**biological venting** *See* bioventing.

**biological waste** Liquid or solid waste from decomposing organic matter (e.g., organic foods and human and animal bodies) and hazardous wastes from hospitals and similar institutions. They may transmit diseases through flies, mosquitoes, and rodents. *See also* hospital waste.

**biological wastewater treatment** A treatment technology that uses bacteria and other microorganisms to consume organic matter in wastewater. *See* biological treatment.

**biology** (1) The science of life and living organisms in all aspects: origin, structure, growth, reproduction, etc.; subdivided into zoology (animals), botany (plants), and microbiology (microorganisms). (2) The living organisms in a stream, lake, or other defined area.

**biomagnification** Same as biological magnification.

**biomanipulation** A lake restoration technique, against the effects of eutrophication, based on the theory that increasing the population of predators will reduce the abundance of planktivores, increase the zooplankton population, and eventually reduce algal biomass and improve water clearness.

**biomarker** A biological indicator used in assessing exposure to toxic and hazardous materials. Common biomarkers include blood enzyme level (e.g., cholinesterase), blood lead level, DNA adducts, hair mercury level, organochlorine fat level, protein adducts, and urinary phenol level (Freeman, 1998).

**biomass** (1) All of the material that is or was a living organism or was excreted from an organism in a given area or volume; it sometimes refers to vegetation generated photosynthetically from carbon dioxide ($CO_2$) and water ($H_2O$) and represented by the formula $CH_2O$:

$$CO_2 + H_2O + E \rightarrow CH_2O + O_2 \qquad \text{(B-56)}$$

where E represents solar energy. (2) The total weight of all biological matter, including extracellular polymeric materials. *See also* microbial biomass. (3) In wastewater treatment, the biomass is the sum total of all living organisms in a reactor or in a process. The consumption and degradation of substrate during biological treatment continuously produces cell biomass, usually represented by the formula $C_{60}H_{87}O_{23}N_{12}P$. *See also* activated sludge composition, active biomass.

**biomass carbon** *See* microbial biomass and Anderson–Domsch equation.

**biomass fuel** A carbon-based, renewable fuel, (e.g., alcohol, methanol, ethanol, wood), derived from biomass, corn, and other grains, as compared to fossil fuel. Also called biofuel.

**biomass growth** In biological treatment, biomass growth is usually measured by volatile suspended solids or particulate COD. Other measures include protein content, DNA, and adenosine triphosphate (ATP). *See* growth pattern.

**biomass solids** The solids in a biological reactor, including the biomass as well as nonbiodegradable volatile suspended solids and inert inorganic suspended solids.

**biomass yield** In biological processes, organic matter is converted to new microorganisms and end products. Bacterial yield or biomass yield is the ratio of the amount of biomass produced to the amount of substrate consumed.

**biomat** In a subsurface-soil absorption system, the layer of biological material formed by microorganisms at the interface between the soil and the effluent particulates. It is a primary factor controlling the soil infiltration capacity, more so than soil permeability. The biomat grows to an equilibrium thickness, serving as an effective mechanical and biological filter. *See* long-term acceptance rate (2).

**BioMatrix® process** A proprietary integrated fixed-film activated sludge method of wastewater treatment that uses a fixed packing in the aeration tank. *See* Bio-2-Sludge®, Captor®, Kaldnes®, Linpor®, moving bed biofilm reactor, Ringlace®, submerged rotating biological contactor.

**Bio Max™** A floating piping apparatus manufactured by Environmental Dynamics, Inc.

**biome** The entire community of living organisms in a single major ecological area, with distinctive flora and fauna, specific habitat and other characteristics, especially a community that has reached a climax. Examples: desert, rain forest, grasslands, tundra. *See also* biotic community.

**biometrics, biometry** (1) Biostatistics; the calculation of the probable duration of human life. (2) The application of statistical methods to biological observations.

**biomineralization** The process by which organisms extract and selectively precipitate inorganic elements from the environment; examples include the formation of silica ($SiO_n$) gels in diatoms or calcium carbonate ($CaCO_3$) shells in mollusks (Stumm & Morgan, 1996).

**Biomizer** A sequencing reactor developed by Environmental Dynamics, Inc.

**bio-Mn** *See* biological iron and manganese removal.

**Bio-Module** A package plant manufactured by Envirex, Inc., using rotating biological contactors for wastewater treatment.

**BioMonitor™** An automated BOD analyzer made by Anatel Corp.

**biomolecules** The molecules that constitute matter in living organisms, often polymers with large molecular masses: carbohydrates, nucleic acids, and proteins.

**biomonitor** A monitor that relies on the responses or properties of living organisms to detect the presence of a contaminant in the environment. Biomonitors are used, for example, to measure changes that result from stresses exerted by toxic materials. They include the canaries used to detect the presence of toxic gases in mines, dynamic fish used in commercial tanks, and the water flea used in the Daphnia test.

**biomonitoring** *See* biological monitoring.

**Biomphalaria** The genus of snails that carry and transmit the pathogen of schistosomiasis, *Schistosoma mansoni.*

**Bio-Net®** A wastewater treatment system developed by NSW Corp., using rotating biological contactors.

**BIONOx™** A submersible apparatus manufactured by Framco Environmental Technologies used for aeration and mixing.

**Bionutre™ process** A version of the Orbal™ process, which is a variation of the oxidation ditch process that uses three concentric channels in series and disk aerators on a horizontal shaft. Developed by Envirex, Inc., it is designed to achieve BOD removal and simultaneous nitrification/denitrification (SNdN). Bionutre™ operates with limited aeration in the first channel to produce anoxic conditions and with or without internal recycle from the third to the first channel. *See* Orbal™ process.

**Bio-Nutri™** A wastewater treatment process developed by Smith & Loveless, Inc. for the removal of nitrogen and phosphorus.

**Bio-Ox™** A reactor manufactured by SRE, Inc. for the biological treatment of wastewater.

**biooxidation** Same as biochemical oxidation.

**Bio-Pac® media** A product of NSW Corp. for use in wastewater treatment by trickling filters.

**Bio Pac plant** A package trickling filter plant manufactured by the FMC Corp.

**Biopaq®** A sludge blanket process used by CBI Walker, Inc. to treat concentrated wastewaters.

**biophysical wastewater treatment** A process that uses activated carbon as an adsorbent and a surface for biological growth in the treatment of wastewater. The flowchart includes an aeration basin, a clarifier, a sand filter, a carbon regeneration unit (using low-pressure wet oxidation), and a thickener for the spent carbon, as well as arrangements for the addition of makeup carbon and the recirculation of the carbon sludge.

**bio-P microorganism** An organism, most commonly a species of the *Acinetobacter* genus, that is active in the process of enhanced phosphorus uptake; it accumulates phosphorus in excess of its growth requirements.

**Biopol** A biodegradable plastic developed around

1990 and marketed by Marlborough Biopolymers Ltd. (an ICI subsidiary in England); built in microbial cells from organic compounds (polyhydroxybutyrate and polyhydroxyvalerate) and water. It constitutes 80% of the cell tissue and the remainder can break down back into carbon dioxide and water after landfill disposal.

**biopolymer** A polymer directly produced by living or once-living cells or cellular components (EPA-40CFR723.250-3).

**BioPro Designer** A process simulator for biochemical processes. *See* BatchPro Designer.

**bioproductivity** The amount of organic matter produced by living organisms in an ecosystem.

**Bio-Pure®** A wastewater treatment system designed by Aqua-Clear Technologies Corp. for reclamation purposes.

**Biopurge** A program developed by Stiles-Kem Division MetPro Corp. of Waukegan, IL for the control of biofilm problems in water treatment installations.

**biopure water** Sterile water that has a very low concentration of total solids.

**bioreactor** (1) A fermentation vat for the production of bacteria, yeast, or other living organisms. (2) A biological reactor, i.e., a vessel, tank, or other container in which biological reactions take place, e.g., the aeration basin of the activated sludge process or a stabilization pond. (3) A sealed vessel used as a growth container in continuous culture, also called a chemostat, in which environmental conditions can be controlled (substrate concentration, flowrate, pH, temperature, oxygen concentration, etc.)

**Bio-Reel™** A wastewater treatment process developed by the Schreiber Corp. using tubing.

**biorefractory organics** Low-molecular-mass compounds of low volatility, such as the organochlorine compounds sometimes found in drinking water and responsible for taste and odor problems. They resist biodegradation and tend to persist and accumulate in the environment. Also called recalcitrant organics.

**biorefractory substance** *See* biorefractory organics.

**bioregeneration** Regeneration of an adsorbent by microorganisms that consume the accumulated contaminants.

**bioremediation** Use of living organisms to clean up oil spills or remove other pollutants from soil, water, or wastewater; use of organisms such as nonharmful insects to remove agricultural pests or counteract diseases of trees, plants, and garden soil. The organisms use the waste materials in their metabolism and convert them to harmless end products. Bioremediation has achieved some success because it uses a natural process, destroys or immobilizes contaminants instead of transferring them from one place to another, and is often cost-effective and less time-consuming than other alternatives. *See also* ex situ bioremediation, in situ bioremediation, intrinsic bioremediation or natural attenuation, remedial action.

**bioretention** (1) A stormwater management practice that uses native plants and soil conditioning to capture and treat sheet flow from impervious areas. It includes a ponding area over the root zone of the plants, a sandbed to drain and aerate the root zone, and an organic layer on the surface of the soil. (2) Same as bioaccumulation.

**Bio-S®** A surfactant produced by Medina Products and used in bioremediation.

**Bio/Scent®** A liquid product of Hinsilblon, Ltd. for neutralizing odors.

**Bioscrub** An apparatus manufactured by the CMS Group for the control of odors and volatile organic compounds in rotating biological contactors.

**bioscrubber** A device that combines biodegradation and absorption in one system to control odors at wastewater treatment plants and pumping stations. *See* AroBIOS™ bioscrubber, Bioscrub, Bioscrubbers™.

**Bioscrubbers™** Equipment manufactured by WRc Process Engineering for odor removal by microorganisms.

**Bio-Separator** A flow diversion baffle manufactured by ThermaFab, Inc. for use in wastewater treatment lagoons.

**Bio-Sock** *See* Biosock.

**Biosock™** An apparatus designed by Sybron Chemicals, Inc. for the introduction of bacterial cultures into flows, including a fabric sock or Bio-Sock.

**biosolids** Primarily organic, semisolid by-products of water or wastewater treatment that have been treated for beneficial use or that can be beneficially recycled, as compared to solids in general or sludge, which are treatment residuals that have not been processed. *See also* bulk biosolids, chemical sludge, Class A biosolids, Class B biosolids, exceptional quality biosolids, monofill, primary sludge, process to further reduce pathogens, process to significantly reduce pathogens, secondary sludge, waste activated sludge, vector attraction reduction.

**biosolids pelletization** A sludge handling method using heat drying to produce sludge particles of 1 to 4 mm in diameter that can be used as a fertilizer. *See also* heat drying-pelletization.

**biosorption process** Commercial name of the equipment of the contact stabilization variation of the activated sludge process.

**biosparging** The injection of nutrients and oxidants mixed in a compressed air stream into the saturated zone to enhance microbial activity for the destruction of contaminants; sometimes used in conjunction with soil-vapor extraction. The air is injected through a sparger or porous bubbler. *See also* air sparging and bioventing (Freeman, 1998).

**biosphere** The space between Earth and its atmosphere that can support life; the mass of living organisms on the Earth's surface.

**BioSpiral** A rotating biological contactor manufactured by Walker Process Equipment Co.

**Bio-Sponge** *See* JAWS Bio-Sponge™.

**biostabilizer** A machine that converts solid waste into compost by grinding and aeration.

**Biostart™** A liquid concentrate of Advanced Microbial Systems, Inc..

**biostat** A substance that inhibits the growth of organisms but does not destroy them.

**biostatistics** The application of statistics to biological and medical data; medical statistics; biometrics or biometry.

**biostimulant** An essential substance or element that promotes the growth of organisms, like the major nutrients nitrogen and phosphorus that stimulate the growth of algae and the eutrophication of receiving waters.

**biostimulation** An in situ biological treatment process in which oxygen and nutrients are added to promote the growth of indigenous microorganisms that can degrade the responsible compounds in the contaminated soil or groundwater (Freeman, 1998).

**biostorage** *See* biomagnification and biological phosphorus storage.

**Biostyr® process** A wastewater treatment process of Kruger, Inc. developed in Denmark for the removal of BOD, nitrogen, and suspended solids. It uses an upflow reactor packed with 2–4 mm polystyrene beads that have a specific density lower than water. *See* submerged attached growth process for related methods, Figure B-15.

**Bio-Surf** A wastewater treatment process developed by Envirex, Inc. using rotating biological contactors.

**biosurfactant** A water-soluble, low-molecular-weight, surface active agent produced in situ by microorganisms and used to enhance the biological treatment of organic-contaminated soils; effective in the treatment of hydrocarbons and in the complexation of heavy metals.

**biosurvey** Biological survey.

**biosynthesis** The production of new cells by microorganisms during wastewater treatment or other biological processes. Also called biological synthesis.

**biota** The animal and plant life of a given region or system; flora and fauna.

**Biotac™** An apparatus manufactured by Davis Industries using bacteria for bioremediation in wet wells.

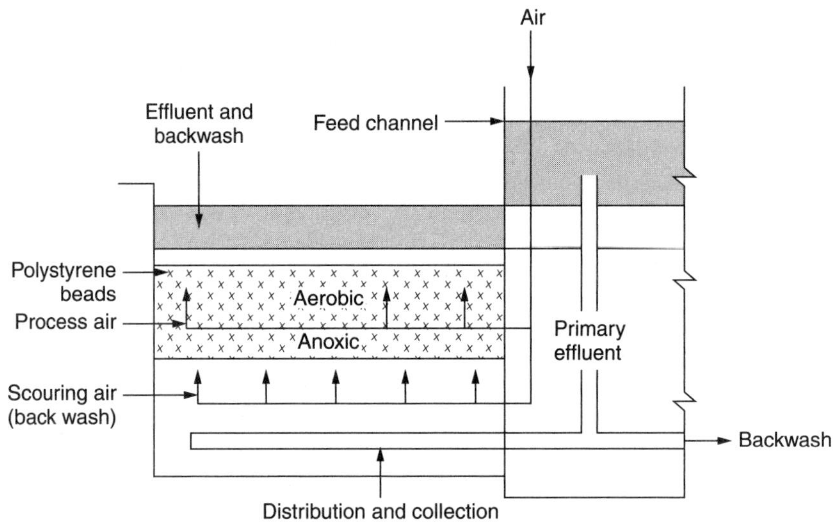


**Figure B-15.** Biostyr® process.

**biotechnology** Applied biological science, i.e., techniques that use living organisms or parts of organisms to produce a variety of products (from medicines to industrial enzymes) to improve plants or animals, or to develop microorganisms to remove toxics from bodies of water, or act as pesticides. *See also* bioengineering.

**Bio-terrorism Act** A U.S. federal law passed in 2002, requiring water supply systems that serve more than 3300 people to conduct vulnerability assessments and have emergency response plans.

**Biothane®** An anaerobic treatment process developed by Biothane Corp.

**biotic** Living or biological.

**biotic community** A naturally occurring assemblage of plants and animals that live in the same environment and are mutually sustaining and interdependent. Also called biocenosis. *See* biome.

**biotic environment** The plants, animals, and microorganisms.

**biotic factor** An influence on the environment or a variable affecting an ecosystem through the activities of living organisms, as compared to climatic or other physical influences.

**biotic index** A rating used to assess environmental quality in general, e.g., species diversity; the greater the diversity, the lower the degree of pollution. For example, a water body can have an index that reflects the amount of dissolved oxygen present or the type of organisms (pollution-tolerant or pollution-sensitive) it supports. *See also* biological score.

**biotic influence** *See* biotic factor.

**biotic potential** The capacity of a biotic community or a population of organisms to grow under optimum environmental conditions.

**biotite [$KMgFe_2AlSi_3O_{10}(OH)_2$]** A common potassium–iron, mica mineral, occurring in black, dark brown, or dark green sheets or flakes in igneous rocks. It dissolves in water to form gibbsite [$Al(OH)_3$] and ferric hydroxide [$Fe(OH)_3$]:

$$KMgFe_2AlSi_3O_{10}(OH)_2 + \tfrac{1}{2}O_2 + 3\ CO_2 + 11\ H_2O = Al(OH)_3 + 2\ Fe(OH)_3 + K^+ + Mg^{2+} + 3\ HCO_3^- + 3\ H_4SiO_4 \quad \text{(B-57)}$$

**Bioton®** An apparatus manufactured by PPC Biofilter for the control of volatile organic compounds and odors.

**biotope** A uniform portion of a habitat in terms of physical and biotic characteristics; the smallest geographical unit of the biosphere or of a habitat, characterized by its biota, that can be defined by convenient boundaries.

**biotower** A trickling filter that uses plastic packing, allowing taller devices, higher loading rates, and less land area. Also called biotrickling filter, or biological filter. *See also* nitrification tower.

**Biotox®** A process developed by Biothermica International for membrane regeneration using thermal oxidation.

**biotoxin** A toxic product of biological activity, e.g., aflatoxin, and plant and marine toxins.

**biotransfer factor (BTF)** The ratio of the concentration of a substance in tissue (in mg/kg) to its daily intake by an organism (in mg/day). It is used as a measure of bioaccumulation from food and water by land animals and correlates well with octanol–water partition coefficients.

**biotransformation** Metabolism of a substance, i.e., its conversion into other compounds by organisms; includes biodegradation.

**biotreatment** Same as biological treatment.

**biotrickling filter** A biological filter or biological stripping tower, particularly a unit filled with plastic packing for the support of microbial growth and designed to remove odors. Same as biofilter. *See also* biotower.

**bioturbation** Contraction of biological perturbation; the phenomenon that occurs at wastewater treatment plant discharges where benthic organisms facilitate the dispersion of pollutants, nutrients, and oxygen between the sediment and water.

**biotype** A group of organisms having the same genotype (genetic makeup), or a distinguishing feature of the genotype; e.g., the classical and El Tor biotypes of the cholera vibrio. *See* serotype.

**bioventing** An in situ bioremediation technique that involves sinking wells into relatively porous soils and drawing air down; it is applicable when the vadose or unsaturated zone is contaminated. *See also* biosparging.

**biowall concept** The idea of containing groundwater contamination by establishing a biological, permeable treatment barrier in the flow path of a contaminant plume. *See also* iron redox barrier.

**Biox™** A package plant manufactured by Bioscience, Inc. for water treatment.

**Bioxide®** An apparatus manufactured by Davis Industries, using a nitrate-based solution for the control of odors in sewer collection systems.

**biphenyl ($C_{12}H_{10}$)** A compound of two phenyl groups; a colorless powder or white scale, insoluble in water. Also called diphenyl or phenylbenzene.

**biradical** An atom or molecule that has two unpaired electrons; a diradical. *See also* free radical.

**Birm™** A medium produced by the Clack Corp. and used in granular filters for the removal of iron and manganese from water.

**bisulfate** A common radical that indicates a salt of sulfuric acid containing the $HSO_4^-$ group, e.g., sodium bisulfate ($NaHSO_4$).

**bisulfite** A common radical that indicates a salt of sulfurous acid ($H_2SO_3$) containing the $HSO_3^-$ group, e.g., sodium bisulfite ($NaHSO_3$).

**bitter lake** A lake that is rich in magnesium sulfate ($MgSO_4$); an example of an extreme environment. *See also* borax lake and soda lake.

**bittern** (1) The saturated brine solution remaining after precipitation of sodium chloride (NaCl) in the solar evaporation process (EPA-40CFR415.161-c). It consists mainly of calcium and magnesium chlorides. *See also* mother liquor. (2) A solution substantially freed from undissolved matter by a solid/liquid separation process such as decanting or filtration.

**bituminous-based activated carbon** Granular or powdered activated carbon produced from bituminous coal.

**bituminous coal** Solid fossil fuel classified as bituminous coal by ASTM Designation D38877 (EPA-40CFR60.251-b). It is high in carbonaceous matter (volatile hydrocarbons) and yields much volatile waste matter. Bituminous coal is one of the materials that can be used to make activated carbon for the treatment of water or wastewater.

**bivalent ion** (1) An ion having a valence charge of 2, like calcium ($Ca^{2+}$) or sulfate ($SO_4^{2-}$). (2) An ion with two valences, like chromium ($Cr^{3+}$ or $Cr^{6+}$) or iron ($Fe^{2+}$ or $Fe^{3+}$).

**bivalve** A mussel, clam, or any other mollusk with two shells hinged together and a soft body.

**bivinyl** Another name of butadiene.

**black alum** Alum that contains some activated carbon, used as a coagulant.

**blackbirds** Char emanating from chimneys, flues, and stacks; the combustible particles in flue gas. *See also* fly ash.

**black-box model** (1) Same as mass balance model. In general, a black box is an electronic component with known input and output that can be easily inserted into or removed from a larger system without knowledge of the component's internal structure. (2) A representation of mixing phenomena or other physical processes by conceptual, empirical or statistical relationships. Also called an input–output model.

**Black Death** *See* bubonic plague.

**black deposit** A mixture of bacteria, iron sulfide, and manganese oxide that forms in dead-end water distribution lines where sulfate compounds are reduced chemically; it causes odors.

**blackfly** The common name of the members of the *Simulium* genus (*S. damnosum, S. neavei*) that carry the pathogen *Onchocerca volvulus,* responsible for river blindness. This mosquito breeds in fast-flowing and well-aerated streams. Also called buffalo gnat.

**blackfoot disease** A disease of the peripheral blood vessels caused by arsenic exposure. It constricts arteries, diminishes blood flow, and makes the feet look black.

**black lead** Another name for graphite.

**black liquor** A highly concentrated organic waste resulting from the kraft pulping process; it is the mixture of cooking liquor and dissolved lignin that remains after digestion. The liquor is usually processed to recover chemicals, concentrated to about 70% solids, and then combusted. *See also* brownstock washing and white liquor.

**blackout dosage** The dosage of activated carbon used to shut out sunlight in sedimentation basins and small reservoirs, and thereby control the growth of algae; it is 0.2–0.5 pound per 1000 square feet of water surface.

**black sand** A coarse, discolored mixture of heavy minerals such as magnetite, ilmenite, and hematite; used as a filter medium.

**black smoker** (1) A buildup of nutrients and chemical precipitates surrounding a warm hydrothermal vent (water temperature less than 300°C). (2) A hydrothermal vent that emits very hot water and minerals.

**black water** (1) Water that contains human, animal, or food waste, or, in general, the wastewater generated through toilet use, as compared to greywater or sullage. (2) Water with a certain dark coloration caused by the presence of manganese oxides (e.g., more than 0.10 mg/L of Mn) and making it unsuitable for laundering and for some industrial processes such as dyeing and papermaking. Black water may also result from the corrosion of iron, steel, copper, and galvanized piping by hydrogen sulfide. *See also* brown water, red water. (3) Blackwater (one word) is a human or animal disease resulting in the breakdown of red blood cells and the production of dark urine.

**blackwater fever** A severe form of malaria resulting in kidney damage and hemoglobin pigment in red or black urine, an indication of massive intravascular hemolysis.

**bladder pump** A device for lifting water in a well, consisting of a submerged bladder that is filled and then squeezed.

**bladder tank** A tank equipped with a bladder precharged to a given pressure to provide protection against pressure surges in piping systems by maintaining a desired air volume in the vessel under normal operating conditions. Other surge vessels include compressor vessels and hybrid tanks.

**bladed-surface aerator** A rotating, bladed device used to inject air into a body of water.

**Blake–Kozeny equation** A formula that expresses the rate of head loss in a granular filter in terms of the bed and fluid characteristics:

$$H/L = v(K/g)(\mu/\rho)(1-f)^2(A/V)^2/f^3 \quad \text{(B-57)}$$

where $H$ = head loss; $L$ = depth of bed; $v$ = face or approach velocity; $K$ = coefficient of permeability, also called Kozeny constant; $g$ = gravitational acceleration; $\mu$ = viscosity of the fluid; $\rho$ = density of the fluid; $f$ = bed porosity; $A$ = total area of the medium; $V$ = total volume of the medium. The formula was proposed by C. F. Blake in 1922 and J. Kozeny in 1927; it is sometimes referred to simply as the Kozeny equation (Fair et al., 1971; AWWA, 1999). *See also* Coakley's, Carman–Kozeny, Ergun, and Rose equations, and the Fair–Hatch and Kozeny formulas.

**blank** A bottle containing only dilution water or distilled water; the analyte being tested is not added. Tests are frequently run on a sample and a blank and the results are compared.

**blanket filtration** *See* sludge-blanket filtration (also called contact filtration).

**blast furnace** A furnace, constructed from refractory bricks, in which hot blast air flows upward through the raw materials and exits at the furnace top; used for the production of pig iron or cast iron from iron ore. Blast furnace effluent contains flue-dust solids, a mixture of iron oxide (70%, $Fe_2O_3$), alumina, silica, carbon, and magnesia.

***Blastocystis*** An intestinal protozoan parasite transmitted through contaminated food or water.

**bleach** (1) A dry or liquid, strong oxidizing and disinfecting compound, usually containing chlorine combined with calcium or sodium. (2) A 5.25% solution of household bleach (sodium hypochlorite, NaOCl). *See also* Javel water. (3) Dry bleach, i.e., dry calcium hypochlorite (99.2%, $[Ca(OCl)_2]$.

**bleaching powder** A compound of chlorine and slaked lime used in water or wastewater disinfection. It is an unstable white powder (e.g., HTH) that liberates chlorine when reacting with an acid; the fresh product contains about 30% of available chlorine but loses strength rapidly. *See* calcium hypochlorite for more detail.

**bleed** To draw accumulated liquid or gas from a line or container.

**bleeder mechanism** A mechanism to drain accumulated water in a container, e.g., a pond. In urban stormwater management, some regulatory agencies do not allow bleeder mechanisms to reduce the retained volume of runoff during a storm event.

**bleed through** The passage of untreated substrate through an aeration basin during peak flows.

**blended phosphate** A mixture of polyphosphates and orthophosphates, used as chemical inhibitor to control corrosion.

**blending** (1) The practice, current during wet weather, of diverting excess wastewater flows around the biological treatment component of a plant to avoid upsets and mixing the diverted flow with the biological treatment effluent. Usually the total flow receives primary treatment and the blended effluent is disinfected before discharge. Blending is regulated by the USEPA. (2) The use of water from different sources, with different characteristics, in the same water supply system. *See* water blending, interface zone. (3) The mixing of sludges from primary, secondary, and advanced wastewater treatment. *See* sludge blending.

**blending plant** *See* water blending.

**Blendmaster** A sludge mixer manufactured by McLanahan Corp.

**Blendrex™** A motionless mixer manufactured by the LCI Corp.

**blind flange** A pipe flange with a blind end (or dead end) used to close the end of a pipeline.

**blinding** (1) The reduction or cessation of flow through a water filter, vacuum filter, or membrane because of fouling by solids. In pressure filter presses, a precoat system helps reduce blinding. (2) Covering a drain tile, after installation, with a permeable layer of loose topsoil, sand, or gravel to prevent misalignment and breakage during backfilling.

**blinds** Water samples that contain a chemical of known concentration with a fictitious name, inserted in the sample flow to test the impartiality of the laboratory staff.

**blind spot** A place in a filter medium or membrane where filtration does not occur.

**blind staggers** An acute livestock disease characterized by impaired vision, wandering in circles, staggering gait, respiratory failure, and death; caused by an excess of selenium (Se). Also called staggers. *See also* alkali disease.

**blind study** An experimental epidemiology study in which it is not known whether a participant belongs to the control group or the study group.

**block** A group of membranes and pressure vessels under the same control in a pressure-driven treatment system.

**block carbon** *See* activated carbon block.

**block rate** A water rate structure in which the rate is the same within a relatively large quantity taken as a unit, e.g., 10,000 gallons. *See* declining, inclining, and inverted block rates.

**Blom's plotting position** *See* Blom's transformation.

**Blom's transformation** A formula that represents the percent or frequency of observations equal to or less than a given value (ranked M) in a series of N measurements (Metcalf & Eddy, 2003):

$$P = 100(M - 3/8)/(N + 1/4) \qquad \text{(B-59)}$$

**bloodborne pathogen** A pathogen like the hepatitis and immunodeficiency viruses carried in blood materials.

**blood fluke** A pathogenic schistosome that is easily destroyed by chlorine, iodine, or storage; copper sulfate is effective against its snail host. *See also Schistosoma.*

**blood products** Any product derived from human blood, including but not limited to blood plasma, platelets, red or white corpuscles, and derived licensed products such as interferon.

**blood-urea nitrogen (BUN)** A measure of renal damage caused by chemicals, as shown by an increase of urea in blood.

**bloodworm** The freshwater larva (*Chironomus* or chironomid larva) of the midge fly, an unwanted predator that may destroy the organisms of activated sludge. (2) A red or red-blooded earthworm.

**bloom** A proliferation of algae, diatoms, and/or higher aquatic plants in a body of water, causing a sudden green or red discoloration; often related to pollution; more common in spring or early summer, especially when pollutants accelerate growth.

**BLOOM II model** A phytoplankton model that can be combined with chemical and water quality models to predict algal growth, depending on nutrient levels.

**bloom of algae** *See* algal bloom.

**blow** The upward flow of water and sediment into an excavation caused by unbalanced hydrostatic pressure. This outside pressure may result from a nearby stream or from the removal of overburden. Also called a boil (3).

**blowby** (1) In membrane filtration processes, a procedure used to recycle concentrate back to the feed. (2) A leakage of fuel and gases between a piston and a cylinder wall or a leakage of contaminants through a water or wastewater treatment device.

**blowdown** (1) The minimum discharge of recirculating water to remove materials contained in the water, the further buildup of which would cause concentrations that exceed limits established by best engineering practice (EPA-40CFR401.11-p). *See also* concentrate. (2) The continuous or intermittent discharge of process water or concentrated water from boilers and cooling towers to maintain accumulated solids at an acceptable level. *See also* cooling tower blowdown water, drift, makeup water.

**blowdown apparatus** The device that allows the regulation of blowdown in boilers and cooling towers.

**blower** Air-conveying equipment that generates pressures up to about 100 kPa (15 pounds per square inch), commonly used for wastewater aeration systems. *See* centrifugal boiler, positive-displacement boiler, inlet guide vane-variable diffuser.

**blower method** The use of an air blower to apply copper sulfate to reservoirs. The chemical is blown over the water for a dose of 0.3 mg/L. The application of copper sulfate controls the growth of plankton (algae) in reservoirs. *See also* burlap bag, continuous-dose, and spray methods.

**blower power requirement** The power requirement of each blower, $P_w$ (hp) is a function of absolute inlet temperature $T$ (°$R$); efficiency $E$ (usually $0.7 < E < 0.9$); absolute inlet and outlet pressures, respectively $P_i$ and $P_o$ (psi); and some constants (Metcalf & Eddy, 2003):

$$P_w = (\text{WRT}/550\,NE)\,[(P_o/P_i)^{0.283} - 1] \qquad \text{(B-60)}$$

where $W$ = weight of the flow of air (lb/sec); $R$ = engineering gas constant = 53.3 ft.lb/lb of air/°$R$; $N$ = 0.283 for air.

**blowfly** A dipterous insect that deposits its eggs or larvae on excreta or in the wounds of living animals.

**blowoff branch** A short section of conduit equipped with a valve and installed at a low point of a pipeline to allow drainage and sediment removal.

**blowoff valve** A small valve with gated takeoffs, installed at a low point in a pressure conduit or at a depression in a pipeline to allow drainage or flushing of the line. Also called a scour valve or washout valve.

**blowout** A severe escape of oil, gas, or water, e.g., a bursting of a conduit, dam, or canal under hydrostatic pressure.

**blue baby** An infant born with a congenital heart or lung defect that causes cyanosis.

**blue-baby disease** A pathological condition caused by an increased blood concentration of methemoglobin, a stable compound that does not yield up its oxygen to tissues. Symptoms, including lethargy, shortness of breath, and a bluish skin color, affect mostly bottle-fed infants under one year of age. *See* methemoglobinemia for detail.

**blue-baby syndrome** *See* methemoglobinemia.

**blue copperas** A common name for copper sulfate ($CuSO_4$). *See* blue vitriol for detail.

**blue-green algae** Actually not algae, but a type of photosynthetic bacteria, now called cyanobacteria, some of which produce toxic metabolites (algal toxins); *see,* e.g., *Microcystis,* geosmin. They are very common and among the oldest organisms on earth. In the absence of available nitrogen, they can abstract it from the atmosphere in free-living state or in symbiosis with other organisms. Their potential health effects include acute respiratory failure and other nerve damages, damages to organs, and malignant tumors. They also cause an earthy or musty taste or odor in water through their nonvolatile metabolic products; *see* geosmin, methylisoborneol. Blue-green algae species of the genera *Anabaena* and *Oscillatoria* are active in the oxidation of organic matter in stabilization ponds, utilizing the carbon dioxide produced by bacteria and releasing oxygen. Blue-green algae produce floating blooms that turn warm water green when the weather is calm and high levels of nutrients are available; blooms that last long decrease dissolved oxygen levels and cause fish kills.

**blue-green bacteria** Same as blue-green algae or cyanobacteria.

**blue stone** (1) A common name for copper sulfate ($CuSO_4$). *See* blue vitriol. (2) A bluish, argillaceous sandstone used in the construction industry.

**blue tide** An overgrowth of blue-green algae that secrete harmful toxins. *See also* red tide.

**blue vitriol** A common name for copper sulfate ($CuSO_4 \cdot 5\,H_2O$) in the water supply field; used in algae control. It is found as a natural mineral in deep-blue crystals or in anhydrous form as a white powder. Also called blue copperas, blue stone, chalcanthite, cupric sulfate.

**blue water** (1) Water discolored by a fine dispersion of copper-corrosion products that can react with soap scum to stain plumbing fixtures and lead to consumer complaints. Blue water may also result when microbial slimes are dislodged. *See also* green water. (2) The open water.

**B lymphocyte** Same as B cell.

**B metal** *See* B-type metal.

**BMP** (1) Acronym of best management practice. (2) Acronym of biochemical methane production.

**BMP procedure** *See* biochemical methane potential test, which indicates the maximum potential methane production from a wastewater.

**BMP test** *See* biochemical methane potential test.

**BMTS** Acronym of Burns and McDonnell treatment system.

**BNC** Acronym of base-neutralizing capacity.

**$B_3O_3N_6$** Chemical formula of borazine.

**BNR** Acronym of biological nitrogen removal and biological nutrient removal.

**$B_2O_3$** Chemical formula of boric oxide.

**Boat Clarifier®** An intrachannel clarifier developed for oxidation ditch wastewater treatment plants and manufactured by United Industries, Inc. (Baton Rouge, LA) in the shape of a boat. *See also* Burns and McDonnell treatment system, Carrousel Intraclarifier™, pumpless integral clarifier, side-channel clarifier, side-wall separator.

**BOC** Acronym of biodegradable organic carbon.

**BOD** Acronym of biochemical oxygen demand or biological oxygen demand.

**$BOD_5$** The amount of dissolved oxygen consumed in five days by biological processes breaking down organic matter; the five-day carbonaceous or nitrification-inhibited oxygen demand.

**$BOD_5$ input** The biochemical oxygen demand of the materials entered into a process. It can be calculated by multiplying the fats, proteins, and carbohydrates by factors of 0.890, 1.031, and 0.691 respectively. Organic acids (e.g., lactic acid) should be included as carbohydrates. Composition of input materials may be based on either direct analyses or generally accepted published values (EPA-40CFR405.11-b).

**$BOD_7$** The biochemical oxygen demand as determined by incubation at 20°C for a period of 7 days using an acclimated seed. Agitation employing a magnetic stirrer set at 200 to 300 rpm may be used (EPA-40CFR417.151-g).

**BOD analysis method** A method used to analyze $BOD_5$ data and determine the BOD reaction rate constant (K) and the ultimate or 20-day BOD. *See,* e.g., the Fujimoto, least-squares, and Thomas BOD methods.

**$BOD_e$** The BOD of the final effluent of a treatment plant.

**BOD exertion** The BOD of a waste is assumed to follow a first-order reaction:

$$BOD_t = L(1 - e^{-Kt}) \qquad \text{(B-61)}$$

where $BOD_t$ = BOD exerted at time t, $L$ = BOD at time t = 0, and $K$ = a rate constant that varies with temperature. *See* Figure B-16.

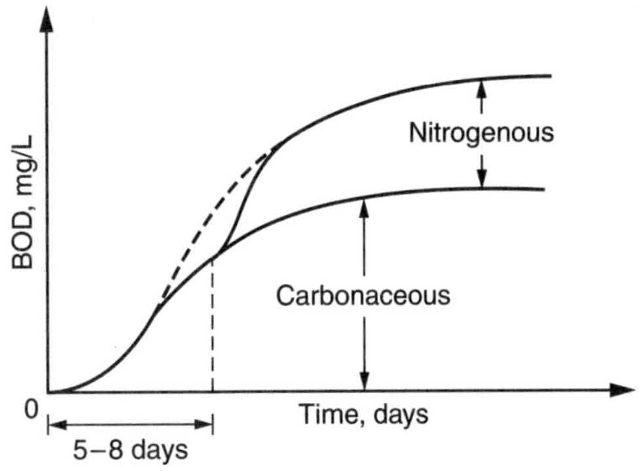


**Figure B-16.** BOD exertion.

**BOD F/M ratio** A parameter used in the design of anoxic tanks on the basis of the specific denitrification rate:

$$F/M_b = QS_0/(V_{nox}X_b) \quad \text{(B-62)}$$

where $F/M_b$ = BOD F/M ratio based on active biomass, g BOD/g biomass/day; $Q$ = influent flow rate, m$^3$/day; $S_0$ = influent BOD concentration, mg/L; $V_{now}$ = volume of anoxic tank, m$^3$; $X_b$ = biomass concentration in the anoxic zone, mg/L.

**BOD$_L$** Symbol for ultimate carbonaceous biochemical oxygen demand; also denoted UBOD or BODu.

**BOD** load The amount of biochemical oxygen demand of a wastewater passing through a treatment unit or discharged into a receiving stream, in weight per time unit, e.g., pounds per day.

**BOD/N/P ratio** A parameter used to indicate the relative weight or concentration of the nutrients nitrogen (N) and phosphorus (P) with respect to the organic content (BOD) of a waste or the nutrient requirements of a biological treatment process. For example, domestic wastewater has BOD/N/P ratio of about 100/17/3, whereas the theoretical aerobic treatment requirement is about 100/(3 – 6)/(0.7 – 1.5), depending on the process and the availability of the nutrients.

**BOD population equivalent** The population equivalent of a wastewater based on an average BOD of 0.17 pound per person per day. Also called organic population equivalent. *See* population equivalent for detail.

**BOD-temperature effect** The effect of temperature on the rate of BOD exertion is usually based on the Arrhenius equation:

$$K_T = K_{20}\,\theta^{(T-20)} \quad \text{(B-63)}$$

where $K_T$ = rate constant at temperature T (°C), $K_{20}$ = rate constant at temperature 20°C, $\theta$ = a parameter that varies with the waste but often taken as 1.047.

**BOD test** A test to determine the oxygen consumed by heterotrophic bacteria of a waste in the dark at 20°C over a period of time, usually 5 days. It is the difference between the dissolved oxygen concentrations in a sample at time 0 and after five days. To obtain only carbonaceous BOD, a nitrification inhibitor may be added. The test measures only biodegradable organic matter and requires a high concentration of acclimated seed bacteria. *See also* COD test.

**BOD to nitrogen to phosphorus ratio** *See* BOD/N/P ratio.

**BOD$_u$** Acronym of ultimate carbonaceous BOD. Also denoted as UBOD or BOD$_L$.

**body burden** The amount or concentration of a substance in an organism, particularly a substance that accumulates, heavy metals, nonpolar chemicals, radioactive or toxic chemicals. It is an indication of chronic toxicity.

**body-contact recreation** Any recreational activity such as bathing or swimming that involves contact with surface water. It is a source of contamination (pathogenic bacteria, viruses, and protozoa) of drinking water sources.

**body feed** Slurried coating material added continuously to the influent of precoat filters during the filtering cycle to maintain or improve the porosity of the cake as the solids removed clog the precoat. It is usually a slurry of the same material as the precoat (e.g., diatomaceous earth or perlite). *See also* diatomaceous earth filtration.

**body feeding** The third and final step in the precoat filtration process, consisting of addition of medium (between 1 and 10 mg/L of body feed per mg/L of suspended solids) to the incoming flow to compensate for the reduction in porosity caused by the compression of the precoat cake.

**body water** The portion of the human body that is water, about 80% by weight.

**body weight** The weight of a person or an animal. In the determination of safe or toxic doses of chemicals, body weights of 70 kg (154 lb) and 10 kg (22 lb), respectively, are sometimes assumed for an adult and a child. *See* AADI, LOAEL, NOAEL, RMCL.

**bog** A type of wetland with a surface layer of living vegetation (typically sphagnum moss) and a deep layer of accumulated peat deposits on poorly drained ground, with a lack of oxygen in the waterlogged soil. Bogs depend primarily on precipitation for their water source; they are usually acidic and rich in plant residue, with a conspicuous mat of living green moss. The sphagnum moss, commonly known as peat moss, is harvested for use as a gardening additive.

**BogenFilter™** A belt filter press manufactured by Klein America, Inc.

**bog lake** A lake that is undergoing severe eutrophication, with bogs already started on the periphery.

**$B(OH)_3$** Chemical formula of boric acid. Also written as $H_3BO_3$.

**$B(OH)_4^-$** Chemical formula of the borate ion.

**boil** (1) An area of a rapids having bubbling and swirling water. (2) A rise in water surface caused by the turbulent upward movement of water. (3) The upward flow of water and sediment into an excavation caused by unbalanced hydrostatic pressure. This outside pressure may result from a nearby stream or from the removal of overburden. Also called a blow. (4) *See* sand boil.

**boiler** A closed vessel or arrangement of vessels and tubes with a heat source, in which water is continually vaporized into steam by the application of heat.

**boiler blowdown** The water removed from the recirculating system of a boiler and replaced by makeup water to maintain impurities at an acceptable level. *See* blowdown and makeup water.

**boiler chemical-cleaning wastes** Wastes generated by the removal of deposits and corrosion products collected on boiler tubes, including dense magnetite ($Fe_3O_4$) deposits, leakage of cooling water, and chemical contaminants. Also called waterside cleaning wastes. *See also* on-line boiler cleaning.

**boiler feedwater** Water supplied to a boiler from a tank or a condenser for steam generation; usually softened, demineralized, heated, and deaerated. Also called boiler-water feed or feedwater.

**boiler-flue gas** Waste gas from a boiler, a residue sometimes used in the neutralization of alkaline wastes. *See* waste boiler flue gas treatment.

**Boilermate®** A deaerator manufactured by Cleaver-Brooks as a packed column.

**boiler scale** An encrustation of precipitated minerals on the heating surfaces of a boiler.

**boiler water** A representative sample of the water or steam condensate circulating in a boiler system, taken after the generated steam has been separated and before the incoming feedwater or any added chemical has become mixed with the sample to change its composition.

**boiler-water feed** Same as boiler feedwater.

**boiling (disinfection)** One of the earliest water treatment methods, conducted primarily in household containers. *See also* heat disinfection.

**boiling heat transfer** One of two modes of operating an evaporator; solvents evaporate on the surface of the device by application of condensing steam or other heat source. *See* flash evaporation.

**boiling point** The temperature at which a component's vapor pressure equals atmospheric pressure or the pressure acting on the liquid; a relative indicator of volatility, it generally increases with increasing molecular weight. Organic compounds have lower boiling points than inorganics. High-boiling liquids usually enter the body through contact, whereas low-boiling liquids usually enter through inhalation. Also called normal boiling point. *See also* melting point.

**boiling-point elevation (BPE)** The difference between the boiling point of a substance and that of pure water under the same pressure; a critical parameter in the design of evaporators. Also called boiling-point rise.

**boiling-point rise** Same as boiling-point elevation.

**boiling temperature** Same as boiling point.

**boiling water reactor** A nuclear reactor that uses enriched uranium as fuel and ordinary water as coolant and moderator.

**boil out** A process using boiled water to remove scale deposits from evaporators.

**boil-water advisory** *See* boil-water order.

**boil-water order** A precautionary order or advisory issued by health and water authorities to boil or otherwise disinfect drinking water that may have been microbiologically contaminated (e.g., because of inadequate treatment or failure in the distribution system) and thus may present a threat to public health and safety. Holding water in a rolling boil for one minute normally kills pathogens at sea level. Boiling water does not protect against thermally resistant bacterial spores and may actually result in greater inhalation health hazards.

**bolson** A desert valley raised by aggradation and draining toward the interior into a playa.

**Boltzmann constant** Boltzmann's constant.

**Boltzmann, Ludwig** Austrian physicist (1844–1906); his concept of a molecular constant, which bears his name, is included in the general gas law.

**Boltzmann's constant** The ratio of the universal gas constant ($R = 8.314 \times 10^7$ ergs per mole degree) to Avogadro's number ($N = 6.022 \times 10^{23}$ molecules per mole), i.e., $1.38 \times 10^{-16}$ erg per degree molecule. *See* Einstein's equation for diffusion, general gas law, Stefan–Boltzmann constant.

**bolus dose** A chemical dose administered to an animal in a single volume as compared to a drinking water contaminant that is absorbed over a period of time. A bolus is a medicine in a round mass larger than an ordinary pill.

**BOM** Acronym of (a) background organic matter or (b) biodegradable organic matter.

**bomb calorimeter** An apparatus used to determine the heat content or fuel value of materials such as water or wastewater sludges. It measures the heat produced by the combustion reaction in a closed vessel. *See also* calorific value, Dulong formula, and sludge fuel value.

**bomb-calorimeter test** A test conducted with a bomb calorimeter to determine the fuel value of a sludge.

**bomb calorimetry** The measurement of heat using a bomb calorimeter.

**bond** (1) The attraction between atoms in a molecule or compound; *see* chemical bond for more detail. (2) A long-term (e.g., 20–30 years) debt instrument used by government to finance such public improvements as water and wastewater services. Bonds sold by publicly owned utilities are secured by the revenue of the utilities or by the resources of the issuing government. Corporations can also issue bonds. *See* general obligation bond, revenue bond.

**bond anticipation note** A short-term interest-bearing note that a government issues in anticipation of a long-term forthcoming bond.

**bond covenant** A formal document requiring an issuer to meet the conditions of a bond.

**bond discount** The excess of the face value of a bond over the acquisition or sale price, corresponding to an actual yield that is higher than the stated interest rate.

**bonded debt** The total of the value of the outstanding bonds of an agency or enterprise.

**bond fund** (1) The total amount of the proceeds of a bond issue. (2) A mutual fund of bonds.

**bond indenture** The legal instrument adopted by the bond issuer, defining its responsibilities toward the bond holders.

**bonding** The adsorption of substances due to electrostatic forces as strong as a chemical bond between the sorbate molecule and the surface of the adsorbent. *See* chemical adsorption.

**bond ordinance** Same as bond ordinance.

**bond premium** The excess of the acquisition or sale price of a bond over its face value, corresponding to an actual yield that is lower than the stated interest rate.

**bond resolution** Same as bond indenture.

**bonds payable** The face value of outstanding bonds.

**bone char** Pulverized bone heated without air or burnt with air. It is a black pigment substance containing 10% carbon, used in water defluoridation at a loading of about 1250 mg $F^-$/L; it can be regenerated with a 1% caustic soda solution. It is also used for arsenic removal and for decolorizing sugar.

**bone marrow** A soft, fatty, vascular tissue in the interior cavities of bones that is a major site of blood cell production. Benzene and other chemicals can affect this material and cause serious or even fatal disorders such as anemia (suppression of red blood cell formation) or leukemia (loss of control over white blood cell production).

**bonnet** The cover on a gate valve.

**BonoZon** An ozone generator manufactured by ProMinent Fluid Controls, Inc.

**BOO** Acronym of build–own–operate. *See also* BOOT.

**book value** The value of a business property, assets of a utility, etc., as stated in a book of accounts, which may be different from their actual or market value.

**boom** A series of connected floating timbers designed to protect the face of a hydraulic structure, deflect floating materials, contain oil spills, etc.

**booming** The accumulation and sudden release of water in placer mining.

**booster chlorination** Additional chlorination of drinking water at some point in the distribution system to assure a residual at the far ends.

**booster pump** *See* booster station.

**booster station** A station housing booster pumps, which raise the pressure of water or wastewater on the discharge side, or which lift the liquid to a higher pressure plane. *See* pumping station.

**BOOT** Acronym of build–own–operate-transfer. *See also* BOO.

**Boothwall™** A cartridge filter manufactured by Dustex Corp. for dust removal.

**borane** Any of the compounds of boron and hydrogen, from $B_2H_6$ to $B_{20}H_{16}$. Also called boron hydride.

**borate** A salt or ester of boric acid or any compound containing boron; e.g., borax.

**borate buffer** A solution of boric acid [$H_3BO_3$ or $B(OH)_3$] and caustic soda (NaOH) used to maintain a constant pH during testing. *See* simulated distribution system test.

**borate ion [$B(OH)_4^-$]** *See* borate.

**borax ($Na_2B_4O_7 \cdot 10H_2O$)** Hydrated sodium borate, a white powder or crystals, soluble in water, used as a cleansing agent. Also called sodium borate, sodium pyroborate, sodium tetraborate.

**borax lake** A lake that is rich in sodium borate ($Na_2B_4O_7$), an example of an extreme environment. *See also* bitter lake and soda lake.

**borax pentahydrate ($Na_2B_4O_7 \cdot 5H_2O$)** A white crystalline, water-soluble solid, used as a herbicide, in water softening, as a disinfectant, and as a deodorant.

**borazine ($B_3N_3H_6$)** A colorless liquid that hydrolyzes to form boron hydrides; it has properties similar to those of benzene ($C_6H_6$).

**Borda, Jean Charles** French military engineer (1733–1799).

**Borda loss** The head loss caused by the sudden expansion or enlargement of a conduit. Unlike other transition or appurtenance head losses, the Borda loss (BL) may not be expressed in terms of velocity head ($kV^2/2$ g) or equivalent conduit length (Fair et al., 1971):

$$BL = (V_1 - V_2)^2/2\ g \qquad \text{(B-64)}$$

where $V_1$ and $V_2$ are the velocities in the original and expanded conduits, respectively, and g is the gravitational acceleration.

**borderline ion** Any metal cation with characteristics between those of type A or hard and type B or soft ions, as to polarizability, toxicity, and preferences for forming complexes. *See* transition metal cation for detail.

**borderline trace metal** A micronutrient metal that is essential to algal growth, e.g., cobalt, copper, iron, manganese, molybdenum, nickel, zinc. *See also* class A ion, class B ion.

**border spring** A spring that issues from the boundary between two formations, at a point where the water table slope equals the surface gradient. Also called alluvial-slope spring or boundary spring.

**border strip** A grassed or vegetated strip placed around a field, along channels, or other places to control erosion.

**border-strip flooding** One of two main types of surface application of wastewater, using ridges in the direction of the slope, a lined ditch with slide gates at the head of each strip, and underground or gated piping (Metcalf & Eddy, 1991). Also called border-strip irrigation. *See also* ridge-and-furrow irrigation.

**border-strip irrigation** Same as border-strip flooding.

**bore** (1) A hole made or enlarged by boring; a borehole or a tunnel. (2) An abrupt rise of water moving rapidly upstream from the mouth of an estuary as a result of high tide or moving downstream as a result of a cloudburst or a sudden release from a reservoir. Also called tidal bore. (3) A standing wave advancing upstream in an open channel.

**bored-hole latrine** *See* borehole latrine.

**bored-hole privy** *See* borehole latrine.

**bored tubewell** A well sunk by hand with an auger.

**bored well** A shallow (10–100 ft), large-diameter well (8–36 in) constructed by a hand-operated or power-driven auger.

**borehole latrine** A latrine that has a hand-augered borehole, approximately 20 ft deep and 16 in diameter, instead of the customary dug pit.

**borehole privy** *See* borehole latrine.

**boric acid [$H_3BO_3$ or $B(OH)_3$]** (1) A white crystalline acid, occurring naturally or prepared from borax; used in the water chemistry laboratory (e.g., in testing for ammonia nitrogen), in industry, and in medicine. Also called orthoboric acid. (2) Any of a number of acids containing boron.

**boric anhydride** Same as boric oxide.

**boric oxide ($B_2O_3$)** A colorless crystalline compound used in metallurgy. Also called boric anhydride or boron oxide.

**boride** A compound of boron plus a less electronegative element.

**boring sample** Material from a boring analyzed to determine the characteristics of the formation.

**boron (B)** A brown powder or a yellow crystalline trivalent element, found naturally in borax ($Na_2B_4O_7 \cdot 10H_2O$) or boric acid [$H_3BO_3$ or $B(OH)_3$]. Atomic number = 5. Atomic weight = 10.811. Specific gravity = 2.34 (crystals) or 2.37 (powder). An inorganic contaminant found in trace concentrations in natural waters; regulated by the USEPA: both the MCLG and MCL between 0.6 and 1.0 mg/L. At sufficiently high concentrations, boron may interfere with enzyme production and cause physiological effects in humans (digestive difficulties, nerve disorders). Boron is an important chemical parameter related to the quality of irrigation water, with a recommended limit of 1.25 mg/L for sensitive crops and 4.0 mg/L for tolerant crops (Cairncross and Feachem, 1993).

**boron oxide** Same as boric oxide.

**borosilicate** A salt of boric and silicic acids.

**borosilicate glass** A highly heat- and shock-resistant glass, containing at least 5% of boric oxide ($B_2O_3$), used in making cookware and labware.

***Borrelia duttoni*** The pathogen that causes tick-borne relapsing fever.

***Borrelia recurrentis*** The spirochaete that causes louse-borne relapsing fever.

**borrow pit** A pit from which earth is removed for use in construction or for other purposes.

**Bosker** An apparatus designed by Brackett Green, Ltd. to clean trash racks.

**botanical pesticide** A pesticide whose active ingredient is a plant-produced chemical such as nicotine or strychnine. Also called a plant-derived pesticide.

**bottled artesian water** Bottled water from a well drilled in a confined aquifer, a labeling regulated by the U.S. Food and Drug Administration. Also called artesian well water.

**bottled fluoridated water** Bottled water containing natural or added fluoride, at least 0.8 mg/L of fluoride ion.

**bottled groundwater** Bottled water from a well drilled in an aquifer that is under a pressure at least equal to 1.0 at and not under the influence of surface water; a labeling regulated by the U.S. Food and Drug Administration.

**bottled mineral water** Bottled water from a protected underground source and from boreholes or springs, containing at least 250 mg/L of total dissolved solids; a labeling regulated by the U.S. Food and Drug Administration.

**bottled purified water** Bottled water that has been treated by a suitable process to meet the definition of purified water; a labeling regulated by the U.S. Food and Drug Administration.

**bottled sparkling water** Bottled water that has been treated by a suitable process and contains the same concentration of carbon dioxide as the source; a labeling regulated by the U.S. Food and Drug Administration. *See also* natural sparkling water.

**bottled spring water** Bottled water that was collected from an underground formation that flows naturally to the surface, through a spring or a borehole tapping the formation of the spring; a labeling regulated by the U.S. Food and Drug Administration.

**bottled sterile water** Bottled water that meets the sterility requirements of the U.S. *Pharmacopeia,* 23rd edition, of January 1, 1995; a labeling regulated by the U.S. Food and Drug Administration.

**bottled water** Drinking water sealed in bottles or other containers, often from a spring, sometimes carbonated, with no additives except for disinfection; a labeling regulated by the U.S. Food and Drug Administration.

**bottled water plant** Any place or establishment in which bottled water is prepared for sale.

**bottled well water** Bottled water from a well or borehole that taps an aquifer; a labeling regulated by the U.S. Food and Drug Administration.

**bottle point technique** A common method of collecting adsorption isotherm data, using at least 10 bottles containing selected amounts of adsorbate and adsorbent, the results being expressed in terms of the equation

$$Q_e = Q_i + (C_0 - C)\,V/M \qquad \text{(B-65)}$$

where $Q_e$ = equilibrium solid phase concentration of contaminant (adsorbate) on the adsorbent, mass or moles per mass; $Q_i$ = initial solid phase concentration of adsorbate, mg/L; $C_0$ = initial concentration of adsorbate in solution, mg/L; $C$ = equilibrium concentration in solution, mg/L; $V$ = volume of sample; $M$ = mass of adsorbent in the bottle.

**bottom ash** The nonairborne combustion residue from burning pulverized coal in a boiler; the material that falls to the bottom of the boiler and is removed mechanically; a concentration of the noncombustible materials, which may include toxics. *See also* fly ash, black birds.

**bottom deposit** A diffuse source of contamination of water resources, e.g., soluble and suspended matter in thicknesses varying from a thin layer to a heavy sludge bank. Also called benthal or benthic deposit.

**bottom ice** Ice formed below the surface of a body of water, attached either to a submerged structure or to the bed. Also called anchor ice, ground ice.

**bottoms** The least volatile fraction of petroleum or any other substance, e.g., a residual liquid, left behind in distillation after more volatile fractions are driven off.

**bottom scour** The transport of particles in traction along the bottom of a sedimentation basin. *See also* auxiliary, scouring velocity.

**bottom sediment** Sediment constituting the bed of a water body.

**bottom water level** The lower level in a tank, reservoir, etc. at which to turn on a pump or open a valve.

**boundary layer** When a fluid flows over a solid surface, the thin layer of fluid in contact with the surface is the boundary layer. Velocity is reduced in the vicinity of the boundary because of the forces of adhesion and viscosity. Boundary-layer flow may be laminar or turbulent; it is prevalent in streams and is described in one-dimensional models.

**boundary-layer diffusion** The transfer of molecules or ions across a boundary between two media, e.g., a particle and a solution. *See also* diffusion and dispersion.

**boundary-layer turbulence** The condition of the flow that develops at the surface of a settling particle and reduces considerably the drag force when the Reynolds number $R_e$ exceeds 200,000, with the drag coefficient $C_D = 0.1$.

**boundary spring** A spring that issues from the boundary between two formations at a point where the water table slope equals the surface gradient. Also called alluvial-slope spring or border spring.

**bound water** (1) Water held on the surface or interior of colloidal particles, particularly the diffuse layer of water in which ions of opposite charge produce a drop in potential. *See* Guoy–Stern colloidal model. (2) The fixed layer of water attached to the slime layer that develops on the surface of the support medium in attached growth processes; it serves to transfer oxygen and nutrients to the slime and waste products from the slime. *See* fixed growth process. (3) Water of hydration of a crystalline compound. (4) Water that is bound to sludge solid particles and that only the thermochemical destruction of the particles can release. *See* bulk water, interstitial water, hydration water, and vicinal water.

**bound water layer** *See* bound water (2).

**bovine serum albumin** A blood substance prepared from the serum of cattle for media enrichment in culturing microorganisms and tissue cells.

**bowl** (1) *See* pump bowl. (2) *See* turbine pump (2).

**Boyle, Robert** British chemist and physicist (1627–1691) who stated in 1662 the thermodynamic principle named after him.

**Boyle's law** At constant temperature and relatively low pressures, the volume of a gas varies inversely with its pressure (or the pressure varies inversely with the volume), or, the product of volume by pressure is constant:

$$P_1V_1 = P_2V_2 = \text{constant} \qquad \text{(B-66)}$$

where $V_1$ is the volume corresponding to pressure $P_1$ and $V_2$ the volume corresponding to pressure $P_2$. Also called Mariotte's law. *See also* general gas law.

**Boythorp** A steel tank coated with glass materials fabricated by Klargestor.

**BPACl** Acronym of basic polyaluminum chloride.

**BPE** Acronym of boiling point elevation.

**BPEO** Acronym of best practicable environmental option.

**BPM** Acronym of best practicable means.

**BPR** Acronym of biological phosphorus removal.

**BPT** *See* best practicable control technology.

**Bq** Abbreviation of Becquerel, an international unit of radioactivity.

**Br** Chemical symbol of bromine.

**$Br_2$AA** Abbreviation of dibromoacetic acid, a regulated disinfection by-product.

**BrAA** Abbreviation of monobromoacetic acid, a regulated disinfection by-product.

**brackets** In chemical formulas or notations, square brackets [. . .] are sometimes used to denote the molar concentration of a species, and curly brackets or braces {. . .} to denote the activity of that species in moles/L; e.g., [A] = molar concentration of A, {A} = activity of A at concentration [A].

**brackish** Mixed fresh and saltwater.

**brackish water** Water with a mineral content between 1000 and 10,000 mg/l (concentration of dissolved solids), as compared to freshwater, saltwater or saline water, seawater or ocean water, and brine. More generally, brackish water (e.g., in an estuary) is more saline than freshwater and less saline than seawater. Treatment processes commonly considered for brackish and saline waters include distillation, electrodialysis, freezing, ion exchange, and reverse osmosis. *See* salinity.

**brackish water conversion** *See* saline water conversion.

**braided river** *See* braided stream.

**braided stream** A wide and shallow stream that divides successively between alluvium islets such as bars, sandbars, islands, or shoals, which change with time or with each runoff event. *See also* sedimentation terms.

**brake horsepower (bhp)** The actual power, expressed in watts, required of a motor to drive a pump; the power delivered to the pump shaft, expressed in horsepower; equal to the ratio of the water horsepower to the efficiency of the pump; it can be measured by a dynamometer applied to the shaft or flywheel. Also called shaft horsepower. *See* horsepower, power terms.

**Bramah's press** A hydrostatic machine consisting of large and a small cylinders connected together and equipped with watertight pistons for magnifying an applied force. Also called hydrostatic press. Named after Joseph Bramah (1748–1814), English engineer and inventor.

**branched network** As opposed to looped networks, a branched sewer network consists of branch sewers collecting flow from subareas or subcatchments and feeding into main lines or trunk sewers. A branched water supply system is a tree-like network, where water flows into the branches from a single trunk. Branched networks are common in irrigation but rare in potable water supply. *See also* dendritic network.

**branch sewer** A sewer collecting wastewater or stormwater from a relatively small area and discharging into another line serving more than one subarea or subcatchment; it receives wastewater from both laterals and house connections.

**Brandol®** A diffuser manufactured by Schumacher Filters, Inc. and producing fine air bubbles.

**brass** A copper alloy with 40% or less zinc, and usually some lead. Zinc adds strength and lead contributes malleability and ductility.

**brass and bronze corrosion** The dissolution of brass and bronze fixtures caused by dezincification or other problems, a potential cause of mechanical failure of valves and faucets and of metal (copper, lead, zinc) leaching into the drinking water.

**$BrCH_2CH_2Br$** Chemical formula of ethylene dibromide or 1,2-dibromomethane.

**BrCN** Chemical formula of cyanogen bromide.

**breakaway point** The flow rate at which a water meter starts to register.

**breakbone fever** A common name of dengue fever.

**breakdown maintenance** A reactive, unplanned maintenance approach that consists of fixing equipment when it breaks down, equivalent to crisis management, as opposed to proactive maintenance. *See also* conditional or indicative maintenance and developmental maintenance.

**breakeven analysis** An examination of projected costs and benefits to determine the payback period of a project, i.e., when the original project investment will be recovered. *See* financial indicators for related terms.

**breakpoint** (1) In water chlorination, the point at which the chlorine demand by ammonia ($NH_3$) is satisfied, the point at which the chlorine residual curve changes from a trend of decreasing residual to a trend of increasing residual, according to the overall reaction

$$2\ NH_3 + 3\ HOCl \rightarrow N_2(gas) + 3\ H_2O + 3\ H^+ + 3\ Cl^- \qquad (B\text{-}67)$$

*See also* the equation and figure under breakpoint chlorination, and peakpoint. (2) The chlorine dose that corresponds to this point, i.e., the breakpoint dosage. Theoretically (under ideal conditions), the molecular weight ratio of chlorine (in the HOCl) to ammonia nitrogen is:

$$Cl_2/N = 3\ (2 \times 35.45)/14 = 7.6 \qquad (B\text{-}68)$$

In practice, the molecular weight ratio varies between 8.0 and 10.0. (3) The phenomenon observed during the chlorination of ammonia-containing water: when the molar ratio ($R$) of chlorine to ammonia is less than 1.0, most of the residual oxidizing chlorine is monochloramine; breakpoint occurs at about $R = 2.0$. In actual experience, the breakpoint occurs at $R > 2$ with some residual chlorine in solution.

**breakpoint chlorination** Addition of chlorine to water until the chlorine demand (from ammonia and readily oxidizable organics) has been satisfied. At this point, further additions of chlorine will result in a free residual chlorine that is directly proportional to the amount of chlorine added beyond the breakpoint. Available chlorine is primarily combined chlorine (chloramines) below the breakpoint and free chlorine above the breakpoint. Breakpoint chlorination of wastewater converts more than 90% of ammonia to nitrogen gas, but usually requires dechlorination to remove the excess residual in the effluent and it is practical only as an effluent polishing method, not for removal of high levels of influent nitrogen. The breakpoint chlorination reaction has been presented under several forms, e.g.: (a) the equation listed above (*see* breakpoint) between ammonia and hydrochlorous acid;

$$\text{(b)}\quad 2\ NH_4^+ + 3\ HOCl \rightarrow N_2(gas) + 3\ H_2O + 5\ H^+ + 3\ Cl^- \qquad (B\text{-}69)$$

$$\text{(c)}\quad 2\ NH_3 + 3\ Cl_2 \rightarrow N_2(gas) + 6\ H^+ + 6\ Cl^- \qquad (B\text{-}70)$$

$$\text{(d)}\quad 2\ NH_3 + 3\ Cl_2 \rightarrow N_2(gas) + 6\ HCl \qquad (B\text{-}71)$$

*See* Figure B-17, chloramination, free chlorination.

**breakpoint curve** A graphical representation of the breakpoint phenomenon, particularly at pHs between 6.5 and 8.5. *See* breakpoint chlorination and Figure B-17. The figure shows the following features: breakpoint B, a hump at point A, an initial demand OI, an irreducible residual BC, and the three forms of residual: free, combined, and total. Also called hump and dip curve.

**breakpoint dosage** Same as breakpoint (2).

**breakpoint phenomenon** *See* breakpoint (3).

**breakpoint reactions** Breakpoint chlorination results in the oxidation of ammonia, the formation of chloramines, and the reduction of chlorine. *See* chloramination.

**break tank** A tank designed to hold water or wastewater before repumping or subsequent flow by gravity to a treatment process.

**breakthrough** (1) A crack or break in a filter bed that allows the passage of floc or particulate matter through the filter; will cause an increase in effluent turbidity. A breakthrough can occur (a) when a filter is first placed in service, (b) when the effluent valve suddenly opens or closes, and (c) during periods of excessive head loss through the filter (including when the filter is exposed to negative heads). (2) The point reached during fil-

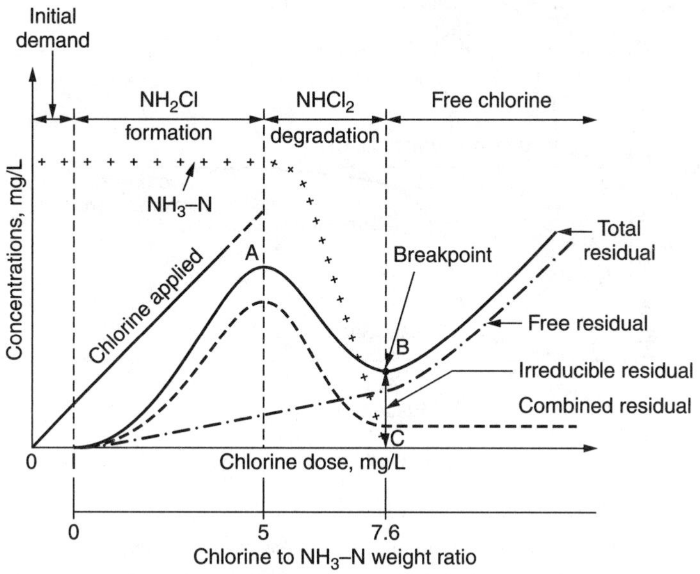


**Figure B-17.** Breakpoint curve.

tration at which the turbidity of the filtrate starts to increase or the headloss reaches a predetermined limit because the pores of the filter bed are filled with solids. The filter run is then terminated and the filter is backwashed. (3) In the adsorption process, the point at which the adsorbate reaches a predetermined concentration in the effluent, e.g., 5% of the influent concentration. *See also* adsorption bed exhaustion. (4) A similar quality deterioration in an ion-exchange resin.

**breakthrough adsorption capacity** The actual performance of an adsorption column, a percentage of the theoretical capacity (also called adsorptive capacity) based on isotherm data. For a single column, the breakthrough capacity is between 25 and 50% of the adsorptive capacity; the percentage increases for multiple columns. *See also* time to breakthrough.

**breakthrough capacity** (1) The amount of adsorbate or filtrate removed or the volume of flow treated when a filter, an adsorption unit, or an ion exchange unit reaches breakthrough. *See* ion-exchange breakthrough capacity, bed volumes, carbon usage rate, column utilization. (2) The capacity of an ion exchange column, expressed in kilograms per cubic meter or cubic foot of ion exchange resin.

**breakthrough concentration** The maximum acceptable concentration ($C_b$) of adsorbate in the effluent selected on the basis of water quality requirements. It is sometimes defined as the minimum detectable concentration. *See also* exhaustion concentration, critical depth, empty bed contact time, bed volumes, column utilization, carbon usage rate.

**breakthrough curve** In adsorption and similar processes, an S-shaped curve plotting the solute concentration in the effluent ($C_e$) versus the treated effluent volume, the time of treatment, or the number of bed volumes treated; sometimes the ratio $C_e/C_0$ of concentrations is used instead of just the effluent, $C_0$ being the feed concentration. The curve is used to determine the breakthrough capacity, carbon usage rate, and column utilization. *See* Figure B-18. *See also* mass transfer zone.

**breakthrough index** A parameter that indicates the point at which effluent quality becomes inadequate in the operation of a rapid sand filter. The breakthrough index ($B$) is computed by Hudson's formula as follows:

$$B = 60\,QD^3H/[L(T + 10)] \qquad \text{(B-72)}$$

where $Q$ = filtration rate (e.g., gpm/ft$^2$), $D$ = sand size (cm), $H$ = terminal headloss (ft), $L$ = depth of sand bed (in), $T$ = temperature (°F). The breakthrough index depends on the degree of pretreatment and on how well the water responds to coagulation, varying from 0.0004 for poor coagulation and average pretreatment to 0.0060 for average coagulation and excellent pretreatment (Fair et al., 1971).

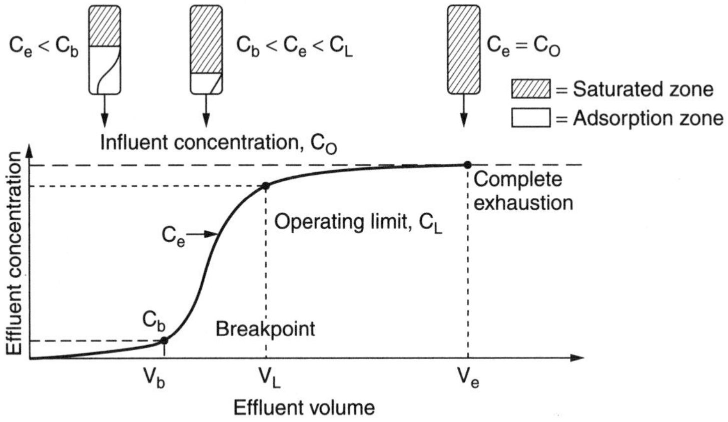


**Figure B-18.** Breakthrough curve.

**breakthrough point** In the operation of an ion exchange column, the point at which the medium becomes saturated with the ions being removed and these start appearing in the effluent.

**breakthrough time** *See* time to breakthrough.

**breakwater** A coastal hydraulic structure built of piles, concrete blocks, large loose rock, etc., from the shore into the littoral zone and designed to protect the littoral zone, as well as shipping and marine structures, from the action of incoming waves. Similar to breakwaters, jetties mainly prevent drift deposit in navigable channels. Other coastal hydraulic structures are seawalls, groins, and revetments. Fixed or floating breakwaters are also used to protect stabilization ponds against erosion of the interior slopes.

**Breeze®** A compact air stripping apparatus manufactured by Aeromix Systems, Inc.

**brewer's yeast** A yeast of the genus *Saccharomyces,* used as a ferment for making beer and wine. Also a strain of *Saccharomyces cerevisiae* used in combination with *Arthrobacter* to biodegrade petroleum hydrocarbons.

**brewery/distillery/winery wastes** Major wastewater generating sources in breweries, distilleries, and wineries include grain steeping and pressing, alcohol distillation residues, and condensate from stillage evaporation. Such wastes are characterized by high dissolved organic matter, high nitrogen, and fermented starches. Trickling filtration is an appropriate treatment method, in addition to by-product recovery and reuse.

**bridging** (1) A coagulation mechanism observed in water treatment whereby the long-chain, high-molecular-weight polymers bridge the distance between the colloids and form larger particles that settle by gravity; also called bridging effect. *See also* adsorption–destabilization, double-layer compression, restabilization, and sweep-floc coagulation. (2) The condition of water treatment units (e.g., softening, filtration, chemical feed) in which the dry products cake, fill pores, or otherwise do not flow as expected. It occurs when salt sticks together to form one large solid mass of pellets, or when salt cakes in a dry-salt brine tank, causing failure of the liquid or brine beneath the dry salt to become saturated. *See also* mushing. (3) The ability of particles to form a crustlike film over void spaces within a filter medium or a membrane.

**bridging effect** Same as bridging (1).

**Briggsian logarithm** Same as common logarithm.

**Briggs logarithm** Same as common logarithm.

**bright field** The illuminated area around the object of a microscope.

**bright-field microscope** A common optical microscope with multiple lenses, a bright field, and a source light to illuminate the specimen. It produces a dark image on a bright background. Maximum magnification = 2000 times. Resolution = 0.2 μm (200 nm). *See also* electron-beam, optical or light, and ultraviolet-ray microscopes.

**bright-field microscopy** The use of a bright-field microscope in water and wastewater microbiology, e.g., in performing Gram stains.

**brimstone** Sulfur.

**brine** (1) A liquid with a dissolved salt concentration higher than that of ordinary seawater; water saturated or strongly impregnated with salt, usually with a mineral content higher than 36,000 mg/l (concentration of dissolved solids), as compared to freshwater, brackish water, salt or saline water, and seawater or ocean water. *See* salinity. (2) The waste stream from such desalination techniques as ion exchange and reverse osmosis, which may be discharged to deep wells, saline surface waters or by evaporation through lagooning. Sometimes called concentrate, reject, reject brine, waste brine. (3) A concentrated solution of potassium chloride (KCl) or sodium chloride (NaCl) used in the regeneration of ion-exchange resins. Also called potassium chloride brine or sodium chloride brine. (4) A salt and water solution for pickling.

**Brinecell** An electrolytic generation device manufactured by Brinecell, Inc.

**brine collector** A device, piping, or system used to collect brine from a brine tank, in an ion exchange or similar desalination unit.

**brine concentrator** A falling-film evaporator, with scale control, used to concentrate cooling tower blowdown, brine, and other wastes to about 20% solids; used with other units or devices (e.g., solar ponds or crystallizer) to solidify liquid wastes. *See also* zero discharge plant, seed slurry, demister or entrainment separator, and MVR evaporator.

**brine consumption factor (BCF)** A measure of the performance of an ion-exchange unit, equal to the ratio of regenerant equivalents required to the equivalents of the target ion removed. Also called brine use factor.

**brine disposal** The ultimate discharge of the waste stream from saline water conversion processes by such methods as controlled thermal evaporation, deep-well injection, evaporating (or evaporation) ponds, land application, ocean disposal, surface water discharge, discharge to wastewater collection system.

**brine draw** The process of drawing a brine solution into an ion exchanger during regeneration.

**brine eductor** A component of a brine tank's control valve designed to educt brine from the tank into an ion exchanger.

**brine ejector** A component of a brine tank's control valve designed to draw brine from the tank into an ion exchanger.

**brine electrolysis** A process used to produce chlorine industrially or adapted to produce disinfectants electrochemically in small, rural drinking water plants: the chlorine gas ($Cl_2$), produced by the oxidation of chloride ($Cl^-$) at the anode of an electrochemical cell, hydrolizes to hypochlorous acid (HOCl), and at the cathode water ($H_2O$) is reduced to hydrogen gas ($H_2$) and hydroxyl ion ($OH^-$), with the formation of short-lived oxidants (chlorine dioxide, hydrogen peroxide, ozone) as well as inorganic by-products (bromate, chlorate, chlorite) (AWWA, 1999):

$$2\,Cl^- \rightarrow Cl_2 + 2\,e^- \quad \text{(B-73)}$$

$$Cl_2 + H_2O \rightarrow HOCl + Cl^- + H^+ \quad \text{(B-74)}$$

$$2\,H_2O + 2\,e^- \rightarrow H_2 + 2\,OH^- \quad \text{(B-75)}$$

**brine heater** The part of a multistage flash evaporator for heating the feedwater.

**brine-injection well** A well used for the underground injection of waste brine from saline water conversion plants, a potential source of groundwater contamination.

**brinelling** Tiny indentations high on the shoulder of a water pump's bearing; a type of bearing failure.

**brine mud** Waste material, often associated with well drilling or mining, composed of mineral salts or other inorganic compounds.

**brine staging** A configuration designed to improve the performance of the reverse osmosis process by using the reject stream from one stage as feedwater for a succeeding stage. Also called reject staging.

**brine tank grid** A perforated platform in the bottom section of a brine tank of home water softeners that creates a zone where water can come in contact with the lower side of the dry salt stored above. As the water reaches up to the salt layer, it creates the brine makeup for regeneration.

**brine use factor (BUF)** A measure of the performance of an ion exchange unit, equal to the ratio of regenerant equivalents used to the equivalents of the target ion removed. Also called brine consumption factor.

**brine waste** The residuals stream resulting from the treatment of seawater or brackish water by such methods as reverse osmosis, electrodialysis, and electrodialysis reversal.

**brining** Regeneration of an ion exchange resin using a concentrated salt solution.

**briquette composting** A composting method in which a homogenized mixture of dewatered sludge and processed municipal refuse is placed in molds, compacted by vibration, and removed in the form of briquettes approximately 15 × 9 × 6 inches, which are stacked for aerobic and anaerobic decomposition. *See also* windrow process and sludge composting.

**bristle worm** A grazing organism that feeds on the film of a trickling filter.

**British degree** A unit of hardness, equivalent to approximately 14.3 mg/L as $CaCO_3$. *See also*

American degree, French degree, German degree, Russian degree.

**British manual formulation** A formula proposed in 1988 for the design of trickling filters, based on a multiple regression analysis (WEF & ASCE, 1991):

$$L_e/L_0 = 1/[1 + K\theta^{(T-15)}(As^m/Q^n)] \quad \text{(B-76)}$$

where $L_e$ = $BOD_5$ of settled filter effluent, mg/L; $L_0$ = influent $BOD_5$, mg/L; $K$ = first-order rate coefficient; $\theta$ = temperature coefficient; $T$ = wastewater temperature, °C; $As$ = media surface area and coefficient, $m^2/m^3$; $m$ = constant characteristic of filter medium; $Q^n$ = volumetric hydraulic rate and coefficient, $m^3/m^3$/day. Other trickling filter design formulas include Eckenfelder, Galler and Gotaas, Germain, Howland, Kincannon and Stover, Logan, modified Velz, NRC, Rankin, Schulze, and Velz.

**British thermal unit (BTU, btu, or Btu)** The quantity of heat required to raise the temperature of one pound of water one degree Fahrenheit at 39°F; used as the standard for comparing heating values of fuels. 1 BTU = 0.252 calorie = 1.055 joules.

**broad area treatment** A treatment method to stabilize a disturbed area by covering it with topsoil and an appropriate seed.

**broadcast chemicals** Chemicals that are spread over entire areas, e.g., pesticides, herbicides, and fertilizers.

**broad-crested weir** A weir with a substantial crest width in the direction of flow over it and no appreciable bottom contraction of the nappe; all weirs having a crest thickness more than 60% of the nappe thickness are considered broad-crested. Also called long-based or wide-crested weir.

**broad irrigation** An operation that applies wastewater and/or sludge to agricultural lands. Its primary purpose is wastewater disposal, whereas a similar operation, called sewage or wastewater farming, serves a dual purpose: wastewater disposal as well as land irrigation and fertilization.

**broad-screen analysis** A chemical analysis that looks for unknown substances.

**Broad Street pump** The pump in London identified by Dr. John Snow as being the source of the 1854 cholera outbreak that killed 700 people in 17 weeks (including 500 fatal attacks in 10 days). Most of the victims used water from the pump's well, adjacent to a leaky sewer that served the house at No. 40 Broad Street where lived the original cholera case. Dr. Snow's pioneering analysis has been cited as a classic epidemiological study.

**broke** A waste generated in the production of paper.

**bromamine** A compound of hypobromous acid (HOBr) and ammonia ($NH_3$), formed, for example, during water chloramination; it is similar to chloramines. Monobromamine and dibromamine are, respectively, $NH_2Br$ and $NHBr_2$.

**bromate** A salt of bromic acid, containing the bromate ion ($BrO_3^-$). Ozonation of bromide-containing water may result in the formation of bromate, an emerging inorganic contaminant, highly soluble in water, a probable human carcinogen with a current MCL of 10 μg/L (which may soon be revised to 5 μg/L). One pathway of bromate formation in water is through the formation of hypobromite ($OBr^-$) and bromite ($BrO_2^-$), and another through the formation of the hypobromous acid (HOBr) and hypobromite radical (BrO):

$$\text{(unbalanced)}\ O_3 + Br^- \rightarrow OBr^- \rightarrow BrO_2^- \rightarrow BrO_3^- \quad \text{(B-77)}$$

$$\text{(unbalanced)}\ O_3 + Br^- \rightarrow HOBR/OBr^- \rightarrow BrO \rightarrow BrO_3^- \quad \text{(B-78)}$$

Lowering the pH of the water may reduce bromate formation. Also, activated carbon may convert the bromate ion to bromide ($Br^-$).

**bromate ion ($BrO_3^-$)** The highest oxidation state of bromide ($Br^-$).

**bromcresol green** *See* bromocresol green.

**bromic acid ($HBrO_3$)** An acid produced from the reaction of barium bromate with sulfuric acid, stable only in dilute solutions, and used in pharmaceutical products.

**bromide ($Br^-$)** A salt of hypobromous acid, e.g., sodium bromide (NaBr), or any compound of bromine such as methyl bromide ($CH_3Br$). An inorganic contaminant, highly soluble in water; acceptable daily intake recommended by the World Health Organization: 1.0 mg/kg for total inorganic bromide or about 7.0 mg/L in drinking water. It finds its way into drinking water sources from saltwater intrusion, connate water, or industrial pollution. *See also* bromate.

**bromide–FAC ratio ($Br^-/Cl^+$)** The molar ratio of the bromide ($Br^-$) concentration to the free available chlorine (FAC) in water, which indicates the relative amounts of hypobromous acid (HOBr) and hypochlorous acid (HOCl), and affects bromine substitution and trihalomethane formation. Also called initial [$Br^-$]/average [$Cl^-$] molar ratio.

**bromide–organic carbon ratio ($Br^-/TOC$)** The ratio of the weight of bromides ($Br^-$) to the weight of total organic carbon (TOC) in water at the point of chlorination, an indication of the potential for bromine substitution and trihalomethane formation.

**bromide utilization** The ratio of the molar sum of bromine in individual trihalomethanes to the molar amount of bromide in water, an indication of actual bromide substitution.

**brominated alcohol** An alcohol that contains bromine; see bromohydrin.

**brominated organic** An organic compound containing bromine. Also called bromine-substituted organic.

**bromination** The use of bromine to destroy pathogens, in particular to inactivate viruses in drinking water of low turbidity. Bromine reactions in water are similar to but not as fast as those of chlorine. *See also* chlorination and iodination.

**bromine (Br)** A brownish-red, fuming halogen used in water disinfection as a chlorine-bromide mixture (bromine chloride), used in pharmaceuticals, dyestuffs, photographic materials, etc. Atomic number = 35. Relative atomic mass = 79.904. It is the only liquid nonmetallic element; relatively soluble in water, toxic at about 60 mg/kg, corrosive, irritating to mucous membranes, with an unpleasant odor. Free bromine ($Br_2$) or the oxidized form of bromide ion ($Br^-$) is a strong oxidant, not found naturally; concentrated brines or saline deposits are commercial bromine sources. Alone, it is an effective disinfectant (particularly of swimming-pool water), more expensive than chlorine; it also forms trihalomethanes, hypobromous acid (HOBr), hypobromite ion ($OBr^-$), and bromamines.

**bromine atom (Br)** *See* bromine.

**bromine chloride** A halogen compound combining bromine and chlorine (BrCl):

$$Br_2 + Cl_2 \rightarrow 2\ BrCl \qquad \text{(B-79)}$$

It remains in equilibrium with the components in solution or in a gas mixture. Used as a disinfectant, bromine chloride is a more powerful germicide than either gas alone. Also called chlorine bromide.

**bromine disinfection** Same as bromination.

**bromine incorporation factor** A measure of the degree of substitution of bromine in haloacetic acids (HAA) or trihalomethanes (THM) equal to (a) the ratio of the molar sum of bromine in the HAAs to the molar amount of HAAs or (b) the ratio of the molar sum of bromine in the THMs to the molar amount of THMs.

**bromine-substituted byproduct** A disinfection by-product that contains bromine.

**bromine-substituted organic** An organic compound that contains bromine. Also called brominated organic.

**bromine-substituted trihalomethane** Same as total trihalomethane bromine.

**brominism** Same as bromism.

**bromism** A pathological condition caused by excessive bromide and characterized by skin eruptions or depression of the central nervous system; unlikely to be related to drinking water.

**bromoacetic acid** A compound of acetic acid of the general formula $CX_3COOH$ in which bromine replaces at least one of the Xs and hydrogen replaces the others.

**bromoacetone ($CH_2BrCOCH_3$)** A colorless and toxic liquid compound of bromoacetic acid used in tear gas and chemical warfare gas.

**bromoacetonitrile** An acetonitrile that contains bromine; general formula: $CX_3C{\equiv}N$ in which bromine replaces at least one of the Xs and chlorine or hydrogen replaces the others.

**bromobenzene ($C_6H_5Br$)** A synthetic organic compound used as a solvent, sometimes found in treated water, possibly a by-product of chlorination. It causes skin irritation and may affect the nervous system. Also called monobromobenzene and phenyl bromide.

**bromochloroacetic acid (BCAA) (CHBrClCOOH)** A haloacetic acid, by-product of water disinfection using chlorine.

**bromochloroacetonitrile (BCAN)** A haloacetonitrile, by-product of water disinfection using chlorine.

**bromochloroiodomethane (CHClBrI)** A trihalomethane and a disinfection by-product.

**bromochloromethane ($CH_2ClBr$)** A clear, colorless, volatile, nonflammable liquid with a sweet, chloroform-like odor, used in organic synthesis and in fire extinguishing agents. Same as chlorobromomethane. Also called methylene chlorobromide.

**bromocresol green** A reagent used in water analyses as an acid–base indicator, with a pH transition range of 3.8 (yellow)–5.4 (blue).

**bromocresol green–methyl red indicator** A chemical substance used as an indicator in the laboratory determination of alkalinity: it changes from bluish-gray at pH = 4.8 to light pink at pH = 4.6. *See also* methyl orange alkalinity, phenolphthalein alkalinity.

**bromodichloroacetic acid (MDCAA) ($CHBrCl_2$-COOH)** A haloacetic acid; a disinfection by-product.

**bromodichloromethane (BDCM) ($CHCl_2Br$)** A common trihalomethane and a disinfection by-product. Bioaccumulation of BDCM is a primary water quality concern; the USEPA recommends a bioconcentration factor of 3.75 in receiving waters. Commercially produced BDCM is used in organic synthesis and as a laboratory reagent. Also called dichlorobromomethane.

**bromodiiodomethane ($CHBrI_2$)** A trihalomethane and a disinfection by-product. Also called diiodobromomethane.

**bromoform ($CHBr_3$)** A colorless, heavy liquid; a trihalomethane and a disinfection by-product. Commercially produced bromoform is used in pharmaceuticals, fire-resistant chemicals, and solvents. Also called tribromomethane, methenyl tribromide, methyl tribromide.

**bromohydrin** A brominated alcohol that is a possible by-product of ozonation; e.g., 3-bromo-2-methyl-2-butanol.

**bromoketone (X–CO–Y)** A haloketone containing at least one bromine atom in each of the radicals X and Y.

**bromomethane ($CH_3Br$)** Another name of methyl bromide, a colorless, poisonous gas used as a solvent, refrigerant, and fumigant. Also called methyl bromide.

**bromopicrin ($CBr_3NO_2$)** A by-product of chlorination and ozonation formed during the disinfection of waters containing bromide. Also called tribromonitromethane.

**bromotrifluoromethane ($CBrF_3$)** A colorless gas with an ether-like odor. Also called bromofluoroform, trifluorobromomethane.

**bromphenol blue** A reagent used in water analyses as an acid–base indicator, with a pH transition range of 3.0 (yellow)–4.6 (purple).

**bromthymol blue** A reagent used in water analyses as an acid–base indicator, with a pH transition range of 6.0 (yellow)–7.6 (blue).

**Brønsted acidity** The property of a substance that can donate a proton to another substance, in accordance with the Brønsted–Lowry theory. *See also* Lewis acidity.

**Brønsted-Lowry theory** A concept developed in 1923 to define an acid as a proton donor and a base as a proton acceptor. This is a more comprehensive definition than the usual reference to the production of hydrogen ions (H+ → acid) and hydroxyl ions ($OH^-$ → base). *See also* Lewis concept, hydrated metal ion, conjugate acid–base pair, polyprotic acid.

**Brønsted's concept** Same as Brønsted–Lowry theory.

**Brønsted theory** Same as Brønsted–Lowry theory.

**bronze** A copper alloy, usually with tin or lead as the main alloying element; a commonly used material in water meters and valves.

**brood parasite** A parasite that entrusts the rearing of its young to members of a different species, e.g., cuckoo.

**brook trout** A common North American trout, an intolerant species that requires cold water, high dissolved oxygen concentrations, and low turbidity.

**brown algae** Algae of the primarily marine group Phaeophyta that have brown pigments in addition to chlorophyll. They include *Macrocystis pyrifera, Sargassum,* and *Turbinaria.*

**brown coal** A common name for lignite.

**brownfield** A site actually or potentially contaminated by hazardous wastes, sometimes in an old urban commercial area or in a recently developed industrial area. The presence of hazardous pollutants or contaminants may complicate the expansion, redevelopment, or reuse of the site. Also called a brownfield site.

**brownfield site** *See* brownfield..

**brownfields program** A program that permits the reclamation and reuse of former industrial properties under relatively lenient standards if public health and safety is not at risk.

**Brownian diffusion** The random motion of small particles caused by thermal effects, a mass transport process in flocculation and coagulation whose intensity depends on the fluid's thermal energy (the product of its absolute temperature by the Boltzmann's constant). Also called molecular diffusion. *See* Brownian movement.

**Brownian diffusivity** A characteristic ($D$) of Brownian movement that determines the rate of transport of particles to the surface of floc (AWWA, 1999):

$$D = 0.106\, BT/(\mu d) \qquad \text{(B-80)}$$

where $B$ = Boltzmann's constant, $T$ = absolute temperature, $\mu$ = absolute viscosity, and $d$ = diameter of the diffusing particle.

**Brownian flocculation** The aggregation of solid particles into larger but still microscopic particles, brought about by Brownian movement and contributing to enhance the agglomeration and removal of particles, particularly those in the size range of 0.001 to 1 micron. Brownian flocculation is significant in ozone treatment, which results in favorable surface chemistry changes. It is not important in the bulk fluid when mixing and convection are significant. Also called microflocculation, perikinetic flocculation. *See also* macroflocculation,

tion or orthokinetic flocculation, and flocculation concepts.

**Brownian motion** Same as Brownian movement.

**Brownian movement** or Brownian motion The random movement of colloidal or other small particles that results from the impact of molecules and ions dissolved or suspended in a solution or in a gas. Brownian motion contributes to the attachment mechanism of rapid rate filtration by bringing the suspended particles into contact with the media grains. It is also responsible for particle removal in perikinetic flocculation (also called microflocculation). (After Robert Brown, 1773–1858, Scottish botanist who first observed it in 1827).

**Brownie Buster** Equipment manufactured by Enviro-Care for the separation of organic solids.

**brownstock washing** The controlled process in the pulp and paper industry that separates the dark brown pulp from the black liquor, a step toward the recovery of chemicals from the liquor.

**brown tide** A discoloration of coastal surface waters caused by the proliferation of brown algae. *See also* red tide.

**brown water** Water with a certain brown coloration caused by the presence of manganese oxides and making it unsuitable for laundering and for some industrial processes such as dyeing and papermaking. *See also* black water, red water.

**brucellosis** A bacterial infection that often causes spontaneous abortions in animals and remittent fever in humans. Also called undulant fever, Malta fever, Mediterranean fever, Rock fever.

**brucite** The mineral magnesium hydroxide, $Mg(OH)_2$, found naturally in foliated crystals, used in magnesia refractories.

***Brugia malayi*** The species of mosquitoes that transmit Malayan filariasis.

**Brunauer–Emmet–Teller method** A laboratory method for determining the required surface area of activated carbon based on the Brunauer–Emmett–Teller or BET isotherm.

**Brunauer–Emmett–Teller (BET) isotherm** An extension of the Langmuir isotherm to multilayer adsorption proposed in 1938:

$$Q = AB \cdot C/\{(C_s - C)[1 + C(B - 1)/C_s]\} \quad \text{(B-81)}$$

where $Q$ = quantity adsorbed by unit mass of adsorbent at equilibrium, $A$, $B$ = two constants that are usually determined empirically, $C$ = the equilibrium concentration of adsorbate in the dilute solution, and $C_s$ = saturation concentration of adsorbate in water. *See also* BET surface area.

**Bruner-Matic™** An apparatus manufactured by the Bruner Corp. for the control of water and wastewater treatment operations.

**Brune's trap efficiency curves** A set of curves that relate the percentage of sediment trapped by a detention basin ($E$, %) to the capacity–inflow ratio ($r$) of the basin (e.g., in acre-feet/acre-feet per year). The capacity–inflow ratio is actually a measure of detention time. There are three separate curves for fine, medium, and coarse solids. For capacity–inflow ratios between 0.0035 and 0.0795, the efficiency varies from 20% to 85% on the median curve and may be estimated from the equation

$$E\,(\%) = 100\,[1 - 1/(1 + 100\,r)]^{1.2} \quad \text{(B-82)}$$

**brush aeration** The extended aeration variation of the activated sludge process operating in an oxidation ditch that provides an aeration time of 24 hours.

**brush aerator** A surface aeration device consisting of several brushes attached to a rotary drum, half-submerged in the center of a tank or basin. It is often used in oxidation ditch wastewater treatment plants. Also called a rotor.

**brush barrier** A structure used in the control of sediment on construction sites. It consists of tree limbs, weeds, vines, root mat, soil, rock, and other cleared materials placed at the bottom of a slope.

**brush discharge** A corona discharge between two electrodes, with long luminous streamers of ionized particles.

**BSA** Acronym of bovine serum albumin.

**bsCOD** Abbreviation of biodegradable soluble COD.

**BTEX** Acronym of benzene, toluene, ethylbenzene, xylene.

**BTF** Acronym of biotransfer factor.

**BTU, btu, or Btu** Acronym of British thermal unit. 1 BTU = 0.252 calorie = 1.055 Joule.

**BTU-Plus®** Filter media produced by Alar Engineering Corp.; it burns to inert ash.

**BTX** Acronym of benzene–toluene–xylene.

**B-type metal** Any of the atmophile metals, including arsenic (As), lead (Pb), mercury (Hg), selenium (Se), and tin (Sn), whose transport through the atmosphere exceeds their transport in streams. They are volatile, have low boiling points, and can be methylated or released to the atmosphere as vapors, e.g., from coal burning. They are potentially hazardous to health. Also called soft Lewis acid. *See also* A-type metal and class B ion.

**B-type metal cation** Same as class B ion.

**bubble** A system under which existing emissions sources can propose alternate means to comply with a set of emissions limitations; under the bubble concept, sources can control more than required at one emission point where control costs are relatively low in return for a comparable relaxation of controls at a second emission point where costs are higher.

**bubble aeration** The process of contacting gas bubbles with a liquid to transfer the gas (e.g., carbon dioxide, oxygen, ozone) to the liquid or to remove substances such as volatile organic compounds by stripping. *See* diffused aeration for detail.

**bubble generator** A device used in sludge digesters to supply gas to draft tubes.

**bubble policy** *See* emissions trading.

**bubble point** The gas pressure required to displace liquid from the pores of a wetted filtration membrane. *See* bubble-point pressure.

**bubble-point pressure** In membrane filtration, wet membrane capillaries are impermeable to bulk flow of a test gas until a pressure high enough to evaluate the liquid from the largest pore is reached. This pressure is commonly referred to as the bubble-point pressure and is defined as the minimum pressure at which a steady stream of bubbles is observed from the downstream side of a wet membrane. The bubble-point pressure $P$ (Pa or psi) is a function of the surface tension of the liquid $\Phi$ (N/m or lbf/ft), the contact angle between the membrane and the liquid ($\alpha$), the diameter of the largest pore ($d$, m or ft), and a shape correction factor $C$ (Farahbakhsh et al., 2003):

$$P = 4(C/d)\Phi \cos \alpha \qquad \text{(B-83)}$$

Also called bubbling pressure, gas entry pressure, or, simply, bubble point. *See also* bubble-point test, diffusive airflow test, and pressure decay test.

**bubble-point test** A test used to determine the integrity of a wetted membrane by submitting it to pressures below the bubble point (e.g., 2 bars or 29.4 psi) and identifying the fibers that emit a steady stream of bubbles. A low-concentration surfactant is used to identify the damaged fibers. *See also* bubble-point pressure, pressure decay test, and diffusive airflow test.

**bubbler** (1) A device for bubbling gas through a liquid. (2) A sprinkler head that delivers a relatively large flow of water for the irrigation of trees and shrubs. (3) A fountain that spouts drinking water.

**bubble rise velocity** The velocity of a rising air bubble in water or wastewater, similar to the settling velocity of a particle. *See* terminal rise velocity.

**bubbler system** (1) A simple, easy-to-maintain, apparatus for measuring liquid levels based on the principle that air bubbled through water at a pressure slightly above the static head produces a back pressure equal to the static head. (2) A common terminology for pneumatic-type differential level controller.

**bubbler-tube level indicator** An instrument that monitors water level using a bubbled air supply.

**bubbler-tube level-sensing transmitter** An electronic device that records water levels based on a signal from a monitoring gauge.

**bubble size** The average diameter of the gas (e.g., air, oxygen, carbon dioxide, or ozone) bubbles discharged from a diffuser at the bottom of the contactor in an aeration or ozonation system. Generally, the finer the bubble size and the longer the bubble residence (contact) time within the water, the greater the transfer of gas to the water.

**bubbling bed** The low-velocity mode of operation of a fluid bed furnace; because of the formation of combustion gas bubbles, like boiling water, the bed moves, expands, shoots upward, and falls back. The bubbles disappear with increasing velocities. *See also* circulating bed.

**bubbling-bed combustor** A fluidized-bed incinerator consisting of a vertical refractory-lined vessel, an air distributor, a fluidized bed of sand or similar inert material, and a freeboard. Waste is introduced and quickly dispersed into the hot bubbling bed. *See also* tuyère, circulating bed combustor.

**bubbling-bed furnace** A fluid bed furnace operating under bubbling-bed conditions.

**bubbling-bed incinerator** A fluidized-bed incinerator that operates under bubbling-bed conditions; commonly used for the incineration of wastewater sludges. Same as bubbling-bed combustor. *See also* circulating-bed incinerator.

**bubbling pressure** The gas pressure required to displace liquid from the pores of a wetted filtration membrane. *See* bubble-point pressure.

**bubonic plague** An acute, sometimes fatal infection caused by the bacterium *Yersinia pestis* and transmitted to humans by rat fleas. Symptoms include fever, weakness, and swelling of armpit, groin, and other lymphatic glands. A form of the disease, called Black Death, reportedly killed a quarter of the European population in the 14th century.

**Büchner funnel** A laboratory instrument consisting of a funnel with a perforated bottom and a disposable filter paper to evaluate wastewater and sludge dewaterability.

**Büchner funnel test** A laboratory test conducted to determine the sludge characteristics (drainability or filterability, specific resistance, chemical dosage requirement); the moisture reduction observed in a given time period is calculated by placing a given quantity of sludge and various conditioning agents on filter paper in a Büchner funnel and measuring the volume of filtrate. Also called specific resistance (SR) test. *See also* Atterberg test, capillary suction time (CST) test, compaction density, filterability constant, filterability index, shear strength, specific resistance, standard jar test, time to filter (TTF) test.

**Büchner funnel test apparatus** The apparatus used to determine the specific resistance of a sludge sample, consisting of a Büchner funnel fitted in a glass adapter and on a graduated volumetric cylinder. The adapter has a side arm connected to a vacuum pump and equipped with a vacuum gage. *See* Figure B-19.

**bucket dredge** A continuous chain of buckets, or a drag bucket, mounted on a scow for lifting excavated materials. Also called chain bucket, ladder dredge.

**bucket elevator** A conveying apparatus consisting of an endless chain or belt with attached buckets. *See also* belt conveyor.

**bucket latrine** A latrine that consists of a seat or squatting slab installed on a bucket or other container. The latter has sufficient capacity for a few days' excreta from a household, is placed outside the dwelling, and emptied by a scavenger or sweeper into a wheelbarrow or cart collector for transportation to a depot for treatment and reuse, or to a disposal site.

**bucket latrine system** The system including bucket latrines and a service for the collection and disposal of excreta; one of the least likely systems to provide health benefits because of the risks involved.

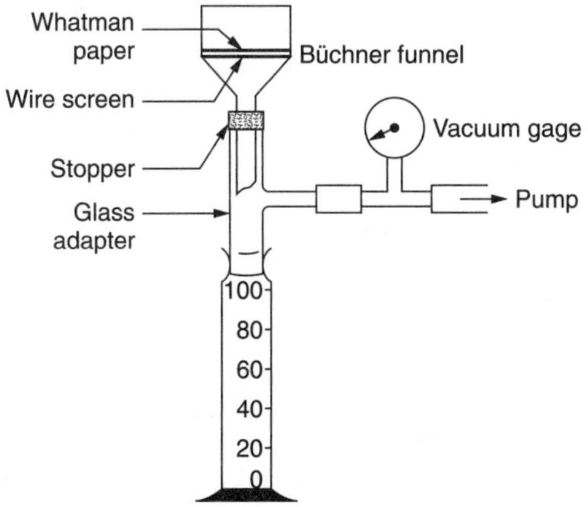


**Figure B-19.** Büchner funnel apparatus (adapted from Crittenden et al., 2005).

**bucket pump** A device that uses buckets or similar containers (e.g., attached to a conveyor belt) to lift water: the container is filled, raised to the desired level, and emptied.

**budding** The process of asexual reproduction of yeasts, free-living protozoans, and other organisms; a mother cell produces a small bubble that grows until it reaches the same size as the mother cell and is then separated by a crosswall. Also called schizogony. *See also* gemmation.

**budget** A financial plan based on an itemized estimate of expected income and expenses for a given time period in the future.

**budgeting** The process of preparing a budget.

**BUF** Acronym of brine use factor.

**buffalo gnat** *See* blackfly.

**Buffalo green monkey (BGM) cell** *See* Buffalo green monkey kidney cell.

**Buffalo green monkey kidney (BGMK) cell** A common cell line used for cultures ("tissue cultures") to grow human enteric viruses in the laboratory. When this cultured cell is inoculated with a sample that contains viruses, they develop and destroy the cell in about 12 days.

**buffer** A solution or liquid whose chemical makeup neutralizes acids or bases without a great change in pH; for example, the combination of a weak acid (such as acetic acid, $CH_3COOH$) or base and one of its salts (such as sodium acetate, $CH_3COONa$). *See also* common ion effect. Buffers are used to stabilize the pH of solutions. Carbon dioxide ($CO_2$) and its related species [carbonic acid ($H_2CO_3$), bicarbonate ($HCO_3^-$), carbonate ($CO_3^{2-}$)] constitute a significant buffering capacity of natural waters. *See* heterogeneous buffer system, homogeneous buffer system.

**buffer action** The effect of some ions or substances in resisting a change in pH of a solution or in neutralizing an acid or base.

**buffer capacity** (1) The capacity of a solution or liquid to neutralize acids or bases; e.g., the capacity of water to resist changes in pH. It is a measure of the quantity of acid or base that can be added to a solution before its pH changes significantly, e.g., the number of moles of buffer to change the pH by one unit. Buffer capacity is also measured by the buffer index, an important parameter in the design of corrosion control systems. (2) The pH change caused by the addition of a given amount of acid or base to a solution. The rate of pH change is minimum when (Droste, 1997)

$$pH = pK_A \quad \text{or} \quad pH = pK_B \qquad \text{(B-84)}$$

and the salt molar concentration equals the acid molar concentration for a basic buffer or the base molar concentration for an acidic buffer. $K_A$ and $K_B$ are equilibrium constants. Also called buffering capacity or buffer intensity.

**buffered** Said of a solution that can resist changes in pH.

**buffer index (β)** A measure of buffer capacity or buffer intensity, often taken as the reciprocal of the slope of the titration curve:

$$\beta = dC_B/d\text{pH} = -dC_A/d\text{pH} \qquad \text{(B-85)}$$

where $C_A$ and $C_B$ are the quantities of strong acid or strong base in moles/liter required to produce a change of dpH.

**buffering capacity** Same as buffer capacity.

**buffer intensity** Same as buffer capacity.

**buffer solution** A compound composed of a salt and a acid or a base that is used to stabilize the pH of a solution by producing or consuming hydrogen ions.

**buffer strip** A strip of grass or other erosion-resisting vegetation between or below cultivated strips or fields, or separating a waterway (ditch, stream, creek) from an intensive land use (e.g., a farm, a residential or commercial subdivision). It is intended to remove sediment, organic matter, and other pollutants from runoff or wastewater. Also called filter strip, grassed buffer, and vegetated filter strip.

**buffer stripping** The cultivation of narrow strips of land across a slope, rather than parallel to it, to reduce soil erosion.

**buffer zone** (1) In stormwater management, a vegetation zone that can be used to spread flows and trap sediment in the vicinity of water bodies. (2) An area maintained around a wastewater spray irrigation field, in addition to the treatment and other requirements, to protect the population against the potential transmission of disease by the spread of droplets through the air.

**bug** (1) A virus or any other microorganism. (2) A defect or imperfection in a computer program or in the hardware.

**builder** An alkaline phosphate or other chemical added to a soap, detergent, or other cleaning agent to increase its effectiveness (e.g., adjust its alkalinity and soil-suspending power). Builders inactivate water hardness by sequestration, precipitation, or ion exchange; they include complex phosphates, sodium carbonate, and sodium aluminosilicate. Builders also (a) supply alkalinity to assist cleaning, (b) provide buffering, (c) aid in keeping removed soil from redepositing during washing, and (d) emulsify oily and greasy soils. *See also* built detergent, built soap.

**building footprint** The surface occupied on the ground by a building; the dimensions and area of the horizontal projection of the building. *See also* small-footprint technology.

**building subdrain** The portion of the plumbing system of a building that does not drain by gravity into the sewer system. Also called house subdrain.

**building trap** A trap in a building drain to prevent air from the sewer system from flowing back into the building. Also called house trap.

**build-own-operate (BOO)** A project implementation approach in which privately owned companies design, finance, construct, operate public works such as water and wastewater facilities. *See also* privatization.

**build–own–operate–transfer (BOOT)** A project implementation approach in which privately owned companies design, finance, construct, and operate public works such as water and wastewater facilities, and transfer such facilities to a municipality or other governmental agency after a definite period. *See also* privatization.

**buildup** The increase in water table elevation around a recharge well or spreading basin. *See also* drawdown.

**buildup of emulsified grease** A common problem in the operation of depth filters used in wastewater treatment, resulting in an increase in head loss and a reduction of filter run. Control measures include air scour, surface wash, and steam cleaning.

**built detergent** A cleaning product that contains both a surfactant and a builder; sometimes referred to as a heavy-duty laundry detergent.

**built soap** A cleaning product that contains both a soap and a builder, designed for general purposes, especially laundering.

***Bulinus*** The principal snail genus infected by the pathogens *Schistosoma haematobium* and *S. intercalum.* It is found mainly in still or very slowly moving water, small pools, and water holes.

**bulk biosolids** Biosolids that are hauled in bulk, as compared to biosolids distributed in individual bags. The USEPA regulates the application of bulk biosolids.

**bulk density** (1) The dry mass of deposited sediment, powdered or granulated solid material per unit of volume; mass is measured after all water has been extracted, but total volume includes the volume of the material and the volume of air space between the grains. The bulk density of water and wastewater treatment chemicals relates to the material as shipped by the supplier. Also called dry density. (2) The bulk density of a granular bed is the mixed density of the grains and surrounding water, calculated as follows:

$$P_b = (1 - \varepsilon) P_s + P \cdot \varepsilon \qquad \text{(B-85)}$$

where $P_b$ = bulk density, $\varepsilon$ = bed porosity, $P_s$ = solids density, and $P$ = density of water. *See* intermixing.

**bulked activated sludge** Same as bulked sludge.

**bulked sludge** Activated sludge that has undergone the bulking phenomenon; the sludge does not settle well because of its excessive volume. *See* more detail under sludge bulking.

**bulk flow** The displacement of a liquid as a volume by the phenomena of convection and advection, as compared to diffusion and dispersion. Bulk flow of substances is common in water and wastewater treatment reactors.

**bulking** An operational problem of activated sludge plants whereby the sludge does not settle well because of its excessive volume. *See* sludge bulking for more detail.

**bulking activated sludge** Activated sludge that is undergoing the bulking phenomenon, which usually occurs when the plant operates with excessive aeration or with insufficient organic loading. *See also* sludge bulking.

**bulking agent** (1) An organic or inorganic material added to sludge (or other feed substrate) in a composting operation to provide structural support and/or supplemental carbon, increase the porosity of the mixture, and promote uniform air flow while reducing the bulk weight and the moisture content. Common bulking agents used in sludge composting include bark, brush (chipped), compost (recycled), leaves, lumber waste (ground), peanut hulls, rice hulls, rubber tires (shredded), sawdust, solid waste (shredded), wood chips, and yard waste. *See also* amendment (2). (2) In the lime stabilization (of sludge) process, excess lime addition results in the precipitation of calcium carbonate ($CaCO_3$) and unreacted lime [$Ca(OH)_2$] that serve as bulking agents to increase porosity and resistance to compression.

**bulking material** Dry material added to improve the porosity of compost substrate; bulking agent.

**bulking microorganisms** Microorganisms that impart poor settling and thickening characteristics to (activated) sludge. They are mostly fungi and filamentous bacteria. *See Sphaerotilus* or sewage fungus, *S. natans, H. hydrossis, Thiothrix, Beggiatoa, Microthrix parvicella, Nocardia* sp. type 021N bacteria, and filamentous sulfur bacteria. Less often, some microorganisms cause viscous or zoogleal bulking. The opposite microorganisms are called floc formers. Nonfilamentous bulking is rare, caused by zooglea-like organisms.

**bulking sludge** A low-density, poorly settling activated sludge that results from the predominance of filamentous organisms or the presence of entrained water in individual cells. It sometimes results in poor plant performance due to high concentrations of suspended solids in the effluent. Remedial measures include modifying the dissolved oxygen concentration, increasing the food-to-microorganism ratio, modifying the sludge recycle and waste ratios, and chlorinating the influent or the return sludge. *See also* filamentous bulking, hydrous bulking, and viscous bulking.

**bulk media filter** A filtering device used for the biological treatment of odorous air, consisting of a bed of soil, compost, or other media. *See* Figure B-20, compost filter, soil filter.

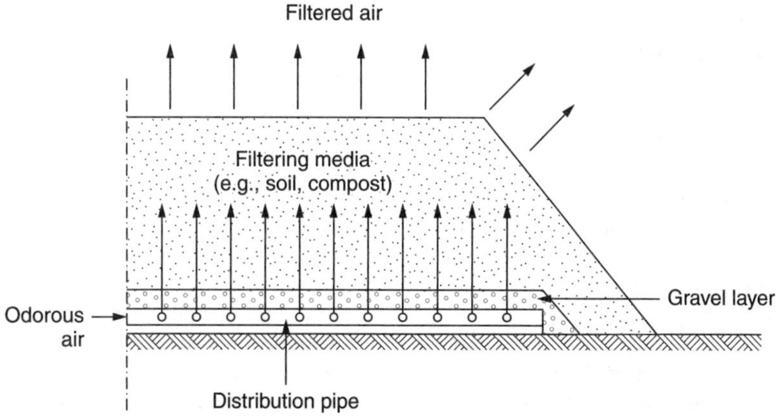


**Figure B-20.** Bulk media filter (adapted from Water Environment Federation, 1991).

**bulk modulus** The coefficient of elasticity of a substance; the ratio of a pressure on a volume to the resulting fractional change in volume:

$$K = \Delta P/(\Delta V/V) \qquad \text{(B-86)}$$

where $K$ = bulk modulus, $\Delta P$ = change in pressure, $\Delta V$ = change in volume, and $V$ = initial volume. At normal temperature and pressure, the bulk modulus of water is 2.15 kN/m$^2$. Also called bulk modulus of elasticity.

**bulk modulus of elasticity** Same as bulk modulus.

**bulk sewage sludge** Sewage sludge that is not sold or given away in a bag or other container for application to the land (EPA-40CFR503.11-e).

**bulk solution transport** The first step in the adsorption process, in which, by advection and dispersion, organic materials to be adsorbed (adsorbates) move through the bulk liquid to the boundary layer of fixed film of liquid around the adsorbent particles. *See also* film diffusion transport, pore (or internal) transport, and adsorption (or adhesion).

**bulk specific gravity, saturated surface dry** The specific gravity of a sample of granular material, starting with a reproducible saturated dry surface. *See* apparent specific gravity, bulk density.

**bulk water** In a sludge suspension, water that is not bound to the sludge particles and does not move with them; water that is relatively easy to remove by such dewatering methods as centrifugation and vacuum filtration. Also called free water. *See also* bound water.

**bulrush** One of the tall, emergent, grasslike vegetation species (*Scirpus* spp. and *Typha* spp.) commonly found in natural wetlands.

**bump** The release of nitrogen gas accumulated in a denitrification filter, through backwashing with water for 1–5 minutes, to reduce head loss across the filter. Bumping is required 5–16 times per day. Also called backwash bump.

**bumping** (1) Releasing accumulated nitrogen gas in denitrification filters through backwashing. Also called backwash bumping or filter bumping. (2) A flow surge in a rapid filter.

**bump joint** A joint that connects large riveted or welded steel pipe sections with one flared end and the other end adequately shaped; it is used in bends of up to five degrees per section.

**BUN** Acronym of blood-urea nitrogen.

**Bunsen burner** A burner commonly used in chemical laboratories that produces a hot, nonluminous flame from a mixture of air and gas. Named after the German chemist Robert Wilhelm Bunsen (1811–1899).

**buoyancy** The ability, tendency or power of a body to float or rise in a fluid; the force supporting a floating body. The upward pressure exerted by a fluid on an immersed body results from the buoyant force, which is also called buoyancy force or hydrostatic uplift. Archimedes' principle states that the buoyant force ($F$), equal to the weight of the displaced fluid, acts vertically upward at the buoyancy center or center of the displaced mass:

$$F = g(\rho'/\rho - 1) \qquad \text{(B-87)}$$

where $F$ is the buoyant force per unit mass, $g$ is the acceleration of gravity, $\rho'$ and $\rho$ are, respectively, the density of the buoyed fluid and that of the surrounding fluid. In the atmosphere, a buoyant force results from a local temperature increase:

$$F = g(T/T' - 1) \qquad \text{(B-88)}$$

where $T$ and $T'$ are, respectively, the temperature of the heated air and that of the surrounding environment, and the expression $(T/T' - 1)$ is called the buoyancy factor. A body immersed in a liquid will rise to the surface, float in the liquid, or sink to the bottom, depending on whether its weight is less than, equal to, or larger than the hydrostatic uplift. The design of ships, the operation of submarines, and the design of pipelines buried in waterlogged soils use buoyancy principles. Buoyancy also explains the floating of solids that were originally heavier than a surrounding fluid, e.g., in dissolved-air flotation.

**buoyancy factor** *See* buoyancy.

**buoyancy force** *See* buoyancy.

**buoyant force** *See* buoyancy.

**buoyant weight** In an expanded granular bed, the buoyant weight of the media grains is the pressure drop after fluidization:

$$\Delta P = H\rho g = L(\rho_s - \rho)g(1 - \varepsilon) \qquad \text{(B-89)}$$

where $\Delta P$ = fluidization pressure drop or buoyant weight, $H$ = head loss in depth of bed, $\rho$ = fluid density, $g$ = gravitational acceleration, $L$ = depth of bed, $\rho_s$ = mass density of media grains, and $\varepsilon$ = bed porosity.

**burden** (1) In utility accounting, burden is overhead, the indirect costs incurred (e.g., clerical support and maintenance) and applied to such direct costs as labor, equipment, and chemicals. (2) The accumulation of a substance in an organism. *See* body burden. (3) The intensity of a helminthic infection. *See* worm burden.

**burette** A glass tube with fine graduation, fitted with a bottom stopcock, used to accurately dispense small quantities of solutions, e.g., during a titration.

**Burger Press** A belt filter press manufactured by EMO France.

**burial ground** A disposal site for radioactive waste materials that uses earth or water as a shield. Also called graveyard.

**buried gate valve** A gate valve installed in the ground in a box that extends to the ground surface.

**burlap bag method** The use of a burlap bag to apply copper sulfate to reservoirs. The bag is dragged through the water from a rowboat or a motor boat with enough chemical to provide a dose of 0.3 mg/L. The application of copper sulfate controls the growth of plankton (algae) in reservoirs. *See also* blower, continuous-dose, and spray methods.

**burned dolomitic stone** *See* dolomitic stone.

**burned lime** *See* burnt lime.

**burner** A heating device using natural gas, bottled gas, or any other fuel; the part of a gas fixture from which flame is produced. Burners are used to provide auxiliary fuel in multiple hearth furnaces. *See* Bunsen burner.

**burner turndown** The ratio of the maximum to the minimum capacity of the burner of a multiple hearth furnace, an important design parameter. A typical range is between 4:1 and 10:1. Also called turndown ratio or burner turndown ratio.

**burner turndown ratio** Same as burner turndown.

**burnout** The intermittent or periodic application of free chlorine to drinking water distribution systems, sometimes coupled with intense flushing, to overcome accumulated chlorine demand and control nitrifying organisms in a chloraminated system (Bryant et al., 1992). Also called free chlorine burnout.

**Burns and McDonnell treatment system (BMTS)** An intrachannel clarifier (ICC) developed for oxidation ditch wastewater treatment plants. It includes an aeration zone, a mixer, upstream and downstream baffles, submerged orifice discharge pipes, and bottom panels. *See* boat clarifier, Carrousel Intraclarifier™, pumpless integral clarifier, side-channel clarifier, side-wall separator.

**burnt lime** Another name for lime, a white or grayish-white, slightly soluble in water, calcined oxide (CaO). It may be obtained from ground limestone (calcite) or calcium carbonate ($CaCO_3$) and has many applications in water treatment, mortars, cements, etc. *See* quicklime for detail.

**burping** The periodic discharge of sludge from small package wastewater treatment plants that use a biological process. It was first thought that these plants could achieve complete oxidation, but excess sludge does build up and occasionally discharge in the effluent.

**bursting** *See* hydraulic bursting, pipe bursting, pneumatic bursting, static bursting.

**bushing** (1) A replaceable, short threaded tube or sleeve that screws into a pipe fitting to reduce its size. (2) The bearing surface for pin rotation when a chain revolves around a sprocket.

**Buswell–Boruff–Sykes relationship** A relationship proposed by A. W. Buswell and C. B. Boruff in 1932 and modified by R. M. Sykes in 2001 to estimate the amount of ammonia ($NH_3$), carbon dioxide ($CO_2$), hydrogen sulfide ($H_2S$), and methane ($CH_4$) produced under anaerobic conditions, knowing the composition ($C_vH_wO_xN_yS_z$) of the organic matter in the wastewater being treated (Metcalf & Eddy, 2003):

$$\begin{aligned} &C_vH_wO_xN_yS_z + (v - w/4 + x/2 + 3y/4 + z/2)H_2O \\ &\rightarrow (v/2 + w/8 + x/4 + 3y/8 + z/4)CH_4 \\ &+ (v/2 - w/8 + x/4 + 3y/8 + z/4)CO_2 \\ &+ yNH_3 + zH_2S \end{aligned} \quad \text{(B-90)}$$

**Buswell–Hatfield equation** An equation proposed by A. W. Buswell and W. D. Hatfield in 1939 for the conversion of organic matter in industrial wastes to carbon dioxide and methane:

$$\begin{aligned} &C_vH_wO_x + (v - w/4 - x/2)H_2O \\ &\rightarrow (v/2 - w/8 + x/4)CO_2 \\ &+ (v/2 + w/8 - x/4)CH_4 \end{aligned} \quad \text{(B-91)}$$

**Butachlor** An organic contaminant, included in USEPA's Phase II (unregulated).

**butadiene ($CH_2CHCHCH_2$ or $C_4H_6$)** A major hydrocarbon derived from butane or butene, used in the production of synthetic rubber, paint, plastics, etc.; a colorless and flammable gas at room temperature, which liquefies at –4.4°C. Also called biethylene, bivinyl, divinyl erythrene, pyrrolylene, vinylethylene.

**butanal** An aldehyde on the USEPA list for quarterly monitoring.

**butane ($CH_3CH_2CH_2CH_3$ or $C_4H_{10}$)** A colorless, flammable gas, used as fuel or to manufacture rubber. Also called diethyl, methylethylmethane.

**butane number** The volume of butane adsorbed per unit weight of activated carbon after passage of butane-saturated air through the material at a given temperature and pressure. *See also* phenol number, molasses number, iodine number.

**butanoic acid ($CH_3CH_2CH_2COOH$ or $C4H_8O_2$)** *See* butyric acid.

**butanol** Same as butyl alcohol.

**butanol wheel** An instrument used to measure the intensity of odors against a scale of various concentrations of butanol. *See also* scentometer and triangle olfactometer.

**butterfly gate** A gate that acts on a shaft inside a pipe, similar to a butterfly valve.

**butterfly valve** A valve that uses a rotating disk to regulate fluid flow in pipes or ducts; the stem-operated disk is parallel to the direction of flow when opened and perpendicular to the flow when closed; a shut-off valve usually found in pipes larger than 4 inches, and used for noncritical flow control.

**butter of arsenic** Another name of arsenic trichloride.

**butyl alcohol ($C_4H_9OH$)** Any of four flammable liquid alcohols used as solvents and in organic synthesis. Also called butanol.

**butylbenzene, *n*-butylbenzene,, *sec*-butylbenzene, *tert*-butylbenzene** Volatile organic chemicals on USEPA's list of principal organic contaminants for monitoring by water supply systems.

**butylbenzyl phtalate** An organic compound sometimes found in water supply wells. Benchmark concentration established by the USEPA is 3.0 mg/L.

**butyric acid ($CH_3CH_2CH_2COOH$ or $C_4H_8O_2$)** A rancid liquid, as in spoiled butter, whose esters are used as flavorings. It is one of three organic acids produced during anaerobic sludge digestion. Also called butanoic acid. *See also* acetic and propionic acids.

**BV** Abbreviation of bed volume.

**$BV_{50}$** Bed volumes to 50% breakthrough.

**BVF®** An anaerobic method of ADI Systems, Inc. for wastewater treatment. *See* ADI-BVF®.

**BVT** Acronym of bed volumes treated.

**$BVT_{50}$** Acronym of bed volumes treated (BVT) to 50% breakthrough. Same as $BV_{50}$.

**bw** Acronym of body weight.

**BWL** Acronym of bottom water level.

**bypass** A system of pipes, conduits, channels, gates, valves, etc. designed to divert flow from the main path or around a treatment unit, structure, device, fixture, or obstruction.

**bypass channel** A channel designed to carry excess floodwater from a stream or to divert water from a main channel. Also called flood-relief channel. *See also* floodway.

**bypass feeder** A device that feeds controlled quantities of chemicals for water or wastewater treatment. It consists of a tank containing the chemicals and through which a small stream is diverted for mixing and then returned to the main line. Also called diffusion feeder.

**bypass flow** The water that bypasses a hydraulic or other structure.

**bypassing** The diversion of water or wastewater around a structure or treatment unit.

**bypass line** *See* pump-bypass line.

**bypass valve** (1) A valve used to divert fluid and avoid exceeding a pressure limit. (2) A small pilot valve used to equalize pressure on both sides of a larger valve.

**by-product** (1) Material, other than the principal product, generated as a consequence of an industrial process. Examples are process residues such as slag from the production of iron in a blast furnace, distillation column bottoms, or compost from the fines of a waste-derived-fuel plant. The term does not include a co-product that is produced for the general public's use and is ordinarily used in the form in which it is produced by the process (EPA-40CFR261.1-3). *See also* disinfection by-product, oxidation by-product, ozonation by-product. (2) The concentrated solution of contaminants rejected into the feedwater of a membrane system. Also called concentrate.

**by-product material** (1) Any radioactive material (except special nuclear material) yielded in, or made radioactive by, exposure incident to the process of producing or utilizing special nuclear material. (2) The tailings or wastes produced by the extraction or concentration of uranium or thorium from ore processed primarily for its source material content, including discrete surface wastes resulting from uranium solution extraction processes. Underground ore bodies depleted by these solution extraction operations do not constitute by-product material within this definition (EPA-10 CFR 20.1003).

**by-product recovery** The recovery of exhausted materials used in an industrial process (including wastewater treatment) for reuse there or elsewhere. Examples: phosphoric acid, copper, nickel, and chromium through ion exchange in the metal-plating industries; caustic soda from cooking liquors through multiple-effect evaporation in specialty paper mills; methane gas from sludge digesters in wastewater treatment plants; sulfite-waste-liquor by-products from paper mills; waste blood from packing houses and slaughterhouses.

# C

**C** Chemical symbol of carbon.

**$C_8$** Another name of perfluorooctanoate or PFOA.

**$^{12}C$** Symbol of the carbon-12 atom. *See* atomic mass unit.

**$^{13}C$ or $C_{13}$** Chemical symbol of carbon-13.

**$^{14}C$** Symbol of carbon-14.

**CA** Acronym of cellulose acetate.

**Ca** Chemical symbol of calcium.

**$C_a$** Symbol of Cauchy's number.

**$Ca^{2+}$** Calcium ion.

**CAA** Acronym of Clean Air Act.

**$CaAl_2Si_2O_8$** Chemical formula of anorthite.

**CAAA** Acronym of Clean Air Act Amendments.

**CAAS™** *See* cyclic activated sludge system.

**$Ca_3(AsO_4)_2$** Chemical formula of calcium arsenate.

**CAB** *See* cellulose acetate (blend) membrane.

**cable-driven screen** A device used in wastewater treatment that includes a cable-driven mechanism to move the rake teeth through the screen openings.

**CableTorq** A sludge thickener of Dorr-Oliver, Inc., with an apparatus for automatic load response.

**$CaC_2$** Chemical formula of calcium carbide.

**$CaC_6H_{10}O_4$** Chemical formula of calcium propionate.

**$CaC_{12}H_{22}O_{14}$** Chemical formula of calcium gluconate.

**$CaCl_2$** Chemical formula of calcium chloride.

**$CaCl_2 \cdot 8NH_3$** Chemical formula of calcium chloride ammoniate.

**$Ca(ClO)_2$** Chemical formula of calcium hypochlorite.

**CaCl(OCl)** Chemical formula of calcium chloride hypochlorite or calcium oxychloride.

**$CaCN_2$** Chemical formula of calcium cyanamide.

**$CaCO_3$** Chemical formula of calcium carbonate.

**$CaCO_3 \cdot MgCO_3$** Chemical formula of dolomite.

**CAD** Acronym of computer-aided design or computer-aided drafting. Program or programs that enable engineers and designers to sketch and draft technical designs, mechanical parts, and illustrations on the computer. *See* AutoCAD.

**CAD/CAM** Computer-aided design/computer-aided manufacturing.

**cadastral map** A map showing such features as boundaries of public and private lands.

**CADD** Acronym of computer aided design and drafting.

**Cadmium (Cd)** (1) A heavy-metal element that accumulates in the environment. It is a soft, silvery-white element found as a sulfide or carbonate in copper, lead, and zinc ores. Atomic weight = 112.40. Atomic number = 48. Specific gravity = 8.65. Valence = 2. The metal itself, its carbonate and hydroxide are insoluble in water, but its chloride, nitrate, and sulfate are water-soluble. It has many industrial applications, e.g., electroplating, batteries, insecticides; it is also used to protect other metals against oxidation. At sufficiently high concentrations, cadmium may interfere with enzyme production and cause physiological effects in humans (high blood pressure, kidney damage, destruction of testicular tissue and red blood cells); *see* itai-itai. At 15 mg/L it may cause nausea and vomiting. It is a trace element that can accumulate in crops from sludge; there is a limit on the amount of sludge containing over 2.0 mg/L of cadmium that can be applied to land. Both the USEPA and WHO recommend a limit of 0.005 mg/L in drinking water. (2) The total cadmium present in the process wastewater stream exiting the wastewater treatment system (EPA-40CFR415.451).

**cadmium bronze** An alloy of copper containing 1% cadmium.

**cadmium cell** A cell having cadmium and mercury electrodes and cadmium sulfate as electrolyte.

**cadmium green** A strong green mixture of cadmium sulfide and hydrated chromium oxide used as a pigment in paints.

**cadmium red** A strong red or reddish mixture of cadmium salts used as a pigment in paints.

**cadmium reduction method** A method used to determine the concentration of nitrate in a sample of water or wastewater, based on the fact that nitrate is reduced to nitrite in the presence of cadmium.

**cadmium sulfide (CdS)** A light yellow or orange powder, insoluble in water, used as a pigment.

**cadmium yellow** Cadmium sulfide used as a pigment.

**CADRE®** A process developed by Vara International for the destruction of volatile organic compounds.

**CAE** Acronym of computer-aided engineering.

**caesium (Cs)** *See* cesium.

**$CaF_2$** Chemical formula of calcium fluoride.

**CAF index** *See* combined lamp aging fouling index.

**$Ca_5F(PO_4)_3$** Chemical formula of calcium fluorophosphate.

**CAG** Acronym of carcinogen assessment group.

**cage screen** A screen built in the form of a cage with bars, rods, or mesh for use in water or wastewater treatment.

**$Ca(HCO_3)_2$** Chemical formula of calcium bicarbonate.

**Cairox®** A proprietary form of potassium permanganate ($KMnO_4$) produced by Carus Chemical Co.

**caisson** A large watertight structure or chamber, open at the bottom, that is usually sunk or lowered by digging from the inside; used to gain access to the bottom of a stream or other body of water.

**cake** Short for sludge cake, i.e., dewatered sludge with a solid concentration sufficient to allow handling as a solid material.

**caked media/medium** Filter bed material that is larger than usual after years of continuous service. The grains with rounded or worn edges have coalesced into larger-sized particles and have a high uniformity coefficient. *See also* media caking.

**cake filtration** Filtration mechanism for filters by which solids are removed by straining at the entering face of the granular material, as in precoat filters and membrane filters. It also occurs in slow sand filters in conjunction with depth filtration. Cake filtration does not require chemical coagulation and sedimentation for a source water of good quality (e.g., low turbidity). The process is also used in the treatment of hazardous wastes. Solids accumulate as a cake of increasing thickness, with increasing resistance to flow, and the cake becomes the filter medium.

**cake filtration equation** A semi-empirical equation used to interpret test data in the design of rotary vacuum filters, assuming negligible flow resistance and constant pressure differential (Freeman, 1998):

$$Y = K[P^{(1-S)}/(\mu R)]^{0.5}(C^M/T^N) \qquad \text{(C-01)}$$

where $K$ = empirical constant, e.g., $2^{0.5} = 1.4142$ $Y$ = cake yield = weight of dry solids per unit filter area per unit time; $P$ = pressure differential across filter cake; $S$ = compressibility coefficient, from 0.2 to 0.6 for industrial waste sludge; $\mu$ = liquid viscosity; $R$ = constant of specific resistance in the cake; $C$ = Solids concentration in feed slurry; $M$ = empirical constant, e.g., 0.5; $N$ = empirical constant, e.g., 0.5; $T$ = filtration time.

**cake formation** The result of concentration polarization and a cause of fouling in membrane processes; it occurs when the majority of solids in the feed water are larger than the pores or molecular weight cutoff of the membrane. Also called gel formation. *See also* fouling indexes, pore narrowing, pore plugging.

**CakePress** A sludge dewatering press manufactured by Parkson Corp. in modular sections.

**cake space** The volume available in a filter for the formation of a sludge cake.

**cake yield** In vacuum filtration and similar filtration mechanisms, the weight of solids per unit filter area per unit time. *See* cake filtration equation.

**calandria** A series of vertical tubes used as the heating element of an evaporator.

**calcareous** Made of or containing calcium carbonate ($CaCO_3$) or other calcium compounds.

**calcareous spring** A spring whose water contains a significant concentration of dissolved calcium carbonate ($CaCO_3$).

**calcify** To make calcareous or become stone-like or chalky due to deposition of calcium salts.

**calcination** (1) The conversion of precipitated calcium carbonate ($CaCO_3$) into calcium oxide (CaO) by heating. *See* sludge calcination. (2) The thermal degradation (pyrolysis) of predominantly inorganic materials; in general, the conversion into calx by heating or burning.

**calcined baryta (BaO)** Also called barium monoxide, barium protoxide, barium oxide; *see* baryta (1).

**calciner** An apparatus used to lower the moisture and organic content of phosphate rocks. Also called a nodulizing kiln.

**calcining** Processing of an inorganic compound through heat or other procedure to alter its form and drive off a substance that was originally part of the compound. For example, in lime softening treatment, lime (CaO) can be recovered by addition of carbon dioxide ($CO_2$) to recalcine calcium carbonate ($CaCO_3$) sludge and remove magnesium; heating the sludge at high temperature (e.g., 900–1200°C) drives off the carbon dioxide, leaving quicklime.

**Calciquest** A liquid polyphosphate product of Calciquest, Inc.

**calcite** (1) The most common form of natural calcium carbonate ($CaCO_3$) found in a variety of crystalline forms; a major constituent of limestone, marble, chalk; a major mineral contributing hardness to natural waters; used as an agricultural lime. Also called whiting and unburned lime. (2) A white solid, essentially powdered calcium carbonate, used instead of clay as a coagulant aid (with alum) in the treatment of cold, soft waters of low turbidity. Also used in water treatment for neutralization, stabilization, or prevention of corrosion.

**calcite contactor** A device that consists of a bed of crushed limestone used to increase such characteristics of water as pH, alkalinity, and calcium carbonate concentration.

**calcite saturation index (CSI)** A saturation index for calcium carbonate ($CaCO_3$):

$$\text{CSI} = \log_{10}([\text{Ca}^{2+}][\text{CO}_3^{2-}]/K) \qquad \text{(C-02)}$$

where $[Ca^{2+}]$ = molar concentration of calcium ions, moles/L; $[CO_3^{2-}]$ = molar concentration of carbonate ions, moles/L; $K$ = solubility constant corrected for ionic strength. A solution is undersaturated, in equilibrium, or supersaturated, respectively, if CSI < 0, CSI = 0, or CSI > 0. *See also* Langelier index.

**calcium (Ca)** A soft, silver-white, chemically active metallic element of the alkaline earth group, with many industrial applications; abundant in the earth's crust and found most commonly as calcium carbonate ($CaCO_3$) in limestone, gypsum, and fluorite; an essential macronutrient for plants and animals. Atomic weight = 40.08. Atomic number = 20. Valence = 2. Specific gravity = 1.55. Melting point = 840°C. Boiling point = 1,484°C. Its compounds are widely used in water treatment (quicklime, slaked lime, bleaching powder) and in construction (gypsum). Calcium and magnesium salts are the principal constituents of hardness in natural waters; they contribute to the formation of scale and insoluble soap curds.

**calcium arsenate [$Ca_3(AsO_4)_2$]** A white, somewhat water-soluble, poisonous powder, used as an insecticide and germicide.

**calcium bicarbonate [$Ca(HCO_3)_2$]** A hypothetical compound of calcium and carbonic acid, often used to represent calcium alkalinity.

**calcium carbide ($CaC_2$)** A common, dark gray to black, industrial compound of calcium, derived from coke or anthracite by reaction with limestone or quicklime; it may combine with water to produce flammable and explosive gases (acetylene, $C_2H_2$) in sewers:

$$\text{CaC}_2 + 2\,\text{H}_2\text{O} \rightarrow \text{C}_2\text{H}_2 + \text{Ca(OH)}_2 \qquad \text{(C-03)}$$

The hydroxide by-product of this reaction is also called carbide lime.

**calcium carbonate ($CaCO_3$)** A white or transparent, chalky, crystalline, solid substance that is the most abundant calcium compound, (e.g., calcite, chalk, limestone, marble); only sparingly soluble in water, in which it is the principal cause of hardness and scales; used in dentifrices, polishes, paints, rubber, plastics, lime, cement, and in medicine. Also called unburned lime.

**calcium carbonate equivalent (mg/L as $CaCO_3$)** A convenient expression of the concentration of specified constituents in water and wastewater in terms of their equivalent value to calcium carbonate, which has a molecular weight of 100 and an equivalent weight of 50. For example, the hardness in water that is caused by calcium, magnesium and other ions is usually described as calcium carbonate equivalent. A calcium (atomic weight 40) concentration of 90 mg/L corresponds to a calcium hardness of 90 × 100/40 = 225 mg/L as calcium carbonate. An alkalinity due to 122 mg/L of bicarbonates ($HCO_3$, molecular weight 61) corresponds to a bicarbonate alkalinity of 122 × 100/61 = 200 mg/L as calcium carbonate. *See also* as calcium carbonate.

**calcium carbonate precipitation potential (CCPP)** The theoretical approximate amount of calcium carbonate that could precipitate on a pipe surface from a given water or dissolve in the water. It may be determined from the Rossum–Merrill formula, based on the correlation of equivalents of calcium precipitated to equivalents of alkalinity precipitated:

$$\text{CCPP} = 50{,}045\,(\text{TALK}_i - \text{TALK}_{eq}) \quad \text{(C-04)}$$

where all variables are in mg/L as $CaCO_3$, $TALK_i$ and $TALK_{eq}$ are, respectively, the initial alkalinity and the alkalinity at equilibrium. At equilibrium, $TALK_i = TALK_{eq}$, CCPP = 0, and there is no precipitation or dissolution of $CaCO_3$. When $TALK_i > TALK_{eq}$, there is oversaturation and CCPP indicates the concentration of calcium carbonate that should precipitate. When $TALK_i < TALK_{eq}$, there is undersaturation and CCPP indicates the concentration of calcium carbonate that should dissolve. *See* corrosion index for a list of corrosion control parameters.

**calcium carbonate precipitation potential index** An index that indicates the potential for precipitation or dissolution of calcium carbonate. *See* calcium carbonate precipitation potential.

**calcium carbonate saturation index** An index used in corrosion control and to evaluate the potential of water for forming and dissolving scales in piping, water heaters, and other equipment. *See* aggressiveness index, buffer intensity indexes, calcite saturation index, calcium carbonate precipitation potential, Ryznar index, Langelier saturation index.

**calcium carbonate solubility** An important consideration in the treatment of water to remove calcium (Ca) by precipitating calcium carbonate ($CaCO_3$). Theoretically, 0.14 meq/L of carbonate ($CO_3$) is required for complete calcium removal, but, in practice, more than 0.70 meq/L is provided through reaction with bicarbonate alkalinity ($HCO_3$) or the addition of soda ash (sodium bicarbonate, $Na_2CO_3$). Calcium carbonate solubility is also important in corrosion control; *see* Langelier index. *See also* common ion effect.

**calcium chloride ($CaCl_2$)** A common, white, lumpy solid, soluble salt of calcium; a drying agent and deicing salt, used in laboratories to remove water gas streams. It is produced as a byproduct of commercial processes or derived from the reaction of calcium carbonate ($CaCO_3$) with hydrochloric acid (HCl):

$$CaCO_3 + 2\,HCl \rightarrow CaCl_2 + H_2CO_3 \quad \text{(C-05)}$$

**calcium chloride hypochlorite [CaCl(OCl)]** One of the earliest chlorine products used as a bleaching powder in water disinfection. Also called chloride of lime. *See* chlorinated lime for detail.

**calcium cyanamide ($CaCN_2$)** A common compound of calcium; a gray-black, often lumpy powder, unstable in water, used in fertilizers and herbicides, produced by heating calcium carbide ($CaC_2$) and nitrogen ($N_2$).

**calcium cycle** The circulation of calcium as a plant and animal nutrient between land, sea, and the atmosphere, involving living organisms and powered by solar energy. Calcium taken up from the soil by plants passes through the food chain, leaks into or is captured by aquatic or terrestrial ecosystems, and eventually returns to the soil.

**calcium fluoride ($CaF_2$)** A common, white, crystalline compound of calcium, insoluble in water, found naturally as the mineral fluorite; used in metallurgy and in dentifrices. Also called fluorspar. When used in water fluoridation, it is often first dissolved in an alum solution that also serves for coagulation.

**calcium fluorophosphate [$Ca_5F(PO_4)_3$]** A mineral mixture of calcium, fluoride, and phosphate. *See* apatite.

**calcium gluconate ($CaC_{12}H_{22}O_{14}$)** A white powder, soluble in water, used as a dietary calcium supplement.

**calcium hardness** Hardness caused by calcium compounds.

**calcium hydrate** Same as calcium hydroxide or hydrated lime.

**calcium hydroxide [$Ca(OH)_2$]** A stable compound of calcium obtained by slowly adding water to quicklime (calcium oxide, CaO):

$$CaO + H_2O \rightarrow Ca(OH)_2 \quad \text{(C-06)}$$

Slightly soluble in water, it is used in water treatment, in wastewater treatment (phosphate precipitation or pH adjustment), in construction, as agricultural lime to raise the soil pH, and in the chemical industries. Also called calcium hydrate, hydrated lime, milk of lime, slaked lime, or whitewash. *See* hydrated lime for detail.

**calcium hydroxyphosphate** Another name of apatite, a mineral that is used as a specific ion-exchange medium for the removal of excess fluoride from drinking water.

**calcium hypochlorite [$Ca(OCl)_2$]** A salt of hypochlorous acid; a white, crystalline compound containing up to 70% available chlorine; frequently used as a bleaching agent or a water or wastewater disinfectant. As a dry powder, it does not conserve well and loses strength rapidly (about 0.013% per day). In water it forms hypochlorous acid, a strong oxidizing/disinfecting agent, according to this overall reaction:

$$Ca(OCl)_2 + 2\ H_2O \rightarrow 2\ HOCl + Ca^{2+} + 2\ OH^- \quad \text{(C-07)}$$

Also called chlorinated lime or bleaching powder. *See also* high-test hypochlorite, hypochlorination.

**calcium magnesium carbonate** A common mineral consisting of the carbonates of calcium and magnesium, occurring naturally in crystals or in masses. *See* dolomite for detail.

**calcium metasilicate [$CaSiO_3$ or $CaO.SiO_2$]** A white powder, insoluble in water, used in antacids and in paper making.

**calcium nitrate [$Ca(NO_3)_2$]** A common compound of calcium, a white solid used in fertilizers, fireworks, and explosives.

**calcium oxalate** A white crystalline powder, insoluble in water.

**calcium oxide** A common compound of calcium and another name for lime; a white or grayish-white, slightly soluble in water, calcined oxide (CaO). It may be obtained from ground limestone (calcite) or calcium carbonate ($CaCO_3$) and has many applications, e.g., in water treatment, mortars, cements, etc. *See* quicklime for detail.

**calcium oxychloride (CaClOCl)** The basic constituent of chlorinated lime. *See* chlorinated lime for detail.

**calcium permanganate [$Ca(MnO_4) \cdot 4\ H_2O$]** A violet, crystalline, solid compound of calcium and manganese, used as a disinfecting or deodorizing agent.

**calcium phosphates** Compounds of calcium and phosphorus, found in rocks and in animal bones, and used in fertilizers, dentifrices, food supplements, and baking powder.

**calcium propionate ($CaC_6H_{10}O_4$)** A white powder, soluble in water, used as a fungal inhibitor in baking products.

**calcium silicates** Compounds of calcium and silica: calcium metasilicate, dicalcium silicate, tricalcium silicate.

**calcium sulfate** A white solid known as the mineral anhydrite ($CaSO_4$) and as gypsum ($CaSO_4 \cdot 2H_2O$), a potential source of scaling in saline water conversion systems. It is used in construction, as a fertilizer, in industry, and often as seed material in brine concentrators. *See* seed slurry.

**calcium sulfate saturation** The point beyond which any further addition of calcium sulfate ($CaSO_4$) to a given solution will cause precipitation. $CaSO_4$ precipitation occurs in cation exchange when resin is regenerated with too strong a sulfuric acid ($H_2SO_4$) solution; it also occurs in electrodialysis systems.

**calcium sulfide (CaS)** A common compound of calcium; a yellow to light gray powder that smells like rotten eggs; it is used in paints, to produce hydrogen sulfide, and as a depilatory.

**Caldicot Screen** A self-cleaning bar screen made by Advanced Wastewater Treatment, Ltd.

**Caldwell–Lawrence approach** A method proposed in 1953 to quantify the potential for deposition of calcium carbonate ($CaCO_3$) in water piping and containers. *See also* Langelier saturation index.

**Caldwell–Lawrence (C–L) diagram** A graphical representation of a number of equilibrium relationships for calcium carbonate ($CaCO_3$), with acidity as ordinate and the excess of alkalinity over calcium as abscissa, all expressed in mg/L as ($CaCO_3$), based on Langelier's stability diagrams. Saturation conditions (pH, soluble calcium concentration, and alkalinity) can be determined at any point on the diagram. C–L diagrams can be used to estimate required dosages in water softening. Also called water conditioning diagram. *See also* calcium carbonate precipitation potential.

**Calgon®** A brand of sodium phosphate glass used in water softening. *See* sodium hexametaphosphate for detail.

**caliber** The inner diameter of a hollow cylinder, e.g., a tube or a pipe.

**calibration (instrument)** The process of verifying the precision of an instrument by comparison with a standard or reference.

**calibration (modeling)** Generally speaking, model calibration is the iterative process of comparing simulation results to measured data, sometimes collected in independent sets (e.g., flows, pressures, velocities, areas, depths, water surface elevations, concentrations, temperatures), and making modifications to assumed data so that the model simulates the system more accurately. Calibration of a water quality model, for example, may involve running the model and comparing the results to collected data in order to determine BOD, dispersion, reaeration, and other rate constants.

**caliche** (1) A surface deposit of sand, clay, or gravel impregnated with crystalline salts like sodium chloride (NaCl) and sodium nitrate ($NaNO_3$). (2) A hard deposit of calcium carbonate ($CaCO_3$), gravel, or sand in the subsoil of semiarid regions. Also called petrocalcic horizon, hardpan.

**caliciviruses** Waterborne and foodborne, single-stranded RNA viruses that cause acute gastroenteritis through the fecal–oral route, e.g., Hawaii agent, Montgomery County agent, Norwalk agent, and Snow Mountain agent. *See also* hepatitis E virus (HEV).

**calk** A material or substance (e.g., lead, oakum, plastic) used to fill or close seams or crevices in windows, tanks, etc. to make them watertight or airtight. Also spelled caulk or called calking.

**calking** (1) Same as calk. (2) The process of inserting a calking to make a structure waterproof or airtight. *See,* e.g., lead calking.

**callable bond** A bond subject to redemption prior to maturity.

**calomel** A common name of mercurous chloride ($Hg_2Cl_2$), a common salt of mercury, highly soluble in water. It is a white, tasteless powder used as a purgative and as a fungicide.

**calorie** The quantity of heat required to raise the temperature of one gram of water one degree C at 15°C; used as the standard for comparing heating values of fuels: 1 calorie = 4.184 joules = 3.966 BTU.

**calorific content** Same as calorific value.

**calorific value** The quantity of heat that can be obtained from a material per unit mass, i.e., the number of units of heat obtained by the complete combustion of a unit mass of the material. It is used in the comparison of different fuels as well as the design of sludge processing units and incineration facilities for industrial or municipal solid wastes. Examples of typical calorific values:

| | |
|---|---|
| Landfill gas (primarily methane and carbon dioxide) | 500 BTU/ft$^3$ |
| Coal | 10,000–14,000 BTU/lb |
| Petroleum products | 19,000–20,000 BTU/lb |
| Wastewater treatment sludge | 4,000–12,500 BTU/lb |

Also called calorific content, fuel value, heating value, heat content, heat value, thermal value. *See also* bomb calorimeter, Dulong formula, heat capacity, and sludge fuel value.

**Calver** A line of chemicals produced by Hach Company and used in the determination of calcium concentrations in aqueous solutions.

**calx** (1) Another name for lime; a white or grayish-white, slightly soluble in water, calcined oxide (CaO). It may be obtained from ground limestone (calcite) or calcium carbonate ($CaCO_3$) and has many applications, e.g., in water treatment, mortars, cements, etc. *See* quicklime for detail. (2) The oxide or ashy substance remaining after a mineral has been thoroughly burned. *See* calcinations.

**Cam-Centric®** A line of plug valves manufactured by Val-Matic. *See* plug valve (Cam-Centric).

**CA membrane** *See* cellulose acetate membrane.

**$CaMg(CO_3)_2$ or $CaMg(CO_3)_2$ (s)** Chemical formula of dolomite or calcium magnesium carbonate.

**Camp Nozzle** A plastic nozzle manufactured by Walker Process Equipment Co. for use in filter underdrainage systems.

**Camp sedimentation theory** The ideal sedimentation basin has four zones and only the overflow rate determines its overall removal efficiency of discrete particles. *See* inlet zone, settling zone, sludge zone, outlet zone, sedimentation zone,

**Camp–Shields equation** An equation proposed in 1942 to estimate the scouring velocity required to resuspend organic particles. *See* scouring velocity.

**Camp, Thomas R.** American sanitary engineer, author of articles on hydraulic analysis of water distribution, sewer design, velocity gradients, grit chamber design, lateral spillways.

***Campylobacter*** A genus of Gram-negative, microaerophilic, waterborne bacteria causing diarrhea in humans; only recently recognized (early 1970s), with a reservoir in animals.

***Campylobacter* enteritis** An enteric infection caused by the waterborne bacterial agent *Campylobacter fetus* ssp. *jejuni,* with varying symptoms that may include profuse and watery diarrhea accompanied by fever, abdominal pains, headaches, blood and mucus in stools, etc.

***Campylobacter fetus* ssp. *jejuni*** *See Campylobacter jejuni.*

**campylobacteriosis** Same as *Campylobacter* enteritis.

***Campylobacter jejuni*** The genus of bacteria causing *Campylobacter* enteritis, characterized by a high infective dose, a persistence of 7–10 days, an incubation period of 2–5 days, and a typical concentration of 10 million per gram of feces. Conventional chlorination is effective against this pathogen. In May 2000, an outbreak of this pathogen and *E. coli* 0157:H7 through contaminated drinking water killed seven people and made more than 2000 people sick in Walkerton, Ont. (Canada).

***Campylobacter pylori*** An organism identified in 1982 and now renamed *Helicobacter pylori.*

**$CaNa_2(SO_4)_2$** Chemical formula of glauberite.

**cancer** A disease characterized by the rapid and uncontrolled growth of aberrant cells into malignant tumors. Some limited data seem to indicate that (1) cancer is less prevalent in municipalities with a water supply system than those without a system and (2) municipalities supplied from polluted rivers have a higher cancer death rate than those supplied from underground sources.

**cancer potency slope** A parameter, derived from animal studies using multistage analysis, that indicates the magnitude of cancer threat from a substance or other agent. Also called q* slope.

**candela (Cd)** The Système International (SI) unit of luminous intensity, equal to that of a source of monochromatic radiation of frequency $540 \times 10^{12}$ Hz that has a radiant intensity of 1/683 watt per steradian.

***Candida albicans*** A yeast species proposed, instead of coliforms, as an indicator of swimming water pollution associated with infections of the respiratory tract, skin, and eyes. *See also Staphylococcus aureus.*

***Candida parapsilosis*** A yeast species proposed as an indicator of organisms that resist disinfection.

**candle** A unit of luminous intensity used prior to the adoption of the candela in 1979; it corresponded to the intensity of a wax candle or a fraction of the intensity of 45 carbon-filament lamps. *See also* candela.

**candlepower** A former measure of luminous intensity, expressed in candles. *See also* candela.

**Candy tank** A pyramidal sedimentation tank with large sloping sides, first designed in India around 1932 for water treatment; similar to the Imhoff tank used in wastewater treatment from 1906. Modifications of the Candy tank for application in the floc-blanket process include the introduction of multiple hoppers or troughs, a flat bottom, polyelectrolytes, and floc ballasting (AWWA, 1999). *See also* Spaulding Precipitator.

**cane sugar waste** Wastewater generated during the production of cane sugar, from such sources as cooling and condenser effluent, solid waste from the filter cake, spillage, scum leaks, washings, cleanings, and machinery operations. This waste, characterized by variable pH and high carbonaceous soluble BOD, is usually processed by neutralization, chemical treatment, and selective aerobic oxidation, with some recycling.

**canicola fever** A water-related disease caused by the spirochete *Leptospira canicola,* characterized by an acute fever, jaundice, and inflammation of the stomach and intestines. It affects humans and dogs. *See also* canine leptospirosis and Weil's disease.

**canine leptospirosis** A severe intestinal disease affecting dogs caused by spirochetes of the genus of water-related bacteria *Leptospira.* Also called canine typhus, Stuttgart disease, or Stuttgart's disease. *See also* leptospirosis.

**canine typhus** Same as canine leptospirosis.

**cannery waste** Wastewater originating from the seasonal processing of fruits and vegetables (trimming, culling, juicing, and blanching), characterized by high suspended, colloidal, and dissolved organic solids. Appropriate treatment: screening, lagooning, soil absorption or spray irrigation.

**canning industry** *See* cannery waste.

**Cannon™** A mixing device manufactured by Infilco Degremont, Inc. for use in digesters.

**Cannonball 2** A portable device manufactured by Biosystems, Inc. for the detection of multiple gases.

**$Ca(NO_3)_2$** Chemical formula of calcium nitrate.

**Cansorb** An activated carbon adsorber manufactured by the TIGG Corp.

**CaO** Chemical formula of calcium oxide or quicklime.

**$Ca(OCl)_2$ or $CaO_2Cl_2$** Chemical formula of calcium hypochlorite.

**$Ca(OCl)_2 \cdot 4H_2O$** Chemical formula of commercial calcium hypochlorite.

**$Ca(OH)_2$** Chemical formula of calcium hydroxide or slaked lime.

**$Ca_5(OH, F, Cl)(PO_4)_3$** Chemical formula of apatite.

**$[Ca(OH)_2]_{0.6}[Mg(OH)_2]_{0.4}$** Chemical formula of dolomitic hydrated lime.

**$Ca_5OH(PO_4)_3$** Chemical formula of hydroxyapatite.

**$(CaO)_{0.6}(MgO)_{0.4}$** Chemical formula of dolomitic quicklime.

**$CaO \cdot SiO_2$** Chemical formula of calcium metasilicate.

**$(CaO)_2 \cdot SiO_2)$ or $Ca_2SiO_4$** Chemical formula of dicalcium silicate.

**$(CaO)_3 \cdot SiO_2)$ or $Ca_3SiO_5$** Chemical formula of tricalcium silicate.

**CAOX** Abbreviation of carbon adsorbable organic halogen.

**cap** (1) A layer of clay or other impermeable material installed over the top of a closed landfill to prevent entry of rainwater and minimize leachate. (2) A fitting for the spigot or screw end of a pipe.

**capacitance** The part of impedance that resists the flow of electrical current.

**capacitive deionization (CDI)** A saline water conversion process in which the feedwater passes between low-voltage, parallel plates of porous carbon, serving as electrodes that attract ions of opposite charge. Mechanisms responsible for the removal of charged constituents include physisorption, chemisorption, electrodeposition, and/or electrophoresis. Capacitive deionization is classified as an emerging concentrate minimization and management method. *See also* carbon aerogel.

**capacitor** A device for accumulating a charge of electricity, consisting of two conductive plates of opposite charge, separated by a nonconductor. Also called a condenser.

**capacity** The ability of an installation to perform or provide a service, such as the ability of a basin to hold water (volume), the ability of a pump to raise wastewater (flow or power), or the ability of a treatment plant to process water or wastewater (flow in mgd, e.g., or pollutant removal in pounds per day). Also the ability of a utility to meet the demand of its customers. The capacity curve of a reservoir (or its storage-capacity curve) shows the relationship between its volume of water and the elevation of its water surface. The maximum discharge of a hydraulic structure or the maximum flow that a treatment plant can process is sometimes called its carrying capacity.

**capacity charge** A one-time charge assessed by a water, sewer, or other utility against new users, developers, or applicants for new service to recover part or all of the costs of existing or additional facilities. It is a means of charging the cost of development impact to those who cause it. Also called impact fee, capital recovery charge, or facility expansion charge. *See also* connection charge development impact fee and system development charge.

**capacity curve** (1) The storage-capacity curve of a reservoir. *See* capacity. (2) A graph of the capacity of an ion exchanger vs. regenerant levels.

**capacity factor** (1) The ratio of the output of an electrical generating plant to the total rating of its various generating components. Also called plant factor. *See also* load factor. (2) In an acid–base system, the number of protons required to reach a certain pH, i.e., the alkalinity or acid-neutralizing capacity (ANC) and acidity or base-neutralizing capacity (BNC) of a solution. Both ANC and BNC are quantitative factors as opposed to pH, which is considered an intensity factor.

**capacity–inflow ratio** A parameter used in the Brune's trap efficiency curves. It is the ratio of the storage capacity of a reservoir, basin, or pond to the annual runoff, or, essentially, the detention time in the unit.

**capacity restoration** The process of returning a component to its original capacity, e.g., by backwashing a filter or regenerating activated carbon beds, ion exchange membranes, and the like.

**CAPDET** A computer program prepared by the USEPA for evaluating the cost of wastewater treatment alternatives.

**CAP grade water** Water that meets the standards established by the College of American Pathologists. The standards cover three types of laboratory grade water: clinical, cell or tissue, and cultural.

**capillarity** A result of surface tension and adhesion between a liquid and a solid surface that draws the liquid upwards.

**capillary** A capillary tube.

**capillary action** Movement of water through small spaces due to capillary forces.

**capillary capacity** The moisture content of soil in the field some time after reaching saturation and after free drainage has practically ceased, or the quantity of water held in a soil by capillary action. *See* field capacity for detail.

**capillary condensation** The phenomenon of water vapor condensation in the microspores of the carbon adsorber of an air stripping unit, which begins when the relative humidity reaches a certain level. It reduces the performance of the unit because the condensed water competes with the contaminants for available adsorption sites.

**capillary electrophoresis (CE)** A technique that separates analytes (particularly ionic species in water) through differences in electrophoretic mobility. *See also* SPE/CE method.

**capillary flow** (1) The ascent or descent of a liquid in a tube of small diameter. (2) The movement of water through soil spaces above the water table as a result of pore surface attraction. *See* capillary rise. (3) The movement of suspended water in the aeration zone through interconnecting films on the surface of soil particles or on the walls of fractures. Also called capillary movement and film flow. *See also* capillary migration.

**capillary force** The molecular force that causes adhesion of a liquid to a solid surface, and causes the liquid to rise against the surface; for example, the movement of water through very small spaces.

**capillary fringe** The part of the vadose zone overlying the water table; a porous material that contains capillary water. The capillary fringe may extend or rise to the root zone and be subject to evapotranspiration. *See* subsurface water.

**capillary head** *See* capillary rise.

**capillary interstice** An interstice that holds water above the water table in the capillary fringe.

**capillary ion electrophoresis** A method used for the rapid analysis of ions (e.g., bromide, chloride, nitrate, organics) with low concentrations in solution.

**capillary migration** The underground movement of water in the zone of aeration by molecular attraction of the soil material.

**capillary moisture** *See* capillary water.

**capillary movement** The movement of suspended water in the aeration zone through interconnecting films on the surface of soil particles or on the walls of fractures. Also called capillary flow and film flow. *See also* capillary migration.

**capillary opening** A small opening that allows capillary flow.

**capillary percolation** Percolation through a capillary interstice.

**capillary rise** The result of capillary action, as may happen in small pores of soils or in small-diameter glass tubes. The following equation expresses the height of the capillary rise ($h$) as a function of the diameter ($d$) of the glass tube or soil pore, the surface tension of the liquid ($\sigma$), the angle of adhesion ($\alpha$) between the liquid and the tube or soil, and the specific weight ($\gamma$) of the liquid (Figure C-01). The capillary rise in soils varies from 3 cm in sand of 2 mm to more than 30 m in clay particles smaller than 0.002 mm.

$$h = 4\sigma \cos \alpha / d\gamma \qquad \text{(C-08)}$$

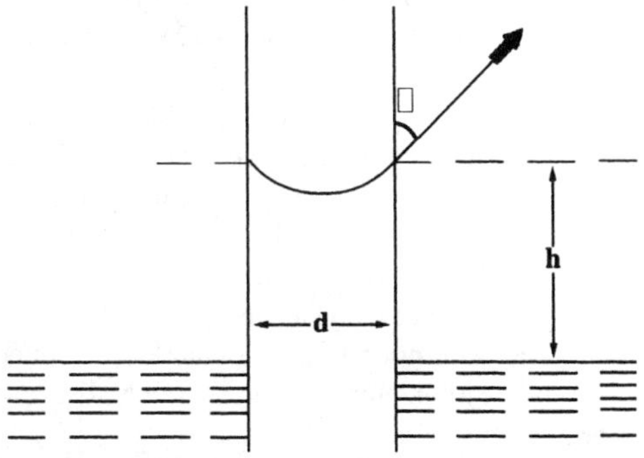


**Figure C-01.** Capillary rise (courtesy CRC Press).

The capillary head is the elevation difference between the final position of the meniscus and its position at a given time.

**capillary seepage trench** A trench used in subsurface disposal of septic tank effluents, having the bottom and lower part of the sides sealed so that water flows horizontally in a large area, thereby reducing the load in the immediate vicinity.

**capillary suction** The process whereby water rises above the water table into the void spaces of a soil due to tension between the water and soil particles.

**capillary suction test** *See* capillary suction time test.

**capillary suction time** The time, in seconds, required for conditioned sludge to yield a small volume of filtrate under the capillary suction pressure of dry filter paper.

**capillary suction time (CST) test** A fast and simple test designed to determine the optimum dose of conditioning chemical for sludge based on its capillary suction time. It measures the time required for free water to travel one centimeter from a 5–7 mL sludge sample on filter paper. *See also* Atterberg test, Büchner funnel test, compaction density, filterability constant, filterability index, shear strength, specific resistance, standard jar test, time-to-filter (TTF) test.

**capillary tube** A tube with a small bore; also called capillary.

**capillary water** (1) Water held by surface tension forces in the pores around soil particles; also called capillary moisture. (2) Subsurface water in the capillary fringe of the vadose zone. It comes from the intermediate zone by gravity and is held by capillary forces while in transit to the water table. *See* subsurface water.

**capillary zone** The zone that holds capillary water. *See* subsurface water.

**capillary zone electrophoresis** A method for separating charged particles by differential migration in an electric field; used, e.g., in the separation of amino acids, carboxylic acids, and proteins.

**capital** The net worth, or assets of an enterprise, a utility, etc., after deducting its liabilities.

**capital assets** Fixed assets; assets of a long-term or relatively permanent nature, e.g., land, buildings, patents.

**capital costs** Costs (usually long-term debt) of financing construction and equipment. Capital costs are usually fixed, one-time expenses.

**capital expenditure** An addition to capital or fixed assets, e.g., the acquisition of a new building.

**capital recovery charge** A one-time charge assessed by a water, sewer, or other utility against new users, developers, or applicants for new service to recover part or all of the costs of existing or additional facilities. It is a means of charging the cost of development impact to those who cause it. Also called capacity charge, impact fee, or facility expansion charge. *See also* connection charge development impact fee, and system development charge.

**capital recovery factor (*A/P, i, n*)** A factor used to convert a capital cost into equivalent periodic (e.g., annual) costs. It is the reciprocal of the present worth factor:

$$(A/P, i, n) = i(1 + i)^n/[(1 + i)^n - 1] \quad \text{(C-09)}$$

where $A$ = periodic (annual) cost, $P$ = present worth, $i$ = interest rate per period, and $n$ = number of periods. *See* financial indicators for related terms.

**Capitox** A modular wastewater treatment plant manufactured by Simon-Hartley, Ltd.

**$Ca_5(PO_4)_3F$ or $Ca_{10}(PO_4)_6F_2$** Chemical formula of fluoroapatite.

**$Ca_{10}(PO_4)_6(OH)_2$** Another formula for hydroxyapatite.

**Capozone®** A system engineered by Capital Controls Co., Inc. for the generation of ozone.

**caproic acid ($C_5H_{11}COOH$)** The common name of hexanoic acid, an oily, colorless or yellow liquid of the carboxylic group, obtained from fatty animal tissue or coconut oil, used in flavoring agents.

**capsid** The shell-like coat of protein that surrounds the nucleic acid core of a virus. This outer structure, composed of up to several hundred proteins, protects the viral core; it harbors receptor sites (for virus attachment) as well as electrical charges and hydrophobic groups. Also called protein coat.

**Capsular** A pumping station manufactured by Smith & Loveless, Inc.

**capsule** A rigid layer of protein that surrounds the wall of a bacterial cell; also called glycocalyx. *See also* slime layer.

**capsule phage** Bacteriophage that infect hosts through the host's polysaccharide or other outer layer. *See also* appendage phage, somatic phage

**Captivated Sludge Process** A wastewater treatment process developed by Waste Solutions, using a fixed film medium.

**Captor®** A proprietary, integrated, fixed-film activated sludge method of wastewater treatment developed by Waste Solutions, using free floating, 30 mm × 25 mm × 25 mm foam pads in the aeration tank. *See also* Bio-2-Sludge®, BioMatrix®, Kaldnes®, Linpor®, moving bed biofilm reactor, Ringlace®, submerged rotating biological contactor.

**capture** A measure of the performance of a centrifuge (used in sludge thickening and dewatering), defined as the percentage of thickened dry solids with respect to feed solids. *See* recovery for detail.

**Capture Flow** A three-stage filtering catch basin designed by Carson Industries to remove solid particles transported by stormwater.

**capture of rejected recharge** *See* rejected recharge.

**capture zone** The part of an aquifer that drains into a well.

**CAR™** An aerobic wastewater treatment process developed by ADI Systems, Inc., using a covered reactor.

**carbacryl** Acrylonitrile.

**Carball** Equipment manufactured by Walker Process Equipment Co. for the generation of carbon dioxide.

**carbamates** Salts or esters of carbamic acid ($NH_3CO_2$), organic compounds used in herbicides, fungicides, or insecticides. They persist in soils and have the potential to affect human skeletal muscles and nervous system.

**carbamate pesticide** A pesticide containing a carbamate as active agent. *See*, e.g., aldicarb, carbaryl, carbofuran, oxamyl.

**carbamic acid ($NH_3CO_2$)** A hypothetical compound known by its salts or esters.

**carbamide [$CO(NH_2)_2$ or $CH_4ON_2$]** A soluble nitrogen compound found in urine (about 24,000 mg/L); an amide derivative of carboxylic acid. *See* urea for detail.

**carbaryl** An unregulated organic chemical, used to kill earthworms and insects, and for thinning in apples.

**carbide** (1) A compound of carbon with a more electropositive element or group. (2) Same as calcium carbide ($CaC_2$).

**carbide lime** A by-product of acetylene production from the reaction of calcium carbide with water; see calcium carbide; sometimes used in the alkaline stabilization of sludge as a substitute for quicklime or for hydrated lime.

**carbinol** Same as methyl alcohol.

**carbocyclic compound** An organic compound with only carbon atoms in the ring, e.g., benzene or cyclopropane.

**Carbo Dur™** Granular activated carbon produced by U.S. Filter Corp.

**Carbofilt** Anthracite filter media produced by International Filter Media.

**carbofuran ($C_{12}H_{15}NO$)** A white-to-grayish, odorless, crystalline solid, synthetic organic chemical (2,3-dihydro-2,2-dimethyl-7-benzouranol-methylcarbamate), with a USEPA-set MCLG of 0.04 mg/L and MCL of 0.004 mg/L; used as an insecticide and nematocide. Solubility in water: 700 mg/L. It is one of the rare pesticides that can be removed by ordinary water treatment methods; it hydrolizes at common pH levels; it can also be adsorbed on powdered or granular activated carbon. It has potential neurological effects, causing headaches and nausea.

**carbohydrate** A compound containing only carbon (C), hydrogen (H), and oxygen (O), with a general formula of $(CH_2O)_n$ or $C_x(H_2O)_y$; for example, glucose ($C_6H_{12}O_6$), a monosaccharide. Carbohydrates also include disaccharides like sucrose (household sugar) and polysaccharides. The latter contribute to cell maintenance (energy) or cell integrity. The ratio of a wastewater's carbohydrates and fats (another source of energy) to nutrients is an important consideration in biological treatment.

**carbol fuchsin** A special stain used to identify mycobacteria, which do not stain with common dyes. This acid-fast procedure uses a solution of ethanol (95%) and hydrochloric acid (3%) to wash organisms, which eliminates the carbol fuchsin from all organisms except the acid-fast bacteria.

**carbolic acid ($C_6H_5 \cdot OH$)** Same as phenol, a monohydroxy derivative of benzene.

**carbon (C)** (1) An element present in all materials of biological origin. Atomic weight = 12.011. Atomic number = 6. Specific gravity varies from 1.8 (amorphous carbon) to 3.5 (diamond). Valence = 4. It forms innumerable compounds, particularly the carbohydrates and chains containing carbon and other atoms close to it in the periodic table. It is the primary element of organic matter, but carbon dioxide ($CO_2$) and its compounds are inorganic. It occurs in a pure state as diamond or graphite or in an impure state as charcoal. (2) Same as activated carbon.

**carbon 12** or **carbon-12 ($^{12}C$)** The atom that comprises 99% of natural carbon, used as a standard of atomic weight with a unit of 12.00000.

**carbon 13** or **carbon-13 ($^{13}C$)** The stable isotope of carbon, atomic mass number = 13; used as a tracer.

**carbon-13 ($^{13}C$) nuclear magnetic resonance** A spectrometric method used to analyze organic compounds.

**carbon 14** or **carbon-14 ($^{14}C$)** A naturally occurring radioactive isotope of carbon that emits beta particles when it undergoes radioactive decay. Mass number = 14. Half-life = 5730 years. It is widely used in the dating of organic materials. Also called radiocarbon.

**carbon absorber** An add-on control device that uses activated carbon to absorb volatile organic compounds (VOCs) from a gas stream. The VOCs are later recovered from the carbon.

**carbonaceous** Containing carbon; rich in organic compounds.

**carbonaceous biochemical oxygen demand (CBOD or carbonaceous BOD)** The portion of biochemical oxygen demand in which oxygen consumption is due to oxidation of carbon, usually measured after a sample has been incubated for 5 days. Also called carbonaceous oxygen demand or first-stage BOD. *See also* nitrogenous oxygen demand, total organic carbon, total oxygen demand, ultimate biochemical oxygen demand.

**carbonaceous BOD** Same as carbonaceous biochemical oxygen demand or CBOD.

**carbonaceous BOD removal** The conversion by microorganisms of the organic matter in wastewater into cell tissue and end products. Simultaneously, nitrogenous compounds are converted to ammonia.

**carbonaceous constituents** The various components of organic matter in wastewater as measured by the BOD and COD tests. *See* COD fractionation for detail.

**carbonaceous exchanger** A cation exchanger of limited capacity, prepared by the sulfonation of coal, lignite, or peat, and often containing both strong acid and weak acid groups.

**carbonaceous oxygen demand** Same as carbonaceous biochemical oxygen demand.

**carbon activation** The process of making a carbon material active or more reactive by oxidation to develop the internal pore structure. It consists of carbonization (i.e., preparing a char by heating the material) and then exposing the char to oxidizing gases. *See also* chemical activation, physical activation.

**carbon adsorbable organic halogen (CAOX)** A surrogate measurement for the total amount of organic compounds that contain one or more halogen atoms in a sample of raw or treated water. *See* halogen-substituted organic material, total organic halide analysis, and total organic halogen for more detail.

**carbon adsorbable organic halogen (CAOX) method** A commonly used procedure to estimate the total organic halide as a surrogate for the total amount of halogenated compounds in a water sample. It uses a nitrate solution to remove residual inorganic halogen salts, and activated carbon to adsorb the organically bound bromine, chlorine, and iodine, which are then converted to inorganic compounds by heat application and then measured by titration. *See also* total organic carbon.

**carbon adsorption** A treatment system that removes contaminants (such as refractory or other organic matter as well as taste and odor compounds and residual inorganics like nitrogen, chlorides, sulfides, and heavy metals) from water or wastewater by forcing it through tanks containing powdered or granular activated carbon treated to attract the contaminants. It is also used to recover organic solvents in metal finishing plants.

**carbon adsorption dechlorination** The removal of free and combined residual chlorine from wastewater using gravity or pressure flow in a granular activated carbon column, according to the following reactions:

Free chlorine: $$C + 2\,Cl_2 + 2\,H_2O \rightarrow 4\,HCl + CO_2 \quad (C\text{-}10)$$

Monochloramine: $$C + 2\,NH_2Cl + 2\,H_2O \rightarrow 2\,NH_4^+ + CO_2 + 2\,Cl^- \quad (C\text{-}11)$$

Dichloramine: $$C + 4\,NHCl_2 + 2\,H_2O \rightarrow 2\,N_2 + CO_2 + 8\,H^+ + 8\,Cl^- \quad (C\text{-}12)$$

**carbon aerogel** An electrode material of high electrical conductivity, high effective surface area, and high selectivity for the removal of iodide and other ions. The Lawrence Livermore National Laboratory has developed an optimized version of carbon aerogel for use in the desalination process of capacitive deionization.

**carbon analyzer** An instrument that can be used online to detect continuously very low concentrations of total organic carbon (TOC), e.g., in the effluents of microfiltration and reverse osmosis treatment units.

**carbonatation** Saturation or reaction with carbon dioxide ($CO_2$). *See also* carbonation.

**carbonate ($CO_3^{2-}$)** (1) The divalent, negatively charged anion $CO_3^{2-}$; it is one of the ions contributing to alkalinity and total dissolved solids in drinking water. (2) A compound containing the anion radical $CO_3^{2-}$; a salt or ester of carbonic acid ($H_2CO_3$); for example, calcium carbonate ($CaCO_3$) or sodium carbonate ($Na_2CO_3$).

**carbonate alkalinity** Alkalinity caused by the bicarbonate ($HCO_3^-$) and carbonate ($CO_3^{2-}$) ions, usually expressed as milligrams per liter of equivalent calcium carbonate ($CaCO_3$).

**carbonated water** *See* soda water.

**carbonate equilibrium** *See* carbonic acid equilibria.

**carbonate hardness** Hardness in water caused by the bicarbonates ($HCO_3^-$) and carbonates ($CO_3^{2-}$) of calcium and magnesium. Also called temporary hardness because it can be removed by boiling the water to precipitate the carbonates. Carbonate hardness is sometimes defined as the hardness caused by the alkalinity ions, which include not only carbonates and bicarbonates, but also the hydroxyl ion ($OH^-$). *See also* noncarbonate or permanent hardness and negative hardness.

**carbonate mineral** A mineral that contains the carbonate group ($CO_3^{2-}$), including the following as well as intermediates between them: ankerite [$CaMg(CO_3)_2$], aragonite ($CaCO_3$), calcite ($CaCO_3$), dolomite [$FeMg(CO_3)_2$], magnesite ($MgCO_3$), rhodocrosite ($MnCO_3$), siderite ($FeCO_3$), and smithsonite ($ZnCO_3$). Malachite [$CuCO_3 \cdot Cu(OH)_2$] contains both the carbonate and hydroxide of copper.

**carbonate rock aquifer** An underground formation of carbonate minerals; water from such a formation is typically alkaline, with high levels of total dissolved solids, and high calcium and/or magnesium hardness.

**carbonate saturometer** An instrument used to determine the calcium carbonate ($CaCO_3$) saturation condition of a sample. It is a modification of the pH meter that operates on the same principle as the marble test: the pH change upon the introduction of calcium carbonate causes a potential difference that can be related mathematically to the oversaturation or undersaturation (AWWA, 1999). *See* corrosion index for a list of corrosion control parameters.

**carbonate species** The chemical species that make up the carbonate system: gaseous carbon dioxide [$(CO_2)_g$], aqueous carbon dioxide [$(CO_2)_{aq}$], carbonic acid [$H_2CO_3$], bicarbonates [$HCO_3^-$], and carbonates [$CO_3^{2-}$]. The carbonate system also includes the hydrogen ($H^+$) and hydroxyl ($OH^-$) ions as well as calcium carbonate ($CaCO_3$).

**carbonate species distribution** *See* carbonic species diagram and carbonic acid system.

**carbonate system** *See* carbonate species.

**carbonation** (1) Saturation with carbon dioxide ($CO_2$), as in soda water. (2) The process of diffusing carbon dioxide through a liquid to stabilize it with respect to the precipitation or dissolution of alkaline constituents; recarbonation. (3) Reaction with carbon dioxide to remove lime in sugar refining. (4) Same as carbonization.

**carbonatite** An igneous intrusion or extrusion of ankerite, calcite, or dolomite.

**carbonator** A device used to carbonate or recarbonate water.

**carbon bisulfide ($CS_2$)** A clear, slightly yellow, poisonous, flammable liquid, used in cellophane, pesticides, and solvents. Same as carbon disulfide.

**carbon black** A finely divided, amorphous carbon from partially burned hydrocarbons or from charred wood, bones, etc. used in pigments, rubber products, and thermoplastics (to prevent their degradation by ultraviolet light). Also used as a clarifying or filtering agent.

**carbon block** *See* activated carbon block.

**carbon chloroform extract (CCE)** The residue from a carbon chloroform extraction test.

**carbon chloroform extraction** A method used to assess organic material in water by first adsorbing it on activated carbon, extracting the activated carbon with chloroform ($CHCl_3$), and then evaporating the chloroform to leave a residue.

**carbon contactor** A pressure or gravity, downflow or upflow reactor filled with granular activated carbon for the treatment of water or wastewater. It may be an expanded-bed, moving-bed, or pulsed-bed contactor.

**carbon cycle** The global circulation of carbon through living organisms, land, water, and the atmosphere, as represented graphically in Figure C-02. Carbon is present in the atmosphere as gaseous carbon dioxide [$(CO_2)_g$], in groundwater and surface water as bicarbonate ($HCO_3^-$) or aqueous carbon dioxide [$(CO_2)_{aq}$], in rocks as limestone ($CaCO_3$), or in organic matter, generally represented by ($CH_2O$). By photosynthesis, plants convert inorganic carbon [$(CO_2)_g$] into complex organic compounds consumed by other organisms, which, upon respiration and decay, return it to the atmosphere. Carbon is also leaked from underground reservoirs and returns underground by chemical precipitation and as mineral carbon in microbial shells. It is the carbon cycle that transfers solar energy to other biological systems.

**carbon dioxide ($CO_2$)** A colorless, odorless, tasteless, noncombustible gas formed in animal respiration, the combustion or decomposition of organic matter, and subterranean geochemical activity. It is heavier than air [$(CO_2)_g$ represents less than 1% of air] and moderately soluble in surface water (less than 10 mg/L), with higher concentrations in groundwater; it combines with water to form carbonic acid (a weak acid) and can contribute to the corrosion of metals and concrete. In water, it is present mostly as molecular $(CO_2)_{aq}$ and often represented as [$H_2CO_3^*$] to denote the inclusion of carbonic acid. Because of its relative safety, easy handling, and antiplugging action, carbon dioxide is sometimes used to stimulate well performance. Carbon dioxide influences pH, corrosivity, alkalinity, coagulation, and iron removal; it is used for recarbonation in the water softening process. In water treatment, carbon dioxide is neutralized by lime or removed by aeration (water droplets exposed to air lose about 80% of their $CO_2$ in seconds). Also called carbonic-acid gas, carbonic anhydride. *See also* total carbon dioxide.

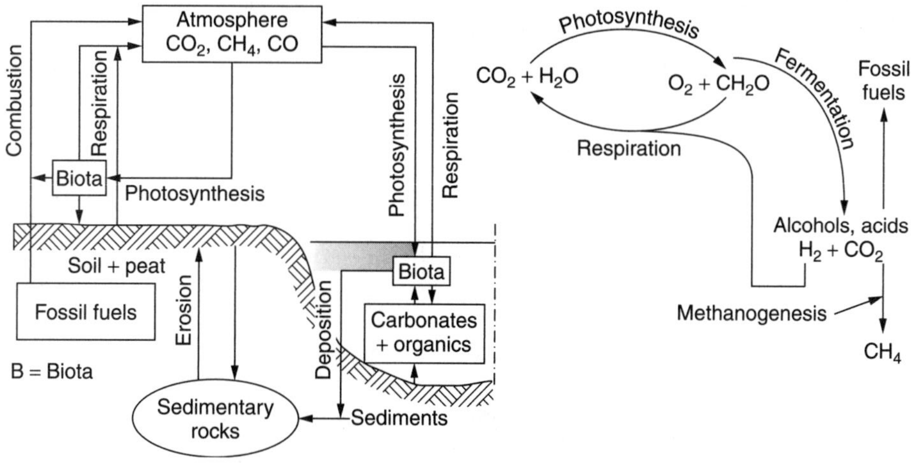


**Figure C-02.** Carbon cycle (adapted from Maier et al., 2000, and Howard, 1998).

**carbon dioxide acidity** *See* $CO_2$ acidity.

**carbon dioxide–calcium equilibria** The interaction between calcium and carbon species in natural waters, an important factor affecting such parameters as alkalinity, pH, and hardness, as shown in the reversible reactions:

$$CaCO_3(s) + (CO_2)_{aq} + H_2O \leftrightarrow Ca^{2+} + 2\ HCO_3^- \quad \text{(C-13)}$$

$$Ca^{2+} + 2\ HCO_3^- \leftrightarrow CaCO_3(s) + (CO_2)_g + H_2O \quad \text{(C-14)}$$

**carbon-dioxide treatment** The use of bottled carbon dioxide ($CO_2$) to neutralize alkaline industrial wastewater. The $CO_2$ combines with water to form carbonic acid ($H_2CO_3$). The process is expensive compared to other neutralization methods.

**carbon disulfide ($CS_2$)** A clear, slightly yellow, poisonous, flammable liquid, used in cellophane, pesticides, and solvents. Same as carbon bisulfide.

**carbon fixation** The process of photosynthesis in which autotrophic organisms, using sunlight energy, convert atmospheric carbon dioxide gas ($CO_2$) into organic matter ($CH_2O$).

**carbon footprint** A measure of the contribution of human activities to the production of greenhouse gases, in units of carbon dioxide ($CO_2$). The primary footprint or direct footprint measures direct $CO_2$ emissions from the combustion of fossil fuels (in cars, planes, etc.), whereas the secondary footprint or indirect footprint measures $CO_2$ emissions from the manufacture, use, and disposal of products. One source estimates individual American production of 20 tons of carbon dioxide and related gases per year.

**carbon hexachloride** Same as hexachloroethane.

**carbonic acid ($H_2CO_3$)** A weak acid formed by the combination of carbon dioxide and water:

$$CO_2 + H_2O = H_2CO_3 \quad \text{(C-15)}$$

It adds to the potential of water to dissolve subsurface minerals and corrode metals and concrete.

**carbonic acid endpoint** One of two carbon dioxide ($CO_2$) titration endpoints, corresponding to pH = 4.6, often used in the determination of acidity and alkalinity of water. *See also* bicarbonate end point.

**carbonic acid equilibria** The equilibrium reactions of carbonic acid ($H_2CO_3$) or carbon dioxide ($CO_2$) in water. *See* carbonic acid system and carbonic species distribution.

**carbonic acid gas ($CO_2$)** Same as carbon dioxide.

**carbonic acid system** A series of chemical reactions among various carbon species that is generally assumed to control the pH of natural waters:

$$CO_2 + H_2O \leftrightarrow H_2CO_3^* \quad \text{(C-16)}$$

$$H_2CO_3^* \leftrightarrow H^+ + HCO_3^- \quad \text{(C-17)}$$

with $K_1 = 4.45 \times 10^{-11}$ at 25°C

$$HCO_3^- \leftrightarrow H^+ + CO_3^{2-} \quad \text{(C-18)}$$

with $K_2 = 4.69 \times 10^{-11}$ at 25°C

$$C_T = [H_2CO_3^*] + [HCO_3^-] + [CO_3^{2-}] \quad \text{(C-19)}$$

where $[H_2CO_3^*]$ denotes the combined concentrations of carbon dioxide ($CO_2$, which hydrolizes only slightly) and carbonic acid ($H_2CO_3$), $C_T$ is the total carbonic species in solution, and $K_1$ and $K_2$ are equilibrium constants.

**carbonic anhydride ($CO_2$)** Same as carbon dioxide.

**carbonic species distribution diagram** A graphical representation of the concentrations of the various carbonic species (carbonic acid, bicarbonate, carbonate) as a function of pH. *See* Figure C-03.

**carbon isotopes** The two stable isotopes C-12 and C-13, and the radioactive isotope C-14. The carbon element occurs mostly as C-12 (98.89%). C-13 represents about 1.11%. There is very little C-14, on the order of $10^{-10}$ %.

**Carbonite** Anthracite filter media produced by Carbonite Filter Corp.

**carbonization** The formation of carbon from organic matter; the first step in the preparation of activated carbon; it consists in heating wood, coal, or any other carbonaceous material in the absence of air to liberate the carbon and convert the material to a char.

**carbon monoxide (CO)** A colorless, odorless, poisonous gas produced by incomplete combustion of many natural and synthetic products, e.g., gasoline, oil, wood, and cigarettes. In the body, carbon monoxide combines with chemicals in the blood,

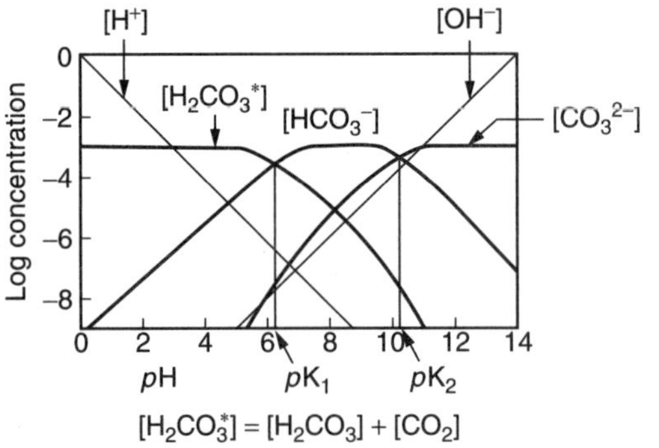


**Figure C-03.** Carbonic species distribution ($10^{-3}$ M solution at 25°C).

prevents the blood from bringing oxygen to the cells, tissues, and organs, and can cause serious health effects and even death. Carboxyhemoglobin is formed when iron is bound to carbon monoxide instead of oxygen. Carbon monoxide is lethal to humans at concentrations exceeding 5000 mg/L.

**carbon nanotube** A hollow cylinder formed by carbon molecules, with a diameter ranging from 1.4 to 2.0 nanometers, about 1/50,000 the size of a human hair; it is used in the manufacture of a special desalination membrane. *See* nanotube membrane filtration.

**carbon/nitrogen ratio** Usually abbreviated as C/N ratio, it is the ratio of the carbon content to the nitrogen content of a compostable or other organic compound, e.g., on a dry basis: 0.8 for urine, 6–10 for nightsoil, and 500 for sawdust. A C/N ratio of about 35 is recommended for raw compost.

**carbon oil** Another name of benzene.

**carbon reactivation** Same as activated carbon reactivation.

**carbon regeneration** Same as activated carbon regeneration.

**carbon regeneration unit** Any enclosed thermal treatment device used to regenerate spent activated carbon (EPA-40CFR260.10).

**carbon reservoirs** Sources or sinks of carbon include the atmosphere (carbon dioxide, $CO_2$), the oceans (biomass, carbonates, solids), and the land (biota, humus, fossil fuel, Earth's crust). The atmospheric reservoir is the smallest, but the most actively cycled of the three.

**carbon respiration** As part of the carbon cycle, the consumption by animals and heterotrophic organisms of the organic compounds produced as a result of photoautotrophic activity. Respiration produces carbon dioxide ($CO_2$) and new cell mass.

**carbon species diagram** *See* carbonic species diagram.

**carbon steel** Steel that contains up to 2.0% of carbon.

**carbon tet** Same as carbon tetrachloride.

**carbon tetrachloride ($CCl_4$)** A colorless, nonflammable, toxic liquid chemical organic compound, insoluble in water, with a strong odor resembling ether; used in the manufacture of chlorofluoromethanes, solvents, and cleaning agents. Acute or chronic exposure to this chemical may lead to renal damage, and gastrointestinal and nervous disorders. It is not a disinfection by-product. MCL = 0.005 mg/L. MCLG = 0. Also called carbon tet, fasciolin, methane tetrachloride, perchloromethane, tetrachlorocarbon, tetrachloromethane.

**carbon tetrachloride activity** A parameter used to describe the adsorptive capacity of activated carbon; the maximum percentage increase in weight of an activated carbon after the passage of tetrachloride-saturated air. *See also* decolorizing index, iodine number, methylene blue number, molasses number, and phenol adsorption value.

**carbon usage rate (CUR)** One of six common parameters used to define the performance of granular activated carbon contactors (and, sometimes, powdered activated carbon units) in the treatment of water and wastewater. It represents the capacity of the activated carbon to treat a given liquid or to remove a specified contaminant and is often defined as the quantity (weight) of activated carbon used per unit volume treated, i.e.,

$$\mathrm{CUR} = M/Q \cdot t \qquad \text{(C-20)}$$

where $M$ is the mass of activated carbon used in grams, $Q$ is the volumetric flow rate in liters/hour, and $t$ = contact time or operation time in hours. The CUR may also be expressed in terms of the activated carbon's apparent density ($\rho_{GAC}$, g/L) and the bed volumes to breakthrough (BVB):

$$\mathrm{CUR\ (g/L)} = \rho_{GAC}/\mathrm{BVB} \qquad \text{(C-21)}$$

The other common performance parameters are activated carbon density, bed life, empty bed contact time, specific throughput, and volume of water treated.

**carbonyl (C═O)** (1) One of three major groups of organic compounds, characterized by the double carbon–oxygen bond C═O and found in such compounds as aldehydes, ketones, organic acids, sugars, and many synthetic organics. (2) A compound of carbon monoxide (CO) and a metal, e.g., nickel carbonyl, $Ni(CO)_4$.

**carbonyl group** *See* carbonyl (1).

**carboxyhemoglobin (COHb)** A stable compound of carbon monoxide (CO) and hemoglobin, which reduces the oxygen-carrying capacity of the blood.

**carboxyl** Containing the carboxyl group.

**carboxylate** A salt or ester of a carboxylic acid.

**carboxyl group** One of three major groups of organic compounds, characterized by the link —COOH, present in many organic acids, e.g., acetic acid ($CH_3COOH$). The presence of the carboxyl group in water is a possible activating factor of trihalomethanes.

**carboxylic acid** Any organic acid containing one or more carboxyl groups, including the following: acetic, benzoic, butyric, formic, lauric, myristic, palmitic, phthalic, propionic, stearic, terphtalic, and valeric acids. Although a component of naturally occurring organic matter, carboxylic acids do not have a high potential for the formation of disinfection by-products. They contribute to the cation exchange capacity of some resins. A carboxylic acid represents the highest oxidation state of an organic radical, beyond which there is formation of carbon dioxide and water, e.g., formic acid (HCOOH) from methane ($CH_4$):

$$CH_4 \rightarrow CH_3OH \rightarrow H_2C{=}O \rightarrow HCOOH \rightarrow CO_2 + H_2O \quad \text{(C-22)}$$

**carboxyl methyl cellulose** *See* carboxymethylcellulose.

**carboxymethylcellulose (CMC)** A white, water-soluble polymer produced from cellulose and used in paper making, in foods, in textile-finishing mills, and in general-purpose synthetic detergents. Also called cellulose gum, CMC, or sodium cellulose glycolate.

**carboxymethyl cellulose** *See* carboxymethylcellulose.

**carboy** A large glass container protected by basket work or a wooden box, used to store or transport liquid chemicals or water samples.

**carcinogen** Any physical, chemical, or microbial substance or agent that can, directly or indirectly, cause or aggravate cancer or tumors in an organism, e.g., benzo[a]pyrene, cutting oils, radioactive substances, and vinyl chloride monomer. In setting standards for drinking water, the USEPA classifies carcinogens in three categories depending on the strength of available evidence. For example, there is strong evidence for arsenic, asbestos, and cadmium; equivocal evidence for lindane and styrene; and inadequate evidence for barium, cyanide, and toluene. Same as carcinogenic agent. *See also* mutagen, known probable human carcinogen via ingestion, and possible human carcinogen via ingestion.

**Carcinogen Assessment Group** A group of the USEPA that develops and promulgates methods for the identification of risks related to carcinogens found in the environment.

**carcinogenesis** The origin or production of cancer, very likely a series of steps. The carcinogenic event so modifies the genome and/or other molecular control mechanisms in the target cells that these can give rise to a population of altered cells. Chemical carcinogenesis relates to the role of xenobiotic chemicals in causing cancer.

**carcinogenesis bioassay** Any in vivo or in vitro test of cancer-causing properties of substances.

**carcinogenic** Cancer producing; causing or inducing uncontrolled growth of aberrant cells into malignant tumors. Also called oncogenic. *See also* genotoxic, initiator, mutagenic, promoter, teratogenic.

**carcinogenic agent** An agent that can produce cancer, such as (a) chemicals (nitrosamines, disinfection by-products, some hydrocarbons), (b) biological agents (e.g., retroviruses, hepatitis B virus), (c) ionizing radiation (e.g., X-rays), and (d) genetic factors. Same as carcinogen or carcinogenic substance.

**carcinogenic compound** A substance that can produce cancer; e.g., the trihalomethanes formed during water disinfection; a carcinogen or carcinogenic substance.

**carcinogenic effect** For radiation, carcinogenic effects relate to bone marrow, blood-forming tissues, and lymph nodes.

**Carcinogenicity** The ability of an agent or substance to cause cancer.

**carcinogenic substance** A substance or preparation that may induce, or increase the incidence of, cancer in humans, if inhaled or ingested or if it penetrates the skin. Same as carcinogenic agent or carcinogen.

**cardiovascular disease** A disease that affects the heart or blood vessels, e.g., angina, hypertension, stroke. Sudden death syndrome and stroke are reportedly associated with drinking water quality. There seems to be some (unexplained) correlation between cardiovascular disease and the consumption of soft water as compared to moderately hard and fluoridated water. There may also be an association between chlorination and cardiovascular disease.

**cardiovascular system** The heart and blood vessels.

**CA•RE™** A program developed by U.S. Filter Corp. for the recovery of spent filter cartridges.

**Carela™** A line of products of R. Spane GmbH used in water disinfection.

***Carex* spp.** A common emergent aquatic sedge found in natural wetlands.

**Carlson index** *See* trophic state index.

**Carman–Kozeny equation** A formula developed by P. C. Carman in 1937 for head loss of fluid flow through a granular bed. C. F. Blake in 1922 and J. Kozeny in 1927 proposed a similar formula. The Carman–Kozeny equation is widely used for flow through porous media (e.g., granular filters, pebble bed flocculators, vacuum filters). Carman's equation is based on the Darcy–Weisbach formula:

$$H/L = 3f(1 - e)v^2/(4gSDe^3) = F(1 - e)v^2/(gSDe^3) \quad \text{(C-23)}$$

where $H$ = head loss, $L$ = depth of bed, $e$ = bed porosity, $f$ = friction factor, $v$ = face or approach velocity, $g$ = gravitational acceleration, $S$ = shape factor or sphericity, $D$ = the diameter of the sphere equivalent in volume to the granular media, and $F = 3f/4$.

The Ergun equation is commonly used to estimate the parameter $F$ (Droste, 1997), which modifies the Carman–Kozeny equation. For other formulas proposed to compute the clean-water head loss through a granular porous medium, see the Blake–Kozeny, Fair–Hatch, Hazen, and Rose equations. *See also* Coakley's equation, Kozeny formula, and porosity factor.

**Carman, P. C.** Author of often quoted articles on filtration mechanisms (1933) and fluid flow through granular beds (1937).

**carnallite ($KCl \cdot MgCl_2 \cdot 6\ H_2O$)** A mineral compound of potassium and magnesium chlorides, and an essential plant nutrient.

**Caro's acid** Same as persulfuric acid.

**carousel ion-exchange process** Water treatment process consisting of 10–20 parallel ion-exchange columns such that some columns are being exhausted while others are being rinsed, regenerated, or in standby. *See also* merry-go-round system.

**carp** A family of hardy fish, some species of which survive well in stabilization ponds used for aquaculture.

**carrier** (1) A person or an animal that harbors a specific pathogen and serves as a potential source of infection but does not show any discernible sign of disease. A healthy cholera carrier, e.g., may excrete up to one million vibrios per gram of feces. The carrier state may last the duration of the illness or may persist for months or even a lifetime. Carriers may be more important in disease transmission than the persons who are actually ill. *See* Typhoid Mary, asymptomatic carrier. (2) The inert liquid or solid material added to an active ingredient in a pesticide.

**carrier state** The condition of a pathogen carrier. For typhoid fever and paratyphoid fever, e.g., a carrier state may follow an infection and a chronic carrier state is most common among females infected during middle age.

**Carrobic** An aerobic sludge digestion and thickening unit of the oxidation ditch wastewater treatment system developed by Eimco Process Equipment Co.

**Carrousel®** Same as Carrousel™ oxidation ditch.

**Carrousel intraclarifier™** An intrachannel clarifier (ICC) developed for oxidation ditch wastewater treatment plants. Same as Carrousel™ oxidation ditch. *See also* boat clarifier, Burns and McDonnell treatment system, pumpless integral clarifier, side-channel clarifier, side-wall separator.

**carrousel ion-exchange process** *See* carousel ion-exchange process.

**Carrousel™ oxidation ditch** A biological wastewater treatment system using complete mix activated sludge in oxidation ditches, developed by DHV Water BV and licensed to Eimco Process Equipment Co.

**carrousel system** A biological wastewater treatment plant that includes (a) a deep, vertical-wall oxidation ditch operating in the extended aeration mode to provide nitrification (in an aerobic zone) and denitrification (in an anoxic zone) and (b) a separate final clarifier.

**carrying capacity** (1) In recreation management, the amount of use a recreation area can sustain without loss of quality. (2) In wildlife management, the maximum number of animals an area can support during a given period. (3) The maximum discharge of a hydraulic structure or the maximum flow that a treatment plant can process is sometimes called its carrying capacity. (4) The maximum level of microbial activity that can be expected in a given environmental setting, taking into consideration biological, physical, and chemical conditions.

**carryover or carry-over** (1) The entrainment of particles in the vapor of a boiling liquid; the particles entrained. (2) The drizzle, droplets, or mist that are entrained in the circulating air around the top of a cooling tower that is not equipped with spray eliminators. *See* cooling tower precipitation for detail.

**cartage** The vehicular or manual nightsoil removal; e.g., the use of containers or vehicles to take away the contents of latrines or vaults.

**cartage system** The combination of an individual sanitation system such as a bucket latrine and a nightsoil collector who periodically empties the bucket and carries away the contents; a system that presents great environmental health risks, unless adequately supervised (Feachem et al., 1983).

**Cartermix** Equipment manufactured by Ralph B. Carter Co. for use as a mixer in anaerobic sludge digestion.

**cartridge filter** A device used in small plants or in point-of-use applications for the treatment of high-quality water. It consists of a pressure vessel containing replaceable elements or cartridges of ceramic or polypropylene materials that retain impurities larger than their pores. It is not effective against viruses and is subject to frequent clogging by algae and high turbidity. When the filter run terminates, the cartridges are replaced by new ones. Disinfection should follow cartridge filtration. *See also* bag filter.

**cartridge microfilter** A sieving filter with pores of 5–20 micrometers commonly used as pretreatment for reverse osmosis and nanofiltration water treatment units to protect the membrane elements against particulate fouling agents.

**Carulite®** A line of chemicals produced by Carus Chemical Co. for use as catalysts in the treatment of volatile organic compounds.

**Carver–Greenfield process** A proprietary, multiple-effect evaporation process used in sludge dewatering. The sludge is first mixed with oil, the mixture treated by multiple-effect evaporation followed by oil–solid separation and condensate–oil separation by centrifugation. The oil is recycled.

**CaS** Chemical formula of calcium sulfide.

**cascade** (1) A stretch of stream that includes a series of shallow waterfalls. (2) A sudden drop in a water producing agitation and aeration of the liquid. (3) A repetitive process for purification, in which components of a mixture are gradually separated. (4) *See* enriching section.

**Cascade** A biological filter manufactured by Mass Transfer, Inc. using synthetic media.

**cascade aeration** An economical aeration system in which a thin film of wastewater falls on a series of steps. *See also* Barrett equation, cascade reoxygenation.

**cascade aerator** A device that serves to inject air into water or wastewater through a series of steps, the splashing of the water creating turbulence and bringing water droplets in contact with air. *See* gravity aerator. *See* Figure C-04.

**cascade air stripping** A water and wastewater treatment method that introduces fresh air along the full packing depth of a stripping column, and not just at the bottom, to remove organic constituents of low volatility. *See also* cascade packed tower, packed column air stripping.

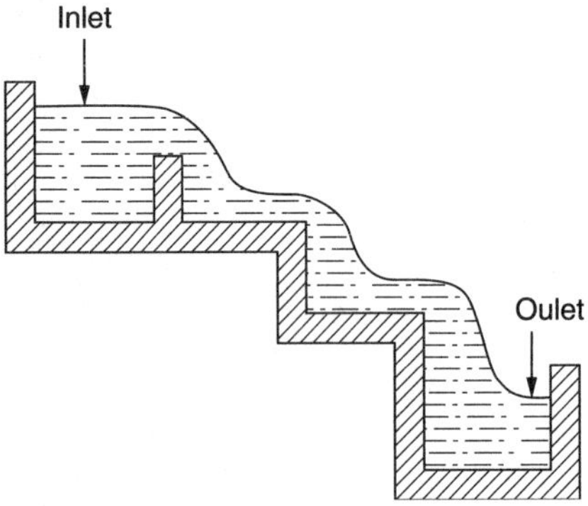


**Figure C-04.** Cascade aerator.

**cascade control** An automatic process control system that uses a primary feedback controller to calculate the set point of a secondary controller, allowing the apparatus to control two variables such as dissolved oxygen (DO) and air flow: the DO controller determines the set point for airflow. *See also* the following types of control: feedback, feed-forward, integral, on–off, proportional, proportional–derivative, proportional–integral, proportional–integral–derivative.

**cascade packed tower** A device used to strip volatile organic compounds, carbon dioxide, and other gases from water or wastewater. It consists of a packed tower, with fresh air introduced at various points along the tower and flowing countercurrent to the liquid. *See also* cascade air stripping, cocurrent packed tower, crossflow packed tower, end effect, low-profile air stripper, minimum air-to-water ratio, packed tower design equation, sieve tray column, wall effect.

**cascade reaeration** *See* cascade reoxygenation.

**cascade reoxygenation** The introduction of oxygen into a treated waste effluent through an air–water interface on a series of steps or at such hydraulic structures as weir overflows, flumes, and spillways. *See also* mechanical or diffused air reoxygenation.

**CASE** Computer-aided software engineering.

**case-control epidemiologic study** *See* case-control study.

**case-control study** An epidemiologic study that looks back in time at the exposure history of individuals who have a health effect (cases) and at a group who do not (controls), to ascertain whether they differ in proportion exposed to the chemical under investigation. *See also* retrospective study.

**case-referent epidemiology study** *See* case-control study.

**cash basis** A method of accounting that records revenues and expenditures when they are actually received or paid, as opposed to the accrual basis.

**cash crop** Any marketable crop such as wheat or cotton; a crop for direct sale as compared to a crop for livestock feed. Wastewater is sometimes reused agriculturally to grow cash crops.

**cash for grass** A rebate program used by some utilities to encourage consumers to use less water by replacing turf grass with xeric plants and groundcovers.

**Casil index (CI)** A measure of water corrosiveness, which increases when anions increase or cations decrease (Symons et al., 2000):

$$CI = (\text{calcium ions} + \text{magnesium ions} + \text{silica ions} - \text{anions})/2 \quad \text{(C-24)}$$

where all ions are expressed in milliequivalents per liter (meq/L).

**casing** A solid pipe or tube, usually of steel, placed in a well or borehole to support it in unstable materials and prevent other fluids from entering or leaving the hole. In a string of casing, the casing head is the top heavyweight element that receives the driving blows, whereas the casing shoe is a rigid annular fitting with a cutting edge at the lower end.

**casing head** *See* casing.

**casing shoe** *See* casing.

**$CaSiO_3$** Chemical formula of calcium metasilicate.

**$CaSO_4$** Chemical formula of calcium sulfate, found in such minerals as anhydrite and gypsum.

**$CaSO_4$ (s)** Chemical formula of anhydrite.

**$CaSO_4 \cdot 2H_2O$ (s)** Chemical formula of gypsum.

**CAS registration number** A number assigned by the Chemical Abstracts Service to identify a chemical. For example, benzene has No. 71-43-2 and chloroform has 67-66-3.

**CASS™** A wastewater treatment system developed by Transenviro, Inc. using sequencing batch reactors.

**cassiterite** The principal tin ore, essentially an oxide ($SnO_2$).

**casting** An article cast in a mold; the shaping of cast iron or ductile iron tees, valves, etc. in a sand mold.

**cast iron** A group of iron–carbon–silicon metallic products obtained by reducing iron ore with carbon at temperatures high enough to render the metal fluid and cast it in a mold. Used in piping, valves, and fittings.

**cast-iron pipe** A pipe made of pig iron (iron tapped from a blast furnace) in a cylindrical sand mold. Corrosion of cast-iron pipe produces tubercles of rust above pits of the metal.

**Cast Iron Pipe Research Association (CIPRA)** A predecessor of the Ductile Iron Pipe Research Association, a trade association of manufacturers of ductile iron pipe and fittings that publishes useful information about water supply and wastewater management. The association changed its name in 1980.

**CastKleen** A filter underdrain system developed by Eimco Process Equipment Co. for in-place casting.

**catabolic reaction** The breakdown of cell constituents and metabolites by organisms for energy and growth; catabolism.

**catabolism** The breaking down of complex molecules into simpler ones by organisms to obtain energy; also called dissimilation. *See also* metabolism and anabolism.

**catalase** An enzyme that contains iron and decomposes hydrogen peroxide ($H_2O_2$) into water ($H_2O$) and oxygen ($O_2$).

**catalyse** Same as catalyze.

**catalysis** A process in which an added substance (a catalyst) alters the rate of a reaction but does not itself change.

**catalyst** A substance that changes the speed or yield of a chemical or biological reaction without being consumed or chemically changed by the reaction; e.g., dissolved metals, enzymes. The catalyst participates in the reaction at one stage and is reconstituted at a later stage. Positive catalysts accelerate reactions while negative or retarding catalysts (also called inhibitors) slow them down.

**catalyst bed** In a catalytic incinerator, the catalyst bed generally consists of a metal mesh mat, ceramic honeycomb, or similar structure that maximizes the surface area of the catalyst. Also called matrix.

**catalyst filter** A granular filter whose bed contains a catalyst medium. It is used to render dissolved molecules [e.g., ferrous iron ($Fe^{2+}$), manganous manganese ($Mn^{2+}$), anionic sulfur ($S^{2-}$)] insoluble and filterable by altering their valence. Also called oxidizing filter.

**catalyst media** Filter media that can cause certain reactions in water treatment, such as calcite, dissimilar metal alloys, magnesium oxides, manganese greensands, pumicites, and zeolites. *See also* catalyst filter, catalytic activated carbon.

**catalytic activated carbon** Activated carbon with modified surface properties that enhance the functionality of the activated carbon in converting the oxidation state of various elements. For example, with hydrogen sulfide ($H_2S$), the sulfide ion ($S^-$) is adsorbed and then converted on the catalytic carbon to elemental sulfur ($S^0$) and sulfate ion ($SO_4^{2-}$). Once the sulfide is adsorbed and converted, it is desorbed and the site is restored. For these reactions to occur, excess dissolved oxygen is required in the water and a minimum empty bed contact time of five minutes may be necessary.

**Catalytic Extraction Processing™** A proprietary process that uses a molten metal bath as a catalyst to dissociate hazardous compounds into their elements, which are then recombined to form marketable products such as ceramics, industrial gases, and metal alloys. For example, a feed of chlorobenzene ($C_6H_5Cl$) contaminated by nickel oxide (NiO) can be converted to carbon monoxide gas [CO(g)], hydrogen gas [$H_2$(g)], a ceramic or gas product, and a metal alloy (Freeman, 1998).

**catalytic filtration** Water treatment using a catalyst filter.

**catalytic incineration** A process in which a catalyst (e.g., a metal oxide, palladium, platinum, rubidium) accelerates the oxidation of volatile organic compounds in an emission stream.

**catalytic incinerator** A control device that oxidizes volatile organic compounds (VOCs) by using a catalyst to promote the combustion process. Catalytic incinerators require lower temperatures than conventional thermal incinerators, thus saving fuel and other costs.

**catalytic oxidation** A flameless oxidation process that uses a catalyst (e.g., a metal oxide, palladium, platinum, rubidium) to destroy volatile organic compounds at temperatures of 600–800°F.

**catalytic oxidizer** An off-gas posttreatment unit for the control of organic compounds. Gas enters the unit and passes over a support material coated with a catalyst (commonly a noble metal such as platinum or rhodium) that promotes oxidation of the organics. Catalytic oxidizers can also be very effective in controlling odors. High moisture content and the presence of chlorine or sulfur compounds can adversely affect the performance of a catalytic oxidizer.

**catalytic ozone destructor** A device that destroys ozone ($O_3$) using a catalyst to enhance the performance of the off-gas treatment system..

**catalytic photochemical oxidation** A process that accelerates the destruction of chemical compounds using ultraviolet absorbance and an oxidant such as ozone ($O_3$).

**catalytic reaction** A chemical or biological reaction involving a catalyst; catalysis.

**catalytic scrubber** A device that uses an aqueous chelated iron catalyst to remove hydrogen sulfide ($H_2S$) from digester gas. Elemental sulfur is produced and an oxidizer unit reactivates the catalyst with air.

**catalyze** To act as a catalyst, or to speed up a chemical reaction. Also spelled catalyse.

**catalyzed oxidation** Chemical or biological oxidation that is enhanced by a catalyst.

**cataphoresis** *See* electrophoresis.

**catch-can method** A method to measure precipitation from a sprinkler system using containers installed throughout an irrigation field.

**catch sample** A single air, water, or wastewater sample, representative of the composition of the flow at a particular time and place. *See* grab sample for detail.

**catechol ($C_6H_6O_2$)** A colorless, crystalline, dihydroxyl derivative of benzene ($C_6H_6$) used in photography, dyeing, and as a reagent. Catechol is a phenolic compound that may produce mutagenic agents during chlorination. Also called pyrocatechol, pyrocatechin.

**categorical industrial user (CIU)** A small industrial wastewater facility that is subject by federal regulations to both discharge standards based on its industrial category (e.g., metal finishing) and standards applying to larger industrial users regardless of the volumes discharged. *See also* significant industrial user.

**categorical industry** A commercial or industrial establishment subject to the USEPA's categorical standards regarding the discharge of certain pollutants to publicly owned treatment works. *See* priority pollutant.

**categorical pretreatment standard** A technology-based effluent limitation for an industrial facility discharging into a municipal sewer system. Analogous in stringency to best available technology (BAT) for direct dischargers.

**categorical standard** A standard used by the USEPA to control pollutant discharges to publicly owned treatment works by certain industrial and commercial establishments. *See also* priority pollutant and prohibited discharge standard.

**category I contaminant** A substance that the USEPA has classified as a known or probable human carcinogen because there is sufficient evidence that it can cause cancer in humans or animals via ingestion.

**category 1 carcinogen** Any substance for which sufficient evidence of carcinogenicity in humans and animals exists, based on epidemiological studies, according to the International Agency for Research on Cancer. *See* known probable human carcinogens via ingestion for detail.

**category II contaminant** A substance that the USEPA has classified as a possible human carcinogen because there is limited evidence that it can cause cancer in humans or animals via ingestion.

**category 2 carcinogen** Any substance for which limited evidence of carcinogenicity in animals exists, based on epidemiological studies, according to the International Agency for Research on Cancer. *See* possible human carcinogens via ingestion for detail. Subdivided into 2A for limited evidence and 2B for no evidence.

**category III contaminant** A substance for which the USEPA has insufficient or no evidence that it can cause cancer via ingestion.

**category 3 carcinogen** Any substance for which sufficient evidence of carcinogenicity in experimental animals exists, according to the International Agency for Research on Cancer.

**category A carcinogen** Same as category I contaminant or known probable human carcinogens via ingestion.

**category a carcinogen** A classification of the National Toxicology Program; same as category I contaminant or known probable human carcinogens via ingestion.

**category B carcinogen** Same as category II contaminant or possible human carcinogens via ingestion; subdivided into B1 or B2 according to whether the evidence is from human studies or animal studies.

**category b carcinogen** A classification of the National Toxicology Program; same as category II contaminant or possible human carcinogens via ingestion.

**category C carcinogen** Same as category III contaminant.

**category D carcinogen** A substance that the USEPA has not classified because of insufficient evidence that it can cause cancer via ingestion.

**category E carcinogen** A substance about which the USEPA has no evidence of carcinogenicity in at least two animal tests in different species or in both animal and epidemiologic studies that it can cause cancer via ingestion.

**catenary bar screen** Same as catenary screen, a mechanical screening device using revolving chain-mounted rakes to clean a stationary bar rack.

**catenary screen** A device used to remove coarse materials from wastewater ahead of other treatment units. It consists of a front-cleaned, front-return chain-driven screen, with the weight of the chain holding the rake against the rack and with all sprockets, shafts, bearings located out of the liquid to reduce wear and corrosion. Also called catenary bar screen. *See also* chain-driven screen, continuous-belt screen, reciprocating-rake screen.

**Cat Floc®** A cationic polymer produced by Calgon Corp. for use in solids separation processes.

**cathode** (1) The negative pole or electrode where the current leaves an electrolytic cell or system and where reduction takes place. The cathode attracts positively charged particles or ions (cations). (2) The negatively charged electrode of an electrodialysis cell. (3) The negative electrode of a vacuum tube.

**cathode ray tube (CRT)** A phosphorescent screen used in the display of computer data.

**cathodic corrosion** An alkaline condition corrosive to aluminum, lead, and zinc, and created by the transport of electrons to the cathodic zone. Normal corrosion occurs at the anodic zone.

**cathodic current** The current that results from the reduction of metal ions in an electrochemical cell, corresponding to a gain of electrons in the reverse reaction:

$$\text{Metal} \leftrightarrow \text{Metal}^{z+} + ze^- \qquad \text{(C-25)}$$

**cathodic inhibitor** A substance that stops the flow of hydrogen or oxygen to cathodic sites by reacting with them and forming a protective film against corrosion; e.g., calcium carbonate ($CaCO_3$).

**cathodic protection (CP)** An electrical system used to prevent or reduce rust, corrosion, and pitting of a metal surface in water or soil. A low-voltage current is made to flow through a liquid (water) or a soil in contact with the metal in such a manner that the external electromotive force renders the metal structure cathodic. This concentrates corrosion on auxiliary anodic parts that are deliberately allowed to corrode instead of letting the structure corrode. This method of corrosion control is achieved through galvanic protection (sacrificial anode) or impressed-current (electrolytic cathodic) protection. Other methods include anodic protection, chemical coating, inhibition, inert materials, and metallic coatings. *See also* passivation.

**cathodic reaction** The transfer of electrons that accompanies metallic corrosion, from the anodic zone to the cathodic zone. Two simultaneous reactions occur, e.g., the oxidation of iron and the reduction of dissolved oxygen (McGhee, 1991):

$$\text{Fe} \rightarrow \text{Fe}^{2+} + 2\,\text{e}^- \qquad \text{(C-26)}$$

$$\tfrac{1}{2}\,\text{O}_2 + 2\,\text{e}^- + \text{H}_2\text{O} \rightarrow 2\,\text{OH}^- \qquad \text{(C-27)}$$

**catholyte** The portion of the electrolyte in the vicinity of the cathode of an electrodialysis unit. *See also* anolyte.

**cation** A positively charged ion or radical in an electrolyte solution, attracted to the cathode under the influence of a difference in electrical potential. Sodium ($Na^+$), hydrogen ($H^+$), zinc ($Zn^{2+}$), and ammonium ($NH_4^+$) are cations. *See also* anion.

**cation exchange** A reversible exchange of positively charged ions (cations) between an aqueous solution and a mineral. (1) More specifically, a water treatment process that transfers a more desirable cation (e.g., hydrogen or sodium) from a synthetic, porous medium or resin to the liquid phase in exchange for a less desirable cation (e.g., calcium or magnesium). A concentrated solution is used to regenerate resins whose exchanging capacity has been exhausted. A common resin used in water softening consists of small polystyrene spheres, about 0.5 mm in diameter. The cation exchange and resin regeneration reactions are:

$$Ca^{2+} + Na_2R \rightarrow CaR + 2\ Na^+ \quad (C\text{-}28)$$

$$Mg^{2+} + Na_2R \rightarrow MgR + 2\ Na^+ \quad (C\text{-}29)$$

$$CaR + 2\ NaCl \rightarrow CaCl_2 + Na_2R \quad (C\text{-}30)$$

$$MgR + 2\ NaCl \rightarrow MgCl_2 + Na_2R \quad (C\text{-}31)$$

where R represents the anionic component of the resin. Also called base exchange. *See also* zeolite softening. (2) Cation exchange is also used to improve soil fertility by adding deficient elements. When the concentration of a cation (e.g., potassium, $K^+$) is increased in a soil solution, this displaces another cation (e.g., magnesium, $Mg^{2+}$) on a charge equivalent basis. *See also* adsorption affinity and cation exchange capacity.

**cation-exchange capacity (CEC)** (1) The sum of exchangeable cations a soil can absorb expressed in milliequivalents per 100 grams of soil as determined by sampling the soil to the depth of cultivation or solid waste placement, whichever is greater, and analyzing by the summation method for distinctly acid soils or the sodium acetate method for neutral, calcareous or saline soils (EPA-40CFR257.3.5-3). In land applications, the organic matter of sludge enhances the capacity of soil to retain such nutrients as calcium, magnesium, and potassium. (2) The charge associated with clay and organic particles in a porous medium; a factor that affects the bioavailability of metals in soils. *See also* isomorphic substitution, ionization.

**cation-exchange material** A material that picks up hardness-causing metals or other cations in water in exchange for nontroublesome ions.

**cation exchanger** A resin or similar material with a negatively charged framework and pores containing positively charged ions (cations), used in water treatment (e.g., softening, desalination, production of boiler-feed water). Strongly acidic cation exchangers have functional groups derived from strong acids (e.g., $SO_3^-$ from $H_2SO_4$), and weakly acidic exchangers derive their groups from weak acids (e.g., $COO^-$ from acetic acid and $PO_3H^-$ from phosphoric acid).

**cation-exchange resin** A synthetic material that is used in water treatment to remove such cations as calcium and magnesium but also iron and manganese. *See also* anionic-exchange resin, hydrogen-cycle resin. Also called a cation exchanger or base exchanger.

**cation-exchange softening** Water softening using the cation-exchange process. Note that this process increases the concentration of sodium in the finished water, a consideration for consumers on a sodium-restricted diet.

**cation-exchange water softener** A device that reduces water hardness through the cation-exchange process.

**cationic** Pertaining to cations; having a positive ionic charge; characteristic of polymers that have a net positive charge.

**cationic coagulant** *See* cationic polyelectrolyte, cationic polymer.

**cationic detergent** A detergent containing quaternary ammonium compounds; found in dishwashing products.

**cationic exchange** *See* cation exchange.

**cationic flocculant** A cationic polyelectrolyte.

**cationic metal** A metal such as lead or calcium that strongly interacts with negatively charged surfaces as opposed to an anionic metal such as arsenic or its compound arsenate ($AsO_4^{3-}$) that is attracted to positively charged surfaces. The nature and intensity of the charge of a metal strongly affect its fate and bioavailability. *See also* adsorption affinity (Maier et al., 2000).

**cationic polyelectrolyte** A natural or synthetic chemical used in the treatment of water and wastewater, where it dissolves to form positively charged ions.

**cationic polymer** A polymer having positively charged groups of ions; often used as a coagulant aid or to improve the filtration of negatively charged particles. In small concentrations (less than 1.0 mg/L), cationic polymers are sometimes added to aeration basins to improve the settleability of the mixed liquor. *See also* ampholytic polymer, anionic polymer.

**cationic surfactant** A positively charged ionic surface-active agent used in cleaning products, mostly for its disinfecting and fabric softening properties; its hydrophilic group is usually a quaternary ammonium salt, e.g., cetyl ammonium bromide. *See also* anionic surfactant, nonionic surfactant.

**cation membrane** A membrane used in electrodialysis to allow the passage of anions but not cations; it is impermeable to water and nonelectrolytes. Also called anion-selective membrane or cation-transfer membrane. An example is a polystyrene–divinylbenzene membrane functionalized with an ammonium group (—$NR^{3+}$).

**cation-selective membrane** A membrane used in electrodialysis to retain anions but not cations; it is impermeable to water and nonelectrolytes. Also called anion transfer membrane or anion membrane. An example is a polystyrene–divinylbenzene membrane functionalized with a sulfonyl group (—$SO_3^-$).

**cation transfer membrane** Same as cation membrane.

**cat liver fluke** The common name of opistorchiasis, an excreted helminth infection caused by the pathogens *Opistorchis felineus* and *Opistorchis viverrini.* It is found mainly in Poland, Ukraine, northern Turkey, Thailand, and southern Laos. Often associated with fish farming in excreta-enriched ponds, it is transmitted from the feces of infected persons or cats through aquatic snails and fish as intermediate hosts. It is similar to Chinese liver fluke.

**Cat-Ox™** A catalytic oxidation system developed by Catalytic Combustion Corp.

**cattail** An aquatic plant often found in ponds and constructed wetlands and with roots attached to the bottom.

**cat-urine odor** A urine odor sometimes observed in water that has been treated with chloride dioxide ($ClO_2$). It results from the reaction of a byproduct (chlorite), free chlorine, and volatile organic matter from carpeting.

**Cauchy number** or **Cauchy's number (Ca)** A dimensionless number used in the study of water hammer; it is equal to the ratio of inertia forces to compression (or elastic) forces and usually computed as the ratio of a characteristic velocity of the system to its kinematic elasticity (Symons et al., 2000):

$$C_a = \rho V^2/E \qquad \text{(C-32)}$$

where $\rho$ = density, $V$ = characteristic velocity (e.g., mean, surface, or maximum velocity), and $E$ = bulk modulus elasticity of the liquid.

**caulk** A material or substance (e.g., lead, oakum, plastic) used to fill or close seams or crevices in windows, tanks, etc. to make them watertight or airtight. Also spelled calk or called caulking.

**caulking** (1) Same as caulk. (2) The process of inserting a caulking to make a structure waterproof or airtight.

***Caulobacter*** One of the few bacteria groups that can survive and even achieve limited growth in pure, distilled, or low-nutrient spring waters by exchanging nutrients with the atmosphere. *See also Pseudomonas aeruginosa* and *P. fluorescens.*

**caustic** (From the Greek work kaustikos meaning burning.) (1) Alkaline or basic; usually applied to some bases. (2) Caustic soda (NaOH) or any similar substance. (3) Any substance that can destroy, burn, or corrode living flesh or tissue.

**caustic alcohol** Same as sodium ethylate.

**caustic alkalinity** The alkalinity caused by hydroxyl ions ($OH^-$). Also called hydroxide alkalinity. When titration is used to determine the acid-neutralizing capacity of water, hydroxide alkalinity, represented by [$OH^-$ alk], is:

$$[OH^- \text{ alk}] = [OH^-] - [HCO_3^-] - 2\,[H_2CO_3^*] - [H^+] \qquad \text{(C-33)}$$

*See also* mineral acidity and p-alkalinity.

**caustic baryta** Another name for barium hydroxide. *See* baryta (2).

**caustic corrosion** Corrosion of aluminum, brass, bronze, tin, zinc, and other materials by sodium hydroxide or caustic soda (NaOH).

**caustic embrittlement** The failure of a steel boiler caused partially by the action of a concentrated caustic solution.

**caustic lime** Another name for lime, a white or grayish-white, slightly soluble in water, calcined oxide (CaO). It may be obtained from ground limestone (calcite) or calcium carbonate ($CaCO_3$) and has many applications, e.g., in water treatment, mortars, cements, etc. *See* quicklime for detail.

**caustic potash** Same as potassium hydroxide.

**caustic soda (NaOH)** Common term for sodium hydroxide, a white, strong alkaline compound, highly soluble in water, used as a concentrated solution or a flake in treatment processes to raise the pH of water or wastewater, neutralize acidity, and increase alkalinity. It is sometimes used to replace lime and soda ash in water softening. *See* sodium hydroxide for more detail.

**caustic-soda softening** A water treatment process that uses caustic soda (NaOH) instead of lime and soda ash to remove hardness ions: the caustic reacts with alkalinity and produces a carbonate ($CO_3$) ion to react with the divalent ions causing hardness. It is used when the raw water does not contain enough carbonate to react with lime. Calcium carbonate ($CaCO_3$) is precipitated in a fluidized bed from the bottom of which it is eventually removed.

**caustic-soda treatment** The addition of concentrated solutions of caustic soda (NaOH) to neutralize acid wastes. Other neutralization techniques include limestone treatment, lime-slurry treatment, and waste mixing (Nemerow & Dasgupta, 1991).

**caustic-soda treatment equations** The reactions that take place in two steps when caustic soda (sodium hydroxide, NaOH) is used to neutralize carbonic acid ($H_2CO_3$) or sulfuric acid ($H_2SO_4$). Representative overall reactions are:

$$2\ NaOH + H_2CO_3 \rightarrow Na_2CO_3 + 2\ H_2O \quad (C\text{-}34)$$

$$2\ NaOH + H_2SO_4 \rightarrow Na_2SO_4 + 2\ H_2O \quad (C\text{-}35)$$

**caustic wash** (1) A solution of 10% sodium hydroxide (NaOH) used to control scale formation on the walls of heat exchangers and reactors in thermal sludge conditioning. (2) A caustic solution of less than 20% used to restore the capacity of caked filter media. *See* media caking.

**cavern** A large underground opening in limestone or other rock resulting from dissolution by water. Caverns can also be formed in igneous rock by large gas bubbles.

**cavern flow** The movement of water through caves or other large openings, or through coarse granular materials.

**cavitation** (1) The action of a centrifugal pump attempting to deliver more water than allowed by suction. Then the water pressure decreases below vapor pressure, which turns the water into vapor. When the vapor bubbles move to points of higher pressure, they collapse violently, damaging the pumps and the pipes. Cavitation results also from the formation and collapse of gas pockets or bubbles on the gate of a valve, causing pitting of the gate or valve surface. Cavitation is accompanied by loud noises that sound like someone is pounding on the impeller or gate with a hammer. Cavitation may also occur under the nappe of a spillway when the actual head over the spillway exceeds the design head by 50% or more. The cavitation parameter is a proportionality factor ($\sigma$) equal to the ratio of the net positive suction head ($H'$) to the total pump head ($H$), and varying from 0.05 to 1.0. The pump manufacturer usually specifies it. *See* the formula for the net positive suction head (NPSH).

$$\sigma = H'/H \quad (C\text{-}36)$$

(2) A mechanical aeration system used to draw air into an industrial waste disposal well. It consists of a vertical draft tube connected to the influent and a rotor assembly mounted on a shaft, both supported by a steel bridge. (3) Same as cavitation process.

**cavitation parameter** *See* cavitation.

**cavitation process** The use of high-frequency sound waves to improve the performance of aqueous cleaners. *See* ultrasonic cleaning.

**CBOD** Acronym of carbonaceous biochemical oxygen demand.

**$CBOD_5$** The five-day measure of the pollutant parameter carbonaceous biochemical oxygen demand.

***C. bovis*** *See Cysticercus bovis.*

**$CBr_2ClCOOH$** Chemical formula of dibromochloroacetic acid.

**$CBr_3NO_2$** Chemical formula of bromopicrin.

**$CBrF_3$** Chemical formula of bromotrifluoromethane.

**CBrNX** Chemical formula of bromoacetonitrile, with X = Br, Cl, or H. Also written as $CBrX_2C{\equiv}N$.

***C. cayatensis*** Abbreviation of *Cyclospora cayatensis*.

**CCA** Acronym of critical component analysis.

**CCAS™** *See* countercurrent aeration system.

**CCC** (1) Acronym of (a) central control center, (b) criterion continuous concentration, (c) critical coagulation concentration, (d) current coagulation control. (2) A streaming current coagulation control center designed by Milton Roy Co.

**CCE** Acronym of carbon chloroform extract.

***C. cellulosae*** *See Cysticercus cellulosae.*

**CCL** Acronym of (drinking water) Contaminant Candidate List.

**CCL1** Acronym of First Contaminant Candidate List.

**CCL2** Acronym of Second Contaminant Candidate List.

**$CCl_2{:}CHCH_3$** Chemical formula of 1,1-dichloropropene.

**$CCl_2F_2$** Chemical formula of dichlorodifluoromethane.

**CCL3** Acronym of Third Contaminant Candidate List.

**$CCl_3CH_2$** Chemical formula of chloral.

**$CCl_3CH(OH)_2$** Chemical formula of chloral hydrate.

**$CCl_3F$** Chemical formula of trichlorofluoromethane.

**$CCl_3NO_2$** Chemical formula of chloropicrin or nitrochloroform.

**$CCl_4$** Chemical formula of carbon tetrachloride.

**$C_2Cl_4$** Chemical formula of tetrachloroethylene. Also written as $Cl_2C{=}CCl_2$.

**$C_4Cl_6$** Chemical formula of hexachlorobutadiene; also written as $Cl_2C{:}CClCCl{:}CCl_2$.

**$C_5Cl_6$** Chemical formula of hexachlorocyclopentadiene.

**$C_6Cl_4(COOCH_3)_2$** Chemical formula of dimethyl-2,3,5,6-tetrachloroterephtalate.

**$C_6Cl_5OH$** Chemical formula of pentachlorophenol.

**$C_6Cl_6$** Chemical formula of hexachlorobenzene.

**CCML method** Abbreviation of Cohen censored maximum likelihood method.

**CC-PCR** Acronym of cell culture-polymerase chain reaction.

**CCPP** Acronym of calcium carbonate precipitation potential.

**CCPP index** Acronym of calcium carbonate precipitation potential index.

**C curve** A tracer response curve, i.e., a graphical representation of the distribution of a tracer versus time of flow. It is used in studying short-circuiting, retardation, and longitudinal mixing in sedimentation basins and other reactors. *See* time–concentration curve for more detail.

**CD** Acronym of current density.

**Cd** (1) The chemical symbol of cadmium. (2) Abbreviation of candela, the international unit of luminous intensity; also written cd.

**CDA membrane** Abbreviation of cellulose diacetate membrane.

**CDC** Acronym of Centers for Disease Control and Prevention.

**CDI** Acronym of (a) capacitive deionization; (b) chronic daily intake; (c) continuous deionization.

**CDI™** A continuous deionization process developed by the U.S. Filter Corp. for electrical resin regeneration.

**CD/N** Acronym of current density to normality.

**CDROM or CD-ROM** Acronym of compact disk read-only memory. A computer storage disk that uses laser optics instead of magnetic means to read data. *See* ROM.

**CDS** Acronym of continuous deflection separator.

**CDTA** Acronym of cyclohexylenediaminetetraacetic acid.

**CE** Acronym of capillary electrophoresis.

**CEBW** Acronym of chemically enhanced backwash.

**CEC** Acronym of cation-exchange capacity.

**Cecarbon®** Granular activated carbon produced by Elf Atochem North America, Inc.

**Cecasorb®** Canisters of activated carbon manufactured by Elf Atochem North America, Inc.

**CELdek®** Synthetic media produced by Munters for evaporative cooling systems.

**cell** (1) The smallest structural part of living matter (protoplasm) capable of functioning as an independent unit; bounded by a thin membrane; divided into a nucleus and the cytoplasm. (2) In solid waste disposal, holes where waste is dumped, compacted, and covered with layers of dirt on a daily basis.

**cell biomass** *See* biomass, cell composition.

**cell composition** A microbial cell (for prokaryotes) consists of about 80% water, 18% dry organic material, and 2% dry inorganic material. Common empirical formulas used for prokaryote cells are $C_5H_7O_2N$ and $C_{60}H_{87}O_{23}N_{12}P$. These elements (C, H, O, N, P) represent about 95% of the weight of the cells, which also contain sulfur (S), potassium (K), sodium (Na), calcium (Ca), magnesium (Mg), chlorine (Cl), iron (Fe), and other trace elements. Algal cell or algal protoplasm is also represented by the equivalent formulas $C_{106}H_{263}O_{110}N_{16}P$ and $(CH_2O)_{106}(NH_3)(H_3PO_4)$.

**cell culture** Growth and replication of cells in vitro.

**cell culture–polymerase chain reaction (CC-PCR)** A testing method recently developed to detect the presence of live, infectious *Cryptosporidium* oocysts in water by targeting the DNA of the pathogen that grows in human cells.

**cell debris** The cell wall or nonbiodegradable material remaining after a cell dies, while the other portion is released for consumption by other microorganisms. Cell debris, measured among volatile suspended solids (VSS), represent 10–15% of original cell weight and are produced at a rate $r$, g VSS/m³/day (Metcalf & Eddy, 2003):

$$r = fk_dX \qquad \text{(C-37)}$$

where $f$ = fraction of biomass remaining as cell debris, 10–15%; $k_d$ = endogenous decay rate, g VSS/g VSS/day; and $X$ = biomass concentration, mg/L. *See* cell lysis.

**cell fusion** The merging of two or more cells into a single unit.

**cell granule** Food reserve stored intracellularly by bacteria, e.g., carbon as glycogen or poly-$\beta$-hydroxybutyrate, elemental sulfur, and phosphorus as polyphosphates. *See also* endospore, gas vesicle.

**cell lysis** Same as lysis, i.e., the decomposition or destruction of a cell by antibodies (lysins), e.g., the cell wall disintegration caused by ozonation. *See* cell debris.

**cell-mediated immunity** Immunity that depends not on antibody but on the ability of T cells to recognize and destroy antigen-bearing cells; e.g., in the resistance to infections, autoimmune diseases, and allergic reactions.

**cell membrane** The semipermeable double layer of lipid and protein that encloses the cytoplasm of a cell and serves, among other functions, as a protective barrier against toxic substances. It controls the passage of substances into the cell and the release of waste materials and metabolic by-products out of the cell.

**cell pair** The combination of an anion membrane, a cation membrane, and two separators or spacers to form an electrodialysis cell. *See* electrodialysis cell pair, membrane stack or module.

**cell pair resistance** The electrical resistance of a cell pair.

**cell potential** The driving force of an overall redox reaction or the difference between the standard potentials of the two half-reactions:

$$A + e^- \rightarrow A^- \qquad E^0_A \qquad \text{(C-38)}$$

$$B^- - e^- \rightarrow B \qquad -E^0_B \qquad \text{(C-39)}$$

$$A + B^- \rightarrow A^- + B \qquad E^0_A - E^0_B \qquad \text{(C-40)}$$

The cell potential ($E_C = E^0_A - E^0_B$) is also called couple potential.

**cell residence time (CRT)** The average time, in days, that microorganisms spend in a biological waste treatment system, a leading design and operation parameter: e.g., nitrifiers' CRTs are longer than CRTs for organisms that stabilize carbonaceous matter. *See* solids residence time for detail.

**cell tissue** Tissue, i.e., an aggregate of similar cells and cell products, often represented by the formula $C_5H_7O_2N$, which shows the main elements present (carbon, hydrogen, oxygen, and nitrogen). *See also* cell composition.

**cell transformation** The process whereby a cell acquires a carcinogenic property.

**cellular adenylate** The sum of the three coenzymes ADP, AMP, and ATP often used as an indication of microbial biomass. *See also* adenylate energy charge.

**cellular macromolecule** A high-molecular-weight biological molecule of proteins and nucleic acids. Such molecules play a role in reactions leading to the development of cancer and chronic diseases.

**cellular uptake** *See* luxury uptake.

**cellulose** An abundant plant polymer (with molecular weight averaging $4.0 \times 10^5$ but going up to $1.8 \times 10^6$), insoluble in water, and a structural carbohydrate, i.e., it serves to maintain the integrity of cells as compared to a carbohydrate that provides energy. Microorganisms produce extracellular enzymes that initiate the degradation of cellulose. It is also defined as a polysaccharide consisting of units joined together; it has the same general formula as starch: $(C_6H_{10}O_5)_n$. It constitutes 60% of wood and 90% of cotton. *See also* hemicellulose.

**cellulose acetate (CA)** An acetic ester of cellulose, used to make water treatment membranes, yarns, textiles, etc.

**cellulose acetate blend (CAB) membrane** A membrane used in water treatment applications and consisting of a blend of cellulose diacetate and cellulose triacetate that has 2.5 acetyl groups per repeating carbon ring.

**cellulose acetate (CA) membrane** An asymmetric membrane used in reverse osmosis and other water treatment processes; it consists of acetic groups attached to cellulose. *See also* polyamide (PA) membrane.

**cellulose diacetate (CDA)** A diacetic ester of cellulose that resists solvents and is used in the manufacture of water treatment membranes, textiles, etc. A CDA membrane has two acetyl groups per repeating carbon ring.

**cellulose gum** A white, water-soluble polymer produced from cellulose and used in paper making, in foods, in textile-finishing mills, and in general-purpose synthetic detergents. Also called carboxymethylcellulose, CMC, or sodium cellulose glycolate.

**cellulose ion exchanger** A cellulose-based material combined with anion or cation groups for selective ion exchange.

**cellulose triacetate (CTA)** A triacetic ester of cellulose that resists solvents and is used in the manufacture of water treatment membranes, textiles, etc. A CTA membrane has three acetyl groups per repeating carbon ring.

**cellulosic membrane** A membrane made from cellulose-based materials.

**cell wall** The component that maintains the cell shape and protects the membrane.

**cell yield** A constant that relates growth to substrate utilization. *See also* bacterial yield or biomass yield, Monod equation.

**cell yield coefficient** The amount of cell mass produced per unit amount of substrate consumed. *See also* bacterial yield or biomass yield, Monod equation.

**Celsius, Anders** The Swedish astronomer (1701–1744) who devised the Celsius temperature scale.

**Celsius degree** Unit of temperature interval equal to the kelvin; formerly known as degree centigrade. Celsius scale is the international name for the centigrade scale, where freezing point is at zero degree (0°C), and boiling point at 100 degrees (100°C), at a barometric pressure of 760 mm Hg or 101.325 kilopascals. The Celsius and

Kelvin scales are related as follows: temperature in Celsius = temperature in Kelvin minus 273.15. To convert to the Fahrenheit scale, multiply degrees centigrade by 1.8 and add 32.

**Celsius temperature scale** *See* Celsius degree.

**CEM** Acronym of continuous emissions monitoring.

**CEMcat™** A device manufactured by Advanced Sensor Devices, Inc. for monitoring continuous emissions.

**cement** (1) A powder that, mixed with water, binds a stone and sand mixture into strong concrete when dry. Cement contains quicklime (CaO), silica ($SiO_2$), alumina ($Al_2O_3$), ferric oxide ($Fe_2O_3$), magnesium oxide (MgO), and gypsum. *See also* sludge cement. Portland cement is the commonly used name. (2) Binding mineral matter in sedimentary rock; it consists of silica, calcite, dolomite, hematite, limonite, siderite, and gypsum. (3) Cement rock is a natural rock that contains the above constituents.

**cementation** The deposition of a metal when its ions react with a more readily oxidized metal, i.e., the displacement and precipitation of a metal from solution by a metal higher in the electromotive series, an electrochemical process that can be used for copper (Cu) removal by reaction with iron (Fe) or for the precipitation of cadmium (Cd) with a more active metal (zinc, Zn):

$$Cu^{2+} + Fe \text{ (iron scrap)} \rightarrow Cu + Fe^{2+} \quad \text{(C-41)}$$

$$Cd^{2+} + Zn \rightarrow Cd + Zn^{2+} \quad \text{(C-42)}$$

*See also* electrodeposition.

**cement grout** A mixture of Portland cement and water that is fluid enough to be pumped through a small-diameter pipe.

**cement industry waste** Wastewaters originating from cement manufacture operations (grinding, blending, kiln, dust leaching and control, cooling) are high in suspended solids and inorganic salts. Common treatment methods include segregation of dust-contact streams, neutralization, and sedimentation.

**cement kiln** A rotary chamber used to calcine a slurry to produce cement.

**cement kiln dust (CKD)** Alkaline material produced during the manufacture of cement; it may be used as a substitute for lime in the stabilization of wastewater sludge. The material is 10–30% quicklime (CaO) equivalent, has a specific gravity of 2.75, a bulk density of 50–65 lb/ft$^3$, and an angle of repose of 70°. *See* advanced alkaline stabilization.

**cement kiln dust lime** A lime product used in the alkaline stabilization of wastewater sludge. *See* cement kiln dust.

**cement-lined pipe** A cast-iron, steel, or other pipe whose interior is lined with a layer of cement mortar for smoothness and corrosion control.

**cement rock** *See* cement (3).

**CEMS** Acronym of continuous emissions monitoring system.

**census tract** A standard subdivision of a county or city used for population enumeration; it is usually a homogeneous area of 2500–8000 people.

**census-tract analysis** An analysis based on census-tract characteristics; it is sometimes used in water and wastewater planning studies to determine water use and wastewater production parameters.

**Censys™** A line of products manufactured by the U.S. Filter Corp. for use in water and wastewater treatment.

**Centaur™** Activated carbon produced by the Calgon Carbon Corp.

**center-feed clarifier** *See* center-feed settling tank.

**center-feed settling tank** A type of circular tank used in primary and secondary sedimentation; it includes a pipe suspended from a bridge to carry the wastewater to a circular well in the center of the tank and a revolving mechanism for sludge transport and removal. The bottom of the tank slopes toward the central hopper. *See also* peripheral-feed or rim-feed settling tank.

**center of buoyancy** The center of gravity of the volume occupied by an object floating in a liquid.

**center of flotation** The centroid of the water plane area of a floating body.

**center of gravity** The point in a body through which the resultant of gravitational forces on the particles of the body passes and from which the resultant force of attraction of the body on other bodies emanates. *See also* center of mass, centroid.

**center of mass** The point at which the entire mass of a body is considered concentrated; the first moment of an object about any line through the center of mass is zero. *See also* center of gravity, centroid.

**center of pressure** The point of application of the resultant of the normal pressures acting on a surface.

**center pivot sprinkler system** *See* center-pivot system.

**center-pivot system** A sprinkler irrigation system commonly used in wastewater distribution; it consists of a rotating lateral suspended by wheel supports. *See also* sprinkler distribution, furrow irrigation, graded-border irrigation, drip or trickle irrigation, gated pipe irrigation.

**Centers for Disease Control and Prevention (CDC)** An agency of the U.S. Department of Health responsible for protecting public health through disease surveillance, control and prevention procedures, and health education.

**Center-Slung** A basket centrifuge manufactured by Ketema, Inc.

**centigrade** *See* Celsius degree.

**centigrade temperature scale** *See* Celsius degree.

**centimeter, gram, second (CGS or cgs) system** The metric system of measurements with the fundamental units of length (centimeter), mass (gram), time (second). *See also* Système International.

**centipoise** The unit of absolute viscosity. One centipoise is the absolute viscosity of water at room temperature or about 20.2°C. Abbreviation: cp.

**Centrac** A metering pump manufactured by Milton Roy Co.

**Centra-flo™** A sand filter of the continuous flow type manufactured by Applied Process Equipment.

**central atom** The metal cation that combines with ligands (anions or molecules) to form a coordination compound. *See also* chelate.

**central control center (CCC)** The control room of a digital control system of a wastewater treatment plant. It functions on a continuous basis, with personnel having control or oversight over the entire facility.

**central tendency** The tendency of a dataset to cluster around some central value, such as the mean, median, or mode.

**centrate** The water or dilute stream leaving or remaining in a centrifuge after the removal of a dense cake of solids. It contains fine, low-density, nonsettling solids and is commonly returned to the influent of the wastewater treatment plant, which may result in the discharge of fine solids with the effluent. Ammonium phosphate scales (struvite) sometimes build up in centrate lines. *See also* sludge cake.

**Centric®** A series of overload clutch products manufactured by Zurn Industries, Inc.

**Centri-Cleaner®** A liquid cyclone produced by Andritz-Ruthner, Inc.

**Centridry™** A process developed by Humboldt Decanter, Inc. for sludge drying and dewatering.

**CentriField®** A wet scrubber manufactured by Entoleter, Inc.

**centrifugal blower** An air conveying device commonly used in wastewater aeration systems, particularly where the required capacity exceeds 15,000 cfm of air; it is similar in operation to a low-specific-speed centrifugal pump.

**centrifugal dewatering** The use of a centrifuge to reduce the water content of water or wastewater treatment residuals. The fast rotation of a cylindrical bowl or basket develops a force that separates sludge into a solid cake and a liquid centrate that are discharged separately.

**centrifugal drying** The partial drying of a sludge or slurry by a device that uses centrifugal action (e.g., a centrifuge).

**centrifugal force** An outward force on a body that rotates about an axis or the inertial force acting away from the center of the curved path of a body; assumed equal and opposite to the centripetal force. Due to the centrifugal force, a moving body tends to continue in the same direction. In water and wastewater engineering, applications of the centrifugal force include the use of centrifugal pumps and centrifugation.

**centrifugal mixed-flow pump** A type of centrifugal pump, with an impeller intermediate between the radial and axial flow propellers, used in dry well installations of wastewater treatment plants.

**centrifugal nonclog pump** A low-cost, high-volume, high-efficiency centrifugal mixed-flow pump used primarily in activated sludge handling operations. *See also* nonclog pump.

**centrifugal pump** A continuous-flow pump that moves water and other liquids by accelerating them radially outward in an impeller fixed on a rotating shaft and enclosed in a casing. The pump imparts pressure to the liquid through the centrifugal force created by the impeller. A centrifugal screw pump has a screw-type impeller. It is one of the most widely used in water supply, drainage, wastewater disposal, and irrigation. Also called radial-flow pump. There are two other types of continuous-flow pumps: axial-flow and mixed-flow. Noncontinuous flow pumps are called positive-displacement pumps. *See also* grinder pump, vortex pump, airlift pump, screw pump.

**centrifugal screw pump** *See* centrifugal pump.

**centrifugal thickening** The use of a centrifuge to thicken or dewater sludge from water or wastewater treatment; centrifugation. *See also* cosettling thickening, dissolved air flotation, gravity belt thickening, gravity thickening, rotary drum thickening, solid-bowl centrifuge, capture, TSS recovery.

**centrifugation** The use of centrifugal force (i.e., in a centrifuge) to separate liquids of different densities, thicken slurries, or separate solids from liquids based on density differences. In wastewater sludge dewatering, centrifuging is a shallow-depth settling process that results in a sludge liquor as by-product, the centrate, which is returned to the head of a water treatment plant or to the primary clarifier. The sludge cake is further processed for final disposal. Addition of chemical agents improves efficiency.

**centrifugation thickening** Centrifugation; centrifugal thickening.

**centrifuge** A mechanical device that uses centrifugal or rotational forces to separate solids from liquids. It is similar to a clarifier with accelerated settling. *See* the two common types of centrifuge used in water and wastewater treatment sludge dewatering: solid-bowl and basket centrifuges. *See also* high-solids or high-torque centrifuges. In wastewater treatment, centrifuges used to thicken activated sludge and dewater digested or chemically conditioned sludges achieve a solids recovery of 90–95%.

**centrifuge classification** The separation by a centrifuge of normal microorganisms and decay products from filamentous organisms, the latter being selectively recycled with the centrate. This phenomenon occurs when there are significant differences between the characteristics of the various solids of the sludge being dewatered and can be minimized by maintaining a high rate of solids recovery (85–90%) during centrifugation.

**centrifuge dewatering** Centrifugation; centrifugal dewatering.

**centrifuge skimming** A step in the operation of an imperforate basket centrifuge for sludge dewatering, which consists in the removal of soft sludge (about 5–10% of the volume) from the inner wall of sludge in the basket; the sludge so removed is returned to the wastewater treatment system. Skimming takes place after filling the centrifuge and before initiating plowing.

**centrifuging** *See* centrifugation.

**centripetal drainage** Radial drainage inward toward a central point.

**centripetal force** The force on a body that rotates about an axis or moves along a curved path. It is directed toward the center of rotation or the center of curvature and is assumed equal and opposite to the centrifugal force.

**Centripress™** A solid bowl centrifuge manufactured by Humboldt Decanter, Inc.

**centroid** In a figure, the point that is such that the sum of the displacements from all points of the figure is zero; the point of intersection of the medians of a triangle. *See also* center of gravity, center of mass.

**centroid of flow** The midpoint of a channel with uniform flow.

**CenTROL®** A gravity sand filter manufactured by General Filter Co.

**Centrox®** A line of aspirating aerators manufactured by Hazelton Environmental Products, Inc.

**CEP** Acronym of catalytic extraction processing.

**CEPT** Acronym of chemically enhanced primary treatment.

**CEQ** Acronym of Council on Environmental Quality.

**Cerabar** A pressure transmitter manufactured by Endress+Hauser.

**Ceraflo®** A line of ceramic membrane filters manufactured by the U.S. Filter Corp.

**ceramic membrane** A membrane consisting of a ceramic layer and an alumina–zirconia layer, used in water filtration.

***Ceratium*** A group of unicellular and flagellated phytoplankton whose blooms contribute to taste and odor problems in reservoirs and lakes. They impart a fishy or septic odor and a bitter taste.

**cercaria** The larval stage of a trematode worm emerging from a snail host; more specifically, the highly mobile, final stage of a schistosome emerging from the aquatic snail through the skin of a human host. Sedimentation and filtration do not effectively remove cercariae; chlorination and storage are more effective means of treatment. *See also* egg, miracidium, metacercaria.

**cercariae** Plural of cercaria.

**cercarial dermatitis** A skin disease caused by snail hosts and schistosome larvae, found in some waters of the United States and transported by infected water fowl; common name: swimmer's itch Copper sulfate is effective against the snail host.

***Ceriodaphnia dubia*** Scientific name of the water flea, a test organism used in toxicity bioassays.

***Cerrhina molitcrella*** A hardy fish species of the carp family that survives well in stabilization ponds used for aquaculture.

**certificate of participation** A financial instrument, issued by a nonprofit entity and equivalent to a tax-exempt bond for the construction of public works, such as water and wastewater facilities.

**certificate of water right** An official document serving as court evidence of a perfected water right.

**certification** A program of a recognized agency to document the capabilities of personnel through education and experience. Examples in the water and wastewater engineering field include state certification for treatment plant operators and certification as a Diplomate of the American Academy of Environmental Engineers.

**certified backflow-prevention assembly technician** A technician qualified by training and experience to install, maintain, repair, and test backflow prevention devices.

**certified laboratory** A laboratory that meets specific requirements and is approved by a state or the Environmental Protection Agency to conduct analyses related to compliance with the Safe Drinking Water Act.

**certified reference material** A stable and well-defined substance that is physically and chemically similar to a given sample and can be analyzed to represent the sample.

**certified utility safety administrator** An individual certified by the National Safety Council Utilities Division to analyze the safety of a specific utility, e.g., drinking water and wastewater management.

**cerussite** A common lead compound, essentially lead carbonate, $PbCO_3(s)$, found in masses or in transparent crystals. Also, a product of lead precipitation resulting from the corrosion of lead piping. *See also* hydrocerussite.

**cesium (Cs)** A soft, silvery-white, ductile, alkali, highly reactive, liquid metallic element, used in photoelectric cells and as a catalyst, with two radioactive isotopes: $^{134}Cs$ and $^{137}Cs$; the most electropositive of all elements. Atomic weight = 132.905. Atomic number = 55. Specific gravity = 1.873. Melting point = 28.4°C. Boiling point = 670°C. Valence = 1. Cesium constitutes a dangerous waste product of the nuclear industry. Also spelled caesium.

**cesium-137** A radioisotope of cesium, used in cancer research and radiation treatment. Its ionizing radiation can also inactivate microorganisms. Half-life reported variously in the literature, from 30 to 37 years. Abbreviation: Cs-137 or $^{137}Cs$.

**cesium-141** A radionuclide of cesium, half-life = 33 days.

**cesspit** A subsurface container for the retention of wastewater until it is removed by a vacuum tanker or other means.

**cesspool** (1) An underground, sealed chamber, without overflow, that stores liquid wastes and sludge and is emptied periodically, similar to a large vault system. (2) An underground pit or covered tank with open joints constructed in permeable soil to receive raw domestic wastewater and allow partially treated effluent to seep into the surrounding soil, while solids are contained and undergo digestion. Sometimes called leaching cesspool or soakaway. *See also* septic tank. (3) A covered, sealed, underground chamber that receives domestic wastewater and has an outlet pipe; a septic tank. (4) A cistern, pit, or well for retaining sediment from a drain.

**Cestoda** A class of flatworms that includes the cestode or tapeworm. They consist of a head (scolex) and a chain of helminth segments (proglottides), each segment having reproductive organs. They pass by a larval stage in an intermediate host before infecting the definitive host as an adult through fecal contamination.

**cestode** A flat or tapelike worm of the class Cestoidea or Cestoda and phylum Platyhelminthes, lacking an alimentary canal; parasitic when adult in the digestive tract of humans and other vertebrates. Also called tapeworm. Cestodes include three parasites of the human intestine: the genera *Diphyllobothrium, Hymenolepis,* and *Taenia.* Some cestodes like *Taenia saginata* exist as cysts in one host (cow) and adults in another (man). *See also* trematode (fluke).

**$(C_2F_4)_n$** Chemical formula of polyfluoroethylene.

**C factor** (1) Pipe roughness coefficient used in the Hazen–Williams formulas. The C factor indicates the smoothness of the interior of a pipe. The higher the C factor, the smoother the pipe, the greater the carrying capacity, and the smaller the friction or energy losses from water flowing in the pipe. To calculate the C factor, measure the flow ($Q$), the pipe diameter, distance between two pressure gages, and the friction or energy loss of the water between the gages:

$$C = Q/[193.75\, D^{2.63} S^{0.54}] \qquad \text{(C-43)}$$

(2) The friction factor in the Chezy equation. (3) A proportionality coefficient in the orifice, rational, and weir equations.

**CFC** Acronym of chlorofluorocarbon.

**CFC-113** A chlorinated solvent used for cleaning in the PCB assembly process; it is being phased out and replaced by less toxic products such as the hydrofluorocarbons.

**$CFCl_3$.** Chemical formula of trichlorofluoromethane or Freon 11.

**CFD** Acronym of computational fluid dynamics.

**cfm** Abbreviation of cubic foot (feet) per minute, a common unit of air flow equal to one cubic foot in one minute.

***C. fetus* ssp. *jejuni*** Abbreviation of *Campylobacter jejuni.*

**CFR** Acronym of Code of Federal Regulations.

***C. fruendii*** Abbreviation of *Citrobacter fruendii.*

**cfs** Abbreviation of cubic foot (feet) per second, a common unit of flow equivalent to one cubic foot in one second, for the measurement of liquids and gases. 1 cfs = 448.8 gpm = 0.646 mgd = 28.32 liters per second = 2446.6 $m^3$/day. Another abbreviation is cusec.

**CFSTR** Acronym of continuous-flow stirred-tank reactor.

**CFU** or **cfu** Acronym of colony forming unit(s).

**CGR** Acronym of coliform growth response.

**CGR test** Coliform growth response test.

**CGS** or **cgs system** Abbreviation of centimeter, gram, second system.

**CH** For chemical formulas of some organic compounds, see the table under organic compound formulas.

**$C_2H_2n$** Chemical formula of cycloparaffin.

**Chabelco** A series of chain products of Envirex, Inc.

**Chadwick, Edwin (1800–1890)** Author of *Report of the Poor Law Commissioners on an Inquiry into the Sanitary Condition of the Labouring Population of Great Britain* (1842). He recommended the use of small pipe sewers and separate sewer systems: "the rain to the river and the sewage to the soil." *See also* the Great Sanitary Awakening.

**Chagas' disease** The common name of American trypanosomiasis, a disease caused by the protozoa *Trypanosoma cruzi* and transmitted by blood-sucking bugs that breed in wall cracks and furniture. The disease affects people in poor rural areas of Central and South America. *See also* Gambian sleeping sickness and Rhodesian sleeping sickness.

**chagoma** A small granuloma in the skin caused by early multiplication of the pathogen of Chagas' disease.

**chain-and-bucket elevator** A device used to remove grit from horizontal flow and aerated grit chambers. *See also* clamshell bucket, inclined screw.

**chain-and-flight mechanism** A mechanism used to scrape and/or collect grit from grit chambers or sludge from clarifiers. *See* chain-and-flight sludge collector, chain-and-bucket elevator, chain-and-flight sludge collector, screw conveyor.

**chain and flights** A mechanism or device used in large wastewater treatment plants to scrape the grit from a horizontal flow grit chamber to a hopper at the chamber's inlet end for removal by other devices. *See also* screw auger, chain-and-bucket elevator.

**chain-and-flight sludge collector** A scraping device or mechanism used to remove sludge from rectangular clarifiers or sedimentation basins. It consists of two endless loops of cast iron or plastic chains with wood or fiberglass flights (scrapers) attached to the chains. Sludge collected into a hopper is discharged into processing units by a screw conveyor. The chains move along the bottom and across the top of the basin. *See also* screw conveyor, traveling bridge.

**chain bucket** A continuous chain of buckets, or a drag bucket, mounted on a scow for lifting excavated materials. Also called bucket dredge, ladder dredge.

**chain-driven bar screen** *See* chain-driven screen.

**chain-driven screen** A device used in wastewater treatment to remove coarse materials from the liquid ahead of other treatment units. It consists of a chain-driven mechanism to move the rake teeth through the screen openings. The screen can be raked from the front/upstream side or the back/downstream side, and the rakes can return to the bottom from the front or from the back. Also called chain-driven bar screen. *See also* catenary screen, continuous belt screen, reciprocating rake screen.

**Chainsaver Rim** A sludge collection mechanism manufactured by Jeffrey Chain Co.

**chalcanthite** A common name for copper sulfate ($CuSO_4$). *See* blue vitriol for detail.

**chalcogen** Any of the elements constituting Group VIA in the periodic table: oxygen (O), sulfur (S), selenium (Se), tellurium (Te), and polonium (Po).

**chalcogenide** (1) The combination of a chalcogen with a more electropositive element or radical. (2) Any of the elements of Group VIA of the periodic table: oxygen (O), sulfur (S), and selenium (Se). *See also* chalcogen.

**chalk** A common compound of calcium, as calcium carbonate ($CaCO_3$).

**challenge water** A water sample prepared specifically to test water treatment equipment or products according to established standards. Challenge water for each type of equipment is specifically defined in the individual equipment testing standards such as those established by the Water Quality Association and the National Sanitation Foundation International.

**chalybeate spring** (From the Greek word chalibe, meaning iron.) A spring whose water contains a high concentration of iron sulfate [$Fe_2(SO_4)_3$] and other iron salts in solution.

**chamber** A space enclosed by walls, e.g.: a compartment, the space in a channel lock between the upper and the lower gates, a diversion, grit, or junction chamber. *See* the following chambers used in water and wastewater treatment: air, chlorine contact, digestion, diversion, dosing, grit, junction, mixing, scum, settling, sludge digestion.

**chance contact straining** A major mechanism of removal of particulate matter within a granular filter, whereby pore spaces trap smaller particles by chance contact. *See* filtration mechanisms.

**change of state** (1) The transformation of a substance from one to another of the three states of matter: gas, liquid, or solid, with changes in the substance's physical properties and molecular structure. (2) A change in the state variables (density, pressure, temperature).

**channel** A natural or artificial waterway that contains flowing water periodically or continuously. It has definite, although variable, geometric characteristics such as bed, banks, and slope. A channel may also be a ditch or drain excavated for the flow of water. There may be channels within large bodies of water, for example, the deep areas used for navigation. Channel geometric properties commonly used in hydraulics and hydrology include the area (A), the wetted perimeter (P), the top width (W), the hydraulic radius (R), and the hydraulic mean depth ($D_m$). *See* open channel flow.

**Channelaire™** A submersible aerator/mixer manufactured by Framco Environmental technologies.

**Channel Flow** A wastewater disintegrator manufactured by C&H Waste Processing.

**channeling** The condition that exists in a filter, ion exchange or other packed bed where water or regenerant flows through furrows or channels without effectively contacting the medium. This condition results from plugging and other causes. A nonuniform medium increases the tendency for channeling and short-circuiting during backwashing of a granular filter. Also called channelization. *See also* short circuiting.

**channelization** (1) Straightening and deepening streams so water will move faster, a marsh-drainage tactic that can interfere with waste assimilation capacity, disturb fish and wildlife habitats, and aggravate flooding. (2) Same as channeling.

**Channel Monster®** A wastewater grinder manufactured by JWC Environmental.

**chapelet** (1) A device consisting of a continuous chain of buckets between rotating sprocket wheels, used for raising water or for dredging. (2) A chain pump with buttons or discs along the chain; sometimes called paternoster pump.

**char** The product of the carbonization of organic matter, used as a fuel or in other applications; charcoal. A char is the first step in the preparation of activated carbon. It is produced by heating the base material (almond, coconut, and walnut hulls, wood, bone, coal, etc.) to 700°C to drive off the hydrocarbons but without enough oxygen for combustion.

**characteristic heating value** The amount of heat released by a fuel in a combustion chamber minus the dry flue gas and the moisture losses. *See* available heat for detail.

**characteristics** A group of biological, chemical, or physical constituents of water or wastewater. *See also* contaminant, pollutant.

**characteristic speed** The speed in revolutions per minute (rpm) at which a turbine would operate to develop one horsepower (1.341 kilowatts) under a head of one foot. It is used in selecting the proper turbine for a particular condition. *See* specific speed for detail. It is also called characteristic type for a waterwheel.

**characteristic type** *See* characteristic speed.

**charcoal** The carbonaceous material obtained by heating wood or another organic substance with little or no oxygen. Sometimes confused with activated carbon, it is used as a fuel and as an adsorbent (e.g., to remove odor and color) but has only 30–40% of the surface of activated carbon.

**charge** The excess or deficiency of electrons in a body or substance. *See also* electric charge.

**charge balance** The sum of the positive and negative charges of the elements of a chemical reaction, which must be neutral or equal to 0. It is sometimes used to determine the formula of a compound or to balance a reaction. For example, in the general COD reaction using dichromate:

$$C_nH_aO_b + c \cdot Cr_2O_7^{2-} + d \cdot H^+ \rightarrow 2c \cdot Cr^{3+} + n \cdot CO_2 + (b + 7c - 2n) \cdot H_2O \quad \text{(C-44)}$$

the charge balance equation is:

$$-2c + d = 6c, \qquad \text{i.e., } d - 8c = 0 \quad \text{(C-45)}$$

The charge balance and electron balance equations are equivalent. (The electron balance equation states that, for a balanced reaction, the number of electrons gained is equal to the number of electrons lost.)

**charge conservation law** One of two fundamental rules applied in balancing chemical reactions; same as charge balance. *See also* law of mass balance.

**charge density** The mole ratio of charged monomers to noncharged monomers of polyelectrolytes.

**charged polysulfonate membrane** A membrane that has been chemically sulfonated to create the ability to reject dissolved ions. The sulfonation process permanently affixes sulfonate ($SO_3^-$) groups on the membrane surface the same way that cation-exchange resins receive their charge characteristics.

**charge for conditional water service** The amount charged for readiness to supply a quantity of water to a customer in the event that its primary source of supply is not available.

**charge neutralization** (1) The process of adding a positively charged material to neutralize or destabilize a negatively charged particle suspension. It is one of two common methods of mixing chemicals for coagulation: coagulants are dispersed rapidly (e.g., in less than one second) and mixed at high intensity, forming hydrolysis products that will contact the colloidal particles for precipitating hydroxides. The other method is sweep coagulation. *See also* surface charge neutralization and floc volume concentration. (2) The mechanism of neutralization and destabilization in coagulation. *See also* interparticle bridging (polymer bridge formation).

**charge neutralization destabilization** The method or mechanism of charge neutralization.

**charge reversal** A change in the charge of a particle resulting from the addition of chemicals to adjust water or wastewater pH and optimize the performance of coagulants. Charge reversal can also result from overdosing of coagulants beyond a zero point of charge and cause particle restabilization.

**Charles–Gay-Lussac law** *See* Charles' law and Gay-Lussac law.

**Charles, J. A. C.** The French physicist (1746–1823) who stated in 1787 the thermodynamic principle known as Gay-Lussac law or Charles' law.

**Charles' law** The volume of a gas ($V$) at constant pressure varies in direct proportion to its absolute temperature ($T$), or, the ratio of volume to temperature is constant:

$$V_1/T_1 = V_2/T_2 = \text{constant} \qquad \text{(C-46)}$$

where $V_1$ is the volume corresponding to temperature $T_1$ and $V_2$ is the volume corresponding to temperature $T_2$. It is sometimes expressed as the Gay-Lussac law in terms of density: the density of an ideal gas at constant (and relatively low) pressure varies inversely with its absolute temperature. *See also* general gas law.

**check dam** A structure installed in a small stream (e.g., a gully) to decrease flow velocity, minimize channel erosion, promote sedimentation, and divert water. *See also* gully reclamation.

**check sample** A water sample collected to verify the results of a previous test; e.g., after the analysis of a sample indicates the presence of bacteria by the membrane filter method or the fermentation tube technique.

**check valve** A semiautomatic device that limits flow in a piping system to a single direction. Its hinged disc or flap opens in the direction of normal flow under the influence of pressure and closes when flow ceases to prevent flow reversal. Examples of its installation are in the discharge pipe of a centrifugal pump and as an internal valve of a diaphragm metering pump. When installed at the end of a suction line, it is called a foot valve; it prevents draining the line when pumping stops. A double check valve is a backflow-preventing device that closes when flow reverses. Also called nonreturn valve or backpressure valve.

**check valve slam** The reversal of forward flow toward a pump before a valve fully closes to stop the pump. When the valve closes, the reverse flow stops and causes a loud water-hammer noise in the pipe.

**Check Well** An instrument manufactured by Drexelbrook Engineering Co. to measure water levels in wells.

**cheese waste** Waste generated in a cheese factory, which processes whole milk, cream, or separated milk in a cheese vat, along with various additives, such as lactic-acid bacterial culture and souring agents. *See* dairy waste for characteristics and treatment methods.

**chelate** (1) A heterocyclic chemical compound whose molecule consists of a central metallic ion attached by covalent bonds to two or more nonmetallic ions in a ring structure; a chelating agent or chelator. (2) Pertaining to a chelate or chelating agent. (3) To react to form a chelating agent or to combine an organic compound with a metallic ion to form a chelate.

**chelate effect** The increase in stability of complexes formed with multidentate ligands as compared to those with unidentate ligands and also the fact that the degree of complexation decreases with dilution more strongly for monodentate complexes. The chelate effect results from the entropy difference between the two complex formations.

**chelating agent** A compound (a ligand) that is soluble in water and combines with metal ions to keep them in solution, forming a heterocyclic ring. As compared to a unidentate (or monodentate) ligand, a chelating agent is a multidentate or polydentate ligand, i.e., it has more than one atom that can bond with a central metal ion at one time to form a ring structure. EDTA and other chelating agents may be incorporated into synthetic resins to produce exchangers for specific metal ions. Another chelating agent, nitrilotriacetic acid (NTA), is used in detergents to combine with the hardness ions. Also called a chelator. *See also* complex, monodentate ligand, polydentate ligand, sequestering agent.

**chelating ion exchanger** An ion-exchange resin that selectively adsorbs one metal ion to the exclusion of other metal ions present.

**chelating reagent** *See* chelating agent.

**chelating resin** An ion-exchange resin, similar to a weak-acid resin, that has an iminodiacetate functional group and exhibits a high degree of selectivity for copper, mercury, lead, nickel, and other toxic metals. Chelating resins can be used to remove metals from waste streams.

**chelation** The chemical complexing (forming or joining together) of metallic cations (such as copper) with certain organic compounds (such as EDTA or ethylene diamine tetraacetic acid); chelation is the formation of a complex by a metal ion with a multidentate ligand. Chelation can be used to prevent the precipitation of metals, to treat heavy metal poisoning, or to remove metals from waste streams. Chelation may also be used for water softening without actually removing the hardness ions from solution. For example, excess EDTA anions (represented by $Y^{4-}$) can bind with and reduce the concentration of free hydrated calcium (Ca) ions:

$$Ca^{2+} + Y^{4-} \rightarrow CaY^{2-} \qquad \text{(C-47)}$$

*See also* complexation, sequestration.

**chelator** A chelating agent.

**Chem-Feed®** A diaphragm-metering injection pump manufactured by Blue-White Industries of Huntington Beach, CA.

**Chem-Flex®** A portable holding tank manufactured by Aero Tec Laboratories, Inc.

**Chem-Gard®** A line of centrifugal pumps manufactured by Vanton Pump & Equipment Co.

**chemical** (1) An association of atoms into a molecule or compound with specific characteristics. (2) A substance produced or used in a chemical process or reaction (e.g., in water and wastewater treatment). *See also* important chemicals, for some of those that relate to water chemistry, water and wastewater treatment, or to sludge processing. (3) Related to or concerned with chemistry or chemicals.

**Chemical Abstracts Service numbers (CAS #)** Numbers used by the Chemical Abstracts Service of the American Chemical Society to identify chemical substances. The USEPA compiles a list of CAS #s for all toxic substances available in the United States. Table C-01 lists the reference numbers of some substances.

**Table C-01.** Chemical Abstracts Service numbers

| Substance | CAS # | Substance | CAS # |
|---|---|---|---|
| Acetic acid | 64-19-7 | Formaldehyde | 50-00-0 |
| Ammonia | 7664-41-7 | Lindane | 121-75-5 |
| Benzene | 71-43-2 | Phenol | 108-95-2 |
| Chloroform | 67-66-3 | Tetrachloroethylene | 127-18-4 |
| Dieldrin | 60-57-1 | Vinyl chloride | 75-01-4 |

**chemical actinometer** An instrument that uses a photochemical reaction to determine the intensity of solar or other radiation, e.g., the intensity of ultraviolet light. The measurement is based on the decomposition of certain substances, e.g., potassium ferric oxalate [$K_3Fe(C_2O_4).3H_2O$], or oxalic acid ($H_2C_2O_4$) in the presence of uranyl $(UO_2)^{2+}$ sulfate. *See also* actinometer.

**chemical activation** The use of a chemical process to produce activated carbon from such materials as wood, peat, lignite, and coal. It combines carbonization (the conversion of the raw material to a char) and activation (oxidation of the char to develop pore structure). Chemical activation may consist of heating the raw material to a temperature of 400–500°C in the presence of phosphoric acid, zinc chloride, or another dehydrating agent. *See also* physical activation, high-temperature steam activation (Bryant, 1992 and AWWA, 1999).

**chemical activity** Same as activity (1).

**chemical adsorption** The adsorption of substances due to electrostatic forces as strong as a chemical bond between the sorbate molecule and the surface of the adsorbent. Also called bonding or chemical interaction. *See* chemisorption for detail.

**chemical agent** An element, compound or mixture that coagulates, disperses, dissolves, emulsifies, foams, neutralizes, precipitates, reduces, solubilizes, oxidizes, concentrates, congeals, entraps, fixes, makes the pollutant mass more rigid or viscous, or otherwise facilitates the mitigation of deleterious effects or the removal of the pollutant from the water (EPA-40CFR300.5).

**chemical analysis** The quantitative and/or qualitative examination of a sample to determine its chemical constituents.

**chemical and allied products** A heavy manufacturing sector including artificial fibers (except glass), biological products, botanical products, cellulosic materials, cosmetics, detergents, enamels, fertilizers, industrial inorganic chemicals, industrial organic chemicals, inks, lacquers, medicinal chemicals, paints, perfumes, pesticides, pharmaceuticals, plastics, soaps, synthetic resins, synthetic rubber, and varnishes. Facilities in this sector are required to conduct quarterly visual examinations of stormwater discharges.

**chemical antagonism** The condition of the reaction of a chemical substance that neutralizes the effects of the substance; e.g., the limitation of the toxicity of mercury and selenium, which form insoluble complexes in tissues.

**chemical–biological phosphorus removal** *See* chemical–biological treatment.

**chemical–biological treatment** The combination of chemical precipitation of phosphorus with biological treatment for removal of organic matter. It involves the addition of aluminum or iron salts ahead of, in, or following primary clarification or aeration, or prior to secondary clarification.

**chemical bond** The attraction between atoms or the force that holds them together in a molecule or compound, e.g., during a chemical reaction. *See also* the following types of bond: coordinate, covalent, electrovalent, hydrogen, ionic, metallic, polar.

**chemical carcinogen** A chemical substance that can cause cancer in humans or animals.

**chemical carcinogenesis** The development of cancer from the action of natural or synthetic chemicals; the role of xenobiotic chemicals in causing cancer.

**chemical cartridge** A cartridge containing activated carbon or other filtering material to remove hazardous substances from a respirator.

**chemical closet** A receptacle used as a toilet and containing a chemical disinfectant to liquefy fecal matter, retard decomposition, and reduce odors. It consists of a commode chair and a container placed beneath the seat with the chemical. Also called chemical toilet.

**chemical coagulant** A chemical compound such as lime and iron or aluminum salts used in water or wastewater treatment to promote the flocculation of finely divided solids into more readily settleable flocs. *See* coagulant, flocculant.

**chemical coagulation** The use of inorganic chemicals such as aluminum sulfate and ferric chloride to promote the formation of settleable flocs, as compared to biological flocculation. Coagulation causes the destabilization and aggregation of colloidal and fine suspended solids. In wastewater treatment, chemical coagulation increases the removal of biochemical oxygen demand, suspended solids, and phosphorus, but it produces sludge that is more voluminous and more difficult to thicken and dewater. *See* coagulation, perikinetic flocculation for more detail.

**chemical coatings** Such materials as asphalt, cement, coal tar preparations, epoxy, and paints used to protect metals against corrosion by isolating metallic surfaces from water. Other methods of corrosion control are anodic protection, cathodic protection (galvanic or impressed-current protection), inhibition, inert materials, and metallic coatings.

**chemical composition** The name and percentage by weight of each compound in an additive and the name and percentage by weight of each element in an additive (chemical agents) (EPA-40CFR79.2-h).

**chemical conditioning** The application of chemicals to water treatment sludge or digested wastewater treatment sludge to improve its dewatering characteristics, or to raw sludge to reduce its nuisance potential as well. Common chemicals used include lime, alone or with ferric chloride, and polymeric coagulants. The objective is to neutralize the charges on the sludge particles, allow the water to drain, and prevent the decay of organic material. Other sludge conditioning methods include elutriation, chlorination, heat treatment, freezing, and bulking agents.

**chemical control** The use of sodium azide, sodium arsenite, copper sulfate, or other chemicals to control the growth of aquatic weeds and plants.

**chemical dehydration** *See* chemical activation.

**chemical disinfecting agent** Any of a number of chemicals used in the practice of water and wastewater disinfection, including acids, alcohols, alkalies, bromine, chlorine and its compounds, dyes, heavy metals and their compounds, hydrogen peroxide, iodine, ozone, peracetic acid, phenol and phenolic compounds, quaternary ammonium compounds, soaps and synthetic detergents.

**chemical disinfection** The application of chemicals, such as chlorine or metal ions, to disinfect water or wastewater, as compared to disinfection by heat, ultraviolet light, or other nonchemical agents.

**chemical dissolving box** A box used to dissolve chemicals before they are introduced to the water or wastewater to be treated. *See also* chemical solution tank.

**chemical dose** The quantity of a chemical substance added to water or wastewater for a specific purpose; e.g., a dose of X mg/L of alum for coagulation or Y mg/L of chlorine for disinfection.

**chemical drinking water standards** Limits established by the USEPA or other regulatory agency for the concentrations of chemical contaminants in drinking water. *See* maximum contaminant level.

**chemical element** *See* element.

**chemical elements table** A table listing all known chemical elements, usually with some characteristics: symbols, atomic numbers, and atomic weights. *See* element.

**chemical equation** A representation of a chemical reaction, showing the reactants on the left and the products of the right, with an arrow between the two sides indicating the direction of the reaction. A bidirectional arrow indicates a reversible reaction. Reactants and products are represented by their chemical formulas. For example, the interaction of carbon dioxide ($CO_2$) with solid calcium carbonate [calcite, $CaCO_3$ (s)] may be represented by:

$$CaCO_3 \text{ (s)} + CO_2 \text{ (g)} + H_2O \leftrightarrow Ca^{2+} + 2\ HCO_3^- \tag{C-48}$$

**chemical equilibria** *See* chemical equilibrium, solubility equilibria.

**chemical equilibrium** The condition of a reversible chemical reaction when there is no transfer of weight or energy either way, or when the rate of transfer is the same in both directions. *See also* equilibrium constant, heterogeneous equilibrium, homogeneous equilibrium, mass-action relationship, solubility equilibria, stoichiometry.

**chemical equivalent** The weight in grams of a substance that combines or displaces one gram of hydrogen; equal to the formula weight divided by the valence. For example, calcium (Ca, molecular weight 40 and valence +2) has a chemical equivalent of 20 grams per mole, whereas sodium (Na, molecular weight 23 and valence +1) has a chemical equivalent of 23. *See also* gram equivalent and equivalent per liter.

**chemical extraction** In hazardous waste treatment, chemical extraction is the removal of a constituent by chemical reaction with a substance (extractant) in solution, as shown in the extraction of lead from lead carbonate ($PbCO_3$, a poorly soluble salt) by reaction with hydrogen ions ($H^+$):

$$PbCO_3 + H^+ \rightarrow Pb^{2+} + HCO_3^- \tag{C-49}$$

Also called leaching (Manahan, 1997).

**chemical feeder** A mechanical device used to measure and dispense chemicals at a predetermined rate. Feeder types include dry feeders with vibrators, gravimetric or volumetric dry feeders, and wet feeders. *See also* ammoniator, chemical gas feeder, chlorinator, plunger pump, slaker, sulfonator.

**chemical feeding** The complete apparatus that serves to handle, store, measure, and dispense the proper amounts of water and wastewater treatment chemicals, including such accessories as air pollution control and worker protection. Chemical feeding varies with the products and the type of plant, e.g., bulk, dry, gas, solution, automatic, manual, etc. *See also* chemical handling and feeding.

**chemical feed pump** A pump used as a chemical feeder in water or wastewater treatment, usually to a conduit or container under pressure or to facilitate measurement of the chemicals. *See also* packed-plunger metering pump.

**chemical fixation** The transformation of a toxic chemical compound to a new, nontoxic form, implying the chemical bonding of the contaminant to the additive (e.g., a polymeric inorganic silicate containing calcium or aluminum); sometimes used to denote solidification, stabilization, or their combination.

**chemical fixation/chemical solidification** The process of stabilizing liquid or dewatered sludge, scum, grit, or incinerated sludge ash by mixing them with chemically reactive or encapsulating agents such as cement, fly ash, lime, pozzolana, and sodium silicate. The process yields a product that is similar to natural clay or a soil-like material. It is used (a) in the treatment of industrial sludge and hazardous wastes to immobilize undesirable constituents and (b) in the stabilization of municipal sludge for disposal as a landfill cover or for land reclamation projects. Other terms related to solidification and stabilization include encapsulation, vitrification, and sorption.

**chemical formula** A chemical formula is an expression of the composition of a compound in terms of its elements and the ratios of their combination; e.g., the formula of calcium carbonate ($CaCO_3$) indicates that one atom of calcium combines with one atom of carbon and three atoms of oxygen. *See also* chemical symbol, Hill's molecular formula arrangement..

**chemical gaging** A method used to measure streamflow by adding a chemical of known strength at a given rate and determining its concentration in the stream after complete mixing. The flow ($Q$) can be computed from:

$$Q = Q_0(C_0 - C)/C \tag{C-50}$$

where $Q_0$ = volume of chemical added, $C_0$ = initial concentration of the chemical, and $C$ = concentration of the mixed stream.

**chemical gas feeder** A device that feeds a gas through a pressure regulator or controlled orifice for water or wastewater treatment. It may be linked to the on-site gas production apparatus in the case of carbon dioxide, chlorine dioxide, and ozone.

**chemical grout** A chemical compound used to seal joints or cracks.

**chemical half-life** (1) The length of time required for the mass, concentration, or activity of a chemical or physical agent to be reduced by one-half through decay. *See* decay processes and half-life. (2) The time required for the elimination of one-half of a total dose from the body.

**chemical handling and feeding** The activities involved in the handling and feeding of chemicals used in water and wastewater treatment: receiving, unloading, storing, feeding, and, sometimes, on-site production. Special equipment is used, depending on the size of the plant and the chemicals applied; e.g., day tanks, storage tanks, feed pumps, flowmeters, and valves.

**chemical hygrometer** *See* absorption hygrometer.

**chemical industry waste** Wastewater produced by plants that manufacture chemical products such as acids, bases, cornstarch, detergents, fertilizers, pesticides, plastics, powder and explosives, resins, etc. Chemical wastes may be toxic, high in BOD and color, but low in suspended solids. *See* acid wastes, chloralkali wastes, organic chemical wastes, pesticide wastes, phosphate-industry wastes, polymer wastes.

**chemical inhibitor** A chemical compound added to water to control corrosion by forming a protective film on the surface of water pipes and a barrier between the water and the pipes. Common chemical inhibitors include orthophosphates, polyphosphates, and their mixtures, as well as sodium silicates. *See* corrosion inhibitor for detail.

**chemical inorganic compound** *See* inorganic chemical and inorganic compound.

**chemical interaction** A minor mechanism contributing to the removal of particulate matter within a granular filter. *See* chemical adsorption.

**chemical ionization** A mass spectrometry procedure for analyzing molecules and determining molecular weights by bombarding the molecules with methane ($CH_4$) or another gas.

**chemical kinetics** The subject dealing with the rate of chemical reactions. *See* zero-order, first-order, and second-order reactions, temperature effect, Arrhenius equation.

**chemical lime** A common compound of calcium; another name for lime, a white or grayish-white, slightly soluble in water, calcined oxide (CaO). *See* quicklime for detail.

**chemically enhanced backwash (CEBW)** In membrane processes, the addition of chemicals to the permeate to enhance the effectiveness of backwash in removing fouling materials and in inactivating biological growth.

**chemically enhanced primary treatment (CEPT)** An advanced primary wastewater treatment that includes the addition of chemicals such as ferric chloride to improve the removal of BOD and suspended solids. Also called ballasted flocculation when a ballast such as microsand is also added. *See also* A-B process (Reardon et al., 2004).

**chemically pure water** Water that is essentially hydrogen and oxygen ($H_2O$), without any material in solution or suspension.

**chemically reactive media** Filtration materials in granular or bead form that can react chemically with constituents in water and serve to modify water quality, such as calcite in pH modification; or serve as a catalyst to initiate chemical reactions such as manganese greensand, pyrolusite, activated carbon, and dissimilar metal alloy products. *See also* catalyst media.

**chemically stabilized water** Water that has a pH, alkalinity, and calcium ($Ca^{2+}$) concentration very close to calcium carbonate ($CaCO_3$) saturation, and thus does not deposit nor dissolve $CaCO_3$.

**chemical microconstituent** *See* microconstituent.

**Chemical Monitoring Reform** A program carried out by the USEPA Office of Ground Water and Drinking Water from 1992 to 1996 to reduce the chemicals monitoring requirements and enhance the monitoring process, while satisfying the provisions of the Safe Drinking Water Act.

**Chemical Monitoring Revisions** A program of the USEPA Office of Ground Water and Drinking Water to reduce the chemicals monitoring requirements and enhance the monitoring process.

**chemical neutralization** The removal of excess acidity or excess alkalinity by treating water or wastewater with a chemical of the opposite composition. *See* neutralization for more detail. *See also* chemical stabilization.

**chemical nitrogen removal** The conversion of nitrogen to a form that is insoluble or only slightly soluble in water, e.g., nitrogen gas ($N_2$) or ammonia ($NH_3$), by such processes as breakpoint chlorination, ion exchange, or air stripping.

**chemical odor control** The elimination or reduction of odors in water or wastewater by chemical oxidation, chemical precipitation, and/or pH control, using such chemicals as chlorine, hydrogen peroxide, oxygen, ozone, potassium permanganate, and sodium hypochlorite.

**chemical organic compound** *See* organic chemical and organic compound.

**chemical oxidation** The process of oxidizing substances in water or wastewater using a chemical such as chlorine, hydrogen peroxide, ozone, or potassium permanganate to remove electrons. Chemical oxidation is used in such operations as color removal, destruction of taste and odor-causing materials, and the removal of synthetic organic chemicals. *See also* oxidation-reduction reaction, advanced oxidation.

**chemical oxygen demand (COD)** An indirect measure of the oxygen required to oxidize all compounds, both organic and inorganic, in water; a widely used parameter to characterize the organic strength of wastewaters and pollution of natural waters. The measure is from a laboratory test based on a chemical oxidant and, therefore, does not necessarily correlate with biochemical oxygen demand (BOD), but, unlike BOD, it can be determined in a few hours instead of 5 days. *See* COD test for detail. Other measures of organic content are BOD, ThOD, TOC, and TOD.

**chemical phosphorus removal** Precipitation of phosphorus compounds using chemicals such as aluminum, calcium, ferric iron, and ferrous iron salts. The phosphorus precipitates as phosphate in complex reactions, which can be approximated by:

$$Fe^{3+} + PO_4^{3-} \rightarrow FePO_4 \qquad \text{(C-51)}$$

$$3\ Fe^{2+} + 2\ PO_4^{3-} \rightarrow Fe_3(PO_4)_2 \qquad \text{(C-52)}$$

$$Al^{3+} + PO_4^{3-} \rightarrow AlPO_4 \qquad \text{(C-53)}$$

*See also* hydroxyapatite and.

**chemical–physical treatment** Same as physical–chemical treatment.

**chemical polishing** The use of chemical storage and feeding equipment on a regular or standby basis to supplement biological phosphorus removal and ensure compliance with effluent requirements. Common supplements include acetate or another low-molecular weight carboxylic acid as a source of readily biodegradable COD, or a metal salt (e.g., aluminum, calcium, iron).

**chemical pollution** Air, water, or soil pollution by liquid, solid, or gaseous chemicals.

**chemical potential ($\mu$)** The partial molal free energy of a chemical species, a function of its standard chemical potential ($\mu^0$, at a temperature of 25°C and pressure of 1 atm), the universal gas constant ($R$ = 1.987 cal/°K/mole), its temperature ($T$, °K), and its activity (a).

$$\mu = \mu^0 + RT(\ln a) \qquad \text{(C-54)}$$

*See also* fugacity, reference state, standard state.

**chemical precipitation** The application of soluble chemicals to throw dissolved ions out of solution so that they can be removed by such physical operations as sedimentation and filtration. Chemical precipitation is used in iron and manganese removal, water softening, fluoride removal, and other treatment processes. In wastewater treatment, chemical precipitation is used to enhance the removal of total suspended solids, biochemical oxygen demand, and nutrients (e.g., nitrogen and phosphorus). The process is used in hazardous waste treatment to remove heavy metal ions from water or aqueous wastes; heavy metals are usually precipitated as hydroxides, sulfides, or carbonates. For detail, see alum precipitation, lime precipitation, lime/soda ash softening, ferrous sulfate and lime precipitation, ferric chloride precipitation, ferric chloride and lime precipitation, ferric sulfate, lime precipitation, cementation, and chemical–biological treatment. *See also* chemical equilibrium, carbonic acid system.

**chemical precipitation sludge** The slimy, dark, red or grayish-brown sludge resulting from chemical precipitation with metal salts.

**chemical precipitation softening** A water treatment process that uses chemicals to precipitate and remove hardness-causing ions. *See* lime softening, and lime–soda ash softening for details.

**chemical proportioner** A device used in water and wastewater treatment to dispense chemicals proportionally to flow. *See also* chemical feeder.

**chemical reaction** The combination of atoms of two or more elements to form molecules, or the breakdown of molecules into individual atoms. The process occurs as a result of heating, as a result of a chemical addition, or spontaneously. *See also* chemical equation.

**chemical reagent** A chemical that is added to produce a chemical reaction.

**chemicals** *See* chemical and important chemicals for some of those that relate to water chemistry, water and wastewater treatment, or sludge processing.

**chemical scrubber** A treatment unit used to remove odors in wastewater treatment plants; it allows contact between air, water, and chemicals for the oxidation or entrainment of odorous compounds. Common chemical scrubbers include countercurrent packed towers, spray chamber absorbers, and cross-flow scrubbers. *See* scrubbing fluid, chemical scrubbing.

**chemical scrubbing** An odor control process used to remove hydrogen sulfide ($H_2S$) and other airborne pollutants generated by wastewater treatment. These pollutants are treated for short periods of 1.5 to 2.0 seconds by contact with appropriate chemicals (e.g., sodium hypochlorite, potassium permanganate, hydrogen peroxide) in the scrubber's packing. Existing chemical scrubbers can be cost-effectively converted to biotrickling filters with a special medium for the same purpose. Other wastewater odor treatment methods or units include activated-carbon adsorbers, biofilters, biotrickling filters, and thermal processing.

**chemicals handling and feeding** *See* chemical handling and feeding.

**chemical-similitude theory** The interpretation of the dose–response relationship of disinfection as a chemical undergoing a kinetic reaction, i.e., on the basis of the Chick–Watson model. *See also* disinfection kinetics, vitalistic theory.

**chemical sludge** Sludge resulting from chemical processes of inorganic wastes that are not biologically active or from the chemical treatment of water or wastewater. *See also* coagulation sludge, precipitation.

**chemical sludge conditioning** *See* chemical conditioning.

**chemical solidification** *See* chemical fixation.

**chemical solution tank** A tank used to dissolve chemicals before they are introduced to the water or wastewater to be treated. *See also* chemical dissolving box.

**chemical-specific approach** An approach to setting water quality standards for toxic compounds; it uses individual criteria for each toxic chemical detected in wastewater. Corresponding criteria can be developed in laboratory experiments. *See also* whole-effluent approach.

**chemicals recovery** The recovery of chemicals used in coagulation or sludge conditioning. *See* recalcination and froth flotation.

**chemical stability** Resistance to the effects of chemical action, e.g., the ability of ion-exchange resins to resist breakdown by aggressive solutions. *See also* the following terms related to the stability of a solution: stability, stability constant, stability index, stability ratio, stabilization.

**chemical stabilization** (1) The use of chemicals to stabilize the organic matter in sludge and reduce its nuisance potential, usually by chemical oxidation (e.g., chlorination) or pH adjustment (e.g., lime stabilization). *See* chemical conditioning and chemical fixation. (2) Chemical stabilization of water consists in adjusting its pH, calcium concentration, and alkalinity to its calcium carbonate ($CaCO_3$) saturation to protect pipes against corrosion and incrustation. *See also* Langelier index.

**chemical stabilization/fixation** *See* chemical fixation.

**chemical substance** Any organic or inorganic substance of a particular molecular identity, including any combination of such substances occurring in whole or in part as a result of a chemical reaction or occurring in nature, and any chemical element or uncombined radical (EPA-40CFR720.3-e or 761.3).

**chemical symbol** A single capital letter or a capital letter followed by a lower case letter used to represent an atom or a chemical element, e.g., C for carbon, Cl for chlorine, N for nitrogen, and Na for sodium. An ion is represented by the chemical symbol of the corresponding element or radical, followed by the number of unit charges on the ion, e.g., $Ca^{2+}$ and $OH^-$ for the calcium and hydroxyl ions, respectively. *See* chemical formula.

**chemical tank** A tank that stores chemicals for use in a water or wastewater treatment plant.

**chemical thermodynamics** The branch of thermodynamics that deals with chemical changes such as reactions and phase transfers. *See* thermodynamics laws, chemical equilibria, enthalpy, entropy, Gibbs free energy, chemical potential.

**chemical toilet** A receptacle used as a toilet and containing a chemical disinfectant to liquefy fecal matter, retard decomposition, and reduce odors. It consists of a commode chair and a container placed beneath the seat with the chemical. Also called chemical closet.

**chemical treatment** A water or wastewater treatment scheme that uses chemicals to achieve such objectives as solids and liquid removal, disinfection, or odor control. Chemical treatment processes include coagulation, disinfection, flocculation, precipitation, and sludge conditioning. *See also* unit operations, unit processes, biological treatment, physical treatment, and physical–chemical treatment.

**chemical unit process** A water or wastewater treatment method in which chemical reactions bring about the removal or modification of constituents, as compared to physical unit operations and biological unit processes. A wastewater treatment plant usually incorporates two or all three treatment methods. Examples of chemical unit processes: advanced oxidation, chemical coagulation, chemical disinfection, chemical neutralization, chemical oxidation, chemical precipitation, chemical scale control, chemical stabilization, ion exchange.

**chemical waste** *See* chemical industry waste.

**chemical weathering** The breakdown of the minerals in a rock brought about by chemical changes in the rock and producing solutes, soils, sediments, and sedimentary rocks. Chemical weathering processes include carbonation, hydrolysis, hydration, and oxidation.

**chemihydrometry** The use of chemical gaging to determine the rate of flow of water. *See also* salt-velocity method.

**Cheminjector-D®** A series of diaphragm pumps manufactured by Hydroflo Corp.

**chemisorption** The adsorption of substances due to electrostatic forces as strong as a chemical bond between the sorbate molecule and the surface of the adsorbent. This adsorption mechanism involves electronic interactions between surface sites and adsorbates that cause the formation of the bonds. Same as chemical adsorption. *See also* desorption, electrostatic adsorption, exchange adsorption, hydrophobic bonding, physical adsorption, polarity, solution force.

**chemistry** The science that deals with the composition, properties, and reactions of substances and elements of matter. *See* water chemistry for details.

**Chemix** A device manufactured by Semblex, Inc. for mixing and feeding dry polymers in water treatment.

**chemoautotroph** An organism that obtains energy from the oxidation of reduced inorganic compounds such as ammonia ($NH_3$), nitrite ($NO_2$), ferrous iron (Fe), and sulfide (S).

**chemoautotrophic** Related to an organism that uses inorganic compounds to produce organic material and energy; e.g., *Thiobacillus* oxidizes hydrogen sulfide ($H_2S$) to sulfur while some iron bacteria oxidize ferrous salts for energy. Also called chemosynthetic.

**chemoautotrophic theory** One of three main hypotheses about the source of primary organic carbon and the process for its introduction into deep-sea hydrothermal vents: bacteria use hydrogen ($H_2$), hydrogen sulfide ($H_2S$), and methane ($CH_4$) as electron donors and carbon dioxide ($CO_2$) as carbon source. *See also* organic thermogenesis and advective plume hypothesis.

**chemocline** A zone or layer of a lake, reservoir, or other impoundment where there is a rapid change with depth in the concentration of dissolved substances. *See* thermocline.

**chemoheterotroph** An organism that obtains energy from the oxidation of organic compounds via respiration. Chemoheterotrophs include most pathogens and are active in biogeochemical cycles, wastewater treatment, and bioremediation.

**chemokinesis** The increased activity of an organism caused by a chemical substance.

**chemolithotroph** An organism that can chemically oxidize inorganic substances to obtain energy and fix carbon dioxide ($CO_2$) exclusively as a carbon source.

**Chemomat** A water treatment system developed by Ionics, Inc., using electrochemical membranes for solids separation.

**chemonite** A wood preservative consisting of acetic acid, ammonia, arsenic trioxide, copper hydroxide, and water.

**chemoorganotroph** An organism that oxidizes organic compounds for energy and carbon.

**chemoprevention** Same as chemoprophylaxis.

**chemoprophylaxis** Disease prevention using chemicals, drugs (including antibiotics), or food nutrients. Also called chemoprevention.

**chemosmosis** Chemical action between substances through a semipermeable membrane. *See also* osmosis, reverse osmosis.

**chemosphere** The region of the atmosphere where chemical, mostly photochemical, activity takes place.

**chemostat** A sealed vessel, also called a bioreactor, used as a growth container in continuous culture, in which environmental conditions can be controlled (substrate concentration, flow rate, pH, temperature, oxygen concentration, etc.)

**chemosterilant** A chemical that causes permanent sterility in animals.

**chemosynthesis** The synthesis of organic matter by an organism, using the energy from chemical reactions. For example, the autotrophic bacteria *Beggiatoa, Nitrosomonas,* and *Nitrobacter* decompose sulfur and nitrogen compounds by chemosynthesis (Fair et al., 1971):

*Beggiatoa:* $$2\ H_2S + O_2 \rightarrow 2\ S + 2\ H_2O + \text{energy} \qquad \text{(C-55)}$$

*Beggiatoa:* $$2\ S + 3\ O_2 + 2\ H_2O \rightarrow 2\ H_2SO_4 + \text{energy} \qquad \text{(C-56)}$$

*Nitrosomonas:* $$2\ NH_3 + 3\ O_2 \rightarrow 2\ HNO_2 + 2\ H_2O + \text{energy} \qquad \text{(C-57)}$$

*Nitrobacter:* $$2\ NO_2 + O_2 \rightarrow 2\ NO_3 + \text{energy} \qquad \text{(C-58)}$$

*See also* photosynthesis.

**chemosynthesizer** Same as chemotroph; see also photosynthesizer.

**chemosynthetic** *See* chemoautotrophic.

**chemosynthetic bacteria** Bacteria that derive energy from the oxidation of organic or inorganic compounds for the decomposition of organic matter. They include heterotrophic bacteria that reduce nitrates and sulfates in the absence of free oxygen and unpigmented autotrophic bacteria. *See* chemosynthesis.

**chemotaxis** The movement of microorganisms toward beneficial substances or away from inhibitory substances, using their flagella over short distances.

**chemotherapy** Treatment of diseases using chemicals (including antibiotics) that destroy the responsible germs or cancerous tissues.

**chemotroph** Any organism that obtains energy from the oxidation or reduction of inorganic or organic compounds, as opposed to an organism that obtains energy from the sun. Also called chemosynthesizer. *See also* chemosynthesis, phototroph.

**chemotrophic** Applied to chemotrophs. *See also* chemo-autotrophic, heterotrophic, phototrophic.

**Chem-Scale™** A weighing scale manufactured by Force Flow Equipment for vertical chemical tanks.

**ChemScan** A series of instruments manufactured by Biotronics Technologies, Inc. for process analysis.

**ChemSensor®** An instrument manufactured by ORS Environmental Systems to monitor volatile organic compounds.

**Chemtact™** Equipment manufactured by Quad Environmental Technologies Corp. for scrubbing atomized mist odors.

**Chem-Tower®** Equipment manufactured by Smith & Loveless, Inc. to feed bulk chemicals.

**Chemtrac®** A device manufactured by Chemtrac Systems, Inc. to monitor streaming currents.

**Chemtube®** A diaphragm metering pump manufactured by Wallace & Tiernan, Inc.

**Chevron™** A system of clarifier tube settlers manufactured by the U.S. Filter Corp. This proprietary design features an approximate V-shaped cross section. Also called chevron tube settler. *See also* tube settler.

**chevron tube settler** *See* Chevron™.

**Chézy, Antoine** French civil engineer and instructor (1718–1798).

**Chézy coefficient** The friction coefficient ($C_z$) used in the Chézy formula.

**Chézy formula** A basic hydraulic formula for flow of water in open channels, developed by the French engineer A. Chézy while designing a canal for the Paris water supply in 1768. It expresses the velocity of flow $V$ as a function of the hydraulic radius $R$, the bed slope $S_0$, and a friction coefficient $C_z$, i.e.,

$$V = C_z(RS_0)^{0.5} \qquad \text{(C-59)}$$

The friction coefficient depends on the Reynolds number, on boundary roughness, and on channel or conduit geometry. These factors can be expressed by the drag coefficient ($C_D$):

$$C_D = (2g/C_D)^{0.5} \qquad \text{(C-60)}$$

where g is the gravitational acceleration. In 1889 Manning presented a formula for this coefficient, which was independently derived by Strickler in 1923 and by others:

$$C_z = R^{1/6}/n \qquad \text{(C-61)}$$

where $n$ is the Manning's "n" factor or coefficient. *See also* the Bazin and Kutter's formulas.

**Chézy resistance factor** The friction coefficient ($C_z$) used in the Chézy formula.

**Chicago Pump** A line of water and wastewater products of the Yeomans Chicago Corp.

**chicane** An obstacle used on a belt thickener or belt filter press to mix sludge and facilitate its dewatering.

**Chick, Harriet** English biochemist who developed in 1908 the relationship between disinfectant efficiency and contact time that bears her name, either as she was working in England or while she was a Fellow at the Pasteur Institute, Paris.

**Chick's law** A relationship based on observed data, between the efficiency of a disinfection process and the time of contact; it indicates that the number of organisms inactivated per unit time is proportional to the number of organisms remaining. It is usually expressed as

$$N/N_0 = e^{-kt} \qquad \text{(C-62)}$$

where N is the number of organisms remaining at time $t$, $N_0$ is the initial number of organisms, and $k$ is a rate constant. The ratio $N/N_0$ represents the surviving fraction of microorganisms at time $t$. *See also* the CT concept. For various reasons, Chick's law is only an approximation; see, for example, the general purification equation.

**Chick–Watson disinfection model** *See* Chick–Watson equation.

**Chick–Watson equation** An equation that combines Chick's law on disinfection kinetics and Watson's expression of the dieoff coefficient as an exponential function of disinfectant concentration:

$$\ln (N/N_0) = -\alpha C^n \cdot T \qquad \text{(C-63)}$$

where $N$ = number of microorganisms remaining at time T; $N_0$ = initial number of microorganisms; $\alpha$ = inactivation constant; $C$ = disinfectant concentration; $n$ = constant of dilution, commonly assumed to be 1; and $T$ = time. This equation does not account well for the variable, heterogeneous characteristics of wastewater. *See also* Chick–Watson law deviations. The USEPA's

Long-term 2 Enhanced Surface Water Treatment Rule uses a linear form of the Chick–Watson equation in developing *Cryptosporidium* $C \cdot T$ tables for ozone and chlorine dioxide. It states that the logarithm of the activation rate (log kill) equals the C.T product times a parameter ($k_{10}$) that depends on the disinfectant, the target microorganism, and water temperature:

$$\log \text{kill} = k_{10}\,(C \cdot T) \qquad \text{(C-64)}$$

[$C \cdot T$ is the product of residual disinfectant concentration ($C$) in mg/L, determined before or at the first customer, and the corresponding disinfectant contact time ($T$) in minutes, i.e., "$C$" times "$T$."]

**Chick–Watson law** *See* Chick–Watson equation.

**Chick–Watson law deviations** Deviations from the predictions of the Chick–Watson equation, due to, e.g., (a) a dilution constant significantly different from 1, and (b) reactions (particularly with organics) that diminish the bactericidal action of a chemical disinfectant. Shoulders (time lags until the onset of disinfection) and tailing off (the inactivation rate decreases progressively) are often observed as shown in Figure C-05.

**Chick–Watson model** *See* Chick–Watson equation.

**chilled water loop** Any closed cooling water system that transfers heat from air handling units or refrigeration equipment to a refrigeration machine or chiller (EPA-40CFR749.68-2).

**chiller** A heat exchanger or similar device used to cool or refrigerate gases or liquids.

**chimney effect** The tendency of air or gas in a vertical passage to rise when it is heated because of a lower density than that of the surrounding gas.

**china clay** Same as kaolin or white clay.

**Chinese liver fluke** The pathogen that causes the Chinese liver fluke disease; a shorter name for the disease itself. Also called oriental liver fluke.

**Chinese liver fluke disease** Common name of clonorchiasis, a helminthic disease caused by the excreted worm *Clonorchis sinensis*. This infection of the bile ducts is found mainly in Southeast Asia and is transmitted from the feces of infected persons or animals through aquatic snails and fish as intermediate hosts; contracted by eating the raw or partly cooked intermediate hosts, but also occasionally by ingestion of drinking water containing the metacercariae from decomposed fish. Symptoms are sometimes absent or vague, but may include diarrhea, abdominal discomfort, liver enlargement, gall bladder colic, etc. *See also Clonorchis sinensis* life cycle and opisthorchiasis. Also called Asiatic liver fluke disease and Oriental liver fluke disease.

**chironomid larva** *See* bloodworm.

***Chironomus*** *See* bloodworm.

**chisel plowing** Cropland preparation by a special implement (chisel) that avoids complete inversion of the soil (as occurs with conventional moldboard plowing). Chisel plowing can leave a protective cover or crop residues on the soil surface that helps prevent erosion and improve infiltration.

**chitin** (1) A nitrogen-containing polysaccharide that is the principal constituent of the outer covering of insects, crustaceans, and arthropods; the organic skeletal substance of crustacean shells. (2) The common polymer chitosan.

**chitosan** A biodegradable, nontoxic cationic polymer produced by acidifying chitin; used as a flocculation aid with alum coagulation in wastewater treatment.

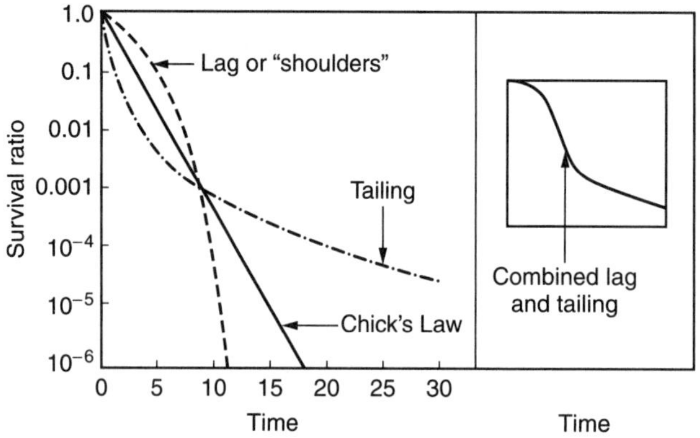


**Figure C-05.** Chick–Watson law deviations (adapted from AWWA, 1999, and Metcalf & Eddy, 2003).

**Chi-X®** A product of the NuTech Environmental Corp. used for odor control.

**chlamydia** A group of eubacteria that live inside other cells and cause eye, lung, sexually transmitted, and other infections.

***Chlamydomonas* genus** A group of green algae active in the stabilization of organic matter in waste stabilization ponds; they are sensitive to adverse conditions (e.g., temperature extremes) and may be found in the pond effluent.

**chloracetic acid** *See* chloroacetic acid.

**chloracne** A severe form of skin irruption caused by exposure to dioxin (2,3,7,8-tetrachlorodibenzodioxin) and some other chlorinated compounds.

**chloral ($CCl_3CH_2$)** The common name of trichloroacetaldehyde, also called trichloroacetic acid aldehyde; a colorless, oily liquid with a pungent odor that combines with water to form chloral hydrate.

**chloral hydrate [$CCl_3CH(OH)_2$]** The common name of trichloroacetaldehyde, a disinfection by-product regulated by the USEPA, which requires its quarterly monitoring. MCLG = 0.04 mg/L. Used as a sedative or a hypnotic, it is a dry, crystalline solid. Also called chloral.

**chlor-alkali process** Production of chlorine by electrolysis of brine. *See* chlor-alkali waste.

**chlor-alkali waste** Wastewater generated during the production of chlorine and caustic soda in an electrolytic cell with a mercury cathode. The effluent from such a process is high in total solids and contains iron, calcium, magnesium, chromium, mercury, and other metals. However, in-plant control of mercury waste and effluent reuse may reduce these constituents. *See also* Minamata disease.

**chloramination** A process used for water disinfection by combining, sequentially or simultaneously, chlorine and ammonia to mass ratios between 3:1 and 5:1; monochloramine, dichloramine, and trace levels of trichloramine are formed. Note that chloramines may induce hemolytic anemia in kidney dialysis patients. *See also* free chlorination, breakpoint chlorination, prechlorination, preammoniation.

**chloramination reaction sequence** A theoretical sequence of four reactions that represent the chemistry of chloramination, i.e., the chlorine–ammonia reactions in water: formation of hypochlorous acid (HOCl), formation of monochloramine ($NH_2Cl$), formation of dichloramine ($NHCl_2$), and the degradation of dichloramine into nitrogen gas and chloride:

$$3\ Cl_2 + 3\ H_2O \rightarrow 3\ HOCl + 3\ Cl^- + 3H^+ \qquad (C\text{-}65)$$

$$2\ HOCl + 2\ NH_3 \rightarrow 2\ NH_2Cl + 2\ H_2O \qquad (C\text{-}66)$$

$$2\ HOCl + 2\ NH_2Cl \rightarrow 2\ NHCl_2 + 2\ H_2O \qquad (C\text{-}67)$$

$$2\ NHCl_2 + H_2O \rightarrow N_2 + HOCl + 3\ H^+ + 3\ Cl^- \qquad (C\text{-}68)$$

$$3\ Cl_2 + 2\ NH_3 \rightarrow 6\ H^+ + 6\ Cl^- + N_2 \qquad (C\text{-}69)$$

Hypochlorous acid may also react with dichloramine to form trichloramine:

$$HOCl + NHCl_2 \rightarrow NCl_3 + H_2O \qquad (C\text{-}70)$$

*See also* breakpoint chlorination.

**chloramine** A disinfecting compound formed by the reaction of hypochlorous acid (or aqueous chlorine, HOCl) with ammonia or organic nitrogen; general formula: $NH_xCl_y$, with x = 0, 1, or 2 and y = 1, 2, or 3. Chloramines, which include monochloramine ($NH_2Cl$), dichloramine ($NHCl_2$), and trichloramine or nitrogen trichloride ($NCl_3$), are more effective than chlorine for maintaining a residual in water distribution systems; their total concentration is called combined chlorine residual. Nondisinfectant, organic chloramines may also be formed when organic nitrogen is present. A potential health effect of chloramines is hemolytic anemia in dialysis patients; they also cause mutations in bacteria and toxic effects in fish. *See also* bromamine, chloramination reaction sequence, culpric chloramine.

**chloramine disinfection** *See* chloramination.

**chloramine formation** *See* chloramination, chloramination reaction sequence.

**chloramine inactivation** Inactivation of microorganisms due to reactions of chloramines with enzymes, particularly interference with the sulfhydryl groups (—SH) of these enzymes.

**chloramine residual** The chlorine remaining as a compound with ammonia in water during breakpoint chlorination; same as combined chlorine residual.

**chloramino acid** The product of the reaction of hypochlorous acid with an amino acid or another amine.

**chlorate ($ClO_3^-$)** A salt of chloric acid ($HClO_3$), i.e., containing the chlorate ion; the term chlorates may include other ions: hypochlorite ($ClO^-$), chlorite ($ClO_2^-$), and perchlorate ($ClO_4^-$), all strong oxidizing, disinfecting, and bleaching agents. They also can form explosive mixtures with other chemicals. Chlorate is a potential contaminant of drinking water when chlorine dioxide ($ClO_2$) is used for disinfection; it is formed from the hypochlorite ion or hypochlorous acid (HClO):

$$3\ HClO \rightarrow 2\ HCl + HClO_3 \quad (C\text{-}71)$$

Chlorate ions are also formed during the decomposition of bleach or sodium hypochlorite (NaOCl) or in ozonated water that contains a chlorine residual. Chlorates may cause hemolysis, methemoglobinemia, renal failure, or even death at high concentrations, e.g., 71 mg/kg of body weight. *See also* perchlorate.

**chlorate ion ($ClO_3^-$)** *See* chlorate.

**Chlor-A-Vac™** A gas induction system manufactured by Capital Controls Co., Inc.

**chlordane ($C_{10}H_6Cl_8$)** A persistent synthetic organic chemical (1,2,4,5,6,7,8,8-octachloro-4,7-methano-3a,4,7,7a-tetrachloroindane), with an MCLG of 0 (zero) and an MCL of 0.002 mg/L, widely used as an insecticide or fumigant (e.g., for termite control); may cause adverse effects on the neurological system and on the liver.

**chlorella** Any freshwater, unicellular green alga of the genus *Chlorella.* A group of algae commonly found in stabilization ponds.

***Chlorella* genus** A group of green algae abundant in waste stabilization ponds and in trickling filters; resistant to anaerobiosis and to temperature extremes. They do not participate directly in waste degradation, but they contribute oxygen during daylight hours. They may also cause clogging of the filter surface and odors.

***Chlorella vulgaris*** A species of freshwater, unicellular, green algae; sometimes used in laboratory analyses because it is easy to grow. *See also Chlorella* genus.

**chloric acid ($HClO_3$)** A hypothetical acid that exists only in solution or by its salts. *See* chlorate.

**chloride ($Cl^-$)** (1) A water-soluble chlorine compound; a salt of hydrochloric acid (HCl), e.g., sodium chloride (NaCl). Higher chloride concentrations than in adjacent waters may be a sign of pollution. Some natural waters are unpalatable because of an excess of chlorides (e.g., 3000 mg/L), from the leaching of chloride-containing rocks and soils; they may have a salty taste and laxative effects on new users. The USEPA sets a limit of 250 mg/L of chloride in drinking water. Chlorides may react with metals in solution and cause increased corrosion. Chloride concentration is used to determine the suitability of wastewater or treated effluent for agricultural reuse. Chlorides may be toxic to freshwater fish at a concentration of 4000 mg/L. Human excreta contain an average of 6 grams of chlorides per person per day. (2) A major anion commonly found in water and wastewater. (3) A compound containing chloride, e.g., methyl chloride ($CH_3Cl$).

**chloride index** The ratio of the quantity of chlorides in wastewater to that in the water supply, an indication of the presence of wastewater from other sources.

**chloride of lime [CaCl(OCl)]** One of the earliest chlorine products used as a bleaching powder in water disinfection. Also called calcium chloride hypochlorite. *See* chlorinated lime for detail.

**chloride reduction process** A process used to generate chlorine dioxide ($ClO_2$) on the site of water treatment plants, with sodium chlorate ($NaClO_3$), sodium chloride (NaCl), and sulfuric acid ($H_2SO_4$):

$$2\ NaClO_3 + 2\ NaCl + 2\ H_2SO_4 \rightarrow 2\ ClO_2 + Cl_2 + 2\ Na_2SO_4 + 2\ H_2O \quad (C\text{-}72)$$

*See also* the Jazka–CIP, Mathieson (or sulfur dioxide, $SO_2$), Solvay (or methanol, $CH_3OH$) processes.

**chlorinated** (1) Pertaining to water, wastewater or other substance that has been treated with chlorine for disinfection or oxidation. (2) The condition of an organic compound that contains chlorine atoms. *See also* chlorine-substituted.

**chlorinated aromatics** Any of the chlorinated derivatives of benzene, toluene, phenol, naphthalene, and biphenyl, and other compounds containing at least one benzene ring; used in medicine, agricultural chemicals, and paints.

**chlorinated by-product** *See* disinfection by-product.

**chlorinated copperas ($FeSO_4Cl$)** A solution of ferrous sulfate ($FeSO_4$) and ferric chloride ($FeCl_3$); the product that results from the addition of chlorine ($Cl_2$) to copperas ($FeSO_4$), which also produces ferric sulfate [$Fe_2(SO_4)_3$] and ferric chloride. The latter then reacts with calcium bicarbonate [$Ca(HCO_3)_2$], if there is sufficient alkalinity, or with any added lime [$Ca(OH)_2$]:

$$3\ FeSO_4 + 3/2\ Cl_2 \rightarrow Fe_2(SO_4)_3 + FeCl_3 \rightarrow 3\ FeSO_4Cl \quad (C\text{-}73)$$

$$2\ FeCl_3 + 3\ Ca(HCO_3)_2 \rightarrow 2\ Fe(OH)_3\ (s) + 3\ CaCl_2 + 6\ CO_2 \quad (C\text{-}74)$$

$$2\ FeCl_3 + 3\ Ca(OH)_2 \rightarrow 2\ Fe(OH)_3\ (s) + 3\ CaCl_2 \quad (C\text{-}75)$$

Chlorinated copperas is effective in prechlorination and for color removal; also used in chemical sludge conditioning.

**chlorinated DBP** A by-product of chlorination when bromine is absent from the source water. It results from the combination of chlorine with natural organic matter. *See also* mixed DBP and DBP shift.

**chlorinated ferrous sulfate** *See* chlorinated copperas.

**chlorinated hydrocarbons** A class of persistent, broad-spectrum insecticides that linger in the environment and accumulate in the food chain, including DDT, aldrin, dieldrin, heptachlor, chlordane, lindane, endrin, mirex, hexachloride, and toxaphene. Another example is TCE, used as an industrial solvent. Some are carcinogenic to laboratory animals. Also called organochlorines.

**chlorinated isocyanurate** A chlorinated potassium or sodium salt of cyanuric acid used in swimming pool disinfection.

**chlorinated ketone** A substance formed when ketones combine with chlorine compounds, e.g., as a by-product of water chlorination. Trichloroacetone, tetrachloroacetone, pentachloroacetone, and hexachloroacetone cause mutations in strains of *Salmonella.*

**chlorinated lime [$CaCl(OCl) \cdot 4\ H_2O$]** A combination of hydrated lime [$Ca(OH)_2$] and chlorine; a relatively unstable chemical used as a source of chlorine to disinfect water. Chlorinated lime is approximately 62.5% calcium oxychloride and contains about 35% available chlorine. It is also called hypochlorite of lime and has as basic active ingredient calcium oxychloride [$CaCl(OCl)$], the formula of which is sometimes confused with that of calcium hypochlorite [$Ca(OCl)_2$]. Calcium chloride hypochlorite or chloride of lime has the same formula as calcium oxychloride and is an earlier product used in water disinfection. Also called calcium hypochlorite, bleach, bleaching powder..

**chlorinated organic** An organic molecule in which chlorine has replaced another element. Chlorinated organic compounds are less biodegradable than their parent constituents and may be toxic to aquatic life. Also called chlorine-substituted organic, chloroorganic. *See also* chlorinated hydrocarbon, trihalomethane.

**chlorinated organic compound** *See* chlorinated organic.

**chlorinated paraffins** Dry chlorinated paraffins, which are mainly straight-chain, saturated hydrocarbons. The wastewater category includes production of chlorinated paraffins by passing gaseous chlorine into a paraffin hydrocarbon or by chlorination using solvents, such as carbon tetrachloride, under reflux (EPA-40CFR63.191).

**chlorinated phenoxy acid ($Cl_2C_6H_3OCH_2COOH$)** A herbicide. MCLG = MCL = 0.07 mg/L. *See* chlorophenoxy.

**chlorinated polyvinyl chloride (CPVC)** A chlorinated form of polyvinyl chloride used in industrial piping, the extra chlorine providing increased heat resistance. It is a high-strength thermoplastic polymer, practically inert toward water, inorganic reagents, hydrocarbons, and alcohols over a broad temperature range.

**chlorinated solvent** An organic solvent containing chlorine atoms, e.g., methylene chloride and 1,1,1-trichloromethane, used in aerosol spray containers and in highway paint.

**chlorination** The application of chlorine or its compounds to drinking water, wastewater, or industrial waste for disinfection or to oxidize undesirable compounds. Chlorine is applied as chlorine gas ($Cl_2$), chlorine dioxide ($ClO_2$), chloramine, or hypochlorite. When added to water, chlorine gas forms hypochlorous acid (HOCl) and hydrochloric acid (HCl):

$$Cl_2 + H_2O \rightarrow HOCl + HCl \qquad \text{(C-76)}$$

One hypothesis for chlorine's germicidal effects is by the destruction of extracellular enzymes and intercellular systems of microorganisms: cells die when enzymes essential to metabolic processes are inactivated; common chlorine doses do not destroy cells by oxidation. Temperature and pH affect the rate and efficiency of chlorination; other important factors include the time of contact, chlorine concentration, and the concentration of organisms. Chlorination is also used to remove incrustations from well screens, as an alternative to well treatment by dry ice, hydrochloric acid, or polyphosphates. *See also* alkaline chlorination, breakpoint chlorination, Chick's law, Chick–Watson equation, chloramination, chlorine disinfection mechanism, dechlorination, disinfection byproducts, hypochlorination, marginal chlorination, plain chlorination, postchlorination, prechlorination, superchlorination.

**chlorination by-product** *See* disinfection byproduct.

**chlorination chamber** A detention basin for the diffusion of chlorine compounds through water or wastewater. Also called chlorine contact chamber.

**chlorination of wells** The mixing of chlorine and well water for periodical well disinfection.

**chlorinator** A metering device (e.g., an orifice flowmeter or a dosimeter with an adjustable opening) that adds chlorine at a predetermined rate, in gas or liquid form, to water or wastewater to kill infectious organisms. *See* Figure C-06. *See also* chemical feeder.

**chlorinator room** *See* chlorine room.

**chlorine ($Cl_2$)** (1) A greenish-yellow gas of the halogen group, with a characteristic odor, soluble in water; 2.5 times heavier than air, extremely toxic; atomic weight = 35.457; molecular weight = 70.914. It occurs naturally as halite or sodium chloride (NaCl) or as chlorides of other metals, and is commonly produced by the electrolysis of brine. Chlorine is a strong oxidant commonly used in water and wastewater treatment as a disinfectant and for the oxidation of taste- or odor-causing organic materials as well as the removal of ammonia and grease; a highly toxic irritant that requires careful handling. For water treatment applications, it is available as a gas (elemental chlorine, $Cl_2$), a concentrated aqueous solution (sodium hypochlorite, NaOCl), or a solid [calcium hypochlorite, $Ca(OCl)_2$]. Other uses of chlorine include the improvement of coagulation, stabilization of sludge in settling tanks, the prevention of anaerobic conditions in wastewater, the conversion of cyanide to cyanate in industrial wastewaters, the destruction of hydrogen sulfide, the reduction of the oxygen demand of return activated sludge, and the preparation of chlorinated copperas. Utilities purchase chlorine gas in pressurized cylinders of 100 to 2200 pounds and ship them on railroad tank cars. Chlorine residuals are toxic to aquatic life even at low levels. *See also* available chlorine, breakpoint, colicidal efficiency, combined available chlorine, chlorine demand, chlorine equivalent, chlorine residual, combined chlorine, free available chlorine, free chlorine, Javelle water. (2) An amber liquid, 1.44 times as heavy as water; unconfined, it vaporizes rapidly at standard temperature and pressure. (3) The total residual chlorine present in the process wastewater stream exiting the wastewater treatment system (EPA-40CFR415.661-d).

**chlorine 36** The radioactive isotope of chlorine, used as a tracer; mass number = 36; half-life = 440,000 years.

**chlorine–ammonia process** The treatment of water using chlorine and ammonia as in chloramination or by adding chlorine to water that already contains ammonia.

**chlorine–ammonia treatment** Same as chlorine–ammonia process.

**chlorine bromide** A halogen compound combining bromine and chlorine (BrCl). It remains in equilibrium with the components in solution or in a gas mixture. Used as a disinfectant, chlorine bromide is a more powerful germicide than either gas alone. Also called bromine chloride.

**chlorine burnout** The intermittent or periodic application of free chlorine to drinking water distribution systems, sometimes coupled with intense flushing, to overcome accumulated chlorine demand and control nitrifying organisms in a chloraminated system. Also called free chlorine burnout or, simply, burnout (Bryant, 1992).

**chlorine–chlorite process** The production of chlorine dioxide ($ClO_2$) by the oxidation of a solid powder or a concentration of sodium chlorite ($NaClO_2$) with chlorine gas ($Cl_2$). *See* chlorine dioxide generation.

**chlorine, combined** *See* combined chlorine.

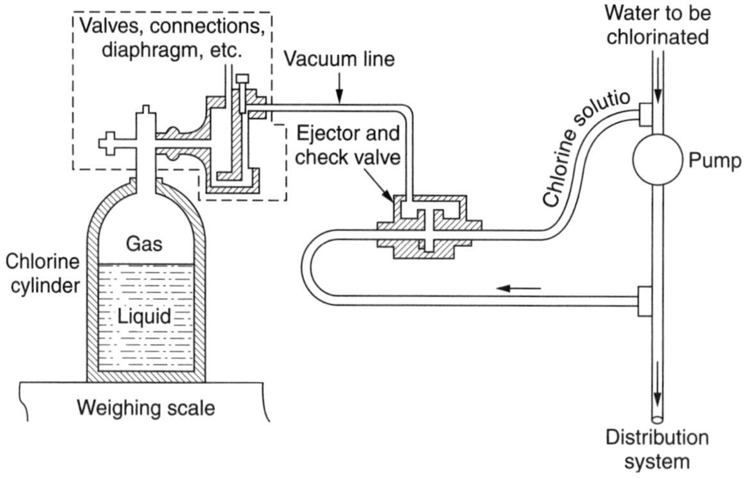


**Figure C-06.** Chlorinator (adapted from Fair et al., 1971, and McGhee, 1991).

**chlorine contact basin** Same as chlorine contact chamber.

**chlorine contact chamber** A detention basin in a water or wastewater treatment plant where effluent is disinfected by chlorine addition; provides adequate contact time. Also called chlorination chamber, chlorine contact basin.

**chlorine cylinder** A container, typically made of steel, for storing chlorine gas under pressure.

**chlorine decay** The degradation of free chlorine in drinking water in the absence of ammonia, commonly assumed to depend on total organic carbon (TOC) and two parameters (*k*, *K*) that vary with the initial chlorine dose and the TOC concentration (AWWA, 1999):

$$C_t = KX \ln (C_0/C_t) - kXt + C_0 \qquad \text{(C-77)}$$

where $C_t$ is the free chlorine concentration in mg/L of $Cl_2$ at time *t*, and *X* is the TOC concentration in mg/L. *See also* chlorine demand, residual chlorine decay.

**chlorine demand** (1) The difference between the amount of chlorine added to water or wastewater and the amount of residual chlorine remaining after a given contact time, usually 15 minutes. Chlorine demand may change with time, dosage, temperature, pH, and nature and amount of the impurities in the water:

$$\text{chlorine demand (mg/L)} = \text{chlorine applied (mg/L)} - \text{residual (mg/L)} \qquad \text{(C-78)}$$

Sometimes called immediate chlorine demand, it is the amount of chlorine required to immediately react with impurities such as reducing substances (iron, manganese, ammonia, hydrogen sulfide, nitrites, living and dead organic matter). Chlorine demand of wastewater may be roughly estimated as 0.5 mg/L of chlorine per mg/L of $BOD_5$. Based on kinetics, two equations have been proposed for chlorine demand exertion:

$$D = kt^n \qquad \text{(C-79)}$$

$$D = C_0\{1 - [xe^{(-k_1t)} + (1 - x)e^{(-k_2t)}]\} \qquad \text{(C-80)}$$

Where $D$ = chlorine demand; $k$, $n$ = empirical constants; $t$ = time, hours; $C_0$ = chlorine dose, mg/L; $x$ = empirical parameter, between 0.4 and 0.6; $k_1$, $k_2$ = rate constants, typically, 1.0/min and 0.003/min, respectively. *See also* breakpoint chlorination, chlorine decay, chlorine dose, combined residual, free residual. (2) The amount of chlorine needed for water disinfection; same as chlorine applied in the above equation.

**chlorine demand—free water** A condition in which most of the chlorine applied to a water becomes residual because of very low oxidant demand.

**chlorine detector** A device installed in a chlorine room that automatically triggers an alarm when it detects a given concentration of chlorine in the air.

**chlorine dioxide ($ClO_2$)** A greenish-yellow, water-soluble, very reactive, unstable gas with an irritating odor, but a relatively stable free radical. It is used as a substitute for chlorine gas in water disinfection because it does not produce chloramines or trihalomethanes and is not affected by normal pH ranges, but a degradation product (chlorate, $ClO_3^-$) may cause hemolytic anemia. Chlorine dioxide is a more effective virocide than chlorine; it can meet the USEPA CT criteria for *Giardia* at less than half the CT level of chlorine. It decomposes rapidly and is usually prepared on-site; *see* chlorine dioxide generation. It is also used to oxidize iron and manganese:

$$Fe^{2+} + ClO_2 + 3\ H_2O \rightarrow Fe(OH)_3 + ClO_2^- + 3\ H^+ \qquad \text{(C-81)}$$

$$Mn^{2+} + 2\ ClO_2 + 2\ H_2O \rightarrow MnO_2 + 2\ ClO_2^- + 4\ H^+ \qquad \text{(C-82)}$$

Industrial uses of chlorine dioxide include bleaching (pulp and paper, flour, oils), and cleaning and tanning of leather. *See also* chlorine dioxide disproportionation, gas-phase chlorine dioxide.

**chlorine dioxide by-products** Chlorite ($ClO_2^-$), chlorate ($ClO_3^-$), and oxidized organics resulting from the oxidation reactions of chlorine dioxide ($ClO_2$) with organic matter. These chlorinated or nonchlorinated by-products include aldehydes and quinines.

**chlorine dioxide demand** The loss of chlorine dioxide during disinfection due to its reaction with water, interconversion to chlorite and chloride, and, to a lesser extent, reaction with material present in water.

**chlorine dioxide disproportionation** The conversion of chlorine dioxide ($ClO_2$) into chlorite ($ClO_2^-$) and chlorate ($ClO_3^-$) under alkaline conditions:

$$2\ ClO_2 + 2\ OH^- \rightarrow H_2O + ClO_2^- + ClO_3^- \qquad \text{(C-83)}$$

**chlorine dioxide generation** Chlorine dioxide is usually prepared on-site by combining chlorine ($Cl_2$) and sodium chlorite ($NaClO_2$) in a ratio of one mole chlorine to two moles chlorite (the "chlorine-chlorite" process), or by acidification of sodium chlorite (the "acid–chlorite" process):

$$2\,NaClO_2 + Cl_2 \rightarrow 2\,ClO_2 + 2\,NaCl \quad (C\text{-}84)$$

$$5\,NaClO_2 + 4\,H^+ \rightarrow 4\,ClO_2 + 5\,Na^+ + Cl^- + 2\,H_2O \quad (C\text{-}85)$$

*See* acid–chlorite, chloride reduction, Jazka–CIP, Mathieson, and Solvay processes.

**chlorine dioxide inactivation** Mechanisms of microorganism inactivation by chlorine dioxide are similar to chlorine disinfection mechanisms.

**chlorine disinfection mechanisms** The mechanisms by which chlorine eliminates the pathogenic effects of microorganisms, e.g., reactions adversely affecting respiratory, transport, and nucleic acid activity in bacteria, or disruption of the nucleic acid or the viral capsid proteins of viruses. One hypothesis for chlorine's germicidal effects is by the destruction or impairment of extracellular enzymes [particularly, interference with the sulfhydryl groups (—SH) of these enzymes] and intercellular systems of microorganisms: cells die when enzymes essential to metabolic processes are inactivated. Common chlorine doses do not destroy cells by oxidation.

**chlorine disinfection models** Models developed to describe the disinfection process and particularly the rate of microorganism inactivation. *See* Chick–Watson equation, Collins–Selleck's model, modified Collins-Selleck's model, CT concept, Gard's model, generalized disinfection model, Hom model.

**chlorine disinfection pathways** *See* chlorine disinfection mechanisms.

**chlorine disproportionation** The reaction of dissolved aqueous chlorine [$Cl_2(aq)$] with water to form hypochlorous acid (HOCl), chloride ions ($Cl^-$), and protons ($H^+$):

$$Cl_2\,(aq) + H_2O \rightarrow HOCl + Cl^- + H^+ \quad (C\text{-}86)$$

This reaction is essentially complete within a few seconds at ordinary water temperatures. Also called chlorine hydrolysis (AWWA, 1999 and Fair et al., 1971).

**chlorine distribution diagram** A graphical representation of the relative amounts of three chlorine species in an aqueous solution, as a function of pH: elemental chlorine ($Cl_2$), hypochlorous acid (HOCl), and hypochlorite ion ($OCl^-$). *See* Figure C-07.

**chlorine dose** The amount (pounds per million gallons) or concentration (mg/L) of chlorine ($Cl_2$) applied to disinfect water or wastewater, corresponding to the sum of the chlorine demand and the desired free chlorine residual.

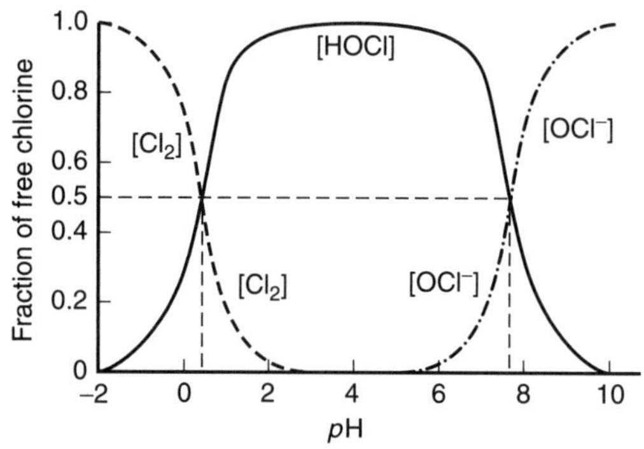


**Figure C-07.** Chlorine distribution diagram ($[Cl^-] = 10^{-3}$ M). (Adapted from AWWA, 1999.)

**chlorine equivalent** An expression of the oxidizing capacity of chlorine compounds. *See* available chlorine for more detail.

**chlorine feeder** *See* chlorinator.

**chlorine feed rate** The rate at which chlorine is fed (e.g., in pounds per hour) to correspond to a desired dosage.

**chlorine feed room** A room in a water or wastewater treatment plant designed specifically for chlorine storage and feed, designed with appropriate safety measures. *See also* chlorine room.

**chlorine, free** *See* free available chlorine.

**chlorine hydrate ($Cl_2 \cdot 8H_2O$)** A yellowish ice resulting from the combination of chlorine with water below 49.2°F. *See* chlorine ice.

**chlorine hydrolysis** The equilibrium reaction of elemental chlorine dissolved in pure water. *See* chlorine disproportionation for detail.

**chlorine ice** A crystalline chlorine hydrate ($Cl_2 \cdot 8\,H_2O$) formed when chlorine gas liquefies in contact with a cold surface and combines with moisture; this phenomenon may cause plugging of a chlorinator's orifice or control valve.

**chlorine inactivation** The mechanisms by which chlorine eliminates the pathogenic effects of microorganisms, e.g., reactions adversely affecting respiratory, transport, and nucleic acid activity in bacteria, or disruption of the nucleic acid of viruses. *See* chlorine disinfection mechanisms.

**chlorine ionization** An instantaneous reverse reaction of chlorine hydrolysis, essentially the decomposition of hypochlorous acid (HOCl) into the hydrogen ($H^+$) and hypochlorite ions ($OCl^-$):

$$HOCl \rightarrow H^+ + OCl^- \quad (C\text{-}87)$$

**chlorine nitride** *See* trichloramine.

**chlorine reactions** *See* chloramination reaction sequence, chlorine disproportionation, chlorine ionization.

**chlorine removal** The removal of chlorine from solution using sulfur dioxide or other means. *See* dechlorination.

**chlorine requirement** The amount of chlorine that is needed for a particular purpose. Some reasons for adding chlorine to water or wastewater are reducing the number of coliform bacteria, obtaining a particular chlorine residual, or oxidizing some substances. In each case, a definite dosage of chlorine is necessary. This dosage is the chlorine requirement.

**chlorine residual** The amount or concentration of chlorine remaining in water or wastewater after application and satisfaction of the chlorine demand. Chlorine residual protects drinking water against contamination and growth of aesthetically significant microorganisms in the distribution system, but it increases the corrosion of copper, iron, and steel. *See also* free chlorine residual.

**chlorine residual decay** *See* residual chlorine decay.

**chlorine room** The separate room, in a water or wastewater treatment plant, that contains all chlorine cylinders, chlorinator, and piping carrying gaseous chlorine under pressure. The room is heated to about 15°C and properly ventilated (one room volume per minute), with the air outlet near the floor and an automatic chlorine detector connected to an alarm. Also called chlorinator room. *See also* chlorine feed room.

**chlorine solution** A solution that contains about 1% of available chlorine by weight, used for water disinfection; e.g., the proprietary disinfectants and bleaches Chloros®, Chlorox®, Domestos®, Javel® water, Milton®, Parazone®, Voxsan®, and Zonite®.

**chlorine sources** Chlorine is used in water and wastewater treatment as a gas, bleaching powder, chlorinated lime, chlorine solution, high-test hypochlorite, or in tablets.

**chlorine-substituted** Characteristic of a substance that results from the substitution of a chlorine atom in another substance.

**chlorine-substituted by-product** A disinfection by-product containing chlorine.

**chlorine-substituted organic** An organic substance that results from the substitution of a chlorine atom in another substance.

**chlorine tablet** Common name of calcium hypochlorite and other solid chlorine compounds used for water disinfection.

**chlorine taste** A common complaint from consumers of drinking water disinfected by chlorine, whose odor threshold may be 0.3 mg/L, whereas the chlorine residual may be higher.

**chlorine-to-ammonia ratio** *See* chlorine-to-ammonia nitrogen ratio.

**chlorine-to-ammonia nitrogen ratio** The ratio ($Cl_2/NH_3$-N) of the chlorine and ammonia doses applied during chloramination of water, typically 7.6:1 milligrams $Cl_2$ per milligram $NH_3$ at the breakpoint, calculated as $6 \times 35.5/2 \times 14 = 7.6$ from the reaction:

$$3\,Cl_2 + 2\,NH_3 \rightarrow 6\,H^+ + 6\,Cl^- + N_2 \quad \text{(C-88)}$$

It is an important factor in the formation of chloramines and the performance of the chloramination process. Lower ratios than the breakpoint ratio, e.g., 3:1–4:1, may be considered to limit the undesirable formation of dichloramines, trichlorides, and trihalomethanes (Bryant et al., 1992).

**chlorine toxicity** The detrimental effects of chlorine residuals on the biota of a receiving stream.

**chlorinity** The quality, state, or degree of being chlorinous, as expressed, for example, by the relative concentration or weight of chlorine in salt water; sometimes applied to the concentration or weight of all halides in water. It is sometimes defined, independently of atomic weights, as 0.3285233 times the weight of silver equivalent to all the halides. *See also* chlorosity, salinity.

**chlorite ($ClO_2^-$)** (1) A salt of chlorous acid ($HClO_2$), i.e., containing the chlorite ion ($ClO_2^-$), a potential contaminant of drinking water when chlorine dioxide ($ClO_2$) is used for disinfection: chlorine dioxide degrades to chlorite, chloride, and chlorate. Chlorite ion can oxidize blood hemoglobin, causing methemoglobinemia (particularly in infants) or hemolytic anemia (in dialysis patients). Chlorites are relatively stable, degrade slowly to chloride, and can be removed from water by the application of ferrous iron salts or reduced sulfur compounds; they are used to produce chlorine dioxide on-site. *See also* chlorate. (2) A group of silicate minerals related to micas and found in green crystals or scales of metamorphic rocks, e.g., the hydrous silicates of aluminum, ferric iron, and magnesium.

**chlorite ion ($ClO_2^-$)** *See* chlorite.

**ChlorMaster™** An installation engineered by Pepcon Systems, Inc. for the generation of sodium hypochlorite.

**chloroacetaldehyde** A disinfection by-product produced during the chlorination of water containing organic matter.

**chloroacetic acid ($C_2H_3ClO_2$)** A colorless, crystalline, water-soluble powder, produced from acetic acid and chlorine, used in dyes. Also called chloracetic acid, monochloroacetic acid.

**chloroacetonitrile** A disinfection by-product produced during the chlorination of water containing organic matter.

**chlorobenzene ($C_6H_5Cl$)** A volatile organic compound almost insoluble in water, used as a solvent and in chemical manufacturing. This substance, which affects the respiratory and nervous systems, is regulated by the USEPA: MCLG = MCL = 0.1 mg/L. *See* monochlorobenzene.

**chlorobromomethane** *See* bromochloromethane.

***Chlorobium*** A species of strictly anaerobic, photoautotrophic, green, sulfur-oxidizing bacteria that use hydrogen sulfide ($H_2S$) instead of water ($H_2O$) for photosynthesis.

**chlorodibromomethane ($CHBr_2Cl$)** A trihalomethane/volatile organic compound, regulated by the USEPA. *See* dibromochloromethane.

**3-chloro-4-(dichloromethyl)-5-hydroxy-2(5H)-furanone (mutagen X or MX)** A bacterial mutagen found in chlorinated effluents of pulp and paper mills and in drinking water samples.

**chlorodiiodomethane ($CHClI_2$)** A trihalomethane/volatile organic compound, regulated by the USEPA. Also called diiodochloromethane.

**6-chloro-n,n-diethyl-1,3,5-triazine-2,4-diamine** *See* simazine.

**1-chloro-2,3-epoxypropane ($CH_2OCHCH_2Cl$)** *See* epichlorohydrin.

**chloroethane** A volatile organic compound regulated by the USEPA.

**chloroethene** Same as vinyl chloride.

**chloroethylene** Same as vinyl chloride.

**chlorofluorocarbon (CFC)** A family of inert, nontoxic, and easily liquefied chemicals used in refrigeration, air conditioning, packaging, insulation, or as solvents and aerosol propellants. Because CFCs are not destroyed in the lower atmosphere, they drift into the upper atmosphere where their chlorine components affect the ozone layer, which protects the Earth's surface from harmful effects of solar radiation. Two common propellants are $CFCl_3$ and $CF_2Cl_2$; two refrigerants are $CFHCl_2$ and $CF_2HCl$. Also called Freon.

**chloroform ($CHCl_3$)** A colorless, volatile, nonflammable, pungent, sweet-tasting liquid; a trihalomethane formed by the reaction of chlorine and natural organic material in water, i.e., methane ($CH_4$) with three chlorine atoms replacing hydrogen atoms; used as a general solvent, refrigerant, propellant, and grain fumigant. It can cause liver damage and cancer. It is regulated by the USEPA as a drinking water contaminant among the trihalomethanes. Also called trichloromethane.

**chloroformbutanol** A reagent used in the laboratory analysis of organic acids.

**chlorogenic acid ($C_{16}H_{18}O_9$)** A colorless, crystalline acid responsible for the browning of cut fruits and vegetables.

**chlorohydrin** An organic substance that contains a chlorine atom and a hydroxyl group.

**Chloromatic** Equipment manufactured by Brinecell, Inc. for the electrolytic generation of chlorine.

**chloromethane ($CH_3Cl$)** A volatile organic compound regulated by the USEPA. Also called methyl chloride.

**chloroorganics** Organic compounds containing chlorine and formed during water or wastewater chlorination. *See* chlorinated organic.

**Chloropac®** Equipment manufactured by ELCAT Corp. for the generation of hypochlorite.

**chlorophenol ($C_6H_5ClO$)** Any of three isomers of a compound of phenol and chlorine that may be formed during water chlorination and cause problems of taste and odor; general formula: $C_6H_xCl_{5-x}OH$, where x varies from 0 to 4. Mono-, di-, and tri-chlorophenols are produced and used industrially as biocides and in dyes. A limit of 0.001 mg/L of phenol is recommended for drinking water. Also spelled chlorphenol. *See also* pentachlorophenol.

**chlorophenolic** Characteristic of chlorophenols, products derived from the chlorination of phenolic compounds.

**chlorophenolic compound** A phenolic compound (carbolic acid) combined with chlorine.

**chlorophenol red ($C_{19}H_{12}Cl_2O_5S$)** A reagent used in water analyses as an acid–base indicator, with a pH transition range of 4.8 (yellow)–6.4 (purple).

**chlorophenothane** *See* DDT.

**chlorophenotone** *See* DDT.

**chlorophenoxy** A class of herbicides that may be found in domestic water supplies and cause adverse health effects (potential damage to the liver and kidney); general formula: $(CH_3)_yCl_xH_{5-x-y}OR$-COOH, where x = 1, 2, or 3, and y = 0 or 1. Two widely used chlorophenoxy herbicides are 2,4-D (2,4-dichlorophenoxy acetic acid, chemical formula: $Cl_2C_6H_3OCH_2COOH$) and 2,4,5-TP (2,4,5-trichlorophenoxy propionic acid or silvex). Regulated by the USEPA: MCL = MCLG = 0.07 mg/L. Also called chlorinated phenoxy acid.

**chlorophenoxy herbicide** *See* chlorophenoxy.

***Chlorophyceae*** *See* chlorophyta.

**chlorophyll** The green coloring matter of leaves and plants, a pigment that enables them to use radiant energy for the conversion of carbon dioxide ($CO_2$) into carbohydrates. Among other elements, it contains magnesium and iron. Also spelled chlorophyl. *See* photosynthesis.

**chlorophyll a ($C_{55}H_{72}MgN_4O_5$)** The bluish-black form of chlorophyll, which is responsible for the green color of algae and essential for leaves and plants to capture light during photosynthesis. It is also found in some bacteria.

**chlorophyll b ($C_{55}H_70MgN_4O_6$)** A dark green form of chlorophyll that assists plants, algae, and some bacteria in photosynthesis.

***Chlorophyta*** A large division of algae (the green algae) found in freshwater, saltwater, and damp places on land. Also called chlorophyceae.

**chloropicrin (CP) ($CCl_3NO_2$)** A colorless, oily, poisonous liquid that causes headaches and other ailments; used as an insecticide, in tear gas, and in other industrial applications. It is a disinfection by-product whose monitoring by water supply systems is required quarterly. Also spelled chlorpicrin. Also called nitrochloroform, trichloronitromethane.

**chloroplast** A plastid that contains chlorophyll and is a photosynthesis site.

**chloroplatinate** A salt of chloroplatinic acid; e.g., potassium chloroplatinate ($K_2PtCl_6$), used in preparing solutions to analyze water for color.

**chloroplatinate unit (cpu or CPU)** A unit used to express the color of a water sample. *See* color unit.

**chloroplatinic acid ($H_2PtCl_6 \cdot 6\ H_2O$)** A red-brown, crystalline solid used in glass, metal, and ceramic work. Also called platinic chloride.

**chloropropylene oxide** Another name of the organic contaminant epichlorohydrin.

**chlororganic** Organic compounds combined with chlorine. These compounds generally originate from, or are associated with, life processes such as those of algae in water.

**Chloros®** A proprietary bleach that contains about 4% of available chlorine; it can be used to make a 1% stock solution for water disinfection. Such a solution is unstable under warm conditions.

**chlorosis** (1) Discoloration or abnormally yellow color of normally green plant parts caused by failure to develop chlorophyll, disease, lack of nutrients, or various air pollutants. (2) Iron-deficiency anemia resulting in a yellow-green complexion; also called greensickness.

**chlorosity** The product of chlorinity and the density of a solution at 20°C.

**chlorosulphonic acid ($HClO_3S$)** A colorless, yellowish, corrosive liquid.

**chlorotene** Another name of trichloroethane.

**chlorothiazide ($C_7H_6lN_3O_4S_2$)** A white, crystalline powder used as a diuretic.

**chlorotrifluoroethylene ($C_2H_2ClF_3$)** A colorless flammable gas that can form oils, grease, and waxes.

**chlorotrifluoromethane ($CClF_3$)** A colorless gas used as a refrigerant and in pharmaceutical products. Also called trifluorochloromethane.

**chlorous acid ($CHClO_2$)** A hypothetical acid that exists only in solution or by its salts. *See* chlorite.

**Chlorox®** A proprietary bleach that contains about 4% of available chlorine; it can be used to make a 1% stock solution for water disinfection. Such a solution is unstable under warm conditions.

**chlorphenol** Same as chlorophenol.

**chlorpicrin** Same as chloropicrin.

**chlorpyrifos ($C_9H_{11}Cl_3NO_3PS$)** An insecticide for lawns and ornamental plants.

**Chlor-Scale™** A scale manufactured by Force Flow Equipment for weighing chlorine ton containers.

**ChlorTainer™** A total containment vessel manufactured by TGO Technologies, Inc. for the storage of chlorine and sulfur dioxide. It meets all the safety requirements of the Uniform Fire Code and does not require mechanical devices to function.

**Chlortrol** An instrument manufactured by Fisher & Porter for the analysis of chlorine residual.

**$C_5H_7NO_2$** Common simplified empirical formula of prokaryote cell biomass. *See* cell composition for detail.

**$C_{60}H_{87}N_{12}O_{23}P$** Common simplified empirical formula of cell biomass. *See* cell composition for detail.

**$C_nH_aO_b$** General chemical formula of organic matter.

**cholera** (From the Greek word chole, meaning bile.) An acute, highly infectious, water-related and excreta-related disease of the gastrointestinal tract caused by the bacteria *Vibrio cholerae,* which produce an exotoxin, a substance that causes excessive fluid losses. It has worldwide distribution but is endemic in some parts of Asia (e.g., China and India). It is transmitted via the fecal–oral route and is characterized by profuse diarrhea, vomiting, dehydration, cramps and other symptoms. Sometimes called Asiatic cholera. *See* classical cholera, El Tor cholera.

**cholera vibrio** The organism that causes cholera. *See Vibrio cholerae.*

**cholestasis** Partial or complete stoppage of the flow of bile that can cause jaundice.

**cholesterol ($C_{23}H_{49}O_3COOH$)** A fat-like sterol (steroid alcohol) found in animal tissues as well as some higher plants and algae; a protective agent in the skin and a detoxifier in the bloodstream, but also a factor in atherosclerosis.

**cholic acid ($C_{27}H_{45}OH$)** A bile acid related to sex hormones and cholesterol; obtained commercially from beef bile; used to form bile salts with glycine ($NH_2CH_2COOH$) or taurine ($NH_2CH_2CH_2SO_3H$).

**cholinesterase** A group of enzymes (e.g., acetylcholinesterase) found in animals; they hydrolyze esters of choline and regulate nerve impulses. Cholinesterase inhibition is associated with poisoning by organophosphorus pesticides and a variety of acute symptoms such as nausea, vomiting, blurred vision, stomach cramps, and rapid heart rate.

**cholinesterase inhibition** *See* cholinesterase.

**cholinesterase inhibitor** An organophosphorus pesticide or other toxic compound that inhibits cholinesterase.

**chopper–grinder pump** A special combination of a centrifugal pump and a grinder, sometimes used to recirculate digested sludge.

**Chopper Pump** A pump manufactured by Vaughn Co., Inc. for used in wastewater conveyance and treatment; it chops solids between the impeller and the fixed bar.

**chopper pump** A wastewater conveyance device equipped with a cutter knife attached to a nonclog impeller to break up large incoming solids. A chopper pump may eliminate the need for a grinder or comminutor but has a relatively low efficiency, in the 40%–60% range.

**chopper-type pump** Same as chopper pump.

**C horizon** The unaltered soil layer underlying the B-horizon containing unweathered or partially weathered material and a minimum of soil fauna and flora.

**chromate** A salt of chromic acid and an important compound of chromium; e.g., sodium chromate ($Na_2CrO_4$) and potassium chromate ($K_2CrO_4$). Anion exchange is a common method used for chromate removal from water.

**chromatin** The substance of a cell nucleus consisting of DNA, RNA, and proteins that forms chromosomes during cell division.

***Chromatium*** A species of strictly anaerobic, photoautotrophic, green or purple sulfur-oxidizing bacteria.

**chromatographic movement** Movement of chemical components through a sorption bed as through a chromatograph.

**chromatographic peaking** A phenomenon observed in adsorption and ion-exchange columns: some species concentrate in the column and later exit in concentrations exceeding the influent concentration, a potentially dangerous situation in the case of toxic ions or substances. In some special resins, the selectivity sequence can be reversed to eliminate this problem (AWWA, 1999).

**chromatography** A technique that allows the separation of a mixture into its component compounds according to their relative affinity for a solvent system or column media. It is used to identify substances based on their rate of movement across a base material compared to the rates of known substances. Gas or liquid chromatography is used to analyze organic compounds in water.

**Chromaver** A line of chemical reagents produced by Hach Co. to detect the presence of chromates.

**chrome waste** Wastewater that contains chromates and chromic acid, originating from such industries as chromium plating and leather tanning.

**chromic acetate [$Cr(C_2H_3O_2)_3 \cdot H_2O$]** A grayish-green powder used in dyes and textiles. Also called chromium acetate.

**chromic acid** (1) A hypothetical acid ($H_2CrO_4$) known by its salts and in solution. (2) A bright orange caustic material that reacts with organic matter ($CrO_3$), used as a cleaning agent and in industrial applications; actually not an acid.

**chromite ($FeCr_2O_4$)** (1) A cubic mineral, basically ferrous chloride; the principal ore of chromium. (2) A salt of bivalent chromium.

**chromium (Cr)** A hard, grayish-white, crystalline, corrosion-resistant, heavy metal that can damage living things at low concentrations and tends to accumulate in the food chain. Under the Clean Water Act, the total chromium present in the process wastewater stream exiting a wastewater treatment system is considered (EPA-40CFR415.661-c). Atomic weight = 51.996. Atomic number = 24. Specific gravity = 7.2. Valence = 2, 3, or 6. It is used in electroplating, in alloys (e.g., 12% or more in stainless steel), as a corrosion inhibitor, and other applications such as plating pump impellers and other parts in wastewater treatment equipment. Regulated by the USEPA in drinking water: MCLG = MCL = 0.1 mg/L. Hexavalent chromium compounds are carcinogenic and also have long-term effects on the skin and kidneys. *See also* chromium (trivalent) III and chromium (hexavalent) VI.

**chromium III** Same as trivalent chromium.

**chromium 51** The radioactive isotope of chromium with a mass number of 51 and a half-life of 27.8 days, used as a tracer.

**chromium acetate** *See* chromic acetate.

**chromium pollution** Water pollution caused by the discharge of chromium wastes.

**chromium VI** Same as hexavalent chromium.

***Chromobacterium* genus** Heterotrophic organisms active in biological denitrification.

**chromogen** (1) A substance of organic fluids that forms colored compounds when oxidized. (2) A colored compound that can be converted to a dye.

**chromogenic** Producing color.

**chromogenic coliform analysis** *See* chromogenic medium test.

**chromogenic medium test** One of the common laboratory procedures used to detect the presence of coliforms in multiple 10-mL or single 100-mL water supply samples. Also called chromogenic coliform analysis. *See also* multiple-tube fermentation and membrane filter technique.

**chromogenic substrate technology** A bacteriological laboratory technique that uses a substrate linked to a chromogen to identify bacteria that have the specific enzymes that can hydrolyze the substrate. *See also* fluorogenic substrate technology.

**chronic** Occurring over a long period of time, continuously or intermittently; used to describe ongoing exposures and effects that develop only after a long exposure, e.g., the effects of asbestos. *See also* acute.

**chronic daily intake (CDI)** A parameter used in the determination of the risk of lifetime exposure to a substance; calculated in milligrams per kg per day as the total dose ($D$, mg) divided by the body weight ($W$, kg) and by the number of days in the lifetime ($L$, days):

$$CDI = D/(WL) \qquad \text{(C-89)}$$

Total dose is usually taken as the product of the substance concentration ($C$, mg/L), the intake rate ($R$, L/day), the exposure duration ($T$, days), and an absorption factor ($f$, dimensionless):

$$D = CRTf \qquad \text{(C-90)}$$

Commonly used values in daily intake calculations include a lifetime of 70 years, a body weight of 70 kg for an adult or 10 kg for a child, and daily water ingestion rates of 2.0 liters for adults and 1.0 liter for children. *See also* lifetime risk, potency factor, slope factor, dose–response assessment, reference dose (Metcalf & Eddy, 2003).

**chronic effect** An adverse effect on a human or animal in which symptoms recur frequently or develop slowly over a long period of time. *See also* health hazard. In studies of chemical exposure, chronic health effects on rodents are considered for periods of 1–2 years, and subchronic effects for periods of 2–13 weeks.

**chronic exposure** Long-term, low-level exposure to a toxic chemical; multiple exposures occurring over an extended period of time, or a significant fraction of the animal's or the individual's lifetime.

**chronic health effect** *See* chronic effect.

**chronicity index** The ratio of the daily dose that produces toxicity to the dose that produces acute toxicity; an indication of the potential for a substance to produce toxic effects.

**chronic toxicity** (1) The capacity of a substance to cause long-term poisonous human health effects: the injury persists because it is irreversible or progressive or because its rate is greater than the rate of repair during the exposure. Chronic toxicity results in sublethal response over 1/10 of lifespan or longer. *See also* acute toxicity and health hazard. (2) The condition of adverse effects persisting for an extended period.

**chronic toxicity test** A method used to determine the concentration of a substance that causes chronic toxicity.

**chronic toxic unit** Same as toxic unit—chronic. *See* toxic units.

**chronic value (ChV)** The geometric mean of the no-observed-effect concentration (NOEC) and lowest-observed-effect concentration (LOEC) from partial-cycle, full-cycle, and early-lifestage tests:

$$\text{ChV} = (\text{NOEC} \times \text{LOEC})^{0.5} \qquad \text{(C-91)}$$

Sometimes used interchangeably with maximum acceptable toxicant concentration. *See also* inhibiting concentration and toxicity terms.

**chronic virus infection** A viral infection that shows no immediate sign, although the host cell can shed the virus. *See* oncogenic virus, acute virus infection.

**chronotropic action** An action, e.g., a chemical effect, that affects the rate or timing of heartbeats or other physiologic processes.

**chrysolite** *See* olivine.

***Chrysops*** A genus of mosquitoes, also called mangrove flies, that breed in shaded areas in the vicinity of slow-moving streams. In Central and West Africa, they carry the pathogen *Loa loa* that causes loiasis.

**chrysotile [$Mg_3Si_{12}O_5(OH)_4$]** One of the minerals classified as asbestos; a compound of silica and magnesium oxide.

**Churchill–Buckingham method** An empirical approach, proposed in the *Sewage and Industrial Wastes Journal* ("Statistical Method for Analysis of Stream Purification Capacity," *28*:517, 1956), using multiple correlation between physical, chemical, and biological processes to assess the effect of pollutants on a receiving water. Accordingly, the dissolved oxygen drop (mg/L) depends on the sag conditions: BOD load (mg/L), temperature (°C), and flow (cfs). Other stream sanitation models include the Streeter–Phelps formulation and the Thomas method.

**Churchill–Elmore–Buckingham's equation** An empirical relationship proposed by M. A. Churchill, H. L. Elmore, and R. A. Buckingham to calculate the reaeration rates in streams. Two forms of the equation were presented in *J. Sanitary Engineering Division, ASCE, 88,* SA4, pp. 1–46 ("The Prediction of Stream Reaeration Rates") and in Advances in Water Pollution Research ("The prediction of Stream Reaeration Rates," Vol. 1, Pergamon Press, London, 1964, p. 89):

$$K_2 = 5.03\ U^{0.969}/H^{1.673} \quad \text{(metric units)} \qquad \text{(C-92)}$$

$$K_2 \approx 5\ U/R^{5/3} = 7.5\ S^{0.5}/(Rn) \quad \text{(U.S. units)} \qquad \text{(C-93)}$$

where $K_2$ = reaeration rate, $U$ = mean flow velocity in the reach under consideration, $H$ = stream depth, $R$ = mean hydraulic radius, $S$ = head loss or drop of water surface in the reach, and $n$ = Kutter's coefficient of roughness. *See also* reaeration constant.

**Churchill method** *See* the Churchill–Buckingham method.

**chute spillway** The overall structure that allows water to drop rapidly through an open channel without causing erosion; usually constructed near the edge of a dam.

**ChV** Abbreviation of chronic value.

**Ci** Abbreviation of curie, a unit of radioactivity.

**CID** Acronym of (1) critical initial dilution; (2) *Cryptosporidium* inactivation device.

**Cide-Trak™** An apparatus manufactured by the Microbics Corp. for monitoring biocides.

**ciénaga** (From Spanish cieno, meaning mud, slime.) A swamp or marsh formed and fed by springs, with standing water in depressions covered by heavy vegetation and occasionally with small outlet streams. Also spelled ciénega.

**ciénega** *See* ciénaga.

**cilia** Plural of cilium.

**ciliate** A highly developed heterotrophic protozoa that moves (and feeds) by the action of hairlike structures, cilia, on its body; e.g., *Balantidium coli, Paramecium,* and the Sporozoa *Plasmodium, Isospora,* and *Toxoplasma.* Ciliated, predatory protozoa are found on wastewater treatment attached films and suspended flocs. *See also* flagellate.

**Ciliophora** A phylum of protozoa in the kingdom Protista; the ciliates.

**cilium (plural: cilia)** A short thread of cytoplasm that projects from the surface of a cell. Many organisms (free-swimming protozoa, rotifers, mollusks, humans, etc.) use cilia for locomotion, for feeding, or in the respiratory passages. *See also* flagellum.

**cinnabar** The main source of mercury in nature, essentially mercuric sulfide (HgS), a mineral occurring in red crystals or masses. Mercuric sulfide is used as a pigment.

**cinnamene** Another name of styrene.

**CIP** Acronym of (a) capital improvement plan, (b) cast iron pipe, (c) clean in place.

**Cipolletti weir** A trapezoidal weir with notch side slope of one horizontal to four vertical; used to measure flow in open channels, particularly at contractions in streams. The flaring of the sides compensates for the effect of the end contraction. The Cipolletti weir's discharge formula relates the discharge ($Q$, cfs) to the base width or weir length ($L$, ft) and the weir head or the water depth ($H$, ft) on the upstream side (at a point where the surface curvature does not affect water level):

$$Q = 3.37\ LH^{1.5} \qquad \text{(C-94)}$$

*See also* the Francis formula.

**CIPRA** Acronym of Cast Iron Pipe Research Association.

**circle of influence** The circular outer edge of a depression produced in the water table by pumping water from a well. *See also* cone of depression and cone of influence.

**CirculAire™** An aspirating aerator manufactured by Aeration Industries, Inc.

**circular clarifier** A circular sedimentation tank or basin with flow usually fed from the center and directed radially to weirs at the periphery. It is equipped with a scraper mechanism at the bottom. *See* center-feed settling tank.

**circulating bed** The high-velocity (10–25 fps) mode of operation of a fluid bed furnace: the particulate bed is entrained over the freeboard into a particle separator and recirculates to the bottom of the combustion chamber, with continuous ash removal. On turndown or when velocities decrease sufficiently, the circulating bed becomes a bubbling bed.

**circulating-bed combustor** A fluidized-bed incinerator operated in the circulating-bed mode; a circulating-bed furnace or incinerator. *See also* bubbling-bed combustor.

**circulating-bed furnace** A fluid-bed furnace operating under circulating-bed conditions.

**circulating-bed incinerator** A fluidized-bed incinerator that operates under circulating-bed conditions; commonly used for the incineration of wastewater sludges and very efficient in the destruction of PCBs and other chlorinated compounds (e.g., up to 99.9999%). Same as circulating-bed combustor or circulating-bed furnace. *See also* bubbling-bed incinerator (Metcalf & Eddy, 2003).

**circulation loop** The part of a membrane treatment unit in which concentrate is blended with the influent to increase crossflow velocity.

**circulation zone** The upper layer of water in a thermally stratified lake or reservoir. This layer, under the influence of surface currents, consists of the warmest water and has a fairly uniform temperature. The layer is readily mixed by wind action. *See* epilimnion for more detail.

**Circuline** A line of circular sludge collectors manufactured by FMC Corp.

**Circumfed** A dissolved air flotation apparatus manufactured by Tenco Hydro, Inc.

**circumferential flow** The flow of a liquid parallel to the circumference of a circular structure. Also called peripheral flow.

**CIS** Acronym of customer information system.

***cis-*** A prefix in chemical nomenclature denoting a geometric isomer having two identical atoms or groups attached to the same side of a double bond. Examples: *cis*-1,2-dichloroethylene, *cis*-2-heptene, *cis*-2-pentene. *See also* geometric isomer, *trans-*. In this text, entries are listed alphabetically as though these prefixes did not exist.

***cis-* isomer** A geometric isomer.

**cistern** A small tank or reservoir that stores water (e.g., rainwater) for a home or a farm.

**cistern-flush toilet** A toilet (or latrine) that uses small quantities of water from a cistern to flush the excreta; a common type of communal latrine. The mixture of excreta and flushing water is usually discharged into a soakaway. *See also* pour-flush latrine.

**CitectSCADA™** A centralized system developed by Citect Pty, Ltd. for remote monitoring and control of water and wastewater treatment facilities.

**citrate** A salt or ester of citric acid. *See* sodium citrate.

**citric acid ($C_6H_8O_7 \cdot H_2O$) or [$C_3H_4(OH)(COOH)_3 \cdot H_2O$]** A white, crystalline, water-soluble powder with a strong acidic taste; present in citrus fruits; commercially available as crystals or powder, it is used in water treatment for pH adjustment.

**citric acid cycle** A series of enzyme-catalyzed reactions occurring in the aerobic metabolism of carbohydrates, proteins, and fatty acids, with the conversion of pyruvic acid into carbon dioxide and water, the buildup of ATP, and the reduction of oxygen. Also called Krebs' cycle, tricarboxylic acid cycle. *See also* glycolysis.

***Citrobacter*** A genus of gram-negative bacteria that ferment lactose very slowly or not at all.

***Citrobacter freundii*** A bacteria species that can cause biofilm problems in potable water distribution systems.

***Citrobacter* genus** A group of bacteria that includes species of total (nonfecal) coliforms; they are commonly found in feces but they also occur naturally in soils and water.

***Citromonas*** A genus of floc-forming bacteria commonly found in activated sludge.

**CIU** Acronym of categorical industrial user.

**CIX™** A water treatment system manufactured by Kinetico Engineered Systems, Inc. using ion exchange.

***C. jejuni*** *See Campylobacter jejuni.*

**CKD** Acronym of cement kiln dust.

**Cl** Chemical symbol of chlorine.

**$Cl_2$** Chemical formula of chlorine gas.

**ClAA** Abbreviation of monochloroacetic acid, a regulated disinfection by-product.

**$Cl_2AA$** Abbreviation of dichloroacetic acid, a regulated disinfection by-product.

**$Cl_3AA$** Abbreviation of trichloroacetic acid, a regulated disinfection by-product.

**cladding** An inorganic surface-finishing technique that combines heat and pressure to mechanically bond a coating to the surface of a part, e.g., the molten nickel alloy placed on the copper core of a coin. It is somewhat similar to electroplating, but produces a thicker coating and less hazardous waste.

***Cladosporium*** A waterborne fungus, pathogenic to humans. It has been identified in unchlorinated groundwater as well as surface water and service mains.

***Cladosporium resinae*** A fungus species that produces organic acids and causes the corrosion of nonferrous metals, e.g., the aluminum of fuel tanks or the protective coatings of other containers.

**CLAM®** Acronym of cleansimatic liquid analysis meter, a device manufactured by Monitek Technologies, Inc.

**clamp** A thin metal strip used to repair or strengthen a leaking pipe.

**clamshell** A clamshell bucket.

**clamshell bucket** A device or mechanism used to remove accumulated grit from the hoppers of a grit chamber. It is equipped with a dredging bucket opening at the bottom, consisting of two similar pieces hinged together at the top, and moved by overhead monorail tracks. *See also* chain-and-bucket elevator, inclined screw, or tubular conveyor (WEF & ASCE, 1991).

**Clara cell** A nonciliated cell of the lung that is easily damaged by chemicals.

**Claraetor** Equipment manufactured by Dorr-Oliver, Inc. combining aeration with a circular clarifier.

**ClarAtor** A clarifier design of Aero-Mod, Inc.

**ClariCone™** A solids contact clarifier manufactured by CBI Walker, Inc.

**clarification** Clearing action that occurs during water or wastewater treatment when solids settle out. This is often aided by centrifugal action and chemically induced coagulation in wastewater. Same as settling and sedimentation, which are now more commonly used.

**clarification failure** The excessive loss of solids to the effluent of a clarifier. *See* the following types of clarification failure: dispersed growth, filamentous bulking, pin floc, and viscous bulking.

**clarified wastewater** Settled wastewater; wastewater that has undergone sedimentation for the separation of settleable solids.

**clarifier** A quiescent, circular or rectangular basin/tank for the settling of suspended (settleable) solids by gravity and their subsequent removal as sludge, usually by a motor-driven rake mechanism. This mechanism used to be considered the difference between clarifier and settling tank but this distinction no longer applies and the two terms are used synonymously. *See* sedimentation basin/tank for more detail. *See also* upflow clarifier.

**clarifier effluent launder** A trough that discharges water or wastewater from a clarifier into the effluent piping. Water elevation in the launder is calculated from the equation for a side-overflow weir. *See also* clarifier hydraulics, diversion dam, diverting weir, effluent weir, skimming weir.

**clarifier/gravity thickener** A circular (or, rarely, rectangular) sedimentation basin incorporating a deepened central section that functions as a gravity thickener. *See,* e.g., Clarigester and Clarithickener®.

**clarifier hydraulics** The various water elevations between a clarifier and an effluent manhole, calculated on the basis of head losses from pressure flow, side-overflow weir, and V-notch weir equations. *See* Figure C-08 and hydraulic profile.

**clarifier overflow rate** *See* overflow rate.

**clarifier weir** The main device constituting the outlet of a sedimentation tank, designed to skim the clarified liquid from the surface and maintain deposited solids in the tank. Typically, it is a 90° V-notch weir with a loading (called weir overflow rate) of 10,000–30,000 gallons per foot per day. *See also* outlet weir and effluent weir.

**Clari-Float®** A package wastewater treatment plant manufactured by Tenco Hydro, Inc. using dissolved air flotation.

**Clarifloc®** A polyelectrolyte produced by Polypure, Inc. for use in solids separation processes.

**Clariflocculator** Equipment manufactured by Dorr-Oliver, Inc. combining clarification and flocculation in one unit.

**ClariFlow** Upflow clarification equipment manufactured by Walker Process Equipment Co.

**Clarigester** Equipment manufactured by Dorr-Oliver, Inc. providing clarification and solids digestion in a two-story tank.

**Clarion** A line of absorption media produced by Colloid Environmental Technologies Co.

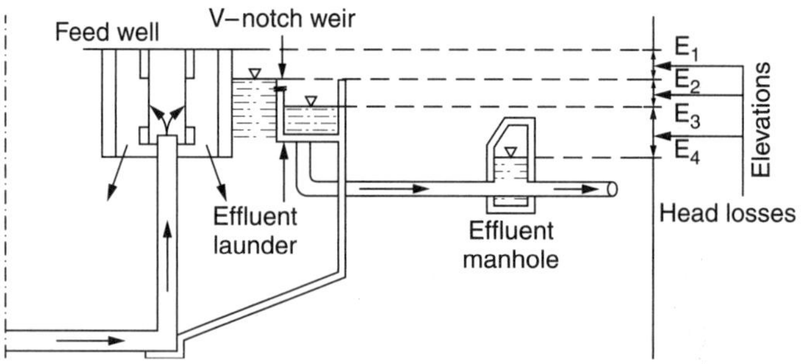


**Figure C-08.** Clarifier hydraulics (adapted from Water Environment Federation, 1991).

**Clar+Ion®** A line of cationic chemicals produced by General Chemical Corp. and used in water treatment for coagulation and flocculation.

**Claripak** An upflow clarifier manufactured by Aerators, Inc. using inclined plates.

**ClariShear™** A floating sludge collector manufactured by Techniflo Systems.

**ClariThickener™** Equipment manufactured by Eimco Process Equipment Co. combining clarification and sludge thickening in one unit.

**Clari-Trac®** A sludge removal apparatus designed by F. B. Leopold Co., Inc. as a siphon mounted on a track in rectangular tanks.

**clarity** State or quality of being clear; clearness.

**Clari-Vac®** A sludge removal apparatus manufactured by F. B. Leopold Co., Inc. as a floating-bridge siphon in rectangular tanks.

**Clar-i-vator®** A solids contact clarifier manufactured by Smith & Loveless, Inc.

**Clar-Vac** An air flotation system manufactured by Dontech, Inc.

**class (pipe and fittings)** The working pressure rating of a specific pipe for use in water distribution systems, including allowances for surges. This term is used for cast iron, ductile iron, asbestos cement, and some plastic pipes.

**class 1 (or I) hazardous waste injection well** *See* class I injection well.

**class 1 (or I) injection well** An EPA classification for injection wells used for the subsurface emplacement of hazardous, industrial, or municipal wastes beneath the lowermost formation that contains an underground source of drinking water within one-quarter (0.25) mile of the well. Wells used for the disposal of water treatment plant residuals are in class I.

**class 1 (or I) sludge management facility** Any publicly owned treatment works (POTW) identified as being required to have an approved pretreatment program (including such POTWs located in a state that has elected to assume local program responsibilities) and any other treatment works treating domestic sewage classified as a class I sludge management facility by the USEPA Regional Administrator, or, in the case of approved state programs, the Regional Administrator in conjunction with the State Director, because of the potential for its sludge use or disposal practices to adversely affect public health and the environment (EPA-40CFR122.2, or 501.2122.2, or 503.9-c).

**class 1 (or I) substance** Any substance so designated by the EPA, such as chlorofluorocarbons, halons, carbon tetrachloride, and methyl chloroform.

**class 1 (or I) treatment works** A USEPA classification, based on reliability and water use, for wastewater treatment facilities that can unacceptably or permanently damage navigable waters (e.g., drinking water sources, shellfish waters, or waters used for contact recreation) by an inadequate effluent discharge even for a few hours.

**class 2 (or II) injection well** A USEPA classification for injection wells used for the subsurface emplacement of nonhazardous fluids from the production and processing of oil and natural gas.

**class 2 (or II) substance** Any substance so designated by the USEPA, such as hydrochlorofluorocarbons.

**class 2 (or II) treatment works** A USEPA classification, based on reliability and water use, for wastewater treatment facilities that can unacceptably or permanently damage navigable waters (e.g., recreational waters) by the continued discharge of an inadequate effluent discharge for a few days but not by a short-term discharge.

**class 3 (or III) injection well** A USEPA classification for wells used in the injection of fluids in the extraction of such minerals as sulfur and metals, including uranium.

**class 3 (or III) treatment works** A USEPA classification, based on reliability and water use, for wastewater treatment facilities that are not designated as class I or class II treatment works.

**class 4 (or IV) injection well** A USEPA classification for injection wells used for the subsurface emplacement of hazardous or radioactive wastes into or above a formation that contains an underground source of drinking water within one-quarter (0.25) mile of the well.

**class 5 (or V) injection well** A USEPA classification for injection wells other than those defined under classes I through IV.

**class A biosolids** Solids or sludge from wastewater treatment that have been processed to reduce enteric viruses, pathogenic bacteria, and viable helminth ova below detectable levels. They have a fecal coliform density below 1000 MPN per gram of total dry solids or a *Salmonella* density less than 3 MPN per 4 grams of total dry solids. Alternatively, they can receive some specific heat or pH treatment. Class A biosolids can be used by the general public, in nurseries, gardens, and golf courses. Also called clean sludge. *See* process to further reduce pathogens (PFRP).

**class A ion** A metal cation characterized by spherical symmetry and low polarizability, with a hard electron sheath not easily deformed by electronic fields. Type A cations form complexes preferentially with fluorides and ligands that have oxygen as donor atom. They attract water more strongly than they attract ammonia or cyanide and they are less toxic than the other types of cations (Stumm and Morgan, 1996). They include: hydrogen ($H^+$), beryllium ($Be^{2+}$), lithium ($Li^+$), calcium ($Ca^{2+}$), aluminum ($Al^{3+}$), potassium ($K^+$), sodium ($Na^+$), magnesium ($Mg^{2+}$), silicon ($Si^{4+}$), strontium ($Sr^{2+}$), titanium ($Ti^{4+}$), scandium ($Sc^{3+}$), lanthanum ($La^{3+}$), thorium ($Th^{4+}$), and zirconium ($Zr^{4+}$). All macronutrient metals belong to this group. Also called A-type metal cation or hard acid. *See also* class B and transition type ions.

**class A sludge** *See* Class A biosolids.

**class B biosolids** Solids or sludge from wastewater treatment that have been processed to reduce enteric viruses, pathogenic bacteria, and viable helminth ova to levels unlikely to pose a threat to public health and the environment under specific use conditions (agricultural land application or landfill disposal). Such biosolids have a fecal coliform density less than $2.0 \times 10^6$ MPN or CFU per gram of total solids. Class B biosolids cannot be sold or given away for lawn or home garden applications. *See* process to significantly reduce pathogens (PSRP).

**class B ion** A metal cation characterized by a deformable electron sheath, with a complexation preference for bases containing iodine (I), nitrogen (N), or sulfur (S). They attract ammonia or cyanide more strongly than they attract water or the hydroxyl ion ($OH^-$) and they are more toxic than the other types of cations (Stumm and Morgan, 1996). They include: copper ($Cu^+$), silver ($Ag^+$), gold ($Au^+$), thallium ($Tl^+$), gallium ($Ga^+$), zinc ($Zn^{2+}$), cadmium ($Cd^{2+}$), mercury ($Hg^{2+}$), lead ($Pb^{2+}$), tin ($Sn^{2+}$), thallium ($Tl^{3+}$), gold ($Au^{3+}$), indium ($In^{3+}$), and bismuth ($Bi^{3+}$). Also called B-type metal cation or soft acid (except $Bi^{3+}$, $Pb^{2+}$, and $Zn^{2+}$). *See also* class A and transition type ions.

**class B sludge** *See* Class B biosolids.

**classical cholera** A pandemic waterborne disease caused by one of the two very similar biotypes of Vibrio cholerae. Classical cholera has a higher fatality rate than El Tor cholera; more than 60% when untreated.

**classical typhus** A water-washed disease caused by the pathogen *Rickettsia prowazeki* and transmitted by lice; also called epidemic typhus. *See* louse-borne typhus.

**classic sag curve** A plot of the dissolved oxygen concentration in a stream as a function of distance from a point of waste discharge, according to the Streeter–Phelps formulation.

**classic Streeter–Phelps formulation** *See* Streeter–Phelps formulation.

**classification** *See* biological classification, biological classification conventions, environmental classification, hydraulic classification.

**classifier** A device for separating elements according to such characteristics as size or density.

**clast** A constituent fragment of a rock formation, e.g., a grain of sand, sediment, silt, or gravel.

**clastic** Pertaining to sediments or rocks composed of fragments or particles of older rocks, or of fragments deposited by mechanical transport. *See also* bioclastic.

**Claus process** A method used for the production of elemental sulfur (S) from sulfur dioxide ($SO_2$) and hydrogen sulfide ($H_2S$), which can be released from some soils by heating them to 1100°C in a vacuum. The dioxide and sulfide are then passed through a heated catalyst bed (e.g., bauxite, $Al_2O_3$) at 320 °C:

$$SO_2 + 2\ H_2S \rightarrow 2\ H_2O + 3\ S \qquad \text{(C-95)}$$

**clay** An inorganic soil material, mainly hydrous aluminum silicate, with grains smaller than 0.005 mm in equivalent diameter, plastic when wet, rigid when dried, and vitrified when fired to high temperatures; used as coagulant aid in water treatment, as lining material in waste disposal, or for the fabrication of sewer pipes. Clay in the form of bentonite or kaolinite preparation may also be used as a weighting agent to enhance settling by supplementing the solids content of a suspension. A clay pipe or tile consists of clay baked in a kiln. Clays make poor aquifers, despite their high porosity; they have low permeability. They may constitute confining layers or leak water to other aquifers.

**clay liner** A layer of clay soil added to the bottom and sides of an earthen basin for use as a waste disposal site. *See also* bentonite layer.

**clay minerals** A group of hydrous aluminum silicate minerals that constitute the major portion of clay. Examples are stable secondary minerals such as illite, kaolinite, and montmorillonite resulting from the weathering of primary minerals.

**claypan** (1) A strongly compacted soil layer underlying soft materials and cemented with clay. *See* hardpan for other similar layers. (2) A shallow, dry depression that holds water after heavy rains.

**clay particle** One of the three size constituents of a porous medium, along with sand and silt. Clay particles, which have about the size of bacterial cells, are important in determining the characteristics of the medium, particularly surface area and average pore size.

**clay pipe** *See* clay.

**clay soil** (1) A mixture of clay, sand, and silt containing less than 40% of any these materials. *See also* soil classification. (2) A soil containing more than 40% clay, but less than 45% sand and less than 40% silt.

**CLB** Cyanobacteria-like body.

**$Cl_2C$:CCl** Chemical formula of perchloroethylene.

**$Cl_2C{=}CCl_2$** Chemical formula of tetrachloroethylene

**$Cl_2C:CH_2$** Chemical formula of 1,1-dichloroethylene.

**$Cl_3C{-}CH_3$** Chemical formula of trichloroethane.

**$Cl_3CCH(C_6H_4OCH_3)_2$** Chemical formula of methoxychlor.

**$Cl_2CF_2$** Chemical formula of dichlorodifluoromethane.

**$(ClC_6H_4)_2C:CCl_2$** Chemical formula of dichlorodiphenyl dichloroethylene.

**$(ClC_6H_4)_2CH(CCl_3)$** Chemical formula of dichlorodiphenyl trichloroethane.

**$(ClC_6H_4)_2CHCHCl_2$** Chemical formula of dichlorodiphenyl dichloroethane.

**$ClCH_2{-}CH_2Cl$** Chemical formula of 1,2-dichloroethane.

**$Cl_2C_6H_3OCH_2COOH$** Chemical formula of 2,4-dichlorophenoxyacetic acid.

**$Cl_2C_6H_3OH$** Chemical formula of 2,4-dichlorophenol.

**C–L diagram** *See* Caldwell–Lawrence diagram.

**Clean Air Act (CAA)** An act originally passed in 1963, revised in 1970, and amended in 1977 and 1990; it establishes air pollutant emission standards.

**Clean Air Act Amendments (CAAA)** The 1990 amendments of the Clean Air Act with further restrictions on air emissions.

**Clean-A-Matic** A self-cleaning basket strainer manufactured by GA Industries, Inc.

**Clean chemicals** A line of high-purity laboratory products of Hach Co.

**clean closure** The closure of a hazardous waste surface impoundment that includes decontaminating all waste residues, impoundment components, contaminated subsoil, and all structures or equipment contaminated with the waste or leachate.

**cleaner production** An approach to comprehensive control of industrial environmental hazards, e.g., the continuous application of an integrated preventive environmental strategy to processes, products, and services.

**cleaning in place** A method used to clean membrane modules (or other devices) in situ by direct application of a chemical solution, typically in a recirculatory mode. The cleaning system may include tanks, pumps, and other devices for storage, mixing, and recirculation.

**cleaning system** *See* cleaning in place.

**cleaning water** Process water used to clean the surface of an intermediate or final plastic product or to clean the surfaces of equipment used in plastics molding and forming that contact an intermediate or final plastic product. It includes water used in both the detergent wash and rinse cycles of a cleaning process (EPA-40CFR463.2-d).

**cleaning wye** A wye fitting with a long branch in a water main for the insertion of cleaning pigs.

**clean in place** *See* cleaning in place.

**clean-in-place system** *See* cleaning in place.

**cleanout** A structure or device providing access for inspection and cleaning purposes, e.g., a pipe fitting containing a removable plug for access to a pipe run.

**clean-out port** A piece of pipe installed for cleaning purposes at the end of a river diffuser outfall, with its opening just above the bottom.

**clean river** A stream that shows no apparent sign of pollution.

**Clean Shot** A pneumatic apparatus manufactured by Wheelabrator Engineered Systems, Inc. for solids delivery.

**clean sludge** Sludge from wastewater treatment that has been treated to reduce enteric viruses, pathogenic bacteria, and viable helminth ova below detectable levels. *See* Class A biosolids for detail.

**Clean Water Act (CWA)** The 1972 U.S. Federal Water Pollution Control Act regulating discharges into national waters; later renamed Clean Water Act in 1977; updated in 1987. Important sections concern the National Pollution Discharge Elimination System (NPDES), lists of pollutants and limits on their discharge, stormwater regulations, and best available technology requirements.

**clean-water reservoir** In a water supply system, it is a reservoir that holds treated water before it is distributed to customers. Also called finished-water reservoir. *See also* clear well.

**Clean Water Restoration Act** A U.S. law of 1966 authorizing a matching federal grant program of 30% for the construction of wastewater treatment facilities.

**clean water standard** Any enforceable limitation, control, condition, prohibition, standard, or other requirement that is established pursuant to the Clean Water Act (CWA) or contained in a permit issued to a discharger by the USEPA, or by a state under an approved program, or by a local government to ensure compliance with pretreatment regulations as required by section 307 of the CWA.

**clear acrylic** A generic name for the thermoplastic polymer Plexiglas®.

**clearance** The rate at which chemicals are cleared from the blood, expressed in milliliters of plasma cleared per unit time.

**Clearcon** Circular clarifier equipment manufactured by Vulcan Industries, Inc.

**Clearflo** Cylindrical clarifier equipment manufactured by Roberts Filter Manufacturing Co.

**Clear View™** An apparatus manufactured by Goal Line Environmental Technologies to monitor emissions continuously.

**clear-water basin or reservoir** In a water supply system, it is a basin or reservoir that holds treated water before it is distributed to customers. Also called finished-water basin or reservoir. *See also* clear well.

**clear-water iron** Same as ferrous iron.

**clear-water pump** A centrifugal pump used in a wastewater installation for such applications as flushing water for general cleaning, spray water to disperse foam, pre- and postchlorination, and water for chemical mixing.

**clear-water reservoir** *See* clear-water basin.

**clear well** A tank or reservoir of filtered water for filter backwashing; sometimes used as a chlorine contact chamber or simply for storage of filtered water of sufficient quantity to prevent the need to vary the filtration rate with variations in demand. *See also* clear-water basin or clear-water reservoir.

**cleated belt conveyor** Same as cleated conveyor.

**cleated conveyor** A device resembling a flat belt conveyor with a series of overlapping cleats or pockets on the carrying side to allow steep incline angles. It is used in wastewater treatment plants to move solid materials that cannot be easily pumped, e.g., grit, screenings, dewatered sludge cakes, and other solids. *See also* belt conveyor, screw conveyor.

***Clematis vitalba*** A perennial plant growing in hedgerows and woodland edges, whose presence is indicative of calcareous soil.

**cleptoparasite** A parasite that steals food caught by another species; also spelled kleptoparasite.

**$ClHC{=}CCl_2$** Chemical formula of trichloroethylene.

**ClHC:CHCl** Chemical formula of 1,2-dichloroethylene, *cis*-1,2-dichloroethylene, and *trans*-1,2-dichloroethylene.

**$Cl_2 \cdot 8\,H_2O$** Chemical formula of chlorine ice.

**$Cl_4H_8O_4$** Chemical formula of alizarin.

**cliff spring** A spring that occurs at the base of a cliff.

**climate** The composite of the weather conditions of a region, i.e., temperature, air pressure, precipitation, humidity, sunshine, cloudiness, and winds, averaged daily, monthly, seasonally, and yearly.

**climatic** Pertaining to climate; due to climate and not to soils or topography.

**climatic cycle** Recurrence of weather-related phenomena such as floods and droughts at regular intervals.

**climatic province** An area with the same general climatic characteristics.

**climatic variation** The gradual change in weather conditions of a region over long periods of time.

**climatology** The science that deals with the phenomena of climate or atmospheric conditions; often used interchangeably with meteorology.

**climax** The ultimate, stable, and self-perpetuating stage in the development of a plant and animal community, given a set of environmental conditions.

**Climber®** A bar screen of Infilco Degremont, Inc. in the form of a reciprocating rake.

**climber screen** Same as reciprocating rake screen.

**ClimbeRack™** A screening rack of Infilco Degremont, Inc. that does not require lubrication.

**cline** The gradual change in morphological or physiological characteristics exhibited by members of adjacent populations of the same species or by a species or other natural group along an environmental or geographic transition.

**clinical studies** Studies of human suffering from symptoms induced by chemical exposure.

**clinical wastes** Wastewaters generated at a clinic or hospital, which may contain human and animal tissue or excretions, drugs, medicinal products, etc. *See* hospital wastes.

**clinophlolite** Same as clinoptilolite.

**clinoptilolite** A natural zeolite (clay) that is used for ammonia removal by ion exchange; in water and wastewater treatment, it can exchange the ammonium ion ($NH_4^+$) for calcium, magnesium, or sodium ions. Lime [$Ca(OH)_2$], sodium chloride (NaCl), or sodium hydroxide (NaOH) are used to regenerate the ion exchange medium, producing ammonium hydroxide ($NH_4OH$) as waste.

**$ClO_2$** Chemical formula of chlorine dioxide.

**$ClO_4^-$** Chemical formula of perchlorate.

**cloaca maxima** The great sewer of Rome, built during the Roman Empire to drain the Forum and the valleys between the hills to the Tiber River in the ancient city, parts of which may still be in operation. A section of it is vaulted and paved with blocks of lava.

**clogging** Flow blockage due to particles or other causes in filters, pipes, or other water and wastewater units.

**clogging head loss** In water filtration design and operation, the gradual increase in head loss due to the removal of particulates from the influent and their accumulation on top of and within the filter bed.

**clonal expansion** An increase in the number of cells with the same genotype, sometimes induced by carcinogenic chemicals.

**clone** Any one cell or organism descended from and genetically identical to a single common ancestor.

**cloning** In biotechnology, obtaining a group of genetically identical cells from a single cell; making identical copies of a gene.

**clonorchiasis** A helminthic disease caused by the excreted worm *Clonorchis sinensis*. *See* its common name, Chinese liver fluke, for more detail.

***Clonorchis sinensis*** A parasitic worm that causes the Chinese liver fluke disease; latency = 6 weeks, persistence = life of intermediate host, infective dose relatively low; often associated with fish farming in excreta-enriched ponds. Also called *Opistorchis sinensis* because of its close similarity with *Opistorchis felineus* and *Opistorchis viverrini*.

**Cloromat** A package chlorine/caustic plant manufactured by Ionics, Inc.

**close-coupled pump** A pump that is connected directly to its motor by a coupling or clutch without any gearing, shafting, or belting. Also called direct-connected pump.

**closed centrifugal pump** A centrifugal pump whose impeller has vanes enclosed in disks.

**closed conduit** A pipe or other closed duct for conveying fluids.

**closed-conduit flow** Flow in a closed conduit, i.e., flow under pressure, as opposed to open-channel flow.

**closed-conduit system** A water or sewer system where all flows are under pressure, i.e., in closed conduits. *See also* manifolded system.

**closed cooling-water system** Any configuration of equipment in which any heat is transferred by circulating water that is contained within the equipment and not discharged to the air; chilled water loops are included (EPA-40CFR749.68-3).

**closed-cycle concept** An approach to industrial production that uses maximum recycling and minimizes water consumption as well as effluent generation.

**closed-cycle cooling system** A cooling system that transfers heat by water recirculation within the system with little blowdown of solids.

**closed fire line** An unmetered connection to the distribution system for fire protection without a hydrant.

**closed impeller** An impeller with sidelines extended to the vane tips.

**closed-loop reactor** *See* CLR.

**closed-loop recycle** A reuse system that treats wastewater from a process stream and returns it to the same process along with some makeup water. *See also* water reuse applications.

**closed-loop stripping analysis** An extraction procedure used to separate volatile organics by bubbling gas into a sample and recycling the head space through an adsorbent trap.

**closed system** A water distribution system that provides pressure through pumps, without elevated storage.

**closing dike** A structure built across a branch of a river to reduce flow to the river.

***Clostridia*** Plural of *Clostridium*.

***Clostridium*** A genus of anaerobic, rod-shaped, Gram-positive, nitrogen-fixing bacteria that form spores to survive in adverse conditions; found in soil and in the intestinal tract of animals. They are sometimes used as indicators of fecal pollution. *See also Bacillus* and *Clostridium spp.*

***Clostridium acetobutylicum*** A species of obligate anaerobic, acidophilic bacteria that ferment sugar.

***Clostridium botulinum*** The anaerobic microorganism that causes botulism, a sometimes fatal disease contracted from spoiled food.

***Clostridium pasteurianum*** A species of anaerobic nitrogen-fixing bacteria active in unmanured grassland.

***Clostridium perfringens* (*C. Perfringens*)** A species of Gram-positive, rod-shaped bacteria exclusively of fecal origin; the predominant anaerobic spore-forming bacteria in water and a prominent member of the intestinal flora: one gram of human feces contains approximately 1600 such organisms. It causes wound infections and food poisoning. It is used, albeit on a limited basis, as a fecal contamination indicator because of its ability to form spores and to tolerate small concentrations of oxygen. The resistant and long-lasting characteristics of its spores favor its use as an indicator of past pollution. Chlorination and granular filtration can remove about 90% of its cells from water. *See also* sulfite-reducing *Clostridium*, anaerobic sporeformer.

***Clostridium spp.*** A group of nonmethanogenic bacteria responsible for hydrolysis and fermentation in anaerobic digestion.

***Clostridium tetani*** The species of bacteria that cause tetanus. *See* tetanus bacillus for detail.

**Cloth-Media Disk Filter® (CMDF)** A surface filtration device consisting of a number of disks mounted vertically in a tank, with a polyester-needle felt cloth or a synthetic-pile fiber cloth. Feedwater enters the tank and flows through the cloth into a central collection tube or header. Backwashing is accomplished by flow reversal. *See also* surface filtration, Discfilter® (Metcalf & Eddy, 2003).

**cloud** A visible mass of water droplets that are too small to fall to the earth and remain suspended at a considerable distance. A cloudburst is a very intense rainstorm of short duration over a small area; a storm of an intensity of 4 inches/hour (100 mm/hr) or 10 times the intensity of a heavy storm as defined by the U.S. Weather Bureau. Cloud detection radars are weather radars that detect clouds instead of precipitation. Cloud seeding is the introduction of silver iodide crystals, dry ice (solid carbon dioxide), sea salt or other chemicals into clouds to induce rain.

**cloudburst** *See* cloud.

**cloud detection radar** *See* cloud.

**cloud seeding** *See* cloud.

**CLR** Acronym of closed-loop reactor. The CLR Process is a CLR oxidation ditch process developed by Lakeside Equipment Corp.

**CLR Process** *See* CLR.

**club soda** Soda water to which salts have been added.

**clumping** The formation of small, close groups of resin or media grains in such treatment units as ion-exchange or filter beds, as a result of organic fouling or electrostatic charges.

**cluster sewer system** A system that provides sewer service to more than one structure, from two to hundreds of structures, e.g., when the extension of existing systems is impractical or traditional onsite treatment systems are difficult to locate on each lot. It incorporates a wide range of methods for collecting, treating, and discharging wastewater, e.g., from large-scale septic systems to sand filters to membrane bioreactors. Also called community sewer system, shared sewer system.

**CMAS** Acronym of complete-mix activated sludge.

**CMAS process** *See* complete-mix activated sludge process.

**CM basin** Complete mixing basin.

**CMB reactor** *See* completely mixed batch reactor.

**CMC** (1) Carboxymethylcellulose; a white, water-soluble polymer produced from cellulose and used in paper making, in foods, in textile-finishing mills, and in general-purpose synthetic detergents. Also called sodium cellulose glycolate. (2) Acronym of criterion maximum concentration.

**CMDF** Acronym of Cloth-Media Disc Filter®.

**CMF** Acronym of coagulation–microfiltration. *See also* iron CMF process.

**CMF reactor** *See* completely mixed flow reactor.

***C. muris*** Abbreviation of *Cryptosporidium muris.*

**CN** Chemical formula of the cyanide radical.

**CNCl** Chemical formula of cyanogen chloride.

**C/N ratio** *See* carbon/nitrogen ratio.

**$CO_2$** Chemical formula of carbon dioxide.

**$CO_2$ acidity** The base neutralizing capacity of an aqueous solution due to the presence of carbon dioxide ($CO_2$) and undissociated carbonic acid ($H_2CO_3$), the sum of which is represented by ($H_2CO_3^*$). Carbon dioxide acidity ($CO_2$ Acy) is calculated as:

$$[CO_2\,Acy] = [H_2CO_3^*] + [H^+] - [CO_3^{2-}] - [OH^-] \quad \text{(C-96)}$$

where $[H^+]$, $[CO_3^{2-}]$, and $[OH^-]$ represent, respectively, the concentrations of hydrogen ion, carbonates, and hydroxyl ions.

**Co-60** Abbreviation or symbol of cobalt 60.

**coaction** The interaction of living organisms upon each other.

**Coagblender** A line of turbine-like mixers manufactured by Aerators, Inc. for installation in-line or in open channels.

**coagulant** A chemical that causes very fine particles to destabilize and clump together into larger particles. This makes it easier to separate the solids from water by settling, skimming, draining, or filtering. Common coagulants used in water and wastewater treatment include alum, chlorinated copperas, ferric chloride, ferric sulfate, ferrous sulfate and lime, polyaluminum chloride, and sodium aluminate. *See also* flocculant.

**coagulant aid** Any chemical or substance, usually a patented polyelectrolyte, used to assist or modify coagulation by providing nuclei for floc formation. Some common coagulant aids are oxidizing agents (e.g., chlorine, ozone) that reduce the interference of organic compounds, and weighting agents (e.g., activated carbon, activated silica, bentonite clay, limestone) that increase particle density as well as adsorptive surface. *See also* chitosan, sodium alginate.

**coagulant carryover** The carryover of some coagulant from the settling basin to a gravity filter to assist in the flocculation and removal of microscopic particulate matter that would otherwise pass through the filter.

**coagulant/flocculant aid** *See* coagulant aid, flocculant aid.

**coagulant recovery** A method used in sludge handling to extract aluminum or iron coagulants from the waste stream and minimize the quantity of waste for final disposal. Acidification is used to put the metal back into solution. Recalcination is used to dewater the lime residual and burn it for the production of quicklime (CaO).

**coagulant sludge** *See* coagulation sludge.

**coagulate** To cause colloidal particles to combine or aggregate into larger masses or clumps by addition of an electrolyte or by physical means.

**coagulated–settled water** Water that has been treated by coagulation and sedimentation before filtration.

**coagulation** Clumping of particles in water or wastewater to settle out impurities, often induced by chemicals such as lime, alum, and iron salts. The chemicals neutralize the electrical charges of the fine particles and cause destabilization and initial aggregation of the particles. This clumping together makes it easier to separate the solids from the water by settling, skimming, draining, or filtering. Coagulation with metallic salts involves a complex series of reactions (dissolution, hydrolysis, and polymerization) that result in the formation of large insoluble particles. For aluminum sulfate [$Al_2(SO_4)_3$], the overall reaction produces aluminum hydroxide [$Al(OH)_3$]:

$$Al_2(SO_4)_3 + 6\ H_2O \rightarrow 2\ Al(OH)_3 + 3\ H_2SO_4 \qquad \text{(C-97)}$$

Temperature, pH, and alkalinity affect the process. Chemical coagulation is used in wastewater treatment to increase the removal of suspended solids, phosphorus, and algal cells in stabilization pond effluents, and for sludge conditioning. Coagulation may also result from a biological process. *See* Schultze–Hardy rule. *See also* coagulation and flocculation for the difference between the two processes.

**coagulation aid** *See* coagulant aid.

**coagulation basin** A basin in which a liquid is mixed gently, with or without chemical addition, to induce the agglomeration of particles and facilitate their subsequent settling.

**coagulation/filtration plant** A traditional form of water treatment combining coagulation and filtration to reduce turbidity, pathogens, color, taste, and odor. It usually includes sedimentation and other unit operations and processes.

**coagulation–microfiltration (CMF)** A treatment process that consists of the addition of a coagulant such as ferric chloride ($FeCl_3$) to water followed by microfiltration; it is used for arsenic removal—the arsenate [As(V)] ion is adsorbed on the particles of ferric hydroxide [$Fe(OH)_3$] that are formed. Chlorine can be used to oxidize any arsenite [As(III)] present to arsenate. The microfiltration step does not require flocculation. *See also* iron CMF process, transmembrane pressure.

**coagulation region** An area where a specific coagulation mechanism predominates on a diagram of the process as a function of pH and coagulant dose or turbidity and coagulant dose. Coagulation mechanisms include adsorption–destabilization, bridging, double-layer compression, restabilization, and sweep-floc coagulation. Also called coagulation zone. *See* Figure C-09.

**coagulation sludge** The semisolid residuals produced by the treatment of water or wastewater using such coagulants as aluminum or iron salts. In addition to the microbial, organic, and inorganic contaminants from the liquid, coagulation sludges contain the metallic or polymeric products used, with a total solids content of about 9%. The quantity of sludge produced ($S$, pounds per day) depends on the flow ($Q$, mgd), the dose of alum ($A$ in mg/L as 17.1% $Al_2O_3$) or iron ($F$ in mg/L as Fe), the concentration of suspended solids ($X$, mg/L), and the concentration of additional solids from polymers or activated carbon ($Y$, mg/L):

$$S = 8.34\ Q(0.44\ A + X + Y) \qquad \text{(C-98)}$$

$$S = 8.34\ Q(2.90\ F + X + Y) \qquad \text{(C-99)}$$

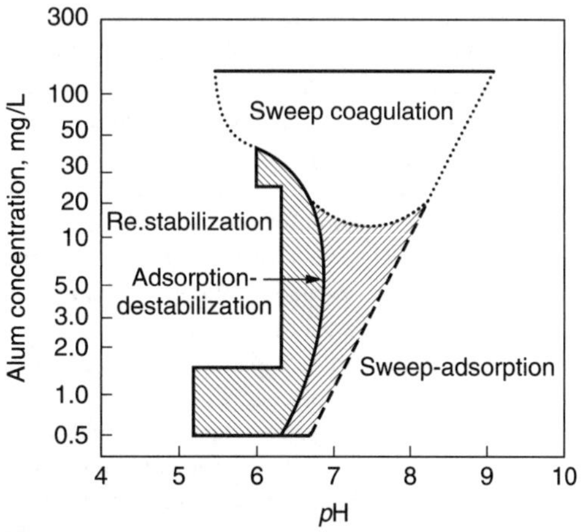


**Figure C-09.** Coagulation regions (adapted from Bryant et al., 1992).

**coagulation zone** Same as coagulation region.

**Coakley's equation** A modification of the Blake–Kozeny equation for application to the filtration of waste solids, proposed by Peter Coakley in 1965 (Fair et al., 1971):

$$t/V_w = SV_w + B \quad \text{(C-100)}$$

where: $t$ = time, and $V_w$ = volume of filtrate collected in time $t$ after cake formation.

$$S = (\mu I_s C_s)/2A^2\,\Delta p \quad \text{(C-101)}$$

$$B = \mu I_m/(A\Delta p) \quad \text{(C-102)}$$

where $\mu$ = absolute viscosity of displaced water.

$$I_s = I/\gamma_s = \text{specific resistance of the sludge solids} \quad \text{(C-103)}$$

$C_s$ = initial dry-weight concentration of sludge solids, $\Delta p$ = pressure difference in a sludge cake of thickness $L$, $I_m$ = initial resistance of a unit area of filtering surface, $I$ = resistance of a unit volume of sludge cake, and $\gamma_s$ = specific weight of sludge solids.

Given a set of pairs of values of time ($t$) and filtrate volume ($V_w$), the straight-line plot of the logarithm of ($t/V_w$) versus the logarithm of ($V_w$) allows the determination of $B$ as the intercept at $V_w$ = 1 and of $S$ as the slope of the line.

**coal** A solid, carbon-based material used as a filter medium, alone (anthracite) or in combination with aluminum. Coal is also used in activated carbons. Pulverized coal is used as a sludge conditioning agent. *See also* anthracite coal.

**coalesce** To grow together to form one mass. *See* coalescence.

**coalescence** The action of two or more droplets, colloidal particles, or the like that merge to form a larger unit as a result of molecular attraction. In water and wastewater treatment, particles that coalesce, as a result of coagulation, can be more easily removed by sedimentation.

**coal naphta** Another name of benzene.

**coal pile runoff** Runoff from or through a coal storage pile; it is subject to an NPDES stormwater discharge permit, including effluent limitations of 50 mg/L TSS and a pH range of 6.0–9.0.

**coal-preparation waste** Wastewater associated with the production of coal, through such operations as crushing, screening, classifying, and washing to remove impurities and separate coal solids smaller than 3.25 inches. The principal constituent of coal-washing water is suspended solids; the wastewater may also contain calcium sulfate, magnesium sulfate, and iron. Treatment methods used for coal-preparation wastes include combinations of settling and impounding basins, froth flotation to remove the ultrafines, and recovery of fine coal.

**coal-processing waste** *See* coal preparation waste and acid mine-drainage waste.

**coal–sand dual-media filter** A water filtration device consisting of a relatively coarse anthracite layer (effective size 0.9–1.0 mm, uniformity coefficient 1.5–1.7, and specific gravity 1.4–1.6) over a finer sand layer (0.45–0.55 mm, uniformity coefficient 1.7, and specific gravity about 2.65).

**coal tar** A thick, black, viscous liquid derived from the distillation of bituminous coal, used in waterproofing and pipecoating.

**coal tar naphta** Another name of benzene.

**Coanda effect** The tendency of a liquid coming out of a nozzle or orifice to travel close to the wall contour even if the wall curves away from the jet's axis.

**Coanda, Henri Marie** French engineer and inventor (1885–1972).

**coarse aggregate** An aggregate measuring 6 to 18 mm.

**coarse-bed filtration** A water pretreatment process located ahead of depth filtration and consisting in the partial removal of flocculated particles.

**coarse-bubble aeration** Water or wastewater aeration with diffusers that produce relatively large bubbles. *See also* fine-bubble aeration.

**coarse-bubble diffuser** An old classification of devices used in the diffused aeration of wastewater according to the size of the bubbles produced. The current tendency is to divide diffusers into porous (or fine-pore) diffusers, nonporous diffusers, and other diffusers.

**coarse fish** Fish, like mullet, that can tolerate low oxygen levels. *See also* tolerant fish species, game fish.

**coarse-grained filter** A trickling filter or a water filter of such coarse material as broken stone, gravel, and slag, instead of sand.

**coarse rack** A rack with more than one inch or 2.5 centimeters between bars.

**coarse sand** Sediment particles with an equivalent diameter between 0.5 and 1.0 millimeter. *See also* sand and soil classification.

**coarse screen** A screening device usually having openings greater than 25 mm or one inch. Same as bar rack. *See also* coarse rack.

**coarse-screen head loss** The head loss through coarse screens ($H_L$):

$$H_L = (V^2 - v^2)/(2gC) \quad \text{(C-104)}$$

where the headloss is expressed in meters and $V$ = flow velocity through the openings of the screen, m/s; $v$ = approach velocity in the upstream channel, m/s; $g$ = gravitational acceleration = 9.81 $m/s^2$; and $C$ = an empirical discharge coefficient varying from 0.7 from a clean screen to 0.6 for a clogged screen.

**coarse screening** A wastewater unit operation for the removal of coarse solids (e.g., rags and sticks) and other debris by surface straining. The device used is a bar rack or coarse screen.

**coarse screenings** Materials removed from wastewater by coarse screening and consisting of rags, sticks, leaves, food particles, bones, plastics, bottle caps, rocks, cigarette butts, etc. Coarse screenings from a separate sewer system average 3.5–35 L/1000 $m^3$, 10–20% solids content, 600–1100 $kg/m^3$, and 70–95% volatile solids content.

**coarse solids reduction** The interception of coarse solids by such units as comminutors and macerators that also grind or shred them, and return them to the flow stream for removal in downstream treatment units. This step, when used, is an alternative to coarse screening.

**CoAsS** Chemical formula of cobaltite.

**coastal** Pertaining to any body of water landward of the territorial seas or any wetlands adjacent to such waters (EPA-40CFR435.41-e).

**coastal high-hazard area** An area subject to high-velocity waters, including hurricane wave wash or tsunamis.

**coastal waters** (1) Waters of the coastal zone. (2) Under the Coastal Zone Management Act, coastal waters include (a) in the Great Lakes area, the waters within the territorial jurisdiction of the United States, consisting of the Great Lakes, their connecting waters, harbors, roadsteads, and estuary-type areas such as bays, shallows and marshes; and (b) in other areas, those waters adjacent to the shorelines, which contain a measurable quantity or percentage of seawater, including sounds, bays, lagoons, bayouts, ponds, and estuaries.

**coastal zone** Lands and waters adjacent to the coast that exert an influence on the uses of the sea and its ecology, or whose uses and ecology are affected by the sea.

**coating** A layer of paint or other material applied to the inside or outside surface of a pipe, valve, etc., to protect it against corrosion.

**cobalt (Co)** A hard, silvery-white, magnetic, brittle, metallic element that occurs in combination with arsenic and sulfur and is used as an alloy, in electroplating, and other applications. Atomic weight = 58.93. Atomic number = 27. Specific gravity = 8.9. Valence = 2 or 3. It is an essential nutrient for development but, at sufficiently high concentrations (e.g., above 1.0 mg/kg of body weight), cobalt may interfere with enzyme production and cause physiological effects in humans. Cobalt in trace amounts promotes methanogenesis during anaerobic digestion. Manufacturers commonly add cobalt (as cobalt chloride) as a catalyst to accelerate the deoxygenation of water during the testing of aeration equipment.

**cobalt-60** A radioactive, artificial isotope of cobalt that provides a source of gamma rays; half-life = 5.25 years. It is used as a tracer and can also be used for sludge irradiation.

**cobaltite (CoAsS)** A major ore mineral of cobalt; essentially cobalt arsenic sulfide, found in hydrothermal veins.

**cobaltous chloride ($CoCl_2$)** One of two chemicals used to dissolve in a mixture to approximate the color of natural waters. The other is potassium chloroplatinate.

**cobaltous hydroxide ($Co_2O_3 \cdot 3\ H_2O$)** A rose-red, amorphous powder used to prepare cobalt salts and paints.

**cobalt potassium nitrite** *See* potassium cobaltinitrite.

**cobalt thiocyanate active substance (CTAS)** A classification of nonionic surfactants used in analytical methods. These surfactants react with CTAS to yield a cobalt-containing product that can be extracted and measured. *See also* MBAS.

**cobalt yellow** A common name for potassium cobaltinitrite or cobalt potassium nitrite.

**cocci** (1) Plural of coccus; bacteria of circular or spherical shape. Other shapes include curved (vibrio), rod (bacillus), and spiral (spirillum). Cocci survive desiccation better than rods and spirals. (2) A common name of coccidioidomycosis.

**Coccidia** Plural of coccidium; an order of sporozoans.

**coccidian cyst or parasite** An intestinal sporozoan that can reproduce sexually or asexually; generally found in larger numbers in anaerobically digested sludge than in raw wastewater.

***Coccidioides*** A genus of fungi, generally called *Coccidia,* including the pathogen of coccidioidomycosis.

***Coccidioides immitis*** The species of fungi responsible for coccidioidomycosis.

**coccidioidomycosis** An acute or progressively chronic respiratory infection caused by inhaling spores of *Coccidioides immitis* fungi; characterized by fever and reddish bumps on the skin; common in hot, semiarid regions. Also called cocci, desert fever, desert rheumatism, San Joaquin Valley fever, valley fever.

**coccidiosis** An infectious disease caused by protozoan parasites that affect the intestines of birds, domestic animals, and dogs.

**coccus** *See* cocci.

**cock** A hand-operated device, such as a valve or faucet, for regulating flow in a pipe; also called stopcock.

**cocktail** A mixture concocted of various substances; e.g., an organic solution used in the analysis of radioisotopes in water, i.e., a scintillation cocktail.

**cocktail effect** The combined effects of two or more toxic chemicals taken together, whereby one ingredient makes the body more sensitive to another.

**$CoCl_2$** Chemical formula of cobaltous chloride.

**COCODAF®** A proprietary device of Thames Water and Paterson Candy Ltd (Isleworth, United Kingdom) that provides countercurrent flotation in an upflow floc-blanket clarifier.

**cocomposting** The combined, static or agitated, composting of sludge and municipal solid waste (MSW) or segregated yard waste, in a 2 to 1 mixture of MSW/sludge.

**coconut-based activated carbon** Activated carbon produced from coconut shells and used for adsorption in water or wastewater treatment.

**cocurrent** In the same direction as water flows. *See also* countercurrent.

**cocurrent centrifuge** One of two basic configurations of solid-bowl centrifuges, whereby the solids and the liquid move parallel to each other over the length of the bowl. *See also* countercurrent centrifuge (WEF & ASCE, 1991).

**cocurrent flow** The downflow pattern in the ion-exchange process; the flow pattern of a separation process in which all contacting elements flow in the same direction, e.g., liquid and gas, water and activated carbon. *See also* countercurrent flow.

**cocurrent operation** The operation of two or more components of a system in the same direction, e.g., in heat exchange or other transfer processes. *See also* countercurrent operation.

**cocurrent packed tower** A device used to strip volatile organic compounds, carbon dioxide, and other gases from water or wastewater. It consists of a packed tower, with both fresh air and the liquid introduced at the top of the tower, flowing through the packing, and exiting at the bottom. *See also* cascade air stripping, cascade packed tower, crossflow packed tower, end effect, low-profile air stripper, minimum air-to-water ratio, packed tower design equation, sieve tray column, wall effect (AWWA, 1999).

**cocurrent settler** An inclined plate sedimentation tank in which the suspension is fed above the inclined surfaces and flows down through the channels; settled solids flow in the same direction as the liquid.

**cocurrent settling** The mode of operation of a cocurrent settler. The condition for particle removal is (AWWA, 1971):

$$V \geq WV'/(L \cos \theta - W \sin \theta) \qquad \text{(C-105)}$$

where $V$ = settling velocity of particles removed, $W$ = perpendicular distance between surfaces, $V'$ = liquid velocity between the inclined surfaces, $L$ = required length of surface, and $\theta$ = the angle of surface inclination from the horizontal. *See also* Figure C-10, countercurrent settling, crossflow settling.

**COD** Acronym of chemical oxygen demand.

**COD analysis** *See* COD fractionation, COD test.

**CodeLine™** A pressure vessel manufactured by Advanced Structures, Inc. to house membranes.

**Code of Federal Regulations (CFR)** A publication of the U.S. government that contains the rules issued by the USEPA and other federal agencies.

**codisposal** The joint disposal of two or more types of waste, e.g., the codisposal of septage and solid wastes as landfilling or composting or the codisposal of wastewater residuals with municipal solid wastes.

**codisposal of sludge** (1) The joint disposal of sludges generated from two separate processes; e.g., the joint disposal of sludges from water treatment and wastewater treatment, either by discharging the water treatment sludge into sanitary sewers or transporting it to the wastewater treatment plant. (2) Joint disposal of water or wastewater treatment sludges with municipal refuse or other organically derived wastes such as lawn clip-

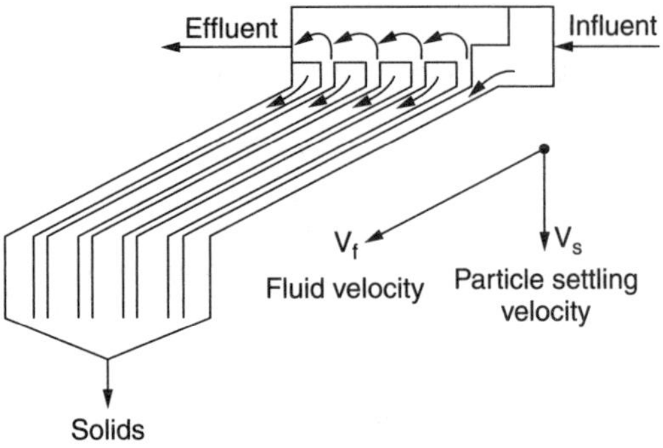


**Figure C-10.** Cocurrent settling.

pings and leaves, using composting, incineration, or landfilling.

**COD fractionation** The division of the total chemical oxygen demand of wastewater into various forms of carbonaceous constituents, according to whether they are biodegradable or nonbiodegradable, readily biodegradable or slowly biodegradable, and soluble or particulate (Figure C-11). This analysis is used in the design of activated sludge and other biological wastewater treatment processes. Other carbonaceous constituents include total, soluble, and ultimate BOD. Wastewater also contains nonbiodegradable volatile suspended solids and inert suspended solids. *See* Table C-02.

**COD:N:P** The ratio of the concentrations of oxygen demand, nitrogen, and phosphorus concentrations, often used to determine the nutrient requirements in biological treatment.

**codon** A sequence of three adjacent nucleotides in DNA or RNA that specifies a particular amino acid.

**COD reflux unit** The instrument used in the dichromate reflux method for measuring the chemical oxygen demand of a wastewater sample. It consists of an Erlenmeyer flask fitted with a condenser and a hot plate.

**COD test** A laboratory test conducted to measure the chemical oxygen demand, i.e., the oxygen equivalent of the organic matter in wastewater that can be oxidized chemically by dichromate in an acid solution. The test is conducted by boiling for two hours a mixture of the sample and potassium dichromate, acidified with sulfuric acid ($H_2SO_4$). After cooling, the remaining dichromate ($Cr_2O_7^{2-}$) is measured by titration with ferrous ammonium sulfate [$FeSO_4 \cdot (NH_4)_2SO_4 \cdot 6H_2O$]. COD values are usually higher than BODs, because they include materials that are only slowly or not at all biodegradable. The reaction for cell biomass ($C_5H_7O_2N$), for example, is:

$$C_5H_7O_2N + 10/3\ Cr_2O_7^{2-} + 83/3\ H^+ \rightarrow 5\ CO_2 + 46/3\ H_2O + NH_4^+ + 20/3\ Cr^{3+} \quad \text{(C-106)}$$

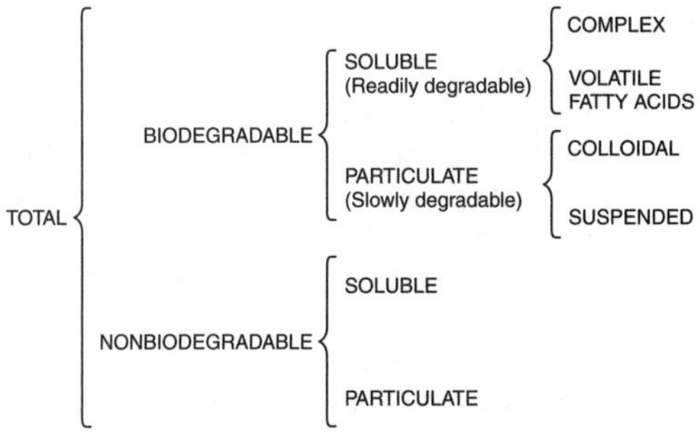


**Figure C-11.** COD fractionation.

**Table C-02.** COD fractionation

| Constituent | Abbreviation |
|---|---|
| Chemical oxygen demand (Total) | COD |
| Soluble COD | sCOD |
| Particulate COD | pCOD |
| Biodegradable COD | bCOD |
| Biodegradable soluble COD | bsCOD |
| Biodegradable particulate COD | bpCOD |
| Readily biodegradable COD | rbCOD |
| Slowly biodegradable COD | sbCOD |
| Nonbiodegradable COD | nbCOD |
| Nonbiodegradable soluble COD | nbsCOD |
| Nonbiodegradable particulate COD | nbpCOD |
| Complex soluble COD | csCOD |
| Volatile fatty acids | VFA |
| Nonbiodegradable volatile suspended solids | nbVSS |

**COD test interferences** Substances that can interfere with the COD test include ammonia, hydrogen peroxide, and any compounds that are oxidized by dichromate.

**coefficient** A quantity, usually represented in formulas or equations by a letter, but determined analytically or empirically.

**coefficient of absorption** *See* absorption coefficient.

**coefficient of area** *See* area coefficient.

**coefficient of attenuation** *See* attenuation coefficient.

**coefficient of axial dispersion** *See* axial dispersion coefficient.

**coefficient of compressibility** The relative decrease ($C_{co}$) of the volume ($V$) of a system with increasing pressure ($P$) in an isothermal process:

$$C_{co} = -(1/V)(\partial V/\partial P) \quad \text{(C-107)}$$

Also called compressibility. *See* bulk modulus, sludge compressibility.

**coefficient of deposition** *See* deposition coefficient.

**coefficient of determination** A dimensionless parameter used in regression analysis to measure the degree of correlation between two sets of data. It approaches 1 for a high degree of correlation, but approaches 0 when the data are not correlated. For example, in hydraulic modeling, for a set of $N$ observations ($O_i$ through $O_N$) and $N$ simulated values

($S_i$ through $S_N$), the coefficient of determination ($r$) is

$$r = D_1/[(D_2) \cdot (D_3)]^{0.5} \quad \text{(C-108)}$$

where

$$D_1 = N\Sigma O_i S_i - \Sigma O_i \Sigma S_i \quad \text{(C-109)}$$

$$D_2 = N\Sigma O_i^2 - (\Sigma O_i)^2 \quad \text{(C-110)}$$

$$D_3 = N\Sigma S_i^2 - (\Sigma S_i)^2 \quad \text{(C-111)}$$

and $\Sigma$ indicates the summation from $i = 1$ through $i = N$. The correlation coefficient is the square ($r^2$) of the coefficient of determination.

**coefficient of diffusion** *See* diffusion coefficient.

**coefficient of drag** *See* drag coefficient.

**coefficient of extinction** *See* extinction coefficient.

**coefficient of fineness** The ratio of suspended solids to turbidity of a water or wastewater sample, an indication of the size of particles causing turbidity.

**coefficient of gas transfer** *See* gas transfer coefficient.

**coefficient of kurtosis** *See* kurtosis coefficient.

**coefficient of molecular diffusion** *See* molecular diffusion coefficient.

**coefficient of nonuniformity** *See* nonuniformity coefficient.

**coefficient of permeability** *See* permeability coefficient.

**coefficient of regime** The ratio of the maximum daily flow to the minimum daily flow in a given year.

**coefficient of reliability (COR)** A dimensionless parameter used to determine a mean value ($m_d$) for a variable in the design of water or wastewater treatment processes, and determined as follows:

$$\text{COR} = [(V_x^2 + 1)^{1/2}]\exp\{-Z_1 - \alpha[\ln (V_x^2 + 1)^{1/2}\} \quad \text{(C-112)}$$

$$V_x = \sigma_x/m_x \quad \text{(C-113)}$$

where $V_x$ = coefficient of variation of the existing distribution, $\sigma_x$ = standard deviation of performance values from an existing treatment process, $m_x$ = mean of performance values from an existing treatment process, $Z_{1-\alpha}$ = number of standard deviations away from the mean of a normal distribution, and $1 - \alpha$ = cumulative probability of occurrence (reliability test). *See also* COR method.

**coefficient of retardance** *See* retardance coefficient.

**coefficient of rigidity** One of two constants that describe a Bingham plastic and a characteristic of sludge. It is a function of the percent solids in the sludge and is expressed in pounds per foot per second (or in kilograms per meter per second). It is related to the dimensionless Reynolds number ($R_e$):

$$R_e = \rho VD/\eta = \gamma VD/\eta \quad \text{(C-114)}$$

Where $\rho$ = sludge density, kg/m$^3$; $V$ = average velocity, m/sec or fps; $D$ = pipe diameter, m or ft; $\eta$ = coefficient of rigidity, kg/m/sec or lb/ft/sec; and $\gamma$ = specific weight of sludge, lb/ft$^3$. *See also* Hedstrom number, yield stress.

**coefficient of self-purification** *See* self-purification coefficient.

**coefficient of skewness** *See* skewness coefficient.

**coefficient of storage** *See* storage coefficient.

**coefficient of tension** The relative increase ($C_{te}$) of the pressure ($P$) of a system with increasing temperature ($T$) under conditions of constant volume:

$$C_{te} = (1/P)(\partial P/\partial T) \quad \text{(C-115)}$$

**coefficient of thermal conduction** A measure of the ability of a substance or material to conduct heat; the rate of heat transfer by conduction through a material or substance of unit thickness per unit area and per unit temperature gradient. *See* thermal conductivity for detail.

**coefficient of thermal expansion** The relative increase ($C_{th}$) of the volume ($V$) of a system or substance with increasing temperature ($T$) under conditions of constant pressure:

$$C_{th} = (1/V)(\partial V/\partial T) \quad \text{(C-116)}$$

**coefficient of transmissivity** *See* transmissivity coefficient.

**coefficient of uniformity** *See* uniformity coefficient.

**coefficient of viscosity (μ)** A measure of the internal resistance of a fluid to flow; it is equal to the ratio of the viscous shearing stress ($\tau$) to the velocity gradient ($\partial V/\partial s$). Also called absolute viscosity or dynamic viscosity.

**Coex Seal™** A containment liner manufactured by National Seal Co.

**coffee processing waste** Wastewater generated during the pulping and fermenting of coffee beans. It is high in BOD and suspended solids. Common treatment method includes screening, sedimentation, and trickling filtration. The effluent can be reused in irrigation.

**cofferdam** A temporary damlike structure built around a site to exclude water and provide access. It may consist of single or double sheet piling, or an earth embankment.

**cofiring** Burning of two fuels in the same combustion unit, e.g., coal and natural gas, oil and coal, or the thermal destruction of hazardous waste with conventional fuels in a boiler.

**CogBridge** A sludge collector manufactured by Walker Process Equipment Co. in the form of a traveling bridge.

**cogen** Abbreviation of cogeneration.

**cogeneration** The use of an energy source for more than one purpose; e.g., the simultaneous production of electricity with natural gas and the use of the thermal energy from the waste heat. Also, the generation of electricity from alternative fuels, the waste heat of solid waste incineration, or of industrial processes. Cogeneration is sometimes practiced with sludge digestion by producing thermal and mechanical energy from digester gas.

**Cog Rake** A bar screen manufactured by the FMC Corp. as a reciprocating rake.

**Cohen censored maximum likelihood method** A statistical method used to determine the parameters of a normal distribution for a sample that is biased toward the left or the right.

**cohesion** Molecular attraction that holds two particles together. *See also* adhesion.

**cohesive soil** Soil with a high proportion of clay; plastic when moist and hard to break when dry.

**COHNS** An acronym used sometimes to represent the major elements of organic waste: carbon (C), oxygen (O), hydrogen (H), nitrogen (N), and sulfur (S). *See also* cell tissue.

**cohort** (1) A group of individuals that share a particular statistical or demographic characteristic (e.g., age, birth period, exposure to a chemical). (2) An individual in a population of the same species.

**cohort epidemiologic study** An epidemiological study of individuals selected for their exposure to a given factor.

**cohort study** An epidemiological study that observes subjects in differently exposed groups and compares the incidence of symptoms. Although ordinarily prospective in nature, such a study is sometimes carried out retrospectively, using historical data. *See also* prospective study.

**Coilfilter** A rotary vacuum belt filter manufactured by Komline-Sanderson Engineering Corp. The filtering medium is made up of two layers of alloy steel, coiled endless springs.

**coincident maximum loading** The loading on a water or wastewater treatment unit when both the hydraulic load (or flow) and the organic or solids concentration peak simultaneously, a condition that is not typical because of storage and other factors (Metcalf & Eddy, 2003). For example, the maximum total suspended solids concentration to a dewatering unit may be twice the average concentration while the actual peak solids loading is only 1.5 times the average loading.

**coincineration** (1) The incineration of wastewater sludges with municipal solid wastes to reduce the combined cost, in a ratio of 1 pound of dry wastewater solids to 4.6 pounds of solid wastes without heat recovery, or a ratio of 1 pound of dry wastewater solids to 7 pounds of solid wastes with heat recovery. (2) Burning of industrial wastes as hazardous waste fuel in industrial furnaces and boilers for energy recovery, or in wastewater sludge incinerators (Metcalf & Eddy, 2003 and Manahan, 1997).

**co-ion** Any one of a group of ions or particles having the same charge. *See also* counterion, Donnan potential, electrical double layer.

**coke** The solid carbon residue resulting from the distillation of coal or petroleum; it can be burnt to generate carbon dioxide for use in water treatment plants. It is also used as a catalyst to promote the oxidation of Mn (II) in iron and manganese removal by aeration.

**coke-tray aerator** An aerator in which water passes or is sprayed over coked-filled trays.

**Colby's method** A graphical method proposed by B. R. Colby in 1964 to determine the total bed-material discharge in a sand-bed stream.

**cold and gel trap** An adsorbent bed used in the cryogenic air separation process to remove the final traces of carbon dioxide ($CO_2$) and hydrocarbons from the feed air.

**cold lime–soda softening** A water treatment process that uses lime and soda ash at ambient temperature. *See also* lime–soda ash softening.

**Cold Plasma®** An apparatus manufactured by Cimco Lewis Ozone Systems, Inc. for water disinfection using ozone.

**cold spring** A nonthermal spring whose water temperature is significantly lower than the ambient temperature; any nonthermal spring in an area of thermal springs. Cold springs are often fed by melted snow and ice in mountainous regions.

**cold sterilization** Disinfection of a fluid using a submicrometer-pore filter.

**cold vapor** A method for testing water for the presence of mercury.

**cold water effect** The upflow rate required to expand a sand or anthracite filter decreases with temperature, because at colder temperatures water is more dense and has a higher viscosity than at warmer temperatures. Filter operation must, therefore, be adjusted accordingly; for example, from an upflow rate of 13.5 gpm/ft$^2$ at 70°F to 7.6 gpm/ft$^2$ at 40°F (Bryant, 1992).

**cold weather operation** The adjustment of plant operations (e.g., chemical doses, detention times) to take cold weather conditions into consideration.

**cold windbox** A fluid-bed furnace in which air is supplied at ambient temperature; air cools the orifice plate to prevent its excessive expansion. *See also* hot windbox.

**coli-aerogenes bacteria** A group of bacteria found mostly in the intestines of humans and warm-blooded animals.

**coliform** A group of bacteria found in the intestines of warm-blooded animals (including humans), plants, soil, air, and water. Fecal coliforms are a specific class of bacteria that only inhabit the intestines of warm-blooded animals. The presence of coliform in water is an indication that the water is polluted and may contain pathogenic organisms, although the coliforms themselves are normally nonpathogenic. Coliforms are Gram-negative, aerobic and facultatively anaerobic, nonsporeforming bacteria that produce acid and gas in the fermentation of lactose at 35–37°C. They are readily identified and enumerated in the laboratory. *See also* fecal coliform, total coliform, presumptive indicator. Also called coliform bacillus, colon bacillus.

**coliform bacillus** *See* coliform.

**coliform bacteria** Rod-shaped bacteria living in the intestines of humans and warm-blooded animals and used as an indication of fecal contamination. *See* coliform.

**coliform count** The number of presumptive coliform bacteria in a 100-mL sample of water or wastewater. The presence of coliforms is indicative of fecal contamination. Drinking water should have a coliform count of zero. Higher counts are allowable for other uses.

**coliform density** The number or concentration of coliform organisms in a volume of water, commonly 100 milliliters. *See* coliform count.

**coliform detection methods** Methods used to identify coliforms in water. *See* most probable number, membrane filtration technique, presence–absence test.

**coliform group** The coliform group includes not only *E. coli*, but also *Citrobacter, Enterobacter aerogenes,* and *Klebsiella pneumoniae,* commonly associated with the intestinal tracts of warm-blooded animals and found in their fecal material. It does not include such pathogens as *S. typhi.*

**coliform group bacteria** *See* coliform.

**coliform growth response (CGR)** A test developed to determine the potential of a sample of treated water to support the growth of coliform bacteria, using a strain of *Enterobacter cloacae* as the bioassay organism. The test measures changes in viable densities of this organism over five days at 20°C, which are used to calculate the coliform growth response index (CGR), an indication of nutrient availability for coliform growth (Maier et al., 2000):

$$\text{CGR} = \log (N_5/N_0) \qquad \text{(C-117)}$$

where $N_5$ = number of colony-forming units (CFU) per milliliter at day 5 and $N_0$ = number of CFU/mL initially (day 0). *See also* assimilable organic carbon, coliform regrowth potential.

**coliform growth response index** *See* coliform growth response.

**coliform index** A rating of the purity of water based on a count of fecal bacteria.

**coliform organisms** Microorganisms found in the intestinal tract of humans and animals. Their presence in water indicates fecal pollution and potentially adverse contamination by pathogens.

**coliform regrowth potential** The potential of a drinking water sample to develop coliform bacteria. Coliform group bacteria can regrow in natural surface and drinking water distribution systems, depending on the amount and type of organic matter present and on water temperature. *See* the coliform growth response test.

**Coliform Rule** A set of sampling requirements established by the USEPA in 1989 for the sampling of water supply systems, based on the population they serve; e.g., a minimum of 100 samples per month is required for a system serving a population of 100,000. *See* Total Coliform Rule for detail.

**coliform standard** The maximum contaminant level goal for coliform bacteria is zero, with an allowance for inadvertent sample contamination by a dirty faucet or from the hands of the person collecting the sample.

**coliform/streptococci ratio** A parameter once proposed, arguably, to determine whether contamination of a surface water is predominantly from human or animal feces, based on the observation that in the United States, human feces contain a coliform/streptococci ratio larger than 4.0, whereas animal feces have a coliform/streptococci ratio lower than 0.7.

**coliform test** A laboratory test conducted to detect the presence of coliform bacteria in a sample of water, which may indicate possible fecal contamination. Depending on the testing method used, the results are reported in most probable numbers (MPN) of organisms per 100 ml or as coliform organisms per 100 ml. *See also* the serial-dilution and membrane filter techniques.

**coliform test deficiencies** As a water quality indicator, the coliform test (a) does not account for the regrowth of microorganisms in the water distribution system or in an aquatic environment, (b) does not account for the suppression by high background bacterial growth, (c) is not indicative of a health threat, and (d) is not applicable for enteric viruses and protozoan parasites, which have greater resistance to disinfectants.

**Colilert®** A reagent produced by IDEXX Laboratories, Inc. for the identification of total coliforms and *E. coli.*

**Colilert test** An inexpensive, reliable, commercial method for the simultaneous detection of total coliforms and *E. coli*; a special substrate containing dyes for the specific organisms is added to a water sample. The hydrated medium is first colorless, but then changes color if the target organisms are present. *See also* presence–absence test.

**coliphage** A bacteriophage (bacterial virus) that is specific to the host *E. coli*; a potential indicator of the presence of pathogens and of the effectiveness of water treatment. Coliphages are common in wastewater and treated effluents. *See* somatic phage and male-specific (F-specific) phage.

**ColiSure** A medium produced by Millipore and used in the presence–absence test for coliforms.

**Coliwasa** *See* composite liquid-waste sampler.

**collagen** An extracellular protein abundant in the bone, skin, teeth, tendon, etc. of higher animals, forming insoluble fibers and connecting tissue between cells. It yields gelatin upon boiling. Ascorbic acid (vitamin C) is essential for its synthesis.

**collapse pulsing** A condition observed in filter backwashing, characterized by the formation and collapse of air pockets within the filter bed and resulting from the combination of air and subfluidization water flow. This condition develops when (AWWA, 1991)

$$0.45\,Q^2 + 100\,V/V' = 41.9 \qquad \text{(C-118)}$$

where $Q$ = airflow rate in standard cubic feet per minute per square foot and $V/V'$ = ratio of superficial water velocity to the minimum fluidization velocity based on the $d_{60}$ grain size of the medium.

**collapse-pulsing backwash** A common and efficient filter backwash method, developed in the 1970s and 1980s, using simultaneous air and subfluidization of water; it provides more abrasion than full fluidization for the removal of retained particles. *See also* subcritical fluidization backwash.

**collar** (1) The upper rim of a borehole, shot hole, or mine shaft. (2) A rim around a hydraulic structure to prevent seepage. (3) A concrete, watertight joint between cast-iron and vitrified-clay pipe sections. (4) A short sleeve used to join two smaller pipe sections.

**Collectaire** An airlift apparatus manufactured by the FMC Corp. for the removal of activated sludge.

**collecting agent** An additive that promotes the flotation of particles by increasing their interfacial tension with air while decreasing their interfacial tension with the liquid. Also called a collector or collecting reagent. Common collectors include fatty acids, soaps, or other surface-active agents (e.g., some alkyl compounds). The combination of collecting agents with activators and depressants enhances their effects. However, collecting and foaming agents may counteract each other.

**collecting reagent** Same as collecting agent.

**collection main** (1) A public sewer receiving flows from building services or individual systems. (2) A main sewer in a collection system, i.e., receiving flows from branch and submain sewers; also called a trunk sewer.

**collection system** In wastewater or stormwater management, a system of conduits and pumping stations that receive and convey flows to a treatment plant or a disposal facility.

**collector** (1) Same as collecting agent. (2) Same as collector sewer. (3) A device for collecting backwash water from a filter or an ion exchange bed. (4) A device used as an upper distributor of water in a downflow column. (5) In packed-bed filtration, collectors are previously deposited particles that form a micropacked bed that serves to intercept and remove other particles that enter the cake. (6) A freshwater organism that feeds on finely divided organic matter that it collects by filtering and gathering mechanisms. Also called fine particulate detritovore. *See also* shredder, gouger, grazer, piercer.

**collector chain** A chain used to convey the sludge scraper in a rectangular sludge collector.

**collector sewers** Pipes used to carry wastewater from individual sources to an interceptor sewer that will carry it to a treatment plant.

**collector well** (1) A vertical vault connecting a series of horizontal shafts for the collection of groundwater near a river. (2) A well near a surface water supply used to lower the water table and induce infiltration of surface water to the well.

**College of American Pathologists (CAP) grade water** Water that meets the clinical, cell, and cultural standards of the College of American Pathologists.

**collimated-beam bioassay** A technique that uses a collimated beam in a small reactor to develop dose–response curves and determine the ultraviolet (UV) light dose required to inactivate microorganisms in a sample. The collimated-beam device consists of a column for adjusting the lines of sight of the instrument and a low-pressure, low-intensity UV lamp connected to a power supply. *See also* Beer's law, transmittance.

**collimated light** Electromagnetic radiation with parallel waves.

**Collins–Selleck model** Based on batch experiments conducted in the early 1970s, the reduction of coliform organisms in a chlorinated primary effluent is plotted as a linear relationship on log-log paper, according to the following equation:

$$S = N/N_0 = 1/(1 + 0.23\ Ct)^3 \qquad \text{(C-119)}$$

where $S$ = mean effluent survival ratio of organisms, $N$ = number of organisms remaining after disinfection at time $t$, $N_0$ = initial number of organisms (before disinfection), $C$ = chlorine residual remaining at time $t$, and $t$ = contact time, min. *See* Figure C-12, chlorine disinfection models, disinfection kinetics, modified Collins–Selleck model (Metcalf & Eddy, 2003 and WEF & ASCE, 1991).

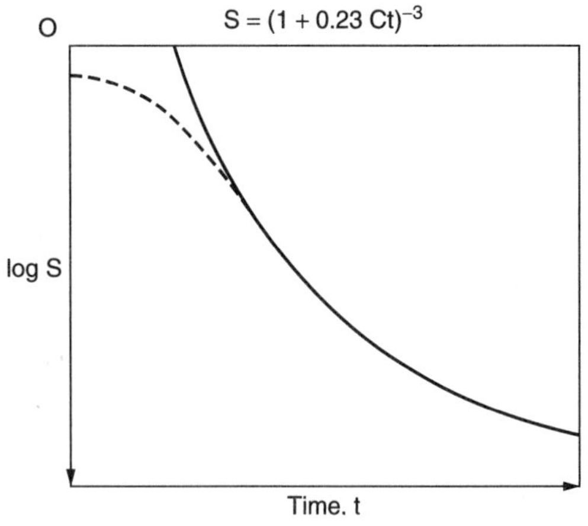


**Figure C-12.** Collins–Selleck model; C = concentration.

**collision efficiency** A dimensionless factor, ranging from 0 to slightly more than 1.0, that indicates the fraction of successful collisions among destabilized colloidal or suspended particles, i.e., the fraction of collisions that lead to actual agglomeration in the coagulation process; defined as follows for fluid motion (Stumm & Morgan, 1996 and AWWA, 1999):

$$\alpha_a = \text{(rate at which particles attach)/(rate at which particles collide)} \qquad \text{(C-120)}$$

More specific equations proposed for orthokinetic flocculation ($\alpha_o$) and for equal-size particles in turbulent flow ($\alpha_t$) are:

$$\alpha_o = 0.61[A/\pi\mu d^3(du/dz)]^{0.18} \qquad \text{(C-121)}$$

$$\alpha_t = 0.61\ [A/\pi\mu d^3(\varepsilon/\nu)^{0.5}]^{0.18} \qquad \text{(C-122)}$$

where $A$ = the Hamaker constant, in the range of $10^{-20} - 10^{-19}$; $d$ = diameter of the interacting particles; $du/dz$ = magnitude of the velocity gradient; $\varepsilon$ = local rate of turbulent energy dissipation per unit mass of fluid; and $\nu$ = kinematic viscosity. Also called flocculation rate correction factor. *See also* attachment probability, colloid stability, flocculation concepts.

**collision efficiency factor** *See* collision efficiency.

**Collision Scrubber™** A scrubber manufactured by Monsanto Enviro-Chem Systems, Inc. for the control of air pollution.

**collocation** The practice of locating seawater reverse osmosis (SWRO) facilities on the same site as water-cooled coastal power plants. The SWRO plant uses as feed a portion of the power plant's cooling water and discharges its concentrate into the power plant cooling water outflow, thereby reducing substantially the cost of desalination and diluting the brine before discharge to the ocean (Alspach & Watson, 2004).

**colloid** *See* colloids.

**Colloidair Separator™** A dissolved air flotation apparatus manufactured by the U.S. Filter Corp. for installation in open basins.

**colloidal dispersion** A mixture resembling a true solution but containing one or more substances that are finely divided but large enough to prevent passage through a semipermeable membrane. It consists of particles that are larger than molecules, settle out very slowly with time, scatter a beam of light, and are too small for resolution with an ordinary light microscope (EPA-40CFR796.1840-i); same as colloidal suspension.

**colloidal gel** A material with pores small enough to retain the dispersed particles of a colloidal suspension but large enough to allow the dispersion medium to pass through; often used as an ultrafilter.

**colloidal material** Very fine particles of solids, gases, or immiscible liquids suspended in water.

**colloidal matter** Same as colloids.

**colloidal particle** *See* colloids.

**colloidal solids** Same as colloids.

**colloidal solution** Small solids that are uniformly dispersed in solution; also called sol or colloidal suspension. *See* colloidal dispersion.

**colloidal stability** The characteristic of solid particles due to their surface charges that causes them to repel each other instead of coalescing and settling out of a suspension. Coagulation promotes destabilization of colloidal particles, whereas flocculation promotes agglomeration of destabilized particles). *See* colloid stability and colloid stability ratio.

**colloidal state** Colloids are characterized by their large specific surface (thousands of square meters per cubic centimeter) and their consequent adsorptive properties and tendency to develop electrical charges with respect to their surrounding medium. *See* the following terms: actinide colloid, association colloid, attachment probability, charge density, charge neutralization, co-ion, collision efficiency, colloid stability ratio, colloidal dispersion, colloidal stability, constant capacitance model, counterions, destabilization, diffuse layer, DLVO theory, electrical double layer, electrokinetic property, electroosmosis, electrophoresis, electrophoretic mobility, electrostatic stabilization, energy hill, Guoy–Chapman model, Guoy–Stern model, hydrophilic particle, hydrophobic colloid, interparticle bridging, isomorphic replacement, isomorphous replacement, Nernst potential, plane of shear, point of zero charge, point of zero net proton charge, polymer bridging, potential energy, rigid layer, Shultz–Hardy rule, steric interaction, steric stabilization, Stern layer, Stern potential, sticking coefficient, surface charge, surface complexation model, surface of shear, surface potential, van der Waals forces, zeta potential.

**colloidal suspension** A system of small (colloidal) particles dispersed by molecular motion in a surrounding medium; same as colloidal dispersion. *See also* mechanical suspension.

**colloiders** Trays or other contact units installed in the settling compartment of two-story septic tanks.

**colloidor** A tray or other surface installed in a wastewater treatment unit to facilitate the coagulation and settling of colloidal particles.

**colloids (colloidal matter, colloidal particles, colloidal solids)** Finely divided, discrete solids that do not dissolve and will not settle by gravity, but remain dispersed in a liquid for a long time due to their very small size (between 1 and 1000 millimicrons), their very large surface/volume ratio, and negative electrical charge, which prevents them from clumping together and settling out. Coagulation promotes the destabilization of colloids and flocculation promotes the agglomeration of destabilized particles, thereby facilitating their settling. *See* association colloid, colloidal dispersion, colloidal gel, dissolved solids, Guoy–Stern model, hydrophilic colloid, hydrophobic colloid, imbibition, lyophilic colloid, lyophobic colloid, solids, suspended solids, Tyndall effect.

**colloids critical mass** The mass of colloidal particles necessary and sufficient for their effective coagulation. The use of colloids as coagulant aids speeds up and improves the process.

**colloid stability** The property of colloidal particles that prevents them from joining into larger particles that can flocculate and precipitate. It can be characterized by the rate of particle aggregation, which is a function of the mass transport coefficient and a sticking coefficient called collision efficiency and Stumm & Morgan, 1996). It can also be determined experimentally (colloid stability ratio). *See* hydrophilic colloid, hydrophobic colloid, zeta potential.

**colloid stability ratio (*W*)** The reciprocal of the collision efficiency ($\alpha_a$), when this parameter is measured experimentally under conditions of a constant mass transport coefficient (Stumm & Morgan, 1996):

$$W = 1/\alpha_a \qquad \text{(C-123)}$$

**colmatage** The reduction in the flux of water through a membrane caused by the reversible accumulation of fouling agents.

**colocidal efficiency** The concentration of aqueous or free available chlorine required to destroy 99% of *E. Coli* in 30 minutes at 2–5°C. It varies from about 0.005 mg/L at pH = 6 to about 0.4 mg/L at pH = 11.

**colon bacillus** *See* coliform.

**colonization** (1) The spread of a species into a new habitat. (2) The multiplication of a microorganism after attachment to host tissues or other surfaces.

**colony** (1) A discrete clump of microorganisms growing together in a nutrient medium as the descendants of a single cell. (2) A group of organisms of the same kind living in close association.

**colony counter** An instrument used in the standard plate count test to count bacterial colonies.

**colony-forming unit (CFU or cfu)** A term used to report the concentration of heterotrophic bacteria present in a sample; the number of visible bacteria colony units present is counted in a laboratory plate count test and the result is expressed as CFU/mL. The report also indicates the method used (e.g., membrane filter), the incubation temperature, and the medium.

**color** An objectionable characteristic of drinking water resulting from the presence of colloidal or dissolved materials (e.g., humic and fulvic acids); it is measured by visual comparison with laboratory-prepared standards. Color in source water is usually caused by iron, manganese, humus, peat, plankton, and weeds. Color of natural waters may also come from wastewater discharges from textile and paper mills, tanneries, slaughterhouses, and other industries. *See* apparent color and true color. Adsorption on activated carbon or activated alumina is an effective color removal method.

**color comparator** An instrument used to determine such water parameters as chlorine residual and pH, based on the color of a sample compared to standards.

**colorimeter** A photoelectric instrument that measures the amount of light of a specific wavelength absorbed by a solution. It is used in several water and wastewater analyses, e.g., the determination of iron and manganese. *See* spectrophotometer.

**colorimetric analysis** A laboratory method used to determine the quantity of a specific constituent in a sample by measuring the amount of light of a specific wavelength absorbed by the sample. *See also* absorbance, Beer–Lambert law, spectrophotometer, atomic absorption spectroscopy, transmittance.

**colorimetric indicator** A chemical that changes color at a specific pH value and is used to determine end points of titrations.

**colorimetric measurement** A means of measuring unknown chemical concentrations in water by measuring a sample's color intensity. The specific color of the sample, developed by addition of a chemical reagent, is measured with a photoelectric colorimeter or is compared with "color standards" using, or corresponding with, known concentrations of the chemical.

**colorimetry** The measurement of color in terms of the wavelength of light absorbed.

**Color-Katch™** A chemical produced by Kem-Tron for use as a coagulant or flocculant in water treatment.

**color removal** The elimination, reduction, or transformation of the natural organic substances causing color in water by adsorption, oxidation, or physical-chemical processes, e.g., activated carbon adsorption, chlorination, coagulation, ozonation. Slow sand filtration alone removes only 30% of natural color.

**color throw** Color contributed to the effluent of an ion exchanger or filter by a component of the system. Color throw occurs after an extended standing period, which allows slowly soluble colored matter to accumulate in the water. It may result from the leaching of color bodies from an ion-exchange resin into the water.

**color unit (cu)** The unit used to report the results of analyses of water for color. The color of a sample of water is determined by comparing it to standards prepared from solutions of potassium chloroplatinate ($K_2PtCl_6$), cobaltous chloride ($CoCl_2 \cdot 6 H_2O$), and hydrochloric acid (HCl) in distilled water. For example, 1246 grams of chloroplatinate, one gram of cobaltous chloride, and 100 milliliters of the acid in 1000 milliliters of distilled water make up a solution having 500 color units.

**colour** British spelling of color.

**ColOx™** A biological reactor unit manufactured by TETRA Technologies, Inc. for the aerobic treatment of wastewater using a fixed film.

**column** A vertical, typically cylindrical vessel, with an inlet and an outlet used in water or wastewater filtration, adsorption, and other treatment processes.

**column settling test** An analysis conducted in a 1- or 2-liter graduated cylinder to determine the theoretical surface overflow rate of a clarifier to treat a given flocculent suspension; the percent reduction in suspended solids is measured for samples drawn along the column and plotted versus time (McGhee, 1991 and Droste, 1997). *See* Figure C-13.

**column utilization** In the adsorption of a substance from water or wastewater, column utilization is the ratio of the mass of a substance adsorbed at breakthrough to the mass adsorbed at complete saturation. Other parameters related to mass transfer and breakthrough in a packed bed reactor include: breakthrough concentration, critical depth, empty-bed contact time, bed volumes, breakthrough capacity, and carbon usage rate (AWWA, 1991).

**Combi-Guard** A compact screening apparatus manufactured by Andritz Sprout-Bauer S.A.

**combination acid pickling** Under the Clean Water Act, those operations in which steel products are immersed in solutions of more than one acid to chemically remove scale and oxides, and those rinsing steps associated with such immersions (EPA-40CFR420.91-c).

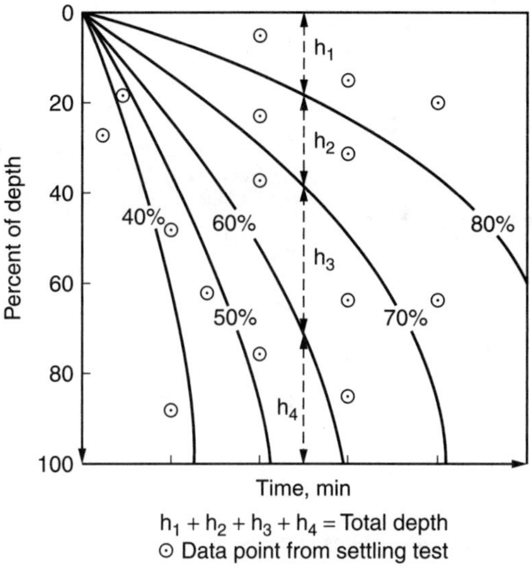


**Figure C-13.** Column settling test (adapted from McGhee, 1991).

**combination bond** A bond backed by the credit of a government but payable from the revenue of a water/wastewater or other utility.

**combination furnace** A device that combines a multiple-hearth dryer for drying dewatered sludge and a fluid-bed furnace for autogenous sludge combustion. The use of hot flue gases from the fluid-bed furnace minimizes the need for auxiliary fuel. Also called multiple-hearth fluid-bed furnace.

**combination well** An open well connected to an infiltration gallery or to another well.

**combined available chlorine** The concentration of chlorine and ammonia as chloramines still available for disinfection and the oxidation of organic matter. It is usually the sum of monochloramine ($NH_2Cl$) and dichloramine ($NHCl_2$), as trichloramine or nitrogen trichloride ($NCl_3$) is insignificant. As a germicidal agent, combined available chlorine is more stable but much less efficient than free available chlorine. Note that dichloramine and trichloramine may produce unpleasant taste and odors even at low concentrations, which can be controlled by keeping pH below certain limits.

**combined available residual chlorine** The concentration of residual chlorine that is combined with ammonia ($NH_3$) and/or organic nitrogen in water as a chloramine (or other chloro derivative), yet is still available to oxidize organic matter and utilize its bactericidal properties.

**combined biological process** A wastewater treatment method that uses more than one process, in one or more tanks, e.g., combined filtration–aeration, combined aerobic–anaerobic–anoxic, or combined fixed film/suspended growth.

**combined chlorine** The sum of the concentrations of the three chloramine species, usually expressed in mg/L as $Cl_2$: monochloramine ($NH_2Cl$), dichloramine ($NHCl_2$), and trichloramine or nitrogen trichloride ($NCl_3$). They result from the combination of hypochlorous acid (HOCl) with ammonia ($NH_3$). *See also* free chlorine, total chlorine.

**combined chlorine residual** The combined concentration of the three forms of chloramines produced during chlorination of water containing ammonia. *See also* free chlorine residual, total chlorine residual.

**combined filtration–aeration process** A method of biological wastewater treatment that combines a trickling filter or biological tower with an activated sludge unit. Various names are used, sometimes confusingly, to designate the corresponding flowsheet variations of Figure C-14 (see, e.g., Hammer et al., 1996; McGhee, 1991; Metcalf & Eddy, 2003; and Water Environment Federation and American Society of Civil Engineers, 1991). This combination is better able to absorb hydraulic or organic shock loads than conventional activated sludge and produces a sludge with improved settleability compared to the trickling filter sludge. In the simple roughing filtration–plain aeration (RF–PA) process, a trickling filter with direct recirculation functions as a roughing filter ahead of a plain aeration basin, which is followed by a final clarifier and usually has a reduced size. If settled biological floc is returned from the final clarifier to the aeration basin, the combination may be called a roughing filtration–activated sludge (RF–AS) process. Sometimes existing activated sludge plants are upgraded to a series trickling filtration–activated sludge (STF–AS) process by adding a trickling filter and intermediate clarifier with solids recycled ahead of the aeration basin. Four other combinations result from the variation of the aeration period (normal aeration = NA and decreased aeration = DA) and the point of return of the recycled sludge (return to filter = RF and return to aeration basin = RB). The four flowsheets, which all include a trickling filter, an aeration basin, and a final clarifier, are called and defined as follows. (a) Activated biological filtration (ABF) with DA and RF; also called activated biofiltration. (b) Biofiltration–activated sludge (BF–AS) with NA and RF. (c) Trickling filtra-

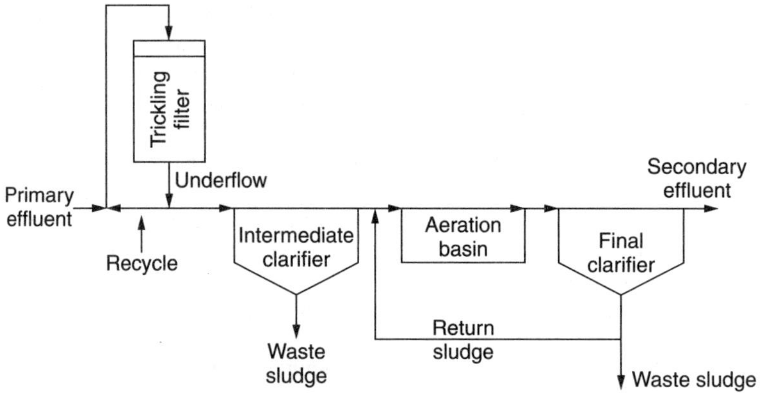


**Figure C-14.** Combined filtration–aeration.

tion–solids contact (TF–SC) with DA and RB; sometimes including a reaeration unit for the return sludge and a flocculating well in the clarifier. (Note the difference with the roughing filter activated sludge, which includes direct recirculation around the filter). (d) Trickling filtration–activated sludge (TF–AS) with NA and RB. Combined filtration–aeration is also called combined biological, combined trickling filter–activated sludge, coupled trickling filter – activated sludge, dual process, etc. The combined processes most used currently are the ABF, BF–AS, RF–AS, TF–AS, and TF–SC. *See also* dual process.

**combined fixed-film/suspended-growth process** A method of biological wastewater treatment that combines the fixed film and suspended growth mechanisms in one reactor by placing inert support media in an activated sludge basin. This arrangement reportedly results in increased biomass, reduced sludge production, and more efficient oxygen transfer (WEF & ASCE, 1991). Current designs are mostly proprietary. Some media used include plastic particles of about 8 mm in diameter, sheets of plastic modules, and even activated carbon for the removal of toxic organics. *See also* contact aeration. Also called a hybrid process.

**combined lamp aging fouling (CAF) index** A parameter that indicates the extent of aging and fouling of ultraviolet (UV) lamps, expressed as a percentage of the expected intensity of new lamps ($S_0$):

$$\mathrm{CAF} = 100\, S/S_0 \qquad \text{(C-124)}$$

where $S$ is the measured UV intensity.

**combined loading rate** The hydraulic loading rate ($L_{wn}$, m/yr) used in the design of slow-rate land treatment systems and based on a combination of water balance and nitrogen balance equations (WEF & ASCE, 1991):

$$L_{wn} = [C_p(P - E) + U/10]/[(1 - F)C_n - C_p] \qquad \text{(C-125)}$$

where $C_p$ = concentration of nitrate nitrogen in percolate, mg/L; $P$ = precipitation, m/yr; $E$ = crop evapotranspiration rate, m/yr; $U$ = uptake of nitrogen by crop, kg/ha/yr; $F$ = fraction of applied nitrogen lost by denitrification, volatilization, and soil storage; and $C_n$ = concentration of nitrogen in wastewater, mg/L. *See also* hydraulic loading rate, nitrogen loading rate, total percolation, limiting loading rate.

**combined moisture** Moisture that is combined with organic or inorganic matter.

**combined nitrification** Nitrification that occurs in the same reactor as that for the satisfaction of carbonaceous oxygen demand, with a BOD/TKN ratio greater than 5.0, instead of a separate nitrification that follows secondary treatment, with a BOD/TKN ratio less than 3.0. Also called single-stage nitrification. *See also* separate nitrification, separate-stage nitrification).

**combined oxidants disinfection** The use of more than one oxidant to inactivate pathogens in water or wastewater. *See also* simultaneous disinfection, sequential disinfection.

**combined process** A wastewater treatment method that incorporates two different processes; also called a dual process. Most plants commonly using dual processes combine a fixed film unit with a suspended growth reactor. *See also* combined filtration–aeration process and dual biological treatment system.

**combined procurement approach** The combination of architect/engineer and turnkey procurement approaches to contract for the design and construction of desired facilities.

**combined residual** Chloramine, the combination of chlorine with ammonia. In general, the combination of an additive with something else.

**combined residual chlorination** Application of chlorine to water or wastewater to produce combined available residual chlorine. The residual can be made up of monochloramine, dichloramine, and nitrogen trichloride, which result from the combination of chlorine with natural or added ammonia or with certain organic nitrogen compounds.

**combined sewer** A sewer designed to carry both wastewater and stormwater or surface water. Combined sewers sometimes overflow during a heavy storm, by design and at selected locations, when the combined flow exceeds the capacity of the sewers. The excess is frequently discharged directly to a receiving stream without treatment, or to a holding basin for subsequent treatment and disposal. *See also* interceptor sewer. (A separate sewer system has separate sanitary and storm sewers.)

**combined sewer overflow** Discharge of a mixture of stormwater and untreated wastewater when the flow capacity of a sewer system is exceeded during rainstorms. Contaminants from combined sewer overflows include bacteria, nutrients, solids, BOD, and metals. Stormwater runoff may also carry toxic chemicals from industrial areas or streets. *See also* first-flush effect.

**combined sewer overflow (CSO) control** For technologies of combined sewer overflow control, see source control, collection system control, CSO storage. *See also* combined sewer overflow treatment.

**combined sewer overflow (CSO) outlet** The place where or the device by which a combined sewer system discharges combined wastewater/stormwater into a receiving water body, frequently near the intersection of a combined sewer with an interceptor sewer. Sometimes, a backwater gate, tide gate, or elastomeric check valve is used to prevent the receiving water from entering the sewer.

**combined sewer overflow (CSO) storage** A method of control of combined sewer overflows, which provides flow equalization, peak flow reduction, and some level of treatment through settling, skimming, and diversion to treatment plants after flows recede. *See* in-system storage, surface storage, off-line storage.

**combined sewer overflow treatment** Plants serving combined sewer systems regularly treat dry-weather flows plus a portion of the wet-weather flows. In some cases, facilities near combined sewer outlets provide storage and/or treatment to some of the excess flows before discharge to receiving waters or pumping back into the sewer system during dry weather. Technologies used in the treatment of combined sewer overflows include disinfection, biological treatment, physical-chemical treatment, horizontal reciprocating screens, high-rate clarification, swirl concentrators, and vortex separators.

**combined sewer system (CSS)** A sewer system that conveys wastewater to a treatment plant during dry weather, but conveys both wastewater and stormwater to treatment or overflow works during wet weather. In addition to the contributing drainage area and wastewater sources, a CSS includes the sewer pipe network, interceptors, regular and diversion structures, and combined sewer overflow outlets.

**combined treatment** (1) Joint treatment of municipal wastewater and other wastes, e.g., commercial and industrial wastes. (2) Treatment provided through the combination of two or more processes, e.g., activated sludge and trickling filtration. *See* dual treatment.

**combined trickling filter/activated sludge process** *See* combined filtration–aeration process.

**combined wastewater** Municipal or industrial wastewater that is mixed with surface runoff. *See also* sanitary wastewater, stormwater.

**combined water** Chemically bound water in soil, which remains after the evaporation of hygroscopic water. Only heating can drive off combined water. *See also* capillary water.

**combining weight** (1) The ratio of the atomic weight of an atom to its valence. *See also* equivalent weight. (2) The formula weight of a substance divided by its electrical charge.

**Combu-Changer®** A regenerative chemical produced by ABB Air Preheater, Inc. for the control of emissions of volatile organic compounds.

**combustible** Pertaining to materials that can catch fire, burn, and release heat in the presence of air; or such materials.

**combustible-gas indicator** A device that measures the concentration of explosive fumes, based on the catalytic oxidation of a combustible gas on a filament incorporated in a Wheatstone bridge. Also called explosimeter.

**combustion** (1) Burning, or rapid oxidation, accompanied by release of energy in the form of heat and light. Many important air pollutants (sulfur dioxide, nitrogen oxides, particulates) are products of the combustion of coal, gas, oil, wood, and other materials. (2) The controlled burning of waste, in which heat chemically alters organic compounds, converting them into stable inorganics such as carbon dioxide and water. *See* sludge combustion or thermal destruction.

**combustion air** The air introduced through the fuel bed and into the primary chamber of an incinerator.

**combustion air requirement** *See* combustion oxygen requirement.

**combustion chamber** (1) The actual compartment where waste is burned in an incinerator. (2) The chamber of a boiler, engine, etc. where combustion occurs.

**combustion gases** The gases and vapors that result from combustion.

**combustion oxygen requirement** The amount of oxygen required for the combustion of organic matter in sludge, represented by the formula $C_aH_bO_cN_d$ for the main constituents carbon, hydrogen, oxygen, and nitrogen, and assuming that carbon and hydrogen are oxidized to carbon dioxide ($CO_2$) and water ($H_2O$), respectively, and that nitrogen is liberated as nitrogen gas ($N_2$):

$$C_aH_bO_cN_d) + (a + 0.25\ c - 0.5\ b)\ O_2 \rightarrow a\ CO_2 + 0.5\ c\ H_2O + 0.5\ d\ N_2 \qquad \text{(C-126)}$$

The theoretical air requirement is 4.35 times the oxygen requirement (because air is about 23% oxygen on a mass basis). Commonly, a 50% excess is added for complete combustion.

**Combustrol®** A treatment technology developed by Wheellabrator Air Systems, Inc. for waste conditioning using fly ash.

**Comet** A rotary distributor manufactured by Simon-Hartley, Ltd. For use in fixed-film reactors.

**cometabolic degradation** A type of biological degradation in which the compound being degraded is not part of the microorganism's metabolism, but degradation results from the action of nonspecific enzymes and does not contribute to cell growth.

**cometabolism** The simultaneous metabolism of two compounds, in which the degradation of the second compound (the secondary substrate) depends on the presence of the first compound (the primary substrate). For example, in the process of degrading methane, some methanotrophic bacteria possess a nonspecific enzyme (methane monooxygenase) that can catalyze the degradation of hazardous chlorinated solvents (e.g., trichloroethylene or TCE) that the bacteria would otherwise be unable to attack. The enzyme transforms TCE into an oxide that subsequently degrades hydrolytically, killing the bacteria in the process. Thus, a constant supply of methane and nutrients is required. Cometabolism is also presented as a type of incomplete or partial oxidation in which the energy derived does not support microbial growth.

**commensalism** A close association between organisms of different species that is beneficial to at least one species and does not harm the others. *See also* parasitism, symbiosis.

**commensal organisms** Organisms that live together without injury to any of them, e.g., commensal bacteria found in human feces include enterobacteria, enterococci, lactobacilli, clostridia, bacteroides, bifidobacteria, and eubacteria.

**comment period** Time provided for the public to review and comment on a proposed USEPA action or rulemaking after publication in the *Federal Register.*

**commercially dry sludge** Sludge that contains at most 5% moisture for use as a fertilizer or 10% moisture for other applications.

**commercial waste** All solid waste emanating from business establishments such as stores, markets, office buildings, restaurants, shopping centers, and theaters.

**commercial water use** The use of drinking water by such business customers as offices, gas stations, laundries, car washes, hotels, shopping centers, and stores.

**commingled discharges** Stormwater that is mixed with process wastewater.

**comminuted solids** Wastewater solids that have been divided into fine particles.

**comminuter** Same as comminutor.

**comminuting screen** A screen equipped with a cutting edge or with knives for grinding or shredding coarse materials in wastewater. The screenings are not removed but they are too small to clog pumps. Also called cutting screen; same as comminutor, grinder.

**comminution** Mechanical shredding or pulverizing of waste; used in solid waste management. In wastewater treatment, in-stream grinding of coarse solids to reduce their size is used as an alternative to screening.

**comminutor** A machine (e.g., a fixed circular screen with moving cutters or a movable slotted drum with knives) used in wastewater treatment downstream of a grit chamber to shred, grind, or pulverize large solids (such as rags, sticks, paper) into smaller settleable particles. New plants generally do not include comminutors because of problems experienced, e.g., accumulation of plastics in sludge digestion tanks and rags on air diffusers. Also called cutting screen or comminuting screen; also spelled comminuter. *See also* barminutor, Barminutor®, grinder, macerator, rag ball, rag rope.

**comminutor balls** *See* rag ball.

**comminutor rope** *See* rag rope.

**common-ion effect** A phenomenon of water chemistry whereby the solubility of one ion decreases in the presence of an increased concentration of another ion. For example, the solubility of calcium (Ca) decreases when the carbonate ($CO_3$) concentration increases. The common-ion effect also explains the buffering capacity of natural waters and aqueous solutions. An example is the combined effect of acetic acid and its salt sodium acetate, which dissociate as follows to maintain a pH balance:

$$CH_3COOH \leftrightarrow CH_3COO^- + H^+ \quad \text{(C-127)}$$

$$3COONa \rightarrow CH_3COO^- + Na^+ \quad \text{(C-128)}$$

*See also* the Henderson–Hasselbach equation, LeChâtelier's principle.

**common logarithm (log)** Logarithm of base 10; the exponent in the representation of a number as a power of 10. For example, for $N = 446.6836 = 10^{2.65}$, $\log N = 2.65$. Also called Briggsian logarithm or Briggs logarithm. *See also* natural logarithm.

**common salt** Sodium chloride (NaCl), a white or colorless, crystalline compound that occurs abundantly in nature, for example, as a mineral or a constituent of seawater, and in animal fluids. *See* sodium chloride for detail.

**common source epidemic** A disease outbreak that affects a group of individuals exposed to the same source.

**common-wall construction** An arrangement of treatment tanks, towers, or other units having a common wall to minimize space requirements and reduce costs.

**communal latrine** A latrine used by more than one household or by the public in general as in a market, school, or other institution. Communal latrines are often of the pour-flush type and sometimes combined with showers and laundry facilities. Also called a public latrine.

**communicable disease** A disease caused by an infectious agent or its products and transmitted from one host or through the environment to another host.

**communicable period** The time during which an infectious agent may be transmitted directly or indirectly from a source (an infected person or animal) to another person or animal, including arthropods.

**community** In ecology, a group of interacting populations in time and space. Sometimes, a particular subgroup may be specified, such as the fish community in a lake or the soil arthropod community in a forest.

**community component** Any portion of a biological community. It may pertain to the taxonomic group (fish, invertebrates, algae), the taxonomic category (phylum, order, family, genus, species), the feeding strategy (herbivore, carnivore, omnivore), or organizational level (individual, population, community association) of a biological entity within the aquatic community.

**community sewer system** *See* cluster sewer system.

**community water supply** Same as community water system.

**Community Water Supply Study** A survey conducted in 1969 by the U.S. Public Health Service (USPHS) to determine whether public water supplies met the 1962 USPHS standards.

**community water system (CWS)** A public water supply system of a certain size, e.g., serving at least 15 connections or 25 year-round residents. Also called a public water system.

**Compact CDI™** A line of continuous deionization products of the U.S. Filter Corp.

**compaction** (1) Reduction of the bulk of solid waste by rolling and tamping. *See also* sludge compaction. (2) The physical reduction in membrane thickness and/or deformation of a membrane, due to pressure, temperature effect, or other causes. *See also* membrane flux retention coefficient.

**compaction curve** A curve that shows the relationship between density and moisture content of sludge. *See* sludge compaction and Proctor compaction.

**compaction density** The density of compacted sludge or the achievable dry density of a given sludge, an important characteristic for landfill or monofill disposal as well as for sludge reuse in backfill or as a soil substitute. *See also* Atterberg test, Büchner funnel test, capillary suction time (CST) test, filterability constant, filterability index, shear strength, specific resistance, standard jar test, time-to-filter (TTF) test.

**Compact RO** A compact reverse osmosis unit manufactured by the U.S. Filter Corp.

**compact technique** *See* compact technology.

**compact technology** A water or wastewater treatment process or operation that requires less space than conventional processes or operations. Compact techniques are design and construction methods that result in installations that are smaller than usual. A compact process may be defined as reducing land requirement by a certain percentage, for example, 30 or 50%. *See also* small footprint technology.

**comparative risk assessment** The comparison of risks associated with various problems or issues. In the United States, e.g., it is estimated that the lifetime risk of mortality from a motor vehicle accident is 12.5 times lower than that of cancer from smoking one pack of cigarettes daily but 200 times higher than that of diarrhea from rotavirus.

**comparator** An instrument used to determine the color of a water sample by visually comparing it to standards prepared with specified amounts of potassium chloroplatinate ($K_2PtCl_6$) and cobaltous chloride ($CoCl_2$) or to standard colored disks.

**compartment** A subdivision of another space, larger than a manhole or a vault, but smaller than a chamber.

**compartmental model** A pharmacokinetic model that represents the body as compartments whose volumes are expressed in terms of concentrations of fluids (blood, urine) and effects of chemicals.

**compensated hardness** A measure of hardness used in the design of cation-exchange softeners to compensate for the reduction in performance due to some substances; it is based on total hardness, the magnesium-to-calcium ratio, as well as the concentrations of iron, manganese, and sodium.

**compensation depth** The depth of the compensation point (2). *See* freshwater profile for related terms.

**compensation point** (1) The light intensity or carbon dioxide ($CO_2$) concentration at which the rates of photosynthesis and respiration in a plant are equal. (2) The point in a lake or ocean profile at which plants use as much organic matter (respiration) as they produce (photosynthesis). This point corresponds to a light intensity of about 1% of sunlight. *See* freshwater profile for related terms. *See also* light compensation point.

**competent formation** In relation to the design and construction of injection wells, a competent formation is limestone, dolomite, or consolidated sandstone that will stand unsupported in a borehole (Freeman, 1998). *See also* incompetent formation, fluid-seal completion.

**competition** The struggle among organisms for water, food, space, and other vital requirements, when supply is limited.

**competitive adsorption** The condition that exists when the presence of a substance hinders the adsorption of the target substance.

**CompGro™** Trademark of the compost produced at the East Bay Municipal Utility District of Oakland (CA) from anaerobically digested, chemically conditioned, dewatered sludge. It is sold to landscape contractors, homeowners, and other customers.

**Completaire** A package plant manufactured by the FMC Corp. for wastewater treatment using the complete mix activated sludge process.

**complete combustion** The complete oxidation of the various elements in organic matter. *See* combustion, combustion oxygen requirements.

**completed test** The third and final phase in the microbiological analysis of water using the multiple-tube fermentation or MPN technique; it is usually conducted for samples collected after final disinfection or from the distribution system. A continuation of the presumptive test, it uses agar cultures for 24–72 hours at 35°C to determine the presence of Gram-negative, nonspore-forming, rod-shaped bacteria in the positive tubes. *See also* confirmed test.

**completely mixed activated sludge** *See* completely mixed biological process.

**completely mixed batch (CMB) reactor** One of two modes of design and operation of powdered activated carbon (PAC) reactors; the solution is introduced into a tank, PAC is added, and the slurry is mixed and held until the desired concentration is reached. *See also* completely mixed flow reactor, complete mixing, complete mixing mass balance for detail (Freeman, 1998).

**completely mixed biological process** A wastewater treatment process that uses a completely mixed biological reactor in which the influent and the return sludge are uniformly distributed. Hence, oxygen demand is uniform and peak loads are reduced.

**completely mixed flow** *See* complete mixing, plug flow. Also called mixed flow.

**completely mixed flow (CMF) reactor** One of two modes of design and operation of powdered activated carbon (PAC) reactors: wastewater and PAC are continuously added to a tank, mixed, and withdrawn at the same rate. *See also* completely mixed batch reactor, complete mixing, complete mixing mass balance for detail.

**completely mixed reactor** A reactor in which flow occurs under complete mixing conditions, implying uniform characteristics throughout the contents of the reactor. *See also* complete mixing and complete mixing mass balance for detail.

**completely mixed state** *See* complete mixing.

**completely mixed system** A wastewater treatment system operating under conditions of complete mixing.

**completely mixed unit** A basin, tank, reactor, or other container operated under complete mixing conditions. *See also* complete mixing and complete mixing mass balance for detail.

**completely stirred tank reactor** Same as completely mixed reactor.

**complete-mix activated sludge (CMAS) process** The activated sludge process using a completely mixed reactor as aeration tank. It is characterized by thorough mixing of the contents and uniform concentrations of the parameters (mixed liquor suspended solids, BOD, oxygen requirements, etc.), simplicity of operation, ability to withstand shock loads, and tendency to promote the growth of filamentous organisms.

**complete-mix anaerobic digester** A completely mixed reactor used for anaerobic digestion, with or without recycle, the solids retention time varying from 15 to 30 days.

**complete mixing (CM)** A flow or treatment condition that assumes that all incoming material to a unit is distributed instantly and uniformly, thus creating uniform pollutant or substrate concentrations throughout the unit, which are also the same concentrations in the effluent. Under complete mixing conditions, the fate of pollutants may be represented by a first-order decay equation, i.e., the change in pollutant mass in the unit equals the influent mass, minus the effluent mass, minus the first-order decay of pollutant (see, e.g., Nix, 1994):

$$d(C \cdot V)/dt = QC_i - QC - K_d \cdot C \cdot V \quad \text{(C-129)}$$

where $C$ = effluent pollutant concentration ($ML^{-3}$), $C_i$ = influent pollutant concentration ($ML^{-3}$), $K_d$ = decay coefficient ($T^{-1}$), $Q$ = inflow rate ($L^3\ T^{-1}$), $t$ = time ($T$), and $V$ = volume of water in the unit ($L^3$). The symbols $M$, $L$, and $T$ represent, respectively, units of mass, length, and time. *See also* continuous stirred reactor and plug flow.

**complete mixing mass balance** The relationship between the masses of a substance going into and out of a completely mixed reactor, as well as any quantities generated or accumulated. *See* complete mixing. Assuming steady-state conditions, the effluent concentration, which is the same as the concentration in the reactor, is:

$$C = C_i/(1 + K_d\theta) \quad \text{(C-130)}$$

where $C$ = effluent pollutant concentration ($ML^{-3}$), $C_i$ = influent pollutant concentration ($ML^{-3}$), $K_d$ = decay coefficient ($T^{-1}$), $Q$ = inflow rate ($L^3\ T^{-1}$), $\theta$ = $V/Q$ = liquid detention time in the reactor ($T$), and $V$ = volume of water in the unit ($L^3$). The symbols $M$, $L$, and $T$ represent, respectively, units of mass, length, and time.

**complete-mix reactor** ***See*** complete mixing and complete mixing mass balance for detail.

**complete treatment** (1) A method of treating water that consists of the addition of coagulant chemicals, flash mixing, coagulation–flocculation, sedimentation, and filtration. Also called conventional filtration or conventional water treatment. (2) A level of wastewater treatment that implies a high percentage removal of such pollutants as suspended, colloidal, and dissolved organic matter to meet regulatory requirements. Such treatment may include primary, secondary, and advanced treatment steps.

**complete waste treatment system** All the treatment works necessary to meet the requirements of Title III of the Clean Water Act, involving the following: (a) the transport of wastewater from individual homes or buildings to a plant or facility where treatment of the wastewater is accomplished; (b) the treatment of the wastewater to remove pollutants; and (c) the ultimate disposal, including recycling or reuse, of the treated wastewater and residues that result from the treatment process (EPA-40CFR35.2005-12).

**complete wastewater treatment** An advanced wastewater treatment method used, e.g., for water reclamation. It is similar to the conventional water treatment flowsheet applied to a secondary wastewater effluent, i.e., including coagulation, flocculation, sedimentation, filtration, and disinfection. Also called full tertiary wastewater treatment.

**CompleTreator** A package trickling filter manufactured by Dorr-Oliver, Inc.

**complex** (1) A compound that consists of a central atom or ion surrounded by ligands, which are other atoms, ions, or molecules arranged in a definite pattern. Important ligands include ammonia ($NH_3$), chloride ($Cl^-$), fluoride ($F^-$), and the hydroxyl ion ($OH^-$). Also called complex ion, coordination compound, coordination ion, and inner-sphere complex. *See also* chelate, chelating agent, ion pair, ligand, organometallic compound. (2) A compound formed by the combination of two or more salts, e.g.:

$$FeSO_4 + 6\ KCN \rightarrow K_4Fe(CN)_6 + K_2SO_4 \quad \text{(C-131)}$$

**complexation** (1) A reaction in which a metal ion and one or more anionic ligands chemically bond. Complex formation affects the distribution of a metal and other species in solution. For example, copper ($Cu^{2+}$) forms complexes with ammonia ($NH_3$), and cadmium ($Cd^{2+}$) with cyanide ($CN^-$):

$$Cu^{2+} + NH_3 \leftrightarrow CuNH_3^{2+} \quad (C\text{-}132)$$

$$CuNH_3^{2+} + NH_3 \leftrightarrow Cu(NH_3)_2^{2+} \quad (C\text{-}133)$$

$$Cu(NH_3)_2^{2+} + NH_3 \leftrightarrow Cu(NH_3)_3^{2+} \quad (C\text{-}134)$$

$$Cu(NH_3)_3^{2+} + NH_3 \leftrightarrow Cu(NH_3)_4^{2+} \quad (C\text{-}135)$$

$$Cd^{2+} + CN^- \leftrightarrow CdCN^+ \quad (C\text{-}136)$$

$$CdCN^+ + CN^- \leftrightarrow Cd(CN)_2 \quad (C\text{-}137)$$

$$Cd(CN)_2 + CN^- \leftrightarrow Cd(CN)_3^- \quad (C\text{-}138)$$

$$Cd(CN)_3^- + CN^- \leftrightarrow Cd(CN)_4^{2-} \quad (C\text{-}139)$$

*See also* sequestration. (2) A reaction between an element and an ion or radical, resulting in the formation of a compound that contributes to the solubility of the element. Such is the case of calcium hydroxide and magnesium carbonate:

$$Ca^{2+} + OH^- \rightarrow CaOH^+ \quad (C\text{-}140)$$

$$Mg^{2+} + CO_3^{2-} \rightarrow MgCO_3^0 \quad (C\text{-}141)$$

(3) The inactivation of an ion by a reagent that combines with it and prevents it from participating in other reactions. Complexation with microorganisms is also a biological mechanism for heavy metals removal from wastewater, but complexes often prevent the precipitation of metals. (4) The reversible dissociation of a metal species (ML) to a metal ion ($M^{2+}$) and an organic complexing species ($H_2L$, the acidic form of the ligand $L^{2-}$), depending on the hydrogen ion ($H^+$) concentration:

$$ML + 2\,H^+ \leftrightarrow M^{2+} + H_2L \quad (C\text{-}142)$$

**complexation model** *See* surface complexation model.

**complex cover** The combination of soil, crops, vegetation, and tillage that characterize a site or area. Also called complex surface.

**complex formation** Same as complexation.

**complex ion** Same as complex.

**complex ion formation** Same as complexation.

**complex organic waste constituent** Any of the following materials that are biodegraded during anaerobic digestion: alcohol, fatty acid, protein, starch.

**complex surface** The combination of soil, crops, vegetation, and tillage that characterize a site or area. Also called complex cover.

**compliance cycle** The nine-year calendar cycle, beginning January 1, 1993, during which public water systems must monitor. Each cycle consists of three three-year compliance periods.

**compliance monitoring** Collection and evaluation of data, including self-monitoring reports and verification, to show whether pollutant concentrations and loads contained in permitted discharges are in compliance with the limits and conditions specified in the permit.

**compliance period** A three-year calendar period within a compliance cycle. Each compliance cycle has three compliance periods. The first compliance period runs from January 1, 1993. *See also* monitoring period and start date.

**compliance schedule** A negotiated agreement between a pollution source and a government agency that specifies data and procedures by which a source will reduce emissions and, thereby, comply with a regulation.

**Compmaster™** A computerized control system manufactured by Wheelabrator Clean Water Systems, Inc., Bio Gro Division, for composting units.

**component** A chemical entity such as a salt, mineral, molecule, ion, or electron considered in the solution of chemical equilibrium equations. *See* phase rule for detail.

**Component Clarifier** A group of clarifier components designed by Eimco Process Equipment Co. for use as required.

**Composite Correction Program** A program of the USEPA to improve the performance of a water treatment plant in two steps: a comprehensive performance evaluation followed by comprehensive technical assistance.

**composite liquid-waste sampler (Coliwasa)** A glass, metal, or plastic tube equipped with an end closure that can be opened for sampling free-flowing liquids and slurries in drums, pits, shallow tanks, and similar containers. *See also* auger, bailer, dipper, thief, trier, weighted bottle.

**composite membrane** A semipermeable membrane used in water treatment by reverse osmosis and consisting of more than one polymer, as opposed to an anisotropic or asymmetric membrane. It has a rejecting layer of one polymer (e.g., polyamide, a thin cellulose acetate, about 0.20 micron) supported by layers of thicker, different porous materials. Also called thin-film composite membrane.

**composite sample** The flow-weighted average of several individual aliquots (or samples) taken over a given period of time (usually every one or two hours during a 24 hour time span) and proportional to the flow at the time of sampling, as compared to a grab sample, which is a single sample, representative of the composition of the flow at a particular time and place. The composite or proportional sample represents the average conditions during the sampling period. For example, in stormwater monitoring, flow-weighted composite samples are taken for the first three hours following a discharge, and the volume of an individual aliquot is determined as follows (Dodson, 1999):

$$V = V_{max}(Q/Q_{max}) \qquad \text{(C-143)}$$

where $V$ = volume of individual aliquot added to composite sample container, $V_{max}$ = volume of aliquot obtained at the time of maximum flow rate, $Q$ = flow rate measured at the time individual aliquot is obtained, and $Q_{max}$ = maximum flow rate observed during storm event. Sometimes called weighted composite sample. *See also* continuous sample, random sample.

**composite sample multiplier** The ratio ($M$) of the volume of composite sample desired ($V$) to the product of the average flow rate ($Q$) and the number of individual samples ($N$):

$$M = V/(NQ) \qquad \text{(C-144)}$$

**composite sampling** *See* compositing (1).

**compositing** (1) The formation of a composite sample from discrete individual samples to obtain a sample that is representative of the source. Same as composite sampling. (2) The construction of a composite unit hydrograph from hydrographs of subareas.

**compost** The relatively stable, humus-like material that is produced from an aerobic or anaerobic composting process in which bacteria in soil, nightsoil, or sludge, mixed with garbage, sawdust, and degradable trash, break down the mixture into organic fertilizer or soil conditioner. The organic matter in compost is not 100% degraded, but it is sufficiently stabilized to reduce the potential for odor generation. Composted sludge solids are usually dark brown to black, with an inoffensive odor similar to that of garden soil conditioners. *See also* *Aspergillus fumigatus,* bulking agent, organic amendment.

**compostable** Same as compostable material.

**compostable material** Any waste that is readily degradable biologically, e.g., garbage and wastewater treatment sludge.

**Compost-A-Matic** Equipment manufactured by Farmer Automatic of America, Inc. for sludge composting.

**compost bioaerosol** Biological agents dispersed aerially from composting operations, including pathogenic microorganisms and microbial enzymes. The greatest risks are from *Aspergillus fumigatus* and endotoxin.

**compost biofilter** A device, typically 6 ft deep, designed to treat odorous gases from wastewater management facilities by passing them through a biologically active bed of compost. As packing material, the compost removes the odor-causing particles by absorption/adsorption and microbial oxidation. A bulking agent is added to the compost in a one-to-one ratio by volume, with periodic addition of compost to make up for biological conversion. Also called compost filter (Metcalf & Eddy, 2003).

**compost briquette** The product of briquette composting.

**composted sludge solids** *See* compost.

**compost filter** Same as compost biofilter.

**composting** The controlled biological decomposition of organic material to form a humus-like material. Controlled methods of composting include mechanical mixing and aerating, ventilating the materials by dropping them through a vertical series of aerated chambers, or placing the compost in piles out in the open air and mixing it or turning it periodically. Composting yields an environmentally sound, nuisance-free product of potential value as a soil conditioner. The process may be aerobic or anaerobic. *See also* Bangalore method, carbon/nitrogen ratio, sludge composting, windrow.

**composting amendment** Manure and other substances used in a composting operation to promote uniform air flow while reducing the bulk weight and the moisture content. *See* amendment and bulking agent.

**composting cooling stage** The last stage in the composting process, following the mesophilic and thermophilic stages and resulting in the production of a stable and mature compost: there is a reduction in microbial activity, in temperature (from thermophilic back to mesophilic range), and in moisture content. Curing takes place during the cooling stage and may occur in a separate container.

**composting latrine** *See* composting toilet.

**composting methods** The principal methods of composting for sludge stabilization are aerated static pile, in-vessel or mechanical, and windrow.

**composting microbiology** Bacteria, actinomycetes, and fungi carry out the destruction of organic matter, the formation of humic acid, and the production of a stable compost. *See also* composting stages.

**composting operations** A typical flowchart of sludge composting includes (a) preparation of the mixture of dewatered sludge, amendments, and bulking agents, (b) aeration and decomposition of the compostable mixture, (c) recovery of bulking agents, (d) curing and storage, (e) screening out of nonbiodegradables, and (f) final disposal.

**composting performance factors** These factors include aeration level, bed porosity, detention time, moisture content, organic/nutrient ratio, and temperature.

**composting privy** *See* composting toilet.

**composting stages** Three different stages of activity in the composting process: mesophilic, thermophilic, cooling.

**composting toilet** A toilet in which excreta are mixed with such other materials as ash, garbage, grass, sawdust, and straw under conditions that permit the batch or continuous production of an inoffensive compost that can be removed by householders for use as an agricultural fertilizer and soil conditioner. Same as composting latrine or composting privy. *See multrum* toilet.

**compost meister** In Germany, the expert specialist or technician in the operation of a composting unit.

**compost mulch** Compost that has no visible water or dust during handling, used as a temporary soil stabilization or erosion control material and placed on the soil surface. The compost is applied at a minimum thickness of two inches and a rate of 100 tons/acre.

**compost row** Same as windrow.

**Compost Storm Water Filter** A unit manufactured by CSF Treatment Systems, Inc. for the filtration of stormwater.

**compost toilet** Same as composting toilet.

**compound** A substance composed of two or more independent elements whose composition is constant and that can be separated only by a chemical reaction. For example, table salt or sodium chloride (NaCl) is a compound of sodium and chlorine.

**Compound 146** Polyurethane material made by the FMC Co. for use in sprocket tooth inserts.

**compound meter** The combination of two devices to measure a varying range of flows.

**compound microscope** An optical instrument that magnifies small objects using two or more lenses mounted in the same tube.

**compound pipe** A pipeline or a loop made up of two or more pipes.

**compound tube** A tube made up of two or more tubes of various lengths and diameters.

**compound weir** A weir made up of two or more sections; e.g., a trapezoidal or rectangular section with a V-notch weir at the bottom. The lower V-notch allows better accuracy in measuring low flows, whereas the full section can measure larger peak flows than a triangular weir.

**comprehensive coliform monitoring plan** A requirement of the Total Coliform Rule for public water supplies to develop state-approved monitoring plans.

**comprehensive performance evaluation** A thorough review of the performance of a water or wastewater treatment plant.

**comprehensive technical assistance** A set of corrective measures taken to optimize the performance of a water or wastewater treatment plant.

**compressibility** (1) *See* coefficient of compressibility. (2) The degree of change in suspended solids or filter cake caused by the application of pressure. *See* sludge compressibility.

**compressibility coefficient** *See* sludge compressibility.

**compression curve** In the study of sludge compressibility, the compression curve plots the void ratio (the ratio of the volume of voids to the volume of the solid phase) versus the logarithm of the consolidation pressure.

**compression index** A parameter used in studying the compressibility of sludge disposed in a landfill and determined from the slope of the compression curve. It indicates the degree of settlement of the landfill. *See also* sludge compressibility and swelling index.

**compression point** In the theory of sedimentation, the point at which all the suspension has passed into Type IV or compression settling; the point where the interface of the accumulating solids reaches the top of the settling suspension.

**compression settling** The settling of particles in a concentrated suspension in which further settling can occur only by compression of the existing structure of settled particles. It is characterized by increased contact between particles, layer compression, squeezing out of water, and very high solids concentration. Compression settling is the primary mechanism in sludge thickening. *See* sedimentation type IV for more detail.

**compressor** A device that increases the pressure of a gas or vapor. Two types of compressor are used in diffused aeration systems: centrifugal and positive displacement rotary lobe. *See also* blower.

**compressor vessel** A vessel equipped with a compressor to provide protection against pressure surges in piping systems by maintaining a desired water level and air volume in the vessel under normal operating conditions. Other surge vessels include bladder tanks and hybrid tanks.

**Com PRO™** Trademark of the compost produced at the Montgomery County (MD) Regional Composting Facility from chemically conditioned, vacuum-filtered sludge. It is sold in bulk and in 18-kg bags to nurseries, homeowners, and other customers.

**computational fluid dynamics (CFD)** A modeling technique that uses the finite element method to simulate flow patterns numerically. CFD incorporates computer-based methods to solve the fundamental equations (continuity, momentum, energy) of fluid dynamics. It is used to optimize the design and operation of water and wastewater treatment plants in such applications as vortex separators, mixing tanks, sedimentation basins, chlorine contact chambers, and ultraviolet disinfection systems.

**computational fluid dynamics (CFD) model** A three-dimensional representation of mixing, flow patterns, heat transfer, or other hydraulic phenomena using mathematical (generally, partial differential) equations.

**computer-aided design and drafting (CADD)** An automated computer program used in mapping, drafting, and architectural and engineering design.

**computer-aided engineering** The use of computer programs and tools in the solution of engineering problems. Acronym: CAE.

**computer-assisted mapping** The use of CADD-based programs and tools in standard mapping operations. Acronym: CAM. It is one of the three current tools used for developing comprehensive mapping programs for such infrastructure elements (e.g., a sewer system). *See also* AM/FM and GIS.

**computer-based modeling** Simulation using a computer program.

**computer mapping program** A comprehensive program of data collection, mapping, and data management services, based on all available, up-to-date information from personnel, records, and field investigations. Acronym: CMP. Currently, three technologies are used to develop a CMP: computer-aided mapping (CAM), geographic information system (GIS), and automated mapping and facilities management (AM/FM).

**computer model** A model designed to be solved using a computer, as opposed to an analytical model, for example. Because of their capacity to carry out mathematical operations at great speed, computers are an excellent tool to perform complex simulations. There are currently a variety of computer models that simulate or assist in biological treatment design, cost estimation for water and wastewater facilities, process optimization, sewer design, stormwater flow, water distribution analysis, water quality, etc. Not all models are computerized. In urban stormwater management, e.g., SWMM Level I is a simple, noncomputerized model for estimating annual runoff and pollutant loading, whereas the full SWMM is a complex hydraulic model that cannot be solved without a computer.

**concentrate** In general, the waste stream containing the pollutants removed by a treatment process. In particular, the portion of the feed stream that contains the retained, dissolved or suspended constituents in membrane filtration processes: dialysis, electrodialysis, hyperfiltration, microfiltration, nanofiltration, pervaporation, reverse osmosis, and ultrafiltration. Also called concentrate stream, permeate stream, reject, reject stream, reject water, residual stream, retained phase, retentate, or waste stream. *See also* blowdown, brine or brine stream, centrate, feed stream or feedwater, permeate.

**concentrated brine disposal** Common methods applicable to the disposal of the concentrated waste stream from a membrane filtration process include controlled thermal evaporation, deep-well injection, discharge to a sewer system, evaporation ponds, land application, ocean discharge, and surface water discharge.

**concentrate staging** The arrangement of a pressure-driven membrane system such that the concentrate from one stage is processed in another stage.

**concentrate stream** Same as concentrate.

**concentration** (1) Amount of a substance dissolved or suspended in a unit volume of solution, or applied to a unit weight of solid; usually expressed in mg/L. *See also* molality, molarity, mole fraction, normality, ppm, standard solution. (2) The process of increasing the amount of a substance per unit volume of solution, e.g., by evaporation. (3) The process of increasing the solids content of sludge by dewatering or thickening.

**concentration as** The expression of the concentration of a compound in terms of its key element, e.g., [$NH_3$-N] for ammonia ($NH_3$) as nitrogen (N). *See also* as calcium carbonate.

**concentration as $CaCO_3$** *See* as calcium carbonate.

**concentration cell** A galvanic cell made up of two electrodes of the same metal in different concentrations of a salt of the same metal.

**concentration cell corrosion** Localized corrosion caused by differences in the total or type of mineralization of the environment, such as in a concentration cell, and resulting in the formation of pits and tubercles. *See also* differential oxygenation corrosion.

**concentration condition** An equation that expresses the mass balance for a chemical reaction at equilibrium, i.e., for a given substance, the sum of the concentrations of the various species is equal to the analytical concentration (the total number of moles) of the substance. For example, for an aqueous solution of C = $10^{-3}$ M boric acid $B(OH)_3$, in addition to $H_3O^+$ and $OH^-$, the species present are $B(OH)_3$ and $B(OH)_4^-$, abbreviated respectively as HB and $B^-$, and the concentration condition is:

$$[HB] + [B^-] = C \qquad \text{(C-145)}$$

where the square brackets indicate molar concentrations. *See also* distribution diagram (Stumm & Morgan, 1996).

**concentration cycle** *See* cycle of concentration.

**concentration factor** (1) A measure of the increased flow available from pondage for electric power generation. It is the ratio of the average flow during pondage to the average daily flow. (2) The degree to which a membrane retains a contaminant, i.e., the factor by which to multiply the concentration of a constituent in the feed to obtain the concentration in the concentrate. The concentration factor (CF) depends on the rate of water recovery and the rate of contaminant rejection:

$$CF = 1/(1 + R)^r \qquad \text{(C-146)}$$

where $R$ is the fractional system recovery and $r$ is the fractional rejection for a given constituent. The concentration factor increases with recovery, as shown on Figure C-15. (3) An expression of the concentration resulting from sample preparation, e.g., the ratio of solute concentration in the final extract to that in the original sample.

**concentration gradient** The change in solute concentration per unit distance, causing decreases from zones of higher concentrations to zones of lower concentrations, a dominant process in groundwater movement and a driving force for mass transfer.

**concentration point** A point that receives runoff from various parts of a drainage area; e.g., the point at which flows are computed for the design of storm sewers. *See also* retardance, time of concentration.

**concentration polarization** (1) The formation of a dense polarized layer of solutes near a membrane surface, or the buildup of ions on either side of the membrane, resulting in a flow restriction and membrane fouling. At the extreme, concentration polarization leads to the formation of a gel or cake layer. It can be controlled by increasing turbulence, decreasing system recovery, using hydrophilic or cross-flow membranes, periodic backflushing, and solubilization with surfactants or cleaning agents. *See also* pore narrowing, pore plugging, membrane fouling control, Figure C-16. (2) The ratio of the salt concentration in the membrane boundary layer to the salt concentration in the bulk stream. The most common and serious problem resulting from concentration polarization is the increasing tendency for precipitation of sparingly soluble salts and the deposition of particulate matter on the membrane surface. Adding sequestering agents to the feed water or using pH control may prevent such precipitation. (3) In corrosion studies, concentration polarization indicates a depletion of ions near an electrode. The polarization in a corroding metal system relates to the concentration differences between the surface and the bulk solution. (4) The basis for chemical analysis by a polarograph. *See also* activation polarization.

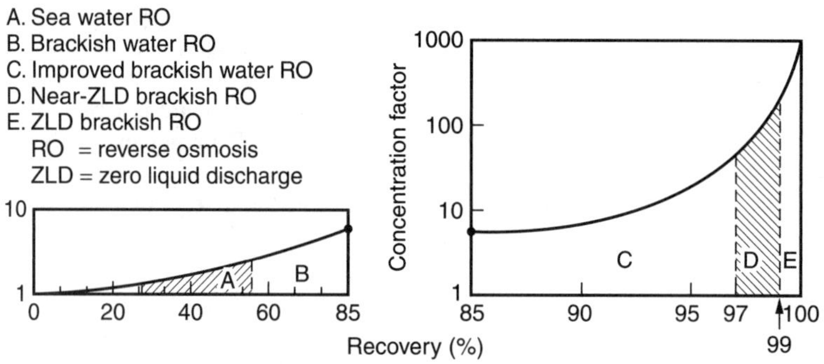


**Figure C-15.** Concentration factor (adapted from Sethi et al., 2006).

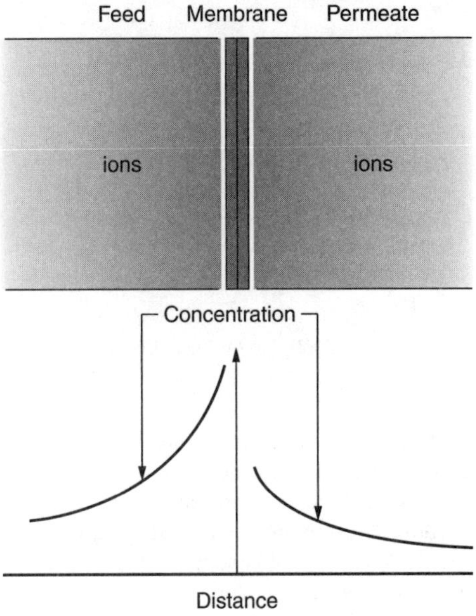


**Figure C-16.** Concentration polarization (adapted from Droste, 1997).

**concentration polarization layer** In a membrane treatment system, the flowing concentration boundary layer where all chemical species reach higher concentrations near the membrane surface, often in advance of the formation of cake and gel layers by particles and macromolecules.

**concentration quotient** *See* equilibrium concentration quotients.

**concentration ratio** The ratio of solids concentration in a water system to the solids concentration in the makeup water.

**concentration ratio diagram** A diagram that shows the ratio of the concentration of a particular metal species to the concentration of free metal ions as a function of pH or some other variable. It is a useful indicator of major species and trends.

**concentration tank** A basin of short detention time used for the concentration of sludge by sedimentation or flotation before further processing.

**concentration time** *See* time of concentration.

**concentration-time concept** *See* CT concept.

**concentration-time relationship** A relationship based on the Chick–Watson law and used in the design of disinfection systems: the disinfectant concentration ($C$) and contact time ($t$) required to produce a fixed percent inactivation are such that:

$$C^n t = \text{constant} \quad \text{(C-147)}$$

where $n$ is the Chick–Watson coefficient, which is often close to 1.0. *See* Figure C-17.

**concentrator** A solids-contact tank used to concentrate wastewater sludge or slurries.

**concentric piping** A new, smaller pipe installed inside a storm or combined sewer to carry wastewater exclusively, thereby separating it from the stormwater.

**concrete pressure pipe** A steel-reinforced concrete pipe.

**concrete sewer corrosion** The corrosion of concrete sanitary sewer pipes from the inside out, resulting from the activity of sulfate-reducing and sulfur-oxidizing bacteria. Under anaerobic conditions, sulfate-reducing bacteria in the wastewater generate volatile hydrogen sulfide ($H_2S$) that sulfur-oxidizing bacteria (*Thiobacillus thiooxidans*) in the moist crown convert to sulfuric acid ($H_2SO_4$). This strong acid reacts with the calcium hydroxide [$Ca(OH)_2$] binder in the concrete to form calcium sulfate, which has no binding capability (Maier et al., 2000):

$$2\,CH_2O + SO_4^{2-} + 2\,H^+ \rightarrow H_2S + 2\,CO_2 + 2\,H_2O \quad \text{(C-148)}$$

$$H_2S + 2\,O_2 \rightarrow H_2SO_4 \quad \text{(C-149)}$$

$$H_2SO_4 + Ca(OH)_2 \rightarrow CaSO_4 + 2\,H_2O \quad \text{(C-150)}$$

where ($CH_2O$) represents the organic matter in wastewater. This phenomenon is sometimes called crown corrosion or crown rot. *See* Figure C-18.

**condensate** (1) The liquid that separates from a vapor during evaporation and condensation; con-

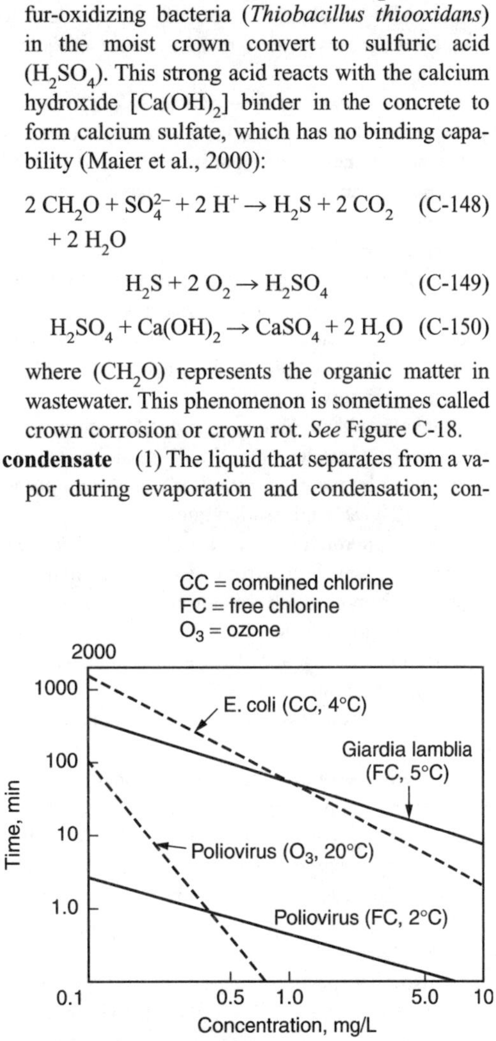


**Figure C-17.** Concentration–time relationship (99% inactivation). (Adapted from AWWA, 1999.)

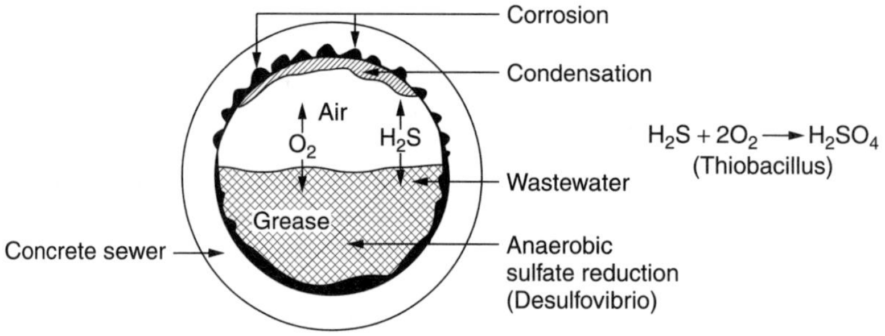


**Figure C-18.** Concrete sewer corrosion (adapted from Maier et al., 2000, and McGhee, 1991).

densed steam from a heat exchanger. (2) The product of the distillation of saline water.

**condensate polishing** Treatment of condensate water before disposal or reuse to attain a required quality.

**condensate water** A type of wastewater from industrial tanks and processes that usually contains low concentrations of organics and total dissolved solids. It can be recycled as makeup water for boilers.

**condensation** (1) The process by which water or any substance changes state from vapor to liquid; the opposite of evaporation. *See also* deposition, dewpoint, sublimation. (2) A chemical reaction, such as polymerization, between molecules to form a complex compound.

**condensation heat** *See* heat of condensation.

**condensation nuclei** (1) Small particles of salt produced by the evaporation of water from ocean spray. (2) *See* condensation nucleus.

**condensation nucleus** A particle of hygroscopic dust or gases (e.g., from sea salt, combustion products) upon which condensation of water vapor occurs to form water drops or ice crystals. Also called nucleus (4). *See also* freezing nucleus.

**Condense-A-Hood** Hoods manufactured by Bedminster Bioconversion Corp. for the collection of odors.

**condensed phosphate** *See* polyphosphate.

**condenser** (1) A device for accumulating a charge of electricity, consisting of two conductive plates of opposite charge, separated by a nonconductor. Also called a capacitor. (2) Any heat transfer device for converting gases or vapors to the solid or liquid phase.

**condensor** Same as condenser.

**conditional maintenance** An unplanned, reactive maintenance approach that consists of inspections to detect problems early, repairs based on intuitive recognition of experienced workers, and effective apprentice programs. Also called indicative maintenance. *See also* breakdown maintenance, proactive maintenance, developmental maintenance.

**conditioned water** Water treated to reduce undesirable substances or conditions such as hardness, iron, color, taste, and odor.

**conditioner selection test** A test conducted to evaluate the effectiveness of chemicals for sludge conditioning. *See* standard jar test, filter leaf test, capillary suction time test, standard shear test, Büchner funnel test.

**conditioning** The pretreatment of water, wastewater or sludge, mostly by chemical, but also biological or physical, means, to facilitate subsequent treatment such as solids removal, thickening, or dewatering. More specifically, the addition of a chemical to water or wastewater treatment residuals or their physical alteration to facilitate dewatering. *See* more detail under sludge conditioning, chemical conditioning, and physical conditioning.

**conditioning chemicals** *See* sludge conditioning chemicals.

**conductance** (1) A measure of the electrical conductivity or conducting power of a solution, equal to the reciprocal of the resistance. Natural and drinking waters have a specific conductance between 50 and 500 micromhos/cm. *See also* Ohm's law. (2) The capacity of a water sample to carry an electrical current; it is related to the concentration of ionized substances in the water. Its measurement is a rapid method of estimating the dissolved solids content of the sample. *See* specific conductance for more detail. *See also* conductivity.

**conductance units** Units used to express conductance include mhos in U.S. customary units and siemens in Système International units. The corresponding conductivity units are micromhos per centimeter and siemens per meter, respectively.

**conduction** (1) The act of conducting water, e.g., through a pipe. (2) The transfer of heat from one body to another, caused by a temperature difference. In heat drying of sludge, the conductive heat transfer rate ($R$, Btu/h) may be calculated as the product of the conductive heat-transfer coefficient ($C_{cond}$, Btu/ft$^2$/h/°F), the area of wetted surface exposed ($A$, ft$^2$), and the temperature difference between the heating medium ($T_m$, °F) and the sludge ($T_s$, °F):

$$R = C_{cond} A(T_m - T_s) \qquad \text{(C-151)}$$

*See also* convection, radiation.

**conduction dryer** A device used to dry wastewater sludge. *See* indirect dryer for detail.

**conduction drying** A method of drying sludge using heat to evaporate water and reduce the moisture content of the solids. The device used has a solid retaining wall between the wet sludge and the steam or other hot fluid. Also called indirect sludge drying. *See* conduction, convection drying, thermal drying.

**conduction sludge drying** *See* conduction drying.

**conductivity** (1) A measure of the ability of water and solutions to carry an electric current; it is directly related to the mineral content of water. Electrical conductivity (EC) is used as a surrogate measure of the total dissolved solids (TDS) content of water and to express the ionic strength ($I$) of solutions. EC is expressed in millisiemens per meter (mS/m), decisiemens per meter (dS/m), micromhos per centimeter (μmhos/cm), or millimhos per centimeter (mmhos/cm):

$$1 \text{ dS/m} = 100 \text{ mS/m} = 1000 \text{ (μmhos/cm)} = 1 \text{ mmho/cm} \qquad \text{(C-152)}$$

$$\text{TDS (mg/L)} = a \text{ EC (dS/m or mmhos/cm)} \qquad \text{(C-153)}$$

where $550 < a < 700$

$$I = 1.6 \times 10^{-5} \times \text{EC (dS/m or mmhos/cm)} = 2.5 \times 10^{-5} \times \text{TDS} \qquad \text{(C-154)}$$

*See also* conductance, ionic strength, salinity. (2) A coefficient of proportionality describing the rate at which a fluid (e.g., water or gas) can move through a permeable medium. Conductivity is a function of both the intrinsic permeability of the porous medium and the kinematic viscosity of the fluid that flows through it. *See* hydraulic conductivity.

**conductivity bridge** An instrument used to measure conductivity, consisting of a Wheatstone bridge supplied with alternating current. The arms of the bridge include a conductivity cell, a fixed resistance, and the two end coils of a calibrated resistance.

**conductivity detector** A device used to determine the concentration of an analyte by measuring the conductivity of the solution.

**conductivity unit** *See* conductance unit.

**conductor** A substance, body, device or wire that readily conducts or carries electrical current.

**conduit** In general, a conduit is a natural or artificial channel, open or closed, for the conveyance of water or other fluids. In a closed conduit, water flows under pressure; open conduit flow is by gravity. In a sewer system model such as SWMM, conduits or links are pipes and channels as opposed to nodes. *See* link and link-node network.

**cone of depression** The depression, roughly conical in shape, produced in the water table by the pumping of water from a well. Also defined as the area around a discharging well where pumping has lowered the potentiometric surface. In an unconfined aquifer, the cone of depression is a cone-shaped depression in the water table where the medium has actually been dewatered. Also called cone of influence. *See also* circle of influence.

**cone of influence** Same as cone of depression.

**Cone Screen** A rotary fine screen manufactured by Andritz-Ruthner, Inc.

**cone valve** A valve with a conical moving plug; it is opened by lifting and rotating the tapered section, and closed by placing it back. Also called conical plug valve.

**confidence interval** *See* confidence limits.

**confidence level** The probability that an interval between two confidence limits will contain an event or a population value.

**confidence limits** The limits within which, at a certain confidence level, the true value of a result lies. For example, if the estimated value of a variable is $X$, the standard error of estimate $E$, and the known Student's $t$ statistic $t$, then the confidence limits are $(X - E \cdot t)$ and $(X + E \cdot t)$. The range $[(X - E \cdot t) - (E + E \cdot t)]$ is called the confidence interval.

**Configurator®** A computerized procedure developed by the U.S. Filter Corp. for preliminary design and equipment selection in water and wastewater treatment.

**confined aquifer** An aquifer in which groundwater is confined under pressure that is significantly greater than atmospheric pressure. It is a fully saturated formation of porous rock or soil, overlain by a confining layer. The potentiometric surface (or hydraulic head) of the water in a confined aquifer is at an elevation that is equal to or higher than the base of the overlying confining layer. A discharging well in a confined aquifer lowers the potentiometric surface, which then forms a cone of depression, but the saturated medium is not dewatered. Same as artesian aquifer. *See* aquifer and Figure A-18.

**confined space** As defined by the National Institute of Occupational Safety and Health, "any space that by design has limited openings for entry and exit and unfavorable natural ventilation that could contain or produce dangerous air contaminants, and that is not intended for continuous employee occupancy." Examples of confined spaces in wastewater treatment facilities include manholes, large conduits, and digesters.

**confining layer** A geologic formation characterized by low permeability that inhibits the flow of water (*see also* aquitard).

**confirmed test** The second step in a coliform (multiple-tube fermentation) test; when the first step (presumptive test) is positive, a small aliquot is further incubated in brilliant green lactose bile broth at 35°C for 48 hours. Gas production in the broth indicates a positive confirmed test. *See also* completed test, confirming test.

**confirming test** Same as confirmed test.

**confluence** *See* confluent.

**confluent** As an adjective, running or flowing together, or blending into one. As a noun, one of two or more confluent streams. The junction of these streams is their confluence.

**confluent growth** A continuous bacterial growth covering the entire filtration area of a membrane filter, or a portion thereof, in which bacterial colonies are not discrete (EPA-40CFR141.2).

**confluent stream** A stream that flows together and blends with another of about the same size. When one stream is much smaller than the other, it is called an affluent or a tributary.

**confounding factor** A variable other than chemical exposure level that can affect the incidence or degree of a parameter being measured.

**congeal** To thicken or solidify, usually by cooling or freezing.

**$CO(NH_2)_2$** Chemical formula of urea. *See also* ammonium cyanate.

**conical plug valve** A valve with a conical moving plug; it is opened by lifting and rotating the tapered section, and closed by placing it back. Also called cone valve.

**conjugate** A metabolite of normal biological constituents; e.g., glucuronide, glycine, sulfate, taurine. *See also* conjugation reaction.

**conjugate acid** A chemical species that can donate a proton and is related to its conjugate base by the difference of that proton, e.g., hydrochloric acid (HCl) is the conjugate acid for the chloride ion ($Cl^-$). *See* conjugate acid–base pair.

**conjugate acid–base pair** A pair of compounds such that the acid (HA) becomes a potential $H^+$ acceptor (the conjugate base, $A^-$) by losing an $H^+$, and the base becomes a potential $OH^-$ acceptor (the conjugate acid, $B^+$) by losing an $OH^-$:

$$HA \leftrightarrow H^+ + A^- \quad K_A \qquad (C\text{-}155)$$

$$BOH \leftrightarrow B^+ + OH^- \quad K_B \qquad (C\text{-}156)$$

$$K_A \cdot K_B = K_W = [H^+] \cdot [OH^-] \qquad (C\text{-}157)$$

where $K_A$, $K_B$, and $K_W$ are equilibrium constants, and the quantities between brackets are molar concentrations.

**conjugate base** A chemical species that can accept a proton and is related to its conjugate acid by the difference of that proton, e.g., the chloride ion ($Cl^-$) is the conjugate base for hydrochloric acid (HCl). *See* conjugate acid–base pair.

**conjugated proteins** Compounds that contain amino acids and carbohydrates, lipids, or other groups, whereas simple proteins contain only amino acids.

**conjugation** The process of passing genetic information between bacteria by transferring chromosomal material.

**conjugation reaction** A detoxification reaction in an organism that is exposed to a chemical, the solubility of which has been increased by the biochemical addition of a conjugate.

**conjunction kinetics** A concept applied in water and wastewater treatment to study the factors that influence the contacts among ions, molecules, colloids, and suspensions. It is an important consideration in the coagulation of colloidal particles, which is the result of perikinetic motion (the random movement of particles known as Brownian motion) and orthokinetic motion (their systematic transport by hydraulic forces). Parameters related to conjunction kinetics include power dissipating function, shear gradient, and velocity gradient (Fair et al., 1971).

**conjunction opportunity** Same as contact opportunity.

**conjunctive management** Integrated management and use of two or more water resources.

**conjunctive use** A water resources management approach involving the use of all water resources in a region (surface, ground, etc.) for optimal results, e.g., a larger and more dependable yield. An important aspect of conjunctive water management is the use of recharged aquifers as underground storage, which eliminates the need for surface reservoirs and pipelines, and reduces evaporation losses.

**conjunctive water management** *See* conjunctive use.

**conjunctivitis** An inflammation of the conjunctiva (the mucous membrane that lines the exposed portion of the eyeball and inner surface of the eyelid); an eye infection, often transmitted by flies, that may lead to blindness. It is prevalent in tropical areas where poor hygiene is associated with a lack of water. *See also* trachoma, water-washed disease.

**connate seawater** Seawater trapped in sediments at the time of their deposition; a source of bromide ions.

**connate water** Interstitial water in the zone of saturation but entrapped in sediments or igneous rocks at the time of their deposition; it is usually highly mineralized. *See* subsurface water and fossil groundwater.

**connection band** A collar joining two pipe sections and holding them together by friction or mechanical bond.

**connection charge** A one-time service charge assessed by a water, sewer, or other utility against new customers to recover the direct and indirect costs of connecting them to existing facilities. *See also* capacity charge, capital recovery charge, facility expansion charge, impact fee, development impact fee, and system development charge.

**connection fee** Same as connection charge.

**Conoscreen** A rotating disc manufactured by Purator Waagner-Brio for use in microscreening.

**consecutive digestion** Digestion of organic matter under thermophilic and then mesophilic conditions.

**consecutive systems** Two systems that are interconnected.

**consent agreement** A written agreement, signed by the parties, containing stipulations or conclusions of fact or law and a proposed penalty or proposed revocation or suspension acceptable to both complainant and respondent (EPA-40CFR22.03).

**consent decree** A legal document approved by a judge that formalizes an agreement reached between EPA and potentially responsible parties (PRPs) through which PRPs will conduct all or part of a cleanup action at a Superfund site; cease or correct actions or processes that are polluting the environment; or otherwise comply with EPA-initiated regulatory enforcement actions to resolve the contamination at the Superfund site involved. The consent decree describes the actions PRPs will take and may be subject to a public comment period. Similar decrees may be signed regarding water pollution control actions. In general, a consent decree is a binding agreement settling all questions in a dispute between two parties.

**consent order** A legal document issued by a regulatory agency to a water or wastewater utility for remedial actions. *See* consent decree and consent agreement.

**consequential damages** Losses sustained as a result of a water or wastewater facility failing to work as specified, including, e.g., the costs to install an adequate system. Also called performance damages. *See also* liquidated or delay damages, performance specification, prescriptive specification.

**conservancy system** The practice of waste disposal by earth closets and privies instead of using water carriage.

**conservation (of resources)** A resource management approach that discourages loss or waste and preserves and renews, when possible, human and natural resources. Conservation is also the use, protection, and improvement of natural resources according to principles that will assure their highest economic or social benefits. A conservation district is a governmental agency that implements soil and water conservation measures. A utility conservation rate is a rate established to discourage waste or reduce use of water, electricity, etc.; it is usually an increased rate or a rate structure such as marginal cost pricing, which charges more for additional units. Conservation storage is storage of water for future use.

**conservation district** *See* conservation.

**conservation laws** Principles derived from Newtonian mechanics stating that basic properties (such as energy, mass, momentum) cannot be created nor destroyed, but can be transferred or transformed. Examples of the application of the laws of conservation are: conservation of energy (first law of thermodynamics; Bernoulli's equation; equivalence between energy, work, heat), conservation of mass (continuity equation, mass balance models, storage equation, materials balances), and conservation of momentum (Newton's first law of motion, momentum equation). *See also* Saint-Venant equations.

**conservation master plan** A water supply master plan that emphasizes water conservation.

**conservation rate** A water use rate that increases over a certain quantity to encourage conservation. *See* conservation.

**conservation storage** *See* conservation.

**conservative** (1) In the case of a contaminant, one that does not degrade and the movement of which is not retarded; unreactive. (2) In the case of an assumption, one that leads to a worst-case scenario, one that is most protective of human health and the environment.

**conservative anion** The conjugate base of a strong acid, e.g., the chloride ($Cl^-$), nitrate ($NO_3^-$), and sulfate ($SO_4^{2-}$) ions corresponding, respectively, to hydrochloric (HCl), nitric ($HNO_3$), and sulfuric ($H_2SO_4$) acids.

**conservative cation** The conjugate cation of a strong base, e.g., calcium ($Ca^{2+}$) and potassium ($K^+$) ions corresponding, respectively, to calcium hydroxide [$Ca(OH)_2$] and potassium hydroxide (KOH).

**conservative constituent/element/pollutant/substance** (1) A constituent, element, pollutant, or substance whose concentration does not change through a water or wastewater treatment process; e.g., bromide ion through coagulation. (2) An element or material that is not created or destroyed, except to a minor extent, during a treatment or other unit process or operation, e.g., inert suspended solids, phosphorus, nitrogen, metals, and biologically resistant organics. (3) A pollutant that is not changed by chemical or biological action. Compliance standards for receiving streams sometimes specify the reaction kinetics of nonconservative pollutants, although they establish limitations for both conservative and nonconservative pollutants. (4) A substance that does not degrade or react over time, or a substance such as chloride that is not affected by any process except dilution.

**conservative element** *See* conservative constituent.

**conservative pollutant** *See* conservative constituent.

**conservative substance** *See* conservative constituent.

**conservative tracer** A nonabsorbing tracer; a substance that passes through a process without any reaction, accumulation, or change in its total mass; e.g., the chloride ion passing through a water filtration unit. *See also* conservative constituent.

**consolidated formation** A geologic formation that has been transformed to stone, e.g., at the edges of a borehole. Sometimes used interchangeably with bedrock.

**consolidation curve** In studying the compaction behavior of sludge, the consolidation curve plots the void ratio versus the consolidation pressure. *See also* zero-air-void curve.

**consortium** A heterogeneous group of microorganisms associated with a biochemical process, e.g., bacteria and fungi in the degradation of crude oil.

**constant capacitance model** A surface complexation model that uses a simple linear relationship between charge and surface potential to express quantitatively the fundamental concept that adsorption occurs at defined coordination sites.

**constant diffusivity model** A mathematical model developed to define the relationships between the breakthrough curve for large and small carbon columns in a rapid small-scale column test. This model assumes that the high hydraulic loading rate renders dispersion negligible in the test and that film diffusion is responsible for mass transfer. *See also* proportional diffusivity model.

**constant displacement velocity condition** A mathematical relationship derived by T. R. Camp in 1942 for outlet control devices designed to maintain constant velocities in a grit chamber (Fair et al., 1971):

$$W = NKH^{N-1}/V \qquad \text{(C-158)}$$

where $W$ = width of the chamber, $N$, $K$ = numerical constants, $H$ = channel depth or head, and $V$ = displacement velocity.

**constant-flow stirred-tank reactor** *See* continuous-flow stirred-tank reactor.

**constant-flux operation** One of three operating modes of a membrane process, which keeps the flux rate fixed but allows the transmembrane pressure (TMP) to increase over time. *See also* constant-TMP operation, variable membrane operation.

**constant-level grit chamber** A shallow, circular tank used in older wastewater treatment plants; it is equipped with a mechanism to scrape settling solids while separating grit from suspended organic matter.

**constant of gravitation** *See* law of gravitation.

**constant-pressure filtration** One of three basic modes of filter operation, whereby filtration begins at a high rate, which then declines as head loss develops; this operation requires considerable storage capacity (WEF & ASCE, 1991). Sometines referred to as true declining-rate filtration. Other common methods of filter operation include constant rate and declining rate (also called variable declining rate).

**constant-rate filter** A filter that produces water at a constant rate, the operation being controlled by an adjustable device on the influent or effluent side. *See also* declining-rate filter.

**constant-rate filtration** Filter operation in which water production is maintained at a constant rate by adjustable influent control pumps or weirs, or effluent control modulating valves. Other common methods of filter operation include constant pressure and declining rate (also called variable declining rate).

**constant-rate fixed-head filtration** Filter operation in which water production is maintained at a constant flow rate and a fixed head through effluent control by manual or mechanical modulating valves.

**constant-rate stage** The second stage of operation of thermal sludge drying, in which moisture evaporates from the surface of the solids and is replaced by internal moisture. It follows the warm-up stage and precedes the falling-rate stage.

**constant-rate variable-head filtration** Filter operation in which water production is maintained at a constant flow rate and a variable head through influent control via pumps or weirs.

**constant-speed pump** A pump designed to operate at a constant speed, and discharge at a constant rate, given the head loss and the pump characteristic curve. The opposite is a variable-speed pump.

**constant spring** A spring whose discharge does not vary much from the average discharge; for example, the variation may not exceed 33%.

**constant-TMP operation** One of three operating modes of a membrane process, which keeps the transmembrane pressure (TMP) fixed but allows the flux rate to decrease over time. *See also* constant flux operation, variable membrane operation.

**constituent** (1) An essential part or component of a system or group; e.g., an ingredient of a chemical mixture: benzene is one constituent of gasoline. (2) An individual compound or element of water or wastewater, e.g., ammonia ($NH_3$) or sulfate ($SO_4^{2-}$). *See also* characteristic, contaminant, pollutant. *See* wastewater constituents for detail.

**constituents of concern** Wastewater constituents that are not easily removed by ordinary treatment methods and that require special consideration for health, reuse, or other reasons. They include biodegradable organics, dissolved inorganics, heavy metals, nutrients, pathogens, priority pollutants, refractory organics, and suspended solids (Metcalf & Eddy, 2003).

**constitutive enzyme** An enzyme that a cell produces continuously, as compared to an inducible enzyme.

**constructed wetland** A wetland built and operated as an aquatic treatment system to minimize point and nonpoint source pollution prior to its discharge into receiving waters. In a constructed wetland the roots of aquatic plants such as cattails, reeds, and rushes use wastewater that is applied above or below the soil surface. The minimum level of preapplication treatment for constructed wetlands is the equivalent of primary clarification or short-detention aeration ponds. *See* free water surface system, subsurface flow system, rock–reed filter, vegetated submerged bed, floating aquatic plant system, preapplication treatment, emergent plant.

**construction activities** Activities related to the construction of water and wastewater facilities include funding and financing, bidding, contractor selection, resident inspection and contract administration, operation and maintenance manuals, warranty testing, acceptance, and performance certification. *See also* facilities planning activities, design activities.

**construction grants program** A former U.S. Government program that provided grants for the construction of wastewater facilities over 30 years; now replaced by state revolving loan fund programs.

**consumptive use** (1) A water use that removes water from the available supply and does not return it to a water resource system (e.g., manufacturing, agriculture, food preparation), as compared to an instream use, which takes place within the stream channel (e.g., hydroelectric power generation, navigation, water quality improvement, fish propagation, recreation). (2) Also used to designate the loss of irrigation water due to evapotranspiration. *See* potential or reference evapotranspiration. Consumptive waste is water that is returned to the atmosphere without further use. *See also* the Blaney–Criddle and the Penman–Monteith methods to estimate consumptive use. (3) More generally in agriculture, consumptive use includes not only evapotranspiration, but also a small quantity of water contained in plant tissue.

**consumptive waste** *See* consumptive use.

**consumptive water use** The use of water by plants; evapotranspiration. *See* consumptive use.

**Contaclarifier** An upflow clarifier manufactured by Roberts Filter Manufacturing Co. and including a buoyant medium.

**Contac-Pac** A package wastewater treatment plant manufactured by the FMC Corp. using the contact aeration process in a circular steel container.

**contact aeration** A precursor of the combination of fixed-film and suspended-growth processes for wastewater treatment in which flat asbestos sheets were installed in aeration basins. *See also* combined fixed-film/suspended-growth process (WEF & ASCE, 1991).

**contact aerator** A biological treatment unit that consists of an aeration tank containing stones, plastic sheets, or asbestos-cement sheets on the surface of which microorganisms grow for wastewater stabilization. *See* aerated contact bed for detail.

**contact bed** A biological treatment unit that consists of (a) a tank, about 5 ft deep, containing layers of slate supported by bricks, or filled with broken stones; (b) sprays that discharge the incoming wastewater over the biomass accumulated on the contact surfaces; and (c) an underdrainage system to keep the bed aerobic. Originally, the system did not use sprays but operated on a fill-and-draw basis. This precursor to the trickling filter was used in England as an alternative to intermittent sand filters. *See* Figure C-19. *See also* aerated contact bed (or contact aerator or submerged contact aerator), bacteria bed, nidus rack, percolating filter.

**contact chamber** A large tank designed for water or wastewater to mix with disinfecting agents or other chemicals.

**contact clarifier** A circular or rectangular water or wastewater treatment unit in which the solids-contact process occurs. *See* solids-contact clarifier and sludge blanket for more detail.

**contact condenser** An apparatus that condenses steam by direct contact with a cooling liquid.

**contact dermatitis** An inflammatory reaction of the skin, often an allergic response to a chemical.

**contact filter** (1) A rapid sand filter, or other type of filter, used in a water treatment plant to reduce turbidity ahead of final filtration. *See* more detail under preliminary filter. (2) A precursor of the trickling filter used in England at the end of the nineteenth century to treat wastewaters: a watertight basin filled with 2–4 in rocks operated on a fill-and-draw cycle followed by resting periods of approximately 6 hours each.

**contact filtration** (1) A water treatment process that combines flocculation and sedimentation in an upflow clarifier. *See* Figure C-20. Operation is controlled hydraulically through power dissipation and residence time for adequate floc growth, floc shear, and floc-blanket surface. Also called sludge-blanket filtration. (2) A water or wastewater treatment process that improves filtration efficiency by promoting the attachment of solid particles to the filter grains. For example, to enhance phosphorus removal in the activated sludge process, chemicals may be added ahead of the aeration basin and to the effluent of the secondary clarifier, which is then routed to a rapid sand filter. *See* Figure C-21. (3) One of three wastewater treatment processes that produce reclaimed water for unrestricted urban reuse. As shown on Figure C-22, it involves on-line coagulation followed by filtration of a secondary effluent. *See also* full tertiary wastewater treatment and direct filtration. (4) Same as preliminary filtration, i.e., water treatment using a contact filter (1). (5) Wastewater treatment using a contact filter (2).

**contact flocculation** A treatment process that uses a coarse-media bed as a flocculation unit for coagulated water to reduce the time necessary to produce a settleable and filterable floc.

**contact freezing** A technique for the desalination of saline and brackish waters using two heat-transfer circuits of recycling hydrocarbons; one to transfer heat from the influent to the treated water and the waste brine, and the other to vaporize the hydrocarbon, freeze the influent, compress the vapor, and melt the ice.

**contact load** *See* contact-sediment discharge.

**contact opportunity** A concept proposed for the conversion of loadings of water and wastewater treatment processes into loading intensities. Such intensities would permit one to relate loadings to performance and compare them. The dimensionless product *GT* has been proposed to measure contact opportunity, which is also called conjunction opportunity (Fair et al., 1971):

$$GT = (PV/M)^{0.5}/Q \qquad \text{(C-159)}$$

where

$$G = (P/VM)^{0.5} = \text{velocity gradient (generally, sec}^{-1}\text{)} \qquad \text{(C-160)}$$

$$T = V/Q = \text{time of displacement (sec)} \qquad \text{(C-161)}$$

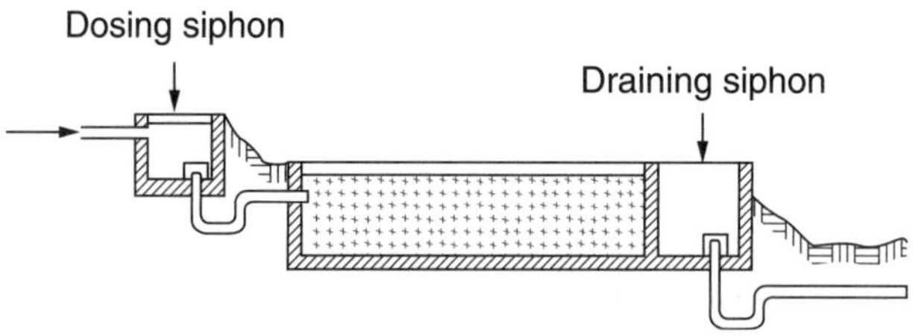


**Figure C-19.** Contact bed (adapted from Fair et al., 1971).

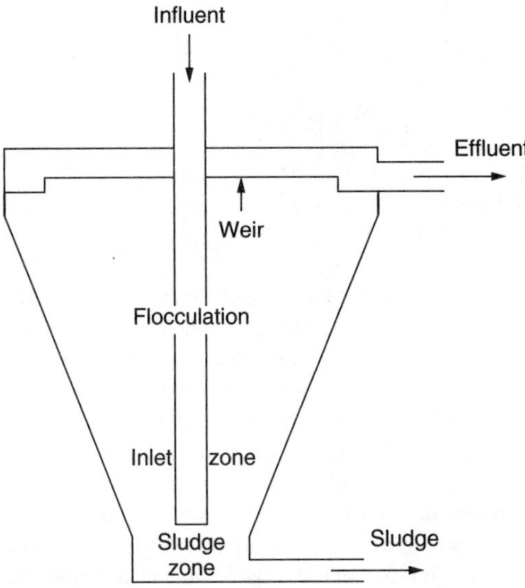


**Figure C-20.** Contact filtration (1) (sludge filtration).

and $P$ = useful power input relative to the volume $V$, $V$ = volume of fluid, $M$ = a proportionality factor having the same dimensions as the absolute viscosity, $Q$ = flow rate. From the first equation, contact opportunity is regarded as the ratio of power-induced flow to displacement-induced flow. The parameter may be used in the design of flocculation and other treatment units, and could replace less adequate parameters such as hours of aeration or volume of air per pound of BOD, pounds of BOD per acre per day, pounds of BOD per daily volume (acre-feet, 1000 cubic feet, etc.), pounds of BOD per unit surface or unit length, etc.

**contactor** (1) An electrical switch, usually magnetically operated. (2) A vertical vessel containing an activated carbon bed. *See* activated carbon contactor for detail. (3) *See* rotating biological contactors.

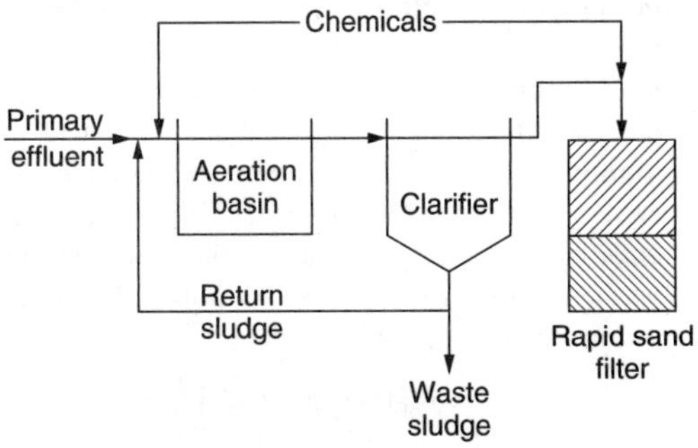


**Figure C-21.** Contact filtration (2).

**contact period** Same as contact time in water or wastewater chlorination.

**contact precipitation** A mechanism of ion transfer between two substances that is different from ion exchange; e.g., the precipitation of iron and manganese on manganese zeolite and the regeneration of the latter with potassium permanganate. Also called surface precipitation.

**contact process** (1) A wastewater treatment process in which diffused air is bubbled over fixed media surfaces. (2) A catalytic method for producing sulfuric acid ($H_2SO_4$) from sulfur dioxide ($SO_2$) and oxygen ($O_2$), using a catalyst to oxidize the dioxide to trioxide.

**contact recreation** An activity that involves the risk of ingestion of water, e.g., swimming, water skiing, diving, surfing, and wading by children.

**contact roughing filter** Same as contact filter (1); a rapid sand filter, or other type of filter, used in a water treatment plant to reduce turbidity ahead of final filtration. *See* more detail under preliminary filter.

**contact sediment, contact-sediment discharge** Sediment transported in a stream by rolling, sliding, or skipping along in contact with the bed. Also called bed load or contact load. *See also* sedimentation terms.

**contact spring** A spring that emerges at the contact between an upper permeable material and an impermeable or less permeable unit.

**contact stabilization** A modification of the conventional, plugflow activated sludge process in which raw wastewater and activated sludge are aerated for a short period (20–60 minutes) prior to solids removal and continued aeration (3–6 hours) in a second aeration unit or stabilization tank. Sludge return is to the stabilization unit, which may also receive the digester supernatant. Contact time is an important operating parameter; still, effluent quality from contact-stabilization is usually somewhat lower than from conventional activated sludge. This variation requires less aeration volume than conventional activated sludge, but it produces little or no nitrification. Also called biosorption process, originally developed as a high-rate biological-oxidation process in Austin (TX). *See* Figure C-23.

**contact tank** A tank designed for water or wastewater to mix with disinfecting agents or other chemicals. *See* contact chamber.

**contact time** (1) The time that constituents and chemicals are reacting with one another in a basin,

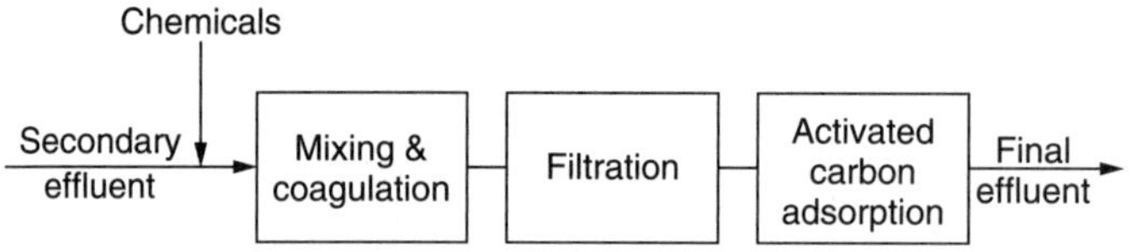


**Figure C-22.** Contact filtration (3).

tank, vessel, etc. (2) Under the Surface Water Treatment Rule, the contact time for adequate disinfection is taken as $t_{10}$, i.e., the residence time exceeded by 90% of the fluid; usually evaluated from tracer studies. (3) The length of time that wastewater remains in a biological reactor (e.g., trickling filter, aeration tank) in contact with microorganisms. For the activated sludge process, it is calculated as the hydraulic detention time ($\theta$), i.e., the ratio of the aeration tank volume ($V$) to the influent flow rate ($Q$):

$$\theta = V/Q \qquad \text{(C-162)}$$

For detention time in trickling filters, see Howland's equation.

**contagious** Capable of being transmitted to a person from a contaminated person or object; carrying or transmitting a contagious disease.

**container valve** A valve mounted on a container, such as a chlorine cylinder.

**containment** A technique used to indefinitely isolate contaminated zones of an aquifer, using such hydraulic barriers as clay caps, interceptor trenches, and slurry walls. Two other groundwater protection methods, extraction and in situ treatment, are preferable when feasible.

**contaminant** In general, any foreign substance present in another substance. More specifically, any biological, chemical, physical, or radiological substance or matter that has an adverse effect on air, soil, or water. Also, a substance that causes the concentration of that substance in groundwater to exceed the specified maximum contaminant level. Or, a substance that causes an increase in the concentration of that substance in the groundwater where the existing concentration of that substance exceeds the specified maximum contaminant level. Water contaminants include bacteria, viruses, and other microorganisms, dissolved minerals, soluble organics, suspended solids, etc. *See also* pollutant.

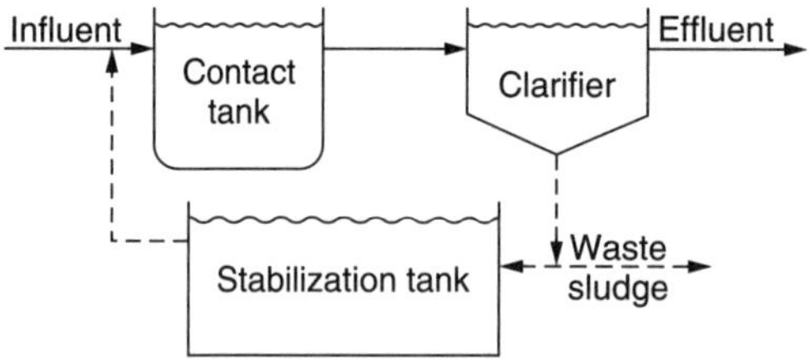


**Figure C-23.** Contact–stabilization process.

**contaminant accumulation** The build-up of such contaminants as dust and dirt prior to a given storm; it is one of the variables included in SWMM and other wastewater or stormwater models. *See also* dust and dirt build-up, contaminant washoff.

**contaminant aging** The process of a contaminant diffusing into micropores that exclude bacteria, thereby becoming unavailable for biodegradation, with an increase in contact time between contaminant and medium (Maier et al., 2000). *See also* micropore exclusion.

**contaminant buildup** *See* contaminant accumulation.

**Contaminant Candidate List (CCL or CCL1)** A list of unregulated contaminants published by the USEPA and presenting potential public health risks in drinking water. The 1998 list contains 50 (mostly organic) chemicals and 10 microbes. *See also* Contaminant Candidate List 2, Drinking Water Contaminant Candidate List.

**Contaminant Candidate List 2 (CCL2)** A list of nine microbial agents and 42 chemicals that the USEPA published in 2004 for possible regulation as drinking water contaminants. This list is part of the first Contaminant Candidate List. *See* Table C-03.

**Contaminant Selection List** *See* Drinking Water Contaminant Candidate List.

**contaminant source inventory** A record of activities that affect water quality within a watershed.

**contaminant wash-off** The load ($L$, pounds) of contaminants (e.g., BOD, suspended solids, nutrients, heavy metals, fecal coliforms) that a storm washes off over a given period ($t$, hours). It may be estimated as:

$$L = L_0[1 - \exp(-CR^y t] \qquad \text{(C-163)}$$

where $L_0$ = initial amount of contaminant, lb; $C$ = wash-off coefficient, per inch; $R$ = runoff rate,

**Table C-03.** Contaminant Candidate List 2 (CCL2)

| |
|---|
| *Chemical Contaminants* |
| Acetochlor |
| Alachlor ESA and other acetanilide pesticide products |
| Aluminum |
| Boron |
| Bromobenzene |
| DCPA di-acid degradate |
| DCPA mono-acid degradate |
| DDE |
| Diazinon |
| 1,1-Dichloroethane |
| 2,4-Dichlorophenol |
| 1,3-Dichloropropane |
| 2,2-Dichloropropane |
| 1,1-Dichloropropene |
| 1,3-Dichloropropene |
| Dinitrophenol |
| 2,4-Dinitrotoluene |
| 2,6-Dinitrotoluene |
| 1,2-Diphenylhydrazine |
| Disulfoton |
| Diuron |
| EPTC |
| Fonofos |
| Isopropyltoluene |
| Linuron |
| Methyl tertiary butyl ether (MTBE) |
| Methylbromide |
| 2-Methyl-phenol |
| Metolachlor |
| Molinate |
| Nitrobenzene |
| Organotins |
| Perchlorate |
| Prometon |
| Royal demolition explosive (RDX) |
| Terbacil |
| Terbulos |
| 1,1,2,2-Tetrachloethane |
| Triazines and their degradation products |
| 2,4,6-Trichlorophenol |
| 1,2,4-Trimethylbenzene |
| Vanadium |
| *Microbiological Contaminants* |
| Adenoviruses |
| *Aeromonas hydrophila* |
| Caliciviruses |
| Coxsakieviruses |
| Cyanobacteria, other freshwater algae, and their toxins |
| Echoviruses |
| *Helicobacter pylori* |
| Microsporidia (Enterocytozoon and Septata) |
| *Mycobacterium acium intracellulare* (MAC) |

inches/hour; and $y$ = wash-off exponent. *See also* dust and dirt build-up (Metcalf & Eddy, 1991).

**contaminate** To introduce a substance that would cause (a) the concentration of that substance in groundwater to exceed the maximum contaminant level specified, or (b) an increase in the concentration of that substance in the groundwater where the existing concentration of that substance exceeds the maximum contaminant level specified (EPA-40CFR257.3.4-2).

**contaminated nonprocess wastewater** Any water that, during manufacturing or processing, comes into incidental contact with any raw material, intermediate product, finished product, by-product, or waste product by means of runoff, accidental spills, accidental leaks, and discharges from safety showers and related personal safety equipment (EPA-40CFR415.91-f).

**contaminated runoff** Runoff that comes into contact with any raw material, intermediate product, finished product, by-product, or waste product located on petroleum refinery property (EPA-40CFR419.11-g).

**contamination** The introduction of a contaminant into water, air and soil, in a concentration that renders the medium unfit for its next intended use. It also applies to surfaces of objects and buildings, and various household and agricultural use products.

**Cont-Flo™** A bar screen manufactured by John Meunier, Inc., including a reciprocating rake for back cleaning.

**Continental®** A line of products and equipment manufactured by the U.S. Filter Corp. for use in water treatment.

**contingency plan** A document setting out an organized, planned, and coordinated course of action to be followed in case of a fire, explosion, or other accident that releases toxic chemicals, hazardous waste, or radioactive materials that threaten human health or the environment. For a water or wastewater utility, a contingency plan defines actions to be undertaken under specified adverse conditions.

**contingency planning** A process that defines specific alternatives to follow in the event of unforeseen occurrences. In water utility risk management, contingency planning aims at ensuring service continuity between the impact of a problem and the return to normal operation.

**continuous-backwash filter** A low-head, shallow-bed, rapid granular filter with multiple compartments and a traveling-bridge backwash system that allows backwashing of the cells successively.

Solids are removed near the top of the bed and there is no need for a filter gallery. *See also* deep-bed continuous-backwash filter.

**continuous-belt filter press** A device for sludge dewatering, consisting of a belt filter press with continuous feeding of sludge.

**continuous-belt screen** A device used to remove fine and coarse materials from wastewater ahead of other treatment units. It consists of a continuous, self-cleaning belt of plastic or stainless steel elements that are pulled through the wastewater, with relatively small openings. Also called continuous self-cleaning screen. *See also* catenary screen, chain-driven screen, reciprocating rake screen.

**continuous composter** A composting toilet based on the Swedish "multrum"; usually a double humus vault underneath a superstructure; see Figure C-24. It requires the addition of a carbon source such as garbage, vegetable leaves, and sawdust to adjust the carbon/nitrogen ratio. As the compost is applied to agricultural land, a minimum retention of three months is recommended. *See also* continuous composter.

**continuous composting toilet** Same as continuous composter.

**continuous control** A combination of three types of process control. Also called proportional–integral–derivative control. *See* proportional control mode, integral control, proportional–derivative control.

**continuous culture** A growth medium and substrate that is continuously replenished to remain constant for the study of an organism or group of organisms. *See also* batch culture.

**continuous-culture technique** *See* continuous-flow test. Also called continuous-flow test or steady-state test. *See also* batch test.

**continuous-culture test** A test conducted at the laboratory to study the behavior of a biological treatment plant. Also called steady-state test. *See also* batch test.

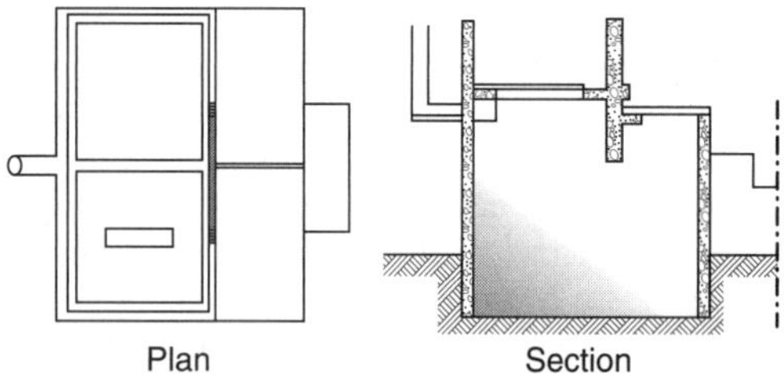


**Figure C-24.** Continuous composter (adapted from Feachem et al., 1983).

**continuous deflection separator (CDS)** A device used for solids separation from mixtures of wastewater and stormwater, particularly to capture more than 90% of particles less than 900 microns from a first flush. It is similar to a vortex separator but uses a filtration mechanism without relying on flow currents induced by vortex action. *See* Figure C-25.

**continuous discharge** A routine release to the environment that occurs without interruption, except for infrequent shutdowns for maintenance, process changes, etc. The one-dimensional model for a continuous discharge is (Metcalf & Eddy, 1991):

$$C = \{M/[A(U^2 + 4\,KE_L)^{1/2}]\}$$
$$\exp\{[(xU/2)[1 \pm (1 + 4\,KE_L/U^2)^{1/2}]\} \quad \text{(C-164)}$$

where $C$ = constituent concentration in liquid, mg/L; $M$ = constituent loading or discharge rate, mass per unit time; $A$ = stream cross-sectional area, ft$^2$; $U$ = cross-section-averaged velocity, fps; $K$ = decay constant; $E_L$ = longitudinal dispersion coefficient; and $x$ = longitudinal distance along river or estuary, ft.

**continuous-discharge pond** A wastewater treatment pond that discharges the effluent at the same rate as the influent, taking into account evaporation and seepage losses. Regarding effluent discharge, there are also controlled-discharge ponds,

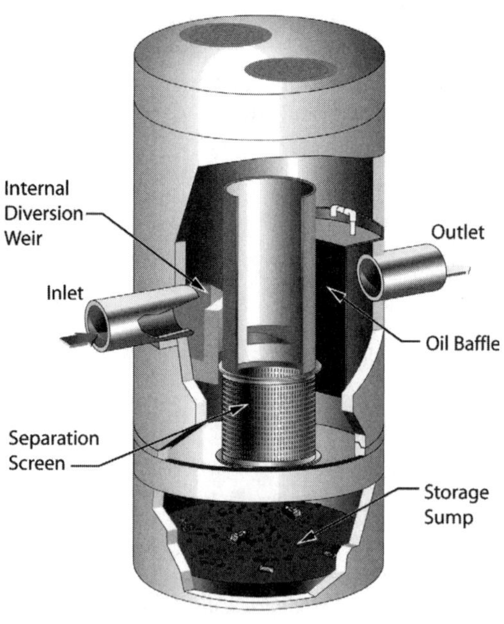


**Figure C-25.** Continuous deflection separator (courtesy Contech Stormwater Solutions, Inc.).

hydrograph-controlled-release ponds, and total-containment ponds. *See also* lagoons and ponds.

**continuous distillation** The continuous introduction of a feed stream into a distillation column, as opposed to a batch operation. The column provides an interface between the liquid and the vapor through plates or packing. *See also* batch distillation and flash distillation.

**continuous-dose method** The use of dosing equipment to apply copper sulfate continuously to reservoirs. The equipment is calibrated to provide a dose of 0.3 mg/L. The application of copper sulfate controls the growth of plankton (algae) in reservoirs. *See also* blower, burlap bag, and spray methods.

**continuous draft** A type of water supply that uses a surface source without an impounding reservoir, i.e., water is pumped directly from the source to the treatment works. Continuous draft is used when available flow is high at all seasons compared to the demand. On the other hand, selective draft diverts and stores enough clean flood waters to meet the demand during low-flow seasons.

**continuous-feed method** A method of disinfection of a new water main before it is placed in service: a solution-feed chlorinator or a hypochlorite feeder injects 50 mg/L of chlorine into water that fills the main and disinfects the associated valves and hydrants for at least 24 hours; it leaves a residual of at least 25 mg/L. Finally, the main is flushed to waste using potable water. *See also* slug method, tablet method.

**continuous filter** A filter in which filtration and backwash occur simultaneously. *See also* semi-continuous filter.

**Continuous-Flo®** A traveling bridge filter manufactured by Zimpro Environmental, Inc.

**continuous-flow operation** A mode of operating a water or wastewater treatment unit with a steady flow, instead of intermittent, fill-and-draw, or batch operation.

**continuous-flow pump** A pump made up of a rotating element called an impeller, and enclosed in a casing that connects to the pipeline. Radial-flow or centrifugal, axial-flow or propeller, and mixed-flow pumps are of the continuous-flow category and are also called rotodynamic pumps. The other broad category is positive displacement pumps.

**continuous-flow stirred-tank reactor (CFSTR)** A theoretical reactor with uniform concentration resulting from contents being completely mixed continuously. Also called complete-mix reactor. *See also* continuous stirred reactor.

**continuous-flow tank** A water or wastewater treatment tank that is operated continuously, as compared to an intermittent, fill-and-draw or batch operation.

**continuous-flow test** A test conducted to study the actual behavior of a treatment plant, i.e., collecting data during the operation of the plant.

**continuous-flow thickener** A circular settling basin operated continuously for the gravity thickening of sludge. Sludge is fed near the center of the basin and distributed radially; settled water exits over a peripheral weir.

**continuous immunomagnetic collection** The capture of target protozoa and viruses by specific antibodies attached to a magnetic head.

**continuous injection** The introduction and mixing of a quantity of tracer into a reactor until the effluent concentration equals the influent concentration. Most commonly a dye is injected into the reactor to characterize tracer response curves and describe the hydraulic performance of reactors. Also called step input or continuous-step input. *See also* pulse input, residence time distribution (RTD) curve, time-concentration curve.

**continuous liquid–liquid extraction** Simple liquid–liquid extraction is a selective isolation process using an organic solvent to separate a liquid constituent from wastewater or other solutions. In continuous liquid–liquid extraction, the solvent is recycled to enhance the performance of the process. It is a commonly used technique to prepare samples for the analysis of organic compounds.

**continuously stirred tank reactor** *See* continuous-flow stirred-tank reactor.

**continuous mixing** The type of mixing used in a reactor or holding tank whose contents must be kept in suspension, as in activated sludge aeration basins, aerated lagoons, aerobic digesters, equalization tanks, and flocculation chambers. *See also* pneumatic mixing and mechanical aerators.

**continuous mode** *See* continuous-flow operation.

**continuous-move sprinkler system** A type of sprinkler system used to distribute wastewater in land applications.

**continuous rapid mixing** The type of mixing used to mix one substance with another, e.g., the blending of coagulation or flocculation chemicals with water or wastewater or the addition of chemicals to sludge. Mixing times vary from less than one second (high-speed induction mixers, in-line mixers, pressurized water jets, pumps, static in-line mixers) to 10–20 seconds (turbine mixers, propeller mixers, and other hydraulic devices).

**continuous reactor** *See* continuous stirred reactor.

**continuous sample** (1) A flow of water from a particular place in a plant to the location where samples are collected for testing. This continuous stream may be used to obtain grab or composite samples. Frequently, several taps (faucets) will flow continuously in the laboratory to provide test samples from various places in a water treatment plant. (2) A fraction of the total flow diverted over a period of time, usually at a constant rate.

**continuous self-cleaning screen** Same as continuous-belt screen.

**continuous sludge-removal tank** A sedimentation tank with a mechanism that allows continuous sludge removal.

**continuous step input** The introduction and mixing of a quantity of a tracer into a reactor until the effluent concentration equals the influent concentration. Most commonly, a dye is injected into the reactor to characterize tracer response curves and describe the hydraulic performance of reactors. Also called continuous injection or step input. *See also* pulse input, residence time distribution (RTD) curve, time–concentration curve.

**continuous stirred reactor (CSTR)** A reactor whose contents are under continuous stirring and satisfy complete mixing conditions; a convenient simplification for such complex phenomena as watershed thermal processes. *See* continuous-flow stirred-tank reactor.

**contour or contour line** (1) A line connecting points of the same elevation above mean sea level (MSL) or another datum. The difference in elevation between two adjacent contours is the contour interval. A contour basin is a basin whose boundaries (e.g., levees) follow contour lines. A contour map or contour plot shows surface configuration by means of contour lines labeled according to their elevations. (2) Sometimes used to designate on a map or a graph points having the same characteristics such as pressure or rainfall.

**contour farming** A conservation-based method of farming in which all farming operations (for example, tillage and planting) are performed across (rather than up and down) the slope. Ideally, each crop row is planted at a right angle to the ground slope. Cultivations take place in lines parallel to the contours to reduce loss of topsoil from erosion, increase the capacity of the soil to retain water, and reduce water pollution by sediments.

**contour plowing** Soil tilling method that follows the shape of the land to discourage erosion.

**contour strip cropping** Crop growing on strips parallel to contours to reduce erosion.

**contour strip farming** A kind of contour farming in which row crops are planted in strips between alternating strips of close-growing, erosion resistant forage crops (grass, grain, hay).

**contract demand** The quantity of water stipulated in a contract between a water purveyor and a large customer.

**contract operation** The operation and maintenance of municipal water and wastewater facilities by a private enterprise. *See also* privatization.

**contracted weir** A rectangular weir whose notch opening is smaller than the width of the upstream channel, thus causing both a horizontal and a vertical contraction of the nappe. The opposite is a full-width or suppressed weir. *See* the weir equations. For rectangular contracted weirs whose length ($L$) and side contraction both exceed three times the head ($H$), the discharge ($Q$) may be estimated as:

$$Q = 3.27(L - 0.2\ H)H^{1.5} \qquad \text{(C-165)}$$

**contraction** The reduction of the cross-sectional area of a conduit, pipe, channel, nappe, or stream. Also, same as constriction. The contraction coefficient in a discharge formula accounts for the effect of the constriction (e.g., weir, orifice); it is taken as the ratio of the smallest cross-sectional area in the constriction to the nominal area of the constriction. A contraction loss is one of several minor head losses, due to the reduction of the size of a conduit or channel, considered in hydraulic modeling studies.

**contraction coefficient** *See* contraction.

**contraction loss** *See* contraction.

**Contraflo®** A solids contact clarifier manufactured by General Filter Co.

**Contraflux®** An activated carbon unit made by Graver Co. for countercurrent operation.

**Contra-Shear®** Screening equipment manufactured by Andritz-Ruthner, Inc.

**Contreat®** A package plant manufactured by EnviroSystems Supply, Inc. for aerobic wastewater treatment.

**control** (1) A section of a conduit or stream where the water level correlates well with the discharge. (2) A sample used to assess the quality of a laboratory procedure. (3) A standard of comparison in experimentation. *See also* toxicity terms. (4) The cross section of a waterway that determines the energy head corresponding to a given discharge.

**control efficiency** The mass of a pollutant in the wastewater sludge fed to an incinerator minus the mass of that pollutant in the exit gas from the incinerator stack divided by the mass of the pollutant in the sludge fed to the incinerator (EPA-40CFR503.41-c). *See also* destruction and removal efficiency.

**control float** A floating device installed in a body of water to control the operation of pumps.

**control flume** *See* control structures.

**control group** A group of subjects observed in the absence of agent exposure or, in the instance of a case/control study, in the absence of an adverse response.

**controlled discharge** Release of wastewater treatment effluent based on streamflow fluctuations and at a rate to maintain an established water quality in the receiving stream.

**controlled-discharge pond** A wastewater treatment pond that discharges effluent at a rate that depends on the level of the receiving stream and does so infrequently, e.g., once or twice a year; it is usually a facultative pond with high detention times (e.g., 120 days). A special case is the hydrograph-controlled pond. Regarding effluent discharge, there are also continuous-discharge ponds and total-containment ponds. *See also* lagoons and ponds.

**controlled evaporation** The use of heat to reduce the volume of the concentrated waste streams produced by membrane processes, an expensive and energy-intensive method used where other alternatives are not available, e.g., in inland locations.

**controlled filtration** A wastewater treatment method that uses a deep (18–24 ft) trickling filter that achieves the same BOD removal as a high-rate filter but at much higher hydraulic and organic loading rates, thus at a lower cost.

**controlled reaction** A chemical reaction under temperature and pressure conditions maintained within safe limits to produce a desired product or process.

**controlled storage** The volume of a storage reservoir from the top of the gates to the bottom of the outlet, i.e., the portion that the operator can control.

**controlled thermal evaporation** *See* controlled evaporation.

**controlled tipping** (1) A method of solid waste disposal, known also as a sanitary landfill. *See also* tip, tipping. (2) A method of solid waste disposal practiced in Germany; the equivalent of a sanitary landfill for milled refuse but without the daily cover or final cover (Sarnoff, 1971). *See also* milling, milled refuse.

**controlled volume pump** Same as metering pump.

**controlled waste** A British legal classification for household, commercial, and industrial wastes.

**controller** A device that controls the starting, stopping, or operation of a device or piece of equipment.

**control loop** The path through the control system between the sensor, which measures a process variable, and the controller, which controls or adjusts the process variable.

**control-loop tuning** The process of determining the proper values of two parameters (the controller proportional gain coefficient and the controller integration gain factor) used in the equation of proportional–integral control.

**control point** A point in a channel where the relationship between discharge and head or depth is known, e.g., spillway, flow measuring devices (weir, flume, gate), point of critical depth. Control points are used in the establishment of surface profiles. *See* control structures.

**control reach** *See* control structures.

**control section** *See* control structures.

**control structures/control works** Hydraulic structures or devices through which a fluid may flow and the flow rate can be measured; also called outlet structures. Control structures may be used as diversion works, at the head or diversion point of a conduit or canal. Control structures include weirs, gates, manholes, etc., as well as reservoirs and other works for flood control. They may affect the quantity and timing of the releases, or simply the manner of operation. Sharp-crested weirs are the most common controls in water supply and wastewater engineering. A control flume has a constriction with minimum head to measure the flow. Similarly for a control reach in an open channel. In general, a control section is a cross section where discharge is uniquely related to depth of flow or to energy head; e.g., at the point of critical depth in open channels.

**control system** A system that senses and controls its own operation on a close, continuous basis in what is called a proportional or modulating control.

**control works** Same as control structures.

**convection** (1) The motion and mixing caused in a fluid by gravity and temperature-induced density differences. In groundwater, convection causes the transport of contaminants by bulk motion. *See also* bulk flow, advection. (2) Heat transfer through the movement of air masses. *See also* conduction, radiation.

**convection dryer** A sludge processing device that brings the feed material into direct contact with heated air. The most common are rotary dryers. Also called direct dryer. *See also* thermal dryer, flash (or pneumatic conveyor) dryer, fluid-bed dryer, indirect dryer, combined direct–indirect dryer, and infrared dryer.

**convection drying** A sludge drying process that brings the wet sludge in direct contact with hot gases or other heat transfer medium. Also called direct drying. The convective heat transfer rate ($Q_{conv}$, Btu/hr) is:

$$Q_{conv} = H_{conv}A(T_g - T_s) \qquad \text{(C-166)}$$

where $H_{conv}$ = convective heat transfer coefficient (Btu/hr/ft$^2$/°F), usually provided by equipment manufacturers or determined from pilot studies; $A$ = wetted area exposed to the heat transfer medium (ft$^2$); $T_g$ = temperature of transfer medium, °F; and $T_s$ = temperature of sludge, °F. *See also* conduction or indirect drying (WEF & ASCE, 1991).

**convection heat-transfer coefficient** *See* convection drying.

**convection sludge drying** *See* convection drying.

**convective diffusion** *See* convection.

**convective heat-transfer rate** *See* convection.

**convective sedimentation** The type of mass transport mechanism that occurs for flocculation when a moving particle collides with a stationary particle, as in a packed bed filter, under the influence of gravity (Stumm & Morgan, 1996). The flocculation rate constant depends on the specific gravity and diameters of the particles, and on the kinematic viscosity of the fluid. This is a special case of differential settling. *See also* flocculation concepts.

**conventional activated sludge** The activated sludge process, as originally configured, i.e., including (a) a relatively long and narrow, rectangular, plug-flow aeration basin with a food-to-microorganism ratio of 0.15–0.40 per day and a mixed liquor suspended solids concentration of 1500–3000 mg/L; and (b) a clarifier, the influent being mixed with return sludge and fed at the head of the basin. Air is supplied through porous diffusers at the bottom of the basin for oxygen and mixing and excess solids are wasted from the sludge return line. Removal of 5-day BOD is within the range of 85–95%. Alternatively, a continuous-flow, stirred-tank reactor is used instead of the plug-flow basin.

**conventional aeration** (1) The addition of air to a water or wastewater treatment basin using surface aerators, mechanical aerators, or submerged diffusers. (2) The conventional activated sludge process.

**conventional anaerobic digestion** Same as conventional anaerobic treatment or low-rate anaerobic digestion.

**conventional anaerobic treatment** Anaerobic digestion, in a well-mixed reactor without solids recycle, with a detention time of 10–30 days. *See* low-rate anaerobic digestion for detail.

**conventional atomic absorption spectrophotometry (or spectroscopy)** Same as flame atomic absorption spectrophotometry.

**conventional concentrate disposal** The disposal of residuals from membrane processes using common methods such as surface water discharge, deep injection wells, drainfields, boreholes, evaporation ponds, land application, sanitary sewers, and thermal evaporation toward zero liquid discharge.

**conventional digester figure** A covered tank used for anaerobic digestion of wastewater treatment sludge. It is equipped with a mixer and outlets for gas collection, digested liquid, and waste sludge.

**conventional disease classification** The classification of diseases according to the characteristics of the pathogens that cause them. Also called generic or biological taxonomy.

**conventional disinfectant** A chemical or physical agent commonly used in the disinfection of water or wastewater. Conventional disinfection processes include chlorination and coagulation/sedimentation/granular media filtration. More recently, ozonation, membrane filtration, and ultraviolet light irradiation have been used. *See also* emerging disinfection technologies.

**conventional downflow depth filter** *See* conventional downflow filter.

**conventional downflow filter** One of five types of depth filter commonly used in advanced wastewater treatment. It consists of a single bed of sand or anthracite, a dual bed of anthracite over sand, or combinations of these media with activated carbon, garnet, or ilmenite. *See* depth filtration.

**conventional extended aeration process** *See* extended aeration.

**conventional filtration** A method of treating water to remove particulates. The method consists of the addition of coagulant chemicals, flash mixing, coagulation–flocculation, sedimentation, and filtration. Also called complete treatment, complete filtration treatment, or conventional treatment. *See also* direct filtration and inline filtration.

**conventional filtration treatment** Same as conventional filtration.

**conventional German digester** One of three commonly used forms of anaerobic digester; a deep cylindrical vessel with top and bottom cones, as shown in Figure C-26. *See also* cylindrical digester, egg-shaped digester.

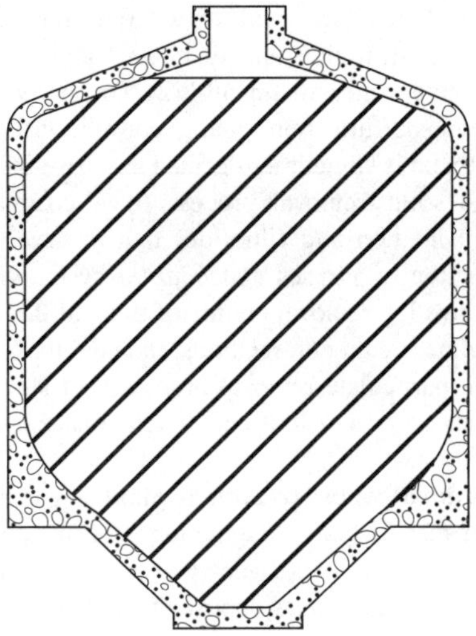

**Figure C-26.** Conventional German digester.

**conventional lagoon** A wastewater treatment lagoon that has a top aerobic layer and an anaerobic bottom layer. *See* aerobic–anaerobic lagoon for more detail. *See also* lagoons and ponds.

**conventional membrane treatment** A membrane treatment system consisting of conventional pretreatment for scaling control, membrane filtration, chlorination, and corrosion control. *See* Figure C-27.

**conventional municipal wastewater treatment** *See* conventional wastewater treatment, as applied to municipal wastes.

**conventional pit latrine** A simple on-site excreta and night-soil disposal unit consisting of a hole in the ground, covered by a seat, squatting slab, or pour-flush bowl, usually with a removable cover and a superstructure. The pit is filled in when two-thirds to three-quarters full and a new pit is dug. Most conventional pit latrines are malodorous, promote insect breeding, and constitute a groundwater pollution hazard where the water table is high. *See also* ventilated improved pit (VIP) latrine, composting toilet, Reed Odorless Earth Closet.

**conventional plug-flow activated sludge** *See* conventional activated sludge.

**conventional pollutant** A statutorily listed pollutant understood well by scientists. It may be in the form of organic waste, sediment, acid, bacteria, viruses, nutrients, oil and grease, and heat. Section 304.a.4 of the Clean Water Act lists the following conventional pollutants: BOD, TSS, pH, fecal coliform bacteria, and oil and grease. *See also* emerging constituent, nonconventional constituent, wastewater constituent.

**conventional pollution** (1) The discharge of a conventional pollutant into the environment. (2) *See* natural pollution.

**conventional pretreatment** In a pressure-driven membrane system, pretreatment consists of an acid or antiscalant addition to prevent the precipitation of salts during filtration. *See also* posttreatment and advanced pretreatment.

**conventional sand drying bed** A sludge dewatering device used in small- and medium-sized communities. *See* sand drying bed.

**conventional septic-tank disposal-field system** An individual wastewater treatment and disposal system consisting of a septic tank and a soil absorption system, used where local site conditions are appropriate. Exceptions include shallow soil cover, extreme percolation rates (too slow or too rapid), high groundwater table, and steep slopes.

**conventional sludge drying** A unit operation designed to reduce the moisture content of sludge by evaporation to the atmosphere through vapor pressure differences. *See also* sludge drying equation and mechanical drying.

**conventional system** A system that has been traditionally used to collect municipal wastewater in gravity sewers and convey it to a central primary or secondary treatment plant prior to discharge to surface waters.

**conventional technology** (1) Wastewater treatment processes and techniques involving centralized treatment by means of biological or physical/chemical unit processes followed by direct point source discharge to surface waters (EPA-40CFR35.2005-14). (2) Wet flue gas desulfurization (FGD) tech-

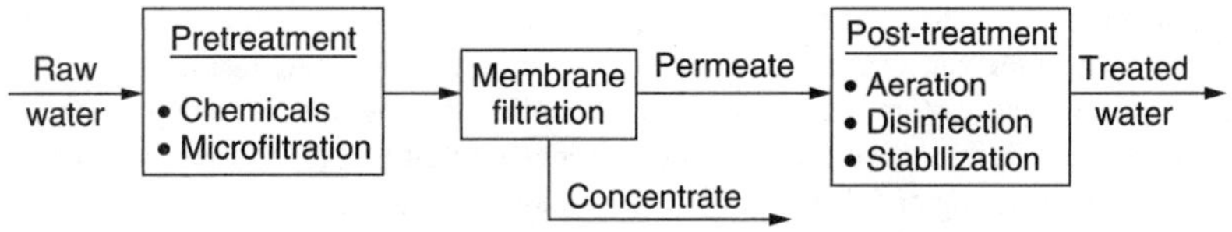


**Figure C-27.** Conventional membrane treatment.

nology, dry FGD technology, atmospheric fluidized bed combustion technology, and oil hydrodesulfurization technology under the Clean Air Act (EPA-40CFR60.41-b). *See also* emerging technology.

**conventional tillage** The traditional method of farming in which soil is prepared for planting by completely inverting it with a moldboard plow. Subsequent working of the soil with other implements is usually performed to smooth the soil surface. Bare soil is exposed to the weather for some varying length of time depending on soil and climactic conditions.

**conventional tilling** Tillage operations considered standard for a specific location and crop and that tend to bury the crop residues; usually considered as a base for determining the cost-effectiveness of control practices.

**conventional treatment** *See* conventional filtration, conventional wastewater treatment, conventional water treatment.

**conventional treatment limitations** Constituents that are insufficiently removed from wastewater by conventional treatment include nitrogen, phosphorus, pathogens, toxins, and soluble nonbiodegradable compounds.

**conventional treatment plant** A plant that includes the unit operations and processes that provide conventional treatment. *See* conventional filtration, conventional wastewater treatment, conventional water treatment.

**conventional treatment train** *See* conventional water treatment.

**conventional trickling filter** A secondary wastewater treatment unit designed to operate without recirculation at loadings of 0.3–1.5 kg BOD/day/$m^3$ and 45–90 gpd/$ft^2$. Also called low-rate trickling filter or standard trickling filter. *See also* high-rate trickling filter, intermediate-rate trickling filter, roughing filter, super-rate filter, biofilter, biotower, oxidation tower, dosing tank, filter fly.

**conventional wastewater constituent** *See* conventional pollutant.

**conventional wastewater treatment** A combination of physical and biological processes designed to remove organic matter from wastewater, e.g., screening, grit removal, sedimentation, activated sludge, or trickling filtration. *See also* primary treatment, secondary treatment, tertiary treatment, advanced wastewater treatment, physical-chemical treatment, sludge handling and disposal.

**conventional water treatment** (1) The water treatment train that includes coagulation, flocculation, and sedimentation, with or without pretreatment, ahead of filtration and disinfection. Also called complete water treatment. (2) For purposes of meeting disinfection goals, conventional water treatment is sometimes defined as a baseline, two-stage solids-removal process (e.g., coagulation/sedimentation and filtration) that achieves some reduction of viruses and *Giardia* cysts, with respective log removal credits of 2.0 and 2.5. Other baseline treatment methods include the single-stage particulate removal processes of slow sand filtration, direct filtration, and diatomaceous earth filtration.

**conventional water treatment plant** A treatment plant that includes coagulation, flocculation, and sedimentation, with or without pretreatment, ahead of filtration, as opposed to a direct filtration plant.

**Convertofuser®** A coarse-bubble diffuser manufactured by FMC Corp.

**conveyance loss** In irrigation, conveyance losses from a canal, a reservoir, or a conduit consist of losses due to leakage, seepage, evaporation, transpiration, and operational waste, or, in general, losses between the point of diversion and the point of delivery.

**conveyor** A machine used to move wet or dry solid materials such as grit, screenings, and dewatered sludge cakes that cannot be easily pumped. *See* belt conveyor and screw conveyor.

**conveyor centrifuge** A device consisting of a rotating helical screw conveyor inside a rotating bowl, used for sludge dewatering. *See* solid-bowl centrifuge for detail.

**CO · OH** Chemical formula of the carboxyl radical.

**$Co_2O_3 \cdot 3\,H_2O$** Chemical formula of cobaltous hydroxide.

**cooking snow** Snow that yields a greater amount of water than average upon melting. Also called water snow.

**cooling** The use of evaporation to reduce water or wastewater effluent temperature to facilitate treatment or prevent thermal pollution.

**cooling agent waste** *See* cooling water.

**cooling coil** A coil of pipe or tubing that is used for cooling a hot fluid that it carries (or that surrounds it). Heat is transferred from the hot fluid, or to the cold fluid, inside the pipe or tubing.

**cooling degree-day** *See* degree-day.

**cooling pond** (1) A pond or other outside depression designed to receive hot process water for cooling by evaporation, convection, and radiation, before water reuse or discharge. (2) A large water tank that stores irradiated fuel elements from nuclear reactors to allow the decay of their fission products.

**cooling tower** A hollow, vertical structure that helps remove heat from water used as a coolant by exposing it to air; e.g., in electric power generating plants. Also defined as an open water recirculating device that uses fans or natural draft to draw or force ambient air through the device to cool warm water by direct contact (EPA-40CFR749.68-6). Also called an evaporative cooling tower. *See also* mechanical draft tower and natural draft tower.

**cooling tower blowdown** Concentrated water removed regularly from a cooling tower and replaced by fresh makeup water to control the concentrations of dissolved solids and other contaminants. It can be recycled where acceptable, e.g., as wash water, pump coolant, scrubber water makeup, utility water, drum seal water, and tank field waste. Same as blowdown (2).

**cooling tower blowdown water** Same as cooling tower blowdown.

**cooling tower makeup water** *See* makeup water.

**cooling tower precipitation** The drizzle, droplets, or mist that are entrained in the circulating air around the top of a cooling tower that is not equipped with spray eliminators; it may contain salt, dissolved materials, and any constituents of the water circulating in the tower, including the pathogens of Legionnaires' disease. It represents about 0.005% of the recirculating water. Also called carry-over, drift. *See also* blowdown, make-up water.

**cooling water** (1) Fresh, saline or brackish water used by industries, usually in a condenser, to reduce the temperature of liquids or gases. Cooling water is subject to more lenient requirements than drinking water or process water, but it should not produce scales, slimes, or otherwise interfere with the operation of plant equipment. It may be recycled with appropriate blowdown or used only once (once-through cooling). Cooling water may become contaminated by small leaks, corrosion products, or heat, but generally contains little, if any, organic matter. *See also* boiler-feed water. (2) Water used in the condensation step of a distillation system.

**coontail** A submerged plant that has most of its foliage beneath the water surface.

**coordinate bond** A chemical bond between two atoms, one of which provides the bonding electrons.

**coordinated phosphate treatment** The use of phosphate buffers to prevent hydroxyl alkalinity in boilers.

**coordination** Any combinations of cations with molecules or anions containing free pairs of electrons (bases). Also called complexation or complex formation.

**coordination compound** A complex compound consisting of a central metal ion attached to a group of surrounding molecules or ions by coordinate covalent bonds. Examples are the hydrolysis products of aluminum and other trivalent metal salts. Also called complex, complex ion, or complex compound. *See also* ligand and ligand donor atom.

**coordination number** The number of ligands usually coordinated to a central atom or ion.

**coordination sphere** The space around a central molecule, atom or ion, that is filled by the ligand species.

**COP** Acronym of clarifier optimization program.

**COP™** A clarifier designed or manufactured by WesTech Engineering, Inc. of Salt Lake City, UT.

**CopaClarifier** Secondary clarifier equipment manufactured by Hydro-Aerobics; includes filter brushes.

**Copa-NILL** A flushing unit manufactured by Hydro-Aerobics as a tipping bucket tank.

**Copasacs** A fine screen sack manufactured by Hydro-Aerobics.

**CopaScreen** A combination screening and dewatering apparatus manufactured by Longwood Engineering Co.

**Copasocks** A sock-like screening unit manufactured by Hydro-Aerobics.

**Copatrawl** A sock-like screening unit manufactured by Hydro-Aerobics.

**Copawash** An apparatus manufactured by Hydro-Aerobics for washing stormwater tanks.

**CopaWets** A coagulation–flocculation process designed by Copa Group for wastewater treatment.

**Copenhagen water** A specially standardized water, prepared by the Hydrographical Laboratories of Copenhagen (Denmark), widely accepted as a control sample for the analysis of seawater salinity. Copenhagen water, also called normal water or standardized seawater, has a chlorinity between 19.30 and 19.50 grams per kg, determined with an accuracy of ±0.001. (The water near the city of Copenhagen has a total dissolved solids content of approximately 30,000 mg/L.)

**copepod** An order of the class crustacea and the phylum arthropoda; a member of the order Copepoda, including *Cyclops*. Copepods are tiny marine or freshwater crustaceans that lack compound eyes or carapace and have six pairs of limbs on the thorax.

**copepoda** A group of tiny freshwater or marine crustaceans, including some plankton and some parasites, e.g., *Cyclops,* commonly found in ponds.

**copiotroph** An organism that can survive only in nutrient-rich conditions; also called r-strategist or zymogenous organism.

**copiotrophic organism** Same as copiotroph.

**Coplastix®** A proprietary composite of Ashbrook Corp. for use in the manufacture of sluice gates and stop logs.

**copper** (Cu, from its Latin name cuprum.) A reddish-brown, malleable, ductile, metallic element that conducts electricity and heat well; found in ores as carbonates, oxides, and sulfates. *See* azurite, bornite, chalcopyrite, cuprite, malachite. Atomic weight = 63.546. Atomic number = 29. Specific gravity = 8.96. Valence = 1 or 2. Copper is used in water pipes, electrical wiring, and utensils. It does not corrode easily, but oxidizes in continuous contact with the atmosphere and corrodes rapidly in the presence of chlorine, ozone, hydrogen sulfide, and other oxidizing agents. Its concentrations in natural waters are normally low; e.g., 20 μg/L. Copper is an essential dietary element (e.g., 2.0 mg daily), but can cause poisoning through drinking water at concentrations much higher than its taste threshold, which is below 8.0 mg/L. USEPA-recommended limits: MCLG = MCL = 1.3 mg/L. At high doses it can cause gastrointestinal disturbances, damage to the liver, and anemia. Dissolved copper can cause staining of laundry and the rapid corrosion of galvanized steel piping. Effective methods of removal: lime softening, reverse osmosis, electrodialysis. *See also* fluoride and molybdenum (for their effects on copper in water), green stain, Wilson's disease.

**copper arsenite ($CuHAsO_3$)** A yellow-green, water-insoluble powder; a highly toxic compound of arsenic trioxide and copper acetate; used as a pigment and in insecticides. Also called acidic copper arsenite, cupric hydrogen arsenite, Scheele's green, Schloss green. *See also* Paris green.

**copperas ($FeSO_4 \cdot 7H_2O$)** Common name for ferrous sulfate heptahydrate, a greenish-white, crystalline solid; a granular acid compound used with lime in water treatment for the coagulation of iron and manganese. It is also available as a liquid from processed spent pickle liquor. *See* ferrous sulfate for more detail. *See also* chlorinated copperas.

**copper complexes** Copper tends to make inorganic and organic complexes such that $[Cu^{2+}]/[Cu(II)]_{inorganic} = 0.01$ and $[Cu^{2+}]/[Cu(II)]_{dissolved} = 0.0000001 - 0.00001$.

**copper–copper surface electrode** A copper rod immersed in a solution of copper sulfate and constituting the cathode of a galvanic cell, which is used in field measurements of pipe-soil potentials.

**copper cyanide** Same as cuprous cyanide.

**copper hydroxide [$Cu(OH)_2$]** A blue, water-insoluble powder used in rayon and for copper salts. Also called cupric hydroxide.

**copper–nickel** A copper alloy that resists corrosion and cracking because of the addition of 10% to 30% of nickel. Also called cupronickel.

**copper pipe** A hollow cylindrical conduit made of copper, used in plumbing; copper tubing.

**copper pollution** The release of wastes containing copper compounds from such industries as copper plating, copper pickling, and rayon manufacture.

**copper salts** Copper sulfate and other copper compounds, used in the control of aquatic weeds and algae. Copper is highly toxic to fish and freshwater invertebrates; it may also accumulate in sediments.

**copper service** A water service line made of copper.

**COPPERSOL** A computer program that develops solubility diagrams for copper.

**copper sulfate ($CuSO_4$)** A chemical used since the early 1900s in water pretreatment for algae control at a concentration of 0.12–0.32 mg/L; higher dosages may be toxic to fish (whose tolerance varies from 0.15 mg/L for trout to 2.10 mg/L for bass). The solubility of the commercial product ($CuSO_4 \cdot 5\ H_2O$) varies between 20% and 31% by weight, depending on temperature. It is applied by dragging sacks of the chemical throughout the water reservoir or by blowing the crystals onto the surface from a boat. Chlorine is a supplemental algicide; the destruction mechanism is similar for both agents. In alkaline waters, sodium citrate is added to prevent or reduce precipitation of copper carbonate. The presence of carbon dioxide also affects dosage. Algae destruction by copper sulfate may lead to an intensification of odors, an increase in the number of saprophytic bacteria, and a decrease in dissolved oxygen. Also called blue copperas, blue stone, blue vitriol, chalcanthite, cupric sulfate. *See also* cupric chloramine.

**copper sulfating** Using copper sulfate to control algae in reservoirs.

**copper tubing** A hollow cylindrical conduit made of copper, used in plumbing.

**coprecipitation** (1) A water or wastewater treatment process that uses chemical precipitation to remove additional contaminants in addition to the usual targets. For example, chemical precipitation usually focuses on hardness elements (calcium, magnesium) and organic contaminants, but can also remove heavy metals, radionuclides, and viruses. Four forms of coprecipitation have been identified; *see* adsorption (4), inclusion, occlusion, solid-solution formation (AWWA, 1999). (2) In wastewater treatment, the addition of chemicals to the primary effluent, to the mixed liquor, or before secondary sedimentation, to form phosphorus and other precipitates that are removed with the waste activated sludge. *See also* preprecipitation, postprecipitation (Metcalf & Eddy, 2003).

**Co–Pt unit** Abbreviation of chloroplatinate unit or cobalt–platinum unit. *See* color unit.

**COR** Acronym of coefficient of reliability.

**coracidium** The ciliated larva that develops in freshwater from one unsegmented egg of a tapeworm such as *Diphyllobothrium latum.* It is a free-swimming stage. Plural: coracidia.

***Corbicula fluminea*** Scientific name of the Asiatic clam, a freshwater mollusk from Southeast Asia, now found in many waters of the United States; it causes problems of clogging, taste, and odor in drinking water sources and long transmission lines; chlorine is an effective control agent. *See also Dresissena polymorpha.*

**core** The uranium-containing heart of a nuclear reactor, where energy is released.

**core sample** A sample of the medium obtained to represent the entire bed when the bed is being analyzed for capacity or usefulness. A hollow tube is sent down through the bed to extract the sample.

**COR method** A statistical approach that uses a dimensionless coefficient (coefficient of reliability, COR) to determine a mean value ($m_d$) for a variable or parameter in the design of water or wastewater treatment processes (Metcalf & Eddy, 2003):

$$m_d = (\text{COR})\, X_s \qquad \text{(C-167)}$$

where $X_s$ is a fixed standard to be met at a specified reliability level (same unit as $m_d$)

**CORMIX1** A simple mixing model developed for the USEPA and used to establish effluent limits.

**CORMIX2** A simple mixing model developed for the USEPA and used to establish effluent limits.

**corner sweep** Tool used for scraping settled solids from the corners of rectangular or square sedimentation tanks.

**corona** (1) A colored circle or set of concentric circles of light around a luminous body, color being attributable to diffraction from thin clouds or dust. (2) A corona discharge.

**corona discharge** A luminous discharge of electricity at the surface of a conductor or between two conductors, causing the ionization of oxygen and the formation of ozone. Also called corona, electric glow, St. Elmo's fire. *See also* brush discharge, ozone generator.

**corona discharge cell** A device consisting of two electrodes separated by a discharge gap and a dielectric material across which a high voltage potential is maintained for the production of ozone. *See also* ozone generator.

**corona discharge method** A procedure for the generation of ozone from a clean, moisture-free air or oxygen gas stream using a corona discharge.

**Coronaviridae** *See* coronavirus.

**coronavirus** A single-stranded RNA, spherical virus of the Coronaviridae family, including some members that cause acute respiratory infections in humans, including the common cold. Some also cause acute gastroenteritis in piglets and calves. These viruses have petal-shaped or coronalike spikes projecting from the capsid.

**Corosex®** Magnesia produced by the Clack Corp. for neutralizing acidity in filters.

**Cor-Pak®** A catalytic oxidizer manufactured by ABB Air Preheater, Inc.

**corporation cock** Same as corporation stop.

**corporation stop** A water service shutoff valve located at a street water main, between the service pipe and the meter. This valve cannot be operated from the ground surface because it is buried and there is no valve box. Also called a corporation cock, corporation valve, ferrule.

**corporation valve** Same as corporation stop.

**corrasion** The mechanical erosion of soil and rock by rock fragments carried by wind, water, ice, or gravity.

**correlation coefficient** The square of the coefficient of determination. A dimensionless parameter used in linear regression analysis to measure the goodness of fit of a data series. A coefficient of 1.0 indicates perfect fit, whereas a coefficient of 0 indicates no correlation at all.

**correlative right** The doctrine according to which landowners are entitled to an amount of water proportional to their land areas during a drought. Also interpreted as meaning that rights of landowners over a common ground water basin are coequal, i.e., any one owner cannot take more than his or her share even if the rights of others are impaired.

**corrode** To eat or wear away by electrochemical action.

**corrosion** The gradual decomposition, dissolution, or destruction of a material by chemical action, often due to a complex electrochemical reaction. Corrosion may be caused by: (a) stray current electrolysis (electrolytic corrosion), (b) galvanic action of dissimilar metals (galvanic corrosion), (c) anaerobic bacteriological action, or (d) differential concentration cells. Corrosion starts at the surface of a material and moves inward. In water works, corrosion may affect well casings and screens, and metallic and concrete pipes. It may reduce their hydraulic capacities, cause structural failures, reduce water quality, etc. Metallic corrosion converts a metal to a salt or oxide and reduces its strength; it normally occurs at an anodic zone, with a transfer of electrons from that zone to a cathodic zone. Corrosion products in drinking water distribution piping can shield microorganisms from the action of disinfectants. *See* acid attack, aggressiveness index, biological corrosion, blue water, buffer index/intensity/capacity, calcium carbonate precipitation potential, cathodic current, cathodic protection, concentration cell corrosion, concrete corrosion, corrosion control, coupon weight-loss method, crevice corrosion, cuprosolvency, deactivation, dealloying, differential aeration cell, differential concentration cell, differential oxygenation corrosion, disequilibrium index, dissimilar metals, Eh–pH diagram, electrochemical cell, electrochemical rate measurement, electrolytic cathodic protection, EMF series, erosion corrosion, ferricalcic deposit, galvanic corrosion, galvanic series, graphitization, green water, immunity, impingement, inhibitors (anodic, cathodic), iron bacteria, Langelier saturation index, Larson ratio, loop system weight-loss method, marble test, microbial corrosion, microbiologically influenced corrosion, Nernst equation, noble metal, oxidation–reduction reactions, passivity, pitting/pitting corrosion, plumbosolvency, polarization, potential-pH diagram, Pourbaix diagram, protecting scale, red water, Riddick index, rusty water, Ryznar index, sacrificial anode, saturation index, saturometry, scale, selective leaching, sewer corrosion, solubility diagram, standard electromotive force, stray current corrosion, sulfate-reducing bacteria, sulfur oxidizer, tuberculation/tubercle, uniform corrosion, weight-loss method, X-ray diffraction.

**corrosion cell** A cell that can cause corrosion by generating electrons and positive ions at an anodic area that migrate to a cathodic area. A conducting solution (e.g., water) can connect the two areas to complete the circuit. A piece of iron immersed in water can form the corrosion cell shown in Figure C-28, with both oxidation and reduction reactions taking place:

$$\text{Anodic:}\quad Fe + 2\,H_2O \rightarrow Fe(OH)_2 + 2\,H^+ + 2\,e^- \qquad (C\text{-}168)$$

$$\text{Anodic:}\quad Fe + HCO_3 \rightarrow FeCO_3 + H^+ + 2\,e^- \qquad (C\text{-}169)$$

$$\text{Cathodic:}\quad 2\,H^+ + 2\,e^- \rightarrow H_2 \qquad (C\text{-}170)$$

$$\text{Cathodic:}\quad \tfrac{1}{2}\,O_2 + 2\,H^+ + 2\,e^- \rightarrow H_2O \qquad (C\text{-}171)$$

*See also* differential-aeration cell, dissimilar metals, galvanic cell, sacrificial anode.

**corrosion control** (1) Measures taken to keep the corrosivity of drinking water below levels that can constitute health hazards. Particular contaminants of interest include metals (cadmium, copper, iron, lead, zinc) and microorganisms that thrive in corrosion products. (2) Measures taken to protect metal piping against internal or external corrosion, e.g., anodic/cathodic/galvanic protection, corrosion-resistant materials, deactivation, select backfill, chemical/metallic coatings and linings, inert materials, inhibition, insulation from electrical current. Corrosion control is difficult in the operation of such processes as wet-air oxidation of sludge. *See also* polyethylene encasement.

**corrosion coupon** *See* coupon.

**corrosion index** Any one of the indexes that predict the corrosivity of water, i.e., the tendency of a water to dissolve or precipitate calcium carbonate or other solids. *See also* aggressive index, calcium carbonate precipitation potential, carbonate saturometer, disequilibrium index, Langelier index, Larson index, marble test, McCauley's driving force index, Pisigan–Singley equation, Riddick's corrosion index, Ryznar stability index, saturation index, saturometry.

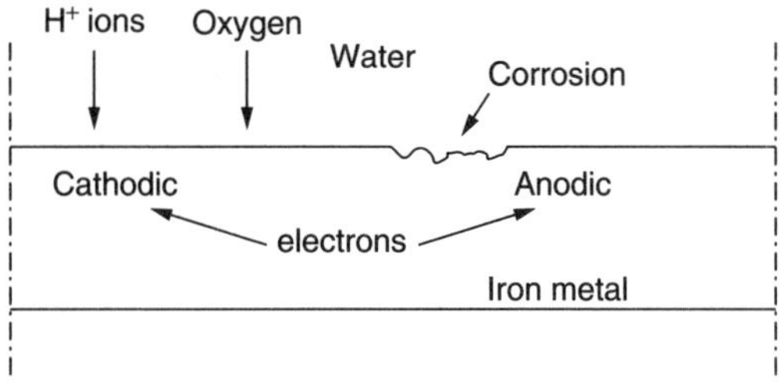


**Figure C-28.** Corrosion cell (adapted from Fair et al., 1971).

**corrosion inhibition** A corrosion control mechanism whereby the deposition or adsorption of insoluble or less soluble compounds protects a metallic surface. *See* inhibition for more detail.

**corrosion inhibitor** A substance that slows the rate of corrosion of metal plumbing materials by water, especially lead and copper materials, by forming a protective film on the interior surfaces of those materials and inhibiting the transfer of electrons and the diffusion of reactants ($H_2O$, $H^+$, $OH^-$). For example, corrosion-inhibition products containing orthophosphates are used to control lead, copper, or even iron uptake, by forming insoluble phosphates. Nonoxidizing inhibitors that promote the formation of passive protective films include bicarbonate, calcium carbonate, metaphosphate, and silicate. However, polyphosphates and silicates may also promote corrosion in some circumstances. *See also* blended phosphates, chemical inhibitor, hydroxypyromorphite, stannous chloride.

**corrosion pitting** One of two types of corrosion-related failures; it occurs when corrosion creates a pit in the pipe wall, which can grow until the pipe is fully penetrated, resulting in a leak.

**corrosion prevention** The elimination or reduction of corrosion. *See* corrosion control.

**corrosion rate** A parameter that expresses the progression of corrosion, for example, the units of length over time, as in mils per year (mpy or thousandths of an inch per year):

$$\text{mpy} = K/RA \qquad \text{(C-172)}$$

where $K$ is an electrochemical constant that depends on the metal and environmental conditions, $R$ is the electric resistance of the metal–soil specimen to linear polarization (ohms), and $A$ is the total area of the corroding specimen (square inches). The corrosion rate may also be expressed as a weight loss in milligrams per square decimeter per day.

**corrosion-resistant materials** (a) Corrosion-resistant metals or alloys that have electromotive forces close to those of noble metals or that produce a protective oxide coating against corrosion, such as copper, stainless steel, and tin; (b) coatings and linings (bituminous materials, chromium, paints, zinc) that prevent anodic and cathodic reactions. They resist corrosion even after prolonged exposure to the appropriate environment.

**corrosion scale** Product of the corrosion or corrosion control treatment of unlined iron and steel pipes in water distribution systems, resulting from such factors as pH variations, decreases in alkalinity, or increases in dissolved solids. Corrosion scales consist of such iron phases as magnetite, maghemite, hematite, green rust, siderite, and vivianite.

**corrosion treatment** Water treatment to eliminate or reduce corrosion. *See* corrosion control.

**corrosive** (1) Pertaining to a chemical agent that reacts with the surface of a material, causing it to deteriorate or wear away, or to a condition that promotes corrosion. (2) Pertaining to a chemical that destroys or irreversibly alters living tissue by direct action at the site of contact.

**corrosive hazardous waste** As defined by the USEPA, any aqueous waste having a pH less than or equal to 2.0, or greater than or equal to 12.5; or any liquid that corrodes steel at a rate greater than 6.35 mm or 0.25 inch per year at 55°C.

**corrosiveness** *See* corrosivity and coupon.

**corrosive substance** As defined by a United Nations Committee on the Transport of Dangerous Goods, a substance that, by chemical action, will cause severe damage when in contact with living tissue, or, in the case of leakage, will materially damage, or even destroy, other items or means of transport; it may also cause other hazards (Nemerow & Dasgupta, 1991).

**corrosive water** Water that is undersaturated with respect to calcium carbonate ($CaCO_3$), i.e., water with a Langelier saturation index LSI < 0; somewhat of an incorrect use of the term because the LSI only relates to the presence or absence of calcium carbonate scale.

**corrosivity** An indication of the corrosiveness of a water. The corrosiveness of a water depends on the water's pH, alkalinity, hardness, temperature, total dissolved solids, dissolved oxygen concentration, and the Langelier index. Other water quality parameters involved in corrosion include chlorides, hydrogen sulfide, sulfates, and bacteria (particularly at dead ends in the distribution system).

**corrugated plate interceptor (CPI)** A device made of corrugated plates to separate nonemulsified oil from water based on their density difference.

**Corten** A corrosion-resistant steel fabricated by U.S. Steel Corp.

***Corynebacterium* genus** A group of heterotrophic organisms active in biological denitrification.

***Corynebacterium spp.*** A group of heterotrophic organisms active in biological denitrification, responsible for hydrolysis and fermentation.

**cosettling thickening** One of six common methods of sludge thickening used in wastewater treatment. The method consists of allowing solids to thicken in a primary clarifier that incorporates a sludge blanket and has retention times of 12 to 24 hours. For example, one "thickening clarifier" in a bank of several clarifiers may be used for cosettling thickening. *See also* centrifugal thickening, dissolved air flotation, gravity belt thickening, gravity thickening, and rotary drum thickening.

**cosmic pollution** Pollution that comes from the sun in the form of radioactivity: alpha, beta, and, particularly, gamma, rays.

**cost–benefit analysis** A technique used to compare various alternatives of a project or various projects. It is a quantitative evaluation of the overall benefits to society of a proposed action versus the costs that would be incurred, including not only monetary factors but also intangible benefits, which are often difficult to assess. *See* benefit-cost analysis for more detail.

**cost/benefit ratio** *See* its reciprocal, benefit/cost ratio.

**cost-effective alternative** An alternative control or corrective method identified after analysis as being the best available in terms of reliability, performance, and cost. *See* financial indicators for related terms.

**cost-effectiveness** The characteristic of a project or an action that produces the optimum result for the cost. Also called cost-efficiency.

**cost-effectiveness analysis** An analysis to determine among several alternatives which one yields the optimum effect at the most reasonable cost, i.e., which one is the most effective in meeting economic, financial, technical, environmental, and other objectives. The cost-effective solution is not necessarily the least-cost solution. *See* financial indicators for related terms.

**cost-efficiency** Same as cost-effectiveness.

**cost sharing** A publicly financed program through which society, as the beneficiary of environmental protection, shares part of the cost of pollution control with those who must actually install the controls.

**cothickening** An old practice consisting of the discharge of biological sludge to the influent end of a primary clarifier for combined settling and consolidation with the primary sludge. Separate sludge thickening is now more common.

**cotreatment** The joint treatment of two different wastes at a single facility, e.g., municipal and industrial wastewaters, or septage with wastewater at a treatment plant in which a pumper truck can unload the septage into a holding tank.

**cotton sizing waste** Wastewater generated during the manufacture of cotton cloth; adsorption is an appropriate treatment method.

**coulomb** A measurement of the amount of electrical charge conveyed in one second by an electric current of one ampere. One coulomb equals about $6.25 \times 10^{18}$ electrons.

**Coulomb, Charles Augustin de** French physicist and inventor (1736–1806).

**Coulombic correction factor** *See* Coulombic factor.

**Coulombic effect** *See* Coulombic factor.

**Coulombic factor** A number that expresses the interdependence of Coulombic interaction energy with pH in surface phenomena. It depends on surface charge ($\Psi$), the Faraday constant ($F$), the molar gas constant ($R$), and the absolute temperature ($T$):

$$C = \exp(-F\Psi/RT) \qquad \text{(C-173)}$$

Coulombic effects may contribute as many as 20 kJ/mol or 200 mV to surface reactions (Stumm & Morgan, 1996).

**Coulombic force** The electrostatic force of attraction or repulsion between two point charges. *See* Coulomb's law.

**Coulomb's law** In an environmental system, the Coulombic force or force of attraction or repulsion ($F$) between a pair of charges $Q_1$ and $Q_2$ separated by a distance $D$ is

$$F = kQ_1Q_2/(\varepsilon D^2) \qquad \text{(C-174)}$$

where $k$ is a proportionality constant and $\varepsilon$ is the dimensionless dielectric constant.

**coulometric cell** A sealed vessel containing two electrodes and a water or wastewater sample, used in conducting coulometric titrations. One of the electrodes produces a substance that reacts with the sample.

**coulometric titration** A method of measuring the quantity of electricity (number of coulombs) generated by an exchange of electrons between a water or wastewater sample and a substance produced by an electrochemical cell.

**coulometry** A quantitative analysis method used to determine the amount of a substance that is liberated or deposited by electrolysis by measuring the quantity of electricity (number of coulombs) passing through the substance.

**Council on Environmental Quality** A council appointed by the president of the United States to advise on national environmental policies.

**count** *See* aerobic plate count, bacterial count, heterotrophic plate count, plate count, total bacterial count, total count, water plate count.

**Counter Current®** An aeration process developed by the Schreiber Corp. using rotating diffusers.

**countercurrent adsorption** A mode of operation of activated carbon adsorption units in the treatment of wastewater with large concentrations of impurities. It uses two (or sometimes more) carbon beds. Wastewater is treated by once-used carbon and then by virgin or regenerated carbon. Carbon used twice is discarded or regenerated (WEF & ASCE, 1991).

**countercurrent aeration system** A proprietary aeration system (CCAS™) used to provide aeration and mixing in the activated sludge process. It incorporates diffusers mounted at the bottom of a revolving bridge that is moving faster than the contents of the circular basin.

**countercurrent centrifuge** One of two basic configurations of solid-bowl centrifuges, whereby, after entering at the junction of the conical and cylindrical sections, the solids move to the conical end and the liquid moves towards the other end. *See also* cocurrent centrifuge.

**countercurrent efficiency** A feature of a granular activated carbon adsorber that allows a partially spent carbon to adsorb contaminants before the feedwater comes in contact with fresh carbon.

**countercurrent elutriation** Elutriation of sludge (washing it out in water to reduce its alkalinity) in multiple tanks, one after another, the entire volume of washwater being introduced into the last tank and moving toward the first tank while the sludge moves in the opposite direction. The washwater from the first tank is wasted or returned to the plant for treatment. Assuming that the tanks are identical, the alkalinity of the elutriated sludge and the washwater requirement may be calculated as follows (Fair et al., 1971):

$$X = [A(R-1) + WR(R^N - 1)]/(R^{N+1} - 1) \quad \text{(C-175)}$$

$$(R^{N+1} - 1)/(R - 1) = (A - W)/(E - W) \quad \text{(C-176)}$$

where $X$ = alkalinity of the elutriated sludge, $A$ = alkalinity of the sludge moisture before elutriation, $R$ = ratio of washwater to sludge volumes, $W$ = initial alkalinity of washwater, and $N$ = number of tanks. *See also* countercurrent elutriation.

**countercurrent flow** Flow in opposite direction. *See* countercurrent operation.

**countercurrent operation** The operation of a granular carbon adsorber or any similar treatment unit with the adsorbent continuously input into the unit in the opposite direction of the fluid movement. More generally, countercurrent operation applies to two components of a unit that flow in opposite directions. Other examples of countercurrent operation include air strippers, heat exchangers, and ozonizing towers. *See also* cocurrent, fixed bed, fluidized bed, recirculation.

**countercurrent settler** An inclined plate or tube settler in which the suspension is fed below the settling modules and flows upward while the solid particles settle onto the lower surface of the channels. *See also* cocurrent settlers, crossflow settlers.

**countercurrent settling** The mode of operation of a countercurrent settler. The condition for particle removal is (AWWA, 1999):

$$V \geq WV'/(L \cos \theta + W \sin \theta) \quad \text{(C-177)}$$

where $V$ = settling velocity of particles removed, $W$ = perpendicular distance between surfaces, $V'$ = liquid velocity between the inclined surfaces, $L$ = required length of surface, and $\theta$ = the angle of surface inclination from the horizontal. *See also* Figure C-29, Lamella® Gravity Settler, cocurrent settling, crossflow settling.

**counterion** Any one of a group of ions or particles having opposite charges; counterions attract each other and are held by electrostatic and van der Waals forces. *See also* co-ion, Donnan potential, electrical double layer.

**coupled biological treatment system** A wastewater treatment method that incorporates two different processes; also called a dual process. Most plants commonly using dual processes combine a fixed film unit with a suspended growth reactor. *See also* combined filtration–aeration process and dual biological treatment system.

**coupled trickling filter/activated sludge system** A wastewater treatment system that combines features of the two processes. *See* combined filtration–aeration process.

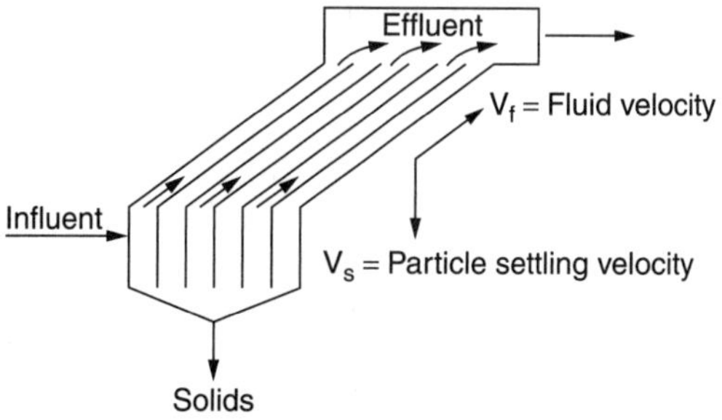


**Figure C-29.** Countercurrent settling (adapted from Crittenden et al., 2005).

**couple potential** The driving force of an overall redox reaction or the difference between the standard potentials of the two half-reactions. *See* cell potential for more detail.

**coupling action** Corrosion caused by a galvanic cell. Also called galvanic corrosion. *See also* electrolytic corrosion.

**coupling model** The combination of a free energy model and a homogeneous or exponential solution-diffusion model to represent mass transfer processes in a multisolute system. It is either the linear solution diffusion model or the film theory model, with the mass transfer coefficient ($K_i$) determined as follows:

$$\ln K_i = \ln (C^*) + \Sigma(-\Delta G/RT) \qquad \text{(C-178)}$$

where $C^*$ = membrane specific constant, $\Delta G$ = difference of coupled ion free energy at the interface and the bulk, $R$ = universal gas constant, and $T$ = temperature.

**coupon** A steel specimen (coupon) inserted into water to measure the corrosiveness of the water. The rate of corrosion is equal to the loss of weight of the coupon (in milligrams) per surface area (in square decimeters) exposed to the water per day.

**coupon test** A method used to determine the rate of corrosion or scale formation by placing metal strips, or coupons, of a known weight in a tank or pipe.

**coupon weight-loss method** A method of measuring corrosion rates; see coupon test. *See also* loop system weight-loss method, electrochemical rate measurement.

**covalent bond** The chemical bond between two atoms that share a pair of electrons; e.g., atoms that form a radical; the bond between carbon atoms, the C—H bond in methane ($CH_4$), and the N—H bond in ammonia ($NH_3$). *See also* coordinate bond.

**cover crop** A crop that provides temporary protection for delicate seedlings and/or provides a canopy for seasonal soil protection and improvement between normal crop production periods. Except in orchards, where permanent vegetative cover is maintained, cover crops usually are grown for one year or less. When plowed under and incorporated into the soil, cover crops are also referred to as green manure crops.

**covered anaerobic lagoon** *See* anaerobic covered lagoon.

**covered sludge drying bed** A sludge drying bed in a glass enclosure to keep precipitation out and facilitate evaporation.

**covered storage** A completely enclosed water reservoir to prevent contamination.

**Covertite** A cover manufactured by Thermacon Enviro Systems, Inc. for wastewater treatment tanks.

**$C_3OX_6$** Chemical formula of haloketone. Also written as $CX_3COCX_3$, with X = $Cl^-$, $Br^-$, or $H^+$.

**coxsackievirus** An enterovirus consisting of single-stranded RNA, classified as a picornavirus, excreted in the feces of infected persons and occasionally found in drinking water, particularly in the summer, with few if any enteric symptoms in humans. These viruses cause a variety of infections and symptoms, including the common cold, fever, aseptic meningitis, hepangina, heart disease (myocarditis), and gastrointestinal problems.

**coxsackievirus A** A nonpolio enterovirus, more resistant to chlorination (i.e., having a higher CT value) than *E. coli*. It causes fever, respiratory disease, meningitis, and herpangina.

**coxsackievirus B** A prevalent nonpolio enterovirus, found in water and wastewater; it causes an infection that is usually more severe in newborns than other age groups, and that may lead to fever, rash, respiratory disease, encephalitis, hepatitis, myocarditis, congenital heart anomalies, meningitis, pleurodynia, and even death.

**CP** Acronym of cathodic protection.

***C. parvum*** *See Cryptosporidium parvum.*

**CPE** Acronym of (a) cytopathic effect, (b) cytopathogenic effect.

**CPI** Acronym of corrugated plate interceptor.

***C. pipiens*** Abbreviation of *Culex pipiens.*

***C. p. molestus*** Abbreviation of *Culex pipiens molestus.*

***C. p. pallens*** Abbreviation of *Culex pipiens pallens.*

***C. p. quinquefasciatus*** Abbreviation of *Culex pipiens quinquefasciatus.*

**cpu** or **CPU** Acronym of chloroplatinate unit.

**CPVC** Acronym of chlorinated polyvinyl chloride.

***C. quinquefasciatus*** Abbreviation of *Culex quinquefasciatus.*

**Cr** Chemical symbol of the metallic element chromium.

**cracking** A primary endothermic reaction of the pyrolysis of hydrocarbon compounds, in which species of low volatility cleave to more volatile species, e.g.:

$$C_xH_y \rightarrow C_dH_d + C_mH_n \qquad \text{(C-179)}$$

where $x = c + m$ and $y = d + n$. *See also* oxidative pyrolysis.

**cracks and contraction of filter bed** A problem encountered in the operation of a depth filter, whereby, as a result of improper cleaning, the grains of the filtering medium become coated and cracks develop at the sidewalls of the filter. Mudball formation also results sometimes. Other depth filtration operation problems include turbidity breakthrough, emulsified grease buildup, filter medium loss, and gravel mounding. *See also* filter crack.

**cradle to grave** A concept in hazardous waste management to track waste from its generation to its ultimate disposal.

**crank-and-flywheel pump,** A steam-driven reciprocating pump with a flywheel mounted on a crankshaft to store energy that is then used to raise a fluid.

**Crane®** A line of products of Cochrane Environmental Systems used in water treatment.

**$Cr(C_2H_3O_2)_3 \cdot H_2O$** Chemical formula of chromic acetate.

**creamery** A plant that processes whole milk, sour cream, and sweet cream into butter and other products.

**created wetland** An upland shallow body of water converted into a wetland. *See also* artificial wetland.

**creep** *See* saltwater creep.

***Crenothrix*** A genus of autotrophic bacteria with unbranched attached filaments having a gelatinous sheath. They can oxidize and precipitate iron as hydroxide [$Fe(OH)_3$], which may cause staining and capacity reduction in wells and pipelines.

***Crenothrix polyspora*** A species of filamentous, iron-oxidizing bacteria that can deposit gelatinous ferric hydroxide.

**cresol ($C_7H_8O$)** Any of three isomeric compounds derived from coal or wood tar and used as a disinfectant. Also called methyl phenol. *See also* tricresol.

**cresol red** A reagent used in water analyses as an acid–base indicator, with pH transition ranges of 7.0 (yellow)–8.8 (purple), and 0.2 (red)–1.8 (yellow).

**crest** (1) The upper edge of a weir or weir plate. (2) The top of a dam, dike, or spillway that water must reach before passing over the structure. (3) The highest point of a wave. (4) The highest elevation reached by floodwaters flowing in a channel.

**crevice corrosion** Localized corrosion in narrow crevices filled with liquid, at gaskets, lap joints, rivets, and surface deposits; caused by changes in acidity, oxygen depletion, and dissolved ions.

**crib weir** A diversion weir of log cribs filled with rock.

**crisis management** An approach used in water utility risk management; a series of measures are taken to solve critical problems or reduce their impacts.

**criteria** Descriptive factors taken into account by USEPA in setting standards for various pollutants. These factors are used to determine limits on allowable concentration levels, and to limit the number of violations per year. When issued by USEPA, the criteria provide guidance to the states on how to establish their standards.

**criteria pollutant** Any pollutant for which the USEPA has established national ambient air quality standards to protect human health and welfare: ozone, carbon monoxide, total suspended particulates, sulfur dioxide, lead, and nitrogen oxide. The term "criteria pollutant" derives from the requirement that EPA must describe the characteristics and potential health and welfare effects of these pollutants. It is on the basis of these pollutants that standards are set or revised (EPA-40CFR51.852).

**criterion continuous concentration (CCC)** A parameter designed to protect aquatic organisms against long-term/chronic effects. *See* toxic units for detail.

**criterion maximum concentration (CMC)** A parameter designed to protect aquatic organisms against short-term/acute effects. *See* toxic units for detail.

**criterion of detection** One of a few parameters used to measure the precision or the detection level of a method of analysis, stated as 1.645 times the standard deviation of blank analyses. Also called critical detection level.

**critical bed depth** (1) In a granular activated carbon bed, the depth that defines the zone where adsorption occurs and lies between the fresh carbon and spent carbon. (2) The minimum depth of an ion exchange unit required to contain the mass transfer zone. *See also* mass transfer zone.

**critical coagulation concentration (CCC)** The concentration of an electrolyte needed to destabilize a colloidal suspension. As electrolytes are added, the thickness of the diffuse electrical layer and the zeta potential are reduced, and so are the surface charges.

**critical component analysis (CCA)** A frequently used approach for analyzing the mechanical reliability of a water or wastewater treatment plant; developed by the USEPA to determine which mechanical components of the plant will have the most immediate effect on performance in case of failure. The analysis is based on the assessment of four factors: expected time before failure, inherent availability, mean time before failure, and operating availability.

**critical deficit ($D_c$)** The maximum dissolved oxygen deficit, corresponding to the lowest point on the oxygen sag curve, i.e., the point of minimum dissolved oxygen. *See* critical point for detail.

**critical depth ($y_c$)** (1) In open channel flow, the depth for which the specific energy ($E$) is minimum for a given discharge ($Q$) or the discharge is maximum for a given specific energy. Given the critical flow equation, critical depth in wide channels occurs, in general, when

$$Q^2 W = gA^3 \qquad \text{(C-180)}$$

For a rectangular channel,

$$y_c = (q^2/g)^{1/3} \qquad \text{(C-181)}$$

the discharge per unit of surface width of channel is

$$q = Q/W \qquad \text{(C-182)}$$

where $W$ is the surface width, $A$ is the cross-sectional area, and $g$ the gravitational acceleration. This flow characteristic serves to determine the flow regime: on a mild slope, the critical depth is smaller than the normal depth ($y_n$), and the flow is subcritical; on a steep slope, the critical depth is greater than the normal depth and the flow is supercritical. (2) Same as critical bed depth.

**critical detection level** One of a few parameters used to measure the precision or the detection level of a method of analysis, stated as 1.645 times the standard deviation of blank analyses. Also called criterion of detection.

**critical dilution rate** In a chemostat or a continuously stirred biological reactor, the critical dilution rate ($D_c$, /time) is the rate below which steady state can be achieved and maintained:

$$D_c = \mu \qquad \text{(C-183)}$$

where $\mu$ is the specific growth rate as given in the Monod equation. The chemostat operates efficiently when the dilution rate is about equal to the maximum specific growth rate ($\mu_{max}$). Above the critical rate, cells are washed out and efficiency declines.

**critical dilution ratio (CDI)** The dilution achieved under the worst-case ambient conditions, e.g., in the zone of ocean discharge under the effects of initial momentum and buoyancy or at the boundary of a mixing zone for a river discharge. *See* toxicity terms, toxic units.

**critical dissolved oxygen** The minimum concentration of dissolved oxygen to maintain in an aeration basin for the biota to be able to metabolize organic matter. The recommended design minimum is 2.0 mg/L, however, some basins operate satisfactorily with as little as 0.5 mg/L.

**critical drawdown period** In the operation of a lake or reservoir, it is the time between the beginning of drawdown and the lowest useful water surface elevation.

**critical flow** In open channel flow, critical flow is a transition between two varied-flow conditions: subcritical and supercritical flows. Critical flow occurs in flow measurement devices such as weirs and at or near free discharges. It corresponds to a minimum specific energy for a given discharge or a maximum discharge for a given specific energy. Specific energy ($E$) is defined as the sum of the depth of flow ($y$) and the velocity head ($V^2/2g$), where $V$ is the average velocity and $g$ the acceleration of gravity. *See* the critical flow equation. Some other characteristics of critical flow follow. (1) The mean velocity is equal to the celerity of a gravity wave, i.e., $V = (gy)^{0.5}$. (2) The Froude number equals 1, i.e., $F_r = 1$. When the Froude number is close to 1, the flow is unstable and subject to wave formation. (3) The velocity head is half of the depth, i.e., $V^2/2\,g = y/2$, for a rectangular channel. (4) Under critical or supercritical flow conditions, a wave cannot move upstream as the water velocity exceeds the wave celerity, i.e., downstream conditions or controls do not affect upstream sections hydraulically. (5) Flow is supercritical if (a) $F_r > 1$, or (b) the mean velocity is greater than the celerity of the gravity wave, or (c) the normal depth ($y_n$) is less than the critical depth ($y_c$), or (d) the normal slope ($S_n$) exceeds the critical slope ($S_c$). (6) Flow is subcritical if (a) the mean velocity is smaller than the celerity of the gravity wave, or (b) $F_r < 1$, or (c) $y_n > y_c$ or (d) $S_n < S_c$; then disturbances travel upstream and downstream.

**critical flow equation** In open channel flow, the relationship between specific energy ($E$), depth of flow ($y$), average velocity ($V$), and the gravitational acceleration ($g$):

$$E = y + V^2/2\,g \qquad \text{(C-184)}$$

It may be written as

$$E = y + q^2/2\,gy^2 \qquad \text{(C-185)}$$

for a rectangular channel, or as equation (C-180) for any wide channel. The unit discharge ($q$) is the ratio of the discharge ($Q$) to the surface width ($W$): $q = Q/W$. *See* the specific energy equation.

**critical flux** *See* critical permeate flux.

**critical groundwater area** An area with groundwater problems, e.g., declining water levels, so that development and use of this resource are limited.

**critical habitat** An area designated as critical for the survival and recovery of threatened or endangered species.

**critical initial dilution (CID)** The dilution of waste discharges achieved under worst-case ambient conditions. *See* toxic units.

**critical load** In general, the maximum load that does not cause harmful effects to the receptor. With respect to weathering and acid deposition phenomena, the critical load (CL) is the tolerable acid deposition, i.e., the amount in milliequivalents per square meter per year ($meq/m^2/yr$) that must not be exceeded to avoid acidification of forest soils and the release of aluminum and hydrogen ions to the soils. It may be estimated as the difference between the weathering rate ($W$, $meq/m^2/yr$) and the amount of alkalinity that leaches from the soil ($Alk_L$, $meq/m^2/yr$):

$$CL = W - Alk_L \quad (C\text{-}186)$$

**critical micelle concentration** The surfactant concentration that triggers the formation of micelles by monomers.

**critical permeate flux** The maximum flux at which a low-pressure membrane filtration system can operate effectively without fouling. Critical flux has also been defined as (a) the flux below which transmembrane pressure (TMP) does not increase with time of operation; (b) the flux that does not cause deposition of colloids on the membrane; (c) the minimum flux that, when successively increased and decreased, does not cause TMP hysteresis.

**critical pitting temperature** A parameter used to define resistance to pitting corrosion.

**critical point** (1) The point at which a substance has the same density, pressure, and temperature in two phases; e.g., the combination of density, pressure and temperature at which point a gas and liquid become indistinguishable. (2) The point on the dissolved oxygen sag curve at which the deficit is maximum. Based on the classical dissolved-oxygen sag equation, the coordinates of the critical point are (Fair et al., 1971):

$$\text{Critical time: } t_c = 2.3 \log\{f[1-(f-1)(D_a/L_a)]\}/[k(f-1)] \quad (C\text{-}187)$$

$$\text{Critical deficit: } D_c = L_a/f\{f[1-(f-1)(D_a/L_a)]\}^{1/(f-1)} \quad (C\text{-}188)$$

where $f$ = self-purification or oxygen-recovery ratio, $D_a$ = initial dissolved-oxygen deficit, $L_a$ = initial first-stage BOD (at the point of discharge or point of reference), and $k$ = deoxygenation rate.

**critical point drying** A procedure used to eliminate distortions of specimens for analysis by a scanning electron microscope.

**critical pressure** The minimum pressure necessary to liquefy a gas that is at critical temperature.

**critical scour velocity** The maximum horizontal velocity in a sedimentation tank to avoid resuspension (scouring) of settled particles. It may be calculated from the same Camp–Shields equation used for scouring velocity.

**critical solids retention time (critical SRT)** In biological treatment, the critical solids retention time (SRT, days) is the SRT value below which waste stabilization does not occur; also called minimum solids retention time (SRTmin), i.e., the residence time at which microbial cells are washed out of the system or wasted faster than they can reproduce (Metcalf & Eddy, 2003):

$$1/SRT_{min} = [YkS_0/(K_s + S_0)] - k_d \quad (C\text{-}189)$$

When $S_0$ is much greater than $K_s$:

$$1/SRT_{min} \approx Yk - k_d = \mu_m - k_d \quad (C\text{-}190)$$

where $Y$ = synthesis yield coefficient, g VSS/g bsCOD; $k$ = maximum specific substrate utilization rate, g/g/day; $S_0$ = influent BOD concentration, mg/L; $K_s$ = half-velocity constant, substrate concentration at one-half the maximum specific substrate utilization rate, mg/L; $k_d$ = endogenous decay coefficient, g VSS/g VSS/day; and $\mu_m$ = maximum specific biomass growth rate, g new cells/g cells/day. *See also* process safety factor.

**critical SRT** *See* critical solids retention time.

**critical stratification temperature difference ($|\Delta T|$)** The absolute value of the difference between the temperature of the inflow and the temperature of the contents of a storage tank that can cause stratification, estimated as follows (Grayman et al., 2004):

$$|\Delta T| = CQ^2/(D^3H^2) \quad (C\text{-}191)$$

where $C$ is a coefficient, $Q$ (cfs) the inflow rate, $D$ (ft) the diameter of the inlet of the tank, and $H$ (ft) the depth of water in the tank. *See also* mixing time.

**critical temperature** The temperature above which a gas cannot be liquefied solely by an increase in pressure.

**critical velocity** (1) In open-channel flow, Belanger's critical velocity ($V_c$) is that velocity corresponding to the critical flow condition, i.e., the point of minimum specific energy and maximum discharge. In a rectangular channel, it is such that the velocity head is one half of the critical depth ($y_c$), i.e.,

$$V_c^2/2\,g = 0.5\,y_c \qquad \text{(C-192)}$$

In general, the critical velocity is the velocity of the gravity wave; it may be computed as the ratio of the critical discharge to the critical flow area, or, as for the critical slope, from the Manning formula. Belanger's critical velocity is different from Reynolds critical velocity. *See also* open channel flow. (2) Critical velocity may also be defined as the minimum velocity that will prevent silting in a channel, computed as

$$V_c = 1.17\,(fR)^{0.5} \qquad \text{(C-193)}$$

with a friction factor ($f$):

$$f = 8\,d^{0.5} \qquad \text{(C-194)}$$

$d$ being the diameter (in inches) of the predominant type of soil transported, and $R$ the hydraulic radius in feet. *See also* critical scour velocity. (3) The velocity above which erosion will occur. *See* erosion threshold flow. (4) There is also a critical velocity for liquid turbulence to prevent polarization.

**$CrO_3$** Chemical formula of chromic acid (2).

**$Cr_2O_7$** Chemical formula of the dichromate ion.

**Cromaglass® SBR** A proprietary sequencing batch reactor tank system of Cromaglass Corp., constructed of lightweight fiberglass, for the treatment of wastewater.

**crooksite** One of the rare minerals that contain the metallic element thallium.

**crop coefficient** A factor ($K_c$) that multiplies the evapotranspiration ($ET$) related to a reference crop such as tall fescue grass or alfalfa (reference $ET$ or $ET_r$) to obtain the evaporation ($ET_a$) applicable to a given vegetation (Ruiz et al., 2006):

$$K_c = ET_a/ET_r \qquad \text{(C-195)}$$

The crop coefficient is used in the design or scheduling of irrigation systems with wastewater effluents.

**crop consumptive use** The amount of water transpired during plant growth plus what evaporates from the soil surface and foliage in the crop area.

**crop for direct human consumption** Any crop that is consumed by humans without processing to minimize pathogens prior to distribution to the consumer.

**crop irrigation system** A natural, slow-rate treatment system designed primarily for water and nutrient reuse through crop production or landscape irrigation. *See* slow-rate system type 2 for detail.

**cropland** Land that is suitable for or actually used for the cultivation of crops, except forest crops and permanent pasture.

**cropped area** The actual area cultivated in a land application system, exclusive of land required for preapplication treatment facilities, buffer zones, service roads, and storage reservoirs. Also called field area and determined as follows (Metcalf & Eddy, 1991):

$$A_w = (365\,Q + \Delta V_s)/[L_w/12)(43{,}560)] \qquad \text{(C-196)}$$
$$= 0.000275\,(365\,Q + \Delta V_s)/L_w$$

where $A_w$ = field area, acres; $Q$ = average daily wastewater flow, ft³/day; $\Delta V_s$ = net loss or gain in stored water volume due to precipitation, evaporation, and reservoir seepage, ft³/yr; and $L_w$ = design hydraulic loading rate, in/yr. The numbers 365, 12, and 43,560 correspond, respectively, to the number of days in a year, the number of inches per foot, and the number of square feet in one acre. There is a slight confusion in the literature between this term and field application area.

**crop rotation** A system of farming in which a regular succession of different crops are planted on the same land area; e.g., a 7-year rotation as opposed to planting the same crop time after time (monoculture). Crop rotation helps maintain soil fertility by reducing the risk of depleting the soil of particular nutrients and prevent the buildup of insect and fungal pests.

**crop salinity threshold** The maximum permissible electrical conductivity of the soil saturation extract for a given crop. Also called vegetation salinity threshold or soil salinity threshold. *See also* leaching requirement.

**cross** A pipe fitting with two pairs of branches at right angles.

**cross-collector** A mechanism used to collect sludge in sedimentation tanks and convey it to a point of removal.

**cross-connection** An actual or potential connection, usually unintended or undesirable, between two water conduits or bodies; for example, (a) between a potable water supply and a wastewater or otherwise polluted source or (b) between sanitary and storm sewers. *See also* backflow, backsiphonage, and interconnection. Cross-connections are responsible for more than 20% of recent waterborne disease outbreaks in the United States. Protected and regularly inspected cross-connections, including backflow preventers, are sometimes allowed to supply industrial firefighting flow.

**cross-connection control** A series of measures taken by health authorities and public health engineers to eliminate the causes of drinking water contamination (mostly pathogens, but also chemicals) through back pressure or backsiphonage. Such measures include survey of potential sources, installation of preventing devices, and administration of a control program.

**cross contamination** (1) Contamination by cross-connection; the intermixing of two water streams that results in unacceptable water quality for a given purpose. (2) Accidental mixing of anion and cation resins following regeneration in a mixed bed deionizer.

**Cross/Counteflo** An inclined plate clarifier manufactured by Zimpro Environmental, Inc.

**Cross-Flo** An inclined static screen manufactured by the Kason Corp.

**crossflow** (1) Flow that occurs at right angles to an inclined plate settler. (2) Same as crossflow filtration (1). (3) Same as crossflow media.

**crossflow filtration** (1) The operating condition of a pressure-driven membrane unit in which the permeate flows through the membrane perpendicularly to the feedwater and concentrate. The feedwater flows at a high velocity to suspend contaminants and reduce fouling; water that does not pass through the membrane may be recirculated through the membrane after blending with new feedwater or directed to a storage reservoir. Also called recirculation filtration. *See* direct feed filtration, transmembrane pressure (TMP), recovery rate, rejection rate, constant flux operation, constant TMP operation. *See also* cake filtration, depth filtration, Figure C-30. (2) A type of filtration in which the feedwater flows parallel to the surface of the medium.

**Crossflow Fouling Index™** An index devised by Argo Scientific to define membrane fouling.

**crossflow media** A plastic material of molded or extruded shape, used as medium in trickling filters and in packed biological reactors.

**crossflow packed tower** A device used to strip volatile organic compounds, carbon dioxide, and other gases from water or wastewater. It consists of a packed tower, with fresh air introduced across the tower packing at right angles to the direction of the liquid flow. *See also* cascade air stripping, cocurrent packed tower, cascade packed tower, end effect, low-profile air stripper, minimum air-to-water ratio, packed tower design equation, sieve tray column, wall effect.

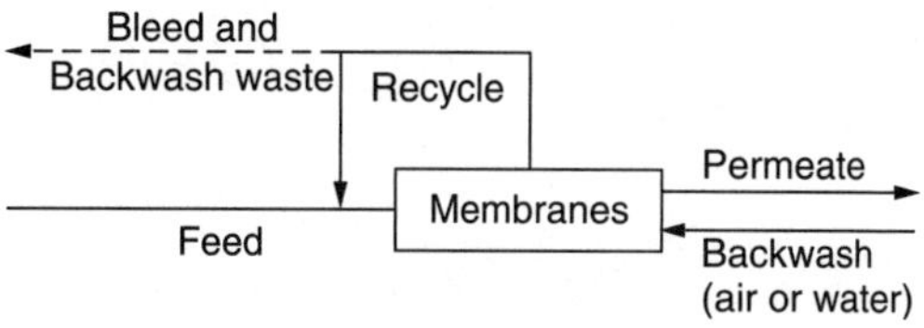


**Figure C-30.** Crossflow filtration.

**crossflow pervaporation system** A membrane separation technique that removes volatile organic compounds from liquids. It is used in groundwater remediation, leachate treatment, and the treatment of wastewaters that contain solvents, degreasers, and gasoline. *See* pervaporation.

**crossflow settler** An inclined plate or tube settler in which the suspension flows horizontally between the inclined surfaces, without affecting the vertical settling velocity, while the solid particles move downward. *See also* cocurrent settler, countercurrent settler.

**crossflow settling** The mode of operation of a crossflow settler. The condition for particle removal is (Metcalf & Eddy, 2003 and AWWA, 1999):

$$V \geq WV'/L \cos \theta \qquad \text{(C-197)}$$

where $V$ = settling velocity of particles removed, $W$ = perpendicular distance between surfaces, $V'$ = liquid velocity between the inclined surfaces, $L$ = required length of surface, and $\theta$ = the angle of surface inclination from the horizontal. *See also* Figure C-31, countercurrent settling, cocurrent settling.

**cross leakage** In an electrodialysis unit, leakage between the demineralized and concentrate streams, or permeation of water from the concentrate to the demineralized stream.

**cross-link** A bond, atom, or group that links the chains of atoms in a polymer or other complex organic molecule.

**cross-linkage** (1) The degree of bonding of a monomer or set of monomers (e.g., with divinylbenzene or acrylic) to form an insoluble, three-dimensional resin matrix. It is one of the factors that determine the strength of a resin and its ability to withstand chemical oxidation. (2) A comparatively short connection composed of either an element, a chemical group, or a compound that bridges between neighboring chains of atoms in a complex chemical molecule (especially a polymer). Cross-linking changes a plastic from ther-

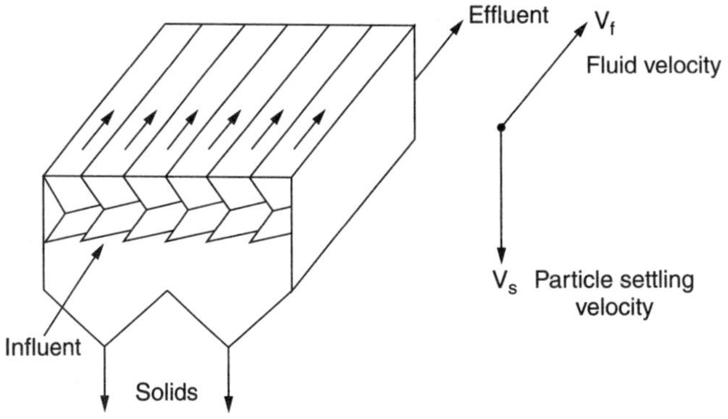


**Figure C-31.** Crossflow settling (adapted from Crittenden et al., 2005).

moplastic to thermosetting, and it increases strength, durability, heat, electrical resistance, and resistance to solvents and other chemicals.

**cross-linked** Characteristic of polymers attached by a cross-link.

**cross-linked polyethylene (PEX)** A recently developed plastic material used in the fabrication of residential piping and plumbing accessories. PEX may contribute an odor threshold number (TON) of up to TON = 5, which exceeds the secondary maximum contaminant level.

**cross-linking** *See* cross-linkage.

**cross section** In a stream, conduit, pipe, or channel, the cross section is the intersection with a vertical plane normal to the direction of flow and bounded by the wetted perimeter (P) and the surface width (W). The cross-sectional area is the surface area of the cross section.

**cross-sectional area** *See* cross section.

**cross-sectional bed area** The area of a bed of activated carbon perpendicular to the direction of flow.

**cross-sectional epidemiologic study** An epidemiological study that deals with disease data related to the same period of time.

**crotyl mercaptan ($CH_3$—CH—CH—$CH_2$—SH)** A skunklike odor-causing compound found in wastewater, with an odor threshold of 0.00029 ppm by volume.

**Crouzat™** A line of products of the U.S. Filter Corp. used in water treatment.

**crown** The inside top of the arch of a pipe, sewer, covered channel, or conduit. Also called soffit.

**Crown™** A line of self-priming pumps manufactured by Crane Pumps & Systems for wastewater conveyance.

**crown corrosion** The corrosion of concrete sanitary sewer pipes from the inside out resulting from the activity of sulfate-reducing and sulfur-oxidizing bacteria. *See* concrete sewer corrosion for detail.

**Crown Press™** A testing instrument manufactured by Neogen Corp. for use in sludge dewatering operations.

**crown rot** The corrosion of concrete sewer pipe flowing partly full by sulfuric acid ($H_2SO_4$) resulting from the biological oxidation of hydrogen sulfide ($H_2S$) accumulated at the head space of the sewer. *See* concrete sewer corrosion for detail.

**CRP®** Continuous recirculation process, developed by the FMC Corp. for sludge mixing in anaerobic digesters.

**CRT** Acronym of cell residence time.

**$C_R$t concept** Same as Ct concept. Here, $C_R$ designates the residual concentration of disinfectant and t represents contact time.

**crude mortality rate** *See* death rate.

**crude rate** A measure of morbidity and mortality for an entire population, without taking into account such characteristics as age, gender, or race.

**crude wastewater** Untreated wastewater; raw wastewater.

**crustacea** A class of aquatic arthropods with a hard-shell body (crust); includes barnacles, crabs, crayfish, lobsters, shrimps, water fleas, and wood lice.

**crustacean** (1) Any member of the class crustacea, including crayfish, lobsters, crabs, shrimps, and water fleas. Sometimes referred to as "water insects." (2) An organism that is active in biological wastewater treatment, at a stage higher than bacteria, protozoa, and fungi; their presence indicates that treatment has proceeded well. *See also Cyclops.*

**cryogenic** Of or pertaining to the production or use of very low temperatures.

**cryogenic air separation (CAS)** A process used to generate high-purity, liquid oxygen. It involves the liquefaction of air and fractional distillation to separate it into oxygen, nitrogen and other components. CAS is generally more efficient than pressure-swing adsorption, the other common process of production of pure oxygen on the site of a wastewater treatment plant. *See also* cold and gel traps, pressure-swing adsorption.

**cryogenic oxygen** Pure oxygen produced by the cryogenic air separation process.

**cryogenic process** A process used by large wastewater treatment plants to generate oxygen at very low temperatures. *See* cryogenic air separation.

**cryogenic system** A system that produces, or operates at, a local temperature that is lower than the surrounding temperature; e.g., refrigerated transport.

**cryolite ($Na_3AlF_6$)** One of the minerals that contain fluorine; it is essentially a compound of this halogen with sodium and aluminum, occurring in white masses and used in the electrolytic production of aluminum. Also called sodium aluminum fluoride and Greenland spar.

**cryophilic digestion** A variation of the aerobic sludge digestion process that uses temperatures lower than 10°C or 50°F for better operational control at small treatment plants. The product $\theta T$ is used as a performance control parameter within the range $250 < \theta T < 300°$ days, with $\theta$ = solids retention time (days) and $T$ = temperature (°C).

**crypto** Abbreviation of *Cryptosporidium.*

**cryptosporidiosis** A gastrointestinal disease caused by the ingestion of water- and excreta-related *Cryptosporidium parvum,* often resulting from drinking water contaminated by runoff from pastures or farmland. Its symptoms include profuse diarrhea, abdominal cramps, nausea, vomiting, low fever; an episode lasts about 12 days; it affects particularly the sensitive population (e.g., people with AIDS or severely weakened immune systems); median infective dose of about 130 oocysts (range of 10–500); incubation period of 2–12 days. There is currently (2005) no particular drug or treatment that is effective. The first significant outbreak occurred in April 1993 in Milwaukee, WI, when 400,000 people became ill and approximately 100 died.

***Cryptosporidium*** A waterborne and foodborne member of the protozoa genus; intracellular parasites of the intestinal tracts of various species, measuring about 3–5 micrometers and prevalent in surface waters. It usually does not display a high degree of host specificity. Some species have been detected in humans and are associated with diarrhea (cryptosporidiosis). It resists disinfection more than other pathogens. Ozone is a more effective disinfectant than chlorine and slow sand filtration can remove 99.9% of the cysts. Membrane filtration is also effective. USEPA proposed protective measures against this pathogen in the Enhanced Surface Water Treatment Rule.

***Cryptosporidium* inactivation device** An ultraviolet device used to inactivate Cryptosporidium oocysts.

***Cryptosporidium muris*** A secondary species of *Cryptosporidium* that infects mammals.

***Cryptosporidium* oocyst** A stage of development of Cryptosporidium in which a fertilized ovum produces a surrounding wall. Oocysts are commonly found in particulate matter from river bottoms and storm drains.

***Cryptosporidium parvum*** The primary species of *Cryptosporidium* infective to humans; found especially in lake and river water contaminated with wastewater and animal wastes. It is an emerging pathogen first identified as a human pathogen in 1976 and an opportunistic pathogen that significantly affects individuals with weak or compromised immune system. Immunofluorescence is currently used to detect its presence in a water sample.

***Cryptosporidium spp.*** Several species of the protozoa causing cryptosporidiosis.

***Cryptosporidium* surrogate** In drinking water treatment processes, potential surrogates for the removal of *Cryptosporidium* oocysts include turbidity, heterotrophic plate counts, *Bacillus subtilis,* microspheres, and dissolved organic carbon.

**crystal** A solid, homogeneous chemical substance that has a definite geometric shape and a characteristic internal structure, with fixed angles between its faces and distinct edges or faces.

**crystal growth** One of the two phases of the physical aspect of precipitation formation; it consists of the dissolution of ions from the surrounding solution to the surfaces of the solid particles until the elimination of supersaturation and the establishment of equilibrium. *See also* nucleation (AWWA, 1999).

**crystalline** Having a regular molecular structure evidenced by crystals; clear, transparent, formed by crystallization.

**crystallization** (1) The development or processing of slurries to concentrate solids in a suspension, e.g., in the treatment of water treatment brines for zero liquid discharge or the extraction of crystals of pure water from seawater. *See also* evaporation. (2) The formation of high-purity crystalline solids from liquid or vapor to recover minerals from by-products.

**crystallization water** Water that combines with salts when they crystallize, i.e., a part of the crystalline compound. Also called water of hydration.

**crystallizer** A device used (a) to concentrate salts for the production of crystalline solids, (b) to treat residuals in zero-discharge wastewater treatment plants, and (c) for concentrate disposal in small-scale desalting systems. *See also* forced-circulation evaporator.

**$CS_2$** Chemical formula of carbon bisulfide.

**Cs-137** Symbol or abbreviation of cesium-137.

**CSF™** A stormwater filter manufactured by CSF Treatment Systems, Inc.

**CSF processes** A water treatment train that includes coagulation, sedimentation, and filtration. *See also* conventional treatment.

***C. sinensis*** *See Clonorchis sinensis.*

**CSO** Acronym of combined sewers overflow(s).

**CSO control** *See* combined sewer overflow control.

**CSO outlet** *See* combined sewer overflow outlet.

**CSO storage** *See* combined sewer overflow storage.

**CST** Acronym of capillary suction time.

**CSTR** Acronym of completely stirred tank reactor or continuously stirred tank reactor.

**CST test** *See* capillary suction time test.

**CT, C.T, C.t, CTcalc,** or **CxT** The product of residual disinfectant concentration (C) in mg/L determined before or at the first customer, and the corresponding disinfectant contact time (T) in minutes measured at peak hourly flow, i.e., $C \times T$. If a public water system applies disinfectants at more than one point prior to the first customer, it must determine the CT of each disinfectant sequence before or at the first customer to determine the total percent inactivation or total inactivation ratio. In determining the total inactivation ratio, the public water system must determine the residual disinfectant concentration of each disinfection sequence and corresponding contact time before any subsequent disinfection application point(s). CT99.9 is the inactivation ratio or CT value required for 99.9% (3-log) inactivation, usually applied as a requirement for *Giardia lamblia* cysts. Similarly, CT99.99 (4 log) is usually applied to the inactivation of viruses. *See also* CT table, Chick– Watson law, Surface Water Treatment Rule.

**CT99.9** *See* CT.

**CT99.99** *See* CT.

**CT approach** *See* CT concept.

**CTAS** Acronym of cobalt thiocyanate active substance. *See also* MBAS.

**CTcalc** *See* CT.

**CT concept** The concept based on the Chick–Watson law: as disinfectant concentration (C) and contact time (T) increase, microorganism inactivation also increases. The CT product is considered an important measure of germicidal efficiency or effectiveness; e.g., the State of California requires a CT value of 450 mg.min/L with a modal contact time of 90 min for wastewater reuse applications. It plays an important role in the Surface Water Treatment Rule. Also called $C_Rt$ concept. *See also* contact time, CT credit, CT table.

**CT credit** The disinfection credit awarded to a given process by the USEPA's Surface Water Treatment Rule, based on the CT concept. *See also* baseline process.

**CT criteria** Criteria developed by the USEPA to regulate drinking water disinfection practices, based on the CT concept. *See also* CT table, log removal, sequential pathogen removal, primary disinfection, secondary disinfection, Surface Water Treatment Rule.

**CT disinfection criteria** *See* CT criteria.

***Ctenopharyngodon idella*** A hardy fish species of the carp family that survives well in stabilization ponds used for aquaculture.

**CT product** The product of disinfectant concentration (C) and contact time (T). *See* CT.

**C.t product** *See* CT.

**CT table** A table that shows the disinfection requirements (as a CT value) for unfiltered water, depending on inactivation goal (percent), pH, temperature, and free chlorine residual. For example, Table C-04 presents CT values for 99.99% (4-log) inactivation, a temperature of 20°C, pH from 6.0 to 9.0, and free chlorine residuals from 0.4 to 3.0 mg/L. The USEPA published this and other CT tables in a final surface water treatment rule dated June 29, 1989. The agency also proposed two CT tables for 99.9% (3-log) inactivation of *Giardia lamblia* cysts by chlorine dioxide and ozone, and by chloramines.

**CxT** Same as CT.

**Cu** Chemical symbol of the metallic element copper.

**$Cu_3(AsO_3)_2$** Chemical formula of Paris green.

**Table C-04.** CT values for a temperature of 20°C

| Free chlorine residual, mg/L | pH < 6.0 | pH = 6.5 | pH = 7.0 | pH = 7.5 | pH = 8.0 | pH = 8.5 | pH < 9.0 |
|---|---|---|---|---|---|---|---|
| < 0.4 | 36 | 44 | 52 | 62 | 74 | 89 | 105 |
| 0.6 | 38 | 45 | 54 | 64 | 77 | 92 | 109 |
| 0.8 | 39 | 46 | 55 | 66 | 79 | 95 | 113 |
| 1.0 | 39 | 47 | 56 | 67 | 81 | 98 | 117 |
| 1.4 | 41 | 49 | 58 | 70 | 85 | 103 | 123 |
| 1.8 | 43 | 51 | 61 | 74 | 89 | 108 | 129 |
| 2.0 | 44 | 52 | 62 | 75 | 91 | 110 | 132 |
| 2.4 | 45 | 54 | 65 | 78 | 95 | 115 | 138 |
| 2.8 | 47 | 56 | 67 | 81 | 99 | 119 | 143 |
| 3.0 | 47 | 57 | Z68 | 83 | 101 | 122 | 146 |

**$CuCO_3 \cdot Cu(OH)_2$** or **$Cu_2(CO_3)(OH)_2$ (s)** Chemical formula of malachite.

**$CuHAsO_3$** Chemical formula of copper arsenite.

**$Cu_3(OH)_2(CO_3)_2$(s)** Chemical formula of azurite.

**$Cu_{12}Sb_4S_{13}$** Chemical formula of tetrahedrite.

***Culex*** A widespread genus of mosquitoes that maintain the body parallel to the feeding or resting surface; many species are vectors of water-related diseases.

***Culex fatigans*** Previous name of *Culex quinquefasciatus.*

***Culex pipiens*** The species of mosquitoes that are vectors of a number of water- and excreta-related diseases (e.g., filariasis) and constitute a major nuisance, particularly at night. Distributed worldwide, these mosquitoes breed in septic tanks, latrines, wastewater, sullage, and other polluted water.

***Culex pipiens fatigans*** Previous name of *Culex quinquefasciatus.*

***Culex pipiens molestus*** The major member of the *Culex pipiens* species of mosquitoes that transmit Bancroftian filariasis in Egypt.

***Culex pipiens pallens*** The major member of the *Culex pipiens* species of mosquitoes that transmit Bancroftian filariasis in China and Japan.

***Culex pipiens quinquefasciatus*** Previous name of *Culex quinquefasciatus.*

***Culex quinquefasciatus*** The major member of the *Culex pipiens* species of mosquitoes that transmit Bancroftian filariasis in urban areas of Asia, coastal Brazil, and coastal East Africa. Previously known as *Culex pipiens fatigans, Culex pipiens quinquefasciatus,* or *Culex fatigans.*

***Culex tarsalis*** A species of mosquitoes that transmit the virus of Western equine encephalitis; it breeds mostly in irrigated lands and wastewater stabilization ponds.

**culicine** (1) Mosquito of the genus Culex or related genera. (2) Pertaining to the subfamily of mosquitoes that includes the *Aedes, Culex,* and *Mansonia* genera. Culicine mosquitoes are bloodsucking arthropods that can transmit arboviral infections.

**Cullar®** An activated carbon filter manufactured by Culligan International Corp.

**cullet water** Water that is exclusively and directly applied to molten glass in order to solidify the glass (EPA-40CFR426.11-b). ("Cullet" means broken glass generated in the manufacturing process or waste glass recycled to a glass meting furnace.)

**Cullex** A resin manufactured by Culligan International Corp. for use in water softening.

**Cullsorb** A greensand filter manufactured by Culligan International Corp.

**culturable cells** Cells that can grow on a culture medium as opposed to dormant or nonculturable cells that may become viable over time but cease to grow.

**cultural eutrophication** The increase in the growth of aquatic animal and plant life caused by overfertilization of water bodies through human activities, specifically water and land uses associated with population growth, industrial development, and intensified agriculture. *See also* natural eutrophication.

**culture** The growth of live microorganisms or tissues, developed by furnishing sufficient nutrients in a suitable medium.

**culture medium (media)** A liquid or solidified nutrient material used to support the growth of microorganisms; e.g., broth and agar-based solid substances. Also called medium.

**culture tube** A slender glass tube, with open top and rounded closed bottom, used in the multiple-tube fermentation and other bacteriological tests.

**cumulative exposure** The summation of exposures of an organism to a chemical over a period of time.

**cumulative impact** The impact on the environment resulting from the incremental impact of an action when added to other past, present, and reasonably foreseeable future actions, regardless of what agency (federal or non-federal) or person undertakes such other actions. Cumulative impacts can result from individually minor but collectively significant actions taking place over a period of time (EPA-40CFR1508.5).

**cumulative pollutant loading rate** The maximum amount of an inorganic pollutant that can be applied to an area of land, under the Clean Water Act (EPA-40CFR503.11-f).

**cumulative residence time distribution curve** A tracer response curve, i.e., a graphical representation of the distribution of a tracer versus time of flow, for a step input, with output concentrations normalized by dividing the measured concentrations by such a factor that the area under the curve equals 1. Also called an *F* curve or an *F*(*t*) curve. The *F* curve may be defined as (Metcalf & Eddy, 2003):

$$F(t) = \int E(t)\, dt \qquad \text{(C-198)}$$

where the symbol $\int$ indicates integration from zero to infinity (0 to ∞) and $E(t)$ is the residence time distribution function. $F(t)$, as a function of time ($t$), represents the fraction of tracer that has arrived at the sampling point and is such that:

$$1 - F(t) = (t/T - p^{1-m})(e^{-A}) \qquad \text{(C-199)}$$

where $T$ = theoretical residence time or hydraulic detention time, $p$ = fraction of flow acting as plug flow, $1 - p$ = fraction of flow acting as mixed flow, $m$ = fraction of dead space of basin, and

$$A = 1/[(1 - p)(1 - m)] \qquad \text{(C-200)}$$

*See also* exit age curve or E curve, time-concentration curve.

**cumulative risk** The risk posed by the combination of multiple contaminants. It is commonly considered in formulating remediation plans for hazardous waste cleanup but not usually in setting maximum contaminant levels for drinking water.

**cumulative toxicity** The toxic effects on an organism caused by successive exposure to a substance or other risk. *See also* toxicity terms.

**cumulonimbus** A cloud or group of clouds indicative of thunderstorm conditions; resembling large towers of high altitude, with ice crystals at the top.

**cumulose deposit** An accumulation of organic matter and small amounts of rock waste.

**cumulus** A cloud of dense individual elements in the form of mounds or towers, with a flat base and top like a cauliflower.

**cunette** A trough at the bottom of a combined sewer; introduced in Europe, the cunette was designed to carry dry-weather flow at an appropriate velocity. *See* Figure C-32.

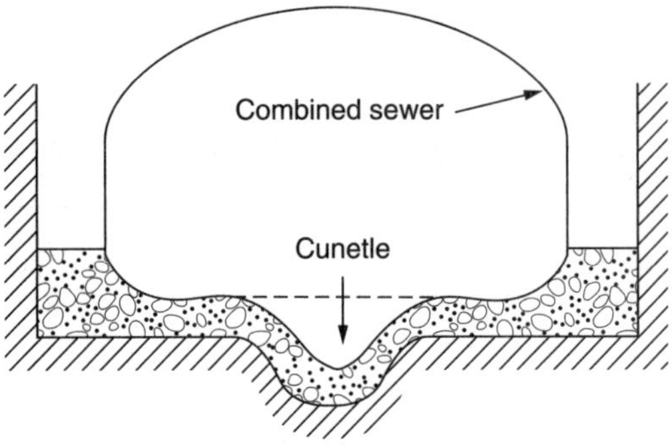


**Figure C-32.** Cunette.

**CuO(s)** Chemical formula of tenorite.

**cupric** Of or containing copper in general, but particularly in the bivalent state; e.g., cupric oxide, CuO.

**cupric copper** Copper in the bivalent state, $Cu^{2+}$.

**cupric hydrogen arsenite** *See* copper arsenite.

**cupric hydroxide** Same as copper hydroxide.

**cupric sulfate** *See* copper sulfate and blue vitriol.

**cuprichloramine** A chemical formed by the combination of ammonia ($NH_3$), chlorine ($Cl_2$), and copper sulfate ($CuSO_4 \cdot 5\ H_2O$); used to control algal blooms in reservoirs.

**cuprite** A copper ore, the mineral cuprous oxide ($Cu_2O$) found in red crystals and brown-black granular masses.

**cupronickel** A copper alloy that contains 10% to 30% nickel. *See* copper–nickel.

**cuprosolvency** (1) The ability of or condition for copper to dissolve, in drinking water, for example. (2) The phenomenon of uniform corrosion whereby soluble copper is released from copper pipes and brass fixtures. Cuprosolvency can be reduced by providing an appropriate buffer intensity through pH adjustment. *See* Lead and Copper Rule, cerussite, plattnerite, scrutinyite.

**cuprous** Of or containing univalent copper, $Cu^+$; e.g., cuprous oxide $Cu_2O$ or cuprous cyanide CuCN.

**cuprous copper** Copper in the univalent state, $Cu^+$.

**cuprous cyanide (CuCN)** A creamy-white, poisonous powder used in electroplating. Also called copper cyanide.

**cuprum** Latin name of copper (Cu).

**cup screen** A drum screen with single entry but double exit.

**CUR** Acronym of carbon usage rate.

**curb cock** Same as curb stop.

**curb stop** A water service shutoff valve near the curb, between the water main and a building connection; it is usually operated by a wrench or valve key to start or stop water service to the building. Also called curb cock. *See* curb stop and box for detail.

**curb stop and box** A metallic or concrete box, housing, or vault installed near the curb, containing and providing access to a shutoff valve in a customer's water service line, with a cover to keep out dirt and debris. It is usually located on the property line, between the main water line and the meter, or between the curb and the sidewalk. The valve may be operated to start or stop water service to the customer. This receptacle is also called a curb cock or curb stop, cutoff, meter stop, stop box, valve box, or water stop. *See also* valve vault.

**cured-in-place lining** (1) A trenchless water main rehabilitation method using a polyester woven hose or a nonwoven felt bag, impregnated with a resin that is cured in place by hot water or steam. *See* the following types of cured-in-place liners: felt-based liner, membrane liner, woven-hose liner. *See also* trenchless technology. (2) A cement–mortar mixture installed in a steel or iron pipe.

**curie (Ci)** A measure of radioactivity, equivalent to $3.7 \times 10^{10}$ nuclear disintegrations per second or $3.7 \times 10^{10}$ becquerel (Bq). A common unit used in environmental measurements is the picocurie (pCi), which is equal to 0.037 disintegration per second or 0.037 Bq = $1/10^{12}$Ci.

**curing** (1) Process of keeping concrete or mortar damp for at least the first week after it is cast so that the cement always has enough water to harden. (2) In composting, the process of stabilization of organic matter by bacteria and turning over of the piles in the windrows.

**curing time** In the method of freeze-thaw sludge conditioning, curing time is the time at which the block of frozen sludge is kept at subfreezing temperatures; it allows extra time for the last frozen portion to completely dehydrate.

**curing yard** An open area where compost windrows are formed.

**curly brackets {...}** A symbol used to denote the activity of a chemical species, as compared to square brackets [...] for concentration.

**current** (1) A movement or flow of electricity. Electric current is measured by the number of coulombs per second flowing past a certain point in a conductor. A flow of one coulomb per second is called one ampere, the unit of the rate of flow of current. (2) A movement or flow of water. Water flowing in a pipe is measured in gallons per second or similar units, not by the number of molecules of water going past a point.

**current density ($E_c$)** The current (amperage) per unit area that flows perpendicularly through a membrane, expressed in milliamperes per square centimeter ($mA/cm^2$).

**current-density-to-normality ratio** In an electrodialysis unit, the ratio of the current density (CD, in $mA/cm^2$) to the normality of the solution (N, in eq/L). Common CD/N ratios range from 500 to 800. *See* polarization (2).

**current diagram** A graphic representation of current velocity and strength over a stretch of tidal waterway.

**current efficiency** The efficiency (expressed as a fraction) of an electrodialysis unit in using the current supplied. *See* electrodialysis current.

**Currie Clarifier** A circular tank manufactured by Dorr-Oliver, Inc., combining clarification and aeration.

**curtain drain** A drain installed to intercept surface or ground water at the upper end of an area and carry the flow away. Also called intercepting drain.

**curtain wall** (1) A wall that extends down below the surface of the water to prevent floating objects from entering a screen forebay. (2) A deep cutoff wall used to protect an overflow masonry dam against seepage greater than the hydrostatic head.

**Curveco** A mechanically cleaned bar screen manufactured by Eco Equipment.

**$Cu_{12}Sb_4S_{13}$** Chemical formula of tetrahedrite.

**$CuSO_4 \cdot 5\ H_2O$** Chemical formula of commercial copper sulfate.

**$CuSO_4 \cdot 4\ NH_3$** Chemical formula of copper sulfate ammoniate.

**cutaneous** Related to the skin. A cutaneous irritation is an inflammatory reaction of the superficial cells of the skin.

**cutback** An uncoated area left at the end of a section of steel pipe so that the coating on the sections is not damaged during welding and construction. *See also* shrink sleeve.

**cutoff** Same as curb stop and box.

**cut pipe** A tubular pipe section, longer than one foot and threaded at both ends. *See also* nipple.

**Cutrine-Plus** An algicide/herbicide produced by Applied Biochemists, Inc.

**cutting screen** A screen equipped with a cutting edge or with knives for grinding or shredding coarse materials in wastewater. The screenings are not removed but they are too small to clog pumps. Also called comminuting screen; same as comminutor, grinder.

**Cuver** A chemical produced by Hach Co. for the detection of waterborne copper.

***C. vulgaris*** Abbreviation of *Chlorella vulgaris.*

**$C_v$ value** Same as valve capacity coefficient.

**CWA** Acronym of Clean Water Act.

**cwt** Abbreviation of hundredweight.

**$CX_3{\equiv}N$ (with X = $Cl^-$, $Br^-$, or $H^+$)** Chemical formula of haloacetonitrile.

**CYANAMER®** or **Cyanamer®** Active scale antiprecipitants, inhibitors, and dispersants produced by Cytec Industries, Inc. for use in water treatment chemical formulations.

**cyanamid** Same as cyanamide.

**cyanamide ($CH_2N_2$)** A white, crystalline, unstable solid produced from the reaction of ammonia and cyanogen chloride or sulfuric acid and calcium cyanamide. Also spelled cyanamid.

**cyanate** A cyanide compound that is formed in water at pH above 8.0 and that decomposes to nitrogen gas ($N_2$) and carbon dioxide ($CO_2$); a salt or ester of cyanic acid (HOCN).

**cyanazine [$C_2H_5NH\text{-}C_3N_3(Cl)\text{-}NHC(CH_3)_2CN$]** a potentially carcinogenic organic chemical used to control broadleaf and grassy weeds. It may be found in surface waters receiving agricultural runoff and is not removed effectively by conventional water treatment processes. It is one of several derivatives of cyanuric chlorides. Its MCLG or MCL may be set at 0.001 mg/L. *See also* atrazine, propazine, simazine, and s-triazine.

**cyanic acid (HOCN)** An unstable, poisonous liquid, isomeric with fulminic acid.

**cyanide ($CN^-$)** A deadly poisonous compound that contains a cyanogen or CN group in combination with another element or radical and behaves like halides; it is used in such industrial processes as electroplating. Its presence is rare in natural waters and indicates pollution. Cyanic salts such as calcium cyanide [$Ca(CN)_2$], potassium cyanide (KCN), and sodium cyanide (NaCN) form aqueous hydrogen cyanide or hydrocyanide acid (HCN) when added to water. HCN, a volatile, weak, very toxic acid (used in gas chambers), can then dissociate into the hydrogen ($H^+$) and cyanide ($CN^-$) ions. In the human body, cyanide dissolved in the blood stream interferes with respiration. At a high pH, cyanide is oxidized to cyanate in water. When absorbed through the lungs and skin, it reacts with heavy metals, affects the nervous, cardiovascular, and other systems; cyanide poisoning occurs very rapidly. For drinking water, the USEPA sets a maximum contaminant level (MCL) and a maximum contaminant level goal (MCLG) of 0.2 mg/L for cyanide; however, the European Community and WHO recommended lower limits. In industrial waste treatment, alkaline chlorination converts cyanide compounds to the less soluble cyanogen chloride gas and the less toxic cyanate ion. Chemical oxidation is an effective removal process, followed by electrodialysis and reverse osmosis.

**cyanide A** A cyanide amenable to chlorination and determined by a certain method.

**cyanide pollution** The release of industrial wastewaters containing cyanides from the manufacture of heating gas, case hardening metals, and from the plating or cleaning of metals.

**cyanide process** A process to extract gold and silver from ore by dissolution in alkaline solution of sodium cyanide (NaCN) and then precipitation from the solution.

**cyanobacteria** Formerly called blue-green algae, a type of photosynthetic and nitrogen-fixing bacteria, some of which produce toxic metabolites (algal toxins) and cause taste-and-odor problems in drinking water sources from lakes and impoundments. Potential health effects include acute respiratory failure and other nerve damages, damages to organs, and malignant tumors. *See* blue-green algae, *Anabaena flos aquae, Aphanizomenon flos aquae, Microcystis.*

**cyanobacteria bloom** Same as cyanobacterial bloom.

**cyanobacterial bloom** Bloom of cyanobacteria. *See* algal bloom, geosmin, MIB.

**cyanobacteria-like bodies (CLB)** Organisms that, upon analysis, appear to be cyanobacteria, i.e., 8–10 micrometers in size, staining red with modified acid-fast stains, and autofluorescing under ultraviolet light. Cyclospora oocysts have been sometimes confused with cyanobacteria-like bodies in microorganism analyses.

**cyanoethylene** Acrylonitrile.

**cyanogen bromide (CNBr)** A colorless, poisonous, volatile, crystalline solid used in insecticides; a by-product of water ozonation and chloramination.

**cyanogen chloride (CNCl)** A toxic gas with limited solubility in water, used in chemical synthesis, fumigants, and tear gas; a disinfection by-product that is often analyzed in raw or treated water. In 1993 the WHO issued a drinking water guideline of 70 μg/L as CN for this contaminant.

**cyanogen halide (CNX)** A by-product of chloramination; cyanogen bromide, cyanogen chloride.

**cyanohydrin** An organic compound that contains the cyanide (CN) and hydroxyl (OH) groups.

**Cyanophyta** A group of primitive, unicellular prokaryotes formerly called blue-green algae. *See* cyanobacteria.

**cyanosis** Skin bluish/purplish discoloration or lividness caused by deficient blood oxygenation. *See* methemoglobinemia.

**cyanuric chloride** *See* cyanogen chloride and derivatives (atrazine, cyanazine, propazine, simazine).

**Cybreak™** An emulsion breaker produced by American Cyanamid Co.

**Cycle-Let®** Equipment manufactured by Thetford Systems, Inc. for the treatment and recycling of wastewater.

**cycle of concentration** (1) The ratio of total dissolved solids (TDS) concentration in a recirculating stream to the TDS concentration of makeup water. (2) The ratio of the concentration of solids in a liquid whose volume has been reduced (e.g., by evaporation) to the initial solids concentration.

**cycle of fluctuation** The period of time of rise and a succeeding decline of a water table; e.g., daily cycle, annual cycle. Also called phreatic cycle.

**Cyclesorb®** A treatment system manufactured by Calgon Carbon Corp. using adsorption on granular activated carbon.

**cycle time** *See* pump cycle time.

**cyclic activated sludge system (CAAS™)** A variation of the activated sludge process using three baffled aeration zones in the approximate volumetric proportions of 1:2:20, with mixed liquor recycling from zone 3 to zone 1, continuous influent feed, and batch effluent release. *See* Figure C-33.

**cyclic aliphatic compound** A compound consisting of carbon atoms bonded together in a ring; e.g., cyclohexane, $C_6H_{12}$. *See* Figure C-34.

**cyclic depression** Water withdrawal that exceeds the supply over a given cycle.

**cyclic recovery** The gradual rise in elevation of the water table following a period of cyclic depression.

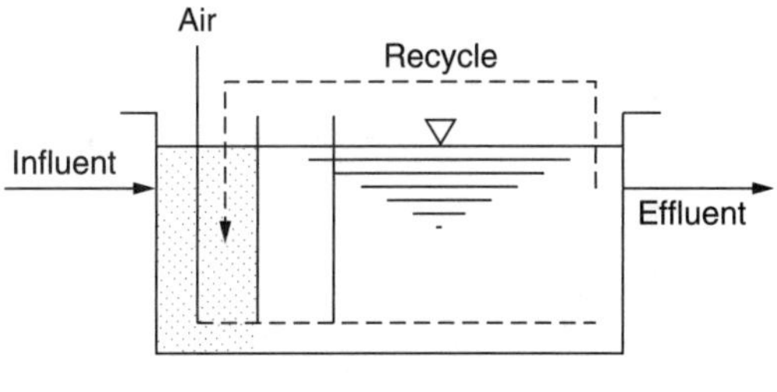


**Figure C-33.** Cyclic activated sludge.

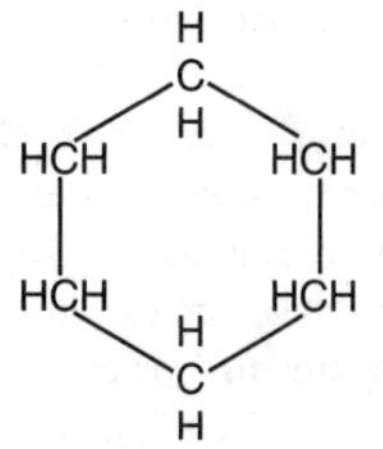


**Figure C-34.** Cyclic aliphatic compound (cyclohexane).

**cyclic storage** The accumulation of water in a storage reservoir during periods of surplus (or of greater than average supply) for use during periods of deficit (or of smaller than average supply). Sometimes called overyear storage, seasonal storage.

**Cyclo Blower** An air blower made by Gardner-Denver.

**CycloClean™** A hydrocyclone separator manufactured by Krebs Engineers.

**cyclodiene** An insecticide of the group of polycyclic chlorinated hydrocarbons. *See* endrin and heptachlor.

**Cyclofloc** A proprietary process developed by Biwater-OTV, Ltd to increase the rinse rates of clarifiers. It uses and recycles fine sand as a ballast. The same process is also known as Simtafier. *See* ballasted floc.

**Cy-Clo-Grit** A cyclonic grit collector manufactured by Jones & Attwood, Inc.

**Cyclo Grit Washer** A wastewater treatment unit manufactured by Eimco Process Equipment Co. and combining grit washing and dewatering.

**Cyclo-Hearth** A multiple hearth furnace manufactured by Zimpro Environmental, Inc..

**cyclohexane** *See* cyclic aliphatic compound.

**cyclohexylenediaminetetraacetic acid (CDTA)** A buffering solution of high ionic strength.

**cyclone** A type of separator for removal of larger particles from an exhaust gas stream. Gas laden with particulates enters the cyclone and is directed to flow in a spiral, causing the entrained particulates to fall out and collect at the bottom. The gas exits near the top of the cyclone.

**Cyclone** A coarse-bubble diffuser manufactured by Aeromix Systems, Inc.

**cyclone degritter** An inclined device that uses centrifugal force to (a) separate grit from primary sludge in wastewater treatment plants that do not have grit chambers or (b) wash grit otherwise collected. It consists of a volute feed chamber, a vortex finder, cylindrical and conical sections, apex valve, and accessories. *See also* cyclone separator (2).

**cyclone precipitator** Same as cyclone separator.

**cyclone scrubber** An air pollution control device that uses water or another liquid to remove up to 90% of particulates, fumes, dusts, mists, and vapors from gas streams; for example, in sludge incinerators and fluidized bed reactors. The device exerts centrifugal forces to increase contact between particles and liquid droplets. *See also* ejector–venturi scrubber, spray tower, and Venturi scrubber.

**cyclone separator** (1) A conical device that separates particles from water or wastewater using centrifugal force; used in air pollution and odor control during heat drying of sludge. Also called cyclone precipitator or hydroclone. (2) A similar device that washes and dewaters grit slurries drawn from grit chambers. *See also* cyclone degritter.

**cyclonic water jet scrubber** *See* cyclone scrubber.

**cyclonic wet scrubber** *See* cyclone scrubber.

**cycloparaffin ($C_nH_{2n}$)** A homogeneous, saturated, alicyclic hydrocarbon.

**Cyclo/Phram®** A metering pump manufactured by Leeds & Northrup.

**cyclopoid** Same as cyclops.

**cyclops (plural: cyclopes)** A member of the *Cyclops* genus; a very small crustacean living in small water bodies and serving as an intermediate host for the development of *Dracunculus medinensis,* the pathogen of Guinea worm, which infects humans through drinking water. It is also an intermediate host for the cestode *Diphyllobothrium latum.* Chlorination is an effective treatment process against cyclopes. *See* Guinea-worm disease.

***Cyclops* genus** A group of freshwater copepods with a single, median eyelike spot.

***Cyclospora*** A genus of waterborne and foodborne, seasonal protozoans of the order *Coccidia* and phylum Apicomplexa that cause diarrhea, anorexia, and other symptoms in humans; opportunistic pathogens that significantly affect individuals with weak or compromised immune system.

***Cyclospora cayetanensis*** A round or ovoid, coccidian protozoan parasite, similar to *Cryptosporidium;* distributed mainly in Asia, the Caribbean, Mexico, and Peru. It is a newly recognized pathogen that was originally thought to be a blue-green alga. Its infective form is a sporulated oocyst causing cyclosporiasis.

**cyclosporiasis** An infection caused by the waterborne or foodborne protozoa *Cyclospora cayetanensis,* and characterized by severe diarrhea, stomach cramps, nausea, fatigue, and vomiting.

**Cyclotherm** A heat exchanger manufactured by FMC Corp. for use in sludge processing.

**Cyclo-Treat™** A cyclone separator manufactured by Envirex, Inc.

**Cygnet** A rotary distributor manufactured by Simon-Hartley for use in fixed-film reactors.

**cylinder infiltrometer** (1) A device used in the field to measure percolation rates, infiltration rates, or hydraulic conductivity. (2) A small-scale test using this device.

**cylinder valve** Same as container valve.

**cylindrical sludge digestion tank** One of three commonly used forms of anaerobic digester; a shallow vertical cylinder with a conical floor sloped 1 vertical to 6 horizontal, as shown in Figure C-35, or with a waffle bottom. *See also* conventional German digester, egg-shaped digester.

***Cylindrospermopsis*** A genus of cyanobacteria found in algal blooms and capable of producing toxins that can damage the liver.

***Cyprinus carpio*** A hardy fish species of the carp family that survives well in stabilization ponds used for aquaculture.

**cyst** A resting, nonvegetative stage formed by some bacteria and protozoa (e.g., *Entameba histolitica* and *Giardia lamblia*) in which the whole cell is surrounded by a protective layer to resist unfavorable conditions. Cysts evolve from trophozoites; they are relatively large and are removed from water by coagulation, sedimentation, and sand filtration, but they are more resistant to disinfection than bacteria and viruses. Stabilization ponds are an effective wastewater treatment process. *See also* resistant form. (2) A bladder, sac, or vesicle containing fluid or semifluid matter.

**cysteine ($C_3H_7NO_2S$)** A crystalline sulfur-containing amino acid, found in almost all proteins and produced by the reduction of cystine.

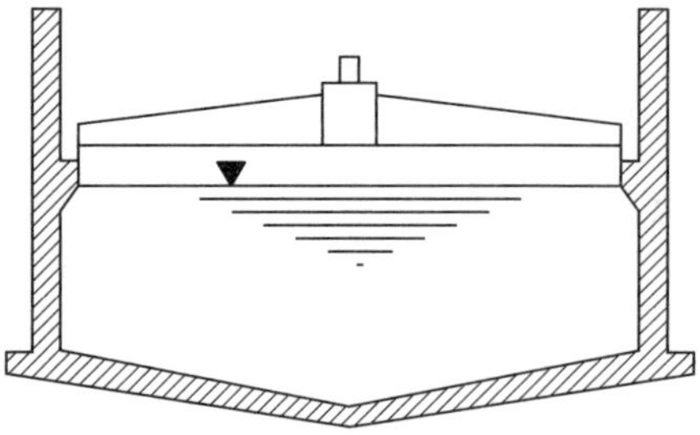

**Figure C-35.** Cylindrical sludge-digestion tank.

**cysteine conjugate lyase** An enzyme that catalyzes reactions that may be responsible for the toxic and carcinogenic effects of tetrachloroethylene ($Cl_2C{:}CCl_2$) and trichloroethylene ($CHCl{:}CCl_2$).

**cysticercosis** An infestation of the intestines by the larval stage of the beef tapeworm (*Cysticercus bovis*) or pork tapeworm (*Cysticercus cellulosae*), which affects the muscles, the brain, and the heart.

***Cysticercus bovis*** The larval stage of the beef tapeworm (*Taenia saginata*), a pathogenic agent of cysticercosis.

***Cysticercus cellulosae*** The larval stage of the pork tapeworm (*Taenia solium*), a pathogenic agent of cysticercosis.

**cystine ($C_6H_{12}N_2O_4S_2$)** A crystalline amino acid found in most proteins and in the bladder. *See also* cysteine.

**cystitis** A urinary infection; inflammation of the bladder, which may increase the concentration of coliforms and other bacteria in the urine.

**cytolysis** Dissolution or degeneration of cells.

**cytopathic** Pertaining to or characterized by cell death, morphological changes, or functional changes within the cell.

**cytopathic effect (CPE)** A morphological change in a cell that leads to its death; the destruction of the tissue culture cell sheet without agar overlay in the most-probable-number method. *See also* cytopathogenic effect. (Cythopathic means of, pertaining to, or characterized by a pathological change in the function or form of a cell that leads to its death.)

**cytopathic unit** A unit that measures cell alterations caused by infection.

**cytopathogenic effect (CPE)** The effect of a substance or microorganism pathologic for or destructive of cells, particularly the observable changes caused in a host cell by virus replication. CPE is used to detect and quantify viruses in cell cultures. For example, adenoviruses cause the formation of grapelike clusters, whereas enteroviruses cause the rounding of the cells. *See also* plaque-forming unit method, serial dilution endpoint or $TCID_{50}$ method, most probable number, cytopathic effect. (Cythopathogenic means of or pertaining to a substance or microorganism that is pathologic or destructive to cells.)

**cytopathogenicity assay** A test based on the observation by microscope of morphologic changes of living cells infected by a virus.

**cytopathogenic virus** A virus that kills the cell in which it replicates.

**cytoplasm** A solution containing the DNA and other essential biomolecules for bacterial metabolism; a cell component that contains the material necessary for cell functions, including water, nutrients, enzymes, ribosomes, and small organic molecules.

**cytoplasmic inclusions** A cell component that contains storage material providing carbon, nutrients, or energy; e.g., carbohydrates, polyphosphates, lipids, or sulfur granules.

**cytosis** Cell activity.

**cytosol** The water-soluble component of cell cytoplasm

**cytotoxic** Pertaining to a substance that is toxic to cells.

**cytotoxicity** Cell destruction by a cytotoxic substance.

**cytotoxin** Any material toxic to cells.

# D

**D** Chemical symbol of deuterium.

**2,4-D** *See* 2,4-dichlorophenoxyacetic acid.

**$D_2O$** Chemical formula of deuterium oxide or heavy water.

**$d_{10}$** Diameter of the 10th percentile sand (or other media) size; the particle diameter for which 10% of the particles in a sample have a smaller or equivalent diameter. It is used to define filter media. Also called $ED_{10}$. *See also* effective diameter, $d_{90}$.

**D-20** A filter underdrain device manufactured by Infilco Degremont, Inc.

**$d_{60}$** Diameter of the 60th percentile sand (or other media) size.

**$d_{90}$** Diameter of the 90th percentile sand (or other media) size. For media of log-normal distribution:

$$d_{90} = d_{10} \cdot U^{1.67} \qquad \text{(D-01)}$$

$d_{10}$ is the diameter of the 10th percentile size and $U$ is the uniformity coefficient. The 90% size is used to determine the backwash rate of depth filters.

**Dac Floc®** A polyelectrolyte manufactured by Dacar Chemical Co. to enhance liquid/solid separation.

**dacthal** A stable dichloropropionic acid herbicide, frequently detected in surface and ground water, which has been shown to damage organs of laboratory animals.

**DAF** Acronym of dissolved air flotation.

**DAF** A treatment system designed and manufactured by WesTech Engineering, Inc. of Salt Lake City, UT.

**DAFT** Acronym of dissolved air flotation thickener.

**daily cover** Cover material (e.g., six inches of sandy loam) spread and compacted on the top and side slopes of a sanitary landfill at the end of each day to control fire, moisture, and erosion, and to ensure an aesthetic appearance. *See also* final cover.

**daily discharge** Discharge of a pollutant measured during a calendar day or any 24-hour period that reasonably represents the calendar for purposes of sampling. For pollutants with limitations expressed in units of mass, the "daily discharge" is calculated as the total mass of the pollutant discharged over the day. For pollutants with limitations expressed in other units of measurement, the "daily discharge" is calculated as the average measurement of the pollutant over the day (EPA-40CFR122.2).

**daily flow** The volume of water that passes through a water or wastewater treatment plant, a pipe, or any hydraulic structure in one day, i.e., 24 hours.

**daily maximum limitation** For the subcategories for which numerical limitations are given, it is a value that should not be exceeded by any one effluent measurement. The 30-day limitation is a value that should not be exceeded by the average of daily measurements taken during any 30-day period (EPA-40CFR429.21-j).

**daily values** Daily measurements used to assess compliance with the maximum for any one day, as applied to produced water effluent limitations and new source performance standards of the Clean Water Act (EPA-40CFR435.11-c).

**dairy waste** Wastewater generated during the processing of dairy products (whole milk, separated milk, buttermilk, whey); characterized by a high concentration of protein, fats, lactose, and other dissolved organics. Common treatment method is a biological process such as activated sludge or trickling filtration.

**Dakota** A belt filter press manufactured by Hydro-Cal Co.

**dalapon ($CH_3CCl_2COOH$)** A herbicide used to kill certain weeds in sugarcane, other crops, and in watercourses. USEPA-issued limit in drinking water is MCLG = MCL = 0.02 mg/L. Also called di-alpha-dichloropionic acid or 2,2-dichloropropionic acid.

**dalton (D)** (1) The unit of atomic mass, used to measure the mass of atomic and smaller particles: one dalton equals one-twelfth the mass of the carbon-12 atom. Also called atomic mass unit: AMU or amu. (2) The unit of measure of molecular weight cutoff (MWCO) in quantifying the rejection characteristics of a membrane. (3) A former unit of mass (or weight) equivalent to one-sixteenth the mass (or weight) of one atom of oxygen-16, or the mass (or weight) of one hydrogen atom, i.e., $1 \times 10^{-24}$ gram.

**Dalton, John** English chemist and physicist (1766–1844). *See* Dalton's law of partial pressures and dalton.

**Dalton's law** Same as Dalton's law of partial pressures.

**Dalton's law of partial pressures** In a mixture of gases, the molecules of each gas exert the pressure they would exert if they occupied the volume alone at the same temperature, that is, independently of the other gases. The partial pressure of each gas is proportional to the amount of that gas in the mixture. The total pressure is the sum of the partial pressures:

$$pV = (p_1 + p_2 + p_3 + \ldots)V \quad \text{(D-02)}$$

where $p$ = absolute (total) pressure of the mixture, $p_i$ = partial pressure of gas ($i$), and $V$ = volume of the gas mixture. Dalton's law, formulated in 1891, is applied in water chemistry, along with other gas laws; sometimes called simply Dalton's law or law of partial pressures. *See* general gas law.

**dam** *See* dike.

**damping effect** The effect of storage in collection lines, different timing of tributary areas and other factors that contribute to reduce the peaks and increase the valleys in wastewater flows. *See*, e.g., the Gifft formulas.

**Danjes system** A tank containing wastewater that is treated by injection of fine air bubbles; this method is used in Munich, Germany.

**Dano Drum™** A rotating vessel manufactured by Reidel Smith Environmental, Inc. and used for composting.

***Daphnia*** A genus of microscopic crustaceans found in water; used in research and as food for tropical fish. A water flea.

**darcy** A unit for the measure of intrinsic permeability; it is the permeability of a medium whose pore is completely filled by a fluid of one centipoise viscosity that flows through the medium at a rate of one cubic centimeter per second per square centimeter of cross-sectional area under a pressure gradient of one atmosphere per centimeter of length. For water at 60°C, it corresponds to 18.2 gpd/ft$^2$ for water under a hydraulic gradient of 1.0 ft/ft.

**Darcy, Henri Philibert Gaspard** A French engineer (1803–1858) who, in a book on municipal water supplies, proposed the relationship that is widely used in the study of groundwater flow. In 1856 he conducted experiments for the design of sand filters at the city of Dijon. *See* Darcy's law. He suggested a form of the Darcy–Weisbach formula for head losses in pipes in 1857.

**Darcy's law** An empirical relationship between hydraulic gradient and viscous flow in the saturated zone of a porous medium under laminar conditions. The flux of vapors through the voids of the vadose zone can be related to a pressure gradient through the air permeability by Darcy's law. The generalized Darcy's law, also called Darcy's law of seepage, states that the flow velocity $V$ (actually the flux) at any point in a permeable material is proportional to the rate of hydraulic energy loss ($\Delta h/\Delta L$) at that point, i.e.,

$$V = K \cdot \Delta h/\Delta L = k(\Delta h/\Delta L) \cdot \gamma/\mu \quad \text{(D-03)}$$

where $K$ is the coefficient of permeability or hydraulic conductivity, $k$ is the intrinsic permeability and $\gamma$ and $\mu$ are, respectively, the specific weight and dynamic viscosity of the fluid. (In a sand filter, for example, $h$ could be the head loss and $L$ the thickness of the filter.) For flow in porous media, Darcy's law is valid up to Reynolds numbers between 1 and 10. *See also* Blake–Kozeny equation, effective porosity, effective velocity.

**Darcy–Weisbach equation** A turbulent flow equation, established in 1850 on the basis of experiments. It expresses the energy loss or head loss $h_f$ for flow in pipes as a function of pipe length ($L$) between two sections, pipe internal diameter ($D$), the velocity head, and a resistance coefficient or friction factor of the pipe ($f$):

$$h_f = f(L/D)(V^2/2\,g) \qquad \text{(D-04)}$$

where $V$ is the average velocity and $g$ the gravitational acceleration. Under laminar-flow conditions, the friction factor

$$f = 64/R_e \qquad \text{(D-05)}$$

where $R_e$ is the Reynolds number. For turbulent flow in smooth pipes and with $R_e > 4000$, the friction factor may be estimated from:

$$f^{0.5}[\log(R_e f^{0.5}) - 0.8] = 1 \qquad \text{(D-06)}$$

*See* the Chézy and Manning equations for more traditional methods of open channel design. *See also* the Blasius equation for an empirical formula of the friction factor. The Darcy–Weisbach equation can be used to derive the Kozeny equation for head loss in flow through porous media.

**Darcy–Weisbach roughness coefficient** The dimensionless roughness coefficient ($f$) in the Darcy–Weisbach equation.

**dark-field microscope** An optical device that uses light as a source of illumination to increase the contrast of a transparent, usually live and unstained, specimen and produce a bright image against a dark background. Maximum magnification = 2000 times. Resolution = 200 nm. *See also* light microscope, electron microscope, ultraviolet ray microscope.

**dark-field microscopy** The use of a dark-field microscope to study microorganism, e.g., the motility and growth of bacteria and protozoa.

**Darwin calibrator** A technique that uses an algorithm to identify the best parameters to calibrate a model. It mimics the principle of natural selection by minimizing the differences between observed and simulated variables.

**data** Generally, any type of information, for example, observations and measurements of physical facts, occurrences and conditions, presented in numbers, words, or graphics.

**data collection unit** A component of some meter reading devices used to organize data for billing and analysis.

**database** An organized, indexed collection of related data, i.e., a program for storing, retrieving and managing information. A set of data files on one or more related subjects.

**database management system or DBMS** A program designed to create and maintain databases. Some current developers of DBMS are Informix, Microsoft, and Oracle. *See* Table D-01 for an example of database management.

**data comparability** The characteristics (e.g., detection limit precision, accuracy, bias) that allow information from many sources to be of definable or equivalent quality so that it can be used to address program objectives not necessarily related to those for which the data were collected.

**DataGator™** An apparatus manufactured by TN Technologies, Inc. for measuring flows in sewers.

**data quality objectives** In the context of water quality monitoring, the characteristics or goals that are determined by a monitoring or interpretive program to be essential to the usefulness of the

**Table D-01.** Database management system (example)*

The sewer modeling study used Microsoft Access® to perform the data management tasks. Due to the magnitude and complexity of the project, traditional data processing tools like spreadsheets were inadequate to perform the data compilation and analysis. The model development required planning and creation of extensive databases to handle the input–output (I/O) operations for the model. The consultant developed several databases in Access®, which proved to be a powerful but relatively simple-to-use tool. These databases served as pre- and postprocessors for the VSC model. They include: (a) Hydrograph generation database—A set of tables that translate the hydrograph development process into sequential queries and creation of the hydrograph import files. (b) Pump dimensions/pump curves database—Another set of tables to aid in pump station data import. The pump import file includes the pump name, pump discharge–dynamic head relationship (pump curve), pump station on/off elevations, number of pumps per station, and initial wet well depth. (c) Minor losses (K factors) database—Designed to compute the overall minor loss coefficient for each of the 350 pump stations in the model. (d) Pump stage–storage database—Designed to generate the stage–storage input in four steps.

*Adapted from Adrien, 2004

data, e.g., limits of precision and bias of measurements, completeness of sampling and measurements, representativeness of sites with respect to objectives, and validity of data.

**DataRam™** A monitoring device manufactured by MIE, Inc. that provides continuous readout of airborne particles.

**dateometer** A small calendar disc attached to motors and equipment to indicate the year in which the last maintenance service was performed.

**datum** (1) A level surface, line, or point used as a reference in surveying. (2) Any quantity or set of quantities used as a reference or base.

**datum level** Same as datum plane.

**datum plane** A permanently established horizontal plane, surface, or level used as a reference to compute heights, depths, soundings, ground and water surface elevations, and tidal data. Mean sea level is the most generally used datum. However, mean low water and lower low water are also used on the Atlantic coast and Pacific coast of the United States, respectively. *See also* NGVD. Also called chart datum, datum level, reference level, or reference plane.

**Davies equation** An equation proposed to calculate the activity coefficient ($\gamma$) of a chemical species as a function of its charge or valence ($c$) and the ionic strength ($I$) of the solution at 25°C, when $I > 0.1$ M (Metcalf & Eddy, 2003):

$$\log \gamma = -0.51\, c^2[I^{0.5}/(1 + I^{0.5}) - 0.3\, I] \quad \text{(D-07)}$$

*See also* Guntelberg approximation.

**day-degrees** The sum of the degrees above a reference temperature over a certain period. *See* degree-day for detail.

**day tank** A tank used to store a chemical solution of known concentration for feed to a chemical feeder. A day tank usually stores sufficient chemical solution to properly treat the water for at least one day. Also called an age tank.

**2,4-DB** A pesticide with a WHO-recommended guideline value of 0.09 mg/L for drinking water.

**DBAN** Acronym of dibromoacetonitrile.

**DBC** Acronym of doubleflow bubble contactor

**DBCM** Acronym of dibromochloromethane.

**DBCP** Acronym of dibromochloropropane.

**DBC Plus®** Microbial culture produced by Enviroflow, Inc. for use in wastewater treatment.

**DBP** Acronym of disinfection by-product(s).

**DBP formation** *See* disinfection by-product formation.

**DBPFP** Acronym of disinfection by-product formation potential.

**DBP precursor** *See* disinfection by-product precursor.

**DBP shift** It is the change in DBP species observed when source water containing bromide is chlorinated or ozonated. Both chlorination and ozonation can lead to the formation of hypobromous acid (HOBr) and brominated as well as mixed DBPs. In the absence of bromide in the source water, chlorination results in the formation of chlorinated DBPs only.

**DBS™** A series of drive units of DBS Manufacturing, Inc. for use in clarifiers and sludge thickeners.

**DCA** Acronym of dichloroacetic acid.

**DCAN** Acronym of dichloroacetonitrile.

**1,1-DCE** *See* 1,1-dichloroethylene.

**D-Chlor™** An apparatus manufactured by Eltech International Corp. for dechlorination with sodium sulfite tablets.

**DCI System Six™** An apparatus manufactured by Fischer & Porter for water disinfection using ultraviolet light in an open channel.

**DCS** Acronym of digital control system, distributed control system, distribution control system, or distributor control system.

**DD** Acronym of (a) diffusion dialysis, (b) Donnan dialysis.

**D/DBP** Acronym of disinfectant/disinfection by-product(s).

**D/DBP Rule** *See* Disinfectant/Disinfection By-Products Rule.

**DDE** Acronym of dichlorodiphenyl dichloroethylene.

**DDF** Acronym of downflow denitrification filter.

***D. dentricum*** *See Diphyllobothrium dentricum.*

**DDS** Acronym of dual digestion system.

**DDT** {$[(ClC_6H_4)_2CH(CCl_3)]$ or $C_{14}H_9Cl_5$} The first chlorinated hydrocarbon insecticide (chemical name: dichloro-diphenyl-trichloroethane); a white, crystalline, water-insoluble solid, formed by the reaction of chloral with chlorobenzene in the presence of sulfuric acid. It has a half-life of 15 years and can collect in fatty tissues of certain animals. EPA banned registration and interstate sale of DDT for virtually all but emergency uses in the United States in 1972 because of its persistence in the environment and accumulation in the food chain, under the Toxic Substances Act. Adsorption on granular activated carbon is an effective treatment process for the removal of DDT and other organic chemicals (e.g., chloroform, dieldrin, lindane).

**DE** Acronym of diatomaceous earth.

**$d_e$** Symbol for effective size of filter media.

**deactivation** (1) The protection of a metallic surface by deaeration (removal of oxygen by vacuum, heating or other methods) and removal of corrosion products by filtration. *See also* passivation. (2) The loss of activity by an adsorbent or a catalyst.

**dead-burned lime** The lumps of inert, semivitrified material obtained during the calcining process by heating limestone beyond the point of quicklime, which is soft, porous, and highly reactive. Dead-burned lime, also called overburned lime, is used in refractory materials. *See* quicklime for detail.

**dead end** The end of a water main that is not connected to other parts of the distribution network by means of a connecting loop or pipe.

**dead-end filtration** (1) A filtration operating configuration with two continuous flow streams through the membrane or medium (feedwater and permeate) and periodic backwashing. *See also* crossflow filtration, direct filtration. (2) The operating condition of a pressure-driven membrane unit in which there is no crossflow; all the water applied passes through the membrane. *See* direct-feed filtration for detail.

**dead-end flow** A filtration operating condition in which all the feedwater passes through the medium.

**dead storage** The portion of a pond or reservoir below the lowest outlet of the structure, thus not available for use, except for sediment collection. *See* reservoir storage.

**dead volume** The unused portion of a basin, tank, reactor, etc., which reduces the volume effectively available for treatment, because of obstructions or short-circuiting.

**dead water** Standing or still water, or water that fails to circulate for the proper operation of a structure or equipment. Different from dead well.

**dead well** A well through an impermeable layer to drain water to a permeable one. Sometimes called absorbing well, drain well, negative well. Different from dead water.

**dead zone** (1) A section of a stream or river in which flow is reduced or even reversed, where there may exist relatively stagnant pools with little inflow or outflow, and where complete, dispersive mixing dominates. In a natural water body, a dead zone is characterized by further reduction and retention of a pollutant cloud. *See also* aggregate dead zone model. (2) An area in the Atlantic Ocean in the approach to New York harbor that is (or was) relatively free of marine plants and animal life because of the sludge that is (or was) being dumped there. The sludge dumping causes high BOD and low dissolved oxygen levels.

**deaeration** The process or the act of eliminating or reducing the concentration of oxygen or air in water, other liquids, or granular activated carbon. Deaeration reduces the corrosivity of water.

**deaerator** A device that removes oxygen, air or other dissolved gases from a solution. *See also* degasifier.

**dealkalization** A process that eliminates or reduces the alkalinity of water, usually by chemical treatment or ion exchange.

**dealkalizer** A strong anion exchange unit used to reduce bicarbonate alkalinity.

**dealkylation** The replacement, through microbial metabolism, of alkyl groups by hydrogen (H) in an organic compound that contains such groups; e.g., the O-dealkylation of methoxychlor insecticides ($R—O—CH_3 \rightarrow R—OH$) and the S-dealkylation of dimethyl mercaptan ($R—S—CH_3 \rightarrow R—SH$).

**dealloying** A form of corrosion that consists in the preferential removal of one metal from an alloy; e.g., dezincification or the removal of zinc from brass or the removal of disseminated lead from brass and soldered joints. Also called selective leaching.

**DeAmine™** A chemical produced by NuTech Environmental Corp. for odor control.

**death phase** The last phase in the growth patterns of bacteria in a batch reactor; no growth is occurring as the substrate is depleted and changes in biomass concentration are exponential, and due to cell death.

**debenture** An unsecured long-term debt instrument.

**debris** Solid material exceeding a 60 mm particle size that is intended for disposal and that is: a manufactured object; or plant or animal matter; or natural geologic material; except such materials as lead acid batteries, cadmium batteries, radioactive lead solids; process residuals such as smelter slag, and residues from the treatment of waste, wastewater, sludges, or air emission residues (EPA-40CFR268.2-g).

**debt service** The amount of money necessary annually to cover interest and principal on a debt or to provide a sinking fund to redeem a bond.

**Debye–Huckel limiting law** *See* Debye–Huckel theory.

**Debye–Huckel theory** A theory proposed to determine the mean activity coefficient ($\gamma$) of a solute as a function of its ionic charge or valence ($c$), the ionic strength of the solution, the dielectric constant, temperature, and an adjustable parameter corresponding to the size of the ion. This theory is used in the Davies and Guntelberg equations.

**decant** (1) To draw off the upper layer of liquid (water) after the heavier material (a solid or another liquid) has settled. (2) To pour a liquid gently from one container to another so as not to disturb the contents. (3) The liquid sidestream resulting from decanting, e.g., the high-strength sidestream from solubilization during thermal sludge conditioning, which is a major source of odors.

**decantation** Separation of a liquid from solids or from a heavier liquid by drawing off the supernatant without disturbing the lower layer.

**decanter centrifuge** A device consisting of a rotating helical screw conveyor inside a rotating bowl, used for sludge dewatering. *See* solid-bowl centrifuge for detail.

**decanting paved drying bed** A paved bed used in sludge dewatering as an alternative to sand drying beds, particularly in warm, arid, or semiarid climates. As shown in Figure D-01, the device incorporates a soil–cement mixture as paving material, drawoff pipes for decanting the supernatant, and a sludge feed pipe at the center. *See also* drainage paved drying bed.

**decant phase** The fourth of the five steps in the operation of sequencing batch reactors, whereby effluent is removed using floating or adjustable weirs or other mechanisms. Also called draw phase. *See also* fill phase, react phase, settle phase, and idle phase.

**decant tank** A container of high-strength sidestreams from thermal sludge conditioning, a major source of odor in wastewater treatment plants. It also functions as storage for conditioned sludge.

**decarbonate** To remove or reduce carbon dioxide ($CO_2$) from water or another liquid.

**decarbonator** A unit that removes carbon dioxide ($CO_2$) from a solution; also a unit that converts alkalinity to carbon dioxide before air stripping.

**decationize** To remove cations by ion exchange; to exchange cations for hydrogen ions using a strong acid cation exchanger.

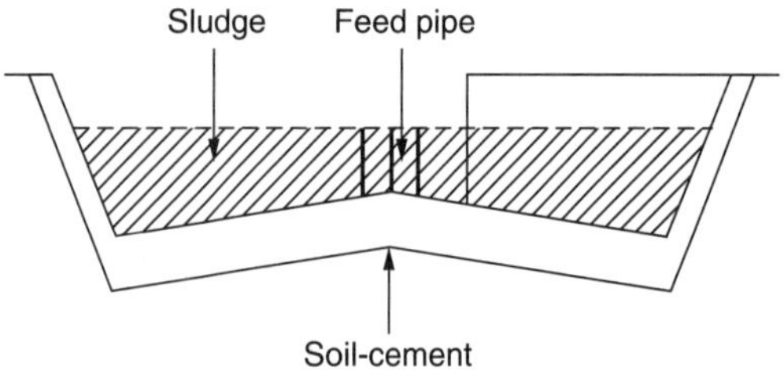


**Figure D-01.** Decanting paved drying bed.

**decay** Disintegration of a substance; disappearance of a pollutant as a result of absorption, chemical reaction, removal, or transformation into another substance.

**decay processes** The hazardous ranking system (HRS) lists five decay processes: biodegradation, hydrolysis, photolysis, radioactive decay, and volatilization. *See also* half-life.

**decay product** An isotope formed by the radioactive decay of some other isotope. This newly formed isotope possesses physical and chemical properties that are different from those of its parent isotope, and may also be radioactive (EPA-40CFR300-AA).

**decay rate** The number of nuclei disintegrating in a unit time. *See* half-life.

**DeCelerating Flo** A gravity sand filter manufactured by CBI Walker, Inc.

**decentralized sewer service** Wastewater treatment and disposal systems used in small communities include (a) systems that treat wastewater close to sources that generate it and maximize the use of soil dispersal and reuse opportunities, and (b) onsite systems that use soil-based methods similar to septic systems.

**Dechlor Demon™** A device manufactured by Hydro Flow Products, Inc. for dechlorinating water while fire flow testing and flushing.

**dechlorinating agent** A reducing agent used to remove chlorine from water. *See* dechlorination (2).

**dechlorination** (1) The deliberate removal of chlorine from water or any other substance by chemically replacing it with hydrogen or hydroxide ions in order to detoxify the substance. (2) The partial or complete reduction of residual chlorine ($Cl_2$, HOCl, $OCl^-$) by any chemical or physical process for various purposes: the reduction of potential toxic effects on aquatic life, taste and odor control, and the protection of a downstream process such as zeolite softening. Reducing agents commonly used in dechlorination include sulfur dioxide ($SO_2$), sodium bisulfite ($NaHSO_3$), sodium sulfite ($Na_2SO_3$), sodium metabisulfite ($Na_2S_2O_5$), and sodium thiosulfate ($Na_2S_2O_3$). Ammonium chloride ($NH_4Cl$) and ascorbic acid ($C_6H_8O_6$) may also be used. The possible overall reactions are:

$$SO_2 + HOCl + H_2O \rightarrow Cl^- + SO_4^{2-} + 3\ H^+ \quad \text{(D-08)}$$

$$SO_2 + Cl_2 + 2\ H_2O + \rightarrow H_2SO_4 + 2\ HCl \quad \text{(D-09)}$$

$$NaHSO_3 + Cl_2 + H_2O \rightarrow NaHSO_4 + 2\ HCl \quad \text{(D-10)}$$

$$2\ Na_2S_2O_3 + Cl_2 \rightarrow Na_2S_4O_6 + 2\ NaCl \quad \text{(D-11)}$$

Chlorine is also removed by aeration (for taste and odor control) or adsorption on activated carbon, although the latter may occur as an undesirable side effect:

$$C + 2\,HOCL \rightarrow CO_2 + 2\,H^+ + 2\,Cl^- \quad (D\text{-}12)$$

$$C + 2\,Cl_2 + 2\,H_2O \rightarrow CO_2 + 4\,HCl \quad (D\text{-}13)$$

**dechlorination agent** Same as dechlorinating agent.

**dechlorination by substitution** *See* substitutive dehalogenation.

**decibel** The unit of the intensity of sound waves.

**decimal reduction time** The time required to thermally destroy 90% of the microorganisms in a sample of water or wastewater. Also called D value. *See also* thermal death time and z value.

**declining block pricing** Pricing of utility services based on the declining rate concept.

**declining block rate or declining rate** A water rate structure in which the unit price decreases for each successive block; e.g., $0.015 per gallon for the first 1000 gallons per month, $0.013 per gallon for the next 5000 gallons, etc. It is the opposite of inclining and inverted block rates.

**declining growth** The phase in the reproduction of bacteria (and microorganisms in general) where the rate of growth slows down following a period of logarithmic growth, assuming suitable pH, temperature, and other environmental conditions. The declining-growth phase begins when food is almost depleted and becomes a limiting factor. A stationary phase follows declining growth before the onset of the endogenous phases. Declining growth prevails in most biological wastewater treatment processes. The Monod equation describes growth in the declining phase.

**declining-rate filter** A filter that produces water at a declining rate. It is sometimes designed as a system of several units such that each unit is backwashed by the effluent from the others without pumping; each filter's rate depends on its head loss.

**declining-rate filtration** A filter operating method whereby (a) the rate of flow through the filter declines and the level of the liquid above the filter bed rises throughout the length of the filter run; or (b) the flow declines through the filter units, the cleanest unit carrying the largest flow; or (c) the water level decreases in a common underdrain as the individual units become plugged. Sometimes called variable declining-rate filtration. *See also* constant-pressure filtration, constant-rate filtration, true declining-rate filtration, uniform-flow filtration.

**declining-rate operation** *See* declining-rate filtration.

**decolorization** The removal of color-causing substances such as tannins or humic acid from water, using such treatment methods as adsorption, coagulation–filtration, ion exchange, and oxidation.

**decolorizing index** One of three parameters used by manufacturers to express the relative capacity of an activated carbon by testing it against a standard molasses solution for the adsorption of large molecules. *See* molasses number for detail.

**decomposers** Bacteria, fungi, and other organisms that break down dead organic matter into simpler compounds. Also called reducers. *See also* mineralization, detritovores, grazers.

**decomposition** (1) The conversion of chemically unstable materials to more stable forms by chemical or biological action. If organic matter decays when there is no oxygen present (anaerobic conditions or putrefaction), undesirable tastes and odors are produced. Decay of organic matter when oxygen is present (aerobic conditions) tends to produce much less objectionable tastes and odors. (2) An event in which a polymerization reactor advances on the point where the polymerization reaction becomes uncontrollable, the polymer begins to break down (decompose), and it becomes necessary to relieve the reactor instantaneously in order to avoid catastrophic equipment damage or serious adverse personnel safety consequences (EPA-40CFR60.561).

**decomposition of wastewater** The chemical, aerobic or anaerobic transformation of organic or inorganic matter in wastewater.

**deconcentrator** A device used to remove accumulated solids from boiler feedwater.

**decontamination** The process of reducing, eliminating, or making harmless biological or chemical agents, so as to reduce the likelihood of disease transmission from those agents.

**decontamination factor** The reciprocal of the surviving fraction of a pathogen that is exposed to a disinfecting agent; it is sometimes used to compare the effectiveness of different disinfectants. For example, at a concentration of 0.5 mg/L, chlorine and ozone have about the same decontamination factor for *E. coli*: 1000 (i.e., a surviving fraction of 0.001), whereas, at 0.3 mg/L, the factor decreases to 100 for chlorine and close to 1.0 for ozone.

**decreasing block rate** Same as declining block rate.

**decross-linking** The physical or chemical destruction of the cross-linked polymer (e.g., divinylbenzene) of an ion-exchange resin, resulting in increased moisture in the resin and swelling of the beads. Destructive agents include heat, chlorine, ozone, and hydrogen peroxide.

**dedicated capacity** The portion of the capacity of a water, wastewater, or other utility that is reserved for an individual customer or group of customers.

**dedicated land disposal (DLD)** (1) A method of treating and disposing of water or wastewater treatment residuals in soils used exclusively for that purpose. The site must be environmentally acceptable and be capable of accepting a high rate of application. Treatment occurs through soil microorganisms, aeration, and oxidation. *See also* disturbed land reclamation. (2) More generally, the surface application of sludge to land for disposal purposes and not for growing crops, usually at rates higher than agronomic rates.

**Dee Fo™** A chemical additive produced by Ultra Additives, Inc. for foam control.

**DeepAer** An aeration apparatus manufactured by Walker Process Equipment Co.

**DeepBed** A fixed-film reactor manufactured by TETRA Technologies, Inc. consisting of a filter bed of granular media.

**deep-bed denitrification filter** Same as denitrification filter.

**deep-bed downflow filter** A 4–8 ft-deep filter packed with relatively large media (e.g., anthracite) for wastewater treatment.

**deep-bed filter** A sand, anthracite, or dual-media filter having a depth of 4–6 feet, used especially in direct filtration; sometimes combined with larger media to allow deeper penetration of particles or with ozonation to remove biodegradable by-products.

**deep-bed filter denitrification** A wastewater treatment method using a 4–6 ft deep filter to remove nitrogen from, and further polish, a nitrified effluent. The granular medium provides support for heterotrophic microorganisms that use an exogenous carbon source such as methanol ($CH_3OH$). *See also* bump.

**deep-bed filtration** *See* deep-bed filter.

**deep-bed upflow continuous backwash filter** A 3–4 ft-deep filter packed with sand for wastewater treatment. The liquid is introduced at the bottom and flows upward through riser tubes and through downward-moving sand. Filtered water exits from the sand bed at the same time that an airlift pipe draws trapped solids downward. *See also* continuous-backwash filter.

**Deep Bubble™** Multistaged aeration systems manufactured by Lowry Aeration Systems, Inc. for the removal of radon, carbon dioxide, and volatile organic compounds.

**Deep Draw** Airlift diffusers manufactured by Wilfley Weber, Inc. for installation in lagoons.

**deep injection well** A well used to discharge water or wastewater into a deep underground stratum. Some wells are more than 1000 ft deep. Hazardous wastewater or toxic chemicals can also be discharged through boreholes that penetrate saline aquifers. It is then desirable that the stratum be overlain by impervious layers to prevent upward migration into freshwater zones.

**deep lake** For seasonal density or thermal stratification considerations, a lake that is deeper than 20 feet.

**deep reservoir** For seasonal density or thermal stratification considerations, a reservoir that is deeper than 20 feet.

**deep shaft activated sludge process** A variation of the conventional activated sludge process, developed in England and used where land is scarce or expensive. The aeration basin is replaced by a shaft of two compartments, up to 500 ft deep into the ground, with wastewater flowing downward in the center and upward in the outer compartment. A flotation tank is used to separate the effluent from the recycle and waste activated sludges. Oxygen transfer is three times as efficient as in the conventional process. *See* Figure D-02.

**deep shaft reactor** The reactor used in the deep shaft activated sludge process.

**deep vertical shaft process** Same as deep shaft activated sludge process.

**deep water (DW)** A mass of water found at great depth, such as the North Atlantic DW, Indian DW, Pacific DW, Japan Sea DW, Mediterranean DW, Greenland Sea DW, and Norwegian Sea DW.

**deep-water habitat** Permanently flooded land that lies below the deep-water boundary of wetlands.

**deep well** A drilled well that draws water or minerals from below shallow impermeable strata, or injects wastewater below such strata. Well depth

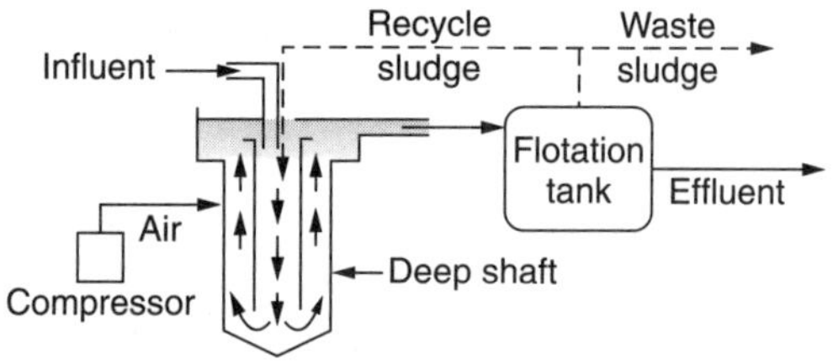


**Figure D-02.** Deep shaft activated sludge process.

varies from a few hundred feet to several thousand feet.

**deep-well disposal** The disposal of wastewater into confined underground strata. *See* deep injection well, underground injection, class I hazardous waste injection well.

**deep-well injection** The discharge, by gravity or pumping, of raw or treated wastes for containment in the pores of a permeable formation that is not used for water supply. *See also* deep well, deep injection well.

**deep-well pump** A centrifugal or reciprocating pump for lifting water from a deep well. *See* deep-well turbine pump and submersible pump.

**deep-well submersible pump** A pump for use in a deep well. The impeller is installed near the water surface to avoid cavitation, whereas the motor is at the bottom, just below the impeller.

**deep-well turbine pump** A multistage centrifugal pump for use in a deep well. The impeller is installed close to the water surface to avoid cavitation, whereas the motor is at ground level. Also called vertical turbine pump.

**deep-well WAO process** A variation of the wet-air oxidation process that uses 5000-ft deep vertical concentric tubes inside a drilled well for the thermal treatment of macerated sludge. Sludge flows down the annular space between the inner tubes (called downcomer); oxygen and steam flow into the central tube; and the oxidized sludge exits upward in the outer annular space (the upcomer) and into a solids–liquid separator.

**defecate** To have a bowel movement; to excrete waste matter from the intestines.

**deferrization** The removal of soluble iron compounds from water, usually by chemical precipitation using dissolved oxygen, sometimes with chlorine or potassium permanganate. The aeration process is also effective in the removal of manganese (demanganization) and odor-producing gases.

**deficit ratio** *See* Barrett equation.

**defilement** Pollution.

**DE filter** Abbreviation of diatomaceous earth filter.

**DE filtration** Abbreviation of diatomaceous earth filtration.

**Defined Substrate Technology™** A reagent system developed by Environetics, Inc. to enhance the growth of specified microorganisms.

**defined-substrate technology (DST)** A microbial testing procedure that uses a special medium to detect simultaneously total coliforms and *E. coli*. *See* Colilert test.

**definitive host** The host in which reproduction of a parasite occurs.

**deflagration** The explosion that may occur in the storage vessels of dried wastewater treatment residuals when organic dust in the air catches fire and the heat of combustion increases the pressure beyond the strength of the vessel.

**Deflectofuser®** An air diffuser manufactured by the FMC Corp. for aeration by coarse bubbles.

**deflocculant** A chemical used to increase fluidity, e.g., to decrease the tendency of colloidal particles in a liquid to clump together.

**deflocculating agent** A material added to a suspension to prevent settling.

**defluoridation** The removal of excess fluoride in drinking water to prevent the mottling (brown stains) of teeth. Fluoride removal methods include (a) ion exchange on such media as bone char, apatite, or activated alumina, all of which can be regenerated by caustic soda; and (b) adsorption on or coprecipitation with magnesium hydroxide (McGhee, 1991).

**defoamant** A substance that has low foam compatibility and low surface tension, used to control foaming in wastewater treatment and sludge processing, even at low concentrations; usually based on silicone, polyamide, vegetable oils, or stearic acid. Upon contact, the droplets of defoamant break the foam bubbles one after another.

**defoamer** Same as defoamant.

**defoaming agent** Same as defoamant.

**defoliant** A chemical, such as 2,4-dichlorophenoxyacetic acid, applied to trees and growing plants that causes them to shed their leaves.

**deformable solids** Suspended solids that get into the interstices of a filter cake and cause plugging.

**degasification** A water treatment process that removes dissolved gases from the water, for water quality enhancement or to improve plant operation. The gases may be removed by either mechanical or chemical treatment methods or a combination of both.

**degasifier** A device for removing carbon dioxide, hydrogen sulfide, and other dissolved gases from solution, usually by air stripping. *See also* deaerator.

**degasify** To reduce the concentration of gases and volatile substances from solution.

**degassing** (1) Under the Clean Water Act, the removal of dissolved hydrogen from molten aluminum prior to casting. Chemicals are added and gases are bubbled through the molten aluminum. Sometimes a wet scrubber is used to remove excess chlorine gas (EPA-40CFR467.02-e). (2) Same as degasification.

**degradable material** Material that can be broken down by biological, chemical, or physical processes.

**degradation** (1) Chemical, physical, or biological breakdown of a complex compound into simpler compounds; same as decomposition. (2) The geological process of wind and water wearing down and carrying away parts of the surface of the earth. (3) A decrease in water quality or in the performance of a water or wastewater treatment unit. (4) The reduction in ion exchange capacity of a resin due to swelling or size reduction of the resin particles.

**degradation potential** The degree to which a substance is likely to be reduced to a simpler form by bacterial or chemical activity.

**degrade** To convert complex materials into simpler compounds.

**degraded water** Water that is polluted and unsuitable for its primary purpose.

**degraded wetland** A wetland that has been impaired physically or chemically by human activities to the point that its habitat value, food storage, or other functions are significantly reduced.

**degreasing** The removal of grease, oil, and dirt from wastewater, sludge, or machine parts.

**degree** Any of several units of temperature, density, concentration, angular measure, or viscosity.

**degree Baumé** A measure of density or concentration of chemicals, e.g., aqua ammonia. *See* Baumé degree.

**degree Celsius** A unit of temperature on the centigrade or Celsius scale equal to 1/100 of the interval between the freezing and boiling points of water under standard conditions.

**degree-day** (1) A parameter that indicates the heating requirement of a building during the winter or its cooling requirement (air conditioning and refrigeration) during the summer. It is the departure of one degree, in one day, of the mean daily temperature from a given standard. The standard temperatures are 24°C = 75°F for the cooling degree-day and 19°C = 66°F for the heating degree-day. *See also* day-degrees. (2) The parameter is also used in relation to plant growth. *See* growing degree-day. *See also* freezing index.

**degree Engler** A measure of the viscosity of a fluid, determined by the Engler viscometer.

**degree Fahrenheit** A unit of temperature on the Fahrenheit scale equal to 1/180 of the interval between the freezing and boiling points of water under standard conditions.

**degree of freedom** An independent variable such as concentration condition, temperature, and pressure in solving chemical equilibrium problems. *See* phase rule for detail.

**degree of purification** The extent to which a natural or artificial process destroys or removes objectionable impurities from water or wastewater. *See also* contaminant, self-purification, treatment processes.

**degree of treatment** A measure of the efficiency or performance of a water or wastewater treatment process with respect to the removal of impurities or contaminants.

**degritter** A square tank designed and operated with a constant level and a short detention time (1.0 min or less) for the removal of grit from wastewater. *See* detritus tank for more detail. *See also* cyclone degritter, DorrClone®.

**degritting** *See* sludge degritting.

**dehalogenation** A microbial reaction in which a halogen atom [bromine (Br), chlorine (Cl), fluorine (F), and iodine (I)] is replaced by the hydroxyl radical (OH). *See also* oxidative dehalogenation, reductive dehalogenation, substitutive dehalogenation.

**dehalogenation by oxidation** *See* oxidative dehalogenation.

**dehalogenation by reduction** *See* reductive dehalogenation.

**dehalogenation by substitution** *See* substitutive dehalogenation.

**dehydrate** To remove, chemically or physically, water in combination with other matter; or to give up water geologically and form a new mineral compound.

**dehydration** A chemical, physical, or geologic process whereby water is removed or lost.

**DeHydro®** A sludge drying bed manufactured by Infilco Degremont, Inc.

**dehydrogenase** An oxydoreductase enzyme that catalyzes the removal of hydrogen from a substance; a dehydrogenase enzyme.

**dehydrogenase enzyme** An organic compound that can be reduced or oxidized to facilitate metabolic reactions. It is used to measure the activity of viable cells in biological treatment. *See also* adenosine triphosphate.

**deicing chemical** A chemical substance used to remove ice or prevent ice formation on roads, airplane wings, etc. Deicing salts may constitute a water quality problem by contributing a variety of pollutants to surface water sources, e.g., sodium, chloride, BOD, chromium, copper, lead, nickel, zinc, solids, and phosphorus.

**deicing salt** *See* deicing chemical.

**deionization (DI)** The process of removing ions (all ionized minerals and salts, both organic and inorganic) from a solution, most commonly through ion exchange. The two-phase batch process removes positively charged ions by a cation exchange resin and negatively charged ions by an anion exchange resin. *See*, e.g., LINX®. *See also* desalination and electrodeionization.

**deionization water** Bottled water that has been treated by deionization to meet the definition of purified water, a labeling regulated by the U.S. Food and Drug Administration. *See* bottled purified water.

**deionize** To remove organic and inorganic minerals and salts from water. *See* deionization.

**deionized water** In general, water produced by deionization, i.e., neutral water that is about equal to distilled water in purity. Also, reagent water used in laboratories and prepared there using ion-exchange cartridges. *See also* demineralized water, bottled purified water.

**deionizer** A device or unit that deionizes water.

**dekSPRAY** Nozzles manufactured by Munters for cooling towers.

**Delaval Filter** A filter manufactured by IDRECO, Ltd. for polishing precoat condensate.

**delay damages** Penalties imposed as a result of a manufacturer or supplier of water or wastewater goods or services failing to complete a project on time, including, e.g., the costs of additional engineering, utility expenses, or regulatory penalties. Also called liquidated damages. *See also* consequential or performance damages, performance specification, prescriptive specification.

**delayed Chick–Watson model** A modification of the Chick–Watson model that includes an initial lag period ($CT_{lag}$) for the $CT$ parameter ($C$ being the disinfectant concentration and $T$ the contact time):

$$N/N_0 = e^{-kCT} \quad \text{for } CT > CT_{lag} \qquad \text{(D-14)}$$
$$= 1 \quad \text{for } CT \leq CT_{lag}$$

where $N$ = number of surviving organisms, $N_0$ = number of live (viable, infectious) organisms before addition of disinfectant, and $k$ = constant. *See also* disinfection kinetics.

**delayed hypersensitivity** An allergic reaction to a biological or chemical substance that develops only some considerable time after exposure.

**delayed inflow** Stormwater that takes some time (e.g., several days) to reach the sanitary sewer system, including discharges of sump pumps from cellar drainage, surface water through manholes in ponded areas, and runoff associated with the melt of an accumulated snow cover. *See also* direct inflow, steady inflow, total inflow, infiltration/inflow.

**delayed start** A procedure used in the operation of water filtration units to reduce the negative effect of the filter ripening sequence on the filter effluent; the filter is kept out of service for a certain period of time after it is refilled, for example, until the next filter is stopped for backwashing. *See also* slow start and extended terminal subfluidization wash.

**delayed toxicity** An adverse chemical effect that becomes manifest only after some minimum latent period.

**delegated state** A state (or other governmental entity such as a tribal government) that has received authority to administer an environmental regulatory program in lieu of a federal counterpart. As used in connection with NPDES, UIC, and PWS programs, the term does not connote any transfer of federal authority to a state.

**deliquescence** The ability of a dry solid to absorb water from the air and become saturated.

**delivered water** The finished product water from a public or private utility water plant that is carried through a water main network and arrives at the point of use (e.g., a home, institution, or business facility).

**del operator** *See* Laplace equation.

**DelPAC** Polyaluminum coagulants produced by Delta Chemical Corp.

**Delrin** Acetyl resins of high molecular weight produced by E. I. DuPont de Nemours & Co.

**delta** An alluvial deposit made of sediments and debris dropped by a stream as it enters a body of water.

**Delta-G®** A parallel-plate separator manufactured by Smith & Loveless, Inc.

**delta P (ΔP)** The pressure drop (in psi) by flowing water in a pressurized system as the result of the velocity and turbulence of the water, conduit restrictions, and surface roughness.

**Deltapilot** An instrument manufactured by Endress + Hauser for measuring hydrostatic levels.

**Delta-Stak®** An inclined-plate clarifier manufactured by Eimco Process Equipment Co.

**DeltΔ™** A traveling chain manufactured by Envirex, Inc.; used in water screening operations.

**deluge shower** A safety apparatus for washing chemicals off quickly.

**Delumper®** A line of products of Franklin Miller for disintegrating and crushing solids.

**demand** (1) The quantity of water required or used by customers during a certain period. The term applies similarly to other services such as electricity or gas. *See also* the following: fire demand, flow demand, peak demand. (2) Any of the common chemical or biological requirements in water and wastewater treatment. *See* the following: benthal demand, biochemical oxygen demand, chemical oxygen demand, chlorine demand, first-stage biochemical oxygen demand, immediate biochemical oxygen demand, oxygen demand, second-stage biochemical oxygen demand, standard biochemical oxygen demand, ultimate biochemical oxygen demand.

**demand charge/rate** A special charge, surcharge, or rate levied by a water utility for seasonal peak demands, particularly on wholesale customers to encourage them to reduce their peak demands via equalization storage. Also called peak-flow charge/rate.

**demand curve** A graphical representation of customers' willingness to pay for water, i.e., a curve showing the amount of water they are willing to buy at different rates.

**demand factor** The ratio of the peak or maximum demand for water (or electricity, gas, etc.) to the average demand.

**demand initiated regeneration (DIR)** A method of automatically initiating regeneration or recycling in filters, deionizers, or softeners after a predetermined, metered volume of water has been processed. In a softener or deionizer, regeneration may be triggered automatically, based on an electrical or mechanical signal. All operations, including bypass, backwashing, brining, rinsing, and returning the unit to service, are initiated and performed automatically in response to the demand for treated water.

**demand management** An approach to produce economies of water use and eliminate, reduce, or delay capacity expansion. Measures to reduce demand on a water supply system include leakage reduction, rate policy (e.g., conservation-oriented pricing), water-saving taps and fittings, water-efficient landscaping, and consumer education.

**demand-side management** An approach used by public utilities to increase capacity through reducing demand instead of installing additional units.

**demand-side waste management** An approach by consumers who use purchasing decisions to communicate to product manufacturers that they prefer environmentally sound products packaged with the least amount of waste, made from recycled or recyclable materials, and containing no hazardous substances.

**demand uncertainty** A safety factor included in demand forecasting and capacity projections to account for the variability of the factors that affect forecasts.

**demanganization** The removal of manganese from water, usually by chemical precipitation using dissolved oxygen, sometimes with chlorine or potassium permanganate. The aeration process is also effective in the removal of iron (deferrization) and odor-producing gases.

**demineralization** A treatment process that removes dissolved minerals (salts) and inorganic constituents from brackish and saline waters, usually through ion exchange or another desalting process, chemical coagulation, or lime softening for heavy metals. *See* desalination and saline-water conversion classification for more detail.

**demineralize** To reduce the concentration of minerals in water. *See* demineralization.

**demineralized stream** A stream with a reduced mineral concentration, e.g., as a result of electrodialysis or other electrical-driven membrane process.

**demineralized water** Water produced by demineralization. *See also* bottled purified water and deionized water.

**demineralizer** A device or unit that demineralizes water.

**demineralizing** Removing dissolved mineral salts and other dissolved solids from water, using such processes as adsorption on a carbon\ electrode, electrodialysis, ion exchange, osmionic, and thermal diffusion, usually to make water drinkable or usable as process or cooling water; demineralization. *See* desalination and saline-water conversion classification for more detail. Same as desalting.

**demister** A device that is used to eliminate or reduce entrainment, i.e., to remove entrained droplets of water from a vapor stream produced during evaporation (e.g., during desalting). Also called demister entrainment device, entrainment separator, mist eliminator, or simply separator. *See also* spacer.

**Demister®** A device manufactured by Otto H. York Co. to eliminate mists.

**demister entrainment device** Same as demister.

**demonstration of performance (DOP)** *See* DOP credit.

**denaturation** Irreversible destruction of a macromolecule, e.g., destruction of protein by heat.

**dendritic distribution system** A tree-like or branched network, having no loops. All flow in a dendritic network converges in a single sink, as in a sewer or drainage system, or originates from a single source, as in an irrigation or a branched water supply system.

**dendritic drainage** A drainage network having channels with many branches.

**dengue** A short name for dengue fever.

**dengue fever** One of the diseases caused by viruses transmitted by the water-related mosquito *Aedes aegyptii.* Also called breakbone fever or simply dengue, it is an acute febrile disease of sudden onset with fever, headache, joint and muscle pains, and other symptoms. *See also* hemorrhagic fever and dengue hemorrhagic fever.

**dengue hemorrhagic fever (DHF)** A severe form of dengue fever, found mainly in urban areas of Southeast Asia. *See also* hemorrhagic fever.

**dengue virus** The virus that causes dengue fever. Transmitted by mosquitoes *Aedes aegyptii* and other *Aedes* species, it is found worldwide.

**Denite®** A process developed by TETRA Technologies, Inc. for denitrification using a bed of granular medium in a fixed-film reactor.

**denitrification** The biological conversion of nitrate nitrogen ($NO_3$-N) and nitrite nitrogen ($NO_2$-N) to gaseous or molecular nitrogen ($N_2$), nitrogen dioxide ($NO_2$), and other end products under anaerobic conditions. Common treatment configurations used for denitrification include (a) nitrification followed by a suspended-growth unit, including methanol addition; (b) nitrification followed by a packed-bed column or a fluidized-bed reactor, with methanol addition; (c) nitrification–denitrification in a suspended-growth unit using wastewater or an endogenous carbon source. Methanol requirements are estimated as follows:

$$C_m = 2.47\,(NO_3) + 1.53\,(NO_2) + 0.87\,(DO) \quad \text{(D-15)}$$

where $C_m$ = methanol requirement, mg/L; $NO_3$ = nitrate concentration, mg/L; $NO_2$ = nitrite concentration, mg/L; and DO = dissolved oxygen concentration, mg/L. Substrate sources other than methanol include acetate, ethanol, and molasses. When denitrification occurs in a secondary clarifier, it causes the problem of rising sludge. Denitrification also occurs in ponds, land application systems, and wetlands. *See* biological denitrification for a list of related terms.

**denitrification filter** A 4–6.5 ft deep filter used to remove nitrogen (postanoxic nitrate removal) and suspended solids from a nitrified wastewater effluent. The granular media provides support for heterotrophic microorganisms that use an exogenous carbon source such as methanol ($CH_3OH$). Proprietary systems include Davco filter from U.S. Filter Davco Products, elimi-NITE from F. B. Leopold, Inc., and Tetra Denite from Severn Trent's Tetra Process Div. Also called deep-bed denitrification filter. *See also* downflow denitrification filter, upflow continuous backwash filter, bump, Tetra® filter.

**denitrification filter bump** Same as bump.

**denitrifiers** Mostly heterotrophic bacteria that use the respiratory pathway of metabolism to reduce nitrates ($NO_3$) to nitrogen gas ($N_2$); there are also some autotrophic and some fermentative denitrifiers as well as some nitrogen-fixing bacteria; e.g., the following genera: *Alcaligenes, Bacillus, Chromobacterium, Flavobacterium, Nitrosomonas, Paracoccus, Pseudomonas, Rhizobium, Rhodopseudomonas, Thiobacillus, Thiomicrospira.*

**Denitri-Filt™** A filter manufactured by Davis Water & Waste Industries, Inc. for biological denitrification.

**denitrifying bacteria** *See* denitrifiers.

**Densadeg®** A proprietary water and wastewater processing unit manufactured by Infilco Degremont, Inc. combining lamella zone solids settling with sludge recirculation and thickening. *See also* Actiflo®, Fluorapide®.

**Densadeg system** *See* Densadeg®.

**Densator®** A high-density solids contact clarifier manufactured by Infilco Degremont, Inc., including primary and secondary mixing zones.

**dense granulated sludge** A dense sludge floc with a solid concentration of up to 100,000 mg/L that is a key feature of the upflow sludge blanket reactor process. Favorable conditions for the formation of this dense sludge are high carbohydrate or sugar content, near neutral pH, plugflow hydraulics, and a COD:N:P ratio around 300:5:1 at start-up. The formation of granulated sludge has also been observed in the anaerobic baffled reactor process.

**dense membrane** A nanofiltration or reverse osmosis membrane that adsorbs a water layer that has rejected small particles in the liquid being treated; a nonporous membrane.

**dense nonaqueous phase liquid (DNAPL)** A toxic compound that is immiscible with and heavier than water, such as perchloroethylene, trichloroethene (TCE), and other chlorinated solvents, produced at dry cleaning operations, military bases or industrial sites. Groundwater contaminated by DNAPLs can be remediated by such a technique as density-modified displacement.

**dense-sludge process** A proprietary high-rate clarification process used in the treatment of stormwater. As shown in Figure D-03, it includes ferric sulfate injection ahead of a grit chamber that serves also for coagulation, addition of a polymer in a two-stage flocculation tank, and a clarifier with a presettling zone and lamella-plate

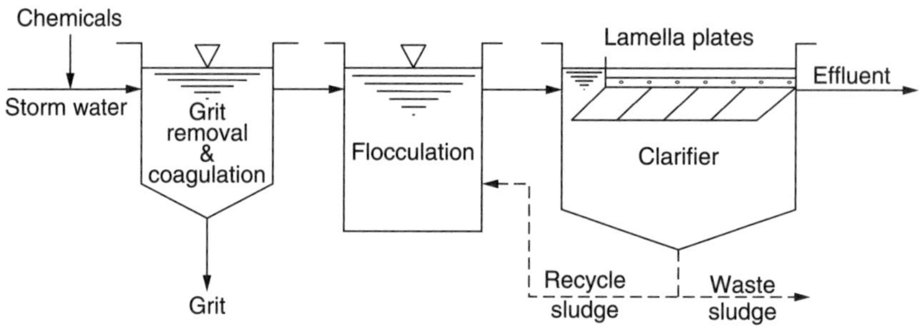


**Figure D-03.** Dense sludge process.

settlers. A portion of the settled sludge is returned to the flocculation tank and the remainder is wasted. *See also* ballasted flocculation and lamella-plate clarification.

**density ($\rho$)** A measure of how heavy a substance (solid, liquid, or gas) is for its size. Density is expressed as the mass of the substance contained in a unit volume, that is, grams per cubic centimeter or pounds per cubic foot. The density of water reaches a maximum of 1000 kg/m$^3$ = 1 kg/liter = 1.941 slugs/ft$^3$ at about 4°C or 39°F. It decreases to 958 kg/m$^3$ or 1.860 slugs/ft$^3$ at 100°C or 212°F. *See also* bulk density, probability density, specific gravity, specific weight.

**density current** Water flowing through a larger body of water and retaining its identity because of its different density. In water and wastewater treatment, density currents due to temperature differences may cause short-circuiting. Inlet baffles help reduce the undesirable effect of density currents in clarifiers. On the other hand, such currents increase receiving water quality by dispersing waste loads and reducing locally high concentrations of pollutants. *See also* density flow and density stratification.

**density flow** Water movement at or near the surface of a reservoir due to an inflow of a different density. *See also* density current and density stratification.

**density-modified displacement** A remediation technique used for cleanup of groundwater contaminated by dense nonaqueous-phase liquids (DNAPLs). It consists of the injection of (a) an emulsion of alcohol and surfactants into the zone of contamination to give the DNAPL about the same density as water, and (b) a second surfactant solution to reduce interfacial tension between the contaminants and the soil to facilitate their separation and the downstream removal of the contaminants.

**density of coliforms** The number or concentration of coliform organisms in a volume of water, commonly 100 milliliters.

**density stratification** The formation of layers of different densities in a body of water. *See also* density current and density flow.

**density wetted in water** In granular activated carbon adsorption systems, the particle density wetted in water is the mass of activated carbon including the water required to fill the internal pores per unit volume of particle, typically 1300–1500 kg/m$^3$.

**Densludge** A sludge digestion unit manufactured by Dorr-Oliver, Inc., including a primary thickening compartment.

**dental** *See* dentated sill or denticle.

**dental fluorosis** *See* fluorosis.

**dentated sill** In general, a small tooth or toothlike part; also called dental, denticle. (1) A notched sill at the end of an apron to reduce water velocity and scour, promote aeration, or improve navigation. (2) A baffle resembling the toothlike projection of an apron used to deflect or reduce the velocity of flowing water.

**denticle** *See* dentated sill.

**Dentrol** A device manufactured by Walker Process Equipment Co. to control sludge density.

**denudation** The action of erosive processes (rain, frost, wind, running water, etc.) that expose or lay bare the solid matter (rock) of the earth.

**Deox/2000™** A device manufactured by Wallace & Tiernan, Inc. to monitor dechlorination.

**deoxygenation** The loss of dissolved oxygen in a stream or other liquid under natural conditions, caused by the oxidation of organic matter (e.g., exertion of BOD) or by chemical reducing agents. Oxygen demand is exerted not only by waste discharges but also by bottom deposits (benthal demand).

**deoxygenation constant** A parameter that expresses the rate of dissolved oxygen depletion (usually per day) in a stream or other liquid; it varies with temperature and is one of the important parameters used in the study of water quality of receiving waters. Also called deoxygenation rate. *See* dissolved-oxygen analysis.

**deoxygenation rate** Same as deoxygenation constant.

**deoxyribonucleic acid (DNA)** The double-stranded, helix-shaped macromolecule that contains the hereditary material vital to reproduction and transmitted between generations of cells. It consists of molecules of long unbranched chains of nucleotides, which are combinations of phosphoric acid ($H_3PO_4$), the monosaccharide deoxyribose, and one of the four nitrogenous compounds: adenine ($C_5H_5N_5$), cytosine ($C_4H_5N_3O$), guanine ($C_5H_5N_5O$), and thymine ($C_5H_6N_2O_2$). *See also* plasmid DNA, ribonucleic acid.

**dependable capacity** The capacity of a water or wastewater facility that can provide the expected level of service except under unusual circumstances. *See also* safe yield for the concept of dependable draft.

**depletion curve** In hydraulics, a graphical representation of water depletion from storage-stream channels, surface soil, and groundwater. A depletion curve can be drawn for baseflow, direct runoff, or total flow.

**Deplution** A line of wastewater treatment products of Ralph B. Carter Co.

**Depolox®** A device manufactured by Wallace & Tiernan, Inc. for analyzing pH, chlorine, and fluoride.

**deposit** Solid material transported to a new position by human activity, gravity, wind, water, or ice. *See* the following types of deposit: alluvial deposit, benthal deposit, fluvial deposit, marine deposit, organic deposit, sludge deposit.

**deposition** (1) The process of solid material settling from a suspension like wastewater or sludge. *See also* reservoir deposition. (2) The material that collects on the inside surface of water pipes or that forms on utensils and machinery as a result of heating. (3) The accumulation of rock materials and other debris resulting from human or natural activities. *See* deposit. (4) Any process in which a solid forms directly from its vapor.

**deposition coefficient** A parameter used in the analysis of dissolved-oxygen sag to reflect the effect of benthal oxygen demand and benthal decomposition. The coefficient of deposition depends on characteristics (composition, quiescence or turbulence) of the wastewater discharged and of the receiving stream.

**depressant** Same as depressing agent.

**depressing agent** A surface-active additive that combines with another foaming agent (a collector) to promote the flotation of solid particles at the surface. Also called a depressant or a depressing reagent. *See also* collecting agent.

**depressing reagent** Same as depressing agent.

**depression landfill** An abandoned quarry or pit, or any other hole or depression, used as a landfill site.

**depth filter** A device used in the treatment of drinking water or the advanced treatment of wastewater. *See* depth filtration and the following types of depth filters: conventional downflow, deep-bed upflow continuous-backwash, deep-bed downflow, pulsed-bed, traveling-bridge, synthetic-medium, DSS Environmental two-stage, Fuzzy Filter, two-stage.

**depth filtration** A classification for filters in which solids are removed within and on the surface of the granular or compressible media (e.g., rapid granular filters), through complex transport and attachment mechanisms. Depth filtration requires chemical pretreatment because the particles removed are smaller than the interstices between the filter grains. Depth filtration also occurs in multimedia filters, which have progressively smaller pore spaces. *See also* surface filtration, multimedia filtration, membrane filtration.

**depth filtration problems** Problems associated with the operation of depth filters include turbidity breakthrough, mudball formation, emulsified grease buildup, cracks and contraction of filter bed, filter medium loss, and gravel mounding.

**depth-integrated sample** A composite sample that represents the characteristics of a body of water at various depths; e.g., a mixture of samples taken at various points such that they are proportional to the velocities at these points.

**depth of flotation** The vertical distance between the surface of a liquid and the lowest point on a floating body.

**depth of submergence** *See* submerged weir.

**depth profile** A graphical representation of the depth of a body of water versus distance or time or the plot of a physical or chemical characteristic of the water versus depth.

**depuration** (1) The process of making something (e.g., shellfish) free from impurities. (2) The process by which the concentration of xenobiotics in tissue decreases by loss from an organism placed in an uncontaminated environment. The loss occurs through such mechanisms as desorption, diffusion, egestion, or excretion.

**Depurator** A flotation unit manufactured by Envirotech Co. using induced air.

**deq** Symbol for equivalent spherical diameter, i.e., diameter of an equal-volume sphere, a concept used in defining grain shape and roundness in filtration theory.

**derivatization** The conversion process of one substance into another, especially to a form more susceptible to analysis; e.g., the conversion of an organic acid into an ester or the conversion of haloacetic acids to methyl esters, through the addition of diazomethanes, before analysis by gas chromatography.

**Derjaguin–Landau–Verwey–Overbeck (DLVO) theory** A theory of colloid stability based on repulsive forces such as electrostatic stabilization, caused by the interaction of electrical double layers of particle surfaces, and attractive forces such as the London–van der Waals force, arising from electrical and magnetic polarizations within and around the particles. *See also* Hamaker constant, stability ratio.

**dermal exposure** Contact between a chemical and the skin.

**dermal toxicity test** A test to determine the potential of a substance to cause allergic reactions in the skin.

**dermatitis** Inflammation and redness of the skin. *See also* cercarial dermatitis.

**dermatosis** Any skin disease, especially one without inflammation.

**DESAL™** A line of desalination products of Desalination Systems, Inc.

**desalinate** To desalt; to remove the salt from seawater and make it potable.

**desalination** (1) The removal of dissolved salts (such as sodium chloride, NaCl) and other dissolved solids from ocean or brackish water by natural means (leaching) or by specific water treatment processes such as adsorption on carbon electrodes, electrodialysis, ion exchange, osmionics, and thermal diffusion, usually to make water drinkable or usable as process or cooling water. Another group of processes that extract water (instead of salts) are commonly included under desalination. *See* saline-water conversion classification. (2) Any mechanical procedure or process to remove all or some of the salt from lake water and return the freshwater portion to the lake (EPA-40CFR35.1605-7). (3) The removal of salts from soil by artificial means, usually leaching. Same as demineralization, desalinization, desalting. *See also* deionization.

**desalinization** Same as desalination.

**Desal membrane** A proprietary spiralwound membrane. *See* spiral-wound module.

**desalt** Same as desalinate.

**desalting** The removal of dissolved mineral salts and other dissolved solids from water. *See* desalination and saline-water conversion classification for more detail.

**desander** A device for removing sand from well drilling fluids.

**DeSander®** A hydroclone separator manufactured by Krebs Engineers.

**descaling** The mechanical or chemical removal of scales and encrustations from pipes.

**descending block rate** Same as declining block rate.

**desert fever** An acute or progressively chronic respiratory infection characterized by fever and reddish bumps on the skin. *See* coccidioidomycosis for detail.

**desert rheumatism** *See* coccidioidomycosis.

**desertification** The process of an area becoming a desert, with rapid depletion of plant life and loss of topsoil caused by drought, overexploitation, and other factors.

**desiccant** A chemical agent that can remove or absorb moisture from the atmosphere in a small enclosure. Some desiccants can dry out plants or insects, causing death.

**desiccation** A process used to thoroughly dry air.

**desiccator** (1) A closed container into which heated weighing or drying dishes are placed to cool in a dry environment. The dishes may be empty or they may contain a sample. Desiccators contain a substance, such as anhydrous calcium chloride, that absorbs moisture and keeps the relative humidity near zero so that the dish or sample will not gain weight from absorbed moisture. (2) An apparatus for absorbing moisture from a chemical substance or for cooling heated items and preventing them from picking up moisture.

**design** Engineering design, as related to water and wastewater facilities, is the development of plans and specifications for their construction and implementation. *See also* facilities plans.

**design activities** Design activities for a water or wastewater project include design concept, P&ID, detailed calculations, plans, specifications, and plans of operation. *See also* facilities planning activities, construction activities.

**design analysis** A step that precedes engineering design and includes the presentation of physical data, project requirements, and design criteria.

**designated project area** The portion of the waters of the United States within which the permittee or permit applicant plans to confine a designated species, using a method or plan or operation (including physical confinement) that, on the basis of scientific evidence, is expected to ensure that specific individual organisms comprising an aquaculture crop will enjoy increased growth attributable to the discharge of pollutants, and be harvested within a defined geographic area (EPA-40CFR122-25.2).

**designated uses** Water uses identified in state water quality standards that must be achieved and maintained as required under the Clean Water Act; e.g., public water supply, irrigation, and cold water fisheries. A designated use definition describes the chemical, physical, and biological attributes covered by the use, i.e., its narrative criteria.

**design–build** A method of project implementation in which one entity is responsible for both design and construction of a facility.

**design capacity** The average daily flow that a treatment plant or other facility is designed to accommodate.

**design criteria** Guidelines for the preparation of final engineering documents, including construction details and materials, processes, unit sizes, etc., in order for a facility or process to achieve its intended function.

**design discharge** In hydraulic studies, the maximum capacity that a hydraulic structure is designed to handle. It affects the cost of the structure and depends on such criteria as the acceptable risk level and corresponding return period. *See also* design flow and design storm.

**designer bug** Popular term for microbes developed through biotechnology and capable of degrading specific toxic chemicals at their source in toxic waste dumps or in groundwater.

**design flow** In water quality studies such as waste load allocations, a low-flow criterion of the form $pQt$ is the minimum average flow $Q$ over the period $p$ expected to occur once in $t$ years. For example, the *7Q10* flow is the minimum seven-day average flow expected once in 10 years. Also defined as the seven-day, consecutive low flow with a 10-year return period or the lowest streamflow for seven consecutive days that would be expected to occur once in 10 years. The *7Q20* and *30Q20* flows are the average streamflows equaled or exceeded on the average once in every 20 years (i.e, 5% of the time), over periods of 7 days and 30 days, respectively. *See also* design discharge. The probability ($p$) that the flow will exceed $Q$ in any given year is

$$p = 1 - 1/t \qquad \text{(D-16)}$$

and the probability ($p'$) that the flow $Q$ will occur at least once in $N$ years is

$$p' = 1 - (1 - 1/t)^N \qquad \text{(D-17)}$$

**design life** The estimated number of years of service of an installation or any component or equipment thereof: The number of years of service of an installation, component, or equipment, as estimated by the design engineer. It is equivalent to useful life at start-up.

**design loading** For a water or wastewater treatment facility, the design loading is the loading (hydraulic or constituent load per an appropriate unit such as flow, surface, or volume) selected for the desirable performance of the facility. *See also* hydraulic loading, organic loading.

**design parameter** A quantity, variable, characteristic, etc., used in the design of a facility. Examples of design parameters for water and wastewater facilities include depth, food-to-microorganism ratio, hydraulic retention time, organic loading, overflow rate, sludge age, solids retention time, velocity gradient, and weir overflow rate.

**design period** The length of time for which water and wastewater facilities, or individual elements thereof, are designed, i.e., the time period that their capacity is adequate to meet the project demand. This parameter, which determines the size of the facilities, depends on a variety of factors such as economic life, costs, ease of expansion, risk of obsolescence. Table D-02 shows some typical design periods.

**design standard** A design procedure, the choice of a design parameter, etc. that reflects common engineering practice. *See also* rule of thumb.

**design velocity** *See* flushing velocity.

**desiliconization** An ion exchange process designed for the reduction of silica from water.

**Table D-02.** Design period (typical)

| Component | Design Period, years |
|---|---|
| Water well | 5–10 |
| Water source with impoundment | 50 |
| Water treatment plant | 10–15 |
| Distribution storage | 10–15 |
| Sewers | 50 |
| Wastewater treatment plant | 15–20 |
| River intake | 20 |
| Transmission pipeline | 25 |
| Water pumping station | 10–15 |
| Distribution lines | 50 |
| Wastewater pump station | 10–15 |

**desilting area** An area covered with vegetation, used for silt and sediment control in flowing water.

**desilting basin** A basin for removing silt and sediment from streamflow or runoff, e.g., just below the diversion structure of a canal. Also called desilting works.

**desilting works** *See* desilting basin.

**desludging** The process of removing accumulated sludge from septic tanks, pits, pour-flush toilets, aqua-privies, facultative ponds, and similar waste treatment units.

**desorption** (1) The reverse of adsorption, i.e., the release of an adsorbed solute from an adsorbent; e.g., the regeneration of a carbon adsorber by removal of the perchloroethylene adsorbed on the carbon (EPA-40CFR63.321). Desorption may be caused by displacement of one substance by another or by a decrease in the concentration of a substance in the influent. It relates more to physical adsorption than to chemical adsorption. *See also* chemical adsorption or chemisorption, electrostatic or exchange adsorption, exchange adsorption, hydrophobic bonding, polarity, solution force, solvophobic force, van der Waals force. (2) The reverse of absorption, e.g., the transfer of a gas out of the liquid phase into the gas phase. *See* two-film theory.

**desorption of gases** Desorption of gases by water is governed by the general gas law and, in particular, the following concepts or principles: absorption coefficient, Avogadro's' hypothesis, Dalton's law of partial pressures, gas absorption, gas dispersion, gas precipitation, gas solubility, gas solution, gas-transfer coefficient, Henry's law, two-film theory.

**desorption rate** The rate at which a gas is transferred from a liquid to the gas phase. It is proportional to its degree of oversaturation in the liquid and similar to the gas absorption rate:

$$R_{de} = K_{de}A(C - C_s) \qquad \text{(D-18)}$$

where $R_{de}$ = rate of desorption, mass per volume per unit time; $K_{de}$ = coefficient of desorption, length per unit time; $A$ = area of interface per unit volume of liquid; $C$ = gas concentration at time $T$; and $C_s$ = saturation concentration at a given temperature. *See also* two-film theory, absorption rate.

**desoxyribonucleic acid** Same as deoxyribonucleic acid.

**destabilization** The process in which the particles in a stable suspension are modified to attract one another or to attach to a stationary surface. Destabilization mechanisms include adsorption, double-layer compression, heterocoagulation, interparticle bridging, and surface charge neutralization. *See also* charge reversal, coagulation zones, sweep floc, restabilization.

**destratification** Elimination or reduction of density or thermal stratification (as well as separate layers of plant or animal life) by vertical mixing within a lake or reservoir. The vertical mixing can be achieved by mechanical means (pumping) or through the use of forced air diffusers that release air into the lower layers of water.

**destruction and removal efficiency (DRE)** (1) An expression of incineration efficiency in destroying hazardous waste; a percentage that represents the number of molecules of a compound removed or destroyed in an incinerator relative to the number of molecules that entered the system:

$$\text{DRE} = 100(W_0 - W_e)/W_0 \qquad \text{(D-19)}$$

where $W_e$ and $W_0$ are, respectively, effluent and influent mass flow rates of a compound or hazardous constituent. For example, a DRE of 99.99% means that 9,999 molecules are destroyed for every 10,000 that enter; 99.99% is known as "four nines." For some pollutants, the RCRA removal requirement may be as stringent as "six nines" or 99.9999%. (2) A performance measure used to determine the effectiveness of a biofilter in the treatment of hazardous wastes (Freeman, 1998):

$$\text{DRE} = 1 - C_e/C_0 = 1 - e^{-kt} \qquad \text{(D-20)}$$

where $C_e$ and $C_0$ are, respectively, effluent and influent concentrations, $k$ is an empirical reaction rate (biodegradation rate), and $t$ is the residence time in the biofilter.

**destructive distillation** (1) The process of heating organic matter in the absence of air and under pressure to boil off, condense, and collect the volatile products, leaving the nonvolatile products as a residue. (2) The use of pyrolysis to destroy wastewater sludge by converting it to a carbon residue and liquid and gaseous products that can be recycled. The term is often used synonymously with pyrolysis.

**Destrux®** A heavy-duty knife manufactured by Franklin Miller for cutting scrap materials.

***Desulfobacter*** A genus of sulfur-reducing bacteria that use sulfate as a terminal electron acceptor and hydrogen as the electron donor; they have been identified as nuisance organisms in anaerobic digesters. *See* dissimilatory sulfate reduction for detail.

***Desulfobulbus*** A genus of sulfur-reducing bacteria that use sulfate as a terminal electron acceptor and hydrogen as the electron donor. *See* dissimilatory sulfate reduction.

***Desulfococcus*** A genus of sulfur-reducing bacteria that use sulfate as a terminal electron acceptor and hydrogen as the electron donor. *See* dissimilatory sulfate reduction.

***Desulfonema*** A genus of sulfur-reducing bacteria that use sulfate as a terminal electron acceptor and hydrogen as the electron donor. *See* dissimilatory sulfate reduction.

***Desulfosarcina*** A genus of sulfur-reducing bacteria that use sulfate as a terminal electron acceptor and hydrogen as the electron donor. *See* dissimilatory sulfate reduction.

***Desulfotomaculum*** A genus of sulfur-reducing bacteria that use sulfate as a terminal electron acceptor and hydrogen as the electron donor. *See* dissimilatory sulfate reduction.

***Desulfovibrio*** A genus of sulfur-reducing bacteria that use sulfate as a terminal electron acceptor and hydrogen as the electron donor; they have been identified as nuisance organisms in anaerobic digesters. *See* anaerobic heterotrophic respiration, dissimilatory sulfate reduction for detail.

***Desulfovibrio desulfuricans*** The species of sulfate-reducing bacteria most responsible for the production of hydrogen sulfide in groundwater.

**desulfurization** The process of lowering the sulfur content of fossil fuels.

***Desulphovibrio spp.*** *See Desulfovibrio.*

**detection criterion** *See* criterion of detection.

**detection level** One of a few parameters used to measure the precision or the detection level of a method of analysis, stated as a result that produces a signal 2 × 1.645 times the standard deviation above the mean of blank analyses. Also called lower level of detection or simply level of detection.

**detection limit** A measure of the accuracy and/or precision of an analytical method indicating at which concentration a reporting result is valid or significant. *See* level of detection.

**detector check** A check valve that has a component to indicate passage of water through the valve.

**detention** In the management of stormwater or combined sewer overflows, detention is the slowing, attenuating, or dampening of flows by temporarily holding the water on a surface area, in a storage basin, or within the sewer system.

**detention basin, detention pond, detention reservoir, detention tank, holding pond, retarding basin, retarding reservoir, storage basin, storage reservoir** These are structures built, with or without outlet control, to temporarily hold excess water such as streamflow, stormwater, and runoff. Detention or retarding basins, and detention or retarding reservoirs, used for flood control, temporarily store streamflow until the stream has sufficient capacity for the ordinary flow plus the release from the basins or reservoirs; they have no control gates. Storage basins or storage reservoirs, also designed and operated for flood mitigation, are equipped with gates and valves for adequate regulation. Ponds are natural or artificial bodies of water of limited size; they are used in stormwater or wastewater management. The EPA defines a holding pond as a pond or reservoir, usually made of earth, built to store polluted runoff. It is current practice for municipalities to require land developers to install detention basins or holding ponds to keep the peak runoff from an area at the same level as before development. Detention tanks are used mainly in wastewater management, to even out surges and allow biological, chemical, or physical treatment. *See also* dry/wet detention systems, and forebay.

**detention lag** The time period between the moment a change is made and the moment such a change is finally sensed by the associated measuring instrument.

**detention pond** *See* detention basin.

**detention reservoir** *See* detention basin.

**detention slaker** A mechanical device that adds enough water to hydrate quicklime (CaO) into slaked or hydrated lime that contains approximately 10% calcium hydroxide [$Ca(OH)_2$]. *See also* paste or pugmill slaker.

**detention tank** *See* detention basin.

**detention time** (1) The calculated time required for a small amount of water to pass through a tank, basin, reservoir, pond, etc., at a given flow rate. (2) The actual time that a small amount of water is in a settling basin, flocculating basin, or rapid-mix chamber. (3) In storage reservoirs, the length of time water will be held before being used. (4) The time that stormwater remains in a BMP; for a runoff event, the theoretical detention time is the average time that stormwater remains in the BMP during release. (5) Same as retention time, hydraulic retention time or retention period, i.e.,

$$t = v/Q \qquad \text{(D-21)}$$

or the time ($t$) that water, wastewater, or stormwater remains in a reactor, unit process, storage basin, or any similar facility of volume ($v$) at a discharge rate ($Q$). This formula and the above definitions correspond to an average; actual residence times vary widely, particularly in stormwater systems. *See also* detention/retention.

**detention time multiplier** A factor ($M$) recommended to multiply the hydraulic detention time in a primary sedimentation tank to correct for the effect of cold weather, when wastewater temperature is below 20 °C (WEF & ASCE, 1991):

$$M = 1.82\, e^{-0.03\,T} \tag{D-22}$$

where $T$ is wastewater temperature, °C.

**detergent** Synthetic washing agent (made from aromatic sulfonates, alkyl sulfates, etc.) that helps remove dirt and oil; sometimes called surfactant or surface active agent, synthetic detergent or syndet. Some detergents contain compounds that kill useful bacteria and encourage the growth of algae when they are in wastewater that reaches receiving waters. The presence of detergents in drinking water sources is an indication of pollution. Detergents may also cause foaming in wastewater aeration units. *See also* alkyl benzene sulfonate, cobalt thiocyanate active substance, methylene-blue active substance.

**deterioration pollution** The increase in waste generation caused by the deterioration of equipment.

**determinant** The specific site on an antigen where an antibody can bind, the chemical structure of the site determining which antibody. Also called antigenic determinant.

**determinant of water quality** A water quality parameter.

**determination** A quantitative analysis to define a substance.

**determination coefficient** *See* coefficient of determination.

**detoxification** Treatment to remove, modify, or reduce toxic materials. Also, metabolism of a substance to a less toxic form; e.g., the enzymatic conversion of the toxic organophosphate insecticide paraoxon ($CH_3$—$CH_2$—$PO_4$—$CH_2$—$CH_3$—■—$NO_2$) to *p*-nitrophenol (HO—■—$NO_2$). The symbol ■ represents a cyclical ring.

**Detritor** Grit removal equipment manufactured by Dorr-Oliver, Inc. and featuring a reciprocating rake. *See* Figure D-04.

**detritovore** An organism that feeds on detritus (nonliving plant and animal matter). *See also* producer, decomposer, grazer, or reducer.

**detritus** (1) Coarse debris in wastewater. (2) Decaying organic matter such as root hairs, stems, and leaves usually found on the bottom of a water body. (3) Grit or fragments of rocks; mineral debris in the bedload of a watercourse.

**detritus tank** A square tank grit chamber incorporating a revolving rake to scrape settled grit to a sump for removal and disposal. It is designed and operated with a constant level and a short detention time (1.0 min or less) for the removal of grit from wastewater. The difference between an ordinary grit chamber and a detritus tank is that the former removes only inert materials, whereas the latter removes both inert and organic materials. *See* Figure D-04. Also called degritter, grit clarifier, square tank degritter, square horizontal-flow grit chamber.

**deuterated internal standard** A compound having a known proportion of an isotope after replacing a number of hydrogen atoms with deuterium; it is used for calibrating mass spectrophotometers.

**deuteride** A hydride in which deuterium replaces the ordinary hydrogen atoms.

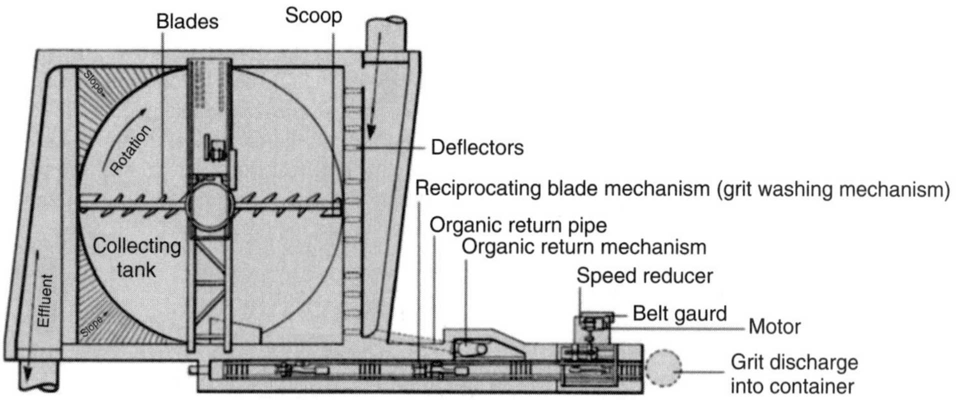


**Figure D-04.** Detritor (courtesy Dorr-Oliver Eimco).

**deuterium (D)** An isotope of hydrogen, with twice the mass of ordinary hydrogen, an atomic number of 1, and an atomic weight of 2.0147. Also called heavy hydrogen.

**deuterium background correction** An adjustment for interferences in the results of atomic absorption spectrophotometry, using a deuterium arc lamp and a second continuum source for visible wavelengths.

**deuterium oxide ($D_2O$)** Water in which deuterium replaces the hydrogen atoms; also called heavy water; used mainly as a coolant in nuclear reactors.

**developed length** The length of a line of pipe along the centerline of the pipe and fittings.

**developed water** (1) Groundwater that is artificially withdrawn and would otherwise be wasted. (2) Artificially induced surface water.

**developmental maintenance** A maintenance approach that focuses on bringing equipment to a level that exceeds its original standards, as compared to reactive maintenance and proactive maintenance, whose goal is to bring a unit to, or maintain the unit at, the as-built level.

**developmental toxicity** The study of adverse effects on the developing organism (including death, structural abnormality, altered growth, or functional deficiency) resulting from exposure prior to conception (in either parent), during prenatal development, or postnatally up to the time of sexual maturation.

**development impact fee** A one-time charge assessed by a water, sewer, or other utility against new users, developers, or applicants for new service as well as existing customers to recover part or all of the costs of additional facilities. It is a means of charging the cost of development impact to those who cause it. *See also* capacity charge, capital recovery charge, impact fee, facility expansion charge, connection charge, and system development charge.

**Developure** A depth filter manufactured by Osmonics, Inc.

**devil's grip** *See* pleurodynia.

**dew** Water droplets that form on cool surfaces following condensation of atmospheric water vapor and condensation of water from the inside of the blades of grass, leaves, etc. *See* dew point.

**dewater** (1) To remove or separate a portion of the water present in a sludge or slurry. To dry sludge so it can be handled and disposed. (2) To remove or drain the water from a tank, a trench, a riverbed, a structure, a cofferdam, an excavation, etc; to unwater.

**dewaterability** The ability of sludge to reach a higher solids concentration by losing water.

**dewatered cake** Biosolids or residuals that have been dewatered to more than 20% dry solids to produce a material of soil-like consistency.

**dewatered sludge** Sludge with a higher solids concentration after losing part of its water through draining or filtering. Dewatered sludge differs from thickened or dried sludge.

**dewatering** (1) Removal or draining of the water from a container such as a tank, a basin, or a trench. (2) An open-air or mechanical liquid–solids separation process used in sludge handling to reduce water content and increase solids concentration to 10–20% by weight. Common dewatering methods include air drying (sand drying bed, freeze-assisted sand bed, solar drying bed, vacuum-assisted drying bed, wedgewire bed), dewatering lagoon, flotation, and mechanical dewatering (belt filter press, centrifuge, pressure filter, vacuum filter). *See also* thickening and drying. *See* more detail under sludge dewatering. (3) A method used to remove and discharge excess water from a construction site, sediment traps, and basins, e.g., by pumping water out or by lowering the groundwater table. This is called groundwater dewatering when it is associated with excavated trenches to provide a stabilized area for construction. Dewatering discharges usually have a high sediment concentration. In sewer construction, excavations may be drained by letting water run along the trench bottom to a sump and then pumping it out. *See also* quick condition and well point.

**dewatering centrifuge** A centrifuge used for sludge dewatering as compared to a thickening centrifuge.

**dewatering lagoon** A large basin, excavated or with earthen berms above ground, used to concentrate solids from water or wastewater treatment sludge. Similar to sand drying beds but with higher initial loading and longer drying times, dewatering lagoons are often equipped with a sand underdrain bottom, liners, leachate collection systems, and monitoring wells to prevent groundwater pollution.

**dewatering of reservoirs** The partial or complete draining of a reservoir to allow aquatic plants to die.

**dewatering processes** For sludge dewatering, see vacuum filter, centrifuge, belt filter press, recessed plate filter press, drying bed, and lagoon.

**dewatering table** The horizontal belt used to dewater sludge in a belt thickener.

**dew point** (1) The temperature to which air, with a given quantity of water vapor and at a given pressure, must be cooled to cause condensation of the vapor in the air and begin to form dew; also defined as the temperature at which water vapor saturates air and below which dew formation begins. Dew point is a moisture reduction indicator in the air preparation step of ozone generation. Also called dew point temperature. *See also* frost point. (2) The temperature at which air becomes saturated with water vapor, and below which the water vapor condenses as droplets. (3) *See* water dew point. (4) *See* acid dew point.

**dew point temperature** Same as dew point.

**dewvaporation** A process of humidification–dehumidification used for the desalination of seawater and brackish water: the feedwater is evaporated by heated air and freshwater is deposited as dew on the opposite side of the heat transfer device. Dew formation supplies the energy for evaporation. Dewvaporation is classified as an emerging concentrate minimization and management method.

**dextran blue** A tracer of high molecular weight used in measuring dispersion in submerged biological filters.

**dezincification** The removal or loss of zinc from an alloy, e.g., by corrosion. This corrosion process removes zinc from brass but leaves the copper in place; the brass valve or fitting retains its original dimensions but is severely weakened and is prone to structural failure, leaks, or seepage. It occurs mostly in waters with high chlorides and pH greater than 8.0. *See also* dealloying.

**DF** Acronym of Discfilter®.

**DFR** Acronym of dynamic fixed-film reactor; also a unit manufactured by the Schreiber Corp.

**D horizon** The massive bedrock underlying the soil layers. Also called R horizon.

**DI** Acronym of deionization.

**diagenesis** The physical and chemical changes affecting sediments during and after burial.

**diagnostic feasibility study** A two-part study to determine a lake's current condition and to develop possible methods for lake restoration and protection. The diagnostic portion of the study includes information and data gathering to determine the characteristics of the lake and its watershed. The feasibility portion analyzes this information, determines procedures to improve lake quality, develops a technical plan and schedule for implementation, and, if necessary, conducts pilot-scale evaluations (EPA-40CFR35.1605-8).

**diagnostic organisms** Excreted bacteria used as indicators of environmental fecal pollution of water sources.

**diagonal-flow pump** A type of pump that directs the fluid diagonally—instead of radially or axially—through the impeller.

**dialdehyde** An organic compound that has two aldehyde (HC═O) functional groups, e.g., glyoxal; an ozonation by-product.

**dialogite** Same as rhodochrosite.

**dialysate** Same as dialyzate.

**dialysis** The separation of substances (solutes) from solution on the basis of molecular size or concentration difference by diffusion through a semipermeable membrane. In electrodialysis, application of a voltage to the membrane accelerates the diffusion rate. *See* electroosmosis, reverse osmosis, saline water conversion classification.

**dialyzate** Also spelled dialyzate. (1) The solution or the crystalline material passing into it through a semipermeable electrodialysis membrane. Also called diffusate. (2) In general, the remaining or colloidal portion of a solution.

**4,6-diamidino-2-phenylindole ($C_{16}H_{15}N_5 \cdot 2\,HCl$)/propidium iodide ($C_{27}H_{34}N_4I_2$)** Acronym: DAPI/PI. Fluorogenic vital dyes that indicate whether *Cryptosporidium* oocysts are viable or dead.

**diamine** (1) A chemical compound that contains two amino groups. Diamines smell like decayed flesh, one of the major categories of offensive odors associated with untreated wastewater. (2) Another name of hydrazine ($N_2H_4$).

**Diamite Series™** Liquid products of King Lee Technologies for removing foulants from membranes.

**Diamond Gate** Equipment manufactured by Andritz-Ruthner, Inc. for the compaction of wastewater treatment screenings.

**Diamond Seal™** A metering gate manufactured by TETRA Technologies, Inc.

**diaphragm** A thin sheet placed between parallel parts of a device to increase its rigidity.

**diaphragm filter press** A modification of the standard filter, a sludge dewatering device. The diaphragm press filters sludge through a cloth for about 20 minutes and then applies compressed air through an expandable diaphragm to squeeze more water out. This operation results in a cake of variable thickness, avoids hanging cakes, and does not require prior sludge conditioning. Also called diaphragm press or variable-volume recessed-plate filter press. *See also* sludge pressing.

**diaphragm float** A float mounted on a truck that runs along a straight channel section to gauge water velocity.

**diaphragm metering pump** A device developed in the 1960s to deliver a controlled volume of liquid in water and wastewater treatment processes as well as other applications. It uses a diaphragm actuated mechanically or by a closed volume of oil between the diaphragm and plunger. *See also* metering pump, lost motion.

**diaphragm press** Same as diaphragm filter press.

**diaphragm pump** (1) A positive-displacement, low-capacity, low-head pump that uses a flexible diaphragm attached to a vertical cylinder to move liquids. The movement of the diaphragm successively creates suction and discharge. It is used in wastewater treatment to pump slurries and sludge (primary, thickened, or digested). (2) A metering pump that uses a diaphragm to separate the operating parts from the pumped fluid or from hydraulic fluid.

**diaphragm-type metering pump** A diaphragm pump used for measuring liquid discharges.

**diarrhea** A symptom or common name for several water-related or excreta-related diseases; an intestinal disorder characterized by frequent and fluid fecal evacuations, sometimes accompanied by vomiting and fever. Diarrhea may be caused by bacteria, viruses, and other pathogenic agents. *See also* amebiasis, amebic dysentery, bacillary dysentery, bacterial enteritis, balantidiasis, *Campylobacter* enteritis, cholera, cryptosporidiosis, *E. coli* diarrhea, gastroenteritis, giardiasis, rotavirus diarrhea, salmonellosis, shigellosis, viral diarrhea, yersiniosis.

**diarrheal agent** A pathogen or substance that causes diarrhea, e.g., *Vibrio cholerae, Entamoeba histolytica. See* diarrheal disease.

**diarrheal disease** A disease causing or indicated by the frequent evacuation of watery, loose stools. *See* Table D-03 for a list of some potential causes of diarrhea.

**diarrhoea** British spelling of diarrhea.

**diatom** Microscopic, unicellular, yellowish brown, marine or freshwater algae that contain silica in their cell walls. Some diatoms can cause taste and odor problems in drinking water. *See also* diatomaceous earth.

**Table D-03.** Diarrheal diseases (selected examples)

| | |
|---|---|
| **Infectious** | Schistosomiasis |
| Amebic dysentery | Shigellosis |
| Balantidiasis | Tularemia |
| Cholera | Typhoid fever |
| Clonorchiasis | Yersiniosis |
| Giardiasis | **Noninfectious** |
| Guinea worm | Cancer |
| Norwalk virus | Chemical poisoning |
| Paratyphoid fever | Endocrine disorders |
| Salmonellosis | Food poisoning |

**diatomaceous earth (DE)** A fine, siliceous, chalk-like material, made up of fossilized diatoms, used to filter out solids in water and wastewater treatment or as an active ingredient in some powdered pesticides. Also called diatomite or fuller's earth, it is most commonly used in precoat filtration, as body feed, and in filter precoating..

**diatomaceous earth filter (DE filter)** A pressure filter having diatomaceous earth of mean pore 5–17 microns as medium, built on a porous septum, which serves as a drainage system; used in the treatment of drinking water, swimming-pool water, and secondary wastewater treatment plant effluent. *See also* body feed, precoat.

**diatomaceous earth filtration (DE filtration)** A filtration method resulting in substantial particulate removal, using a process in which: (a) a "precoat" cake of diatomaceous earth is deposited on a support membrane (septum), and (b) while the water is filtered by passing through the cake on the septum, additional filter medium, known as "body feed," is continuously added to the feed water to maintain the permeability of the filter cake (EPA-40CFR141.2). Diatomite filters also remove *Giardia* cysts but are not suitable for waters of high turbidity (e.g., > 30 NTU) or secondary effluents with more than 13 mg/L of suspended solids. Typical precoat applications are 0.3 $kg/m^2$ for water and 0.5 $kg/m^2$ for secondary effluent, while body coat is added at the rates of 2–3 mg/L and 5–6 mg/L per turbidity unit, respectively, for water and secondary effluent. These filters are used mainly as mobile units for drinking water in the field or to treat swimming pool water.

**diatomic molecule** A molecule that contains two atoms, [e.g., hydrogen ($H_2$) and oxygen ($O_2$)], or two replaceable atoms or groups; binary molecule.

**diatomic oxygen** The common form of molecular oxygen ($O_2$), consisting of two atoms in one molecule. *See also* free radical.

**diatomite** Same as diatomaceous earth.

**diatomite filtration** Same as diatomaceous earth filtration.

**diazine ($C_4H_4N_2$)** Any of three isomeric compounds that contain a ring of four carbon atoms and two nitrogen atoms.

**diazinon $\{[(CH_3)_2CHC_4N_2H(CH_3)O]PSO(C_2H_5)_2\}$** The generic name for 0,0-diethyl-0-(2-isopropyl-4-methyl-6-pyrimidinyl) phosphorothioate, a common organophosphate insecticide, banned in 1986 by EPA for use on open areas such as sod farms and golf courses because it poses a danger to migratory birds. The ban does not apply to agricultural, home lawn, or commercial establishment uses.

**diazomethane ($H_2C{=}N^+{=}N^-$ or $CH_2N_2$)** A yellow, odorless, toxic, explosive gas, used in organic synthesis and as a methylating agent, e.g., to derivatize haloacetic acids and convert them to methyl esters for analysis by gas chromatography.

**diazotroph** An organism that can use atmospheric nitrogen as sole source of nitrogen gas for growth.

**dibasic sodium phosphate** Same as disodium phosphate.

**dibenzofurans** A group of highly toxic organic compounds.

**dibenzo-p-dioxin** *See* dioxin.

**dibromamine ($NHBr_2$)** A brominated compound similar to dichloramine and formed during chloramination when the source water contains bromine.

**dibromide** A compound containing two bromine atoms, e.g., ethylene dibromide ($C_2H_4Br_2$).

**dibromoacetic acid ($CHBr_2COOH$)** Acronym: DBAA or $Br_2$AA. A haloacetic acid containing two bromine atoms; a regulated disinfection byproduct whose quarterly monitoring in drinking water is required by the USEPA.

**dibromoacetonitrile ($CHBr_2C{\equiv}N$)** Acronym: DBAN. A haloacetonitrile containing two bromine atoms; a byproduct of water chlorination and ozonation.

**dibromochloroacetic acid ($CBr_2ClCOOH$)** Acronym: DBCAA. A haloacetic acid containing two bromine and one chlorine atoms.

**dibromochloromethane ($CHClBr_2$)** Acronym: DBCM. A trihalomethane/volatile organic compound containing two bromine and one chlorine atoms; used in fire-extinguishing products, aerosol propellants, refrigerants, and pesticides; regulated by the USEPA. Also called chlorodibromomethane. Bioaccumulation of DBCM is a primary water quality concern; the EPA recommends a bioconcentration factor of 3.75 in receiving waters.

**dibromochloropropane ($CH_2BrCHBrCH_2Cl$)** Abbreviation: DBCP. MCLG = 0. MCL = 0.0002 mg/L. *See* 1,2-dibromo-3-chloropropane.

**1,2-dibromo-3-chloropropane ($CH_2BrCHBrCH_2$-$Cl$)** Acronym: DBCP. The contaminant in dibromochloropropane, a volatile organic compound, soluble in water (about 1.0 g/L), used as a fumigant in agriculture for nematode control. Banned by the USEPA, it may adversely affect kidney and testicular functions.

**dibromoiodomethane ($CHBr_2I$)** A trihalomethane/volatile organic compound, regulated by the USEPA.

**1,2-dibromomethane ($BrCH_2CH_2Br$)** The contaminant in ethylene dibromide, a colorless, volatile, heavy, liquid pesticide with a characteristic odor. Also called ethylene bromide.

**DIC** Acronym of dissolved inorganic carbon.

**dicalcium silicate [$(CaO)_2 \cdot SiO_2$) or $Ca_2SiO_4$]** A compound of lime and silica, used in cement and as an acid neutralizer.

**dicamba** A derivative of dichlorobenzoic acid, a pesticide used to control broadleaf weeds in agriculture, regulated by the USEPA: MCLG = MCL = 0.2 mg/L.

**dichloramine ($NHCl_2$)** An unstable product of the reaction of ammonia ($NH_3$) and hypochlorous acid (HOCl). *See* breakpoint chlorination, chloramination.

**dichloroacetaldehyde** A haloacetaldehyde mutagenic in one *Salmonella* strain but not in others.

**dichloroacetic acid ($CHCl_2COOH$)** Acronym: DCAA or $Cl_2$AA. A haloacetic acid containing two chlorine atoms, used as a chemical intermediate and in pharmaceuticals; a regulated disinfection by-product.

**dichloroacetonitrile ($CHCl_2C{\equiv}N$)** Acronym: DCAN. A haloacetonitrile containing two chlorine atoms; a disinfection by-product whose monitoring in drinking water is required quarterly by the USEPA.

**dichlorobenzene ($C_6H_4Cl_2$)** Any of three isomers of benzene in which two hydrogen atoms are replaced by chlorine atoms. A synthetic organic chemical with low vapor pressure used in the production of dyes and pesticides. It may affect the nervous system, kidney, liver, and lungs. Two isomers, *ortho-* and *para*-dichlorobenzene, are regulated by the USEPA in drinking water: MCLG = MCL = 0.6 mg/L.

**1,2-dichlorobenzene ($C_6H_4Cl_2$)** Common name: *ortho*-dichlorobenzene. A clear, colorless to pale yellow liquid with a pleasant odor; a volatile organic compound used as a solvent, a fumigant, etc.; regulated by the USEPA: MCLG = MCL = 0.6 mg/L.

**1,3-dichlorobenzene ($C_6H_4Cl_2$)** Common name: *meta*-dichlorobenzene. A colorless liquid used as a fumigant, an insecticide, and in organic synthesis; regulated by the USEPA: MCLG = MCL = 0.6 mg/L.

**1,4-dichlorobenzene ($C_6H_4Cl_2$)** Common name: *para*-dichlorobenzene. A colorless to white, crystalline, volatile, synthetic organic compound, with a penetrating mothball odor, produced by the chlorination of benzene, soluble in water, with various applications as solvent, insecticide, germicide, fumigant, and deodorant. May cause anorexia, nausea, and atrophy of the liver; regulated in drinking water by the USEPA: MCLG = MCL = 75 μg/L.

***m*-dichlorobenzene** Abbreviation of *meta*-dichlorobenzene.

***meta*-dichlorobenzene** One of the contaminants in dichlorobenzene. *See* 1,3-dichlorobenzene.

***o*-dichlorobenzene** Abbreviation of *ortho*-dichlorobenzene.

***ortho*-dichlorobenzene** One of the contaminants in dichlorobenzene. *See* 1,2-dichlorobenzene.

***p*-dichlorobenzene** Abbreviation of *para*-dichlorobenzene.

***para*-dichlorobenzene** One of the contaminants in dichlorobenzene. *See* 1,4-dichlorobenzene.

**dichlorobromomethane** A trihalomethane and a disinfection by-product; chemical formula: $CHBrCl_2$. *See* bromodichloromethane.

**1,1-dichloro-2,2-bis-(*para*-chlorophenyl) ethane** *See* dichlorodiphenyl dichloroethane.

**dichlorodiethyl sulfide ($C_4H_8Cl_2S$)** An oily liquid with a mustard odor, used in chemical warfare; it can cause blindness or even death and damage the skin and the lungs. Also called mustard gas.

**dichlorodifluoromethane ($CCl_2F_2$)** A colorless, nonflammable, relatively inert, gas or liquid, volatile organic chemical, insoluble in water, used as a refrigerant. Also called Freon 12.

**dichlorodiphenyl dichloroethane [$(ClC_6H_4)_2CHCHCl_2$]** Acronym: DDD. The generic name for 1,1-dichloro-2,2-bis-(*para*-chlorophenyl) ethane, a discontinued insecticide. Also called tetrachlorodiphenylethane.

**dichlorodiphenyl dichloroethylene [$(ClC_6H_4)_2C{:}CCl_2$]** Acronym: DDE. A degradation product of dichlorodiphenyl trichloroethane.

**dichlorodiphenyl trichloroethane [$(ClC_6H_4)_2CH(CCl_3)$]** *See* DDT.

**1,1-dichloroethane ($CH_3CHCl_2$)** A volatile organic compound that the USEPA requires water supply systems to monitor. It is used as a solvent.

**1,2-dichloroethane ($ClCH_2—CH_2Cl$)** A volatile organic compound with a pleasant odor and sweet taste, somewhat soluble in water and with limited biodegradability, used as a solvent and in the production of insecticides, paint, etc. Its vapor may cause eye, nose and throat irritations, and affect the nervous system and some organs. Regulated in drinking water by the USEPA: MCLG = 0 and MCL = 0.005 mg/L. Also called ethylene dichloride.

**1,1-dichloroethene** Same as 1,1-dichloroethylene.

**dichloroethylene** A solvent used in the electronics industry and as a dry-cleaning fluid, suspected of causing cancer in humans.

**1,1-dichloroethylene ($Cl_2C{:}CH_2$)** Abbreviation: 1,1-DCE. A volatile organic liquid, almost insoluble in water; used in food packaging. It may be damaging to the kidney and liver at high doses. Regulated by the USEPA: MCL = MCLG = 0.007 mg/L. Also called vinylidene chloride and 1,1-dichloroethene.

**1,2-dichloroethylene ($ClHC{:}CHCl$)** *See* acetylene dichloride.

***cis*-1,2-dichloroethylene ($ClHC{:}CHCl$)** One of the two contaminants of acetylene dichloride. Regulated by the USEPA: MCLG = MCL = 0.07 mg/L in drinking water.

***trans*-1,2-dichloroethylene ($ClHC{:}CHCl$)** One of the two contaminants of acetylene dichloride. Regulated by the USEPA: MCLG = MCL = 0.1 mg/L in drinking water.

**dichlorofluoromethane ($CHCl_2F$)** A colorless liquid or gas with an ether-like odor, used in fire extinguishers, solvents, and refrigerants.

**dichloroiodomethane** A trihalomethane and a disinfection by-product; chemical formula: $CHCl_2I$.

**dichloromethane ($CH_2Cl_2$)** A volatile organic compound, used in paint removers, urethane foam, pesticides, etc.; regulated by the USEPA as a drinking water contaminant: MCLG = 0, MCL = 0.005 mg/L. Also called methylene chloride.

**dichlorophenol** A phenol compound having two atoms of chlorine. *See* chlorophenol and 2,4-dichlorophenol.

**2,4-dichlorophenol ($C_6H_4Cl_2O$ or $Cl_2C_6H_3OH$)** An organic compound, colorless to yellow crystals with a sweet, musty or medicinal odor, used in chemical synthesis; a disinfectant by-product, regulated by the USEPA.

**2,4-dichlorophenoxyacetic acid ($Cl_2C_6H_3OCH_2COOH$ or $C_8H_6O_3Cl_2$)** Abbreviation: 2,4-D. An odorless, white to pale yellow crystalline powder; a hormonal herbicide used to control weeds on lawns and in cereals; a drinking water contaminant regulated by the USEPA: MCLG = MCL = 0.070 mg/L. *See* chlorophenoxy.

**1,2-dichloropropane ($CH_3CHClCH_2C$ or $C_3H_6Cl_2$)** A clear, colorless, liquid volatile organic chemical with a chloroform-like odor, used as an industrial solvent and as a soil fumigant, metal degreaser, lead scavenger, or in dry-cleaning fluids; regulated by the USEPA as a drinking water contaminant: MCLG = 0, MCL = 0.005 mg/L. It may affect the kidney and liver. Effective removal methods include granular activated carbon and packed tower aeration. Also called propylene dichloride.

**1,3-dichloropropane ($CH_2ClCH_2CH_2Cl$)** A volatile organic chemical regulated by the USEPA as a drinking water contaminant.

**2,2-dichloropropane ($CH_3CCl_2CH_3$)** A volatile organic chemical regulated by the USEPA as a drinking water contaminant.

**1,1-dichloropropene ($CCl_2$:$CHCH_3$)** A volatile organic chemical that water supply systems are required to monitor.

**1,3-dichloropropene (CHCl:$CHCH_2Cl$)** An organic chemical that water supply systems are required to monitor; used as a soil fumigant.

**dichlorprop** A pesticide (chlorophenoxy herbicide) with a WHO-recommended guideline value of 0.1 mg/L for drinking water.

**dichromate ($Cr_2O_7^{2-}$)** A chromium compound used in the COD test to oxidize organic matter to the end products of water and carbon dioxide.

**dichromate reflux method** The determination of chemical oxygen demand (COD) using an Erlenmeyer flask to reflux (vaporize and condense) a mixture consisting of a wastewater sample, standard potassium dichromate ($K_2Cr_2O_7$), and sulfuric acid containing silver sulfate.

**DIC microscope** *See* differential interference contrast microscope.

**dicofol** A pesticide used on citrus fruits.

**diderate** A common ore of iron, essentially a carbonate $FeCO_3$.

**dieldrin ($C_{12}H_{10}Cl_6O$)** A light tan, crystalline, insoluble, poisonous solid, no longer produced or used as an insecticide in the United States. It is the generic name for the persistent insecticide 1,2,3,4,10,10-hexachloro-6,7-epoxy-1,4,4a,5, 6,7,8,8a-octahydro-1,4-endo,exo-5,8-dimethanonapthalene, and a stereoisomer of endrin. Also known as Illoxol, Octalox, Panoram D-31, and Quintox.

**dielectric** A nonconductive substance, an insulator, or a substance that can maintain an electric field with minimum loss of power.

**dielectric constant (D)** (1) The ratio of the flux density of an electric field in a dielectric to its flux density in a vacuum. Also called permittivity, relative permittivity, specific inductive capacity. (2) An electrical property of water and other media related to their polarization. Changes in the dielectric constant of groundwater with respect to a baseline indicate the presence of pollutants. *See* zeta potential.

**dielectric fitting** A plumbing fitting made of an electrical nonconductor, such as plastic, and used to control galvanic corrosion when joining pipes of dissimilar metals, e.g., copper and galvanized steel. When it is used in a water main, a bypass strap may be necessary to maintain continuity of existing electrical grounding.

**diene** A chemical compound that contains two double bonds; e.g., 1,3-butadiene ($H_2C{=}CH{-}HC{=}CH_2$). *See also* aliphatic hydrocarbon.

***Dientamoeba fragilis*** A waterborne species of amebas that cause an intestinal infection with much milder symptoms than those of *Entamoeba histolytica.*

**die-off constant** *See* die-off rate constant.

**die-off rate constant** The constant ($k$) commonly used in first-order decay models to express the rate at which microorganisms die off, particularly when they are taken from their natural habitat (Droste, 1997 and Feachem, 1983):

$$k = (k_d + k_p + k_r + k_s)\theta^{(T-20)} \qquad \text{(D-23)}$$

where $k$ = overall dieoff rate constant, $k_d$ = dieoff coefficient in the dark, $k_p$ = predation coefficient, $k_r$ = coefficient of dieoff from radiation, $k_s$ = settling coefficient, $\theta$ = Arrhenius constant, and $T$ = temperature in °C. *See also* $T_{90}$.

**dietary $LC_{50}$** A statistically derived estimate of the concentration of a test substance in the diet that would cause 50% mortality to the test population under specified conditions (EPA-40CFR152.161-a).

**diethylaminoethanol ($C_6H_{15}NO$)** A colorless, hygroscopic, soluble liquid used for chemical synthesis, in antirust compounds, and photographic products.

**diethylene** Another name of butane.

**diethyl ether [$(C_2H_5)_2O$ or $C_4H_{10}O$]** Abbreviation: DEE. A colorless, highly volatile, flammable liquid, with an aromatic odor and sweet, burning taste, resulting from the action of sulfuric acid ($H_2SO_4$) on ethyl alcohol ($C_2H_5OH$); used as a solvent in analytical chemistry to extract polar organics. Also called diethyl oxide, ether, ethyl ether, ethyl oxide, sulfuric ether.

**diethylhexyl adipate** or **di(2-ethylhexyl)adipate {[$CH_2CH_2COOCH_2CH(C_2H_5)C_4H_9]_2$}** A synthetic organic compound of adipic acid ($C_6H_{10}O_4$); a white, crystalline, slightly soluble solid. Adipates have been found in treated wastewater effluent from meat processing plants. MCLG = MCL set by the USEPA: 0.5 mg/L in drinking water.

**diethylhexyl phtalate** An organic chemical regulated by the USEPA as a drinking water contaminant: MCLG = 0; MCL = 0.006 mg/L.

**diethyl oxide** *See* diethyl ether.

**N,N-diethyl-p-phenylenediamine [$(C_2H_5)_2NC_6H_4NH_2$]** Abbreviation: DPD. A reagent used in the determination of chlorine residuals. *See* DPD method.

**N,N-diethyl-p-phenylenediamine-ferrous ammonium sulfate** Abbreviation: DPD-FAS. A reagent used in the laboratory for the determination of chlorine residuals in water. It consists of a solution of N,N-diethyl-p-phenylenediamine [$(C_2H_5)_2NC_6H_4NH_2$] with a titration of ferrous ammonium sulfate [$Fe(SO_4) \cdot (NH_4)_2SO_4 \cdot 6H_2O$].

**diethyl sulfide ($C_2H_5SH_5C_2$)** A mercaptan, a compound derived from hydrogen sulfide ($H_2S$).

**differential-aeration cell** The corrosion cell that forms on the surface of a metal in water, with a cathodic area where there is an oxygen concentration and an anodic area with less access to oxygen (Fair et al., 1971). Also called a differential-concentration cell.

**differential-concentration cell** *See* differential-aeration cell.

**differential distillation** A noncontinuous distillation operation in which a discrete quantity or batch of liquid feed is charged into a distillation unit and distilled at one time. *See* batch distillation operation for detail.

**differential enrichment medium** A medium that contains a substance that reacts in a specific manner to indicate the presence or absence of a given group of organisms, or to promote their growth relative to other organisms. If it also inhibits the growth of the competitive species, it is a differential and selective enrichment medium. *See also* Colilert test and selective enrichment medium.

**differential gage** *See* differential pressure gage.

**differential interference contrast microscope** An optical device that uses two separate visible light beams as a source of illumination: a reference beam and a beam through the specimen, giving the perception of a three-dimensional image. Maximum magnification = 2000 times. Resolution = 200 nm. *See also* light microscope, electron microscope, and ultraviolet ray microscope.

**differential medium** *See* differential enrichment medium.

**differential oxygenation corrosion** Corrosion caused by differences in dissolved oxygen concentration between two metal surfaces, e.g., under rivets/washers/debris or in crevices. *See also* concentration cell corrosion.

**differential plunger pump** A reciprocating pump with a plunger to draw the liquid upward; a double-action pump.

**differential pressure cell** A device used to determine the pressure difference between two sites, e.g., head loss in a filter.

**differential pressure gage** An instrument that measures the relative difference in pressure between two points of a fluid in a pipe or other container. *See also* differential manometer.

**differential pulse polarography** A method used in the electrochemical analysis of metal cations and anions.

**differential sedimentation** *See* differential settling.

**differential settling** The type of flocculation that occurs when particles of different settling velocities collide under the influence of gravity. The flocculation rate constant depends on the specific gravity and diameters of the particles and on the kinematic viscosity of the fluid. *See also* convective sedimentation and flocculation concepts.

**differential speed** The difference in the rotation speeds (in revolutions per minute or RPM) of the conveyor and the bowl of a solid-bowl centrifuge; the factor that causes sludge to separate into a liquid layer and a solid cake.

**differential stain** A stain consisting of a primary dye and a counterstain used in preparing water or wastewater samples for microscopic observation; *see,* e.g., Gram stain. *See also* simple stain, smear, acidic dye, basic dye, acid-fast stain.

**differential surge tank** A surge tank used to dampen flow and pressure fluctuations in a water distribution system.

**difficult waste** As used in the United Kingdom, waste that can be harmful to the environment or whose properties present handling problems; e.g., alkalis, biocides, dyes and pigments, fuel/oil/grease, inorganic acids, organic acids, polymetallic materials and precursors, sludge, and toxic metal compounds.

**Diffusadome®** A coarse-bubble air diffuser manufactured by Amwell, Inc.

**Diffusair** An apparatus manufactured by Walker Process Equipment Co. for the diffusion of carbon dioxide. *See also* Diffuserator.

**diffusate** The solution or the crystalline material passing into it through a semipermeable electrodialysis membrane. Also called dialyzate.

**diffused aeration** The process of contacting gas bubbles with a liquid to transfer the gas (e.g., carbon dioxide, oxygen, ozone) to the liquid or to remove substances such as volatile organic compounds by stripping. The diffusing system includes diffusers, header pipes, air mains, blowers, and appurtenances. Also called bubble aeration, diffused-air aeration. *See also* diffusivity exponent, minimum air-to-water ratio, Stanton number.

**diffused aerator** A device commonly used to transfer oxygen in aerobic wastewater treatment or in air stripping of volatile organics. *See* diffuser.

**diffused air** (1) Compressed air broken into small bubbles below the surface of water or wastewater to transfer oxygen to the liquid. (2) A type of aeration that forces oxygen into wastewater by pumping air through perforated pipes inside a holding tank.

**diffused-air aeration** The introduction of compressed air into water by means of submerged diffusers or nozzles. *See* diffused aeration for detail.

**diffused-air reoxygenation** The injection of oxygen into a treated wastewater effluent, using diffused aeration, to increase its dissolved oxygen content. *See also* cascade reoxygenation, mechanical reoxygenation.

**diffuse layer** The counterbalancing cloud of ions that forms around a negatively charged particle in the theory of electrostatic stabilization. *See* electrical double layer.

**diffuser** (1) A porous plate or tube through which air or another gas is forced and divided into fine or coarse bubbles for diffusion in a liquid, as used in water or wastewater treatment. *See also* air diffuser, jet aeration, oxygen transfer efficiency, U-tube aeration, and the following types of diffuser: coarse-bubble, disk, dome, fine-bubble, fixed-orifice, flexible, membrane, nonporous, orifice, panel, porous, slotted-tube, sparger, static-tube. (2) A structure used, instead of a single outlet, on an ocean or river outfall to discharge wastewater effluents. Typically, it consists of several small ports or holes over one-third of the length of the outfall and discharges in one or two directions. *See* Figure D-05.

**Diffuserator** An apparatus manufactured by Walker Process Equipment Co. for the diffusion of carbon dioxide. *See also* Diffusair.

**diffuser fouling** The reduction of the oxygen transfer efficiency of a porous diffuser due to (a) internal impurities in the compressed air that the air filters do not remove or (b) external accumulation of biological slimes and inorganic precipitants.

**diffuser plate** A porous plate installed in an aeration basin to diffuse air into water or wastewater.

**diffuser tube** An air tube installed in an aeration basin to diffuse air into water or wastewater.

**diffuser vane** A vane in the casing of a centrifugal pump designed to convert velocity head to pressure head. Also called diffusion vane.

**diffuser wall** Another name for baffle wall: a wall used as a baffle to equalize flow distribution and prevent short-circuiting, for example, in a paddle flocculator. Baffle walls may be used in a reactor to approximate plug-flow conditions.

**diffuse source of pollution** A source of pollutants dispersed on land by human activities; a phrase sometimes applied to urban stormwater runoff because of both its point and nonpoint characteristics. *See* background pollution, nonpoint source, point source.

**diffusing well** Same as diffusion well.

**diffusion** (1) The movement of suspended or dissolved molecules or particles from a more concentrated to a less concentrated area as a result of the random movement of individual particles. The process tends to distribute them uniformly throughout the available volume. *See also* advection, boundary-layer diffusion, diffusion and dispersion, eddy (or turbulent) diffusion, longitudinal diffusion, molecular diffusion. (2) The spreading or scattering of a fluid, e.g., the process by which molecules in a single phase equilibrate to a zero concentration gradient by random molecular or thermal motion (Brownian motion). The flux of molecules is from regions of high concentration to regions of low concentration and is governed by Fick's Second Law. The theory of diffusion is used in many applications in the field of water and wastewater treatment, e.g., gas transfer (aeration) and reverse osmosis.

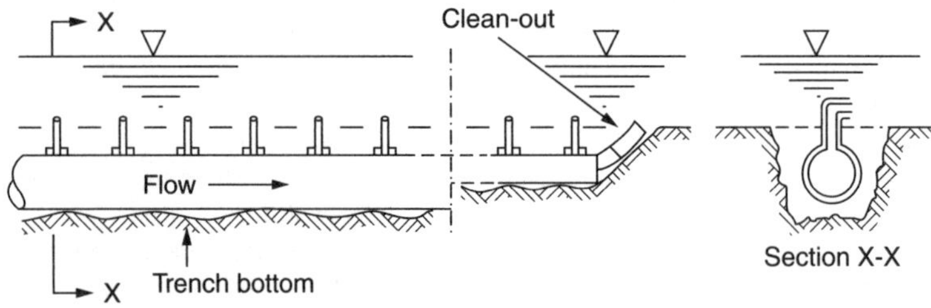


**Figure D-05.** Diffuser (river outfall). (Adapted from Metcalf & Eddy, 1991.)

**diffusion aerator** A low-pressure aerator that blows air through submerged plates, pipes, or similar devices to form rising bubbles through water or wastewater.

**diffusion coefficient** *See* dispersion coefficient, eddy diffusion coefficient, molecular diffusion coefficient.

**diffusion dialysis (DD)** A process that uses ion-exchange membranes to separate and recover acids from spent pickling liquor in the steel-finishing industry. The unit consists of a stack of membranes with spent liquor fed into one compartment and water fed countercurrently into another compartment. *See also* Donnan dialysis.

**diffusion and dispersion** In all three states, water has space between its molecules for other ions and molecules to penetrate (or diffuse into). Liquid water has $H_2O$ molecules as well as hydrogen ($H^+$) and hydroxyl ($OH^-$) ions. Water molecules in the vapor state occupy a larger volume than in the two other states. Gases and liquid substances can diffuse into water and ice. When a substance is introduced into water, mass transfer occurs in reverse direction of the concentration gradient, as diffusion or dispersion. Molecular diffusion is the slow concentration equalization process that occurs under quiescent conditions, similar to the phenomenon of heat conduction. Mass transfer is called eddy diffusion or turbulent diffusion when it is caused by turbulence without flow; when it involves mechanical mixing, it is referred to as mechanical dispersion or kinematic dispersion. Hydrodynamic dispersion represents the combined effect of molecular diffusion and mechanical dispersion. Longitudinal diffusion is the longitudinal spreading of a substance in a fluid moving under turbulent conditions, faster or slower than the velocity of the fluid. Advection is also longitudinal spreading but at the same speed as the host fluid. Dispersion includes all forms of diffusion: molecular, mechanical, and longitudinal. When a slug of contaminant enters an aquifer or wastewater is discharged into a surface stream, the net result of dispersion is an increase of the volume of contaminant in the receiving medium. Fick's laws of diffusion apply to each term, with an appropriate diffusion coefficient. *See also* advection–dispersion equation. Boundary-layer diffusion is the transfer of molecules or ions across a boundary between two media, e.g., a particle and a solution. The literature on water and wastewater treatment cites mostly molecular diffusion or simply diffusion.

**diffusion feeder** A device that feeds controlled quantities of chemicals for water or wastewater treatment. It consists of a tank containing the chemicals, through which a small stream is diverted for mixing and then returned to the main line. Also called bypass feeder.

**diffusion layer** The stagnant liquid layer that separates the biofilm from the bulk liquid in a biological attached-growth treatment process. Through this layer, substrate, oxygen, and nutrients diffuse to the biofilm, while biodegradation products diffuse to the bulk liquid. Also called bound water layer, stagnant liquid layer, or fixed water layer. *See* fixed-growth process.

**diffusion limitation** A characteristic of attached-growth processes, as compared to a suspended-growth process like activated sludge, whereby process performance is limited by the depth of the biofilm, within which occur substrate removal and electron donor utilization. This characteristic is often exploited for nitrification and denitrification. Because of this limitation, attached-growth processes require higher dissolved oxygen concentrations in the bulk liquid than suspended-growth processes.

**diffusion-limited concept** *See* diffusion limitation.

**diffusion-limited process** Any biological process, the performance of which depends on the rate of diffusion and consumption of substrate and nutrients within the biofilm. The concentration of these constituents is lower in the biofilm than in the bulk liquid. *See also* surface flux, substrate flux, substrate utilization rate.

**diffusion vane** A vane in the casing of a centrifugal pump designed to convert velocity head to pressure head. Also called diffuser vane.

**diffusion well** A well used to add water to an aquifer. Also called a recharge well. *See also* diffusing pit and diffusing well.

**diffusive airflow test** A test used to determine the integrity of a wetted membrane by submitting it, offline, to an airflow below the bubble point pressure. The airflow through the membrane is measured and compared to airflow through an intact membrane. *See also* bubble-point test and pressure decay test.

**diffusivity** Same as molecular diffusion coefficient.

**diffusivity exponent** A dimensionless parameter ($m$) that affects the liquid diffusivity term in the equation used to calculate the overall mass transfer coefficient (the "$K_La$" factor) for bubble and surface aeration; e.g., $m = 0.6$ for bubble aeration and $m = 0.5$ for surface aeration.

**Digesdahl** A digestion apparatus manufactured by Hach Co. used to prepare laboratory samples to be analyzed for proteins and minerals.

**digested sludge** Wastewater treatment sludge that has undergone aerobic or anaerobic decomposition and now contains relatively nonputrescible solids.

**digester** (1) In wastewater treatment, a closed tank or vessel used for sludge digestion, including systems for gas collection, sludge heating, and solids recirculation. It operates continuously or intermittently. Digester liquor is returned to the plant influent or treated for disposal in the final effluent. Digester design considerations include size (see digester capacity), gas collection, heating system, and solids circulation. Also called sludge digester, digestion tank, hydrolytic tank. (2) In solid waste conversion, a unit in which bacterial action is induced and accelerated in order to break down organic matter and establish the proper carbon-to-nitrogen ratio.

**digester capacity** The design capacity of a digestion tank depends on the sludge load, digestion time, loss of sludge moisture, and temperature. The following formula is sometimes used (Fair et al., 1971):

$$V = C[F - 2/3(F - D)]T \qquad \text{(D-24)}$$

where $V$ = capacity of the digester, in cubic feet per capita; $C$ = a factor of safety to account for sludge liquor, scum, and gas in addition to the basic space requirements; $F$ = daily per capita volume of fresh solids, cubic feet; $D$ = daily per capita volume of digested solids, cubic feet; and $T$ = digestion time, days.

**digester coils** Hot water or steam pipes installed in a digester for heating.

**digester gas** The gas mixture (specific gravity = 0.86) released during the anaerobic decomposition of organic matter, more specifically, during sludge digestion. Saturated with water vapor, it contains by volume approximately 60–70% methane ($CH_4$), 25–30% carbon dioxide ($CO_2$), and small amounts of nitrogen ($N_2$), hydrogen ($H_2$), and hydrogen sulfide ($H_2S$). The fuel value of digester gas is usually used for heating, power production, etc.; because of its water vapor content and the gases other than methane, this fuel value is about 600 BTU/ft$^3$, i.e., 60% of that of natural gas.

**digester loading** A measure of the mass of sludge processed by an anaerobic digester per unit volume (pounds of volatile solids per day per ft$^3$) or per unit weight of volatile solids in the digester (pounds of volatile solids added per day per pound of digester volatile solids). Typical design parameters for mesophilic anaerobic digesters include a solids loading rate of 100–300 lb VSS/1000 ft$^3$/day and a solids retention time of 15–20 days. (VSS = volatile suspended solids.)

**digester sludge age** The solids retention time (SRT) of a digester.

**digester stratification** The formation of separate layers in sludge digesters that are not vigorously stirred, e.g., from top to bottom: an inactive scum layer, a layer of tank liquor or supernatant, an active layer of digesting solids (which usually receives incoming solids), and a bottom layer of digested solids, which are withdrawn for further treatment and disposal.

**digestion** The biochemical decomposition of organic matter, resulting in stabilization, i.e., partial gasification, liquefaction, mineralization of pollutants, and volume reduction. Some pathogens are also destroyed during digestion. Digestion is one metabolic mechanism that organisms use to process energy and materials: they break down complex molecules to simpler substances that they can use. Other mechanisms include photosynthesis and respiration. *See also* sludge digestion and the following types of digestion: aerobic, anaerobic, high-rate, mesophilic, single-stage, thermophilic, two-stage.

**digestion chamber** (1) The sludge digestion compartment of an Imhoff tank. (2) A sludge digestion tank.

**digestion tank** A digester.

**Digichem** A programmable instrument manufactured by Ionics, Inc. for use in titration analysis.

**digital control system (DCS)** An instrumentation and control system of a water or wastewater treatment plant that uses electronic (digital, instead of analog) devices. A DCS consists of a number of building blocks, such as remote terminal units, programmable logic controllers, distributed process controllers, area control centers, central control centers, media-connecting devices, single-loop controllers, and multiloop controllers. Also called distributed control system, distribution control system, or distributor control system. *See* distributed process controller and programmable logic controller.

**digital dosing pump** A recent improvement of the diaphragm metering pump that uses robotic technology to improve accuracy and lower costs in delivering a controlled volume of liquid in water and wastewater treatment processes as well as other applications. Operation is controlled by software. This device focuses on control and measure of the output instead of on flow inducement. *See also* metering pump.

**digital particle image velocimetry (DPIV)** A procedure using laser beam illumination to photograph the movement of neutrally buoyant fluorescent particles and explain fluid movement in mixing devices. *See also* computational fluid dynamics, laser Doppler anemometry, laser-induced fluorescence.

**digital readout** Use of numbers to indicate the value or measurement of a variable. The readout of an instrument by a direct numerical reading of the value.

**dihaloacetonitrile ($CHX_2C{\equiv}N$, with X = Br and/or Cl)** Abbreviation: DHAN. A haloacetonitrile with two halogen atoms.

**dihydrate** A hydrate that contains two molecules of water; e.g., potassium sulfite ($K_2SO_3 \cdot 2\ H_2O$).

**dihydrogen monoxide ($H_2O$)** Another name for water.

**dihydrogen oxide ($H_2O$)** Another name for water.

**dihydroxy** Containing two hydroxyl groups.

**diiodobromomethane ($CHBrI_2$)** A trihalomethane and a disinfection by-product. Also called bromodiiodomethane.

**Dijbo** A hydraulic trash rake manufactured by Landustrie Sneek BV.

**dike** A low wall, embankment, or ridge that can act as a barrier to prevent the movement of liquids, e.g., to prevent a spill from spreading. When is a dike a levee or a dam? According to one common definition: a levee retains water temporarily during peak flood stages and a dam impounds a permanent pool. Thus, safety criteria are more stringent for dams than for levees. Also, the U.S. Congress authorizes levees as navigational structures to protect property, not people.

**diketone** A class of organic compounds containing two ketone functional groups; general formula [R—(C=O)—R]. Some diketones are ozonation or other disinfection by-products, e.g., dimethyl glyoxal ($CH_3COCOCH_3$) and dimethyl ketone or acetone ($CH_3COCH_3$).

**diluent** Any fluid or solid material used to dilute or carry an active ingredient.

**dilute and disperse** An approach to industrial waste disposal relying on its absorption onto domestic refuse or other waste material in the unsaturated zone.

**diluted sludge volume index** The sludge volume index (SVI) determined using a sample diluted with process effluent until the settled volume after 30 minutes is 250 ml/L or less, which allows meaningful comparisons of SVI results for different sludges. *See also* stirred sludge volume index (Metcalf & Eddy, 2003).

**dilute-phase bed** The phase in fluidized-bed combustion at which the bubbling of the reactor bed becomes so great that the boundary between the bed and the gas above it becomes indistinct.

**dilute solution** A solution that has been made weaker, usually by the addition of water.

**diluting water** Distilled water that has been stabilized, buffered, and aerated, as used in the BOD test.

**dilution** (1) The process of lowering the concentration of a solution by adding more solvent; the dispersal of a fluid in a larger volume of another fluid. *See* simple dilution, serial dilution. (2) A natural method of liquid disposal in which wastewater or treated effluent is discharged into a stream or body of water. When dilution by lakes or oceans is used as a treatment method, it requires a large area and a long outfall to prevent the contamination of beach waters.

**dilution bottle** A glass bottle used for diluting bacteriological samples. Also called French square, milk dilution bottle.

**dilution constant** The exponent ($n$) of concentration ($C$) in Watson's formula for the dieoff coefficient ($k$) in the theory of disinfection kinetics:

$$k = \alpha C^n \quad \text{or} \quad n = \log (k/\alpha)/\log C \qquad \text{(D-25)}$$

where $\alpha$ is an activation constant.

**dilution disposal** Disposal by dilution, i.e., discharge of collected wastewater into a receiving body of water with or without suitable treatment.

**dilution factor** (1) The ratio (at the point of disposal or at a point downstream) of the flow of receiving water to the flow of wastewater, treated effluent, or stormwater discharged. Also called available dilution. *See also* dilution rate, dilution ratio. (2) *See* simple dilution, serial dilution.

**dilution gaging** A flow measuring method involving the injection of a solution of known concentration at one point in a stream and determination of the concentration or dilution at another point downstream. *See* dilution method.

**dilution method** A method used to evaluate a stream discharge ($Q$) by injecting a tracer of known concentration ($C_1$) and flowrate ($Q_1$), and measuring its concentration ($C_2$) at a point downstream. *See also* salt method.

$$Q = Q_1(C_1 - C_2)/C_2 \qquad \text{(D-26)}$$

**dilution rate (*D*)** (1) A measure of the hydraulic load in a flow-through reactor, equal to the reciprocal of the flow rate ($Q$):

$$D = 1/Q \qquad \text{(D-27)}$$

See loading intensity. (2) The reciprocal of detention time ($T$) or the ratio of flow rate ($Q$) to velocity ($V$):

$$D = 1/T = Q/V \qquad \text{(D-28)}$$

**dilution ratio** The relationship between the volume of water in a stream and the volume of incoming water. It affects the ability of the stream to assimilate waste.

**dilution requirement** The minimum dilution ratio or the minimum streamflow that will prevent objectionable conditions or nuisances in a receiving water. In the past, for combined sewer discharges, it was estimated to be between 2.5 and 10 cfs per 1000 population. The required streamflow is actually a function of the first-stage BOD of the wastewater ($L$, pounds per capita per day) and the permissible loading ($L_a$, mg/L); it may be determined from the following equation (Fair et al., 1971):

$$Q = 185.5\, L/L_a \qquad \text{(D-29)}$$

where $Q$ is in cfs per 1000 population.

**dilution to threshold (D/T)** A measure of detectable odor concentration in a sample, reported as the number of dilutions to the minimum detectable threshold odor concentration (MDTOC). If $N$ volumes of diluted air are required to reduce an odor to its MDTOC, then $D/T = N$. *See also* $ED_{50}$, threshold odor, threshold odor number (Metcalf & Eddy, 2003).

**dilution water** The water to which the test substance is added and in which the test species is exposed (EPA-40CFR797.1600-5).

**dilution weight** Under CERCLA, a unitless parameter in the hazard ranking system (HRS) that adjusts the assigned point value for certain targets subject to potential contamination as a function of the flow or depth of the water body at the target (EPA-40CFR300-AA).

**dilution zone** A limited area or volume of water where initial dilution of a discharge takes place, and where numeric water quality criteria can be exceeded but acutely toxic conditions are prevented from occurring (EPA-40CFR131.35.d-8). Same as mixing zone. *See also* zone of initial dilution.

**dimension** A basic quantity or characteristic of physical systems such as force, mass, temperature, length, and time. Most hydraulics problems use one of two sets of dimensions: the FLT (force–length–time) system or the MLT (mass–length–time) system.

**dimensional analysis** A method for deriving dimensionless relationships within physical systems or for comparing the quantities occurring in a problem without actually solving the problem. This comparison may be made through the use of dimensionless numbers.

**dimensionless numbers** In hydraulics, a series of fundamental parameters derived from governing equations or by dimensional analysis: Courant, densimetric Froude, dimensionless variance, dispersion number, Euler, Froude, Galileo, gradient Richardson, kinematic wave, Lewis, Mach, orifice/soil, Peclet, Prandtl, Reynolds, Rossby, Schmidt, Weber.

**dimensionless variance** A parameter ($\nu$) related to the Peclet number ($P_e$), used in the formulation of normalized density function for residence times, and calculated as follows (AWWA, 1999):

$$\nu = (2/P_e) + 2[1 - \exp(-P_e)]/P_e^2 \qquad \text{(D-30)}$$

**dimer** A molecule composed of two identical molecules or a polymer derived from two identical monomers, e.g., the hydroxo ferric complex $2[Fe(H_2O)_5(OH)]^{++}$.

**dimeric species** Same as dimer.

**dimethylarsine [$(CH_3)_2AsH$]** A toxic methyl derivative of arsenic.

**dimethylarsinic acid [$(CH_3)_2AsO(OH)$]** A toxic methyl derivative of arsenic.

**1,2-dimethylbenzene [$C_6H_4(CH_3)_2$]** One of three compounds that make up the organic contaminant xylene, the others being 1,3- and 1,4-dimethylbenzene.

**dimethylcarbinol isopropanol ($C_3H_8O$ or $CH_3$—$CH_3$—CHOH)** Same as isopropyl alcohol.

**1,1-dimethyl-4,4-dipyridylium** *See* paraquat.

**dimethyl glyoxal ($CH_3COCOCH_3$)** A diketone formed during water ozonation.

**dimethyl ketone ($CH_3COCH_3$)** A common household solvent of fats and cleaning agent for laboratory glassware. *See* acetone for detail.

**dimethylnitrosamine ($C_2H_6N_2O$)** Abbreviation: DMN or DMNA. A yellow, water-soluble, carcinogenic liquid found in tobacco smoke and some foods. It is assumed that nitrites, secondary amines, and bacteria may combine to produce this carcinogen in the stomach or bladder of some individuals.

**dimethylpolysulfide** A bacterially produced sulfur compound that causes swampy and fishy taste and odor problems in drinking water distribution systems.

**dimethyl sulfate [$(CH_3)_2O_2S$ or $CH_3SO_2CH_3$]** A colorless or yellow poisonous liquid, used in organic synthesis. Also called methyl sulfate.

**dimethyl-2,3,5,6-tetrachloroterephtalate [$C_6Cl_4(COOCH_3)_2$]** Abbreviation: DTCT. A herbicide.

**dimethyl-2,3,5,6-tetrachloroterephtalate (DTCT) di-acid degradate** A degradation product of the herbicide DTCT.

**dimethyl-2,3,5,6-tetrachloroterephtalate (DTCT) mono-acid degradate** A degradation product of the herbicide DTCT.

**dimictic** Pertaining to lakes and reservoirs that freeze over and normally go through two stratification and two mixing cycles a year.

**Dimminutor®** An open-channel comminutor manufactured by Franklin Miller.

**dinitrobenzene ($C_6H_4NO_2$)** Any of three isomeric benzene derivatives used in dyes.

**dinitrogen ($N_2$ or N—N)** Nitrogen in the form of an inert gas; atmospheric nitrogen, consisting of two atoms.

**dinitrogen pentoxide ($N_2O_5$)** An unstable compound formed by the reaction of the nitrate radical ($NO_3$) and nitrogen dioxide ($NO_2$). It hydrolyzes to nitric acid ($HNO_3$):

$$NO_3 + NO_2 \rightarrow N_2O_5 \quad \text{(D-31)}$$

$$N_2O_5 + H_2O \rightarrow 2\ HNO_3 \quad \text{(D-32)}$$

**2,4-dinitrophenol [$C_6H_3OH(NO_2)_2$]** One of six isomers of phenol in which nitrogen groups replace two hydrogen atoms; a synthetic organic chemical used in dyes, wood preservatives, and other industrial products.

**2,6-dinitrophenol** A reagent used in water analyses as an acid–base indicator, with a pH transition range of 1.7 (colorless)–4.7 (yellow).

**dinitrotoluene** A mixture of two synthetic organic chemicals: 2,4-dinitrotoluene and 2,6-dinitrotoluene. Regulated by the USEPA as a drinking water contaminant: MCL = 0.003 and MCLG = 0.

**2,4-dinitrotoluene [$C_6H_3CH_3(NO_2)_2$]** A synthetic organic chemical used industrially.

**2,6-dinitrotoluene [$C_6H_3CH_3(NO_2)_2$]** A synthetic organic chemical used industrially.

***Dinobrion*** A genus of filter-clogging algae (diatoms) sometimes found in drinking water sources; they may cause short direct filtration runs.

**dinoflagellates** Single-celled aquatic organisms with both plant and animal characteristics, e.g., photosynthesis and motility, sometimes classified as algae and other times as protozoa. They may produce red tides under certain favorable conditions.

**dinoseb [$CH_3(C_2H_5)CHC_6H_2(NO_2)_2OH$]** A synthetic organic chemical, slightly soluble in water, used as a herbicide, a fungicide and an insecticide. It was banned by the USEPA in 1986 because it posed the risk of birth defects and sterility. Regulated as a drinking water contaminant: MCL = MCLG = 0.007 mg/L. Also called 2-*sec*-butyl-4,6-dinitrophenol.

**dioxin ($C_{12}H_4Cl_4O_2$)** Any of a family of chlorinated hydrocarbon compounds known chemically as dibenzo-p-dioxins, which have as a nucleus a triple-ring structure consisting of two benzene rings connected through a pair of oxygen atoms, e.g., 2,3,7,8-tetrachlorodibenzo-p-dioxin (TCDD). Concern about them arises from their potential toxicity and contaminants in commercial products. Tests on laboratory animals indicate that dioxin is one of the most toxic man-made products. At high concentrations, it may cause chloracne. USEPA set an MCLG of 0 and an MCL of $3 \times 10^{-8}$ for dioxin in drinking water, but its solubility in water is very low: 0.2 ppb.

**Dioxytrol** Centrifugal aeration equipment manufactured by Hazleton Environmental Products, Inc.

**Dipair™** An aerator manufactured by Infilco Degremont, Inc. using a static tube.

**diphenamid ($C_{16}H_{17}NO$)** A selective herbicide used for weed control on lawns and crops.

**diphenyl** Another name of biphenyl.

**diphenylamine ($C_{12}H_{11}N$)** A colorless, crystalline benzene derivative used in dyes, in propellants, and in analytical chemistry.

**1,2-diphenylhydrazine ($C_6H_5NHNHC_6H_5$)** A synthetic organic chemical.

**Diphonix™** Ion exchange resins manufactured by Eichrom Industries, Inc.

**diphtheria** A febrile, infectious disease of the air passage caused by the bacillus *Corynebacterium diphtheriae.*

**diphyllobothriasis** A helminthic excreted infection caused by *Diphyllobothrium latum. See* its common name, fish tapeworm, for more detail.

***Diphyllobothrium dentricum*** A tapeworm of the gull that occasionally infects humans in Siberia. Like *Diphyllobothrium latum,* it uses fish as an intermediate stage.

***Diphyllobothrium latum*** The species of helminthic pathogens that cause fish tapeworm disease. It is widely distributed, mainly in lakeside areas of temperate regions. Its mode of transmission is from humans or animals to copepods to fish to humans; period of latency: 2 months.

***Diphyllobothrium pacificum*** A tapeworm of the fur seal that occasionally infects humans in Peru. Like *Diphyllobothrium latum,* it uses fish as an intermediate stage.

***Diplocystis*** A genus of blue-green algae or cyanobacteria that can cause off-tastes and odors in drinking water sources.

***Diplogonoporus grandis*** A tapeworm of the whale that occasionally infects humans in Japan. Like *Diphyllobothrium latum,* it uses fish as an intermediate stage.

**dipole** A molecule that has a definite separation between its center of positive charge and center of negative charge. Also called a polar molecule, which is more attracted to charged species than nonpolar molecules.

**dipole array** A particular arrangement of electrodes used to conduct electrical resistivity surveys, including the dipole–dipole, Schlumberger, and Wenner arrays.

**dipole–dipole interaction** One of the forces that hold adsorbates on the surface of adsorbents. *See also* hydrogen bond and van der Waals force.

**dipper** A hazardous waste sampling device consisting of an aluminum or fiberglass pole equipped at one end with a glass or plastic beaker; it is used for liquids and free-flowing slurries. *See* auger, bailer, composite liquid-waste sampler, thief, trier, weighted bottle.

**DIPRA** Acronym of Ductile Iron Pipe Research Association.

**Di-Prime™** An automatic pump manufactured Goodwin Pumps for handling trash.

**Diptera** A group of two-winged flies; they are among the biting insects that transmit leishmaniasis, loiasis, onchocerciasis, and trypanosomiasis in tropical areas.

**diquat** $[(C_5H_4NCH_2)_2Br_2]$ A herbicide and plant growth regulator, harmful to mammals, and regulated by the USEPA in drinking water: MCLG = MCL = 0.02 mg/L.

**DIR** Acronym of demand-initiated regeneration.

**diradical** An atom or molecule that has two unpaired electrons; a biradical. *See also* free radical, diatomic oxygen.

**direct-acting reciprocating pump** A steam-driven reciprocating pump with a piston rod connecting the steam piston to the plunger.

**direct additive** Chlorine, alum, lime, polymer, or any other chemical product that is added to treat drinking water or to maintain its quality in the storage and distribution system. *See also* indirect additive.

**direct benefit** *See* primary benefit.

**direct-connected pump** A pump that is connected directly to its motor by a coupling or clutch without any gearing, shafting, or belting. Also called close-coupled pump.

**direct-contact membrane distillation** Membrane distillation is a desalination method using a temperature-driven, hydrophobic membrane to separate water, in the form of condensed vapor, from contaminants. In the direct-contact configuration, a cool solution contacts the membrane directly and condenses the vapor that passes through the membrane.

**direct costs** Costs associated with the construction and operation of facilities, mainly capital expenditures (buildings, equipment, utility connections, equipment installation, project engineering, financing, etc.) and operation and maintenance expenses (labor, waste disposal, utilities, etc.).

**direct count** (1) One of four common methods of enumeration and identification of bacteria using the Petroff–Hauser counting chamber or the electronic particle counter. *See also* the pour-and-spread plate method, membrane-filter technique, multiple-tube fermentation, and heterotrophic plate count (Metcalf & Eddy, 2003). (2) The enumeration of all viable cells in an environmental sample, including culturable and nonculturable cells. *See also* direct viability count, dormant cells.

**direct current** Electrical current flowing in one direction only and essentially free from pulsation.

**direct discharge** The discharge of a pollutant (EPA-40CFR122.2). *See* direct discharger.

**direct discharger** A municipal or industrial facility that introduces pollution through a defined conveyance or system such as outlet pipes; a point source.

**direct dryer** A sludge processing device that brings the feed material into direct contact with heated air. The most common are rotary dryers. Also called convection dryer. *See also* thermal dryer, flash (or pneumatic conveyor) dryer, fluid-bed dryer, indirect dryer, combined direct–indirect dryer, and infrared dryer.

**direct drying** A sludge drying process that brings the wet sludge in direct contact with hot gases or other heat transfer media. *See* convection drying for detail.

**direct-feed filtration** The operating condition of a pressure-driven membrane unit in which there is no crossflow, all the water applied passes through the membrane, and raw feedwater is used to flush accumulated material from the membrane. Also called dead-end filtration. *See* Figure D-06, crossflow, transmembrane pressure (TMP), recovery rate, rejection rate, constant flux operation, constant TMP operation. (3) Same as crossflow media.

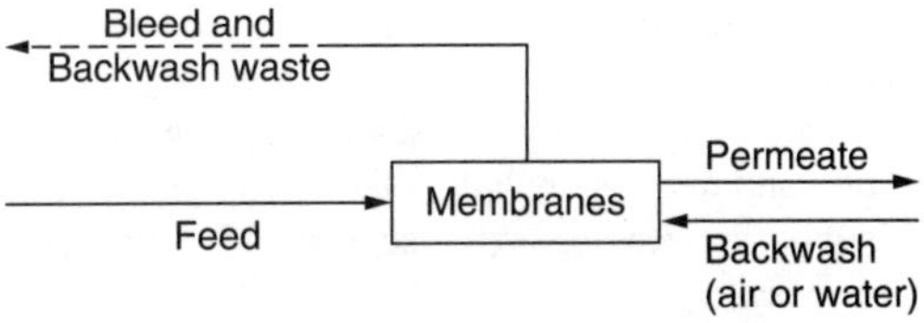


**Figure D-06.** Direct-feed filtration.

**direct filtration** (1) A water treatment method that consists of the addition of coagulant chemicals, flash mixing, coagulation, minimal flocculation, and filtration. The flocculation step may be omitted, but the physical-chemical reaction will occur to some extent. The sedimentation process is omitted. *See also* conventional filtration and inline filtration. Direct filtration is usually applied to waters of very low turbidity and excellent bacteriological quality, e.g., a surface water with color < 40 units, turbidity < 5 NTU, algae < 2000 areal standard units per mL, iron < 0.3 mg/L, and manganese < 0.05 mg/L. It differs from direct in-line filtration. *See also* direct wastewater filtration. (2) A membrane filtration operating configuration with two continuous flow streams: feedwater and permeate. Also called dead-end filtration.

**direct footprint** *See* carbon footprint.

**direct halogenation** The formation of a compound by substituting a halogen atom for a hydrogen atom in another compound, e.g., the formation of the monomer of polyvinyl chloride ($CH_2{=}CHCl$) from ethylene ($CH_2{=}CH_2$).

**direct–indirect dryer** A jacketed trough or vessel that uses large volumes of heated gas to dry wastewater treatment sludge; this operation improves heat transfer between solids and heating surfaces and reduces energy consumption.. *See also* thermal dryers classification, Milorganite (WEF & ASCE, 1991).

**direct inflow** Inflow to the sanitary sewer system from a source directly connected to stormwater runoff, e.g., from roof leaders, yard drains, manhole covers, cross connections with catch basins, and combined sewers (Metcalf & Eddy, 2003). The effect of direct inflow on wastewater flow rates is immediate. *See also* delayed inflow, steady inflow, total inflow, infiltration/inflow.

**direct influence** *See* groundwater under the direct influence of surface water.

**direct injection** The direct conveyance and injection of highly treated reclaimed water into a saturated groundwater zone such as a well-confined aquifer for recharge. This recharge method is used, usually, where groundwater is deep and surface spreading is impractical. It is effective in creating freshwater barriers in coastal aquifers against saltwater intrusion. Also called direct subsurface recharge.

**direct in-line filtration** The addition of chemical coagulants directly to the filter inlet pipe. The chemicals are mixed by the flowing water. Flocculation and sedimentation facilities are eliminated, but coagulation and flocculation occur within the piping. This treatment method is commonly used in pressure filter installations and for waters of low turbidity. Also called in-line filtration. *See* conventional filtration and direct filtration.

**direct-input respirometer** An instrument for measuring the character and extent of respiration (e.g., oxygen consumption and carbon dioxide production), usually equipped for data collection and processing by computer. It delivers pure oxygen to a sample through a metering device to satisfy the demand detected by minute pressure differences. *See* respirometric method, Gilson respirometer, Warburg respirometer.

**direct irrigation** Land application of wastewater, by spraying or piping, for treatment and disposal and not for raising crops. *See* land filtration, land application, spray irrigation.

**directly influenced by surface water** Characteristic of springs, infiltration galleries, wells, and other collectors or water sources that are subject to contamination from surface waters. *See* groundwater under the direct influence of surface water.

**Director™** A baffle manufactured by Environetics, Inc. for use in flow diversion.

**direct oxidation** The direct combination of oxidants and chemicals without the mediation of living organisms; e.g., chlorination.

**direct photolysis** The direct absorption of light by a chemical followed by a reaction that transforms the parent chemical into one or more products (EPA-40CFR796.3700-xii).

**direct potable-water reuse** The incorporation of reclaimed water directly into a potable water supply system, often by blending with another source of potable water. Windhoek (Namibia) operates intermittently, according to needs, a water treatment plant that uses reclaimed wastewater as intake water. *See also* emergency potable-water reuse, water reuse applications.

**direct reuse** The use of reclaimed water that has been conveyed from a wastewater reclamation plant to the reuse site, without discharging it to a natural body of water. *See also* water reuse applications.

**direct runoff** or **storm runoff** (1) Water that flows over the ground surface or through the ground directly into streams, rivers, and lakes. (2) The portion of precipitation that reaches a stream shortly after rainfall and remains in the river basin for only a few days. It is the sum of surface runoff and interflow, and the difference between stream flow and base flow. It corresponds to precipitation excess. Also called direct stream flow or quick-response runoff.

**direct service area** The area receiving complete sewer service from a central sewer authority, as opposed to other utilities that provide only collection service within their own boundaries but deliver their flows to the central agency for transmission, treatment, and final disposal.

**direct sludge drying** A sludge drying process that brings the wet sludge in direct contact with hot gases or other heat transfer medium. *See* convection drying for detail.

**direct subsurface recharge** Same as direct injection.

**DirecTube** A gas mixing device manufactured by Walker Process equipment Co. as an eductor tube used in anaerobic treatment.

**direct viability count (DVC)** A laboratory procedure consisting in the incubation of a sample with nalidixic acid to inhibit DNA synthesis, thus allowing the determination of the number of viable and actively growing cells. The difference between the total direct count with acridine orange and the direct viability count is a rough estimate of viable but nonculturable microorganisms.

**direct wastewater filtration** An advanced wastewater treatment method consisting of coagulation, flocculation, and filtration of a secondary effluent. *See* complete wastewater treatment, full tertiary wastewater treatment, contact filtration, direct filtration.

**direct water filtration** *See* direct filtration.

**dirty filter** A rapid filter that has been backwashed inadequately or whose auxiliary scour system (surface washers or air scour) operates improperly, resulting in filter cracks and mudballs (AWWA, 1999). Another manifestation is the deposition of minerals (e.g., calcium carbonate, aluminum oxide, iron oxide) on the grains of the filter medium, which minerals can later leach into the filtered water, increase bed depth, and impair filtration efficiency.

**dirty skin** *See* schmutzdecke, mud blanket.

**disaccharide** A carbohydrate formed by the combination of two monosaccharides, e.g., sucrose or household sugar, which combines fructose and glucose.

**disaster** An event that causes widespread destruction and distress, including the disruption of such public services as water supply and wastewater disposal.

**disaster preparedness** Readiness for events that may cause widespread destruction and distress.

**Disc Accelerator™** A precision-formed stainless steel mechanism designed by Val-Matic® to close valve discs rapidly and avoid slamming by flow reversal and yet allow the disc to be stabilized under flow conditions. It is fully enclosed within the valve and completely out of the flow path. *See also* SurgeBuster®.

**disc dryer** A device used to dry wastewater treatment sludge by circulating steam through a stationary horizontal vessel or trough, which has a jacketed shell and contains a number of discs in a rotating assembly. The discs transport the sludge through the device and provide the heat transfer surface. *See also* indirect or conduction dryer (WEF & ASCE, 1991).

**Discfilter® (DF)** A surface filtration device consisting of a number of disks mounted vertically in a tank and supporting a filter cloth in polyester or stainless steel, each disk connected to a central feed tube. The unit operates in continuous or intermittent backwash mode. *See also* surface filtration, Cloth-Media Disk Filter®.

**DiscFuser®** A coarse-bubble air diffuser manufactured by the FMC Corp.

**discharge** (1) Flow of surface water in a stream or canal, or the outflow of groundwater from a flowing artesian well, ditch, or spring. The word also applies to discharge of liquid effluent from a facility or of chemical emissions into the air through designated venting mechanisms. In various regulations, the USEPA defines discharge as including any spilling, leaking, pumping, pouring, emitting, emptying, or dumping. (2) In general, in hydraulics, hydrodynamics, or hydrology, discharge, flow or flow rate is the flux of water through an area, from a stream, canal, conduit, pump, tank, etc. By definition, discharge ($Q$) is the product of the flow area ($A$) by the velocity ($V$) of the flowing water, or

$$Q = AV \tag{D-33}$$

an expression of the conservation of mass principle. It may also be computed by formulas (Darcy–Weisbach, Chézy, Manning), or from field measurements using the stage–discharge relationship or the velocity–area method. *See* discharge measurement, and open channel flow. (3) The quantity of sediment passing a given cross section of stream in a unit of time. *See also* sedimentation terms.

**discharge allowance** The amount of pollutant (mg per kg of production unit) that a plant will be permitted to discharge. The allowance is specific to a given manufacturing operation (EPA-40CFR461.2-e).

**discharge area or flow area (*A*)** The cross-sectional area normal to the direction of flow; used in the computation of the discharge (*Q*) of a stream, a pipe, or a conduit by multiplying it by the flow velocity (*V*). Also used in the determination of the hydraulic radius (*R*), which is the ratio of the flow area to the wetted perimeter. *See* open channel flow, and channel section.

**discharge capacity** The maximum flowrate that can pass through a hydraulic structure, such as a conduit, a channel, or a treatment plant.

**discharge coefficient** A coefficient used in the weir, orifice, and other formulas or in electromagnetic current meters. To obtain the actual flow through an orifice, weir, nozzle, or current meter, the theoretical discharge is multiplied by this coefficient. Also called flow coefficient. *See* orifice flow equations, velocity coefficient, and weir equations.

**discharge consent** In the United Kingdom, an authorization to discharge effluents to a sewer system or to a receiving stream, specifying such conditions as flow rates and characteristics (BOD, COD, TSS, etc.) *See also* discharge permit, discharge standard.

**discharge curve** A graphical representation of the relationship between discharge and a pertinent hydraulic property such as depth of flow or gage height. Same as rating curve and discharge rating curve. *See* Figure D-07.

**discharge farfield** A region far from the port of discharge of wastewater or effluent into the ocean or a large lake, beyond the nearfield and the transition regions. In the farfield, ambient currents carry the wastewater away and further dilute it by diffusion. *See also* initial dilution, wastewater plume (Metcalf & Eddy, 1991).

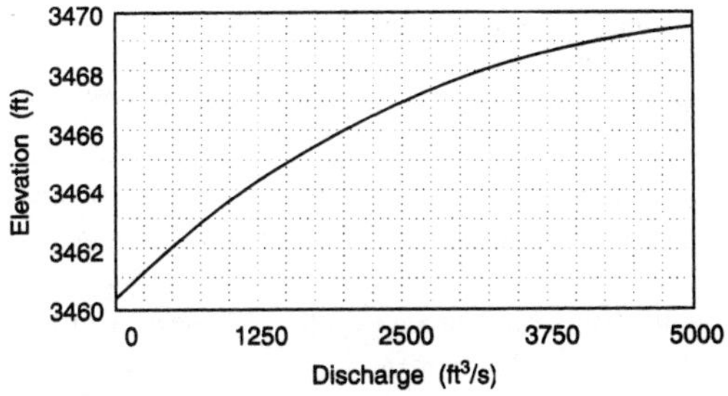


**Figure D-07.** Discharge curve (courtesy CRC Press).

**discharge head** A measure of the pressure exerted by a fluid at the point of discharge; for example, the vertical distance between the intake level of a water or wastewater pump and the level of discharge. Also called dynamic discharge head. *See* dynamic head.

**discharge measurement** or **flow measurement** The determination of the discharge in a canal, conduit, stream, orifice, etc., using such devices as a current meter, weir, or Pitot tube. Direct determinations use measurements of volume or weight in a given time, as in a positive displacement meter. In indirect determinations, a flow characteristic such as pressure or water surface elevation is measured and used in a formula to compute flow rate.

**discharge nearfield** A region close to the port of discharge of wastewater or effluent into the ocean or a large lake, where the effluent forms a buoyant, rapidly rising plume that entrains ambient water for dilution until the plume density is about equal to that of ambient water. Also called initial mixing region. *See also* initial dilution, discharge farfield, transition region, wastewater plume (Metcalf & Eddy, 1991).

**discharge of a pollutant** Any addition of any pollutant to navigable waters from any point source; any addition of any pollutant to the waters of the contiguous zone or the ocean from any point source other than a vessel or other floating craft that is being used as a means of transportation (EPA-40CFR122.2)

**discharge permit** A formal, written authorization by the USEPA or a state regulatory agency to discharge wastewater effluents into the environment. *See* NPDES.

**discharge piping** The piping or header through which water discharges at the outlet side of a pump or reservoir.

**discharge rate** Flow rate; the volume of water or wastewater that flows out of a pump, structure, etc. over a given period of time, expressed in gallons per day, cubic feet per second, million gallons per day, or corresponding metric units.

**discharge standard** *See* effluent standard.

**discharge valve** A valve installed on the outlet side of a pump or hydraulic structure to control the release of water.

**disc-nozzle centrifuge** A device that uses centrifugal or rotational forces to separate solids from liquids, e.g., in sludge thickening or dewatering. Because of its clogging potential, wear problems, and inefficient solids capture, this device (as well as the basket centrifuge) is now installed less often than solid-bowl centrifuges (WEF & ASCE, 1991).

**Discor** A disc dryer manufactured by Andritz-Ruthner, Inc.

**Discostrainer** Fine screen manufactured by Hycor Corp.

**Discotherm** A thermal sludge processor manufactured by LIST, Inc.

**discount rate** The interest used in cost-effectiveness, benefit–cost, and similar analyses to discount future benefits and costs and obtain their present worth. It usually includes an inflation component and a real return-on-investment component. *See* financial indicators for related terms.

**Disc-Pak** A cartridge filter manufactured by Alsop Engineering Co.

**Discreen®** A screening apparatus manufactured by Ingersoll-Dresser Pump as a rotating disc.

**discrete dynode detector** An array of electrodes or dynodes, used increasingly as an electron multiplier in mass spectrometers instead of channel electron multiplier detectors.

**discrete particles** Suspended solids such as grit, sand, or suspended metal scales or particles that settle easily and independently of one another.

**discrete particle settling** The phenomenon referring to sedimentation of particles in a suspension of low solids concentration; the settling of discrete, nonflocculent particles, as separate, unhindered units. *See* sedimentation type I for more detail.

**discrete sedimentation** *See* sedimentation type I.

**discrete settling** The settling of discrete, nonflocculent particles, as separate, unhindered units. *See* sedimentation type I for more detail.

**disc screen** A circular disc equipped with wire mesh, rotating on a horizontal axis for wastewater screening.

**Disc-Tube™** A reverse osmosis apparatus manufactured by Rochem Separation Systems.

**disease** A disorder in or incorrect functioning of an organ, part, or structure of the body, often with a characteristic train of symptoms; a change from a state of health resulting from infections, poisons, toxicity, unfavorable environmental factors, etc.

**disease-causing bacteria** Pathogenic bacteria that can cause disease by infection.

**disease reservoir** An active, passive, or inanimate host in which a pathogen is found.

**disease vectors** Rodents, flies, and mosquitoes capable of transmitting disease to humans.

**Di-sep** Water treatment units manufactured by Smith & Loveless, Inc. using membrane filtration.

**disequilibrium index** An index reflecting the equilibrium pH of a water with respect to calcium and alkalinity. This index is used in stabilizing water to control both corrosion and the deposition of scale. *See* saturation index for detail (AWWA, 1999).

**dishpan hand** A hand that is rough and irritated, as from dishwashing products and strong cleaning agents; this causticity condition affecting the skin of sensitive persons, sometimes chronic, may be caused by mineralized or hard waters that require much soap (Fair et al., 1971).

**disinfectant** Any physical or chemical agent that is added to water or wastewater in any part of the distribution or treatment process, and is intended to kill or inactivate pathogenic organisms. Disinfectants are also used in wells, swimming pools, and other bathing waters. *See* disinfecting agent.

**disinfectant contact time ("T" in CT calculations)** The time in minutes that it takes for water to move from the point of disinfectant application or the previous point of residual measurement to a point before or at the point where residual disinfectant concentration (C) is measured. In pipelines, the time is calculated by dividing the internal volume of the pipe by the maximum hourly flowrate. Within mixing basins and storage reservoirs it is determined by tracer studies or an equivalent demonstration.

**disinfectant/disinfection by-product** A disinfectant or disinfection by-product regulated by the USEPA.

**Disinfectant/Disinfection By-products Rule** A primary regulation issued by the USEPA for disinfectants and disinfection by-products in drinking water. *See also* Disinfection By-products Rule.

**disinfectant effectiveness** The percent reduction or inactivation of microorganisms produced by a concentration (C) of a disinfectant in a contact time (T). *See* C · T product, disinfecting efficiency, log reduction, log survival.

**disinfectant performance variables** Factors that affect the effectiveness of disinfection include contact time, concentration of a chemical disinfectant, intensity and nature of a physical agent, temperature, type of organisms, and nature of suspending liquid (e.g., natural water, distilled water, wastewater).

**disinfectant-resistant pathogen** Same as disinfection-resistant pathogen.

**disinfectant time** The time it takes water to move from the point of disinfectant application (or the previous point of residual disinfectant measurement) to a point before or at the point where the residual disinfectant is measured. In pipelines, the time is calculated by dividing the internal volume of the pipe by the maximum hourly flowrate; within mixing basins and storage reservoirs it is determined by tracer studies of an equivalent demonstration. *See also* disinfectant contact time.

**disinfected wastewater** Wastewater that has been treated by a disinfecting agent such as chlorine or ultraviolet light.

**disinfecting agent** Any chemical or physical agent that is added to water or wastewater in any part of the distribution or treatment process, and is intended to kill or inactivate pathogenic organisms. Disinfectants are also used in wells, swimming pools, and other bathing waters. Disinfecting agents include oxidizing chemicals (chlorine, chlorine dioxide, chloramines, bromine, iodine, ozone, potassium permanganate, hydrogen peroxide, Javelle water, surfactants), metal ions (copper, silver), alkalis, acids, surfactants, and physical agents (heat, radiation, and ultraviolet light). *See* Table D-04 for a comparison of five common disinfecting agents.

**disinfecting chemical** Common disinfecting chemicals include chlorine, chloramines, chlorine dioxide, ozone. *See* disinfecting agent.

**disinfecting efficiency** The degree to which a disinfecting agent reduces the number of indicator organisms (e.g., coliforms) present in the water being treated. *See also* disinfectant effectiveness.

**disinfection** A chemical or physical process designed to kill most microorganisms in water or wastewater, essentially all pathogenic bacteria, viruses, fungi, and protozoa cysts. For surface waters, the USEPA also requires 99.9 and 99.99% inactivation of *Giardia* cysts and viruses, respectively. Chlorine is one of the most frequent disinfectants in water treatment. Disinfection is also used in wastewater treatment. Other disinfection methods commonly used in water and wastewater treatment include ozonation, ultraviolet irradiation, heat, ultrasonic waves, metallic ions, and oxidation by potassium permanganate. In addition to the inactivation of pathogens, disinfection may be construed to include their physical removal, as recognized by the USEPA through credit for solids separation processes. *See also* boil water order, Chick–Watson equation, chlorination, dilution constant, disinfection kinetics, inactivation constant, log survival, ozonation, sterilization, ultraviolet radiation.

**disinfection alternatives** Alternatives to the common disinfecting methods or agents (chlorine and ozone, e.g.) include ultraviolet light, peracetic acid, peroxone, and combined chemical disinfection (i.e., the sequential or simultaneous use of two or more disinfectants). *See also* interactive disinfection.

**disinfection by-product (DBP)** (1) Under the Safe Drinking Water Act, a compound formed by the reaction of a disinfectant (such as chlorine, chlorine dioxide, chloramines, and ozone) with natural organic material (NOM) in the water supply at very low concentrations, expressed in micrograms per liter or even nanograms per liter. Halogen-substituted by-products are produced by the substitu-

**Table D-04.** Disinfecting agents

| Disinfectant | Advantages | Disadvantages |
|---|---|---|
| Chloramines | No current disinfection by-products (DBPs); sustainable residual; possible inactivation of *Cryptosporidium.* | Possible concerns with nitrification, some DBPs, NDMA, CNCl; ineffective against viruses and *Cryptosporidium.* |
| Chlorine (gas) | Facultative onsite generation; removal of iron and manganese. | Formation of DBPs; safety concern for gas storage; ineffective against *Cryptosporidium.* |
| Chlorine dioxide | No DBPs; inactivation of viruses and *Giardia* cysts; iron and manganese removal | Concerns with chlorite and chlorate; potential for taste and odor problems at customer's tap. |
| Ozone | No DBPs; effective inactivation of viruses, *Cryptosporidium,* and *Giardia;* oxidation of taste, odor, color, synthetic organic chemicals. | Concerns with microbial regrowth and bromate formation; no measurable or sustained residual; *Cryptosporidium* inactivation requires long contact time at low temperatures. |
| UV light | No regulated DBPs; inactivation of viruses, *Cryptosporidium,* and *Giardia.* | No measurable or sustained residual; limited inactivation of adenovirus. |

Adapted from Long et al. (2005).

tion of chlorine or bromine for hydrogen atoms. Other DBPs, e.g., aldehydes, result from the oxidation of NOM. Some DBPs may cause cancer and other toxic effects. DBPs are hydrophilic and have low molecular weights, which makes them difficult to remove by physicochemical processes. Common organic DBPs include (a) trihalomethanes (chloroform, bromodichloromethane, dibromochloromethane, and bromoform); (b) haloacetic acids (particularly dichloroacetic acid and trichloroacetic acid); (c) haloacetaldehydes (chloroacetaldehyde, dichloroacetaldehyde, and trichloroacetaldehyde or chloral hydrate); (d) formaldehyde; (e) haloketones (tri-, tetra-, penta-, and hexa-chloroacetones); (f) haloacetonitriles (chloroacetonitrile, trichloroacetonitrile, bromochloroacetonitrile, and dibromoacetonitrile); (g) chloropicrin (also called trichloronitromethane or nitrochloroform); (h) cyanogen chloride; (i) chlorophenols (monochlorophenol, dichlorophenol, and trichlorophenol); and (j) 3-chloro-4-(dichloromethyl)-5-hydroxy-2(5h)-furanone (MX). (2) Disinfectants also form inorganic byproducts, including residual free chlorine, chloramines, chlorine dioxide, chlorite, chlorate, bromide, and bromate.

**disinfection by-product (DBP) formation potential** An indirect measure of the amount of DBP precursors present in a water sample, obtained by dosing the sample with a large amount of disinfectant and measuring the DBPs formed under conditions favorable to maximum production (pH, temperature, contact time). The formation potential is the difference between the levels of DBP after and before the test.

**disinfection by-product precursor (DBP precursor)** Natural organic matter in source water that may combine with chlorine and other agents to form disinfection by-products. DBPs include trihalomethanes, haloacetic acids, haloacetonitriles, haloketones (HKs), chloral hydrate (CH), and chloropicrin (CP). Precursor materials can be reduced by conventional treatment or such innovative approaches as riverbank filtration.

**disinfection by-product (DBP) regulatory assessment model** An empirical computer model developed by the USEPA to assess DBP formation under the Disinfectant/Disinfection By-products Rule.

**Disinfection By-products (DBP) Rule** A primary regulation issued by the USEPA for disinfection by-products in drinking water. Phase 1, finalized in 1998, required enhanced coagulation as a strategy for removal of natural organic matter, provided alternatives to this treatment technique, and established maximum contaminant levels (MCL) for DBPs. Phase 2 revised these recommendations. *See also* Disinfectant/Disinfection By-products Rule.

**disinfection goals** The removal or inactivation of pathogenic agents (primary disinfection) and the provision of a disinfectant residual to prevent the regrowth of pathogens in the distribution system (secondary disinfection).

**disinfection kinetics** The rate or efficiency of disinfection depends on the time of contact ($t$), the concentration or intensity of the disinfecting agent ($C$), the number or concentration of organisms ($N$), and temperature ($T$). *See* chemical-similitude theory, Chick–Watson model, Collins–Selleck model, delayed Chick–Watson model, dilution constant, Gard model, Hom model, inactivation constant, microbial diversity concept, modified Chick–Watson model, modified Collins–Selleck model, modified Gard model, threshold CT, vitalistic theory.

**disinfection mechanisms** The various ways by which disinfecting agents destroy or inactivate microorganisms: (a) damage to or destruction of the cell wall, resulting in lysis and death; (b) alteration of the permeability of the cytoplasmic membrane, allowing vital nutrients to escape; (c) alteration of the nature of the protoplasm; (d) alteration of the DNA or RNA and disruption of the reproduction process; and (e) inhibition and inactivation of enzymes.

**disinfection-resistant pathogen** A bacterium or other microorganism that is not inactivated by ordinary disinfectant doses, e.g., bacterial spores, *Mycobacterium, Giardia* and *Cryptosporidium* cysts and oocysts, and spores of fungi.

**disinfestation** The elimination or reduction of insects, rodents, etc. from a person, clothing, or in the environment of an individual, or on domestic animals.

**disintegration** Decay; the breaking down of a substance into its components.

**disk aerator** A submerged horizontal-axis aerator, similar to a surface aerator but with a disk attached to a rotating shaft to agitate the liquid. *See also* Kessener brush aerator, paddle aerator, mechanical aerator.

**disk diffuser** A rigid ceramic disk mounted on an air distribution pipe near the floor of an aeration basin. Other types of porous diffusers used in wastewater aeration include dome, membrane, and panel. *See also* nonporous diffuser.

**disk filter** *See* Cloth-Media Disk Filter®.

**disk meter** A positive displacement water meter used to register flow in small pipes by a nutating disk.

**disk screen** A circular disk that rotates about a perpendicular, central axis.

**disk-type meter** *See* disk meter.

**disodium ethylenediaminetetraacetate ($Na_2EDTA$)** A chemical compound that forms stable complex ions with calcium, magnesium, and other hardness-causing cations. It is used as a reagent in testing for hardness in water.

**disodium phosphate** Sodium phosphate that occurs (a) as a white, soluble, anhydrous powder ($Na_2HPO_4$), used in ceramics, baking powder, etc., or (b) as clear colorless, hydrated crystals ($Na_2HPO_4 \cdot x\ H_2O$) used in detergents, fertilizers, and pharmaceutical products. Also called dibasic sodium phosphate. *See also* sodium phosphate and trisodium phosphate.

**dispersant** (1) A chemical agent used to break up concentrations of organic material such as spilled oil, or to prevent the agglomeration of particles. Some antiscalants can surround particles of suspended salts or organic solids and serve as dispersants. (2) A surface-active agent or any admixture used to promote and stabilize a dispersion.

**dispersed-air flotation** A treatment process used to remove materials with hydrophobic surfaces (such as emulsified oil, suspended solids, low-density particles) from process or high-volume wastewaters by forming large, dispersed gas bubbles (500–1000 micron diameter) in the solution. These attach to the oil and solid particles and cause them to rise as dense froth to the surface, where they are removed by skimming paddles. The gas bubbles are generated by mechanical shear of propellers, diffusion of gas through porous media, or by homogenizing a gas and liquid stream. Because of their compact size, dispersed-air flotation units have low capital costs, but they have high power requirements. Sometimes called induced-air flotation. Two widely used forms of dispersed flotation used in the mineral industry are foam flotation and froth flotation. *See also* dissolved air flotation, Ozoflot, vacuum flotation.

**dispersed growth** A type of clarification failure characterized by a turbid effluent and no distinct solids–liquid interface in the clarifier; bioflocculation is insufficient because of an excess of nonflocculating or filamentous organisms, the presence of surfactants and toxic compounds, and short solids retention times. *See also* dispersed growth, filamentous bulking, pin floc, and viscous bulking.

**dispersed-growth aeration** A biological process used to treat industrial wastes containing a high concentration of dissolved organic matter in the absence of flocculent growths. After aeration and settling, a portion of the supernatant liquor is retained for seeding incoming wastewater and the settled sludge from the secondary clarifier is processed for disposal (Nemerow & Dasgupta, 1991).

**dispersed hepatitis A virus** A preparation of homogeneously dispersed, instead of clumping, virus particles.

**dispersed remnant and filter media conditioning** A stage of the filter-ripening sequence in which newly attached particles improve filtration performance by collecting other particles.

**dispersing agent** A material that increases the stability of particles in a liquid; a dispersant.

**dispersion** (1) The process by which a substance or chemical spreads and dilutes in flowing water or gas. It is the net result of a number of mass transfer phenomena. The scattering and mixing are caused by turbulence, molecular diffusion, and by differences in pore velocity and in flow paths. *See also* diffusion and dispersion. (2) The separation of colloidal or other small particles. (3) The phenomenon of short-circuiting in a water or wastewater treatment tank.

**dispersion coefficient** (1) A parameter that accounts for the effects of dispersion due to differences in fluid velocity, turbulent eddies, and molecular diffusion. In wastewater treatment reactors, the coefficient of dispersion varies between $10^2$ and $10^6$ $cm^2/sec$. *See also* eddy diffusion coefficient, molecular diffusion coefficient. (2) For the one-dimensional modeling of wastewater disposal in estuaries and rivers, the coefficient of longitudinal dispersion ($D_L$, $ft^2/sec$) may be estimated as (Metcalf & Eddy, 1991):

$$D_L = 0.011\, U^2B^2/HU^* \qquad \text{(D-34)}$$

where $U$ = stream velocity averaged over the cross section, fps; $B$ = stream width, ft; $H$ = stream depth, ft;

$$U^* = \text{shear velocity, fps} = (gHS)^{1/2}; \qquad \text{(D-35)}$$

$g$ = gravitational acceleration, ft/sec; and $S$ = stream slope, ft/ft.

**dispersion factor** (1) The ratio of the increase in the ground level ambient air concentration for a pollutant at or beyond the property line of the site where the sewage sludge incinerator is located to the mass emission rate for the pollutant from the incinerator stack (EPA-40CFR503.41-d). (2) Same as dispersion number.

**dispersion force** *See* London-van der Waals force.

**dispersion index** A measure of short-circuiting of a liquid through a continuous-flow tank. *See* Morrill index, volumetric efficiency, time–concentration curve.

**dispersion number (*d*)** A dimensionless number that gives an idea about the degree of axial dispersion; the reciprocal of the Peclet number ($P_e$):

$$d = 1/P_e = D/u \cdot L \qquad \text{(D-36)}$$

where is $D$ a coefficient of axial dispersion in square feet per second or square meters per second, $u$ is the fluid velocity in fps or meters per second, and $L$ is a characteristic length in feet or meters. At one extreme, there is no dispersion in an ideal plug-flow reactor ($d = 0$); at the other, dispersion is extremely high, as in a complete-mix basin ($d \to \infty$). Dispersion is considered low or high when $d < 0.05$ or $d > 0.25$, and moderate in between, i.e., $0.05 < d < 0.25$. Also called a dispersion factor. *See also* dispersion coefficient.

**dispersion rate** A parameter used in the analysis of diffusion of plumes or effluents.

**dispersion symmetry index (DSI)** A parameter that indicates whether the dispersion curve for a reactor is symmetrical about the mode (Droste, 1997):

$$\text{DSI} = (T_{90} - T_P)/(T_P - T_{10}) \qquad \text{(D-37)}$$

where $T_P$ is the time of maximum concentration (mode), $T_{90}$ is the time of passage of 90% of the tracer, and $T_{10}$ is the time of passage of 10% of the tracer. The index is equal to 1 for a symmetrical curve.

**displacement ejector** An ejector that lets a liquid accumulate in a pressure chamber and displaces it by compressed air when the liquid level reaches a certain elevation.

**displacement meter** A water meter used to measure low flows based on the filling and emptying of a container of known volume.

**displacement period** Same as displacement time.

**displacement pump** A type of pump that moves water through the alternate filling and emptying of the pump chamber by the movement of a piston or similar device. *See also* positive displacement pump.

**displacement time** The average length of time a moving liquid is detained or held in a tank, channel, or other structure; it is equal to the ratio of active volume to flow rate. Also called displacement period. *See also* detention time.

**displacement velocity** The rate at which the incoming liquid displaces the contents of a sedimentation tank.

**disposable component** A component designed to be thrown away after use instead of being repaired or reused, e.g., a cartridge filter element or a disposable resin.

**disposable IX resin** A resin that is used in the ion-exchange process and that is disposed of as a solid waste without regeneration. It is effective in the removal of perchlorate and other difficult contaminants.

**disposal** Final placement or destruction of toxic, radioactive, or other wastes; surplus or banned pesticides or other chemicals; polluted soils; and drums containing hazardous materials from removal actions or accidental releases. Disposal may be accomplished through the use of secure landfills, surface impoundments, land farming, deep-well injection, ocean dumping, or incineration. *See also* deep-well disposal, ocean disposal, ultimate disposal, reuse, injection, percolation, evaporation.

**disposal bed** A disposal field with a bottom width greater than 3–4 ft. Also called seepage bed. *See* soil absorption system for detail.

**disposal by dilution** The discharge of treated or untreated wastewater into a receiving body of water, involving natural purification and dispersion.

**disposal facility** A facility or part of a facility at which hazardous waste is intentionally placed into or on any land or water, and at which waste will remain after closure (EPA-40CFR260.10 & 270.2).

**disposal field** A subsurface area containing perforated pipes laid in gravel-lined, looped or lateral trenches or in a bed of clean stones, through which treated wastewater may seep into the surrounding soil for further treatment and disposal. *See* soil absorption system for detail.

**disposal field biomat** In a subsurface-soil absorption system, the layer of biological material formed by microorganisms at the interface between the soil and the effluent particulates. It is a primary factor controlling the soil infiltration capacity, more so than soil permeability. The biomat grows to an equilibrium thickness, serving as an effective biological filter.

**disposal field dosing** The periodic application of septic tank effluent to a soil absorption system, using a pump or a dosing siphon.

**disposal pit** A covered excavation that is used to introduce, by seepage through the bottom and sides, effluents of septic tanks or other partial treatment devices from small installations. *See* seepage pit for detail. *See also* soil absorption system.

**disposal SI** Disposal surface impoundment.

**disposal site** (1) That portion of the "waters of the United States" in which specific disposal activities are permitted and consist of a bottom surface area and any overlying volume of water. In the case of wetlands on which surface water is not present, the disposal site consists of the wetland surface area (EPA-40CFR230.3-i). (2) An interim or finally approved and precise geographical area within which ocean dumping of wastes is permitted under conditions specified in issued permits. Such sites are identified by corner coordinates, or by center point coordinates of the radius from that point (EPA-40CFR228.2-a). (3) The region within the smallest perimeter of residual radioactive material (excluding cover materials) following completion of control activities (EPA-40CFR192.01-d).

**disposal surface impoundment (SI)** An excavated or diked area used to contain liquid wastes. *See also* storage SI, treatment SI.

**disposal well** (1) A well used for the disposal of waste into a subsurface (EPA-40CFR146.3). (2) A shallow well used for the disposal of surface runoff and treated water into an aquifer, which purifies the liquid before subsequent withdrawal. *See also* aquifer storage and recovery well.

**disposition of complaint** The conclusion of the investigation of a customer complaint.

**Disposorbs** Disposable adsorption units manufactured by Calgon Carbon Corp.

**disproportionation** The simultaneous oxidation and reduction of a substance reacting with itself and forming two different molecules, e.g., the disproportionation of chlorine dioxide ($ClO_2$) into chlorite ($ClO_2^-$) and chlorate ($ClO_3^-$) under alkaline conditions, that of ethylene ($C_2H_4$) into acetylene ($C_2H_2$) and ethane or dimethyl ($C_2H_6$), and that of aqueous chlorine [$Cl_2^0$ (aq)] into hypochlorous acid (HOCl) and chloride ($Cl^-$):

$$2\,ClO_2 + 2\,OH^- \rightarrow ClO_2^- + ClO_3^- + H_2O \quad \text{(D-38)}$$

$$2\,C_2H_4 \rightarrow C_2H_2 + C_2H_6 \quad \text{(D-39)}$$

$$Cl_2^0\,(aq) + H_2O \rightarrow HOCl + Cl^- + H^+ \quad \text{(D-40)}$$

(2) The biological conversion of a substrate into its oxidized form (carbon dioxide, $CO_2$) and its most reduced form (methane, $CH_4$):

$$2\,CH_2O \rightarrow CO_2 + CH_4 \quad \text{(D-41)}$$

**Dissanayake's model** An empirical, first-order equation proposed in 1980 to describe the removal of fecal bacteria in stabilization ponds (Feachem et al., 1983). The first-order rate constant ($K$, /day) is expressed in terms of temperature ($T$, °C), the average concentration of algae in the pond ($C$, mg/L), and the organic loading on the pond ($P$, kg COD/ha/day):

$$K = 0.7716(1.0281)^T(1.0016)^c(0.999)^P \quad \text{(D-42)}$$

**dissimilar metals** Two different metals in contact, e.g., through such a conducting solution as water, may constitute a galvanic cell, with the possibility of corrosion of the anodic metal (the one with the higher electromotive force). The same result may occur in a single metal that has areas of different oxidation potentials. *See* corrosion cell, sacrificial anode.

**dissimilating nitrate reduction** Same as dissimilatory nitrate reduction.

**dissimilation** The breakdown of reduced compounds into simpler substances by organisms to obtain energy. Same as catabolism.

**dissimilatory nitrate reduction** The use of nitrate by microorganisms as a terminal electron acceptor in the oxidation of organic matter. In one pathway, facultative organisms produce energy to drive the oxidation, with ammonium as end product. The other pathway is biological denitrification, with nitrogen gas as end product. Also called dissimilating nitrate reduction (Maier et al., 2000 and Metcalf & Eddy, 2003)

**dissimilatory sulfate reduction** One of two forms of anaerobic sulfur reduction whereby sulfate-reducing bacteria use sulfate ($SO_4^{2-}$) as terminal electron acceptor, hydrogen ($H_2$) as electron donor and a low-molecular-weight compound such as acetate ($CH_3COOH$) or methanol ($CH_3OH$) as a carbon source (Maier et al., 2000):

$$4\,CH_3OH + 3\,SO_4^{2-} \rightarrow 4\,CO_2 + 3\,S^{2-} + 8\,H_2O \quad \text{(D-43)}$$

*See also* assimilatory sulfate reduction and sulfur respiration.

**dissociation** (1) The reversible breaking up of a chemical combination into simpler constituents without a change in valence, e.g., the decomposition of water into hydrogen and oxygen. *See* ionization. (2) *See* electrolytic dissociation.

**dissociation constant** A quantity that describes the tendency of a molecule to ionize in a solution or the rate at which a compound dissociates into its constituent species, e.g., the rate of dissociation of an acid (HA) into its carboxylate anion ($A^-$) and the hydrogen ion ($H^+$):

$$HA_{aq} \leftrightarrow H^+ + A^- \quad \text{(D-44)}$$

In an aqueous solution at equilibrium, the dissociation constant for this reaction ($K_a$) is the ratio of the product of the molar concentrations ([$H^+$] and

[$A^-$]) of the products (ions) to the molar concentration [HA] of the reactant (nonionized electrolyte):

$$K_a = [H^+][A^-]/[HA] \quad \text{(D-45)}$$

The dissociation constant increases with increased ionization and is larger for strong acids (weak bases) than weak acids (strong bases). The constant is usually expressed in logarithmic terms:

$$pK_a = -\log_{10} K_a \quad \text{(D-46)}$$

*See* Table D-05 for some dissociation constants. *See also* affinity constant, apparent dissociation constant, association constant, ionization constant, equilibrium constant, Michaelis constant.

**dissociation degree** A number that indicates to what extent a substance dissociates in water. It is the ratio of the molar concentration of the substance produced to the analytical concentration (total number of moles of the pure substance). It may be computed by considering the concentration condition and the equilibrium constant (Stumm, 1996). For example, for an aqueous solution of $C = 10^{-3}$ M boric acid $B(OH)_3$, in addition to $H_3O^+$ and $OH^-$ the species produced is $B(OH)_4^-$, abbreviated $B^-$, and the degree of dissociation ($\alpha_1$) is

$$\alpha_1 = [B^-]/C = K/(K + [H^+]) \quad \text{(D-47)}$$

where the square brackets indicate molar concentrations and K is the equilibrium constant. Also called ionization fraction or degree of protolysis. *See* distribution diagram.

**dissociation in water** Ionization; the separation of molecules in water into ions.

**dissolution** Dissolving of a substance in a liquid solvent (e.g., water).

**dissolution reaction** The reversible reaction of the separation of a molecule into positive and negative ions; generally, for a solid compound [$A_xB_y(s)$]:

$$A_xB_y(s) = x\,A^{y+} + y\,B^{x-} \quad \text{(D-48)}$$

The solubility product constant ($K_{sp}$), similar to the dissociation constant, is based on this equation:

$$K_{sp} = [A^{y+}]^x[B^{x-}]^y \quad \text{(D-49)}$$

**dissolved** Characteristic of elements that pass through a 0.45 μm filter.

**dissolved-air flotation (DAF)** A process used in the treatment of wastewater (removal of suspended solids and grease) and in sludge processing. Flocculated material in contact with small gas bubbles (50–100 micron diameter) in a supersaturated solution results in the air/floc mass being buoyed to the surface, leaving behind a clarified water. It is also used in the clarification of flocculated water, e.g., surface waters with significant algal blooms or with low-density solids. There are three types of DAF, with bubbles produced by reducing the pressure of an air-saturated water stream: microflotation, pressure flotation, and vacuum flotation. Concepts related to dissolved-air flotation include Henry's law, bulk density, Archimedes' principle, buoyancy, and Stokes' law. *See also* dispersed air flotation, electrolytic flotation or electroflotation, flotation thickening. *See also* Figures D-08 and D-09.

**dissolved-air flotation thickening** The use of dissolved-air flotation to thicken sludge from wastewater treatment. *See* flotation thickening for detail.

**dissolved chemical oxygen demand (sCOD)** The soluble portion of the chemical oxygen demand of wastewater, as compared to the particulate fraction (which includes colloidal and suspended solids). Also called soluble COD. *See* COD fractionation.

**dissolved constituent** *See* dissolved solids, dissolved gases.

**dissolved gases** (1) Gases to which water is exposed, particularly nitrogen ($N_2$), oxygen ($O_2$), carbon dioxide ($CO_2$), hydrogen sulfide ($H_2S$), and methane ($CH_4$), all of which can be removed by aeration or other processes. (2) The sum of dissolved gaseous components. High concentrations

**Table D-05.** Dissociation constants at 25°C

| Substance | Equilibrium reaction | pK | Importance in water chemistry |
|---|---|---|---|
| Acetic acid | $CH_3COOH \leftrightarrow H^+ + CH_3COO^-$ | 4.7 | Organic wastes, anaerobic digestion |
| Carbonic acid | $H_2CO_3^* \leftrightarrow H^+ + HCO_3^-$ | 6.4 | Various |
| Hydrochloric acid | $HCl \leftrightarrow H^+ + Cl^-$ | –3 | Analyses |
| Hydrogen sulfide | $H_2S \leftrightarrow H^+ + HS^-$ | 7.0 | Odor, corrosion, anaerobic digestion |
| Hypochlorous acid | $HOCl \leftrightarrow H^+ + OCl^-$ | 7.5 | Disinfection |
| Phosphoric acid | $H_3PO_4 \leftrightarrow H^+ + H_2PO_4^-$ | 2.1 | Buffer, nutrient |
| Sodium hydroxide | $NaOH \leftrightarrow Na^+ + OH^-$ | — | Analyses, neutralization |
| Sulfuric acid | $H_2SO_4 \leftrightarrow H^+ + HSO_4^-$ | –3 | Coagulation, pH adjustment |

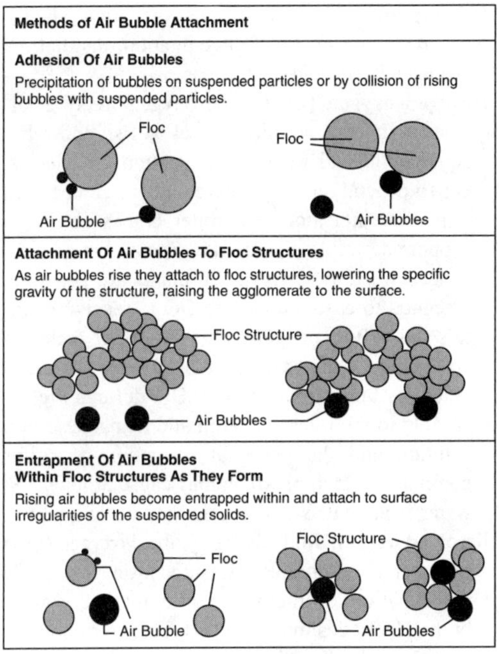


**Figure D-08.** Dissolved-air flotation (bubble attachment). (Courtesy Eimco Water Technologies.)

of such gases may cause operational problems in water and wastewater facilities, e.g., air binding in filters and cavitation in pumps.

**dissolved inorganic carbon (DIC)** The sum of all dissolved carbonate-containing species. When the concentrations of complexes like Ca and Mg are negligible,

$$DIC = [CO_3^{2-}] + [HCO_3^-] + [H_2CO_3^*] \quad \text{(D-50)}$$

where the concentrations are in mol/L and $[H_2CO_3^*]$ represents the sum of dissolved carbon dioxide gas and carbonic acid (AWWA, 1999). DIC can also be determined as a function of pH and alkalinity with a given temperature and ionic strength. Figure D-10.

**dissolved inorganics removal** The application of an advanced treatment method to remove ammonia ($NH_3$), nitrate ($NO_3$), phosphorus (P), and total dissolved solids (TDS) from water or wastewater. Common methods include chemical precipitation, distillation, electrodialysis, ion exchange, reverse osmosis, and ultrafiltration.

**dissolved load** The portion of a stream load that consists of weathered rock constituents carried in chemical solution by moving water.

**dissolved matter** The portion of total matter that passes through a 0.45-micrometer-pore-diameter filter. *See also* particulate matter.

**dissolved metal** A metallic element of an unacidified sample that passes through a 0.45-μm membrane filter. *See also* acid extractable metal, suspended metal, total metal.

**dissolved minerals** The portion of dissolved solids that is made up of inorganic salts in solution in water or wastewater.

**dissolved nitrogen flotation** A variation of the dissolved flotation process that uses nitrogen instead of air to cause the suspended particles to rise to the surface.

**dissolved organic bromine** A surrogate measurement of the bromine-substituted organic matter in water or wastewater; it is caused by synthetic organic chemicals in raw water or by disinfection by-products and similar substances in treated water.

**dissolved organic carbon (DOC)** The fraction of total organic carbon that is dissolved in a water sample; operationally, the organic carbon that passes through a 0.45-μm filter; a measure of total dissolved organic matter and an indication of the disinfection by-product precursors in a water sam-

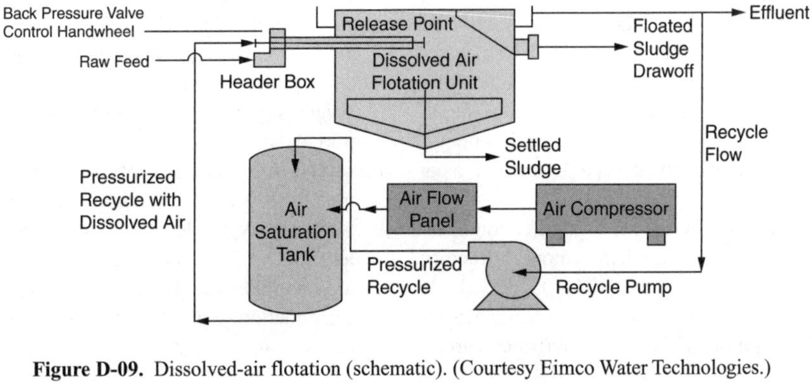


**Figure D-09.** Dissolved-air flotation (schematic). (Courtesy Eimco Water Technologies.)

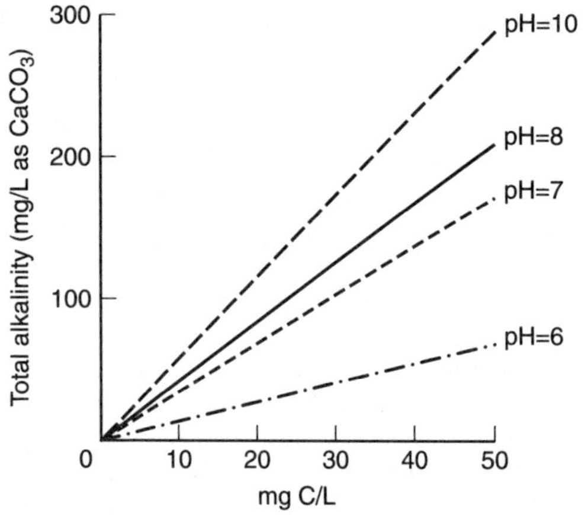


**Figure D-10.** Dissolved inorganic carbon (adapted from AWWA, 1999).

ple. Aquatic natural organic matter is the main DOC source in natural waters. On the average, groundwater has a DOC less than 2.0 mg C/L and surface water has a DOC between 2 and 10 mg C/L. The USEPA proposes enhanced coagulation as an appropriate technology for DOC removal. *See also* particulate organic carbon, UV254 and SUVA254.

**dissolved organic chlorine (DOCl)** A surrogate measurement of the dissolved chlorine-substituted organic matter in water or wastewater; it is caused by synthetic organic chemicals in raw water or by disinfection by-products and similar substances in treated water.

**dissolved organic halogen (DOX)** A surrogate measurement for the total amount of organic compounds that contain one or more halogen atoms in a sample of raw or treated water. *See* halogen-substituted organic material, total organic halide analysis, and total organic halogen for more detail.

**dissolved organic halogen formation potential** *See* total organic halogen formation potential.

**dissolved organic matter (DOM)** The fraction of total organic matter that is dissolved in a water sample; operationally, the organic matter that passes through a 0.45-μm filter. It consists of amino acids, carbohydrates, organic acids, and nucleic acids. *See* dissolved organic carbon as a practical measure.

**dissolved organics removal** The application of an advanced treatment method to remove refractory organics, total organic carbon (TOC), and volatile organic compounds (VOC) from water or wastewater. Common methods include activated carbon adsorption, advanced chemical oxidation, chemical oxidation, chemical precipitation, distillation, electrodialysis, and reverse osmosis.

**dissolved oxygen (DO)** The oxygen freely available in aqueous solutions, vital to fish and other aquatic life and for odor prevention. DO levels, which depend on temperature and pressure, are an important indicator of a water body's ability to support desirable aquatic life. Secondary and advanced wastewater treatment plants are generally designed to ensure adequate DO in receiving waters. Dissolved oxygen is an important factor in corrosivity; it increases the rate of many corrosion reactions but inhibits others. DO deficits are favorable to the release of iron and manganese into solution and the potential formation of sulfur compounds, both of which may cause problems in water treatment. *See also* irrigation water quality.

**dissolved oxygen analysis** Dissolved oxygen (DO) in water is commonly measured using the azide modification of the iodometric method. Membrane electrodes may also be used; see dissolved-oxygen probe. The DO test is used in the determination of biochemical oxygen demand.

**dissolved oxygen deficit** The difference, at a given temperature, between the saturation concentration of dissolved oxygen (DO) and the actual DO concentration in water. *See also* oxygen transfer rate and dissolved-oxygen sag analysis.

**dissolved oxygen probe** A device consisting of two metal electrodes, with one end containing a potassium chloride solution and covered with a polyethylene membrane; the other end is in contact with a salt solution. It is used to measure the dissolved oxygen concentration in a water sample separated from the salt solution by a selective membrane.

**dissolved-oxygen sag** The decline in the concentration of dissolved oxygen (DO) downstream from a point of pollution (e.g., discharge of treatment plant effluent or untreated wastewater) in a stream until the DO deficit reaches a maximum; subsequently, the DO level recovers. The decline and recovery result from the interplay of microbial activity, deoxygenation, and reaeration. Same as oxygen sag. *See* dissolved-oxygen sag analysis for more detail.

**dissolved-oxygen sag analysis** When wastewater or effluent is discharged into a receiving stream, it exerts an oxygen demand along with other oxygen demands, at the same time that atmospheric reaeration supplies oxygen to the stream. The oxygen sag equation represents the interplay of deoxygenation and reaeration. The classic sag curve, derived by H. W. Streeter and E. B. Phelps in 1925, neglects the effects of photosynthesis, respiration, and sediment oxygen demand:

$$D = [KL_a/(R - K)](e^{-KT} - e^{-RT}) + D_a e^{-RT} \quad \text{(D-51)}$$

where $D$ = dissolved oxygen deficit at a distance $x$ from the point of pollution or at a time $T$ corresponding to that distance, $K$ = the decay constant or deoxygenation rate, $L_a$ = initial first-stage BOD of the mixture of stream and discharge, $R$ = surface reaeration rate, $T$ = time, and $D_a$ = initial dissolved-oxygen deficit (at time $T = 0$). The coordinates of the critical point or point of maximum deficit ($T_c$, $D_c$) and the point of inflection or point of maximum recovery ($T_i$, $D_i$) are:

$$T_c = [1/(R - K)] \ln \{(R - K)[1 - D_a(R - K)/KL_a]\} \quad \text{(D-52)}$$

$$D_c = (K/R)(L_a e^{-KTc}) \quad \text{(D-53)}$$

$$T_i = [1/(R - K)]\ln\{(R/K)(R - K)[1 - D_a(R - K)/KL_a]\} \quad \text{(D-54)}$$

$$D_i = K(R + K)(L_a e^{-KTi})/R^2 \quad \text{(D-55)}$$

To complete the analysis, the effects of photosynthesis, respiration, and sediment oxygen demand must be taken into account. The combined effect ($PRS$) may be added to the deficit:

$$PRS = (1 - e^{-RT})\,[(N + S/H - P)/R] \quad \text{(D-56)}$$

where $N$ = rate of oxygen consumption by respiration per unit time per unit volume, $S$ = rate of oxygen uptake by sediments per unit area per unit time, $H$ = stream depth, and $P$ = rate of oxygen production by photosynthesis per unit time per unit volume. The use of an oxygen-recovery ratio or self-purification $F = R/K$ may somewhat simplify the above equations. Figure D-11 is an example of the classic sag curve.

**dissolved-oxygen sag curve** A spoon-shaped curve that represents the profile of dissolved-oxygen (DO) deficit along a receiving water downstream from a point of pollution such as a wastewater discharge. The curve, vs. time or distance of travel, is characterized by (a) an initial DO deficit ($D_a$) in a zone of relatively clean water, (b) a maximum deficit ($D_c$, also called critical deficit or minimum DO) between a zone of degradation and a zone of recovery, and (c) a point of inflection between recovery and relatively clean water. Same as oxygen sag curve. *See* dissolved-oxygen sag analysis.

**dissolved oxygen saturation** Saturation values of dissolved oxygen in water exposed to the atmosphere (20.9% oxygen) and under atmospheric pressure (760 mm of mercury) depend on temperature and salt content; see Table D-06.

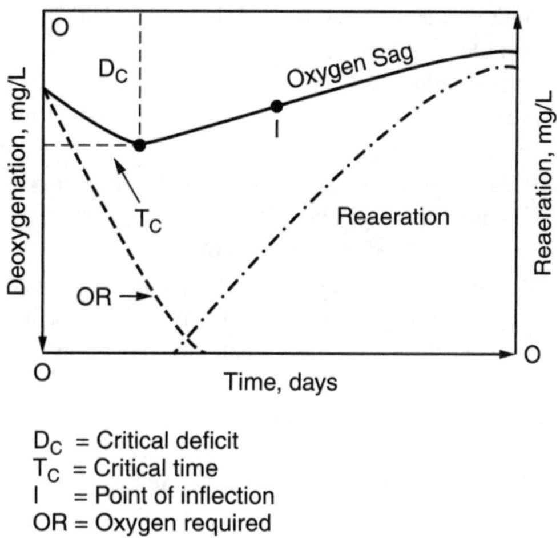


**Figure D-11.** Dissolved-oxygen sag curve.

**dissolved oxygen saturation equation** An equation used to calculate the dissolved oxygen saturation concentration $C'$ (mg/L) at a barometric pressure $P'$ (mm) and temperature $T$, given the saturation concentration $C$ (mg/L) at the barometric pressure of 760 mm and the pressure of saturated water vapor $P$ (mm) at the same temperature:

$$C' = C(P' - P)/(760 - P) \quad \text{(D-57)}$$

**dissolved particle** *See* dissolved solids.

**dissolved solids** Disintegrated organic and inorganic matter in water that cannot be removed by filtration: they pass through a 0.45-micrometer-pore-diameter filter; but if a measured volume of the filtrate is evaporated until dry, the residue represents the dissolved solids content. *See also* col-

**Table D-06.** Dissolved oxygen saturation values (mg/L)

| | Chloride content, mg/L | | |
|---|---|---|---|
| Temperature, °C | 0 | 10,000 | 20,000 |
| 0 | 14.7 | 13.0 | 11.3 |
| 5 | 12.8 | 11.4 | 10.0 |
| 10 | 11.3 | 10.1 | 8.9 |
| 15 | 10.0 | 9.0 | 8.0 |
| 20 | 9.0 | 8.1 | 7.3 |
| 25 | 8.2 | 7.4 | 6.5 |
| 30 | 7.4 | 6.7 | 6.0 |

loids, filterable residues, solids, suspended solids, and total dissolved solids (TDS).

**dissolved zinc (Zn)** The portion of total zinc in a water sample that passes through a 0.45-micrometer-pore-diameter filter.

**distillate** In the distillation of a substance, a portion is evaporated; the distillate is the product, the part that is condensed afterward from vapors.

**distillation** The act of purifying liquids through boiling, so that the steam condenses to a pure liquid and the pollutants remain in a concentrated residue. The oldest method of producing freshwater from saline and brackish waters, distillation may also be used to destroy sludge; *see* destructive distillation. Various types of evaporators are used for distillation, including multiple-effect evaporator, multistage flash evaporator, and vapor-compression still, all of them characterized by the recovery of the latent heat of vaporization. Distillation may be classified as a water recovery process as compared to demineralizing or desalting processes; *see* desalination and saline water conversion classification. *See also* flash evaporation, binary distillation, batch distillation, simple continuous distillation, differential distillation.

**distillation bottoms** By-products of distillation, often hazardous pollutants, consisting of unevaporated solids, semisolid tars, and sludges. Also called still bottoms.

**distillation operation** A batch or continuous operation separating one or more feed streams into two or more exit streams, each exit stream having component concentrations different from those in the feed streams. The separation is achieved by the redistribution of the components between the liquid and vapor phases by vaporization and condensation as they approach equilibrium within the distillation unit (EPA-40CFR264.1031 & 60.661).

**distillation process** *See* distillation.

**distillation unit** A device or vessel in which distillation operations occur, including all associated internals (such as trays or packing) and accessories (such as reboiler, condenser, vacuum pump, steam jet, etc.) plus any associated recovery system (EPA-40CFR 60.661, 63.101, & 63.111).

**distillation water** Bottled water that has been treated by distillation to meet the definition of purified water, a labeling regulated by the U.S. Food and Drug Administration. *See* bottled purified water.

**distilled water** Water formed by the condensation of steam or water vapor and from which dissolved salts, colloidal particles, and other impurities have been removed. It is water of great purity, chemically pure water, i.e., $H_2O$.

**distillery waste** *See* brewery/distillery/winery wastes.

**distilling** The distillation process, i.e., heating a mixture to separate its components by condensation of the volatile elements.

**distributed control system (DCS)** *See* digital control system.

**distributed process controller (DPC)** A computer whose control logic, memory, and input/output control are connected to a central computer that stores all the data related to a treatment plant. *See also* digital control system, programmable logic controller.

**distributed source** The opposite of a point source, that is, runoff, drainage from agricultural lands, or other sources that discharge pollutants over a wide area. Also called nonpoint source.

**distributing reservoir** Same as distribution reservoir.

**distribution and distribution function** *See* probability density.

**distribution box** A small structure used to distribute flow from one source to two or more tanks or pipes. *See also* Y junction.

**distribution coefficient ($K_d$)** (1) A measure of the extent of partitioning of a substance between geologic materials (for example, soil, sediment, rock) and water. The distribution coefficient (in mL/g) is used in the hazard ranking system (HRS) in evaluating the mobility of the substance for the groundwater migration pathway (EPA-40CFR300-AA). (2) In the adsorption process, the distribution coefficient or distribution ratio is the ratio of the quantity of adsorbate associated with the adsorbent to the total chemical concentration of the adsorbate in solution:

$$K_d = C_s/C_w \qquad \text{(D-58)}$$

or

$$K_d = SM_s/M_L \qquad \text{(D-59)}$$

where $K_d$ = distribution coefficient or distribution ratio, $L$/kg; $C_s$ = quantity of adsorbate, mol/kg; $C_w$ = adsorbate concentration, mol/L; $S$ = solids concentration; $M_s$ = mass of constituent in the solid phase; and $M_L$ = mass of constituent in the liquid phase. (3) The ratio of the concentration of a solute in one phase to its concentration in another phase at equilibrium, when the ratio varies with the initial concentration. It indicates the ability of a method to extract a component from a matrix. *See also* distribution constant, linear isotherm, partition coefficient, ozone distribution coefficient, retardation factor.

**distribution constant** The ratio of the concentration of a solute in one phase to its concentration in another phase at equilibrium, when the ratio is independent of the initial concentration. It indicates the ability of a method to extract a component from a matrix. *See also* distribution coefficient.

**distribution control system (DCS)** *See* digital control system.

**distribution diagram** In analyzing the distribution of an acid (HB) and its conjugate base (B) in a solution, the distribution diagram is a plot of the degree of acid formation ($\alpha_0$) versus pH. The distribution diagram, also called the formation function, is related to other terms as follows:

Concentration condition $C = [HB] + [B]$ (D-60)

Acidity constant $K = [H^+][B]/[HB]$ (D-61)

Degree of acid formation $\alpha_0 = [HB]/C = [H^+]/(K + [H^+])$ (D-62)

Ionization fraction $\alpha_1 = [B]/C = K/(K + [H^+])$ (D-63)

with $\alpha_0 + \alpha_1 = 1$ (D-64)

The brackets indicate molar concentrations. The ionization fraction is also called degree of dissociation or degree of protolysis. *See also* dissociation degree.

**distribution graph** A type of unit hydrograph with percentages of the total runoff volume (instead of flows) as ordinates. These percentages are called distribution ratios, which are constant for all unit hydrograhs of a given drainage area. This distribution graph assumedly represents all storms of a given duration in the basin. *See also* dimensionless unit hydrograph.

**distribution point** A point at which water quality is representative of a distribution system.

**distribution ratio** (1) Same as distribution coefficient. (2) *See* distribution graph. (3) For a trickling filter, the ratio of the flow applied to a fictitious high-rate filter to the flow of the actual filter. The fictitious filter would operate at a rate equal to the instantaneous peak rate of the actual filter.

**distribution reservoir** A reservoir in the distribution network of a water supply system providing local storage in case of an emergency and to respond to daily fluctuations in demand. Also called distributing reservoir. *See also* service storage.

**distribution sample** A water sample taken from a pipe or site in the distribution system.

**distribution system** The system of conduits and appurtenances used to deliver municipal or irrigation water to consumers.

**distribution system characteristic curve** A graphical representation of the total dynamic head in a water distribution system as a function of total discharge.

**distribution works** Pipes, reservoirs and appurtenances used in distributing water to consumers for all uses: domestic, firefighting, industrial, etc.

**distributor** (1) A fixed device (e.g., perforated pipe or sprinkler nozzle) or a movable device (e.g., rotating or reciprocating perforated pipe) used to apply water or wastewater to the surface of a filter or contact bed. *See also* revolving distributor, rotary distributor. A typical trickling filter distributor consists of two or more hollow arms, containing nozzles, installed on a central pivot, and revolving horizontally. (2) A fitting installed at the top and bottom of an ion exchanger or filter bed to retain the media and distribute the flow evenly through all sections.

**distributor control system (DCS)** *See* digital control system.

**district metering** The monitoring of water or wastewater flows by hydraulically separate sections of a water distribution network or a sewer collection system.

**disturbance** A factor that causes a change in the performance of a treatment process. *See* external process disturbance and internal process disturbance.

**disturbed land reclamation** The one-time application of biosolids to land at a high rate (e.g., 50–100 dry tons per acre) to correct adverse soil conditions and facilitate revegetation of disturbed land. *See also* dedicated land reclamation.

**disulfate** A salt of pyrosulfuric acid, e.g., sodium disulfate ($Na_2S_2O_7$).

**disulfide** A compound containing two atoms of sulfur, e.g., carbon disulfide ($CS_2$) or the bivalent group —SS—, e.g., diethyl disulfide ($C_4H_{10}S_2$).

**disulfoton [$C_8H_{19}O_2PS_3$ or $(C_2H_5O)_2P(S)SCH_2CH_2SCH_2CH_3$)]** A pale-yellow, toxic liquid, used as an insecticide.

**ditch oxidation** A biological wastewater treatment method similar to the activated sludge process but using a ring-shaped or oval ditch instead of the conventional aeration basin. *See* oxidation ditch.

**diuresis** Increased discharge of urine.

**diurnal** Occurring during a 24-hour period, daily, during the daytime as opposed to nighttime, or during a tidal day.

**diurnal demand curve** A graphical representation of water demand over a 24-hour period.

**diurnal fluctuation** The cyclic rise and fall of a variable during a 24-hour period. For example, water use and wastewater production of a community have definite diurnal patterns during each season of the year. Water table, stream flow, water and wastewater characteristics (BOD, suspended solids, etc.) also fluctuate daily. Photosynthesis is a diurnal process, with significant daily variations in the concentrations of dissolved oxygen and carbon dioxide, which affect the performance of stabilization ponds and the oxygen sag in receiving waters.

**diurnal variations** *See* diurnal fluctuation.

**diurnal wastewater variation** Figure D-12 shows typical variations of domestic or municipal wastewater during a 24-hour period.

**diuron [$C_6H_3Cl_2NHCON(CH_3)_2$]** A white, crystalline weed killer.

**divalent** Having a valence of 2, e.g., ferrous ion ($Fe^{2+}$ or $Fe^{++}$), sulfate ion ($SO_4^=$ or $SO_4^{2-}$).

**divergence factor** A correction factor applied to incident irradiance to obtain the average UV irradiance used in bench-scale experiments. It accounts for the light that strays outside the testing apparatus.

**diverging tube** A tube whose diameter increases from one end.

**diversion** The redirection of stream, wastewater, or storm flows using a channel, canal, pipeline, or other conduit. The flows are diverted to protect a site from flooding, overflow, or erosion, for use or safe disposal in another site, or for use as a water supply. Also called terrace or diversion terrace when built across a slope to control surface runoff and soil erosion. A diversion bank or diversion berm is the part of the diversion that prevents water from flowing onto lower lying areas.

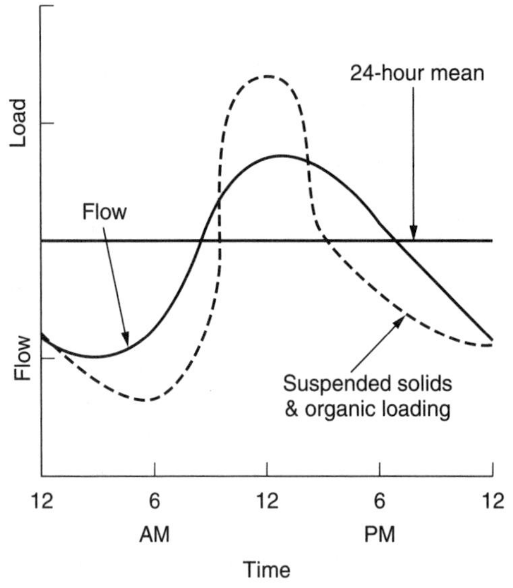


**Figure D-12.** Dirurnal wastewater variation.

**diversion bank** *See* diversion.

**diversion berm** *See* diversion.

**diversion box** A small diversion chamber.

**diversion chamber** A chamber designed to divert all or part of a stream or other flow.

**diversion manhole** A manhole that functions as a diversion chamber.

**diversion rate** The percentage of waste materials diverted from traditional disposal such as landfilling or incineration to be recycled, composted, or reused.

**diversion structure** A flow regulator, installed at the intersection of a combined sewer and an interceptor, used to divert combined sewer flows in excess of wastewater treatment plant capacity to receiving water via the combined sewer overflow outlet, thus preventing flooding and treatment plant overloading. Typical diversion structures, also called regulators or regulator structures, include automatic regulators, drop inlets, high outlet regulators, Hydro-Brake regulators, leaping weirs, mechanical regulators, orifices, relief siphons, reverse taintor gates, side weirs, tipping-plate regulators, and transverse weirs.

**diversion terrace** *See* diversion.

**diversion weir** Same as diverting weir.

**diversion works** Dams, pump stations, conduits, tunnels, weirs, etc., and appurtenances used to divert water to protect a construction site or for other purposes.

**diversity** The distribution and abundance of different kinds of plant and animal species and communities in a specified area.

**diversity index** A mathematical expression that depicts the diversity of a species in quantitative terms.

**diverting weir** A weir over which excess water, stormwater or wastewater is allowed to flow. Diverting weirs are used in wastewater treatment plants, for example, as a bypass to prevent surcharging of upstream units in case of electrical or mechanical failure of screening devices. Same as overfall weir or overflow weir. *See also* diversion dam, effluent launder, effluent weir, and skimming weir.

**divinylbenzene [$C_{10}H_{10}$ or $C_6H_4(CH{:}CH_2)_2$]** Abbreviation: DVB. A clear liquid, a polymerization monomer used in the manufacture of rubbers, cation and macroporous ion-exchange resins, and other products.

**divinylethylene** Another name of butadiene.

**division box** (1) A device, equipped with weirs or control valves, that divides incoming flow into two or more streams, e.g., to distribute flow from primary clarifiers equally among aeration tanks. Also called splitter box, splitting box. (2) A structure in an irrigation canal that controls and measures the water delivered to a customer.

**DLD** Acronym of dedicated land disposal.

**DLO™** Acronym of dynamic light obscuration, a technique developed by Chemtrac Systems, Inc. to monitor particles.

**DLVO theory** Short for Derjaguin, Landau, Verwey, and Overbeek theory.

**DMDT** Another name of methoxychlor.

**DNA** Acronym of deoxyribonucleic acid.

**DNA microarray** An advanced molecular technology that can detect thousands of genes on a single slide and is being considered for application in virulence investigations in environmental microbiology.

**DNAPL** Acronym of dense nonaqueous-phase liquid.

**DNRA** Acronym of dissimilatory nitrate reduction to ammonia.

**DNF** Acronym of dissolved nitrogen flotation.

**DO** Acronym of dissolved oxygen.

**$D_2O$** Chemical formula of deuterium.

**Dobbs et al. equation** An equation proposed in 1989 to estimate partition coefficients for wastewater treatment processes and therefore the distribution of an organic compound between the solid biomass and the liquid phases (Metcalf & Eddy, 2003):

$$K_p = 0.58 \log K_{ow} + 1.14 \quad \text{(D-65)}$$

where $K_p$ is the partition coefficient of the compound in L/g and $K_{ow}$ is the dimensionless octanol–water partition coefficient.

**DOC** Acronym of dissolved organic carbon.

**doctor blade** A scraping mechanism that removes or regulates materials on a belt, roller, or other moving or rotating surface.

**Dokwed®** A bar screen manufactured by Hubert Savoren BV.

**dolomite [$CaMg(CO_3)_2$ or $CaCO_3 \cdot MgCO_3$]** (1) A common mineral consisting of the carbonates of calcium and magnesium, occurring naturally in crystals or in masses; chemically altered limestone containing a high fraction of magnesium. It is a primary hardness-contributing mineral in natural waters. Also called calcium magnesium carbonate. It becomes dolomitic lime by calcination. (2) The rock dolomite stone or dolostone, consisting essentially of the mineral dolomite.

**dolomite stone** Same as dolomite (2).

**dolomitic hydrated lime** A mineral or chemical product consisting of calcium and magnesium hydroxides, e.g., corresponding to the formula $\{[Ca(OH)_2]_{0.6}\,[Mg(OH)_2]_{0.4}\}$; approximate specific gravity = 2.5; bulk density = 30–40 pounds per cubic foot. It is used in wastewater treatment for pH adjustment or as an acid-neutralizing agent, with a basicity factor of 0.912 as compared to 1.306 for magnesium oxide (MgO). Also called pressure dolomitic hydrated lime. *See also* normal dolomitic hydrated lime.

**dolomitic lime** A mixture or compound of calcium and magnesium oxides or hydroxides; e.g., lime that contains 30–50% magnesium oxide (magnesia, MgO) and 70–50% calcium oxide (CaO). It is sometimes used in water or wastewater treatment for acid neutralization or pH adjustment. *See* dolomitic hydrated lime and dolomitic quicklime.

**dolomitic limestone** A rock that is essentially dolomitic lime, i.e., calcium and magnesium oxides. It is used in wastewater treatment for pH adjustment or as an acid-neutralizing agent, with a basicity factor of 0.564 as compared to 0.71 for high-calcium hydrated lime.

**dolomitic quick lime** A mineral or chemical product consisting of calcium and magnesium oxides, e.g., corresponding to the formula $[(CaO)_{0.6}\,(MgO)_{0.4}]$; approximate specific gravity = 3.35; bulk density = 55–60 pounds per cubic foot. It is used in wastewater treatment for pH adjustment or as an acid-neutralizing agent, with a basicity factor of 1.110 as compared to 0.687 for caustic soda (NaOH) that is 78% $Na_2O$.

**dolomitic stone** Same as dolomite (2). Burned dolomitic stone, used to neutralize nitric and sulfuric acid wastes in the lime-slurry treatment method, consists essentially of the oxides of calcium and magnesium; it contains approximately 48% CaO, 34% MgO, and 2.0% $CaCO_3$ (Nemerow & Dasgupta, 1991).

**dolostone** Same as dolomite (2).

**DOM** Acronym of dissolved organic matter.

**dome diffuser** A porous, dome-shaped ceramic device mounted on an air distribution pipe for the aeration of wastewater.

**domestic** Of or pertaining to the home or the household. Domestic use of water and wastewater services is distinguished from commercial, industrial, and institutional uses. *See also* municipal.

**domestic consumption** The quantity of water used by households in a given period. It may include not only such uses as cooking and hygiene but also unaccounted for water.

**domestic filter** A small filter used to purify water in a household or any other unit of comparable size. Also called home treatment filter, household filter, point-of-use filter.

**domestic flow** (1) The flow of water necessary to meet the demand of residential customers; it may include commercial usage but excludes industrial, agricultural, and fire protection needs. (2) *See* domestic wastewater.

**domestic meter** A service meter for a single domestic consumer.

**domestic or other nondistribution system plumbing problem** A coliform contamination problem in a public water system with more than one service connection that is limited to the specific service connection from which the coliform positive sample was taken (EPA-40CFR141.2).

**domestic septage** Solid or liquid material removed from a septic tank, cesspool, portable toilet, Type III marine sanitation device, or similar treatment works that receives only domestic sewage. Domestic septage does not include liquid or solid material removed from a septic tank, cesspool, or similar treatment works that receives either commercial wastewater or industrial wastewater and does not include grease removed from a grease trap at a restaurant (EPA-40CFR503.9-f).

**domestic service** The piping that is necessary to provide water service to a domestic consumer. Also called house service.

**domestic sewage** *See* domestic wastewater. Wastewater is now the generally accepted term for the spent or used water from all sources. Sewage, as distinguished from industrial and institutional wastewaters, is produced by households and commercial establishments.

**domestic use** Water used by households. *See* domestic flow.

**domestic wastewater** Untreated liquid wastes that flow directly to a sewer system from the kitchens, bathrooms, lavatories, toilets, and laundries of households; a combination of human feces, urine, and graywater. Also called sanitary sewage or sanitary wastewater, which may include similar contributions from commercial and institutional facilities. *See also* domestic sewage, industrial wastewater, infiltration/inflow, stormwater.

**domestic wastewater flow rate** The flow rate of wastewater from domestic sources; sometimes used interchangeably with interior water use because of the correlation between the two.

**Domestos®** A proprietary bleach that contains about 4% of available chlorine; it can be used to make a 1% stock solution for water disinfection. Such a solution is unstable under warm conditions.

**dominance** The condition of the most abundant species in a vegetation stratum.

**Donaldson, Wellington** A consultant and then chief of wastewater treatment operations of New York City; he proposed the use of the sludge density index in 1932 as a measure of sludge compactibility.

**Donnan dialysis** A variation of diffusion dialysis; a process that uses ion exchange membranes to separate and recover acids from spent pickling liquor in the steel-finishing industry. The unit consists of a stack of membranes with the metals containing solution fed into one compartment and a mineral acid (e.g., sulfuric acid, $H_2SO_4$) fed countercurrently into another compartment (Freeman, 1995).

**Donnan potential** The condition created in a membrane in which charged functional groups attract ions of opposite charge (counterions) and create a deficit of like-charged ions (co-ions) in the membrane and an accumulation of co-ions in the concentrate (AWWA, 1999).

**Don-Press** A screw-type press manufactured by Dontech, Inc. to process solids and screenings from wastewater treatment.

**DOP credit** A numerical value of the credit granted a surface water treatment facility for the protection provided against *Cryptosporidium,* based on demonstration of performance (DOP). It is calculated as

$$\text{DOP credit} = \log (R/T) \qquad \text{(D-66)}$$

where $R$ = mean, median, or geometric mean concentration of raw water; and $T$ = respectively, mean, median, or geometric mean concentration of treated water. *See also* log removal ratio.

**Doppler current meter** An acoustic instrument that uses a single ultrasonic transducer to measure current velocities. The procedure is to project a sound signal of known frequency into the water, i.e., a continuous beam of ultrasonic waves that are reflected to a receiver by suspended solids or gas bubbles in the liquid, and to measure the Doppler shift, which is proportional to the water velocity past the meter. (The Doppler effect or Doppler shift is the difference between the frequency of the signal and that of the reverberation due to interaction with particles carried by the flow.) Also called ultrasonic Doppler velocity meter. *See also* ultrasonic transit time velocity meter.

**Doppler effect** *See* Doppler current meter; same as Doppler shift.

**Doppler shift** *See* Doppler current meter; same as Doppler effect.

**dormant cells** Cells, often rounded and relatively small, that are in a state of minimal metabolic activity, not growing, in reaction to adverse conditions or as part of their normal annual rhythm. Over time they may become viable but nonculturable. *See also* culturable count, direct count.

**DorrClone** Equipment manufactured by Dorr-Oliver, Inc. to classify grit. It may be used to wash grit or to separate grit from the clarifier underflow. *See* Figures D-13 and D-14.

**DorrCo** A line of products of Dorr-Oliver, Inc.

**Dortmund tank** An upward flow sedimentation tank used in treating wastewater, with sludge removal at the bottom.

**DO sag curve** *See* dissolved oxygen sag curve.

**dosage** The actual quantity of a chemical administered to an organism or to which it is exposed. Also, the quantity of a substance added to a specified quantity or volume of solution to reach a specified effect; dose.

**dose** The amount of test substance administered or the level of exposure to a substance; expressed as weight or volume of test substance per unit weight of test animal, for example mg/kg, or as weight of test substance per unit weight of food or drinking water (EPA-40CFR795.260-2, 798.1100-2, 798.1175-2, 798.2250-2, 798.2450-5, & 2650-2). *See* absorbed dose, toxicity terms.

**dose-effect curve** A graphical representation of the relationship between a dose of chemical and the frequency or intensity of its effect.

**dose equivalent** The product of the absorbed dose from ionizing radiation and such factors as account for differences in biological effectiveness due to the type of radiation and its distribution in the body as specified by the International Commission on Radiological Units and Measurements. The unit of dose equivalent is the rem or millirem and the sievert in SI units (EPA-40CFR141.2, 190.02-h, & 191.12).

**dose of radioactivity** A quantity of ionizing radiation, expressed in roentgens.

**dose rate** The amount of a substance administered, or the level of exposure to a substance, per unit time.

**dose response** A quantitative relationship between the dose or concentration of a chemical and an effect caused by the chemical. The dose response may also be defined as the quantitative shift of an organism's response to changes in exposure level; e.g., a small dose of carbon monoxide may cause drowsiness, whereas a large dose can be fatal.

**dose–response assessment** (1) The estimation of the potency of a chemical. (2) A quantitative determination of what effects (e.g., toxic injury, disease) a constituent may cause at different doses; a component of risk assessment. Exposure or dosage is expressed in relative weight (e.g., mg of a substance per kg of body weight per day), in concentration of the substance in drinking water (mg/L), or similar units. *See also* chronic daily intake, dose–response models, exposure assessment, hazard identification, lifetime risk, potency factor, risk characterization, slope factor, total dose, toxicity terms.

**dose–response curve** A graphical representation of the relationship between the dose of a chemical and an effect caused by the chemical.

**dose–response evaluation** See dose–response assessment.

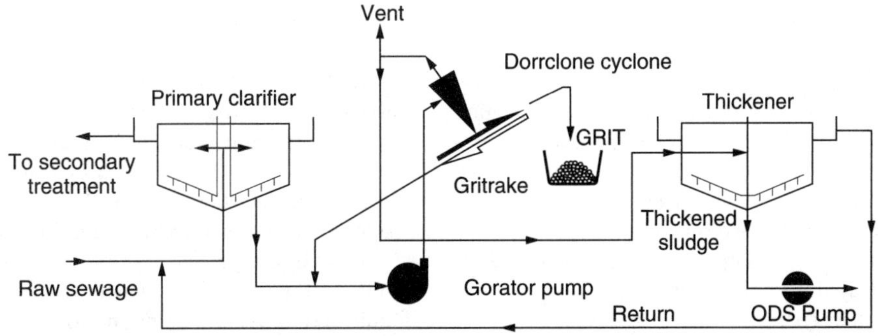


**Figure D-13.** Dorrclone® (degritter). (Courtesy Dorr-Oliver Eimco.)

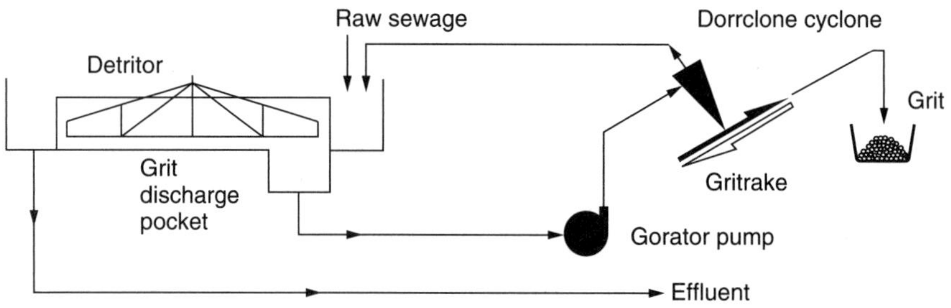


**Figure D-14.** Dorrclone® (grit washing). (Courtesy Dorr-Oliver Eimco.)

**dose–response experiment** A biological or chemical experiment to obtain dose–response data, e.g., jar tests for optimal coagulation dosage.

**dose–response model** A mathematical formula or set of such formulas that describe the dose–response relationship for a particular agent. Common models proposed for human exposure include (a) beta-Poisson model, (b) exponential model, (c) linear (or linearized) multistage model, (d) logit model, (e) log-probit model, (f) multihit model, (g) multistage model, one-hit (or single-hit) model, (h) probit model, and (i) Weibull model. *See also* annual risk, lifetime risk, margin of exposure, margin of safety.

**dose–response relationship** The quantitative relationship between the amount of exposure to a substance (administered, absorbed or believed to be effective) and the changes in certain aspects of a biological system (e.g., the extent of toxic injury or disease produced).

**Dosfolat®** A solution of folic acid produced by Bioprime.

**dosimeter** (1) An instrument for measuring and recording radiation dosage or exposure to radiation. (2) A device, operated under pressure or under vacuum, designed to measure and feed chemicals (e.g., ammonia, chlorine, sulfur dioxide) at a predetermined rate for water or wastewater treatment.

**dosing** The periodic application of septic tank effluent to a soil absorption system.

**dosing apparatus** A tank, siphon, chamber, any other device or arrangement, or combination thereof, used for regulating the application of water, wastewater, or chemicals to treatment units. *See also* multiple dosing tanks.

**dosing chamber** *See* dosing tank.

**dosing pump** (1) Same as metering pump. (2) A device used to apply septic tank effluent to a soil absorption system.

**dosing rate** (The volume of wastewater discharged on the packing of a trickling filter for each pass of the distributor, expressed in depth of liquid ($D_r$, mm/pass). The dosing rate depends on the BOD loading, e.g., $D_r$ = 40–120 mm/pass for a BOD of 2.0 kg/m$^3$/day. It affects the number of filter flies as well as biofilm thickness and odors. Also called operating dose, SK value, or spülkraft. *See also* flushing dose, rotational speed.

**dosing ratio** The ratio of the maximum application rate to the average application rate of water or wastewater to a filter or any other treatment unit.

**dosing siphon** A siphon that automatically discharges liquid onto a trickling filter bed, soil absorption system, or other wastewater treatment device.

**dosing tank** (1) A tank that receives and holds raw or partly treated wastewater for subsequent, intermittent treatment and discharge at a constant rate. It is used when a trickling filter (or other wastewater treatment system) experiences poor distribution and inadequate velocity at low flows. A dosing chamber may be defined similarly. *See* Figure D-15. (2) Any tank used to apply a liquid to subsequent units.

**dot-blot** A method of virus detection by immobilizing the virus' nucleic acid and hybridizing it to a probe of a complementary substance. Also called dot blot hybridization. *See also* nucleic acid probe, in situ hybridization, fluorescent in situ hybridization.

**dot blot hybridization** *See* dot-blot.

**double-action device/mechanism** A device/mechanism acting in two directions, such as a reciprocating piston in a cylinder with a chamber at each end. A double-action pump admits water on both sides of the piston, thereby assuring a more or less constant discharge.

**double-action pump** *See* double-action device.

**double-action reciprocating pump** *See* double-action device.

**double check-valve assembly** *See* check valve.

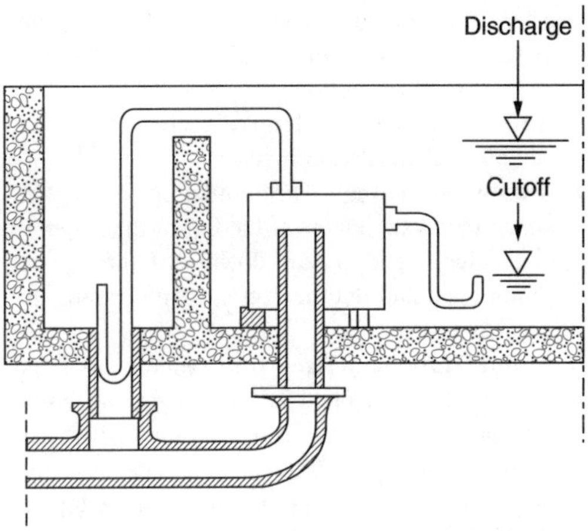


**Figure D-15.** Dosing tank (adapted from Fair et al., 1971).

**double coagulation** The application of coagulants to water or wastewater at two different points of the treatment train. This is different from dual coagulation.

**double-compartment tank** A septic tank with two compartments. The first compartment, with twice the volume of the second compartment, receives both toilet effluent and household sullage. Its settled effluent flows to the second compartment, which may also receive sullage directly.

**Double Dish** Underdrain manufactured by U.S. Filter Corp. for pressure filters.

**Double Ditch** A wastewater treatment system developed by Kruger, Inc. using the oxidation ditch process.

**double filtration** The use of two slow sand filters in series or a rapid sand filter followed by a slow sand filter, or two rapid sand filters in series to treat water.

**DoubleGuard™** A screening apparatus manufactured by Envirex, Inc. incorporating two sets of bar racks.

**double layer** The charged surface of a colloidal particle and its surrounding layer of counterions. *See* electrical double layer for detail.

**double-layer compression** A coagulation mechanism that reduces the repulsive forces of similar particles and promotes their destabilization, subsequent agglomeration, and settling. It is not practical to use this phenomenon in drinking water treatment, but it is an important destabilization factor in the formation of estuaries or deltas where rivers flow into seawater. *See also* adsorption–destabilization, bridging, restabilization, and sweep-floc coagulation.

**double-layer repulsion** The repulsive potential energy that develops between two similar colloidal particles when they approach each other and their diffuse layers begin to overlap; it tends to slow the motion of the particles and inhibit their collision. *See also* hydrodynamic retardation and van der Waals force.

**double-layer theory** *See* electrical double layer for detail.

**double offset** The combination of two plumbing fittings to realign a pipe.

**double-pan balance** A balance with one pan for the material to weigh and the other for counterbalancing weights.

**double-strength pipe** A vitrified clay sewer pipe that has at least a strength of 1000 lb/ft.

**double-stroke deep-well pump** A reciprocating pump with two sets of rods connecting the power head to the cylinder plungers.

**double-suction impeller** An impeller with two suction inlets.

**double-suction pump** A centrifugal pump with a suction pipe on both ends.

**double-vault composting toilet** A variation of the composting toilet that has a vault with two compartments used alternatingly.

**double-vault latrine** A double-vault composting toilet or a pit latrine with two alternating pit sites, each pit being dug out a year or two after closing and the contents used as fertilizer.

**doubling time** The time it takes for a microbial cell division to occur; the time needed for a population to double. It can be determined from the linear portion of semilog plot of growth versus time, which is defined by:

$$X = X_0 e^{\mu t} \qquad \text{(D-67)}$$

where $X$ = number or mass of cells at time $t$, $X_0$ = initial number or mass of cells (at time $t = 0$), and $\mu$ = specific growth rate. *See also* generation time.

**Dow Chemical formula** A formula developed to determine the packing head loss ($N_p$), expressed in terms of velocity heads, in a vertical trickling filter or tower (Metcalf & Eddy, 1991):

$$N_p = 10.33\, De^{0.0000136L/A} \qquad \text{(D-68)}$$

where $D$ = packing depth, m; $L$ = liquid loading rate, kg/hr; and $A$ = tower cross section area, m$^2$. The total tower head loss also includes inlet, underdrain, and minor losses, representing 30–50% of the packing loss ($N_p$). Another correction factor may be applied for nonvertical trickling filters, depending on the specific surface area of the packing. *See also* tower resistance.

**Dowex®** An ion-exchange resin produced by Dow Chemical Co.

**downcomer** (1) Same as downspout. (2) The inner annular space between the concentric vertical tubes of a deepwell WAO process.

**downflow** The conventional flow pattern in water and wastewater treatment units; the downward flow of the liquid or treatment chemicals from the top of the units. Sometimes called cocurrent flow. *See also* upflow, crossflow, countercurrent.

**downflow anaerobic filter** *See* anaerobic downflow attached-growth process.

**downflow attached-growth anaerobic process** See anaerobic downflow attached-growth process.

**downflow bubble contactor** A cone-shaped device that provides prolonged oxygen bubble contact time and high rates of oxygen transfer to aerate water or wastewater. It has special application in reservoir reaeration, temperature destratification, and postaeration. Also called Speece cone. *See also* U-tube contactor.

**downflow carbon adsorption** An activated carbon adsorption process consisting of two or more columns operated in series and in countercurrent mode, with the liquid moving vertically downward, allowing the adsorption of organics and the filtration of suspended solids.

**downflow column** A sand filter, anaerobic digester, or any similar device in which liquid flows by gravity from top to bottom; the opposite of an upflow column.

**downflow contactor** A downflow carbon adsorption unit.

**downflow denitrification filter (DDF)** A 4–6.5 ft deep filter, consisting of a sand and gravel bed supported by underdrains, used to remove nitrogen (postanoxic nitrate removal) and suspended solids from a nitrified wastewater effluent. The granular media provides support for heterotrophic microorganisms that use an exogenous carbon source such as methanol ($CH_3OH$). Also called downflow packed-bed denitrification filter. *See also* bump, Tetra® filter, upflow continuous backwash filter.

**downflow filter** A water or wastewater filtration configuration whereby the liquid is introduced at the top of the unit and the effluent withdrawn at the bottom; sometimes called conventional, because it is the mode of operation of most wastewater filters in the United States. *See also* biflow filter, upflow filter.

**downflow fixed film reactor** A reactor developed for anaerobic wastewater treatment; it provides a vertical, porous surface (e.g., clay or fibrous polyester) for the anaerobic bacteria to grow and wastewater flows from the top while the effluent is withdrawn at the bottom.

**downflow packed-bed denitrification filter** *See* downflow denitrification filter.

**downflow softening** The softening process in which raw water enters at the top of the softening bed column and passes downward through the cation resin and out the bottom, with brining in a cocurrent direction.

**downflow stationary fixed-film reactor** A fixed-film reactor developed in Canada for wastewater treatment using bacteria grown on vertical surfaces of porous, rough media (e.g., clay or fibrous polyester), with influent introduced at the top and withdrawn at the bottom.

**downgradient** The direction of groundwater flow, i.e., in the direction of decreasing static head; the same as downstream for surface water.

**downspout** A vertical pipe leading from a roof drain or gutter down to the ground, a cistern, a storm drain or other means of disposal. Also called conductor, downcomer, or leader.

**downstream** With or in the direction of the current of a stream.

**downstream disinfectant** The disinfectant applied second or last in the sequential use of two or more disinfecting agents. *See also* synergistic inactivation and upstream disinfectant.

**downwash** The drawdown of a plume after emission due to a low-pressure area downwind of the stack.

**dowsing** The process of searching for or detecting underground water or other substances using a divining rod.

**DOX** Abbreviation of dissolved organic halogen.

**d-part®** A chemical produced by Medina Products for waste processing.

**DPD** Acronym of N,N-diethyl-p-phenylene-diamine.

**DPD indicator** DPD or N,N-diethyl-p-phenylene-diamine is a colorimetric redox indicator. When oxidized by chlorine, it can be measured by titration or spectrophotometrically.

**DPD method** A method of measuring the chlorine residual in water by adding the reagent DPD to a sample, which turns the water red. The residual may be determined by either titrating or comparing a developed color with color standards. *See also* amperometric titration, iodometric method, orthotolidine test.

**DPD test** *See* DPD method.

**DPIV** Acronym of digital particle image velocimetry.

**DR** Acronym of dosing rate.

**drab** A conduit that carries surplus groundwater, stormwater, or surface water, but not wastewater (Symons et al., 2000).

**dracontiasis** Same as dracunculiasis.

**dracunculiasis** A helminthic infection caused by the pathogen *Dracunculus medinensis.* Also called dracontiasis. *See* its common name, guinea worm, for more detail.

***Dracunculus medinensis*** The helminth species that causes dracunculiasis; also called guinea worm, distributed mostly in West and Central Africa and in India. The mature female worm is about one foot and a half long, usually creating a blister under the skin of the leg of an infected person. *See also* cyclopoid, gravid.

***Dracunculus medinensis* life cycle** The route followed by the guinea worm from an infected host to another host: on contact with water, the guinea worm blister of an infected host releases thousands of microscopic larvae that become infective within cyclopoids as intermediate hosts. When a human host consumes water containing infected cyclopoids, the larvae develop further and a fertilized female worm eventually reaches the legs of the host to restart the cycle, which lasts 12–15 months.

**draft** (1) The process of withdrawing water or the quantity of water drawn from a tank, reservoir, stream, etc. *See also* yield. (2) The pressure head resulting from temperature and moisture differences, e.g., in the operation of a trickling filter. *See* natural draft for detail.

**draft environmental impact statement** A document prepared under the guidelines of the National Environmental Policy Act to describe the beneficial and adverse effects of a major project.

**draft tube** (1) A centrally located vertical cylinder, usually with flared ends, used to promote mixing as part of a mechanical aerator; it is used, e.g., to extend the inlet of an aerator in a deep pond or a deep tank. Also, a draft tube aerator or draft tube turbine aerator. (2) One of the four basic parts of reaction hydraulic turbines such as the Francis and the Kaplan turbines. The other parts are the scroll case, the wicket gates, and the runner. The draft tube is a conical tube designed to decelerate the flow from the runner. (3) A part of the mixing apparatus of a high-rate sludge digester.

**draft tube aerator** A mechanical aerator that includes a draft tube; sometimes called simply a draft tube.

**Draft Tube Channel** A wastewater treatment process developed by Lightin using oxidation ditches.

**draft tube loss** The energy lost in a draft tube through eddies and friction.

**drag** The resistance of a liquid to the sedimentation or flotation of suspended particles. The drag coefficient is a measure of this resistance.

**drag coefficient ($C_D$)** (1) A dimensionless coefficient that expresses the nature of flow and channel characteristics, as in the Chézy coefficient. (2) Newton's drag coefficient is a dimensionless measure of the resistance to the settling or flotation of suspended solids, depending on their characteristics (size, shape, density, terminal settling velocity), and defined as the ratio of the force per unit area to the stagnation pressure:

$$C_D = F_D/(0.5\, \rho V^2 A) \qquad \text{(D-69)}$$

where $F_D$ is the drag force, $A$ the surface area on which it acts, $V$ the average velocity, and $\rho$ the fluid density. It may also be computed as a function of the Reynolds number ($R_e$):

$$C_D = (24/R_e) + (3/R_e^{0.5}) + 0.34 \qquad \text{(D-70)}$$

*See also* Chézy coefficient, friction factor, and Stoke's law.

**drag equation** The equation of frictional drag force, as given under drag coefficient.

**drag force ($F_D$)** One of the forces acting on a particle traveling in a resistant fluid, as used to determine the terminal velocity of a settling particle. *See* drag coefficient.

**Drag-Star™** A cable and scraper apparatus manufactured by Smith & Loveless, Inc. to remove sludge from rectangular sedimentation tanks.

**drag tank** A rectangular sedimentation basin that uses a chain and flight collector mechanism to remove dense solids.

**Draimad** A sludge dewatering system developed by Aero-Mod, Inc.

**drain** A channel, buried pipe, lined or unlined ditch, or other conduit that carries off liquids by gravity. Drains that carry stormwater, wastewater, or a combination of both are called storm, sanitary, or combined sewers, respectively. Relief drains are used to dewater a construction site with a high water table. *See also* building drain, French drain, underdrain.

**drainable sludge** Sludge that can be dewatered by gravity.

**drainage** (1) Same as drainage water. (2) Removal of excess ground, surface, or storm water from structures or from an area, by gravity or pumping, to prevent inconvenience or protect against losses. (3) Same as drainage area or watershed. (4) The drainage system: channels, ditches, drains, pump stations, manholes, and other appurtenances. (5) Water lost from the soil by percolation. Soil drainage refers to the frequency and duration of periods when the soil is not saturated; water is removed fast from well-drained soils, but slowly from poorly drained or waterlogged soils.

**drainage basin** (1) The area of land that drains water, sediment, and dissolved materials to a common outlet at some point along a stream channel. (2) The area from which a single drainage system carries surface runoff away; also called catchment area, watershed, drainage area. (3) The largest natural drainage area subdivision of a continent; e.g., the United States has been divided at one time or another, for various administrative purposes, into some 12 to 18 drainage basins. *See* catchment, drainage area, watershed.

**drainage channel** A semicircular or rectangular channel that carries the flow from a trickling filter's underdrains at a velocity of 2–3 fps and admits air for their ventilation.

**drainage coefficient** The discharge of an underdrainage system designed to remove excess rainfall water, for salinity control or other purposes. The coefficient is expressed as inches of depth of water per 24 hours or as a percentage of the mean annual rainfall. Also called drainage modulus.

**drainage density** The relative density of natural drainage channels in a given area, i.e., the ratio of miles (or kilometers) of channel to square miles (or square kilometers) of area.

**drainage field** Same as drainfield.

**drainage paved drying bed** A paved bed used in sludge dewatering as an alternative to sand drying beds, with a front-end loader to improve sludge removal. The bed consists of a concrete or bituminous concrete lining overlaying a sand or gravel base of 8–12 inches. *See also* decanting paved drying bed.

**drainage swale** A channel lined with vegetation, riprap, asphalt, concrete, or other material, installed to convey surface runoff and control erosion.

**drainage system** A network of pipes, pumps, structures, and appurtenances designed and operated to effect drainage.

**drainage trench** A trench that contains a perforated pipe surrounded by gravel or a similar material and used as a soakaway for the dispersion of liquids.

**drainage water** Incidental surface waters from such sources as rainfall, permafrost melt, or snowmelt (EPA-EPA-40CFR440.141-3). Ground, surface, or storm water collected by a drainage system and discharged into a natural waterway. Different from drain well and drainage well.

**drainage well** (1) In general, it is a well constructed to receive runoff as a stormwater management structure. *See also* drywell (2). (2) A well drilled to carry excess water off agricultural fields. Because it acts as a funnel from the surface to the groundwater, a drainage well can contribute to groundwater contamination. (3) A vertical shaft in a masonry dam designed to divert seepage away from the downstream face. (4) A well to carry water or wastewater into underground strata. Different from drain well and drainage water.

**Drain-Dri®** A pump manufactured by Yeomans Chicago Corp. for installation with a wet well.

**drainfield** An open area, the soil of which absorbs the contents of a septic tank. More specifically, it is a system of trenches containing open-jointed pipes connected together and laid on a bed of gravel or similar materials. *See* soil absorption field for more detail.

**drain lines** The trenches of a subsurface soil absorption system. Also called leach lines.

**drainpipe** A large pipe that carries away the discharge of waste pipes, soil pipes, etc.

**drain tile** A perforated drain used at the bottom of a building foundation, in trenches of short lengths, to collect and carry off excess groundwater. Also used, in an envelope of gravel, for underground disposal of wastewater. *See also* tile drainage.

**draw** (1) A natural depression or swale; the dry bed of a stream. (2) A small natural drainageway with a shallow bed; a gully. (3) A coulee or ravine (in the Western U.S.).

**draw-and-fill digester operation** One of two commonly used procedures to fill a sludge digester, whereby digested sludge is withdrawn before feeding. The other procedure is fill-and-draw.

**drawback tank** A tank that supplies permeate to the membranes of a reverse osmosis unit to prevent damage during depressurization.

**drawdown** (1) The drop in surface elevation or in water level resulting from the withdrawal of water, as when a well is pumped or when a pond loses water by infiltration and evaporation. (2) The quantity of water drawn from a tank or reservoir, or the drop of water level of the tank or reservoir. *See also* drop (2), and depression head.

**draw phase** The fourth of the five steps in the operation of sequencing batch reactors, whereby effluent is removed using floating or adjustable weirs or other mechanisms. Also called decant phase. *See also* fill phase, react phase, settle phase, and idle phase.

**DRE** Acronym of destruction and removal efficiency.

**dredge** To clear out with a machine, remove sediment or sludge from rivers or estuaries to maintain navigation channels.

**dredging** Removal of mud from the bottom of water bodies. This can disturb the ecosystem and cause silting that kills aquatic life. Dredging of contaminated muds can expose biota to heavy metals and other toxics. Dredging activities may be subject to regulation under Section 404 of the Clean Water Act.

***Dreissena polymorpha*** The scientific name of zebra mussels.

**drenching shower** A device used to wash chemicals off the body (Symons et al., 2000).

**D-R equation** *See* Dubinin–Raduskevich equation.

**Dresser coupling** A device used to join iron pipe sections, consisting of two external rings clamped together against a central ring.

**Dresser/Jeffrey** Wastewater screening equipment of Jones & Attwood, Inc.

**Drewgard** A product of Ashland Chemical used as a corrosion inhibitor.

**drift** Water lost from the top of a cooling tower as mist or droplets entrained in the circulating air. Drift represents about 0.005% of the recirculating water. *See* cooling tower precipitation for detail. *See also* blowdown, makeup water.

**drifting-sand filter** An obsolete rapid filter whose sand drifts continuously to the point of withdrawal for washing and recycling with the raw water.

**Driftor™** A product of Kimre, Inc. used to eliminate drift.

**driller's well log** A drilling log showing the depth, thickness, characteristics of penetrated strata, location of water-bearing strata, depth, size, and characteristics of casing installed.

**drilling fluid** The circulating fluid (mud) used in the rotary drilling of wells to clean and condition the hole and to counterbalance formation pressure. A water-based drilling fluid is the conventional drilling mud in which water is the continuous phase and the suspending medium for solids, whether or not oil is present. An oil-based drilling fluid has diesel oil, mineral oil, or some other oil as its continuous phase with water as the dispersed phase (EPA-40CFR435.11-1). *See also* drilling mud.

**drilling foam** A bubbly fluid produced by injecting a solution of surfactant into an airstream and used when drilling in unconfined aquifers. The fluid carries the cuttings along with water to the surface and helps stabilize the borehole.

**drilling machine** A tool attached to a water main for cutting and tapping a hole for the installation of a corporation cock; same as a tapping machine. Also called pressure-tapping machine if there is flow in the main, or dry-tapping machine otherwise.

**drilling mud** (1) A heavy suspension used in drilling an injection well, introduced down the drill pipe and through the drill bit (EPA-40CFR144.3). (2) A water-based or oil-based suspension of clays introduced into an oil well during drilling. Same as drilling fluid.

**drilling rig** A trailer- or truck-mounted machine used for drilling a borehole.

**drinking fountain** A fountain that ejects a jet of water for drinking without a cup.

**drinking water** Water intended for human consumption or for use in the preparation of food or beverages, or for cleaning articles used in the preparation of food or beverages. Often used interchangeably with potable water, which implies that the water is safe, i.e., the water delivered to the consumer meets the established physical, chemical, bacteriological, radionuclide, and pressure standards.

**drinking water compliance rate** A key indicator that water utilities can use to assess their performance compared to established standards. It measures the percentage of days in a reporting year during which the water supply system is in compliance with the primary drinking water regulations—maximum contaminant levels and mandated treatment methods.

**drinking water contaminant candidate list** A list of contaminants regulated under the Safe Drinking Water Act, published by the USEPA in March 1998, with scheduled revisions every 5 years.

**drinking water equivalent level (DWEL)** Estimated exposure (in mg/L) that is interpreted to be protective for noncarcinogenic endpoints of toxicity over a lifetime of exposure. DWEL was developed for chemicals that have a significant carcinogenic potential. It provides risk managers with evaluation information on noncancer endpoints, but infers that carcinogenicity should be considered the toxic effect of greatest concern. It can be estimated as follows:

$$\text{DWEL (mg/L)} = R \times \text{body weight (kg)}/\text{drinking water volume (L/day)} \quad \text{(D-71)}$$

where $R$ is the reference dose (RfD). Common assumptions are a body weight of 70 kg and a daily consumption of 2.0 liters. Then,

$$\text{DWEL (mg/L)} = 35\ \text{RfD} \quad \text{(D-72)}$$

**drinking water priority list** A list of drinking water contaminants that may pose a health risk and warrant regulation; developed by the USEPA in 1988 under the Safe Drinking Water Act. *See also* priority pollutant.

**drinking water quality standards** Rules or recommendations issued by governmental agencies and international organizations for the protection of water supply sources and the assurance of the bacteriological, chemical, and physical quality of drinking water.

**drinking water regulations** Federal or state regulations applicable to public water systems. Current (2007) USEPA drinking water regulations relate to volatile organic chemicals, synthetic organic chemicals, inorganic chemicals, trihalomethanes, surface water treatment, total coliform, lead and copper, consumer confidence reports, and radionuclides.

**drinking water standard** Numerical limits established by federal, state, or local agencies for contaminants in drinking water, e.g., limits on radioactivity and the concentrations of coliforms, chemicals, and physical parameters. *See also* Primary Drinking Water Regulations, Secondary Drinking Water Regulations, maximum contaminant level, maximum contaminant level goal.

**drinking water standards, primary** *See* Primary Drinking Water Regulations.

**drinking water standards, secondary** *See* Secondary Drinking Water Regulations.

**drinking water supply** Any raw or finished water source that is or may be used by a public water system (as defined in the Safe Drinking Water Act) or as drinking water by one or more individuals (EPA-40CFR300.5).

**drinking water treatment facility** All the installations, equipment, structures, etc. that are necessary to treat and distribute potable water to consumers.

**drinking water treatment plant** Same as water treatment plant.

**drip distribution** An onsite disposal system that distributes wastewater through flow-regulating emitters. The tubing can be installed at various depths below the ground surface. The system consists of a pretreatment device, a pump tank, a pump and controls, a flow metering device, a filtering device, and a drip distribution field.

**drip-feed chlorinator** A device that is used to apply a chlorine solution at a more or less constant rate, usually to water at low pressure. It may consist of a ceramic filter candle with a 3-mm drip aperture, the rate of feed being controlled by the head on and the size of the aperture.

**drip irrigation** An irrigation method using perforated plastic pipes and small emitters or applicators at the base of the plants. It realizes water savings (due to a reduction of evaporation and percolation) and sometimes economy of nutrients as well as salinity reduction. Also called trickle irrigation. Other basic irrigation methods include: flooding, furrow, sprinkler, and subirrigation.

**drip leg** A piece of pipe installed on a chlorine container to keep collected moisture out of the container.

**dripstone** A deposit of calcium carbonate that includes stalactites, stalagmites, columns, and cave pearls.

**drive shoe** The protective end at the bottom of a drive pipe and casing.

**driven tubewell** A type of well used for water supply in small communities: it is constructed by driving a well point (a perforated or slotted tube) into the ground.

**driving force** In membrane filtration processes, any of the forces that induce the transport of materials through the membranes, e.g., concentration gradient, electrical potential, hydraulic pressure difference, pressure gradient, temperature gradient, vacuum.

**driving head** (1) The adjusted head ($h'$) across a weir to account for the restriction on flow imposed by a tide gate. *See* effective driving head. (2) In Darcy's law, the driving head is used to determine the hydraulic energy loss.

**drop** (1) A hydraulic structure used to drop the water level and dissipate energy. Same as drop structure. *See also* chute, drop spillway, and fall. (2) The magnitude of the actual drop in water level or the difference in water surface elevations upstream and downstream of an obstruction. *See also* drawdown (2) and fall.

**drop inlet** An overfall structure consisting of a vertical riser connected to a discharge pipe; sometimes used in conjunction with an orifice in a combined sewer overflow diversion structure.

**drop manhole** A manhole used to match the grades of two sewer lines with a considerable difference (e.g., more than 2 ft). The wastewater from the incoming line falls through a vertical pipe to the elevation of the outgoing line. *See* Figure D-16.

**drop pipe** The suction pipeline of a deep-well pump. It is common practice to disinfect the drop pipe and cylinder of a new or repaired well before lowering the assembly into the casing.

**drop structure** A structure, such as a drop inlet, drop manhole, drop spillway for dropping water, stormwater, or wastewater to a lower level. The

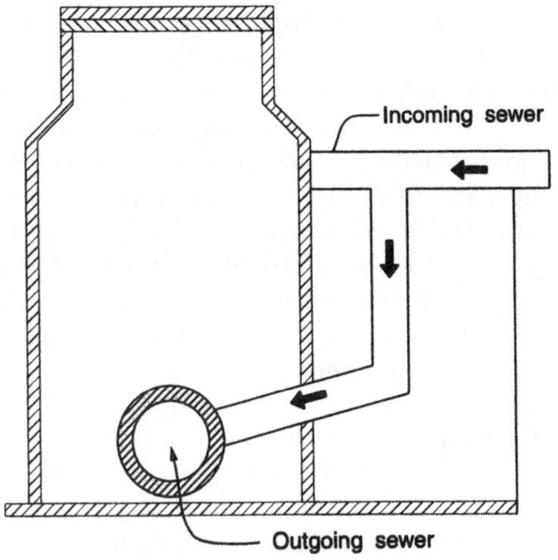


**Figure D-16.** Drop manhole (courtesy CRC Press).

drop, vertical or inclined, is for energy dissipation or other purposes. (The advanced interconnected pond routing model allows for the simulation of drop structures.)

**drop test** A test used to estimate water leakage by shutting down the facilities and measuring the drop in downstream pressure or the water level drop in storage facilities.

**dropping head** The head reduction that occurs on a reservoir's orifice or outlet pipe when outflow exceeds inflow.

**drought** An extended, indefinite, period of dry weather or deficient precipitation resulting in an inability to meet normal domestic, industrial, or agricultural water demands. Drought is sometimes defined in terms of a particular project by determining low-flow frequency curves or flow-duration curves. Drought conditions refer to an extended period of dry weather.

**drought condition** *See* drought.

**drought price** A special inclining block price applied by a water during a drought to encourage conservation.

**drowned sluice gate** A hydraulic structure such that a downstream obstacle or control causes the depth of water below the gate to exceed the width of the gate. *See* Figure D-17.

**drowned weir** Same as submerged weir.

**drug manufacturing wastewater** *See* pharmaceutical waste.

**drug residual** A residue of pharmaceutical products found at very low concentrations in wastewater after their use and excretion by water consumers. These products are predominantly organic, highly soluble compounds that are not easily degradable in the gut. Also called pharmaceutically active compound, pharmaceutical residual, or pharmaceutical residue. *See also* xenobiotic.

**Drumm-Scale™** A weighing scale manufactured by Force Flow Equipment.

**drum screen** A cylindrical screening device used in water or wastewater treatment to remove floating and suspended solids. The screening or straining medium is mounted on a rotating cylinder.

**Drumshear** A rotating fine screen manufactured by Aer-O-Flo Environmental, Inc.

**Drumstik** A tube manufactured by Stranco, Inc. for chemical feed in drum units.

**drum submergence** A dimensionless parameter in the estimate of a vacuum filter cake yield based on the specific resistance equation; it is defined as the fraction of the drum circumference below the sludge surface in the pan.

**drum-type screen** *See* drum screen.

**dry absorption** A method of controlling acids in flue gas emissions by injection of dry calcium.

**dry accumulation rate** The rate of accumulation (exceptionally the rate of decrease) of sediment from runoff events during dry weather. *See also* wet accumulation rate.

**Dry-All** A vacuum belt press manufactured by Baler Equipment Co.

**drybed** A shallow, artificial pond in which sewage sludge is dumped and dried under the influence of solar heat. Also called an evaporation pond. *See also* drying bed.

**dry-bulb temperature** The temperature of air when measured by a conventional thermometer.

**dry chemical feed** The introduction of treatment chemicals into water or wastewater using a dry chemical feeder.

**dry chemical feeder** A mechanical device that dispenses chemicals (or other fine materials) in

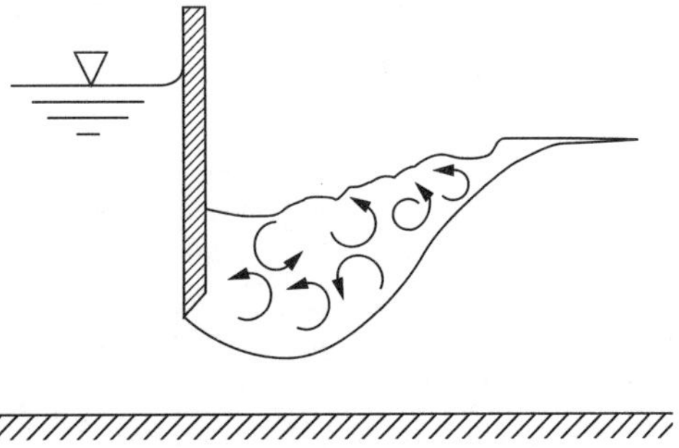

**Figure D-17.** Drowned sluice gate.

dry form to water or wastewater. It controls the dosage by the rate of volumetric or gravimetric displacement of the chemicals. A typical dry-chemical-feed system consists of a storage hopper, a dry chemical feeder, a dissolving tank, and a pumped or gravity distribution system. *See* Figure D-18. Examples of chemicals that are fed dry, in granular or powdered forms, are alum and ferrous sulfate. *See* liquid-chemical-feed system.

**dry chemical feeding** *See* dry chemical feed.

**dry-chemical-feed system** *See* dry chemical feeder.

**dry cleaning wastes** Wastewater from laundries that use nonaqueous solvents in their operations. *See* laundry wastes.

**dry combustion** The ignition and incineration of heat-dried wastewater treatment sludges at high temperature, with or without additional fuel.

**dry connection** Service connection to a water or wastewater line when the line is empty or otherwise not in service.

**dry day** *See* wet day.

**dry-day, dry-period flow** The average wastewater flow during a dry day of the dry season, i.e., without any influence of rain; it may, however, contain a certain quantity of groundwater infiltration. Same as dry-weather flow, the sum of base wastewater flow and infiltration.

**dry density** The mass of deposited sediment or powdered or granulated solid material per unit of volume. Also called bulk density.

**dry deposition** Material deposited from the atmosphere on the earth's surface in the absence of precipitation. That includes particles and such gases as sulfur dioxide ($SO_2$), nitric acid ($HNO_3$), and ammoniac ($NH_3$). In forests, sulfate ($SO_4^{2-}$), nitrate ($NO_3^-$), and hydrogen ($H^+$) ions are more prevalent. *See also* wet deposition.

**dry detention basin** A basin that detains temporarily a portion of stormwater runoff, releases it slowly to reduce flooding and pollutant concentration, and dries out between rain events. *See also* extended detention basin.

**dry excreta disposal system** A disposal system that does not use water to carry excreta; a night-soil system as compared to a wet or sewer system.

**dry feed** *See* dry chemical feed.

**dry feeder** *See* dry chemical feeder.

**dry feeding** *See* dry chemical feeding.

**dry ice** A refrigerant, solidified carbon dioxide, that vaporizes directly without passing through the liquid phase. It is used to remove incrustations from a well screen. The product is placed above screen level and agitated intermittently for a few hours. Subsequent pumping of the well removes the dry ice and loosened incrustations. *See also* polyphosphate treatment, acid treatment, chlorination.

**drying** An open-air or mechanical liquid–solids separation process used in sludge handling to increase solids concentration to more than 35% by weight, or even 50% for landfilling. Common dewatering methods include solar drying, lagooning, and filter pressing. Thermal drying is also feasible; e.g., the Carver Greenfield process. *See also* flash drying, thickening and dewatering.

**drying agent** A substance that removes water or moisture, e.g., calcium oxide and silica gel.

**drying bed** A water or wastewater treatment unit that separates water and solids in residuals. It consists of confined, underdrained shallow layers of sand, gravel, or other natural or artificial porous

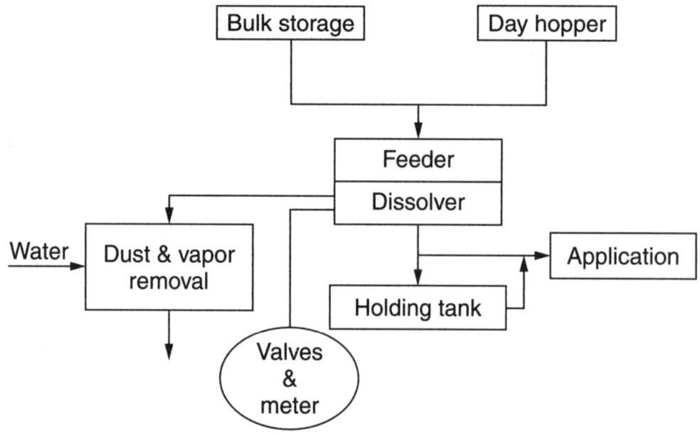


**Figure D-18.** Dry chemical-feed system.

materials on which wet sludge is distributed for draining and evaporation. Drying beds may be open to the atmosphere or covered as in a greenhouse. They are simple to operate but require large land areas. After several weeks (or months, depending on climate), the sludge cake formed on the surface is removed manually or mechanically for landfill disposal or use as a soil conditioner. The term also applies to underdrained, shallow, dyked, earthen structures used for drying sludge. Also called sludge bed, sludge drying bed. *See also* air drying, (conventional) sand drying bed, drybed, evaporation pond, paved drying bed, artificial-media drying bed, vacuum-assisted drying bed, wedge-wire drying bed, and solar drying bed.

**drying lagoon** A lagoon used for dewatering digested sludge, as an alternative to a drying bed, particularly in areas of high evaporation rates. It includes facilities for supernatant decanting and recycling to the treatment units.

**drying period** A resting period provided in the operation of rapid wastewater infiltration systems for soil reaeration and biological reactions between applications. Drying periods vary from 4 to 16 days, depending on the type of wastewater applied, the season, and the application period. *See also* operating or loading cycle, and average application rate.

**drying process** A sludge or other waste treatment process that produces a residue that is generally solid, as compared to the more viscous residue of evaporation.

**dry intake** A towerlike structure built to draw water from a lake or a river; it consists of an outer dry well, a central wet well, and gated pipes that bridge them and transfer water to the intake conduit through the wet well. *See also* wet intake.

**dry pit** An enclosure in which a pump is installed, providing access for maintenance without contact with the liquid, which is in an adjacent wet well. *See also* dry well.

**dry-pit pump** A pump operated without a wet well, the liquid being conducted to and from the unit by piping.

**dry polymer** A polymer in powder, granule, flake, or microbead forms with 90–95% of active constituents and a shelf life of several years. Polymers are also available in emulsion, gel, liquid, and mannich forms.

**dry rendering** A method used in rendering plants to process meat; unlike wet rendering, it does not produce a waste liquor but results in some drainage and press liquor that can be recycled. *See* rendering pollution, meat packing waste.

**dry-salt saturator tank** A brine tank, usually full of undissolved salt and with saturated brine below the undissolved salt; a type of brine tank used with most automatically regenerated home softeners because it reduces the frequency of refilling the tank with salt.

**dry scrubber** A device or method used to reduce the emission of acid gases (hydrogen bromide, hydrogen chloride, hydrogen fluoride, sulfur oxides) during sludge combustion. It uses a reagent such as lime to precipitate chlorides, sulfates, and sulfides. *See also* fluid bed combustion, wet scrubber.

**dry season** The period of the year when atmospheric precipitation is less intense than the average. For example, the dry season in southeastern Florida is between November and April, when about one third of the annual precipitation of 60 inches falls.

**dry solids** A measure of the solids content of a water or wastewater sample on a dry basis.

**dry suspended solids** The weight of suspended matter in a sample of water or wastewater after drying for a specified period of time and a specified temperature.

**dry-tapping machine** *See* drilling machine.

**dry unit weight** The dry weight of a unit of sludge or other semisolid residual, expressed in pounds per cubic foot or $kN/m^3$. *See also* zero-air-void curve and standard Proctor compaction.

**dry vent** A vent that does not carry any liquid.

**dry-weather flow** (1) The flow of wastewater in a sanitary or combined sewer system during dry weather. It is not affected by recent or current rain; it includes domestic, industrial, commercial, and institutional wastes as well as groundwater infiltration, but excludes any storm or surface water. Same as dry-day, dry-period flow. (2) Stream flow during dry weather, consisting usually of groundwater. *See* base flow.

**dry weight** The weight of a compound or body without its moisture or water content; a basis used to express the concentration of a substance in living organisms.

**dry weight basis** Calculated on the basis of having been dried at 105°C until reaching a constant mass, i.e., essentially 100% solids content (EPA-40CFR503.9-h).

**dry-weight capacity** (1) The dry weight of a chemical that can be stored in a given container. (2) The capacity of an ion-exchange resin, expressed in milliequivalents per gram of dry resin. *See also* wet-volume capacity, operating capacity.

**dry well** (1) The dry compartment of a pumping station where the pumps are located, as opposed to the wet well. *See also* dry pit. (2) Same as drainage well (1). An infiltration device for stormwater disposal, sometimes installed on residential properties to receive water from roof drains. (3) A well that has been completely drained or a well that produces no water.

**DS** Abbreviation of dissolved solids.

**DSFF reactor** *See* downflow stationary fixed-film reactor.

**DSS Environmental two-stage filter** An advanced wastewater treatment device that uses two deep-bed upflow, continuous backwash filters in series for the removal of turbidity, total suspended solids, and phosphorus. *See also* depth filter for other similar devices.

**DST™** Acronym of defined substrate technology, a process developed by Environetics, Inc. to achieve the growth of specific microorganisms.

**DSVI** Acronym of diluted sludge volume index.

**D/T** Abbreviation of dilution to threshold.

**DTCT** *See* dimethyl-2,3,5,6-tetrachloroterephthalate.

**D Tech™** A series of portable kits manufactured by EM Industries, Inc. for environmental testing.

**Dual** The commercial name for the chemical metolachlor. *See also* Primext.

**Dualator®** A gravity sand filter manufactured by Tonka Equipment Co.

**dual biological treatment system** A wastewater treatment system that incorporates elements of both fixed film and suspended growth methods, e.g., trickling filtration and activated sludge (Metcalf & Eddy, 2003 and WEF & ASCE, 1991). Sometimes called combined, coupled, hybrid, or series system. Some dual systems are described under combined filtration–aeration process. Other, less common, systems include combinations of a roughing filter with an aerated lagoon, a facultative lagoon, a pure oxygen activated sludge unit, or a rotating biological contactor.

**dual chamber** A water processing system composed of two separate tanks or compartments vertically connected, one above the other, and operated by one common set of valve controls.

**dual coagulation** The coagulation process used in water treatment for two purposes, e.g., removal of both particles and natural organic matter. This is different from double coagulation.

**dual digestion system (DDS)** Sludge digestion in two stages: aerobic thermophilic digestion using air or high-purity oxygen, followed by mesophilic anaerobic digestion. It is usually applied in pure-oxygen activated sludge plants.

**Dual Disc® Cushion Swing** A line of check valves manufactured by Val-Matic. *See* Figure D-19. *See also* Swing-Flex®, Titled Disc® Cushion Swing, and Silent Check.

**dual distribution system** (1) A water distribution system that provides water at normal pressures of 150–300 kPa (20–40 lb/in$^2$) for ordinary household use and at a pressure of 2100 kPa for fire fighting in high-value districts. (2) A water supply that has two separate piping networks: one for drinking water and the other for nonpotable uses (e.g., reclaimed water for lawn watering and untreated surface supply for industrial fire fighting). Also called dual supply, two-main system. *See also* dual-main grid, dual plumbing system, and water reuse applications.

**dual flow** A combination of radial flow and axial flow.

**dual-flow mixer** A combination mixer used to disperse and suspend slurries such as powdered activated carbon. It consists of an upper, radial flow turbine and a lower, axial flow impeller mounted on a single vertical shaft. *See* Figure D-20.

**dual-flow screen** A moving screen installed in a channel with water entering through two panels and leaving through the center.

**dual-flush device** *See* dual-flush toilet (2).

**dual-flush toilet** (1) A toilet that can be flushed at either of two flow rates, e.g., 1.6 and 0.8 gpf. *See also* high-efficiency, high-flow, and low-flow toilets. (2) A toilet with a special retrofit device that allows the use of a portion of the tank for an ordinary flush or the full tank to flush solids.

**dual-function media** Any filter or ion exchange media that is used to perform two treatment steps in one application; e.g., activated carbon adsorp-

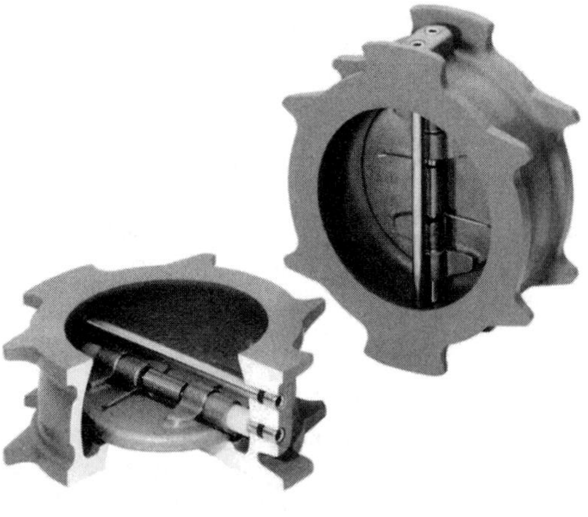

**Figure D-19.** DualDisc® (courtesy Val-Matic).

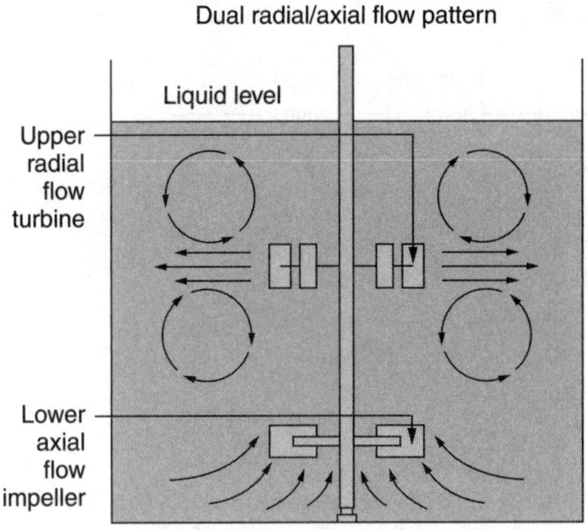


**Figure D-20.** Dual-flow mixer.

tion and filtration, or cation resin softening and dissolved iron removal.

**dual-main grid** A water distribution system consisting of two lines on each street to serve customers on both sides, with appropriately placed service headers, as compared to a single-main grid. The dual-main system is more useful against breaks and avoids dead ends. It is different from the dual distribution system.

**dual-media bed** A bed of two layers of different granular media, each with its own specific gravity, which prevents bed stratification and promotes longer filter runs. Typical dual-media beds commonly used in water and wastewater treatment include (a) activated carbon–sand, (b) anthracite–sand, (c) resin–sand, and resin–anthracite. Garnet and ilmenite are often used in multimedia beds.

**dual-media filter** A deep-bed filter that uses two types of granular media one on top of the other, usually silica sand and anthracite coal. Compared to a monomedium bed, a dual-media filter encourages penetration of solids in the bed and reduces the rate of head loss development while lengthening filter runs. *See also* triple-media or mixed-media filter.

**dual-media filtration** Water or wastewater filtration using a dual-media filter.

**dual-membrane system** A membrane filtration plant that includes two processes, e.g., the combination of a low-pressure unit such as ultrafiltration or microfiltration and a high-pressure unit such as reverse osmosis or nanofiltration, for the removal of multiple contaminants such as particles and dissolved materials. Also called integrated membrane system.

**dual oxidation control** A procedure that monitors residuals and measures oxidation–reduction potential to control disinfection and dechlorination of wastewater.

**dual plumbing system** The use of two sources of water by a residence or other consumer: one plumbing set for drinking and the other for nonpotable uses such as washing, laundering, toilet flushing, and landscape watering. *See also* dual distribution system.

**dual pond** Same as retention/detention pond.

**dual-powered flow-through lagoon system** A wastewater treatment system that consists of (a) a complete-mix lagoon for bioconversion of the substrate and flocculation of the biomass and (b) two or three separate or contiguous facultative lagoons that provide solids separation, stabilization, and storage. Both the complete-mix and facultative lagoons require a power input. *See also* lagoons and ponds.

**dual process** A wastewater treatment method that incorporates two different processes. Most plants commonly using dual processes combine a fixed film unit with a suspended growth reactor. *See also* combined filtration–aeration process and dual biological treatment system.

**dual-purpose plant** A plant that serves two purposes, e.g., an electricity generation plant that recovers waste heat to operate a water desalination facility on the same site.

**dual-sludge process** A wastewater treatment system designed for nutrient removal and including two biological reactor–clarifier sets in series, one of the reactors being aerobic and the other anoxic. There are various flowsheets depending on which reactor precedes the other and on bypass and recycling arrangements. Short-term aeration is normally provided before clarification when the aerobic basin precedes the anoxic basin. *See* Figure D-21.

**dual supply** *See* dual distribution system.

**dual system** *See* dual distribution system.

**dual vent** A vent installed at the junction of lines from two fixtures to serve both.

**dual water system** *See* dual distribution system.

**Dubinin–Raduskevich equation** An equation used in the design of adsorption of air-stripping off-gases to estimate the single-solute gas-phase capacity for a volatile organic compound (AWWA, 1999):

$$q = \rho W \exp(-B\varepsilon^2/\alpha^2) \qquad \text{(D-73)}$$

where $q$ = solid-phase volatile organic compound (VOC) equilibrium concentration, $\rho$ = liquid den-

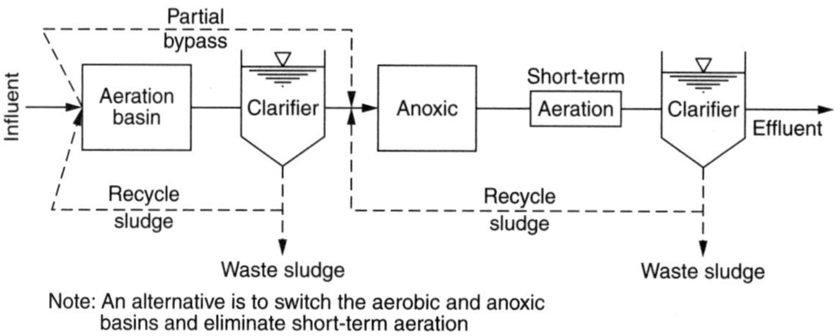


**Figure D-21.** Dual-sludge process.

sity of the VOC, $W$ = maximum adsorption space or pore volume, $B$ = microporosity constant,

$$\varepsilon = \text{adsorption potential} = RT \ln (P_s/P), \quad \text{(D-74)}$$

$\alpha$ = polarizability of the VOC (see the Lorentz–Lorentz equation), $R$ = universal gas constant, $T$ = temperature in °K, $P_s$ = saturation vapor pressure of the VOC at temperature $T$, and $P$ = partial pressure of the VOC in the gas.

**Dubl-Safe** A measuring device manufactured by Culligan International Corp. for use in the brine regeneration of softening installations.

**Duckbill®** A device manufactured by Markland Specialty Engineering for wastewater sampling.

**duckbill gate** A backwater gate used as a combined sewer overflow outlet and consisting of a durable rubber sleeve shaped like a duck's bill. *See* elastomeric check valve for detail.

**duckweed** A floating aquatic plant used in wetlands wastewater treatment systems, and to remove algae from lagoon and stabilization pond effluents. Compared to some other aquatic plants, duckweed is less sensitive to cold climates but is sensitive to wind and has a shallow root system. *See* lemnaceae and water hyacinth.

**duct** A tube, canal, pipe, or conduit for the conveyance of a fluid.

**ductile iron** A hard, nonmalleable ferrous metal strengthened by the addition of graphite nodules instead of flakes and containing other materials such as magnesium or cerium. This material is more flexible than ordinary cast iron; it is somewhat resistant to corrosion but reacts with hydrogen sulfide to form ferrous sulfide. Also called nodular cast iron. Ductile iron is used in the water and wastewater field to manufacture piping (e.g., force mains), pumps, gear, shafts, etc.

**ductile iron pipe (DIP)** A water pressure pipe made from ductile iron.

**Ductile Iron Pipe Research Association** A trade association of manufacturers of ductile iron pipe and fittings that publishes useful information about water supply and wastewater management. The association changed its name in 1980 from Cast Iron Pipe Research Association.

**Dulong formula** A formula proposed to estimate the total fuel value of a sludge ($Q$, BTU/lb):

$$Q = 14{,}544\ \text{C} + 62\ 208\ (\text{H} - \text{O}/8) + 4050\ \text{S} \quad \text{(D-75)}$$

where C, H, O, and S represent respectively the sludge percent content of carbon, hydrogen, oxygen, and sulfur. The sludge fuel value determined by the Dulong formula (WEF & ASCE, 1991) is always lower than that obtained with a bomb calorimeter, by as much as 10%. Note that combustion does not ordinarily oxidize the nitrogen in sludge. *See also* sludge fuel value.

**dummy plating** The selective plating of a trace contaminant metal out of an electroplating bath in the presence of other metals. The higher the potential of a metal in the electrochemical series, the higher the selectivity; e.g., $Au^{3+} > Ag^{+} > Cu^{2+} > Fe^{3+} > Pb^{2+} > Sn^{2+} > Ni^{2+} > Cr^{3+}$.

**dump valve** A valve at the bottom of a tank or container to empty it quickly in case of an emergency.

**Dunbar filter** A trickling filter developed in 1901; it consists of a layer of coarse sand over successively coarser broken stones.

**Dunkers** An apparatus manufactured by Munters for balancing flows.

**Duo-Clarifier** A clarifier of Dorr-Oliver, Inc.

**Duo-Deck** A floating cover manufactured by Envirex, Inc. for anaerobic sludge digesters.

**Duo-Filter** A two-stage trickling filter of Dorr-Oliver, Inc.

**Duo-Flo** A belt screen manufactured by Dontech, Inc., incorporating a traveling mesh.

**Duolite A-7** An anion-exchange resin made of hydrophilic matrix of weak-base phenol formaldehyde, regenerable with sodium hydroxide (NaOH). This resin is more effective in adsorption than ion-exchange removal of dissolved organic carbon.

**Duo-Pilot®** An apparatus manufactured by Wheelabrator Engineered Systems, Inc. for control of coagulants using pilot filters.

**DuoReel** A power cable reel manufactured by Walker Process Equipment Co.

**DuoSparj** A coarse-bubble diffuser manufactured by Walker Proccess Co.

**Duo-Vac** An inclined screen manufactured by Bowser-Briggs Filtration Co.

**DuoVAL** A gravity sand filter manufactured by General Filter Co. with automatic backwash.

**duplex pump** A reciprocating pump consisting of two side-by-side cylinders connected to the same suction and discharge lines, one exerting suction and the other exerting discharge.

**duplex station** A pumping station having two pumps.

**duplicate samples** Two or more identical samples obtained by splitting one sample.

**DuPont equation** An equation developed for the temperature correction factor (TCF) for a DuPont membrane but used for normalized flux:

$$TCF = 1.03^{(T-25)} \quad \text{(D-76)}$$

where $T$ = temperature (°C). *See also* DuPont Permasep®, Hagen–Poiseuille equation (AWWA, 1999).

**DuPont Permasep®** A manufacturer of membranes for water filtration, such as the hollow fine fiber (HFF) membranes. These membranes contain 650,000 HFFs, 4-ft long, folded in a U form, and sealed into a tube; total surface area: 1500 square feet.

**Dura-Disc** A fine-bubble membrane diffuser manufactured by Wilfley Weber, Inc.

**Dura-Fuser™** An aeration piping system designed by Davis Water and Waste Industries, Inc.

**Dura-Mix** A mixing apparatus manufactured by Wilfley Weber, Inc. for use in aeration units.

**Dura-Trac™** A streaming current sensor manufactured by Chemtrac Systems, Inc.

**Durbin–Watson test** A test used to determine serial correlation in a time series.

**Durco** Filtration equipment manufactured by Duriron Co., Inc.

**Durex®** Material produced by Envirex, Inc. for the fabrication of roller chains.

**duricrust** Same as duripan.

**duripan** In semiarid climates, a strongly compacted soil layer underlying soft materials and cemented with silica. Also called duricrust or silcrete. *See* hardpan for other similar layers.

**DuroFlow®** An air blower manufactured by Gardner-Denver.

**Duroy** A water screen manufactured by Envirex, Inc.

**dust** Fine particles suspended in air and generally smaller than 100 millimicrons; particles below 1 millimicron are categorized as fumes or smoke.

**dust and dirt buildup** An indication of the contaminants accumulated prior to a given storm; it is one of the variables considered in the assessment of water pollution from runoff. Three linear, exponential, and power functions are commonly used to estimate dust and dirt build-up (DD, pounds per 100 ft of curb) as a function of time ($t$):

$$DD = at + b \quad \text{(D-77)}$$

$$DD = c(1 - e^{-dt}) \quad \text{(D-78)}$$

$$DD = mt^f \quad \text{(D-79)}$$

where $a$, $b$, $c$, $d$, $f$, and m are empirical coefficients. *See also* contaminant washoff (Metcalf & Eddy, 1991).

**Dustube®** A fabric filter manufactured by Wheelabrator Clean Water Systems, Inc.

**duty of water** The amount of water required to satisfy the irrigation demand of land, based on careful use and management and on ordinary crops for such land. *See also* diversion duty, farm duty of water, field duty of water, gross duty, head-gate duty, high duty, low duty, net duty.

**D value** In a disinfection process, the time required to reach one log inactivation, that is, the time for a disinfectant to inactivate 90% of the pathogens. Also called decimal reduction time. *See also* thermal death time and z value.

**DVC** Acronym of double-vault composting.

**DVB** Acronym of divinylbenzene.

**dwarf tapeworm** The excreted helminth pathogen, *Hymenolepis nana,* or the infection that is causes. Common name of hymenolepiasis. It is distributed worldwide and is transmitted from the feces of infected persons or rodents.

**DWCCL** Acronym of Drinking Water Contaminant Candidate List.

**DWEL** Acronym of drinking water equivalent level.

**DWF** Acronym of dry weather flow.

**DWPL** Acronym of drinking water priority list.

**dye** A conservative, nonreactive coloring material; a liquid containing coloring matter, used as a tracer in water and wastewater applications. For example, the following dyes are used in laboratory analyses: Eriochrome cyanine R and Solochrome cyanine R-200 for aluminum; fluorogenic dyes are used in the immunofluorescence assay for detecting *Cryptosporidium* oocysts and *Giardia* cysts. Dyes are also used in stream discharge measurements. *See also* stains.

**dye dilution method** The use of a fluorescent or other type of dye in the dilution method of discharge measurements.

**Dynablend™** An apparatus manufactured by Fluid Dynamics, Inc. for mixing and feeding polymers.

**DynaClear®** A package plant manufactured by the Parkson Corp. for treating water by contact filtration.

**DynaCycle®** A regenerative oxidation apparatus manufactured by Enviro-Chem Systems, Inc.

**DynaFloc Feedwell** A clarifier feed designed by Dorr-Oliver, Inc. for mixing flocculants.

**Dyna-Grind** A grinder manufactured by FMC Corp. for screenings and wastewater solids.

**dynamic composting** One of two variations of in-vessel composting: the material to be composted (e.g., a mixture of dewatered sludge and a bulking agent) is agitated periodically in an enclosed, long rectangular or circular container or vessel for aeration, temperature control, and mixing. Curing takes place in an outside static pile. Also called agitated-bed composting. *See* sludge composting for detail (Metcalf & Eddy, 2003).

**dynamic discharge head** Same as discharge head.

**dynamic equilibrium** The condition of a system in which outflow equals inflow, e.g., an aquifer whose natural discharge matches its recharge.

**dynamic head** When a pump is lifting or pumping water, it is the vertical distance (in feet) from the elevation of the energy gradeline on the suction side of the pump to the energy gradeline on the discharge side of the pump. Also called total dynamic head (TDH) or total head (H). In general, TDH is the total energy that a pump must impart to a fluid to move it from one point to another, or the head against which the pump works. The pump head (*H*) represents the net work done on a unit weight of fluid from the inlet or suction flange to the discharge flange:

$$H = (p/\gamma + V^2/2\,g + z)_{\text{discharge}} - (p/\gamma + V^2/2\,g + z)_{\text{suction}} \quad \text{(D-80)}$$

where $H$ = pump head or pump total head, or total dynamic head (TDH); $\gamma$ = specific weight of fluid; $p$ = static pressure; $V$ = average fluid velocity; $g$ = gravitational acceleration; $z$ = elevation above or below a datum, also called elevation or potential head; $p/\gamma$ = pressure head or flow work; $V^2/2\,g$ = velocity head or kinetic energy; $(p/\gamma + V^2/2\,g + z)_{\text{discharge}}$ = discharge head, dynamic discharge head, or total dynamic discharge head; and $(p/\gamma + V^2/2\,g + z)_{\text{suction}}$ = suction head or dynamic suction head when it is positive, suction lift or dynamic suction lift when it is negative. Note that internal pump losses are not added to the suction lift nor subtracted from the suction head. Note also that acceleration head in reciprocating pumps is not a head loss. *See also* pump-head terms.

**dynamic membrane** A membrane with a layer of deposited particles that also acts as a filtration barrier.

**dynamic model** A physical model whose characteristics are proportional to the full-size object. Also called dynamic similarity model.

**dynamic pressure** The pressure exerted by a fluid in motion, as opposed to static pressure. When a pump is operating, the dynamic pressure or dynamic head is the vertical distance (in feet) from a reference point (such as the pump centerline) to the hydraulic gradeline.

**dynamic pressure head** The vertical distance from a reference point to the hydraulic gradeline for a pump in operation.

**Dynamic Probe™ Expert Transmitter** Modular or built-in equipment manufactured by Roberts Filter Group of Darby, PA. It is an intelligent sensing system that optimizes filtration and backwash, for use in rehabilitation or new construction projects.

**dynamic random access memory (DRAM)** The storage location on a memory chip for data, which are updated at very high speeds, measured in nanoseconds.

**dynamic settling** Settling of particles in moving water as opposed to quiescent settling. (In extended detention wet ponds, particle settling is assumed to be quiescent in the permanent pool between runoff events and dynamic in the active storage zone.)

**dynamic similarity model** A physical model whose characteristics are proportional to the full-size object. Also called dynamic model.

**dynamic suction head** The vertical distance from the source of supply, when pumping proceeds at required capacity, to the center of the pump, minus the velocity head, and the entrance and friction losses. Dynamic suction lift is similarly defined, but with the addition of the velocity head and entrance and friction losses. Internal pump losses are not added or subtracted in either case. *See* dynamic head and pump head terms.

**dynamic suction lift** *See* dynamic suction head.

**dynamic viscosity (μ)** A measure of the internal resistance of a fluid to tangential or shear stress, and thus to flow; equal to the ratio of the viscous shearing stress ($\tau$) to the velocity gradient ($\partial V/\partial s$); expressed in Newton seconds per square meter ($N \cdot s/m^2$) in SI units. Also called absolute viscosity or coefficient of viscosity.

**DynaSand®** A sand filter manufactured by Parkson Corp. and incorporating a moving bed that is continuously backwashed. It achieves denitrification and suspended solids removal.

**Dynasieve®** A rotary, self-cleaning screen manufactured by Andritz-Ruthner, Inc.

**Dynatherm** An in-vessel composting unit manufactured by Compost Systems, Inc.

**Dynatrol®** Devices manufactured by Automation Products, Inc. for measuring flows and liquid levels.

**DynaWave®** A scrubbing apparatus manufactured by Monsanto-Enviro-Chem Systems, Inc. using a reverse jet.

**dyne** The unit of force in the CGS (centimeter–gram–second) system: it is the force that imparts an acceleration of one cm/sec/sec to a mass of one gram.

**dysentery** Disease of the gastrointestinal tract, involving abdominal pain, inflammation of the stomach and intestinal linings, and the passing of bloody stools; usually resulting from poor sanitary conditions and transmitted by contaminated food and water; also called diarrhea or gastroenteritis. Two major pathogenic agents of dysentery are the bacterial genus *Shigella* and the protozoon *Entamoeba histolytica*. *See also* amebic dysentery, shigellosis.

**dysphothic zone** The part of the photic zone of a water body that receives only dim light. It is between the euphotic and aphotic zones.

**Dystor** An apparatus manufactured by Envirex, Inc. for holding the gas from anaerobic digesters.

**dystrophic** Pertaining to a freshwater that has a very low concentration of dissolved nutrients and plankton, is rich in colloidal humic materials, and does not support abundant plant life.

**dystrophic lake** An acidic (low pH), shallow body of water that contains much humus and/or other organic matter in suspension or on the bottom, but little calcium, phosphorus or nitrogen; it contains many plants but few fish. When it is stratified, the deep layers lack dissolved oxygen. Algal blooms are infrequent in dystrophic lakes, which may eventually develop into peat bogs or marshes. *See also* eutrophic lake, oligotrophic lake, and wetland.

**dystrophy** Disorder caused by defective nutrition or metabolism; the state of being dystrophic.

# E

**$E^0$** Symbol of electromotive force (EMF).

**E.A. Aerator** A package plant manufactured by Lakeside Equipment Corp. for wastewater treatment.

**EAD** Acronym of electroacoustical dewatering, characteristic of a press manufactured by Ashbrook Corp.

**Eadie–Hoftsee plot** A graphical representation of the Michaelis–Menten equation such as to magnify departures from linearity. It is a plot of the rate of uptake of a substance ($K$) versus the ratio ($K/C$) of this rate to the concentration ($C$) of the substance (Droste, 1997):

$$K = -C^*(K/C) + K^* \qquad \text{(E-01)}$$

where $K^*$ = the maximum rate of uptake, and $C^*$ = the concentration at which the rate is half the maximum rate. *See* Figure E-01 and Lineweaver–Burke plot.

**EAFD** Acronym of electric arc furnace dust.

**early closure device** A toilet retrofit device that saves water by closing the flapper early.

**early warning system (EWS)** An apparatus set up by drinking water utilities to detect changes in source water quality and implement appropriate protective measures. EWSs may include a combination of elements such as monitoring, detection, event characterization, communication, and responses.

**earthen reservoir** Same as earth reservoir.

**earth flow** A flow of slippery earth materials from clay terraces. Also called landslide flow.

**earth-ground** The principal grounding conductors in an electrical installation.

**earth reservoir** A lined or unlined reservoir with embankments made from excavated materials; an earthen reservoir.

**Earth's gravitational constant** *See* gravitational acceleration or acceleration of gravity.

**Earth's water distribution** *See* hydrologic cycle and Table E-01.

**earth tank** An impoundment consisting of an excavation and an earthen dam to provide water for livestock. Also called a stock pond.

**EarthTec®** A low-pH algicide/bactericide containing a high concentration of biologically active cupric ions, produced by Earth Sciences Laboratories, Inc.

**earthworm** *See* red worm.

**earthworm conversion** An emerging technology of sludge stabilization and disposal that uses earthworms to consume the organic matter in municipal wastewater treatment sludge. After screen-

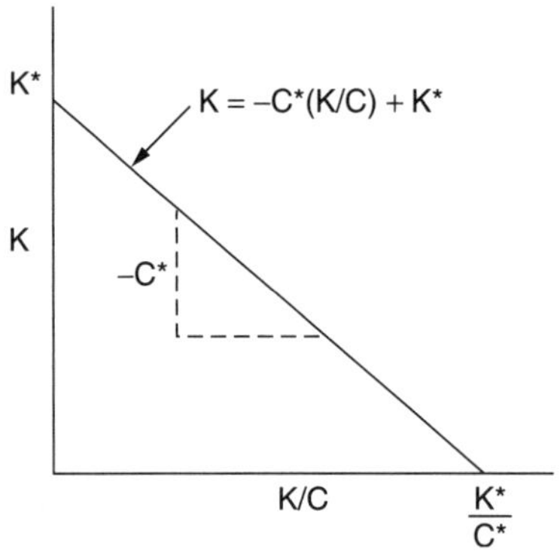


**Figure E-01.** Eadie–Hoftsee plot.

ing, the worms are used as animal feed and the worm castings as soil conditioner. Also called vermiculture or vermilization.

**earthy–musty odor** The smell of soil (dirt), caused in water by chemicals produced by blue-green algae. *See* geosmin, 2-methylisoborneol.

**easement** The legal right held by one person, usually without profit, to use the land owned by another person for a limited purpose, for example, for the construction, operation, and maintenance of water and sewer facilities. *See also* leasehold and rights of way.

**EaseOut** An apparatus manufactured by Walker process Equipment Co., incorporating an air heading and a drop pipe.

**Table E-01.** Earth's water distribution (all volumes in $10^{12}$ $m^3$)

| | | | |
|---|---|---|---|
| Total: | 1,433,378 | (100.0%) | |
| Saltwater in oceans: | 1,400,000 | (97.7%) | |
| Other than ocean water: | 33,378 | (2.3%) | |
| Snow, snowpack, ice, glaciers: | 29,000 | (86.88%) | |
| Snow, ice: | | | 86.9% |
| Groundwater: | 4,000 | (11.99%) | |
| Groundwater, soil moisture: | | | 12.2% |
| Freshwater lakes, reservoirs, ponds: | 125 | (0.37%) | |
| Fresh surface water: | | | 0.4% |
| Saline lakes: | 104 | (0.31%) | |
| Saline surface water: | | | 0.3% |
| Soil moisture: | 65 | (0.19%) | |
| Moisture in living organisms: | 65 | (0.19%) | |
| Atmospheric moisture: | 13 | (0.04%) | |
| Atmospheric and living organism moisture: | | | 0.2% |
| Wetlands: 4 (0.01%) | | | |
| Rivers, streams, canals | 2 | (< 0.01%) | |
| Total (other than ocean) | 33,378 | (100%) | |

**Eastern Econoline** A line of mixers manufactured by EMI, Inc.

**EasyLogger®** A stormwater data logger made by Omnidata International, Inc. of Logan, UT.

**eau de Javelle** A solution of sodium hypochlorite (NaOCl) in water, used as a bleach, antiseptic, etc. *See* Javelle water and Javel® water for detail. After the former town of Javel, now a part of Paris, France.

**EBC** Acronym of equivalent background compound.

**EBCT** Acronym of empty bed contact time.

**EBDC** Acronym of ethylene bisdithiocarbamate.

**eberthella group** A group of bacteria that includes the species *Salmonella typhosa,* which transmits typhoid fever.

**EBIC** Acronym of Environmentally Balanced Industrial Complexes.

**EBPR** Acronym of enhanced biological phosphorus removal.

**ebullition** The act or process of boiling of a liquid, with the formation, rise, and escape of bubbles.

**EC** Acronym of (1) effective concentration, (2) electrical conductivity.

**$EC_{50}$** Symbol of median effective concentration. (1) The experimentally derived concentration of a test substance in dilution water that is calculated to affect 50% of a test population during continuous exposure over a specified period of time, the effect measured being immobilization (EPA-40CFR141.43). (2) The experimentally derived concentration of a chemical in water that is calculated to induce shell deposition 50% less than that of the controls in a test batch of organisms during continuous exposure within a specified period (EPA-40CFR797.1800-2 & 1830-6). *See also* toxicity terms and EC-X.

**$EC_{dw}$** Symbol for the salinity of the drainage water percolating below the root zone, a parameter used to assess the potential effects of reclaimed water, used in irrigation, on crop yield and on groundwater quality:

$$EC_{dw} = EC_w/LF \qquad \text{(E-02)}$$

where $EC_w$ is the salinity of the irrigation water and LF is the leaching fraction (the fraction of applied water that passes through the root depth and percolates below). A conservative estimate for $EC_{dw}$ is $EC_e$, the salinity of the saturation extract of the soil sample; a more reasonable estimate for most soils is $EC_{dw} = 2\ EC_e$. *See also* salinity, electrical conductivity

**$EC_e$** Symbol for the salinity of the saturation extract of a soil sample. *See* $EC_{dw}$.

**ECH** Abbreviation of epichlorohydrin.

**echinococcosis** A disease transmitted by the ingestion of infective eggs in water or food contaminated by dog feces. Also called hyadatid disease.

**Echo-Lock** A compensative device manufactured by Marsh-McBirney, Inc. for flow meters.

**echovirus** The acronym of enteric cytopathogenic human orphan virus, a potentially dangerous enterovirus occasionally found in drinking water and pathogenic to humans; (orphan, because it was not known originally to cause any disease). It may cause diarrhea, meningitis, respiratory disease, rash, and fever.

**Eckenfelder's formulas** Two formulas proposed in 1963 for the design of trickling filters, to include the effects of specific surface area, temperature, and flow rate:

Without recirculation: $S_e/S_0 = \exp[(-K_s A_s^{1+m} D)/q^n]$ (E-03)

With recirculation: $S_e/S_0 = [\exp(-k_s D/q^n)]/[(1 + R) - \exp(-k_s D/q^n]$ (E-04)

where $S_e$ = effluent soluble $BOD_5$, mg/L; $S_0$ = influent soluble $BOD_5$, mg/L; $K_s$ = treatability coefficient based on soluble BOD, gpm $^{0.5}$/ft$^2$; $A_s$ = clean surface area, ft$^2$/ft$^3$; $m$ = coefficient accounting for surface loss with increasing area; $D$ = depth of media, feet; $q$ = influent rate, gpm/ft$^2$; $n$ = hydraulic coefficient; and $k_s = K_s A_s$; $R$ = recycle ratio. Other trickling filter design formulas include British Manual, Galler–Gotaas, Germain, Howland, Kincannon and Stover, Logan, modified Velz, NRC, Rankin, Schulze, Velz.

**ECL** Acronym of electrochemiluminescence, a technology used in chemical and biological analyses.

**Eclipse®** A dispersion polymer produced by Calgon Corp.

***E. cloacae*** *See Enterobacter cloacae.*

**EC medium** A culture medium for the detection of coliform bacteria and *E. coli* in water, milk, shellfish, and other material. It contains peptone, lactose, sodium chloride, bile salts, and a strong potassium phosphate buffer. Fecal coliforms from feces of humans or warm-blooded animals produce lactic acid and gas in EC medium after about 24 hours of incubation at about 45°C. *See also* P-A broth, lauryl tryptose broth, M-endo medium.

**EcoCare™** A product of Nature Plus, Inc. to control odors.

**Ecochoice** An apparatus manufactured by Eco Purification Systems USA, Inc. for the catalytic oxidation of dissolved organics.

**Ecodenit** A process developed by Biwater-OTV, Ltd. for biological nitrate removal.

**EcoDry** A sludge drying system of Andritz-Ruthner, Inc.

**ecokinetics** The study of rates of uptake and metabolism of substances by organisms in an ecosystem.

***E. coli*** Abbreviation of *Escherichia coli.* Same as *B. coli.*

***E. coli* O157:H7** Abbreviation of *Escherichia coli* O157:H7.

***E. coli* diarrhea** An enteric disease caused by *E. coli* bacteria, with symptoms varying from a mild form to those of cholera and dysentery.

**Ecolo-Chief** A package plant manufactured by Chief Industries, Inc. for wastewater treatment.

**EcoLogic™** A wastewater treatment scheme developed by Atlantic Ultraviolet Corp. using aeration, ultraviolet light, and ozonation.

**ecological amplitude** The range of environmental conditions in which an organism can function.

**ecological engineering** The use of ecological processes in natural or man-made systems to achieve engineering objectives; e.g., composting systems and constructed wetlands.

**ecological impact** The effect that a man-made or natural activity has on living organisms and their nonliving (abiotic) environment.

**ecological indicator** (1) A characteristic of the environment that, when measured, quantifies the magnitude of stress, habitat characteristics, degree of exposure to a stressor, or ecological response to exposure. It is a collective term for response, exposure, habitat, and stress indicators. (2) A community or organism whose presence is indicative of a set of environmental conditions; a plant or animal species, community, or special habitat with a narrow range of ecological tolerance.

**ecologically sustainable development** Development at such a rate that allows an ecosystem to sustain itself. *See* sustainable development.

**ecological model** One of three types of model developed to represent the biological, chemical, and physical processes of water quality. As an extension of kinetic models, ecological models focus on specific portions of the aquatic food chain in addition to the reaction-rate constants. *See also* biocoenotic models and kinetic models.

**ecological niche** The position, function, or status of an organism in a community of plants and animals that determines its activities and interactions with other organisms. Also called niche.

**ecological refugees** Populations fleeing environmental disaster areas or destitution. *See* environmental refugees.

**ecological risk assessment** The application of a formal framework, analytical process, or model to estimate the effects of human actions on a natural resource and to interpret the significance of those effects in light of the uncertainties identified in each component of the assessment process. Such analysis includes problem formulation, initial hazard identification, exposure and dose–response assessments, risk characterization, and risk management. *See also* stressor, toxicity terms.

**ecology** The relationship of living things to one another and their environment, or the study of such relationships. Applied ecology deals with the management of habitats and the consequences of pollution.

**Ecomachine** A belt filter press manufactured by WesTech Engineering, Inc.

**Econ-Abator®** Catalytic oxidation equipment manufactured by Wheelabrator Clean Air Systems, Inc.

**Econex** Ion exchange equipment manufactured by Ionics, Inc., incorporating counterflow regeneration.

**Econ-NOx™** Selective catalytic reduction equipment manufactured by Wheelabrator Clean Air Systems, Inc.

**econometric forecasting** The use of water prices, household income, and other economic variables to predict water demand.

**economic development rate** A water pricing schedule designed to entice commercial and industrial customers to remain or locate in the utility's service area.

**economic groundwater yield** The maximum rate at which groundwater can be withdrawn without depleting the aquifer or impacting its water quality.

**economic life** The number of years after which structures, equipment, and other capital goods should be replaced or abandoned to minimize their life cycle costs (capital, operation, maintenance, repair). Sometimes called project life.

**economic yield** *See* economic groundwater yield.

**Economixer** A solid bowl centrifuge manufactured by Centrisys Corp.

**economy** In thermal desalination, the efficiency of distillate production, e.g. in pounds per 1000 BTU input.

**economy of scale** The reduction in unit capital or fixed cost resulting from an increase in the size of production. Water and wastewater facilities usually exhibit such a reduction, so that their capital or construction cost [$C(Q)$] is sometimes represented as a function of their capacity ($Q$) modified by a coefficient ($a$) and the economy-of-scale parameter ($b$):

$$C(Q) = a \cdot Q^b \qquad \text{(E-05)}$$

**economy-of-scale parameter** *See* economy of scale.

**Econopure** A reverse osmosis unit of Osmonics, Inc.

**ecoparasite** A parasite that feeds on a specific host or group of hosts.

**Ecopure** A series of products of Durr Environmental Division for the control of volatile organic compounds and particulates.

**ecorefugees** Same as ecological refugees, i.e., populations fleeing environmental disaster areas or destitution. *See* environmental refugees.

**ecoregion** A homogeneous area defined by similarity of climate, landform, soil, potential natural vegetation, hydrology, or other ecologically relevant variables. Regions of ecological similarity help define the potential designated use classifications of specific water bodies.

**ecorock** A hard material produced from the residues of the incineration of municipal solid waste and sludge; it is used as an aggregate for road construction.

**Ecorock** A process developed in 1982 by the Philadelphia Water Department; it produces a hard, dense rock from the ashes of an incinerated, combined material consisting of ground municipal solid waste and dewatered sludge. The rock, produced by fusing the ash in a furnace at 2000°F, is marketed as a road material.

**Ecosorb®** A chemical product of Odor Management, Inc. for neutralizing odors.

**ecosphere** The "bio-bubble" that contains life on Earth, in surface waters, and in the air. *See also* biosphere.

**ecosystem** The interacting system of a biological community and its nonliving surroundings. Examples of ecosystems include the Earth, the atmosphere, a freshwater pond, a forest, etc.

**ecotoxicology** The study of environmental responses to the effects of toxic substances.

**ecotoxic substance** A substance that, if released, presents or may present immediate or delayed adverse impacts to the environment by means of bioaccumulation and/or toxic effects upon biotic systems.

**ecotype** A local variety, subspecies, or race that resulted from and is especially adapted to a particular set of environmental conditions.

**EcoVap™** An apparatus manufactured by Wheelabrator Clean Air Systems, Inc. for the control of volatile organic compounds.

**ectoparasite** A parasite that lives on the surface of a host organism, or for long or short periods on the outside of the host. *See also* endoparasite.

***Ectothiorhodospira*** A genus of anaerobic phototrophic, sulfur-oxidizing bacteria.

**E curve** A tracer response curve, i.e., a graphical representation of the distribution of a tracer versus time of flow, for a pulse input, with output concentrations normalized by dividing the measured concentrations by such a factor that the area under the normalized curve is equal to 1. *See* exit age curve for more detail.

**$EC_w$** Symbol for the salinity of irrigation water. *See* $EC_{dw}$.

**EC-X** The experimentally derived chemical concentration that is calculated to effect X percent of the test criterion (EPA-40CFR797.1075-1). *See also* $EC_{50}$.

**ED** Acronym of (a) effective dose and (b) electrodialysis.

**$ED_{10}$** Diameter of the 10th percentile sand (or other media) size; the particle diameter for which 10% of the particles in a sample have a smaller or equivalent diameter. It is used to define filter media. Also called $d_{10}$. *See also* effective diameter, $d_{90}$.

**$ED_{10}$** or **10 percent effective dose** An estimated dose, expressed in milligrams toxicant per kilogram body weight per day (mg/kg-day), associated with a 10% increase in response over control groups. For hazard ranking system (HRS) purposes, the response considered is cancer (EPA-40CFR300-AA).

**$ED_{50}$** A measure of odor strength in a sample of water or wastewater, the median effective dilution, or the number of times an odorous air sample must be diluted for the average person to barely detect an odor. *See also* threshold odor number, minimum detectable threshold odor concentration, dilution to threshold.

**EDB** Abbreviation of ethylene dibromide.

**EDC** Acronym of endocrine-disruptive chemical.

**ED cell** *See* electrodialysis cell.

**eddy** A vortex-like motion of a fluid running contrary to a main current; a small whirlpool; random movement of turbulent air.

**eddy current clutch** One of two common variable-torque transmission systems used to drive wastewater pumps at varying speeds. Its main parts are a constant-speed input shaft close-coupled to a synchronous or induction motor and a variable-speed output shaft connected to the pump. It also includes an electromagnet that produces eddy currents and develops torque when excited. *See also* liquid clutch.

**eddy diffusion** Mass transfer is called eddy diffusion or turbulent diffusion when it is caused by turbulence without flow. *See also* diffusion and dispersion.

**Eddyflow** An upflow clarifier manufactured by Gravity Flow Systems, Inc.

**edetic acid** Same as ethylenediaminetetraacetic acid (EDTA).

**Edge Track** A mechanism designed by Eimco Process Equipment Co. for the alignment of drum filter cloths.

**EDI** Acronym of electrodeionization.

**EDR** Acronym of electrodialysis reversal.

**EDR process** Electrodialysis reversal process.

**EDTA** Abbreviation of ethylenediaminetetraacetic acid.

**EDTA titration method** A standard method used in the laboratory to measure hardness. *See* EDTA titrimetric method.

**EDTA titrimetric method** A common method used for hardness testing, using the Eriochrome Black T dye and the property of disodium EDTA ($Na_2EDTA$) to form stable complex ions with hardness-causing divalent cations (e.g., calcium and magnesium). At pH = 10, the solution turns wine red in the presence of hardness or blue in the absence of hardness. Same as EDTA titration method.

**EDU** Acronym of equivalent dwelling unit.

**eduction** The removal of used filter sand from a filter using a water slurry.

**Eductogrit** An aerated grit chamber manufactured by Aerators, Inc.

**eductor** A hydraulic device used to create a negative pressure (suction) by forcing a liquid through a restriction, such as a venturi. An eductor or aspirator (the hydraulic device) may be used in the laboratory in place of a vacuum pump. As an injector, it is used to produce vacuum for chlorinators.

***Edwardsiella*** A waterborne bacterial agent (a genus of enterobacteriaceae) causing gastroenteritis.

**EEC** Acronym of emerging environmental contaminant.

**EED** Acronym of estimated exposure dose.

**effect** One of several units or steps in the evaporation–condensation operations of water distillation, each unit operating at a different pressure. Also called stage (of an evaporator). *See also* multiple-effect distillation and multiple-effect evaporator.

**effective acidity (EA)** The acid content of hydrolyzing metal salts used in coagulation. It depends on the percent basicity of prehydrolyzed products ($B$), the weight percent of 93% sulfuric acid solution in acid-supplemented products ($A$), the atomic weight of the metal ($W$ = 27 for aluminum and 55.9 for iron), and the percent weight metal concentration ($C$) in the product solution (AWWA, 1999):

$$EA \text{ (meq/mg metal)} = [(300 - 3\,B)/100\,W] + (A/52.7\,C) \quad \text{(E-06)}$$

For example, commercial alum with a metal concentration of 4.3% in the product solution has an effective acidity of 0.11 meq/mg, whereas ferrous sulfate with a basicity of 7% and a metal concentration of 12% has an effective acidity of 0.05 meq/mg.

**effective concentration (EC)** The concentration estimated for a constituent to cause a specified effect in a specified time period, for example the 48-hour $EC_{50}$. *See also* toxicity terms.

**effective corrosion inhibitor residual** A concentration of corrosion inhibitor sufficient to form a protective coating on the interior walls of a pipe, reducing its corrosion (EPA-40CFR141.2).

**effective depth** For a floc-blanket clarifier, the effective depth is the ratio of the total volume of blanket to the area of its upper interface with the supernatant; for a hopper it is about one-third of the actual depth.

**effective diameter (ED)** A parameter used to define soil or filter media. *See,* e.g., $ED_{10}$.

**effective discharge area** The nominal or calculated area of flow through a pressure-relief valve or similar device; it is used to determine the flow capacity of the device.

**effective dose (ED)** The sum over specified tissues of the products of the dose equivalent received following an exposure of, or intake of radionuclides into, specified tissues of the body, multiplied by appropriate weighting factors. This allows the various tissue-specific health risks to be summed into an overall health risk (EPA-40CFR191.12).

**effective dose equivalent** The sum of the products of the absorbed dose and appropriate factors to account for differences in biological effectiveness due to the quality of radiation and its distribution in the body of reference man. The unit of the effective dose equivalent is the rem (EPA-40CFR61.101-b, 61.21-b, and 61.91-a).

**effective grain size** *See* effective size.

**effective groundwater velocity** The ratio of the volume of groundwater through a formation in unit time to the product of the cross-sectional area by the effective porosity. *See* actual groundwater velocity for more detail.

**effective head** The head available for the production of hydroelectric power after the deduction of frictional, entrance, and other losses, except turbine losses. Also called net head.

**effective heating value** The amount of heat released by a fuel in a combustion chamber minus the dry flue gas and the moisture losses. *See* available heat for detail.

**effective height** The total head on a pump.

**effective MCRT** The mean cell residence time (MCRT) or solids retention time (SRT) of a dual wastewater treatment process calculated only for the suspended-growth component, neglecting the residence time in the fixed-film reactor. Effective sludge age.

**effective opening** The minimum cross-sectional area of a discharge, expressed as the diameter of an equivalent circle.

**effective plug-flow reactor** As defined by the USEPA, a plugflow reactor that has a Morrill dispersion index (MDI) equal to or less than 2.0. For an ideal plug-flow reactor, MDI = 1.0. *See also* dispersion index, volumetric efficiency.

**effective porosity** The ratio (or percentage) of the volume of liquid obtained from a saturated volume of rock or soil, under a given hydraulic condition, to the total volume of soil or rock. Also defined as the amount of interconnected pore space in a soil or rock through which fluids can pass, expressed as a percent of bulk volume. Some of the voids and pores in a rock or soil will be filled with static fluid or other material, so that effective porosity is always less than total porosity. *See also* specific yield, effective velocity.

**effective precipitable water** The maximum precipitation by convective action.

**effective rain** A rainstorm lasting a definite time unit (e.g., a day or an hour) and producing surface runoff after deduction of various losses.

**effective range** That portion of the design range (usually above 90%) in which an instrument has acceptable accuracy. *Also see* range and span.

**effective size (E.S.)** A method of characterizing filter sand: the effective size is the diameter (in mm) of the particles in a granular sample (filter medium) for which 10% of the total grains are smaller and 90% larger on a weight basis; usually designated as $d_{10}$. Effective size is obtained by passing granular material through sieves with varying dimensions of mesh and weighing the material retained by each sieve. The effective size is also approximately the average size of the grains (e.g., for sand, the 10% size by weight corresponds to the 50% size by count). Compared to coarse materials, fines produce higher head losses, retain more small particles, and require lower backwash velocities, but are more likely to produce mudballs. Also called Hazen's effective size or 10th percentile, $P_{10}$. *See also* manufacturer's rating and uniformity coefficient.

**effective size relationship** An equation that relates the effective sizes ($d_1$, $d_2$) of two filter mediums to their densities ($\rho_1$, $\rho_2$) and the density of water ($\rho_w$):

$$d_1/d_2 = [(\rho_2 - \rho_w)/(\rho_1 - \rho_w)]^{0.667} \quad \text{(E-07)}$$

It is used to determine appropriate sizes for dual- and multimedium filters (Metcalf & Eddy, 2003).

**effective sludge age ($\theta_e$)** The actual sludge residence time ($\theta_T$) in sequencing batch reactors, adjusted to take into account the fractions of the cycle time ($\theta_c$) in the fill and draw periods; it represents the time during which the biomass is active:

$$\theta_e = \theta_T (\theta_f + \theta_r)/\theta_c \quad \text{(E-08)}$$

where $\theta_f$ and $\theta_r$ represent, respectively, the fill and react times. *See also* effective MCRT.

**effective SRT** *See* effective MCRT and effective sludge age.

**effective storage** The volume of water available in a reservoir for a given purpose, excluding, e.g., dead storage. There are special considerations for power generation and flood control.

**effective surface area** In granular media filtration, the effective surface area is the actual surface area adjusted to take into account that particles shield each other.

**effective velocity** The actual velocity of groundwater, as compared to the hydraulic conductivity or coefficient of permeability or to the theoretical velocity of Darcy's law. The effective velocity ($V$) is the ratio of the theoretical velocity to the effective porosity ($P$):

$$V = KI/P \quad \text{(E-09)}$$

Where $K$ is a constant of proportionality (hydraulic conductivity or coefficient of permeability) and $I$ is hydraulic gradient, dimensionless.

**effervescence** The escape of gas bubbles from a liquid.

**efficiency** The ratio of work done or energy developed to energy supplied, expressed as a percentage; or, more generally, a measure of the effectiveness of an operation, defined as the ratio of output to input. *See* wire-to-water efficiency.

**effluent** (1) Treated or untreated wastewater that flows out of a treatment plant, sewer, or industrial outfall. Generally refers to wastes discharged into surface waters. (2) Water or some other liquid (raw, partially or completely treated) flowing from a reservoir, basin, treatment process, or treatment plant. Also, dredged material or fill material, including return flow from confined sites (under CWA) (EPA-40CFR232.2).

**effluent charge** A charge levied against a polluter for effluent discharge.

**effluent concentrations consistently achievable through proper operation and maintenance** For a given pollutant parameter, (a) the 95th percentile value for the 30-day average effluent quality achieved by a treatment works in a period of at least two years, excluding values attributable to upsets, bypasses, operational errors, or other unusual conditions; or (b) a seven-day average value equal to 1.5 times the value derived under (a) (EPA-40CFR133.101-f).

**effluent data** Information necessary to determine the characteristics of any pollutant discharged or resulting from a discharge, and the characteristics of the source of the discharge (EPA-40CFR2.302-2).

**effluent disposal** The release of treated wastewater into the environment and its reentry into the hydrologic cycle, usually by discharge and dilution into ambient waters or by land application.

**effluent drain** A drain of small diameter (e.g., 3–4 inches) and nominal gradient (e.g. 0.5%) that carries the effluent from septic tanks. Also called a small-bore sewer.

**effluent exchange** The use of wastewater discharge as compensation for upstream diversions.

**effluent guidelines** Technical USEPA documents that set effluent limitations for given industries and pollutants.

**effluent launder** A trough that discharges water or wastewater from a treatment basin (e.g., a clarifier) into the effluent piping. *See also* clarifier hydraulics, diversion dam, diverting weir, effluent weir, skimming weir.

**effluent limitation** Any restriction established by a State or the USEPA Administrator on quantities, rates, and concentrations of chemical, physical, biological, and other constituents that are discharged from point sources into navigable waters, the waters of the contiguous zone, or the ocean, including schedules of compliance. Or, simply, restrictions established on quantities, rates, and concentrations in wastewater discharges. Also called effluent standard. *See also* National Pollutant Discharge Elimination System (NPDES), water quality standard, chemical-specific approach, stream standard, whole-effluent approach.

**effluent limitation guideline** A regulation published by the USEPA under section 304(b) of the Clean Water Act to adopt or revise effluent limitations (EPA-40CFR122.2 & 401.11-j).

**effluent-limited pollutant** A pollutant whose concentration is limited by technology-based criteria that are more stringent than limits based on water quality criteria. *See also* water quality limited impact.

**effluent polishing** Further treatment of a secondary effluent, e.g., to upgrade it to tertiary standards using such methods as microscreening or filtration.

**effluent quality** The set of parameters that define the physical, chemical, and biological conditions of water or wastewater from a basin, reservoir, or pipe. For example, the effluent quality achievable by a single-stage activated sludge/nitrification process may be defined as:

| | |
|---|---|
| 5-day biochemical oxygen demand | $BOD_5$ = 5–15 mg/L |
| Total suspended solids | TSS = 10–25 mg/L |
| Chemical oxygen demand | COD = 20–45 mg/L |
| Total nitrogen | N = 20–30 mg/L |
| Ammonia nitrogen | $NH_3$-N = 1–5 mg/L |
| Phosphate (as phosphorus) | $PO_4$-P = 6–10 mg/L |
| Turbidity | 5–15 NTU |

**effluent seepage** The discharge of groundwater to the surface or a surface water body; leakage.

**effluent standard** Same as effluent limitation, including the prohibition of any discharge.

**effluent trough** A channel into which effluent flows from a wastewater treatment tank.

**effluent weir** A dam or a weir at the outflow end of a watercourse, sedimentation basin or other hydraulic structure. *See also* clarifier weir, outlet weir, diversion dam, diverting weir, effluent launder, and skimming weir.

**effluent weir loading** The hydraulic loading on the weir of a sedimentation basin, i.e., the average daily overflow divided by the total weir length, in gallons per day per linear foot.

**efflux tube** A tube inserted into an orifice for water discharge.

***E. fishelsoni*** *See Epulopiscium fishelsoni.*

**egg** A resistant form taken by a helminth to survive under adverse conditions. *See also* cyst, endospore, oocyst, spore.

**egg-shaped digester** One of three commonly used forms of anaerobic digester; a vessel similar to an upright egg with a steep cone at the bottom to minimize grit accumulation, as shown in Figure E-02. *See also* cylindrical sludge digestion tank, conventional German digester.

**EGL** Acronym of energy gradeline.

**$E_H$** The electric potential in an oxidation–reduction reaction; the redox potential; the electrode potential measured against the hydrogen electrode. *See also* standard redox potential, Nernst equation.

**EHEC** Acronym of enterohemorrhagic *E. coli.*

**E horizon** A layer on a soil profile, beneath the A horizon and above the B horizon; it contains nutrients and inorganics removed from the A horizon (by elutriation).

**$E_H$–pH diagram** *See* potential diagram or Pourbaix diagram.

**EHRC** Acronym of enhanced high-rate clarification.

**EIA** Acronym of environmental impact assessment.

***Eichhornia*** A genus of tropical freshwater plants.

***Eichhornia crassipes*** A species of prolific aquatic plants, commonly called water hyacinth, that colonizes artificial lakes and may serve as a habitat for the vector of Malayan filariasis. It is commonly used in aquatic treatment systems of wastewater and can be used as fertilizer and food for livestock.

**EIEC** Acronym of enteroinvasive *E. coli.*

**eighteen-megohm water** A high-quality water with an approximate electrical resistivity of 18 megohm-cm and a conductivity of 0.0556 micromhos/cm at a temperature of 25°C. Because it conducts electricity very poorly, it is used in the

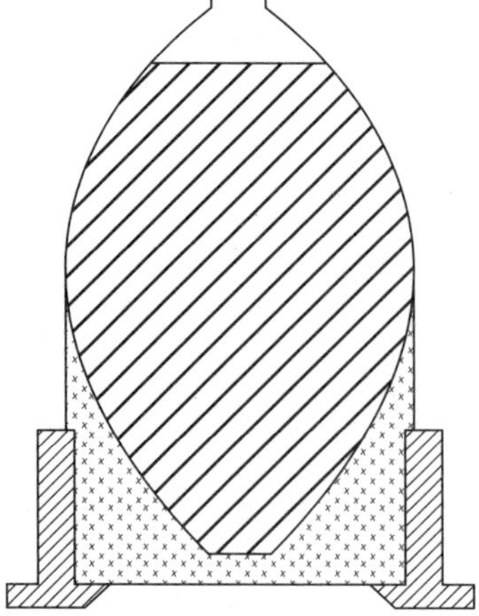

**Figure E-02.** Egg-shaped digester.

semiconductor industry as rinse water. *See also* ultrapure water, pharmaceutical-grade water, and reagent-grade water.

**EI ionization** Electron impact ionization.

**EimcoBelt®** A vacuum filter manufactured by Eimco Process Equipment Co. including a continuous belt.

**EimcoMet** A line of polypropylene components manufactured by Eimco Process Equipment Co.

**Einstein's equation** An equation for the diffusion coefficient of suspended particles, incorporated into the single collector efficiency (*E*) model for diffusive transport (AWWA, 1999):

$$E = 0.9(KT/\mu CPV)^{2/3} \quad \text{(E-10)}$$

where $K$ = Boltzman's constant, $T$ = absolute temperature, $\mu$ = absolute viscosity, $C$ = diameter of the collector, $P$ = particle diameter, and $V$ = approach velocity of flow.

**Eintein's method** A procedure proposed by H. A. Einstein in 1950 to determine total bed material and total sediment discharge in sand-bed streams.

**ejector** A device for dispersing chemical solutions into water being treated. More generally, a device that moves fluids or solids by entrainment in air or water under pressure. Also called injector.

**ejector pump** A jet pump; a pump operating on the venturi principle.

**ejector scrubber** *See* ejector-venturi scrubber.

**ejector-venturi scrubber** An air pollution control device that uses a high-pressure water or other liquid jet to remove up to 90% of particulates, fumes, dusts, mists, and vapors from gas streams; for example, in sludge incinerators. Also called ejector scrubber. *See also* cyclone scrubber, spray tower, and venturi scrubber.

**elapsed-time meter** A meter that records the duration of operation of a pump or other device. It is sometimes used in lieu of, or in addition to, a flowmeter to monitor the operation of a pumping station. If the pump operates at a constant rate per hour, the elapsed time can be used to determine the flow or volume pumped in any desired period. Some agencies place a limit on the nominal daily average pump operating time (NAPOT) of fixed-speed pumps or the nominal average power consumption (NAPC) of variable-speed and multiple-speed pumps. The elapsed-time meter serves to monitor NAPOT.

**elastic demand** A demand (for goods or services such as water supply) whose percent change is greater than a corresponding percent price change. For a water utility, elastic demand results in a decrease in revenue for a rate increase. *See also* price elasticity and inelastic demand.

**elasticity** *See* price elasticity.

**Elasti-Liner®** Material fabricated by KCC Corrosion Control Co. for containment liners.

**Elastol®** A chemical product of Lemacon Techniek B.V. used to improve oil/water separation.

**elastomer** An elastic, resilient, natural or synthetic material similar to rubber.

**elastomeric check valve** A backwater gate used as a combined sewer overflow outlet and consisting of a durable rubber sleeve shaped like a duck's bill. It discharges at small head differentials and seals around debris caught in the gate. Also called duckbill gate. *See* flap gate.

**Elastox®** A membrane air diffuser manufactured by Eimco Process Equipment Co.

**Elbac** A product of Eltech International Corp. used for wastewater bioaugmentation.

**Elbow Rake** A hydraulic trash rake manufactured by Acme Engineering Co., Inc.

**elective enrichment medium** A substance that combines nutritional or physiological elements to promote the growth of a specific or limited type of organism so that this organism can be more easily detected or enumerated. *See also* selective enrichment medium.

**Electraflote** A sludge thickener manufactured by Ashbrook Corp. using electrolysis to generate bubbles.

**electrical actuator** A device that uses an alternative current motor or a stepping motor to render a valve or a gate automatic. This type of device is replacing hydraulic-operated gates and valves in many water and wastewater applications. *See also* pneumatic actuator.

**electrical conductivity** The reciprocal of electrical resistance. *See* conductivity for detail. *See also* irrigation water quality.

**electrical double layer** The charged surface of a colloidal particle and its surrounding layer of counterions. *See* Figure E-03. This is a concept used in the theory of coagulation, ion exchange, and other water and wastewater treatment processes. *See* the following related terms: diffuse layer, counterion, co-ion, Nernst potential, Zeta potential, rigid layer, Stern layer, plane of shear, potential energy, Gouy–Chapman model, van der Waals forces, electrostatic stabilization, surface complexation model, constant capacitance model.

**electrically regenerable ion exchange** *See* LINX®.

**electrical resistance** *See* electrical conductivity.

**electrical stage** A group of membrane cell pairs in an electrodialysis unit.

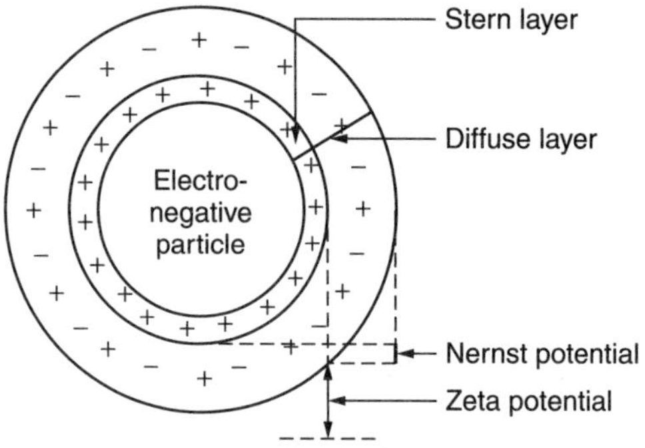


**Figure E-03.** Electrical double layer (adapted from AWWA, 1999).

**electric arc furnace dust (EAFD)** Residues containing heavy metals and resulting from the production of steel in electric furnaces.

**electric corona discharge** Corona discharge.

**electric furnace** A type of incinerator sometimes used for sludge combustion; it is a rectangular, horizontally oriented, refractory-lined shell equipped with a moving wire belt that carries the sludge beneath infrared heating elements. It includes feeding, drying/combustion, and ash discharge zones. Also called infrared furnace.

**electric glow** A luminous discharge of electricity at the surface of a conductor or between two conductors, causing the ionization of oxygen and the formation of ozone. Also called corona, corona discharge, St. Elmo's fire. *See also* brush discharge.

**electric incineration** The use of an electric furnace for sludge incineration.

**electric timer** An electrical device used to control the operation of water softeners, filters, or other water treatment units.

**electrochemical cell** A device used to measure the potential of half-reactions and consisting of two different metals in contact with each other through a conductive solution. The more resistant metal becomes the cathode and the less resistant metal corrodes as the anode. Corrosion requires the presence of an electrochemical cell. *See* Figure E-04. *See also* electrolytic cell.

**electrochemical corrosion** Corrosion caused by electrochemical reactions.

**electrochemical gaging** Using the concentration and electrical conductivity of a salt solution as a tracer to measure the flow of water.

**electrochemical generation** The development of an electric current from the reaction between anions and cations. Lightning or cathodic protection devices may trigger electrochemical processes in water storage tanks with a resulting production of undesirable chlorine species such as chlorate ($ClO_3$) and perchlorate ($ClO_4$).

**electrochemical potential** A quantity indicative of the tendency of an oxidant to take up an electron or a reductant to give up an electron; it is the driving force for initiating a chemical reaction such as corrosion. *See also* standard cell potential, Gibbs free energy, Faraday's constant.

**electrochemical rate measurement** A method for determining the rate of corrosion based on electrical resistance, linear polarization, corrosion current, or galvanic current.

**electrochemical reaction** Chemical changes produced by electricity (electrolysis) or the production of electricity by chemical changes (galvanic action). In corrosion, a chemical reaction is accompanied by the flow of electrons through a metallic path. The electron flow may come from an external force and cause the reaction, such as the electrolysis caused by a direct current electric railway, or the electron flow may be caused by a chemical reaction as in the galvanic action of a flashlight dry cell.

**electrochemical separation process** A process, such as electrodialysis, that removes ions from a feedstream, as compared to pressure driven processes that remove water from the process stream.

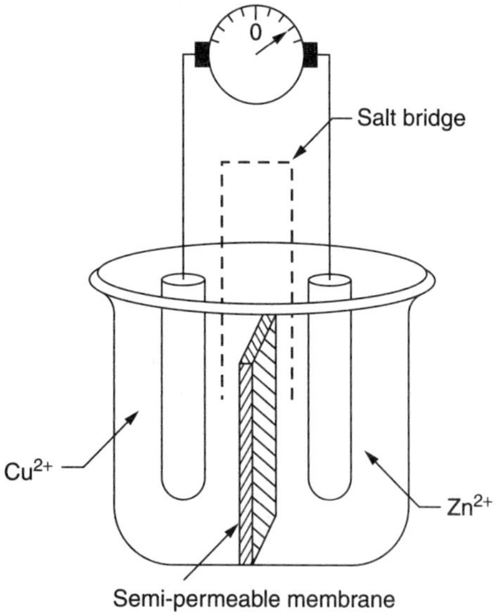


**Figure E-04.** Electrochemical cell (adapted from Droste, 1997).

**electrochemical series** A list of metals with their standard electrode potentials (in volts), which indicate how easily these elements will take on or give up electrons. Hydrogen is conventionally assigned a value of zero. *See* electromotive series, galvanic cell, and noble metal.

**electrochemistry** The branch of chemistry that studies the chemical changes produced by electricity or the chemical production of electricity.

**electrode** A conductor through which electrical current enters or leaves a nonmetallic medium such as a cell or a liquid electrolyte.

**electrodeionization (EDI)** A variation of the electrodialysis process used to remove ions from solution, as in water treatment for drinking purposes. It uses a specialized membrane containing a mixed bed resin. It is a continuous process as compared to the batch operation of ion exchange (e.g., the LINX® process).

**electrode method** An analytical procedure that uses an electrode and a volt meter to measure concentrations in water or wastewater.

**electrodeposit** Electrodeposition.

**electrodeposition** A deposit of metal or other element by electrolysis; the reduction of metal ions to metal by electrons at an electrode, a method of metal removal from solution. *See also* ion exchange, solvent extraction, cementation, and adsorption.

**electrode potential** The difference (E) in charge between metals in an electrolyte, which can generate a current and metal corrosion. It is an indication of electron activity and of the relative oxidizing or reducing tendency of a water solution: high electron activity → reducing conditions and low activity → oxidizing conditions. *See also* $E_H$, pE.

**electrodialysis (ED)** A process that uses electromotive force (direct electrical current) applied to ion-selective, cationic or anionic, semipermeable membranes, positioned between positively and negatively charged plates, to separate a solution's ionic components (e.g., to remove minerals from water, in softening or demineralization). Often used to desalinize salty or brackish water, but ineffective for pathogen and organics removal. *See also* electrodialysis reversal, electrolysis, Faraday's laws of electrolysis, anolyte, catholyte, current density, current-density-to-normality ratio, polarization, Ohm's law, makeup water, and membrane fouling.

**electrodialysis cell pair** The combination of an anion membrane, a cation membrane, and two separators or spacers to form an electrodialysis cell. *See* Figure E-05, membrane stack or module.

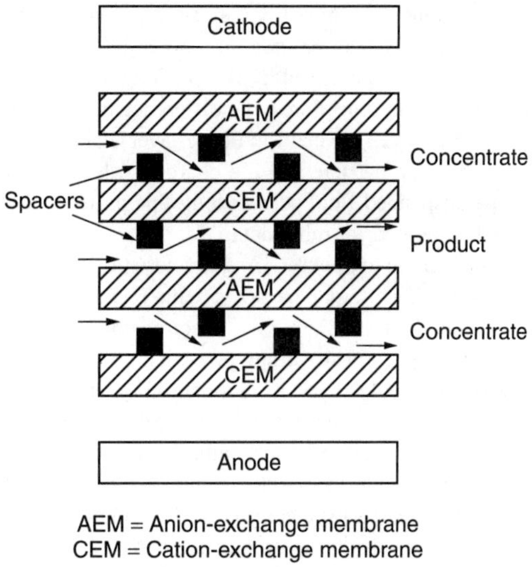


**Figure E-05.** Electrodialysis cell pair.

**electrodialysis current** The current flow ($I$, amp) required by an electrodialysis unit to treat a solution of flowrate ($Q$, L/s) and normality ($N$, eq/L) with $n$ membranes in a stack:

$$I = FQN\eta/(nE_c) \qquad \text{(E-11)}$$

where $F$ = Faraday's constant = 96,485 amp · s/eq, $\eta$ = fraction of electrolyte removal, and $E_c$ = fraction current efficiency.

**electrodialysis reversal (EDR)** A variation of the electrodialysis process in which the electrical polarity of the electrodes is reversed cyclically (e.g., every 15–20 minutes) to automatically clean membrane surfaces and reduce their scaling and fouling.

**electroextraction** The recovery of metal from metallic salts by electrolysis. *See* elctrowinning for detail.

**electroflotation** A wastewater and sludge treatment process that generates hydrogen and oxygen bubbles in a dilute aqueous solution through a direct current between two electrodes to remove suspended solids. Also called electrolytic flotation.

**electrohydraulic discharge** The rapid release of an electrical charge across electrodes submerged in a liquid, resulting in the production of ultraviolet radiation, thermal shocks, chemical radicals, and reactions that can be used for microbial inactivation and oxidation of organic compounds.

**electrokinetic measurement** A quantity used to monitor the effects of coagulants and changes in solution chemistry on the stability of particles. *See* electrophoretic mobility or zeta potential and streaming current or streaming potential.

**electrokinetic potential (ζp)** In coagulation and flocculation procedures, the difference in the electrical charge between the dense layer of ions surrounding a particle and the charge of the bulk of the suspended fluid surrounding this particle. *See* zeta potential for more detail.

**electrokinetic property** A characteristic of the behavior of colloidal particles with respect to the surrounding medium. *See* electroosmosis, electrophoresis, and streaming potential.

**electrolysis** (1) The decomposition of materials (electrolytes) by an outside electrical current: positive ions migrate to the cathode while negative ions migrate to the anode. In this process, a metal ion is reduced by electrons at the cathode while another species gives up electrons and is oxidized at the anode. It is widely used to recover metals such as cadmium (Cd), copper (Cu), gold (Au), lead (Pb), silver (Ag), and zinc. For example, for the electrolysis of a copper solution:

At the cathode: $Cu^{2+} + 2\,e^- \rightarrow Cu$ (E-12)

At the anode: $H_2O \rightarrow \frac{1}{2}\,O_2 + 2\,H^+ + 2\,e^-$ (E-13)

Net reaction: $Cu^{2+} + H_2O \rightarrow Cu + \frac{1}{2}\,O_2 + 2\,H^+$ (E-14)

(2) Sometimes used synonymously with stray current corrosion, which use is discouraged by the USEPA. *See also* Faraday's law.

**electrolysis cell respirometer** A laboratory instrument used in determining the BOD value and rate constant of a one-liter wastewater sample: it produces oxygen through an electrolytic reaction and replaces the oxygen used by the microorganisms to maintain constant the oxygen pressure over the sample. The BOD is determined from the length of time to generate oxygen and its correlation to the amount of oxygen produced. It has replaced the Gilson and Warburg respirometers in the respirometric BOD determination. Also called electrolytic respirometer. *See also* respirometric method.

**electrolyte** An ionic conductor, i.e., a substance that dissociates into two or more ions when it is dissolved in water; e.g., simple electrolytes like the coagulants and polyelectrolytes like the flocculants. Acids, bases, and salts are electrolytes. *See* Arrhenius theory.

**electrolyte solution** One of the three components of an electrochemical cell; a solution containing an electrolyte and connecting the anode and cathode of cell.

**electrolytic analysis** The use of electrolysis to determine the quantities of chemicals in solution, by measuring the potential at an electrode, the quantity of material deposited at an electrode, or the volume of gas released at an electrode.

**electrolytic cathodic protection** A method of protecting a pipe or a structure against corrosion by using a rectifier to pump electrons into them; a graphite anode becomes surrounded by ions that are oxidized. Other methods of corrosion control include anodic protection, impressed-current protection, chemical coatings, inhibition, inert materials, and metallic coatings.

**electrolytic cell** A device in which the chemical decomposition of material causes an electric current to flow; sometimes simply called cell, it consists essentially of a cathode, an anode, and an electrolyte in a container. Also, a device in which a chemical reaction occurs as a result of the flow of electric current. Chlorine and caustic soda (NaOH) are made from salt (NaCl) in electrolytic cells. *See also* electrochemical cell.

**electrolytic chlorine** Chlorine that is produced by the dissociation of hydrochloric acid (HCl) or one of its salts, such as sodium chloride (NaCl), in an electrolytic cell.

**electrolytic conductivity detector** An instrument used to detect organic substances by gas chromatography.

**electrolytic corrosion** One of two basic forms of metal corrosion, involving the action of direct electrical current from an external source, such as electrified transit, direct current welding equipment, and crane operations. *See also* galvanic corrosion.

**electrolytic corrosion cell** An electrochemical cell in which an external direct current generates corrosion.

**electrolytic dissociation** The separation of a molecule of an electrolyte into its constituent atoms.

**electrolytic flotation** A wastewater and sludge treatment process that generates hydrogen and oxygen bubbles in a dilute aqueous solution through a direct current between two electrodes to remove suspended solids. Also called electroflotation.

**electrolytic metal recovery** (1) The recovery of metal from metallic salts by electrolysis. *See* electrowinning for detail. (2) The use of special electroplating equipment to reduce the concentration of dissolved metals in rinses and concentrated-rinse tanks.

**electrolytic method** The recovery of metals from solutions through electrolysis. *See* electrowinning and dummy plating.

**electrolytic recovery** *See* electrolytic metal recovery (2).

**electrolytic respirometer** An instrument that measures oxygen uptake by microorganisms in a sample as incremental changes in gas volume while maintaining pressure constant within a reaction vessel; usually equipped for data collection and processing by computer. Also called electrolysis cell respirometer. *See* respirometric method.

**electromagnetic conductivity** The inverse of electrical resistivity; the same as electric conductivity, i.e., the ability of the earth to conduct electricity. Also called terrain conductivity.

**electromagnetic flowmeter** *See* magnetic flowmeter.

**electromagnetics** The study of the phenomena associated with electric and magnetic fields. Also called electromagnetism.

**electromagnetism** Electromagnetics.

**Electromat** Electrodialysis equipment manufactured by Ionics, Inc.

**Electromedia®** Mineral filter media produced by Filtronics, Inc.

**Electromedia® 1** A water treatment system manufactured by Filtronics, Inc. of Anaheim, CA for the removal of iron, manganese, arsenic, radium, and hydrogen sulfide.

**electrometric method** A method used in the laboratory to determine the pH of a solution. Also called glass electrode method.

**electrometric titration** The titration of a base or an acid, using a pH meter to measure the endpoints and observing the change of potential of an electrode.

**electromotive force (EMF** or **emf;** symbol: **$E^0$)** The difference in volts between the electrode potentials of two half-reactions whose combination results in no gain or loss of electrons; or the electrical pressure available to cause a flow of current (amperage) when an electrical circuit is closed. *See* voltage. A reaction tends to proceed as indicated when EMF > 0. The standard electromotive force is the potential of a cell in which a reaction occurs among substances of unit activity.

**electromotive force (EMF) series** Same as electrochemical series.

**electromotive series** A list of metals and alloys presented in the order of their tendency to corrode (or go into solution). Also called galvanic series. This is a practical application of the theoretical electrochemical series. *See* Table E-02.

**electron** An extremely small, negatively charged particle (charge = –1); the part of an atom that determines its chemical properties; mass = 0.0005 u (atomic mass units), compared to a mass of 1.007 u for a proton.

**electron acceptor** A molecule or ion, any chemical entity that accepts electrons transferred to it from another compound. It is an oxidizing agent that, by virtue of its accepting electrons, is itself reduced in the process. Also called an electrophile. *See also* covalent bond, electron donor, Lewis acid, oxidant, oxidation–reduction, and terminal electron acceptor.

**electron balance** The sum of electrons gained and electrons lost in a chemical reaction. For a balanced equation, the electron balance is neutral, i.e., 0. *See also* charge balance.

**electron beam microscope** *See* electron microscope.

**electron beam process** A disinfecting process that uses ionizing radiation (e.g., gamma radiation from cobalt-60) but no radioactive source.

**electron capture** A method used to detect trace organics in water or wastewater samples that contain compounds of high electron affinities; the effluent from a gas chromatographic column passes over a radioactive emitter (an electron capture detector, ECD), resulting in the ionization of the carrier gas and production of electrons, which are captured by electrophilic molecules. Captured electrons are collected through the application of a short voltage pulse; the compound concentration is related to the change in pulse rates. *See*

**Table E-02.** Electromotive and electrochemical series (examples)

| Element | Corrosion reaction | SEP | Electrode reaction* | SEP |
|---|---|---|---|---|
| Magnesium | $Mg(s) + 2\,H^+ = Mg^{2+} + H_2(g)$ | +2.4 | $Mg^{2+} + 2\,e^- = Mg\,(s)$ | –2.4 |
| Zinc | $Zn(s) + 2\,H^+ = Zn^{2+} + H_2(g)$ | +0.8 | $Zn^{2+} + 2\,e^- = Zn\,(s)$ | –0.8 |
| Iron | $Fe(s) + 2\,H^+ = Fe^{2+} + H_2(g)$ | +0.4 | $Fe^{2+} + 2\,e^- = Fe\,(s)$ | –0.4 |
| Copper | $Cu(s) + 2\,H^+ = Cu^{2+} + H_2(g)$ | –0.3 | $Cu^{2+} + 2\,e^- = Cu\,(s)$ | +0.3 |
| Gold, Silver | $Au(s) + 3/2\,H_2O + 3/4\,O_2(g) = Au(OH)_3(s)$ | –0.2 | $Ag^+ + e^- = Ag\,(s)$ | +0.8 |

*SEP = Standard electrode potential, volts at 25°C.

*also* flame ionization, mass spectrometry, and thermal conductivity.

**electron capture detector (ECD)** A device used in gas chromatography for the analysis of organic compounds. *See* electron capture.

**electron charge** The charge of one electron, equal to $1.602 \times 10^{-19}$ coulomb.

**electron donor** A molecule or ion, any chemical entity that donates electrons to another compound. It is a reducing agent that, by virtue of its donating electrons, is itself oxidized in the process, e.g., organic compounds undergoing aerobic oxidation or ammonia ($NH_3$), nitrite ($NO_2$), or other reduced inorganic substance for autotrophic nitrifying bacteria. Also called a nucleophile. *See also* covalent bond, electron acceptor, Lewis base, reducing agent, oxidation–reduction.

**electronegative atom** An atom that attracts electrons and tends to migrate to the positive pole in electrolysis. Fluorine (F), nitrogen (N), and oxygen (O) are highly electronegative. *See* hydrogen bond.

**electronegative filter** A microporous device made of cellulose esters or fiberglass with organic resin binders, used for collecting viruses from large volumes of water. *See also* electropositive filter, virus adsorption elution.

**electronegativity** The ability of an atom to pull electrons away from a hydrogen atom and to migrate to the positive pole. *See also* polarity.

**electroneutrality** The characteristic of a reaction, compound, atom or molecule that has no net electrical charge.

**electroneutrality balance** A diagram that shows the requisite concentrations of chemicals in the lime–soda ash softening method, based on water characteristics and the assumed chemical reactions.

**electroneutrality condition** The relationship that expresses the fact that a solution in equilibrium is electrically neutral, i.e., the total number of positive charges per unit volume equals the total number of negative charges. *See also* concentration condition, proton condition.

**electroneutrality equation** *See* electroneutrality condition.

**electroneutrality principle** The sum ($S_c$, meq/L) of the equivalent concentrations of positive ions (cations) equals the sum ($S_a$ meq/L) of the equivalent concentrations of negative ions (anions) in a solution. It is used in checking the accuracy of analytical measurements (cation–anion balances). Such measurements are considered accurate if the following analytical acceptance criteria are met (Metcalf & Eddy, 2003):

$0 < S_a < 3.0$: $S_c - S_a \leq \pm 0.2$ meq/L

$3.0 < S_a < 10$: $D \leq \pm 2.0\%$

$10 < S_a < 800$: $D \leq \pm 5.0\%$

where

$$D = 100\,(S_c - S_a)/(S_c + S_a) \qquad \text{(E-15)}$$

**electronic-grade water** Water that meets special requirements (e.g., resistivity, particle count, and silica concentration) for use in the production of microelectronic devices. Also called microelectronic water.

**electronic particle counter** A device used to obtain a direct count of bacteria in a sample; as the sample passes through an orifice, the electrical conductivity of the fluid decreases and the reduction is correlated to the number of bacteria. *See also* direct count, Petroff–Hauser counting chamber.

**electronic particle size counting** A procedure used to study and quantify particles in wastewater by diluting a sample and passing it through a calibrated orifice or past laser beams. *See* electronic particle counter, microscopic counting, and serial filtration.

**electronic path** The movement of electrons through an electrical circuit.

**electron impact ionization** A mass spectrometry method used to identify unknown compounds by bombarding the analyte with an electron beam.

**electron micrograph** An image taken by an electron microscope.

**electron microscope** An optical device that uses electromagnets as lenses and electrons instead of light rays to form images and achieve a very high magnification (100,000–1,000,000 times) and resolution (0.5–10 nm). Also called electron beam microscope. *See also* scanning electron microscope, transmission electron microscope, light or optical microscope, and ultraviolet ray microscope.

**electron microscopy** The use of an electron microscope to obtain a gray or computer-colored image of a stained specimen.

**electron spectroscopy chemical analysis (ESCA)** A precise technique using an X-ray beam to determine the type and amount of residue on a surface by exciting the contaminant atoms in a sample and causing the emission of electrons. Also called X-ray photoelectron spectroscopy.

**electron spin resonance** A spectroscopic technique that uses the absorption of microwave radiation by electrons to study the reactions of free radicals.

**electron-volt** or **electron volt (eV or ev)** A unit of

energy equal to the energy of one electron accelerating through a potential difference of one volt; it is equivalent to $1.602 \times 10^{-19}$ joule.

**electroosmose treatment** Same as electroosmosis (2).

**electroosmosis** (1) The movement of a solvent with respect to a stationary charged particle caused by an electric field. The liquid moves in a direction opposite to the one the colloidal particles would follow. *See also* electrophoresis, sedimentation potential, streaming potential, surface potential. (2) A treatment process that produces freshwater from saline or brackish waters by separating the water from the salts or other dissolved solids, using an electric field. The feedwater passes from one electric cell to another. Also called electroosmose treatment. *See also* saline-water conversion classification.

**electrophile** (1) An electron acceptor; same as Lewis acid. (2) A positively charged ion or molecule containing atoms without full octets, e.g.. aluminum trichloride ($AlCl_3$), bromine (molecular, $Br_2$), carbon dioxide ($CO_2$), chlorine (molecular, $Cl_2$), hydrogen ion ($H^+$), hydronium ion ($H_3O^+$), hypochlorous acid (HOCl), nitrogen dioxide ($NO_2$), nitrogen oxide (NO), ozone ($O_3$), and sulfite ($SO_3$). *See also* nucleophile, hydrolysis (3), octet.

**electrophilic intermediate** A substance that readily accepts electrons from nucleophiles in covalent bonding to form macromolecules.

**electrophoresis** The movement of dissolved, suspended, or colloidal, charged particles toward the pole of opposite charge, under the influence of an electric field. The solution is considered stationary and the rate of migration is proportional to the applied potential gradient. Also called cataphoresis. *See* electroosmosis, microelectrophoresis, electrophoretic mobility, sedimentation potential, streaming potential, surface potential, zeta potential.

**electrophoretic mobility (EPM)** A parameter that indicates the tendency of ions or colloidal particles to migrate in solution under the influence of an electrical potential. It can be used as a control technique of the coagulation process by correlating its values or changes with particle removal efficiency. When used in the calculation of the zeta potential, it leads to a value lower than the surface potential from the double-layer theory. Many organic and microbial particles in natural waters ($5.0 < pH < 9.0$) are negatively charged, as indicated by their EPM. *See also* electrokinetic potential, streaming current.

**electrophotometer** An instrument that uses colored glass filters to obtain wavelengths for analysis. Also called filter photometer.

**electroplating** Plating or coating with a metal by electrolysis, e.g., in the manufacture of strategic and consumer products that include printed circuit boards and automotive parts. It uses a series of water-based solutions that contain chemicals.

**electroplating process wastewater** Liquid wastes generated during such electroplating operations as deburring, polishing, solvent cleaning, alkaline cleaning, pickling, etching, chromating, phosphating, paint coating, and anodizing. They contain residues of the metals processed and the chemicals used, particularly chromium, cyanides, copper, nickel, and cadmium, some of which are recovered and recycled. *See also* evaporative recovery (atmospheric and vacuum), electrolytic recovery, ion exchange, reverse osmosis, electrodialysis, and carbon adsorption.

**electropositive filter** A microporous filter used to concentrate viruses from large volumes of water or wastewater (e.g., seawater or water with high amounts of organic matter and turbidity); it consists of a bed of fiberglass or cellulose–diatomaceous earth. Also called zeta plus filter. *See also* electronegative filter, virus adsorption elution.

**electroselectivity** The property of an ion-exchange resin whose preference for higher-valence ions decreases with increasing ionic strength of the solution. The phenomenon is called selectivity reversal or electroselectivity reversal when ion-exchange isotherm changes from favorable (convex) to unfavorable (concave). *See* ion-exchange equilibria for related terms.

**electroselectivity reversal** The phenomenon of an ion-exchange isotherm that changes from favorable (convex) to unfavorable (concave) when the ionic solution increases and the selectivity for higher-valence ions decreases.

**electrospray ionization** A liquid chromatography–mass spectrometry technique that forms ions from analyte molecules by evaporation of charged droplets.

**electrostatic adsorption** One of the three conditions that cause the adsorption of substances onto adsorbents. Similar to ion exchange, it results from the electrical attraction or coulombic attractive forces between ions and charged functional groups. *See also* chemical adsorption or chemisorption, desorption, hydrophobic bonding, physical adsorption, polarity, solution force, van der Waals attraction. Also called exchange adsorption.

**electrostatic filter** A device that applies static

electricity to a filter to improve its efficiency in collecting small particles.

**electrostatic interactions** Interactions between charged particles; one of the factors influencing microbial transport. *See also* Coulomb's law, van der Waals force, hydrophobic interactions.

**electrostatic precipitator (ESP)** An air pollution control device that removes particles from a gas stream (smoke) after combustion occurs. The ESP imparts an electrical charge to the particles, causing them to adhere to metal plates inside the precipitator. Rapping on the plates causes the particles to fall into a hopper for disposal.

**electrostatic repulsion forces** Forces developed by positive ions adsorbed onto the surfaces of particles and holding these particles apart. *See* zeta potential.

**electrostatic stabilization** The most important repulsive force that tends to keep particles away from one another in a suspension; it is caused by the interaction of the electrical double layers of the particle surfaces. It is an important concept in the theory of coagulation and flocculation. *See also* Derjaguin–Landau–Verwey–Overbeck theory, electrical double layer, hydrodynamic retardation, London–van der Waals force, polymer bridging, seric stabilization, surface charge.

**electrovalence** (1) *See* electrovalent bond. (2) The valence of an ion, i.e., the number of charges acquired by an atom when losing or gaining electrons. Also called electrovalency or polar valence.

**electrovalency** Same as electrovalence (2).

**electrovalent bond** The electrostatic bond formed by the transfer of electrons between two ions. Also called ionic bond. Same as electrovalence (1).

**electrovitrification** A reuse technology for the waste products of the power generation industry, involving an electric arc to burn organics and fuse inorganic by-products into an amorphous mass that is then cooled, crushed, and used as construction aggregate.

**electrowinning** The recovery of metal from metallic salts by electrolysis. It uses plating cells to recover metal from rinsewaters, wastewaters, and even process streams. Its purpose is to recover a valuable metal such as gold or silver, for pollution control, or to prepare the wastewater for recycling. Also called electroextraction or electrolytic metal recovery. *See also* dummy platting.

**element** (1) A substance that cannot be separated into its constituent parts and still retain its chemical identity; for example, sodium (Na) or chlorine (Cl). Table E-03 is a list of common elements. Important chemical elements found in organisms related to water and wastewater treatment include (a) the major constituents of organic matter (C, H, O, N, P, S), (b) the elements found above trace levels (Ca, Cl, K, Mg, Na), and (c) the trace elements (Co, Cu, Fe, Mn, Zn). (2) A fundamental, ultimate part of a system. (3) A cartridge. (4) Membrane elements are individual fibers or sheets, the assemblage of which forms a module with other parts such as feed, permeate, and concentrate ports.

**elemental analysis** A laboratory method for analyzing the basic elements of a sample: carbon (C),

**Table E-03.** Elements (common)

| Element | Symbol | Atomic Weight | Common Valence | Element | Symbol | Atomic Weight | Common Valence |
|---|---|---|---|---|---|---|---|
| Aluminum | Al | 27.0 | 3+ | Lead | Pb | 207.2 | 2+, 3+ |
| Arsenic | As | 74.9 | 3+ | Magnesium | Mg | 24.3 | 2+ |
| Barium | Ba | 137.3 | 2+ | Manganese | Mn | 54.9 | 2+, 4+, 7+ |
| Boron | B | 10.8 | 3+ | Mercury | Hg | 200.6 | 2+ |
| Bromine | Br | 79.9 | 1– | Nickel | Ni | 58.7 | 2+ |
| Cadmium | Cd | 112.4 | 2+ | Nitrogen | N | 14.0 | 3–, 5+ |
| Calcium | Ca | 40.1 | 2+ | Oxygen | O | 16.0 | 2– |
| Carbon | C | 12.0 | 4–, 4+ | Phosphorus | P | 31.0 | 5+ |
| Chlorine | Cl | 35.5 | 1– | Potassium | K | 39.1 | 1+ |
| Chromium | Cr | 52.0 | 3+, 6+ | Selenium | Se | 79.0 | 6+ |
| Copper | Cu | 63.5 | 2+, 1+ | Silicon | Si | 28.1 | 4+ |
| Fluorine | F | 19.0 | 1– | Silver | Ag | 107.9 | 1+ |
| Gold | Au | 197.0 | 3+ | Sodium | Na | 23.0 | 1+ |
| Hydrogen | H | 1.0 | 1+ | Sulfur | S | 32.1 | 2–, 6+ |
| Iodine | I | 126.9 | 1– | Tin | Sn | 118.7 | 2+ |
| Iron | Fe | 55.8 | 2+, 3+ | Zinc | Zn | 65.4 | 2+ |

hydrogen (H), nitrogen (N), oxygen (O), sulfur (S), silicon (Si), and metals. It is used in thermal processes, for example, to determine the characteristics of the ash that exits the kiln and the quantity of the waste that will enter the gas stream in vapor and particle forms.

**elemental chlorine (Cl)** The gas chlorine as an element or atom. *See* chlorine and molecular chlorine for more detail.

**elemental composition** The composition of a waste in terms of its individual elements. *See also* elemental analysis.

**elemental oxygen (O)** The element or atom of oxygen, which can be obtained from molecular oxygen ($O_2$) by application of energy, as is the case of the formation of ozone:

$$O_2 + \text{energy} \rightarrow O + O \quad \text{(E-16)}$$

Also called nascent oxygen.

**elemental phosphorus plant** A facility that processes phosphate rock to produce elemental phosphorus. A plant includes all buildings, structures, operations, calciners and nodulizing kilns on one contiguous site (EPA-40CFR61.121-a).

**elementary reaction** A chemical reaction that involves one, two, and, rarely, three molecules, and occurs exactly as written, without any intermediate step. *See also* reaction order and molecularity.

**elephantiasis** A swelling of the limbs caused by mosquito-borne worms that obstruct lymph ducts; a symptom of filariasis.

**elevated storage** Water storage at an elevation, usually in a tank supported on a tower, to provide pressure in the distribution system. *See also* ground storage. Elevated storage may also be provided by earthen, steel, or concrete reservoirs installed on high ground, or by standpipes or tanks raised above the ground surface.

**elevated tank** A tank supported above ground by a tower, posts, or columns. It stores water for local distribution, where a ground-level reservoir is inadequate. *See* water tower.

**elevation charge** A surcharge to a basic water rate for customers at certain elevations to compensate for the additional pumping costs.

**elevation energy** Same as elevation head.

**elevation head or elevation energy** The element of the total dynamic head of a pump that represents the elevation of the fluid above or below a datum. Also called potential head or potential energy. *See also* hydraulic energy, dynamic head, and pump head terms.

**Elf/Anvar** An oil–water separator produced by Graver Co.

**elimination capacity** A parameter ($E_c$, g/m$^3$/hr) used in the design and analysis of biofilters and other odor control devices. It is the product of the volumetric flow rate ($Q$, m$^3$/hr) and the difference in concentrations of the influent gas ($C_0$, g/m$^3$) and effluent gas ($C_e$, g/m$^3$) divided by the volume ($V$, m$^3$) of the filter bed contactor:

$$E_c = Q(C_0 - C_e)/V \quad \text{(E-17)}$$

**elimination half-life** The time required for one-half of a substance to be eliminated from the body.

**elimination rate** Same as elimination capacity.

**elimination rate constant** The rate at which a substance is eliminated from the body.

**ELISA** Acronym of enzyme-linked immunosorbent assay.

**EloxMonitor™** An on-line COD monitor manufactured by Anatel Corp.

**El Tor cholera** A waterborne disease cause by the El Tor vibrio.

**eltor (El Tor) vibrio** One of two biotypes of *Vibrio cholerae,* the causative organism of cholera. It causes a high fatality rate in untreated cases. It was identified first at a camp in El Tor, Egypt, and later in the 1937 outbreak as well as the pandemics of the 1960s and 1970s. *See also* classical cholera, *Vibrio cholerae.*

**eluent** A solution used to remove particles absorbed in or adsorbed on solid particles, e.g., a slightly alkaline proteinaceous fluid of 1.5% beef extract for the adsorption of viruses.

**elute** To remove by dissolving an absorbate (or adsorbate) from an absorbent (or adsorbent), e.g., by application of a chemical that reverses the absorption (or adsorption) process.

**elution** (1) The use of a solution to separate absorbates (or adsorbates) from absorbents (or adsorbents). (2) The stripping of contaminants from an ion exchange medium using a concentrated solution.

**elution curve** The graphical representation of the concentration of ions as a function of bed volumes treated to determine the regeneration of ion exchange resins. *See also* breakthrough curve.

**elution order** The order in which ions are stripped from an ion-exchange resin; it depends on the selectivity sequence and on the total ionic concentration of the feedwater.

**elutriated sludge** Wastewater treatment sludge, the water alkalinity of which has been reduced by dilution, sedimentation, and decantation.

**elutriation** A sludge conditioning process that consists of washing sludge with water of low alkalinity (e.g., freshwater or the effluent of the treatment plant) to remove organic and inorganic components, reduce the alkalinity of the sludge and, consequently, the chemical dosages (lime, ferric chloride) required for additional treatment, and improve filtration. Sludge conditioning using polyelectrolytes does not require elutriation. *See also* countercurrent elutriation, multiple elutriation.

**elutriator** A device that thickens the solids slurry in an evaporator.

**eluviation** (1) The transport of insoluble soil particles (e.g., nutrients, clay and other inorganics) in water from upper to lower layers. *See also* illuviation, leaching. (2) The removal of dissolved or suspended, organic or inorganic, materials by percolating water.

**EM** Acronym of (a) electrophoretic mobility; (b) enhanced monitoring.

**EMAP data** Environmental monitoring data collected under the auspices of the Environmental Monitoring and Assessment Program. All EMAP data share the common attribute of being of known quality, having been collected in the context of explicit data quality objectives and a consistent quality assurance program.

**EMB** Eosin-methylene blue (agar plate).

**Embden–Meyerhof–Parnas pathway** The process of glucose fermentation into lactic and pyruvic acids. Also called Embden–Meyerhof pathway or EMP pathway. *See* glycolysis for more detail.

**Embden–Meyerhof pathway** *See* Embden–Meyerhof–Parnas pathway.

**EMBR process** *See* extractive membrane bioreactor process.

**embryotoxic** Poisonous to embryos; pertaining to a substance that poses a greater risk to an embryo than to a pregnant woman or animal.

**EMC** Acronym of event mean concentration.

**emergency** An unexpected serious situation requiring immediate action to avoid or mitigate severe consequences.

**emergency management** Special measures taken to prepare for or respond to an emergency.

**emergency potable-water reuse** The direct potable reuse of treated municipal wastewater on a short-term, emergency basis, as happened in Chanute (Kansas) for five months in 1956–1957 when the water source, the Neosho River, ceased to flow: the chlorinated secondary effluent was mixed with water stored in the river channel behind the water treatment dam and used as intake water. *See also* direct potable-water reuse.

**emergency preparedness plan** *See also* disaster preparedness.

**emergency source** A source of water developed during an emergency for temporary use.

**emergent or emerging** (1) Coming into view or notice, as from concealment or obscurity. In the fields of water supply and wastewater disposal, emerging issues are those that have been identified recently, are potentially significant, but have not been regulated. Often, appropriate approaches or technologies to deal with emerging issues are only at the testing stage, as opposed to conventional issues, which are well understood and for which there are well-established solutions. (2) Same as emergent plant.

**emergent plant** An aquatic plant rooted in the soil or granular support medium and whose stem and leaves extend above water; e.g., bulrush, cattail, reeds, sedges. Such plants promote wastewater treatment in both natural and constructed wetlands by providing a surface for the attachment of bacteria, adsorbing wastewater constituents, transferring oxygen to the water, and controlling algal growth through the restriction of sunlight penetration.

**emerging** *See* emergent (1).

**emerging clinical pathogen** A pathogen for which the incidence of infection is growing, the evolution of known species causes new infections, and infections are spreading to previously unaffected areas. An example is adenoviruses.

**emerging compound** *See* emerging wastewater constituent.

**emerging concentrate disposal** A method of concentrate disposal that enhances overall recovery and reduces the volume of concentrate. Current (2007) emerging concentrate minimization and management methods include capacitive deionization, dewvaporation, forward osmosis, high-efficiency reverse osmosis (RO), membrane distillation, salt solidification and sequestration, seeded slurry precipitation and recycle, two-pass nanofiltration, two-phase RO—biological, two-phase RO—chemical, and wind-aided intensified evaporation. *See also* conventional concentrate disposal.

**emerging constituent** *See* emerging wastewater constituent.

**emerging disinfection technology** A disinfection method that is recent or being investigated. Common methods include chlorination and granular filtration, and, more recently, ozonation, membrane filtration, and ultraviolet light irradiation. Technologies under development (2007) include pulsed arc electrohydraulic discharge, acoustic cavitation, and combined oxidants disinfection.

*See also* bank filtration and computational fluid dynamics.

**emerging environmental contaminant (EEC)** *See* emerging wastewater constituent.

**emerging environmental problem** A problem that may be new and/or becoming known because of better monitoring and use of indicators.

**emerging infectious disease** A disease that has been documented in humans over the past 20 years or that may pose a threat in the near future. Current (2007) emerging waterborne diseases concern viruses, bacteria, and protozoans; e.g., (a) the Norwalk and similar viruses, the astroviruses, coronaviruses, and adenoviruses; (b) the bacteria *E. Coli* 0157:H7, *A. hydrophyla, M. avium,* and *H. pylori;* (c) the protozoan parasites *Nosema, Pleistophora, Enterocytoozoon, Encephalitozoon, Microsporidum, C. cayetanensis, T. gondii* and, previously, *Giardia* and *Cryptosporidium. See also* Surface Water Treatment Rule (SWTR) and Enhanced SWTR.

**emerging inorganic contaminant** Any of a number of inorganic constituents that may pose long-term health risks. Such contaminants usually appear at low concentrations and may be found in groundwater sources that do not receive any treatment beyond disinfection. Currently (2007) the list of inorganic contaminants of concern includes arsenic, high salinity, inorganic disinfection by-products, iron, manganese, nitrate, and perchlorate (a rocket fuel component).

**emerging organic compound** An organic substance that has been identified in water supply sources or in treated wastewater effluents but for which there is no established standard or requirement. The most common sources of these compounds are veterinary and human antibiotics, prescription and nonprescription drugs, industrial and household products, sex and steroidal hormones, personal care products, other endocrine disruptors, and methyl tertiary butyl ether (MTBE, introduced into groundwater via leaking underground storage tanks). Drinking water disinfection by-products are also included in emerging contaminants. *See also* emergent (1).

**emerging pathogens** Recently discovered or recently recognized microbes that have or may have a significant effect on human health, including calicivirus, *Cryptosporidium, Cyclospora cayetanesis, E. coli* 0157:H7, *Helicobacter pylori, Hepatitis E, Legionella,* microspora, rotavirus, and *Toxoplasma gondii. See* emerging infectious disease.

**emerging pollutants of concern (EPOC)** Chemical substances that pollute water sources from wastewater discharges, including endocrine disrupters, industrial and household products, pharmaceuticals, and personal-care by-products. *See also* wastewater constituents.

**emerging technology** Any sulfur dioxide control system that is not defined as a conventional technology, and for which the owner or operator of the facility has applied to the USEPA and received approval to operate as an emerging technology (EPA-40CFR60.41-b). This definition can be adapted to apply to wastewater treatment. *See also* conventional technology (1).

**emerging viral pathogen** *See* emerging pathogen and emerging infectious disease.

**emerging wastewater constituents** Constituents found in municipal wastewater and meeting the definitions of emerging inorganic contaminant or emerging organic compound. Their concentration is usually in the microgram/L or nanogram/L range; they may pose health and environmental concerns. Some are difficult to remove from wastewater, even with advanced treatment methods. *See also* conventional wastewater constituents, nonconventional wastewater constituents, emerging pollutants of concern, emerging infectious disease.

**Emerzone®** Ozone generating equipment manufactured by Emery-Trailigaz Ozone Co.

**EMF** or **emf** Acronym of electromotive force.

**EMF series** *See* electromotive force series.

**eminent domain** The power of any level of government to take private property for public use with payment of fair compensation to the owner. The right of eminent domain is sometimes exercised by water and wastewater authorities to acquire sites needed for the construction of reservoirs and other waterworks. In addition to fair compensation, the authorities often foster good will by providing relocation assistance.

**emission spectroscopy** A laboratory technique that measures characteristic radiations to analyze metals in water and wastewater.

**emission standard** A limitation on the maximum level of an airborne contaminant, based on applicable regulations.

**emissivity** A dimensionless coefficient characteristic of a surface used to dry sludge by radiation heat transfer. *See* radiation drying.

**emollient** An ingredient for making skin soft or supple, or for soothing the skin; e.g., fatty acids and lanolin included in toilet bars and skin preparation products.

**empirical** Relying upon or gained from experiment or observation.

**empirical collision efficiency factor ($\alpha$)** In the theory of filtration mechanisms, $\alpha$ is a factor that affects the overall collector efficiency to take into account the fact that the particles are not completely destabilized. The overall efficiency is the sum of efficiencies for sedimentation transport, interception, and diffusive transport. *See* fundamental filtration model.

**empirical equation** A mathematical expression among variables and parameters, based on observations. *See* empirical model.

**empirical formula or empirical method** (1) A formula or method based more on observations and practice than on theoretical considerations, e.g., the plotting position in frequency analysis, the Creager and Myers–Jarvis enveloping curves in peak runoff discharge computations, and the Churchill and Buckingham technique for water quality modeling. (2) An empirical formula in organic chemistry is a molecular formula that may represent more than one compound; e.g., the molecular formula $C_2H_6O$ may represent ethyl alcool ($CH_3$—$CH_2$—OH) or dimethyl ether ($CH_3$—O—$CH_3$).

**empirical method** Same as empirical formula (1).

**empirical model** A representation of chemical or physical processes by generalities and simplifications based on observations, measurements, or practical experience rather than solely on principles or theory; e.g., the Darcy and Manning equations. Most analytical and numerical models include empirical elements. Also called lumped-parameter model, statistical or black-box model. *See also* deterministic model, stochastic model.

**EMP pathway** *See* Embden–Meyerhof–Parnas pathway.

**empty bed contact time (EBCT)** The residence time of water in an empty packed-bed reactor (PBR) or, more generally, the time during which water or wastewater to be treated is in contact with the treatment medium; equivalent to the ratio of bed volume ($v$) to flow rate ($Q$) or media depth ($D$) to superficial velocity ($V$):

$$\text{EBCT} = v/Q = D/V \qquad \text{(E-18)}$$

EBCT is a principal design variable; e.g., about 12–15 minutes for a PBR in water treatment, 2–8 minutes for ion exchange resins. In adsorption, the minimum contact time ($\text{EBCT}_\text{m}$) is related to the critical depth of the column ($L_\text{c}$):

$$\text{EBCT}_\text{m} = L_\text{c}/(Q/A) \qquad \text{(E-19)}$$

where $A$ is the cross-sectional area of the column. This minimum is related to the empty bed contact time ($\text{EBCT}_\text{MTZ}$) that is long enough to contain the mass transfer zone (MTZ). *See also* bed-depth-service time, bed life, carbon usage rate, empty bed residence time, exhaustion rate, service flow rate, specific throughput.

**empty bed residence time** The ratio of the total volume of a filter bed contactor to the volumetric flow rate of gas, similar to the empty bed contact time of a packed-bed reactor. *See also* actual residence time and elimination capacity.

**empty-tank maintenance cleaning** A relatively new (2007) cleaning method used to remove accumulated solids from the surface of wastewater treatment membranes: chemicals are fed back through the membranes into the empty tank. *See also* air scour, backpulse, maintenance cleaning, and recovery cleaning.

**EM-Pure™** A membrane system manufactured by Filtronics, Inc. of Anaheim, CA for use in municipal drinking water applications; it incorporates the proprietary design "Hyper-Filtration™."

**EMR™** A metal recovery apparatus of Kinetico Engineered Systems, Inc..

**Emscher filter** A biological wastewater treatment unit similar to the contact aerator.

**Emscher tank** A deep two-story wastewater treatment tank originally designed and patented by Karl Imhoff, providing sedimentation in the upper compartment and anaerobic sludge digestion in the lower compartment. It is an improvement of the Travis tank carried out in the Emscher District of Germany. *See* Imhoff tank for more detail.

**emulsified grease buildup** An operational problem of depth filters when emulsified grease accumulates within the medium bed, increases head loss, and reduces filter run. Air scour and water surface wash are effective means of control. Other depth filtration operation problems include turbidity breakthrough, mudball formation, filter medium loss, cracks and contraction of filter bed, and gravel mounding.

**emulsifier** An emulsifying agent.

**emulsifying agent** A substance that modifies the surface charge of droplets to prevent their coalescence and create or maintain an emulsion; e.g., surface-active agents, soaps, certain proteins and gums.

**emulsion** A stable, heterogeneous dispersion of two immiscible liquids, maintained as one suspended in another by agitation or by emulsifying agents; for example, an oil and water mixture used in the aluminum forming and nonferrous metals forming industries.

**emulsion polymer** A dispersion of polymer particles in oil, with the addition of surface-active agents, when used in water and wastewater treatment to prevent separation from the liquid. Polymers are also available in dry, gel, liquid, and mannich forms.

**emulsoid** (1) A sol that has a liquid-dispersed phase, i.e., a dispersion of colloidal liquid particles in a liquid. *See also* suspensoid. (2) It is also defined as "a colloid that is readily dispersed in a suitable medium and may be redispersed after coagulation" (APHA, 1981; Symons et al., 2000).

**encapsulate** To enclose or seal off.

**encapsulation** The complete enclosure, sealing, coating, or jacketing of waste or other material with a solidification/stabilization additive, binder, polymer, asphalt, etc.; e.g., the treatment of asbestos-containing materials with a liquid that covers the surface with a protective coating or imbeds fibers in an adhesive matrix to prevent their release into the air. *See also* macroencapsulation, microencapsulation, and solidification/stabilization.

**encased membrane system** A low-pressure membrane system consisting of (a) pressure vessels mounted on racks and encasing modules of hollow-fiber membranes and (b) a pump feeding water to the modules under positive pressure. Residuals are withdrawn by intermittent backwashing or continuous wasting of concentrate. Also called pressurized membrane system.

**encephalitis** (1) An inflammation of the brain. (2) Encephalitis lethargica, a form of encephalitis caused by a virus (e.g., arthropod-borne, influenza, herpes, chickenpox) and characterized by apathy and abnormal sleepiness. Also called sleeping sickness.

**encephalitis lethargica** *See* encephalitis (2).

***Encephalitozoon cuniculi*** A parasite having as reservoirs laboratory rodents, rabbits, and dogs; the spores are transmitted through feces and urine.

***Encephalitozoon intestinalis*** One of two significant species of microsporidia pathogenic to humans and common in people affected by AIDS; the other is *Enterocytozoon bieneusi*. Its infective form is a spore, transmitted through feces, urine, and orally.

**enclosed trickling filter** A covered, sometimes domed, trickling filter.

**enclosed-reactor composting** Composting that takes place in an enclosed container under environmentally controlled conditions. *See* in-vessel composting for detail.

**encrustation** A covering or crust on the surface of an object; e.g., the build-up of calcium carbonate on the inside of a water pipe or the tubercle resulting from the corrosion of an exposed metal. Pipe encrustation reduces the capacity of the pipe.

**encystment** The transformation of a protozoan into an inactive, nonmotile, resistant cyst in response to adverse environmental conditions. Encystment changes include volume reduction; loss of cilia, flagella, and feeding vacuoles; and formation of a multilayered cyst wall.

**endangered species** Animals, birds, fish, plants, or other living organisms threatened with extinction by man-made or natural changes in their environment. Requirements for declaring a species endangered are contained in the Endangered Species Act.

**Endangered Species Act** A federal law of 1973 (Public Law 93-205), amended in 1984 (PL 98-327) and 1988 (PL 100-478), that creates a program to protect endangered species.

**endangerment assessment** (1) A site-specific risk assessment of the actual or potential danger to human health or welfare and the environment from the release of hazardous substances or wastes. The endangerment assessment document is prepared in support of enforcement actions under federal regulations. (2) A study to determine the nature and extent of contamination at a site on the National Priorities List and the risks posed to public health or the environment. The USEPA or a state conducts the assessment when legal action is to be taken to direct potentially responsible parties to clean up a site or pay for it. The endangerment assessment supplements a remedial action.

**end-around baffles** Baffles installed in water or wastewater treatment basins to force the liquid to flow around the end of the baffles, thereby reducing short-circuiting and encouraging mixing.

**end bell** A device used to hold the stator and the rotor of a motor in position.

**end contraction** The condition of a rectangular weir when its ends project inward from the sides of the channel. *See* weir end contractions for detail.

**end effect** In a packed tower used for stripping volatile organic chemicals (VOC) from water or wastewater, end effect refers to the VOC removal that occurs as the liquid contacts air above the packing at the top of the tower and air below the packing at the bottom. *See also* cascade air stripping, cocurrent packed tower, crossflow packed tower, cascade packed tower, low-profile air stripper, minimum air-to-water ratio, packed tower design equation, sieve tray column, wall effect.

**endemic** Something peculiar to a particular people or locality, such as a disease or a pathogen that is always present in a given geographic area or community. For example, cholera is endemic in some parts of Asia. *See also* epidemic.

**endemic disease** A disease or infection that is constantly present within a given geographic area or population group.

**endemic goiter** A disease or syndrome caused by the consumption of iodine-deficient water or water containing goitrogens.

**endemic typhus** A disease transmitted by rats and often associated with poor refuse disposal. *See also* epidemic or louse-borne typhus.

**endergonic** Pertaining to a chemical reaction, a process or an exchange (e.g., photosynthesis) that requires an input of energy.

**endergonic reaction** A reaction that results in a positive change in free energy release; it requires an additional energy input to proceed in the direction indicated. *See also* bioenergetics, exergonic reaction, Gibbs free energy.

**endobiotic habitat** The tissues of fish and other large organisms that support smaller organisms in the marine environment. *See also* neuston layer and epibiotic habitat.

**endocrinal** *See* endocrine (1).

**endocrine** (1) Of or pertaining to an endocrine gland or its internal secretion into the blood or lymph; also called endocrinal, endocrinic, endocrinous. (2) An internal secretion; hormone. (3) Endocrine gland.

**endocrine disruption** A physiological disorder caused by a variety of organic compounds (pesticides, dioxin, PCB, etc.), e.g., disruption of the reproductive function by substances that affect the thyroid and steroid hormones.

**endocrine-disruptive chemical (EDC)** Same as endocrine disruptor.

**endocrine disruptor** An exogenous substance that can adversely impact the functions of endocrine glands. A notable example is the group of persistent organic pollutants (DDT, etc.), organic compounds that have long half-lives and undergo slow physical, chemical, and biological degradation. They can disrupt the endocrine system by mimicking the function of hormones and other steroid compounds. *See also* emerging pollutant of concern, xenobiotic.

**endocrine gland** Any of the glands that secrete substances directly into the blood or lymph: e.g., adrenal gland, pancreas, pituitary gland, thyroid gland. The latter can be damaged by radioactive iodine.

**endocrine system disruption** *See* endocrine disruption.

**endocrinic** *See* endocrine (1).

**endocrinous** *See* endocrine (1).

**end-of-line treatment** A treatment system in which all wastewater in an area is collected and brought to treatment at one plant. *See also* satellite treatment.

**end-of-pipe alternative** A stormwater management strategy that usually consists of a traditional drainage system for runoff collection and transmission to an end-of-pipe facility (e.g., a wet pond or a constructed wetland) for treatment and disposal. Other strategies are source or downstream control, or a combination of source controls and end-of-pipe facilities.

**end-of-pipe facility** *See* end-of-pipe alternative.

**endogenous** Growing from within, inside an organism.

**endogenous decay** (1) The phase in the growth of microorganisms in which starvation, death, predation, and autooxidation become significant because of limited substrate. It is commonly represented by a first-order reaction. In biological treatment (e.g., activated sludge, aerobic digestion), endogenous decay increases at high solids retention times. *See also* endogenous respiration. (2) The actual decrease in cell mass resulting from this process; it is proportional to the concentration of organisms present.

**endogenous decay coefficient** A parameter that accounts for the decrease in cell mass due to various factors such as death, predation by higher organisms in the food chain, and oxidation of storage products for cell maintenance; usually represented by the symbol kd and expressed in g VSS/g VSS per day or as the reciprocal of time. (VSS = volatile suspended solids). *See also* specific growth rate, lysis–regrowth model.

**endogenous growth** Growth from within.

**endogenous growth phase** A period in the growth of microorganisms during which their metabolism decreases, with an eventual decrease in biomass and number of viable cells. *See* endogenous respiration.

**endogenous metabolism** Growth of microorganisms on dead cells as a source of carbon and energy.

**endogenous phase** *See* endogenous decay and endogenous respiration.

**endogenous respiration** A growth phase, also called endogeny, in which microorganisms metabolize their own protoplasm without replacement due to low concentration of available food. Cell multiplication may continue for some time, the starving cells feeding on their own stored metabolites or on dead cells undergoing lysis, but, eventually, the population reaches a stationary phase, followed by a declining phase. Endogenous respiration is prevalent in such biological treatment processes as activated sludge and aerobic digestion. A generalized endogenous respiration reaction, using $C_5H_7NO_2$ for the formula of cell tissue, is:

$$C_5H_7NO_2 + 5\ O_2 \rightarrow 5\ CO_2 + NH_3 + 2\ H_2O \quad \text{(E-20)}$$

See endogenous decay, growth pattern.

**endogeny** Litterally, growth from within, i.e., microbial growth based on the products of metabolic processes when the food supply or substrate is below the required level and toxic waste products accumulate in the system; the endogenous respiration phase.

**endoparasite** A parasite that lives internally within a host organism. *See also* ectoparasite.

**endospore** A spore (an inclusion or granule) formed within a cell, coated with layers of protein, containing all the information necessary for reproduction, and extremely resistant to heat, desiccation, chemicals, and other harmful agents; a dormant stage in the life cycle of the *Bacillus, Clostridium,* and other groups of bacteria. Endospores are about 0.5 × 1.0 × 1.5 micrometers, are commonly found in drinking water sources, and are removable by coagulation–flocculation-filtration. Also called spore. *See* aerobic sporeformer, anaerobic sporeformer, resistant form.

**endospore detection method** A simple and quick laboratory procedure to detect the presence of endospores in a water sample: the sample is heated to 80°C for 12 minutes, rapidly cooled in an ice bath to 35°C, membrane-filtered, placed in a Petri dish containing an appropriate nutrient, and incubated for 20–22 hours. The colonies arising on the membrane filter are sporeformers.

**endothall ($C_8H_{10}O_5$)** The common name for 7-oxalobicyclo-[2.2.1]-heptane-2,3-dicarboxylic acid, a synthetic organic chemical used as a defoliant or herbicide, and regulated by the USEPA: MCLG = MCL = 0.1 mg/L.

**endothermic** Pertaining to a process, exchange, or chemical reaction that takes place with absorption of heat, e.g., melting ice.

**endotoxin** A heat-resistant toxin or poisonous substance from the cell wall of Gram-negative bacteria that is released during cell lysis; it is a bioaerosol that poses a health risk to sensitive people during sludge composting. Algal endotoxins may also be toxic to aquatic life and wildlife.

**end point** (1) The point at which the titration of a sample is stopped, which is when the color of the sample changes (e.g., from blue to clear). In addition to a color change, an end point may be reached by the formation of a precipitate or the reaching of a specified pH. An end point may be detected by the use of an electronic device such as a pH meter. (2) A response measure in a toxicity study.

**endrin ($C_{12}H_8Cl_6O$)** The generic name of 1,2,3,4,10,10-hexachloro-6,7,epoxy-1,4,4a,5,6,7,8,8a-octahydro-1,4-endo-endo-5,8-dimethanophthalene, a stereoisomer of dieldrin; a white, odorless, crystalline solid pesticide and rodenticide that is toxic to freshwater and marine aquatic life and produces adverse health effects in domestic water supplies; almost insoluble in water. Symptoms of endrin poisoning include dizziness, headache, insomnia, nausea, and sweating. MCLG = MCL = 0.002 mg/L.

**Endurex** A coarse-bubble diffuser manufactured by the Parkson Corp.

**Enelco** A line of water treatment products of Infilco Degremont, Inc.

**energy** The capacity to do work, usually expressed as the capacity per unit mass of fluid, as compared to power, which is the work done per unit of time. Hydraulic energies include potential (or elevation) energy, pressure energy, and kinetic energy. Energy loss (or head loss) occurs when friction converts part of the hydraulic energy into heat energy. Total energy is the sum of these three energies. *See* Bernoulli's law.

**energy audit** A technical investigation conducted for a manufacturing plant, treatment plant, etc., to assess energy uses, identify and evaluate energy conservation options as well as an implementation plan.

**energy barrier** The electrostatic force between two particles that prevents their flocculation and aggregation. Also called energy hill. *See also* electrostatic stabilization, London–van der Waals force, DLVO theory, steric interaction, polymer bridging.

**energy budget** A quantitative account of energy inputs, transformations, and outputs in an ecosystem, industrial process, or a wastewater treatment plant.

**energy conservation** Measures taken to reduce energy use from all sources (e.g., electricity, fuel oil, natural gas, methane, gasoline). In water supply and wastewater works, the greatest potential for energy conservation is in pumping (water distribution pumping, or pumping in wastewater treatment).

**energy conservation law** The principle according to which energy cannot be destroyed but can be converted. *See* Bernoulli's equation.

**energy content** *See* fuel value, sludge energy content.

**energy dissipation** The conversion of the mechanical energy of flowing water into heat energy by hydraulic jumps, baffles, buckets, and aprons in spillways and stilling basins.

**energy equation** Bernouilli's equation applied to two points along a conduit or channel, showing the relationship between velocity head, pressure head, elevation head, and frictional head losses.

**energy flow** The passage of energy from sunlight to autotrophic organisms and so on through the trophic levels of a food chain, with much dissipation from one level to another.

**energy gradeline (EGL)** The energy gradeline represents the elevation of the energy head of water flowing in a pipe, conduit or channel. The line is drawn above the hydraulic gradeline a distance equal to the velocity head of the water flowing at each section or point along the pipe or channel. The total energy (H) is the sum of the water surface elevation ($h$) and the velocity head ($V^2/2g$), i.e.,

$$H = h + V^2/2g \qquad \text{(E-21)}$$

In steady uniform flow, the EGL is parallel to the water surface and to the bottom of the channel or conduit. The hydraulic gradeline corresponds to the water surface in open-channel flow or the line to which water would rise in pressurized flow if the conduit were freely vented. The energy gradient is the slope of the energy gradeline; the hydraulic gradient is the slope of the hydraulic gradeline. The energy head is the height of the hydraulic gradeline above the centerline of a conduit plus the mean velocity head. The energy and hydraulic gradelines slope downward in the direction of flow (because of friction losses), and rise or drop abruptly with the introduction or withdrawal of energy by pumps or turbines. *See* Figure E-06. Sometimes called energy line or energy gradient line.

**energy gradient** The difference in total (energy) head per unit horizontal distance measured in the direction of flow. *See* energy gradeline.

**energy gradient line** Same as energy gradeline.

**energy head** *See* energy gradeline.

**energy hill** The electrostatic force between two particles that prevents their flocculation and aggregation. *See* energy barrier.

**energy line** A line joining the elevations of the energy heads; same as energy gradeline.

**energy loss** The difference in total (energy) head between two sections of a conduit or channel. Same as headloss.

**Energy Mix** Rapid mixing equipment manufactured by Walker Process Equipment Co.

**energy of activation** *See* activation energy.

**energy recovery** The production of energy from waste through combustion and other processes or the retrieval of waste energy for a beneficial purpose. For example, devices such as impulse turbines and turbopumps recover residual pressure from membrane concentrate streams in seawater desalination by reverse osmosis.

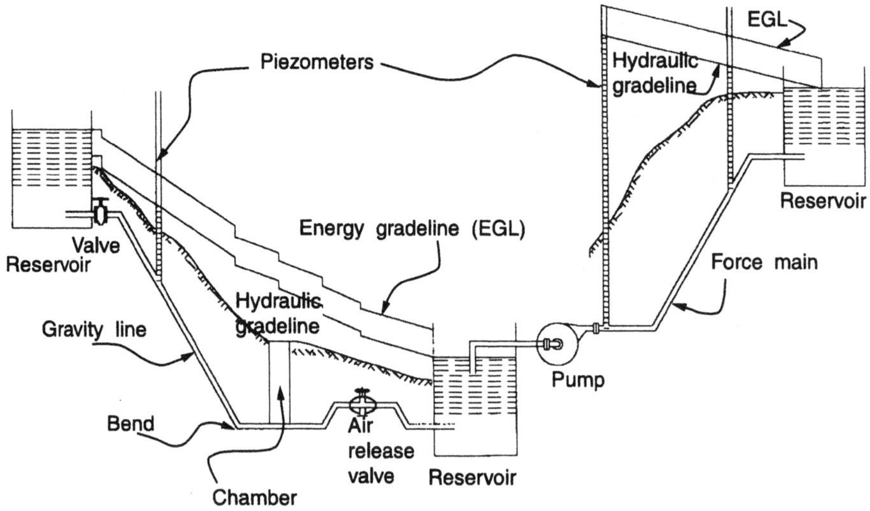


**Figure E-06.** Energy grade line (courtesy CRC Press).

**energy-intensive industries** Nonferrous smelting and refining, cement production, utilities, pulp and paper.

**enforceable requirements** Conditions or limitations in permits issued under the Clean Water Act, Section 402 or 404 that, if violated, could result in the issuance of a compliance order or initiation of a civil or criminal action under federal or applicable state laws. If a permit has not been issued, the term includes any requirement that, in the EPA Regional Administrator's judgment, would be included in the permit when issued. Where no permit applies, the term includes any requirement that the Regional Administrator determines is necessary for the best practical waste treatment technology to meet applicable criteria.

**enforcement** Federal, state, or local legal actions to obtain compliance with environmental laws, rules, regulations, or agreements and/or obtain penalties or criminal sanctions for violations. Enforcement procedures may vary, depending on the requirements of different environmental laws and related implementing regulations. Under CERCLA, for example, USEPA will seek to require potentially responsible parties to clean up a Superfund site, or pay for the cleanup, whereas under the Clean Air Act the agency may invoke sanctions against cities failing to meet ambient air quality standards that could prevent certain types of construction or federal funding. In other situations, if investigations by USEPA and state agencies uncover willful violations, criminal trials and penalties are sought.

**engineering economics** *See* engineering economy.

**engineering economy** The analysis of economic consequences of engineering decisions, i.e., application of economic principles, techniques, and data to the planning and evaluation of engineering projects, e.g., cost-effectiveness analysis, cost/benefit analysis, value engineering, cashflow equivalence, present worth analysis, replacement analysis, etc.

**engineering geology** The application of geologic principles, techniques, and data to mining, construction, petroleum engineering, materials science, groundwater resources.

**engineering hydrology** The subject that covers the quantitative aspects of the rainfall–runoff relationship, particularly the variation of runoff over time for the prediction of design events such as floods and droughts.

**engineering structure** A structure designed to withstand reasonable physical forces or loads applied against or on its various members or parts.

**Engler viscometer** An instrument designed to determine the degree Engler, a measure of the viscosity of a fluid.

**enhanced biological phosphorus removal (EBPR)** *See* enhanced phosphorus uptake.

**enhanced coagulation** A water treatment process used to remove dissolved organic carbon (DOC, mainly natural organic matter) and reduce the formation of disinfection by-products during coagulation with iron or aluminum salts. Enhanced coagulation uses higher coagulant doses than are required for the removal of color and turbidity. USEPA proposes enhanced coagulation as an appropriate technology for DOC removal, which can be modeled using the following equation:

$$(\mathrm{DOC_s} - \mathrm{DOC_f})/D = ab \cdot \mathrm{DOC_f}/(1 + b \cdot \mathrm{DOC_f}) \quad \text{(E-22)}$$

where $\mathrm{DOC_f}$ = the final concentration of sorbable DOC in the water (mg/L), $\mathrm{DOC_s}$ = the sorbable DOC in the water before coagulant addition (mg/L), $D$ = coagulant dose (milliequivalent per liter, meq/L), $a$ = the maximum DOC sorption capacity (mg DOC/meq metal), and $b$ = the sorption coefficient in L/mg. Enhanced coagulation has also been considered for arsenic removal. *See also* enhanced softening.

**enhanced high-rate clarification (EHRC)** An advanced primary treatment process in which a chemical such as ferric chloride and a ballast such as microsand are added to screened, degritted wastewater to improve the removal of BOD and suspended solids. Also called ballasted flocculation.

**enhanced particle flocculation** A component of high-rate clarification processes used in water and wastewater treatment, consisting in the addition of silica sand, recycled chemically conditioned sludge, or another ballasting agent and a polymer to a partially coagulated and flocculated suspension, and then gently stirring the mixture to promote the formation of large floc particles. *See also* ballasted flocculation.

**enhanced phosphate removal** *See* enhanced phosphorus uptake.

**enhanced phosphorus uptake** A wastewater treatment process that achieves both BOD and phosphate removal by promoting the growth of so-called bio-P microorganisms that accumulate phosphorus well in excess of their growth requirements, up to 18% P as opposed to a typical 1.5% P in activated sludge biomass. The main feature of the process is an anaerobic reactor followed by an aerobic reactor of the same size with sludge recirculation and wastage from a secondary clarifier. Bacteria species of the *Acinetobacter* genus are the microorganisms most commonly implicated in the process. Figure E-07 illustrates a possible removal mechanism. Also called enhanced biological phosphorus uptake, excess biological phosphorus removal. *See* biological phosphorus removal model, polyhydroxybutyrate (PHB), and the A/O, A/O with nitrification, modified Bardenpho, OWASA Nutrification, and Phostrip processes.

**enhanced primary sedimentation** A wastewater treatment method that uses preaeration or chemical coagulation to promote the flocculation of finely divided solids and increase the removal of suspended solids in a primary clarifier.

**enhanced rhizosphere biodegradation** A series of effects that plants have on the microbial population in the immediate area surrounding the root system of a plant; a mechanism used in phytoremediation (the use of plants to clean contaminated sites). The roots provide moisture and nutrients for microorganisms to grow and metabolize or destroy the pollutants. *See also* phytodegradation.

**enhanced sedimentation** *See* enhanced primary sedimentation.

**enhanced softening** A complex water treatment process that removes natural organic matter (NOM) during softening to reduce the formation of disinfection by-products (DBP). It uses, for example, lime dosages higher than that required for precipitative softening but not high enough to induce magnesium precipitation. *See also* enhanced coagulation.

**Enhanced Surface Water Treatment Rule (ESWTR)** A revision of the Surface Water Treatment Rule issued by the USEPA in July 1994 to be effective in 1999: it requires water systems that serve at least 10,000 people to provide additional protection against the waterborne pathogens *Cryptosporidium, Giardia,* and viruses.

**enhanced volatilization** A technique used for removing volatile organic compounds in unsaturated soils; see soil-vapor extraction.

**enhanced wetland** A wetland with values increased by human activity. *See* wetland value.

**enlargement coefficient** *See* enlargement loss.

**enlargement loss** The headloss caused by the sudden change in water velocity at an enlargement of the cross section of a conduit or channel. The enlargement coefficient in a discharge formula accounts for the effect of the change. *See also* expansion and contraction.

**enmeshment** The phenomenon that occurs when coagulation is practiced with chemical dosages in excess of coagulant demand, resulting in the formation of an unstable precipitate and the rapid formation of visible floc. *See also* sweep flocculation.

**Enning ESD®** An enhanced, egg-shaped anaerobic digester manufactured by CBI-Walker, Inc. under license from the German firm Enning.

**enol** An organic compound containing a hydroxyl group attached to a doubly-linked carbon atom as in C=C=OH.

**enolate** A metallic derivative of an enol.

**enolization** The conversion of a carbonyl (CH—C=O) into an enol under acidic conditions (C=C=OH) or into an enolate ion (C=C—O⁻) under basic conditions.

**enriching section** The cascade or set of plates above the feed introduction point of a continuous distillation column that involves only one feed stream. Also called rectifying section. *See also* stripping section.

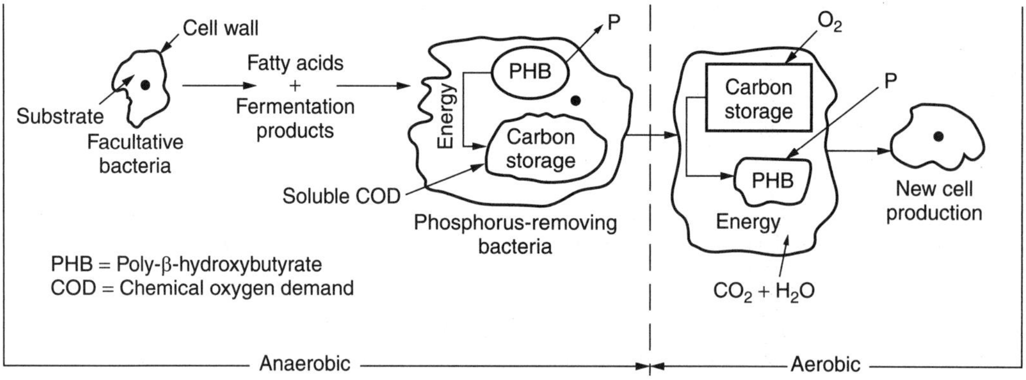


**Figure E-07.** Enhanced biological phosphorus removal.

**enrichment** The increase in the growth potential for algae and other aquatic plants resulting from the addition of nutrients (e.g., nitrogen, phosphorus, carbon compounds) from wastewater effluents or agricultural runoff to surface water.

**enrichment culture** A selective culture medium used to isolate microorganisms directly from nature under special incubation conditions.

**enrichment medium** A liquid substance that promotes the growth of a type of organism and inhibits the growth of other organisms that may be present in a mixture so that the former can be more easily detected or enumerated. *See also* elective enrichment medium and selective enrichment medium.

***Ent.*** Abbreviation of (a) *Entamoeba,* (b) *Enterococcus.*

***Entamoeba coli*** An ameba closely related to *Entamoeba hystolitica,* but nonpathogenic. Abbreviation: *Ent. coli,* not *E. coli.*

***Entamoeba hartmanni*** An ameba closely related to and smaller than *Entamoeba hystolitica,* but nonpathogenic.

***Entamoeba histolytica*** A parasite of the large intestine, the pathogenic, waterborne and foodborne protozoan pathogen of amebiasis, the third most common cause of parasitic death worldwide. Causes amebic dysentery and amebic hepatitis. Its cysts are resistant to chemical disinfection. Persistence = 25 days; low median infective dose, < 100. *See also Dientamoeba fragilis.*

***Entamoeba moshkovskii*** A free-living organism found in wastewater, with trophozoites and cysts similar to those of *Entamoeba hystolitica.*

***Ent. avium*** Abbreviation of *Enterococcus avium.*

***Ent. coli*** Abbreviation of *Entamoeba coli.*

***Ent. durans*** Abbreviation of *Enterococcus durans.*

**enteric** Of intestinal origin, especially applied to wastes or bacteria; pertaining to the small intestine.

**enteric adenovirus** Same as adenovirus.

**enteric bacteria** Rod-shaped, Gram-negative, nonsporeforming, facultative anaerobic bacteria, including the genera *Citrobacter, Escherischia, Enterobacter, Klebsiella, Salmonella,* and *Shigella* of the Enterobacteriaceae family, found in the gastrointestinal tract of warm-blooded animals; some of them are pathogenic. *Escherichia coli* is a widely used indicator of fecal pollution. Enteric bacteria also include the genus *Erwinia,* found in plants. Also called enterics or enterobacteria. *See also* fecal coliform and total coliform.

**enteric cythopathogenic human orphan virus** *See* echovirus.

**enteric disease** A disease of fecal–oral origin manifested in the intestinal tract, diarrhea being a very common symptom; e.g., cholera and typhoid.

**enteric fever** A water-related and excreta-related disease, also called typhoid fever or simply typhoid, caused by the bacteria *Salmonella typhi.* It is transmitted via the fecal–oral route and has worldwide distribution. *See* typhoid fever for more detail.

**enteric organism** Any organism of intestinal origin, particularly the enteric pathogens: bacteria, fungi, protozoa, viruses, and worms.

**enteric pathogen** A pathogen that infects the gastrointestinal tract and is transmitted through the fecal–oral route. Common enteric pathogens include protozoan parasites, *Ascaris,* enteroviruses, rotavirus, adenovirus, *Salmonella,* and *Shigella.*

**enterics** Same as enteric bacteria.

**enteric virus** A small virus, ranging from 20 to 80 nanometers, found in the gastrointestinal tract and transmittable through drinking water contaminated by excreta. Many are pathogenic (e.g., coxsackievirus, echovirus, poliovirus) and may cause disorders of the central nervous system. The waterborne Norwalk virus and rotavirus cause diarrheal disease. Human enteric viruses also include adenovirus, astrovirus, calicivirus, hepatitis A virus, parvovirus, and reovirus. Note that enteric viruses include the genus *Enterovirus.*

**enteritis** An enteric infection, very prevalent in tropical areas, transmitted through contaminated food or drinking water; an inflammation of the intestine.

***Enterobacter*** A genus of rod-shaped, lactose-fermenting bacteria of the family Enterobacteriaceae; one of four genera of total coliform bacteria. *See also* enterobacteria.

***Enterobacter aerogenes*** A species of short, rod-shaped, gas-producing, Gram-negative coliform bacteria of the genus *Enterobacter,* widely found in nature (in soil and on plants, e.g.) but also in the intestinal tracts of humans and other animals. Also called *Aerobacter aerogenes.*

***Enterobacter cloacae*** A species of lactose-fermenting bacteria of the Enterobacteriaceae family; a coliform used as an indicator of fecal pollution in water. *See* coliform growth response test.

**enterobacteria** Rod-shaped, Gram-negative bacteria, including the genera *Escherischia, Salmonella,* and *Shigella* of the Enterobacteriaceae family, found in the intestinal tracts of humans and other animals. They also include the genus *Erwinia,* found in plants. *See* more detail under enteric bacteria.

**Enterobacteriaceae** A family of bacteria that includes the coliforms, salmonellae, and other groups.

***Enterobacter species*** A classification used for bacteria of the *Enterobacter* genus whose species has not been identified.

**enterobiasis** An excreted helminth infection caused by the pathogen *Enterobius vermicularis. See* its common name, pinworm, for more detail.

***Enterobius vermicularis*** The helminth species that causes pinworm infections.

***Enterococci*** Plural of *Enterococcus.*

***Enterococcus avium*** *See Enterococcus* genus.

***Enterococcus durans*** *See Enterococcus* genus.

***Enterococcus faecalis*** *See Enterococcus* genus. This species is specific to the human gut.

***Enterococcus faecium*** *See Enterococcus* genus. This species is specific to the human gut.

***Enterococcus gallinarium*** *See Enterococcus* genus.

***Enterococcus* (*Enterococci*) genus** A group of fecal streptococci, Gram-positive Lancefield group D bacteria, often of nonfecal origin, distinguished from other streptococci by the ability to grow in 6.5% sodium chloride at pH 9.6 and 45°C; including the species *Ent. Avium, Ent. faecium, Ent. durans, Ent. faecalis, Ent. gallinarium. See also Streptococcus genus.*

**enterocolitis** Inflammation of the small intestine and of the colon. Symptoms include diarrhea, low-grade fever, and abdominal cramps. *See* yersiniosis.

***Enterocytozoon bieneusi*** One of two significant species of microsporidia pathogenic to humans and common in people affected by AIDS; the other is *Encephalitozoon intestinalis. E. bieneusi* is transmitted via the fecal–oral route; its infective form is a spore.

**enterohemorrhagic *Escherichia coli* (EHEC)** A strain of *Escherichia coli* associated with foodborne and waterborne hemorrhagic colitis. *See Escherichia coli* 0157:H7 and hemolytic-uremic syndrome.

**enterohepatic circulation** The process of absorption of metabolites, drugs, etc. from the intestines into the blood and then into the liver.

**enterohepatitis** Inflammation of the intestines and liver.

**enteroinvasive** Capable of invading the mucosal surface and deeper tissues of the bowel.

**enteroinvasive *E. coli* (EIEC)** Strains of *Escherichia coli* that invade the colonic mucosa and cause a bloody diarrhea in humans, with symptoms like those of shigellosis.

**Enterolert™** A reagent produced by IDEXX Laboratories, Inc. for the detection of enterococci.

**enteropathogenic *E. coli*** Strains of *Escherichia coli* that do not produce heat-labile or heat-stable enterotoxins; they cause infantile diarrhea in developing countries. This group includes some enteroinvasive and some enterotoxigenic strains.

**enterotoxic *E. coli*** *See* enterotoxigenic *E. coli.*

**enterotoxigenic** Pertaining to microorganisms that produce toxic substances affecting the cells of the intestinal mucosa.

**enterotoxigenic *E. coli*** Strains of *Escherichia coli* that cause diarrhea through heat-labile and/or heat-stable enterotoxins, a common cause of foodborne or waterborne travelers' diarrhea with cholera-like symptoms: cramping, vomiting, diarrhea, prostration, dehydration. With an incubation period of 10–72 hours, the illness lasts 3–5 days.

**enterotoxin** A toxic substance produced by some bacteria that causes dysfunction in the human gastrointestinal tract (e.g., violent vomiting and diarrhea).

**enterovirus** Any of several picornaviruses of the genus *Enterovirus,* infecting the human intestinal tract and causing diseases of the nervous system, and including the polio, echo, and coxsackie viruses. Note that enterovirus is distinguished from enteric virus. *See also* excreted virus.

***Enterovirus genus*** *See* enterovirus.

**Enterprise** A floating aerator manufactured by Air-O-Lator Corp.

**Enterprise™** A regenerative product of Grace TEC Systems using thermal oxidation.

***Ent. faecalis*** Abbreviation of *Enterococcus faecalis.*

***Ent. faecium*** Abbreviation of *Enterococcus faecium.*

***Ent. gallinarium*** Abbreviation of *Enterococcus gallinarium.*

**enthalpy** The total heat content of a liquid, vapor, system, or body; or the heat released or taken up by a chemical reaction. Enthalpy ($H$, Btu or kilojoules) is usually calculated as the sum of internal energy ($U$, Btu or kilojoules) and the product of pressure ($P$, psi or kilopascals) and volume ($V$, cubic feet or cubic meters):

$$H = fPV \qquad \text{(E-23)}$$

$f$ being a conversion factor, 144/778 = 0.1851 for U.S. customary units and 1 for SI units.

***Ent. hartmanni*** Abbreviation of *Entamoeba hartmanni.*

***Ent. histolytica*** Abbreviation of *Entamoeba histolytica.*

***Ent. moshkovskii*** Abbreviation of *Entamoeba moshkoskii.*

**entrain** To trap bubbles in water either mechanically through turbulence or chemically through a reaction.

**entrained** Characteristic of particulates or vapor transported along with flowing gas or liquid.

**entrainment** (1) The incorporation of small organisms, including eggs and larvae of fish and shellfish, into an intake system. (2) The carryover of droplets of water with vapor produced during evaporation. (3) The mechanical or chemical trapping of bubbles in water.

**entrainment separator** A device that is used to eliminate or reduce entrainment, i.e., to remove entrained droplets of water from a vapor stream produced during evaporation (e.g., during desalting). Also called demister, demister entrainment device, mist eliminator, or, simply, separator. *See also* spacer.

**entrance head** The head required for water to flow into a conduit, channel, or hydraulic structure.

**entrance loss** The head required to overcome resistance to entrance of water into a conduit, channel, or hydraulic structure. Also called entry loss or inlet loss. *See also* exit loss.

**entrapment** One of two principal mechanisms responsible for solids separation in rapid rate filtration, whereby the filter medium, like a sieve or a strainer, retains the coagulated particles in pore openings. *See also* attachment.

**entropy** A measure of unavailable energy during a thermodynamic process, depending on temperature, pressure, and other variables. It is a thermodynamic term, with units of Btu per degree Rankine or joules per degree Kelvin, expressing the degree of disorder in a system, as in the second law of thermodynamics. *See also* free energy.

**entry loss** Same as entrance loss.

**entry point** A point of entry of water into the distribution system, often selected for collecting samples.

**entry portal** The point of entrance of microorganisms to the human body. In the case of waterborne or foodborne pathogens, they are most often carried through the mouth to the gastrointestinal tract. *See* portal of entry for detail.

**Envessel Pasteurization™** A process developed by RDP Co., using lime stabilization and pasteurization to further reduce pathogens in sludge.

**Enviro-Blend™** A line of chemicals produced by American Minerals, Inc. for the removal of heavy metals from wastewater.

**Envirodisc** A rotating biological contactor manufactured by Walker Process Equipment Co. for wastewater treatment.

**ENVIROFirst™** Granules of sodium carbonate peroxyhydrate produced by Solvay Interox for use in water treatment.

**EnviroGard™** A test kit from Millipore for screening chemicals.

**Enviromat** A wastewater treatment system designed by Ionics, Inc.

**environment** The sum of all external conditions (physical, chemical, biotic) affecting the life, development and survival of an organism. The environment includes water, air, land, and all plants, man and other animals living therein, and interrelationships that exist among these. *See also* ecosystem.

**environmental alteration** A corrosion control method using an electrolytic backfill material to protect buried pipes. Also called trench improvement.

**environmental assessment (EA)** An environmental analysis prepared pursuant to the National Environmental Policy Act to determine whether a federal action would significantly affect the environment and thus require a more detailed environmental impact statement. *See also* environmental review.

**environmental audit** An independent assessment of the current status of a party's compliance with applicable environmental requirements or of a party's environmental compliance policies, practices, and controls.

**environmental chemistry** The study of natural and manmade substances in the environment; their detection, monitoring, transport, and chemical transformation in air, water, and soil. *See also* aquatic chemistry.

**environmental classification** The classification of diseases according to the environmental factors (e.g., transmission routes and life cycles) that can be altered to affect the pathogens that cause the diseases, as compared to the generic or biological classification. *See also* biological classification, biological classification convention, phylogenetic classification, organic nomenclature, taxonomic classification.

**environmental engineering** An offshoot of civil/sanitary engineering focusing on the development of physical systems to control waste streams. Civil/sanitary engineers started to refer to themselves as environmental engineers around 1970 with the great boom in new environmental regulations, such as the National Environmental

Policy Act and the Water Pollution Control Act. Environmental engineering deals with the enhancement of human life, the protection of public health, and the protection of environmental resources. It applies principles from not only engineering but also such disciplines as hydraulics, chemistry, microbiology, ecology, and epidemiology for the planning, design, construction, administration, and operation of facilities for water supply, wastewater disposal, solid waste disposal, air pollution control, abatement of noise pollution, and water quality. *See also* environmental health engineering, public health engineering, sanitary engineering.

**environmental epidemiology** The application of epidemiological data and principles to the assessment of the effects of environmental contaminants on human health.

**environmental equity** Equal protection from environmental hazards of individuals, groups or communities regardless of race, ethnicity, or economic status; a major challenge for sustainable development.

**environmental exposure** Human exposure to pollutants originating from facility emissions. Threshold levels are not necessarily surpassed, but low-level chronic pollutant exposure is one of the most common forms of environmental exposure. *See* threshold level.

**environmental geochemistry** The branch of geochemistry that studies the interactions among rock, water, air, and human systems that determine the characteristics of the surface environment.

**environmental geology** The application of geological data and principles to the solution of environmental problems, e.g., the effects of mineral extraction and erosion.

**environmental health** The subfield of public health that deals with the impacts of people on their environment and the impacts of the environment on them. According to the World Health Organization "Environmental health comprises those aspects of human health, including quality of life, that are determined by physical, biological, social, and psychological factors in the environment. It also refers to the theory and practice of assessing, correcting, controlling, and preventing those factors in the environment that can potentially affect adversely the health of present and future generations."

**environmental health engineering** A branch of engineering that aims at the improvement of public health by focusing on reducing the transmission of infectious diseases through safe practices in domestic water supply, excreta and wastewater disposal, refuse disposal, drainage, housing, and other areas. *See also* environmental engineering, public health engineering, sanitary engineering.

**environmental impact** A change in an environmental condition or parameter resulting from human or natural activities, for example, the impoundment of a surface stream, the discharge of emissions into the air, discharge of inadequately treated effluents into a stream, or the withdrawal of water from an aquifer. Impacts can be beneficial or adverse, primary (or direct) or secondary (indirect).

**environmental impact assessment (EIA)** A method of analysis that attempts to predict probable repercussions of a proposed development on the social and physical environment of the surrounding area. *See* environmental assessment.

**environmental impact statement (EIS)** A document required of federal agencies by the National Environmental Policy Act for major projects or legislative proposals significantly affecting the environment. A tool for decision making, it describes the positive and negative effects of the undertaking and cites alternative actions.

**environmental indicator** A measurement, statistic, or value that provides a proximate gauge or evidence of the effects of environmental management programs or of the state or condition of the environment.

**environmental justice** The fair treatment of all races, cultures, incomes, and educational levels with respect to the development, implementation, and enforcement of environmental laws, regulations, and policies. Fair treatment implies that no population of people should be forced to shoulder a disproportionate share of the negative environmental impacts of pollution or environmental hazards due to a lack of political or economic strength. There are cases of landfills and hazardous waste sites located in minority communities.

**environmental laws** Laws passed to protect natural resources, including environmental health; e.g., Clean Air Act, Clean Water Act, National Environmental Policy Act, CERCLA, SARA, RCRA, Toxic Substances Control Act, Surface Mining Control and Reclamation Act.

**environmentally balanced industrial complex (EBIC)** A group of compatible industrial plants selected and located together in such a way as to minimize environmental impacts and production costs.

**environmentally related measurements** Any data collection activity or investigation involving the assessment of chemical, physical, or biological factors in the environment that affect human health or the quality of life; e.g., pollutant concentrations in the ambient environment, pollutant effects on health, and a determination of the environmental impact of cultural and natural processes (EPA-40CFR30.200).

**environmentally sensitive** (1) Aware of the effects that human actions can have on the environment. (2) Pertaining to environmental resources that are more susceptible to impacts. Examples of environmentally sensitive areas commonly discussed in an environmental impact assessment or statement include surface waters, marshes, flood plains, forests, wildlife habitats, groundwater recharge areas, steeply sloping lands, prime agricultural areas, and archaeologic/historic/cultural sites.

**environmentally transformed** Under the Toxic Substances Control Act, a chemical is environmentally transformed when its chemical structure changes as a result of the action of environmental processes on it (EPA-40CFR271.3).

**environmental microbiology** A branch of microbiology that applies our understanding of microbes in the environment to the benefit of public health and human society in general. It initially focused on water quality but currently interfaces with additional subjects, including aeromicrobiology, agriculture and soil microbiology, biogeochemistry, bioremediation, biotechnology, food quality, and wastewater treatment. *See also* microbial ecology.

**environmental overload** The introduction into the environment of more pollutants than the environment can handle or dispose of on its own.

**Environmental Performance Index (EPI)** A parameter developed by the Niagara Mohawk Corporation to measure progress toward the major objectives of its corporate environmental policy. The EPI is a weighting and rating index in the categories of waste emission reductions, compliance, and enhancement.

**environmental planning** A subdivision of public health engineering or environmental engineering dealing with the preliminary approval of facilities for water supply, wastewater disposal, solid waste management, etc. Environmental planning includes such activities as facilities planning and environmental impact assessments.

**environmental pollution** "The unfavorable alteration of our surroundings, wholly or largely as a by-product of man's action, through direct or indirect changes in energy patterns, radiation levels, chemical and physical constitution, and abundance of organisms . . ." (President's Advisory Committee, as cited in Sarnoff, 1971).

**Environmental Protection Act (EPA)** A United Kingdom legislation enacted in November 1990 for pollution control and other environmental protection goals.

**Environmental Protection Agency (EPA or USEPA)** An agency of the U.S. government primarily responsible for enforcing federal environmental laws. It also conducts research in the field, issues helpful guidelines, and develops tools such as the SWMM model.

**environmental refugees** "People who have been forced to leave their traditional habitat, temporarily or permanently, because of a marked environmental disruption—either natural or human-induced—that jeopardized their existence and/or seriously affected their quality of life." Such environmental disruption often involves several factors such as soil erosion, deforestation, desertification and water shortage (World Health Organization, 1997). Also called ecological refugees, ecorefugees.

**environmental resistance** The limiting effect of ecological conditions on the growth of a population.

**environmental review** The process whereby the USEPA undertakes an evaluation to determine whether a proposed Agency action may have a significant impact on the preparation of an environmental impact statement. *See also* environmental assessment (EPA-40CFR6.101-c).

**environmental risk analysis** A qualitative or quantitative evaluation of the environmental and/or health risk resulting from exposure to a chemical or physical agent; it combines exposure assessment results with toxicity assessment results to estimate risk. It includes four components: hazard identification, dose–response assessment, exposure assessment, and risk characterization. Also called risk assessment.

**environmental science** The study of the interactions among terrestrial, aquatic, atmospheric, soil, and living ecosystems, including all disciplines that affect these interactions (e.g., ecology, chemistry, biology, sociology, technology, and government).

**Environmental Simulation Program (ESP)** A set of modeling and simulation tools developed by OLI Systems, Inc. (Morris Plains, NJ) for environmental and other applications, including biological treatment, sedimentation, neutralization, precipitation, etc.

**environmental stress** A condition that is likely to cause an adverse change in an ecosystem.

**environmental taxonomy** *See* environmental classification.

**Environmental Technology Verification (ETV)** A program of the USEPA and the NSF International to validate the efficacy of water treatment technologies.

**environmental water reuse** *See* recreational/environmental water reuse.

**Enviropac** A wastewater treatment system developed by Walker Process Equipment Co. using rotating biological contactors.

**Enviropax** Tube settlers manufactured by Enviropax, Inc.

**Enviropress** A piston-type press manufactured by Environmental Engineering, Ltd. for use on screenings.

**Enviro-Seal™** A valve packing apparatus manufactured by Fisher Controls International, Inc. to control emissions.

**Envirosorb™** A material manufactured by Geosource, Ltd. to absorb oil spills.

**Envirovalve** A telescopic valve manufactured by EnviroQuip, Inc.

**enzootic disease** An endemic infectious disease transmissible to humans by animals.

**enzymatic assay** A test used to detect the presence of total coliform bacteria and *E. coli* in a water or wastewater sample; the sample is added to bottles or MPN tubes containing salts and specific enzymes as carbon source. The color of the sample changes to yellow if coliforms are present and fluoresces when exposed to long-wave ultraviolet illumination if *E. coli* are detected. *See also* MPN test, presence–absence test.

**enzymatic hydrolysis** In anaerobic digestion, the conversion by enzymes of complex organic constituents (starches, fatty acids, proteins, etc.) to lower-molecular-weight soluble intermediates (e.g., sugars, amino acids).

**enzymatic reactions** Biochemical changes (e.g., fermentations) resulting from the catalytic action of proteins from living cells; enzymatic reactions occur in the endogenous respiration phase of microbial growth patterns.

**enzyme** An unstable organic substance (such as a protein or conjugated protein produced by living organisms) that causes or speeds up chemical reactions; an organic and/or biochemical catalyst in the biochemical transformation of compounds. Enzymes control chemical reactions of metabolism, each enzyme being applicable only to a limited range of reactions. Examples: adenosine triphosphate, dehydrogenase, oxidase. *See* constitutive enzyme, inducible enzyme.

**enzyme complexing** One of the mechanisms of removal of suspended and colloidal solids from wastewater by trickling filtration.

**enzyme destruction** A primary mechanism of destruction of microorganisms by chemical and physical agents (e.g., chlorine, heat): they inactivate enzymes that are essential to the metabolic processes of the cells.

**enzyme induction** The synthesis of a new enzyme to increase the activity of cells; e.g., for the metabolism of chemicals foreign to the body.

**enzyme-linked immunosorbent assay (ELISA)** A sensitive laboratory technique using specific antibodies to detect the presence of antigens and viruses. The antigen of interest is concentrated, solubilized, and incubated for its capture (i.e., binding with the antibody). A secondary antibody, associated with a signal molecule is then captured (i.e., bound to the antigen). The addition of a substrate causes a color change in proportion to the amount of antigen present.

**enzyme–substrate complex** The intermediate compound produced during a reaction catalyzed by an enzyme, which recognizes the compound by its molecular structure and binds to it. The complex then breaks down into energy and the product of the reaction, regenerating the enzyme in the process for additional reactions.

**$E^0$** Symbol of electromotive force (EMF).

**eosinophilia** The presence in the blood of an abnormally high number of substances containing eosin (a red, crystalline, water-insoluble solid, $C_{20}H_8Br_4O_5$) and other acid stains; a symptom of diphyllobothriasis.

**EOX** Acronym of extractable organohalogen.

**EPA** Acronym of (a) (United States) Environmental Protection Agency and (b) (United Kingdom) Environmental Protection Act.

**EPC** Acronym of engineer, procure, and construct.

**EPCO™** Rotating biological contactors manufactured by U.S. Filter Corp. for wastewater treatment.

**EPDM** Acronym of ethylene propylene diene monomer.

**EPEC** Acronym of enterohepathogenic *E. coli.*

**EPER** Acronym of European Pollutant Emissions Register.

**E-pH diagram** Same as Pourbaix diagram.

**EPI** (1) Abbreviation of epichlorohydrin. (2) Acronym of Environmental Performance Index.

**epibiotic habitat** The surfaces in the marine environment on which attached communities develop. *See also* neuston layer and endobiotic habitat.

**EPIC™** An apparatus manufactured by Norchem Industries to improve polymer control.

**epichlorohydrin ($C_3H_5OCl$ or $CH_2OCHCH_2Cl$)** Abbreviation: ECH or EPI. A highly volatile liquid with a chloroformlike odor, used as a solvent and in the production of glycerin, resins, agricultural chemicals, and flocculents for food preparation or water treatment. It is a strong mutagen; its metabolite, alphachlorohydrin, can produce infertility in laboratory animals. Other names: chloropropylene oxide, halogenated alkyl epoxide.

**epichlorohydrin dimethylamine (epiDMA)** A cationic polyelectrolyte widely used in water treatment, sold as an aqueous solution and classified as a quaternary amine.

**epidemic** Widespread outbreak of a disease, or a large number (clearly in excess of normal expectancy) of cases of a disease in a single community, region, or a relatively small area, during the same time period. Disease may spread from person to person, and/or by the exposure of many persons to a single source, such as a contaminated water supply.

**epidemic curve** A graph of the number of cases of a disease versus time, providing information about the incubation period and other characteristics.

**epidemic hepatitis** *See* hepatitis A.

**epidemic jaundice** *See* hepatitis A.

**epidemic pleurodynia** *See* pleurodynia.

**epidemic typhus** A water-washed disease caused by the pathogen *Rickettsia prowazeki* and transmitted by lice; also called classical typhus. *See* louse-borne typhus.

**epidemiological features of excreted pathogens** Characteristics of excreta-related pathogens include excreted load, latency, persistence, median infective dose, immunity, reservoir, intermediate host.

**epidemiological risk analysis** The establishment of cause-and-effect relationships between exposure to a chemical and disease. *See* risk assessment.

**epidemiologic study** Study of human populations to identify causes of disease. Such studies often compare the health status of a group of persons who have been exposed to a suspect agent with that of a comparable nonexposed group.

**epidemiology** A branch of medicine that studies epidemics to determine the factors that cause the diseases and how to prevent them. It is a study of the distribution of disease or other health-related states and events in human populations, as related to age, sex, occupations, ethnic, and economic status in order to identify and alleviate health problems and promote better health.

**epidemiology study** The investigation of factors contributing to disease or toxic effects in the general population.

**epiDMA** *See* epichlorohydrin dimethylamine.

**epifluorescence microscope** A microscope that uses ultraviolet light to illuminate the specimen, which reflects visible light to the ocular.

**epilimnion** The upper layer of water in a thermally stratified lake or reservoir. This layer, under the influence of surface currents, consists of the warmest water and has a fairly uniform temperature. The layer is readily mixed by wind action and contains relatively high levels of dissolved oxygen during daylight hours. It is also called circulation zone. *See* thermal stratification.

**epilimnion zone** The epilimnion.

**epilithic** Pertaining to organisms that grow on stones, rocks, or other hard, inorganic substances.

**epipelagic zone** The upper part of an ocean or other water body through which sufficient sunlight can penetrate to enable photosynthesis. Its depth may reach 200 meters depending on water turbidity. *See* photic zone for detail.

**epiphyte** A nonparasite plant that grows above ground on other plants; also called air plant or aeroplant.

**episode** An air pollution incident in a given area caused by a concentration of atmospheric pollutants under meteorological conditions that may result in a significant increase in illness or deaths. Episode may also apply to a water pollution event or a hazardous material spill.

***Epistylis*** One of the groups of ciliate protozoa active in trickling filters, feeding on the biological film and maintaining its high growth rate.

**epizootic** An epidemic disease affecting an animal population.

**epizootic disease** An epidemic disease transmissible to humans by animals.

**epm** Acronym of equivalent(s) per million.

**epoxidation** The formation of an oxygen bridge across a double bond; a chemical reaction that results in the formation of an epoxide. It is a means of metabolic attack on aromatic rings.

**epoxide** An organic chemical containing a group consisting of an oxygen bound to two connected carbon or other atoms; general formula: $R_2C—O—CR_2$, where R is hydrogen or a carbo-containing group. It is an ozonation by-product.

**epoxy lining** A trenchless method of water main rehabilitation involving the application of a thin layer of epoxy (a dielectric insulator) resin and hardener to the pipe wall; the mixture cures in approximately 16 hours. *See also* trenchless technology, spray-applied liner.

**EPS** Acronym of extended period simulation.

**epsomite** The natural form of Epsom salt, found in caves and lake deposits.

**Epsom salt ($MgSO_4 \cdot 7H_2O$)** Small colorless crystals of hydrated magnesium sulfate having cathartic properties and laxative effects; also used in fertilizers, leather tanning, and textile dyeing. (Epsom is a town in Surrey, southeast England.)

**EP toxic** *See* extraction procedure.

**EP toxicity test** *See* extraction procedure toxicity test.

**EP toxicity threshold** *See* extraction procedure toxicity test threshold.

***Epulopiscium fishelsoni*** A giant marine bacterium (0.60 mm × 0.08 mm), visible with the naked eye; isolated from the gut of Red Sea surgeonfish. It does not reproduce by bacterial division but through two vegetative offspring grown inside the cytoplasm of the mother cell.

**eq** Abbreviation of equivalent.

**eq/L** Abbreviation of equivalent(s) per liter.

**equalization** The process of dampening hydraulic or organic variations in a flow to achieve nearly constant conditions, e.g., the provision of sufficient storage capacity to equalize pumping rate or to equalize demand over a long period.

**equalization basin** A basin or tank used to provide flow equalization; same as equalizing basin.

**equalization storage** *See* equalizing basin.

**equalizing basin** A basin used in a hydraulic system to render uniform the volume and composition of the flows. It is often used in wastewater and water supply systems to provide stable influent characteristics ahead of treatment units or to facilitate water distribution to customers. Also called equalization basin or balancing basin. An equalizing reservoir is defined similarly. Equalization storage or equalizing storage refers to the volume of water in the basin or reservoir.

**equalizing reservoir** Also called balancing reservoir. *See* equalizing basin.

**equalizing storage** *See* equalizing basin. Also called operating storage.

**equation of state** (1) An empirical relationship among some physical properties of water such as density, temperature, salinity, and turbidity; e.g., the relationship between water density ($\rho$) and (i) the density ($\rho_T$) of pure water at temperature $T$, (ii) the changes in density due to turbidity or suspended solids ($\Delta\rho_{SS}$) and to salinity or dissolved solids ($\Delta\rho_{DS}$):

$$\rho = \rho_T + \Delta\rho_{SS} + \Delta\rho_{DS} \quad \text{(E-24)}$$

(2) Same as the ideal gas equation. *See also* Boyle law, Charles law, state variable.

**equations and formulas** *See* Section IV: Illustrations.

**equilibrium** (1) In relation to radiation, the state at which the radioactivity of consecutive elements within a radioactive series is neither increasing or decreasing. (2) The state in which the concentrations of all species are constant in a chemical reaction. *See also* chemical equilibrium. (3) A state of balanced partitioning of a gas or organic contaminant between air and water in aeration and air stripping applications to water or wastewater treatment. *See* Henry's law equilibrium.

**equilibrium analysis of wells** The evaluation of aquifer characteristics using the Theim and Dupuit equations, which make a number of assumptions, including constant flow rate and a uniform aquifer of infinite extent.

**equilibrium concentration quotients** (1) The equilibrium conditions derived from ion-exchange reactions and characterizing the selectivity of an exchanger for a specific ion. *See* ion-exchange process. (2) *See* equilibrium constant.

**equilibrium constant** A number that describes the quantitative relationship between the species of a reversible reaction that is in equilibrium. For the chemical reaction

$$aA + bB + \ldots \leftrightarrow cC + dD + \ldots \quad \text{(E-25)}$$

the equilibrium constant is

$$K = [C_C]^c \cdot [C_D]^d \ldots / [C_A]^a \cdot [C_B]^b \ldots \quad \text{(E-26)}$$

where $[C_A]$, $[C_B]$ . . . denote the concentrations of substances A, B . . . . The quantities a, b, c, and d are called stoichiometric coefficients. *See also* the van't Hoff equation. The equilibrium constant for the dissociation of water is:

$$K_w = [H^+] \cdot [OH^-] \quad \text{(E-27)}$$

which varies with temperature as follows:

$$K_w = -6.0875 + 4470.99/T + 0.01706\,T \quad \text{(E-28)}$$

where $T$ is temperature in °K. For a given reaction, the equilibrium constant varies with the temperature and the ionic strength of the solution. *See also* activity product, common-ion effect, crystal growth, dissociation constant, dissolution reaction, equilibrium concentration quotient, equilibrium condition, ionization constant, ion product, law of mass action, LeChatelier's principle, molar concentration, reaction constant, saturated solution, solubility product constant, solubility product, supersaturated solution.

**equilibrium contact operation** A mechanism of the mass transfer process that contributes to equilibrium concentrations between two phases through a series of contacts that result in a progressive change of concentration from one phase to another.

**equilibrium contact separation** The mechanism of mass transfer whereby material is transferred from one homogeneous phase to another under a pressure or concentration gradient. The transfer of a constituent stops when equilibrium is reached. Also called equilibrium phase separation.

**equilibrium line** A straight line that represents equilibrium in gas stripping operations, based on Henry's law. *See* stripping tower mass balance.

**equilibrium phase separation** Same as equilibrium contact separation.

**equilibrium quotient** *See* equilibrium concentration quotient.

**equilibrium shift** A change in the relative concentrations of reacting substances that causes a different reaction or a change in the reaction rate. For example, a change in the relative concentrations of sodium and calcium ions will dictate both the exchange rate and the selection of which ions will be adsorbed to and released from ion-exchange resin beads.

**equilibrium vapor pressure** The pressure in equilibrium with its liquid or solid phase. *See also* saturation vapor pressure, vapor tension.

**equipment** *See* facilities.

**equivalence data** Chemical data or biological test data intended to show that two substances or mixtures are equivalent (EPA-40CFR790.3).

**equivalence point** (1) On a titration curve, it is the pH at which the equivalents of titrant added are equal to the equivalents in solution; for example, in the titration of strong acids and strong bases (e.g., HCl by NaOH) or in the redox titration of one metal by another (e.g., $Ce^{4+}$ by $Fe^{3+}$). (2) The point at which the equivalent concentrations of two reactants are equal.

**equivalency** Any body of procedures and techniques of sample collection and/or analysis for a parameter of interest that has been demonstrated in specific areas to produce results not statistically different from those obtained from a reference method.

**equivalent (eq)** *See* equivalent weight.

**equivalent available chlorine (EAC)** A measure of the disinfecting or oxidizing power of chlorine dioxide ($ClO_2$), based on the oxidation half-reaction:

$$ClO_2 + 5\,e^- + 4\,H^+ \rightarrow Cl^- + 2\,H_2O \qquad \text{(E-29)}$$

and the percent weight of chlorine in chlorine dioxide (52.6%) and on the number of electrons changed (5), i.e., EAC = 5 × 52.6% = 263% or 2.63 times the oxidizing power of chlorine; 1.0 mg/L of chlorine dioxide is equivalent to 2.63 mg/L of chlorine.

**equivalent background compound** A hypothetical single compound equivalent to the background organic matter that competes with a given trace organic in adsorption applications. *See also* ideal adsorbed solution theory.

**equivalent calcium carbonate** A concentration expressed as calcium carbonate. *See also* chemical equivalent.

**equivalent conductance** A characteristic ($\lambda$, micromhos/cm) of a substance or a solution proportional to its specific conductance ($\kappa$, micromhos/cm) and inversely proportional to the normality (N = number of equivalents per liter) of the solution (or the equivalent weight of the substance):

$$\lambda = 1000\,\kappa/(kN) \qquad \text{(E-30)}$$

where $k$ is an adjustment factor (Droste, 1997).

**equivalent conduit length** *See* equivalent pipes.

**equivalent customer** The ratio of the water consumption of a large consumer to the consumption of a single-family residence through a 5/8-inch meter (Symons et al., 2000). *See also* equivalent dwelling unit, equivalent meter, and population equivalent.

**equivalent depth** The concept of equivalent depth of filter media is based on the assumption that the ratio ($L/d$) of bed depth ($L$) to grain diameter ($d$) is constant and results in equal filtrate quality for the same influent suspension and the same filtration rate. Another ratio, $L/d^{\beta}$, has also been proposed instead of $L/d$, with $1.5 < \beta < 1.67$.

**equivalent diameter** For noncircular conduits flowing full or partly full, the equivalent diameter is equal to four times the hydraulic radius. It may be used in the Darcy–Weisbach formula to calculate the head loss.

**equivalent dwelling unit (EDU)** A convenient concept used in water, wastewater and similar studies. It has the size, water use, wastewater production, etc. of the average household. Also called equivalent residential unit. *See also* population equivalent.

**equivalent length of fittings** *See* equivalent pipes.

**equivalent meter** A means of expressing the size of a water meter ($X$) in terms of a standard meter ($x$); it is the ratio of the squares of the two sizes, i.e., $X^2/x^2$. For example, for $X = 2''$ and $x = 5/8''$, the equivalent meter is $4/(25/64) = 10.24$ times the standard meter. *See also* equivalent customer.

**equivalent method** Any method of sampling and analyzing for air pollution that has been demonstrated to the USEPA Administrator's satisfaction to be, under specific conditions, an acceptable alternative to normally used reference methods.

**equivalent(s) per liter (eq/L)** A unit of concentration equal to normality, used in volumetric analysis; it is equal to concentration divided by the equivalent weight. *See also* equivalent per million.

**equivalent(s) per million (epm)** A unit of ionic concentration or chemical equivalent weight equal to the concentration in parts per million (ppm) divided by the equivalent weight. *See also* equivalent per liter.

**equivalent pipes** Two pipes or two systems of pipes are said to be equivalent when their head losses are equal for equal discharges. Similarly, one pipe may be equivalent to a system of pipes. The equivalent length of a pipe fitting is the length of pipe that would yield the same head loss as the fitting for a given discharge. It is expressed as a multiple of the pipe diameter, or as the ratio of the length ($L$) to the diameter ($D$). From the Darcy–Weisbach formula, the head loss (hf) in the fitting is

$$h_f = f(L/D)(V^2/2g) \qquad \text{(E-31)}$$

or

$$h_f = k(V^2/2g) \qquad \text{(E-32)}$$

with

$$k = f(L/D) \qquad \text{(E-33)}$$

where $f$ is the friction factor, $V$ is the average velocity, and $g$ is the gravitational acceleration. The coefficient $k$, called the minor loss coefficient (sometimes referred to as the $k$-factor), is constant for a given fitting and does not vary with the Reynolds number ($R_e$). $k$ is not a friction factor but is a function of changes in direction, obstructions, or changes in velocity. For air piping, minor losses can be computed as equivalent length:

$$L = 55.4\, CD^{1.2} \qquad \text{(E-34)}$$

where $C$ = a dimensionless resistance factor.

**equivalent population** The estimated population that would contribute a given load of water, wastewater, solid waste, etc. *See* population equivalent for detail.

**equivalent residential unit** *See* equivalent dwelling unit.

**equivalent solids reservoir (ESR)** A method used to model the washoff and transport of pollutants in stormwater management studies.

**equivalent spherical diameter** The diameter of an equal volume sphere, a concept used in defining the characteristics of filter media. *See also* sphericity.

**equivalent weight (eq wt)** The weight of a substance that contains one gram atom of hydrogen or proton, i.e., the weight of a substance that is associated with one hydrogen ion, proton, or hydroxyl ion. The equivalent weight of a simple element (e.g., sodium, Na) is its atomic weight divided by its number of reactive protons (i.e., 23/1 = 23); *see* element. The equivalent weight of a compound (e.g., calcium carbonate, $CaCO_3$) is its molecular weight divided by its number of reactive protons (i.e., 100/2 = 50). The number of reactive protons relates to the substance's valence and oxidation state. For compounds that do not donate or take up protons, the equivalent weight is based on their reaction with water. For example, carbon dioxide ($CO_2$) combines with water ($H_2O$) to form carbonic acid ($H_2CO_3$), which can donate two $H^+$ ions; thus, its eq wt is its molecular weight (44 g) divided by 2, i.e., 22 g. Also called combining weight. *See also* equivalent per liter, equivalent per million, milliequivalent weight, normality.

**eq wt** Abbreviation of equivalent weight.

**ergonomics** An applied science that considers the capacities and requirements of workers in designing devices, machines, systems, and physical working conditions. Water and wastewater treatment plants are normally designed to prevent or minimize injuries, fatigue, stress, etc. to operation and maintenance personnel. Also called human engineering, human-factors engineering.

**Ergun equation** (1) An equation developed by S. Ergun in 1952 to compute the applicable friction factor (F) in the Carman–Kozeny equation (AWWA, 1991 and Droste, 1997):

$$F = 1.75 + 150(1 - e)/R_e \qquad \text{(E-35)}$$

where $e$ = porosity of the filter bed and $R_e$ = Reynold's number. (2) An extension of the Carman–Kozeny equation proposed by Ergun in 1952 to account for kinetic energy loss:

$$H/L = 150\, \mu(1 - e)^2 v/[\rho g e^3 (SD)^2] + 1.75(1 - e)\, v^2/(gSDe^3) \qquad \text{(E-36)}$$

where $H$ = head loss, $L$ = depth of bed, $\mu$ = absolute viscosity of fluid, $v$ = face or approach velocity, $\rho$ = density of the fluid, $g$ = gravitational acceleration, $S$ = shape factor or sphericity, $D$ = the diameter of the sphere equivalent in volume to the granular media, and $\mu$ = viscosity of the fluid.

**Erlenmeyer flask** A conical flask with a wide base and a narrow neck used in laborartories to swirl and mix liquids, e.g., in the dichromate reflux method for the determination of COD.

**erosion** The wearing away of the land surface by running water, waves, moving ice, or wind, or by such processes as mass wasting. Geologic erosion refers to natural erosion processes occurring over long time spans. Accelerated erosion generally refers to erosion in excess of what is presumed or estimated to be naturally occurring levels, and which is a direct result of human activities (e.g., farming, residential or industrial development, road building, or logging) (Hawley and Parsons in Dodson, 1999).

**erosion control** The application of measures to reduce land erosion.

**erosion corrosion** The deterioration of a material or surface by a moving fluid, resulting from the combined effect of corrosion and erosion of suspended solids or gas bubbles. This phenomenon is often observed at points in the distribution system or in building plumbing where flow velocity is high or there is an abrupt change in direction. *See also* impingement attack.

**erosion threshold flow** The flow that is expected to result in a velocity above which erosion of a given channel at a specific location will occur.

**error** In wastewater process control applications, the difference between the measured value of a control parameter and the set point under non-steady-state operating conditions.

**ES** Acronym of effective size.

**ESCA** Acronym of electron spectroscopy chemical analysis.

**escape coefficient** A parameter used in a formula proposed for the surface reaeration rate or reaeration constant, $C_e$ = 0.054/ft = 0.177/m at 20°C. It is lower for large streams and varies with temperature (correction factor $\theta$ = 1.022).

**escarpment** The topographic expression of a fault.

***Escherichia coli* (*E. coli*)** The main species of fecal coliforms, bacteria of fecal origin used as an indicator organism in the determination of water pollution. Facultatively anaerobic, Gram-negative, nonsporeforming, *E. coli* are highly present in the intestine (up to $4 \times 10^{10}$ organisms excreted per person per day) and, although not all species are pathogenic, they are associated with pathogens. They can ferment sugars to produce organic acid and gas. *See also* coliform group. Ultraviolet light is an effective disinfectant against *E. coli*, producing 99.99 kill at 3000 mw-sec per cm$^2$.

***Escherichia coli* O157:H7** A foodborne or waterborne serotype of *Escherichia coli,* O-antigen 157, H-antigen 7, that produces enterotoxins (verotoxins I and II), a prime cause of bloody diarrhea in infants. In May 2000, an outbreak of this pathogen and *Campylobacter jejeuni* through contaminated drinking water killed 7 people and made more than 2000 people sick in Walkerton, Ont. (Canada). *See also* enterohemorrhagic *E. coli.*

***Escherichia* genus** A genus of fecal coliform and total coliform bacteria.

**ESD™** Acronym of enhanced sludge digestion, an egg-shaped anaerobic digester manufactured by CBI-Walker, Inc.

**ESP** Acronym of (a) electrostatic precipitator; (b) Environmental Simulation Program.

**ESP®** Equipment manufactured by Wheelabrator Clean Water Systems, Inc. for drying and pelletization of wastewater sludge.

**ESPA™** A membrane manufactured by Hydranautics using a low-pressure polyamide.

**espundia** Latin American word for a mouth and nostril deformity transmitted by sandflies. *See also* leishmaniasis.

**ESR** Acronym of equivalent solids reservoir.

**essential elements** Essential elements for agricultural crops and other plants: (a) the basic elements of carbon, oxygen, and hydrogen, available from water and air; (b) the fertilizing elements of nitrogen, phosphorus, potassium, calcium, magnesium, sulfur, and iron available from the soil and, to a certain extent, from wastewater and sludge solids; and (c) the trace elements such as boron, copper, manganese, and zinc.

**essential metal** *See* essential metal cation.

**essential metal cation** Any of the macronutrients and micronutrients required for the growth of organisms and having known biological functions: sodium (Na), potassium (K), magnesium (Mg), calcium (Ca), manganese (Mn), iron (Fe), cobalt (Co), nickel (Ni), copper (Cu), zinc (Zn), and molybdenum (Mb). Chromium (Cr), vanadium (V), and tungsten (W) are also thought to have essential functions in some organisms.

**essential minerals** Nutrients required for the growth of active biomasses are usually present in wastewater except, sometimes, nitrogen and phosphorus. *See also* essential elements.

**ester** An organic compound formed by the reaction between an inorganic or organic acid and an alcohol with the elimination of a molecule of water; it is similar to a salt in inorganic chemistry. General formula: R—COO—R′, where R and R′ denote, respectively, the radicals for an acid and an alcohol, e.g.:

$$RCO{-}OH + H{-}OR' \rightarrow H_2O + R{-}COO{-}R' \quad (E\text{-}37)$$

Esters are also derived from an organic acid by the replacement of the hydrogen by an organic radical, e.g., ethyl acetate ($CH_3COOC_2H_5$) from acetic acid ($CH_3COOH$).

**esterification** The process of formation of an ester; e.g., derivitization.

**estimated exposure dose (EED)** The measured or calculated dose to which humans are likely to be exposed considering exposure by all sources and routes.

**estuarine discharge/disposal** Discharge/disposal of wastewater or effluents in an estuary. Differences with stream disposal include tidal action and less pronounced currents (thus more opportunity for sedimentation near the discharge point).

**estuarine habitat** A tidal habitat and adjacent tidal wetlands that are usually semienclosed by land but have open, partly obstructed, or sporadic access to the open ocean and in which ocean water is at least occasionally diluted by freshwater runoff from the land.

**estuarine waters** Deepwater tidal habitats and tidal wetlands, usually enclosed by dry land but with access to the ocean, and occasionally diluted by freshwater runoff; e.g., bays, lagoons, river mouths, salt marshes.

**estuarine zone** The area near the coastline, consisting of estuaries and coastal saltwater wetlands.

**estuary** A region of interaction between rivers and near-shore ocean waters, where tidal action and river flow mix fresh and saltwater into brackish water. Estuaries include bays, river mouths, salt marshes and lagoons. These brackish water ecosystems shelter and feed marine life, birds, and wildlife. More simply, an estuary is a coastal water body at the mouth of a river, where the sea tide meets the river current. *See also* partially mixed estuary, salt wedge estuary, well-mixed estuary.

**ESWTR** Acronym of Enhanced Surface Water Treatment Rule.

**ET** Acronym of evapotranspiration.

**ETA system** Acronym of evapotranspiration absorption system.

**ETBF** Acronym of expected time before failure.

**etching** The chemical deterioration of the surface of glassware caused by high temperatures and detergents; it is more prevalent or intensified in soft or softened water systems.

**ETEC** Acronym of enterotoxigenic *E. coli.*

**ethanal ($CH_3CHO$ or $C_2H_4O$)** A volatile, colorless liquid, soluble in water, with a pungent, apple-like flavor. It is an oxidation product of ethanol ($C_2H_5OH$) and can be oxidized to acetic acid ($CH_3COOH$). It may be formed during water disinfection, particularly with ozone. It is used industrially and also as a food additive. Also called acetaldehyde.

**ethane ($C_2H_6$)** A colorless, odorless, flammable gas used in chemical synthesis and as a fuel. Also called dimethyl.

**ethaneperoxide acid** A commercial name of peracetic acid. *See* acetyl hydroxide for detail.

**ethanoic acid** The proper name of acetic acid.

**ethanol ($C_2H_5OH$ or $CH_3CH_2OH$)** An alternative automotive fuel derived from grains, molasses, starch, etc.; usually blended with gasoline to form gasohol. The basis of alcoholic drinks, it is an inflammable organic compound resulting from sugar fermentation. Proper name of ethyl alcohol.

**ethene** *See* ethylene.

**ether** (1) A colorless, highly volatile, flammable liquid, with an aromatic odor and sweet, burning taste, resulting from the action of sulfuric acid ($H_2SO_4$) on ethyl alcohol ($C_2H_5OH$); used as a solvent in analytical chemistry to extract polar organics. Also called diethyl ether, diethyl oxide, ethyl ether, ethyl oxide, sulfuric ether. (2) A hydrocarbon derivative in which two alkyl groups (R) are attached to an oxygen atom (O); used widely as an industrial solvent. General formula: R—O—R. Ethyl ether ($CH_3CH_2OCH_2CH_3$) and isopropylether [$CH_3)_2CHOCH(CH_3)_2$] are potential odor-causing compounds.

**ethinyl trichloride ($ClHC{=}CCl_2$)** Another name of trichloroethylene.

**ethyl** (1) Containing the ethyl group ($-C_2H_5-$), e.g., ethyl ether ($C_4H_{10}O$). (2) An antiknock fluid, acetic ether, containing tetraethyl lead.

**ethyl acetate ($C_4H_8O_2$)** A colorless, volatile, flammable liquid used as a scent in perfumes. Also called acetic ester.

**ethyl acrylate ($C_5H_8O_2$), ($CH_2$:$CHCOOCH_2CH_3$)** A foul-smelling, colorless liquid used in the plastics industry.

**ethyl alcohol** *See* ethanol.

**ethylbenzene ($C_6H_5C_2H_5$)** A colorless, flammable, water-soluble liquid, one of many constituents of gasoline and fuel oil, used to produce styrene; also used as paint solvent and diluent. This volatile organic chemical may cause damage to organs and the nervous system. MCLG = MCL = 0.7 mg/L.

**ethyl butyrate ($C_6H_{12}O_2$)** A colorless, volatile liquid used in flavoring extracts.

**ethyl carbamate** Same as urethane.

**ethyl chloride ($C_2H_5Cl$)** A flammable gas used as a refrigerant, solvent, etc. Also called chloroethane.

***S*-ethyl di-*N,N*-propylthiocarbamate [$C_2H_5SC(O)N(C_3H_7)_2$]** Abbreviation: EPTC. A herbicide.

**ethylenation** The introduction of the ethylene group ($C_2H_4$—) into a compound.

**ethylene** (1) A substance containing the ethylene group ($C_2H_4$—). (2) Same as ethanol. (3) A colorless, flammable gas ($C_2H_4$) of the ethylene series used in agriculture (citrus fruit coloring) and in chemical synthesis. Also called ethene, olefiant gas.

**ethylene alcohol** *See* glycol.

**ethylene bisdithiocarbamate (EBDC)** An organic substance used in fungicides that break down into ethylene thiourea.

**ethylene bromide (BrCH2CH2Br)** *See* ethylene dibromide.

**ethylenediaminetetraacetic acid [$(HOOCCH_2)_2NCH_2CH_2N(CH_2COOH)_2$]** Abbreviation: EDTA. An organic ligand that can form up to six bonds with a metal ion (e.g., calcium, magnesium, and other divalent ions); commonly used to titrate metals, chloride, or cyanide. Also called edetic acid.

**ethylene dibromide ($BrCH_2CH_2Br$)** Abbreviation: (EDB). A colorless liquid used as an agricultural fumigant and in certain industrial processes. Extremely toxic and found to be a carcinogen in laboratory animals, EDB has been banned for most agricultural uses in the United States. Also called ethylene bromide. It is a colorless, volatile, and heavy liquid. MCLG = 0, MCL = 0.00005.

**ethylene dichloride ($C_2H_4Cl_2$ or $ClCH_2$—$CH_2Cl$)** A volatile organic chemical used as a solvent and in the production of insecticides, paint, vinyl chloride, etc.; it can cause eye, throat, and nose irritations as well as damage to organs and the nervous system. MCLG = 0, MCL = 0.005 mg/L. Also called 1,2-dichloroethane.

**ethylene dioxide** A gas that is sometimes used in sterilizing laboratory apparatus that cannot withstand autoclaving or heating.

**ethylene glycol** Same as glycol.

**ethylene group (—$C_2H_4$—)** The bivalent group derived from ethylene or ethane. Same as ethylene radical.

**ethylene oxide ($C_2H_4O$)** A colorless, odorless, toxic gas from the oxidation of ethylene; used in synthetizing glycol.

**ethylene radical (—$C_2H_4$—)** The bivalent group derived from ethylene or ethane. Same as ethylene group.

**ethylene series** *See* alkene series.

**ethylene thiourea ($NHCH_2CH_2NHCS$)** Abbreviation: ETU. The common name for 2-imidazolidinethione, a breakdown product of the pesticide EBDC that may cause cancer. It is regulated by the USEPA: MCLG = 0; MCL = 0.025 mg/L.

**ethylene propylene diene monomer (EPDM)** A synthetic rubber used in the manufacture of pipe joint seals. *See* internal joint seal lining.

**ethyl ether** *See* diethyl ether and ether.

**ethyl group** *See* ethyl and ethane.

**ethylic acid** Same as acetic acid.

**ethyl mercaptan ($CH_3CH_2$—SH)** Same as mercaptan.

**ethyl methyl ketone ($CH_3CH_2COCH_3$)** *See* ketone.

**ethyl nitrate ($C_2H_2NO_2$)** A colorless, sweet, insoluble, explosive liquid used in manufacturing organic products.

**ethyl oxide** *See* diethyl ether and ether.

**ethyl sulfide ($C_4H_{10}S$)** A colorless, oily liquid, used as a solvent. Also called ethylthioethane, ethylthioether.

**ethylthioethane** Same as ethyl sulfide.

**ethylthioether** Same as ethyl sulfide.

**ethyl urethane** Same as urethane.

**ethyne** Same as acethylene.

**ethynyl group** The univalent group HC≡C; also called the ethynyl radical.

**ethynyl radical** The univalent radical HC≡C; also called the ethynyl group.

**etiologic agent** The agent that is the cause or the origin of a disease or an infection; pathogen.

**etiological agent** Etiologic agent.

**etiology** The study of the causes and origins of diseases. *See also* pathogenesis.

**ETSW** Acronym of extended terminal subfluidization wash.

**ET system** Acronym of evapotranspiration system.

**ETU** Acronym of ethylene thiourea.

**ETV** Acronym of Environmental Technology Verification.

**eubacteria** "True" bacteria, the principal microorganisms that carry out the decay of organic matter in wastewater treatment, often organized in branches of gelatinous mass (*zooglea ramigera*). The most common of these organisms are the cyanobacteria or blue-green algae, purple bacteria, Gram-positive bacteria, chlamydia, and spirochetes.

***Eubacterium spp.*** Any of the nonsporulating eubacteria found in feces.

**eucaryote** *See* eukaryote.

**eucaryotic** *See* eukaryotic.

***Euglena* genus** A group of green algae active in the decomposition of organic matter in waste stabilization ponds; sensitive to adverse conditions (e.g., temperature extremes), they may cause high BOD levels in the pond effluent.

**eukaryote** An organism having specialized organelles in the cytoplasm and one or more cells with well-defined nuclei enclosed within nuclear membranes; a higher stage of living than a prokaryote. Eukaryotic cells are one order of magnitude larger than bacteria. This group includes algae, fungi, and protozoa. Also spelled eucaryote.

**eukaryotic** Pertaining to cells that have a nuclear membrane separating the cytoplasm from the nuclear material or to organisms composed of such cells. Also spelled eucaryotic.

**eukaryotic cell** *See* eukaryote.

**eukaryotic organism** *See* eukaryote.

**euphotic** Pertaining to the upper layer of a lake or impoundment with sufficient sunlight for photosynthesis.

**euphotic zone** The upper layer of water in a natural water body through which sufficient sunlight can penetrate for active photosynthesis; it is up to 100 m deep. *See also* dysphotic zone, photic zone, Secchi depth.

**Euroform** A product of Munters used to reduce or eliminate mists.

**European Pollutant Emissions Register (EPER)** An agency established by the European Union Commission in 2000 to which Member States must report every three years on the emissions of industrial facilities to air and water. It will be replaced in 2009 by the European Pollutant Releases and Transfer Register.

**eurytopic** Pertaining to organisms that tolerate a broad range of habitats.

**eutectic** Pertaining to or characteristic of a material that is easily melted; having a very low melting point.

**eutectic freezing** A treatment process used to produce freshwater from saline and brackish waters by bringing them to the eutectic point (the lowest melting point possible). Ice crystals are formed, which separate from and float above the salts and other materials. *See also* freezing, contact freezing, saline water conversion classification.

**eutectic point** *See* eutectic freezing.

**eutrophic** (1) Pertaining to reservoirs and lakes that are rich in nutrients and very productive in terms of aquatic animal and plant life. (2) Characteristic of a lake with a lower layer of cold water that is depleted of oxygen in the summer by the decomposition of organic matter.

**eutrophication** The increase in the nutrient levels of a body of water; this usually causes an increase in the growth of aquatic animal and plant life. Eutrophication is a slow, natural aging process during which a lake, estuary, or bay evolves into a bog or marsh and eventually disappears. During the later stages of eutrophication, the water body is choked by abundant plant life due to higher levels of nutritive compounds such as nitrogen and phosphorus; the latter is often the limiting nutrient. Human activities can accelerate the process. *See* cultural eutrophication, natural eutrophication.

**eutrophication prevention** Any measure taken to eliminate the causes of overfertilization of receiving streams, lakes, ponds, etc. and prevent the development of nuisance growth. Such measures include the use of copper sulfate to control the growth of algae and aquatic plants in reservoirs and the removal of nutrients, particularly phosphorus, from effluents by chemical precipitation. Even low levels of nutrients (e.g., 0.30 mg/L of inorganic nitrogen and 0.05 mg/L of inorganic phosphorus at spring turnover) may produce an algal bloom in a stratifying lake. Another measure is the diversion of nutrient-rich effluents from lakes and other impoundments. However, eutrophication of large, deep lakes is not reversed immediately when pollution stops.

**eutrophication remedial measures** *See* eutrophication prevention.

**eutrophic dimictic lake** A eutrophic lake that goes through two stratification and two mixing periods each year, with a period of dissolved oxygen concentration reduction during the winter stratification because of prolonged snow and ice cover.

**eutrophic lake** A lake that exhibits any of the following characteristics: (a) excessive biomass accumulations of primary producers; (b) rapid organic and/or inorganic sedimentation and shallowing; or (c) seasonal and/or diurnal dissolved oxygen deficiencies that may cause obnoxious odors, fish kills, or a shift in the composition of aquatic fauna to less desirable forms (EPA-40CFR35.1605-5). Algal blooms are common in eutrophic lakes. *See also* dystrophic lake, oligotrophic lake.

**eutrophic reservoir** *See* eutrophic lake.

**eutrophic stage** The stage in the natural life cycle of a lake or reservoir in which nutrient levels and microbiological activity are high, associated with very low dissolved oxygen, high turbidity, color, and formation of disinfection by-product precursors; eutrophy.

**eutrophy** The stage in the aging process of a lake during which the lake is rich in nutrients and has an abundance of algae and other microscopic life that causes it to eventually dry up; eutrophic stage.

**Ev** Symbol of bulk modulus of elasticity.

**eV** Symbol of electron volt.

**Eva** A bar screen manufactured by Brackett Green, Ltd. with provision for back-raking.

**evaporating dish** A glass or porcelain dish used to evaporate water or wastewater samples by heating.

**evaporation** (1) The process by which water is converted into a vapor; the opposite of condensation, it depends on temperature, wind velocity, and relative humidity. For measuring evaporation from surface water, see evaporation pan. Evaporation from land surfaces is less than plant transpiration, and the two are usually lumped into evapotranspiration. (2) The quantity of water that evaporates. (3) A process used to concentrate solids in a suspension; water or wastewater is brought to its boiling point and then converted to vapor. (4) The recovery of a volatile solvent as a vapor from a solution containing nonvolatile dissolved solutes, with the vapor and residue as by-products. It is used to extract proteins or concentrate metals from a solution or slurry. (5) *See* thin-film evaporation. *See* actual evaporation, boiling heat transfer, brine concentrator, crystallization, distillation, drying, flash evaporation, flash vaporization, latent heat of vaporization, pan evaporation, vaporization.

**evaporation area** The surface area that loses water to the atmosphere by evaporation, including a body of water and any adjacent moist land.

**evaporation heat** *See* heat of evaporation.

**evaporation pond** A pond for water evaporation under the influence of solar heat, for example, (a) an area where sewage sludge is dumped and dried (also called drybed), (b) a wastewater disposal pond (also called a total-containment pond) in a region where annual evaporation exceeds precipitation, (c) a pond for the disposal of waste concentrate from water desalination, (d) an enclosed body of seawater that becomes a salt deposit after evaporation.

**evaporation rate** (1) The quantity or depth (in, mm, cm) of water that evaporates from a given surface per unit of time (day, month, year). (2) The rate at which a chemical changes into a vapor. A chemical that evaporates quickly can be a more dangerous fire or health hazard.

**evaporative concentration** A laboratory technique used to concentrate a solute by vaporizing the solvent.

**evaporative cooler** A device that pumps hot air through water from an area to be cooled and the cooled air back to the area. The air is cooled by water evaporation.

**evaporative cooling tower** A hollow, vertical structure that helps remove heat from water used as a coolant by exposing it to air; e.g., in electric power generating plants. Also simply called a cooling tower. *See also* mechanical draft tower and natural draft tower.

**evaporative recovery** The use of evaporation to recover chemicals and materials from electroplating and other industrial operations.

**evaporator** A device that heats water to create a phase change from liquid to vapor and to distill fresh water from brackish or saline waters. *See also* gain output ratio.

**evaporator design parameters** Critical liquid properties that include heat capacity, sensible heat, heat of vaporization, density, thermal conductivity, boiling-point rise, and heat transfer coefficient.

**evaporator load** The free moisture that is removed in the form of a gas from sludge solids by evaporation and discharged to the atmosphere, before the solids can be incinerated or used as a soil additive.

**evaporite** (1) The mineral residue resulting from evaporation. (2) A sedimentary rock (such as gypsum or rock salt, anhydrite, halite, oolitic limestone or oolite, carnallite, sylvite) formed by precipitation of minerals and evaporation of seawater.

**evapotranspiration (ET)** All water lost to the atmosphere, mainly the combined water withdrawal by evaporation (from water surfaces, soil, vegetation, and subsurface sources) and by transpiration from plants growing in the soil, that is about two-thirds of the precipitation over land surfaces. *See* the Penman–Monteith method for the evaluation of potential evapotranspiration. *See also* consumptive use and water budget.

**evapotranspiration–absorption (ETA) system** An unlined evapotranspiration bed used for wastewater treatment and disposal where soils are highly impermeable, such as heavy clays, thus using not only evaporation and transpiration, but also absorption.

**evapotranspiration bed** A bed of 2.0–2.5 ft of fine sand covered with topsoil and vegetation. It is used for the treatment of wastewater, which is spread over the bed and drawn by the vegetation from the sand bottom. Water is lost by evaporation from the soil and by transpiration from growing plants. A liner (natural clay, synthetic material, or concrete) is required if the surrounding soil is very permeable.

**evapotranspiration mound** Same as soakaway mound.

**evapotranspiration (ET) system** Same as evapotranspiration treatment system.

**evapotranspiration treatment system** A wastewater treatment system that combines surface evaporation with plant transpiration. *See* evapotranspiration bed and evapotranspiration–absorption system.

**event mean concentration (EMC)** The arithmetic mean concentration of a pollutant during a storm event, from grab or composite sampling. Because of first-flush phenomena, short-duration storms usually have higher EMCs than average or long-duration storms.

***E. vermicularis*** *See Enterobius vermicularis.*

**EVT** A belt filter press manufactured by Eimco Process Equipment Co.

**Eweson®** A compartmentalized rotary digester manufactured by Bedmintser Bioconversion Corp.

**EWS** Acronym of early warning system.

**examination of wastewater** The analysis of raw or treated wastewater to determine its composition, condition, possible effects on receiving streams, treatment requirements or treatment plant performance with respect to regulations. *See* standard tests.

**examination of water** The analysis of raw or treated water to determine its safety, palatability, or treatment requirements. *See* standard tests.

**exceedance** Violation of the pollutant levels permitted by environmental protection standards.

**Excel®** A cationic flocculant produced by Cytec Industries, Inc.

**exceptional-quality biosolids** Solids from wastewater treatment that meet the Class A requirements for pathogen reduction as well as the vector reduction standards and restrictions on heavy metals.

**excess activated sludge** The sludge produced in a biological wastewater treatment process in excess of what is needed for process operation. *See* excess sludge for more detail.

**excess chlorine** An obsolete term for residual chlorine. *See* combined available residual chlorine, free available residual chlorine.

**excessive encrustation** The accumulation of precipitates on a pipe or container and the corresponding capacity reduction, commonly caused by the precipitation of calcium carbonate ($CaCO_3$). *See also* tuberculation.

**excessive infiltration** As per USEPA guidelines, infiltration into a sewer system is excessive if the flow to the wastewater treatment plant in nonrunoff conditions, adjusted for nondomestic use, is greater than 120 gpcd. *See* allowable infiltration, excessive inflow.

**excessive infiltration/inflow** The quantities of infiltration/inflow (I/I) that can be economically eliminated from a sewerage system by rehabilitation, as determined in a cost-effectiveness analysis that compares the costs for correcting the I/I conditions to the costs for transportation and treatment of the I/I (EPA-40CFR35.905). *See also* nonexcessive infiltration and nonexcessive inflow.

**excessive inflow** As per USEPA guidelines, storm-induced inflow into a sewer system is excessive if the flow to the wastewater treatment plant is greater than 275 gpcd. *See also* allowable infiltration, excessive infiltration.

**excessive mineralization** A property of water that contains more dissolved solids than desirable for drinking purposes. Some highly mineralized waters, e.g., those containing magnesium and sulfate ions, have laxative properties. Hard water is another example. *See also* dishpan hand, winter chapping.

**excess-lime method** Same as excess-lime treatment.

**excess-lime treatment** A modification of the lime–soda ash softening process that feeds excess lime to remove calcium and magnesium hardness to a practical limit of 35–40 mg/L. A typical two-stage flow scheme includes lime addition to the raw water, flocculation–clarification, recarbonation, addition of soda ash, stage-2 flocculation–clarification, stage-2 recarbonation, filtration, sludge handling, and disposal. Also called excess-lime method, excess lime-soda softening, and railway softening.

**excess lime–soda softening** Same as excess lime treatment.

**excess-lime softening** Same as excess lime treatment.

**excess sludge** In the activated sludge or any other biological treatment process, it is the sludge that is produced in excess of the required solids recirculation and that does not escape in the effluent. It is withdrawn from the system and wasted. Also called excess activated sludge, waste sludge, sludge excess, or waste-activated sludge. *See also* return sludge.

**exchangeable sodium percentage (ESP)** The relative amount of sodium in a soil that can be exchanged for other ions. Sodic soils have high

ESPs, which can reduce their permeability. *See* sodium adsorption ratio.

**exchange adsorption** One of the three conditions that cause the adsorption of substances onto adsorbents. Similar to ion exchange, it results from the electrical attraction or coulombic attractive forces between ions and charged functional groups. *See also* chemical adsorption or chemisorption, desorption, hydrophobic bonding, physical adsorption, polarity, solution force, van der Waals attraction. Also called electrostatic adsorption.

**exchange capacity** An ion-exchange unit's or ion-exchange resin's limited capacity for storage of ion equivalents while maintaining electroneutrality in the solution and the ion-exchange medium, taking into account the number of charges displaced, e.g., one calcium ion ($Ca^{++}$) displaces two sodium ions ($Na^+$); usually expressed in eq/L, eq/kg, meq/L, meq/kg, or in grams $CaCO_3$ per cubic meter of resin:

$$1 \text{ meq/L} = 1 \text{ eq/m}^3 = 50 \text{ g CaCO}_3\text{/m}^3 \quad \text{(E-38)}$$

*See also* resin exchange capacity, matrix, functionality, porosity.

**exchanger** An ion-exchange unit. *See* ion exchanger.

**exchange site** A location on a bead of ion-exchange resin that holds mobile ions available for exchange with other ions in the solution that passes through the entire resin bed. Also called functional group. *See* functionality.

**exchange tank** *See* portable exchange tank.

**exchange velocity** The rate of displacement of ions in an ion-exchange operation.

**excimer** An excited dimer (a molecule composed of two identical, simpler molecules or a polymer derived from two identical monomers), formed by the collision of an electron from a corona discharge with a molecule of gas (e.g., chlorine, bromine, fluorine). In water and wastewater treatment, excimers participate in the inactivation of pathogens and the destruction of organic pesticicdes. Same as excited dimer.

**excited dimers** Two monomers, e.g., two gas molecules joined together by a corona discharge. *See* excimer.

**excited oxygen atom** The single oxygen atom, represented by the symbol $O(^1D)$, as opposed to molecular oxygen ($O_2$) or ozone ($O_3$). It is formed in the intermediate reaction of ozone photolysis:

$$O_3 + UV \rightarrow O_2 + O(^1D) \quad \text{(E-39)}$$

where UV denotes ultraviolet light. Also called singlet oxygen (Metcalf & Eddy, 2003).

**excitotoxin** A molecule that overstimulates nerve cells and kills or damages them, e.g., glutamate.

**exclusionary ordinance** Zoning that excludes classes of persons or businesses from a particular neighborhood or area.

**exclusion chromatography** Same as high-performance liquid chromatography.

**exclusively human infection** An infection that is spread only from one person to another, e.g., cholera, shigellosis, typhoid.

**excreta** Excreted matter; feces, urine, vomit, sweat. Excreta contribute approximately 41 g of $BOD_5$ daily per adult, which comes 68% from fecal material, 25% from urine (about 1.2 kg/day), and 7% from anal cleansing materials.

**excreta disposal disease** An excreta-related disease.

**excreta-related disease** A disease caused by pathogens that are transmitted in human excreta. *See also* water- and excreta-related diseases, fecal–oral disease, soil-transmitted helminth, tapeworm, excreta-related insect vector, water-related disease.

**excreta-related infection** *See* excreta-related disease.

**excreta-related insect vectors** Two groups of insects (*Culex pipiens,* flies, cockroaches) that breed in excreta or polluted waters and transmit pathogens that cause filariasis and various fecal–oral infections.

**excreted bacteria** Common pathogenic bacteria released in excreta include *Campylobacter, E. coli, leptospira, Salmonella, Shigella, Vibrio cholerae,* and *Yersinia.*

**excreted helminths** A number of parasitic worms that are released in the excreta of infected hosts. *See* the following infections (and their pathogens and common names): ancylostomiasis, ascariasis, clonorchiasis, diphyllobothriasis, enterobiasis, fasciolopsiasis, hymenolepiasis, hookworm, opisthorchiasis, paragonimiasis, schistosomiasis, strongyloidiasis, taeniasis, trichuriasis.

**excreted infection** *See* excreta-related disease.

**excreted load** The number, concentration or mass of pathogens passed by an infected person. It is related to other key factors that affect the transmission of an infection: infective dose, latency, persistence, and multiplication.

**excreted pathogen output** *See* excreted load.

**excreted protozoa** Protozoans carried in excreta and responsible for human infections include *Balantidium coli, Cryptosporidium, Entamoeba histoyitica,* and *Giardia lamblia. See also* protozoal disease, vector-borne protozoa.

**excreted viral pathogen** *See* excreted virus.

**excreted virus** Common viruses that infect the alimentary canal and are shed in very large numbers in excreta. They cause water-related diseases and include hepatitis A, Norwalk agent, echovirus, poliovirus, rotavirus, and other viruses.

**excyst** To emerge from a cyst.

**excystation** The release of sporozoites or trophozoites from a protozoan cyst or oocyst (e.g., *Cryptosporidium* or *Giardia*). Parasitic protozoa enter a host as cyst and then bile salts help break down cyst walls as temperature increases to a more tolerable level.

**exempted aquifer** An aquifer or its portion that meets the criteria in the definition of "underground source of drinking water" but that has been exempted (EPA-40CFR144.3 or 146.3). Underground bodies of water defined in the Underground Injection Control program as aquifers that are potential sources of drinking water though not being used as such, and thus exempted from regulations barring underground injection activities.

**exemption** A waiver granted by a primacy state to a public water system, from a requirement concerning a maximum contaminant level and/or a treatment technique, if certain conditions exist. These conditions are: (a) the system cannot comply with the requirement due to economic or other compelling factors; (b) the system was in operation on the effective date of the requirement; and (c) the exemption will not result in an unreasonable public health risk. *See also* variance.

**exergonic** Pertaining to a chemical reaction, a process or an exchange, e.g., respiration, that gives out energy.

**exergonic reaction** A reaction that results in a negative change in free energy release; it proceeds in the direction indicated ("spontaneously") without additional energy input; e.g., biological oxidation in autothermal aerobic digestion increases liquid temperatures by as much as 60°C. *See also* bioenergetics, endergonic reaction, Gibbs free energy.

**exfiltration** (1) Leakage of water from a water main, or wastewater or stormwater from a sewer to the surrounding ground from unintentional openings. Exfiltration is often considered in the analysis of infiltration/inflow problems in sewer systems; the analysis is thus called infiltration/exfiltration/inflow or I/E/I analysis. (2) The flow of groundwater into ditches and channels. (3) Intentional leakage of stormwater from French drains over the duration of a storm event. (4) The quantity of water that leaks.

**exhaustion** The condition of activated carbon, ion-exchange resin, or other adsorbents that have depleted their capacity by using all available sites; the effluent concentration then matches the influent concentration. *See also* adsorption bed exhaustion, breakthrough curve.

**exhaustion concentration** The effluent solute concentration in a fixed bed adsorber at exhaustion; it is commonly chosen at 90–99% of the initial concentration, depending on the shape of the breakthrough curve. *See also* breakthrough concentration, adsorption zone, empty bed contact time.

**exhaustion rate** The rate at which the capacity of an ion-exchange bed is being exhausted; the reciprocal of the empty bed contact time. It is commonly designed in the range of 1.0–5.0 gpm/cubic foot. *See* service flow rate for detail.

**exhaustion wave front** The boundary between the absence and presence of a contaminant as it passes through a media bed.

**exit age curve** A tracer response curve, i.e., a graphical representation of the distribution of a tracer versus time of flow, for a pulse input, with output concentrations normalized by dividing the measured concentrations by such a factor that the area under the normalized curve is equal to 1. Also called an E curve or normalized residence time distribution curve. It is used in studying short-circuiting, retardation, and longitudinal mixing in sedimentation basins and other reactors. *See* Figure E-08. For a complete-mix reactor the exit

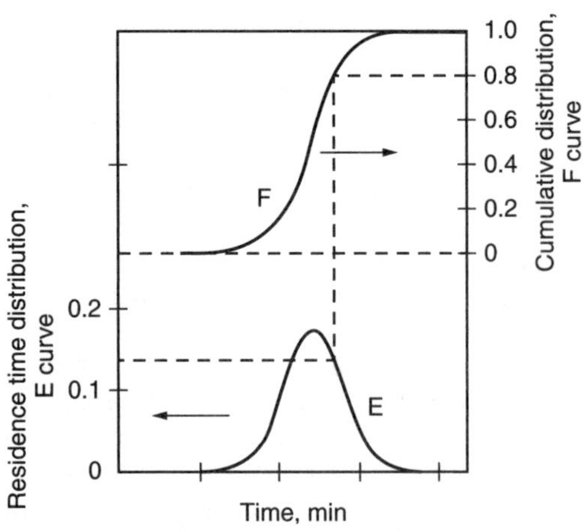


**Figure E-08.** Exit age (E) and F curves (adapted from Metcalf & Eddy, 2003).

age curve, as a function of time ($t$) is (Metcalf & Eddy, 2003):

$$E(t) = e^{-t/\tau/\tau} \quad \text{(E-40)}$$

where $\tau$ is the theoretical hydraulic residence time. *See also* average retention time index, continuous injection, cumulative residence time distribution curve (or F curve), F(t) curve, mean retention time index, modal retention time index, Morrill dispersion index, pulse input, short-circuiting index, time-concentration curve (also called flow-through curve, fluid residence time distribution curve, residence time distribution curve), and volumetric efficiency.

**exon** A portion of an interrupted gene represented in the DNA or RNA.

**exopolymer** A macromolecular, polymeric material released by microorganisms outside their cells. *See also* capsule, glycocalyx, sheath, slime.

**exopolysaccharide** A complex molecule of monosaccharides released by microorganisms outside their cells; an exopolymer.

**exothermal** Same as exothermic.

**exothermic** Pertaining to a chemical change, reaction, or process in which heat is released; e.g., combustion. Also called exothermal. *See also* endothermic, exergonic.

**exothermic reaction** A reaction that releases heat; e.g., in alkaline sludge stabilization, the addition of quicklime (CaO) generates approximately 27,500 BTU per pound · mole or 64 kJ/gram · mole by reaction of the chemical with water ($H_2O$) to form hydrated lime [$Ca(OH)_2 \cdot x\ H_2O$]. The reaction between lime and carbon dioxide ($CO_2$) also releases heat, approximately 78,000 BTU per pound mole.

**Exotoc®** An instrument manufactured by Neotronics for the detection of gases.

**exotoxin** A soluble toxin or poisonous substance excreta by algae, bacteria, or other microorganisms, which may be toxic to aquatic life, wildlife, and humans; e.g., the exotoxin produced by the cholera vibrios. *See also* endotoxin.

**expanded bed** A bed of granular materials, such as sand or activated carbon, that is expanded for backwashing or for normal operation. *See* fluidized bed for more detail.

**expanded-bed carbon contactor** An activated carbon column that functions like a filter during backwashing, i.e., with the influent introduced at the bottom and the bed allowed to expand. Exhausted carbon is removed at the bottom and new carbon added at the top.

**expanded bed equation** *See* backwash velocity equations.

**expanded-bed reactor** A wastewater treatment technique that uses bacteria grown on a suspended medium to stabilize organic matter. *See* fluidized-bed reactor, anaerobic expanded-bed reactor; expanded beds are sometimes considered to be "less expanded" than fluidized beds and have little particle movement. *See also* wastewater treatment reactors for other types used in wastewater treatment.

**expansion coupling** A coupling that allows two joined pipe sections to move.

**expansion flow** Flow in a conduit with decreasing velocity and increasing pressure because of an increase in cross-sectional area.

**expansion joint** A joint installed in a structure or pipeline to allow for thermal expansion or contraction (i.e., fluctuations in length or size resulting from temperature changes).

**expansion loss** One of several minor head losses, due to the augmentation of the size of a conduit or channel. *See also* enlargement loss.

**expected time before failure (ETBF)** A statistical measure of the reliability of mechanical equipment, as determined by the ratio of elapsed time (hours) to the number of failures; it is one of the factors used in critical component analysis. *See also* mean time before failure, inherent availability, operating availability.

**expert system** A computer program that responds to specific situations, based on available information, and using procedures that attempt to duplicate the thought processes and apply the knowledge of an expert. SCADA is an example of an expert system that uses artificial intelligence in the water and wastewater utilities. *See also* artificial intelligence and intelligent system.

**explosimeter** A device that measures the concentration of explosive fumes, based on the catalytic oxidation of a combustible gas on a filament incorporated in a Wheatstone bridge. Also called combustible-gas indicator.

**explosion proof (XP)** Characteristic of a motor or electrical enclosure indicating its capacity to withstand explosions or prevent ignition of gases or vapors.

**explosive** A chemical that causes a sudden, almost instantaneous release of pressure, gas, and heat when subjected to sudden shock, pressure, or high temperatures.

**explosive limit** The amounts of vapor in the air that form explosive mixtures; limits are expressed as lower or upper limits and give the range of vapor concentrations in air that will explode if an ignition source is present. *See also* flash point, lower explosive limit, upper explosive limit.

**explosive substance** A liquid or solid substance or mixture of substances that can react chemically and produce gas at such a temperature and pressure and at such a speed as to cause damage to the surroundings.

**exponential growth phase** A phase in the growth pattern of microorganisms in batch culture in which their number and mass increase exponentially: the rate of increase of cells in the culture is proportional to the number or mass ($X$) of cells present at any particular time:

$$dX/dt = \mu X \quad \text{(E-41)}$$

where $\mu$ is the specific growth rate. *See also* generation time, doubling time.

**exponential infection model** A distribution function used to describe the probability of human infection by enteric microorganisms. Thus, the probability ($P$) of infection from a single exposure is:

$$P = 1 - e^{-rN} \quad \text{(E-42)}$$

where $N$ is the number of organisms ingested in the exposure and $r$ is the fraction of ingested microorganisms that overcome the host defenses. *See* dose–response model for related distributions.

**exponential phase** *See* exponential growth phase.

**exposure** (1) The process by which an organism comes into contact with a chemical or physical agent, e.g., inhalation of air, ingestion of food or water, or absorption through the skin. (2) The amount of radiation or pollutant present in a given environment that represents a potential health threat to living organisms.

**exposure assessment** A component of risk assessment: the determination or estimation of the magnitude, frequency, duration, route, and extent of exposure to a chemical. *See also* hazard identification, dose–response assessment, risk characterization.

**exposure coefficient** A term that combines information on the frequency, mode, and magnitude of contact with contaminated medium to yield a quantitative value of the amount of contaminated medium contacted per day.

**exposure indicator** A characteristic of the environment measured to provide evidence of the occurrence or magnitude of a response indicator's exposure to a chemical or biological stress.

**exposure level** The amount or concentration of a chemical at the absorptive surfaces of an organism.

**exposure pathway** The course taken by a hazardous agent from a source to a receptor via environmental carriers or media, e.g., air for volatile compounds and water for soluble substances. *See also* exposure route or intake pathway.

**exposure route** Ingestion, inhalation, dermal contact, or any other mechanism by which a hazardous agent is transferred to a receptor. Also called intake pathway. *See also* exposure pathway.

**exposure scenario** A set of conditions or assumptions about sources, exposure pathways, concentrations of toxic chemicals and populations (numbers, characteristics and habits) that aid the investigator in evaluating and quantifying exposure in a given situation.

**exposure time** (1) The mean time of exposure of water to purification processes, equal to the ratio of the volume of a treatment unit to the rate of flow. (2) The time period during which a test organism is exposed to a test constituent or other risk. *See also* toxicity terms.

**ExpressClean™** A service provided by Coster Engineering for cleaning reverse osmosis units.

**Expressor** A belt filter press manufactured by Eimco Process Equipment Co.

**ex situ** Moved from its original place; excavated; removed or recovered from the subsurface.

**ex situ bioremediation** The clean up or treatment of soil or water away from the contaminated site, using microorganisms to degrade the contaminants.

**ex situ treatment** Processing and treating wastes after removing them from a contaminated site; term often applied to the solidification, stabilization, and encapsulation of hazardous wastes. *See also* in situ treatment, in-drum processing, in-plant processing, mobile-plant processing.

**extended aeration** A variation of the activated sludge process with increased detention time (24 hr) and sludge age (solids retention time = 20–30 days), and low organic loading to allow endogenous respiration. Some excess solids are stabilized aerobically in the aeration unit, but sludge accumulates in the system and is withdrawn approximately every six months. Its operation requires little supervision; the plant can withstand shock and toxic loads. Typically, screened and degritted wastewater is introduced to the aeration basin, without primary clarification. Most extended aeration plants are small prefabricated, package units used in small communities. Two operating problems of extended aeration are the loss of pinpoint floc and mixed liquor suspended solids following a period of low influent loading.

**extended-aeration package plant** A preengineered plant that uses the extended aeration process to treat wastewater; usually a circular configuration with an aeration tank and an aerobic digester around a secondary clarifier, without primary settling.

**extended-detention basin** A stormwater detention basin that dries more slowly than a dry detention basin and may retain a permanent pool.

**extended period simulation (EPS)** A water distribution model that allows the user to identify situations other than worst cases, which are the focus of steady-state distribution models. For example, Haestad Methods' WaterCAD® and Cybernet® can evaluate pumping and piping capacity, tank volume and elevation, effects on pressures and flows, movement of contaminants and chemical constituents, etc.

**extended terminal subfluidization wash (ETSW)** A backwashing technique proposed to reduce the effects of filter ripening. ETSW is a procedure that extends filter backwash duration at subfluidization rates sufficiently to displace the entire contents of the filter box. Subfluidization refers to the reduction or elimination of filter bed expansion. *See also* filter ripening.

**Extendor** A detention tank designed by Semblex, Inc.

**extent of chlorination** The percent by weight of chlorine for each isomer (ortho, meta, and para) (EPA-40CFR704.45-2).

**external corrosion** Corrosion that occurs outside a pipe or container, caused by soil, surface, or bedding characteristics. *See* cathodic protection.

**externality** A cost or benefit attributable to an economic activity that is not reflected in the price of goods or services; a cost or benefit accruing to a group as a result of actions by others or events over which such group has no control; e.g., pollution or other environmental damages. *See also* polluter pays principle.

**external process disturbance** An external factor that causes a change in the operation or performance of a treatment process, e.g., variations in the flow and/or characteristics of the influent to a wastewater treatment plant or the source water. Equalization basins and collection-system storage are sometimes used to minimize the effects of such disturbances. *See also* internal process disturbance.

**external water treatment** Treatment of boiler feedwater or boiler makeup water outside the boiler.

**extinction** Complete disappearance of a species that fails to adapt to environmental changes.

**extinction coefficient** A parameter used in evaluating the germicidal effectiveness of ultraviolet light or other forms of radiant energy. The extinction coefficient ($k_a$) is related to the Secchi depth ($d_s$) and to radiation intensity. *See* attenuation coefficient for more detail.

**extracellular polymeric substance (EPS)** A substance released by a microorganism outside its cell. *See* exopolymer.

**extracellular slime** Exopolymer that allows microorganisms to adhere to solid surfaces.

**extracellular water** Water found in an organism outside its cell membrane.

**extractable metal** The metal that can be extracted from a water or soil sample by treatment with mineral acid and filtration.

**extractable organohalogen (EOX)** The fraction of adsorbable organic halogens that is extractable by a nonpolar organic solvent. It provides a good indication of the amount of organic halogens susceptible to lipophilic absorption.

**extraction** (1) A process used to concentrate a solute or reduce interferences from a matrix. It consists of dissolving and separating out constituents of a liquid treated with specific solvents. Examples include the extraction of chlorophyll, oil and grease, and disinfection by-products. (2) One of three common methods of groundwater protection; it involves either excavation and disposal of the contaminating materials, or pumping out and treating ("pump-and-treat") the contaminated water, which may be reinjected upgradient. *See also* containment and in situ treatment.

**extraction procedure** (1) A specific protocol for extracting a given substance. (2) A procedure to determine toxicity by simulating leaching; if a certain concentration of a toxic substance can be leached from a waste, that waste is considered hazardous, i.e., "EP toxic."

**extraction procedure (EP) toxicity test** A test that uses a leaching procedure to determine if a waste has the characteristics of a toxic chemical for land disposal purposes. *See also* toxicity characteristics leaching procedure.

**extraction procedure (EP) toxicity threshold** A limit set by the USEPA for the concentration of a waste constituent, based on the EP toxicity test, above which a waste is considered hazardous. Thresholds have been established for pesticides, metals, and herbicides; e.g., 10.0 mg/L for 2,4-D, 1.0 mg/L for cadmium, and 0.02 mg/L for endrin.

**extraction well** (1) A well employed to extract fluids (either water, gas, free product, or a combination of these) from the subsurface. Extraction is usually accomplished by either a pump located within the well or suction created by a vacuum pump at the ground surface. (2) A well that recovers percolated flow from a rapid infiltration wastewater treatment system for reuse or to control the level of the water table. Also called recovery well.

**extractive membrane bioreactor (EMBR) process** A wastewater treatment process in which membranes extract degradable soluble organic molecules from inorganic constituents such as acids, bases, and salts for subsequent biological treatment. It consists of hollow-fiber membranes with attached biomedium (for the organic extraction) and a suspended solids bioreactor.

**Extractor** A horizontal belt press manufactured by Eimco Process Equipment Co.

**Extractoveyor** A conveyor apparatus manufactured by Compost Systems Co. for handling composted sludge.

**extraintestinal amebiasis** An infection by amebas outside the intestine, the parasites migrating from ulcers via the hepatic portal vein to the liver and other organs, where they form an amebic abscess.

**Extreme Duty™** Sludge mixing equipment manufactured by WesTech Engineering, Inc.

**extreme environment** (1) An environment in which conditions like pH, temperature, salinity, pressure, and nutrients are either too high or too low. (2) An environment that favors extremely low microbial diversity. *See* extremophile, neuston, ultrapure water.

**extreme environmental conditions** Exposure to any or all of the following: ambient weather conditions, temperature consistently above 95°C (203°F), detergents, abrasive and scouring agents, solvents, or corrosive atmospheres (EPA-40CFR52.741).

**extremely hazardous chemicals** Chemicals on a USEPA list that requires special handling for the protection of the public and of workers.

**extremely hazardous substance** Any of more than 400 chemicals identified by USEPA as toxic, and listed under SARA Title III. The list is subject to periodic revision.

**extremophile** An organism that has adapted to an environment where it is difficult or impossible for other organisms to survive. Examples of extremophiles include thermotolerant bacteria, halotolerant organisms, barophiles, psychrophilic organisms, and acidophiles. *See* extreme environment.

**extrinsic property** A characteristic of a body that can be measured, e.g., temperature, flow, pollutant concentration, density, velocity, but is not conserved; it cannot be used in accordance with the law of conservation to predict changes in the body. *See also* intrinsic property.

**eye of a spring** The point where water emerges from a water-bearing formation over an impervious stratum. *See also* silt trap, spring box, spring capping.

**eyewash** (1) A solution applied locally to the eye for irrigation, e.g., to wash chemicals or debris out. (2) A safety device used to wash chemicals from the eyes with a spray of water.

**eyewash station** A flushing fountain for washing chemicals or debris for the eyes.

**E-Z™** A batch centrifuge manufactured by Western States Machine Co.

**E-Z Tray®** An air stripper manufactured by QED Environmental Systems.

# F

F F F F F F F F F F F F F F F F F F F F F F

**F** Farad, the SI unit of electrical capacitance.

**F** Faraday's constant.

**F** Fahrenheit degree.

**FAB gas chromatography–mass spectrometry** Abbreviation of fast atom bombardment gas chromatography–mass spectrometry.

**fabricated metals products industry** A standard industrial classification that covers the manufacture of such metal products as cans, cutlery, forgings, general hardware, hand tools, ordnance, stampings, and structural parts. Its major water pollutants are volatile organic compounds, acids, heavy metals, and oils. Also called metal fabrication industry.

**fabricated self-supporting underdrain** A filter underdrain system made of vitrified clay or plastic blocks and grouted to the filter floor. *See also* underdrain, manifold-lateral underdrain, false-floor underdrain with nozzles.

**fabric filter** An air pollution control device used to remove particulates in the thermal processing of sludge at wastewater treatment plants. Also called bag house. It bypasses the dust-containing gas stream through a fabric medium that retains the particulates. Periodically, the accumulated dust is removed and the bag is cleaned. *See also* mechanical collectors, wet scrubbers, and electrostatic precipitators.

**FAC** Acronym of free available chlorine.

**facilities or equipment** Buildings, structures, process or production equipment or machinery that form a permanent part a new pollution source and that will be used in its operation, if these facilities or equipment are of such value as to represent a substantial commitment to construct. It excludes facilities or equipment used in connection with feasibility, engineering, and design studies regarding the source or water pollution treatment for the source (EPA-40CFR122.29-5). *See also* facility.

**facilities planning activities** Activities necessary for the development of a facilities plan for a water or wastewater project. *See also* design activities, construction activities.

**facilities plans** (1) Plans and studies related to the construction of treatment works necessary to comply with the Clean Water Act. A facilities plan provides information on current conditions and problems, needs, alternate solutions, the cost-effectiveness of alternatives, a recommended plan, an environmental assessment of the recommendations, descriptions, preliminary design and costs of the treatment works, and a completion schedule. (2) Similar documents for any water supply or sewer project. (3) *See* state facility plan.

**facility** All or any portion of buildings, structures, equipment, roads, walks, parking lots, rolling stock or other conveyances, or other real or personal property (EPA-40CFR12.103).

**facility expansion charge** A one-time charge assessed by a water, sewer, or other utility against new users, developers, or applicants for new service to recover part or all of the costs of existing or additional facilities. It is a means of charging the cost of development impact to those who cause it. Also called capacity charge, capital recovery charge, or impact fee. *See also* connection charge development impact fee, and system development charge.

**facility plan** *See* (a) facilities plans and (b) state facility plan.

**factor of safety** *See* safety factor.

**facultative** Pertaining to bacteria and other organisms that are able to grow in either the presence or absence of a specific environmental factor; e.g., bacteria that can use either molecular (dissolved) oxygen or oxygen obtained from food material such as sulfate or nitrate ions. In other words, facultative bacteria can live under aerobic or anaerobic conditions.

**facultative aerated lagoon** *See* facultative aerated pond.

**facultative aerated pond** A shallow earthen basin used for wastewater treatment. The aeration is sufficient for aerobic treatment but not for mixing. At the bottom it functions like a facultative lagoon. *See* facultative partially mixed lagoon for more detail.

**facultative aerobe** A normally anaerobic microorganism that can also grow in the presence of oxygen.

**facultative aerobic bacteria** Bacteria that can use nitrate or nitrite as an electron acceptor when free oxygen is not available.

**facultative anaerobe** A normally aerobic microorganism that can also grow in the absence of molecular oxygen or air. In the absence of oxygen, these microorganisms can utilize another compound (e.g., sulfate, nitrite, or nitrate) as a terminal electron acceptor. *See also* aerotolerant anaerobe and true facultative anaerobe.

**facultative anaerobic autotrophic respiration** *See Thiobacillus denitrificans.*

**facultative anaerobic heterotrophic respiration** *See Pseudomonas denitrificans.*

**facultative bacteria** Bacteria that can live under aerobic or anaerobic conditions, i.e., with or without the presence of dissolved oxygen. Also called microaerophilic bacteria.

**facultative lagoon/pond** A lagoon or pond in which stabilization of wastewater occurs as a result of aerobic (near the surface), anaerobic (near the bottom), or facultative microorganisms. It is the most common type of stabilization pond, usually the largest in a pond system. *See* aerobic–anaerobic lagoon for more detail.

**facultative organism** An organism that can live in the presence or absence of free oxygen. Most of the bacteria that carry out the decomposition of organic matter in wastewater treatment are facultative as opposed to strict or obligate organisms. Exceptions include nitrification and methanogenesis.

**facultative parasite** An organism that is a parasite under certain conditions, but can also live as a saprophyte. *See also* obligate parasite.

**facultative partially mixed lagoon** A shallow earthen basin, with a depth of 6 to 15 feet, aerated by mechanical devices on floats or fixed platforms, and used for wastewater treatment. The aeration is sufficient for aerobic treatment but not for mixing, e.g., 1.0–1.5 kg of oxygen per kg of BOD. Thus solids that settle at the bottom undergo anaerobic decomposition, and there this lagoon functions like a facultative lagoon. Also called aerated facultative lagoon, facultative aerated pond, or partial-mix aerated lagoon or pond. *See also* lagoons and ponds.

**facultative plant** A plant species likely to occur in a wetland or a nonwetland.

**facultative pond** *See* facultative lagoon, and lagoons and ponds.

**facultative pond maximum loading** The maximum organic load ($L_{max}$, pounds/acre/day of $BOD_5$) that can be applied to a facultative pond without creating complete anaerobic conditions throughout, estimated as follows (Droste, 1997):

$$L_{max} = k(1.054)^{(1.8T+32)} \qquad \text{(F-01)}$$

where $T$ is temperature in °C and $k$ a conversion factor that varies with location and climate.

**facultative pond performance** The percent removal of organic matter ($BOD_5$) effected by a facultative pond or series of ponds. *See* complete mixing mass balance, plug flow mass balance for standard models relating pond performance to a reaction rate constant ($k$) and detention time ($t$). The rate constant is a function of temperature as indicated in the Arrhenius equation, with an Arrhenius constant $\theta$ = 0.987 for complete mixing and $\theta$ = 0.962 for plug flow.

**facultative pond volume** The required volume ($v$, $m^3$) of a pond, based on desired BOD removal efficiency of 80–90%, algal toxicity factor ($f_a$), sulfide oxygen demand ($f_s$), flow ($Q$, $m^3/d$), temperature ($T$, °C), temperature correction factor $\theta$ = 1.085, and influent ultimate BOD ($L_a$):

$$v = 0.035 f_a \cdot f_s \cdot Q \cdot L_a \cdot \theta^{(35-T)} \quad \text{(F-02)}$$

**facultative pond with aeration** A wastewater stabilization pond whose upper layer is aerated mechanically to maintain aerobic conditions while the bottom layers undergo anaerobic decomposition.

**facultative process** A biological wastewater treatment process in which microorganisms can function in the presence or absence of molecular oxygen. *See also* aerobic process, anaerobic process, anoxic process.

**facultative sludge lagoon** A method used to store sludge during the off-seasons of a dedicated land disposal site. It can store sludge for several months with limited odor.

**facultative upland plant** A plant species that occurs at least two-thirds of the time outside wetlands.

**facultative wetlands plant** A plant species that occurs at least two-thirds of the time in wetlands.

**FAD** Acronym of filtration avoidance determination

**faecal** Same as fecal.

**faecal–oral** Same as fecal–oral.

**faeces** Same as feces.

**faeces decomposition** Same as feces decomposition.

**faeces volume** Same as feces volume.

**Fahrenheit** The Fahrenheit scale.

**Fahrenheit, Gabriel Daniel** The German physicist who devised the Fahrenheit scale (1686–1736).

**Fahrenheit degree** A degree on the Fahrenheit scale. Also called degree Fahrenheit.

**Fahrenheit scale** A temperature scale with the freezing point at 32°F and the boiling point of water at 212°F at a barometric pressure of 1 atmosphere (760 mm Hg).

**fail-safe** Characteristic of an installation that is unlikely to fail because of the minor malfunction of a single element. In water and wastewater systems, e.g., important elements such as pumps usually are designed with backup units and ring water distribution is preferable to dendritic networks.

**Fairfield** An in-vessel composting apparatus manufactured by Compost Systems Co.

**Fair, Gordon M.** A prominent sanitary engineer (1894–1970), professor at Harvard University, investigator, mentor; author of a well-known text, *Water and Wastewater Engineering,* and of several related articles.

**Fair–Hatch equation** An equation proposed by G. M. Fair and L. P. Hatch in 1933 to estimate the head loss in a clean, stratified filter bed (Metcalf & Eddy, 2003 and Fair et al., 1971):

$$H/L = (K/g)\nu V[(1 - F)^2/F^3](6/\Psi)^2 \sum_{i=1}^{n} (P_i/D_i^2) \quad \textbf{(F-03)}$$

where $H$ = head loss; $L$ = bed depth; $K$ = the dimensionless Kozeny coefficient, with $K$ = 5 under most conditions; $g$ = gravitational acceleration; $\nu$ = kinematic viscosity; $V$ = approach velocity; $f$ = bed porosity; $\Psi$ = sphericity; $n$ = number of layers of the stratified bed; $P_i$ = the fractional weight of layer $i$; and $D_i$ = the sieve size of layer $i$.

**Fair's formula** A formula defining the oxygen-recovery or self-purification ratio of a stream as the ratio ($f$) of the reaeration rate ($r$) to the deoxygenation rate ($k$):

$$f = r/k \quad \text{(F-04)}$$

*See* self-purification coefficient (Nemerow & Dasgupta, 1991).

**falciparum malaria** Malaria caused by the protozoal parasite *Plasmodium falciparum*, the most serious form of malaria, prevalent in humid areas of the tropics.

**fall** (1) the amount by which one point is lower than a higher point along the slope of a pipe, channel, floor, etc. (2) A sudden drop in water surface elevation or the difference in the water surface elevations of two points. Also the required drop to facilitate proper flow in a pipe.

**falling-film evaporation** The multiple-effect evaporation technique used in the Carver–Greenfield process to extract water from sludge; a fluidized sludge-oil slurry is pumped to the first-stage evaporator with steam added to the side of its shell; the vaporized water is then removed and the slurry rolls down in a thin film.

**falling-film evaporator** A device with vertical heat transfer surfaces in which a film of liquor falling down the surfaces is heated to boiling by steam condensing on the other side of the surface; vapor is driven off. *See also* forced-circulation evaporator and rising-film evaporator.

**falling-rate stage** The third and final stage in the operation of a thermal sludge dryer, in which moisture evaporates from the surface of the solids faster than it can be replaced by internal moisture. *See also* warm-up stage, constant-rate stage.

**fall overturn** Same as fall turnover.

**fall overturning** Same as fall turnover; also called great overturning.

**fall turnover** *See* turnover; also called great overturn because it lasts longer than the spring overturn.

**Fallova™** Equipment manufactured by International Shredder, Inc. and used for shredding wastewater solids.

**fall velocity** The velocity of a particle falling alone in quiescent, distilled water of infinite extent. *See also* terminal settling velocity.

**false bottom** *See* false filter bottom, filter floor.

**false filter bottom** A horizontal partition above the actual bottom of a container; e.g., a porous or perforated floor suspended above the true bottom of a filter, serving as underdrainage. *See* Figure F-01.

**false floor** *See* false filter bottom, filter floor.

**false-floor underdrain** A filter underdrain system consisting of a false floor or steel plate 1–2 ft above the bottom, with nozzles every 5–8 inches. *See* Figure F-02. *See also* underdrain, fabricated self-supporting underdrain, manifold–lateral underdrain.

**false positive** A positive result from noncoliform bacteria in the presumptive test. *See* completed test, confirmed test.

**false positive reaction** The reaction during a standard coliform test in which coliforms of nonfecal origin mimic the truly fecal coliforms by fermenting lactose at 44°C (particularly in warm tropical waters).

**Fanning's formula** A formula proposed to estimate flood flows in New England (Fair et al., 1971):

$$Q = CA^{5/6} \quad \text{(F-05)}$$

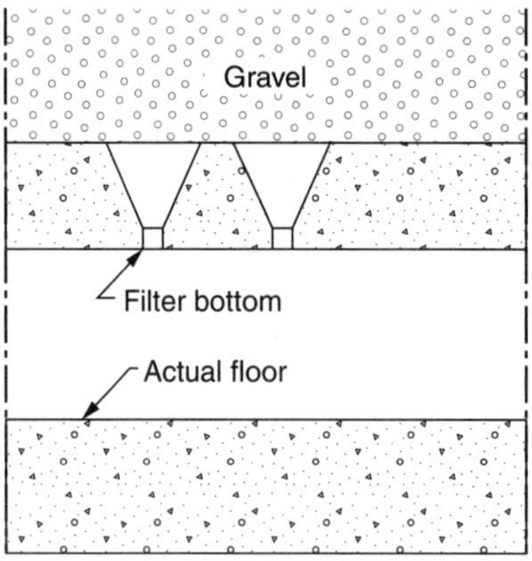


**Figure F-01.** False filter bottom (adapted from Fair et al., 1971).

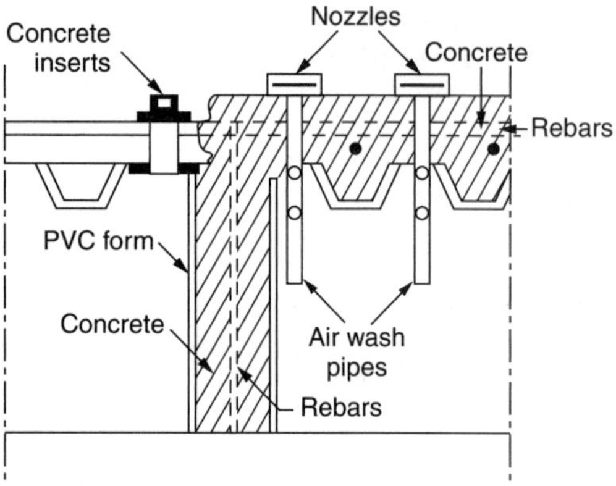


**Figure F-02.** False-floor underdrain (adapted from AWWA, 1999).

where $Q$ is the peak flood flow in cfs; $C$ is a coefficient that accounts for the effects of storm frequency, runoff–rainfall relationship, and maximum rainfall; and $A$ is the drainage area. For $A$ in square miles, $C = 200$.

**farad (F)** The SI unit of capacitance, equal to the capacitance of a capacitor between the plates of which there is a potential difference of one volt when the capacitor is charged by 1 coulomb.

**faraday** A unit of electricity used in electrolysis, equal to 96,500 coulombs.

**Faraday, Michael** English physicist and chemist (1791–1867), discoverer of electromagnetic induction.

**Faraday's constant or Faraday constant** A proportionality constant F = 96,493 coulombs/equivalent, used in some formulas, e.g., electrodialysis current, Nernst equation.

**Faraday's law** An electrochemical principle stating that the amount of chemical change of any substance undergoing electrolysis is proportional to the total electrical charge passed. In electrodialysis theory, the law is often expressed as one Faraday of electricity causes one gram equivalent (gr eq) of a substance to migrate from one electrode to another, and is translated by the following formula:

$$\text{gr eq/unit time} = QNF \quad \text{(F-06)}$$

where $Q$ = flowrate, $N$ = solution normality in eq/L, and $f$ = fraction of electrolyte removal.

**Faraday's law of electrolysis** Same as Faraday's law.

**farfield** A region far from the port of discharge of wastewater or effluent into the ocean or a large lake, beyond the nearfield and the transition regions. *See* discharge farfield.

**farm tank** A tank located on a tract of land devoted to the production of crops or raising animals, including fish, and associated residences and improvements. A farm tank must be located on the farm property (EPA-40CFR280.12).

**farmyard manure** A mixture of straw and animal excrement used after composting as a fertilizer and soil conditioner.

**FAS** Acronym of ferrous ammonium sulfate.

***Fasciola*** A genus of parasitic, excreted helminthes infecting primarily cattle and sheep, found in wet pastures and small streams, and using intermediate host vector snails.

***Fasciola gigantica*** A species of parasitic flatworms related to *Fasciola hepatica* and measuring up to 7.5 cm × 1.2 cm. It is a parasite of cattle, camels, and other herbivores in Africa, Asia, and some Pacific Islands.

***Fasciola hepatica*** A species of visible (up to 3 cm in length and 1.3 cm in breadth), parasitic flatworms that use the water snail (*Limnaea truncatula*) as intermediate host. Latency = 2 months. Persistence = 4 months. Median infective dose > $10^8$. Also called sheep liver fluke or liver fluke.

**fascioliasis** The water-related infection caused by *Fasciola hepatica* and *Fasciola gigantica,* parasitic worms that affect mainly sheep and cattle. Humans acquire the disease (an infection of the bile ducts with inflammation and fibrosis) by eating watercress and other aquatic plants infested with encysted metacercariae (larvae of the parasites). Human infection is reported in South America, North America, Cuba, and some parts of Europe. Also called sheep liver fluke and liver rot.

**fasciolin** Same as carbon tetrachloride.

**fasciolopsiasis** An excreted helminth infection of the duodenum (small intestine) caused by the pathogen *Fasciolopsis buski. See* its common name, giant intestinal fluke, for detail.

***Fasciolopsis buski*** The helminth species that causes the giant intestinal fluke disease, in Southeast Asia, mainly in central and south China; found in humans, pigs, and dogs, with snails as intermediate hosts, it may reach almost two inches in length. It is transmitted through metacercariae encysted on water caltrop, water chestnut, water bamboo, water hyacinth, and other aquatic plants. Excreted load = $10^3$/gram of feces. Latency = 2 months. Median infective dose > $10^6$.

**Fasflo** Equipment manufactured by Vikoma International for skimming floating oil.

**FAST®** A wastewater treatment plant engineered by Scienco/Fast Systems using the activated sludge system.

**fast atom bombardment gas chromatography–mass spectrometry** A laboratory technique using ionization, liquid chromatography, and capillary electrophoresis to analyze biological molecules and other substances of high molecular weight.

**Fastek™** A thin membrane manufactured by Osmonics for use in reverse osmosis installations.

**fast fluidization** A condition or phase in the fluidized-bed incineration process in which all solid particles begin to move out of the bed and maintain a vertical density gradient. *See also* incipient fluidization.

**fast rinse** The accelerated application of rinse water to a water softener at the end of brine regeneration.

**fast-track design** The preparation of plans and specifications for a project in such a way that construction can begin before design is complete, thereby shortening the implementation schedule.

**fat** *See* fats; fats and oils; fats, oil, and grease.

**fatality rate** The percentage of persons diagnosed with a disease who die as a result of that episode of the disease.

**fate and transport model** A mathematical or graphical model designed to assess the impact on water quality from watershed activities. More specifically, it predicts contaminant loads affecting water quality. *See also* reservoir loading model.

**fathead minnow** A freshwater organism sometimes used in toxicity bioassays. Scientific name: *Pimephales promelas.*

**fatigue corrosion** *See* stress corrosion.

**fats** A variety of biochemical substances that are soluble in organic solvents but sparingly soluble in water; triglyceride esters of fatty acids; e.g., glycerol oleobutyropalmitate. *See also* grease.

**fats and oils** One of three major categories of natural organic compounds (with carbohydrates and proteins), only sparingly soluble in water, and sometimes odoriferous. General formula: $C_nH_{2n+1}COOH$.

**fats, oil, and grease (FOG)** The oil and grease content of wastewater, determined by extraction with trichlorotrifluoroethane and including animal fats, vegetable oils, waxes, water- and petroleum-based oils and grease, etc.

**fatty acid** Any of a chain of aliphatic acids, consisting of a long chain of hydrocarbons ending in a carboxyl group; an important energy compound for microbial metabolism, richer than carbohydrates per unit weight. General formula of saturated fatty acids is $[CH_3(CH_2)_xCOOH_x$, with $x = 0 - 20]$, and that of unsaturated fatty acids is [$CH_3(CH_2)_xCH=CH(CH_2)_yCOOH$, with x + y ≥ 14]. For example, stearic acid [$CH_3(CH_2)_{16}COOH$] and oleic acid [$CH_3(CH_2)_7CH=CH(CH_2)_7COOH$].

**fatty acid methyl ester (FAME)** A fatty acid in which the methyl group of methanol replaces the active hydrogen ion.

**fatty acid profile** The array of fatty acids contained in an organism grown in pure culture.

**fatty acid profiling** The generation of a fatty acid profile to identify an organism.

**faucet aerator** A device attached to a faucet head to reduce water flow and add air to the flow. It increases the rinsing power of water and concentrates flow. *See also* flow-reduction devices.

**faucet restrictor** A device attached to a faucet to restrict flow; e.g., combined with a faucet aerator.

**fault protection** *See* flexible coupling.

**fauna** Animal life of an area, region, or period, considered as a whole. *See also* flora.

**Favair®** Equipment manufactured by the U.S. Filter Corp. for us in dissolved-air flotation units.

**favorable isotherm** *See* ion-exchange isotherm.

**FBBR** Acronym of fluidized-bed bioreactor.

**FBC** Acronym of fluidized bed combustion.

**FBR** Acronym of (a) floc blanket reactor; (a) fluidized-bed reactor.

**FBR/PAC/UF process** A water treatment process that includes flocculant and coagulant chemicals addition ahead of a floc blanket reactor (FBR), followed by treatment with fresh powdered activated carbon (PAC) and ultrafiltration (UF) membranes. The concentrate from the membranes is returned to the activated carbon unit and the used PAC from the membranes is returned to the influent to the FBR.

**FBS** Acronym of fine bar screen, wastewater treatment equipment manufactured by Jones & Attwood, Inc.

**FC/FS ratio** *See* fecal coliform/fecal streptococci ratio.

**F+ coliphage** Bacteriophage that infect strains of *E. coli* and related bacteria through the $F^+$ or sex pili. *See* male-specific coliphage for detail.

**F curve** A graphical representation of a cumulative data function. *See* cumulative residence time distribution curve.

**FDA** Acronym of Food and Drug Administration.

**FDS** Acronym of fixed dissolved solids.

**Fe** Chemical symbol of iron.

**Fe (II)** Symbol of ferrous iron.

**Fe(III)** Symbol of ferric ion.

**Fe3®** Liquid ferric sulfate produced by FE3, Inc.

**feasibility determination** A determination by the USEPA under the Safe Drinking Water Act whether a treatment method is feasible and best available technology, considering efficiency and cost.

**feasibility study** (1) Analysis of the practicality of a proposal; e.g., a description and analysis of potential cleanup alternatives for a site on the National Priorities List. The feasibility study usually recommends selection of a cost-effective alternative. It usually starts as soon as the remedial investigation is underway. (2) A small-scale investigation of a problem to ascertain whether a proposed research approach is likely to provide useful data.

**fecal or faecal** Of or pertaining to feces.

**fecal bacteria reduction in ponds** The reduction of fecal bacteria (and other organisms) in a pond depends on the number of bacteria in the influent ($N_i$, number in 100 mL), the retention time ($t$, days) and a rate constant ($K$), usually expressed as a function of temperature ($T$, °C). For fecal coliforms (Cairncross & Feachem, 1993):

$$N_e = N_i/(1 + K \cdot t) \qquad \text{(F-07)}$$

$$K = 2.6\,(1.19)^{T-20} \qquad \text{(F-08)}$$

where $N_e$ is the number of bacteria per 100 mL in the effluent.

**fecal coliform** A coliform present in the feces of warm-blooded animals, a subgroup of total coliforms (96% in human feces) used as an indicator of fecal contamination. *E. coli* and some *Klebsiella pneumoniae* strains are the predominant members of the group. Fecal coliforms are aerobic and facultative, Gram-negative, nonsporeforming, rod-shaped bacteria that ferment lactose at 44.5°C. They provide a better indication of recent fecal contamination than total coliforms, but do not distinguish between human and animal sources.

**fecal coliform bacteria** Bacteria found in the intestinal tracts of animals; they produce gas at high incubation temperatures. Their presence in water or sludge is an indicator of pollution and possible contamination by pathogens. *See* fecal coliform.

**fecal coliform/fecal streptococci (FC/FS) ratio** A ratio that may indicate whether a recent (24 hours) fecal water contamination was from human or animal wastes: e.g., FC/FS > 4 → human origin and FC/FS < 0.7 → animal origin.

**fecal coliform procedure** A test that uses fermentation of lauryl tryptose broth to indicate the presence of total coliforms from feces, soil, or other origin; the test detects the production of acid and gas at 35°C or 95°F. Alternatively, an EC medium can be used to detect the production of lactic acid and gas.

**fecal composition (chemical)** On a dry-weight basis, human feces contain approximately 4.5% calcium as CaO, 50% carbon, 6.0% nitrogen, 93% organic matter, 4.2% phosphorus as $P_2O_5$, and 1.75% potassium as $K_2O$. *See also* urine composition.

**fecal detection** A determination of water contamination by fecal matter using fecal indicators.

**fecal indicator** An organism that indicates contamination by fecal matter from humans or other warm-blooded animals. *See* fecal coliform, total coliform.

**fecal indicator bacteria** *See* fecal indicator.

**fecal indicator group** *See* fecal indicator.

**fecal matter** Feces.

**fecal–oral** Pertaining to a disease or pathogen that is transmitted through fecal material that reaches the mouth.

**fecal–oral bacterial disease** A disease caused by bacteria transmitted through the fecal–oral route, e.g., gastroenteritis, cholera, typhoid, and bacillary dysentery).

**fecal–oral disease** A disease that is transmitted through fecal pathogens that reach the mouth, usually an enteric disease, i.e., a disease of the intestinal tract, with diarrhea as the most common symptom. *See* bacterial diseases, excreted viruses, excreted protozoa, insect-borne disease, and water- and excreta-related diseases.

**fecal–oral infection** *See* fecal–oral disease.

**fecal–oral nonbacterial disease** Fecal–oral disease caused by a pathogen other than bacteria, e.g., by helminthes (e.g., roundworm), protozoa (e.g., giardiasis), or viruses (e.g., infectious hepatitis).

**fecal–oral pathway** Same as fecal–oral route.

**fecal–oral route** The mode of transmission of many water- and excreta-related diseases, whereby fecal material reaches the mouth and then the intestinal tract. Also called fecal–oral, anal–oral, or anus-to-mouth pathway. *See also* portal of entry.

**fecal–oral route of exposure** Same as fecal–oral transmission.

**fecal–oral transmission** Direct or indirect, waterborne or foodborne transmission of a disease or pathogen through human or animal excreta that reach the mouth.

**fecal pollution indicator** *See* fecal indicator.

**fecal streptococci (FS)** A group of Gram-positive bacteria of the genera *Enterococcus* and *Streptococcus* (Lancefield Group D), abundant in the intestinal tracts of warm-blooded animals, used as an indicator of fecal contamination, consisting mainly of seven strains: *Ent. avium, Streptococcus bovis, Ent. durans, S. equinus, Ent. faecalis, Ent. faecium,* and *Ent. gallinarium.* Also called enterococci.

**fecal streptococcus** Singular of fecal streptococci.

**fecal weight** The weight of wet feces produced per person per day varies with diet, climate, and other factors, e.g., 100–200 grams in Europe and North America to 130–520 grams in developing countries.

**Fecascrew** A screw press manufactured by Hydropress Wallender & Co. for handling screenings from wastewater treatment.

**Fecawash** A washing and conveying unit manufactured by Hydropress Wallender & Co. for handling screenings from wastewater treatment.

**feces or faeces** The excrement from the gastrointestinal tract of humans and animals, consisting of residue from food digestion and bacterial action, whose composition and weight vary with diet and climate; wet weight varies from 100 g to 500 g and bacterial population is equivalent to $2.0 \times 10^{12}$ *E. coli* cells daily per person.

**$FeCl_3$** Chemical formula of ferric chloride.

**$FeCO_3$** Chemical formula of ferrous carbonate.

**$FeCO_3$ (s)** Chemical formula of siderite.

**$FeCr_2O_4$** Chemical formula of a compound of iron and chromium, chromite, the principal ore of the latter.

**fecundity index** The percentage of copulations resulting in pregnancy. *See also* fertility index.

**Federal Emergency Management Agency (FEMA)** A department of the U.S. government that responds to disasters.

**Federal Insecticide, Fungicide and Rodenticide Act (FITRA)** The federal law of 1972 that requires the registration and toxicity testing of pesticides.

**Federalloy®** Lead-free alloys manufactured by Federal Metal Co. of Bedford, OH and used in water supply piping.

**Federal Register** A daily publication of the U.S. government (Office of Federal Register) providing official public notices on proposed and final laws, rules, regulations, and other actions. *See also* Code of Federal Regulations (CFR).

**Federal Water Pollution Control Act (FWPCA) Amendments of 1972** U.S. water pollution control legislation of 1972 (Public Law 92-500), amended in 1977. It included goals for the attainment of water quality in U.S. waterways, enforcement powers under the National Pollutant Discharge Elimination System (NPDES), and grants for the construction of water pollution control facilities. *See also* Clean Water Act.

**feed** (1) In pressure-driven membrane processes, the input stream to the membrane array. Also called feedstream or feedwater. (2) The wastewater, process water, or sludge delivered to a dissolved air flotation unit; its concentration is measured in mg/L suspended solids and its flow usually in gpm.

**feed and bleed** The use of multiple stages of ultrafilters operated with recirculation of the reject water from an initial stage to subsequent stages.

**feedback** The circulating action between a sensor measuring a process variable and the controller that controls or adjusts the process variable.

**feedback control** In wastewater process control applications, a control method in which the value of an input variable is based on adjustments of a controlled parameter. *See also* the following types of control: cascade, feed-forward, integral, on-off, proportional, proportional–derivative, proportional–integral, and proportional–integral–derivative

**feed crop** A crop produced primarily for consumption by animals.

**feeder reservoir** Any impounding reservoir.

**feed-forward control** In wastewater process control applications, a control method in which the value of an input variable is based on measured disturbances. *See also* static feed-forward control and the following types of control: cascade, feedback, integral, on–off, proportional, proportional–derivative, proportional–integral, and proportional–integral–derivative

**feedlot** A confined area for the controlled feeding of animals. Crop or forage production or growth is not sustained in the area of confinement. Feedlots tend to concentrate large amounts of animal wastes that cannot be absorbed by the soil and, hence, may be carried to nearby streams or lakes by runoff (EPA-40CFR412-11.b & 412.21-b).

**feedlot pollution** *See* feedlot wastes.

**feedlot wastes** Solid and liquid wastes from a facility where cattle or other animals are raised for market. On modern feedlots where cattle can gain more than two pounds per day per head, the animals produce huge concentrated piles of manure and excreta that the soil cannot handle as on an open range. Feedlot wastes are high in organic suspended solids and BOD; common treatment methods include anaerobic lagoons and land disposal. One dairy cow is the BOD equivalent of 20–25 persons.

**Feedpac** An apparatus manufactured by Nalco Chemical Co. for coagulant feeding in water or wastewater treatment.

**feed pipe/feed pump** A pipe/pump used to supply water to a boiler.

**feed pressure** The pressure of the feedwater to a treatment unit.

**feed pump** (1) *See* feed pipe. (2) A pump used to supply chemicals accurately to a water or wastewater treatment unit.

**feed stream** In pressure-driven membrane processes, the input stream to the membrane array. Also called feed or feedwater.

**feed tank** A device used to control pressure surges in piping systems. It consists of an open or closed tank that admits water into a pipe after a downsurge to prevent initial low pressures. When equipped with a check valve to allow only inflow, it functions as a one-way surge tank. *See also* other types of pressure surge control devices.

**feedwater** (1) Water supplied to a boiler from a tank or a condenser for steam generation; it is usually softened, demineralized, heated, and deaerated. Also called boiler-water feed or boiler feedwater. (2) In general, the water to be treated, fed to a given treatment system. In pressure-driven membrane processes, the input stream to the membrane array. Also called feedstream or simply feed.

**feedwater heater** A device used to preheat boiler feedwater with steam from a turbine.

**feedwater treatment** (1) Special treatment of water to be used in boilers or industrial processes, including, e.g., demineralization and corrosion control. (2) Pretreatment of the feedwater to a membrane process.

**Feigenbaum index** A corrosion index ($Y$) developed in Israel for hard-saline waters (Symons et al., 2000):

$$Y = AH + B[Cl^- + SO_4^{2-}]e^{-1/AH} + C \quad (F\text{-}09)$$

$$H = [Ca^{2+}][HCO^-]^2/[CO_2] \quad (F\text{-}10)$$

where $A = 0.00035$, $B = 0.34$, and $C = 19.0$. All concentrations are in milligrams bicarbonate per liter. The higher the $Y$ value, the less corrosion and vice versa, with moderate corrosion for $200 < Y < 500$.

**feldspar** Any of a group of minerals—aluminosilicates of potassium, sodium, calcium. Important constituents of igneous rocks. *See also* albite, anorthite, orthoclase. Also spelled felspar.

**felspar** *See* feldspar.

**felt-based lining** A trenchless water main rehabilitation method using a nonwoven polyester felt, impregnated with a resin (elastomer) that is cured in place by hot water or steam. *See* the following types of cured-in-place liners: cured-in-place lining, woven-hose lining, membrane lining. *See also* trenchless technology.

**FEMA** Acronym of Federal Emergency Management Agency.

**female end** The inside threaded connection of a pipe section, which connects with the spigot end of another section. Also called bell end.

**female fertility index** *See* fertility index.

**F.E.M.S.®** Acronym of fugitive emissions management system, a relational database developed by EnviroMetrics.

**fen** A type of wetland that accumulates peat deposits. Fens are less acidic than bogs, deriving most of their water from groundwater rich in calcium and magnesium. *See* wetlands.

**fenoprop** A pesticide with a WHO-recommended guideline value of 0.009 mg/L for drinking water.

**Fenton mechanism** The use of Fenton's reagent to inactivate bacteria or viruses, remove color, or oxidize toxic chemicals.

**Fenton's reaction** *See* the first reaction listed under Fenton's reagent. An alternative representation of Fenton's reaction is:

$$Fe^{2+} + H_2O_2 + H^+ \rightarrow Fe^{3+} + OH^{\cdot} + H_2O \quad \text{(F-11)}$$

**Fenton's reagent** A mixture of ferrous sulfate $[Fe(SO_4)]$ or other ferrous salt and hydrogen peroxide $(H_2O_2)$ used to produce the hydroxyl radical $(^{\cdot}OH)$ for the advanced oxidation of toxic organic chemicals or for the removal of color. Fenton's mechanism may also be used to inactivate macromolecules (proteins or nucleic acids): a metal ion binds to a biological target, is reduced and subsequently reoxidized by $H_2O_2$ while generating hydroxide radicals. Actually, divalent metal ions other than iron $(Fe^{2+})$, e.g., copper $(Cu^{2+})$ and manganese $(Mn^{2+})$, can be used for Fenton's reaction, the first of a series of reactions (Freeman, 1998):

$$Fe^{2+} + H_2O_2 \rightarrow Fe^{3+} + OH^- + {}^{\cdot}OH \quad \text{(F-12)}$$

$$Fe^{3+} + H_2O_2 \rightarrow Fe^{2+} + HO_2^{\cdot} + H^+ \quad \text{(F-13)}$$

$$Fe^{2+} + {}^{\cdot}OH \rightarrow Fe^{3+} + OH^- \quad \text{(F-14)}$$

$${}^{\cdot}OH + H_2O_2 \rightarrow H_2O + HO_2^{\cdot} \quad \text{(F-15)}$$

$$Fe^{3+} + O_2^- \rightarrow Fe^{2+} + O_2 \quad \text{(F-16)}$$

**$Fe_2O_3$** Chemical formula of ferric and in iron rust oxide; the iron compound found in the ores hematite and limonite.

**$\alpha$-$Fe_2O_3$** Chemical formula of the iron ore hematite.

**$Fe_2O_3 \cdot 3H_2O$** Chemical formula of the iron ore limonite.

**$Fe_3O_4$ or $FeFe_2O_4$** Chemical formula of ferrous oxide, the iron compound of the ore magnetite.

**$Fe(OH)_3$** Chemical formula of ferric hydroxide, the result of the oxidation of ferrous iron by oxygen or chlorine.

**$Fe_5OH_8 \cdot 4H_2O$** Chemical formula of ferrihydrite.

**$\alpha$-FeOOH** Chemical formula of goethite.

**$\gamma$-FeOOH** Chemical formula of lepidocrocite.

**$Fe_3(PO_4)_2$** Chemical formula of vivianite.

**Feripac** A water treatment plant manufactured by Vulcan Industries, Inc. for iron and manganese removal.

**fermentation** The anaerobic conversion of organic matter to carbon dioxide $(CO_2)$, methane $(CH_4)$, and other low-molecular-weight compounds; anaerobic organisms extract chemical energy from substrates in the absence of molecular oxygen. Fermentation is used in wastewater treatment, sludge treatment (e.g., anaerobic digestion), as well as commercially (e.g., wine making, beer brewing, production of ethyl alcohol and vinegar, cheese making). The following reactions represent the production of ethyl alcohol $(CH_3CH_2OH)$ from carbohydrates $(C_6H_{12}O_6)$ and the anaerobic decomposition of organic matter $(CH_2O)$:

$$C_6H_{12}O_6 \rightarrow 2\ CH_3CH_2OH + 2\ CO_2 \quad \text{(F-17)}$$

$$2\ CH_2O \rightarrow CH_4 + CO_2 \quad \text{(F-18)}$$

*See also* acidogenesis, digestion, anaerobic composting.

**fermentation industry waste** *See* brewery/distillery/winery waste.

**fermentation tube** A glass tubular container with a screw cap or a slip-on stainless steel closure, containing smaller tubes upside down and used in the multiple-tube fermentation technique.

**fermentation tube procedure** A method of assessing the microbial quality of a sample of water through the detection of indicator organisms; it does not include anaerogenic and lactose-negative coliforms. *See also* membrane filter technique.

**fermentation tube technique** *See* multiple-tube fermentation technique, membrane filter technique.

**fermentative metabolism** The use of heterotrophic organisms and an internal electron acceptor for energy production in biological treatment. *See also* respiratory metabolism.

**fermentative respiration** Anaerobic respiration in which products of catabolic reactions serve as electron acceptors; respiratory metabolism.

**ferrate** A salt of the hypothetical ferric acid ($H_2FeO_4$) that can potentially be used in water or wastewater disinfection.

**ferredoxin** A red-brown protein containing iron and sulfur that acts as an electron carrier during photosynthesis, nitrogen reduction, or redox reactions.

**ferric** Pertaining to iron that is trivalent [Fe(III)] or in a higher oxidation state, as in ferric sulfate ($Fe_2SO_4$) and ferric chloride ($FeCl_3$). *See also* ferrous.

**ferric acid** *See* ferrate.

**ferricalcic deposit** A passivating film layer consisting of a mixture of calcium carbonate ($CaCO_3$), ferrous carbonate ($FeCO_3$), and ferric oxide ($Fe_2O_3$) that forms in the presence of dissolved oxygen on unlined iron piping, providing protection against internal corrosion.

**ferric chloride ($FeCl_3$)** A black-brown, soluble, solid iron salt, commonly used as a coagulant in water and wastewater treatment (e.g., phosphorus or suspended solids removal) or as a sludge conditioner; it forms ferric hydroxide floc [$Fe(OH)_3$] by combining with alkalinity, natural or added as lime. It begins to form crystals ($FeCl_3 \cdot 6\ H_2O$) at about 7°C. In hydrated form, $FeCl_3 \cdot x\ H_2O$, it occurs as yellow-orange crystals.

**ferric chloride–lime precipitation** The same as ferric chloride precipitation, but with lime [$Ca(OH)_2$] added to supplement natural alkalinity:

$$2\ FeCl_3 + 3\ Ca(OH)_2 \rightarrow 2\ Fe(OH)_3 + 3\ CaCl_2 \quad \text{(F-19)}$$

**ferric chloride precipitation** A water or wastewater treatment method using ferric chloride ($FeCl_3$) for the chemical precipitation of dissolved and suspended solids. The chemical combines with soluble calcium bicarbonate [$Ca(HCO_3)_2$] to form insoluble ferric hydroxide [$Fe(OH)_3$]:

$$2\ FeCl_3 + 3\ Ca(HCO_3)_2 \rightarrow 2\ Fe(OH)_3 + 3\ CaCl_2 + 6\ CO_2 \quad \text{(F-20)}$$

**ferric coagulants** Iron-based chemicals used in water treatment, including chlorinated copperas, copperas, ferric chloride, and ferric sulfate.

**ferric hydroxide [$Fe(OH)_3$]** A coagulant precipitate, formed when ferrous iron is oxidized by oxygen or chlorine, e.g., for the removal of iron from groundwater or the removal of turbidity, natural organic matter, and arsenic:

$$4\ Fe^{++} + O_2\ (aq) + 10\ H_2O \rightarrow 4\ Fe(OH)_3(s) + 8\ H^+ \quad \text{(F-21)}$$

$$2\ Fe^{++} + HOCl + 5\ H_2O \rightarrow 2\ Fe(OH)_3(s) + Cl^- + 5\ H^+ \quad \text{(F-22)}$$

$$Fe^{++} + ClO_2 + 3\ H_2O \rightarrow Fe(OH)_3(s) + ClO_2^- + 3\ H^+ \quad \text{(F-23)}$$

Red-brown ferric hydroxide sometimes precipitates from groundwater on fixtures and laundry.

**ferric iron ($Fe^{3+}$) or Fe(III) or $Fe^{+++}$** (1) A relatively stable insoluble ion used in the coagulation process, its effect resulting from the formation of hydrolysis products and from the tendency of hydroxo ferric complexes to polymerize (Fair et al., 1971):

$$[Fe(H_2O)_6]^{+++} + H_2O \rightarrow [Fe(H_2O)_5(OH)]^{++} + H_3O^+ \quad \text{(F-24)}$$

$$[Fe(H_2O)_5(OH)]^{++} + H_2O \rightarrow [Fe(H_2O)_4(OH)_2]^+ + H_3O^+ \quad \text{(F-25)}$$

*See also* dimeric species. (2) Small solid particles of trivalent iron suspended in water as ferric oxide [$Fe_2O_3$] or ferric hydroxide [$Fe(OH)_3$] and removable by filtration. Also called precipitated iron. *See also* rusty water.

**ferric iron salt** Any of the salts of the hypothetical ferric acid, some of which are used as coagulants.

**FerriClear®** Ferric sulfate produced by Eaglebrook, Inc.

**ferric oxide ($Fe_2O_3$)** A dark-red, crystalline, insoluble solid, occurring naturally as rust or hematite, used as a pigment. Also called iron oxide.

**ferric salt** *See* ferric chloride, ferric sulfate.

**ferric sulfate [$Fe_2(SO_4)_3 \cdot x\ H_2O$]** A coagulant commonly used in water and wastewater treatment (e.g., phosphorus and suspended solids removal); it reacts with the natural alkalinity of water or added lime to form a ferric hydroxide floc that is heavier than alum floc. Commercially available as granules that contain two or three molecules of water, and 21% or 18.5% Fe, respectively; e.g. Ferri-Floc, FerriClear®. Also called iron sulfate. Ferric sulfate reacts with calcium bicarbonate [$Ca(HCO_3)_2$], if there is sufficient alkalinity, or with any added lime [$Ca(OH)_2$]:

$$Fe_2(SO_4)_3 + 3\ Ca(HCO_3)_2 \rightarrow 2\ Fe(OH)_3\ (s) + 3\ CaSO_4 + 6\ CO_2 \quad \text{(F-26)}$$

$$Fe_2(SO_4)_3 + 3\ Ca(OH)_2 \rightarrow 2\ Fe(OH)_3\ (s) + 3\ CaSO_4 \quad \text{(F-27)}$$

**ferric sulfate–lime precipitation** A water or wastewater treatment method using ferric sulfate and lime for the chemical precipitation of dissolved and suspended solids. *See* ferric sulfate and equation F-27.

**ferricyanic acid [$H_3Fe(CN)_6$]** A brown, crystalline, soluble solid.

**ferricyanide** A salt of ferricyanic acid, e.g., potassium ferricyanide [$K_3Fe(CN)_6$].

**Ferri-Floc** Ferric sulfate produced by Boliden Intertrade, Inc.

**ferrihemoglobin** Same as methemoglobin.

**ferrihydrite ($5Fe_2O_3 \cdot 9\ H_2O$ or $Fe_5HO_8 \cdot 4\ H_2O$)** A poorly crystalline compound formed during the corrosion of iron pipes; also called an iron phase or a corrosion scale. It can be transformed to more stable forms: hematite and goethite.

**ferrite** (1) A compound of ferric oxide and a more basic oxide, e.g., sodium ferrite ($NaFeO_2$). (2) The pure iron constituent of ferrous metals. *See also* iron carbide.

**ferroalloy** An alloy of iron with another element other than carbon, used in steel.

**ferroaluminum** A ferroalloy with up to 80% aluminum.

***Ferrobacillus*** A genus of bacteria that can catalyze the oxidation of iron (II) to iron (III) and obtain energy for metabolic needs.

**ferrocement** Cement mortar reinforced by layers of steel mesh; sometimes used in the fabrication of slabs for latrines.

**ferrocene [$(C_5H)_2Fe$]** An orange, crystalline, insoluble compound used as an antiknock additive in gasoline and as a catalyst. Also called dicyclopentadienyliron.

**ferrochrome** Same as ferrochromium.

**ferrochromium** A ferroalloy containing up to 70% chromium. Also called ferrochrome.

**ferroconcrete** Reinforced concrete.

**ferrocyanic acid [$H_4Fe(CN)_6$]** A white, crystalline, soluble solid resulting from the combination of a ferrocyanide and an acid.

**ferrocyanide** A salt of ferrocyanic acid, e.g., sodium ferrocyanide [$Na_4Fe(CN)_6$].

**ferromanganese** A ferroalloy containing up to 90% manganese.

**ferromolybdenum** A ferroalloy containing up to 60% molybdenum.

**ferronickel** A ferroalloy containing up to 46% nickel.

**ferrosilicon** A ferroalloy containing up to 95% silicon.

**ferrotitanium** A ferroalloy containing up to 45% titanium.

**ferrotungsten** A ferroalloy containing up to 80% tungsten.

**Ferrosand®** Granular media produced by Hungerford & Terry, Inc. for use in iron and manganese removal by filtration.

**Ferrosand Filter** A treatment process designed by Hungerford & Terry, Inc. of Clayton, NJ for the removal of iron, manganese, arsenic, radium, and hydrogen sulfide from well waters.

**ferrous** Pertaining to iron that is divalent [Fe(II)] or in a lower oxidation state, e.g., ferrous sulfate ($FeSO_4$) or ferrous oxide (FeO). *See also* ferric.

**ferrous ammonium sulfate (FAS) [$FeSO_4 \cdot (NH_4)_2SO_4 \cdot 6H_2O$]** A chemical used to determine the presence of chlorine and to titrate potassium dichromate in the COD test. *See also* N,N-diethyl-p-phenylethylene diamine, DPD test.

**ferrous carbonate ($FeCO_3$)** An iron compound that may dissolve in groundwater, causing iron to precipitate, when carbon dioxide is present and oxygen is lacking:

$$FeCO_3 + CO_2 + H_2O \rightarrow Fe^{++} + 2\ HCO_3^- \quad \text{(F-28)}$$

**ferrous chloride ($FeCl_2$)** A soluble iron salt used as a coagulant.

**ferrous iron ($Fe^{2+}$) or Fe(II) or $Fe^{++}$** A soluble, reduced form of iron that can act as a reducing agent or be oxidized to ferric hydroxide for iron removal from groundwater.

**ferrous oxide (FeO)** A black, insoluble powder. Also called iron monoxide.

**ferrous salt** Such compounds as ferrous chloride ($FeCl_2$) and ferrous sulfate ($FeSO_4$) used in water treatment as coagulants or reducing agents.

**ferrous sulfate ($FeSO_4 \cdot x\ H_2O$)** A bluish-green, saline-tasting, coagulant commonly used in water and wastewater treatment (e.g., odor control, phosphorus removal, sludge conditioning, suspended solids removal) and for other industrial purposes (fertilizer, ink, medicine, etc.); it reacts with the natural alkalinity of water or the added lime to form a ferrous hydroxide floc, which itself may form ferric hydroxide:

$$FeSO_4 \cdot 7\ H_2O + Ca(OH)_2 \rightarrow Fe(OH)_2 + CaSO_4 + 7\ H_2O \quad \text{(F-29)}$$

$$4\ Fe(OH)_2 + O_2 + 2\ H_2O \rightarrow 4\ Fe(OH)_3(s) \quad \text{(F-30)}$$

Commercially available as granules, crystals, powder, and lumps that contain seven molecules of water and 20% Fe. Also called copperas or ferrous sulfate heptahydrate, green vitriol, iron vitriol, and iron sulfate. *See also* chlorinated copperas.

**ferrous sulfate heptahydrate ($FeSO_4 \cdot 7\ H_2O$)** Same as ferrous sulfate, but including seven molecules of water; copperas.

**ferrous sulfate-lime precipitation** A treatment method using ferrous sulfate and lime to precipitate solids and improve plant performance. Lime is added when alkalinity is insufficient. The reactions can occur as shown above for ferrous sulfate or as follows:

$$FeSO_4 \cdot 7\,H_2O + Ca(HCO_3)_2 \rightarrow Fe(HCO_3)_2 + CaSO_4 + 7\,H_2O \quad \text{(F-31)}$$

$$Fe(HCO_3)_2 + 2\,Ca(OH)_2 \rightarrow Fe(OH)_2 + 2\,CaCO_3 + 2\,H_2O \quad \text{(F-32)}$$

**ferrous sulfide (FeS)** A dark metallic, crystalline, insoluble compound used in ceramics and to produce hydrogen sulfide ($H_2S$). It is responsible for the blackening of wastewater and sludge in which it forms by the combination of hydrogen sulfide and iron.

**ferrovanadium** A ferroalloy containing up to 55% vanadium.

**Ferrover** Chemical reagent produced by Hach Co. for iron analysis.

**Ferrozine** Spectrophotometric reagent produced by Hach Co. for the analysis of iron and iron compounds.

**ferruginous** Iron-bearing; of the color of iron rust.

**ferrum** Latin name of iron (Fe).

**Ferr-X** A process developed by Aquatro ferr-X Corp. for iron removal.

**fertility index** The percentage of males that impregnate fertile, nonpregnant females (the male fertility index) or the percentage of females that conceive after impregnation by a fertile male (the female fertility index). *See also* fecundity index.

**fertility plasmid** A circular piece of DNA outside the chromosome, originally found in *E. coli.*

**fertilizer** Material usually containing nitrogen (N), phosphorus (P), and potassium (K), or NPK, added to soil to provide essential nutrients for plant growth, including organic fertilizers (e.g., farmyard manure, crop residue, compost, bonemeal, blood, fishmeal) and inorganic fertilizers in the form of chemical mixtures. Excessive use of fertilizers may cause water pollution (e.g., nitrates in groundwater). *See also* soil conditioner.

**fertilizer industry waste** Wastewater originating from the production of fertilizers: chemical reactions of basic elements, spills, cooling waters, boiler blowdowns, product washing, and other operations. It contains sulfuric, phosphorous and nitric acids, suspended solids, and various minerals. Except for the blowdown and cooling water, fertilizer industry waste is classified as ammonia ($NH_3$) effluent or phosphoric acid ($H_3PO_4$) effluent. Common treatment methods include neutralization, sedimentation, ammonia stripping, lime precipitation. Reuse is also practiced. *See also* phosphate industry waste.

**fertilizer value** The weight or percentage of the total content of nitrogen (N), phosphorus (as $P_2O_5$), and potassium (as potash, $K_2O$) in a fertilizer. *See* Table F-01 for the fertilizer value of sludge.

**$FeS_2$** Chemical formula of iron sulfide or the iron ore pyrite; also iron disulfide, marcasite.

**FeSAs** Chemical formula of arsenopyrite or mispickel, a common arsenic ore.

**$FeSO_4Cl$** Chemical formula of chlorinated copperas.

**$Fe_2(SO_4)_3$** Chemical formula of ferric sulfate.

**$FeSO_4 \cdot 7\,H_2O$** Chemical formula of green vitriol, iron sulfate.

**$FeSO_4 \cdot x\,H_2O$** Chemical formula of copperas or commercial ferrous sulfate; usually with $x = 7$ (ferrous sulfate heptahydrate), but also sometimes $x = 9$.

**$Fe_2(SO_4)_3 \cdot 2H_2O$** A chemical formula of commercial ferric sulfate.

**$Fe_2(SO_4)_3 \cdot 3H_2O$** A chemical formula of commercial ferric sulfate.

**$FeSO_4 \cdot (NH_4)_2SO_4 \cdot 6H_2O$** Chemical formula of ferrous ammonium sulfate.

**$FeTiO_3$** Chemical formula of ilmenite.

**FFG equation** *See* Frumkin–Fowler–Guggenheim equation.

***F. gigantica*** *See Fasciola gigantica.*

**Fibercone Press** A conical press manufactured by Black Clawson for sludge dewatering.

**FiberFlo™** A cartridge filter manufactured by Fibercor.

**fiberglass** (1) A material made of extremely fine filaments of glass, used for insulation and to reinforce plastic materials in the fabrication of components of water and wastewater facilities (e.g., manholes, weirs, launders, and other light structures). It resists to chemicals, corrosion, and solvents. The fibers can penetrate the skin and have adverse effects on the lungs if inhaled. (2) Hardened plastic material reinforced with glass fibers.

**fiberglass-reinforced plastic (FRP)** A corrosion-resistant material consisting of glass fibers imbed-

**Table F-01.** Fertilizer value of sludge (typical, in % of total solids)

| | Raw primary sludge | Digested primary sludge | Raw activated sludge |
|---|---|---|---|
| Nitrogen (as N) | 2.5 | 3.0 | 3.7 |
| Phosphorus (as $P_2O_5$) | 1.6 | 2.5 | 6.9 |
| Potassium (as potash, $K_2O$) | 0.4 | 1.0 | 0.6 |

ded in vinyl esters or other thermosetting resins; it is used to manufacture pipes, tanks, and fasteners. It is less strong than metals and subject to thermal expansion.

**fiber optics** The branch of optics that deals with the transmission of light through transparent fibers or the transmission of signals by optical light.

**fiber optics transmission** The use of transparent glass fibers to transmit data.

**fibreglass** *See* fiberglass.

**fibrosarcoma** A sarcoma derived from fibroblast cells that produce collagen.

**fibrosis** The development of excessive fibrous tissue in an organ. Liver fibrosis ultimately results in cirrhosis.

**Fick, Adolph** German physiologist (1829–1901).

**Fickian diffusion** Spreading of a solute under the influence of a concentration gradient, from regions of higher concentrations to regions of lower concentrations, according to Fick's laws of diffusion.

**Fick's first law of diffusion** *See* Fick's laws of diffusion.

**Fick's laws of diffusion** Two physical chemistry laws proposed to explain the process of mass transfer of a substance by diffusion in a medium, e.g., the diffusion and transport of contaminants in water. The first law, similar to Darcy's law, states that the diffusive flux ($M$) is proportional to the concentration gradient ($\partial C/\partial x$, the partial derivative of concentration $C$ with respect to distance $x$). According to the second law, the rate of change ($\partial C/\partial t$, partial derivative of concentration with respect to time $t$) is proportional to the rate of change of the concentration gradient ($\partial^2 C/\partial x^2$). ($D$) is a proportionality factor called the coefficient of diffusion or coefficient of molecular diffusion (in area per unit time) in both equations:

$$\text{First law: } M = -D(\partial C/\partial x) \quad \text{(F-33)}$$

$$\text{Second law: } \partial C/\partial t = D(\partial^2 C/\partial x^2) \quad \text{(F-34)}$$

*See also* molecular diffusion, turbulent or eddy diffusion, coefficient of dispersion, dispersion number, dispersion factor, Peclet number, coefficient of axial dispersion.

**Fick's second law of diffusion** *See* Fick's laws of diffusion.

**fictive component** A fictitious substance used to represent different natural organic compounds having similar adsorption properties, particularly the Freundlich isotherm $n$ and $K$ parameters, in modeling the competitive adsorption of organic compounds in natural waters. *See also* ideal adsorbed solution theory and equivalent background compound.

**fictive natural organic matter** *See* fictive component.

**FID** Acronym of flame ionization detection.

**field application area** The actual area ($A$, ha) cultivated in a wastewater application, exclusive of roads, buffer zones, and storage areas (WEF & ASCE, 1991):

$$A = 0.0001\,(365\,Q + V_s)/L_h \quad \text{(F-35)}$$

where $Q$ is wastewater flow ($m^3$/day); $V_s$ is the net loss or gain of stored wastewater volume due to evaporation, seepage, or precipitation ($m^3$/year); and $L_h$ is the limiting hydraulic loading rate ($m^3$/year). There is a slight confusion in the literature between this term and field area. *See also* cropped area.

**field area** The actual area cultivated in a land application system, exclusive of land required for preapplication treatment facilities, buffer zones, service roads, and storage reservoirs. *See* cropped area for detail.

**field blank** A sample of highly purified water or of a wastewater of definite composition used to verify the integrity of a laboratory's sampling and analysis procedures.

**field capacity or field moisture capacity** A parameter related to the water content of a soil or rock. It is the moisture content of soil in the field some time after reaching saturation and after free drainage has practically ceased, or the quantity of water held in a soil by capillary action after the gravitational or free water has drained; expressed as moisture percentage on a dry weight basis. It is sometimes called capillary capacity, field carrying capacity, maximum water-holding capacity, moisture-holding capacity, normal moisture capacity. *See also* specific retention.

**field carrying capacity** Same as field capacity.

**field groundwater velocity** The ratio of the volume of groundwater through a formation in unit time to the product of the cross-sectional area by the effective porosity. *See* actual groundwater velocity for more detail.

**field moisture capacity** Same as field capacity.

**field oxygen transfer rate** *See* field rate of oxygen transfer.

**field rate of oxygen transfer** The rate of oxygen transfer ($R_{ww,T}$) to wastewater at temperature $T$ (°C), taking into account aeration efficiency and mass transfer factors (WEF & ASCE, 1991):

$$R_{ww,T} = 0.11\,\alpha F R_{w,T}\,(\beta C_s - C_l)\,\theta^{T-20} \quad \text{(F-36)}$$

where $\alpha$ = a correction factor; $F$ = fouling factor, accounting for performance impairment due to fouling or material deterioration; $R_{w,T}$ = the standard oxygen transfer rate (SOTR), i.e., the transfer rate in tap water at temperature $T$; $\beta$ = a correction factor; $C_s$ = saturation concentration based on partial pressure; $C_l$ = gas concentration in the liquid bulk phase; and $\theta$ = empirical temperature correction factor, usually = 1.024. The coefficient 0.11 is the reciprocal of 9.09, which is the dissolved oxygen concentration in tap water at 20°C.

**field survey** An investigation conducted in the field to determine the conditions of prospective water sources and anticipate their probable composition. Field surveys include industrial waste surveys, pollutional surveys, sanitary surveys, observation of the gross quality of water sources, and collection of samples.

**field testing** Practical and generally, small-scale testing of innovative or alternative technologies directed to verifying performance and/or refining design parameters not sufficiently tested to resolve technical uncertainties that prevent the funding of a promising improvement in innovative or alternative treatment technology (EPA-40CFR35.2005-17).

**FIFRA** Acronym of Federal Insecticide, Fungicide, and Rodenticide Act.

**FIFRA pesticide** An ingredient of a pesticide that must be registered with USEPA under the Federal Insecticide, Fungicide, and Rodenticide Act, and may be subject to labeling and use requirements.

**filamentous bacteria** *See* filamentous organisms.

**filamentous bulking** An operational problem of activated sludge plants in which the sludge does not settle well because of its excessive volume, as indicated by the sludge volume index (SVI). It is a type of clarification failure characterized by a large irregular floc, a large SVI (> 150) with a clear supernatant, poor settling and compaction, and a rising sludge blanket in the clarifier. Filamentous strands formed of single-cell filaments attached end-to-end protrude out of sludge flocs. Conditions that promote excessive filaments include low dissolved oxygen concentrations, low food-to-microorganism ratios, large solids retention times, the presence of sulfide, readily biodegradable organics, septic wastewater, nutrient deficiency, and low pH. *See also* dispersed growth, hydrous bulking, pin floc, selector tank, and viscous bulking.

**filamentous growth** *See* filamentous organisms.

**filamentous microorganism** *See* filamentous organisms.

**filamentous organisms** Intertwined, hairlike biological growths of some species of bacteria, algae, and fungi that grow under adverse conditions, which result in poor sludge settling and dewatering, a condition called filamentous bulking or sludge bulking. Filamentous growth occurs in wastes deficient in nutrients, at low pH, with high sludge ages or low F/M ratios. Filamentous species include Type 021N bacteria, sulfur bacteria (*Beggiatoa, Leucothrix, Thiotrix*), *Nocardia, Sphaerotilus natans, Microthrix parvicella, Halsicomenobacter hydrossis, Nostocoida limicola,* etc. *See also* pinpoint floc.

**filamentous sludge** Activated sludge that contains excessive amounts of filamentous organisms and settles poorly as a result. *See also* sludge bulking and floating sludge.

**filamentous sulfur bacteria** A type of bacteria that grow in filamentous sheaths and in activated sludge systems that have low or variable nutrient concentrations. They use organic acids as a carbon source for growth and do not develop well at low pH. They include the genera *Beggiatoa, Leucothrix,* and *Thiotrix.*

***Filaria*** A genus of nematodes, some of which cause serious diseases in humans. Filariae are small threadlike roundworms carried as larvae by mosquitoes; adult worms are blood and tissue parasites of vertebrates. *See Wuchereria bancrofti, Culex fatigans, Dracunculus medinensis, Cyclops.*

**filariasis** A helminthic disease transmitted by mosquito-borne pathogens that develop in the lymphatic system and release larvae in the blood, causing lymph obstruction and elephantiasis. *See* Bancroftian filariasis, Malayan filariasis.

**Filawound®** A pressure vessel manufactured by Spaulding Composites for use with reverse osmosis membranes.

**fill** (1) Same as backfill; any material (earth, stones, etc.) used to build up the level of an area of ground such as a channel, valley, sink, or other depression. (2) The volume of material used. (3) Same as embankment.

**fill and draw** A method of operation whereby a container is filled for a reaction to occur and then emptied. An example is the wastewater treatment process called sequencing batch reactors. *See* batch process and continuous-flow system.

**fill-and-draw digester operation** One of two commonly used procedures to fill a sludge digester, whereby sludge is fed to the reactor and mixed for a short period before the digested sludge is withdrawn. The other procedure is draw and fill.

**filled electron shell** The configuration characteristic of the atom of a noble gas, which has no tendency to lose, gain, or share electrons with other elements, and is thus chemically unreactive. Also called satisfied outer shell. *See* noble gas outer electron shell, Lewis symbol, octet rule, free radical.

**fill material** (1) Same as fill (1). (2) Material used to replace an aquatic area with dry land, excluding discharged pollutants.

**fill phase** The first of the five steps in the operation of sequencing batch reactors, whereby raw wastewater or primary effluent are added to fill the reactor, the contents of which are mixed and aerated. *See also* draw phase, react phase, settle phase, and idle phase.

**film diffusion** The second step in the adsorption process; film diffusion transport.

**film diffusion transport** One of four steps in the adsorption process, in which materials to be adsorbed (adsorbate) overcome resistance and move by molecular diffusion through a stationary liquid film ("hydrodynamic boundary layer") to the entrance of pores of the adsorbing particles. *See also* bulk solution transport, pore transport, and adsorption.

**film flow** The movement of suspended water in the aeration zone through interconnecting films on the surface of soil particles or on the walls of fractures. Also called capillary flow or capillary movement. *See also* capillary migration.

**FilmShear** Air diffusers manufactured by Aerators, Inc. using fine and coarse bubbles.

**Filmtec®** Membranes manufactured by Dow Chemical Co. for use in reverse osmosis.

**Filmtec membrane** A proprietary spiralwound membrane. *See* spiral-wound module.

**Filmtec NF70 membrane** A proprietary spiral-wound membrane containing four envelopes with 90 square feet in a sheet of $3' \times 3.75'$. *See* spiral-wound module.

**film theory model** A modification of the linear solution diffusion model used in sizing arrays of membrane elements. It assumes that the solute concentration increases exponentially from the center of the feed stream channel toward the surface of the membrane and diffuses back into the bulk stream (AWWA, 1999):

$$C_p = C_f K_i \exp(J/K_b)/[K_w(\Delta P - \Delta\Pi)(2 - 2R)/(2 - R) + K_i \exp(J/K_b)] \qquad \text{(F-37)}$$

where $C_p$ = permeate stream solute concentration, $C_f$ = feed stream solute concentration, $K_i$ = solute mass transfer coefficient, $J$ = water flux, $K_b$ = back-diffusion coefficient (solute diffusion from the membrane surface to the bulk in the feed stream), $K_w$ = solvent mass transfer coefficient (solute diffusion through the membrane to the permeate stream), $\Delta P$ = pressure gradient, $\Delta\Pi$ = osmotic pressure, and $R$ = recovery. *See also* linear solution diffusion model, coupling model.

**Filox®** Filter media produced by Matt-Son, Inc. for the removal of iron, manganese, and hydrogen sulfide.

**FiltaBand** Fine screening equipment manufactured by Longwood Engineering Co., including a self-cleaning feature.

**filter** (1) A device utilizing a granular material, woven cloth, or other medium to remove suspended solids from water, wastewater, or air. (2) A porous device through which perchloroethylene is passed to remove contaminants in suspension; for example: lint filter (button trap), cartridge filter, tubular filter, regenerative filter, prefilter, polishing filter, and spin disc filter (EPA-40CFR63.321). (3) A porous layer of paper, glass fiber, or other material used in the laboratory to remove particulate matter from water samples and chemical solutions. (4) To screen or remove pollutants from air, water, etc.

**filterability constant** An important measure of sludge dewaterability, which indicates that the water released from sludge depends on solids concentration and viscosity, as determined, e.g., by the capillary suction time. The filterability constant is linearly related to the specific resistance of the sludge (AWWA, 1999):

$$\chi = 10^{-6}\,\Phi\mu S/T \qquad \text{(F-38)}$$

$$\chi = 0.04\,R + 0.25 \qquad \text{(F-39)}$$

where $\chi$ = filterability constant, $(\text{kg/s/m}^2)^2$; $\Phi$ = dimensionless instrument constant; $\mu$ = viscosity, centipoises (cP); $S$ = solids concentration, mg/L; $T$ = capillary suction time, seconds; and $R$ = specific resistance, m/kg. Also called filterability index. *See also* Atterberg test, Büchner funnel test, capillary suction time (CST) test, compaction density, filterability index, shear strength, specific resistance, standard jar test, time-to-filter (TTF) test.

**filterability index** Same as filterability constant.

**filterable constituent** A dissolved constituent of water or wastewater; a constituent that can pass through a 0.45 micrometer pore-diameter filter.

**filterable residues** Solid particles that pass through the filter during water filtration; they are mostly dissolved solids, but also include some colloids; a small portion of the dissolved solids are retained on the filter material. *See also* solids.

**filterable residue test** *See* total dissolved solids test.

**filter–adsorber** A device, such as a biological filter or a post-filter adsorber, that uses granular activated carbon (GAC) to remove suspended particles and adsorb organic compounds, particularly those responsible for taste and odor in drinking water. It is sometimes constructed simply by substituting GAC for all or a portion of the granular media of a rapid sand filter. *See also* mudball, a common operating problem in GAC filter adsorbers.

**Filter AG®** A granular filter medium produced by Clack Corp. for the adsorption of tastes and odors.

**filter agitation** An effective method for cleaning granular filters by removing attached materials from the medium, using water or a mixture of air and water at high pressure to agitate the media and loosen the materials. *See* surface washing for more detail.

**filter aid** (1) An agent added to the suspension to be filtered or placed on the filter media to improve the effectiveness of the filtration process by enhancing particle retention or increasing media permeability. Filter aids include fine materials such as diatomaceous earth (see precoat filter) and polymers. Same as filtration aid. (2) A flocculant in water treatment. *See also* coagulant/flocculant aid. (3) An inorganic agent used in wastewater sludge conditioning associated with pressure filtration. Conditioning agents used as a precoat in filter presses include fly ash, cement kiln dust, and pulverized coal.

**filter aiding polymer** Same as filter aid (1), usually added just ahead of filtration.

**filter air wash** *See* air wash.

**filter alum** The commercial, granular, powdered, or lump form of aluminum sulfate, a readily soluble coagulant used in water treatment. Also called alum or sulfate of alumina; formula: $Al_2(SO_4)_3 \cdot 14H_2O$ or $Al_2(SO_4)_3 \cdot 18H_2O$.

**filter appurtenances** Proper operation and control of a filter require a number of appurtenances: sluice gates and valves on all lines, venturi meters or other measuring devices, rate controllers, flow and headloss gages, sand-expansion and washwater controllers, etc. *See also* filter underdrain system, washwater trough, filter surface washing system, surface washers.

**filter area** The effective area through which water enters the filter media.

**filter background level** In reference to asbestos testing, the concentration of structures per square millimeter of filter that is considered indistinguishable from the concentration measured on a blank (a filter through which no air has been drawn) (EPA-40CFR763-AA-13)

**filter backwash** The process of removing particles retained on a filter media. *See* backwashing.

**filter backwashing** *See* backwashing.

**filter backwash rate** The rate at which flow is reversed through a filter to remove particles retained on the media, usually expressed in gpm/square foot of filter area.

**filter bag** A bag made of fabric used to improve the effluent quality of on-site wastewater treatment units by filtration; it reduces the suspended solids content to less than 30 mg/L.

**filter bed** (1) The layer of granular material through which a liquid passes for treatment. Typically, a granular filter bed is 2 ft deep, supported by a graded gravel layer over underdrains, in a 9 ft concrete box. (2) A pond or tank with a sand bedding, or a slow sand filter. (3) The media of a trickling filter.

**filter bed cracks and contractions** *See* cracks and contraction of filter bed.

**filter bed expansion** The flow of water through a filter underdrain to remove impurities retained on the media, thereby expanding the stationary bed by about 50%. The ratio of expanded bed depth ($L$) to the fixed-bed depth ($L_0$) is (AWWA, 1999):

$$L/L_0 = (1 - \varepsilon_0)/(1 - \varepsilon) \qquad \text{(F-40)}$$

where $\varepsilon_0$ and $\varepsilon$ are the porosities of the fixed bed and expanded bed, respectively.

**filter bed expansion formula** *See* filter bed expansion.

**filter belt** *See* belt filter.

**filter blanket** A layer of sand and gravel to retain fines.

**filter blinding** *See* blinding (1).

**filter block** A concrete or vitrified-clay block, usually rectangular and with openings in the upper face, used as underdrain in stone-media filters.

**filter bottom** The filter underdrainage system, collecting filtered water and distributing the backwash water. *See* underdrain (2) and (3) for more detail.

**filter box** The rectangular box containing the media and underdrains of a filter. *See* filter bed (1) and filter tank.

**filter breakthrough index** *See* breakthrough index.

**filter bumping** Releasing accumulated nitrogen gas in denitrification filters through backwashing. *See* bump and bumping for detail.

**filter cake** (1) The dewatered solids retained on the surface of a mechanical filter. (2) The layer of fine solids retained on a diatomaceous earth filter or any similar unit. Filter cake is sometimes called mud cake.

**Filter Cel®** Diatomaceous earth produced by Celite Corp. for use as filter medium.

**filter clogging** The effect of fine particles or biological growths filling the voids of a rapid sand filter or a biological bed. Clogged sand filters can be restored by removing and washing the sand, adding a detergent to the bed, or other methods.

**filter cloth** A fabric used as medium stretched around the drum of a vacuum filter.

**filter coefficient** A proportionality constant ($\lambda$) used in the phenomenological model of filtration, such that (AWWA, 1999):

$$\delta C/\delta z = -\lambda C \qquad \text{(F-41)}$$

$$\delta \sigma/\delta \theta = \lambda VC \qquad \text{(F-42)}$$

where $C$ = concentration of particles in suspension (volume of particles/volume of suspension liquid), $z$ = depth into the bed in the direction of flow, $\sigma$ = specific deposit (volume of deposited solids/volume of filter), $\theta$ = corrected time, and $V$ = approach velocity of flow.

**filter contactor** A water filtration device in which the conventional medium has been replaced by granular activated carbon for the dual purpose of adsorption/biodegradation and removal of particles.

**filter crack** Bed cracking is one of the difficulties of filter operation. Short-circuiting water pushes the granular material away from the walls and creates shrinkage cracks. Other cracks result from pressure differential or from inadequate cleaning, which leaves a thin layer of compressible matter around medium grains. Other operational problems include mud balls, jetting and sand boils, and sand leakage. *See also* dirty filter, cracks and contractions of filter bed.

**filter crib** A water supply intake consisting of a wooden crib in the bed of a stream, filled with gravel and covered with sand to the streambed level. Water is pumped through this device for further treatment and distribution.

**filter crops** Crops planted across a slope to slow runoff and reduce erosion.

**filter cycle** The operating time of a filter between backwashes. Same as filter run (1).

**filter drawdown** The lowering of free water above the media filter before backwashing.

**filtered wastewater** Wastewater effluent treated by granular filtration.

**filtered-water reservoir** A basin used to store enough filtered water to allow filtration at a constant rate even during peak hours. *See also* clearwell, clear-water reservoir.

**filter efficiency** The performance of a filter as measured by such parameters as turbidity or pathogen reduction. *See also* log reduction.

**filter element** The flat or tubular structure that supports the septum and medium of a precoat filter. A leaf is a rectangular or round, flat element.

**filter equations** *See* trickling filter equations.

**filter fabric** A pervious sheet of burlap or a synthetic material (polypropylene, nylon, polyester, or polyethylene yarn) with ultraviolet ray inhibitors and stabilizers, used to line the outlet area of a temporary sediment trap or for the protection of a storm drain inlet.

**filter feeder** An animal that feeds on particles and organisms strained out of water, e.g., barnacle, clam, coral, oyster, sponge.

**filter feeding** The feeding method of oysters and other bivalve mollusks, which concentrates pathogenic organisms from their surrounding water into their tissues.

**filter fence** A temporary fence used in construction sites for sediment control in shallow overland flow.It consists of a geotextile fabric stretched across a series of posts and supported by a wire fence, with a backfilled trench at the lower edge. Also called a silt fence.

**filter flooding** The filling of a trickling filter above the medium to control filter flies.

**filter floor** A false bottom or false floor that supports the filter bed and serves to collect filtered water as well as dispense washwater. It may be made of porous blocks, a porous plate, other material. *See also* false filter bottom.

**filter fly** A small, dark grey fly that is a nuisance around trickling filter beds, particularly during warm weather. These flies are mainly an irritation to operating personnel, but wind can carry them a long distance and they can pass through window screens. Their larvae breed in the zoogleal film on the filter media and on the inside retaining walls of the filter. Effective control measures include spraying the filter with insecticide and periodically submerging it to drown the larvae. Smooth plastic media do not commonly support many filter flies. Also called Psychoda, *Psychoda alternata,* or Psychoda fly.

**filter gallery** In a treatment plant or filtration installation, a gallery or passageway that provides access for the installation and maintenance of conduits and accessories. *See also* pipe gallery.

**filter head loss** Head loss through the filter medium indicates when to backwash; excessive head loss causes negative head and air binding. For formulas proposed to compute the clean-water head loss through a granular porous medium see the Blake–Kozeny, Carman–Kozeny, Ergun, Fair–Hatch, Hazen, and Rose equations. *See also* Coakley's equation, Kozeny formula, and porosity factor.

**filter head loss development** The determination of head losses in a clogged filter using one of the formulas proposed for clean water or an alternative formula that relates head loss to the progression of filtration or to the amount of solids removed (Metcalf & Eddy, 2003):

$$H_t = H_0 + \sum_{i=1}^{n} (h_i)_t \qquad \text{(F-43)}$$

$$(h_i)_t = a(q_i)_t^b \qquad \text{(F-44)}$$

where $H_t$ = total head loss at time $t$, ft; $H_0$ = total initial clear-water head loss, ft; $(h_i)_t$ = head loss in the $i$th layer of the filter at time $t$, ft; $(q_i)_t$ = amount of material deposited in the ith layer at time $t$, mg/cm$^3$; and $a$, $b$ = constants.

**filtering crucible** A small container of porcelain or other heat resistant material, with a filtering bottom, used in the total suspended solids test. Also called a Gooch crucible.

**filtering materials** *See* filter media

**filtering medium** *See* filter media.

**filtering to waste** *See* filter to waste.

**Filterite®** A cartridge filter manufactured by Memtec America Corp.

**filter jetting** *See* jetting (2).

**filter layers intermixing** The phenomenon that occurs when media grains of adjacent layers mix in a multimedia filter; e.g., the upper sand grains move into the lower anthracite grains of a dual-media bed. Intermixing tendency depends on the comparative bulk densities of adjacent layers. *See* bulk density (2), sand boil, gulf streaming, jet action.

**filter leaf** *See* filter element. Same as leaf.

**filter leaf test** A test conducted on a sludge sample to determine conditioner requirements for vacuum filtration. *See also* Atterberg test, Büchner funnel test, capillary suction time (CST) test, filterability constant, filterability index, shear strength, specific resistance test, standard jar test, standard shear test, time-to-filter (TTF) test.

**filter loading** (1) The hydraulic load on a granular filter or on a trickling filter, equal to the volume of water or wastewater applied per unit time per unit surface area, usually expressed in gpm/square foot or in mgd/acre. (2) The organic load on a trickling filter, equal to the weight of organic matter per unit time per unit surface area, expressed, for example, in pounds of BOD per acre per day.

**filter loading rate** Same as filter loading (1).

**filter loading, hydraulic** The flow applied per unit area of the filter, e.g., in gallons per square foot per day.

**filter loading, organic** The weight of organic matter applied per unit area of the filter per unit time, e.g., in pounds of BOD per square foot per day.

**filter mat** The layer of solids trapped on a step screen and enhancing the performance of the device.

**Filtermate** A filtration coagulant produced by Argo Scientific.

**filter media** (1) Materials used in filter beds include natural silica sand, crushed anthracite, crushed magnetite, garnet sand, ilmenite, crushed glass, slag, metallic ores, shredded coconut husks, burned rice husks, diatomaceous earth, and granular activated carbon. Common selection and design criteria are effective size and uniformity coefficient. *See also* area–voids volume ratio, manufacturer's rating, shape factor, and sphericity. (2) A cartridge that contains membranes, fabric, or other filtering materials. (3) A cloth or other suitable material used to intercept solids in sludge filtration.

**filter medium** *See* filter media.

**filter medium loss** The loss of filter material during backwashing. Other depth filtration operation problems include turbidity breakthrough, emulsified grease buildup, cracks and contraction of filter bed, and gravel mounding.

**filter mud** *See* mudball.

**FilterNet™** A fabric composite fabricated by SLT North America, Inc. for filters.

**filter operating table** A table set on the floor of a filtration plant to support the filter head loss and flow rate gauges.

**filter operation performance** *See* filter run and solids breakthrough.

**filter pack** A cylinder containing filtering materials installed around the intake screen of a well to retain fines, protect pumping equipment, and prevent screen clogging.

**filter packing** The material used in a trickling filter to serve as support for the development of the biofilm. Typical materials currently used include river rocks and plastics, all having a high surface area per unit volume, high durability, and high porosity.

**Filterpak** Plastic media produced by Mass Transfer, Inc. for biological filters.

**Filter-Pak™** A gravity sand filter of Graver Co.

**filter paper** Porous paper used in filtering, e.g., a paper used in wastewater analysis to measure suspended solids.

**filter photometer** An instrument that uses colored glass filters to obtain wavelengths for analysis. Also called electrophotometer.

**filter plant** *See* filtration plant.

**filter ponding** The formation of ponds on the surface of a trickling filter by medium degradation or other causes.

**filter pooling** The formation of wastewater pools on a clogged filter.

**filter press** A device that applies pressure between two plates or belts to dewater sludge from water or wastewater treatment, leaving a cake and a filtrate. *See* sludge pressing for names commonly used. *See also* plate-and-frame filter press, diaphragm filter press, hanging cake.

**filter-pressed sludge** Sludge dewatered in a filter press.

**filter rate** *See* filter loading.

**filter rating** A parameter used to define the sizes of particles that will be retained on a filter medium. For a given rating, the medium will retain a percentage of solid particles that are larger than the rating number in micrometers. This percentage is commonly taken as 85% for a nominal rating and 99.9% for an absolute rating. Also called micrometer rating or micron filter rating.

**filter resanding** The restoration of the original depth of a slow sand filter by replacing the sand lost after several scraping operations, which remove the schmutzdecke along with some sand. *See also* filter scraping.

**filter ripening** A phenomenon observed in water filtration, whereby the concentration of particles increases in the filter effluent immediately following backwash. The filter ripening period and extent of the phenomenon depend on influent characteristics, backwashing procedure, and other factors. Ripening coincides with the development of the biological surface coat called schmutzdecke. Filter ripening may be responsible for 90% of the particles passing through a filter. As shown in Figure F-03, the filter-ripening sequence (FRS) comprises five stages: a lag phase, media disturbance and intramedia remnant stage, upper filter remnant stage, influent mixing and particle stabilization stage, and the dispersed remnant and filter media conditioning stage. *See also* biological ripening period.

**filter-ripening sequence (FRS)** *See* filter ripening.

**filter run** (1) The operating time of a filter between backwashes; same as filter cycle. The length of filter runs is usually based on a limit on head losses or on the turbidity of the product water. *See also* air-bound filter, backwashing, floc breakthrough. (2) The time the medium of a sludge-dewatering filter is used until it is changed.

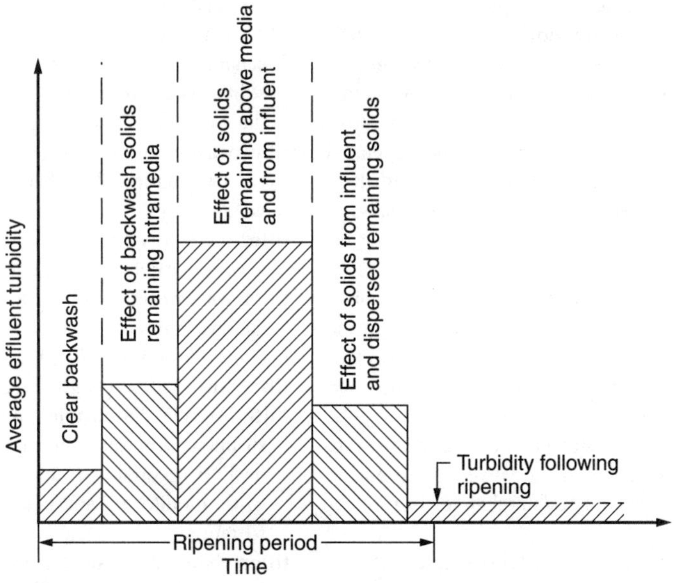


**Figure F-03.** Filter ripening (adapted from Amburgey et al., 2003).

**filters** Filtration plants are sometimes called "the filters."

**filter sand** Sand prepared specifically for use in a filter.

**filter sand boil** *See* sand boil.

**filter scraping** The cleaning of a slow sand filter by removing the schmutzdecke along with a small amount of sand, which is washed and stockpiled for later reuse. *See also* filter resanding.

**filter's efficiency** A measure of the performance of a filter or the capacity of a filter to retain solids, expressed as the mass of suspended solids removed per unit area of filter medium per unit of head loss developed. *See also* filtration efficiency and backwashing efficiency.

**filter septum** The filtering material in a surface filter, e.g., a woven metal fabric or cloth fabric. *See also* septum.

**filter shock treatment** The application of potassium permanganate ($KMnO_4$) as a solid or a concentration solution directly to the surface of a filter for the control of biological growths in a unit used for iron or manganese removal.

**FilterSil™** Sand from Unimin Corp. for filter beds.

**filter strainer** A perforated device installed in a rapid sand filter for the flow of filtered water and backwash water. Also called strainer head.

**filter strip** Strip or area of vegetation used for removing sediment, organic matter, and other pollutants from runoff and wastewater. Also, a stormwater pollution control measure consisting of a strip of permanent, close-growing vegetation that retards the flow of runoff and reduces the pollutant load to the receiving water. Used as an outlet or pretreatment device, as well as above dams, diversions, and other hydraulic structures. Also called buffer strip, grassed buffer, and vegetated filter strip. *See also* dry detention basin, infiltration basin, infiltration, trench, catch basin, sand filter, grassed swale, constructed wetland, wet retention basin

**filter surface washing system** A filter appurtenance used to remove attached material from the filter medium. Common surface washers are single-arm and dual-arm agitators, fixed or mounted on rotary sweeps.

**filter surging** The immediate increase in flow rate observed when changing the valve setting during backwash or to accommodate changes in production.

**filter tank** The concrete or steel structure that contains the media, gravel bed, underdrain, and other components. *See also* filter bed and filter box.

**filter-to-waste connection** A valved pipe stub or similar device that branches from the filter effluent so as to waste the product water after washing the bed with detergents or after backwashing. Also called rewash connection in older filters that are washed with raw water.

**filter-to-waste period** *See* filter-to-waste procedure.

**filter-to-waste procedure** A procedure during which, immediately after backwash, the operator diverts the filtrate (filter effluent) from the finished water reservoir to improve the quality of the water supplied. The filter-to-waste procedure usually lasts 5 to 10 minutes or until the quality of filtered water is acceptable, but may extend to 2 days when *Giardia lamblia* cysts are of concern. It is one technique used to reduce the effects of filter ripening (the phenomenon that causes an increase of particles in filter effluent immediately following backwashing). The filter-to-waste procedure is also used after cleaning the bed with detergents. *See also* slow start, delayed start, filter ripening.

**filter troughs** *See* filter washtroughs.

**filter underdrain** The filter underdrainage system, supporting the filtering materials, collecting filtered water, and distributing the backwash water and air. *See* underdrain (2) and (3) for more detail.

**filter unloading** The periodic sloughing of biological growth from the medium of a trickling filter.

**filter wash** Backwash; the reversal of flow through a rapid filter to remove impurities retained on the media.

**filter washbox** A baffled compartment incorporated in a moving-bed filter to allow cleaned sand to separate by gravity from the concentrated waste solids. The sand returns to the top of the bed and the solids are moved for disposal.

**filter washtroughs** Opened conduits installed on a filter media to collect backwash water. *See* washwater trough.

**filter-wash waste tank** A basin used to hold filter washwater before treatment and disposal.

**filter washwater** The wastewater that results from backwashing a filter, characterized by a solids concentration of 100–1000 mg/L.

**filter washwater trough** *See* washwater trough.

**FilterWorx™ Automatic Control System** An instrumentation and control package manufactured by F. B. Leopold of Zelienople, PA for the automatic monitoring and control of municipal water filtration.

**filter yield** A parameter used to describe the performance of a vacuum filter press; it is the mass of dry cake solids removed from the filter media per hour per square foot of filter. It is computed as follows, based on the specific resistance equation (AWWA, 1999):

$$Y = (2\ PWD)/UST)^{0.5} \quad \text{(F-45)}$$

where $Y$ = filter cake yield (kg/m$^2$/s); $P$ = pressure (N/m$^2$); $W$ = feed solids concentration, mg/L; $D$ = drum submergence (the fraction of the drum circumference below the sludge surface in the pan), dimensionless; $U$ = viscosity (Ns/m$^2$); $S$ = specific resistance, m/kg; and $T$ = cycle time (time for a complete revolution of the drum, s.

**Filtomat®** Cooling water filter designed by Orival, Inc. to be self-cleaning.

**Filtra-Matic™** A pressure leaf filter fabricated by U.S. Filter Corp.

**FiltraPak** A package wastewater treatment manufactured by Diagenex, Inc.

**Filtrasorb®** Granular activated carbon produced by Calgon Carbon Corp.

**filtrate** (1) The liquid passing through a filter or the liquid remaining after removal of solids through filtration. (2) The product water from a low-pressure membrane filtration process; *see also* permeate.

**filtration** (1) A process for removing particulate matter from water or wastewater by passage through porous media such as sand or a man-made filter; often used to remove particles containing pathogens. Following coagulation, when water or wastewater flows through a bed of granular materials, the flocculated solids are transferred onto the grains by straining, sedimentation, and interfacial contact, and then removed for disposal. *See also* air binding, backwashing, biological filtration, breakthrough, constant-rate filtration, declining-rate filtration, depth filtration, direct filtration, effective size, effective surface area, filter bed expansion, filtration mechanisms, filtration plant, fluidized bed, Galileo number, incipient fluidization, in-line filtration, minimum fluidization condition, minimum fluidization velocity, porosity, pressure filtration, rapid filtration, Richardson–Zaki equation, ripening filter, schmutzdecke, shape factor, slow sand filtration, sphericity, underdrainage, uniformity coefficient. (2) The natural passage of water through permeable formations to recharge groundwater or the passage of wastewater through natural soils for land treatment or land disposal. (3) Membrane filtration.

**filtration aid** *See* filter aid.

**filtration avoidance criteria for surface water** Requirements established by the USEPA under the Surface Water Treatment Rule for a public water system to avoid providing filtration.

**filtration avoidance determination (FAD)** A waiver granted by the USEPA, allowing a water supply agency to avoid the installation of filtration facilities. For example, to inactivate *Cryptosporidium* and other pathogens, New York City was allowed to provide ultraviolet disinfection instead of a more costly filtration plant.

**filtration driving force** Gravity or applied pressure through pumping. *See* gravity filter, pressure filter.

**filtration efficiency** A measure of the performance of a filtration operation, expressed in terms of turbidity or suspended solids remaining after treatment rather than perecent of solids removed. *See also* filter's efficiency, backwashing efficiency.

**filtration kinetics** During a filter run, filtration proceeds at a rate that depends on the initial rate, the porosity of the bed, and the rate of solids deposition. *See* Ives equation.

**filtration mechanisms** Mechanisms and phenomena that are responsible for or contribute to the removal of particulate matter within a granular filter include adhesion, biological growth, bonding, chance contact straining, chemical interaction, flocculation, inertial impaction, interception, mechanical straining, physical adsorption, sedimentation, transport, and attachment.

**filtration model** A mathematical representation of the straining and sorption processes of contaminant removal through a porous medium, assuming that the contaminant is particulate:

$$\partial C/\partial x = \lambda C \quad \text{(F-46)}$$

where $C$ = concentration of particulate contaminant (mass/volume, e.g., mg/L), $x$ = distance traveled through the porous medium (length, e/g/, m or ft), $\lambda$ = filtration coefficient (reciprocal of length, e.g.,/m or/ft), and $\partial C/\partial x$ = partial derivative of concentration with respect to distance.

**filtration plant** A water treatment plant that includes filtration and other processes. Although the plant includes other processes, to underscore the importance of filtration, it is often called by the type of filtration used, such as conventional, direct, slow sand, rapid sand, pressure, or diatomaceous earth. Also called a filter plant, or the filters.

**filtration pretreatment** A treatment process installed ahead of the filters to enhance filterability. *See* pretreatment (2) for more detail.

**filtration process variables** The following factors are considered in the design of filtration units: fluid characteristics, bed charge, media characteristics (grain size, shape, density, head loss characteristics), allowable head loss, bed depth, filtration rate, filtration pretreatment, removal efficiency, and characteristics of influent solids (concentration, composition, particle size).

**filtration rate** The rate of application of water to a filter, expressed as the ratio of the flow rate to the surface area of the filter, e.g., in gpm/ft$^2$ or in mgd/acre. *See also* standard or traditional filtration rate.

**filtration rate equation** An equation proposed to represent the rate of wastewater filtration as a function of eight variables and constants. *See* generalized filtration rate equation for detail.

**filtration spring** A spring that occurs where the water table meets the ground surface: water flows by gravity from numerous small openings in permeable materials or from openings in a rock formation. Also called a gravity spring or a seepage spring.

**filtration wastes** Wastes generated during filtration operations, including backwash water from rapid and precoat filters and dirty sand scrapings from slow sand filters.

**Filtroba®** Helical elements fabricated by Ketema, Inc. for pressure filters.

**Filtromatic** A traveling-bridge filter manufactured by Biwater Treatment, Ltd.

**Filtros®** Fine-bubble diffusers manufactured by Ferro Corp.

**fimbria** Singular of fimbriae.

**fimbriae** Small protein, hairlike appendages to bacterial cells involved in conjugation and other functions such as sticking to surfaces or attaching to each other. Also called pili, which are longer. *See also* cilia, flagella.

**final clarifier** Same as final sedimentation tank. Also called secondary clarifier in a secondary treatment plant.

**final clarifier criteria** Criteria commonly used in the design and operation of final (or secondary) clarifiers include overflow rate, solids loading rate, and weir loading rate.

**final cover** A 24–36 inch layer of earth deposited over a sanitary landfill upon closure and permanently exposed to the elements. *See also* daily cover.

**final disinfection** Disinfection of water as the last treatment step to maintain a disinfectant residual throughout the distribution system. *See* secondary disinfection for detail.

**final effluent** The effluent from the last unit of a wastewater treatment plant.

**final environmental impact statement** An environmental impact statement prepared, reviewed, and approved in accordance with the National Environmental Policy Act.

**final sedimentation** Settling of suspended solids in the last sedimentation basin of a wastewater treatment plant.

**final sedimentation tank** The last sedimentation tank in a wastewater system, e.g., following a trickling filter or an aeration basin. It corresponds to a secondary clarifier in a secondary treatment plant. Also called final clarifier or final settling tank.

**final settling tank** Same as final sedimentation tank.

**final treatment** A stage of water treatment, beyond sedimentation and filtration, that includes such processes as disinfection, fluoridation, and pH control.

**financial indicators** *See* benefit, benefit–cost analysis, benefit/cost ratio, break-even analysis, capital recovery factor, cost–benefit analysis, cost/benefit ratio, cost-effective alternative, cost-effectiveness analysis, discount rate, internal rate of return, life-cycle costing, net present value, payback period, present value (or worth) factor, profitability index, return on investment, sustainability.

**finding of no significant impact (FNSI)** A document prepared by a federal agency showing why a proposed action would not have a significant impact on the environment and thus would not require preparation of an environmental impact statement. An FNSI is based on the results of an environmental assessment.

**fine aggregate** Sand.

**FineAir** A fine-bubble diffuser fabricated by Parkson Corp. of ceramic.

**fine-bubble aeration** Diffused aeration of water or wastewater using fine bubbles. *See also* coarse-bubble aeration.

**fine-bubble diffuser** An old classification of devices used in the diffused aeration of wastewater according to the size of the bubbles produced. The current tendency is to divide diffusers into porous (or fine-pore) diffusers, nonporous diffusers, and other diffusers.

**fine-pore diffuser** A device made from ceramic, plastic, rubber, or cloth materials for the diffusion of air to water or wastewater. Also called porous diffuser. *See* disk, dome, membrane, panel diffusers. *See also* nonporous diffuser, jet aeration, U-tube aeration, coarse-bubble diffuser, fine-bubble diffuser.

**fineness coefficient** *See* coefficient of fineness.

**fine particulate detritovores** *See* collector (6).

**fine rack** A rack with clear spaces of one inch or less between bars.

**fines** (1) The smallest or finer-grained particles of a mass of soil, sand, or gravel; the solid material that settles last in a body of water. (2) Crushed stone or ore that passes through a given screen.

**fine sand** Sediment particles with diameters ranging from 0.10 mm to 0.25 mm. *See also* soil classification and sand.

**fine screen** (1) A screen with openings generally smaller than one inch (25 mm). Fine screens are used in preliminary or primary wastewater treatment, ahead of trickling filters, or for the treatment of combined sewer overflows. Water intake screens have openings smaller than ¼ inch (6 mm). Very fine screens have openings of 0.2 to 1.5 mm. *See* drum screen, horizontal reciprocating screen, static wedgewire screen, step screen, tangential flow screen. (2) In wastewater treatment, (rotating drum or fixed surface) fine screens have openings smaller than 1/16 inch. They are sometimes used for primary treatment in lieu of sedimentation, but not with the same efficiency, or to upgrade primary clarification units. *See also* coarse screen, microscreen.

**fine screen head loss** The head loss through fine screens depends on the size and amount of solids in the influent wastewater as well as aperture size and cleaning frequency and method. For a given screen, the clear-water head loss is provided by the equipment manufacturer or estimated from the following equation (WEF & ASCE, 1991):

$$H_L = (\tfrac{1}{2} g)(Q/CA)^2 \qquad \text{(F-47)}$$

where $H_L$ = head loss, ft; $g$ = gravitational acceleration = ft/sec$^2$; $Q$ = discharge through the screen, cfs; $C$ = discharge coefficient for screens, typically 0.6 for clean screens; and $A$ = effective open area of submerged screen, ft$^2$.

**fine screening** A wastewater treatment process designed to remove small particles.

**fine screenings** Small particles removed from wastewater on screens with openings less than 6 mm or 0.25 in. They include rags, paper, plastics, grit, food waste, fecal materials, etc. Typical materials from a fine rotary drum screen have the following characteristics: 80–90 % moisture content, specific weight of 900–1100 kg/m$^3$, volume of 6 ft$^3$/MG (million gallons).

**fine sediment discharge** *See* fine sediment load.

**fine sediment load** The portion of the suspended solid load composed of smaller particles than those generally found in the streambed. These smaller particles are transported without deposition. Typically in a sand-bed stream, the fine sediment load consists of particles finer than sand (0.062 mm). Also called fine-sediment discharge, washload, or washload discharge. *See also* sedimentation terms.

**finger channel** A long effluent channel extending into a sedimentation basin to increase weir length and reduce the vertical velocity as the liquid rises up to the overflow channel.

**finger structure** A typical structural profile of membranes used in water filtration showing well-defined cavities in the shape of fingers. *See also* sponge structure.

**finished drinking water** *See* finished water (1).

**finished water** (1) Water that has received proper treatment and official certification by health authorities, and is ready for delivery to consumers; potable water. Also called product water. *See also* raw water. (2) Processed water used to remove waste plastic material generated during a finishing process or to lubricate a plastic product during a finishing process. It includes water used to machine or assemble intermediate or final plastic products (EPA-40CFR463.2-e). Also called finishing water.

**finished-water reservoir** In a water supply system, it is a reservoir that holds treated water before it is distributed to customers.Also called clean-water reservoir. *See also* clear well.

**finishing water** Processed water used to remove waste plastic material generated during a finishing process or to lubricate a plastic product during a finishing process. It includes water used to machine or assemble intermediate or final plastic products (EPA-40CFR463.2-e).

**fire algae** A group of unicellular algae, most of which have flagella. They constitute a large part of plankton in seas and freshwaters, including the Dinophyceae that can cause red water when abundant. Also called Pyrrophyta.

**fire cistern** A cistern that stores water for fire-fighting as a backup or a supplement to the distribution system.

**fire demand** The amount of water required for fighting fires, representing a small quantity annually but a high rate (McGhee, 1991):

$$F = 18\, C(A)^{0.5} \qquad \text{(F-48)}$$

where $F$ = required fire flow, gallons per minute; $C$ = a coefficient depending on the type of construction, varying from 0.6 for fire-resistant materials to 1.5 for wood frame; and $A$ = total floor area in square feet, excluding basements. The fire demand may also specify the duration (in hours) for which the flow is required.

**fire demand rate** Same as fire demand, but with an indication of the pressure required at a certain location.

**fire flow** The flow that a water distribution system can deliver to fire-fighting pumpers at a specified pressure, e.g., 20 psi.

**fire flow demand** The fire demand determined by a fire department.

**Fire Flow Tester™** A device manufactured by Hydro Flow Products, Inc. for use in fire flow testing and flushing.

**fire hydrant** A hydrant used in firefighting. Also called a fire plug.

**fire line service charge** A charge levied for providing a water line for fire protection, separate from the domestic water charge. *See also* inch-foot charge for water service.

**fire plug** (1) A fire hydrant. (2) A wooden plug that stops a hole in a wooden pipe that is removed to obtain water for fire fighting.

**fireside wash** In the power generation industry, a fireside wash consists of washing the heat transfer surfaces with large amounts of water.

**fireside waste** In the power generation industry, fireside wastes include (a) the impurities that form gaseous and solid combustion by-products and (b) the solid residues from the incomplete combustion of fuels and from the corrosion of boiler internals.

**firm pumping capacity** The capacity of a pumping station or other system with the largest unit out of service.

**firm yield** *See* safe yield.

**firn** Granular snow on top of a high mountain that is compacted into glacial ice as a result of melting, refreezing, and other processes; a field of such snow. Also called névé.

**first-class water quality** The quality required of water that is used for drinking and cooking, a total annual average of approximately 8 liters per person per day or less than 2% of domestic water consumption. *See also* second-class water quality, third-class water quality.

**First Contaminant Candidate List (CCL1)** Same as Contaminant Candidate List (CCL).

**first draw** The water that immediately comes out when a tap is first opened; it is likely to have the highest level of lead contamination from plumbing materials.

**first-draw residential lead sample** A first-draw sample as required by the USEPA's Lead and Copper Rule.

**first-draw sample** A one-liter sample collected in accordance with CFR Section 141.86(b)(2) from tap water that has been standing in plumbing pipes at least 6 hours, without flushing the tap (EPA-40CFR141.2). Such a sample is used to determine whether prolonged exposure to a pipe or faucet causes high levels of a contaminant. Also called first-draw residential lead sample, first-draw tap sample, standing sample. *See also* flowing sample.

**first-draw tap sample** Same as first-draw sample.

**first flush** In stormwater and combined sewer overflow management studies, the first flush refers to the action of the first storm of the rainy season (and less often to the first part of any storm). The first flush usually carries an unusually high pollution load, consisting of sediment from soil erosion; oil, grease, and heavy metals from automobiles; and nitrates and phosphates from fertilizers. For the first flush of a storm event, pollutant concentration decreases as the flow duration increases. *See* continuous deflection separator as a device that captures the first flush.

**first-flush effect** The result of the first-flush phenomenon, i.e., higher pollutant concentrations in runoff at the beginning of a storm event or at the beginning of the rainy season. For example, concentrations of $BOD_5$ and TSS may drop by two-thirds within the first hour of a storm event. First-flush loads are a primary target in modeling studies for the abatement of stormwater pollution; they are affected by combined sewer slopes, design and cleaning frequency of streets and catch basins, rainfall intensity and duration, and surface buildup of contaminants and debris. *See also* event mean concentration.

**first in time, first in right** An expression of the doctrine that older water rights have priority over more recent rights.

**first law of thermodynamics** Energy cannot be created or destroyed, but it can be changed from one form to another (electrical, chemical, thermal, nuclear, etc.). Also called law of conservation of energy. *See also* zeroth law, second law, third law of thermodynamics.

**first-order consumer** In an aquatic food chain, first-order consumers are animals, e.g., fly nymphs, copepods, and water fleas, which consume plants and are in turn eaten by second-order consumers.

**first-order decay** Decay according to a first-order equation.

**first-order equation** An equation representing a rate of change proportional to the variable itself;

commonly used for microorganism survival in water as:

$$dC/dt = -kC \quad \text{(F-49)}$$

or

$$C = C_0 e^{-kt} \quad \text{(F-50)}$$

where $C$ = concentration of microorganisms at time t (e.g., number per 100 mL), $k$ = first-order decay or die-off rate constant (reciprocal of time, e.g.,/hour), and $C_0$ = initial concentration of microorganisms (number per 100 mL at time t = 0). *See also* first-order reaction.

**first-order half-life ($t_{1/2}$)** The time required for the concentration of a substance to be reduced to one-half its initial value (EPA-40CFR796.3780-vii).

**first-order kinetics** The characteristic of a first-order reaction.

**first-order reaction** A reaction in which the rate of disappearance of a chemical is directly proportional to the concentration of the chemical and is not a function of the concentration of any other chemical present in the reaction mixture (EPA-40CFR796.3700-v). It is the type of reaction that occurs most frequently in chemical and biological processes of water and wastewater. For example, in stormwater modeling studies, a first-order reaction is sometimes assumed for the removal efficiency of the settling process:

$$C = C_0 \cdot e^{-kt} \quad \text{(F-51)}$$

where $C_0$ and $C$ are pollutant concentrations initially and at time $t$, respectively, and $k$ is the reaction rate constant. Actually, the rate of change ($r$) is:

$$r = dC/dt = -kC \quad \text{(F-52)}$$

or

$$r = -k(C - C_s) \quad \text{(F-53)}$$

where $C_s$ is a saturation concentration. First-order kinetics is assumed for other processes such as radioactive decay, mixing, flocculation, activated sludge, and extended aeration.

**first-order reaction kinetics** *See* first-order reaction.

**first-order retarded reaction** A first-order reaction in which the rate constant ($r$) changes with time ($t$), distance, or degree of treatment, and depends on a retardation factor:

$$r = \pm kC/(1 + \alpha t)^n \quad \text{(F-54)}$$

where $k$ = reaction rate constant, $C$ = concentration at time $t$, $\alpha$ = rate factor due to concentration change, and $n$ = an exponent related to particle size distribution in wastewater treatment. The quantity $[1/(1 + \alpha t)^n]$ is a retardation factor, acting to reduce reaction rate over time. The first-order retarded reaction has been proposed to express the change of concentration due to depth filtration, the rate $r$ decreasing with distance $z$ from the top of the filter bed (instead of time $t$). *See also* retardant reaction.

**first-pass effect** The reduction of the concentration of a chemical, i.e., the fraction cleared from the blood, the first time the chemical passes through the liver.

**first-stage biochemical oxygen demand** The oxygen demand exerted by carbonaceous matter, which usually occurs before the oxidation of nitrogenous matter, or second stage, begins. Also called ultimate biochemical oxygen demand or ultimate carbonaceous biochemical oxygen demand (UBOD, $BOD_L$, or $BOD_u$). *See also* carbonaceous biochemical oxygen demand, nitrification.

**first-stage BOD** *See* first-stage biochemical oxygen demand.

***Fischerella*** A genus of cyanobacteria that produce geosmin, an odor-causing substance in drinking water sources.

**FISH** Acronym of fluorescent in situ hybridization.

**fish bioassay** Determination of the biological activity or potency of a substance by testing its effect on the growth of fish.

**fish elevator** A structure that allows fish to pass upstream around over a dam. *See also* fish ladder.

**fishery** A place where fish, shellfish, and other living aquatic resources are bred, processed, or sold. A capture fishery exploits resources whose production is controlled by natural events but may be impacted by human actions. In aquaculture or fish farming, an operator controls the production of fish and other aquatic organisms.

**fisheyes** The lumps of improperly mixed polymer in water or the resulting condition.

**fish farm** A facility in which fish are bred for commercial purposes; same as fish ranch.

**fish farming** The production of fish for commercial purposes, one of three important forms of aquaculture; same as fish ranching. Breeding fish in ponds enriched by wastewater or nightsoil has been or was practiced for centuries in such places as China, Egypt, and Europe. The addition of wastes causes an increase in the population of bacteria, algae, and zooplankton, on which a variety of fish (carp, tilapia, etc.) feed. This practice raises health issues, however, related to the transfer of animal pathogens and helminths through fish and other intermediate hosts. *See* aquaculture, algal culture, macrophyte production.

**fish kill** The sudden destruction of large quantities of fish caused by insufficient dissolved oxygen, excessive temperature, toxic discharges, or other conditions. Sudden algal blooms or massive deaths of algae often cause fish kills. Some effective weed killers are also toxic to fish.

**fish ladder** An inclined structure that allows fish to swim upstream of a dam to reach their spawning grounds; e.g., a series of progressively lower baffled chambers that reduce the velocity of the water. *See also* fish elevator.

**fish or wildlife** Any member of the animal kingdom, including any mammal, fish, bird, amphibian, reptile, mollusk, crustacean, arthropod or other invertebrate, and including any part, product, egg, or offspring thereof, or the dead body or part thereof (Endangered Species Act).

**fish pond** A small natural or artificial pond containing fish raised for commercial purposes.

**fish processing waste** Wastewater generated during the processing of fish and originating from centrifuges, fish pressing, evaporators, and other wash waters. It is very high in BOD, total organic solids, and odor. This waste stream is usually treated by evaporation and the residue barged to sea.

**fish pump** A special pump with sufficient clearance between the vanes or a bladeless impeller to pass fish and other objects; sometimes used as a nonclog pump for untreated wastewater.

**fish ranch** A facility in which fish are bred for commercial purposes; same as fish farm.

**fish ranching** The production of fish for commercial purposes. *See* fish farming for more detail.

**fish screen** (1) A screen installed at the head of or inside an intake pipeline or canal to prevent fish from entering. *See also* velocity cap. (2) A moving screen designed to remove impinged fish and return them to the water body.

**fish tapeworm** An excreted helminth infection caused by the pathogen *Diphyllobothrium latum.* Distributed worldwide, it is transmitted from the feces of infected persons or animals through intermediate hosts (copepod and fish). It is the common name of diphyllobothriasis.

**fish toxicity** Impairment of fish development or immediate death of fish caused by a toxic substance.

**fishway** A structure that allows fish to swim upstream and downstream of a dam. It may consist of a series of ponds of low falls, one above the other. *See also* fish elevator and fish ladder.

**fission** (1) The spontaneous or induced splitting of the nucleus of an atom into two or more nuclei of lighter elements, with a concurrent release of a relatively large amount of energy and, often, production of neutron and gamma rays. Also called nuclear fission. *See also* fusion. (2) A form of asexual reproduction in which the parent organism splits into two independent organisms.

**Fitch Feedwell** A horizontal feedwell manufactured by Dorr-Oliver, Inc. for clarifiers.

**fittings** Pipe fittings are connections in a piping system that modify the size or the direction of the conduits. Velocity or direction changes cause minor (head) losses proportional to the velocity head. Besides valves, orifices, nozzles, and venturi meters that also cause minor head losses, examples of fittings include elbows, bends, standard Ts, basket strainers, couplings, unions, reducers, and increasers.

**five-day BOD ($BOD_5$)** The oxygen demand exerted in five days and usually associated with the biochemical oxidation of carbonaceous matter. *See also* first-stage BOD.

**five-day BOD test** *See* BOD test.

**five hundred series methods** Analytical methods published by the USEPA for drinking water.

**five hundred (500-) year floodplain** The area, including the base floodplain, that is subject to inundation from a flood having a 0.2% chance of being equaled or exceeded in any given year. Also called the 0.2% chance floodplain.

**five-stage Bardenpho™ process** A wastewater treatment process designed to accomplish carbonaceous BOD removal as well as enhanced phosphorus uptake and nitrification–denitrification. *See* modified Bardenpho™ process for detail.

**fix (a sample)** To add chemicals to a sample in the field to prevent the water quality indicators of interest in the sample from changing before final measurements are performed later in the laboratory.

**fixatif** Same as fixative.

**fixation** (1) Conversion of a soil nutrient essential for plant growth from a soluble, available form to an insoluble, unavailable form. (2) Conversion of nitrogen into a useful compound such as nitrate fertilizer. *See also* nitrogen fixation. (3) Reduction from a volatile or fluid to a stable or solid form; stabilization or solidification of a waste material by involving it in the formation of a stable solid derivative. *See* chemical fixation.

**fixative** Formaldehyde, ethanol, or other chemical substance used to preserve cells or other material for analysis. Also called fixatif.

**fixed assets** Assets of a long-term or relatively permanent nature, e.g., land, buildings, patents.

**fixed bed** A steel or concrete vessel that contains stationary filtration material such as sand, activated carbon, or ion-exchange resin. Water or wastewater flows downward and the bed is periodically backwashed to remove particulates accumulated at the top of the medium, which is supported by a layer of sand and gravel on a filter block. *See also* expanded bed, moving bed.

**fixed-bed adsorber** A granular activated carbon (GAC) adsorber, used in water or wastewater treatment, with a stationary medium; it operates in a downflow or upflow mode. A conventional rapid sand filter may be converted to a fixed-bed adsorber by replacing the sand with GAC as granular medium. *See also* fluidized-bed adsorber, pulsed-bed adsorber, upflow expanded-bed mode.

**fixed-bed adsorption** A process for the removal of taste and odor compounds using a fixed-bed adsorber.

**fixed-bed column** A vessel containing stationary media for the treatment of water or wastewater in a downflow mode (e.g., gravity filters, ion-exchange columns, activated carbon columns), with periodic media expansion to remove accumulated impurities. Fixed-bed columns (or contactors) of granular activated carbon in series or in parallel are sometimes used downstream of granular-medium filters to remove suspended solids from secondary effluents.

**fixed-bed porosity** The ratio of void volume to total bed volume of a granular media filter, expressed as a decimal fraction or a percentage. It depends on grain sphericity and affects backwash rate, head loss, and the capacity of the medium to retain solids.

**fixed-bed treatment** A process that uses a fixed bed to treat water or wastewater, e.g., for the removal of perchlorate.

**fixed capital** The investment in fixed assets.

**fixed charges/fixed costs** Charges/costs that do not depend on the level of production or that cannot be avoided or shifted, e.g., rent, interest, amortization, executive salaries.

**fixed cost** *See* fixed charges.

**fixed cover** A stationary cover that maintains the volume of an anaerobic digester constant, including a free space between the roof and the liquid surface. Gas storage is provided separately so that air will not be drawn into the digester and cause an explosion when the liquid volume changes. *See also* floating cover, membrane gas cover.

**fixed digester cover** *See* fixed cover.

**fixed dissolved solids (FDS)** The residue remaining after total dissolved solids (TDS) are ignited to 500 ± 50°C.

**fixed distributor** A stationary apparatus of perforated pipes, notched troughs, or sprinkler nozzles used to apply water or wastewater to the surface of a filter or contact bed. *See* distributor for detail.

**fixed film** *See* fixed growth.

**fixed-film activated sludge** An aeration basin that includes a fixed medium for microorganisms to attach to and grow.

**fixed-film process** *See* fixed-growth process.

**fixed-film process efficiency** The treatment performance of a fixed-film anaerobic reactor (e.g., an upflow anaerobic filter) in terms of COD removal (Freeman, 1998):

$$E = 100\,(1 - T^{-m}) \qquad \text{(F-55)}$$

where $E$ = percent COD removal, $T$ = hydraulic retention time (days), and $m$ = coefficient varying from 0.5 to 1.0 depending on the process.

**fixed-film reactor** A reactor that uses a fixed medium as a supporting surface for bacterial growth, with a resulting high solids retention time (e.g., > 100 days). It can be an upflow reactor (also called a biofilter), downflow reactor, or fluidized or expanded bed reactor. *See also* fixed-film activated sludge and fixed-growth process.

**fixed groundwater** Water held in saturated material and not available for water supply.

**fixed growth** One of two types of microbial development used in biological wastewater treatment; microorganisms grow attached to natural or artificial media and use the contaminants of the liquid stream as substrate. The other type is suspended growth. Some treatment techniques use both types of growth in series or in the same reactor. Also called attached growth or fixed film. *See also* biofilm and fixed-growth process.

**fixed-growth process** A biological wastewater treatment process in which the microorganisms are attached to rock, slag, sand, redwood, plastic or other inert packing materials. When wastewater is sprayed over the fixed media, it produces a biological slime or biofilm consisting mainly of bacteria, protozoa, and fungi that use oxygen, organic matter, and other waste constituents for food and energy, releasing carbon dioxide. These transfers occur mainly by diffusion. *See* Figure F-04. Fixed-film processes are characterized by high solids retention times, resistance to shock loads, energy efficiency, and low maintenance requirements. Examples: trickling filter, intermittent sand filter, rotating biological contactor (RBC), packed-bed

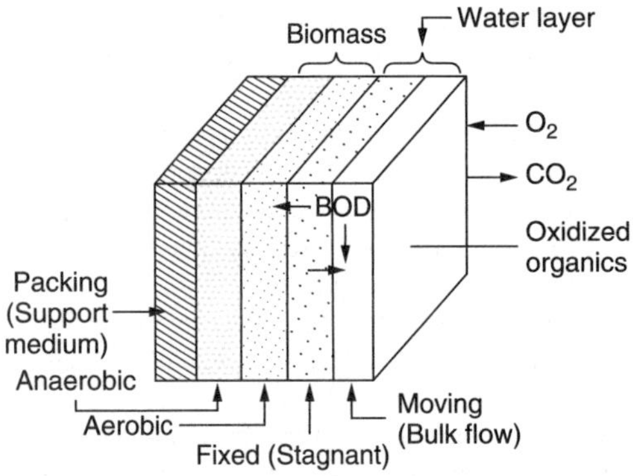


**Figure F-04.** Fixed-growth process (adapted from McGhee, 1991, and Metcalf & Eddy, 2003).

reactor, biological activated filter, and anaerobic packed or fluidized bed. Also called attached-growth process, biofilm process, or fixed-film process. *See also* stagnant liquid layer, suspended-growth process, hybrid processes.

**fixed-growth system** A wastewater treatment system that uses a fixed-growth process.

**fixed liabilities** A liability or debt that does not mature in the short term, e.g., a mortgage or debenture.

**fixed matter** The residue of particulate and dissolved materials remaining in a water or wastewater sample after heating the sample to 600°C or burning it to drive off the volatile solids. Also called fixed solids.

**fixed-media filter** A device used for the biological treatment of wastewater; a trickling filter.

**fixed nozzle** A stationary nozzle used to spray water or wastewater, e.g., nozzles for backwash water or surface wash.

**fixed-nozzle distributor** A device consisting of fixed nozzles for the continuous dosing of wastewater over a trickling filter medium, as compared to a rotary distributor that provides intermittent dosing.

**fixed-orifice diffuser** A nonporous device used for the aeration of wastewater. *See* orifice diffuser and slotted tube.

**fixed sample** A sample is fixed in the field by adding chemicals to prevent water quality indicators of interest in the sample from changing before laboratory measurements are made.

**fixed screen** A fine-to-medium screen that uses a stationary, inclined deck as a sieve to remove solids from wastewater, following coarse bar screens in small or industrial waste treatment plants. *See* static screen for more detail.

**fixed solids** The residue of particulate and dissolved materials remaining in a water or wastewater sample after heating the sample to 600°C or burning it to drive off the volatile solids. Also called fixed matter.

**fixed suspended solids** The inorganic content of suspended solids in water or wastewater determined after ignition or heating of total suspended solids at 500 ± 50°C. *See* fixed matter.

**fixed-volume filter press** A sludge dewatering device consisting of plates or trays installed rigidly in a frame and pressed together between a fixed end and a moving end. Each plate or tray is covered with a filter cloth as medium.

**fixed-volume, recessed-plate filter press** A device that applies pressure to dewater sludge. It consists of a series of two-sided recessed plates, supported vertically on a frame and held together by powered screws, with a filter cloth fitted over each frame. *See also* sludge pressing.

**fixed water layer** The stagnant liquid layer that separates the biofilm from the bulk liquid in a biological attached growth treatment process. Through this layer, substrate, oxygen, and nutrients diffuse to the biofilm, while biodegradation products diffuse to the bulk liquid. Also called bound water layer, diffusion layer, stagnant liquid layer. *See* fixed-growth process.

**fixture** *See* plumbing fixture and fixture unit.

**fixture branch** The pipe that supplies water to a fixture from a distribution pipe.

**fixture count** The total number of plumbing fixtures or water outlets in a building; used to estimate peak flow rates and size equipment.

**fixture unit** A parameter used to estimate domestic and institutional water uses and corresponding wastewater generation. The fixture unit method assumes a flow per plumbing fixture. This term includes installed receptacles, devices, or appliances either supplied with water or receiving on discharge liquids or liquidborne wastes, or both. Examples of plumbing fixtures are toilets, sinks, bathtubs, dishwaters, and drinking fountains. The flow rate per fixture varies widely, e.g., from 17 gallons per hour for a restaurant kitchen sink to 150 gallons per hour for a public park shower.

**flagella** Plural of flagellum.

**flagellate** A microorganism (e.g., *Giardia, leishmania,* and other protozoans) that moves by the action of tail-like projections called flagella. Flagellates are one of the groups active in the decomposition of organic matter in wastewater. *See also* ciliate.

**flagellum** (plural: **flagella**) A hairlike protein

thread of cytoplasm that projects from the surface of a cell. Many organisms (gametes, spores, rod-shaped bacteria) use flagella for locomotion. *See also* cilium.

**flame arrester** A screen or tube installed in a vent or pipe to protect it against flames.

**flame atomic absorption spectrophotometry** A highly sensitive instrumental technique for measuring trace quantities of elements in water when the concentrations are on the order of milligrams per liter; the element being measured is atomized in a flame. Also called conventional atomic absorption spectrophotometry. *See also* atomic absorption spectrophotometry and graphite furnace atomic absorption spectrophotometry.

**flame ionization** A method used to detect trace organic chemicals in water or wastewater samples. *See also* electron capture, mass spectrometry, thermal conductivity.

**flame ionization detector (FID)** A device widely used with gas chromatographic columns to identify and quantify organic compounds. It consists of a gas diffusion flame burning at the end of a jet. Organic compounds that enter the flame from a column form charged intermediates that are collected and measured through voltage application. Other detectors use electrolytic conductivity, electron capture, or photoionization.

**flameless atomic absorption spectrophotometry (or spectroscopy)** Same as graphite furnace atomic absorption spectrophotometry.

**flameless thermal oxidation** A technology patented by Thermatrix, Inc., using an inert ceramic medium in a refractory-lined reactor, for the destruction of organic gases and vapors from industrial operations, including the treatment of hazardous wastes.

**flame polished** Melted by a flame to smooth out irregularities. Sharp or broken edges of glass (such as the end of a glass tube) are rotated in a flame until the edge melts slightly and becomes smooth.

**flame temperature** *See* theoretical flame temperature.

**flaming** Using a flame to disinfect an object; e.g., flaming the end of a faucet before collecting a water sample for analysis.

**flammable, inflammable** (1) Pertaining to objects that are easily set on fire, combustible, inflammable, or that burn easily, strongly and at a rapid rate. (2) A flammable substance, which can cause fires and explosions.

**flammable liquid, inflammable liquid** (1) A liquid with a flash point below 100°F or about 38°C. (2) A liquid, mixture of liquids, or liquids containing solids in suspension or solution that give off an inflammable vapor at a temperature of not more than 60.5°C in a closed-cup test or not more than 65.6°C in an open-cup test.

**flange** A projecting rim, edge, lip, or rib used for attachment to another object.

**flanged joint** Flanges bolted together to form a pipe joint.

**flanged pipe** A pipe with flanged ends.

**flap gate** A backwater gate consisting of a flap hung against an inclined seat; it opens or closes by rotation around hinges at the top for a single leaf or at the side for a double leaf. It is used as a combined sewer overflow outlet discharging below the high-water level of the receiving water. *See* Figure F-05, elastomeric check valve.

**flap valve** A device that rotates about a hinged flap to open in the direction of flow or to close when flow reverses.

**flare** A control device that burns hazardous materials to prevent their release into the environment; may operate continuously or intermittently, usually on top of a stack. In the field of wastewater treatment, steam- or air-assisted and pressure-head flares are used for the disposal of waste digester gas and the destruction of volatile organic compounds found in off-gas streams.

**flash** The portion of a fluid that vaporizes when its pressure drops below the saturation pressure.

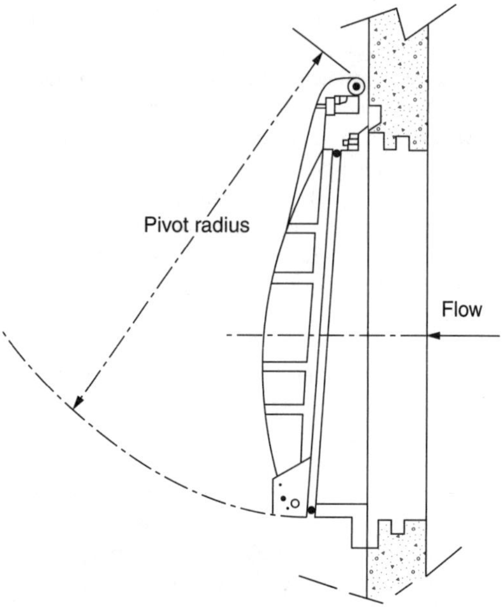


**Figure F-05.** Flap gate (adapted from Metcalf & Eddy, 1991).

**flash distillation** A continuous distillation process that uses a single stage, with the liquid and vapor leaving in equilibrium. For example, hot, but not boiling, water at high pressure is introduced into a lower-pressure chamber, causing some of the water to evaporate quickly into steam. Also called simple continuous distillation. *See also* batch distillation, binary distillation.

**flash dryer** A device that uses hot air or superheated vapor for vaporizing water from sludge pulverized in a cage mill or by an atomized suspension technique. It consists of a furnace, mixer, cage mill, cyclone separator, and vapor fan. Also called pneumatic conveyor dryer. *See also* direct dryer, fluid bed dryer, rotary dryer, indirect dryer.

**flash drying** A method of sludge drying; sludge is dispersed in a stream of hot, dry air or other gas to separate the solids, which may be used to provide part of the required heat. Moisture is transferred from the sludge to the gases. Cyclone separators separate the dried sludge from the moisture-carrying gases. The dried sludge is used as a fertilizer or soil conditioner, or is incinerated autogenously with wet sludge. *See also* pelletization.

**flash-drying incineration** A heat treatment method for sludge; flash drying.

**flash evaporation** Vaporization of a solvent by introducing the solution into an enclosed vessel pressurized below the saturation vapor pressure of the solvent, the heat for evaporation coming from the cooling of the liquid; one of two modes of evaporator operation. Also called flash vaporization. Used in gas chromatography to convert volatile constituents into gases. *See also* boiling heat transfer.

**flash evaporator** A device used in flash distillation, by which saline water is vaporized under vacuum through pressure reduction.

**flashing** The conversion of a portion of a fluid into vapor by pressure reduction instead of temperature elevation. *See* multistage flash evaporation.

**flashing point** *See* flash point.

**flash mix** (1) Same as flash mixing. (2) A method used for initial breaking of emulsion polymers in chemical sludge conditioning. It consists of rapidly combining the polymer with water, bringing the mixture first into a high energy state, and then releasing it at a lower velocity. It produces a uniformly blended solution in one second.

**flash mixer** A device used in flash mixing, i.e., to quickly mix or disperse chemicals throughout a liquid. One common type is a concrete tank equipped with an impeller, e.g., a turbine with flat or curved blades connected to a vertical or horizontal shaft. Important design parameters of flash mixers include detention time and circulation capacity.

**flash mixing** A method of mixing or dispersing coagulants, softening or other chemicals instantly or rapidly (e.g., in one minute), using mechanical means such as motor-driven devices. In the flocculation process, rapid mixing is followed at once by rapid agitation and initiates the mechanism of particle aggregation. Also called rapid mixing. *See also* flash mix (2) and velocity gradient.

**flash point** The lowest temperature at which a substance (e.g., a gas or volatile liquid) ignites momentarily on application of a flame. A substance with a flash point of 100°F or less is considered dangerous; with a flash point above 200°F, the substance is said to have low flammability. Also called flashing point. *See* upper explosive limit, lower explosive limit.

**flash range** The difference between the maximum brine temperature and the temperature of the brine in the last stage of a flash evaporator.

**flash tank** A component of an evaporator system consisting of a vertical cylinder enclosed at both ends, with an orifice in the feed line to cause flashing.

**Flashvap** A flash evaporator manufactured by Licon, Inc.

**flash vaporization** *See* flash evaporation.

**flask** A glass bottle having a rounded body and narrow neck used in the laboratory; e.g., Erlenmeyer flask.

**flat fee** A fee charged to water and wastewater customers, based on size of service pipe, number of fixtures or other factors, but regardless of the amount of water use or wastewater discharge. Also called uniform rate or uniform block rate.

**flat rate** *See* flat fee.

**flat-sheet membrane** A membrane manufactured as a flat sheet and used as a single layer, particularly in laboratory separations. *See* spiral-wound membrane, plate-and-frame membrane.

**flatworm** A parasitic worm having a flat body and sometimes found in wastewater treatment processes; also called a trematode.

***Flavobacterium*** A genus of heterotrophic, floc-forming organisms active in biological denitrification, activated sludge, microbial regrowth, and in trickling filters, including some opportunistic pathogens.

**flavor** A distinctive taste experienced in the mouth.

**flavor profile** The characteristics of taste and odor of a water sample.

**flavor profile analysis (FPA)** A method proposed for the identification of the character as well as intensity of tastes and odors in a water or wastewater sample; it uses a group of trained panelists to establish the matrix of tastes and odors in the sample. *See also* threshold odor number (TON).

**FlexAir™** A fine-pore membrane manufactured by Environmental Dynamics, Inc. for use as an air diffuser.

**Flex-A-Tube®** A fine-bubble diffuser manufactured by Parkson Corp..

**FlexDisc** A membrane manufactured by EnviroQuip International Corp. for use as a fine-bubble diffuser.

**FlexDome** A membrane manufactured by EnviroQuip International Corp. for use as a fine-bubble diffuser.

**Flexflo®** A peristaltic metering pump manufactured by Blue-White® Industries of Huntington Beach, CA.

**flexible coupling** A joint or connection that allows movement in a pipeline without failure. Also called fault protection.

**flexible diffuser** A device made of polymeric materials that supplies dust-free (filtered), fine air bubbles for the aeration of water or wastewater; unlike ordinary ceramic diffusers, the flexible diffuser is not prone to plugging. *See also* sparger, jet aeration.

**flexible joint** A joint between two pipe sections, such that one can be deflected without affecting the other; e.g., the mechanical pipe joint, which includes lugs and bolts.

**flexible-joint pipe** A cast iron pipe that can be installed underwater and can be deflected without damage or risk of leakage.

**Flexi-Fabric** A fabric manufactured by Eimco Process equipment Co. for filter presses.

**Flexiflo** A high-speed surface aerator manufactured by Aerators, Inc.

**Flexi Jet** A mixer manufactured by Aerators, Inc. using air sparging.

**Flexi-Jet** Spray nozzles manufactured by F. B. Leopold Co. for rotary surface skimmers of sand filters.

**Flex-i-liner®** Sealless self-priming rotary pumps manufactured by Vanton Pump & Equipment Co.

**Fleximix** A high-speed surface aerator manufactured by Aerators, Inc.

**flexiring** A supporting medium used in upflow fixed-film reactors, providing a void volume of 95%

**Flexishaft™** A progressing cavity pump manufactured by MGI Pumps, Inc.

**FlexKlear** An inclined-plate settler manufactured by Eimco Process Equipment Co.

**FlexKleen™** Nozzles fabricated by Eimco Process Equipment Co. for filter underdrains.

**FlexLine™** A tubular diffuser manufactured by EnviroQuip International Corp.

**Flexmate®** A skid-mounted vessel manufactured by U.S. Filter Corp. for use in water treatment.

**Flexofuser®** A fine-bubble diffuser manufactured by FMC Corp.

**Flexoplate™** A membrane manufactured by FMC Corp. as a medium-bubble air diffuser.

**FlexRO Mobile®** A reverse osmosis unit manufactured by U.S. Filter Corp. and mounted on a skid.

**FlexRO™** A mobile reverse osmosis unit manufactured by U.S. Filter Corp. and mounted on a skid.

**FlexScour®** A filter underdrain manufactured by Eimco Process Equipment Co. that can use simultaneously air and water for backwash. It features 18 ga stainless steel lateral ducts axially partitioned into upper and lower channels for air and water. *See* Figure F-06.

**flight** (1) The helical blade of a screw pump. (2) The horizontal scraper on a rectangular sludge or grit collector. *See* chain and flights, chain-and-flight sludge collector.

**Flight Guide** Nonmetallic shoes manufactured by Trusty Cook, Inc. for sludge collector flights.

**Flint Rim** A cast sprocket with hardened rim manufactured by FMC Corp.

**float** The sludge that accumulates on the surface of a flotation tank; it can be removed by flooding or mechanically; its concentration is measured in percent solids.

**floatable** A constituent that can float at the surface of a water or wastewater treatment unit, e.g., foam, oil and grease.

**FloatAll** Dissolved-air flotation equipment manufactured by Walker Process Equipment Co.

**float control** A float device used to control the operation of a piece of equipment.

**float gage** An instrument that includes a buoyant float resting on the surface of a liquid whose elevation is to be measured.

**float gaging** Using a float gage to measure the velocity of flowing water.

**floating aquatic plant** For floating aquatic plants used in wastewater treatment systems, see duckweed, pennywort, and water hyacinth.

**floating aquatic plant system** A free water surface wetland, 1.5–6.0 ft deep, using floating plant species, usually duckweed and water hyacinth. It may be aerated, nonaerated, or facultative. The

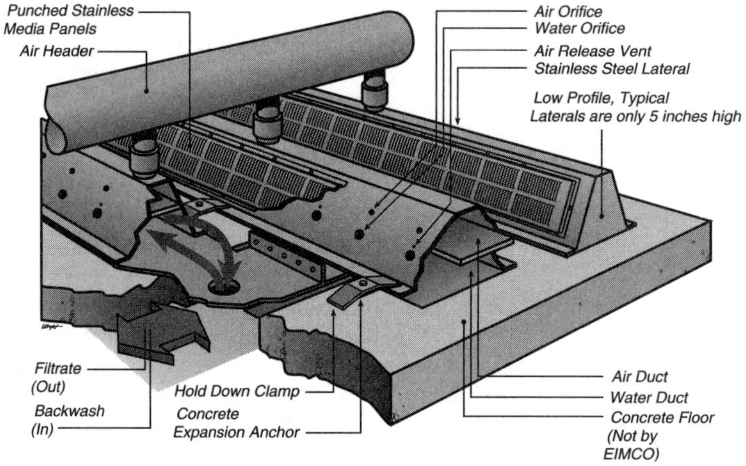


**Figure F-06.** FlexScour® (courtesy Eimco Water Technologies).

plants are used to remove algae from lagoon and stabilization pond effluents. The minimum preapplication treatment for a floating aquatic plant system is the equivalent of primary clarification or short-detention aerated ponds.

**floating compressor** An aerating device used, especially during the summer, to raise the oxygen content of receiving waters and destroy their stratification.

**floating cover** An airtight metal cover on an anaerobic sludge digestion tank that can move up or down as the sludge level changes. The cover floats on the surface of the digester contents, thus allowing the volume of the unit to fluctuate without letting air in, which would create a risk of explosion from the mixing of air and gas. *See also* fixed cover, membrane gas cover.

**floating digester cover** *See* fixed cover.

**floating gas holder** A floating cover on an anaerobic sludge digestion tank that moves up or down as the gas volume changes.

**floating liquid** *See* floating materials.

**floating materials** Oil, grease, and other materials that float on the surface of water or wastewater; often, they are unsightly, restrict the passage of light, and interfere with plant growth, with treatment processes, or with reaeration.

**floating on the system** A mode of operating a water supply storage system such that daily inflow matches daily demand.

**floating sludge** Sludge that floats at the surface of a clarifier as a result of nitrogen gas bubbles from denitrification. This phenomenon occurs in suspended growth processes operated with high sludge age (e.g., more than 10 days), high dissolved oxygen (more than 2.0 mg/L), and long solids retention in the clarifier.

**floating solid** *See* floating materials.

**floating solids** *See* floating sludge.

**floating weed** An aquatic plant that floats on the surface of a pond or other water body.

**float run** The distance over which a float is used.

**float switch** An electrical or pneumatic switch operated by a float in response to changing liquid levels in a tank or reservoir.

**Float-Treat®** Dissolved-air flotation equipment manufactured by Envirex, Inc.

**float tube** (1) An intake pipe fixed at the bottom, the submerged mouth rising and falling with the water surface. (2) A tube the mouth of which is held below the water surface by a float that rises and falls with the water level. (3) A vertical pipe that contains a float that controls the operation of a pumping station.

**float valve** A valve that is actuated by a float.

**Flo-Buster™** Equipment manufactured by Enviro-Care for the agitation and separation of organic solids.

**floc** (1) Clumps of bacteria and particulate impurities that have come together and formed a cluster as a result of chemical coagulation or biological activity in wastewater treatment. (2) A fine, fluffy mass formed by the agglomeration of suspended and colloidal particles during water treatment; it is usually induced by coagulation and flocculation.

**floc age** Similar to sludge age or solids retention time in the activated sludge process, floc age is the ratio of the amount (weight or volume) of floc in circulation to the weight or volume of floc formed in a unit time. The older the floc, the more dead cells it contains and the less active it is in oxidizing organic matter.

**floc ballasting** The use of ballasted floc to enhance flocculation and sedimentation. *See* ballasted flocculation.

**Floc Barrier®** Inclined settling tubes manufactured by Graver Co. for installation in clarifiers.

**floc blanket** A fluidized bed of aggregated particles that have reached the maximum rate of accumulation in a settling tank.

**floc-blanket clarification** Water clarification using the floc-blanket concept. The principal mechanism of clarification within the blanket involves flocculation, entrapment, and sedimentation, but settling, entrainment, and particle elutriation occur above the blanket.

**floc-blanket clarifier** A basin designed to accomplish clarification using the sludge blanket concept. Several designs can be used for a floc-blanket clarifier, including solids contact clarifiers, hopper-bottomed tanks like the Candy tank, flat-bottomed tanks, inclined settling tanks, and ballasted floc systems. Also called floc-blanket tank. *See also* Candy tank, Spaulding Precipitator.

**floc-blanket process** *See* floc-blanket clarification.

**floc blanket reactor/powdered activated carbon/ ultrafiltration process** *See* FBR/PAC/UF process.

**floc-blanket sedimentation** *See* floc-blanket clarification.

**floc-blanket tank** *See* floc-blanket clarifier.

**floc breakthrough** A condition in the operation of a filter when the turbidity of the product water increases; it is often used as an indication to terminate the filter run.

**floc breakup** The disaggregation of flocs by stirring and agitation, intended to increase the number of particles. Floc breakup occurs with high mixing intensities (velocity gradient $G > 100$/sec) and large flocs. Also called floc disaggregation or floc shear, the opposite of floc buildup or floc growth.

**floccose** Flocculent.

**flocculant** (1) A water-soluble organic polyelectrolyte that is used alone or in combination with metal salts to form clumps of solids that settle rapidly in water or wastewater. Examples include aluminum or iron salts such as alum $[Al_2(SO_4) \cdot 14\ H_2O]$, ferrous sulfate ($FeSO_4$), and lime $[Ca(OH)_2]$. Same as flocculating agent or flocculation agent. *See also* coagulant, flocculent polymer. (2) A chemical used to improve the plasticity of clay for ceramic work. (3) Sometimes used as an adjective, but the preferred adjective form is flocculent. *See also* coagulation and flocculation.

**flocculant aid** Any chemical or substance, usually a patented polyelectrolyte, used to assist or modify flocculation. *See* coagulant aid for detail.

**flocculant settling** *See* flocculent settling.

**flocculate** To form flocculent masses.

**flocculating agent** Same as flocculant.

**flocculating tank** Same as flocculation tank.

**flocculation** (1) The gathering together of fine particles in water by gentle mixing after the addition of coagulant chemicals to form larger particles. Important flocculation mechanisms include Brownian motion, differential settling, and differential velocity gradients. *See* macroflocculation and microflocculation. (2) The water and wastewater treatment process that consists of chemical addition and gentle stirring to promote the agglomeration of destabilized particles that can then be removed by sedimentation, flotation, and coarse bed filtration. *See also* flocculation concepts. (3) *See* bioflocculation. (4) The formation of lumps that settle as silt, and sometimes mudbanks, when colloidal clays from river water mix with solution particles in seawater. (5) A mechanism of removal of particulate matter within a granular filter: the larger particles resulting from flocculation are then removed by another mechanism such as straining or interception. *See* filtration mechanisms. *See* coagulation and flocculation for the difference between these two processes and flocculation concepts for associated terms.

**flocculation agent** Same as flocculant.

**flocculation aid** Same as flocculant aid.

**flocculation basin** *See* flocculation tank.

**flocculation concepts** *See* Brownian diffusion, Brownian diffusivity, Brownian flocculation, Brownian motion, Brownian movement, charge neutralization destabilization, collision efficiency, conjunction opportunity, contact clarifier, contact flocculation, contact opportunity, convective sedimentation, differential settling, flash mix, flash mixing, floc blanket, floc breakup, flocculant, flocculation aid, flocculation limit, flocculation mixer, flocculation rate constant, flocculation rate correction factor, floc disaggregation, floc shear, floc volumetric concentration, G value, G value concept, Hamaker constant, hydraulic flocculation, hydrodynamic retardation, Kolmorogoff microscale, limiting size, macroflocculation, microflocculation, mixers, orthokinetic flocculation, orthokinetic motion, perikinetic flocculation, perikinetic motion, power dissipation function, rapid mix, rapid mixing, shear gradient, sludge blanket, stability ratio, Stokes' law, velocity gradient, volumetric concentration.

**flocculation devices** *See* mixing and flocculation devices.

**flocculation limit** The water content of a deflocculated sediment or soil. *See also* flocculation ratio.

**flocculation mixer** For mixers used for flocculation in water and wastewater treatment, see paddle flocculator, turbine/propeller flocculator, static mixer.

**flocculation rate constant** A second-order constant used in characterizing the aggregation mechanisms involved in flocculation. Along with the collision efficiency factor and the concentrations of particles, it determines the rate of collision between two particle sizes ($i$) and ($j$). The flocculation rate constant ($K_{ij}$) depends on the thermal energy of the fluid (the product of Boltzmann's constant, $B$, and the absolute temperature, $T$), the absolute viscosity ($\mu$), and the diameters of the particles ($D_i$ and $D_j$). Specific equations proposed for Brownian (or perikinetic) flocculation in an infinite stagnant fluid ($K_p$), for differential settling ($K_d$), and for isotropic turbulent flow ($K_t$) are (AWWA, 1999):

$$K_p = (2/3)(BT/\mu)(D_i + D_j)/(D_iD_j) \qquad \text{(F-56)}$$

$$K_d = [\pi g(S-1)/72\ \nu](D_i + D_j)^3(D_i - D_j) \qquad \text{(F-57)}$$

$$K_t = [(D_i + D_j)^3/6.18](\varepsilon/\nu)^{1/2} \qquad \text{(F-58)}$$

where $g$ = gravitational acceleration, $S$ = specific gravity of the particles, $\nu$ = kinematic viscosity, and $\varepsilon$ = local rate of turbulent energy dissipation per unit mass of fluid. *See also* flocculation concepts.

**flocculation rate correction factor** A dimensionless factor, ranging from 0 to slightly more than 1.0, indicating the fraction of successful collisions among destabilized colloidal or suspended particles. *See* collision efficiency for more detail.

**flocculation ratio** The void ratio of a deflocculated sediment or soil. *See also* flocculation limit.

**flocculation tank** A tank in which liquid suspensions are gently agitated to promote the formation of large settleable flocs. Sometimes a flocculant aid is added. A flocculator.

**flocculator** A tank or other apparatus for enhancing the formation of floc in water or wastewater, mixing energy being provided mechanically or hydraulically. *See also* Alabama flocculator, Flocsillator®, baffled channel, flash mixer, paddle flocculator, pebble bed flocculator, propeller, useful power input.

**flocculator–clarifier** A circular or rectangular water or wastewater treatment unit where the solids-contact process occurs. *See* sludge blanket and solids-contact clarifier for more detail.

**floccule** A piece of flocculent matter in a liquid.

**flocculent** (The preferred adjective form, instead of flocculant.) Consisting of or pertaining to flocs.

**flocculent polymer** An anionic or nonionic polyelectrolyte, with a molecular weight 10 times that of a typical primary coagulant polymer; used after flocculation to increase floc size and strength. *See also* polyacrylamide polymer.

**flocculent precipitate** A wooly-looking substance, aluminum hydroxide [$Al(OH)_3$], resulting from the combination of ammonia ($NH_3$) with an aluminum-salt solution.

**flocculent sedimentation** Same as flocculent settling.

**flocculent settling** The settling of particles in a dilute suspension as a result of flocculation. *See* sedimentation type II for more detail. *See also* Stokes' law, and other types of settling: I or free, III or hindered or zone, IV or compression.

**flocculent suspension** An agglomeration of particles or chemical precipitates formed during coagulation and flocculation.

**floc density** The mass of floc per unit volume, a characteristic that affects floc volume, floc settling velocity, and turbidity removal.

**floc disaggregation** The disaggregation of flocs by stirring and agitation intended to increase the number of particles. *See* floc breakup for detail.

**floc/filtration method** A procedure recommended to estimate the readily biodegradable chemical oxygen demand (rbCOD) and determine the true soluble COD of a wastewater sample. It consists of vigorously mixing 1 mL of a 100 g/L solution of zinc sulfate ($ZnSO_4$) and a 100 mL sample of wastewater for 1 minute, raising the pH by addition of sodium hydroxide (NaOH), gentle mixing for floc formation, settling for about 15 min, withdrawing and filtering the supernatant on a 0.45-μm membrane, and measuring the COD of the filtrate. The rbCOD is taken as the difference between the COD of the wastewater sample and the COD of the filtrate, based on the assumption that treatment eliminates the rbCOD.

**floc-forming bacteria** *See* floc-forming microorganisms.

**floc-forming microorganisms** A mixture of microorganisms that settle and thicken well during biological wastewater treatment or sludge processing, as compared to bulking organisms or filamentous bacteria. Common floc-forming organisms found in activated sludge include bacteria groups (*Achromobacter, Acinetobacter, Alcaligenes, Arthrobacter, Citromonas, Flavobacterium, Pseudomonas, Zooglea*) and protozoa (*Aspidisca, Paramecium, Vorticella*). These organisms have higher growth rates at high soluble substrate concentrations, a characteristic that is used in the design of selector processes. *See* selector concept.

**floc growth** During flocculation, the aggregation of discrete particles into larger particles that can be removed more easily after sedimentation; also called floc buildup, the opposite of floc breakup or floc shear.

**Flocide®** A sanitizing additive produced by FMC Corp.

**Floclean** A chemical product of FMC Corp. for cleaning membranes.

**Floclear** A solids contact clarifier manufactured by Aerators, Inc.

**Flocon®** A feedwater additive produced by FMC Corp. for use in reverse osmosis units.

**Flo-Conveyor™** A conveyor manufactured by Enviro-Care Co. for handling screenings.

**FloCor®** A PVC medium manufactured by Gray Engineering Co. (Imperial Chemical Industries in Great Britain) for use in trickling filters. It allows a void space of about 97% and an exposed surface area of 27–37 square feet per cubic foot, or more than twice the capacity of rocks and fieldstones. *See also* Koroseal® and Surfpac®.

**Floc-Pac™** Flocculant packages produced by American Cyanamid Co.

**Flocpress®** A belt filter press manufactured by Infilco Degremont, Inc.

**Flocsettler** A flocculator–clarifier manufactured by Amwell, Inc.

**floc shear** The disaggregation of flocs by stirring and agitation intended to increase the number of particles. *See* floc breakup for detail.

**Flocsillator** A horizontal oscillating flocculator manufactured by Eimco Process Equipment Co. Its mechanism consists of a wedge-shaped paddle array suspended from a horizontal drive shaft that spans the basin above water level.

**FlocTreator** Coagulation and flocculation equipment manufactured by PWT Americas.

**Floctrol** A paddle flocculator manufactured by Envirex, Inc. with a horizontal shaft.

**floc volumetric concentration** *See* volumetric concentration.

**floe ice** Thick ice broken into pieces and floating on the surface of a body of water.

**Flofilter™** A water treatment unit of Purac Engineering, Inc. using flotation and filtration.

**Flo-Lift™** A vertical device manufactured by Enviro-Care Co. for lifting wastewater treatment screenings.

**Flo-Line** A screening device manufactured by Derrick Corp. with fine mesh.

**FloMag™** Water treatment products of Martin Marietta Specialties, Inc., containing magnesium in granular, powder, or slurry form.

**FloMaker™** A submersible mixer manufactured by ITT Flygt Corp.

**Flo-Mate®** A portable flowmeter manufactured by Marsh-McBirney, Inc. of Frederick, MD.

**flood or flooding** (1) A general and temporary condition of partial or complete inundation of normally dry land areas from the overflow of inland and/or tidal waters, and/or runoff of surface waters from any source, or flooding from any other source (EPA-40CFR6-AA-c). (2) "A general and temporary condition of partial or complete inundation of two or more acres of normally dry land area or of two or more properties from (i) overflow of inland or tidal waters; (ii) unusual and rapid accumulation or runoff of surface waters from any source; (iii) mudflow; (iv) collapse or subsidence of land along the shore of a body of water as a result of erosion or undermining by waves or current" (as defined by the National Flood Insurance Program). (3) Excessive streamflow resulting from precipitation or snowmelt and causing a watercourse to overtop its banks and flow onto the floodplain. (4) A relatively high discharge or any discharge exceeding a designated value.

**flooded fixture** A plumbing fixture overflowing at the top or rim.

**flooded-suction station** A type of wastewater pumping station, as opposed to submersible or suction lift stations.

**flood-flow formulas** Formulas derived from empirical evaluations and hydrological factors to estimate flood flows. *See* the following formulas: Fanning's, Kinnison–Colby's, McMath's, Potter's, and rational.

**flood fringe** The portion of the floodplain that is outside of the floodway. Also called floodway fringe.

**flood hazard boundary map** An official map of a community showing the boundaries of the flood, mudslide (i.e., mudflow), and related erosion areas.

**flooding** *See* flood.

**flooding basin technique** A method used to measure the hydraulic conductivity of soils.

**flooding of powdered activated carbon** *See* powdered activated carbon flooding.

**flood insurance rate map (FIRM)** An official map of a community that shows the special hazard areas and the risk-premium zones applicable to the community.

**flood insurance study** (1) A study that establishes floodway boundaries to accommodate a base flood of a given frequency. (2) An examination, evaluation and determination of flood hazards with, if appropriate, water surface elevations; or an examination, evaluation and determination of mudslide (i.e., mudflow) and/or flood-related erosion hazards.

**flood irrigation** (1) An old irrigation method that uses natural flood waters through canals and ditches. (2) The application of wastewater through distributors on land surrounded by low embankments and the collection of the percolated water in underdrains discharging into ditches. This method of wastewater treatment and disposal is also called land filtration. *See* Figure F-07. (3) A method of wastewater treatment and disposal that consists in applying the wastewater through ditches across beds of farmland that are kept moist by seepage. Also called bed irrigation, land filtration. *See also* direct irrigation, spray irrigation.

**flood mitigation** (1) The use of hydraulic structures and other means to reduce flood damage to an acceptable minimum. *See also* flood control. (2) All steps necessary to minimize the potentially adverse effects of a proposed flood protection action, to restore and preserve the natural and beneficial floodplain values, and to preserve and enhance natural values of wetlands. (Definition from the code of federal regulations on floodplain management and the protection of wetlands.)

**flood polygon** *See* floodplain.

**floodplain** (1) The flat or nearly flat, lowland area adjacent to inland and coastal waters and within the perimeter subject to inundation by floods of a given frequency. More generally, the floodplain or flood polygon is the part of a river valley that water covers when the river overtops its banks. The USEPA (EPA-40CFR6-AA-d) further defines a base floodplain (the 100-year floodplain or 1% chance floodplain) and the critical-action floodplain (the 500-year floodplain or the 0.2% floodplain). (2) Definition from the code of federal regulations on floodplain management and the protection of wetlands; the lowland and relatively flat areas adjoining inland and coastal waters including, at a minimum, that area subject to a 1% or greater chance of flooding in any given year. When a critical action (an action for which even a slight chance of flooding is too great) is involved, the floodplain is the area subject to inundation from a flood having a 0.2% chance of occurring in any given year (the 500-year floodplain). Floodplain does not include areas subject only to mudflow.

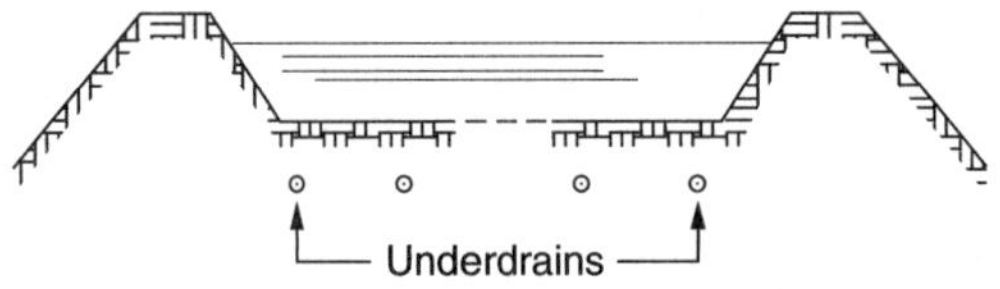


**Figure F-07.** Flood irrigation.

**floodplain values** *See* natural values of floodplains and wetlands.

**floodproofing** Modification of structures and facilities, their sites, and their contents, including individual measures (such as levees and floodwalls) and nonstructural measures, to protect the contents or the upper floors of buildings against the effects of floodwaters.

**flood rim** The edge of a plumbing fixture or other receptacle from which water may overflow.

**FloodSafe™ Inflow Preventer** A patented system that provides vault protection from contamination. It is designed to help prevent the entry of contaminated water or other fluids from entering the air valve outlet and, subsequently, the pipeline, while allowing the air valve or vent to exhaust and admit air to the system. *See* Figure F-08.

**Flo-Poke®** A portable instrument to measure flowrate, manufactured by ISCO, Inc. of Lincoln, NE.

**flora** (1) Plants, listed by species and considered as a whole, and plant life of a particular region or period. *See also* fauna. (2) The aggregate of microorganism in a human body or animal, e.g., intestinal flora.

**Flo-Screen™** A bar screen manufactured by Enviro-Care Co. with a reciprocating rake.

**flotation** A unit operation by which particles less dense than water rise to the surface for removal; gas bubbles or other flotation agents introduced to water (by aeration, injection, or vacuum flotation) attach to solid particles, decreasing their density and creating bubble–solid agglomerates that float to the surface, from which they are removed. Flotation is used in the removal of grease, oil, emulsified substances, and finely divided suspended or colloidal solids and in sludge thickening. *See also* air/solids ratio, collector, dispersed-air flotation, dissolved-air flotation, elec-

**Figure F-08.** FloodSafe™ inflow preventer (courtesy Val-Matic).

trolytic flotation (electroflotation), float, flotation thickener, foam flotation, froth flotation, frothing agent, heterogeneous flotation model, microflotation, natural flotation, pressure flotation, Stokes' law, vacuum flotation, white-water collector model.

**flotation agents** Fine-air bubbles or hydrophobic wetting and foaming chemical compounds that are added to water or wastewater to make suspended substances lighter so that they rise to the surface and can be skimmed off. Chemical flotation agents include detergents (anionic, cationic, neutral), glues, greases, oils, and resins. *See also* flotation reagents.

**flotation aid** Any chemical that produces coagulation, breaks an emulsion, and/or aids in the absorption of air bubbles by the liquid or particles to be removed.

**flotation–filtration plant** A water (or wastewater) treatment plant that includes a unit in which flotation is carried out over a filter, a compact arrangement attractive for package plants or where land is at a premium.

**flotation reagent** Surface-active compounds that promote air flotation of particles heavier than water by attaching to the particles and altering the conditions of the suspension. *See* activating, collecting, depressing, and foaming reagents

**flotation thickener** A tank or similar unit used to concentrate sludge solids by flotation. Appurtenant units include an air saturation tank, a mixing chamber, a recycle pump, and an air compressor. The common design parameters of flotation thickeners include air/solids ratio between 0.005 and 0.060, surface overflow rate between 250 and 1000 gallons per square foot per day, detention time between 30 and 60 min, and recycle ratio. *See also* gravity thickener.

**flotation thickening** A process designed to concentrate low-density particles in water and wastewater treatment residuals. It uses air bubbles to absorb particles and cause them to float to the surface for removal. Three common flotation thickening techniques are dispersed air flotation, dissolved air flotation, and vacuum flotation. Thickening reduces the volume of sludge prior to digestion or dewatering; solids content increases from 2–3% to approximately 4–6%. Dissolved-air flotation is particularly effective with waste activated sludge and other relatively light sludges. Flotation thickening is also used (in Europe) for water treatment instead of sedimentation. *See also* air-to-solids ratio, sludge volume index, centrifugal thickening, cosettling thickening, gravity belt thickening, gravity thickening, rotary drum thickening.

**flotator** A circular or rectangular dissolved air flotation unit used to remove suspended solids from liquids in industrial and municipal applications, particularly where oils, fats, fibers, greases, or colloids are present.

**Flo-Tote®** An open-channel flow meter equipped with a computer, manufactured by Marsh-McBirney, Inc. of Frederick, MD.

**flotsam** Debris from natural sources or human activities floating on water bodies, e.g., wreckage of ships and their cargoes. *See also* jetsam.

**flow** The movement of a fluid stream from one point to another; the fluid itself or the quantity of fluid per unit time, i.e., the same as discharge or flowrate. (The word flow applies equally to the forward movement of solids). The two types of fluid flow are open-channel or gravity flow and pressurized flow. *See* discharge, flowrate, and open-channel flow.

**flowage** The act of flowing, e.g., the flow of water, wastewater, or sludge through the various units of a treatment plant. Also called flow line, as represented on a process flow diagram or a hydraulic profile.

**Flo-Ware** A software of Marsh-McBirney, Inc. of Frederick, MD for logging data from flow meters.

**flow augmentation** The increase of natural streamflow by water from an impoundment, which normally increases the dissolved-oxygen content and promotes reaeration. *See also* low-flow augmentation.

**flow-balancing tube** A special tubing or pipe in a reverse osmosis plant that provides the required pressure adjustment to equalize concentrate flow rates from individual permeators.

**flow cell** An electrode or other sensing element immersed in a fluid to measure a property like pH, dissolved oxygen, or electrical conductivity.

**flowchart** (1) A detailed graphical representation of the operations necessary for the execution of a task such as manufacturing or data processing. Symbols are used to represent operations, equipment, and materials. Also called flowsheet or flow diagram. *See also* process flow diagram. (2) A description of logical steps in a computer algorithm.

**flow coefficient** Same as discharge coefficient.

**flow control device** A device that controls the rate of flow of a fluid manually or automatically. Called rate-of-flow controller when automatic. *See* regulator.

**flow control valve** A valve that controls flowrate by differential pressure across an orifice.

**flow cytometry** The measurement and analysis, through a light beam, of signals emitted by particles while flowing in a liquid. It is used in the detection of bacteria and parasites, e.g., *Cryptosporidium* and *Giardia. See also* particle counter.

**flow demand** (1) The quantity of water required to meet the demand of a system or customer, e.g., fire demand or industrial demand. (2) The natural or regulated discharge necessary for a stream to meet all requirements of a hydroelectric power plant.

**flow diagram** *See* flowchart.

**flow diffuser** A device or fitting that distributes flow; a distributor, a diffuser.

**flow equalization** An operation that consists in storing water or wastewater in a reservoir to dampen flow and/or concentration fluctuations on a daily or longer-term basis. In wastewater treatment, e.g., an equalization basin or tank may be installed between the preliminary and primary treatment units or between an overflow structure and primary treatment. Equalization achieves a relatively constant flow to treatment units, the size of which can be reduced, or to receiving streams.

**flow equalization weir** A weir included in the control structure ahead of such units as clarifiers to ensure equal flow distribution.

**flowers of zinc** Zinc oxide.

**flow factors** Factors that are used to multiply an average flow to obtain other water or wastewater flow parameters such as instantaneous peak flow, maximum one-hour flow, minimum one-hour flow.

**flowing sample** A sample collected from a well-flushed line in a water distribution system for bacteriological analysis and determination of general water quality characteristics. *See also* first-draw sample.

**flow injection analysis (FIA)** An automated water and wastewater examination method whereby a sample is injected into a flowing carrier stream, forming a concentration gradient. A color reaction or other detector is used to measure the analytes. FIA has been used for the analysis of inorganics, from bromide to sulfide.

**flow level control** The use of a barrier such as a gate or dam to control the water level in a channel, stream, or pipe.

**flow-limiting shower head** A shower head that restricts and concentrates water passage through orifices that direct the shower flow for optimum use. *See also* flow-reduction devices.

**flow line** (1) The path of fluid particles in a system such as a well or a stream; a contour line around a body of water. (2) The hydraulic grade line of an open channel.

**FlowLogger** A flow-measuring device manufactured by ISCO, Inc. of Lincoln, NE.

**flow measurement** Same as discharge measurement.

**flowmeter** A gage indicating the velocity of wastewater moving through a treatment plant or of any liquid moving through various industrial processes. More generally, any instrument for measuring flow rate, velocity, or pressure of a fluid. Also called fluid meter. *See also* manometer, magnetic flow meter, pitometer, Parshall flume, orifice, weir, and venturi meter.

**flowmeter accuracy** The closeness of the measurement made by a flowmeter to the reference value of the flow being measured, expressed as the difference between the measurement and the reference value (EPA-40CFR72.2).

**Flowminutor™** A comminutor manufactured by Enviro-Care Co. with a horizontal shaft.

**flow-nozzle meter** A water meter in which flow creates a pressure difference in a nozzle that a float tube uses to determine the discharge.

**flow-paced feed system** A device that feeds chemicals proportionally to the flow being treated.

**flow path study** A study designed to monitor hydrologic and water quality parameters along the path followed by groundwater, for example, in the induced infiltration and natural filtration processes. Flow path wells are drilled at increasing depths from a surface water source toward the production wells. Parameters monitored along the flow path may include temperature, river stage, streambed permeability, groundwater elevation, *Giardia, Cryptosporidium,* aerobic spores, turbidity, total coliforms, and heterotrophic plate counts.

**flow path well** *See* flow path study.

**flow-proportional composite** A flow-proportional composite sample.

**flow-proportional composite sample** A composite of grab samples collected continuously or discretely in proportion to the total flow at time of collection or to the total flow since collection of the previous grab sample. The grab volume or frequency of grab collection may be varied in proportion to flow (EPA-40CFR471.02-rr).

**flow-proportional control** A method of control of chemical feed rates. *See* flow-paced feed system.

**flow rate** The volume per time unit given to the flow of gases or other fluid substance that emerges from an orifice, pump, or turbine, or passes along a conduit or channel (EPA-40CFR146.3). The rate, expressed in gallons or liters per hour, at which a fluid escapes from a hole or a fissure in a tank. Such measurements are also made of liquid waste, effluent, and surface water movement. The term flow rate applies as well to the volume or mass of solid material that passes through a cross section of conduit in a given time, measured, for example, in kg/hr or $m^3$/day. Common flow rate units in the water and wastewater field are acre-feet/day, gpd, gpm, mgd, cfs, $m^3$/sec, liters/sec, and liters/day. *See also* dilution rate.

**flow recording** The process, usually automatic, of documenting flow rates past a given point in a process.

**flow-reduction devices and appliances** Devices and appliances that reduce interior water use and the corresponding wastewater flow. *See* bathroom retrofit kit, faucet aerator, flow-limiting shower head, low-flow toilet, pressure-reducing valve, pressurized shower, toilet dam, toilet leak detector, vacuum toilet.

**flow regulator** A device or structure used to control the flow or the level of water or wastewater in a canal, conduit, channel, basin, or treatment unit. *See* regulator, flow control device.

**flowsheet** Same as flowchart.

**FlowSorb™** Canisters of granular activated carbon fabricated by Calgon Carbon Corp.

**flow splitter** A box or chamber that splits incoming flow into two or more streams.

**flow test** The measurement of the rate of flow of water at a point in the distribution system, e.g., in a pipe, at a pump discharge, or at a fire hydrant.

**flow through** A continuous or intermittent passage of test solution or dilution water through a test chamber, culture tank, or holding or acclimation tank with no recycling (EPA-40CFR795.120, 797.1300-4, 797.1400-6, 797.1600-8, & 797.1970-2).

**flow-through aerated lagoon** A shallow earthen basin, aerated by mechanical devices and used for wastewater treatment. The aeration is sufficient for aerobic treatment and for maintaining only a portion of the solids in suspension. A sedimentation unit is used to remove the effluent solids. Also called aerobic flow-through partially mixed lagoon. *See also* aerated facultative lagoon, and lagoons and ponds.

**flow-through chamber** The upper compartment of an Imhoff tank or a two-story sedimentation tank.

**flow-through curve** A tracer response curve, i.e., a graphical representation of the distribution of a tracer versus time of flow. It is used in studying short-circuiting, retardation, and longitudinal mixing in sedimentation basins and other reactors. *See* time-concentration curve for more detail.

**flow-through pond** *See* flow-through system.

**flow-through process tank** A tank that forms an integral part of a production process through which there is a steady, variable, recurring, or intermittent flow of materials during the operation of the process. Flow-through process tanks do not include tanks used for the storage of materials prior to their introduction into the production process or for the storage of finished products or by-products from the production process (EPA-40CFR280.12).

**flow-through system** A treatment system that does not include recirculation; the opposite of a recirculating system; e.g., a flow-through pond. Also called once-through system. *See also* flow-through aerated lagoon.

**flow-through test** An enhanced bioassay method in which a 24-hour composite sample of wastewater effluent is continuously supplied to laboratory containers to determine toxicity to aquatic organisms (e.g., fathead minnow and water flea). The sample is filtered, aerated, and stabilized at 20°C. The containers outflow continuously.

**flow-through time** The time required for a volume of liquid to pass through a basin from inlet to outlet; equal to the ratio of the volume of the basin to the flow rate, or estimated from tracer recovery curves. *See also* hydraulic retention time.

**flow totalizer** A type of flow meter that shows the cumulative volume at all times; to obtain the volume for a given period, subtract the volume for the previous period from the cumulative volume. The average flow in that period is the volume divided by the length of the period. The totalizer is operated from the movement of a manometer transducer or from the movement of a chart pen arm. Also called an integrator.

**Flowtrex** A pleated cartridge filter manufactured by Osmonics, Inc.

**flow tube** A calibrated flow-measuring device (similar to a venturi tube) made for a specific range of flow velocities and fluids. It measures velocity based on pressure drop.

**flow valve** A valve that automatically closes when the flow rate reaches a certain level.

**flow-weighted composite sample** A mixture of aliquots collected at a constant time interval, with the volume of each aliquot proportional to the flow rate. *See also* aliquot.

**flow-weighted concentration** For *N* observations of flows and concentrations, the flow-weighted concentration is the ratio of the sum of the products of flow rate and concentration for each observation to the sum of the flow rates:

$$C_w = \left(\sum_{i=1}^{N} C_i Q_i\right) \Bigg/ \left(\sum_{i=1}^{N} Q_i\right) \quad \text{(F-59)}$$

where $C_w$ = flow-weighted average concentration, mass/unit volume; $C_i$ = concentration at *i*th observation, mass/unit volume; and $Q_i$ = flowrate at *i*th observation, volume/unit time.

**flow weighting** The adjustment of pollutant concentrations for the effect of flow in a series of measurements.

**flow work** The ratio of static pressure ($p$) to the specific weight of the fluid ($\gamma$) in the formula of dynamic head. Same as pressure head.

**fluctuation cycle** The cycle of fluctuation, i.e., the period of time of rise and a succeeding decline of a water table; e.g., daily cycle, annual cycle. Also called phreatic cycle.

**flue** The chimney leading from an incinerator; a passage or duct for smoke in a chimney.

**flue gas** The air coming out of a chimney after combustion in the burner it is venting. It can include nitrogen oxides, carbon oxides, water vapor, sulfur oxides, aerosols, particles, and many chemical pollutants. It also contains carbon dioxide (about 14% when well burned). Also called incinerator effluent or incinerator emission, stack gas. *See also* waste boiler-flue gas treatment.

**FluePac™** Powdered activated carbon produced by Calgon Carbon Corp.

**fluid** Any material or substance that flows or moves, whether in a semisolid, liquid, sludge, gas, or any other form or state (EPA-40CFR144.3 or 146.3). More generally, an inelastic substance that assumes the shape of its container, can flow and deform continuously under the effect of a shear force without returning to its original disposition. Fluids include liquids and gases. Fluid properties include bulk modulus (force/surface), density or mass density (mass/volume), mass, specific gravity or relative density (dimensionless), specific weight (force/volume), surface tension (force/length), viscosity (mass/length/time), volume, and weight (force).

**Fluidactor** A fluid bed incinerator manufactured by Walker Process Equipment Co. for sludge processing.

**fluid bed combustion** A method used to reduce the emission of acid gases (hydrogen bromide, hydrogen chloride, hydrogen fluoride, sulfur oxides) during sludge combustion. It uses a bed of granular limestone to form solid calcium compounds with the gases and these solids are removed with the ash. *See also* dry scrubber, wet scrubber.

**fluid bed dryer** Same as fluidized-bed dryer. *See also* direct dryer.

**fluid bed furnace** Same as fluidized-bed furnace.

**fluid-bed suspension** The concept applied in fluidized-bed incinerators: solid wastes or sludge solids are shredded and blown into the furnace.

**Fluid Dynamics®** A filter system manufactured by Memtec America Corp.

**fluidization** The upward flow of a fluid through a granular bed at sufficient velocity to suspend the grains. In water treatment, filter beds are fluidized (expanded) for cleaning purposes, using water, air, or a combination of the two. Fluidization is similar to hindered or type III settling; it results in stratification of the finer grains of each layer in a multimedia filter. *See also* air scour, backwash velocity, buoyant weight, fluidization pressure, Galileo number, incipient fluidization, minimum fluidizing velocity.

**fluidization point** Same as incipient fluidization point. On a log-log plot (Figure F-09) of pressure drop ($\Delta P$) across a packed bed versus backwash velocity as represented by the Reynolds number ($R_e$), $\Delta P$ first varies linearly (line OA). At F, the fluidization point, the medium grains begin to move freely as in hindered settling.

**fluidization pressure drop** In an expanded granular bed, the buoyant weight of the media grains is

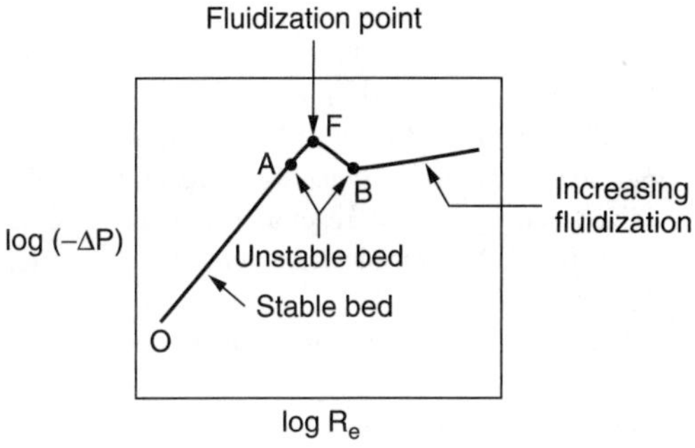


**Figure F-09.** Fluidization point (F).

the pressure drop after fluidization. *See* buoyant weight for detail.

**fluidization velocity** *See* minimum fluidization velocity.

**fluidized** Pertaining to a mass of solid particles that is made to flow like a liquid by injection of water or gas. In water treatment, a bed of filter media is fluidized by backwashing water through the filter.

**fluidized** (1) A bed of granular materials, such as sand or activated carbon, that is expanded for backwashing or for normal operation; the particles do not rest on each other but are supported by the drag forces from the liquid. Fluidized beds require less pumping and less downtime than fixed beds. Also called expanded bed. *See also* fixed or packed bed. (2) A device for burning coal, organic refuse, or other solid carbonaceous fuel, mixed with inert materials, at low temperatures, thereby minimizing emissions.

**fluidized-bed adsorber** A granular activated carbon (GAC) adsorber, used in water or wastewater treatment, with the medium held in suspension by the flow of liquid. *See also* fixed-bed adsorber, pulsed-bed adsorber, upflow expanded-bed mode.

**fluidized-bed biological reactor** *See* fluidized-bed bioreactor.

**fluidized-bed biological treatment** An aerobic biological wastewater treatment process that uses microorganisms attached to the surface of fine media such as sand for the stabilization of organic matter. The very high surface area/volume ratio of the media (hence a very high concentration of active biological solids and reduced detention time in the bed) is an advantage but its small openings can plug easily. *See also* anaerobic fluidized-bed process.

**fluidized-bed bioreactor (FBBR)** A reactor packed with 0.4–0.5 mm sand or activated carbon for the biological treatment of wastewater (e.g., post denitrification) or contaminated groundwater. Part of the effluent is aerated for recycling. *See* submerged attached-growth process for related methods. *See also* Figure F-10 and fluidized-bed reactor.

**fluidized-bed combustion** A method of burning particulate fuel, such as powdered or granular coal, by which fuel is burned after injection into a rapidly moving gas stream. The emission of sulfur dioxide ($SO_2$) can be eliminated or reduced by burning the fuel in a bed of finely divided limestone or dolomite.

**fluidized-bed denitrification** A wastewater treatment process using a fluidized-bed reactor fed at the bottom as a denitrification filter. A soluble carbon source (e.g., methanol, $CH_3OH$) is usually added. A smaller reactor, shaped as an inverted cone, is installed inside the large reactor for sludge collection and wastage.

**fluidized-bed dryer** A device that uses heat to dry sludge, similar to a rotary dryer but with a steam boiler and fluid-bed reactor instead of a hot gas furnace and rotary drum. It consists of a stationary, vertical chamber with a perforated bottom through which a hot gas is forced. Also called fluid bed dryer. *See* Figure F-11. *See also* heat drying for other types.

**fluidized-bed furnace** A vertical, cylindrical, refractory-lined steel shell used for sludge incineration by heated air through a fluidized sand bed. Also called fluid-bed furnace.

**fluidized-bed GAC adsorber** *See* fluidized-bed adsorber.

**fluidized-bed incineration** Sludge incineration using a fluidized-bed incinerator. *See also* fluidization, fast fluidization, fluidized-bed suspension, incipient fluidization.

**fluidized-bed incinerator** A vertical, cylindrical shell of refractory-lined steel that contains a bed of hot sand or other granular material, fluidized by the

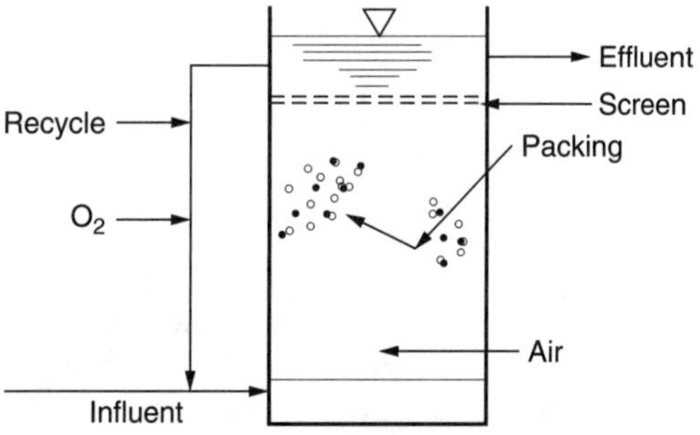


**Figure F-10.** Fluidized-bed bioreactor.

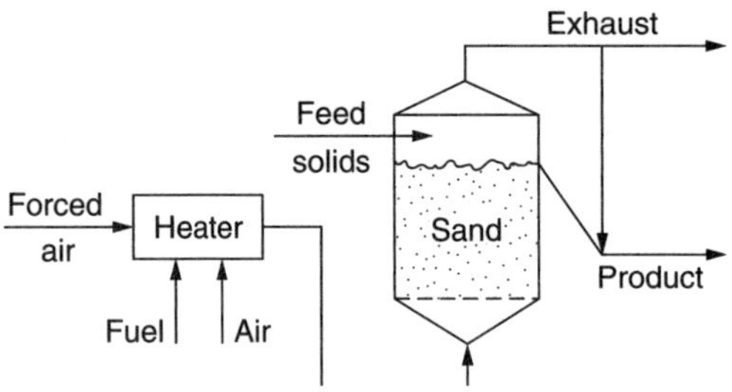


**Figure F-11.** Fluidized-bed dryer.

upward flow of air and waste, to transfer heat directly to the waste; used mainly for destroying municipal sludge. The granular material, preheated to 800°C, conserves the heat even when the unit is idle. The air flow carries out ash for removal by an air pollution control system. *See also* bubbling-bed incinerator (combustor), circulating-bed incinerator (combustor), tuyere, incineration, combustion, multiple-hearth incineration, coincineration.

**fluidized-bed pellet reactor** A reactor designed to remove water hardness while minimizing sludge production. The process includes the introduction of silica sand or other pellets into a column reactor, directing the mixture of water and chemicals through the column, and letting the crystallized pellets grow and settle at the bottom.

**fluidized-bed porosity function ($F_\varepsilon$)** A dimensionless expression that combines (a) the pressure drop in a fixed granular bed correlated to a modified Reynolds number and (b) the constant head-loss equation of a fluidized bed (AWWA, 1999):

$$F_\varepsilon = [\varepsilon^3/(1-\varepsilon)^2][g\rho(\rho_s - \rho)/S_s^3\mu^2] \quad \text{(F-60)}$$

$$\log F_\varepsilon = 0.56543 + 1.09348\,X + 0.17971\,X^2 - 0.00392\,X^4 - 1.5\,(\log \Psi)^2 \quad \text{(F-61)}$$

$$X = \log R_s^* \quad \text{(F-62)}$$

where $\varepsilon$ = porosity, $g$ = gravitational acceleration, $\rho$ = mass density of the fluid, $\rho_s$ = mass density of the grains, $S_s$ = specific surface of the grains, $\mu$ = absolute viscosity of the fluid, $R_e^*$ = modified Reynolds number (which uses the interstitial velocity—the ratio of velocity to porosity—as characteristic velocity), and $\Psi$ = sphericity. *See also* filter bed expansion, modified Reynolds number, interstitial velocity.

**fluidized-bed reactor** (1) A pressurized vessel designed for the separation of solid particles from a fluid, which flows upward and keeps the particles in suspension. Such reactors are used in ion exchange, granular activated carbon adsorption, biological contactors, and sludge blanket processes, as well as in furnaces and kilns. Also called expanded-bed reactor, although expanded beds are sometimes considered to be "less expanded" than fluidized beds and have little particle movement. *See also* anaerobic fluidized-bed reactor and fluidized-bed biological reactor. (2) Such a vessel used for the anaerobic treatment of wastewater. *See* Figure F-12.

**fluidized-bed softening** *See* caustic soda softening.

**fluidized-bed unit** One of three types of continuous flow, granular activated carbon adsorbers, in which water or wastewater is introduced at the bottom with sufficient velocity to expand the medium particles. *See also* batch treatment, countercurrent operation, fixed-bed adsorber.

**Fluidizer-Minor** A flash drying unit manufactured by Centrico, Inc.

**fluidizing** *See* fluidization.

**fluid mechanics** A branch of engineering science that studies the aspects of fluid behavior of interest to civil engineers, particularly hydraulics, which concentrates on the study of water and other liquids.

**fluid meter** Same as flowmeter.

**fluid ounce** A unit for the measurement of liquid volumes, equal to 1/16 liquid pint or 231/128 cubic inches. (The United Kingdom fl oz is 1/20 pint.)

**fluid parcel** A continuous mass of fluid.

**fluid potential (Φ)** The quantity of energy per unit mass or the product of the gravitational acceleration ($g$) by the hydraulic head ($H$):

$$\Phi = gH \quad \text{(F-63)}$$

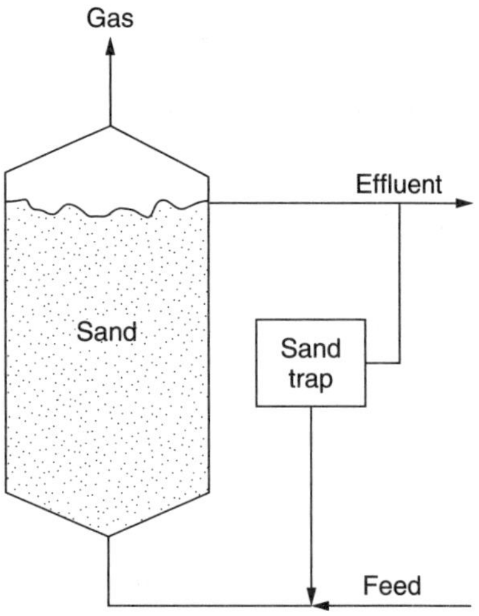


**Figure F-12.** Fluidized-bed reactor.

**fluid residence time distribution** A tracer response curve, i.e., a graphical representation of the distribution of a tracer versus time of flow. It is used in studying short-circuiting, retardation, and longitudinal mixing in sedimentation basins and other reactors. *See* time–concentration curve for more detail.

**fluid-seal completion** A method of construction of wastewater injection wells with tubing inside a long casing string and an open annulus at the bottom of the casing. The annulus is filled with a light fluid that floats on the aqueous waste. *See also* competent formation, incompetent formation.

**Fluid Systems membrane** A proprietary spiral-wound membrane. *See* spiral-wound module.

**Fluidyne ISAM®** A proprietary sequencing batch reactor system that includes three chambers for the anaerobic/anoxic/aerobic treatment of wastewater.

**fluke** A schistosome or other parasitic flatworm of the class Trematoda that usually develops in a snail as intermediate host; e.g., the blood fluke (*Schistosoma*), the sheep liver fluke (*Fasciola hepatica*); a trematode. *See also* the water snail (*Limnaea truncatula*).

**flume** (1) A deep and narrow channel, especially that of a mountain stream. (2) An open channel of wood, masonry, metal, or reinforced concrete, elevated or on a grade, used to carry water for power, transport, etc. across valleys, and minor depressions, or over obstructions. *See also* aqueduct and stream for the differences among various watercourses. (3) A flow measuring device in which the flow is locally accelerated by a streamlined lateral contraction (then called a venturi flume) or the lateral contraction and a hump in the invert. Critical flow usually occurs in the throat (or narrowest section). *See* Parshall flume, an example of a venturi flume. *See also* Figures F-13 (measuring flume) and F-14 (flume over a depression).

**fluorapatite [$Ca_5(PO_4)_3F$] or [$Ca_{10}(PO_4)_6F_2$]** A crystalline compound of fluoride and hydroxyapatite [$Ca_5(PO_4)_3(OH)$ or $Ca_{10}(PO_4)_6(OH)_2$] that has a hardening effect on teeth and bones; this mineral may also leach toxic substances into water sources. Also spelled fluoroapatite. *See* hydroxyapatite.

**Fluorapid®** A process developed by Biwater-OTV, Ltd. to increase the rate of clarifier rise in water and wastewater treatment. It combines floc-blanket settling and sand ballasting with inclined settling. *See also* Actiflo® and Densadeg®.

**Fluorapide system** *See* Fluorapid®.

**fluorescein ($C_{20}H_{12}O_5$)** An orange-red, crystalline, insoluble solid compound that produces an orange color and an intense green fluorescence in alkaline solutions; a common dye used as a tracer to assess dispersion and mixing regimes in water and wastewater treatment reactors.

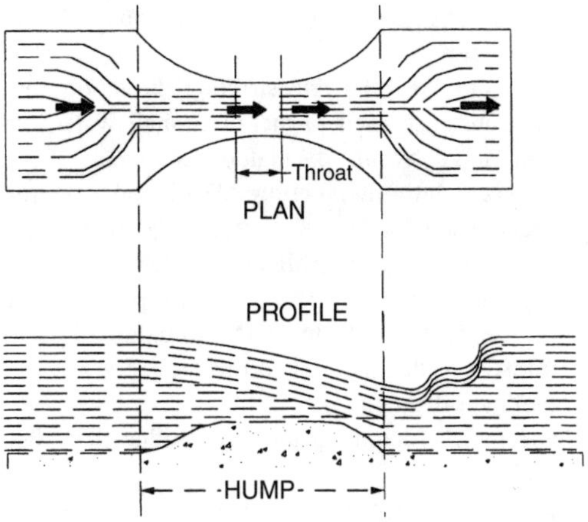


**Figure F-13.** Flume (measuring) (courtesy CRC Press).

**fluorescence** The emission of visible light by a substance when exposed to external radiation (e.g., X-rays or ultraviolet light), a phenomenon used to detect the presence of certain organic compounds in water or wastewater; the property to produce such emission or the radiation so produced. *See also* immunofluorescence, phosphorescence, fluorspar.

**fluorescence detector** An instrument that detects trace organic compounds in water.

**fluorescence in situ hybridization** *See* fluorescent in situ hybridization.

**fluorescence method** *See* fluorescent in situ hybridization.

**fluorescence microscope** An optical device that uses ultraviolet light as a source of illumination to obtain bright color images of specimens stained with fluorescent dyes (e.g., acridine orange or fluorescein). Maximum magnification = 2000 times. Resolution = 200 nm. Also called ultraviolet ray

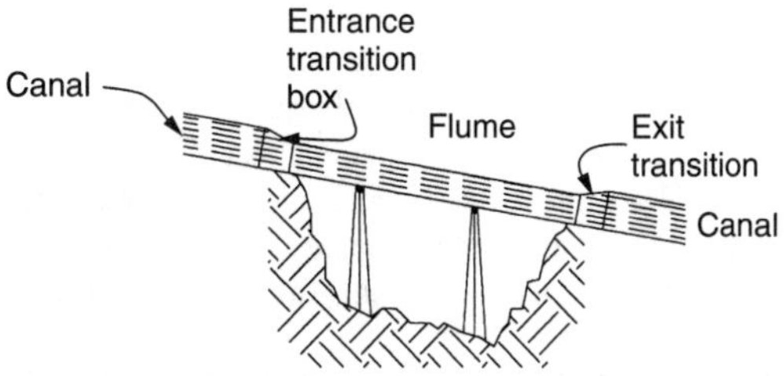


**Figure F-14.** Flume over a depression (courtesy CRC Press.).

microscope. *See also* light microscope, electron microscope.

**fluorescent dye** A stable, strikingly bright or glowing substance used to measure small water currents and determine the path of sediments.

**fluorescent immunolabeling** *See* immunofluorescence assay.

**fluorescent in situ hybridization (FISH)** The use of fluorescently labeled nucleic acid probes to target the RNA of cells in the identification of bacteria. Also called in situ hybridization. *See also* dot blot hybridization.

**Fluoretrak** Liquid tracing dye produced by Formulabs, Inc.

***Fluoribacter*** A genus of bacteria of the family *Legionellaceae,* sometimes implicated in pneumonia-like diseases. *See* legionellosis.

**fluoridation** The addition of a chemical to increase the concentration of fluoride ions in drinking water to a predetermined optimum limit (e.g., between 0.8 and 1.2 mg/L) to reduce the incidence of dental caries in children. Chemicals used include sodium fluoride (NaF), sodium silicofluoride ($Na_2SiF_6$), and hydrofluosilicic acid ($H_2SiF_6$); all three are corrosive and require tanks and piping in fiberglass, plastic, or other resistant materials. Defluoridation is the removal of excess fluoride in drinking water to prevent the mottling of tooth enamel. *See also* apatite, cryolite, dental fluorosis, fluorination, fluorosis, fluorspar, mottled enamel index, mottling.

**fluoridation notification** A requirement of the USEPA that water supply systems that exceed the secondary MCL for fluorides (2.0 mg/L) must notify their users of the actual fluoride concentration and of the risk of fluorosis.

**fluoride ($F^-$)** A gaseous, solid, or dissolved compound of fluorine with another organic or inorganic element; sometimes resulting from industrial processes, or found naturally in the minerals apatite, cryolite, and fluorspar; a salt of hydrofluoric acid (HF) such as sodium fluoride (NaF) or any compound containing fluorine such as methyl fluoride ($CH_3F$). Excessive amounts in food or water can lead to dental fluorosis or mottling (a brownish discoloration); fluoride can be removed from drinking water using adsorption on activated alumina. The USEPA recommends an upper limit of 0.6–2.4 mg/L of fluoride in drinking water, depending on air temperature. *See also* fluoride deficiency, Fluoride Rule, Genu Valgum, sodium fluoride.

**fluoride deficiency** Insufficient fluoride intake (in food and water) may result in dental caries and poor growth of bones and teeth. Fluoride (e.g., 1.0 mg/L) is commonly added to public water supplies.

**fluoride effect on health** *See* dental caries, fluorosis (mottled enamel), Genu Valgum.

**fluoride excess** In drinking water, a fluoride concentration of more than 2.0 mg/L may cause mottling of tooth enamel (fluorosis) and prolonged consumption of water with more than 4.0 mg/L may cause joint stiffness and skeletal deformities. *See also* Genu Valgum.

**fluoride ion ($F^-$)** *See* fluoride.

**fluoride removal** Effective fluoride removal processes include reverse osmosis, electrodialysis, and adsorption on granular activated carbon, activated alumina, bone char, etc. *See also* Nalgonda process.

**Fluoride Rule** A regulation of the USEPA setting the limit on fluoride concentration in drinking water.

**fluoridone** A herbicide applied directly to water to control submerged weeds.

**fluorimeter** An instrument that measures fluorescence, e.g., such fluorescent molecules as polyaromatic hydrocarbons. *See* fluorometer for detail.

**fluorination** The process of treating or combining (e.g., a hydrocarbon) with fluorine (F). *See also* fluoridation.

**fluorine (F)** A pale greenish-yellow, corrosive, noxious, very reactive gas of the halogen group, occurring naturally as fluorite and cryolite, abundant in the Earth's crust; with a characteristic odor, detectable at very low concentrations. The most electronegative of all elements, it cannot be oxidized to a positive state. It combines easily with all elements, forming fluorides. Atomic weight = 18.998. Atomic number = 9. Density = 1.696. Valence = 1. *See* fluorosis, fluorspar (calcium fluoride), cryolite.

**fluorite** A common compound of calcium (calcium fluoride, $CaF_2$), found in green, blue, purple, yellow, or colorless crystals of sedimentary and igneous rocks; it is used in ceramics and for the manufacture of hydrofluoric acid. Also called fluorspar.

**fluoroapatite** *See* fluorapatite.

**fluorocarbons (FCs)** Any of a number of organic compounds analogous to hydrocarbons in which one or more hydrogen atoms are replaced by fluorine. Once used in the United States as a propellant for domestic aerosols, they are now found mainly in coolants and some industrial processes. FCs containing chlorine are called chlorofluorocarbons (CFCs). They are believed to be modifying the ozone layer in the stratosphere, thereby allowing more harmful solar radiation to reach the Earth's surface.

**fluorochrome** Any of a group of fluorescent dyes used to label biological materials; e.g., fluorescein (resorcinolphtalein, $C_{20}H_{12}O_5$) and rhodamine B ($C_{28}H_{31}ClN_2O_3$).

**fluoroelastomer** An elastic, rubberlike polymer that contains fluorine.

**fluorogen** A fluorescent substance, i.e., a substance capable of emitting visible light or other form of radiation when exposed to external variation.

**fluorogenic substrate technology** A bacteriological laboratory technique that uses a substrate linked to a fluorogen to identify bacteria that have the specific enzymes that can hydrolyze the substrate. *See also* chromogenic substrate technology.

**fluorometer** An instrument that measures fluorescence to determine the nature of the material that emits the fluorescence. It measures the light emitted by a fluorescent material irradiated by ultraviolet light; sometimes used in tracing studies to determine the movement of currents and sediment in water. The fluorometer can detect very low concentrations of some dyes and chemicals used in tracer studies (e.g., fluorescein, rhodamine WT). Also called fluorimeter.

**fluorosilicic acid ($H_2SiF_6$)** A common substance used in the fluoridation of water supplies; molecular weight 144.08, available commercially as a straw-colored, transparent, fuming, corrosive, 20–35% aqueous solution.. It may contain dissociation products as impurities like hydrofluoric acid (HF) and silicon tetrafluoride ($SiF_4$):

$$H_2SiF_6 \rightarrow 2\ HF + SiF_4 \qquad \text{(F-64)}$$

Also known as hexafluorosilicic acid, silicofluoric acid, hydrofluorosilicic acid.

**fluorosis** (1) An abnormal condition caused by excessive intake of fluorine (e.g., more than 2.0 mg/L of fluoride), characterized chiefly by discoloration and pitting of the teeth (dental fluorosis), but also detrimental bone changes. However, a primary MCL of 4.0 mg/L is set to prevent skeletal fluorosis, which is crippling. (2) The changes in tooth enamel that indicate fluorosis; also called mottled enamel.

**fluorotrichloromethane** *See* trichlorofluoromethane.

**fluorspar ($CaF_2$)** Natural form of calcium fluoride, a common source of fluoride compounds used in water fluoridation when it is dissolved by a solution of alum. Also called fluorite.

**fluosilicic acid ($H_2SiF_6$)** Same as fluorosilicic acid.

**Fluosolids** A fluid bed reactor manufactured by Dorr-Oliver, Inc. for sludge processing.

**flush** (1) To open a water tap to clear out all the water that may have been sitting in the pipes for a long time. In new homes, to flush a system means to send large volumes of water through the unused pipes to remove loose particles of solder and flux. (2) To force large amounts of water through liquid to clean out piping, tubing, or storage or process tanks.

**flush hydrant** A hydrant that has outlets at or below grade.

**flushing** A method used to clean water distribution and sewer lines. Hydrants are opened and water with a high velocity flows through the pipes, and flows out of the hydrants. Water from hydrants is also used to remove deposits of grit from sewers.

**flushing chamber** *See* flush tank.

**flushing dose** In the operation of a trickling filter, the flushing dose is an intermittent high dose used daily to control the biofilm thickness and solids inventory. The flushing dose, expressed as the amount of liquid applied for each pass of a distributor arm (in mm/pass), is a function of the BOD loading and operating dose (or dosing rate) of the filter. *See also* rotational speed, dosing rate.

**flushing intensity** A measure of the hydraulic loading or dosing of a trickling filter, equal to the depth of wastewater applied for each pass of an arm of the distributor; it is sometimes referred to as the SK factor (in mm/pass of an arm), from the German word spülkraft, for dosing rate. Also called instantaneous dosing intensity. *See* SK concept.

**flushing manhole** A manhole, at the upper end of a sewer line, that can be occasionally filled with water to flush the line. Such an upper sewer section normally does not receive enough flow to be self-cleaning.

**flushing program** A preventative maintenance program for periodically flushing dead-end water mains, other parts of a water distribution system, manholes, and sewer lines.

**flushing tank** *See* flush tank.

**flushing velocity** The recommended velocity for flushing solids deposited in channels and pipelines through wastewater treatment plants, e.g., 5.0–6.0 fps for raw wastewater. Common values of the design velocity through a treatment plant for raw wastewater and other internal plant flows are in the range of 2.0–3.0 fps, but entrance velocities to clarifiers or other units may be reduced to 1.0 fps. *See also* minimum velocity, scouring velocity.

**Flush-Kleen®** A nonclogging ejector manufactured by Yeomans Chicago Corp. for wastewater pumping.

**flush tank or flush chamber** (1) A tank or chamber that stores water or wastewater for flushing a sewer line or manhole. (2) A tank or chamber storing water to flush a toilet.

**flush valve** A self-closing valve used to release water to flush a urinal or toilet.

**fluvial** Pertaining to streams or rivers. *See also* sedimentation terms.

**flux** (1) A flowing or flow; the rate of movement of mass through a unit cross-sectional area per unit time in response to a concentration gradient or some advective force. Also, heat transfer rate per unit area. (2) In a pressure-driven membrane process, the mass or volume rate of transfer through the membrane surface, i.e., the permeate flow per unit of membrane surface area (e.g., gallons per square foot per day); used to express the membrane's rate of water production. *See also* permeate flux.

**flux curve** A curve that represents the variation of solids flux as a function of solids concentration in studying sludge thickening. Solids flux is the product of solids concentration by the velocity of the sludge–water interface. The curve is used in the design of thickeners: it allows the determination of limiting variables once an underflow concentration is selected. *See* solids flux.

**fluxing lime** Another name for lime, a white or grayish-white, slightly soluble in water, calcined oxide (CaO). It may be obtained from ground limestone (calcite) or calcium carbonate ($CaCO_3$) and has many applications in water treatment, mortars, cements, etc.

**flux rate** The volume or mass of permeate (water, solvent) processed by a membrane per unit area per unit time, expressed, e.g., in gpd/square foot. *See* flux.

**flux rate decline** A measure of hydraulic performance of a membrane, equal to the decline in flux rate over a period of time.

**fly ash** (1) Noncombustible residual particles expelled by flue gas; e.g., suspended particles, charred paper, dust, soot, and other partially oxidized matter carried in the products of combustion (EPA-40CFR240.101-h). *See also* bottom ash, blackbirds. (2) The component of coal that results from the combustion of coal and is the finely divided mineral residue that is typically collected from boiler stack gases by electrostatic precipitators, mechanical collection devices, and/or fabric filters. Economizer ash is included when it is collected with fly ash (EPA-40CFR249.04-e and 423.11-e). (3) Fly ash weighs about 30 pounds per cubic foot, constitutes about 10% of the coal burned, has a 10–25% content of quicklime (CaO), and contains oxides of aluminum, iron, and silicon, as well as oxides of other metals, soot, carbon black, and pozzolanic materials. A properly equipped stack collects some of the fly ash, but the portion that escapes consists of smaller particles that do the most damage to human health and plants. Fly ash is sometimes used as an additive in sludge conditioning, resulting in a relatively dry product after dewatering; *see* chemical fixation. It may contain incompletely burned fuel and other pollutants. The addition of fly ash, instead of lime, to sludge is considered an advanced alkaline stabilization technology. *See* electrostatic precipitator.

**fly-borne disease** Any of a number of diseases that are caused by pathogens carried by flies (biting insects other than mosquitoes), e.g., leishmaniasis, loiasis, river blindness, and sleeping sickness.

**Flygt** Pump and other products of the ITT Flygt Corporation of Trumbull, CT.

**F:M** *See* food-to-microorganism ratio.

**F:M$_b$ (or F/M$_b$) ratio** The food-to-microorganism ratio based on active biomass concentration, expressed in grams of BOD per gram of biomass. This parameter is used in the design of anoxic tanks (Metcalf & Eddy, 2003):

$$F/M_b = QS_0/(V_{nox}X_b) \qquad \text{(F-65)}$$

where $Q$ = influent flowrate, m/day; $S_0$ = influent BOD concentration, mg/L; $V_{nox}$ = volume of anoxic reactor, m; and $X_b$ = biomass concentration in anoxic reactor, mg/L.

**F:M ratio** or **F/M ratio** Abbreviation of food-to-microorganism ratio.

**FNSI** Acronym of finding of no significant impact.

**FO** Acronym of forward osmosis.

**foam** The frothy substance, composed of minute bubbles formed by agitation, aeration, fermentation, etc., on the surface of a liquid containing soap, detergents, and other surface active materials and solid particles. In wastewater treatment, *Nocardia* is the organism responsible for foaming and sludge bulking; the resulting froth is slippery, difficult to remove when dried, and contains also grease and many bacteria. *See also Nocardia* foam, frothing agent, scum.

**Foam Ban™** An additive produced by Ultra Additives, Inc. for foam control.

**foam flotation** One of two forms of dispersed air flotation; a process widely used in the mineral industry, developed in 1914 to separate suspended particles in a suspension by introducing air bubbles through porous diffusers. The process is unsuitable for treating drinking water (because of the formation of large bubbles and the use of undesirable chemicals) but applicable to wastewater treatment. *See also* froth flotation.

**foam fractionation** Same as foam separation.

**foam fractionator** *See* foam separation.

**foaming** A nuisance condition affecting some biological treatment systems, caused by the development of the *Microthrix* and *Nocardia* genera of bacteria, which attach to and stabilize air bubbles. Foaming may be controlled by adjusting the mixed liquor suspended solids or by spraying a defoaming agent.

**foaming agent** Any of a classification of anionic surfactants (e.g., alkyl or amyl alcohols in the $C_5$ to $C_{12}$ range, pine oil) that reduce surface tension (and increase the stability of the particle–air mixture) just enough to promote the formation of bubbles and foam at the surface to support rising particles. *See* MBAS. Also called frother or frothing agent. Other types of flotation chemicals include activating, collecting, and depressing agents.

**foaming reagent** Same as foaming agent.

**foam phase separation** Same as foam separation.

**foam separation** A treatment process consisting of the injection of fine air bubbles into wastewater for the removal of suspended matter and excess surface active agents. The device used in this process, the foam fractionator, consists of a column of wastewater into which a sparger introduces bubbles that rise through the liquid with the solids and surfactants. This forms a foam at the surface, which is forced into a foam breaker, collapsed, and discharged. The treated effluent, or foamed product, exits at the bottom of the column. With a few simplifying assumptions, the following equation represents a mass balance for the process (Nemerow & Dasgupta, 1991):

$$C_f - C_b = 1000\,(G/F)E_sS \qquad \text{(F-66)}$$

where $C_f$ = solute concentration in the feed, mg/L; $C_b$ = solute concentration in the foamed or bottom product, mg/L; $G$ = volumetric gas rate, L/min; $F$ = liquid feed rate, L/min; $E_s$ = solute surface excess corresponding to $C_b$, mg/cm$^2$; and $S$ = specific surface of the bubbles in the foam phase, cm$^2$/cm$^3$. *See* Figure F-15. Also called fractionation.

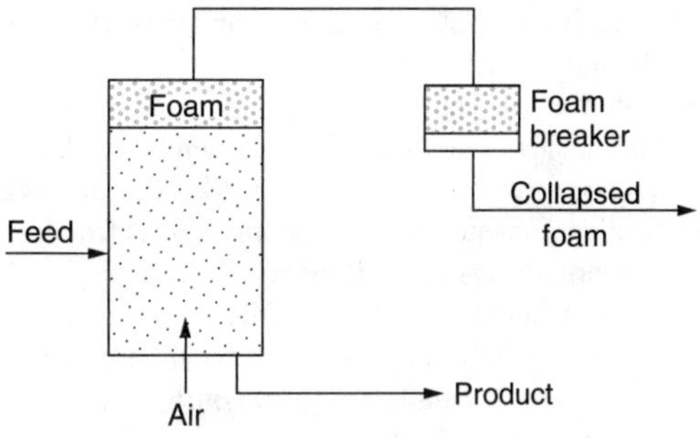


**Figure F-15.** Foam separation.

**Foamtrol™** An additive produced by Ultra Additives, Inc. for foam control.

**foci** Plural of focus.

**focus of infection** Initial site of a cell culture from which infection spreads.

**fodder crops** Forages such as corn, sorghum, and other crops (including leaves, stalks, and grain) grown principally for animal feed. Treated effluents, sludge, and nightsoil are often applied to fields of fodder crops. *See also* food, food chain, and forage crops.

**FOG** Acronym of fats, oils, and grease.

**fog** Droplets formed in the water-saturated atmosphere by condensation on aerosol particles.

**fold-and-form lining** A modified version of sliplining, a trenchless method of water main rehabilitation; a HDPE or PVC liner pipe is folded into a "C" or "U" shape at the factory to facilitate insertion in the field, attached to a cable, pulled into the damaged pipe, and then reformed into a round shape by an appropriate method. *See* trenchless technology, modified sliplining, tight-fit lining.

**follow-up monitoring** Monitoring for such drinking water quality parameters as copper and lead after the completion or rehabilitation of a treatment plant.

**fomes** Singular form of fomites.

**fomites** Inanimate objects that can be contaminated and transmit disease, e.g., toys, dirty clothes, door handles, dishes, hypodermic needles, bedding.

**fonolos ($C_{10}H_{15}OPS_2$)** Common name of the insecticide ethylphosphonodithioic acid O-ethyl S-phenyl ester.

**FONSI** Acronym of finding of no significant impact.

**Font'n Aire®** A floating aerator manufactured by Air-O-Lator Corp. that is also used as a decorative fountain.

**Food and Drug Administration (FDA)** A division of the U.S. Department of Health and Human Services that regulates food, drugs and cosmetics, including bottled water.

**food chain** A sequence of organisms, each of which (as predator) uses the next, lower member of the sequence (prey) as a food source. Also defined as a feeding sequence in which energy passes from primary producers to primary consumers to secondary consumers, i.e., from plants to herbivores to carnivores. An example of a food chain is unicellular algae → *Daphnia* → dragonfly nymph → smooth newt → grass snake, with energy dissipation accompanying every transfer. The food chain sometimes serves as a mechanism for concentrating certain toxic chemicals in tissues of predators. *See also* bioconcentration and biomagnification.

**food chain crop** Tobacco and crops grown for human consumption and feed for animals whose products are consumed by humans (EPA-40CFR257.3.5-4). *See also* fodder, food, and forage crops.

**food crop** A crop grown mainly for human consumption; e.g., fruits, vegetables, tobacco (EPA-40CFR503.9-1). *See also* fodder, food chain, and forage crops.

**food cycle** A series of interconnected food chains; a food web.

**food: microorganism ratio** or **food/microorganism ratio** *See* food-to-microorganism ratio.

**food poisoning** A gastrointestinal illness caused by ingestion of food containing pathogenic organisms and their toxic products. *See also* salmonellosis.

**food pyramid** The succession of predation levels in a food chain, with a decreasing number of predators at each higher level.

**food to biomass ratio** Same as food-to-microorganism ratio.

**food-to-microorganism ratio (F/M, F:M)** An expression of the organic loading rate of a biological waste treatment system, equal to the ratio of the influent BOD or COD to the volatile suspended solids (VSS) in the mixed liquor, i.e., pounds of BOD or COD per pound of VSS per day; also called food-to-biomass ratio, it is an important design and operation control parameter:

$$F/M = U = S_0/\theta \cdot X_v \qquad \text{(F-67)}$$

where $S_0$ is the concentration of BOD or COD in the influent, $\theta$ is the hydraulic detention time, and $X_v$ is the concentration of VSS in the mixed liquor. This definition is valid when the BOD or COD ($S_e$) of the effluent is negligible. Otherwise the correct equation from a mass balance is:

$$F/M = U = (S_0 - S_e)/\theta \cdot X_v \qquad \text{(F-68)}$$

Also called process loading factor, which is related to the solids retention time ($\theta_x$) as follows:

$$(\theta_x)(Y \cdot U - k_c) = 1 \qquad \text{(F-69)}$$

where $Y$ is a yield factor (mass of microorganisms produced per mass of substrate removed and $k_c$ is a rate constant (mass removed through endogenous respiration per mass present per day).

**food waste** The organic residue generated by the handling, storage, sale, preparation, cooking, and serving of foods; commonly called garbage (EPA-40CFR243.101-1).

**food web** A series of interconnected food chains in a natural community; a food cycle.

**fool's gold** A mineral that is sometimes mistaken for gold, e.g., iron or copper pyrites. Iron pyrite ($FeS_2$) is a sulfide mineral used in the production of sulfuric acid ($H_2SO_4$); it is also a source of cobalt. *See also* iron sulfide.

**foot-pound** Unit of measure of the work done by a force of one pound over a distance of one foot.

**footprint** *See* building footprint, small-footprint technology.

**foot-candle** A unit of illuminance or illumination; the illuminance of a luminous flux of one lumen uniformly distributed over a surface of one foot square, i.e., 1 lx = 1 lumen/ft$^2$ = 1 steradian-candela/ft$^2$ = 10.7639 meter-candles.

**foot valve** A special type of check valve located at the bottom end of the suction pipe on a pump. This valve opens when the pump operates to allow water to enter the suction pipe but closes when the pump shuts off to prevent water from flowing out of the suction pipe.

**forage crop** A crop that can be used as feed for domestic animals, by being grazed or cut for hay. *See also* fodder crop, food crop, and food chain crops.

**Forager™** A polymer sponge fabricated by Dynaphore, Inc. for the removal of heavy metals.

**forbay** The water behind a dam.

**forced aeration** The use of diffusers to apply pressurized air below the surface of a liquid.

**forced-aeration composting** One of three common methods of sludge composting, using a grid of aeration or exhaust piping supporting a mixture of dewatered sludge and bulking agent, and equipped with a blower. *See* aerated static pile sludge composting for detail.

**forced-circulation evaporator** A device that evaporates liquor by flash vaporization caused by the flow through heated tubes; circulation is maintained by pumping the liquid through the heating element with relatively low evaporation per pass. *See also* crystallizer, natural-circulation evaporator, falling-film evaporator, batch pan, and rising-film evaporator.

**forced draft deaerator** A device for removing dissolved gases from water or other solutions, using a countercurrent of air or another gas through a packed column. Also called a forced draft degasifier.

**forced draft degasifier** Same as forced draft deaerator.

**force main** A pressurized sewer line that conveys wastewater or stormwater from a pumping station to another force main, a manhole, a treatment plant or a point of disposal. A gravity sewer flowing full is said to be under surcharge conditions, but is not called a force main. Velocities in force mains are usually greater than in gravity sewers. A force main may also be a pipeline supplying water from pumps, as opposed to a gravity main, but a pressurized water supply pipe is usually called a water main. *See also* pumping main. Also called pressure main.

**forested wetland** A wetland dominated by vegetation greater than 20 ft.

**forfeited water right** A water right that has been canceled as a result of several consecutive years of nonuse.

**forfeiture** Cancellation of a water right for several consecutive years of nonuse. Same as abandonment.

**formaldehyde ($CH_2O$ or HCHO)** A colorless, toxic, pungent, irritating, biodegradable, soluble, and potentially carcinogenic gas, used chiefly as a disinfectant and preservative and in synthesizing other compounds like resins. It is formed during water disinfection by oxidants, particularly ozone. The USEPA requires water supply systems to monitor formaldehydes quarterly. *See* Figure F-16, urea formaldehyde. Also called methanal.

**formalin** A clear, colorless, aqueous solution of 40% formaldehyde. Also called formol.

**formalin-fixed cysts or oocysts** Cysts or oocysts of protozoans preserved in formalin, which renders them noninfectious.

**formation** A group of similar consolidation rocks or unconsolidated minerals.

**formation constants** Groundwater parameters that characterize the hydraulic properties of aquifers: hydraulic conductivity or coefficient of permeability, hydraulic diffusivity, specific yield, storage coefficient, and transmissivity.

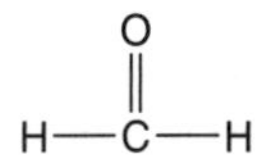


**Figure F-16.** Formaldehyde.

**formation fluid** Fluid present in a formation under natural conditions as opposed to introduced fluids, such as drilling mud (EPA-40CFR144.3).

**formation function** *See* distribution diagram.

**formation heat** *See* heat of formation.

**formazin turbidity unit (FTU)** A standard measure of turbidity based on a chemical reaction with a formazin solution. It is equivalent to the nephelometric turbidity unit (NTU) when a nephelometric turbidimeter is used. *See also* JTU.

**formic acid (HCOOH or $CH_2O_2$)** A colorless, irritating, fuming, water-soluble liquid, obtainable from ants or synthetically; used in dyeing, tanning, and in medicine; the common name of methanoic acid, a carboxylic acid.

**form loss of head** Loss of head caused by the change in shape of a conduit or waterway; distinct from loss of head by friction.

**formol** A clear, colorless, aqueous solution of 40% formaldehyde; another name for formalin.

**form resistance** Resistance to fluid flow causing head losses in transitions and appurtenances; such losses in conduits are significant for short distances but negligible compared to friction losses over long distances. *See also* Borda loss, equivalent pipe length, minor losses, velocity head.

**formula** (1) An algebraic equation or other expression that represents a variable or parameter in terms of other variables and constants. *See,* for example, the Bazin, Chézy, and Hazen–Williams formulas. (2) An equation representing a statement or other reasoning. (3) A chemical formula is an expression of the composition of a compound in terms of its elements and the ratios of their combination; e.g., the formula of calcium carbonate ($CaCO_3$) indicates that one atom of calcium combines with one atom of carbon and three atoms of oxygen. *See also* Hill's molecular formula arrangement.

**formulation** The substances comprising all active and inert ingredients in a pesticide.

**formula unit** The chemical formula of an ionic compound that does not form molecules, with the least number of elements possible. *See also* empirical formula, molecular formula, structural formula.

**formula weight** (1) Same as molecular weight for a molecule. (2) The sum of the atomic weights of the atoms in the formula unit of an ionic compound.

**forsterite ($Mg_2SiO_4$)** The magnesium end member of the olivine group; named after the German naturalist J. R. Forster (1729–1798). It weathers primarily by hydrolysis:

$$Mg_2SiO_4 + 4\,H_2O \rightarrow 2\,Mg^{2+} + 4\,OH^- + H_4SiO_4 \qquad \text{(F-70)}$$

**forward osmosis (FO)** A seawater or brackish water desalination technique similar to reverse osmosis but using an osmotic pressure gradient instead of hydraulic pressure as driving force. Simple and reliable, with a low fouling potential and low energy consumption, FO is classified (2007) as an emerging concentrate minimization and management method.

**forward reaction rate constant** The rate constant for a chemical reaction that occurs as written (as opposed to a reverse reaction). *See* kinetic law of mass action.

**Fossil Filter™** A filter manufactured by KriStar Enterprises for the treatment of stormwater runoff.

**fossil fuel** Fuel derived from ancient organic remains, e.g., peat, coal, crude oil, and natural gas, or any gaseous fuel derived from these materials.

**fossil groundwater** (1) The portion of a groundwater reservoir that has been in storage for a very long time as opposed to the water that is renewed through the current hydrologic cycle. Its use constitutes groundwater mining and cannot continue indefinitely. However, withdrawals of fossil groundwater may occur on a temporary and limited basis. The Ogallala aquifer underlying the High Plains area of the United States is considered fossil water from the melting of glaciers after the last Ice Age; it has been overexploited. (2) The term fossil water is sometimes used, synonymously with connate water, to denote interstitial water that was buried at the same time as the original sediment.

**fossil salt** Another name of halite, a soft white, red, yellow, blue, or colorless evaporite mineral that is at least 95% sodium chloride (NaCl) in cubic crystals; it occurs together with other minerals such as gypsum and sylvite. Also called native salt, rock salt.

**fossil water** *See* fossil groundwater.

**Fotovap** An evaporator manufactured by Licon, Inc. for the treatment of wastewater from photoprocessing laboratories.

**fouling** In general, an undesirable condition caused by bacterial growth and accumulation of materials on surfaces or within pores. (1) A slimy deposit of organisms on concrete, masonry, and metal surfaces in contact with water. (2) The accumulation of solids in the media of filters and ion-exchange units, thus increasing their maintenance and reducing their capacity. *See also* breakthrough, resin fouling, tailing. (3) The reversible or irreversible deposition and accumulation of colloidal or suspended matter, microorganisms, etc. from the feed stream on the surface and in the pores of membranes. *See also* colmatage. (4) The accumulation of solids on the adsorptive surface of activated carbon. (5) The formation of biological slimes or inorganic precipitants on the outside of porous diffusers. (6) The attachment and growth of deposits in cooling tower recirculation systems. *See also* biofouling.

**fouling factor** A correction factor used to allow for some variation of aeration equipment performance due to fouling. It depends on diffuser type and on wastewater characteristics. *See* actual oxygen transfer rate. For fine-pore diffusers, it varies between 0.5 and 1.0. *See also* fouling rate, apparent alpha.

**fouling index** A parameter that indicates the fouling potential (amount of particulate and colloidal matter) of a feedwater and corresponding pretreatment requirements of a membrane, as determined from simple tests (filtration of a sample of feedwater through a 0.45 micron Millipore filter at 30 psi). Common indexes, resembling mass transfer coefficients, include the silt density index, the modified fouling index, and the mini plugging factor index. *See also* plugging factor.

**fouling potential** The amount of suspended and colloidal solids, metal oxides, microbial matter, organics, and other solutes contained in a feedwater and capable of fouling a membrane unit. *See* fouling index.

**fouling rate** The rate at which the fouling factor changes.

**fountain** (1) A spring or source of water; the head of a stream of water. (2) A drinking fountain. (3) A jet or stream of water from an opening or structure to cool the air or to serve as an ornament. (4) A type of spray aerator consisting of towers that spray onto the surface of raw water reservoirs; a fountain aerator. Also called spray tower.

**fountain aerator** A spray aerator commonly used in water treatment, consisting of a device in the form of a fountain in several sections whose diameters decrease from the bottom; it sprays onto the surface of a body of water. *See* Figure F-17. Also called fountain, fountain spray aerator, spray tower.

**fountain flow** The type of flow from a vertical pipe open at the top; the fluid rises and then spreads like an umbrella.

**fountain head** (1) A fountain or spring from which a stream flows; the head or source of a spring. (2) The head in a saturated confined aquifer.

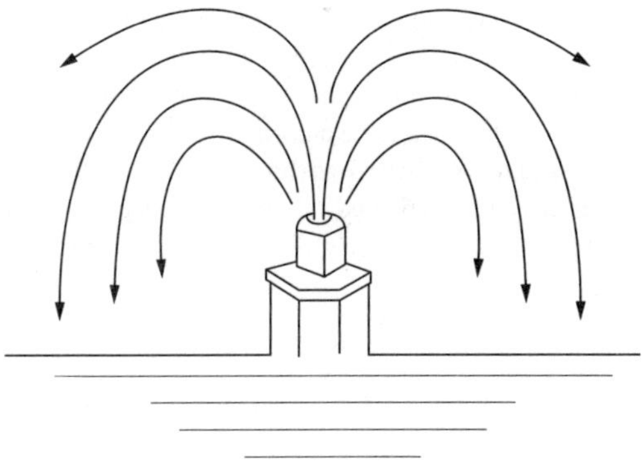

**Figure F-17.** Fountain aerator.

**fountain spray aerator** *See* fountain aerator.

**Four-Beam™** An instrument manufactured by BTG, Inc. to measure turbidity and suspended solids concentrations.

**Fourier transform infrared (FTIR) analysis** A spectroscopic method that uses continuous emissions monitoring in combination with such devices as gas and liquid chromatographs to analyze the molecular structure of a substance, e.g., to identify organic and inorganic compounds in liquids, solids, and gases.

**Fourier transform infrared (FTIR) spectrometry** *See* Fourier transform infrared analysis.

**four-log removal** The removal of 99.99% of pathogens or other contaminants from water or wastewater. Note that 99.99% = 0.9999 corresponds to a residual of $R = 1 - 0.9999 = 0.0001 = 10^{-4}$. Therefore, $-\log R = 4$. *See* four nines DRE.

**four-methylumbelliferyl-β-D-glucuronide** *See* methylumbelliferyl-β-D-glucuronide.

**four nines DRE** An expression of the efficiency of an incinerator that achieves a DRE (destruction and removal efficiency) of 99.99%. *See* four-log removal.

**four-stage Bardenpho process** Same as Bardenpho®.

**four-way valve** A valve that has a movable element to allow flow passage in either pair of two directions.

**Fox-Pac** A sanitation unit manufactured by Red Fox Environmental, Inc. for marine uses.

**FPA** Acronym of flavor profile analysis.

**F+ phage** *See* male-specific phage.

**fpm** Acronym of foot (feet) per minute.

**fps** Acronym of foot (feet) per second.

**fractal index** A parameter that indicates the pattern of solids deposition on a surface; used to predict the volume and density of sludge (Symons et al., 2000). (Fractal = geometrical or physical structure of irregular shape; fractal dimensions = dimensions of such a structure.)

**fractional distillation** The separation of volatile components in a mixture by gradually increasing its temperature and collecting the components as they reach their different boiling points; used, e.g., in cryogenic air separation and the production of paraffin hydrocarbons from petroleum.

**fractionate** To separate a mixture into its components, fractions or parts, based on differences between their boiling points or other properties. *See* fractionation, fractional distillation, crystallization.

**fractionation** (1) A distillation method for separating successively the volatile components of a mixture based on the differences in their boiling points. (2) *See* COD fractionation. (3) *See* foam fractionation.

**fragipan** A strongly compacted soil layer underlying soft materials and cemented with an acid; it is a platey structure between deposited materials and a parent rock. *See* hardpan for other similar layers.

***Francisella tularensis*** A small, nonmotile, bacterial agent that causes the intestinal disease tularemia; usually transmitted through insect bites, with a reservoir in small mammals. It has a high mortality ratio and is used in biological warfare.

**Francis formula** An expression of flow ($Q$) over a rectangular, sharp-crested, suppressed, or Cipolletti weir of length ($L$) and head ($H$), where $H \leq L/3$:

$$Q = \alpha L H^{1.5} \tag{F-71}$$

with $\alpha = 3.33 - 3.37$.

**frazil** Ice crystals formed in swift streams or rough seas.

**frazil ice** Fine ice crystals that form when water is turbulent and has a temperature far below freezing. When extended to the bottom of a stream, frazil ice may impede flow and cause damage. In waterworks, frazil ice may form in intakes or plug screens.

**Fred Hervey Plant** A 10.0-mgd water reclamation plant constructed by the city of El Paso (Texas) for direct groundwater recharge. It includes the following unit operations or processes: preliminary treatment, primary sedimentation, suspended-growth BOD removal and nitrification, methanol addition, suspended-growth denitrification, lime coagulation, recarbonation (carbon dioxide addition), sand filtration, ozonation, adsorption on granular activated carbon, chlorination before and after storage, and deep well injection. The effluent meets drinking water standards. *See also* Water Factory 21.

**free acid form** The condition of a regenerated weak acid cation exchanger.

**free ammonia** Ammonia that is free in solution, as compared to the ammonia bound in chloramines.

**free available chlorine (FAC)** (1) The value obtained using the amperometric titration method for free available chlorine described in "Standard Methods for the Examination of Water and Wastewater" (EPA-40CFR423.11-1). (2) Same as free chlorine; the sum of the three species in water: molecular chlorine ($Cl_2$), hypochlorous acid (HOCl), and hypochlorite ion ($OCl^-$), with the concentrations expressed in mg/L as $Cl_2$. *See also* combined available chlorine.

**free available chlorine residual** *See* free chlorine residual.

**free available residual chlorine** *See* free chlorine residual.

**free base form** The condition of a regenerated weak base anion exchanger.

**freeboard** The vertical distance between the top of a hydraulic structure and the normal maximum liquid level, provided to prevent overflows due to liquid movement. For example: (1) Thc vertical distance between the top of a tank or surface impoundment dike, and the surface of the waste contained therein. (2) The vertical distance from the normal water surface to the top of the confining wall. (3) The vertical distance from the sand surface to the underside of a trough in a sand filter; also called available expansion (EPA-40CFR260.10).

**freeboard height** The wall height of an open-top vapor cleaning tank above the vapor layer. The freeboard ratio is this height divided by the width of the tank opening; this ratio is usually set at 0.75.

**freeboard ratio** *See* freeboard height.

**free carbon dioxide** The carbon dioxide that is present in water as gas ($CO_2$) or as carbonic acid ($H_2CO_3$), but not in the compounds carbonates and bicarbonates. $[H_2CO_3^*]$ denotes the concentration of free carbon dioxide in eq/L:

$$[H_2CO_3^*] = [CO_2] + [H_2CO_3] \qquad \text{(F-72)}$$

*See also* acidity.

**free chlorination** Chlorination beyond the break point, as distinguished from chloramination.

**free chlorine** The amount of chlorine available as dissolved gas ($Cl_2$), hypochlorous acid (HOCl), or hypochlorite ion ($OCl^-$) that is not combined with ammonia or other compounds. Free chlorine is toxic to pathogenic microorganisms and may also serve as an oxidizing agent. Chloride, with an oxidation state of minus one (–1), is the stable form. Same as free available chlorine.

**free chlorine-activated carbon reactions** The reactions of activated carbon in water with hypochlorous acid (HOCl) and hypochlorite ion ($OCl^-$):

$$HOCl + C^* \rightarrow C^*O + H^+ + Cl^- \qquad \text{(F-73)}$$

$$OCl^- + C^* \rightarrow C^*O + Cl^- \qquad \text{(F-74)}$$

where C* and C*O represent the carbon surface and a surface oxide, respectively. Such reactions result in the production of organic by-products.

**free chlorine burnout** The intermittent or periodic application of free chlorine to drinking water distribution systems, sometimes coupled with intense flushing, to overcome accumulated chlorine demand and control nitrifying organisms in a chloraminated system. Also simply called burnout.

**free chlorine residual** That portion of the total available residual chlorine composed of dissolved chlorine gas ($Cl_2$), hypochlorous acid (HOCl), and/or hypochlorite ion ($OCl^-$) remaining in water after chlorination, i.e., after satisfaction of all the chlorine demand; i.e., at the end of the specified contact period. This does not include chlorine that has combined with ammonia nitrogen or other compounds. Also called free available chlorine residual, free available residual chlorine, free residual chlorine, or simply free residual.

**free chlorine species** The various products of the reactions of chlorine in water that are not associated with ammonia or other compounds. Major chlorine species are, depending on pH, molecular chlorine ($Cl_2$), hypochlorous acid (HOCl), and hypochlorite ion ($OCl^-$). The reactions may result also in transient chlorine species ($H_2OCl^+$, $Cl^+$, and $Cl_3^-$). *See also* combined chlorine, chloramines, and chlorine residual.

**free convection** Motion in a fluid due to density differences. Also called gravitational convection.

**free cyanide** The cyanide ion (CN or HCN) in wastewater, as compared to the cyanide bound in metal cyanide complexes, e.g., copper cyanide, $Cu(CN)_4^{3-}$.

**freedom degree** *See* degree of freedom.

**free energy** *See* Gibbs' free energy.

**free fall** The condition of discrete particles settling unhindered in a liquid.

**free-falling weir** or **free-fall weir** Same as free weir.

**free-fall weir** Same as free weir.

**free flow** (1) Gravity flow or open-channel flow in sewers, pipes, streams, etc. (2) A condition of flow that is not affected by the submergence of a structure (e.g., a weir) or the existence of a tailwater downstream.

**Free-Flow™** Ceramic diffuser plates manufactured by Davis Water & Waste Industries, Inc. for use in fine-bubble aeration.

**free-flow condition** *See* free flow (2).

**free groundwater** Same as free water.

**free ligand** The fraction of a chemical dissociated from its binding site on a protein or chelating agent.

**free liquid** A liquid that readily separates from the solid portion of a waste under ambient temperature and pressure (EPA-40CFR260.10).

**free mineral acid** A strong acid such as sulfuric acid ($H_2SO_4$) or hydrochloric acid (HCl) in acid mine water or other solution.

**free mineral acidity (FMA)** Same as mineral acidity.

**free moisture** Liquid that will drain freely by gravity from solid materials.

**free oil** Nonemulsified oil that readily separates from water.

**free product** A petroleum hydrocarbon or other regulated substance in the liquid ("free" or non-aqueous) phase. *See also* nonaqueous phase liquid (NAPL).

**free radical** A very reactive and unstable chemical, atom, or molecule, bearing an unpaired (or unshared) electron, that can destabilize other molecules and generate more free radicals through rapid chain reactions. Most free radicals are short-lived and can be damaging to living organisms; sometimes responsible for toxic responses in the body and even degenerative conditions, from natural aging to Alzheimer's disease. A dot (˙) next to an element or radical indicates the unpaired electron phenomenon, as in ($O^{\cdot}$) and ($OH^{\cdot}$). Chlorine dioxide ($ClO_2$) is also known to have an unpaired electron in its molecular structure, thus being in an unstable and reactive condition. *See also* diradical, lipid peroxidation.

**free residual** The dissolved chlorine gas ($Cl_2$), hypochlorous acid (HClO), or chlorite ions ($OCl^-$) that remain in solution after the chlorine demand of a water is satisfied. *See* free chlorine residual.

**free residual chlorination** The application of chlorine to water to produce a free available chlorine residual equal to at least 80% of the total residual chlorine (sum of free and combined available chlorine residual).

**free residual chlorine** *See* free chlorine residual.

**free settling** The settling of discrete, nonflocculent particles in a dilute suspension, without flocculation or any interaction between particles. *See* sedimentation type I for more detail. *See also* Stokes' law, and other types of settling: II or flocculent, III or hindered or zone, IV or compression.

**Free-Slide** Traveling water screens manufactured by Envirex, Inc. as a wire mesh basket.

**free surface** The boundary of a liquid in contact with the atmosphere.

**free-surface flow** Same as open-channel flow.

**free vortex** The condition created by a rotating flow within a cylindrical vessel, such that the product of the tangential velocity ($V$, fps) and the radius ($R$, ft) is constant:

$$VR = \text{constant} \qquad \text{(F-75)}$$

This condition is applied in the design of devices that use accelerated gravity separation; *see,* e.g., Teacup.

**free water** (1) Water occurring just below the water table, moving according to the slope of the latter, and extending down to the first confining bed. Also called free groundwater or mobile water, but sometimes confused with gravitational water. *See* subsurface water. (2) In a sludge suspension, water that is not bound to the sludge particles and does not move with them, water that is relatively easy to remove by such dewatering methods as centrifugation and vacuum filtration. Also called bulk water. *See also* bound water.

**free-water content** *See* water content.

**free water knockout (FWKO)** A vessel used in an oil field to separate water from oil by gravity.

**free water surface system** One of two types of constructed wetlands used for wastewater treatment: it consists of level, shallow, parallel basins or channels with a relatively impermeable bottom, in which emergent vegetation grows. Bacteria attached to the plants and vegetative litter decompose the organic matter and remove BOD from the wastewater, whereas suspended solids are entrapped in the vegetation or are removed by sedimentation. The required surface area of an FSF wetland ($A$, ha) depends on its depth ($D$, m), the volume of wastewater ($Q$, $m^3$/day), and the detention time ($t$, days):

$$A = 0.0001\, Qt/D \qquad \text{(F-76)}$$

*See also* subsurface flow system.

**free water surface wetland** A wetland with its water surface exposed to the atmosphere. *See* free water surface system.

**free water system** *See* free water surface system.

**free weir** A weir that is not submerged and that falls over the crest without any tailwater interference. Also called a free-fall weir, a free-falling weir, or a friction weir.

**freeze-assisted drying** *See* freezing (4).

**freeze-assisted sand bed** A device used to dewater water or wastewater treatment sludge, similar to the sand drying bed, but with the sludge being mechanically or naturally frozen and then thawed to facilitate water separation from the solids. *See also* air drying.

**freeze distillation** Production of water by freezing a saline solution and removing the salts from the crystals before they melt. *See also* freezing.

**freeze drying** A method of preservation of heat-sensitive substances (foods, blood, antibiotics, microorganisms, etc.) by freezing them and then subliming the ice in a vacuum. Freeze drying is used in the treatment of sludge and hazardous wastes. Also called lyophilization.

**freeze–thaw bed** A device used to dewater sludge using the freeze–thaw process and designed with the following features: (a) sludge application in several thin layers instead of a single thick layer; (b) the bed is covered to keep snow and rain out; (c) the sides of the housing are left open to allow free air circulation; (d) the depth ($D$, m) of sludge that can be frozen is (AWWA, 1999):

$$D = Tt/(11{,}371 + 19{,}294\,Z) \qquad \text{(F-77)}$$

where $T$ = average ambient temperature (°C), $t$ = freezing time (hours), and $Z$ = thickness of sludge layer (m). Also called a freezing bed or a residuals freezing bed.

**freeze–thaw conditioning** *See* freezing (2).

**freeze–thaw cycles** Same as freeze–thaw process.

**freeze–thaw process** A procedure used in sludge handling to improve the effectiveness of such dewatering methods as sand drying bed and lagoon dewatering.

**freeze–thaw sludge dewatering** Same as freeze–thaw process. *See also* freeze–thaw bed and freeze (2).

**freezing** A process used, under natural or artificial conditions, to produce fresh water from brackish and saline waters, to condition sludge for further processing, or to reduce the water content of sludge. Freezing is a water recovery process as compared to demineralization or desalination. *See* contact freezing, eutectic freezing, latent heat of vaporization, saline water conversion classification. (1) Water to be desalinated is cooled by heat exchange until ice crystals are produced, which are removed and melted, leaving a brine for disposal. Some of the product water is used to wash the ice crystals. (2) Natural freezing is used to condition wastewater sludge by converting this jellylike material into a granular material that easily drains upon thawing; the process is called freeze–thaw conditioning. Freezing, which destroys the water-binding capacity of the material, must extend through the entire depth ($X$, cm) of the sludge, which depends on the duration of freeze ($t$, days) and the difference ($\Delta T$, °C) between the actual temperature and the freezing temperature (McGhee, 1991):

$$X = 2.04\,(\Delta T \cdot t)^{0.5} \qquad \text{(F-78)}$$

*See also* bound water, free water, interstitial water, surface water, curing time. (3) Alum coagulation sludge from water treatment can be concentrated by freezing before disposal to landfills or through land application. (4) Freeze-assisted drying of sludge involves the application of sludge in 75-mm layers to accelerate its dewatering under freezing temperatures; each layer is completely frozen before the application of the next. The total depth ($Y$, cm) of sludge that can be applied depends on the maximum depth of frost penetration ($F_p$, cm) in the area (McGhee, 1991):

$$Y = 1.76\,F_p - 101 \qquad \text{(F-79)}$$

**freezing bed** Same as freeze–thaw bed.

**freezing index** A parameter used to determine the extent of frost penetration, as it affects the depth of cover over water and sewer lines. It is the range of maximum positive cumulative departure minus maximum negative cumulative departure of daily mean temperatures from 0°C or 32°F, beginning the first day a freezing temperature is recorded. *See* Figure F-18 An approximate value of the freezing index ($F$) is (in North America) (Fair et al., 1971):

$$F = (32\,M - \Sigma T_m)N \qquad \text{(F-80)}$$

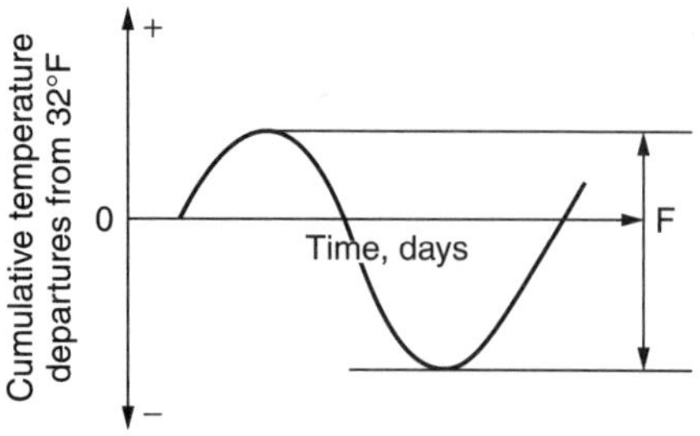


**Figure F-18.** Freezing index (F).

where $M$ = number of months with temperature below freezing, $N$ = average number of days during these months (e.g., $N$ = 30.2 for December through March), and $T_m$ = sum of the mean temperatures during these months. *See also* frost depth, degree-day.

**freezing nucleus** A particle of clay, dry ice ($CO_2$), silver iodide (AgI), or other mineral serving as a core for the formation of ice crystals. *See also* condensation nucleus.

**freezing point** The temperature at which a liquid freezes to a solid at standard pressure. *See* melting point for detail.

**freezing process** *See* freezing.

**Fre-Flo** A cement underdrain manufactured by Infilco Degremont, Inc.

**French degree** A unit of hardness, equivalent to 10 mg/L as $CaCO_3$. *See also* American degree, British degree, German degree, Russian degree.

**French square** A glass bottle used for diluting bacteriological samples. Also called dilution bottle, milk dilution bottle.

***Frenothrix*** One of three groups of iron bacteria, filamentous organisms that use iron compounds and cause a number of problems in water supply systems, e.g., taste and odor, color ("red water"), and clogging of well screens. *See also Gallionella* and *Leptothrix.*

**frequency factor** A parameter of the Arrhenius equation; also called preexponential factor.

**fresh-air inlet** An opening with a perforated cover for ventilation in a wastewater line.

**fresh–saltwater interface** The coastal area where fresh groundwater meets saline groundwater. Sometimes called bad water line.

**fresh sludge** Sludge with little decomposition.

**fresh wastewater** Relatively recent wastewater, still with some dissolved oxygen.

**freshwater** Water with a low mineral content, as expressed by a concentration of dissolved solids (DS) of less than 1000 mg/l as compared to sea or ocean water, brackish water, saltwater or saline water, brine, and saline estuarine waters. Inland freshwater sources include ponds, lakes, and streams. *See* salinity.

**freshwater barrier** A hydraulic structure, e.g., a line of recharge wells or trenches, installed between sea and land to delay or stop saltwater intrusion. *See* Ghyben–Herzberg principle.

**freshwater composition** *See* Table F-02 for the composition of a typical stream, lake, or groundwater. *See also* soda lake, seawater constituents.

**freshwater environment** A body of freshwater and all its living things, inorganic matter, and dead organic matter. *See also* running-water habitat, standing-water habitat.

**Table F-02.** Freshwater composition (typical)

| Constituent, unit | River | Lake | Groundwater |
|---|---|---|---|
| Bicarbonate ($HCO_3$), $pHCO_3$ | 3.5 | 2.7 | 2.5 |
| Calcium (Ca), pCa | 3.7 | 3.0 | 3.0 |
| Chlorine (Cl), pCl | 5.6 | 3.6 | 3.6 |
| Hydrogen ion (H), pH | 7.3 | 7.7 | 7.5 |
| Magnesium (Mg), pMg | 4.6 | 3.4 | 3.2 |
| Potassium (K), pK | 4.9 | 4.3 | 4.1 |
| Sodium (Na), pNa | 4.3 | 3.4 | 3.0 |
| Sulfate ($SO_4$), $pSO_4$ | 4.2 | 3.6 | 3.4 |

*Note:* The notation p indicates the negative logarithm (base 10) of the concentration of a constituent.

**freshwater lake** Any inland pond, reservoir, impoundment, or other similar body of water that has recreational value, exhibits no oceanic or tidal influences, and has a total dissolved solids concentration of less than 1% (EPA-40CFR35.1604-2).

**freshwater lens** The body of underground freshwater supported by saltwater where these two water sources are in contact. *See* Ghyben–Herzberg principle.

**freshwater profile** *See* the following terms: benthic zone, compensation depth, compensation point, epilimnion, hypolimnion, lake stratification, limnetic zone, littoral zone, neuston layer, overturn, profundal zone, thermocline. *See also* Figure F-19, marine profile.

**fretting corrosion** *See* stress corrosion.

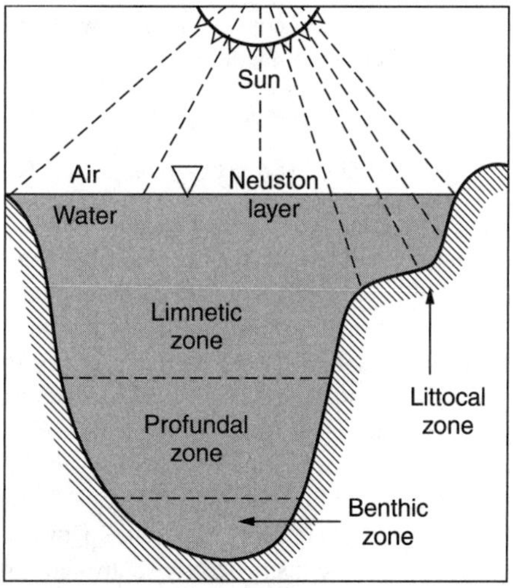


**Figure F-19.** Freshwater profile (adapted from Maier et al., 2000).

**Table F-03.** Freundlich adsorption isotherm parameters for granular activated carbon

| Compound | $K$ | $1/n$ |
|---|---|---|
| Heptachlor | 9,320 | 0.92 |
| Endrin | 666 | 0.80 |
| Alachlor | 482 | 0.26 |
| Lindane | 299 | 0.43 |
| Chlordane | 194 | 0.50 |
| Aldicarb | 133 | 0.40 |
| Chlorobenzene | 101 | 0.35 |
| Bromoform | 93 | 0.66 |
| Chloral hydrate | 27 | 0.05 |
| Phenol | 21 | 0.54 |

**Freundlich adsorption constant** Same as Freundlich constant.

**Freundlich adsorption isotherm** *See* Freundlich isotherm.

**Freundlich adsorption isotherm parameters** The coefficient $K$ and exponent $1/n$ in the Freundlich isotherm. *See* Table F-03 for some parameters related to granular activated carbon adsorption.

**Freundlich capacity coefficient** Same as Freundlich constant.

**Freundlich constant** *See* Freundlich isotherm and Freundlich adsorption isotherm parameters.

**Freundlich equation** *See* Freundlich isotherm.

**Freundlich equilibrium adsorption equation** *See* Freundlich isotherm.

**Freundlich isotherm** A graphical representation of the equilibrium relationship between adsorbate, adsorbent, and solution at a given temperature; derived from empirical considerations. Two forms of the Freundlich equation are:

$$Q = KC^{1/n} \quad \text{(F-81)}$$

$$\log Q = \log K + (1/n) \log C \quad \text{(F-82)}$$

where $Q$ = quantity adsorbed by unit mass of adsorbent at equilibrium; $K$ = an empirical constant (called the Freundlich constant, Freundlich adsorption constant, or Freundlich isotherm constant), indicative of the adsorptive capacity of an adsorbent for an adsorbate. It may be determined graphically by plotting $\log Q$ vs. $\log C$. $C$ = the equilibrium concentration of adsorbate in the dilute solution; and $n$ = another empirical constant usually greater than unity. It is one of two commonly used isotherms to describe activated carbon adsorption in water and wastewater applications. These equations can be used to determine adsorption coefficients of a given adsorbent or to describe the adsorption of tastes and odors in terms of threshold values, using:

$$Q = (C_0 - C)/M \quad \text{(F-83)}$$

$C_0$ being the threshold taste or odor concentration, $C$ the residual concentration, and $M$ the mass of adsorbent. The Freundlich isotherm is also called van Bemmelan equation. *See also* modified Freundlich isotherm, BET isotherm, Langmuir isotherm, phenol value.

**friability** Characteristic of a material that is easily crumbled or reduced to powder. The ability of ion-exchange beads to resist hydrostatic pressure.

**friable** Capable of being crumbled, pulverized or reduced to powder by hand; easily crumbled, not cohesive or sticky.

**frictional resistance** Resistance to fluid flow causing head losses by friction over surfaces; such losses in conduits are significant for long distances but negligible compared to form losses over short distances. Also called surface resistance. *See* the Darcy-Weisbach and Hazen–Williams equations.

**friction coefficient or friction factor** A coefficient used in hydraulic formulas to reflect the energy gradient caused by friction, i.e., a measure of the resistance to fluid flow. It depends on the Reynolds number of the flow and on the roughness of the conduit wall. *See also* the Darcy–Weisbach and the Manning formulas.

**friction factor** Same as friction coefficient.

**FrictionFlex®** A process developed by SLT North America, Inc. for the production of containment liners.

**friction head, friction head loss, friction loss** The head (also called pressure or energy) lost by a fluid flowing in a pipe or channel as a result of turbulence caused by velocity, the roughness of the pipe and channel walls, and restrictions caused by fittings. Water flowing in a conduit loses pressure or energy as a result of friction losses. The hydraulic grade line drops as a result of friction head loss.

**friction head loss** Same as friction head.

**friction loss** Same as friction head.

**friction slope** The slope of the energy line at a cross section of channel or conduit. It may be determined from the Manning formula if the other variables are known (discharge, water surface elevation, roughness coefficient, and hydraulic radius). The HEC-2 and HEC-RAS models use four approximations of friction slope between two cross sections: average friction slope, average conveyance, geometric mean friction slope, and harmonic mean friction slope.

**friction weir** Same as free weir.

**Fridgevap®** A package plant manufactured by Licon, Inc. using the distillation process.

**fringe water** Water in the capillary fringe above the water table.

**Fritsche's formula** A formula for the calculation of the Darcy–Weisbach friction factor ($f$) for circular air pipes smaller than 10 inches in diameter (Fair et al., 1971):

$$f = 0.048\, D^{0.027}/Q^{0.148} \quad \text{(F-84)}$$

where $D$ is the pipe diameter in inches and $Q$ is the rate of airflow in cubic feet per minute. *See also* air piping.

**front clean/back return screen** A mechanically cleaned bar screen used to remove coarse materials from wastewater ahead of other treatment units. It consists of a chain-driven mechanism to move the rake teeth through the screen openings. The screen is raked to clean from the front/upstream side, and the rakes return to the bottom from the back. *See* chain-driven screen.

**front clean/front return screen** A mechanically cleaned bar screen used to remove coarse materials from wastewater ahead of other treatment units. It consists of a chain-driven mechanism to move the rake teeth through the screen openings. The screen is raked to clean from the front/upstream side, and the rakes return to the bottom from the front. *See* chain-driven screen.

**Frontloader** A bar screen manufactured by Schreiber Corp. with a reciprocating rake.

**Frontrunner** A bar screen manufactured by Jones & Attwood, Inc., with a reciprocating rake.

**frost depth** The depth of frozen of frozen soils, a consideration in the design and construction of water and sewer lines. An approximate value of frost depth ($D$, in) may be determined from the following equation (Fair et al., 1971):

$$D = 1.65\, F^{0.468} \quad \text{(F-85)}$$

where $F$ is the freezing index.

**frost point** A temperature below 0°C to which air, with a given quantity of water vapor and at a given pressure, must be cooled to cause condensation of the vapor in the air and begin to form dew. *See also* dew point.

**FrostSafe™ Two-way Air Damper** A corrosion-resistant valve manufactured by Val-Matic to help minimize the flow of cold air into vaults through vent piping and help prevent the freezing of the vault components. *See* Figure F-20.

**froth** A collection of bubbles, as on an agitated liquid; *see* foam for detail.

**Figure F-20.** FrostSafe™ (courtesy of Val-Matic).

**frother** Same as frothing agent.

**froth flotation** One of two forms of dispersed air flotation; a process widely used in the mineral industry, developed in 1905 to separate suspended particles by agitating finely divided ore in water with entrained air and a small amount of oil. The process is unsuitable for treating drinking water (because of the formation of large bubbles and the use of undesirable chemicals) but applicable to wastewater treatment. Froth flotation is applicable to the recovery of lime from calcium carbonate sludge or, more specifically, the separation of clay particles or other colloidal impurities. *See also* foam flotation.

**frothing** A foaming action that results in a layer of froth on a wastewater aeration tank.

**frothing agent** Any of a classification of anionic surfactants (e.g., alkyl or amyl alcohols in the $C_5$ to $C_{12}$ range, pine oil) that reduce surface tension (and increase the stability of the particle–air mixture) just enough to promote the formation of bubbles and foam at the surface to support rising particles. *See* MBAS. Also called frother or foaming agent. Other types of flotation chemicals include activating, collecting, and depressing agents.

**frothing reagent** Same as frothing agent.

**Froude number ($F_r$)** A dimensionless number equal to the ratio of the average flow velocity ($V$) to the square root of the product of the gravitational acceleration ($g$) by a characteristic length ($L$) such as the hydraulic mean depth, i.e.,

$$F_r = V/(gL)^{0.5} \quad \text{(F-86)}$$

When the characteristic length equals the depth of flow ($y$), as in a rectangular channel, the Froude number becomes the ratio of the average velocity to the celerity of the gravity wave, i.e.,

$$F_r = V/(gy)^{0.5} \quad \text{(F-87)}$$

The Froude number determines the regime of flow, which is critical, subcritical, or supercritical if $F_r$ is, respectively, equal to, less than, or greater than 1.0. Flow conditions tend to become unstable where $F_r$ is close to 1.0 and result in wave formation. In a plug flow sedimentation tank, The Froude number increases with the length/width ratio, which increases velocity and decreases the characteristic length. The increase in $F_r$ is associated with improved flow stability. The Froude number is also an element of the general equation of gradually varied flow.

**FRP** Acronym of fiberglass-reinforced plastic.

**FRS** Acronym of filter-ripening sequence.

**fructose ($C_6H_{12}O_6$)** A yellowish-to-white, crystalline, water-soluble sugar, sweeter than sucrose, found in honey and many fruits, used in foodstuffs and in medicine. Also called fruit sugar.

**fruit sugar** Fructose.

**Frumkin equation** Same as Frumkin–Fowler–Guggenheim equation.

**Frumkin–Fowler–Guggenheim (FFG) equation** An expression comparable to the Langmuir isotherm and developed to take lateral interactions into account in describing adsorption phenomena (Stumm & Morgan, 1996):

$$\theta/(1 - \theta)e^{-2a\theta} = BX \quad \text{(F-88)}$$

where

$$\theta = [SA]/[S_T] \quad \text{(F-89)}$$

where $SA$ = adsorbate on surface sites, $[S_T]$ = maximum concentration of surface sites, $a$ = interaction coefficient (for $a$ = 0, the FFG equation is equivalent to the Langmuir isotherm), $B$ = adsorption constant, and $[X]$ = equilibrium (bulk) concentration of adsorbate.

**fry** A young fish.

**FS** Acronym of fixed solids.

**FSL** Acronym of facultative sludge lagoon.

**F-specific phage** Bacteriophage that infect hosts through the sex pili or flagella. Also called appendage phage. *See also* capsule phage, somatic phage

**F-specific RNA coliphage** Bacteriophage that infect strains of *E. coli* and related bacteria through the $F^+$ or sex pili. It is one of the groups of bacteriophage considered as possible indicators of fecal contamination. Also called male-specific coliphage. *See also* somatic coliphage.

***F*(*t*) curve** A graphical representation of a cumulative data function. *See* cumulative residence time distribution curve and dispersion index.

**FTIR** Acronym of Fourier transform infrared (spectrometry).

**FT procedure** *See* fermentation tube procedure.

**FTU** Acronym of formazin turbidity unit.

**Fuchs ATAD** An autothermal aerobic digestion (ATAD) system developed by Kruger, Inc.

**fuel value** The quantity of heat that can be obtained from a material per unit mass, i.e., the number of units of heat obtained by the complete combustion of a unit mass of the material. *See* calorific value for detail.

**fugacity** An ideal pressure that characterizes the tendency of a constituent to escape from a phase, as compared to the activity of the constituent in a solution (Stumm & Morgan, 1996):

$$\mu_i = \mu_i^0 + RT \ln (f_i/f_i^0) \quad \text{(F-90)}$$

where $\mu_i$ = chemical potential of species $i$, $\mu_i^0$ = standard chemical potential of species $i$; $R$ = universal gas constant, $T$ = absolute temperature, $f_i$ = fugacity of species $i$, and $f_i^0$ = standard fugacity of species $i$.

**fugitive emission** Emissions not caught by a capture system in chimneys, stacks, or vents.

**fugitive water** Water that leaks from a reservoir or an irrigation system.

**Fujimoto BOD analysis** A method proposed in 1961 to analyze BOD data: BOD at time $t$ + 1 ($BOD_{t+1}$) is plotted vs. BOD at time $t$ ($BOD_t$). The intersection of the curve with the line of slope 1 determines the ultimate BOD ($BOD_u$) and the following equation is used to determine the BOD rate constant (Metcalf & Eddy, 2003):

$$BOD_r = (BOD_u)\, e^{-kt} \quad \text{(F-91)}$$

where $BOD_r$ = amount of waste remaining at time $t$ (days), in oxygen equivalents, mg/L; and $k$ = first-order reaction rate constant, per day. *See also* least-squares BOD analysis.

**full-cone nozzle** One of two types of pressurized nozzles in spray aerators commonly used in water treatment; it delivers a uniform spray of droplets. *See also* hollow-cone nozzle.

**full-cost pricing** The application of full long-term costs to customers of water, wastewater, or other utilities, e.g., by using asset management plans over a century or the life expectancy of the longest-lived assets.

**Fuller, George Warren** Designer of the first sizable rapid filtration plant for treating municipal drinking water in 1909 (Little Falls, NJ). A decade earlier, he conducted feasibility studies for coagulation-rapid filtration at Louisville, KY, including the use of aluminum and iron salts and the effective sand size.

**fuller's earth** (1) A fine, clay-like substance, used as a coagulant aid in waters of low turbidity; it reduces alum dosage and absorbs taste and odor-producing substances. (2) *See* diatomaceous earth. (3) An absorbent clay used to remove grease from fabrics and as a dusting powder.

**Fuller's formula** A formula proposed by Weston E. Fuller (1879–1935) in 1914 to estimate flood flows in the United States (Fair et al., 1971):

$$Q = CA^{0.8}(1 + 0.8 \log T)(1 + 2A^{-0.3}) \quad \text{(F-92)}$$

where $Q$ is the peak flood flow in cfs; $C$ is a coefficient between 25 and 200 to account for the effects of storm frequency, runoff–rainfall relationship, and maximum rainfall; $A$ is the drainage area in square miles; and $T$ is the number of years in the period considered.

**Full-Fit™** A membrane separator manufactured by Osmonics, Inc.

**full-flow pressure flotation** One of three basic modes of operation of pressure (dissolved-air) flotation. *See also* recycle-flow pressure flotation and split-flow pressure flotation.

**full fluidization backwash** A filter backwashing method in which the media grains separate completely from one another, supported only by the liquid.

**full tertiary wastewater treatment** An advanced wastewater treatment method used, e.g., for water reclamation. It is similar to a conventional water treatment flowsheet applied to a secondary wastewater effluent, i.e., including coagulation, flocculation, sedimentation, filtration, and disinfection. Also called complete wastewater treatment.

**full-width weir** *See* suppressed weir.

**fulminic acid (HOCN)** An unstable, poisonous liquid, isomeric with cyanic acid.

**fulvic acid** A complex organic compound formed by the decomposition of plant matter in soil or water; soluble in both acidic and alkaline conditions. Fulvic acids are, along with humic acids, the major organic constituent of natural waters, a source of color, and major precursors of disinfection by-products; they can be removed by coagulation or modified by oxidation (e.g., by chlorine dioxide or ozone). *See also* humic acid, aquatic fulvic acid, aquagenic organic matter, natural organic matter, tannin, disinfection by-product.

**fume** Tiny solid particles trapped in vapor in a gas stream; they result from the condensation of gas particles and from oxidation or other reactions. *See also* dust and smoke.

**fume hood** An enclosed cabinet with a fan to vent fumes in a laboratory.

**fumigant** A pesticide or disinfectant vaporized to kill pests and pathogens in soil, buildings, and greenhouses. *See,* e.g., dibromochloropropane.

**fumigrain** Acrylonitrile.

**functional antagonism** The condition in which the effect of one substance offsets the effect of another substance.

**functional assessment** The evaluation of the vital processes and attributes of a wetland.

**functional equivalency** The characteristic of a restored or artificial ecosystem that functions as a natural ecosystem with respect to some parameters such as flood control.

**functional equivalent** Term used to describe USEPA's decision-making process and its relationship to the environmental review conducted under the National Environmental Policy Act (NEPA). A review is considered functionally equivalent when it addresses the substantive components of a NEPA review.

**functional groups** *See* organic functional groups, exchange sites.

**functionality** The characteristic of an ion exchange resin that relates to the chemical composition of the fixed-charge sites. Common functional groups include (a) strongly acidic (e.g., sulfonate, $—SO_3^-$), (b) weakly acidic (e.g., carboxylate, $—COO^-$), (c) strongly basic [e.g., quaternary amine, $—N^+(CH_3)_3$], and (d) weakly basic [e.g., tertiary amine, $—N(CH_3)_2$]. *See also* resin exchange capacity, matrix, porosity.

**functional tests** Tests performed on samples of water, wastewater, or sludge to evaluate specific performance, e.g., bioassay (sediment, sludge), BOD, chlorine demand, chlorine residual, COD, filterability (sludge), heat content (sludge), jar test, settleability, stability, threshold odor, and Warburg determination (Fair et al., 1971).

**fundamental filtration model** A model that attempts to predict particle removal in a filter, through the mechanisms of particle transport and attachment. For example, the following equation describes particle removal in a single collector (AWWA, 1999):

$$\ln(C/C_0) = -1.5(1 - \varepsilon)\,\alpha\eta_0 L/d_c \quad \text{(F-93)}$$

where $C$ = effluent particle concentration, $C_0$ = influent particle concentration, $\varepsilon$ = porosity, $\alpha$ = empirical collision efficiency factor, $\eta_0$ = overall collector efficiency, $L$ = bed depth, and $d_c$ = collector diameter. Also called macroscopic filtration model. *See also* phenomenological (macroscopic) filtration model.

***Fundulus*** A genus of fish used in bioassays of toxic wastes in saltwater.

**fungal, fungous** Of, pertaining to, caused by, of the nature of, or resembling fungi.

**fungi** Plural form of fungus. Aerobic, multicellular, heterotrophic, nonphotosynthetic microorganisms (such as mushrooms, yeast, molds, mildews, rusts, and smuts) that are small non-chlorophyll-bearing plants lacking roots, stems and leaves. Most fungi are saprophytes, obtaining their nourishment from dead organic matter. Along with bacteria, fungi are the principal organisms responsible for the decomposition of carbon in the biosphere. Fungi have some ecological advantages over bacteria: they can grow in low-moisture areas, in low pH environments, and in wastes that are deficient in nitrogen. They are important in sludge composting and an important cause of filamentous growth and bulking sludge. They occur in natural waters and grow best in the absence of light. Their decomposition may cause objectionable tastes and odors in water. They may also cause the degradation of gasket and joint materials. Certain fungi, e.g., *Aspergillus,* are pathogenic to humans. *See also* budding, hyphae, lichen, mycorrhizae.

**fungicide** A chemical that is used to control, deter, or destroy fungi.

**fungistat** A chemical that keeps fungi from growing.

**fungous** *See* fungal.

**fungus** Singular form of fungi.

**funnel** A utensil, usually cone-shaped with a tube at the apex, used in the laboratory for pouring liquids into containers.

**furan ($C_4H_4O$)** Any of the five members of the colorless, liquid, unsaturated heterocyclic, toxic chlorinated organic compound present in minute amounts in the air emissions from hazardous waste incinerators; used in organic synthesis. Also called furfuran.

**furfural ($C_4H_3OCHO$)** A colorless, oily liquid used in the manufacture of plastics and as a solvent. Also called furfurol.

**furfuran** Same as furan.

**furfurol** Same as furfural.

**furnace** (1) The chambers of the combustion train where drying, ignition, and combustion of waste material and evolved gases occur (EPA-40CFR240.101-1). (2) A solid-fuel-burning appliance that is designed to be located outside ordinary living areas and that warms places other than the space where the appliance is located, by the distribution of air heated in the appliance through ducts (EPA-40CFR60.531). (3) Any container in which materials are heated to very high temperatures.

**furrow irrigation** Irrigation method in which water travels through the field by means of small channels between each row or group of plants. It can also be used to apply reclaimed wastewater or sludge by gravity to row crops, which may cause an odor problem. *See also* sprinkler distribution, gated pipe irrigation, graded-border irrigation, drip or trickle irrigation.

***Fusazium*** A genus of fungi identified in trickling filters, active in waste stabilization but possibly causing problems of clogging and restriction of ventilation.

**fusion** The process, or the resulting nuclear reaction, when nuclei of light elements (such as hydrogen and its isotopes) combine to form a heavier element and release considerable energy. *See also* fission.

**fusion heat** *See* heat of fusion.

**fusion latent heat** Same as heat of fusion.

**futile call** The case where a junior water right is allowed to draw water in spite of a downstream senior right because there would not be more water available to the latter if the former were curtailed.

**Futura-Thane** A lining manufactured by Futura Coatings for drinking water tanks.

**future capacity** Capacity built into a water or wastewater facility in excess of current demand, in anticipation of future requirements.

**fuzzy filter** An innovative filter consisting of compressible, almost spherical polymeric media of about 1.25 inches in diameter placed between two porous plates.

**Fuzzy Filter®** An upflow filter manufactured by the Schreiber Corp.; it uses a highly porous, synthetic medium made of polyvaniladene.

**fuzzy inference** The formulation or mapping of a system from a given input to an output on the basis of "if–then" statements or rules.

**FWKO** Acronym of free water knockout.

**FWPCA** Acronym of Federal Water Pollution Control Act.

**FWS system** Abbreviation of free water surface system.

**FWS wetland** *See* free water surface wetland.

**FYM** Acronym of farmyard manure.

# G

**G** Symbol for velocity gradient.

**g** Symbol for the acceleration of gravity; g = 32.2 ft/sec$^2$ = 9.81 m/sec$^2$.

**Ga** Notation for Galileo number.

**GAC** Acronym of granulated activated carbon.

**GAC Sandwich™ filter** A modification of the slow sand filter developed by the Thames Water Utilities, Ltd. of the London metropolitan area to remove trace pesticides and other organics from water. It consists of 6 in of granular activated carbon between 12 in of sand at the bottom and 18 in of sand at the top.

**gage (gauge)** A device for measuring the elevation of the water surface above some datum.

**gage pressure** The pressure within a closed container or pipe as measured with a gage. In contrast, absolute pressure is the sum of atmospheric pressure and gage pressure. Most pressure gages read in gage pressure or psig (pounds per square inch gage).

**Gaia hypothesis** or **Gaia theory** "Living organisms and their material environment are tightly coupled. The coupled system is a superorganism, and as it evolves there emerges a new property, the ability to self-regulate climate and chemistry" (James Lovelock, British scientist, as quoted in Maier et al., 2000). In essence, living organisms respond to and modify the environment, regulate global climate, and drive biogeochemical cycles. From Gaea, the ancient Greek goddess of the earth.

**gain output ratio (GOR)** The ratio of mass of distillate to steam output, used to assess the performance of an evaporator.

**galactose ($C_6H_{12}O_6$)** A white, crystalline, soluble, six-carbon sugar, obtained from the hydrolysis of milk sugar; a constituent of lactose.

**β-D-galactosidase** An enzyme that hydrolyses lactose into glucose and galactose. It is used in testing for coliform group bacteria using the defined-substrate technology. Also called lactase.

**β-D-galactoside** A glycoside derived from galactose and used as substrate for the enzyme $\beta$-D-galactosidase in the coliform bacteria test.

**4-β-galactosidase-glucose** Same as lactose.

**galena** A common lead compound, essentially lead sulfide, PbS; the principal ore of lead. It is an important source of silver, which it contains as an impurity. Also called galenite.

**galenite** Same as galena.

**gal/flush** Abbreviation of gallon(s) per flush.

**gal/ft$^2$** Abbreviation of gallon(s) per square foot.

**Galileo number** A dimensionless number ($G_a$) relating fluid density (ρ), particle diameter (d), acceleration of gravity (g), particle density (ρ′), and absolute viscosity (μ). It is the reciprocal of the sedimentation number and is used in the study of sedimentation and filter backwashing. *See* minimum fluidization condition.

**Galler-Gotaas formula** An equation developed in 1964 from a multiple regression analysis for the design of stone media trickling filters (WEF & ASCE, 1991):

$$L_e = [K(QL_0 + RL_e)^{1.19}]/[(Q + R)^{0.78}(1 + D)^{0.67}a^{0.25}] \quad \text{(G-01)}$$

where $L_e$ = $BOD_5$ of settled filter effluent at 20°C, mg/L.

$$K = [0.464\ (43{,}560)/\pi)^{0.13}]/[Q^{0.28}t^{0.15}] \quad \text{(G-02)}$$
$$= 3.127/[Q^{0.28}t^{0.15}]$$

where $Q$ = influent flow, mgd; $L_0$ = $BOD_5$ of filter influent at 20°C, mg/L; $R$ = recirculation flow, mgd; $D$ = dilter depth, feet; $a$ = filter radius, feet; $t$ = wastewater temperature, °C. Other trickling filter design formulas include British Manual, Eckenfelder, Germain, Howland, Kincannon and Stover, Logan, modified Velz, NRC, Rankin, Schulze, Velz.

**gallery** (1) A small tunnel or passageway in a dam, mine, rock, or water or wastewater treatment plant for access, inspection, or drainage, or to carry pipe or house machinery, or for some other purpose. (2) An underground structure that collects subsurface water.

***Gallionella*** One of three groups of rod-shaped iron bacteria, filamentous organisms that use iron compounds and cause a number of problems in water supply systems, e.g., taste and odor, color ("red water"), clogging of well screens. *See also Frenothrix* and *Leptothrix.*

***Gallionella ferruginea*** A species of bacteria that metabolize iron and deposit a gelatinous ferric hydroxide [$Fe(OH)_3$].

**gallon** A unit of volume for the measurement of liquids, equal to 231 cubic inches or 128 fluid ounces in the United States. Thus, one gallon = 0.1336806 $ft^3$ = 3.7854118 liters. Abbreviation: gal. Note that the United Kingdom gallon or imperial gallon is equal to 160 fluid ounces or 4.54609 liters, i.e., about 1.2 U.S. gallons.

**gallon per flush** A unit for measuring the flow required to flush a toilet. Abbreviations: gpf, gal/flush). *See* dual-flush, high-efficiency, high-flow, and low-flow toilets

**galvanic anode** A metal that is higher in the galvanic series than the metal to be protected. The presence of the two dissimilar metals creates a galvanic corrosion cell and the galvanic anode (or sacrificial element) is consumed to protect the other metal.

**galvanic cathodic protection** The use of a sacrificial anode to protect another metal from corrosion. *See* galvanic protection for detail.

**galvanic cell** An electrolytic cell capable of producing electrical energy by electrochemical action; a battery. The decomposition of materials in the cell causes an electric current to flow from cathode to anode. *See also* corrosion cell, electrochemical cell.

**galvanic corrosion** One of two basic forms of metallic corrosion. It involves the action of a direct electrical current generated within the galvanic cell and connecting two dissimilar metals. The metal serving as anode deteriorates. Galvanic corrosion increases with the potential difference between the two metals. It is a problem in water supply systems, particularly where brass, bronze, or copper is in direct contact with aluminum, galvanized iron, or iron. Sometimes called coupling action. *See also* electrolytic corrosion.

**galvanic coupling** A galvanic cell of two dissimilar metals.

**galvanic protection** One of two methods of providing cathodic protection (the conversion of an entire metal surface into a cathode; corrosion normally occurs only at anodic areas). Galvanic protection consists of a sacrificial anode, i.e., a metal of a higher corrosion potential than the one to be protected. The sacrificial anode corrodes and is periodically replaced. For example, in water works magnesium is usually used to protect iron and steel; aluminum and zinc may be used elsewhere. Other methods of corrosion control include anodic protection, impressed-current protection, chemical coatings, inhibition, inert materials, and metallic coatings.

**galvanic series** A list of metals and alloys presented in the order of their tendency to corrode in a certain environment; a practical application of the electrochemical series. *See* electromotive series for more detail. *See also* noble metals.

**galvanization** The electrolytic or hot dipping process of coating a metal with zinc to increase its resistance to corrosion.

**galvanize** To coat a metal (especially iron or steel) with zinc by immersion in a bath or by deposition from a solution as protection against corrosion.

**galvanized iron** Iron or steel coated with zinc to prevent corrosion (rust).

**galvanized pipe** A pipe consisting of a base steel layer underlying layers of iron/zinc alloys, with the zinc content increasing progressively toward the interior surface of the pipe. It has been used extensively in water distribution systems but is being replaced by copper and plastic pipes.

**galvannealing** The condition in which the iron content of a galvanized pipe extends all the way to the surface of the pipe, thus compromising protection against corrosion.

**Gambian sleeping sickness** Common name of one variety of African trypanosomiasis, a disease caused by the protozoa *Trypanosoma gambiense* and transmitted by the riverine tsetse fly (*Glossina* spp.). It is found mainly in Central and West Africa. *See also* Rhodesian sleeping sickness and Chagas' disease.

***Gambusia affinis*** A species of fish used in stocking ponds to control mosquitoes. Also called mosquito fish.

**game fish** Edible species like trout, salmon, or bass, also caught for sport. Many of them show more sensitivity to environmental change and pollution (e.g., BOD) than "rough" fish. *See also* intolerant species, tolerant fish.

**gamete** A germ cell; a mature sexual reproductive cell that can unite with another cell to form an organism, e.g., a motile sperm or a nonmotile egg or ovum.

**gametogenesis** The development of male and female sex cells.

**gamma (γ) -benzene hexachloride ($C_6H_5Cl_6$)** The chlorinated hydrocarbon contaminant of lindane. Abbreviation: γ-BHC.

**γ-BHC** Abbreviation of gamma-benzene hexachloride. *See* lindane.

**gamma (γ)** decay A nuclear reaction in which the nucleus of an atom emits gamma rays.

**γ-$Fe_2O_3$** Chemical formula of maghemite.

**γ-FeOOH** Chemical formula of lepidocrocite.

**gamma (γ) -glutamyltranspeptidase** *See* glutamyltranspeptidase.

**gamma (γ) -HCH** Abbreviation of gamma (γ) -hexachlorocyclohexane.

**gamma (γ) -hexachlorocyclohexane** Another name of the insecticide lindane. Abbreviation: gamma (γ) -HCH.

**gamma (γ) irradiation** A possible method of water disinfection using gamma rays from a radioactive source.

**gamma (γ) radiation** A type of ionizing radiation, consisting of electromagnetic, wave-type energy, e.g., X-rays, that can be used as an effective water disinfecting agent.

**gamma (γ) ray** A form of electromagnetic radiation of short wavelength and high energy, usually between 0.01 and 10 megaelectron volts (Mev), emitted sometimes with alpha and beta rays but always with fission. Lead or depleted uranium may be used to protect the body against the penetrating effects of gamma radiation. Sometimes called nuclear X-ray or radiant energy.

**gamma (γ) -ray irradiation** A technique that uses doses > 1.0 Mrad of gamma rays from cobalt-60 or cesium-135 at 20°C as a process to further reduce pathogens.

**Ganguillet, E.** Joint author of the Kutter–Ganguillet formula (1888) for the roughness coefficient in the Chézy formula.

**Gantt chart** A bar chart or bar graph, i.e., a chart using parallel bars to illustrate comparative elements such as costs, birth rates, and treatment performances.

**garbage** Animal and vegetable waste resulting from the handling, storage, sale, preparation, cooking, and serving of foods, and consisting of putrescible organic matter and its moisture.

**garbage fish** A fish like carp or catfish that can tolerate lower water quality (e.g., higher BOD) than the game fish.

**Gard** A rotary distributor manufactured by General Filter Co. for trickling filters.

**Gard's disinfection model** A relationship proposed in 1957 to define the decreasing rate constant of microorganism inactivation during the disinfection wastewater treatment of effluents (Metcalf & Eddy, 2003):

$$S = N/N_0 = 1/(1 + aCt)^{k/a} \qquad \text{(G-03)}$$

where $N$ = mean effluent survival ratio of organisms; $N$ = number of organisms remaining after disinfection at time $t$; $N_0$ = initial number of organisms (before disinfection); $a$ = rate coefficient; $C$ = chlorine residual remaining at time $t$; $t$ = contact time, min; and $k$ = initial first-order inactivation rate constant (at $t$ = 0). *See* chlorine disinfection models, disinfection kinetics.

**Gar-Dur** Chain and flight sludge collectors fabricated by Garland Manufacturing Co. with plastics of high molecular weight.

**garnet** A group of hard, reddish, glassy mineral sands made up of silicates of base metals (calcium, magnesium, iron, and manganese). Garnet has a higher density (specific gravity of 3.6–4.2) than sand; it is often used as the bottom layer of a multimedia filter or as one element of a mixed-media filter.

**garnet sand** Same as garnet.

**gas** A state of matter; a substance characterized by low density, low viscosity, perfect molecular mobility, and the ability to expand indefinitely; a mixture of such substances. Common dissolved gases in water include carbon dioxide ($CO_2$), hydrogen sulfide ($H_2S$), methane ($CH_4$), nitrogen ($N_2$), and oxygen ($O_2$). Dangerous gases in sewers and manholes include those from gasoline leaks or such industrial chemicals as calcium carbide, $CaC_2$, or those resulting from biological activity (hydrogen sulfide and methane); they pose a risk of explosion or they may be toxic.

**gas absorption** The general gas law governs the absorption or desorption of gases in water and wastewater. The absorption (or desorption) rate is assumed to be proportional to the undersaturation of the gas in the liquid (Fair et al., 1971; Metcalf & Eddy, 2003):

$$(C - C_0)/(C_s - C_0) = 1 - e^{-AKT} \quad \text{(G-04)}$$

or

$$R_{ab} = K_{ab}A(C_s - C) \quad \text{(G-05)}$$

and

$$R_{de} = K_{de}A(C - C_x) \quad \text{(G-06)}$$

where $C$ = gas concentration at time $T$; $C_0$ = initial concentration ($T$ = 0); $C_x$ = saturation concentration at a given temperature; $T$ = time; $A$ = the area of interface per unit volume of liquid; $K$ = gas transfer coefficient; $R_{ab}$ = rate of absorption, mass per volume per unit time; $K_{ab}$ = coefficient of absorption, length per unit time; $R_{de}$ = rate of desorption, mass per volume per unit time; and $K_{de}$ = coefficient of desorption, length per unit time. *See also* two-film theory, desorption rate.

**gas absorption rate** The rate at which a gas is transferred to a liquid. *See* gas absorption.

**GA salt** *See* ground alum salt.

**gas chlorination** The use of chlorine gas ($Cl_2$) to disinfect water or wastewater.

**gas chlorinator** A device used to add chlorine gas ($Cl_2$) to water or wastewater. *See* chlorinator for detail.

**gas chromatograph** An instrument for the separation of volatile organic compounds. It includes a sample injector, a chromatography column in an oven, and a detector.

**gas chromatograph–mass spectrometer** A highly sophisticated instrument that identifies the molecular composition and concentrations of various chemicals in water and soil samples. It consists of the combination of a gas chromatograph and a mass spectrometer. *See also* gas chromatography–mass spectrometry.

**gas chromatography (GC)** An analytical method for the analysis of organic compounds at trace concentrations (e.g., THMs) in a water sample. It consists of the volatilization of the substances to be analyzed and their identification and concentration based on their retention time in the chromatography column. *See also* stationary phase, liquid chromatography.

**gas chromatography–electron capture detector (GC–ECD)** The combination of two techniques to analyze organic compounds. *See* gas chromatography and electron capture detector.

**gas chromatography–mass spectrometry (GC–MS)** The combination of gas chromatography and mass spectrometry for the analysis of contaminants in water.

**gas constant (*R*)** A proportionality constant used in the universal (or ideal) gas equation: $R$ = 0.082057 atm · L/mole · 0°K = 8.3145 J/mole · 0°K. Also called universal gas constant or universal gas law constant.

**gas desorption** *See* gas absorption.

**gas desorption rate** Same as desorption rate.

**gas dispersion** The combined effect of molecular diffusion and eddy diffusion (convection and agitation).

**gas dome** A steel cover floating partially or entirely on the liquid sludge of a digestion tank.

**gas entry pressure** The gas pressure required to displace liquid from the pores of a wetted filtration membrane. *See* bubble-point pressure.

**gaseous emission** Same as gaseous waste.

**gaseous waste** Volatile or uncondensed residuals discharged into the atmosphere at a wastewater treatment plant. Also called gaseous emission. *See also* smoke.

**gas equation of state** *See* equation of state.

**gas feed** A system used to feed chemicals in gaseous form in water or wastewater treatment processes (e.g., chlorination, dechlorination). *See* chlorinator, dry chemical feed, wet chemical feed.

**gas-holder cover** The component of an anaerobic sludge digester under which gas is collected. *See* fixed cover, floating cover, membrane gas cover.

**gasification** (1) The conversion of solid material such as coal into a gas for use as fuel. (2) The conversion of organic matter into gas during fermentation or other waste decomposition processes. The amount of gas produced ($y$) in a time ($t$) may be modeled as a first-order reaction or as logistic growth, as a function of a saturation value ($L$) and coefficients $K_1$ and $K_2$ (Fair et al., 1971):

$$dy/dt = K_1(L - y) + K_2(L - y)^2 \qquad \text{(G-07)}$$

$$dy/dt = K_1(L - y) + K_2y(L - y) \qquad \text{(G-08)}$$

**gasket** A rubber, metal, or rope ring used to pack a piston or make a joint or connection watertight.

**gas laws** *See* general gas law.

**GasLifter** Equipment manufactured by Walker Process Equipment Co. to provide mixing in anaerobic digesters.

**gas membrane** Same as gas separation membrane.

**gasohol** Mixture of unleaded gasoline and ethanol derived from fermented agricultural products containing at least 9% ethanol. Gasohol emissions contain less carbon monoxide than those of gasoline.

**gas-phase chlorine dioxide ($ClO_2$)** An oxidant produced by the reaction of hydrochloric acid (HCl) and potassium chlorate ($KClO_3$) for use as a water disinfectant. *See* chlorine dioxide for detail.

**gas production** The formation of a gas by a chemical or biological reaction, e.g., during sludge digestion. *See* digester gas.

**gas separation membrane** A membrane used to increase the oxygen content of air or in other gas separation processes.

**gas solids separator (GSS)** A device used in the anaerobic sludge blanket process to collect biogas, prevent solids washout, promote the separation of solids and gas, and prevent the return of the solids to the sludge blanket.

**gas solubility** The extent to which a gas will dissolve or the amount of gas that dissolves in water, wastewater, or other solvents. For water and wastewater, the solubility of a gas depends on its partial pressure in the liquid, the liquid temperature, and the concentration of impurities in the liquid. *See also* gas dispersion, gas precipitation, Dalton's law of partial pressures, Henry's law, general gas law, absorption coefficient, Avogadro's hypothesis, gas-transfer coefficient, two-film theory.

**gas solubility coefficient** The quantity of a gas absorbed by a unit volume of water or wastewater at a given temperature and under a barometric pressure of 1 atmosphere, usually expressed in mg/L; also called solubility coefficient.

**gas stripping** The mass transfer of a gas from the liquid phase to the gas phase. Gas stripping applications to water and wastewater treatment include the removal of ammonia ($NH_3$), carbon dioxide ($CO_2$), hydrogen sulfide ($H_2S$), oxygen ($O_2$), and volatile organic compounds (VOCs).

**gas stripping tower** *See* stripping tower.

***Gasterosteus aculeatus*** A three-spine stickleback used in fish bioassays for acute toxicity.

**gas transfer** The exchange of gas molecules between a gas such as oxygen or carbon dioxide and a liquid (water or wastewater); it is often accompanied by biological, biochemical, or chemical effects. Gas transfer principles are applied in many operations and processes of water or water treatment. Examples include activated sludge aeration, aerobic digestion, ammonia stripping, carbon dioxide release from sludge, carbon dioxide removal, chlorination, deferrization and demanganization, deoxygenation, hydrogen sulfide removal, methane release from sludge, odor removal, ozonation, recarbonation, and removal of volatile substances. *See also* gas absorption, oxygen transfer.

**gas transfer coefficient** A coefficient in the gas absorption equation. *See also* oxygen transfer coefficient.

**gas trap** A device installed in a drain line to retain water and prevent the counterflow of gases. *See also* gas vent.

**gastric** Pertaining to the stomach.

**gastric lavage** The washing out of the stomach by pumping a fluid in and out, a procedure often used in the case of recent poisoning.

**gastritis** An inflammation of the mucous membrane of the stomach. Gastritis may be caused by the infectious agent *Helicobacter pylori;* if left untreated for a long time, it may lead to ulcers and stomach cancer.

**gastrodisciasis** Gastrodiscoidiasis.

***Gastrodiscoides hominis*** A pathogen transmitted from pigs to humans through aquatic snails and aquatic vegetation; found mainly in Bangladesh, India, the Philippines, and Vietnam. Latency = 2 months. Median infective dose < 100.

**gastrodiscoidiasis** An infection caused by the excreted helminth *Gastrodiscoides hominis.*

**gastroenteric illness** An illness of the gastrointestinal system, caused by an infectious or other agent. *See,* e.g., gastritis, gastroenteritis.

**gastroenteritis** An inflammation of the mucous membrane of the stomach and intestines resulting in diarrhea, with vomiting and cramps when irritation is excessive. When caused by an infectious agent—waterborne or foodborne—it is often associated with fever. Also called bacterial enteritis or, simply, diarrhea. When the infectious agent is not identified, the illness is designated as an "acute gastroenritis of undetermined etiology." *See also* viral enteritis. It may also be caused by chemicals.

**gastroenteritis (acute)** *See* acute gastroenteritis.

**gastroenteritis virus** A Norwalk-type agent that causes fever and epidemic vomiting and diarrhea.

**gastroenterology** Study of the stomach, intestines, and their diseases.

**gastrointestinal** Pertaining to the stomach or intestines.

**gastrointestinal illness** An illness of the stomach and intestines, e.g., diarrhea, often caused by such waterborne pathogens as *Salmonella typhi* and *S. paratyphi. See also* salmonellosis.

**gastrointestinal tract** The portion of the alimentary tract that is in the stomach and the intestines. *See also* intestinal tract.

**gas vent** A passage that lets gases escape, e.g., an opening in the sludge chamber of an Imhoff tank for digester sludge to escape to the atmosphere without passing through the settling chamber. *See also* gas trap.

**gas vesicle** A protein inclusion within a bacterial cell that can provide buoyancy. *See* cell inclusion, endospore.

**gate** A door or other movable watertight barrier for controlling the passage of materials through a pipe, channel, or other waterway.

**gate chamber** A structure that houses a valve or a similar regulating device while providing access for maintenance and repair. Same as gate vault. *See also* gate house.

**gated pipe irrigation** A type of distribution system used in surface applications of municipal wastewater. It consists of a pressure or gravity flow (from a storage basin or a lagoon) and a perforated distribution pipe installed level across the upper end of a field that is sloped 0.2–5.0 %. *See also* furrow ieeigation.

**gate headloss** *See* hydraulic grade differential.

**gate house** The superstructure of a gate chamber or the superstructure at the headworks of a dam, powerhouse, etc.

**gate valve** A valve whose closing element is a disk fitting tightly over an opening; very common in water and wastewater works due to its efficiency and low cost

**gate vault** Same as gate chamber.

**gauge** A number that defines the thickness of the sheet used to make steel pipe. The larger the number, the thinner the pipe wall.

**Gaussian probability curve** Same as normal distribution.

**Gauss, Karl Friedrich** German astronomer and mathematician (1777–1855).

**gauze filtration** The recovery of viruses from water by filtering them through an open, meshlike woven material.

**gauze number** The number of openings per inch in a wire mesh; a number used when specifying screens for wells and filters.

**gavage** A type of exposure in which a substance is administered to an animal through a stomach tube.

**Gay-Lussac, Joseph Louis** French chemist and physicist (1778–1850).

**Gay-Lussac's law** The density of an ideal gas at constant (and relatively low) pressure varies inversely with its absolute temperature. *See also* Charles' law and general gas law.

**GBT** Acronym of gravity belt thickener.

**GC** Acronym of gas chromatography.

**GC–ECD** Acronym of gas chromatography–electron capture detector.

**GC–MS** Acronym of gas chromatography–mass spectrometry.

**GDD** Acronym of growing degree-day

**GDT Process™** GDT is the acronym of gas–degas treatment, a process developed by Mazzei Injector Corp. to remove volatile organic compounds.

***G. duodenalis*** *See Giardia duodenalis.*

**Ge** Chemical symbol of germanium.

**gear pump** A positive rotary pump using two meshing gear wheels to move the fluid from suction to discharge.

**GEHO®** A piston pump manufactured by Envirotech Co. for handling heavy sludges.

**Geiger counter** A hand-held instrument for the detection of ionizing radiation; it consists of a gas-filled tube that produces electric-current pulses when the gas is ionized by radiation plus a device to measure the pulses. Also called Geiger–Muller counter.

**Geiger–Muller counter** Same as Geiger counter.

**Geiger–Muller threshold** Same as Geiger threshold.

**Geiger–Muller tube** *See* Geiger tube.

**Geiger threshold** The lowest voltage in a Geiger counter that produces pulses, the size of which are independent of the number of ions produced.

**Geiger tube** The gas-filled tube serving as an ionization chamber in the Geiger counter. Also called Geiger–Muller tube.

**gel** (1) A jellylike, viscous, colloidal material that contains as little as 0.5% solid matter, that does not dissolve, remains suspended, and does not precipitate without physical or chemical agents (e.g., heat, electrolyte). (2) A borehole lubricant similar to drilling mud. (3) A semirigid polymer, such as cellulose acetate or polyacrilamyde, used in the electrophoretic separation of proteins and nucleic acids. (4) A compound injected into an aquifer to reduce its permeability locally and change the direction of flow.

**gelatinous** Jellylike; pertaining to, having the nature of, or resembling jelly or slimy suspended solids that can cause filter clogging.

**gel electrophoresis** A biochemical technique that uses an electric field to separate protein molecules in a mixture by moving them through a gel of agarose or polyacrylamide and observing that the smaller molecules move faster than the larger ones.

**Gelex** Standardization procedures designed by Hach Co. for turbidimeters.

**gel formation** The result of concentration polarization and a cause of fouling in membrane processes; it occurs when the majority of solids in the feed water are larger than the pores or molecular weight cutoff of the membrane. Also called cake formation. *See also* fouling indexes, pore narrowing, pore plugging.

**gel permeation chromatography** A technique used to determine the molecular weight of a substance or to separate substances based on their molecular weights compared to the pore size of a gel such as dextran.

**gel polymer** A wastewater treatment chemical consisting of high-molecular-weight monomers with about 30% active solids. Polymers are also available in dry, emulsion, liquid, and mannich forms.

**gel resin** A jellylike material used as an ion-exchange resin.

**gel zeolite** A synthetic sodium aluminum silicate used as a cation-exchange product in residential water softeners. Also called siliceous gel zeolite. *See also* aluminosilicate.

**gel zone** A material formed during the corrosion of lead piping; it conducts both electrons and ions, and combines with lead oxides to form a hydrated surface layer, $PbO(OH)_2 \cdot H_2O$.

**Gemco** A filter manufactured by Gauld Equipment Sales Co. for the treatment of spent liquor.

**Gemini** (1) A granular activated carbon contactor made by Roberts Filter Manufacturing Co. (2) A self-cleaning basket strainer manufactured by S. P. Kinney Engineers, Inc.

**Gemini Polymaster** An apparatus manufactured by Komax Systems, Inc. for blending polymers in solution or emulsion.

**gemma (plural: gemmae)** A bud or cluster of cells that separate from a parent plant to form a new organism.

**gemmation** Asexual reproduction of plants in which new individuals develop from groups of cells that grow on and detach from the parent's body; e.g., in mosses and liverworts. When applied to animals, gemmation is called budding.

**GEMS** Acronym of Global Environmental Monitoring System.

**GEMS/Water** A program of the United Nations Environment Program (UNEP) and the World Health Organization (WHO), in cooperation with UNESCO (UN Educational, Scientific and Cultural Organization) and WMO (World Meteorological Organization) aimed to generate data for a global assessment of freshwater quality. *See* Table G-01 for basic monitoring data measured at four types of monitoring stations: stream, headwater lake, groundwater, global river flux.

**Gen2®** Chemical feed equipment manufactured by Stranco, Inc.

**gene** The basic unit of heredity; a linear sequence of nucleotides in the DNA providing instructions for the synthesis of RNA, which leads to the hereditary traits.

**gene mapping** The process of constructing a model of the linear sequence of genes of a chromosome or a map of genes on a chromosome. *See also* genetic mapping.

**genera** Plural of genus.

**general equation of state** Same as general gas law or ideal gas equation.

**general gas law** An equation that expresses the relationship between the absolute pressure ($P$), the volume ($V$), the number of moles ($N$), and the absolute temperature ($T$) of a gas in water:

$$PV = NRT \qquad \text{(G-09)}$$

where $R$ is the universal gas constant = 8.3136 × $10^7$ dyne-cm per gram-mole and °C absolute, or 1546 lb-ft per lb-mole and °F absolute. This law, also called ideal gas law or universal gas law, combines the Boltzmann's constant and a series of re-

**Table G-01.** GEMS/Water (basic monitoring variables)

| | |
|---|---|
| alkalinity | ammonia |
| calcium | chloride |
| chlorophyll *a* | coliforms (fecal) |
| dissolved oxygen | electrical conductivity |
| fluoride | magnesium |
| nitrate | nitrite |
| pH | phosphorus (dissolved) |
| phosphorus (total) | potassium |
| silica (reactive) | sodium |
| sulfate | temperature |
| total suspended solids | transparency |
| water discharge or level | |

lated hypotheses by Avogadro, Boyle, Charles, and Gay-Lussac. It is a good approximation for real gases and applies to many situations in the water and wastewater treatment field. *See also* Dalton's, Graham's, and Henry's laws.

**generalized disinfection model** A formulation proposed to represent the disinfection process in secondary and filtered wastewater treatment effluents. It is an extension of the model developed for ultraviolet disinfection, incorporating two different portions of the inactivation curve for free-swimming (dispersed) organisms and those associated with particles (Metcalf & Eddy, 2003):

$$N(t) = N_d(0)\, e^{-kCt} + N_p(0)(1 - e^{-kCt})/(kCt) \quad \text{(G-10)}$$

where $N(t)$ = number of organisms remaining after disinfection at time $t$; $N_d(0)$ = initial number of dispersed organisms (before disinfection, time $t$ = 0); $k$ = inactivation rate constant; $C$ = chlorine residual remaining at time $t$; $t$ = contact time, min; and $N_p(0)$ = initial number of particles containing organisms (before disinfection, time $t$ = 0). *See* chlorine disinfection models.

**generalized filtration rate equation** An equation proposed to represent the rate of wastewater filtration (the change in the concentration $C$ of suspended solids with respect to the distance $z$ from the top of the filter bed) as a function of the initial removal rate ($r$), the quantity of suspended solids accumulated in the filter ($q$), and the ultimate quantity ($q_u$) of solids that can be deposited in the filter (Metcalf & Eddy, 2003):

$$dC/dz = [rC/(1 + az)^n](1 - q/q_u)^m \quad \text{(G-11)}$$

where $a$ and $n$ are constants and $m$ = a parameter indicative of floc strength.

**generalized purification equation/function** Same as purification function.

**general obligation bond** A debt obligation backed by the full faith and credit of the issuing government or community, with funds for repayment drawn from ad valorem taxes on real property. Such bonds are sometimes used to finance improvements to water and wastewater systems. *See also* revenue bond and special assessment bond.

**general permit** (1) A permit applicable to a class or category of dischargers; an NPDES or "404" permit authorizing a category of discharges or activities under the Clean Water Act within a geographical area (EPA-40CFR124.2). (2) A permit authorizing a category of discharges of dredge or fill material; general permits are for categories of discharge that are similar in nature, will cause only minimal adverse environmental effects when performed separately, and will have only minimal cumulative adverse effects on the environment (EPA-40CFR232.2).

**generation time** The time ($T$) for an organism to reproduce, e.g., the time for a bacterial cell to divide into two daughter cells, with a rate constant $k$:

$$k = (1/T) \ln 2 \quad \text{(G-12)}$$

*See also* doubling time.

**generator** (1) A facility or mobile source that emits pollutants into the air or releases hazardous waste into water or soil. (2) Any person, by site, whose act or process produces regulated medical waste or whose act first causes such waste to become subject to regulation. In a case where more than one person (e.g., doctors with separate medical practices) is located in the same building, each business entity is a separate generator. (3) A rotating device that produces electricity.

**generic classification** The conventional classification of diseases according to the characteristics of the pathogens that cause them, as compared to an environmental classification. Also called biological classification. *See also* biological classification convention.

**generic taxonomy** *See* generic classification.

**Generox™** A chlorine dioxide generator manufactured by the Drew Division of Ashland Chemical.

**Genesis** (1) A wastewater shredder manufactured by International Shredder, Inc. (2) A package treatment plant manufactured by Roberts Filter Co. using gravity filtration.

**genetic damage** The detrimental effect of radiation on reproductive elements (genes and chromosomes).

**genetic engineering** The process of inserting new genetic information into existing cells in order to modify an organism for the purpose of changing one of its characteristics.

**genetic mapping** Arranging genes on a chromosome. *See also* gene mapping.

**genetic material** A quantity of DNA.

**genetic toxicology** The study of the effects of chemicals and physical agents on hereditary processes.

**genome** The total DNA of an organism; the nucleic acid molecule that is surrounded by a protein coat to form a virus.

**genotoxic** Capable of causing alteration or damage to the genetic material (DNA) in living cells. Many carcinogens and mutagens are also genotoxic.

**genotoxic activity** The production of mutations in a cell that are transmitted to daughter cells; it may result in cancer, birth defects, or other damages.

**genotoxic carcinogen** A substance or agent that induces cancer by altering or damaging the DNA directly or indirectly.

**genotoxin** A toxin that affects fetal development.

**genotype** The genetic make up of an organism or group of organisms with respect to one or several traits transmitted from parent to offspring. *See also* biotype and serotype.

**Gen-Ozi** An ozone generator manufactured by Matheson Gas Products.

***G. enterica*** *See Giardia enterica.*

**Genter's formula** An expression of the final alkalinity of elutriated sludge, as a fraction ($F$) of the alkalinity of digested sludge and the number ($N$) of volumes of clean water used to wash the sludge (Nemerow & Dasgupta, 1991):

$$F = 1/(N + 1)^2 \quad \text{(G-13)}$$

Alkalinity is sometimes used as an indication of the degree of chemical fouling from sludge digestion.

**genus** One of the subdivisions of the plant and animal kingdom. A genus (e.g., *Salmonella*) is a direct subdivision of a family (e.g., Enterobacteriaceae) and comprises a number of species (e.g., *Salmonella typhi*). Plural: genera. *See* biological classification convention.

**Genu Valgum** A deformity of the skeleton associated with a high intake of molybdenum, present in some areas of India that are also prone to fluorosis.

**geochemistry** The science that deals with the composition of and changes in the substances and pore water of the earth's crust. *See also* environmental geochemistry.

**geodetic head** Static head or total head, including velocity head and losses.

**geographic data** Graphic features and their attributes related to a physical location.

**geographic information system or GIS** A computer system designed for storing, analyzing, manipulating, and displaying data in a geographic context. A system of computer hardware, software, and procedures designed to support the capture, management, manipulation, analysis, modullary, and display of spatially referenced data for solving complex planning and management problems (Federal Interagency Coordinating Committee). For example, in one sewer modeling project, the consultant used ArcInfo and ArcView, both developed by the Environmental Systems Research Institute. The advantages of using GIS in addition to AutoCAD maps are two-fold. First, by overlaying the pump station area boundaries on the gravity elements of the conveyance system, one can derive the stage–storage relationship. Second, with the overlay and spatial query capabilities of GIS, one can easily generate inventories of all elements of a sewer system.

**Geoguard®** A device manufactured by American Sigma for groundwater sampling.

**geohydrology** Same as hydrogeology.

**geologic erosion** Natural erosion caused by geological processes over long periods and resulting in the wearing away of mountains, the building up of floodplains and coastal plains, etc.

**geology** The science that deals with rocks, minerals, water, and soils of the Earth, including their physical, chemical, and biological changes.

**geomembrane** A material used as a liner in wastewater treatment ponds to minimize percolation.

**geometric isomers** Compounds that have the same molecular formula but a different structural formula, including a different spatial arrangement of the carbon–carbon double bond.

**geometric mean** The $N$th root of the product of $N$ numbers; e.g., the geometric root of 4, 6.5, 83.6, and 250 is $(4 \times 6.5 \times 83.6 \times 250)^{1/4} = 27.151$.

**geomorphology** The study of the characteristics, origin, and development of land forms (excluding ocean basins and mountain ranges), and the changes that take place as a result of the forces of nature. *See also* sedimentation terms.

**geopressured reservoir** A geothermal reservoir consisting of porous sands that contain water or brine at high temperature or pressure.

**geopurification** A method of water reclamation used in hydrogeological conditions favorable for groundwater recharge with spreading basins. *See* soil–aquifer treatment for detail.

**geosmin ($C_{12}H_{22}O$)** An earthy-smelling organic contaminant produced by certain blue-green algae and *Actinomycetes;* it may be converted to aliphatic and aromatic compounds that cause a sweet, fruity odor in water even at very low concentrations, e.g., 0.2 ppb. It can be removed from drinking water by ozonation or adsorption on activated carbon. Also called *trans*-1, 10-dimethyl-*trans*-9-decalol. *See also* methylisoborneol.

**geosynthetic** A synthetic material (e.g., geomembrane, geonet, geotextile) used in erosion control, channel stabilization, lining, soil filtration, leachate control, road improvement, and similar applications.

**Geothane®** A lining manufactured by Futura Coatings, Inc. for wastewater containment.

**geothermal** Pertaining to the heat transferred from the interior of the Earth to the surface by hot water or steam, as in geothermal energy.

***Geotrichum*** A genus of fungi identified in trickling filters, active in waste stabilization but susceptible to causing problems of clogging and restriction of ventilation.

**germ** A disease-causing microorganism.

**germ cell** Any stage of a sexual reproductive cell; e.g., sperm, egg.

**Germain's formula** A 1965 application of the Schulze formula to plastic media in the design of trickling filters:

$$L_e/L_0 = e^{-kD}/q^n \qquad \text{(G-14)}$$

where $L_e$ = $BOD_5$ of settled filter effluent, mg/L; $L_0$ = influent $BOD_5$, excluding recirculation, mg/L; $k$ = rate constant determined experimentally, also called wastewater treatability and packing coefficient; $D$ = filter depth, feet; $q$ = hydraulic application rate, gpm/ft$^2$; and $n$ = constant characteristic of filter medium. Other trickling filter design formulas include British Manual, Eckenfelder, Galler and Gotaas, Germain, Howland, Kincannon and Stover, Logan, modified Velz, NRC, Rankin, Schulze, and Velz.

**German degree** A unit of hardness equivalent to approximately 10 mg/L of CaO or 17.9 mg/L of $CaCO_3$. *See* American degree, British degree, French degree, Russian degree.

**German digester** *See* conventional German digester.

**germicidal band** Same as germicidal range.

**germicidal effectiveness** The degree to which a disinfection process kills or inactivates pathogens, measured usually by the survival of discrete bacteria or by the inactivation rate (in log units) and dependent on the product of the residual disinfectant concentration ($C$) and the contact time ($t$). Used interchangeably with germicidal efficiency. *See Ct* concept.

**germicidal efficiency** *See* germicidal effectiveness.

**germicidal range** The range of wavelengths within which ultraviolet light is germicidal, i.e., 200–300 nanometers. Also called short-wave ultraviolet band.

**germicidal treatment** Treatment to destroy or inactivate microorganisms, particularly germs; disinfection.

**germicidal ultraviolet** Ultraviolet light of such a wavelength that it has germicidal properties. The germicidal wavelength range is between 200 and 300 nanometers with a peak at about 254 nanometers. Also called shortwave ultraviolet.

**germicidal unit** A measure of radiant energy equal to the intensity of 100 mw per square centimeter for radiations of wavelength 2537 Å. *See also* coefficient of attenuation.

**germicide** A substance formulated to kill germs or microorganisms. The germicidal properties of chlorine make it an effective disinfectant.

**gestation index** The percentage of pregnancies that result in a live birth.

**Gewe** An inclined-plate settler manufactured by Purac Engineering, Inc.

**Geyer, J. C.** A prominent sanitary engineer (1906–1995), professor at Johns Hopkins University, investigator, mentor; coauthor of the well-known text, *Water and Wastewater Engineering,* and of several related articles.

**geyser** A periodic thermal spring resulting from the expansive force of superheated steam.

**GFAA** Acronym of graphite furnace atomic absorption.

***G. fusca*** Abbreviation of *Glossina fusca.*

**Ghyben–Herzberg lens** The lens of saltwater that forms beneath freshwater when these two waters of different densities are in contact, e.g., on islands, peninsulas, spits, or in artesian aquifers outcropping in the sea.

**Ghyben–Herzberg principle** When freshwater and saltwater are in contact, because of the difference in densities, the freshwater column must balance the pressure at the bottom of the saltwater lens (Fair et al., 1971):

$$H_s(S_s - S_f) = S_f(H_f - H_s) \qquad \text{(G-15)}$$

where $H_s$ = depth of saltwater interface below sea level, $S_s$ = specific gravity of saltwater, $S_f$ = specific gravity of freshwater, and $H_f$= thickness of the freshwater lens. For the common conditions of $S_f$ = 1.0 and $S_s$ = 1.025,

$$H_s = 40\ (H_f - H_s) \qquad \text{(G-16)}$$

This means that, theoretically, the depth of the freshwater lens $H_s$ below sea level is 40 times the elevation of the water table above sea level. In other words, if pumping lowers the level of freshwater by 1.0 foot, saltwater will rise 40 feet, with a risk of salt contamination. *See* Figure G-01, freshwater barrier.

**GI** Acronym of gastrointestinal.

**giant intestinal fluke** The excreted helminth pathogen *Fasciolopsis buski* or the disease that it causes, mainly in China and other parts of Southeast Asia. Its eggs are transmitted from the feces of infected persons or pigs through aquatic snails as intermediate hosts. The adult trematode, fleshy and elongated, measures up to 7.5 cm × 2.0 cm and lives attached to the small intestine. The dis-

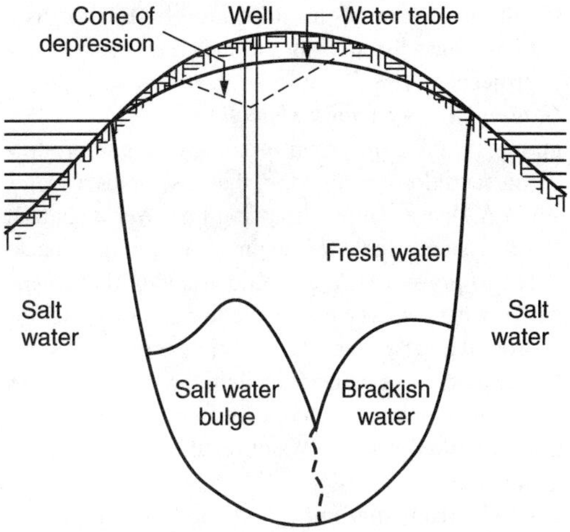


**Figure G-01.** Ghyben–Herzberg principle (adapted from Fair et al., 1968).

ease, fasciolopsiasis, is usually mild and symptomless, but occasionally characterized by abdominal pain, diarrhea, nausea, fever, and even intestinal obstruction. Human infection occurs by eating raw, infective water chestnuts, water caltrop, and other aquatic plants. Also called intestinal fluke.

**giant roundworm** A common soilborne, but occasionally waterborne, parasite causing ascariasis.

***Giardia*** A genus of single-celled, flagellated, pathogenic protozoans, existing as trophozoites or as cysts, that cause gastroenteritis and resist disinfection. Coagulation–sedimentation removes 65–90% of *Giardia* cysts. At low temperatures, cysts may persist in water for many months. Slow sand filtration achieves essentially complete removal of cysts. In-line filtration is an effective alternative for waters of low turbidity.

***Giardia* cyst** A stage in the life cycle of the genus *Giardia.*

***Giardia enterica*** Another name of *Giardia lamblia.*

**Giardia enteritis** Same as giardiasis.

***Giardia intestinalis*** A species of protozoans found as trophozoites in the small intestine of human and animal hosts, or as cysts, which average about 13 micrometers in length. They resist adverse environmental conditions and cause infections through water or food contaminated by feces. Another name of *Giardia lamblia.*

***Giardia lamblia*** Flagellate protozoan shed in the feces of humans and animals mostly during its cyst stage. It is the species of *Giardia* that can cause giardiasis, a severe gastrointestinal disease with an infective dose of as few as 10 viable cysts, when contaminated drinking water containing the cysts is ingested; an opportunistic pathogen that significantly affects individuals with weak or compromised immune system. Filtering-to-waste is an effective treatment procedure against this pathogen. Cysts measure approximately 9 micrometers by 12 micrometers. Also called *Giardia enterica, Giardia intestinalis, Lamblia intestinalis.*

***Giardia muris*** A species of *Giardia* found in rodents; sometimes used as a proxy for the human pathogen in water quality testing.

**giardiasis** A gastrointestinal infection caused by ingestion of waterborne (and less often, foodborne) *Giardia lamblia.* The disease has an average incubation period between 7 and 10 days, is often asymptomatic, but sometimes causes chronic diarrhea, abdominal cramps, flatulence, weight loss, and fatigue. Infection often results from surface water contaminated by wild and domestic animals. The disease may be transmitted from person to person, e.g., in institutions and day-care centers. Also called beaver fever in the western United States.

**Gibbs equation** A relationship between interfacial tension and adsorption, for a given species, at temperature $T$ and pressure $P$ (Stumm & Morgan, 1996):

$$S_c = -(1/RT)(\partial S_t/\partial \ln A) \qquad \text{(G-17)}$$

where $S_c$ = surface concentration (mol/m$^2$), $R$ = gas constant, $T$ = absolute temperature, $S_t$ = surface tension or interfacial tension, J/m$^2$), and $A$ = activity (or concentration) of the species.

**Gibbs' free energy (*G*)** The property of a thermodynamic system that determines the spontaneity of processes such as chemical reactions. It is equal to the enthalpy of the system minus the product of its entropy and absolute temperature. Each reaction corresponds to a net change in free energy ($\Delta G$), the amount of useful energy available from the equation or the work required for the reaction to proceed. For the reaction

$$\text{aA} + \text{bB} \leftrightarrow \text{cC} + \text{dD} \qquad \text{(G-18)}$$

$$\Delta G = \Delta G^0 + R \cdot T \cdot \ln[\{\text{C}\}^\text{c} \cdot \{\text{D}\}^\text{d}/\{\text{A}\}^\text{a} \cdot \{\text{B}\}^\text{b}] \qquad \text{(G-19)}$$

$$\Delta G^0 = -R \cdot T \cdot \ln K \qquad \text{(G-20)}$$

where $\Delta G^0$ is the standard free energy change, obtained at standard conditions (pH = 7.0, pressure of 1.0 atm, and temperature of 25°C); $R$ is the universal gas constant = 1.987 cal/°K/mole; $T$ is temperature in °K; {X} indicates the activity of substance X; and *K* is the equilibrium constant. Named after the American physicist Josiah Willard Gibbs (1839–1903). The German physiologist and physicist Hermann Ludwig Ferdinand von Helmholtz (1821–1894) deserves a similar credit. Also simply called free energy, Gibbs' free energy function, Gibbs' function, or thermodynamic potential. Other related thermodynamic terms include: standard cell potential, electrochemical potential, Faraday's constant, equilibrium constant, oxidation–reduction reaction.

**Gibbs' free energy function** Same as Gibbs' free energy.

**Gibbs' function** Same as Gibbs' free energy.

**gibbsite** A whitish or grayish mineral, essentially hydrated aluminum oxide ($Al_2O_3 \cdot 3\ H_2O$) or aluminum hydroxide [$Al(OH)_3$], an important constituent of bauxite ore and laterite. It is also a result of the dissolution process of silicates; see biotite for detail. Named after the American mineralogist George Gibbs (1776–1833); also called hydrargillite.

**Gifft formulas** Three equations proposed in 1945 to account for the damping effects in wastewater flows (Fair et al., 1971):

$$Q_{max}/Q_{ave} = 5.0\ P^{-1/6} \quad \text{(G-21)}$$

$$Q_{min}/Q_{ave} = 0.2\ P^{1/6} \quad \text{(G-22)}$$

$$Q_{max}/Q_{min} = 25.0\ P^{-1/3} \quad \text{(G-23)}$$

where $Q_{max}$ = maximum daily flow, $Q_{ave}$ = average daily flow, $Q_{min}$ = minimum daily flow, and $P$ = population in thousands.

**Gilson manometer** Same as Gilson respirometer.

**Gilson respirometer** A device that measures oxygen consumption, e.g., the biochemical oxygen demand (BOD) of a wastewater sample. It is similar to the Warburg respirometer (i.e., a manometer connected to a flask that contains the sample and seed culture and an alkali solution to trap the carbon dioxide produced), except that it holds the gas pressure (instead of the volume) constant. Currently (2007), the electrolysis cell respirometer is preferred for the determination of BOD. *See also* respirometric method.

***G. intestinalis*** *See Giardia intestinalis.*

**Girasieve®** A rotating drum screen manufactured by Andritz Sprout-Bauer S.A., including an external feed.

**GIS** Acronym of geographic information system.

**glacial acetic acid** *See* acetic acid.

**glacier** A large mass of land ice that consists of recrystallized snow; it moves slowly. *See also* firn.

**Gladiator** A pump manufactured by Ejector Systems, Inc. for use in groundwater remediation projects.

***G. lamblia*** *See Giardia lamblia.*

**gland** (1) A cell, group of cells, or organ producing secretions or an organ that resembles a gland. (2) A sleeve within a stuffing box over a shaft or valve stem, tightened against compressible packing to prevent leakage while allowing movement. (3) A device used to retain and compress packing around a valve stem. *See* packing gland.

**GLASdek** A synthetic medium manufactured by Munters for use in evaporative cooling applications.

**glass-distilled water** Water distilled in an all-glass apparatus.

**glass electrode method** A method used in the laboratory to determine the pH of a solution. Also called electrometric method.

**glass enclosure** A greenhouse-like enclosure used to protect and improve the performance of sludge drying beds.

**glass fiber filter (GFF)** A water treatment device made of glass fibers and used to filter fine particles and algae.

**glassification** A process that consists in imbedding wastes in a glass material. *See* vitrification for detail.

**glass washer** A plumbing fixture installed in older buildings with inlets below their water level and thus a potential source of contamination by cross-connection.

**glauberite [$CaNa_2(SO_4)_2$]** A soluble salt mixture of calcium and sodium sulfates, sometimes found in the brine resulting from desalination processes.

**Glauber salt ($Na_2SO_4 \cdot 10H_2O$)** An anhydrous (decahydrate) form of sodium sulfate; a colorless, crystalline, soluble solid, used in textile dyeing and as a purgative. Named after the German chemist J. R. Glauber (1604–1668).

**Glauber's salt** Same as Glauber salt.

**glauconite** A natural, dull-green, micacious hydrous potassium iron aluminum silicate, used as a filter medium (with affinity for the adsorption of manganese) and as an ion exchanger; approximate formula: [$K_{15}(Fe, Mg, Al)_{4-6}(Si, Al)_3O_{20}(OH)_4$]. *See* greensand.

**Global Environmental Monitoring System (GEMS)** One of four components of Earthwatch, a part of the United Nations Environment Program (UNEP). The purpose of GEMS is to provide early warning of impending natural or man-induced environmental changes or trends that pose direct or indirect harm to human health or well-being. *See* GEMS/WATER.

**global positioning system** or **GPS** A new technology to determine elevations and locations using three-dimensional signals from satellites. It is a system of satellites orbiting the earth twice daily to transmit precise time and position signals. With GPS, water and sewer system elements (hydrants, valves, inverts, wetwells, etc.) can be surveyed with excellent accuracy. *See also* modeling software.

**global rejection** The total flux of the substances that do not pass through a filtration membrane, as opposed to the recovery of the permeating stream. Rejection is a measure of the membrane's ability to resist the passage of solutes. *See also* local (or apparent) rejection, rejection rate, steric rejection.

**globe valve** A valve consisting of a movable disk as closing element and a matching ring seat in a spherical body; used mainly in household plumbing. The valve has a globe-shaped plug that rises or falls vertically when the stem handwheel is rotated.

**globulin** A protein in plant or animal tissue, or a blood plasma protein, insoluble in water but soluble in a salt solution.

**glomerulonephritis** An inflammatory disease of the kidney characterized by the presence of albumin in the urine, edema, and hypertension; caused by mercury, cadmium, or lead poisoning.

***Glossina fusca*** One of two species of tsetse fly that transmit the protozoan *Trypanosoma gambiense,* which causes the Gambian sleeping sickness in Central and West Africa. The fly breeds in riverine and forested areas.

***Glossina longipennis*** A species of tsetse fly, vector of sleeping sickness.

***Glossina morsitans*** A species of tsetse fly that transmits the protozoa *Trypanosoma rhodesiense,* which causes Rhodesian sleeping sickness in East Africa. The fly breeds in open woodlands and feeds on large game animals.

***Glossina palpalis*** One of two species of tsetse fly that transmit the protozoan *Trypanosoma gambiense,* which causes the Gambian sleeping sickness in Central and West Africa. The fly breeds in riverine and forested areas.

**glucose ($C_6H_{12}O_6$)** A simple sugar, a monosaccharide that may be produced by photosynthesis, with release of oxygen:

$$6\,CO_2 + 6\,H_2O = C_6H_{12}O_6 \quad + 6\,O_2 \quad \text{(G-24)}$$

*See also* fructose and sucrose.

**glucuronic acid [$COOH(CHOH)_4CHO$ or $C_6H_{10}O_7$]** An oxidation product of glucose found in the blood and urine. Also called glycuonic acid.

**β-glucuronidase** An enzyme that hydrolyzes compounds of glucuronic acid and chemicals containing the hydroxyl group.

**glucuronide** A glycoside that forms glucuronic acid upon hydrolysis.

**GLUMRB** Acronym of Great Lakes Upper Mississippi River Board.

**glutaraldehyde [$OHC(CH_2)_3CHO$ or $C_5H_8O_2$]** A nonflammable, toxic, soluble liquid used in leather tanning and as a fixative for electron microscope samples.

**glutathione ($C_{10}H_{17}O_6N_3S$)** A crystalline, soluble compound of glutamic acid, cysteine, and glycine found in blood and tissues, important for the activation of some enzymes.

**glutathione peroxidase** An enzyme that inactivates peroxides using glutathione.

**glutathione reductase** An enzyme that reduces glutathione that has been oxidized.

**glutathione-S-transferase** An enzyme that transfers glutathione to chemical metabolites.

**glycine ($H_2NCH_2COOH$ or H—CH—COOH—$NH_2$)** The simplest amino acid; a colorless, crystalline, sweet, water-soluble solid used in organic synthesis, in biochemical research, and as a constituent of an alkaline buffer used as an eluent liquid in testing for enteric viruses. Also called aminoacetic acid, glycocoil. *See also* cholic acid.

**glycocalyx** A coating of macromolecules (carbohydrates or proteins) that surrounds the wall of a bacterial cell. It protects the cell against predation, disinfection, and other adverse conditions. Also called either a slime layer or a capsule.

**glycocoil** Glycine.

**glycogen [$(C_6H_{10}O_5)_n$]** A white, tasteless polysaccharide similar to starch; an important carbohydrate material in the liver, muscles, fungi, and yeasts. Also called animal starch.

**glycol ($C_2H_6O_2$)** A colorless, readily biodegradable, sweet liquid used as antifreeze, as a solvent, and a primary ingredient of deicing compounds at airports. Also called ethylene alcohol, ethylene glycol.

**glycolysis** The process of carbohydrate (e.g., glucose or glycogen) fermentation into lactic or pyruvic acid; the first stage in the anaerobic liberation of energy from food during respiration. Also called Embden–Meyerhof–Parnas pathway.

**glycoside** Any carbohydrate compound that hydrolyzes to a sugar and an aglycon (a nonsugar product). Some cyanogenic glycosides contain hydrogen cyanide, which is poisonous.

**glycuronide** Same as glucuronide.

**Glydaseal** A sluice gate manufactured by Rodney Hunt Co.

**glyoxal (OHCCHO)** A dialdehyde formed during disinfection, particularly ozonation, of water containing natural organic matter. The USEPA requires water supply systems to monitor it quarterly. Biologically active carbon filtration is an effective method of glyoxal removal.

**glyoxylic acid (OHCCOOH or $C_2H_2O_3$)** A soluble crystalline compound, active in plant photorespiration; it is also formed during disinfection, particularly ozonation, of water containing natural organic matter. Biologically active carbon filtration is an effective method of removal of glyoxylic acid.

**glyphosate [HO—CO—$CH_2$—NH—$CH_2$—PO—$(OH)_2$]** The common name of the herbicide N-(phosphonomethyl)glycine; an organic contaminant regulated in drinking water by the USEPA: MCL = MCLG = 0.7 mg/L.

***G. morsitans*** Abbreviation of *Glossina morsitans.*

**$G^0$** Symbol of standard free energy.

**G/O** A coarse-bubble diffuser manufactured by G-H Systems, Inc.

**go-devil** (1) A flexible, jointed apparatus with self-adjusting blades, used to free a pipeline from such obstructions as accumulations and tuberculations. (2) A dart dropped into a well to explode a charge of dynamite.

**goethite (α-FeOOH)** A compound formed during the corrosion of iron pipes or a fine particle resulting from the hydrolysis and polymerization of ferric iron; also called an iron phase or a corrosion scale. *See also* hematite.

**goiter** An enlargement of the thyroid gland around the neck; an endemic disease that results from an iodine deficiency (in diet as well as in drinking water). This deficiency is usually remedied by the distribution of iodide tablets instead of adding iodide to drinking water; *see* fluoridation. Also spelled goitre.

**goitre** Goiter.

**goitrogen** A substance that causes goiter, e.g., thiouracil.

**gold (Au** from its Latin name *aurium*) A bright, yellow, soft, precious metallic element, not subject to oxidation or corrosion. Atomic number 79. Atomic weight 196.967. Specific gravity 19.3 at 20°C.

**goldfish** A common ploy of home water treatment filter sales persons is to place a goldfish in a prospective customer's tap water. The goldfish dies within an hour from chlorine. That is not necessarily a threat to the health of humans, who do not breathe through gills.

**Golfwater®** Aeration equipment manufactured by Aeration Industries, Inc. for use in golf course ponds.

**Gooch crucible** A small container in porcelain or other heat resistant material, with a filtering bottom, used in laboratory tests, e.g., the total suspended solids test and the measurement of organic acid concentrations. Also called a filtering crucible.

**Goodrich formula** A formula used to predict the maximum water consumption for a moderate residential community in the absence of actual water use data:

$$p = 180\, t^{-0.10} \qquad \text{(G-25)}$$

where $p$ is the percentage of the annual average consumption corresponding to a period t in days from 1/12 to 360. Thus, the maximum daily rate would be 180% of the annual average.

**gooseneck** Generally, a curved object resembling the neck of a goose; in water supply, a portion of a service connection between the water distribution main and a meter, providing some flexibility for relative displacement between the connection and the main. Also called a pigtail.

**GOR** Acronym of gain output ratio.

**Gore-Tex®** Material made by Gore & Associates, Inc. for the fabrication of microporous membranes.

**gouger** A freshwater organism that burrows into larger particles of wood. *See also* shredder, collector, grazer, piercer.

**Gouy–Chapman model** A mathematical presentation of the electrical double layer theory, with electrostatic attraction and diffusion being the processes responsible for the formation of the diffuse layer and ions being treated as point charges without any other properties. It is a qualitative model of electrostatic stabilization and not a predictive tool for coagulant requirements and coagulation rates.

***G. palpalis*** Abbreviation of *Glossina palpalis.*

**GPC** Acronym of gel permeation chromatography.

**gpcd** Abbreviation of gallon(s) per capita per day, a rate of water use or wastewater production.

**gpd** Abbreviation of gallon(s) per day, a unit of flow or discharge.

**gpd/ft²** Abbreviation of gallons per day per square foot, a unit of flux or hydraulic loading.

**gpd/in-mi** Abbreviation of gallon(s) per day per inch diameter per mile. A measure of the rate of infiltration/inflow (I/I) in a gravity sewer. For example, if a 15″ sewer section of 1100 feet has an I/I flow of 10,000 gallons per day, its I/I rate is:

10,000 gpd/(15 in × 1100 ft/5280 ft/mi) = 3200 gpd/in-mi. Some regulatory agencies specify 5000 gpd/in-mi as a maximum I/I rate for sewer performance or as a guideline for cost-effectiveness analysis of sewer system rehabilitation.

**gpf** Abbreviation of gallon(s) per flush.

**gpg** Abbreviation of grain(s) per gallon, a dosage or a unit of concentration.

**gpm** Abbreviation of gallon(s) per minute, a unit of flow; often used to define the discharge capacity of a pump. One gpm = 0.00144 mgd = 0.002228 cfs = 0.0631 liter/second = 5.4510 $m^3$/day.

**gpm/ft²** Abbreviation of gallons per minute per square foot, a unit of flux or hydraulic loading.

**GPS** Acronym of global positioning system.

**gpy** Acronym of gallons per year.

**grab sample** A single air, water or wastewater sample, representative of the composition of the flow at a particular time and place as compared to a composite sample, which is a flow-weighted average of several samples taken over a given period of time. For example, in stormwater monitoring, a first-flush grab sample is taken within the first 30 minutes of a discharge from a designated storm event. Also called a catch sample or a spot sample.

**grab sampling** The monitoring of air, water, or wastewater characteristics by collecting a single sample to represent flow composition. Grab sampling sometimes leads to grossly misleading information.

**Grabber-1** Equipment manufactured by Franklin Miller for catching heavy objects.

**Grabber-2** A bar screen manufactured by Hycor Corp. with a reciprocating rake.

**gradation coefficient** *See* pebble count method.

**grade** (1) The elevation of the invert or the bottom of a pipeline, canal, culvert or similar conduit. (2) The inclination or slope of a pipeline, conduit, stream channel, or natural ground surface; usually expressed in terms of the ratio or percentage of number of units of vertical rise or fall per unit of horizontal distance. A 0.5% grade would be a drop of one-half foot per hundred feet of pipe. (3) The finished surface of a structure such as a canal bed, roadbed, top of an embankment, or bottom of an excavation.

**graded-border irrigation** One of two main surface application methods of distributing wastewater in land application systems, using low, narrow or wide, parallel soil ridges or borders constructed in the direction of the slope. *See also* furrow irrigation, drip (or trickle) irrigation, gated pipe irrigation, sprinkler system.

**gradient** The change in quantity over distance; the rate of change of any characteristic per unit length; e.g., bottom slope, which may be expressed in ft/ft, m/m, or as a percentage; e.g., elevation gradient, hydraulic gradient, pressure gradient, and velocity gradient.

**gradient ion chromatography** An analytical technique that mixes solvents to increase the strength of a constituent during the separation of ion-exchange analytes.

**graduated cylinder** A tall cylinder of glass or plastic, with a pouring lip and fine graduations, used in laboratories for rapid liquid measurements.

**Graetz–Leveque correlation** A set of adjustable parameters proposed for the Sherwood number equation under particular conditions. *See* Sherwood number for detail.

**Graham's law of diffusion** A thermodynamic principle that states that at a given temperature and pressure, the rate of diffusion of a gas is inversely proportional to the square root of its density, i.e.,

$$K_1/K_2 = (\rho_2/\rho_1)^{0.5} \quad \text{(G-26)}$$

where $K_1$ is the rate of diffusion corresponding to density $\rho_1$, and $K_2$ corresponds to $\rho_2$. *See also* general gas law.

**Graham, Thomas** British chemist (1805–1869)

**grain density** The mass of a grain of filter media per unit volume; a characteristic that affects backwash rates.

**grain hardness** The ability of a mineral to scratch or be scratched by another; it affects the durability of a filter medium. *See* MOH hardness number.

**grain per gallon (gpg)** A measure of the concentration of a solution equal to 17.1 mg/L, used mainly in water hardness calculations.

**grain roundness** A measure of how closely a grain of filter medium approximates a sphere. It affects bed porosity, head loss through the medium, backwash flow requirements, and treatment efficiency. *See also* equivalent spherical diameter, sphericity.

**grain shape** *See* grain roundness.

**grain size and distribution** An important characteristic of a filter medium, determined by sieve analysis. It affects filtration efficiency and backwashing requirements. An appropriate filter medium is neither too coarse nor too fine. *See also* effective size, uniformity coefficient.

**grain specific gravity** *See* grain density.

**gram** A unit of mass equivalent to one milliliter of water at 4°C; one pound = 454 grams.

**gram atomic weight** The quantity of an element in grams that is equal to its atomic weight or the quantity of a compound that contains one Avogadro number ($6.023 \times 10^{23}$) of each of its atoms.

**gram equivalent weight** The equivalent weight in grams. A 1 normal (1 N) solution contains 1 gram equivalent weight of the substance or reagent per liter of solution. *See also* chemical equivalent and equivalent per liter.

**Gram, Hans Christian J.** Danish bacteriologist (1853–1938) who differentiated bacteria as Gram-positive or Gram-negative based on differences in dye uptake.

**gram-milliequivalent** The equivalent weight in grams divided by 1000.

**gram-mole** *See* gram molecular weight.

**gram molecular weight** The molecular weight of a compound in grams; e.g., the gram molecular weight of sodium hydroxide (NaOH) is 40 grams. Also called gram-mole or gram molecule.

**gram molecule** Same as gram molecular weight.

**Gram-negative** (or **gram-negative**) The characteristic of a cell wall that remains red (does not retain the violet dye) after completion of the Gram-staining procedure. Yeasts, some bacteria (e.g., *Shigella dysenteriae*), and a few molds are Gram-negative.

**Gram-negative** (or **gram-negative**) **rods** Coliforms that, in the multiple-tube fermentation technique, indicate a positive completed test.

**Gram-positive** (or **gram-positive**) The characteristic of a cell wall that remains blue (retains the violet dye) after completion of the Gram-staining procedure; most organisms have Gram-positive cells.

**Gram-positive** (or **gram-positive**) **bacteria** Bacteria that have a Gram-positive cell wall. *See Clostridium, Lactobacillus,* and *Staphyloccocus.*

**Gram-positive** (or **gram-positive**) **rods** Microorganisms that, in the multiple-tube fermentation technique, indicate the absence of the coliform group.

**Gram reaction** or **gram reaction** A bacteriological staining technique. *See* Gram's stain.

**Gram's method** or **gram's method** A method of staining and identification of bacteria. *See* Gram's stain.

**Gram's solution** or **gram's solution** A solution of iodine ($I_2$), potassium iodide (KI), and water ($H_2O$) that is used to stain and identify bacteria.

**Gram's stain** or **gram's stain** A common staining procedure for differentiating and categorizing bacteria as Gram-positive and Gram-negative. It involves two dyes (purplish and red colors), iodine, and a solvent. At the completion of the procedure, Gram-positive organisms are blue or violet and Gram-negative organisms are red (pink) or violet. *See also* coliform bacteria.

**Gram stain** or **gram stain** *See* Gram's stain.

**Gram staining** (or **gram staining**) *See* Gram's stain.

**Gran plot** A diagram used to determine titration end points more precisely than on a pH vs. acid or base curve. The Gran procedure is based on the principle that added increments of mineral acid increase hydrogen ion concentration [$H^+$] and decrease hydroxyl ion concentration [$OH^-$], and vice versa for increments of a strong base.

**granular activated carbon (GAC)** A granular form of activated carbon with a diameter larger than 0.1 mm (approximately 140 sieve) and a large surface area per unit mass, used for air, water, or wastewater treatment in filters, fixed-bed contactors, or as support to a biological population. GAC, particularly effective against organics, can be produced from coal, peat, wood, and agricultural by-products. Its characteristics include a total surface area of 700–1300 $m^2/g$, a bulk density of 400–500 $kg/m^3$, an effective size of 0.6–0.9 mm, an iodine number of 600–1100, and a uniformity coefficient $\leq 1.9$. *See also* powdered activated carbon.

**granular activated carbon capped filter** A multimedia filter with granular activated carbon as the top layer, used mainly for the removal of organic matter.

**granular activated carbon–sand filter** A dual-media filter with sand at the bottom and granular activated carbon as the top layer, sometimes replacing anthracite.

**granular activated carbon sandwich filter** *See* GAC Sandwich™ filter.

**granular activated carbon (GAC) treatment** A filtering system often used in small water plants and individual homes to remove organics. GAC can be highly effective in removing taste and odor compounds, synthetic organic chemicals, and elevated levels of radon from water.

**granular bed filter** A water treatment device using filtration through a bed of granular material, commonly a substantial depth of sand, anthracite coal, granular activated carbon, or a combination of these materials. *See also* precoat filter.

**granular-bed filter equation** A semiempirical equation developed to interpret depth filtration data and design granular-bed filters (Freeman, 1998):

$$t = KH/(CQ) \qquad \text{(G-27)}$$

where $t$ = total run time, $K$ = an empirical constant, $H$ = operating head available beyond the headloss of a clean bed, $C$ = suspended solids concentration in the feedwater, and $Q$ = flow velocity.

**granular carbon** *See* granular activated carbon.

**granular filtering material** *See* filter media.

**granular filtration** A treatment process that uses a bed of granular media to remove suspended solids and colloidal matter from water or wastewater. *See* granular media filter for the types of device used. *See also* membrane filtration.

**granular hypochlorite** Calcium hypochlorite [$Ca(OCl)_2$] or lithium hypochlorite (LiOCl) used in water disinfection after being dissolved in batch solution near the point of application.

**granular media** Granular materials such as sand, anthracite coal, and activated carbon used in water filtration.

**granular media filter** A tank or vessel filled with sand or other granular media to remove suspended solids and colloids from a water or wastewater that flows through it. Granular media filters are classified according to the direction of flow, type and number of media, driving force, and flow control method. *See* biflow filter, depth filter, downflow filter, dual-media filter, gravity filter, mixed-media filter, multimedia filter, pressure filter, rapid sand filter, single-medium filter, slow sand filter, surface filter, upflow filter.

**granular media filtration** *See* granular filtration.

**granular-media gravity filter** *See* gravity filter.

**granular media properties** *See* grain density, grain hardness, fixed-bed porosity, grain shape and roundness, grain size and distribution.

**granular medium filtration** *See* granular filtration.

**granulated dense sludge** A dense sludge floc with a solids concentration of up to 100,000 mg/L that is a key feature of the upflow sludge blanket reactor process. *See* dense granulated sludge.

**granulated sludge** *See* dense granulated sludge.

**granulated sludge solids** *See* dense granulated sludge.

**granule** *See* cell granule.

**graphite** A black or dark grey, natural or manufactured, soft carbon, with metallic luster and greasy feel, used as a lubricant, as pencil lead, and for making crucibles and other refractories. Graphite is one of the materials that can be used to make activated carbon for the treatment of water or wastewater. Also called black lead.

**graphite furnace atomic absorption (GFAA) spectrophotometry** A highly sensitive instrumental technique for measuring trace quantities of elements in water when the concentrations are on the order of micrograms per liter; the element being measured is atomized in a graphite tube. *See also* atomic absorption spectrophotometry and flame atomic absorption spectrophotometry.

**graphitization** One of two types of corrosion-related failures observed in steel and iron pipes. It occurs when the iron content of the pipe is leached away, leaving behind a carbon flake matrix. Graphitization is common in highly mineralized or low-pH waters that cause the removal of the iron–silicon metal alloy and leave a hard mass of graphite. *See also* corrosion pitting.

**grapple dredge** A floating derrick equipped with a grab bucket for removing materials from deep waters.

**GRAS** Acronym of generally regarded as safe, a materials designation of the U.S. Food and Drug Administration.

**grassed waterway** Natural or constructed watercourse or outlet that is shaped or graded and established in suitable vegetation for the disposal of runoff without erosion.

**grating** A fixed frame of bars used to exclude coarse materials while admitting air or water; a screening device used in water or wastewater treatment.

**graupel** Snow covered with tiny, white, granular ice particles; a material resembling small soft hail. Also called snow pellet. *See also* rime.

**Gravabelt** A gravity-belt thickener manufactured by Komline-Sanderson Engineering Corp. for use in sludge processing.

**gravel** Rock fragments measuring 2–75 mm (0.008–3.0 inches) and used as support material in granular filter media and in concrete. Gravels and sands constitute the best water-producing materials of aquifers, with good storage and transmission characteristics.

**gravel filter berm** A temporary ridge constructed of loose gravel or crushed rock to capture sediment from runoff by slowing and filtering flow, and diverting it from an exposed traffic area.

**gravel-less pipe** An 8 or 10 inch pipe made of corrugated, perforated polyethylene, wrapped with geotextile fabric.

**gravel-less pipe system** An on-site disposal system that distributes treated wastewater into the soil, consisting of (a) a treatment device, e.g., a septic tank; (b) gravel-less pipes; (c) 2-ft wide pipe trenches.

**gravel mounding** An operational problem of depth filtration, whereby excessive flow rates or uncontrolled air application during backwashing cause the disruption of the support gravel. Other depth filtration operational problems include turbidity breakthrough, emulsified grease buildup, filter medium loss, and cracks and contraction of the filter bed.

**gravel pack** (1) Gravel placed around a well intake screen and perforated well casing, thereby filtering out sediments and increasing the effective well diameter of the well. (2) A similar packing around a landfill gas extraction well.

**gravel support bed** Layers of different size gravel and coarse sand installed above the underdrainage system of a filter or ion exchanger. It supports the bed and contributes to the distribution and collection of finished water as well as the dispersal of backwash water.

**Gravex** A zeolite softener produced by Graver Co.

**graveyard** *See* burial ground.

**gravid** As a noun: a stage in the life cycle of the guinea worm, when it migrates to superficial tissues of the host.

**gravid proglottid** A segment or joint of a tapeworm containing a complete, male and female, reproductive system. Its recovery from feces or the perianal region is used to diagnose taeniasis, diphyllobothriasis, and hymenolepiasis.

**Gravilectric** Equipment manufactured by Patterson Candy International, Ltd. for wasting excess sludge using load cells.

**Gravi-Merik™** A gravimetric belt feeder manufactured by Merrick Industries.

**gravimetric** Of or pertaining to the measurement of unknown concentrations of water quality indicators in a sample by weighing a precipitate or residue of the sample.

**gravimetric analysis** Analysis by weight, e.g., the determination of the total solids content of a water or wastewater sample. *See also* volumetric analysis.

**gravimetric dry feeder** Same as gravimetric feeder.

**gravimetric feeder** A feeder that provides a specific (sometimes constant) weight of dry chemical over a determined period of time. Also called a gravimetric dry feeder. *See also* volumetric feeder.

**gravimetric measurement** Measurement of weight.

**gravimetric method** A quantitative method of analysis. *See* gravimetric analysis.

**gravimetric procedure** A procedure using weight or density to determine concentration.

**gravimetry** The measurement of weight or density; gravimetric analysis.

**Gravipak** Equipment manufactured by Aerators, Inc. for inclined-plate clarifiers.

**Gravi-Pak** An oil–water separator manufactured by Bowser-Briggs Filtration Co.

**Gravisand™** Filter components manufactured by Davis Water & Waste Industries, Inc.

**gravitational acceleration or acceleration of gravity** The acceleration of a free-falling body in a vacuum, a factor used in several fluid mechanics formulas. It varies between 9.78 and 9.82 m/sec$^2$ on the surface of the Earth. In engineering applications, it is usually taken as g = 32.2 ft/sec$^2$ or 9.81 m/sec$^2$.

**gravitational constant** *See* law of gravitation.

**gravitational convection** Motion in a fluid due to density differences. Also called free convection.

**gravitational force** The force due to gravity ($F_g$, kg.m/sec$^2$), exerted by a particle falling in water or wastewater is a function of the densities of the particle ($\rho$, kg/m$^3$) and the liquid ($\rho'$, kg/m$^3$), the volume of the particle ($V$, m$^3$), and the gravitational acceleration ($g$ = 9.81 m/sec$^2$):

$$F_g = (\rho - \rho')gV \qquad \text{(G-28)}$$

The gravitational force is used in Newton's law and the theory of sedimentation to derive terminal velocity.

**gravitational phenomena** *See* gravity separation.

**gravitational potential** (1) The energy required to move a mass of water from the surface of a saturated water column to a higher point against the force of gravity. (2) The potential of groundwater or soil moisture above a certain datum. Also called gravity potential.

**gravitational settling** Same as gravity settling.

**gravitational water** Water that moves by gravity from the soil and pellicular subzones through the capillary fringe and the water table. Sometimes called vadose water or lumped with pellicular water as intermediate groundwater or intermediate vadose water. *See* subsurface water. Sometimes confused with free water and gravity water.

**Gravitator™** A clarifier–thickener designed by DAS International, Inc.

**gravity** The gravitational attraction at the surface of the planet.

**gravity aerator** A device that serves to inject air into water or wastewater through (a) a series of steps (cascade aerators), (b) an inclined plane studded with riffle plates (inclined-plane or inclined-apron aerator), (c) a vertical tower or vertical stacks (tower aerator) with liquid and air flowing in countercurrent, or (d) a stack of perforated trays (tray aerator). The time of exposure ($t$, sec) depends on the available head ($H$, ft) and the number ($N$) of descents, steps, or trays:

$$T = (2\,NH/g)^{0.5} \qquad \text{(G-29)}$$

where g = gravitational acceleration = 32.2 ft/sec$^2$.

**gravity belt thickener (GBT)** A sludge dewatering unit that consists of a porous filter belt that allows water to drain by gravity.

**gravity belt thickening** A process designed to concentrate solid particles in water and wastewater treatment residuals, using a horizontal, porous screen through which water is removed by gravity. *See* belt-press dewatering, centrifugal thickening, cosettling thickening, flotation thickening, gravity thickening, rotary drum thickening.

**gravity collection** Wastewater or stormwater collection involving only open-channel flow (no pumping station and no force main).

**gravity dewatering** A sludge dewatering method that produces a cake by thickening and compression. *See* gravity belt thickener.

**gravity drainage zone** The second area of a belt filter press used to dewater sludge. It consists of a slightly inclined belt for the drainage of free water by gravity.

**gravity filter** (1) A rapid granular media filter that is open for operation at atmospheric pressure; the filter operating level is near the hydraulic gradeline of the influent, which flows by gravity through the bed. It is used to remove nonsettleable floc from water. *See also* pressure filter. (2) A similar device used in wastewater treatment; a deep box containing not only the granular media but also the underdrain system, wash troughs, etc.

**gravity flow** Flow of a fluid with a free surface open to the atmosphere, in an open channel or in a closed conduit flowing partly full. *See* open-channel flow for detail.

**gravity flux** The product ($N_g$) of the hindered settling velocity ($V_h$) of a solids suspension and the initial solids concentration ($C_i$) of the suspension:

$$N_g = C_i V_h \qquad \text{(G-30)}$$

*See also* solids flux.

**gravity grit chamber** *See* gravity separator.

**gravity potential** Same as gravitational potential (2).

**gravity separation** A unit operation used in water and wastewater treatment to remove suspended and colloidal solids by gravity. Applicable gravitational phenomena include accelerated gravity settling, ballasted flocculent settling, compression settling, discrete particle settling, flocculent settling, flotation, hindered settling. *See also* inclined plate settler, tube settler.

**gravity separator** A device designed for separating grit from wastewater by gravity. Grit particles settle and are scoured at the same velocities as suspended solids, in accordance with Newton's law; *see* settling velocity, scouring velocity. *See also* catch basin, settling tank.

**gravity settling** *See* gravity separation.

**gravity sewer** A sewer with open-channel flow, as opposed to a force main where flow is under pressure.

**gravity sludge thickener** *See* gravity thickener.

**gravity spring** A spring that occurs where the water table meets the ground surface; water flows by gravity from numerous small openings in permeable materials or from openings in a rock formation. Also called a filtration spring or a seepage spring.

**gravity survey** A geophysical investigation conducted to locate groundwater sources, indicating the location of fractures and intrusions, and the depth to alluvial deposits. *See also* resistivity survey, seismic survey.

**gravity system** A network of gravity sewers.

**gravity thickener** A sedimentation basin or tank used to concentrate solids in water, wastewater, or sludge. Gravity thickeners operate at high solids loading rates, and have rotating arms to facilitate thickening and scrapers to remove the sludge. Design parameters are surface overflow rate between 350 and 850 gallons per square foot per day and a detention time between 3 and 4 hours. Good performance also depends on aerobic conditions and a high sludge blanket. *See also* flotation thickener.

**gravity thickening** A process designed to concentrate solid particles in water and wastewater treatment residuals, prior to digestion or dewatering, using a batch or continuous, usually circular, concrete thickening tank operated at a flow rate that allows the solids to settle by gravity. Gravity thickening also serves an equalization and storage purpose, ahead of other sludge processing operations. In wastewater treatment, the supernatant resulting from thickening is returned to the influent or the primary settling tank. *See also* centrifugal thickening, cosettling thickening, flotation thickening, gravity belt thickening, rotary drum thickening, sludge blanket.

**gravity water** Water conveyed by gravity as opposed to pumping. *See also* gravitational water.

**gravity water supply** A water supply that operates by gravity, i.e., without pumping.

**gray** A unit of the Système International for measuring absorbed doses of ionizing radiation, equal to 1 joule per kilogram (1.0 J/kg) or 100 rads. *See also* roentgen equivalent man (rem).

**grayling** Any of a group of freshwater fish related to the trout and having a brilliantly colored fin. Like the rainbow trout, it is an intolerant species that requires cold water, high dissolved oxygen content, and low turbidity.

**gray wastewater** *See* gray water.

**gray water** All nontoilet household wastewater from sinks, basins, washing machines, dishwashers, baths, and showers, that is, domestic wastewater that does not contain excreta and is expected to contain considerably fewer pathogenic microorganisms than sewage. Gray water is often separated from black water to reduce the loading on onsite systems. Gray water ponding on the ground is a health risk because it can promote the breeding of such vectors as *Culex pipiens* and *Anopheles* mosquitoes. Also called grey water, gray wastewater, sullage. *See also* black water, sanitary water, sullage soakaway.

**gray water system** A wastewater disposal system designed to handle only household wastewater from the kitchen and other sources except toilets. *See also* on-site nitrification/denitrification.

**grazer** A heterotrophic organism that feeds on living organic matter. *See also* detritovore, gouger.

**grazing food chain** The food chain consisting of phytoplankton as primary producers, consumed by zooplankton, which in turn are consumed by other organisms.

**grease** A common term for fats, oily matter, waxes, and related substances that are found in wastewater and usually lumped with oil. *See* grease and oil.

**grease and oil** A group of water-insoluble organic substances found in wastewater, including rendered animal fats, waxes, free fatty acids, soaps, mineral oils, oily matter, thick lubricants, and some nonfatty materials. Grease and oil as an aggregate can be determined analytically by extraction with trichlorotrifluoroethane and is usually removed by flotation skimming. Also called oil and grease.

**GreaseBurn** An incinerator manufactured by Walker Process Equipment Co. for grease and skimmings.

**Grease Grabber** A device manufactured by Abanaki Corp. for skimming oil and grease.

**grease interceptor** A device designed to keep grease out of a sewer system. *See also* grease trap.

**grease-removal tank** A tank used to remove oil and grease by flotation.

**grease separator** Same as grease trap.

**grease skimmer** A device for removing floating grease or scum from a grease-removal tank or any wastewater treatment tank.

**grease-skimming tank** A grease-removal tank that has a grease skimmer.

**grease trap** An apparatus, like a small settling, skimming, or holding tank, for the collection and separation of grease from wastewater; it is usually required for effluents from some commercial and industrial establishments such as restaurants, hotels, garages, packing houses, bakeries, and creameries.

**Great Lakes Initiative** A set of guidelines proposed to establish uniform water quality criteria for the U.S. portion of the Great Lakes basin.

**Great Lakes–Upper Mississippi River Board** A board of 10 member states that sets standards for the design of water and wastewater facilities. *See* Ten-States Standards.

**Great Lakes–Upper Mississippi River Board of State Sanitary Engineers** *See* Great Lakes–Upper Mississippi River Board.

**great overturning** Same as fall overturning.

**Great Sanitary Awakening** A reaction in Great Britain and the United States to the dangers to public health created by poor drinking water quality and inadequate wastewater disposal in the nineteenth century, particularly in the urban areas. Measures taken as a result of this awakening led to a significant decrease in the prevalence of diseases associated with poor sanitation and crowded environments even without immunizations and curative medicine. Also called Sanitary Revolution. *See* Adams (Julius), Bazalgette (John), Chadwick (Edwin), Kirkwood (James P.), Mills (Hiram F.), Rawlison (Robert), Shattuck (Lemuel), Simpson (James).

**green algae** A group of organisms active in the decomposition of organic matter in stabilization ponds, utilizing carbon dioxide produced by bacteria and releasing oxygen. Common species are from the genera *Chlamydomonas, Chlorella,* and *Euglena*.

**greenhouse effect** The warming of the Earth's atmosphere attributed to a build-up of carbon dioxide or other gases; some scientists think that this build-up allows the sun's rays to heat the Earth, while infrared radiation makes the atmosphere opaque to a counterbalance loss of heat. This phenomenon is analogous to what happens in a glass greenhouse.

**greenhouse gases** Gases that cause the greenhouse effect, including carbon dioxide, chlorofluorocarbons, methane, and nitrous oxide.

**Greenland spar ($Na_3AlF_6$)** One of the minerals that contain fluorine; it is essentially a compound of this halogen with sodium and aluminum, occurring in white masses and used in the electrolytic production of aluminum. Also called sodium aluminum fluoride and cryolite.

**Greenleaf Filter Control** A rapid sand gravity filter manufactured by Infilco Degremont, Inc. with multiple cells, including an apparatus to control operation and backwashing.

**green liquor** A solution of smelt from the kraft recovery furnace in water.

**green rust** A compound formed during the corrosion of iron pipes in the presence of oxygen or when ferric oxyhydroxides are dissolved under reducing conditions; also called an iron phase or a corrosion scale. The formula of its carbonate form is $Fe_4^{(II)}Fe_2^{(III)}(OH)_{12}(CO_3)$.

**greensand** A naturally occurring sandstone with a greenish hue caused by its large content of glauconite, a hydrous potassium iron silicate. Because of its ion-exchange properties, it is used extensively in water softening units and in manganese greensand zeolites; regeneration is by potassium permanganate. Greensand is sometimes used to remove iron and manganese from water. *See* greensand filter. Also known as glauconite.

**greensand filter** A filtering device consisting of sand coated with manganese dioxide ($MnO_2$) for the adsorption of soluble iron and manganese. The medium is regenerated by the addition of chlorine or potassium permanganate ($KMnO_4$) to oxidize the iron and manganese removed.

**greensickness** Same as chlorosis; iron-deficiency anemia resulting in a yellow-green complexion.

**green sludge** Fresh sludge that has not undergone advanced decomposition.

**green stain** Discoloration due to a reaction between soap and copper at concentrations higher than 1.0 mg/L in water.

**greenstone** A common name for various dark-green igneous rocks containing chlorite.

**green sulfur bacteria** A group of bacteria that evolved in the absence of oxygen; found in mud, stagnant water, sulfur springs, and saline lakes, they fix carbon using light energy and oxidize sulfide to sulfur:

$$CO_2 + H_2S \rightarrow S^0 + \text{fixed carbon} \qquad \text{(G-31)}$$

**green tide** A profuse algal bloom in a water body, caused by excessive nutrient concentrations. *See also* red tide, brown tide.

**green vitriol ($FeSO_4 \cdot 7\ H_2O$)** A bluish-green, saline-tasting, coagulant commonly used in water treatment and for other industrial purposes (fertilizer, ink, medicine, etc.); a common name for copperas. *See* ferrous sulfate for more detail.

**green water** Water discolored by a fine dispersion of copper-corrosion products that can react with soap scums to cause green staining of plumbing fixtures and lead to consumer complaints. *See also* blue water.

**grey water** Same as gray water or sullage.

**Griductor®** A comminutor manufactured by Infilco Degremont, Inc.

**Griffin, A. E.** Engineer who discovered the breakpoint chlorination phenomenon in 1939.

**Griffin Generator** An ozone generator manufactured by Ozonia North America.

**Grifter®** A compact unit manufactured by Ingersoll-Dresser Pump as a wastewater grinding and pumping station.

**Grind Hog™** A shredding device manufactured by G. E. T. Industries, Inc.

**grinder** A device, usually located in piping or in channels, used to reduce the size of wastewater solids for further processing by chopping them into particle sizes of 6–9 mm. *See also* barminutor, comminutor, grinder pump, hammermill, rag ball, rag rope, screenings grinder.

**grinder pump** A mechanical device that shreds solids and raises wastewater to a higher elevation through pressure sewers; usually a submersible, centrifugal pump, installed in a small wet well serving up to five residences; commonly used to pump septage, holding-tank waste, sludge, and slurries. *See also* pressure sewerage, septic tank effluent pumping (STEP), chopper–grinder pump.

**grinder pump system** A pressure sewer system that includes a holding tank and a grinder pump that reduces the size of the solids and discharges the wastewater into a pressure main.

**grinding** *See* sludge grinding.

**grit** Sand, gravel, cinders, and other suspended, inert, inorganic solid matter with a nominal diameter of 0.12–0.20 mm and settling velocities and specific gravities substantially greater than those in the organic solids of wastewater. Grit, usually removed in a grit chamber to protect subsequent treatment units from abrasion, is used as fill material or disposed of as solid waste (e.g., in sanitary landfills). Grit volumes depend on sewer conditions and wastewater characteristics, with an average volume of about 0.5 $m^3$ per 1000 $m^3$ of wastewater. Grit that settles in sewer lines with low flows or low slopes may be removed by flushing or by appropriate sewer cleaning equipment.

**grit catcher** A chamber usually placed at the upper end of a depressed sewer or at other points on combined or storm sewers where wear from grit is possible; the chamber is sized and shaped to reduce the velocity of flow through it and thus permit the settling out of grit. Also called sand catcher. *See also* grit chamber and sand interceptor.

**grit chamber** A wastewater treatment unit, like a detention chamber or a grit channel, designed to separate grit from organic solids by differential sedimentation or through air-induced spiral agitation. An important aspect of grit chambers is flow or outlet control by such devices as a proportional-flow weir, a vertical throat, or a Parshall flume. *See also* aerated grit chamber, constant-level grit chamber, detritus tank, gravity separator, grit catcher, grit removal, horizontal-flow grit chamber, grit separator, grit washer, sand catcher, sand interceptor, vortex grit chamber.

**grit channel** The enlarged portion of a sewer designed to reduce the velocity of flow so that grit can settle.

**grit clarifier** A grit removal device with influent and effluent weirs on opposite sides. *See* detritus tank for detail.

**grit classifier** A mechanical device using an inclined screw or reciprocating rake to wash putrescible organics from grit and raise the separated grit to a point of discharge. It consists of an inlet box, settling compartment, and screw- or rake-type conveyor. It is sometimes equipped with water sprays to facilitate cleansing. *See also* grit washer, hydrocyclone separator, cyclone degritter.

**grit collector** A device in a grit chamber that conveys settled grit to the point of collection.

**grit compartment** The part of a grit chamber where grit collects before it is removed.

**grit dewatering** The concentration of grit materials by reducing its water content, often conducted sequentially with grit washing in a hydrocyclone separator and a grit classifier.

**grit disposal** The conveyance of grit removed from wastewater to landfills, dumpsters, or storage hoppers. Clean grit may be used as fill material.

**Grit King™** A grit removal unit manufactured by H. I. L. Technology, Inc.

**GritLift** An airlift pump manufactured by Walker process Equipment Co. for grit handling.

**Gritmeister™** A grit separator equipped with a screw conveyor, manufactured by Hycor Corp.

**Gritreat** An aerated grit chamber by Envirex, Inc.

**grit removal** A preliminary wastewater treatment operation that is designed to separate grit from organic solids, and prevent it from reaching subsequent treatment units such as primary clarifiers or aeration basins. *See also* aerated grit chamber, vortex grit removal, detritus tank, hydrocyclone, horizontal-flow grit chamber, grit washing.

**grit removal mechanism** *See* chain and bucket elevator, chain-and-flight mechanism, clamshell bucket, horizontal screw conveyor, inclined screw, tubular conveyor.

**grit separator** A device used to separate grit from organic solids in water or wastewater.

**Grit Snail™** An apparatus manufactured by Eutek Systems, Inc. for the removal of fine grit.

**grit trap** A device designed for grit settling and collection.

**grit washer** A device used for washing organic matter from grit. *See* grit classifier for detail.

**grit washing** A preliminary wastewater treatment operation that separates grit from lighter organic material captured during grit removal or from sludge when grit is removed in primary clarifiers or in detritus tanks. Grit washing facilitates the storage and disposal of the materials removed, while preventing odors and nuisance conditions caused by putrescible matter. *See* cyclone degritter, hydrocyclone separator.

**gross alpha activity** The total radioactivity due to alpha particle emission as inferred from measurements on a dry sample. The USEPA has set an MCL of 15 pCi/L (a practical quantification level) for gross alpha particle activity in drinking water. The World Health Organization recommends more detailed radionuclide analysis when gross alpha activity exceeds 0.1 Bq/L.

**gross alpha/beta particle activity** The total radioactivity due to alpha or beta particle emission as inferred from measurements on a dry sample.

**gross alpha emitter** A method proposed by the USEPA to identify and measure gross alpha activities.

**gross alpha (particle) activity** *See* gross alpha activity.

**gross alpha radiation** *See* gross alpha activity.

**gross beta activity** The total radioactivity due to beta particle emission as inferred from measurements on a dry sample. The World Health Organization recommends more detailed radionuclide analysis when gross beta activity exceeds 1.0 Bq/L. The USEPA's practical quantification level for this contaminant is 30 pCi/L.

**gross beta emitter** A method proposed by the USEPA to identify and measure gross beta activities.

**gross beta (particle) activity** *See* gross beta activity.

**gross calorific value** Same as sludge heating value. *See also* available heat.

**gross concentration** A measure of the aggregate organic matter in water or wastewater greater than 1.0 mg/L, as determined currently by such laboratory methods as biochemical oxygen demand

(BOD), chemical oxygen demand (COD), and total organic carbon (TOC), or calculated as the theoretical oxygen demand (ThOD) from the formula of the organic matter. Previous methods include albuminoid nitrogen and oxygen consumed. Total, organic, and ammonia nitrogen are still analyzed, mainly to determine nitrogen availability for biological processes. *See also* trace concentration.

**gross heating value** Same as sludge heating value. *See also* available heat.

**ground air** Same as soil air.

**ground alum (GA) salt** A coarse salt the size of ground alum, formerly used to regenerate zeolite in water softeners; also used as a secondary coagulant.

**ground garbage** Kitchen waste shredded or ground by an apparatus and discharged to the sewer system through the sink.

**ground ice** Ice formed below the surface of a body of water, attached either to a submerged structure or to the bed. Also called bottom ice, anchor ice.

**ground key valve** A valve installed on water service pipes and operated by a turn of the plug.

**groundwater** The supply of freshwater found beneath the Earth's surface, usually in aquifers, which supply wells and springs. Because groundwater is a major source of drinking water, there is growing concern over contamination from leaching agricultural or industrial pollutants or leaking underground storage tanks (under the Safe Drinking Water Act).

**groundwater age** The period of time it takes water to travel from the point of recharge to a given point underground. *See also* water age.

**groundwater contamination** The introduction of contaminants into groundwater by natural or manmade processes, via surface water or percolation through the soil (e.g., leaking sewers, malfunctioning septic tanks, latrines, abandoned wells, landfills, dump sites, industrial leaks and spills). These contaminants move slowly, are carried over long distances, and are difficult to remove; often, contaminated aquifers are simply abandoned as drinking water sources. Contaminants of particular concern include nonaqueous-phase liquids, cosolvents, other organics, and microorganisms.

**groundwater dewatering** A method used to remove and discharge excess water from a construction site by lowering the groundwater table. Dewatering discharges usually have a high sediment concentration. *See also* dewatering.

**groundwater discharge** (1) Discharge of subsurface water where the water table intersects the land surface, e.g., through springs and seepage outcrops, or to the atmosphere by evaporation. (2) Groundwater entering near coastal waters, which has been contaminated by landfill leachate, deep well injection of hazardous wastes, septic tanks, etc. (under the Safe Drinking Water Act).

**Groundwater Disinfection Rule (GWDR)** A regulation of the USEPA setting the disinfection requirements for systems using groundwater, with an emphasis on the risk of viral diseases. *See also* Groundwater Rule, natural disinfection, groundwater under the influence of surface water.

**groundwater exploration** An investigation conducted to locate a suitable source of groundwater, including the examination of maps and aerial photographs, and the completion of geophysical surveys. *See* gravity, resistivity, and seismic surveys.

**groundwater hydrology** The study of the origin, nature, and occurrence of subsurface water, as well as its movement through and seepage from the underground formations. The general hydrologic equation relates the various recharge factors ($\Sigma R$) to discharges ($\Sigma D$) and change in storage ($\Delta S$):

$$\Sigma R = \Sigma D + \Delta S \qquad \text{(G-32)}$$

*See also* hydrography (or surface water hydrology) and hydrometeorology.

**groundwater infiltration** (1) Same as infiltration (1). (2) Water that enters a treatment facility as a result of the interception of natural springs, aquifers, or runoff that percolates into the ground and seeps into the treatment facility's tailings pond or wastewater holding facility and that cannot be diverted by ditching or grouting the tailings pond or wastewater holding facility (under the Clean Water Act) (EPA-40CFR440.132-d). *See also* infiltration/inflow.

**groundwater law** The common law doctrine of riparian rights and prior appropriation as applied to groundwater.

**groundwater mining** Withdrawal of groundwater over a period of time in excess of the rate of aquifer recharge. *See also* fossil groundwater.

**groundwater mound** The rise in groundwater level that occurs when wastewater is applied to a spreading basin. Its height and extent depend on basin geometry, application rate, existing level and slope of water table, hydraulic conductivity, porosity, and other factors.

**groundwater recharge** Replenishment of groundwater naturally or artificially, including infiltration from rainfall and snowmelt or from surface water, leakage through confining layers, and water from diffusion or spreading operations. Artificial recharge is undertaken to restore water withdrawn from an aquifer, to protect the aquifer against such sources of pollution as saltwater intrusion, or to control subsidence. *See also* water reuse applications.

**groundwater renovation** The restoration of the quality of a contaminated groundwater or the reduction of the extent of its contamination. *See* containment, extraction, and in situ treatment.

**groundwater reservoir** An aquifer system in which water is stored naturally or artificially.

**Ground Water Rule (GWR)** An anticipated regulation of the USEPA under SDWA section 1412(b)(8). It is to require the use of disinfectants for groundwater systems and to establish multiple protective measures against bacteria and viruses, including sanitary surveys, hydrologic sensitivity measurements, microbial monitoring of source water, and corrective actions

**groundwater table** The level of groundwater; the upper surface of the zone of saturation of groundwater above an impermeable layer of soil or rock (through which water cannot move) as in an unconfined aquifer. This level can be near the surface of the ground or far below it. In an unconfined aquifer, the fluid at the water surface is at atmospheric pressure. Sometimes simply called water table.

**groundwater under the direct influence of surface water (GWUDI)** Any water beneath the surface of the ground with: (a) significant occurrence of insects or other macroorganisms, algae, or large-diameter pathogens such as *Giardia lamblia;* (b) significant and relatively rapid shifts in water characteristics such as turbidity, temperature, conductivity, or pH that closely correlate to climatological or surface water conditions. The State must determine direct influence for individual sources in accordance with its criteria. The State determination of direct influence may be based on site-specific measurements of water quality and/or documentation of well construction characteristics and geology with field evaluation (EPA-40CFR141.2). GWUDI sources are considered to be subject to contamination by *Giardia lamblia* and *Cryptosporidium parvum.* They fall under the USEPA's Surface Water Treatment Rule (SWTR), the Interim Enhanced Surface Water Treatment Rule (IESWTR), and the Long Term 2 Enhanced Surface Water Treatment Rule (LT2ESWTR). These regulations provide guidance for filtration and disinfection of GWUDI sources.

**Group D streptococci** *See* fecal streptococci.

**grout** A watery mixture of cement (and, commonly, bentonite) without aggregate that is used to seal the annular space around well casings to prevent infiltration of water or short-circuiting of vapor flow. Grouting is also used in joints of brickwork or masonry.

**growing degree-day (GDD)** (1) A parameter that indicates the growth and development potential of crops, defined as an accumulation of heat units (Ruiz et al., 2006):

$$GDD = 0.5\,(T_{max} - T_{min}) - T_{base} \quad \text{(G-33)}$$

where $T_{max}$ = daily maximum temperature at which heat units accumulate, $T_{min}$ = daily minimum temperature at which heat units accumulate, and $T_{base}$ = the temperature below which growth stops for a specific crop. GDD is used in the design and scheduling of land application systems of wastewater effluents. (2) A parameter used in relation to plant growth and representing one degree-day above the standard of 5°C = 41°F. Sometimes called degree-day.

**growth factor** Organic nutrients, i.e., compounds required by an organism for growth but which it cannot synthesize from other carbon sources: amino acids, nitrogen bases, and vitamins. Microorganism growth depends also on environmental factors such as temperature, salinity, and pH. *See also* purine, pyrimidine, minor nutrient, inorganic nutrient, mesophilic organism, psychrophilic organism, thermophilic organism.

**growth pattern** In the presence of adequate food and environmental conditions, the number of microorganisms in a medium varies at various rates: first slowly (lag phase), then logarithmically (log growth phase) and at a declining rate until the number is constant (stationary phase) and starts decreasing (increasing death, log death, and death phases). The actual mass of microorganisms varies in a similar fashion, with an endogenous phase corresponding to the death phases. *See* Figures G-02 and G-03. *See also* basal metabolism, endogeny, lysis, protoplasm.

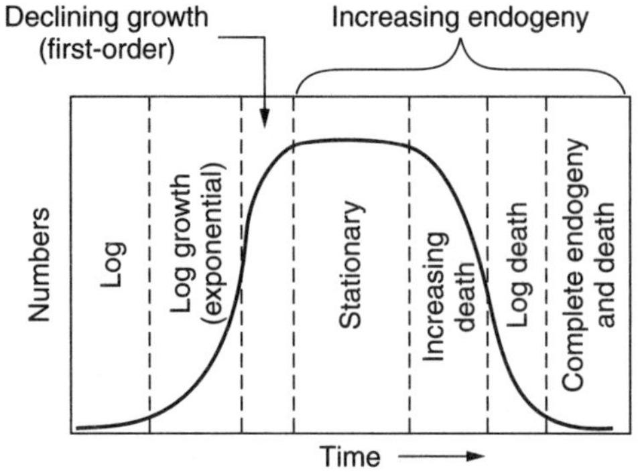


**Figure G-02.** Growth pattern (numbers).

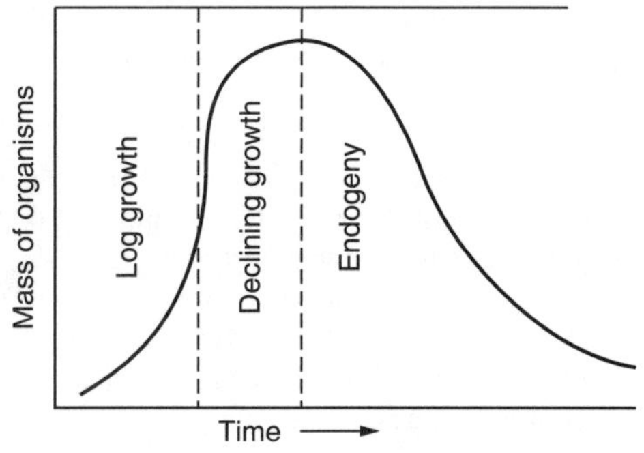


**Figure G-03.** Growth pattern (mass).

**growth yield** The increase in biomass from metabolism of an incremental amount of substrate.

**growth yield coefficient** A dimensionless parameter used in the theory of biological waste treatment to relate cell yield to the mass of the critical nutrient metabolized. This coefficient is often assumed to be $Y = 0.65$.

**GSS** Acronym of gas solids separator.

**GT** The dimensionless product of velocity gradient and detention time, two parameters used in the design of mixing and flocculation basins; for example, a GT value of 200,000 is recommended for optimum performance of rapid mixers. *See also* contact opportunity or conjunction opportunity.

**Guardian™** Strainer products of Tate Andale, Inc.

**Guardian Blue™** An early warning system manufactured by the Hach Company that includes continuous monitoring of water quality parameters and compares recorded data against a proprietary library of contaminant fingerprints.

**Guggenheim process** A wastewater treatment process that uses chemical flocculation of dispersed biological growth and sludge recycle for the removal of organic matter, suspended solids, and phosphorus.

**guinea worm** Common name of *Dracunculus medinensis,* the excreted helminth species that causes dracunculiasis. It is a long, slender roundworm that is common in Africa and India. Also a short name for guinea-worm disease. *See also Cyclops.*

**guinea-worm disease** Common name of dracunculiasis, a helminthic infection caused by the pathogen *Dracunculus medinensis.* The disease is transmitted from an infected person through *Cyclops* as intermediate hosts. Its final stage is a 1.5 ft long female worm forming a painful blister under the skin of the leg of an infected person. Guinea worm is recognized as the only disease exclusively transmitted by drinking water and its eradication is sometimes adopted as an index to measure success in providing safe water supplies. Filtration (e.g., through a cloth) can remove the cyclopoids (and the pathogens) from drinking water.

**Gujer–Boller equation** An equation proposed for the design of nitrification biotowers (WEF & ASCE, 1971):

$$R_{ZT} = e^{0.044(T-10)}\, N/(K_N + N) \qquad \text{(G-34)}$$

where $R_{ZT}$ = surface removal rate or nitrification rate (grams per square meter per day) at depth $Z$ and temperature $T$; $T$ = temperature, °C; $N$ = bulk liquid $NH_4$-N concentration, mg/L; $K_N$ = half-velocity constant (saturation concentration), mg/L. *See also* the Parker et al. formula.

**gulf streaming** The phenomenon of stratification by size of the grains of a single-medium filter during backwash with fluidization, the finer grains resting on top and the coarser grains at the bottom. The tendency to stratify depends on the backwash rate and the differences in bulk densities of the various grain sizes. Also called jet action, sand boil.

**gullet** (1) In general, an open channel for water, a gully, a ravine, or a cut in an excavation. (2) An open channel that collects backwash water in a filter for subsequent discharge, with backwash troughs or a weir.

**gully** A small, elongated, and deep channel created by the eroding action of running water, usually dry except after a rainstorm, icemelt, or snowmelt. Gullies are similar to but smaller than ravines, and deeper than rills or rivulets. *See* stream for the difference between various watercourses. Gully erosion is severe erosion in which trenches are cut to a depth greater than one foot or 30 centimeters. Generally, ditches that can be crossed with farm equipment are considered gullies.

**gully erosion** *See* gully.

**gully reclamation** The use of a small dam of straw, earth, stone, or concrete to collect silt and gradually fill in channels of eroded soil. *See also* check dam.

**gum ammoniac** A brownish-yellow gum resin, with an acrid taste, extracted from a plant and used in ceramics and in medicine. Sometimes simply called ammoniac.

**Gundline®** A containment liner manufactured by Gundle Lining Systems, Inc.

**Gundnet** A drainage net manufactured by Gundle Lining Systems, Inc.

**Gundseal** A geocomposite liner manufactured by Gundle Lining Systems, Inc.

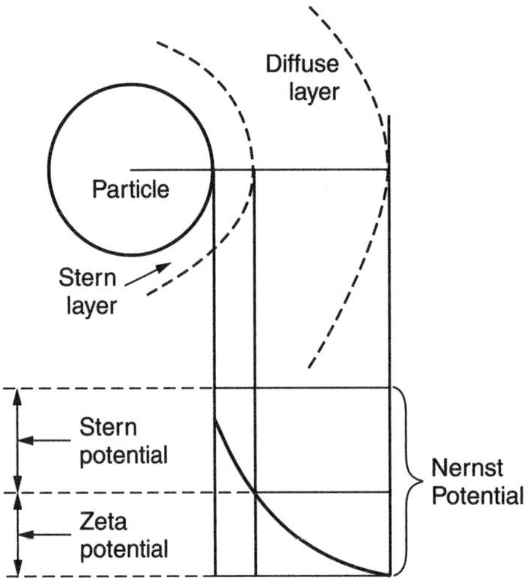


**Figure G-04.** Guoy–Stern colloidal model.

**Guntelberg approximation** An equation proposed to calculate the activity coefficient ($\gamma$) of a chemical species as a function of its charge or valence ($c$) and the ionic strength ($I$) of the solution, when $I < 0.1$ M (Metcalf & Eddy, 2003):

$$\log \gamma = -0.5\, c^2 I^{0.5}/(1 + I^{0.5}) \qquad \text{(G-35)}$$

*See also* Davies equation.

**Guoy–Stern colloidal model** A schematic representation of the surface phenomena around colloidal particles, used in the theory of coagulation and flocculation; a rigid layer (Stern layer), positively charged, is attached to the electronegative particle and is surrounded by a diffuse or boundwater layer. Ions of opposite charge, drawn from the solution, create a drop in Nernst or thermodynamic potential, which is the sum of the potentials across the rigid layer (Stern potential) and the diffuse layer (zeta potential). *See* electrical double layer and Figure G-04.

**gutter flow** The flow carried by gutters during a storm and coming from the adjoining streets; usually determined using Manning's formula applied to a triangular cross section (McGhee, 1991):

$$Q_0 = \delta(z/n)S^{0.5}y^{8/3} \qquad \text{(G-36)}$$

where $Q_0$ = gutter flow (m$^3$/s or ft$^3$/s); $\delta$ = a unit conversion constant = 0.38 for SI units and 0.56 for English units; $z$ = reciprocal of the cross slope of the gutter, dimensionless; $n$ = roughness coefficient, dimensionless; $S$ = gutter slope (or hydraulic gradient, parallel to the longitudinal slope of the street), dimensionless; and $y$ = depth of flow at the curb (m or ft). Gutter flow ($Q_0$) and the actual flow ($Q$, cfs, with $Q \leq Q_0$) from a gutter of wedge-shaped cross-section into curb or gutter inlets may be calculated as follows (Fair et al., 1971):

$$Q = 0.00482\, Ly(gy)^{0.5} \qquad \text{(G-37)}$$

$$Q = 1.87\, LI^{0.579}(nQ_0/S^{0.5})^{0.563} \qquad \text{(G-38)}$$

where $L$ is the length of the inlet opening (ft), $g$ is the gravitational acceleration, and $I$ is the crosswise slope of the gutter within the width of flow (dimensionless).

**G value** *See* velocity gradient, a parameter usually represented by the letter "G." *See also* conjunction opportunity.

**G value concept** *See* velocity gradient concept.

**GWDR** Acronym of Groundwater Disinfection Rule.

**GWI** Acronym of groundwater infiltration.

**GWR** Acronym of Ground Water Rule.

**GWUDI** Acronym of groundwater under the direct influence (of surface water).

**GWUI** Acronym of groundwater under the direct influence of surface water.

**gypsum ($CaSO_4 \cdot 2\,H_2O$)** (1) A mineral consisting primarily of fully hydrated calcium sulfate, or a rock containing calcium sulfate, used as a soil amendment or in the fabrication of gypsum cement, wallboard, plaster. (2) A scale of insoluble calcium sulfate formed during water desalination.

**Gyrazur™** A clarifier manufactured by Infilco Degremont, Inc. with softening capability.

# H

**H-3** Chemical symbol of tritium. Also represented by $H^3$, $^3H$, $H^c$.

**$H^3$ or $^3H$** Chemical symbol of tritium. Also represented by $H^c$, H-3.

**ha** Abbreviation of hectare(s).

**HAA** Acronym of haloacetic acid.

**HAA5** The sum of the concentrations of five haloacetic acids (monochloroacetic, dichloroacetic, trichloroacetic, monobromoacetic, and dibromoacetic acids), regulated under the Disinfectant/Disinfection By-Products Rule.

**HAA6** Symbol representing the sum of the six haloacetic acids considered in the formation of disinfection by-products: trichloroacetic, dichloroacetic, chloroacetic, bromochloroacetic, monobromoacetic, and dibromoacetic acids.

**Haber–Bosch process** *See* Haber process.

**Haber process** An industrial process for producing ammonia ($NH_3$) from atmospheric nitrogen ($N_2$) and hydrogen ($H_2$) at high temperature and pressure through a catalyst bed:

$$N_2 + 3\,H_2 \rightarrow 2\,NH_3 \qquad \text{(H-01)}$$

Developed by the German chemist and Nobel prize laureate Fritz Haber (1868–1934).

**Haber's law** As a rule of thumb: the equivalent effective dose is equal to the product of the concentration of a toxin and the time of exposure (Symons et al., 2000).

**habitat** The place and its living as well as nonliving surroundings where a population (e.g., human, animal, plant, microorganism) lives; the environment where an organism naturally breeds and prospers. A habitat indicator is a physical attribute of the environment measured to characterize conditions necessary to support an organism, population, or community in the absence of pollutants, e.g., salinity of estuarine waters or substrate type in streams or lakes.

**habitat capability** The existing or future capacity of an area to support wildlife, fish, or a sensitive plant population; expressed in numbers of animals, pounds of fish, or acres of plants.

**habitat indicator** *See* habitat.

**habitat protection** Measures taken for the well-being of the organisms that live in a place.

**Hach One** A pH electrode manufactured by Hach Co.

**H-acidity** Mineral acidity, the characteristic of a natural water that contains more protons than indicated by the proton condition or carbonic acid ($H_2CO_3$) equivalent point. Abbreviation or symbol: H-Acy:

$$\text{H-Acy} = [H] - [OH^-] - [HCO_3^-] - 2\,[CO_3^{2-}] \quad \text{(H-02)}$$
$$= -\,[\text{Alk}]$$

where the brackets indicate molar concentrations of the species and Alk represents alkalinity. *See also* caustic alkalinity.

**H-Acy** Abbreviation or symbol of H-acidity or mineral acidity.

***Haemagogus spp.*** A group of mosquitoes that transmit the yellow fever virus.

**haematite** British spelling of hematite.

**haemoglobin** British spelling of hemoglobin.

**Hagen, Gotthilf Heinrich Ludwig** German physicist (1797–1874) who proposed that the velocity of liquids through capillary tubes is proportional to the hydraulic gradient in an 1839 article, "On the flow of water in narrow cylindrical tubes."

**Hagen–Poiseuille equation** (1) A formula used to calculate the flow of water ($Q$) through a membrane as a function of viscosity ($\mu$), membrane porosity ($\rho$), pore radius ($r$), applied pressure ($\Delta P$), and membrane thickness ($\delta$):

$$Q = \rho r^2 \Delta P/(8\,\mu\delta) \quad \text{(H-03)}$$

This equation is used to derive a temperature correction factor by assuming that all the variables are constant at a given temperature. *See also* sludge flow, based on Poiseuille's equation for viscous liquids (AWWA, 1999). (2) An equation for flow through porous media, also applied to pipe flow:

$$K = kg/\nu \quad \text{(H-04)}$$

where $K$ is hydraulic conductivity, $k$ is intrinsic permeability, $g$ is the gravitational acceleration, and $\nu$ is the kinematic viscosity.

**hail** Precipitation in the form of balls or lumps of ice over half a centimeter in diameter, formed by alternate freezing and melting as precipitation from convective clouds is carried up and down by turbulent air currents. *See also* graupel, snow pellets, rime, soft hail.

**hailstone** An individual unit of hail.

**Halberg®** A digester draft tube manufactured by SIHI Pumps, Inc. for sludge mixing.

**half-cell reaction** An oxidation or reduction half-reaction. Two such reactions may be combined to indicate the overall corrosion of a metal in drinking water:

$$2\,\text{Pb (metal)} \leftrightarrow 2\,Pb^{2+} + 4\,e^- \quad \text{(H-05)}$$

$$O_2 + 2\,H_2O + 4\,e^- \leftrightarrow 4\,OH^- \quad \text{(H-06)}$$

$$2\,\text{Pb (metal)} + O_2 + 2\,H_2O \leftrightarrow 2\,Pb^{2+} + 4\,OH^- \quad \text{(H-07)}$$

**half-life ($t_{1/2}$)** (1) The length of time required for the mass, concentration, or activity of a chemical or physical agent to be reduced by one-half through decay. *See* decay processes. The half-life of a radioactive material is the time it takes for half of the material to radiate energetic particles and rays and transform to new materials; e.g., the half-life of cesium (Cs-137) is 30 years, after which time half of it decays to a nonradioactive stable nuclide, barium (Ba-137). Starting with 100 kg of Cs-137, one obtains 50 kg of Cs-137 and 50 kg of Ba-137 after 30 years; after 30 more years only 25 kg of Cs-137 remain, and so on. The biochemical half-life of DDT in the environment is 15 years, of radium 1580 years, of carbon-14 5730 years, of plutonium-239 24,000 years, and of uranium-238, billions of years. Chemical half-life and radioactive half-life refer, respectively, to chemical and radioactive substances; ten half-lives are required to eliminate 99.9% of the activity of a radionuclide, but 5.5 half-lives are considered sufficient for a pesticide to be completely degraded. For a first-order reaction of rate $k$ or a radioactive decay constant of $\lambda$:

$$t_{1/2} = 0.693/k = 0.693/\lambda \quad \text{(H-08)}$$

The mean lifetime or mean life expectancy of a radioactive atom ($\tau$) is the reciprocal of the decay constant:

$$\tau = 1/\lambda \quad \text{(H-09)}$$

(2) The time required for the elimination of one-half of a total dose from the body. *See also* half-time.

**half-reaction** The oxidation or the reduction part of an overall redox reaction.

**half-saturation constant** (1) In the Monod equation, the half-saturation constant ($K_s$) is the substrate concentration at which microbial growth occurs at one-half the maximum specific growth rate. *See also* half-velocity constant. (2) In the interaction of a trace metal with phytoplankton, $K_s$ is the metal ion concentration at which half the transport molecules are bound in complexes with ligands. Usually, $K_s$ is much larger than the molecular concentration $[M]$ of the metal and the uptake rate ($V$) is:

$$V = V_{max}[M]/K_s \quad \text{(H-10)}$$

where $V_{max}$ is the maximum uptake rate when the transport ligands are fully saturated (Maier et al., 2000 and Stumm & Morgan, 1996).

**half-time** (1) A parameter used to compare die-off rates of bacteria; equal to the time required for a 50% reduction in the initial population. (2) When an organism contaminated by a substance is placed in clean water, the half-time is the period required for half of the substance to be eliminated from the organism or for its tissue concentration to be halved. Same as half-life of the substance.

**half-velocity constant** A parameter of the Monod equation (or Michaelis–Menten equation), representing the substrate concentration at one-half the maximum growth rate, expressed in mass per unit volume. *See also* half-saturation constant.

**halide ($X^-$)** The ionic form of a halogen atom, sometimes represented by the symbol $X^-$; e.g., chloride ($Cl^-$), bromide ($Br^-$), or fluoride ($F^-$), or a chemical compound containing a halogen, e.g., aluminum trichloride ($AlCl_3$), which reacts in water ($H_2O$) to produce aluminum hydroxide [$Al(OH)_3$] and hydrochloric acid (HCl):

$$AlCl_3 + 3\ H_2O \rightarrow Al(OH)_3 + 3\ HCl \quad \text{(H-11)}$$

*See also* halogen.

**halite** A soft white, red, yellow, blue, or colorless evaporite mineral that is at least 95% sodium chloride (NaCl) in cubic crystals; it occurs together with other minerals such as gypsum and sylvite. Also called native salt, rock salt, fossil salt.

**Hall detector** An instrument used to detect specific groups of chemical compounds or the electrolytic conductivity of a substance. Also called Hall electrolytic conductivity detector.

**Hall electrolytic conductivity detector** Same as Hall detector.

**halo** Prefix from the Greek word halos, meaning salt.

**haloacetaldehyde** The combination of a halogen and an aldehyde, sometimes formed as a disinfection by-product. *See* chloroacetaldehyde, chloral hydrate, and dichloroacetaldehyde.

**haloacetic acid ($CX_3COOH$,** with X = $Cl^-$, $Br^-$, or H) Acronym: HAA. Any one of a group of nine contaminants regulated by the USEPA in drinking water under the Disinfectants and Disinfection By-products Rule. The most prominent members are dichloroacetic acid ($Cl_2AA$) and trichloroacetic acid ($Cl_3AA$). Also found in chlorinated drinking water: monochloroacetic acid (ClAA), monobromoacetic acid (BrAA), bromochloroacetic acid (BrClAA), dibromoacetic acid ($Br_2AA$), bromodichloroacetic acid ($BrCl_2AA$), chlorodibromoacetic acid ($ClBr_2AA$), and tribromoacetic acid ($Br_3AA$). Five of them (HAA5) are regulated by the USEPA with MCL = 60 μg/L: ClAA, BrAA, $Cl_2AA$, $Br_2AA$, and $Cl_3AA$.

**haloacetonitrile ($CX_3{\equiv}N$,** with X = $Cl^-$, $Br^-$, or $H^+$) Abbreviation: HAN. A disinfection by-product formed during the chlorination of water that contains natural organic matter. The USEPA requires water supply systems to monitor quarterly the following disinfection by-products: dichloroacetonitrile, trichloroacetonitrile, bromochloroacetonitrile, and dibromoacetonitrile.

**haloamine** An aqueous combination of hypochlorous acid (HOCl) with an amino acid or other amine.

***Haloanaerobium*** A genus of halotolerant bacteria; they require salt concentrations substantially higher than found in seawater.

***Halobacterium*** A group of heterotrophic halotolerant organisms active in biological denitrification; they require high concentrations of salt for growth.

**halocarbon** The chemical compounds $CFCl_2$ and $CF_2Cl_2$ and such other halogenated compounds that may reasonably be anticipated to contribute to reductions in the concentration of ozone in the stratosphere.

**halocline** The boundary between two masses of water of different salinities; a well-defined salinity gradient in an ocean.

**haloform ($CHX_3$)** Any of the trihalomethanes that contain one carbon and three atoms of a halogen. In the general formula $CHX_3$, X represents a halogen; e.g., chloroform ($CHCl_3$) and bromoform ($CHBr_3$).

**haloform reaction** The reaction of a methyl ketone with a halogen, with a base as a catalyst. It initially produces a trihaloketone, but the final products are a haloform and a carboxylic acid. It is a common reaction pathway for the formation of trihalomethanes. For example, acetone ($C_3H_6O$) reacts with bromine to form bromoform ($CHBr_3$).

**halogen** One of the chemical elements of group VIIA of the periodic table: astatine, bromine, chlorine, fluorine, or iodine. Chlorine and its compounds as well as bromine and iodine are effective germicidal agents used in water and wastewater disinfection; they are also oxidizing agents. Halogens are very reactive elements; they combine with metals to form salts called halides.

**halogenated aliphatics** Aliphatic compounds that contain halogen atoms, such as trichloroethylene (TCE) and other one- or two-carbon halogenated compounds commonly used as industrial solvents. They degrade more slowly than nonhalogenated aliphatics. *See* substitutive dehalogenation, oxidative dehalogenation, reductive dehalogenation.

**halogenated alkyl epoxide** *See* epichlorohydrin.

**halogenated by-product** A disinfection by-product containing a halogen such as chlorine or bromine. Same as halogen-substituted by-product.

**halogenated chloro-organic compound** A derivative of methane, in which chlorine or another halogen replaces one or more hydrogen atoms during water chlorination. *See* trihalomethane.

**halogenated compound** A compound that contains a halogen, e.g., the disinfection by-products such as the trihalomethanes and haloacetic acids.

**halogenated hydrocarbon** Any compound of a hydrocarbon with a halogen, in particular the trihalomethanes formed during water disinfection. *See also* chlorinated hydrocarbon.

**halogenated nitrile** An organic compound and disinfection by-product containing the cyanide group (—CN) and a halogen such as chlorine or bromine. Same as halogen-substituted nitrile. *See also* haloacetonitrile.

**halogenated organic compound (HOC)** A compound containing at least one halogen—carbon bond, i.e., an organic that has undergone halogenation. HOCs include disinfection by-products and synthetic organic chemicals, both of which may pose a health risk even in low concentrations in drinking water.

**halogenated organic material** Same as halogenated organic compound.

**halogenated organics** Organic molecules that contain one or more halogen atoms. Same as halogenated organic compounds.

**halogenating species** A chemical compound such as hypochlorous acid (HOCl) or hypobromous acid (HOBr) that can cause halogenation by combining with an organic substance.

**halogenation** The replacement of a hydrogen atom by an atom of chlorine, bromine, or another halogen in a chemical compound.

**halogen-demand-free phosphate buffer saline** A stock buffer solution prepared for disinfection experiments, which does not lose more than 10% or 0.1 mg/L of its halogen content after a holding period of one hour.

**halogen-substituted by-product** Same as halogenated by-product.

**halogen-substituted nitrile** Same as halogenated nitrile.

**halogen-substituted organic material** Same as halogenated organic compound.

**halogen substitution** Same as halogenation.

**haloketone ($CX_3COCX_3$, with X = $Cl^-$, $Br^-$, or $H^+$)** Acronym: HK. A compound of a halogen and a ketone; a by-product of the disinfection of water containing natural organic matter. The USEPA requires the monitoring of 1,1-dichloropropanone and 1,1,1-trichloropropanone in drinking water. Haloketones are also found as intermediate industrial products. There are indications that some chloroacetones are mutagenic to strains of *Salmonella*.

**halomethane** A substance formed during water disinfection when a halogen like chlorine or bromine combines with natural humic substances, chloromethane ($CH_3Cl$), chloroform ($CHCl_3$), bromoform ($CHBr_3$), and methyl iodide ($CH_3I$). *See also* trihalomethane and organohalogen.

**haloperoxidase** *See* organohalogen.

**halon** A bromine-containing compound with long atmospheric lifetimes whose breakdown in the stratosphere causes depletion of ozone. Halons are used in firefighting.

**halo-obligatory** Characteristic of an organism that requires salt concentrations above 12% for growth (Droste, 1997).

**halo-organic** A compound formed by the substitution of a halogen for hydrogen; it may be toxic or carcinogenic.

**halophile** An organism that requires high salt concentrations for growth. *See also* halotolerant.

**halophilic** Pertaining to organisms that thrive in a salt environment, e.g., bacteria found in unsanitary or incompletely dried salts.

**halophilic bacteria** *See* halophilic, halophile.

**halophyte** A plant that can grow in soils that contain high concentrations of salts of chlorine, bromine, iodine, and other minerals.

**halotolerant** Pertaining to organisms that can grow in the presence of salt but do not require it. *See also* halophile.

***Halsicomenobacter hydrossis*** A species of filamentous bacteria responsible for activated sludge bulking when dissolved oxygen concentration is low or nutrients are insufficient.

**Hamaker constant** A parameter used in calculating interaction energy, collision efficiency factor, and other variables related to colloid stability in the theory of adsorption, flocculation, and coagulation. It depends on the density and polarizability of the dispersed particles but is independent of ionic strength. It usually varies between $10^{-20}$ and $10^{-19}$, but may be less than for organic colloids (Stumm & Morgan, 1996). *See* collision efficiency.

**Hamburg Rotor** A surface aerator manufactured by Hellmut Geiger GmbH & Co.

**Hammer-Head™** A groundwater pump manufactured by QED Environmental Systems.

**hammermill** A high-speed rotating assembly that uses hammers and cutters to crush, grind, chip, or shred solids for further treatment and disposal.

**HAN** Acronym of haloacetonitrile.

**HAN4** Symbol representing the sum of the four haloacetonitrile species considered in the formation of disinfection by-products: trichloroacetonitrile, 1,1-dichloroacetonitrile, bromochloroacetonitrile, and dibromoacetonitrile.

**hand-cleaned coarse screen** A coarse screen used in small wastewater pumping stations or at the headworks of small- or medium-size wastewater treatment plants, often as standby units for use when the mechanically cleaned screens are being repaired.

**hand-dug well** A common and simple method of groundwater extraction; it is constructed using hand tools and local materials. It is sometimes equipped with a pump, but a bucket and rope can also be used to draw water from it.

**handpump** A device designed to raise groundwater from wells, using hand power. It consists mainly of a piston that moves up and down inside a cylinder. It is usually simple enough for repair by a village mechanic. *See* VLOM.

**hanging cake** A dewatered sludge cake that refuses to leave the filtering cloth, a problem sometimes encountered in plate-and-frame filter presses.

**Hankin™** Water treatment products containing ozone, from Wheelabrator Clean Water Systems, Inc.

**Haptenization** The formation of an antigenic substance from the combination of two compounds, one having a low molecular weight and the other having a high molecular weight.

**Harbor Bosun** Dye tablets produced by Formulabs, Inc.

**Harborlite®** Perlite material produced by Celite Corp.

**hard detergent** A detergent that resists biological degradation, e.g., alkylbenzene sulfonate.

**HaRDE®** An electrostatic precipitator manufactured by Wheelabrator Clean Water Systems, Inc.

**hardness** (1) A characteristic of water caused mainly by the salts of calcium and magnesium, such as bicarbonate ($HCO_3^-$), carbonate ($CO_3^{2-}$), sulfate ($SO_4^{2-}$), chloride ($Cl^-$), and nitrate ($NO_3^-$); usually expressed as calcium carbonate ($CaCO_3$) equivalent. Other polyvalent metallic ions may also be included in hardness, e.g., iron, aluminum, manganese, strontium, and zinc, as well as hydrogen ions. Excessive hardness in water is undesirable because it causes the formation of soap curds (calcium or magnesium oleates), increased use of soap, deposition of scale in boilers, damage in some industrial processes, and sometimes objectionable tastes in drinking water. However, hardness does not affect modern synthetic detergents. *See also* hard water, EDTA titration method. Hardness is removed or reduced by softening processes (e.g., lime–soda, excess lime, split treatment, and cation-exchange methods). *See also* the following types of hardness: carbonate, negative, noncarbonate, permanent, temporary. Calcium and magnesium hardness can be calculated as follows:

$$\text{Hardness (mg equivalent } CaCO_3/L) = 2.497\ (\text{Ca, mg/L}) + 4.12\ (\text{Mg, mg/L}) \quad \text{(H-12)}$$

(2) Resistance of a mineral to deformation. *See* grain hardness, Moh's hardness scale.

**hardness as calcium carbonate ($CaCO_3$)** The usual way of reporting water hardness in the United States. *See* hardness units, calcium carbonate equivalent.

**hardness bar graph** A diagram used in solving water softening problems and showing the concentrations in meq/L of the species involved at various stages of the process. It provides an easy check on the total cation and anion charge balance of a water sample. Also called hardness distribution diagram. *See also* ion balance, ionic proportion diagram.

**hardness–cardiovascular link** There may be an inverse relationship between hardness in drinking water and cardiovascular disease, perhaps due to the protective effects of calcium and magnesium in hard water or the negative effects of higher concentrations of cadmium, copper, lead, and other trace elements in soft water (Droste, 1997 and AWWA, 1999).

**hardness classification** A scale that indicates whether water is soft or hard. *See* hard water.

**hardness distribution diagram** A graphical representation of the contributions of the major anions and cations to the total hardness of a water sample. Also called hardness bar graph.

**hardness ions** The minerals responsible for scaling in boilers, deposits on pipes, and excessive consumption of soaps made of natural animal fats; the divalent cations such as calcium and magnesium, but also aluminum, iron, and other metals.

**hardness (permanent)** *See* permanent hardness.

**hardness range** The range of concentrations of total hardness as $CaCO_3$, according to which a water is considered moderately hard, hard, or very hard. *See* hard water.

**hardness removal** The removal of ions causing hardness. *See* softening.

**hardness (temporary)** *See* temporary hardness.

**hardness (total)** *See* total hardness.

**hardness units** Units used to express water hardness include American degree, British degree (Clark scale), calcium equivalent, calcium carbonate equivalent, French degree, German degree, and Russian degree, where:

| | | |
|---|---|---|
| American degree | = 1 grain per U.S. gallon | = 17.24 mg/L of $CaCO_3$ |
| British degree | = 1 grain per British gallon | = 14.29 mg/L of $CaCO_3$ |
| French degree | | = 10 mg/L of $CaCO_3$ |
| 1 German degree | = 10 mg/L of CaO equivalent | = 17.9 mg/L of $CaCO_3$ |
| 1 Russian degree | = 1 mg/L of Ca | = 2.5 mg/L of $CaCO_3$ |

**hardpan** *See* caliche (2).

**hard piping** Pipe or tubing that is manufactured and properly installed using good engineering judgment and standards (EPA-40CFR63.111 or 63.161).

**hard water** Alkaline water containing dissolved salts that interfere with some industrial processes and prevent soap from lathering; the opposite of soft water. Hard water also produces scales in hot-water pipes, boilers, and heaters. On the other hand, it is less corrosive and may taste better than soft water. Water may be considered hard if it has a hardness greater than the typical hardness of water from the region. Some textbooks define hard water as water having a hardness of more than 100 mg/L as calcium carbonate. More generally, the degree of water hardness may be interpreted as follows:

| | |
|---|---|
| 0–50/75 mg/L as $CaCO_3$ | = soft water |
| 50/75–150 mg/L as $CaCO_3$ | = moderately hard water |
| 150–300 mg/L as $CaCO_3$ | = hard water |
| more than 300 mg/L as $CaCO_3$ | = very hard water |

*See also* hardness.

**hard water scale** The scale deposited on the inside of pipes and appliances by water containing a high concentration of carbonates and bicarbonates of calcium and magnesium. *See also* scale.

**harmful industrial ingredients** A variety of constituents of industrial wastewater that can cause malfunctions of collection and treatment facilities, including toxic metals, feathers, rags, acids and alkalis, flammables, noxious gases, fats, detergents, and phenols.

**harmful inorganics** Inorganic substances that are found in drinking water and can cause metabolic disturbances and other undesirable effects; e.g., heavy metals and salts.

**harmful organics** Organic compounds that are toxic, carcinogenic, or that cause objectionable taste and odor in drinking water, e.g., pesticides, polynuclear aromatic hydrocarbons, and trihalomethanes.

**harmful substances** Any substances or preparations that represent limited health risks when they are inhaled or ingested or if they penetrate the skin.

**harmonic feedback** A problem experienced by equipment that uses variable-frequency drives: some energy is released in the form of the harmonics of the output frequency, which can affect the electrical power bus protection relays.

**harmonic filtering** The use of filters, in wastewater treatment plant equipment having variable-frequency drives, to prevent harmonic feedback problems by isolating the control units from the plant's electrical system.

**harmonic function** Any solution to the Laplace equation.

***Hartmannella*** A genus of free-living amebas that produce round cysts, including the species *Hartmannella vermiformis* that is often associated with the agent of Legionnaires' disease in hot-water systems.

**harvesting** A method of control of aquatic plants. *See* plant harvesting.

**Hasachlor®** A 12.5% sodium hypochlorite solution produced by HASA, Inc. for the reduction of metals, suspended solids, and chlorate formation.

**$H_3AsO_4 \cdot ½\ H_2O$** Chemical formula of arsenic acid.

**Hastelloy®** A nickel alloy fabricated by Haynes International, Inc.

**HAV** The acronym of hepatitis-A virus.

**Hawaii agent** A calicivirus first identified after an acute, waterborne gastrointestinal outbreak in Hawaii.

**hazard** A potential source of harm or damage, often associated with a disaster.

**hazard evaluation** A component of risk assessment that involves gathering and evaluating data on the type of health injury or disease (e.g., cancer) that may be produced by a chemical and on the conditions of exposure under which injury or disease is produced.

**hazard identification** The first step in the risk assessment procedure, consisting in determining if a chemical can cause adverse health effects in humans and what those effects might be.

**hazardous chemical** A USEPA designation for any hazardous material requiring an MSDS (material safety data sheet) under OSHA's Hazard Communication Standard. Such substances can produce fires and explosions or adverse health effects like cancer and dermatitis. Hazardous chemicals are distinct from hazardous waste.

**hazardous class** A group of materials, as designated by the U.S. Department of Transportation, that share a common major hazardous property such as radioactivity or flammability.

**hazardous constituent** A constituent listed as such by the USEPA and regulated under the Resource Conservation and Recovery Act (RCRA). Sludge from wastewater treatment may contain such hazardous constituents as heavy metals or toxic materials (arsenic, beryllium, cadmium, chromium, copper, lead, mercury, molybdenum, nickel, selenium, zinc) and organic compounds.

**hazardous material** *See* hazardous substance.

**hazardous ranking system (HRS)** The principal screening tool used by USEPA to evaluate risks to public health and the environment associated with abandoned or uncontrolled hazardous waste sites. The HRS calculates a score based on the potential of hazardous substances spreading from the site through the air, surface water, or ground water, and on other factors such as density and proximity of human population. This score is the primary factor in deciding if the site should be on the National Priorities List and, if so, what ranking it should have compared to other sites on the list. Also called hazard ranking system.

**hazardous substance** (1) Any material that poses a threat to human health and/or the environment. Typical hazardous substances are toxic, corrosive, ignitable, explosive, or chemically reactive. (2) Any substance designated by USEPA to be reported if a designated quantity of the substance is spilled in the waters of the United States or otherwise released into the environment.

**hazardous waste** A subset of solid wastes, hazardous wastes are by-products of society that can pose a substantial or potential hazard to human health or the environment when improperly managed. They possess at least one of four characteristics (ignitability, corrosivity, reactivity, or toxicity) or appear on special USEPA lists. *See also* hazardous chemical.

**hazard ranking system (HRS)** Same as hazardous ranking system.

**hazards analysis** A series of procedures used to (a) identify potential sources of release of hazardous materials from fixed facilities or transportation accidents; (b) determine the vulnerability of a geographical area to a release of hazardous materials; and (c) compare hazards to determine which present greater or lesser risks to a community.

**hazard summary** A list of hazards to which a community, business, or employee may be exposed, e.g., chemicals, equipment, natural disasters.

**Hazen, Allen** A prominent American water supply engineer (1870–1930) who made several contributions related to the development of principles of probability to determine the safe yield of impounding reservoirs, piping networks, the theory of sedimentation, and research on water filtration at the Lawrence Experiment Station in Massachusetts (effective size and uniformity coefficient). *See also* the Hazen–Williams formula.

**Hazen equation** A formula proposed by A. Hazen in 1905 to compute the clean-water headloss through a granular porous medium (Metcalf & Eddy, 2003):

$$H = (1/C)[60/(T + 10)](LV_h/d_{10}^2) \quad \text{(H-13)}$$

where $H$ = headloss, ft; $C$ = coefficient of compactness, dimensionless, e.g., from 600 for closely packed sand to 1200 for very uniform clean sand; $T$ = temperature, °F; $L$ = depth of filter bed or layer, ft; $V_h$ = superficial (approach) filtration velocity, ft/day; and $d_{10}$ = effective grain size diameter, mm. *See* filter head loss for other similar equations.

**Hazen method** An empirical procedure to determine hydraulic conductivity of a sediment from its grain size.

**Hazen number** A unit of measurement of color in water, one unit corresponds to the color produced by one mg/L of platinum in a cobalt-based compound (Porteous, 1992). *See* chloroplatinate unit or Pt–Co color unit, and Hazen unit.

**Hazen's effective size** Same as effective size.

**Hazen's law** Under ideal settling conditions (discrete particle settling), settling efficiency is independent of the depth ($H$) of the sedimentation tank but depends on the tank plan area (AWWA, 1999):

$$V^* = Q/L^*W = Q/A^* \quad \text{(H-14)}$$

where $V^*$ = overflow rate of the ideal tank, $Q$ = flow rate, $L^*$ = length of the ideal tank, $W$ = tank width, and $A^*$ = plan area of the tank. However, depth is important in inclined or horizontal settlers and in limiting scour.

**Hazen's settling equation** A formula presented by Allen Hazen in 1904 to relate the efficiency of a sedimentation basin, under conditions of discrete and unhindered settling, to settling velocity of the particles ($V$) and the surface loading or overflow rate ($L$). Efficiency ($E$) is the proportion of particles removed (Fair et al., 1971):

$$E = V/L = V/Q/A \qquad \text{(H-15)}$$

where $Q$ is the flow rate and $A$ the surface area of the settling zone. *See* ideal sedimentation basin.

**Hazen's sieve rating** A sieve size or sieve number as determined by a method proposed by Allen Hazen in 1892. The Hazen's rating is about 10% larger than the corresponding sieve manufacturer's rating.

**Hazen unit** A unit of color in water, as defined by Allen Hazen around 1890 and based on the platinum–cobalt standard. *See* chloroplatinate unit or Pt–Co color unit, and Hazen number.

**Hazen–Williams coefficient** Same as the Hazen–Williams roughness coefficient.

**Hazen–Williams formula** A widely used empirical formula, proposed by Allen Hazen and Gardner Williams in 1905, to calculate pressure pipe friction for water flowing under turbulent conditions. It yields comparable results to the Darcy–Weisbach formula at moderately high Reynolds numbers. It expresses flow ($Q$) in a pipe as a function of its diameter ($D$), slope ($S_0$) and a friction coefficient ($C$) between 100 and 150, depending on the material and age of the pipe:

$$Q = a \cdot C \cdot D^{2.63} S_0^{0.54} \qquad \text{(H-16)}$$

with $a$ = 0.432 for English units (cfs and ft), and $a$ = 0.278 for metric units (cubic meters per second and meters). Equivalent formulas for head loss ($h_f$) and average velocity ($V$) are:

$$V = b \cdot C \cdot R^{0.63} S_0^{0.54} \qquad \text{(H-17)}$$

and

$$h_f = c \cdot V^{1.85} L/(C^{1.85} C^{1.165}) \qquad \text{(H-18)}$$

with $b$ = 1.318 for English or 0.849 for metric units, $c$ = 3.02 for English or 6.79 for metric units, $R$ being the hydraulic radius, and $L$ being the length of the pipe. *See also* the Manning's and Kutter–Ganguillet formulas.

**Hazen–Williams roughness coefficient A coefficient ($C$)** expressing the influence of the material and age of a pipe on the flow velocity in the pipe. Used in the Hazen–Williams formula, it varies from 100 to 150.

**hazmat** Abbreviation of hazardous material.

**$H_3BO_3$** Chemical formula of boric acid, orthoboric acid. Also written as $B(OH)_3$.

**H-bond** Abbreviation of hydrogen bond.

**$HBrO_3$** Chemical formula of bromic acid.

**$H^c$** Chemical symbol of tritium. Also represented by $H^3$, $^3H$, or H-3.

HCB Acronym of hexachlorobenzene.

**HCFC** Acronym of hydrochlorofluorocarbon.

**HCl** Chemical formula of hydrochloric acid and hydrogen chloride.

**HClO** Chemical formula of the hypochlorite ion.

**$HClO_3$** Chemical formula of chloric acid.

**$HClO_4$** Chemical formula of perchloric acid.

**$HClO_3S$** Chemical formula of chlorosulphonic acid.

**HCN** Chemical formula of hydrocyanic acid or hydrogen cyanide.

**$H_2CO_3$** Chemical formula of carbonic acid.

**$H_2CO_3^*$** Chemical formula of the sum of dissolved carbonic acid ($H_2CO_3$) and carbon dioxide ($CO_2$).

**$H_2C_2O_4$** The chemical formula of oxalic acid.

**$H_2CrO_4$** Chemical formula of the hypothetical chromic acid.

**HCR pond** *See* hydrograph controlled release pond.

**HDD** Acronym of horizontal directional drilling.

**HDD** or **2,3,7,8-HDD** A dibenzo-p-dioxin, totally chlorinated or totally brominated.

**HDF** or **2,3,7,8-HDF** A dibenzofuran, totally chlorinated or totally brominated.

**HDFPBS** Acronym of halogen-demand-free phosphate buffer saline.

***H. diminuta*** *See Hymenolepis diminuta.*

**HDPE** Acronym of high-density polyethylene.

**HDPE pipe** A pipe manufactured with high-density polyethylene, a material that is flexible and easy to install; used, e.g., in leachate collection systems and in water and wastewater facilities.

**HDRPP** Acronnym of hydraulically driven reciprocating piston pump.

**head** (1) The vertical distance (in feet) equal to the pressure (in psi) at a specific point. The pressure head is equal to the pressure in psi times 2.31 ft/psi. (2) the depth of water above the crest of a weir or similar structure. (3) The kinetic or potential energy of each unit weight of a liquid, expressed as the vertical height through which a unit weight would have to fall to release the average energy possessed. (4) The source or upper end of a system, as in headwall, headwater, and headworks. *See also* acceleration head, discharge head, dynamic discharge head, dynamic head, dynamic suction head, dynamic suction lift, elevation head, head loss, kinetic head, potential head, pressure head, static head, suction head, suction lift, total dynamic discharge head, total dynamic head or TDH, total head, velocity head, weir head.

**headbox** A chamber used for the addition of chemicals to water or wastewater

**header** A manifold or other pipe, conduit, or chamber fitted with several smaller outlet pipes to distribute fluid.

**headgate** The gate that controls water flow into irrigation canals and ditches; regulated by a watermaster during water distribution, with appropriate notices for official regulations.

**head increaser** A device used to reduce the tailwater pressure of a draft tube and increase its discharge.

**head loss** The head, pressure, or energy lost in a pipe or channel as a result of turbulence caused by the velocity of the flowing fluid and the roughness of the pipe and channel walls, or restrictions caused by fittings. Water flowing in a pipe loses head, pressure, or energy as a result of friction losses. Head loss does not include changes in the elevation of the hydraulic gradeline unless the hydraulic and energy lines are parallel to each other. Formulas for the calculation of head losses include those developed by Darcy–Weisbach, Hazen–Williams, and Manning. Head loss is an important consideration in the design of water and wastewater treatment units, e.g., comminutors, filter appurtenances, filter backwash, filter launders, and screens. *See also* Ergun equation, Kozeny equation, negative head. Also called lost head, energy loss.

**head loss development** *See* filter head loss development.

**headspace** The space between the level of a liquid and the cover in a container. It may contain volatile chemicals such as disinfection by-products.

**headspace analyzer** An instrument used to analyze volatile organics, which accumulate in the space above a sample.

**headspace BOD analysis** A BOD determination procedure conducted in a small bottle or test tube with sufficient volume for air and based on the application of Henry's law, knowing the atmospheric pressure of the sealed sample, the volume of the sample, and the volume of headspace.

**headwater** (1) The source of a river, including any emerging groundwater; the waters from which it rises, or its upper reaches. (2) The water(s) upstream of hydraulic works.

**headworks** The initial structures and devices at the head or diversion point of a conduit, canal, water, or wastewater treatment plant.

**health advisory** An estimate of acceptable drinking water levels for a chemical substance based on health effects information; a health advisory is not a legally enforceable federal standard, but serves as technical guidance to assist federal, state, and local officials. For example, USEPA classifies metolachlor as a possible human carcinogen and issues a health advisory of 0.1 mg/L for lifetime exposure.

**health advisory level** A nonregulatory health-based reference level of chemical traces (usually in mg/L) in drinking water at which there are no health risks when ingested over various periods of time. Such levels are established for one day, 10 days, long term, and lifetime exposure periods. They contain a large margin of safety.

**health education** A recommended component of water supply and sanitation programs, particularly in developing countries, that is designed to ensure that the greatest health benefits are achieved. Health education, often regarded as a dialogue between officials and users instead of a lecture about good hygiene, focuses on the relationship between health, water, excreta, sullage, solid waste, etc. Also called hygiene education.

**health hazards** Various risks to human health associated with exposure to toxic agents, including the following. (1) Acute toxicity. The older term used to describe immediate toxicity. Its former use was associated with toxic effects that were severe (e.g., mortality), in contrast to the term "subacute toxicity" that was associated with toxic effects that were less severe. The term "acute toxicity" is often confused with that of acute exposure. (2) Allergic reaction. Adverse reaction to a chemical resulting from previous sensitization to that chemical or to a structurally similar one. (3) Chronic toxicity. The older term used to describe delayed toxicity. However, the term "chronic toxicity" also refers to effects that persist over a long period of time whether or not they occur immediately or are delayed. The term "chronic toxicity" is often confused with that of chronic exposure. (4) Idiosyncratic reaction. A genetically determined abnormal reactivity to a chemical. (5) Immediate versus delayed toxicity. Immediate effects occur or develop rapidly after a single administration of a substance, whereas delayed effects are those that occur after the lapse of some time. These effects have also been referred to as acute and chronic, respectively. (6) Reversible versus irreversible toxicity. Reversible toxic effects are those that can be repaired, usually by a specific tissue's ability to regenerate or mend itself after chemical exposure, whereas irreversible toxic effects are those that cannot be repaired. (7) Local versus systemic toxicity. Local effects refer to those that occur at the site of first contact between the biological system and the toxicant; systemic effects are those that are elicited after absorption and distribution of the toxicant from its entry point to a distant site. (2) *See* cross-connection.

**health risk assessment** *See* risk assessment.

**heart disease** Heart disease has been statistically associated with water softness in some industrialized countries but this association has not been explained.

**hearth** The lower part of a blast furnace where molten metal collects or the bottom of a furnace where waste materials are burned.

**Heatamix** An apparatus manufactured by Simon-Hartley, Ltd. to provide heating and recirculation in anaerobic digesters.

**heat balance** An accounting of the heat gains and heat losses in a system.

**heat budget** The quantity of heat required to raise a body of water from its minimum temperature to its maximum temperature on a yearly basis.

**heat capacity** The quantity of energy that must be supplied to raise the temperature of a substance one degree, or the amount of heat released by a decrease of one degree in the temperature of the substance, expressed in calories/°C or BTUs/°F. For contaminated soils, heat capacity is the quantity of energy that must be added to the soil to volatilize organic components. The typical range of heat capacity of soils is relatively narrow; therefore, variations are not likely to have a major impact on application of a thermal desorption process. Same as thermal capacity. Sometimes used synonymously with specific heat. *See also* calorific value, sensible heat.

**heat conductivity** A measure of the ability of a substance or material to conduct heat; the rate of heat transfer by conduction through a material or substance of unit thickness per unit area and per unit temperature gradient. *See* thermal conductivity for detail.

**heat content** The quantity of heat that can be obtained from a material per unit mass, i.e., the number of units of heat obtained by the complete combustion of a unit mass of the material. *See* calorific value for detail.

**heat dryer** A device that uses heat to reduce the moisture content of sludge. *See* direct or convection dryers (flash dryers, and rotary dryers, including kiln and cage-mill types), indirect dryers, combined direct–indirect dryers, and infrared dryers.

**heat drying** A sludge processing method using heat to drive off moisture and concentrate the solids to more than 90% content. It also reduces pathogens and produces Class A biosolids that are suitable as a soil amendment, conditioner, or fertilizer. The USEPA regulations promulgated in 1993 (40 CFR Part 503) define heat drying as a process to significantly reduce pathogens: "Sewage sludge is dried by direct or indirect contact with hot gases to reduce the moisture content of the sewage sludge to 10 percent or lower. Either the temperature of the sewage sludge particles exceeds 80 degrees Celsius or the wet bulb temperature of the gas in contact with the sewage sludge as the sewage sludge leaves the dryer exceeds 80 degrees Celsius." *See* conduction, convection, radiation, heat dryers.

**heat drying–pelletization** A sludge handling and disposal method that consists of heating the feed sludge in a rotary dryer, cooling to 120°F the desirable particles (typically between 1 and 4 mm in diameter), and recycling the other particles to the feed material. The biosolids pellet thus produced is marketed as a fertilizer.

**heater treater** A treatment unit that is used in oil fields to separate water from oil emulsions.

**heat exchange** The exchange of heat between two bodies by such phenomena as convection or radiation. Same as heat transfer.

**heat exchanger** A device that is used for the transfer of heat from one substance or body to another; e.g., a coiled pipe carrying hot water through a tank of cooler water.

**heat-exchanger tank** A tank equipped with coils for heating, water, wastewater, or sludge.

**Heath Aqua-Scope** A leak-detection device manufactured by Heath Consultants, Inc.

**heating degree-day** *See* degree-day.

**heating value** The quantity of heat that can be obtained from a material per unit mass, i.e., the number of units of heat obtained by the complete combustion of a unit mass of the material. *See* calorific value for detail.

**heating, ventilating, and air conditioning (HVAC)** The component of a building or its plans and specifications that deals with air circulation, including heating, ventilation, and air conditioning. In a water or wastewater treatment building, HVAC provides a comfortable and safe working environment while protecting sensitive equipment.

**heat of adsorption** The energy released by the adsorption of molecules or particles.

**heat of condensation** The amount of heat released when a vapor changes state to a liquid; specifically, the heat released by a unit mass of gas at the boiling point when it condenses completely to a liquid. It is equal, in absolute value, to the heat of vaporization.

**heat of evaporation** Same as heat of vaporization.

**heat of formation** The amount of heat required to form one gram-molecule of a compound from its constituent elements.

**heat of fusion** The amount of heat absorbed during fusion; specifically, the heat taken up by a unit mass of a solid at the melting point when it changes completely to a liquid at the same temperature, e.g., 144 BTU per pound (or about 333 J/g) at 0°C when ice melts. It is equal, in absolute value, to the heat of solidification. Sometimes called latent heat of fusion or heat of melting, it is an important parameter in the design of freezing units for water or wastewater treatment. Water has the second highest heat of fusion of all substances (after ammonia); this characteristic means that additional energy is required to freeze or thaw it and prevents rapid temperature changes around 0°C (actually, water at the bottom of deep ice-covered lakes is above the freezing temperature).

**heat of melting** Same as heat of fusion.

**heat of solidification** The heat released by a unit mass of liquid at its freezing point as it changes to a solid; it is equal, in absolute value, to the heat of fusion.

**heat of solution** The amount of heat that is absorbed or released when a substance is dissolved in a solvent.

**heat of sublimation** The heat absorbed by a unit mass of a solid as it changes to the gaseous state at constant temperature and pressure; the direct conversion of ice to vapor requires 1,222 BTU per pound (2,843 kJ/kg).

**heat of vaporization** The amount of energy required to change a volume of liquid to vapor; specifically, the energy absorbed when a unit mass of a given material at the boiling point changes completely to a gas at the same temperature; it is equal, in absolute value, to the heat of condensation. Sometimes called the latent heat of vaporization. The heat of vaporization of water affects the evaporation rate; it is an important factor of the water cycle and an important parameter in the design of evaporators and distillation units for water or wastewater treatment. Water has the highest heat of vaporization of all substances: about 2.26 kJ/g at 100°C and 1.0 atmosphere, helping to reduce the loss of water and heat to the atmosphere. All water, wastewater, or sludge treatment processes involving evaporation require some heat input to vaporize the volatile solvent. The heat of vaporization of water ($H_v$), also called heat of evaporation, may be estimated from (Fair et al., 1971):

$$H_v = 1094 - 0.56\,T \qquad \text{(H-19)}$$

where $H_v$ is in BTU per pound and $T$ is the temperature in °C. For combustion calculations, the temperature is usually taken as $T = 0°C = 32°F$ or $15.55°C = 60°C$.

**heat pollution** *See* thermal pollution.

**heat requirement (sludge digestion)** The amount of heat that is necessary for the proper operation of a sludge digestion tank, i.e., to raise the temperature of incoming sludge to that of the tank and to compensate for all heat losses (through influent piping, cover, walls, bottom). The heat ($Q$) required for a surface ($A$) and a temperature difference ($\Delta T$) is (Fair et al., 1971):

$$Q = CA(\Delta T) \qquad \text{(H-20)}$$

$C$ being a coefficient of heat flow, e.g., 0.30 BTU/square foot/hour for wet earth and 0.15 BTU/square foot/hour for air.

**heat sensor** A device that opens and closes a switch in response to changes in temperature. This device might be a metal contact, or a thermocouple that generates a minute electrical current proportional to the difference in heat, or a variable restor whose value changes in response to changes in temperature. Also called temperature sensor.

**heat-shrinkable sleeve** A corrosion protective coating in the form of a wraparound or tubular sleeve, consisting of (a) a polyethylene or polypropylene backing to which an adhesive is applied and (b) an epoxy primer. Sleeves are attached onto the cutbacks between sections of pipelines during welding and construction. Also called shrink sleeve.

**heat transfer** The exchange of heat between two bodies by such phenomena as conduction, convection, or radiation. Same as heat exchange.

**heat transfer coefficient** A measure of the rate of heat transfer between steam and liquid per unit surface area and per unit temperature gradient; a critical parameter in the design of evaporators. In thermal sludge drying, the heat transfer coefficient includes the effects of the surface films and the medium; it is usually provided by the equipment manufacturer or determined from pilot studies. *See also* conduction drying, convection drying, aerated lagoon temperature.

**heat treatment** The treatment of sludge with heat to condition it for further processing. *See* more detail under thermal conditioning.

**heat value** The quantity of heat that can be obtained from a material per unit mass, i.e., the number of units of heat obtained by the complete combustion of a unit mass of the material. *See* calorific value for detail.

**HeatX** A gas heating apparatus manufactured by Walker Process equipment Co. for anaerobic digesters.

**heavy Baumé scale** *See* Baumé scale.

**heavy hydrogen (D)** Deuterium or another heavy isotope of hydrogen.

**heavy liquid** A dense liquid used to separate heavy minerals, e.g., bromoform, specific gravity (sp) 2.87, compared to quartz (sg = 2.65) and calcite (sp = 2.71).

**heavy metal** A metallic element having a high atomic weight or a specific gravity greater than 5, particularly those most associated with metal pollution: arsenic (As), cadmium (Cd), copper (Cu), mercury Hg), chromium (Cr), lead (Pb), and zinc (Zn), but also manganese (Mn), tin (Sn), thallium (Th), selenium (Se), and nickel (Ni). Common hazardous wastes, heavy metals can damage living things at low concentrations and tend to accumulate in the food chain; e.g., they accumulate in the soil and can be taken up by food crops. Cadmium, lead, mercury, and zinc are not effectively eliminated by conventional wastewater treatment and are a concern in effluent and sludge applications. They can be precipitated by hydrogen sulfide in an acid solution. Heavy metals, usually added to wastewater from commercial and industrial sources, are present in trace concentrations in sludge and biosolids, which may limit the land application of residuals.

**heavy metal inactivation** Water disinfection using heavy metals, particularly copper (Cu) and silver (Ag) in swimming pools and hot tubs.

**heavy-metal selective chelating resin** A chelating resin with the general formula R—EDTA—Na; it behaves like a weak-acid cation resin but has a high selectivity for heavy-metal cations

**heavy metals removal** For methods applicable to the removal of heavy metals from wastewater, see carbon adsorption, chemical precipitation,, ion exchange, and reverse osmosis.

**heavy mineral** The detrital accessories of sedimentary rocks that have high specific gravities (e.g., higher than that of bromoform, 2.87).

**heavy sludge** Sludge with a low moisture content (i.e., a relatively high solids concentration).

**heavy spar** The mineral barium sulfate, principal ore of barium; it is found in white, yellow, or colorless crystals called barytes. Also called barite.

**heavy water ($D_2O$)** Water in which deuterium (an isotope of hydrogen with an atomic weight of 2) replaces the hydrogen atoms; also called deuterium oxide; used mainly as a coolant in nuclear reactors.

**hectare (ha)** A unit of area in the metric system equal to 100 ares or 10,000 square meters. Also, one hectare equals 2.471 acres or 107,639 square feet.

**Hedstrom number ($H_e$)** One of the two dimensionless numbers used to determine the pressure drop due to friction for sludge (Metcalf & Eddy, 2003):

$$H_e = D^2 S_y g\gamma/\eta^2 \qquad \text{(H-21)}$$

where $D$ = pipe diameter, ft; $S_y$ = yield stress, lb/ft$^2$; $g$ = gravitational acceleration = 32.2 lb · ft/lb · s$^2$; $\gamma$ = specific weight of sludge, lb/ft$^3$; and $\eta$ = coefficient of rigidity, lb/ft/s. The other number is the Reynolds number. *See also* rheology.

**Hela-Flow®** Plastic products of Liquid-Solids Separation Corp. for use in water and wastewater treatment.

**Helaskim** A surface skimmer manufactured by Walker Process Equipment Co.

**helical cell** A bacterium (or other microorganism) shaped like a long spiral. Also called a helix.

HeliCarb A carbon dioxide contactor manufactured by CBI Walker, Inc.

**Heliclean®** An open-channel screen manufactured by Hycor Corp., with a screenings washer. *See also* Helisieve®.

**Helico** A press manufactured by Infilco Degremont, Inc. for handling screenings.

***Helicobacter pylori (H. pylori)*** A Gram-negative, spiral-shaped, waterborne and foodborne bacterium that causes duodenum and stomach ulcers as well as other gastric ailments. Initially, it was identified erroneously as *Campylobacter pylori.* Chlorination is an effective treatment method.

**heliothermometer** An instrument for measuring the intensity of solar radiation (e.g., ultraviolet light) using a thermoelectric couple.

**Heli-Press** Equipment manufactured by Vulcan Industries, Inc. to compact screenings from wastewater treatment.

**Helisieve®** An open-channel screen manufactured by Hycor Corp. *See also* Heliclean®.

**Helisieve Plus™** A unit manufactured by Hycor Corp. for handling septage.

**HeliSkim** A helical surface skimmer manufactured by Walker Process Equipment Co.

**HeliThickener** A flight screw conveyor manufactured by Walker Process Equipment Co. for sludge processing.

**helix** A bacterium (or other microorganism) shaped like a long spiral. Also called a helical cell.

**Helixor** A combined aerator and mixer manufactured by Polcon Sales, Ltd. for subsurface installation.

**Helixpress** Equipment manufactured by Hycor Corp. for dewatering and conveying screenings.

**Helmholtz free energy** *See* Gibbs' free energy.

**helminth** A free-living or parasitic worm such as *Ascaris, Dracunculus medinensis, Onchocerca volvulus, Schistosoma,* or *Taenia;* a multicellular complex organism containing organs and tissue. Some helminths depend on aquatic intermediate hosts to cause infections in humans. Most helminthes multiply outside the human host, *Strongyloides* being an exception. In addition to *Ascaris lumbricoides* (the stomach worm), those transmitted by drinking water include hookworms, threadworms, and whipworms. Helminths refer to worms collectively, most of which are in the phyla Nematoda (roundworms), Platyhelminthes (flatworms), and Annelida (segmented worms). *See also* hookworm, tapeworm, fluke.

**helminthic disease** *See* helminth infection.

**helminth infection** A long lasting, widespread, and insidious infection or disease caused by a helminth or parasitic worm. Helminth infections are excreta-related, mosquito-borne, and, in one case, water-related. *See* the following infections (and their pathogens and common names): ascariasis, clonorchiasis, diphyllobothriasis, dracunculiasis, enterobiasis, fasciolopsiasis, filariasis, hymenolepiasis, hookworm, onchocerciasis, loiasis, opisthorchiasis, paragonimiasis, schistosomiasis, strongyloidiasis, taeniasis, trichuriasis. Also called helminthiasis, helminthic disease, or worms. *See also* cyclopoid.

**helminth life cycle** See *Ascaris lumbricoides, Clonorchis sinensis, Dracunculus medinensis,* hookworm, schistosome, and *Taenia saginata* life cycles.

**HELP model** *See* Hydrologic Evaluation of Landfill Performance model.

**hemacytometer** An instrument for counting microorganisms under a microscope; it consists of a glass chamber and a coverslip under which a water sample flows.

**hematite ($\alpha$-$Fe_2O_3$)** (1) An iron oxide mineral and important iron ore, found in igneous rocks, in hydrothermal veins, and in ooliths; used as a pigment and in anticorrosion paints. (2) A compound formed during the corrosion of iron pipes or a fine particle resulting from the hydrolysis and polymerization of ferric iron; also called an iron phase or a corrosion scale. Also spelled haematite. *See also* goethite.

**hematolysis** Same as hemolysis.

**hematopoiesis** The production of blood and blood cells; hemopoiesis.

**heme ($C_{34}H_{32}N_4O_4Fe$)** A pigment in red blood cells (hemoglobin) containing reduced iron; heme iron.

**heme iron** Organically bound iron, sometimes causing a pinkish coloration in water sources.

**hemicellulose** A common plant polymer, molecular weight of about 40,000; a gummy polysaccharide, intermediate between sugar and cellulose, that readily hydrolyzes to monosaccharides.

**hemiparasite** A parasite that obtains only part of its food from its host; also called kleptoparasite.

**hemochromatosis** A health condition characterized by an excessive accumulation of iron in the body and causing dysfunction or even failure of the liver, pancreas, and heart. It may be hereditary (idiopathic hemochromatosis) or caused by blood transfusion or by excessive iron consumption. The nutritional iron requirement is 10–15 mg per day; the secondary standard for iron in drinking water is 0.3 mg/L.

**hemodialysis** Blood dialysis, using, e.g., a membrane to purify a kidney patient's blood.

**hemoglobin** A substance that is 94% protein and 6% heme ($C_{34}H_{32}N_4O_4Fe$); in red blood cells, it carries oxygen ($O_2$) from the lungs to the tissues and carbon dioxide ($CO_2$) from the tissues to the lungs.

**hemolysis** The release of hemoglobin that accompanies a breakdown of red blood cells, e.g., during hemodialysis; it may be caused by chlorate poisoning or the presence of chloramines in dialysis water.

**hemolytic anemia** A disease characterized by a decrease of hemoglobin or red cells in the blood; it may result from a concentration of chlorite ions ($ClO_2^-$, e.g., from water disinfection by chlorine dioxide), which can oxidize hemoglobin.

**hemolytic–uremic syndrome (HUS)** An occasional complication of enterohemorrhagic *E. coli* infections of the very young, which may result in renal failure and hemolytic anemia.

**hemopoiesis** Same as hematopoiesis.

**hemorrhagic fever** An arboviral infection (such as dengue) characterized by fever, chills, hemorrhages, and other symptoms.

**hemorrhagic jaundice** A water-related bacterial infection; same as Weil's disease. *See also* leptospirosis.

**Henderson–Hasselbach equation** An equation that relates the pH of a solution to the dissociation constant pK of an acid and the molar concentrations of the acid and its salt:

$$\begin{aligned} pH &= pK + \log([salt]/[acid]) \\ &= pK - p([salt]/[acid]) \end{aligned} \quad (H\text{-}22)$$

where the brackets denote molar concentrations and the notation "p" indicates the logarithm of the reciprocal of a quantity. *See also* common-ion effect.

**Henry equation** Same as linear isotherm. *See also* Henry's law.

**Henry, Joseph** American physicist (1797–1888).

**Henry's constant** Same as Henry's law constant.

**Henry's law** The relationship between the partial pressure of a compound and the equilibrium concentration in the liquid through a proportionality constant known as the Henry's law constant. Henry's law states that, at constant temperature, the weight of any gas dissolved in water is directly proportional to the pressure exerted by the gas above the water. This law is used in water and wastewater treatment to determine the extent of concentrations of gases (saturation) in the liquid phase and the extent of volatilization from the liquid phase. It is written in several forms in the literature:

$$C_{aq} = K \cdot C_g \quad (H\text{-}23)$$

$$C_g = H \cdot C_{aq} \quad (H\text{-}24)$$

$$C_g = H_u C_s \quad (H\text{-}25)$$

$$H = 4.559\ TH_u \quad (H\text{-}26)$$

$$H = RTH_u \quad (H\text{-}27)$$

where $C_{aq}$ = gas concentration in the liquid phase (mole fraction of gas in water), moles, mg/L, or mL/L; $K$ = Henry's law constant, moles/atm, mg/L/atm, or mL/L/atm; $C_g$ = gas concentration in the gas phase (partial pressure, mole fraction of gas in air), atm; $H$ = Henry's law constant at temperature $T$, °K, atm, atm/mole, or atm · m$^3$/mole; $H_u$ = unitless Henry's law constant; $C_s$ = saturation concentration of constituent in liquid, mg/L; $T$ = temperature, °K = 273.15 + °C; and $R$ = universal gas law constant = 0.000082057 atm · m$^3$/mole. °K. For an illustrated presentation on Henry's law, see Metcalf & Eddy (2003), pp. 66–69 or Droste (1997), pp. 21–23. *See also* absorption coefficient, fugacity, activity, Henry equation, linear isotherm.

**Henry's law coefficient** Same as Henry's law constant.

**Henry's law constant** The ratio of the concentration of a compound in air (or vapor) to the concentration of the compound in water under equilibrium conditions; e.g., a Henry's law constant greater than 1000 atm is recommended for the removal of semivolatile compounds by diffused aeration. The constant depends on the gas, its temperature, and nature of the liquid. Its numerical value varies with the definition of Henry's law. *See* Table H-01, absorption coefficient, salting-out effect. *See also* air–water portioning coefficient for more detail.

**hepatic** Pertaining to the liver.

**hepatitis** An inflammation and necrosis of the liver usually caused by an acute viral infection or by chemical contaminants in drinking water. Yellow jaundice is one symptom of hepatitis; other symptoms are fever, weakness, nausea, vomiting, and diarrhea. Milder cases of hepatitis are reversible, but the fulminant form can lead to death. *See* the various forms of the virus: hepatitis A, B, C, D, and E.

**Table H-01.** Henry's law constants

| Gas | $K$, mg/L/atm, 0°C | 25°C | $H$, atm 20°C | $H_u$, unitless 20°C |
|---|---|---|---|---|
| Ammonia, $NH_3$ | | | 0.75 | 0.000561 |
| Carbon dioxide, $CO_2$ | 3480 | 1450 | 1420 | 1.06 |
| Chlorine, $Cl_2$ | 1460 | | 579 | 0.43 |
| Chlorine dioxide, $ClO_2$ | | | 1500 | 1.12 |
| Hydrogen, $H_2$ | 1.91 | 1.56 | 68,300 | 51.10 |
| Hydrogen sulfide, $H_2S$ | 6640 | 3505 | 483 | 0.36 |
| Methane, $CH_4$ | 3968 | 2152 | 37,600 | 28.13 |
| Nitrogen, $N_2$ | 29.1 | 18.0 | 80,400 | 60.16 |
| Oxygen, $O_2$ | 69.6 | 39.3 | 41,100 | 30.75 |
| Ozone, $O_3$ | 1375 | 584 | 5300 | 3.97 |

**hepatitis A** (1) One of the five forms of hepatitis; commonly called infectious hepatitis, epidemic hepatitis, epidemic jaundice, or short-incubation hepatitis; an inflammation and necrosis of the liver caused by the hepatitis-A virus (HAV). It is a water- and excreta-related disease transmitted via the fecal–oral route, distributed worldwide, and recorded since the fifth century BC. Symptoms include jaundice, fatigue, abdominal pain, loss of appetite, nausea, diarrhea, and fever. (2) The hepatitis-A virus.

**hepatitis-A virus (HAV)** An RNA, human-excreted virus, of the Picornaviridae family, transmitted by contaminated food or water, by person-to-person contact, or by contaminated blood products. Very resistant to acids, disinfectants, and heat. It is communicable during the incubation period of 10–50 days. An effective treatment process is coagulation–flocculation–sedimentation–filtration followed by chlorination for 30 minutes.

**hepatitis B** The most severe of the five forms of hepatitis; also called serum hepatitis. Symptoms include jaundice, fatigue, abdominal pain, loss of appetite, nausea, joint pain, and vomiting.

**hepatitis-B virus (HBV)** A DNA virus causing hepatitis and transmitted through contaminated needles, blood products, tissue fluids, or syringes.

**hepatitis-C virus (HCV)** A hepatitis virus transmitted through contaminated blood, e.g., through blood transfusion. Symptoms of hepatitis C include jaundice, fatigue, abdominal pain, loss of appetite, nausea, and dark urine. Four out of five cases do not show any symptoms.

**hepatitis delta agent** The hepatitis-D virus.

**hepatitis-D virus (HDV)** An RNA virus that causes hepatitis and is transmitted in combination with the hepatitis-B virus. Also called hepatitis delta agent.

**hepatitis E** A form of hepatitis similar to hepatitis A, with mild symptoms (except in pregnant women) that include fever, abdominal pain, and loss of appetite.

**hepatitis-E virus (HEV)** A waterborne virus that causes hepatitis through the fecal–oral route; formerly known as non-A, non-B hepatitis virus; clinically indistinguishable from hepatitis A. Symptoms include jaundice, malaise, anorexia, abdominal pain, arthralgia, and fever. It resembles the Norwalk agent and caliciviruses.

**hepatitis type A** *See* hepatitis A.

**hepatitis virus** *See* hepatitis.

**hepatocyte** A liver cell that can be damaged by specific waterborne chemicals and microbial agents.

**hepatomegaly** Enlargement of the liver.

**hepatotoxin** An algal toxin that can cause damage to the liver. The most prevalent hepatotoxins in U.S. drinking water sources are the microcystins produced by the genera *Microcystis, Anaboena,* and *Oscillatoria.* (Westrick, 2003). Conventional water treatment does not remove hepatotoxins; activated carbon adsorption is more effective.

**heptachlor ($C_{10}H_5Cl_7$)** A synthetic organic chemical of the group of cyclodienes, containing seven chlorine atoms; a generic name for 1,4,5,6,7,8,8-heptachloro-3a,4,7,7a-tetrahydro-4,7-methanoindene. This persistent contaminant, which rapidly oxidizes to epoxide, was previously used as an insecticide in agriculture and for termite control. May affect the nervous system. Regulated by the USEPA as a drinking water contaminant: MCLG = 0 and MCL = 0.0004 mg/L.

**heptachlor epoxide ($C_{10}H_5Cl_7O$)** An oxidation product of heptachlor in soils and on crops. It is regulated by the USEPA.

**herbicide** A compound, usually a man-made organic chemical, used to kill or control plant growth. Herbicides are regulated by the USEPA as drinking water contaminants under the Phase II contaminants rule; *see* pesticide. Common herbicide contaminants of water sources include alachlor, metolachlor, atrazine, cyanazine, prometon, simazine, 2,4-D, paraquat, diquat, and 2,4,5-T.

**Hercules** A pressure filter manufactured by Liquid–Solids Separation Corp..

**Hermann–Gloyna formula** An expression proposed for the determination of the detention time (td, days) in a waste stabilization pond (Fair et al., 1971):

$$t_d = 0.49\ HY_0/(PS\theta^{T-20}) \qquad \text{(H-28)}$$

where 0.49 = a factor of safety related to the ability of light energy to produce the required quantity of oxygen; $H$ = pond depth, ft; $Y_0$ = pond loading in mg/L of $BOD_5$; $P$ = the efficiency of conversion of light energy to chemical energy, in the range of 0.02–0.06; S = insolation, cal/cm²/day; $\theta$ = van't Hoff–Arrhenius temperature factor, e.g., = 1.072; and $T$ = pond temperature, °C.

**HERO™** Acronym of high-efficiency reverse osmosis, a patented technology used for the desalination of brackish water and seawater; originally developed to produce extra-pure water for the electronics industry. It includes components for hardness and alkalinity removal, degasification (removal of carbon dioxide, $CO_2$), pH adjustment through caustic addition, and a two-phase reverse osmosis system. It is classified as an emerging concentrate minimization and management method (Sethi et al., 2006).

**herpangina** An infectious disease caused by the human excreted Coxsackievirus A, affecting mostly children and characterized by a sudden occurrence of fever, loss of appetite, and throat ulcerations.

**hertz (Hz)** The number of complete electromagnetic cycles or waves in one second of an electrical or electronic circuit. Also called the frequency of the current.

**Hervey, Fred** *See* Fred Hervey Plant.

**heteroaerobic pond** A wastewater treatment pond in which organic matter is decomposed by microorganisms that function facultatively as aerobic or anaerobic, depending on diurnal or seasonal factors such as photosynthesis changes. Also called amphiaerobic pond. *See also* facultative pond, and lagoons and ponds.

**heterocoagulation** A charge neutralization mechanism of destabilization of colloidal particles in which particles are deposited on other particles of opposite charge, as, for example, in the destabilization of large negatively charged particles by depositing smaller positively charged particles. *See also* surface charge neutralization, interparticle bridging.

**heteroecious parasite** A parasite that is not host-specific or that needs more than one host for its entire life cycle; the opposite of an autoecious parasite. Also called heteroxenous parasite.

**heterogeneous** Varying in structure or composition at different locations in space.

**heterogeneous buffer system** A solid mineral that can directly or indirectly stabilize the pH of a solution by consuming hydrogen $[H^+]$ or hydroxyl $[OH^-]$ ions; e.g., kaolinite $[Al_2Si_2O_5(OH)_4]$, magnesium hydroxide $[Mg(OH)_2]$, and calcium carbonate $(CaCO_3)$.

**heterogeneous equilibrium** A condition of chemical equilibrium between substances in two or more physical states, e.g., the equilibrium reached by calcium carbonate $(CaCO_3)$ with the calcium $(Ca^{2+})$ and carbonate $(CO_3^{2-})$ ions in solution at high pH. *See also* homogeneous chemical equilibrium, heterogeneous reaction.

**heterogeneous flocculation model** A set of equations used to describe collisions among bubbles and particles in the theory of dissolved-air flotation (AWWA, 1999):

$$dN_{f,i}/dt = -kN_b(\alpha_{f,i}N_{f,i} - \alpha_{f,i-1}N_{f,i-1}) \quad (i = 1 - m_f) \quad \text{(H-29)}$$

$$dN_{f,0}/dt = -k\alpha_f N_b N_{f,0} \quad (i = 0) \quad \text{(H-30)}$$

$$k = aG(d_b + d_f)^3 = a(\varepsilon/\mu)^{1/2}(d_b + d_f)^3 \quad \text{(H-31)}$$

where $N_{f,i}$ = number concentration of floc particles with $i$ attached air bubbles, $T$ = time; $k$ = kinetic coefficient based on the collision mechanism, $N_b$ = number concentration of air bubbles, $\alpha_f$ = collision attachment factor, $m_f$ = maximum number of attached air bubbles to floc particles, $N_{f,0}$ = number concentration of floc particles without attached air bubbles, $\alpha_0$ = collision attachment factor without attached air bubbles, $a$ = a constant, $G$ = mean velocity gradient, $d_b$ = bubble diameter, $d_f$ = floc diameter, $\varepsilon$ = mean energy dissipation rate, and $\mu$ = absolute water velocity. *See also* white water collector model.

**heterogeneous nucleation** The formation of very small particles in a supersaturated solution in the presence of solid substrates, such as recycled lime sludge, acting as seeds for crystal growth.

**heterogeneous reaction** (1) A chemical reaction that involves more than one phase, the reaction usually occurring at or near the interface between the phases. *See also* heterogeneous equilibrium, homogeneous reaction, reversible reaction, irreversible reaction. (2) A reaction that involves constituents associated with specific sites, as on an ion-exchange resin, or a reaction that requires the presence of a solid-phase catalyst, as in packed- and fluidized-bed reactors.

**heterolactic fermentation** *See* lactic acid fermentation.

***Heterophyes heterophyes*** An excreted hermaphroditic flat trematode transmitted from dogs or cats → brackish-water snails → brackish-water fish → humans. Latency = 6 weeks. Median infective dose < 100.

**heterophyiasis** A snail vector disease prevalent in the Nile delta but also in the Balkans, Far East, Middle East, Spain, and Turkey; caused by the water-related pathogen *Heterophyes heterophyes*. It is a minor, usually asymptomatic infection of the small intestine; occasionally characterized by nausea, diarrhea, fever, and abdominal pain.

**heterotroph** An organism that uses reduced organic compounds for energy and as a carbon source; sometimes called saprophyte. Heterotrophs include animals, fungi, most bacteria, and a few flowering plants. *See also* autotroph.

**heterotrophic** Designating or typical of organisms that derive carbon for the manufacture of cell mass from organic matter; they depend on other organisms for food. *See also* autotrophic.

**heterotrophic activity** A measure of the extent of utilization of natural organic matter as a carbon source by microorganisms. Also called carbon respiration.

**heterotrophic bacteria** *See* heterotrophic microorganism.

**heterotrophic microorganism** Bacteria and other microorganisms that use organic matter synthesized by other organisms for energy and growth. Most pathogenic bacteria are heterotrophic, including some opportunistic species, but some heterotrophic bacteria are innocuous. *See also* autotrophic organism.

**heterotrophic organisms** Species that depend on organic matter for food, e.g., denitrifying bacteria, certain sulfur-reducing bacteria; heterotrophs. They dominate the microbial population of activated sludge.

**heterotrophic plate count (HPC)** A laboratory method of bacterial enumeration in a water or other sample. It indicates the number of colonies of heterotrophic (pathogenic and innocuous) bacteria grown on selected solid media at a given temperature and incubation period (i.e., the number of bacteria originally in the sample that will develop into colonies), usually expressed in number of bacteria per milliliter of sample. Colonies are derived from pairs, chains, clusters, or single cells. HPC, an improvement of the traditional standard plate count, is conducted using one of three recognized procedures (see membrane filter method, pour plate method, spread plate method). HPC enumerates aerobic and facultatively anaerobic bacteria that include Gram-negative organisms of the genera *Acinetobacter, Aeromonas, Alcaligenes, Citrobacter, Flavobacterium, Enterobacter, Klebsiella, Moraxella, Proteus, Pseudomonas,* and *Serratia.* A high HPC is not a direct indicator of fecal contamination but may indicate a change in drinking water quality and a potential for microbial regrowth in the distribution system. An HPC limit of 500/mL has been recommended for tap water. *See also* colony-forming units.

**heterovalent exchange** An ion exchange reaction between two ions of unequal valences, e.g., a sodium ($Na^+$) for calcium ($Ca^{2+}$) exchange. *See* ion-exchange equilibria for related terms.

**heteroxenous parasite** *See* heteroecious parasite.

**HEV** Acronym of hepatitis-E virus.

**Hevi-Duty** Replacement parts manufactured by Envirex, Inc. for traveling water screens.

**hexachloroacetone** An organic disinfection by-product that is mutagenic to strains of *Salmonella.*

**hexachlorobenzene ($C_6Cl_6$)** Abbreviation: HCB. A white solid, synthetic organic chemical, only slightly soluble in water, that is not produced any more in the United States; previously used in industrial applications and as a fungicide in agriculture. Prolonged exposure to HCB may cause porphyria cutanea tarda (cutaneous lesions and hyperpigmentation).

**hexachlorobutadiene ($Cl_2C{:}CClCCl{:}CCl_2$)** A synthetic organic chemical used as a solvent; regulated by the USEPA as a drinking water contaminant: MCL = MCLG = 0.001 mg/L. Acronym: HCBD.

**1,2,3,4,5,6-hexachloro-cyclohexane** *See* lindane.

**hexachlorocyclopentadiene ($C_5Cl_6$)** Abbreviation: HEX. A synthetic organic chemical, used industrially and as a pesticide or fungicide; regulated by the USEPA as a drinking water contaminant: MCL = MCLG = 0.05 mg/L.

**hexafluorosilicic acid** Another name of fluorosilicic acid.

**Hex-Air** Aeration equipment manufactured by Dunlop, Ltd. using fine-bubble diffusers.

**hexametaphosphate** A high-molecular-weight compound of phosphorus that is converted to orthophosphate by acid hydrolysis of boiling water. Of the general formula $(MPO_3)_6$, where M represents a metal, these compounds are used as sequestering agents: the phosphate combines with calcium to form a stable suspension and avoid incrustations. *See also* sodium hexametaphosphate.

**hexanoic acid ($C_5H_{11}COOH$)** An oily, colorless or yellow liquid of the carboxylic group, obtained from fatty animal tissue or coconut oil, used in flavoring agents. Commonly called caproic acid

**hexavalent chrome** *See* hexavalent chromium.

**hexavalent chromium** A toxic form of chromium used in plating operations, usually reduced to the trivalent form and precipitated as hydroxide. It causes dermatitis, respiratory disorders, and hemorrhagic effects on the organs. It is commonly found as the chromate ion (e.g., $Na_2CrO_4$) or the dichromate ion (e.g., $K_2Cr_2O_7$). Regulated by the USEPA as a drinking water contaminant: MCL = MCLG = 0.1 mg/L. Also called chromium VI.

**hexose** *See* monosaccharide.

**HF** Chemical formula of hydrofluoric acid.

**HFC** Acronym of hydrofluorocarbon.

**$H_4Fe(CN)_6$** Chemical formula of ferrocyanic acid

**HFF membrane** Hollow fine fiber membrane.

**$HgCl_2$** Chemical formula of mercuric chloride.

**$Hg_2Cl_2$** Chemical formula of mercurous chloride or calomel.

**$Hg(ONC)_2$** Chemical formula of mercury fulminate.

**HGPRT** Acronym of hypoxanthine guanine phosphoribosyl transferase.

**HgS** Chemical formula of mercuric sulfide, found in the ore cinnabar.

***H. hydrossis*** Abbreviation of *Halsicomenobacter hydrossis.*

***H. hydrossis* type 0041** *Halsicomenobacter hydrossis* type 0041, a type of filamentous bacteria responsible for activated sludge bulking and associated with nutrient deficiency.

***H. hydrossis* type 0675** *Halsicomenobacter hydrossis* type 0675, a type of filamentous bacteria responsible for activated sludge bulking and associated with nutrient deficiency.

**Hi-Capacity** A treatment unit manufactured by Eimco Process Equipment Co. combining clarification and thickening.

**Hidrostal** A centrifugal pump manufactured by Envirotech Co.

**Hi-Flo®** Products fabricated by Culligan International Corp. for use in depth filters and water softeners.

**high blood pressure** High dietary intake of sodium (e.g., as sodium chloride) has been linked with high blood pressure, but the contribution from drinking water is relatively low.

**high-calcium hydrated lime** Lime that contains 95–98% hydrated lime or calcium hydroxide [$Ca(OH)_2$]; approximate specific gravity = 2.35; bulk density = 25–35 pounds per cubic foot. It is used in wastewater treatment for pH adjustment or as an acid-neutralizing agent, with a basicity factor of 0.71 as compared to 0.687 for caustic soda (NaOH) that is 78% $Na_2O$.

**high-calcium lime** Lime that contains 95–98% quicklime (calcium oxide, CaO) or hydrated lime [calcium hydroxide, $Ca(OH)_2$].

**high-calcium limestone** Limestone that contains a high fraction of quicklime or calcium oxide (CaO). It is used in wastewater treatment for pH adjustment or as an acid-neutralizing agent, with a basicity factor of 0.489 as compared to 0.564 for dolomitic limestone.

**high-calcium quick lime** Lime that contains 95–98% quicklime (calcium oxide, CaO); approximate specific gravity = 3.3; bulk density = 55–60 pounds per cubic foot. It is used in wastewater treatment for pH adjustment or as an acid-neutralizing agent, with a basicity factor of 0.941 as compared to 0.507 for sodium carbonate ($Na_2CO_3$) that is 58% $Na_2O$.

**high-capacity filter** (1) A water filtration device that operates at a rate higher than average. (2) A high-rate trickling filter.

**high-density polyethylene (HDPE) [$(H_2C{:}CH_2)_x$]** A synthetic organic material that produces toxic fumes when burned. It is used to make plastic bottles, landfill liners (because of its low permeability), and other products. HDPE is sometimes used in pipes and porous diffusers for water supply and wastewater disposal. *See also* styrene acrylonitrile.

**high-density polyurethane drying bed** A sludge drying device consisting of a sloped slab supporting interlocking high-density polyurethane panels. Each panel has a dewatering area and an underdrain system. *See also* wedgewire drying bed.

**high-efficiency reverse osmosis** *See* HERO™.

**high-efficiency RO** *See* HERO™.

**high-efficiency toilet** A toilet that uses 1.3 gallons per flush (gpf) or less. High-flow toilets use 3.5 gpf or more. Older toilets used 5.0 to 7.0 gpf. The 1992 U.S. Energy Policy Act allows the sale of only low-flow toilets, which use 1.6-gpf. Some dual-flow toilets can use 1.6 or 0.8 gpf.

**high-energy electron beam irradiation** The use of an electron beam to convert water contaminants into less harmful products.

**high-flow toilet** A toilet that uses 3.5 gallons per flush (gpf) or more. Older toilets used five to seven gpf. The 1992 U.S. Energy Policy Act allows the sale of only low-flow toilets, which use 1.6 gpf. High-efficiency toilets use 1.3 gpf or less. Some dual-flow toilets can use 1.6 or 0.8 gpf.

**High Flux Series™** Products of King Lee Technologies for cleaning membranes.

**High-Flux TF™** A chemical produced by King Lee Technologies for oil control in membranes.

**high F/M selector** A bioreactor designed to promote the growth of floc-forming instead of filamentous bacteria, on the basis of high substrate concentrations as indicated by the food-to-microorganism ratio (F/M). For example, a selector may consist of a series of three reactors loaded, respectively, at 12, 6, and 3 grams of COD per gram of mixed liquor suspended solids per day.

**high-frequency ozonation** The generation of ozone for use in water or wastewater treatment at a frequency of at least 1000 cycles per second (1000 Hz). *See also* high-voltage electrode.

**high-intensity flushing** The use of high dosing rates in the operation of a trickling filter to control fly larvae. *See* SK concept.

**high-intensity pulsed ultraviolet light treatment** A method of water or wastewater disinfection using ultraviolet light generated through a high-speed switch acting on a direct current capacitor.

**high-level radioactive waste (HLW)** Waste generated in the core fuel of a nuclear reactor or by nuclear fuel reprocessing; it is a serious threat to anyone who comes near the waste without shielding. HLW is also a mixed waste because it has highly corrosive components or has organics or heavy metals. HLW may include other highly radioactive material that requires permanent isolation according to the Nuclear Regulatory Commission. *See also* low-level radioactive waste.

**high-level waste** High-level radioactive waste.

**High-lime Combi** A rotating drum screen manufactured by Hercules Systems, Ltd.

**high-line jumper** A pipe or hose connected to a fire hydrant and laid on top of the ground to provide emergency water service to an isolated portion of a distribution system.

**high-outlet regulator** An orifice-like device used to regulate combined sewer overflows; the invert of the overflow pipe is at a higher elevation than the crown of the combined sewer.

**high-performance liquid chromatography (HPLC)** A technique that uses ultraviolet adsorption to separate and measure trace organics in water, based on their affinity for the stationary phase.

**high-pressure liquid chromatography** Same as high-performance liquid chromatography.

**high-pressure membrane** A semipermeable membrane used in such high-pressure filtration systems as nanofiltration and reverse osmosis for the separation of dissolved or ionic constituents from water by a solution–diffusion mechanism. *See* low-pressure membrane.

**high-pressure minicolumn (HPMC)** A small-scale test developed to simulate the operation and performance of a full-scale activated carbon reactor. It uses a high-pressure liquid chromatography column to evaluate the capacity of activated carbon to remove volatile organic compounds. *See also* rapid small-scale column test, constant diffusivity model, proportional diffusivity model.

**high-pressure minicolumn technique** A small-scale test that uses a high-pressure liquid chromatography column to generate performance data of granular activated carbon contactors. *See also* rapid small-scale column test, proportional diffusivity design.

**high-pressure oxidation (HPO)** A method of feeding sludge to a wet air oxidation unit; the sludge reacts with 100 cubic feet of air per pound of dry solids at a reactor temperature of 500°F and a pressure of 1000–1500 psig. *See also* intermediate-pressure oxidation.

**high-pressure piston pump** A pump used to pump sludge under high pressure and over long distances. It incorporates separate power pistons or membranes that separate the drives from the sludge.

**high pressure zone** The fourth and last area of a belt filter press used to dewater sludge from water or wastewater treatment. This is the area where the two porous plates pass between rollers of decreasing diameters to squeeze out water.

**high-purity oxygen** A gas stream that contains at least 90% oxygen by weight, produced by selective liquefaction or adsorption of other constituents from a gas stream.

**high-purity oxygen activated sludge** A variation of the activated sludge process that uses high-purity oxygen instead of air in the aeration basin. *See* high-purity oxygen process for detail.

**high-purity oxygen aerobic digestion** A variation of the conventional aerobic sludge digestion process that uses high-purity oxygen instead of air. It is particularly applicable in temperate climates because it is not affected by changes in ambient temperatures.

**high-purity oxygen generation** *See* pressure-swing adsorption system, cryogenic air separation process.

**high-purity oxygen process** A variation of the activated sludge process that uses high-purity oxygen instead of air in an enclosed aeration basin, which is divided into three or four compartments, the first of which receives the influent, the return activated sludge, and the oxygen. It requires a relatively small aeration volume, produces sludge with good settling characteristics, and emits a smaller volume of volatile organic compounds and off-gas. However, it has limited capability for nitrification, may produce *Nocardia* foaming, and does not handle peak flows well.

**high-quality groundwater** From the point of view of bacteriological quality, groundwater that contains less than 1 coliform per 100 mL and a heterotrophic bacterial population less than 10 organisms per mL at a temperature of 15°C or more. Other contaminants of concern in groundwater include nitrate, iron, and sulfur.

**high-quality sludge** Sludge that contains no more than 25 mg/kg cadmium, 1000 mg/kg lead, and 10 mg/kg PCBs on a dry weight basis, for land application purposes, as defined in a 1991 joint publication of the USEPA, the U.S. Food and Drug Administration, and the U.S. Dairy Association on land application of sludge for the production of fruits and vegetables.

**high-rate activated sludge** Same as high-rate aeration process.

**high-rate aeration process** A variation of the complete mix activated sludge process characterized by a low mixed liquor suspended solids concentration, a high volumetric BOD loading, short solids retention time, and high sludge recycle ratio. It produces an effluent with lower quality than conventional activated sludge. This process is not the same as high-rate aerobic treatment. Also called high-rate activated sludge or short-term aeration.

**high-rate aerobic treatment** A method of treatment for organic wastes, particularly in small installations, consisting of comminution, long-period aeration (1–3 days), final settling, and high sludge recycle rate (100–300% of flow). Also called total oxidation process, but different from high-rate aeration process.

**high-rate anaerobic digester** A reactor of cylindrical shape for the anaerobic digestion of thickened sludge, including an apparatus for heating, mixing, and uniform feeding. High-rate anaerobic digestion systems are designed as single-stage or two-stage installations, and usually operated in the mesophilic range.

**high-rate clarification** A wastewater treatment method that uses special flocculation and sedimentation to achieve rapid settling. *See* enhanced particle flocculation, ballasted flocculation, lamella plate clarification, dense-sludge process.

**high-rate clarifier** A sedimentation basin designed and operated at a hydraulic loading significantly higher than average. *See* plate settler, tube settler.

**high-rate digester** Same as high-rate anaerobic digester.

**high-rate digestion** The anaerobic stabilization of organic matter (e.g., in wastewater sludge) enhanced by thorough mixing and by operation in the thermophilic range. Detention time is reduced to 10 days from 30 days for standard rate digestion. *See also* high-rate anaerobic digester.

**high-rate filter** (1) A trickling filter operating at a rate greater than 10 mgad, including effluent recycle. (2) A rapid sand filter.

**high-rate filtration** Water or wastewater treatment using a high-rate filter.

**high-rate infiltration** A method of land application of wastewater at loading rates higher than 50 cm per week. The wastewater is treated by percolation through coarse-textured soil and then used for groundwater recharge or recovered for irrigation. Secondary treatment is recommended before application for recharge. *See also* geopurification, high-rate sprinkling, low-rate irrigation, overland flow, rapid infiltration extraction, rapid infiltration, soil aquifer treatment, and surface spreading.

**high-rate pond** A shallow aerobic stabilization pond used mainly for the rapid treatment of soluble organic wastes or for nutrient removal; it is designed to handle a load of 80 to 160 pounds of BOD per acre per day and to produce 100 to 260 mg/L of algal cell tissue (Metcalf & Eddy, 1991). Waste decomposition in a high-rate pond may be represented by this reaction (Nemerow & Dasgupta, 1991):

$$C_{11}H_{29}O_7N + 14\,O_2 + H \rightarrow 11\,CO_2 + 13\,H_2O + NH_4 \qquad \text{(H-32)}$$

*See also* low-rate and maturation ponds, and lagoons and ponds.

**high-rate process** A water or wastewater treatment process that operates at a hydraulic or surface loading rate significantly higher than average. *See* ballasted floc, coarse-bed filtration, dissolved-air flotation, flotation–filtration, high-rate filtration, inclined settling.

**high-rate sedimentation** An alternative to conventional rapid rate filtration of water in which the conventional sedimentation basin is replaced by an upflow blanket clarifier, solids contact unit, plate or tube settler, dissolved air flotation unit, or adsorption clarifier.

**high-rate sprinkling** Application of pretreated wastewater to vegetated land at a high rate through sprinklers instead of a spreading basin. *See also* geopurification, high-rate infiltration, low-rate irrigation, overland flow, rapid infiltration extraction, rapid infiltration, soil aquifer treatment, and surface spreading.

**high-rate trickling filter** A secondary wastewater treatment unit consisting of a rock- or plastic-packed filter, designed to operate with recirculation around the filter at loadings of 1.5–18.7 kg BOD/day/$m^3$ and 230–690 gpd/$ft^2$ (10.0–30.0 mgad). *See also* intermediate-rate trickling filter, low-rate trickling filter, roughing filter, super-rate filter, biofilter, biotower, oxidation tower, dosing tank, filter fly.

**high-recovery array** In pressure-driven membrane processes, an array (of multiple interconnected stages in series) whose feed is the concentrate from the previous array and whose concentrate is the feed to the succeeding array, thus increasing recovery from one array to the next.

**high salting** A procedure using a high dosage of salt, e.g., 15 pounds of sodium chloride (NaCl) or potassium chloride (KCl) per cubic foot, to regenerate cation resins treating water with high concentrations of hardness and dissolved heavy metals.

**high-service district** A section of a community that is served by piping and storage independent from the main distribution system because it is too high with respect to the main sections.

**high-service pump** (1) A pump that serves a high-service district, using the main supply and boosting it to the required pressure. (2) A device that pumps water from the clearwell of a conventional water treatment plant, as compared to the low-service pump that conveys source water to the plant.

**high-solids centrifuge** A modification of the solid-bowl centrifuge, designed with a longer bowl and a lower differential bowl speed, to produce a dryer solids cake (e.g., 30% solids). Also called high-torque centrifuge.

**high-speed induction mixer** A device used for instantaneously mixing chemicals (e.g., alum, chlorine, ferric chloride, cationic polymer) in water or wastewater treatment. An example is a proprietary mixer consisting of a propeller that creates a vacuum in the mixing chamber and induces the chemical from the storage container. *See also* mixing and flocculation devices.

**high-temperature incineration** Incineration at a temperature higher than 2000°F, which transforms the material into a molten residue that flows like a fiery liquid.

**high-temperature steam activation** The selective oxidation of a carbonaceous material to produce activated carbon. The raw material is heated to 800–1000°C under controlled conditions and in the presence of steam as an oxidizing gas. *See also* chemical activation.

**high-test calcium hypochlorite [$Ca(OCl)_2$]** A calcium hypochlorite product containing 50%, 70% or more available chlorine and used in water treatment as well as in the disinfection of water tanks, water lines, and swimming pools.

**high-test hypochlorite (HTH)** Same as high-test calcium hypochlorite. High Test Hypochlorite® and HTH® are also registered names.

**high-to-low-dose extrapolation** The process of predicting low exposure risks to rodents from the measured high exposure–high risk data.

**high-torque centrifuge** A modification of the solid-bowl centrifuge, designed with a longer bowl and a lower differential bowl speed, to produce a dryer solids cake (e.g., 30% solids). Also called high-solids centrifuge.

**high-velocity wash** A technique used in filter operation to scour by increasing the rate or velocity of backwash. *See also* scour (2).

**high-viscosity sludge** A dense material that consists of a mixture of wastewater treatment residuals (primary sludge, waste activated sludge, clarifier scum, digested sludge) that has been thickened by gravity. It requires positive-displacement pumps.

**high-voltage electrode** The outlet post on a transformer used in the production of ozone at high voltage. *See also* high-frequency ozonation.

**Hi-Iron** A process developed by Aquatrol Ferr-X Corp. to remove iron by contact aeration.

**Hi-Lift** A bar screen manufactured by Longwood Engineering Co., including a reciprocating rake.

**Hill's molecular formula arrangement** A chemical indexing system proposed in 1900 and adopted by the U.S. Patent Office; organic compounds are arranged by carbon (C), hydrogen (H), and remaining elements in alphabetical order.

**Hi-Lucid** A treatment process developed by Hitachi Metals America, Ltd. using high-rate coagulation and sedimentation.

**hindered sedimentation** *See* hindered settling.

**hindered settling** The sedimentation of closely spaced particles in a suspension of intermediate concentration (e.g., sludge or wastewater) so that settling is hindered by interparticle forces and is about the same for an entire zone. *See* sedimentation type III for more detail.

**HIP** Acronym of high-intensity press, a belt press manufactured by Andritz–Ruthner, Inc.

**Hi-pass** A fixed in-line mixer manufactured by Komax Systems, Inc.

**Hi-Rate Thickener** A circular unit manufactured by Dorr-Oliver, Inc. for gravity thickening of sludge.

**histidin** or **histidine [$C_3H_3N_2CH_2CH(NH_2)COOH$]** An essential amino acid constituent of proteins and the iron-binding site of hemoglobin, used in the Ames test.

**histogram** A graph of a frequency distribution as a series of rectangles with bases on the horizontal axis representing class intervals and heights representing frequencies. Histograms are used in the analysis of wastewater treatment plant data. *See* Figure H-01 and Pareto chart.

**histology** The study of the structure of cells and tissues; usually involves microscopic examination of tissue.

**historical average wastewater flow** A yardstick of 120 gpcd established by the USEPA to measure total dry-weather base flow and indicate that infiltration into a sewer system is not excessive. It includes a domestic flow of 70 gpcd, a commercial and small industrial flow of 10 gpcd, and infiltration of 40 gpcd.

**Hi-Tork®** A portable mixer manufactured by the Philadelphia Mixer Corp.

**HK** Acronym of haloketone.

**HK2** Symbol representing the sum of two haloketone species considered in the formation of disin-

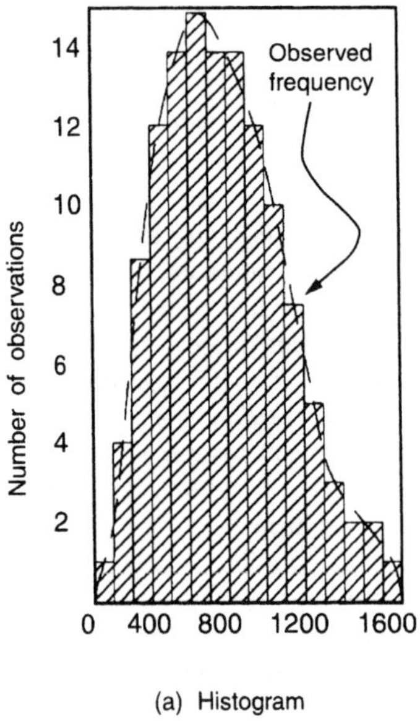


**Figure H-01.** Histogram (courtesy CRC Press).

fection by-products: 1,1-dichloropropanone and 1,1,1-trichloropropanone.

**HMS coagulant** *See* hydrolyzing metal salt coagulant.

**$H_2NNH_2$** Chemical formula of hydrazine; also written as $N_2H_4$.

***H. nana*** *See Hymenolepis nana.*

**$HO^{\cdot}$** A hydroxyl radical that has an unpaired electron. It may be produced by the action of ultraviolet (UV) light on ozone in water or in wet air or in the peroxone process. The overall reaction for the photolysis of ozone is:

$$O_3 + UV + H_2O \rightarrow O_2 + HO^{\cdot} + HO^{\cdot} \quad (H\text{-}33)$$

*See also* the equation under peroxone.

**$HO_2$** Chemical formula of the hydroperoxyl radical.

**$H_2O$** Chemical formula of water, dihydrogen monoxide, dihydrogen oxide.

**$H_2O$ Express** A belt dewatering press manufactured by Magnet Machinery, Inc.

**$H_2O_2$** Chemical formula of hydrogen peroxide.

**$H_3O^+$** Chemical formula of hydronium or hydroxonium ion.

**HOBr** Chemical formula of hypobromous acid.

**HOC** Acronym of halogenated organic compound.

**hockey puck** Informal for the solid tablet form of calcium hypochlorite used in a swimming pool chlorinator.

**HOCl** Chemical formula of hypochlorous acid.

**HOCN** The chemical formula of cyanic acid and fulminic acid.

**HOD-WACX** A clarifier designed by Dean Wacx including filtration media.

**HOG™** Acronym of halogenated organic gas, a destruction process developed by Quantum Technologies, Inc.

**HOI** Chemical formula of hypoiodous acid.

**holding pond** A pond or reservoir, usually made of earth, built to store polluted runoff or other liquids until they can be treated to meet water quality standards or be used in some other way. *See also* detention basin.

**holding tank** A tank that stores wastewater before final disposal.

**Hollosep** A reverse osmosis unit manufactured by Toyobo Co., Ltd. using hollow fiber membranes.

**hollow-cathode lamp** A lamp used in the atomic absorption spectrophotometric analysis of a metal, which makes up the cathode, with an argon-filled glass anode.

**hollow-cone nozzle** One of two types of pressurized nozzles in spray aerators commonly used in water treatment; it delivers a circular spray of droplets concentrated at the circumference. *See also* full-cone nozzle.

**hollow fiber** A hollow-fiber membrane.

**hollow-fiber membrane** A membrane consisting of an outer wall (shell), small-diameter tubes, and a hollow central bore or channel (lumen), used in reverse osmosis and ultrafiltration pressure-driven units. Also called a hollow fiber. These membranes are arranged in modules or elements in single pressure vessels.

**hollow-fiber membrane configuration** Same as hollow-fiber module.

**hollow-fiber module** The arrangement of a bundle of hundreds or thousands of hollow fibers in single pressure vessels, with feed application to the inside or outside of the fiber. It allows a relatively large membrane surface area per unit volume, compared to other configurations (plate-and-frame, spiral-wound, and tubular).

**hollow fine fiber membrane** A membrane consisting of a bundle of hollow fine fibers folded in a U form and sealed in a tube, used in water treatment by reverse osmosis or nanofiltration. *See also* Dupont Permasep® and spiral wound membrane.

**hollow-flight dryer** A device used to dry wastewater treatment sludge by circulating steam through a stationary horizontal vessel or trough, which has a jacketed shell and contains a number of flights in a rotating assembly. The flights transport the sludge through the device and provide the heat transfer surface. *See also* indirect or conduction dryer.

**home aeration unit** *See* aerobic treatment unit.

**homeostasis** A condition of balance within an organism between functions and chemical composition; a constant internal state with respect to pH, temperature, salt concentration, and blood sugar levels.

**homeowner water system** Any water system that supplies piped water to a single residence.

**home treatment filter** A small filter used to purify water in a household or any other unit of comparable size. Also called domestic filter, household filter, point-of-use filter.

**home water treatment unit** A granular activated carbon filter, a membrane filtration unit, a water softener, or other water treatment device attached to the service connection, a sink, or any other point of a dwelling.

**Hom model** A flexible and empirical modification of the Chick–Watson model proposed by Leonard C. Hom in 1972 to represent the rate ($r$) of inactivation of microorganisms in disinfection processes:

$$r = -k' m N t^{m-1} C^n \qquad \text{(H-34)}$$

$$\ln (N/N_0) = -k' t^m C^n \qquad \text{(H-35)}$$

where

$$k' = k/C^n = \text{a rate constant independent of disinfectant concentration} \qquad \text{(H-36)}$$

$k$ = a rate constant; $C$ = disinfectant concentration; $n$ = the coefficient of dilution in the Watson equation; $m$ = a constant, which may indicate whether or not there is a shoulder or tailing; $N$ = concentration of viable organisms (at time $t$); $t$ = time of exposure; and $N_0$ = initial concentration of viable organisms (at time $t$ = 0). *See also* Chick–Watson model, chlorine disinfection models, modified Hom model, shoulder effect, tailing effect.

**homogeneity** The characteristic of a porous medium whose permeability does not change from point to point. It is, along with isotropy, a common assumption in the analysis of groundwater problems.

**homogeneous** Uniform in structure or composition at all locations in space.

**homogeneous buffer system** A compound or mixture consisting of aqueous species that combine to stabilize the pH of a solution.

**homogeneous chemical equilibrium** A condition of chemical equilibrium in which all reactants and products occur in the same physical state, e.g., all in solution.

**homogeneous nucleation** The formation of very small particles in a supersaturated solution in the absence of solid substrates acting as seeds for crystal growth.

**homogeneous (porous) medium** A porous medium whose permeability does not vary from point to point.

**homogeneous reaction** (1) A chemical reaction that involves reactants and products in only one phase. *See also* heterogeneous equilibrium, heterogeneous reaction. (2) A reversible or irreversible reaction between constituents distributed uniformly throughout a fluid, e.g., the reactions that take place in batch, complete-mix, or plug-flow reactors.

**homogeneous surface diffusion model** A model that is widely used to predict the performance of adsorption units, particularly small-scale columns and pilot plants, in the removal of organic compounds from water or wastewater. It has limited applicability to the removal of trace compounds. *See also* solution diffusion model.

**homolactic fermentation** *See* lactic acid fermentation.

**Homomix** A propeller-type mixer manufactured by Amwell, Inc.

**homovalent exchange** An ion exchange reaction between two ions of the same valence, e.g., a chloride-for-nitrate exchange. For such an exchange, the separation factor equals the selectivity coefficient. *See* ion-exchange equilibria for related terms.

**honey wagon** A vacuum truck used to remove and transport septage.

**hookworm** (1) Members of the species *Ancylostoma duodenale* and *Necator americanus* that cause the hookworm disease. Excreted load = 100 per gram of feces. Latency = 7 days. Persistence = 3 months. Median infective dose < 100. (2) A short name for the disease itself.

**hookworm disease** An excreted helminth infection caused by the pathogens *Ancylostoma duodenale* and *Necator americanus*. Distributed worldwide but mainly in warm, wet climates, it is transmitted from soils contaminated by the feces of an infected person and occasionally in water. The pathogens inhabit the small intestine and secrete an anticoagulant that causes blood loss and anemia.

**hookworm life cycle** The mode of transmission of the hookworm, from the feces of an infected host through maturation in the soil to penetration (usually through the feet) and development in a new host.

**Hoover–Porges formula** A widely used empirical formula proposed in 1952 to represent the organic fraction of cells or sludge in terms of their content of carbon (C), hydrogen (H), oxygen (O), and nitrogen: $C_5H_7O_2N$. It may be modified as $C_{60}H_{87}O_{23}N_{12}P$ or $C_5H_7O_2NP_{0.074}$ to include phosphorus (P). *See also* COD:N:P.

**Hoover process** A two-stage process designed to recover lime from the sludge produced in lime softening. First, lime is added just to precipitate calcium and is recovered; the lime added for magnesium precipitation is wasted with the sludge. Carbon dioxide is also recycled for recarbonation. *See also* Lykken–Estabrook process.

**Horg** A collective term for all organic acids in a natural water.

**horizon** *See* soil horizon.

**horizontal centrifugal pump** *See* horizontal pump.

**horizontal directional drilling (HDD)** A trenchless process used in the replacement of water or sewer pipes as an alternative to open cut: a pilot bore is installed by a drilling rig and enlarged to an adequate size by pulling back reamers, and then the new pipe is installed in the bore. HDD is used in busy or crowded areas or to avoid disturbing environmentally sensitive areas. *See* trenchless technology, open cut.

**horizontal-flow grit chamber** A square or rectangular device that uses a proportional weir or a rectangular control section (e.g., a Parshall flume) to vary the depth of flow and keep the velocity of a wastewater stream constant (about 1.0 fps), thereby allowing grit particles to settle and lighter organic particles to remain suspended and be carried out of the channel. The chamber is provided with appropriate equipment to remove the collected grit. *See also* detritus tank, chain and bucket elevator, chain and flights, screw auger. Aerated grit chambers and vortex-type grit chambers are more common in new wastewater treatment plants.

**horizontal-flow tank** A sedimentation tank or other type of basin, with or without baffles, in which water flows (horizontally) from one end to the other, as opposed to vertical- (upward or downward) and spiral-flow tanks.

**horizontal media trickling filter** A trickling filter that uses flat pieces of wood, about 6 × 25 mm, as media instead of rocks or plastics. The device sometimes combines attached and suspended growth processes to treat wastewater.

**horizontal pump** A reciprocating pump whose piston or plunger moves horizontally or a centrifugal pump with a horizontal shaft.

**horizontal reciprocating screen** A medium-mesh screen designed for the treatment of combined sewer overflows. It consists of (a) parallel bars normal to the direction of flow and mounted on a weir and (b) a cutting mechanism that moves horizontally along the bars. *See also* tangential flow screen.

**horizontal screw conveyor** A device or mechanism installed in the trough of a grit chamber to collect and pull removed materials to a hopper at the head of the chamber. *See also* chain-and-flight mechanism, inclined screw, clamshell bucket, chain and bucket elevator.

**horizontal screw pump** A pump that has a radial-blade runner in a horizontal casing, often used in drainage. Also called wood screw pump.

**horizontal submerged aerator** A mechanical aerator consisting of disks or paddles attached to a rotating shaft and designed to entrain oxygen into a liquid from the atmosphere or from air or pure oxygen introduced at the bottom of the basin. *See also* vertical submerged aerator.

**horizontal surface aerator** A mechanical device that is designed to transfer oxygen from the atmosphere into a liquid by surface renewal and exchange, and to provide mixing of the contents of the basin or container. It consists of a horizontal cylinder rotated by a motor, with bristles, angle steel, plastic bars, or blades just above the liquid surface. *See also* Kessener brush aerator, vertical surface aerator, submerged aerator.

**horizontal velocity** A parameter considered in the design of grit chambers, such that only grit (inorganic) and not organic matter settles in the unit. It is selected close to but less than the scouring velocity of the grit. It is also one of the design parameters of sedimentation basins.

**horsepower (HP)** A unit of power equivalent to 550 foot-pounds per second or 745.7 watts. For example, it is used to express the theoretical power requirement, or water horsepower (also called hydraulic horsepower, $P_w$), of an electric motor to drive a pump, and is calculated as

$$P_w = \gamma QH/550 \quad \text{(H-37)}$$

where $\gamma$ = specific weight of the fluid in pounds per cubic foot, $Q$ = fluid discharge in cfs, and $H$ = total dynamic head in ft. To obtain the actual power needed, or brake horsepower ($P_b$), one must consider the efficiency ($e$) of the pump, which also depends on the discharge and the head:

$$P_b = P_w/e \quad \text{(H-38)}$$

The brake horsepower ($P_b$), also called shaft horsepower, is the energy going to the pump, whereas the water horsepower ($P_w$) is the energy used by the motor, the difference between the two being the power consumed by mechanical losses, noise, heat, viscous drag, and internal recirculation. *See also* motor efficiency, power terms, wire-to-water efficiency.

**hose barb** A twist-type connector used for connecting a small-diameter hose to a valve or faucet.

**hose bib** The point of connection of a hose in a water line.

**Hose Monster®** A device manufactured by Hydro Flow Products, Inc. for use in fire flow testing and flushing.

**Hosepump** A peristaltic pump manufactured by Waukesha Pumps, Inc.

**hospital-acquired infection** *See* nosocomial pathogen.

**hospital waste** Liquid or solid wastes from hospitals, laboratories, and similar establishments are considered hazardous as they may contain human pathogens and other dangerous substances. Incineration is a safe disposal method for such wastes.

**host** (1) In genetics, the organism, typically a bacterium, into which a gene from another organism is transplanted. (2) In medicine, an animal infected or parasitized by another organism. (3) In general, a plant or animal that provides nourishment or protection to an agent (infectious or not). A parasite may use more than one host, e.g., primary or definitive hosts, secondary or intermediate hosts, and transport hosts.

**host resistance** The ability of an individual to resist infection.

**hot-lime–cation softening** A complete variation of the hot-process softening method that combines hot-lime softening with cation exchange softening.

**hot-lime–soda softening** A variation of the hot-process softening method that includes addition of lime and soda ash to water at about 100°C. Calcium, carbon dioxide, iron, magnesium, and silica are removed. This method achieves partial softening.

**hot-lime softening** A variation of the hot-process softening method that includes (a) addition of a lime slurry to water at a temperature between 100 and 125°C, (b) precipitation and removal of calcium and magnesium by sedimentation and filtration. This method achieves partial softening.

**hot-process softening** A water softening and clarifying method used for boiler feedwater and in sulfur mining. It consists of the addition of lime alone, with soda ash, or with cation softening to water at about 100°C, to remove carbon dioxide, magnesium and silica. Partial and complete variations of this process include hot-lime softening, hot-lime–cation softening, and hot-lime–soda softening.

**hot spring** A spring with a temperature higher than the temperature of the human body (37°C or 98.6°F). *See also* thermal spring, warm spring.

**hot well** A container for collecting condensate as in a steam condenser serving a vacuum-jet or steam-jet ejector (EPA-40CFR264.1031).

**hot windbox** A fluid bed furnace with a brick dome or a refractory-lined plate, in which the air supply is heated by burners or heat exchangers. *See also* cold windbox.

**house cistern** A cistern used to store rainwater to supply a household.

**house connection** The pipe carrying wastewater from a house drain to a common or public sewer, or to a point of immediate disposal. Also called building drain, building sewer, service connection, or house sewer.

**household detergent** A detergent produced and marketed for use by households; generally, an anionic or nonionic surfactant containing phosphates and other constituents.

**household disposal system** A system used for the treatment and disposal of wastewater from a household, e.g., a septic tank and soil absorption system. *See also* on-site disposal, individual sewage disposal.

**household filter** A small filter used to purify water in a household or any other unit of comparable size. Also called home treatment filter, domestic filter, point-of-use filter.

**household waste** (1) Solid waste, composed of garbage and rubbish, that normally originates in a private home or apartment house. Domestic waste may contain a significant quantity of toxic or hazardous waste. (2) Any material (including garbage, trash, and sanitary wastes in septic tanks) derived from households (including single and multiple residences, hotels and motels, bunkhouses, ranger stations, crew quarters, campgrounds, picnic grounds, and day-use recreation areas) (EPA-40CFR258.2). (3) Wastewater from kitchens, lavatories, toilets, and laundries.

**house service** The piping that is necessary to provide water service to a domestic consumer. Also called domestic service.

**house sewer** Same as house connection.

**house subdrain** The portion of the plumbing system of a building that does not drain by gravity into the sewer system. Also called building subdrain.

**house trap** A trap in a building drain to prevent air from the sewer system to flow back into the building. Also called building trap.

**Howland formula** *See* Howland's equation.

**Howland's equation** A formula proposed in 1957 to approximate the detention time ($\theta$) of clean water in a trickling filter (Fair et al., 1971):

$$\theta = CH(\nu/g)^{1/3}(S/Q)^{2/3} \qquad \text{(H-39)}$$

where $C$ = a coefficient that depends on film buildup and medium structure, $H$ = filter depth; $\nu$ = kinematic viscosity of the liquid, $g$ = gravitational acceleration, $S$ = specific surface of the contact medium = the ratio of its surface area to its volume, and $Q$ = hydraulic load = flow per unit area and time. The Howland formula is also written as (Metcalf & Eddy, 2003):

$$\theta = CH/(Q/A)^n \qquad \text{(H-40)}$$

where $A$ = filter cross section area and $n$ = hydraulic constant for the packing material used. A third formulation assumes flow through a trickling filter as steady down an inclined plane under laminar conditions (Nemerow, 1991):

$$\theta = (3\,\nu/gs)^{1/3}(L/q^{2/3}) \qquad \text{(H-41)}$$

where $s$ is the sine of the angle between the inclined plane and the horizontal, $L$ is the length of the plane, and $q$ is the rate of flow per unit width of the plane. Other trickling filter design formulas include British Manual, Eckenfelder, Galler and Gotaas, Germain, Kincannon and Stover, Logan, modified, NRC, Rankin, Schulze, Velz. *See also* trickling filter contact time.

**HP** Water horsepower, as opposed to brake horsepower (hp).

**hp** Abbreviation of horsepower, a standard unit of power, equal to 550 foot-pounds per second or approximately 745.7 watts. Also used to designate brake horsepower, as opposed to water horsepower (HP).

**HPC** Acronym of heterotrophic plate count.

**HPD** Products of Wheelabrator Engineered Systems, Inc. for use in evaporators.

**HP-Hybrid** An automatic filter press manufactured by Heinkel Filtering Systems, Inc.

**HPLC** Acronym of high-performance liquid chromatograph or chromatography.

**HPMC** Acronym of high-pressure minicolumn.

**HPO** Acronym of high-pressure oxidation.

**$H_2PtCl_6$** Chemical formula of chloroplatinic acid.

***H. pylori*** Abbreviation of *Helicobacter pylori.*

**HRB®** A baler manufactured by Harris Waste management Group, Inc. for use in solid waste management.

**HRGC** Acronym of high-resolution gas chromatography.

**HRMS** Acronym of high-resolution mass spectrometry.

**HRS** Acronym of hazard (or hazardous) ranking system.

**HRT** Acronym of hydraulic residence time.

**$H_2S$** Chemical formula of hydrogen sulfide or sulfurated hydrogen.

**$H_2S^+$** Chemical formula of sulfonium.

**HSC™** A high-pressure centrifugal pump manufactured by Pump Engineering, Inc. for use in reverse osmosis units.

**HSDM** Acronym of homogeneous surface diffusion model.

**$H_2SiF_6$** Chemical formula of hydrofluosilicic acid.

**$HSO_3$** Formula of a radical characteristic of bisulfite. *See,* e.g., sodium bisulfite.

**—$HSO_3$** *See* sulfonic acid. Also written —$SO_2OH$.

**$H_2SO_3$** Chemical formula of sulfurous acid.

**$H_2SO_4$** Chemical formula of sulfuric acid.

**$H_2S_2O_3$** Chemical formula of thiosulfuric acid.

**$HSO_3NH_2$** Chemical formula of sulfamic acid.

**HT** Acronym of heat treatment.

**HTC™** Acronym of hydraulic turbocharger, a device manufactured by Union Pump Co. for reverse osmosis units.

**HTH** (1) Acronym of high-test calcium hypochlorite. (2) A calcium hypochlorite (chemical formula: $Ca[OCl]_2$) produced by Olin Corp.

**HTP** Acronym of heat-treated peat, a cleanup product of American Products for use in polymer spills.

**Hudson's formula** A formula proposed by H. E. Hudson in 1959 to predict the point of breakthrough of inadequately treated water in rapid filters. *See* breakthrough index.

**Huisman** A wastewater treatment unit manufactured by Envirex, Inc. using oxidation ditches.

**human carcinogenic potency factor** The risk resulting from a lifetime exposure of 1.0 mg/day of a substance per kg of body weight. *See* slope factor for detail.

**human consumption** The use of water by individuals for drinking, cooking, dishwashing, bathing, or showering.

**human engineering** An applied science that considers the capacities and requirements of workers in designing devices, machines, systems, and physical working conditions. Water and wastewater treatment plants are normally designed to prevent or minimize injuries, fatigue, stress, etc. to operation and maintenance personnel. Also called ergonomics, human-factors engineering.

**human enteric virus** *See* adenovirus, enterovirus, gastroenteritis virus, hepatitis-A virus, parvovirus, reovirus, rotavirus.

**human equivalent concentration** Exposure concentration for humans that has been adjusted for dosimetric differences between experimental animal species and humans to be equivalent to the exposure concentration associated with observed effects in the experimental animal species. If occupational human exposures are used for extrapolation, the human equivalent concentration represents the equivalent human exposure concentration adjusted to a continuous basis.

**human equivalent dose** A dose that, when administered to humans, produces an effect equal to that produced by a dose in animals.

**human excreta** Feces, urine, and vomitus from humans; the principal vehicle of transmission and spread of many communicable diseases. Worldwide, human beings produce an average of 1150 grams of urine and 200 grams of feces per day.

**human excreta-related disease** *See* excreted bacteria, excreted helminths, excreted protozoa, excreted virus.

**human-excreted infection** *See* excreted bacteria, excreted helminths, excreted protozoa, excreted virus.

**human exposure evaluation** A component of risk assessment that involves describing the nature and size of the population exposed to a substance and the magnitude and duration of their exposure. The evaluation could concern past, current, or anticipated exposures.

**human-factors engineering** *See* human engineering.

**human health risk** The likelihood that a given exposure or series of exposures may have damaged or will damage the health of individuals experiencing the exposures.

**human immunodeficiency virus (HIV)** *See* acquired immune deficiency syndrome.

**human pollution** Pollution of surface water by human feces. *See* coliform/streptococci ratio.

**human viruses** Viruses that infect human hosts, including those that cause the following ailments: colds, dengue fever, diarrhea, encephalitis, eye infections, hepatitis, herpes, influenza, measles, meningitis, mumps, rabies, smallpox, warts, yellow fever, etc.

**humic acid** A brown compound formed by decomposing organic matter and found in water, soils, lignite, and peat. It is a mixture of several substances. In water, it is a source of color (a yellowish-brown hue) and a major precursor of disinfection by-products. It is sometimes characterized as the portion of organic matter that is soluble in alkaline solutions but precipitated in acidic solutions. *See also* fulvic acid, aquatic fulvic acid, humin, aquagenic organic matter, natural organic matter, tannin, disinfection by-product.

**humic material** Same as humic substance.

**humic matter** Humic substance.

**humic organic substance** Humic substance.

**humic substance** A product of the partial decomposition of organic matter in soils, and, to a lesser extent, in water and sediments; a major component of natural organic matter in water sources. Aquatic humic substances are nonvolatile, have molecular weights between 500 and 5000 grams, and, like clay, have a net negative charge and a large surface area; they include humin, humic acid, and fulvic acid.

**humic water** Water that contains humic substances.

**humin** One of three forms of humic substances; the part of the organic matter in soil that does not dissolve under the action of hot alkali or that cannot be extracted by acid or base. It is not considered a significant THM precursor. *See also* humic acid, fulvic acid.

**hump and dip curve** Another name for the breakpoint curve, a graphical representation of the breakpoint phenomenon, particularly at pHs between 6.5 and 8.5. *See* breakpoint chlorination.

**humus** The dark, amorphous, colloidal residue in soil remaining after prolonged microbial decomposition of all animal and plant tissues. Also, the decomposed vegetable matter remaining after composting. Humus is a complex and stable molecule that improves soil texture and helps retain water. Activated sludge and well-digested sludges contain a similar material. *See also* trickling-filter humus.

**humus sludge** Sludge that resembles humus or that deposits in the clarifier following trickling filtration; trickling-filter humus.

**humus tank** A sedimentation tank that follows an aeration basin, a trickling filter, or any other biological treatment reactor; a secondary clarifier.

**Hurricane** (1) A mixer manufactured by Franklin Miller. (2) A unit manufactured by Harmsco Industrial Filters including a centrifuge and a cartridge filter.

**HUS** Acronym of hemolytic–uremic syndrome.

**hutchinsonite** A rare mineral containing the metallic element thallium.

**HVAC** Acronym of heating, ventilating, and air conditioning.

**HVAC/R** Acronym of heating, ventilating, air conditioning, and refrigeration.

**hyacinth** A floating aquatic plant, the roots of which provide a habitat for a diverse culture of aquatic organisms that metabolize organics in wastewater.

**hyacinth system** A floating aquatic plant system that uses water hyacinth in a series of rectangular basins similar to stabilization ponds, with recycle and step feed, to treat wastewater.

**hybrid aerobic process** The combination of aerobic suspended and attached growth processes such as activated sludge and trickling filtration for the removal of carbonaceous BOD and nitrification.

**hybrid anaerobic process** The combination of anaerobic suspended growth, attached growth, and sludge blanket processes, such as upflow sludge blanket and trickling filtration, for the removal of carbonaceous BOD.

**hybrid plant** A water or wastewater treatment plant that combines two or more processes to improve performance, e.g., distillation and reverse osmosis or activated sludge and trickling filtration.

**hybrid system** A system that includes more than one process or technology, for example, a wastewater treatment method that incorporates two different processes; also called a dual process. Most plants commonly using dual processes combine a fixed film unit with a suspended growth reactor. *See also* combined filtration–aeration process and dual biological treatment system.

**hybrid tank** A vessel equipped with a compressor or a dipping tube to provide protection against pressure surges in piping systems by maintaining a desired air volume in the vessel and controlling the air pressure above atmospheric pressure under normal operating conditions. Other surge vessels include bladder tanks and compressor vessels.

**hydatid** The cyst of a tapeworm.

**hydatid disease** Echinococcis, a disease transmitted through the ingestion of infective eggs in water or food contaminated by dog feces.

**Hydecat™** A product of ICI Katalco designed to destroy hypochlorite.

**HYDRA** A sludge removal unit manufactured by Hazleton Environmental Products, Inc. using a hydraulic rake.

**HYDRA®** Acronym of Hydrologic Data Retrieval and Alarm System, a computer modeling program.

**Hydradenser** An inclined screw thickener manufactured by Black Clawson.

**hydragyrum** The metallic element mercury (Hg).

**Hydra-Mix** A hydraulic mixer manufactured by Air-O-Lator Corp.

**Hydranautics membrane** A proprietary spiral-wound membrane. *See* spiral-wound module.

**hydrant** An upright pipe or other device equipped for drawing water from a water main into a hose for fire fighting, street washing, or for flushing out the water main. Also called fire hydrant or fire plug.

**Hydrapaint** A spiral membrane manufactured by Hydranautics for use in ultrafiltration.

**Hydra-Press™** A hydraulic compactor manufactured by Vulcan Industries, Inc. for handling screenings from wastewater treatment.

**hydrargyrum** Greek name of mercury (Hg).

**Hydrasand®** A sand filter manufactured by Andritz-Ruthner, Inc., including a continuously cleaned moving bed.

**Hydrasieve®** A fine screen manufactured by Andritz-Ruthner, Inc.

**hydrate** A substance formed by the combination of water with another substance. Water is loosely bound in some hydrates (e.g. washing soda, $Na_2CO_3 \cdot 10H_2O$), but strongly bound in others (e.g., $SO_3 \cdot H_2O = H_2SO_4$, sulfuric acid).

**hydrated calcium sulfate** *See* gypsum, $CaSO_4 \cdot 2H_2O$.

**hydrated hydrogen ion** The hydronium ion, $H_3O^+$ or $H^+$, associated with a number of $H_2O$ molecules in an aqueous solution.

**hydrated lime** Limestone that has been burned and treated with water under controlled conditions until the calcium oxide (CaO) portion has been converted to calcium hydroxide [$Ca(OH)_2$]. Hydrated lime is quicklime combined with water:

$$CaO + H_2O \rightarrow Ca(OH)_2 \qquad \text{(H-42)}$$

Also called slaked lime or slaked hydrated lime; it is used in water treatment, directly in mixing tanks or in dry-feeding equipment, to increase alkalinity. *See* lime–soda process. *Note.* Lime is (a) calcium oxide, commonly called quicklime, but also burned lime, calx, carbide lime, caustic lime, chemical lime, dead-burned lime, overburned lime, unslaked lime; or (b) calcium hydroxide, commonly called hydrated lime, but also calcium hydrate, milk of lime, slaked lime, slaked hydrated lime, or whitewash. Related materials include chlorinated lime, dolomitic lime, dolomitic limestone, high-calcium limestone, and limestone. General uses in water and wastewater treatment: heavy metals removal, odor control, pH adjustment, phosphorus removal, sludge conditioning, sludge stabilization, softening, suspended solids removal.

**hydrated metal ion** A hydrate; the coordination of a metal ion with a number of $H_2O$ molecules, which is considered an acid, e.g., hydrated aluminum ion, $Al(H_2O)_6^{3+}$.

**hydrated proton** The product of the strong reaction of a proton with a water molecule, the hydronium or hydroxonium ion, $H_3O^+$.

**hydration** The chemical combination of a substance with water to form another substance. *See also* hydrate.

**hydration radius** The radius of the complex formed by a cation in soil and its associated water molecules. It depends on the surface charge density and affects the extent of the diffuse double layer as well as the soil structure; e.g., magnesium (Mg) has a higher charge density than sodium (Na) and attracts water molecules more strongly. *See* Figure H-02.

**hydration water** (1) Water that is chemically bound to solid particles in a sludge suspension. *See also* bound water, bulk water, interstitial water, and vicinal water. (2) Water that combines with salts when they crystallize, i.e., a part of the crystalline compound. Also called water of crystallization.

**Hydra-Tracker™** A self-tracking device manufactured by Dontech, Inc. for sludge, grease, and oil removal.

**Hydraucone** A diffuser plate manufactured by Amwell, Inc. for mechanical aerators.

**hydraulically driven reciprocating piston pump (HDRPP)** A device consisting of a twin-screw auger feeder, a pumping assembly, and a hydraulic power unit; developed for pumping concrete and used to pump sludge cake instead of transporting the cake by conveyors.

**hydraulic burster** *See* hydraulic bursting.

**hydraulic bursting** A trenchless process used in the replacement of water or sewer pipes: a bursting head is pulled through the excavation to break the existing pipe, push the pieces into the surrounding soil, and install the new pipe that is attached at the rear of the head. The hydraulic burster consists of a head with leaves that expand under hydraulic pressure to burst the existing pipe. *See* trenchless technology, open cut, pneumatic bursting, static bursting.

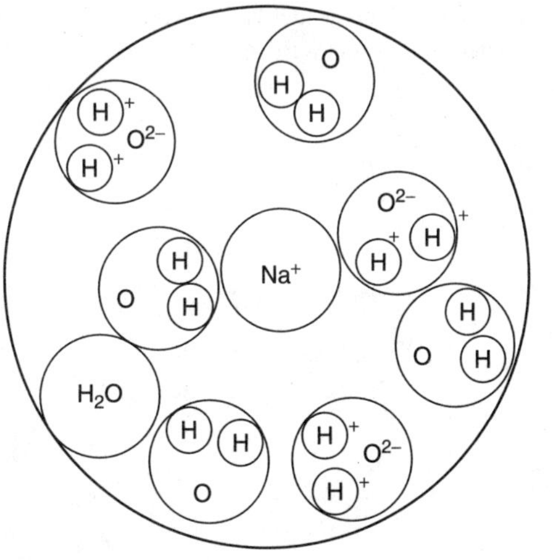


**Figure H-02.** Hydration histogram radius (sodium, $Na^+$).

**hydraulic classification** The stratification of ion exchange or granular media during backwashing, with the smallest particles at the top.

**hydraulic coefficient ($h_c$)** A coefficient that reflects the influence of the slope ($S_0$) and roughness factor ($n$) of a channel, conduit, or sewer on its discharge by the Manning equation. It is

$$h_c = S_0^{0.5}/n \tag{H-43}$$

**hydraulic conductivity ($K$)** A coefficient of proportionality describing the rate at which water can move through a permeable medium. Hydraulic conductivity is a function of both the intrinsic permeability of the porous medium and the kinematic viscosity of the water that flows through it. Also referred to as the coefficient of permeability. It is used in the Darcy formula for groundwater flow and has the same unit as velocity. It is expressed as the product of the density ($\rho$) of the fluid and the gravitational acceleration ($g$) divided by the dynamic viscosity ($\mu$):

$$K = \rho \cdot g/\mu \tag{H-44}$$

$K$ values greater than 10 m/day are considered large, whereas $K < 0.10$ m/day indicates limited water flow. Hydraulic conductivity can be measured by laboratory analysis of aquifer samples, in the field (tracer technique or aquifer test), or examination of water-level maps. It is an important factor in the evaluation and selection of sites for the slow-rate application of wastewater to land. *See also* effective porosity, effective velocity. Also called vertical permeability.

**hydraulic detention time** The theoretical average time that a liquid or suspended particles pass through a container during treatment, calculated as the ratio of volume to flow. *See* hydraulic residence time for detail.

**hydraulic diffusivity** A groundwater parameter ($D$) equal to the ratio of transmissivity ($T$) to storage coefficient ($S$) or the ratio of permeability ($K$) to unit storage ($s$):

$$D = T/S = K/s \tag{H-45}$$

See formation constant for other measures of hydraulic properties of aquifers.

**hydraulic dredge** A scow that carries a centrifugal pump for pumping sand. Also called sand-pump dredge or suction dredge.

**hydraulic drop** A drop in the hydraulic gradeline or in the water surface elevation as a result of a structure (e.g., a weir) or a transition (e.g., when the bottom slope changes from mild to steep). *See also* hydraulic jump.

**hydraulic elements** The variables, parameters, or quantities included in such flow formulas as Manning's or Darcy–Weisbach's: depth, velocity, discharge, hydraulic radius, flow area, friction factor, mean hydraulic depth, wetted perimeter, etc. For circular conduits flowing partially filled, a hydraulic elements graph (also called partial flow diagram) shows how the ratios of the partial elements to full elements vary with the depth ratio. *See* Figure H-03 and Table H-02. Note that a pipe flowing partially filled may carry a discharge larger than the full discharge: for example, at a depth ratio of 90%, the discharge ratio is approximately 107%.

**hydraulic elements graph** *See* hydraulic elements.

**hydraulic energy** The energy stored in fluids in one of three forms: kinetic, potential, or pressure energy. *See* energy and Bernoulli's law.

**hydraulic flocculation** In the dissolved-air flotation method, the use of the flowing water to provide the energy required for flocculation, e.g., using a baffled tank with a higher G value than when mechanical energy is used.

**hydraulic flocculator** A device for mixing or distributing chemicals in water or wastewater treatment using the turbulence of the liquid. *See* Alabama flocculator, upflow solids contact clarifier, hydraulic mixer.

**hydraulic friction** The resistance to flow of a fluid in a channel or conduit caused by the loss of energy that surface roughness induces.

**Table H-02.** Hydraulic elements of a circular conduit*

| Depth of flow | Area of flow | Hydraulic radius | Velocity | Discharge | Roughness |
|---|---|---|---|---|---|
| 1.00 | 1.00 | 1.00 | 1.00 | 1.00 | 1.00 |
| 0.90 | 0.95 | 1.19 | 1.12 | 1.07 | 0.94 |
| 0.80 | 0.86 | 1.22 | 1.14 | 0.99 | 0.88 |
| 0.50 | 0.50 | 1.00 | 1.00 | 0.50 | 0.81 |
| 0.30 | 0.25 | 0.68 | 0.78 | 0.20 | 0.78 |
| 0.10 | 0.05 | 0.25 | 0.40 | 0.02 | 0.82 |

*Ratios of partial elements to full elements.

**hydraulic friction coefficient** Same as roughness coefficient.

**hydraulic grade** In any water body the hydraulic grade of a section is the piezometric level of the water at that section, i.e., the elevation to which water would rise in a freely vented pipe under atmospheric pressure at that section. In open-channel flow, the hydraulic grade is at the free water surface. The hydraulic grade ($H_g$) is such that

$$P = \rho \cdot g(H_g - z) \quad \text{(H-46)}$$

where $P$ is the pressure, $\rho$ is the density of the fluid, $g$ is the acceleration of gravity, and $z$ is the elevation above or below a datum. *See* hydraulic gradeline, hydraulic gradient.

**hydraulic grade differential** The difference in the elevations of the hydraulic gradeline between two points along a channel. In the case of ports or gates used to distribute flows evenly to treatment process reactors from a common header pipe or channel, the hydraulic grade differential ($\Delta h$, m or ft), the minimum headloss through the port or gate ($h_p$, m or ft), the actual headloss through a gate or port ($H$, m or ft), and the header head loss ($h_L$, m or ft) are related as follows (WEF & ASCE, 1991):

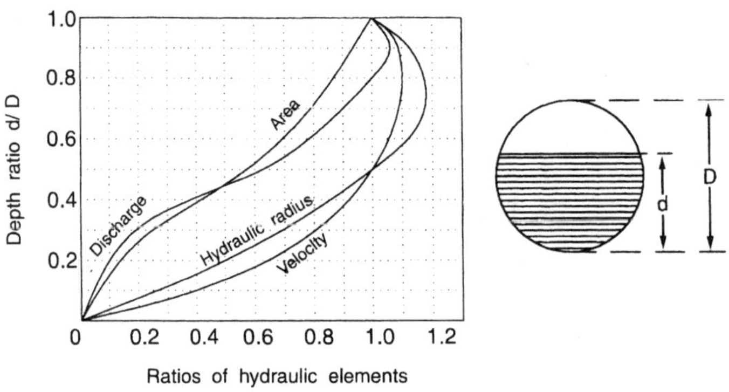


**Figure H-03.** Hydraulic elements.

$$h_p = \Delta h/(1 - m)^2 \quad \text{(H-47)}$$

$$H = Q^2/[(CA)^2\, 2\, g] \quad \text{(H-48)}$$

$$\Delta h = V_0^2/2g - h_L \quad \text{(H-49)}$$

where $m$ = ratio of flow in the first port to that through the last port, $Q$ = flow, $m^3$/sec or cfs, $C$ = dimensionless gate or port coefficient, $A$ = area of gate or port opening, $m^2$ or $ft^2$, $g$ = gravitational acceleration = 9.8 m/sec² or 32.2 ft/sec², and $V_0$ = header inlet velocity, m/sec or fps.

**hydraulic gradeline or HGL** In any water body the hydraulic gradeline is a profile of the piezometric level of water at all points along a line. In a closed conduit flowing full, it is a line joining the elevations to which water would stand in risers or vertical pipes connected to the conduit at their lower end and open at their upper end. In open-channel flow, the hydraulic gradeline is the free water surface. In general, the hydraulic gradeline is one velocity head ($V^2/2\,g$) below the energy gradeline, and the elevation ($H_g$) of the hydraulic gradeline is $H_g = z + P/\rho g$ which is the same as Equation (H-46) ($P$ is the pressure, $\rho$ is the density of the fluid, $g$ is the acceleration of gravity, and $z$ is the elevation above or below a datum). The hydraulic gradeline is one of the outputs of hydraulic models. It is also used in the engineering design of water and wastewater facilities; see hydraulic profile. *See also* energy gradeline, hydraulic grade, hydraulic gradient. Sometimes called hydraulic gradient.

**hydraulic gradeline differential** *See* hydraulic grade differential.

**Hydraulic gradeline profile** The proper term is either hydraulic gradeline or hydraulic profile.

**hydraulic gradient** (1) In general, the direction of groundwater flow due to changes in the depth of the water table. (2) The change in total potentiometric (or piezometric) head between two points divided by the horizontal distance separating the two points. (3) The slope of the hydraulic gradeline; also called hydraulic slope. This is the slope of the water surface in an open channel, the slope of the groundwater table, or the slope of the water pressure for pipes under pressure. (4) The hydraulic gradeline itself. *See* Figure E-06. In the groundwater field, hydraulic gradient (a) is a factor in Darcy's formula for groundwater flow, (b) is a driving force in advection and in contaminant travel, and (c) follows more or less the ground topography in the absence of significant sinks and sources. *See also* Darcy's law, energy gradient and hydraulic gradeline.

**hydraulic gradient test** A test conducted, usually during a period of constant peak flow, to determine if the pressure of a water distribution system is adequate. A plot of pressure observations vs. distance along a line indicates the head losses between points and steeper hydraulic gradients that correspond to sections of greater friction losses.

**hydraulic head** or, simply, **head** The height of the free surface of a body of water above a given point beneath the surface. *See* static head or lift, velocity head, friction head.

**hydraulic horsepower** Same as water horsepower, i.e., the difference between brake horsepower and the power consumed by various losses. It is the output of a pump or the liquid horsepower delivered by the pump. *See* horsepower, power terms.

**hydraulic load** Same as hydraulic loading.

**hydraulic loading** The flow (volume per unit time) or the flow per unit surface area applied to a water, wastewater, or stormwater treatment or storage facility, including recirculation; commonly expressed in liters per square meter per second, cubic meters per square meter per hour, gallons per minute per square foot, cubic meters per hectare per day, or million gallons per day per acre. *See also* dilution rate, organic loading.

**hydraulic loading intensity** The flow rate or hydraulic load of a water or wastewater treatment unit, expressed in such a way as to indicate a performance aspect of the unit. The detention time ($T$) is the ratio of the volume ($V$) of the unit to the flow rate ($Q$). The dilution rate ($R$) is a measure of the hydraulic load in a flow-through reactor, equal to the inverse of the flow rate ($1/Q$). The reciprocal detention time ($1/T$) is sometimes used in the design of trickling filters and expressed in million gallons per acre-foot per day. The velocity factor ($v = Q/A$), the ratio of the flow rate ($Q$) to the applicable surface area ($A$), is used as million gallons per day per acre for trickling filters, land treatment, intermittent sand filters and stabilization ponds, and as gallons per day per square foot (surface overflow rate) for settling tanks.

**hydraulic loading rate** The volumetric flow rate per unit area in a slow rate land application system, determined from a mass balance on the receiving soil. It can be based on (a) permeability and other factors that affect the percolation rate, (b) nitrogen limits, or (c) irrigation water requirements (WEF &ASCE, 1991 and Metcalf & Eddy, 1991):

Permeability: $L = ET - P + F_a \cdot F_c \cdot K$ (H-50)

Nitrogen limit: $L = C_p[(P - ET) + 4.4\, U]/[(1 - f)(C_n - C_p)]$ (H-51)

Irrigation water: $L = (ET - P)/(1 - \lambda/100)$ (H-52)

where $L$ = hydraulic loading rate, depth per unit time, e.g., in/yr; $ET$ = rate of evapotranspiration, depth per unit time, e.g., in/yr; $P$ = rate of precipitation, depth per unit time, e.g., in/yr; $K$ = soil permeability, the lowest value measured at the site, depth per unit time, e.g., in/month; $F_a$ = a dimensionless coefficient accounting for periods of nonapplication; $F_c$ = a dimensionless coefficient accounting for soil clogging and reflecting a long-term acceptance rate; $C_p$ = total nitrogen concentration in percolating water, mg/L; $U$ = nitrogen uptake by crops, lb/acre/yr; $f$ = fraction of applied total nitrogen removed by denitrification and volatilization; $C_n$ = total nitrogen concentration in applied wastewater, mg/L; $R$ = net irrigation water requirement, depth per unit time, e.g., in/yr; and $\lambda$ = leaching requirement, %.

**hydraulic losses** The friction and minor head losses.

**hydraulic mean depth ($D_m$)** The ratio

$$D_m = A/W \qquad \text{(H-53)}$$

of flow area ($A$) to surface width ($W$). It is different from the hydraulic radius, although sometimes the two terms are used interchangeably. Also called hydraulic depth or average depth. *See also* open channel flow characteristics and hydraulic radius.

**hydraulic mixer** A device for mixing or distributing chemicals in water or wastewater treatment using the turbulence of the liquid. It usually includes pumps, but weirs and baffled chambers are also used. *See also* hydraulic flocculator.

**hydraulic mixing** The use of hydraulic devices such as turbines, hydraulic jumps, weirs, Parshall flumes, and pressurized water jets for rapid mixing in water or wastewater treatment when the flow rate is sufficient to provide the required energy. Injecting chemicals into the inlet of a centrifugal pump is also considered hydraulic mixing. *See also* mixing and flocculation devices.

**hydraulic model** A physical or mathematical representation of fluid flow. Physical hydraulic models are full- or reduced-scale prototypes used to conduct experiments. For mathematical models, see flow model. Physical models must have the properties of (a) dynamic similarity (similarity between the inertia forces of the model and prototype); (b) geometric similarity (all lengths of the model are in the same ratio to the corresponding lengths of the prototype); (c) kinematic similarity (the ratio of the velocities in the model and prototype is constant throughout the system).

**hydraulic performance of reactors** Terms used to describe the hydraulic performance of reactors and related to tracer response curves include average retention time index, hydraulic residence time, mean retention time index, modal retention time index, Morrill dispersion index, short-circuiting index, and volumetric efficiency.

**hydraulic population equivalent** The number of persons that would contribute a given flow of wastewater, based on average rate, e.g., 120 gpcd. *See* population equivalent.

**hydraulic profile** A graphical summary presentation of the hydraulic calculations performed for the various trains in a water or wastewater treatment facility (liquid treatment, residuals processing, effluent disposal). The profile starts at the downstream control point (e.g., the receiving water) and goes back to the head end of the facility. It includes, for average and peak flow rates, water surface elevations, hydraulic control devices such as valves and weirs, and sometimes the ground surface and structure elevations. The hydraulic profile serves various purposes: to ensure adequate hydraulic gradient, to determine required pump heads, and to prevent flooding of the facilities or backup under peak flow conditions. *See* energy gradeline, hydraulic gradeline, sewer profile.

**hydraulic radius ($R$)** A widely used hydraulic property defined as the ratio of the cross-sectional area of flow ($A$) to the wetted perimeter ($P$):

$$R = A/P \qquad \text{(H-54)}$$

The wetted perimeter is the length of the line of intersection of the flow area with the wetted surface of the channel, i.e., it does not include the surface width $W$ in an open channel. *See also* hydraulic mean depth. The hydraulic radii of some common cross sections follow. For a rectangle of width $B$ and depth of flow $y$:

$$R = yB/(2\,y + B) \qquad \text{(H-55)}$$

For a trapezoid of base $B$, water depth $y$, side angle $\theta$, and $x = \cot\theta$:

$$R = y(B \cdot \sin\theta + y \cdot \cos\theta)/(B \cdot \sin\theta + 2\,y) = (B + xy)/[B + 2\,y/(1 + x^2)^{0.5}] \qquad \text{(H-56)}$$

For a triangle of side angle $\theta$, and $x = \cot\theta$:

$$R = 0.5\,y\cos\theta = xy/2(1 + x^2)^{0.5} \qquad \text{(H-57)}$$

For a semihexagonal section (side angle $\theta = 60°$):

$$R = y(y + 1.732\,B)/(1.732\,B + 4\,y) \qquad \text{(H-58)}$$

For a semihexagonal section (side angle $\theta = 60°$) with the depth of flow such that the base B is equal to the side:

$$R = y/2 = 0.433\,B \qquad \text{(H-59)}$$

For a parabola when $0 < 4\,y/W < 1$, where W is the surface width, a reasonable approximation is

$$R = 2\,yW^2/(3\,W^2 + 8\,y^2) \qquad \text{(H-60)}$$

For wide channels, e.g., when $B \geq 100\,y$, a convenient assumption for overland flow analysis is:

$$R \cong y \qquad \text{(H-61)}$$

The hydraulic radius appears in several hydraulic formulas: e.g., the Reynolds number (as the characteristic length), channel conveyance, critical slope, equivalent roughness, kinematic wave approximation, and Chézy and Manning equations. For a channel or conduit of given flow area, roughness coefficient, and bottom slope, the best hydraulic cross section is defined as the section that yields the maximum discharge. According to the Manning or Chézy equation, this corresponds to the maximum hydraulic radius and the minimum wetted perimeter. For a rectangular section, this requires that the depth of flow equal half the surface width or $B = 2\,y$, and the hydraulic radius is $R = B/4 = y/2$. For a trapezoid, the section of maximum discharge is half a regular hexagon, with $\theta = 60°$, and $R = y/2 = 0.433\,B$. In open-channel flow, the half-circle is the cross section with the highest efficiency; its hydraulic radius is

$$R = D/4 = (A/4\,\pi)^{1/2} \qquad \text{(H-62)}$$

$D$ being its diameter. *See* hydraulic efficiency and open-channel flow characteristics.

**hydraulic ram** A device for lifting water using the impulse of larger masses of (driving) water in coordination with the pressure waves created by water hammer. The driving water may or may not be from the same source as the water supplied, and the ratio of the two masses is usually between 2:1 and 6:1. Also called water ram or hydraulic pump. In a simple form, the device uses the energy of a descending mass of water to raise part of the mass to a higher level.

**hydraulic residence time (HRT)** Hydraulic residence time, time of flow, or time of travel, an important parameter in water quality studies, is the time water spends in a water body. In stormwater management models, the time of concentration is an example of hydraulic residence time. If the volume ($V$) and the flow ($Q$) are known, the hydraulic residence time ($t$) is their ratio:

$$t = v/Q \qquad \text{(H-63)}$$

The time of travel in streams may be computed as the ratio of distance ($x$) to velocity ($V$), or the ratio of length of channel ($L$) to wave celerity ($c$):

$$t = x/V \qquad \text{(H-64)}$$

and

$$t = L/c \qquad \text{(H-65)}$$

In tracer studies the centroid time of travel ($t_c$) is

$$t_c = \Sigma C \cdot t/\Sigma C \qquad \text{(H-66)}$$

where $C$ is the dye or tracer concentration at time $t$. $\Sigma$ indicates the summation at various times and concentrations.

**hydraulics, hydrodynamics, hydrology** These three branches of science and engineering treat the properties and movement of water and other fluids, and the distinction between the three is not always clear, particularly regarding water. They are all common to several fields or professions that involve water resources. Hydraulics and hydrodynamics deal with the motion of water and other fluids. Hydraulics studies the laws governing the motion of water and other liquids as well as their engineering applications. Hydrodynamics (also called hydromechanics) is the branch of physical science (specifically fluid dynamics) that studies the motion of fluids (particularly liquids) and the forces acting on them and on immersed bodies. Hydrology is concerned only with water, not the other fluids, and is in fact the study of the hydrologic cycle. Simply put, hydraulics is the study of fluid flow, whereas hydrology is mainly the study of runoff resulting from precipitation. Another related branch of physical science is fluid mechanics, which deals with the flow of fluids and the way they respond to and exert forces; the subject of hydraulics is often included in college courses on fluid mechanics.

**hydraulic slope** The slope of the hydraulic gradeline; hydraulic gradient.

**hydraulic staging** A procedure to improve the performance of a water treatment device by allowing multiple passes of the flow through the treatment components of such processes as electrodialysis, granular filtration, and pressurized membrane filtration.

**hydraulic structure** Any of a number of engineering works to control the flow, quality, or distribution of water, wastewater, or stormwater, or to maintain water levels in streams and channels. Hydraulic structures may range under the following classifications: collection, conveyance, diversion, energy dissipation, flow control, flow measurement, hydraulic machinery, quality control, river training, sediment control, shore protection, storage, waterway stabilization. Examples of hydraulic structures: breakwater, bridge, canal, conduit, culvert, dam, dike, drain, embankment, flume, gate, groin, headworks, hydraulic ram, infiltration gallery, inlet, intake, jetty, levee, lock, open channel, orifice, outfall, outlet, pier, pipe, pump station, reservoir, revetment, seawall, sewer, sluiceway, spillway, stilling basin, surge tank, tank, treatment plant, turbine, valve, well.

**hydraulic surging** The sudden increase in the filtration rate, e.g., at the end of a filter run or occasionally during short periods. Hydraulic surging may cause a small increase in turbidity but a significant breakthrough of *Giardia* cysts.

**hydraulic transient** A hydraulic phenomenon lasting only a short duration, e.g., water hammer or surge. It is caused by the sudden closure or opening of valves or when pumps start or stop. *See also* pressure surge, pressure transient.

**hydraulic valve** A valve operated by means of a hydraulic device.

**hydrazine ($H_2NNH_2$ or $N_2H_4$)** A fuming, liquid amino acid compound used as an oxygen scavenger in boiler feedwater and cooling water, as a reducing agent for some metals, in the production of explosives, and as a rocket fuel.

**hydride** A compound containing hydrogen and another element, e.g., water ($H_2O$), hydrogen sulfide ($H_2S$), ammonia ($NH_3$), and lithium aluminum hydride ($LiAlH_4$).

**hydriodic acid** A colorless, corrosive liquid, an aqueous solution of hydrogen iodide (HI).

**hydroblast** The use of a hydraulic jet to remove debris.

**Hydro-Brake regulator** A patented device that uses a centrifugal motion to control combined sewer overflows by reducing the rate of discharge into the interceptor and increasing storage in the combined sewer.

**Hydroburst** An air backwash device manufactured by Wheelabrator Engineered Systems, Inc. for use with wastewater screens.

**hydrocarbon (HC)** A chemical compound derived from petroleum, coal tar, and plant sources. Hydrocarbons consist entirely of carbon and hydrogen; they are the simplest organic compounds. The simplest hydrocarbon is methane, $CH_4$. Other examples are benzene ($C_6H_6$), ethane ($C_2H_6$), and acetylene ($C_2H_2$). Hydrocarbons in wastewater are included in the collective term of oil and grease. *See also* alkane, alkene, aromatic hydrocarbon, saturated hydrocarbon, unsaturated hydrocarbon.

**HydroCeal** A mixed-bubble diffuser manufactured by Hydro-Aerobics, Inc.

**hydrocele** The swelling of the scrotum caused by an infection of mosquito-borne worms; a symptom of filariasis. *See also* elephantiasis.

**HydroCell®** Separators manufactured by the U.S. Filter Corp. for use in induced-air flotation units.

**hydrocerussite [$Pb_3(CO_3)_2(OH)_2$(s) or $2PbCO_3 \cdot Pb(OH)_2$(s)]** A form of basic lead carbonate, a compound of lead that resembles calcium carbonate and is often found as scales on water distribution lines made of this metal. *See also* cerussite.

**Hydro-Chek** A coarse-bubble diffuser manufactured by Pollution Control, Inc.

**hydrochloric acid (HCl)** A colorless or faintly yellow aqueous solution of hydrogen chloride that is poisonous, phytotoxic, a strong acid and a highly corrosive agent; also used to regenerate cation resins. It is formed at pH > 4, along with hypochlorous acid (HOCl), by the combination of chlorine ($Cl_2$) with water ($H_2O$):

$$Cl_2 + H_2O \rightarrow HCl + HOCl \qquad \text{(H-67)}$$

Also called muriatic acid, sometimes used in dilute form to remove calcium carbonate scale from water or wastewater treatment components or in well stimulation. At ordinary temperatures, it may be stored in steel tanks or conveyed in PVC piping.

**hydrochloric acid pickling** An operation in which steel products are immersed in hydrochloric acid solutions to chemically remove oxides and scale, and those rinsing operations associated with such immersions (EPA-40CFR420.91-b).

**hydrochloride** A salt formed by hydrochloric acid (HCl) and an organic base.

**hydrochlorination** Hypochlorination; the application of hypochlorite compounds to water for disinfection.

**hydrochlorinator** A chlorine pump, chemical feed pump, or other device used to dispense chlorine solutions made from hypochlorite, such as bleach (sodium hypochlorite) or calcium hypochlorite, into the water being treated. Also called hypochlorinator.

**hydrochlorofluorocarbon (HCFC)** A substance used in refrigeration as a temporary substitute for a chlorofluorocarbon, with lower ozone depletion potential.

**Hydro-circ®** A nonmechanical sludge apparatus designed by Graver Co. for sludge recirculation.

**Hydrocleaner** Dissolved-air flotation equipment manufactured by Envirotech Co.

**Hydro Clear®** A gravity sand filter manufactured by Zimpro Environmental, Inc., including a pulsed bed.

**hydroclone** A device that uses centrifugal force to separate solids from liquids. *See* cyclone separator.

**HydroClor-Q™** A test kit manufactured by Dexsil Corp. for organic chlorine.

**hydrocolloid** A substance that forms a colloid with water.

**Hydro-Cone** An underdrainage system manufactured by Leeds & Northrup for sand filters.

***Hydrocotyle ranunculoides*** A species of floating aquatic plant used in constructed wetlands to treat wastewater. *See* pennywort.

***Hydrocotyle umbellata*** A species of floating aquatic plant used in constructed wetlands to treat wastewater. *See* pennywort.

**hydrocyanic acid (HCN)** A colorless, highly poisonous liquid; an aqueous solution of hydrogen cyanide (HCN). Also called prussic acid. *See also* cyanide.

**hydrocyclone** A conical-shaped device that utilizes centrifugal force to separate grit from organics or to remove water, slime, or grit from primary sludge. *See also* Teacup™.

**hydrocyclone separator** A device that uses centrifugal force to concentrate grit slurries at a pressure between 5 and 20 psi. Separation is often combined with grit washing. *See also* grit classifier, cyclone degritter.

**Hydrodarco®** Activated carbon manufactured by Norit Americas, Inc.

**HydroDoc™** Trickling filter equipment designed or manufactured by WesTech Engineering, Inc. of Salt Lake City, UT.

**HydroDri™** A screenings press manufactured by Conveyor Corp.

**hydrodynamic boundary layer** The stationary layer of liquid immediately surrounding an adsorbing particle. *See* film diffusion transport. Also called surface film.

**hydrodynamic dispersion** A mass transfer phenomenon resulting from the combined effect of molecular diffusion and mechanical dispersion. *See also* diffusion and dispersion.

**hydrodynamic retardation** The phenomenon caused by viscous fluid flow acting on two particles that move close together in a suspension; it tends to slow the motion of the particles and inhibit their collision. *See also* double-layer repulsion and van der Waals force.

**hydrodynamics** The branch of physics that studies the motion of fluids and the forces acting on them, including hydrostatics and hydrokinetics. Also called hydromechanics.

**hydroelectric plant** A power plant using the energy of falling water to turn a turbine generator and produce electricity.

**hydroextraction** A procedure used in the Carver–Greenfield process in conjunction with centrifugation to separate the carrier oil from the sludge solids.

**HydroFlo™** Disposable in-line filters manufactured by Schlicher & Schuell for monitoring groundwater.

**Hydro-Float** An apparatus manufactured by HydroCal Co. for removing grease, oil, and solids by flotation.

**Hydrofloc** A polyelectrolyte made by Aqua Ben Corp. to improve solids separation from liquids.

**HydroFloc™** A thickener manufactured by Klein America, Inc. with a rotating screen.

**Hydrofluor Combo** Reagents produced by Meridian Diagnostics, Inc. to identify *Cryptosporidium* and *Giardia* cysts.

**hydrofluoric acid (HF)** A colorless, fuming, corrosive, poisonous liquid; a strong acid; an aqueous solution of hydrogen fluoride (HF); an impurity or an intermediate dissociation product of fluorosilicic acid. An aqueous form of hydrogen fluoride.

**hydrofluorocarbon (HFC)** A substitute for chlorofluorocarbons and hydrochlorofluorocarbons; it poses no threat to the ozone layer as it contains no chlorine.

**hydrofluorosilicic acid ($H_2SiF_6$)** A strongly acidic chemical used to add approximately 1.0 mg/L of fluoride to drinking water after conventional treatment. *See also* ammonium silicofluoride, sodium fluoride, sodium silicofluoride. Same as fluorosilicic acid.

**Hydroflush** A bar screen manufactured by Beaudrey Corp. and operated by cable.

**hydrogel** A gel with water as liquid constituent..

**hydrogen ($H_2$)** A nonmetallic element occurring in water ($H_2O$), hydrocarbons, and other organic or inorganic compounds. It is a colorless, odorless, flammable gas, and the lightest of known elements. Atomic weight = 1.0079. Atomic number = 1. Density = 0.0899 g/L at standard temperature and pressure. Used in the synthesis of ammonia ($NH_3$), methanol ($CH_3OH$), hydrazine ($N_2H_4$), and other compounds, and in laboratories (e.g., in gas chromatography).

**hydrogen abstraction** An oxidation reaction in which the abstraction of a hydrogen atom from an organic compound leads to the formation of hydroxyl radicals ($^{\cdot}OH$). *See also* radical reaction, ionic reaction.

**hydrogenate** To combine or treat with hydrogen. Also to hydrogenize.

**hydrogenation** The addition of hydrogen gas to an unsaturated organic compound under controlled conditions, e.g., in the production of solid fats from oils or the production of hydrocarbon from coal.

**hydrogen bond (H-bond)** A weak but significant secondary chemical bond where a hydrogen atom has a covalent link with one electronegative atom (e.g., F, N, O) and an electrostatic link with another electronegative atom within the same molecule or another molecule. It occurs, for example, in water and affects adsorption equilibria (in the same energy range as van der Waals interactions), boiling/melting point, and other properties of substances. *See also* covalent bond, polar bond, dipole–dipole interaction, van der Waals forces.

**hydrogen bonding** A characteristic of water such that individual molecules can join through a hydrogen atom. *See* Figure H-04. Through hydrogen bonding water can dissolve alcohols, salts, sugars, and most other substances.

**hydrogen bromide (HBr)** A colorless gas with a pungent odor; the anhydride of hydrobromic acid (HBr).

**hydrogen chloride (HCl)** A colorless, water-soluble, poisonous, corrosive, phytotoxic gas with a choking odor; the anhydride of hydrochloric acid (HCl); used in the production of PVC, ferric chloride, and silicones. It is produced by burning hydrogen and chlorine together. It is also a by-product of the chlorination of organic compounds.

**hydrogen cyanide (HCN)** A colorless, poisonous gas with a bitter almond-like odor; formed by the reaction of sodium cyanide (NaCN) with dilute sulfuric acid ($H_2SO_4$); used for fumigation. It dissolves in water to form hydrocyanic acid (HCN).

**hydrogen cycle** A step in cation exchange that includes the regeneration of the medium and the exchange of cations for hydrogen ions.

**hydrogen-cycle resin** A resin, of the general formula $H_2X$ (where X is a hydrogen exchange radical), that exchanges the hydrogen ion for all cations present in water, producing an acid in the treated water and an exhausted bed. For water containing calcium sulfate:

$$CaSO_4 + H_2X \rightarrow H_2SO_4 + CaX \quad \text{(H-68)}$$

A strong acid, like hydrochloric acid (HCl) is used for bed regeneration. *See also* anionic-exchange resin.

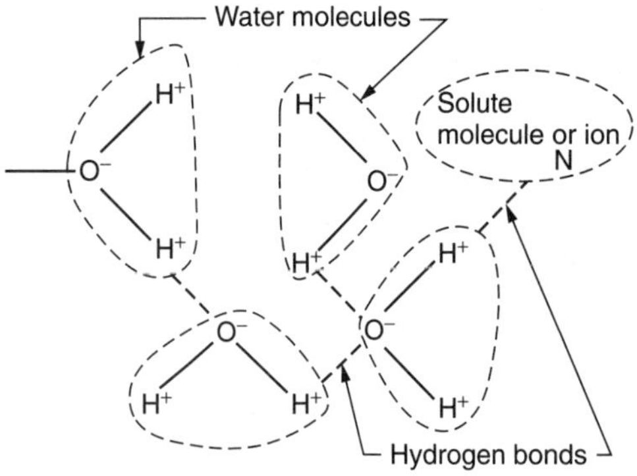


**Figure H-04.** Hydrogen bonding.

**hydrogen electrode** An electrode consisting of a platinum-black surface covered with hydrogen bubbles, used in pH and standard electrode potential measurements. *See* standard hydrogen electrode for detail.

**hydrogenesis** The condensation of air moisture in surface soils or rocks.

**hydrogen exchange** A mechanism used in the cation-exchange process whereby hydrogen is exchanged for a metallic ion, e.g.:

$$H_2R + Ca(HCO_3)_2 \rightarrow CaR + 2H_2O + 2CO_2 \quad \text{(H-69)}$$

where R is a hydrogen cation exchange radical.

**hydrogen fluoride (HF)** A colorless, fuming, toxic, corrosive gas, the anhydride of hydrofluoric acid (HF), used as a catalyst and in hydrocarbon processing.

**hydrogen halide (HX)** An inorganic acidic compound containing a halogen, e.g., hydrochloric acid, hydrofluoric acid.

**hydrogen iodide (HI)** A colorless gas with a suffocating odor, the anhydride of hydriodic acid (HI), used as a catalyst and in hydrocarbon processing.

**hydrogen ion** The single proton, $H^+$ or its hydrated form, $H_3O^+$, the latter being the combination of a water molecule ($H_2O$) and a proton ($H^+$). *See also* hydrated proton, hydronium ion, hydroxonium ion.

**hydrogen-ion concentration** The concentration of hydrogen ions in a solution, usually expressed in moles per liter or as its negative logarithm. *See* pH for detail.

**hydrogen ion-exchange softening** The process of exchanging hydrogen for hardness-causing ions using a strong-base or a weak-base cation exchanger instead of an acid-form resin when a sodium-free water is desirable.

**hydrogenize** *See* hydrogenate.

**hydrogen peroxide ($H_2O_2$ or H—O—O—H)** A clear and colorless, odorless, unstable liquid, a strong oxidizing agent and a weak reductant that is used for bleaching, odor control, dechlorination, disinfection, sludge-bulking control, phenol and cyanide oxidation, and to increase the dissolved oxygen content of groundwater to stimulate aerobic biodegradation of organic contaminants. Hydrogen peroxide is infinitely soluble in water, but rapidly dissociates to form a molecule of water [$H_2O$] and one-half molecule of oxygen [½ $O_2$]:

$$H_2O_2 \rightarrow H_2O + \tfrac{1}{2}\, O_2 \qquad \text{(H-70)}$$

Dissolved oxygen (DO) concentrations of greater than 1,000 mg/L are possible using hydrogen peroxide, but high levels of DO can be toxic to microorganisms. Hydrogen peroxide is used in combination with ozone (peroxone) or with ultraviolet light in advanced oxidation processes, or in solution as a scrubbing fluid in chemical scrubbers to destroy odorous compounds (e.g., hydrogen sulfide, $H_2S$):

$$H_2S + H_2O_2 \rightarrow S^0 \downarrow + 2\, H_2O \qquad \text{with pH} < 8.5 \qquad \text{(H-71)}$$

*See* Fenton's reagent.

**hydrogen peroxide treatment** The enhancement of bioremediation measures by the addition of hydrogen peroxide ($H_2O_2$) to the subsurface as an oxygen source, as indicated by its quick decomposition, which can be mediated by the enzyme, catalase. *See* equation (H-70) above.

**hydrogen peroxide/visible–ultraviolet light process** An advanced oxidation process that produces two hydroxyl radicals from the hydrogen peroxide ($H_2O_2$) using ultraviolet light.

**hydrogen sulfide ($H_2S$)** A dense, colorless, inflammable, toxic, water-soluble gas emitted during anaerobic decomposition of organic matter or produced by the reduction of sulfates and the dissolution of pyrites. Also a by-product of oil refining and burning. Industrial wastes may contain hydrogen sulfide (e.g., from chemical plants, paper mills, tanneries, and textile mills). It smells like rotten eggs. It is explosive and very toxic; in heavy concentration, it can kill or cause illness. Symptoms of exposure to the gas include dizziness, headache, and nausea. It contributes to the corrosion of metal and concrete in water works; *see* black water, crown corrosion or crown rot, pitting. In drinking water, a hydrogen sulfide concentration of 0.05 mg/L is likely to give rise to taste and odor complaints by consumers; it can be removed by aeration, but less effectively than methane ($CH_4$) and carbon dioxide ($CO_2$). In wastewater works, hydrogen sulfide causes odor problems in sewers and at treatment headworks.

**hydrogen sulfide in sewers** *See* crown corrosion.

**hydrogen sulfide speciation** The distribution of the three species of sulfur in wastewater as a function of pH: below pH 6.0, it is mostly hydrogen sulfide ($H_2S$), at pH 7.1 it is about 50% $H_2S$ and 50% $HS^-$, above pH 9.0 it is essentially $HS^-$ and $S^=$, without $H_2S$.

**hydrogen-utilizing methanogen** An organism that uses hydrogen (H) as electron donor and carbon dioxide ($CO_2$) as electron acceptor to produce methane during the anaerobic oxidation of organic waste. *See also* aceticlastic methanogen.

**hydrogeochemistry** The study of the chemical properties of water in a geological context.

**hydrogeologic condition** A condition stemming from the interaction of groundwater with the surrounding soil and rock.

**hydrogeologic cycle** The natural process recycling water from the atmosphere down to and through the earth and back to the atmosphere. *See* hydrologic cycle.

**hydrogeologist** A person who studies and works with groundwater.

**hydrogeology** The geology of groundwater, with particular emphasis on the chemistry and movement of water. Hydrogeology actually concerns all aspects of subsurface water, although engineering applications deal mostly with groundwater movement and distribution. Also called geohydrology. *See also* groundwater hydraulics and groundwater hydrology.

**hydrograph** Basically, a plot of a hydraulic property (usually the discharge) versus time. The ordinate of the graph represents discharge or flow, and the area under the curve represents volume (of runoff, stream flow, etc.) Examples of hydrograph are: (a) A graphical representation of a stream discharge (stream flow) at a single location. (b) A graph showing, for a given point on a stream or conduit, the discharge, stage, velocity or other hydraulic property as a function of time. (c) A time sequence of runoff discharge versus duration time of a storm. (d) A unit hydrograph for stormwater runoff analysis. (e) A graph of average or peak wastewater flows, during dry or wet weather, over the 24 hours of the day (called diurnal hydrograph). (f) Flood and stage hydrographs showing the variation of a flood discharge ($Q$) and elevation ($h$) with time, according to a stage–discharge relationship or rating curve. A common form of rating curve is

$$Q = a \cdot h_b \qquad \text{(H-72)}$$

where a and b are constants for a given stream location. (g) An inflow hydrograph. *See* clean hydrograph, net hydrograph, stage–discharge relationship, and unit hydrograph.

**hydrograph-controlled-release (HCR) pond** A wastewater treatment pond (usually a facultative pond) that discharges the effluent based on a hydrograph correlated to the level or flow of the receiving stream and only when stream flow is above a specified limit. Regarding effluent discharge, there are also continuous-discharge ponds, controlled-discharge ponds, and total-containment ponds. *See also* lagoons and ponds.

**Hydrogritter** An apparatus manufactured by Envirotech Co. for washing grit.

**HydroGuard™** A device used for the automatic flushing of water distribution systems.

**Hydro-Lift** Steel lift stations prefabricated by Hydro-Aerobics, Inc.

**hydroliquefaction** A process used for the complete destruction of wastewater sludge while producing some oil. *See also* Tubingen process, pyrolytic water, sludge-to-oil process.

**Hydro-lite** Medium fabricated by Hydro-Aerobics, Inc. for biological filters.

**hydrological cycle** Same as hydrologic cycle.

**hydrologic budget** An accounting of all inflows to and outflows from a hydrologic unit. *See* water budget for detail.

**hydrologic cycle** Movement or exchange of water between the atmosphere and the earth. Also called water cycle. The hydrogeologic cycle is similarly defined: the natural process recycling water from the atmosphere down to (and through) the earth and back to the atmosphere again. The hydrologic cycle comprises the unending processes controlling the distribution and movement of water on the Earth's surface, in the soil, and in the atmosphere: evaporation from the oceans and the earth, transport over the land masses, condensation of the water vapor, fog or cloud formation, precipitation, evapotranspiration, depression storage, infiltration, percolation, runoff ultimately to the oceans, etc. Evaporation and transpiration account for two-thirds of the precipitation over land, and runoff to oceans for the remaining third. *See also* rainfall–runoff relationships and Figure H-05.

**Hydrologic Evaluation of Landfill Performance (HELP) model** A two-dimensional model developed by the USEPA to estimate surface runoff, subsurface drainage, and leachate from active and closed landfills, using hydrologic processes.

**hydrologic unit** A geographic area representing a surface drainage basin or a distinct hydrologic feature.

**hydrology** The science that treats the waters of the earth, their occurrence, circulation, and distribution, their chemical and physical properties, and their reaction with their environment, including their relation to living things. *See also* hydraulics, hydrologic cycle, hydrometeorology, groundwater hydrology, and hydrography.

**hydrolysate** Any compound formed by hydrolysis.

**hydrolysis** (1) A chemical reaction in which a compound is converted into another compound by taking up water. It is one of the reactions that occur during water treatment by coagulation; for aluminum sulfate (alum), the complex hydrolysis reactions are simplified as follows:

$$Al_2(SO_4)_3 + 6\ H_2O \rightarrow 2\ Al(OH)_3 + 3\ H_2SO_4 \quad \text{(H-73)}$$

Hydrolysis is also explained by the strong bonds formed in aqueous solutions between highly posi-

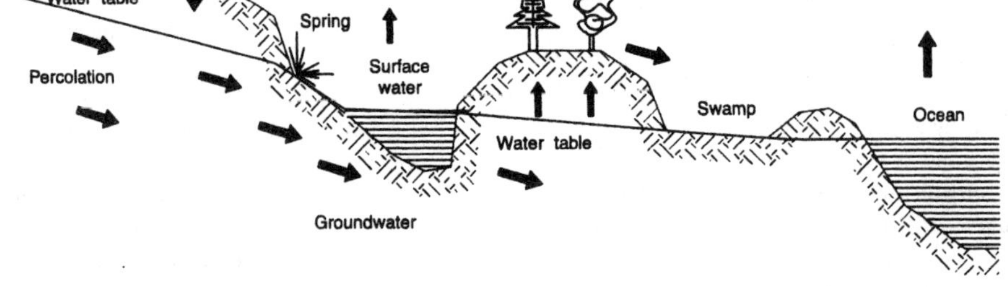

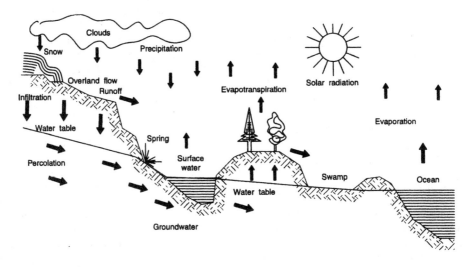


**Figure H-05.** Hydrologic cycle (courtesy of CRC Press).

tive metal ions (e.g., $Al^{3+}$ and $Fe^{3+}$) and the oxygen atoms of six water molecules. *See* aquo aluminum ion deprotonation. (2) The formation of an acid and a base from a salt by the ionic dissociation of water, or the formation of alcohols and acids from esters. For example, (a) hydrochloric acid (HCl) and caustic soda (NaOH) from sodium chloride (NaCl) and (b) ethyl alcohol ($C_2H_5OH$) and acetic acid ($CH_3COOH$) from ethyl acetate ($CH_3COOC_2H_5$):

$$NaCl + H_2O \rightarrow HCl + NaOH \quad \text{(H-74)}$$

$$CH_3COOC_2H_5 + H_2O \rightarrow C_2H_5OH + CH_3COOH \quad \text{(H-75)}$$

(3) The decomposition or liquefaction of solid matter as a result of biological activity. It is the first stage in the simplified representation of the anaerobic digestion process (followed by acid formation and methanogenesis), whereby hydrolytic organisms convert complex organics such as proteins, cellulose, and lipids to soluble substances. (4) A reaction in which a water molecule ($H_2O$) or the hydroxyl ion ($OH^-$) replaces an atom or group of atoms in an organic molecule; "a nucleophilic substitution at a saturated carbon atom" or the reaction of an organic electrophile with a nucleophile (Stumm and Morgan, 1996). For example,

$$R{-}X + H_2O \rightarrow R{-}OH + X^- + H^+ \quad \text{(H-76)}$$

and

$$RCH_2{-}X + HS^- \rightarrow RCH_2{-}SH + X^- \quad \text{(H-77)}$$

with X = Br, Cl, or I. (5) A method of treatment of inorganic and organic chemicals, whereby they are made to react with water and to form safe or less toxic products. For example, by hydrolysis, silicon tetrachloride ($SiCl_4$) is converted to silica ($SiO_2$) and hydrochloric acid (HCl), and toxic acetic anhydride ($CH_3{-}CO{-}O{-}CO{-}CH_3$) is converted to acetic acid ($CH_3{-}CO{-}OH$):

$$SiCl_4 + 2\,H_2O \rightarrow SiO_2 + 4\,HCl \quad \text{(H-78)}$$

$$CH_3{-}CO{-}O{-}CO{-}CH_3 + H_2O \rightarrow 2\,CH_3{-}CO{-}OH \quad \text{(H-79)}$$

(6) Weathering caused by the chemical alteration of certain minerals as they react with water, e.g., the formation of kaolin (white clay or china clay) from the hydrolysis of feldspar in granite.

**hydrolysis product** Any of the various metallic species formed during the hydrolysis of metal salts used in coagulation and responsible for particle aggregation. *See also* aluminum hydrolysis product, amphoteric compound, coordination compound, ligand.

**hydrolysis reaction** The reaction of a salt constituent with water. *See also* hydrolysis, protolysis.

**hydrolysis reaction pathway** One of the possible reactions occurring during the hydrolysis of metal coagulants. Floc formation results from the sorption of natural organic matter or mineral particles, and from precipitation and/or stabilization processes. Hydrolysis may also result in the formation of soluble metal species.

**hydrolytic bacteria** One of two main groups of bacteria that carry out anaerobic digestion; they are saprophytic bacteria that reproduce rapidly and convert complex organic substances into simpler compounds. *See* acid-forming bacteria.

**hydrolytic tank** A vertical-flow vessel in which the organic matter in wastewater or sludge undergoes biochemical degradation, including liquefaction and gas production; a digester.

**hydrolyzing enzyme** An enzyme that brings about the breakdown of biological compounds by the addition of water, e.g., a gastric enzyme for the breakdown of proteins (gastric proteinase) or a pancreatic enzyme for lipids (pancreatic lipase).

**hydrolyzing metal salt (HMS) coagulants** Metal salts, particularly the sulfates and chlorides of aluminum and iron, used in water treatment. The metal ions form strong bonds with the oxygen atoms of surrounding water molecules, release hydrogen atoms to the aqueous solution, and form metal hydroxides.

**Hydromaster** A compact water treatment plant manufactured by Gladwall Engineering Services, Ltd.

**hydrometallurgy** A metal extraction process using liquid solvents for ore leaching at ordinary temperatures, including the separation or recovery of heavy or noble metals by ion exchange.

**hydrometer** *See* hydrometry.

**hydrometry** (1) The determination of the specific gravity of liquids, using a hydrometer, i.e., a graduated tube weighted to float upright in the liquid. (2) The measurement and analysis of the flow of water.

**hydronium ion ($H_3O^+$)** The hydrated hydrogen ion, a positively charged combination of a proton with a water molecule. Also called hydroxonium ion.

**Hydroperm®** A microfiltration unit manufactured by Wheelabrator Engineered Systems, Inc., with crossflow operation.

**hydroperoxyl radical ($HO_2$)** An intermediate product in the chain of reactions of ozone autodecomposition:

$$OH^- + O_3 \rightarrow HO_2 + O_2^- \qquad \text{(H-80)}$$

*See also* odd hydrogen, superoxide ion.

**hydrophilic** Having a strong affinity for water or capable of dissolving in water; soluble or miscible in water; the opposite of hydrophobic.

**hydrophilic acid** Any dissolved organic acid that is not retained by nonionic resins, including single aliphatic acids, uronic acids, and polyuronic acids.

**hydrophilic colloid** A large-molecule, stable, water-loving colloidal particle; a particle attracted by or to water, e.g., starch, protein, soap, synthetic detergent, blood serum, or a similar substance that is stabilized by a layer of water. Biocolloids (bacteria and viruses) are hydrophilic. *See also* association colloid, hydrophobic colloid.

**hydrophilic radical** A radical, such as hydroxyl (OH) and acetate (COOH), that has a strong affinity for water, a property that may be used in the control of evaporation.

**hydrophobic** Having a strong aversion for water; tending not to combine with water, or incapable of dissolving in water. A property exhibited by nonpolar organic compounds, including the petroleum hydrocarbons; the opposite of hydrophilic.

**hydrophobic binding** A sorption mechanism in which humus sorbs nonpolar solutes from a soil solution. *See* hydrophobic bonding.

**hydrophobic bonding** The phenomenon observed in aqueous solutions in which nonpolar organic molecules readily adsorb to any available solid surface, thereby resulting in adsorption rates higher than those associated with surface reactions alone. Hydrophobic bonding results from the strong attraction between $H_2O$ molecules, not because of the attraction between nonpolar groups. Also called hydrophobic effect or hydrophobic sorption. *See also* chemical adsorption or chemisorption, desorption, electrostatic or exchange adsorption, micelle, physical adsorption, polarity, solution force, solvophobic force, van der Waals attraction.

**hydrophobic colloid** (1) A colloidal particle that has relatively little attraction for water; stable because of its positive or negative charge; e.g., clay particles or petroleum droplets. *See also* association colloid, hydrophilic colloid. (2) A charged particle, e.g., metal or salt, that repels a similar particle, thus causing their stability in water.

**hydrophobic compound** A compound that does not readily dissolve in water (e.g., a hydrocarbon), but may be soluble in nonpolar solvents. *See also* lipophile, octanol–water partition coefficient.

**hydrophobic effect** *See* hydrophobic bonding.

**hydrophobic interaction** The tendency of nonpolar groups to associate in an aqueous environment, not because of high affinity with one another, but because of the strong affinity between polar water molecules. *See also* bacterial adherence to hydrocarbon test.

**hydrophobic particles** *See* hydrophobic colloids.

**hydrophobic sorption** *See* hydrophobic bonding.

**hydrophobic substance** *See* hydrophobic compound.

**hydrophone** An underwater microphone that is used to detect leaks in underground water lines, based on the acoustic signal produced by the leakage. The hydrophone was used originally to locate enemy submarines.

**hydrophyte** (1) A plant that grows in water or saturated soil; a hygrophyte. (2) A herbaceous plant whose surviving parts lie in water.

**Hydropillar™** An elevated tank manufactured by Pitt-Des Moines, Inc. for water storage.

**hydropneumatic** A small water system, in which a pump is automatically controlled by the air pressure in a compressed air tank. Same as hydropneumatic tank.

**hydropneumatic system** A system that uses air and water, e.g., a compressed-air tank to control the operation of a water pump.

**hydropneumatic tank** A compressed-air tank that stores liquids under pressure, thus using pressure head instead of elevation head; usually installed in small water or wastewater systems because of its high capital cost. Also called hydropneumatic.

**hydropower** Electrical energy produced by falling water. *See also* hydroelectric plant.

**Hydropress** A belt filter press manufactured by Clow Corp.

**Hydro-Press** A belt filter press manufactured by HydroCal Co.

**HydroPunch®** A device manufactured by QED Environmental Systems for groundwater sampling.

**hydropyrolysis** The thermal degradation of organic matter in the presence of hydrogen. *See also* oxidative pyrolysis.

**Hydrorake** A trash rake manufactured by Hercules Division/Atlas Polar Co.

**HydroRanger** An ultrasonic device manufactured by Milltronics for level measurement.

**Hydro-Rotor** A brush aerator manufactured by Amwell, Inc.

**Hydro-SAFe** A biological filter manufactured by Hydro-Aerobics, Inc.

**hydroscopic** Sometimes used for hygroscopic.

**Hydro Scour** A backwashing apparatus manufactured by Zimpro Environmental, Inc. for pulsed bed filters.

**Hydroscreen** A static fine screen manufactured by Hycor Corp.

**HydroSeal®** A cover manufactured by Eimco Process Equipment Co. for anaerobic digestion gas holders.

**Hydrosep** Grit removal equipment manufactured by Aerators, Inc. for installation in shallow basins.

**Hydroseparator** A gravity thickener manufactured by Dorr-Oliver, Inc. for use in industrial wastewater treatment.

**Hydroshear** An aeration tank manufactured by the FMC Corp. used in package treatment plants with low flows.

**Hydro-Shear** A rotary fine screen manufactured by Dontech, Inc. with internal feed.

**Hydrosil** A static screen manufactured by JDV Equipment Corp., including a cleaning apparatus that uses brushes and water jets.

**Hydro-Sock** An upflow cartridge filter manufactured by Hydro-Aerobics, Inc.

**hydrosphere** (1) The portion of the surface of the Earth that is covered with water, i.e., oceans, seas, lakes, streams, rivers, swamps, glaciers, and ice caps. *See also* atmosphere, lithosphere, biosphere, geosphere. (2) Sometimes extended to mean the entire aqueous environment of the Earth, including also water in the atmosphere and underground.

**hydrostatic approximation** *See* quasihydrostatic approximation.

**hydrostatic equation** (1) The equation expressing hydrostatic pressure ($P$) as the product of the fluid density ($\rho$) by its depth ($y$) and the acceleration of gravity ($g$):

$$P = \rho g y \tag{H-81}$$

Also called hydrostatic law or basic hydrostatic equation. (2) As applied to the atmosphere, the vertical component of the equation of motion, considering only the forces of vertical pressure and gravity:

$$\partial P / \partial z = -\rho g \tag{H-82}$$

where $z$ is the geometric height.

**hydrostatic equilibrium** The condition of a fluid with a complete balance between the forces of pressure and gravity; the surfaces of constant pressure and constant mass are horizontal and coincident. *See also* hydrostatic equation, quasihydrostatic approximation.

**hydrostatic head** The height of a column of water that would produce a given pressure; often simply called head.

**hydrostatic joint** A bell-and-spigot joint made of lead placed into the bell under hydraulic pressure.

**hydrostatic law** *See* hydrostatic equation.

**hydrostatic level** The level to which water from an artesian aquifer or from a conduit under pressure would rise in an open tube. *See also* hydraulic gradeline.

**hydrostatic press** A hydrostatic machine for magnifying an applied force. *See* Brahma's press for detail.

**hydrostatic pressure** The pressure exerted by a body of water due to depth alone, or by the weight of groundwater at higher levels in the same saturation zone.

**hydrostatic sludge removal** A method of sludge removal from settling tanks using the hydrostatic pressure above the sludge outlet.

**hydrostatic test** *See* hydrotest.

**hydrosulfite ($Na_2S_2O_4 \cdot 2\ H_2O$)** A sodium salt used in the reduction of chlorine residuals in water or wastewater. *See also* the following sodium salts: bisulfite, hydrosulfite, hyposulfite, metabisulfite, sulfite, thiosulfate.

**hydrotest** A test conducted under pressure to verify the integrity of tanks, piping, etc. containing or conveying water; the vessel or pipe is filled with water, purged of air, sealed, and observed for leaks, distortion, or failure. Also called hydrostatic test.

**hydrothermal** Pertaining to the action of hot, aqueous solutions or gases within or on the surface of the Earth; applied to geological processes that involve superheated water, specifically alteration and deposition.

**hydrothermal oxidation** A treatment process that oxidizes organic materials in wastewater sludge and hazardous wastes at high temperature and pressure above the fluid's critical point. *See* supercritical water oxidation for more detail.

**hydrothermal vents** Cracks and fissures on the basaltic ocean floor through which flows mineral-rich hot water driven by hydrothermal energy from magma and rich in metal sulfides and hydrogen sulfide. They include warm vents having exit temperatures of 5–25°C, intermediate vents (white smokers) with temperatures up to 300°C, and hot vents (black smokers) up to 400°C. Very productive areas surround these vents, despite the lack of direct photosynthetic food sources. *See also* photic zone, mesopelagic zone, bathypelagic zone, benthic region.

**hydrous bulking** An operational problem of activated sludge plants in which an excess of extracellular hydrophilic biopolymer produces a slimy, jellylike sludge that has a low density and does not settle well. Hydrous bulking is associated with nutrient deficiencies and very high organic loading. Also called viscous bulking

**Hydrovex™** Flow control products manufactured by John Meunier, Inc.

**hydroviscous drive** One of two common variable-torque transmission systems used to drive wastewater pumps at varying speeds. It transmits energy through the strength of its oil. Also called liquid clutch. *See also* eddy current clutch.

**Hydrowash** A recycling pump and aerator apparatus manufactured by Amwell, Inc. for grit removal.

**hydroxide** A chemical compound containing the hydroxyl group (OH), e.g., sodium hydroxide (NaOH) and calcium hydroxide $[Ca(OH)_2]$. Also, a negatively charged ion consisting of a hydrogen atom and an oxygen atom; *see* hydroxide ion, hydroxyl ion.

**hydroxide alkalinity** Alkalinity caused by the hydroxyl ion ($OH^-$). Sometimes called hydroxyl alkalinity. *See* caustic alkalinity for detail.

**hydroxide ion ($OH^-$)** The negatively charged ion formed by the combination of one atom of oxygen and one atom of hydrogen and responsible for the hydroxide alkalinity of a solution. It corresponds to the hydroxyl radical.

**hydroxonium ion** *See* hydronium ion.

**hydroxyapatite $[Ca_5(OH)(PO_4)_3]$ or $[Ca_{10}(OH)_2(PO_4)_6]$** A compound precipitated during phosphorus removal from wastewater; it is formed by the reaction between lime, phosphate, and water:

$$5\ CaO + 5\ H_2O + 3\ PO_4^{3-} \rightarrow Ca_5(OH)(PO_4)_3 + 9\ OH^- \quad \text{(H-83)}$$

It is the principal form of storage of phosphorus and calcium in bones. Also spelled hydroxylapatite. *See also* fluorapatite. Hydroxyapatite can be used to remove fluoride by selective ion exchange; it is converted into fluoroapatite and $OH^-$ ions:

$$Ca_{10}(PO_4)_6(OH)_2 + 2\ F^- = [Ca_{10}(PO_4)_6F_2] + 2\ OH^- \quad \text{(H-84)}$$

and also, in the alkaline pH range, $Ca^{++}$ ions can precipitate phosphate:

$$Ca_{10}(PO_4)_6(OH)_2(s) = 10\ Ca^{++} + 6\ PO_4^{3-} + 2\ OH^- \quad \text{(H-85)}$$

**hydroxybenzene** Same as phenol.

**hydroxyl** Containing the hydroxyl group (OH).

**hydroxyl alkalinity** Same as hydroxide alkalinity, i.e., the alkalinity caused by hydroxyl ions ($OH^-$). *See* caustic alkalinity for detail.

**hydroxylamine ($NH_3O$)** An unstable, weakly basic, crystalline compound used as a reducing agent or laboratory reagent, e.g., in the phenanthroline method of iron and manganese analysis.

**hydroxylapatite** *See* hydroxyapatite.

**hydroxylation** The attachment of —OH groups to hydrocarbon chains or rings.

**hydroxyl-cycle resin** Same as anionic-exchange resin.

**hydroxyl group (—OH)** A univalent group consisting of one hydrogen atom and one oxygen atom, as in the hydroxide ion ($OH^-$); it is commonly found in many inorganic compounds and in alcohols.

**hydroxyl radical ($^{\bullet}OH$ or $OH^{\bullet}$)** A strong oxidizing agent for organic compounds, used in advanced oxidation processes in water and wastewater treatment. It is an intermediary product of the decomposition of ozone and an important factor in ozone's oxidative properties. It can also be produced by the ultraviolet (UV) irradiation of hydrogen peroxide ($H_2O_2$):

$$H_2O_2 + UV \rightarrow 2\ OH \quad \text{(H-86)}$$

*See also* odd hydrogen.

**hydroxyl radical-based process** A process in which a pH increase or a reaction between ozone and an oxidant (e.g., hydrogen peroxide or UV light) promotes the formation of the hydroxyl radical, a highly reactive agent. It is used to remove organic contaminants from drinking water or synthetic organic chemicals from wastewater. *See also* advanced oxidation process.

**hydroxypyromorphite** A solid compound that is formed when phosphates are used to control lead corrosion; as a protective scale layer on pipe walls, it is less soluble than lead carbonate.

**Hydro-Zap®** An apparatus manufactured by Hydro-Aerobics, Inc. for wastewater disinfection using ultraviolet light.

**Hydrozon®** An ozone system manufactured by Carus Chemical Co.

**hydrozone** An area of landscape containing plants with similar water and irrigation needs.

**Hydrymax** Sludge drying equipment manufactured by D. R. Sperry & Co.

**HY-FLEX®** An internal pipe joint seal fabricated by Miller Pipeline Corp. of Indianapolis, IN for self-installation by contractors.

**Hy Flo Super-Cel®** Diatomaceous earth media fabricated by Celite Corp. for use in filtration units.

**Hygene** Bacteriostatic filter media manufactured by Ionics, Inc.

**hygiene education** Same as health education.

**hygrophyte** A plant that thrives in water, wet ground or moist ground and is sensitive to dry conditions; an aquatic plant. Also called hydrophyte.

**hygroscopic** Pertaining to a substance, such as sodium chloride (NaCl), that tends to absorb moisture from air; (the prefix hygro indicates moisture). Similarly, hydroscopic is sometimes used to indicate a substance that readily absorbs or attracts water.

**hygroscopic moisture** Same as hygroscopic water.

**hygroscopic nuclei** Dust or other atmospheric particles around which water condenses into tiny droplets that collide and coalesce to form a raindrop.

**hygroscopic water** A thin film of moisture held in the soil by surface tension forces. It cannot evaporate or drain by gravity because of its equilibrium with atmospheric water vapor, and is not available to plants.

**hymenolepiasis** An excreted helminth infection caused by the pathogen *Hymenolepis nana. See* its common name, dwarf tapeworm, for more detail.

***Hymenolepis diminuta*** A common parasite of rodents; rat tapeworm, occasionally causing infections in children.

***Hymenolepis nana*** The helminth species that causes dwarf tapeworm infections; distributed worldwide, with humans and rodents as reservoirs. Persistence = 1 month. Median infective dose = 100.

**Hymergible®** A hydraulically driven submersible pump manufactured by Crane Pumps and Systems, Inc.

**hyperfiltration (HF)** A treatment process that uses a very dense semipermeable membrane to separate water from ionic or organic impurities of molecular mass 100–500. It is similar to ultrafiltration, except for a different range of molecular masses; it retains dissolved and suspended materials on the basis of size, shape, and molecular flexibility. Sometimes used synonymously with reverse osmosis at a pressure of 800–1200 psi.

**Hyper-Filtration™** A proprietary design of Filtronics, Inc. of Anaheim, CA used in the EM-Pure™ membranes.

**HyperFlex®** A high-density polyethylene liner manufactured by SLT North America, Inc.

**Hyperfloc®** A polyelectrolyte produced by Hychem, Inc. to enhance solids/liquid separation.

**Hyper+Ion™** *See* Hyper Plus Ion™.

**hyperkalemia** A disease caused by an excess of potassium, e.g., a dosage of 10 grams per day.

**hyperkinetic syndrome** A hyperactivity disorder in children, often associated with excessive exposure to heavy metals (e.g., lead in drinking water).

**HyperNet™** A fabric composite manufactured by SLT North America, Inc.

**hyperparasite** A parasite that feeds on other parasites. Also called a superparasite.

**Hyper Plus Ion™** or **Hyper+Ion™** A cationic coagulant produced by General Chemical Corp.

**Hyperpress™** A combined belt filter plate and frame press manufactured by Klein.

**Hypersperse®** An antiscalant produced by Argo Scientific for use with reverse osmosis units.

**hypertension** Elevation of the blood pressure, which may require a sodium-restricted diet for parts of the population, including any contribution from drinking water.

**hyphae** (1) Plural of hypha; the threadlike elements of the mycelium of a fungus. (2) A similar element of the elongated cells of actinomycetes.

**hypo** A common name for sodium hyposulfite or sodium thiosulfate.

**hypobromite** Same as hypobromite ion.

**hypobromite ion ($OBr^-$)** The ionized form of hypobromous acid (HOBr); it forms bromate ($BrO_3^-$) during water disinfection by ozone.

**hypobromous acid (HOBr)** A bromine species that is a disinfectant but tends to dissociate into hypobromite ions ($OBr^-$). In aqueous solutions of neutral pH, it is the principal bromine species, formed by the combination of bromine gas with water, but ionizing to the hypobromite and hydrogen ions:

$$Br_2 + H_2O \rightarrow HOBr + HBr \qquad \text{(H-87)}$$

$$HOBr \rightarrow H^+ + OBr^- \qquad \text{(H-88)}$$

Hypobromous acid may form disinfection by-products with natural organic matter.

**hypochlorhydria** A lower than normal gastric acidity, a condition that may lower the infective doses of cholera.

**hypochlorination** Water disinfection using sodium hypochlorite or calcium hypochlorite, usually under emergency conditions and in small plants. Also called hydrochlorination.

**hypochlorinator** A chlorine pump, chemical feed pump, or other device used to dispense chlorine solutions made from hypochlorites such as bleach (sodium hypochlorite) or calcium hypochlorite into the water being treated. Also called hydrochlorinator.

**hypochlorite** (1) A chemical compound containing available chlorine, e.g., calcium hypochlorite and sodium hypochlorite [$Ca(OCl)_2$ and NaOCl]; used for disinfection (hypochlorination) and available as a liquid (bleach) or solid (granules, powder, and pellets). A salt of hypochlorous acid. *See also* high-test hypochlorite. (2) A short name for the hypochlorite ion ($OCl^-$).

**hypochlorite ion ($OCl^-$)** The ionized form of hypochlorous acid (HOCl) or the product of the dissociation of calcium hypochlorite or sodium hypochlorite when they are added to water:

$$Ca(OCl)_2 \rightarrow Ca^{2+} + 2\ OCl^- \qquad \text{(H-89)}$$

$$NaOCl \rightarrow Na^+ + OCl^- \qquad \text{(H-90)}$$

**hypochlorite of lime** A combination of hydrated lime and chlorine having calcium oxychloride as active ingredient for disinfection. *See* chlorinated lime for detail.

**hypochlorite reactions** Both calcium hypochlorite [$Ca(OCl)_2$] and sodium hypochlorite (NaOCl) hydrolyze to form hypochlorous acid (HOCl) and lime [$Ca(OH)_2$] or caustic soda (NaOH):

$$Ca(OCl)_2 + 2\ H_2O \rightarrow 2\ HOCl + Ca(OH)_2 \qquad \text{(H-91)}$$

$$NaOCl + H_2O \rightarrow HOCl + NaOH \qquad \text{(H-92)}$$

**hypochlorite solution** An aqueous solution of a salt of hypochlorous acid (HOCl), a chlorine-based disinfectant used in water treatment plants, particularly in small or remote towns. Some proprietary disinfectants and bleaches can also be used, e.g., Chloros, Chlorox, Domestos, Javel water, Milton, Parazone, Voxsan, Zonite.

**hypochlorous acid (HOCl)** A chlorine species that is an effective disinfectant but tends to dissociate into chlorite ions at high pH. It is a weak acid, formed by the combination of chlorine gas with water, but ionizes to the hypochlorite and hydrogen ions:

$$Cl_2 + H_2O \rightarrow HOCl + HCl \qquad \text{(H-93)}$$

$$HOCl \rightarrow H^+ + OCl^- \qquad \text{(H-94)}$$

It is also used as a bleaching and oxidizing agent. It forms disinfection by-products with natural organic matter in water containing bromide. *See* free residual and the Morris correlation.

**hypochlorous acid ionization** The decomposition of hypochlorous acid (HOCl) into hypochlorite ion ($OCl^-$) and hydrogen ion ($H^+$).

**Hypo-Gen®** A unit manufactured by Capital Control Co., Inc. for the production of sodium hypochlorite.

**hypoiodous acid (HOI)** One of the forms of iodine in aqueous solution, with properties similar to those of hypobromous acid or hypochlorous acid.

**hypokalemia** Potassium deficiency; it may be caused by an adult daily intake of less than 2 grams.

**hypolimnetic aeration** Aeration of the bottom layer (hypolimnion) of a lake.

**hypolimnion** The lowest layer in a thermally stratified body of water. It consists of colder, more dense water, has a constant temperature and no mixing occurs; water is almost stagnant. The hypolimnion of a eutrophic lake is usually low or lacking in oxygen. Also called the stagnation zone. *See* thermal stratification.

**hypolimnion zone** The hypolimnion.

***Hypomicrobium* genus** A group of heterotrophic organisms active in biological denitrification.

***Hypophthalmichthys molitrix*** A hardy fish species of the carp family that survives well in stabilization ponds used for aquaculture.

**hyposulfite** Another name for sodium thiosulfate or sodium hyposulfite.

**hypoxanthine guanine phosphoribosyl transferase (HGPRT) test** A bioassay using a special enzyme to indicate the mutations caused by a substance in drinking water. *See also* the Ames test.

**hypoxemia** Inadequate oxygenation in the blood, due to a lack of oxygen for breathing, interference of chemicals (carbon monoxide, nitrate, chloride), or other causes. *See also* hypoxia.

**hypoxia** (1) The condition of a water that has a very low dissolved oxygen concentration, e.g., less than 2.0 mg/L. (2) Lack of oxygen in the blood; same as hypoxemia.

**hypoxic** A condition of low oxygen concentration, below that considered aerobic.

**Hypress** A press manufactured by Hycor Corp. for dewatering screenings from preliminary wastewater treatment.

**Hy-Q** A sluice gate closure manufactured by the Rodney Hunt Co. with a flush bottom.

**Hysep®** A decanter centrifuge manufactured by Centrico, Inc.

**Hy-Speed®** Mixing equipment manufactured by Alsop Engineering Co.

**Hytrex®** Cartridge prefilter systems manufactured by Osmonics, Inc.

**Hyveyor** A conveyor manufactured by Hycor Corp.

# I

**$I_2$, $I^-$** Chemical formula of iodine, iodide.
**I/A** Acronym of innovative and alternative.
**IAF** Acronym of induced-air filtration.
**IAST** Acronym of ideal adsorbed solution theory.
**IC** Acronym of inhibiting concentration.
**I&C** Acronym of instrument and control.
**ICA** Acronym of inclined cascade aeration.
**ICC** Acronym of intrachannel clarifier.
**ICEAS™** Acronym of intermittent cycle extended aeration system, a wastewater treatment process developed by Austgen Biojet Wastewater Systems.
**Ice-away®** Equipment manufactured by Air-O-Lator Corp. for melting ice.
**ichthyology** The branch of zoology that studies fishes. *See* intolerant, tolerant, game, and rough fishes.
**ICPAES** Acronym of inductively coupled plasma atomic emission spectroscopy.
**ICR** Acronym of Information Collection Rule.
**icterus** A health condition caused by an excess of bile pigments and a yellow surface skin coloration. *See also* jaundice.
**ID** Acronym of inside diameter or infectious dose.
**$ID_{50}$** *See* infectious dose 50.
**IDDF** Acronym of integrated disinfection design framework.
**ideal adsorbed solution theory (IAST)** A model used to describe competitive adsorption of contaminants from aqueous solutions, based on any of the common single-solute adsorption isotherms and on a limited number of fictive components to reduce the number of different compounds in natural organic matter (AWWA, 1991):

$$C_{i,0} - Cq_i - X_iY_i = 0 \qquad \text{(I-01)}$$

where $C_{i,0}$ = initial liquid-phase concentration of compound $i$, $C$ = carbon dose, and $q_i$ = equilibrium solid-phase concentration of compound $i$.

$$X_i = q_i/\Sigma q_i \qquad \text{(I-02)}$$

$\Sigma$ indicates summation from $i = 1$ to $N$.

$$Y_i = [(\Sigma n_iq_i)/n_iK_i]^a \qquad \text{with } a = n_i \qquad \text{(I-03)}$$

$N$ = number of components in the solution, and $n_i$, $K_i$ = single-solute Freundlich isotherm parameters for compound $I$. *See also* equivalent background compound.
**ideal adsorption** Same as physical adsorption.
**ideal combustion** A process in which all consumable elements are burned or oxidized in accordance with the law of energy conservation. Also called stoichiometric combustion.

**ideal fluid** A fictitious fluid that would be inviscid (i.e., without viscosity) and incompressible, and would have uniform density and no surface tension. The assumption of ideal flow is convenient and produces results that are fairly accurate for actual fluids. Also called perfect fluid.

**ideal gas** A gas whose molecules are subject only to the forces of collision with one another and with the walls of its container; this gas obeys the ideal gas law. One mole of an ideal gas occupies a volume of 22.414 liters at standard temperature (0°C) and pressure (1.0 atm). Also called a perfect gas.

**ideal gas equation** The mathematical expression of the ideal gas law. Also called general equation of state, ideal gas equation of state, or, simply, equation of state.

**ideal gas equation of state** Same as ideal gas equation.

**ideal gas law** The principle that results from the combination of Avogadro's law, Boyles's law, and Charles' law; the product of the pressure ($P$) and volume ($V$) of one gram molecule of an ideal gas equals the product of its absolute temperature ($T$, 0°K) and the universal gas constant ($R$). For a number of moles ($n$),

$$PV = nRT \quad \text{(I-04)}$$

This equation can be combined with Dalton's law to compute the concentration of a gas when its partial pressure in a mixture is known. Also called general gas law, perfect gas law, universal gas law, or sometimes, simply, gas law.

**ideal indicator organism** An organism that is found in the intestinal microflora of warm-blooded animals and that is present in a fecally contaminated sample, with appropriate characteristics relative to the target pathogens (e.g., survival, reproduction outside the host, speed of isolation and quantification). *See also* surrogate organism.

**ideal sedimentation basin** A concept used for the theoretical design of sedimentation units where solid particles have a horizontal velocity equal to the velocity of the fluid (the ratio of flow to surface area) and a vertical velocity equal to its terminal velocity as determined from Stokes' law. Particles with a resultant velocity such that they reach the bottom before the outlet zone are removed. *See* Figure I-01.

**ideal settling basin efficiency** Same as Hazen's settling equation.

**idiosyncratic reaction** An uncommon, genetically based reaction of an individual to exposure to a chemical or biological agent.

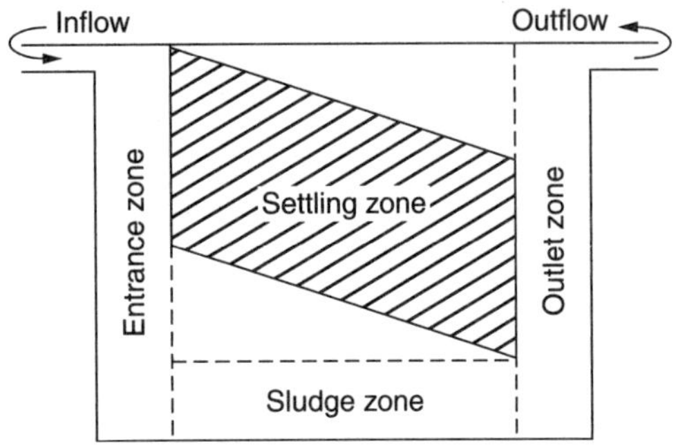


**Figure I-01.** Ideal sedimentation basin.

**IDL** Acronym of instrumental detection level.

**idle phase** The last of the five steps in the operation of multitank sequencing batch reactors, providing time for one reactor to complete the fill phase before switching to another one. *See also* fill phase, react phase, settle phase, and decant phase.

**IDLH** Acronym of immediately dangerous to life and health.

**Idrex** A pressure leaf filter manufactured by Zimpro Environmental, Inc.

**IDS Drumshear** A rotating fine screen manufactured by Aer-O-Flo Environmental, Inc.

**IE** Abbreviation of ion exchange.

**i.e.** Abbreviation of the Latin expression id est, meaning that is.

**IFA** Acronym of immunofluorescent antibody.

**IFAS** Acronym of integrated fixed-film activated sludge.

**IFR-6000** Tube settlers manufactured by Brendwood Industries..

**IFU** A fine-bubble membrane diffuser manufactured by Envirex, Inc.

**igneous** Pertaining to rocks produced under intense heat, like rocks of volcanic origin or crystallized from molten magma.

**igneous rock** A type of rock formed from cooled and solidified magma; classified as acid, intermediate, or basic, depending on its silica content. Extrusive or volcanic rocks are usually more productive aquifers than fresh intrusive rocks. *See also* metamorphic and sedimentary rocks.

**ignitability** The characteristic of a substance that has a low flash point, 60°C or less.

**ignitable hazardous waste** Any of the following substances: an oxidizer, an ignitable compressed gas, a liquid with a flash point below 60°C, or a waste that can cause fire, under standard conditions of temperature and pressure, through friction, absorption of moisture, or spontaneous chemical change.

**ignitable waste** Waste that can cause a fire that burns vigorously and persistently under standard conditions of temperature and pressure.

**I/I** Acronym of infiltration/inflow.

**I/I analysis** An engineering and economic analysis demonstrating possible excessive or nonexcessive infiltration/inflow.

**I/I ordinance** A public regulation concerning infiltration/inflow into sanitary sewers. For example, a Miami-Dade County (Florida) ordinance requires all publicly owned or operated sanitary sewer collection systems of the county to participate in a computerized model to optimize transmission capacity and to facilitate the evaluation of the impact of I/I rehabilitation programs (Adrien, 2004).

**I/I performance standard** *See* infiltration/inflow.

**II-PLP®** A reverse osmosis unit manufactured by U.S. Filter Corp. with a double pass and chemical feed device for pH adjustment.

**illite** A clay mineral having (a) a three-layer mica-like structure, (b) a gray, light green, or yellowish-brown color, and (c) permanent surface charges due to isomorphic substitutions; e.g., hydrous potassium aluminosilicate. Illites make up some of the colloids found in water.

**illuminance** Same as illumination.

**illumination** An expression of the intensity of light; the luminous flux incident per unit area. *See* lumen, lux. Also called illuminance or intensity of illumination.

**illuviation** The accumulation of dissolved or suspended matter deposited by percolating water, e.g., from the E horizon into the B horizon of a soil profile.

**ilmenite** The crystalline or massive, black mineral iron titanate ($FeTiO_3$), a dense iron–titanium ore, associated with hematite and magnetite; a mineral often used as a component of a multimedia filter; approximate specific density = 4.2, and porosity = 0.40–0.55.

**imbibition** The process of assimilating or taking into solution or the characteristic of a substance that takes in water, e.g., gels.

**Imhoff cone** A clear, cone-shaped, graduated container, used to measure the volume of settleable solids in a specific volume (usually one liter) of water or wastewater.

**Imhoff, Karl** A well-known German wastewater engineer (1876–1965) who developed the Imhoff tank while managing two river authorities, the Emscher Genossenschaft and the Ruhrverband.

**Imhoff tank** A deep two-story wastewater treatment tank originally designed and patented by Karl Imhoff, based on the concept of the Travis tank. It provides sedimentation in the upper compartment and anaerobic sludge digestion in the lower compartment. Settled solids slide down a steeply sloped trough into the lower chamber, which has gas vents and from the bottom of which digested sludge is removed. Also called Emscher tank because it was designed in the Emscher District of Germany. *See* Figure I-02 and two-story tank.

**immediate biochemical oxygen demand** *See* immediate oxygen demand.

**immediate BOD** *See* immediate oxygen demand.

**immediate chemical oxygen demand** *See* immediate oxygen demand.

**immediate chlorine demand** The amount of chlorine required to immediately react with readily oxidizable impurities in water or wastewater, such as iron ($Fe^{2+}$), manganese ($Mn^{2+}$), hydrogen sulfide ($H_2S$), and organic matter; it may be roughly estimated as 0.5 mg/L of chlorine per mg/L of $BOD_5$. *See* chlorine demand for more detail.

**immediately dangerous to life and health (IDLH)** The maximum level to which a healthy individual can be exposed to a chemical for 30 minutes and escape without suffering irreversible health effects or impairing symptoms. Used as a "level of concern."

**immediate oxygen demand** The oxygen demand exerted through the direct and rapid action of reducing agents such as ferrous iron ($Fe^{2+}$) or sulfides ($S^{2-}$) or of the end products of previous biochemical action. It is common procedure to subtract the BOD of 15 minutes to account for the immediate BOD:

$$\text{immediate BOD} = (\text{15-min DO depletion}) \times (\text{dilution factor}) \qquad \text{(I-05)}$$

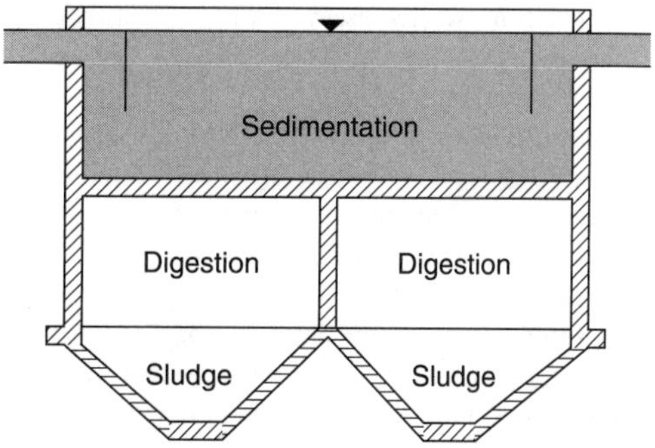


**Figure I-02.** Imhoff tank.

The immediate oxygen demand is a factor considered in determining the rate of oxygen consumption in biological wastewater treatment. The immediate COD of a waste is calculated by supplying a known quantity of oxygen to an unseeded sample of the waste.

**immersed membrane system** A pressure-driven system using hollow-fiber membranes mounted on racks and submerged in a feedwater tank. Vacuum is used to draw water from the membranes and residuals are removed by intermittent backwashing or continuous bleeding of concentrate from the tank. Also called submerged membrane system.

**immiscible** Pertaining to substances that cannot be mixed, e.g., oil in water.

**immune metal** A metal that is protected from corrosion by being thermodynamically stable, e.g., copper in anoxic groundwater or a metal in a water having a cathodic protection system. *See also* passivation, Pourbaix diagram.

**immunity** (1) The inherited or acquired, active or passive capacity of an organism to resist infection by a specific pathogen. (2) The property of a metal that is protected from corrosion over a certain range of pH and oxidation–reduction potential.

**immunization** The process of increasing, naturally or artificially, the resistance of an individual to infectious disease.

**immunoassay** A laboratory test used for the detection and/or quantification of a substance based on its capacity to act as an antigen. *See* immunofluorescence, enzyme-linked immunosorbent assay (ELISA).

**immunofluorescence** Any technique that detects an antigen or antibody in a sample by associating its respective antibody or antigen with a fluorescent compound and observing the reaction under an ultraviolet-light microscope. The technique is used, e.g., to detect the protozoan parasites *Cryptosporidium* and *Giardia* in water.

**immunofluorescence assay** A blood test that uses the immunofluorescence technique. Also called fluorescent immunolabeling.

**immunofluorescent antibody (IFA)** A test that is used to detect *Cryptosporidium* in filtered drinking water; it does not determine the species or infectivity of the oocysts. *See also* cell culture–polymerase chain reaction.

**immunology** The branch of biology that studies the components of the immune system of higher organisms, immunity from disease, cell-mediated and antibody-mediated responses, and techniques of immunologic analysis.

**Impac™** Packing media fabricated by Lantec Products, Inc. for use in air stripping towers.

**impact** A change in the biological, chemical, or physical quality or condition of a water body caused by external sources.

**impact fee** A one-time charge assessed by a water, sewer, or other utility against new users, developers, or applicants for new service to recover part or all of the costs of existing or additional facilities. It is a means of charging the cost of development impact to those who cause it. Also called capacity charge, capital recovery charge, or facility expansion charge. *See also* connection charge development impact fee, and system development charge.

**impact head** A device used in irrigation to rotate nozzles and distribute water over a relatively large area and with low water application rates.

**impaction** *See* inertial impaction, a major filtration mechanism.

**impact loss** Head loss in a flowing stream due to water particles striking one another or striking against a boundary or an obstacle.

**impact-type sprinkler system** A system used to distribute wastewater in land applications.

**impairment** A detrimental effect on the integrity of a water body that prevents attainment of a designated use.

**impeller** The rotating set of vanes in a blower, axial or centrifugal pump, turbine, or mixing apparatus, designed to impel rotation to a fluid mass. *See also* rotor.

**impeller mixer** A common, efficient mixing device driven by a rotating impeller and used in water and wastewater treatment. It is available in three types: propellers and turbines used in rapid mixing, and paddles.

**impeller mixer power requirement** The power input ($P$, in N-m/sec or ft-lb/sec) necessary for an impeller mixer to maintain adequate turbulent conditions in a basin (corresponding to a Reynolds number $R_e > 100{,}000$):

$$P = \rho K N^3 D^5 / g \qquad \text{(I-06)}$$

where $\rho$ = mass density of the fluid ($\rho$ = 1000 kg/m$^3$ = 62.4 lb/ft$^3$ for water); $K$ = constant depending on the characteristics of the basin and impeller as well as those of the fluid; $N$ = impeller revolutions per minute; $D$ = diameter of the impeller, meters or feet; $g$ = gravitational acceleration = 19.8 (metric) = 32.2 (English). *See also* mean temporal velocity gradient, mixing loading parameter, mixing opportunity parameter, power dissipation function, rapid mixing parameters.

**impeller pump** A device that uses a mechanical medium to move water.

**imperforate basket centrifuge** A centrifuge consisting of a vertically mounted spinning bowl, operated on a batch basis to remove moisture in sludge thickening or dewatering operations. The solids accumulate on the wall of the bowl, while the centrate is decanted. Dewatering includes a skimming step before initiating plowing; *see* centrifuge skimming. New dewatering installations tend to use solid-bowl centrifuges instead of imperforate basket centrifuges.

**impermeability** *See* impermeable.

**impermeable or impervious** Not easily penetrated. The property of a material or soil that does not allow, or allows only with great difficulty, the movement or passage of water at ordinary hydrostatic pressure; e.g., impervious surfaces include pavement and rooftops. Permeability is an important factor in runoff studies. *See also* permeable.

**impervious** Same as impermeable.

**impingement attack** Same as impingement corrosion.

**impingement corrosion** The deterioration of a material or surface by a moving fluid, resulting from the combined effect of corrosion and erosion of suspended solids or gas bubbles; e.g., the removal of the protective wall coating of copper pipes by the flow of water of high velocity. Also called erosion corrosion and impingement attack.

**impingement separator** An air pollution control device used to remove particulates larger than 10 μm in the thermal processing of sludge at wastewater treatment plants. It directs the dust-containing gas stream against obstacles so that the particles lose momentum and drop out of the gas. *See also* mechanical collectors, cyclone separator, settling chamber.

**important chemicals** Table I-01 lists some important chemical substances related to water chemistry, water and wastewater treatment, or sludge processing.

**impoundage** Same as impoundment.

**impoundment** To impound is to gather and enclose a liquid, especially water, for irrigation, flood control, water supply, hydropower, or similar purpose. (1) An impoundment is a natural or man-made pond, lake, reservoir, basin, tank, or similar space used for the storage, regulation, and control of water. An artificial impoundment is created by such engineering structures as dams, levees, or dikes. Impounding dams for water supply may incorporate intakes, spillways, and diversion conduits. Also called impoundage. For terms related to the water quality of impoundments see circulation zone, dystrophic lake, epilimnion, eutrophic lake, eutrophication, fall turnover, great turnover, hypolimnion, metalimnion, oligotrophic lake, spring circulation, spring turnover, stagnation zone, summer stagnation, thermocline, transition zone, winter stagnation. (2) The act of impounding or the condition of being impounded.

**impressed current** An external source of direct current used for cathodic protection, with the positive terminal connected to the anode and the negative terminal connected to the metal to be protected. *See also* galvanic anode.

**impressed-current protection** One of two methods of providing cathodic protection (the conversion of an entire metal surface into a cathode; corrosion normally occurs only at anodic areas). In this method, an impressed current is established so that the metal receives the electrons emitted at a rate that prevents corrosion. Other methods of corrosion control include anodic protection, galvanic protection, chemical coatings, inhibition, inert materials, and metallic coatings.

**impressed-voltage cathodic protection** A procedure to protect a metal against corrosion by making it the cathode of an impressed current. Same as impressed-current protection.

**Impulse®** A countercurrent water softener manufactured by U.S. Filter Corp.

**impulse pump** A device, like the hydraulic ram, that lifts water using the impulse principle.

**impurity** Any element or substance that is (a) unintentionally present with another, or (b) produced coincidentally with a primary product, or present in a raw material. An impurity does not serve a useful purpose (EPA-40CFR63.101 & 761.3), but is not necessarily harmful. Water impurities include living organisms and solid or dissolved organics and inorganics. *See also* constituent, contaminant, pollutant.

**IMS** Acronym of (1) integrated membrane system, (2) ion mobility spectrometry.

**IMS®** A support cap manufactured by F. B. Leopold Co. for sand filters.

**in** Abbreviation of inch (inches), a unit of length. One inch = one-twelfth of a foot = 2.54 cm.

**inactivate** To render inactive, inert, or harmless; to stop the activity of (biological substances, antigens, or microorganisms).

**inactivation** (1) The process of making a virus unable to reproduce or infect, using a chemical or physical method, e.g., chlorination, heating, UV disinfection. The term also applies to the destruction (killing) of other microorganisms. Heat inac-

**Table I-01.** Important chemicals.

| Substance | Formula | Remarks: applications, processes, etc. |
|---|---|---|
| Activated carbon | C | Adsorption, dechlorination, denitrification, odor removal, organics removal, sludge stabilization |
| Activated silica | $SiO_2$ | |
| Alumina (aluminum oxide) | $Al_2O_3$ | |
| Aluminum ammonium sulfate | $AlNH_4(SO_4)_2 \cdot 12H_2O$ | Coagulation |
| Aluminum hydroxide | $Al(OH)_3$ | Coagulation |
| Aluminum phosphate | $AlPO_4$ | Phosphate removal |
| Aluminum sulfate (alum) | $Al_2(SO4)_3 \cdot xH_2O$ | Coagulation, phosphorus removal, suspended solids removal, sludge conditioning |
| | $Al_2Si_2O_5(OH)_4$ | |
| | $Al_4(Si_4O_{10})_2(OH)_4 \cdot xH_2O$ | |
| Ammonium chloride | $NH_4Cl$ | Dechlorination |
| Ammonium fluosilicate | $(NH_4)_2SiF_6$ | Fluoridation |
| Ammonium hydroxide | $NH_4OH$ | Chloramination |
| Ammonium nitrate | $NH_4NO_3$ | Nutrient in industrial waste treatment |
| Ammonium silicofluoride | $NH_4SiF_6$ | Fluoridation |
| Ammonium sulfate | | Chloramination |
| Anhydrous ammonia | $NH_3$ | Chloramination |
| Aqua ammonia | $NH_3 \cdot xH_2O$ | Chloramination |
| Ascorbic acid | $C_6H_8O_6$ | Dechlorination |
| Calcium bicarbonate | | |
| Calcium carbide | $CaC_2$ | Explosive gases in sewers |
| Calcium carbonate | $CaCO_3$ | Softening, corrosion control |
| Calcium chloride hypochlorite (chloride of lime) | $CaCl(OCl)$ | Chlorination |
| Calcium fluoride | $CaF_2$ | Fluoridation |
| Calcium hydroxide (hydrated lime) | $Ca(OH)_2$ | Softening. *See also* lime |
| Calcium hypochlorite | $Ca(OCl)_2 \cdot xH_2O$ | Disinfection |
| Calcium oxide (quicklime) | CaO | *See* lime |
| Calcium phosphate | $Ca_3(PO_4)_2$ | Softening, phosphate cycle |
| | $CaHPO_4$ | Phosphate cycle |
| Calcium sulfate | $CaSO_4 \cdot xH_2O$ | Hardness, scale formation |
| Carbon dioxide | $CO_2$ | pH, corrosivity, alkalinity, coagulation, and iron removal |
| Carbonic acid | $H_2CO_3$ | Dissolution of subsurface minerals, corrosion |
| Chlorate | $ClO_3^-$ | Potential drinking water contaminant |
| Chlorinated copperas | | Prechlorination, color removal |
| Chlorine | $Cl_2$ | Ammonia removal, disinfection, filter fly control, grease removal, odor control, prechlorination, sludge-bulking control, sludge stabilization |
| Chlorine dioxide | $ClO_2$ | Chlorination |
| Chlorine hydrate, chlorine ice | $Cl_2 \cdot 8H_2O$ | Chlorinator plugging |
| Chlorite | $ClO_2$- | Potential drinking water contaminant |
| Citric acid | $C_6H_8O_7 \cdot H_2O$ | pH adjustment |
| Copperas | $FeSO_4 \cdot 7H_2O$ | Coagulation of iron and manganese |
| Copper hydroxide | $Cu(OH)_2$ | Algae control |
| Copper sulfate | $CuSO_4 \cdot x\ H_2O$ | |
| Ferric chloride | $FeCl_3$ | Phosphorus removal, sludge conditioning, suspended solids removal |
| Ferric hydroxide | $Fe(OH)_3$ | Coagulation, deferrization |
| Ferric sulfate | $Fe_2(SO_4)_3 \cdot xH_2O$ | Phosphorus removal, suspended solids removal |
| Ferrous ammonium sulfate | $FeSO_4 \cdot (NH_4)_2SO_4 \cdot 6H_2O$ | COD test |
| Ferrous carbonate | $FeCO_3$ | Iron cycle |
| Ferrous hydroxide | $Fe(OH)_2$ | Corrosion, deferrization |
| Ferrous phosphate | $FePO_4$ | Phosphate cycle |
| | FeS | Anaerobic corrosion |

**Table I-01.** *Continued*

| Substance | Formula | Remarks: applications, processes, etc. |
|---|---|---|
| Ferrous sulfate (copperas) | $FeSO_4 \cdot xH_2O$ | Odor control, phosphorus removal, sludge conditioning, suspended solids removal |
| Fluosilicic acid | $H_2SiF_6$ | Fluoridation |
| Greensand | | |
| Hydrochloric acid | HCl | Cation resin regeneration, removal of calcium carbonate scale |
| Hydrogen peroxide | $H_2O_2$ or H—O—O—H | Advanced oxidation, bleaching, disinfection, odor control, sludge-bulking control |
| Hydrogen sulfide | $H_2S$ | |
| Hydroxyapatite | $Ca_5(OH)(PO_4)_3$ or $Ca_{10}(OH)_2(PO_4)_6$ | Phosphorus removal, fluoride removal |
| Hypochlorous acid | HOCl | Disinfection |
| Lime | *See* calcium hydroxide, calcium oxide | Heavy metals removal, odor control, pH adjustment, phosphorus removal, sludge conditioning, sludge stabilization, softening, suspended solids removal |
| Magnesium carbonate | $MgCO_3 \cdot xH_2O$ | Hardness |
| Magnesium fluoride | $MgF_2$ | Defluoridation |
| Magnesium hydroxide | $Mg(OH)_2$ | Softening |
| Magnetite compound | $Fe_3O_4$ | Product of iron pipe corrosion |
| Manganese carbonate | $MgCO_3$ | Manganese cycle |
| Manganese hydroxide | $Mg(OH)_2$ | Demanganization |
| Methanol | $CH_3OH$ | Carbon supplement in postanoxic denitrification |
| Methyl orange | $C_{14}H_{14}N_3SO_3Na$ or $(CH_3)_2NC_6H_4NNC_6H_4SO_3Na$ | Color indicator in acid–base (pH) titration |
| Monochloramine | $NH_2Cl$ | Disinfection |
| Ozone | $O_3$ | Disinfection, odor control, sludge-bulking control |
| Phenol | $C_6H_5 \cdot OH$ | Taste and odor problems |
| Phenolphthalein | $C_{20}H_{14}O_4$ | Acid–base color indicator |
| Phosphoric acid | $H_3PO_4$ | |
| Polymers | | Coagulation/flocculation, sludge conditioning, suspended solids control |
| Potassium permanganate | $KMnO_4$ | Odor control |
| Siderite | $Fe_2CO_3$ | Product of iron pipe corrosion |
| Silica | $SiO_2^-$ | Filter medium |
| Sodium aluminate | $Na_2Al_2O_4$ | Phosphorus removal, softening agent and secondary coagulant, suspended solids removal |
| Sodium bicarbonate | $NaHCO_3$ | Reagent used in the laboratory |
| Sodium bisulfite | $NaHSO_3$ | Dechlorination, disinfection, oxidation |
| Sodium carbonate (soda ash) | $Na_2CO_3$ | Softening, pH adjustment |
| Sodium chlorate | $NaClO_3$ | |
| Sodium chloride | NaCl | |
| Sodium chlorite | | Disinfection |
| Sodium citrate | | Prevention of carbonate precipitation in algal control by copper hydroxide |
| Sodium fluoride | NaF | Fluoridation |
| Sodium hexametaphosphate | $(NaPO_3)_n$ | Sequestering, dispersing, or deflocculating agent |
| Sodium hydroxide | NaOH | pH, acidity, and alkalinity adjustment |
| Sodium hypochlorite | NaOCl | Disinfection, odor control, oxidation |
| Sodium metabisulfite | $Na_2S_2O_5$ | Dechlorination |
| Sodium silicate | $Na2O(SiO_2)_{3.25}$ | |
| Sodium sulfite | $Na_2SO_3$ | Dechlorination |
| Sodium thiosulfate | $NaS_2O_3 \cdot 5H_2O$ | Dechlorination |
| Stannous chloride | $SnCl_2$ | Corrosion inhibition |
| Sulfanilamide | $C_6H_8N_2O_2S$ | Reagent in the laboratory determination of nitrite |
| Sulfur dioxide | $SO_2$ | Dechlorination |
| Sulfuric acid | $H_2SO_4$ | Ion exchange regeneration, COD test, pH adjustment |
| Zinc hydroxide | $Zn(OH)_2$ | Corrosion |

tivates by denaturation of enzymes and other essential proteins, whereas chemicals destroy membranes, cell walls, and enzymes, and impair the ability to replicate nucleic acids. UV radiation inhibits biological activity indirectly. (2) The reduction of the toxicity of a chemical or the activity of a biomolecule.

**inactivation constant** The coefficient $k'$ in Watson's disinfection model:

$$K = k'C^n \qquad \text{(I-07)}$$

where $k$ is the inactivation rate constant (or dieoff coefficient), $C$ is the disinfectant concentration, and $n$ is the dilution constant. *See also* inactivation rate.

**inactivation rate** In a disinfection method, the number of organisms killed per unit volume and per unit time. In the Chick's law formulation, the inactivation rate ($r$) is a function of the concentration of viable microorganisms ($N$) and the inactivation rate constant ($k$):

$$r = -kN \qquad \text{(I-08)}$$

**inactivation rate constant** *See* inactivation rate and Chick–Watson law.

**inactivation rate tailing** A deviation from the Chick–Watson law in which the rate of inactivation of microorganisms decreases progressively. *See* Chick–Watson law deviations.

**inactive mine** A site where there have been in the past extraction, beneficiation, or processing of mining materials but that is not currently subject to active mining. Inactive mines are still required to obtain a stormwater discharge permit.

**inapparent infection** Infection present in a host that does not show any recognizable clinical signs or symptoms. It may be detected by laboratory testing. Also called asymptomatic, occult, or subclinical infection.

**incentive rate** A rate charged customers to encourage conservation; an inclining block rate.

**inch-foot (inch-feet)** The characteristic of a sewer line obtained by multiplying its diameter in inches by its length in feet. Similarly, inch-mile(s) is obtained by multiplying its diameter in inches by its length in miles. Used in the definition of I/I performance standard and gpd/in.-mi.

**inch-foot charge for water service** A charge levied for providing a water line for fire protection, separate from the domestic water charge. It is proportional to the product of the diameter (D, in) and the length (L, ft) of the water mains.

**inch-mile** *See* inch-foot.

**inchoate water right** A nonperfected water right.

**incidence** The number of new cases of a disease or infection in a defined population within a specified period of time. *See also* incidence rate, prevalence.

**incidence rate** The rate at which people develop a disease or an infection, i.e., the ratio of the number of new cases over a period of time to the population at risk; usually expressed as the number of cases per 100,000 population per year.

**incident irradiance ($E_0$)** The intensity of radiant energy as it strikes an object; it can be measured with a radiometer. *See* average UV irradiance.

**incineration** A treatment technology involving destruction or volume reduction of waste by controlled burning at high temperatures, e.g., burning sludge to remove the water and reduce the remaining residues to safe, nonburnable ash that can be disposed of safely on land, in some waters, or in underground locations. Heat drying usually precedes incineration, whose by-products are stack gases and quenching waters. For details on sludge incineration techniques see Carver–Greenfield process, coincineration, combustion, electric furnace, flash drying, fluidized-bed incinerator, infrared incineration, multiple-hearth incinerator, pyrolysis, rotary dryer, sludge incineration, solvent extraction, starved-air incineration, wet oxidation.

**incineration at sea** Disposal of waste by burning at sea on specially designed incinerator ships.

**incinerator** A furnace for combustion of solid matter or semiliquid or dewatered materials, such as sludge, under controlled conditions. Incinerators are used for destroying volatile and other organic compounds, do not extract energy in the form of steam or process heat, and do not rely on the heating value of the waste to sustain combustion. Auxiliary fuel may be used to heat waste gas. Incinerators are different from boilers, sludge dryers, and carbon regeneration units (EPA-40CFR260.10, 61.301 & 63.111). *See* fly ash, flue gas.

**incinerator effluent** Same as flue gas.

**incinerator emission** Same as flue gas.

**incipient fluidization** The condition of a filter bed just before fluidization begins; it is still a fixed bed. The point of incipient fluidization or fluidization point is the intersection of the lines representing fixed bed and fluidized bed head losses on a curve of head loss vs. superficial velocity, as shown on Figure I-03. *See also* backwash velocity, minimum fluidization velocity, fast fluidization.

**incipient fluidization point** *See* incipient fluidization, fluidization point.

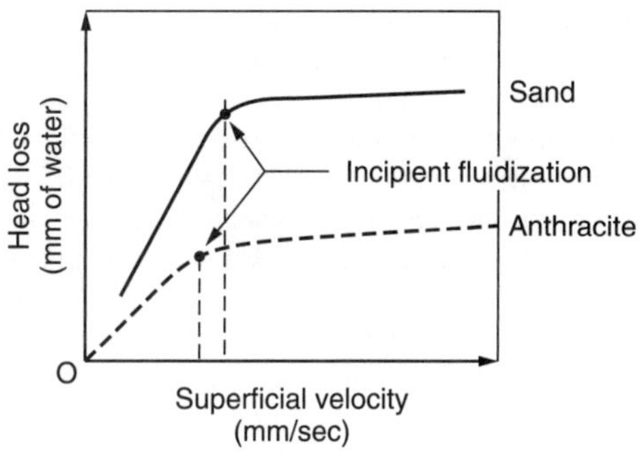


**Figure I-03.** Incipient fluidization.

**inclined-apron aerator** *See* gravity aerator.

**inclined cascade aeration (ICA)** An oxygenation method originally developed to treat wastewater from sugar refineries by pumping the liquid to the top of an inclined corrugated surface and letting it flow by gravity.

**inclined-plane aerator** *See* gravity aerator.

**inclined plate separator** A shallow settling unit consisting of inclined parallel plates installed at the end of a clarifier or gravity thickener in a countercurrent, cocurrent, or crossflow configuration. The plates improve settling performance by increasing the settling area and reducing the distance the solids have to travel before being captured. Also called plate settler or inclined plate settler. *See also* lamella clarifier, inclined tube settler or tube settler, and ballasted flocculation.

**inclined plate settler** Same as inclined plate separator.

**inclined screw** A sloped mechanism or device used in large wastewater treatment plants to lift grit out of a grit chamber, and wash and dewater the materials removed. Also called tubular conveyor. *See also* horizontal screw conveyor, chain-and-flight mechanism, clamshell bucket, chain-and-bucket elevator.

**inclined screw press** A device used to dewater sludge; it consists of a screw conveyor inclined about 20 degrees and rotating inside a stainless steel, wedge-shaped wire screen that narrows upward to create pressure and cause filtrate to flow out. *See also* sludge pressing.

**inclined settling** A settling method that uses inclined plates or tubes in a shallow unit. *See* inclined plate separator and inclined tube settler.

**inclined tube settler** A shallow settling unit defined similarly to the inclined plate separator, but with bundles of small plastic tubes instead of stacked trays.

**inclining block rate or inclining rate** A water rate structure in which the block price increases for each successive block; e.g., \$1.00 for the first 1000 gallons per month, \$1.50 for the next 1000 gallons up to 5000 gallons, etc. It is similar to inverted block rate and the opposite of declining block rate.

**included gas** Gas in the aeration or saturation zone of subsurface water.

**inclusion** (1) A form of coprecipitation of secondary elements (e.g., heavy metals, radionuclides, and viruses) along with the main targets of chemical precipitation (e.g., calcium, magnesium, and organic contaminants) during water treatment. It involves the mechanical entrapment of a portion of the solution around a growing particle. *See also* adsorption, occlusion, solid-solution formation. (2) *See* cell inclusion.

**incompetent formation** In relation to the design and construction of injection wells, incompetent formations are unconsolidated sands and gravels that will cave into the borehole if they are not artificially supported. *See also* competent formation, fluid-seal completion.

**incomplete combustion** A process that involves a combustion reaction that is not carried to completion, leaving, for example, some organic matter and ammonia ($NH_3$) in the end products, in addition to carbon dioxide ($CO_2$) and water ($H_2O$). *See* wet combustion, wet-air oxidation.

**incomplete combustion products** Hazardous products of thermal reactions detected in stack emissions but not present in the waste feed at a concentration of 100 ppm or larger. *See* product of incomplete combustion (PIC) for detail.

**incompressible fluid** A fluid having zero compressibility (a coefficient of compressibility of 0); a fluid with constant density for isothermal pressure changes.

**increasing block rate** Same as inclining block rate.

**increasing death phase** A phase in the growth of microorganisms in which their death rate exceeds the production of new cells. It follows the stationary growth phase. *See* growth pattern.

**incremental cost** The cost of producing the next unit of a product, e.g., the cost of the next increment of water supply, which may be one gallon, 1000 gallons, or one million gallons per day. Also called marginal cost.

**incremental cost pricing** *See* marginal cost pricing.

**incrustants** A solid crust on the inside surface of pipe resulting from water hardness.

**incrustation** Calcium carbonate ($CaCO_3$) precipitates on well screens resulting from an increase in pH caused by the loss of carbon dioxide ($CO_2$) when the well is pumped. Iron and manganese may also cause incrustations by precipitating or promoting the growth of oxidizing bacteria. For methods of incrustation removal *see* acid treatment, chlorination, dry ice, and polyphosphate treatment.

**incubation** (1) The maintenance of microorganisms or cultures under conditions of growth and reproduction (e.g., appropriate nutrients and temperature). (2) The process of a fertilized egg reaching full development or an infectious agent producing symptoms.

**incubation period** The period of time it takes an infectious agent to produce symptoms of a disease from the time of initial contact or exposure; it ranges from 6–12 hours for the Norwalk virus of diarrhea to 30–60 days for the hepatitis-A virus. *See also* latency period.

**incubation time** Same as incubation period.

**incubator** A container used to maintain a constant temperature for the development of microbial cultures.

**Incutrol** An apparatus manufactured by Hach Co. for BOD incubation with temperature control.

**index** A numerical or other indicator of the relation between two phenomena, processes, parameters, etc. *See* the following indexes: antecedent-precipitation, average retention time, chloride, dispersion, flow, fluvial, infiltration, inverse sludge, Langelier saturation, mean retention time, modal retention time, moisture, Morrill dispersion, plasticity, pollutional, rainfall, saturation, sludge density, sludge volume, short-circuiting, Thornthwaite moisture, wetness, zero moisture.

**index of average retention time** *See* average retention time index.

**index of dispersion (tracer)** *See* Morrill dispersion index.

**index of mean retention time** *See* mean retention time index.

**index of modal retention time** *See* modal retention time index.

**index of short-circuiting** *See* short-circuiting index.

**index organism** Same as indicator organism.

**index period** The sampling period during which selection is based on the temporal behavior of the indicator and practical sampling considerations.

**indicated number** The reciprocal of the smallest quantity of sample, in a decimal series, that yields a positive result in the dilution method for microbial density.

**indicative maintenance** An unplanned, reactive maintenance approach that consists of inspections to detect problems early, repairs based on intuitive recognition of experienced workers, and effective apprentice programs. Also called conditional maintenance. *See also* breakdown maintenance, proactive maintenance, developmental maintenance.

**indicator** (1) In biology, an organism, species, or community whose characteristics show the presence of specific environmental conditions. *See* indicator organism and indicator species for detail. (2) In chemistry, a substance (e.g., litmus, orthotolidine reagent) that shows a visible change (usually of color), which indicates the presence of a constituent or the desired point in a chemical reaction; see, e.g., acid–base indicator and redox indicator. (3) A device that indicates the result of a measurement, e.g., a pressure gauge or a movable scale.

**indicator bacteria** Bacteria that are present in large numbers in the excreta of healthy or sick warm-blooded animals. Their presence in a water sample indicates fecal contamination of the sample, which may also contain pathogens and represent a health hazard. *See also* fecal coliforms.

**indicator (chemical)** *See* indicator (2).

**indicator gage** A gage equipped with an index or the like to show instantaneously a flow characteristic such as stage, discharge, or velocity.

**indicator (instrument)** *See* indicator (3).

**indicator microorganism** A microorganism whose presence indicates that pathogenic microorganisms may also be present. *See* indicator (1), indicator organism, coliform group.

**indicator organism** An organism that can be detected easily and whose presence indicates the presence or absence of a specific condition or contaminant. For example, the coliform test is used as a reliable indicator of the possible fecal contamination and the possible presence of pathogens in water. Also called index organism. *See* acid-fast bacteria, alternative indicator, biotic index, fermentation tube procedure, heterotrophic plate count, indicator (1), membrane filtration technique, saprobic classification, and the following organisms: *A. hydrophyla,* bacteroides, bifidobacteria, *Clematis vitalba, Clostridium perfringens, E. coli,* enterococci, fecal coliform bacteria, fecal streptococci, *Klebsiella, P. aeruginosa,* presence–absence test, presumptive test, total coliform bacteria.

**indicator species** Species whose presence indicates certain environmental conditions, such as soil type (e.g., alkaline or acid), high level of pollution, or low dissolved oxygen levels. *See* indicator (1) and indicator organism.

**indicator test** A test conducted for a specific constituent (or organism) to detect the presence of something else, e.g., coliform test.

**indigenous** Living or occurring naturally in a specific area or environment; native.

**indigenous species** A species that originally inhabited a particular area.

**indirect additive** Any of a number of products (paints, coatings, and other chemicals) that enter a water supply after being used in maintenance operations.

**indirect benefit** A benefit that accrues to people who do not use directly the product or service resulting from an action. For example, industrial wastewater treatment contributes to the quality of the receiving water, which may be used by a downstream community for water supply or recreation. Also called secondary benefit. *See also* primary benefit, intangible benefit.

**indirect charges** Overhead and other expenses that are necessary but not directly related to production.

**indirect-contact condenser** A condenser in which there is no direct contact between the heating medium and the vapor stream, but a partition divides the two. The water vapor is removed separately from the heating medium.

**indirect cost** Overhead and other cost items not specifically allocated to a production process or product, e.g., compliance (permitting, reporting, monitoring, insurance, utilities, pollution control, and on-site waste management.

**indirect discharge** Introduction of pollutants from a nondomestic source into a publicly owned waste treatment system. Indirect dischargers can be commercial or industrial facilities whose wastes enter local sewers.

**indirect discharger** A nondomestic discharger of pollutants into a publicly owned treatment works; e.g., a water treatment plant.

**indirect dryer** A sludge processing device that uses a rotating dehydration tube that is heated on the outside to about 212°F; the sludge moves through the tube without directly contacting the heat source. The moisture removed is collected as steam. Indirect dryers include the paddle, hollow-flight, disc, and multiple-effect evaporation types. Also called conduction dryer. *See also* direct dryer, conduction, thermal drying.

**indirect drying** A method of drying sludge using heat to evaporate water and reduce the moisture content of the solids. The device used has a solid retaining wall between the wet sludge and the steam or other hot fluid. Also called conduction sludge drying.

**indirect footprint** *See* carbon footprint.

**indirect impedance approach** A water analysis method using an impedance whose electrodes do not contact the sample directly.

**indirect photolysis** The chemical reaction of a substance caused by other substances that absorb solar energy.

**indirect potable reuse** The indirect reuse of reclaimed water or treated wastewater as a drinking water source, by incorporating the reclaimed water into a raw water source (before treatment) such as a reservoir, surface water, or groundwater. *See also* water reuse applications.

**indirect reuse** The use of reclaimed water or treated wastewater after it has been discharged and diluted into a surface or ground water source. *See also* water reuse applications.

**indirect sludge drying** A method of drying sludge using heat to evaporate water and reduce the moisture content of the solids. The device used has a solid retaining wall between the wet sludge and the steam or other hot fluid. Also called conduction sludge drying. *See* conduction, convection drying, thermal drying.

**indirect water reuse** *See* indirect reuse.

**individual drain system** The system used to convey wastewater from a process unit, product storage tank, feed storage tank, or waste management unit to a waste management unit. The term includes all process drains and common junction boxes, together with their associated sewer lines and other junction boxes, manholes, sumps, and lift stations, down to the receiving waste management unit. A segregated stormwater sewer system designed and operated solely for the collection of runoff and separated from all other individual drain systems is excluded from this definition (EPA-40CFR63.111).

**individual risk** The probability that an individual person will experience an adverse effect. This is identical to population risk unless specific population subgroups can be identified that have different (higher or lower) risks.

**individual sewage disposal** A sewage disposal system serving only one property, e.g., a septic tank–soil absorption system for a single family. *See also* household disposal system.

**individual water supply** A water supply serving only one property, e.g., a well for a single family. *See also* private water supply.

**indole ($C_8H_6NH$)** A colorless to yellowish solid with a characteristic fecal, nauseating odor, associated with untreated wastewater; a putrefaction product from animals' intestines; used as a reagent and in perfumery. It turns red on exposure to light and air. Some fecal and nonfecal coliforms produce indole from the amino acid tryptophan [($C_8H_6N$)$CH_2CH(NH_2)COOH$] at 44°C.

**indoxyl-β-D-glucuronide ($C_{14}H_{15}NO_2$—$C_6H_{13}N$)** A chemical substance used to detect and count *E. coli* in a bacterial culture.

**in-drum processing** Treatment of wastes in a drum or other container, including the addition of reagents, mixing, and other processes; term often applied to the solidification, stabilization, and encapsulation of hazardous wastes. *See also* in situ treatment, ex situ treatment, in-plant processing, mobile-plant processing.

**induced-air flotation (IAF)** A treatment process used to remove materials with hydrophobic surfaces (such as emulsified oil and suspended solids) from process or high-volume wastewaters by forming dispersed air bubbles in the liquid. These attach to the oil and solid particles and cause them to rise as dense froth to the surface where they are removed by skimming paddles. Because of their compact size, induced-air flotation units have low capital costs, but they have high power requirements. Often called dispersed-air flotation.

**induced-draft cooling tower** A cooling tower with an electric fan to induce air flow.

**induced infiltration** Artificial aquifer recharge by injection of water at various points; the process of sustaining well yield by taking advantage of a surface water body as a source of groundwater recharge. When vertical wells, infiltration galleries, and other collection devices are located near a surface water, pumping these devices creates a hydraulic gradient into the aquifer. *See also* natural filtration, riverbank filtration, and groundwater under the direct influence of surface water. Also called induced recharge.

**induced radioactive nuclide** A radioactive nuclide formed by induced nuclear reactions in nature, which result in transmutations. Such a nuclide has a short geological life.

**induced recharge** *See* induced infiltration.

**induced surface water** Groundwater resulting from induced infiltration.

**inducible enzyme** An enzyme that a cell produces in response to the presence of a particular substance. *See also* constitutive enzyme.

**induction flowmeter** An instrument for measuring the flow of a liquid using a tube in a magnetic field.

**induction mixer** A device used for instantaneously mixing chemicals (e.g., alum, chlorine, ferric chloride, cationic polymer) in water or wastewater treatment. *See* high-speed induction mixer for detail.

**induction period** The time between the initial exposure to a toxic or injurious agent and the manifestation or detection of a health effect or a biological response. *See* latency period for detail.

**induction pump** A pump operated by electromagnetic induction.

**induction valve** A valve that draws fluid into the cylinder of a positive-displacement pump, engine, or compressor. Same as an inlet valve.

**inductive capacity** The ratio of the flux density of an electric field in a dielectric to its flux density in a vacuum. *See* dielectric constant for detail.

**inductively coupled argon plasma spectroscopy** Same as inductively coupled plasma spectroscopy.

**inductively coupled plasma atomic emission spectroscopy (ICPAES)** Same as inductively coupled plasma spectroscopy.

**inductively coupled plasma mass spectrometry** Same as inductively coupled plasma spectroscopy.

**inductively coupled plasma spectroscopy** An atomic emission method that uses inductively coupled plasma instruments to analyze metals in water or wastewater, based on the light emitted from metal atom electrons that are excited by argon plasma.

**industrial consumption** The water used by industries as compared to domestic, commercial, or institutional uses. *See* industrial water use for detail.

**industrial cost recovery** A grantee's recovery from industrial and commercial users of a treatment works of the grant amount allocable to the treatment of waste from such users under sections 204(b) and 204(h) of the Clean Water Act (EPA-40CFR35.905). *See also* joint treatment.

**industrial cost recovery period** The period over which industrial cost recovery takes place.

**industrial detergent** A heavy-duty cleaning product sold in bulk for industrial uses; it usually contains anionic or nonionic surfactants.

**industrial effluents** Wastewater and emissions emerging from industrial processes. *See* industrial wastewater.

**industrial fixation** Fixation of nitrogen industrially. *See* Haber process.

**industrial hygiene** The subfield of environmental health related to work environments.

**industrial nitrogen fixation** Fixation of nitrogen industrially. *See* Haber process.

**industrial pollution** Discharge of pollutants and toxic substances into the air, water, and soil by industrial sources.

**industrial pollution prevention** The combination of industrial source reduction and toxic chemical use substitution.

**industrial pretreatment** Pretreatment of industrial wastewater before its discharge into publicly owned treatment works. *See* pretreatment (1).

**industrial prevention programs** *See* industrial pollution prevention and pollution prevention.

**industrial recycling** The reuse of process water by industry, e.g., for cooling, as boiler feed or process water, and in heavy construction. *See also* water reuse applications.

**industrial reuse** The use of reclaimed water or treated wastewater by industry for nonpotable purposes; industrial recycling. Potential industrial uses of reclaimed water from external sources include evaporative cooling water, boiler feedwater, process water, and landscaping and maintenance of grounds. *See also* water reuse applications.

**industrial source** Any source of nondomestic pollutants regulated under the Clean Water Act that discharges into a publicly owned treatment works (EPA-40CFR125.58-i).

**industrial source reduction** Practices that reduce the amount of any hazardous substance, pollutant, or contaminant entering any waste stream or otherwise released into the environment. These practices also reduce the threat to public health and the environment associated with such releases. The term includes equipment or technology modifications, substitution of raw materials, and improvements in housekeeping, maintenance, training, or inventory control.

**industrial use of water** *See* industrial water use.

**industrial user** A source of indirect discharge under the Clean Water Act. More specifically, a nongovernmental, nonresidential user of a publicly owned treatment works that is identified in the Standard Industrial Classification Manual under one of the following divisions: agriculture, forestry and fishing; mining; manufacturing; transportation, communications, electric, gas, and sanitary services (EPA-40CFR35.2005.-19).

**industrial waste** Unwanted materials from an industrial operation; they may be liquid, sludge, solid, or hazardous waste. Their composition, constituent concentration, and other characteristics are usually very different from those of municipal wastes. Also called manufacturing waste. *See also* industrial wastewater. Industrial waste includes waste from (a) land, water, or air transport services, (b) gas, water, sewer, or electricity services, and (c) postal or telecommunications services.

**industrial waste pretreatment** *See* industrial pretreatment and pretreatment (1).

**industrial waste proportioning** Releasing industrial wastewater into a municipal sewer in proportion to the nonindustrial flow of the sewer, which may eliminate the need for flow equalization at the treatment plant. Proportioning may also apply to the discharge of an industrial effluent to a receiving stream.

**industrial waste segregation** The separation of industrial wastewater into two wastes: a strong waste of small volume and a weaker stream of a much larger volume, e.g., the segregation of process waste from cooling water and stormwater, making it easier to apply specific treatment methods to each stream. Sometimes, it is even easier to treat a more concentrated than a dilute waste; that is the case for dye wastes. Segregation is commonly practiced in textile mills and in metal-finishing plants.

**industrial waste stream** Wastewater or, more generally, all liquid, solid, and gaseous wastes created during industrial operations.

**industrial waste survey** An investigation conducted to determine the sources, volumes, and characteristics of the wastewaters produced by a given industry or within a sewer district. Other field investigations include pollutional and sanitary surveys.

**industrial wastewater** Wastewater generated in an industrial process or practice (chemical industry, food processing, manufacturing, materials industry), as opposed to domestic or sanitary wastewater. Also called trade waste. Table I-02 lists some of the industries that have developed specific treatment for their wastewaters. *See also* industrial waste, joint treatment, pretreatment.

**industrial wastewater classification** The three main classifications of wastewater generated at an industrial plant are process, cooling, and sanitary wastes. *See also* process generated wastewater, process waste, process wastewater, process water, sanitary wastewater.

**industrial wastewater surcharge** A charge imposed on industries that use a municipal sewer system above the domestic user fee to compensate for the quantity and strength of their wastes and to encourage them to practice waste reduction and materials recycling.

**industrial water** Water used for an industrial purpose only.

**Table I-02.** Industrial wastewater characteristics

| Industry | Parameter of concern |
|---|---|
| Acids | Low pH |
| Breweries | Nitrogen, carbohydrate |
| Canneries | Solids, BOD |
| Dairies | Fat, protein, carbohydrate |
| Detergents | BOD, phosphate |
| Explosives | Nitrogen, organics |
| Fish | BOD, odor |
| Foundries | Solids |
| Insecticides | Organics, toxics |
| Meat processing | Protein, fat |
| Oil | Solids, BOD |
| Paper | Solids, variable pH |
| Pickles | BOD, solids, low pH |
| Plating | pH, heavy metals, CN, toxics |
| Rubber | BOD, solids, odor |
| Soft beverages | BOD, solids |
| Steel | Low pH, solids, phenols |
| Sugar | Carbohydrate |
| Textiles | pH, solids, BOD |

**industrial water conservation** Measures taken to minimize industrial water use, including reducing water use and recycling process water.

**industrial water requirement** *See* industrial water use.

**industrial water reuse** *See* industrial reuse.

**industrial water use** The use of water in such industrial activities as manufacturing, processing, and power generation. Industrial water requirement is usually expressed in gallons of water actually used to produce a unit of product. In the United States, on average, industries recycle 60% of their water use and consume only 7% of their water intake, i.e., they return 93% to surface or ground waters. Table I-03 lists the water requirements of specific industries per unit production.

**industry water requirement** *See* industrial water use.

**inelastic demand** A demand (for goods or services such as water supply) whose percent change is less than a corresponding percent price change. For a water utility, inelastic demand results in an increase in revenue for a rate increase. *See also* price elasticity and elastic demand.

**Table I-03.** Industrial water use

| Industry | Gallons per unit | Unit |
|---|---|---|
| Automobile | 10,000 | Vehicle |
| Canned vegetables | 125 | Case |
| Leather | 32 | Square foot of hide |
| Meat | 2 | Pound live weight |
| Paper and pulp | 50,000 | Ton |
| Petroleum | 2,000 | Barrel of crude oil |
| Steel | 26,000 | Ton |

**inert** Inactive, lacking the ability to react with other substances, e.g., uncombined atmospheric nitrogen.

**inert gas** *See* noble gas.

**inertial impaction–sedimentation** A major mechanism of removal of particulate matter within a granular filter, whereby heavy particles settle on the filtering medium instead of following the flow streamlines. *See also* filtration mechanisms.

**inert ingredient** Pesticide components such as solvents, carriers, dispersants, and surfactants that are not active against target pests. Not all inert ingredients are innocuous.

**inert materials** (1) Materials that do not react or are subject only to minimal biological transformation or dissolution, e.g., fixed inorganic solids remaining after sludge combustion. *See also* nonbiodegradable, refractory. (2) Materials that are not subject to corrosion and are used in water works instead of metal; for example, asbestos cement and plastic in pipes, weirs, launders, etc. For methods of corrosion control, see anodic protection, cathodic protection (galvanic or impressed-current protection), chemical coatings, inhibition, and metallic coatings.

**inert media** Nonreactive resin beads or granular materials used in water and wastewater treatment.

**inert solids** Fixed inorganic solids that remain after combustion. In biological treatment, inert solids undergo little transformation and dissolution; they tend to accumulate in or pass through the system.

**inert suspended solids** *See* inert solids.

**inert-tracer method** A method used to measure indirectly the rate of oxygen transfer of operating wastewater aeration devices by determining the transfer rate of a radioactive or stable inert gas tracer. *See also* mass balance method and oxygen uptake rate.

**inert total suspended solids (iTSS)** Nonvolatile suspended solids found in raw wastewater and in the mixed liquor of a biological treatment process.

**inert wastes** Wastes that do not react chemically, biologically, or physically, e.g., demolition waste.

**Infa Screen** A bar screen with multiple rakes manufactured by Biwater Treatment, Ltd.

**infantile cyanosis** *See* methemoglobinemia.

**infecting dose** Same as infective dose.

**infection** A condition, symptomatic or not, that is caused by the entry and development of a pathogen in the body and that can be transmitted to another person. *See* carrier.

**infection dose** See the more commonly used expression, infective dose.

**infectious agent** Any organism, such as a virus or a bacterium, that is pathogenic and capable of being communicated by invasion and multiplication in body tissues (EPA-40CFR259.10-b). *See also* pathogenic organism.

**infectious disease** A disease that is caused by a pathogenic organism and can be transmitted from one person to another. *See also* infection, water-related disease, excreta-related disease.

**infectious dose** Same as infective dose.

**infectious dose 50 ($ID_{50}$)** The dose of an infective agent that will infect 50% of the population exposed to it. *See* median infective dose for detail.

**infectious hepatitis** One of the five forms of hepatitis; also called hepatitis-A or jaundice. It is an acute, viral, water- and excreta-related disease transmitted via the fecal–oral route and distributed worldwide. *See* hepatitis-A for more detail.

**infectious substance** A substance that contains viable microorganisms or their toxins, which are known, or suspected, to cause disease in humans or animals.

**infectious waste** Hazardous waste with infectious characteristics, including: contaminated animal waste; human blood and blood products; isolation waste, pathological waste; discarded sharps (needles, scalpels or broken medical instruments); and equipment, pathological specimens, or other wastes from persons with a communicable disease. The majority of infectious wastes originate within hospitals, clinics, laboratories, and similar facilities; in general they have a high organic content. Also called red bag waste and medical waste.

**infective dose** The number of viable pathogens that can cause an infection (disease or colonization) when entering a host simultaneously; e.g., $10^3$–$10^9$ pathogenic cells, 10–500 (median of 130) *Cryptosporidium* oocysts, or as few as 10 *Giardia* cysts per person, depending on such factors as age, previous exposure, and organism survival in water. Also called infectious dose. *See also* median infective dose ($ID_{50}$), minimum infectious dose, excreted load, latency, persistence.

**infectivity** The likelihood of an organism causing an infection. Infectivity is inversely related to the median infective dose ($ID_{50}$); e.g., the probability (and infectivity) of a given number of *Shigella* bacteria ($ID_{50}$ = 10,000) causing an infection is greater than that for *Salmonella* bacteria ($ID_{50}$ > 1,000,000). *See also* latency, persistence.

**infectivity assay** A method of virus detection involving the inoculation of a sample onto host cells and allowing replication to the point of visible signs of infection. Viral plaques are then enumerated. *See also* molecular virus detection, integrated virus detection.

**infectivity titer** A laboratory method using dilution to determine infective dose.

**infestation** Attack or persistence of a parasitic insect on the skin, hair, or clothing.

**infiltration** (1) Water other than wastewater that enters a sewer system (including sewer service connections and foundation drains) from the ground through such means as defective pipes, pipe joints, connections, or manholes. Infiltration does not include, and is distinguished from, inflow (EPA-40CFR35.2005-20 or 40CFR35.905). (2) The technique of applying large volumes of wastewater to land to penetrate the surface and percolate through the underlying soil. (3) The movement of water through the soil surface and into the soil. Infiltration capacity is the maximum rate at which water will enter the soil in a given condition. Infiltration coefficient is the ratio of infiltration to precipitation. Infiltration rate is the actual rate of absorption of water by the soil during a storm; it is equal to the smaller of the infiltration capacity and the rainfall rate.

**infiltration allowance** The maximum infiltration rate permitted in sewer construction specifications; e.g., 200–500 gallons per day per mile of sewer length and inch of pipe diameter (gpd/mi/in). *See also* I/I performance standard.

**infiltration basin or pond** (1) A retention basin or pond excavated into permeable soil for the temporary storage of stormwater runoff, which drains by infiltration or percolation but is not discharged directly into a receiving water. In addition to storage, infiltration basins provide a certain level of stormwater treatment by retaining settleable solids. Design parameters include basin volume, bottom area subject to infiltration, and hydraulic conductivity of the underlying soil. *See also* stormwater retention. Other stormwater infiltration units are drywells and infiltration trenches. (2) Same as spreading basin.

**infiltration capacity** The ability of a soil to process wastewater; hydraulic conductivity, as measured by such tests as flooding basin and sprinkling infiltrometer.

**infiltration gallery** (1) A subsurface groundwater collection system, typically shallow, constructed with open-jointed or perforated, horizontal or vertical, pipes that discharge collected water into a water-tight chamber from which the water is pumped to treatment facilities and into the distribution system. Usually located close to streams or ponds. Infiltration galleries may also consist of single gravel-filled trenches, a network of trenches, or a combination of trenches and pipes. (2) A horizontal underground conduit into a water-bearing formation—often under a river bed—to collect percolating water.

**infiltration/inflow** The total quantity of water from both infiltration and inflow without distinguishing the source (EPA-40CFR35.905). Often abbreviated as I/I or I & I. Excessive I/I is the quantity of I/I that can be economically eliminated from a sewerage system by rehabilitation, as determined in a cost-effectiveness analysis that compares the costs for correcting the infiltration/inflow conditions to the total costs for transportation and treatment of the I/I (EPA-40CFR35.905). Basically, infiltration is groundwater flow during the dry or wet season, whereas inflow is stormwater during the wet season or surface water, but both use up capacity in the conveyance and treatment facilities. In general, infiltration is larger on an average basis and affects both treatment and collection facilities, whereas inflow affects peak flows and transmission lines. I/I may be expressed as total flow in such units as gallons per day or million gallons per day, or as rate in gallons per day per inch diameter per mile of sewer (gpd/in.-mi). For example, some agencies have established an I/I performance standard of 5000 gpd/in-mi, i.e., it is presumed that it is not cost-effective to remove I/I below this rate.

**infiltration pond** *See* infiltration basin.

**infiltration rate** *See* infiltration.

**infiltration sump** A deep chamber, with perforated walls, filled with sand or gravel, used for the infiltration of runoff into the ground and thereby reduce inflow to a combined sewer system. A sedimentation chamber usually precedes the sump to remove suspended solids.

**infiltration water** Water that permeates through the earth. *See also* infiltration and infiltration/inflow.

**infiltrometer** An instrument used to measure the infiltration rate (in/hr) as the difference between the amount of water applied and the amount of runoff divided by the period of percolation. It is also used to measure hydraulic conductivity. *See* cylinder infiltrometer, sprinkling infiltrometer.

**Infinity™ Underdrain** Modular or built-in equipment manufactured by Roberts Filter Group of Darby, PA; a jointless lateral underdrain for use in rehabilitation or new construction projects.

**inflammable** *See* flammable.

**inflammable liquid** *See* flammable liquid.

**inflatable dam** An inflatable device, operated by low-pressure air and installed in a combined sewer system to divert dry-weather flows to an interceptor and store wet-weather flows

**inflow** (1) Water other than wastewater that enters a sewer system (including sewer service connections) from sources such as roof leaders, cellar drains, yard drains, area drains, drains from springs and swampy areas, manhole covers, cross connections between storm sewers and sanitary sewers, catch basins, cooling towers, storm waters, surface runoff, street wash waters, or drainage. Inflow does not include, and is distinguished from, infiltration (EPA-40CFR35.2005-21). (2) In stormwater modeling, net inflow is the difference between precipitation intensity and the rates of infiltration or evaporation. *See also* infiltration/inflow.

**inflow pollutant load** *See* pollutant load.

**influence** *See* area of influence, circle of influence, cone of depression, cone of influence, and zone of influence.

**influent** Water, wastewater, stormwater or other liquid flowing into a reservoir, basin, pumping station, or treatment unit.

**influent mixing and particle stabilization** A stage in the filter-ripening sequence whereby filter influent mixes with backwash remnant water.

**influent weir** A weir at the inflow end of a sedimentation basin, channel, or other hydraulic structure.

**Information Collection Rule (ICR)** A regulation of May 1996 of the USEPA that provides for the collection, by water utilities serving more than 10,000 people, of data necessary to assess risk trade-offs between the control of pathogens and the control of disinfection by-products (DBP) in water treatment. The ICR also requires large utilities to conduct the feasibility of using granular activated carbon adsorption and membrane filtration to remove DBP precursors.

**infrared dryer** An infrared furnace or a multiple-hearth furnace used to reduce the moisture content of municipal sludge. Also called a radiant dryer or radiation dryer.

**infrared drying** The use of infrared lamps, electric resistance elements, or gas-heated incandescent refractories to transfer heat to wet sludge and evaporate its moisture content. *See* radiation drying for detail.

**infrared furnace** A type of incinerator sometimes used for sludge combustion. *See* electric furnace for detail.

**infrared incineration** Combustion of wastewater sludge in an electric furnace using infrared heating elements. *See* electric furnace.

**infrared radiation** Low-energy radiation corresponding to the portion of the electromagnetic spectrum between visible light and microwave; used in the laboratory to identify organic molecules. *See* infrared spectrophotometry and infrared spectroscopy.

**infrared rays** Electromagnetic radiation of a wavelength between 0.8 and 1000 microns, invisible to the human eye.

**infrared sludge drying** The use of infrared lamps, electric resistance elements, or gas-heated incandescent refractories to transfer heat to wet sludge and evaporate its moisture content. *See* radiation drying for detail.

**infrared spectrophotometry** A laboratory method that measures the absorption of light in the infrared range to analyze a substance.

**infrared system** An apparatus that generates intense infrared radiation by passing electricity through silicon carbide resistance heating elements for the combustion of sludge or hazardous wastes.

**infrastretching** Stretching "the performance of existing infrastructure through applied research programs" (Metcalf & Eddy, 2003).

**infrastructure** The basic framework or features of a system or organization; the basic facilities, installations, equipment, and utility systems needed for the functioning of a country or an area.

**ingestion** Exposure to an agent by swallowing, i.e., through the mouth.

**inhalation** Exposure through the lungs.

**inherent availability (AVI)** A statistical measure of the reliability of mechanical equipment, as determined by the fraction of calendar time a component or unit was operating; it is one of the factors used in critical component analysis. *See also* expected time before failure, mean time before failure, operating availability.

**inhibiting concentration (IC)** The estimated concentration of a constituent that causes a specified percentage inhibition or impairment in a qualitative function. *See also* toxicity terms.

**inhibition** A corrosion control mechanism whereby the deposition or adsorption of insoluble or less soluble compounds protects a metallic surface. Inhibition is accomplished using (a) chemicals such as chromates, nitrates, etc.; (b) precipitation of calcium carbonate on the metallic surface by adjusting water chemistry; and (c) passivation to form a less soluble metal–carbonate–hydroxide compound; *see* passivating film. Other methods of corrosion control include anodic protection, cathodic protection (galvanic or impressed-current protection), chemical coatings, inert materials, and metallic coatings.

**inhibitor** (1) Any substance that interferes with a process or a chemical reaction such as precipitation or corrosion. *See* corrosion inhibitor. (2) A substance that decreases the speed or yield of a chemical or biological reaction without being consumed or chemically changed by the reaction; e.g., dissolved metals, enzymes. Also called a negative or retarding catalyst.

**inhibitory toxicity** The inhibitory effect of a substance on a living organism.

**initial [$Br^-$]/average [$Cl^-$] molar ratio** The molar ratio of the bromide ($Br^-$) concentration to the free available chlorine (FAC, $Cl^-$) in water. *See* bromide-to-FAC ratio for detail.

**initial compliance period** The first full three-year compliance period that begins at least 18 months after the promulgation of USEPA drinking water regulations.

**initial depth** *See* sequent depths.

**initial dilution** (1) The dilution ($D_i$) achieved in the vicinity of a river diffuser outfall, e.g., within one diffuser length (Metcalf & Eddy, 1991):

$$D_i = (UHL/2\,Q)\{1 + [(1 + 2\,QV \cos \alpha)/U^2LH]^{0.5}\} \quad \text{(I-09)}$$

where $U$ = river velocity, ft; $H$ = water depth, ft; $L$ = diffuser length, ft; $Q$ = discharge flowrate, fps; $V$ = discharge velocity through each port of the diffuser, fps; and $\alpha$ = orientation angle of the ports above the horizontal. (2) The dilution achieved in the initial mixing region (or discharge nearfield) of an ocean or large lake outfall. For a vertical single-port discharge, the average plume dilution ($D_0$) is:

$$D_0 = 0.13\, B^{1/3}Q^{-2/3}H^{5/3} \quad \text{(stagnant ambient) (I-10)}$$

$$D_0 = 0.29\, U'H^2/Q \quad \text{(flowing ambient) (I-11)}$$

where

$$B = \text{discharge buoyancy} = g\Delta\rho/\rho, \text{ ft/sec}^2 \quad \text{(I-12)}$$

$g$ = gravitational acceleration; $\rho$ = ambient water density, g/L; $\Delta\rho$ = discharge density difference, g/L; and $U'$ = ambient current speed, fps.

**initial dilution zone** *See* initial mixing region.

**initial disinfection byproduct concentration** *See* instantaneous disinfection by-product concentration.

**initial mixing region** A region close to the port of discharge of wastewater or effluent into the ocean or a large lake, where the effluent forms a buoyant, rapidly rising plume. Also called zone of initial dilution. *See* discharge nearfield for detail.

**initial total trihalomethane concentration** *See* instantaneous total trihalomethane concentration.

**initiator** A chemical substance or physical agent, often a mutagen, that initiates a series of DNA changes that lead to carcinogenesis. The organism response is related to the total dose and independent of the time frame of exposure. *See also* promoter or promotor.

**injection aerator** Same as air diffuser.

**injection technique** A method of injection of a substance into a gas chromatograph for analysis.

**injection test** The introduction of clean water into an injection well to monitor pressure and water quality.

**injection well** (1) A well into which fluids are injected for waste disposal, improving the recovery of crude oil, solution mining, etc. (2) A bored, drilled, or driven shaft used for the underground injection of fluids such as brines and residuals of water and wastewater treatment. A dug well whose depth exceeds its largest surface dimension may also be used. For toxic chemicals, the well is preferably more than 1000 ft deep and into a saline aquifer overlain by impervious layers. *See* class I, II, II, IV, and V injection wells. Also called an underground injcction well.

**injection zone** The geological formation that receives fluids through an injection well. *See also* underground injection.

**injector** A device for dispersing chemical solutions into water being treated; e.g., injectors are used to feed chlorine gas directly into water or wastewater. More generally, a device that moves fluids or solids by entrainment in air or water under pressure. Also called ejector.

**inland freshwater wetland** A swamp, marsh, bog, etc. found inland beyond the coastal saltwater wetlands.

**inlet** In general, a connection, entrance, or orifice for the admission of a fluid. In drainage or stormwater collection networks, inlets are connections between the catchment area and drains or sewers for the admission of surface or storm water. Inlets are also structures at the entrance or diversion ends of a conduit. *See also* catch basin, curb inlet.

**inlet guide vane variable diffuser** An air-conveying device used to supply air in wastewater aeration systems; it is designed as a single-stage centrifugal blower operable under variable temperature, pressure, and discharge conditions.

**inlet zone** One of the four zones of the Camp sedimentation theory. The inlet zone of a sedimentation basin is characterized by uniform distribution of wastewater solids and flows over the cross-sectional area of the basin. *See also* settling, sludge, and outlet zones.

**inlet well** (1) A well, surface opening, compartment, or chamber in which water or wastewater is collected and to which the suction pipe of a pump is connected. Also called wetwell. (2) A well through which a fluid (liquid or gas) is allowed to enter the subsurface under natural pressure. (3) A well in the center of a circular sedimentation tank behind which the outlet ports of a riser pipe discharge the influent.

**in-line equalization** A configuration in which all influent water or wastewater passes through an equalization basin to realize considerable dampening of flows and concentrations. *See also* off-line or sideline equalization.

**in-line equipment** Equipment installed within a process line, a network, or thc like; e.g., an in-line pump, flowmeter, or nozzle mixer. In sanitary engineering, in-line storage is the storage of stormwater or wastewater within an existing sewer system. Obstructions such as dams or weirs are used to block the flow and create a backup and storage, with a resulting attenuation of peak flows. *See also* off-line storage.

**in-line filtration** The addition of chemical coagulants directly to the filter inlet pipe. The chemicals are mixed by the flowing water. Flocculation and sedimentation facilities are eliminated, but coagulation and flocculation occur within the piping. This treatment method is commonly used in pressure filter installations and for waters of low turbidity. Sometimes called direct in-line filtration. *See also* conventional filtration and direct filtration.

**in-line mixer** A device that creates turbulence to instantaneously mix chemicals (e.g., alum, ferric chloride, chlorine, cationic polymers) in water or wastewater using fixed sloping baffles or vanes without the application of power. *See also* mixing and flocculation devices.

**in-line rotameter** A flow measurement device for liquids and gases that uscs a flow tube and specialized float. The float device is supported by the flowing fluid in the clear glass or plastic flow tube. The vertical-scaled flow tube is calibrated for the desired flow volumes/time.

**in-line static mixer** An in-line mixer that does not have a moving part; a device used in water or wastewater treatment for mixing and blending chemicals instantaneously, in one second or less (e.g., alum, ferric chloride, chlorine, cationic polymers). *See also* mixing and flocculation devices.

**in-line storage** *See* in-line equipment.

**inner-sphere complex** A stable compound resulting from the covalent bond between a metal ion and a ligand. *See* complex for detail.

**inner-sphere surface charge** The surface charge density of a complex formed by two colloidal particles. It depends on the concentrations of the metal ions or ligands on the surface of the particles.

**innova-Tech™** A wastewater treatment process developed by Innova-Tech, Inc. using oxidation ditches.

**innovative control technology** Any system of air pollution control that has not been adequately demonstrated in practice, but would have a substantial likelihood of achieving greater continuous pollutant reduction than any control system currently used or of achieving at least comparable reductions at lower cost in terms of energy, economics, or non-air-quality environmental impacts (EPA-40CFR51.166-19 & 52.21-19). *See* innovative technology

**innovative technology** (1) A new or inventive method to effectively treat hazardous waste and reduce risks to human health and the environment. (2) A developed wastewater treatment process or technique that has not been fully proven under the circumstances of its contemplated use and that represents a significant advancement over the state of the art in terms of significant reduction in life-cycle cost or significant environmental benefits through the reclaiming and reuse of water, otherwise eliminating the discharge of pollutants, utilizing recycling techniques such as land treatment, more efficient use of energy and resources, improved or new methods of waste treatment management for combined municipal and industrial systems, or the confined disposal of pollutants so that they will not migrate and cause water or other environmental pollution (EPA-40CFR35.2005-23).

**inoculate** To implant microorganisms onto or into a culture medium.

**inoculum** (1) Bacteria placed in compost to start biological action. (2) A medium containing organisms that are introduced into cultures or living organisms, e.g., a bacterial culture from domestic wastewater added to an industrial waste sample to conduct the BOD test.

**inorganic** Material such as sand, silt, iron, calcium salts, and other minerals. Inorganic substances are of mineral origin, whereas organic substances are usually of animal or plant origin.

**inorganic arsenic** The oxides and other noncarbon compounds of the element arsenic included in particulate matter, vapors, and aerosols (EPA-40CFR61.161).

**inorganic carbon** The carbon content of inorganic compounds; e.g., oxides, sulfides, carbon dioxide ($CO_2$), and metallic bicarbonates ($HCO_3^-$) and carbonates ($CO_3^{2-}$). *See also* organic carbon.

**inorganic chemical (IOC)** A chemical substance of mineral origin, not of basically carbon structure. Inorganic chemical water quality parameters of particular concern include arsenic, asbestos, barium, cadmium, chromium, copper, fluoride, lead, mercury, nitrate, nitrite, and selenium. The inorganic chemicals industry has the potential to contaminate water with trace elements, particularly the producers of chlor–alkali, chrome pigments, hydrofluoric acid, hydrogen cyanide, titanium dioxide, and various salts of aluminum, copper, nickel, and sodium. *See also* inorganic compound.

**inorganic chemical conditioning** The use of ferric chloride ($FeCl_3$), lime [$Ca(OH)_2$], or other inorganic chemicals used in association with sludge dewatering by vacuum filters, centrifuges, and belt filter presses.

**inorganic chloramine** A compound formed during the chlorination of ammonia-containing waters. *See* chloramines.

**inorganic compound (IOC)** A compound that does not include organic carbon; a substance that is not derived from a hydrocarbon. Some IOCs are regulated under amendments to the Safe Drinking Water Act. Membrane filtration processes effective in IOC removal include electrodialysis reversal, nanofiltration, and reverse osmosis.

**inorganic constituent (IOC)** Inorganic parameters of importance in water and wastewater treatment include heavy metals, particularly those that can come from distribution and plumbing systems (copper, lead, zinc), alkalinity, aluminum, arsenic, asbestos, barium, cadmium, chromium, cyanide, fluoride, gases, hardness, iron, manganese, mercury, molybdenum, nickel, nitrogen (nitrate and nitrite), nutrients, phosphorus, pH, selenium, sodium, and sulfate.

**inorganic contaminant (IOC)** An inorganic constituent or inorganic compound regulated by the USEPA under the Safe Drinking Water Act. *See* inorganic chemical, emerging inorganic contaminants.

**inorganic DBP** Same as inorganic disinfection by-product.

**inorganic disinfection by-product** Inorganic by-products of water disinfection. Bromide-containing water that is treated with ozone may produce bromates and iodates, whereas disinfection by chlorine may result in the formation of chlorites and chlorates. Cyanogen chloride (CNCl) is a potentially harmful inorganic by-product. These constituents are often discussed under the category of emerging inorganic contaminants (as of 2007) in the literature and in regulatory documents.

**inorganic dissolved solid** Minerals dissolved in water or wastewater, e.g., chlorides, nitrates, phosphates, and sulfates. They can be removed by advanced treatment methods such as evaporation, dialysis, ion exchange, demineralization, and reverse osmosis.

**inorganic fouling** Fouling of a water or wastewater treatment membrane caused by the formation and precipitation of struvite ($MgNH_4PO_4$) when pH rises and carbon dioxide ($CO_2$) escapes. *See also* organic fouling.

**inorganic material** *See* inorganic matter.

**inorganic matter** Any chemical substance that is of mineral origin and contains no hydrocarbons or compounds of a basic carbon structure. Inorganic substances are usually not volatile, combustible, or biodegradable.

**inorganic nutrient** Major nutrients required by microorganisms are nitrogen (N), sulfur (S), phosphorus (P), potassium (K), magnesium (Mg), calcium (Ca), iron (Fe), sodium (Na), and chlorine (Cl). Minor nutrients are zinc (Zn), manganese (Mn), molybdenum (Mo), selenium (Se), cobalt (Co), copper (Cu), and nickel (Ni). *See also* organic nutrients.

**inorganics** *See* inorganic constituent.

**inorganic salt** An inorganic compound found in nature or resulting from human activities. *See* inorganic compound, inorganic constituent, inorganic nutrient. Inorganic salts may cause hardness in water and induce the growth of algae in surface waters.

**inorganics in drinking water** *See* inorganic constituent.

**inorganic solute** Inorganic dissolved matter in water; a class of contaminants examined for the revision of drinking water regulations.

**inorganics removal** Common water treatment methods used to remove some inorganics are listed in Table I-03.

**inorganic substance** *See* inorganic compound.

**inorganic waste** Discarded mineral matter, e.g., sand and silt, that is relatively unaffected by microbiological action.

**in-parallel flow** A piping arrangement that directs flow through two or more units. *See also* in-series flow.

**in-place closure** The closure of a hazardous waste surface impoundment similar to a landfill closure, including the elimination of free liquids, the stabilization of remaining wastes to a sufficient bearing capacity for the final cover, and installation of the cover. *See also* clean closure.

**in-plant processing** The treatment or processing of wastes in a dedicated plant located on the site where the wastes are generated or an outside site; term often applied to the solidification, stabilization, and encapsulation of hazardous wastes. *See also* in situ treatment, ex situ treatment, in-drum processing, mobile-plant processing.

**in-plant study** A full-scale investigation of a water or wastewater treatment method.

**in-plant training** On-the-job training of water or wastewater treatment operators.

**in-process wastewater** Any water that during

**Table I-03.** Inorganics removal processes

| Contaminant | MCL | Recommended process | Required pretreatment |
|---|---|---|---|
| Arsenic, $HAsO_4^{2-}$, As(V) | 0.002–0.02 mg/L | Anion exchange | Iron removal by oxidation and filtration |
| Barium, $Ba^{2+}$ | 2.0 mg/L | $H^+$ ion-exchange softening | Iron removal |
| Chromium, $CrO_4^{2-}$, Cr(VI) | 0.10 mg/L | Anion exchange | — |
| Fluoride, $F^-$ | 4.0 mg/L | Activated alumina | pH adjustment |
| Hardness, (a) $Ca^{2+}$, (b) $Mg^{2+}$ | — | Ion-exchange softening, (a) $Na^+$, (b) $H^+$ | (a) Iron removal, (b) None |
| Nitrate-N, $NO_3^-$ | 4.0 mg/L | Anion exchange | — |
| Perchlorate, $ClO_4^-$ | — | Anion exchange | Turbidity reduction |
| Radium, Ra-226 or Ra-228 | 20 pCi/L | Ion exchange | Iron removal |
| Selenium, $HSeO_3$, Se(IV) | 0.05 mg/L | Adsorption on activated alumina | pH adjustment |
| Uranium, $UO_2(CO_3)_3^{4-}$ | 0.02 mg/L | Anion exchange | Turbidity reduction |

manufacturing or processing comes into direct contact with an industrial product (e.g., vinyl chloride or polyvinyl chloride) or results from the production or use of any raw material, intermediate product, finished product, byproduct, or waste product containing the industrial product (e.g., vinyl chloride or polyvinyl chloride) but that has not been discharged in a wastewater treatment process or discharged untreated as wastewater. Gasholder seal water is not in-process wastewater until it is removed from the gasholder (EPA-40CFR61.61-j).

**input horsepower** The total power used in operating a pump and motor. It is equal to the brake horsepower multiplied by the motor efficiency. *See* power terms.

**input–output model** Same as black-box model.

**in-receiving water flow balance method** A technique that uses floating pontoons and flexible curtains to create an in-water facility to store combined sewer overflows, which displace the clean water during wet weather and are pumped to the collection system following the storm.

**insanitary** Unhealthy.

**insect-borne disease** A disease that is caused by a pathogen that uses an insect as a vector. Some insect-borne diseases are also water-related or excreta-related in that the vectors breed in water or in excreta, or bite near water, e.g., dengue, filariasis, loiasis, malaria, river blindness, sleeping sickness, and yellow fever. *See also* water-related disease and other categories of excreta-related diseases: fecal–oral, soil-transmitted, tapeworm, and water-based.

**insecticide** Any substance or chemical formulated to kill or control insects.

**insect vector route** A mode of transmission of water-related diseases (e.g., dengue, malaria, river blindness, yellow fever); insects that breed in or near water carry the pathogen from one host to another.

**in-series flow** A piping arrangement that directs the entire flow from one unit to another unit. *See also* in-parallel flow.

**insert valve** A shutoff valve that can be installed in a pressurized pipeline during operation.

**inside-out filtration** A water treatment process in which feedwater is applied inside a hollow-fiber membrane and permeate collected outside.

**inside-threaded connection** The bell end of a pipe section, which connects with the spigot end of another section. Also called female end.

**in situ** In place, in its original location, in the natural environment; unmoved; unexcavated; remaining in the subsurface. Pertaining to treatment or disposal methods that do not require movement of the waste.

**in situ air stripping** Same as in-situ stripping.

**in situ bioremediation** The in-place treatment or cleanup of a contaminated site; e.g., the introduction of nutrients and oxygen into contaminated soil and groundwater to promote the growth of microorganisms that will degrade the contaminants.

**in situ biotreatment** The use of microorganisms to degrade hazardous, toxic, and other wastes without excavating or pumping out the contaminated soil or water. *See* in situ treatment, bioaugmentation.

**in situ degradation** *See* in situ treatment.

**in situ hybridization** Same as fluorescent in situ hybridization.

**in situ natural filtration** Same as riverbank filtration.

**In-Situ Oxygenator™** A mechanical floating aerator manufactured by Praxair, Inc.

**in situ stripping** A treatment system that removes or "strips" volatile organic compounds from contaminated ground or surface water by forcing an air stream through the water and causing the compounds to evaporate. Also called air sparging and in situ air stripping.

**in situ treatment** (1) One of three common methods for the short-term renovation of contaminated groundwater, using such biological or chemical processes as addition of oxygen and trace nutrients, hydrolysis, oxidation–reduction reactions, and polymerization. *See also* containment, extraction. Also called in situ degradation. (2) The treatment of wastes in a lagoon, area of contaminated soil, or subsurface of a contaminated site, including the addition of reagents, mixing, and processing; term often applied to the solidification, stabilization, and encapsulation of hazardous wastes. *See also* ex situ treatment, in-plant processing, mobile-plant processing.

**in situ vitrification** A method of treatment of contaminated soil in place by heating it to 1000–2000°C using electrical current through electrodes in the soil surface, with graphite and glass frit between the electrodes to increase conductivity. The top layer melts and transmits heat to deeper layers for the decomposition of organic contaminants.

**in situ volatilization** A technique used for removing volatile organic compounds in unsaturated soils; see air sparging and soil-vapor extraction.

**insoluble** Pertaining to a substance that has a low solubility or is incapable of being dissolved or liquefied. In this book, when no solvent is mentioned for a constituent or substance, insoluble means water-insoluble.

**inspection and maintenance (I&M)** Activities to assure that vehicles' emissions controls work properly. The term also applies to wastewater treatment plants and other antipollution facilities and processes.

**Instant Ocean®** A chemical product of Aquarium System used to simulate seawater salinity.

**instantaneous disinfection byproduct concentration ($DBP_0$)** The initial concentration of disinfection by-products in a sample that is dechlorinated to capture the by-products at the time of sampling, as compared to a sample that is analyzed with a chlorine residual after a period of time and yields a terminal concentration ($DBP_t$). The difference ($DBP_t - DBP_0$) is the DBP formation potential.

**instantaneous dosing intensity** A measure of the hydraulic loading or dosing of a trickling filter, equal to the depth of wastewater applied for each pass of an arm of the distributor; it is sometimes referred to as the SK factor (in mm/pass of an arm), from the German word spülkraft, for dosing rate. Also called flushing intensity. *See* SK concept.

**instantaneous grab sample** A grab sample representative of conditions at the time of sampling, as compared to a composite sample.

**instantaneous point source** A single event contributing pollutants, e.g., a sneeze or a rainfall episode, as compared to continuous point sources.

**instantaneous total trihalomethane concentration ($TTHM_0$)** The initial concentration of total trihalomethanes (TTHM) in a sample that is dechlorinated to capture the TTHM at the time of sampling, as compared to a sample that is analyzed with a chlorine residual after a period of time and yields a terminal concentration ($TTHM_t$). The difference ($TTHM_t - TTHM_0$) is the TTHM formation potential.

**institutional use (pesticide)** Any application of a pesticide in or round any property or facility that functions to provide service to the general public or to public or private organizations, including: (a) hospitals and nursing homes; (b) schools other than preschool and day care facilities; (c) museums and libraries; (d) sports facilities; and (e) office buildings (EPA-40CFR152.3-n).

**institutional wastewater** Wastewater from hospitals, prisons, schools, and similar establishments.

**InstoMix** Mixing equipment manufactured by Walker Process Equipment Co. for installation in lines or in channels.

**in-stream use** A water use that takes place within the stream channel (e.g., hydroelectric power generation, navigation, water quality improvement, fish propagation, recreation), as compared to a consumptive use, which removes water from the available supply (e.g., manufacturing, agriculture, food preparation).

**Instream Water Temperature Model** A model of the U.S. Fish and Wildlife Service that predicts stream water temperatures from hydrological and meteorological conditions, and from stream geometry. Acronym: SNTEMP.

**in-stream water use** *See* in-stream use.

**instrumental detection level** One of a few parameters used to measure the precision or the detection level of a method of analysis, stated as a result that produces a signal greater than five times the signal-to-noise ratio of the instrument. *See also* criterion of detection and critical detection level.

**instrumentation** The set of instruments used in the water and wastewater field for control, monitoring, or analysis.

**instrumentation and control diagram** A simplified process flow diagram that shows the major mechanical equipment, instrumentation, and control elements. *See also* process and instrumentation diagram.

**instrumentation system** A group of equipment components used to condition and convey a sample of the process fluid to analyzers and instruments for the purpose of determining process operating conditions (e.g., composition, pressure, flow, etc.). Valves (0.5 inch and smaller) and connectors (0.75 inch and smaller) are the predominant type of equipment used in instrumentation systems (EPA-40CFR63.161).

**insufficient mineralization** A low concentration of some minerals in water, which may be associated with a health problem, e.g., fluoride deficiency (dental caries), iodide deficiency (goiter), and soft water (cardiovascular disease).

**in-system storage** Complete or partial storage of wet-weather flows within a combined sewer system to control overflows. *See* automatic sluice gate, bascule gate, inflatable dam.

**intake** The works, devices, or structure at the head of a raw-water treatment or any other water supply system. Also called intake structure, intake works, or waterworks intake. Ground or surface water intakes include wellpoints, aqueducts, intake cribs, intake towers, pumping stations, tunnels, pipelines, storage facilities. Intakes may consist of a simple submerged pipe or an elaborate structure. *See also* dry intake, wet intake.

**intake area** The surface area that receives and absorbs the water that reaches a groundwater formation or basin. *See also* recharge area.

**intake chamber** A chamber that gradually narrows to a tunnel at the head end of some hydraulic structures to minimize undesirable effects of currents.

**intake pathway** Ingestion, inhalation, dermal contact, or any other mechanism by which a hazardous agent is transferred to a receptor. Also called exposure route. *See also* exposure pathway.

**intake pipe** The head pipe that receives water in a pipeline system, or a pipeline that conveys water by gravity to a well.

**intake screen** A screen used to prevent fish from entering into an intake pipeline or canal.

**intake section** The bottom part of a well where water flows in.

**intake structure** *See* intake.

**intake tower** A structure that rises above the water surface of a river or an impoundment to draw water, including such appurtenances as gates, openings, screens, meters, and pumps.

**intake works** *See* intake.

**intangible assets** Nonphysical assets such as patents or goodwill.

**intangible benefit (cost)** A benefit (cost) that is not easily defined or quantified in monetary terms. *See also* primary benefit, secondary benefit.

**integral control** In wastewater process control applications, the value of a variable parameter based on a linear relationship with the accumulated error of a controlled parameter. *See also* the following types of control: cascade, feedback, feed-forward, on–off, proportional, proportional–derivative, proportional–integral, proportional–integral–derivative.

**integrated biological nutrient removal** "A wastewater treatment method that combines biological and chemical or physical unit operations and processes to reduce the concentrations of nitrogen and phosphorus in the plant effluent below the levels that would be attainable solely by synthesis in a typical secondary treatment facility" (WEF & ASCE, 1991). *See also* biological nutrient removal.

**integrated disinfection** The sequential application of disinfecting agents to enhance their effectiveness while limiting disadvantages such as the formation of disinfection by-products. Two examples of integrated disinfection currently under investigation (in 2007) are the use of chlorine dioxide before ozonation and the addition of ozone before ultraviolet light disinfection. *See also* disinfecting agent.

**integrated disinfection design framework (IDDF)** A design approach for chlorine disinfection units that improves on the concentration × time (C × T) concept. The IDDF algorithm translates disinfectant decay and microbial inactivation interactions from an ideal batch reactor model to a nonideal continuous-flow model.

**integrated exposure assessment** A summation over time, in all media, of the magnitude of exposure to a toxic chemical.

**integrated fixed-film activated sludge process (IFAS)** A variation of the activated sludge process that combines fixed and suspended biological growth in one reactor: e.g., small plastic sponges or rings in the aeration basin enhance biomass nitrification without increasing suspended solids retention time. IFAS is a small-footprint technology that can operate at loading rates twice those of conventional activated sludge (e.g., 40 pounds of $BOD_5$ per day per 1000 cubic feet). IFAS is used extensively in Germany for nitrification/denitrification, and in Broomfield, Colorado for upgrading the rated capacity of an existing plant.. *See* red worm, submerged biological contactors, and the following proprietary processes: Bio-2-Sludge®, BioMatrix®, Captor®, Kaldness®, Linpor®, and Ringlace®.

**integrated membrane bioreactor** *See* membrane biological reactor.

**integrated membrane system (IMS)** The combination of a membrane process with another water treatment process; e.g., coupling microfiltration with nanofiltration, or nanofiltration with a conventional coagulation–sedimentation–filtration system. This does not include required pretreatment or posttreatment processes. Also called a dual-membrane system.

**Integrated Pollution Control** An approach advocated in the United Kingdom Environmental Protection Act for the simultaneous, instead of isolated, consideration of all major emissions to land, air, and water in order to reduce overall environmental pollution.

**Integrated Risk Information System (IRIS)** A database, developed and maintained by the USEPA, on the adverse health effects of hundreds of substances.

**integrated sample** (1) Same as depth-integrated sample. (2) A composite wastewater sample.

**integrated virus detection** A combination of the infectivity assay and molecular virus detection methods. Integrated cell culture PCR, e.g., involves inoculating a sample onto cells, allowing viral replication to occur, extracting the viral DNA from the cells, and analysis using polymerase chain reaction (PCR).

**integrated waste management** Using a variety of practices to handle municipal solid waste, including source reduction, recycling, incineration, and landfilling.

**integrated water use** The coordinated and planned utilization of all water resources in an integrated manner. *See* conjunctive use.

**integrator** A device or meter that continuously measures and calculates total flows in gallons, million cubic feet, or some other unit of volume. Also called a totalizer.

**intelligent system** A program, commonly computer-based, that can respond to specific situations and propose valid options for action, e.g., in the management of water and wastewater works. Intelligent systems use such methods as expert systems, neural networks, fuzzy logic, genetic algorithms, or Bayesian networks. *See also* artificial intelligence.

**IntensAer** A surface aerator manufactured by Walker Process Equipment Co.

**intensity factor** *See* capacity factor (2).

**interaction** The additive, inhibitory, or synergistic effects of one substance or organism on another.

**interaction coefficient** A coefficient introduced in the Langmuir isotherm to account for lateral interactions at the surface of an adsorbent. *See* the Frumkin equation.

**interactive disinfection** The sequential or simultaneous application of two or more chemical agents to disinfect water or wastewater with a synergistic effect (i.e., the overall result exceeds the added effects of the agents applied separately).

**interbasin transfer** Transfer of water from one basin (watershed) to another.

**intercepting sewer** Same as interceptor.

**intercepting drain** A drain installed to intercept surface or ground water at the upper end of an area and carry the flow away. Also called curtain drain.

**interception** (1) The portion of precipitation retained on buildings and vegetation without reaching the ground because it eventually evaporates; the process itself. *See also* rainfall–runoff relationship. (2) The process of diverting wastewater from a main sewer (usually a combined sewer) for conveyance to a treatment plant. (3) A mechanism that contributes to the process of attachment in rapid rate filtration, whereby suspended particles flowing with the fluid contact or intercept the media grains. Interception is more significant with larger particles. *See also* Brownian motion, filtration mechanisms, mass transport process.

**interceptor** A sewer that intercepts wastewater and sometimes predetermined quantities of stormwater from a combined sewer system for conveyance to a treatment plant or to an outfall. Also called intercepting sewer or interceptor sewer.

**interceptor sewer** (1) Same as interceptor. (2) A major sewer line used to collect and convey flows from smaller lateral sewers in a separate sewer system.

**intercondenser** A condenser used between the stages of an evaporator to reduce steam consumption.

**interconnected ponds** Ponds in series are interconnected if a downstream pond affects the hydraulics of an upstream pond or if the water surface fluctuations at the outfall affect the hydraulics of the downstream pond. Otherwise, the series of ponds is standard. *See also* AdICPR and pond routing.

**interconnection** The physical connection between two networks, e.g., two water supply systems or two sewer networks. *See also* backflow connection and cross-connection.

**interface** The common boundary layer between two adjacent phases, e.g., between a liquid and a solid; or between two fluids such as water and a gas; or between a liquid (water) and another liquid (oil). Also called phase boundary, an important consideration in the study of kinetics of purification and treatment processes. *See* conjunction, diffusion, ion exchange, sorption.

**interface settling velocity** The settling velocity of the sludge/water interface at the beginning of a sludge settleability test. *See* zone settling velocity for detail.

**interface zone** In a water supply distribution system, an interface zone is an area in which waters from different sources, with different hydraulic and chemical characteristics, work against one another and remain longer than desired. *See also* water blending.

**interfacial Gibbs free energy** As a factor in adsorption mechanisms, the value of surface tension or interfacial tension ($\gamma$) when temperature ($T$), pressure ($P$), and the number of moles ($N$) are held constant. It is represented by the partial derivative of the Gibbs free energy ($G$) with respect to the interfacial area ($A$):

$$\gamma = (\partial G/\partial A)_{T,P,N} \qquad \text{(I-13)}$$

**interfacial sloughing** An important feature of trickling filters whereby the interface serves as the locus for the transfers between liquid and film, floc, or biomass as well as for the capture of the trickling filter humus.

**interfacial tension** The tension that occurs at the interface between two fluids or a liquid and a solid; the surface tension between water and air. The higher the interfacial tension of a compound with water, the easier it is to separate after mixing and the less likely to form emulsions. Interfacial tensions of organic compounds range from 0 for a completely miscible liquid (e.g., methanol, $CH_3OH$) to 72 dyn/cm for water at 25°C.

**interference** The discharge of sulfides in quantities that can result in human health hazards and/or risks to human life, and an inhibition or disruption of publicly owned treatment works (EPA-40CFR425.02pj).

**interference substance** A substance that reacts with chlorine and interferes with the determination of the chlorine residual.

**interflow** (1) The portion of precipitation that reaches a stream after infiltration into the soil and after lateral flow, without reaching the water table. Also called subsurface stormflow or throughflow, as opposed to the other part of subsurface runoff that reaches the saturated zone. *See also* rainfall– runoff relationship. (2) The movement of water between layers of different densities in a lake or reservoir. (3) Lateral movement of water in the upper layer of soil. *See also* unsaturated zone and upper zone storage.

**intergranular** Between the individual grains in a rock or sediment.

**Interim Primary Drinking Water Regulations** Regulations initially issued under the Safe Drinking Water Act of 1974. *See also* Primary Drinking Water Regulations.

**interior water use** The water used inside residences through various devices and appliances; sometimes used interchangeably with domestic wastewater rate because of the correlation between the two.

**intermediate clarifier** A sedimentation tank between two trickling filters or between a trickling filter and an aeration basin, as compared to a primary clarifier or a final clarifier.

**intermediate chlorination** The application of chlorine somewhere in the middle of the water or wastewater treatment train, as opposed to prechlorination and final chlorination.

**intermediate ozonation** The application of ozone to settled water. *See also* postozonation, preozonation.

**intermediate-pressure oxidation** One of two current methods of feeding sludge to a wet air oxidation unit: the sludge reacts with 45 cubic feet of air per pound of dry solids at a reactor temperature of 450°F and a pressure of 500–600 psig. *See also* high-pressure oxidation.

**intermediate-rate trickling filter** A trickling filter designed to operate between the low-rate and high-rate ranges, i.e., in the range of 4.0–10.0 mgd/ac. (million gallons per day per acre). *See also* high-rate trickling filter, low-rate trickling filter, roughing filter, super-rate filter, biofilter, biotower, oxidation tower, dosing tank, filter fly.

**intermediate technology** A technology developed in the industrialized world but designed to use materials, assembly, and maintenance resources available in a developing country. *See also* appropriate technology.

**intermediate treatment** (1) Wastewater treatment somewhere between primary treatment and advanced treatment, in terms of the removal of suspended solids and biochemical oxygen demand. (2) The removal of colloidal solids (intermediate in size between suspended and dissolved solids) from wastewater by chemical coagulation or other methods to reduce oxygen demand and turbidity.

**intermittent aeration** A nitrogen removal process carried out in one basin (single-sludge activated sludge process or oxidation ditch) in which the (aerobic) nitrification and the (anoxic) denitrification steps occur alternately. This requires timing the aeration equipment for the two cycles, adequate basin volume for the two steps, and a wastewater feeding arrangement for denitrification. Also called alternating aeration. *See also* Nitrox™, ORP knee, specific denitrification rate.

**intermittent cycle extended aeration system (ICEAS™)** A modification of the sequencing batch reactor process developed in Australia for wastewater treatment. It uses a batch decant reactor, separated by a baffle into a prereact compartment and a main chamber for the react, settle, and decant steps. The influent is fed continuously but withdrawal is intermittent.

**intermittent filter** A natural or artificial bed of granular materials (e.g., sand) used for the aerobic filtration of wastewater applied in intermittent flooding doses.

**intermittent sand filter** (1) A biological wastewater treatment device consisting of a bed of sand 18–30 inches deep underlain by 12 inches of coarse stones and graded gravel of 6–50 mm, with an underdrainage system of perforated or unjointed pipes to collect the effluent. Two applications of intermittent sand filters are solids separation for the effluent of small stabilization ponds or treatment of septic tank effluents. The sand has an effective size of 0.2–0.5 mm and a uniformity coefficient of 2–5. *See* Figure I-04. (2) A natural sand deposit converted into an intermittent filter for the biological treatment of wastewater.

**intermittent sand filtration** A wastewater treatment method, first tried at Medford (MA) in 1887, using filtration through a natural or artificial bed of sand, the flow being applied intermittently with or without primary or secondary pretreatment. Doses are applied daily every 10–20 minutes to a depth of 1–4 inches, with occasional surface scraping of accumulated solids and some sand replacement. Long resting periods between applications help air penetrate the bed and maintain aerobic conditions. Common hydraulic loadings vary from 3,000 gpd/acre for the irrigation of cultivated soils to 25,000 gpd/acre for grasslands, and from 80,000 gpd/acre for the treatment of raw wastewater to 500,000 gpd/acre for the treatment of a secondary effluent. Intermittent sand filtration achieves an average performance of 90% removal of suspended solids, 92% removal of BOD, and 96% removal of bacteria. Intermittent sand filtration is sometimes used to treat septic tank effluents where impermeable soils do not allow soil absorption systems.

**Inter-Mix®** A slow-speed mixer manufactured by Air-O-Lator Corp.

**intermixing** The phenomenon that occurs when media grains of adjacent layers mix in a multimedia filter. *See* filter layer intermixing for detail.

**internal friction** Friction within a fluid due to cohesive forces.

**Internalift®** A screw pump manufactured by Wheelabrator Engineered Systems, Inc.

**internal joint seal** A trenchless water main rehabilitation method using a synthetic rubber (EPDM) material to seal leaking joints; stainless steel bands are used to keep the seal in place. It is used in pipes that are large enough to allow worker access or the use of robots. *See also* trenchless technology.

**internal process disturbance** An internal factor that causes a change in the operation or performance of a treatment process, e.g., intermittent operation of a unit or changes in some operating parameters. *See also* external process disturbance.

**internal rate of return** The discount rate that equates the present worth of a project's estimated revenues to the present value of the project's estimated costs. It is calculated by setting the net present value formula to zero. *See* financial indicators for related terms.

**internal recycle (IR) rate** The recycle flow rate divided by the influent flow rate in any system with recirculation. It is an important factor in the determination of the specific denitrification rate. Also called internal recycle ratio.

**Internal recycle ratio** Same as internal recycle rate.

**internal transport** The third step in the adsorption process, in which the materials to be adsorbed (adsorbate) move through the pores of the adsorbing particles (adsorbent) by molecular diffusion and surface diffusion to adsorption sites. Also called pore transport. *See also* bulk diffusion transport, film diffusion transport, and adhesion.

**internal water** The second broad category of subsurface water. Unlike interstitial water, internal water is chemically combined with the rock formations, at enormous depths and under great pressures. *See* subsurface water.

**internal water treatment** Chemical treatment of boiler makeup water to prevent scale buildup or corrosive pitting of the boiler components.

**international (drinking water) standards** Standards established by such agencies as the World Health Organization and the European Community.

**international system of units** The version of the metric system that has been established by the International Bureau of Weights and Measures and is administered in the United States by the National Institute of Standards and Technology. The ab-

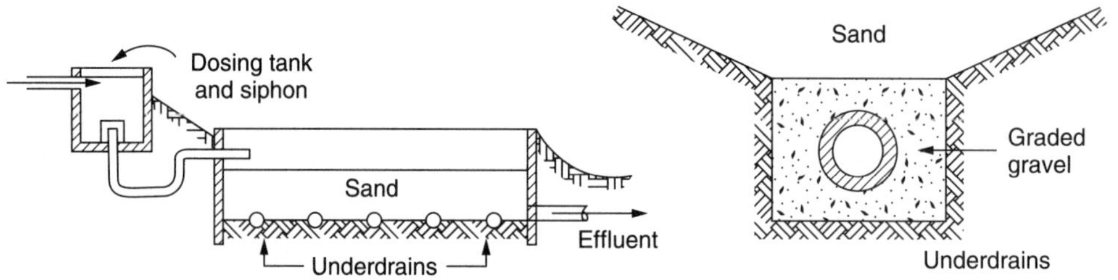


**Figure I-04.** Intermittent sand filter.

breviation for this system is "SI" (EPA-40CFR191.12). *See* Système International.

**interparticle bridging** A mechanism of destabilization in the coagulation process, whereby high-molecular-weight, anionic and nonionic polymers adsorb on and link several colloidal particles. *See also* double-layer compression, surface charge neutralization.

**Inter-Sep™** A rotary screen manufactured by Dontech, Inc.

**interspecies dose conversion** The process of extrapolating from animal doses to equivalent human doses.

**interspecies extrapolation model** Model used to extrapolate from results observed in laboratory animals to humans.

**interspecies hydrogen transfer** The utilization by methanogenic organisms of the hydrogen produced by acidogenic and other anaerobes during the fermentation process.

**interstate carrier** Any vehicle or transport that conveys passengers in interstate commerce.

**interstate carrier water supply** A source of water for drinking and sanitary use on planes, buses, trains and ships operating in more than one state. Such a source is federally regulated.

**interstate waters** Waters that flow across or form part of state or international boundaries, e.g., the Great Lakes, the Mississippi River, or coastal waters.

**interstice** A very small open space in a rock or granular material. Also called a void or void space. *See also* pore.

**interstitial velocity** A parameter ($V_i$) used to determine the Reynolds number of a fluidized bed and equal to the ratio of the superficial backwash velocity ($V_s$) to the porosity ($e$) of the bed:

$$V_i = V_s/e \qquad \text{(I-14)}$$

*See* modified Reynolds number, fluidized-bed porosity function.

**interstitial water** (1) Water contained in rock fractures (voids or interstices) comprising vadose water just below the ground surface and a deeper layer of phreatic or groundwater. *See* subsurface water. (2) Water contained in the interstices within flocs and cells of sludge. It flows with the floc or is held by capillary forces between the solid particles. *See also* bound water, bulk water, free water, hydration water, surface water, vicinal water.

**intervention value** In European practice, a level of groundwater contamination that requires remediation; it is several times larger than the natural or background concentration of a given contaminant. *See also* reference or target value.

**intestinal fluke** *See* giant intestinal fluke and fasciolopsiasis.

**intestinal nematode infection** *See* ancylostomiasis, ascariasis, enterobiasis, strongyloidiasis, and trichuriasis.

**intestinal tract** The portion of the alimentary canal following the stomach. *See also* gastrointestinal tract.

**intolerant fish species** Choice edible fish species that require cold water, high dissolved oxygen concentration, and low turbidity, e.g., grayling, white fishes, brook trout, rainbow trout. *See also* game fish, tolerant fish.

**intrachannel clarifier (ICC)** A proprietary clarifier developed for oxidation ditch wastewater treatment plants, incorporating a clarifier in the channel of the ditch or adjacent to the ditch with a common wall. Sludge is wasted from the aeration channel or the intrachannel clarifier. ICCs are less expensive and require less land than conventional oxidation ditches. However, they are subject to sludge deposits and they lack operational flexibility. *See also* boat clarifier, Burns and McDonnell treatment system, Carrousel Intraclarifier™, pumpless integral clarifier, side-channel clarifier, sidewall separator.

**Intracid®** A tracing dye produced by Crompton & Knowles Corp.

**intraparticle diffusion** The mechanism that carries the bulk of the adsorbate into and through the pores of the adsorbent during adsorption on granular activated carbon and similar materials with relatively small exterior surfaces. Also called intraparticle transport. By contrast, solution transport is assumed to be the controlling mechanism for powdered activated carbon and similar adsorbents with relatively large exterior surfaces

**intraparticle transport** Same as intraparticle diffusion.

**intrinsic bioremediation** The normal degradation of contaminants by indigenous microorganisms without any stimulation or treatment. Also called natural attenuation.

**intrinsic permeability** A coefficient ($k$) that expresses the ability of a porous medium to transmit a given fluid; a measure of the relative ease with which a permeable medium can transmit a fluid (liquid or gas). Intrinsic permeability is a property only of the medium and is independent of the nature of the fluid:

$$k = \mu K/\gamma = q\nu/gi \qquad \text{(I-15)}$$

where $\mu$ and $\nu$ = the dynamic and kinematic viscosities of the fluid, respectively; $K$ = hydraulic conductivity; $\gamma$ = specific weight of the fluid; $q$ = unit (or specific) discharge = discharge per unit cross-sectional area; $g$ = gravitational acceleration; and $i$ = hydraulic gradient. Also called specific permeability, sometimes expressed in darcys.

**intrinsic property** A characteristic of a body that cannot be measured directly, e.g., energy, mass, and momentum, but may be derived from its relationships with extrinsic properties such as weight, temperature, concentration, density, and flow velocity, which are measurable but not conserved. Intrinsic properties are conserved, i.e., they can be used to predict changes in the body in accordance with the laws of conservation. Examples of fundamental relationships between intrinsic and extrinsic properties are: (a) heat content or heat energy = the product of density, specific heat, temperature, and volume; (b) mass = the product of density and volume; (c) momentum = the product of mass and velocity, i.e., the product of density, volume, and velocity. *See also* the difference between fundamental and derived quantities.

**intrinsic resistance** The reciprocal of intrinsic permeability, as applied to a porous medium or to a sludge cake.

**intrinsic water** Pure water that has essentially no ion or mineral content. Also called polished water. *See also* distilled water.

**intrusion** A backflow situation in which outside contaminated water enters a potable water supply through leaks, submerged valves, faulty seals, or other openings. *See also* backsiphonage, pressure transient.

**intrusive rock** An igneous rock formation with very low porosity and permeability; thus, a nonproductive aquifer unless it is fractured. *See also* extrusive, metamorphic, and sedimentary rocks.

**inverse miscibility** Pertaining to a substance that is fully miscible with water at low temperatures (e.g., < 40°F) but only slightly miscible above that temperature; e.g., aliphatic amines used to dewater sludge and partially remove organic contaminants. *See* B.E.S.T.® amine extraction process.

**inverse sludge index** The reciprocal of the sludge volume index times 100. *See* sludge density index for detail.

**inverse solubility** The characteristic of a substance whose solubility decreases with increasing temperature.

**invert** The floor, bottom, or lowest point of the internal surface of a conduit, aqueduct, sewer, tunnel, manhole, or canal, at any cross section.

**invertebrate** An animal that does not have a backbone.

**inverted block rate or inverted rate** A water rate structure in which the unit price increases for each successive block; e.g., \$0.01 per gallon for the first 1000 gallons per month, \$0.015 per gallon for the next 5000 gallons, etc. It is similar to inclining block rate and the opposite of declining block rate.

**inverted siphon** A section of a gravity sewer constructed lower than adjacent sections. It is not a siphon but a U- or V-shaped section dropped below the hydraulic grade line to pass under an obstruction or obstacle (railway, highway cut, stream, subway, valley, tidal estuary, or other depression). The inverted siphon flows full under positive pressure. Also called depressed pipe, depressed sewer, dive culvert, sag line, or sag pipe. Depressed sewers are usually designed as multiple pipes, one of which carries the minimum flow and maintains self-cleansing velocities. Besides the pipe, the inverted siphon includes inlet and outlet chambers. An inverted siphon may also be used (instead of a flume) when a canal must cross a depression. *See* Figure I-05.

**in-vessel composting** A system where sludge composting occurs in a partially or completely enclosed container with equipment for odor control, aeration, mixing, and material handling. There are two types of in-vessel composting: plug flow or static bed and agitated bed or dynamic. Also called enclosed-reactor composting, mechanical composting. *See also* other composting systems: aerated static pile and windrow.

**investor-owned water utility** A public water system owned by shareholders.

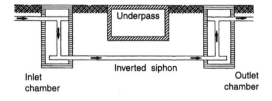

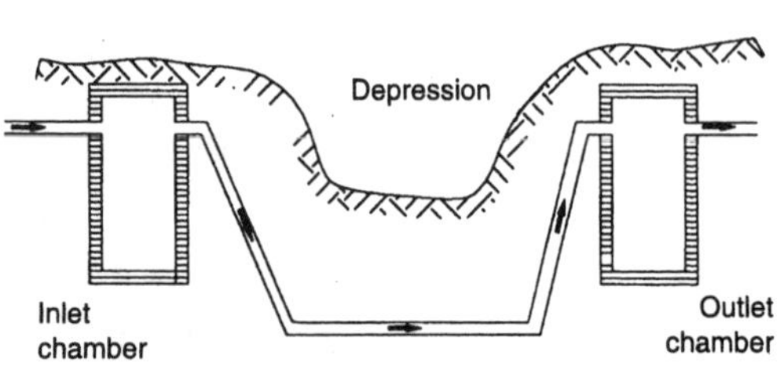


**Figure I-05.** Inverted siphons (courtesy CRC Press).

**in vitro** In glass; a laboratory experiment performed in a test tube or other vessel.

**in vitro study** A study of chemical effect conducted in tissues, cells, or subcellular extracts from an organism (i.e., not in the living organism).

**in vitro test** A toxicity test conducted in a glass petri dish or in a test tube. *See also* toxicity terms.

**in vivo** Within a living organism; a laboratory experiment in which the substance under study is inserted into a living organism.

**in vivo study** The study of chemical effects conducted in intact living organisms.

**in vivo test** A toxicity test conducted using a whole living organism. *See also* toxicity terms.

**$IO_3^-$** The radical representing an iodate.

**IOC** Acronym of inorganic chemical, inorganic compound, inorganic constituent, inorganic contaminant.

**iodamine** A hypothetical compound of iodine and ammonia.

**iodate** A common compound of iodine identified by the radical $IO_3^-$, e.g., potassium and sodium iodates ($KIO_3$, $NaIO_3$); one of the forms of iodine in aqueous solutions.

**iodide ($I^-$)** A common compound of iodine, e.g., potassium and sodium iodides (KI, NaI); a form very often used to represent iodine. Iodides in natural waters originate from industrial wastes, natural brines, and seawater intrusion. *See* goiter.

**iodination** (1) The use of iodine to destroy pathogens, in particular to inactivate viruses in drinking water of low turbidity. Iodine is also used to disinfect swimming pools, which leaves a free residual of 0.2 to 0.6 mg/L. *See also* bromination and chlorination. (2) The addition of iodine salts to water to prevent goiter. Also called iodine treatment.

**iodine ($I_2$)** A bluish-black, nonmetallic chemical element, slightly soluble in water. Atomic weight = 126.90. Atomic number = 53. Specific gravity = 4.93. Gas density = 11.27 g/L. Valence = 1, 3, 5, or 7. Its blue-violet vapor is dangerous to the eyes. Used in medicine and in the treatment of swimming pool water (iodination). Iodine or iodide is an effective disinfectant (e.g., against the cercariae of pathogenic schistosomes, cysts, and spores) and has a relatively low level of toxicity. Unlike chlorine and bromine, it does not react with ammonia or organic nitrogen to any appreciable extent (i.e., no iodoamines). On the other hand, iodine deficiency may cause goiter; potassium iodide (KI) is added to table salt at the rate of 0.1 mg per gram of sodium chloride (NaCl) to prevent it. Iodine is not widely used in water treatment because of its high cost and its negative effects on thyroid activity; *see* tetraglycine hydroperiodide.

**iodine-127** A stable isotope of iodine.

**iodine-131** An artificial radioisotope of iodine, a fission product of nuclear explosions and from nuclear reactors in power plants; half-life = 8 days.

**iodine deficiency** Insufficient iodine intake (in food and water) may result in goiter, which is sometimes endemic in isolated rural villages.

**iodine number** One of three parameters used by manufacturers to express the relative capacity of an activated carbon by testing it against a standard solution for the adsorption of small molecules. It measures the amount of iodine that the sample of activated carbon will adsorb under specified conditions (e.g., milligrams of iodine per gram of carbon from a 0.02 normal solution). For example, AWWA recommends a minimum iodine number of 500 for water treatment by granular activated carbon, but commercial products have higher numbers (AWWA, 1999). *See also* molasses number, phenol number. Other related parameters include carbon tetrachloride activity and methylene blue number.

**iodine treatment** *See* iodination (2).

**iodism** Chronic iodide poisoning.

**iodoform** A trihalomethane and a disinfection by-product; chemical formula $CHI_3$. Also called triiodomethane.

**iodometric chlorine test** A test conducted to determine the chlorine residual in water or wastewater by addition of potassium iodide and titration of the iodine released using sodium thiosulfate and a colorimetric indicator.

**iodometric methods** Two of a few methods available for the determination of chlorine residuals in water and wastewater; iodide added to a sample is oxidized and then titrated with thiosulfate or the reducing agent is added first and then titrated with iodine or iodate. *See also* amperometric titration, azide modification, DPD method.

**ion** (1) An electrically charged, unstable atom or radical (e.g., $SO_4^{2-}$) that is formed by the loss or gain of one or more electrons and can bond electrostatically to an oppositely charged species; ions can be removed from water or wastewater by ion exchange or electrodialysis. *See also* cation (a positive ion) and anion (a negative ion). In water, ions tend to form hydrates instead of remaining free. (2) Any of several electrically charged submicroscopic particles found in the atmosphere.

**Ionac®** Ion-exchange resins produced by Sybron Chemicals, Inc.

**ion balance** A calculation used to verify the accuracy of the laboratory analysis of a water sample.

It shows an equality of the concentrations of cations and anions expressed in milliequivalents per liter or in mg/L as calcium carbonate. It is used in the determination of carbonate, noncarbonate, and total hardness. *See also* hardness distribution diagram, ionic proportion diagram.

**ion charge** In the computation of the equivalent weight of a compound, the molecular weight (in grams) is divided by the ion charge, which is also the number of hydrogen ($H^+$) or hydroxyl ($OH^-$) ions a species can react with or yield in an acid–base reaction. It is also defined as the absolute value of the change in valence that occurs in a redox reaction.

**ion chromatography** A laboratory method using ion exchange to separate chemical substances and analyzed ions in water.

**ion exchange (IE** or **IX)** (1) A reversible chemical process involving the exchange of ions between an insoluble solid medium with ions in a surrounding solution, the direction of the exchange depending on the selectivity of the medium and the ion concentrations. Both anion exchange and cation exchange are used in water treatment. Common media include zeolites or greensands, bentonite clay, and synthetic organic resins. Ion exchange is currently used for many purposes: desalination, nitrogen removal, iron and manganese removal, and softening (exchange of calcium and magnesium ions for sodium ions). *See also* acid retardation, anionic exchange, cationic exchange, deionization, hydrogen cycle, ion exchange treatment, macroporous resin matrix, regeneration, saline-water conversion classification, sodium cycle, strong-acid cation exchange, strong-base anion exchange, weak-acid cation exchange, and weak-base anion exchange. (2) More generally, any replacement of an ion in a solid phase by another ion in a solution, e.g.,

$$CaCO_3(s) + Sr^{2+} \rightarrow SrCO_3(s) + Ca^{2+} \quad \text{(I-16)}$$

**ion-exchange breakthrough capacity** The volume of effluent or total quantity of ions passed through an ion-exchange bed or column before saturation and regeneration. It varies with flow rate, column dimensions, temperature, solution composition, and particle size of the medium. *See also* breakthrough capacity.

**ion-exchange capacity** (1) The quantity of the target ions that can be replaced by a unit volume or weight of the exchange material; usually expressed in milliequivalents per milliliter of wet resin (meq/mL) or in milliequivalents per gram of dry resin (meq/g). It is usually higher for synthetic than natural resins: e.g., 4.0–8.0 me/mL vs. 0.1–1.0 me/mL. *See also* operating capacity. (2) "The ion exchange capacity of a soil (or of soil-minerals in waters or sediments) is the number of moles of adsorbed ion charge that can be desorbed from unit mass of soil, under given conditions of temperature, pressure, soil solution composition, and soil-solution mass ratio" (Sposito quoted in Stumm & Morgan, 1996).

**ion-exchange chromatography** Same as high-performance liquid chromatography.

**ion-exchange demineralization (IXDM)** A water treatment process that completely removes both anions and cations, e.g., for water softening or for the production of pure and ultrapure water for laboratory and industrial purposes. Ion-exchange demineralization is not widely applied for drinking water because of its high costs compared to alternative membrane processes.

**ion-exchange equilibria** The equilibrium relationships that govern the distribution of ions between the phases. *See* binary separation factor, concentration quotient, electroselectivity, electroselectivity reversal, equilibrium concentration quotient, favorable isotherm, heterovalent exchange, homovalent exchange, ion-exchange isotherm, ion-exchange process, selectivity, selectivity coefficient, selectivity quotient, selectivity reversal, selectivity sequence, selectivity series, separation factor, unfavorable isotherm, unit isotherm.

**ion-exchange isotherm** A constant-temperature equilibrium plot of resin-phase concentrations (y-coordinates) versus aqueous-phase concentrations (x-coordinates). In the exchange of ion $i$ (e.g., $Cl^-$) for ion $j$ (e.g., $NO_3^-$), the isotherm is favorable, i.e., convex to the x-axis or above the line y = x. The isotherm is unfavorable if it is concave to the x-axis, as in the case of an exchange of bicarbonate ($HCO_3^-$) for chloride ($Cl^-$). It is a unit isotherm if equivalent fractions ($x_i$) and ($y_i$) are used instead of concentrations (AWWA, 1999). *See* ion exchange equilibria for related terms.

**ion-exchange membrane** A highly selective membrane consisting of a thin film of ion-exchange resins cast in flat sheets and used as a semipermeable barrier that lets some constituents (ions) pass while retaining others (e.g., water). It has a polymeric support structure with fixed anion or cation sites and water-filled passages that retain ions while passing counterions. Also called electrodialysis membrane; used in electrodialysis and electrodialysis reversal. *See also* electrodialytic separation.

**ion-exchange process** The ion-exchange operation involves two reversible reactions, governed by the mass law, a charge balance, and equilibrium concentration quotients. The pairs of reactions representing cation and anion exchange are:

$$H^+R^- + Na^+ \leftrightarrow Na^+R^- + H^+ \quad \text{(I-17)}$$

$$2\,Na^+R^- + Ca^{2+} \leftrightarrow Ca^{2+}R_2^{2-} + 2\,Na^+ \quad \text{(I-18)}$$

$$R^+Cl^- + OH^- \leftrightarrow R^+OH^- + Cl^- \quad \text{(I-19)}$$

$$2\,R^+OH^- + SO_4^{2-} \leftrightarrow R_2^{2+}SO_4^{2-} + 2\,OH^- \quad \text{(I-20)}$$

where $R^+$ and $R^-$ represent, respectively, the anion and cation exchangers. The equilibrium concentration quotients or (selectivity coefficients) derived from these reactions characterize the selectivity of an exchanger for a specific ion (e.g., $Ca^{++}$). Each pair of reactions is sometimes summarized in one reaction to show the net result of ion exchange. For example, the following represent the overall reactions for a strong-acid cation exchanger and a weak-base anion exchanger, respectively:

$$2\,RSO_3H + 2\,Na^+ + Ca^{2+} \leftrightarrow (RSO_3)_2Ca + 2\,Na^+ + 2\,H^+ \quad \text{(I-21)}$$

$$2\,RNH_3OH + 2\,Cl^- + SO_4^{2-} \leftrightarrow (RNH_3)_2SO_4 + 2\,Cl^- + 2\,OH^- \quad \text{(I-22)}$$

**ion exchanger** An insoluble, natural or synthetic, material with sufficient space in its porous structure for ions to pass freely in and out. Examples include soil materials (clay, humus, etc.), alumina, greensand zeolite, and resins. *See* anion exchanger, cation exchanger.

**ion-exchange reactions** *See* ion-exchange process.

**ion-exchange regeneration** *See* ion-exchanger regeneration.

**ion-exchange resin** A cross-linked, insoluble polymer matrix supporting bead-like materials by covalent bonding. Most resins are polystyrene-based with divinylbenzene as a cross-linking agent. These materials represent functional groups that can exchange acceptable anions or cations for less desirable ions. Ion-exchange resins are used in water treatment, e.g., in ion-exchange softening. *See also* acid retardation, anion exchanger, cation exchanger, chelating resin, functionality, porosity, resin exchange capacity, strong-acid resin, strong-base anion resin, weak-acid resin, and weak-base anion resin.

**ion-exchanger regeneration** The periodic restoration of the properties of an exhausted ion-exchange bed by filling it with a nearly saturated solution (of sodium chloride, NaCl, for example) for about 15 minutes, draining it, and wasting enough product water to obtain the desired quality. Regeneration cycle is defined by the volume that can be treated or the time between the cycles. *See also* empty bed detention time.

**ion-exchange selectivity coefficient** *See* selectivity coefficient.

**ion-exchange softener** A device that uses ion exchange to remove hardness constituents (mainly calcium and magnesium) from water.

**ion-exchange softening** The most important application of ion exchange to drinking water treatment; a process that replaces hardness ions (mainly $Ca^{2+}$ and $Mg^{2+}$, but also some $Fe^{2+}$, $Mn^{2+}$, and $Sr^{2+}$) by sodium ions. It is used both at the household and municipal scales. Sodium chloride (NaCl) is used for regeneration. Split treatment allows partial removal of hardness.

**ion-exchange treatment** (1) A common water softening method often found on a large scale at water purification plants that remove some organics and radium by adding calcium oxide or calcium hydroxide to increase the pH to a level at which the metals will precipitate out. (2) More generally, the use of the ion-exchange process to remove undesirable ions from water or wastewater by substituting more acceptable ions.

**ion-exchange water softener** *See* ion-exchange softener.

**ion-exclusion chromatography** A laboratory method that separates chemical constituents on the basis of their molecular sizes; smaller molecules are retained longer in the chromatography column than larger molecules.

**Ion Grabber** A treatment apparatus manufactured by Atlantes Chemical Systems, Inc. using electrolysis.

**ionic bond** The electrostatic bond formed by the transfer of electrons between two ions; a weaker bond than a covalent bond. Also called electrovalent bond. Same as electrovalence (1).

**ionic concentration** The concentration of any ion in solution, usually expressed in moles per liter. *See also* ionic strength.

**ionic constant** A parameter that indicates to what extent a substance dissociates in a solution. *See also* ionization constant, ion product.

**ionic product** *See* ion product.

**ionic product of water** *See* ion product of water.

**ionic proportion diagram** A graphical representation of the percentages of the major anions and cations in a water sample; e.g., bicarbonate ($HCO_3^-$), carbonate ($CO_3^{2-}$), chloride ($Cl^-$), sulfate

($SO_4^{2-}$), calcium ($Ca^{2+}$), magnesium ($Mg^{2+}$), and potassium ($K^+$), with all concentrations expressed in nilliequivalents per liter or as calcium carbonate ($CaCO_3$). *See also* hardness distribution diagram, ion balance..

**ionic reaction** An oxidation reaction that involves species with paired electrons. The five major types of ionic reaction in aqueous solutions are (a) addition of hypobromous (HOBr) and hypochlorous (HOCl) acids to carbon–carbon double bonds to form halohydrins; (b) formation of organohalides by activated aromatic substitution; (c) formation of a N-halo compound by transfer of a halogen onto a nitrogen atom; (d) formation of an inorganic halide ion with oxygen transfer; and (e) oxidation with electron (and no atom) transfer (AWWA, 1999). *See also* radical reactions.

**ionic strength** A measure of strength of a solution or chemical potential of its electrolytes, based on their concentrations and valences, as defined by the following equation, called the Lewis–Randall correlation:

$$\text{ionic strength} = 0.5\ \Sigma(c_i z_i^2) \qquad \text{(I-23)}$$

where $\Sigma$ represents summation for all ions and $c_i$ and $z_i$ represent, respectively, the concentration (in moles per liter) and the valence of the *i*th ion. Ionic strength is a measure of the interionic effect caused by electrical attractions and repulsions between the ions. The ionic strength of a soil solution influences transport mechanisms by altering the size of the diffuse double layer and the soil structure. *See also* hydration radius, ionic concentration, Langelier's ionic strength approximation, Russell's ionic strength approximation, conductivity.

**ionic weight** The sum of the atomic weights of the components of an ion.

**ionizable compound** A compound that can be separated or changed into ions, e.g., acids, amines, and phenols.

**ionization** (1) The splitting or dissociation of molecules into negatively and positively charged ions. For example, water ($H_2O$) ionizes to hydrogen ($H^+$) and hydroxyl ($OH^-$) ions. Ionization of carboxyl and amino groups is responsible for the development of surface charges on proteins and microorganisms. Ionization occurs in the hydroxyl group of metallic compounds exposed on the surface of clay particles; it increases with pH. *See also* Arrhenius theory, water ionization, cation exchange capacity. (2) The process of an atom acquiring a net electrical charge by gaining or losing an electron.

**ionization constant** A value indicating the tendency of a molecule to ionize in solution at a given temperature. The ionization constant of water ($K_w$) at 25°C is approximately $1.008 \times 10^{-14}$. Hence,

$$K_w = [H^+][OH^-] = 1 \times 10^{-14} \qquad \text{(I-24)}$$

$$pH + pOH = 14 \qquad \text{(I-25)}$$

where the square brackets represent molar concentrations and pH and pOH are the negative logarithms of the hydrogen and hydroxyl ion concentrations, respectively. The ionization constant of acetic acid ($CH_3COOH$) in water at 25°C is $1.76 \times 10^{-5}$. *See also* affinity constant, apparent dissociation constant, association constant, dissociation constant, equilibrium constant, ion product, Michaelis constant, solubility product.

**ionization fraction** A number that indicates to what extent a substance dissociates in water. It is the ratio of the molar concentration of the substance produced to the analytical concentration (total number of moles of the pure substance). *See* dissociation fraction for more detail.

**ionization potential** The energy required to remove an electron from the outer shell of the molecule of a compound, expressed in electron volts (eV = 23,053 cal/mol). It is used in determining which compounds will ionize under a given ultraviolet lamp.

**ionizing radiation** Radiation with wavelengths in the range 0.001–100 nm that can strip electrons from atoms and produce ions in materials, i.e., alpha, beta, and gamma radiation. Exposure to ionizing radiation (ionization) causes biological damage by producing broken chemical bonds. Cosmic rays from outer space are a source of ionizing radiation less significant environmentally than the radioactive substances produced by humans. Ionizing radiation is a powerful disinfectant but does not leave any residual.

**ionizing radiation inactivation** Inactivation of microorganisms using free radicals through the ionizing radiation generated by radioactive materials such as cesium-127 or cobalt-60 and high-energy electron beams. This method is used for sludge treatment.

**ion-mobility spectrometry** A technique used to measure pollutants in gaseous mixtures.

**ionography** A standard method in the electronic industry to measure the amount of ionic concentration on a surface.

**ion pair** Two particles of opposite charge or two ions of opposite charge joined by electrostatic attraction in an aqueous solution; e.g., a metal ion and a ligand. The two constituents form a temporary compound and are separated by one or more water molecules. Also called outer-sphere complex. *See also* complex.

**ion-pair extraction** A technique used to extract a constituent or compound (e.g., a metal or an organic substance) from a solution by forming an ion pair with another constituent of opposite charge. An example is the extraction of methylene blue substances with an organic solvent by forming an ion pair with an ionic substance.

**ion product** The solubility product of a dissociation reaction: the product of the molar concentrations of the reactants raised to powers equal to their respective numbers of ions. For example, the ion product ($K_{sp}$) of the dissolution of the solid $A_xB_y$ is:

$$K_{sp} = [A]^x \cdot [B]^y \qquad \text{(I-26)}$$

*See also* ionization constant.

**ion product of water** The product of the molecular concentrations of the constituent ions of water, $[H^+]$ and $[OH^-]$:

$$K_w = [H^+][OH^-] \qquad \text{(I-27)}$$

which may also be written as

$$pK_w = pH + pOH \qquad \text{(I-28)}$$

where the "p" operator indicates the negative common logarithm of the constituent. $K_w$ depends on temperature: at 25°C, $K_w = 1.008 \times 10^{-14}$, $pK_w = 14.0$, and a neutral solution has pH = pOH = 7.0. Also called ionic product of water. *See also* acidity constant, basicity constant.

**Ionpure®** Water treatment products and services of U.S. Filter Corp.

**ion-selective electrode (ISE)** An electrochemical probe that has a high degree of selectivity for one ion over other ions in a sample; used in chemical analysis to measure the electrode potential and concentration of ammonia ($NH_4^+$), dissolved oxygen, fluoride ($F^-$), hydrogen ion ($H^+$), etc., and some metals: arsenic ($As^{3+}$), cadmium ($Cd^{2+}$), lead ($Pb^{2+}$), potassium ($K^+$), and sodium ($Na^+$). *See also* atomic absorption spectroscopy, inductively coupled plasma atomic emission.

**ion-selective membrane** A membrane used in electrodialysis that lets specific ions pass while retaining others.

**ion selectivity** The capacity of an ion-exchange resin to let certain ions pass and retain others.

**Ion Stick®** A water treatment unit manufactured by York Energy Conservation to control scale formation and fouling.

**ion toxicity** *See* specific ion toxicity.

**ion transfer** A unit operation of water or wastewater treatment that moves ions into or out of the fluid. Ion transfer processes include adsorption, chemical coagulation, chemical precipitation, and ion exchange.

**Iopor** An ultrafiltration unit manufactured by Dorr-Oliver, Inc. with low pressure.

**IPO** Acronym of intermediate-pressure oxidation.

**IQS/3™** A programmable device manufactured by Bruner Corp. to control water treatment operations.

**IQ-Tox Test™** A kit prepared by Kingwood Diagnostics to test for water threat contaminants selected by the USEPA: aldicarb, colchicines, cyanide, dicrotophos, thallium sulfate, botulinum toxin, ricin, soman, and VX.

**IRA 458** A hydrophilic, polyacrylic resin manufactured by Rohm and Haas Company and used in ion exchange.

**IRIS** Acronym of Integrated Risk Information System.

**iron (Fe from its Latin name, ferrum)** A white, malleable, ductile, metallic element, with very abundant ores in the earth's crust (see diderate, hermatite, limonite, magnetite, pyrite). Atomic weight = 55.85. Atomic number = 26. Specific gravity = 7.87. Valence = 2, 3, 4, and 6. Common compounds include oxides, carbonates, silicates, chlorides, sulfates, and sulfides. Ferric and ferrous salts are very soluble in water; colloidal and suspended or settled, ferric and ferrous compounds are also found in water. Precipitation of oxidized iron compounds causes water to have a metallic taste and (at a concentration greater than 0.3 mg/L) to stain clothes, utensils, and plumbing fixtures. On the other hand, like manganese and zinc, iron may combine with the cement matrix of asbestos–cement pipe to form a protective coating against corrosion. The metal corrodes rapidly. Trace amounts of iron are essential for animal and plant growth. Iron is the inorganic element that affects ultraviolet absorption most. *See also* hemochromatosis, red water.

**iron-55** A radionuclide produced by high-energy neutrons acting on iron in weapons hardware. Half-life = 2.7 years.

**iron and manganese removal** Iron and manganese are released in solution when dissolved oxygen concentration is low. When oxidized to insoluble ferric hydroxide [$Fe(OH)_3(s)$] and manganese dioxide [$MnO_2(s)$], they precipitate as reddish-brown or black deposits. They can be removed from water by aeration or adsorption on activated carbon. Wetlands are also used for the treatment of acid mine drainage. *See also* deferrization, demanganization, biological iron and manganese removal.

**iron and manganese sequestering** Use of sequestering agents (such as sodium silicate, phosphates, and polyphosphates) to bind iron and manganese.

**iron arsenic (FeAsS)** The common white-to-gray mineral, arsenopyrite, an arsenic ore. Also called mispickel.

**iron bacteria** Bacteria that can metabolize reduced iron or cause the dissolution of ferrous iron and deposition of ferric oxide in or on their secretions. In water supply systems they may be a source of turbidity, clog well screens, cause taste and odor problems, or increase the rate of corrosion by interfering with the development of passivating scales. *See also* black deposit, *Crenothrix, Ferrobacillus, Frenothrix, Gallionella, Leptothrix, Sphaerotilus,* red water. Measures against iron bacteria and black deposits in water supplies include a dissolved oxygen of 2.0 mg/L, a free chlorine residual of 0.2 mg/L, and a pH above 7.2; treatment of open reservoirs with copper sulfate is also effective.

**iron CMF process** A water treatment process that uses coagulation by iron salts and direct microfiltration to remove arsenic (As) from groundwater. It does not include the customary flocculation step. *See also* coagulation–microfiltration.

**iron coagulant** *See* iron salts.

**iron coagulation–microfiltration process** *See* iron CMF process.

**iron complexing** The formation of stable iron and manganese complexes of less than 1.0 mg/L in water by the addition of polyphosphates or silicates that sequester the metals.

**iron disulfide ($FeS_2$)** *See* iron sulfide and marcasite.

**iron fouling** The reduction of the capacity of an ion-exchange resin or granular filter medium by the accumulation of iron on or within the resin bed.

**iron monoxide** Same as ferrous oxide.

**iron oxide** Same as ferric oxide. The precipitation of iron oxide is a cause of turbidity in treated corrosive water

**iron oxide adsorption** The adsorption of contaminants, e.g., natural organic matter, on the surface of synthetic ferric oxide ($Fe_2O_3$) within a granular medium.

**iron-oxide-coated olivine** An iron–magnesium silicate [$(Mg, Fe)_2SiO_4$] coated with synthetic ferric oxide ($Fe_2O_3$); used to adsorb organic matter from water.

**iron-oxide-coated sand** Sand used in fixed-bed reactors for water or wastewater treatment; a granular material coated with a synthetic ferric oxide ($Fe_2O_3$) prepared from a heated mixture of sodium hydroxide (NaOH) and ferric nitrate [$Fe(NO_3)_3 \cdot 6 H_2O$].

**iron-oxidizing bacteria** Bacteria that oxidize iron, causing the corrosion of water pipes at the tubercles that they form. *See also* iron bacteria.

**iron pan** A strongly compacted, thin soil layer underlying soft materials and cemented with ferric oxide. *See* hardpan for other similar layers.

**iron perchloride** *See* ferric chloride.

**iron phases** Various compounds resulting from the corrosion of iron pipes in water, including: ferrihydrite, goethite, green rust, hematite, lepidocrocite, maghemite, magnetite, siderite, and vivianite. *See also* corrosion scale.

**iron pyrite ($FeS_2$)** *See* fool's gold, iron sulfide, marcasite.

**iron redox barrier** An oxidation–reduction gradient and sorption capacity established to immobilize a hydrocarbon plume and contain the contamination of an aquifer. *See also* biowall concept.

**iron removal** Dissolved iron ($Fe^{2+}$) concentration in drinking water can be reduced by aeration, which renders the metal insoluble and removable by sedimentation or filtration. *See also* iron and manganese removal.

**Iron Remover** A contact bed unit manufactured by Walker Process Equipment Co. for iron removal.

**iron salts** Compounds of iron used for coagulation in water and wastewater treatment include ferrous sulfate ($FeSO_4$), chlorinated copperas ($FeCl_3$), and ferric sulfate [$Fe_2(SO_4)_3$].

**iron sludge** The residuals from the treatment of water or wastewater using iron salts as coagulants, which settle to the bottom of sedimentation basins along with the impurities removed. *See also* alum sludge, brine, concentrate, lime sludge, polymeric sludge.

**ironstone** Any iron-bearing mineral with silicate impurities.

**iron sugar** A common name of ferrous sulfate ($FeSO_4$).

**iron sulfates [$Fe_2(SO_4)_3 \cdot x\ H_2O$] or ($FeSO_4 \cdot 7 H_2O$)** Coagulants commonly used in water treatment; they react with the natural alkalinity of water or the added lime to form a ferric or ferrous hydroxide floc. *See* ferric sulfate and ferrous sulfate for more detail.

**iron sulfide ($FeS_2$)** An iron and sulfur compound that causes black deposits in dead-end water supply lines. Also called iron disulfide, iron pyrite, marcasite. *See also* iron bacteria.

**iron trichloride** Same as ferric chloride.

**iron vitriol ($FeSO_4 \cdot 7\ H_2O$)** A bluish-green, saline-tasting, coagulant (iron sulfate) commonly used in water treatment and for other industrial purposes (fertilizer, ink, medicine, etc.); see ferrous sulfate for more detail.

**irradiance** Same as irradiation (3).

**irradiation** (1) Exposure to radiation of wavelengths shorter than those of visible light (gamma, X-ray, or ultraviolet), for medical purposes, to sterilize milk or other foodstuffs, or to induce polymerization of monomers or vulcanization of rubber. (2) Treatment by exposure to radiation, as of ultraviolet light; a method of food preservation that may produce free radicals. (3) A measure of the intensity of radiant energy per unit area; same as irradiance.

**irreducible benefit** A benefit that is real but cannot be expressed in exact monetary terms, e.g., water pollution control.

**irreducible residual** The total chlorine residual at the breakpoint. *See* breakpoint curve and Figure B-17.

**irreducible wastes** Wastes generated in petroleum refining by the separation of unwanted impurities, e.g., salts in desalter brine.

**irregular weir** A weir that has a nonstandard or irregular crest.

**irreversible effect** Effect characterized by the inability of the body to partially or fully repair injury caused by a toxic agent.

**irreversible reaction** A reaction that proceeds in only one direction and continues until exhaustion of the reactants.

**irrigation** Artificial application of water or wastewater to land areas to supply the water and nutrient needs of plants and grass.

**irrigation efficiency** (1) The amount of water stored in the crop root zone compared to the amount of irrigation water applied. (2) The percentage of soil moisture increase for consumptive use in terms of the water applied that can be accounted for.

**irrigation return flow** Surface and subsurface water that leaves the field following irrigation; it is often contaminated with dissolved salts, fertilizers, and pesticides. In litigation, it may be limited to measurable water returning to the stream from which it was diverted. Also called irrigation return water.

**irrigation return water** Same as irrigation return flow.

**irrigation scheduling** A procedure designed to save water and energy by determining how much water a field requires and when to apply it. It maximizes irrigation efficiency and restores soil moisture depending on evapotranspiration.

**irrigation water quality** Parameters used to determine the suitability of water and treated effluents for crop irrigation include the total dissolved solids (or electrical conductivity), sodium adsorption ratio, nematode eggs, and fecal coliforms.

**irritant** A substance that can cause irritation of the skin, eyes, or respiratory system. Effects may be acute from a single high level exposure, or chronic from repeated low-level exposures to chlorine or such compounds as nitrogen oxide and nitric acid.

**Irving–Williams order** The sequence of complex stability for metal ions. It shows the following order for transition metal cations and other bivalent metals (Stumm & Morgan, 1996):

$$\text{Transition metals: } Mn^{2+} < Fe^{2+} < Co^{2+} < Ni^{2+} < Cu^{2+} > Zn^{2+} \quad \text{(I-29)}$$

$$\text{Bivalent metals: } Be > Mg > Ca, Ba, Sr > Sn > Pb > Zn > Cd > Hg \quad \text{(I-30)}$$

Also called Irving–Williams series.

**Irving–Williams series** Same as Irving–Williams order.

**ISE** Acronym of ion-selective electrode.

**ISEP®** A contactor manufactured by Advanced Separation Technologies for continuous adsorption and desorption.

**ISO** Acronym of International Organization for Standardization.

**ISO 14000** A standard of the ISO concerning environmental management in the manufacturing and services sectors.

**isobaric** Having or showing equal or constant pressure.

**isochlor** An imaginary line connecting on a map all the points of equal chloride concentration, e.g., in groundwater or estuaries.

**isochoric** Having equal or constant volume.

**isodrin** The pesticide 1,4:5,8-dimethanonaphtalene,1,2,3,4,10,10-hexachloro-1,4,4a,5,8,8a-hexahydro (EPA-40CFR704.102-3).

**isoelectric point (pI)** The pH at which the net charge on a substance is neutral; an important factor in such water treatment processes as protein precipitation, membrane filtration, and coagulation. For bacteria, pI = 2.5 – 3.5; they are negatively charged around pH = 7.0. Viruses have pIs ranging from 2.5 to 8.5. As the pH of a surrounding solution decreases from the neutral range, viruses shift their surface proteins from a negative to a positive charge. *See also* amphoteric compound, electrophoretic separation, point of zero charge, point of zero net proton charge.

**isogram** Same as isoline.

**isoline** A line joining points of equal value with respect to a given variable. Also called isogram, isopleth.

**ISOLUX™ Arsenic Treatment Systems** Water treatment systems manufactured by MEI for the removal of heavy metals (arsenic, cadmium, chromium, and lead).

**isomeric ratio** The ratios of ortho-, meta-, and para-chlorinated terphenyls, under the Toxic Substances Control Act (EPA-40CFR704.45-3).

**isomerism** The state, condition, or phenomenon characterized by the relation between two or more compounds, radicals, or ions that share the same kinds and numbers of atoms but have different configurations, structural formulas, or properties. Isomerism results from the multibonding and chaining ability of carbon and other atoms or groups of atoms. Also called structural isomerism, as compared to nuclear isomerism and optical isomerism.

**isomers** (1) Two or more compounds having the same molecular formula, but different chemical structures and different properties, due to different arrangements of the atoms. For example, there are 35 isomers of $C_9H_{20}$. *See* Figure I-06 for the structural formulas of two isomers of dichloroethane, $C_2H_4Cl_2$. (2) Two or more atomic nuclei having the same atomic number and mass number, but different energy states and half-lives. Also called nuclear isomers. *See also* isotope.

**isomorphic replacement** Same as isomorphic substitution.

**isomorphic substitution** One of the three common processes that impart a charge to mineral and organic particles in water, in addition to reactions with the water or its solutes. In isomorphic substitution the surface charge arises from imperfections in the structure of a particle, particularly in clay minerals. For example, the substitution of an aluminum (Al) atom for a silicon (Si) atom in a silica ($SiO_2$) tetrahedron results in the formation of a negatively charged framework. Also called isomorphic replacement, isomorphous replacement.

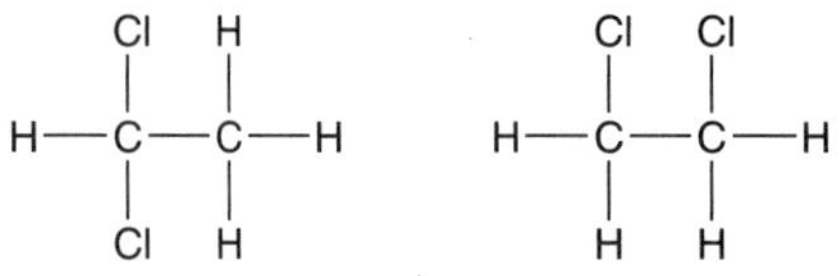

**Figure I-06.** Isomers (dichloroethane).

**isomorphous replacement** Same as isomorphic substitution.

**isopleth** *See* isoline.

**isopropanol ($C_3H_8O$ or $CH_3$—$CH_3$—CHOH)** Same as isopropyl alcohol.

**isopropyl** Containing the univalent isopropyl group or isopropyl radical ($C_3H_7$ or $CH_3$—$CH_3$—CH), an isomer of the propyl group.

**isopropyl alcohol ($C_3H_8O$ or $CH_3$—$CH_3$—CHOH)** Common name of 2-propanol, a colorless, flammable, water-soluble liquid produced from propylene by the action of sulfuric acid and hydrolysis; used as a solvent and in the production of antifreeze and rubbing alcohol. Also called dimethylcarbinol, isopropanol, secondary propyl alcohol.

**isopropyl group** *See* isopropyl.

**isopropyl radical** *See* isopropyl.

***p*-isopropyltoluene [$CH_3C_6H_4CH(CH_3)_2$]** The solvent para-cymene.

**isopycnic** Of equal or constant density; *see* isosteric.

***Isospora belli*** The protozoal species that causes isosporiasis.

**isosporiasis** A rare and little-known protozoal infection of the human intestinal tract, caused by the pathogen *Isospora belli.*

**ISO standard** A standard developed by the International Organization for Standardization and adopted by many countries.

**ISO standard day conditions** A temperature of 288°K, a relative humidity of 60%, and a pressure of 101.3 kilopascals (EPA-40CFR60.331-g).

**isosteric** Of equal or constant specific volume; see isopycnic.

**isotherm** (1) A line connecting points of equal temperature (real or at mean sea level), e.g., to illustrate regions on a weather map or temperature-induced density currents in a water body. (2) A graphical representation of the equilibrium relationship, based on experimental measurements, between adsorbate, adsorbent, and solution at a given temperature. *See also* adsorption isotherm, BET, Freundlich, and Langmuir isotherms. (3) *See* ion-exchange isotherm.

**isothermal process** Any thermodynamic change of state of a system under conditions of constant temperature. *See* isothermy.

**isothermy** The condition of an isothermal body of water, i.e., one that is mixed and maintains the same temperature throughout; e.g., in some lakes in spring and fall.

**Table I-04.** Isotopes

| Isotope | Radiation | Half-life | Isotope | Radiation | Half-life |
|---|---|---|---|---|---|
| Bromine-78 | β, γ | 6.4 min | Radium-226 | α, γ | 1600 yr |
| Carbon-14 | β | 5800 yr | Radon-222 | α, γ | 3.8 days |
| Cesium-137 | β, γ | 30 yr | Sodium-24 | β, γ | 15 hours |
| Cobalt-60 | β, γ | 5.3 yr | Strontium-90 | β | 29 yr |
| Iodine-129 | β, γ | $1.7 \times 10^7$ yr | Sulfur-35 | β | 88 days |
| Iodine-131 | β, γ | 8.1 yr | Thorium-232 | α, γ | 3.3 days |
| Iron-59 | β | 45 days | Uranium-233 | α, γ | 160,000 yr |
| Phosphorus-32 | β | 11 yr | Uranium-235 | α, γ | $7.1 \times 10^8$ yr |
| Plutonium-239 | α | 24,000 yr | Uranium-238 | α, γ | $4.5 \times 10^9$ yr |
| Potassium-40 | β, γ | $1.3 \times 10^9$ yr | Zinc-65 | β, γ | 245 days |

**isotope** Any of two or more forms of a chemical element that have the same number of protons in the nucleus, the same atomic number, but a different number of neutrons and different atomic weights. Isotopes are represented by the chemical symbol of the element preceded by its mass number in superscript, e.g., hydrogen ($^1H$), deuterium ($^2H$), tritium ($^3H$), uranium-233 ($^{233}U$), uranium-235 ($^{235}U$), and uranium-238 ($^{238}U$). Most elements have two or more isotopes; various isotopes of the same element have the same chemical behavior, but different physical or radioactive behaviors; some are highly unstable. Table I-04 lists the type of radiation and half-life of some important isotopes. *See also* isomer (2), radioisotope.

**isotropic** Having the same physical properties in all directions or along all axes; applicable to a medium's elasticity, heat conduction, light transmission, and, in the case of an aquifer, permeability. The opposite is anisotropic.

**isotropic (porous) medium** A porous medium whose permeability is the same in all directions. The opposite is an anisotropic medium.

**isotropy** The characteristic of a porous medium whose permeability does not change with direction. It is, along with homogeneity, a common assumption in the analysis of groundwater problems, whereas the opposite, anisotropy, is often the case, particularly in sedimentary deposits.

**itai-itai** Bone deterioration caused by cadmium poisoning. (From a Japanese word meaning ouch-ouch).

**iTSS** Acronym of inert total suspended solids.

**Ives equation** A semiempirical equation proposed by K. J. Ives in 1960 and based on the purification function to express the progressive changes in filtration rate during a filter run (Fair et al., 1971):

$$k = k_0 + c\sigma - \sigma\varphi^2/(f_0 - \sigma) \qquad \text{(I-31)}$$

where $k$ = rate constant, which changes during the filter run; $k_0$, $c$, $\varphi$ = initial rate constant and two coefficients that describe a specific unit, respectively; $\sigma$ = specific deposit or volume of deposited matter per unit volume, the principal variable as the filter run progresses; $f_0$ = porosity of the clean filter bed. The rate coefficient reaches a maximum at a critical specific deposit.

**IWA model** Same as ASM1 model.

**IWT®** Acronym of Illinois Water Treatment, a series of products manufactured by the U.S. Filter Corp.

**IX** Abbreviation of ion exchange.

**IXDM** Acronym of ion-exchange demineralization.

# J

**Jackbolt™** Aluminum covers manufactured by Enviroquip, Inc. for clarifiers.

**jacketed pump** A pump with jackets around cylinders and other parts for handling materials that melt when heated.

**jacking and boring** A method used to install sewers below highways and railroads without interfering with traffic; hydraulic jacks on one side drive the pipes through a hole opened by a cutting head while an auger draws out the excavated materials.

**JackKnife** An apparatus manufactured by Walker Process Equipment Co., consisting of an air header and a drop pipe.

**Jackson candle turbidimeter** An instrument used to determine the turbidity of a water sample. It consists of a candle and a calibrated glass tube on a supporting frame. Because of its limitations for measuring low turbidities, it is rarely used now.

**Jackson turbidity unit (JTU)** A standard measure of turbidity based on the observation of the outline of a candle; also called the visual method. The FTU (formazin turbidity unit, based on a known chemical reaction) and the NTU (nephelometric turbidity unit, measured by an instrument) are more reproducible and more currently used; however, the JTU is sometimes preferred for the examination of highly turbid waters.

**Jacob's equation** *See* storage coefficient.

**JAC Oxyditch** A wastewater treatment plant manufactured by Chemineer, Inc. using oxidation ditches.

**jar test** (1) A laboratory procedure that simulates a water treatment plant's coagulation/flocculation units with differing chemical doses, energy of rapid mix, energy of slow mix, and settling time. The purpose of this procedure is to estimate the minimum or ideal coagulant dose required to achieve certain water quality goals. Samples of water to be treated are commonly placed in six jars. Various amounts of chemicals are added to each jar, and the settling of solids is observed. The dose of chemicals that provides satisfactory removal of turbidity and/or color is the dose used to treat the water being taken into the plant at that time. When evaluating the results of a jar test, the operator should also consider the floc quality in the flocculation area and the floc loading on the filter. (2) The jar test applied to sludge conditioning. *See* standard jar test, filter leaf test, capillary suction time test, standard shear test, Büchner funnel test.

**jar test apparatus** An automatic stirring apparatus with an arm and six test tubes corresponding to the beakers that contain the water to be treated. It is used to conduct jar tests; chemicals (coagulants, coagulant aids, alkalinity agents) are added simultaneously to the beakers. Also used to determine the appropriate dose of powdered activated carbon.

**Jaszka–CIP process** A process used to generate small quantities of chlorine dioxide ($ClO_2$) on the site of water treatment plants, using sodium chlorate ($NaClO_3$) and sulfur dioxide ($SO_2$):

$$2\ NaClO_3 + SO_2 \rightarrow 2\ ClO_2 + Na_2SO_4 \quad \text{(J-01)}$$

*See also* the chloride reduction, Mathieson (or sulfur dioxide), Solvay (or methanol, $CH_3OH$) processes.

**jaundice** (1) A condition caused by an excess of bile pigment deposition in the skin giving a yellow appearance to the patient, especially in the eyes. *See also* icterus, hemorrhagic jaundice or Weil's disease. (2) Common name of hepatitis-A or infectious hepatitis.

**Javelle water** (1) A solution of sodium hypochlorite (NaOCl) in water, used as a bleach, antiseptic, etc. Also called eau de Javel, Javel water, Javelle water. After the former town of Javel, now a part of Paris, France. (2) Chlorine gas dissolved in an alkaline potassium solution, used for waste treatment in France in the 19th century and as a cholera prophylactic agent in Europe in 1831.

**Javel® water** A proprietary disinfectant that contains about 1% of available chlorine by weight; it can be used directly as a stock solution for water disinfection. Such a solution is unstable under warm conditions.

**Javex-12™** Sodium hypochlorite produced by Javex Manufacturing Corp.

**JAWS Bio-Sponge™** An odor control apparatus manufactured by Jacobs Air Water Systems of Tampa, FL, including an organic medium to react with hydrogen sulfide ($H_2S$) and mercaptans.

**Jayfloc** A polyelectrolyte produced by Exxon Chemical Co. for use as a flocculant in water and wastewater treatment.

**Jeffrey-1** Screening equipment manufactured by Jones & Attwood, Inc.

**Jeffrey®-2** Chain, sprocket and other equipment manufactured by Jeffrey Chain Corp. for use in treatment plants.

**jejunum** The part of the small intestine extending from the duodenum to the ileum.

**JelClear™** A granular filtration medium with a bonded coagulant produced by Argo Scientific.

**JETA** A vortex grit collector manufactured by Jones & Attwood, Inc. *See* Figures J-01 and J-02.

**jet action** The phenomenon of stratification by size of the grains of a single-medium filter during backwash with fluidization, the finer grains resting on top and the coarser grains at the bottom. The tendency to stratify depends on the backwash rate and the differences in bulk densities of the various grain sizes. Also called gulf streaming, sand boil.

**jet aeration** A wastewater aeration system using floor-mounted nozzle aerators that entrain air in the liquid in the form of very fine bubbles; unlike diffusers, it is not prone to plugging. Jet aeration combines pumping with air diffusion and is particularly applied in deep aeration basins (> 25 ft). *See also* flexible diffuser, sparger, porous diffuser, nonporous diffuser, U-tube aeration.

**Jeta-Matic®** Spray jet apparatus manufactured by U.S. Filter Corp. for cleaning pressure leaf filters.

**Jet Breaker™** Equipment manufactured by Mahr for washing and compacting screenings from wastewater treatment.

**Jet-Chlor®** An apparatus manufactured by Jet, Inc. for water disinfection with chlorine tablets.

**jet diffuser** *See* jet aeration.

**jet ejector** *See* jet pump.

**JetIII®** A fabric filter manufactured by Wheelabrator Clean Water Systems, Inc.

**JetMix** A unit manufactured by A. O. Smith Harvestore products, Inc. to store and mix sludge.

**jet mixing** A rapid mixing technique used in water and wastewater treatment. *See* pressurized water jet mixer.

**Jet Plant** A package plant manufactured by Jet, Inc. for wastewater treatment.

**jet pump** A pump using an accelerating jet of air to entrain another fluid and deliver it at a higher

**Figure J-01.** JETA grit removal system (view). (Courtesy Jones & Attwood.)

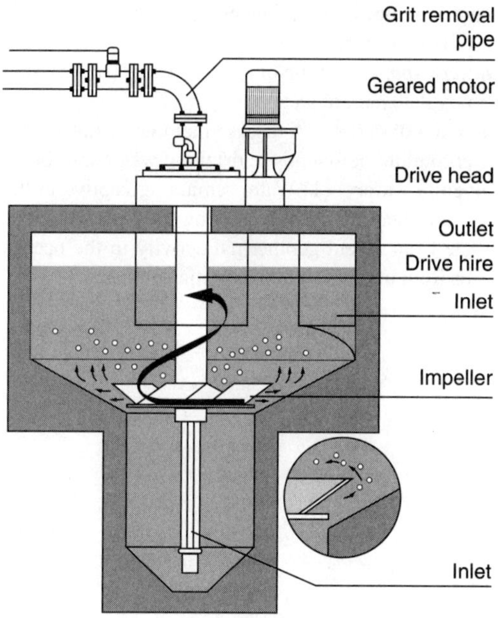


**Figure J-02.** JETA grit removal system (basic elements). (Courtesy Jones & Atwood.)

pressure. Jet pumps and jet ejectors are used in wells and for dewatering.

**jetsam** Material deliberately discarded overboard, which may sink, float, or be washed ashore. *See also* flotsam.

**Jet Shear** Continuous-mixing equipment manufactured by Flo-Trend Systems, Inc.; uses jet nozzles.

**jetted tubewell** A small well in unconsolidated deposits, constructed by sinking a pipe into soil loosened by water that is pumped through the pipe. The flow washes out the material to let the casing advance with the hole. *See also* palm and sludger method.

**Jet-Tex®** A fabric manufactured by Jet, Inc. for leach bed filters.

**jetting** (1) The process of constructing a jetted tubewell. (2) A difficulty sometimes encountered in the operation of a filter, when porosity and permeability differences between sand and gravel cause backwash water to break through at various points on the surface. Slow opening of backwash valves and surface washing before fluidization may help reduce the extent of this phenomenon. *See also* filter crack, mud ball, sand boil, and sand leakage.

**Jet Tray** A deaerator manufactured by Cochrane Environmental Systems.

**Jet-Wet™** An apparatus manufactured by Fluid Dynamics, Inc. for dry polymer feed.

**Jigrit** A screw grit washer manufactured by Jeffrey Division/Indresco.

**J-Mate** A sludge heating and shearing apparatus manufactured by JWI, Inc.

**jogging** The frequent starting and stopping of an electric motor.

**Johannesburg process** A biological nutrient (nitrogen and phosphorus) removal process that consists of four biological reactors in the following sequence: anoxic-1, anaerobic, anoxic-2, aerobic, followed by a final clarifier. The influent enters the anaerobic reactor; activated sludge is recycled as a nitrate source from the mixed liquor to anoxic-2 and from the clarifier underflow to anoxic-1. *See also* the VIP process, the UCT process, and the modified UCT process.

**John Snow** *See* Snow, Dr. John.

**Johnson Screen** A wedgewire screen manufactured by Wheelabrator Engineering Systems, Inc.

**joint** The connection between two pipe sections, often with a third part that may include nuts, bolts, joint compound, etc. Typically, iron pipe joints for water supply are of the following varieties: ball, bell-and-spigot, Dresser coupling, flanged, mechanical, push-on, threaded, and Victaulic coupling. Concrete pipes used in sewer work typically have joints of rubber gasket, mortar or mastic packing, O-ring gasket, or spigot groove with O-ring. *See also* flexible joint.

**joint compound** A synthetic compound used instead of lead in pipe joints for water tightness.

**joint treatment** The treatment of industrial and municipal wastewaters in the same plant, with or without industrial pretreatment. It is usually done in a municipal plant, with the costs assigned according to flow, BOD, suspended solids, and other parameters; *see* industrial cost recovery. Joint treatment normally results in cost savings through economies of scale but also may improve treatability if one waste supplies the nutritional deficiencies of another.

**joule** A measure of energy, work, or quantity of heat; an SI unit. One joule is the work done when a force of one Newton is displaced a distance of one meter in the direction of the force. 1 joule = 0.948 BTU = 0.239 calorie.

**J-Press®** A plate-and-frame filter press manufactured by JWI, Inc.

**J-Track™** A chain-and-flight device manufactured by FMC Corp. for use in sludge collectors.

**JTU** Acronym of Jackson turbidity unit.

**JUD** A tracking device manufactured by Klein America, Inc. for belt filter presses.

**junction box** A protective enclosure for the connection or termination of electrical wires and cables. In sewer networks, a manhole or other access point to a sewer line or a pump station.

**junction chamber** A chamber or large conduit section for the junction of two or more conduits.

**junction invert** The invert of the lowest conduit connecting to a junction. Such information may be available on as-built drawings or may be determined from sewer line length and slope.

**junction manhole** A manhole at the intersection of two or more sewers.

***Juncus spp.*** A group of aquatic plants (rushes) that populate natural wetlands.

**juvenile diabetes** A disease that may result from a chromium deficiency in drinking water and food.

**juvenile water** (1) Water remaining captive in the lithosphere. (2) Water entering the hydrologic cycle from recent geothermal activity in the oceans or from underground magmatic sources.

# K

**K** Chemical symbol of the metallic element potassium, from its latin name kalium.

**kala azar** A visceral, sometimes fatal, disease; a form of leishmaniasis, transmitted by sandflies.

**Kaldnes® process** A proprietary integrated fixed-film activated sludge method of wastewater treatment of Purac Engineering, Inc. of Wilmington, Delaware, developed by the Norwegian company Kaldnes Miljoteknologi. It uses film growth supported by small cylindrical polyethylene elements in the aeration tank. *See also* Bio-2-Sludge®, BioMatrix®, Captor®, moving-bed biofilm reactor, Ringlace®, submerged rotating biological contactor.

**kalium** Latin name of potassium (K).

**$K_2Al(Si_6Al_2)O_{20}(OH, F)_4$ or $KAl_3Si_3O_{10}(OH)_2$** Chemical symbol of muscovite.

**$KAlSi_3O_8$** Chemical symbol of orthoclase.

**kame** A ridge, hill, or mound of stratified drift left by a retreating ice sheet.

**KAMET** A system developed by Krofta Engineerirng Corp. to treat municipal wastewater effluents.

**Kan-Floc™** A chemical produced by Kem-Tron for use as a coagulant/flocculant in wastewater treatment.

**kaolin** A fine, white clay material, high in aluminum, used in the manufacture of porcelain; sometimes used in water treatment as a weighting agent to enhance settling. Also called china clay or white clay. It is a decomposition of feldspar and is essentially hydrated aluminum disilicate or kaolinite. *See also* hydrolysis (5).

**kaolinite [$Al_2Si_2O_5(OH)_4$]** The mineral from which kaolin is extracted, essentially hydrated aluminum disilicate, formed by the alteration of feldspar and other minerals.

**karst** A geologic formation with characteristics of relief and drainage arising from a high degree of rock solubility in natural waters. The majority of karst occurs in limestones, but karst may also form in dolomite, gypsum, and salt deposits. Features associated with karst terrains include irregular topography, sinkholes, vertical shafts, abrupt ridges, caverns, abundant springs, and/or disappearing streams (EPA-40CFR300-AA).

**Katec®** A thermal oxidizer produced by Grace TEC Systems.

**Kat-Floc™** A chemical produced by Kem-Tron for use as a coagulant/flocculant in water and wastewater treatment.

**Katox** A catalytic oxidizer produced by Adwest Technologies, Inc.

**KCl** Chemical formula of potassium chloride; sylvite.

**$KCl \cdot MgCl_2 \cdot 6\ H_2O$** Chemical formula of carnallite.

**$K_2Cr_2O_7$** Chemical formula of potassium dichromate.

**$K_2CrO_4$** Chemical formula of potassium chromate.

**KD-HF™** A deionization unit fabricated by Kinetico Engineered Systems, Inc.

**Kebab™** A disc oil skimmer manufactured by Vikoma International.

**Kedem–Kactchalsky equation** A relationship that describes the irreversible transport of solutes across a reverse osmosis membrane. It is similar to a coefficient ($L$) that relates the flux of a solute to the applicable driving force (AWWA, 1999):

$$L = 1/(\mu R_m) \qquad \text{(K-01)}$$

where $\mu$ is the absolute viscosity of water and $R_m$ (reciprocal of length) is the hydraulic resistance of a clean membrane.

**kelvin** or **Kelvin (K)** The SI unit of absolute or thermodynamic temperature equal to 1/273.16 of the absolute temperature of the triple point of water (ice, liquid, and vapor). Also called degree Kelvin. *See also* Celsius degree.

**Kemmerer sampler** A vertical device for collecting samples of water with suspended sediments.

**Kenics®** Static mixers manufactured by Chemineer, Inc.

**Kenite®** Diatomite product of Celite Corp.

**Kessener brush** Same as Kessener brush aerator.

**Kessener brush aerator** A rotating horizontal cylinder with submerged bristles or a cylindrical metal brush used as a horizontal surface aerator, providing aeration and mixing in oxidation ditches or relatively shallow tanks.

**ketoacid** Any of a number of organic compounds having the ketone and carboxylic acid groups in their structure, including some disinfection by-products.

**ketone** A hydrocarbon derivative in which the carbonyl group (C : O) is linked to two carbon atoms: R · CO · R. An example is acetone; Figure K-01. Ketones are used as solvents and some are formed as disinfection by-products. *See also* haloketone.

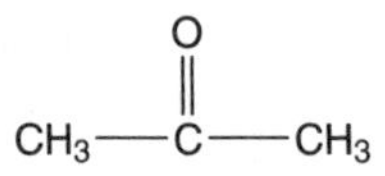

**Figure K-01.** Ketone (acetone).

**Key-Tech** Products of Keystone Engineering & Treatment Technology Co. for use in water and wastewater treatment.

**KF** Chemical formula of potassium fluoride.

**k-factor, K-factor** (1) A factor used in the computation of minor or local head losses caused by geometric changes, obstructions, or fittings. It serves to express these minor head losses as a function of the kinetic energy or velocity head, or as an equivalent length of pipe. In streams, the minor loss coefficient varies from 0.1 for a gradual contraction to 0.6 for an abrupt contraction, and from 0.3 to 0.8 for similar expansions. For valves and fittings in pipes, the k-factor may vary widely, for example, from 0.2 for a wide-open gate valve, to 5.6 for a half-open gate valve, to 10 for a wide-open globe valve. *See* equivalent pipes and minor losses. (2) A parameter of the rate of oxygen transfer from air to wastewater, depending on aeration equipment and wastewater characteristics.

**$K_3Fe(C_2O_4) \cdot 3H_2O$** Chemical formula of potassium ferric oxalate.

**K-Floc™** A flocculant/coagulant produced by Kem-Tron for use in wastewater treatment.

**K-Floor** A monolithic component manufactured by PWT Americas for use as a suspended floor in filters.

**$KH_2PO_4$** The chemical formula of potassium dihydrogen phosphate.

**kg** Abbreviation of kilogram.

**kiln** A furnace, oven, or any heated enclosure for processing a substance by baking, drying, or burning bricks, pottery, and refractory products. Kilns are used in thermal processing of sludge and hazardous wastes. *See* rotary kiln for detail.

**kiln dust** *See* cement kiln dust.

**kiln residence time** A design parameter for rotary kilns, indicative of the average time ($T$, min) that the materials spend in the unit and defined by the following equation (Freeman, 1998):

$$T = 106.2\ L\theta^{0.5}/(SDN) \qquad \text{(K-02)}$$

where $L$ = kiln length, ft; $\theta$ = dynamic angle of repose, degrees from horizontal; $S$ = kiln slope, ft/ft; $D$ = inside diameter, ft; and $N$ = rotational speed of kiln, revolutions per minute.

**kilo** A prefix meaning "thousand" used in the metric system and other scientific systems of measurement. It is also used as an abbreviation for kilogram or kilometer.

**kilogram** The unit of mass in the MKS (meter-kilogram-second) system, abbreviated, kg, equal to 1000 grams or approximately 2.205 pounds.

**Kincannon–Stover model** A relationship between substrate utilization rate and organic loading developed in 1982 for the design of trickling filters, based on the Monod equation (WEF & ASCE, 1991):

$$A_s = [(8.34\ QS_0/\mu_{max}S_0)/(S_0 - S_e)] - K_b \quad \text{(K-03)}$$

where $A_s$ = total media surface area, 1000 ft²; $Q$ = influent flow rate, mgd; $S_0$ = influent soluble $BOD_5$, mg/L; $\mu_{max}$ = maximum specific substrate utilization rate of $A_s$, lb $BOD_5$/day/1000 ft²; $S_e$ = effluent soluble $BOD_5$, mg/L; and $K_b$ = proportionality constant, lb/day/1000 ft². Other trickling filter design formulas include British Manual, Eckenfelder, Galler and Gotaas, Germain, Howland, Logan, NRC, modified Velz, Rankin, Schulze, and Velz.

**kinematic dispersion** The mass transfer caused by mechanical mixing. Also called mechanical dispersion. *See also* diffusion and dispersion.

**kinematic viscosity ($\nu$)** The ratio of dynamic viscosity to mass density. Kinematic viscosity is a measure of a fluid's resistance to gravity flow; the lower the kinematic viscosity, the easier and faster the fluid will flow. For liquids, it decreases with increasing temperature. Also see viscosity and Reynolds' number.

**kinematic viscosity coefficient** Same as kinematic viscosity.

**kinetic coefficient** A coefficient used in kinetic models; e.g., the synthesis yield coefficient ($Y$), specific substrate utilization rate ($U$), maximum specific substrate utilization rate ($k$), half-velocity constant ($K_s$), and endogenous decay coefficient ($k_d$) used in modeling the suspended growth treatment processes. Kinetic coefficients are used in the design, operation, and performance evaluation of wastewater treatment units, along with such other parameters as hydraulic detention time, food-to-microorganism ratio, and solids retention time.

**kinetic constant** *See* kinetic coefficient.

**kinetic energy** Kinetic energy is the energy of a moving body as a result of its motion. The kinetic head is the element of total dynamic head of a pump that represents the kinetic energy or velocity head $V^2/2\ g$, $V$ being the average fluid velocity and $g$ the gravitational acceleration. Same as velocity head. *See also* dynamic head.

**kinetic head** *See* kinetic energy.

**kinetic law of mass action** The thermodynamic principle according to which the formation of products in a chemical reaction depends on the concentration of the reactants and products, among a number of factors (AWWA, 1999). In the reaction between hypochlorous acid (HOCl) and bromide ($Br^-$), for example:

$$HOCl + Br^- \rightarrow HOBr + Cl^- \quad \text{(K-04)}$$

The following equation expresses the kinetic law of mass action:

$$d[HOBr]/dt = k_f[HOCl][Br^-] \quad \text{(K-05)}$$

where the concentrations are in molar units and $k_f$ is the forward reaction rate constant in liters per mole per unit time. *See also* reaction rate constant, rate law.

**kinetic model** A common type of model developed to represent the biological, chemical, and physical processes of water quality. Kinetic models focus on the transformation of constituents, their concentrations, and the reaction-rate constants. Examples of kinetic models are nutrient cycles and stream oxygen balance. *See also* ecological models and biocoenotic models.

**kinetic pump** A rotodynamic pump, e.g., the types commonly used in sludge handling (non-clog mixed-flow pump, recessed-impeller pump, grinder pump), as compared to a positive-displacement pump.

**kinetic rate coefficient** A number that describes the rate at which a water constituent, such as biochemical oxygen demand or dissolved oxygen, rises or falls.

**kinetics** (1) The branch of mechanics that studies the motion of material bodies under the action of given forces; the dynamics of material bodies. (2) The study of the rate of change in physical, chemical, or biological processes. Kinetic concepts applied to water and wastewater treatment concern the rate of chemical and biological reactions and the transfer of substances across boundaries (e.g. adsorption, aeration, diffusion, ion exchange). *See* zero-order reaction, first-order kinetics, first-order reaction, second-order reaction, kinetic coefficient.

**kingdom** The highest category in the taxonomic classification.

**Kinnison–Colby's formula** A formula proposed to estimate flood flows in New England (Fair et al., 1971):

$$Q = (0.000036\ H^{2.4} + 124)\ A^{0.85}/(RL^{0.7}) \quad \text{(K-06)}$$

where $Q$ is the peak flood flow in cfs, $H$ = median altitude of the basin in feet above the outlet, $A$ is the drainage area in square miles, $R$ is the percent of impoundment surface in the area (lakes, ponds, reservoirs), and $L$ is the average distance in miles to the outlet.

**$KIO_3$** Chemical formula of potassium iodate.

**Kirkwood, James P.** Constructor of the first sizable water filters for the Hudson River water of Poughkeepsie, N.Y., one of the significant achievements of the Great Sanitary Awakening.

**Kirschmer relationship** An empirical equation proposed by the German engineer O. Kirschmer in 1926 to compute the head loss through racks and screens in preliminary wastewater treatment (Fair et al., 1971):

$$H = B(M/W)^{4/3}\ V \sin A \qquad \text{(K-07)}$$

where $H$ = head loss, ft; $B$ = a bar-shaped factor (from 0.76 for a bar with semi-circular face to 2.42 for a sharp-edged rectangular bar); $M$ = maximum width of the bars facing the flow, ft; $W$ = minimum width of the clear openings between pairs of bars, ft; $V$ = velocity head of the liquid approaching the bar, ft; and $A$ = angle of the bar with the horizontal.

**Kirschmer's formula** Same as Kirschmer relationship.

**kitchen wastes** Liquid wastes originating from the kitchen, mainly related to food, e.g., garbage grindings and dishwahing waste.

**Kjeldahl method** A laboratory method used to determine the combined concentration of ammonia nitrogen and organic nitrogen in a water sample, involving digestion and distillation of the sample and analysis of the distillate. For organic nitrogen, first the sample is boiled to drive off the ammonia ($NH_3$) and then digested to convert the organic nitrogen to ammonium ($NH_4^+$) through the action of heat and acid. For total Kjeldahl nitrogen (the sum of organic nitrogen and ammonia), the entire sample is digested without driving off the ammonia. The method does not account for nitrogen in the form of azide, azine, azo, hydrazone, nitrate, nitrite, nitro, nitroso, oxime, and semi-carbazone.

**Kjeldahl nitrogen** Nitrogen in the form of organic proteins or their decomposition product, ammonia, as measured by the Kjeldahl method; it includes ammonia nitrogen and organic nitrogen. Also called total Kjeldahl nitrogen. *See also* inorganic nitrogen.

**Kjeldahl nitrogen test** A standard laboratory test used to determine the Kjeldahl nitrogen, i.e., the sum of ammonia nitrogen and organic nitrogen, of a water sample. *See* Kjeldahl method, total Kjeldahl nitrogen.

**$K_La$** Volumetric mass transfer coefficient. *See* two-film theory.

**Klampres®** A belt filter press manufactured by Ashbrook Corp.

***Klebsiella* genus** A genus of lactose-fermenting, thermotolerant bacteria of the family of Enterobacteriaceae, included in the total coliform groups and used as an indicator of fecal contamination. They are opportunistic pathogens commonly found in finished water and on pipe biofilms.

***Klebsiella oxytoca*** A species of lactose-fermenting bacteria of the coliform group, used as an indicator of fecal contamination of drinking water.

***Klebsiella pneumoniae*** A member of the group of coliform bacteria, an early indicator of fecal contamination of water sources. However, this organism is also found in uncontaminated sites.

**Kleer Flow** Spiral-wound membranes manufactured by Great Lakes International, Inc. for use in reverse osmosis.

**Klenphos-300** A corrosion inhibitor containing zinc phosphate produced by Klenzoid, Inc.

**Klensorb** A chemical produced by Calgon Carbon Corp. to absorb oil and grease.

**KL Series™** Powders produced by King Lee Technologies for cleaning fouling membranes.

**$KMgFe_2AlSi_3O_{10}(OH)_2$** Chemical formula of biotite.

**$KMnO_4$** Chemical formula of potassium permanganate.

**K2Modular™** A screw feeder manufactured by K-Tron North America.

**known probable human carcinogen via ingestion** Any contaminant for which sufficient evidence of carcinogenicity in humans and animals exists, based on epidemiological studies. For example, there is strong evidence for arsenic, asbestos, and cadmium. Also called category I contaminant or category A carcinogen. This USEPA classification corresponds more or less to category 1 of the International Agency for Research on Cancer and category A of the National Toxicology Program. The MCLG for such substances is set at zero. *See also* carcinogen, mutagen, and possible human carcinogen via ingestion.

**$K_2O$** Chemical formula of potash or potassium oxide.

**Koagulator** A solids contact clarifier manufactured by Zimpro Environmental, Inc.

**Koch, Robert** German physician (1843–1910), bacteriologist, and 1905 Nobel prize winner who established the importance of water filtration to control waterborne diseases based on observations in Altona and Hamburg in Germany in the 1880s. He identified the vibrio responsible for cholera. *See also* Pasteur (Louis) and Snow (John).

**Koflo®** An in-line static mixer manufactured by Koflo Corp.

**Kolmogoroff microscale** Turbulent fluid flow creates eddies that transfer their kinetic energy from large ones to smaller and smaller eddies until the energy is dissipated as heat. The Kolmogoroff microscale is the difference between the inertial eddy size (energy transfer with little dissipation) and the viscous eddy size (energy dissipated as heat). This parameter is used in the theory of flocculation, e.g., to show that the velocity gradient (G value) is not an effective concept for the performance of microflocculation.

**Kolmogoroff microscale length** The relationship developed by Kolmogoroff in 1941, relating the length of the Kolmogoroff microscale ($L_K$, m) to the kinematic viscosity ($\nu$, m$^2$/s) and the power input per unit mass ($P = G^2\nu$, kg · m$^2$/s$^3$/kg), where $G$ is the velocity gradient (per sec):

$$L_K = (\nu^3/P)^{1/4} = (\nu/G)^{1/2} \qquad \text{(K-08)}$$

**Komara™** A device manufactured by Vikoma International for skimming floating oil.

**Kompress®** A belt filter press manufactured by Komline-Sanderson Engineering Corp.

**Koroseal®** A PVC medium manufactured by B. F. Goodrich Industrial Products Co. for use in trickling filters. It allows a void space of about 97% and an exposed surface area of 27–37 square feet per cubic foot, or more than twice the capacity of rocks and fieldstones. *See also* FloCor® and Surfpac®.

**Koro-Z** Biological filter medium produced in PVC by B. F. Goodrich Co.

**$K_{OW}$ or $k_{ow}$** *See* octanol/water partition coefficient, bioaccumulation, lipophilicity, bioconcentration.

**Kozeny constant** The dimensionless coefficient $K$ in the Blake–Kozeny equation, which defines the head loss in a granular medium. For most filtration conditions, $K = 5$.

**Kozeny equation** A formula that expresses the rate of head loss in a granular filter in terms of the bed and fluid characteristics. *See* Blake–Kozeny equation (B-57) for more detail.

**Kozeny formula** A formula for determining the specific capacity of a partially penetrating well (Fair et al., 1971):

$$Q/Q' = K\{1 + 7\,[R/(2\,KB)]^{0.5}\cos(\pi K/2)\} \qquad \text{(K-09)}$$

where $Q$ = specific capacity of a partially penetrating well, gpm/ft; $Q'$ = specific capacity of a fully penetrating well, gpm/ft; $K$ = ratio of the screen length to the thickness of the saturated aquifer; $R$ = effective well radius, ft; and $B$ = aquifer thickness, ft. *See also* Blake–Kozeny equation, Carman–Kozeny equation. The formula is applicable under steady-state conditions, for small values of $B$ and large values of $K$ and $R$.

**$K_2PtCl_6$** Chemical formula of potassium chloroplatinate.

**kraft** A strong brown paper manufactured from pulp; used in bags and as wrapping paper.

**kraft process** A chemical process for making wood pulp, using an alkaline liquor [a mixture of sodium sulfate ($Na_2SO_4$) or sodium sulfite ($Na_2S$) and caustic soda (NaOH)] to digest wood chips. It uses 25,000–50,000 gallons of water per ton of pulp, depending on whether the kraft is unbleached or bleached. Kraft mill wastes have average concentrations of 1200 mg/L total solids, 175 mg/L $BOD_5$, 150 mg/L suspended solids, 8.2 pH, and 175 mg/L alkalinity. Also called sulfate process. *See also* pulping process, sulfite process

**kraft pulp mill** Any stationary source that produces pulp from wood by cooking (digesting) wood chips in a water solution of sodium hydroxide and sodium sulfite (white liquor) at high temperature and pressure. Regeneration of the cooking chemicals through a recovery process is also considered part of the kraft pulp mill (EPA-40CFR60.281-a).

**Kraus-Fall** A clarifier with peripheral feed manufactured by Graver Co.

**Kraus process** A variation of the activated sludge process designed to improve sludge settleability and increase oxygen through nitrates by recycling aerobically treated supernatant from anaerobic digestion to the aeration basin. The process includes separate sludge reaeration and joint aeration of some digested sludge, digester supernatant, and activated sludge. It is used to treat nitrogen-deficient industrial wastes.

**KrCl** Formula of krypton chloride.

**Krebs' cycle** A series of enzyme-catalyzed reactions occurring in the aerobic metabolism of carbohydrates, proteins, and fatty acids, with the conversion of pyruvic acid into carbon dioxide and water, the buildup of ATP, and the reduction of oxygen. Also called citric acid cycle, tricarboxylic acid cycle. *See also* glycolysis.

**Krebs, Sir Hans Adolf** German-born, naturalized British biochemist; winner of the 1953 Nobel Prize for medicine and pharmacology; discoverer of the nitric acid cycle.

**Kruger/Fuchs** An ATAD (autothermal thermophilic aerobic digestion) system manufactured by Kruger, Inc.

**krypton (Kr)** An inert gaseous element found in very small amounts in the atmosphere, used in tungsten-filament light bulbs and in narrow-band excimer lamps with possible applications to wastewater disinfection. Atomic weight = 83.80. Atomic number = 36.

**krypton-85** A radionuclide sometimes found in water; half-life = 10.3 years.

**krypton chloride (KrCl)** A compound of krypton and chlorine used in narrow-band excimer lamps with possible applications to wastewater disinfection.

**$K_2S_2O_5$** Chemical formula of potassium persulfate. Also written as $K_2S_2O_8$.

**$K_2SO_4 \cdot Al_2 (SO_4)_3 \cdot 24H_2O$** Chemical formula of potassium aluminum sulfate.

**K-strategist** Same as an oligotroph or an autochthonous organism.

**KUBE³** A belt filter press manufactured by Klein America, Inc.

**Kubota bioreactor** The bioreactor used in the Kubota system, similar to that of the Zenogem® process.

**Kubota MBR process** *See* Kubota system.

**Kubota membrane** The membrane used in the Kubota system.

**Kubota system** A wastewater treatment method using slightly loose plate membranes submerged in an aerobic basin, called a membrane bioreactor. It includes cartridges of fine porous membranes mounted on both sides of a supporting plate. The cartridges can be removed individually for cleaning, or the membranes can be cleaned in place, using a dilute bleach solution (e.g., a 0.5% solution of hypochlorite), without taking them out of service. *See also* membrane biological reactor.

**Kuch mechanism** The mechanism of oxidation of iron (from $Fe^{2+}$ to $Fe^{3+}$) in corrosion scales and its precipitation as ferric hydroxide [$Fe(OH)_3$] to cause a red discoloration of water.

**kurtosis** (1) The quality, state or tendency of the curve of a frequency distribution to be very flat or very tall about its mode (i.e., in the region of peak frequency). (2) Same as kurtosis coefficient.

**kurtosis coefficient** A statistical parameter sometimes used in the analysis of water and wastewater management data. It measures the peakedness or flatness of the frequency curve about the mode. It is the fourth moment ($\mu_4$) of a frequency distribution about the mean and the origin. The first moment is the mean, the second moment is the variance, and the third moment is the coefficient of skewness. The normal distribution has a kurtosis of 3; the parameter is $\mu_4 > 3$ for peaked curves and $\mu_4 < 3$ for flat curves.

**Kutter–Ganguillet formula** An empirical formula proposed in 1869 by W. R. Kutter and E. J. Ganguillet to evaluate the velocity or discharge coefficient (Cz) in the Chézy formula (Fair et al., 1971):

$$C_z = (K + 1.811/n)/[1 + K(n/R^{0.5})] \quad \text{(K-10)}$$

where $n$ is a coefficient of roughness, $s$ is the invert slope, $R$ is the hydraulic radius, and

$$K = 41.65 + 0.00281/s \quad \text{(K-11)}$$

*See also* Manning's formula.

**$K_v$ value** Same as valve capacity coefficient.

**kW** Abbreviation of kilowatt.

**kWh** Abbreviation of kilowatt-hour.

# L

**labile** Unstable; inactivated by high temperature.

**laboratory operation wastewater** Wastewater from laboratories may contain volatile organic compounds (e.g., benzene, chloroform, methylene chloride, and toluene) and metal-bearing inorganic chemicals. Their discharge into publicly owned treatment works may be regulated by federal or local agencies.

**laboratory water** Purified water used for making up solutions or dilutions in the laboratory.

**lactase** An enzyme that hydrolyzes lactose into glucose and galactose. It is used in testing for coliform group bacteria using the defined-substrate technology. Also called $\beta$-D-galactosidase.

**lactic acid bacteria** A group of Gram-positive, nonsporeforming bacteria that carry out lactic acid fermentation of sugars, e.g., the genera *Lactobacillus* and *Streptococcus.*

**lactic acid fermentation** The complete or partial conversion of lactose, glucose, or other sugars by certain bacteria to lactic acid ($C_3H_6O_3$)—homolactic fermentation—or a mixture of lactic acid and other products—heterolactic fermentation.

**lactobacilli** Plural of lactobacillus.

**lactobacillus** A group of long, slender, rod-shaped, nonmethanogenic bacteria responsible for hydrolysis and fermentation in anaerobic digestion. They produce a considerable amount of lactic acid ($C_3H_6O_3$) in the fermentation of milk and other carbohydrates.

**lactose ($C_{12}H_{22}O_{11}$)** A disaccharide, present in milk, that is easily metabolized by some genera of coliform bacteria. It forms glucose and galactose upon hydrolysis by the enzyme β-galactosidase. Also called milk sugar.

**lactose bile broth** A culture medium that contains lactose, oxgall, peptone, and brilliant green dye, used in the multiple-tube fermentation technique. *See also* lauryl tryptose broth.

**lactose negative** Pertaining to a substance that cannot hydrolyze or ferment lactose.

**lactose positive** Pertaining to a substance that can hydrolyze or ferment lactose.

**lacustrine** Pertaining to materials that are present or formed in lakes or lake beds; e.g., sedimentary deposits.

**lacustrine habitat** A wetland or deep-water habitat of at least 20 acres, located in a depression or dammed river channel, with more than 30% of the area lacking trees, shrubs, persistent emergents, emergent mosses, or lichens.

**lacustrine plain** A lake bed from which water has disappeared.

**lacustrine zone** The deep-water zone of a lake or reservoir.

**ladder dredge** A continuous chain of buckets, or a drag bucket, mounted on a scow for lifting excavated materials. Also called bucket dredge, chain bucket.

**Ladder Gate® Climb Preventive Shield** A device that controls access to fixed ladders on tanks, towers, buildings, or other structures. Its angled sides prevent reaching around the shield to gain access to the ladder.

**lag** (1) Drainage lag in the formula of hydrograph time to peak; estimated as the distance on the time scale between the centroid of the rainfall and the peak discharge. It is one of two basic parameters used in the Snyder unit hydrograph method, the other being the storage coefficient. (2) The time difference between peak inflow and peak outflow as a result of storage. *See* reservoir storage routing. Also called lag time or basin lag.

**Lagco** A Parshall flume manufactured by F. B. Leopold Co., Inc.

**lag coefficient** An empirical parameter used in the estimation of mechanical sieving, i.e., rejection of particles at the surface of a membrane.

**lag effect** The phenomenon observed in wastewater chlorination according to which initially there is little reduction in the number of organisms, before a phase of straight-line inactivation (log inactivation vs. chlorine dose times contact time). Also called shoulder effect. *See also* tailing effect, Gard's disinfection model, Collins' disinfection model.

**lag growth phase** *See* lag phase.

**lagoon** (1) An artificial, shallow pond where sunlight, bacterial action, and oxygen work to purify wastewater or sludge; also used for storage of wastewater or spent nuclear fuel rods. *See also* lagoons and ponds. (2) A shallow body of water, often separated from the sea by low sandy dunes, coral reefs, or sandbars; also called a laguna. (3) A small, pondlike body of water; also called a lagune.

**lagoon dewatering** The use of a lagoon to dewater sludge. *See* dewatering lagoon.

**lagooning** The use of a lagoon, pond, basin, reservoir, or artificial impoundment for the disposal, storage or treatment of wastewater, sludge, or other material. In wastewater or sludge treatment, solids settle to the bottom and undergo microbial decomposition while the liquor evaporates or is removed for reuse or disposal. *See also* lagoon process and lagoons and ponds.

**lagoon process** A biological wastewater treatment or sludge processing method that uses a lagoon or a pond to effect carbonaceous BOD removal, nitrification, or waste stabilization; same as lagooning. *See also* lagoons and ponds.

**lagoons and ponds** There is some confusion in the literature on the use of lagoons and ponds:

Aerobic pond = maturation pond (Droste, 1997)

Facultative pond = oxidation pond (Water Environment Federation, 1991)

Facultative pond = aerobic–anaerobic pond (Metcalf & Eddy, 1991)

Lagoon = large holding and detention pond (APHA et al., 1981)

Lagoon = oxidation pond = stabilization pond = mechanically aerated pond (Droste, 1997)

Lagoon = oxidation pond = stabilization pond (Metcalf & Eddy, 1991)

Lagoon = oxidation pond = shallow pond without mechanical aeration (Sarnoff, 1971)

Lagoon = aerated pond (Water Environment Federation, 1991)

Stabilization lagoon = a shallow storage pond (APHA et al., 1981)

Stabilization pond = nonaerated lagoon (Metcalf & Eddy, 2003)

Stabilization pond = a type of oxidation pond (APHA et al., 1981)

For wastewater treatment, this book makes no difference between a lagoon and a pond, except that a lagoon may be considered as smaller and shallower than a pond. Definitions proposed for the above terms are listed alphabetically. Ponds, lagoons, and other waste treatment systems are not included in waters of the United States. *See also* aerated facultative lagoon, aerated lagoon, aerobic lagoon with solids recycle, amphiaerobic pond, anaerobic pretreatment pond, aquaculture, continuous-discharge pond, controlled-discharge pond, conventional lagoon, dual-powered flow-through lagoon system, evaporation pond, facultative partially mixed lagoon, flow-through aerated lagoon, heteroaerobic pond, high-rate pond, hydrograph-controlled-release pond, lagooning, lagoon process, lagoon temperature, low-rate pond, maturation lagoon, maturation pond, nominal complete-mix regime, partial-mix aerated lagoon, partial-mix aerated pond, photosynthetic pond, polishing pond, pond systems, ponds in dual processes, sewage lagoon, suspended-growth aerated lagoon, tertiary pond, total-containment pond, waste-oxidation basin.

**lagoon temperature** *See* aerated lagoon temperature. *See also* lagoons and ponds.

**lag phase** (1) The first phase in the growth of microorganisms in the presence of adequate food and environmental conditions, during which the number of microorganisms increases slowly as they acclimate to the new environment. The actual mass of microorganisms varies in a similar fashion. Also called lag growth phase. *See* growth patterns, basal metabolism, endogeny, lysis, protoplasm. (2) A stage in the filter-ripening sequence that results from the clean water remaining in the underdrain system at the end of backwashing.

**lag time** (1) Same as lag, one of a few parameters used in runoff analysis. *See also* antecedent precipitation index, attenuation constant, peak flow, plotting time width, standard duration of rainfall, time base, time of concentration, and time of equilibrium. (2) Same as acclimation period, i.e., the time necessary for a process or a system to reach design operation and performance.

**laguna** Same as lagoon (2).

**lagune** Same as lagoon (3).

**laid length** The total length of pipe or pipeline measured after its actual placement, including joints, gaskets, spaces between pipe ends, etc.

**lake** An inland body of fresh or salt water of a certain size, e.g., at least 50 acres. (The deepest lake in the world, Lake Baikal in Russia's southern Siberia, is larger than all five Great Lakes combined). Besides reservoirs, lakes are classified as autotrophic, bog, dystrophic, eutrophic, mesotrophic, oligotrophic, and oxbow lakes. *See also* pond.

**lake discharge** *See* lake disposal.

**lake disposal** The discharge of treated effluent or, less often, raw wastewater into a lake. Design considerations include poor natural dispersion due to low currents and little mixing, tendency of the solids to settle in the vicinity of the discharge; tendency of the warmer (and less dense) wastewater to overrun the receiving water and form a plume; low dissolved oxygen content at the bottom level of a deep lake; risk of polluting water intakes, beaches, and lakefront properties. *See also* ocean disposal.

**lake life cycle** See the three principal stages or trophic levels of a lake or reservoir: oligotrophic, mesotrophic, and eutrophic stages.

**lake overturn** The turnover and recirculation of materials that occur in a deep lake in the spring and fall as a result of temperature and density changes. *See* turnover, fall turnover, spring turnover, for stratification for detail.

**lake pollution** Lake pollutants come from discharges of raw wastewater and treated effluent, thermal effluent, offal, wind-blown seeds and grasses, surface runoff, and snowmelt. *See* eutrophication, cultural eutrophication, limnological considerations.

**lake profile** *See* the following terms: benthic zone, epilimnion, hypolimnion, limnetic zone, littoral zone, neuston layer, overturn, profundal zone, thermocline. *See also* marine profile.

**lake stratification** The formation of layers of different temperatures in a lake, reservoir, or other body of water. *See* Figure L-01, stratification, thermal stratification.

**lake turnover** Same as lake overturn.

**lakewide management plan** A written document that embodies a systematic and comprehensive ecosystem approach to restoring and protecting the beneficial uses of the open waters of each of the Great Lakes in accordance with the Great Lakes Water Quality Agreement (Clean Water Act).

**Lakos IPC** A self-cleaning screen manufactured by Claude Laval Corp. for pump intakes.

**LAL test** Short for limulus amebocyte lystate test.

**Lambert's law** *See* Beer–Lambert law.

***Lamblia intestinalis*** The name given in Eastern Europe to the species of parasitic flagellate protozoon that causes giardiasis in humans. *See Giardia lamblia.*

**Lamella®** A proprietary gravity settler and thickener manufactured by Parkson Corp., including inclined plates and countercurrent flow. *See* lamella clarifier.

**lamella clarifier** A lamella (plural, lamellae) is (a) a thin plate, scale, membrane, or layer, or (b) in a building, a member joined in a crisscross pattern with other lamellae to form a vault. A lamella clarifier includes uniformly spaced, inclined, parallel trays of plastic, rawhide, etc., to improve performance by increasing the surface area of the basin and reducing the surface overflow rate, while suppressing wind currents and reducing turbulence. It may be designed for counterflow (upward), cocurrent, or cross flow operation. It is an efficient gravity settling device but subject to clogging as a secondary clarifier. Lamellae are sometimes installed to upgrade existing clarifiers. Also called

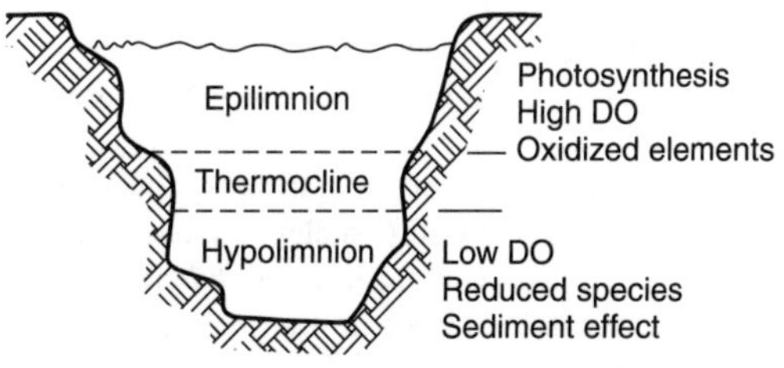


**Figure L-01.** Lake stratification.

inclined plate separator, inclined plate settler, lamella plate clarifier, lamella separator, lamella settler. *See also* multiple-tray clarifier, plate settler, tube settler.

**Lamella® Gravity Settler** *See* Lamella®.

**lamella plate** *See* lamella clarifier.

**lamella plate clarification** (1) In general, any use of a lamella clarifier to separate settleable solids from water, wastewater, or stormwater. (2) A high-rate clarification process that includes addition of chemicals for flocculation and coagulation, three-stage flocculation, and lamella clarification. *See also* ballasted flocculation and dense-sludge process.

**lamellar clarification** The use of lamella clarifiers in water or wastewater treatment.

**lamella separator** Same as lamella clarifier.

**lamella settler** Same as lamella clarifier.

**LamGard** An apparatus manufactured by Lamson Corp. for automatic oxygen control.

**laminar boundary layer** An interfacial region having smooth and nonturbulent flow. *See also* laminar sublayer.

**laminar flow** The smooth and orderly flow of a viscous fluid in parallel layers. Each layer moves with respect to adjacent layers with a constant velocity. Exchanges between layers are limited to molecular and thermal diffusion and molecular transfer of momentum. The opposite of laminar flow is turbulent flow. The Reynolds number ($R_e$) determines whether flow is laminar ($R_e < 2100$) or turbulent ($R_e > 4000$). Between these two numbers, the flow may be laminar or turbulent, depending on other factors. Laminar flow occurs when average velocity is relatively low and energy head is lost mainly through viscosity. Under laminar flow conditions, the friction factor ($f$) in the Darcy–Weisbach formula is $f = 64/R_e$. In most cases, water or wastewater flow in conduits is turbulent. For example, in a pipe of diameter 12 inches, laminar flow cannot exist for velocities exceeding 0.05 fps. Laminar flow is sometimes called streamline flow or viscous flow. *See* open-channel flow.

**laminar range** The range of Reynolds numbers ($R_e$) within which flow is laminar and the drag coefficient ($C_D$) is equal to 24 divided by the Reynolds number. *See* Stokes range for more detail.

**laminar shear** One of the transport mechanisms that cause destabilized particles to move and collide during flocculation; in a laminar flow field, differences in fluid velocities cause the particles to come into contact. *See also* orthokinetic flocculation and flocculation rate correction factor.

**laminar sublayer** A layer with smooth, nonturbulent flow, between a surface and a turbulent layer above. *See also* laminar boundary layer.

**Lam-Pak®** A package plant manufactured by Graver Co. for wastewater treatment.

**LAN** Acronym of local area network.

**Lancom™** An apparatus manufactured by Land Combustion, Inc.

**Lancy™** Wastewater treatment products of U.S. Filter Corp.

**land** Any surface or subsurface land that is not part of a disposal site and is not covered by an occupiable building (EPA-40CFR192.11-b).

**land application** (1) The spraying or spreading of sewage sludge onto the land surface; the injection of sewage sludge below the land surface, or the incorporation of sewage sludge into the soil so that the sludge can either condition the soil or fertilize crops or vegetation grown in the soil (EPA-40CFR503.11-h). Current regulations concern heavy metals concentrations, pathogen densities, and vector attraction. *See* Class A sludge, Class B sludge. (2) The discharge of wastewater onto the ground for treatment or reuse. *See also* land disposal and land treatment.

**land application limit (sludge)** *See* land application loading rates.

**land application loading rates** Limiting rates imposed on the land application of sludge to satisfy various criteria; they also depend on land use, e.g., agricultural, forest, land reclamation, or a dedicated disposal site. The maximum application rate ($A$, tons/acre/year) for pollutants is:

$$A = K \cdot L/C \tag{L-01}$$

where $L$ is the constituent limitation in pounds/acre/year, $C$ is the pollutant concentration in the sludge in mg/kg, and $K$ is a conversion factor based on 1 ton = 2000 pounds and 1 kg = 1,000,000 mg. The formula is applied for each pollutant and the lowest limit applies. *See also* nitrogen-limiting application rate and phosphorus-limiting application rate.

**land application pollutant loading** *See* land application loading rates.

**land application pollutants** Accumulation in the soil is a major consideration for the following heavy metals: arsenic (As), cadmium (Cd), chromium (Cr), cobalt (Co), copper (Cu), lead (Pb), mercury (Hg), molybdenum (Mo), nickel (Ni), selenium (Sn), and zinc (Zn). Pathogens are also a concern.

**land application system** *See* land treatment (of wastewater).

**land application unit** An area where wastes are applied to or incorporated into the soil for treatment, disposal, or reuse, exclusive of operations of manure spreading.

**land burial** A method of disposal of dewatered sludge by putting it in the ground and covering it with earth. *See also* landfilling, land disposal.

**land disposal** The placement of wastewater, sludge, or other wastes in or on the land, except in a corrective action management unit, including but not limited to placement in a landfill, surface impoundment, waste pile, injection well, land treatment facility, salt dome formation, underground mine or cave, or placement in a concrete vault or bunker intended for disposal purposes (EPA-40CFR268.2-c). *See also* land application and land treatment. Land disposal and land treatment systems include overland flow, rapid infiltration, slow rate or irrigation, subsurface, and wetlands. Some design restrictions concern storage requirements for unfavorable weather conditions (freezing, precipitation), buffer zone, nutrients, organic compounds, and heavy metals.

**land disposal restrictions** Restrictions, mandated by the 1984 amendments to the Resource Conservation and Recovery Act, prohibiting the disposal of hazardous wastes into or on the land unless the wastes meet treatability standards of lower toxicity.

**land drainage** The removal of water from the superficial layer of soil in the root zone of plants.

**land farming (of waste)** A disposal process in which hazardous wastes or biosolids deposited on or in the soil are degraded naturally by microbes. The soil and waste mixture is then periodically tilled and watered to promote biological activity. *See also* land treatment.

**land fill** Open dumping of solid wastes; often used synonymously with landfill or sanitary landfill.

**landfill** A facility in which municipal and/or industrial solid waste is disposed in or on land, according to accepted engineering practices to minimize environmental impacts. The term may apply to the disposal site or to the method of disposal (landfilling). Sludges from water or wastewater treatment are sometimes codisposed in landfills with municipal solid waste. According to federal regulations, a landfill cannot be a waste pile, a land treatment facility or land application unit, a surface impoundment, an underground injection well, a salt dome formation, a salt bed formation, an underground mine or a cave (EPA-40CFR259.10-a and 260.10). *See also* sanitary landfill, secure chemical landfill, secure landfill.

**landfill cap and liner systems** The top portion of a completed landfill (cap system) consists of an impermeable layer (e.g., bentonite, clay, or synthetic membrane), a sand drainage layer, and a layer of topsoil to support vegetation. There may be layers of intermediate soil cover and daily soil cover between the cap and the deposited wastes. The liner system consists of two sand drainage layers separated by a membrane and placed on top a bentonite liner. Leachate collection pipes are installed in both sand drainage layers of the liner. *See* Figure L-02.

**landfill cell** A discrete volume of a hazardous landfill that uses a liner to provide isolation of wastes from adjacent cells or wastes; for example, trenches and pits (EPA-40CFR260.10).

**landfill closure** *See* in-place closure.

**landfill gas** A colorless mixture of methane ($CH_4$), carbon dioxide ($CO_2$), and other gases produced by the anaerobic decomposition of wastes in a landfill; it has an offensive odor and a heating value of approximately 500 Btu/cubic foot.

**landfilling** The disposal of water or wastewater treatment residuals (biosolids, grit, screenings, and other solids) in a monofill or in a sanitary landfill. The presence of hazardous or toxic constituents in the sludge and the risk of groundwater pollution by leachate are major concerns. *See* landfill.

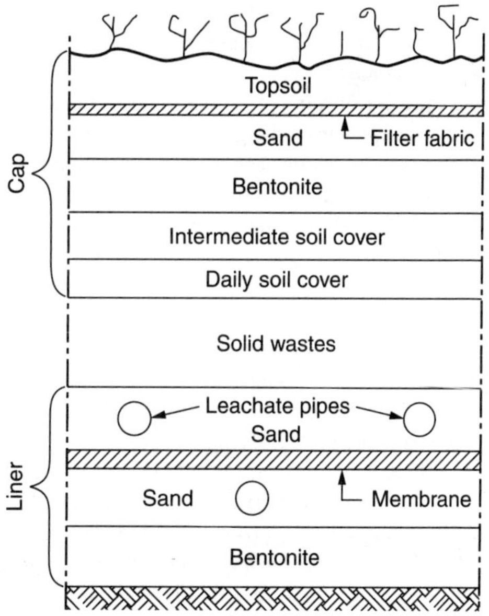


**Figure L-02.** Landfill cap and liner (adapted from Water Environment Federation, 1991).

**landfill leachate** The liquid produced in a landfill as a result of the infiltration and percolation of rainfall, groundwater, runoff, or flood water into and through a landfill; often a cause of groundwater contamination. Such leachate contains organic matter with high BOD and COD, heavy metals, alkaline metals, bicarbonates, chlorides, and sulfates. Leachate formation can be minimized by intercepting surface and ground waters. Leachate can also be collected and recirculated through the landfill or treated by aerobic, anaerobic, advanced treatment, or physical-chemical methods.

**landfill water balance** *See* water balance (2).

**land filtration** A method of wastewater treatment and disposal that consists in applying the wastewater through ditches across beds of farmland that are kept moist by seepage. Also called bed irrigation. *See* flood irrigation for detail and also direct irrigation, spray irrigation.

**landscape ecology** The study of the distribution patterns of communities and ecosystems, the ecological processes that affect those patterns, and changes in patterns and processes over time.

**landscape impoundment** A body of water used for aesthetic or other purposes, excluding contact recreation.

**landscape indicator** A measurement of the landscape, calculated from mapped or remotely sensed data, used to describe spatial patterns of land use and land cover across a geographical area. Landscape indicators may be useful as measures of certain kinds of environmental degradation such as forest fragmentation.

**landscape irrigation** The use of reclaimed water on parks, playgrounds, golf courses, freeway medians, school yards, cemeteries, greenbelts, and landscaped areas around residential, commercial, or industrial developments. *See also* water reuse applications.

**landslide flow** A flow of slippery earth materials from clay terraces. Also called earth flow

**land treatment** (1) The use of microorganisms in the upper soil layers and the filtrating capacity of the soil matrix to treat wastewater, residuals, or even hazardous wastes, and, more ordinarily, treated effluents, before their discharge to surface water. The soil microorganisms most active in biodegradation are *Agrobacterium, Arthrobacteri, Bacillus, Flavobacterium, Pseudomonas, Actinomycetes,* and fungi. Beside pretreatment, a land treatment system requires facilities to store the wastewater during unfavorable weather conditions. *See also* land application, land disposal, overland flow treatment, rapid infiltration system, slow rate system. (2) Irrigation of land with water or wastewater treatment sludge, in which the soils, microorganisms, and crops provide additional treatment, including nutrient removal. (3) Land farming.

**Landy™** Surface aeration equipment designed or manufactured by WesTech Engineering, Inc. of Salt Lake City, UT.

**Langelier index, Langelier's index,** or **Langelier saturation index (LSI)** An index, widely used in the water treatment and distribution field, proposed by W. F. Langelier in 1936 to reflect the equilibrium pH of a water with respect to calcium and alkalinity; sometimes called simply saturation index (SI). This index is used in stabilizing water to control both corrosion and the deposition of scale.

$$\text{Langelier index} = \text{LSI} = \text{pH} - \text{pH}_s \quad \text{(L-02)}$$

where pH = actual pH of the water, and $pH_s$ = pH at which the water having the same alkalinity and calcium content is just saturated with calcium carbonate. The Langelier index is also considered as a measure of the degree of saturation of calcium carbonate ($CaCO_3$) in water with respect to pH, alkalinity, and hardness; at equilibrium pH = $pH_s$ and LSI = 0. If LSI is positive, water is supersaturated with $CaCO_3$ and potentially scale-forming. If LSI is negative, water is undersaturated or aggressive and will potentially dissolve $CaCO_3$ deposits. Determination of $pH_s$ requires the knowledge of the concentrations of calcium and alkalinity as well as the activity coefficient for monovalent ions, and solubility and equilibrium constants. The USEPA requires water supply systems to monitor the LSI among various corrosivity parameters. Water supply systems tend to maintain a positive LSI, e.g., by increasing pH, temperature, alkalinity, or calcium content. *See* corrosion index for a list of corrosion control parameters.

**Langelier saturation index (LSI)** Same as Langelier index.

**Langelier's ionic strength approximation** A relation between the total dissolved solids content (TDS, mg/L) and the ionic strength ($\mu$) of a solution:

$$\mu = \text{TDS}/40{,}000 \quad \text{(L-03)}$$

*See also* specific conductance and Russell's ionic strength approximation.

**Langelier's stability diagram** A graphical representation of the stability of water and a method of determining the quantity of chemical to use in water treatment. *See also* Caldwell–Lawrence diagram.

**Langmuir equation** *See* Langmuir isotherm.

**Langmuir, Irving** American chemist (1881–1957), winner of Nobel Prize in 1932 for work on surface chemistry. Also known for experiments on cloud seeding.

**Langmuir isotherm** A graphical representation of the equilibrium relationship between adsorbate, adsorbent, and solution at a given temperature; the first adsorption isotherm, derived in 1918, based on the Michaelis–Menten equation. Two forms of the equation developed by I. Langmuir are:

$$Q = AB \cdot C/(1 + A \cdot C) \qquad \text{(L-04)}$$

$$1/Q = (1/B)[1 + (1/A)(1/C)] \qquad \text{(L-05)}$$

where $Q$ = quantity adsorbed by unit mass of adsorbent at equilibrium, $A$, $B$ = two constants that are usually determined empirically, and $C$ = the equilibrium concentration of adsorbate in the dilute solution. It is one of two commonly used isotherms to describe activated carbon adsorption in water and wastewater applications. When $A \cdot C \ll 1$ (at low concentrations), the first equation is linear in $C$. When $A \cdot C \gg 1$ (at high concentrations), the equation indicates a limiting capacity of adsorption, $Q = B$. This isotherm can be used to determine the adsorption coefficients of a given adsorbent or to describe the adsorption of tastes and odors in terms of threshold values, using

$$Q = (C_0 - C)/M$$

which is the same as equation (F-84), $C_0$ being the threshold taste or odor concentration, $C$ the residual concentration, and $M$ the mass of adsorbent. *See also* Freundlich isotherm, modified Freundlich isotherm, BET isotherm, Frumkin equation, phenol value.

**Lanpac** Packing media manufactured by Lantec Products, Inc. for air stripping towers.

**lantern ring** A bronze or other metal ring installed in the packing over the shaft of a pump to improve lubricant distribution.

**Laplace equation** The second-order partial differential equation,

$$\nabla^2 \phi = 0 \qquad \text{(L-06)}$$

where $\nabla$ is the del operator and $\nabla^2$ the Laplace operator or Laplacian operator. In rectangular Cartesian coordinates x, y, and z,

$$\nabla = \partial/\partial x + \partial/\partial y + \partial/\partial z \qquad \text{(L-07)}$$

$$\nabla^2 = \partial^2/\partial x^2 + \partial^2/\partial y^2 + \partial^2/\partial z^2 \qquad \text{(L-08)}$$

$$\partial^2\phi/\partial x^2 + \partial^2\phi/\partial y^2 + \partial^2\phi/\partial z^2 = 0 \qquad \text{(L-09)}$$

A solution to the Laplace equation is called a harmonic function. The Laplace equation is used to describe steady, two-dimensional groundwater flow, with $\phi$ as velocity gradient:

$$\phi = -KH + C \qquad \text{(L-10)}$$

where $K$ is a constant that depends on the fluid and the solid medium, $H$ is the drop in groundwater table between two points, and $C$ is an integration constant.

**Laplace operator** *See* Laplace equation.

**Laplacian operator** *See* Laplace equation.

**lapse rate** The rate of decrease in atmospheric temperature with increasing altitude above a given location; the rate is not the same for wet as for dry air masses.

**large and medium size municipal separate storm sewer system** A municipal separate storm sewer that is (a) located in an incorporated place of population 100,000 or more at the latest decennial census; or (b) located in a county with unincorporated urbanized population of 100,000 or more, except the systems located in incorporated places; or (c) owned or operated by a municipality other than those included in (a) or (b) and excluded by the regional administrator of the USEPA.

**large meter** A water meter larger than 2 inches.

**large water system** A water supply system serving more than 50,000 people.

**L*ARO** A reverse osmosis unit manufactured by PWT Americas.

**Larson index** Same as Larson ratio.

**Larson ratio** or **Larson's ratio** An index proposed in 1957 to characterize the corrosion potential of a water source, based on factors other than calcium carbonate solubility, e.g., chloride and sulfate concentrations, the effects of which can be overcome by an increase in alkalinity:

$$\text{LR} = ([\text{Cl}^-] + 2\,[\text{SO}_4^{2-}])/[\text{HCO}_3^-] \qquad \text{(L-11)}$$

where the square brackets indicate molar concentrations (mol/L) of chlorides, sulfates, and bicarbonates. To prevent enhanced corrosion of unlined iron, originally it was recommended that LR equal approximately 0.2 or 0.3, but other researchers extend the ratio up to 1.0. Others use the inverse form of the index, i.e., the ratio of bicarbonates to the sum of chlorides and sulfates. Also called Larson index. *See* corrosion index for a list of corrosion control parameters.

**larva** A worm-like developmental stage whereby helminths, insects, and other organisms can move and secure their own nourishment. Plural: larvae.

**larvae** Plural of larva.

**LAS** Acronym of linear alkyl sulfonate.

**Lasaire®** Aeration equipment manufactured by A. B. Marketech, Inc. for installation in lagoons.

**laser doppler anemometry** A technique to study turbulence and collect data on the mean velocity of particles in a mixing chamber: a particle reflects light as it passes through the intersection of two laser beams; its velocity is determined by correlation with the wavelength of the reflected light. *See also* computational fluid dynamics, digital particle image velocimetry, laser-induced fluorescence.

**laser-induced fluorescence** A method used to measure the degree of mixing of a solution: laser light of a given wavelength induces fluorescence of dyes or similar materials, and the degree of mixing is indicated by the degree of light scattering. *See also* computational fluid dynamics, digital particle image velocimetry, laser doppler anemometry.

**laser particle counter** An instrument equipped with a laser as lighting source for counting particles.

**Laser Profiler** A stand-alone, snap-on tool designed by CUES, Inc. for use with a CCTV survey system and camera to determine internal pipeline conditions before and after rehabilitation. It collects survey data and creates reports, including measurements of internal faults, pipe sizes, laterals, and water levels, as well as performing automatic analysis of pipe shape and capacity up to 30 times per second. *See* Figure L-03.

**laser scanning cytometry (LSC)** An effective method of microsphere enumeration, using fluorescent dyes to examine microbial cells. It detects microspheres captured on a slide or membrane. Also called solid-phase cytometry.

**latency** (1) Time from the first exposure to a chemical until the appearance of a toxic effect; same as latency period. (2) The time between the release of a pathogen and the infection of a human host; i.e., some excreted pathogens must first develop in an intermediate medium (soil) or host (animals) before infecting man. Table L-01 shows latency of some excreted pathogens. Excreted bacteria, viruses, and protozoa have no latent period. *See also* persistence, infective dose, incubation period, latent virus.

**latency period** The time between the initial exposure to a toxic or injurious agent and the manifestation or detection of a health effect or a biological response; crudely estimated as the time (or some fraction of the time) from first exposure to detection of the effect; same as latency (1). Sometimes called latent period or induction period. *See also* incubation period.

**latent** Potential; not manifest.

**latent energy** The energy necessary to effect a change of state (e.g., melting, vaporization, condensation) in a substance at constant temperature. *See also* latent heat.

**latent excreted infection** *See* excreted helminths.

**latent heat** The amount of heat absorbed or radiated during a change of state at constant temperature; e.g., heat of fusion, heat of solidification, heat of sublimation, heat of vaporization, and the heat required to evaporate moisture from sludge. The loss or gain of latent heat does not affect the temperature of the substance changing state. *See also* latent energy, sensible heat.

**latent heat of evaporation** Same as heat of evaporation.

**latent heat of fusion** Same as heat of fusion.

**latent heat of vaporization** Same as heat of vaporization.

**latent period** Same as latency period.

**latent virus** A virus that is present or even integrated in the host chromosome but does not produce progeny virions until appropriately stimulated. The resulting initial disease remains dormant

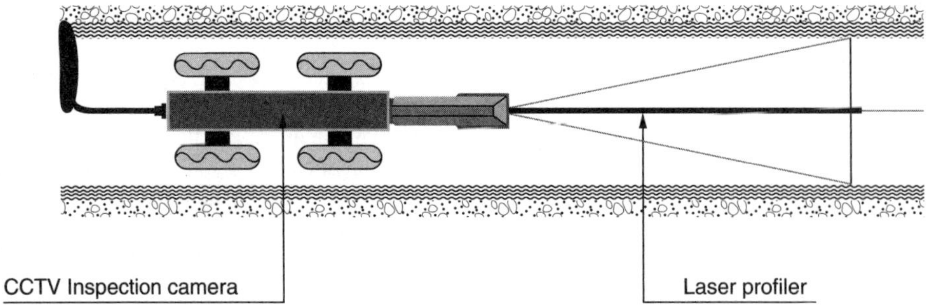


**Figure L-03.** Laser Profiler (courtesy Cues, Inc.).

**Table L-01.** Latency

| Pathogen | Latency (days) |
|---|---|
| *Ascaris lumbricoides* | 10 |
| Hookworm | 7 |
| *Paragonimus westermani* | 120 |
| *Schistosoma haematobium* | 50 |
| *Strongyloides stercoralis* | 3 |
| *Taenia saginata* | 60 |

and noninfectious until reactivated. *See also* latency.

**lateral** (1) A secondary pipe extending from a main water pipe or header. (2) A lateral sewer.

**lateral diffusion coefficient** A parameter ($E_y$, $ft^2$/sec) used in the formulation of the distribution of a constituent concentration in the mixing zone, equal to the product of the stream depth ($H$, ft) by the shear velocity ($U$, fps):

$$E_y = HU = (gH^{3/2}S)^{1/2} \qquad \text{(L-12)}$$

where $g$ is the acceleration due to gravity (ft/sec$^2$) and $S$ is the stream slope (ft/ft).

**Lateral Flow Sludge Thickener™** A gravity sludge thickener manufactured by Gravity Flow Systems, Inc.

**lateral sewer** A pipe that runs under streets and receives wastewater from homes and businesses, as opposed to domestic feeders and main trunk lines. Also, a sewer discharging into a branch or other sewer and having no tributary. *See also* lateral.

**laterite** A strongly compacted soil layer underlying soft materials and cemented with ferric oxide. Also called plinthite. *See* hardpan for other similar layers.

**lithium amide ($LiNH_2$)** A compound of lithium (Li) and nitrogen (N) that ignites spontaneously on contact with water:

$$LiNH_2 + H_2O \rightarrow LiOH + NH_3 \qquad \text{(L-13)}$$

**latrine** A place or building for defecation and urination, usually not within a house or another building; sometimes used synonymously with toilet. *See* the following types of latrines or toilets: borehole, bucket, chemical, cistern-flush, communal, composting, double-pit, offset-pit, overhung, pour-flush, raised-pit, single-pit, twin-pit pour-flush, vault, ventilated double-pit, ventilated-pit, VIP.

**latrine (borehole)** *See* borehole latrine.

**latrine (bucket)** *See* bucket latrine.

**latrine (composting)** *See* composting toilet.

**launder** Sedimentation basin and filter discharge channel, consisting of overflow weir plates (in sedimentation basins) and conveying troughs; sometimes applied to the trough itself.

**laundering weir** The overflow weir of a sedimentation basin. A plate with V-notches along the top to assure a uniform flow rate and avoid short-circuiting.

**launder trough** A trough in a treatment basin or tank to stabilize inflow or outflow.

**laundry waste** Wastewater from industrial laundries that may be characterized by high alkalinity, turbidity, and organic solids, and by the presence of lint, fibers, oils, and greases. *See also* dry cleaning waste. Common treatment methods include screening, chemical precipitation, flotation, and adsorption.

**lauroyl** or **lauryl** Containing the monovalent organic lauroyl (or lauryl) group ($C_{12}H_{23}O$—) derived from lauric acid ($C_{12}H_{24}O_2$), which is a white, crystalline, water-insoluble powder; a fatty acid found in vegetable fats.

**lauryl tryptose broth** A culture medium that contains tryptose, lactose, and inorganic nutrients, used in the multiple-tube fermentation technique.

**lawn** In the laboratory detection of male-specific coliphages, a lawn is a large number of host bacteria grown in a semisoft agar medium and on which the viruses form plaques by infecting the bacteria. The number of plaques is recorded as plaque-forming units (PFU) per volume of sample.

**lawn sprinkling** Water use ($Q_s$, gpd) for lawn and garden watering in metered and sewered residential areas may be estimated as follows (Fair et al., 1971):

$$Q_s = 13{,}100\,(E - P)\,D^{-1.26} \qquad \text{(L-14)}$$

where $E$ = potential evapotranspiration, in; $P$ = average daily precipitation, in; and $D$ = gross housing density in dwelling units per acre.

**law of gravitation** Two masses attract each other with a force ($F$) equal to the product of their masses ($M_1$, $M_2$) and the constant of gravitation ($G$), divided by the square of the distance ($D$) between them:

$$F = GM_1M_2/D^2 \qquad \text{(L-15)}$$

The constant of gravitation, also called gravitational constant or universal gravitational constant, is equal to $3.434 \times 10^{-8}$ pounds force per square foot per slug squared.

**law of mass action** One of two fundamental rules applied in balancing chemical reactions: any element has the same number of atoms on the left-hand side as on the right-hand side. *See* mass action law for more detail.

**law of partial pressures** Same as Dalton's law of partial pressures.

**law of the biggest pump** *See* right of free capture.

**Lawrence–McCarty model** A kinetic model proposed in 1970 for the plug-flow activated sludge process and based on two assumptions: (a) the concentration of organisms in the influent to the aeration basin is approximately the same as that in the basin effluent and (b) the rate of substrate utilization in the basin is of the form of the Michaelis–Menten equation (see substrate utilization model). The Lawrence–McCarty model is:

$$1/\mathrm{SRT} = Yk(S_0 - S)/[(S_0 - S) + (1 + R)K_s \ln (S_i/S)] - k_d \quad \text{(L-16)}$$

where SRT = solids retention time, days; $Y$ = synthesis yield coefficient, g VSS/g bsBOD (VSS = volatile suspended solids, bsBOD = biodegradable soluble BOD); $K$ = maximum specific substrate utilization rate, g/g/day; $S_0$ = influent concentration, mg/L; $S$ = effluent concentration, mg/L; $R$ = recycle ratio; $k_d$ = endogenous decay coefficient, g VSS/g VSS/day; $K_s$ = half-velocity constant or substrate concentration at one-half the maximum specific substrate utilization rate, mg/L; and $S_i$ = influent concentration after dilution with recycle flow, mg/L.

**laws of thermodynamics** *See* thermodynamics laws.

**laxative effect** Drinking water with chloride and sulfate concentrations may produce laxative effects, particularly in new consumers.

**layered bed** The bed of a multimedia filter, with layers of different filtering materials or an ion-exchange bed consisting of resins of different bead sizes.

**lb** Abbreviation of pound(s).

***L. biflexa*** *See Leptospira biflexa.*

**lb/mil cu ft** Abbreviation of pound(s) per million cubic feet, a unit of concentration.

**lb/mil gal** Abbreviation of pound(s) per million gallons, a unit of concentration.

**LC** Acronym of lethal concentration.

**$LC_{50}$** Acronym of lethal concentration–50% of median lethal concentration, a median level concentration used as a measure of toxicity. It is the concentration of a chemical in air or water that is expected to cause death in 50% of test animals or organisms living in that air or water, upon a specified exposure period such as 96 hours. The concentration is typically expressed in micrograms per cubic meter of air or micrograms per liter of water (EPA-40CFR116.3, 797.1350-3, 797.1400-8, 300-AA, etc.). *See also* $LD_{50}$, lethal concentration, toxicity terms.

**LCA** Acronym of life-cycle assessment.

***L. canicola*** *See Leptospira canicola.*

**LC-MS** Acronym of liquid chromatography–mass spectrometry.

**LC-PB-MS** Acronym of liquid chromatography–particle beam–mass spectrometry.

**LD** Acronym of lethal dose.

**$LD_{50}$** Acronym of lethal dose–50% or median lethal dose–50; i.e., the empirically derived single dose of a chemical, taken by mouth or absorbed by the skin, that is expected to cause death in 50% of the test animals so treated. The $LD_{50}$ is used in the hazardous ranking system (HRS) to assess acute toxicity, expressed in milligrams of toxicant per kilogram of body weight. The lower the $LD_{50}$, the more toxic the compound. *See also* $LC_{50}$ and lethal dose.

**LDA** Acronym of laser doppler anemometry.

**LDF** Acronym of limiting design factor.

**leach** To extract a soluble substance, lose a solute through a filtering medium, or dissolve through a percolating liquid.

**leachate** Water that collects contaminants or other soluble, suspended, or miscible materials as it drains or trickles through soil, wastes, pesticides, or fertilizers. Leachate may occur in farming areas, feedlots, and landfills, and may result in hazardous substances entering surface water, groundwater or soil (EPA-40CFR257.2). Leachate usually contains sodium chloride (NaCl), nitrate ($NO_3^-$), trace metals, and various organic materials. *See also* landfill leachate, leaching.

**leachate collection system** A system that gathers leachate and pumps it to the surface for treatment.

**leachfield** (1) An area of land in which a septic tank drains or wastewater is discharged. Also called leaching field. (2) A soil absorption system or disposal field.

**leaching** (1) The process by which soluble substances are dissolved and transported down through the soil by recharge. (2) The removal of a constituent by chemical reaction with a substance (extractant) in solution. *See* chemical extraction for detail. (3) The dissolution of substances into a liquid through a porous medium. (4) The extraction of solutes by percolating water from landfills, mine wastes, sludge, or soils. *See also* acid mine drainage. (5) The removal of valuable metals from ores by microbial action. (6) The disposal of effluents through porous soils or rocks. *See also* soil absorption field.

**leaching cesspool** An underground pit or covered tank with open joints constructed in permeable soil to receive raw domestic wastewater and allow partially treated effluent to seep into the surrounding soil, while solids are contained and undergo digestion. *See* cesspool for detail.

**leaching chamber** A plastic, dome-shaped chamber with a solid top, louvered sides, and open bottom to allow water flow.

**leaching chamber system** An onsite wastewater treatment and disposal system that includes (a) a septic tank or other treatment device; (b) a leaching chamber; and (c) leaching trenches. A pipe carries the effluent from the treatment device to the chamber, which stores the liquid until it enters the soil through the trenches.

**leaching field** Same as leachfield.

**leaching fraction** A parameter (LF) used in irrigation management; the fraction of applied water that passes through and percolates below the root zone. It is calculated as the ratio of depth of water below the root zone ($D_d$, mm or in.) to the depth of water applied at the surface ($D_i$, mm or in.):

$$LF = D_d/D_i = (D_i - ET_c)/D_i \qquad \text{(L-17)}$$

where $ET_c$ represents crop evapotranspiration (mm or in.).

**leaching pit** A kind of soakaway.

**leaching requirement** In irrigation, the extra water needed to prevent the excessive accumulation of salts in the root zone; usually expressed as a fraction or percentage of the total volume of water applied to the soil and calculated as:

$$LR = 100\ C_i/(5\ C_e - C_i) \qquad \text{(L-18)}$$

where LR = leaching requirement, percent; $C_i$ = electrical conductivity of the irrigation water, deciSiemens/m; and $C_e$ = maximum permissible electrical conductivity of the soil saturation extract (or soil salinity threshold of the crop). LR is one of the parameters considered in the design and scheduling of irrigation systems using wastewater effluents.

**leach lines** The trenches of a subsurface soil absorption system. Also called drain lines.

**lead** (**Pb**, from its Latin name plumbum) A soft, blue-gray, highly malleable heavy metal that is hazardous to health if breathed or swallowed; sometimes found as a trace element and cumulative poison in air or water; the final stable product of uranium decay. In drinking water, lead occurs from corrosion of lead plumbing or leaded brass faucets. It is associated with acid mine drainage. Its use in gasoline, paints, and plumbing compounds has been sharply restricted or eliminated by federal laws. It melts at low temperatures, is easily used to form pipes, and resists acid corrosion. Atomic weight = 207.2. Atomic number = 82. Specific gravity = 11.35. Valence = 2 or 4. Common lead compounds include anglesite ($PbSO_4$), cerussite ($PbCO_3$), galena (PbS), and minim ($Pb_3O_4$). At sufficiently high concentrations, lead ingestion or inhalation may interfere with enzyme production and cause physiological effects in humans such as kidney damage, anemia, birth defects, elevated blood pressure, delayed neurological development, and, perhaps, cancer. The USEPA regulates lead in water under the Lead and Copper Rule (MCLG = 0 and MCL = 0.015 mg/L).

**lead-210 ($^{210}Pb$)** An isotope of lead, with a half-life of 21 years; a product of the decay of uranium-226.

**lead acetate [$Pb(C_2H_3O_2)_2 \cdot 3\ H_2O$]** A white, crystalline, soluble, poisonous solid used in dyes, textiles, and paints. Also called sugar of lead.

**Lead and Copper Rule (LCR)** A regulation established in June 1991 by the USEPA to protect the public from lead and copper exposure in drinking water. It requires water purveyors to monitor for lead; if its concentration exceeds 15 μg/L, the water must be treated or lead service connections must be replaced. It sets an action level of 1.3 mg/L for copper, i.e., at higher concentrations corrosion control or other corrective measures must be implemented.

**lead arsenate [$Pb_3(AsO_4)_2$]** A toxic compound formerly used in pesticides.

**lead calking** An operation that consists in driving lead into a pipe joint.

**lead carbonate ($PbCO_3$)** A compound of lead that is often found as scales on water distribution lines made of this metal. *See also* hydrocerussite.

**Lead Contaminant Control Act (LCCA)** Public Law 100-572 of 1988, which establishes a program to eliminate the risk of lead poisoning from water coolers in schools.

**lead dioxide ($PbO_2$)** A product of the corrosion of lead piping. *See* plattnerite and scrutinyite.

**lead-free** Pertaining to fittings and pipes that contain less than 8.0% lead or flux and solder that contain less than 0.2% lead, according to the Safe Drinking Water Act.

**lead joint** Lead wool or molten lead, poured into the annular space of a bell-and-spigot joint, used as the connection between two pipe sections. *See also* joint and joint compound.

**lead load** The amount of lead found in a sample of water, air, or soil.

**lead line** A rope or cable attached to a weight and used for measuring depths of water during hydrographic surveying. Also called sounding line.

**lead pipe** A service line or a small water main made of lead. Such lines are no longer used or are being replaced in the United States.

**lead poisoning** An illness caused by the accumulation of lead in the body from drinking water, metal vessels, or other sources. Lead accumulates in the body; e.g., 0.6 mg/day over a long period may be dangerous.

**lead service** Short for lead service line.

**lead service line** A service line made of lead that connects the water main to the building inlet and any lead pigtail, gooseneck or other fitting that is connected to such lead line (EPA-40CFR141.2). *See also* lead pipe and lead poisoning.

**LEADSOL** A FORTRAN computer program developed by the USEPA to investigate lead solubility in drinking water.

**lead sugar** Same as lead acetate.

**Leadtrak** Test kits manufactured by Hach Co. to measure lead concentrations in water.

**leaf** A rectangular or round, flat element that supports the septum and medium of a precoat filter. *See also* membrane leaf.

**leaf filter** A precoat filter with leaves.

**leakage** (1) The uncontrolled, natural or man-made, loss of water from hydraulic structures, or from one aquifer to another, caused by hydrostatic pressure. (2) The magnitude of flow leaked. (3) The passage of ions into the effluent of an ion exchange unit due to bed exhaustion. (4) In an electrodialysis or electrodialysis reversal unit, the diffusion of ions from the concentrate to the product water, caused by a high concentration gradient between these two streams. It can be controlled by keeping the ratio of concentrations below 150. (5) *See* sand leakage.

**LeakfinderRT** A personal computer-based device manufactured by Echologics Engineering, Inc. for correlating leak noise with signal processing.

**leak survey** An investigation conducted, when excessive leakage is suspected, to determine the extent and location of leaks in a water distribution system. *See also* water waste survey.

**Leakwise®** An apparatus manufactured by Agar Corp. to monitor the presence of oil on water.

**leaky aquifer** An aquifer such that water leaks in or out of it either naturally because of a semi-pervious overlying or underlying bed, or through a man-made encased opening. Also called a semi-confined aquifer. A leaky confined aquifer is an aquifer underlain by a confining layer but overlain by a semipervious layer and a phreatic aquifer.

**leaky confined aquifer** *See* leaky aquifer.

**leaping weir** A device consisting of an adjustable steel weir plate and a gap or an opening in the invert of a combined sewer to let the dry-weather flow fall to a sanitary sewer; wet-weather flows leap the opening into the overflow outlet. *See* Figure L-04. Also called a separating weir.

**least-squares BOD analysis** A method used to analyze a time series of BOD measurements by fitting a curve through a set of data points so that the difference between the observed value and the fitted value is a minimum. The least-squares method is used to determine the first-order reaction rate constant and the ultimate BOD. *See also* the Fujimoto BOD analysis method.

**L'Eau Claire®** Upflow filters manufactured by the U.S. Filter Corp.

**LeChâtelier's principle** The application of stress to a system in equilibrium causes the system to tend to relieve the stress and restore equilibrium under different conditions. *See,* e.g., common ion effect.

**Lectra/San** Marine sanitation system developed by Eltech International Corp.

**LED** Acronym of light-emitting diode.

***Legionella*** A genus of gram-negative, aerobic, nonspore-forming bacteria of the family *Legionellaceae,* some species of which have caused a type of pneumonia called legionnaires' disease. Their natural habitat is fresh surface water, but they may be found in water distribution systems and heating units. Transmission is through the inhalation of contaminated aerosols. *Legionella* is also responsible for the milder disease, Pontiac fever. *See also* legionnaires' disease, Pontiac fever.

***Legionellae*** Plural of *Legionella.*

***Legionellaceae*** *See* Legionella.

***Legionella pneumophila*** The species of *Legionella* that causes most of the cases of legionnaires'

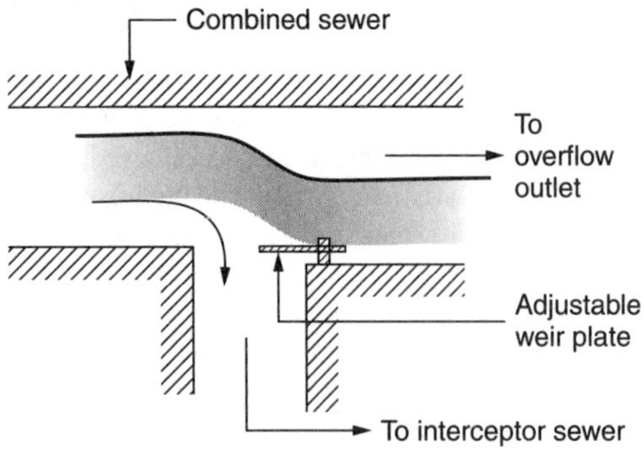


**Figure L-04.** Leaping weir.

disease. It is commonly found in rivers, lakes, and groundwater, and sometimes in biofilms on water mains and plumbing materials, in wastewater and reclaimed water. It also causes Pontiac fever. Potassium permanganate ($KMnO_4$) is an effective inactivating agent at a pH of 6.0.

**legionellosis** Legionnaires' disease (pneumonia-like) or milder infection by the agent *Legionella* (a nonpneumonia form of the disease). *See also* Pontiac fever.

**Legionnaires' disease** A pneumonia-like disease, sometimes fatal, caused by bacteria of the genus *Legionella,* contracted most probably by inhaling contaminated aerosols and characterized by malaise, myalgia, fever, headache, and respiratory illness. (A prominent outbreak occurred at a 1976 Legionnaires' Convention in Philadelphia.) The bacteria are transported through the air and find favorable growth conditions in or near air conditioners, cooling towers, and hot water systems. Five percent of the persons exposed to the bacteria contract the disease, which may kill up to 40% of the persons infected. Also called legionellosis. *See also* Pontiac fever.

***Leishmania*** The species of flagellate protozoon transmitted by sandflies and causing leishmaniasis.

**leishmaniasis** Any of various tropical infections caused by the flyborne pathogen *Leishmania.* They cause cutaneous sores or mouth and nostril deformities. *See also* espundia, kala azar, Oroya fever, papatasi fever, phlebotomine sandfly.

**LEL** Acronym of lower explosive limit.

**Lemna®** A biological process developed by Lemna Corp. for wastewater treatment using duckweed.

**Lemnaceae** A family of floating aquatic plants that provide habitat for organisms that metabolize organics in wastewater. The genus *Lemna* includes duckweed; it is a genus of small, free-floating aquatic weeds with short roots of about 12 mm, sometimes planted in constructed wetlands.

***Lemna spp.*** *See* Lemnaceae.

**Lenard effect** The separation of electric charges when water drops break up aerodynamically; after the German physicist Phillip Lenard (1862–1947). Also called spray electrification, waterfall effect.

**length to mixing** The distance, $L_m$ (in ft), from the point of waste discharge into a stream to the point of complete mixing. *See* mixing length for detail.

**length-to-width ratio** The ratio of the length to the width of a basin, a parameter used in the design of rectangular sedimentation tanks (commonly between 3:1 and 5:1) and chlorine contact chambers, and to define the prevailing mixing conditions: a ratio of at least 20 is required for plug flow. The ratio also affects the dispersion index in these tanks.

**lentic** Pertaining to or living in still or stagnant water.

**lentic habitat** One of two types of freshwater environment. *See* lentic water.

**lentic water** A relatively calm, nonflowing or standing water body such as a lake, a reservoir, a marsh, or a pond, as opposed to a lotic or running water body such as a river or a canal.

**Leo-Lite** A fiberglass trough manufactured by F. B. Leopold Co. for effluent and scum collection.

**LeoVision** A computerized graphic display designed by F. B. Leopold Co. to monitor operating conditions at water and wastewater treatment plants.

**lepidocrocite (γ-FeOOH)** A compound formed during the corrosion of iron pipes; also called an iron phase or a corrosion scale.

**lepidolite** An important lithium ore.

**leptomitus** A fungus that forms unsightly masses on the walls and baffles of wastewater treatment structures.

***Leptospira*** A genus of aerobic bacteria of a spiral shape; some species are pathogenic to humans. Sometimes found in natural bathing waters. They are excreted in the urine of animal carriers and reach new hosts through skin abrasions or mucous membranes. Plural: leptospirae or leptospiras.

***Leptospira canicola*** Older name of the species of bacteria that causes leptospirosis (canicola fever) in humans and animals; now called *Leptospira interrogans* serovar *canicola.*

**leptospirae** Plural of leptospira.

***Leptospira icterohaemorrhagiae*** The species of bacteria that causes leptospirosis (hemorrhagic jaundice or Weil's disease) in humans.

***Leptospira interrogans*** The species of bacteria that causes leptospirosis (canicola fever) in humans and animals.

**leptospiras** Plural of leptospira; bacteria of the genus *Leptospira.*

**leptospire** Leptospira.

**leptospirosis** A water-related infection caused by the spirochaete *Leptospira interrogans* and affecting humans and domestic animals. Its symptoms include fever, jaundice, and muscle pain. It is transmitted from the urine of rodents and other animals through a person's skin, eyes, or mouth. Related diseases include canicola fever, canine leptospirosis, Weil's disease or hemorrhagic jaundice.

***Leptothrix*** One of three groups of iron bacteria, filamentous organisms that use iron compounds and cause a number of problems in water supply systems, e.g., taste and odor, color ("red water"), and clogging of well screens. *See also Gallionella* and *Frenothrix.*

**lesion** A pathological or traumatic discontinuity of tissue or loss of function of a part.

**LET** Acronym of linear energy transfer.

**lethal** Deadly; fatal.

**lethal concentration (LC)** The concentration of a test material or other substance in air or water that is fatal to a specified number of test organisms in a specified time period. The lethal concentration may be related to a percentage of the population, as in the 96-hour $LC_{50}$. *See also* lethal dose, toxicity terms.

**lethal concentration–50 percent or lethal concentration–50%** *See* $LC_{50}$.

**lethal dose (LD)** The amount of a test material or other substance in air or water that is fatal to an organism. The lethal concentration may be related to a percentage of the population, as in the $LD_{50}$. *See also* lethal concentration, toxicity terms.

**lethal dose–50 percent or lethal dose–50%** *See* $LD_{50}$.

**lethal time** The time period required for a given dose to be fatal. *See* latency period.

***Leucothrix* genus** A genus of filamentous sulfur bacteria, suspected to be among the organisms responsible for bulking and foaming sludge.

**leukemia** A progressive disease of the bone marrow and organs that produce blood cells, associated in humans with exposure to benzene ($C_6H_6$).

**levee** *See* dike.

**Level Bed Agitator** An agitator manufactured by Wheelabrator Clean Water Systems, Inc. for use in composting units.

**level control** A float device (or pressure switch) that senses changes in a measured variable and opens or closes a switch in response to that change. In its simplest form, this control might be a floating ball connected mechanically to a switch or valve such as is used to stop water flow into a toilet when the tank is full.

**Level Mate™** A device manufactured by Ametek for measuring liquid levels.

**level of concern (LOC)** The concentration in air of an extremely hazardous substance above which there may be serious immediate health effects to anyone exposed to it for short periods. The term can also be applied to substances in water.

**level of detection** One of a few parameters used to measure the precision or the detection level of a method of analysis, stated as a result that produces a signal 2 × 1.645 times the standard deviation above the mean of blank analyses. Also called lower level of detection or, simply, detection level.

**level of quantification** One of a few parameters used to measure the precision or the detection level of a method of analysis, stated as a result that produces a signal 10 times the standard deviation above the blank signal.

**Lévêque's correlation** An equation indicating that, in the theory of pervaporation, the liquid-boundary-layer mass transfer coefficient ($K$) depends on the superficial velocity of the fluid in the feed flow ($U$), the diffusivity ($D$), the hydraulic diameter of the cell ($D_H$), and the length ($L$) of the cell (Freeman, 1998):

$$K = 1.6\,(UD^2/D_H L)^{1/3} \qquad \text{(L-19)}$$

**Lewatit** Ion-exchange resins manufactured by Miles.

**Lewis acid** A substance, molecule, or ion, that can accept and share a lone pair of electrons donated by a Lewis base. That includes metal ions, acidic oxides, or atoms.

**Lewis acidity** The property of a substance that can accept a pair of electrons from another substance and form a covalent bond with it. *See also* Bronsted acidity.

**Lewis acid site** A surface site that can receive a pair of electrons from an adsorbate serving as a Lewis base.

**Lewis–Whitman theory** *See* two-film theory.

**Lewis base** Any substance that can form a covalent bond with an acid by transferring a pair of electrons to it; e.g., amine functional groups. Also called a nucleophile. Lewis bases are also Bronsted bases. *See also* ligand.

**Lewis concept** A broad concept according to which acids and bases combine to form a covalent bond. *See* Lewis acid, Lewis base, Bronsted concept.

**Lewis, Gilbert Newton** American chemist (1875–1946).

**Lewis–Randall correlation** An expression used to estimate the magnitude of the ionic strength of an aqueous solution in terms of the molar concentrations and charges of its ionic species. *See* ionic strength.

**Lewis structure** A compound or ion based on the Lewis concept: in acquiring or giving up electrons, atoms in a compound tend to attain the configuration of the nearest inert element. *See also* resonance.

**Lewis symbol** A chemical element or formula with

dots around the symbol to represent outer electrons, e.g., H· for hydrogen and He: for helium. *See also* noble gas outer electron configuration.

**Lewis–Whitman two-film theory** *See* two-film theory.

**Leydig cell** The endocrine cell of the testis, susceptible to damage by haloacetic acids.

**LF** Acronym of leaching fraction.

**LI** Acronym of (1) Langelier index and (2) liquidity index.

**Li** Chemical symbol of lithium.

**lice-carried infection** A disease caused by pathogens carried by lice, e.g., typhus and relapsing fever, which affect people who do not regularly wash their clothes.

**lichen** A greenish, gray, yellow, or blackish, sponge-like plant growing on wood, stone, or soil; it is a symbiotic combination of a fungus and a green alga (cyanophyta) or blue-green alga (cyanobacterium); often used as an air pollution indicator species, being very sensitive to sulfur dioxide ($SO_2$). Lichens play an important role in the weathering process that converts rocks to soil.

***L. icterohaemorrhagiae*** *See Leptospira icterohaemorrhagiae.*

**Liebig, Baron Justus von** German chemist (1803–1873), well known in agricultural chemistry.

**Liebig's law of the limiting nutrient** Same as Liebig's law of the minimum.

**Liebig's law of the minimum** The principle that, in a given community or ecosystem, the nutrient in shortest relative supply is the limiting or growth-determining factor. Thus, phosphorus is often regarded as the limiting nutrient for the control of eutrophication in receiving waters because carbon can be obtained from atmospheric carbon dioxide or decaying organic matter and algae can abstract enough nitrogen from the atmosphere.

**LIF** Acronym of laser-induced fluorescence.

**life cycle** The sequence of changes or stages in the existence of an organism (from fertilization through reproduction to death) or in the operation of a manufactured product (from development through use to final disposal).

**life-cycle analysis** *See* life-cycle costing.

**life-cycle assessment** *See* life-cycle costing.

**life-cycle cost** The total cost of equipment, facilities, and other capital goods over their economic life, including capital, operation, maintenance, and repair.

**life-cycle cost analysis** *See* life-cycle costing.

**life-cycle costing** A method of project evaluation to determine the most economical among the technically feasible alternatives by estimating their life-cycle costs, including capital as well as operation and maintenance. The life-cycle analysis of a manufactured product, for example, may include the following stages: (a) raw material acquisition, (b) material manufacturing, (c) product manufacturing, (d) product use, and (e) product disposal. *See* financial indicators for related terms.

**lifeline rate** A special rate applied by public utilities to lower-income customers so that they can afford a certain level of service.

**lifeline water rate** A rate intended to provide affordable water service to consumers of limited income. It is a minimum bill corresponding to a limited monthly flow charged to all customers and usually based on recovery of some basic costs such as billing and meter reading. The lifeline rate is less than the average cost of service, thus, to compensate, use in excess of the minimum is charged above the average cost of service.

**lifer** A mixing device installed in a rotary kiln to prevent the burning material from sliding down and promote uniform heat transfer throughout the unit.

**Lifeserver™** Wastewater treatment plants constructed by Davis Water and Waste Industries, Inc.

**life span** (1) The mean value of the length of existence of an organism or a species, a variable used in risk assessment studies. Human life span is sometimes assumed to be 70 years. *See* lifetime exposure. (2) The longest period over which the life of an individual or species may extend. (3) Longevity.

**lifetime exposure** The total amount of exposure to a substance that a human would receive in a lifetime (usually assumed to be 70 years). *See* life span.

**lifetime risk** (1) The risk ($P_L$) of contracting one or more infections during a lifetime (taken as 70 years = 25,550 days), a function of the probability ($P$) of infection from a single exposure. Assuming daily exposure to a constant concentration and a Poisson distribution of the contamination in water, for example,

$$P_L = 1 - (1 - P)^{25{,}550} \qquad \text{(L-20)}$$

*See* dose–response model for related distributions. (2) As defined by the USEPA, the product of the chronic daily intake (CDI) over a 70-year lifetime in mg/kg/day and a potency factor (PF) in kg.day/mg:

$$P_L = \text{CDI} \times \text{PF} \qquad \text{(L-21)}$$

*See also* toxicity terms.

**lift** In a sanitary landfill, a compacted layer of solid waste and the top layer of cover material.

**lift-and-turn valve** A manual control valve used in the regeneration of water softeners and filters.

**lifting station** *See* lift station and pumping station.

**lift pump** A pump for lifting fluid to the pump's own level, e.g., a pump that lifts wastewater in a sewer to allow gravity flow.

**Lift Screen** A bar screen manufactured by Envirex, Inc. with a reciprocating rake.

**lift station** *See* pumping station.

**ligand** (1) Any of a group of atoms, ions, or small molecules that surround the central atom or ion of a complex; for example, ammonia ($NH_3$) is a ligand in the complexation of copper ($Cu^{2+}$). Other ligands include carbonates ($CO_3^{2-}$), chlorides ($Cl^-$), cyanides ($CN^-$), hydroxides ($OH^-$), phosphates ($PO_4^{3-}$), and water ($H_2O$). The molecules or ions surrounding the central metal ions in hydrolysis products are ligands; they donate an electron pair to form the complex and are called Lewis bases. Ligands can be unidentate, bidentate, tridentate, or multidentate, according to their number of ligand atoms. Also called complexing agent. *See also* chelating agent, coordination compound. (2) An antibody, hormone, drug, or other molecule that binds to a receptor.

**ligand atom** An atom attached directly to the central metal ion of a coordination compound; in a ligand composed of several atoms, the ligand atom is responsible for the basic nature of the ligand.

**ligand compound** Same as ligand.

**ligand binding site** A site on a compound (e.g., a protein or an adsorbent) with affinity for another compound or group of compounds (e.g., drugs, toxins).

**ligand donor atom** Same as ligand atom.

**ligand exchange** The mechanism by which contaminant anions are exchanged for surface hydroxides on activated alumina. *See* activated alumina adsorption for detail.

**light Baumé scale** *See* Baumé scale.

**light-blocking sensor** Same as light-obscuration sensor.

**light-compensation point** The depth in a sea or lake below which, because of low light intensity, plants use up as much or more organic matter in respiration than they make during photosynthesis. *See* compensation point.

**light microscope** A microscope that uses visible light as a source of illumination. Common types provide a maximum magnification of 2000 times and a resolution of 0.2 μm or 200 nm. Also called optical microscope. *See also* bright-field, dark-field, differential interference, phase-contrast, electron-beam and ultraviolet-ray microscopes.

**light-obscuration sensor** A device used in counting particles by measuring the decrease in intensity of the light they block from a known source. Also called light-blocking sensor.

**Lightspeed** A digital flowmeter manufactured by Presto-Tek Corp.; uses fiber optics.

**light water** Ordinary water, as compared to heavy water or deuterium oxide.

**light-water reactor** A nuclear reactor that uses light water (instead of deuterium or heavy water) as a coolant and moderator. *See also* boiling-water reactor, pressurized-water reactor.

**lignin** (1) A complex, noncarbohydrate, organic substance constituting, with cellulose, the main part of woody tissue, imparting strength and rigidity to plants. It is a phenolic polymer resistant to biological degradation. *See also* fulvic acid, humic acid, and tannin. (2) The part of wood pulp discarded to make paper.

**lignite** A type of coal, soft and usually dark brown, with a woodlike structure and a low energy content; used to make activated carbon. Also called brown coal.

**lignite-based activated carbon** Activated carbon produced from the combustion of lignite for use in water purification, wastewater treatment, and other applications.

**Limberflo** Precast components manufactured by Aerators, Inc. for use as filter bottoms.

**lime** Lime is (a) calcium oxide (CaO), commonly called quicklime, but also burned lime, calx, carbide lime, caustic lime, chemical lime, dead-burned lime, fluxing lime, overburned lime, and unslaked lime; or (b) calcium hydroxide [$Ca(OH)_2$], commonly called hydrated lime, but also calcium hydrate, slaked lime, slaked hydrated lime, or milk of lime. Related materials include chlorinated lime, dolomitic lime, dolomitic limestone, high-calcium limestone, and limestone. General uses in water and wastewater treatment: heavy metals removal, odor control, pH adjustment, phosphorus removal, sludge conditioning, sludge stabilization, softening, and suspended solids removal. (1) A white or grayish-white, slightly soluble in water, calcined oxide; very caustic in solution. It may be obtained from ground limestone (calcite) or calcium carbonate ($CaCO_3$) and has many applications, e.g., in water treatment, sludge stabilization, mortars, cements, etc. Average bulk density = 1000 kg/m$^3$. (2) Any of a number of chemicals consisting essentially of calcium hydroxide made from limestone or slaked lime; average bulk density approximately half that of water. (3) A calcium compound used in soil amendment. (4) A fruit or a tree.

**lime alkalies** Lime-containing products used for acid neutralization or pH adjustment in water and wastewater treatment, including quicklime (CaO), slaked lime [$Ca(OH)_2$], and dolomitic lime. *See also* sodium alkalies.

**lime–alum technique** *See* Nalgonda process.

**lime-and-settle** A technology that uses chemical precipitation, usually with lime, followed by sedimentation.

**lime and soda ash process** Same as lime–soda ash softening.

**lime and soda ash softening** Same as lime–soda ash softening.

**lime chloride [CaCl(OCl)]** One of the earliest chlorine products used as a bleaching powder in water disinfection. Also called calcium chloride hypochlorite.

**lime clarification** A process that uses lime to precipitate hardness and reduce turbidity in the form of metal carbonate and hydroxide from water. *See* lime softening for detail.

**lime deterioration** The formation of calcium carbonate ($CaCO_3$) when quicklime (CaO) or hydrated lime [$Ca(OH)_2$] reacts with atmospheric carbon dioxide ($CO_2$):

$$CaO + CO_2 \rightarrow CaCO_3 \quad \text{(L-22)}$$

$$Ca(OH)_2 + CO_2 \rightarrow CaCO_3 + H_2O \quad \text{(L-23)}$$

Quicklime and hydrated lime deteriorate within 6–12 months.

**lime kiln** A furnace for the production of lime.

**lime milk** Another name for calcium hydroxide. *See* milk of lime.

**lime neutralization treatment** The use of lime to neutralize acid wastes. *See* limestone treatment and lime-slurry treatment. Other neutralization methods for acid wastes include caustic-soda treatment and waste mixing.

**lime posttreatment** A form of alkaline stabilization; the addition of hydrated lime [$Ca(OH)_2$] or quicklime (CaO) to sludge after dewatering to stabilize the material. A lime posttreatment device consists of a dry lime feeder, a dewatered sludge cake conveyor, and a lime–sludge mixer. *See* Figure L-05. Also called postlime stabilization.

**lime precipitation** The addition of lime or calcium hydroxide [$Ca(OH)_2$] to water or wastewater to facilitate the removal of dissolved and suspended solids. Clarification occurs as a result of the reactions between lime and carbonic acid ($H_2CO_3$) or alkalinity in the form of calcium bicarbonate [$Ca(HCO_3)_2$]:

$$Ca(OH)_2 + H_2CO_3 \leftrightarrow CaCO_3 + 2\,H_2O \quad \text{(L-24)}$$

$$Ca(OH)_2 + Ca(HCO_3)_2 \leftrightarrow 2\,CaCO_3 + 2\,H_2O \quad \text{(L-25)}$$

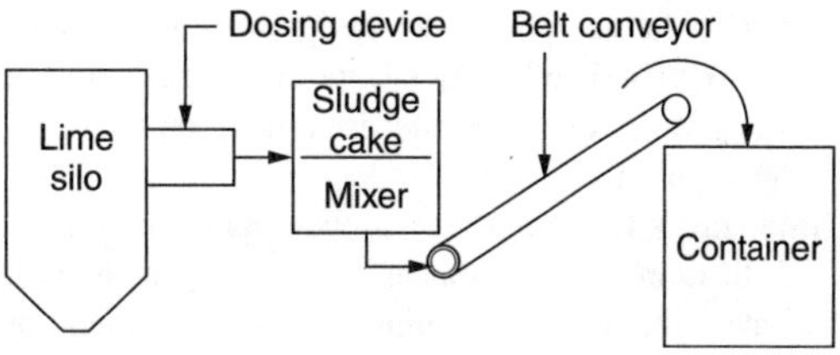


**Figure L-05.** Lime posttreatment.

**lime pretreatment** A form of alkaline stabilization; the addition of hydrated lime [$Ca(OH)_2$] or quicklime (CaO) to sludge before dewatering to stabilize the material.

**lime recalcining** The recovery of lime from water or wastewater treatment sludge by heating the sludge in a multiple-hearth or other furnace to drive off the water and carbon dioxide. The term is sometimes used to mean not only the recovery but also the reuse of lime. This leaves the lime (CaO) and some inert material that is periodically wasted:

$$CaCO_3 \rightarrow CaO + CO_2 \quad \text{(L-26)}$$

**lime recovery** The recovery of lime from lime sludge by (a) solubilization of magnesium through carbonation, (b) mechanical solids separation, and (c) recalcination of the calcium carbonate sludge. *See also* lime recalcining.

**lime scale** Scale formed by the precipitation of metal carbonates and other salts in pipes and containers; it is mostly the carbonates of calcium ($CaCO_3$) and magnesium ($MgCO_3$).

**lime slaker** A device used to hydrate quicklime (CaO) into hydrated or slaked lime by adding enough water to form a slurry. A paste (or pugmill) slaker produces a mixture that contains about 35% calcium hydroxide [$Ca(OH)_2$], whereas the slurry of a detention slaker has about 10% calcium hydroxide.

**lime slaking** *See* slaking.

**lime sludge** Residuals produced during the treatment of water or wastewater with lime. Lime sludge is easily dewatered by centrifugation and calcium carbonate ($CaCO_3$) is recovered by recalcining.

**lime slurry [$Ca(OH)_2$]** A solution containing approximately 5% calcium hydroxide, prepared in a slaker. Quicklime in granular form is at least 90%

CaO, whereas powdered, hydrated lime contains approximately 68% CaO. Lime slurries are used in many water or wastewater treatment applications. *See also* milk of lime.

**lime-slurry treatment** A process used in the neutralization of acid wastes by mixing them with lime slurries containing hydrated lime or dolomitic stone. The process, usually slow, can be hastened by heating or oxygenating the mixture. Other neutralization processes for acid wastes include caustic soda treatment, limestone treatment, and waste mixing. *See also* sulfation.

**lime soap** A substance consisting of the insoluble calcium and magnesium salts resulting from the reaction of hard water minerals with fatty acids in soap. Also called soap curd.

**lime–soda ash method** Same as lime–soda ash softening.

**lime–soda ash softening** A water treatment process that uses hydrated lime [$Ca(OH)_2$] and soda ash (sodium carbonate, $Na_2CO_3$) to reduce carbonate and noncarbonate hardness. A lime–soda ash plant includes mixing, flocculation, sedimentation, and filtration. Calcium carbonate and magnesium hydroxide precipitate and are removed as sludge (see lime–soda process equations). Some hardness remains in the effluent, depending on the quantity of chemicals added and the resulting pH. The process produces a large volume of sludge and requires proper pH adjustment to avoid problems in further treatment (filtration) or in the distribution system. *See also* lime softening, hot lime–soda softening, and excess lime treatment.

**lime–soda method** Same as lime–soda ash softening.

**lime–soda process** Same as lime–soda ash softening.

**lime–soda process equations** The reactions that occur when water is treated with lime [$Ca(OH)_2$] and soda ash ($Na_2CO_3$) to remove hardness, which is precipitated as calcium carbonate ($CaCO_3$) and magnesium hydroxide [$Mg(OH)_2$]. Two possible sequences follow.

*Sequence 1*

$$Ca(OH)_2 + CO_2 \rightarrow CaCO_3 \downarrow + H_2O \quad (L\text{-}27)$$

$$Ca(OH)_2 + Ca(HCO_3)_2 \rightarrow 2\ CaCO_3 \downarrow + 2\ H_2O \quad (L\text{-}28)$$

$$Na_2CO_3 + CaCl_2 \rightarrow CaCO_3 \downarrow + 2\ NaCl \quad (L\text{-}29)$$

$$2\ Ca(OH)_2 + Mg(HCO_3)_2 \rightarrow 2\ CaCO_3 \downarrow + Mg(OH)_2 \downarrow + 2\ H_2O \quad (L\text{-}30)$$

$$Ca(OH)_2 + MgCl_2 + Na_2CO_3 \rightarrow CaCO_3 \downarrow + Mg(OH)_2 \downarrow + 2\ NaCl \quad (L\text{-}31)$$

*Sequence 2*

$$Ca(OH)_2 + H_2CO_3 \rightarrow CaCO_3 \downarrow + 2\ H_2O \quad (L\text{-}32)$$

$$Ca(OH)_2 + Ca^{2+} + 2\ (HCO_3) \rightarrow 2\ CaCO_3 \downarrow + 2\ H_2O \quad (L\text{-}33)$$

$$Ca^{2+} + Na_2CO_3 + (SO_4^{2-} \text{ or } 2\ Cl^-) \rightarrow CaCO_3 \downarrow + 2\ Na + (SO_4^{2-} \text{ or } 2\ Cl^-) \quad (L\text{-}34)$$

$$2\ Ca(OH)_2 + Mg^{2+} + 2\ (HCO_3) \rightarrow 2\ CaCO_3 \downarrow + Mg(OH)_2 \downarrow + 2\ H_2O \quad (L\text{-}35)$$

$$Mg^{2+} + Ca(OH)_2 + (SO_4^{2-} \text{ or } 2\ Cl^-) \rightarrow Mg(OH)_2 \downarrow + Ca^{2+} + (SO_4^{2-} \text{ or } 2\ Cl^-) \quad (L\text{-}36)$$

$$Ca^{2+} + Na_2CO_3 + (SO_4^{2-} \text{ or } 2\ Cl^-) \rightarrow CaCO_3 \downarrow + 2\ Na + (SO_4^{2-} \text{ or } 2\ Cl^-) \quad (L\text{-}37)$$

**lime–soda softening** Same as lime–soda ash softening.

**lime-softened** Pertaining to water that has undergone lime softening.

**lime softening** A water treatment process that uses lime and other chemicals to precipitate hardness (and enhance clarification, when included) prior to filtration. The residuals of the process include particulate matter in addition to the hardness precipitates, calcium carbonate ($CaCO_3$), and magnesium hydroxide [$Mg(OH)_2$]. The lime-softening process can be designed in several configurations, e.g., single-stage, two-stage, or split softening. It produces larger amounts of residuals than coagulation and may eventually be replaced by membrane processes. *See also* lime–soda ash softening.

**lime stabilization** The addition of quicklime (CaO) or hydrated lime [$Ca(OH)_2$], alone or with ferric chloride, to untreated or digested sludge to raise the pH to 12 for a minimum of 2 hours, chemically inactivate microorganisms, and improve dewatering characteristics. *See* lime pretreatment, lime posttreatment.

**lime stabilization chemicals** Quicklime (CaO), hydrated lime [$Ca(OH)_2$], fly ash, and cement kiln dust.

**lime stabilization process** *See* lime stabilization.

**lime stabilization reactions** The quicklime (CaO) or hydrated lime [$Ca(OH)_2$] added for sludge stabilization reacts with the alkalinity (in the form of bicarbonate, $HCO_3^-$), inorganic constituents (e.g., calcium, carbon dioxide, phosphorus), and organic constituents (e.g., acids, fats) of the sludge. *See also* saponification.

**limestone** A sedimentary rock consisting mainly of calcium carbonate ($CaCO_3$) in the form of calcite, but also some magnesium carbonate ($MgCO_3$).

Calcium carbonate is a white or transparent solid, insoluble in water. Although their porosity and permeability are variable, limestone formations are the second most common source (after sandstone) of groundwater. Limestone is used as building stone, as aggregate, in the production of lime, and in making cement. It is also used as a weighting agent in the treatment of high-color, low-turbidity waters. *See also* lime.

**limestone contactor** A bed of limestone through which water passes to dissolve calcium carbonate and decrease corrosivity.

**limestone scrubbing** Use of a limestone and water solution to remove gaseous stack-pipe sulfur before it reaches the atmosphere.

**limestone treatment** The neutralization of acid wastes by passing them through a bed of limestone (calcium carbonate, $CaCO_3$). For sulfuric acid ($H_2SO_4$), e.g., the reaction is

$$H_2SO_4 + CaCO_3 \rightarrow CaSO_4 + H_2CO_3 \quad \text{(L-38)}$$

This process sometimes results in the problems of disposal of used limestone beds and foaming. *See also* caustic-soda treatment, lime-slurry treatment, waste mixing.

**lime treatment** Water treatment using lime (CaO), often in combination with soda ash ($NaCO_3$) to remove the hardness ions, calcium (Ca), and magnesium (Mg), from solution. *See* lime–soda ash softening.

**limited degradation** An environmental policy permitting some degradation of natural systems but terminating at a level well beneath an established health standard.

**limited water-soluble substance** A chemical that is soluble in water at less than 1000 mg/L. *See also* readily water-soluble substance.

**limiting current density (LCD)** As electrodialysis proceeds, electrical resistance and current density increase exponentially; the LCD is the point at which polarization occurs and water is dissociated into protons and hydroxide ions.

**limiting design factor (LDF)** A factor used in the design of slow-rate, land-treatment systems of wastewater, such as soil permeability and the allowable loading for nitrogen or another constituent. *See* hydraulic loading rate, limiting loading rate, slow infiltration or Type 1 system.

**limiting factor** A condition or environmental factor (e.g., temperature, pH, nutrient concentration) whose absence or excess is incompatible with the needs or tolerance of a species or population and that may have a negative influence on their ability to thrive or survive; i.e., a condition that inhibits a biological, chemical, or social process.

**limiting flux** The maximum quantity of solids that can reach the bottom of a clarifier. The thickening component of settling fails and the sludge blanket rises when the solids loading rate of a clarifier exceeds the limiting flux. For a solids suspension, the limiting flux corresponds to the minimum ($N_L$) on the curve of total flux versus solids concentration; *see* solids flux. In the design of sedimentation units, the limiting flux is used to determine the minimum surface area ($A$) of a clarifier:

$$A \geq QC_0/N_L \quad \text{(L-39)}$$

where $Q$ and $C_0$ are, respectively, the influent flow rate and solids concentration. *See also* gravity flux, total flux, underflow flux.

**limiting loading rate** The loading rate selected for the design of a slow-rate land-treatment system; the lowest of the rates calculated to satisfy soil permeability and nitrogen limits.

**limiting nutrient** A nutrient, such as carbon (C), nitrogen (N), or phosphorus (P), that allows microorganisms to grow in prolific amounts. It is usually phosphorus in freshwater bodies and nitrogen in estuaries.

**limiting salt** In water treatment by a membrane process, the limiting salt determines the appropriate dosage of antiscalant or acid used for scaling control. It is identified from the solubility products of candidate salts and the characteristics of the feedwater. It is the salt that allows the least recovery.

**limiting size** In flocculation theory, the size at which flocs start to break as they grow larger at the same time as the velocity gradients across them. *See also* floc breakup and flocculation concepts.

**limiting solids flux** *See* limiting flux.

**limiting solids retention time** A parameter used to calculate the design solids retention time (SRT) of a sludge digester by application of a safety factor. The limiting SRT ($SRT_{min}$, days) is selected on the basis of a desired digester performance, expressed as the fraction of reduction of the concentration of biodegradable substrate ($E$). For a complete-mix digester (WEF & ASCE, 1991),

$$SRT_{min} = \{[YKS_e/(K_c + S_e)] - B\}^{-1} \quad \text{(L-40)}$$

where $Y$ = growth yield of anaerobic organisms, g VSS/g COD destroyed; VSS = volatile suspended solids; $K$ = maximum specific substrate utilization rate, g COD/g VSS/day; $S_e$ = concentration of biodegradable substrate in digested sludge and in digester, g COD/L:

$$S_e = S_0 (1 - E) \quad \text{(L-41)}$$

$S_0$ = concentration of biodegradable substrate in feed sludge, g COD/L; $K_c$ = half saturation concentration of biodegradable substrate in feed sludge, g COD/L; and $B$ = endogenous decay coefficient,/day. The design SRT or actual SRT is $SRT_{min}$ multiplied by a safety factor.

**limit of detection (LOD)** The minimum concentration of a constituent that can be detected, using a certain analytical method; a measure of the accuracy and/or precision of an analytical method indicating at which concentration a reported result is valid or significant. It is used in determining such important parameters as maximum contaminant level (MCL) but is variously defined in the regulations, for example: (a) 0.5 μg/L for volatile organic compounds, (b) the smallest concentration within a 95% confidence interval, (c) limits determined by the USEPA for synthetic organic chemicals, and (d) different limits for toxics. *See also* method detection limit or MDL, practical quantification limit.

**limit of quantification (LOQ)** The minimum concentration of a constituent of a water sample for an analytical result to be reliable.

**limit stop** A diaphragm valve that can be adjusted to control flow rates in a water or wastewater treatment process.

**limnetic** Living in or relating to the open water of lakes, marshes, or ponds, i.e., beyond the littoral zone and to a depth where sunlight can penetrate.

**limnetic community** The aggregate of organisms (fish, phytoplankton, zooplankton) that live in the limnetic zone.

**limnetic zone** The surface layer of a body of freshwater, beyond the littoral zone, where sunlight can penetrate sufficiently for significant photosynthesis but not to the bottom. It is populated by fish, phytoplankton, and zooplankton. *See* freshwater profile for related terms.

**limnobion** The living organisms in a lake.

**limnological considerations** The effects of temperature, density, and wind govern the response of lakes and impoundments to pollution, including wastewater discharges. *See also* density stratification, epilimnion, fall turnover, great turnover, hypolimnion, spring circulation, spring turnover, summer stagnation, thermal stratification, thermocline, transition zone, winter stagnation, zonal differentiation.

**limnology** A subdivision of hydrology; the study of freshwater, particularly lakes and other inland waters, their flora and fauna, as well as their chemical, physical, geographical, and other features. *See* surface water hydrology, microlimnology, lentic habitat, lotic habitat, oceanography.

**limonite** A common iron ore: $Fe_2O_3 \cdot 3\,H_2O$.

**Limulus amebocyte lystate test (LAL test)** A test that detects the presence of endotoxins in treated water intended for pharmaceutical uses.

**lindane ($C_6H_6Cl_6$)** The common name for $\gamma$-isomer of 1,2,3,4,5,6-hexachlorocyclohexane, a white, crystalline, water-insoluble powder; a persistent pesticide that causes adverse health effects in domestic water supplies; it is also harmful to bees and livestock, and toxic to freshwater and marine aquatic life. It is mostly a chlorinated hydrocarbon, benzene hexachloride, or gamma-benzene hexachloride. Regulated as a drinking water contaminant by the USEPA: MCL = MCLG = 0.0002 mg/L.

**linear alkyl sulfonate (LAS)** A synthetic detergent that is biodegradable; general formula: R-$C_6H_4$-$SO_4$M, with R = $C_{10+}$ and M = a sodium or another salt. LAS has largely replaced alkylbenzene sulfonate as a detergent.

**linear correlation** A statistical method of evaluating data from historical records such as flows, concentrations, and chemical costs.

**linear energy transfer (LET)** Average amount of energy lost by an ionizing particle or photon per unit length of track in matter.

**linear isotherm** A special case of the Freundlich isotherm, used to relate the equilibrium concentration of a constituent in the liquid and solid phases:

$$C_S = K_d C_L \quad \text{(L-42)}$$

where $C_S$ = mass of constituent per unit mass of solid, mass/mass; $C_L$ = constituent concentration in the liquid, mass per unit volume, e.g., mg/L; and $K_d$ = distribution coefficient, volume per unit mass, e.g., L/mg. It indicates that the affinity of the adsorbent for the adsorbate remains the same for all levels of $C_S$. The linear isotherm is sometimes called the Henry equation, because of its similarity with Henry's law for the absorption of gases.

**linearity** An indication of how closely an instrument measures actual values of a variable through its effective range; a measure used to determine the accuracy of an instrument.

**linearized multistage model (LMM)** The modified form of the multistage model that assumes that the basic dose–response data are linear at low doses. *See* multistage model for detail.

**linearized multistage procedure** *See* linearized multistage model.

**linear-motion valve** Linear-motion valves, as defined by the type of throttling element movement, include diaphragm, gate, and globe valves. *See also* rotary-motion valve.

**linear multistage model** A model used to assess nonthreshold effects of toxic constituents that is linear at low doses with a constant of proportionality that yields less than a 5% chance of underestimating risk; it is a modification of the multistage model.

**linear polarization** A technique that uses a probe to measure the potential of a fluid to corrode metals based on its polarization resistance. The probe measures the flow of electrical current, which is compared to metal corrosion rates.

**linear solution diffusion model** A model used in the design of membrane arrays, assuming a homogeneous solution in which solute flow is controlled by diffusion and solvent flow is controlled by pressure (AWWA, 1999):

$$C_p = K_i C_f / [K_w(\Delta P - \Delta\Pi)(2 - 2R)/(2 - R)] = Z_i C_f \quad \text{(L-43)}$$

where $C_p$ = permeate stream solute concentration; $K_i$ = solute mass transfer coefficient; $C_f$ = feed stream solute concentration; $K_w$ = solvent mass transfer coefficient (solute diffusion through the membrane to the permeate stream); $\Delta P$ = pressure gradient; $\Delta\Pi$ = osmotic pressure; $R$ = recovery; and $Z_i$ = combined mass transfer term, including the effects of mass transfer, pressure, and recovery. *See also* coupling model, film theory model.

**liner** Often used interchangeably with lining, which is more common in the water, wastewater, and waste disposal fields.

**line swabbing** A technique used to develop a well with a tightly fit brushlike device.

**Lineweaver–Burke plot** The graphical representation of a rearrangement of the Michaelis–Menten equation:

$$1/K = (1/K_m) + (C_h/K_m)(1/C) \quad \text{(L-44)}$$

where $K$ = rate of uptake of the substance, $C$ = concentration of the substance, $K_m$ = the maximum rate of uptake, and $C_h$ = the concentration at which the rate is half the maximum rate. It plots $1/K$ versus $(1/C)$, which allows the graphical determination of the constants $K_m$ and $C_h$. *See also* Eadie–Hofstee plot.

**lining** In general, a material, such as clay, asphalt, and plastic membranes, used or suitable for covering the inner surface of something. (1) A replaceable tubular sleeve inside a hydraulic or pump-pressure cylinder. (2) A protective layer attached or bonded to the inside of a tank, reservoir, pond, or conduit to protect it against corrosion, resist erosion, prevent seepage, or reduce friction losses. (3) A string of casing in a borehole. (4) A layer of plastic, dense clay, or other impermeable material that prevents leachate from reaching the groundwater. (5) An insert or sleeve for sewer pipes to prevent leakage or infiltration.

**Link-Belt®** A line of products manufactured by FMC Corp. for use in water and wastewater treatment.

**Linpor® process** A proprietary, integrated fixed-film activated sludge method of wastewater treatment developed by Waste Solutions, using free-floating, 10–13 mm cube foam pads in the aeration tank. *See also* Bio-2-Sludge®, BioMatrix®, Captor®, Kaldnes®, moving bed biofilm reactor, Ringlace®, submerged rotating biological contactor.

***L. interrogans*** *See Leptospira interrogans.*

**Linton–Sherwood correlation** A set of adjustable parameters proposed for the Sherwood number equation under particular conditions. *See* Sherwood number for detail.

**LINX®** Electrically regenerable ion exchange, an emerging technology developed by the Pionetics Corp. of San Carlos, CA for point-of-use drinking water systems. The LINX® apparatus consists of one or more electrochemical cells (each having two electrodes, a water-splitting membrane between two electrodes, an inlet, and outlet), a flow control device, and a power source. The application of a voltage accelerates the extraction of undesirable ions and the formation of water component ions at the boundary of the membrane layers, thereby replacing the salts with water molecules in the solution.

**LiOCl** Chemical formula of lithium hypochlorite.

**lipid** Any of a group of organic compounds, insoluble in water but soluble in alcohol and ether. Lipids are long-chain fatty acids and one of three main constituents of living cells. Raw wastewater often contains lipids in the form of fatty acids and their salts, animal fats, and mineral and vegetable oils; they form a dispersed emulsion that is not removed by quiescent settling. *See also* grease and oil.

**lipid peroxidation** The formation of reactive compounds through the oxidation of double-bond fatty acids, particularly in the presence of some metals.

**lipid solubility** The maximum concentration of a chemical that will dissolve in fatty substances. Lipid soluble substances are insoluble in water. They will very selectively disperse through the environment via uptake in living tissue.

**lipophilic** Having a strong affinity for lipids or enhancing the solution or absorption of lipids.

**lipophilicity** The nonpolar character of a substance, its strong affinity for lipids or its ability to promote the dissolution or absorption of lipids. Lipophilicity is an indication of the ability of an organism to accumulate organic substances, often measured by the octanol–water partition coefficient ($K_{ow}$) and used as a predictive parameter for bioaccumulation or biomagnification.

**lipophite** A substance that is soluble in fats and lipids, but often hydrophobic.

**lipoprotein** A protein that contains a lipid associated with a simple protein.

**liposome** A lipid layer around an aqueous particle; a microscopic sac of fatty substances around a cell used in research.

**Liquaclone®** A water treatment unit manufactured by Sanborn Environmental Systems to remove granular solids.

**Liquapac™** A water treatment unit manufactured by Sanborn Environmental Systems to remove solids from spent coolant and oil clarification streams.

**liquefaction** (1) The conversion of a gas or solid into a liquid using changes in temperature or pressure; the term melting is more commonly used for solids. (2) The conversion of organic matter in wastewater from a suspended to a soluble form. (3) A change in the condition of subsurface soil and other materials below the water table, whereby they become fluidlike during seismic activities, which may cause underground structures to float and above-ground structures to tilt and settle.

**liquid** A substance that has a definite volume, flows freely but takes the shape of its container. *See also* fluid, which includes both liquids and gases.

**Liquid A™** A sludge stabilization process developed by RDP Co.

**liquid alum** An aqueous solution of alum that is easy to transport and feed; used in medium-size and large water treatment plants.

**liquidated damages** Penalties imposed as a result of a manufacturer or supplier of water or wastewater goods or services failing to complete a project on time, including, e.g., the costs of additional engineering, utility expenses, or regulatory penalties. Also called delay damages. *See also* consequential or performance damages, performance specification, prescriptive specification.

**liquid bleach** Sodium hypochlorite (NaOCl).

**liquid-chemical-feed system** A system designed to dilute and distribute chemicals in liquid form in water or wastewater treatment; it consists of a solution storage tank, transfer pump, day tank equipped with a mixer, and feed pump. *See* Figure L-06. Also called wet chemical feed or solution feed. *See also* dry chemical-feed system.

**liquid chlorine** (1) Dry, purified molecular chlorine gas converted to a liquid under pressure and refrigerated; shipped under pressure in steel drums and cylinders for various uses, including water and wastewater disinfection. (2) A hypochlorite solution, e.g., bleach or sodium hypochlorite (NaOCl).

**liquid chromatography (LC)** A laboratory method used to analyze organic compounds by separating them into a liquid mobile phase and a porous stationary phase packed inside a tubular column. *See also* gas chromatography.

**liquid chromatography–mass spectrometry (LC–MS)** A laboratory method that combines the use of liquid chromatography and mass spectrometry to analyze organic compounds.

**liquid chromatography–particle beam–mass spectrometry (LC–PB–MS)** The combination of liquid chromatography, mass spectrometry, and an interface between them to analyze organic compounds.

**liquid clutch** One of two common variable-torque transmission systems used to drive wastewater pumps at varying speeds. It transmits energy through the strength of its oil. Also called hydroviscous drive. *See also* eddy current clutch.

**liquid crystal display (LCD)** A device consisting of crystals between glass to display information on computers, instruments, and the like.

**liquid disposal** The disposal of liquid wastes. *See* wastewater disposal.

**liquid feed** The addition of chemicals to water or wastewater as liquids or solutions instead of solids or gases. *See* liquid chemical-feed system for detail. *See also* dry feed.

**liquid feeder** A device that applies chemicals (e.g., in water or wastewater treatment) in solu-

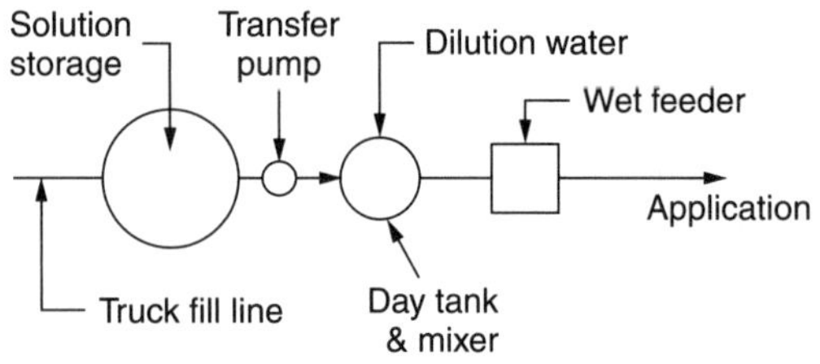


**Figure L-06.** Liquid-chemical-feed system.

tions or suspensions. *See also* solution feeder, dry feeder.

**liquid glass** Same as sodium silicate.

**liquid-injection incinerator** Commonly used system that relies on high pressure to prepare liquid wastes for incineration, breaking them up into tiny droplets to allow easier combustion.

**liquidity index (LI)** A quantitative value used to assess whether a soil will behave as a brittle solid, semisolid, plastic, or liquid. LI is equal to the difference between the natural moisture content of the soil and the plastic limit (PL) divided by the plasticity index (PI).

**liquid limit (LL)** One of the three Atterberg limits: the lower limit for viscous flow of a soil; the water content (as a percentage of the dry weight) of a brittle soil when it passes from the liquid to the plastic state. *See also* plastic limit, shrinkage limit, plasticity index.

**liquid–liquid extraction (LLE)** A selective extraction process using an organic solvent to separate a liquid constituent from wastewater or other solutions. It is a commonly used technique to prepare samples for the analysis of organic compounds. Also called solvent extraction. *See also* continuous liquid–liquid extraction.

**Liquidow®** Calcium chloride produced by the Dow Chemical Co.

**liquid oxygen (LOX)** Oxygen that has been liquefied by decreasing its temperature to facilitate storage, transport, and delivery.

**liquid polymer** A polymer with low to medium molecular weight, active solids of 10–50%, and shelf life of 2–12 months. Polymers are also available in dry, gel, emulsion, and mannich forms.

**LiquidPure®** An activated carbon adsorber manufactured in a small drum by American Norit Co.

**liquid scintillation cocktail** A common fluid used in medical laboratories to analyze DNA and proteins. It often uses radioactive tracers and listed hazardous materials such as Toluene and Xylene. The combination of the two makes it a mixed waste. By volume, it is the most common form of commercially generated mixed waste.

**liquid scintillation counter** An instrument used in the analysis of radionuclides, particularly radon ($^{222}Rn$) in water sources.

**liquid sludge** Water or wastewater treatment sludge that can flow by gravity or pumping.

**liquid water content** *See* water content (2).

**Liqui-Fuge™** A rotary fine screen manufactured by Vulcan Industries, Inc., with internal feed.

**Liqui/Jector®** A gas coalescer manufactured by Osmonics, Inc.

**LiQuilaz®** A device manufactured by Particle Measuring Systems, Inc. for measuring solids concentration in-line.

**Liquiphant** A device manufactured by Endress+Hauser to monitor liquid levels.

**Liqui-pHase®** An apparatus manufactured by Liquid Carbonic to neutralize carbon dioxide.

**Liqui-Strainer** A rotating drum screen manufactured by Vulcan Industries, Inc. with external feed.

**LiquiTrak™** A device manufactured by TSI, Inc. to monitor nonvolatile residues.

**Liquitron®** A device manufactured by Liquid Metronics, Inc. to control pump operation.

**liquor** An aqueous solution of one or more chemical compounds in water; any combination of water or wastewater, as in mixed liquor and mother liquor.

**list** Shorthand term for USEPA's list of violating facilities or firms debarred from obtaining government contracts because they violated certain sections of the Clean Air or Clean Water Acts. The list is maintained by the Office of Enforcement and Compliance Monitoring.

**listed hazardous waste** A waste that is on the USEPA's list of specific hazardous wastes or hazardous waste categories; a listed waste.

**listed species** A species of fish, wildlife, or plant officially designated by an agency as an endangered species or threatened species.

**listed waste** A waste that is listed as hazardous under the Resource Conservation and Recovery Act but that has not been subjected to the Toxic Characteristics Listing Process because the dangers it presents are considered self-evident. Examples: acutely hazardous wastes (arsenic acid, cyanides, and many pesticides) and toxic wastes (benzene, creosote, phenols, and toluene).

**liter (L or l)** A metric system unit of volume for the measurement of liquids and gases, equal to one cubic decimeter or one thousandth of a cubic meter. Also one liter equals 0.264 gallon or 0.0353 cubic foot. Abbreviation: L or l.

**lithium (Li)** A relatively rare, soft, alkali, silver-white, metallic element, the lightest of all metals, used in nuclear armament and other industrial applications; chemically similar to, but less active than, sodium; found most commonly in amblygonite, lepidolite, petalite, and spodumene. Atomic weight = 6.94. Atomic number = 3. Specific gravity = 0.53. Melting point = 181°C. Boiling point = 1342°C. Average concentration in natural waters is 0.05 mg/L. It is not regulated as a drinking water contaminant by USEPA or WHO.

**lithium aluminum hydride ($LiAlH_4$)** A compound that reacts violently with water to produce lithium and aluminum hydroxides and hydrogen gas:

$$LiAlH_4 + 4\,H_2O \rightarrow LiOH + Al(OH)_3 + 4\,H_2 \quad \text{(L-45)}$$

**lithium hypochlorite (LiOCl)** A lithium salt of hypochlorous acid that is applied in chlorination in dry or liquid form. The dry powder form dissolves in water to release approximately 35% chlorine.

**lithology** (1) The science that studies the mineral composition and structure of rock or rock types, and the character of a rock formation. (2) The description of rocks on the basis of their physical and chemical characteristics: structure, composition, color, and texture. It affects the quality of natural groundwater.

**lithophile element** A chemical element that has an affinity for or is concentrated in the Earth's crust. *See also* atmophile element, A-type metal.

**lithosphere** The solid part of the Earth below the surface, composed of rocks, soils, and subsoils; it includes any groundwater contained within it. *See also* atmosphere, hydrosphere.

**lithotrophic** Pertaining to organisms that use inorganic compounds as electron donors in their energetic processes.

**Litmustik®** A portable device manufactured by Omega Engineering, Inc. for pH testing.

**Little, Arthur D.** The American chemical engineer (1863–1935) who first proposed the term "unit operations," which was later applied to the theory of water and wastewater treatment.

**Little Fox** A modular plant manufactured by Red Fox Environmental, Inc. to treat wastewater at sea.

**littoral current** A current produced by waves and moving parallel to the shore, within the stretch where waves break.

**littoral drift** The movement of marine and other sediments parallel to the contour of a beach, due to the action of waves and littoral currents. Also called beach drift, longshore drift.

**littoral zone** The part of the coastal area where waves and currents can move sediment and sunlight can penetrate to the bottom, or practically the area along the shoreline between the high and low water levels. The littoral zone may also include a permanent shallow area where aquatic vegetation grows. In seawater, the littoral zone corresponds to the area between the high-tide and low-tide marks.

**liver fluke** An excreted helminth or the disease that it causes. *See* Asiatic liver fluke (clonorchiasis), cat liver fluke (opisthorchiasis), Chinese liver fluke, fascioliasis.

**liver fluke disease** *See* liver fluke.

**liver rot** A water-related disease of sheep and cattle but occasionally affecting humans. *See* fascioliasis for detail.

**live storage** In a reservoir or other impoundment, the volume between the outlet and the full-pond line, i.e., the sum of the active storage and inactive storage. Also called useful storage, it can be used for any principal or secondary purpose. *See* reservoir storage.

**live vaccine** A vaccine that contains a mixture of weakened strains of a pathogen, which causes an infection leading to resistance.

**LL** Acronym of liquid limit.

**LLD** Acronym of lower level of detection.

**LLE** Acronym of liquid–liquid extraction.

**LLRW** Acronym of low-level radioactive waste.

**LME®** An inclined-plate separator manufactured by Zimpro Environmental, Inc.

**LMM** Acronym of linearized multistage model.

**load** Any quantity relating to the work performed by an installation. *See* loading and the following types of load: base, bed, BOD, hydraulic, organic, peak, pollutional, sediment, suspended, wash.

**load allocation** The portion of a receiving water's loading capacity that is attributed either to one of its existing or future nonpoint sources of pollution or to natural background sources. Load allocations are best estimates of the loading, which may range from reasonably accurate estimates to gross allotments, depending on the availability of data and appropriate techniques for predicting the loading. Wherever possible, natural and nonpoint source loads should be distinguished (EPA-40CFR130.2-g).

**load cell** A device that is used to weigh the shaft of rotating biological contactors in place and to monitor biomass growth or accumulation.

**load curve** A graphical representation, versus time, of the hydraulic or organic load on a treatment plant, a treatment unit, or a pumping station.

**load factor** The ratio of average load of an operation to its maximum load during a given period. *See also* capacity factor.

**loading** The rate at which liquids or pollutants are introduced to a unit, expressed as flow, flow per unit area, mass per unit time, or mass per unit area, e.g., mgd, kg/day, or $lb/ft^2/day$.

**loading capacity** The greatest amount of loading that a water can receive without violating water quality standards (EPA-40CFR130.2-f).

**loading cycle** The sum of wastewater application and drying periods in rapid infiltration systems. It is selected to maximize the infiltration rate, nitrogen removal, or nitrification. It varies from 5–8 days for the application of secondary effluents in the summer to 21–28 days for winter applications. Also called operating cycle. *See also* average application rate.

**loading intensity** A parameter proposed to relate hydraulic and pollutant loadings to the performance of biological treatment units (Fair et al., 1971). The general loading intensity parameter is weight of removable substance in the influent applied to a unit contact surface in a unit time. The parameter can be determined for entire processes or for each component thereof. The area of the contact surface may be measured directly in some cases or indirectly in others. *See also* contact opportunity. For hydraulic loading intensities, see velocity factor, reciprocal detention time, detention time, dilution rate.

**loading rate** The hydraulic, organic, or any other load per unit area or unit length, e.g., gallons per minute per square foot of filter, gallons per day per foot of weir, or pounds of BOD per cubic foot of air.

**Load Limitor** A chain tensioning device manufactured by Envirex, Inc. for traveling water screens.

**LOAEL** Acronym of lowest observed adverse effect level.

**loaiasis** *See* loiasis.

***Loa loa*** The nematode worm that causes loiasis in Central and West Africa. *See also* mangrove fly.

**loam soil** Soil consisting of organic material, sand, silt, and clay.

**Lobe-Aire®** A rotary lobe blower manufactured by Spencer Turbine Co.

**Lobeflo™** A rotary lobe pump manufactured by MGI Pumps, Inc.

**Lobeline™** A positive displacement pump manufactured by Positive Flow Systems, Inc.

**local head losses** Same as minor losses.

**localized corrosion** Corrosion limited to an area exposed to the corrosive action.

**local rejection** The inherent rejection rate ($R^*$) of a membrane for a specific contaminant, taking into consideration flow and concentration variations on either side of the membrane at a specific location, as opposed to the global rejection rate ($R$) that is based on the average feed rate and permeate concentrations. The two rates are related as follows (AWWA, 1999):

$$R^* = 1 - C^*_p/C^*_m \quad \text{(L-46)}$$

$$R = 1 - C_p/C_f = 1 - [1 - (1 - r)^{P(1-R^*)}]/r \quad \text{(L-47)}$$

where $C^*_p$ = permeate concentration of a specific contaminant,

$$C^*_m = PC^*_b \quad \text{(L-48)}$$

= concentration of a specific contaminant at the membrane surface, $C_p$ = the permeate concentration; $C_f$ = the feedwater concentration, $r$ = single-pass recovery of the membrane; $P$ = polarization factor, and $C^*_b$ = bulk concentration of a contaminant at a given location. Also called apparent rejection. *See also* global rejection, polarization factor, lag coefficient, mechanical sieving (steric rejection), recovery rate.

**local toxic effect** An effect occurring at the site of exposure to a toxic substance.

**Lo-Cat®** A process developed by Wheelabrator Clean Air Systems, Inc. to oxidize hydrogen sulfide in anaerobic bio-gas units.

**Lockett, W. T.** One of the two reported originators of the activated sludge process. The other is Edward Ardern. They described their studies in a 1914 article entitled "Experiments on the Oxidation of Sewage without the Aid of Filters." *See also* activated sludge process.

**LOD** Acronym of level of detection.

**lodestone** An iron mineral that possesses magnetic polarity and attracts iron. *See* magnetite for detail.

**LOEC** Acronym of lowest-observed-effect concentration.

**LOEL** Acronym of lowest observed effect level.

**log** (1) Abbreviation of common logarithm. (2) A measure of the performance of a water or wastewater disinfection process. *See* log removal efficiency.

**Logan model** A computer model developed in 1987 to design plastic media trickling filters based on the media geometry. Numerical equations are used to determine the rate of removal of soluble BOD. Other trickling filter design formulas include British Manual, Eckenfelder, Galler and Gotaas, Germain, Howland, Kincannon–Stover, modified Velz, NRC, Rankin, Schulze, Velz.

**logarithm** (1) The exponent that indicates the power to which a number must be raised to produce a given number. For example, if $B^2 = N$, the 2 is the logarithm of $N$ to the base $B$, or $10^2 = 100$ and log 100 (base 10) = 2. Abbreviation of log base 10 or common logarithm is "log." Natural logarithm is log base e. (2) Same as common logarithm.

**log boom** A floating structure of logs or timber used to protect hydraulic structures (dams, intakes, etc.) from wave damage, to deflect floating material, or to control spills.

**log death phase** A phase in the growth of microorganisms in which their death rate exceeds the production of new cells, and the number of microorganisms decreases at a logarithmic rate. *See* growth pattern.

**LogEasy™** A device manufactured by Hach Co. to count particles.

**log growth** Growth at a logarithmic rate.

**log growth phase** The period during which microorganisms grow logarithmically with time, when food is not a constraint and other growth factors are adequate. The growth rate depends on the ability of the cells to process food, or on cell division time in the case of bacteria. The population doubles over a regular time interval called the generation time. *See* growth pattern.

**logistic growth curve** The graphical representation of the first-order, logistic equation, sometimes used in forecasting the population to be served by water and wastewater facilities:

$$y = L/[1 + e^{(\ln P - Qt)}] \quad \text{(L-49)}$$

where $y$ = population at time $t$, and $L$ = saturation or maximum population,

$$P = (L - y_0)/y_0 \quad \text{(L-50)}$$

$$Q = KL \quad \text{(L-51)}$$

$y_0$ = initial population (at time $t = 0$), and $K$ = a growth constant (per time). The most characteristic point of this curve is the point of maximum growth rate, which occurs at time $t^*$:

$$t^* = (-\ln P)/ Q \quad \text{(L-52)}$$

Then $y = 0.5\ L$, i.e., half the saturation population. *See* Figure L-07. The logistic model may also be used to predict the growth of microorganisms.

**logit dose–response model** A model that relates the proportion of test animals infected to the number of infecting agents inoculated, e.g., proportion of mice infected by *Cryptosporidium parvum* oocysts. *See also* response logit.

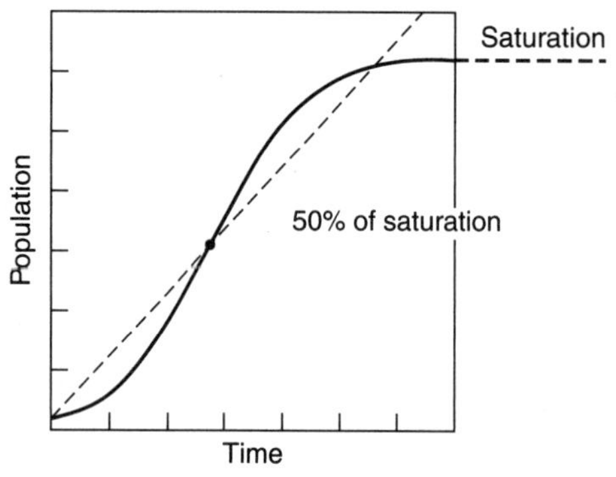


**Figure L-07.** Logistic growth curve.

**logit model** A dose–response or tolerance distribution model of the form

$$P(d) = 1/[1 + e^{-(a + b \log d)}] \quad \text{(L-53)}$$

where $P(d)$ is the probability of toxic effects from a continuous dose rate $d$, and $a$ and $b$ are constants. The logit model is represented graphically by a logistic curve and gives results similar to those of the log-probit model except at low doses. *See* dose–response model for related distributions.

**log-mean feed concentration (LMFC)** The average concentration of feedwater in a concentrate-staged, pressure-driven membrane treatment unit (AWWA, 2000):

$$\text{LMFC} = C_f \ln (C_c/C_f)/[1 - (C_f/C_c)] \quad \text{(L-54)}$$

where $C_f$ and $C_c$ represent, respectively, the feed concentration and the concentrate concentration, both in milligrams per liter.

**log-normal** (or **lognormal) distribution** A special case of the log Pearson Type III distribution (when the skew coefficient is equal to zero), used in the frequency analysis of floods and other extreme events. It is obtained from the normal distribution by substituting the natural logarithm (log $x$) for the random variable ($x$).Thus, the lognormal probability $P(x)$ of an event ($x$) is:

$$P(x) = \exp \{-0.5\ [\log x - \mu)/\sigma^2]^2\}/[x \cdot \sigma(2\ \pi)^{0.5}] \quad \text{(L-55)}$$

for the two-parameter distribution and

$$P(x) = \exp\{-0.5[\log (x - x_0) - \mu]/\sigma^2]^2\}/[(x - x_0)\sigma(2\pi)^{0.5}] \quad \text{(L-56)}$$

for the three-parameter distribution, where $\mu$ is the mean and $\sigma$ is the standard deviation. These two parameters are estimated for a sample from the population. The third parameter $x_0$ is the location parameter. The number $\pi = 3.1416$.

**log phase** *See* log growth phase.

**log-probit model** A normal distribution used to describe the probability of human infection by enteric microorganisms or of adverse effects by a toxic substance, based on the assumption that each individual has a threshold dose or exposure for infection or toxic effect. The data are transformed to log base 10. A probit is a normal equivalent deviate (the difference between a variable and the mean) increased by a whole integer to avoid negative numbers. *See* dose–response model for related distributions.

**log reduction** *See* log removal efficiency.

**log reduction unit** Same as log unit.

**log removal** *See* log removal efficiency.

**log removal efficiency (LRE)** An expression of the removal efficiency of a treatment process, the negative logarithm of the ratio of effluent concentration ($C$) to influent concentration ($C_0$):

$$\text{LRE} = -\log C/C_0 \qquad \text{(L-57)}$$

LRE is often used in relation to the Surface Water Treatment Rule and water treatment designed to remove, kill, or inactivate pathogens, but it applies to other impurities as well. For example, a 3-log efficiency implies a removal of 99.9% of impurities, a 4-log efficiency implies a removal of 99.99%, and so on, corresponding to remaining fractions of 0.001 (log = –3) and 0.0001 (log = –4), respectively. Conventional wastewater treatment processes have an LRE of 1 or 2 (90–99% removal) for enteric bacteria, whereas a pond system with 30-day detention in a warm climate may achieve an LRE of 6 (99.9999% removal). *See also* fundamental filtration theory, log unit.

**log removal ratio (LRR)** A parameter that defines the performance of a surface water treatment facility in terms of *Cryptosporidium* spore removal:

$$\text{LRR} = \log (R/F) \qquad \text{(L-58)}$$

where $R$ and $F$ represent in spores/liter the concentration in the raw and finished water respectively. *See also* DOP credit.

**log survival** *See* log removal efficiency.

**log unit** A unit of measure of pathogen reduction by wastewater treatment, equal to the inverse power of 10 that expresses the percent of pathogens remaining after treatment. For example, a treatment process that reduces coliforms by 99% leaves 1% or $0.01 = 10^{-2}$ and has an efficiency of 2 log units; 99.99995 corresponds to a residual of $0.0000005\% = 10^{-6.3}$ or 6.3 log units. If $p$ is the percent reduction, the corresponding log removal is:

$$\text{log removal} = -\log_{10}[(100 - p)/100] \text{ units} \qquad \text{(L-59)}$$

**log unit reduction** *See* log removal efficiency.

**log weir** A triangular weir made of logs placed in the direction of flow with butt ends downstream.

**Lo-Head™** A traveling bridge filter manufactured by Agency Environmental, Inc.

**loiasis** A mosquito-borne helminthic infection, the eye worm disease, caused by the pathogen *Loa loa.* Distributed near the equatorial rain forest in Central and West Africa, it is transmitted by *Chrysops* species or mangrove flies. The worm may migrate through the body and cause transient swelling, pain, and itching. Also spelled loaiasis.

**London–van der Waals dispersion interaction** *See* van der Waals dispersion interaction.

**London–van der Waals force** *See* van der Waals force.

**long-based weir** Same as broad-crested weir.

**longitudinal diffusion** The longitudinal spreading of a substance in a fluid moving under turbulent conditions, faster or slower than the velocity of the fluid. *See also* diffusion and dispersion.

**longitudinal dispersion** Movement of a substance in the direction of flow, forward or backward. *See also* diffusion and dispersion.

**longitudinal flow** Movement of a water up or down in a cartridge or loose-medium filter. Also called axial flow.

**longitudinal mixing** The dispersion of a substance lengthwise in a stream or a treatment unit. Longitudinal mixing is limited in tidal estuaries because of density differences that keep the freshwater over the saltwater. *See* saltwater creep.

**longitudinal study** A study in which subjects are followed forward in time from initiation of the study. This is often called a prospective or cohort study.

**long-launder approach** One of two common methods of designing weirs and launders for clarifiers: it assumes that the placement and length of the weir are as important for a rectangular primary clarifier as for a secondary clarifier, hence, a weir length equal to 33–50% of the length of the sedimentation basin. *See also* short-launder approach.

**Longopac** An apparatus manufactured by Spirac Engineering for handling wastewater treatment screenings.

**long pipe** A pipeline the length ($L$) of which is much greater than its diameter ($D$), e.g., with $L > 500\,D$. For such a pipe, minor (or local) head losses are negligible.

**longshore current** A current generated by the action of waves, winds, and tides, and moving parallel to the coast. *See also* littoral current.

**longshore drift** The movement of marine and other sediments parallel to the contour of a beach, due to the action of longshore currents. Also called beach drift, littoral drift.

**long-term acceptance rate (LTAR)** (1) One of two factors affecting the parameter soil permeability ($K$) in the formula of hydraulic loading of a slow rate land application system. It accounts for soil clogging and is represented by the symbol $F_c$. (2) The long-term hydraulic capacity of the biomat of a subsurface-soil absorption system, i.e., the flow rate that can be applied to a given soil over a long period of time before application must stop to allow the soil–wastewater interface to dry and restore infiltrative capacity. Typically, LTAR = approximately 16 liters per square meter per day.

**Long-term 2 Enhanced Surface Water Treatment Rule (LT2ESWTR)** A regulation proposed by the USEPA to enhance the protection of drinking water against pathogens. It requires most water supply systems to monitor *Cryptosporidium* initially and sets the stage for the use of ultraviolet (UV) light disinfection.

**LoNox™** A combustion burner manufactured by John Zink Co.

**LOOP** A package plant manufactured by Smith & Loveless, Inc. for wastewater treatment using the oxidation ditch process.

**Loop Chain** A nonmetallic chain manufactured by Envirex, Inc. for sludge collection.

**loop system weight-loss method** A corrosion assessment method that uses actual pipe sections to study the effects of water quality parameters on materials in the distribution system as well as the effectiveness of corrosion control techniques. *See also* coupon weight-loss method and electrochemical rate measurement.

**loose media filter** A filter using a loose medium.

**loose medium** A medium that can be expanded during backwashing and rinsing, e.g., in a filter, ion-exchange unit, or adsorption column, as compared to a fixed medium.

**loose RO** Another name for nanofiltration.

**loping** A problem sometimes encountered in the operation of rotating biological contactors, characterized by uneven shaft rotation of air-driven units.

**Lo-Pro™** An air stripper manufactured by ORS Environmental Systems.

**LOQ** Acronym of level of quantification.

**lorandite** One of the rare minerals that contain the metallic element thallium.

**Lorentz-Lorentz equation** An equation used to determine the polarizability ($\alpha$) of a volatile organic compound:

$$\alpha = (\eta^2 - 1)M/[\rho(\eta^2 + 2)] \qquad \text{(L-60)}$$

where $\eta$ = refractive index of the volatile organic compound (VOC), $M$ = molecular weight of the VOC, and $\rho$ = liquid density of the VOC. *See also* Dubinin–Raduskevich equation.

**Loschmidt, Joseph** Austrian chemist, 1821–1895 who calculated the number of moles in one cubic centimeter of an ideal gas.

**Loschmidt's number** The number of molecules of an ideal gas, at standard temperature (0°C) and pressure (760 mm Hg), contained in one cubic centimeter of the gas, equal to $2.687 \times 10^{19}$. *See also* Avogadro's number.

**loss of head** Same as head loss.

**loss-of-head gage** A gage that indicates (and may record) head losses, e.g., filtration head loss.

**loss on ignition (LOI)** The percent weight change of a 50-gram sample of sand heated at 1800°F in a muffle furnace for two hours (Freeman, 1995). Although the heating causes calcium carbonate ($CaCO_3$) to dissociate into calcium oxide and carbon dioxide ($CO_2$), the latter being as a gas, the change may actually be a weight gain due to the oxidation of metallic iron to ferric oxide ($FeO_3$).

**lost motion** A mechanism introduced in a diaphragm metering pump to control its capacity by adjusting the quantity of actuating fluid or the degree of mechanical actuation.

**lotic** Pertaining to or living in rivers or other flowing water bodies. *See also* lentic.

**lotic habitat** A moving freshwater habitat such a spring, stream, or river.

**lotic water** A running watercourse such as a river or a canal, as opposed to a lentic or standing water body such as a reservoir or a lake.

**louse-borne fever** *See* relapsing fever.

**louse-borne typhus** A water-washed, rickettsial disease caused by the pathogen *Rickettsia prowazeki* and transmitted from a person to another person via lice. Acute and sometimes fatal, it is found mostly in high-altitude areas of Africa. Also called epidemic typhus.

**low chromium content** *See* juvenile diabetes.

**low-density polyethylene (LDPE)** A thermoplastic polymer or copolymer comprised of at least 50% ethylene by weight and having a density of 0.940 g/cm$^3$ or less (EPA-40CFR60.561).

**low-DO oxidation ditch process** An oxidation ditch operated under dissolved oxygen conditions that promote both nitrification and denitrification. *See* Sym-Bio™ process.

**low dose** One tenth of the high dose (EPA-40CFR795. 232-4).

**lower alternate stage** *See* alternate stages or alternate depths.

**lower explosive limit (LEL)** (1) The concentration of a compound in air below which the mixture will not catch on fire. *See also* flash point, upper explosive limit. (2) The concentration of a gas below which the concentration of vapors is insufficient to support an explosion. LELs for most organics are generally 1 to 5% by volume.

**lower heating value** In the evaluation of gas production in anaerobic digestion, the lower heating value of digester gas is the difference between the heat of combustion and the heat of vaporization of any water vapor present, i.e., approximately 600 Btu/ft$^3$.

**lower level of detection** One of a few parameters used to measure the precision or the detection level of a method of analysis, stated as a result that produces a signal 2 × 1.645 times the standard deviation above the mean of blank analyses. Also called level of detection or, simply, detection level.

**lowest-effect level (LEL)** Same as lowest observed adverse effect level.

**lowest observed adverse effect level (LOAEL)** (1) The lowest dose in an experiment that produced an observable adverse effect. (2) The lowest exposure level at which there are statistically or biologically significant increases in frequency or severity of adverse effects between the exposed population and its appropriate control group. Sometimes used as a target of acceptable daily intake of a substance in milligrams per kilogram (mg/k) of body weight per day; it is thus equal to the NOAEL multiplied by a factor of safety. *See also* acceptable daily intake, benchmark dose, reference dose, and no observed adverse effect level (NOAEL).

**lowest observed effect concentration (LOEC)** The lowest concentration of a constituent for which the measured values are statistically different from the control. *See also* toxicity terms.

**lowest observed effect level (LOEL)** The lowest dose or exposure that produces a change in an organism, not necessarily an adverse effect as in the LOAEL.

**low flow** (1) Stream flow during the driest period of the year. Also called minimum flow. *See also* design flow. (2) A flow condition in flow routing problems; *see* combination flow.

**low-flow augmentation** An addition to stream flow during periods of low flow, e.g., to meet required dilution ratios. For example, a dam and reservoir on a stream or on one of its tributaries can store water during periods of high flows and release it when needed. *See also* pumped storage, flow augmentation.

**low-flow criterion** A criterion defining the minimum flow that is acceptable for a given purpose. For example, in water quality studies, the 7Q10 flow is the minimum seven-day average flow expected once in ten years. *See* design flow.

**low-flow stream augmentation** *See* low-flow augmentation.

**low-flow toilet** A toilet that uses 1.6 gallons per flush (gpf) or less, as allowed by the 1992 U.S. Energy Policy Act. Older toilets used 5.0 to 7.0 gpf. High-efficiency toilets use 1.3 gpf or less. High-flow toilets use 3.5 gpf or more. Some dual-flow toilets can use 1.6 or 0.8 gpf. Also called low-flush toilet, ultra-low-flush toilet. *See also* flow-reduction device and water saver.

**low-flush toilet** Same as low-flow toilet.

**low-head, continuous-backwash filter** *See* continuous-backwash filter.

**low-level mixed waste (LLMW)** A waste that contains low-level radioactive waste and hazardous waste.

**low-level radioactive waste (LLRW)** A waste less hazardous than most of those associated with nuclear reactors; generated by hospitals, research laboratories, and certain industries. The U.S. Department of Energy (DOE), Nuclear Regulatory Commission, and USEPA share responsibilities for managing LLRW. Its specific definition per the DOE is "waste that contains radioactivity and is not classified as high-level waste, transuranic waste, or spent nuclear fuel. Test specimens of fissionable material irradiated for research and development only, and not for the production of power or plutonium, may be classified as low-level waste, provided the concentration of transuranics is less than 100 nanocuries per gram of waste." *See also* high-level radioactive waste.

**low-level waste** Low-level radioactive waste.

**low-pressure filtration** *See* low-pressure membrane filtration.

**low-pressure, high-intensity UV lamp** A mercury–indium amalgam–argon lamp that generates monochromatic radiation at a wavelength of 254 nm, a lamp wall of 40°C, an internal pressure of 0.001–0.01 mm Hg, and an output of 50–100 W for an input of 70–80 W. In water and wastewater disinfection applications, quartz sleeves isolate the lamp from direct contact with the liquid. The average useful life of the high-intensity lamp is about 13,000 hours but the sleeves last 4–8 years. *See also* low-pressure, low-intensity UV lamp; medium-pressure, high-intensity UV lamp; narrow-band excimer lamp; pulsed UV lamp.

**low-pressure, low-intensity UV lamp** A mercury–argon lamp that generates monochromatic radiation at a wavelength of 254 nm, a lamp wall of 40°C, an internal pressure of 0.007 mm Hg, and an output of 25–27 W for an input of 70–80 W. In water and wastewater disinfection applications, quartz sleeves isolate the lamp from direct contact with the liquid. The average useful life of the low-intensity lamp is about 10,000 hours but the sleeves last 4–8 years. *See also* low-pressure, high-intensity UV lamp; medium-pressure, high-intensity UV lamp; narrow-band excimer lamp; pulsed UV lamp.

**low-pressure membrane** A porous membrane used in such low-pressure filtration systems as microfiltration and ultrafiltration for the removal of contaminants from water by a size-exclusion mechanism. *See also* high-pressure membrane.

**low-pressure membrane filtration** A water filtration technique, such as microfiltration and ultrafiltration, that removes nearly all particles larger than average pore size, including turbidity, bacteria, protozoa, algae, and some viruses.

**low-pressure oxidation (LPO)** One of two basic modes of thermal sludge conditioning used in wastewater treatment. An LPO system consists mainly of a storage/blending/feed apparatus pumping macerated sludge to a heat exchanger that also receives sludge from a reactor heated by steam, a solids settling unit, and a dewatering unit. *See also* heat treatment.

**low-pressure zone** The third area of a belt filter press where the two porous belts come together with the sludge solids in between as in a sandwich. Also called wedge zone.

**low-profile air stripper** A type of packed tower stripper used to remove volatile organic chemicals (VOC) from contaminated waters. It consists of a countercurrent, 10-ft high column with a series of perforated trays along the column, each tray ending with inlet and outlet channels. The liquid flows from the top and horizontally across the trays. Also called a sieve tray column. *See also* cascade air stripping, cocurrent packed tower, cascade packed tower, end effect, crossflow packed-tower, minimum air-to-water ratio, packed-tower design equation, wall effect.

**low-rate anaerobic digestion** A single-stage anaerobic sludge stabilization process operated in the mesophilic range (30–38°C), including digestion, thickening, and supernatant formation in one tank; used mostly in small installations. One of four basic configurations of the anaerobic digestion process, it provides anaerobic treatment in a well-mixed reactor without solids recycle, with a detention time of 10 to 30 days. Also called conventional or standard-rate digestion. *See* Figure L-08. *See also* anaerobic contact, high-rate anaerobic digestion, phase separation, secondary digester, separate sludge digestion, single-stage high-rate digestion, thermophilic digestion, two-stage digestion.

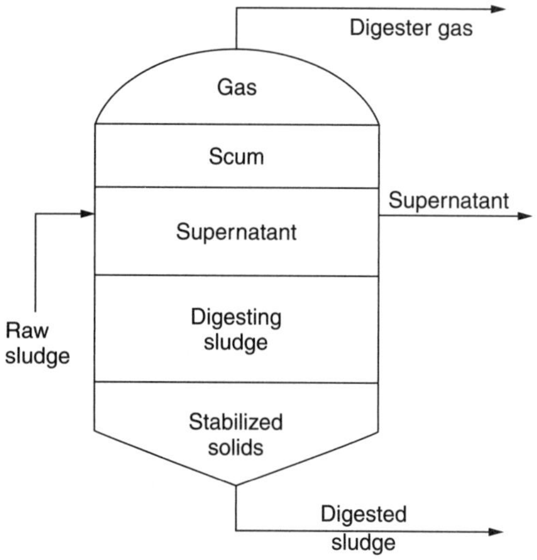


**Figure L-08.** Low-rate anaerobic digestion.

**low-rate digestion** *See* low-rate anaerobic digestion.

**low-rate filter** A trickling filter that is designed and operated at a low hydraulic and/or organic loading rates, e.g., 2.0–4.0 mgd/acre and/or 5–25 pounds of BOD per 1000 cubic feet. *See* low-rate trickling filter.

**low-rate irrigation** The land application of wastewater by sprinkling or surface application at a rate of 1.5 to 10 cm per week. *See also* geopurification, high-rate infiltration, high-rate sprinkling, overland flow, rapid infiltration extraction, rapid infiltration, soil aquifer treatment, and surface spreading.

**low-rate pond** An aerobic stabilization pond used mainly for the treatment of soluble organic wastes or secondary effluents; it is designed to handle a load of 60 to 120 pounds of BOD per acre per day and to produce 40 to 100 mg/L of algal cell tissue. *See also* high-rate and maturation ponds, and lagoons and ponds.

**low-rate trickling filter** A secondary wastewater treatment unit designed to operate without recirculation at loadings of 0.3–1.5 kg BOD/day/$m^3$ and 45–90 gpd/$ft^2$ (or 2.0–4.0 mgad); the only type of trickling filter installed in the United States before 1936. It is a circular or rectangular unit, generally fed from a dosing tank and consistently producing a nitrified effluent with relatively low BOD. Also called conventional trickling filter or standard-rate trickling filter. *See also* high-rate trickling filter, intermediate-rate trickling filter, roughing filter, super-rate filter, biofilter, biotower, oxidation tower, dosing tank, filter fly.

**Lowry's concept** A concept developed in 1923 to define an acid as a proton donor and a base as a proton acceptor. *See* Brønsted-Lowry theory for detail.

**low-service district** The portion of a community that is served by the main distribution system, without the use of energy-boosting pumps.

**low-service pump** In a conventional water treatment plant, a device that pumps source water through the plant, as compared to the high-service pumps that convey water from the clear well to the distribution system.

**low-solids flux (LSF)** A new soldering technique proposed for use in the electronics industry, whereby a fluxing material contains only 1–5% solids as compared to 20–35% solids in a conventional flux. LSF eliminates the cleaning step.

**Lowther Plate ozonator** An ozone generator that consists of a series of air-cooled aluminum blocks, dielectrics, and stainless-steel electrodes between which air or oxygen is passed. *See also* Otto Plate ozonator.

**LOX** Abbreviation of liquid oxygen.

***L. pneumophila*** *Legionella pneumophila.*

**LPO** Acronym of low-pressure oxidation.

**LR** Acronym of Larson ratio.

**LSC™** A compact spray deaerating heater manufactured by Graver Co.

**LSI** Acronym of Langelier saturation index.

**LT2ESWTR** Acronym of Long-term 2 Enhanced Surface Water Treatment Rule.

**LTAR** Acronym of long-term acceptance rate.

**Ludzack–Ettinger process** A biological nitrogen removal process, first proposed in 1962, that includes an anoxic reactor followed by an aerobic reactor and a final clarifier, with recirculation to the anoxic stage from either or both the mixed liquor and the clarifier underflow, as shown in Figure L-09. The return activated sludge (RAS) rate is an important design and operation parameter because RAS provides the only nitrate source to the anoxic tank. *See also* modified Ludzack–Ettinger process, Wuhrmann process.

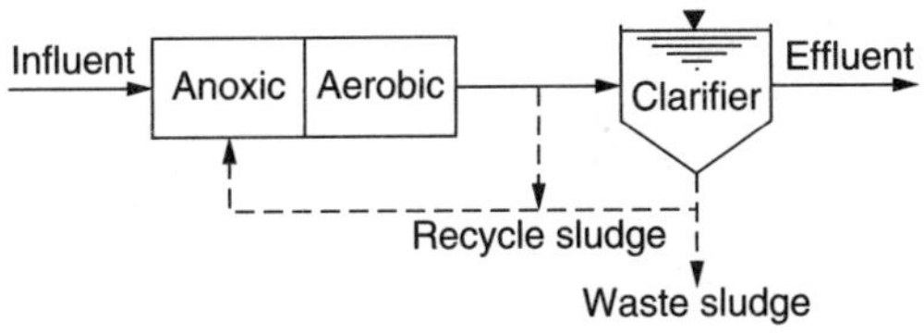


**Figure L-09.** Ludzack–Ettinger process.

**lumen** (1) The Système International (SI) unit of luminous flux; the luminous flux emitted in one steradian by a point source of intensity one candela, i.e., 1 lumen = 1 steradian-candela. (2) The canal, duct, or cavity of a tubular organ. (3) The interior of a hollow fiber membrane, within which solid materials can accumulate and cause fouling.

**luminous intensity** *See* candela.

**luminous flux** *See* lumen.

**lung fluke** The excreted helminth *Paragonimus westermani,* or the infection that it causes (lung fluke disease).

**lung fluke disease** The common name for paragonimiasis, a helminthic disease caused by the pathogen *Paragonimus westermani,* found in East Asia, Africa, and South America. It is transmitted to human by pigs and domestic animals through aquatic intermediate hosts (e.g., crab, crayfish, snail).

**lux (lx)** The Système International (SI) unit of illuminance; the illuminance of a luminous flux of one lumen uniformly distributed over a surface of one meter square, i.e., 1 lx = 1 lumen/$m^2$ = 1 steradian-candela/$m^2$ = 0.0929 foot-candle. Also called meter-candle. Plural = luces.

**luxury uptake** The removal of phosphorus in wastewater treatment beyond the theoretical bacterial requirements. This mechanism results from the growth of a particular bacterial population that can store large quantities of phosphorus within the cell (cellular uptake) in addition to the metabolic uptake. *See also* activated algae process, bacterial assimilation.

**L'vov platform** A graphite tube adapted for use in atomic absorption spectrophotometers.

**LWL** Acronym of low water level.

**lx** Abbreviation of lux.

**Lyco™** A line of wastewater treatment products of the U.S. Filter Corp.

**lye** A highly concentrated aqueous solution of potassium hydroxide (KOH) or sodium hydroxide (NaOH).

**Lykken–Estabrook process** A process designed to recover lime from the sludge produced in lime softening; the precipitate is added to 12% of the influent water and this overtreated stream is mixed with the main stream. Sludge from the second stage is wasted. *See also* the Hoover process.

**Lyme disease** An acute inflammatory disease caused by a tick-borne spirochete, *Borrelia burgdorferi,* first recognized in Lyme (CT) and affecting humans as well as animals. Symptoms include a large rash at the biting site, acute headache, backache, chills, and fatigue.

***Lymnaea*** A genus of amphibious snails that are intermediate hosts for *Fasciola hepatica.*

**lymphatic system** A network of small ducts that drain excess fluids from tissues of the body to the bloodstream.

**lymphocyte** A type of white blood cell, with a nongranular cytoplasm surrounding a spherical nucleus, involved in inflammatory reactions to toxic substances.

***Lyngbya*** A genus of cyanobacteria that produce geosmin, a substance that causes odor in water sources.

**lyophilic** Characteristic of colloids, the particles of which have a strong affinity for the liquid in which they are dispersed. The viscosity of the suspension increases parabolically with colloids concentration.

**lyophilization** A method of preservation of heat-sensitive substances (foods, blood, antibiotics, microorganisms, etc.) by freezing them and then subliming the ice in a vacuum. Also called freeze-drying.

**lyophobic colloidal suspension** A suspension in which the colloidal particles have little or no affinity for the surrounding liquid. The viscosity of the suspension is only slightly higher than that of the dispersing medium.

**lyse** To undergo lysis.

**lysin** An antibody that causes the disintegration or dissolution of bacterial cells.

**lysine** A crystalline, basic amino acid produced by hydrolysis of proteins, essential to human and animal nutrition.

**lysis** Death of a living microorganism causing the rupture of the cell and the release of protoplasm for resynthesis by other organisms. During the endogenous growth phase, lysis decreases the number of cells and total biomass. *See* lysin, cell debris, cell lysis.

**lysis–regrowth model** A model according to which microorganisms release biomass particulate matter during endogenous decay. The biodegradable portion of this release is hydrolyzed and becomes a source of readily biodegradable COD; the other portion remains as cell debris. This approach, used in the ASM1 and ASM2 models, results in a higher endogenous decay coefficient than the decay coefficient traditionally used for the activated sludge process.

**lysogen** A prokaryote that contains a prophage. *See also* temperate virus.

**lysogenic phage** A temperate phage; a phage in which the nucleic acid is integrated with the chromosome of the host, which transmits it to its descendants and daughter cells.

**lysogeny** The condition of a bacterial carrier host that is harboring a temperate virus as a prophage (a stable, noninfectious form of the virus) and is still sensitive to other phage populations.

# M

**m** Abbreviation of meter(s), a metric or SI unit equal to 3.281 ft.

**MABR process** Acronym of membrane aeration bioreactor process.

**MAC** Acronym of (a) maximum allowable concentration and (b) *Mycobacterium avium* complex. *See also* Mai.

**MacConkey-PA** A simple medium proposed to conduct the presence–absence test for multiple indicators in drinking water.

**maceration** Chopping or tearing. Maceration is a unit operation of wastewater treatment consisting of the side-stream grinding of coarse solids.

**macerator** A low-speed device, consisting of two counter-rotating assemblies with blades, used to intercept coarse solids in wastewater and grind or shred them. Macerators are installed in pipelines or in channels. *See also* comminutor, bar screen.

**macerator–grinder** *See* sludge macerator–grinder.

**Mach, mach, Mach number, mach number** The ratio of the speed of an object to the speed of sound in the undisturbed medium through which the object is moving. After the Austrian physicist/psychologist/philosopher Ernst Mach (1838–1916).

**Macpac** A chemical feed device manufactured by Milton Roy Co.

**macrobenthos** Animals and plants larger than microscopic that live on the bottom of a body of water.

**Macro-Cat** An ion-exchange resin produced by Sybron Chemicals, Inc.

**macroencapsulation** The embedding of an agglomerate of waste particles in a material that isolates them from the surroundings. *See also* encapsulation, microencapsulation, solidification/stabilization.

**macrofloc** A floc particle that is too large to go through a filter bed.

**macroflocculation** The aggregation of particles larger than 1 or 2 micrometers due to induced velocity gradients or differential settling; it is not effective for viruses. Also called orthokinetic flocculation. *See also* flocculation concepts.

**macrofouling** The fouling of a filter bed, membrane, or other water treatment unit by macroorganisms (e.g., clams and mussels).

**macroinvertebrate** An animal that is larger than microscopic but has no backbone or spinal column. Macroinvertebrates are among the organisms that settle in sedimentation basins.

**Macrolite®** A proprietary ceramic filtration medium manufactured by Kinetico, Inc. of Newbury,

OH for municipal water treatment at a loading rate of 4–10 gpm/ft$^2$.

**macromolecule** A large molecule formed by the combination of several small molecules.

**Macronet®** An adsorbent resin made of macroporous polystyrene and acrylic by the Purolite Co. of Bala Cynwyd, PA and used in ion exchange.

**macronutrient** Any of the principal inorganic nutrients needed for microbial cell synthesis in relatively large amounts. Some macronutrients belong to class A ions, e.g., calcium ($Ca^{2+}$), iron ($Fe^{2+}$), magnesium ($Mg^{2+}$), potassium ($K^+$), and sodium ($Na^+$). Other macronutrients are carbon (C), hydrogen (H), oxygen (O), nitrogen (N), sulfur (S), phosphorus (P), and chlorine (Cl). Cell biomass is often represented as $C_{60}H_{87}O_{23}N_{12}P$. *See also* micronutrients, trace metals.

**macroorganism** Any plant, animal, or fungus that is larger than microscopic, i.e., visible to the naked eye. *See also* macroinvertebrate.

**macrophage** A large, phagocytic cell found in the walls of blood vessels, body fluids, and connective tissues; derived from monocytes, and activated by inflammation. Macrophages remove dead cells and foreign particles (including bacteria, viruses, and toxins) by engulfing and biochemically processing them.

**macrophyte** A plant large enough to be visible to the naked eye, especially a marine or aquatic plant (e.g., crowfoot, water lily), as opposed to microphytes (such as phytoplankton). Macrophytes often grow in stabilization ponds, with roots suspended or attached to the bottom.

**macrophyte production** The production of large aquatic plants for human or animal consumption, one of three important forms of aquaculture. The plants include water spinach, water chestnut, water bamboo, and water hyacinth. The latter may also be used in wastewater treatment to remove nutrients, metals, and other pollutants. *See also* algal culture and fish farming.

**macropore** (1) A pore (of an adsorbing material) having a diameter larger than 100 nanometers (1000 Å) or even a pore larger than 50 nanometers (500 Å). Macropores are normally associated with the adsorption of large molecules. *See also* mesopore and micropore. (2) A relatively large opening in a soakage or land application site, e.g., root channels, structural voids, rodent burrows, and solution channels or fissures, through which wastewater can drain down and pollute groundwater.

**macroporous resin** An anionic or cationic ion-exchange resin that has large pores and resists oxidation, organic fouling, and swelling.

**macroporous resin matrix** A polymeric resin of polystyrene cross-linked with divinyl benzene with functional groups impregnated into the resin.

**macroreticular resin** An anionic or cationic ion-exchange resin that retains its pore structure even after drying and is less subject to fouling than other resins. *See also* macroporous and fixed-pore resins.

**macroscopic** Visible to the naked eye.

**macroscopic filtration model** *See* phenomenological filtration model.

**macroscopic organism** An organism big enough to be seen by the naked eye (i.e., without the aid of a microscope).

**macrotransport** One of the three steps of the adsorption process; the movement of organic matter or other adsorbate through water or wastewater to the solid–liquid interface by advection and diffusion. *See also* microtransport, sorption.

**MACT** Acronym of maximum achievable control technology.

**maghemite ($\gamma$-$Fe_2O_3$)** A compound formed during the corrosion of iron pipes; also called an iron phase or a corrosion scale.

**Magicblock™** A device manufactured by Osmonics, Inc. for fluid control.

**magmeter** *See* magnetic flowmeter.

**Magna** A rotor aerator manufactured by Lakeside Equipment Corp.

**Magna Cleaner™** A liquid cyclone manufactured by Andritz-Ruthner, Inc.

**MagneClear™** Magnesium hydroxide chemicals produced by Martin Marietta Specialties, Inc. for use in water treatment.

**magnehelic gauge** A sensitive differential pressure or vacuum gauge manufactured by Dwyer Instrument Co. that uses a precision diaphragm to measure pressure differences. This gauge is manufactured in specific pressure or vacuum ranges such as 0 to 2 inches of water column. Magnehelic gauges are typically used to measure SVE system vacuums.

**magnesia (MgO)** A white, tasteless substance; processed magnesium oxide, a chemical used in medicine and in water treatment for pH adjustment. *See also* milk of magnesia.

**magnesia milk** *See* milk of magnesia.

**magnesian calcite** Calcite with extensive magnesium ($Mg^{2+}$) substitution, a seawater mineral of biogenic origin; abbreviated Mg-calcite; distinct from both calcite and magnesite.

**magnesite** A mineral occurring in white masses; the principal ore of magnesium, essentially magnesium carbonate ($MgCO_3$). *See also* magnesian calcite.

**magnesium (Mg)** A light, silver-white, malleable, ductile, abundant, metallic element, generally found as magnesium carbonate (magnesite) or as a carbonate in combination with calcium (dolomite). Atomic weight = 24.3. Atomic number = 12. Specific gravity = 1.74. Valence = 2. Melting point = 648°C. Boiling point = 1090°C. Used in photography, pyrotechnics, other industrial applications, and in medicine (Epsom salt and milk of magnesia), it is also an essential nutritional element. Magnesium tarnishes easily in air and burns with an intense flame (magnesium light) to form magnesium oxide (MgO). Its salts are soluble in water, nontoxic to humans at natural concentrations, but may produce laxative and diuretic effects. Along with calcium, magnesium is one of the main contributors to water hardness.

**magnesium ammonium phosphate scale** *See* struvite.

**magnesium carbonate ($MgCO_3$)** A white powder, insoluble in water and alcohol, soluble in acid; used in dentifrices, medicine, cosmetics, and refractories.

**magnesium dioxide ($MgO_2$)** Magnesium peroxide.

**magnesium hardness** Water hardness caused by the carbonate, sulfate, or other compounds of magnesium; it can be precipitated as magnesium hydroxide.

**magnesium hydroxide [$Mg(OH)_2$]** A white, crystalline, somewhat soluble powder, used in medicine. The product that is precipitated during magnesium removal, it is formed by the reaction of magnesium with the hydroxyl ion:

$$Mg^{2+} + 2\ OH^- \rightarrow Mg(OH)_2 \qquad \text{(M-01)}$$

*See also* lime–soda softening, milk of magnesia.

**magnesium light** The white light produced when magnesium is burned; used in photography, pyrotechnics, etc.

**magnesium oxide (MgO)** Magnesia.

**magnesium peroxide ($MgO_2$)** A white, tasteless, insoluble powder used in medicine, as an oxidant, and as a bleaching agent. Also called magnesium dioxide.

**magnesium silicate ($MgSiO_3$)** A white powder, insoluble in water and alcohol, used as a bleaching agent. The commercial product is variably hydrated, e.g., $3MgSiO_3 \cdot 5H_2O$. *See also* forsterite.

**magnesium sulfate ($MgSO_4$)** A white, water-soluble salt, used in medicine and industrially. *See* Epsom salt.

**magnesium trisilicate ($Mg_2Si_3O_8 \cdot xH_2O$)** A white, odorless, and tasteless powder used industrially and pharmaceutically.

**magnetic flowmeter** A device used to measure fluid flow through a pipe. It consists of an electromagnet placed perpendicularly to the pipe and a galvanometer to measure the intensity of the electromotive force created, which is proportional to the flow velocity. Magnetic meters can also be used to measure velocities in rivers and lakes. Also called magnetic meter, magnetic inductive flowmeter, electromagnetic flowmeter, or magmeter.

**magnetic inductive flowmeter** *See* magnetic flowmeter.

**magnetic ion exchange resin** A resin specifically developed for the removal of dissolved organic carbon from drinking water. *See* MIEX®.

**magnetic meter** *See* magnetic flowmeter.

**magnetic separation** Use of a magnetic field (through magnets) to separate ferrous materials from mixed municipal waste streams or magnetic suspended solids from a liquid.

**magnetic stirrer** A laboratory device used for mixing chemicals.

**magnetite ($Fe_3O_4$ or $FeFe_2O_4$)** A black, crystalline ferrous–ferric iron mineral whose grains can be separated using a magnet; sometimes occurs with another metal (e.g., magnesium or zinc) substituting for the ferrous form. Magnetite is an important source of iron and vanadium. Also, a compound formed during the corrosion of iron pipes, called an iron phase or a corrosion scale. Magnetite buildup is a major cause of reduced thermal efficiency of boilers; at the same time, it prevents excessive corrosion of the metal. Also called lodestone. *See also* spinel, waterside cleaning wastes.

**Magnifloc®** A polyelectrolyte produced by American Cyanamid Co. to improve the separation of solids from liquids.

**Magnum** Equipment manufactured by Atlantic Ultraviolet Corp. for water disinfection by ultraviolet light.

**Magnum®** A belt filter press manufactured by the Parkson Corp.

**Magnum™** A catalytic oxidizer manufactured by Grace TEC Systems.

**Magox®** Abbreviation of magnesium oxide, a product of Premier Services Corp.

**Mai** Acronym of *Mycobacterium avium intracellulare.* Also called MAC.

**Mai complex** Same as *Mycobacterium avium intracellulare.*

**main** A principal pipe, conduit, or line in a water, gas, or electricity distribution system; a duct or pipe that supplies or drains ancillary branches; a trunk sewer that collects wastewater from smaller lines. *See also* arterial main, force main, secondary main.

**maintenance** Activities required to keep a facility in as-built condition, i.e., the upkeep and repair of facilities for their efficient operation, excluding their replacement or retirement. In engineering economics, maintenance costs are usually included with operation costs under O&M. *See* the following maintenance approaches: reactive (breakdown and conditional or indicative), proactive (preventive and predictive or reliability), and developmental.

**maintenance cleaning** A method used to remove accumulated solids from the surface of wastewater treatment membranes; as in backpulsing, flow is reversed to dislodge the particles, but with the addition of a solution of sodium hypochlorite or citric acid. *See also* air scour, backpulse, recovery cleaning, and empty-tank maintenance cleaning.

**maintenance cost** Labor, materials, and similar expenses associated with the upkeep of a facility; often associated with operating expenses under the category operation and maintenance costs, as compared to capital cost or debt service.

**maintenance energy** The energy required to maintain a cell in a nongrowth condition, i.e., without producing new cells, when the substrate or some other nutrient is limiting.

**maintenance hatch** Same as manhole.

**maintenance wastewater** Wastewater generated by the draining of process fluid from components in the chemical manufacturing process unit into an individual drain system prior to or during maintenance activities. Maintenance wastewater can be generated during planned and unplanned shutdowns and during periods not associated with a shutdown. Examples of activities that can generate maintenance wastewaters include descaling of heat exchanger tubing bundles, cleaning of distillation column traps, draining of low legs and high-point bleeds, draining of pumps into an individual drain system, and draining of portions of the chemical manufacturing process unit for repair (EPA-40CFR63.101).

**major industrial wastes** *See* the following terms for the major industrial water users and generators of wastewater: acid waste, agricultural waste, bakery waste, beet sugar waste, bottling waste, brewery/distillery/fermentation waste, cane sugar waste, cannery waste, cement industry waste, chloralkali waste, coal processing waste, coffee pressing waste, dairy waste, detergent waste, dry cleaning waste, feedlot waste, fish processing waste, laundry waste, meat and poultry processing waste, metal plating waste, organic chemical waste, palm oil waste, pesticide waste, petrochemical waste, pharmaceutical waste, phosphate industry waste, photographic waste, pickles and olives waste, polymer waste, pulp and paper mill waste, radioactive waste, refinery waste, rubber waste, soft drink waste, steam power plant waste, steel mill waste, tannery waste, textile waste, water production waste.

**major ions** The anions and cations that usually have the highest concentrations in natural water: bicarbonate ($HCO_3^-$), calcium ($Ca^{2+}$), carbonate ($CO_3^{2-}$), chloride ($Cl^-$), magnesium ($Mg^{2+}$), potassium ($K^+$), sodium ($Na^+$), and sulfate ($SO_4^{2-}$). *See also* ionic proportion diagram.

**major municipal separate storm sewer outfall** A municipal separate storm sewer outfall that discharges from a single pipe with an inside diameter of 36 inches or more or its equivalent (discharge from a single conveyance other than circular pipe that is associated with a drainage area of more than 50 acres); or, for municipal separate storm sewers that receive stormwater from lands zoned for industrial activity (based on comprehensive zoning plans or its equivalent), an outfall that discharges from a single pipe with an inside diameter of 12 inches or more or from its equivalent (discharge from other than a circular pipe associated with a drainage area of 2 acres or more) (EPA-40CFR122.26-5).

**major outfall** Same as major municipal storm sewer outfall.

**major pathogenic agents** Major microbial disease-causing agents related to water and wastewater include (a) enteric bacteria (*Salmonella, Shigella, E. coli, Campylobacter jejuni, Yersinia, Leptospira, Legionella, Mycobacteria, Vibrio cholerae*); (b) enteric viruses (enteroviruses, hepatitis-A virus, adenoviruses, reoviruses, rotaviruses, Norwalk-type viruses); (c) helminths (nematodes, hookworms, cestodes, trematodes); and (d) protozoans (*Acanthamoeba, Balantidium coli, Cryptosporidium, Entamoeba histolytica, Giardia lamblia*).

**majors** Larger publicly owned treatment works (POTW) with flows equal to at least one million gallons per day (1.0 mgd) or servicing a population equivalent of 10,000 or more persons. Also, certain other POTWs having significant water quality impacts. *See also* minors.

**makeup** Fluid or solid material added to a process to replace lost products or to maintain a parameter in equilibrium (e.g., temperature, concentration). *See also* makeup carbon and makeup water.

**makeup carbon** Fresh activated carbon added to an adsorption column after the regeneration cycle.

**makeup water** (1) Water added to the circulating stream of a cooling tower to replace water lost by evaporation, drift, leakage, or blowdown, and maintain a proper salt balance. (2) Water or recycled wastewater added to a gravity sludge thickener to adjust the surface overflow rate. (3) Water added to an electrodialysis unit, about 10% of the feed volume, to continuously wash the membranes. (4) Water added to the recirculating loop of a boiler to replace steam leaks; like the feedwater, it must be free of actual or potential insoluble constituents and meet stringent requirements to prevent scaling, fouling, and corrosion.

**malachite [$CuCO_3 \cdot Cu(OH)_2$ or $Cu_2CO_3(OH)_2$]** A green mineral, basic copper carbonate; an ore of copper, used for ornamental objects.

**malaria** A debilitating and sometimes fatal disease caused by the protozoa *Plasmodium,* transmitted by female mosquitoes of the genus *Anopheles;* characterized by fever, often periodic, chills, sweating, anemia, enlargement of the spleen, and various complications. It is found in most of the warm areas of the world.

**malathion ($C_{10}H_{19}O_6PS_2$)** A common organophosphate insecticide with low toxicity for mammals but harmful to fish and bees.

**Malayan filariasis** A mosquito-borne helminthic infection caused by the pathogen *Brugia malayi.* Distributed worldwide but mostly in India and Southeast Asia, it is transmitted by *Aedes, Anopheles, Mansonia,* and other mosquitoes. The pathogens develop in the lymphatic system and release larvae in the blood, causing lymph obstruction, elephantiasis, and hydrocele. *See also* Bancroftian filariasis.

**male bacteria** Bacteria that have appendages called pili on the cell surface.

**male end** The outside threaded connection of a pipe section, which connects with the bell end of another section. Also called spigot end.

**male fertility index** *See* fertility index.

**male-specific bacteriophage** *See* male-specific coliphage.

**male-specific coliphage** Bacteriophage that infect strains of *E. coli* and related bacteria through the $F^+$ or sex pili. It is one of the groups of bacteriophage considered as possible indicators of fecal contamination. Also called F-specific coliphage or $F^+$ coliphage. *See also* somatic coliphage.

**male-specific phage** *See* male-specific coliphage.

**malignant** Very dangerous or virulent, tending to become progressively worse, causing or likely to cause death.

**malignant tertian malaria** Malaria caused by *Plasmodium falciparum.*

**M alkalinity** or **M-alkalinity** Short for methyl orange alkalinity.

**Mallard** An apparatus manufactured by the Copa Group to remove scum from clarifiers.

**Mammoth®** A brush aerator manufactured by Zimpro Environmental, Inc.

**managed competition** A method of selecting consultants or contractors for a project by selecting candidates on the basis of prequalifying criteria.

**managed wetland** A wetland that operates with human intervention.

**managerial control** A method of nonpoint source pollution control that is derived from managerial decisions, such as changes in application times or rates for agrochemicals.

**manganese (Mn)** A hard, brittle, grayish-white metallic element, commonly found as oxides and carbonates (pyrolusite, rhodochrosite); used in alloys, chemicals, batteries, and agricultural products. Atomic weight = 54.94. Atomic number = 25. Specific gravity = 7.3. Valence = 1, 2, 3, 4, 6, or 7. Melting point = 1244°C. Boiling point = 1962°C. Manganese is an essential nutritional element; its concentrations in natural waters (average 0.06 mg/L) are nontoxic to humans. However, above 0.15 mg/L it may cause taste and color problems as well as black stains on laundry and plumbing fixtures. The USEPA and WHO recommend a limit of 0.05 mg/L in drinking water. In water chemistry, manganese resembles iron in many respects; precipitation of their oxidized compounds causes water to have a metallic taste and to stain clothes, utensils, and plumbing fixtures. They may also combine with the cement matrix of asbestos–cement pipe to form a protective coating against corrosion. Iron and manganese can be removed from water by aeration or adsorption. *See also* iron and manganese, demanganization, manganese zeolite process, manganism.

**manganese-54** A radionuclide from nonfission neutron reactors, sometimes found in water sources; half-life = 310 days

**manganese bacteria** Bacteria that can use dissolved manganese for energy and convert it to manganic hydroxide [$Mn(OH)_2$], which precipitates.

**manganese bronze** An alloy that is 55% copper, 40% zinc, and up to 3.5% manganese.

**manganese dioxide ($MnO_2$)** A dark- or light-brown insoluble compound of manganese, found naturally as the mineral pyrolusite; used in water treatment as an oxidizing agent or in the production of potassium permanganate ($KMnO_4$). It may also be produced by the reduction of permanganate:

$$MnO_4^- + 4\,H^+ + 3\,e^- \rightarrow MnO_2(s) + 2\,H_2O \quad \text{(M-02)}$$

Manganese dioxide can also adsorb ferrous iron, manganous manganese, trace inorganics, and natural organic materials (DBP precursors).

**manganese dioxide-coated pumicite** An aluminum silicate mineral coated with manganese dioxide ($MnO_2$) used as a catalyst to oxidize iron and manganese.

**manganese greensand** Greensand to which higher oxides of manganese have been added for use in water treatment to remove iron, manganese, and hydrogen sulfide; it is usually regenerated by a solution of potassium permanganate ($KMnO_4$).

**manganese greensand process** A historic process in which a filter medium is coated with manganese dioxide ($MnO_2$) for the removal of iron, manganese, and radium. Permanganate is added to the filter backwash or to the influent to oxidize the adsorbed materials.

**manganese greensand zeolite** Same as manganese zeolite.

**manganese removal** Manganese concentration in drinking water can be reduced by aeration (which renders the metal insoluble and removable by sedimentation), water softening (e.g., using lime and soda ash), oxidation, or adsorption on activated carbon.

**manganese sulfate** *See* manganous sulfate.

**manganese zeolite** A synthetic gel zeolite, sodium alumino silicate, to which higher oxides of manganese have been added for use in water treatment to remove iron, manganese, and hydrogen sulfide; it is usually regenerated by a solution of potassium permanganate ($KMnO_4$).

**manganese zeolite process** A process used in water treatment to remove iron and manganese from solution by adsorption on granular pyrolusite and periodic oxidation with permanganate.

**manganic hydroxide** *See* manganite.

**manganic manganese** Trivalent or tetravalent manganese ($Mn^{3+}$ or $Mn^{4+}$).

**manganism** A slow, irreversible neurotoxic disease, caused by occupational inhalation of manganese or exposure to the metal through other sources. The USEPA has identified a LOAEL of 2.0 mg/L and NOAEL of 0.17 mg/L in drinking water.

**manganite [MnO(OH)]** (1) A gray to black, high-density manganese ore that is essentially manganic hydroxide, used in water treatment to control iron, manganese, and hydrogen sulfide ($H_2S$) in filters. *See also* manganese dioxide ($MnO_2$) or pyrolusite. (2) Any of various salts containing tetravalent manganese ($Mn^{4+}$) and derived from the acids $H_4MnO_4$ or $H_2MnO_3$.

**manganous manganese** Bivalent manganese ($Mn^{2+}$).

**manganous sulfate ($MnSO_4 \cdot 4H_2O$)** A pink, water-soluble salt used in fertilizers. Also called manganese sulfate.

**mangrove fly** A mosquito of the genus *Chrysops* that breeds in shaded areas in the vicinity of slow-moving streams. In Central and West Africa, mangrove flies carry the pathogen *Loa loa* that causes loiasis.

**Manhattan Process** A high-rate filtration process developed by Roberts Filter Manufacturing Corp.

**manhead** *See* manhole.

**manhole (MH)** (1) A hole through which a person may enter a sewer, tank, boiler, or similar structure. (2) A gravity sewer appurtenant structure, usually where two or more sewer sections meet. Manholes are provided every 300 or 400 feet for inspection or whenever the sewer changes size or slope. Also called maintenance hatch, access hole, access port, manway, or manhead. *See* terminal manhole.

**manifold** A large pipe with several apertures or branches for multiple connections, or to permit flow diversions from one of several sources or to one of several discharge points. Also called a header.

**manifolded system** A sewer system consisting of pump stations and force mains, without any element open to the atmosphere (a manhole or a gravity line). *See also* closed-conduit system.

**manifold-lateral underdrain** A filter underdrain system consisting of perforated pipe laterals along a manifold. *See* Figure M-01. *See also* underdrain, fabricated self-supporting underdrain, false-floor underdrain with nozzles.

**manmade lake** An artificial body of water larger than a swimming pool, used for runoff control, recreation, cooling, sports, etc.

**manmade wetland** Same as artificial wetland.

**mannich polymer** A very high-viscosity wastewater treatment chemical that contains 4–7% active polymer and has a relatively short shelf life.

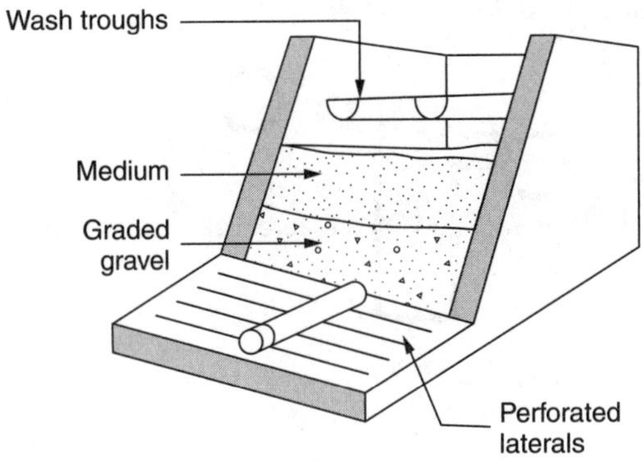


**Figure M-01.** Manifold—lateral underdrain.

Polymers are also available in dry, emulsion, gel, and liquid forms.

**Manning formula** An empirical equation published by Manning in 1890 for open-channel flow. He derived the formula by curve fitting to observations in large rivers and channels. It expresses the average longitudinal velocity ($V$) as a function of the hydraulic radius ($R$) of the channel, the channel slope ($S_0$), and a roughness coefficient or retardance factor of the channel lining ($n$):

$$V = (\delta/n)R^{2/3}S_0^{1/2} \qquad \text{(M-03)}$$

where $\delta$ is a unit conversion constant = 1.00 for SI units and 1.49 for English units. It is similar to the formula established theoretically by Chézy in 1775; in fact, the two formulas are identical if the Chézy roughness factor ($C_z$) is taken as

$$C_z = (\delta/n)R^{1/6} \qquad \text{(M-04)}$$

It is also equivalent to the kinematic wave equation (which assumes equilibrium between gravitational and frictional forces) under steady, uniform flow conditions. Known in Europe as the Strickler formula, the Manning equation is widely used in the United States to determine discharges and flow velocities, to estimate the effects of friction in the momentum equation, and to solve for depth of flow ($y$) using the Newton–Raphson method, for example. It can also be used to determine the Chézy coefficient. In the SWMM model, the RUNOFF, EXTRAN, and TRANSPORT blocks use the Manning equation (Nix, 1994) to simulate surface runoff or to estimate the friction slope ($S_f$), in the forms

$$dy/dt = i_e - [(\delta W)/(An)](y - y_p)^{5/3}S_0^{1/2} \qquad \text{(M-05)}$$

and

$$S_f = Q^2/[(\delta/n)^2A^2R^{4/3}] \qquad \text{(M-06)}$$

where, in addition to the variables and parameters defined above, $A$ = drainage area, $i_e$ = rainfall excess, $Q$ = runoff flow rate, $t$ = time, $W$ = width of overland flow, and $y_p$ = depth of maximum depression storage.

**Manning, Robert** Irish engineer (1816–1897).

**Manning roughness coefficient** The empirical bottom roughness coefficient $n$ in the Manning formula, which reflects the effect of channel or conduit roughness on the velocity of flow; roughness retards the flow, increases the potential for infiltration, and decreases erosion. The roughness coefficient varies from 0.025–0.033 for natural, clean, straight, full-stage channels without ripples to 0.070–0.150 for weedy reaches or floodways with heavy underbrush. Where field measurements are not possible, the Manning coefficient may be estimated from an empirical formulation such as

$$n = 0.031\, d^{1/6} \qquad \text{(M-07)}$$

$d$ being the size of channel particles. For open-channel flow in pipes, it may vary from 0.009 to 0.017. *See also* equivalent roughness coefficient.

**manometer** An instrument for measuring fluid pressure, particularly water pressure in a pipe or other container. The simple manometer is a piezometer tube bent into a U-shaped loop, containing a manometer fluid such as mercury or oil, with one end connected to a tap in the pipe and the other end open to the atmosphere. *See* Figure M-02. The fluid pressure at the tap ($P_t$) is determined from the formula

$$P_t = \gamma' \cdot z - \gamma \cdot h \qquad \text{(M-08)}$$

where $\gamma'$ and $\gamma$ represent, respectively, the unit weights of the manometer fluid and the fluid in the pipe, $h$ is the elevation difference between the tap and the lower manometer fluid column, and $z$ the elevation difference between the two manometer columns. A differential manometer serves to measure the pressure difference between two sources or two taps, each end of the tube being connected to a tap. *See* Figure M-02. If the fluid specific weight is the same ($\gamma$) for both sources and $h'$ is the elevation difference between the two taps, the pressure difference ($\Delta P$) is

$$\Delta P = \gamma(h' - z) + \gamma \cdot z \qquad \text{(M-09)}$$

An inverted manometer is an inverted U-shaped differential manometer for use with manometer fluids lighter than the fluids in the sources.

**manometric respirometer** An instrument for measuring the character and extent of respiration

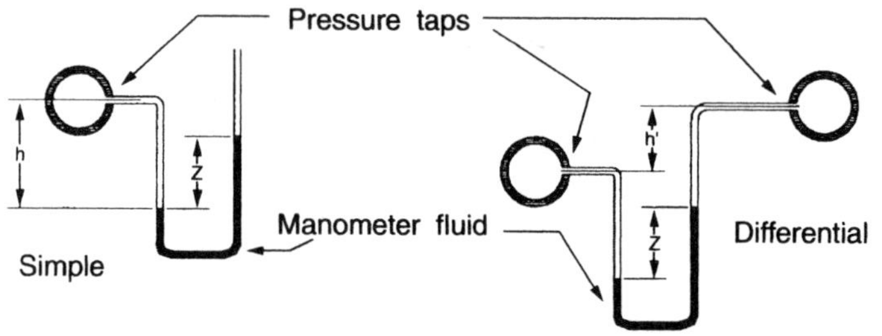


**Figure M-02.** Manometer.

(e.g., oxygen consumption and carbon dioxide production), usually equipped for data collection and processing by computer. It measures oxygen uptake by relating it to the pressure change caused by oxygen consumption while maintaining constant the gas volume. *See* respirometric method, Gilson respirometer, Warburg respirometer.

**manometric technique** A procedure that quantifies microbial respiration by measuring oxygen depletion from the atmosphere within a sealed container, based on the ideal gas law; it monitors the change in gas volume ($V$) or gas pressure ($P$) while holding one of these two variables constant.

**Manor®** A filter press manufactured by Simon-Hartley, Ltd.

***Mansonia*** A genus of mosquitoes that carry the pathogen *Brugia malayi,* which causes Malayan filariasis in Asia. Their larvae breed on the roots and leaves of floating aquatic plants (e.g., water fern, water hyacinth, water lettuce).

**mantle** *See* soil mantle.

**manually cleaned bar screen** A wastewater screening device, with bars inclined 30–45 degrees to the vertical and having 1–2 in. openings. The screenings are raked manually.

**manual solution feed** A method of feeding chemicals to a small water treatment basin by dissolving them in a plastic container, transferring them to a day tank, and then to the basin by using a metering pump.

**manufacturer's rating** A U.S. standard sieve size or sieve number as determined by direct measurement of the clear dimensions of a representative number of screen openings. *See also* sieve size, Hazen's sieve rating.

**manufacturing wastes** Wastes from industrial operations. *See* industrial waste and industrial wastewater for detail.

**Manu-Matic** An apparatus manufactured by U.S. Filter Corp. to clean pressure leaf filters manually.

**manure** Animal excrement or other refuse used as fertilizer.

**Manver** A chemical product of Hach Co. used to analyze water hardness.

**manway** *See* manhole.

**marble** A common compound of calcium, in the form of calcium carbonate ($CaCO_3$), a white or transparent solid, insoluble in water.

**marble test** A practical laboratory test used to determine the saturation concentration of calcium carbonate ($CaCO_3$) and the stability condition of water. It measures the alkalinity of a sample before and after 24-hr contact with powdered calcium carbonate. The water is depositing (supersaturated), stable, or corrosive (undersaturated) if, respectively, alkalinity has decreased, is unchanged, or has increased. The test is particularly helpful in analyzing waters treated with corrosion inhibitors or when high concentrations of magnesium, zinc, and sulfate significantly affect calcium carbonate precipitation. *See* corrosion index for a list of corrosion control parameters.

**marcasite ($FeS_2$)** Iron disulfide, a common mineral chemically similar to pyrite.

**MARD** Acronym of motor-actuated rotary distributor, a flow distributor manufactured by General Filter Co. for trickling filters.

**marginal chlorination** The application of just enough chlorine for disinfection but not for oxidation as effected, for example, by breakpoint chlorination. Marginal chlorination may also result in the formation of chlorophenol, phenolic taste, and the intensification of odors and tastes caused by certain algae such as *Synura.*

**marginal cost** The cost of producing the next unit of a product, e.g., the cost of the next increment of water supply, which may be one gallon, 1000 gallons, or one million gallons per day. Also called incremental cost.

**marginal cost pricing** A method used to determine the rate of water or wastewater service by setting it at the marginal cost, i.e., the cost of the next increment of supply. In practice this method is used to charge a higher price for each increment of consumption. Also called incremental cost pricing.

**margin of exposure (MOE)** The ratio of the no observed adverse effect level (NOAEL) to the estimated exposure dose (EED); a means of expressing the degree of safety for a compound that does not follow the linearized multistage model. *See* margin of safety.

**margin of safety (MOS)** The maximum amount of exposure producing no measurable effect in animals (or studied humans) divided by the actual amount of human exposure in a population; the older term used to describe the margin of exposure.

**mariculture** Cultivation of fish and shellfish in estuarine and coastal waters. *See also* aquaculture.

**marine environment** A shoreline, wetland, or any area with a salinity greater than 5000 mg/L and in contact with water reaching the ocean (AWWA, 2000).

**marine profile** *See* the following terms: aphotic zone, bathypelagic zone, benthic region, benthopelagic zone, benthos, compensation depth, compensation point, dysphotic zone, endobiotic habitat, epibiotic habitat, epipelagic zone, euphotic zone, hydrothermal vent, mesopelagic zone, neuston layer, pelagic zone, photic zone. *See also* freshwater profile.

**marine salt** Same as salt.

**marine sanitation device** Any equipment or process installed on board a vessel to receive, retain, treat, or discharge wastewater.

**marine water** Water characterized by a salinity of 33–37% and a depth of up to 11,000 m. *See* the following terms related to marine water: advective plume hypothesis, algal bloom, aphotic zone, benthopelagic zone, brackish water, chemoautotrophic theory, endobiotic habitat, epibiotic habitat, epipelagic zone, estuary, hydrothermal vent, neuston, organic thermogenesis, pelagic zone, photic zone, red tide, salinity.

**Mariotte, Edme** French physicist (1620–1684).

**Mariotte's law** Same as Boyle's law; at constant temperature and relatively low pressures, the volume of a gas varies inversely with its pressure.

**Marox** A wastewater treatment plant designed by Zimpro Environmental, Inc. based on the use of pure oxygen.

**MARS** A biological wastewater treatment process developed by Kruger, Inc. and controlled by membranes.

**marsh** Periodically wet or continually flooded area with the surface not deeply submerged. A type of wetland that does not accumulate appreciable peat deposits and is dominated by herbaceous vegetation, sedges, cattails, rushes, or other hydrophytes. Marshes may be either fresh or saltwater, tidal or nontidal. *See* wetlands.

**Marsh funnel viscosity** The time in seconds that it takes a quarter gallon of a drilling fluid to flow through a Marsh funnel.

**marsh gas** The gaseous product of the anaerobic degradation of organic matter, as in a swamp or other wetland area, consisting primarily of methane ($CH_4$), but it contains also ammonia ($NH_3$) and sulfur-containing compounds. Also called swamp gas and used as a common name for methane. *See also* biogas.

**marshland** (1) An area or region of marshes, swamps, bogs, or the like. (2) A soft wetland vegetated by grasses and reeds.

**masking agent** A perfume scent sprayed in fine mists near a wastewater treatment unit that emits offensive odors, to mask or overpower such odors. *See also* neutralizing agent.

**Maspac®** Plastic media produced by Clarkson Controls & Equipment Co.

**mass** The quantity of matter in a body as measured by its inertia; a quantitative measure of a body's resistance to acceleration. Also defined as the inverse of the ratio of the body's acceleration to the acceleration of a standard mass under otherwise identical conditions. The gravitational force on an object is proportional to its mass.

**mass-action equilibrium** *See* mass-action relationship.

**mass-action law** One of two fundamental rules applied in balancing chemical reactions. Any element has the same number of atoms on the left-hand side as on the right hand side; electrons given up or taken up do not have any mass. Expressions of the law of mass action include equilibrium constant equations, Henry's law, and solubility products. *See also* charge conservation law, forward reaction rate.

**mass-action relationship** The relationship between reactants and products of a chemical reaction that indicates that the reaction is in true equilibrium, e.g.,

$$aA + bB \leftrightarrow cC + dD \quad \text{(M-10)}$$

$$[C]^c[D]^d/([A]^a[B]^b) = K \quad \text{(M-11)}$$

where A and B are the reactants; C and D are the products; a, b, c, and d are the stoichiometric coefficients; the symbol ↔ indicates a reversible reaction; the brackets [] indicate molar concentrations; and K is the equilibrium constant. The reaction is said to be in mass-action equilibrium.

**mass balance** An inventory of all identified materials (fluids, solids) entering, leaving, or accumulating in a system (e.g., a basin, a reservoir, a wastewater treatment unit), or a quantitative analysis of the changes occurring in the system. A simple mass balance for a specific compound or constituent is based on the principle of mass conservation:

$$\text{Mass in} = \text{Mass out} - \text{Generation} + \text{Consumption} + \text{Accumulation} \quad \text{(M-12)}$$

When the inventory concerns only flows, e.g., in a water or wastewater system, it is called a water balance or a flow balance. A basic hydrologic balance includes such components as rainfall, runoff, evapotranspiration, infiltration, groundwater recharge, base flow, and direct surface discharge. *See* mass curve or Rippl diagram, and mass diagram. Also called material(s) balance. *See also* solids balance.

**mass-balance analysis** An analysis that is based on a mass balance.

**mass-balance equation** (1) The general expression that represents a mass balance:

$$\text{accumulation} = \text{inflow} - \text{outflow} + \text{generation} \quad \text{(M-13)}$$

*See also* steady-state simplification. (2) A relationship that represents a mass balance; e.g., an expression between the total solubility of a chemical constituent and the concentrations of various species, as shown in the following simplified case for the solubility of lead (II) from aqueous orthophosphate species:

$$S_T\,[\text{Pb(II)}, PO_4] = [PbHPO_4^0] + [PbH_2PO_4^+] \quad \text{(M-14)}$$

where $S_T$ is total lead solubility.

**mass-balance method** A method used to measure the rate of oxygen transfer of operating wastewater aeration devices by striking a mass balance for the gas between the influent and effluent flows. *See also* inert-tracer method and oxygen uptake rate.

**mass concentration** The mass or weight of a constituent per unit volume of solution, e.g., mg/L. *See* concentration, density.

**mass flow rate** The mass of a constituent in a wastewater stream, determined by multiplying the average concentration of that constituent in the wastewater stream by the annual volumetric flowrate and density of the wastewater stream (EPA-40CFR63.111).

**mass loading** The loading on a treatment unit based on the total mass of the constituent of interest, expressed usually in pounds per day or kilograms per day and calculated as:

$$\text{Loading, kg/day} = (\text{concentration, mg/L}) \times (\text{flowrate, m}^3\text{/d})/1000 \quad \text{(M-15)}$$

$$\text{Loading, lb/day} = 8.34 \times (\text{concentration, mg/L}) \times (\text{flowrate, mgd}) \quad \text{(M-16)}$$

**mass number** The nearest integer to the atomic weight of an atom; equal to the number of nucleons (protons and neutrons) in the nucleus of the atom. Represented by the symbol A and commonly used to denote isotopes and subatomic particles.

**mass per unit volume** An expression of concentration, e.g., milligram of constituent per liter of solution (mg/L).

**mass ratio** An expression of concentration, e.g., milligram of constituent per million milligrams of solution (ppm).

**mass spectrometer (MS)** An instrument that allows the determination of molecular structure. It is used in water analysis for the detection of organic materials separated by gas or liquid chromatography. MS detects ions according to their masses and electrical charges. *See also* gas chromatography–mass spectrometer.

**mass spectrometer interface** A device serving as a link between a mass spectrometer and a liquid chromatograph or other instrument. *See,* e.g., liquid chromatography–particle beam–mass chromatography.

**mass spectrometer–mass spectrometer (MS–MS)** Two mass spectrometers used in series. Also called tandem mass spectrometer.

**mass spectrometry** A laboratory method of separating ions according to their masses, using a mass spectrometer. It is used to detect trace organic chemicals in water or wastewater samples. *See also* electron capture, flame ionization, thermal conductivity.

**mass transfer** (1) The transfer of material from one homogenous phase to another, an important mechanism in the removal of volatile substances, in membrane filtration, and flow through porous media in general. *See* Fick's laws of diffusion, Stokes–Einstein law of diffusion, two-film theory, absorption, adsorption, desorption, ion exchange. (2) The amount of a substance passing through an interface from one phase to another.

**mass-transfer coefficient (MTC)** A constant of proportionality, determined experimentally to establish equilibrium conditions between two phases. It is related to the diffusion coefficient of the target compound through water and through air. In pressure-driven membrane processes, it is the mass or volume unit transfer through a membrane based on the driving force. *See also* Henry's law coefficient.

**mass-transfer model** An equation that applies the two-film theory to the absorption of a gas in a liquid:

$$(C_s - C_t)/(C_s - C_0) = \exp(-tK_L a) \quad \text{(M-17)}$$

where $C_s$ = concentration of the constituent in equilibrium according to Henry's law, mg/L; $C_t$ = concentration of the constituent in the liquid phase at time $t = t$, mg/L; $C_0$ = initial concentration of the constituent in the liquid phase, i.e., at time $t = 0$, mg/L; $K_L a$ = volumetric mass transfer coefficient,/time.

**mass-transfer zone (MTZ)** The zone in an adsorption or ion-exchange column where most of the adsorption actually takes place; the concentration of adsorbate in the liquid decreases to a small value. As the operation progresses, the zone moves downward toward the breakthrough condition. *See* adsorption zone for detail. *See also* breakthrough curve.

**mass-transport coefficients** Coefficients that characterize the aggregation and deposition of particles in coagulation and filtration, and that can be derived from physical considerations of transport mechanisms. The aggregation coefficient ($A$, $m^3$/sec) and the deposition coefficient ($D$, dimensionless) are such that

$$K_A = A\alpha_A \quad \text{(M-18)}$$

$$K_D = D\alpha_D \quad \text{(M-19)}$$

where $K_A$ and $K_D$ are rate constants that depend on the chemical and physical characteristics of the systems, and are sticking coefficients called collision efficiency factor for aggregation and attachment probability for deposition.

**Master-Flo** A bladder pump manufactured by American Sigma.

**master plan** A general plan or program for achieving an objective; e.g., a document that establishes the present and future development conditions of a community. It may relate to development in general or to a specific area such as water supply or sewerage. *See also* facilities plans.

**MATC** Acronym of maximum allowable toxicant concentration.

**material balance (or materials balance)** Same as mass balance.

**material safety data sheet (MSDS)** Product safety information sheets prepared by manufacturers and marketers of products containing toxic chemicals. These sheets can be obtained from the manufacturer or marketer. Some stores, such as hardware stores, may have material safety data sheets on hand for products they sell.

**materials-balance equation** Same as mass balance equation. *See also* batch reactor, complete-mix reactor, plug-flow reactor.

**materials exchange** An activity used in waste minimization or pollution prevention programs; a public or private organization matches the waste or by-product of one manufacturer with the need of another manufacturer, thus reducing disposal costs and promoting reuse. Also called waste exchange.

**materials mass-balance principle** The fundamental basis for the analysis of unit operations and processes in water or wastewater treatment, accounting for all quantities of materials before and after reactions and conversions occur. *See also* mass-balance analysis, mass-balance method.

**materials recovery facility (MRF)** A central facility for the processing of recycled materials.

**mathematical model** A set of mathematical relationships representing the behavior of a system. Open-channel flow models are generally partial differential or integral equations that cannot be solved analytically. Such models are also used to describe groundwater flow and transport processes. Mathematical models are deterministic (or mechanistic) or stochastic. Examples of simple mathematical models are the rational formula to determine the peak runoff discharges and the Streeter–Phelps formulation of the dissolved oxygen sag in a receiving stream. Mathematical models exist also for the activated sludge process, groundwater flow, risk assessment, sanitary and storm sewers, water distribution systems, water quality, etc. A mathematical model may also consist of thousands of relationships.

**Mathieson process** A process used to generate chlorine dioxide ($ClO_2$) on the site of water treatment plants, using sodium chlorate ($NaClO_3$), sulfur dioxide ($SO_2$), and sulfuric acid ($H_2SO_4$):

$$2\,NaClO_3 + SO_2 + H_2SO_4 \rightarrow 2\,ClO_2 + Na_2SO_4 + H_2SO_4 \quad \text{(M-20)}$$

Also called the sulfur dioxide process. *See also* the chloride reduction, Jazka–CIP, and Solvay (or methanol) processes.

**matrix** (1) The polymer backbone of an ion-exchange resin to which the negatively charged exchange sites, (—$SO_3^-$) or (—$COO^-$), are fixed. (2) In a catalytic incinerator, the matrix generally consists of a metal mesh mat, ceramic honeycomb, or similar structure that maximizes the surface area of the catalyst. Also called catalyst bed.

**matrix modifier** A reagent used to delay atomization of the metal being analyzed by graphite furnace atomic absorption spectrophotometry.

**maturation** The intermediate step in ballasted flocculation, between coagulation and sedimentation. After addition of the ballasting agent to the flocculated suspension, the mixture is gently stirred in a maturation compartment or a separate maturation tank to allow the floc particles to grow into larger particles that settle more easily.

**maturation compartment/tank/zone** The compartment, tank, or area in which flocculated particles and ballast grow into larger particles during ballasted flocculation.

**maturation lagoon** *See* maturation pond, and lagoons and ponds.

**maturation pond** A shallow, aerobic waste stabilization pond used for polishing wastewater effluents from an activated sludge plant, a trickling filter plant, or a facultative pond. The maturation pond is usually 2–4 ft deep and provides a detention time of 10 to 15 days. In addition to BOD, suspended solids and some ammonia, this pond reduces considerably the concentration of fecal organisms and produces an effluent that may be essentially free of pathogens (e.g., protozoal cysts, and helminth eggs). Maturation ponds are designed to handle a BOD load less than or equal to 15 pounds per acre per day and to produce between 5 and 10 mg/L of algal cell tissue. They are sometimes used in fish farming. Also called polishing pond, tertiary pond. *See* the Wehner and Wilhelm's equation for the design of aerobic ponds. *See also* fecal bacteria in ponds, high-rate pond, low-rate pond, lagoons and ponds.

**maturation tank** *See* maturation compartment.

**maturation zone** *See* maturation compartment.

**mature filter** A granular filter (e.g., sand and anthracite) that has become coated with hydrous oxides so that it can effectively remove iron and manganese.

***M. avium*** *Mycobacterium avium.*

**MaxAir™** A coarse-bubble diffuser manufactured by Environmental Dynamics, Inc.

**Maxi-Flo®** Pressurized sand filters manufactured by U.S. Filter Corp.

**Maxim®** An evaporator manufactured by Beaird Industries, Inc. for seawater conversion.

**maximally tolerated dose** The maximum dose of a substance to which a laboratory animal can be exposed without an observable toxic reaction. *See* maximum tolerated dose.

**MaxiMizer®** A solid bowl centrifuge manufactured by Alfa Laval Separation, Inc.

**maximum abrasion** The maximum effect on filter backwash velocity resulting from interparticle contacts, estimated to occur when the minimum fluidization velocity ($V_f$) is one-tenth of the terminal or washout velocity ($V_t$). *See* backwash velocity.

**maximum acceptable toxicant concentration (MATC)** The maximum concentration of a constituent that does not cause significant harm to productivity or other uses of a receiving water. Sometimes used interchangeably with chronic value. *See also* toxicity terms.

**maximum allowable concentration (MAC)** (1) The legal standard for a substance in air, water, or food. (2) The maximum level of radionuclides to which an employee may be exposed during an 8-hour work day. Also called maximum permissible concentration.

**maximum allowable level (MAL)** The maximum acceptable concentration of a constituent in drinking water according to the American National Standards Institute.

**maximum allowable toxicant concentration (MATC)** The concentration of a constituent that does not cause significant harm to productivity or other uses of a receiving water. Sometimes called chronic value. *See also* toxicity terms.

**maximum contaminant level (MCL)** The maximum permissible level of a contaminant in water that is delivered to the free-flowing outlet of the ultimate user of a public water system, except in the case of turbidity, for which the maximum permissible level is measured at the point of entry to the distribution system. Contaminants added to the water under circumstances controlled by the user are excluded from this definition, except those contaminants resulting from the corrosion of piping and plumbing caused by the water quality (EPA-40CFR142.2). A primary MCL is an enforceable standard of the USEPA established to protect public health. A secondary MCL is a nonenforceable standard for such contaminants as taste, odor, and color, established by the USEPA to protect public welfare.

**maximum contaminant level goal (MCLG)** Under the Safe Drinking Water Act, a nonenforceable concentration of a drinking water contaminant, set at a level at which no known or anticipated adverse effects on human health occur and that allows an adequate safety margin. The MCLG is usually the starting point for determining the regulated maximum contaminant level. Also called a recommended maximum contaminant level (RMCL). *See also* risk assessment, the research conducted by the USEPA before establishing an MCLG.

**maximum contaminant level (primary)** *See* maximum contaminant level.

**maximum contaminant level (secondary)** *See* maximum contaminant level.

**maximum deficit** *See* critical deficit.

**maximum discharge** The largest volume in unit time that can pass through a hydraulic structure.

**maximum exposure limit (MEL)** Maximum level of exposure to a harmful substance allowed for a worker or any individual.

**maximum flow** The largest volume of water through a stream, unit, or structure within a given time period.

**maximum infiltration rate** The limit on infiltration permitted by sewer construction specifications. *See* infiltration allowance.

**maximum permissible concentration (MPC)** *See* maximum allowable concentration.

**maximum residual disinfectant level (MRDL)** The maximum permissible level of a disinfectant residual in drinking water at the consumer's tap. The MRDL is a USEPA-enforceable standard, analogous to MCL, established to protect public health.

**maximum residual disinfectant level goal (MRDLG)** A nonenforceable concentration limit of the USEPA on disinfectant residuals in drinking water, set at a level at which no known or anticipated adverse effects on human health occur and that allows an adequate safety margin. The MRDLG is usually the starting point for determining the regulated maximum residual disinfectant level.

**maximum specific growth rate** A constant in the Monod equation that reflects intrinsic physiological properties of a particular microorganism or culture; usually designated by $\mu_{max}$.

**maximum thirty-day average** The maximum average of daily values for 30 consecutive days.

**maximum tolerated dose (MTD)** The dose that an animal species can tolerate for a major portion of its lifetime without significant impairment or toxic effects other than carcinogenicity. Same as maximally tolerated dose.

**maximum total trihalomethane potential (MTTP)** The maximum concentration of total trihalomethanes produced in a given water containing a disinfectant residual, after 7 days at 25°C or above (EPA-40CFR141.2).

**maximum trihalomethane formation potential (MTHMFP)** The amount of trihalomethanes formed when a water sample is dosed with a large amount of chlorine or chlorine compound.

**maximum trihalomethane potential** *See* maximum total trihalomethane potential.

**maximum water density** The density of water at 4°C (more precisely 3.98°C): 1.0 gram per milliliter.

**maximum water-holding capacity** The moisture content of soil in the field some time after reaching saturation and after free drainage has practically ceased, or the quantity of water held in a soil by capillary action. *See* field capacity for detail.

**Maxipress** A belt filter press manufactured by Envirex, Inc.

**Maxi-Rotor** A rotary brush aerator manufactured by Kruger, Inc.

**Maxi-Strip®** An apparatus manufactured by Hazleton Environmental products, Inc. to remove volatile organics.

**Maxi-Yield™** A device manufactured by Wallace & Tiernan, Inc. to blend polymers.

**Max-Pak™** Plastic media produced by Jaeger Products, Inc.

**Maz-O-Rator** A solids grinder manufactured by Robbins & Myers, Inc.

**MBAS** Acronym of methylene-blue active substances, which are used in detergents or surfactants.

**MBBR** Acronym of moving-bed biofilm reactor.

**MBR** Acronym of membrane biological reactor or membrane bioreactor.

**MBR process** *See* membrane bioreactor process.

**MCB** Acronym of monochlorobenzene.

**McCauley's driving force index** or, simply, **McCauley index** An index used in corrosion control, based on the solubility of calcium carbonate ($CaCO_3$). It predicts the potential weight of precipitate (DeZuane, 1997). *See* corrosion index for a list of corrosion control parameters.

**McCauley index** Same as McCauley's driving force index.

**MCL** Acronym of maximum contaminant level.

**MCLG** Acronym of maximum contaminant level goal.

**McMath's formula** A formula proposed to estimate flood flows in St. Louis, Missouri (Fair et al., 1971):

$$Q = CIA^{4/5}S^{1/5} \quad \text{(M-21)}$$

where $Q$ is the peak flood flow in cfs; $C$ is a coefficient that accounts for the effects of storm frequency, runoff-rainfall relationship, and maximum rainfall; $I$ is the rainfall intensity in inches per hour; $A$ is the drainage area; and $S$ is slope in feet per 1000 feet. For $A$ in square miles, $C = 480$; and for $A$ in acres, $C = 0.75$.

**m-ColiBlue24™** A broth produced by Hach Co. for use in coliform testing.

**MCRT** Acronym of mean cell residence time.

**MDI** Acronym of Morrill dispersion index.

**MDL** Acronym of method detection limit.

**MDTOC** Acronym of minimum detectable threshold odor concentration.

**me** Abbreviation of milliequivalent.

**ME** Acronym of multiple effect (distillation).

**mean cell-residence time (MCRT)** A basic design, control, and operation parameter of activated sludge and other suspended growth processes. It is the average time that a microbial cell (cell mass) remains in the system (usually limited to the aeration basin) and calculated as the mass of cells (total mixed liquor suspended solids in the aeration tank) divided by the rate of cell wasting from the system (wastage). *See* solids retention time for more detail.

**mean depth** Same as hydraulic mean depth.

**meandering stream** A stream with a sinuous flow pattern. *See also* sedimentation terms.

**mean flow** The arithmetic average of the discharge at a point on the flow line.

**mean life expectancy** *See* half-life.

**mean lifetime** *See* half-life.

**mean retention time (MRT) index** One of the terms used to characterize tracer response curves and describe the hydraulic performance of reactors; it is the ratio of the time at which 50% of the tracer has passed through the reactor ($T_{50}$) to the theoretical hydraulic residence time ($\tau$):

$$\text{MRT} = T_{50}/\tau \quad \text{(M-22)}$$

*See also* residence time distribution (RTD) curve, time–concentration curve. The MRT index measures the skewness of the RTD curve; the curve is skewed to the right or to the left according whether MRT > 1.0 or MRT < 1.0. For an ideal plug-flow reactor, the RTD follows a normal distribution and MRT = 1.0.

**mean solids retention time** *See* mean cell residence time or solids retention time.

**mean temporal velocity gradient** A measure of the mixing intensity or mechanical agitation in a water or wastewater treatment process such as flocculation. *See* velocity gradient for detail.

**mean time before failure (MTBF)** A statistical measure of the reliability of mechanical equipment, as determined by the ratio of operating hours to the number of failures; it is one of the factors used in critical component analysis. *See also* expected time before failure, inherent availability, operating availability.

**mean velocity** The average velocity of a fluid flowing in a channel, pipe, etc. It is equal to the ratio of the discharge ($Q$) to the cross-sectional area ($A$) for a section, or to the average cross-sectional area for a reach.

**mean velocity curve** A plot of velocity versus depth of water flowing in an open channel, at a given point and along a vertical line. Also called vertical velocity curve.

**mean velocity gradient** A measure of the mixing intensity or mechanical agitation in a water or wastewater treatment process such as flocculation. It is used in the design of mixing and flocculation units. *See* velocity gradient for detail.

**measured variable** A characteristic or component part that is sensed and quantified by a primary element or sensor.

**meat packing waste** The waste resulting from meat packing operations, mainly from stockyards (where the animals are kept before they are killed), slaughterhouses or abattoirs (where the killing, dressing, and some by-product processing take place), and the packinghouses (for further processing and production of salable products). Poultry processing differs somewhat from the processing of other meats. Stockyard wastewater is lower in BOD (< 100 mg/L) and suspended solids (< 200 mg/L) than domestic waste because of its high hydraulic population equivalent. Slaughterhouse wastewater has high concentrations of BOD (about 2000 mg/L) and nitrogen (about 500 mg/L) as well as a reddish-brown color and a considerable amount of hair, dirt, etc. For a packinghouse, the wastewater characteristics vary with the specific source; the most concentrated sources are the blood and tank water, the scalding tub, and the gut washer, with concentrations as high as 32,000 mg/L BOD, 5400 mg/L organic nitrogen, and 15,000 mg/L suspended solids. *See also* rendering waste, paunch manure, packers pollution.

**meat processing waste** Same as meat packing waste.

**mechanical aeration** (1) The use of mechanical energy—through such means as brushes, blades, paddles, paddle wheels, spray mechanisms, or turbines—to inject air into a liquid; e.g., to promote mixing of water with atmospheric air or to cause a wastewater stream to absorb oxygen. (2) The use of mechanical means to mix air and mixed liquor in the activated sludge process.

**mechanical aerator** A device used to transfer oxygen from air to water or wastewater; in the activated sludge process, mechanical aerators also keep mixed liquor suspended solids in suspension. *See* submerged aerator, surface aerator, Kessener brush aerator, mixing and flocculation devices.

**mechanical agitation** (1) Same as mechanical aeration (1). (2) The use of air under pressure in filter backwashing to clean the media.

**mechanical agitator** *See* mechanical surface agitator.

**mechanical analysis** A procedure for determining the size distribution of grains of a granular material, using sieves. *See also* uniformity coefficient.

**mechanical collector** An air pollution control device used to remove particulates in the thermal processing of sludge at wastewater treatment plants, usually as a precollector ahead of more efficient devices. *See* the three types of mechanical collectors: cyclone separator, impingement separator, and settling chamber, and also fabric filter, wet scrubber, and electrostatic precipitator.

**mechanical composting** A system in which sludge composting occurs in an enclosed container with equipment for odor control, aeration, mixing, and material handling. There are two types of mechanical composting: plug flow or static bed, and agitated bed or dynamic. Also called in-vessel composting. *See also* other composting systems: aerated static pile and windrow.

**mechanical coupling** A treadless pipe coupling.

**mechanical dewatering** Dewatering of water or wastewater treatment sludge using such equipment as a belt filter press, centrifuge, pressure filter, or vacuum filter. *See also* air drying and lagoon dewatering.

**mechanical disease transmission** Disease transmission by a mechanical vector.

**mechanical disinfection** The removal of bacteria and other organisms in such unit operations as coarse screening, fine screening, grit removal, and plain sedimentation.

**mechanical dispersion** The mass transfer caused by mechanical mixing. Also called kinematic dispersion. *See also* diffusion and dispersion.

**mechanical draft cooling tower** Same as mechanical draft tower.

**mechanical draft tower** An evaporative cooling tower, generally built in standard size, that uses a fan to induce and circulate the cooling air. *See also* natural draft tower.

**mechanical filter** A filter that removes suspended solids from water or wastewater by physical means. *See also* mechanical filtration.

**mechanical filtration** A former name of rapid sand filtration.

**mechanical joint** An example of a flexible joint that uses a gasket compressed by lugs and bolts to join pipes or fittings.

**mechanically cleaned bar screen** *See* mechanically cleaned screen.

**mechanically cleaned screen** A water or wastewater screening device that is cleaned mechanically of retained solids. For common types of mechanically cleaned screens, see catenary screen, chain-driven screen, continuous-belt screen, and reciprocating rake screen.

**mechanical pipe joint** *See* mechanical joint.

**mechanical pump** A pump that conveys fluid by direct contact of the fluid with a moving part of the pump.

**mechanical rake** A device operated by a machine to remove debris from racks at intakes of water conduits or conduits to convey wastewater to pumping stations.

**mechanical regulator** A device used to regulate flow in combined or interceptor sewers: water level is controlled by float and gate travels. *See* Figure M-03. Also known as automaatic regulator or reverse taintor gate.

**mechanical reliability** The reliability of the mechanical equipment in a water or wastewater facility, including any standby items. *See also* critical component analysis, expected time before failure, inherent availability, mean time before failure, operating availability.

**mechanical reoxygenation** The injection of oxygen into a treated wastewater effluent, using mechanical aeration, to increase its dissolved oxygen content. *See also* cascade reoxygenation, diffused-air reoxygenation.

**mechanical scour** A technique used in filter operations to intensify scour by stirring the fluidized bed mechanically. *See also* scour (2).

**mechanical seal** A device or apparatus installed between a casing and a shaft to prevent leakage.

**mechanical sieving** The principal mechanism of rejection of particles by filtration membranes, which may be affected by electrostatic interactions, dispersion forces, and hydrophobic bonding. Mechanical sieving may be defined by the

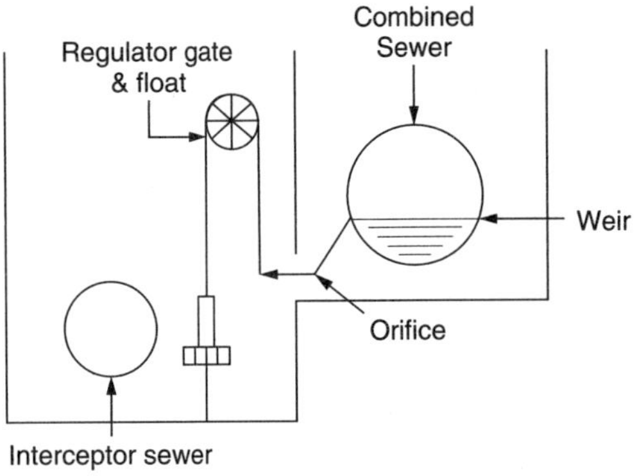


**Figure M-03.** Mechanical regulator.

fraction of particles ($P$) passing through a cylindrical pore or the fraction of particles ($1 - P$) rejected by the membrane (AWWA, 1999):

$$P = G(1 - \lambda)^2[2 - (1 - \lambda)^2] \quad \text{(M-23)}$$
for $\lambda \leq$ and $P = 1$ for $\lambda > 1$

where $G$ = an empirical lag coefficient:

$$G = \exp(-0.7146\,\lambda^2) \quad \text{(M-24)}$$

and $\lambda$ = a dimensionless variable equal to the ratio of the particle radius to the pore radius. Also called steric rejection. *See also* rejection, global rejection, local (apparent) rejection, polarization factor.

**mechanical sludge dewatering** *See* mechanical dewatering.

**mechanical sludge drying** A sludge processing operation in which an apparatus provides auxiliary heat to increase the vapor-holding capacity of ambient air and the latent heat for evaporation. *See also* conventional sludge drying.

**mechanical sparger** *See* sparger.

**mechanical straining** A major mechanism of removal of particulate matter within a granular filter, whereby particles larger than the pore spaces of the medium are strained out. *See also* filtration mechanisms.

**mechanical surface aerator** Same as surface aerator or mechanical aerator.

**mechanical surface agitator** A device developed to improve filter backwashing by increasing the scrubbing action in the expanded bed. It consists of a rotating arm installed above the media and beneath the wash troughs, with nozzles directing the water jets to the bed surface.

**mechanical suspension** A system of small particles dispersed by agitation in a surrounding medium. *See also* suspension.

**mechanical vapor recompression (MVR) evaporator** A device that uses a compressor to produce vapor for reuse in evaporation by raising the temperature and pressure of evaporated water. MVR evaporators are often used in brine concentrators. Also called vapor compressor.

**mechanical vector** An agent involved in the transmission of an infection. Contrary to a biological vector, the mechanical vector is not actually infected by the pathogen; it serves to transport the pathogen on or in its body from one site to another. For example, flies and other insects may carry excreta-related pathogens from latrines to kitchens. Flies are also implicated in the transmission of eye infections (e.g., conjunctivitis and trachoma).

**mecoprop** A persistent, soluble, chlorophenoxy pesticide with a WHO-recommended guideline value of 0.01 mg/L for drinking water. At high doses, it may affect the kidneys.

**Mectan™** A grit chamber manufactured by John Meunier, Inc.

**MED** Acronym of multiple-effect distillation.

**media** (1) Specific environments (air, water, soil) that are the subject of regulatory concern and activities. (2) *See* medium.

**media caking** A condition of the media of a filter bed whose grains have coalesced into larger-size particles, causing filtration and backwashing flow restrictions. Media caking typically results from poor size matching of a dual-medium filter, inadequate backwashing, and the coalescing effects of alum floc or powdered activated carbon. A caked medium can be restored by a caustic solution that dissolves the substance holding the grains together and by agitating the bed with an air wand. *See also* mudball.

**media disturbance and intramedia remnant** A stage in the filter-ripening sequence whereby particles are dislodged from the media and remain in the pore water.

**MEDIA G2®** An adsorption system engineered by ADI International, Inc. for the reduction of arsenic in water treatment.

**media gradation** The undesirable phenomenon observed during the operation of a granular filter, whereby the media settle after backwashing, with the finest particles on top and the coarsest at the bottom. It causes higher head loss resulting from a concentration of the solids removed in the finer strata. It affects mixed-media filters less than single-medium filters.

**median** The midmost term of a distribution; in an array of data, the term having as many items larger as items smaller in value. *See also* mean, average.

**median diameter** The particle diameter of which 50% of the particles are coarser and 50% are finer; usually represented by $D_{50}$. *See also* sedimentation terms.

**median effective concentration ($EC_{50}$)** The constituent concentration at which at least 50% of test organisms do not survive for a specified period of time and a specified confidence limit. It can be determined using such methods as binomial, moving averages, and probit. *See also* median lethal concentration, median tolerance limit, chronic toxicity, toxicity terms.

**median effective dilution** *See* $ED_{50}$.

**median infective dose** Also called infectious dose 50 ($ID_{50}$), the dose of an infective agent that will infect 50% of the population exposed to it. *See* Table M-01 for the median infective doses of some water- and excreta-related pathogens. *See also* minimum infectious dose, infectivity.

**median lethal concentration ($LC_{50}$)** The constituent concentration at which at least 50% of test organisms show a sublethal effect such as immobilization, fatigue in swimming, and avoidance. A fish swimming chamber, e.g., is used in conducting acute toxicity tests to assess sublethal effects. *See also* median effective concentration, median tolerance limit, chronic toxicity, toxicity terms.

**median lethal dose fifty** The empirically derived single dose of a chemical, taken by mouth or absorbed by the skin, that is expected to cause death in 50% of the test animals so treated. *See* $LD_{50}$ for detail.

**median tissue culture infective dose ($TCID_{50}$)** A measure of the infective dose of a virus based on observed cell death following inoculation with varying dilution of sample. Also called medium tissue culture infective dose. *See* serial dilution endpoint method.

**Table M-01.** Median infective dose ($ID_{50}$)

| Pathogen | Disease | $ID_{50}$ |
|---|---|---|
| *Ascaris lumbricoides* | Ascariasis | < 100 |
| *Entamoeba histolytica* | Amebic dysentery | < 100 |
| *Giardia lamblia* | Giardiasis | < 100 |
| Hepatitis-A virus | Infectious hepatitis | < 100 |
| Hookworm | Hookworm | < 100 |
| *Salmonella typhi* | Typhoid | > 1,000,000 |
| *Schistosoma* | Schistosomiasis | < 100 |
| *Shigella* | Shigellosis | 10,000 |
| *Taenia saginata* | Teniasis | < 100 |
| *Vibrio cholerae* | Cholera | > 1,000,000 |

**median tolerance limit** In toxicity assays, the median tolerance limit (TLm) represents the concentration of a substance that kills 50% of the test organisms after a specified exposure time. It is determined graphically from a semilogarithmic plot of concentration vs. percent survival or from the equation

$$\log TL_m = \log C_2 + (\log C_1 - \log C_2)(P_2 - 50)/(P_2 - P_1) \quad \text{(M-25)}$$

where $C_1$ and $C_2$ are the substance concentrations closest to the median survival and $P_1$ and $P_2$, the corresponding percentage survivals, respectively. Median effective concentration and median lethal concentration are currently used instead of median tolerance limit.

**median toxic dose** A characteristic of the potency of a substance for adverse effects; a dose such that 50% of laboratory animals show an adverse response.

**medical waste** Any waste generated in the diagnosis, treatment, or immunization of human beings or animals, in research pertaining thereto, or in the production or testing of biologicals, excluding hazardous waste identified or listed under CFR Part 261 or any household waste. Medical wastes from hospitals, laboratories, and doctors' offices may contain pathogens. Also called red bag waste and infectious waste.

**medicinal spring** A spring whose water contains constituents having curative or remedial properties.

**Medina®** Bioremediation products of Medina Products.

**medium** (plural: **media**) (1) A liquid or solidified nutrient material used to support the growth of microorganisms; e.g., broth and agar-based solid substances. Also called culture medium. (2) A material used in filtering devices to retain suspended or dissolved solids or molecules. Sometimes used in the plural form, media, for granular filtration materials, ion-exchange resins, or adsorbents.

**medium-pressure, high-intensity UV lamp** A mercury lamp that generates polychromatic radiation with a lamp wall of 600–800°C, an internal pressure of 100–10,000 mm Hg, and an output of 1000–2000 W. Only about 35% of the output is in germicidal UV-C wavelength range and 10% near 254 nm. In water and wastewater disinfection applications, quartz sleeves isolate the lamp from direct contact with the liquid. *See also* low-pressure, high-intensity UV lamp; low-pressure, low-intensity UV lamp; narrow-band excimer lamp; pulsed UV lamp.

**medium sand** Sediment particle; granular material having a diameter between 0.25 and 0.50 mm.

**medium-size water system** A public water system serving between 3300 and 50,000 persons.

**medium-specific concentration (MSC)** A concentration limit specified for a constituent in a given medium and a type of exposure.

**medium tissue culture infective dose ($TCID_{50}$)** *See* serial dilution endpoint method.

**meg** A procedure used for checking the insulation resistance on motors, feeders, buss bar systems, grounds, and branch circuit wiring. *See also* megger and megohm.

**Megatron™** Water purification plant manufactured by Atlantic Ultraviolet Corp. that uses ultraviolet light.

**megger** An instrument used for checking the insulation resistance on motors, feeders, buss bar systems, grounds, and branch circuit wiring.

**megohm** One million ohms, the unit used in meggers.

**Megos®** An apparatus manufactured by Capital Controls Co., Inc. for the production of ozone.

**meiosis** Cell division whereby daughter nuclei receive only half the original number of chromosomes in the parent cell. *See also* mitosis.

**MEL** Acronym of maximum exposure limit.

**Mellafier** An inclined-plate clarifier manufactured by Industrial Filter & Pump Manufacturing Co.

**melting heat** *See* heat of melting.

**melting point** The temperature at which a solid changes to a liquid at the standard pressure. Organic compounds have lower melting points than inorganics. At this temperature, both the liquid and the solid forms of a pure substance may exist in equilibrium. It is equal to the freezing point of a given substance. *See also* boiling point.

**melting temperature** Same as melting point.

**melt water** Icemelt and snowmelt, i.e., water derived from the melting of ice and snow.

**Membio®** An aerobic digester manufactured by Memtec America Corp.

**Membralox®** Ceramic membrane filters manufactured by U.S. Filter Corp.

**membrane** A thin, solid, natural or synthetic material that is selectively permeable ("permselective") to dissolved, colloidal, or suspended particles depending on such characteristics as size, diffusion rate, solubility, or electrical charge. Synthetic organic polymer and pressure-driven membranes are commonly used in water and wastewater treatment, e.g., in reverse osmosis, ion exchange, advanced filtration, and electrodialysis. They represent an excellent option to meet existing and future drinking water regulations on pathogens and inorganics. *See also* advanced pretreatment, apparent rejection, array, asymmetric membrane, back-diffusion coefficient, bank, Brownian diffusion, bubble-point test, composite membrane, concentrate, concentrate stream, concentration–polarization layer, conventional pretreatment, driving force, duPont equation, DuPont Permasep®, feed, Filmtec NF70, finger structure, flux, fouling, fouling index, Graetz–Leveque correlation, Hagen–Poiseuille equation, ion-exchange membrane, isoelectric point, Kozeny equation, lag coefficient, leakage, Linton–Sherwood correlation, mass transfer coefficient, mechanical sieving, membrane element, modified fouling index, molecular weight cutoff, osmotic pressure, permeate, permeate stream, phase-inversion membrane, polarization factor, posttreatment, pressure vessel, pressure-driven membrane process, product, reject, rejection, residual stream, retentate, scaling, Sherwood number, sieving, silt density index, solute, solvent, sponge structure, stage, steric rejection, symmetrical membrane, thin-film composite membrane, train.

**membrane aeration bioreactor process** A wastewater treatment process that uses plate-and-frame, tubular, and hollow membranes to transfer pure oxygen to the biomass attached to the outside of the membrane

**membrane air stripping** A water treatment process using microporous, hydrophobic membranes to remove volatile substances, based on a concentration gradient between air and water; contaminated water is passed through the inside of the hollow fibers while air flows countercurrently outside the fibers. *See also* cascade air stripping, inclined cascade aeration, packed tower aeration, steam stripping.

**membrane area** The active surface area of a membrane.

**membrane autopsy** The disassembly of a membrane element or module to examine its conditions after its use in a filtration run. This procedure may include visual inspection of membrane elements for foulants and/or scales, measurement of internal membrane area and other element characteristics, and scanning electron microscopy of a cross section of the membrane. It is used in evaluating membrane performance. *See also* membrane destructive analysis, membrane flux, net operating pressure, rejection, specific flux, temperature correction factor.

**membrane backflush** A procedure in which flow of water and/or air is reversed through a microfiltration or ultrafiltration membrane to remove retained materials; similar to granular filter backwashing.

**membrane-based technology** A process that uses membranes to separate, fractionate, and concentrate contaminants. *See* reverse osmosis, ultrafiltration, microfiltration, nanofiltration, electrodialysis, pervaporation, liquid membranes, membrane reactors.

**membrane biofilm reactor (MBfR)** A vessel containing hollow-fiber membranes and using hydrogen gas to remove such water contaminants as perchlorate and nitrates. The hydrogen fed into the hollow fiber serves as an electron donor to the microorganisms in the biofilm that forms on the outside. These reduce the contaminants to nitrogen gas and chlorine ion:

$$5\,H_2 + 2\,NO_3^- \rightarrow N_2 + 4\,H_2O + 2\,OH^- \quad \text{(M-26)}$$

$$2\,H_2 + ClO_4^- \rightarrow Cl^- + 2\,H_2O + O_2 \quad \text{(M-27)}$$

**membrane biological reactor or membrane bioreactor (MBR)** A variation of the activated sludge process that uses microfiltration or ultrafiltration membranes for enhanced removal of BOD and suspended solids from wastewater. The installation of the membranes in the aeration basin eliminates the need for settling tanks, return sludge facilities, and tertiary filters, while avoiding the formation of filamentous bacteria and bulking sludge; this is called an integrated membrane bioreactor. MBR is a small-footprint process, often used in satellite treatment; it can be operated at loading rates four times higher than those of conventional activated sludge (e.g., 70 pounds per day of $BOD_5$ per 1000 cubic feet). MBR can also provide nitrogen control and biological phosphorus removal. Another configuration is the recirculated membrane bioreactor, in which the membrane separation unit is installed externally to the bioreactor. *See* Figure M-04, membrane aeration bioreactor, membrane flux rate, extractive membrane bioreactor process, Kubota system, Zenogem®, and ZeeWeed®.

**membrane bioreactor (MBR)** Same as membrane biological reactor.

**membrane bioreactor (MBR) process** *See* membrane biological reactor.

**membrane bioreactor (MBR) system** A biological treatment system consisting of a bioreactor and a membrane separation unit installed inside or outside the reactor.

**membrane bundle** A group of hollow-fiber membranes in a pressure vessel.

**membrane cartridge** A spiral-wound membrane element.

**membrane cleaning** The use of backflushing or cleaning chemicals to remove in-place materials retained on a membrane.

**membrane cleaning techniques** Methods used to remove accumulated particles from the surface of water or wastewater treatment membranes and restore the capacity of the membranes. *See* air scour, backpulse, maintenance cleaning, recovery cleaning, and empty-bed maintenance cleaning.

**membrane cleaning solution** A chemical used for cleaning membranes of carbonate scales, metal hydroxides, and organic foulants. Examples in-

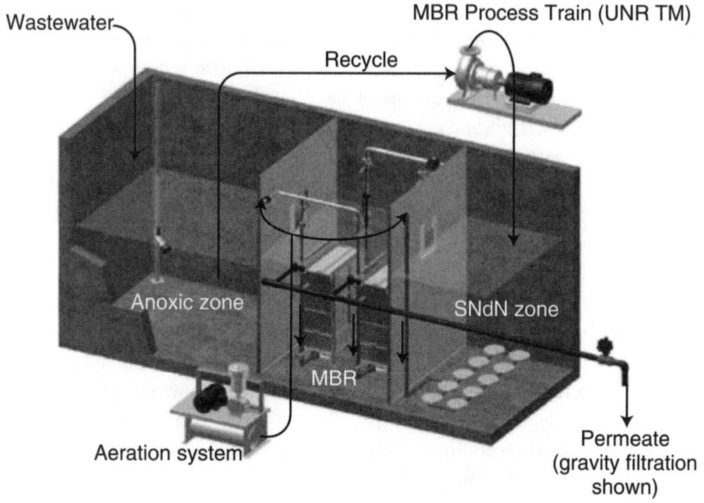


**Figure M-04.** Membrane biological reactor (courtesy Enviroquip, Inc.).

clude acidic or alkaline solutions, disinfectants, and chelating agents.

**membrane compaction** *See* compaction (2) and membrane flux retention coefficient.

**membrane concentrate** The concentrated solution of contaminants rejected into the feedwater of a membrane system or ultimately disposed of when contaminant concentrations are too high. Also called membrane by-product or, simply, concentrate.

**membrane configuration** The arrangement of membranes in a module, which varies with processes and may affect performance. For example, reverse osmosis and nanofiltration membranes have spiral wound or hollow fine fiber configurations, whereas microfiltration and ultrafiltration use hollow fiber membranes. Other configurations include tubular and plate-and-frame arrangements. Configurations also relate to the combination of flow and pressure types: outside-in or inside-out flow, and negative pressure or positive pressure.

**membrane destructive analysis** The disassembly of a membrane to analyze its conditions. Such a membrane cannot be reused. *See* membrane autopsy.

**membrane diffuser** A porous, perforated flexible plastic membrane mounted on an air distribution grid for the aeration of wastewater with fine bubbles.

**membrane distillation** The driving force for desalination is the difference in vapor pressure of water across the membrane. *See* the following membrane distillation arrangements: air gap, direct-contact, sweep-gas, and vacuum. As of 2007, membrane distillation is classified as an emerging concentrate minimization and management method.

**membrane electrode** A selective, polyethylene or Teflon™, membrane used in a device for measuring dissolved oxygen (DO) without chemical treatment of the sample. The membrane electrode is calibrated using a water sample of known DO concentration as determined by the iodometric method. *See* dissolved oxygen probe.

**membrane element** A single unit containing a group of spiral-wound or hollow fine-fiber membranes and spacers. *See also* module.

**membrane element autopsy** *See* membrane autopsy.

**membrane feedwater pretreatment** Treatment provided to the feedwater to nanofiltration and reverse osmosis units to prevent their fouling by colloidal matter and other constituents. Such pretreatment may consist of one or more of the following steps: cartridge filtration, removal of colloidal and suspended solids, removal of iron and manganese, disinfection, pH adjustment, dechlorination, deoxygenation.

**membrane filter (MF)** (1) A small filtration device made of artificial material and to specific pore sizes; it is used to retain microorganisms and other solid particles for the examination of water or wastewater in the laboratory. (2) Larger filtration units are used in water treatment processes such as electrodialysis, microfiltration or ultrafiltration. *See also* membrane filtration (2).

**membrane filter analysis** *See* membrane filtration technique.

**membrane filter coliform test** *See* membrane filtration technique.

**membrane filter method** (1) One of three laboratory procedures used in the heterotrophic plate count method of bacterial enumeration in a water or other sample. It indicates the number of colonies of heterotrophic (pathogenic and innocuous) bacteria grown on selected solid media at a given temperature and incubation period: m-HPC agar or R2A agar, for 72 hours at 35°C. The other two procedures are pour plate and spread plate. (2) *See* membrane filtration technique.

**membrane filter procedure** *See* membrane filtration technique.

**membrane filter technique** *See* membrane filtration technique.

**membrane filtration** (1) *See* membrane filtration technique. (2) A water treatment process using a membrane filter; classified as cake filtration, i.e., the solids are retained on the face of the granular material and not within it.

**membrane filtration technique** A method used to detect the presence of coliforms and fecal coliforms in a water sample. Water is filtered through a cellulose membrane, which is then placed on a nutrient medium and incubated. The retained bacteria take on a particular color depending on the medium; their colonies can be counted and expressed as a number per 100 mL. When the density exceeds 200 colonies per 100 mL, the result is reported as too numerous to count (TNTC). It has largely replaced the MPN technique because of its simplicity and speed. *See also* agar, broth, heterotrophic plate count, most probable number, multiple tube fermentation, standard plate count.

**membrane flow equations** Simple equations used to characterize a membrane filtration system or predict its performance (Bryant, 1992):

$$Q_w = K_w (P - \Pi) A/T \qquad \text{(M-28)}$$

$$Q_s = K_s SA/T \qquad \text{(M-29)}$$

$$Y = 100\, Q_p/Q_f \quad \text{(M-30)}$$

$$E = 100\, C_p/C_f \quad \text{(M-31)}$$

where $Q_w$ = flow rate of liquid through the membrane, $K_w$ = membrane permeability coefficient, $P$ = hydraulic pressure differential across the membrane, $\Pi$ = osmostic pressure differential across the membrane, $A$ = active membrane surface area, $T$ = membrane thickness, $Q_s$ = quantity of salt bleed through the membrane per unit time, $K_s$ = membrane permeability coefficient for a given salt, $S$ = differential salt concentration across the membrane, $Y$ = productivity or percent recovery; $E$ = percent salt passage, $Q_p$ = flow rate of permeate or treated water, $Q_f$ = flow rate of feed, $C_p$ = salt concentration in permeate stream, and $C_f$ = salt concentration in feed stream. *See also* normalization.

**membrane flushing** Using air or water to dislodge and remove substances accumulated in a membrane module.

**membrane flux** The flow rate of permeate through a membrane per unit area, i.e., $Q/A$, where $Q$ is the flow of permeate produced and $A$ is the membrane area. It is an important design, operation, and performance evaluation parameter for membrane processes. *See also* normalized specific flux, net operating pressure, rejection, specific flux, temperature correction factor.

**membrane flux rate** Same as membrane flux.

**membrane flux retention coefficient** A dimensionless parameter equal to the remaining fraction of productivity of a pressure-driven membrane after a period of operation and indicative of membrane compaction.

**membrane fouling** The accumulation, reversible or irreversible, of foulants on the membrane surface or within the membrane pores, which reduces membrane flux and performance of the membrane system. Serious fouling sometimes occurs in nanofiltration and reverse osmosis membranes treating highly organic feedwaters for control of total organic carbon or disinfection by-products. Foulants include inorganic colloidal or suspended particles, inorganic scales from the precipitation of supersaturated salts, organic matter, and biological materials. Related factors include electrostatic repulsion, osmotic pressure. On the other hand, fouling considerations affect membrane system design and operation as well as pretreatment requirements. Chemical cleaning (or cleaning in place) or backflushing can restore performance. *See also* concentration–polarization layer, fouling index.

**membrane fouling control** Measures taken to control the deposition and accumulation of foulants in membrane filtration processes include membrane feedwater pretreatment, membrane backflushing, and chemical cleaning.

**membrane gas cover** A gas-holder cover for cylindrical sludge digesters consisting of a support structure for a small center gas dome and flexible air and gas membranes, with an air blower to pressurize the space between the two membranes. *See also* fixed cover, floating cover.

**membrane-introduction mass spectrometry (MIMS)** A high-speed technique for the direct analysis of environmentally significant compounds of low concentrations (e.g., ppb) without isolation or derivitization.

**membrane leaf** An envelope-like device consisting of two back-to-back, flat membrane sheets sealed on three sides, with the fourth side open for permeate water collection. Leafs with spacers between them constitute elements or modules.

**membrane life** The period of time a membrane performs satisfactorily and remains in service.

**membrane lining** A trenchless gas and water main rehabilitation method using a thin elastomeric membrane impregnated with a resin that is cured in place by hot water or steam. It is used to restore the integrity of a main that is damaged by breaks, corrosion, small pinholes, or joint gaps. *See* cured-in-place lining, felt-based lining, woven-hose lining. *See also* trenchless technology.

**membrane materials** Current raw materials used in the manufacture of membranes: natural cellulose acetate and such synthetic materials as polyamides, vinyl polymers, polyfuran, polyolefins, and polysulfone.

**membrane module** *See* module.

**membrane packing density** The number of membranes or the total membrane area per unit volume of a module.

**membrane permeability coefficient** A parameter that indicates the capacity of a water treatment membrane in flow rate per unit membrane area per unit driving force.

**membrane poisoning** The incapacitation of a membrane by low-mobility ions. *See also* membrane fouling.

**membrane posttreatment** *See* posttreatment.

**membrane pretreatment** The use of a membrane process, instead of a conventional process, as pretreatment to another membrane process, e.g., microfiltration before nanofiltration, or ultrafiltration before reverse osmosis. *See also* pretreatment.

**membrane process** A water or wastewater treatment technique that uses membranes to remove dissolved solids, colloidal materials, or other undesirable constituents. In water treatment, common membrane processes include electrically driven processes: electrodialysis (ED) and ED reversal, low-pressure processes that use porous membranes under less than 40 psig (e.g., microfiltration and ultrafiltration), and high-pressure processes that use semipermeable membranes under more than 75 psig—nanofiltration (NF) and reverse osmosis (RO). Other membrane processes are dialysis, electroosmosis, osmosis, pervaporation, piezodialysis, and thermoosmosis. Membrane processes are classified according to the membrane materials, driving force, separation mechanisms, and pore sizes.

**membrane recovery rate** The percent of feedwater that a membrane system converts into product water; e.g., a system with a recovery rate of $P$ % produces a concentrate stream of (100 – $P$) % of the feedwater flow:

$$P, \% = 100\,(Q_p/Q_f) \qquad \text{(M-32)}$$

where $Q_p$ and $Q_f$ represent, respectively, the permeate and feedwater flowrates. *See also* concentration factor, membrane rejection rate.

**membrane reject** *See* reject.

**membrane rejection** *See* rejection.

**membrane rejection rate** The percent ($R$) of solute that a membrane system rejects:

$$R, \% = 100 - 100\,(C_p/C_f) \qquad \text{(M-33)}$$

where $C_p$ and $C_f$ represent, respectively, the permeate and feedwater concentrations. *See also* concentration factor, membrane recovery rate

**membrane selectivity** *See* selectivity.

**membrane separation** A water treatment method that uses membranes to remove undesirable constituents. *See* membrane process.

**membrane separation anaerobic treatment** A wastewater treatment method that uses membranes for solids separation in conjunction with an anaerobic reactor. As in the membrane biological reactor, the membrane may be installed inside or outside the anaerobic reactor, with internal or external recirculation.

**membrane separation process** *See* membrane process.

**membrane separation technique** *See* membrane process.

**membrane shock treatment** The disinfection of a membrane using a biocide for a short period.

**membrane softening** A water treatment process that uses semipermeable membranes to remove hardness ions (mainly calcium and magnesium). Nanofiltration is commonly used, alone or in combination with reverse osmosis.

**membrane stack** Same as module.

**membrane treatment** *See* membrane process.

**Membrastill™** A processing unit manufactured by U.S. Filter Corp. for the production of pharmaceutical-grade water.

**Memcor®** A unit manufactured by Memtec America Corp. for water treatment using continuous microfiltration.

**Memcor®** A membrane treatment system manufactured by U.S. Filter.

**Memcor®** Membrane System A water and wastewater filtration system manufactured by U.S. Filter Corp.

**Memory-Flex™** A check valve manufactured by Val-Matic Corp.

**Mem*Recon™** A product of King Lee Technologies for reconditioning reverse osmosis membranes.

**Memstor** A product of King Lee Technologies for use in membrame storage.

**Memtrex™** Pleated filters manufactured by Osmonics, Inc.

**m-endo medium** A selective culture used to detect the presence of coliform bacteria. It consists of a mixture of tryptose, thiopeptone, casitone, yeast extract, lactose, inorganic nutrients, and basic fuchsin. *See also* EC medium, lauryl tryptose broth, P-A broth.

**meniscus** The curved top of a column of liquid in a small tube. When the liquid wets the sides of the container (as with water), the curve forms a valley. When the confining sides are not wetted (as with mercury), the curve forms a hill or upward bulge. *See also* capillarity.

**Mensch™** A bar screen manufactured by Vulcan Industries, Inc., including a reciprocating rake.

**MEP** Acronym of multiple extraction procedure.

**meq** Abbreviation of milliequivalent.

**meq/L** Abbreviation of milliequivalent per liter.

**mercaptan** Any of a number of aliphatic organic compounds that contain sulfur and have a disagreeable garlicky odor noticeable at low concentrations; they are found in certain industrial wastewaters and are formed during anaerobic decomposition. Also called a thiol or thioalcohol. General formula: RSH, e.g., diethyl sulfide $C_2H_5SC_2H_5$ [or butylmercaptan $CH_3(CH_2)_3SH$], crotylmercaptan $CH_3CHCHCH_2SH$, ethyl mercaptan $CH_3CH_2SH$, methylmercaptan $CH_3SH$, phenylmercaptan $C_6H_5SH$, and propylmercaptan $CH_3CH_2CH_2SH$.

**mercapto** Containing the mercapto group; sulfhydryl; thiol.

**mercapto group** Containing the univalent group —SH. Also called mercapto radical. *See* sulfhydryl group for detail.

**mercapto radical** *See* mercapto group.

**Merco®** A centrifuge manufactured by Dorr-Oliver, Inc.

**mercuric chloride ($HgCl_2$)** A common salt of mercury and a corrosive sublimate, highly soluble in water.

**mercuric sulfide (HgS)** The principal ore of mercury and a compound used as paint pigment; it is highly soluble in water.

**mercurous chloride ($Hg_2Cl_2$)** A common salt of mercury, also called calomel, highly soluble in water.

**mercury (Hg,** from its Greek name, hydragyrum) A heavy, silver-white metal, liquid at ordinary temperatures, usually found in nature in combination with sulfur in the ore cinnabar or in organic compounds (e.g., methyl mercury). Atomic weight = 200.59. Atomic number = 80. Specific gravity = 13.546. Valence = 1 or 2. Mercury has many industrial applications: it is used in amalgams, instruments, mercury vapor lamps, boilers, etc. It is excreted very slowly from the body, can accumulate in the environment and is highly toxic, even lethal at low concentrations, if breathed, swallowed or adsorbed through the skin. It is insoluble in water but many of its salts are highly water-soluble, e.g., its chlorides (calomel), fulminate, and sulfide. At sufficiently high concentrations, mercury can interfere with enzyme production and cause physiological effects in humans (neurologic and renal disturbances). It is regulated by the USEPA as a drinking water contaminant: MCL = MCLG = 0.002 mg/L. WHO and European Community's guidelines are lower: 0.001 mg/L. Toxic compounds include methyl mercury and mercuric chloride. A combination of conventional and advanced treatment processes can effectively remove mercury from water and wastewater. *See* heavy metals, methyl mercury.

**mercury fulminate [$Hg(ONC)_2$]** A common salt of mercury, highly soluble in water and a detonator.

**mercury vapor ultraviolet light** The light resulting from an electron flow through ionized mercury vapor in an ultraviolet lamp, used in water or wastewater disinfection at a wavelength of 254 nanometers.

**Merlin®** A progressive cavity pump manufactured by MGI Pumps, Inc.

**Mer-Made** Filter leaves manufactured by Mer-Made Filter, Inc. for installation in vacuum diatomite filter units.

**meromictic** Pertaining to a permanently stratified lake.

**merozoite** A stage in the asexual reproduction of certain sporozoans; a cell, developed from a shizont, that can reproduce sexually.

**merry-go-round** A sequence of three or more granular activated carbon columns operated in series as lead column, roughing column(s), and finisher(s). The lead is taken out of service when exhausted, replaced by the first roughing column, and regenerated to become the last finisher, and so on.

**merry-go-round ion exchange** Same as merry-go-round system.

**merry-go-round system** A water treatment plant that consists of three or more ion-exchange columns in series, with one being regenerated while the others are operating as roughing and polishing columns. When put back into service, the newly regenerated column becomes the polishing column, and the former polishing column becomes the roughing column. *See also* carousel ion-exchange process.

**Mesa-Line®** A portable submersible pump manufactured by Crane Pumps & Systems.

**mesh** (1) One of the openings or spaces in a screen or woven fabric. (2) The mesh number, that is, the number of openings per inch, measured from the center of one wire or bar to a distance of one inch, including the diameter of the wire or fabric.

**mesh screen** A woven fabric of different materials.

**mesh size** The diameter of a particle as determined from the U.S. Sieve Series; the diameter of a screen opening.

**mesolimnion** *See* metalimnion.

**mesopelagic zone** The layer of the ocean just below the photic zone and above the bathypelagic zone. It is dimly illuminated and inhabitated by sharks and other sea animals. *See also* benthic region and hydrothermal vent.

**mesophile** (1) Any of a group of organisms that grow best within a moderate temperature range (from the prefix meso, meaning middle or mid). The literature reports different temperatures for this range; they generally fall between 10°C (50°F) and 50°C (122°F); the optimum range for growth is between 27 and 38°C (AWWA, 2000; Pankratz, 1996; Manahan, 1997; Metcalf & Eddy, 1991; Metcalf & Eddy, 2003; Hammer & Hammer, 1996; Droste, 1997; Maier, 2000, WEF,

1991). *See also* psychrophile and thermophile. (2) Same as mesophilic.

**mesophilic** Pertaining to a mesophile. Mesophilic bacteria active in wastewater treatment have an optimum temperature of 27–38°C. Anaerobic digestion tanks are usually heated for operation near 35°C (Hammer & Hammer, 1996). *See also* psychrophilic and thermophilic.

**mesophilic anaerobic digestion** Same as mesophilic digestion.

**mesophilic bacteria** Bacteria that grow at moderate temperatures. *See* mesophile and mesophilic.

**mesophilic composting range** Same as mesophilic stage.

**mesophilic composting stage** Same as mesophilic stage.

**mesophilic digestion** Anaerobic sludge digestion within the mesophilic range of 27–38°C. With a mean retention time of 15 days it yields reductions in the order of 1–2 $\log_{10}$ in total coliforms, fecal coliforms, and fecal streptococci. Mesophilic digestion facilities are designed and operated in various modes. *See also* single-stage, high-rate, two-stage, separate, and standard-rate or low-rate digesters, thermophilic digestion.

**mesophilic (micro)organism** A (micro)organism that grows at moderate temperatures, e.g., 20–50°C, with an optimum range of 25–40°C. *See* mesophile, mesophilic, psychrophilic, thermophilic.

**mesophilic range** (1) The range of moderate temperatures over which some organisms can exist. *See* mesophile. (2) The optimum temperature range to operate waste digestion units so as to emhance the growth of mesophilic organisms; reported as between 27 and 38°C (between 81 and 100°F).

**mesophilic saprophytes** A group of bacteria that thrive in moderate temperatures while carrying out sludge digestion.

**mesophilic stage** The first of three stages of the sludge composting process, during which temperature of the composting material increases to about 40°C (104°F), with a preponderance of fungi and acid-producing bacteria. *See also* cooling stage and thermophilic stage.

**mesophyte** A plant that grows under usual or moderate conditions of atmospheric moisture. *See also* xerophyte.

**mesopore** A pore (of an adsorbing material) having a diameter between 2 and 50–100 nanometers (20 and 500–1000 Å). Mesopores are normally associated with the adsorption of large molecules. *See also* macropore and micropore.

**mesotrophic** Characteristic of reservoirs and lakes that contain moderate quantities of nutrients and are moderately productive in terms of aquatic animal and plant life. *See also* oligotrophic, eutrophic.

**mesotrophic lake** A lake that is moderately productive and still aerobic.

**mesotrophic stage** A stage in the life cycle of a lake or reservoir, associated with a moderate amount of nutrients and moderate biological activity.

***Metabacterium polyspora*** A species of marine bacteria that produce multiple endospores by asymmetric division at both ends of the cell and by symmetric division of the endosopores.

**metabolic** Pertaining to the chemical change processes (e.g., digestion and respiration) that occur within living organisms.

**metabolic nutrient requirement** The amount of essential elements that organisms require for growth. For example, the cell mass of wastewater treatment microorganisms contains approximately 12% of nitrogen and 2% of phosphorus by weight, corresponding to nutrient removal by ordinary biological treatment. *See also* biological nutrient removal.

**metabolic pathway** One of the routes that may contribute to a biological process; for example, in anaerobic digestion one pathway is the conversion of 30% of the organic matter to propionic acid, 60% of which is converted to acetic acid and then to methane.

**metabolism** The sum of the chemical reactions occurring within a cell or a whole organism to convert organic matter, including the energy-releasing breakdown of molecules (catabolism) and the synthesis of new molecules (anabolism). Metabolism keeps the organism alive and productive; it contributes to accelerating the rate of such chemical reactions as sulfide oxidation. Metabolism applies to the biotransformation of both substrates and synthetic chemicals. *See also* basal metabolism, photosynthesis, respiration, digestion, biotransformation.

**metabolite** (1) A product of metabolism, especially of the biotransformation of chemicals by an organism. Examples are the algal metabolites geosmin and methyl-isoborneol that cause taste and odor problems in drinking water even at very low concentrations. Some endotoxins and exotoxins are also toxic to humans and aquatic life. (2) A substance involved in metabolism; an essential nutrient. *See also* antimetabolite.

**metacercariae** Plural of metacercaria, the encysted larva of a trematode, usually found in or on an aquatic intermediate host; the last stage in the development of some water-related parasites such as *Fasciolopsis, Clonorchis,* and *Paragonimus.*

**metacresol purple** A reagent used in water analyses as an acid–base indicator, with pH transition ranges of 1.2 (red)–2.8 (yellow) and 7.4 (yellow)–9.0 (purple).

**meta-dichlorobenzene ($C_6H_4Cl_2$)** *See* dichlorobenzene.

**metagonimiasis** A snail vector disease prevalent in the Far East and Siberia, usually asymptomatic; caused by the excreted pathogen *Metagonimus yokogawai.* Occasionally, symptoms occur, e.g., nausea, diarrhea, fever, and abdominal pain.

***Metagonimus yokogawai*** A fluke that causes minor infections of the small intestine; transmitted from dogs and cats to humans through aquatic snails and freshwater fish as intermediate hosts. Latency = 6 weeks. Median infective dose < 100.

**metal** An element that possesses most or all of the following characteristics: crystallic solids with metallic luster, malleability, ductility, conductivity of heat and electricity, high specific gravity, high chemical reactivity. As an exception, mercury is a liquid. Metals easily lose electrons to form positive ions; metals are also electron pair acceptors or Lewis acids. Metallic oxides combine with water to form hydroxides rather than acids. Some metals are highly toxic to humans and aquatic life, but trace amounts of metals are necessary for biological treatment, whereas high metals contents may render wastewater unfit for reuse. *See also* complex, metal contamination, metalloid, heavy metals, essential metal, toxic metal, noble element, tramp element.

**metal alkoxide** Same as alkoxide. *See also* sodium methoxide ($NaOCH_3$).

**metal alkyl** An organometallic compound formed by the substitution of a metallic ion for the hydroxyl group in methyl alcohol ($CH_3OH$), e.g., methylmercury ($CH_3Hg^+$).

**metal amide** A compound resulting from the combination of a metal and ammonia. *See,* e.g., lithium amide ($LiNH_2$).

**metal contamination** The contamination of water sources by metals of concern for public health and aquatic life, resulting from industrial activities, wastewater discharges, erosion of contaminated sediments and soils, as well as nonpoint source runoff and natural geologic processes. Examples of metals important in the field of water and wastewater treatment, most of which are regulated by the USEPA are antimony, arsenic, barium, beryllium, cadmium, chromium, copper, lead, mercury, selenium, and thallium. *See also* heavy metals.

**Metal-Drop™** A chemical product of Kem-Tron used as a flocculant or coagulant.

**metal fabrication industry** A standard industrial classification that covers the manufacture of various metal products. *See* fabricated metals products industry for detail.

**metal-finishing plant** Wastewater from such a plant contains chromium, cyanides, and other metals. These plants usually practice waste segregation.

**metal finishing wastes** Wastewater from the metal-finishing industry, which originates from such operations as electroplating and galvanizing, and may contain acids, cautic materials, and metal particles.

**metalimnion** The middle layer in a thermally stratified lake or reservoir. In this layer, there is a rapid decrease in temperature with depth, e.g., 0.5°F or more per foot. *See* thermocline for more detail.

**metallic bond** A chemical bond between atoms of a metallic element.

**metallic coatings** Materials applied to form a film on metallic surfaces for protection against corrosion. Common application methods include electroplating, metal spraying, and hot dipping. Other methods of corrosion control include anodic protection, cathodic protection (galvanic or impressed-current protection), chemical coatings, inhibition, and inert materials.

**metallic sulfide** A compound of sulfur and a metal that forms under anaerobic conditions in wastewater and imparts to the liquid a gray, dark gray, or black color.

**metalloid** (1) An element that possesses some but not all of the characteristics of a metal, or that has properties intermediate between those of metals and nonmetals, e.g., antinomy (Sb), arsenic (As), boron (B), germanium (Ge), selenium (Se), silicon (Si), and tellurium (Te). Also called a semimetal or semiconductor. (2) A nonmetal that forms alloys with a metal.

**metal-plating waste** Wastewater generated during the manufacture of metal-plated products, originating mainly from the stripping of oxides and cleaning and plating of metals. It contains acids, metals, and toxic mineral materials. Common treatment methods include alkaline chlorination of cyanide, reduction and precipitation of chromium, and lime precipitation of other metals.

**metal pollution** The discharge of concentrated metal wastes into the environment from such activities as mining, ore refinement, nuclear processing, and the manufacture of batteries, metal alloys, electrical components, paints, etc.

**metal release** A product of the corrosion of a metal pipe released to the water, carried in dissolved, colloidal, or suspended form.

**metal salt coagulants** Metallic salts used as coagulants in the treatment of water and wastewater; commonly salts of aluminum and iron such as aluminum sulfate (alum), ferric chloride, and ferric sulfate.

**metal solvency** The property of a metal that dissolves in water or other solvents.

**metals removal** Common wastewater treatment processes used to remove heavy metals include chemical precipitation, activated carbon adsorption, ion exchange, and reverse osmosis. The metals of interest are arsenic (As), barium (Ba), cadmium (Cd), copper (Cu), mercury (Hg), nickel (Ni), selenium (Se), and zinc (Zn). Metal finishing wastes are often treated by addition of a base to precipitate the hydroxides. *See also* amphoteric compound.

**metal toxicity** Toxic metals affect microbial cells by binding to cellular ligands and displacing native essential elements, by disturbing proteins, and disturbing nucleic acids.

**metal uptake** The absorption of a metal by water from a corroding surface or the incorporation of a metal in an organism's cells or tissues.

**MetalWeave®** Baffles manufactured by Eimco Process Equipment Co. in stainless steel for use in water or wastewater treatment tanks.

**metamorphic rock** One of three broad categories of geologic formations constituting aquifers; created by intense heat, pressure, or chemical action on igneous or sedimentary materials. Such formations are compact, crystalline, and poor sources of groundwater, compared to the igneous and sedimentary rocks.

**metastasis** The transfer of disease from one organ or part to another not directly connected with it.

**metastatic** Pertaining to the transfer of disease from one organ or part to another not directly connected with it.

**meteoric water** Water fallen as rain and percolating to the water table.

**meteorological measurements** Measurements of wind speed, wind direction, barometric pressure, temperature, relative humidity, and solar radiation (EPA-40CFR58.1-z).

**meteorology** The study of the chemical, physical, and dynamic aspects of atmospheric phenomena; the science of weather and weather forecasting.

**meterage** The practice of billing utility customers based on actual use as indicated by metering, which encourages conservation.

**meter box** A container enclosing and protecting a water meter, while allowing access for meter reading.

**meter-candle** Another name of lux, the Système International (SI) unit of illuminance; the illuminance of a luminous flux of one lumen uniformly distributed over a surface of one meter square, i.e., 1 lx = 1 lumen/$m^2$ = 1 steradian-candela/$m^2$ = 0.0929 foot-candle.

**metering pump** A pump used to feed measured volumes of chemical solutions, e.g., a positive-displacement pump that can adjust its capacity to deliver a controlled volume of liquid. It is used in the chemical, pharmaceuticals, foods, and other industries. Water and wastewater treatment applications include chemicals dosing for pH control, coagulation, and disinfection. Also called dosing pump, controlled volume pump. *See also* packed-plunger metering pump, diaphragm metering pump, step valve, chemical feed pump, lost motion, solenoid-driven metering pump, digital dosing pump

**meter rate** The charge levied for water service based on the quantity consumed, as measured by a meter.

**meter stop** Same as curb stop and box.

**methaemoglobinaemia** *See* methemoglobinemia.

**methanal** Same as formaldehyde.

**methane ($CH_4$)** A colorless, odorless, nonpoisonous but asphyxiating, flammable gaseous hydrocarbon (of the paraffin group) of high fuel value, created by anaerobic decomposition of organic matter; somewhat soluble in water; often found in natural gas, marshes, swamps, sewers, landfill gas, etc. It is usually not found in large quantities in untreated wastewater because oxygen is toxic at low concentrations to the methanogenic organisms. It may also be produced naturally through volcanic activity or by heating carbon monoxide (CO) and hydrogen ($H_2$) over a catalyst. It is a greenhouse gas produced mainly by bacterial breakdown of cellulose in the guts of termites and ruminants. Methane forms an explosive mixture with air at a volume concentration of 5.5–13.5%. It is easily removed from water by aeration. Also called marsh gas or methyl hydride, it is one of two primary constituents of biogas. *See also* trihalomethane.

**methane bacteria** A group of obligate anaerobic bacteria that form methane while decomposing organic matter. *See* methanogen and methanogenesis for detail.

**methanecarboxylic acid** Same as acetic acid.

**methane fermentation** The sequence of reactions that produce methane during the anaerobic decomposition of organic matter. *See* methanogen and methanogenesis for detail.

**methane formation** The third and final stage in the simplified representation of the anaerobic digestion process, following hydrolysis and acid formation. *See* methanogen and methanogenesis for detail.

**methane former** Same as methanogen.

**methane-forming bacteria** *See* methanogen.

**methane monooxygenase** A nonspecific enzyme that incorporates oxygen into a substrate. *See* cometabolism (of trichloroethylene).

**methane potential** The potential of a waste to produce methane under anaerobic conditions. It depends on the concentration of organic matter (for example, as measured by the COD or the 5-day BOD) and on treatment efficiency. *See also* methane yield, methane production, biochemical methane potential test, and anaerobic toxicity assay.

**methane production** The quantity of methane ($CH_4$) produced per unit time ($Q_m$) by anaerobic decomposition of an organic waste. It depends on the waste flow ($Q$), the efficiency of substrate removal ($E$, between 0 and 1), the methane yield of the waste ($M$) and the influent COD ($X$):

$$Q_m = QEMX \quad \text{(M-34)}$$

The following formulas are often used in digester design for the volume ($Q_m$ in cubic meters per day) of methane produced at standard conditions of temperature and pressure:

$$Q_m = 0.35\,[Q(S_0 - S) - 1.42\,P_x] \quad \text{(M-35)}$$

$$P_x = YQ(S_0 - S)/(1 + \theta k_d) \quad \text{(M-36)}$$

where $Q$ = waste flowrate, $m^3/d$); $S_0$ = influent biodegradable COD, mg/L; $S$ = effluent biodegradable COD, mg/L; $P_x$ = net mass of cell tissue produced, kg/d; $Y$ = yield coefficient, g volatile suspended solids/g biodegradable COD; $\theta$ = solids retention time, d; and $k_d$ = endogenous decay coefficient, /d.

**methane production from nightsoil** The generation of methane, or biogas, by mixing nightsoil with animal wastes, straw or other crop wastes, and water in a anaerobic digester. The liquid effluent is used as a fertilizer.

**methane tetrachloride** Same as carbon tetrachloride.

**methane yield** The quantity of methane produced per unit of organic matter removed, usually expressed as a volume $M$ per COD. The theoretical yield is $M_t$ = 5.6 cubic feet of methane per pound of COD removed. Toxicity, refractory organics, and other reasons will reduce the biochemical methane potential of the waste, as determined by the BMP test. Environmental conditions and various treatment losses further reduce the actual yield to about $M$ = 3.2 cubic feet per pound of COD removed.

***Methanobacillus*** A group of archaebacteria, strict obligate anaerobes responsible for methane ($CH_4$) production during fermentation, and active in anaerobic digesters under mesophilic conditions. They oxidize hydrogen ($H^+$) with carbon dioxide ($CO_2$) as the electron acceptor to produce methane.

***Methanobacterium*** A group of thermotolerant archaea, strict obligate anaerobes responsible for methane ($CH_4$) production during fermentation, and active in anaerobic digesters under mesophilic conditions. They oxidize hydrogen ($H^+$) with carbon dioxide ($CO_2$) as the electron acceptor to produce methane.

***Methanocarcina*** A group of archaea, strict obligate anaerobes responsible for methane production during fermentation, and active in anaerobic digesters under mesophilic conditions. They use acetate to produce methane ($CH_4$) and carbon dioxide ($CO_2$). Also called *Methanosaeta.*

***Methanococcus*** A genus of spherical, mesophilic bacteria identified from a mixed culture and responsible for methane production during anaerobic digestion.

**methanogen** Methanogens are a group of anaerobic bacteria that produce methane gas and carbon dioxide by anaerobic decomposition of organic acids. They are found in anaerobic digesters, landfills, swamps, and lake sediments. *See also* aceticlastic methanogen, acetogen, hydrogen-utilizing methanogen, *Methanosaeta, Methanosarcina,* and *Methanothrix.* Some autotrophic methanogens produce methane by using hydrogen to reduce carbon dioxide or carbon monoxide.

**methanogenesis** The formation of methane gas ($CH_4$) as the ultimate product of anaerobic digestion. (To a much lesser extent methane also results from volcanic activity.) Microbial formation of methane occurs in two major processes: fermentation of acetic acid ($CH_3COOH$) and other organic acids with concomitant production of carbon dioxide ($CO_2$), or reduction of a carbon compound such as carbon dioxide or carbon monoxide (CO). Neglecting the intermediate steps, the following reactions illustrate these processes:

$$CH_3COOH \rightarrow CH_4 + CO_2 \quad \text{(M-37)}$$

$$CO_2 + 4H_2 \rightarrow CH_4 + 2H_2O \quad \text{(M-38)}$$

$$CO + 3H_2 \rightarrow CH_4 + H_2O \quad \text{(M-39)}$$

*See also* hydrolysis and fermentation, the other two basic steps that precede methanogenesis in the anaerobic oxidation of organic matter.

**methanogenic** Referring to the formation of methane by certain anaerobic bacteria during the process of anaerobic fermentation.

**methanoic acid (HCOOH or $CH_2O_2$)** A colorless, irritating, fuming, water-soluble liquid, obtainable from ants or synthetically; used in dyeing, tanning, and in medicine. Common name: formic acid, a carboxylic acid.

**methanol ($CH_3OH$)** An alcohol that can be used as an alternative fuel, as a solvent, or as a gasoline additive. It can be produced by the oxidation of methane or from wood by dry distillation. It is less volatile than gasoline; when blended with gasoline, it lowers the carbon monoxide emissions but increases hydrocarbon emissions. When it is used as pure fuel, its emissions are less ozone-forming than those from gasoline. In wastewater treatment, methanol (a rapidly degradable organic matter) is often used a supplemental source of carbon in postanoxic denitrification:

$$5\,CH_3OH + 6\,NO_3^- \rightarrow 3\,N_2\uparrow + 5\,CO_2\uparrow + 7\,H_2O + 6\,OH^- \quad \text{(M-40)}$$

Also called methyl alcohol or wood alcohol.

**methanol process** A process used to generate chlorine dioxide ($ClO_2$) on the site of water treatment plants, with sodium chlorate ($NaClO_3$), methanol ($CH_3OH$), and sulfuric acid ($H_2SO_4$). *See* Solvay process for more detail.

***Methanomonas* genus** A group of heterotrophic organisms active in biological denitrification.

***Methanosaeta*** One of two groups of bacteria that can produce methane and carbon dioxide from acetate (or acetic acid). It was previously identified as *Methanothrix. See* acetoclastic bacteria.

***Methanosarcina*** genus One of two groups of bacteria that can produce methane and carbon dioxide from acetate (or acetic acid). *See* acetoclastic bacteria.

***Methanothrix*** A genus of bacteria identified from a mixed culture and originally considered a methanogen.

***Methanothrix sohengenii*** A species of bacteria originally thought to produce methane and carbon dioxide from acetate. *See Methanosaeta.*

**methanotrophs** Methanotrophic bacteria.

**methanotrophic bacteria** A group of aerobic, heterotrophic soil bacteria that use methane for energy and as a sole source of carbon; they can cometabolize trichloroethylene using the enzyme methane monooxydase. Also called methanotrophs. *See* cometabolism.

**methemoglobin** The oxidized form of hemoglobin, i.e., its iron has been oxidized to the ferric state by such chemicals as nitrite ($NO_2^-$) and chlorite ($ClO_2^-$); it does not carry oxygen in the blood. *See* methemoglobinemia.

**methemoglobinemia** A pathological condition caused by an increased blood concentration of methemoglobin, a stable compound that does not yield up its oxygen to tissues. Methemoglobinemia may lead to asphyxiation in bottle-fed infants under one year of age and in ruminants when they ingest nitrates from food or water, which are bacterially converted to nitrites in the digestive tract; nitrite oxidizes blood hemoglobin to methemoglobin. Symptoms like lethargy, shortness of breath, and a bluish skin color start when methemoglobin concentration reaches about 5 or 10%. For this reason, there is a limitation of 45 mg/L on nitrate concentration in drinking water. By oxidizing hemoglobin, chlorite ion (e.g., from water disinfection with chlorine dioxide) may also cause methemoglobinemia; at higher dosages, chlorate may also cause the disease. Also called infantile cyanosis, blue baby syndrome, or blue baby disease.

**methenyl tribromide** Another name of bromoform.

**method 18** A USEPA test method that uses gas chromatographic techniques to measure the concentration of volatile organic compounds in a gas stream.

**method 24** AUSEPA reference method to determine the density, water content, and total volatile content (water and VOC) of coatings.

**method 25** A USEPA reference method to determine the VOC concentration in a gas stream.

**method blank** Laboratory grade water taken through an entire analytical procedure to determine if samples are being accidentally contaminated in a laboratory.

**method detection level** Same as method detection limit.

**method detection limit (MDL)** One of a few parameters used to measure the precision or the detection level of a method of analysis, stated as a result that produces a signal with a 99% probability that it is different from the blank. Also defined as the minimum concentration of a constituent that can be measured and reported with 99% confidence that the constituent concentration is greater than zero. Sometimes erroneously called minimum detection limit. *See also* limit of detection, minimum reporting limit, practical quantification level.

**method of standard addition** An atomic absorption technique. *See* standard addition method.

**methoxychlor [$Cl_3CCH(C_6H_4OCH_3)_2$]** A pesticide that causes adverse health effects in domestic water supplies and is also toxic to freshwater and marine life. Its chemical name is 2,2-bis (p-methoxyphenyl)-1,1,1-trichloroethane. It is regulated by the USEPA as a drinking water contaminant: MCL = MCLG = 0.04 mg/L.

**methoxy DDT** Another name of metoxychlor.

**2,2-bis-(p-methoxyphenyl)-1,1,1-trichloroethane** Another name of metoxychlor.

**methyl** The radical —$CH_3$ from methane ($CH_4$), occurring in the names of some organic compounds, e.g., methylamine ($CH_5N$). (2) A compound formed from methylene.

**methyl alcohol** Common name of the primary alcohol methanol ($CH_3OH$).

**methylamine** *See* amine. Also called aminomethane.

**methylarsinic acid [$CH_3AsO(OH)_2$]** A toxic methyl derivative of arsenic.

**methylation** The attachment of an alkyl group (—$CH_3$) to a metal or metalloid ion, which results in the formation of an organometal, increases the availability of the metal to organisms, and may lead to bioaccumulation and increased toxicity in the food webs. For example, methyl mercury ($CH_3Hg^+$ and $CH_3 \cdot Hg \cdot CH_3$) is available as $CH_3HgCl$ in saline water and as $CH_3HgOH$ in freshwater. Arsenic is methylated by fungi into methylarsines. Same as alkylation (2).

**methyl benzene** An aromatic solvent that is only slightly soluble in water; used in the production of benzene derivatives, perfumes, and in other industrial applications. Acute exposure to this product may cause damages to the nervous system and certain organs. Regulated by the USEPA as a drinking water contaminant: MCL = MCLG = 1.0 mg/L. Also called phenylmethane and toluene.

**methyl bromide $CH_3Br$)** Another name of methyl bromide, a colorless, poisonous gas used as a solvent, refrigerant, and soil and space fumigant. Also called bromomethane.

**methyl chloride** Same as chloromethane.

**methyl chloroform** Another name of 1,1,1-trichloroethane, a chemical widely used as an industrial cleaner, degreaser, solvent, and in adhesives, coatings, and drain cleaners.

**methylene blue** A cationic dye that reacts with certain (methylene blue active) substances to form an ion pair. It is used in the laboratory analysis of anionic surfactants and in the photodynamic inactivation of such microorganisms as polioviruses.

**methylene blue active substance (MBAS)** A substance that reacts with methylene blue to form a soluble complex as part of the procedure to determine the concentration of anionic surfactants such as linear alkyl sulfonate and alkyl sulfate in water and wastewater. When the resulting product is extracted into a solvent (e.g., chloroform, $CHCl_3$), the concentration of MBAS is proportional to the intensity of the blue color of the product. Also called a foaming agent. *See also* the cobalt thiocyanate active substances (CTAS) test.

**methylene blue number** A parameter used to describe the adsorptive capacity of activated carbon; the number of milligrams of methylene blue (MB) adsorbed by one gram of the activated carbon sample from a solution of 1.0 mg/L of MB. *See also* carbon tetrachloride activity, decolorizing index, iodine number, molasses number, and phenol adsorption value.

**methylene chloride ($CH_2Cl_2$)** Same as dichloromethane, a volatile organic compound, used in paint removers, urethane foam, pesticides, etc.; regulated by the USEPA as a drinking water contaminant: MCLG = 0, MCL = 0.005 mg/L.

**methylene chlorobromide** Same as bromochloromethane.

**methyl ester** Any of a group of fatty esters derived from vegetable oils; they can be formed by derivatizing haloacetic acids. *See* diazomethane.

**methyethylmethane** Another name of butane.

**methyl glyoxal ($CH_3COCHO$)** An aldehyde (or aldoketone) formed during water ozonation; the USEPA requires water supply systems to monitor this disinfection by-product.

**methyl group** A group of organic compounds derived from methane ($CH_4$) and containing the group —$CH_3$ in their formulas.

**methyl hydride** Another name for methane.

**methylisoborneol ($C_{11}H_{20}O$)** Abbreviation: MIB. The common name for 2-exo-2-methylbornane (I), a substance produced by blue-green algae and actinomycetes, and responsible for taste and odor problems in drinking water sources, even at very low concentrations (in the order of 10–30 ng/L). Effective water treatment methods include activated carbon adsorption and ozonation. Also written as 2-methylisoborneol.

**methyl isobutyl ketone [$(CH_3)_2CHCH_2COCH_3$)]** Acronym: MIBK. A solvent for paints and varnishes.

**methyl isocyanate ($CH_3NCO$)** A highly toxic compound that reacts with water to produce a methylamine solution and a carbonate:

$$CH_3NCO + 2\,H_2O \rightarrow CH_3NH_2 + H_2CO_3 \quad \text{(M-41)}$$

**methyl mercaptan ($CH_3SH$)** One of the sulfur compounds produced by some bacteria; it is responsible for swampy, fishy, or decayed cabbage tastes and odors in drinking water distribution systems; odor threshold = 0.0021 ppm per volume.

**methylmercury** A volatile, toxic organic compound of mercury, formed by bacteria in sediment and acidic waters; it accumulates in the body (of humans and fish) because it is produced at a faster rate than it is degraded. It can cause damage to organs and the nervous system; see Minamata disease. Formulas: $CH_3Hg^+$ and $CH_3HgCH_3$. The stable species are $CH_3HgOH$ in freshwater and $CH_3HgCl$ in seawater. Also called alkylmercury, a highly toxic compound used as a fungicide and seed dressing.

**methyl orange [$C_{14}H_{14}N_3SO_3Na$ or $(CH_3)_2NC_6H_4NNC_6H_4SO_3Na$]** A dye used in the laboratory as a pH color indicator in acid–base titrations. When an aqueous solution has a pH less than 3.1, its color is red; when the pH is greater than 4.4, the color is yellow. *See also* phenolphthalein and other acid–base indicators.

**methyl orange acidity** The acidity that exists in an aqueous solution with a pH below 4.5 as measured by the methyl orange test. It is caused mainly by strong inorganic acids. *See also* phenolphthalein acidity and total acidity.

**methyl orange alkalinity (M alkalinity)** A measure of the total alkalinity in a water sample. The alkalinity is measured by the amount of standard sulfuric acid ($H_2SO_4$) required to lower the pH of the water to a pH level of 4.5, as indicated by the change in color of methyl orange from orange to pink. Methyl orange alkalinity is expressed in milligrams per liter equivalent calcium carbonate ($CaCO_3$).

**methyl orange indicator** *See* methyl orange.

**methylotroph** An organism that can oxidize organic compounds that do not contain carbon–carbon bonds. *See also* methanotroph.

**methylphenol ($C_7H_8O$)** (1) Any of three isomeric compounds derived from coal or wood tar and used as a disinfectant. Also called cresol. *See also* tricresol. (2) 2-methylphenol ($CH_3C_6H_4OH$) or orthocresol is synthetic organic chemical, used as a disinfectant or an insecticide.

**methyl red** A reagent used in water analyses as an acid–base indicator, with a pH transition range of 4.4 (red)–6.2 (yellow).

**methyl sulfate [$(CH_3)_2SO_2$]** A colorless or yellow, poisonous liquid, used in organic synthesis. Also called dimethyl sulfate.

**methyl-tert-butyl-ether** Same as methyl tertiary butyl ether.

**methyl tertiary butyl ether [$(CH_3)_3COCH_3$)]** Acronym: MTBE. A low-cost organic chemical used as a highly soluble oxygen additive to gasoline to improve air quality. As a water contaminant, it has a low taste-and-odor threshold, migrates quickly through subsurface water, and is difficult to remove from drinking water by conventional methods. Promising treatment processes include air stripping, activated carbon adsorption, and advanced oxidation processes. It is also used as a solvent. Also spelled methyl-*tert*-butyl-ether. *See also* xenobiotic.

**methyl tribromide** Another name of bromoform.

**methylumbelliferyl-β-D-glucuronide (MUG)** A fluorogenic substance used as the substrate for the enzyme β-glucoronidase in cultures to detect the presence of *E. coli* in a water sample. Also spelled 4-methylumbelliferyl-β-D-glucuronide.

**metoecious parasite** A parasite that is not host-specific; the opposite of ametoecious parasite. Also called metoxenous parasite.

**metolachlor [$(CH_3)C_6H_3(C_2H_5)$-N($COCH_2Cl$)CH($CH_3$)$CH_2OCH_3$]** The common name for 2-chloro-*N*-(2-ethyl-6-methylphenyl)-*N*-(2-methoxy-1-methyethyl)acetamide, a chloracetanilide herbicide that is used to control grasses in fields of corn, sunflower, and ornamental plants. It is biodegradable (half-life of 6–10 weeks) but has a potential for bioaccumulation in fatty tissues. USEPA classifies metolachlor as a possible human carcinogen and issues a health advisory of 0.1 mg/L for lifetime exposure (which is both the MCL and MCLG). In addition, it can cause skin irritation and perhaps, at very high concentrations, damage the liver. Activated carbon adsorption and reverse osmosis are recommended treatment processes to remove metolachlor. *See also* Dual and Primext.

**metoxenous parasite** Same as metoecious parasite.

**metribuzin ($C_8H_{14}N_4OS$)** The common name for 4-amino-6-(1,1-dimethylethyl)-3-(methylthio)-1,2,4-triazin-5(4*H*)-one, a triazine herbicide regulated by the USEPA as a drinking water contaminant: MCL = MCLG = 0.2 mg/L. It is one of the rare pesticides that can be removed by ordinary water treatment methods; it can be oxidized by chlorine or chlorine dioxide or adsorbed on granular or powdered activated carbon. Long-time exposure to metribuzin causes damages to the thyroid, kidney, and liver of laboratory animals.

**metric** A biological attribute, feature, or characteristic of a community that reflects ambient conditions.

**metric ton** A unit of mass or weight of 1000 kg, or about 2205 pounds. Also called a tonne.

**metric units** A decimal system of units based on the meter for length, the second for time, and the kilogram for mass. One meter (m) = 100 cm = 1000 mm = 0.001 km = 3.28 feet. One kilogram (kg) = 1000 g = 0.001 metric ton = 2.205 pounds. *See also* Système International.

**METROGRO™** The trademark of three sludge products of Metropolitan Denver Sewage Disposal District No. 1: an aerated windrow sludge compost, a polymer-conditioned and dewatered sludge cake containing 17% solids and a liquid 7%-solids sludge.

**MeV** Abbreviation of million electron volts.

**MF** Acronym of (a) membrane filter, membrane filtration; (b) microfiltration.

**MF analysis** *See* membrane filter method, membrane filtration technique.

**MFI** Acronym of modified fouling index.

**MFI plot** A curve that plots, for a membrane filtration system, the inverse of flow vs. the cumulative volume of filtrate. *See* modified fouling index.

**MF method** *See* membrane filter method, membrane filtration technique.

**MF technique** *See* membrane filter method, membrane filtration technique.

**MF test** *See* membrane filter test.

**MG** Abbreviation of million gallons. 1 MG = 133,681 $ft^3$ = 3,785.4 $m^3$.

**mg** Abbreviation of milligram(s).

**Mg** Chemical symbol of the metallic element manganese.

**mgad** Abbreviation of million gallons per acre per day, a unit of hydraulic loading.

**Mg-calcite** Magnesian calcite.

**$MgCO_3$** Chemical formula of magnesium carbonate or magnesite.

**$MgCO_3 \cdot 3H_2O$** Chemical formula of nesquehonite.

**$Mg_4(CO_3)_3(OH)_2 \cdot 3H_2O$** Chemical formula of hydromagnesite.

**mgd** Abbreviation of million gallons per day, a unit used to measure discharges or flow capacities of treatment plants, pump stations, sewer lines, and similar facilities. One mgd = 1.5472 cfs = 43.8125 liters/second = 3785.4 $m^3$/day.

**$(Mg, Fe)_2SiO_4$** Chemical formula of olivine.

**mgid** Acronym of million gallon(s) imperial per day. *See* gallon. 1 mgid = 1.2 mgd.

**mg/L or mg/l** Abbreviation of milligram(s) per liter, a unit of concentration.

**mg/L as $CaCO_3$** A convenient expression of the concentration of specified constituents in water and wastewater in terms of their equivalent value to calcium carbonate. *See* calcium carbonate equivalent.

**$MgNH_4PO_4$ or $MgNH_4PO_4 \cdot 6H_2O$** Chemical formula of struvite.

**MgO** Chemical formula of magnesium oxide or magnesia.

**$MgO_2$** Chemical formula of magnesium dioxide or magnesium peroxide.

**$Mg(OH)_2$** Chemical formula of magnesium hydroxide or brucite.

**$MgSiO_3$** Chemical formula of magnesium silicate.

**$Mg_2Si_3O_8 \cdot xH_2O$** Chemical formula of magnesium trisilicate.

**$Mg_2SiO_4$** Chemical formula of fosterite or magnesium silicate.

**$Mg_3Si_{12}O_4(OH)_4$** Chemical formula of chrysotile.

**$MgSO_4$** Chemical formula of magnesium sulfate.

**$MgSO_4 \cdot 6H_2O$** Chemical formula of Epsom salt.

**MH** Abbreviation of manhole or maintenance hatch.

**MHF** Acronym of multiple-hearth furnace.

**mho** The unit of conductivity (electrical conductance), the reciprocal of resistivity (mho is the backward spelling of ohm, the unit of resistivity).

**MHT®** Magnesium hydroxide produced by Dow Chemical Co.

**mi** Abbreviation of mile(s). One mile = 5280 ft = 1.609 km.

**MIB** Abbreviation of methylisoborneol.

**MIBK** Acronym of methyl isobutyl ketone.

**MIC** Acronym of microbiologically induced corrosion or microbiologically influenced corrosion.

**mica** A group of silicate minerals with a single cleavage that produces flakes; used as a thermal and electrical insulator. *See also* biotite and muscovite.

**micelle** A negatively charged colloidal particle surrounded by positively charged, hydrophobic and hydrophilic ions or molecules. It can increase or decrease in size without a structural change. For example, the molecule typical of soap, sodium stearate [$NaC_{18}H_{35}O_2$ or $CH_3(CH_2)_{16}COO^-Na^+$], has a long hydrophobic tail [$C_{17}H_{35}$ or

$CH_3(CH_2)_{16}$] and a charged ionic, hydrophilic head ($COO^-$). In water, it can form a micelle of up to 100 stearate anions with their tails inside a colloidal particle and their heads in the surrounding water, in contact with the $Na^+$ counterions. *See* Figures M-05 and M-06. Also called an association colloid.

**Michaelis constant** An apparent dissociation constant related to the effects of substrate concentration on the rate of substrate transformation into product (Symons et al., 2000). *See also* apparent dissociation constant, association constant, dissociation constant, equilibrium constant, ionization constant.

**Michaelis–Menten constant** The overall constant or half-velocity constant ($C_h$) used in the Michaelis–Menten equation. It combines the effects of all the rate constants involved in enzyme kinetics.

**Michaelis–Menten equation** A saturation-type equation proposed by L. Michaelis and M. Menten in 1913 for enzyme kinetics, but since applied to various reactions, such as the rate of uptake of a substance (e.g., phosphate in a lake, substrate removal) in relation to its concentration:

$$K = CK_m/(C + C_h) \quad \text{(M-42)}$$

where $K$ = rate of uptake of the substance, $C$ = concentration of the substance, $K_m$ = the maximum rate of uptake, and $C_h$ = the concentration at which the rate is half the maximum rate. *See also* the Eadie–Hofstee plot, Langmuir isotherm, Lineweaver–Burke plot, Monod equation, substrate utilization model.

**Michaelis–Menten formulation** Same as Michaelis–Menten equation.

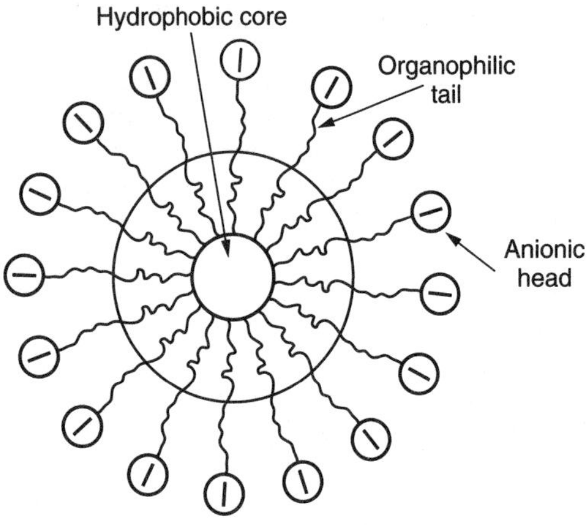


**Figure M-05.** Micelle (association colloid).

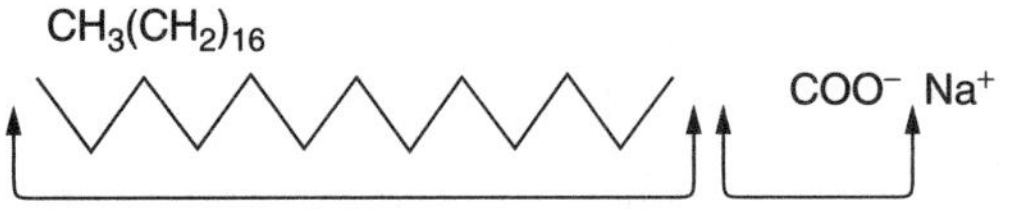


**Figure M-06.** Micelle (sodium stearate, $NaC_{18}H_{35}O_2$).

**Micrasieve™** A static fine screen manufactured by Andritz-Ruthner, Inc. with a pressure feed.

**Micro/2000®** A device manufactured by Wallace & Tiernan, Inc. for the analysis of chlorine dioxide, potassium permanganate, and chlorine residual.

**microaerobic** Requiring a partial pressure of oxygen less than that of atmospheric oxygen for growth. *See also* microaerophile.

**microaerobic condition** A condition of oxygen deficit, i.e., oxygen demand exceeds oxygen supply. Autothermal thermophilic aerobic digestion (ATAD) systems usually operate under microaerobic conditions for the fermentation of proteinaceous matter.

**microaerophile** A microaerophilic organism.

**microaerophilic bacteria** *See* microaerophilic organisms.

**microaerophilic organisms** (1) Obligate aerobes that function best under conditions of low oxygen concentration (e.g., less than 1.0 mg/L) or, more generally, facultative microorganisms that can switch metabolism from aerobic to anaerobic, depending on the availability of free oxygen. Such organisms constitute a good portion of the biomass of wastewater treatment. (2) Organisms that grow in the presence of oxygen but below the level in the atmosphere (21%). Some such organisms are also able to respire anaerobically with electron acceptors other than oxygen.

**microaerotolerant anaerobe** An organism that grows in an anaerobic system and microaerophilic conditions (less than 5% oxygen), but not in a carbon dioxide ($CO_2$) incubator (15% oxygen) or in air (21% oxygen).

**microbe** A microscopic organism; a plant or animal not visible to the naked eye, observable only through a microscope. Same as microorganism.

**microbial biomass** A parameter that indicates soil condition and potential microbial activity. A common measure of microbial biomass is biomass carbon, the amount of respiration corrected for basal respiration (Maier et al., 2000):

$$\text{BioC} = (F - U)/K \quad \text{(M-43)}$$

where BioC is the carbon trapped in the microbial biomass, $F$ is the carbon dioxide ($CO_2$) produced by a fumigated soil sample, $U$ is the $CO_2$ produced by the nonfumigated soil sample, and $K$ is the fraction of biomass carbon mineralized to $CO_2$. *See also* Anderson–Domsch equation.

**microbial cell composition** *See* cell composition.

**microbial cell formula** *See* cell composition.

**microbial corrosion** *See* microbiologically induced corrosion.

**microbial diversity concept** The diversity of a microbial population in its resistance to a disinfectant explains the decline in the number of surviving organisms with increased exposure time and concentration (CT). This concept is proposed as an alternative to the kinetics of a chemical reaction, which has been implied in the Chick–Watson and similar models (Najm, 2006). *See also* disinfection kinetics, vitalistic theory.

**microbial ecology** The study of the interactions of microorganisms with air, water, or soil. *See also* environmental microbiology.

**microbial film** The gelatinous film of microbial growth covering the surface of a medium or spanning the interstices of a granular bed; e.g., the biofilm that is produced when wastewater is sprayed over fixed media, as in a trickling filter, or when water flows through a slow sand filter. *See* slime for detail.

**microbial fuel cell** A technology currently in the experimental stage (as of 2007), intended to use bacteria to generate electricity while treating wastewater: the bacteria oxidize organic matter and transport electrons from the cell surface to the anode. *See also* air cathode single chamber.

**microbial growth** The activity and growth of microorganisms such as bacteria, algae, diatoms, plankton, and fungi.

**microbial loop** A pathway in the aquatic food web in which bacterioplankton mineralize a portion of the organic carbon into carbon dioxide ($CO_2$) and assimilate the remainder to produce new biomass or secondary production.

**microbial mat** An extreme case of interfacial aquatic habitat in which microorganisms are laterally compressed into a thin, biologically active, largely self-sufficient mat of up to 1 cm deep. Mats may be found in hot springs, saline lakes, and marine estuaries. Typically, they have a bottom layer of precipitated iron sulfide, covered by layers of purple sulfur bacteria, oxidized iron, cyanobacteria, and sand.

**microbial pesticide** A microorganism that is used to control a pest, but of minimum toxicity to man.

**microbial products** Materials produced in large batches and in large-scale fermentations, including antibiotics, vitamins, aminoacids enzymes, yeast, vinegar, and alcoholic beverages.

**microbial quality of drinking water** Drinking water must be free of all pathogenic organisms (bacteria, viruses, protozoa, and helminths).

**microbial regrowth** *See* regrowth.

**microbial removal credit** The credit allowed by regulatory agencies for the performance of baseline water treatment processes in reducing microbial contaminants.

**microbial risk assessment** The process of quantitatively estimating responses to contaminating agents in terms of risk of infection. It usually includes the following steps: hazard identification, exposure assessment, dose–response, and risk characterization.

**microbials** Contaminants from microorganisms.

**microbial transport factors** Factors affecting the transport of microorganisms through soils, vadose zone materials, sediments, porous media, etc. include micropore exclusion, microbial filtration, contaminant aging, adhesion, medium characteristics, water flow rate, predation, and physiological state of the cells.

**microbiocide** A substance capable of controlling or destroying living microorganisms.

**microbiofouling** The development of biological growth, e.g., a biofilm, which impairs the performance of membrane filters, other water treatment units, condensers, cooling towers, etc. *See* biofouling for detail.

**microbiological** Pertaining to microorganisms.

**microbiological analysis** An examination of water or wastewater to determine their microbiological quality or characteristics, e.g., the number and density of microorganisms. *See also* microbiological test, microbiological standard, sanitary survey.

**microbiological drinking water standard** *See* microbial quality of drinking water.

**microbiologically induced corrosion (MIC)** Corrosion caused by microbial activity, i.e., corrosion resulting from a reaction between microorganisms and/or their metabolic by-products on the one hand and pipe material on the other, including changes in such water quality parameters as pH and dissolved oxygen. MIC may also be associated with taste and odor problems. Also called microbial corrosion or microbiologically influenced corrosion. *See also* iron bacteria, sulfate-reducing bacteria.

**microbiologically influenced corrosion (MIC)** Same as microbiologically induced corrosion or microbial corrosion.

**microbiological quality** An expression of the absence or degree of pollution of a water or wastewater sample in terms of the concentration and frequency of occurrence of particular species of pathogens, most commonly the indicator bacteria of the coliform group.

**microbiological standards** Guidelines established by an agency to ensure an acceptable microbiological quality for drinking water. *See* treated water standards, untreated water standards.

**microbiological test** An examination of a water sample to detect the presence of fecal coliforms, total coliforms, and excreted protozoa. The two commonly used microbiological tests are the membrane filtration technique and most probable number (MPN) or multiple tube method.

**microbiology** The study of microorganisms and their processes. The part of aquatic microbiology that concerns drinking water and wastewater treatment deals with bacteria, viruses, algae, protozoa, and helminths.

**microbiota** Microorganisms; microscopic plants and animals, e.g., bacteria, viruses, fungi, algae, and other microfauna.

**Microbloc** An apparatus manufactured by Wheelabrator Clean Air Systems, Inc. to control volatile organic compounds in carbon beds.

**Microcat®** A microbial product of Bioscience, Inc. for use in biological wastewater treatment processes.

***Microccocus*** A genus of bacteria that can degrade hydrocarbons, e.g., for the elimination of petroleum wastes from water and soil.

**Microchem™** A tablet chlorinator produced by Mooers Products, Inc.

**microchemical** Of or pertaining to chemical reactions between small quantities of compounds.

**microconstituent** A natural or man-made microscopic compound found in water or wastewater. Of concern in drinking water are some chemical microconstituents (alkylphenols, antibiotics, disinfection by-products, endocrine disruptors, estrogens, ethers, heavy metals, NDMA, personal care products, pharmaceuticals, and polybrominated biphenyl) and some biological microconstituents (pathogenic bacteria, protozoa, and viruses).

**microcosm** A diminutive, representative system analogous to a larger system in composition, development, or configuration. As used in biodegradation treatability studies, microcosms are typically constructed in glass bottles or jars.

**microcosmic salt** A common name of sodium ammonium phosphate ($NaNH_4HPO_4 \cdot 4\ H_2O$).

**microcrystallite** A body made up of microscopic crystals; a stack of graphitic planes in the structure of active carbon.

**microcurie** A measure of radioactivity equal to one millionth or $10^{-6}$ of a curie.

**microcystin** The toxin produced by the cyanobacterium *Microcystis.*

**microcystin-LR** A hepatotoxin from *Microcystis aeruginosa,* commonly found in water sources and responsible for human and animal poisoning.

***Microcystis*** A group of cyanobacteria (blue-green algae), some strains of which are toxic; they sometimes produce floating blooms on warm water when the weather is warm and nutrients are available, and release odorous sulfur compounds when they decay.

***Microcystis aeruginosa*** A species of cyanobacteria (blue-green algae) commonly found in algal blooms and responsible for the production of hepatotoxins and neurotoxins.

**MicroDAF** A dissolved-air flotation unit manufactured by Princeton Clearwater.

**microelectronic water** Water that meets special requirements (e.g., resistivity, particle count, and silica concentration) for use in the production of microelectronic devices. Also called electronic-grade water.

**microelectrophoresis** A technique used in water treatment to determine electrophoretic mobility (EM or zeta potential): an electric field is applied to a flocculent suspension in a tube equipped with electrodes at one end. The EM is measured by correlation with the velocity of particle migration and the voltage gradient.

**microencapsulation** (1) The process of enclosing a substance in microcapsules (tiny capsules used for the slow release of drugs, pesticides, or the like). (2) Isolation of individual, fine particles of a waste by mixing with another material, which then forms an encapsulated product. *See also* encapsulation, macroencapsulation, solidification/stabilization.

**microenvironment** The immediate surroundings of a microorganism.

**microfauna** Microscopic animals or the fauna of a microhabitat.

**MicroFID™** An apparatus manufactured by Photovac, Inc. for the detection of volatile organic compounds.

**microfiltration (MF)** A filtration process that removes suspended and colloidal solids larger than the pore size of the low-pressure membranes used (nominal diameter of 0.1–0.2 micron). These particles include total suspended solids, turbidity, coliforms, protozoan cysts and oocysts, *Cryptosporidium*, *Giardia,* and some, but not all, bacteria and viruses. The driving force of the process is hydraulic pressure difference or vacuum in open vessels. Microfiltration is used in wastewater treatment to condition the liquid for effective disinfection or as pretreatment for reverse osmosis.

**Micro Fine** An ultrafiltration unit manufactured by Memtec Americas Corp.

**Microfloat™** An aspirating aerator manufactured by Aeration Industries, Inc. for surface installation in dissolved-air flotation units.

**microfloc** Destabilized floc particle formed immediately after coagulation, consisting of small clumps of small particles in the size range of 0.001 to 1 micron. Microflocs penetrate the depth of a granular filter bed and contribute to enhance the filter's performance.

**Microfloc®** A line of products of Wheelabrator Engineered Systems, Inc.

**Microfloc®** A preassembled treatment plant manufactured by U.S. Filter Corp.

**microflocculation** The aggregation of solid particles into larger but still microscopic particles, brought about by Brownian movement and contributing to enhancement of the agglomeration and removal of particles. *See* Brownian flocculation for more detail.

**microflora** Microscopic plants or the flora of a microhabitat.

**microflotation** A type of dissolved-air flotation applicable to water treatment. Another type, pressure flotation, is more commonly used. *See also* vacuum flotation.

**microfouling** Fouling of a water treatment unit caused by microbial activity.

**microgram (μg)** A unit of mass equal to one-millionth of a gram or about 0.000000035 ounce.

**microgram per liter (μg/L)** A unit of concentration of trace constituents; one microgram of a substance dissolved in each liter of water. This unit is about the same as parts per billion (ppb) since one liter of water is equal in weight to one billion micrograms.

**micrograph** A reproduction of an image formed by a microscope. *See* photomicrograph and electron micrograph.

**microirrigation** A water management technique that tries to minimize water use with microsprinklers or drip irrigation.

**microlimnology** The study of freshwater microorganisms.

**Micro-Matic®** A rotating microscreen manufactured by U.S. Filter Corp.

**Micromesh Strainer** A microscreen manufactured by Lakeside Equipment Co.

**micrometer (μm)** A unit of length equal to one-millionth of a meter or one-thousandth of a mm. Also called micron.

**micrometer rating** Same as micron rating.

**micromho** A unit used to measure conductivity, equal to one-millionth of a mho.

**micromonospora** A group of moldlike bacteria usually associated with algae and responsible for taste and odor problems in surface waters. They may also be found in drinking water treated by flocculation, sedimentation, and filtration.

**micron** Same as micrometer, or one-millionth of a meter or approximately 0.00004 inch.

**micron filter rating** *See* micron rating.

**Micronizer™** A dissolved-air flotation unit manufactured by Microlift Systems, Inc. and using fine bubbles.

**micron rating** A parameter used to define the sizes of particles that will be retained on a filter medium. *See* filter rating for more detail. Also called micrometer rating or micron filter rating.

**micronucleus test** A test used to determine whether a chemical can damage the chromosomes.

**micronutrient** A beneficial element that is required in very small amounts for the growth of microorganisms. Micronutrient metals belong to the borderline group, e.g., iron (Fe), manganese (Mn), molybdenum (Mo), copper (Cu), cobalt (Co), nickel (Ni), and zinc (Zn). Other micronutrients include boron (B), chloride (Cl), chromium (Cr), lead (Pb), magnesium (Mg), selenium (Se), and vanadium (V). Some are toxic at higher concentrations. Some metal micronutrients are catalysts for enzymes; others may be involved in photosynthesis. *See also* trace element, vitamin, macronutrient.

**microorganism** A microscopic organism, i.e., a living plant or animal that can be seen individually only with the aid of a microscope. Microorganisms include bacteria, protozoans, yeast, fungi, mold, viruses, and algae. Also called microbe.

**microorganism composition** *See* cell composition.

**microorganism growth factors** Organic nutrients and environmental conditions required for a microorganism to grow. *See* growth factor for detail.

**microorganism regrowth** *See* regrowth.

**microphyte** A bacterium, microscopic plant, or small alga.

**Micro-Pi™** A rotary screen manufactured by Andritz-Ruthner, Inc., including pressure feed.

**Micro-Polatrol®** Power devices manufactured by Cathodic Protection Services Co. for corrosion control.

**micropollutant** A pollutant that occurs in water or air in very small concentrations, on the order of micrograms per liter (μg/L) instead of milligrams per liter (mg/L) for ordinary pollutants. Of particular importance are the harmful organics found in drinking water.

**micropore** A pore (of an adsorbing material) having a diameter smaller than 2 nanometers (20 Å); however, pores smaller than 1 nanometer (10 Å) are generally considered inaccessible to most solutes. Micropores are normally associated with the adsorption of small molecules. *See also* mesopore and macropore.

**micropore exclusion** A filtration mechanism indicating that bacteria and other microorganisms may be excluded from the microporous domain of aggregated, macroporous media. *See also* microbial filtration, contaminant aging.

**microporosity constant** *See* Dubinin–Radushkevich equation.

**microporous filter** A cellulose nitrate or fiberglass filter of very small pores used in testing water samples for enteric viruses.

**microporous membrane** A membrane of very small pores used in the treatment of water by microfiltration or microfiltration.

**microporous resin** Ion-exchange resin of low porosity, with little strength or resistance to degradation and swelling.

**MICROQL** A BASIC language computer program for modeling chemical equilibria. *See also* MINEQL and REDEQL.

**microsand** A material consisting of microscopic sand particles (usual diameter range = 100–150 μm and specific gravity = 2.6) that is added to water or wastewater as a weighting agent or ballast to enhance flocculation and coagulation. *See* ballasted flocculation.

**microsand ballasted floc particle** A floc particle attached onto a microsand grain (as ballast) by a polymer layer serving as a glue.

**microsand enhanced coagulation** *See* ballast and ballasted flocculation.

**microscope** An instrument used to visually magnify objects and organisms that are too small to be observed by the naked eye. Modern laboratory microscopes have various lenses: ocular and objective, low- or high-power, and oil immersion. *See also* specialized microscopes: light or optical (brightfield, dark-field, differential interference, phase-contrast), electron beam (transmission electron, scanning electron), and ultraviolet-ray (fluorescent).

**microscope (brightfield)** *See* brightfield microscope.

**microscope (electron)** *See* electron microscope.

**microscope (fluorescence)** *See* fluorescence microscope.

**microscope (light)** *See* light microscope.

**microscope (phase contrast)** *See* phase-contrast microscope.

**microscope (scanning electron)** *See* scanning electron microscope.

**microscope (transmission electron)** *See* transmission electron microscope.

**microscopic** Very small, not visible to the naked eye, visible only when magnified. Microscopic organisms (microorganisms) have dimensions between 0.5 and 100 micrometers.

**microscopic analysis** (1) The examination of water or wastewater to determine the presence of microscopic organisms (algae, bacteria, crustacean, diatoms, protozoa) and other microscopic solids, their density, and the source of contamination. Also called microscopic examination. (2) The examination of microorganisms in activated sludge and similar suspensions.

**microscopic counting** The use of a particle counting chamber to enumerate the individual microorganisms in a sample. *See also* optical imaging.

**microscopic examination** *See* microscopic analysis.

**microscopic filtration model** *See* fundamental filtration model.

**microscopic organism** *See* microorganism.

**microscopic particulate analysis (MPA)** The examination of water for the presence of a number of particles that may indicate the direct influence of surface water on groundwater and to determine the effectiveness of water filtration plants. Particles of concern are isolated on a one-micrometer cartridge filter and include amoebas, crustaceans, *Cryptosporidium* and other coccidia, diatoms and other algae, *Giardia* cysts, nematodes, plant debris, pollen, and rotifers.

**microscopic reversibility** The principle that chemical reactions are reversible and, in equilibrium, products are formed at the same rate as reactants (Droste, 1997).

**microscopy** The use of a microscope to examine water and wastewater or for other purposes.

**microscreen** A variable, low-speed, continuously backwashed, screen used for the removal of fine solids from water, from wastewater, and often from treated effluents. It consists of a motor-driven, rotating drum in stainless steel or polyester with a fine-mesh screen of 20 to 30 microns on its periphery. The liquid flows in the drum and a pressure wash removes the solids retained by the mesh. *See also* microstrainer.

**microscreening** A treatment unit operation that uses a microscreen to remove fine solids, floatable matter, and algae from wastewater under gravity-flow conditions.

**Micro-Sieve** A microscreen manufactured by Passavant Corp.

**microsphere** A microscopic inert particle similar in size and density to microorganisms and used as a surrogate (e.g., for *Cryptosporidium* oocysts) in studying the exposure of contaminants to ultraviolet light.

***Microspora* phylum** *See* Microsporidia.

**Microsporidia** Plural of microsporidium; the nontaxonomic name for the phylum *Microspora,* a large group of spore-forming protozoan parasites (more than 100 genera and 1000 species) that live inside cells of insects, fish, rodents, and other animals. They cause gastrointestinal infections (microsporidiosis), particularly in immunodeficient patients. Among the five species identified as pathogenic to humans, the most significant are *Enterocytozoon bieneusi, Encephalitozoon hellem, Encephalitozoon council, Encephalitozoon intestinalis, Pleistophora* spp, *Nosema corneum,* and *Microsporidium* spp.

**microsporidiosis** An opportunistic gastrointestinal infection by microsporidia infecting mainly AIDS and other immunodeficient patients. In addition to diarrhea, weight loss, and gastric pain, symptoms may include illness of the respiratory tract, urogenital tract, eyes, muscles, and organs.

***Microsporidium*** The genus of unidentified microsporidia.

**microstrainer** A water or wastewater pretreatment unit used for the removal of suspended solids, including algae but not bacteria to any significant degree. It consists of a rotating, variable low-speed drum in stainless steel or polyester with a fine-mesh screen on its periphery (usually 20 to 30 microns, but sometimes as small as 1 micrometer). The liquid flows in the drum and a pressure wash removes the solids retained by the mesh. Common design parameters include drum diameter, head loss (average), head loss (maximum), hydraulic loading, peripheral drum speed, screen mesh, submergence, and washwater flow. *See also* microscreen.

**microstraining** The use of a microstrainer in water or wastewater treatment.

**Micro-T** A turbidimeter manufactured by HF Scientific, Inc., with a remote control.

***Microthrix*** One of two genera of nuisance bacteria that have hydrophobic surfaces and attach to air bubbles to cause foaming problems in activated sludge wastewater treatment units. *See also* foaming, *Nocardia.*

***Microthrix parvicella*** A species commonly found among filamentous organisms responsible for foaming in wastewater treatment.

***Microthrix parvicella* type 0041** A common filamentous organism responsible for foaming and bulking problems in activated sludge, particularly under conditions of low food-to-microorganism ratio.

***Microthrix parvicella* type 0092** A common filamentous organism responsible for foaming and bulking problems in activated sludge, particularly under conditions of low food-to-microorganism ratio.

***Microthrix parvicella* type 0675** A common filamentous organism responsible for foaming and bulking problems in activated sludge, particularly under conditions of low food-to-microorganism ratio.

***Microthrix parvicella* type 1701** A common filamentous organism responsible for foaming and bulking problems in activated sludge, particularly under conditions of low dissolved oxygen concentration.

***Microthrix parvicella* type 1851** A common filamentous organism responsible for foaming and bulking problems in activated sludge, particularly under conditions of low food-to-microorganism ratio.

**Microtox®** assay A rapid bacterial assay for acute toxicity developed by Microbics Corp. The test causes the luminescent marine bacteria *Photobacterium phosphoreum,* very sensitive to toxicants, to produce a light that is compared to a standard.

**microtransport** One of the three steps of the adsorption process; the diffusion of organic matter or other adsorbate through the macropores of the solid adsorbent (e.g., activated carbon) to the adsorption sites in the micropores of the adsorbent. *See also* macrotransport, sorption.

**microtunneling** A trenchless process used in the installation of deep, new water mains, but also in the rerouting of existing mains. It combines a remotely controlled boring machine with a jacking tool. *See* trenchless technology, open cut.

**microturbine** A device that turns digester gas into electricity to power pumps, fans, and blowers while producing hot water for digester operation.

**midge fly** Any of numerous minute insects resembling mosquitoes, sometimes found in water supply systems. Their larvae feed on some aquatic microorganisms and on decomposing vegetation. *See also* bloodworm.

**Midi-Rotor** A rotary brush aerator manufactured by Kruger, Inc.

**midnight dumping** Deliberate and illegal disposal of sludge or other waste materials down storm drains or at an unauthorized, nonpermitted location.

**MIEX** Acronym of magnetic ion exchange.

**MIEX®** Membrane ion exchange system designed or manufactured by WesTech Engineering, Inc. of Salt Lake City, UT for the removal of dissolved organics.

**MIEX®** Process A continuous process that uses a magnetized ion-exchange resin to remove negatively charged organic acids, which constitute the majority of dissolved organic carbon (DOC) found in natural waters and a precursor to the formation of trihalomethanes. It is also effective in removing malodorous compounds such as dimethyltrisulfide, which has an odor threshold of 10 ng/L.

**MIEX®** resin A resin developed in Australia for the removal of dissolved organic carbon to minimize the formation of disinfection by-products in drinking water; it is a microporous, microsize, strong-base resin that removes negatively charged organic ions in a stirred reactor. Resin separation from the water occurs in conventional upflow settlers.

**MightyPure™** A water purification apparatus manufactured by Atlantic Ultraviolet Corp.; uses ultraviolet light.

**migration** The movement of contaminants, water, oil, gas, etc. through porous and permeable rock.

**MIL** A series of agitators manufactured by Denver Equipment Co.

**mil** A unit of length equal to one thousandth of an inch (0.001 in.). The diameter of wires and tubing is measured in mils, as is the thickness of plastic sheeting. The rate of metallic corrosion is measured in mils per year (mpy).

**mildly brackish water** A mixture of saline or brackish water and freshwater, such as some inland waters with a mild mineral content as expressed by a concentration of dissolved solids between 1,000 and 2,000 mg/L. Also called mildly saline water. *See* salinity.

**mildly saline water** *See* mildly brackish water.

**milk dilution bottle** A glass bottle used for diluting bacteriological samples. Also called dilution bottle.

**milk of lime** Another name for slaked lime or calcium hydroxide [$Ca(OH)_2$]. The product is used in water or wastewater treatment as a wet suspension or slurry in a water-to-lime ratio between 2:1 and 6:1 by weight. *See* lime for detail.

**milk of magnesia** A milky, white suspension of magnesium hydroxide [$Mg(OH)_2$], used as an antacid or laxative. *See also* magnesia.

**milk sugar** *See* lactose.

**milled refuse** Refuse that has been pulverized or ground into small particles that hardly have any odor and can be reused in roadbeds or disposed of in a sanitary landfill. *See also* controlled tipping.

**Miller's fumigrain** Acrylonitrile.

**millicurie** A measure of radioactivity equal to one thousandth or $10^{-3}$ of a curie.

**milliequivalent (me** or **meq)** A unit of weight, one-thousandth the equivalent weight of a substance. Also called milliequivalent weight.

**milliequivalent per liter (meq/L)** A unit of concentration of a substance based on its equivalent weight. It is equal to the concentration of the substance in mg/L divided by its equivalent weight.

**milliequivalent weight** Same as milliequivalent.

**milligram (mg)** A unit of mass, equal to one-thousandth of a gram.

**milligram per liter (mg/L)** A common unit of measurement of concentrations of dissolved or suspended materials. Abbreviation: mg/l or mg/L. *See also* parts per million (ppm).

**milliliter (ml** or **mL)** A unit of volume equal to one-thousandth of a liter or one cubic centimeter.

**millimicron (mμ)** A unit of length, equal to one-thousandth of a micron. Same as nanometer.

**millimole (mmol, m*M*)** A unit of weight, equal to one-thousandth of a mole.

**milling** The process of converting solid waste into milled refuse.

**million gallons per acre per day (mgad)** A unit used to measure hydraulic loading on trickling filters and other treatment units.

**million gallons per day** A unit used to measure water flow or plant capacity. *See* the abbreviation mgd.

**millirem** *See* mrem.

**Milliscreen™** A rotary fine screening device, internally fed, manufactured by Andritz-Ruthner, Inc. and Contra-Sheer Engineering, Ltd.

**millisiemen per meter (mS/m)** The SI unit of electrical conductivity: 1 mS/m = 10 μmho/cm.

**mill scale** A coating of oxide that forms on heated steel.

**Mills, Hiram F.** Member of the Massachusetts Board of Health who made the Lawrence Experiment Station available for research on water and wastewater treatment in 1886, one of the significant achievements of the Great Sanitary Awakening.

**Mills–Reincke phenomenon** The correlation between the improvement of the quality of drinking water in a community and the decrease in mortality associated with nonwaterborne diseases; first noted independently by two observers in 1893 and 1894.

**Milorganite™** The trade name of heat-dried activated sludge from the Jones Island wastewater treatment plant in Milwaukee, WI, sold as a fertilizer. The facility uses a direct–indirect rotary dryer.

**mils** Plural of mil.

**mils per year (mpy)** The expression of the loss of metal due to corrosion. One mpy = 0.001 inch per year.

**Milton®** A proprietary disinfectant that contains about 1% of available chlorine by weight; it can be used directly as a stock solution for water disinfection. Such a solution is unstable under warm conditions.

**MIMS** Acronym of membrane-introduction mass spectrometry.

**Minamata disease** A disease of the central nervous system, first diagnosed in 1956 and resulting from mercury poisoning; named after the bay and town of Minamata, Japan. Symptoms include numbness in fingers and lips, and speech and hearing difficulties. Inorganic mercury used in an industry was discharged with wastewater in a small bay and biologically converted to the highly toxic dimethylmercury, which accumulated in fish and other seafood consumed in large amounts by the local residents. That caused the death of 43 persons (officially) and the incapacitation of many more, including some after the discharge had stopped.

**mine drainage** Drainage from mining operations, rendered acidic by the presence of sulfur-bearing minerals. *See* acid mine drainage for detail.

**MINEQL** Acronym of Mineral Equilibrium, a series of computer programs used for modeling chemical equilibria. *See also* MICROQL and REDEQL.

**mineral** A natural substance, usually inorganic, of definite chemical composition and crystal structure. The dissolved substances most commonly encountered in natural waters result from the combination of the cations sodium (Na), potassium (K), calcium (Ca), magnesium (Mg), iron (Fe), and manganese (Mn) with the anions bicarbonate ($HCO_3$), carbonate ($CO_3$), sulfate ($SO_4$), and chloride (Cl). *See also* rock, soil.

**mineral acid** An inorganic acid such as the strong hydrochloric (HCl), nitric ($HNO_3$), and sulfuric ($H_2SO_4$) acids or the weak carbonic acid ($H_2CO_3$).

**mineral acidity** Acidity cause by the presence of mineral acids. *See* H-acidity for detail.

**mineral cycles** The cycles of such elements as nitrogen and sulfur that are taken up in the form of mineral salts and circulated in a series of oxidations and reductions, e.g., nitrogen can be found in various states from –3 in ammonium ($NH_4^+$) to +5 in nitrate ($NO_3^-$).

**mineral-free water** Deionized or distilled water.

**mineral impurities** *See* inorganic contaminants.

**mineralization** (1) The conversion (or stabilization) of a compound from an organic form to an inorganic or mineral form as a result of microbial decomposition or combustion. Microorganisms convert various elements in organic matter—carbon (C), hydrogen (H), nitrogen (N), phosphorus (P), and sulfur (S)—to carbon dioxide ($CO_2$), water ($H_2O$), nitrate ($NO_3^-$), phosphate ($PO_4^=$), and sulfate ($SO_4^=$), respectively. Very little mineralization occurs in water treatment processes. However, the aerobic or anaerobic decay of organic wastes releases inorganic chemicals that become available for new cell growth. *See also* biodegradation. (2) The mineral content of water, measured as total dissolved solids (TDS). Drinking water with excessive or insufficient mineralization may be associated with such problems as cardiovascular disease, dental caries, dishpan hands, Genu Valgum, goiter, laxative effects, tooth mottling, and winter chapping. In drinking water, high concentrations of chlorides and sulfates may impart taste and odor problems and cause laxative effects. High TDS also affect cooling, boiler feed, and industrial processes.

**mineralization factor** The fraction of organic nitrogen that is converted to a form available to plants during sludge application; it is generally between 0.2 and 0.4. *See* plant available nitrogen.

**mineral lead** A compound material containing 50–65% sulfur used to seal bell-and-spigot pipe joints.

**mineral naphtalene** Another name of benzene.

**mineral pollution** Pollution of surface or ground water that comes in contact with mineral deposits or veins.

**mineral salt** The product of the reaction between a mineral acid and a base. Also a mineral dissolved from rock by water; e.g., salts of sodium, calcium, magnesium, etc. picked up by rainwater as it falls on the earth and seeps through soil. An excess of mineral salt in water may be harmful to human health.

**mineral spring** A spring containing water with a high concentration of mineral salts.

**mineral wastes** Slag heaps and mill tailings from mines and metal smelting plants; they eventually reach water sources with stormwater runoff and snowmelt.

**mineral water** Carbonated water, soda water, or water that comes from a mineral spring or from a protected groundwater source, containing at least 250 mg/L of dissolved solids.

**miners salt** Mined rock salt.

**mine wastes** Wastes originating from mining operations; mineral wastes.

**mine water** Same as mine drainage.

**MiniBUS™** A membrane cartridge manufactured by Cuno Separation Systems Division for water treatment by ultrafiltration.

**MiniChamp** A chemical induction unit manufactured by Gardiner Equipment Co., Inc.

**Minigas®** A device manufactured by Neotronics for gas detection.

**minim** A common lead compound, essentially lead oxide, $Pb_3O_4$.

**Mini-Magna®** A rotor aerator manufactured by Lakeside Equipment Corp.

**minimal medium** A culture used in coliform tests required under the Safe Drinking Water Act. It consists of ortho-nitrophenol-$\beta$-D-galactopyranoside and 4-methylumbelliferyl-$\beta$-D-glucuronide.

**Minimax** A pressure filter manufactured by Larox, Inc. for dewatering.

**Mini-Maxi** A dissolved-air flotation unit manufactured by Tenco Hydro, Inc.

**Mini-Miser™** An apparatus manufactured by Recra Environmental for sludge dewatering using multiple feeds.

**minimization** A comprehensive program to minimize or eliminate wastes, usually applied to wastes at their point of origin. *See* waste minimization.

**Mini Monster®** A low-flow wastewater grinder manufactured by JWC Environmental.

**minimum air-to-water ratio** In the design of a packed tower as well as diffused or bubble aeration systems, the minimum air-to-water ratio $[(V/Q)_{\min}]$ is the lowest ratio of volumetric airflow rate (V) to the water flow rate ($Q$) that meets a given treatment objective:

$$(V/Q)_{\min} = (C_0 - C_e)/(HC_0) \qquad \text{(M-44)}$$

where $C_0$ = influent contaminant concentration; $C_e$ = treatment objective, i.e., the bulk-water phase contaminant concentration at the bottom of the tower; $H$ = Henry's law constant. *See also* cascade air stripping, cocurrent packed tower, cascade packed tower, end effect, crossflow packed tower, minimum air-to-water ratio, packed tower design equation, wall effect.

**minimum bactericidal concentration (MBC)** The minimum amount of antimicrobial agent required to yield a 3-log (99.9%) reduction in viable colony-forming units of a bacterial or fungal culture.

**minimum basic freshwater need** *See* basic water requirement.

**minimum billing** *See* lifeline rate.

**minimum-day (hour, month) demand** The least volume of water required by consumers per day (hour, month) in a year.

**minimum detectable threshold odor concentration (MDTOC)** The minimum odor level that a panel of human subjects can detect. *See* dilution to threshold.

**minimum detection limit** The correct expression is method detection limit.

**minimum flow** (1) The smallest volume of fluid through a stream, unit, or structure within a given period. (2) Streamflow during the driest period of the year; also called low flow. *See also* design flow. (3) The specific amount of water reserved to support aquatic life, to reduce pollution, or for recreation.

**minimum fluidization condition** The relationship that exists between the Reynolds number (Re) and the Galileo number ($G_a$) at minimum fluidization during filter backwashing. It can be used to determine the minimum fluidization velocity ($V_f$, in m/min):

$$R_e = (1135.69 + 0.0408\, G_a)^{1/2} - 33.7 \qquad \text{(M-45)}$$

$$G_a = d^3\rho(\rho' - \rho)g/\mu^2 \qquad \text{(M-46)}$$

$$R_e = \rho d V_f/\mu \qquad \text{(M-47)}$$

where $d$ = grain diameter of sphere of equal volume, $\rho$ = mass density of the fluid; $\rho'$ = density of medium particles, $g$ = gravitational acceleration, and $\mu$ = absolute viscosity of the fluid.

**minimum fluidization velocity** The superficial velocity of water ($V_f$, in m/min) that just begins to fluidize a granular bed during backwash but is insufficient to cleanse the filter of the suspended matter removed. It is used in determining minimum backwash rate. On a curve of head loss versus superficial velocity, it is the intersection of the fixed-bed and fluidized-bed head loss lines. It can also be calculated as indicated under minimum fluidization condition. *See also* backwash velocity, incipient fluidization.

**minimum fluidizing velocity** Same as minimum fluidization velocity.

**minimum-hour demand** *See* minimum-day demand.

**minimum infectious dose** The minimum dose or number of pathogens that can cause an infection. *See* infectious dose 50, infectivity.

**minimum inhibitory concentration (MIC)** The minimum concentration of antimicrobial agent required to prevent visually discernible bacterial or fungal growth.

**minimum-month demand** *See* minimum-day demand.

**minimum reporting level (MRL)** The lowest concentration of a constituent that a laboratory can report with some level of confidence. One author defines the MRL as three or more times the method detection limit (MDL), which corresponds to the lowest level at which the method can measure an analyte with a ± 50% accuracy.

**minimum reproduction time (MRT)** The minimum period of time required for the reproduction of microorganisms in waste treatment, e.g., 3–5 days at 35°C for methane formers. Below this minimum, the microorganisms are washed out and the treatment process fails. The design of biological treatment systems usually includes a safety factor of 3–20 times the MRT. *See also* solids retention time.

**minimum sludge age** Same as minimum solids retention time.

**minimum solids retention time ($SRT_{min}$)** In suspended growth treatment processes, the minimum solids retention time is the residence time at which the microorganisms are washed out of the system faster than they can reproduce. *See* critical SRT for detail.

**minimum streamflow** *See* minimum flow.

**minimum velocity** The velocity required to prevent solids deposition in channels and pipelines of a treatment plant; e.g., 2.0 fps for raw wastewater. *See also* flushing velocity, scouring velocity.

**mining** *See* mining of an aquifer.

**mining of an aquifer** Withdrawal of groundwater over a period of time in excess of the rate of aquifer recharge. *See also* fossil groundwater.

**Mini Osec** An apparatus manufactured by Wallace & Tiernan to provide electrolytic chlorination.

**mini plugging factor index (MPFI)** An empirical measure of the fouling tendency or plugging characteristics of water or wastewater based on the time it takes 500 mL of filtrate to pass through a 0.45 micron pore membrane filter at a constant pressure of 30 psi at the beginning and at the end of the test period. The MPFI is used to assess the treatability of a given water or wastewater by nanofiltration and reverse osmosis. Pretreatment is recommended if the index exceeds certain limits. Two other indexes are the silt density index and the modified fouling index.

**Minipure™** A water treatment apparatus manufactured by Atlantic Ultraviolet Corp. that uses ultraviolet light.

**Mini-Ring** Plastic media produced by Mass Transfer, Inc. for use in biological filters.

**Mini-San** An apparatus manufactured by Eltech International Corp. to feed disinfection tablets.

**Miniseries™** A package water treatment plant manufactured by Matrix Desalination, Inc. using desalination technology.

**minor intestinal flukes** *See Gastrodiscoides hominis, Heterophyes heteropyes,* and *Metagonimus yokogawai.*

**minor loss coefficient** or **k-factor** A coefficient used in the computation of minor or local head losses. It serves to express these minor losses in function of the kinetic energy or velocity head, or as an equivalent length of pipe. In streams, the minor loss coefficient varies from 0.1 for a gradual contraction to 0.6 for an abrupt contraction, and from 0.3 to 0.8 for similar expansions. In sewer system modeling, the minor loss coefficient simulates the friction losses contributed by all the fixtures and valves in a pumping station piping. *See* equivalent pipes and k-factor. For example, in a sewer modeling project (Adrien, 2004), the consultant designed a k-factor database to compute the overall minor loss coefficient ($k_{eq}$) for each of the 350 pump stations. The relevant data for computing the k-factors was collected and stored in a pump dimension database. Individual factors used to calculate the overall factor ranged from 0.05 for a cone valve or an 11( elbow, to 0.90 for a side-outlet elbow, and to 2.50 for a fully open check valve

**minor losses** or **local head losses** In addition to friction losses, minor losses or local head losses are energy or head losses due to flow contraction or expansion or to obstacles such as weirs and bridges. Minor losses are caused in pipes by abrupt changes in size, bends, elbows, junctions, valves and fittings. Local or minor losses may be neglected for long pipes, but are significant in a section of 100 ft or shorter. Such losses are usually expressed in function of the kinetic energy or velocity head ($V^2/2\,g$) multiplied by an appropriate minor loss coefficient or k-factor. An alternative method is the equivalent conduit length technique; a length of pipe estimated to cause the same pressure drop as the fitting or geometric change is added to the actual pipe length. It has been found that the equivalent length is almost constant while the k-factor decreases with the size of the fittings. *See* k-factor and equivalent pipes.

**minor nutrient** Any of the less important inorganic nutrients needed for microbial cell synthesis, e.g., cobalt (Co), copper (Cu), manganese (Mn), molybdenum (Mo), nickel (Ni), and zinc (Zn). *See also* macronutrients, micronutrients, trace metals.

**minors** Publicly owned treatment works (POTW) with flows less than one million gallons per day (1.0 mgd). *See also* majors.

**MINTEQ** A series of FORTRAN language computer programs developed by the USEPA and combining features of MINEQL and WATEQ.

**miracidia** Plural of miracidium.

**miracidium** A tiny swimming organism; the embryo or larval stage of a schistosome (or other trematodes) in a snail. Miracidia result from the hatching of schistosome eggs excreted in the feces or urine of infected individuals and in turn produce a great number of cercariae. Adequate excreta disposal can thus contribute to the control of schistosomiasis. Long detention times in wastewater treatment units (e.g., anaerobic waste stabilization ponds) inhibit the production or survival of miracidia. *See also* cercaria, egg, metacercaria.

**miscible** Pertaining to two substances that can be mixed together or dissolved into each other to form a homogeneous substance.

**miscible liquids** Two or more liquids that can be mixed and will remain mixed under normal conditions.

**mispickel (FeSAs)** Another name of arsenopyrite, the common white-to-gray mineral iron arsenic sulfide, an arsenic ore.

**missing value** A missing data point in a statistical array.

**mist** (1) Liquid particles measuring 40 to 500 microns, formed by the condensation of vapor and suspended in air by spraying or splashing; such fine droplets are not easily separated by gravity. By comparison, fog particles are smaller than 40 microns. (2) A suspension of water droplets in the atmosphere, which reduces visibility to 1–2 km.

**mist eliminator** A device that is used to eliminate or reduce entrainment, i.e., to remove entrained droplets of water from a vapor stream produced during evaporation (e.g., during desalting). Also called demister, demister entrainment device, entrainment separator, or, simply, separator. *See also* spacer.

**Mist-Master®** A mesh pad manufactured by ACS Industries, Inc. for use as a mist eliminator.

**mist scrubber system** A modified, wet packed-bed device using small droplets of scrubbing liquid to oxidize odorous gases in wastewater treatment plants. It consists mainly of a contact chamber, chemical storage tank, air compressor, and chemical distribution system.

**mite** A vector of such diseases as typhus affecting mostly animals; a grazing organism found in large numbers in trickling-filter films.

**mitigation** (1) Measures taken to avoid, minimize, or reduce and mitigate for adverse impacts on the environment; limitation of negative consequences. Also, measures taken to reduce the probability of a disaster. (2) *See* flood mitigation (2).

**mitigation banking** The process of creating, restoring, or enhancing wetlands in compensation for future wetland losses due to development.

**mitochondria** Plural of mitochondrion.

**mitochondrion** An organelle in the cytoplasm of aerobic cells that contains enzymes for the metabolism of substrate to produce adenosine triphosphate ($C_{10}H_{16}N_5O_{13}P_3$), an important energy element.

**mitogen** A substance or agent that stimulates normal cell division. Some mitogens can promote cancer.

**mitosis** The usual process of nucleus division in eukaryotes. *See also* meiosis.

**Mixaerator** A static mixing aerator manufactured by Raph B. Carter Co.

**Mixco** A batch mixer manufactured by Lightin.

**mixed bed** A filter or ion exchange bed that contains a mixture of two or more media

**mixed-bed demineralizer** An ion-exchange unit using a mixture of resins of strong bases and strong acids to demineralize water.

**mixed-bed system** A single ion-exchange column that provides nearly complete deionization through the use of two different types of resins, e.g., a mixture of strong-acid cation and strong-base anion resins. *See also* cocurrent fixed bed, countercurrent method.

**mixed culture** A culture of more than one organism in or on the same medium. *See also* pure culture.

**mixed DBP** A disinfection by-product that contains both chlorine and bromine atoms; e.g., dibromochloromethane and bromodichloromethane.

**mixed flow** An ideal flow condition under which the influent to a reactor instantaneously blends with the contents of the tank and the effluent has the same uniform composition as the contents. *See* *also* complete mixing, plug flow, short-circuiting. Also called completely mixed flow.

**mixed-flow impeller** An impeller that has characteristics of both radial- and axial-flow impellers; it is used to pump wastewater or stormwater. *See also* centrifugal mixed-flow pump.

**mixed-flow pump** A pump that combines features of both axial and radial-flow pumps; centrifugal force and the lift of the vanes provide impulse, while flow is admitted axially but discharged radially and axially. It is used for pumping raw wastewater and stormwater, and in sludge recirculation as well as trickling filter recirculation. *See* centrifugal mixed-flow pump, centrifugal nonclog pump.

**mixed liquor** A mixture of raw or settled sludge, living and dead microorganisms, inert suspended solids, colloidal matter, and wastewater containing organic matter undergoing activated sludge treatment in an aeration tank.

**mixed liquor filamentous organisms** An operational problem sometimes encountered in suspended-growth wastewater treatment plants. Recommended remedial actions include increasing sludge wasting and chlorinating the return sludge.

**mixed liquor suspended solids (MLSS)** The suspended solids contained in the mixed liquor, including biomass, nonbiodegradable volatile suspended solids, and inert total suspended solids; sometimes used to represent the concentration of total suspended solids in the aeration tank of an activated sludge plant, including both volatile and nonvolatile solids.

**mixed liquor volatile suspended solids (MLVSS)** The volatile fraction of the mixed liquor suspended solids, often used to represent the concentration of active microorganisms available for decomposition of the organic matter.

**mixed media** A combination of two or more media of different characteristics in a single bed for the chemical or physical treatment of water and wastewater. For example (a) a mixed bed of calcium carbonate ($CaCO_3$) and magnesium oxide (MgO) can be used for pH adjustment, and (b) a mixed-media filter may contain anthracite, silica sand, and garnet or ilmenite of respective specific gravities 1.5, 2.6, and 4.2.

**mixed-media filter** A filter that contains a mixture of two or more media of different characteristics (size, specific gravity, composition). A mixed-media filter differs from a dual- or multimedia filter in that the materials of the former are intermixed while those of the latter are in stratified layers. Mixed media are used to avoid the undesirable gradation of a single medium that occurs after backwashing. *See also* triple-media filter.

**mixed-order reaction** A reaction whose order changes depending on the relative magnitude of the concentrations of the reactants; also called saturation-type reaction. For example:

$$r = kC/(K + C) \qquad \text{(M-48)}$$

where $r$ = reaction rate, $k$ = reaction rate constant, $C$ = concentration, and $K$ = a coefficient. When $C$ is large compared to $K$, the rate r is approximately equal to the constant $k$ and corresponds to a zero-order reaction. When $C$ is low, the rate is approximately $r = kC/K$, corresponding to a first-order reaction.

**mixed oxidant** A mixture of chlorine and oxygen-based compounds produced during an electrochemical process: mainly hypochlorous acid (HOCl), but also chlorine dioxide ($ClO_2$), hydrogen peroxide ($H_2O_2$), hydroxyl ion (OH), and ozone ($O_3$). *See* brine electrolysis.

**mixed-oxidant disinfection** The use of mixed oxidants to disinfect drinking water, particularly in small, rural areas.

**mixed-oxidant generator** A device consisting mainly of an electrochemical cell for producing hypochlorous acid and other short-lived oxidants. *See* brine electrolysis.

**mixed waste** A combination of hazardous and radioactive wastes, jointly regulated by the Nuclear Regulatory Commission and the Environmental Protection Agency. Radioactive waste includes source, special nuclear, or by-product material.

**mixer head loss** The head loss $H$ (m) that is dissipated as water or wastewater passes through a mixer is usually estimated as a function of the approach velocity $V$ (m/s):

$$H = kV^2/2\,g = KV^2 \qquad \text{(M-49)}$$

where $k$ = a dimensionless empirical coefficient characteristic of the mixing device, $g$ = gravitational acceleration = 9.81 m/s$^2$, and $K$ = overall coefficient for the mixing device = $k/2$ g.

**mixers** Devices mounted vertically or horizontally in flocculation basins of older water treatment plants; they rotate slowly to promote agglomeration of particles. Newer plants incorporate flash mixing in a single unit. *See* mixing and flocculation devices, hydraulic mixers, pneumatic mixers, velocity gradient, rapid mixing, contact clarifier.

**mixing** A brief operation in the coagulation/flocculation process of water or wastewater treatment—blending chemicals with the liquid—usually followed by stirring and agitation. Mixing is slow in flocculation, but rapid in coagulation. The key design parameters of a rapid mixing device are related as follows:

$$GT^*C^{1.46} = 5{,}900{,}000 \qquad \text{(M-50)}$$

where $G$ is the velocity gradient (/sec), $T^*$ is the optimum detention time (sec), and $C$ is the concentration of chemical (mg/L). Mixing is also used for homogenizing and maintaining solids in suspension. *See also* velocity gradient, flash mixing.

**mixing and flocculation devices** *See* high-speed induction mixer, hydraulic mixing devices, in-line mixer, mechanical aerator, paddle mixer, pneumatic mixing, pressurized water jet, propeller flocculator, propeller mixer, static in-line mixer, static mixer, turbine flocculator, turbine mixer.

**mixing basin/chamber/channel/tank** A basin/chamber/channel/tank in which liquids and or chemicals are mixed under the action of mechanical or hydraulic energy.

**mixing chamber** *See* mixing basin.

**mixing channel** *See* mixing basin.

**mixing energy** A parameter used in the design and evaluation of coagulation and flocculation units. It is characterized by the velocity gradient ($G$), usually expressed in the inverse of time, e.g., per second or $sec^{-1}$.

**mixing index** A parameter that indicates the degree of mixing and provides an idea about flow conditions in a basin, as determined from a tracer test. It is the ratio of the time of the first appearance of the tracer to the detention time. A common index is the Morrill dispersion index. *See also* time–concentration curve, average retention time index, cumulative residence-time distribution curve, dispersion number, dispersion symmetry index, exit age curve, mean retention time index, modal retention time index, residence-time distribution curve, short-circuiting index, volumetric efficiency.

**mixing length** The distance, $L_m$ (in ft) from the point of waste discharge into a stream to the point of complete mixing. Assuming lateral and vertical homogeneity and plug flow (no longitudinal mixing), the mixing length can be estimated as follows:

$$L_m = 2.6\ UB^2/H \quad \text{for a side-bank discharge} \quad \text{(M-51)}$$

$$L_m = 1.3\ UB^2/H \quad \text{for a midstream discharge} \quad \text{(M-52)}$$

**mixing loading parameter** The hydraulic retention time in a mixing basin or the ratio of the volume of the basin to the design flowrate through it. *See also* mixing opportunity parameter.

**mixing opportunity parameter** The unitless product of the velocity gradient ($G$, /sec) and the hydraulic retention time ($t_d$, sec) of a mixing basin, calculated as the ratio of the power-induced rate of flow to the hydraulic-induced rate of flow ($Q$, cfs):

$$G \cdot t_d = (PV/\mu)^{0.5}/Q \qquad \text{(M-53)}$$

where $P$ = power requirement, ft-lb f/sec; $V$ = mixing chamber volume, $ft^3$; and $\mu$ = absolute viscosity of the fluid, $lb.sec/ft^2$.

**mixing power requirement** The power input ($W$) required for operating a mixer is a function of impeller and flow characteristics:

$$P = N_p \rho n^3 D^5 \qquad \text{(M-54)}$$

where $N_p$ = the dimensionless power number of the impeller; $\rho$ = fluid density, $kg/m^3$; $n$ = revolutions per second, r/s; and $D$ = impeller diameter, m. *See also* impeller mixer power requirement

**mixing ratio** The ratio of the mass of water vapor (or any other atmospheric constituent) to the mass of dry air to which it is associated. Water vapor mixing ratio ($R_m$, dimensionless) may be expressed in terms of the pressure ($P$) and vapor pressure ($P_v$):

$$R_m = 0.622\ P_v/(P - P_v) \qquad \text{(M-55)}$$

For practical purposes, the mixing ratio is often approximated by specific humidity ($H_s$). *See also* absolute humidity, dewpoint, relative humidity.

**mixing tank** *See* mixing basin.

**mixing time (t)** The period of time it takes for a tank to be relatively well mixed. For a cylindrical tank operated in the fill-and-draw mode, mixing time ($t$, seconds) is a function of the volume at the start of fill (v, cubic feet), the inflow rate ($Q$, cfs), and the inflow velocity ($V$, fps):

$$t = 10.2\ v^{2/3}/(VQ)^{1/2} \qquad \text{(M-56)}$$

*See also* critical temperature difference.

**mixing zone** The part of a receiving water where mixing occurs with an effluent from a point source; constituent concentrations vary between those of the effluent and fully mixed values.

**mixolimnion** The upper layer of a meromictic lake.

**mixotroph** An organism that can assimilate organic compounds as carbon sources, using inorganic compounds as electron donors. *See also* autotroph, heterotroph.

**mixture** The combination of two or more elements and/or compounds without a chemical reaction.

**mL or ml** Abbreviation of milliliter.

**MLC** Acronym of multiloop controller.

**MLE process** Abbreviation of Modified Ludzack–Ettinger process.

**MLSS** Acronym of mixed liquor suspended solids.

**MLVSS** Acronym of mixed liquor volatile suspended solids.

**m*M*** Abbreviation of millimole.

**mm** Abbreviation of millimeter.

**MMO** Acronym of minimal medium ortho-nitrophenyl-β-D-galactopy-ransoside.

**mμ** Abbreviation of millimicron.

**Mn** Chemical symbol of manganese.

**$MnC_3$** Chemical symbol of rhodochrosite, a common manganese mineral.

**$MnCO_3(s)$** Chemical formula of manganese carbonate.

**$MnO_2$** Chemical formula of pyrolusite or manganese oxide.

**$MnO_4^-$** Chemical formula of the permanganate ion.

**MnO(OH)** Chemical formula of manganite.

**$MnSO_4 \cdot 4H_2O$** Chemical formula of manganous sulfate.

**Mo** Chemical symbol of molybdenum.

**MobileFlow** A trailer-mounted apparatus manufactured by Ecolochem, Inc. for water treatment.

**mobile phase** The first phase of gas or liquid chromatography in which the constituents of interest are transported for separation in the stationary phase.

**mobile-plant processing** Treatment and processing of wastes in a plant operated on a trailer and easily moved from site to site; term often applied to the solidification, stabilization, and encapsulation of hazardous wastes. *See also* in situ treatment, ex situ treatment, in-plant processing.

**MobileRO** A reverse osmosis unit manufactured by Ecolochem, Inc. and mounted on a trailer.

**Mobius** A fine-mesh belt screen manufactured by Pro-Ent., Inc.

**modal retention time (MoRT) index** One of the terms used to characterize tracer response curves and describe the hydraulic performance of reactors, it is the ratio of the time of peak tracer concentration ($T_p$) to the theoretical hydraulic residence time ($\tau$):

$$\text{MoRT} = T_p/\tau \qquad \text{(M-57)}$$

*See also* residence-time distribution (RTD) curve, time–concentration curve. The MoRT index measures the uniformity of flow in the reactor. MoRT = 1.0 for an ideal plug-flow reactor, MoRT = 0 for a complete-mix reactor. For a typical sedimentation basin, $0 < \text{MoRT} < 0.8$.

**mode** In an array of values, the magnitude that has the highest frequency.

**model** A scaled reproduction of a system, a mental conceptualization, an empirical relationship, or a series of mathematical and statistical equations representing a system. In any case, it is an imperfect representation, but a valuable tool to study or predict a variety of conditions and obtain answers that would be impractical by measuring or observing the actual system. A "mathematical" or "mechanistic" model is usually based on biological or physical mechanisms, and has model parameters that have real-world interpretation. In contrast, "statistical" or "empirical" models are curve-fitting to data, where the mathematical function used is selected for its numerical properties. Extrapolation from mechanistic models (e.g., pharmacokinetic equations) usually carries higher confidence than extrapolation using empirical models (e.g., logit). The advent of high-speed computers (hardware) and development of advanced programs (software) has enhanced all these model properties. Computer models are real electronic laboratories to conduct experiments, evaluate scenarios, and plan and design projects.

**modeling** The quantitative or mathematical simulation of a system (also phenomenon, event, or the like) by a model designed to adequately represent the actual system by replicating its properties, laws, and behavior. The simulation experiment allows the prediction of the behavior of the actual system under various conditions, which cannot be easily studied on the actual system. *See also* simulation. Also spelled modelling.

**modelling** Same as modeling.

**moderately hard water** Water with a hardness between 50/75 and 150 mg/L as calcium carbonate ($CaCO_3$). *See also* hard water.

**moderately saline water** Inland water containing 2000–10,000 mg/L of dissolved solids. Also called brackish water. *See also* salinity.

**MODFLOW** Abbreviation of modular three-dimensional finite-difference groundwater flow model. A program developed by the U.S. Geological Survey to evaluate current steady-state groundwater conditions. *See also* MODPATH.

**modified aeration** A variation of the activated sludge process that uses a short aeration period of 1.5–3.0 hours, a higher food-to-microorganism ratio, and a low mixed liquor suspended solids concentration of 200–500 mg/L, which result in poor sludge settling characteristics and high effluent solids concentration.

**modified Ames test** The Ames DNA test, typically conducted on rats, with the addition of an S-9 liver extract from various animals to simulate mammalian enzyme effects.

**modified Bardenpho™ process** A wastewater treatment process designed to accomplish carbonaceous BOD removal as well as enhanced phosphorus uptake and nitrification–denitrification. As shown in Figure M-07, it includes five stages in series: anaerobic, anoxic-1, aerobic-1, anoxic-2, aerobic-2, and a final clarifier, with mixed liquor recycling from aerobic-1 to anoxic-1 and from the clarifier underflow to the anaerobic reactor. Sludge is wasted only from the underflow. Also called five-stage Bardenpho™ process. *See also* Bardenpho or four-stage Bardenpho™ process and Phoredox process.

**modified Chick–Watson model** Same as Hom model. *See also* disinfection kinetics.

**modified Collins–Selleck model** A refinement of the original Collins–Selleck model for the reduction of coliform organisms in a chlorinated secondary effluent to account for the lag observed at low *Ct* values:

$$S = N/N_0 = 1.0 \quad \text{for } Ct < b \qquad \text{(M-58)}$$

$$S = N/N_0 = (Ct/b)^{-n} \quad \text{for } Ct > b \qquad \text{(M-59)}$$

where $S$ = mean effluent survival ratio of organisms; $N$ = number of organisms remaining after disinfection at time $t$; $N_0$ = initial number of organisms (before disinfection); $C$ = chlorine residual remaining at time $t$; $t$ = contact time, min; $b$ = value of the $x$ – intercept with $S = N/N_0 = 1.0$ or $\log S = \log N/N_0 = 0$; and $n$ = a numerical constant equal to the slope of the inactivation curve. *See* Figure M-08, chlorine disinfection models, Collins–Selleck model.

**modified exponential model** A modified exponential model used to describe the probability of human infection by enteric microorganisms. *See* beta-Poisson model for detail.

**modified fouling index (MFI)** An empirical measure of the fouling tendency or plugging characteristics of water or wastewater based on the volume of filtrate that passes every 30 sec through a 0.45 micron pore membrane filter at a constant pressure of 30 psi over a test period of 15 min. The MFI is defined by the following relationship:

$$1/Q = a + mV \qquad \text{(M-60)}$$

where $m$ = MFI, expressed in seconds per liter square (s/L$^2$); $a$ = constant, s/L; $Q$ = average flow, L/s, and $V$ = throughput volume, L. The MFI may be determined graphically as the slope of the straight line portion of the plot of inverse flow ($1/Q$) against throughput volume ($V$), as shown in Figure M-09. It is used to assess the treatability of a given water or wastewater by nanofiltration (NF) and reverse osmosis (RO). Pretreatment is recommended if MFI exceeds 10 for NF or 2.0 for RO. Two other indexes are the silt density index and the mini plugging factor index.

**modified Freundlich isotherm** The Freundlich isotherm applied to the removal of biochemical oxygen demand (BOD) by activated sludge during a contact period of 20 minutes:

$$(X/M)\, C' = KC^{1/n} \qquad \text{(M-61)}$$

where $X$ = dissolved BOD concentration in influent wastewater, mg/L; $M$ = mass of activated sludge; $C'$ = concentration of activated sludge, mg/L; $K$ = empirical constant similar to the Freundlich constant, typically ranging from 0.0001 to 0.001. $C$ = residual dissolved BOD concentration, mg/L; and $n$ = another empirical constant, typically close to 1.0.

**modified Gard model** The combination of Chick–Watson and Gard models. *See also* disinfection kinetics.

**modified Hom model** A modification of the Hom model to represent the rate ($r$) of inactivation of microorganisms in disinfection processes when the disinfectant decays as a first-order reaction:

$$r = -kmNt'^{m-1} \qquad \text{(M-62)}$$

$$\ln (N/N_0) = -kt^m C_0^n A \qquad \text{(M-63)}$$

where $N$ = concentration of viable organisms (numbers per unit volume at time $t$); $N_0$ = initial

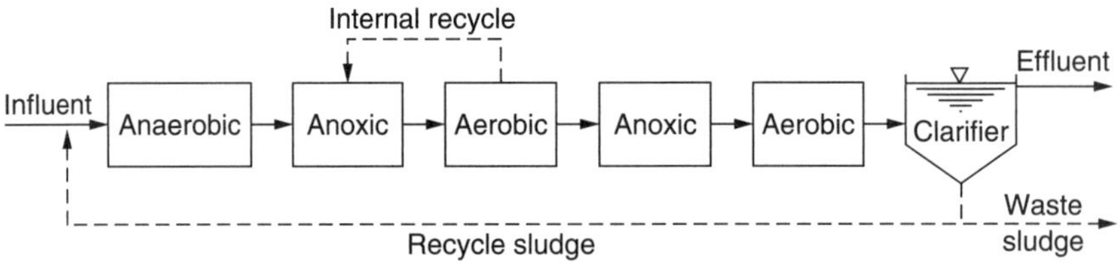


**Figure M-07.** Modified Bardenpho™ process.

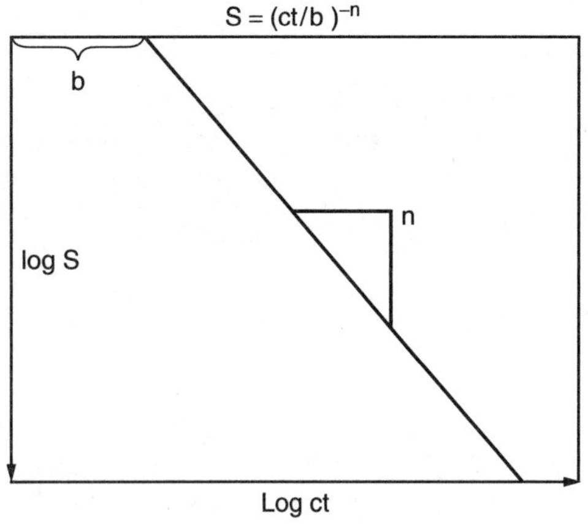


**Figure M-08.** Modified Collins–Selleck model; c = concentration.

concentration of viable organisms (at time $t = 0$); $k$ = the inactivation rate constant (organisms inactivated per unit time); $T$ = time of exposure; $m$ = a constant, which may indicate whether there is a shoulder or tailing; $C_0$ = initial disinfectant concentration, mg/L; $n$ = the coefficient of dilution in the Watson equation;

$$A = m\{[1 - e^{(-nk't/m)}]/(nk't)\}^m \quad \text{(M-64)}$$

$k'$ = first-order rate constant for disinfectant decay. *See also* Chick–Watson model.

**modified Ludzack–Ettinger process (MLE process)** A biological nitrogen removal process that includes an anoxic reactor followed by an aerobic reactor and a final clarifier, with recirculation to the influent line from both the mixed liquor and the clarifier underflow, as shown in Figure M-10. The total solids recycle is an important design and operation parameter because it provides the only nitrate source to the anoxic tank. The process may be designed as one biological reactor that includes an aerobic zone preceded by an anoxic zone, with internal mixed liquor recycle from the aerobic to the anoxic zones. Proposed in 1973, MLE is an improvement over the original Ludzack–Ettinger process. *See also* the Wuhrmann process.

**modified Proctor density** *See* Proctor density.

**modified Reynolds number ($R_e^*$)** A parameter used in the development of a model for predicting the porosity of an expanded bed during backwashing. It includes the interstitial velocity as characteristic velocity and the mean hydraulic radius of the flow channel as characteristic length:

$$R_e^* = (V/\varepsilon)[\varepsilon/S(1-\varepsilon)](\rho/\mu) \quad \text{(M-65)}$$
$$= V\rho/[\mu S(1-\varepsilon)]$$

where $V$ = superficial backwash velocity, $\varepsilon$ = bed porosity, $V/\varepsilon$ = interstitial velocity, $S$ = specific surface of the bed grains, $\rho$ = mass density of the fluid, and $\mu$ = absolute viscosity of the fluid. *See also* fluidized-bed porosity function.

**modified sliplining** *See* fold-and-form liner and tight-fit liner.

**modified UCT process** Abbreviation of Modified University of Cape Town process.

**Modified University of Cape Town process** A wastewater treatment process designed to remove nitrogen and phosphorus beyond the metabolic requirements of the organisms. As shown in Figure M-11, it consists of four biological reactors (in the sequence anaerobic, anoxic-1, anoxic-2, aerobic) followed by a final clarifier, with multiple sludge recycles: (a) internally from anoxic-1 to anaerobic, (b) internally from aerobic to anoxic-2, and (c) from the clarifier underflow to anoxic-1. *See also* combined nutrient removal.

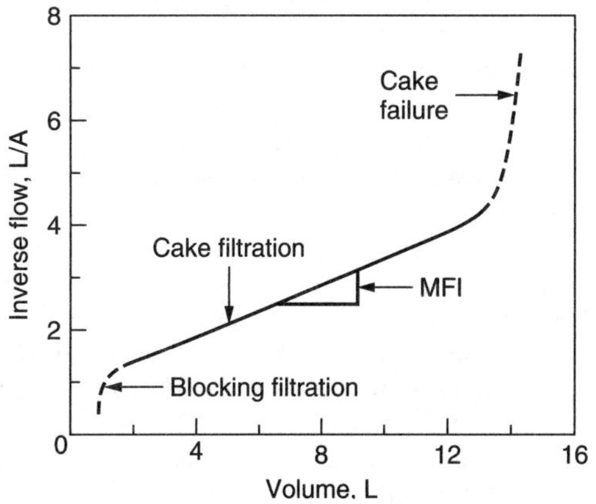


**Figure M-09.** Modified fouling index.

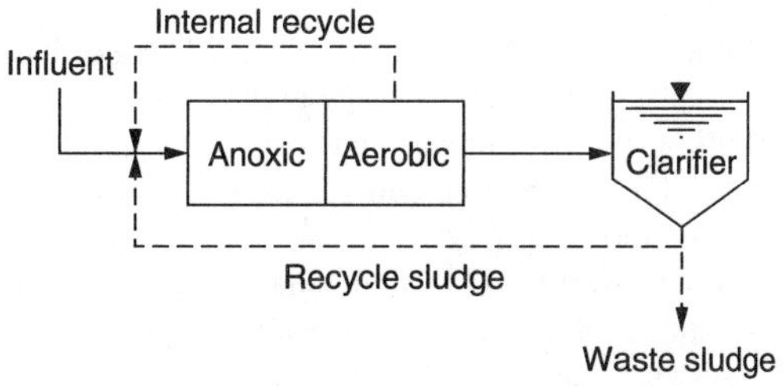


**Figure M-10.** Modified Ludzack–Ettinger process.

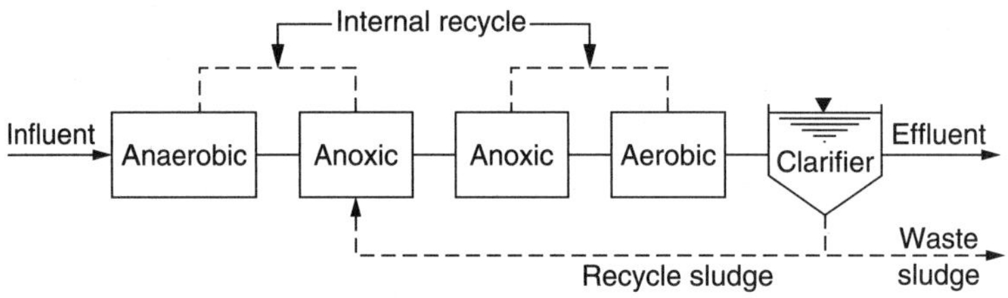


**Figure M-11.** Modified University of Cape Town process.

**modified Velz formula** Another form of the second Eckenfelder formula for the design of trickling filters with recirculation:

$$S_e/S_0 = 1/\Delta \quad \text{(M-66)}$$

$$\Delta = (R + 1)\exp\{(k_{20}A_sD\theta^{T-20})/[q(R + 1)]^n\} - R \quad \text{(M-67)}$$

where $S_e$ = effluent soluble $BOD_5$, mg/L; $S_0$ = influent soluble $BOD_5$, mg/L; $R$ = recycle ratio; $k_{20}$ = filter treatability constant at 20°C; $A_s$ = clean packing specific surface area, $m^2/m^3$; $D$ = depth of media, m; $\theta$ = temperature correction coefficient = 1.035; $q$ = influent rate, $L/m^2/s$; and $n$ = constant characteristic of packing used. Other trickling filter design formulas include British Manual, Eckenfelder, Galler–Gotaas, Germain, Howland, Kincannon and Stover, Logan, NRC, Rankin, Schulze, Velz.

**modifying factor (MF)** An uncertainty factor that is greater than zero and less than or equal to 10; the magnitude of the MF depends upon the professional assessment of scientific uncertainties of the study and database not explicitly treated with the standard uncertainty factors (e.g., the completeness of the overall database and the number of species tested); the default value for the MF is 1.

**MODPATH** Abbreviation of modular path, a computer program used to simulate flow conditions and solute transport in aquifers. *See also* MODFLOW, particle-path model.

**Mod-U-Flo** Clarifier equipment manufactured by Western Filter Co.

**Moduflow** A sand filter manufactured by Smith & Loveless, Inc. in concrete for gravity flow.

**Modulab®** A treatment unit manufactured by the U.S. Filter Corp. to produce high-quality water for laboratories.

**Modular Aquarius®** A modular water treatment plant manufactured by Wheelabrator Engineerred Systems, Inc.

**modular treatment system** (1) Another name for package treatment units. Modular water treatment plants are regulated by the USEPA under "small systems". (2) A membrane filtration system, e.g., microfiltration, consisting of modules joined together.

**module** A complete pressure-driven membrane unit, consisting of the membrane fibers or sheets, pressure support, feed inlet, and product outlet. A module includes one or more membrane elements. Also called a permeator for hollow fiber membranes and hollow fine fiber membranes, and membrane stack for electrodialysis or electrodialysis reversal.

**Modu-Plex** A pumping station manufactured by Smith & Loveless, Inc., including a wet well.

**Moh hardness** Short term for Moh hardness number.

**Moh hardness number** A measure of the hardness of filter bed grains. Moh's hardness scale classifies minerals on their ability to scratch or be scratched by another, from talc (hardness 1) to corundum (9) and diamond (10). Hardness is an important characteristic for the durability of a filter medium. A Moh hardness of 3.0 is commonly specified for anthracite coal and 7.0 for silica sand. Materials with high Moh's hardness are preferable for use in wastewater filtration media. Named after Friedrich Moh (1772–1839) who devised the system.

**Mohlman, F. W.** Director of Laboratories for the Chicago Sanitary District in 1934 when he proposed the use of the sludge volume index.

**Mohr pipette** A graduated pipette used in the laboratory to measure liquids.

**Moh scale hardness** The hardness of a material based on Moh's scale. *See* Moh hardness number.

**Moh's hardness scale** *See* Moh hardness number.

**moiety** An indefinite portion, part, or share of a sample; one of the portions into which something is divided; a portion of a molecule, e.g., the phenolic moiety, the phosphate moieties in the DNA and RNA macromolecules, etc.

**moisture** Water or other liquid causing wetness or dampness on a surface.

**moisture content** The quantity or weight percentage of water in a mass of soil, wastewater, sludge, compost, screenings, or other by-products; for example, the amount of water lost from a soil upon drying to a constant weight, expressed as the weight per unit weight of dry soil or as the volume of water per unit bulk volume of the soil. For a fully saturated medium, moisture content equals the porosity. Also called water content. *See also* sludge moisture content, volumetric moisture content.

**moisture-holding capacity** The moisture content of soil in the field some time after reaching saturation and after free drainage has practically ceased, or the quantity of water held in a soil by capillary action. *See* field capacity for detail.

**moisture–weight–volume relationships** When sludge is dewatered from an initial volume ($V_0$) containing a percentage ($P_0$) of water to a volume ($V$) of water content ($P$), assuming that the specific gravity of the solids ($S$) is about the same as that of the combined weight of water and dry solids:

$$V \approx V_0 (100 - P_0)/(100 - P) \qquad \text{(M-68)}$$

When water no longer fills the voids of the dewatered sludge, the volume ($V$) of the cake is:

$$V = W/[(1 - f)SY] \qquad \text{(M-69)}$$

where $W$ is the weight of dry solids, $f$ the porosity ratio, and $Y$ the specific weight of water.

**mol** Abbreviation of mole (1).

**molal concentration** Same as molality.

**molality** A measure of concentration, equal to the number of moles of a solute per liter of solvent (m or mol/L); sometimes taken as moles of solute per kilogram of solvent (m or mol/kg). Same as molal concentration. A similar measure, molarity, is more commonly used in the field of water and wastewater treatment, particularly in chemical equilibrium equations; however, molal concentrations in mol/kg are normally used for chemical analyses of brine solutions and seawater.

**molar (M or mol/L)** A unit that expresses the molarity of a solution. *See* molar concentration.

**molar concentration** A measure of concentration equal to the number of moles of a solute in one liter of solution (abbreviation M, mol/L); used as a substitute for molal concentrations in chemical equilibrium equations applied to dilute solutions. It is usually represented in square brackets [ ]. Same as molarity.

**molarity** *See* molar concentration. *See also* molality and normality.

**molar solution** A molar solution consists of one gram molecular weight of a compound dissolved in enough water to make one liter of solution. A gram molecular weight is the molecular weight of a compound in grams. For example, the molecular weight of sulfuric acid ($H_2SO_4$) is 98. A 1.0 M solution of sulfuric acid would consist of 98 grams of $H_2SO_4$ dissolved in enough distilled water to make one liter of solution. A molar solution of sodium hydroxide contains 40.0 grams of pure NaOH. *See also* normal solution.

**molasses** A thick, brownish, syrupy by-product of sugar refining, used as a substrate for the preparation of microbial products. It contains mainly sucrose, water, and inorganic components.

**molasses number** One of three parameters used by manufacturers to express the relative capacity of an activated carbon by testing it against a standard molasses solution for the adsorption of large molecules. The number is determined from the ratio of optical densities of the filtrate of the molasses solution to that of the activated carbon sample. Also called decolorizing index. *See also* iodine number, phenol number. Other related parameters include carbon tetrachloride activity and methylene blue number.

**molds** Multicellular, nonphotosynthetic, heterotrophic, aerobic filamentous fungi that resemble higher plants with branched threadlike growths; often observed on the exterior of decaying fruit. They may cause poor settling of activated sludge, e.g., in the treatment of industrial wastes at low pH and with high sugar content.

**mole** (1) The molecular weight of a substance, usually expressed in grams; containing a number of atoms or molecules equal to Avogadro's number; abbreviation: mol. For example, one mole of Argon (Ar) = 40.0 g, one mole of hydrogen ($H_2$) = 2.0 g, one mole of methane ($CH_4$) = 16.0 g. The mole is considered as the SI unit of amount of substance, the amount that contains as many elementary units as there are atoms in 0.012 kg of carbon-12. (2) A device used to clear sewers and pipelines. (3) A massive harbor structure extending from the shore into deep water; breakwater or jetty. (4) A mobile electromechanical device used to agitate and move drying solids in a solar drying bed.

**molecular ammonia** The gas $NH_3$. *See* ammonia, anhydrous ammonia, and aqueous ammonia.

**molecular chlorine ($Cl_2$)** A dense gas that reacts with two electrons ($e^-$) to form inert chloride ($Cl^-$):

$$Cl_2 + 2\,e^- \rightarrow 2\,Cl^- \qquad \text{(M-70)}$$

It is typically provided in pressurized tanks for water and wastewater treatment applications, but it can also be generated electrolytically on-site from brine (sodium chloride, NaCl). *See* chlorine for more detail.

**molecular diffusion** (1) The process whereby molecules of various gases tend to intermingle under the influence of a concentration gradient and eventually become uniformly dispersed. More generally, molecular diffusion applies to all dissolved and suspended substances, whose concentrations in water, for example, eventually become uniform. Molecular diffusion, described mathematically by Fick's laws, is commonly discussed in the theory of water and wastewater treatment (e.g., aeration and reverse osmosis), but is not a major consideration in other environmental conditions. It may be significant in the transport of organic compounds in saturated aquifers of low pore water velocities. *See* diffusion and dispersion for more detail. (2) The random motion of small particles caused by thermal effects, a mass transport process in coagulation. Also called Brownian diffusion.

**molecular diffusion aeration** An aeration technique that uses thin synthetic polymer membranes to transfer air to wastewater without liquid agitation, thereby avoiding foam formation and stripping of volatile organics. A separate apparatus is used to provide mixing.

**molecular diffusion coefficient** A proportionality factor used in the Fick's laws of diffusion, usually expressed in $cm^2/hr$ (or $cm^2/sec$) and representing grams of solute diffusing through one square centimeter in one hour (or one sec) under a concentration gradient of one gram per milliliter per centimeter. *See* Table M-02 for the diffucion coefficients of some gases in water. For organic substances in water (Montgomery, 1996):

$$D^* = 13{,}260/(\mu^{1.14}V^{0.589}) \qquad \text{(M-71)}$$

**Table M-02.** Molecular diffusion coefficient in water, *D*

| Substance | Temperature, °C | *D*, $cm^2/hr$ |
|---|---|---|
| Acetic acid, $CH_3COOH$ | 13.5 | 0.032 |
| Ammonia, $NH_3$ | 15.2 | 0.064 |
| Bromine, $Br_2$ | 12.0 | 0.030 |
| Carbon dioxide, $CO_2$ | 20.0 | 0.065 |
| Chlorine, $Cl_2$ | 12.0 | 0.051 |
| Glucose, $C_6H_{12}O_6$ | 18.0 | 0.020 |
| Hydrochloric acid, HCl | 19.2 | 0.092 |
| Nitrogen, $N_2$ | 20.0 | 0.064 |
| Oxygen, $O_2$ | 20.0 | 0.067 |
| Sodium chloride, NaCl | 15.0 | 0.039 |
| Sulfuric acid, $H_2SO_4$ | 12.0 | 0.047 |

where $D^*$ = molecular diffusion coefficient, $cm^2/sec$; $\mu$ = water viscosity (cps); and $V$ = molar volume of the substance ($cm^3/mol$). Also called diffusivity. *See* Stokes–Einstein law of diffusion for another formula applicable to spherical particles.

**molecular fingerprinting** An emerging (as of 2007) and more accurate testing method than traditional methods; e.g., it is used to isolate such organisms as *Paenibacillus spp.,* which are difficult to identify.

**molecular formula** The formula that represents one molecule of a compound; it shows the number of each constituent atom as a subscript. For example, octane ($C_8H_{18}$) has 8 carbon atoms and 18 hydrogen atoms, whereas commercial alum [$Al_2(SO_4)_3 \cdot xH_2O$] has 2 aluminum atoms, 3 sulfur atoms, 12 oxygen atoms, and x water molecules. *See also* Hill's molecular formula arrangement, isomer, empirical formula, structural formula.

**molecular ion** A compound from which an electron has been removed in a mass spectrometer.

**molecularity** The number of molecules or ions involved in an elementary chemical reaction; it is equal to the overall order of the reaction. (An elementary reaction occurs exactly as written, without intermediate steps, and usually involves just one or two reactants.)

**molecular mass** The sum of the products of the atomic mass of the elements of a compound by their respective numbers of atoms in the compound. For example, for ammonia ($NH_3$), the molecular mass is $(14.0 \times 1) + (1.0 \times 3) = 17.0$. The molecular mass of calcium carbonate ($CaCO_3$) is $40 + 12 + 3 \times 16 = 100$. *See also* molecular weight.

**molecular oxygen ($O_2$)** Oxygen gas. *See* oxygen and elemental oxygen for detail.

**molecular separation** The separation of dissolved contaminants from a solvent by passage through a size-selective membrane under pressure. *See also* membrane processes, reverse osmosis, hyperfiltration, ultrafiltration, electrodialysis.

**molecular size** The size of a material based on its passage through a membrane filter of known size classification. *See also* molecular weight cutoff.

**molecular virus detection** A method used to detect the presence of viral nucleic acid, e.g., through polymerase chain reaction. *See also* infectivity assay, integrated virus detection.

**molecular viscosity** *See* viscosity.

**molecular weight** The molecular weight of a compound in grams is the sum of the atomic weights of the elements in the compound. For example, the molecular weight of sulfuric acid ($H_2SO_4$) is 98 = (2 × 1) + (1 × 32) + (4 × 16). *See also* mole, molecular mass.

**molecular weight cutoff (MWCO)** The molecular weight of the smallest material rejected by a membrane, usually expressed in daltons (D); a common classification of pressure-driven membranes. Also defined as the nominal molecular weight of a known species that would always be rejected in a fixed percentage under specific conditions. It is a classification parameter used to define the contaminant removal rating of a high-pressure process. Reverse osmosis membranes typically have an MWCO between 100 and 200 D, whereas nanofiltration membranes have an MWCO between 200 and 2000. *See also* operating pressure, osmotic pressure, salt rejection capability, and silt density index.

**molecular weight distribution (MWD)** An expression of the percentages of a mixture of organic substances in various ranges of molecular weight cutoffs. MWDs are used in the quantification and characterization of natural organic matter in such methods as high-performance size-exclusion chromatography.

**molecular weight fractionation** The determination of the molecular weight distribution of a heterogeneous liquid mixture by passing it through a series of membrane filters of decreasing molecular weight cutoffs.

**molecule** The smallest division of a compound that still retains or exhibits all the properties of the substance; made up of one or more atoms; e.g., a molecule of oxygen ($O_2$) has two atoms.

**mole drainage** Land drainage using underground tubular channels built with a mole plow.

**mole fraction** The ratio ($x_i$) of the number of moles ($n_i$) of substance ($i$) to the total number of moles in a compound that has ($j$) substances:

$$x_i = n_i/\Sigma n_j \qquad \text{(M-72)}$$

Mole fractions are used in thermodynamic analysis, mass transfer of gases, and solution chemistry.

**molinate ($C_2H_5$—S—CO—$NC_6H_{12}$)** The common name for S-ethyl hexahydro-1H-azepine-1-carbothiolate, pesticide whose WHO-recommended limit in drinking water is 6 micrograms per liter.

**mol/L** Abbreviation of mole per liter.

**molluscicide** A substance that kills or otherwise controls snails and other mollusks. Molluscicide application is used with engineering methods in the control of schistosomiasis.

**mollusk** Any invertebrate, such as a clam or snail, with a soft body in a calcareous shell.

**molten glass process** An advanced waste incinerator design that uses a pool of molten glass to transfer heat to the waste and retain combustion products in a poorly leachable glass. *See also* plasma incinerator, infrared system, molten salt combustion.

**molten salt combustion** An advanced waste incinerator design that uses a bed of molten sodium carbonate ($Na_2CO_3$) to destroy wastes and retain gaseous products. *See also* plasma incinerator, infrared system, molten glass process.

**molten salt reactor** A thermal treatment unit that rapidly heats waste in a heat-conducting fluid bath of carbonate salt.

**mol. wt.** Abbreviation of molecular weight.

**molybdenite** The principal ore of molybdenum, essentially molybdenum sulfide ($MoS_2$).

**molybdenum (Mo)** A lustrous, hard, silver-white metal, found in nature in combination with sulfur (molybdenite) or lead (wulfenite) or a by-product of copper and tungsten mining. Atomic weight = 95.94. Atomic number = 42. Specific gravity = 10.2. Valence = 1, 3, 4, 5, or 6. Melting point = 2617°C. Sublimation at 4507°C. It is used in alloys and in nuclear energy, military, and other applications. It is an essential nutritional element for nitrogen fixation, but at high dietary intake and in association with fluorides, molybdenum may cause the skeletal deformity Genu Valgum and exacerbate the effects of copper deficiencies. Molybdenum in natural waters originates from industrial wastes, natural sources, and cooling-tower water additives.

**Molyver** Reagents produced by Hach Co. for the determination of molybdenum concentrations in the laboratory.

**momentary rate** The volume of wastewater distributed in one second over a surface of one square foot of a trickling filter.

**monel, Monel metal** A trademark corrosion-resistant nickel alloy with 30% of copper; used in waterworks where corrosion prevention is desirable.

**monitoring** (1) Periodical or continuous surveillance, sampling, or testing to determine the level of compliance with statutory requirements and/or pollutant levels in various media or in humans, plants, and animals. *See also* compliance monitoring. (2) Routine observation, sampling, and testing of water constituents or water quality parameters to determine treatment efficiency. (3) Locating and testing radioactive contamination. (4) Collecting and evaluating data to determine exposure levels to various substances.

**monitoring and reporting** Activities conducted to comply with the requirements of the Primary Drinking Water Regulations or similar regulations of the Environmental Protection Agency.

**monitoring well** (1) A well used to obtain samples for analysis or to measure groundwater levels. Also called a sampling well. (2) A well drilled at a hazardous waste management facility or Superfund site to collect groundwater samples for the purpose of physical, chemical, or biological analysis to determine the amounts, types, and distribution of contaminants in the groundwater beneath the site.

**Monkey Screen** A bar screen manufactured by Brackett Green, including a reciprocating rake.

**Mono®** Pumps and accessories manufactured by Ingersoll-Dresser Pump.

**monoaldehyde** An organic compound that has one aldehyde group (HC═O), e.g., formaldehyde ($CH_2O$). Some monoaldehydes are disinfection by-products.

**monoaromatic** An aromatic hydrocarbon containing a single benzene ring.

**monobasic sodium phosphate** Same as sodium phosphate.

**Monoblock** An apparatus manufactured by Wheelabrator Clean Air Systems, Inc. for the control of volatile organic compounds in carbon beds.

**monobromamine ($NH_2Br$)** A product of the combination of hypobromous acid (HOBr) and ammonia ($NH_3$); it is similar to monochloramine.

**monobromoacetic acid ($CH_2BrCOOH$)** A regulated disinfection by-product, one of the six haloacetic acids (HAA6) that water supply systems are required to monitor quarterly. Abbreviations: MBAA or BrAA.

**monobromobenzene** Same as bromobenzene.

**monochloramine ($NH_2Cl$)** One of the products of the combination of hypochlorous acid (HOCl) and ammonia ($NH_3$). Chloramines are sometimes used in water disinfection but are not as effective as chlorine gas (or chlorine dioxide and ozone); however, monochloramine may be more effective than free chlorine in controlling biofilms in distribution systems. *See also* breakpoint chlorination.

**monochloramine formation** During breakpoint chlorination, most of the residual oxidizing chlorine is monochloramine until the breakpoint is reached.

**monochloroacetic acid ($CH_2ClCOOH$)** A haloacetic acid regulated by the USEPA as a disinfection by-product. Abbreviations: MCAA or ClAA.

**monochlorobenzene ($C_6H_5Cl$)** A volatile organic compound almost insoluble in water, used as a solvent and in chemical manufacturing. This substance, which affects the respiratory and nervous systems, is regulated by the USEPA: MCLG = MCL = 0.1 mg/L. Abbreviation: MCB; also called chlorobenzene and phenylchloride.

**monochloroethene** Same as vinyl chloride.

**monochloroethylene** Same as vinyl chloride.

**monochlorophenol** Also called 2-, 3-, or 4-chlorophenol. *See* chlorophenol for detail.

**monochromatic light** A light developed by a prism and grating system to select wavelengths in the visible spectrum. (Monochromatic means having one color or one range of wavelengths).

**monochromator** A device used to isolate monochromatic light or other forms of radiant energy, e.g., in an atomic absorption spectrophotometer for the chemical analysis of trace metals.

**Monocluster** A package plant manufactured by Graver Co. for water treatment.

**monoculture** *See* crop rotation.

**monodentate ligand** An anion or molecule that has only one site that bonds to a metal ion to form a complex, e.g., the cyanide ion ($CN^-$). Also called unidentate ligand. *See also* polydentate ligand.

**Monod equation** A mathematical expression developed by Jacques Monod in 1949 to describe the relationship between biomass production or microbial growth rate (μ, in mg/L/hr) and the concentration (S, mg/L) of a growth-limiting substrate or nutrient:

$$\mu = \mu_{max}S(K_s + S) \qquad \text{(M-73)}$$

where $\mu_{max}$ is, in mg/L/hr, the maximum growth rate when the substrate is not limiting, and $K_s$ is, in mg/L, the half-saturation constant or half-velocity constant, i.e., the value of $S$ when $\mu = \mu_{max}/2$ (also called saturation constant). The parameters μ and $\mu_{max}$ are sometimes called, respectively, specific growth rate and maximum specific growth rate and expressed in the inverse of time, e.g., per hour (1/hr) or per day (1/day). The Monod equation is used in the theory of biological waste treatment, e.g., in deriving expressions of substrate removal, cell yield, cell yield coefficient, and volatile acids production in anaerobic treatment. *See also* the Michaelis–Menten equation.

**Monod formulation** *See* the Monod equation.

**Monod, Jacques** French chemist (1910–1976); winner of Nobel prize in 1965.

**Mono-Ferm** A gravity filter manufactured by Graver Co. for the removal of iron and manganese.

**monofilament** A single continuous fiber used in making filter cloths. *See also* multifilament.

**monofill** A landfill or other waste disposal facility receiving only one type of waste. For example, a monofill may be used to dispose of water or wastewater treatment sludges to the exclusion of municipal solid wastes, or vice versa. The monofill is usually operated as a sanitary landfill, for example, with a lining, collection and treatment of leachate, and appropriate closure when it is filled.

**monofilling** The exclusive use of a landfill for the disposal of treatment plant residuals. *See also* area filling and trench filling.

**Monoflo** A progressing cavity pump manufactured by MGI Pumps, Inc.

**Monoflo®** A sceening grinder manufactured by Ingersoll-Dresser Pump.

**Mono-Floc®** A gravity sand filter manufactured by Graver Co., including a device for coagulant feeding.

**Monoflor®** A filter underdrain manufactured and cast in place by Infilco Degremont, Inc.

**Monolift®** A progressing cavity pump manufactured by Ingersoll-Dresser Pump for groundwater applications.

**monomedia filter, monomedia filtration** *See* monomedium filter and monomedium filtration.

**monomedium filter** A granular filter using a single medium such as sand, anthracite, activated carbon, garnet, or ilmenite. It is usually deeper than other types of filters and uses particles of larger size, e.g., anthracite and granular activated carbon. *See also* dual-media, multimedia, and mixed-media filters.

**monomedium filtration** A water or wastewater treatment method that uses a monomedia filter.

**monomer** The basic or single molecule of a synthetic resin or plastic; a molecule of low molecular weight capable of reacting with identical or different monomers to form polymers. *See also* polymer.

**monomictic** Characteristic of lakes and reservoirs that are relatively deep, do not freeze over during the winter months, and undergo a single stratification and mixing cycle per year. These lakes and reservoirs usually become destratified during the mixing cycle, usually in the fall.

**Mono-Pac** A gravity filter manufactured in concrete by Graver Co.

**Mono-Pilot®** A pilot filter column manufactured by Wheelabrator Engineered Systems, Inc. for use as a coagulant control device.

**Monorake** A traveling bridge raking apparatus manufactured by Dorr-Oliver, Inc. for use in rectangular clarifiers.

**monosaccharide** A simple carbohydrate (a single sugar ring) that may bond with similar compounds to form disaccharides and polysaccharides. Examples of monosaccharides include fructose, galactose, glucose, and sucrose, all having the same molecular formula, $C_6H_{12}O_6$, but different structures. Also called simple sugar, or hexose (because of its six carbon atoms).

**Mono-Scour®** A gravity sand filter manufactured by Graver Co. for flows with high concentrations of solids.

**MonoSparj** A coarse-bubble diffuser manufactured by Walker Process Equipment Co.

**Monosphere™** An ion-exchange resin produced by Dow Chemical Co.

**monovalent** Having a valence of one, positive or negative, such as the cuprous ion (copper, $Cu^+$) or the chloride ion ($Cl^-$).

**monovalent ion** An ion of valence one.

**Monovalve®** Filter A gravity sand filter manufactured by Graver Co.

**monoxenix parasite** A parasite that is host-specific. Also called an ametoecious parasite; the opposite of a metoecious parasite.

**Monozone®** An ozone generation unit manufactured by Capital Controls Co., Inc.

**Montgomery County agent** A calicivirus first identified after an acute, waterborne gastrointestinal outbreak in Montgomery County. *See also* Norwalk virus.

**montmorillonite $[Al_4(Si_4O_{10})_2(OH)_4 \cdot x\ H_2O]$** A clay mineral with a layer lattice structure and variable water content; sometimes used in the laboratory to simulate inorganic turbidity particles of natural waters. *See also* fuller's earth, bentonite.

**Montreal Protocol** An international agreement to phase out the use of chlorofluorocarbons.

***Moraxella*** genus A group of heterotrophic organisms active in biological denitrification.

**morbidity** Rate of disease incidence.

**morbidity rate** The incidence of sickness; the proportion of a population who are ill with a disease in a period of time, an indication of the incidence and prevalence of the disease. *See also* mortality rate.

***Moringa oleifera*** A tropical shade plant, the extract of which contains a dimeric cationic protein that can be used as a coagulant for water treatment, mainly in developing countries where it is less expensive than hydrolyzing metal salts.

**Morrill dispersion index (MDI)** One of the terms used to characterize tracer response curves and describe the hydraulic performance of reactors, it is the ratio of the 90th percentile to the 10th percentile (i.e., the times at which 90% and 10%, respectively, of the tracer has passed through the reactor):

$$MDI = T_{90}/T_{10} \qquad (M\text{-}74)$$

*See also* residence time distribution (RTD) curve, time–concentration curve. The MDI was proposed by A. B. Morrill in 1932 to measure dispersion. For an ideal plug-flow reactor, MDI = 1.0 and for a complete-mix reactor, MDI ≈ 22. For the USEPA, an effective plug-flow reactor has an MDI ≤ 2.0 (Metcalf & Eddy, 2003). A related term is the volumetric efficiency (VE), defined as follows:

$$VE = 100/MDI \qquad (M\text{-}75)$$

**Morrill index** Same as Morrill dispersion index.

**Morris correlation** A relationship established in 1966 by J. C. Morris for the ionization constant ($K_a$) of hypochlorous acid (HOCl) as a function of temperature ($T$, °K):

$$\ln (K_a) = 23.184 - 0.0583\,T - 6908/T \qquad (M\text{-}76)$$

**Morris, J. Carrell** American chemist, professor at Harvard University; special area of research: chlorination.

**mortality** Death rate.

**mortality rate** The incidence of death; the proportion of a population who die of a disease or of all causes in a period of time. *See also* morbidity rate.

**$MoS_2$** Chemical formula of molybdenum sulfide or molybdenite.

**mosquito-borne disease** Any disease that is caused by a pathogen that uses a mosquito as an intermediate host. Mosquito-borne, water- or excreta-related diseases include dengue, filariasis, malaria, and yellow fever. *See also* arboviral infection.

**mosquito-borne helminth** A parasitic worm that is transmitted to man by mosquitoes. *See* the following infections, their pathogenic agents, and their common names: Bancroftian filariasis, loiasis, Malayan filariasis, onchocerciasis.

**mosquito-borne virus** A virus that uses a mosquito as host, e.g., the dengue, yellow fever, encephalitis, and hemorrhagic fever viruses.

**mosquito fish** A species of fish, *Gambusia affinis,* used in stocking ponds to control mosquitoes.

**most probable number (MPN)** The most probable number of organisms (usually coliform-group bacteria) in a 100-mL sample. It is derived from a statistical analysis of the number of positive and negative results of tests on multiple portions of equal volume (based on the Poisson distribution for extreme values). The MPN is a statistical estimate of the concentration of organisms, not the actual concentration. *See also* multiple-tube fermentation technique.

**most probable number (MPN) formula** A simple formula proposed by Harold A. Thomas in 1942 to calculate the MPN based on the number of positive tubes ($N$), the volume of the negative tubes ($V_n$, mL), and the volume of all tubes ($V_a$, mL). *See* Thomas' MPN equation for the formula.

**most probable number (MPN) method** One of two commonly used methods to determine the microbiological quality of a water sample. *See* most probable number technique.

**most probable number (MPN) technique** (1) One of two commonly used methods to test for the presence of total coliforms and fecal coliforms in a water sample; the sample is mixed with a nutrient and incubated in test tubes (e.g., for 24 hours at 37°C for total coliforms and 44°C for fecal coliforms). The production of acid and gas indicates the presence of coliforms and the MPN is determined from a standard statistical analysis of the number of positive tubes; *see* binomial distribution and Poisson distribution. Also called multiple tube method or multiple-tube fermentation technique. *See also* Colilert test, fecal coliform test, confirmed test, microbial test, membrane filtration technique, presence–absence test, presumptive test. (2) One of two basic methods to estimate viable bacteria counts in the environment. *See also* standard plate count. (3) One of two procedures used to quantify viruses based on the cytopathogenic effect (CPE) method; it is similar to the MPN technique for coliforms. *See also* serial dilution endpoint method.

**most probable number test** A test conducted to determine the microbiological quality of a water sample. *See* most probable number technique.

**mother liquor** (1) The residual brine or concentrated solution that remains after evaporation and salt crystallization in a vacuum pan. *See also* bittern. (2) The solution, essentially free of undissolved matter, that remains after filtration, centrifugation, decantation, or any other process of solids separation.

**motile** Capable of self-propelled movement; a term that is sometimes used to distinguish between certain types of organisms found in water.

**motive flow** Flow of water through a device that provides the suction to induce the flow of a regenerant or gas.

**MotoDip** A motorized skimmer pipe manufactured by Walker Process Equipment Co.

**motor benzol** Another name of benzene.

**motor efficiency** The ratio of energy delivered by a motor to the energy supplied to it during a fixed period or cycle. Motor efficiency ratings will vary depending upon motor manufacturer and usually will range from 88.9 to 90.0%. *See also* brake horsepower, water horsepower, wire-to-water efficiency, power terms.

**motor input horsepower** The power input to a motor expressed in horsepower. *See also* power terms.

**motor pumper** A fire truck equipped to provide adequate pressure for fire fighting. Also called a pumper truck.

**mottled enamel** Tooth enamel discolored because of nutritional and developmental defects; it is often caused by dental fluorosis.

**mottled enamel index** A parameter that measures the severity of dental fluorosis and is used to establish the fluoride level in drinking water that does not produce significant enamel discoloration, e.g., 1.0 mg/L.

**mottled tooth enamel** *See* mottled enamel.

**mottling** Damage to the teeth due to an excess of fluoride. Related guidelines or limits established for fluoride concentrations in drinking water vary from 0.7–1.5 mg/L (European Community), to 1.5 mg/L (WHO), and 2.0 mg/L (USEPA).

**mottling of teeth** *See* mottling of tooth enamel.

**mottling of tooth enamel** The staining of teeth, a health problem caused by consuming water with high fluoride concentrations (e.g., more than 2.0 mg/L).

**mound disposal system** A mound system.

**mound system** An intermittent sand filter installed aboveground for the treatment and disposal of septic tank effluent where local site conditions preclude the use of a conventional subsurface disposal field (e.g., high ground water, shallow impervious layer, low permeability soil, shallow soil profile). *See* Figure M-12 and soakaway mound.

**movable-spray aerator** A rotary or straight-line distributor, with nozzles or orifices, used in trickling filters or other wastewater treatment units.

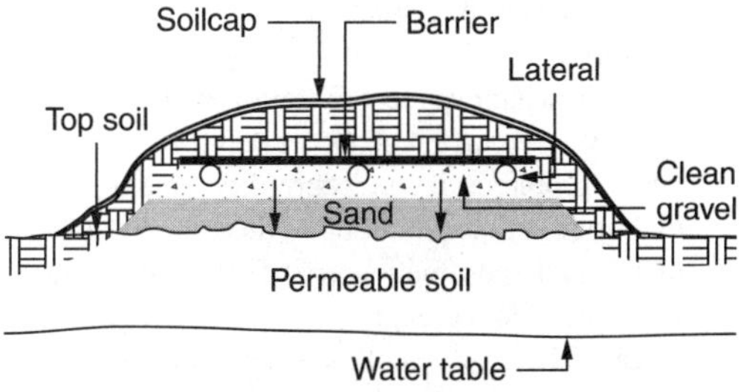


**Figure M-12.** Mound system.

**movable weir** A temporary weir or similar structure that can be removed from and replaced in a channel, or an adjustable overflow weir in a sedimentation tank. *See also* diverting weir.

**moving bed** A filter configuration used in the design and operation of granular filtration and activated carbon adsorption units. Instead of a fixed bed, it uses a moving bed that allows continuous operation of the filter by removing the spent medium at one end while adding fresh medium at the other end. Also called pulse bed. *See also* the following related terms: air mix–pulse mix, downflow moving-bed filter, expanded bed, fixed bed, moving-bed filter, pulse bed filter, upflow moving-bed filter, washbox.

**moving-bed biofilm reactor (MBBR)** An integrated fixed-film activated sludge method of wastewater treatment that uses film growth supported by small cylindrical polyethylene elements suspended in the aeration tank. It can be designed with or without recirculation, for BOD removal and nitrification, or as a type of solids contact process where the MBBR replaces the trickling filter. *See also* Bio-2-Sludge®, BioMatrix®, Captor®, Kaldnes®, moving-bed biofilm reactor, Ringlace®, submerged rotating biological contactor.

**moving-bed filter** A granular media filter that operates continuously using the moving-bed approach instead of a fixed bed. *See also* upflow moving-bed filter, pulse bed filter, washbox.

**moving screen** *See* traveling screen.

**moving water layer** The layer of water that receives waste products in the theory of such attached-growth processes as trickling filtration; attached to the support medium are, successively, an anaerobic zone, an aerobic zone, a fixed water layer, and the moving water layer. *See* fixed-growth process.

**Moyno®** Progressing cavity pumps manufactured by Robbins & Myers, Inc.

**MPA** Acronym of microscopic particulate analysis.

***M. parvicella*** *Microthrix parvicella.*

**MPFI** Acronym of mini plugging factor index.

**MPN** Acronym of most probable number.

**MPN formula** Abbreviation of most probable number formula.

**MPN method** Abbreviation of most probable number method.

**MPN technique** Abbreviation of most probable number technique.

**MPN test** Abbreviation of most probable number test.

**MPVT™** Acronym of multipurpose vertical turbine, a pumping device manufactured by Patterson Pump Co.

**mpy** Acronym of mils per year, a unit used to measure the rate of metallic corrosion.

**MRDL** Acronym of maximum residual disinfectant level.

**MRDLG** Acronym of maximum residual disinfectant level goal.

**mrem** Abbreviation of millirem, a measure of the biological damage due to radiation; it is equal to one thousandth of a rem.

**mrem/yr** A measure of equivalent dose of radiation from any source; it indicates the excess lifetime cancer risk from exposure to radiation. For example, 0.1 mrem/yr corresponds to about a $10^{-6}$ risk excess.

**MRF** Acronym of materials recovery facility.

**MRF Press** A belt filter press manufactured by Idreco USA, Ltd..

**M-roy** Metering pumps manufactured by Milton Roy Co.

**MS2 coliphage** A test virus used commonly as a surrogate for pathogenic viruses, for the validation of ultraviolet (UV) light reactors, and for the development of UV dose–response relationships. MS2 provides reliable and reproducible responses.

**MS4** Acronym of municipal separate storm sewer system. An MS4 permit authorizes storm water discharges from municipal systems, subject to two standards prohibiting nonstormwater discharges into the system and requiring the maximum possible reduction of pollutants.

**MS4 permit** *See* MS4.

**MS Diffuser** A medium-bubble diffuser manufactured by Envirioquip, Inc.

**MSDS** Acronym of material safety data sheet.

**MSF** Acronym of multistage flash distillation or multistage flash evaporation.

**MSF-BR** Acronym of multistage flash evaporation–brine recirculation.

**MSF distillation** Abbreviation of multistage flash distillation.

**MSF-OT** Acronym of multistage flash evaporation, once through.

**MSL** Acronym of mean sea level, a reference used in expressing elevations. *See also* NGVD.

**mS/m** Millisiemens per meter.

**MT®** Products of B. F. Goodrich Co. for cleaning reverse osmosis membranes.

**m T7 medium** A special culture medium used to detect the presence of coliform bacteria in treated water.

**MTBE** Acronym of methyl tertiary butyl ether.

**MTBF** Acronym of mean time between failure.

**MTC** Acronym of mass transfer coefficient.

**MTD** Acronym of maximum tolerated dose.

**MTS®** Acronym of mobile treatment system, a trailer-mounted plant manufactured by Graver Co. for treatment of industrial wastewaters.

**MTTP** Acronym of maximum total trihalomethane potential.

**MTTR** Acronym of mean time to repair.

**MTZ** Acronym of mass transfer zone.

**mucopolysaccharide** A component of bacterial cell walls consisting of N-acetylglucosamine, N-acetylmuramic acid, and a pentapeptide. Also called glycosaminoglycan, murein, or peptidoglycan.

***Mucor*** A genus of fungi identified in trickling filters, active in waste stabilization but susceptible of causing problems of clogging and restriction of ventilation.

**MUD** Acronym of municipal utility district.

**mud** (1) Wet loose mixture of particles less than 60 μm in diameter. *See also* clay, silt. (2) Drilling mud.

**mud accumulation** During a filter cycle, mud accumulates first on the surface of the media, then forms mudballs, which may sink within the bed or to the bottom.

**mudball** (1) A material that is approximately round in shape and varies from a pea size up to two or more inches in diameter. It consists of floc, solids, and filter media. This material forms in a filter bed as a result of inefficient backwash and long filter run times, gradually increases in size, and reduces filtration efficiency. The use of polymers as coagulant aids or filter aids accentuates the formation of mudballs. Mudballs collect at the surface of the filter after backwash and normally represent less than 0.1% of the top 6 inches of sand. They are also a common problem in granular activated carbon filter–adsorbers. *See also* auxiliary air scour, dirty filter, surface wash. (2) A ball of sediment in stream flow or in bottom deposits.

**mudball formation** Mudballing.

**mudballing** An operational problem experienced in filters, which renders filtration and backwashing slow and inefficient. Other depth filtration operation problems include turbidity breakthrough, emulsified grease buildup, filter medium loss, cracks and contraction of filter bed, gravel mounding, slime clogging

**mudbank** Lumps of colloidal solids deposited as silt on the banks of a river as a result of flocculation when river water carrying electrically charged clay particles mixes with seawater containing charged dissolved partricles. *See* flocculation (4).

**mud blanket** A layer of solids and flocculent material that forms on the surface of a sand filter; schmutzdecke.

**mud cake** (1) *See* filter cake. (2) A film of drilling fluid on the wall of a borehole during drilling.

**mud flat** A muddy, flat, low-lying area, formed by the tide or by surface water discharges.

**mud flow** (1) A river of liquid and flowing mud on the surfaces of normally dry land areas, as when earth is carried by a current of water (as defined by the National Flood Insurance Program). *See also* flood. (2) Flow of water heavily loaded with sediments and debris, sometimes carrying large rocks. Also called mud slide, mud spate. *See also* mud wave.

**mud slide** *See* mud flow (2).

**mud spate** *See* mud flow (2).

**mud pot** The mixture of water from a hot spring with clay and other particles; a hot spring filled with boiling mud.

**mud valve** A plug valve used to drain sediment from the bottom of settling tanks or sedimentation basins.

**mud wave** A change in surface elevations of an area caused by variations in soil pressures in the underlying silt formations. *See also* mud flow (2).

**Muffin Monster®** A wastewater grinder manufactured by JWC Environmental.

**MUG** Acronym of 4-methylumbelliferyl-β-D-glucuronide.

**μg/L** Abbreviation of microgram(s) per liter, a unit of constituent concentration.

**mulch** Any substance spread or allowed to remain on the soil surface to conserve soil moisture and shield soil particles from the erosive forces of raindrops and runoff. It may consist of wood chips, compost, sawdust, sludge, or other organic matter.

**multiclone** A device consisting of parallel cyclone separators to remove particulates.

**Multicoil** A sludge drying apparatus manufactured by Kvaerner Eureka, USA.

**Multicone** A cascade aerator manufactured by Infilco Degremont, Inc.

**Multicrete®** A filter underdrain apparatus manufactured by General Filter Co.

**multidentate ligand** An anion or molecule that has more than one site that can bond to a central metal ion at one time to form complexes. For example, oxalate and ethylenediamine are bidentate ligands (2 atoms), citrate is tridentate (3 atoms), and EDTA is hexadentate (6 atoms). Also called polydentate ligand or chelating agent. *See also* unidentate ligand.

**Multidigestion** A two-stage sludge digester manufactured by Dorr-Oliver, Inc.

**MultiDraw** A circular clarifier manufactured by Walker Process Equipment Co., including multiple nozzles for solids removal.

**multieffect distillation** Same as multiple-effect distillation.

**multifilament** A bundle of fiber strands used in making filter cloths. *See also* monofilament.

**Multiflo** Nozzles manufactured by Amwell, Inc. for rotary distributors.

**Multiflo®** A pump column manufactured by Jet Tech, Inc.

**Multiflow** A water treatment plant manufactured by Ecolochem, Inc. and mounted on a skid.

**Multi-Flow** Medium fabricated in PVC by B. F. Goodrich Co. for use in biological filters.

**multifunctional medium** A material that can be used in more than one process or for more than one purpose. For example, activated carbon can be used as a filter medium or as an adsorbent. Activated carbon can also be used to remove solids or to reduce chlorine residual. *See also* activated alumina.

**multihit model** A model that uses age-specific rates to define risk from cancer and assumes that interactions between a carcinogenic agent and a cell follows a Poisson distribution. *See* dose–response model for related distributions.

**multilayered bed** A filter or ion-exchange unit that contains two or more media of different densities and in stratified layers. *See also* multimedia filter.

**multilevel clarifier** Two or more settling tanks, one on top of the other, fed independently but operating on a common water surface. This arrangement saves space and facilitates odor control. Also called stacked clarifier.

**multilocus enzyme electrophoresis** A laboratory technique that uses enzyme mobility to subtype organisms.

**Multilog** A data logger manufactured by Radcom Technologies. It is a multichannel device, with a telemetry option, for the monitoring of flow, pressure, and water quality.

**Multilogger** An instrument manufactured by Stevens Water Monitoring Systems for multiple measurements.

**multiloop controller (MLC)** A single device connected to the control system of a wastewater treatment installation, as part of a distributed control system, to regulate multiple variables. *See also* single-loop controller.

**Multi-Mag™** An electromagnetic flowmeter manufactured by Marsh-McBirney, Inc.

**multimedia filter** A granular filtration apparatus using more than one medium such as silica sand, anthracite, activated carbon, garnet, and ilmenite. The term often applies to filters with three or more different media; the bottom layer usually consists of a medium (garnet or ilmenite) with finer grains than the other layers. The top layer may be of granular activated carbon for the control of taste, odor, and organic compounds. *See also* dual-media, monomedium, and mixed-media filters.

**multimedia filter bed** (1) The various layers of intermixed or segregated media constituting a multimedia filter. (2) The term is sometimes used synonymously with depth filter.

**multimedia filtration** A water or wastewater treatment process using multimedia filters or depth filters.

**multinuclear complex** A complex that has more than one central metal atom. Also called a polynuclear complex.

**multiple chlorination** The application of chlorine for disinfection of water at several points from intake to distribution, usually for economy and/or ease of construction and operation and for a raw water of relatively good quality, e.g., an MPN less than 10/100 mL and a turbidity under 5 NTU.

**multiple dosing tanks** A series of dosing tanks that can be used one at a time or in any combination.

**multiple effect distillation (MED)** A desalting process using several vessels or effects operating at successively lower pressures. When a thin film of heated saline water is introduced in one vessel, some of the water vaporizes and then condenses in another vessel, releasing heat that is used for further vaporization in this vessel, and so on. (In water distillation, an effect is an evaporation–condensation step). Also called multieffect distillation.

**multiple-effect evaporation** A process using multiple-effect evaporators in the distillation of water and wastewater. *See* multiple-effect distillation. Other distillation processes that may be considered for the reclamation of municipal wastewater are multistage flash evaporation and vapor-compression distillation. Multiple-effect evaporation is also used to extract water from wastewater sludge as part of the Carver–Greenfield process.

**multiple-effect evaporation dryer** A sludge drying device that uses steam or other medium to conduct heat indirectly through sludge. *See also* indirect dryer.

**multiple-effect evaporator** A set of single-effect evaporators connected in series, with the vapor from one effect becoming a source of heat for the next effect. (In water distillation, an effect is an evaporation–condensation step.) Some evaporators use geothermal power as the energy source. Some multiple-effect evaporators are used in sludge drying; see Carver–Greenfield process.

**multiple elutriation** Elutriation of sludge (washing it out in water to reduce its alkalinity) in multiple tanks, one after another, without recirculation of the washwater. Assuming that the tanks are identical and each receives an equal amount of washwater, which is wasted after use, the alkalinity of the elutriated sludge and the washwater requirement may be calculated as follows (Fair et al., 1971):

$$EA = \{A + W[(R/N + 1)^N - 1]\}/(R/N + 1)^N \quad \text{(M-77)}$$

$$R = N\{[(A - W)/(E - W)]^{1/N} - 1\} \quad \text{(M-78)}$$

where $EA$ = alkalinity of the elutriated sludge, $A$ = alkalinity of the sludge moisture before elutriation, $R$ = ratio of washwater to sludge volumes, $W$ = initial alkalinity of washwater, and $N$ = number of tanks. *See also* countercurrent elutriation.

**multiple extraction procedure (MEP)** A procedure designed to simulate the transformation of a waste material subject to the precipitation of acid rain.

**multiple-hearth dryer** A multiple-hearth incinerator used for sludge drying and burning.

**multiple-hearth, fluid-bed furnace** A device that combines a multiple-hearth dryer for drying dewatered sludge and a fluid-bed furnace for autogenous sludge combustion. The use of hot flue gases from the fluid-bed furnace minimizes the need for auxiliary fuel. Also called combination furnace.

**multiple-hearth furnace** A furnace or incinerator consisting of several vertically stacked hearths that is used for regenerating granular activated carbon, incinerating or drying organic sludge, recalcinating lime, etc. These materials are fed into the top hearth and are regenerated or dried as they gradually move downward. Oxygen for combustion or an outside fuel source may be required depending on the specific application.

**multiple-hearth incineration** A process that uses a multiple-hearth incinerator to convert dewatered sludge to inert ash; commonly used in large plants because of its complex operation. *See also* starved air incineration.

**multiple-hearth incinerator** An apparatus used to carry out incineration; same as multiple-hearth furnace. More precisely, a countercurrent incinerator used to convert dewatered sludge to an inert ash:

the sludge cake moves downward from the top hearth and is continuously raked while preheated air admitted at the lowest hearth rises through the incoming sludge. A multiple-hearth furnace may be designed and operated to only dry sludge.

**multiple linear regression** A technique using the least squares method to estimate the value of one variable (or parameter) from two or more other variables (or parameters).

**multiple outlets** A network of conduits with separate outlets for the distribution of wastewater or effluent under water.

**multiple-purpose reservoir** Same as multipurpose reservoir.

**multiple sampler** An instrument used to collect several water, wastewater, or sediment samples at the same time.

**multiple-sludge system** A wastewater treatment system designed for nutrient removal and including multiple stages, each in separate tanks and with its own clarifier and sludge recycle lines. *See* dual-sludge process and triple-sludge process.

**multiple-stage flash evaporation** Same as multistage flash evaporation.

**multiple-stage ozonation** A water treatment scheme designed in France and Germany for efficient application of ozone. It includes the following flow sequence: preozonation, flocculation, sedimentation, ozonation, sand filtration, adsorption on granular activated carbon, ground storage, and chlorination. It achieves removal of organics, elimination of THM formation, and maintenance of a disinfectant residual in the finished water.

**multiple-stage sludge digestion** The biochemical decomposition of wastewater sludge in two or more tanks in series, with each tank operated under different mixing conditions. Also called stage digestion.

**multiple-tray aerator** A water aeration device that consists of trays spaced 20 inches apart. The trays, made of perforated plates, screens, or wooden slats, contain open side louvers for air flow, as well as media like ceramic balls, coke, slag, or stones to enhance the oxidation of iron and manganese. *See also* cascade aerator.

**multiple-tray clarifier** A sedimentation basin equipped with parallel trays, and sometimes with individual sludge scrapers, to improve solids collection in water or wastewater treatment. *See also* lamella clarifier, tray clarifiers.

**multiple-tube analysis** *See* multiple-tube fermentation test.

**multiple-tube fermentation (MTF)** *See* multiple-tube fermentation test.

**multiple-tube fermentation technique** *See* multiple-tube fermentation test.

**multiple-tube fermentation test** A microbiological test that uses multiple tubes to grow bacteria (in general coliform bacteria) and determine their density in soil, water, or food samples. *See* most probable number method.

**multiple-tube method** Same as most probable number (MPN) method.

**multiple use** Use of land for more than one purpose; e.g., grazing of livestock, wildlife production, recreation, watershed, and timber production. Also, the use of water bodies for recreational purposes, fishing, and water supply.

**multiple-use reservoir** Same as multipurpose reservoir.

**multiplier (composite sample)** The ratio ($M$) of the volume of composite sample desired ($C$) to the product of the average flowrate ($Q$) by the number of portions ($N$):

$$M = C/(QN) \qquad \text{(M-79)}$$

**Multi-Point** A device manufactured by Drexelbrook Engineering Co. to control liquid levels.

**multiport valve** A master valve used in filters, softeners, or other water treatment units to control such operations as backwashing and regeneration.

**Multiport Valve™** A master valve manufactured by U.S. Filter Corp. to control backwashing and rinsing.

**multipurpose reservoir** A reservoir designed and operated for two or more (sometimes conflicting) purposes such as municipal water supply, flood mitigation, navigation, irrigation, power development, recreation, and pollution abatement.

**MultiRanger** A device manufactured by Milltronics, Inc. for measuring liquid levels and volumes.

**multistage filtration** The use of two or more filters in series to treat water or wastewater. It usually consists of a high-rate filter followed by lower-rate units.

**multistage flash distillation (MSF)** A desalting process that uses distillation in several vessels or stages to separate water from a saline solution or to treat wastewater for reclamation. The process operates at successively lower temperatures and pressures. Flashing refers to the vapor or boiling resulting from the pressure reduction. The incoming water or wastewater is preheated. At each stage, a portion of the liquid is flashed and then condenses while its latent heat is used in the the other portion being returned to the first stage. MSF is less efficient than simple evaporation but has a lower construction cost. Scaling and corrosion are operating problems common to all distillation processes. *See also* flashing, and multiple-effect distillation.

**multistage flash evaporation (MSF)** A variation of the distillation process that operates in countercurrent and at successively lower temperatures and pressures so as to recapture the latent heat of vaporization of water. Same as multistage flash distillation. *See also* multiple-effect evaporation and vapor-compression process.

**multistage model** A mathematical model based on the multistage theory of the carcinogenic process, which yields risk estimates either equal to or less than the one-hit model. It is a dose–response model often expressed in the form

$$P(d) = 1 - \exp\{-[q(0) + q(1)d + q(2)d^2 + \ldots + q(k)q^k]\} \quad \text{(M-80)}$$

where $P(d)$ is the probability of cancer from a continuous dose rate $d$, the $q(i)$ are the constants, and $k$ is the number of dose groups (or, if less, $k$ is the number of biological stages believed to be required in the carcinogenesis process). Under the multistage model, it is assumed that cancer is initiated by cell mutations in a finite series of steps. A one-stage model is equivalent to a one-hit model. A modified form, the linearized multistage model, assumes that the basic dose–response data are linear at low doses. The constant $q(1)$ is forced to be positive (>0) in the estimation algorithm and is also the slope of the dose–response curve at low doses. The upper confidence limit of $q1$ [$q(1)$*] is called the slope factor or human carcinogenic potency factor. *See* dose–response model for related distributions.

**multistage pump** A centrifugal pump using multiple impellers operating in series in the same casing to increase the head of the discharging fluid.

**Multi-Tech®** A water treatment process developed by Cullligan International Corp., including the addition of chemicals, contact flocculation, and filtration.

**Multi-Turi®** A wet scrubber with a venturi manufactured by CMI-Schneible Co.

**multivalent** (1) Pertaining to a chemical element that has a valence greater than 2. (2) Containing several antibodies. (3) Pertaining to an antibody that has many antigen-binding sites or many determinants.

**Multiwash®** A sand filter manufactured by General Filter Co. and combining air and water backwash.

**Multi-Wash®** A wet scrubber manufactured by CMI-Schneible Co.

**Multi-Zone** An apparatus manufactured by Zimpro Environmental, Inc. for anaerobic digestion of sludge.

**multrum composting toilet** Same as multrum toilet.

**multrum continuous-composting toilet** Same as multrum toilet.

**multrum toilet** A continuous composting toilet originated in Sweden and replicated in Tanzania and elsewhere. It consists of a superstructure over the decomposition compartment, which is covered by the squatting plate, and a humus vault.

**Muncher®** Wastewater grinders manufactured by Ingersoll-Dresser Pump.

**Munchpump®** An assembly manufactured Ingersoll-Dresser Pump, including a combined pump and grinder.

**municipal contract operation** A project implementation arrangement whereby a private contractor provides such services as water supply and wastewater disposal to residents of a municipality, which retains ownership of the facilities. The services may be limited to operation of the facilities but may extend to complete operation, maintenance, and administration, including billing. *See also* design–build–operate.

**municipal discharge** Discharge of effluent from wastewater treatment plants that receive flows from households, commercial establishments, and industries, including combined sewer/separate storm sewer overflows.

**municipal separate storm sewer** One of two basic types of public storm sewers. A conveyance or system of conveyances (including roads with drainage systems, municipal streets, catch basins, curbs, gutters, ditches, man-made channels, or storm basins): (i) owned or operated by a state, city, town, borough, county, parish, district, association, or other public body; (ii) designed or used for collecting or conveying stormwater; (iii) which is not a combined sewer; (iv) which is not part of a publicly owned treatment works (EPA-40CFR122.26-8). The other type of public drainage is a combined sewer system (CSS).

**municipal separate storm sewer system (MS4)** *See* municipal separate storm sewer.

**municipal sewage** Wastes (mostly liquid) originating from a community; may be composed of domestic, commercial, and industrial discharges. Same as municipal wastewater.

**municipal softening** Water softening at a municipal treatment plant, as compared to the same operation for industrial use.

**municipal utility district** A legal entity organized to provide such services as water supply, wastewater disposal, stormwater management, electricity, and gas.

**municipal waste** (1) Same as municipal wastewater. (2) Combined solid waste originating from residential, commercial, and industrial sources, including sludges from water and wastewater treatment plants as well as any residual products of solid waste processing. *See also* industrial waste.

**municipal wastewater** Combined liquid waste originating from residential, commercial, and industrial sources, including any infiltration/inflow or street washoff that ends up in the wastewater collection system. Although wastewater characteristics and volumes vary widely with seasons, times of day, and other factors, a typical municipal wastewater in the United States has a flow of 100 gallons per person per day, a BOD of 200 mg/L and a total suspended solids concentration of 240 mg/L. Also called municipal sewage or municipal waste. *See also* industrial or trade waste, domestic wastewater, and sanitary wastewater.

**municipal wastewater reuse** The principal categories of municipal reuse of wastewater or reuse of municipal wastewater, after appropriate treatment, are: agricultural irrigation, groundwater recharge, industrial recycling/reuse, landscape irrigation, nonpotable urban reuse, potable reuse, and recreational/environmental reuse. *See also* water reuse applications.

**municipal wastewater treatment** A plant designed to treat municipal wastewater. *See also* publicly owned treatment works.

**municipal water** Water distributed to municipal households and businesses by public or private agencies.

**Muniflo** A positive displacement pump manufactured by Envirotech Co. for sludge handling, including a rotary lobe.

**Munox** Cultures of selective bacteria produced by Osprey Biotechnics for use in wastewater treatment.

**Munster** A cleaning apparatus manufactured by Landustrie Sneek B. V., including a trash rake.

**Muntz metal** An alloy containing approximately 58% copper and 42% zinc; from the name of its inventor, the English metallurgist G. F. Muntz. Also called alpha-beta brass.

**muriatic acid (HCL)** Another name for hydrochloric acid; used in acid treatment or well stimulation.

***Musca domestica*** The ordinary housefly, a vector of many excreta-related diseases. *See* flyborne disease.

**muscovite** A common, light-colored form of the silicate mineral mica, used as an electrical insulator. Two formulas are found in the literature: $K_2Al(Si_6Al_2)O_{20}(OH,Fe)_4$ and $KAl_3Si_3O_{10}(OH)_2$.

**mushing** The phenomenon of salt pellets breaking down into crystals in a water softener. *See also* bridging.

**mushroom valve** A flat disk that rises or descends without rotation while closing or opening. Also called a poppet valve.

**Mushroom Ventilator** An air diffuser manufactured by Knowles Mushroom Ventilator Co. in cast iron.

**mustard gas ($C_4H_8Cl_2S$)** An oily liquid with a mustard odor, used in chemical warfare; it can cause blindness or even death and damage the skin and the lungs. Also called dichlorodiethyl sulfide.

**mutagen** A substance, agent, or preparation that causes a permanent genetic change (a mutation, a DNA alteration) in a living cell or organism other than the change that occurs during normal genetic recombination; e.g., radionuclide, ultraviolet irradiation, some chemicals. *See also* carcinogen and teratogen.

**mutagenesis** The process or condition in which an inheritable genetic change occurs in a cell's DNA, as compared to a somatic mutation, which is not inheritable; also, the study of a mutation, as to its origin and development. Mutations, which can lead to cancer or other chronic diseases, may result from human exposure to certain pollutants. The Ames test and other assays are used to detect mutation potential.

**mutagenic** Pertaining to a chemical or physical agent that can cause permanent genetic changes. *See* mutation, carcinogenic, genotoxic, teratogenic.

**mutagenicity** The capacity of a chemical or physical agent to cause permanent alteration of the genetic material within living cells. One of the factors considered in establishing no observed adverse effect levels, it is correlated with carcinogenicity and teratogenicity.

**mutation** A random and spontaneous or induced change in the cellular or biochemical characteristics of an organism as a result of changes in genes or other hereditary factors. Some mutations are natural but most mutations are harmful. *See also* mutagen.

**Mutrator®** An apparatus manufactured by Ingersoll-Dresser Pump, including wastewater pumps and grinders.

**mutual aid plan** An agreement between utilities, municipalities, or other organizations to assist each other during disasters or emergencies.

**mutualism** A beneficial interaction between two organisms or organisms of different species. *See also* commensalism, symbiosis, parasitism.

**mutual water company** A cooperative that provides water supply services to its members.

**MVR** Acronym of mechanical vapor recompression.

**MVR evaporator** *See* mechanical vapor recompression evaporator.

**MW** Acronym of megawatt.

**MWC** Acronym of molecular weight cutoff.

**MWCO** Acronym of molecular weight cutoff.

**mycelium** A mass of hyphae forming the vegetative portion of a fungus. Plural: mycelia.

**myco** A prefix meaning mushroom, fungus.

**mycobacteria** Any organism of the genus *Mycobacterium.*

***Mycobacterium*** A genus of unicellular, acid-fast, Gram-negative, pathogenic bacteria found in humans and cattle, responsible for tuberculosis and leprosy; so-called because of the mycolic acids on their cell surface. Mycobacteria grow slowly in soils but can degrade hydrocarbons and many other organic contaminants.

***Mycobacterium avium*** **complex (MAC)** A group of 28 serovars of two opportunistic species of bacterial pathogens, *Mycobacterium avium* and *M. intracellulare,* that cause pulmonary disease, particularly in hospital patients and patients with predisposing factors such as AIDS. They are commonly found in the environment and resist chlorination to some extent; hospital water systems often harbor MAC. (*M. scrofulaceum* was previously included in the group.) Also called *Mycobacterium avium intracellulare* (Mai).

***Mycobacterium avium intracellulare*** **(Mai)** *See Mycobacterium avium* complex.

***Mycobacterium tuberculosis*** A species of bacteria isolated from raw municipal wastewater and suspected in outbreaks of tuberculosis among individuals swimmimg in water contaminated with wastewater (e.g., from institutions treating tuberculosis patients).

**mycology** The study of fungi.

**mycoplasma** A group of very small bacteria without a cell wall.

**mycorrhizae** A symbiotic association of fungi and plant roots; hyphae of the fungi growing on the roots absorb water and nutrients required by the plants.

**mycosis** A disease caused by a fungus. Plural: mycoses.

**mycotoxin** A poisonous substance (e.g., aflatoxin) produced by fungi.

**Myers rating** A method of comparing flood flow characteristics of U.S. drainage basins by plotting extreme peak flows ($Q$, cfs) vs. the drainage area ($A$, mi$^2$, with $A \geq 4.0$):

$$Q = 100\,pA^{0.5} \qquad \text{(M-81)}$$

where

$$p = 100\,Q/Q_u \qquad \text{(M-82)}$$

$$Q_u = 10{,}000\,A^{0.5} \qquad \text{(M-83)}$$

$Q_u$ being the postulated ultimate maximum flood flow. For example, the rating is 25% for the Colorado River basin and 64% for the lower Mississippi River basin (Fair et al., 1971).

**myocarditis** Inflammation of the myocardium (a muscular structure of the heart), one of the symptoms of Coxsackievirus B.

**Mystaire®** An air scrubber manufactured by Misonix, Inc.

**mμ** Abbreviation of millimicron.

# N

**N** Abbreviation of normal.

**Na** Symbol of the metallic element sodium, from its Latin name, natrium.

**NA** (1) Not analyzed. (2) Not applicable. (3) Not available.

**NAA** Acronym of neutron activation analysis.

**$Na_3AlF_6$** Chemical formula of cryolite; also called Greenland spar or sodium aluminum fluoride.

**$NaAlO_2$** A chemical formula of sodium aluminate.

**$Na_2Al_2O_4$** Another chemical formula of sodium aluminate.

**$NaAlSi_3O_8$** Chemical formula of albite.

**$Na_{1/3}Al_{7/3}Si_{11/3}O_{10}(OH)_2$** Chemical formula of montmorillonite.

**$NaAsO_2$** Chemical formula of sodium arsenite or sodium metaarsenite. *See also* $Na_3AsO_3$.

**$Na_3AsO_3$** Chemical formula of sodium arsenite or sodium metaarsenite. *See also* $NaAsO_2$.

**$NaBH_4$** Chemical formula of sodium borohydride.

**$Na_2B_4O_7$** Chemical formula of borax. Also written as $Na_2B_4O_7 \cdot (10H_2O)$.

**$NaBO_2 \cdot 3H_2O$** A chemical formula of sodium perborate. Also written as $NaBO_3 \cdot 4H_2O$.

**$NaBO_3 \cdot 4H_2O$** Another chemical formula of sodium perborate. Also written as $NaBO_2 \cdot 3H_2O$.

**$Na_2B_4O_7 \cdot (5\ H_2O)$** Chemical formula of borax pentahydrate.

**$Na_2B_4O_7 \cdot 10\ H_2O$** Chemical formula of borax.

**NaBr** Chemical formula of sodium bromide.

**NaCl** Chemical formula of sodium chloride.

**$NaClO_2$** Chemical formula of sodium chlorite.

**NaCN** Chemical formula of sodium cyanide.

**$Na_2CO_3$** Chemical formula of soda ash or sodium carbonate.

**$Na_2CO_3 \cdot (H_2O)$** Chemical formula of sodium carbonate monohydrate.

**$Na_2CO_3 \cdot (7\ H_2O)$** Chemical formula of sodium carbonate heptahydrate.

**$Na_2CO_3 \cdot (10\ H_2O)$** Chemical formula of sodium carbonate decahydrate; washing soda.

**$Na_2CrO_7$** Chemical formula of sodium dichromate.

**NAD** Acronym of nicotinamide adenine dinucleotide.

**NADPH-cytochrome P-450 reductase** Abbreviation of nicotinamide adenine dinucleotide phosphate-cytochrome P-450 reductase.

***Naegleria*** A genus of free-living amoebae, sometimes present in freshwater; human protozoan pathogens that can infect nasal passages and invade brain tissues. They are less sensitive to chlorine disinfection than coliform bacteria.

***Naegleria fowleri*** A worldwide species of free-living amoebae that occur in water, soil, and decaying vegetation; they are responsible for a deadly infection, primary amoebic meningoencephalitis.

**NaF** Chemical formula of sodium fluoride.

**NAFCO** A precoat filter aid produced by Liquid-Solids Separation Corp.

**Nagler test** A test that presumptively identifies *Clostridium perfringens.* It uses an antitoxin specific to this and three other *Clostridium* species.

**$NaHCO_3$** Chemical formula of baking soda (sodium bicarbonate).

**$NaH_2PO_4$** Chemical formula of sodium phosphate.

**$Na_2HPO_4$** Chemical formula of disodium phosphate.

**$NaHSO_2$** Chemical formula of sodium bisulfite.

**$NaHSO_3$** Chemical formula of sodium bisulfate.

**Nalgonda process** A simple water treatment process developed in India, using a higher alum dose (e.g., 600 mg/L of hydrated alum) than usual to assist in the removal of fluoride by sedimentation. Lime is added before alum to adjust alkalinity. Also called the lime–alum technique.

**Nalgonda technique** *See* Nalgonda process.

***N. americanus*** *See Necator americanus.*

**$NaNH_2$** Chemical formula of sodium amide.

**$NaNH_4HPO_4$** Chemical formula of sodium ammonium phosphate.

**nano** An SI unit prefix indicating $10^{-9}$, e.g., nanometer = $10^{-9}$ meter or one millionth of a millimeter.

**$NaNO_2$** The chemical formula of sodium nitrite.

**$NaNO_3$** The chemical formula of sodium nitrate.

**nanocurie** A measure of radioactivity equal to one billionth or $10^{-9}$ of a curie.

**nanofiltration** A high-pressure, specialty filtration using semipermeable membranes to remove particles larger than approximately one nanometer (10 angstroms). In nanofiltration, the driving force is a hydrostatic pressure difference; diffusion and exclusion are the separation mechansisms. Nanofiltration is used in water softening (instead of chemical precipitation) to separate nonvolatile organic matter (e.g., synthetic organics and disinfection by-products) having molecular weight cutoffs ranging from 200 to 2000 daltons and multivalent inorganics. It is also applied to the removal of selected dissolved constituents (e.g., metal ions responsible for hardness) from wastewater. Also called loose reverse osmosis. *See also* ultrafiltration, macrofiltration.

**nanogram (ng)** One thousandth of one millionth of a gram, i.e., $10^{-9}$ g or $10^{-6}$ mg.

**nanogram per liter (ng/L)** A unit of small concentrations, equal to one millionth of a milligram per liter (mg/L); sometimes used interchangeably with parts per trillion.

**nanometer (nm)** A measure of length equal to one billionth of a meter, i.e., $10^{-9}$ m or one millionth of a millimeter, i.e., $10^{-6}$ mm.

**nanoparticle** A particle small enough to be measured in nanometers (nm), e.g., synthesized magnetite of 10–20 nm that can be used to adsorb arsenic from water. These particles can then be removed by a magnet with a low magnetic field.

**nanoplankton** The smallest phytoplankton.

**nanoporous silica** A highly porous material (approximately 800 square meters per gram) that can be used in water or wastewater treatment; *see,* e.g., self-assembled monolayers on mesoporous supports.

**nanotube membrane filtration** A water desalination process that uses membranes of carbon nanotubes and produces extremely high gas and water flows compared to reverse osmosis and other conventional methods.

**$Na_2O$** Chemical formula of sodium monoxide.

**$Na_2O_2$** Chemical formula of sodium peroxide.

**NaOCl** Chemical formula of sodium hypochlorite.

**NaOH** Chemical formula of caustic soda or sodium hydroxide.

**$Na_2O(SiO_2)_{3.25}$ approx** Chemical formula of water glass.

**NAPC** Acronym of nominal average power consumption, a parameter used to determine the adequacy of a sewer transmission system's multiple- or variable-speed pumps; the capacity of a pump station is inadequate if its NAPC exceeds a certain value. *See also* elapsed-time meter.

**naphthalene ($C_{10}H_8$)** A white, crystalline, water-insoluble volatile organic compound that water supply systems are required to monitor; it is used as a moth repellent, as a fungicide, and in industry.

**NAPL** Acronym of nonaqueous phase liquid.

**$Na_3P_3O_{10}$** Chemical formula of sodium tripolyphosphate.

**$Na_3PO_4 \cdot 12H_2O$** Chemical formula of trisodium phosphate.

**$(NaPO_3)_n$** Chemical formula of sodium hexametaphosphate.

**NAPOT** Acronym of nominal average pump operating time, a parameter used to determine the adequacy of sewer transmission systems having fixed-speed pumps; if a pump station's NAPOT exceeds a certain value (e.g., 10 hours a day on a monthly basis), the capacity is inadequate. *See also* elapsed-time meter.

**nappe** (From the French word for sheet). The stream of water overflowing a weir, dam, or spillway. The ratio of the nappe thickness to the crest thickness serves to differentiate between sharp-crested and broad-crested weirs.

**Nara** A paddle dryer manufactured by Komline-Sanderson Engineering Corp.

**narrow-band excimer lamp** A device that uses a corona discharge to form excited dimers and produce monochromatic radiation at a wavelength of 172, 222, or 308 nm. *See also* low-pressure, high-intensity UV lamp; low-pressure, low-intensity UV lamp; medium-pressure, high-intensity UV lamp; pulsed UV lamp. (Excited dimmers = excimers, i.e, two molecules of xenon, krypton, or other gases joined together.)

**$Na_2S$** Chemical formula of sodium sulfide.

**nascent oxygen (O)** Same as elemental oxygen.

**$Na_2SiF_6$** Chemical formula of sodium silicofluoride or sodium fluorosilicate.

**$Na_2SiO_3 \cdot 9H_2O$** Chemical formula of aqueous sodium metasilicate.

**$Na_2SO_3$** Chemical formula of sodium sulfite.

**$Na_2SO_4$** Chemical formula of sodium sulfate.

**$Na_2S_2O_3$** Chemical formula of sodium thiosulfate. Also written: $Na_2S_2O_3 \cdot 5H_2O$.

**$Na_2S_2O_4$** Chemical formula of sodium hydrosulfite. Also written: $Na_2S_2O_4 \cdot 2H_2O$.

**$Na_2S_2O_5$** Chemical formula of sodium metabisulfite.

**$Na_2S_2O_7$** Chemical formula of sodium disulfate.

**$Na_2S_2O_4 \cdot 2H_2O$** Chemical formula of sodium hydrosulfite. Also written: $Na_2S_2O_4$.

**$Na_2S_2O_3 \cdot 5H_2O$** Chemical formula of sodium thiosulfate. Also written: $Na_2S_2O_3$.

**Nasty Gas™** A regenerative blower manufactured by EG&G Rotron, Inc. for the removal of noxious gases.

**National Drinking Water Advisory Council** A panel established under the Safe Drinking Water Act and selected from the general public, state and local agencies, and private organizations to advise the USEPA on matters related to potable water.

**National Environmental Policy Act (NEPA)** A United States law passed in 1969 and declaring a national policy of protection and enhancement of the environment. As a result of this act, the Council of Environmental Quality (CEQ) and the federal Environmental Protection Agency (EPA or USEPA) were established. The act requires the preparation of environmental impact assessments or statements for public and private actions that may have major environmental consequences.

**National Estuary Program** A program established under the Clean Water Act Amendments of 1987 to develop and implement conservation and management plans for protecting estuaries and restoring and maintaining their chemical, physical, and biological integrity, as well as controlling point and nonpoint pollution sources.

**National Geodetic Vertical Datum** *See* NGVD.

**National Institute for Occupational Safety and Health (NIOSH)** An organization that tests and approves safety equipment for particular applications. NIOSH is the primary federal agency engaged in research in the national effort to eliminate on-the-job hazards to the health and safety of working people. The NIOSH Publications Catalog contains a listing of documents, mainly on industrial hygiene and occupational health.

**National Interim Primary Drinking Water Regulation (NIPDWR)** A USEPA regulation published in December 1975 and based on 1962 drinking water standards. It contains maximum contaminant levels (MCL) for several organic and inorganic chemicals, physical parameters, radioactivity, and bacteriological factors, with turbidity listed as a health-related instead of an aesthetic parameter.

**National Municipal Plan** A policy created in 1984 by USEPA and the states to bring all publicly owned treatment works into compliance with the Clean Water Act requirements.

**National Oil and Hazardous Substances Contingency Plan** The federal regulation that guides determination of the sites to be corrected under both the Superfund program and the program to prevent or control spills into surface waters or elsewhere.

**National Pollutant Discharge Elimination System (NPDES)** A provision of the Clean Water Act (CWA) that prohibits the discharge of pollutants into waters of the United States unless a special permit is issued by the USEPA, a state, or, where delegated, a tribal government on an Indian reservation. This national program issues, modifies, revokes, reissues, terminates, monitors, and enforces permits, and imposes and enforces pretreatment requirements under sections 307, 318, 402, and 405 of CWA (EPA-40CFR122.2). An amendment of the CWA (the Water Quality Act of 1987) established a phased approach for stormwater discharge regulation. The permits regulate discharges from industries, municipal treatment plants, landfills, and other point sources.

**National Pollution Discharge Elimination System (NPDES)** The commonly used term is National Pollutant Discharge Elimination System.

**national pretreatment standard** Any regulation containing pollutant discharge limits promulgated by the USEPA in accordance with the Clean Water Act, which applies to industrial users of a publicly owned treatment works. It further means any state or local pretreatment requirement applicable to a discharge and that is incorporated into a permit issued to a publicly owned treatment works under the national pollutant discharge elimination system (EPA-40CFR117.1). Sometines called pretreatment standard or, simply, standard.

**National Primary Drinking Water Regulation (NPDWR)** Federally enforceable drinking water regulations that set maximum contaminant levels and applicable treatment techniques. *See* Primary Drinking Water Regulation.

**National Priorities List (NPL)** USEPA's list of the most serious uncontrolled or abandoned hazardous waste sites identified for possible long-term remedial action under Superfund. The list is based primarily on the score a site receives from the Hazard Ranking System. USEPA is required to update the NPL at least once a year. A site must be on the NPL to receive money from the Trust Fund for remedial action.

**National Research Council data** Data collected from trickling filters at military installations and used in the formulation of trickling filter design. *See* NRC formula.

**National Research Council equations** *See* NRC equations.

**National Rivers Authority** The agency set up in 1989 to oversee the activities of regional water companies in England and Wales.

**National Safe Drinking Water Regulation** Same as Primary Drinking Water Regulation.

**National Sanitation Foundation** Former name of NSF International.

**National Secondary Drinking Water Regulation (NSDWR)** *See* Secondary Drinking Water Regulation.

**National Storm Water Program** A two-phase program established under a provision of the 1987 water quality amendment of the Clean Water Act (CWA): (1) Regulation of stormwater discharges associated with industrial activity, stormwater discharges from large and medium-size municipal separate storm sewer systems (MS4), and stormwater discharges contributing to a violation of a water quality standard or contributing pollutants significantly to waters of the United States. (2) Identification of additional sources of stormwater contamination and establishment of procedures and methods to control them.

**native salt** Another name of halite, a soft white, red, yellow, blue, or colorless evaporite mineral that is at least 95% sodium chloride (NaCl) in cubic crystals; it occurs together with other minerals such as gypsum and sylvite. Also called rock salt, fossil salt.

**natrium** Latin name of sodium (Na).

**natural air draft** *See* natural draft.

**natural amino acid** Any of a number of constituent compounds of proteins, e.g., glycine: $CH_2—NH_2—CO_2H$.

**natural attenuation** The normal degradation of contaminants by indigenous microorganisms without any stimulation or treatment. Also called intrinsic bioremediation.

**natural background radiation** Naturally occurring radiation, e.g., cosmic rays from outer space. *See also* natural radioactivity.

**natural chlorine compound** Compounds containing chlorine and found naturally include inorganic chlorides, methyl chloride produced by marine algae, and more than 2000 organic compounds.

**natural-circulation evaporator** A single-pass device similar to a heat exchanger in which liquid flow is driven by density gradients. *See also* batch, falling-film, and rising-film evaporators.

**natural decay** The decay of contaminants in nature as a result of death, photooxidation, and other phenomena. First-order kinetics is usually assumed for natural or radioactive decay:

$$R = k_d N \qquad \text{(N-01)}$$

where $R$ = rate of decay, number/time; $k_d$ = first-order reaction rate coefficient,/time; and $N$ = number of organisms remaining, dimensionless.

**natural disinfection** As defined in USEPA's Groundwater Disinfection Rule, a source water treatment via virus attenuation by natural subsurface processes such as virus inactivation, dispersion (dilution), and irreversible sorption to aquifer-framework solid surfaces.

**natural draft** The mechanism that provides airflow in the operation of a trickling filter to aerate the bed and prevent odors without using forced ventilation. It is driven by the temperature difference between ambient air and the air inside the pores of the bed. The pressure head resulting from the temperature and moisture differences, called draft or natural air draft, $D_{air}$ (mm of water), may be calculated as (Metcalf & Eddy, 2003):

$$D_{air} = 353\, H(1/T_c - 1/T_w) \qquad \text{(N-02)}$$

where $H$ = height of tricking filter, m; $T_c$ = colder temperature, °K; and $T_w$ = warmer temperature, °K.

**natural draft cooling tower** Same as natural tower.

**natural draft tower** An evaporative cooling tower that is designed to use density differences between air inside and air outside the tower to induce and circulate the cooling air. As compared to a mechanical draft tower, it is noiseless but creates mist and fog.

**natural eutrophication** The increase in the growth of aquatic animal and plant life due to overfertilization of water bodies from such natural causes as runoff from uncultivated or undeveloped lands. *See also* cultural eutrophication.

**natural filtration** The use of soils or aquifers as filtering media, as in the riverbank filtration or in situ processes.

**natural flotation** A wastewater treatment process that uses a quiescent tank for substances lighter than water (e.g., grease, oil) to rise, without flotation agents, to the surface where they are removed by skimming. *See also* air flotation.

**natural flow** The flow of water past a given point of a stream that is not subject to diversion by storage, import, export, return flow, or any other man-made modification. *See* virgin flow.

**natural freeze–thaw dewatering** The use of the natural freezing and thawing cycles in cold climates to increase the solids content of sludge to 20–50%. The device used is similar to a sand drying bed or deep lined underdrained trenches.

**natural gas** (1) A natural, combustible mixture of gaseous hydrocarbons and nonhydrocarbons found in porous sedimentary formations and consisting mainly (80%) of methane ($CH_4$). Natural gas also contains small quantities of butane, ethane, propane, and other gases. It is used as a fuel and in industrial applications. There are also natural gases composed maily of carbon dioxide ($CO_2$) and hydrogen sulfide ($H_2S$). (2) Methane gas resulting from the decomposition of wastewater and sludge (sewage gas) or from organic decomposition in a landfill. *See* digester gas.

**natural groundwater** Groundwater that is not subject to contamination by human activities.

**natural logarithm (ln)** A logarithm with base $e$ = 2.718, such that, if $y = e^x$, then $\ln y = x$. Also called Napierian logarithm. *See also* common logarithm.

**naturally soft water** Water that contains relatively low concentrations of hardness-causing salts and does not require softening treatment for municipal or industrial uses.

**natural organic material** *See* natural organic matter.

**natural organic matter (NOM)** A diverse mixture of substances with varying chemical and physical characteristics, present in surface and groundwater, including hydrophobic humic and fulvic acids as well as nonhumic elements. Derived from decaying organic matter and dead organisms, they contribute to the color of water and have the potential to react with chlorine and other disinfectants to form disinfection by-products (DBP) that may have adverse health effects. NOM can also foul ion-exchange resins. *See* activated carbon adsorption, enhanced coagulation, and enhanced softening as treatment processes for NOM reduction. NOM is an important source of dissolved organic matter in natural waters.

**natural percolation** The portion ($NP$) of total percolation attributable to rainfall that exceeds the evaporation rate. It may be calculated as the difference between rainfall (R) and crop evapotranspiration ($ET$):

$$NP = R - ET \qquad \text{where } R > ET \qquad \text{(N-03)}$$

$$NP = 0 \qquad \text{where } ET > R \qquad \text{(N-04)}$$

**natural phosphorus removal** The removal of phosphorus from wastewater using a natural treatment system through such mechanisms as chemical precipitation with calcium and iron or aluminum, adsorption by clay minerals and organic soil constituents, and plant uptake. *See* natural treatment system.

**natural pollutant** A constituent or substance of natural origin that is considered a pollutant when present in excessive concentration, e.g., volcanic dust, sea salt particles, ozone from lightning or photochemical reactions, and products of forest fires.

**natural pollution** Pollution that is not caused directly or indirectly by human activities, but by geologic events such as volcanic eruptions, spontaneous forest fires, and seashore algal blooms. An example is the dissolution of minerals from rocks into ground or surface water. *See also* natural eutrophication.

**natural polyelectrolyte** Any of the polymers of biological origin or made from cellulose derivatives and alginates. They are used to promote particle destabilization and aggregation in water and wastewater treatment. *See also* synthetic polyelectrolyte.

**natural purification** The slow reduction or elimination of contaminants from natural waters by such physical, chemical, and biological phenomena or factors as atmospheric reaeration, bottom deposit, currents, dilution, non-point-source pollution, photosynthesis, runoff, sedimentation, sunlight, and temperature. *See* self-purification for more detail.

**natural radioactivity** Radioactivity generated by radioactive elements in the Earth's crust, from atomic number 84 (polonium) to atomic number 92 (uranium) and radioisotopes of lighter elements such as potassium-40. *See also* natural background radiation.

**natural recharge** Aquifer replenishment from precipitation or other natural sources as compared to artificial recharge.

**natural resources** The natural wealth of a country or any region, consisting of renewable and nonrenewable products and features of the land, water, and air; e.g., trees, fish, farmland, mineral deposits, forests, fossil fuels, and solar energy.

**Natural Resources Conservation Service (NRCS)** New name of the U.S. Soil Conservation Service (SCS).

**natural slope** *See* angle of repose.

**natural sludge dewatering** The use of sand drying beds and other drying beds to reduce the water content of sludge. *See* sand bed dewatering, paved drying bed, wedge-wire drying bed, vacuum-assisted drying bed.

**natural softening** (1) A temporary phenomenon that sometimes occurs in small lakes or reservoirs during the summer: algal growth reduces carbon dioxide, increases pH, and causes the precipitation of calcium carbonate, which can be redissolved when conditions change again. (2) The exchange of hardness-causing cations (e.g., calcium and magnesium) for other ions (e.g., sodium and potassium) through the normal flow of groundwater.

**natural sparkling water** Effervescent water in which the carbon dioxide comes from the same source as the water. *See* natural water and sparkling water.

**natural treatment** Wastewater treatment through the physical, chemical, and biological processes that occur naturally through the interactions of microorganisms, plants, soil, water, and the atmosphere.

**natural treatment system** Systems or processes that use natural treatment for wastewater or sludge include aquaculture, constructed and natural wetlands, and land application. Natural treatment systems often produce water of quality comparable to or better than the effluent of advanced wastewater treatment. *See also* sewage farming, slow-rate system, rapid-infiltration system, high-rate sprinkling, overland flow, percolate.

**natural uranium** Uranium that is approximately 99.27% U-238, 0.7% U-235, and 0.006 U-234, as defined by the USEPA. U.S. water supplies contains 0.3–2.0 pCi/L of natural uranium.

**natural values of floodplains and wetlands** The qualities of or functions served by floodplains and wetlands, including (a) water resources (flood moderation, water quality maintenance, groundwater recharge); (b) living resources (fish, wildlife, plant resources, and habitats); (c) cultural resources (open space, natural beauty, scientific study, outdoor education, archaeological and historic sites, recreation); and (d) cultivated resources (agriculture, aquaculture, forestry). (Definition from the code of federal regulations on floodplain management and the protection of wetlands.)

**natural wastewater treatment** The use of natural components (soil, vegetation, microorganisms, and other animal life) for wastewater treatment and disposal. Natural treatment systems include soil absorption, land treatment, ponds, floating aquatic plants, and constructed wetlands.

**natural water** Water as it occurs in nature, i.e., in the hydrologic cycle, including atmospheric moisture, rainwater, surface water, groundwater, ocean water, all containing solid, liquid, and gaseous particles in suspension or in solution. The quality of a natural water depends on its source: groundwater usually contains some calcium and magnesium carbonates, chlorides, sulfates, iron and manganese, whereas surface water may contain particulates, nitrates and phosphates; seawater contains sodium chloride (NaCl) and other salts.

**natural wetland** A naturally inundated area supporting emergent plants in water depths less than 2 feet, as compared to a man-made or constructed wetland. In water quality regulations, natural wetlands are classified as receiving waters and require that wastewater discharges receive secondary or advanced treatment.

**Nautilus®** An apparatus manufactured by Wheelabrator Engineered Systems, Inc. and consisting of a traveling bridge siphon for sludge removal.

**navigable water** Traditionally, waters sufficiently deep and wide for navigation by all or specified vessels; any stream, lake, arm of sea, etc., declared navigable by the Congress of the United States. Such waters in the United States come under federal jurisdiction and are protected by certain provisions of the Clean Water Act.

**NBOD** Acronym of nitrogenous biochemical oxygen demand.

**nbpON** Acronym of nonbiodegradable particulate organic nitrogen.

**$NBr_3$** Chemical formula of tribromamine.

**nbsCOD** Acronym of nonbiodegradable soluble chemical oxygen demand.

**nbsON** Acronym of nonbiodegradable soluble organic nitrogen.

**$NCl_3$** Chemical formula of trichloramine.

***n*-butane ($CH_3$—$CH_2$—$CH_2$—$CH_3$)** *See* butane.

***n*-butyl ($CH_3$—$CH_2$—$CH_2$—$CH_2$—)** *See* butyl.

***n*-butylbenzene** *See* butylbenzene.

***n*-butyric acid ($CH_3$—$CH_2$—$CH_2$—COOH)** *See* butanoic or butyric acid.

**nbVSS** Acronym of nonbiodegradable volatile suspended solids.

**$NCl_3$** Chemical formula of trichloramine or nitrogen trichloride.

**NCWS** Acronym of noncommunity water system.

**NDMA** Acronym of N-nitrosodimethylamine.

**NDP** Acronym of net driving pressure.

**NDWAC** Acronym of National Drinking Water Advisory Council.

**nearfield** A region close to the port of discharge of wastewater or effluent into the ocean or a large lake, where the effluent forms a buoyant, rapidly rising plume. *See* discharge nearfield for detail.

**near the first service connection** In drinking water regulations, any of the first 20% of water connections from the treatment plant, as measured by flow time in the distribution network.

**neat solution** A full-strength solution.

**nebulizer** In general, a device that reduces liquids to fine sprays. In water laboratories, nebulizers are used to introduce water and wastewater samples into atomic absorption spectrophotometers.

***Necator americanus*** A latent, persistent pathogen excreted in human feces. It causes the hookworm disease in Central America, South America, and the Caribbean. Also called New World hookworm. *See also Ancylostoma duodenale.*

**necrosis** Death of cells caused by exogenous, chemical or biological, factors.

**needle valve** A device that controls flow through a circular outlet by a tapered needle that advances into or retreats from the outlet.

**needle weir** A type of movable-frame weir made of vertical square timbers. *See also* rolling-up curtain weir.

**negative aeration** The condition of an aerated static pile composting system whose blowers draw air down through the pile, allowing it to heat up more quickly and facilitating the collection and treatment of odorous gases. Also called negative ventilation.

**negative catalyst** A substance that decreases the speed or yield of a chemical or biological reaction without being consumed or chemically changed by the reaction; e.g., dissolved metals, enzymes. Also called a retarding catalyst or an inhibitor.

**negative charge** The electrical potential of an electron; the charge carried by anions like arsenate, chloride, chromate, fluoride, fulvate, humate, nitrate, selenate, and sulfate, which have more electrons than protons.

**negative hardness** The difference between alkalinity (Alk) and the total calcium and magnesium compounds as calcium carbonate equivalent (mg/L) when alkalinity exceeds total hardness.

**negative head** (1) A head loss in excess of the static head, as in a partial vacuum. (2) A pressure below atmospheric pressure in a filter bed that experiences clogging or other causes of high head losses, a condition that may result in the formation and release of gases, a condition called air binding. To avoid negative heads, terminate filter runs before the total head loss reaches the submergence depth of the medium. Also called negative pressure.

**negatively charged filter** A filter with a net negative charge, which requires lowering of the pH for virus adsorption.

**negative metal oxide semiconductor (NMOS)** A semiconductor that includes a negatively charged silicon structure.

**negative pressure** Pressure less than atmospheric pressure. *See also* negative head.

**negative pressure valve** A valve that closes automatically before pressure in a pipeline or system drops below atmospheric pressure.

**negative sample** A sample that does not show the presence of a specific organism in a test, e.g., the presence of coliform bacteria in the multiple-tube fermentation technique.

**negative ventilation** Same as negative aeration.

***Neisseria* genus** A group of heterotrophic organisms active in biological denitrification.

**nekton** An animal that can move about entirely on its own power.

**Nemathelminthes** A phylum of worms with elongated, unsegmented, cylindrical bodies, including the nematodes and hair worms.

**nematocide** A chemical agent that destroys nematodes.

**Nematoda phylum** A large group of unsegmented worms, including some free-living in soil and water, and many parasites. *See* nematode.

**nematode** An invertebrate animal of the phylum Nematoda and class nematode, that is, an unsegmented round worm with elongated, fusiform, or saclike body covered with cuticle, and inhabitating soil, water, plants, or plant parts; may also include nemas and eelworms (Federal Insecticide, Fungicide and Rodenticide Act of 1947, or FIFRA). Nematodes cause a number of intestinal infections, e.g., filariasis and loiasis and other water-related diseases: guinea worm, cercarial dermatitis, and schistosomiasis. They are also found in the biomass active in biological treatment. *See Ascaris lumbricoides, Trichuris triciura, Necator Americanus, Ancylostoma duodenale,* hookworm, pinworm.

**nematode worm** A nematode.

**Nemo®** A progressing cavity pump manufactured by Netzsch, Inc.; includes a macerator.

**neonatal** Pertaining to the first four months after birth.

**neoplasm** A new growth of abnormal tissue; tumor.

**neoprene** An oil-resistant synthetic rubber used in linings of tanks and other containers.

**Neosepta®** A stack manufactured by Graver Co. for electrodialysis membranes.

**Neozone™** An ozone generator manufactured by Northeast Environmental Products, Inc.

**NEPA** Acronym of National Environmental Policy Act.

**nephelometer** (From the Greek prefix nephelo, meaning cloud.) (1) A precalibrated instrument for measuring the concentration of a suspension—as of bacteria, suspended solids, or other substances—by its scattering of a beam of light. For example, turbidity of a water sample is measured by passing light through the sample and measuring the amount of light deflected by the particles. Also called nephelometric turbidimeter or, simply, turbidimeter. (2) An apparatus that uses barium chloride standards to measure the number of bacteria in a sample.

**nephelometric** Pertaining to means of measuring turbidity by passing light through a sample and measuring the amount of light deflected.

**nephelometric method** The use of a nephelometer to measure turbidity. It is the preferred method for turbidity measurements, because of its precision and applicability, and the one approved by the USEPA. *See also* the visual method.

**nephelometric turbidimeter** Same as nephelometer.

**nephelometric turbidity** The turbidity of a sample of water as measured by a nephelometer.

**nephelometric turbidity unit (NTU)** A unit of turbidity as determined by a nephelometer: one NTU corresponds to the concentration of 1.0 mg/L of silica ($SiO_2$). Treated drinking water usually has a turbidity lower than 1.0 NTU, whereas highly turbid waters have more than 1000 NTUs. *See also* JTU, FTU, visual method.

**nephelometry** (1) The application of the nephelometric method. (2) The use of a nephelometer to determine the number of bacteria in a sample.

**nephritis** Inflammation of the kidneys, which may be caused by the consumption of lead-containing water, e.g., rainwater collected from lead-clad roofs.

**neritic zone** (1) The portion of the ocean that includes the estuarine zone and the continental shelf. (2) The region lying directly above the sublittoral zone of the sea bottom.

**Nernst equation** The equation, related to free energy change ($\Delta G$), that describes the potential ($E$) of a chemical reaction:

$$E = -(\Delta G)/(n \cdot F) \qquad \text{(N-05)}$$

$$E^0 = -(\Delta G^0)/(n \cdot F) \qquad \text{(N-06)}$$

$$E = E^0 - 2.3\,(RT/n \cdot F) \log (\{\text{red}\}/\{\text{ox}\}) \qquad \text{(N-07)}$$

where $n$ = number of electrons transferred (number of equivalents/mole), $F$ = Faraday constant = 96,493 coulombs/equivalent, $E^0$ = standard potential, $\Delta G^0$ = standard free energy change, {red} = activity of the reduced species in a reduction reaction, and {ox} = activity of the oxidized species in a reduction reaction. The ratio {red}/{ox} is the reaction quotient. The Nernst equation is used to calculate the equilibrium constant ($K$) for oxidation–reduction reactions:

$$\ln K = nFE^0/RT \qquad \text{(N-08)}$$

See oxidation–reduction potential.

**Nernst potential** In the electrical double layer theory of electrostatic stabilization, the maximum electric potential between the surface of a particle and the bulk of the solution is at the particle surface and is called the Nernst potential. *See also* zeta potential.

**nesquehonite** Aqueous magnesium carbonate: $MgCO_3 \cdot 3H_2O$.

**nesslerization** A method used in laboratories to determine the presence of ammonia in a water sample by adding Nessler reagent and comparing the mixture to standard solutions visually or through an absorption spectrophotometer.

**Nesslerizer** A box in which Nessler tubes are placed for comparison to standard or dummy tubes in conducting colorimetric measurements.

**Nessler reagent** An alkaline solution of mercuric iodide (HgI), sodium hydroxide (NaOH), and potassium iodide (KI) used in the colorimetric procedure (nesslerization) to determine the presence of ammonia.

**Nessler tube** A clear-glass tube with a flat bottom used in the laboratory for comparing opacity or color density, or for other colorimetric measurements (e.g., DPD test, nitrite determination, orthotolidine test).

**net available head** In a waterwheel, the difference between the pressure elevation in the power conduit before and after the waterwheel. *See also* net head.

**net biomass yield** The ratio of the net biomass growth rate to the substrate utilization rate. *See also* observed yield, synthesis yield or true yield, solids production or solids yield.

**net calorific value** The amount of heat released by a fuel in a combustion chamber minus the dry flue gas and the moisture losses. *See* available heat for detail.

**net driving force** Same as net driving pressure.

**net driving pressure (NDP)** The net pressure that is available to operate a pressure-driven membrane apparatus (e.g. water treatment by reverse osmosis). It is equal to the hydraulic pressure differential minus the osmotic pressure differential and may be computed as

$$NDP = P_{fh} + P_{po} - P_{ph} - P_{fo} \qquad \text{(N-09)}$$

where $P_{fh}$ is the net feed pressure, $P_{po}$ the permeate osmotic pressure, $P_{ph}$ the permeate line pressure, and $P_{fo}$ the feedwater osmotic pressure.

**net head** The head available for the production of hydroelectric power after the deduction of frictional, entrance, and other losses, except turbine losses. Also called effective head. *See also* net available head.

**net heating value** The amount of heat released by a fuel in a combustion chamber minus the dry flue gas and the moisture losses. *See* available heat for detail.

**net irrigation water requirement** The amount of water required by a crop to replace water lost through evapotranspiration and provide for leaching, germination, etc. The following formula may be used to estimate the net requirement ($R$, in):

$$R = (ET - P)/(1 - L/100) \qquad \text{(N-10)}$$

where $ET$ = evapotranspiration, in; $P$ = precipitation, in; and $L$ = leaching requirement as a percent of applied water. *See also* total irrigation water requirement.

**net operating pressure** In a membrane process, the pressure applied minus the osmotic pressure difference, an important parameter for evaluating membrane filtration performance. *See also* membrane flux, normalized specific flux, rejection, specific flux, temperature correction factor.

**net positive suction head (NPSH)** The minimum suction head ($H'$) required for a pump to operate, or the absolute pressure at the suction intake of the pump. It varies with the absolute gage pressure at the centerline of the pump intake ($p_i$), the vapor pressure ($p_v$) and specific weight ($\gamma$) of the liquid being pumped, the velocity ($V$) at the intake, and gravitational acceleration ($g$):

$$NSPH = H' = (p_i - p_v)/\gamma + V^2/2g \qquad \text{(N-11)}$$

The allowable NSPH may be less than zero for pumps of high specific speeds. The NSPH is used to define the cavitation parameter.

**net present value (NPV)** The algebraic sum of the discounted cash flows over the planning period (time horizon) of a project:

$$NPV = \sum_{i=1}^{N} F_i/(1 + r)^i \qquad \text{(N-12)}$$

where $N$ is the number of time periods (e.g., years, months) in the planning period, $F_i$ is the cash flow in the $i$th period, and $r$ is the discount rate expressed as a decimal. For constant future cashflows, see the present value factor. *See* financial indicators for related terms.

**net proton charge (NPC)** The charge due to the binding of protons ($H^+$ ions), equal to the product of the Faraday constant ($F$) and the difference between the sorption densities of the hydrogen ($\Gamma_H$) and hydroxyl ($\Gamma_{OH}$) ions:

$$NPC = F(\Gamma_H - \Gamma_{OH}) \qquad \text{(N-13)}$$

Also called surface protonation. *See also* point of net proton charge, point of zero salt effect.

**net proton charge density** Same as net proton charge.

**net proton surface charge density** Same as net proton charge.

**net soil evaporation** A parameter ($E_n$) used in determining the design loading rate for a dedicated land disposal site and calculated as a function of the pan evaporation rate ($E_p$) and annual precipitation ($P$), all expressed in mm/yr or in/yr (Metcalf & Eddy, 2003):

$$E_n = 0.7\,E_p - P \qquad \text{(N-14)}$$

**net specific growth rate** For suspended growth treatment processes, the net specific growth rate or specific biomass growth rate ($\mu$) is the reciprocal of the solids retention time ($\theta$) and is related to the specific substrate utilization rate ($U$) and the food-to-microorganism ratio ($F/M$):

$$\mu = 1/\theta = YU - k_d = Y(F/M)E/100 - k_d \quad \text{(N-15)}$$

where $Y$ = yield coefficient (or biomass yield); $E$ = BOD or COD removal efficiency, %.

**Net-Waste** A screw press manufactured by Diemme USA.

**net yield** *See* net biomass yield.

**neurotoxicity** One of the factors considered in establishing no observed adverse effect levels (NOAEL).

**neurotoxin** A toxin affecting the nervous system, produced by certain species of algae and other organisms. It may cause twitching, muscle contraction, convulsions, and even death to humans and livestock and are not effectively removed by conventional water treatment processes. *See also* hepatoxin.

**neuston** (1) The aggregate of microorganisms and small organisms (e.g., mosquito larvae, pond-skaters) floating or swimming in the surface film of a body of water. (2) The neuston layer.

**neuston layer** The air–water interface of a body of fresh or seawater, a thin (1–10 microns), but nutrient-rich biofilm between the air and the limnetic or pelagic zone. The neuston layer is also subject to intense solar radiation, large temperature fluctuations, and toxic substances. *See* freshwater profile and marine profile for related terms.

**neutral** (1) Pertaining to an element or substance that has neither an excess nor a lack of electrons. (2) Pertaining to water that has a pH of 7.0, i.e., an equal number of free hydrogen ($H^+$) and hydroxyl ions ($OH^-$).

**neutral depth** Same as normal depth.

**neutralism** The condition that exists between species or organisms that do not interact much with each other.

**Neutralite** Filter media manufactured by U.S. Filter Corp. for use in the treatment of acidic wastewaters.

**neutralization** (1) A chemical process used to decrease the acidity or alkalinity of a substance by adding alkaline or acidic minerals, respectively, so as to obtain a solution that has a near neutral pH (pH of about 7.0), through the simple reaction,

$$H^+ + OH^- \rightarrow H_2O \quad \text{(N-16)}$$

Neutralization is required for effluents with a pH too high or too low, as well as for scaling control or corrosion control. *See also* lime neutralization treatment, limestone treatment, lime-slurry treatment, caustic soda treatment, waste boiler–flue gas treatment, carbon dioxide treatment, sulfuric acid treatment. (2) A chemical reaction in which water is formed by hydrogen ions from an acid and hydroxyl ions from a base. (3) An acid pickling operation that does not include acid recovery or acid regeneration processes (EPA-40CFR420.91-i). (4) The inhibition of the infectivity of a virus when an antibody binds the cell's protein.

**neutralizer** (1) An alkaline compound used to neutralize an acidic water; e.g., calcium carbonate ($CaCO_3$) and magnesium oxide (MgO). (2) An acidic compound used to neutralize an alkaline water; e.g., hydrochloric acid (HCl) or acetic acid ($CH_3COOH$). (3) A filter medium used for carbon dioxide ($CO_2$) reduction and pH adjustment.

**Neutralizer Plus™** Filter media produced by Matt-Son, Inc. for use in pH adjustment.

**neutralizing agent** A chemical compound sprayed or atomized in fine mists to react with, neutralize, or dissolve odorous compounds in wastewater treatment units that emit objectionable odors. *See also* masking agent.

**neutral molecule** A molecule that does not have a net electrical charge or dipole moment, but has a dynamic distribution of charge that makes interaction between two molecules possible. In adsorption theory, there is a weak attraction between neutral molecules, called dispersion force or London–van der Waals force.

**neutral salt splitting** A typical first step in the reactions that occur during the treatment process of strong acid cation exchange (SACE), in which a neutral salt such as calcium chloride ($CaCl_2$) is split by a solid ion exchange resin [$RSO_3^- H^+(s)$] while exchanging hydrogen ions for calcium and producing hydrochloric acid (HCl):

$$2\ RSO_3^- H^+(s) + CaCl_2 \leftrightarrow (RSO_3^-)_2Ca^{2+}(s) + 2\ HCl \quad \text{(N-17)}$$

SACE is used to remove carbonate and noncarbonate hardness. *See also* basic salt splitting, weak-acid cation exchange.

**neutral solution** A solution of pH = 7, i.e., of equal concentrations of hydrogen ions ($H^+$) and hydroxyl ions ($OH^-$).

**neutral zone** The space between scum and sludge in a clarifier, thickener, or Imhoff tank settling compartment.

**neutron** An unstable, high-speed and deeply penetrating, neutral elementary particle having a mass number of one and a half-life of approximately 12 minutes; found in all atomic nuclei except those of hydrogen. Neutrons are involved in fission and fusion reactions. *See also* electron, proton.

**neutron activation analysis (NAA)** A laboratory technique used to determine the concentrations of chemical elements based on a measure of induced radiation.

**Neva-Clog** Filter media manufactured by Liquid-Solids Separation Corp. as metal plates.

**névé** Granular snow on top of a high mountain, compacted into glacial ice; a field of such snow. Also called firn.

**New England sewage farm** A natural sand bed used as a wastewater filter, a precursor of granular wastewater filters.

**newspaper pulp** An additive sometimes used to condition wastewater sludge to produce a dry material by mechanical dewatering.

**newton** The derived SI unit of force; the force required to give a mass of 1 kilogram an acceleration of 1 meter per second per second. *See also* pascal. Named after Sir Isaac Newton (1642–1727), English philosopher and mathematician.

**Newtonian fluids** Water, oil, and other fluids whose viscous shearing stress ($\tau$) is the product of the coefficient of viscosity ($\mu$) by the velocity gradient ($\partial V/\partial s$) between the fluid layers, according to Newton's law of viscosity:

$$\tau = \mu \cdot \partial V/\partial s \qquad \text{(N-18)}$$

where $V$ is the mean velocity and $s$ is the vertical distance. For Newtonian fluids, a pressure drop is proportional to the velocity and viscosity under laminar flow conditions. *See also* viscosity, Reynolds number, non-Newtonian fluid.

**Newton–Raphson method** An iterative technique used to find an approximate solution to the equation $f(x) = 0$. It is based on the truncation of the Taylor series expansion of $f(x)$ after the first derivative term:

$$x_{i+1} = x_i - f(x_i)/f'(x_i) \qquad \text{(N-19)}$$

where $f'(x)$ is the first derivative of $f(x)$ and $x_i$ the solution at step $i$. The procedure starts with an initial estimate of the solution $x_0$ ($i = 0$) to compute $x_1$, which is used to compute $x_2$ and so on until the difference between two successive solutions meets a specified tolerance. The Newton–Raphson method is often used to solve simultaneous nonlinear flow equations.

**Newton's law** The principle that explains the sedimentation of discrete particles. Newton's law of viscosity relates the unit shear force to the dynamic viscosity and horizontal velocity:

$$F_{yx} = -\mu dV_x/dy \qquad \text{(N-20)}$$

where $F_{yx}$ = shear force in the $x$ direction or momentum in the $y$ direction from motion in the $x$ direction; $\mu$ = dynamic viscosity; $V_x$ = velocity in the $x$ direction; and $y$ = vertical direction. *See* terminal settling velocity.

**New World hookworm** *See Necator Americanus.*

**NF** Abbreviation of nanofiltration.

**n factor** The roughness factor used in the Manning formula or Kutter formula. Also called roughness coefficient or roughness factor.

**ng/L** Nanogram per liter.

**NGVD** National Geodetic Vertical Datum; a reference used for elevations. The U.S. Coast Guard and Geodetic Survey uses the "Sea Level Datum of 1929." *See also* MSL, and sea-level datum.

**$NH_3$** Chemical formula of ammonia.

**$NH_4$** Chemical formula of ammonium.

**$N_2H_4$** Chemical formula of hydrazine.

**$NH_2Br$** Chemical formula of monobromamine.

**$NH_2CH_2CH_2O_3H$** Chemical formula of taurine.

**$NH_2CH_2COOH$** Chemical formula of glycine.

**$NHCl_2$** Chemical formula of dichloramine.

**$NH_2Cl$** Chemical formula of monochloramine.

**$NH_4Cl$** Chemical formula of ammonium chloride or sal ammoniac.

**$NH_4HCO_3$** Chemical formula of ammonium bicarbonate.

**$NH_4HF_2$** Chemical formula of ammonium bifluoride.

**$NH_3$-N** Abbreviation of ammonia nitrogen.

**$NH_3O$** Chemical formula of hydroxylamine.

**$NH_4NO_3$** Chemical formula of ammonium nitrate.

**$NH_4OH$** Chemical formula of ammonium hydroxide or aqua ammonia.

**$NH_4SiF_6$** Chemical formula of ammonium silicofluoride.

**$(NH_4)_2SiF_6$** Chemical formula of ammonium fluosilicate.

**$(NH_4)_2SO_4$** Chemical formula of ammonium sulfate.

**$(NH_4)_2S_2O_3$** Chemical formula of ammonium thiosulfate.

**Ni** Chemical symbol of the metallic element nickel.

**niche** The position, function, or status of an organism in a community of plants and animals, which determines its activities and interactions with other organisms. Also called ecological niche.

**nickel (Ni)** A hard, silver-white, malleable, ductile metal, resistant to oxidation, used in alloys (e.g., in stainless steel), in plating operations, in alkaline storage batteries, as a catalyst, and other industrial applications. Atomic weight = 58.69. Atomic number = 28. Specific gravity = 8.9. Valence = 1, 2, or 3. melting point = 1453°C. Boiling point = 2732°C. Water treatment processes effective for nickel removal include lime softening, cation exchange, reverse osmosis, and electrodialysis. The USEPA regulates nickel in drinking water: MCL = MCLG = 100 mg/L.

**nickel carbonyl [$Ni(CO)_4$]** A volatile, toxic, explosive intermediary compound in the extraction of nickel from the impure metal by the direct action of carbon monoxide (CO); a respiratory irritant and possible carcinogen.

**$Ni(CO)_4$** Chemical formula of nickel carbonyl.

**nicotinamide adenine dinucleotide (NAD)** A common electron carrier associated with dehydrogenase enzymes.

**nicotinamide adenine dinucleotide phosphate-cytochrome P-450 (NADPH-cytochrome P-450) reductase** An enzyme that is associated with the reduction of heme iron to the ferrous state, a crucial step in the metabolism of chemical elements.

**nidus rack** A biological treatment unit that consists of an aeration tank with a partially submerged, rotating mechanism on the surface of which microorganisms grow for wastewater stabilization. *See also* aerated-contact bed, bacteria bed, contact bed, intermittent sand filter, percolating filter.

**nightsoil** Human fecal wastes, with or without anal cleaning material, deposited in a bucket or similar container for manual removal, usually at night; collected and transported without flushing water; used as fertilizer on fallow fields, mixed with earth and used as manure, or spread onto sewage farms.

**nightsoil reuse** Nightsoil is considered a natural resource and often reused in agriculture, aquaculture, and biogas production. *See* agricultural reuse.

**NIMBY** Acronym of not in my backyard.

**NIOSH** Acronym of National Institute for Occupational Safety and Health.

**NIPDWR** Acronym of National Interim Primary Drinking Water Regulations.

**nipple** A short tubular pipe section threaded at both ends. *See also* cut pipe.

**Nitox®** An treatment apparatus manufactured by TIGG Group, using adsorption on activavted carbon.

**Nitra-Select™** Media produced by Matt-Son, Inc. for nitrate removal.

**nitrate ($NO_3^-$)** A stable, oxidized ion of nitrogen and one of the major anions in natural waters; converted from nitrite by nitrifying bacteria; used as fertilizer, a food preservative, and an oxidizing agent. It is also a salt or ester of nitric acid ($HNO_3$), highly soluble in water and prone to leaching from land supporting vegetation. Plant nutrient and inorganic fertilizer (e.g., potassium nitrate [$KNO_3$], sodium nitrate [$NaNO_3$]) that enters water supply sources from mineral deposits, soils, septic systems, latrines, leaching pits, animal feedlots, agricultural fertilizers, manure, industrial wastewaters, sanitary landfills, and garbage dumps. Although not new to the water supply field, nitrate is considered an emerging inorganic contaminant (as of 2007). Above 45 mg/L in drinking water, nitrates may cause methemoglobinemia, particularly in infants (blue baby disease). High nitrate intakes from drinking water or food may also be responsible for gastric cancer. Regulated by the USEPA as a drinking water contaminant: MCL for nitrate/nitrite nitrogen = 10 mg/L. Removal of nitrates from drinking water is expensive; ion exchange, for example, is feasible but the disposal of the waste brine is problematic. *See* membrane biofilm reactor.

**nitrate bacteria** Aerobic bacteria that oxidize nitrogenous material, converting nitrites to nitrates according to the reactions

$$2\,NO_2^- + O_2 \rightarrow 2\,NO_3^- \qquad \text{(N-21)}$$

or

$$2\,HNO_2^- + O_2 \rightarrow 2\,HNO_3^- \qquad \text{(N-22)}$$

Also called nitrate formers or nitro-bacteria. The most commonly noted nitrate bacteria are the genus *Nitrobacter;* there are other autotrophic nitro-bacteria, e.g., *Nitrococcus, Nitroeystis, Nitrospina, Nitrospira. See also* nitrifying bacteria, *Nitrobacter* and *Nitrosomonas.*

**nitrate contamination** *See* nitrate pollution.

**nitrate dumping** The release of nitrate in product water by an ion-exchange column operated beyond nitrate breakthrough; it results from the displacement of nitrates on the resin by sulfates in the feedwater.

**nitrate formers** Same as nitrate bacteria.

**nitrate immobilization** The uptake of nitrate by plants and microorganisms, followed by the reduction of the nitrate to ammonium and its incorporation into the biomasss. *See* assimilatory nitrate reduction for detail.

**nitrate N** *See* nitrate nitrogen.

**nitrate nitrogen ($NO_3$-N)** A stable, oxidized form of nitrogen found in wastewaters. *See* nitrate.

**nitrate pollution** The contamination of freshwater by nitrate from fertilized farmlands or other sources. *See* methemoglobinemia.

**nitrate reduction** A bacterially mediated transformation of the nitrogen ion of nitrate ($NO_3$) to a lower oxidation state. *See also* nitrification, nitrogen fixation, denitrification.

**nitrate removal** The process of reducing the concentration of nitrates in water or wastewater, e.g., by ion exchange in a packed bed or biological denitrification.

**nitration benzene** Another name of benzene.

**Nitraver** Reagents produced by Hach Co. to determine nitrite concentrations.

**nitric acid ($HNO_3$)** A strong mineral acid produced by catalytic oxidation of ammonia with air, with formation of nitric oxide and nitrogen peroxide as intermediate products.

**nitric oxide (NO)** (1) A gas formed by combustion under high temperature and high pressure in an internal combustion engine; it changes into nitrogen dioxide ($NO_2$) in the ambient air and contributes to photochemical smog. (2) An intermediate product in the denitrification process, by which nitrate ($NO_3^-$) is sequentially reduced to nitrite ($NO_2^-$), nitric oxide (NO), nitrous oxide ($N_2O$), and nitrogen ($N_2$).

**nitrification** The biochemical transformation of reduced inorganic nitrogen (ammonia or ammonium nitrogen, $NH_4$-N) to nitrite ($NO_2$) and then nitrate ($NO_3$) nitrogen by nitrifiers, mostly the autotrophic bacteria *Nitrosomonas* and *Nitrobacter:*

$$2\ NH_4^+ + 3O_2 \rightarrow 2\ NO_2^- + 2\ H_2O + 4\ H^+ + \textit{Nitrosomonas}\ \text{cells} \qquad \text{(N-23)}$$

$$2\ NO_2^- + O_2 \rightarrow 2\ NO_3^- + \textit{Nitrobacter}\ \text{cells} \qquad \text{(N-24)}$$

$$2\ NH_4^+ + 4\ O_2 \rightarrow 2\ NO_3^- + 2\ H_2O + 4\ H^+ \qquad \text{(N-25)}$$

The overall nitrification reaction may also be presented as:

$$NH_3 + 2\ O_2 \rightarrow HNO_3 + H_2O \qquad \text{(N-26)}$$

Conditions that promote the process include higher pH, higher dissolved oxygen concentration, and higher temperature. Nitrification can occur in secondary wastewater treatment operating at low organic loading and warm temperature, but it increases oxygen consumption and may cause floating sludge on the following clarifier. Adverse environmental factors include toxicity, metals, and unionized ammonia.

**nitrification biotower** A tall trickling filter, as high as 42 ft, filled with plastic packing, used in tertiary nitrification, i.e., for the oxidation of ammonia nitrogen ($NH_4$-N) to nitrates ($NO_3^-$).

**nitrification–denitrification** The conversion of ammonia nitrogen to nitrites, nitrates, and, finally, nitrogen gas by microorganisms, a process used to remove nitrogen from wastewater. *See* biological nitrogen removal.

**nitrification inhibitor** A chemical that slows down the conversion of ammonium ($NH_4^+$) to nitrate nitrogen ($NO_3$-N), e.g., heavy metals and certain organic compounds that inhibit the growth of nitrifiers. Such wastewater characteristics as alkalinity, dissolved oxygen, pH, solids retention time, and temperature also impact the rate of nitrification.

**nitrification rate** *See* trickling filter nitrification rate.

**nitrification trickling filter design** An empirical approach to the design of trickling filters intended for tertiary nitrification, based on the ammonium nitrogen ($NH_4$-N) surface removal flux:

$$R_n = R_{max} N e^{-rZ}/(K_n + N) \qquad \text{(N-27)}$$

where $R_n$ = $NH_4$-N surface removal rate or nitrification rate at depth $Z$ and temperature $T$, g/m²/day; $R_{max}$ = maximum $NH_4$-N surface removal at temperature $T$, g/m²/day; $N$ = bulk liquid $NH_4$-N concentration, mg/L; $r$ = an empirical constant related to the rate decrease as a function of depth, dimensionless; $Z$ = depth in trickling filter, m; and $K_n$ = half-velocity constant, mg/L.

**nitrifier** An organism that is involved in the biochemical transformation of inorganic nitrogen to nitrite and nitrate. *See* nitrification, nitrifying bacteria.

**nitrifying bacteria** Bacteria involved in the transformation of inorganic nitrogen to nitrite and nitrate, mainly the *Nitrosomonas* genus (*N. europaea, N. eutrophus, N. marina*) and *Nitrobacter* genus (*N. hamburgensis, N. vulgaris, N. winogradskyi*), but also other ammonium oxidizing genera (*Nitrosococcus, Nitrosospira, Nitrosolobus, Nitrosovibrio*) and nitrite oxidizing genera (*Nitrosospina, Nitrococcus, Nitrospira*).

**nitrile** (1) An organic compound that contains nitrogen; a cyanide. (2) Acrylonitrile.

**nitrilotriacetic acid (NTA)** An organic compound that now replaces phosphate in detergents. Guideline value recommended by WHO: 0.2 mg/L in drinking water.

**nitrite ($NO_2^-$)** (1) An unstable ion of nitrogen, an intermediate product in the nitrification process; a salt or ester of nitrous acid, converted from ammonia and organic nitrogen by bacteria; used in food preservation. Nitrite is usually not a significant constituent of natural waters; it occurs as an intermediate oxidation state of nitrogen within a narrow pH range. It is regulated by the USEPA. *See also* methemoglobinemia, nitrate, nitrogen dioxide. (2) An intermediate product in the denitrification process, where nitrate ($NO_3^-$) is sequentially reduced to nitrite ($NO_2^-$), nitric oxide (NO), nitrous oxide ($N_2O$), and nitrogen ($N_2$). (3) Same as sodium nitrite.

**nitrite bacteria** Aerobic bacteria that convert ammonia nitrogen into nitrites according to the reaction

$$2\ NH_4^+ + 3\ O_2 \rightarrow 2\ NO_2^- + 4\ H^+ + 2\ H_2O \quad \text{(N-28)}$$

or

$$2\ NH_3 + 3\ O_2 \rightarrow 2\ HNO_2 + 2\ H_2O \quad \text{(N-29)}$$

Also called nitrite formers or nitroso-bacteria. The most commonly noted nitrite bacteria are the genus *Nitrosomonas,* but there are other nitroso-bacteria, e.g., *Nitrosococcus, Nitrosolobus, Nitrosorobrio, Nitrosospira. See also* nitrifying bacteria, *Nitrobacter* and *Nitrosomonas.*

**nitrite formers** Same as nitrite bacteria.

**nitrite ion** *See* nitrite (1).

**nitrite N** Short for nitrite nitrogen.

**nitrite nitrogen ($NO_2$-N)** A relatively unstable form of nitrogen, which is easily oxidized to nitrate; usually found below 1.0 mg/L in wastewater or 0.1 mg/L in natural waters. *See* nitrite.

***Nitrobacter*** A genus of autotrophic bacteria most commonly noted for the conversion of nitrites to nitrates. *See* nitrate bacteria, nitrification.

**nitro-bacteria** Bacteria that oxidize nitrogenous material, converting nitrites to nitrates. *See* nitrate bacteria and nitroso-bacteria.

**nitrobenzene ($C_6H_5NO_2$)** A pale yellow, toxic, water-soluble, synthetic organic liquid used as a solvent and in the manufacture of aniline. Also called nitrobenzol.

**nitrobenzol** Nitrobenzene.

**nitrochloroform ($CCl_3NO_2$)** *See* chloropicrin.

***Nitrococcus*** A genus of autotrophic bacteria that can oxidize nitrite. *See* nitrification.

***Nitrocystis*** A genus of autotrophic bacteria that can oxidize nitrite. *See* nitrification.

**nitrogen (N)** A colorless, odorless, chemically inactive gaseous element of molecular formula $N_2$ (78% of the atmosphere by volume), essential for microbial and plant growth, available in substantial quantities in the soil, present in water and wastewater as ammonia ($NH_3$), organic nitrogen, nitrite ($NO_2^-$), and nitrate ($NO_3^-$). An important natural source of nitrogen is saltpeter or sodium nitrate ($NaNO_3$). Atomic weight = 21.0067. Atomic number = 7. Concentrations are usually expressed in terms of elemental nitrogen (N). Excess nitrogen in wastewater effluents can contribute to eutrophication of receiving waters. Nitrogen as ammonia is toxic to fish even at low levels (0.01 mg/L). The principal sources of nitrogenous compounds include substances of animal and plant origin, sodium nitrate ($NaNO_3$), and atmospheric nitrogen.

**nitrogenase** The enzyme complex that catalyzes the reaction of nitrogen fixation.

**nitrogen chloride ($NCl_3$)** A compound formed during water chlorination and resulting from the reaction of chlorine with ammonia; it can cause objectionable taste and odor if not removed (e.g., by aeration, dechlorination, or exposure to sunlight). *See* trichloramine.

**nitrogen cycle** A graphical representation of nitrogen's conservation and pathways in nature (a) as it circulates through the soil, water, and air, and (b) as it changes form from living animal matter through decomposition of dead organic matter, to plant life and back to living animal matter. See Figure N-01.

**nitrogen dioxide ($NO_2$)** A reddish-brown gas resulting from combustion or from the combination of nitric oxide with oxygen in the atmosphere. A major component of photochemical smog and a primary air pollutant, it causes respiratory problems at low concentrations. *See also* nitrite.

**nitrogen fertilizer** An essential nitrogen-containing plant nutrient but also a cause of algal blooms in some water bodies.

**nitrogen fixation** (1) The biological or chemical process by which elemental nitrogen ($N_2$, gas) from the air is converted to organic nitrogen or available nitrogen ($NH_4^+$, $NO_3^-$); e.g., when cyanobacteria (or blue-green algae) abstract nitrogen from the atmosphere. (2) The biologically mediated production of nitrogen from the atmos-

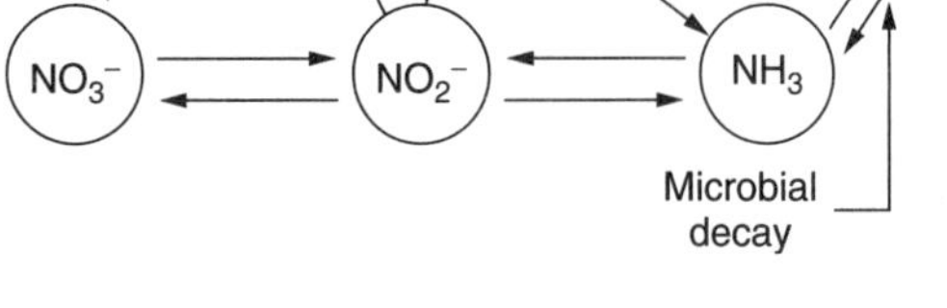

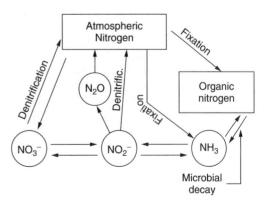


**Figure N-01.** Nitrogen cycle.

phere. (3) The production of nitrogen oxides by lightning. (4) The industrial production of nitrogen in fertilizers, e.g., through the Haber process.

**nitrogen fixation process** The conversion of atmospheric nitrogen to ammonia is represented by various reactions, e.g.:

$$3\ CH_2O + 2\ N_2 + 3\ H_2O + 4\ H^+ \rightarrow 3\ CO_2 + 4\ NH_4^+ \quad \text{(N-30)}$$

$$N_2 + 3\ H_2 \rightarrow 2\ NH_3 \quad \text{(N-31)}$$

$$N_2 + 8\ H^+ + 8\ e^- + 16\ \text{Mg-ATP} \rightarrow 2\ NH_3 + H_2 + 16\ \text{Mg-ADP} + 16\ P_i \quad \text{(N-32)}$$

where ATP = adenosine triphosphate, ADP = adenosine diphosphate, and $P_i$ = inorganic phosphate. *See also* Figure N-01. *See also* biological dinitrogen fixation.

**nitrogen-fixing organisms** Free-living nitrogen-fixing bacteria (e.g., *Azotobacter, Clostridium pasteurianum*), bacteria that live symbiotically in root nodules of plants, actinomycetes, and cyanophyta.

**nitrogen fixing plants** A plant that can assimilate and fix atmospheric nitrogen through the bacteria in its roots.

**nitrogen limiting application rate** *See* nitrogen loading rate.

**nitrogen loading rate** (1) The limiting application rate of sludge to land, based on nitrogen requirements, considering that sludge is a slow-release fertilizer. It may be determined from the following equations:

$$A_N = U/N_P \quad \text{(N-33)}$$

$$N_P = [(NO_3) + K(NH_4) + MN_0]F \quad \text{(N-34)}$$

where $A_N$ = sludge loading rate based on nitrogen, tons N/acre/year; $U$ = crop uptake of nitrogen, pounds/acre/year, e.g., alfalfa = 200 – 600, barley = 110; NP = nitrogen available to crop in sludge, pounds/ton; $(NO_3)$ = fraction of nitrate nitrogen in sludge; $K$ = volatilization factor for ammonia loss, from 0.5 to 1.0; $(NH_4)$ = fraction of ammonia nitrogen in sludge; $M$ = mineralization factor for organic nitrogen = 0.3–0.5 depending on climate; $N_0$ = fraction of organic nitrogen in sludge; and $F$ = factor to convert tons to pounds (1 ton = 2000 pounds). *See also* septage loading rate. (2) From a nitrogen balance, the nitrogen loading rate on (wastewater) land treatment systems is:

$$A'_N = (U' + 0.001\ C_p P_w)/(1 - F') \quad \text{(N-35)}$$

where $A'_N$ = nitrogen loading rate, kg/ha/year; $U'$ = crop uptake of nitrogen, kg/ha/year; e.g., $U'$ = 200–270 for Kentucky blue grass and $U'$ = 115 for oats; $C_p$ = percolate nitrate nitrogen concentration, mg/L; $P_w$ = percolate flow, meters per year; and $F'$ = fraction of applied nitrogen lost to denitrification, volatilization, and soil storage. *See also* hydraulic loading rate, total percolation, combined loading rate, limiting loading rate.

**nitrogen monoxide** *See* nitric oxide, nitrogen oxides.

**nitrogen (nitrate)** *See* nitrate nitrogen.

**nitrogen (nitrite)** *See* nitrite nitrogen.

**nitrogenous** A term used to describe chemical compounds (usually organic) containing nitrogen in combined forms; e.g., proteins and nitrates.

**nitrogenous biochemical oxygen demand (NBOD)** Same as nitrogenous oxygen demand.

**nitrogenous BOD** Same as nitrogenous oxygen demand.

**nitrogenous oxygen demand (NOD)** The quantity of oxygen consumed by the biochemical oxidation of ammonia, organic, and other forms of nitrogen in wastewater, which starts after satisfaction of a good deal of the carbonaceous oxygen demand. Nitrification is a two-step process, in which *Nitrosomonas* bacteria oxidize ammonia to nitrites and *Nitrobacter* convert the nitrites to nitrates. For the average municipal wastewater, the NOD represents about 35% of the total oxygen demand. Also called second-stage biochemical oxygen demand. *See also* first-stage BOD and ultimate BOD.

**nitrogenous wastes** Animal or vegetable residues that contain significant amounts of nitrogen.

**nitrogen oxides ($NO_x$)** Products of the oxidation of nitrogen during the combustion of fossil fuels from transportation and stationary sources and a major contributor to the formation of ozone in the troposphere and to acid deposition. Nitrogen oxides are also produced in the atmosphere under high-energy conditions. They include nitrous oxide or dinitrogen oxide, nitric oxide, and nitrogen dioxide. They react with volatile organic compounds to form smog. The production of nitrogen oxides can be minimized through modifications of the combustion process and treatment of the flue gases.

**nitrogen removal** The removal of nitrogen from wastewater using single or combined biological, chemical, or physical processes, e.g., biological assimilation, nitrification, denitrification, breakpoint chlorination, ion exchange, air stripping.

**nitrogen reservoirs** Major global reservoirs of nitrogen include the atmosphere, the ocean (biomass, dissolved and particulate organics, soluble salts, and dissolved gas), and the land (biota, organic matter, the Earth's crust). The components actively involved in the nitrogen cycle are the ocean biomass, organics and soluble salts (about $9.9 \times 10^{11}$ metric tons) and the land biota (about $2.5 \times 10^{10}$ metric tons).

**nitrogen toxicity** Ammonia nitrogen is toxic to fish and other aquatic organisms, sometimes at levels as low as 0.01 mg/L, its effect depending also on the concentrations of dissolved oxygen and carbon dioxide, on temperature, and on pH.

**nitrogen trichloride ($NCl_3$)** One of the products of the reaction of chlorine with ammonia; also called trichloramine.

**nitrogen uptake** The use of nitrogen for cell growth by microorganisms during waste treatment, a secondary reaction in the biological nitrogen removal processes. *See also* biological nutrient requirement.

**nitrophenols** Synthetic organopesticides that contain carbon, hydrogen, nitrogen, and oxygen.

**nitrosamide** *See* nitrosamine.

**nitrosamine** A product of the reaction of aqueous nitrous acid ($HNO_2$) with secondary amines. Nitrosamines and nitrosamides are suspected of being powerful carcinogens; they can be formed in the stomachs of adults who consume water with high levels of nitrate, converted to nitrite. *See N*-nitrosodimethylamine.

**Nitroseed** A test devised by Polybac Corp. to detect nitrogen toxicity.

**nitroso-bacteria** A group of bacteria that convert ammonia to nitrite. *See also* nitrite bacteria and nitro-bacteria.

***Nitrosococcus*** A genus of autotrophic bacteria that can oxidize ammonia to nitrite. *See* nitrification.

**N-nitrosodimethylamine [O═N—N($CH_3$)$_2$ or $C_2H_6N_2O$]** Acronym: NDMA. A semivolatile organic chemical soluble in water (3978 mg/L), unlikely to bioaccumulate, biodegradate, or volatilize; detected in groundwater and wastewater effluent. It is also a combustible, yellow, oily liquid, with a low octanol–water partitioning coefficient (log $K_{ow}$ = –0.51) and a Henry's law constant of $2.63 \times 10^{-7}$ atm-m$^3$/mol. It is an impurity of rocket fuels. It is used in rubber curing, as an antioxidant, and in other industrial processes. It may be formed in water or wastewater treatment as a chloramination by-product and is classified as a probable human carcinogen. *See also* nitrosamine, xenobiotic.

***Nitrosolobus*** A genus of autotrophic bacteria that can oxidize ammonia to nitrite. *See* nitrification.

***Nitrosomonas*** A genus of nitrite bacteria that typically oxidize ammonia ($NH_3$) to nitrite ($NO_2^-$) under aerobic conditions. *See* nitrification.

***Nitrosomonas europea*** A species of autotrophic bacteria that can use nitrite ($NO_2^-$) to oxidize ammonia ($NH_3$) and produce nitrogen gas ($N_2$) in the absence of oxygen ($O_2$).

***Nitrosorobrio*** A genus of autotrophic bacteria that can oxidize ammonia to nitrite. *See* nitrification.

***Nitrosospira*** A genus of autotrophic bacteria that can oxidize ammonia to nitrite. *See* nitrification.

***Nitrospina*** A genus of autotrophic bacteria that can oxidize nitrite. *See* nitrification.

***Nitrospira*** A genus of autotrophic bacteria that can oxidize nitrite. *See* nitrification.

**nitrous oxide ($N_2O$)** An intermediate product in the denitrification process, by which nitrate ($NO_3^-$) is sequentially reduced to nitrite ($NO_2^-$), nitric oxide (NO), nitrous oxide ($N_2O$), and nitrogen ($N_2$). It can contribute to global warming as a major greenhouse gas and cause depletion of the ozone layer.

**Nitrox™** A patented process developed by United Industries, Inc. that uses oxidation–reduction potential (ORP) measurements to control the operation of oxidation ditch, intermittent aeration plants that achieve biological nitrogen removal. See Figure N-02. *See also* ORP knee, specific denitrification rate.

**NMOS** Acronym of negative metal oxide semiconductor.

**NO** Chemical formula of nitric oxide.

**$NO_2$** Chemical formula of nitrogen dioxide or nitrite.

**$NO_3^-$** Chemical formula of nitrate.

**$N_2O$** Chemical formula of nitrous oxide.

**NOAEL** Acronym of no observed adverse effect level.

**noble** Inert, unreactive chemically, resistant to corrosion or other chemical action.

**noble element** An element in the electrochemical series that has a relatively high standard potential, tends to be cathodic with respect to the surrounding environment, and does not readily oxidize; e.g., in order of decreasing potential: gold ($Au^{3+}$),

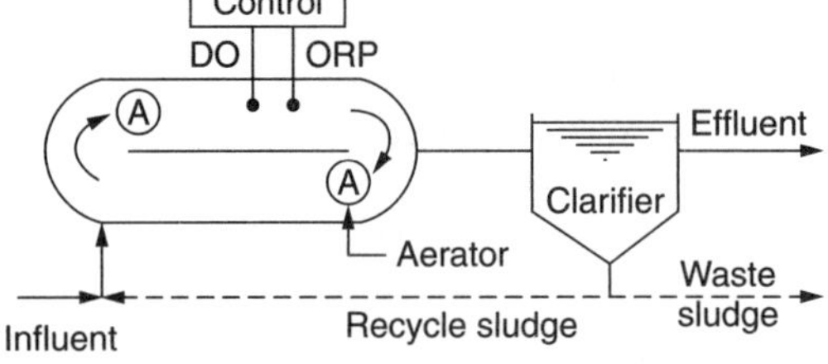


**Figure N-02.** Nitrox™.

platinum ($Pt^{2+}$), silver ($Ag^+$), copper ($Cu^+$), and copper ($Cu^{2+}$). The more noble elements (copper, as well as nickel and tin) used in iron and steel alloys tend to remain completely in the alloy; they are called tramp elements. The more reactive elements (e.g., silicon, aluminum) tend to be taken up by the slag.

**noble gas** Any of the nonreactive gases, listed in Group 0 of the periodic table, including helium (He), neon (Ne), argon (Ar), krypton (Kr), xenon (Xe), and radon (Rn), which volatilize quickly from water. They all have a complete octet (or completely filled atomic orbital), i.e., eight outer electrons in their Lewis symbols, except helium, which has two electrons. Also called inert gas, rare gas.

**noble gas outer electron configuration** The Lewis symbol of an atom showing it with eight outer electrons acquired by being a noble gas or by losing, gaining, or sharing electrons with other atoms. See Figure N-03, octet rule, free radical.

**noble metal** A chemically inactive metal, high on the EMF series; a metal that does not corrode easily and is much scarcer than the so-called useful or base metals. Examples: gold ($Au^{3+}$), platinum ($Pt^{2+}$), silver ($Ag^+$), copper ($Cu^+$), and copper ($Cu^{2+}$).

***Nocardia*** Filamentous, aerobic soil organisms with hydrophobic cell surfaces, whose accumulation creates foaming in diffused-air basins and secondary clarifiers. Nocardia are often associated with algae and responsible for earthy, musty taste and odor problems in surface waters. They can degrade hydrocarbons.

***Nocardia amarae*** A species of filamentous bacteria often associated with the problems of sludge foaming and bulking.

***Nocardia* foam** A thick layer of brown foam formed by bacteria of the genus *Nocardia* by trapping air bubbles that subsequently float on the top of aeration basins and secondary clarifiers; e.g., in pure oxygen activated sludge systems with high solids retention times. *Nocardia* foaming may be controlled by a chlorine solution, a cationic polymer, by the installation of spray nozzles, or by reducing the concentration of oil and grease in the influent. Otherwise, the foaming problem is repeated in the sludge digestion tanks.

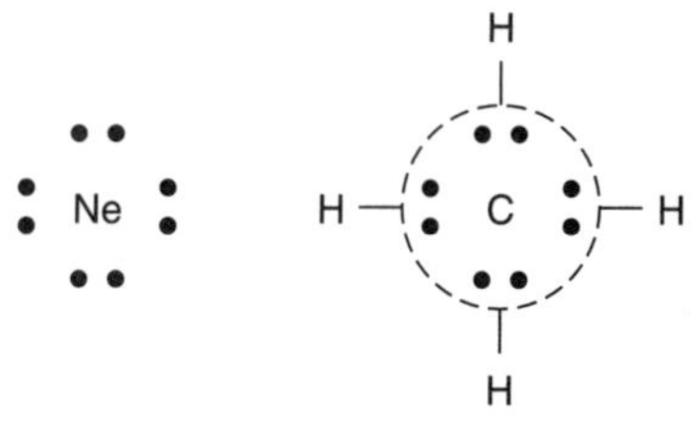


**Figure N-03.** Noble gas outer electron configuration for neon (Ne) and methane ($CH_4$).

***Nocardia* foaming** Formation of *Nocardia* foam.

***Nocardia* sp. Type 0041** A group of bacteria associated with filamentous bulking caused by a low food to microorganism ratio in the activated sludge process.

***Nocardia* sp. Type 0092** A group of bacteria associated with filamentous bulking caused by a low food to microorganism ratio in the activated sludge process.

***Nocardia* sp. Type 021N** A group of bacteria associated with filamentous bulking caused by a low food to microorganism ratio in the activated sludge process.

***Nocardia* sp. Type 0581** A group of bacteria associated with filamentous bulking caused by a low food to microorganism ratio in the activated sludge process.

***Nocardia* sp. Type 0675** A group of bacteria associated with filamentous bulking caused by a low food to microorganism ratio in the activated sludge process.

***Nocardia* sp. Type 0803** A group of bacteria associated with filamentous bulking caused by a low food to microorganism ratio in the activated sludge process.

***Nocardia* sp. Type 0961** A group of bacteria associated with filamentous bulking caused by a low food to microorganism ratio in the activated sludge process.

***n*-octanol [$CH_3(CH_2)_7OH$]** A solvent that serves as a surrogate for many environmental and physiological organic substances and a reference phase for organic–water partitioning. *See* octanol–water partition coefficient.

**NOD** Acronym of nitrogenous oxygen demand.

**Nodak system** (From North Dakota). A wastewater treatment and disposal system developed at North Dakota State University using absorption trenches elevated above the natural soil in a permeable sand fill. *See* mound system.

**no-dig technology** A method of water or sewer pipe repair that does not involve excavation. *See also* trenchless technology.

***Nodularia*** A genus of cyanobacteria that are found in algal blooms and produce hepatotoxins (nodularins).

**nodulizing kiln** An apparatus used to lower the moisture and organic content of phosphate rocks. Also called a calciner.

**NOEC** Acronym of no observed effect concentration.

**no-effect level** *See* no observed adverse effect level.

**NOEL** Acronym of no observed effect level.

**Nokia** Fine-bubble diffusers manufactured by Munters.

**NOM** Acronym of natural organic matter.

**nomenclature** *See* biological nomenclature, environmental classification, generic classification, organic nomenclature.

**nominal average power consumption** *See* NAPC.

**nominal average pump operating time** *See* NAPOT.

**nominal complete mix** The mixing regime prevalent in an aerobic lagoon with solids recycle; the power input (16–20 kW/1000 $m^3$) is sufficient to maintain most of the mixed liquor solids in suspension.

**nominal diameter** Diameter used for general identification, not necessarily the same as the actual diameter. For sand particles and similar materials, the nominal diameter is the diameter of a sphere of the same volume as the given particle. *See also* inside and outside diameters.

**nominal filter rating** Same as nominal rating.

**nominal pipe size** The approximate inside diameter of a pipe, which varies with pipe class and wall thickness; e.g., a 12-inch pipe.

**nominal rating** A parameter used to define the sizes of particles that will be retained on a filter medium. *See* filter rating for more detail, also nominal filter rating.

**nomograph** A diagram with more than two scales to solve problems graphically.

**nomographic method** Using a nomograph to solve formulas or equations.

**non-A non-B hepatitis** A more recent form of viral hepatitis, common in post-transfusion cases.

**nonaqueous phase liquid (NAPL)** (1) A contaminant that remains as the original bulk liquid in the subsurface. *See also* dense nonaqueous phase liquid and free product. (2) A liquid that keeps its distinct properties and boundaries in a mixture with water, e.g., oil.

**nonbiodegradable** For the purpose of wastewater treatment, some organic compounds (e.g., cellulose and long-chain-saturated hydrocarbons) are considered nonbiodegradable because of time and environmental limitations. Synthetic organic compounds are also resistant to biodegradation, whereas some other compounds are toxic to or inhibit biological treatment processes.

**nonbiodegradable carbon** Carbon (or organic matter) that heterotrophic microorganisms cannot convert to an inorganic or mineral form. *See also* biodegradable organic carbon.

**nonbiodegradable chemical oxygen demand (nbCOD)** The soluble and particulate portions of organic matter that microorganisms cannot decompose within the time frame of biological treatment methods. *See also* COD fractionation.

**nonbiodegradable particulate chemical oxygen demand (nbpCOD)** The particulate portion of organic matter in wastewater that is not decomposed by microorganisms; also called nonbiodegradable volatile suspended solids (nbVSS), a portion of the mixed liquor suspended solids (MLSS). This constituent originates in the influent wastewater or is produced during endogenous respiration as cell debris. Assuming a constant ratio of COD to volatile suspended solids (VSS):

$$\text{nbVSS} = \text{VSS}\,(1 - \text{bpCOD}/\text{pCOD}) \quad \text{(N-36)}$$

where bpCOD = biodegradable particulate COD, mg/L, and pCOD = particulate COD, mg/L. *See also* COD fractionation.

**nonbiodegradable soluble chemical oxygen demand (nbsCOD)** The portion of dissolved organic matter that is not removed by biological treatment of wastewater and is found in the effluent. *See also* COD fractionation.

**nonbiodegradable volatile suspended solids (nbVSS)** *See* nonbiodegradable particulate chemical oxygen demand.

**nonbiological surrogate** A chemical or physical constituent used as a surrogate for a biological agent, e.g, turbidity or microspheres for *Cryptosporidium* oocysts.

**non-BNR process** A process that removes nutrients from wastewater by the application of nonbiological methods.

**noncarbonate hardness** The difference between total hardness and carbonate hardness. Caused mainly by chlorides, nitrates, and sulfates of calcium and magnesium, it is calculated as the difference between total hardness and alkalinity (equal to carbonate hardness), all quantities being expressed as calcium carbonate. Precipitation of noncarbonate hardness requires the addition of soda ash ($Na_2CO_3$). Noncarbonate hardness is also called permanent hardness because they are approximately the same. *See also* temporary hardness and negative hardness.

**nonchlorine preoxidant** Ozone ($O_3$), potassium permanganate ($KMnO_4$), or any other oxidizing agent used ahead of water chlorination.

**noncholera vibrio** A bacterial pathogen that causes diarrhea, but not as severe as that of cholera.

**nonclog centrifugal pump** A typical device used to pump sludge and scum in wastewater treatment plants.

**nonclog mixed-flow pump** A device used to pump grit slurries and incinerator ash slurries. *See* mixed-flow pump and nonclog pump.

**nonclog pump** A pump with an impeller that is closed and has at most two or three vanes with sufficient clearance to pass large solids; used for pumping untreated wastewater. Bladeless impellers are also used. *See also* centrifugal nonclog pump, two-port nonclog radial pump, and fish pump.

**nonclogging impeller** An impeller that can pass large solids.

**noncommunity water system** A water system that serves a nonresident population (e.g., a school, factory, campsite, public park); it may be transient or nontransient. *See* community water system or public water system.

**noncompostables** Inert objects (e.g., metals, glass, ceramics) that must be removed from the waste stream before and after the grinding step in composting.

**noncondensable gas** A gaseous compound that does not liquefy when it is combined with water vapor.

**Nonconservative constituent/element/pollutant/substance** A constituent/element/pollutant/substance that is not changed by chemical or biological action during a treatment process or in the environment. *See* conservative constituent.

**nonconsumptive use** The use of water that is not consumed, e.g., for the generation of hydroelectric power or for recreational purposes.

**noncontact cooling water** Water used for cooling that does not come into direct contact with any raw material, product, by-product, or waste.

**noncontact cooling water pollutants** *See* noncontact cooling water.

**noncontatct cooling water system** A once-through collection and treatment system designed and operated for collecting cooling water that does not come into contact with hydrocarbons or oily wastewater and that is not recirculated through a cooling tower (EPA-40CFR60.691).

**noncontact recreation** A recreational activity that does not involve a significant risk of water ingestion, e.g., fishing, boating.

**nonconventional concentrate disposal** The disposal of residuals from membrane processes using evaporation, crystallization, solar ponds, or solar distillation.

**nonconventional pollutant** Any pollutant that is not statutorily listed, is poorly understood by the scientific community, or can be removed or reduced only by advanced wastewater treatment methods, e.g., metals, refractory organics, surfactants, volatile organic compounds, total dissolved solids. *See also* wastewater constituent.

**nonconventional wastewater constituents** *See* nonconventional pollutant.

**nonculturable cell** A cell that is in a nonculturable state; a dormant cell.

**nonculturable state** The condition of an organism that cannot be detected by growth in a laboratory culture.

**nondebt financing** The use by a utility of revenue generated by rates and charges in excess of operation and maintenance needs to finance new construction. Also called pay-as-you-go financing.

**nondegradable** Pertaining to a constituent or a substance that resists decay by biological, chemical, or physical action. *See also* nonbiodegradable.

**nondegradation** An environmental policy that disallows any lowering of naturally occurring water quality regardless of preestablished health standards.

**nondischarging treatment plant** A treatment plant that does not discharge treated wastewater into any surface water body. Most such plants are pond systems that dispose of the total flow they receive by means of evaporation or percolation to groundwater, or facilities that dispose of their effluent by recycling or reuse (e.g., spray irrigation or groundwater recharge).

**noneffluents** Fluid sanitary wastes that are not liquid effluents of wastewater treatment plants; they include nightsoil, the contents of pit latrines and composting toilets, septage, and the sludge from aquaprivies.

**nonexcessive infiltration** The quantity of flow that is less than 120 gallons per capita per day (domestic baseflow and infiltration) or the quantity of infiltration that cannot be economically and effectively eliminated from a sewer system as determined by a cost-effectiveness analysis (EPA-40CFR35.2005-28). *See also* excessive infiltration/inflow.

**nonexcessive inflow** The maximum total flow rate during storm events that does not result in chronic operational problems related to hydraulic overloading of the treatment works or that does not result in a total flow of more than 25 gallons per capita per day (domestic baseflow plus infiltration plus inflow). Chronic operational problems may include surcharging, backup, bypasses, and overflows (EPA-40CFR35.2005-29). *See also* excessive infiltration/inflow.

**nonfilamentous bulking** Sludge bulking problems caused by nonfilamentous organisms, such as zooglea, whose exocellular slime forms voluminous flocs that settle poorly and cause foamimg as well. *See* viscous bulking for detail.

**nonfilterable residues** Solid particles that do not pass through the filter during water filtration; they are mostly suspended solids, but also include some colloids and even a small portion of dissolved solids retained on the filter material. *See also* solids.

**nonflocculent settling** The settling of discrete, nonflocculent particles, as separate, unhindered units; characterized by very low solids concentrations. It is the primary mechanism in grit removal. *See* sedimentation Type I for detail.

**nonfood crop** Any crop that is not used for human consumption, e.g., fodder, fiber, seed, and pasture. Such crops are often irrigated with secondary treatment effluents.

**nongrowth condition** A condition in which the substrate or some other nutrient is limiting; the substrate is used for maintenance energy, i.e., maintenance without the production of new cells.

**nonhardening salt** A salt that contains agents that easily absorb atmospheric moisture and prevent caking. Nonhardening agents include calcium and magnesium chlorides ($CaCl_2$ and $MgCl_2$).

**nonhomogeneous medium** A medium having a permeability that varies from point to point.

**nonionic** (1) Pertaining to a substance that does not have an ionic charge or that does not separate into ions. (2) A short form of a nonionic substance, such as nonionic surfactant.

**nonionic detergent** A polar detergent that does not ionize in water; a combination of ethylene oxide with phenolic compounds or fatty acids (Sarnoff, 1971).

**nonionic flocculant** A flocculant (typically a polyelectrolyte) that does not have a net electrical charge. *See also* nonionic polymer.

**nonionic polyelectrolyte** A polyelectrolyte that does not have a net electrical charge. *See also* nonionic polymer (AWWA, 1999).

**nonionic polymer** A polymer that has no net electrical charge or has a very low tendency to develop charged sites in aqueous solutions; sometimes used as flocculant aid with metal coagulants to promote the formation of larger and tougher flocs. *See also* ampholyte polymer, anionic polymer, cationic polymer.

**nonionic resin** An ion exchange resin that has no exchangeable charged ions such as chloride or hydrogen; it is used to separate and concentrate humic substances or other natural organic matter.

**nonionic surfactant** A surfactant that does not dissociate in aqueous solutions. *See also* cationic surfactant, anionic surfactant.

**nonionizing radiation** Radiation that dos not cause ionization, e.g., sound or radio waves, visible, infrared, or ultraviolet light.

**nonmechanical dewatering** Sludge dewatering without using mechanical devices. *See* dewatering lagoon, freeze-drying bed, sand drying bed, solar drying bed, and mechanical dewatering.

**nonmetal** An element that does not have all the characteristics of a metal. Nonmetals may have a dull appearance or may not be malleable; they tend to gain electrons to form negative ions. Examples: atmospheric oxygen, bromine, sulfur, carbon, nitrogen. *See also* metalloid.

**non-Newtonian fluid** A fluid whose viscosity is not constant and pressure drop under laminar flow conditions is not proportional to flow; e.g., sludge.

**nonpathogenic** Pertaining to an agent that does not cause disease.

**nonpermit confined space** A confined space that, based on appropriate testing, does not actually or potentially contain any risk of death or serious injury.

**nonpersistent pesticide** A phosphate-based pesticide that has a shorter half-life than the chlorinated hydrocarbon pesticides but still poses a health risk to the workers applying it.

**nonpoint pollutant** *See* nonpoint source.

**nonpoint pollution** Man-made or man-induced pollution originating from a nonpoint source.

**nonpoint source** A diffuse pollution source (i.e., without a single point of origin or not introduced into a receiving stream from a specific outlet). The pollutants may be carried off the land by stormwater. Common nonpoint sources are agriculture, forestry, natural mineral springs, urban runoff, mining, construction, dams, channels, highway deicing salts, land disposal, saltwater intrusion, and city streets. Also, a pollution source that generally is not controlled by establishing effluent limitations under sections 301, 302, and 402 of the Clean Water Act. Nonpoint source pollutants are not traceable to a discrete identifiable origin, but generally result from land runoff, precipitation, dry fallout, drainage, or seepage (EPA-40CFR35.1605.4). *See also* agricultural nonpoint source, diffuse source of pollution, point source, and Table P-01.

**nonpoint-source pollutant** *See* nonpoint source.

**nonpolar** Possessing water-repelling properties and not easily dissolved in water.

**nonporous diffuser** A device that produces relatively large bubbles for the aeration of grit chambers, wastewater channels, aerobic digesters, and in other applications. Common nonporous diffusers include orifice diffusers, slotted tubes, and static tubes. *See also* porous diffuser, valved orifice diffuser.

**nonpotable** Unsafe or unpalatable.

**nonpotable reuse** The use of reclaimed water except for direct human consumption.

**nonpotable urban reuse** The use of reclaimed water for air conditioning, fire protection, and toilet flushing. *See also* water reuse applications.

**nonpotable water** Water that is either not safe or not satisfactory for drinking and cooking because it contains objectionable pollution, contamination, minerals, or infective agents.

**nonpotable water reuse** Any use of reclaimed water that does not involve direct or indirect potable application. *See also* water reuse applications.

**nonprotecting scale** A scale or deposit that forms on the surface of a metal and does not reduce the rate of the corrosion of the metal. *See also* protecting scale, erosion corrosion.

**nonpurgeable organic carbon (NPOC)** The portion of total organic carbon (TOC) that cannot be removed by purging a sample with an inert gas. Also called nonvolatile TOC.

**nonputrescible** Pertaining to materials that cannot be readily decomposed biologically. *See also* nonbiodegradable.

**nonreferee method** A quick and practical method of testing water for general information.

**nonreturn valve** A device that limits flow in a piping system to a single direction. Its hinged disc or flap opens in the direction of normal flow and closes to prevent flow reversal. Also called check valve or backpressure valve.

**nonselective medium** A medium that contains all necessary nutrients for the growth of all organisms, without any inhibitory factors. *See also* selective and elective enrichment media.

**nonsettleable matter** Suspended matter that does settle to the bottom or float to the surface of a tank or water body, under quiescent conditions and for a reasonable time period of one or two hours. Nonsettleable solids are fine suspended or colloidal particles less than 0.1 micron in diameter.

**nonsettleable solids** *See* nonsettleable matter.

**nonsignificant categorical industrial user (NSCIU)** A categorical industrial user that discharges less than 100 gallons per day of wastewater.

**nonsignificant noncomplier** A public water supply system that violates the National Drinking Water Regulations but not seriously enough to be a significant noncomplier.

**nonspore forming** Nonsporulating.

**nonsporulating** Pertaining to an organism that does not produce spores; nonspore forming.

**nonstructural alternative** In stormwater management, an alternative that does not involve a sewer system, e.g., initiatives to reduce or eliminate stormwater flows: downspout disconnection, porous pavements, rain barrels, soak pits, etc.

**nonsubmerged attached growth process** A wastewater treatment process that uses a fixed medium as support for the development of a biofilm; *see* biofilter, contact filter, rotating biological contactor, trickling filter. *See also* suspended growth with fixed-film packing, submerged attached growth process.

**nonsulfur purple bacteria** A group of phototrophic prokaryotes that have a low tolerance for hydrogen sulfide ($H_2S$).

**nonthreshold pollutant** A substance that is harmful to a particular organism at any concentration.

**nontransient noncommunity water system** A public water system that regularly serves at least 25 of the same nonresident persons per day for more than six months per year.

**nonuniform flow** The opposite of uniform flow, i.e., a flow whose depth, width, discharge or velocity is not constant. If any characteristic changes, the flow is varied, gradually or abruptly. Gradually varied flow takes place in the vicinity of the transition between subcritical and supercritical flows, e.g., at the intersection of mild and steep bottom slopes. The nonuniform character of flow in sewers is taken into consideration at transitions such as junctions, measuring devices, and changes in gade or size.

**nonuniformity coefficient** *See* uniformity coefficient.

**nonvolatile matter** Mineral material such as sand, salt, or metal that is not affected or that is affected only slightly by the action of organisms, and that is not lost upon ignition of dry solids at 550°C. *See also* fixed matter, fixed solids, inorganic waste, volatile solids.

**nonvolatile organic carbon** *See* nonpurgeable organic carbon.

**nonvolatile residue** *See* nonvolatile matter.

**nonvolatile TOC** *See* nonpurgeable organic carbon.

**no observable adverse effect level (NOAEL)** From long-term toxicological studies of active ingredients, levels that indicate a safe, lifetime exposure level for a given chemical. Used to establish tolerance for human diets.

**no observed adverse effect level (NOAEL)** An exposure level at which there are no statistically or biologically significant increases in the frequency or severity of adverse effects between the exposed population and its appropriate control; some effects may be produced at this level, but they are not considered as adverse, nor precursors to adverse effects. In an experiment with several NOAELs, the regulatory focus is primarily on the highest one, leading to the common usage of the term NOAEL as the highest exposure without adverse effect. *See also* adjusted acceptable daily intake (AADI), lowest observed adverse effect level, reference dose, drinking water equivalent level.

**no observed effect concentration (NOEC)** The highest concentration of a constituent at which the measured effects are no different from the control. *See also* toxicity terms.

**no observed effect level (NOEL)** An exposure level at which there are no statistically or biologically significant increases in the frequency or severity of any effect between the exposed population and its appropriate control.

**NOPOL™** A diffuser apparatus manufactured by Aeration Industries, Inc. for installation at the bottom of aeration basins.

**Nopol** Fine-bubble aeration equipment designed or manufactured by WesTech Engineering, Inc. of Salt Lake City, UT.

**Noramer®** Polymers produced by Rohm & Haas for use in water treatment.

**Norit Roz** An activated carbon produced by Norit Americas, Inc., using peat and steam.

**NORM** Acronym of naturally occurring radioactive materials.

**normal** (1) A normal solution contains one gram equivalent weight of reactant per liter of solution. The equivalent weight of an acid is that weight which contains one gram atom of ionizable hydrogen or its chemical equivalent. For example, the equivalent weight of sulfuric acid ($H_2SO_4$) is 98/2 = 49 because it has two replaceable hydrogen ions. A 1.0 N solution of sulfuric acid would consist of 49 grams of $H_2SO_4$ dissolved in enough water to make one liter. *See also* molar. (2) Pertaining to a hydrocarbon molecule that has a single unbranched chain of carbon atoms.

**normal boiling point** The temperature at which a component's vapor pressure equals atmospheric pressure or the pressure acting on the liquid; a relative indicator of volatility. *See* boiling point for detail.

**normal density, normal distribution** The normal distribution is one of a few distributions used to simulate extreme events. It corresponds to the normal density function $f(x)$, with the mean ($\mu$) and standard deviation ($\sigma$) as parameters:

$$f(x) = \exp[-(x - \mu)^2/2\sigma^2]/[\sigma(2\pi)^{0.5}] \quad \text{(N-37)}$$

Also called Gaussian distribution or Gaussian probability curve. See Figure N-04.

**normal depth ($y_n$)** In open-channel flow, the depth of uniform flow or the hypothetical depth of steady-nonuniform flow, as determined, e.g., by the Manning or Chézy formulas. It corresponds to uniform velocity for a given flow and to a water surface parallel to the channel bottom. Sometimes called neutral depth. Normal depth designates also the depth of water measured perpendicularly to the bed. *See* normal-flow.

**normal distribution** Same as normal density.

**normal dolomitic hydrated lime** A mineral or chemical product consisting of calcium hydroxide [$Ca(OH)_2$] and magnesium oxide, ($MgO$); approximate specific gravity = 2.8; bulk density = 30–40 pounds per cubic foot. It is used in wastewater treatment for pH adjustment or as an acid-neutralizing agent. *See also* pressure dolomitic hydrated lime.

**normal flow** The conceptual open channel flow under uniform or steady-nonuniform conditions. Normal discharge and its characteristics (normal

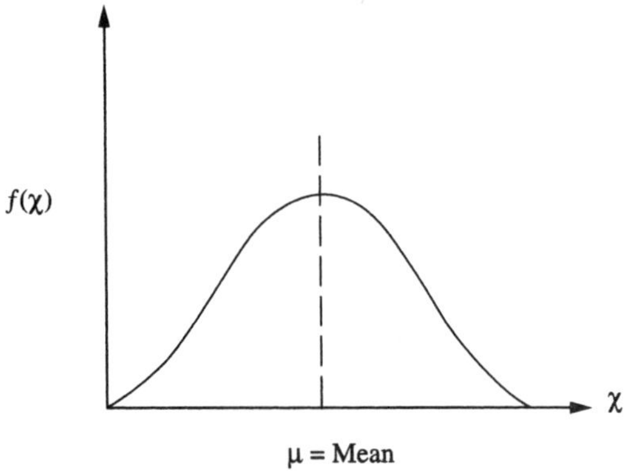


**Figure N-04.** Normal density.

depth, normal slope, and normal velocity) may then be determined from the Chézy or Manning formulas. Normal flow conditions very rarely exist in nature because of changes in channel properties (bottom slope, roughness, cross-sectional area). *See* open channel flow. The EXTRAN Block of SWMM uses a normal flow equation when (a) the channel flow $Q$ computed by the finite-difference form of the Saint-Venant equation is positive, (b) the water surface slope is less than the slope of the conduit, and (c) the normal flow $Q_n$ is less than the channel flow $Q$:

$$Q_n = (gS_0A_{up}R_{up}^{2/3})^{0.5} \quad \text{(N-38)}$$

where $g$ = gravitational acceleration, $S_0$ = bed slope, and $A_{up}$ and $R_{up}$ = cross-sectional area and hydraulic radius at the upstream end of the conduit.

**normal flow filtration** Water filtration with the feed stream flowing perpendicularly to the media surface.

**normality (N)** A measure of the strength or concentration of a solution, as compared to the normal solution. It is equal to the number of gram-equivalent weights (or gram-molecular weights) of solute in one liter of solution, expressed in equivalents per liter (eq/L) or milliequivalents per liter (meq/L). A 1.0 N solution contains 1.0 equivalent weight of the substance per liter. *See also* molality, molarity, and normal solution.

**normalization** The process of adjusting performance data of nanofiltration and reverse osmosis units to correspond to constant conditions (e.g., feedwater characteristics and feed pressure) and facilitate comparison.

**normalized concentration** In adsorption or other mass transfer processes, the normalized concentration ($C^*$) is the ratio of the concentration ($C$) of a constituent at any time or point in the column to the initial concentration ($C_0$):

$$C^* = C/C_0 \quad \text{(N-39)}$$

It is a convenient parameter used as ordinate versus throughput (volume of liquid treated) in drawing a breakthrough curve and mass transfer zone for an activated carbon column.

**normalized flow** (1) A loading relationship ($Q^*$) used in the analysis of individual treatment process performance, equal to the ratio of actual flow ($Q$) to the design flow ($Q_d$):

$$Q^* = Q/Q_d \quad \text{(N-40)}$$

(2) The permeate from a nanofiltration or reverse osmosis unit adjusted to correspond to constant operating conditions (e.g., feedwater characteristics and feed pressure).

**normalized residence time** *See* normalized time.

**normalized residence time distribution curve** A tracer response curve, i.e., a graphical representation of the distribution of a tracer versus time of flow, for a pulse input, with output concentrations normalized by dividing the measured concentrations by such a factor that the area under the normalized curve is equal to 1. *See* exit age curve for detail.

**normalized salt passage** The amount of salt from a nanofiltration or reverse osmosis unit adjusted to correspond to constant operating conditions (e.g., feedwater characteristics and feed pressure).

**normalized specific flux** The ratio of the specific flux at any point to the initial flux in a membrane filtration process. An important parameter for evaluating membrane performance, it is used to evaluate flux decline during the operation of the process. *See also* membrane flux, net operating pressure, rejection, specific flux, temperature correction factor.

**normalized time** A unitless parameter used in modeling flows in reactors; equal to the ratio of any time to the theoretical detention time.

**normalized treatability and packing coefficient** A modification of the treatability and packing coefficient used in the design of trickling filters. The coefficient is normalized with respect to packing depth and influent BOD concentration as follows:

$$k^* = k\,(D/D^*)^{0.5}(S/S^*)^{0.5} \quad \text{(N-41)}$$

where $k^*$ = normalized treatability and packing coefficient; $K$ = coefficient at a depth of 20 ft and an influent BOD of 150 mg/L; $D$ = packing depth of 20 ft; $D^*$ = actual packing depth at the site, ft; $S$ = BOD of 150 mg/L; and $S^*$ = actual influent BOD concentration, mg/L.

**normal moisture-holding capacity** The moisture content of soil in the field some time after reaching saturation and after free drainage has practically ceased, or the quantity of water held in a soil by capillary action. *See* field capacity for detail.

**normal probability curve** *See* normal density.

**normal slope ($S_n$)** and **normal velocity** Respectively, the bottom slope and velocity of flow corresponding to normal-flow conditions. *See* open channel flow.

**normal solution** A solution that has a normality of 1, i.e., contains 1.0 eq/L of a substance. A normal solution contains 1.008 g of replaceable hydrogen ($H^+$) in one liter for an acid or 17.008 g of hydroxyl ion ($OH^-$) in one liter for a base.

**normal velocity** *See* normal slope.

**normal water** A specially standardized water, prepared by the Hydrographical Laboratories of Copenhagen (Denmark), widely accepted as a control sample for the analysis of seawater salinity. *See* Copenhagen water for detail.

**normal year** A year in which hydrologic characteristics of a basin are approximately equal to their arithmetic mean over a long period (e.g., temperature, stream flow, precipitation). Also called an average year.

**Norovirus** The genus name for a group of single-stranded RNA, nonenveloped viruses called Norwalk-like viruses of the family caliciviridae that cause acute gastroenteritis or the stomach flu. Also called calicivirus, small round virus (SRV), or small round structured virus. *See* Norwalk virus.

**Nor-Pac®** A random tower packing manufactured by NSW Corp.

**Nortex** Seals manufactured by Norair Engineering Corp. for water screens.

**North™** Rotating drum screens manufactured by KRC Hewitt, Inc. with internal feed.

**North American** Pressure leaf filters manufactured by Liquid-Solids Separation Corp.

**North Filter** Rotary fine screens manufactured by KRC Hewitt, Inc.

**Norton** Media produced by Aeration Engineering Resources Corp. for use in biological reactors.

**Norvac Systems** Dry feed systems engineered by Northland Industrial Specialists, using a particle dispersion device to dissolve such chemicals as polymers, powdered activated carbon, and bentonite.

**Norwalk agent** Same as Norwalk virus.

**Norwalk virus** An unusually hardy, extremely contagious calicivirus that causes excreta-related diarrhea; named after the location of its first documented outbreak (an elementary school in Norwalk, Ohio in 1968) and one of the most likely agents to cause a waterborne outbreak in the United States; incubation period: one or two days. Transmission through water, food, ice, and contaminated surfaces; of 17 documented outbreaks of Norwalk virus, 13 were traced to contaminated water. Now officially called Norovirus. Recent outbreaks have affected cruise ships and nursing homes. Often not diagnosed, the second most common illness in the United States after the common cold. Symptoms include nausea, vomiting, diarrhea, cramping, chills, muscle aches, and fatigue, for one or two days but it can kill the elderly and weak. People are contagious from the moment they feel ill until 3–15 days after recovery. Outbreaks of similar small round viruses have occurred in Montgomery County, Hawaii, Taunton, Amulree, Otofuke, Sapporo, and Snow Mountain.

**nosocomial** Pertaining to or originating in hospitals.

**nosocomial pathogen** A pathogen acquired in a hospital or other healthcare facility, whether by a patient, visitor, or staff, e.g., *Pseudomonas aeruginosa,* which is responsible for about 10% of nosocomial infections.

***Nostoc*** A genus of cyanobacteria that produce toxins.

***Nostocoida limicola*** A group of bacteria associated with filamentous bulking in the complete-mix activated sludge process.

**not analyzed (NA)** Pertaining to a water or wastewater sample that was collected but not analyzed.

**not applicable (NA)** Pertaining to a laboratory test that was not conducted because it is not required (and no sample was collected for it).

**notch** An opening in a hydraulic structure for the measurement of discharge.

**notched weir** A weir with a notch or notches in the crest. *See also* V-notch weir.

**not detected** Pertaining to a substance or constituent that was not detected in a laboratory analysis at or above the equipment or method detection limit.

**notifiable disease** An infectious disease that must be reported to public health agencies when diagnosed.

**Notim™** Media produced by Matt-Son, Inc. for the removal of organic iron and tannins.

**not in my back yard (NIMBY)** An expression used to indicate the opposition of local inhabitants to the location of an undesirable project in their neighborhood, city, etc. Such a project may be publicly needed but is usually considered a nuisance or an environmental risk. Examples of projects that lead to NIMBY conflicts include waste disposal, treatment plants, jails, drug rehabilitation centers, and nuclear facilities. Where environmental concerns predominate, NIMBY conflicts may be resolved by the use of so-called green technologies (Freeman, 1998).

**not quantitated** Pertaining to a substance or constituent that is detected below a quantitation limit.

**NoVOCs™** A stripping process developed by EG&G Environmental to remove volatile organic compounds.

**Novus®** An emulsion polymer produced by Betz.

**No-Wear™** A traveling bridge mechanism manufactured by Davis Water & Waste Industries, Inc. for use in filter backwashing.

**No-Well** A traveling screen design developed by FMC Corp. without using a channel intake.

**NOx** Abbreviation or formula of nitrogen oxides; used also to represent the amount of ammonia-nitrogen oxidized.

**Noxidizer™** An incineration apparatus manufactured by John Zink Co.

**Noxon** A combined unit manufactured by Purac Engineering, Inc. as a decanter and a centrifuge.

**NoxOut** An apparatus manufactured by Nalco Chemical Co. for the reduction of nitrogen oxide.

**nozzle** A cone-shaped, tubelike device, usually streamlined, for accelerating and directing a fluid whose pressure decreases as it leaves the nozzle. Many of the devices used in the aeration of water and wastewater contain nozzles.

**nozzle aerator** A pressure nozzle that sprays fine droplets of water or wastewater into the air. *See* spray aerator for detail.

**Nozzle Air** An aeration apparatus manufactured by Envirotech Co. for wastewater treatment using dissolved-air flotation.

**NPDES** Acronym of National Pollutant Discharge Elimination System. An NPDES permit is issued pursuant to section 402 of the Clean Water Act. An NPDES State is a state or interstate water pollution control agency with an NPDES permit program approved pursuant to section 402 of the Clean Water Act.

**NPDES permit** *See* NPDES.

**NPDES permit state** *See* NPDES.

**NPDOC** Acronym of nonpurgeable dissolved organic carbon.

**NPDWR** Acronym of National Primary Drinking Water Regulation.

**NPK** The informal name for an artificial fertilizer, from its content of nitrogen (N), phosphorus (P), and potassium (K). *See also* fertilizer value.

**NPL** Acronym of National Priorities List.

**NPSH** Acronym of net positive suction head.

**NPSHA** Acronym of net positive suction head available.

**NPSHR** Acronym of net positive suction head required.

**NPV** Acronym of net present value.

**NRA** Acronym of National Rivers Authority.

**NRC equations** *See* NRC formulation.

**NRC filter equations** *See* NRC formulation.

**NRC formulas** *See* NRC formulation.

**NRC formulation** Empirical formulas developed in 1946 for the design of trickling filters, based on data collected at military bases, mostly in the southern United States, during World War II:

$$E = 100/[1 + 0.0085\,(W/VF)^{0.5}] \qquad \text{(N-42)}$$

where $E = BOD_5$ removal, %; $W = BOD_5$ loading, excluding recycle, pounds/day; $V$ = filter volume (surface area times depth of media), acre-feet; $F$ = recirculation factor = number of passes of organic material through the filter:

$$F = (1 + R)[1 + (1 - P)R]^2 \qquad \text{(N-43)}$$

$R$ = recirculation ratio (recirculated flow divided by influent flow), dimensionless; and $P$ = a weighing factor. Other trickling filter design formulas include British Manual, Eckenfelder, Galler–Gotaas, Germain, Howland, Kincannon and Stover, Logan, modified Velz, Rankin, Schulze, and Velz.

**NRCS** Acronym of Natural Resources Conservation Service.

**NRC trickling filter formulas** *See* NRC formulation.

**NSCIU** Acronym of nonsignificant categorical industrial user.

**NSDWR** Acronym of National Secondary Drinking Water Regulations.

**NSF International** Formerly the National Sanitation Foundation. A nonprofit organization that conducts research and provides educational services in the field of environmental health.

**NTNCWS** Acronym of nontransient, noncommunity water system.

**NTP** Acronym of National Toxicology Program.

**N-Trak** A test kit manufactured by Hach Co. for the analysis of nitrogen in water.

**NTU** or **ntu** Acronym of nephelometric turbidity unit.

**nuclear isomerism** The state, condition, or phenomenon characterized by the relation of two or more nuclides having the same atomic number and mass number but different properties (e.g., energy level and half-life). *See also* structural isomerism.

**nuclear isomers** Isomers; two or more atomic nuclei having the same atomic number and mass number but different energy states and half-lives. *See also* isotope.

**nuclear magnetic resonance (NMR)** An analytical technique used in research laboratories to identify organic compounds based on spectrometry and the nuclear particles in a sample.

**nuclear transmutation** Conversion of one type of atom to another.

**nucleation** (1) The formation of very small particles in a supersaturated solution from which precipitation and crystallization can occur; e.g., the conversion of dissolved calcium and carbonate ions into solid calcium carbonate. In lime precipitation processes, sludge is often recycled to provide seed crystals for nucleation. (2) The initiation of a phase change to a lower thermodynamic state, e.g., from vapor to liquid (condensation) or to solid (deposition), and from liquid to solid (freezing).

**nuclei** Plural of nucleus.

**nucleic acid** A large acidic biological polymer, e.g., DNA or RNA, that contains phosphoric acid ($H_3PO_4$), purine ($C_5H_4N_4$), pyrimidine ($C_4H_4N_2$), cytosine, uracil or thymine, and ribose or deoxyribose. Nucleic acids store and pass on genetic information that controls reproduction and protein synthesis.

**nucleic acid probe** A molecule that has a strong interaction only with a specific organism and can be used to detect that organism. The probe is synthesized to correspond to the RNA or DNA of the organism. *See also* dot blot hybridization, in situ hybridization, fluorescent in situ hybridization.

**nucleic acid probe method** A procedure using a nucleic acid probe to identify bacteria and protozoa.

**nuclein** A constituent of the nuclei of cells containing nitrogen, phosphorus, and sulfur.

**nucleophile** (1) Same as Lewis base or electron donor. (2) A negatively charged ion, a molecule that contains atoms with unshared pairs of electrons, or a molecule with highly polarized or polarizable bonds; e.g., acetate ($CH_3COO^-$), bicarbonate ($HCO_3^-$), chloride ($Cl^-$), cyanide ($CN^-$), fluoride ($F^-$), hydroxide ($OH^-$), iodide ($I^-$), nitrate ($NO_3^-$), orthophosphate ($HPO_4^{3-}$), sulfate ($SO_4^{2-}$), sulfide ($HS^-$), and water ($H_2O$). *See also* electrophile, hydrolysis (3).

**nucleophilic reaction** A reaction in which the reacting species brings an electron pair. *See* substitutive dehalogenation.

**nucleotide** The simplest unit of a nucleic acid polymer, consisting of a nitrogen-containing base, phosphate, and a simple sugar.

**Nuclepore®** A membrane cartridge filter manufactured by Costar.

**nucleus** (1) The mass of protoplasm encased in a membrane of a eukaryotic cell and responsible for various functions of the organism. (2) The small core at the center of an atom, containing positively charged particles (protons) and uncharged particles (neutrons). (3) A fundamental arrangement of atoms in a compound, e.g., the benzene ring. (4) A particle on which water vapor condenses to form a water drop or ice crystal. Also called a condensation nucleus.

**nuclide** A species of atom characterized by the number of protons, neutrons, and energy in the nucleus. An unstable isotope decays through intermediate nuclides before reaching a final form.

**nuisance** A visually or olfactorily undesirable environmental condition.

**nuisance bacteria** *See* nuisance organism.

**nuisance microorganism** *See* nuisance organism.

**nuisance organism** (1) Any of a diverse group of organisms that affect the aesthetic quality of drinking water, e.g., the iron and sulfur bacteria that cause color, taste, and odor problems. (2) Any of a number of organisms that cause operational problems at wastewater treatment plants, for example the bacterial genera *Microthrix* and *Nocardia* that cause sludge bulking and foaming, or the genera *Desulfivibrio* and *Desulfobacter* that reduce sulfate to sulfide, which can be toxic to the methanogens during anaerobic digestion. *See also* bulking microorganisms, filamentous bacteria.

**nuisance rule** A regulation that establishes limits on specific pollutants and prohibits injury, detriment, nuisance, or annoyance to the public.

**number-average molecular weight** The weighted average of the molecular weights of the individual components of natural organic matter.

**Nu-Notch Mushroom** An air diffuser manufactured in cast iron by Knowles Mushroom Ventilator Co.

**NuTralite®** A product of NuTech Environmental Corp. for the control of odors.

**Nu-Treat** A flocculator-clarifier manufactured by Envirex, Inc.

**nutrient** Any substance that is assimilated by organisms and promotes their growth, including water, carbon dioxide, oxygen, and several mineral elements. Nitrogen and phosphorus are major nutrients that promote the growth of algae. There are other essential and trace elements that are also considered nutrients. Trace elements include sulfur, potassium, magnesium, and calcium. All nutrients can lead to undesirable aquatic growth or to the pollution of groundwater. *See also* inorganic nutrients, organic nutrients, growth factor.

**nutrient agar** Solid nutrient broth supplemented with agar.

**nutrient assimilation** (1) The uptake of nutrients (e.g., nitrogen and phosphorus) by microorganisms from solution to satisfy their nutritional requirements. (2) A method used in wastewater treatment for nutrient removal. *See also* algal culture, bacterial assimilation.

**nutrient broth** A liquid basal medium, a mixture of beef extract and peptone, used to grow organisms.

**nutrient budget** In an ecosystem, the nutrient budget refers to the energy intake and output of the biotic components of the ecosystem (the internal budget) and the intake and output of the whole system (external budget).

**nutrient equalization** The use of complete mixing in the activated sludge process for better distribution of the influent load.

**nutrient pollution** Contamination of water resources by excessive inputs of nutrients; in surface waters, excess algal production is a major concern.

**nutrient stripping** The removal of nutrients from wastewater (for control of eutrophication), from food, or any other substance.

**nutrient-type metal** A reactive trace metal, such as zinc or cadmium, that is removed from surface waters by biogenic particles and then remineralized at depth. *See also* scavenged-type metal.

**Nutrification™ process** A sidestream fermentation process developed in North Carolina for the enhanced biological removal of phosphorus from wastewater. *See* OWASA Nutrification™ process for detail.

**Nutrigest®** Clarifier equipment manufactured by Smith & Loveless, Inc.

**nutrition** The process by which organisms obtain nutrients for energy and new structural material.

**NVCU™** A vapor control apparatus manufactured by NAO, Inc.

**N-Viro** A process developed by N-Viro Energy Systems, Inc. to disinfect and stabilize sludge using pasteurization and chemical treatment.

**nylon** A thermoplastic material that has good structural, chemical, and hydraulic properties.

# O

**O** Chemical formula of elemental or nascent oxygen.

**$O_2$** Chemical formula of molecular oxygen.

**$O_2^-$** Chemical formula of superoxide ion.

**$O_2$ Minimizer®** A device manufactured by Schreiber Corp. to control the aeration of mixed liquor.

**$O_3$** Chemical formula of ozone.

**$O_3$-$H_2O_2$** *See* ozone-hydrogen peroxide.

**obligate aerobe** An organism that requires the presence of molecular oxygen ($O_2$) for its metabolism; it can survive only in the presence of dissolved oxygen.

**obligate aerobic microorganism** *See* obligate aerobe.

**obligate anaerobe** An organism for which the presence of molecular oxygen is toxic. Such an organism generates energy by fermentation and derives the oxygen needed for cell synthesis from chemical compounds.

**obligate organisms** Organisms that cannot grow in either the presence or absence of a specific environmental factor; also called strict organisms. Viruses can be classified as obligate parasites or obligate pathogens because they can thrive only in a living host cell. *See also* facultative organisms, obligate aerobe, obligate anaerobe. Most of the bacteria that carry out the decomposition of organic matter in wastewater treatment are facultative as opposed to obligate organisms. Some specific biological reactions such as nitrification and methanogenesis are mediated by strict organisms.

**obligate parasite** An organism that can live only as a parasite. *See also* obligate organism, facultative parasite.

**obligate pathogen** *See* obligate organism.

**$OBr^-$** Hypobromite ion.

**OBS®** Sensors manufactured by D&A Instrument Co. for the analysis of turbidity.

**observation port** An access point that allows an operation and maintenance worker to look at a buried component, e.g., in an on-site wastewater treatment system.

**observed biomass yield** Same as observed yield.

**observed solids yield** Same as observed yield.

**observed yield** A parameter that represents two phenomena of microbial growth: substrate removal and endogenous decay. The net growth of microorganisms ($r_x$) is then the product of the observed yield ($Y_{obs}$) or actual measurements of biomass production and the rate of substrate removal ($r_s$), taking into account concurrent cell losses:

$$r_x = -Y_{obs} \cdot r_s \qquad \text{(O-01)}$$

*See also* net biomass yield, synthesis yield or true yield, solids production or solids yield.

**observed yield factor** Same as observed yield.

**occlude** To cause to become obstructed or closed and thus prevent passage either into or from. In physical chemistry, to incorporate as by absorption or adsorption.

**occlusion** A physical chemistry phenomenon whereby one solid particle adheres to another by absorption or adsorption, sometimes resulting in coprecipitation of secondary elements (e.g., heavy metals, radionuclides, and viruses) along with the main targets of chemical precipitation (e.g., calcium, magnesium, and organic contaminants) during water treatment. *See also* adsorption (4), inclusion, solid-solution formation.

**Occupational Safety and Health Act of 1970 (OSHA)** A law designed to protect the health and safety of industrial workers, including the operators of water supply and wastewater systems and treatment plants. OSHA also refers to the federal and state agencies that administrate the OSHA regulations.

**Occupational Safety and Health Administration (OSHA)** An agency of the U.S. Department of Labor that oversees safety and health in the workplace.

**ocean desalination** The treatment of ocean water to remove dissolved salts (such as sodium chloride, NaCl) and other dissolved solids to produce drinking water.

**ocean discharge** The discharge of wastewater to ocean water, usually through an ocean outfall so as to provide sufficient initial dilution and maximum protection of the marine environment. Ocean discharge is also used for the disposal of concentrated waste streams from membrane filtration processes, sometimes combined with power-plant cooling water. *See* zone of initial dilution.

**ocean discharge waiver** A variance from Clean Water Act requirements for discharges into marine waters.

**ocean disposal** The disposal of effluents and residuals into the ocean by such methods as barging or through an ocean outfall. Because of the significant dilution available, ocean disposal usually requires lower than secondary treatment, but water quality modeling is required to assure adequate mixing. Other considerations include the lack of dissolved oxygen, low temperatures, and slow decomposition rates in deep waters; lower dissolved oxygen saturation concentration in saline waters; and the tendency of wastewater to spread without mixing in saline water because of its lower density. Ocean disposal of sludge is being prohibited in the United States; when it is practiced, a disposal site 100 miles offshore is effective. *See also* ocean dumping, ocean incineration, surface vessel disposal.

**ocean dumping** Discharge of wastewater sludge and other wastes in ocean water or the sea by barge or incineration at sea (ocean incineration). Ocean dumping can cause contamination of the sea floor sediments by copper, lead, organic materials, and particulates.

**Ocean Dumping Act (ODA)** A federal law of 1988 (Public Law 100-688) that regulates ocean disposal. It prohibits dumping by barge after December 31, 1991 and bans ocean incineration.

**ocean incineration** The incineration of sludge and other wastes on ocean-going vessels and far from the shore. Sludge incineration at sea is now banned under the Ocean Dumping Act.

**ocean outfall** A pipeline and appurtenances that discharge stormwater, wastewater, treated effluent, or waste concentrate from membrane filtration into the sea. Secondary treatment is recommended before wastewater discharge, but the minimum requirement is for the removal of all settleable solids and floating matter. Also called submarine outfall. *See* Figure O-01.

**ocean profile** *See* the following terms: aphotic zone, bathypelagic zone, benthic region, benthopelagic zone, benthos, dysphotic zone, endobiotic habitat, epibiotic habitat, epipelagic zone, euphotic zone, hydrothermal vent, mesopelagic zone, neuston layer, pelagic zone, photic zone. *See also* freshwater profile.

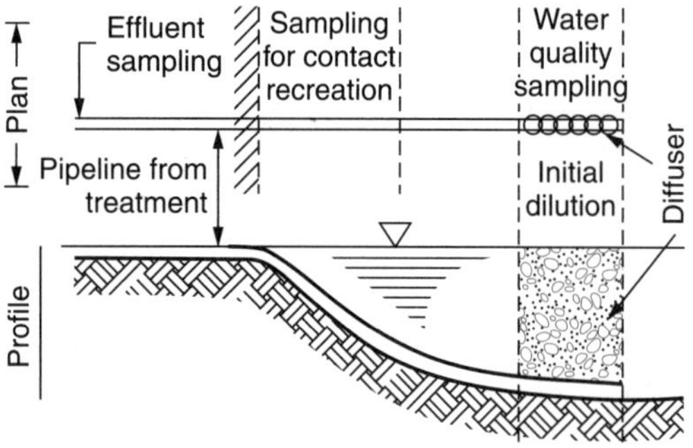


**igure O-01.** Ocean outfall (adapted from Hammer & Hammer, 1996).

**ocean sludge dumping** The disposal of wastewater treatment sludge into the ocean. *See* ocean dumping.

**ocean waters** Coastal waters landward of the baseline of the territorial seas, the deep waters of the territorial seas, or the waters of the contiguous zone. Sea or ocean water has between 30,000 and 36,000 mg/L of dissolved solids (DS) as compared to freshwater, brackish water, saltwater, and brine. *See* salinity.

***Ochrobactrum anthropi*** A species of bacteria effective in the degradation of mineral oils and aliphatic hydrocarbons.

**$OCl^-$** Hypochlorite ion.

**$O_2C_2O_2^{-2}$** Chemical formula of oxalate.

**O'Connor–Dobbins formula** A formula proposed in 1958 to calculate the surface reaeration rate ($K_2$) as a function of depth, current speed, and the molecular diffusion coefficient. *See* reaeration constant for detail.

**octane** Any of the isomeric saturated hydrocarbons represented by the formula $C_8H_{18}$.

**octanol** *See* *n*-octanol.

**octanol–water coefficient** *See* octanol–water partition coefficient.

**octanol–water partition coefficient ($K_{ow}$, $K_{OW}$, or $k_{ow}$)** A dimensionless coefficient representing the ratio of the solubility of a compound in the water-saturated octanol phase (a nonpolar solvent) to its solubility in the octanol-saturated water phase (a polar solvent). The higher the $K_{ow}$, the more nonpolar, the more adsorptive, and the less strippable the compound. log $K_{ow}$ is generally used as a relative indicator of the tendency of an organic compound to adsorb to soil. log $K_{ow}$ values are generally inversely related to aqueous solubility and directly proportional to molecular weights and bioconcentration factors in aquatic organisms. The strippability threshold is considered to be log $K_{ow} = 3.5$ for some and log $K_{ow} = 2.0$ for others. For example, for benzene, trichloromethane, and toluene, log $K_{ow}$ = 2.12, 1.90, and 2.21, respectively. For acids, amines, phenols, and other ionizable compounds, the coefficient is a function of pH. *See also* n-octanol, sorption coefficient, lipophilicity, bioconcentration factor.

**octet** Any group of eight, particularly the group of eight electrons in the noble gas outer electron configuration. All noble gases, except helium (He), have a full octet; chlorine (Cl) has 7 electrons, oxygen and sulfur have 6, nitrogen 5, carbon 4, and so on. *See also* electrophile, free radical, octet rule.

**octet rule** The tendency of chemical elements to acquire a full octet, i.e., an 8-electron outer configuration by reacting with other elements to lose, gain, or share electrons. This rule helps determine the nature of chemical bonding and the formulas of chemical compounds.

**OD** Acronym of outside diameter.

**ODA** Acronym of Ocean Dumping Act.

**odd hydrogen** Any of the three species found in the destruction of ozone ($O_3$) in the stratosphere: hydrogen atom (H), hydroxyl radical (OH), and hydroperoxyl radical ($HO_2$). *See also* ozone decomposition.

**odds ratio** In epidemiology, the relative risk of disease when comparing groups, equal to the ratio of the odds of exposure among the cases of disease to the odds of exposure in the controls.

**Odin** A package plant manufactured by Davis Water & Waste Industries, Inc. for water treatment.

**Odophos** An apparatus manufactured by Davis Water & Waste Industries, Inc. to control hydrogen sulfide and remove phosphorus.

**odor** One of the physical parameters used in the evaluation of drinking water, often discussed along with taste. There are a number of fundamental odor classifications, e.g., (a) sweet or fragrant, (b) acid or sour, (c) burnt or empyreumatic, and (d) goaty or caprylic. Odor problems in drinking water are caused by chemicals (hydrogen sulfide, dissolved iron and manganese, leachate), natural decomposition of organic matter, or microbial activity (e.g., by actinomycetes and algae). *See also* chlorine taste, threshold odor number. Fresh municipal wastewater has a soapy or oily odor, while stale or septic wastewater smells like hydrogen sulfide. Most wastewater treatment and sludge processing techniques, particularly those involving anaerobic decomposition, require some level of odor control. In wastewater treatment plants, inorganic gases and vapors (e.g., hydrogen sulfide ($H_2S$) and ammonia ($NH_3$) are generally responsible for odors, and odorous compounds include mercaptans, organic sulfides, and amines. Odors generated by industrial wastes may pose health and safety risks.

**Odor Buster®** A device manufactured by United Industries, Inc. that uses aeration to control odors at wastewater pumping stations and preliminary treatment units.

**odor control** The prevention, elimination, or reduction of odors in drinking water or wastewater, or at conveyance and treatment facilities. Aeration, algae elimination, oxidation, dechlorination, and adsorption on activated carbon are used to remove or destroy tastes and odors caused by dissolved gases and volatile organics. Oxidizing agents include chlorine, ozone, permanganate, and hydrogen peroxide. A number of procedures help contain odors in treatment plants, e.g., covers on trickling filters and other units, dispersion through tall stacks, wet scrubbing, combustion, and masking agents.

**Odorgard™** An air scrubbing process developed by ICI Katalco Co. using hypochlorite.

**odor index** A parameter (*OI*) used to characterize the odor potential of a substance; defined as the ratio of the vapor pressure (*VP*, in ppm on a volume basis) to the 100% odor recognition threshold, *RT*, in ppm (Droste, 1997):

$$OI = VP/RT \qquad \text{(O-02)}$$

*See* Table O-01 for examples of compounds with high odor potential.

**OdorLok™** An apparatus manufactured by Eaglebrook, Inc. for the control of odor and corrosion due to hydrogen sulfide.

**odor masking** The addition of a chemical agent to wastewater or off-gases to mask an offensive odor.

**odor-masking agent** A perfume scent sprayed in fine mists near a wastewater treatment unit that emits offensive odors, to mask or overpower such odors. *See also* neutralizing agent.

**OdorMaster™** An odor control apparatus manufactured by Pepcon Systems, Inc. as an electrolytic gas scrubber.

**odor neutralization** The addition of a chemical agent to an off-gas so that they cancel each other's odor or produce an odor of lower intensity.

**Odor-Ox** A multistage air scrubber manufactured by Purafil, Inc.

**odor prevention** *See* odor control.

**odor recognition threshold** At the 100% odor recognition level, all the members of an odor panel perceive the odor. *See* odor index.

**odor removal** *See* odor control.

**Table O-01.** Odor index (odor potential)

| Compound | Formula | Odor index, $10^6$ |
|---|---|---|
| Acetaldehyde | $CH_3CHO$ | 4 |
| Ammonia | $NH_3$ | 0.2 |
| 1-Butene | $CH_3CH_2CH{=}CH_2$ | 43 |
| Ethylmercaptan | $CH_3CH_2SH$ | 290 |
| Ethylsulfide | $(CH_3{-}CH_2)_2S$ | 14 |
| Formaldehyde | HCHO | 5 |
| Hydrogen sulfide | $H_2S$ | 17 |
| Isobutene | $(CH_3)_2C{=}CH_2$ | 5 |
| Propionic acid | $CH_3CH_2COOH$ | 0.1 |
| Propylmercaptan | $CH_3CH_2CH_2SH$ | 263 |

**odor puff** A concentrated cloud of odor formed under quiescent conditions over a wastewater treatment unit and transported without breaking up over long distances (e.g., 15 miles) from the source.

**odor threshold** The minimum odor of a water sample that can just be detected after successive dilutions with odorless water. Odor threshold may also be defined as the lowest concentration of the responsible substance that the olfactory sense can detect; it is commonly reported in μg/L or μg/kg. Odor threshold in water varies widely; e.g., 1 μg/L for geosmin and 20 mg/L for chloroform; chlorine taste complaints are often reported by consumers at a concentration as low as 0.2–0.4 mg/L. Also called threshold odor. *See also* dilutions to threshold, minimum detectable threshold odor concentration, odor index, odor threshold concentration, threshold odor number, Weber–Fechner law.

**odor threshold concentration (OTC)** The level at which one or more individuals in a panel can sense the presence of a substance. Also called threshold odor concentration. *See also* odor index.

**odor threshold test** A quantitative test that measures the minimum detectable odor in water, or the minimum quantity of a substance that produces a detectable odor, using a sample of the water and odor-free distilled water for dilution. *See* threshold odor number.

**odor-treatment method** A technique or measure used to treat an odor-producing compound in wastewater or the foul air that results, e.g., adsorption on activated carbon, compost, sand, or soil; dilution with air; masking agents; scrubbing towers; chemical precipitation, chemical oxidation, neutralizing agents; biotrickling filters; and compost filters.

**odor unit** The unit used to report odor intensity; the reciprocal of the dilution ratio with odor-free water at which the odor is just noticeable. *See* odor threshold, threshold odor number.

**odour** British spelling for odor.

***O. felineus*** *See Opisthorchis felineus.*

**off-gas** Air or vapor from a water or wastewater treatment process.

**off-line equalization** A configuration in which only the portion of influent water or wastewater above the daily average flow or another predetermined limit passes through an equalization basin, e.g., to capture the first flush from combined sewer systems. Also called sideline equalization. *See also* in-line equalization.

**off-line equipment** Equipment or subsystem out of service, in standby, maintenance, or mode of operation other than on-line; also equipment installed outside of a process line or outside a network.

**off-line pump station** A pump station with a wetwell, the rate of pumping depending on the volume of water in the wetwell. An off-line station may also pump according to the head difference over the pumps. *See* pumping station and on-line pump station.

**off-line storage** The storage of stormwater or wastewater outside an existing sewer system, usually in a tank, deep tunnel, or abandoned pipeline to which the liquid waste is diverted from the main sewer system. *See also* in-line storage, in-receiving water flow balance method, in-system storage.

**offset** The difference between the actual value and the desired value (or set point); characteristic of proportional controllers that do not incorporate reset action.

**offset manhole** A manhole installed tangentially to a sewer line, rather than on the centerline.

**offset pit** A pit that is entirely or partially displaced from its superstructure.

**offset-pit latrine** A latrine with a pit that is partially or entirely displaced from its superstructure.

**offsite well** A well used for the injection of hazardous wastes generated at other locations, e.g., wells operated by commercial waste disposers.

**offstream use** Withdrawal of surface or ground water at one location for use at another location.

**OFR** Acronym of overflow rate.

**OH** Chemical symbol of the hydroxyl radical.

**$OH^-$** Chemical symbol of the hydroxide ion or hydroxyl ion.

**$OH^•$** A hydroxyl radical that has an unpaired electron. It is produced by the action of ultraviolet light on ozone in water or in wet air, or in the peroxone process. *See* the reactions under peroxone and for the photolysis of ozone under $HO^•$.

**[$OH^-$ alk]** Symbol of caustic alkalinity.

**ohm,** The unit of electrical resistance; the resistance of a conductor in which one volt produces a current of one ampere.

**Ohm's law** For a current that flows through a resistance ($R$, ohms), the voltage drop ($E$, volts) equals the product of the intensity of the current ($I$, amperes) and the resistance:

$$E = RI \qquad \text{(O-03)}$$

Ohm's law is applied in the determination of conductivity and specific conductance, as well as the power required in the electrodialysis process.

**O horizon** The dark, organic-rich surface layer of a soil profile, above the A horizon.

**oil** A viscous substance derived from animals and plant seeds or nuts, or composed of hydrocarbon mixtures. *See also* grease, fats.

**oil and grease** A group of water-insoluble organic substances found in wastewater, including rendered animal fats, waxes, free fatty acids, soaps, mineral oils, oily matter, thick lubricants, and some nonfatty materials. Oil and grease interferes with biological treatment and, if it is not removed during wastewater treatment, it can interfere with biological life and cause aesthetic problems in receiving waters. Oil and grease as an aggregate can be determined analytically by extraction with trichlorotrifluoroethane and is usually removed by flotation skimming. Also called grease and oil; previously called fats, oil, and grease (FOG).

**Oil Grabber** An oil skimming apparatus of Abanaki Corp.

**oil interceptor** A device designed to keep oil out of a sewer system. *See* oil trap.

**OilMaster** An apparatus manufactured by National Fluid Separators, Inc. to separate oil from water.

**Oil-Minder** A submersible pump manufactured by Stancor, inc.

**oil of vitriol** An old and obsolete term for sulfuric acid ($H_2SO_4$).

**Oil Pollution Act (OPA)** A U.S. law of 1990 that requires tank owners and operators to pay for damages caused by oil spills.

**oil production and refinery waste** Wastes produced during the production and refining of oil include drilling mud, salts, oils, natural gas, and acid sludges. They contain high levels of dissolved salts, BOD, odor, phenol, and sulfur. Some of these constituents can be diverted, recovered, or injected. Sludges can be neutralized or incinerated. However, most of the water used in refineries is for cooling only.

**oils** *See* fats and oils.

**oil separation** The removal of oil and grease from wastewater.

**oil separator** *See* oil trap.

**oil skimmer** A device designed to remove oil from water surfaces.

**Oilspin II** A hydroclone separator manufactured by Serck Baker, Inc.

**oil trap** An apparatus, like a small settling, skimming, or holding tank, for the collection and separation of oil from wastewater; it is usually required for effluents from some commercial and industrial establishments such as restaurants, hotels, garages, packinghouses, bakeries, and creameries.

**oil–water separator** Wastewater treatment equipment used to separate oil from water, consisting of a separation tank, which also includes the forebay and other separator basins, skimmers, weirs, grit chambers, and sludge hoppers. This term includes slop oil facilities, storage vessels, and auxiliary equipment located between individual drain systems and the oil–water separator. Examples of oil–water separators include API (American Petroleum Institute) separator, parallel-plate interceptor, and corrugated-plate interceptor with the associated ancillary equipment (EPA-40CFR60.691, 61.341, & 63.111). An organic–water separator is similarly defined to separate organics from water.

**oil well flooding** An underground mining method used to recover secondary oil by injecting water into a formation to force out additional oil into a producing well. *See also* water flooding.

**oily sludge** The sludge resulting from the emulsification of oil with water, usually in the presence of suspended fines, in the petroleum refining industry.

**oily wastewater** Wastewater generated during the refinery process that contains oil, emulsified oil, or other hydrocarbons. Oily wastewater originates from a variety of refinery processes, including cooling water, condensed stripping steam, tank draw-off, and contact process water (EPA-40CFR60.691).

**OKI** A mechanical aerator manufactured by Outomec USA, Inc.

**Old World hookworm** *See Ancylostoma duodenale.*

**olefin** An unsaturated hydrocarbon with multiple bonds between some carbon atoms, e.g., ethane ($H_2C{=}CH_2$) and acetylene ($HC{\equiv}CH$). *See* alkene, paraffin.

**oleic acid** A typical unsaturated, fatty acid: $CH_3(CH_2)_7CH{=}CH(CH_2)_7COOH$.

**Oleofilter™** A filter manufactured by Aprotek, Inc. for the removal of hydrocarbons from water.

**oleum** A solution of sulfuric acid ($H_2SO_4$) and sulfur trioxide ($SO_3$).

**olfactory fatigue** A condition in which a person's nose, after exposure to a certain odor, is no longer able to detect the odor.

**oligo-dynamic action** The effect of small amounts of heavy metals on the growth of bacteria.

**oligosaprobic zone** An area or reach that is slightly polluted with organic wastes in a slow-moving stream, but still contains the mineralized products of self-purification. *See also* saprobic classification.

**oligotroph** A microorganism that requires little organic carbon (e.g., 1.0 mg/L) for growth; same as an autochthonous organism or an r-strategist.

**oligotrophic** Characteristic of reservoirs and lakes that are nutrient poor and contain little aquatic plant or animal life.

**oligotrophic lake** A deep clear lake with few nutrients, little organic matter in suspension or on the bottom, a high dissolved oxygen level, low biochemical oxygen demand, absence of hydrogen sulfide, and very rare algal blooms; e.g., a deep cold-water lake or a spring-fed lake. *See also* dystrophic lake, eutrophic lake.

**oligotrophic organism** An organism that can live in nutrient-deficient environments.

**oligotrophic stage** A period in the natural life cycle of a lake or reservoir associated with low nutrient concentrations, limited algal production, and limited organic production.

**oligotrophic water** Water characterized by a low level of nutrients, little organic production, and high dissolved oxygen.

**oligotrophy** The condition of an oligotrophic water body.

**olivine [$(Mg,Fe)_2SiO_4$]** A group of olive-green to gray-green magnesium iron silicates found in igneous rocks. Also called chrysolite. *See* iron oxide-coated olivine.

**O & M** Acronym of operation and maintenance.

**Omega** A horizontal rotor aerator manufactured by Purestream, Inc.

**Omega®** An apparatus manufactured by Merrick Industries, Inc. for slaking and feeding lime in water treatment.

**Omnichlor** An apparatus manufactured by Eltech International Corp. for the production of sodium hypochlorite.

**Omniflo®** Equipment manufactured by Jet Tech, Inc., using sequencing batch reactors for wastewater treatment.

**Omnipure** Products of Eltech International Corp. for use in marine wastewater treatment.

**once-through cooling** A cooling system that uses water only once, without recirculation; saline or brackish water may be used for this purpose.

**once-through cooling water** Water used for heat removal, passing through the main cooling condensers in one or two passes, without coming into direct contact with any raw material, intermediate, or finished product (EPA-40CFR419-11.e and 423-11.g). *See also* recirculated cooling water.

**once-through operation or system** An operation or system that does not include recycling, reuse, or recirculation; e.g., a once-through cooling system; the opposite of a recirculating system. Also called a flow-through system.

***Onchocerca*** The species of filarial nematode that causes onchocerciasis or river blindness.

**onchocerciasis** A disease caused by the mosquito-borne helminth *Onchocerca volvulus,* which is transmitted by blackflies or members of the *Simulium* species. *See* its common name, river blindness, for more detail.

**oncogene** Any gene that may lead to the formation of cancerous growth because of alterations or mutations. Drinking water may contain substances such as haloacetic acids that may activate oncogenes. (The prefix onco, meaning tumor or mass comes from the Greek onkos = mass, bulk).

**oncogenic** Pertaining to a substance that may cause tumors. *See* carcinogenic for detail.

**oncogenic virus** A virus that does not kill its host cell but directs it to produce malignant cancer cells.

**oncology** Study of cancer.

***Oncomelania*** A genus of snails that the pathogen *Schistosoma japonicum* excreted in feces may infect to transmit schistosomiasis. It is amphibious, resistant to environmental modification and molluscicides, and can survive in dry periods. *See also Tricula.*

**one-hit model** A mathematical model based on the biological theory that a single "hit" of some minimum critical amount of a carcinogen at a cellular target (namely DNA) can initiate an irreversible series of events, eventually leading to a tumor. In other words, a tumor can be induced after a single susceptible target or receptor has been exposed to a single effective dose unit of a substance. The one-hit model is a dose–response model of the form

$$P(d) = 1 - \exp(-bd) \qquad \text{(O-04)}$$

where $P(d)$ is the probability of cancer from a continuous dose rate $d$, and $b$ is a constant. Also called single-hit model. *See* dose–response model for related distributions.

**Onguard®** A device manufactured by Ashland Chemical to monitor corrosion.

**on-line boiler cleaning** A procedure used to remove deposits and corrosion products collected on boiler tubes, consisting of the injection of a sodium polyacrylate solution into the boiler feedwater.

**on-line pump station** or **booster station** A station that pumps according to the level of the water surface at the junction being pumped; it has no wetwell. It may also operate according to the head difference over the pumps. *See also* pumping station and off-line pump station.

**on-line turbidimeter** An instrument that continuously samples a water stream and measures and records its turbidity.

**on-off control** A simple type of intermittent feedback control initiated when a controlled parameter reaches a predetermined value, e.g., the water level in a wetwell can be used to turn a wastewater pump station on or off. *See also* the following types of control: feedback, feed-forward, integral, proportional, proportional–derivative, proportional–integral, and proportional–integral–derivative.

**on-site disposal** The disposal of wastes, in particular excreta or wastewater, on the site where they originate and sludge on the site of a treatment plant.

**on-site disposal system** A system of disposal for wastewater or excreta from individual dwellings and community facilities in unsewered areas. Acceptable disposal systems are capillary seepage trench, extended aeration package plant, mound, pit privy, sand filter, and septic tank and subsurface disposal field. *See* on-site wastewater or excreta disposal.

**on-site excreta disposal** *See* on-site wastewater or excreta disposal.

**on-site maintenance district** *See* onsite wastewater management district.

**on-site nitrification/denitrification** An individual wastewater treatment process designed to remove nitrogen. It includes two separate septic tanks for gray water and black water, an aerobic subsurface sand filter or recirculating granular filter that nitrifies the graywater tank effluent, and an anaerobic media tank for denitrification using the graywater tank effluent. *See* Figure O-02.

**on-site sanitation** *See* on-site wastewater or excreta disposal.

**on-site sewage disposal** *See* on-site wastewater or excreta disposal.

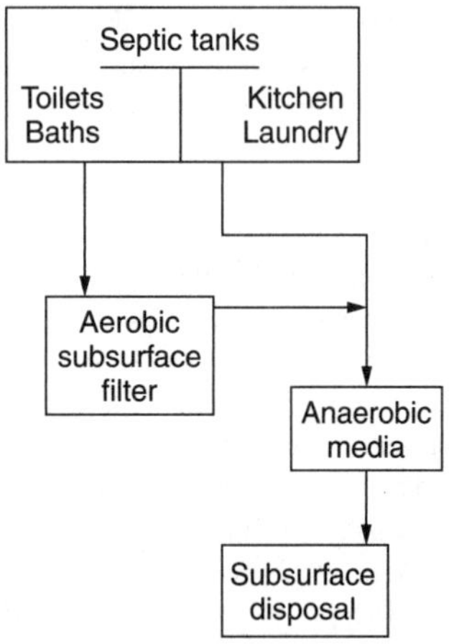


**Figure O-02.** On-site nitrification/denitrification.

**on-site sludge disposal** The disposal of sludge at the site of the water or wastewater treatment plant, practiced by very few communities because of the risk of disease transmission. In such cases, sludge processing may include lime treatment, chlorination, or a thermal process.

**on-site source control** In general, the practice of reducing pollutants at their source. In stormwater management, it is the use of measures designed to retain stormwater at or near the point of rainfall as opposed to conveyance through a sewer system and use of end-of-pipe facilities. Source controls that promote infiltration and reduce runoff include the discharge of roof leaders to pervious areas, reduced lot grading, rear-yard ponding, soakaway pits, and rural road cross sections instead of the urban standard of curb and gutter.

**on-site stormwater retention** The control of peak stormwater discharge within the site producing it. It is a common requirement of local and state regulatory agencies for all new development or redevelopment projects to install detention basins or holding ponds to keep the peak runoff from an area at the same level as before development.

**on-site system** A system of treatment and disposal for wastewater or excreta from individual dwellings and community facilities in unsewered areas; e.g., a septic tank and a subsurface disposal field.

**on-site wastewater management district (OSWMD)** A public or private agency formed to manage on-site wastewater systems in a defined area such as a housing development or a small rural community. Its responsibilities may vary from the approval of the design of individual systems to the scheduling of routine maintenance activities.

**on-site wastewater or excreta disposal** Disposal of wastewater or excreta on the property where they originate, usually in a rural area or the outskirts of an urban area that does not have a sewer system, e.g., within a residential building lot, a commercial site, or a campground.

**$OOCCOO^{-2}$** Chemical formula of oxalate.

**oocyst** A stage in the development cycle of a sporozoan (e.g., *Cryptosporidium*) or the resulting structure (enclosing cyst wall) that protects it in the environment and usually is infectious. Oocysts are excreted in the feces and, once ingested, split open and release sporozoites in the small intestine. *See also* resistant form.

***Oocystis*** A genus of green algae.

**oogenesis** The production of germ cells in a female; the origin and development of the ovum.

**OPA** Acronym of Oil Pollution Act.

**opacity** The amount of light obscured by particulate pollution in the air; clear window glass has zero opacity, whereas a brick wall is 100% opaque. Opacity is an indicator of change in the performance of particulate control systems.

**Opatken–Grady RBC model** A second-order model developed by Opatken in 1985 to determine BOD removal in rotating biological contactors (RBC) and converted to Système International (SI) units by Grady et al. in 1999. The model calculates the soluble BOD at any stage $n$ of the system as $S_n$ (mg/L):

$$S_n = 51.3347\, Q[-1 + (1 + 0.03896\, A_\mathrm{n} S_{n-1}/Q)^{0.5}]/A \qquad \text{(O-05)}$$

where $Q$ = flow rate, $m^3/d$; and $A_\mathrm{n}$ = surface area of disk at stage $n$, $m^2$.

**open centrifugal pump** A centrifugal pump with an impeller that has independent vanes.

**open channel** A natural or artificial waterway or conduit in which liquids flow with a free surface, i.e., under atmospheric pressure. Examples of open channels include all natural streams (rivers, creeks, brooks, and ravines) as well as artificial channels (aqueducts, canals, chutes, culverts, ditches, flumes, partially full conduits, and tunnels). Gravity sewers are designed as open channels.

**open-channel flow** Flow of a fluid with a free surface open to the atmosphere, in an open channel or in a closed conduit flowing partly full; also called gravity flow. The opposite is pressurized flow, such as in a closed conduit flowing full or in a confined aquifer. Free surfaces are subject to atmospheric pressure of one atmosphere, which is equal to 14.7 psi = 101.4 $kN/m^2$ = 29.92 in (760 mm) of mercury = 33.90 ft of water at average sea level under standard conditions. *See* Figures O-03 and O-04 for an open-channel flow profile and cross sections. Common characteristics of open-channel flow include: area ($A$), bottom width ($B$), depth ($y$), discharge or flow rate or, simply, flow ($Q$), velocity ($V$), and Froude Number ($F_r$):

$$F_r = V/(gy)^{0.5} \qquad \text{(O-06)}$$

The hydraulic mean depth ($D_\mathrm{m}$) is

$$D_\mathrm{m} = A/B \qquad \text{(O-07)}$$

and the hydraulic radius ($R$):

$$R = A/P \qquad \text{(O-08)}$$

The Reynolds number is

$$R_e = VL_s/\nu \qquad \text{(O-09)}$$

where $V$, $L_s$, and $\nu$ are, respectively, the mean velocity, a characteristic length, and the kinematic viscosity. Other characteristics are roughness coefficient ($n$), slope ($S_0$), stage ($h$), surface width ($W$), and types of flow (laminar or turbulent, uniform or varied, steady or unsteady; critical, subcritical, or supercritical, stratified or not). The general equation of gradually varied flow is

$$dy/dx = (S_0 - S_f)/(1 - F_r^2) \qquad \text{(O-10)}$$

where $S_f$ is the slope of the total energy line, $S_0$ is the bed slope, $y$ is the depth of flow, $x$ the longitudinal distance, and $F_r$ is the Froude number.

**open cut** An excavation method used in the construction of water mains and sewers. For the replacement of damaged water pipes, a new trench is excavated parallel to the existing one or the old main is dug up to install the new pipe. *See also* trenchless technology.

**open-cut mining** *See* strip mining.

**open cycle cooling system** A system that discharges cooling water without recycling.

**open defecation** The unacceptable practice of defecation in open fields, in areas that have no toilet facilities, or have objectionable facilities.

**open-ended valve** Any valves, except pressure-relief valves, having one side of the valve seat in contact with process fluid and one side open to the atmosphere, either directly or through open piping (EPA-40CFR60.481).

**open impeller** An impeller that has no attached side wall.

**open-impeller pump** A pump that has no attached side wall, usually installed in wastewater and sludge handling facilities because it can pass larger debris than a closed-impeller pump.

**open surge tank** Same as surge tank or standpipe (1).

**operable treatment works** A treatment works that (a) upon completion of construction will treat wastewater, transport wastewater to or from treatment, or transport and dispose of wastewater in a manner that will significantly improve an objectionable water quality situation or health hazard; and (b) is a component part of a complete waste treatment system that, upon completion of construction for the complete waste treatment system (or completion of other treatment works in the system in accordance with an approved schedule), will comply with all applicable statutory and regulatory requirements (EPA-40CFR35.905).

**operating availability (AVO)** A statistical measure of the reliability of mechanical equipment, as determined by the fraction of time a component or a unit is expected to be operational, excluding preventive maintenance; it is one of the factors used in critical component analysis. *See also* expected time before failure, inherent availability, mean time before failure.

**operating capacity** A measure of the actual performance of an ion-exchange resin, depending on feedwater composition, feedwater rate, regeneration level, and other conditions. It is usually lower than the capacity advertised by the resin manufacturer. *See also* dry-weight capacity, wet-volume capacity.

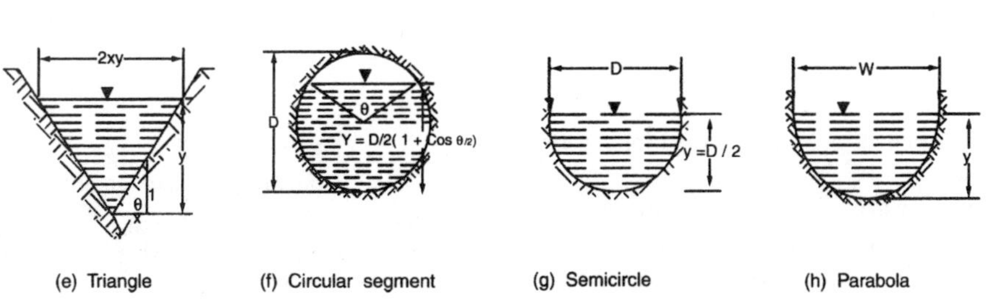

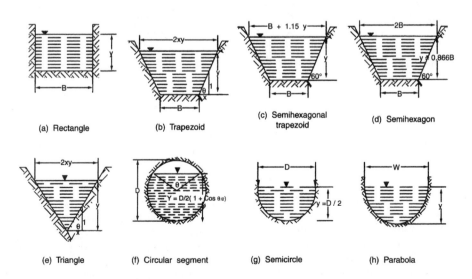


**Figure O-03.** Open-channel flow (cross section) (courtesy CRC Press).

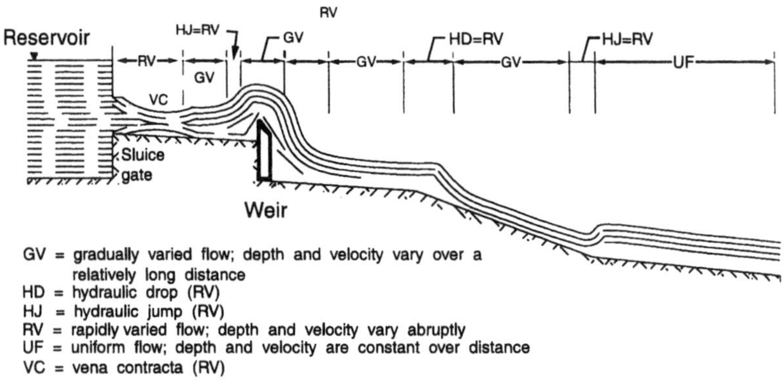


**Figure O-04.** Open-channel flow (profile) (courtesy CRC Press).

**operating cycle** (1) The complete cycle of service run of a water or wastewater treatment unit (e.g., a granular filter, an ion exchanger, or a membrane filter), including such operations as filter service, backwash, regeneration, rinses, membrane cleaning, and return to service. (2) The sum of wastewater application and drying periods in rapid infiltration systems. It is selected to maximize the infiltration rate, nitrogen removal, or nitrification. It varies from 5–8 days for the application of secondary effluents in the summer to 21–28 days for winter applications. Also called loading cycle. *See also* average application rate.

**operating dose** Same as dosing rate of a trickling filter.

**operating line** A straight line that represents actual operating conditions of a stripping tower or any separation process. It is derived from a mass balance relating solute concentrations initially and after contact. For example, for a batch reactor treating an aqueous organic solution by adsorption on powdered activated carbon, the line can be drawn from the following mass balance:

$$q = (V/M)(C_0 - C) \qquad \text{(O-11)}$$

where $q$ = solute concentration to the activated carbon at any time, mg/g; $V$ = volume in liquid in the reactor, L; $M$ = mass of powdered activated carbon, g; $C_0$ = initial solute concentration, mg/L; and $C$ = solute concentration after adsorption, mg/L. *See* Figure O-05. *See also* stripping tower mass balance.

**operating pressure** The pressure range within which a water or wastewater processing device is designed to function, as indicated by its manufacturer, in psi or in any other unit of pressure. Also called working pressure.

**operating pressure (reverse osmosis and nanofiltration)** The pressure at which pressure-driven membrane systems operate, namely the pressure of the feedwater, typically lower for nanofiltration than reverse osmosis. It is used as a parameter for classifying these membranes. *See also* molecular weight cutoff, osmotic pressure, salt rejection capability, and silt density index.

**operating pressure differential** The operating pressure of a hydropneumatic system.

**operating storage** Same as equalizing storage. *See* equalizing basin.

**operational salt efficiency** A measure of the long-term performance of a water softener in terms of its capacity to remove hardness. *See* salt efficiency.

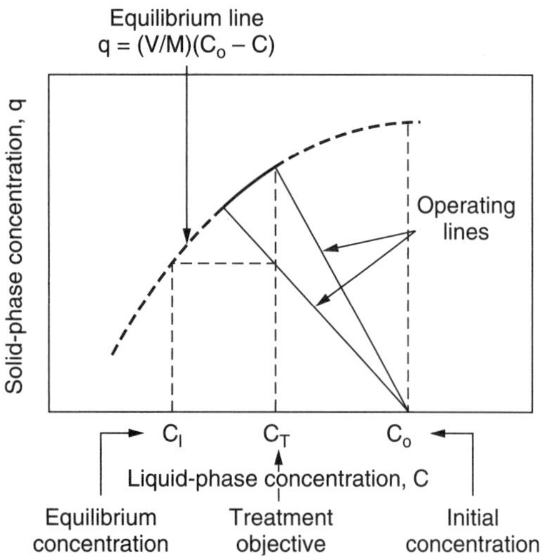


**Figure O-05.** Operating line (adapted from Crittenden et al., 2005).

**operation and maintenance (O&M)** (a) Control of the unit processes and equipment of a facility. This includes personnel and financial management; and records, laboratory control, process control, safety and emergency operation planning. (b) Preservation of functional integrity and efficiency of equipment and structures. This includes preventive maintenance, corrective maintenance and replacement of equipment as needed (EPA-40CFR35.2005-30).

**operation and maintenance costs** The ongoing repetitive costs of operating a water or wastewater system; for example, employee wages and benefits, costs of chemicals, and periodic equipment repairs.

**operation and maintenance manual** A document prepared for a unit, system, plant, etc. to define the equipment and processes included as well as their proper operation, repair, and maintenance.

**operation expense** *See* operation and maintenance costs.

**operation waste** Water wasted during the operation of a hydraulic structure, e.g., through spillways or from irrigation canals.

**operator** A person who participates in the operation of a water or wastewater facility.

**operator training** Instruction provided to water and wastewater operators.

***Opercularia*** One of the groups of ciliate protozoa active in trickling filters, feeding on the biological film and maintaining its high growth rate.

**opisthorchiasis** An excreted helminth infection caused by the pathogens *Opisthochis felineus* (cat liver fluke, found in Poland, Ukraine, and northern Turkey) and *Opisthorchis viverrini* (found in Thailand and southern Laos). It is closely related to clonorchiasis. *See* its common name, cat liver fluke, for more detail.

***Opisthorchis felineus*** The helminth species that causes cat liver fluke disease (opisthorchiasis) in East Europe.

***Opisthorchis sinensis*** Another name of *Clonorchis sinensis,* because of its close similarity to *Opisthorchis felineus* and *Opisthorchis viverrini.*

***Opisthorchis viverrini*** The helminth species that causes cat liver fluke disease (opisthorchiasis) in East Asia.

**O-$PO_4$** Symbol of orthophosphate.

**opportunistic bacterial pathogen** *See* opportunistic pathogen.

**opportunistic infection** An infection caused by an opportunistic pathogen.

**opportunistic pathogen** An organism that is not pathogenic to healthy individuals but can cause disease in immunologically weak individuals, e.g., AIDS patients and other ill persons, burn patients, cancer patients, patients taking antibiotics, pregnant women, the very young, and the elderly. Many hospital-acquired infections are caused by opportunistic waterborne pathogens, mostly heterotrophic bacteria. Some opportunistic pathogens include *Aeromonas* species (e.g., *A. hydrophila*), *Mycobacterium avium intracellulare, Pseudomona species* (e.g., *P. aeruginosa*), and the *Acinetobacter, Flavobacterium, Klebsiella, Proteus,* and *Serratia* genera. Astroviruses, some algae that produce neurotoxins, microsporidia, and the protozoan genus *Toxoplasma* are also considered opportunistic. *See also* frank pathogen.

**opportunistic protozoa** Any of the protozoan pathogens that significantly affect persons with compromised immune systems (young children, elderly, AIDS patients, etc.), including *Cryptosporidium parvum, Cyclospora, Encephalitozoon intestinalis, Enterocytozoon bieneusi, Giardia lamblia,* and *Taxoplasma gondii.*

**optical bleach** A fluorescent organic dye that absorbs ultraviolet light while emitting visible light; used as a brightener in the textile and detergent industries.

**optical imaging** A technique that uses a microscope attached to a video camera to obtain a quantitative assessment of particles in a small wastewater sample. The camera transmits the images of the particles to a computer for such measurements as mean, minimum and maximum diameter, length/width ratio, surface area, and volume of particles. Also called particle imaging. *See also* microscopic counting.

**optical microscope** A microscope that uses visible light as a source of illumination. Common types provide a maximum magnification of 2000 times and a resolution of 0.2 μm or 200 nm. Also called light microscope. *See also* bright-field, dark-field, differential interference, phase-contrast, electron-beam, and ultraviolet-ray microscopes.

**optical-range scanning** A nondestructive, technique that measures the relative amounts of contamination on smooth, transparent substrate such as glass; commonly used in optics-cleaning applications.

**Opti-Core** Media manufactured in PVC by B. F. Goodrich Co. for use in biological filters.

**optimal corrosion control treatment** The corrosion control treatment that minimizes the lead and copper concentrations at users' taps while insuring that the treatment does not cause the water system to violate any national primary drinking water regulations. *See* optimization of corrosion control.

**optimal point of coagulation** A set of operating conditions (pH, chemical dose, mixing) determined by testing and producing the best floc in the shortest time and the most effective removal of target contaminants (e.g., particulates, turbidity, disinfection by-product precursors) in water or wastewater treatment. *See also* optimum flocculation condition.

**Optimem™** Reverse osmosis membranes manufactured by NW Acumem, Inc.

**Optimer™** A flocculant produced by Nalco Chemical Co.

**optimization of corrosion control** The water treatment method that minimizes the concentration of lead and copper at the consumer's tap without violating any drinking water regulations. *See* optimal corrosion control treatment.

**optimized coagulation** *See* optimal point of coagulation.

**Optimum®** A water treatment plant manufactured by BCA Industrial Controls using direct filtration.

**optimum flocculation condition** The combination of flocculation design parameters that produces the best floc, i.e., the velocity gradient ($G$), time ($t$), and volumetric concentration ($\Phi$). In water and wastewater treatment, the optimum condition is sometimes expressed by setting the product of these variables equal to a constant ($C \approx 4000$):

$$Gt\Phi = C \quad \text{(O-12)}$$

*See* volumetric concentration.

**optimum point of coagulation** *See* optimal point of coagulation.

**oral** Of the mouth; through or by the mouth.

**oral–fecal transmission** Same as fecal–oral transmission.

**oral rehydration therapy (ORT)** The administration of a liquid by mouth to treat a diarrheal disease such as cholera.

**Orbal™ oxidation ditch** Same as Orbal™ process.

**Orbal™ process** A variation of the oxidation ditch process that uses three concentric channels in series and disk aerators on a horizontal shaft, such that dissolved oxygen increases from the outer to the inner loops. Developed by Envirex, Inc., it is designed to achieve BOD removal and simultaneous nitrification/denitrification (SNdN). It can achieve nitrogen removals as high as 95%. *See* Figure O-06. *See also* Bionutre™ process.

**Orbal Simpre™** Same as Orbal™ process.

**order of elution** In the ion-exchange treatment process, the order in which ions (contaminants) are stripped from an exchange resin using a concentrated solution. It is determined by the selectivity sequence.

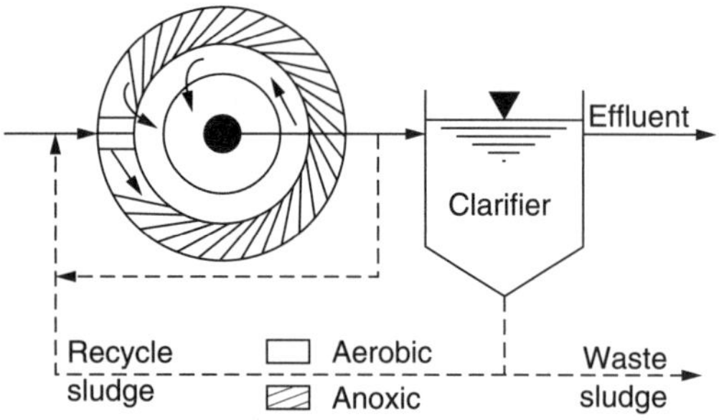


**Figure O-06.** Orbal™ process.

**order of reaction** The order of a reaction with respect to a compound is the stoichiometric coefficient for that compound. The overall order of a reaction is the sum of the stoichiometric coefficients of the various compounds. For example, in the reaction:

$$aA + bB + cC + \ldots \rightarrow \ldots \quad \ldots \quad \text{(O-13)}$$

the reaction order for compound A is a and the overall reaction order is $n = a + b + c + \ldots$. *See also* first-order reaction, first-order retarded (or retarded first-order) reaction, second-order reaction, saturation-type equation, retardation factor, and rate law.

**Orec** An apparatus manufactured by Ozone Research & Equipment Corp. for the production of ozone.

***Oreochromis mossambicus*** A genus of the fish tilapia, which feeds on the eggs and young of snails, and thus can be used in the control of schistosomiasis.

**Organagro®** Agricultural compost produced by Bedminster Bioconversion Corp.

**organelle** A permanent, specialized structure of a microorganism.

**organic** (1) Referring to or derived from living material or organisms. (2) In chemistry, any volatile, combustible compound containing carbon and sometimes biodegradable. Organic matter commonly found in wastewater includes carbohydrates, fats and oils, and proteins. *See also* inorganic.

**organic acid** An acid that contains organic carbon; general formula: R—COOH, where R represents a radical; e.g., acetic acid ($CH_3COOH$), butyric acid ($CH_3CH_2CH_2COOH$), propionic acid ($CH_3CH_2COOH$). In stabilization ponds, acid-forming bacteria decompose benthal deposits of organic matter to produce organic acids that are used by methane-forming bacteria and algae. Organic acid concentration is routinely measured to monitor the anaerobic digestion process.

**organic adsorption** The removal of organic substances from water or wastewater by adsorption on activated carbon or other adsorbents.

**organic agriculture** *See* organic farming.

**organic amendment** An organic material, such as dried manure, sawdust, or straw, added to sludge (or other feed substrate) in a composting operation to promote uniform air flow while reducing the bulk weight and the moisture content. *See also* bulking agent and amendment (2).

**organic carbon** Carbon (C) derived from living organisms. *See also* inorganic carbon.

**organic chemical** A substance produced by animals or plants and containing mainly carbon, hydrogen, nitrogen, and oxygen. Chloroform and other trihalomethanes in drinking water are among the organic chemicals that have recently received considerable attention. *See also* organic compound, organic material, organics and organic matter.

**organic chemicals, plastics, and synthetic fibers (OCPSF)** An industry that is involved in chemical synthesis and the recovery of organic chemicals as by-products of other manufacturing operations. Wastewater generated by this industry has widely varying characteristics, including conventional pollutants (BOD, TSS, oil and grease), metals and other toxic priority pollutants, as well as nonconventional pollutants.

**organic chemistry** The branch of chemistry that deals with the compounds of carbon, mainly in combination with hydrogen, oxygen, or a halogen, but also with nitrogen, phosphorus, and sulfur. *See also* organic compound, inorganic chemistry.

**organic chloramine** An organic compound of nitrogen and chlorine, formed during the chlorination of water containing organic nitrogen; e.g., RNHCl, where R represents the organic group attached to nitrogen. The formation of organic chloramines reduces the disinfection capability of chlorine and can be reduced by increasing the ratio of ammonia nitrogen to organic nitrogen. Also called organochloramine.

**organic chlorine** Chlorine that is chemically bound to an organic compound.

**organic chlorine compound** Any of more than 2000 compounds containing carbon and at least one chlorine atom; they are important building blocks of the chemical industry. *See* organochlorine compound.

**organic compound** Any compound of carbon, excluding carbon monoxide (CO), carbon dioxide ($CO_2$), carbonic acid ($H_2CO_3$), metallic carbides or carbonates, and ammonium carbonate ($NH_4CO_3$) (EPA-40CFR52.741). Organic compounds are usually combustible, high in molecular weight, and sparingly soluble in water. Examples of compounds that contain carbon from living organisms are carbohydrates, ethers, and hydrocarbons. Important industrial organic compounds include agricultural chemicals, biological materials, and synthetic polymers. Organic compounds (normally composed of a combination of carbon, hydrogen, oxygen, and nitrogen in some cases) may be a source of food for bacteria and other microorganisms active in waste treatment. Organic compounds of particular concern in drinking water sources include chlorinated hydrocarbons, trihalomethanes, and volatile organic chemicals, some of which are potentially carcinogenic. The organic content of wastewater typically consists of 40–60% proteins, 25–50% carbohydrates, 10% oils and fats, and synthetic organic molecules; urea is also found in fresh wastewater. *See also* organic chemical, organic material, organics, and organic matter.

**organic compound characteristics** Organic compounds are usually combustible, have low melting point, low boiling point and high molecular weights. They may be natural compounds, products of fermentation, or synthetic products of a manufacturing process. *See also* natural organic matter, synthetic organic chemical.

**organic compound formulas** *See* Table O-02.

**organic contamination** Contamination by an organic compound.

**organic disinfection by-product** Undesirable by-products of water disinfection by oxidizing agents include trihalomethanes and haloacetic acids, as well as chlorophenols, chloropicrin, cyanogen chloride, haloacetonitriles, haloaldehydes, halo ketones, and 3-chloro-4-(dichloromethyl)-5-hydroxy-2(5H)-furanone or MX. More generally, organic disinfection by-products are classified under chlorine by-products, chlorine dioxide by-products, chloramine by-products, and ozone by-products.

**organic farming** Farming using compost, seaweed, manure, other products from living things, and, in some cases, a naturally occurring chemical such as nicotine, instead of artificial fertilizers (e.g., nitrates and phosphates), pesticides (fungicides, herbicides, insecticides), and other agrochemicals (fruit regulators, growth stimulants,

**Table O-02.** Organic compound formulas

| | |
|---|---|
| **$CHBrCl_2$** | bromodichloromethane |
| **$CHBrClI$** | bromochloroiodomethane |
| **$CHBrI_2$** | bromodiiodomethane |
| **$CHBr_2Cl$** | dibromochloromethane |
| **$CHBr_2CN$** | dibromoacetonitrile |
| **$CHBr_2I$** | dibromoiodomethane |
| **$CHBr_3$** | bromoform |
| **$CHClO_2$** | chlorous acid |
| **$CHCl_2F$** | dichlorofluoromethane |
| **$CHCl_3$** | chloroform or trichloromethane |
| **$CHClI_2$** | chlorodiiodomethane |
| **$CHI_3$** | triiodomethane |
| **$CHO$** | monoaldehyde, also written as HC═O |
| **$CHX_3$** | haloform, X representing a halogen |
| **$CH_2BrCl$** | bromochloromethane |
| **$CH_2Cl_2$** | dichloromethane or methylene chloride |
| **$CH_2N_2$** | cyanamide |
| **$CH_2N_2$** | diazomethane, also written as $H_2C{=}N^+{=}N^-$ |
| **$CH_2O$** | simplified organic matter or biomass |
| **$CH_2O_2$** | methanoic acid or formic acid, also written as HCOOH |
| **$CH_3Br$** | bromomethane |
| **$CH_3Cl$** | chloromethane |
| **$CH_3Hg^+$** | methylmercury |
| **$CH_3NO_2$** | carbamic acid, also written as $NH_3CO_2$ |
| **$CH_3NaO$** | sodium methylate |
| **$CH_4$** | methane |
| **$CH_4N_2O$** | ammonium cyanate |
| **$CH_4N_2O$** | urea or carbamide, also written as $CO(NH_2)_2$ |
| **$(CH_4N_2O)_n$** | polyurea, also written as $[CO(NH_2)_2]_n$ |
| **$CH_4N_2S$** | ammonium thiocyanate |
| **$CH_4O$** | methanol, also written as $CH_3OH$ |
| **$CH_4S$** | methyl mercaptan, also written as $CH_3SH$ |
| **$CH_5AsO_3$** | methylarsinic acid, also written as $CH_3AsO(OH)_2$ |
| **$CH_6N_2O_2$** | ammonium carbamate |
| **$C_2HBrO_2X_2$** | bromoacetic acid, also written as $CX_2BrCOOH$, with X = $Br^-$ or H |
| **$C_2HBr_2ClO_2$)** | dibromochloroacetic adid, also written as $CBr_2ClCOOH$ |
| **$C_2HBr_2N$** | dibromoacetonitrile, also written as $CHBr_2C{\equiv}N$ |
| **$C_2HBr_3O_2$** | tribromoacetic acid, also written as $CBr_3COOH$ |
| **$C_2HCl_2N$** | dichloroacetonitrile, also written as $CHCl_2C{\equiv}N$ |
| **$C_2HCl_3$** | trichloroethylene, also written as $ClHC{=}CCl_2$ |
| **$C_2HCl_3O_2$** | trichloroacetic acid, also written as $CCl_3COOH$ or $HC_2Cl_3O_2$ |
| **$C_2HO_2X_3$** | haloacetic acid, also written as $CX_3COOH$, with X = $Cl^-$, $Br^-$, or H |
| **$C_2H_2$** | acetylene |
| **$C_2H_2BrCl_2O_2$** | bromodichloroacetic acid, also written as $CHBrCl_2COOH$ |
| **$C_2H_2Br_2O_2$** | dibromoacetic acid, also written as $CHBr_2COOH$ |
| **$C_2H_2ClF_3$** | chlorotrifluoroethylene |
| **$C_2H_2Cl_2$** | 1,1-dichloroethylene (vinylidene chloride), also written as $CH_2CCl_2$ |
| **$C_2H_2Cl_2O_2$** | dichloroacetic acid, also written as $CHCl_2COOH$ |
| **$C_2H_2FNaO_2$** | sodium fluoroacetate |
| **$C_2H_2NO_2$** | ethyl nitrate |
| **$C_2H_2O_2$** | glyoxal, also written as OHCCHO |
| **$C_2H_2O_3$** | glyoxylic acid, also written as OHCCOOH |
| **$C_2H_2O_4$** | oxalic acid |
| **$C_2H_3Cl$** | vinyl chloride, vinyl chloride monomer, also written as $CH_2CHCl$ |
| **$C_2H_3ClO_2$** | chloroacetic acid or mono-chloroacetic acid, also written as $CH_2ClCOOH$ |
| **$C_2H_3Cl_3$** | 1,1,1-trichloroethane, also written as $CH_3CCl_3$ |
| **$C_2H_3Cl_3$** | vinyl trichloride or 1,1,2-trichloroethane, also written as $CHCl_2CH_2Cl$ |
| **$C_2H_3NaO_2$** | sodium acetate |
| **$C_2H_3NO$** | methyl isocyanate, also written as $CH_3NCO$ |
| **$C_2H_4$** | ethylene |
| **$(C_2H_4)_x$** | high-density polyethylene (HDPE), also written as $[(H_2C{:}CH_2)_x$ |
| **$C_2H_4Br_2$** | ethylene dibromide, also written as $BrCH_2$. Also written $CH_2Br$ |
| **$C_2H_4Cl_2$** | ethylene dichloride, also written as $ClCH_2{-}CH_2Cl$ |
| **$C_2H_4Cl_2$** | 1,1-dichloroethane, also written as $CH_3CHCl_2$ |
| **$C_2H_4O$** | ethanal (acetaldehyde), also written as $CH_3CHO$ |
| **$(C_2H_4O)_x$** | polyvinyl alcohol, also written as $({-}CH_2CHOH{-})_x$ |
| **$C_2H_4O_2$** | acetic acid (ethanoic), also written as $CH_3COOH$ |
| **$C_2H_4O_3$** | peracetic acid, also written as $CH_3CO_3H$ |
| **$C_2H_5Cl$** | ethyl chloride |
| **$C_2H_5NO_2$** | glycine, also written as $H_2NCH_2COOH$ or H—CH—COOH—$NH_2$ |
| **$C_2H_6$** | ethane or dimethyl, also written as $CH_3CH_3$ |
| **$C_2H_6N_2O$** | dimethylnitrosamine |
| **$C_2H_6O$** | ethanol (ethyl alcohol), also written as $CH_3CH_2OH$ or $C_2H_5OH$ |

Table O-02. *Continued*

| | |
|---|---|
| $C_2H_6O_2$ | glycol |
| $C_2H_6O_2S$ | methyl sulfate or dimethyl sulfate, also written as $(CH_3)_2SO_2$ or $CH_3SO_2CH_3$ |
| $C_2H_6S$ | ethyl mercaptan, also written as $CH_3CH_2—SH$ |
| $C_2H_7As$ | dimethylarsine, also written as $(CH_3)_2AsH$ |
| $C_2H_7AsO_2$ | dimethylarsinic acid, also written as $(CH_3)_2AsO(OH)$ |
| $C_2H_7NO_2$ | ammonium acetate, also written as $NH_4(C_2H_3O_2)$ |
| $C_2H_7NO_3S$ | taurine, also written as $H_2NCH_2CH_2SO_3H$ |
| $C_3H_3Cl_3O$ | 1,1,1-trichloropropanone, also written as $CCl_3COCH_3$ |
| $C_3H_4Cl_2$ | 1,3-dichloropropene, also written as $CHCl{:}CHCH_2Cl$ |
| $C_3H_4Cl_2O_2$ | dalapon (di-alpha-dichloropropionic acid), also written as $CH_3CCl_2COOH$ |
| $C_3H_4O_2$ | methyl glyoxal, also written as $CH_3COCHO$ |
| $(C_3H_4O_2)_x$ | polyacrylic acid, also written as $(H_2C{:}CHCOOH)_x$ |
| $C_3H_5BrO$ | bromoacetone, also written as $CH_2BrCOCH_3$ |
| $C_3H_5Br_2Cl$ | dibromochloropropane; 1,2-dibromo-3-chloropropane, also written as $CH_2BrCHBrCH_2Cl$ |
| $C_3H_5ClO$ | epichlorohydrin, also written as $CH_2OCHCH_2Cl$ |
| $C_3H_5NaO_2$ | sodium propionate |
| $C_3H_5NaO_3$ | sodium lactate |
| $C_3H_6$ | propene or propylene, also written as $CH_3CHCH_2$ |
| $(C_3H_6)_n$ | polypropylene |
| $C_3H_6Cl_2$ | propylene dichloride, 1,2-dichloropropane, also written as $CH_3CHClCH_2Cl$ |
| $C_3H_6Cl_2$ | 1,3-dichloropropane, also written as $CH_2ClCH_2CH_2Cl$ |
| $C_3H_6Cl_2$ | 2,2-dichloropropane, also written as $CH_3CCl_2CH_3$ |
| $C_3H_6N_2S$ | ethylene thiourea, also written as $NHCH_2CH_2NHCS$ |
| $C_3H_6O$ | acetone (propanone) or dimethyl ketone, also written as $CH_3COCH_3$ |
| $C_3H_6O_2$ | propanoic (or propionic) acid, also written as $CH_3CH_2COOH$ |
| $C_3H_6O_3$ | lactic acid |
| $C_3H_7—$ | the *n*-propyl radical, also written as $CH_3—CH_2—CH_2—$ |
| $C_3H_7NO_2S$ | cysteine |
| $C_3H_8$ | propane |
| $C_3H_8NO_5P$ | glyphosate, also written as $HO—CO—CH_2—NH—CH_2—PO—(OH)_2$ |
| $C_3H_8O$ | dimethylcarbinol isopropanol, isopropyl alcohol, or 2-propanol, also written as $CH_3—CH_3—CHOH$ |
| $C_3H_8O$ | 1-propanol, also written as $CH_3—CH_2—CH_2—OH$ |
| $C_3H_9NO_3$ | ammonium lactate |
| $C_4H_4N_2$ | diazine and pyrimidine |
| $C_4H_4N_2O_2$ | uracil (*see* RNA) |
| $C_4H_4O$ | furan |
| $C_4H_4O_5$ | oxalacetic acid, also written as $HOOCCOCH_2COOH$ |
| $C_4H_5N_3O$ | cytosine (*see* RNA) |
| $C_4H_6$ | butadiene, also written as $CH_2CHCHCH_2$ |
| $C_4H_6O_2$ | dimethyl glyoxal, also written as $CH_3COCOCH_3$ |
| $(C_4H_8)_n$ | polybutylene |
| $C_4H_8Cl_2S$ | dichlorodiethyl sulfide and mustard gas |
| $C_4H_8O$ | ethyl methyl ketone, also written as $CH_3CH_2COCH_3$ |
| $C_4H_8O_2$ | butanoic acid (butyric acid), also written as $CH_3CH_2CH_2COOH$ |
| $C_4H_8S$ | crotyl mercaptan, also written as $CH_3CHCHCH_2SH$ |
| $C_4H_{10}$ | butane, also written as $CH_3CH_2CH_2CH_3$ |
| $C_4H_{10}O$ | butyl alcohol, also written as $C_4H_9OH$ |
| $C_4H_{10}O$ | diethyl ether, also written as $(C_2H_5)_2O$ |
| $C_4H_{10}S$ | diethyl sulfide, also written as $C_2H_5SC_2H_5$ |
| $C_5H_4O_2$ | furfural, also written as $C_4H_3OCHO$ |
| $C_5H_5N_5$ | adenine (*see* RNA) |
| $C_5H_5N_5O$ | guanine (*see* RNA) |
| $C_5H_6N_2O_2$ | thymine |
| $C_5H_7NO_2$ | Chemical formula commonly assumed for sludge, microorganism composition, cell mass, cell tissue, biomass, organic fraction of cells. |
| $C_5H_7NO_2P_{0.074}$ | Another chemical formula proposed for sludge. |
| $C_5H_7NO_2P_{0.2}$ | Chemical formula that is generally used to represent the composition of activated sludge microorganisms. |
| $C_5H_8O_2$ | ethyl acrylate, also written as $CH_2{:}CHCOOCH_2CH_3$ |
| $C_5H_8O_2$ | glutaraldehyde, also written as $OHC(CH_2)_3CHO$ |
| $(C_5H_8O_4)_x$ | pentosans |
| $C_5H_{10}O_2$ | pentanoic acid or valeric acid, also written as $C_4H_9COOH$ |
| $C_5H_{12}O$ | methyl tertiary butyl ether, also written as $(CH_3)_3COCH_3$ |
| $C_5H_{16}NCl_2$ | atrazine |

(*continued*)

**Table O-02.** *Continued*

| | |
|---|---|
| $C_6H_3Cl_3$ | trichlorobenzene |
| $C_6H_3Cl_3N_2O_2$ | picloram |
| $C_6H_3Cl_3O$ | 2,4,6-trichlorophenol, also written as $C_6H_2Cl_3OH$ |
| $C_6H_4Cl_2$ | dichlorobenzene (1,2-dichlorobenzene, 1,3-dichlorobenzene, and 1,4-dichlorobenzene) |
| $C_6H_4Cl_2O$ | 2,4-dichlorophenol, also written as $Cl_2C_6H_3OH$ |
| $C_6H_4NO_2$ | dinitrobenzene |
| $C_6H_4N_2O_5$ | 2,4-dinitrophenol, also written as $C_6H_3OH(NO_2)_2$ |
| $C_6H_4O_2$ | quinine |
| $C_6H_5Br$ | bromobenzene |
| $C_6H_5Cl$ | monochlorobenzene or phenyl chloride |
| $C_6H_5ClO$ | chlorophenol |
| $C_6H_5NO_2$ | nitrobenzene |
| $C_6H_5Na_3O_7$ | sodium citrate |
| $C_6H_6$ | benzene |
| $C_6H_6Cl_6$ | lindane |
| $C_6H_6O$ | phenol (carbolic acid, hydroxybenzene, oxybenzene, phenylic acid), also written as $C_6H_5 \cdot OH$ |
| $C_6H_6O_2$ | catechol and resorcinol |
| $C_6H_7N$ | aniline |
| $C_6H_7NO_3S$ | sulfanilic acid |
| $C_6H_8N_2O_2S$ | sulfanilamide |
| $C_6H_8O_6$ | ascorbic acid, a dechlorinating agent |
| $C_6H_8O_7$ | citric acid, also written as $C_3H_4(OH)(COOH)_3$ |
| $C_6H_{10}O_5$ | cellulose |
| $(C_6H_{10}O_5)_n$ | starch, glycogen |
| $C_6H_{10}O_7$ | glucuronic acid, also written as $COOH(CHOH)_4CHO$ |
| $C_6H_{12}N_2O_4S_2$ | cystine |
| $C_6H_{12}N_2S_4$ | thiram |
| $C_6H_{12}O$ | methyl isobutyl ketone, also written as $(CH_3)_2CHCH_2COCH_3$ |
| $C_6H_{12}O_2$ | caproic acid (or hexanoic acid), also written as $C_5H_{11}COOH$ |
| $C_6H_{12}O_6$ | fructose, galactose, glucose |
| $C_6H_{15}NO$ | diethylaminoethanol |
| $C_7H_5N_3O_6$ | trinitrotoluene, also written as $CH_3C_6H_2(NO_2)_3$ |
| $C_7H_5NaO_2$ | sodium benzoate |
| $C_7H_6ClN_3O_4S_2$ | chlorothiazide |
| $C_7H_6N_2O_4$ | 2,4-dinitrotoluene, 2,6-dinitrotoluene, also written as $C_6H_3CH_3(NO_2)_2$ |
| $C_7H_8$ | toluene (methyl benzene, phenylmethane), also written as $C_6H_5CH_3$ |
| $C_7H_8O$ | cresol |
| $C_7H_8O$ | orthocresol or 2-methylphenol, also written as $CH_3C_6H_4OH$ |
| $C_7H_9N$ | toluidine |
| $C_7H_{12}ClN_5$ | simazine, also written as $ClC_3N_3(NHC_2H_5)_2$ |
| $C_7H_{13}N_3O_3S$ | oxamyl, also written as $(CH_3)_2N—CO—C(SCH_3){:}N—O—CO—NH—CH_3$ |
| $C_7H_{14}N_2O_2S$ | aldicarb |
| $C_7H_{14}N_2O_3S$ | aldoxycarb |
| $C_8H_4O_4R_2$ | General formula of phthalates, also written as $C_6H_4(COOR)_2$ |
| $C_8H_5Cl_3O_3$ | 2,4,5-trichlorophenoxyacetic acid |
| $C_8H_6Cl_2O_3$ | 2,4-dichlorophenoxyacetic acid, also written as $Cl_2C_6H_3OCH_2COOH$ |
| $C_8H_7N$ | indole, also written as $C_8H_6NH$ |
| $C_8H_8$ | styrene, also written as $C_6H_5CH{=}CH_2$ |
| $C_8H_{10}$ | ethylbenzene, also written as $C_6H_5C_2H_5$ |
| $C_8H_{10}$ | xylene (dimethylbenzene), also written as $C_6H_4(CH_3)_2$ |
| $C_8H_{13}Cl(CN)N_5$ | cyanazine, also written as $C_2H_5NH—C_3N_3(Cl)—NHC(CH_3)_2CN$ |
| $C_8H_{14}ClN_5$ | atrazine desethyl |
| $C_8H_{14}N_4OS$ | metribuzin |
| $C_8H_{18}$ | octane |
| $C_8H_{18}O$ | the solvent *n*-octanol, also written as $CH_3(CH_2)_7OH$ |
| $C_8H_{19}O_2PS_3$ | disulfoton, also written as $(C_2H_5O)_2P(S)SCH_2CH_2SCH_2CH_3$ |
| $C_8H_{20}Pb$ | tetraethyl lead, also written as $Pb(C_2H_5)_4$ |
| $C_9H_7Cl_3O_3$ | silvex, also written as $Cl_3C_6H_2OCH(CH_3)COOH$ |
| $(C_9H_9)_n$ | polystyrene, also written as $(C_6H_5CH{:}CHCH_2)_n$ |
| $C_9H_9N$ | skatole |
| $C_9H_{10}Cl_2N_2O$ | diuron, also written as $C_6H_3Cl_2NHCON(CH_3)_2$ |
| $C_9H_{11}Cl_3NO_3PS$ | chloropyrifos |
| $C_9H_{12}$ | *n*-propylbenzene |
| $C_9H_{12}$ | 1,2,4-trimethylbenzene, also written as $C_6H_3(CH_3)_3$ |
| $C_9H_{13}ClN_6$ | cyanazine. *See also* $C_8H_{13}Cl(CN)N_5$ |
| $C_9H_{17}NOS$ | molinate, also written as $C_2H_5—S—CO—NC_6H_{12}$ |
| $C_{10}H_5Cl_7$ | heptachlor |
| $C_{10}H_5Cl_7O$ | heptachlor epoxide |
| $C_{10}H_5Fe$ | ferrocene, also written as $(C_5H)_2Fe$ |
| $C_{10}H_6Cl_4O_4$ | dimethyl-2,3,5,6-tetrachloroterephtalate, also written as $C_6Cl_4(COOCH_3)_2$ |
| $C_{10}H_6Cl_8$ | chlordane |
| $C_{10}H_7Cl_7$ | heptachlor |
| $C_{10}H_8$ | naphthalene |
| $C_{10}H_{10}$ | divinylbenzene, also written as $C_6H_4(CH{:}CH_2)_2$ |

**Table O-02.** *Continued*

| | |
|---|---|
| $C_{10}H_{10}Cl_8$ | toxaphene |
| $C_{10}H_{12}N_2O_5$ | dinoseb, also written as $CH_3(C_2H_5)CHC_6H_2(NO_2)_2OH$ |
| $C_{10}H_{14}$ | *p*-isopropyltoluene, also written as $CH_3C_6H_4CH(CH_3)_2$ |
| $C_{10}H_{14}NO_5PS$ | parathion |
| $C_{10}H_{15}OPS_2$ | fonolos |
| $C_{10}H_{16}N_2$ | N,N-diethyl-p-phenylenediamine, also written as $(C_2H_5)_2NC_6H_4NH_2$ |
| $C_{10}H_{16}N_2O_8$ | ethylenediaminetetraacetic acid (EDTA), also written as $(HOOCCH_2)_2NCH_2CH_2N(CH_2COOH)_2$ |
| $C_{10}H_{16}N_5O_{13}P_3$ | adenosine triphosphate |
| $C_{10}H_{17}N_3O_6S$ | glutathione |
| $C_{10}H_{19}NO_3$ | A common formula used to represent (1) a typical domestic wastewater and (2) the biodegradable organic matter in wastewater. |
| $C_{10}H_{19}N_5O$ | prometon, also written as $(H_7C_3HN)_2C_3N_3OCH_3$ |
| $C_{10}H_{19}O_6PS_2$ | malathion |
| $C_{11}H_{12}N_2O_2$ | tryptophan, also written as $(C_8H_6N)CH_2CH(NH_2)COOH$ |
| $C_{11}H_{20}O$ | methylisoborneol |
| $C_{12}H_4Cl_4O_2$ | dioxin |
| $C_{12}H_7Cl_3O_2$ | triclosan |
| $C_{12}H_8Cl_6$ | aldrin |
| $C_{12}H_8Cl_6O$ | endrin |
| $C_{12}H_{10}Cl_6O$ | dieldrin |
| $C_{12}H_{10}N_2O$ | 1,10-phenanthroline, also written as $C_{12}H_8N_2 \cdot H_2O$ |
| $C_{12}H_{11}N$ | diphenylamine |
| $C_{12}H_{12}Br_2N_2$ | diquat, also written as $(C_5H_4NCH_2)_2Br_2$ |
| $C_{12}H_{12}N_2$ | 1,2-diphenylhydrazine, also written as $C_6H_5NHNHC_6H_5$ |
| $C_{12}H_{14}N_2 \cdot 2CH_3SO_4$ | paraquat. *See also* $C_{14}H_{20}N_2O_8S_2$ |
| $C_{12}H_{15}NO$ | carbofuran |
| $C_{12}H_{21}N_2O_2PS$ | diazinon, also written as $[(CH_3)_2CHC_4N_2H(CH_3)O]PSO(C_2H_5)_2$ |
| $C_{12}H_{22}O$ | geosmin |
| $C_{12}H_{22}O_{11}$ | lactose, sucrose |
| $C_{12}H_{23}O$— | the monovalent organic lauroyl (or lauryl) group |
| $C_{12}H_{24}O_2$ | lauric acid |
| $C_{14}H_2Cl_2$ | tetrachloroethane, also written as $CHC_{12}CHCl_2$ |
| $C_{14}H_8Cl_4$ | dichlorodiphenyl dichloroethylene, also written as $(ClC_6H_4)_2C{:}CCl_2$ |
| $C_{14}H_9Cl_5$ | dichlorodiphenyl trichloroethane (DDT), also written as $(ClC_6H_4)_2CH(CCl_3)$ |
| $C_{14}H_{10}Cl_4$ | dichlorodiphenyl dichloroethane or tetrachlorodiphenylethane, also written as $(ClC_6H_4)_2CHCHCl_2$ |
| $C_{14}H_{14}N_3NaO_3S$ | methyl orange, also written as $(CH_3)_2NC_6H_4NNC_6H_4SO_3Na$ |
| $C_{14}H_{16}N_2$ | tolidine, also written as $\{[C_6H_3(CH_3)NH_2]_2$ |
| $C_{14}H_{20}N_2O_8S_2$ | paraquat, also written as $CH_3(C_5H_4N)_2CH_3 \cdot 2CH_3SO_4$ or $[C_{12}H_{14}N_2 \cdot 2CH_3SO_4$; *see also* $C_{12}H_{14}N_2 \cdot 2CH_3SO_4$ |
| $C_{15}H_{22}ClNO_2$ | metolachlor, also written as $[(CH_3)C_6H_3(C_2H_5)$—$N(COCH_2Cl)CH(CH_3)CH_2OCH_3$ |
| $(C_{16}H_{14}O_3)_n$ | polycarbonate, also written as [—$C_6H_4C(CH_3)_2C_6H_4OCOO$—$]_n$ |
| $C_{16}H_{17}NO$ | diphenamid |
| $C_{16}H_{18}O_9$ | chlorogenic acid |
| $C_{17}H_{12}O_6$ | aflatoxin B1 |
| $C_{17}H_{34}O_2$ | oleic acid, also written as $CH_3(CH_2)_7CH{=}CH(CH_2)_7COOH$ |
| $C_{18}H_{36}O_2$ | stearic acid, also written as $CH_3(CH_2)_{16}COOH$ or $C_{17}H_{35}COOH$ |
| $C_{18}H_{39}NO_2$ | ammonium stearate |
| $C_{19}H_{12}Cl_2O_5S$ | chlorophenol red |
| $C_{20}H_8Br_4O_5$ | eosin |
| $C_{20}H_{10}O_5$ | fluorescein |
| $C_{20}H_{12}$ | benzo(a)pyrene |
| $C_{20}H_{14}O_4$ | phenolphthalein |
| $C_{20}H_{28}N_2O_2$ | indoxyl-$\beta$-D-glucuronide, also written as $C_{14}H_{15}NO_2$—$C_6H_{13}N$ |
| $C_{22}H_{42}O_4$ | diethylhexyl adipate or di(2-ethylhexyl)adipate, also written as $[CH_2CH_2COOCH_2CH(C_2H_5)C_4H_9]_2$ |
| $C_{24}H_{50}O_5$ | cholesterol, also written as $C_{23}H_{49}O_3COOH$ |
| $(C_{27}H_{22}O_4S)_n$ | polysulfone, also written as (—$C_6H_4SO_2C_6H_4OC_6H_4C(CH_3)_2C_6H_4O$—$)_n$ |
| $C_{27}H_{46}O$ | cholic acid, also written as $C_{27}H_{45}OH$ |
| $C_{34}H_{32}FeN_4O_4$ | heme |
| $C_{36}H_{70}O_4Zn$ | zinc stearate, also written as $Zn(C_{18}H_{35}O_2)_2$ |
| $C_{55}H_{70}MgN_4O_6)$ | chlorophyll b |
| $C_{55}H_{72}MgN_4O_5$ | chlorophyll a |
| $C_{60}H_{87}N_{12}O_{23}P$ | Chemical formula proposed for the organic fraction of microbial cells or cell biomass when phosphorus is included |
| $C_{106}H_{263}N_{16}O_{110}P$ | A formula proposed for the composition of algal cells, algal protoplasm, also written as $(CH_2O)_{106}(NH_3)_{16}(H_3PO_4)$ |
| $C_nH_{2n}$ | cycloparaffin |

hormones), and according to some restrictive procedures.

**organic flocculation** A laboratory method using beef extract at a low pH to concentrate viruses in a water sample and then raising the pH to elute the viruses.

**organic fouling** The clogging of the pores of an anion exchanger or membrane by natural organic matter such as compounds of humic and fulvic acids or by the accumulation of colloidal material and bacteria on a membrane surface. Organic fouling results in the loss of exchange capacity and the deterioration of product water quality. *See also* inorganic fouling.

**organic-free water** A water that does not contain any organic substance.

**organic functional groups** Organic compounds fall into three major functional groups: alcohol (—OH), carbonyl (—C═O), and carboxyl (—COOH).

**organic halogen** A compound in which a halogen such as chlorine (Cl) or bromine (Br) has replaced another element, e.g., the disinfection by-products in drinking water.

**organic iron** Iron that is bound or complexed with fulvic acids, humic acids, humin, tannin, or other organic compounds. Organic iron compounds in water are sometimes colorless, but most often yellow, yellowish-brown, or pink.

**organic-laden waste** Industrial wastewater that contains any of the following organic and potentially toxic chemicals: dyes and pigments, explosives and defoliants, formaldehyde, organic acids, organic nitrogen and sulfur, PCB, paint and varnish residues, phenol, spent solvents, etc.

**organic loading** A commonly used design and operating parameter for biological treatment units: the rate at which organic matter is applied to a unit, expressed as mass of BOD or COD per unit area, mass per unit time, or mass per unit area per unit time, e.g., $kg/m^2$, kg/day, $lb/ft^2/day$, or $kg/m^2/day$. Loading on aeration basins is commonly expressed in pounds of $BOD_5$ per 1000 $ft^3$ of tank per day. Loading on trickling filters for nitrification may be expressed in $g/m^2/day$. *See also* hydraulic loading, volumetric loading.

**organic material** Any chemical compound of carbon, including diluents and thinners that are liquids at standard conditions and that are used as dissolvers, viscosity reducers, or cleaning agents, but excluding methane ($CH_4$), carbon monoxide (CO), carbon dioxide ($CO_2$), carbonic acid ($H_2CO_3$), metallic carbonic acid, metallic carbide, metallic carbonates, and ammonium carbonates (EPA-40CFR52.741). *See also* natural organic matter, organic chemical, organic compound, organic matter, organic substance, and organics.

**organic matter** (1) The organic fraction of the sediment or soil; it includes plant and animal residues at various stages of decomposition, cells and tissues of soil organisms, and substances synthesized by the microbial population (EPA-40CFR796.2750-iii). (2) Carbonaceous waste contained in plant or animal matter and originating from domestic or industrial sources, often represented by the simplified formula $CH_2O$ or more generally $C_nH_aO_b$ and $C_nH_aO_bN_c$. *See also* natural organic matter, humus, humic substance, organic chemical, organic compound, organic material, organic substance, organics, theoretical oxygen demand, carbon content.

**organic matter degradation** The biological or chemical decomposition of organic matter into simpler substances or elements through such processes as hydrolysis, oxidation, and reduction.

**organic monitoring device** A unit of equipment used to indicate the concentration level of organic compounds exiting a recovery device based on a detection principle such as infrared, photoionization, or thermal conductivity (EPA-40CFR63.111).

**organic nitrate** A nitrogen-containing compound formed by the reaction of nitric oxide with an alkyl group (R); general formula: $RNO_3$. The presence of organic nitrates in an air mass is an indication of its age.

**organic nitrogen** Nitrogen that is chemically bound to soluble or particulate organic compounds such as amines, amino acids, or proteins. Nitrogenous organic matter is essential to all living things. Other common forms of nitrogen are ammonia, gaseous nitrogen, nitrate, and nitrite. Organic nitrogen is determined analytically by subtracting total ammonia ($NH_3$ + $NH_4$) nitrogen from total Kjeldahl nitrogen (TKN). *See also* Kjeldahl nitrogen and urea.

**organic nomenclature** A set of rules established by the International Union of Pure and Applied Chemistry (Geneva, Switzerland) to identify and classify the many organic compounds, such as the water contaminants: synthetic organic chemicals and volatile organic chemicals, e.g.,

- The largest carbon chain determines the name of a series: e.g., penthyl for $CH_3$—$CH_2$—$CH_2$—$CH_2$—$CH_2$. The name of each series ends with a characteristic suffix: e.g., "-ane" for the paraffins. *See also* Hill's molecular formula arrangement.

- The methyl group branches off the $CH_3$— group: e.g., methyl from methane, ethyl from ethane, propyl from propane, and so on for benzyl, phenyl, styryl, tolyl, and xylyl.
- Numbers are used, separated by commas if necessary and ending with a hyphen, to denote the carbon atoms of parent names and of branches attached to the same carbon. In this book, the names of organic chemicals are alphabetized without regard to the numbers; e.g., 2,3,4-trimethylpentane as simply trimethylpentane and 3-propyl-1,4-pentadiene as propyl-pentadiene.

**organic nutrient** Compounds required by an organism for growth but that it cannot synthesize from other carbon sources: amino acids, nitrogen bases, and vitamins. Also called growth factor. *See also* purine, pyrimidine, minor nutrient, inorganic nutrient.

**organic peroxide** An organic compound that contains univalent oxygen atoms that form a bivalent O—O group, e.g., the strong oxidizing agents formed between ozone ($O_3$) and natural organic matter during water ozonation.

**organic phosphorus** Phosphorus that is bound to compounds that contain carbon. Wastewater may contain organic phosphate, which is usually lumped under the total phosphate ion ($PO_4^{3-}$) along with orthophosphate and polyphosphate.

**organic polymer** An anionic, cationic, or nonionic long-chain molecule with molecular weight varying from 10,000 to 1,000,000; used in water or wastewater treatment to promote the formation of larger particles by bridging. *See also* cellulose, hemicellulose, starch, chitin, and lignin.

**organic population equivalent** The population equivalent of a wastewater based on an average BOD of 0.17 pound per person per day. Also called BOD population equivalent. *See* population equivalent for detail.

**organic precursor material** The part of organic matter that can react to form chloroform ($CHCl_3$) and other disinfection by-products. *See also* precursor (1), disinfection by-product precursor, trihalomethane formation potential.

**organics** Short for organic compounds; chemical compounds made from carbon molecules, including natural materials from animal or plant sources and man-made materials such as synthetic organics. Besides carbon, they contain hydrogen and oxygen as major constituents, as well as nitrogen, phosphorus, and sulfur as minor elements, and certain metals. *See also* organic chemical, organic compound, organic material, organic matter, refractory organics.

**organics in drinking water** Organic compounds derived mainly from the breakdown of natural organic matter, domestic, commercial, and industrial activities, and reactions during water treatment. Significant classes of organic contaminants include chlorinated hydrocarbons, trihalomethanes, and volatile organic chemicals. Table O-03 lists some of the drinking water organic contaminants regulated by the USEPA.

**organics interference** Interference of organic matter with the removal of iron and manganese. *See* peptization for detail.

**organics in water** *See* organics in drinking water.

**organic solute** One of the five classes of contaminants examined by the National Research Council in 1977 to serve as a basis for revising the drinking water regulations.

**organic substance** Virtually any compound that contains carbon, but particularly an animal or vegetable chemical substance having carbon in its molecular structure. All other substances are inorganic.

**organic thermogenesis** An abiotic process presented as an explanation for the formation of primary organic carbon in hydrothermal vents: organic-free rocks near the vents synthesize sugars and amino acids from paraformaldehyde and urea at high temperatures in the presence of carbonates. *See also* chemoautotrophic theory and advective plume hypothesis.

**organic volumetric loading** *See* organic loading and volumetric loading.

**organic–water separator** *See* oil–water separator.

**organism** Any form of animal or plant life; any individual animal or plant.

**organism composition** *See* cell composition.

**organizational best practices index** A key indicator that water and wastewater utilities can use to assess their performance compared to established standards. It measures the degree to which a utility implements primary management practices: strategic planning, long-term financial planning, risk management planning, optimized asset management, performance measurement, customer involvement, and continuous improvement.

**organobromine compound** An organic compound that contains bromine (Br), e.g., a by-product of the disinfection of water that contains bromine and natural organic matter. Some organobromine compounds are synthetic.

**organochloramine** An organic compound of nitrogen and chlorine, formed during the chlorination of water containing organic nitrogen. Also called organic chloramine.

**organochloride** An organic chemical that contains

**Table O-03.** Organics in drinking water

| | | | |
|---|---|---|---|
| Acenaphthylene | Chloromethane | Endothall | Paraquat |
| Acifluorfen | Chlorophenol | Endrin | Pentachlorophenol |
| Acrylamide | Cynazine | Epichlorohydrin | Phenol |
| Acrylonitrile | 2,4-D | Ethylbenzene | Picloram |
| Adipate | Dacthal | Ethylene dibromide | Polychlorinated byphenyls |
| Alachlor | Dalapon | Ethylene glycol | Prometon |
| Aldicarb | Diazinon | Ethylene thiourea | Propachlor |
| Aldicarb sulfone | Dibromoacetonitrile | Fluorotrichloromethane | Propazine |
| Aldicarb sulfoxide | Dibromochloropropane | Fonofos | Pyrene |
| Aldrin | DibromomethaneDicamba | Formaldehyde | Simazine |
| Ametryn | Dichloroacetic acid | Glyphosate | Styrene |
| Ammonium sulfamate | Dichloroacetonitrile | Heptachlor | Tetrachloroethane |
| Anthracene | Dichlorobenzene | Heptachlor epoxide | Tetrachloroethylene |
| Atrazine | Dichlorodifluoromethane | Hexachlorobenzene | Toluene |
| Baygon | Dichloroethane | Hexachlorobutadiene | Toxaphene |
| Bentazon | Dichloroethylene | Hexachlorocyclopentadiene | Trichloroacetic acid |
| Benzene | Dichloromethane | Hexachloroethane | Trichloroacetonitrile |
| Benzo(a)pyrene | Dichlorophenol | Isopropylbenzene | Trichlorobenzene |
| Bromacil | Dichloropropane | Lindane | Trichloroethane |
| Bromochloromethane | Dichloropropene | Malathion | Trichloroethylene |
| Bromodichloromethane | Dieldrin | Methoxychlor | Trichlorophenol |
| Bromoform | Dimethylphthalate | Methyl ethyl ketone | Trichloropropane |
| Bromomethane | Dinitrobenzene | Methyl parathion | Trihalomethanes |
| Carbaryl | Dinitrotoluene | Methyl *tert* butyl ether | Trinitrotoluene |
| Carbofluran | Dinoseb | Metolachlor | Vinyl chloride |
| Carbon tetrachloride | Dioxin | Metribuzin | Xylenes |
| Chlordane | Diquat | Monochlorobenzene | |
| Chlorodibromomethane | Disulfoton | Naphthalene | |
| Chloroform | Diuron | Oxamyl (vydate) | |

chlorides. *See also* chlorinated organic compound, organochlorine compound.

**organochlorine** An organochlorine compound.

**organochlorine compound** An organic compound that contains chlorine (Cl), e.g., a by-product of the chlorination of water that contains natural organic matter. Some organochlorine compounds are synthetic. *See* organic chlorine compound.

**organochlorine pesticide** A pesticide formed by the substitution of chlorine in a hydrocarbon compound.

**organoclays** An adsorbing material consisting of bentonite clay mixed with amines and anthracite.

**organogenesis** The origin and development of an organ. Also called organogeny.

**organogeny** Organogenesis.

**organohalide** An organic compound containing a halogen, e.g., a trihalomethane. *See also* organohalogen.

**organohalogen** Any chlorinated, brominated, or iodinated organic compound; a natural product of the marine environment, synthesized by algae and other marine organisms as a result of reactions with certain enzymes (haloperoxidases); e.g., the halomethanes: chloromethane ($CH_3Cl$), chloroform ($CHCl_3$), bromoform ($CHBr_3$), and methyl iodide ($CH_3I$). More generally, organohalogens contain, in addition to carbon, elements of the halogen group.

**organohalogen compound** *See* organohalogen.

**organoleptic** Affecting or involving a sense organ as of taste, smell, or sight.

**organoleptic measurement** Odor measurement using the human olfactory system. *See* sensory method.

**organoleptic test** An unscientific test used to determine the minimum detectable levels of taste and odor in drinking water; it depends on the reactions of a panel to the sample being analyzed.

**organometallic compound** A compound of a metal and organic matter in which the organic ligand is linked to the metal by a carbon–metal bond; usually, the ligand is not a stable separate species. Examples of organometallic compounds are tetraethyllead [$Pb(C_2H_5)_4$] and iron pentacarbonyl [$Fe(CO)_5$]. *See also* chelate, complex.

**organophosphate** An organic compound that constains phosphorus; a pesticide that contains phosphorus; short-lived, but some can be neurotoxic when first applied. *See* organophosphorus pesticide.

**organophosphate insecticide** An organic compound that contains phosphate and is used as an insecticide, e.g., diazinon. It can affect the nervous system of humans. *See* organophosphorus pesticide.

**organophosphorus compound** *See* organophosphate.

**organophosphorus pesticide** A pesticide that contains phosphorus and may damage the enzyme required for proper functioning of the nervous system. Common compounds are malathion, parathion, diazinon, and phosdrin, all related to nerve gas.

**organotins** Any of a group of alkyl tin compounds used in the plastics industry.

**organ weighting factor** The ratio of the radiation risk for that organ to the total risk, when the whole body is exposed uniformly.

**Ori-Cast** A cast elastomer material manufactured by Oritex Corp. and used in equipment for rectangular clarifiers.

**Oriental liver fluke disease** A helminthic infection of the bile ducts caused by the worm *Clonorchis sinensis,* found mainly in Southeast Asia and transmitted from the feces of infected persons or animals through aquatic snails and fish as intermediate hosts. *See* clonorchiasis for detail.

**orifice** An opening (hole) in a plate, wall, or partition. An orifice flange placed in a pipe consists of a slot or a calibrated circular hole smaller than the pipe diameter. The pressure difference in the pipe above and at the orifice may be used to determine the flow in the pipe. Orifices are used in a number of flow regulators; *see,* e.g., transverse weir. *See* orifice flow, orifice (large), dropout or sump orifice.

**orifice box** A stilling box with a submerged orifice for measuring the flow of liquids.

**orifice diffuser** A nonporous diffuser made of molded plastic and mounted on an air distribution pipe for the aeration of wastewater.

**orifice feed tank** A small tank used to mix and feed chemicals from a storage container to a water or wastewater treatment process unit.

**orifice flange** *See* orifice.

**orifice plate** A flow measurement device for liquids or gases that uses a restrictive orifice consisting of a machined hole that produces a jet effect. Typically, the orifice meter consists of a thin plate with a square-edged, concentric, and circular orifice. The pressure drop of the jet effect across the orifice is proportional to the flow rate. The pressure drop can be measured with a manometer or differential pressure gauge. The orifice-plate discharge equation expresses the flow ($Q$) through an orifice plate as a function of the energy loss coefficient ($C$), the orifice area ($A$), the ratio ($D/D$) of the diameters of the orifice and pipe, the pressure differential ($\Delta P$), the specific gravity of the fluid ($\gamma$), the flow coefficient ($K$) and the acceleration of gravity ($g$):

$$Q = cA(2\, g\Delta P/\gamma)^{1/2}/[1 - c^2(d/D)^4] \quad \text{(O-14)}$$

or

$$Q = KA(2g\, \Delta P/\gamma)^{1/2} \quad \text{(O-15)}$$

with

$$K = c/[1 - c^2(d/D)^4] \quad \text{(O-16)}$$

The coefficients $c$ and $K$ may be determined experimentally for given orifice sizes and Reynolds numbers. *See also* discharge coefficient. Flow through an orifice plate is similar to flow through a venturi meter but with a lower discharge coefficient.

**orifice/soil number (OS)** The dimensionless ratio of orifice head loss to soil head loss in models that represent a small leak from a buried pipe as flow from an orifice (Walski et al., 2006):

$$\text{OS} = 1 + (KAQ/2\, gL)(1/C_d A_0)^2 \quad \text{(O-17)}$$

where $K$ = hydraulic conductivity, $A$ = cross-sectional area of soil, $Q$ = flow, $g$ = gravitational acceleration, $L$ = length of flow path, $C_d$ = discharge coefficient, and $A_0$ = area of the orifice.

**O-ring** A round rubber or neoprene gasket used to seal the bell and spigot ends of two pipe sections; also used with valves, fire hydrants, and other devices.

**Ori-Plastic** A plastic material manufactured by Oritex Corp. and used in equipment for rectangular clarifiers.

***Ornithodorus moubata*** The species of soft ticks that transmit relapsing fever in East and Southern Africa; it is usually found during the day in earth floors and walls, from which it emerges at night for feeding.

**orographic precipitation** Precipitation that occurs as a result of warm, humid air forced to rise by a mountain or another topographic feature. *See also* rain shadow.

**ORP** Acronym of oxidation–reduction (redox) potential.

**orpiment** Another name of the mineral arsenic trisulfide. *See also* red orpiment.

**ORP knee** The dramatic decline in oxidation–reduction potential observed in the operation of Nitrox™ or any other intermittent aeration unit when the nitrate nitrogen ($NO_3$-N) is depleted.

**Orsat apparatus** An apparatus used in analyzing digester gas, consisting of a water-jacketed gas buret with leveling bulb, absorption pipets for oxygen ($O_2$) and carbon dioxide ($CO_2$), and oxidation assemblies for methane ($CH_4$) and cupric oxide–hydrogen ($CuO–H_2$).

**Orsat method** A method used to analyze mixtures of gases such as carbon dioxide ($CO_2$), carbon monoxide (CO), methane ($CH_4$), and molecular oxygen ($O_2$) in a portable apparatus.

**ORT** Acronym of oral rehydration therapy.

**orthoboric acid [$B(OH)_3$ or $H_3BO_3$]** Same as boric acid.

**orthoclase** A type of feldspar or silicate mineral that contains potassium and aluminum; formula: $KAlSi_3O_8$.

**orthocresol ($CH_3C_6H_4OH$)** A synthetic organic chemical, derived from coal or wood tar and used as a disinfectant or an insecticide. Also called 2-methylphenol. *See also* cresol, tricresol.

***ortho*-dichlorobenzene** An organic chemical with low vapor pressure used in the production of pesticides and dyes. *See* dichlorobenzene ($C_6H_4Cl_2$).

**orthokinetic flocculation** The aggregation of particles larger than 1 or 2 micrometers due to induced velocity gradients or differential settling; it is not effective for viruses. Also called macroflocculation. *See also* flocculation concepts.

**orthokinetic motion** *See* conjunction kinetics.

**orthophosphate (O-$PO_4$)** (1) A salt or ester of phosphoric acid that contains phosphorus as the trivalent group $PO_4^{3-}$; a necessary nutrient for plant and animal growth; available for biological metabolism without further breakdown. One of the three common compounds of phosphorus in wastewater, consisting of four ionic forms, the other two being organic phosphate and polyphosphate. The four ionic forms are $PO_4^{3-}$, $HPO_4^{3-}$ (orthophosphate mono-hydrogen), $H_2PO_4^-$ (orthophosphate di-hydrogen), and $H_3PO_4$ (orthophosphoric acid). (2) An ingredient of corrosion-inhibition products. Orthophosphates reduce the corrosion of steel and the release of iron by forming insoluble phosphates in the corrosion scales and by increasing impermeability and adherence. (3) A product of the hydrolysis of polymeric phosphates. *See* reversion.

**orthophosphate di-hydrogen ($H_2PO_4^-$)** *See* orthophosphate (1).

**orthophosphate mono-hydrogen ($HPO_4^{3-}$)** *See* orthophosphate (1).

**orthophosphoric acid ($H_3PO_4$)** A colorless, crystalline solid, the tribasic acid of pentavalent phosphorus, used in fertilizers, in soft drinks, and as a source of phosphorus salts.

**orthophosphorous acid ($H_3PO_3$)** A white-yellowish, crystalline, soluble solid, used in phosphorus salts.

**orthotolidine {[$C_6H_3(CH_3)NH_2]_2$}** A light-yellow, slightly soluble liquid; the ortho isomer of tolidine used in dyes, organic compounds, textiles, and as a reagent. Also spelled *ortho*-tolidine.

**orthotolidine (or *ortho*-tolidine) arsenite (OTA) test** A test that uses the orthotolidine reagent, sodium arsenite ($NaAsO_2$) and colorimetric methods to detect and differentiate between free available chlorine and combined available chlorine. The test is obsolete and only approximate because of the presence of color and other interfering substances.

**orthotolidine (or *ortho*-tolidine) reagent** A solution of orthotolidine crystals used as a colorimetric indicator of chlorine residual or available chlorine. If chlorine is present, a yellow-colored compound is produced. The reagent is no longer approved for chemical analysis in the United States as it is suspected of causing cancer.

**orthotolidine (or *ortho*-tolidine) (OT) test** A test that uses the orthotolidine reagent to measure the approximate free chlorine residual concentration in water. When the liquid reagent is added to water, it produces a yellow coloring that is compared to standard colors in Nessler tubes to determine the concentration of the residual. Color and other interfering substances affect the accuracy of the test. *See* the more modern DPD test.

***Oscillatoria*** A group of blue-green algae (cyanobacteria) commonly found in waste stabilization ponds; they can produce geosmin, a substance responsible for taste and odor in water sources.

***Oscillatoria agardhii*** A species of cyanobacteria (blue-green algae), commonly found in algal blooms and capable of producing hepatotoxins and neurotoxins.

**OSEC™** An apparatus manufactured by Wallace & Tiernan, Inc. for water treatment by electrolytic chlorination.

**OSHA** Acronym of Occupational Safety and Health Act (or Administration).

***O. sinensis*** *See Opisthorchis sinensis.*

**osmionic separation** A saline-water conversion process that removes dissolved salts by ion-selective membranes. It is similar to electrodialysis, but driven by the dilution of a concentrated salt instead of electromotive force. *See also* saline-water conversion classification.

**Osmo®** A water treatment apparatus manufactured by Osmonics, Inc.

**osmometry** Measurement of the osmotic pressure.

**Osmo® MUNI Series™** A reverse osmosis treatment system engineered by GE Water Technologies using the Osmo MUNI Series membrane elements for municipal water plants.

**osmophilic** Characteristic of organisms that prefer high concentrations of solutes.

**osmosis** The passage of a liquid from a weak solution to a more concentrated solution across a semipermeable (permselective) membrane. The membrane allows the passage of the solvent (water) but not the solute (dissolved solids). This process tends to equalize the conditions on either side of the membrane, so the diffusion proceeds from the side of lower concentration to the side of higher concentration. *See also* reverse osmosis.

**osmosis cell** The conceptual unit in which the phenomenon of osmosis occurs. It includes a semipermeable membrane through which the osmotic pressure causes water to flow from the area of low total dissolved solids (TDS) concentration to the area of high TDS concentration, while diffusion causes ions to flow in the reverse direction. *See also* reverse osmosis cell.

**osmosis (reverse)** *See* reverse osmosis.

**Osmostill** A distillation apparatus manufactured by Osmonics, Inc.

**osmotic pressure** The pressure resulting from the phenomenon of osmosis; also equal to the excess pressure that must be applied to produce or maintain equilibrium and prevent reverse osmosis. A greater pressure causes water to flow in a reverse direction, from the higher to the lower solids concentration. For natural waters, osmotic pressure is about 7 kilopascals or 1.0 pound per square inch (psi) per 100 mg/L of total dissolved solids. Seawater has an osmotic pressure of about 400 psi and requires an operating pressure or feedwater pressure of 800 psi or larger. Actual pressure applied is 5–50 times the osmotic pressure. The osmotic pressure of a solute is inversely proportional to its molecular weight.

**osmotic water transport** The transfer of water under osmotic pressure through a semipermeable membrane, from a zone of lower solute concentration to a zone of higher solute concentration.

**OSWMD** Acronym of onsite wastewater management district.

**Osteosclerosis** A disease characterized by an abnormal hardening of the bones, which may be associated with an excess of fluoride in the diet.

**OTA® Aerator** A rotor aerator manufactured by Scoti-Zahner, Inc.

**OTE** Acronym of (a) orbital tube evaporation, (b) oxygen transfer efficiency.

**OTR** Acronym of oxygen transfer rate.

**Otto Plate ozonator** An early ozone generator that consists of a series of water-cooled aluminum blocks, dielectrics, and stainless-steel electrodes between which air is passed. *See also* Lowther Plate ozonator.

**ounce** A unit of mass equal to 1/16 pound or approximately 28.35 grams. Abbreviation: oz.

**OUR** Acronym of oxygen uptake rate.

**outbreak** An epidemic of a common source. *See also* waterborne disease outbreak.

**outcrop** The exposed portion of a stratum or vein at the surface of the earth; often an aquifer recharge zone.

**outer-sphere complex** Two particles of opposite charge or two ions of opposite charge joined by electrostatic attraction. *See* ion pair for detail.

**outer-sphere surface charge** The surface charge density of an ion pair formed by two colloidal particles. It depends on the concentrations of the metal ions or ligands on the surface of the particles.

**outfall** (1) The place where effluent is discharged into receiving waters; the orifice of a sewer. (2) A point where a municipal separate storm sewer discharges to waters of the United States, excluding open conveyances connecting two municipal separate storm sewers, or pipes, tunnels, or other conveyances that connect segments of the same stream or other waters of the United States and are used to convey waters of the United States (EPA-40CFR122.26-9). (3) An outfall sewer receives wastewater from a sewer network or from a treatment plant and carries it to a point of final disposal. A wastewater outfall is an outlet or a structure for final wastewater disposal. Outfall structures are structures for wastewater or stormwater outfall, such as those that SWMM can simulate: transverse weirs with or without tide gates, sideflow weirs without tide gates, and free outfalls without tide gates. *See also* ocean outfall, submarine outfall.

**outfall sewer** *See* outfall (3).

**outfall structures** *See* outfall (3).

**outhouse** A fixed or portable outbuilding with one or more seats and a pit or vault serving as a toilet to collect human feces. Outhouses are used in rural and other areas that have no or limited access to running water. Also called privy, pit toilet, pit privy. *See also* latrine.

**outlet** (1) The point at which water discharges from a stream, river, lake, tidal basin, pipe, channel or drainage area. (2) Opening near the bottom of a dam for draining the reservoir. (3) The discharge opening from a water distributing system, a boiler, heating system, or any water-operated device or equipment. An outlet channel is a channel or waterway that carries water away from a lake, reservoir, or other body of surface water, or from man-made structures such as terraces, subsurface drains, diversions, or impoundments.

**outlet channel** *See* outlet.

**outlet weir** The main structure or device constituting the outlet of a sedimentation tank, installed at the discharge side of a rectangular or square tank, or around the periphery of a circular tank. *See also* clarifier weir and effluent weir.

**outlet zone** One of the four zones of the Camp sedimentation theory. It is the zone of the sedimentation basin that collects the effluent before it discharges over the weir. *See also* settling, sludge, and inlet zones.

**outside diameter** The outer diameter of a pipe, including the wall thickness. *See also* inside diameter and nominal diameter.

**outside-in filtration** A water treatment process in which feedwater is applied outside hollow-fiber membranes and permeate is collected from the inside.

**outside threaded connection** The spigot end of a pipe section, which connects with the bell end of another section. Also called male end.

**outwash** A deposit of sand and gravel formed by meltwater from a glacier.

**outwash plain** A broad, sloping landform of outwash deposits.

**ova** Plural of ovum.

**ovale malaria** A rare form of malaria found in West Africa and caused by *Plasmodium ovale.*

**overall pump efficiency** The combined efficiency of a pump and motor together. Also called wire-to-water efficiency.

**over-and-under baffles** Baffles arranged in a water or wastewater treatment basin to reduce short-circuiting by forcing the liquid to pass sequentially over and under them.

**over-and-under flow pattern** The flow regime in a basin with over-and-under baffles. This term also applies to basins, chambers, or compartments in series and separated by over-and-under baffles.

**overburden** The unconsolidated material that overlies bedrock in a given location; also the soil and rock that overlies a mineral deposit and is removed before strip mining.

**overburned lime** The lumps of inert, semivitrified material obtained during the calcining process by heating limestone beyond the point of quicklime, which is soft, porous, and highly reactive. Overburned lime, also called dead-burned lime, is used in refractory materials. *See* quicklime for detail.

**Overdraft** The pumping of water from a groundwater basin or aquifer in excess of the supply flowing into the basin; it results in a depletion or "mining" of the groundwater in the basin.

**overfall weir** Same as overflow weir.

**overfertilization** Soil or surface water enrichment with excessive amounts of nutrients.

**overflow** (1) The intentional or unintentional diversion or discharge of untreated wastewater or stormwater to the environment, caused by inadequate conveyance or treatment capacity, electrical or mechanical failure, line blockages or restrictions, or human error. (2) The actual volume of liquid that overflows, i.e., the volume of wastewater or stormwater that exceeds the capacity of the conveyance or treatment facilities. (3) The water that exceeds ordinary limits such as stream banks, spillway crests, or the ordinary level of a container. *See also* reservoir storage and surcharge storage. (4) An overflow structure or device that allows the discharge of the excess liquid, such as (a) overflow channel or spillway channel (an artificial waterway from a reservoir, aqueduct, or canal); (b) overflow manhole, used mostly in combined sewer systems, at pump stations, or at treatment plants; (c) overflow spillway (also called ogee spillway or gravity spillway), i.e., a widely used structure to discharge the overflow of a dam; (d) overflow weir (sometimes called an overfall weir or a diverting weir), i.e., a weir over which excess water, stormwater, or wastewater is allowed to flow. *See also* combined sewer overflow, diversion dam, effluent launder, effluent weir, and skimming weir. In wastewater treatment, the overflow rate, also called the surface loading rate or the rise rate, is a design and operation parameter for settling tanks and clarifiers; expressed in gallons per day per square foot, it is equal to the ratio of the average flow rate (in gpd) to the surface area of tank (in $ft^2$). Operators use the overflow rate to determine the proper hydraulic loading of the tanks. *See also* weir overflow rate (or weir loading).

**overflow manhole** *See* overflow (4).

**overflow rate (OFR)** The upward velocity of water in a clarifier as the solids settle, equal to the ratio of flow ($Q$, gpd or cfs) to the clarifier surface area ($A$, ft$^2$); also called surface loading rate or surface overflow rate, an empirical design parameter for sedimentation basins:

$$V_0 = Q/A \qquad \text{(O-18)}$$

where $V_0$ is the overflow rate, expressed in gpd/sq. ft or in fps. It represents the minimum settling velocity necessary for sedimentation; all particles with a settling velocity higher than the operating overflow rate are removed, but lower settling particles are carried over in the effluent. *See also* solids loading rate and weir loading rate.

**overflow siphon** A siphon installed in a basin to discharge liquid at a higher elevation.

**overflow solids flux** The amount of solids per unit area in an activated sludge clarifier ($F$, lb/ft$^2$/day or kg/m$^2$/day):

$$F = QX/A \qquad \text{(O-19)}$$

where $Q$ = clarifier flow rate, $X$ = mixed liquor suspended solids concentration, and $A$ = clarifier cross-section area.

**overflow standpipe** A standpipe installed in a basin or tank with the top open below the hydraulic grade line and the bottom discharging through the side of the container.

**overflow structure** A weir or other structure to allow water to spill freely to a lower level or into another carrier.

**overflow tower** A device placed on one or more summits of a closed conduit to control the pressure by discharging water. *See also* pressure relief valve.

**overflow weir** *See* overflow.

**overhung latrine** A latrine that is constructed so that the excreta fall directly into a body of water.

**overland flow** (1) Also called excess rainfall, storm flow, or surface runoff, it includes overland runoff and the precipitation that falls directly into streams, brooks, rivulets, ravines, and rills. *See also* rainfall–runoff relationship. (2) A land application technique that treats wastewater by allowing it to flow over a graded, vegetated, sloped surface. As water flows over the surface, contaminants are absorbed and, except for evapotranspiration and minor percolation, the water is collected at the bottom of the slope for reuse. Minimum preapplication treatment requirement is fine screening or, preferably, the equivalent of primary clarification. *See also* geopurification, high-rate infiltration, high-rate sprinkling, low-rate irrigation, rapid infiltration extraction, rapid infiltration, soil aquifer treatment, and surface spreading.

**overland flow design** Design of overland flow systems is based on the following equation:

$$C_x = C_s + C_0 A \exp(-KX/Q^n) \qquad \text{(O-20)}$$

where $C_x$ = effluent BOD concentration at point $x$, mg/L; $C_s$ = residual BOD concentration at the end of the slope of the field, mg/L; $C_0$ = initial (applied) BOD concentration, mg/L; $A$ = constant; $K$ = an empirical constant; $X$ = length of the slope, m or ft; $Q$ = application rate, m$^3$/m/hr or gal/ft/hr; and $n$ = an empirical constant.

**overland flow land application** *See* overland flow.

**overland flow treatment** Wastewater treatment using the overland flow technique.

**overload** *See* environmental overload.

**overpack** An aboveground, concrete container for the placement of smaller containers of special hazardous wastes. The overpack is backfilled with sand and gravel and sealed with a concrete top.

**overrun** To operate a water or wastewater treatment unit (e.g., a filter, ion exchanger, or carbon adsorber) beyond its design exhaustion point.

**oversaturated solution** A solution that contains a greater concentration of a solute than is possible at equilibrium under fixed conditions of temperature and pressure (EPA-40CFR796.1840-v). Also called a supersaturated solution.

**overturn** The almost spontaneous mixing of all layers of water in a reservoir or lake when the water temperature becomes similar from top to bottom, causing density to be uniform and thermal stratification to disappear. This may occur in the fall or winter when the surface water cools to the same temperature as the bottom water, in the spring when the surface water warms after ice melts, or after storms. Overturn results in uniformity of chemical and physical properties of water at all depths, and an increase in biological activity. Also called turnover. *See also* thermocline.

**overturning** Same as turnover.

***O. viverrini*** Abbreviation of *Opistorchis viverrini,* an excreted helminth that causes cat liver fluke or opistorchiasis.

**ovum (plural: ova)** An egg; the female reproductive cell or gamete of plants and animals.

**Owamat®** An oil–water separator manufactured by BEKO Condensate Systems Corp.

**OWASA Nutrification™ process** A sidestream fermentation process developed at the Orange Water and Sewer Authority in Carrboro, North Carolina for the enhanced biological removal of phosphorus from wastewater. The complete flowchart includes preliminary treatment, primary treatment, trickling filter, two aeration basins in series followed by an anoxic tank, another aeration basin, and a final clarifier. Two other units are a fermentation tank for the primary sludge and an anaerobic contactor that processes the volatile fatty acids from the supernatant of the fermentation tank as well as the return activated sludge. The content of the anaerobic contactor is recycled to the first aeration basin. *See* Figure O-07.

**Owens–Edwards–Gibbs correlation** A formula proposed by M. Owens, R. W. Edwards, and J. W. Gibbs in 1964 to estimate the stream reaeration constant (Droste, 1997):

$$K_2 = K_c U^{0.67}/H^{1.85} \quad \text{(O-21)}$$

where $K_2$ = reaeration constant,/day; $K_c$ = a coefficient = 5.34 in SI units and 21.7 in U.S. units; $U$ = stream velocity in m/s or fps; and $H$ = stream depth in m or ft. *See also* reaeration constant.

**oxalacetic acid ($HOOCCOCH_2COOH$)** An organic compound that includes both a carbonyl group (C═O) and a carboxylic acid group in its chemical structure; a disinfection by-product.

**oxalate ($OOCCOO^{2-}$)** A salt or ester of oxalic acid; occurring in spinach, rhubarb, and other vegetables or nuts; a substance used as a source of carbon and energy to grow *Spirillum* NOX and determine a surrogate measure of assimilable organic carbon.

**oxalate-carbon equivalent** A surrogate for assimilable carbon determined by growing *Spirillum* NOX on oxalate ($OOCCOO^{2-}$) instead of on organic carbon.

**oxalic acid ($H_2C_2O_4$)** A white, crystalline, water-soluble, poisonous substance used in a chemical actinometer to react with uranyl sulfate for the measurement of the intensity of ultraviolet light and other forms of radiation.

**oxamyl [$(CH_3)_2N—CO—C(SCH_3):N—O—CO—NH—CH_3$]** The common name of methyl N′—N′—dimethyl—N—[(methylcarbamoyl)oxy]—1 thiooxamimidate], a synthetic organic chemical and nonpersistent carbamate used as a pesticide and regulated by the USEPA as a drinking water contaminant: MCL = MCLG = 0.2 mg/L. It is neurotoxic, but degradable by bacteria, with a half-life of 1–5 weeks in soil and 1–2 days in rivers. Related trade names include Dioxamyl®, Thioxamyl®, and Vydate®.

**Oxazur® system** The proprietary air header used for aeration in the Biofor® process.

**oxbow** (1) A bow-shaped bend in a river, or the land it embraces. (2) An oxbow lake.

**oxbow lake** A bow-shaped lake formed in the former channel of a river, e.g., at the site of a former river meander or by the natural closing of the oxbow bend of an aging river.

**oxic** Pertaining to a liquid or an area that contains dissolved oxygen, such as the oxic zone of a reservoir, lake, or treatment unit.

**oxic process** Aerobic process; a biological treatment process in the presence of oxygen. *See also* anaerobic process, anoxic process.

**Oxidair™** An apparatus manufactured by EPG Companies, Inc. using thermal oxidation for soil remediation and off-gas treatment.

**oxidant** (1) An oxidizing agent; the opposite is a reducing agent or reductant; a chemical substance, such as chlorine, chlorine dioxide, bromine, hydrogen peroxide, iodine, ozone, or potassium permanganate that can oxidize other substances. Oxidants are often used as water or wastewater disinfectants. They destroy or alter organic compounds that may interfere with coagulation and flocculation in water treatment. (2) A substance that contains oxygen and reacts in air to produce a new substance; the primary ingredient of photochemical smog.

**oxidation** (1) A chemical reaction in which an element or compound loses electrons to an oxidant; e.g., the addition of oxygen, removal of hydrogen, or the removal of electrons from an element or compound. Oxidation is the opposite of reduction,

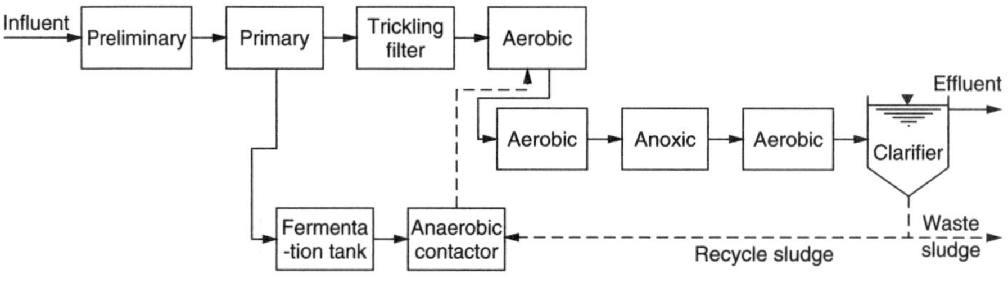


**Figure O-07.** OWASA Nutrification™ process.

and both are part of the oxidation–reduction (redox) reaction. The reaction is called a reduction if the target substance is reduced, and an oxidation if the target substance is oxidized. Water and wastewater treatment uses oxidation in a variety of processes, e.g., for iron and manganese removal. (2) The chemical or biological process whereby organic waste or chemicals such as cyanides, phenols, and organic sulfur compounds are converted (or oxidized) to more stable substances such as carbon dioxide ($CO_2$) and water ($H_2O$). *See also* autoxidation, synthesis, wet oxidation.

**oxidation basin** An earlier name given to lagoons and ponds used for the storage of industrial wastes when it was observed that these basins undergo decomposition; same as oxidation pond.

**oxidation by-product** A secondary compound formed during an oxidation reaction, e.g., the disinfection by-products.

**oxidation ditch** A special configuration of the extended aeration variation of the activated sludge process using an oval channel or ditch, equipped with mechanical aeration and mixing devices. First developed in the Netherlands for small communities, the process is characterized by a long detention time, corresponding to a dilution factor of about 25 times the influent flow, and a low dissolved oxygen concentration toward the exit of the aeration zone, with occasional denitrification. A separate tank or an intrachannel clarifier provides secondary sedimentation. Oxidation ditches may be used to achieve biological nitrogen removal. The process is also called ditch oxidation. *See* Figure O-08, brush aeration, phased isolation ditch, and also the proprietary designs Orbal™, CCAS™, and Biolac™.

**oxidation ditch BNR process** An oxidation ditch designed for BOD reduction and biological nitrogen removal (BNR) in a single tank with anoxic and aerobic zones. *See* low DO oxidation ditch.

**oxidation half reaction** The oxidation half of a redox reaction.

**oxidation lagoon** *See* oxidation pond.

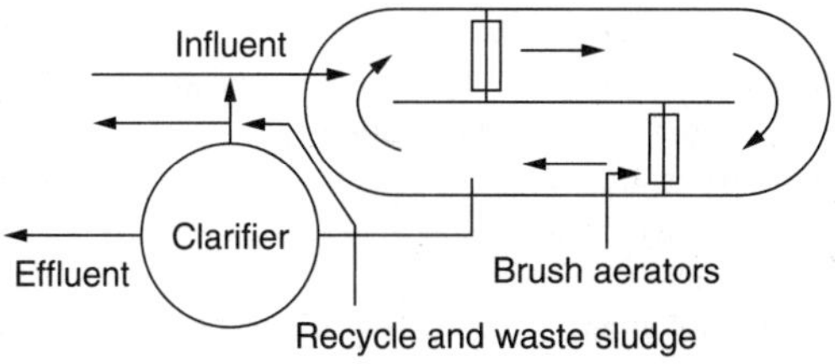


**Figure O-08.** Oxidation ditch.

**oxidation number (ON)** For an atom in a compound, the oxidation number is the number of electrons that are associated with it, i.e., the number of electrons that must be added or subtracted to convert the atom to the elemental form. The algebraic sum of oxidation numbers of a neutral compound is 0. In elemental form, an atom has an oxidation number equal to 0, as in aqueous oxygen [$O_{2(aq)}$] or solid sulfur [$S_{(s)}$]. Hydrogen (H) has an oxidation number of +1, except in metallic hydrides where ON = –1. For oxygen, ON = –2, except in hydrogen peroxide ($H_2O_2$) where it is ON = –1. Carbon (C) has oxidation numbers varying from –4 (in methane, $CH_4$) to +4 (in carbon dioxide, $CO_2$). *See also* oxidation state, valence, ionic strength.

**oxidation pond** A shallow, earthen lagoon or basin used to provide biological treatment of wastewater, including aeration by natural or mechanical means; it is usually part of a series of ponds. Also called a sewage lagoon. *See* stabilization pond for more detail. *See also* lagoons and ponds.

**oxidation pond upgrading** The use of additional processes (e.g., coagulation and sedimentation, dissolved-air flotation, filtration, intermittent sand filtration) to remove algal cells (and suspended solids) from the effluent of stabilization ponds.

**oxidation process** Any wastewater treatment process that oxidizes organic matter biologically or chemically.

**oxidation rate** The rate of destruction or stabilization of organic matter in wastewater treatment.

**oxidation reaction** See oxidation (1).

**oxidation–reduction (redox)** *See* oxidation–reduction reaction.

**oxidation–reduction equation** *See* oxidation–reduction reaction.

**oxidation–reduction chemistry** The chapter of chemistry that deals with the transfer of electrons in redox equations.

**oxidation–reduction couple** The two half reactions that make up an oxidation–reduction reaction.

**oxidation–reduction potential (ORP, pE)** A dimensionless number (pE) that represents the electrical potential required to transfer electrons from one compound or element (the oxidant) to another compound or element (the reductant); used as an indication of the relative strength potential of an oxidation–reduction reaction and as a quantitative measure of the state of oxidation in water and wastewater treatment systems. For a redox half-reaction, ORP can be derived from the Nernst equation (Droste, 1997):

$$ORP = E^0 - (0.059/n) \log ([\text{reduced species}]/[\text{oxidized species}]) \quad (O\text{-}22)$$

where $E^0$ = standard potential, $n$ = number of electrons transferred, and [ ] = concentration. Higher positive pE values correspond to oxidizing solutions (with a predominance of ions like $SO_4^{2-}$, $NO_3^-$, and $Fe^{3+}$), whereas more negative values represent reducing solutions (with the reduced forms $S^{2-}$, $NH_4^+$, and $Fe^{2+}$). Also called redox potential.

**oxidation–reduction (redox) reaction** A chemical reaction consisting of an oxidation reaction in which a substance loses or donates electrons (the oxidized or reducing agent), and a reduction reaction in which a substance gains or accepts electrons (the reduced or oxidizing agent). Redox reactions are always coupled because free electrons cannot exist in solution and electrons must be conserved. For example, the redox reactions for the oxidation of hydrogen sulfide ($H_2S$) to sulfate ($SO_4^{2-}$) by chlorine ($Cl_2$) are:

$$\tfrac{1}{2}\,Cl_2 + e^- = Cl^- \quad (O\text{-}23)$$

$$\tfrac{1}{8}\,SO_4^{2-} + \tfrac{5}{4}\,H^+ + e^- = \tfrac{1}{8}\,H_2S + \tfrac{1}{2}\,H_2O \quad (O\text{-}24)$$

$$\tfrac{1}{2}\,Cl_2 + \tfrac{1}{8}\,H_2S + \tfrac{1}{2}\,H_2O = \tfrac{1}{8}\,SO_4^{2-} + \tfrac{5}{4}\,H^+ + Cl^- \quad (O\text{-}25)$$

The significance of oxidation–reduction reactions in biological wastewater treatment is that all heterotrophic organisms use them to obtain energy. Also called redox equation.

**oxidation state** The net charge on an atom, ion, or molecule; or the hypothetical charge that an atom would have if the ion or molecule were to dissociate; e.g., the oxidation state of the ferric ion ($Fe^{3+}$) is +3 and that of the hypochlorite ($OCl^-$) ion is –1. In all compounds other than peroxides, oxygen has an oxidation state of –2. The concepts of oxidation number and oxidation state are used in balancing chemical reactions. Oxidation states are sometimes represented by roman numerals. *See also* oxidation number.

**oxidation tower** A trickling filter that uses synthetic media. *See also* high-rate trickling filter, intermediate-rate trickling filter, low-rate trickling filter, roughing filter, super-rate filter, biofilter, biotower, dosing tank, filter fly.

**oxidation treatment** The use of an oxidation process to stabilize organic matter.

**oxidative dealkylation** The separation of an alkyl group ($C_nH_{2n+1}$) as an aldehyde from a compound through an oxidation reaction, with the production of a simpler amine or an alcohol.

**oxidative deamination** The separation of a primary amine (R—$NH_2$) from an alkyl chain through an oxidation reaction, with the production of an aldehyde (R—CHO) and ammonia ($NH_3$).

**oxidative deferrization/demanganization** The use of an oxidizing agent (chlorine, dissolved oxygen, ozone, permanganate, for example) to convert iron and manganese ions to insoluble ferric and manganic oxides that are removed by sedimentation and filtration after their precipitation.

**oxidative dehalogenation** (1) The separation of a halogen from an organic compound through an oxidation reaction with production of an aldehyde (R—CHO) and a haloacid ($RX_n$—COOH). (2) An enzyme-mediated reaction in which the halogen atoms are separated from a halogenated aliphatic compound, e.g., the dehalogenation of chloroform (CH—$Cl_3$):

$$CH\text{—}Cl_3 + O_2 + H_2 \rightarrow C(OH)Cl_3 + H_2O \rightarrow CO_2 + 3\,Cl^- + 3\,H^+ \quad (O\text{-}26)$$

*See also* reductive dehalogenation, substitutive dehalogenation.

**oxidative pyrolysis** The partial combustion of a substance in the presence of oxygen, resulting in the formation of carbon monoxide (CO) and carbon dioxide ($CO_2$). It is used in the petroleum industry, electronics manufacturing, and the conversion of biomass and waste materials to energy and fuels. Also called oxygen-starved combustion. *See also* hydropyrolysis.

**Oxidator** A wastewater treatment apparatus manufactured by Eimco Process Equipment Co., combining aeration, flocculation and sedimentation in one unit.

**oxide** A compound of oxygen with one or more electropositive elements, e.g., ferric oxide ($Fe_2O_3$) and calcium oxide or lime (CaO).

**oxidizable salt** A dissolved salt in groundwater that may be oxidized to another form and precipitate in surface water when exposed to dissolved oxygen or air.

**oxidize** To combine with oxygen or, more generally, to remove electrons from an atom, ion, or molecule.

**oxidized sludge** Sludge that has been stabilized by biological or chemical oxidation.

**oxidized wastewater** Wastewater in which the organic matter has been stabilized biologically or chemically.

**oxidizing agent** Any substance, such as oxygen ($O_2$) or chlorine ($Cl_2$), that will readily add electrons to a reaction. *See* oxidant for more detail.

**oxidizing chemical** Same as oxidizing agent or oxidant.

**oxidizing filter** A granular filter whose bed contains a catalyst medium. It is used to render dissolved molecules [e.g., ferrous iron ($Fe^{2+}$), manganous manganese ($Mn^{2+}$), anionic sulfur ($S^{2-}$)] insoluble and filterable by altering their valence. Catalyst media include activated carbon, calcite, manganese greensand, magnesium oxides, dissimilar metal alloys, pumicites, and zeolites. Also called a catalyst filter.

**Oxifree®** A disinfection apparatus manufactured by Capital Controls Co. using ultraviolet light.

**Oxigest®** A cylindrical package plant manufactured by Smith & Loveless, Inc. for wastewater treatment using the extended aeration process.

**Oxigritter** A primary wastewater treatment plant manufactured by Eimco Process Equipment Co.

**oxime** A compound containing the group C=NOH; the product of the reaction of hydroxylamine ($H_2NOH$) with an aldehyde (R—CHO) or a ketone (R—O—R′).

**oxime derivative** Same as oxime.

**Oxitech®** A process developed by U.S. Filter Corp. using resins for the removal of total organic carbon.

**Oxitrace™** An instrument manufactured by Capital Controls Co. for the analysis of oxidants.

**Oxitron™** A wastewater treatment plant manufactured by Kruger, Inc. using a fixed-film biological process.

**oxoacid** An organic compound whose chemical structure includes a carbonyl group (C=O) and a carboxylic acid group (COOH). Some oxoacids are disinfection by-products. *See also* glyoxylic acid, ketomalonic acid, oxalacetic acid, pyruvic acid.

**oxybenzene** Same as phenol.

**Oxycat** A catalyst produced by Met-Pro Corp. for air pollution control.

**OxyCharger** A static aerator manufactured by Parkson Corp.

**Oxychlor** An apparatus manufactured by International Dioxide, Inc. for the production of chlorine dioxide.

**oxychloride** A compound of oxygen, chlorine, and another element, e.g., calcium oxychloride.

**oxychlorine residual** The sum of the residual concentrations of compounds of oxygen and chlorine in water, mainly chlorine dioxide ($ClO_2$), chlorite ion ($ClO_2^-$), and chlorate ion ($ClO_3^-$).

**oxychlorine species** Any of the compounds of oxygen and chlorine, e.g., chlorine dioxide ($ClO_2$), chlorite ion ($ClO_2^-$), and chlorate ion ($ClO_3^-$).

**Oxyditch®** A wastewater treatment plant manufactured by Chemineer-Kenics using the oxidation ditch process.

**Oxy Flo** A mechanical aerator manufactured by Aqua-Aerobic Systems, Inc.

**Oxy-Gard** An apparatus manufactured by Lamson Corp. to monitor the dissolved oxygen level in aeration units.

**oxygen ($O_2$)** A nonmetallic gaseous element that makes up approximately 20% of the atmosphere and is necessary for biological oxidation; the most abundant element in the Earth's crust, occurring in rocks, water, and air. It combines with silicon (Si) to form a large variety of minerals, e.g., quartz ($SiO_2$) and mica [$K_2Al_2O_5[Si_2O_5]_3Al_4(OH)_4$]. Atomic weight = 15.9994. Atomic number = 8. Molecular oxygen has two atoms ($O_2$) and ozone, three atoms ($O_3$). Photosynthesis produces atmospheric oxygen. *See* dissolved oxygen.

**oxygen absorbed** A test that measures the amount of oxygen in a water sample by observing the quantity of oxygen absorbed by permanganate ($MnO_4$) over a period of 4 hours at 27°C.

**oxygenate** A chemical used to add oxygen to a substance; in particular an additive to gasoline such as MTBE to obtain oxygenated fuel where air quality standards are exceeded.

**oxygenate additive** Same as oxygenate.

**oxygenated solvent** An organic solvent containing oxygen as part of its molecular structure. Alcohols and ketones are oxygenated compounds often used as paint solvents.

**oxygenation** Oxidation by dissolved oxygen in water; aeration.

**oxygenation capacity** A measure of the ability of an aeration device to supply oxygen to water or wastewater.

**oxygen balance** The resulting dissolved oxygen concentration of a stream from deoxygenation and reaeration.

**oxygen consumed** A measure of the oxygen used in the stabilization of organic and inorganic matter in wastewater. *See also* oxygen demand, biochemical oxygen demand, chemical oxygen demand.

**oxygen cycle** The interchange of oxygen between the elemental or gaseous form of the atmosphere ($O_2$) and the chemically bound forms in carbon dioxide ($CO_2$), water ($H_2O$), and organic matter ($CH_2O$). Oxygen is released by photosynthesis and chemically bound by combustion and metabolic processes.

**oxygen deficiency** (1) A condition in which an atmosphere has less than the normal or saturation concentration of oxygen, e.g., less than approximately 20% in the Earth's atmosphere. (2) The amount of dissolved oxygen needed to satisfy the requirement of water or wastewater. *See also* oxygen deficit and oxygen demand.

**oxygen deficit** The difference, at a given temperature, between the saturation concentration of dissolved oxygen (DO) and the actual DO concentration in water. *See also* oxygen transfer rate, dissolved-oxygen sag analysis.

**oxygen delignification** Treatment of pulp with oxygen under alkaline conditions before bleaching and after the pulping process. Its purpose is to remove lignin from the pulp, save chemicals, and reduce wastewater constituents such as AOX, BOD, chlorinated compounds, and color. *See also* ozone bleaching.

**oxygen demand** The amount of oxygen that is needed to oxidize the organic matter and reduced inorganics in water or wastewater. It may be used as an indirect measure of the organic load of a wastewater sample. *See* biochemical oxygen demand, biodegradable organic matter, carbonaceous oxygen demand, chemical oxygen demand, nitrogenous oxygen demand, total oxygen demand.

**oxygen demanding waste** Wastewater that contains organic matter, the biological degradation of which requires dissolved oxygen. *See* BOD.

**oxygen depletion** The complete or near complete loss of dissolved oxygen from water or wastewater due to biochemical or chemical demand.

**oxygenic** Capable of producing oxygen.

**oxygen probe** An electrode covered by a gas-permeable membrane combined with a meter to measure the amount of dissolved oxygen in a sample by reference to an oxygen-saturated solution.

**oxygen requirement** (1) The amount of oxygen ($R_o$, kg/day) required for the aerobic degradation of organic matter is the difference between the biodegradable COD removed and the COD of the waste sludge. For the design of a suspended growth process, it is calculated as:

$$R_o = Q(S_0 - S) - 1.42\,P_x \qquad \text{(O-27)}$$

where $Q$ = influent flow rate, m$^3$/day; $S_0$ = influent soluble substrate concentration (BOD or biodegradable soluble COD), mg/L; $S$ = effluent soluble substrate concentration (BOD or biodegradable soluble COD), mg/L; and $P_x$ = volatile suspended solids wasted per day, kg/day. (2) The amount of oxygen required for combustion of organic matter in sludge. *See* combustion oxygen requirement for detail.

**oxygen sag** The decline in the concentration of dissolved oxygen (DO) downstream from a point of pollution (e.g., discharge of treatment plant effluent or untreated wastewater) in a stream until the DO deficit reaches a maximum and the DO level recovers. The decline and recovery result from the interplay of microbial activity, deoxygenation, and reaeration. Same as dissolved-oxygen sag. *See* dissolved-oxygen sag analysis for more detail.

**oxygen sag curve** A spoon-shaped curve that represents the profile of dissolved-oxygen (DO) deficit along a receiving water downstream from a point of pollution such as a wastewater discharge. The curve, vs. time or distance of travel, is characterized by (a) an initial DO deficit ($D_a$) in a zone of relatively clean water, (b) a maximum deficit ($D_c$, also called critical deficit or minimum DO) between a zone of degradation and a zone of recovery, and (c) a point of inflection between recovery and relatively clean water. Same as dissolved-oxygen sag curve. *See* dissolved-oxygen sag analysis.

**oxygen sag equation** An equation that represents the interplay of deoxygenation and reaeration when wastewater or effluent is discharged into a receiving stream. *See* dissolved-oxygen sag analysis for more detail.

**oxygen saturation** The condition of a liquid that contains the maximum concentration of dissolved oxygen in equilibrium with the atmosphere, at a given temperature and pressure. *See* dissolved oxygen saturation.

**oxygen saturation coefficient** The beta ($\beta$) coefficient in the formula for oxygen transfer rate.

**oxygen scavenger** A chemical used for deaeration; mechanical equipment that deaerates.

**oxygen-starved combustion** *See* oxidative pyrolysis.

**oxygen transfer** (1) The exchange of oxygen between a gaseous phase (air) and a liquid phase (water or wastewater). In wastewater treatment, oxygen transfer occurs in two phases: (a) diffused or mechanical aeration dissolves the oxygen from air bubbles in the wastewater and (b) microorganisms use the dissolved oxygen to metabolize waste organic matter. *See* oxygen transfer rate, mass transfer model. (2) The oxygen transfer efficiency (OTE), i.e., the ratio, usually expressed as a percentage, of the quantity of oxygen absorbed by water or wastewater to the amount fed to the liquid; e.g., 5–15% for bubble aeration. OTE depends on many factors such as diffuser characteristics, air flow rate, and tank geometry.

**oxygen transfer coefficient** For an aerated lagoon, the oxygen transfer coefficient or alpha ($\alpha$) coefficient is the ratio of the oxygen transfer coefficient in the wastewater to the transfer coefficient in clean water. *See also* gas transfer coefficient.

**oxygen transfer efficiency(OTE)** *See* oxygen transfer (2).

**oxygen transfer rate** The quantity of oxygen transferred from air bubbles into solution per unit time. *See* actual oxygen transfer rate.

**oxygen uptake** The amount of oxygen used during biochemical oxidation of organic matter, expressed in terms of the concentration of volatile solids or in weight of oxygen per unit volume per day:

$$r_o = -r_{su} - 1.42\, r_g \qquad \text{(O-28)}$$

where $r_o$ is the oxygen uptake rate in grams of oxygen per cubic meter per day; $r_{su}$ is the substrate utilization rate, grams of biodegradable soluble COD per gram of volatile suspended solids; and $r_g$ is the rate of growth of the biomass in grams of volatile suspended solids per cubic meter per day. The constant 1.42 is the g COD per g cell tissue as represented by the formula $C_5H_7NO_2$.

**oxygen uptake rate (OUR)** The rate of dissolved oxygen utilization by microorganisms in wastewater treatment, a function of organic loading (e.g., food-to-microorganism ratio) and temperature. It is about 10 mg/L/hr for extended aeration, 30–50 mg/L/hr average and 60–80 mg/L/hr peak for conventional activated sludge, with lower rates for aeration following a fixed-film unit. *See also* specific oxygen uptake rate and biological uptake.

**oxygen utilization** The oxygen effectively used during aerobic waste treatment or during sludge combustion by incineration or wet air oxidation.

**Oxygun™** A subsurface aerator manufactured by Framco Environmental Technologies.

**oxyhalide** An ion that contains oxygen and a halogen, including such disinfection by-products as bromate ($BrO_3^-$), chlorate ($ClO_3^-$), and chlorite ($ClO_2^-$) ions. Phosphorus oxybromide ($POBr_3$) reacts with water to form phosphoric acid ($H_3PO_4$) and hydrogen bromide gas (HBr):

$$POBr_3 + 3\,H_2O \rightarrow H_3PO_4 + 3\,HBr \qquad \text{(O-29)}$$

**oxypause** The depth at which dissolved oxygen depletion occurs in a facultative pond; it fluctuates, lowering during the daytime and rising at night.

**Oxyrapid** An air diffusion apparatus manufactured by Infilco Degremont, Inc. for the activated sludge process.

**Oxyrotor** A surface brush aerator manufactured by Euroquip Fabrication, Ltd.

**Oxystream™** Oxydation ditch wastewater treatment system designed or manufactured by WesTech Engineering, Inc. of Salt Lake City, UT.

**Oxytrace™** A device manufactured by Capital Controls Co., Inc. for the analysis of chlorine residual.

**Oxytrap™** An aerator manufactured by DAS International, Inc. for use in wastewater treatment.

***Oxyuris vermicularis*** Another name of the nematode species *Enterobius vermicularis,* also known as pinworm, seatworm, or threadworm.

**oz** Abbreviation of ounce.

**OZ** An apparatus manufactured by Ozone Pure Water, Inc. for the generation of ozone.

**Ozat** A compact installation of Ozonia North America for the generation of ozone.

**Ozoflot** A French dispersed-air flotation system that uses ozone-enriched air through diffusers to treat water with high algae concentrations.

**ozonation** The application of ozone to water, wastewater, or air for oxidation of organics, disinfection, or for color, taste, and odor control. Ozone decomposes in water as follows:

$$2\,O_3 \rightarrow 3\,O_2 \qquad \text{(O-30)}$$

Ozonation reduces the concentration of trihalomethane (THM) precursors and contributes to the partial elimination of THMs. However, ozonation does not leave a residual to protect against contamination in the distribution system and, in some circumstances, it may increase the formation of halogenated organics. *See also* multiple-stage ozonation.

**ozonation by-product** (1) A compound formed as a result of disinfection by ozone, namely an epoxide, an organic peroxide, an unsaturated aldehyde (R—CHO), or an aldoacid (HOOC—R—CHO). Specific by-products include formaldehyde, acetaldehyde, glyoxal, methyl glyoxal, acetic acid, formic acid, oxalic acid, succinic acid, and pyruvic acid. (2) In waters containing bromides, ozonation converts bromide to hypobromous acid, which forms bromates and brominated organic compounds, including trihalomethanes, haloacetonitriles, and haloacetic acids. Specific by-products are bromoform, bromopicrin, and cyanogen bromide.

**ozonation environmental impact** Ozone is acutely toxic to aquatic life, and ozonation may produce mutagenic or carcinogenic compounds.

**ozonation plant** An installation set up at a treatment plant for the disinfection of water or wastewater by ozone, including air filters, blowers, air dryers, an ozonator, and an ozone contact chamber.

**ozonation system** An installation that generates ozone on-site and mixes it with water or wastewater for disinfection or other purposes. A typical ozonation system includes the following main elements: air filter, blower, heat exchanger/cooler, refrigerant and desiccant dryers, ozone generator, ozone contactor, and ozone destructor.

**ozonator** (1) An apparatus that produces ozone by passing pure oxygen or air through an electric field. *See also* ozone generator or an ozonizer. (2) A device used to apply ozone to water or wastewater. *See also* ozone contactor.

**ozone ($O_3$)** An unstable, corrosive, pale-blue or colorless gas consisting of three oxygen atoms stuck together into an ozone molecule, whereas molecular oxygen consists of two oxygen atoms. Discovered in 1783 but named around 1840; becomes liquid below −112°C; half-life of 20–30 min depending on temperature and pH; produced electrolytically, chemically, or via ultraviolet light. Ozone has a distinct odor, detectable at low concentrations of 0.01–0.05 ppm; it produces the sharp smell noticed near a lightning strike. Ozone is found in two layers of the atmosphere. In the stratosphere, ozone is a natural form of oxygen that provides a protective layer shielding the Earth from ultraviolet radiation. In the troposphere, ozone is a strong chemical oxidant and major component of photochemical smog. It is poisonous at high concentrations, can seriously impair the respiratory system and is one of the most widespread of all the criteria pollutants for which the USEPA sets standards under the Clean Air Act. Ozone in the troposphere is produced through complex chemical reactions of nitrogen oxides, which are among the primary pollutants emitted by combustion sources; hydrocarbons, released into the atmosphere through the combustion, handling and processing of petroleum products; and sunlight. Ozone is used in water disinfection (in Europe since the early twentieth century), wastewater disinfection, sludge-bulking control, or for advanced oxidation of organics; as such, it may also act as a coagulation–filtration aid. It is a very effective agent for oxidizing such odor-causing substances as geosmin and methylisoborneol. *See also* ozone–hydrogen peroxide.

**ozone-assisted coagulation** An improved coagulation method that uses ozone to alter the surface charge on particles, resulting in reduced coagulant dosages.

**ozone bleaching** A process of the pulp and paper industry using ozone, instead of chlorine, as the primary bleaching agent, following oxygen delignification.

**ozone breakdown** The spontaneous decay of ozone to oxygen; thus, the impracticality of its production off-site and the absence of a residual after application. *See also* ozone decomposition.

**ozone by-product** *See* ozonation by-product.

**ozone contactor** A basin or chamber, usually with several compartments, for ozone to contact water or wastewater; similar to a chlorine contact chamber.

**ozone decomposition** Ozone decomposes to oxygen in water at a rate that depends on the concentration of alkalinity ions, UV radiation, and hydrogen peroxide ($H_2O_2$), according to an autocatalytic model (Droste, 1997 and Metcalf & Eddy, 2003):

$$d[O_3]/dt = -k_1[O_3] - k_2[O_3]^2 \quad \text{(O-31)}$$

where $[O_3]$ is the concentration of ozone and $k_1$ and $k_2$ are constants. The decomposition reactions may be represented as follows:

$$O_3 + H_2O \rightarrow HO_3^+ + OH^- \rightarrow 2\,HO_2 \quad \text{(O-32)}$$

$$O_3 + HO_2 \rightarrow HO + 2\,O_2 \quad \text{(O-33)}$$

$$HO + HO_2 \rightarrow H_2O + O_2 \quad \text{(O-34)}$$

*See* ozone decomposition cycle for another representation of the decomposition reactions. *See also* odd hydrogen.

**ozone decomposition cycle** A graphical representation of the reactions associated with ozone decomposition, including the formation of the superoxide anion ($O_2^-$), the production and release of oxygen ($O_2$), and the production of the hydroxyl radical ($OH^\cdot$) as an intermediate. *See* Figure O-09. The hydroxide ion ($OH^-$) may initiate the process, whereas ozone itself may serve as a

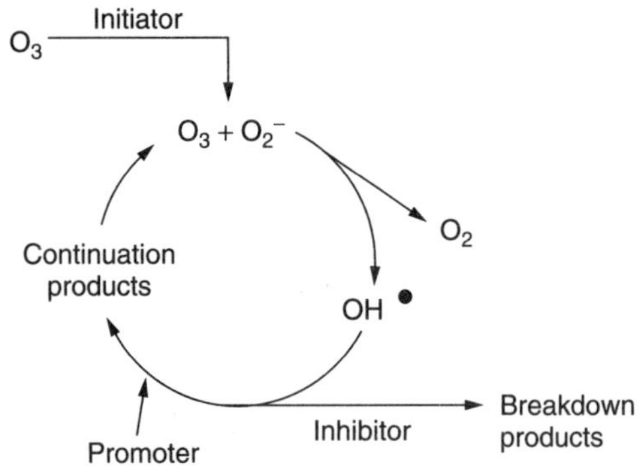


**Figure O-09.** Ozone decomposition cycle (adapted from Bryant et al., 1992).

promoter and carbonates and bicarbonates may be inhibitors.

**ozone demand** The amount of ozone that is required to react with hydroxide ions (to form hydroxyl radicals and organic radicals), bromides, and cyanides. These reactions reduce the half-life of ozone in water and cause the formation of ozonation by-products. *See also* ozone dose.

**ozone destruction** The process of destroying completely or partially the ozone contained in an off-gas.

**ozone disinfection models** Equations that describe the disinfection process with ozone, adapted from the chlorine disinfection models:

$$N/N_0 = 1 \qquad \text{when } U < U_0 \qquad \text{(O-35)}$$

$$N/N_0 = (U/U_0)^{-n} \qquad \text{when } U > U_0 \qquad \text{(O-36)}$$

where $N$ = number of organisms after disinfection; $N_0$ = initial number of organisms (before disinfection); $U$ = ozone dose utilized, mg/L; $N_0$ = initial ozone demand = value of abscissa intercept in a plot of log $N/N_0$ vs. $U$ (i.e., when $N/N_0 = 1$ or log $(N/N_0) = 0$); and $n$ = slope of the dose–response curve. *See also* ozone dose.

**ozone disinfection mechanisms** *See* ozone inactivation.

**ozone distribution coefficient** The ratio of the equilibrium concentration of ozone in water to that in air at the same temperature and pressure. *See also* distribution coefficient or partition coefficient.

**ozone dose** The amount of zone applied for disinfection ($D$, mg/L), including the initial ozone demand, is the ratio of the ozone dose utilized or transferred ($U$, mg/L) to the ozone transfer efficiency ($E$, typically 0.8–0.9):

$$D = U/E \qquad \text{(O-37)}$$

**ozone enrichment** The process of increasing the ozone content of a gas. *See,* e.g., Ozoflot.

**ozone generation** Ozone is generated on-site by the application of a high-voltage electrical discharge into air, oxygen, or a mixture of the two gases, and across the gap of narrowly spaced electrodes. The corona discharge dissociates one oxygen molecule into two atoms that combine with two other oxygen molecules to form two ozone molecules:

$$3\ O_2 \rightarrow 2\ O_3 \qquad \text{(O-38)}$$

*See* Figure O-10.

**ozone generator** An apparatus that produces ozone by passing pure oxygen or air through an electric field. Also called ozonator or ozonizer.

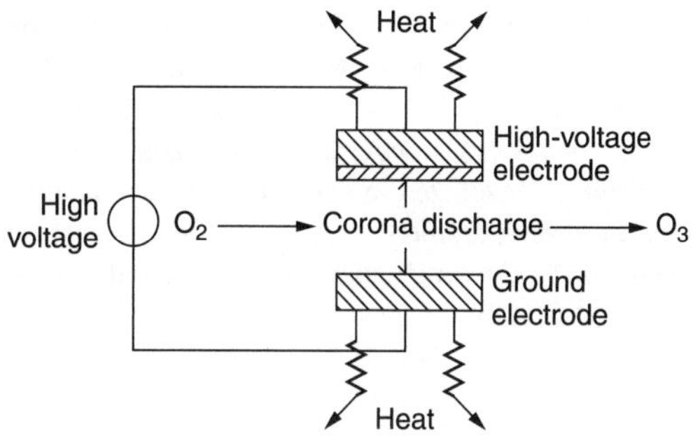


**Figure O-10.** Ozone generation.

**ozone–$H_2O_2$** Same as ozone–hydrogen peroxide.

**ozone half-life** The time required for the decomposition of half of a quantity of ozone at a given temperature and pressure.

**ozone–hydrogen peroxide ($O_3$–$H_2O_2$)** A combination of ozone and hydrogen peroxide used in advanced oxidation processes, e.g., for the removal of xenobiotics and endocrine disruptors from drinking water. *See* peroxone process for detail.

**ozone inactivation** Ozone inactivation mechanisms are similar to those of chlorine: disintegration of cell walls by oxidation, disruption of membrane permeability, impairment of enzyme function and protein integrity by oxidation of sulfhydryl groups, and nucleic acid denaturation.

**ozone residual** The concentration of ozone (mg/L) in treated water or wastewater.

**ozone scavenger** A chemical (such as hydrogen peroxide, sodium bisulfite, or even calcium thiosulfate) used to quench the ozone residual before it leaves the contact chamber of a treatment plant.

**ozone treatment** *See* ozonation.

**ozone–ultraviolet ($O_3$–UV) process** An advanced oxidation process that uses ultraviolet light to catalyze the decomposition of ozone and produce hydroxyl radicals for the oxidation of compounds in water or wastewater

**ozonide** (1) A by-product of ozonation. (2) A compound formed by the reaction of ozone ($O_3$) with another compound. Organic ozonides (e.g., from olefins) are unstable and usually explosives. Inorganic ozonides are explosive, dark-red ionic compounds formed by burning alkali metals in ozone.

**ozonization** Conversion to ozone or treatment with ozone; ozonation.

**ozonized gas** A mixture of ozone and the feed gas (air, high-purity oxygen, or oxygen-enriched air) in an ozonation system.

**ozonizer** An apparatus that produces ozone by passing pure oxygen or air through an electric field. Also called ozonator or ozone generator.

**ozonizing tower** A countercurrent ozonation unit in which ozonized air moves upward against falling water droplets.

**Ozonmat®** An instrument manufactured by Z Polymetron for the analysis of ozone.

**ozonolysis** (1) The decomposition or oxidation of an organic compound (e.g., an unsaturated hydrocarbon) by ozone. With alkenes or alkynes, ozone forms carbonyl compounds. (2) The use of ozone in analytical chemistry to locate double bonds in organic compounds.

# P

**p** A mathematical operator designating the negative logarithm (or the logarithm of the reciprocal) of a quantity; i.e., p = $-\log_{10}$; for example pH = $-\log_{10}$ [H]. *See* pAlk, pE, pH, pK, p$\varepsilon$.

**P2** Pollution prevention.

**$P_{10}$** Symbol for 10th percentile size and for effective size. *See also* uniformity coefficient.

**P17** Abbreviation of *Pseudomonas fluorescens* strain P17.

**$P_{60}$** Symbol for 60th percentile size. *See also* uniformity coefficient.

**PA** Acronym of polyamide.

**PAA** Acronym of (1) peracetic acid; (b) polyacrylic acid.

**P-A broth** A culture used in the presence–absence test (or P–A test), containing beef extract, peptone, lactose, tryptose, inorganic nutrients, and bromocresol purple indicator. *See also* EC medium, lauryl tryptose broth.

**PAC** Acronym of powdered activated carbon.

**PAC** A device manufactured by F. B. Leopold for the positioning and control of valves.

**PAC dose** The quantity of powdered activated carbon per unit volume of gas or liquid treated, expressed in milligrams per liter (mg/L).

**Pace®** An apparatus manufactured by Scienco/FAST Systems for the separation of oil from water.

**Pacer** A package plant manufactured by Roberts Filter Manufacturing Co. for water treatment.

**Pacesetter** An apparatus manufactured by Envirotech Co. for liquid–liquid separation by gravity.

**packaged water system** A package water treatment plant. *See* package plant, modular system, point-of-entry unit, point-of-use unit.

**package plant** A transportable, prefabricated treatment plant that performs one or more processes (extended aeration, sedimentation, flotation) for small flows (e.g., less than 1.0 mgd) usually in a single tank; sometimes used as a temporary measure until the installation of permanent facilities. Besides extended aeration, processes used in package plants include contact stabilization, physical/chemical treatment, rotating biological contactors, and sequencing batch reactors. Also called a compact unit or preengineered plant.

**package plant burping** The periodic discharge of sludge from small package wastewater treatment plants that use a biological process. It was first thought that these plants could achieve complete oxidation, but excess sludge does build up and is occasionally discharged in the effluent.

**package treatment plant** *See* package plant.

**package wastewater plant** *See* package plant.

**packed bed** A treatment unit, such as a filter or ion exchange bed, that does not allow expansion of the medium. *See also* expanded bed.

**packed-bed aeration tower** *See* packed-bed aerator.

**packed-bed aerator** A tower or column packed with a medium (e.g., rocks, plastic materials, Raschig rings) used in air stripping of volatile organic compounds from water or in other aeration operations. The medium provides a large surface and a tortuous path for the liquid and forced air to move in opposite direction. The compounds removed may require further treatment. *See also* air–water ratio, tower aerator.

**packed-bed filter** A water or wastewater treatment filtration unit that uses a packed bed, e.g., a rapid sand filter, an activated carbon adsorber, or an ion exchanger.

**packed-bed reactor** A vessel filled with rock, slag, ceramic, plastic, or some other type of packing material, and operated continuously or intermittently, for water or wastewater treatment, in the upflow or downflow mode. *See also* fluidized-bed reactor.

**packed-bed scrubber** An air pollution control device in which emissions pass through alkaline water to neutralize hydrogen chloride gas. It consists of a contact chamber containing an inert packing material, a chemical solution tank, and a blower, with the necessary ducts and control system. Also called wet packed-bed scrubber.

**packed biological reactor** A basin filled with sand or other small media for the aerobic or anaerobic treatment of wastewater; used, for example, in denitrification. *See* fluidized bed for more detail.

**packed column** A hollow vertical unit filled with packing material and designed to strip gases or transfer contaminants from the liquid to the gas phase. The influent liquid flows over the packing, which provides the contact surface for a countercurrent airflow. Also called a packed tower. *See also* air stripping and degasifier.

**packed-column air stripping** A process that uses a packed column to remove volatile organic compounds from water.

**packed-plunger metering pump** An early version of a metering pump developed by a specialty chemicals company in the 1940s. To deliver a controlled volume of liquid in water and wastewater treatment processes as well as other applications, it uses a packed plunger driven through a gear box by a foot-mounted motor, a slider crank mechanism, and check valves in the pump head. *See also* step valve.

**packed tower** A pollution control device that forces dirty air through a tower packed with crushed rock or wood chips while liquid is sprayed over the packing material. The pollutants in the air stream either dissolve or chemically react with the liquid. Also called a packed column, and used in water treatment to strip volatile organic compounds, carbon dioxide, hydrogen sulfide, and ammonia.

**packed-tower aeration (PTA)** A method of treating water to remove volatile organic chemical (VOC) contaminants. As water is mixed with air, VOCs move from water to air, which then passes through carbon filters to trap the contaminants. The treatment unit is also called a packed-tower stripper or packed-tower air stripper. *See* Figure P-01. *See also* packed-bed aerator, stripping tower.

**packed-tower air stripper** *See* packed-tower aeration.

**packed-tower design equation** An equation that relates the bulk air-phase contaminant concentration ($y_b$) in a packed tower to the bulk water-phase contaminant concentration ($C_b$), based on a mass balance at the bottom of the tower:

$$y_b(Z) = (Q/V)[C_b(Z) - C_e] \qquad \text{(P-01)}$$

where $Z$ = elevation of any point from the bottom of the tower, $Q$ = liquid flow rate to the tower, $V$ = volumetric airflow rate to the tower, $C_e$ = bulk water-phase contaminant concentration at the bottom of the tower. *See also* cascade air stripping, cocurrent packed tower, cascade packed tower, end effect, crossflow packed tower, low-profile air stripper, minimum air-to-water ratio, wall effect.

**packed-tower stripper** *See* packed-tower aeration.

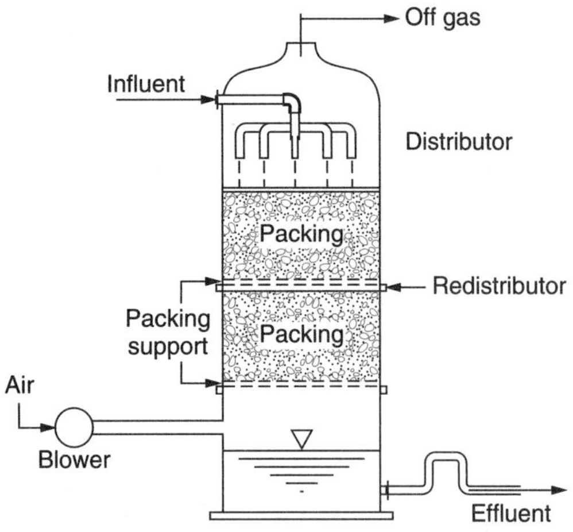


**Figure P-01.** Packed-tower air stripper.

**packed-tower stripping** Same as packed-tower aeration.

**packers pollution** Pollution caused by wastewater from the meatpacking industry, which has a high concentration of common pollutants (BOD, suspended solids, nitrogen), as well as blood, grease, hair, paunch manure, etc. *See* meatpacking waste, rendering pollution.

**packing** (1) The fill material placed in a packed column, a fixed-film reactor, or a stripping vessel to provide a large contact surface area per unit volume and, thus, a high gas transfer rate. Typical trickling filter packing materials include rock, cross-flow plastic, vertical-flow plastic, and horizontal or random-pack redwood. (2) The material placed around a pump shaft to reduce water losses. *See* more detail under pump packing.

**packing factor** A parameter used in the design of stripping towers to characterize the type and size of packing (Metcalf & Eddy, 2003). It is usually provided by the manufacturer. For example, typical ranges are 180–240 for 12.5-mm pall rings and 120–160 for 25-mm Raschig rings.

**packing gland** (1) A device that holds in place the packing installed in the seal cage of a water or wastewater pump. (2) The space between the pump casing and the pump shaft containing the packing that controls flow along the shaft.

**pack joint** A compression fitting or coupling that connects water lines.

**PACl** Abbreviation of polyaluminum chloride.

**PAC/MF system** A water treatment system that combines the use of powdered activated carbon (PAC) with microfiltration (MF) membranes to remove suspended particles and microorganisms (but not dissolved organic matter). *See also* FBR/PAC/UF process, PAC/UF system.

**P-A coliform test** *See* presence–absence test.

**Pacpuri®** An apparatus manufactured by Electrocatalytic, Ltd. for the production of sodium hypochlorite.

**PACT process** A wastewater treatment process developed by DuPont in the 1970s. It integrates adsorption on activated carbon into a conventional activated sludge system. The activated carbon enhances the removal of recalcitrant and toxic contaminants and provides support for microbial growth. PAC is added to the aeration tank so that the equilibrium PAC-mixed liquor suspended solids content ($X_p$, mg/L) is:

$$X_p = (X)(SRT)/HRT \qquad \text{(P-02)}$$

where $X$ is the PAC dosage (between 20 and 200 mg/L), $SRT$ = solids retention time (days), and $HRT$ = hydraulic retention time (days). *See also* biological activated carbon.

**PACT®** system Powdered activated carbon (PAC) treatment, a proprietary process developed by Zimpro Environmental, Inc. *See also* PACT process.

**PAC/UF system** A water treatment system that combines the use of powdered activated carbon (PAC) with ultrafiltration membranes to remove suspended particles and microorganisms (but not dissolved organic matter). *See also* FBR/PAC/UF process, PAC/MF system.

**paddles** Blades attached to horizontal or vertical shafts for providing mechanical mixing and stirring in water or wastewater treatment.

**paddle aerator** (1) A device that looks like a paddle wheel and is used to aerate liquids. Also called paddle-wheel aerator. (2) A submerged horizontal-axis aerator, similar to a surface aerator but with paddles attached to a rotating shaft to agitate the liquid. *See also* Kessener brush aerator, disk aerator, mechanical aerator.

**paddle dryer** A device used to dry wastewater treatment sludge by circulating steam through a stationary horizontal vessel or trough, which has a jacketed shell and contains a number of paddles in a rotating assembly. The paddles transport the sludge through the device and provide the heat transfer surface. *See also* indirect or conduction dryer.

**Paddle Dryer** A sludge dryer manufactured by Komline-Sanderson.

**paddle flocculator** A flocculation device that uses rotating paddles to accomplish mixing of coagulated particles; e.g., wooden paddles in a series of compartments separated by baffles. Because of maintenance problems of paddle flocculators, turbine flocculators are being used more often. *See* Flocsillator®, Floctrol, paddle mixer.

**paddle mixer** A flocculation device that consists of a series of paddles mounted vertically or horizontally on a shaft and regulated by variable-speed drives. *See also* mixing and flocculation devices.

**paddle wheel** A waterwheel equipped with paddles or strips of wood or metal on its periphery, used in power generation, aeration, or mixing.

**paddle-wheel aerator** Same as paddle aerator.

**PADRE™** A low-cost polymeric material developed by the Dow Chemical Company for the adsorption of volatile organic compounds, which are then condensed, transferred to a storage tank, and recovered using proprietary solvents such as PurSorb™ 100 and PurSorb™ 200.

**PAED** Acronymn of pulsed arc electrohydraulic discharge.

***Paenibacillus* spp.** A genus of nonfilamentous organisms that are sometimes significantly present in sludge, causing clarifier failure, but cannot be identified through traditional methods. *See* molecular fingerprinting.

***P. aeruginosa*** *See Pseudomonas aeruginosa.*

**PAH** Acronym of (a) polyaromatic hydrocarbon, (b) polycyclic aromatic hydrocarbon, and (c) polynuclear aromatic hydrocarbon.

**paint filter test** A test that uses a paper filter to determine the free water content of sludge.

**PakTOR** A packed-bed reactor manufactured by General Filter Co.

**palatability test** *See* palatable water.

**palatable water** Water at a desirable temperature that is free from objectionable tastes, odors, colors, and turbidity. In addition to these parameters, palatability tests may include plankton, chloride, CCE, hydrogen sulfide, and natural organic matter.

**pAlk** Representation of the negative logarithm of alkalinity in equivalents per liter. *See* LSI, SDSI.

**p-alkalinity (p-Alk)** Alkalinity of a sample equal to the sum of the concentrations of hydroxyl and carbonate ions, minus the concentrations of carbon dioxide/carbonic acid and hydrogen ions:

$$\text{p-Alk} = [OH^-] + [CO_3^{2-}] - [H_2CO_3^*] - [H^+] \quad \text{(P-03)}$$

**P alkalinity** Short for phenolphthalein alkalinity.

**Pall Aria™** A hollow fiber filtration system developed by Pall Water Processing of East Hills, NY, particularly for the treatment of groundwater.

**pall ring** A common material used as a medium in upflow reactors.

**palm and sludger method** A simple method that uses a lever to move a pipe up and down into the ground to construct jetted tubewells in Bangladesh. *See also* jetting.

**palm oil waste** Wastewater generated during the production of palm oil; originating from the mechanical extraction of crude oil and its refining to edible oil. Such waste is high in BOD, COD, total solids, and fats, and has low pH. Common treatment methods include neutralization, coagulation, flotation, filtration, and biological oxidation.

**Palmer–Bowlus flume** A portable, prefabricated fiberglass, reinforced plastic, and stainless steel type of venturi flume that uses a constricted throat to measure flow in manholes and partially full pipes. Its flat bottom makes it easier (than the Parshall flume) to install in an existing structure.

**palustrine habitat** (1) A nontidal wetland in which the dominant vegetation includes trees, shrubs, persistent emergents, emergent mosses, or lichens. (2) A wetland that lacks such vegetation but is smaller than 20 acres, has a low-water depth less than 6.5 ft in its deepest part, and has an ocean-derived salinity less than 0.5 part per thousand. (3) A tidal wetland with ocean-derived salinity less than 0.5 part per thousand.

**PAM** Acronym of primary amoebic meningoencephalitis.

**PA membrane** *See* polyamide membrane.

**PAN** Acronym of (a) peroxyacetyl nitrate; (b) plant available nitrogen.

**pan** In a pour-flush toilet or cistern-flush toilet, a basin or plate that receives the excreta that are then flushed by water.

**pandemic** (1) Widespread throughout an area, a nation, or the world. (2) An epidemic that affects a large proportion of the population of a wide area.

**panel diffuser** A device used to aerate water or wastewater, consisting of a rectangular panel with a flexible plastic perforated membrane. Other porous diffusers include disk, dome, and membrane diffusers. *See also* nonporous diffusers.

**Pano–Middlebrooks relationship** An equation that relates the extent of nitrification in rotating biological contactors and the soluble BOD loading (Metcalf & Eddy, 2003):

$$F = 1.0 - 0.1\ \text{sBOD} \quad \text{(P-04)}$$

where $F$ is the fraction of nitrification and sBOD (g/m/day) is the soluble BOD loading.

**PAO** Acronym of phosphate-accumulating organism.

**papatasi fever** A viral fever, sometimes called leshmaniasis, transmitted in the Mediterranean and Near East by the sandfly *Phlebotomus papatasii.*

**paper chromatography** A method that uses a strip of paper and a solvent to separate and identify small amounts of mixtures. *See also* chromatography.

**parabolic weir** A weir that has a parabolic notch and a vertical axis.

***Paracoccus* genus** A group of heterotrophic organisms active in biological denitrification.

***Paracoccus pantotropha*** A species of denitrifying bacteria that can carry out simultaneously the oxidation of ammonia and the reduction of nitrate.

**Para Cone** A rotary fine screen manufactured by Andritz-Ruthner, Inc., with internal feed.

***para*-dichlorobenzene ($C_6H_4Cl_2$)** The common name for 1,4-dichlorobenzene; a synthetic organic chemical produced by chlorination of benzene or chlorobenzene, soluble in water, and used as an insecticide and a solvent. The chemical can cause anorexia, nausea, liver atrophy, and blood disorders. Regulated by the USEPA as a drinking water contaminant: MCL = MCLG = 0.075 mg/L.

**paraffin** A saturated hydrocarbon, mainly of the alkane series, containing only carbon and hydrogen atoms, with single bonds between the carbons. Also called paraffin oil. *See* alkanes.

**paraffin hydrocarbon** A paraffin.

**Paraflash** An evaporator manufactured by APV Crepaco, Inc. and equipped with forced circulation.

**Paraflow** A plate heat exchanger manufactured by APV Crepaco, Inc.

**paragonimiasis** An excreted helminth infection caused by the pathogen *Paragonimus westermani. See* its common name, lung fluke, for more detail.

***Paragonimus*** A genus of trematodes that inhabit the lungs and sometimes the brain.

***Paragonimus africanus*** A species of trematodes that cause the lung fluke disease in Africa.

***Paragonimus uterobilateralis*** A species of trematodes that cause the lung fluke disease in Africa (McJunkin, 1982).

***Paragonimus westermani*** A thick, fleshy, ovoid fluke of about 12 × 6 mm, the helminth species that causes the lung fluke disease in Asia. It is transmitted from excreta through snails and crabs or crayfish as intermediate hosts. Latency = 4 months. Median infective dose < 100.

**parallel-finger weir** A type of launder installed on supporting piers as part of traveling-bridge sludge collection equipment in wastewater sedimentation tanks.

**parameter** A variable, measurable property whose value is a determinant of the characteristics of a system; e.g., temperature, pressure, and density are parameters of the atmosphere. Also see model parameter.

**paraquat [$C_{12}H_{14}N_2 \cdot 2CH_3SO_4$ or $CH_3(C_5H_4N)_2CH_3 \cdot 2CH_3SO_4$]** A standard herbicide resistant to microbial degradation but adsorbable on soil particles, used to kill various types of crops, including marijuana; human exposure may cause death. It is the common name of 1,1-dimethyl-4,4-dipyridylium.

**parasite** An organism that lives on or inside a larger host organism and draws nourishment from the host; the presence of parasites is usually harmful and without any use to the host. *See,* e.g., bacteriophage. *See also* ametoecious (monoxenic) parasite, autoecious parasite, brood parasite, ecoparasite, ectoparasite, endoparasite, commensalism, facultative parasite, hemiparasite, heteroecious (heteroxenous) parasite, hyperparasite (superparasite), kleptoparasite (cleptoparasite), metoecious (metoxenous) parasite, obligate parasite, tropoparasite, xenoparasite, mutualism, symbiosis.

**parasitic bacteria** Bacteria that thrive in or on other living organisms. *See* parasite.

**parasitic protozoa** Protozoa that thrive in or on other living organisms, particularly those that cause diseases such as malaria, sleeping sickness, Chagas disease, leishmaniasis, giardiasis, and cryptosporidiosis.

**parasitism** An association between living organisms of two different species whereby the parasite obtains food at the expense of the host. *See also* commensalism, symbiotism, and mutualism.

**parasitology** The study of parasitic organisms and the diseases they cause, which includes a large diversity of eukaryotes, more specifically, protozoa and helminths. Bacteria, viruses, and their associated diseases are studied separately;*see* bacteriology, virology.

**Para-Stat** A static screen manufactured by Dontech, Inc.

**parathion ($C_{10}H_{14}NO_5PS$)** An organophosphate pesticide used in agriculture; toxic to mammals and birds but easily degradable, e.g., by some *Flavobacterium* or *Bacillus* bacteria. Also called thiophos.

**paratyphoid** Same as paratyphoid fever.

**paratyphoid fever** A water-related and excreta-related disease caused by the bacteria *Salmonella paratyphi,* transmitted via the fecal–oral route, distributed worldwide, and similar to typhoid fever in many respects but with milder symptoms. Also called salmonellosis.

**Paravap** An evaporator manufactured by APV Crepaco, Inc. for treatment of waters with high solids concentrations.

**Parazone®** A proprietary bleach that contains about 4% of available chlorine; it can be used to make a 1% stock solution for water disinfection. Such a solution is unstable under warm conditions.

**parent material** The (igneous, sedimentary, or metamorphic) rock from which soil is formed under the action of climate and living organisms (weathering).

**Pareto chart** A histogram showing the relative contributions of small problems that cause one larger problem, thereby helping to establish priorities for action. The Pareto 80–20 principle may well apply to treatment plants, i.e., 80% of the operational problems are due to 20% of the causes; e.g., 80% of tank overflows are caused by sensor and electrical contact problems. Named after the Italian economist and sociologist Vilfredo Pareto (1848–1923).

**Paris green** An emerald-green, poisonous, water-insoluble compound of arsenic trioxide and copper acetate used in pigments and insecticides; also called copper arsenite: $Cu_3(AsO_3)_2$.

**Parker et al. formula** An equation proposed for the design of nitrification biotowers (Metcalf & Eddy, 2003):

$$RZT = 0.23\ EMe^{-RZ}N/(KN + N) \qquad \text{(P-05)}$$

where $RZT$ = surface removal rate or nitrification rate (grams per square meter per day) at depth $Z$ and temperature $T$; $Z$ = depth in tower, m; $T$ = temperature, °C; $E$ = media effectiveness factor; $M$ = maximum surface oxygen transfer rate; $R$ = empirical parameter describing the decrease in rate as a function of depth; $N$ = bulk liquid $NH_4$-N concentration, mg/L; and $KN$ = half-velocity constant (saturation concentration), mg/L. *See also* the Gujer–Boller equation.

**Parkwood** Wastewater treatment equipment manufactured by Longwood Engineering Co.

**Parshall flume** A calibrated device developed by R. L. Parshall for the measurement of small open-channel flows. It consists of a converging section, a throat with a sill (creating critical-flow conditions), and a diverging (or expanding) section. *See* Figure P-02. The discharge ($Q$) is determined as a function of the throat width ($W$) and the upstream water depth ($H$):

$$Q = 4\ W \cdot H^x \qquad \text{with } x = 1.522\ W^{0.226} \quad \text{(P-06)}$$

*See also* Palmer–Bowlus flume.

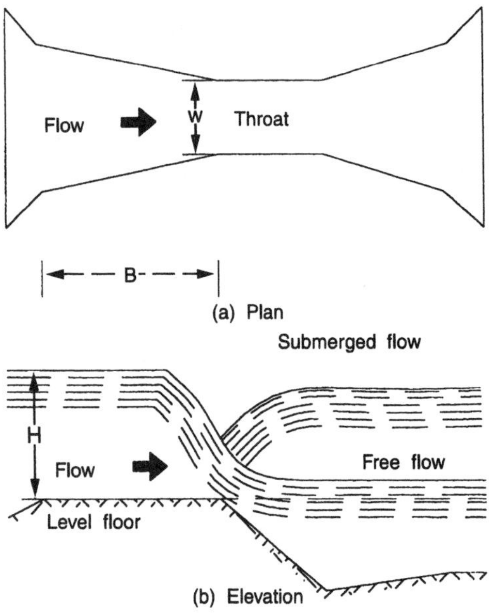


**Figure P-02.** Parshall flume.

**Parshall measuring flume** Same as Parshall flume.

**parthenogenesis** Development of an egg without male fertilization, e.g., by female worms of the *Strongyloides stercoralis* species.

**parthenogenic** or **partenogenetic** Capable of parthenogenesis.

**partial equilibrium** The condition of a system that is in equilibrium with respect to one reaction, but not in equilibrium with respect to others.

**partially mixed estuary** Hydrodynamically speaking, an estuary that has a gradual salinity gradient between the less saline surface layers and the more saline bottom layers. It is usually represented by a two-dimensional or two-layer model to account for the slow upstream movement of saltwater at the bottom and its gradual seaward entrainment by freshwater at the surface. *See also* salt wedge estuary, well-mixed estuary.

**partial-mix aerated lagoon** *See* partial-mix aerated pond.

**partial-mix aerated pond** A shallow earthen basin used for wastewater treatment. The aeration is sufficient for aerobic treatment but not for mixing. At the bottom, it functions like a facultative lagoon. *See* facultative partially mixed lagoon for more detail.

**partial nitrification** The oxidation of ammonia to nitrite in drinking water, without formation of nitrate.

**partial pressure** The portion of total vapor pressure in a system due to one or more constituents and proportional to their amounts (percent by volume) in the vapor mixture. In a mixture of gases, the partial pressure of a component is the pressure the gas would exert if it alone occupied the entire volume at the same temperature. *See* Dalton's law of partial pressures.

**partial pyrolysis** Incomplete combustion because of insufficient oxygen. *See* starved-air combustion.

**particle** A very small solid suspended in water or wastewater; it can vary widely in size, shape, density, and electrical charge. Colloidal and dispersed particles can be gathered together by coagulation and flocculation.

**particle beam mass spectrometry (PBMS)** An improvement of the liquid chromatography–mass spectrometry technique that uses a special interface to identify compounds by comparing them to computerized databases of mass spectra.

**particle concentration measurement** *See* turbidity measurement and particle counting for methods used to determine the performance of particle removal processes in water treatment.

**particle contact theory** An attempt to define the design and performance of water and wastewater treatment processes (particularly flocculation and sedimentation) on the basis of "conjunctions" or contacts between ions, molecules, colloids, and suspensions:

$$J_{ij} = (N_i N_j/6)G(D_i + D_j)^3 \quad \text{(P-07)}$$

where $J_{ij}$ = number of conjunctions of $i$ with $j$ particles, $N_i$ = number of particles of diameter $D_i$, $N_j$ = number of particles of diameter $D_j$,

$$G = [P/(VM)]^{0.5} = \text{shear or velocity gradient} \quad \text{(P-08)}$$

$P$ = useful power input, $V$ = volume of fluid, and $M$ = a proportionality factor with the same dimensions as the absolute viscosity ($\mu$). Equations (P-07) and (P-08) indicate that the number of contacts per unit time and volume increases with the number and size of particles, the power input per unit volume, and the inverse of the viscosity of the fluid ($1/M$) (Fair et al., 1971). *See also* conjunction opportunity, power dissipation function, velocity gradient.

**particle count** The result of a microscopic examination of treated water with a special particle counter that classifies suspended particles by number and size. Conducted with special instruments (electrical sensing, light-blocking, or light-scattering), it is used in evaluating raw and treated water (see particle counting). Also called particle size count or particle count and size distribution. *See also* particle size distribution.

**particle count and size distribution** *See* particle count.

**particle counter** An instrument that measures the number and size of particles within specified ranges; it is used to improve the performance of the filtration of water and wastewater.

**particle counting** A procedure used to evaluate the quality of filtered water and the performance of water filtration plants. The USEPA is considering the use of particle counting in water supply as a surrogate for the monitoring and removal of *Cryptosporidium* and *Giardia* cysts. *See* particle count and particle size distribution.

**particle density** The mass of particles per unit volume. For granular activated carbon, the volume excludes the voids between particles and cracks larger than 0.1 mm; it can be determined by immersion in mercury and measuring the displacement of mercury.

**particle imaging** A technique that uses a microscope attached to a video camera to obtain a quantitative assessment of particles in a small wastewater sample. *See* optical imaging for detail.

**particle-path model** A computer program that uses flow models to determine the path (time and direction) of particles along a flow path. *See* MODFLOW and MODPATH.

**particle size** The average diameter of the particles or grains in a sediment, rock, or bed of material.

**particle size analysis** Measurements and analysis conducted to determine particle size distribution. In wastewater studies, for example, this analysis follows one of three common methods: serial filtration, electronic particle size counting, and direct microscopic observation. *See also* particle counter.

**particle size count** *See* particle count.

**particle size distribution** A graphical or other representation of the size distribution of a mixture of particles. The distribution shows the weight fractions of the mixture larger (or smaller) than or equal to given sizes. Table P-01 is an example of a particle size distribution. Particle size distribution is used in the design of sedimentation units, chlorine and UV disinfection, and filtration. *See also* sieve analysis.

**particulate** (1) As an adjective: of, pertaining to, or composed of particles. (2) As a noun: a separate and distinct particle larger than one micron, removable by filtration, or a material composed of such particles. Examples: clay and silt particles, microorganisms, colloidal and precipitated humic substances, metal precipitates, calcium carbonate precipitates. Particulates are also fine liquid or solid particles such as dust, smoke, mist, fumes, or smog found in air or emissions.

**particulate chemical oxygen demand (pCOD)** A measure of the organic matter of wastewater in the

**Table P-01.** Particle size distribution (example)

| Size, mm | 0.10 | 0.08 | 0.07 | 0.06 | 0.05 | 0.04 | 0.03 | 0.02 | 0.01 |
|---|---|---|---|---|---|---|---|---|---|
| Weight fraction greater than, % | 10 | 15 | 40 | 68 | 85 | 92 | 97 | 99 | 100 |

form of colloidal and suspended solids; also called volatile suspended solids (VSS), a parameter used to measure biomass growth in full-scale biological treatment, although the VSS measure includes other particulate organic matter. *See also* COD fractionation.

**particulate matter** (1) Any airborne, finely divided solid or liquid material with an aerodynamic diameter smaller than 100 micrometers, other than uncombined water (EPA-40CFR51.100-oo, 60.2, or 61.171). (2) Dust, soot and other tiny bits of solid materials that are released into and move in the air. They are produced by many sources: burning of diesel fuels, garbage incineration, road construction, industrial processes, mining operations, agricultural burning. (3) Any minute separate particle of organic or inorganic matter.

**particulate organic carbon (POC)** Organic carbon in particulate form; the fraction of organic matter in water or wastewater that can be removed by a 0.45 micron filter. Except for surface water sources affected by stormwater runoff, it is usually a very small portion of total organic carbon.

**particulate organic matter (POM)** A compound of large molecules that makes up the structural components of plant and animal cells, is suspended in water, and removable by filtration.

**Partisol®** A device manufactured by Rupprecht & Patashnick Co. for air sampling.

**partition coefficient** (1) A parameter that indicates the tendency of a substance to partition between two phases or two media. When applied to the extraction of a solute, it is the ratio of the mass fraction of solute in the extraction phase to the mass fraction in the liquid phase. *See also* Dobbs et al. equation, air–water partitioning coefficient, octanol–water partition coefficient, soil–sediment partition coefficient. (2) A measure of the extent to which a pesticide is divided between the soil and water phases. *See also* distribution coefficient ($K_d$) and ozone distribution coefficient.

**parts per billion (ppb)** *See* parts per million.

**parts per million (ppm), parts per billion (ppb), and parts per million by volume (ppmv)** Units commonly used to express contamination ratios or other concentrations, as in establishing the maximum permissible amount of a contaminant in water, land, or air. One ppm means one part of a substance contained in one million parts of a solid, liquid, or gas, on a weight or volume basis. Similarly for one ppb. In water and wastewater calculations, ppm is approximately equal to, and has been largely replaced by, mg/L or $\mu g/m^3$. The unit ppmv is used mainly for gases; it means a volume/volume ratio that expresses the volumetric concentration of a gaseous contaminant in one million unit volume of air.

**parts per million by volume (ppmv)** *See* parts per million.

**parts per trillion (ppt)** A concentration unit equivalent to one nanogram per liter (ng/L).

**parvoviridae** A family of human excreted viruses, the parvoviruses.

**parvovirus** (1) Any of a number of small DNA viruses of the genus *Parvovirus* and family Parvoviridae that cause respiratory disease in children. (2) An often fatal virus of dogs.

**pascal (Pa)** The pressure or stress of one newton per square meter. 1 psi = 6895 Pa = 6.895 kN/sq.m = 0.0703 kg/sq cm.

**PASS®** A polyaluminum sulfate coagulant produced by the Alumina Co., Ltd.

**Passavant** Wastewater treatment products of Zimpro Environmental, Inc.

**passivating film** A thin layer formed on a metallic surface (e.g., a water supply pipe) by a corrosion inhibitor, or a water quality parameter and consisting of hydroxide and carbonate compounds that are less soluble than the metal itself.

**passivation** (From the verb passivate, to make a surface less chemically active) (1) The passage of the chemically active surface of a metal to a less reactive state. (2) The chemical process of making a metallic surface less reactive and less subject to corrosion. Passivation of stainless steel is accomplished by immersion in acid. Protection of water pipes against corrosion may be achieved by creating a passivating film on the pipe wall. *See also* deactivation.

**passivation dose** The amount of chemical (inhibitor) required to rapidly create a passivating film on a metal and protect it against corrosion, usually taken as 2–3 times the normal inhibitor concentration.

**passive immunity** Immunity of an organism resulting from the inoculation of antibodies from another organism or resulting from the transfer of antibodies to an infant through the placenta. *See also* active immunity.

**passive intrusion** *See* saltwater intrusion.

**passive metal** A metal that is protected against corrosion by application of a chemical to build a stable film on its surface. *See also* immune metal.

**passive screen** A screening device that is not mechanically cleaned.

**passivity** Characteristic of a metal that has become less reactive, i.e., more resistant to corrosion, by forming a protective oxide film either by means of an external electromotive force or by means of an oxidizing solution such as fuming nitric acid.

**pass through** A discharge that exits a publicly owned treatment works (POTW) into waters of the United States in quantities and concentrations that, alone or in conjunction with a discharge or discharges from other sources, is a cause of a violation of any requirement of the POTW's NPDES permit (including an increase in the magnitude or duration of a violation) (EPA-40CFR403.3-n).

**paste slaker** A mechanical device that adds enough water to hydrate quicklime (CaO) into slaked or hydrated lime that contains approximately 35% calcium hydroxide [$Ca(OH)_2$]. Also called pugmill slaker. *See also* detention slaker.

**Pasteur, Louis** French chemist and bacteriologist (1822–1895) who demonstrated the germ theory of disease in the 1880s. *See also* Koch, Robert and Snow, John.

**Pasteur filter** A filtering device that uses a medium of unglazed porcelain.

**pasteurization** A heat treatment process for killing pathogens, originally used to eliminate undesirable bacteria in wine but now used mostly to rid milk of disease-causing organisms; it is also used for water disinfection, but only on a small scale because of its high cost and its lack of a residual. Pasteurization applies heat for a period of time sufficient to eliminate the pathogenic but not all organisms. Sludge pasteurization by the exothermic reaction of quicklime with water to maintain a temperature of at least 70°C for 30 minutes is called an advanced alkaline stabilization process and qualifies as a process to further reduce pathogens; the product may also meet Class A sludge criteria. *See also* sterilization.

**Pasteur pipette** A small glass tube with a tapered tip, used for transferring small quantities of liquids.

**pasture crop** A crop such as legume, grass, grain stubble, and stover that is consumed by animals while grazing (EPA-40CFR 257.3.5).

**paternoster pump** A chain pump with buttons or discs along the chain; another name for chapelet.

**P-A test** *See* presence–absence coliform test.

**pathogen** An infectious organism (usually a microorganism) that can cause disease in other organisms, or in humans, animals, and plants. Principal waterborne pathogens are bacteria, viruses, protozoa, and algae, all found in wastewater, in runoff from animal farms or rural areas populated by domestic and/or wild animals, and in water used for swimming. Fish and shellfish contaminated by pathogens, or the contaminated water itself, can cause serious illnesses. *See also* infectious agent, pathogenic organisms, frank pathogen, opportunistic pathogen, water-related infection, excreta-related infections, latency, and persistence.

**pathogenesis** The mechanism of disease production by an etiologic agent.

**pathogenic bacteria** Bacteria that produce disease in a host. *See* pathogenic organisms.

**pathogenic *Escherischia coli*** See enteroinvasive, enteropathogenic, and enterotoxigenic *E. coli.*

**pathogenic organisms** Organisms, including bacteria, viruses, helminths, and protozoa, capable of causing disease in a host. There are many types of organisms that do not cause disease; they are called nonpathogenic. Unlike composting and waste stabilization ponds, conventional wastewater treatment processes are not effective in removing pathogenic organisms: they remove 90–99% of fecal pathogens, but the effluent still contains 100,000–1,000,000 fecal coliforms per 100 mL. *See also* pathogen, infectious agent.

**pathogenic protozoa** Common protozoa involved in the transmission of waterborne diseases include *Cryptosporidium parvum, Cryptosporidium muris, Entamoeba histolytica,* and *Giardia lamblia. See also* helminths.

**pathogenic viruses** The most significant disease-causing, waterborne viruses are the hepatitis-A agent, Norwalk-like agents, and the enteroviruses (coxsakievirus, echovirus, poliovirus).

**pathogen latency** *See* latency (2).

**pathogen persistence** *See* persistence (2).

**pathogen reduction alternative** Any of a number of sludge handling processes that can be used to meet the requirements for Class A or Class B biosolids, e.g., thermal treatment, high pH-high temperature, processes to further reduce pathogens (PFRP), and processes to significantly reduce pathogens (PSRP).

**pathogen removal** *See* disinfection and sterilization.

**pathological waste** Waste originating from hospitals, laboratories, clinics, and similar establishments; they may contain pathogens. *See also* medical wastes.

**pathology** The study of disease.

**pathway** A history of the flow of a pollutant from source to receptor, including qualitative description of emission type, transport, medium, and exposure route.

**Pathwinder®** A conveyor manufactured by Serpentix Conveyor Corp. for handling screenings.

**Patriot** An apparatus manufactured by SanTech, Inc. for the treatment of fluids recovered from spent coolant and oils.

**paunch manure** The contents of the digestive systems of animals killed in slaughterhouses, including entrails and the viscera. This material is a considerable source of pollution with a high suspended solids content that can interfere with the operation of publicly owned treatment works. It may be carried in tank trucks out of rendering plants to agricultural areas for use as soil fertilizer. *See also* meat packing waste.

**P availability** *See* phosphorus availability.

**paved drying bed** A device that uses evaporation to dewater water or wastewater treatment sludge that is spread 12-in deep and periodically mixed mechanically. It typically consists of an asphalt or concrete pavement overlying a porous gravel layer, with sand drains on the perimeter and in the center of the bed to collect the drainage. Heavy equipment is used for solids removal. Also called solar drying bed. *See also* air drying, sand drying bed, drainage paved drying bed, decanting paved drying bed.

**paved drying bed design** The determination of the required bottom area ($A$, m$^2$) for a drying bed, based on the annual sludge production ($S$, kg/yr of dry solids), the percentage of dry solids in the decanted sludge ($S_d$, as a decimal), the percentage of dry solids after evaporation ($S_e$, as a decimal), the annual precipitation ($P$, m/yr), the reduction factor for evaporation from sludge instead of a free water surface ($K$, unitless), and the free water pan evaporation rate ($E_p$, cm/yr):

$$A = 1.04\, S[(1 - S_d)/S_d - (1 - S_e)/S_e]/(10\, KE_p - 1000\, P) \qquad \text{(P-09)}$$

**pay-as-you-go financing** The use by a utility of revenue generated by rates and charges in excess of operation and maintenance needs to finance new construction. Also called nondebt financing.

**payback** (1) Same as payback period. (2) The return on an investment.

**payback period** The estimated number of years required to recover the original investment of a project; one method of project evaluation. At the end of the period, the project is said to break even. Payback does not take into account the cost of the initial capital and the effect of cashflows beyond the period.

**payback technique** Use of the payback period for project evaluation.

**Paygro** An in-vessel composting apparatus manufactured by Compost Systems Co.

**$Pb_3(AsO_4)_2$** Chemical formula of lead arsenate.

**$Pb(C_2H_5)_4$** Chemical formula of tetraethyl lead or lead tetraethyl.

**$Pb(C_2H_3O_2)_2 \cdot 3H_2O$** Chemical formula of lead acetate.

**$PbCO_{3(s)}$** Chemical formula of lead carbonate and the mineral cerussite.

**$Pb_3(CO_3)_2 \cdot (OH)_{2(s)}$** Chemical formula of hydrocerussite. Also written as $2PbCO_3 \cdot Pb(OH)_2$(s).

**$2PbCO_3 \cdot Pb(OH)_{2(s)}$** *See* $Pb_3(CO_3)_2 \cdot (OH)_2$(s).

**$PbMoO_4$** Chemical formula of wulfenite, a minor ore of molybdenum.

**$PbO_2$** Chemical formula of lead dioxide.

**$Pb_3O_4$** Chemical formula of the mineral minim, essentially lead oxide.

**$PbO(OH)_2 \cdot H_2O$** Chemical formula of the hydrated surface layer formed during the corrosion of lead piping. *See* gel zone.

**PbS** Chemical formula of galena or lead sulfide.

**$PbSO_4$** Chemical formula of lead sulfate and the mineral anglesite.

**PCB** Acronym of polychlorinated biphenyl.

**PCE** Abbreviation of perchloroethylene.

**pCa** Representation of the negative logarithm of the molar concentration of calcium. *See* LSI, SDSI.

**pCi** Abbreviation of picocurie.

**pCi/L** A unit of radioactivity equal to $10^{-12}$ curie per liter.

**PCP** Acronym of (a) pentachlorophenol; (b) progressive cavity pump.

**PCR** Acronym of polymerase chain reaction.

**PDB** *See p*-dichlorobenzene.

**PDC™** Acronym of polymer dosage control, an apparatus manufactured by Andritz-Ruthner, Inc.

**PD control** *See* proportional–derivative control.

**PDM/Roediger®** An anaerobic sludge digester manufactured by Pitt-Des Moines, Inc.

**PE** Acronym of polyethylene.

**pE** A parameter, defined like pH, as the negative logarithm of electron activity:

$$\text{pE} = -\log(\text{A}_{e-}) = \text{E}/0.0591 \text{ at } 25°\text{C} \qquad \text{(P-10)}$$

where E is an electrode potential. A high pE implies a low concentration of electrons and a tendency to oxidize; a low pE indicates a high concentration of electrons and a tendency to reduce. The oxidative or reductive capacity of a solution (like acidity or alkalinity) can be expressed as the quantity of electrons to add or remove from the solution to attain a given pE. Natural waters have an approximate range of $-10 < \text{pE} < +17$, outside of which $H_2O$ is reduced to $H_2$ or oxidized to $O_2$. pE and pH affect the nature of chemical species in aqueous solutions. Also denoted by pε. *See* pε–pH diagram, redox activity, redox potential.

**peak** The maximum quantity occurring over a relatively short period of time such as one hour. Also called peak demand or peak load. Peaks are sometimes defined for longer periods, e.g., peak monthly wastewater flows.

**peak demand** or **peak load** The maximum momentary load on a utility such as an electric generating plant or a water, stormwater, or wastewater pump station or treatment plant. In sanitary engineering, the peak hydraulic demand is expressed in flow per hour or a shorter period of time.

**peak design flowrate** The peak discharge that a facility is designed to accommodate.

**peak discharge** or **peak flow** (1) The maximum instantaneous, hourly, or other flow to a treatment plant, pumping station, or other facility. In stream flow studies, the peak discharge corresponds to the maximum water surface elevation during a given storm event and at a given location. It is a widely used parameter in the hydraulic design of pipes, pumping stations, treatment plants, inlets, culverts, detention facilities, etc. *See* distribution ratio. (2) Peak flow is one of the four parameters of Snyder's method of synthetic hydrograph. *See also* lag time, standard duration of rainfall, and time base. Other parameters used in runoff analysis include antecedent precipitation index (API), attenuation constant, plotting time width, time of concentration, and time of equilibrium.

**peak flow** Same as peak discharge.

**peak-flow charge/rate** A special charge, surcharge, or rate levied by a water utility for seasonal peak demands, particularly on wholesale customers to encourage them to reduce their peak demands via equalization storage. Also called demand charge/rate.

**peak flow study** A study to determine the likelihood of sewer overflows and the cost-effective measures to prevent or alleviate them, including the upgrading of collection and transmission facilities.

**peaking costs** Costs incurred by a utility to meet peak demand.

**peaking factor** The ratio of peak (hourly, daily, etc.) flow rates to long-term average flows. Also called sustained peaking factor. *See also* peak rate factor, sustained loadings.

**peak load** Same as peak demand.

**peak-load pricing** A rate structure that charges utility customers higher prices for use during peak hours.

**peakpoint** In water chlorination the hump on the breakpoint curve, i.e., the point at which the chlorine residual curve changes from a trend of increasing residual to a trend of decreasing residual. *See* breakpoint.

**peak rate factor** A parameter used in the SCS dimensionless unit hydrograph method to relate the peak discharge to the drainage area, the runoff depth, and the time to peak.

**Pearlcomb®** A fine-bubble diffuser manufactured by the FMC Corp.

**Pearth** An apparatus manufactured by Envirex, Inc. for mixing gases from anaerobic digesters.

**peat** A dark brown or black material formed by the partial decomposition and disintegration of vegetation in marshes and other wet places (e.g., sedges and mosses) and having a high moisture content. Peat soil is an organic soil resulting from the accumulation of peat; it contains less than 20% of mineral material. *See also* muck soil.

**peat filter** A wastewater treatment device used in onsite disposal systems and consisting of a prefabricated, fiberglass or high-density polyethylene container of peat material (e.g., sphagnum or coarse fabric).

**peat soil** *See* peat.

**pebble bed flocculator** A vertical tank used for flocculation and filled with gravel media to direct flow through a large number of void openings.

**pebble count method** The determination of the particle size distribution of sediment in the surface layer of a stream based on the weight or length of at least 100 particles. A characteristic of the distribution is the gradation coefficient ($G$) calculated as follows:

$$G = 0.5(D_{50}/D_{16} + D_{84}/D_{50}) \qquad \text{(P-11)}$$

where $D_i$ is the size of the particle such that $i$ percent of the particles are finer.

**Peclet criterion** A limit set on the Peclet number to control numerical dispersion. To ensure numerical stability and minimize dispersion in a groundwater transport model, e.g., a Peclet number less than 2 is recommended.

**Peclet number ($P_e$)** A dimensionless number that indicates when advection dominates ($P_e \gg 1$) or turbulence mixing dominates ($P_e \ll 1$). The Peclet number is the ratio of the advective flux to the dispersive flux, i.e., the product of a characteristic length ($L_s$) and mean velocity ($V$) in the direction of that length, divided by an appropriate mixing or diffusion coefficient ($\varepsilon$). The Peclet number is sometimes used to control the stability of numerical solutions in hydraulic modeling. *See* Peclet criterion. ($\ll$ and $\gg$ mean, respectively, much less than and much more than.)

$$P_e = L_s \cdot V/\varepsilon \qquad \text{(P-12)}$$

It is the reciprocal of the dispersion number, one of the parameters in the dimensionless variance, which is used in defining the density function for the residence time of disinfectants.

***Pediastrum*** A genus of green algae that can contribute to the formation of trihalomethanes during water chlorination.

***Pediculus humanus*** The human louse, the species of lice that transmit *Rickettsia prowazeki,* the pathogen of typhus.

**pedogenic refractory organic matter (PROM)** Refractory organic matter that comes from soils. *See also* aquagenic organic matter, natural organic matter.

**pedology** The study of soils; soil science.

**peds** A secondary structure of surface soils formed by sand, silt and clay particles with microbial cells and their extracellular metabolites, which act as a glue. *See* soil aggregate for detail.

**PEL** Acronym of permissible exposure limit.

**pelagic** Of or pertaining to organisms (plankton, nekton) living or growing in open seas or the surface of the ocean.

**pelagic zone** The water column or planktonic habitat of the oceans, including the epipelagic or photic zone, the mesopelagic zone, the abyssopelagic habitat, and the bathypelagic zone or benthos. *See* marine profile for related terms.

**Pelican** An apparatus manufactured by Copa Group for installation on clarifier walls to remove scum.

**Pelldry** Pellets produced by Sheldahl Industrial Absorbents for the absorption of liquids.

**Pelletech®** An apparatus manufactured by Wheelabrator Clean Water Systems, Inc. used as a sludge dryer and pelletizer.

**pelletization** A sludge handling method using heat drying to produce sludge particles of 1 to 4 mm in diameter that can be used as a fertilizer. *See also* heat drying–pelletization. Pelletization is also used prior to storage to reduce the dust content and associated fire hazards.

**pellet reactor** An upflow water softening device that consists of an inverted conical tank that contains a fluidized bed of fine sand used for the crystallization of calcium carbonate ($CaCO_3$). It is a common type of pellet softener developed and used in Europe.

**pellet softener** A fluidized bed of granular material used in water softening for the crystallization of calcium carbonate ($CaCO_3$).

**pendimethalin** A pesticide; guideline value recommended by the WHO for drinking water: 0.02 mg/L.

**penetration** A parameter that represents the percent amount ($P$, %) of a hazardous compound that is not destroyed by a thermal process. It is determined as follows from the process destruction removal efficiency or DRE (Freeman, 1998):

$$P = 1 - \mathrm{DRE}/100 \qquad \text{(P-13)}$$

**Penfield®** Products of the U.S. Filter Corp. for use in water treatment.

***Penicillium*** A waterborne fungus, pathogenic to humans. It has been identified in unchlorinated groundwater as well as surface water and service mains. It also grows in trickling filters where it can clog ventilation systems.

**Penman–Monteith method** A procedure to determine the reference evapotranspiration for use in the design and scheduling of irrigation systems (e.g., land application of wastewater effluents), based on a mathematical model that incorporates weather data.

**pennywort** Any of several plants having round leaves, generally rooted but sometimes forming hydroponic rafts across water bodies; including the species *Hydrocotyle ranunculoides* and *Hydrocotyle umbellate,* floating aquatic plants used in constructed wetlands to treat wastewater. *See also* water hyacinth, duckweed.

**Penro** A reverse osmosis unit manufactured by Penfield Liquid Treatment Systems.

**penta** Same as pentachlorophenol or PCP.

**pentachloroacetone** An organic compound used in industry, sometimes formed as a disinfection by-product. It causes mutations in strains of *Salmonella.*

**pentachlorophenol ($C_6Cl_5OH$)** A synthetic organic compound, slightly soluble in water, used as a wood preservative, pesticide, and defoliant. The human body easily absorbs and excretes pentachlorophenol, which may damage the organs and central nervous system. Regulated by the USEPA as a drinking water contaminant, but not as a disinfection by-product: MCL = 0.001 mg/L and MCLG = 0. Also called penta and abbreviated PCP.

**pentanal** An aldehyde on the USEPA list for quarterly monitoring.

**pentanoic acid ($C_4H_9COOH$)** The proper name of one of the simplest carboxylic acids (valeric acid).

**PentaPure®** A disinfecting resin produced by WTC Industries, Inc.

**pentasodium tripolyphosphate** Same as sodium polyphosphate.

**pentosans [$(C_5H_8O_4)_x$]** A class of polysaccharides found with cellulose in plants, humus, etc.

**pen waste** Highly concentrated organic waste containing pig excreta, hair, and decayed food.

**pE–pH diagram** *See* pε–pH diagram.

**pE–pH stability field diagram** Same as pE–pH diagram. *See* pε–pH diagram.

**peptide** A compound of two or more amino acids formed by condensing the amino (—$NH_2$) group of one acid and the carboxyl (—COOH) group of another into —NH—CO—.

**peptization** (1) The phenomenon observed during coagulation using hydrolyzing metal salts and resulting in the stabilization and dispersion of microcrystal particles that have a net negative surface charge after adsorbing a significant amount of natural organic matter. It is also called steric stabilization. *See also* sweep flocculation (or enmeshment), heterocoagulation. (2) Interference of organic matter with the removal of iron and manganese through a combination of (a) complexation without complete neutralization of molecular charge, (b) formation of soluble complexes, and (c) reduction of oxidized metals.

***Peptococcus*** A nonsporulating, anaerobic Gram-positive coccus found in large numbers in feces and considered as a possible candidate for the indication of fecal contamination.

***Peptococcus anaerobus*** A species commonly found in anaerobic digesters.

***Peptostreptococcus spp.*** A nonsporulating, anaerobic Gram-positive coccus found in large numbers in feces and considered as a possible candidate for the indication of fecal contamination.

**per** A prefix indicating the maximum amount of an element in an inorganic acid or salt, e.g., percarbonic acid ($H_2C_2O_5$), persulfuric acid ($H_2S_2O_8$), potassium permanganate ($KMnO_4$). *See also* perchlorate, perchloric, peroxide, peroxone, peroxyacetic.

**peracetic acid ($CH_3COOOH$ or $C_2H_4O_3$)** Abbreviation: PAA. A product of the reaction of acetic acid ($CH_3COOH$) and hydrogen peroxide ($H_2O_2$) in water; it can be used as a chemical disinfectant. *See* acetyl hydroxide for detail.

**peracetic acid (PAA) disinfection** The use of a peracetic acid solution as a biocide, due to the presence of the undissociated PAA and of hydrogen peroxide.

**perborax** *See* sodium perborate.

**PERC** Abbreviation of perchloroethylene.

**percent actual chlorine** *See* available chlorine.

**percentage reduction** *See* percent reduction.

**percent available chlorine** *See* available chlorine.

**percent by mass (% by mass)** A unit that expresses analytical results, equal to 100 times the mass of solute divided by the combined mass of solute and solvent.

**percent by volume (% by vol)** A unit that expresses analytical results, equal to 100 times the volume of solute divided by the total volume of solution.

**percent capture** A measure of the performance of a centrifuge (used in sludge thickening and dewatering), defined as the percentage of thickened dry solids with respect to feed solids:

$$C, \% = 100\,[1 - C_R\,(C_C - C_S)/C_S\,(C_C - C_R)] \quad \text{(P-14)}$$

where $C$ is the percent capture; $C_R$, $C_C$, and $C_S$ represent respectively, in mg/L or %, the concentration of solids in the reject wastewater (centrate), the cake, and the sludge feed. *See also* recovery (3).

**percent reduction** The percentage of material removed by water or wastewater treatment, equal to 100 times the ratio of the change in concentration to the initial concentration. Also called percentage reduction. *See also* percent removal.

**percent removal** A percentage expression of the removal efficiency across a treatment plant for a given pollutant parameter, as determined from the 30-day average values of the raw wastewater influent pollutant concentrations to the facility and the 30-day average values of the effluent pollutant concentrations for a given time period (EPA-40CFR133.101-j). *See also* percent reduction.

**percent saturation** The amount of a substance that is dissolved in a solution compared to the amount that could be dissolved the solution, expressed as a percent.

**percent volatile** The percentage of a chemical that will evaporate at ordinary temperatures. A high volatile percentage may mean there is more risk of explosion, or that dangerous fumes can be released. Evaporation rates are a better measure of the danger than the percent volatile.

**percent water extractable phosphorus (PWEP)** The percentage of total phosphorus (P) in fertilizers or biosolids that is soluble in a water extract. Biosolids have less soluble phosphorus than manures or fertilizers. For biosolids, PWEP depends on the solids treatment process (biological, composting, or heat drying).

**perchlorate ($ClO_4^-$)** A salt or ester of perchloric acid; a nonvolatile, inorganic chemical, highly soluble in water; used in the production of solid rocket fuel, highway safety flares, explosives, missiles, fireworks, etc. The primary form of this contaminant in water sources is ammonium perchlorate. Perchlorate is considered as both an endocrine disruptor and a xenobiotic. Even at low concentrations in drinking water, perchlorate is suspected of causing cancer and other afflictions of the thyroid; it has been added to the Candidate Contaminant List.

**perchlorate removal** Advanced water treatment techniques such as membrane processes are the most effective means of perchlorate removal, but they are expensive. Ion exchange, for example, is feasible but the disposal of its waste brine stream is problematic. Perchlorate can also be converted to chloride and oxygen through a fixed-bed biological treatment process, which also eliminates such other contaminants as nitrate, organic matter, and taste and odor compounds. *See also* electrochemical generation, membrane biofilm reactor.

**perchloric acid ($HClO_4$)** A strong, colorless, corrosive, highly oxidized acid of chlorine; a powerful oxidant when hot; used as reagent.

**perchloroethylene ($Cl_2C{=}CCl_2$)** Acronym: PCE. A volatile organic chemical used as a solvent and in other industrial applications, and formerly as a septic tank cleaner. It is classified as a probable human carcinogen. Same as tetrachloroethylene.

**perchloromethane** Same as carbon tetrachloride.

**Perchloron** A brand of calcium hypochlorite.

**Percol** A polyelectrolyte produced by Allied Colloids, Inc. for use in solid–liquid separation.

**percolate** The liquid that passes through a porous body, e.g., the water that percolates through the soil profile in a slow-rate wastewater treatment process and reaches groundwater or surface water, or is recovered.

**percolating filter** Same as trickling filter. *See also* aerated contact bed (or contact aerator or submerged contact aerator), bacteria bed, intermittent sand filter, nidus rack, percolating filter.

**percolating water** Water that passes through rocks or soil under the force of gravity, without a definite channel.

**percolation** (1) The movement of water downward and radially through subsurface soil layers, continuing downward to the groundwater. Can also involve upward movement of the water or the slow seepage of water through a filter (i.e., filtration). (2) Water lost from an unlined conduit through the sides or the bottom.

**percolation field** The area over which septic tank or other effluents are spread for further treatment and disposal. *See* soil absorption field, subsurface disposal field.

**percolation test** A test to determine the percolation rate of water through soil by measuring the drop in water level in a test hole over a period of time. To evaluate the percolation capacity of a proposed subsurface disposal field, a 100-mm hole at least 500-mm deep, with 50 mm of coarse sand or fine gravel at the bottom, is filled with water to a depth of 300 mm. *See* Figure P-03. The drop in water level is measured while observing some detailed procedures. The effluent flow ($Q$, liters per square meter per day) that can be applied to the field is calculated from the following formula, as a function of the time ($t$, min) it takes the water to fall 25 mm:

$$Q = 204/t^{0.5} \qquad \text{(P-15)}$$

The percolation test is often used to determine the required size of the soil absorption system. Also called perc test. *See also* soil profile examination, soil absorption test, shallow trench pump-in test.

**Perc-Rite®** A filtration apparatus manufactured by Waste Water Systems, Inc.

**perc test** Same as percolation test.

**perfected water right** A water right indicating beneficial uses by an applicant under permit; usually irrevocable unless voluntarily canceled or forfeited after several consecutive years of nonuse. *See also* abandoned water right, certificate of water right, forfeited water right.

**perfect fluid** A fictitious fluid that would be incompressible and inviscid (i.e., without viscosity), and would have uniform density and no surface tension. The convenient assumption of perfect flow produces results that are fairly accurate for actual fluids. Also called ideal fluid.

**perfect gas** A gas whose molecules are subject only to the forces of collision with one another and with the walls of its container. *See* ideal gas for more detail.

**perfect gas law** The product of the pressure ($P$) and volume ($V$) of one gram molecule of a perfect

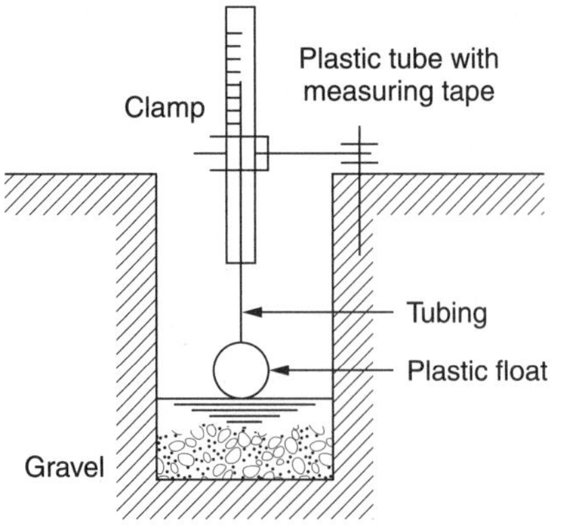


**Figure P-03.** Percolator test.

gas equals the product of its absolute temperature ($T$, 0°K) and the universal gas constant ($R$). *See* ideal gas law for more detail.

**perfluorooctanoate (PFOA)** A substance that is used to manufacture Teflon products and that is suspected of causing adverse health effects in drinking water. Adsorption on granular activated carbon is a possible treatment technology for the removal of PFOA from groundwater supplies. Also called C8.

**perforated baffle wall** A wall that separates water or wastewater treatment units or unit compartments but allows flow through holes in the wall; it promotes uniform flow distribution and reduces short-circuiting.

**perforated casing** A well casing with holes to permit the passage of water.

**perforated-casing well** A well with a perforated casing.

**perforated launder** A trough in a sedimentation basin with holes below the water surface to allow uniform flow.

**perforated pipe** *See* drain tile.

**perforated plate** A plate with holes to allow uniform fluid flow and dissipate energy.

**perforated plate-type jet scrubber** An air pollution control device that uses water or another liquid to remove up to 90% of particulates, fumes, dusts, mists, and vapors from gas streams; for example, in sludge incinerators and fluidized bed reactors. *See also* cyclone scrubber, ejector–venturi scrubber, spray tower, and venturi scrubber.

**performance audit** A procedure to analyze blind samples, the contents of which are known by the USEPA, simultaneously with the analysis of performance test samples in order to provide a measure of test quality data (EPA-40CFR63.2)

**performance damages** Losses sustained as a result of a water or wastewater facility failing to work as specified, including, e.g., the costs to install an adequate system. Also called consequential damages. *See also* liquidated or delay damages, performance specification, prescriptive specification.

**performance evaluation sample** A reference sample provided to a laboratory for the purpose of demonstrating that the laboratory can successfully analyze the sample within specified limits of performance. The true value of the concentration of the reference material is unknown to the laboratory at the time of the analysis.

**performance–loading relationship** An expression of the performance of a biological waste treatment plant or process as a function of loading intensity. The following equation has been proposed for domestic wastewater (Fair et al., 1971):

$$E = 100/(1 + mX^n) \qquad \text{(P-16)}$$

where $E$ is performance as percentage efficiency, $X$ is the loading intensity, $m$ is a proportionality factor depending on the measurement units, and $n$ is a coefficient that depends on the specific process (e.g., 0.03 for activated sludge). The loading intensity may be expressed in pounds of BOD per 1000 square feet per hour of aeration, (b) pounds of BOD per acre-ft, (c) pounds of BOD per 1000 pounds of mixed liquor suspended solids, etc.

**performance ratio** A measure of the performance of an evaporator equal to the weight of distillate produced divided by the energy consumed, e.g., kg/joule.

**performance specification** One of two general types of specifications used in the procurement of goods and services for such municipal projects as water and wastewater facilities. It defines the input conditions and the desired output (i.e., performance). Sometimes, prequalification of manufacturers and/or suppliers is used to reduce risks. *See also* prescriptive specification, liquidated or delay damages, consequential or performance damages.

**performance standards** (1) Regulatory requirements limiting the concentrations of designated organic compounds, particulate matter, and hydrogen chloride in emissions from incinerators. (2) Operating standards established by the USEPA for various permitted pollution control systems, asbestos inspections, and various program operations and maintenance requirements. (3) Specific goals established for replacement wetlands to reach maturity before wetland restoration credits are granted.

**performance–time relationship** An expression of the performance of a biological waste treatment plant or process as a function of time, e.g., removal of BOD vs. hours of aeration by an activated-sludge unit.

**pergelisol** Same as permafrost.

**perikinetic flocculation** The aggregation of solid particles into larger but still microscopic particles, brought about by Brownian movement and contributing to enhance the agglomeration and removal of particles

**perikinetic motion** *See* conjunction kinetics.

**periodic table** An arrangement of all chemical elements according to their symbols, atomic weights, and atomic numbers, in a manner that reflects the recurring behavior of the elements. *See* Figure P-04.

**peripheral-feed clarifier** *See* peripheral-feed settling tank.

**peripheral-feed settling tank** A circular sedimentation basin in which the flow enters around the perimeter (instead of at the center) through a surface launder. This type of tank used in primary and secondary sedimentation includes (a) an annular space formed by a circular baffle into which the influent wastewater discharges and (b) a revolving mechanism for sludge transport and removal. The bottom of the tank slopes toward the central hopper. Also called peripheral-feed clarifier, rim-feed settling tank. *See* Figure P-05. *See also* center-feed settling tank.

**peripheral flow** The flow of a liquid parallel to the periphery of a circular structure. Also called circumferential flow.

**peripheral-flow clarifier** Same as peripheral-feed clarifier.

**peripheral pump** A pump with an impeller that recirculates the fluid through rotating vanes.

**peripheral weir** The outlet weir of a circular settling tank for effluent discharge.

**periphyton** Microscopic plants and animals (e.g., diatoms, algae, and water moss) that are firmly attached to solid surfaces under water such as rocks, logs, pilings, and other structures. Also called aufwuchs.

**peristaltic hose pump** *See* peristaltic pump.

**peristaltic pump** A self-priming, positive displacement pump that uses a rotary head to alternately compress and relax a resilient hose through which the fluid is delivered. This simple pump, which lacks seals, valves, or bearings, is easy to maintain but hose replacement is expensive. It is used mainly in the industrial sector but also to a limited extent in handling wastewater sludges. Also called peristaltic hose pump.

**perlite** A glass of volcanic origin with concentric fractures, used as a medium in precoat filters and for growing plants. It consists of a low-melting-point combination of fernite (a ferric oxide compound, $NaFeO_2$) and cementite (an iron carbide, $Fe_4C$). It contains 2–3 percent of water and expands into a mass of glass bubbles when heated. It serves also as bulking material to maintain the porosity of compost and peat biofilters.

**Perma-buoy** A fiberglass flight manufactured by Jeffrey Division/Indresco for chain-and-flight sludge collectors. *See also* Permalife.

**PermaCare** Chemical products of Houseman, Ltd for reverse osmosis units.

**PermaChem** Chemical products of Hach Co.

**permafrost** The permanently frozen subsurface soil layer in arctic and subarctic regions that experiences freezing temperatures (below 32°F or 0°C) for several years. Also called pergelisol.

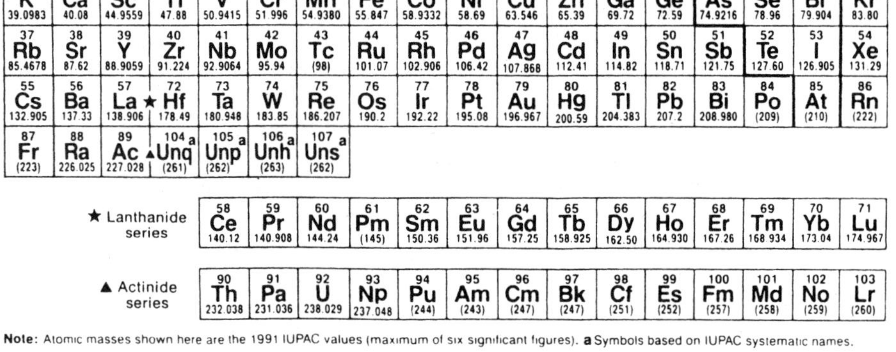


**Figure P-04.** Periodic table.

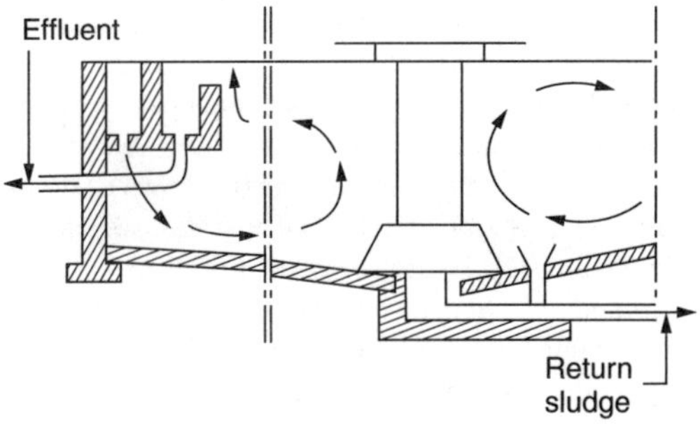


**Figure P-05.** Peripheral-feed settling tank.

**PermaGlas** Coating manufactured by A. O. Smith Harvestore Products, Inc. for storage tanks.

**Permaklip** A seam manufactured by Tetko, Inc. for filter press belts.

**Permalife** Components manufactured by Jeffrey Division/Indresco for chain-and-flight sludge collectors. *See also* Perma-buoy.

**Permalog®** A series of loggers manufactured by Fluid Conservation Systems for use in in-house leak-detection surveys.

**permanent hardness** The portion of total hardness that remains after boiling; it is caused mostly by the chlorides, nitrates, and sulfates of calcium and magnesium. Also called noncarbonate hardness because they are approximately the same. It may be determined as the difference between total hardness and alkalinity (which represents carbonate hardness), all these concentrations being expressed as calcium carbonate. *See also* temporary or carbonate hardness.

**permanent pool** The portion of a pond, tank, or other detention/retention facility that is considered dead storage, as opposed to the active storage zone, which releases water during storm events. Extended-detention dry ponds have only an active storage zone, whereas a wet pond has both an active storage zone and a permanent pool. Also called permanent volume. *See* reservoir storage.

**permanent volume** Same as permanent pool.

**permanganate** (1) A salt of permanganic acid ($HMnO_4$); the latter occurs only in solution. (2) Short for potassium permanganate ($KMnO_4$), a strong oxidant with germicidal properties; used in iron and manganese removal, for taste and odor control, and in the regeneration of greensand.

**permanganate ion ($MnO_4^-$)** *See* permanganate (1).

**permanganate value (PV)** The amount of oxygen absorbed from a standard potassium permanganate ($KMnO_4$) solution in 4 hours at 27°C; used as a rapid and approximate measure of the oxygen demand of an effluent.

**Permasep®** Products of E. I. DuPont de Nemours, Inc. for use in reverse osmosis units.

**permeability** Property of a soil or other material that permits the movement of water through it; a qualitative description of the relative ease with which rock, soil, or sediment will transmit a fluid (liquid or gas); the capacity of a rock to transmit water under a hydraulic gradient. Also, the rate at which liquids pass through soil or other material in a specified direction. Same as perviousness. Often used synonymously for hydraulic conductivity or coefficient of permeability. *See* intrinsic permeability, porosity.

**permeability coefficient** A constant of proportionality used in Darcy's law; see hydraulic conductivity.

**permeate** (1) To penetrate and pass through, as water permeates soil and other porous materials. (2) The liquid product (water) that passes through a semipermeable membrane in a pressure-driven process; it has a lower concentration of total dissolved solids than the feed stream. Same as filtrate in a low-pressure system. Also called product flow stream, permeating stream, product stream, or, simply, product. *See also* feed stream (or feedwater) and concentrate (or reject, retained phase, retentate, waste stream).

**permeate flux** The flow of permeate per unit active membrane area, expressed, for example, in gallons per day per square foot. *See also* permeation flux.

**permeate staging** An arrangement of pressure-driven membrane modules in series so that the permeate from one stage is further treated in the next stage.

**permeate stream** Same as permeate (2).

**permeating stream** Same as permeate (2).

**permeation** The phenomenon of water passing through a porous medium (e.g., a membrane). *See also* dialysis, leaching, osmosis, percolation.

**permeation flux** One of two experimental parameters used to evaluate the performance of pervaporation membranes. For a given membrane material, it is the amount of a substance or component permeating per unit membrane area per unit time; it is proportional to the difference in concentrations between the liquid phase and the bulk vapor phase. *See also* permeate flux and selectivity.

**permeator** A pressure vessel that contains hollow-fiber, semipermeable membranes for water treatment.

**PermeOx®** Solid peroxygen produced by FMC Corp.

**permethrin** A pesticide with a WHO-recommended guideline value of 0.02 mg/L for drinking water.

**per mille (‰)** Per thousand or $10^{-3}$; used to express salinity and chlorinity; e.g., a salinity of 12.56% = 125.6‰

**permissible dose** The dose of a chemical that may be received by an individual without the expectation of a significantly harmful result. *See also* NOAEL.

**permissible exposure limit (PEL)** An exposure limit established by the Occupational Safety and Health Administration for industrial chemicals in the workplace. For each chemical it is the maximum concentration to which a worker may be exposed for 30 years without experiencing adverse health effects. *See also* threshold limit value and recommended exposure limit.

**permissible velocity** The maximum velocity of water that will not cause damage when passing through a hydraulic structure.

**permit** An authorization, license, or equivalent control document issued by the USEPA or an approved state agency to implement the requirements of an environmental regulation; e.g., a permit to operate a wastewater treatment plant or to operate a facility that may generate harmful emissions.

**permitted capacity** The maximum average flow that a treatment plant is allowed to process as determined on its National Pollutant Discharge Elimination System (NPDES) permit.

**permittivity** The ratio of the flux density of an electric field in a dielectric to its flux density in a vacuum. *See* dielectric constant for detail.

**Permofilter** A horizontal pressure filter manufactured by the U.S. Filter Corp.

**permonosulfuric acid** Same as persulfuric acid.

**permselective membrane** A membrane without measurable pores but permeable to small molecules or to molecules of a special nature. It is used in such processes as dialysis and reverse osmosis for the separation of constituents (e.g., colloids) in a fluid based on differences in one or more characteristics, e.g., diffusion rate, solubility, electrical charge, or size and shape. Also called semipermeable membrane.

**Permupak** A package water treatment plant manufactured by the U.S. Filter Corp.

**PermuRO** A reverse osmosis unit manufactured by the U.S. Filter Corp.

**Permutit®** Water treatment products of the U.S. Filter Corp.

**peroxidase** An enzyme that can catalyze the decomposition of peroxides.

**peroxide** (1) A compound that contains two atoms of oxygen linked as a bivalent —O—O— (or —$O_2$—) group; e.g., sodium peroxide ($Na_2O_2$) or dimethyl peroxide ($C_3H_6O_2$) Peroxides are strong oxidants. (2) Hydrogen peroxide ($H_2O_2$). (3) A compound that contains a large amount of oxygen.

**peroxisome** An organelle that contains peroxidase and other enzymes that can oxidize fatty acids.

**peroxisome proliferator** A chemical compound that can increase the number of peroxisomes in tissues.

**peroxone** (1) A combination of hydrogen peroxide ($H_2O_2$) and ozone ($O_3$), used for disinfection or taste and odor control in water and wastewater treatment. (2) The peroxone process. *See also* ozone–hydrogen peroxide.

**peroxone process** An advanced oxidation process that produces hydroxyl radicals ($^{\cdot}OH$ or $HO^{\cdot}$) using peroxone (a combination of hydrogen peroxide, $H_2O_2$, and ozone, $O_3$) in two steps. Ozone is dissolved in the solution and hydrogen peroxide is added to accelerate the decomposition of ozone and increase the formation of hydroxyl radicals according the following overall reaction

$$H_2O_2 + 2\ O_3 \rightarrow HO^{\cdot} + HO^{\cdot} + 2\ O_3 \quad \text{(P-17)}$$

The hydroxyl radical, a strong oxidant, is then used for disinfection or for taste and odor control. Peroxone has been successfully used to oxidize such compounds as trichloroethylene, perchloroethylene, geosmin, and MIB. It is also effective in the control of coliforms.

**Perox-Pure®** An apparatus manufactured by Peroxidation Systems, Inc. for the production of hydrogen peroxide using ultraviolet light.

**Perox-serv™** Services offered by Peroxidation Systems, Inc. for odor control.

**Perox-stor™** Services offered by Peroxidation Systems, Inc. to users of hydrogen peroxide.

**peroxyacetic acid** A product of the reaction of acetic acid and hydrogen peroxide in water; it can be used as a chemical disinfectant. *See* acetyl hydroxide for detail.

**peroxyacetylnitrate (PAN)** A secondary air pollutant found in photochemical smog, formed by the combination of nitrogen oxides and reactive hydrocarbons.

**peroxydisulfuric acid** Same as persulfuric acid.

**peroxymonosulfuric acid** Same as persulfuric acid.

**peroxysulfuric acid** Same as persulfuric acid.

**Perpac** Equipment manufactured by Vulcan Industries, Inc. for the treatment of surface water.

**persistence** (1) The resistance to degradation as measured by the period of time required for complete decomposition of a material. Once introduced into the environment, a compound may persist for less than a second or indefinitely. (2) The tendency of a material to resist change in the environment. (3) The period of time that a pathogen can survive in the environment outside a human host; *see also* latency.

**persistent chemical** A chemical substance that resists removal or transformation in water or soil, e.g., some metals (arsenic, copper, lead), pesticides, hard detergents, and radionuclides.

**persistent excreted infection** An infection caused by an excreted pathogen that survives outside its human host longer than the average pathogen, e.g., the persistence of *Taenia saginatta* is 9 months, whereas that of *Schistosoma mansoni* is 2 days.

**persistent organic pollutant (POP)** An organic compound (a) having a long half-life in the environment and high vapor pressure and (b) undergoing slow physical, chemical, and biological degradation; e.g., aldrin, chlordane, DDT, dieldrin, endrin, furans, heptachlor, polychlorinated biphenyls.

**persistent pesticide** A pesticide that does not break down chemically or that breaks down very slowly and remains in the environment after a growing season.

**personnel access opening** An opening in a vault, caisson, pipeline, etc. to allow access by inspection and maintenance personnel. *See also* manhole.

**person-to-person spread** The transmission of an infection directly from one person to another.

**persulfate** A salt of persulfuric acid ($H_2SO_5$ or $H_2S_2O_8$), e.g., potassium persulfate ($K_2SO_5$ or $K_2S_2O_8$).

**persulfate method** A technique used in the laboratory to determine the concentration of manganese ($Mn^{2+}$) in a sample; the metal is converted to the permanganate ion ($MnO_4^-$) by persulfate ($S_2O_8^{2-}$).

**persulfuric acid** (1) A white, crystalline solid ($H_2SO_5$) used as an oxidizing agent. Also called Caro's acid, permonosulfuric acid, peroxymonosulfuric acid, peroxysulfuric acid. (2) A white, crystalline solid ($H_2S_2O_8$) used to prepare hydrogen peroxide ($H_2O_2$). Also called peroxydisulfuric acid.

**pervaporation (PV)** An industrial treatment process that uses semipermeable membranes to remove volatile organic compounds from aqueous solutions or separate mixtures of dissolved solvents; the analyte is transferred from the solution phase on one side of the membrane into the vapor phase on the other side. It produces a permeate vapor under pressure; the partial pressure difference between solution feed and permeate drives the transport across the membrane. *See also* permeation flux, selectivity, concentration polarization, Lévêque's correlation.

**pervious** Same as permeable. A pervious surface allows water to infiltrate to the subsurface.

**pest** An insect, rodent, nematode, fungus, weed, or other form of terrestrial or aquatic plant or animal life that is injurious to health or the environment.

**pesticide** Any substance or chemical designed or formulated to kill or control weeds or animal pests. Also any substance intended for use as a plant regulator, defoliant, or desiccant. (EPA-40CFR710.2-b). Pesticides in water supplies come from agricultural and household uses. They degrade slowly and are not easily removed by conventional water or wastewater treatment processes, but adsorption on activated carbon is an effective method. The USEPA regulates pesticides, herbicides, and PCBs in drinking water under the Phase II Contaminants (see Table P-02). Pesticides can cause a wide range of health effects, e.g., birth defects, cancer risks, diarrheal symptoms, as well as damages to the immune system, kidney, liver, nervous system, and reproductive system. *See also* algicide, fungicide, herbicide, insecticide, molluscicide, rodenticide, toxic substance.

**pesticide chemical** Any substance that, alone or in chemical combination with or in formulation with one or more other substances, is an econom-

**Table P-02.** Pesticides, herbicides, and PCBs regulated in drinking water

| | | |
|---|---|---|
| Acrylamide | Alachlor | Aldicarb |
| Aldicarb sulfone | Aldicarb sulfoxide | Atrazine |
| Carbofuran | Chlordane | 2,4-D (Formula 40) |
| Dibromochloropropane | Epichlorohydrin | Ethylenedibromide |
| Heptachlor | Heptachlor epoxide | Lindane |
| Metoxychlor | PCBs | Pentachlorophenol |
| Toxaphine | 2,4,5-TP (silvex) | |

ic poison within the meaning of the Federal Insecticide, Fungicide, and Rodenticide Act. The term includes any substance that is an active ingredient, intentionally added inert ingredient, or impurity of such a pesticide (EPA-40CFR177.3 & 180.1-k).

**pesticide product** A pesticide in the particular form it is, or is intended to be, distributed or sold. The term includes any physical apparatus used to deliver or apply the pesticide if distributed or sold with the pesticide (EPA-40CFR152.3-t).

**pesticide related waste** All pesticide-containing wastes or by-products that are produced in the manufacturing or processing of a pesticide and that are to be discarded, but that are not ordinarily a part of or contained within an industrial waste stream discharged into a sewer or the waters of a state (EPA-40CFR165.1-q).

**pesticide residue** A residue of a pesticide chemical or any metabolite or degradation product of a pesticide chemical (EPA-40CFR177.3).

**pesticide tolerance** The amount of pesticide residue allowed by law to remain in or on a harvested crop. The USEPA sets these levels well below the point where the compounds might be harmful to consumers.

**pesticide waste** Wastewater generated during the production of chemicals used to make pesticides, e.g., 2,4-D (dichlorophenoxyacetic acid), DCP (dichlorophenol), and DDT. This waste is acidic, high in organic matter, and toxic to fish. It may be treated by dilution, storage, activated carbon adsorption, or alkaline chlorination.

**PET™** A demineralization apparatus manufactured by U.S. Filter Corp.

**petalite** An important lithium ore.

**petri dish** A shallow, covered glass or plastic dish, containing an agar medium and used in laboratories to grow microbial cultures.

***Petriellidium boydii*** A species of fungi that have been identified as regrowth organisms in water distribution systems.

**petri factor** A correction factor applied to incident irradiance to obtain the average UV irradiance used in bench-scale experiments. It accounts for the nonuniform distribution of light over the surface area of the testing apparatus.

**petrocalcic horizon** A hard deposit of calcium carbonate ($CaCO_3$), gravel, or sand in the subsoil of semiarid regions. Also called caliche. *See* hardpan for other strongly compacted soil layers underlying soft materials.

**petrochemical** A product or compound resulting from the processing of petroleum, coal, natural gas, and miscellaneous sources of carbon; e.g., gasoline, kerosene, petrolatum. Products from the raw materials include first-generation petrochemicals (alkenes, aromatics, butylenes, etc.), basic intermediates (ethylene oxide, ethyl chloride, etc.), and final products such as synthetic detergent bases, solvents, fuel additives, plastics, resins, and synthetic fiber bases.

**petrochemical industry waste** Wastewater generated during the production of petrochemicals has varying characteristics because of the diverse range of products. Generally, such waste is high in COD, TDS, inhibitory compounds, and heavy metals. The industry practices chemicals recovery, water reuse, and wastewater recycling in cooling towers. Waste treatment methods include equalization, pH adjustment, flocculation, coagulation, primary clarification, flotation, trickling filtration, activated sludge, oxidation ponds, activated carbon adsorption, and anaerobic digestion. *See also* oil refinery waste.

**petrochemical waste** *See* petrochemical industry waste.

**Petroff–Hauser counting chamber** An instrument used in the direct microscopic count of bacteria. *See also* acridine orange, electronic particle counting.

**Petro-Flex®** A portable tank manufactured by Aero Tec Laboratories, Inc. for holding petrochemicals.

**petroleum derivative** A substance derived from petroleum, such as a gasoline breakdown product in groundwater.

**Petrolux** A ceramic membrane filter manufactured by the U.S. Filter Corp.

**Petro-Pak** Specialized media produced by McTigue Industries, Inc. for oil removal.

**Petro-Xtractor™** An apparatus manufactured by Abanaki Corp. to skim oil from water wells.

**PEX** *See* cross-linked polyethylene.

**pezodialysis** A desalination process that separates salt by making it pass through a permeable membrane.

**PF** Acronym of (a) phenol–formaldehyde; (b) plug flow; (c) polarization factor; (d) potency factor.

**PF basin** A basin that operates under plug flow conditions.

**PFD** Acronym of process flow diagram.

***Pfiesteria piscidia*** A species of dinoflagellates found in algal blooms, very toxic to fish life; it produces algal toxins that can cause neurological damages and skin lesions in humans.

***P. fluorescens*** *See Pseudomonas fluorescens.*

**PFOA** Acronym of perfluorooctanoate.

**PF resin** Phenol–formaldehyde resin; an adsorbent resin that has no functional groups other than those in the matrix.

**PFRP** Acronym of process to further reduce pathogens.

**PFU** Acronym of plaque-forming unit.

**p-function** A convenient notation used to express concentrations in aqueous solutions:

$$pX = -\log [X] = \log 1/[X] \quad \text{(P-18)}$$

**pH** The negative base-10 logarithm of the hydrogen ion concentration (or the logarithm of the reciprocal of hydrogen ion concentration in moles/liter):

$$pH = -\log [H^+] = \log (1/[H^+]) \quad \text{(P-19)}$$

An expression of the intensity of the basic or acid condition of a liquid; a measure of the acidity of a solution. The pH may range from 0 to 14, where 0 is the most acid, 7 is neutral, and 14 is most alkaline. Natural waters usually have a pH between 6.5 and 8.5; the concentration range suitable for most organisms is 6.0–9.0. *See* colorimetric and electrometric (glass electrode) methods for pH determination. pH is one of the most important characteristics of aqueous solutions, as it affects many parameters, analyses, processes, etc. For example,

- Acidic (low pH) waters are corrosive to distribution system and house plumbing metals.
- pH ranges between 2.0 and 4.0 for soft drinks and between 4.8 and 8.4 for human urine.
- The recommended pH range for drinking water is 6.5–8.5, which is also where the rate of the breakpoint reaction is maximum.
- Chlorination is more effective at a low pH; most disinfection by-products (except THMs) decrease in concentration with increasing pH.
- pH is also related to acidity, adsorption on activated carbon, alkalinity, carbon dioxide stability, coagulation, corrosion, hardness, and solubility in aqueous solutions.
- Both acidic and caustic substances are dangerous to skin and other valuable surfaces.

**PhAC** Acronym of pharmaceutically active compound.

***Phaeophyta*** The genus of brown algae, including the giant kelps.

**phage** Short for bacteriophage.

***Phanerochaete chrysosporium*** A species of fungi (the white rot fungus) that can combine with some bacteria to degrade polychlorinated biphenyls in sandy soils.

**pharmaceutical-grade water** Water that contains no added substance and meets the quality requirements of U.S. Pharmacopeia after purification by an appropriate water treatment process such as distillation, ion exchange, or other USEPA-approved drinking water treatment. *See also* the following water grades defined by U.S. Pharmacopeia: bacteriostatic water for injection, purified water, sterile water for inhalation, sterile water for injection, sterile water for irrigation, and water for injection. Also called USP purified water or USP grade water. *See also* reagent-grade water and ultrapure water.

**pharmaceutically active compound (PhAC)** A residue of pharmaceutical products found at very low concentrations in wastewater after their use and excretion by water consumers. These products are predominantly organic, highly soluble compounds that are not easily degradable in the gut. They include prescription drugs, over-the-counter medications, drugs used in hospitals, and veterinary drugs. They enter drinking water via municipal wastewater and agricultural runoff. Also called drug residual, pharmaceutical residual, or pharmaceutical residue. *See also* xenobiotic.

**pharmaceutical residual** Same as pharmaceutically active compound.

**pharmaceutical residue** Same as pharmaceutically active compound.

**pharmaceutical waste** Wastewater generated during the production of pharmaceutical products and originating mainly from the spent liquors of the fermentation processes, floor washings, and laboratory wastes. It is high in suspended and dissolved organic matter and contains vitamins. It is usually processed by evaporation and drying.

**pharmacodynamics** The quantitative relationship between a chemical dose and the responses it produces in body tissues.

**pharmacokinetics** The dynamic behavior of chemicals inside biological systems, including uptake, distribution, metabolism, and excretion.

**pharmacological antagonism** The condition of two chemicals that produce opposite effects on a receptor.

**pharmakinetics** Same as pharmacokinetics.

**phase** A distinct state of a substance: solid, liquid, and gas are accepted as the principal phases of matter but vapor is also considered a phase and colloids may be referred to as the dispersed phase.

**phase boundary** The common layer between two adjacent phases, e.g., between a liquid and a solid; or between two fluids such as water and a gas; or between a liquid (water) and another liquid (oil). Also called interface, an important consideration in the study of kinetics of purification and treatment processes. *See* conjunction, diffusion, ion exchange, sorption.

**phase change** The passage of water from one phase to another, e.g., in distillation and freezing.

**phase-contrast microscope** A light microscope that contains a series of diaphragms to separate and reconstitute direct versus diffracted light rays. It uses the phase differences of light rays transmitted by different portions of an object to create an image with distinct details. The image shows a contrasted specimen against a gray background. Maximum magnification = 2000 times. Resolution = 0.2 μm (200 nm). Also called phase microscope. *See also* electron-beam, optical or light, and ultraviolet-ray microscopes.

**phase-contrast microscopy** The use of a phase-contrast microscope in water and wastewater microbiology for the examination of microorganisms, particularly fine internal details and bacterial growth rates in biofilms.

**phased ditch** Same as phased isolation ditch.

**phased isolation ditch** A wastewater treatment configuration designed to achieve combined biological nutrient (nitrogen and phosphorus) removal. It uses the oxidation ditch configuration and the sequencing batch reactor operation, i.e., an oxidation ditch operated on a fill-and-draw basis with sequential aerobic, anoxic, and anaerobic conditions in one or more tanks. This technology, developed in Denmark, is applied in several forms:

- A-ditch: one flow-through ditch, no clarifier
- A-ditch with recirculation, no clarifier
- AE-ditch: one ditch, a clarifier, recirculation of the underflow
- D-ditch: two parallel flow-through ditches, no clarifier
- DE-ditch: two parallel ditches, one clarifier, recirculation of the underflow
- T-ditch: three flow-through ditches in parallel, no clarifier
- TE-ditch: three ditches in parallel, a clarifier, underflow recirculation

See the following processes: Bio-Denipho™, Bio-Denitro™, T-Ditch, Trio-Denipho™, VR-Ditch.

**Phase Five Rule** *See* Phase V Rule.

**phase I reaction** The initial reaction of a chemical substance in the body, usually resulting in the production of reactive intermediates. *See also* phase II reaction.

**Phase I rule** A set of standards and monitoring requirements, maximum contaminant level goals, and National Primary Drinking Water Regulations issued by the USEPA in July 1987 for volatile organic chemicals in drinking water. *See* Table P-03. *See also* primacy and vulnerability assessment.

**phase II reaction** A secondary reaction following the introduction of a chemical in the body; it tends to modify the effects of the initial reaction. *See* phase I reaction.

**Phase II Rule** A set of specific sampling and monitoring requirements, maximum contaminant level goals, and National Primary Drinking Water Regulations issued by the USEPA in January 1991 for compliance with drinking water standards. The list includes the nine inorganics and 10 volatile organics shown in Table P-04, as well as 20 pesticides, herbicides and PCBs. Phase II Rule also lists 30 unregulated contaminants (6 inorganics and 24 organics); *see* Table P-05. *See also* primacy and vulnerability assessment.

**phase inversion membrane** A porous material produced from a homogeneous polymer solution.

**Table P-03.** Phase I volatile organic chemicals (VOCs)

| | |
|---|---|
| **VOCs with MCL & MCLG** | **List 1** (*cont.*) |
| benzene | ethylbenzene |
| carbon tetrachloride | styrene |
| *para*-dichlorobenzene | 1,1,1,2-tetrachloroethane |
| 1,2-dichloroethane | 1,1,2,2-tetrachloroethane |
| 1,1-dichloroethylene | tetrachloroethylene |
| 1,1,1-trichloroethane | toluene |
| trichloroethylene | 1,1,2-trichloroethane |
| vinyl chloride | 1,2,3-trichloropropane |
| | *m*-xylene |
| **List 1** | *o*-xylene |
| bromobenzene | *p*-xylene |
| bromodichloromethane | |
| bromoform | **List 2** |
| bromomethane | 1,2-dibromo-3-chloro-propane |
| chlorobenzene | Ethylene dibromide |
| chlorodibromomethane | |
| chloroethane | **List 3** |
| chloroform | bromochloromethane |
| chloromethane | *n*-butylbenzene |
| *o*-chlorotoluene | *sec*-butylbenzene |
| *p*-chlorotoluene | *tert*-butylbenzene |
| dibromomethane | dichlorodifluoromethane |
| *m*-dichlorobenzene | fluorotrichloromethane |
| *o*-dichlorobenzene | hexachlorobutadiene |
| *cis*-1,2-dichloroethylene | isopropylbenzene |
| 1,2-dichloroethane | *p*-isopropyltoluene |
| *trans*-1,2-dichloroethylene | naphtalene |
| dichloromethane | *n*-propylbenzene |
| 1,2-dichloropropane | 1,2,3-trichlorobenzene |
| 1,3-dichloropropane | 1,2,4-trichlorobenzene |
| 2,2-dichloropropane | 1,2,4-trimethylbenzene |
| 1,1-dichloropropene | 1,3,5-trimethylbenzene |
| 1,3-dichloropropene | |

**Table P-04.** Phase II inorganics and volatile organics

| Inorganics | Volatile Organics |
|---|---|
| asbestos | *o*-dichlorobenzene |
| barium | *cis*-1,2-dichloroethylene |
| cadmium | *trans*-1,2-dichloroethylene |
| chromium | 1,2-dichloropropane |
| mercury | ethylbenzene |
| nitrate | monochlorobenzene |
| nitrite | styrene |
| total nitrate-nitrate | tetrachloroethylene |
| selenium | toluene |
| | xylenes |

*See* finger structure, sponge structure, symmetrical membrane, asymmetric membrane, composite membrane, thin-film composite membrane.

**Phase One Rule** *See* Phase I Rule.

**phase rule** A relationship derived from thermodynamics between the number of components (or chemical entities, $C$), the number of phases ($P$), and the degrees of freedom (or independent variables, $F$) in an equilibrium system (Stumm & Morgan, 1996):

$$F = C - P + 2 \qquad \text{(P-20)}$$

**phase separation process** One of the four basic configurations of anaerobic digestion: hydrolysis and acid formation in one reactor and methane formation in a separate reactor. *See also* low-rate digestion, high-rate digestion, and anaerobic contact.

**Phase Six (b) Rule** *See* Phase VIb Rule

**phase transfer** A concentration change resulting from the movement of an ion, molecule, or substance from one phase (solid, liquid, gas, vapor) into another, for example, between the atmospheric and aqueous phases. *See also* solvent extraction, leaching, sorption.

**Phase Two Rule** *See* Phase II Rule.

**Phase V Rule** A set of standards and monitoring requirements, maximum contaminant level goals, and National Primary Drinking Water Regulations issued by the USEPA in July 1992 for six inorganics, three volatile organic chemicals (VOCs), nine pesticides/herbicides, and six other synthetic organic chemicals (SOCs) in drinking water. *See* Table P-06. *See also* primacy and vulnerability assessment.

**Phase VIb Rule** A set of standards and monitoring requirements issued by the USEPA for four inorganics, ten pesticides, five other synthetic organic chemicals (SOCs), and three unregulated substances in drinking water. *See* Table P-07. *See also* primacy and vulnerability assessment.

**PHB** Abbreviation of poly-b-hydroxybutyrate or polyhydroxybutyrate.

**pH buffer** A substance used to help maintain the pH of a solution constant, e.g., the combination of a weak acid and one of its salts, or a weak base and a salt of that base. pH buffers are used in titrations.

**PhD2** A portable device manufactured by Biosystems, Inc. for the detection of multiple gases.

**pH electrode** A specific ion electrode with a membrane made of a special glass that is sensitive to hydrogen ions ($H^+$) only. The electrode, containing a solution of hydrochloric acid (HCl), is used in pH meters.

**Phelps, E. B.** American public health engineer, member of the U.S. Public Health Service; also

**Table P-05.** Phase II unregulated contaminants

| | |
|---|---|
| **Inorganics** | **Organics** (*cont.*) |
| antinomy | dinoseb |
| beryllium | diquat |
| cyanide, nickel | endothall |
| sulfate | glyphosate |
| thallium | hexachlorobenzene |
| | hexachlorocyclopentadiene |
| **Organics** | 2,3,7,8-TCDD (dioxin) |
| aldrin | methomyl |
| benzo(a)-pyrene | metolachlor |
| butachlor | metribuzin |
| carbaryl | oxamyl (vydate) |
| dalapon | picloram |
| dicamba | propachlor |
| dieldrin | simazine |
| di(2-ethylhexyl)adipate | 3-hydroxycarbofuran |
| di(2-ethylhexyl)phthalates | |

**Table P-06.** Phase V contaminants

| | |
|---|---|
| **Inorganics** | **Pesticides, herbicides** (*cont.*) |
| antimony | endrin |
| beryllium | glyphosate |
| cyanide | oxamyl (vidate) |
| sulfate | picloram |
| thallium | simazine |
| **VOCs** | **Other SOCs** |
| dichloromethane | benzo(a)-pyrene |
| 1,2,4-trichlorobenene | di(2-ethylhexyl)adipate |
| 1,1,2-trichloroethane | di(2-ethylhexyl)phtalate |
| | hexachlorobenzene |
| **Pesticides, herbicides** | Hexachlorocyclopentadiene |
| dalapon | HEX) |
| dinoseb | 2,3,7,8-TCDD (dioxin) |
| diquat | |
| endothall | |

Table P-07. Phase VIb contaminants

| | |
|---|---|
| **Inorganics** | **Pesticides** (*cont.*) |
| boron | metribuzin |
| manganese | trifluralin |
| molybdenum | **Other SOCs** |
| zinc | acrylonitrile |
| **Pesticides** | 2,4- & 2,6-dinitrotoluene |
| acifluorfen | hexachlorobutadiene |
| bromomethane | 1,1,1,2-tetrachloroethane |
| cyanazine | 1,2,3-trichloropropane |
| dicamba | **Unregulated Substances** |
| 1,3-dichloropropene | bromacil |
| ethylene thiourea | methyl-*t*-butyl-ether |
| methomyl | prometon |
| metolachlor | |

professor at Columbia University and the University of Florida. Author or coauthor of articles related to water quality published in 1909 through 1942, including the classic oxygen-sag analysis.

**1,10-phenanthroline ($C_{12}H_8N_2 \cdot H_2O$)** A chemical compound used as a colorimetric indicator in determining the concentration of iron in water. The phenanthroline method also involves the prior reduction of iron to $Fe^{2+}$ by boiling with hydrochloric acid (HCl) and hydroxylamine ($NH_2OH \cdot HCl$).

**phenol ($C_6H_5 \cdot OH$)** A colorless, crystalline, organic by-product of petroleum refining, tanning, coal coking, resin manufacturing, and other manufacturing operations; highly soluble in water. A monohydroxy derivative with a hydroxyl (—OH) attached to a carbon atom in the benzene ring, it is also known as carbolic acid. At low concentrations it causes taste and odor problems in water; higher concentrations can kill aquatic life and humans; 1.5 grams may be fatal to humans. It is not volatile enough to be removed by aeration; its boiling point is 182°C. Marginal chlorination intensifies phenolic taste and odor problems in water. Phenol and formalin are used in the manufacture of phenolic resins. Also called carbolic acid, hydroxybenzene, oxybenzene, phenylic acid. *See also* chlorophenol, phenolic compound, phenol index, phenol number.

**phenol adsorption value** Same as phenol number.

**phenol-formaldehyde resin** A three-ring adsorbent resin without any functional groups beyond those in the matrix, used for the removal of organic compounds from water or industrial wastewater.

**phenolic compound** An organic compound that is a derivative of benzene ($C_6H_6$). Phenolic compounds or phenols are formed when the hydroxyl group (OH) replaces one or more hydrogen atoms in the aromatic nucleus. The simplest such compound is hydroxyl benzene ($C_6H_5OH$). *See* phenol index, phenol number.

**phenolics** Phenolic compounds.

**phenol index** The total concentration of phenolic compounds or total phenol content in drinking water. The USPHS and WHO recommend a limit of 0.001 mg/L, whereas the European Union adopts a maximum admissible concentration of 0.0005 mg/L that excludes phenols that do not react with chlorine.

**phenol number** One of three parameters used by manufacturers to express the relative capacity of an activated carbon by testing it against a standard solution for the adsorption of complex molecules. It is the concentration of carbon (mg/L) required to reduce 100 micrograms per liter of pure phenol by 90%. Most activated carbons commercially available have phenol numbers between 15 and 30. Also called phenol value or phenol adsorption value. *See also* butane number, molasses number, iodine number, phenol index.

**phenolphthalein ($C_{20}H_{14}O_4$)** An organic compound used as an acid-base color indicator; a phenolphthalein solution changes from clear or colorless to pink at about pH = 8.3 and from pink to dark red at about pH = 10.0.

**phenolphthalein alkalinity** The alkalinity in a water sample measured by the amount of standard sulfuric acid required to lower the pH to 8.3, as indicated by the change in color of phenolphthalein from pink to clear or colorless. Phenolphthalein alkalinity is expressed as milligrams per liter equivalent calcium carbonate.

**phenol pollution** Phenolic compounds in wastewater originate from resin manufacturing, petroleum refining, textile plants, dye works, tanneries, chemical manufacture, and other industrial operations.

**phenol red** A reagent used in water analyses as an acid–base indicator, with a pH transition range of 6.4 (yellow)–8.2 (red).

**phenols** Phenolic compounds.

**phenol value** Same as phenol number, but different from phenol index.

**phenol wastes** Industrial wastewater that contains phenolic compounds, e.g., from coking or refinery.

**phenomenological model** A model that attempts to predict particle removal in a filter, using a mass balance equation (particles removed from suspension = solids accumulated in the media pores) and an empirical, first-order rate removal expression

(AWWA, 1999). *See also* filter coefficient, fundamental filtration model. Also called macroscopic filtration model.

**phenotype** The observable properties or constitution of an organism. *See also* genotype.

**phenotypic analysis** A method used to identify bacteria based on their physical and metabolic characteristics.

**phenylbenzene** Another name of biphenyl.

**phenyl bromide** Another name of bromobenzene.

**phenyl chloride ($C_6H_5Cl$)** An organic compound almost insoluble in water, used as a solvent and in chemical manufacturing. This substance, which affects the respiratory and nervous systems, is regulated by the USEPA in drinking water: MCLG = MCL = 0.1 mg/L. *See* monochlorobenzene.

**phenylethylene** Same as styrene.

**phenyl hydride** Another name of benzene.

**phenylic acid** Same as phenol.

**phenyl mercury** A moderately toxic compound with a short retention time in the body; it can be transformed rapidly in the environment to release inorganic mercury.

**phenylmethane** An aromatic solvent that is only slightly soluble in water; used in the production of benzene derivatives, perfumes, and in other industrial applications. Acute exposure to this product may cause damages to the nervous system and certain organs. Regulated by the USEPA as a drinking water contaminant: MCL = MCLG = 1.0 mg/L. Also called methyl benzene and toluene.

**1-phenylpropane** The contaminant in *n*-propylbenzene.

**phlebotomine sandfly** An insect that transmits the tropical infection leishmaniasis. It breeds in damp organic debris, animal burrows, and pit latrines.

**pH meter** An instrument, e.g., a voltmeter used to measure the pH of liquids.

**Phoenix Press** A belt filter press manufactured by Phoenix Process Equipment Co.

**pH of saturation ($pH_s$)** The pH at which water is saturated with calcium carbonate. *See* saturation pH and Langelier saturation index.

**PhonRead®** A flowmeter reading system manufactured by Invensys Metering Systems; it uses the Intelligent Communications Encoder. *See also* RadioRead® and TouchRead®.

**pH operating range** Wastewater aeration systems usually operate at $6.5 < pH < 8.5$ because higher pHs inhibit microbial activity whereas lower pHs favor fungi over bacteria. Aerobic digestion has a smaller pH tolerance range, from 6.7 to 7.4.

**Phoredox process** A term used in South Africa to designate any wastewater treatment process designed to accomplish biological phosphorus removal in an anaerobic–aerobic sequence; in particular, a process similar to the A/O process, the Bardenpho™ process, and the modified Bardenpho™ processes. *See also* $A^2O$ process.

***Phormidium*** A genus of cyanobacteria (blue-green algae) abundant in waste stabilization ponds and in trickling filters; resistant to anaerobiosis and to temperature extremes. They do not participate directly in waste degradation, but they contribute oxygen during daylight hours. They may also cause clogging of the filter surface and produce odor-causing substances (geosmin and methylisoborneol or MIB).

**phosdrin** An organophosphorus insecticide used against garden pests.

**phosphatase** A class of compounds that specifically catalyze the hydrolysis of phosphoric esters.

**phosphate ($PO_{43}^-$)** A chemical compound containing phosphorus; a salt or ester of phosphoric acid ($H_3PO_4$); found in the tri-calcium phosphate rock (apatite); an inorganic constituent of bones and teeth; used as a chemical fertilizer and in the production of detergents, glasses, etc. Its ammonium, potassium, and sodium salts are soluble in water; calcium phosphates are less soluble. *See also* orthophosphate, phosphorus, polyphosphate, superphosphate, BOD:N:P ratio, agricultural value, eutrophication.

**phosphate-accumulating organisms (PAOs)** Organisms used in wastewater treatment processes for enhanced biological phosphorus removal; they consume and retain more phosphorus than ordinary organisms. Some treatment processes, such as the University of Cape Town or UCT process, promote the formation of PAOs by special configurations, e.g., anoxic and anaerobic zones.

**phosphate buffer** A phosphate-based substance used to buffer water samples for laboratory analysis; e.g., (a) a solution of $KH_3PO_4$, $K_2HPO_4$, $Na_2HPO_4 \cdot 7H_2O$, and (b) $NH_4Cl$ for the 5-day BOD or a solution of $KH_2PO_4$ and $Na_2HPO_4$ for the amperometric titration of chlorine residual.

**phosphate-buffered saline** Same as phosphate-buffered saline solution.

**phosphate-buffered saline solution** A saline solution that contains a phosphate buffer for use in some laboratory analyses, e.g., (a) a solution of distilled water, solution of sodium chloride (NaCl), potassium chloride (KCl), monobasic potassium phosphate ($KH_2PO_4$), magnesium chloride hexahydrate ($MgCl_2 \cdot 6H_2O$), and dibasic sodium phosphate ($Na_2HPO_4$) for washing bacteria in the mutagenicity test; (b) distilled water, $KH_2PO_4$, $Na_2HPO_4$, NaCl, and KCl in the detection of DNA damage by cell gel electrophoresis; and (c) a similar solution for the male-specific coliphage assay.

**phosphate concentration** The mass of phosphorus compounds per unit volume of a water or wastewater sample, expressed as mg/L of phosphate ion ($PO_4^{3-}$) or phosphorus (P).

**phosphate industry waste** Wastewater generated during mining and processing phosphate rock to extract phosphorus and other chemicals. It originates from washing, screening, and floating operations, and from condenser bleedoff. It has low pH, contains clays, slimes, oils, high suspended solids, phosphorus, silica and fluoride. Common treatment methods include lagooning, coagulation, and sedimentation.

**phosphate removal** Phosphates can be precipitated by multivalent metal ions such as ferric iron [Fe(III)], aluminum [Al(III)], and calcium [Ca(II)]. *See* phosphorus removal for detail.

**phosphate rock** A sedimentary rock rich in apatite; an important source of phosphorus.

**phosphate uptake rate** The rate at which microorganisms use phosphorus compounds for growth; it is generally assumed to follow the Michaelis–Menten equation when considered as the critical nutrient in lake eutrophication.

**phosphoclastic bacteria** Bacteria that produce acidic substances that can dissolve and release inorganic phosphorus from poorly soluble minerals.

**phosphoric** Containing phosphorus, mainly in the pentavalent state.

**phosphoric acid** Any of three acids: metaphosphoric ($HPO_3$), orthophosphoric ($H_3PO_4$), or pyrophosphoric ($H_4P_2O_7$), derived from phosphorus pentoxide ($P_2O_5$) and water ($H_2O$). Orthophosphoric acid may be produced by dissolving the phosphate ($P_2O_5$) portion of phosphate rock in a strong acid (sulfuric, nitric, or hydrochloric); it is sometimes used as a supplemental source of phosphorus in the treatment of nutrient-deficient industrial waste.

**phosphorite** Any phosphorus-rich deposit, particularly lithified phosphate rock. *See* phosphorous.

**phosphorous** Containing trivalent phosphorus, e.g., phosphorous acid, $H_3PO_3$, a colorless, crystalline, water-soluble compound from which phosphorites are derived.

**phosphorous acid** *See* phosphorous.

**phosphorous plant** A facility that uses an electric furnace to produce elemental phosphorus for commercial use, such as high grade phosphoric acid, phosphate-based detergent, and organic chemicals.

**phosphorus (P)** A solid, nonmetallic element, most commonly white, waxy, poisonous, flammable; an essential chemical element and nutrient found as orthophosphate, organic phosphate, or tripolyphosphate, all salts of phosphoric acid, and expressed as elemental phosphorus (P) or phosphate ($PO_4^{3-}$); used in matches and in fertilizers. Dark-red phosphorus is not poisonous and not very flammable. Phosphorus is found naturally only in the combined state, e.g., as phosphate rock. Atomic weight = 30.974. Atomic number = 15. Specific gravity = 1.82–2.69 at 20°C. Phosphorus can contribute to the eutrophication of lakes and other water bodies. Increased phosphorus levels result from discharge of phosphorus-containing materials into the surface waters. The USEPA and WHO do not regulate phosphorus in drinking water but the European Community has a guide value of 0.4 mg/L and a maximum concentration of 5.0 mg/L as $P_2O_5$. Phosphate is regulated in municipal and industrial wastewater discharges to control algal growth and eutrophication (e.g., USEPA has a limit of 0.0001 mg/L of elemental phosphorus for marine and estuarine waters); lime, alum, or other coagulants may be used to precipitate phosphorus.

**phosphorus accumulating organism (PAO)** A specialized aerobic heterotroph that can take up phosphorus from wastewater beyond its metabolic requirement of approximately 2% by weight. *Acinetobacter* and other PAOs are instrumental in enhanced biological phosphorus removal processes. They require a source of rapidly biodegradable organic matter (such as a volatile fatty acid) in an anaerobic zone. *See also* denitrifying phosphorus accumulating organism.

**phosphorus (P) availability** The availability of phosphorus (P) as a nutrient to plants and microorganisms. Phosphorus sources differ in the fraction of total phosphorus available for plant growth; this fraction depends mainly on solubility and release rate. P availability is higher in biosolids from biological phosphorus removal processes than in heat-dried materials. *See also* bioavailability.

**phosphorus coprecipitation** *See* coprecipitation (2).

**phosphorus cycle** The circulation of phosphorus atoms mainly by living organisms, from the phosphate rock through water and soil, and then plants and animals. At the death of living tissues, phosphorus becomes available to plants. *See* Figure P-06.

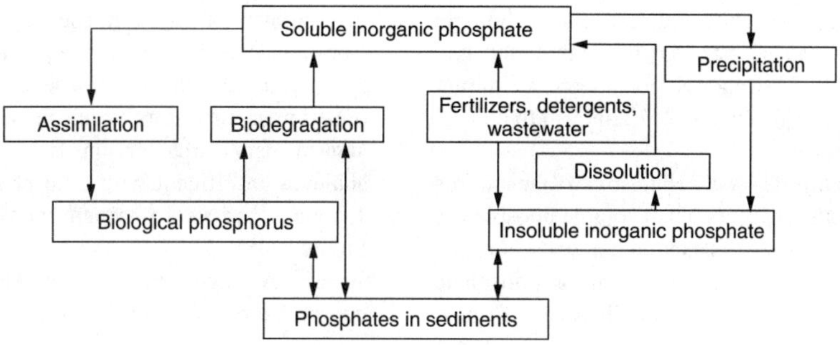


**Figure P-06.** Phosphorus cycle (adapted from Manahan, 1997).

**phosphorus erosion loss** One of three mechanisms of phosphorus loss from soils; it happens when soil particles with attached phosphorus are eroded by and carried in water during a rainstorm. *See also* phosphorus leaching and phosphorus runoff.

**phosphorus fertilization** The addition of phosphorus to soils through fertilizers or other sources. Phosphorus may build up to excess in a soil, which may not be detrimental to crops but has adverse environmental impacts when the excess phosphorus migrates to water bodies.

**phosphorus fertilizer** A plant nutrient containing phosphorus (P).

**phosphorus index** or **P index** A parameter proposed by the U.S. Department of Agriculture's Natural Conservation Service to identify areas with a potential for phosphorus loss. The index is an annual rating for the probability of phosphorus migration from agricultural sources. It depends on transport factors (e.g., soil erosion, runoff, distance from water bodies) and source factors (e.g., solubility, soil background, soil management practices). *See also* phosphorus source coefficient.

**phosphorus in wastewater** Common forms of phosphorus in wastewater include orthophosphate ($PO_4^{3-}$), polyphosphate, and organically bound phosphate. *See also* hexametaphosphate, biological uptake, biological synthesis. Based on 120 gpcd, the average sanitary wastewater contains approximately 4 mg/L of inorganic P and 3 mg/L of organic P. Settling and biological treatment do not change the inorganic concentration but reduce the organic concentration to 2 and 1 mg/L, respectively.

**phosphorus leaching** A mechanism of phosphorus loss from soils; phosphorus applied to the soil for fertilization dissolves in soil water and migrates down to eventually reach groundwater and surface water sources. *See also* phosphorus erosion loss, phosphorus runoff.

**phosphorus limiting application rate ($A_p$)** The maximum quantity of phosphorus that can be applied to land from sludge, in tons per acre per year, assuming that a percentage of the phosphorus is available to plants:

$$A_p = U_p/(PC_p) \qquad \text{(P-21)}$$

where $U_p$ = phosphorus needed by crop, pounds per acre per year; $P$ = phosphorus content of sludge, pounds per ton of dry solids; and $C_p$ = phosphorus availability in sludge, pounds per pound.

**phosphorus loss mechanisms** *See* phosphorus erosion loss, phosphorus leaching, phosphorus runoff.

**phosphorus oxybromide ($POBr_3$)** A compound of phosphorus (P) and bromine (Br) that reacts with water ($H_2O$) to produce phosphoric acid ($H_3PO_4$) and hydrogen bromide gas:

$$POBr_3 + 3\ H_2O \rightarrow H_3PO_4 + 3\ HBr \qquad \text{(P-22)}$$

**phosphorus pentasulfide ($P_2S_5$)** A compound of phosphorus (P) and sulfur (S) that reacts with water ($H_2O$) to produce phosphorus pentoxide and hydrogen sulfide gas ($H_2S$):

$$P_2S_5 + 5\ H_2O \rightarrow P_2O_5 + 5\ H_2S \qquad \text{(P-23)}$$

**phosphorus pentoxide ($P_2O_5$)** A white crystalline powder that absorbs water to form phosphoric acid.

**phosphorus postprecipitation** *See* postprecipitation.

**phosphorus preprecipitation** *See* preprecipitation.

**phosphorus removal** Phosphorus is removed from wastewater by biological and chemical processes: activated algae, bacterial assimilation, biological-chemical processes, chemical processes, luxury uptake, and sequencing batch reactors. Soluble phosphorus may be removed by coagulation and sedimentation. Phosphorus, considered a limiting nutrient in most freshwaters, is removed to control eutrophication.

**phosphorus runoff** A mechanism of phosphorus loss from soils: it occurs when soluble phosphorus is washed off the soil surface into a water body.

**phosphorus salt** Sodium ammonium phosphate ($NaNH_4HPO_4 \cdot 4\ H_2O$); also called salt of phosphorus.

**phosphorus source coefficient (PSC)** A parameter that indicates the solubility of a source of phosphorus; it is used to distinguish between types of phosphorus source (biosolids, fertilizer, manure) and to account for the smaller losses of phosphorus from sources with low solubility. For example average BPR biosolids have a PSC of 0.80, whereas heat-dried biosolids' PSC is about 0.12. Also called PSC value. *See also* phosphorus index.

**phosphorus uptake** *See* phosphate uptake rate.

**PhoStrip™ process** A biological–chemical treatment process developed by TETRA Technologies for the removal of BOD and phosphorus from wastewater. It includes an aeration basin followed by a secondary clarifier. Phosphorus is removed in an anaerobic stripper tank from a portion of the return sludge and precipitated with lime in a reactor–clarifier from the supernatant, which is returned ahead of the plant. Even without filtration, this process can consistently achieve effluent phosphorus below 0.5 mg/L. *See* Figure P-07.

**PhoStrip II™ process** A biological–chemical process developed by TETRA Technologies for the removal of BOD, nitrogen, and phosphorus from wastewater. It includes a primary clarifier, an aeration basin followed by a secondary clarifier, a prestripper, an anaerobic stripper, lime addition, and a final reactor–clarifier. PhoStrip II™ achieves an effluent with total phosphorus below 1.0 mg/L and total nitrogen below 10 mg/L. *See* Figure P-08.

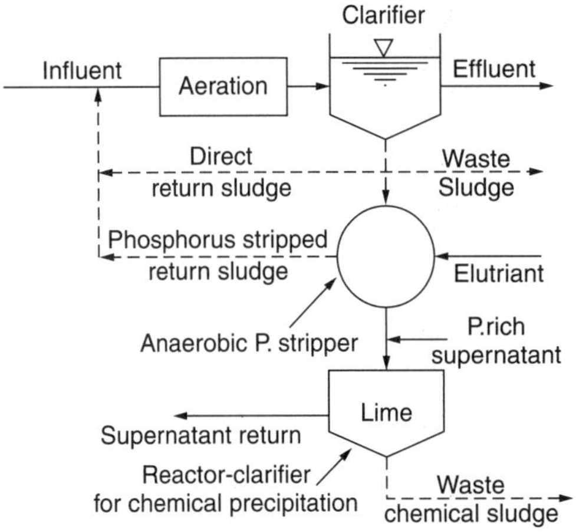


**Figure P-07.** PhoStrip™ process.

**Phosver** A reagent produced by Hach Co. for the determination of phosphates in water or wastewater.

**photic zone** The upper part of an ocean or other water body through which sufficient sunlight can penetrate to enable photosynthesis. Its depth may reach 200 meters depending on water turbidity; subdivided into the dysphotic and euphotic zones. The lower portion is called the aphotic zone, where organisms use the detrital matter that falls from the photic zone. *See* marine profile and freshwater profile for other related terms.

**photoautotroph** Photosynthetic organisms that obtain energy from sunlight and use carbon dioxide as a carbon source, e.g., the cyanobacteria. *See also* chemoautotroph.

***Photobacterium*** A genus of facultative anaerobic bacteria typically found in seawater; they produce energy through dissimilatory nitrate reduction to ammonium.

***Photobacterium phosphoreum*** A species of luminescent marine bacteria used in the Microtox® assay for acute toxicity assessment.

**photocatalytic oxidation** The destruction of organic matter by conversion to carbon dioxide ($CO_2$) and water ($H_2O$) through the irradiation of such photocatalysts as zinc titanate ($Zn_2TiO_2$), zinc oxide (ZnO), and titanium dioxide ($TiO_2$) by sun lamps. *See also* photolysis.

**photocatalytic treatment** An advanced oxidation process that irradiates a catalyst such as titanium dioxide ($TiO_2$) (or another semiconductor) with ultraviolet light. This generates hydroxyl radicals (OH) on the surface of the $TiO_2$ crystals. The process may be used to oxidize chemicals (e.g., MTBE and endocrine disruptors) and pathogens (e.g., cryptosporidium) that are difficult to eliminate by more conventional methods.

**photochemical** Pertaining to the chemical action of sunlight or chemical reactions that depend on light.

**photochemically driven reaction** *See* photochemical reaction.

**photochemical oxidants** Air pollutants formed by the action of sunlight on oxides of nitrogen and hydrocarbons.

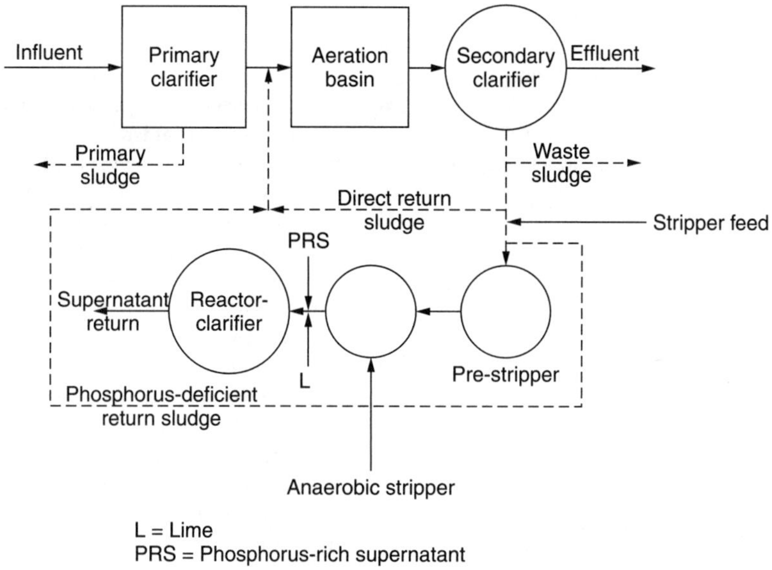


**Figure P-08.** PhoStrip II$^{TM}$ process.

**photochemical oxidation** Chemical change caused by sunlight. *See also* photocatalytic oxidation, photolysis.

**photochemical reaction** A chemical change induced by radiant energy, e.g., photosynthesis, photolysis, photochemical oxidation.

**photodegradation** Decomposition of a substance into simpler components under the action of light or solar radiation.

**photodynamic inactivation** The use of methylene blue, other dyes, titanium dioxide, or other solid materials to make microorganisms more sensitive to inactivation by visible light.

**photoelectric colorimeter** *See* spectrophotometer.

**photoelectron spectroscopy** *See* X-ray photoelectron spectroscopy.

**photographic waste** Wastewater from film-developing and printing operations, containing spent fixer solutions, thiosulfates, silver compounds, and various organic and inorganic reducing agents. Photographic wastes may be discharged to municipal sewers after recovery of silver.

**photoionization detector (PID)** An instrument used in gas chromatography to detect electrical signals from aromatic compounds.

**photolysis** (1) The chemical reaction of a substance caused by direct absorption of solar energy (direct photolysis) or caused by other substances that absorb solar energy (indirect photolysis) (EPA-40CFR300-AA). (2) Photolytic decomposition. *See also* photocatalytic oxidation.

**photolytic decomposition** Decomposition of a substance into simpler components under the action of photons or ultraviolet light.

**photomicrograph** An image taken by a light microscope.

**photon** The basic quantity of radiant energy or electromagnetic radiation, characterized by momentum but having no charge or mass. Light waves, X-rays, etc. consist of photons. Also called light quantum. States may monitor community water supply systems for photon radioactivity.

**photooxidant** An oxidant involved in a photochemical reaction.

**photooxidation** An advanced oxidation process using ultraviolet or other light source to convert or remove such chemical organics as MTBE from contaminated water.

**photoreactivate** To repair DNA damage and become active again when exposed to certain wavelengths of light.

**photoreactivation** The repair of the damage caused to bacteria (e.g., fecal and total coliforms, but not fecal streptococci) by UV light disinfection when they are exposed to visible wavelengths between 300 and 500 nm. Reactivation occurs through the action an enzyme that restores nucleic acid replication. *See also* thymine dimerization.

**photosynthate** A compound formed by photosynthesis.

**photosynthesis** A process in which organisms, with the aid of chlorophyll and associated pigments (as a catalyst), convert carbon dioxide, water, and inorganic substances into oxygen and additional plant material (carbohydrates), using sunlight for energy. For example, pigmented bacteria can use hydrogen sulfide ($H_2S$) and carbon dioxide ($CO_2$) to produce glucose ($C_6H_{12}O_6$) and sulfuric acid ($H_2SO_4$):

$$3\ H_2S + 6\ H_2O + 6\ CO_2 \rightarrow C_6H_{12}O_6 + 3\ H_2SO_4 \quad \text{(P-24)}$$

All green plants grow by photosynthesis. Other reactions representative of photosynthesis include:

$$CO_2 + PO_4 + NH_3 + \text{sunlight energy} \rightarrow \text{new cells} + O_2 \quad \text{(P-25)}$$

$$6\ H_2O + 6\ CO_2 + \text{sunlight energy} \rightarrow C_6H_{12}O_6 + 6\ O_2 \quad \text{(P-26)}$$

$$CO_2 + 2\ H_2O + \text{sunlight energy} \rightarrow CH_2O + O_2 \quad \text{(P-27)}$$

Here, $CH_2O$ represents biomass or a simple carbohydrate. Photosynthesis plays a role in the reoxygenation of polluted waters; when pollution shuts off sunlight to a body of water, it may reduce or eliminate photosynthesis by killing green plants. *See also* chemosynthesis, respiration, oxygen-sag analysis.

**photosynthesizer** *See* phototroph.

**photosynthetic bacteria** Bacteria that use photosynthesis to derive energy for growth.

**photosynthetic organism** *See* primary producer.

**photosynthetic oxygenation** The rate at which photosynthesis adds oxygen to a stream. *See* dissolved-oxygen sag analysis.

**photosynthetic pond** A shallow earthen basin designed to treat wastewater by decomposing organic matter and converting inorganic elements (phosphates and ammonia) to algal cells. It follows a first-stage pond that removes suspended and floating matter in the treatment of industrial wastes. *See also* lagoons and ponds.

**phototoxicity** (1) Damage caused to organisms by photochemical reactions. (2) The ability of a chemical to induce damage through photochemical reactions.

**phototroph** Any organism that uses principally light to generate energy (by photosynthesis) for cellular activity, growth, and reproduction. Phototrophs may be heterotrophic, like certain sulfur-reducing bacteria, or autotrophic like algae and photosynthetic bacteria. Same as phototrophic organism, photosynthesizer. *See also* chemotroph.

**phototrophic** Pertaining to organisms that obtain energy from sunlight. *See also* chemotrophic.

**phototrophic bacteria** A group of bacteria that derive energy from sunlight and use carbon dioxide or organic carbon as a carbon source; e.g., the green bacteria and purple bacteria.

**phototrophic organism** Same as phototroph.

**photovoltaic** Characteristic of materials or devices that convert incident light to electromagnetic force. The photovoltaic effect relates to the phenomenon where incident light or other electromagnetic radiation induces electromotive force on two dissimilar materials, e.g., a metal and a semiconductor.

**photovoltaic cell** A solid-state device that uses crystalline materials to convert light energy from solar radiation directly to electrical energy (electricity): sunlight acting on a semiconductor releases electrons to an electrode.

**photovoltaic effect** *See* photovoltaic.

**pHpznpc** The pH at which the surface charge dependent on pH is zero.

***Phragmites spp.*** Unspecified species of reed found in natural wetlands.

**phreatic** From the Greek phreas meaning artificial well. Denoting or pertaining to groundwater, to the water table or to the layers of soil and rock below the water table but above the zone of rock flowage.

**phreatic cycle** The period of rise and a succeeding decline of a water table; e.g., daily cycle, annual cycle. Also called cycle of fluctuation.

**phreatophyte** A plant having roots long enough to reach the water table.

**pH Redox Equilibrium Equation (pHREEQE)** A FORTRAN computer program developed by the U.S. Geological Survey to perform chemical equilibrium calculations for aqueous solutions.

**pHREEdom™** Chemical products of Calgon Corp. for use in the treatment of cooling water.

**pHREEGE** Acronym of pH redox equilibrium equation.

**pH$_s$** Notation for the pH of saturation.

**pH scale** A scale showing the range of pH values: the acidic range from 0 to 7.0 and the basic range from 7.0 to 14.0.

**pH shock** A temporary and significant reduction of pH levels in a water body caused by a sudden addition of acidic material such as acid runoff. Also called acid shock.

**phthalate** A class of synthetic organic chemicals used industrially; general formula: $C_6H_4(COOR)_2$, e.g., the regulated contaminant di(2-ethylhexyl) phthalate $[C_6H_4(COOCH_2CH(C_2H_5)C_4H_9)_2]$.

**pH tolerance range** *See* pH operating range.

**phyla** Plural of phylum.

**phyletic classification** Phylogenetic classification.

**phyletics** Phylogenetic classification.

**phylogenetic classification** The classification of organisms based on genetic relationships related to evolutionary history. Ribosomal RNA is used to identify microorganisms and determine evolutionary relationships between species. According to the phylogenetic classification, cellular life comprises three basic kingdoms: two prokaryotic (archaea and bacteria) and one eukaryotic (eukarya), with differences not only in RNA composition, but also in cell wall composition, lipid chemistry, and protein synthesis mechanisms. Also called phyletic classification or phyletics. *See also* biological classification, environmental classification, and taxonomic classification.

**phylogeny** The development, evolution or evolutionary history of a group of organisms.

**phylum** (plural: **phyla**) A very broad group of members of the plant and animal kingdoms, e.g., the arthropods, mollusks, and protozoa. A phylum is divided into classes. *See also* biological classification convention, taxonomic classification.

**phys-chem** Abbreviation of physical–chemical treatment.

**physical activation** The manufacturing process using pyrolysis and oxidizing gases to carbonize and activate raw materials in the preparation of activated carbon. *See also* chemical activation, high-temperature steam activation.

**physical adsorption** (1) The accumulation of a substance on the surface of another due to relatively weak interactions produced by the motion of electrons in their orbitals, called van der Waals forces; e.g., solute molecules may be loosely held on the surface of activated carbon particles. Sometimes called ideal adsorption or van der Waals attraction. *See also* chemical adsorption or chemisorption, desorption, electrostatic or exchange adsorption, exchange adsorption, hydrophobic bonding, polarity, solution force, solvophobic force, van der Waals force. (2) A minor mechanism of removal of particulate matter within a granular filter, whereby electrostatic, electrokinetic, and van der Waals forces keep in place particles brought in contact with the surface of the medium. *See also* filtration mechanisms.

**physical analysis** The examination of water or wastewater samples for such physical characteristics as color, conductivity, taste and odor, temperature, and turbidity.

**physical and chemical treatment** Any combination of water or wastewater treatment processes that are not based on microbiological activity. *See also* physical treatment, physical–chemical treatment, and chemical treatment. The term can also refer to treatment of toxic materials in surface and ground waters, oil spills, and some methods of dealing with hazardous materials on or in the ground.

**physical assets** Land, buildings, wells, reservoirs, pumping stations, treatment plants, and other permanent property of a physical nature.

**physical attraction** *See* physical adsorption.

**physical characteristics** Parameters that are considered in the evaluation of water and wastewater, along with bacteriological and chemical parameters. Typical physical characteristics include color, electrical conductivity, odor, pH, taste, temperature, and turbidity. Alkalinity, hardness, and solids (colloidal, dissolved, suspended) are both physical and chemical parameters.

**physical–chemical treatment** A wastewater treatment scheme that was promoted in the 1970s as an alternative to conventional secondary treatment. A typical flow diagram of physical-chemical treatment includes preliminary treatment, rapid mix and flocculation ahead of the primary sedimentation tank, recarbonation (if necessary), optional flow equalization, and filtration, granular carbon adsorption, effluent disinfection, carbon regeneration, and sludge processing and disposal. Backwash wastewaters, scrubber underflow, carbon washwaters as well as the centrate or filtrate from sludge dewatering are returned to the flocculation chamber. Because of high chemicals and sludge handling costs and the difficulty in consistently meeting effluent standards, the physical–chemical treatment is not widely applied for municipal wastewaters. It has found some application in industrial waste treatment. Also called physico-chemical treatment or chemical–physical treatment. *See also* physical and chemical treatment.

**physical conditioning** A process designed to improve the effectiveness of sludge thickening or dewatering by the addition of a precoat or nonreactive product such as diatomaceous earth, or by a unit operation such as freeze–thaw or thermal conditioning. *See also* chemical conditioning.

**physical disinfectant** Physical agents used for water or wastewater disinfection include heat, light, and sound waves. *See also* pasteurization, ultraviolet disinfection, radiation.

**physical stability** A measure of the ability of a treatment unit to resist breakdown due to physical forces.

**physical treatment** A water or wastewater treatment scheme that uses only physical methods of solids–liquid separation such as aeration, air stripping, centrifugation, distillation, filtration, flocculation by agitation, flotation, heat treatment, and sedimentation. *See also* unit operations, unit processes, biological treatment, chemical treatment, and physical–chemical treatment.

**physical unit operation** A water or wastewater treatment method in which physical forces bring about the removal or modification of constituents, as compared to chemical unit processes and biological unit processes. Treatment plants usually incorporate two or all three treatment methods. Examples of physical unit operations are adsorption, filtration, gas transfer, mixing, screening, and sedimentation.

**physical weathering** Breaking down of parent rock by exposure to temperature changes and the action of moving ice, moving water, growing roots, and human activities. *See also* chemical weathering.

**physicochemical method** A quantitative method of analysis used to characterize wastewater. Unlike gravimetric or volumetric methods, it measures properties other than mass or volume. Examples of physicochemical analyses include adsorption spectrometry, colorimetry, fluorometry, polarography, potentiometry, spectroscopy, and turbidimetry.

**physico-chemical treatment** The treatment of effluents or untreated wastewater by nonbiological methods. Same as physical–chemical treatment.

**phytodegradation** The degradation of contaminants by plants through the production of such enzymes as dehalogenase, nitroreductase, and peroxidase. *See also* phytoremediation.

**phytoplankton** Small, usually microscopic plants (such as algae and diatoms), floating or suspended freely in lakes, reservoirs, and other bodies of water. They are photoautotrophic organisms at the bottom of the food chain that include eukaryotes (algae) and prokaryotes (cyanobacteria). *See also* bacterioplankton, zooplankton.

**phytoplankton quotient** The relative ratio of species numbers of two algal groups; once proposed as a biological indicator of a lake trophic state, based on the observation that phytoplankton species composition differs between productive and unproductive lakes.

**phytoremediation** The use of plants or trees to treat contaminated soils and groundwater. In one application, hybrid poplar trees are used to absorb MTBE from a polluted plume. *See* enhanced rhyzosphere biodegradation, phytodegradation. Plants can also accumulate (absorb and store) heavy metals and other inorganic contaminants, and affect the movement of contaminants in groundwater.

**phytotoxic** Harmful to plants.

**phytotoxicity** The characteristic of a substance that inhibits the growth of, or is toxic, to plants. For example, some metals in sludge are phytotoxic even below levels harmful to humans.

**phytotoxin** A poison produced by such algal species as *Anabaena flos* aquae and *Microcystis aeruginosa*. Water contaminated by these algae is repulsive to humans but may cause the death of animals.

**PI** Acronym of plasticity index.

**pI** Isoelectric point.

**PIC** Product of incomplete combustion.

**Picabiol®** An activated carbon process developed by Pica for the production of potable water.

**pickerelweed** An immersed plant attached to the bottom of a water body and extending its foliage above the surface.

**pickets** Vertical paddles of a gravity sludge thickener (Pankratz, 1996).

**pickle** (1) A solution of salt and vinegar for preserving or flavoring foods; any food preserved in such a solution, particularly vegetables. (2) A chemical solution, usually an acid, used to remove oxide scales, rust, and other debris from metal objects.

**pickle and olive waste** Wastewater produced during the processing of vegetables and other foods; it contains lime water, brine, alum, turmeric, syrup, seeds, and pieces of vegetables. It is characterized by a variable pH, high suspended solids, color, and organic matter.

**pickle liquor** Waste acid from the steel pickling process. It contains ferrous iron ($Fe^{2+}$) in a solution of sulfuric acid ($H_2SO_4$) or hydrochloric acid (HCl). Spent pickle liquor is sometimes used to precipitate phosphorus from wastewater at publicly owned treatment works. The resulting ferrous sulfate or ferrous chloride constitutes a good reagent for phosphorus removal. These solutions are usually added to the aeration tank to facilitate iron oxidation. *See* diffusion dialysis.

**pickle liquor removal** Removal of phosphorus from wastewater using pickle liquor.

**pickling** *See* pickling process.

**pickling liquor** Same as pickle liquor.

**pickling process** A method of removing mill scale, rust, dirt, grease, and other debris from steel by washing or immersing in a dilute (sulfuric or hydrochloric) acid solution. As the operation proceeds, the acid becomes weaker and must be renewed; the concentration of the ferrous salt produced increases and the pickling liquor must be spent. Pickling also produces sludges and rinse water. (2) A phenomenon that happens in anaerobic digestion if organic acids produced are allowed to accumulate, lower pH, and inhibit further decomposition.

**PICl** Abbreviation of polyiron chloride.

**picloram ($C_6H_3O_2N_2Cl_3$)** The common name of 4-amino-3,5,6-trichloropicolinic acid, a herbicide sometimes found in groundwater; an unregulated drinking water contaminant listed under the Phase II Rule. It causes liver damage in rodents and dogs.

**pico** Prefix meaning one trillionth ($10^{-12}$)

**picocurie (pCi)** A measure of radioactivity; one pCi is equivalent to one trillionth or $10^{-12}$ of a curie or 0.037 nuclear disintegration per second.

**PI control** *See* proportional–integral control.

**Picornaviridae** A family of viruses that includes the enteroviruses. *See* picornavirus

**picornavirus** The smallest of the RNA viruses, of the Picornaviridae family, including hepatitis-A virus, the enteroviruses, the poliovirus, and the rhinoviruses that cause the common cold.

**PID** Acronym of (a) photoionization detector and (b) proportional integral derivative.

**P&ID** Acronym of process and instrumentation diagram.

**PID control** Proportional integral derivative control.

**Pielkenroad separator** A sedimentation basin partially equipped with parallel corrugated plates to increase surface area and enhance particle separation. *See also* inclined plate separator and tube settler.

**piercer** A freshwater organism that feeds on juices that it sucks from bottom-rooted plants. *See also* shredder, collector, grazer, gouger.

**piezometer** An instrument for measuring pressure head in fluids (in tanks, conduits, or soil) or compressibility of materials. It may consist of a small pipe or tube attached to a conduit or tank and connected to a manometer. A vessel may be used to measure compressibility as the change in volume of a substance in response to hydrostatic pressure.

**pig** A device that is driven by water pressure or pulled by a cable through new drinking water pipes to clean them before they are disinfected. *See also* polypig.

**pigging** The use of a "pig" to clean a new water pipe or to dislodge sand, grit, scale, or other deposited materials from sewer lines.

**pigtail** A portion of a service connection between the water distribution main and a meter, providing some flexibility for relative displacement between the connection and the main.

**pigmented bacteria** A group of autotrophic bacteria that derive energy by photosynthesis, using hydrogen sulfide ($H_2S$), water ($H_2O$), and carbon dioxide ($CO_2$) to produce glucose ($C_6H_{12}O_6$) and sulfuric acid ($H_2SO_4$):

$$H_2S + H_2O + CO_2 \rightarrow C_6H_{12}O_6 + H_2SO_4 \quad \text{(P-28)}$$

**pili** Small protein, hairlike appendages of bacterial cells involved in conjugation and other functions such as sticking to surfaces or attaching to each other. Also called fimbriae, which are shorter. *See also* cilia, flagella.

**pilot filter** A tube filled with granular media, used as a small-scale filter.

**pilot plant** A scaled-down, continuous-flow water or wastewater treatment plant used to simulate, test, and evaluate new or alternative technologies. A pilot test, pilot-scale test, pilot-scale testing, pilot-scale development, or pilot-plant study consists of the operation of a pilot plant, typically with 5–10% of the design flow, to collect basic data for the design and operation of a full-scale plant, including capital and operating costs. Pilot plants are sometimes used to train staff in the operation of new processes. All such pilot activities, although smaller than full-scale, are larger than laboratory-scale or bench-scale. *See also* treatability study.

**pilot plant study** *See* pilot plant.

**pilot-scale development** *See* pilot plant.

**pilot-scale test** *See* pilot plant.

**pilot-scale testing** *See* pilot plant.

**pilot test** *See* pilot plant.

**pilus** Singular of pili.

**Pinch Press®** A device manufactured by Waste Tech, Inc. for high-pressure sludge filtration and dewatering.

**pinch valve** A valve that has flexible elements to stop or start flow.

**P index** *See* phosphorus index.

**pin floc** In water or wastewater treatment, a small floc particle (like the head of a pin) or a large spherical particle, associated with a low sludge volume index. Small pin flocs do not settle well by gravity and contribute to the BOD and suspended solid concentrations of the effluent. Pin floc or pinpoint floc occurs in aerobic biological treatment units operated at a very high solids retention time (SRT), e.g., SRT > 20 days (or high F:M ratio); it is a type of clarification failure caused by too few filaments. Loss of pinpoint floc in the effluent is a common problem in extended aeration plants. *See also* filamentous sludge, dispersed growth, filamentous bulking, and viscous bulking.

**pink water** Water with a reddish-orange coloration because it contains unreacted potassium permanganate ($KMnO_4$) or reduced manganese with excess ozone.

**pinpoint floc** Same as pin floc.

**pinworm** The excreted helminth pathogen *Enterobius vermicularis* or the disease that it causes worldwide.

**pipe (**or **piping)** A hollow cylinder or tubular conduit that is constructed of metal, clay, plastic, wood, asbestos cement, or concrete, and used to conduct fluids or finely divided solids.

**pipe bursting** A trenchless process used in the replacement of water or sewer pipes; a bursting head is pulled through the excavation to break the existing pipe, push the pieces into the surrounding soil, and install the new pipe that is attached at the rear of the head. *See* trenchless technology, open cut, and the following types of bursting methods: hydraulic bursting, pneumatic bursting, static bursting.

**pipe characteristics** Properties of a pipe that affect fluid flow, e.g., roughness coefficient and k-factors.

**pipe class** The working pressure rating of a specific pipe for use in water distribution systems, including allowances for surges. This term is used for cast iron, ductile iron, asbestos–cement, and some plastic pipes.

**pipe corrosion** Internal or external destruction of a pipe as a result of physicochemical interactions with its surroundings (e.g., chemical reactions and stray currents). *See also* corrosion control, pit, tubercle.

**pipe fitting** The work performed by a pipe fitter.

**pipe fittings** Various pieces used in connection with pipes, e.g., bends, bushings, caps, couplings, crosses, diminishers, elbows, joints, nipples, plugs, reducing sockets, tees, unions.

**pipe flow methods** Approaches used in the analysis of flow in closed conduits, e.g., to determine energy losses or required pipe sizes: the empirical, scientific, and conveyance methods are based on the Hazen–Williams, Darcy–Weisbach, and conveyance formulas. The Hardy Cross method is used to solve small pipe-network problems.

**pipe gallery** A gallery, passageway, or conduit (in a treatment plant, for example) for the installation of pipes and accessories, with appropriate space allowance for maintenance and repair.

**pipe gauge** A number that indicates the thickness of the sheet metal used to make the pipe.

**pipe grade** The slope of the pipe in the direction of flow.

**pipe lateral underdrain** A filter underdrain system with orifices or nozzles, a deep or shallow layer of gravel, medium-to-high headloss, and with or without air scour.

**pipeline** Pipes jointed and connected to control devices for conducting fluids or finely divided solids.

**pipeliner** A person or firm who has experience in pipe laying or actually performs such work.

**Pipeliner** A device manufactured by Robbins & Myers, Inc. and installed in wastewater pipes as a grinder/cutter.

**Piper diagram** A graphical representation of the percent equivalents of major ions in a water sample, consisting of two triangles for the anions and cations and a rhombus for the total ionic constituents.

**pipe schedule** A sizing system of arbitrary numbers that specifies the inside and outside diameters of each pipe size. This term is used for steel, wrought iron, and some types of plastic pipe, and to describe the strength of some types of plastic pipe. For example, a plastic pipe schedule 40, common in water and wastewater work, with a nominal diameter of 16 inches, actually has an inside diameter of 15 inches. *See also* pipe class.

**PipeSonde™ probe** An in-pipe monitor manufactured by Hach Co. to analyze seven water quality parameters: conductivity, dissolved oxygen, line pressure, ORP, pH, temperature, and turbidity.

**pipe spool** A prefabricated pipe section.

**pipet** Same as pipette.

**pipe-to-pipe potable reuse** The direct incorporation of reclaimed water into a water distribution system. *See also* water reuse applications.

**pipette** (1) A slender graduated, glass or plastic, tube used in the laboratory to deliver small quantities of liquids, e.g., less than 10 ml, from one container to another. *See also* volumetric pipette or transfer pipette. (2) To measure or transfer liquids using a pipette.

**pipette technique** A method used in determining the size distribution of sediment, particularly for clays and silts.

**pipework** Same as piping (2).

**piping** (1) A phenomenon, directly related to water level, that accompanies seepage underneath a dam, levee, or dike, whereby water carries some of the finer materials away from the structure, which may cause excessive leakage or structural failure. *See also* creep ratio and quicksand. (2) Same as pipe; pipes, collectively, or a system of pipes, fittings, and appurtenances. Also called pipework.

**piping and instrumentation diagram (P & ID)** A component of the process control diagrams established for the design of water and wastewater treatment plants. The P & ID is a graphical representation of the plant, showing all process equipment with associated piping, valves, and control instrumentation. *See also* process flow diagram.

**piping system** The pipes, fittings, and appurtenances of a treatment plant or other hydraulic structure.

**pisciculture** The breeding, rearing, and transportation of freshwater and marine fish by artificial means. *See* fish farming for detail.

**Pisigan–Singley equation** An equation based on laboratory tests on steel, proposed in 1985 to calculate the saturation pH (i.e., pHs) that appears in such corrosion indices as the Langelier index or the Ryznar index (AWWA, 1999):

$$\begin{aligned}\text{pHs} = {} & 11.017 + 0.197 \log(\text{TDS}) \\ & - 0.995 \log(\text{Ca}^{2+}) - 0.016 \log(\text{Mg}^{2+}) \\ & - 1.041\,(\text{TAlk}) + 0.021 \log(\text{SO}_4^{2-})\end{aligned} \quad \text{(P-29)}$$

Total alkalinity (Talk) is expressed in mg/L as calcium carbonate $CaCO_3$ and the other concentrations are in mg/L. *See* corrosion index for a list of corrosion control parameters.

**Pista® Grit** Vortex grit removal equipment manufactured by Smith & Loveless, Inc. Wastewater enters and exits tangentially, while a turbine maintains constant flow velocity and promotes the separation of organics from grit. *See* vortex grit removal, Teacup™.

**piston pump** A displacement pump that moves and imparts pressure to fluids through a reciprocating piston or plunger in a cylinder; a single-action or double-action pump, depending on whether the piston acts in one end or both ends of the cylinder. Same as reciprocating pump.

**Pistra stratiotes** A species of aquatic plants, water lettuce, which may provide suitable breeding habitats for *Mansonia filariasis.*

**pitch** (1) The distance between any two adjacent things in a series, such as screw heads, rivets, gear teeth, pins in a chain, etc. (2) The advance per revolution of an impeller, considering it as a screw; it is a square pitch when it is equal to the diameter.

**Pit Hog®** Sludge pumping equipment manufactured by LWT, Inc.

**pit latrine** A simple and inexpensive excreta disposal unit consisting of a latrine with a pit in which excreta accumulates and decomposes, and the liquid infiltrates into the surrounding soil. *See* VIP latrine.

**pitometer, pitot-static tube, pitot tube** These are all differential-head meters; devices that indicate velocity heads ($V^2/2\,g$) in pipes and are used for calculating the velocity ($V$) of flowing fluids, for establishing velocity profiles, for computing flows, and for investigating waste and leakage. They measure the pressure ($P$) at the center in a Pitot tube and the static pressure ($P'$) in a static tube. With a flow coefficient ($K$), a fluid specific weight ($\gamma$), and a gravitational acceleration ($g$),

$$V = K[2\,g(P - P')/\gamma]^{0.5} \quad \text{(P-30)}$$

However, the distinction between one term and the others is not always clear. *See,* e.g., (a) *Webster's New World Dictionary of American English,* 3rd edition (1991): definitions of Pitot-static tube, Pitot tube, and static tube; (b) APHA et al. (1981): definitions of pitometer and Pitot tube; (c) Linsley et al. (1992, page 36): discussion of Pitot-static tubes, pitometers, and Cole pitometers; (d) Simon and Korom (1997, pp. 77–79): discussion of Pitot, static, and Prandtl tubes; (e) Chadwick and Morfett (1998, pp. 40–41): discussion of velocity measurement. A pitometer survey is a survey of a water distribution system using a pitometer to determine velocities at various points as well as the conditions of the piping system. The Pitot-static tube is a device used to measure the velocity of flowing fluids; it consists of the combination of a Pitot tube and a static tube. Also called Prandtl tube. The Pitot tube is a device for measuring the total pressure of a fluid stream; essentially a tube attached to a manometer at one end and pointed upstream at the other.

**pitometer survey** *See* pitometer.

**pitot-static tube** *See* pitometer.

**pitot tube** *See* pitometer.

**pit privy** A fixed or portable outbuilding with one or more seats and a pit or vault serving as a toilet to collect human feces. Privies are used in rural and other areas that have no or limited access to running water. Also called outhouse, pit toilet, privy. *See also* latrine.

**Pittchlor** High-test sodium hypochlorite (NaOCl) produced by PPG Industries.

**pitting** Localized corrosion causing deep attacks over small surface areas, a cause of failure of steel and copper water piping (one of two types of corrosion-related failures). Pitting results from the creation of concentration cells (pits) on the wall of the pipe by a variety of materials such as sand, dirt, and mill scale. It may be initiated or exacerbated by microorganisms in the presence of hydrogen sulfide ($H_2S$) and oxygen. Pitting is common in cold-water piping carrying hard groundwater with excess carbon dioxide ($CO_2$) and excess dissolved oxygen. The pit can grow until the pipe is fully penetrated, resulting in a leak. Also called pitting corrosion. *See also* graphitization, tuberculation.

**pitting corrosion** Same as pitting.

**pit toilet** A fixed or portable outbuilding with one or more seats and a pit or vault serving as a toilet to collect human feces. Pit toilets are used in rural and other areas that have no or limited access to running water. Also called outhouse, privy, pit privy. *See also* latrine.

**p$K_a$** The negative common logarithm of the equilibrium constant (K) for the dissociation of an acid (or the logarithm of the reciprocal of the equilibrium constant):

$$pK_a = -\log [K] = \log (1/[K]) \qquad \text{(P-31)}$$

**plague** A refuse-related infection that is caused by pathogens transmitted by rats and fleas.

**plain aeration** Wastewater aeration without recycling settled sludge to the aeration basin, as used in the roughing filter-plain aeration process.

**plain filtration** Wastewater treatment that consists in applying directly the biological treatment effluent to granular-media filters, without chemical coagulation. Plain filtration can reduce suspended solids to less than 10 mg/L. *See also* direct filtration.

**plain sedimentation** The settling of suspended particles in water or wastewater by gravity and natural aggregation, without the addition of chemicals for flocculation and coagulation; sometimes used as pretreatment of raw waters that contain a high sediment load, ahead of conventional treatment. *See also* primary treatment.

**plain settling tank** A basin used for plain sedimentation.

**plan** (1) *See* planning and plans. (2) An implementation plan or a plan under the Clean Air Act that establishes emission standards for designated pollutants from designated facilities and provides for the implementation and enforcement of such emission standards (EPA-40CFR58.1-I and 60.21-c).

**Planck's constant** A number used in the formula of radiant energy, $h = 6.62 \times 10^{-27}$ erg-sec, named after the German physicist Max Karl Ernst Ludwig Planck (1858–1947).

**plane of shear** In the electrical double layer theory, the outer surface of the rigid layer, or Stern layer, of a negatively charged particle; or the boundary between the charge that remains with a particle and the charge that does not. *See* electrical double layer.

**Planet** A rotary distributor manufactured by Simon-Hartley, Ltd. for installation in fixed-film reactors (e.g., trickling filters).

**plankton** (1) Small, usually microscopic, plants (phytoplankton) and animals (zooplankton) that float, drift, or swim feebly in aquatic systems; used as food by fish and other higher aquatic organisms. (2) All of the smaller floating, suspended or self-propelled organisms in a body of water. *See also* periphyton.

**planned reuse** The deliberate, direct or indirect, use of wastewater or reclaimed water, e.g., irrigation or other specifically designed reuse as compared to unintentional reuse downstream of an effluent discharge. *See also* water reuse applications.

**planning** *See* planning and plans.

**planning and plans** The planning process identifies problems, defines objectives, collects information, analyzes alternatives and determines necessary activities and courses of action. Plans are reports and drawings, including a narrative operating description, prepared to describe a facility and its proposed operation (EPA-40CFR256.06 and 240.101-t). *See also* wastewater facilities plan.

**plant** (1) A multicellular organism that generates biomass photosynthetically for the benefit of humans and virtually all other organisms. Plants range from single-cell algae to rooted and complex trees. (2) The equipment, fixtures, buildings, machinery, etc. necessary to carry out such an operation as water or wastewater treatment. *See* treatment plant.

**plant available nitrogen (PAN)** The amount of nitrogen available for plant growth from sludge applied to land. Expressed in pounds of nitrogen per acre per dry ton of sludge applied, PAN is used to establish sludge application rates and may be calculated from the following formula (WEF & ASCE, 1991):

$$PAN = 20\,(VN_a + N_n + MN_o) \qquad \text{(P-32)}$$

where $V$ = volatilization factor ($V$ = 1 for sludge injection, $V$ = 0.5 for surface application); $N_a$ = ammonia nitrogen, %; $N_n$ = nitrate nitrogen, %; $M$ = mineralization factor (from 0.2 to 0.4); and $N_o$ = organic nitrogen, %. *See also* available nitrogen for another formula.

**plant factor** The ratio of the output of an electrical generating plant to the total rating of its various generating components. Also called plant operating factor. *See also* load factor.

**plant harvesting** The gathering and removal of aquatic plants from a natural wastewater treatment system to maintain a crop with high metabolic nutrient uptake and for nutrient removal. Harvested plants are dried and landfilled, spread on land, or composted.

**plant hydraulic capacity** The flow or discharge for which the plant is designed and above which it is considered hydraulically overloaded.

**plant operating factor** Same as plant factor.

**plant-scale studies** Full-scale studies, as compared to pilot or laboratory studies.

**plant tap** A sampling tap at the water treatment plant or between the plant and the distribution system.

**plan view** A diagram or photo showing a facility as it would appear when looking down on top of it.

**plaque** A colony of viruses formed by lysis of a host cell on a plate culture, as represented by a clear area in the cell monolayer. *See also* plaque-forming unit.

**plaque assay** A test conducted to detect the areas of lysis in an infected culture; a common method used for the enumeration of viruses. It involves overlaying inoculated tissue cells with agar to localize released viruses and appropriate incubation. Culturable viruses destroy infected cells, which appear as holes or plaques. The concentration is reported in plaque-forming units. *See also* cytopathic effect.

**plaque assay method** *See* plaque assay.

**plaque-forming unit (PFU or pfu)** The number of infectious virus particles, each plaque representing one virus particle or a clump of viruses. *See* plaque assay, colony-forming unit. The virus concentration of a water or wastewater sample is expressed in PFU/L.

**PLASdek** Filling materials produced by Munters for use in cooling towers.

**plasma** (1) The fluid portion of blood containing suspended particles. (2) A discharge of ionized gas containing an approximately equal number of positive ions and electrons; used to analyze metals in spectroscopic instruments. (3) Cytoplasm.

**plasma arc vitrification** A process using a molten bath in a plasma centrifugal furnace to detoxify hazardous materials and turn organic contaminants into innocuous products by vaporizing them at 2000–2500°F. Solids melt and are vitrified at 2800–3000 °F.

**plasma incinerator** An advanced incinerator that injects hot plasma (ionized air) through an electrical arc for the combustion of hazardous and other wastes.

**plasma system** A system that creates very high temperatures to destroy toxic wastes, leaving as residuals gaseous emissions (hydrogen and carbon monoxide), scrubber acid gases, and ashes in scrubber water.

**plasma water** The aqueous portion of blood.

**plasmid** A circular piece of DNA that exists apart from the chromosome and replicates independently of it. Bacterial plasmids carry information that renders the bacteria resistant to antibiotics. Plasmids are often used in genetic engineering to carry desired genes into organisms.

**plasmid DNA** A small circular DNA molecule that provides genetic material; plasmid.

***Plasmodium*** A genus of spore-forming (sporozoa), parasitic and pathogenic protozoans.

***Plasmodium falciparum*** The protozoal parasite that causes falciparum malaria, a serious and sometimes fatal form of the disease in humid tropical areas.

***Plasmodium malariae*** The protozoal parasite that causes quartan malaria in tropical and subtropical areas.

***Plasmodium ovale*** The protozoal parasite that causes ovale malaria in West Africa.

***Plasmodium spp.*** The protozoal pathogen that causes malaria and is transmitted by the *Anopheles* mosquito.

***Plasmodium vivax*** The protozoal parasite that causes vivax malaria.

**plastic** A natural or usually synthetic, high-molecular-weight polymer material, used in many applications in the field of water and wastewater engineering.

**plastic dual-lateral block underdrain** A filter underdrain system that includes a deep layer of gravel and concurrent air-and-water scour; characterized by low head loss.

**plastic fluid** A fluid that can develop a velocity gradient and move only after reaching a threshold shear called yield stress. Wastewater sludge is considered a pseudoplastic fluid obeying the so-called viscosity law:

$$\tau = \eta(dv/dy)^n \qquad \text{(P-33)}$$

where $\tau$ = shear stress, $\eta$ = plastic viscosity, $n$ = a constant, and $dv/dy$ = velocity gradient.

**plasticity index (PI)** The range of water content in which soil is in a plastic state. PI is calculated as the difference between the percent liquid limit and percent plastic limit. *See* Atterberg limits.

**plasticizer** (1) In general, a substance added to make a material flexible, resilient, and easier to handle, e.g., an organic compound added to high-molecular weight polymers in plastic manufacturing. The majority of plasticizers are esters (adipates and phthalates). (2) An admixture for making mortar and concrete work with little water.

**plastic limit (PL)** One of the three Atterberg limits: the lower limit of the plastic state of a soil; the water content (as a percentage of the dry weight) of a brittle soil when it passes from the plastic to the solid state. *See also* liquid limit, shrinkage limit, plasticity index.

**plastic media** Plastic materials used in water and wastewater treatment units such as trickling filters, packed beds, and packed towers.

**plastic nozzle underdrain** A filter underdrain system using plastic nozzles with a shallow or no layer of gravel, air scour or concurrent air-and-water scour; characterized by high head losses.

**plastic pipe** A nonmetallic pipe made of polyvinyl chloride (PVC) $[(CH_2{:}CHCl)_n]$, polyethylene $[(H_2C{:}CH_2)_n]$, polybutylene $[(C_4H_8)_n]$, or similar materials.

**plastics** Nonmetallic chemoreactive compounds molded into rigid or pliable construction materials, fabrics, etc.

**plastic soil** A soil that will deform without shearing (typically silts or clays). Plasticity characteristics are measured using a set of parameters known as Atterberg limits.

**plate** A filter element that has a flat surface.

**plate-and-frame filter press** A large pressure filter, a mechanical device used for batch dewatering of waste chemical sludges. It consists of a series of recessed parallel plates fitted with cloth filters and intervening frames forming a filtering chamber. Chemical sludge, or organic sludge conditioned with ferric chloride and lime, is pumped through the plates, which are then pressed together as in an accordion to force the water out. In addition to filtering and compression, the operation includes discharging the sludge cake and washing of the filter cloths. This device, which is available in two designs, fixed volume and variable-volume diaphragm, is very efficient, achieving a compact, dry cake with solids contents of up to 50%. Also called plate-and-frame press or plate press. *See* sludge pressing for names commonly used.

**plate-and-frame membrane configuration** An arrangement of pressure-driven membranes. *See* plate-and-frame membrane modules.

**plate-and-frame membrane modules** A membrane filtration device, similar to the conventional filter press, consisting of a series of membrane sheets and alternating support plates for water or wastewater treatment; it is used mainly as an electrodialysis module, but also in reverse osmosis. The influent liquid is treated between two adjacent membranes and the permeate exits through the porous support plates. This arrangement allows less membrane surface area per unit volume than other configurations (hollow-fiber, spiral-wound, and tubular).

**plate-and-frame press** Same as plate-and-frame filter press.

**plateau period** A period of above normal but uniform incidence of disease or death in response to a toxic or injurious agent.

**plate clarifier** A clarifier whose treatment capacity has been increased by the addition of inclined parallel plates or trays, which increases the effective surface area of the basin and reduces the surface overflow rate. *See also* lamella clarifier, tube settler.

**plate count** The number of bacteria (or other microorganisms) that grow on a plate of agar from a diluted sample, after a specified incubation period (e.g., 24 to 48 hours) and an appropriate temperature. The count is reported in colony-forming units (CFU) per milliliter and adjusted for dilution. *See also* heterotrophic plate count, petri dish, standard plate count.

**Plate-Pak®** Mist eliminators produced by ACS Industries, Inc.

**plate press** Same as plate-and-frame filter press, a device that uses pressure to dewater sludge from water and wastewater treatment.

**plate screen** A screen that consists of perforated plates for the passage of water or wastewater.

**plate settler** A shallow settling unit consisting of inclined parallel plates installed at the end of a clarifier or gravity thickener in a countercurrent, cocurrent, or crossflow configuration. The plates improve settling performance by increasing the settling area and reducing the distance the solids have to travel before being captured. Also called inclined plate separator or inclined plate settler. *See also* inclined tube settler, Lamella clarifier, tube settler, ballasted flocculation.

**plate tower scrubber** An air pollution control device that neutralizes hydrogen chloride gas by bubbling alkaline water through holes in a series of metal plates.

**Platetube** Porous diffuser plates manufactured by Walker Process Equipment Corp.

**platform furnace** A graphite tube adapted as a furnace to stabilize temperatures in atomic absorption spectrophotometers. Also called L'vov platform.

**plating method** The method used to apply a diluted sample to a growth medium; pour plate method, spread plate method.

**platonic chloride** Same as chloroplatinic acid.

**platinum (Pt)** An expensive, heavy, grayish-white, malleable, ductile, metallic element, resistant to most chemicals, unoxidizable except in the presence of bases. Atomic weight = 175.09. Atomic number = 78. Specific gravity = 21.5 at 20°C.

**platinum black** A black powder of finely divided metallic platinum used as a catalyst in synthesis. *See also* standard hydrogen electrode.

**platinum–cobalt color unit** A unit used to measure color in natural waters, based on standard solutions of potassium chloroplatinate and cobaltous chloride. Also called chloroplatinate unit or Pt–Co unit. *See* color unit for detail.

**platinum–cobalt method** A laboratory procedure used in the analysis of color in drinking water. The platinum–cobalt (Pt–Co) scale, established on the basis of the color produced by a given concentration of chemical, serves to compare color units.

**platinum–cobalt scale** *See* platinum–cobalt method.

**plattnerite** One of two forms of lead dioxide ($PbO_2$), resulting from the corrosion of lead piping. *See also* cerussite and scrutinyite.

**playa** A temporary lake formed after rainfall or its dried-up bed, e.g., in a desert.

**PLC** Acronym of programmable logic controller.

**pleated-cartridge filter** A disposable device used in microfiltration and particularly to concentrate viruses from treated wastewater for analysis. *See also* membrane modules.

**pleomorphic cell** A bacterial cell without a well-defined shape. *See also* coccus, rod, helical cell.

**plerotic water** Subsurface water in the zone of saturation. Also called groundwater or phreatic water.

**pleurodynia** An epidemic disease caused by the human excreted pathogen Coxsackievirus B with such symptoms as sudden chest pain and fever. Also called devil's grip, epidemic pleurodynia.

**plinthite** A strongly compacted soil layer underlying soft materials and cemented with ferric oxide. Also called laterite. *See* hardpan for other similar layers.

**plotting position** An estimate of the probability or return period of an event in the exceedance series to determine its abscissa on the frequency curve. Three commonly used formulas for calculating plotting positions are the Hazen, median, and Weibull formulas. *See also* Blom's transformation.

**plug** (1) A fitting used to close the bell end of a cast iron pipe or to close another fitting; the movable part of a faucet, valve, or similar device. (2) The quantity of water moving under plug flow conditions. (3) A pipe blockage; clogging. (4) Cement, grout, or other material used to fill and seal a hole drilled for a water well.

**plug cock** A shutoff valve that incorporates a rotatable plug.

**plug flow** The type of flow that occurs in tanks, basins, or reactors when a slug of water moves through without ever dispersing or mixing with the rest of the water flowing through. Flow in which fluid and solid particles move as well-mixed discrete parcels or volumes, without mixing between adjacent parcels, and are discharged in the same sequence in which they enter. Under plug flow conditions, the fate of pollutants, in one-dimensional, longitudinal models of lakes and reservoirs, may be represented by a first-order decay equation:

$$\partial C/\partial t = -K_d \cdot C - V \cdot \partial C/\partial x \qquad \text{(P-34)}$$

where $C$ is the constituent concentration, $t$ the time, $K_d$ is a first-order decay coefficient, $V$ is the average longitudinal velocity, and $x$ is the longitudinal coordinate. For plug flow in a rectangular sedimentation basin, a length/width ratio of at 20 is recommended. The Froude number increases with this ratio. *See also* completely mixed flow.

**plug flow activated sludge process** The activated sludge process using a plug flow reactor (aeration basin) with a large length/width ratio, e.g., L/W > 10:1; the most common type of reactor used since its inception in the 1920s until the late 1960s. *See also* complete-mix activated sludge.

**plug flow aeration** *See* plug flow activated sludge.

**plug flow basin** A basin designed and operated under plug flow conditions. *See* plug flow reactor.

**plug flow composting** One of two variations of in-vessel composting, a method of sludge stabilization; there is no mixing between the particles or portions of the composting mass; they leave the enclosed container in the order they enter it. *See* sludge composting for detail.

**plug flow in-vessel composting** Same as plug flow composting.

**plug flow mass balance** The relationship between influent and effluent flow rates and concentrations for a plug flow reactor, taking into account any removal or transformations. Assuming a steady volumetric flow rate:

$$C/C_0 = e^{-kt} \qquad \text{or} \qquad \ln (C/C_0) = -kt \qquad \text{(P-35)}$$

where $C$ and $C_0$ are the concentrations at time $t = t$, and $t = 0$, respectively, and $k$ is the rate coefficient. *See also* complete mixing mass balance.

**plug flow model** A flow model of the activated sludge process that assumes that all particles entering the reactor remain in the reactor the same amount of time. *See* Lawrence–McCarty model.

**plug flow reactor (PFR)** A reactor operating under plug flow conditions; particles flow through the reactor with essentially no mixing and exit in the same sequence they entered, e.g., activated sludge aeration basin, chlorine contact basin, aquatic treatment system. Other types of reactor include batch, complete-mix, fluidized-bed, and packed-bed.

**plug flow sampling** The collection of grab samples so that they come from the same "chunk" of water.

**plug flow system** A wastewater treatment system that uses a plug flow reactor, with or without recycle.

**plugged-flow reactor** *See* plug flow reactor.

**plugging** (1) The act or process of stopping the flow of water, oil, or gas into or out of a formation through a borehole or well penetrating that formation (EPA-40CFR144.3). (2) A phenomenon observed in the operation of aquifer storage and recharge (ASR) and other water wells, whereby precipitates accumulate in the pores and progressively reduce the injection capacity of the wells. (3) The accumulation of solids in a membrane, resulting in a flow restriction and decrease in performance.

**plugging factor** An empirical measure of the fouling tendency or plugging characteristics of water based on the time it takes 500 ml of filtrate to pass through a 0.45 micron pore membrane filter at a constant pressure of 30 psi at the beginning and at the end of the test period. Also called silt density index. *See also* modified fouling index, and mini plugging factor index.

**plug valve** A device consisting of a cylindrical or conical plug installed in a seat in a pipe. The seat has two ports that correspond to the opening of the two pipe sections. When the valve is open, a hole in the plug coincides with the ports and lets the flow through with very little resistance.

**Plug Valve (Cam-Centric®)** A quarter-turn valve with low torque actuation manufactured by Val-Matic® in sizes ½″ to 36″ for shut-off and throttling service in water and wastewater applications. It handles slurries and wastewater without clogging and with minimal head loss due to the valve linear flow path.

**plumbing** All the elements of water supply inside a building and wastewater removal from it: pipes, fixtures, fixture traps, etc.; the installation of these elements.

**plumbing fixture** A sink, toilet, or other receptacle, attached or appended to a house, that receives water or wastewater and discharges it into the drainage system.

**plumbing system** All the elements that handle water supply, wastewater removal, and stormwater in and around a building, including the plumbing, vent pipes, building drains, building sewers, and drainage pipes.

**plumbosolvation** The dissolution of lead into drinking water; plumbosolvency (1).

**plumbosolvency** (1) The ability of or condition for lead to dissolve in drinking water. (2) The phenomenon of uniform corrosion when soluble lead is released from lead pipes, brass fixtures, and lead-based solders. Plumbosolvency can be reduced by providing an appropriate buffer intensity through pH adjustment. *See* Lead and Copper Rule, cerussite, plattnerite, scrutinyite.

**plumbum** Latin name of lead (Pb).

**plume** (1) A visible or measurable discharge of a contaminant from a given point of origin, e.g., a discrete volume of wastewater or drainage floating in estuarial or coastal waters. Can be visible or thermally measurable in water as it extends downstream from the pollution source, or visible in air as, for example, a plume of smoke. *See also* discharge farfield, discharge nearfield, transition region, initial dilution. (2) The area taken up by contaminants in an aquifer. (3) The area of radiation leaking from a damaged reactor. (4) Area downwind within which a release could be dangerous for those exposed to leaking fumes.

**plunger pump** A reciprocating pump whose plungers do not contact the cylinder walls but enter and withdraw from it through packing glands. Simplex, duplex, and triplex pumps have, respectively, one, two, or three plungers. Plunger pumps are used to pump sludge (primary, thickened, digested), slurries, and chemical feed.

**Plus 5** An air diffuser manufactured by Enviro-Quip International Corp.

**Plus 150™** A water treatment apparatus manufactured by U.S. Filter Corp. for laboratories.

**plutonium** A radioactive metallic element chemically similar to uranium.

**plutonium-239** A radioisotope having a half-life of 24,300 years, originating from uranium-235 reacting with a neutron; emits gamma radiation.

***P. malariae*** Abbreviation of *Plasmodium malariae.*

**PMD™** Acronym of pipe-mounted diffuser, equipment manufactured by Environmental Dynamics, Inc.

**PMR** Acronym of proportionate mortality ratio.

**PNA** Acronym of polynuclear aromatic hydrocarbon or simply polynuclear aromatic.

**pneumatic** Of or pertaining to air, gases, wind; filled with compressed air. An airlift pump is used in pneumatic pumping. A pneumatic tank is a holding tank under pressure used in a closed water supply system instead of a pump to avoid surges associated with pump starts and stops. *See also* hydropneumatic tank and pressure tank.

**pneumatic actuator** A device that uses a flexible diaphragm or a piston cylinder to render a valve or a gate automatic. This type of device is replacing hydraulic-operated gates and valves in many water and wastewater applications. *See also* electrical actuator.

**pneumatic burster** *See* pneumatic bursting.

**pneumatic bursting** A trenchless process used in the replacement of water or sewer pipes: a bursting head is pulled through the excavation to break the existing pipe, push the pieces into the surrounding soil, and install the new pipe that is attached at the rear of the head. The pneumatic burster consists of an air-driven hammer contained in a hollow section and impacting on a head to burst the existing pipe; this tool is pulled with a winch and cable. *See* trenchless technology, open cut, hydraulic bursting, static bursting.

**pneumatic conveyor dryer** A device that uses hot air gases or superheated vapor for vaporizing water from sludge pulverized in a cage mill or by an atomized suspension technique. *See* flash dryer for detail.

**pneumatic ejector** A device that raises raw wastewater, scum, or sludge into a pipe through reduced air pressure.

**pneumatic mixers** Diffused aerators used to provide mixing through their rising gas bubbles.

**pneumatic mixing** The injection of a gas, such as air or oxygen, into the bottom of a mixing tank or an aeration basin to create turbulence and mixing via rising gas bubbles. In the activated sludge and other suspended-growth processes, pneumatic mixing maintains the mixed liquor suspended solids in suspension while meeting the oxygen requirements. *See* power dissipation and velocity gradient, two important design parameters for pneumatic mixing. *See also* mixing and flocculation devices.

**pneumatic pumping** *See* pneumatic.

**pneumatic tank** *See* pneumatic.

**p notation** A mathematical operator designating the negative logarithm (or the logarithm of the reciprocal) of a quantity; i.e., $p = -\log_{10}$; for example, $pH = -\log_{10} [H]$. *See* pE, pH.

**$PO_4$** Chemical formula of phosphate.

**$POBr_3$** Chemical formula of phosphorus oxybromide.

**POC** Acronym of (a) particles of complete combustion and (b) particulate organic carbon.

**Pocket Pal** A portable pH meter manufactured by Hach Co.

**PODR** Acronym of point of diminishing return.

**POE** Acronym of point of entry.

**pOH** The negative logarithm of the concentration of the hydroxyl ion ($OH^-$), which may be approximated by the difference 14 – pH.

**POHC** Acronym of principal organic hazardous constituent.

**point dilution method** A method of measuring discharge using a tracer at one point, measuring its concentration at a downstream point, and estimating the dilution ratio between the two points.

**point of concentration** A point that receives runoff from various parts of a drainage area; e.g., the point at which flows are computed for the design of storm sewers. *See also* retardance, time of concentration.

**point of diminishing returns (PODR)** The point on the curve of total organic carbon vs. coagulant dosage where the slope of the curve changes from greater than $S$ to less than $S$, with $S = 0.03$ mg C per mg coagulant. The PODR is a critical factor in determining the feasibility of enhanced coagulation.

**point of disinfectant application** The point at which disinfectant is applied and water downstream of that point is not subject to recontamination by surface water runoff.

**point of entry (POE)** The point at which drinking water or other product enters a delivery system and where there is usually a control apparatus, e.g., a water treatment or water quality control device. *See also* point of use.

**point-of-entry treatment** Treatment applied at the point of entry by such devices as water processing units using distillation, activated carbon filters, reverse osmosis, and ion exchange.

**point-of-entry treatment device** A treatment device applied to the drinking water entering a house or building for the purpose of reducing contaminants in the drinking water distributed throughout the house or building.

**point of fluidization** *See* fluidization point.

**point of incipient fluidization** *See* incipient fluidization point.

**point-of-sale treatment** A small-scale, membrane-based treatment system built on or near existing sewer lines in areas with high potential for reclaimed water. The wastewater collected is given adequate treatment and delivered to the reuse customers. Also called satellite treatment.

**point of use (POU)** The point at which drinking water or other product reaches a consumer (e.g., a faucet in a home) and where a control apparatus (e.g., a water treatment or water quality control device) may be installed. *See also* point of entry.

**point-of-use filter** A small filter used as a point-of-use treatment device. *See also* home treatment filter, household filter, domestic filter.

**point-of-use treatment** Treatment applied at the point of use by such devices as water processing units using distillation, activated carbon filters, reverse osmosis, and ion exchange. *See,* e.g., LINX®.

**point-of-use treatment device** A treatment device applied to a single tap used to reduce contaminants in drinking water at that one tap.

**point of zero charge (pzc)** The pH value ($pH_{pzc}$) at which the net surface charge is 0, corresponding to the zero proton condition. Also called isoelectric point (pI), indicating that particles do not move in the applied electric field. *See also* point of zero net proton charge.

**point of zero net proton charge (pznpc)** A point of zero charge (pzc) or isoelectric point corresponding to a surface charge established solely by the exchange of hydrogen ions [$H^+$] or hydroxyl ions [$OH^-$].

**point of zero net proton condition** Same as point of zero net proton charge.

**point of zero salt effect (pzse)** A point such that the partial derivative of the net proton charge with respect to the ionic strength is zero, i.e., a change in the concentration of an inert background electrolyte does not affect the surface charge.

**point rainfall** The precipitation that occurs at a specific point, e.g., the quantity of rainfall collected or recorded where a gage is located, which usually varies from the records of neighboring gages.

**point source** In general, a stationary location or fixed facility from which pollutants are discharged; any single identifiable source of pollution, e.g., a pipe, ditch, outfall, ship, ore pit, or factory smokestack. Concerning water pollution, a point source is any discernible, confined and discrete conveyance, including but not limited to any pipe, ditch, channel, tunnel, conduit, well, discrete fissure, container, rolling stock, concentrated animal feeding operation, landfill leachate collection system, vessel or other floating craft from which pollutants are or may be discharged. It does not include return flows from irrigated agriculture or agricultural stormwater runoff (EPA-40CFR122.2). Table (P-07) shows some differences between point and nonpoint sources. *See* diffuse source, background pollution, distributed or nonpoint source.

**point source discharge** The discharge of pollutants from a pipe, ditch or any other point source.

**poise** A measure of absolute viscosity, equal to one tenth of a newton-second per square meter ($0.1\ N \cdot s/m^2$), or 100 centipoises. Named after Jean-Louis Poiseuille.

**Poiseuille, Jean Louis** French physician (1799–1869) who investigated the flow of fluids in small tubes (like arteries and veins). *See also* the Hagen–Poiseuille equation.

**Poiseuille's equation** *See* sludge flow. *See also* the Hagen–Poiseuille equation.

**poison** A chemical that, in relatively small amounts, is able to produce injury when it comes in contact with a susceptible tissue. *See also* toxin.

**poisonous waste** A waste that presents a material risk of death, injury or impairment of health of persons or animals, or a waste that threatens the pollution or contamination of any water supply (United Kingdom's 1972 Deposit of Poisonous Waste Act).

**Poisson distribution** The probability distribution function for an event to occur $r$ times with a mean frequency $\mu$:

$$P_r = e^{-\mu}\mu^r/r! \qquad \text{(P-36)}$$

where $P_r$ is the probability of $r$ occurrences, e.g., the probability of finding $r$ bacteria in a water sample. The sign (!) indicates the factorial of the preceding number. The Poisson distribution is

**Table P-07.** Point sources (general differences from nonpoint sources)

| Point sources | Nonpoint sources |
|---|---|
| Origin and cause | |
| Single source, discrete point. | Many locations or sources, diffuse distribution |
| Domestic, industrial, and commercial water uses | Runoff from precipitation or groundwater movement |
| Means of abatement and control | |
| Wastewater treatment before discharge | Reduction or prevention of release of pollutants |
| Compliance with regulatory permits, inspections, and monitoring | Best management practices, change in land use activities, voluntary compliance, or regulation |

Adapted from Adrien (2004), Table P-1, page 260.

used (with the binomial distribution) in the most probable number (MPN) technique.

**polar** *See* polar character (1).

**polar bond** The chemical bond that exists between two different atoms that are covalently bonded, with the formation of secondary bonds (e.g., hydrogen bond) among compounds. Ammonia is an example of a compound formed by polar bonding.

**polar character** (1) The characteristic of a compound that has distinct positively and negatively charged ends; polarity. (2) A characteristic of water that results from its unbalanced nature and allows its individual molecules to join by hydrogen bonding.

**polarity** The property or characteristic producing unequal effects at different points of a system; polar character. Polarity determines the solubility of organic molecules in water and other polar solvents. It explains how a neutral molecule can have positive and negative charges at different locations. Polar compounds are more soluble and less readily adsorbed in water than nonpolar compounds. *See also* chemical adsorption or chemisorption, desorption, electrostatic adsorption, exchange adsorption, hydrophobic bonding, ideal adsorption, physical adsorption, solution force, solvophobic force, van der Waals force.

**polarity reversal** The periodic change of electrode polarity in the electrodialysis reversal process; demineralizing compartments become concentrate compartments and vice versa.

**polarizability** A characteristic of volatile organic compounds used in the design of fixed-bed adsorbers for the control of air stripping off gases. *See also* the Dubinin–Raduskevich and the Lorentz–Lorentz equations.

**polarization** (1) The concentration of ions in the layer adjacent to an ion-exchange membrane. (2) The condition in the electrodialysis process whereby a high current density causes the dissociation of water molecules into $H^+$ and $OH^-$ ions or localized deficiencies of ions on the surface of the membrane. *See also* activation polarization, concentration polarization, current-density-to-normality ratio.

**polarization curve** A graphical representation of the electrochemical mechanism of corrosion in terms of current density, when the corroding metal is used as an electrode and a current is applied through an inert auxiliary electrode. The polarization curve plots the electrode potential vs. the current density.

**polarization factor (PF)** A parameter used to express the phenomenon of concentration polarization in membrane filtration processes. It is equal to the ratio of the concentration of solute at the membrane surface ($C_m$) to the bulk concentration ($C_b$) and expressed as an exponential function of the recovery rate (AWWA, 1999):

$$PF = C_m/C_b = \exp (J\delta_{cp}/D) \qquad \text{(P-37)}$$

$$PF = e^{(Kr)} \qquad \text{(P-38)}$$

where $J$ = permeate flux, $\delta_{cp}$ = thickness of concentration polarization layer, $D$ = diffusivity of the particle species subject to concentration polarization, $K$ = a semiempirical constant = 0.6–0.9 for reverse osmosis membranes, and $r$ = recovery rate.

**polarography** A laboratory technique that measures current flow through an aqueous solution as a function of the potential of an electrolytic cell. A polarographic apparatus automatically registers the current at a dropping mercury electrode. The polarographic method is used in the analysis of chlorine and iodate.

**polar organic by-product** A polar organic compound that is formed during water disinfection.

**polar organic compound** An organic compound having functional groups with an imbalance of electrons and difficult to extract from water because of high solubility.

**polar solvent** A solvent, such as water ($H_2O$) and ethanol ($CH_3OH$), that is electrostatistically positive on one side and electrostatistically negative on the other side.

**polar valence** The valence of an ion, i.e., the number of charges acquired by an atom when losing or gaining electrons. Also called electrovalence or electrovalency.

**Pol-E-Z®** An emulsion polymer produced by Calgon Corp. for use in solids separation processes.

**polio** The common name of poliomyelitis.

**poliomyelitis** A water- and excreta-related, acute disease caused by the poliovirus and affecting the central nervous system. It is transmitted via the fecal–oral route and is distributed worldwide.

**poliomyelitis virus** *See* poliovirus.

**poliovirus** A small, spherical enterovirus of the Picornaviridae family, sometimes found in drinking water and responsible for aseptic meningitis and poliomyelitis. Poliovirus Type 1 is more resistant to chlorination than *E. coli.*

**polished water** Ultrapure water that has essentially no ion or mineral content. Also called intrinsic water. *See also* distilled water.

**polisher** A final water or wastewater treatment step used to improve the quality of the finished water or treated effluent; e.g., ion exchange in deionization plants or filtration of secondary effluents.

**polishing** Pertaining to a water or wastewater treatment process or unit that is applied after conventional ones. *See also* polisher, polishing filter, polishing pond, tertiary treatment, advanced waste treatment.

**polishing filter** (1) An advanced wastewater treatment method used to remove residual solids from a secondary effluent by filtration on granular media. Process variations range from plain filtration (no chemicals, no further sedimentation), to direct filtration (with coagulation and optional flocculation), to traditional filtration. Also called tertiary filter. (2) A water polisher.

**polishing pond** A shallow, aerobic waste stabilization pond used for polishing wastewater effluents from an activated sludge plant, a trickling filter plant, or a facultative pond. *See* maturation pond for more detail.

**pollutant** (or **contaminant**) Any substance introduced into the environment that adversely affects the usefulness of a resource, or, more generally, a constituent added to water (or air) through use. An element, compound, or mixture, an organic substance, an inorganic substance, a combination of organic and inorganic substances, or a pathogenic organism that, after discharge and upon exposure, ingestion, inhalation, or assimilation into an organism either directly from the environment or indirectly by ingestion through the food chain, could cause death, disease, behavioral abnormalities, cancer, genetic mutations, physiological malfunctions (including malfunction in reproduction), or physical deformations in either organisms or their offspring (EPA-40CFR503.9-t and 40 CFR300.5). The term pollutant is often used interchangeably with contaminant and impurity. *See* characteristic, constituent, wastewater constituent.

**pollutant accumulation methods** Two methods used in the Corps of Engineers' STORM model to simulate the concentration of six pollutants (BOD, suspended solids, coliform, etc.) in runoff. In one method, the concentration is a function of dust and dirt accumulation on streets. In the other, pollutant accumulation is simply based on a rate per unit area. In the same model, pollutant availability factor indicates that not all of an accumulated pollutant is washed off with runoff at any given time; it varies with the pollutant itself and with the runoff rate from impervious areas.

**pollutant availability factor** *See* pollutant accumulation method.

**pollutant limit** A numerical value that describes the amount of a pollutant allowed per unit amount of sewage sludge (e.g., milligrams per kilogram of total solids); the amount of a pollutant that can be applied to a unit area of land (e.g., kilograms per hectare); or the volume of a material that can be applied to a unit area of land (e.g., gallons per acre) (EPA-40CFR503.9-u).

**pollutant load** The amount of pollutant carried to a unit (inflow pollutant load) or from that unit (pollutant export), usually expressed in mass (pounds or kilograms). The pollutant loading or pollutant load rate is the mass of pollutant per unit time or the mass of pollutant per unit area, e.g., kg/day, lb/ft$^2$/day. In stormwater management, the yearly export and inflow loads ($L$, lb) of such pollutants as nitrogen and phosphorus may be estimated from the yearly rainfall depth ($P$, inches), two dimensionless factors that adjust the rainfall depth for storms that produce runoff ($C$) and for imperviousness ($I$), the flow-weighted mean concentration of the pollutant ($C_0$, mg/L), and the drainage area ($A$, acres):

$$L = [(P)(C)(I)/12](C_0)(A)(2.72) \qquad \text{(P-39)}$$

**pollutant loading** *See* pollutant load.

**pollutant load rate** *See* pollutant load.

**pollutant standard index (PSI)** A measure of adverse health effects of air pollution levels in major cities.

**polluted stream** A stream that contains an excessive amount or concentration of a specific pollutant or pollutants.

**polluter** A person, corporation, or government agency that produces (and discharges) pollutants in the air, water, or soil.

**polluter-pays principle** The requirement or idea that polluters must pay the cost of avoiding pollution or remedying its effects, e.g., through a system of charges on the person or corporation that causes the pollution.

**polluting material** Pollutant.

**pollution** (1) Generally, the presence of matter or energy whose nature, location, or quantity produces undesired environmental effects. Under the Clean Water Act, e.g., the term is defined as the man-made or man-induced alteration of the physical, biological, chemical, and radiological integrity of water. *See also* natural pollution. (2) The addition of harmful substances to air, soil, or water.

**pollution abatement** The reduction of the daily load of air, water, or air pollution, e.g., through appropriate wastewater treatment before discharge into a receiving stream. *See* pollution control.

**pollution abatement benefit** *See* primary benefit, secondary benefit, intangible benefit.

**pollutional index** *See* pollution index.

**pollutional load** The amount of material that requires treatment in a waste stream or that exerts a detrimental effect in a receiving stream. It is usually expressed in weight of material per unit time, e.g., pounds of BOD per day.

**pollutional survey** A field investigation conducted to determine (a) the quality of each waste stream of an industry or a municipality, and (b) the effects of wastewater discharges on the quality of receiving waters; also called a wastewater survey. Other field investigations include industrial waste and sanitary surveys.

**pollution control** Measures taken to eliminate or reduce pollution, including administrative mechanisms as well as technical processes and devices.

**pollution conversion** The creation of one form or source of pollution while eliminating or reducing another form or source. For example, (a) incineration of solid wastes may eliminate land pollution and create air pollution; (b) land application of sludge may lead to build-up of toxic metals in soils.

**pollution index** A criterion developed to gauge the degree of pollution of a stream or other natural water body, applying weights to various parameters, such as ammonia, bacterial density, benthos level, biochemical oxygen demand, chemical oxygen demand, chloride, dissolved oxygen, iron, manganese, nitrate, pH, plankton level, and suspended solids. Such indices can be used to establish a water quality classification; see Table P-08. Also called pollutional index.

**pollution load** *See* pollutional load.

**pollution plume** A mass of heavily polluted water, often of an elliptical shape with the long axis in the direction of flow.

**pollution prevention** The active process of identifying areas, processes, and activities that create excessive waste by-products for substitution, alteration, or elimination of the process to prevent waste generation. As defined by the USEPA (cited in Freeman, 1995), pollution prevention is "the use of materials, processes, or practices that reduce or eliminate the creation of pollutants or wastes at the source. It includes practices that reduce the use of hazardous materials, energy, water, or other resources and practices that protect natural resources through conservation or more efficient use."

**Table P-08.** Pollution index (surface water quality classification)

| | Water quality classification | |
|---|---|---|
| Parameter | Excellent | Grossly polluted |
| Ammonia, mg/L | 0.1 | 2.7 |
| COD, mg/L | 10 | >80 |
| Dissolved oxygen, % | 90–110 | <20 or >200 |
| Iron, mg/L | 0.1 | >2.7 |
| Nitrate, mg/L | 4 | >100 |
| pH | 6.5–8.0 | <4.0 or >10.0 |
| Suspended solids, mg/L | 20 | >280 |

**pollutograph** A plot similar to a hydrograph, showing the variations of a pollutant concentration or a pollutant load versus time.

**Polly Pig** A device manufactured by Knapp Polly Pig, Inc. for cleaning pipelines.

**polonium (Po)** A radioactive element; atomic number = 84; atomic weight about 210.

**polonium-210** A radioisotope of polonium; one of the radionuclides that have traces in natural water sources.

**poly (3-hydroxyburate)** *See* polyhydroxybutyrate.

**polyacid** A substance that consists of a linear or branched chain of small identical subunits, such as the substances excreted at the surface of microorganisms and used in bioflocculation.

**polyacrylamide [$(C_3H_3ONH_2)_n$]** The most common material found in anionic and nonionic polymers used in water treatment; a monomer acrylamide with a molar mass between 1 and 30 million.

**polyacrylamide polymer** An anionic or nonionic polymer made with polyacrylamide.

**polyacrylate** An anionic polyelectrolyte, compound of polyacrylic acid and consisting of selected monomers. *See* Figure P-09 for its formula.

**polyacrylic acid [$(H_2C{:}CHCOOH)_x$]** Acronym: PAA. A group of chemicals divided in lower molecular weights (from 1000 to 5000) and higher molecular weights (from 6000 to 25,000). Used as scale inhibitor (antiscalant) and antifoulant pretreatment of membrane systems. *See also* polyphosphonate.

**Polyad™** An apparatus manufactured by Weatherly, Inc. for the control of volatile organic compounds in fluidized beds.

**Poly-Alum** An inorganic coagulant produced by Rochester Midland Co.

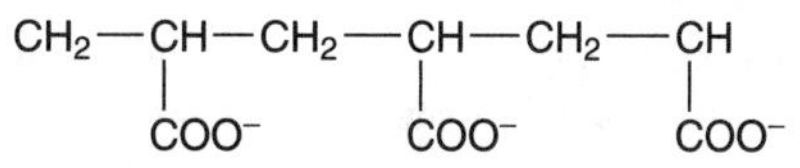


**Figure P-09.** Polyacrylate.

**polyaluminum chloride $[Al(OH)_x(Cl)_y(SO_4)_z]$, with $1.5 \leq y \leq 2.5$ and $0 \leq z \leq 1.5$** Acronym: PACl. A prehydrolyzed form of aluminum chloride ($AlCl_3$); a chemical commonly used in water and wastewater treatment and produced by reacting hydrochloric acid (HCl) with a purified source of aluminum. It may contain variable concentrations of other metals (chromium, sodium, and zinc) as compared to two other water treatment chemicals (low-iron alum and standard alum), but the principal hydrolysis product is $Al_{13}O_4(OH)_{24}^{7+}$. Also called polyaluminum hydroxychloride. *See also* aluminum-based coagulants.

**polyaluminum hydroxychloride (PAHC)** Another name of polyaluminum chloride, including the water of hydration; aluminum/hydroxide ratio = 1.5.

**polyaluminum hydroxysulfate (PAHS)** Another name of polyaluminum sulfate, including the water of hydration.

**polyaluminum sulfate (PAS) $[Al(OH)_x(Cl)_y(SO_4)_z]$, with $1.0 \leq y \leq 2.5$ and $0 \leq z \leq 1.5$** Acronym: PAS. A prehydrolyzed form of aluminum sulfate $[Al_2(SO_4)_3]$, similar to polyaluminum chloride.

**polyamide (PA) membrane** A type of flat-sheet or hollow-fiber membrane used in water treatment; it is made of aromatic or aliphatic polyamide. *See also* cellulose acetate (CA) membrane.

**polyaromatic hydrocarbon (PAH)** An aromatic hydrocarbon found in crude oil and gasoline, containing more than one fused benzene ring. *See* polynuclear aromatic hydrocarbon for more detail.

**poly-b-hydroxybutyrate (PHB)** *See* polyhydroxybutyrate.

**PolyBlend** Equipment manufactured by Stranco, Inc. for handling polymers.

**Poly Boss** An instrument manufactured by WesTech Engineering, Inc. to monitor settling velocities in clarifiers.

**Polybrake** A product of AquaPro, Inc. for removing polymers.

**polybrominated biphenyl (PBB)** A compound similar to polychlorinated biphenyls, persistent in the environment and harmful to humans and other animals; used as a flame retardant.

**polybutylene $[(C_4H_8)_n]$** A thermoplastic polymer used in pipes, tubing, fittings, and other applications.

**polycarbonate $\{[—C_6H_4C(CH_3)_2C_6H_4OCOO—]_n\}$** A high-strength thermoplastic polymer resin and a linear polyester of carbonic acid ($H_2CO_3$); used in windows, tubing, cartridge filters, and piping.

**polychaete worm** A small worm commonly found in oceans and estuaries and used for bioassays of coastal regions.

**Polychem** Plastic products manufactured by Budd Co. for use in rectangular clarifiers.

**polychlorinated biphenyl (PCB)** Any chemical substance that is limited to the biphenyl molecule that has been chlorinated to varying degrees or any combination of substances that contain such substance (EPA-40CFR761.3). PCBs are aromatic compounds having the general formula $C_{12}H_aCl_{10-a}$, where $a \leq 8$. They were used in the manufacture of electrical, heating, and cooling equipment. Because of their high toxicity and persistence in the environment, their use has been discontinued. The USEPA regulates PCBs in drinking water under the Phase II contaminants rule.

**polycyclic aromatic hydrocarbons (PAH)** A group of aromatic compounds containing several, usually fused, atomic rings. PAHs result from industrial operations or the processing of petroleum and coal and include benzo(a)pyrene, dibenzopyrene, and dibenzoacridine, which are carcinogenic. *See* more detail under polynuclear aromatic hydrocarbon.

**polyDADMAC** Abbreviation of polydiallyldimethyl ammonium chloride.

**polydentate ligand** An anion or molecule that has more than one site that can bond to a central metal ion at one time to form complexes. For example, oxalate and ethylenediamine are bidentate ligands (2 atoms), citrate is tridentate (3 atoms), and EDTA is hexadentate (6 atoms). Also called multidentate ligand or chelating agent. *See also* unidentate ligand.

**polydiallyldimethyl ammonium chloride (polyDADMAC)** A quaternary amine, widely used as a cationic polyelectrolyte in water treatment, available commercially as an aqueous solution.

**polydispersity** A parameter that indicates the range of molecular weights of natural organic matter in a water sample; a low polydispersity corresponds to a narrow range.

**polyelectrolyte** A high-molecular-weight substance having points of positive or negative electrical charges that is formed by natural or man-made processes. Natural polyelectrolytes may be of biological origin or derived from starch products and cellulose derivatives. Man-made polyelectrolytes consist of simple substances that have been made into complex substances of high molecular weight. Water-soluble polyelectrolytes are used with other chemical coagulants to aid in binding small suspended particles to larger chemical flocs for their removal from water. Insoluble polyelectrolytes are used in ion exchange. Polyelectrolytes are also used as sludge-conditioning chemicals. Also called a polymer, polymer coagulant, polymeric coagulant, or polyelectrolyte coagulant. *See* ampholytic polyelectrolyte, anionic polyelectrolyte, cationic polyelectrolyte, polyacid.

**polyelectrolyte coagulant** A polymeric organic compound used as a coagulant in water or wastewater treatment.

**polyelectrolyte flocculant** A polymeric organic compound used as a flocculant in water or wastewater treatment or in sludge dewatering.

**polyelectrolyte impurities** Contaminants found in polyelectrolytes from the manufacturing processes, including residual monomers, reactants, and by-products. Most industrialized countries have stringent limits on the application of polyelectrolytes to the treatment of drinking water.

**polyethene** *See* polyethylene.

**polyethylene (PE)** A thermoplastic polymer or copolymer comprised of at least 50% ethylene by weight. It has a low coefficient of friction, resists chemicals, absorbs very little moisture, and is often used for tubing, piping, etc. *See* low-density polyethylene and high-density polyethylene (EPA-40CFR60.561). Also called polyethene, polythene.

**polyethylene encasement** An economical technique used for the protection of gray or ductile cast-iron pipes against corrosion. It consists of a 0.2-mm tube or sheet of polyethylene placed around the pipe before installation in the trench. The encasement constitutes an unbonded and durable film that prevents contact with corrosive soils.

**Poly-Filter** A plate-and-frame filter press manufactured by Clow Corp.

**polyhydroxybutyrate, poly (3-hydroxyburate),** or **poly-b-hydroxybutyrate (PHB)** (1) A carbon compound that is stored by phosphorus-accumulating organisms as a result of their uptake of volatile fatty acids in an anaerobic zone. *See* enhanced biological phosphorus removal. (2) A plastic material digested by soil microorganisms from organic compounds. *See also* Biopol.

**polyhydroxybutyric acid (PHB)** or **poly-β-hydroxybutyric acid** A substance found in many bacteria and containing carbon or energy; a long chain of hydroxybutyric acid molecules.

**polyiron** A chemical substance that contains a significant amount of metal hydrolysis products.

**polyiron chloride (PICl)** A prepolymerized, prehydrolyzed salt of iron; a chemical commonly used as a coagulant in water and wastewater treatment and produced by reacting hydrochloric acid (HCl) with a purified source of iron.

**Polyjet** A flow control valve manufactured by Bailey.

**Poly-Links** A chain manufactured by NRG, Inc. for sludge collectors.

**Polymair** An apparatus manufactured by Acrison, Inc. for handling polymers.

**Polymaster®** An apparatus manufactured by Komax Systems, Inc. for mixing and feeding polymers.

**PolyMax** Products of Semblex, Inc. for mixing and feeding polymers.

**polymer** A natural or synthetic organic compound of high molecular weight, formed by the union of many monomers (molecules of low molecular weight); a basic molecular ingredient in plastic. Polymers are used in chemical coagulation, in water-insoluble ion-exchange resins, in trickling filter media, and in the fabrication of plastic or plastic-lined pipes. A polymer slurry can also be injected into sewer pipes to reduce their friction and increase their capacity significantly. All polyelectrolytes are polymers, but not vice versa. *See also* polyacrylamide, dry polymer, liquid polymer, emulsion polymer, mannich polymer.

**polymer aging** The process of storing and slowly mixing dilute polymer in day tanks until it acquires the desirable characteristics.

**polymerase chain reaction (PCR)** A molecular method that allows exponential amplification of a specific DNA sequence; useful in the clinical arena and potentially to environmental microbiology, e.g., for the rapid detection of potential pathogens. DNA microarrays are also an advanced molecular technology. *See also* primer, RT-PCR.

**polymer bridging** A flocculation mechanism; one of the forces that cause particles in a suspension to attract each other; a polymer connects previously formed particles into a larger particle that settles more readily. *See also* steric interaction, DLVO theory, electrostatic stabilization, London–van der Waals force, hydrodynamic retardation.

**polymer coagulant** A polymer used in water or wastewater treatment to destabilize particles in a suspension so that they can agglomerate into larger particles and settle more readily.

**polymer conditioning zone** The first unit of a belt filter press, consisting of a small tank, a drum, or an injector for the addition of polymer to the sludge to be dewatered.

**polymeric coagulant** A polyelectrolyte used as a coagulant.

**polymeric sludge** The residuals of water or wastewater treatment processes that include the use of polymer.

**polymerization** (1) The formation of a compound (polymer) by the joining of simple basic chemical units (monomers). (2) Synthesis of complex and stable compounds from the partial or incomplete degradation of simpler compounds.

**polymer slurry** *See* polymer.

**PolyMixer** An apparatus manufactured by Atlantes Chemical Systems, Inc. for mixing and feeding polymers.

**Polymizer** A centrifuge manufactured by Alfa Laval Separation, Inc.

**polynuclear aromatic hydrocarbons (PNA or PAH)** A group of aromatic compounds, very insoluble in water, containing several benzene rings ($C_6H_6$), usually fused. Examples: naphthalene (2 rings → $C_{10}H_8$), anthracene (3 rings → $C_{14}H_{10}$), and pyrene (4 rings → $C_{16}H_{12}$). Also called polyaromatic hydrocarbons and polycyclic aromatic hydrocarbons, they are the organic particles of greatest concern in the atmosphere. PAHs result from industrial operations, from the processing of petroleum and coal, or from the pyrolysis or partial combustion of organic compounds. They include benzo(a)pyrene (the most cited example), dibenzopyrene, and dibenzoacridine, which are carcinogenic. Other PAHs, e.g., the BTEX group (benzene, toluene, ethylbenzene, and xylene), are among the top 50 chemicals manufactured. *See also* aryl hydrocarbon.

**polynuclear complex** A complex that has more than one central metal atom. Also called a multinuclear complex.

**polyolefin** *See* polyethylene.

**polyoxyethylene (20) sorbitan monooleate** Same as polysorbate 80.

**Polyozone** Equipment manufactured by Polymetrics to handle ozone.

**Polypak** An apparatus manufactured by Leeds & Northrup for polymer feeding.

**polyphosphate** Any of a number of inorganic compounds with two or more phosphorus atoms, arranged as polymeric chains or rings, often used to prevent the formation of iron, manganese, and calcium carbonate deposits; e.g., sodium polyphosphate [$Na_3(PO_3)_6$], hexametaphosphate ($MPO_3$, where M represents a metal), pyrophosphate ($P_2O_7^{5-}$), and tripolyphosphate ($P_3O_{10}$). Polyphosphates tend to hydrolyze into orthophosphates over time and at high pH and temperature; see reversion. They also constitute nutrients enabling growth of plants and organisms. *See also* phosphate.

**polyphosphate treatment (well)** The addition of 30–40 grams per liter of polyphosphates to remove incrustations from a well screen. The solution is placed above screen level and agitated intermittently for a few hours. Subsequent pumping of the well removes the chemical and loosened incrustations. *See also* acid treatment, dry ice, chlorination.

**polyphosphonate** A relatively inexpensive compound of sodium hexametaphosphate (SHMP) and potassium pyrophosphonate that is used in the control of scales in boilers and in reverse osmosis membranes.

**polypig** A flexible polyurethane device that is driven by water pressure or pulled by a cable through new drinking water pipes to clean and assist in flushing the pipes before they are disinfected. *See also* pig.

**PolyPress** A plate-and-frame filter press manufactured by Star Systems.

**PolyPro** An apparatus manufactured by AquaPro, Inc. for feeding dry polymer.

**polypropylene [$(C_3H_6)_n$]** A thermoplastic polymer of propylene, resistant to chemical attack, used in the manufacture of pipes, tubing, fittings, and in other applications.

**polyprotic acid** An acid having two or more transferable protons (hydrogen ions), e.g., carbonic acid ($H_2CO_3$) and phosphoric acid ($H_3PO_4$). *See also* polyprotonic acid.

**polyprotic base** A base that can accept more than one proton, e.g., the hydroxyl ion ($OH^-$) and carbonate ion ($CO_3^{2-}$).

**polyprotonic acid** An acid that can donate more than one proton or hydrogen ion, e.g., the cation exchangers used in water treatment. *See also* polyprotic acid.

**polysaccharide** (1) A carbohydrate consisting of a chain of monosaccharides, including energy compounds like starch and structural integrity compounds like cellulose. (2) A natural polyelectrolyte excreted at the surface of microorganisms and influential in bioflocculation.

**polysaprobic zone** A grossly polluted area or reach of a slow-moving stream; it contains organic wastes in mostly anaerobic decomposition. *See also* saprobic classification.

**Polyseed®** A bacterial culture prepared by Polybac Corp. for use in BOD tests.

**polysorbate 80** The generic name for polyoxyethylene (20) sorbitan monooleate, a nonionic surfactant used as a dispersant, defoaming agent, and in other applications.

**Poly-Stage™** An air scrubbing apparatus manufactured by Davis Industries.

**polystyrene [$(C_6H_5CH{:}CHCH_2)_n$]** Polymerized styrene; a clear plastic or stiff foam, used as insulator or in the manufacture of ion exchange resin beads.

**polystyrene resin** A common ion-exchange resin made of polystyrene in the form of small spheres of 0.5 mm in diameter.

**polysulfone [$(—C_6H_4SO_2C_6H_4OC_6H_4C(CH_3)_2C_6H_4O—)_n$]** A synthetic thermoplastic polymer used in the manufacture of water treatment membranes.

**polysulfone membrane** A water treatment membrane with a base of polysulfone.

**polytetrafluoroethylene [$(C_2F_4)_n$]** Acronym: PTFE. A polymer of tetrafluoroethylene, with excellent chemical and electrical resistance, used as a coating and for slide bearings, gaskets, and liners for pipes and tanks.

**polythene** *See* polyethylene.

**PolyThickener** A thickener manufactured by Walker Process Equipment Co. for processing waste activated sludge.

**Polytox™** A kit manufactured by Polybac Corp. for conducting biological toxicity tests.

**PolyTube** A tubular air diffuser manufactured by Walker Process Equipment Co.

**polyunsaturated** Pertaining to a chemical substance based on fatty acids that have two or more double bonds per molecule.

**polyurea {[$CO(NH_2)_2]_n$}** A polymer of urea used in the manufacture of plastics, membranes, resins, etc.

**polyurea membrane** A composite membrane with a polyurea base.

**polyurethane** An elastomer or thermoplastic polymer containing the group NHCOO used for padding, insulation, etc., for the production of resins, and in the equipment and structures of water and wastewater facilities.

**polyvalent** Pertaining to a chemical element that has a valence greater than 2. Also called multivalent.

**polyvinyl alcohol [$(—CH_2CHOH—)_x$]** Acronym: PVA. A soluble, synthetic polymer resulting from the combination of an alcohol and polyvinyl acetate {$[—CH_2CH(OOCCH_3)—]_x$}.

**polyvinyl chloride (PVC)** A polymer of vinyl chloride; a tough, environmentally indestructible thermoplastic that releases hydrochloric acid (HCl) when burned; resistant to corrosion, used to fabricate water and wastewater piping, PVC resins, and biological filter media.

**polyvinyl chloride (PVC) pipe** A water distribution pipe manufactured from polyvinyl chloride.

**polyvinylidene fluoride (PVDF)** A tough thermoplastic material, resistant to chemicals, abrasives, and heat.

**polyvinylpyridinium salt** A cationic polyelectrolyte prepared from selected monomers.

**poly wrap** A thin plastic sheet placed around a ductile iron pipe to protect the pipe against corrosion in the surrounding soil.

**poly-β-hydroxybutyrate (PHB)** *See* polyhydroxybutyrate.

**poly-β-hydroxybutyric acid (PHB)** *See* polyhydroxybutyric acid.

**POM** Acronym of particulate organic matter

**pond** A natural or man-made body of freshwater, smaller than a lake and larger than a pool; sometimes artificially formed by damming a stream. *See also* basin, lagoon, oxidation pond, stabilization pond, tank. In wastewater treatment, pond and lagoon are used synonymously, although a lagoon may be considered smaller and shallower than a pond. *See also* lagoons and ponds.

**pond dispersion factor** A parameter used in the design of stabilization ponds to account for the degree of dispersion in a reactor, from $d = 0$ for an ideal plug-flow reactor to infinity for a complete-mix reactor, but within a range of 0.1–2.0 for most stabilization ponds. *See also* Wehner–Wilhem's equation.

**pond *E. coli* removal** *See* Dissanayake's equation.

**ponding** (1) The impoundment of streamflow to create a pond. Ponding also refers to the situation in which water backs up in a channel, ditch, or conduit because of insufficient capacity at a downstream point. The ponding option is an XP-SWMM option that allows water to be stored at a junction without increasing the hydraulic gradeline until there is sufficient hydraulic capacity for the water to rejoin the network. The opposite is the sealed option. (2) Same as pooling. (3) The accumulation of wastewater flow on the biomat at the bottom of a disposal field trench when the application rate exceeds the long-term acceptance rate of the biomat. (4) The filling up of void spaces in a biological filter or in a landfill.

**pond series** A combination of earthen stabilization basins including aerobic, anaerobic, and facultative ponds designed to exploit the advantages of each type and provide complete treatment or complete retention of wastewaters. An example is a system consisting of an anaerobic pond followed successively by an aerated lagoon, a facultative pond, and one or more aerobic (maturation) ponds. *See also* dual process, lagoons and ponds.

**ponds in dual processes** Aerated or facultative lagoons that are used in combination with roughing filters or other fixed-film reactors for wastewater treatment. *See also* lagoons and ponds.

**pond system** *See* pond series. Stabilization ponds may also be operated in parallel, which allows a better distribution of settled solids.

**pond weed** A submerged plant with most of its foliage beneath the water surface.

**Pond-X®** A product of NuTech Environmental Corp. for odor control.

**Pontiac fever** A mild, nonpneumonic form of legionellosis or legionnaires' disease, which heals on its own; caused by the pathogen *Legionella pneumophilia.* It has flu-like symptoms (anorexia, malaise, myalgia, headache) and affects almost all exposed subjects.

**Pontoon** A floating cover manufactured by FMC Corp. for use on anaerobic digesters.

**pool** (1) A small body of standing water, smaller than a pond, or any small collection of liquid; a puddle. (2) A quiet, deep place in a stream.

**pooling** The formation of pools of liquid on the surface of a clogged filter. *See also* ponding.

**POP** Acronym of persistent organic pollutant.

**poppet** A poppet valve.

**poppet valve** A flat disk that rises or descends without rotation while closing or opening; used, e.g., in internal-combustion and steam engines Also called a mushroom valve, poppet.

**population** A group of interbreeding organisms of the same type or species occupying a particular space; the number of humans or other living creatures in a designated area. Population characteristics include birth and death rates, age distribution, density, growth rates, and sex ratio.

**population at risk** A population subgroup that is more likely to be exposed to a chemical or is more sensitive to a chemical than the general population.

**population curve** A graphical representation of the number or mass of bacteria, showing the various growth phases (lag, log, stationary, death) as a function of time.

**population dynamics** The interrelationships between the various microscopic plants and animals involved in waste processing, governed by three major factors: competition for food, predator–prey relationship, and symbiotic association. *See also* growth patterns, primary feeder, biomass, food-to-microorganism ratio, secondary feeder.

**population equivalent** The estimated population that would contribute a given load of water, wastewater, solid waste, etc. It is usually applied to nondomestic users such as industrial, commercial, or institutional establishments, and relates to a specific parameter such as flow, five-day BOD, or total suspended solids (TSS). In wastewater studies, e.g., per-capita daily values commonly used include a flow of 100 gallons, 0.20 pound of TSS, and 0.17 pound of BOD. Thus, a wastewater stream of 50,000 gallons per day carrying 102 pounds of BOD and 105 pounds of TSS would have a hydraulic population equivalent of 500, but population equivalents of 600 with respect to BOD, and 525 with respect to TSS. A related concept is equivalent dwelling unit, which is the average household size.

**population served** A parameter used in setting monitoring and sampling requirements for water supply systems; see, e.g., Coliform Rule.

**Porcupine** A sludge dryer manufactured by Bethlehem Corp.

**pore** A very small open space in a rock, aggregate, or granular material. *See also* interstice.

**pore diffusion** One of the mechanisms of activated carbon adsorption, in which solute molecules migrate through the carbon pores to an adsorption site.

**pore narrowing** A reduction of the pore size of a membrane caused by solid materials attached to the interior surface of the pores; a mechanism that accompanies concentration polarization and contributes to membrane fouling. *See also* fouling indexes, cake or gel formation, and pore plugging.

**pore plugging** The obstruction of the pores of a membrane by solid particles of about the same size that become stuck; a mechanism that accompanies concentration polarization and contributes to membrane fouling. *See also* fouling indexes, cake or gel formation, and pore narrowing.

**pore size** The nominal size of a membrane material's pores, an indication of the size of particles retained on the membrane surface. Usually expressed in microns, it is used to distinguish between low-pressure membranes.

**pore transport** The third step in the adsorption process, in which the materials to be adsorbed (adsorbate) move through the pores of the adsorbing particles (adsorbent) by molecular diffusion and surface diffusion to adsorption sites. Also called internal transport. *See also* bulk diffusion transport, film diffusion transport, and adhesion.

**pore volume** (1) The total volume of pore space in a given volume of rock or sediment; it usually relates to the volume of air or water that must be moved through contaminated material to flush the contaminants. (2) The volume of water or air that will completely fill all of the void space in a given volume of porous matrix. Pore volume is equivalent to the total porosity. The rate of decrease in the concentration of contaminants in a given volume of contaminated porous media is directly proportional to the number of pore volumes that can be exchanged (circulated) through the same given volume of porous media.

**pork tapeworm** (1) The common name of Taenia solium, a parasite that is distributed worldwide and is transmitted to persons who consume insufficiently cooked pork. (2) The excreted helminth infection caused by this parasite and called taeniasis. Cysticercosis is infection by the larval stage of the worm, which affects the muscles, the brain, and the heart. *See also* beef tapeworm.

**Poro-Carbon** An automatic filter manufactured by R. P. Adams Co., Inc. *See also* Poro-Stone.

**Poro-Edge** An automatic water strainer manufactured by R. P. Adams Co., Inc.

**porosity** (1) The degree to which soil, gravel, sediment, or rock is permeated with pores or cavities through which water or air can move. (2) The percentage of such pores or cavities; i.e., the volume fraction of a rock or unconsolidated sediment not occupied by solid material but usually occupied by water and/or air. Porosity differs from perviousness or permeability; clay and chalk have relatively high porosities but are not permeable. Porosity can increase by leaching or decrease by compaction. *See also* specific yield or effective porosity, specific retention, and bulk density. (3) The characteristic of an ion exchange resin that relates to the degree of openness of its polymer structure. *See also* matrix, resin exchange capacity, functionality. *See* microporous resin, gel, macroporous resin.

**porosity factor** The ratio $(1 - f)^2/f^3$, used in the Blake–Kozeny equation for head loss in water filtration, where $f$ is the porosity of the bed. The porosity factor results from the conversion of an approach velocity into an interstitial velocity.

**porosity function** *See* fluidized bed porosity function.

**Poro-Stone** An automatic filter manufactured by R. P. Adams Co., Inc. *See also* Poro-Carbon.

**porous** Pertaining to materials that have pores or cavities through which fluids or light may pass. *See* porosity.

**porous diffuser** A device made from ceramic, plastic, rubber, or cloth materials for the diffusion of air to water or wastewater. Also called fine-pore diffuser. *See* disk, dome, membrane, panel, and nonporous diffusers; jet aeration, U-tube aeration, coarse-bubble diffuser, fine-bubble diffuser.

**porous disk diffuser** A fine-bubble aeration device made of porous plastic or ceramic.

**porous medium** A material that combines three different phases: a solid or inorganic mineral, a liquid or solution, and a gas or the atmosphere. Some characteristics of porous media include texture, structure, cation exchange capacity, isomorphic substitution, and adsorption affinity. *See also* electrical double layer, hydrophobic bonding, and soil water potential.

**porous-media model** A bench-scale model to study the movement of groundwater through porous media; e.g., the studies of Darcy in 1856 led to the equation that defines groundwater flux as a function of hydraulic conductivity and hydraulic gradient. *See* porosity.

**porous membrane** A membrane in which gases travel through by diffusion and separation occurs on the basis of molecular size and membrane pore size.

**porous pavement** A pavement that reduces imperviousness and surface runoff; it is made of asphalt, special concrete (without the finer sediment), or interlocking open-cell cement blocks over a base of coarse gravel.

**porphyria cutanea tarda** A condition caused by prolonged exposure to chemicals such as hexachlorobenzene and characterized by cutaneous lesions and hyperpigmentation.

**Port-A-Berm™** A portable containment apparatus manufactured by Aero Tec Laboratories, Inc.

**portable exchange tank** A small container (up to 2 cubic feet) of ion exchange or filter medium ready for use in household or commercial applications.

**Portacel®** An apparatus manufactured by A. B. Leopold Co., Inc. for disinfection by gaseous chlorine.

**Porta-Cleanse** A mixer manufactured by ITT Flygt Corp. for installation in pumping station wet wells.

**Porta-Feed®** An apparatus manufactured by Nalco Chemical Co. for handling chemicals.

**portal of entry** The point of entrance of microorganisms to the human body; the pathway by which pathogenic agents enter the body. In the case of waterborne or foodborne pathogens, they are most often carried through the mouth to the gastrointestinal tract. The nose, the skin, and the eyes are also portals of entry. *See also* fecal–oral route.

**Porta-Tank®** An apparatus manufactured by Environetics, Inc. for storage of liquids.

**port head loss** *See* hydraulic grade differential.

**Posi-Clean** A straining element manufactured by Tate Andale, Inc.

**Posirake** A bar screen manufactured by Zimpro Environmental, Inc., including a reciprocating rake.

**positive** (1) Having an electrical charge associated with an excess of protons or a deficiency of electrons. (2) Showing the presence of disease or organism.

**positive aeration** The condition of an aerated static pile composting system whose blowers force air up through the pile, allowing the removal of moisture. Also called positive ventilation.

**positive bacteriological sample** A water sample that shows the presence of coliforms according to an approved test. *See* Total Coliform Rule. *See also* positive sample.

**positive catalyst** A substance that increases the speed or yield of a chemical or biological reaction without being consumed or chemically changed by the reaction; e.g., dissolved metals, enzymes.

**positive charge** The electrical potential of a proton; the charge of a cation.

**positive-displacement blower** An air conveying device used in wastewater aeration systems, operating at constant capacity and variable pressure.

**positive-displacement diaphragm pump** A pump used in water and wastewater treatment to feed and meter chemicals, e.g., hypochlorite solutions or fluosilicic acid.

**positive-displacement meter** A flow measuring device that captures definite volumes of water and counts the operations.

**positive-displacement pump** A piston, diaphragm, gear, or screw pump in which liquid is drawn into a space and its pressure is increased for its discharge. It delivers a constant volume with each stroke and is often used for pumping out ground works. Positive-displacement pumps are also used to move wastewater treatment sludges. The other broad category of pumps includes continuous-flow or rotodynamic pumps. *See* diaphragm, plunger, progressive cavity, peristaltic hose pump, reciprocating piston, air lift, Archimedes screw, and rotary lobe pumps, pneumatic ejector.

**positive head** The element of the total dynamic head of a pump that represents the elevation of the fluid above a datum. Also called elevation head, potential head, or potential energy. *See also* hydraulic energy, dynamic head, and pump head terms.

**positive rotary pump** A displacement pump consisting of elements rotating in a case.

**positive sample** A sample that contains bacteria according to the multiple-tube fermentation or the membrane filter tests. *See also* positive bacteriological sample.

**Positive Seal** A rotary distributor manufactured by Walker Process Equipment Co. for use in trickling filters.

**positive ventilation** Same as positive aeration.

**positron** A positively charged particle with the same mass as an electron. *See also* antiparticle.

**possible human carcinogens via ingestion** Any contaminant for which limited evidence of carcinogenicity in animals exists, based on epidemiological studies. For example, there is equivocal evidence for lindane and styrene. Also called category II contaminant or category B carcinogen. This USEPA classification corresponds more or less to category 2 of the International Agency for Research on Cancer and category b of the National Toxicology Program. The MCLG for such substances is based on RfD and safety margins or set within a cancer risk range of 100,000 to 1,000,000. *See also* carcinogen, mutagen, and known probable human carcinogen via ingestion.

**postaeration** The process of aerating the effluent of a treatment plant to increase its dissolved oxygen concentration so that it meets effluent standards and permit requirements. Postaeration methods include cascade aeration, mechanical aeration, and diffused air.

**postammoniation** The addition of ammonia ($NH_3$) to treated water that contains hypochloric acid (HOCl) to produce chloramines.

**postanoxic denitrification** A biological nitrogen removal process using an aerobic/nitrification reactor followed by an anoxic reactor and a final clarifier, with sludge recycle and wastage from the clarifier underflow. Sometimes an exogenous carbon source, e.g., methanol ($CH_3OH$) or acetate ($CH_3COOH$), is added to supplement endogenous respiration.

**postanoxic nitrogen removal process** A biological nitrogen process in which an anoxic/denitrification reactor follows an aerobic/nitrification reactor and a final clarifier. *See* the following processes: postanoxic denitrification, single sludge, Bardenpho™, oxidation ditch, two-sludge, simultaneous nitrification/denitrification, preanoxic denitrification.

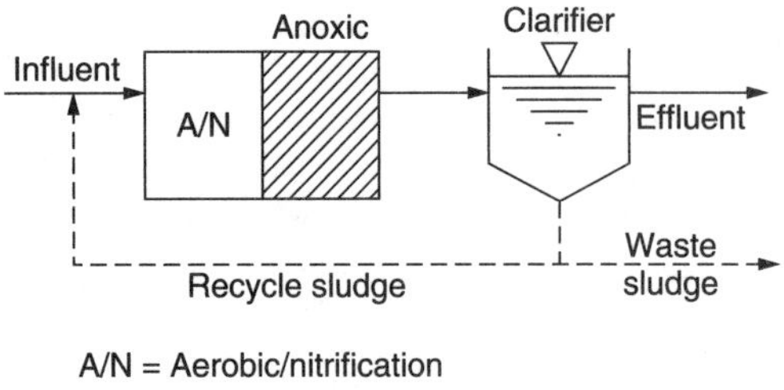


**Figure P-10.** Postanoxic denitrification.

**postchloramination** The addition of chloramines to filtered water as a secondary disinfection step.

**postchlorination** The addition of chlorine to the effluent of a water or wastewater treatment plant (following treatment), for disinfection purposes or to maintain a residual in the distribution system. In water treatment, chlorine is often added before the clear well or immediately before the filter. Also called terminal disinfection. *See also* prechlorination.

**postdenitrification** A single-sludge process designed to remove nitrogen beyond the metabolic requirements of the organisms. *See* Wuhrman process for detail.

**postdisinfection** The addition of a disinfectant to water following filtration or the final treatment process. *See* postammoniation, postchlorination, postchloramination, postozonation.

**postfilter adsorber** A granular activated carbon contactor used for the removal of dissolved organic compounds, particularly those responsible for taste and odor, from water that has passed through a granular media filter. This adsorber does not require backwashing unless it is subject to extensive biological growth. *See also* filter-adsorber, biological filter.

**postlime stabilization** The addition of hydrated lime or quick lime to dewatered sludge to raise the pH of the mixture and stabilize the material. The system consists of a dry lime feeder, a dewatered sludge cake conveyor, and lime–sludge mixer. *See* lime posttreatment.

**postozonation** The application of ozone following the final treatment step (e.g., filtration) to provide disinfection. *See also* intermediate ozonation, preozonation, source water ozonation.

**postprecipitation** A phosphorus removal process that consists of the addition of chemicals to the effluent from a secondary clarifier and the separation of chemical precipitates, usually in a separate basin or filter. *See also* coprecipitation, preprecipitation, biological phosphorus removal.

**posttreatment** An additional step of water or wastewater treatment, for example, corrosion control or dechlorination. In a pressure-driven membrane filtration system, posttreatment consists of aeration (to remove hydrogen sulfide and carbon dioxide), disinfection, stabilization, alkalinity recovery, and other conventional unit operations, as required.

**potability** The overall characteristic of potable water; it is a challenge to establish it and maintain it throughout the distribution system.

**potable** Safe to drink. *See* potable water.

**potable reuse** The use of reclaimed water for potable purposes, e.g., by blending in water supply reservoirs or piping into the distribution system. Issues related to potable reuse include aesthetics, public acceptance, trace organics, and transmission of enteric viruses. *See also* pipe-to-pipe potable reuse.

**potable supply** A potable water supply.

**potable water** Water that is safe *and* satisfactory for drinking and cooking. Safe water does not contain harmful organisms, toxic materials, or chemicals; potable water when delivered to the ultimate consumer, meets the established standards for bacteriological, chemical, physical, and radionuclide parameters. Water that has objectionable odor, color, taste or mineral problems may still be safe for drinking but is not considered potable. Sometimes called drinking water, finished drinking water.

**potable water approval** Approval of a water supply by a regulatory agency as meeting drinking water standards.

**potable water reuse** The direct or indirect augmentation of a drinking water supply by reclaimed water, after appropriate treatment to protect public health. *See also* water reuse applications.

**potamology** The branch of hydrology that studies streams and rivers; surface water hydrology.

**potash** (1) A compound of potassium, essentially the oxide of the metal, $K_2O$. It is one three substances commonly used to rate the fertilizing value of commercial fertilizers as well as the agricultural value of sludge. (2) The crude, impure form of potassium carbonate ($K_2CO_3$) obtained from wood ashes. (3) Potassium hydroxide (KOH).

**potash alum** A metal coagulant used in water treatment; same as potassium aluminum sulfate; see alum (2).

**potassium (K,** from its Latin name kalium**)** An abundant, soft, light, silver-white, waxlike, very reactive, metallic element of the alkali group, easily oxidizable in air, found only in mostly insoluble minerals. Atomic weight = 39.102. Atomic number = 19. Specific gravity = 0.86 at 20°C. It is an essential nutritional element, but may be toxic at very high intake rates; *see* hyperkalemia, hypokalemia. The USEPA and WHO do not regulate potassium in drinking water but the European Community has a guide value of 10 mg/L as K. Its compounds are used as fertilizer and in hard glass. *See also* potash.

**potassium-40** A significant radionuclide in drinking water, representing 0.012% of total potassium and only an intake of 8 pCi per day from a consumption of 2 liters per day. Half-life > 1 billion years.

**potassium acid phthalate** Same as potassium hydrogen phthalate.

**potassium alum** Same as potassium aluminum sulfate; see alum (2).

**potassium aluminum sulfate** *See* alum (2).

**potassium brine** A solution of potassium chloride (KCl) used in the regeneration of ion exchange resins.

**potassium chloride (KCl)** A white or colorless, crystalline solid salt of potassium that is used in the regeneration of cation exchange softeners, manufacture of fertilizers, and mineral water.

**potassium chloroplatinate ($K_2PtCl_6$)** One of the substances used to produce a platinum–cobalt solution that approximates the color of a natural water. The other is cobaltous chloride ($CoCl_2$).

**potassium cycle** The regeneration of cation exchangers with potassium chloride (KCl) instead of sodium chloride (NaCl).

**potassium dichromate ($K_2Cr_2O_7$)** An orange-red, crystalline, water-soluble, poisonous powder used in dyeing, photography, and as a laboratory reagent. Also called potassium bichromate. *See* the (COD) dichromate reflux method.

**potassium dihydrogen phosphate ($KH_2PO_4$)** A chemical used in laboratory experiments as a source of nutrients.

**potassium ferric oxalate [$K_3Fe(C_2O_4) \cdot 3H_2O$]** A substance used in a chemical actinometer for the measurement of the intensity of ultraviolet light and other forms of radiation.

**potassium fluoride (KF)** A white, crystalline, hygroscopic compound of potassium and fluorine that is sometimes used in water fluoridation. It is also a main ingredient of nerve gas.

**potassium hydrogen phthalate** Also called potassium acid phthalate.

**potassium iodide** A white, crystalline, water-soluble powder used as a reagent, in the preparation of Gram's solution, and in medicine.

**potassium permanganate ($KMnO_4$)** A dark purple, crystalline, soluble salt of potassium and manganese; a strong oxidizing agent used in water treatment for taste and odor control, for iron and manganese removal, and for disinfection; a fire and explosion hazard. Also called purple salt. *See also* scrubbing fluid.

**potassium permanganate ($KMnO_4$) oxidation** A treatment process in which permanganate ($MnO_4^-$) is reduced to manganese dioxide [$MnO_2(s)$]:

$$MnO_4^- + 4\,H^+ + 3\,e^- \rightarrow MnO_2(s) + 2\,H_2O \quad \text{(P-40)}$$

**potassium persulfate ($K_2S_2O_5$ or $K_2S_2O_8$)** A salt of persulfuric acid used in the water-soluble carbon test.

**pot chlorinator** A simple and cheap device used in villages to disinfect well water. It consists of a single or double pot that contains a mixture of sand and bleaching powder. The pots hang underwater by a rope. *See* Figure P-11.

**potency** Amount of material necessary to produce a given level of a deleterious effect.

**potency factor (PF)** The risk resulting from a lifetime exposure of 1.0 mg/day of a substance per kg of body weight. *See* slope factor for detail.

**potential** (1) The work required to move a unit mass against a unit force over a given distance. (2) The force or head that makes water move through soil. (3) *See* capillary potential. (4) *See* electrostatic potential. (5) *See* gravitational potential. (6) *See* potential. (7) *See* DBP and THM formation potential. *See also* potential energy.

**potential *Cryptosporidium* surrogate** *See Cryptosporidium* surrogate

**potential-determining ion** A charged ion that determines the attraction or repulsion among particles in a suspension. In chemical coagulation, using strong acids or strong bases, potential-determining ions or counterions may be added to solutions that contain charged colloidal particles to react with the colloid surface and reduce the surface charge. *See also* Shultz–Hardy rule.

**potential energy** (1) Same as potential head. (2) In the electrical double layer theory, the energy ($E$) required to reduce the distance between two particles in a suspension, resulting in an attraction (when $E < 0$) or a repulsion (when $E > 0$).

**potential evapotranspiration** *See* reference evapotranspiration (2).

**potential habitat** A habitat that is suitable for, but currently unoccupied by, a given species or community.

**potential head** The element of the total dynamic head of a pump that represents the elevation of the fluid above or below a datum. Also called elevation head, elevation energy, potential energy. *See also* hydraulic energy, dynamic head, and pump head terms.

**potentially hazardous metal** See cadmium (Cd), chromium (Cr), lead (Pb), mercury (Hg).

**potential-pH diagram** Same as Pourbaix diagram.

**potential surrogate** *See Cryptosporidium* surrogate.

**potential unit infiltration rate (PUIR)** The potential rate of water infiltration in an active streambed zone of one foot square; an infiltration rate parameter expressed in gallons per minute per square foot (gpm/ft$^2$) or in meters per hour (m/hr), it can be estimated from such hydrologic data as river stage, groundwater elevation, surface water

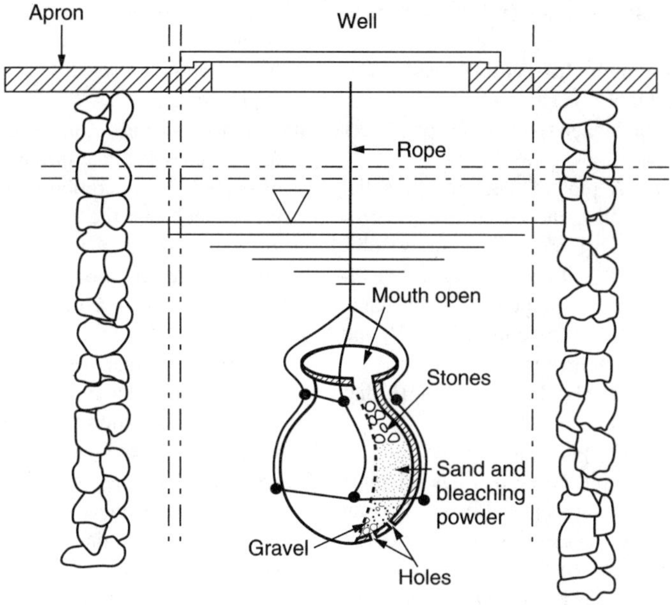


**Figure P-11.** Pot chlorinator (adapted from Cairncross & Feachem, 1992).

temperature, and streambed permeability and thickness, using Darcy's law.

**potentiation** The effect of one chemical to increase the effect of another chemical, e.g., when an inactive substance enhances the action of an active substance. *See also* antagonism, synergism.

**potentiometer** A variable resistor of electricity used as an instrument for measuring the electrical potential of chemical reactions; it opposes a voltage to that of an electrochemical cell so as to block the passage of current and keep concentrations constant.

**potentiometrical procedure** A laboratory procedure that measures the difference in voltage to determine the concentration of a constituent of a water or wastewater sample.

**potentiometric surface** The imaginary surface defined by the water level in wells tapping the confined or semiconfined groundwater zone. Also called piezometric surface.

**potentiometric titration** Titration in which the end point is determined by the voltage of an electric current of a given amperage.

**Potter's formula** A formula proposed to estimate flood flows in the Cumberland Plateau of Kentucky and Tennessee (Fair et al., 1971):

$$Q = CA^{7/6}/(L/S^{1/2}) \qquad \text{(P-41)}$$

where $Q$ is the peak flood flow in cfs; $C$ is a coefficient that accounts for the effects of storm frequency, runoff–rainfall relationship, and maximum rainfall; $A$ is the drainage area; $L$ is the length of the principal waterway in miles; and $S$ is the slope of the waterway in feet per mile. For $A$ in square miles, $C = 1920\ D$; $D$ is a factor that relates the basin to a base station in Columbus, Ohio.

**POTW** Acronym of publicly owned treatment works. The POTW treatment plant is the portion of the POTW that is designed to provide treatment (including recycling and reclamation) of municipal sewage and industrial waste (EPA-40CFR403.3-p).

**POTW treatment plant** *See* POTW.

**POU** Acronym of point of use.

**pound** A unit of mass, abbreviated lb, equal to approximately 0.4536 kg.

**pound per square foot (psf)** A unit of pressure, equal to a force of one pound applied uniformly over an area of one square foot. Pound per square inch (psi) is a unit of pressure equal to a force of one pound applied uniformly over an area of one square inch. Pound per square inch absolute (psia) is the absolute pressure of one pound per square inch. Pound per square inch differential (psid) is the pressure differential in psi between two points in a fluid. Pound per square inch gage (psig) is a gage pressure of one pound per square inch.

**pound per square inch (psi)** *See* pound per square foot.

**pound per square inch absolute (psia)** *See* pound per square foot.

**pound per square inch differential (psid)** *See* pound per square foot.

**pound per square inch gage (psig)** *See* pound per square foot.

**Pourbaix diagram** A plot of electrode potential (E) versus pH indicating the zones where corrosion (as well as passivation and immunity) of a metal (e.g., iron, lead) will occur in water. The diagram shows dominant soluble and insoluble species of the metal, as a function of E, pH, and other water quality parameters. *See* Figure P-12. Also called potential-pH diagram, Eh-pH diagram, or E-pH diagram.

**pour-flushed toilet** *See* pour-flush latrine.

**pour-flush latrine** A pit latrine that has a U-pipe filled with water serving as a seal below the seat or squatting plate to block flies and odors. The mixture of excreta and flushing water is usually discharged into a soakaway. *See also* aqua-privy, cistern-flush toilet, twin-pit pour-flush latrine.

**pour-flush toilet** *See* pour-flush latrine.

**pour plate count** A method of enumerating individual bacteria in water, wastewater or other samples, involving the serial dilution of a sample, mixing each dilution with a warm culture medium, pouring it into a dish and incubating it, and counting the bacterial colonies formed; the pour plate method. *See also* direct count, spread plate count, membrane-filter technique, multiple-tube fermentation, presence–absence test, heterotrophic plate count.

**pour plate method** One of three laboratory procedures used in the heterotrophic plate count method of bacterial enumeration in a water or other sample. It indicates the number of colonies of heterotrophic (pathogenic and innocuous) bacteria grown on selected solid media at a given temperature and incubation period: tryptone glucose extract agar or tryptone glucose yeast, for 48 hours at 35°C. The other two procedures are membrane filter and spread plate. *See* pour plate count.

***P. ovale*** Abbreviation of *Plasmodium ovale.*

**powdered activated carbon (PAC)** One of two size classifications of activated carbon, with an approximate diameter of 0.074 or 200 sieve; the other size is granular activated carbon. PAC provides a large surface area (800–1800 $m^2/g$) for the adsorption of organic particles; it is usually fed as a slurry ahead of filtration, without reuse, for the removal of compounds that cause taste and odor and for the control of trace organic contaminants. Parameters used to characterize powdered activated carbon include a bulk density of 360–740 $kg/m^3$, an iodine number of 800–1200, molasses number, and phenol number. Jar tests are used to determine the dose of powdered activated carbon

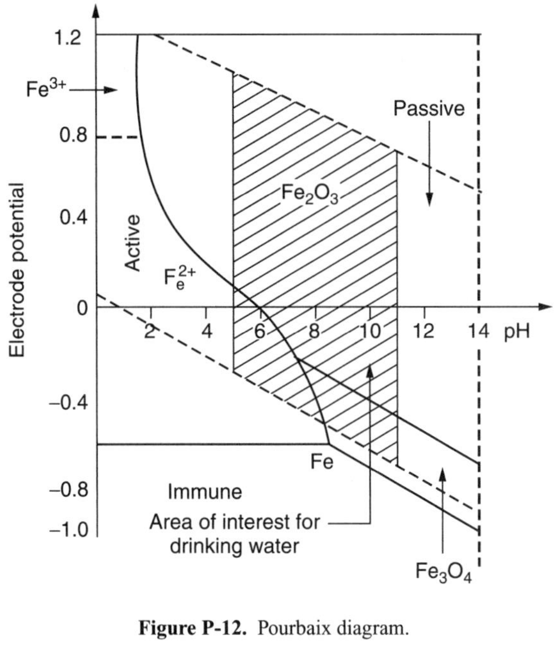


**Figure P-12.** Pourbaix diagram.

required for a given application. *See also* floc blanket reactor/PAC/UF process, PAC/UF system, Roberts–Haberer process.

**powdered activated carbon activated sludge process** The activated sludge process with powdered activated carbon added to the aeration basin to enhance the removal of recalcitrant and toxic substances.

**powdered activated carbon flooding** The uncontrolled running of powdered activated carbon through the dry feeding machines used in water treatment plants.

**powdered activated carbon—microfiltration** *See* PAC/MF system.

**powdered activated carbon—ultrafiltration** *See* PAC/UF system.

**powdered silica** A mineral powder used as a weighting agent or coagulant aid.

**Powder Pop®** A chlorine testing device manufactured by HF Scientific, Inc.

**Powdex** A treatment unit manufactured by Graver Co., combining filtration and ion exchange.

**PoweRake** A bar screen manufactured by Enviro-Fab, Inc., with a reciprocating rake.

**Power Backwash™** A filter backwash device manufactured by TETRA Technologies, Inc., using both air and water.

**Power Brush™** A pipeline cleaner manufactured by Pipeline Pigging Products, Inc.

**power coefficients** *See* power number.

**power dissipation function** A concept used in water and wastewater treatment as a rough measure of mixing effectiveness; it is equal to the ratio of the useful power input ($P$) to the volume ($V$) of the reactor in which treatment takes place. The power dissipation function is related to the velocity gradient or shear gradient ($G$):

$$P = MVG^2 = \rho g Q H_L \qquad \text{(P-42)}$$

where $M$ is a proportionality factor with the same dimensions as the absolute viscosity, $\rho$ is the density of the liquid, $g$ is the acceleration of gravity, $Q$ is the volumetric flow rate, and $H_L$ is the headloss. *See also* contact opportunity, loading intensity.

**power factor** The ratio (KW/KVA) of the true power passing through an electric circuit to the product of the voltage and amperage in the circuit. This is a measure of the lag or load of the current with respect to the voltage. Medium- and high-frequency ozone generators have higher power factors than low-frequency generators. *See also* power terms.

**power head** A mechanism that transmits power to a deep-well pump.

**Powerhouse®** An alkaline cleaner and degreaser produced by Technical Products Corp.

**power input** The electrical input to a motor expressed in kilowatts. *See* useful power input, power terms.

**Powermatic** A bar screen manufactured by Brackett Green, Ltd., with a reciprocating rake.

**Power Mizer™** A multistage centrifugal blower manufactured by Spencer Turbine Co.

**power number** A dimensionless parameter, analogous to the drag coefficient or friction factor, used to express mixing and stirring relationships. There are two equations for the power number ($\Phi$); they both depend on the Reynolds number ($R_e$) of the impeller and the characteristics of impeller and basin:

$$\Phi = 6{,}750\, P/(\rho N^3 R^5) = K R_e^p F_r^q \qquad \text{(P-43)}$$

$$\Phi = K' R_e^{p'} \qquad \text{(P-44)}$$

where $P$ = useful power input; $\rho$ = fluid density; $N$ = number of revolutions per minute; $R$ = effective radius arm of the impeller blade; $K$, $p$, $q$ = coefficients depending on flow conditions; $F_r$ = Froude number; $K'$ = characteristic constant of impeller and basin, called power coefficient, varying over a wide range, e.g., from 0.32 (three-blade, square-pitch, propeller in the turbulent range) to 172.5 (unbaffled, shrouded turbine with stator, viscous range); $p'$ = coefficient depending on flow conditions—$p' = -1$ for laminar flow, $p' = 0$ for turbulent flow. *See also* power input and impeller mixer power requirement.

**PowerPro™** A chlorination unit manufactured by PPG Industries, Inc. of Pittsburgh, PA; it incorporates the Accu-Tab® chlorinator.

**power pump** A pump that uses a form of power other than human energy.

**power requirement** The rate of energy input required to operate a piece of equipment.

**power terms** Terms related to the operation of pumps and motors include:

*Brake horsepower.* The power ($P_b$) delivered to a pump shaft expressed in horsepower. $P_b = E_m$ (1.341 kW).

*Horsepower.* A unit of power equivalent to 550 foot-pounds per second or 745.7 watts.

*Hydraulic horsepower.* The pump output or the liquid horsepower ($P_w$) delivered by the pump. $P_w$ = head (ft) × capacity (gpm)/3960.

*Motor efficiency ($E_m$).* A measure of how effectively the motor turns electrical energy into mechanical energy; the ratio of power input to power output.

*Motor input horsepower.* The power input to the motor expressed in horsepower = 1.341 kW.

*Overall efficiency.* Total efficiency.

*Power factor.* The ratio of the true power to the volt-amperes in an alternative current circuit = 1000 k W/(Amp × Volts × Phase).

*Power input.* The electrical input to a motor expressed in kilowatts (kW); a measure of the rate at which work is done.

*Pump efficiency ($E_p$).* The ratio of the energy delivered by the pump to the energy supplied to the pump shaft = 100 $(P_w)/P_b$.

*Shaft horsepower.* Brake horsepower.

*Total efficiency ($E_t$).* The ratio of the energy delivered by the pump to the energy supplied to the input side of the motor = 100 $(P_w)$/(1.341 kW)

*Water horsepower.* Hydraulic horsepower.

*Wire-to-water efficiency.* Total efficiency.

*Wire-to-wire efficiency.* Total efficiency.

**Power Units™** Pipe treatment device manufactured by Magnetics International, Inc.

**PoweRupp®** An electrical diaphragm pump manufactured by Warren Rupp, Inc.

**Powrclean** A bar screen manufactured by Aerators, Inc., with a multiple-rake.

**Powr-Trols** A control panel manufactured by Healy-Ruff Co. for pump station motors.

**PO*WW*ER™** A treatment process developed by ARI Technologies to remove residual solids from wastewater.

**Poz-O-Lite®** A lightweight aggregate produced by Conversion Systems, Inc.

**Poz-O-Tec®** An additive produced by Conversion Systems, Inc. for sludge stabilization.

**pozzolana** A porous volcanic tuff or ash used in hydraulic cement and mortars.

**pozzolanic** Pertaining to finely divided materials having cementitious properties similar to those of pozzolana; e.g., fly ash.

**pozzolanic admixture** A material like fly ash that has cementitious properties and is sometimes used in concrete to increase resistance to seawater, sulfate-bearing soil, and acidic waters.

**pozzolanic materials** Siliceous and albuminous substances that have some of the characteristics of pozzolana, particularly the cementitious property. Fly ash, which contains pozzolanic materials, is sometimes used in sludge stabilization. It yields a soil-like material with up to 50% solids when dried.

**ppb** Acronym of parts per billion, a measure of concentration approximately equal to μg/L.

**ppm** Acronym of parts per million, a mass/mass (or weight/weight) ratio, sometimes used to express concentration; it is related to mg/L as follows:

$$\text{ppm} = (\text{mg/L})/\text{specific gravity of fluid} \quad \text{(P-45)}$$

**ppmv** Acronym of parts per million by volume.

**ppt** Acronym of parts per trillion, a concentration measure approximately equal to ng/L.

**PQ®** A corrosion inhibitor manufactured by PQ Corp. and based on sodium silicate.

**PQL** Acronym of practical quantification level.

**pQt** *See* design flow.

**practicable** Available and capable of being done after taking into consideration cost, existing technology, and logistics in light of overall project purposes. Also, capable of being used consistent with: performance in accordance with applicable specifications, availability at a reasonable price, availability within a reasonable period of time, and maintenance of a satisfactory level of competition. The test of what is practicable depends upon the situation and includes consideration of the pertinent factors such as environment, community welfare, cost, or technology (EPA-40CFR230.3-q, 248.4-bb, and 6-AA-g).

**practical quantification level (PQL)** The lowest concentration of a water or wastewater constituent that can be quantified within given limits of precision and accuracy in a routine laboratory analysis. The USEPA commonly establishes PQLs that are 5 to 10 times the method detection limits or MDLs. *See also* level of detection, reliable quantification level.

**practical quantification limit** Same as practical quantification level.

**Praestol®** Polymers produced by Stockhausen, Inc.

**Pratt formula** A formula proposed in 1914 to compute the total discharge ($Q_t$) through a Sutro weir, as sometimes used at the end of a grit chamber. *See* Sutro weir for the equation.

**PRD-1 phage** The bacteriophage of *Salmonella typhirmurium.*

**preaeration** (1) A wastewater treatment process using a conventional aeration unit ahead of primary sedimentation to improve treatability by scrubbing out entrained gases, adding dissolved oxygen, and promoting the flotation of grease as well as coagulation. Preaeration is sometimes combined with grit removal. (2) In water supply, preaeration uses packed tower or tray aerators to control taste and odor, oxidize iron and manganese, and facilitate subsequent treatment processes.

**preammoniation** In chloramination practice, the addition of ammonia ($NH_3$) before chlorine ($Cl_2$)

to minimize chlorine contact time and reduce DBP formation. *See also* postammoniation, prechlorination.

**preanoxic denitrification** A biological nitrogen removal process using an anoxic/denitrification reactor followed by an aerobic/nitrification reactor and a final clarifier, with sludge recycle and wastage from the clarifier underflow, and internal nitrate feed from the aerobic zone to the anoxic zone. The nitrification and denitrification reactions can take place in attached- or suspended-growth vessels, or even a combination of the two.

**preanoxic nitrogen-removal process** A biological nitrogen removal process in which an anoxic/denitrification zone precedes an aerobic/nitrification zone and a final clarifier. Nitrate produced in the aeration zone is fed to the anoxic zone and sludge is recycled form the clarifier underflow. *See* the following processes: attached growth denitrification, preanoxic denitrification, modified Ludzack–Ettinger, step-feed, sequencing batch reactor, batch decant, Bio-Denitro™, Nitrox™, Bardenpho™, oxidation ditch, postanoxic denitrification, simultaneous nitrification/denitrification.

**preapplication treatment** Mechanical pretreatment of wastewater before application of a natural treatment technique; e.g., at least fine screening or primary sedimentation to remove gross solids and reduce nuisance conditions. *See also* type 1 slow-rate system, type 2 slow-rate system, rapid infiltration system, overland-flow system, constructed wetlands, floating aquatic plant treatment system.

**precast concrete T-pees underdrain** A filter underdrain system that includes a deep layer of gravel, low head loss, and concurrent air-and-water scour.

**prechloramination** Chloramination of water before other treatment processes. Chloramine addition may occur before coagulation or before filtration. *See also* postchloramination.

**prechlorination** (1) The addition of chlorine or chlorine compounds at the headworks of a plant prior to other treatment processes, mainly for disinfection and control of tastes, odors, and aquatic growths, and to aid in coagulation and settling. In water supply, prechlorination may be applied to the source water, ahead of filtration or any other process, to oxidize iron, manganese, and other metals. It prevents the formation of slime on filters, pipes, and tanks. Because of concerns for the formation of chlorination by-products, ozone and permanganate are being considered as alternative oxidants. In wastewater treatment, chlorine may be added after preliminary treatment to control odors in sedimentation tanks. *See also* postchlorination. (2) The addition of free chlorine to water before ammonia is added to form chloramines. *See also* preammoniation.

**precious metals** Gold, iridium, osmium, palladium, platinum, rhodium, ruthenium, silver, and any of their alloys containing 30% or greater by weight (EPA-40CFR421.261-b, 468.02-x and 471.02-bb)

**precipitant** A chemical that causes precipitation, e.g., the hydrolyzed metal salts of iron and aluminum.

**precipitate** (1) An insoluble, finely divided substance that is the product of a chemical reaction within a liquid. (2) To form such a substance, or to separate from a solution or suspension. (3) The substance that is precipitated.

**precipitated iron** *See* ferric iron.

**precipitation** (1) The process by which atmospheric moisture falls onto a land or water surface as rain, snow, hail, or other forms of moisture. (2) The amount of such precipitation, divided into three main components: direct runoff, river basin recharge, and groundwater accretion. (3) The chemical transformation of a substance in solution into an insoluble form (precipitate); the settling out of small particles. For example, the precipitation of iron and manganese. *See also* chemical precipitation. (4) Formation of calcium carbonate on interior pipe surfaces as protection against corrosion. (5) Removal of hazardous solids from liquid waste to permit safe disposal; removal of particles from airborne emissions.

**precipitation–evaporation ratio** The ratio of rainfall to the amount of water released back to the atmosphere; an indication of the water gain or loss to surface sources.

**precipitation-induced inflow** Inflow of rainwater or surface runoff into a sewer system following precipitation; often characterized by low pH and pollutants from roadways, rooftops, etc. *See* inflow.

**precipitation reaction** A reaction that results in the formation of insoluble precipitates of undesirable substances, e.g., lime (CaO) softening for the precipitation of calcium carbonate ($CaCO_3$) and the reduction of hardness.

**precipitation softening** A water treatment process that removes calcium, magnesium, and other dissolved salts using slaked or hydrated lime to increase pH and form precipitates, which are removed by sedimentation before filtration. Also called precipitative softening, precipitative lime softening. *See also* excess lime treatment, lime treatment, selective calcium carbonate removal, solids contact clarifier, split treatment, two-stage softening, recarbonation, two-stage recarbonation.

**precipitative lime softening** Same as precipitation softening.

**precipitative softening** Same as precipitation softening.

**precipitative softening plant** A variation of the traditional coagulation/filtration plant that includes a process such as lime softening to precipitate hardness-causing constituents from water. *See* precipitation softening.

**Precipitator™** A package treatment plant manufactured by U.S. Filter Corp.

**precision** The standard deviation of replicated measurements. Variation about the mean of repeated measurements, expressed as one standard deviation about the mean (EPA-40CFR53. 23-e and 40CFR86.082.2). *See also* accuracy, level or limit of detection or quantification, practical quantification level or limit, method detection level or limit, minimum detection limit, reliable detection, or quantification level. Precision errors are caused in simulation or monitoring by random deviations from true values, deviations due to limitations of the model or instrument. *See also* bias error.

**Precision** A tube diffuser manufactured by FMC Corp.

**precision error** *See* precision.

**precoat** A 3- to 5-mm layer of diatomaceous earth, perlite, or other fine granular medium built on a porous septum by recirculation of a slurry of the medium. Also called precoat cake. *See also* body feed, diatomaceous earth filtration.

**precoat cake** *See* precoat.

**precoat filter** A water treatment device that uses a thin layer (the precoat) of diatomaceous earth, perlite, or other fine medium to coat the septum before a filter run. It relies on entrapment, straining, and attachment to remove solids. It is not effective in removing color. The medium is usually not reused, but disposed of after each filter cycle. Precoat filters are widely used industrially and in swimming pools. (The septum is a support consisting of a woven wire or synthetic fabric.)

**precoat filter aid** The fine medium that is applied uniformly to the surface of the septum of a precoat filter. Also called filter aid. *See also* body feed.

**precoat filtration** A water treatment process using a precoat filter. It includes three steps: precoating or formation of a precoat layer over the septum, application of raw water for treatment by entrapment and straining, and body feeding or addition of slurried media to the raw water to maintain or improve the porosity of the cake. The filter run ends when pressure drops to the maximum available head; the medium is then disposed of and replaced for another run. Pretreatment by alum conditioning or by ozonation may be used to enhance the removal of colloidal particles associated with color, bacteria, and viruses. *See also* body feed, filter element, septum.

**precoating** The process of creating a precoat on a septum by recirculating a slurry of the medium through the filter until the accumulated cake reaches the desired thickness. Bridging is the principal mechanism of precoat formation as the particles of fine medium are smaller than the openings in the septum. Precoating uses 0.1 to 0.5 pound of medium per square foot of septum. *See* precoat filtration.

**precursor** (1) Natural organic compounds, mainly humic and fulvic acids, found in all surface and ground waters; during disinfection, they may be converted to disinfection by-products. They react with halogens to form trihalomethanes (THM), e.g.,

$$\text{Cl} + \text{precursors} \rightarrow CHCl_3 \text{ (chloroform)} + \text{other THMs} \quad \text{(P-46)}$$

Coagulation, ozonation, reverse osmosis, electrodialysis, and addition of chlorine dioxide are effective in the removal or alteration of precursors. *See* organic precursor material. (2) A chemical substance that is not contaminated due to the process conditions under which it is manufactured, but because of its molecular structure, and under favorable process conditions, it may cause or aid the formation of HDDs/HDFs in other chemicals in which it is used as a feedstock or intermediate (EPA-40CFR766.3). (3) In photochemistry, a compound antecedent to a volatile organic compound. Precursors react in sunlight to form ozone or other photochemical oxidants.

**precursor compound** Same as precursor.

**predator** An animal that kills other animals for food; any organism that preys on other organisms. *See also* parasite.

**predator–prey interaction** A natural control process in a community whereby predators survive by eating prey. Wastewater treatment offers examples of predator-prey relationships: (a) in activated sludge and trickling filters, protozoa consume bacteria; (b) in stabilization ponds, protozoa and rotifers graze on algae and bacteria. Such a relationship also exists in soils between protozoa and bacteria. *See* Figure P-13. *See also* symbiosis, parasitism.

**predator–prey relationship** *See* predator–prey interaction.

**predictive maintenance** A planned, proactive maintenance approach that focuses on predicting

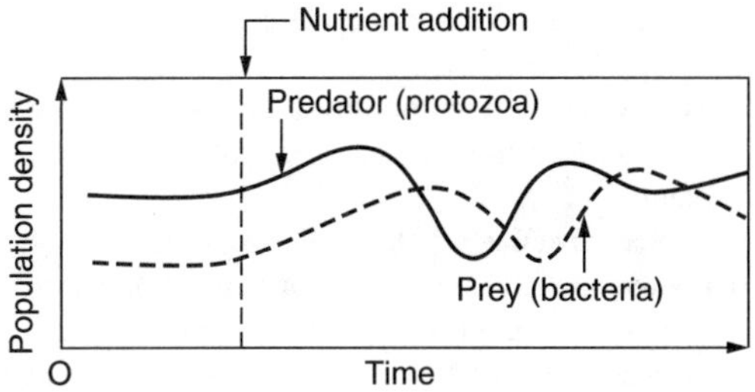


**Figure P-13.** Predator–prey relationship in soil (adapted from Maier et al., 2000).

when equipment will fail and taking timely appropriate corrective actions. It includes inspections, performance monitoring, and predictive analysis. Also called reliability maintenance. *See also* preventive maintenance, reactive maintenance, developmental maintenance.

**predisinfection** The application of a disinfecting agent ahead of a treatment process such as coagulation or filtration. *See also* primary disinfection, postdisinfection, secondary disinfection.

**preexponential factor** A parameter of the Arrhenius equation; also called frequency factor.

**preferential adsorption** A mechanism of surface charge development during chemical coagulation whereby inert substances (e.g., oil droplets, gas bubbles) acquire a negative charge through the adsorption of hydroxyl ions or other anions.

**preferential flow** A phenomenon of hydrologic heterogeneity observed where variations in soil structure (cracks, fissures, channels) create preferred flow paths and affect flow rates and hydraulic conductivity.

**prefiltration process** A water treatment scheme consisting essentially of coagulation and flocculation to eliminate suspended matter particles in preparation for rapid sand filtration. In addition to the coagulation and flocculation units, it may include pretreatment, mixing, chemical feed, and chemical storage.

**PreFLEX®** A pretreatment apparatus manufactured by U.S. Filter Corp. for reverse osmosis and demineralizing units.

**preformed chloramine** A chloramine solution prepared outside of the disinfection unit for direct addition to the water as compared to separate introduction of chlorine and ammonia. *See also* preammoniation.

**prehydrolyzed metal salt** Alum, ferric chloride, or any other metal salt whose acid has been partially neutralized with a base during manufacturing. This is to reduce the reaction of the salt, when used as a coagulant in water treatment, with the alkalinity of the solution. *See also* basicity.

**prelaboratory** Pertaining to monitoring activities such as collection, preparation, and delivery of samples to a place of analysis.

**preliminary assessment** The process of collecting and reviewing available information about a known or suspected waste site or release.

**preliminary filter** A rapid sand filter, or other type of filter, used in a water treatment plant to reduce turbidity and accelerate final filtration. Also called contact filter (1), contact roughing filter, or roughing filter (1).

**preliminary sludge processing operations** Activities intended to provide a constant, homogeneous feed of sludge for further processing. Preliminary sludge processing includes such operations as grinding, degritting, blending, and storage.

**preliminary treatment** (1) Any treatment steps used to prepare water or wastewater for further treatment by removing gross solids, grit, grease, gases, odors, etc. Some preliminary treatment processes include chemical addition, comminution, flotation, flow equalization, flow measurement, grit removal, pH adjustment, preaeration, prechlorination, presedimentation, screening, and skimming. A typical arrangement of preliminary wastewater treatment units is: mechanically cleaned bar screens → wet well/dry well pumping station → Parshall flume → aerated grit chamber → primary treatment. (2) Preliminary treatment is sometimes confused with pretreatment of industrial wastewater to make its characteristics acceptable to publicly owned treatment works.

**preloading** The deliberate or accidental addition of a compound to a treatment unit before the unit is operated, e.g., the preloading of natural organic matter to activated carbon units.

**preloading effect** The phenomenon observed in granular activated carbon adsorption, whereby at the end of an adsorber train the medium adsorbs less trace material per unit mass than at the beginning. This is due to the initial adsorption of large organic molecules that reduces pores or adsorption sites available to the trace compounds. Also called premature exhaustion.

**premature exhaustion** Same as preloading effect.

**premix clarifier** A solids contact clarifier with a central preliminary, mechanically agitated mixing zone, from which water flows to the base of the settling zone. *See also* Spaulding Precipitator, premix-recirculation clarifier.

**premix-recirculation clarifier** A solids contact clarifier with a primary mixing zone from which water flows to the middle of the settling zone. It provides for solids recirculation and secondary mixing. *See* Accelator® clarifier.

**preoxidation** The application of an oxidant such as chlorine or ozone ahead of coagulation, sedimentation, or filtration to oxidize iron, manganese, and other metals or to control microbial growth throughout a treatment plant.

**preozonation** The application of ozone ahead of coagulation, sedimentation, or filtration. Also called source water ozonation. *See* preoxidation, intermediate ozonation, postozonation.

**prepatent period** The interval between infection of an individual by a parasite and the first ability to detect in that host a diagnostic stage of the organism. For example, the median prepatent period of *Entamoeba hystolytica* in humans is 5 days, whereas the mean period of *Ent. coli* is 8–10 days. *See also* incubation period.

**preplumbed building** A building that is ready for the installation of a water treatment device because it has all the necessary plumbing.

**preprecipitation** A phosphorus removal process that consists of the addition of chemicals to raw wastewater and the removal of the precipitate phosphate with the primary sludge. *See also* coprecipitation, postprecipitation, biological phosphorus removal.

**Prerostal®** An apparatus of Envirotech Company for adjusting pumping volume to inflow rate.

**presaturant ion** The ion in the adsorbent or solid phase (e.g., sodium, Na) that is exchanged for an unwanted ion in the liquid phase (e.g., calcium, Ca) of such water treatment processes as ion exchange with synthetic resins or adsorption onto activated alumina.

**prescriptive** Pertaining to water rights that are acquired by diverting water and putting it to use in accordance with specified procedures. These procedures include filling a request to use unused water in a stream, river, or lake with a state agency.

**prescriptive specification** One of two general types of specifications used in the procurement of goods and services for such municipal projects as water and wastewater facilities. It defines explicit criteria for the processes, equipment, or services to be procured, with specific or equal manufacturers and suppliers. *See also* performance specification, liquidated or delay damages, consequential or performance damages.

**presedimentation** A plain sedimentation process that removes gravel, sand, silt, grit, and organic matter from muddy river source water before the main treatment train.

**presedimentation impoundment** A large basin that holds source water ahead of the treatment plant.

**presence–absence broth** *See* P-A broth.

**presence–absence (P-A) coliform test** A modification of the multiple-tube procedure to detect the presence or absence of total coliforms in a water sample. It does not indicate the quantity or type (e.g., fecal or not) of organisms; after a sample is inoculated into a medium and following incubation, a color change to yellow or the production of gas, acid, or both indicates whether or not an organism is present. *See also* Colilert test, enzymatic assay, P-A broth, MacConkey-PA.

**presence–absence method** *See* presence–absence coliform test.

**presence–absence rule** A requirement that a small water supply system should not have more than one positive water sample per month and large systems should not have more than 5% of total coliform samples positive per month, based on the presence–absence test.

**presence–absence technique** *See* presence–absence coliform test.

**presence–absence test** *See* presence–absence coliform test.

**presence indicator** A species whose presence indicates the existence of a certain environmental factor.

**present value factor (*P*/*A*, *r*, *n*)** A factor used to convert a periodic (e.g., annual) cost to a present worth cost. It is the reciprocal of the capital recovery factor:

$$(P/A, r, n) = [(1 + r)^n - 1]/[r(1 + r)^n] \quad \text{(P-47)}$$

where $A$ = periodic cost, $P$ = present worth, $r$ = interest rate per period, and $n$ = number of periods. *See* financial indicators for related terms.

**present worth factor** Same as present value factor.

**preservative** A substance added to a water or wastewater sample to maintain its characteristics, i.e., to keep it stable as well as to keep its compounds and microorganism density from changing.

**presettling** Same as presedimentation.

**pressate** The liquid waste stream from belt filter press thickening of sludge. *See also* centrate, filtrate, leachate, supernatant flow.

**press filter** The name filter press is more commonly used to designate a device that uses pressure to dewater sludge from water or wastewater treatment. *See* sludge pressing for names commonly used.

**PressMaster™** A device manufactured by Eimco Process Equipment Co. for hydraulic sludge dewatering.

**pressure** A type of stress exerted uniformly in all directions; the total load or force per unit area acting on a surface, measured in various units such as pascals and pounds per square inch.

**pressure aerator** An aeration device that sprays water droplets in a whirling motion into the air from stationary or moving orifices or nozzles. *See* spray aerator for detail.

**pressure control** A switch that operates on changes in pressure. Usually, this is a diaphragm pressing against a spring. When the force on the diaphragm overcomes the spring pressure, the switch is actuated.

**pressure decay test** A test used to determine the integrity of a wetted membrane by submitting it, offline, to an airflow below the bubble point pressure. Sharp pressure drops across the membrane are indicative of defects. *See also* bubble point test and diffusive airflow test.

**pressure differential** The difference in pressure between two points of a pipeline or a treatment unit.

**pressure differential meter** An instrument used to determine flow by measuring the pressure difference between two points. *See* flow nozzle, orifice meter, venturi meter.

**pressure dolomitic hydrated lime** A mineral or chemical product consisting of calcium and magnesium hydroxides, e.g., corresponding to the formula $\{[Ca(OH)_2]_{0.6}[Mg(OH)_2]_{0.4}\}$. *See* dolomitic hydrated lime for detail.

**pressure-driven membrane** A membrane that uses pressure differential as the driving force, in such water treatment processes as reverse osmosis, nanofiltration, ultrafiltration, and microfiltration. Such a membrane rejects pathogens and other particles that are too large to pass through its pores.

**pressure-driven membrane process** A process that uses a pressure differential to separate dissolved impurities from water through semipermeable membranes, as opposed to processes that use an electrical potential. A pressure-driven separation process removes water from the feed stream, whereas an electrochemical separation process removes undesirable ions from the feed stream. The basic components of a pressure-driven membrane system include a feedwater pump, the membrane modules in pressure vessels, piping for the permeate, concentrate disposal and concentrate recycling, and appropriate controls. Sew low-pressure membrane process and high-pressure membrane process.

**pressure-driven separation process** Same as pressure-driven membrane process.

**pressure drop** The difference (expressed in kilopascals or another convenient unit) in static pressure measured immediately upstream and downstream of a component, for example, an air filter element.

**pressure energy** Same as pressure head.

**Pressure Exchanger™** An energy recovery device manufactured by Energy Recovery Inc. designed to lower operating costs of seawater and brackish water desalination plants. Abbreviation: PX™.

**pressure filter** (1) A rapid filter that operates in a horizontal or vertical pressure container, with inline application of chemicals from a coagulation tank. Pressure filters are used for iron and manganese removal, in the treatment of swimming-pool water, and in industrial waste treatment. They allow higher hydraulic loading rates and higher head losses than gravity filters. However, because of operational difficulties, pressure filters have been implicated in outbreaks of waterborne diseases and have been excluded by some agencies for the treatment of surface waters. *See* Figure P-14. (2) A similar device for dewatering sludge from water or wastewater treatment. *See* sludge pressing for names commonly used.

**pressure filter press** Filter press is the common name of a device that uses pressure to dewater sludge from water and wastewater treatment. *See* sludge pressing for names commonly used.

**pressure filtration** (1) The use of a pressure filter to treat water or wastewater, usually when there is insufficient elevation for gravity filtration or in

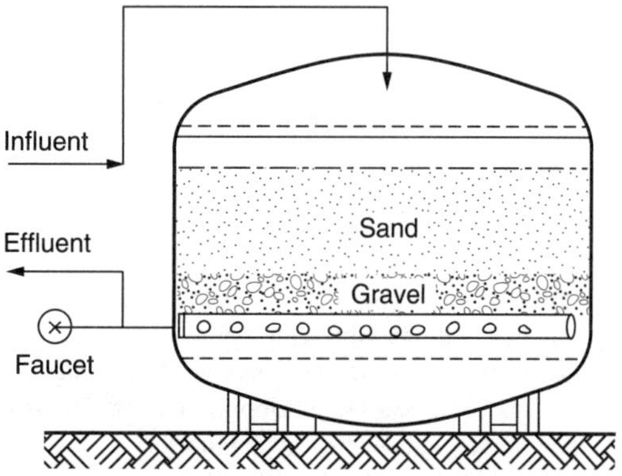


**Figure P-14.** Pressure filter.

treating water for industrial applications. Pressure filtration through a specialty medium such as greensand, with regeneration, is a commonly used technology for the removal of iron and manganese from drinking water. (2) Pressure filtration is also used in sludge dewatering or the thickening of alum sludge. *See* belt filter press, plate-and-frame filter press.

**pressure flotation** A type of dissolved-air flotation in which air dissolved under pressure in water forms bubbles that attach to solid particles and cause them to float to the surface for removal.

**pressure flow** Fluid flow for which energy is provided by a pump or similar device, as compared to flow under the influence of gravity.

**pressure gage** An instrument with a metallic sensing element or a piezoelectric crystal for measuring the pressure of fluids or solids. *See also* piezometer.

**pressure gradient** A pressure differential in a given medium (e.g., water or air) that tends to induce movement from areas of higher pressure to areas of lower pressure.

**pressure granular filter** A granular filter operating under pressure in a steel vessel.

**pressure head** The vertical distance (in feet) equal to the pressure (in psi) at a specific point. The pressure head is equal to the pressure in psi times 2.31 ft/psi. In the expression of total dynamic head of a pump, pressure head is equal to the ratio of the static pressure ($p$) to the specific weight ($\gamma$) of the fluid or $p/\gamma$. Also called pressure energy or flow work. *See also* Bernoulli's law and the dynamic head formula.

**pressure intensity** Same as pressure.

**pressure loss** Same as head loss or pressure drop.

**pressure main** A pressurized sewer line that conveys wastewater or stormwater from a pumping station to another force main, a manhole, a treatment plant or a point of disposal. *See* force main for detail.

**pressure reducer** (1) A control valve that closes and stops flow when downstream pressure is too high. (2) A safety device that allows water to flow from a higher-pressure plane to a lower-pressure plane and maintain a set downstream pressure. A pressure-regulating valve.

**pressure-reducing valve** A device that reduces water pressure in a building below the pressure of the distribution system, thereby decreasing the probability of leaks and dripping faucets. *See also* flow-reducing devices and appliances.

**pressure-regulating valve** Same as pressure reducer (2).

**pressure regulator** A pressure-regulating valve or pump drive-speed device that controls the pressure in a pipeline or a pressurized container; a pressure reducer (2).

**pressure relief** A pressure relief device.

**pressure relief device or valve** A safety device used to prevent operating pressures from exceeding the maximum allowable working pressure of the process equipment. A common pressure-relief device is a spring-loaded pressure-relief valve. Devices that are actuated either by a pressure less than or equal to 2.5 psig or by a vacuum are not pressure-relief devices (EPA-40CFR63.161). Pressure-relief devices are sometimes simply called pressure reliefs. Pressure-relief valves reclose automatically upon return to normal operating conditions. *See also* safety valve, overflow tower, and other pressure surge control devices.

**pressure relief drain** A drain that allows groundwater to enter an empty tank or other buried structure, thereby relieving pressure and avoiding their flotation during periods of high water table.

**pressure relief valve** *See* pressure relief device.

**pressure sewer** *See* pressure sewerage.

**pressure sewerage** A system of pipes in which wastewater or another liquid is pumped to a higher elevation. Wastewater is normally collected in a septic tank or holding tank before being pumped. Pressure sewerage is used to avoid deep gravity sewers in low-density areas; it includes small pipes and grinder pumps that reduce the size of solids to prevent plugging. *See* Figure P-15. *See also* septic tank effluent pumping system, grinder pump system, small-bore sewerage, small-diameter variable-slope sewer.

**pressure sewer system** *See* pressure sewerage.

**pressure surge** The phenomenon that occurs when steady-state flow and pressure conditions change. Sudden changes can cause severe damages to a piping system. Also called hydraulic transient, pressure transient, water hammer.

**pressure surge control device** A device designed

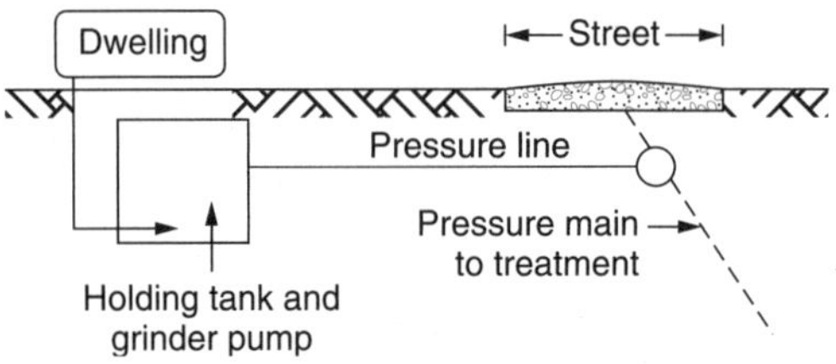


**Figure P-15.** Pressure sewerage.

to smooth the transition between pressure states in piping and reduce the effect of hydraulic or pressure transients (e.g., water hammer) by storing water in a vessel or discharging water from the line. Common surge protection devices include standpipes, surge vessels, feed tanks, pressure-relief valves, surge anticipation valves, air release valves, check valves, and pump-bypass lines.

**pressure survey** A field investigation conducted to determine the hydraulic efficiency of a water supply system in meeting ordinary demands. The survey establishes available pressures and flows at selected points or in specific areas.

**pressure-sustaining valve** A device that maintains the upstream pressure at a set value by limiting the flow.

**pressure-swing adsorption (PSA)** A selective sorption process used by small wastewater treatment plants to generate pure oxygen continuously on-site. It uses a zeolite or other adsorbent in a multibed system to separate individual components of such a gas mixture as air. *See also* cryogenic air-separation process.

**pressure switch** A device that determines whether a line is open or closed by controlling the fluid's pressure or grade.

**pressure tank** An airtight tank that contains water and compressed air, used to pump water for a house, a small installation, or as part of a distribution system. *See also* pneumatic tank, hydropneumatic tank.

**pressure-tapping machine** *See* tapping machine.

**pressure test** The measurement of the static pressure (when there is no flow) or the residual pressure (when water is flowing) at a point in the distribution system.

**pressure transient** A wave having both positive and negative amplitude created in a pipeline by an abrupt change in fluid velocity, and causing low or negative pressures; it is sometimes called surge or water hammer. Pressure transients in a potable water distribution system may result from a number of operational changes (e.g., pump stoppage due to power failures, main breaks, opening and closing of fire hydrants, flushing, etc.) and cause cross connections and contamination. Water pressure may increase by 50–60 psi for every fps of velocity in a sudden stop. *See also* hydraulic transient, intrusion, backflow.

**pressure tube** Same as pressure vessel.

**pressure vessel** A single tube containing several membrane elements in series; designed to support the membranes, separate the feedwater from the product stream, and prevent leaks and pressure losses. Also called pressure tube.

**pressurized membrane system** A low-pressure membrane system consisting of (a) pressure vessels mounted on racks and encasing modules of hollow-fiber membranes and (b) a pump feeding water to the modules under positive pressure. Residuals are withdrawn by intermittent backwashing or continuous wasting of concentrate. Also called encased membrane system.

**pressurized sewer system** *See* pressure sewerage.

**pressurized shower** A device that reduces water use at the shower by mixing water and compressed air. *See also* flow-reducing devices and appliances.

**pressurized water jet mixer** A device used for mixing and blending chemicals in water treatment or in reclaimed water operations; e.g., a flow nozzle in a reactor tube, with an external power source such as a feed pump. *See also* mixing and flocculation devices.

**pressurized water reactor** A nuclear reactor that uses enriched uranium for fuel and light water under pressure as a moderator and coolant in a welded-steel vessel.

**Pressveyor®** A hydraulic screenings press/conveyor manufactured by Hycor Corp.

**Prestex®** A belt manufactured by Tetko, Inc. for belt filter presses.

**prestressed concrete pipe** A reinforced concrete pipe placed in compression by helical wire winding.

**prestripper tank** In the PhoStrip™ II process for phosphorus and nitrogen removal, a prestripper is a tank installed between the secondary clarifier and the anaerobic stripper. It receives the nitrate-rich clarifier underflow and a portion of the phosphorus-deficient sludge from the anaerobic stripper.

**presumptive indicator** An organism used as an indication of fecal pollution of water sources. *See* fecal coliform, total coliform, *E. coli.*

**presumptive phase** The first phase in the multiple-tube fermentation technique; see presumptive test.

**presumptive test** The first phase in the multiple-tube fermentation technique; tubes of a water sample are incubated for one to four periods of 24 hours at 35°C in lauryl tryptose (LT) broth or brilliant green lactose bile (BGLB) broth to determine the number of presumptive or confirmed tubes and positive or negative tubes. A positive tube shows the formation of gas at the end of the incubation period. A presumptive positive shows gas formation in the LT broth; a confirmed positive shows gas formation also in the BGLB broth. The presumptive test is conducted for samples collected prior to final disinfection. *See also* completed test, confirmed test.

**pretreated** Pertaining to water or wastewater that has undergone some form of pretreatment.

**pretreatment** (1) Processes used to reduce, eliminate, or alter the nature of wastewater pollutants from nondomestic sources before they are discharged into publicly owned treatment works (POTWs). The reduction or alteration may be obtained by physical, chemical or biological processes, process changes, or by other means, except as prohibited. Appropriate pretreatment technology includes control equipment, such as equalization tanks or facilities, for protection against surges or slug loadings that might interfere with or otherwise be incompatible with the POTW (EPA-40CFR403.3-q). Pretreatment is required for industrial wastes that may cause problems in the collection system, treatment facilities, or receiving stream, e.g., wastes that contain corrosive, flammable, or explosive materials or materials containing excessive concentrations of grease or solids. *See also* preliminary treatment (2). (2) Pretreatment is used in water supply, e.g., in large tanks or small artificial lakes, or using such processes as aeration, eutrophication control, multiple chlorination, prechlorination, or sedimentation. Coarse filters are sometimes used as pretreatment ahead of slow filtration of high-turbidity waters. Filtration pretreatment of wastewater usually consists of the use of coagulants upstream of sedimentation and/or ahead of the filters. Also called filtration pretreatment. (3) *See* membrane pretreatment.

**pretreatment standard** Same as national pretreatment standard.

**Pretreat Plus™** Chemical products of King Lee Technologies used as antiscalants and dispersants.

**Pretreat SDP™** An antiscalant produced by King Lee Technologies for use in reverse osmosis units.

**prevalence** (1) The number of persons in a population who are sick or affected by a condition at a given time. Prevalence differs from incidence in that the latter includes only new cases in a given period. (2) The prevalence rate, i.e., the number of persons in a population who are sick or affected by a condition at a given time divided by the number of persons in the population.

**prevalence rate** The number of persons in a population who are sick or affected by a condition at a given time divided by the number of persons in the population.

**prevalence study** An epidemiological study that examines the relationships between disease and exposures as they exist in a defined population at a particular point in time.

**prevalent level sample** An air sample taken under normal conditions; also known as ambient background sample.

**preventive maintenance** A planned, proactive maintenance approach that focuses on discovering and remedying conditions that can lead to failures and other adverse consequences. It includes such activities as routine maintenance, comprehensive cleaning, recalibration, and scheduled inspections. *See also* predictive or reliability maintenance, reactive maintenance, developmental maintenance.

**preventive maintenance program** A specific plan to inspect and repair or replace water supply or wastewater management facilities at scheduled intervals, particularly equipment, motors, pumps, valves, and vehicles.

**prey–predator relationship** *See* predator–prey interaction.

**price elasticity** The percentage change in quantity divided by the percentage change in price; normally a negative number, assuming that demand goes up when price decreases and vice versa. It is less than –1.0 when demand is elastic and between 0 and –1.0 when demand is inelastic. *See* elastic demand, inelastic demand.

**primacy** The responsibility for ensuring that a law is implemented; the authority to enforce a law and related regulations. A primacy agency has the primary responsibility for administering and enforcing regulations. For example, under the Safe Drinking Water Act, the states have the authorization to exercise primary enforcement responsibility for all related regulations, provided that the state regulations are at least as stringent as the federal regulations.

**primary amoebic meningoencephalitis (PAM)** A rare but fatal infection caused by the protozoan pathogen *Naegleria fowleri* and contracted more likely by swimming in warm fresh waters than through drinking water. It causes severe, massive headaches and often death in 4–6 days. Permanent brain damage is likely in surviving patients.

**primary benefit** A benefit that accrues to people who use directly the product or service resulting from an action. Pollution abatement is often thought of having no primary or direct benefit. *See also* secondary benefit, intangible benefit.

**primary clarification** A wastewater treatment process designed to remove suspended solids at an approximate rate of 0.5 mm/s and an efficiency of

30–60%. The process also reduces biochemical oxygen demand by 20–40%. Plain primary clarification involves mostly discrete particle settling. With chemical addition, primary clarification is sometimes used for phosphate removal.

**primary clarifier** A rectangular, square, or circular sedimentation basin for the removal of settleable solids, following preliminary treatment and/or preceding secondary treatment. Design parameters for primary clarifiers include retention time (1–2 hr at peak flow) and surface overflow rate (0.3–0.7 mm/s at a depth of 1–5 m). Rectangular tanks, common in most, except small, plants, are equipped with inlet and scum baffles, and have weir overflow rates of 10,000–30,000 gallons per foot per day (120–370 $m^3$ per meter per day). *See also* primary sedimentation.

**primary coagulant polymer** A polyelectrolyte, usually cationic and of low molecular weight, used in water treatment to enhance the agglomeration and deposition of negatively charged particles. Examples: polyDADMAC and epiDMA. *See also* coagulant aid, flocculent polymer, quaternary amine, restabilization.

**primary contaminant** A substance that can have adverse health effects when present in drinking water. *See also* secondary contaminant.

**primary disinfecting agent** Any of the chemical or physical agents used in primary disinfection. A 1980 evaluation recognized only three agents as suitable for primary disinfection: free chlorine ($Cl_2$), chlorine dioxide ($ClO_2$), and ozone ($O_3$). Other agents evaluated include iodine ($I_2$), bromine ($Br_2$), ferrate, ionizing radiation, and UV light.

**primary disinfection** The application of chlorine, ozone, or other disinfecting agent to water before or after filtration but following sedimentation and preceding the clear well, for the primary purpose of inactivating pathogens. *See also* predisinfection, secondary disinfection.

**Primary Drinking Water Regulation (PDWR)** A regulation of public water supply systems specifying contaminant levels that (the USEPA assumes) will not affect human health adversely. Under the Safe Drinking Water Act, the USEPA specifies maximum contaminant levels and applicable treatment methods.

**primary drinking water standard** *See* Primary Drinking Water Regulation.

**primary effluent** The liquid that comes out of a wastewater treatment plant.

**primary element** An instrument that measures a physical condition or variable of interest, e.g., a float or a thermocouple. Also called a sensor.

**primary environmental care** Actions that create supportive physical, social, spiritual, economic, and political conditions in a community. The aim of primary environmental care "is to help meet people's basic livelihood and health needs, ensure optimal use and sustainable management of natural resources, and empower local groups or communities to undertake self-directed sustainable development" (WHO, 1997).

**primary feeder** (1) Same as primary main. (2) In the mixed microorganism population of aerobic and anaerobic wastewater treatment, bacteria are usually the primary feeders for the degradation of organic matter. *See also* secondary feeder, predator–prey relationship.

**primary footprint** *See* carbon footprint.

**primary industry categories** The categories of industrial installations that the Clean Water Act requires to use best available technology to control toxic water pollutants.

**primary main** In a pipe network, one of the pipes that constitute its basic structure, carrying flow to and from storage tanks and usually laid in loops. Also called arterial main. Secondary mains form smaller loops within primary loops, running from one primary line to another.

**primary maximum contaminant level (PMCL)** A maximum contaminant level (MCL), an enforceable numerical limit in drinking water mandated by the USEPA to protect public health. *See also* secondary maximum contaminant level.

**primary MCL** *See* primary maximum contaminant level.

**primary metabolite** A product, e.g., ethanol, in a continuous culture that is generated at a high flow or dilution rate to stimulate microbial growth. *See also* secondary metabolite.

**primary pollutant** A pollutant that exists in the environment in the same form as it was released or that is released directly into the environment, e.g., sulfur dioxide ($SO_2$) and carbon monoxide (CO). *See also* secondary pollutant.

**primary producer** In all food chains, autotrophs are the primary producers, i.e., they synthesize and store all the energy used by the other organisms. Also called photosynthetic organisms, including algae and other green plants, cyanobacteria, some bacteria, and some protozoa.

**primary production** The amount of food and energy synthesized by primary producers within a given community or food chain.

**primary productivity** *See* algal growth rate.

**primary radioactive nuclide** A disintegrating atom that emits alpha or beta rays; it has a half-life larger than 100,000,000 years. *See also* induced and secondary radioactive nuclides.

**primary sedimentation** The principal means of primary wastewater treatment, a widely used unit operation consisting of a clarifier or tank designed to remove readily settleable solids and floating material and to produce an effluent suitable for biological treatment. *See* primary clarification, primary clarifier, primary treatment.

**primary sedimentation tank** *See* primary settling tank.

**primary settling tank** A tank used to provide primary sedimentation, ahead of biological treatment.

**primary sludge** Sludge produced in primary wastewater treatment, a gray and slimy material with an offensive odor and the following average characteristics: easily digested, total dry solids of 7% and 70% volatile, pH 6.5, alkalinity 1000 mg/L as $CaCO_3$. *See also* biosolids, chemical sludge, secondary sludge.

**primary sludge fermentation** A technique used to improve the performance of biological phosphorus removal systems by providing supplemental readily biodegradable chemical oxygen demand in the form of acetate; the fermentation of the primary sludge occurs in a deep primary clarifier or in a separate fermentation reactor; the volatile fatty acids produced are fed to the anaerobic zone of the secondary treatment.

**primary sludge solids** The solid particles in the sludge from primary wastewater treatment as compared to sludge from biological treatment (e.g., activated sludge process or trickling filter). Its dry weight ($W_p$, pounds per day) is proportional to the applied suspended solids ($SS$) load expressed in pounds per day:

$$W_p = F(SS) \qquad \text{(P-48)}$$

where $F$ is the fraction of suspended solids removed, about 0.5 for domestic wastewater.

**primary standard** (1) A pollution limit based on health effects. *See also* secondary standards. (2) A reagent selected and maintained to preserve a high level of accuracy in laboratory analyses. Primary standards are also used to calibrate other reagents.

**primary substrate** *See* cometabolism.

**primary treatment** The first steps in wastewater treatment; fine screens and/or quiescent sedimentation tanks are used, with or without coagulants, to remove most materials that float or will settle, and to produce an effluent suitable for biological treatment. Primary treatment does not significantly affect colloidal or dissolved contaminants; it removes about 50–70% of suspended solids and 25–40% of carbonaceous biochemical oxygen demand from domestic wastewater. The term primary treatment is sometimes restricted to primary clarification or plain sedimentation, whereas preliminary treatment may apply to preceding processes. *See also* preliminary treatment, primary clarification, secondary treatment, biological treatment, tertiary treatment, advanced wastewater treatment.

**primary waste treatment** *See* primary treatment.

**primary wastewater treatment** *See* primary treatment.

**Prima-Sep®** A clarifier manufactured by Graver Co., including tray separators.

**prime** *See* priming.

**prime farmland** Land that historically has been used for intensive agricultural purposes and classified as such by the Secretary of Agriculture on the basis of such factors as moisture availability, temperature regime, chemical balance, permeability, surface layer composition, susceptibility to flooding, and erosion characteristics.

**primer** A cell fragment used in the polymerase chain reaction to amplify the DNA of a microorganism; the primer triggers the formation of millions of copies of the microorganism's DNA. The primer is a short piece of single-stranded DNA, carefully chosen and commercially synthesized.

**Primext** A commercial name for the chemical metolachlor. *See also* Dual.

**priming** The action of starting the flow of fluid in a pump or siphon, e.g., by filling the pump casing with water to remove the air and allow suction by the impeller. Most pumps must be primed before startup.

**priming pipe** A pipe used for priming a siphon spillway.

**Primox®** An apparatus manufactured by BOC Gases to inject oxygen in primary treatment.

**principal isotope** The common form of an element that has more than one form, e.g., hydrogen ($^1H$) as compared to deuterium ($^2H$) and tritium ($^3H$).

**principal source aquifer** An aquifer that supplies a significant portion (e.g., 50%) of the drinking water of an area. *See also* sole source aquifer.

**principal study** The study that contributes most significantly to the qualitative and quantitative risk assessment.

**prion** A proteinaceous infectious particle, similar to a viroid, smaller than a virus, heat-resistant and difficult to detect in infected tissues.

**prior appropriation** A doctrine of water law that allocates the right to use water on a first-come, first-serve basis.

**priority date** The date a water right is established.

**priority list** A list of 39 dangerous substances established in the United Kingdom under two categories: (a) a "red list" of 23 substances (mercury, cadmium, pesticides) that require the application of Best Available Techniques Not Entailing Excessive Costs, and (b) a list of 16 additional substances (heavy metals and pesticides) the discharge of which should be reduced by at least 50%.

**priority pollutant** Any one of 129 toxic pollutants listed as such in federal regulations such as the Clean Water Act. Wastewater contains many of these organic and inorganic substances, which were selected on the basis of their known or suspected carcinogenicity, mutagenicity, teratogenicity, or acute toxicity. Some priority pollutants are volatile organic compounds. The USEPA has established discharge standards for priority pollutants. *See* prohibited discharge standard, categorical standard.

**priority toxic pollutant** *See* priority pollutant.

**priority water quality areas** Specific segments of water bodies, as determined by a state, where municipal discharges have resulted in the impairment of a designated use or significant public health risks, and where the reduction of pollution from such discharges will substantially restore surface or groundwater uses (EPA-40CFR35.2005-34).

**PRISM®** The gas separation membrane system originally developed by Monsanto Corp. in 1929 in a hollow-fiber membrane.

**pristine environment** An environment that contains no man-made contaminants.

**pristine watershed** A clean watershed, unspoiled by human activities.

**privately owned treatment works** A system or device that is not a POTW and that is used to treat wastes from any facility whose operator is not the operator of the treatment works (EPA-40CFR122.2). Or, simply, a waste treatment works that is not owned by a state, unit of local government, or Indian tribe.

**privately owned utility** A public utility (e.g., water supply or wastewater management) owned by shareholders, private individuals, or a corporation. Also called investor-owned utility or private utility.

**private utility** Same as privately owned utility.

**private water supply** A water supply that is not available or accessible to the general public, e.g., a well that serves a household or a commercial building. *See also* individual water supply, privately owned utility.

**privatization** (1) The provision of public services, such as water supply or wastewater management, by a private company under a partnership with a public agency. The private party may own or simply operate and maintain the facilities, i.e., undertake any or all of the following activities: planning, design, financing, construction, operation, and maintenance. Privatization is expected to result in lower life-cycle costs but is sometimes opposed by those who want to preserve local control over utilities. *See also* contract operation and management, build-own-operate, and build-own-operate-transfer. (2) The transfer of equity or control in public utilities from government to private investors.

**privy** A fixed or portable outbuilding with one or more seats and a pit or vault serving as a toilet to collect human feces. Privies are used in rural and other areas that have no or limited access to running water. Also called outhouse, pit toilet, pit privy. *See also* latrine, pit privy, toilet.

**privy vault** The fixed concrete or masonry vault of a privy, equipped with a cleanout opening.

**proactive maintenance** A planned maintenance approach that consists of initiating maintenance activities before adverse consequences such as failure, deterioration, and obsolescence. *See* reactive maintenance, preventive maintenance, predictive or reliability maintenance, developmental maintenance.

**probe method** Same as electrode method.

**Probiotics™** Products of Bio Huma Netics, Inc. for use in sludge oxidation lagoons.

**probit model** A dose–response model expressed in standard deviations ($y$) and having the form

$$P(d) = 0.4 \int \exp[-(y^2)/2]\, dy \qquad \text{(P-49)}$$

where $\int$ denotes the integral from minus infinity ($-\infty$) to $[\log(d - u)]/s$, $P(d)$ is the probability of cancer from a continuous dose rate $d$, and $u$ and $s$ are constants. It assumes that the tolerance of the exposed population follows a log-normal distribution. A probit is a normal equivalent deviate (the difference between a variable and the mean) increased by a whole integer (e.g., 5 in this model) to avoid negative numbers. *See also* log-probit model, linearized multistage model, logit model, multihit model, multistage model, one-hit model, Weibull model.

**ProBlend** An apparatus manufactured by AquaPro, Inc. for use as a feeder of liquid polymers.

**procaryote** A cellular organism in which the nucleus has no limiting membrane. *See* prokaryote.

**process** (1) The preparation of a chemical substance or mixture, after its manufacture, for distribution in commerce (EPA-40CFR372.3, 710.2-t, 761.3). (2) Any source operation including any equipment, devices, or contrivances and all appurtenances thereto for changing any material or for storage or handling of any material, the use of which may cause the discharge of an air contaminant (EPA-40CFR52.1881-viii). (3) *See* water treatment process. (4) *See* wastewater treatment process.

**process and instrumentation diagram (P&ID)** A schematic illustration of how the components of a plant work to implement and control a process. It includes monitoring devices, switches, alarms, etc. The P&ID contains enough information for a reader to understand how the process is monitored and controlled. *See* process flow diagram and instrumentation system diagram for detail.

**process disturbance** A factor, phenomenon, change, etc. that causes a change in the operation or performance of a water or wastewater treatment process. *See* external process disturbance, internal process disturbance.

**process flexibility** The ability of a water or wastewater treatment process to accommodate future modifications of regulations, changes in source water quality, or changes in wastewater characteristics and flows.

**process flowchart** A visual representation of the components of a process; a process flow diagram.

**process flow diagram (PFD)** A diagram developed for a water or wastewater treatment plant, showing all the major process elements as well as the liquid and solids flowpaths, including major equipment involved in each and important operating conditions (e.g., peak flow, minimum flow, etc.) The PFD shows valves, pipe sizes, piping material, equipment, instrumentation and control, basic electrical requirements, electrical signals, and conduits. *See also* instrumentation and control diagram.

**process-generated wastewater** Any water/wastewater directly or indirectly used in any industrial operation, for example, the operation of a feedlot, in the slurry transport of mined material, air emissions control, and processing exclusive of mining. The term includes any other water that becomes commingled with such wastewater in a pit, pond, lagoon, mine, or other facility for treatment (EPA-40CFR412.11-d, 412.21-d, 436.181-e, 436.21-e, and 436-31.e). *See also* process waste, process wastewater, process water.

**processing** Water processing may be defined as the movement of water from a source through treatment and distribution, including the handling and disposal of water treatment residuals. Similarly, wastewater processing involves collection, transmission, treatment, and effluent disposal, including sludge processing and disposal.

**process kinetics** The determination of reaction rates. *See* zero-, first-, and second-order kinetics.

**process loading factor** Same as food-to-microorganism ratio.

**process safety factor** A factor incorporated in the design of a treatment system to protect it against failure. For example, in suspended growth wastewater treatment systems, the selected design solids retention time (SRT) is up to 20 times the required minimum.

**process safety management** Implementation of a program of occupational health and safety to protect the public and workers against hazardous chemicals in a plant.

**process to further reduce pathogens (PFRP)** Any of the processes recommended by the USEPA to meet the pathogen and vector reduction requirements for Class A biosolids. These processes are: beta-ray irradiation, composting, gamma-ray irradiation, heat drying, heat treatment, pasteurization, and thermophilic aerobic digestion.

**process to significantly reduce pathogens (PSRP)** Any of the processes recommended by the USEPA to meet the pathogen and vector reduction requirements for Class B biosolids. These processes are: aerobic digestion, air drying, anaerobic digestion, composting, and lime stabilization.

**process train** An independent treatment process or processes in series.

**process variable** A physical or chemical quantity that is usually measured and controlled in the design and operation of a water treatment, wastewater treatment, or industrial plant.

**process verification** Verifying that process raw materials, water usage, waste treatment processes, production rates and other facts relative to quantity and quality of pollutants contained in discharges are substantially described in the permit application and the issued permit.

**process waste** Any designated toxic pollutant, whether in wastewater or otherwise present, that is inherent to or unavoidably resulting from any manufacturing process, including that which comes into direct contact with or results from the production or use of any raw material, intermediate product, finished product, by-product, or waste product and is discharged into navigable waters (EPA-40CFR129.2-n). *See also* process-generated wastewater, process wastewater, process water.

**process wastewater** Any water that comes into contact with any raw material, product, by-product, or waste during manufacturing or processing. Examples are product tank blowdown or feed tank drawdown, water formed during a chemical reaction or used as a reactant, water used to wash impurities from organic products or reactants, water used to cool or quench organic vapor streams through direct contact, and condensed steam from jet ejector systems pulling vacuum on vessels containing organics (EPA-40CFR63.101). *See also* industrial wastewater classification, process-generated wastewater, process waste, process water.

**process water** Any raw, service, recycled, or reused water that is used in a manufacturing or treatment process, in the actual product, or in materials incorporated in an end product; process water is usually cleaner than cooling water. Various industries such as food processing, laundering, and manufacturing have specific water quality requirements related to turbidity, color, hardness, alkalinity, solids, metals, etc. *See also* process wastewater.

**Prochem®** Agitators and mixers manufactured by Chemics, Inc.

**Proctor density** The density of a soil sample according to one or two laboratory tests conducted to determine the ultimate dry density and optimum moisture content of the sample. The standard Proctor density test is conducted in accordance with ASTM Standard D-698, whereas the modified Proctor density test follows ASTM Standard D-1557. This parameter is used to characterize materials used or found in water and wastewater treatment facilities, e.g., (a) landfill bentonite liners may be compacted to 90% standard Proctor density, and (b) the optimal water content and dry unit weight of the standard Proctor compaction of coagulant sludge are 45% and 72 $lb/ft^3$, respectively.

**procurement approach** The method of contracting for design- and construction-related activities for water, wastewater, and other facilities. *See* details under architect/engineer procurement, combined procurement approach, turnkey procurement approach, privatization.

**produced water** Water resulting from the production of oil and gas.

**producer** An organism that uses solar energy to synthesize organic matter from inorganic materials. *See also* primary producer, secondary producer.

**product** (1) A substance resulting from a chemical reaction and shown on the right-hand side of the corresponding equation. *See also* reactant. (2) The liquid product (water) that passes through a semipermeable membrane in a pressure-driven process; it has a lower concentration of total dissolved solids than the feed stream. Same as filtrate in a low-pressure system. *See* permeate for detail.

**product flow stream** The liquid product (water) that passes through a membrane in a pressure-driven process; same as filtrate in a low-pressure system. Also called permeate.

**production rate** The quantity of finished water produced by a treatment system per unit time, e.g., gpm or gpd in membrane filtration.

**productivity** (1) The rate of biomass production in a biological system per unit area or unit volume. *See also* eutrophication. (2) The ability of a water body to produce living material.

**product of incomplete combustion (PIC)** An organic compound formed by combustion, e.g., carbon monoxide and hydrocarbons. Usually generated in small amounts and sometimes toxic, PICs are heat-altered versions of the original material fed into the incinerator (e.g., charcoal is a PIC from burned wood). Also defined as a hazardous product of a thermal reaction detected in stack emissions but not present in the waste feed at a concentration of 100 ppm or larger.

**product staging** A procedure used in reverse osmosis whereby the product from one stage is used as feedwater for a subsequent stage.

**product stream** *See* product (2).

**product water** The product of a water treatment plant, i.e., water that has passed through all the treatment processes and is ready to be delivered to the consumers. Also called finished water.

**product water recovery** The collection of the effluent in bottom ditches of an overland flow wastewater treatment system.

**proficiency testing** A program of performance evaluation for laboratories conducting water and wastewater analyses.

**profile** A plot of elevations versus distances, on a longitudinal section or a cross section. In particular, a graphical representation of water surface elevation, conduit invert and crown, ground, and hydraulic gradeline. *See also* hydraulic profile, sewer profile, and soil profile.

**Profiler** A device manufactured by Mt. Fury Co., Inc. to monitor the level of sludge blanket and the concentration of suspended solids.

**profile wire** A special wire that has a triangular or trapezoidal cross section. Also called wedgewire.

**profiling** *See* well profiling.

**profitability index** For a given project, the profitability index is the present value of benefits or cash inflows divided by the present value of costs or cash outflows over the planning period; the higher the ratio, the more attractive the project. Also called benefit/cost ratio. *See* financial indicators for related terms.

**profundal zone** In a freshwater body, the open water where biological activity is low and variable because of the lack of photosynthesis, with a light intensity less than the compensation point (1% of sunlight). It serves as a transition for the transmission of detrital matter from the limnetic zone to the benthic zone. *See* freshwater profile for related terms.

**proglottid** *See* gravid proglottid. Plural: proglottides or proglottids. Also spelled proglottis.

**proglottis** *See* proglottid.

**programmable logic controller (PLC)** A device that allows an operator to control several functions or operations at a treatment plant, e.g., water level, chemical feeding, and alarm generation. *See also* SCADA, single-station controller, distributed process controller, digital control system.

**progressing cavity pump (PCP)** A positive displacement pump used for wastewater and sludge and consisting of a single-helix, worm-shaped metal rotor rotating in an elastomeric, double-helix stator. The name of the pump comes from the volume or cavity that moves progressively from suction to discharge when the rotor turns. A PCP does not clog easily, is simple to operate, allows a smooth flow, and can operate over a wide range of flows, but tends to cost more to maintain than plunger pumps. Also called progressive cavity pump.

**progressive cavity pump** Same as progressive cavity pump.

**progressive wave** A wave in which water particles move in circles. *See also* standing wave.

**ProGuard** Backwashable cartridge filters manufactured by ProGuard Filtration Systems.

**prohibited discharge standard** A standard used by the USEPA to control pollutant discharges to publicly owned treatment works by any industrial or commercial establishment. Such standard relates to pollutants that have a pH less than 5.0 or that may create a fire or explosion hazard in sewers or treatment plants, upset treatment processes, or increase the influent wastewater temperature above 40°C. *See also* priority pollutant, categorical standard.

**prokaryote** Any cellular organism in which the nucleus has no limiting membrane, mitochondria, or plastids; a simple life form of a diverse group that includes bacteria, blue-green algae (cyanobacteria), and archaea. Also spelled procaryote. *See also* eukaryote.

**prokaryotic** Pertaining to prokaryotes.

**proliferation** The multiplication of microbial cells by budding or division.

**Promal** Malleable materials manufactured by FMC Corp. for the fabrication of iron chains.

**prometon [$(H_7C_3HN)_2C_3N_3OCH_3$]** The common name for the synthetic organic compound 2,4-bis(isopropylamino)-6-methoxy-S-triazine used as a herbicide. It is often found in surface water and groundwater.

**ProMix™** An apparatus manufactured by BlenTech, Inc. for blending polymers.

**promoter** In studies of skin cancer in mice, an agent that results in an increase in cancer induction when administered after the animal has been exposed to an initiator, which is generally given at a dose that would not result in tumor induction if given alone. A cocarcinogen differs from a promoter in that it is administered at the same time as the initiator. Cocarcinogens and promoters do not usually induce tumors when administered separately. Complete carcinogens act as both initiator and promoter. Some known promoters also have weak tumorigenic activity, and some also are initiators. Carcinogens may act as promoters in some tissue sites and as initiators in others. Same as promotor. *See also* initiator.

**promotor** A chemical substance or physical agent that promotes the growth of cells that leads to DNA changes and carcinogenesis; same as promoter. *See also* initiator.

**propachlor** An unregulated organic contaminant, listed in Phase II Rule.

**ProPack** Random media produced by Gray Engineering Co. for trickling filters.

**propanal** An aldehyde on the USEPA list for quarterly monitoring.

**propane ($C_3H_8$)** A colorless, flammable gas of the alkane series, found in petroleum and natural gas; used as a fuel and in organic synthesis.

**propanil** A pesticide with a WHO-recommended guideline value of 0.02 mg/L for drinking water.

**propanoic acid ($CH_3CH_2COOH$)** The proper name of propionic acid.

**1-propanol ($C_3H_8O$ or $CH_3—CH_2—CH_2—OH$)** Proper name of *n*-propyl alcohol.

**2-propanol ($C_3H_8O$ or $CH_3—CH_3—CHOH$)** Same as isopropyl alcohol.

**propanone ($CH_3COCH_3$)** *See* acetone.

**propazine (2-chloro-4,6-diisopropyl-amino-S-triazine)** A common, potentially carcinogenic organic chemical, derivative of cyanuric chloride, used to control broadleaf and grassy weeds. It may be found in surface waters receiving agricultural runoff and is not removed effectively by conventional water treatment processes. *See also* atrazine, simazine, and s-triazine.

**PRO*PEL** Bead catalysts produced by Prototech Co. for use in air pollution control.

**propeller** The bladed rotor of a pump that drives the fluid axially; an impeller that is curved around its axis like a ship screw.

**propeller blade** A mechanical aerator installed at the bottom of a tube to aspirate air into an aeration basin.

**propeller flocculator** A mixer used in flocculation. *See* turbine/propeller flocculator for detail. *See also* propeller mixer.

**propeller flow pump** Same as propeller pump.

**propeller mixer** A device, with an axial-flow high-speed impeller, used in water or wastewater treatment for mixing or blending thick chemical solutions, for keeping materials in suspension, or for aeration, e.g., in mixing alum for sweep floc flocculation. *See also* mixing and flocculation devices.

**propeller pump** A centrifugal pump that diverts liquids (clean water or treated effluent) in the axial direction of the pipeline in which its propeller-type impeller is installed. Also called axial-flow pump.

**propeller-type impeller** A straight axial-flow impeller.

**propeller-type turbine** An axial-flow turbine that has a runner with blades similar to those of a ship's impeller.

**propenamide** or **2-propenamide** *See* acrylamide.

**propene ($CH_3CH{:}CH_2$)** Common name of propylene.

**propenenitrile** or **2-propenenitrile** Acrylonitrile.

**prophage** A stable, noninfectious form of a virus that replicates in synchrony with the bacterial host. *See* lysogenic phage.

***Propionibacterium* genus** A group of heterotrophic organisms active in biological denitrification.

**propionic acid ($CH_3CH_2COOH$)** The common name of propanoic acid of the group of carboxylic acids; one of the organic acids used to monitor the anaerobic sludge digestion process.

**proportional control mode** In wastewater process control applications, the value of variable control parameter based on a linear relationship with the current error of a controlled parameter. *See also* the following types of control: cascade, feedback, feed-forward, integral, on–off, proportional–derivative, proportional–integral, proportional–integral–derivative.

**proportional counter** An instrument used to detect radiation, particularly alpha activity, but also beta and gamma activity; the strength of the electric pulse generated is proportional to the energy causing the pulse.

**proportional–derivative control** An improvement of the proportional control mode with the addition of a derivative term, e.g., derivative time.

**proportional diffusivity design** An approach used in development of small-scale column tests to generate performance data of granular activated carbon contactors. It is based on the assumption that the internal diffusion coefficient varies linearly with the column diameter. *See also* high-pressure minicolumn technique, rapid small-scale column test.

**proportional diffusivity model** A mathematical model developed to define the relationships between the breakthrough curve for large and small carbon columns in a rapid small-scale column test. This model assumes that the high hydraulic loading rate renders dispersion negligible in the test and that interparticle diffusion is responsible for mass transfer. *See also* constant diffusivity model.

**proportional flowmeter** A flowmeter that measures a specific portion of the total flow.

**proportional-flow weir** Same as proportional weir. *See also* Sutro weir.

**proportional–integral control** A combination of two types of process control. *See* proportional control mode and integral control.

**proportional–integral–derivative control** A combination of three types of process control. Also called continuous control. *See* proportional control mode, integral control, proportional–derivative control.

**proportional sample** Same as composite sample.

**proportional weir** A flow-measuring structure in which the discharge $Q$ is directly proportional to the head $H$. (In contrast, discharge varies with the 1.5 power and the 2.5 power, respectively, of rectangular and triangular weir heads.) The proportional weir has a horizontal crest and at least one side (the Sutro weir) or both sides (the Rettger weir) curved so that the half-width of the notch ($x$) varies with the inverse of the square root of the elevation above the crest ($y$). The curved sides may be shaped according to

$$x = K/2\, y^{0.5} \qquad \text{(P-50)}$$

where $K$ is a coefficient. The proportional weir is used for irrigation diversions, for measuring very small flows, or for regulating wastewater velocity in grit chambers. *See* Figure W-08. Also called proportional-flow weir.

**proportionate mortality ratio (PMR)** The number of deaths from a specific cause and in a specific period of time per 100 deaths in the same time period.

**Proportioneer** A mixing unit manufactured by Lightnin.

**proportioning** *See* industrial waste proportioning.

**proprietary filter** A complete filtration unit provided by a manufacturer, including controls, design criteria, and performance specifications, as compared to a filter that is individually designed by an engineer and built by contractors.

**Propulsair** Aspirating aerator manufactured by Eimco Process Equipment Co.

***n*-propyl ($C_3H_7$ or $CH_3—CH_2—CH_2—$)** A saturated hydrocarbon radical.

***n*-propyl alcohol ($C_3H_8O$ or $CH_3—CH_2—CH_2—OH$)** A primary alcohol; common name of 1-propanol.

***n*-propylbenzene ($C_9H_{12}$)** An organic compound, insoluble in water, used in textile dyeing and in the manufacture of chemicals. It affects the eyes, nose, throat, skin, and nervous system. It is included in List 3 of USEPA's Phase I of volatile organic compounds.

**propylene ($CH_3CH{:}CH_2$ or $C_3H_6$)** A colorless, flammable gas of the alkene series of hydrocarbons; used in organic synthesis. Also called propene.

**propylene dichloride ($CH_3CHClCH_2Cl$)** Another name of 1,2-dichloropropane.

**ProSep** An apparatus manufactured by Advanced Membrane Technology, Inc. for the treatment of industrial wastewater using membranes.

**Prosonic™** An ultrasonic apparatus manufactured by Endress+Hauser for level control.

**prospective study** A study in which subjects are followed forward in time from initiation of the study; for example, an epidemiological study that examines the development of disease in a group of persons determined to be presently free of the disease. Often called a longitudinal or cohort study.

**Prosser/Enpro** Pump products manufactured by Crane Pumps & Systems.

**protean** An insoluble substance from the reaction of a protein with water, a dilute acid, or an enzyme.

**ProTecRO™** A powder produced by King Lee Technologies for use as an antifoulant.

**protecting scale** A scale or deposit that forms on the surface of a metal and reduces the rate of the corrosion of the metal. *See also* nonprotecting scale, erosion corrosion.

**protein** A complex organic compound of high molecular weight made of amino acids; essential for growth and repair of animal tissue. It has no general formula, but all proteins contain carbon, nitrogen, hydrogen, oxygen, phosphorus, sulfur, and trace elements. Many proteins are enzymes; there are also structural and transport compounds. Wastewaters with high protein content also contain significant ammonia concentration and high alkalinity. Proteins represent 40–60% of the organic matter in wastewater. The ratio of carbohydrates and fats to protein is a consideration in aerobic treatment processes. *See also* simple protein, conjugated protein, cytoplasm.

**protein coat** The shell-like coat of protein that surrounds the nucleic acid core of a virus. This outer structure, composed of up to several hundred proteins, protects the viral core; it harbors receptor sites (for the virus attachment) as well as electrical charges and hydrophobic groups. Also called capsid.

**proteolysis** The breaking down (e.g., hydrolysis) of proteins or peptides into simpler compounds, as in digestion.

***Proteus*** A genus of enterobacteriaceae associated with waterborne gastroenteritis; some of them are opportunistic.

***Proteus mirabilis*** A species of bacteria involved in putrefaction and causing diaper rash.

**protist** A member of the kingdom Protista, a class of living organisms including eukaryotic algae, protozoa, slime molds, and other unicellular eukaryotes. (They do not include prokaryotic cells like bacteria. Some protista resemble fungi.) Some protista are pathogenic (e.g., the protozoa *Giardia, Cryptosporidium, Microsporidium*), some are free-living, and others aggregate in colonies. Algae are very significant in aqueous environments (*see,* e.g., cyanobacteria or blue-green algae, photosynthesis).

**Protista** Protists or the kingdom Protista.

**Protista kingdom** The taxonomic kingdom of the protists. Sometimes called Protista phylum or Protista superphylum.

**Protoc** Devices manufactured by Tytronics, Inc. for the analysis of total organic carbon.

**protolysis** The transfer of protons from one substance to another.

**protolysis degree** A number that indicates to what extent a substance dissociates in water. It is the ratio of the molar concentration of the substance produced to the analytical concentration (total number of moles of the pure substance). *See* dissociation fraction for more detail. *See also* distribution diagram.

**proton** (1) A positively charged elementary (subatomic) particle in the nucleus of an atom; designated *p*, $p^+$, or +, having a charge of +1 and a mass of 1.007 u (u = atomic mass unit = 1/12 the mass of carbon-12). The number of protons of an atom determines its atomic number. *See also* electron, neutron. (2) The hydrogen ion ($H^+$).

**protonate** To add a hydrogen ion; e.g., in ion exchange, to add a hydrogen ion ($H^+$) to the carboxylate functional group of a resin and prevent the exchange of hydrogen ions for calcium ($Ca^{2+}$) ions. *See also* split neutral salts.

**protonation** The addition of a hydrogen ion to a substance.

**proton charge** The charge acquired by a particle because of binding or dissociating protons, which affects ion-exchange capacity.

**proton condition** After an aqueous solution reaches equilibrium, the number of excess protons equals the number proton deficiencies. It is equivalent to the electroneutrality condition.

**proton nuclear magnetic resonance (PNMR)** A spectrometric method used to characterize organic compounds, based on the absorption of radiation by protons in a magnetic field.

**proton number** Same as atomic number.

**protoplasm** The material inside the cell membrane that characterizes all living tissue and is responsible for metabolic activity; it consists mainly of proteins, fats, and carbohydrates.

**protoplast** A Gram-positive, membrane-bound cell from which the outer wall has been partially or completely removed. The term is often applied to plant cells. *See also* spheroplast.

**Protozoa** A major group or superphylum of the kingdom Protista, comprising the protozoans. They constitute one of the five categories of parasitic organisms that infect humans. Also, plural of protozoon.

**protozoal disease** A water- and/or excreta-related disease or infection caused by protozoans, which may be excreted (amebiasis, balantidiasis, cryptosporidiosis, giardiasis) or vector-borne (malaria, trypanosomiasis). *See,* for example, amebic dysentery.

**protozoal infection** *See* protozoal disease.

**protozoan** (Plural: protozoans, collectively: protozoa.) (1) As a noun, any member of a diverse group of one-cell, mobile, microscopic eukaryotes of the kingdom Protista that are larger and more complex than bacteria; some are pathogenic, causing, for example, amebic dysentery or malaria. They include amoebae, ciliates, sporozoans, and flagellates; they feed on bacteria, algae, and other protists. Most protozoa reproduce by binary fission or budding, but some also use a kind of sexual reproduction. Most protozoa are free-living although many are parasitic or live in symbiotic relationships. The majority of protozoa are aerobic or facultatively anaerobic heterotrophs. They are present in natural waters, activated sludge, trickling filters, and oxidation ponds; they also survive conventional wastewater treatment as well as sludge digestion and air drying. They often act as polishers of biological treatment effluents by consuming bacteria and particulate organic matter; they prefer higher dissolved oxygen concentrations (> 1.0 mg/L) and larger solids retention times than bacteria. *See also* oocyst, sporozoite, trophozoite, schizont, merozoite, zygote. (2) As an adjective: of, pertaining to, or characteristic of protozoans.

**protozoan cyst** A stable life stage of protozoan parasites, excreted through feces.

**protozoon** Singular form of protozoa. The smallest and simplest animal, which has one cell between 0.002 and 0.5 mm in size; e.g., *Entamoeba hystolitica.*

**protozoon pathogen** A disease-causing protozoon.

**PRO*VOC** Catalyst products of Prototech Co. for use in air pollution control.

**Provox®** sodium percarbonate A chemical produced by OCI Chemical of Shelton, CT for use in water treatment, i.e., oxygen-powered bleaching and algae control.

**prussian blue** A blue paste or liquid (often on a paper like carbon paper) used to show a contact area; it is used to determine if gate valve seats fit properly.

**prussic acid** A colorless, highly poisonous liquid; an aqueous solution of hydrogen cyanide (HCN). Also called hydrocyanic acid. *See also* cyanide.

**$P_2S_5$** Chemical formula of phosphorus pentasulfide.

**PSA** Acronym of pressure swing adsorption.

**PSC** Acronym of phosphorus source coefficient.

**PSC value** Same as phosphorus source coefficient.

**pseudohardness** A characteristic of water that has a high concentration of sodium (e.g., ocean water) and interferes with the action of soap.

**Pseudomonad** or **Pseudomonades** Plural of *Pseudomonas.*

***Pseudomonas*** (Plural: pseudomonad or pseudomonades.) A genus of rod-shaped, Gram-negative, aerobic bacteria, some of which cause disease in plants and animals; they are commonly found in trickling filters and as floc-formers in activated sludge. Some *Pseudomonas* bacteria produce obnoxious sulfur compounds, and others can degrade hydrocarbons.

***Pseudomonas aeruginosa*** A species of Gram-negative, aerobic, nonsporulating, rod-shaped, opportunistic bacteria that transmit various waterborne intestinal infections, as well as eye, ear, and wound infections. Although ubiquitous in the environment, it is used as an indicator of swimming pool and recreational water quality. It produces blue-green and yellow-green pigments, as well as a surfactant that shows specificity for cadmium and lead. This species can survive or even grow in low-nutrient environments, e.g., in tap water and in spring water bottling plants with low-molecular-weight substrates (acetate, amino acid, lactate) at the level of 1 microgram per liter.

***Pseudomonas cepacta*** A species of bacteria that are effective in the biodegradation of creosote extracted from soil.

***Pseudomonas denitrificans*** A species of bacteria that can accomplish denitrification by using nitrate as an electron acceptor to oxidize organic matter in the absence of oxygen. They are said to achieve facultative anaerobic heterotrophic respiration, e.g.,

$$5\ C_6H_{12}O_6 + 24\ KNO_3 \rightarrow 30\ CO_2 + 18\ H_2O + 24\ KOH + 12\ N_2 \qquad \text{(P-51)}$$

***Pseudomonas fluorescens*** A species of Gram-negative, aerobic bacteria that produce a yellow-green fluorescent pigment; commonly found in water but apparently not a human pathogen. It is one of the very few bacteria groups that can survive and even achieve limited growth in pure, distilled, or low-nutrient spring waters by exchanging nutrients with the atmosphere.

***Pseudomonas fluorescens* strain P17** A strain of the bacterial species *Pseudomonas fluorescens* used to determine the concentration of assimilable organic carbon in water.

***Pseudomonas* genus** The most common and widely distributed heterotrophic organisms active in biological denitrification. *See Pseudomonas.*

***Pseudomonas putida*** A species of bacteria that are effective in the biodegradation of some aromatic hydrocarbons (benzene, toluene, ethylbenzene, and xylenes).

**pseudoplastic fluid** A material, like wastewater sludge, whose characteristics vary between those of a Newtonian fluid and a plastic fluid. *See* plastic fluid.

**pseudopodia** Temporary protrusions that certain protozoa (ameboids) use for locomotion; plural of pseudopodium; also called pseudopod. *See also* cilia and flagella.

**PSI** Acronym of pollutant standard index.

**psi** Abbreviation of pound(s) per square inch. A unit of pressure or pressure drop across a flow resistance. One psi is equivalent to the pressure exerted by 2.31 feet of water column. psig is the abbreviation of pound(s) per square inch gauge and psia is pound(s) per square inch absolute such that 0 psig = 14.696 psia = 1.0 atmosphere. psid = pound(s) per square inch differential. psig = pound(s) per square inch gage, i.e., the pressure within a closed container or pipe measured with a gage in psi.

**psia** Acronym of pounds per square inch absolute. *See* psi.

**psid** Acronym of pounds per square inch differential. *See* psi.

**psig** Acronym of pounds per square inch gage. *See* psi.

**P source coefficient** *See* phosphorus source coefficient.

**PSRP** Acronym of process to significantly reduce pathogens.

***Psychoda*** Genus of Psychoda fly.

***Psychoda alternata*** The species of Psychoda or filter flies.

**Psychoda fly** A small, dark grey fly that is a nuisance around trickling filter beds, particularly during warm weather. Psychoda flies are mainly an irritation to operating personnel, but wind can carry them a long distance and they can pass through window screens. Their larvae breed in the zoogleal film on the filter media and on the inside retaining walls of the filter. Effective control measures include spraying the filter with insecticide and periodically submerging it to drown the larvae. Smooth plastic media do not commonly support many filter flies. Higher dosing rates on the filter contribute to reducing the fly population. Also called filter fly. *See also* Anisopus fly.

**psychrometer** An instrument used to determine atmospheric humidity and vapor tension. It consists of two thermometers, one having a wet and ventilated bulb, and the other a dry bulb. *See also* hygrometer.

**psychrometry** (1) Measurement of atmospheric humidity. (2) The thermodynamics of air and water vapor mixtures.

**psychrophile** (1) Any of a group of organisms that grow best within a low temperature range (from the prefix psychro, meaning cold). The literature reports different temperatures for this range; they generally fall between 0°C (32°F) and 30°C

(86°F); the optimum range for growth is between 12 and 18°C (AWWA, 2000; Pankratz, 1996; *Webster's New World Dictionary,* 1991; Manahan, 1997; Metcalf & Eddy, 1991; Metcalf & Eddy, 2003; Hammer & Hammer, 1996; Droste, 1997; Maier, 2000, WEF, 1991). *See also* mesophile and thermophile. (2) Same as psychrophilic.

**psychrophilic** Pertaining to a psychrophile. Psychrophilic bacteria have an optimum temperature range of 12–18°C. Psychrophilic organisms are often found in deep-sea environments because they can also tolerate high pressures. *See also* mesophilic and thermophilic.

**psychrophilic bacteria** Bacteria that grow at low temperatures. *See* psychrophile and psychrophilic.

**psychrophilic (micro)organism** A (micro)organism that grows at low temperatures, e.g., 10–30°C, with an optimum range of 12–18°C. *See* psychrophile and psychrophilic.

**psychrophilic range** (1) The range of low temperatures over which some organisms can exist. *See* psychrophile. (2) The optimum temperature range for the growth of psychrophilic organisms; reported as between 12 and 18°C (between 54 and 64°F).

**Pt-Co** *See* platinum–cobalt color unit.

**PTFE** Acronym of polytetrafluoroethylene.

**public access** A consideration in the design of wastewater reuse systems. An agricultural irrigation system with restricted public access (e.g., for nonfood crops) can use a secondary treatment effluent, whereas an unrestricted public access system (e.g., for food crops and residences) requires advanced treatment for pathogen removal.

**public comment period** The time allowed for the public to express its views and concerns regarding an action by the USEPA (e.g., *Federal Register* notice of proposed rulemaking, a public notice of a draft permit, or a notice of intent to deny).

**public health** A field that encompasses services to improve and protect the physical and mental health of the community; originally, sanitation (for the prevention of such waterborne diseases as cholera and typhoid), immunization, and preventive medicine. Public health is also concerned with exposure to new hazards, risks, or pollutants. *See* environmental health.

**public health engineer** A sanitary engineer that emphasizes the health aspects of environmental protection, e.g., air pollution control, drinking water, environmental planning, housing sanitation, radiation control, recreation area sanitation, solid waste management, and wastewater control.

**public health engineering** The field in which a public health engineer practices.

**public hearing** A formal meeting wherein USEPA officials hear the public's views and concerns about a USEPA action or proposal. USEPA is required to consider such comments when evaluating its actions. Public hearings must be held upon request during the public comment period.

**public latrine** A latrine used by more than one household or by the public in general as in a market, school, or other institution. Public latrines are often of the pour-flush type and are sometimes combined with showers and laundry facilities. Also called a communal latrine.

**Public Law 92-500** *See* federal Water Pollution Control Act Amendments of 1972.

**publicly owned freshwater lake** A freshwater lake that offers public access to the lake through publicly owned contiguous land so that any person has the same opportunity to enjoy nonconsumptive privileges and benefits of the lake as any other person. If user fees are charged for public use and access through state or substate operated facilities, the fees must be used for maintaining the public access and recreational facilities of this lake or other publicly owned freshwater lakes in the state, or for improving the quality of these lakes (EPA-40CFR35.1605-3).

**publicly owned treatment works or POTW** A waste treatment works owned by a state, unit of local government, or Indian tribe, usually designed to treat domestic wastewater. Any devices or systems used in the storage, treatment, recycling, and reclamation of municipal sewage or industrial wastes of a liquid nature, which are owned by a state or municipality. This definition includes sewers, pipes, or other conveyances only if they convey wastewater to a POTW providing treatment (EPA-40CFR122.2 and 40CFR403.3-o). The term also means the municipality that has jurisdiction over the indirect discharges to and the discharges from such a treatment works (EPA-40CFR117.1).

**publicly owned water utility** A water supply system owned by a municipal agency.

**public notice** (1) Notification by the USEPA informing the public of Agency actions such as the issuance of a draft permit or scheduling of a hearing. USEPA is required to ensure proper public notice, including publication in newspapers and broadcast over radio stations. (2) In the safe drinking water program, water suppliers are required to publish and broadcast notices when pollution problems are discovered.

**public notification** Disclosure to the public of a water supply system's violation of the Safe Drinking Water Act.

**public utility** A company that provides a public service such as water supply or wastewater management and is regulated by a federal, state, or local agency.

**public water point** A communal water point from which users collect water (as opposed to an individual connection). *See* standpipe.

**public water supply** Water distributed from a public water system.

**public water system** A system for the provision to the public of piped water for human consumption, if such system has at least fifteen service connections or regularly serves an average of at least 25 individuals daily at least 60 days out of the year. Such term includes: (a) any collection, treatment, storage, and distribution facilities under control of the operator of such system, and (b) any collection or pretreatment storage facilities not under such control that are used primarily in connection with such system (EPA-40CFR142.2).

**puddle** A small pool of water, a few inches deep.

**puddled clay** Clay, sand, and water mixed in layers of 150 mm for use as a watertight seal.

**puff movement** *See* odor puff.

**pugmill** (1) A mill for grinding and mixing materials such as water and soil or the ingredients of cement. Typically, mixing is aided by an internal mechanical stirring/kneading device. (2) A grinder for reducing the size of sludge solids or solid waste for further processing.

**pugmill slaker** A mechanical device that adds enough water to hydrate quicklime (CaO) into slaked or hydrated lime that contains approximately 35% calcium hydroxide [$Ca(OH)_2$]. Also called paste slaker. *See also* detention slaker.

**PUIR** Acronym of potential unit infiltration rate.

**PullUp** An aeration apparatus manufactured by Aerators, Inc., consisting of a removable header and drop pipe.

**pulp** *See* newspaper pulp.

**pulp and paper industry** An industry that uses wood as primary source to manufacture a wide variety of products used in communications, packaging, and many other applications.

**pulp and paper mill waste** Wastewater generated during papermaking activities such as cooking, refining, washing of fibers, and screening of paper pulp. Such wastewater has variable pH, high color, total solids, and inorganic fillers. Characteristics vary with the pulping method (groundwood, soda, sulfate or kraft, and sulfite). Pulp and paper mill waste is usually treated by sedimentation, lagooning, and aeration. The industry also practices product recovery. *See* closed-cycle concept, kraft process, pulping process, sulfite process, white water.

**pulp and paper wastewater** *See* pulp and paper mill waste.

**pulping process** A step in pulp and papermaking in which wood cellulose fiber is separated from the lignin binder. The kraft and sulfite processes are used predominantly in the United States.

**pulsation dampener** A device in which compressible gas absorbs pressure fluctuations and facilitates uniform flow in pumps and discharge pipes.

**Pulsator®** A solids contact clarifier manufactured by Infilco Degremont, Inc.

**Pulsatron®** An electronic metering pump manufactured by Pulsafeeder.

**pulse bed** A filter configuration used in the design and operation of granular filtration and activated carbon adsorption units. *See* moving bed for detail.

**pulse bed filter** *See* pulsed-bed filter.

**pulsed arc electrohydraulic discharge (PAED)** *See* electrohydraulic discharge.

**pulsed-bed adsorber** A device used to remove contaminants from liquid or gaseous streams. It consists of a column of activated carbon operated in an upflow mode, with fresh carbon added at the top and spent carbon withdrawn from the bottom. *See also* adsorption zone, biological fluidized-bed reactor, breakthrough, fixed-bed adsorber, mass-transfer zone, upflow expanded-bed mode.

**pulsed-bed filter** A type of depth filter commonly used in wastewater treatment; consisting of an unstratified shallow layer of fine sand, with an intermittent air pulse mechanism (called air mix–pulse mix) that forces air through the bed to break up the surface mat and renew the sand surface. *See also* conventional downflow filter, deep-bed downflow filter, deep-bed upflow continuous-backwash filter, traveling-bridge filter.

**pulsed UV lamp** A device that converts alternate current to direct current, stores it in a capacitor, and then releases it through a high-speed switch to produce UV radiation 20,000 as intense as sunlight at sea level. *See also* low-pressure, high-intensity UV lamp; low-pressure, low-intensity UV lamp; medium-pressure, high-intensity UV lamp; narrow-band excimer lamp.

**pulse input** The introduction and mixing of a quantity of a tracer into a reactor over a very short period of time compared to the detention time. Most commonly, a dye is injected into the reactor to characterize tracer response curves and describe the hydraulic performance of reactors. Also called slug-dose input. *See also* step input, residence time distribution (RTD) curve, time–concentration curve.

**Pulsemate™** Monitoring devices manufactured by Wallace & Tiernan, Inc., including metering pumps and controls.

**Pulse Mix** A backwashing process developed by Zimpro Environmental, Inc. to regenerate sand filter media.

**pumice** *See* pumicite.

**pumice stone** *See* pumicite.

**pumicite** A natural, porous or spongy form of volcanic glass, containing aluminum silicate, used as an abrasive or as a water filtration medium. Also called pumice or pumice stone. *See also* aluminosilicate and manganese-dioxide-coated pumicite.

**pump** A mechanical device installed in a water or sewer system or other liquid-carrying pipeline that moves the liquid to a higher level. More generally, a pump is a mechanical device that applies pressure to raise or lift a fluid or to cause the fluid to flow. *See* the following types of pumps: airlift, axial-flow, centrifugal, constant-speed, continuous-flow, deep-well, deep-well turbine, diagonal-flow, diaphragm, displacement, displacement ejector, double-action, double-suction, dry-pit, gear, horizontal, jet, mixed-flow, multistage, open-impeller, piston, plunger, positive-displacement, propeller, radial-flow, reciprocating, rotary, rotodynamic, screw, single-action, single-stage, sludge feed, submersible, suction, sump, vacuum, variable-speed, vertical, volute, wind.

**pump affinity laws** Basic relationships that indicate how the capacity or discharge ($Q$), head ($H$), and power ($P$) of a centrifugal pump change with changes in its impeller diameter ($D$) or its rotational speed ($N$), for the same fluid. These relationships are not valid for simultaneous changes in impeller diameter and rotational speed, or for significant changes in pump efficiency. The subscripts 1 and 2 refer to the pump before and after the change in diameter or speed:

$$Q_1/Q_2 = D_1/D_2 = r \quad \text{(P-52)}$$

(assuming constant speed)

$$H_1/H_2 = r^2 \quad \text{(P-53)}$$

$$P_1/P_2 = r^3 \quad \text{(P-54)}$$

$$Q_1/Q_2 = N_1/N_2 = R \quad \text{(P-55)}$$

(assuming constant diameter)

$$H_1/H_2 = R^2 \quad \text{(P-56)}$$

$$Q_1/Q_2 = R^3 \quad \text{(P-57)}$$

**pumpage** The quantity of water, wastewater, or other liquid pumped in a given period of time, e.g., in gallons or million gallons per day, month, or year. Also called water intake.

**Pumpak** Devices manufactured by Healy-Ruff Co. to control the operation of pumps.

**pump-and-treat** *See* extraction as a groundwater renovation technique.

**pump bowl** The submerged part of a pumping unit: housing, impeller, and shaft.

**pump-bypass line** or **(bypass line)** A short pipe segment equipped with a check valve and installed around a pump as a pressure surge control device and to prevent backflow and cavitation.

**pump capacity** The rate of flow that a pump can deliver, depending on head and efficiency.

**pump characteristic curves, pump performance curves,** or, simply, **pump curves** A set of curves, prepared by a pump manufacturer and showing graphically the relationships between the pump discharge capacity, head, power, and efficiency for a given speed. The graph is a plot of efficiency (in %), head (in ft), and brake horsepower versus flow capacity (usually in gpm). For a given speed, to increase the head is to decrease the flow capacity and vice versa, and the rating of a pump corresponds to the combination of head and discharge that yields the maximum efficiency.

**pump characteristics** Certain properties of pumps and pump systems that are used for their selection, comparison, and performance evaluation: capacity, head, power, efficiency, and specific head. They depend on pump size, speed, and design. *See* pump characteristic curves.

**pump curves** Same as pump characteristic curves.

**pump cycle time** In the operation of a pump, cycle time is time elapsed between successive motor starts. Minimum cycle times recommended by manufacturers range from 5 to 30 minutes. Lower times may result in excessive wear of the equipment.

**pumped distribution** A method of distributing drinking water by pumping it directly to consumers without storage. It provides no reserve in case of power failure but increased pressure is available for fire fighting.

**pumped hydroelectric storage** *See* pumped storage.

**pumped storage** A system of upper and lower reservoirs operated as follows: water is pumped from the lower reservoir to the upper reservoir (when power is available, e.g., at night) and then later released to turn turbines and generate electricity. The pumped storage concept could be applied to alleviate water pollution through low-flow augmentation.

**pump efficiency** The ratio of pump output to the pump horsepower, usually expressed as a percentage. This ratio is called wire-to-water efficiency or pump overall efficiency when the motor is included with the pump. *See also* brake horsepower, water horsepower, motor efficiency.

**pumper truck** (1) A fire truck equipped to provide adequate pressure for fire fighting. Also called a motor pumper. (2) A truck used to withdraw the contents of septic tanks and convey it to a point of disposal or treatment. *See* cotreatment.

**pump grinder** Same as grinder pump.

**pump head terms** Common terms related to pumping, used in the design and operation of pumps and pump stations, some of which are illustrated on Figure P-16. *See also* system head curve.

**pump horsepower** The energy delivered to a pump shaft and expressed in horsepower (HP) to obtain a certain rate of flow at a given pressure. *See also* brake horsepower, water horsepower.

**pump house** *See* pumping station.

**pumping head** The sum of static and friction heads. Figure P-17 illustrates head relationships in pumping systems.

**pumping level** (1) The water level elevation in a well that is being pumped at a given rate. (2) The vertical distance in feet or meters from the centerline of a pump discharge to the level of the free pool while water is being drawn from the pool. Also called pumping water level.

**pumping line** The discharge line from a pump or a pumping main, i.e., a pressurized pipeline for the conveyance of water, wastewater, stormwater, or other fluids. *See also* force main.

**pumping main** Same as pumping line.

**pumping station, booster station, lift station, pump station** There exists some confusion in the use of these phrases. Some writers use them interchangeably. *See,* e.g., their definitions in APHA et al. (1981). For water distribution and waste or storm water collection systems, the following definitions can be used. Booster pumps raise the pressure of water or wastewater on their discharge side. The pumps may be installed on pressurized pipelines or force mains, or housed in a booster station. The pressure is added directly to the force main, without a wet well. A lift station contains pumps that elevate wastewater in a sewer to allow gravity flow. Usually, it is the cost-effective alternative to a deeper gravity line. Pump stations (interchangeably, pumping stations) contain water or wastewater pumps and their appurtenances, mechanical devices installed in sewer or water systems or other liquid-carrying pipelines that move the liquids to a higher level. They receive their inflow in a wet well or other appropriate basin and discharge the outflow into a force main. Structures sheltering small water pumps are called pump houses. *See also* off-line and on-line pump stations.

**pumping station area** The area that contributes wastewater to a given pumping station, excluding the areas of any upstream stations. Same as pumping station service area.

**pumping station service area** Same as pumping station area.

**pumping water level** *See* pumping level.

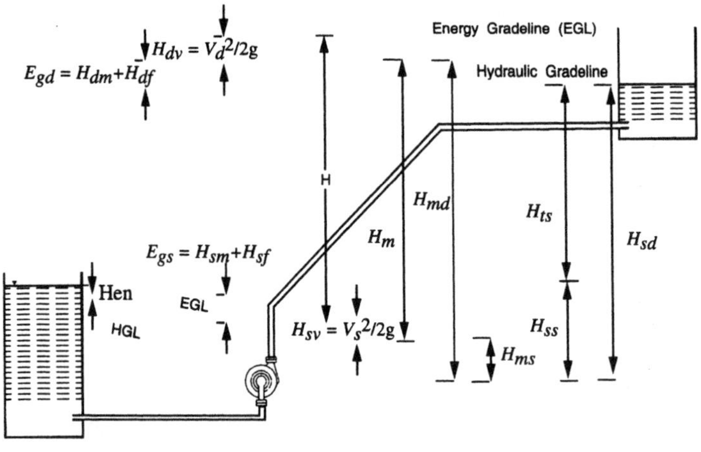


**Figure P-16.** Pump head terms (courtesy CRC Press).

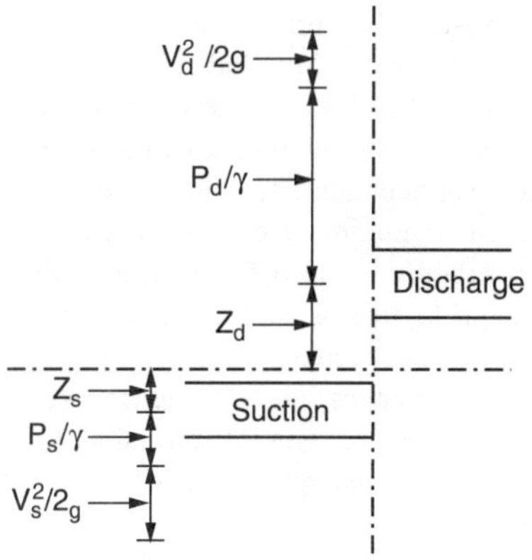


System Head = $(Z_d - Z_s) + (P_d/\gamma - P_s/\gamma) + (V_d^2/2_g - V_s^2/2_g)$

| | Discharge | Suction |
|---|---|---|
| Elevation | $Z_d$ | $Z_S$ |
| Pressure | $P_d$ | $P_S$ |
| Velocity | $V_d$ | $V_S$ |

$\gamma$ = Fluid specific weight

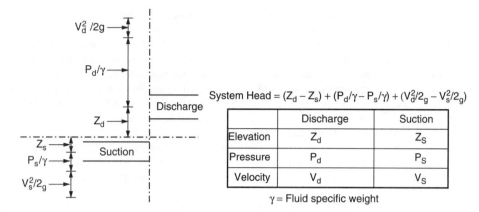

**Figure P-17.** Pumping head relationships.

**pump-in test** A test used to determine the assimilative capacity of a site with regard to on-site disposal of wastewater. *See* shallow trench pump-in test for detail.

**Pumpless Integral Clarifier™** An intrachannel clarifier manufactured by Innova-Tech (Valley Forge, PA) for oxidation ditch wastewater treatment systems. *See also* boat clarifier, Burns and McDonnell treatment system, Carrousel Intraclarifier™, side-channel clarifier, sidewall separator.

**pump operating level** The level of liquid in a wet well that determines when a pump starts or stops; it is based on the net positive suction head requirement of the pump.

**pump operating point** The intersection of a pump head-capacity curve and the characteristic curve of the system in which the pump operates. *See* Figure P-18.

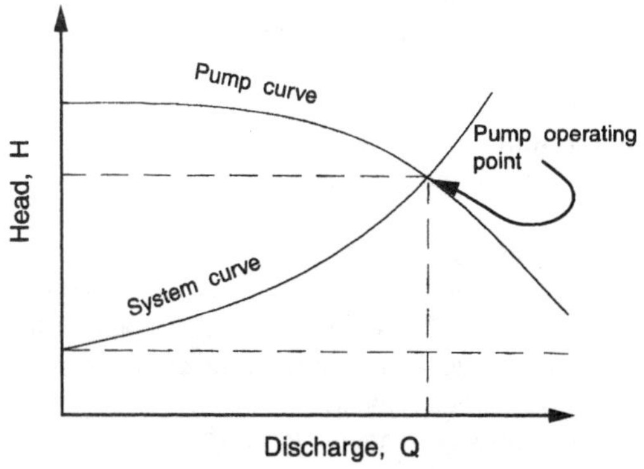


**Figure P-18.** Pump operating point (courtesy CRC Press).

**pump output** The product, expressed in horsepower, of flow rate and total head of a pump at a given pressure. *See also* pump efficiency, pump horsepower, wire-to-water efficiency.

**pump overall efficiency** Also called wire-to-water efficiency. *See* pump efficiency.

**pump packing** Special material placed around a pump shaft to prevent air from entering the pump and thus reduce water losses along the shaft. The packing is a series of rings compressed by a gland; it is kept moist by an annular pool of water so as to prevent overheating and wear of the rotating shaft. *See also* packing.

**pump performance curves** Same as pump characteristic curves.

**pump pit** The dry well or chamber below grade in which a pump is installed.

**pump primer** A device attached to a pump to prime it automatically.

**pump priming** A step consisting of filling the casing of a centrifugal pump with water with the discharge valve closed before starting the pump. Priming replaces air with water and provides a liquid for the pump to act on.

**pump rods** The rods that connect the piston of a pump with the power head.

**pump setting** The height of the discharge column of a well pump.

**pump slip** The proportion of water taken into the suction end of a pump but not discharged.

**pump stage** The number of impellers in a centrifugal pump, e.g., single-stage = one impeller, double-stage or two-stage = two impellers.

**pump station** Same as pumping station.

**pump strainer** A device placed on the inlet end of a pump to retain suspended matter.

**pump stroke** The distance traveled by the piston or plunger of a pump from top to bottom.

**pump submergence** The height of water in the pump pit below the inlet or suction.

**pump-valve cage** A container for the pump valves.

**pump valves** The openings through which water enters or leaves the cylinder of a pump.

**pump well** A nonflowing well; a well that requires a pump or other lifting device to discharge water at the ground surface.

**pupa** The stage in the development of an insect between larva and adult.

**pupae** Plural of pupa.

**Puratex™** A cartridge filter manufactured by Osmonics, Inc.

**PurCycle™** An apparatus manufactured by Purus Corp. for the removal of volatile organic compounds.

**pure compound** A substance that contains elements in the precise proportions represented in its chemical formula. Also called a stoichiometric compound.

**pure culture** A culture in which only one organism grows. Same as axenic culture. *See also* mixed culture.

**pure culture growth** Microbial growth under ideal conditions: availability of substrate and other requirements, absence of inhibitory substances and predators. *See also* growth pattern.

**pure fixed-film process** An industrial wastewater treatment process that uses a tower containing a fixed plastic medium with a specific surface area of about 100 $m^2/m^3$. *See also* biological activated filter.

**pure ion-exchange reaction** An ion-exchange reaction that does not involve weak acid cation resins in the RCOOH form or free-base forms of weak-base resins. Such a reaction proceeds at a more rapid rate than a reaction involving large ions, chelating resins, or the acid or base form of weak resins.

**Pureone®** An apparatus manufactured by U.S. Filter Corp. to produce water for laboratories.

**pure oxygen** A gas that is 90–100% oxygen as opposed to air that contains only approximately 20% oxygen (and 80% nitrogen). In water or wastewater treatment, pure oxygen is produced on-site or carried in liquid form. *See* tonnage oxygen.

**pure oxygen activated sludge** A variation of the activated sludge process using complete-mix reactors in series and pure oxygen with mechanical dispersion instead of air for aeration and mixing. Mixed liquor suspended solids concentration varies from 1000–3000 mg/L for municipal wastewater to more than 10,000 mg/L for some industrial wastes. The process is sometimes credited with a better settling sludge, better handling of peak loads, and lower energy consumption, but *Nocardia* organisms may accumulate in the aeration basin and cause foaming. The basin is often covered to minimize escape of the oxygen-rich gas; carbon dioxide ($CO_2$) then accumulates and cause pH to decrease to 6.0–6.5. *See also* high-purity oxygen activated sludge.

**pure oxygen generation** For on-site production of oxygen at pure oxygen activated sludge plants, *see* cryogenic air separation and pressure swing adsorption.

**pure oxygen process** A variation of the activated sludge process that uses pure (molecular) oxygen instead of air in the aeration tank. *See* pure oxygen activated sludge.

**pure oxygen treatment** *See* pure oxygen activated sludge.

**Puresep®** A solids separation process developed by U.S. Filter Corp.; uses resins.

**pure water** *See* deionized water, distilled water, reagent grade water.

**Purgamix** An apparatus manufactured by Jones & Attwood, Inc. to heat and mix sludge.

**purge** To force a gas through a water sample to release volatile chemicals or other gases.

**purgeable organic carbon (POC)** The portion of total organic carbon removed by gas stripping; volatile organic carbon.

**purgeable organic halogen (POX)** The volatile portion of total organic halogen of a sample, a surrogate measure of the total quantity of purgeable halogen-substituted organic matter.

**purgeable organics** Volatile organic compounds that can be released by forcing a gas through a water sample.

**purged line** A water line having a faucet opened and allowed to run for a specified length of time, e.g., 1–5 minutes.

**Purge Saver™** An instrument manufactured by QED Environmental Systems for groundwater analysis.

**purge water** Water used in cleaning or disinfecting a water line or reservoir.

**Purifax®** A proprietary process developed by Leeds & Northrup for sludge stabilization using wet oxidation by chlorine at high concentrations (e.g., 1200 mg/L for a 1% solids sludge; higher for activated sludge than primary sludge and increasing with solids content).

**Purifax process** Same as Purifax®.

**purification** The process of removing objectionable constituents from water.

**purification function** A general first-order equation that expresses the rate of water purification processes as a function of the remaining property such as the concentration of a substance. It takes one of two forms:

$$y = y_0 e^{\pm kt} \quad \text{(P-58)}$$

$$y = y_0 (1 - e^{\pm kt}) \quad \text{(P-59)}$$

where $y$ = the remaining property (e.g., concentration), $y_0$ = initial concentration (at time $t = 0$), $k$ = reaction rate constant, and $t$ = time of exposure. Several water and wastewater treatment formulations use these equations as a basis; for example, Chick's law, filtration kinetics (Ives equation), gas absorption, and Streeter and Phelps' classic oxygen sag.

**purification works** Treatment facilities that are used to make water potable; same as treatment works, as compared to water collection, transmission, and distribution works.

**purified water** Water that contains no added substance and meets the quality requirements of U.S. Pharmacopeia after purification by an appropriate water treatment process such as distillation, ion exchange, or other USEPA-approved drinking water treatment. *See* pharmaceutical-grade water. Also called USP purified water or USP grade water.

**purine ($C_5H_4N_4$)** A white, crystalline compound with such derivatives as uric acid, caffeine, and fundamental constituents of nucleic acids (adenine, guanine). Purines and pyrimidines constitute one of three major classes of growth factor requirements of organisms.

**Puritan** An apparatus manufactured by SanTech, Inc. for the treatment of fluid recovered from spent coolant and oils.

**Purofine™** A resin produced by Purolite Co. for use in ion-exchange units.

**Purolite A-850** A hydrophilic polyacrylic resin manufactured by the Purolite Co. of Bala Cynwyd, PA and used in ion exchange.

**purple bacteria** A group of Gram-negative pink to purplish-brown bacteria found in anaerobic aquatic environments and capable of carrying out photosynthesis using the pigment bacteriochlorophyll in the absence of oxygen. They include the purple sulfur and purple nonsulfur bacteria.

**purple salt** A common name for potassium permanganate ($KMnO_4$), a dark purple, crystalline salt of potassium and manganese used in water treatment for taste and odor control, for iron and manganese removal, and for disinfection.

**purple sulfur bacteria** A group of bacteria that evolved in the absence of oxygen; they fix carbon using light energy and oxidize hydrogen sulfide to sulfur. *See also* green sulfur bacteria and purple bacteria for detail.

**PurSorb™ 100** A solvent used to reclaim organics. *See* PADRE™.

**PurSorb™ 200** A solvent used to reclaim organics. *See* PADRE™.

**Purveyor (water)** An agency or person that supplies water.

**Pusher™** A screw press manufactured by William R. Perrin, Inc. for sludge dewatering.

**push joint** A gasket used in bell-and-spigot pipe.

**putrefaction** Biological decomposition and incomplete oxidation of organic matter, with the production of ill-smelling and bad-tasting products, associated with anaerobic conditions. *See* anaerobic digestion, fermentation.

**putrescibility** The tendency of organic matter, wastewater, or sludge to degrade in the absence of oxygen.

**putrescible** Able to rot quickly enough to cause odors and attract flies.

**putrescible waste** Animal or vegetable waste degraded bacteriologically.

**Putzmeister** A pump of Asdor used in handling sludge cake.

**PV** Acronym of (a) permanganate value and (b) pervaporation.

**PVC** Acronym of polyvinyl chloride, a material used to fabricate water and wastewater piping.

**PVDF** Acronym of polyvinylidenefluoride.

***P. vivax*** Abbeviation of *Plasmodium vivax.*

***P. westermani*** *See Paragonimus westermani.*

**PWEP** Acronym of percent water extractable phosphorus.

**PWI™ Wheeler Bottom Insert** Modular or built-in equipment manufactured by Roberts Filter Group of Darby, PA for use in rehabilitation or new construction projects. *See also* Wheeler Bottom.

**PWMP®** Acronym of pure water management program, a group of U.S. Filter Corp. that provides outsourced water services.

**PX™** Abbreviation of Pressure Exchanger™.

**pycnometer** A container of known capacity and weight that is used to measure the densities of a liquid or powder.

**Pyramed™** A proprietary filter underdrain system manufactured by WesTech Engineering of Salt Lake City, UT for use in municipal and industrial water filtration.

**pyrene ($C_{16}H_{10}$)** A colorless, polycyclic, aromatic, crystalline hydrocarbon that consists of four fused benzene ($C_6H_6$) rings, found in tar and reportedly carcinogenic.

**pyrheliograph** *See* pyrheliometer.

**pyrheliometer** An instrument for measuring the intensity of direct solar radiation (e.g., ultraviolet light), using a blackened disk or a thermoelectric device. *See also* actinometer and heliothermometer. A pyrheliograph also records the measurement.

**pyridate** A pesticide with a WHO-recommended guideline value of 0.1 mg/L for drinking water.

**pyrimidine ($C_4H_4N_2$)** A heterocyclic compound with important biochemical derivatives such as the fundamental constituents of nucleic acids (e.g., cytosine, uracil). Purines and pyrimidines constitute one of three major classes of growth factor requirements of organisms.

**pyrite ($FeS_2$)** A common iron ore, a sulfur compound, frequently found in coal and a source of cobalt; used in the production of sulfuric acid ($H_2SO_4$). It is involved in the initial reaction leading to the formation of acid mine drainage. Also called fool's gold.

**PyroBatch®** A process developed by Midland Ross, Inc. and marketed by Surface Combustion, Inc. for the destruction of hazardous wastes by pyrolysis at 450–550°C in a rotary kiln. It includes an afterburner for the incineration of volatile products from the pyrolysis chamber.

**pyrobenzol** Another name of benzene.

**pyrocatechin** Same as catechol.

**pyrocatechol** Same as catechol.

***Pyrococcus*** A genus of bacteria adapted to high temperatures (in excess of 100°C).

***Pyrodictium*** A genus of thermolerant bacteria; adapted to temperatures in excess of 100°C.

**pyrogen** A stable substance or cell material from bacteria that produces fevers in mammals; pyrogens are not always destroyed by conditions that kill bacteria.

**pyroligneous acid** Same as acetic acid.

**pyrolusite** A common manganese mineral, essentially manganese dioxide ($MnO_2$); used as a catalytic material to enhance the removal of iron from groundwater by aeration followed by sedimentation and filtration.

**pyrolysis** Decomposition of a chemical by extreme heat in oxygen-deficient conditions; pyrolysis converts organic matter into gases, liquid, and char, through a combination of thermal cracking and condensation. It is a thermal process used in sludge dewatering and in the preparation of activated carbon; *see* carbonization, physical activation. *See also* destructive distillation, partial pyrolysis or starved-air combustion, wet oxidation.

**pyrolytic water** Water produced as a by-product of the Tubingen process.

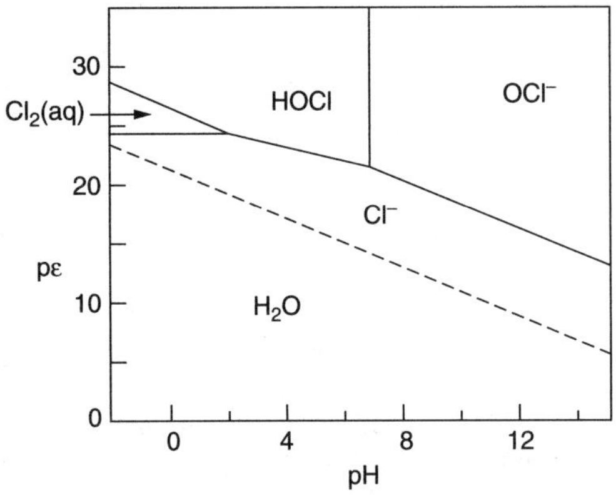


**Figure P-19.** pε–pH diagram (0.04 M total chlorine).

**Pyrospout** An apparatus manufactured by Process Combustion Corp. for sludge processing by fluid bed combustion and incineration.

**Pyroxidizer®** A process developed by Envimid, Inc. for the destruction of medical wastes by pyrolysis at up to 1000°C in a batch autoclave.

**pyrrolylene** Another name of butadiene.

**Pyrrophyta** A group of unicellular algae, most of which have flagella. They constitute a large part of plankton in seas and freshwaters, including the Dinophyceae that can cause red water when abundant. Also called fire algae.

**Python Press** An apparatus manufactured by Waste Tech, Inc. for sludge filtration and dewatering under high pressure.

**PZC** Acronym of point of zero charge.

**PZNPC** Acronym of point of zero net proton charge.

**pε** A parameter, similar to pH, that represents the hypothetical electron activity at equilibrium; the relative tendency of a solution to accept or transfer electrons ($e^-$).

$$p\varepsilon = -\log \{e^-\} \qquad \text{(P-60)}$$

*See* pE for detail.

**pε–pH diagram** A graphical representation that combines equilibria between chemical species as a function of pH and the same equilibria at a given pH as a function of pε. "Such pε–pH stability field diagrams show in a comprehensive way how protons and electrons simultaneously shift the equilibria under various conditions and can indicate which species predominate under any given condition of pε and pH" (Stumm and Morgan, 1996). *See* Figure P-19 for a pε–pH diagram of a chlorine system. *See also* Pourbaix diagram. Also called pE – pH diagram.

# Q

**Q** The usual notation for flow.

**q1*** Upper bound on the slope of the low-dose linearized multistage procedure.

**$Q_{10}$** A temperature coefficient that indicates the increase in the rate of a process that is due to a temperature increase of 10°C.

**QA** Acronym of quality assurance.

**QAPP** Acronym of quality assurance project plan.

**QA/QC** Acronym of quality assurance/quality control.

**QC** Acronym of quality control.

**QL-1™** An apparatus manufactured by Infilco Degremont, Inc. for disinfection by ultraviolet light.

**QLS** Acronym of quality lock sprocket, a device of Budd Co..

**QRA** Acronym of quantitative risk analysis.

**q* slope** A parameter derived from animal studies using multistage analysis and indicating the magnitude of cancer threat from a substance or other agent. Also called cancer potency slope.

**qt** Abbreviation of quart.

**Q-Tracker™** A device manufactured by Badger Meter Co., Inc. to monitor flows in sewers.

**Quadra-Clean™** A vertical basket centrifuge manufactured by Western States Machine Co.

**Quadra-Kleen** An apparatus manufactured by Culligan International Corp. for backwashing sand filters.

**Quadramatic III®** An automatic batch centrifuge manufactured by Western States Machine Co.

**Quadra Press** A plate-and-frame filter press manufactured by Duriron Co.

**Quadrufil** Gravity filter equipment manufactured by Vulcan Industries, Inc.

**quadrupole mass spectrometer** A compact instrument used to detect and identify chemical compounds. It includes four parallel metal rods to filter ions from the mass spectrometer. Only ions of a certain mass/charge ratio pass through the device. This spectrometer is sometimes used in the analysis of odorous compounds such as ammonia, amino acids, and volatile organic compounds.

**quagga mussel** A bivalve that multiplies rapidly in freshwater and can clog intake screens and pipes; present in the Great Lakes, it looks like the zebra mussel. Hypochlorite and permanganate are effective means of control. (A bivalve is a mollusk that has two hinged shells, a soft body, and lamellate gills. The quagga is an extinct mammal of southern Africa resembling the zebra.)

*Processing Water, Wastewater, Residuals, and Excreta for Health and Environmental Protection.* By N. G. Adrien 

**QUAL2E** A model widely used to study river water quality, starting with temperature-dependent BOD–DO relationships and now including 10 constituents as well as the simulation of the impacts of nutrients and photosynthesis on the oxygen balance.

**qualification test** A test or verification procedure conducted to validate the quality of water or wastewater treatment equipment.

**qualified groundwater scientist** A scientist or engineer who has received a baccalaureate or postgraduate degree in the natural sciences or engineering, and has sufficient training and experience in groundwater hydrology and related fields as may be demonstrated by state registration, professional certifications, or completion of accredited university course that enable that individual to make sound professional judgments regarding groundwater monitoring, contaminant fate and transport, and corrective action (EPA-40CFR260.10 and 503.21-1).

**qualitative** Descriptive of kind, type or direction, as opposed to size, magnitude, or degree.

**qualitative assay** A test or measurement that produces a nonnumerical result, such as the presence–absence test. *See also* quantitative assay.

**quality assurance (QA)** A series of planned or systematic actions required to provide adequate confidence that a product or service will satisfy given needs. A quality assurance plan describes an orderly assembly of management policies, objectives, principles, organizational responsibilities, and procedures designed to achieve the quality objectives of a program or project (EPA-40CFR30.200 and 40CFR35.6015-39). A quality assurance statement describes how precision, accuracy, completeness and compatibility will be assessed; it is sufficiently detailed to allow an unambiguous determination of the quality assurance practices to be followed throughout a project (EPA-40CFR30.200). Quality assurance/quality control is a system of procedures, checks, audits, and corrective actions to ensure that technical and reporting activities are of the highest achievable quality.

**quality assurance narrative statement** A description of how precision, accuracy, representativeness, completeness, and compatibility will be assessed; it is sufficiently detailed to allow an unambiguous determination of the quality assurance practices to be followed throughout a research project (EPA-40CFR30.200).

**quality assurance plan** *See* quality assurance.

**quality assurance program plan** A formal document that describes an orderly assembly of management policies, objectives, principles, organizational responsibilities, and procedures by which an agency or laboratory specifies how it intends to (a) produce data of documented quality, and (b) provide for the preparation of quality assurance project plans and standard procedures (EPA-40CFR30.200).

**quality assurance project plan (QAPP)** A written document that represents in specific terms the organization, objectives, functional activities, and specific quality assurance and quality control activities and procedures designed to achieve the data quality objectives of a specific project or continuing operation (EPA-40CFR35.6015-39 and 300.5).

**quality assurance/quality control** *See* quality assurance.

**quality assurance statement** *See* quality assurance.

**quality control** A series of activities to achieve and maintain the desired level of quality in manufacturing or service operations: inspections, analyses, reviews, corrective changes, etc.

**quality control sample** A solution obtained from an outside source having known concentration values to be used to verify the calibration standards (EPA-40CFR136-AC-9).

**QUAL models** A series of water quality models developed by Water Resources Engineers, Inc., starting in the early 1970s. *See* QUAL2E.

**QualServ** A computer program developed by the American Water Works Association and the Water Environment Federation to help water supply and wastewater management utilities improve performance.

**Quanti-Cult™** Organisms produced by IDEXX Laboratories, Inc.

**quantitative** Descriptive of size, magnitude, or degree, as opposed to kind, type, or direction.

**quantitative assay** A test or measurement that produces a numerical result, such as the MPN method. *See also* qualitative assay.

**quantitative risk assessment** The quantitative expression of risks in terms of infection, illness, and death toward the determination of the costs and benefits of related corrective measures.

**Quanti-Tray™** A test kit manufactured by IDEXX Laboratories, Inc. for water samples.

**Quantum®** A floating aerator manufactured by Air-O-Lator Corp.

**quarry water** The moisture content of freshly quarried stone.

**quart** A unit of volume for the measurement of liquids, equal to ¼ gallon, or 2 pints, or 57.75 cubic inches, or approximately 0.9464 liter. Abbreviation: qt.

**quartan fever** *See* quartan malaria.

**quartan malaria** A form of malaria caused by the parasite *Plasmodium malariae,* found in tropical and subtropical areas and characterized by a fever at its paroxysm every three or four days. Also called quartan fever.

**quartz ($SiO_2$)** A crystalline mineral characteristic of acid igneous rocks and abundant in sedimentary and metamorphic rocks; water-clear when pure, but also found in a number of varieties such as amethyst (purple) and rose quartz (pink).

**quartzite** A granular metamorphic rock composed mainly of silica (quartz).

**quartz jacket** A clear, fused quartz tube used to protect ultraviolet lamps. Also called quartz sleeve.

**quartz sleeve** Same as quartz jacket.

**quasihydrostatic approximation** The use of the hydrostatic equation to represent the vertical equation of motion, assuming that vertical accelerations are small. Also called hydrostatic approximation or quasihydrostatic assumption.

**quasihydrostatic assumption** *See* quasihydrostatic approximation.

**quat** Abbreviation of quaternary amine functional group.

**quaternary amine ($R_4N^+$)** A positively charged organic compound of nitrogen (N) derived from the ammonium ion ($NH_4^+$) by substituting alkyl groups for the hydrogen ions; each monomer contains a nitrogen atom bound to four carbon atoms. Quaternary amines used in water treatment include the cationic polyelectrolytes polyDADMAC and epiDMA.

**quaternary amine functional group (quat)** A bacteriostatic long-chain substance included on the surface of special ion exchange resins to prevent bacterial growth.

**quaternary ammonium salt ($R_4N^+X^-$)** An organic nitrogen compound formed by the combination of an alkyl halide with a tertiary amine (an amine with three hydrogen atoms); it produces positive charges in cationic polymers. In the formula, X represents an acid radical and R is an organic group.

**quench** To cool something suddenly or halt a process abruptly by immersion in a liquid.

**quench tank** A water-filled tank used to cool incinerator residues or hot materials during industrial processes.

**quenching agent** A substance used as an inhibitor, e.g., ascorbic acid ($C_6H_8O_6$) as an inhibitor of hypochloric acid (HOCl) or chlorine.

**quick condition** *See* quicksand.

**quicklime** Another name for lime. A white or grayish-white material, slightly soluble in water, produced by calcining limestone to liberate carbon dioxide ($CO_2$); mostly calcium oxide (CaO) or calcium oxide in natural association with a lesser amount of magnesium oxide. Quicklime combines with water to form hydrated lime. It may be obtained from ground limestone (calcite) or calcium carbonate ($CaCO_3$) and has many applications, e.g., in water treatment, mortars, cements, etc. *See* lime for more detail.

**quick-operating valve** A valve with a plug and lever that allow quick opening and closing.

**Quick Purge®** Technology developed by Integrated Environmental Solutions, Inc. for soil and groundwater remediation.

**quicksand** (1) An unstable condition that occurs when the hydraulic gradient of the seepage lines under a structure is higher than unity. It causes the material underneath the structure to have fluid characteristics and lose its bearing capacity. *See also* piping and creep ratio. (2) In sewer construction a quick condition may develop when the excavations extend below the water table and the subsurface strata are permeable, allowing the fluidization of the soils. To prevent this condition, well points along the sides of the trench and extending below its bottom allow depressing the water table by pumping.

**quicksilver** Another name for the metallic element mercury.

**quicksilver water** A solution of mercury nitrate used in gilding.

**quickwater** (1) The part of a river or other stream that has a strong current. (2) A current or bubbling patch of water formed behind a moving boat.

**quiescent impoundment** A surface impoundment used for equalization, settling, storage, biological treatment, and disposal of liquid hazardous wastes. *See also* surface impoundment, aerated impoundment.

**quiescent sedimentation** Same as quiescent settling.

**quiescent settling** The settling of solid particles in still water, i.e., a body of water at rest, without any inflow or outflow, as opposed to dynamic settling. In quiescent settling, the particles first accelerate until the drag or frictional resistance equals their weight, and then settle at a uniform velocity. *See* terminal velocity, flocculent settling, hindered settling, sedimentation types I, II, and III. In extended-detention wet ponds, particle settling is assumed to be quiescent in the permanent pool between runoff events and dynamic in the active storage zone.

**Quik-Clamp** A clamp-on spray nozzle manufactured by FMC Corp.

**quinone ($C_6H_6O_2$)** A yellow, crystalline, cyclic unsaturated diketone; a nonchlorinated by-product of water disinfection by chlorine dioxide; used in dyes.

**quotidian malaria** A type of malaria with fever every 24 hours.

# R

**R** Symbol of roentgen.

**Ra-224** Abbreviation of radium-224.

**Ra-226** Abbreviation of radium-226.

**Ra-228** Abbreviation of radium-228.

**RAASR** Acronym of reservoir-assisted aquifer storage and recovery.

**rabble arm** A rotating metal arm used in a multiple-hearth incinerator to scrape and move materials such as sludge around.

**rabbling process** A procedure during the thermal processing of sludge through a multiple-hearth furnace, whereby the sludge is raked by rabble arms across the hearth toward the center shaft. Rabbling exposes new sludge surfaces to heat for drying.

**race track** A common name for the oval-shaped channel used in the oxidation ditch process of wastewater treatment.

**rack** (1) A manually or mechanically cleaned device, with openings of 25 mm or 1 in, used in preliminary wastewater treatment for the removal of large floating and suspended solids. Screenings accumulated on the racks are periodically raked, ground, and returned to the influent or combined with primary sludge. They may also be processed by other methods (anaerobic digestion, burial, incineration, landfilling). *See also* comminution. Racks and screens are also used to remove leaves, sticks, and other debris from water in dams and at intakes. (2) An assemblage of membrane modules installed in parallel or in series and operated as a single unit. Also called train, unit.

**rack rake** A manual or mechanical rake used to clean racks and screens.

**RACM** Acronym of reasonably available control measure.

**RACT** Acronym of reasonably available control technology.

**RAD** or **rad** Acronym of radiation absorbed dose, a unit that represents the absorption of 100 ergs per gram of material. The unit rad can be used for any kind of radiation and any type of material; it has been replaced by the gray (Gy). The absorbed dose relates to the amount of energy actually absorbed: 1 rad = 0.01 joule of energy per kilogram of material (0.01 J/kg) = 0.01 Gy. *See also* rem.

**Radial Filter** A filtration apparatus manufactured by Aero-Mod, Inc.

**radial flow** Flow of a fluid along the radiuses of a rotating container, that is, from the center to the periphery or vice versa. A radial-flow tank is a circular tank that has radial flow, with the inlets at the center and the outlets on the periphery in radial inward flow, and vice versa in radial outward flow.

**radial-flow mixer** A device that uses a vertical shaft, flat-blade turbine to disperse chemicals, suspended particles, and other substances in a mixing basin. *See* Figure R-01.

**radial-flow pump** A centrifugal device that directs water or wastewater in a volute or turbine casing through the eye of an open or closed impeller. *See also* axial-flow pump, diagonal-flow pump.

**radial-flow tank** *See* radial flow.

**radial inward flow tank** *See* radial flow.

**radial outward flow tank** *See* radial flow.

**Radial Plate Dryer** A sludge drying apparatus manufactured by Envirotech Systems Corp.

**radial gate** A type of spillway crest gate whose face is a section of cylinder that rotates about a horizontal axis on the downstream end of the gate. The gate, widely used in large installations, can be raised or lowered by winches or hoists acting on the bottom; it can also be closed under its own weight. *See also* flashboards, stop logs, sliding gates, bear-trap gates, and drum gates. Also called tainter or taintor gate and canal lock.

**radial to impeller** Perpendicular to the impeller shaft. Material being pumped flows at a right angle to the impeller.

**radial well** A wide, productive and shallow well system in a water-bearing formation, having a central sump or tubewell from which radiate a number of horizontal strainers or cylindrical screens. *See also* strainer well.

**radiant dryer** An infrared furnace or a multiple-hearth furnace used to reduce the moisture content of municipal sludge. Also called an infrared dryer or radiation dryer.

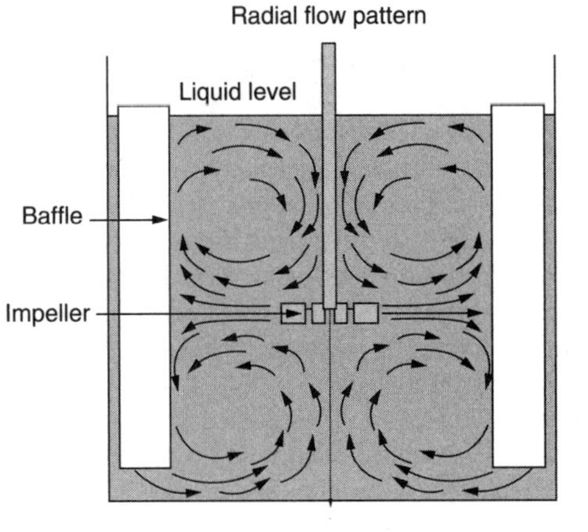


**Figure R-01.** Radial-flow mixer (courtesy Eimco Water Technologies).

**radiation** (1) Transmission of energy through space or any medium, especially by electromagnetic waves or high-speed particles rather than by conduction and convection. (2) The energy emitted or radiated. Also known as radiant energy. Ionizing radiation comprises highly energetic and penetrating X-rays and gamma rays and less penetrating particles. Beta particles are simply energetic electrons and alpha particles are helium nuclei, both arising from the nucleus of a decaying atom. The alpha particle is the easiest of these radiations to stop and the gamma rays are the most difficult to shield against. A piece of paper can stop an alpha particle, but it may take as much as many inches of lead shielding to stop most of the X-rays or gamma rays in a beam. Depending on the dose, kind of radiation, and observed endpoint, the biological effects of radiation can differ widely. Ionizing radiation has been proven to cause cancer at high doses and is assumed to cause cancer and other deleterious effects at low doses.

**radiation absorbed dose (RAD** or **rad)** *See* rad.

**radiation concentration guide** (1) The concentration of a radioisotope that is acceptable in water for lifetime consumption. (2) The maximum level of radionuclides to which an employee may be exposed during an 8-hour workday. Also called maximum allowable concentration or maximum permissible concentration.

**radiation daughter product** *See* radon progeny, radon daughter.

**radiation dose equivalent** A measure of the effect of ionizing radiation on a substance or tissue, an indication of the risk to health from the amount absorbed. *See* sievert, dose equivalent, rem.

**radiation dryer** An infrared furnace or a multiple-hearth furnace used to reduce the moisture content of municipal sludge. Also called a radiant dryer or infrared dryer.

**radiation drying** The use of infrared lamps, electric resistance elements, or gas-heated incandescent refractories to transfer heat to wet sludge and evaporate its moisture content. The rate of radiation heat transfer, $Q_{rad}$, in Btu/hr is (WEF & ASCE, 1991):

$$Q_{rad} = EAS(T_r^4 - T_s^4) \qquad \text{(R-01)}$$

where $E$ = emissivity of the drying surface, a dimensionless coefficient provided by the equipment manufacturer or determined from pilot studies; $A$ = sludge surface area exposed to the radiant source (ft$^2$); $S$ = the Stefan–Blotzman constant =

$1.73 \times 10^{-9}$ Btu/hr/ft²/°R; $T_r$ = absolute temperature of the radiant source, °R; and $T_s$ = absolute temperature of the sludge drying surface, °R. Also called infrared drying. *See also* convection drying, conduction drying.

**radiation equivalent man (rem)** *See* rem.

**radiation protection guide (RPG)** A set of tables showing radiation concentration limits for workers.

**radiation sickness** Sickness that results from overexposure to ionizing radiation, e.g., from X-rays, gamma rays, neutrons, weapon fallout, etc., symptoms of which may include nausea, vomiting, diarrhea, bleeding, hair loss, and death.

**radiation sludge drying** The use of infrared lamps, electric resistance elements, or gas-heated incandescent refractories to transfer heat to wet sludge and evaporate its moisture content. *See* radiation drying for detail.

**radical** A group of atoms that, acting as a single atom, can remain unchanged during a series of chemical reactions. Such combinations exist in the molecules of many inorganic compounds; ethyl ($C_2H_5$) is an organic radical. *See* Table R-01 for a list of radicals commonly found in aqueous chemistry. *See also* free radical.

**radical reaction** An oxidation reaction that involves species with unpaired electrons. The three major types of radical reaction in aqueous solutions are (a) addition of the hydroxyl radical ($^{\cdot}OH$); (b) hydrogen abstraction; and (c) single-electron transfer. *See also* ionic reactions.

**radical scavenger** A substance that, when present in a mixture, consumes or inactivates free radicals; e.g., the bicarbonate ion ($HCO_3^-$) and *tert*-butanol [$(CH_3)_3COH$] are scavengers for the hydroxyl radical.

**Radicator** A solid waste incinerator manufactured by Hitachi Metals America, Ltd.

**radioactive** Characteristic of an unstable substance that undergoes spontaneous decay or disintegration while emitting radiation; also, anything that exhibits or is caused by radioactivity.

**Table R-01.** Radicals (aqueous chemistry)

| | |
|---|---|
| Ammonium, $NH_4^+$ | Nitrate, $NO_3^-$ |
| Bicarbonate, $HCO_3^-$ | Nitrite, $NO_2^-$ |
| Bisulfate, $HSO_4^-$ | Orthophosphate, $PO_4^{3-}$ |
| Bisulfite, $HSO_3^-$ | Orthophosphate- 1H, $HPO_4^-$ |
| Carbonate, $CO_3^{2-}$ | Orthophosphate- 2H, $H_2PO_4^-$ |
| Hydroxyl, $OH^-$ | Sulfate, $SO_4^{2-}$ |
| Hypochlorite, $OCl^-$ | Sulfite, $SO_3^{2-}$ |

**radioactive decay** Spontaneous change in an atom by emission of charged particles and/or gamma rays; also known as radioactive disintegration and radioactivity.

**radioactive disintegration** *See* radioactive decay.

**radioactive half-life** The half-life of a radioactive material. *See* half-life.

**radioactive isotope** *See* radioisotope, radionuclide.

**radioactive material** A substance that does not exist in nature in a stable form and that forms new isotopes or new atomic species during decay.

**radioactive nuclide** An atom that disintegrates and emits corpuscular or electromagnetic radiation such as alpha, beta, and gamma rays. *See* induced, primary, and secondary radioactive nuclides.

**radioactive spring** A spring whose water has a high level of radioactivity.

**radioactive substance** A substance that emits ionizing radiation. Radioactive contaminants in water sources may come from fallout or from the nuclear-energy industry.

**radioactive waste** Unwanted by-products of nuclear fission or any waste that contains radioactive material in concentrations that exceed certain limits; high-level transuranic radioactive waste. Radioactive wastes originate from the processing of uranium ores, laundering of contaminated clothes, research laboratory wastes, hospitals, processing of fuel elements, and power-plant cooling waters. Liquid radioactive wastes are (a) concentrated and stored for subsequent burial or (b) disposed by dilution and dispersion. Sometimes abbreviated as radwaste.

**radioactive water pollution** The introduction of alpha particles, beta particles, gamma rays, and neutrons into water, with possible biological effects on humans (e.g., eye cataracts, leukemia, bone cancer, skin cancer); one possible cause is an accident at a nuclear power plant. Natural radioactivity in water is normally less than 1000 pCi/L, but levels as high as 1,000,000 pCi/L of radon have been recorded in some mineral springs.

**radioactivity** The property of those isotopes of elements (radium, radon, uranium, etc.) that exhibit spontaneous radioactive decay and emit radiation, or the actual spontaneous decay or disintegration accompanied by radiation. Radioactivity is usually measured in curies, becquerels, or picocuries. Recommended radioactivity limits in drinking water are 3 picocuries per liter of radium-226, 10 picocuries per liter of strontium-90, and 1000 picocuries per liter of gross beta. *See also* alpha radiation, beta radiation, gamma radiation, mrem, rad, rem, sievert.

**radioactivity decay** Radioactivity or radioactive decay.

**radioactivity standards** Maximum contaminant level of radioactivity set by the USEPA for drinking water is 5 pCi/L of combined Ra-226 and Ra-228 or about 0.000000005 mg/L of radium. WHO and European Community's standards are five times more stringent. *See also* radium drinking water standards.

**radioactivity unit** *See* radioactivity.

**radiocarbon** A naturally occurring radioactive isotope of carbon that emits beta particles when it undergoes radioactive decay. Mass number = 14. Half-life = 5730 years. It is widely used in the dating of organic materials. Also called carbon-14.

**radiochemical analysis** The examination of radioactive substances.

**radioimmunoassay** The detection of viral antibodies in blood using antibodies labeled radioactively.

**radioisotope** A radioactive isotope of an atom, i.e., undergoing nuclear changes and emitting nuclear radiation. It is a chemical variant of an element with potentially oncogenic, teratogenic, and mutagenic effects on the human body; usually produced artificially for use in research. Used synonymously with radionuclide.

**radiological contaminant** Any type of radiation or radioactive substance regulated by the USEPA.

**radionuclide** Any unstable man-made or natural element, with a distinct atomic weight number, that emits radiation in the form of alpha or beta particles, or as gamma rays. It can have a long life as a soil or water pollutant. Radionuclides of interest in drinking water include radium-226, radium-228, radon-222, uranium-238, strontium-90, potassium-40, and cesium-137. Polonium-210 and radium-224 are also commonly found in water sources. Used synonymously with radioisotope or radioactive isotope.

**Radionuclide Rule** The USEPA's National Primary Drinking Water Regulations regarding radionuclides.

**RadioRead®** A flowmeter reading system manufactured by Invensys Metering Systems; it uses the Intelligent Communications Encoder. *See also* TouchRead® and PhonRead®.

**radium (Ra)** A rare, naturally occurring (Ra-224, Ra-226, or Ra-228) radioactive, alkaline-earth group metallic element that is created in the decay of the uranium and thorium series and that successively emits alpha, beta, and gamma rays, and then forms an isotope of lead; see also radon, radium-224, radium-226, radium-228. Atomic weight 224, 226, or 228; atomic number 88; specific gravity 5; melting point 700°C; boiling point 1140°C; valence 2. It is carcinogenic and accumulates in bones. All radium isotopes cause bone sarcomas. Radium decomposes in water and is found in some groundwaters; it can be removed by cation exchange softening.

**radium-224** Less harmful than radium-226 and radium-228, but also frequently found in water sources.

**radium-226** The most stable isotope of radium, with a radioactivity of 1.0 curie ($3.7 \times 10^{10}$ disintegrations per second) per gram and a half-life of 1620 years. It is frequently found in water sources. It decays to radon and induces head carcinomas. Regulated by the USEPA as a drinking water contaminant: MCL = 20 pCi/L.

**radium-226 plus radium-228 (Ra-226 + Ra-228)** The sum of the concentrations of the natural radioactive isotopes of radium, a drinking water contaminant regulated by the USEPA under the Radionuclide Rule. Effective removal processes: cation exchange, reverse osmosis, electrodialysis, lime softening.

**radium-228** An isotope of radium of concern because of its health effects; frequently found in water sources. Regulated by the USEPA as a drinking water contaminant: MCL=20 pCi/L.

**radium drinking water standards** A screening parameter devised to monitor compliance with radionuclide limits in drinking water. (a) Radium concentration must be determined if gross alpha particle activity exceeds 5 pCi/L. (b) Man-made radionuclides must be determined if gross beta activity exceeds 50 pCi/L. *See also* radioactivity standards.

**radium selective complexer (RSC)** A barium sulfate ($BaSO_4$)-impregnated, strong acid cation resin that can selectively adsorb radium onto the barium sulfate even in the presence of competing ions. The product is manufactured by the Dow Chemical Company. Final disposal of the radium-contaminated waste is a problem.

**radius of hydration** *See* hydration radius.

**radius of influence** The maximum distance away from an air injection or extraction source that is significantly affected by a change in pressure and induced movement of air.

**radon (Rn)** A colorless, odorless naturally occurring, radioactive, inert gas formed with alpha rays by radioactive decay of radium in air, water, soil or other media; 20 known radioisotopes; atomic weight 222 for the most stable isotope; atomic number 86; density 9.73 g/L; melting point –71°C; boiling point –61.8°C; usual valence 9. A so-called noble or nonreactive gas that quickly volatilizes, it may be inhaled while showering, bathing, or cooking with contaminated groundwater. It is estimated that a concentration of 100 pCi/L of radon in drinking water corresponds to a cancer risk of more than 50 per million. Aeration is an effective means of radon removal from water. Regulated by the USEPA as a drinking water contaminant: MCL = 300 pCi/L.

**radon-219** An isotope of radon derived from actinium and corresponding to actinon as the alpha emitter; half-life 3.9 seconds.

**radon-220** An isotope of radon derived from thorium and corresponding to thoron as the alpha emitter; half-life 54.5 seconds.

**radon-222** The most stable isotope of radon, formed by the decay of radium-226, itself a progeny of U-238; half-life 3.8 days; found in surface water at a concentration of 1.0 pCi/L and in groundwater at up to 500,000 pCi/L; may be linked to lung cancer.

**radon daughters or radon progeny** Short-lived radioactive decay products of radon that become longer-lived lead isotopes. The daughter isotopes can attach themselves to airborne dust and other particles, and, if inhaled, damage the lining of the lung. Also known as radon decay products, a term used to refer collectively to the immediate products of the radon decay chain (Po-218, Pb-214, Bi-214, and Po-214), which have an average combined half-life of about 30 minutes.

**radon decay product** *See* radon daughters.

**radon progeny** *See* radon daughters.

**radon rule** A regulation of the USEPA under the Safe Drinking Water Act. It establishes a standard of 300 pCi/L for radon in drinking water with possible exceptions up to 4000 pCi/L under certain conditions.

**radwaste** Abbreviation of radioactive waste.

**raffinate** What remains from oil or other liquid after extraction of soluble elements by a solvent; the waste stream resulting from a solvent extraction process.

**rag** A piece of worn or torn cloth; a fragmentary bit of anything. In wastewater treatment, rags can clog pumps and valves and interfere with the operation of bar screens, comminutors, and grinders.

**rag ball** A material that is approximately round in shape, consisting of fragments of cloth or similar materials found in wastewater and shredded by comminutors or grinders. Rag balls can clog pumps and valves and interfere with the operation of mechanical aerators, mixers, sludge pipelines, and heat exchangers.

**rag rope** A material that has approximately the shape of a rope, consisting of fragments of cloth or similar materials found in wastewater and shredded by comminutors or grinders. Rag ropes can clog pumps and valves and interfere with the operation of mechanical aerators, mixers, sludge pipelines, and heat exchangers.

**railway softening** A modification of the lime–soda ash softening process that feeds excess lime to remove calcium and magnesium hardness to a practical limit of 35–40 mg/L. *See* excess-lime treatment for more detail.

**rainbow trout** A freshwater fish, *Salmo gairdneri,* used as a test organism in acute toxicity bioassays; an intolerant fish that requires cold water, high dissolved oxygen, and low turbidity.

**Rainlogger** A device manufactured by American Sigma for sampling stormwater.

**rainmaking** Rainfall inducement by such techniques as cloud seeding with silver iodide (AgI) crystals from an airplane. *See also* artificial rain, watermaker.

**rainout** The removal of foreign substances (radioactive particles, dissolved and particulate matter) from the atmosphere by rain or snow. Also called washout. *See also* wet deposition, fallout.

**rain shadow** An area on the leeward side of a coastal mountain range, with warm air, water vapor below saturation concentration, and low rainfall (or lower rainfall than the windward area). *See* Figure R-02.

**rainwater garden** A planted basin of loose soil atop a layer of sand, installed on residential or public lots and designed to capture the first por-

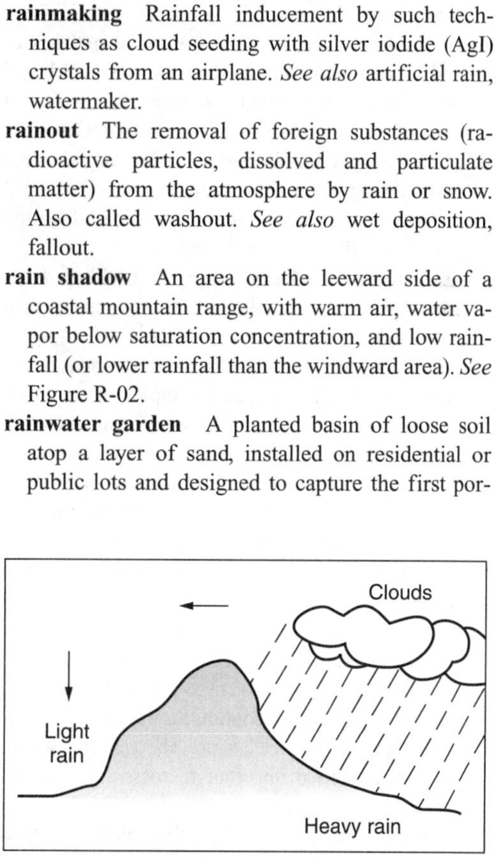


**Figure R-02.** Rain shadow.

tion of a given rainfall, trap stormwater contaminants, and replenish local groundwater.

**rainwater harvesting** A water management technique used mainly in areas where this resource is scarce. It consists of capturing water in such storage systems as tanks, ponds, wells, and rooftops to recharge groundwater that is subsequently used for drinking and other purposes.

**rainwater yield** The quantity of water that is available for water supply in a rainwater collection system. It depends on precipitation, the extent of the collection area, and various losses.

**rakings** Trash removed from bar screens in a wastewater treatment plant.

**Raleigh's law** The scattering of white light by suspended particles depends on the volume ($V$) of particles, the number of particles ($N$), and the wavelength ($\lambda$) of light (Droste, 1997):

$$I_s \sim NV^2/\lambda^4 \qquad \text{(R-02)}$$

where $I_s$ is the intensity of scattered light and the sign "~" means "is proportional to."

**Ram®** A waste compactor manufactured by S & G Enterprises, Inc.

**ramp method** One of three common methods of landfilling: incoming wastes are deposited and compacted on a slope and an excavation is dug in front of the pile to obtain cover material. This forms the slope for the next load. *See also* the area method and the trench method.

**Ram-Rod** A dewatering screw press manufactured by Ketema, Inc.

**random media trickling filter** A trickling filter using media of molded and extruded shapes randomly dumped in the bed.

**random sample** A sample from a statistical population such that all possible samples of the same size have the same probability of being selected. *See also* spot or grab sample, composite sample.

**range** The spread from minimum to maximum values that an instrument is designed to measure. *See also* span and effective range.

**ranking of excreta disposal technologies** Excreta disposal methods ranked in descending order of actual health benefits: (a) flush toilet, sewers, oxidation ponds; (b) septic tank, (c) vault/vacuum truck, pit latrine, or aquaprivy, (d) double-vault batch composter, (e) continuous (multrum) composter, and (f) bucket latrine. The ranking would be different based on other factors, e.g., ease of operation and water needs.

**Rankin's formulas** Empirical formulas developed in 1955 for the design of trickling filters, based on Ten States' Standards and data collected in the Great Lakes area:

$$E = (I - E)(1 + R)/(1 + 1.5\,R) \qquad \text{(R-03)}$$

$$E = (I - E)(1.78)(1 + R)/(2.78 + 1.78\,R) \qquad \text{(R-04)}$$

$$E = (I - E)(1 + R)/(2 + R) \qquad \text{(R-05)}$$

where $E$ = effluent BOD, mg/L; $I$ = influent BOD, mg/L; $R$ = the recirculation ratio, dimensionless. These formulas apply, respectively, to (a) high-rate processes with an organic loading of up to 1.75 kg of BOD per cubic meter per day, (b) higher organic loadings, and (c) first-stage trickling filters without intermediate clarification. Other trickling filter design formulas include British Manual, Eckenfelder, Galler–Gotaas, Germain, Howland, Kincannon–Stover, Logan, modified Velz, NRC, Schulze, Velz.

**Ranney** Products manufactured by Hydro Group, Inc. for use in well screens and caisson intakes. The Ranney Intake is an apparatus consisting of a passive screen and a caisson used as a surface water intake.

**ranney collector** A water collector constructed as a dug well from 12 to 16 feet in diameter that has been sunk as a caisson near the bank of a river or lake.

**Ranney Intake** *See* Ranney.

**R.A.P.I.D.® System** A monitoring system manufactured by Idaho Technology, Inc.

**RaPID Assay™** A reagent kit produced by Ohmicron for conducting soil analyses in the field.

**rapid bench-scale membrane test** A study defined in the USEPA's Information Collection Rule of May 1996 to evaluate membranes of molecular weight cutoffs lower than 1000 daltons at different recovery rates.

**Rapid Decanter®** A solid-bowl centrifuge manufactured by Flottweg.

**rapid filter** A filter using sand or another granular medium (e.g., anthracite coal, crushed magnetite, plastic sphere) to treat water that has undergone coagulation and sedimentation. Typical design filtration rates range from 80 to 160 liters per minute per square meter (2 to 4 gpm/sq ft) of surface area. The unit is backwashed by reversing the flow of previously treated water. *See,* for example, rapid sand filter and pressure filter. Also called rapid rate filter.

**rapid filter breakthrough** The production of water of inadequate quality by a rapid filter. *See* breakthrough index and Hudson's formula.

**rapid filter underdrainage** A pipe grid, floor, or false bottom that is designed (a) to collect the finished water from a rapid filter and convey it to a clear well and (b) to distribute washwater during backwashing.

**rapid filtration** A water treatment method that uses a rapid filter, but also implies the use of coagulation, flocculation, sedimentation, and disinfection. In rapid filtration, solid particles are transported to the surface of the filter medium by a number of mechanisms (diffusion, interception, settling, impingement, entrapment, attachment) and attached to the medium by surface charges. Also called rapid rate filtration.

**rapid granular bed filtration** A new name for rapid sand filtration.

**rapid granular filter** A water treatment device similar to the rapid sand filter but using a medium different from sand, for example, anthracite coal.

**rapid granular filtration** A water treatment method similar to rapid sand filtration but using a medium different from sand.

**Rapid Gravity Dewatering™** An inclined gravity filter manufactured by Wil-Flow, Inc. for sludge dewatering.

**rapid infiltration** A land treatment or disposal method used for wastewater disposal or groundwater recharge, whereby water applied intermittently at high rates percolates laterally or downward in permeable soils. The shallow spreading or infiltration basin is usually covered by appropriate grasses. The effluent is collected in drains or wells. The critical design element is the infiltration capacity of the soils. The minimum preapplication treatment is the equivalent of primary clarification; secondary or advanced treatment may be required in some cases. Oxidation ponds, storage ponds, and chlorination are not recommended. *See also* geopurification, high-rate infiltration, high-rate sprinkling, low-rate irrigation, overland flow, rapid infiltration extraction, soil aquifer treatment, and surface spreading.

**rapid infiltration extraction (RIX)** Same as high-rate infiltration.

**rapid infiltration land application** *See* rapid infiltration.

**rapid infiltration nitrogen removal** The removal of nitrogen from wastewater during the rapid infiltration process, mainly through denitrification, is a function of the total organic carbon concentration (TOC, mg/L):

$$\Delta N = (TOC - 5)/2 \qquad \text{(R-06)}$$

where $\Delta N$ (mg/L) is the maximum amount of nitrogen removed under optimum operating conditions.

**rapidly biodegradable organic matter (RBOM)** A source of carbon and energy used by denitrifying organisms. Examples of RBOM are methanol, volatile fatty acids (e.g., acetic acid), and other end products of fermentation.

**rapid mix** Same as rapid mixing.

**rapid mixing** A physical treatment unit in which chemical solutions (e.g. coagulants or conditioning chemicals) are quickly and uniformly mixed with raw water or wastewater; dissolution and dispersion typically occur in 10–30 seconds. It is the first step in the coagulation process, followed by flocculation. Common methods include hydraulic mixing, in-line mixing, and mechanical mixing. *See* velocity gradient, as a measure of mixing intensity. Also called rapid mix. *See also* flash mix, flash mixing.

**rapid mixing device** A device that provides rapid mixing in water or wastewater treatment. *See* mixing and flocculation devices.

**rapid mixing parameter** A parameter used to describe the performance of rapid mixing devices. *See* mixing loading parameter, mixing opportunity parameter, and velocity gradient.

**Rapidor** A pressure leaf filter manufactured by Liquid-Solids Separation Crop.

**rapid rate filter** Same as rapid filter.

**rapid rate filtration** Same as rapid filtration.

**rapid sand filter** A filter using sand as the granular media to treat water that has undergone coagulation and sedimentation. Typical design filtration rates range from 80 to 320 liters per minute per square meter (2 to 8 gpm/sq ft) of surface area, as compared to 10 times less for slow sand filters. The filter consists of (a) a 24–30 inch layer of sand, with an effective particle size of 0.35–0.80 mm and a uniformity coefficient of 1.3–1.7, on a supporting bed of gravel or on a porous material, and (b) an underdrainage system that removes the filtrate and distributes the wash water. *See also* pressure filter, Fuller (George Warren).

**rapid sand filter rating** The design flow rate of a rapid sand filter.

**rapid sand filter strainer** A perforated device installed in a rapid sand filter for the flow of filtered water and backwash water. Also called filter strainer or strainer head.

**rapid sand filtration** A high-rate method of water treatment using a rapid sand filter, usually following coagulation and sedimentation. Particle removal is primarily through the mechanism of depth filtration. The device operates as a static-bed, downflow, batch or semicontinuous unit. It is periodically backwashed to remove accumulated particles. Also called rapid granular bed filtration.

**rapid small-scale column test (RSSCT)** A bench-scale test used to represent the operation of a full-scale granular activated carbon column; it consists of the application of water to a small-scale column of crushed granular activated carbon of 4–40 mm in diameter and 100–300 mm in length. The test lasts a few weeks. It assumes that the internal diffusion coefficient is constant and dispersion is negligible. *See also* high-pressure minicolumn technique, constant diffusivity model, proportional diffusivity design.

**rare gas** *See* noble gas.

**RAS** Acronym of return activated sludge.

**raschig ring** A material as a medium in upflow fixed film reactors.

**rasp** A machine that grinds waste into a manageable material and helps prevent odor.

**ratchet rate** A water rate schedule that charges the highest rate block reached by a customer during a period.

**rate** (1) The speed of a chemical reaction. (2) Flow rate, i.e., the volume of flow per unit time, expressed, e.g., as gpm, mgd, cfs. (3) The rate charged by a utility for water supply and/or wastewater management services.

**rate base** (1) The number of customers of a water supply or wastewater management utility, or their water use as a basis to establish a rate schedule. (2) The allowable assets of a utility that serve to derive earnings and rate of return.

**rate coefficient** A parameter that indicates the rate at which a chemical or biological reaction proceeds. Rate expressions are available for the following common reactions in wastewater treatment: bacterial conversion, chemical reactions, gas absorption/desorption, natural decay, sedimentation, and volatilization. Rate coefficients are also available for zero-order, first-order, second-order, and saturation reactions. Constants used in water quality evaluation studies include the rate of BOD exertion or deoxygenation rate ($K_1$), the reaeration rate ($K_2$), and the rate of BOD removal by sedimentation and adsorption ($K_3$).

**rate constants** *See* rate coefficient.

**rate covenant** The legal commitment by a water supply or wastewater management utility that issues bonds to maintain rates that are adequate to service the debt.

**RatedAeration®** A wastewater treatment plant manufactured by FMC Corp. in a circular steel package.

**rated capacity** (1) A manufacturer's statement as to the expected performance of its equipment; e.g., expected number of days or volume of product water before backwashing, rinsedown, regeneration, or replacement. (2) *See* permitted capacity.

**rated in-service life** Same as rated capacity (1), but with an emphasis on the length of time before replacement or between servicings.

**rated pressure drop** A manufacturer's statement as to the expected pressure drop across filtration or softening equipment, based on the flow of clean water at 60°F and at the specified flow rate.

**rated service flow** A manufacturer's statement as to the range of flow rates to operate its water or wastewater treatment equipment.

**rated softening capacity** A manufacturer's statement as to the total hardness (e.g., in mg/L as $CaCO_3$) that its equipment will remove, based on flow rate and regenerant dosage.

**rate expression** *See* rate coefficient.

**rate law** A general thermodynamic relationship between the rate of an elementary chemical reaction and the concentrations of the reactants. For example, for the chemical reaction

$$aA + bB \rightarrow cC + dD \qquad \text{(R-07)}$$

the rate law is

$$d[A]/dt = k_{fa}\,[A]^a[B]^b \qquad \text{(R-08)}$$

where the concentrations are in molar units, and $k_{fa}$ is the forward reaction rate constant. *See also* order of reaction.

**rate-limiting step** In a process with units operating in series, the slowest unit is the rate-limiting step, which controls the performance of the process. For example, when the principal adsorption mechanism is physical, diffusion transport is likely to be the rate-limiting step.

**rate making** The process of establishing a rate schedule for water supply and/or wastewater management services, based on a cost-of-service study.

**rate of deoxygenation** *See* deoxygenation rate.

**rate-of-flow controller** An automatic device, like a valve or an orifice plate, that registers and controls the rate of flow of water in a rapid sand filter or, more generally, the flow of a fluid.

**rate-of-flow indicator** A device that indicates the rate of flow of a fluid at any time, but does not necessarily register it.

**rate-of-flow recorder** A device for registering the rate of flow at any time in a rapid sand filter or other hydraulic unit.

**rate of flux** In a membrane process, the rate at which the permeate flows through the membrane, expressed in $kg/m^2$/day or in $gal/ft^2$/day.

**rate of gas absorption** *See* gas absorption rate.

**rate of gas desorption** *See* gas desorption rate.

**rate of phosphate uptake** *See* phosphate uptake rate.

**rate of photosynthetic oxygenation** *See* photosynthetic oxygenation rate.

**rate of reaction** *See* reaction rate.

**rate of reaeration** *See* reaeration rate.

**rate of return** The level of profit realized by a water supply and/or wastewater management utility, equal in percentage to 100 times the ratio of profit to the total allowable rate base.

**rate schedule** A set of rates and charges established by a water supply and/or wastewater management utility, based on customer categories. Rate schedules influence water consumption; *see,* e.g., flat rate, increasing block rate, progressive rate, ratchet rate.

**rate tariff** Rate schedule.

**Ratio** A turbidimeter manufactured by Hach Co.

**RatioFlo™** A valve manufactured by Stranco to control the ratio of flow to polymer in water treatment.

**rational model** A model of inactivation of microorganisms during disinfection (Symons et al., 2000):

$$r = -kCr^nN^x \qquad \text{(R-09)}$$

where $r$ = change in microorganism density (number of microorganisms per unit time), $k$, $n$ = constants, $C$ = residual concentration of disinfectant (mg/L), $N$ = initial microorganism density (number of microorganisms per unit volume), and $x$ = coefficient to account for the shoulders and tailing off effect. Other disinfection models include the Chick–Watson, Hom, and modified Hom models.

**ratio of self-purification** *See* self-purification ratio.

***Rattus norvegicus*** The brown rat or sewer rat, a carrier of the *Leptospira icterohaemorrhagiae* bacteria.

**raw** Pertaining to water or wastewater that has not undergone any treatment, or to sludge that has not been processed.

**raw data** Any laboratory worksheets, records, memoranda notes, or exact copies thereof, that are the result of original observations and activities of a study and are necessary for the reconstruction and evaluation of the report of that study. In the event that exact transcripts of raw data have been prepared (e.g., tapes that have been transcribed verbatim, dated and verified accurate by signature), the exact copy or exact transcript may be substituted for the original source as raw data. Raw data may include photographs, microfilm or microfiche copies, computer printouts, magnetic media, including dictated observations, and recorded data from automated instruments (EPA-40CFR160.3).

**Rawlison, Robert** First chief engineer of England's National Board of Health, studied the water and wastewater needs of cities and towns; one of the significant achievements of the Great Sanitary Awakening.

**raw sanitary wastewater** Sanitary wastewater that has not undergone any treatment. Typically in the United States, such wastewater has an approximate BOD/N/P ratio of 100/17/3 as compared to 100/5/1 as required for biological treatment and 100/23/5 for a settled effluent, concentrations being expressed in mg/L for biochemical oxygen demand (BOD), nitrogen (N), and phosphorus (P).

**raw sewage** Untreated wastewater and its contents.

**raw sludge** Settled sludge that has been removed recently from a sedimentation unit before any processing and before significant decomposition. Also called undigested sludge.

**raw wastewater** Untreated wastewater and its contents.

**raw water** Intake water prior to any treatment or use. Also a water supply source from a spring, stream, aquifer, river, or lake, potentially usable for human consumption. Also called source water; its quality determines what kind of treatment is used to produce drinking water.

**Raymond** A sludge incinerator manufactured by Dorr-Oliver, Inc.

**Raypro** An air diffuser manufactured by Ray Products, Inc.

**Raysorb** An apparatus manufactured by RaySolv, Inc. to control volatile organic substances using activated carbon.

**RBC** Acronym of rotating biological contactor.

**RBC model** A second-order equation developed for the USEPA in 1985 to estimate RBC surface area requirements, based on data collected from full-scale operating plants:

$$S_n = \{[1 + (4)(0.00974)(A_s/Q)S_{n-1}]^{0.5} - 1\}/[(2)(0.00974)(A_s/Q)] \qquad \text{(R-10)}$$

$$= [2635 + 104.8\,(A_s/Q)S_{n-1}]^{0.5} - 51.33\}/(A_s/Q)$$

where $S_n$ = soluble BOD concentration in stage $n$, mg/L; $A_s$ = disk surface area in stage $n$, m$^2$; and $Q$ = flow rate, m$^3$/day.

**RBC operational problems** Performance below design expectations, structural problems with shafts and media, loping, excessive buildup of biomass.

**RBC performance** Removal of 5-day biochemical oxygen demand ($BOD_5$) by rotating biological contactors (RBC), based on an empirical equation developed for the USEPA in 1977. The model is similar to the Schulze and Germain equations for the design of trickling filters:

$$S_e = S_i \exp[-K(V/Q)^{0.5}] \qquad \text{(R-11)}$$

where $S_e$ = total $BOD_5$ of secondary clarifier effluent, mg/L; $S_i$ = RBC influent total $BOD_5$, mg/L; $K$ = reaction constant = 0.30; $V$ = volume of media, $ft^3$; and $Q$ = hydraulic loading, gpm.

**RBC process** A method of wastewater treatment using rotating biological contactors.

**RBC staging** The arrangement of rotating biological contactors (i.e., the compartmentalization of the RBC disks) to form independent cells and optimize treatment performance. Two to four stages are recommended for BOD removal and more than six for nitrification, with flow parallel or perpendicular to the shaft and with step feed flow or tapered feed flow.

**rbCOD** Readily biodegradable chemical oxygen demand.

**RBE** Acronym of relative biological effectiveness.

**RBF** Acronym of riverbank filtration.

**RCM** Acronym of reliability-centered maintenance.

**R&D** Acronym of research and development.

**RDF** Acronym of reuse-derived fuel.

**RDT** Acronym of rotary drum thickener.

**REACH** registration, evaluation, and authorization of chemicals, a policy of the European Union.

**Reacher** A bar screen manufactured by Schloss Engineered Equipment, including a reciprocating rake.

**reactant** Any of the substances that take part in a chemical reaction and are shown on the left-hand side of the equation. *See also* product, catalyst.

**reaction** *See* chemical reaction, homogeneous reaction, heterogeneous reaction, reversible reaction, irreversible reaction.

**reaction order** An indication of the relationship between the rate of a chemical or biological reaction and the concentrations of the reactants. The order of a chemical reaction for a compound is the stoichiometric coefficient of that compound; the overall order of the reaction is the sum of its stoichiometric coefficients. *See* zero-order reaction, first-order reaction, first-order retarded reaction, second-order reaction, saturated or mixed-order reaction. *See* order of reaction, rate law.

**reaction pathway** *See* hydrolysis reaction pathway.

**reaction quotient** In a redox reaction, the ratio of the activity of the reduced species to the activity of the oxidized species. *See also* Nernst equation.

**reaction rate** The rate at which a reaction proceeds; it increases with temperature. For a given substance, it is the decrease or increase in the number of moles per unit volume per unit time. Reaction rates are an important consideration in water or wastewater treatment, where many reactions do not go to completion. *See also* kinetics, rate, reaction order.

**reaction-rate constant** A parameter used in determining reaction rates, usually considered a function of temperature. *See* Arrhenius equation, forward reaction rate constant, rate law, kinetic law of mass action.

**reaction site** The point or place where an enzyme comes into contact with a contaminant substrate in a degradation reaction. The presence of branches or functional groups at the reaction site causes a steric effect, i.e., a reduction of the reaction rate.

**reaction tank** A basin in which water or wastewater is mixed and reacts with chemicals.

**reactivate** To restore by thermal or chemical means a spent adsorbent's (e.g., activated carbon's) porous structure and adsorption capacity by removing the adsorbed materials.

**reactivation** The process of restoring a spent adsorbent's capacity, usually similar to the initial activation method. Also called regeneration. *See* more detail under activated carbon regeneration.

**reactivation by-product** A secondary product of the reactivation or regeneration of activated carbon, e.g., the particulate and gaseous emissions, which can be reduced through the use of scrubbers and afterburners.

**Reactivator®** A solids contact clarifier manufactured by Graver Co.

**reactive industrial waste** A category of waste that, according to the USEPA, (a) is unstable or explosive, or (b) reacts violently, generates toxic gases/fumes/vapors, or forms explosive mixtures with water.

**reactive intermediate** An unstable metabolite of a chemical that reacts with biological molecules and may produce a mutation after cell division.

**reactive maintenance** An unplanned maintenance approach, including such activities as troubleshooting and firefighting when breakdown and failure occur. *See* breakdown maintenance, conditional or indicative maintenance, proactive maintenance, developmental maintenance.

**reactive metabolite** *See* reactive intermediate.

**reactive substance** A reactant.

**reactive wall** A permeable vertical wall of a reactive mixture, installed below the ground surface to treat water that flows through it.

**reactive waste** An unstable waste that undergoes abrupt change accompanied by toxic products or explosions.

**reactor** A container, tank, vat, vessel, or other device in which chemical or biological reactions or processes take place. These reactions and processes include oxidation, reduction, solubilization, coagulation, flocculation, adsorption, recarbonation, etc. Unit operations and processes of water and wastewater treatment take place in reactors and separators. (Generally, a separator separates a feed into two products.)

**reactor–clarifier** A basin in which both flocculation and sedimentation take place; typically, a center-feed basin with a central, conical flocculation compartment. *See also* sludge blanket clarifier, absorption clarifier.

**Reactor-Clarifier™** A solids-contact unit manufactured by EIMCO®, combining slow-speed turbine flocculation and high-volume internal recirculation to achieve mixing, flocculation, and solids contact. *See* Figure R-03.

**reactor classification** Common reactor types are batch, complete-mix, complete-mix in series, complete-mix with recycle, continuous stirred tank, fixed-bed, fluidized-bed, heterogeneous, homogeneous, packed-bed, plug-flow, plug-flow with recycle, semicontinuous stirred tank, suspended-bed, and tubular flow.

**Reactor-Thickener** A sludge thickener manufactured by Ralph B. Carter Co., including mixers and dewatering screens.

**React-pH™** An activated carbon produced by Calgon Carbon Corp.

**react phase** The second of the five steps in the operation of sequencing batch reactors, whereby the biomass consumes the substrate under controlled environmental conditions. *See also* fill phase, decant phase, settle phase, and idle phase.

**readily accessible equipment** In water and wastewater treatment works, accessible equipment can be inspected and cleaned without stoppage and with the use of only simple tools. Readily accessible equipment does not require any tools for cleaning and inspection.

**readily biodegradable chemical oxygen demand (rbCOD)** The portion of organic matter in wastewater that is quickly assimilated by the biomass because it is soluble, whereas particulate matter must first be dissolved by extracellular enzymes. It is measured using the oxygen uptake rate or the floc/filtration method. *See also* COD fractionation.

**readily water-soluble substance** A chemical that is soluble in water at a concentration equal to or greater than 1000 mg/L (EPA-40CFR797.1060-5). *See* limited water-soluble substance.

**ready biodegradability** An expression used to describe those substances that, in certain biodegradation test procedures, produce positive results that are unequivocal and that lead to the reasonable assumption that the substance will undergo rapid and ultimate biodegradation in aerobic aquatic environments (EPA-40CFR796.3100-iii).

**reaeration** (1) Introduction of air into the lower layers of a reservoir. As the air bubbles form and

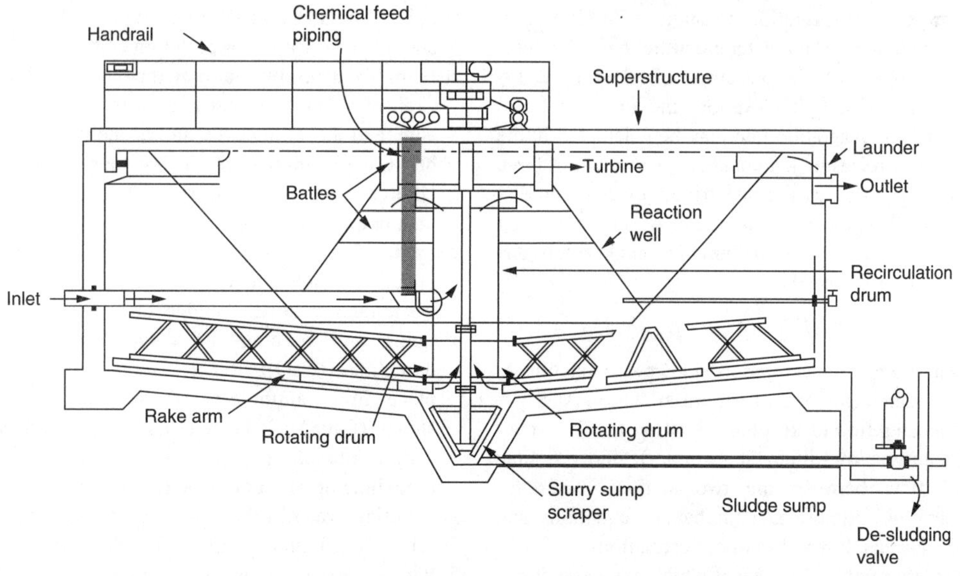


**Figure R-03.** Reactor Clarifier™ (Courtesy Eimco Water Technolgies).

rise through the water, the oxygen from the air dissolves into the water and replenishes the dissolved oxygen. The rising bubbles also cause the lower waters to rise to the surface where they take on oxygen from the atmosphere. (2) The absorption of oxygen from the atmosphere and/or from photosynthesis by green plants. This is sometimes called surface reaeration. (3) The introduction of air or oxygen into return sludge as practiced in the contact-stabilization variation of the activated sludge process. (4) The injection of air or oxygen into a treated wastewater effluent to increase its dissolved oxygen concentration as required by permits. It is usually the last step in the treatment flowchart; it is not advisable ahead of filtration. Also called reoxygenation. *See* cascade reoxygenation and mechanical or diffused air reoxygenation.

**reaeration coefficient** *See* reaeration constant.

**reaeration constant ($K_2$)** A parameter used in water quality modeling to express the rate at which oxygen dissolves in water; it varies with stream turbulence and with temperature. It may also be determined from energy dissipation and travel time considerations:

$$K_2 = (D_m U/H^3)^{0.5} \quad \text{(R-12)}$$

(this is the O'Connor–Dobbins formula)

$$K_{2(T)} = K_{2(20)}(1.025)^{(T-20)} \quad \text{(R-13)}$$

$$K_2 = (C_e)(\Delta H)/t_f \quad \text{(R-14)}$$

where $D_m$ = molecular diffusion coefficient = 0.00002037 $cm^2/s$ at 20°C; $U$ = average velocity of the stream, cm/s; $H$ = average stream depth, cm; $K_{2(20)}$ = reaeration constant at 20°C; $K_{2(T)}$ = reaeration constant at temperature $T$°C; $T$ = temperature in °C; $C_e$ = escape coefficient,/m; $\Delta H$ = change in surface elevation, m; and $t_f$ = travel time. The reaeration rate may be calculated in the field by releasing a homogeneous mixture of three tracers (a fluorescent dye, tritiated water, and dissolved krypton-85) at a station A, and measuring the concentrations of tritium (Tr) and krypton (Kr) at station B (McGhee, 1991):

$$K_2 = \ln [(Kr_a/Tr_a)/(Kr_b/Tr_b)]/0.83\ t \quad \text{(R-15)}$$

where $Kr_a$ = concentration of krypton at station A; $Tr_a$ = concentration of tritium at station A; $Kr_b$ = concentration of krypton at station B; $Tr_a$ = concentration of tritium at station B; and $t$ = time of flow between the two stations. *See also* Churchill–Elmore–Buckingham's equation and the Owens–Edwards–Gibbs correlation.

**reaeration rate** The rate at which oxygen is transferred from the atmosphere to the surface of a body of water. *See* reaeration constant, Churchill–Elmore–Buckingham's equation, dissolved-oxygen sag analysis.

**reagent** A pure chemical substance or a solution that is used to make new products or is used in chemical tests to measure, detect, or examine other substances. *See also* flotation reagent (activating, collecting, depressing, and foaming or frothing reagents).

**reagent blank** A volume of deionized, distilled water containing the same acid matrix as the calibration standards carried through the entire analytical scheme (EPA-40CFR136-AC-12).

**reagent bottle** A glass bottle fitted with a stopper for storing standard chemical solutions.

**reagent-grade water** High-purity water produced for use in the preparation of reagents or in laboratory testing. Such water meets the standards of the American Society for Testing and Materials (ASTM), including a limit on the total bacterial count. *See also* analytic grade, pharmaceutical-grade water.

**reagent water** Reagent-grade water.

**real fluid** A fluid that does not possess one or more of the characteristics of ideal fluids, particularly the absence of viscosity.

**realgar ($As_2S_2$)** An arsenic ore, essentially arsenic disulfide, found as orange-red mineral; used in pyrotechnics. Also called red orpiment.

**reasonable maximum exposure** The maximum exposure reasonably expected to occur in a population.

**reasonably available control measure (RACM)** A broadly defined term referring to technological and other measures for pollution control.

**reasonably available control technology (RACT)** Control technology that is both reasonably available, and technologically and economically feasible. Usually applied to existing sources in nonattainment areas; in most cases, it is less stringent than new source performance standards.

**recalcination** An operation common in treatment plants in which lime is recovered from water and wastewater treatment sludge for reuse, by dewatering and burning the residuals to produce quicklime (CaO). Recalcination usually takes place in a multiple-hearth furnace; the lime sludge ($CaCO_3$) is heated at 1850°C to drive off carbon dioxide ($CO_2$) and water and recover quicklime or calcium oxide (CaO) and some inert material. Water is then added to the quicklime to form slaked lime [$Ca(OH)_2$]. Lime sludge contains a certain amount of magnesium hydroxide, which is normally wasted before recalcining, and colloidal impurities such as clay from surface water.

**recalcining** Same as recalcination.

**recalcitrant** Unreactive, resistant to microbial attack, nondegradable; refractory.

**recalcitrant organic compounds** Stable organic compounds that resist conventional biological treatment and potentially toxic to the environment and human health. They are mostly synthetic organic chemicals or naturally occurring substances such as those found in petroleum products. *See also* refractory organics, xenobiotic compounds, biorefractory organics.

**recalcitrant substance** Same as refractory compound.

**recarbonation** (1) An operation in which carbon dioxide ($CO_2$, from flue gas or from carbon dioxide generators) is bubbled into the water being treated to lower the pH to about 8.5. The pH may also be lowered by the addition of acid. Recarbonation is the final stage in the lime–soda ash softening process, ahead of filtration. This process converts carbonate ions to bicarbonate ions and stabilizes the solution against the precipitation of carbonate compounds on sand grains and in pipelines:

$$Ca^{2+} + CO_3^{2-} + CO_2 + H_2O \rightarrow Ca^{2+} + 2\,HCO_3^- \quad \text{(R-16)}$$

Single-stage recarbonation is used in softening low-magnesium waters without excess lime; otherwise, two-stage recarbonation is applied. Same as recarbonization; sometimes called secondary carbonation. *See also* excess-lime treatment, and sodium hexametaphosphate. (2) The addition of carbon dioxide to wastewater treatment plant effluents with high pH due to such processes as ammonia air stripping.

**recarbonization** An operation in which carbon dioxide is bubbled into water being treated to lower the pH. Same as recarbonation.

**receiver** A device that indicates the result of a measurement. Most receivers in the water utility field use either a fixed scale and movable indicator such as pressure gage, or a movable scale and movable indicator such as those used on a circular flow recording chart. Also called an indicator.

**receiving manhole** The manhole that receives the flow of a gravity sewer ahead of a pump station, as opposed to the discharge manhole, which receives the discharge from a pump station. *See also* key manhole.

**receiving stream** *See* receiving water

**receiving water** A body of water (river, lake, ocean, estuary, stream) that receives treatment plant effluents, untreated wastewater, stormwater or combined wastes, and nonpoint source discharges. Such discharges undergo aerobic decomposition unless the receiving water oxygen supply cannot beep pace with their biochemical oxygen demand. *See also* dissolved-oxygen sag analysis.

**receiving water classification** The classification of a receiving water according to its current or intended uses, with appropriate water quality criteria and effluent standards to protect such uses. *See* water body classification.

**receptor** (1) In biochemistry, a specialized molecule in a cell that binds a specific chemical with high specificity and high affinity. (2) An area on the surface of a cell for the attachment of viruses. (3) In exposure assessment, an organism that receives, may receive, or has received environmental exposure to a chemical.

**recessed-impeller pump** A centrifugal pump that has the impeller at the end of the casing and with vanes only on one side, so that large solids and grit can pass between the impeller and the inlet. It can also be used for sludge of up to 4% solids. Also called vortex pump, shear-lift pump, or torque flow pump.

**recharge** The natural or artificial process of adding water to the saturated zone of an aquifer from precipitation, by percolation from the soil surface, infiltration from surface streams, deep-well injection, or other means. Recharge area is a land area in which water reaches the zone of saturation or an artesian aquifer, naturally or artificially, e.g., where rainwater soaks through the earth to reach an aquifer. A major recharge area is an area where a major part of the recharge of an aquifer occurs. *See* aquifer. Recharge basin is a basin constructed in sandy material to recharge an aquifer from streamwater or stormwater, thus capturing flood and other flows that would otherwise be lost to the ocean. Other means of groundwater recharge include check dams, recharge wells, seepage ponds, and underground leaching systems. Recharge rate is the volume of water per unit of time that replenishes or refills an aquifer. A recharge well or diffusion well allows water to flow into an aquifer for groundwater recharge purposes and sometimes as a barrier to saltwater intrusion.

**recharge area** *See* recharge.

**recharge basin** *See* recharge.

**recharge pretreatment** The treatment of wastewater before its use to recharge groundwater. Pretreatment requirements depend on the sources of reclaimed water, the recharge methods, and location.

**recharge rate** *See* recharge.

**recharge well** *See* recharge.

**recharging** The periodic restoration of the properties of an exhausted material such as an ion-exchange resin or a catalyst medium. *See* regeneration for more detail.

**reciprocal detention time (1/*T*)** The reciprocal of the detention time (*T*), sometimes used in the design of trickling filters and expressed in million gallons per acre-foot per day. *See* hydraulic loading intensity.

**reciprocating flocculator** (1) A conical device mounted on rods to produce slow mixing of water by the alternating raising and lowering of the rods. (2) A flocculation apparatus consisting of mixing paddles attached to an oscillating horizontal beam. Also called a walking beam flocculator.

**reciprocating piston pump** A positive displacement device used in many sludge applications.

**reciprocating plunger pump** *See* plunger pump, reciprocating pump.

**reciprocating positive-displacement pump** A device that combines the features of reciprocating pumps and positive-displacement pumps.

**reciprocating pump** A displacement pump that moves and imparts pressure to fluids through a reciprocating piston or plunger in a cylinder; called single-action or double-action depending on whether the piston acts in one end or both ends of the cylinder. Also called a piston pump.

**reciprocating-rake bar screen** Same as reciprocating-rake screen.

**reciprocating-rake screen** A device used to remove gross pollutants from wastewater to facilitate further treatment and protect downstream equipment. The reciprocating-rake-type screen has a single rake that moves up and down to pull the screenings to the top for removal. Also called climber screen. Other types of screen include chain-driven, catenary, and continuous belt.

**recirculated cooling water** Water that passes through the main condensers to remove waste heat, passes through a cooling device to remove such heat from the water, and then passes again, except for blowdown, through the main condenser (EPA-40CFR423.11-h). *See also* once-through cooling water.

**recirculated membrane bioreactor** *See* membrane biological reactor.

**recirculating granular-medium filter** A common type of on-site system for the management of wastewater from individual residences and other establishments. It is similar to an intermittent sand filter with (a) septic tank (or other pretreatment unit) effluent recicrculated through the filter, (b) medium of larger effective size, and (c) larger hydraulic loading. *See* Figure R-04.

**recirculating sand filter** *See* recirculating granular-medium filter.

**recirculating system** A treatment system that uses recirculation, e.g., trickling filter or activated sludge; the opposite is called a flow-through system or once-through system.

**recirculation** (1) The return of a portion of the effluent of a wastewater treatment plant to the incoming flow; also called recycling. Recirculation uniformizes hydraulic and organic loading and increases biological mass in wastewater treatment; it also reseeds trickling filters with sloughed bacteria. *See also* recycle. (2) In the activated sludge process, the return of a portion of the settled solids to the aeration basin. (3) The return of effluent from a process to the incoming flow to minimize makeup water.

**recirculation factor (*F*)** The average number of passes of influent organic matter through a trickling filter; calculated as

$$F = (1 + R)/(1 + 0.1\ R)^2 \qquad \text{(R-17)}$$

where $R$ is the recycle ratio and the term $0.1\ R$ accounts for the decrease in recirculation benefits as the number of passes increases. The recirculation factor is used to evaluate the performance of trickling filters. For a two-stage trickling filter,

$$E_2 = 100/[1 + 0.4432(W_2/VF)^{0.5}/(1 - E_1)] \qquad \text{(R-18)}$$

where $E_2$ = BOD removal for the second-stage filter at 20°C, percent; $W_2$ = BOD loading to the second-stage filter, kg/day; $V$ = volume of filter packing, m$^3$; and $E_1$ = fraction of BOD removed in the first-stage filter. *See also* NRC formulation, recycle ratio or recirculation ratio.

**recirculation filtration** Same as crossflow filtration.

**recirculation ratio** *See* recycle ratio.

**reclaim** To treat or process wastewater or sludge and reuse them as appropriate, e.g., for recreation, irrigation, recharge, or agriculture.

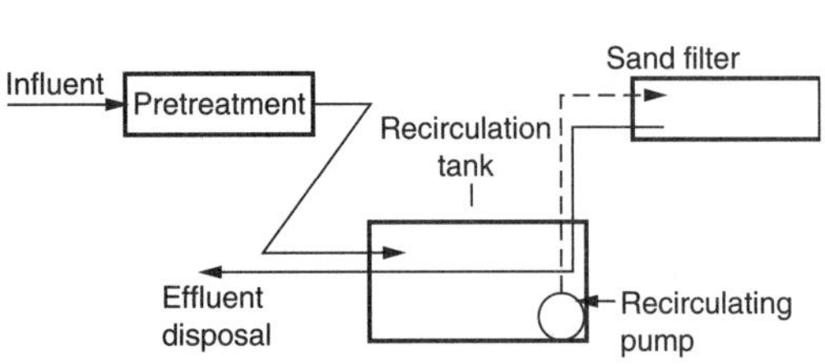


**Figure R-04.** Recirculating granular-medium filter.

**reclaimed** A material is reclaimed if it is processed to recover a usable product, or if it is regenerated. Examples are recovery of lead from spent batteries and regeneration of solvents (EPA-40CFR261.1-4).

**reclaimed brine** Brine from a desalination process, at least 30% saturated and with low hardness, used to regenerate a cation resin.

**reclaimed wastewater** Wastewater that has been recovered for some useful purpose, usually after some treatment to meet the requirements of the intended use.

**reclaimed water** Wastewater or stormwater that has been adequately treated for reuse in irrigation, industrial cooling, groundwater recharge, or for other beneficial purposes. Sometimes called reclamation water or recycled water. *See also* reclaimed wastewater, reclamation, water reuse applications.

**Recla-Mate®** A modular wastewater treatment plant manufactured by Wheelabrator Engineered Systems, Inc.; uses the physical–chemical process.

**reclamation** The process of recovering a resource such as land or water (in the form of wastewater or stormwater, e.g.) and improving it so that it can be used or reused instead of being wasted. Land can be reclaimed by irrigation, drainage, or flood protection. Wastewater can be renovated by advanced waste treatment and reused as cooling water, irrigation water, for groundwater recharge, or for other purposes. *See also* wastewater reuse.

**reclamation water** Same as reclaimed water.

**Recla-Pac** A package plant manufactured by Wheelabrator Engineered Systems, Inc. and using biological wastewater treatment.

**Recoflo™** A proprietary ion-exchange technique developed by Eco-Tec, Inc. for wastewater treatment and recovery of heavy metals and waste acid. It uses countercurrent regeneration and packed resin beds.

**recombinant bacteria** A microorganism whose genetic makeup has been altered by deliberate introduction of new genetic elements. The offspring of these altered bacteria also contain these new genetic elements.

**recommended contaminant level** *See* maximum contaminant level, maximum contaminant level goal, and recommended maximum contaminant level.

**recommended exposure limit** An exposure limit established by NIOSH for industrial chemicals in the workplace; similar to OSHA's permissible exposure limit. *See also* threshold limit value.

**recommended maximum contaminant level (RMCL)** Under the Safe Drinking Water Act, a nonenforceable concentration of a drinking water contaminant, set at a level at which no known or anticipated adverse effects on human health occur and that allows an adequate safety margin. Also called a maximum contaminant level goal (MCLG). *See also* adjusted acceptable daily intake.

**Recommended Standards for Water Works** A publication generally referred to as 10 State Standards and containing "Policies for the Review and Approval of Plans and Specifications for Public Water Supplies—A Report of the Committee of the Great Lakes–Upper Mississippi River Board of Sanitary Engineers."

**recorder** A device that creates a permanent record, on a paper chart, magnetic tape, or other media of the changes in some measured variable.

**recovery** (1) In emergency management, a phase of stabilization between the initial response and the return to normal operations. (2) One of two phenomena characteristic of pressure-driven membrane processes such as microfiltration and ultrafiltration. Recovery refers to the flux of permeate that passes through the membrane and contains lower total dissolved solids than the feed stream, as opposed to the reject stream. *See also* rejection. (3) A measure of the performance of a centrifuge (used in sludge thickening and dewatering), defined as the percentage of thickened dry solids with respect to feed solids:

$$r, \% = 100\ T_P(T_F - T_C)/[T_F(T_P - T_C)] \quad \text{(R-19)}$$

where $r$ is the percent recovery; $T_P$, $T_F$, and $T_C$ represent, respectively, the percent by weight of total suspended solids concentration in the thickened solids product, total suspended solids concentration in the feed, and total suspended solids concentration in the centrate. Also called capture or total suspended solids recovery. *See also* percent capture.

**recovery cleaning** A method used to remove accumulated particles from the surface of wastewater treatment membranes; a membrane is removed from service and allowed to soak in a solution of sodium hypochlorite or citric acid before being replaced in service. *See also* air scour, backpulse, maintenance cleaning, and empty-tank maintenance cleaning.

**recovery rate** (1) Percentage of usable recycled materials that have been removed from the total amount of municipal solid waste generated in a specific area or by a specific business. (2) *See* recovery (3). (3) One of two parameters used to express the performance of a pressure-driven membrane process used in water treatment. Recovery rate is the net percentage of the permeate stream with respect to the feed stream (after deducting any water used for backwash, e.g., in the case of low-pressure membranes):

$$Y, \% = 100(Q_p)/Q_f \quad \text{(R-20)}$$

where $Y$ is the recovery rate, and $Q_p$ and $Q_f$ are, respectively, the flow of permeate stream and the flow of feed stream. A system with a recovery rate of $Y$% produces a concentrate stream of $(100 - Y)$ % of the feedwater flow. Also called water recovery. *See also* concentration factor, rejection rate.

**recovery well** A well that recovers percolated flow from a rapid infiltration wastewater treatment system for reuse or to control the level of the water table. Also called extraction well.

**recreation area sanitation** A series of activities designed to maintain health conditions in recreational facilities; e.g., sampling, monitoring, classification of swimming pools, beach areas, camps, and marinas.

**recreational water** A natural or man-made water body that is used for boating, fishing, swimming, or other recreational activities. Recreation is often included in the objectives of multipurpose impoundment or reservoir projects.

**recreational water reuse/environmental water reuse** A nonpotable use of reclaimed water in recreational lakes, marsh enhancement, streamflow augmentation, fisheries, lakes, and ponds. *See also* water reuse applications.

**Rectangulaire** A package wastewater treatment plant manufactured by FMC Corp., including an aeration tank and a clarifier.

**rectangular weir** A weir with a rectangular notch. The rectangular weir equation estimates the head over the crest ($H$, m or ft) as follows:

$$H = [Q/(C_w L)]^{0.67} \quad \text{(R-21)}$$

where $Q$ = flow (in m³/sec or in cfs); $C_w$ = weir coefficient; and $L$ = length of weir, m or ft. The weir coefficient accounts for the approach velocity; it ranges from 1.8 to 2.3 in metric units and from 3.2 to 4.2 in English units. *See also* weir-end contraction, submerged weir.

**rectifying section** The cascade or set of plates above the feed introduction point of a continuous distillation column that involves only one feed stream. Also called enriching section. *See also* stripping section.

**rectilinear distributor** A distributor carried on a truss that spans a rectangular trickling filter bed and moves back and forth on its side walls.

**recuperative oxidizers** An emissions control device using heat to oxidize volatile organic compounds in odor management. In thermal oxidation, the odorous gases are preheated before going into the combustion chamber. The recuperative oxidizer consists of thin-walled tubes that transfer heat from exhaust gases to the incoming air.

**recuperative thermal oxidation** A process designed to reduce energy consumption in thermal methods of odor management by preheating the incoming air. *See* recuperative oxidizer and regenerative thermal oxidation.

**recycled water** Reclaimed water used beneficially or, simply, reclaimed water. *See also* water reuse applications.

**recycle-flow pressure flotation** One of three basic modes of operation of pressure, dissolved-air flotation; the entire influent goes through the flotation tank, but part of the clarified effluent is pressurized, saturated with air, and recycled *See also* full-flow pressure flotation and split-flow pressure flotation.

**recycle flow pressurization** *See* recycle-flow flotation. The recycle flow is the percentage of the influent flow that is recycled.

**recycle or reuse** (1) Minimizing waste generation by recovering and reprocessing usable products that might otherwise become waste; i.e., recycling of aluminum cans, paper, and bottles. *See also* recirculation. (2) The return of water after appropriate treatment for further use.

**recycle rate** Same as recycle ratio.

**recycle ratio** The ratio ($r$) of recycled flow ($R$) to the influent flow ($I$), a parameter used in the design of conventional activated sludge, trickling filter, and other systems:

$$r = R/I \quad \text{(R-22)}$$

$$N = 1 + r \quad \text{(R-23)}$$

where $N$ is the average number of passes of the influent or any portion thereof through a treatment unit such as an aeration basin. If the performance of a treatment process decreases when the number of passes (or the recirculation ratio) increases, as expressed by a weighting factor ($f$), the number of effective passes ($n$) and the optimum recycle ratio ($r^*$) are (Fair et al., 1971):

$$n = N/[1 + (1 - f)r]^2 \quad \text{(R-24)}$$

$$r^* = (2f - 1)/(1 - f) \quad \text{(R-25)}$$

Also called recirculation ratio. *See also* recirculation factor, sludge volume index (SVI).

**recycle treatment system** A self-contained system that collects wastewater from buildings, then treats and returns the treated effluent for flushing toilets and urinals. Treatment involves three or more steps; all materials removed are returned to the first step for further processing.

**recycling** (1) The process by which recovered materials are transformed into new products (EPA-40CFR244.101-j and 245.101-j), for example, the conversion of solid waste into usable materials or energy and the reuse of wastewater or biosolids in industry or agriculture. (2) A feature of a process in which a substance passes through pipes, tanks, etc. more than once; e.g., the solids recycling in the activated sludge process.

**RED** Acronym of reduction equivalent dose.

**red algae** A large, mostly marine group of algae that owe their coloration to a red or purple pigment that masks their chlorophyll and other pigments; red seaweeds. Some contribute to the formation of coral reefs. Other large groups of algae are brown and green. Also called *Rhodophyta.*

**red bag waste** A category of wastes that includes infectious waste and medical waste.

**red bed** A layer of sedimentary materials with iron in the ferric state.

**Red-B-Gone®** A product of Technical Products Corp. for use in removing rust and iron stains.

**RED bias** In ultraviolet (UV) light disinfection validation testing, the RED bias is a factor that accounts for the effects of test microbe inactivation kinetics on reduction equivalent dose (RED), i.e., the difference between the RED measured by the test microbe during validation and the actual RED delivered to a pathogen. RED bias also depends on the design of the UV reactor. It is used to determine the validation factor.

**red clay** A brown to red, soft, deep marine deposit, rich in iron oxides and manganese, consisting of volcanic particles and other wind-blown dust, insoluble organic remains, etc.

**REDEQL** Acronym of Redox Equilibrium, a computer program used for modeling precipitation and dissolution reactions. *See also* MICROQL and MINEQL.

**Redfield model** A stoichiometric formula for algal protoplasm, based on the Redfield ratios for macronutrients and trace elements, which can be used for the biological control of these elements: $C_{106}H_{263}O_{110}N_{16}P_1Si_xFe_aMn_bZn_cCu_dCd_eNi_f$.

**Redfield ratio** The ratio of nitrogen to phosphorus to carbon (N:P:C) in protoplasm. For algae, it is about N:P:C = 16:1:106. A nutrient that is present in a smaller proportion tends to be a growth limiting factor; e.g., phosphorus in lakes. *See also* cell composition.

**Redfield stoichiometry** The interaction between phytoplankton and trace metals as well as other elements. *See* Redfield model and Redfield ratio.

**Red Fox** An apparatus manufactured by Red Fox Environmental, Inc. for marine wastewater treatment.

**Red List** A list of 23 dangerous substances, designated by the United Kingdom, whose discharge to water should be minimized. *See* priority list.

**red orpiment ($As_2S_2$)** An arsenic ore, essentially arsenic disulfide, found as an orange-red mineral; used in pyrotechnics. Also called realgar.

**redox** Abbreviation of oxidation–reduction.

**redox chemistry** The chemistry of oxidation–reduction.

**redox couple** Oxidation–reduction couple.

**redox equation** *See* oxidation–reduction reaction.

**Redox Equilibrium** The computer program REDEQL.

**redox potential** A dimensionless number (pE) that represents the electrical potential required to transfer electrons from an oxidant to a reductant. Redox potentials are used to determine if a redox equation is thermodynamically feasible. *See* more detail under oxidation–reduction potential.

**redox reaction** An oxidation–reduction reaction.

**redox titration** A laboratory method used to determine the concentration of a dissolved substance; for a given oxidant, known volumes of an appropriate reductant are added to the solution until it reaches an endpoint, and similarly for the titration of a reductant by an oxidant. It is similar to an acid–base titration, except that the potential of the solution changes, instead of its pH.

**Red Rubber™** Urethane rake parts manufactured by Rubber Millers, Inc. for use in bar screens.

**red tide** A brownish-red discoloration of coastal surface waters caused by the proliferation of a red-pigmented marine plankton that produces a strong toxin, often fatal to fish and rendering mollusks toxic to humans. This natural phenomenon is perhaps stimulated by the addition of nutrients. *See also* dinoflagellate, brown water, green tide, blue tide, red water, algal bloom.

**reduced monitoring** A provision of the Lead and Copper Rule that allows small water supply systems to reduce the number of required samples under certain conditions.

**reduced-positive-pressure backflow preventer** Same as reduced-pressure-principle valve.

**reduced-pressure backflow preventer** A device used to prevent water supply contamination through backflow. It consists of a pressure-regulated relief valve between two pressure-reducing check valves; water discharges from the relief port when any of the valves malfunctions.

**reduced-pressure principle backflow-prevention assembly (RPBA)** A pressure-reducing assembly that consists of two independent check valves and one relief valve installed between two shutoff valves. *See* reduced-pressure backflow preventer.

**reduced-pressure-principle valve** A valve used as a backflow preventer; it closes when pressure drops. *See* reduced-pressure backflow preventer.

**reducer** (1) A pipe or fitting larger at one end than at the other, threaded inside or flanged, used to join two pipes of different sizes. A reducing coupling is used to connect a smaller and a larger pipe. A reducing tee connects a branch to a main pipe, with at least one of the outgoing sections smaller than the main. (2) An organism that breaks down dead organic matter into simpler compounds. Reducers consist mainly of bacteria and fungi, a subclass of detritovores; also called decomposers.

**reducing agent** A substance that can give up electrons in a reaction and decrease the positive valence of an ion. *See also* reductant, oxidant, redox reaction.

**reducing coupling** *See* reducer.

**reducing tee** *See* reducer.

**reductant** A substance that causes a reduction while being oxidized; a reducing agent. In water and wastewater treatment, reductants use free or combined oxygen for stabilization. *See also* oxidant, redox reaction.

**reduction** A chemical reaction involving the addition of hydrogen, removal of oxygen, or addition of electrons to an element or compound, thus decreasing its valence. Reduction is the opposite of oxidation, and both are part of the oxidation–reduction (redox) reaction. The reaction is called a reduction if the target substance is reduced, and an oxidation if the target substance is oxidized. Under anaerobic conditions, sulfur compounds are reduced to odor-producing hydrogen sulfide ($H_2S$) and other compounds.

**reduction equivalent dose (RED)** The ultraviolet (UV) dose that corresponds to the inactivation of a test microbe on a UV dose–response curve. Assuming first-order kinetics,

$$\text{RED} = -(1/k) \ln [\Sigma P_i \exp(-kD_i)] \quad \text{(R-26)}$$

where RED is expressed in $mJ/cm^2$, $k$ is the first-order inactivation coefficient in $cm^2/mJ$, and $P_i$ is the probability of delivering the $i$th dose $D_i$ in the dose distribution [$\exp(-kD_i)$]. *See also* RED bias, validation factor.

**reduction half reaction** The reduction half of a redox reaction.

**reduction intensity parameter** A parameter ($R_H$) proposed in 1921 to indicate the reducing intensity of an aqueous solution and equal to the negative logarithm of the partial pressure ($p_{H2}$) of hydrogen:

$$R_H = \log(1/p_{H2}) = -\log p_{H2} \quad \text{(R-27)}$$

It is not currently used.

**reduction reaction** A reaction that involves the removal of oxygen or the addition of electrons. Same as reduction half-reaction. *See also* oxidation–reduction reaction.

**reductive dechlorination** Removal of chlorine as chloride ($Cl^-$) by reducing the carbon atom from C—Cl to C—H.

**reductive dehalogenation** (1) The scission of a carbon–halogen bond and the formation of a halogen radical as well as a carbon free radical. (2) The biodegradation of a halogenated organic compound, mediated by a reduced transition metal complex. Thus, tetrachlroethane ($CHCl_2$—$CHCl_2$) can be reduced to trichloroethane ($CHCl_2$—$CH_2Cl$) or to dichloroethylene (CHCl—CHCl). *See also* oxidative dehalogenation, substitutive dehalogenation.

**redundancy** The condition of a water or wastewater facility that has units, equipment, or processes in excess of the bare minimum required to meet demand or rated capacity, e.g., a backup pump in a pumping station or two medium-size clarifiers instead of one large clarifier. Redundancy allows flexible and efficient operation during equipment failures or periods of reduced as well as peak loading. In a wastewater treatment plant, unit process redundancy means that the plant effluent will not be adversely affected with one or more reactors out of service under average flow conditions.

**red water** (1) Water whose reddish or rusty color is due to the precipitation of ferric iron ($Fe^{3+}$) salts or the presence of microorganisms that depend on iron or manganese. A cause of customer complaint, red water is unsuitable for laundering as well as papermaking and other manufacturing processes. Also called rusty water when it is caused by the corrosion of an iron pipe attached to a bronze faucet, which creates a situation of dissimilar metals in which iron is anodic (corrodible) to copper (the main constituent of bronze). Maintaining a saturation index near zero can help mitigate (but not eliminate) the problem of corrosion and red water. Polyphosphates are also used commonly to control the formation of red water. The increase in apparent color in the distribution system ($\Delta C$, in color units) has been proposed as an indicator for the potential release of corrosion products causing red water:

$$\Delta C = (Cl)^{0.485} \cdot (Na)^{0.561} \cdot (SO_4)^{0.118} \cdot (DO)^{0.967} \cdot (T)^{0.813} \cdot (HRT)^{0.836}/(Alk)^{0.912} \quad \text{(R-28)}$$

where Alk is alkalinity in mg/L as CaCO3, HRT is the hydraulic retention time of treatment in days, and $T$ is the temperature in degrees Celsius; the other parameters are concentrations in mg/L of chlorides, sodium, sulfates, and dissolved oxygen, respectively. Red water is sometimes called yellow water. *See also* black or brown water, calcium carbonate precipitation potential. (2) Same as red tide.

**redwood** A reddish or red-colored wood, naturally rot-resistant, formerly used as trickling filter medium and for clarifier flights.

**red worm** An annelid worm that burrows in soil and feeds on soil nutrients and decaying organic matter; an earthworm. Red worms are found in biological wastewater treatment systems where they graze on attached biomass and can compromise nitrification in the integrated fixed-film activated sludge system if the food/microorganism ratio is low and dissolved oxygen high.

**reed** A tall slender grass that grows in wet areas such as natural or constructed wetlands used for wastewater treatment. As emergent plants, reeds usually grow on shorelines or in water depths of 5 ft or less. They provide support for bacteria films, adsorb wastewater constituents, and help control the growth of algae by restricting sunlight penetration. *See also* reed bed.

**reed bed** An apparatus (similar to a subsurface-flow wetland) used for wastewater treatment or sludge processing. It consists of sand- or rock-filled trenches supporting reeds or other emergent vegetation. Reed beds are used as a tertiary treatment device to polish secondary effluents before disposal or for sludge dewatering, treatment, and storage.

**Reed odorless earth closet (ROEC)** An on-site disposal device designed for 15–30 years for human excreta, featuring a superstructure, a concrete squatting plate from which an inclined chute leads to a completely offset pit, and a ventilation pipe with a fly screen.

**REEF®** A fine-pore diffuser manufactured by Environmental Dynamics, Inc. for installation on the floor of the reactor.

**ReelAer** A surface aerator manufactured by Walker Process Equipment Co., including a horizontal cage.

**reference** A physical or chemical quantity whose value is known exactly, and thus is used to calibrate or standardize instruments.

**reference concentration (RfC)** An estimate of a continuous inhalation exposure to the human population (including sensitive subgroups) that is likely to be without an appreciable risk of deleterious noncancer effects during a lifetime, with uncertainty spanning perhaps an order of magnitude.

**reference dose (RfD)** The daily dose or exposure level that, during an entire lifetime of a human (including sensitive subgroups), appears to be without appreciable risk of deleterious effects on the basis of all facts known at the time; also referred to as ADI, or acceptable daily intake. *See also* benchmark dose, LOAEL, NOAEL, toxicity terms.

**reference electrode** An electrode wired to a voltmeter and inserted into a solution to complete an electrochemical cell and maintain a constant potential. It is used in redox titrations. *See also* pH electrode, specific ion electrode.

**reference evapotranspiration ($ET_0$)** (1) The evapotranspiration rate corresponding to a reference crop such as alfalfa or tall fescue grass. It is used in the design and scheduling of land application systems of wastewater effluents. *See* crop coefficient and Penman–Monteith method. (2) The rate of evapotranspiration from an extended surface of well-watered, full-cover short grass. It is often used as a basis to estimate actual evapotranspiration (ET). For example, (a) for evergreen trees, ET = 1.1 – 1.3 $ET_0$; (b) for water surfaces and moist bare soil, evaporation = 1.05 – 1.15 $ET_0$. Also called potential evapotranspiration. *See also* water deficit.

**reference state** The condition of a solution in which the activity coefficient of a species tends to unity. *See also* standard state.

**reference value** In European practice, the expected natural or background concentration of a given contaminant in groundwater, which normally does not require remediation. Also called target value. *See also* intervention value.

**refinery waste** *See* oil production and refinery waste.

**reflectance FTIR spectrometry** A nondestructive form of Fourier transform infrared (FTIR) spectrometry used to analyze organic contaminants without removing them from the substrate.

**reflection factor (*r*)** A correction factor applied to incident irradiance to obtain the average ultraviolet (UV) irradiance used in bench-scale experiments. It accounts for the reflection of UV light at the air–water interface; commonly, $r = 0.975$.

**reflux** In the flash or differential distillation process, reflux is the return of some of the condensed overhead product to the still. A constant reflux ratio is sometimes used to keep the overhead product from becoming progressively richer in less volatile components.

**reflux valve** A device used in a rising pipeline to prevent ascending water from flowing back.

**Refotex** A fine-bubble diffuser manufactured by Refraction Technologies Corp.

**Refractite** A ceramic filter membrane manufactured by Refraction Technologies Corp.

**refractive index** A characteristic of volatile organic compounds used in the Lorentz–Lorentz equation.

**refractory** (1) A stable, organic or inorganic, material that is difficult to convert or to remove entirely from wastewater. *See* refractory organics. In a sense, inorganic refractory pollutants include nitrogen and phosphorus, 70% of which remain in the effluent of conventional biological treatment. (2) A highly heat-resistant brick or other material used as a liner in furnaces and incinerators.

**refractory index** A measure of the ability of a substance to be biodegraded by bacterial activity. The lower the refractory index, the greater the biodegradability.

**refractory materials** Inert or nonbiodegradable materials found in wastewater; same as refractory (1).

**refractory organic compound** *See* refractory organics.

**refractory organic matter** *See* refractory organics.

**refractory organics** (1) Stable organic materials, mostly dissolved solids, that are difficult or impossible to remove by conventional wastewater treatment methods; e.g., fats, lignin, cellulose, surfactants, phenols, and agricultural pesticides. They represent approximately the difference between BOD and COD or TOC. Adsorption on granular activated carbon is an effective method of removal of refractory organics; membrane filtration and evaporation can also be used. *See also* recalcitrant organic compounds and biorefractory organics. (2) A stable by-product of the bacterial decomposition of cells of dead organisms.

**refractory pollutant** Same as refractory (1).

**refuse** Rubbish or garbage. *See* solid waste.

**Refuse Act** *See* Rivers and Harbors Appropriations.

**refuse-derived fuel (RDF)** Fuel obtained from municipal solid waste using pyrolysis or other methods.

**Regal™** A device manufactured by Chlorinators, Inc. for use in water disinfection.

**regenerant** A chemical product used to restore the capacity of an exhausted ion exchange resin or catalyst medium.

**regenerate** To restore the properties of an exhausted material such as an ion exchange resin or a catalyst medium so that it can be reused for water or wastewater treatment.

**regenerated water** Water diverted for irrigation but returned unused to surface or groundwater; same as return water.

**regeneration** (1) The periodic restoration of the properties of an exhausted material. Ion exchange resins are regenerated with a solution (regenerant) that provides the ions removed. This operation is also called recharging or rejuvenation of the resin. A catalyst medium can similarly be regenerated. (2) The periodic removal of adsorbed materials from spent activated carbon. *See* activated carbon regeneration for more detail. (3) Manipulation of cells to cause them to develop into whole plants.

**regeneration cycle** (1) The period between two consecutive regenerations, e.g., between 12 and 24 months for activated carbon beds. (2) The sequence of steps in the regeneration of ion-exchange resins, activated carbon beds, oxidizing filters, etc.

**regeneration efficiency** The extent of regenerant ion used in ion-exchange regeneration, equal to the ratio of regeneration level to the breakthrough capacity of the ion-exchange bed.

**regeneration level** The relative quantity of regenerant used in ion exchange, expressed, e.g., in pounds of regenerant per cubic foot of ion-exchange bed. Also called salt dosage.

**regeneration rate** The flow rate of regeneration solution per unit area of an ion-exchange bed.

**regeneration tank** A tank used for the regeneration of ion-exchange resins from portable units.

**regeneration water** Water used in the regeneration of ion-exchange resins.

**regenerative oxidizer** An emissions-control device using heat to oxidize volatile organic compounds in odor management. In thermal oxidation, the odorous gases are preheated before going into the combustion chamber. The regenerative oxidizer consists of a ceramic packing material that transfers heat from exhaust gases to the incoming air.

**regenerative pump** A centrifugal pump with rotating vanes that converts mechanical impulse and centrifugal force into pressure to raise low volumes of liquid at high heads. *See also* turbine pump.

**regenerative thermal oxidation (RTO)** A process designed to reduce energy consumption in thermal methods of odor management by preheating the incoming air. *See* regenerative oxidizer and recuperative thermal oxidation.

**regenerative thermal oxidizer (RTO)** Same as regenerative oxidizer.

**Re-Gensorb™** An apparatus manufactured by M & W Industries, Inc. for the removal of volatile organic compounds and other air pollutants.

**regime coefficient** *See* coefficient of regime.

**regrowth** An increase in the concentration of microorganisms (mainly heterotrophic bacteria) in treated water, receiving water bodies, and reclaimed water storage reservoirs. In drinking water pipes or storage tanks with low residuals, bacteria in the biofilm on the pipe wall or in sediments may cause regrowth. Some other distribution system conditions are favorable to regrowth, e.g., low velocity, dead ends, and corrosion-related tuberculation. Regrowth also results in wastewater reuse transmission lines where organic matter remaining after treatment is sufficient for bacterial growth. Regrowth may be controlled by an adequate disinfectant residual or even additional disinfection at intermediate points, and by limiting the availability of biodegradable organic matter. Also called aftergrowth or microbial regrowth. *See also* secondary disinfection, regrowth organisms, bacterial aftergrowth, bacterial regrowth, bacteriological aftergrowth.

**regrowth organisms** The organisms often found in microbial regrowth in water distribution systems, including opportunistic bacteria (*Aeromonas, Bacillus subtilis, Flavobacterium, Legionella, Mycobacterium, Pseudomonas*) and fungi (*Aspergillus fumigatus, Aspergillus niger, Petriellidium boydii*). Currently, Legionnaires' disease, caused by the *Legionella* bacteria, is the greatest concern of regrowth organisms.

**Reg-U-Flo®** A valve manufactured by H. I. L. Technology, Inc. for vortex control.

**regulated flow** Stream flow that has been affected by such hydraulic structures as reservoirs, diversions, and other controls; the opposite of natural flow. Weirs or orifices may also regulate conduit flow. *See also* flow regulator.

**regulation** *See* flow regulator.

**regulator** (1) a structure or device that maintains a quantity at a desired value or varies it according to a predetermined plan, for example, regulators to control the flow or level of water or wastewater at intakes, in canals, channels, reservoirs, or treatment units. (2) A sewer regulator diverts stormwater or wastewater from one line to another. (3) Specifically, a flow regulator, used to divert combined sewer flows in excess of wastewater treatment plant capacity to receiving water via the combined sewer overflow outlet. *See* diversion structure for detail.

**regulator structure** Same as diversion structure or regulator.

**regulatory floodway** The area regulated by federal, state, or local requirements to provide for the discharge of the base flood so the cumulative increase in water surface elevation is no more than a designated amount, not to exceed one foot as set by the National Flood Insurance Program. (Definition from the code of federal regulations on floodplain management and the protection of wetlands.)

**regulatory dose (RgD)** The daily exposure to the human population reflected in the final risk management decision; it is entirely possible and appropriate that a chemical with a specific reference dose (RfD) may be regulated under different statutes and situations through the use of different RgDs.

**Rehydro-Floc™** A flocculent solution produced by Reheis, Inc. and containing aluminum chlorohydrate.

**reject** (1) In general, the waste stream containing the pollutants removed by a treatment process. In particular, the portion of the feed stream that contains the retained, dissolved or suspended constituents in membrane filtration processes: dialysis, electrodialysis, hyperfiltration, microfiltration, nanofiltration, pervaporation, reverse osmosis, and ultrafiltration. Also called concentrate, concentrate stream, reject solution, reject stream, reject water, residual stream, retained phase, retentate, or waste stream. *See also* blowdown, brine or brine stream, feed stream or feedwater, permeate. (2) The solids removed for disposal from the washbox of a moving-bed filter, after gravity separation from cleaned sand.

**reject brine** The waste stream containing the solids removed or separated from saline feedwater by reverse osmosis or electrodialysis. Disposal of reject brine is a major problem at inland desalting plants, which may use such methods as evaporating ponds, deep-well injection, and ocean outfall. *See also* waste brine.

**rejected recharge** The amount of water that is available to recharge an aquifer but cannot enter the aquifer because the latter is full. The rejected recharge may be captured later if a drawdown causes a dewatering in a recharge area.

**rejection** One of two phenomena characteristic of pressure-driven membrane processes such as microfiltration and ultrafiltration. Rejection refers to the flux of the substances that do not pass through the membrane, as opposed to the recovery of the permeating stream. Rejection is a measure of the membrane's ability to resist the passage of solutes through the membrane. It is expressed, in percent, as the ratio of salts removed to the initial salt concentration. *See* rejection rate. Also called salt rejection. *See also* global rejection, local (apparent) rejection, lag coefficient, mechanical sieving (steric rejection), polarization factor.

**rejection coefficient** Same as rejection rate.

**rejection rate** One of two parameters that express the performance of a pressure-driven membrane process used in water treatment. It is the percentage of solute concentration reduction of the permeate stream with respect to the feed stream:

$$R, \% = 100(C_f - C_p)/C_f \qquad \text{(R-29)}$$

where $R$ is the rejection rate, and $C_f$ and $C_p$ are, respectively the feedwater concentration and the permeate concentration. *See also* recovery rate. Also called rejection rate or solute rejection coefficient.

**reject solution** Same as reject.

**reject staging** A configuration designed to improve the performance of the reverse osmosis process by using the reject stream from one stage as feedwater for a succeeding stage. Also called brine staging.

**reject stream** Same as reject.

**reject water** Same as reject.

**rejuvenated stream** A stream that has entered a new cycle of erosion following a geological uplift of the adjoining area.

**rejuvenated water** Water returned to the terrestrial supply by a geological process; water of compaction, metamorphic water.

**rejuvenation** The periodic restoration of the properties of an exhausted material such as an ion-exchange resin or a catalyst medium. *See* regeneration for more detail.

**REKO** Products of Kopcke Industries B. V. for use in bar screens.

**relapsing fever** A louse-borne, water-related disease caused by the spirochaete *Borrelia recurrentis,* found worldwide but particularly in mountainous area of poor sanitation. Also called louse-borne fever. *See also* louse-borne typhus.

**relative biological effectiveness (RBE)** An adjustment factor that converts rads to rems, depending on the type of radiation, dose rate, dose total, and type of tissue; e.g., 1 RBE for beta particles and gamma rays, 10 RBE for alpha particles, neutrons, and protons.

**relative density or specific gravity** The dimensionless ratio of the density of a substance to a standard density. For liquids, the standard density is that of water at 4°C, i.e., 1000 $kg/m^3$ or 1 kg/liter; in U.S. customary units it is 1.941 $slugs/ft^3$.

**relative error** A measure of model accuracy, equal to the average error divided by the average observation. (The average error is the average difference between observations and simulations.) *See* Martin & McCutcheon (1999). In a modeling application, the mean relative error (*RE*) for any variable (*X*) is

$$RE = |X_m - X_s|/X_m \qquad \text{(R-30)}$$

where $X_m$ and $X_s$ are, respectively, the measured (or observed) and simulated values. *See also* average error.

**relative humidity** The water vapor content of the air at a given temperature, expressed as a percentage of the maximum amount that the air could hold at that temperature if saturated with water vapor. It is also defined as the ratio of mixing ratio to saturation mixing ratio. *See also* dew point, sublimation, condensation nuclei.

**relative permittivity** The ratio of the flux density of an electric field in a dielectric to its flux density in a vacuum. *See* dielectric constant for detail.

**relative risk** (1) The ratio of incidence or risk among exposed individuals to incidence or risk among nonexposed individuals; sometimes referred to as risk ratio. (2) The ratio of the risk of occurrence of a disease in one group as compared to the risk of occurrence in another group.

**relative source contribution** The contribution of drinking water to the total exposure to a contaminant.

**relative stability** The ratio of oxygen available in dissolved form, in nitrites, and nitrates to the total oxygen required to satisfy the biological oxygen demand.

**release** Any spilling, leaking, pumping, pouring, emitting, emptying, discharging, injecting, escaping, leaching, dumping, or disposing into the environment of a hazardous or toxic chemical or extremely hazardous substance.

**reliability** The complement of risk, i.e., the probability that a system will perform as designed and will reach its objective. A reliability analysis is an exercise to evaluate the reliability of a simulation, prediction, assessment or the like to achieve a desired result, considering the effects of uncertainty, randomness, and other factors. *See also* risk of failure. The reliability index is a relative measure of reliability and risk. For a given performance parameter, the reliability index ($\rho$) may be defined as the ratio of its mean ($\mu$) to its standard deviation ($\sigma$) :

$$\rho = \mu/\sigma \qquad \text{(R-31)}$$

where $\mu$ and $\sigma$ are, respectively, the mean and standard deviation of the measurements. Reliability also expresses the degree of variability of a statistical parameter as a function of its standard deviation ($\sigma$) and the size of the sample ($n$); e.g., $[\sigma/n^{1/2}]$ and $[1.25\ \sigma/n^{1/2}]$, respectively, for the arithmetic mean and the median. *See also* inherent availability, mechanical reliability, treatment plant reliability, critical component analysis, mean time before failure, expected time before failure, operating availability.

**reliability analysis** *See* reliability.

**reliability-centered maintenance (RCM)** A maintenance strategy designed to ensure that physical assets continue to perform as intended by their users. RCM focuses on identifying and managing equipment failure. *See also* preventive maintenance.

**reliability class** A characteristic of wastewater treatment works based on their degree of reliability as defined in USEPA's Construction Grants program. *See* Class I treatment works, Class II treatment works, and Class III treatment works.

**reliability coefficient** *See* coefficient of reliability.

**reliability concept** As related to water and wastewater treatment, the concept of reliability is a based on a statistical analysis of treatment plant performance to determine relationships between results (e.g., effluent concentrations) and probable frequencies of occurrence.

**reliability index** *See* reliability.

**reliability maintenance** A planned, proactive maintenance approach that focuses on predicting when equipment will fail and taking timely appropriate corrective actions. It includes inspections, performance monitoring, and predictive analysis. Also called predictive maintenance. *See also* preventive maintenance, reactive maintenance, developmental maintenance.

**relief damper** A device in a multiple hearth furnace that ensures that the flue gas automatically exits through the emergency vent stack when exhaust gas temperature reaches a set point.

**relief drain** A drain used to dewater a construction site with a high groundwater table. *See also* subsurface drain. A relief sewer is intended to carry flows in excess of the capacity of a district or of another sewer. The relief sewer may or may not rejoin the original sewer downstream; it is sometimes called relief line or relief sewer line.

**relief line** *See* relief drain.

**relief sewer** *See* relief drain.

**relief sewer line** *See* relief drain.

**relief siphon** A flow-regulating device consisting of a siphon having its inlet as far below the top water level as possible. *See* Figure R-05.

**relief valve** A pressure-relief valve or a valve used to relieve unplanned, nonroutine discharges. Also a valve that automatically introduces air into a line to raise internal pressure above atmospheric pressure. *See also* safety valve.

**relift station** A pumping station that lifts wastewater to avoid excessive trench depths; a lift station.

**reliquefaction** The return of a gas to the liquid state.

**REM** or **rem** Acronym of Roentgen equivalent man or radiation equivalent man; the unit of dose equivalent (or equivalent dose) from ionizing radiation to the total body or any internal organ or organ system. The unit rem relates the absorbed dose (in rads) to the effective biological damage of the radiation or relative biological effectiveness (rbe):

$$1 \text{ rem} = 1 \text{ rad in tissue} \times \text{rbe} \qquad \text{(R-32)}$$

Equivalent dose is often expressed in millirem (mrem), which is one thousandth of a rem. The rem has been replaced by the sievert (Sv): 1 Sv = 100 rem. *See also* becquerel, curie.

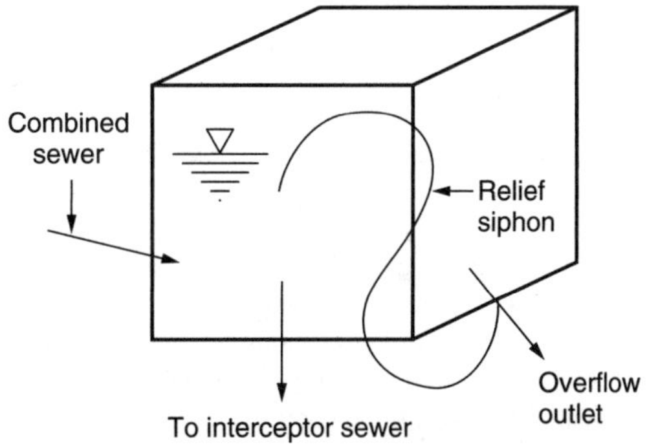


**Figure R-05.** Relief siphon.

**remediation** (1) Cleanup or other methods used to remove or contain a toxic spill or hazardous materials from a Superfund site. (2) For the Asbestos Hazard Abatement Program, abatement methods, including evaluation, repair, enclosure, encapsulation, or removal of greater than 3 linear feet or square feet of asbestos-containing materials from a building.

**remote sensing** A process of data collection, mapping, analysis, or interpretation from a distance, e.g., aerial photogrammetry, GPS, or satellite imaging. It uses such equipment as broadband sensors, satellite and aircraft scanners.

**remote terminal unit (RTU)** A field device equipped with a microcomputer and linking a SCADA master station with instrumentation and control relays. It can perform such tasks as data transmission (e.g., from a lift station to a wastewater treatment plant) and command execution.

**removal** A reduction in the amount of a pollutant in a publicly owned treatment works (POTW)'s effluent or alteration of the nature of a pollutant during treatment at the POTW. The reduction or alteration can be obtained by physical, chemical, or biological means and may be the result of specifically designed POTW capabilities or may be incidental to the operation of the treatment system. Also, excluding the dilution of a pollutant in the POTW (EPA-40CFR403.7(a)(1)-i).

**renal** Pertaining to the kidney.

**Renalin** Hydrogen peroxide ($H_2O_2$) commercially available and diluted 24 to 1.

**rendering plant** A plant that processes livestock carcasses and grease into tallow, hides, fertilizer, fats, oils, etc. for industrial use. A part of the meatpacking industry, a rendering plant may process edible or inedible products. Rendering may be a wet or dry process, carried out at low or high temperatures.

**rendering pollution** The pollution resulting from rendering plant operations, where the main sources of water pollution are the tank water (the process water without the grease and residue) and the stick (waste washwater in which bones are cooked). Effluents from rendering plants contribute a great deal of grease and odor and may cause clogging of treatment plant nozzles. The liquor waste from wet rendering contains high concentrations of nitrogenous and other organic compounds. For discharge into publicly owned treatment works, recommended pretreatment measures include baffled grease traps, dry rendering at low temperatures, blood recovery, and evaporation ponds for tank water. Besides in-plant recovery of blood, grease, and other materials, common treatment processes used for meatpacking waste are screening, flotation, sedimentation, chemical precipitation, trickling filtration or activated sludge. *See also* packers pollution and meatpacking waste.

**renewable resource** Any natural resource that is theoretically inexhaustible, like biomass, hydroelectric, solar, tidal, wave, and wind energy; farmland; forests; fisheries; and surface water.

**Reo-Pure** A reverse osmosis unit manufactured by Great Lakes International, Inc.

**Reoviridae family** A family of viruses, including the rotaviruses, that consist of RNA segments surrounded by a two-layer protein capsid. *See* reovirus.

**reovirus** Any of a group of respiratory, nonlatent, low-effective-dose, double-stranded RNA viruses that produce enteric symptoms and eye infections in humans. Reoviruses have been isolated in wastewater.

**reoxygenation** The introduction of air or oxygen into water or sludge. *See* reaeration for more detail.

**repeat compliance period** Any subsequent compliance period after the initial compliance period.

**repeat sample** A sample that is collected and tested to confirm a previous test result.

**replacement cost** The capital needed to purchase and install all the depreciable components in a facility (EPA-40CFR60.481).

**replicate sample** Any of a number of samples collected simultaneously or sequentially.

**replication** The conversion of one double-stranded DNA molecule into two identical double-stranded molecules, each original strand synthetizing a complementary strand.

**Reporter™** A multiprobe device manufactured by Hydrolab Corp. for monitoring water conditions.

**repose** *See* angle of repose.

**representative sample** A portion of material or water that is as nearly identical in content and consistency as possible to that in the larger body of material or water being sampled. A representative grab sample, for example, cuts across the entire section of flow. *See also* scoop sampler, tube sampler.

**repurification** Very advanced treatment of wastewater to make it suitable for indirect or even direct potable reuse.

**resampling** A requirement for a water supply system to follow a routine sample that yields positive results for total coliform in 3 or 4 repeat samples within 24 hours, including at least one from the same tap as the original sample.

**resanding** Filter resanding is the restoration of the original depth of a slow sand filter by replacing the sand lost after several scraping operations, which remove the schmutzdecke along with some sand. *See also* filter scraping.

**resazurin** A substance ($E_0 = -0.051$ V at pH = 7.0) that is colorless when it is reduced and pink when it is oxidized; used as an indicator for the presence of oxygen in the biochemical methane potential test.

**reserve capacity** The capacity in excess of current demand built into such installations as water and wastewater facilities to accommodate future population, commercial, or industrial growth.

**reserved water right** A water right created by a federal reservation of land for Indian reservations or other purposes.

**reservoir** (1) Any natural or artificial holding area used to store, regulate, or control water or other liquids. A pond, lake, tank, basin, etc. created by the construction of an engineering structure such as a dam to store water, for recreation, flood mitigation, water supply, or hydroelectric power production. (2) A person, animal, plant, soil, etc. in which pathogens can live and breed; e.g., humans are the reservoirs of typhoid and tuberculosis pathogens (*Salmonella typhi* and *Mycobacterium tuberculosis*).

**reservoir-assisted aquifer storage and recovery (RAASR)** *See* attenuation pond.

**reservoir deposition** The accumulation of sediment in a reservoir, which reduces its storage capacity.

**reservoir life cycle** The natural life cycle of lakes and reservoirs includes three stages defined by their nutrient levels, biological activity, and other factors. *See* oligotrophic stage, mesotrophic stage, eutrophic stage.

**reservoir lining** An impervious material installed on the bottom and sides of a reservoir to prevent or reduce water loss through seepage. Common materials used include plastic, concrete, and clay.

**reservoir loading model** A mathematical or graphical model designed to assess the impact on water quality from watershed activities. More specifically, it defines the trophic level of a lake or reservoir from a number of parameters: nutrient input, algae production, transparency, etc. *See also* fate and transport model.

**reservoir of disease** A live, passive or inanimate source of a pathogen; the host in which a disease (or infection) producing parasite is principally found and that accounts for the continuous maintenance of the parasite. Examples of disease reservoirs: persons, animals, soil, arthropods, plants, substances.

**reservoir of infection** *See* reservoir of disease.

**reservoir sanitation** The practice of healthy environmental measures for the protection of the quality of impounding reservoirs intended for drinking water supply, e.g., the regulation of recreational

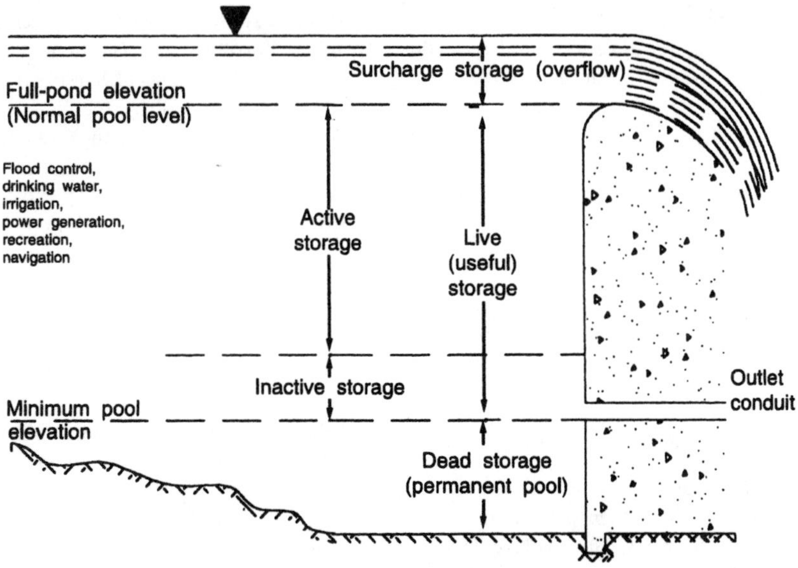


**Figure R-06.** Reservoir storage (courtesy CRC Press).

use of the reservoir and residential development within the watershed.

**reservoir storage** The total volume of water in a reservoir at any time. *See* Figure R-06. Its various elements may include: active storage (also called working volume), inactive storage, live storage (also called useful storage), surcharge storage or overflow, dead storage (or permanent pool), and total storage.

**residence time** The average length of time that a solid particle or a volume of liquid remains in a process, a tank or other unit, or a system. *See also* hydraulic residence time. Under steady-state conditions, the reciprocal of residence time ($1/R_E$) of an element E in a body of water can be used to determine its removal from the water column by sedimentation and by outflow:

$$1/R_E = 1/R_w + 1/R_s \qquad \text{(R-33)}$$

where $R_w$ and $R_s$ are, respectively, the residence time of water and the residence time of the element with respect to sedimentation.

**residence time distribution (RTD) curve** A tracer response curve, i.e., a graphical representation of the distribution of a tracer versus time of flow. It is used in studying short-circuiting, retardation, and longitudinal mixing in sedimentation basins and other reactors. *See* time–concentration curve for more detail. *See also* flow-through curve or fluid residence time distribution.

**residual** (1) Amount of a pollutant remaining after a natural or technological process has taken place, e.g., the particulates remaining in air after it passes through a scrubbing or other process. *See also* desalting residuals. (2) Residuals include all solid, liquid, or gaseous wastes produced by water and wastewater treatment. *See* water treatment plant residuals, free water, interstitial water, bound water, surface water (2). (3) Wastes produced by membrane treatment processes, including backwash, concentrate, chemical cleaning wastes. (4) *See* chlorine residual, combined residual, free residual.

**residual alkalinity** The difference ($Alk_R$) between the concentration of bicarbonate $[HCO_3^-]$ and double the concentration of calcium (2 $[Ca^{2+}]$) of an aqueous solution; caused by the weathering of silicate minerals:

$$Alk_R = [HCO_3^-] - 2\,[Ca^{2+}] \qquad \text{(R-34)}$$

Evaporation of natural waters having residual alkalinity tends to increase their pH and may eventually lead to the formation of soda lakes. Evaporation of waters with negative residual alkalinity tends to increase their calcium hardness.

**residual chlorine** The amount of free and/or available chlorine remaining in water or wastewater after a given contact time under specified conditions. Forms of residual chlorine include hypochlorous acid (HOCl), hypochlorite ion ($OCl^-$), trichloramine (also called nitrogen trichloride, $NCl_3$), dichloramine ($NHCl_2$), and monochloramine ($NH_2Cl$). *See also* chlorine demand, free chlorine, available chlorine.

**residual chlorine decay** The degradation of total residual chlorine of treated effluent in receiving streams, commonly assumed to follow a first-order model, whose overall rate coefficient ($K_T$, at temperature $T$ in °C) depends on turbulence, evaporation, photolysis (effect of sunlight), and temperature (Droste, 1997):

$$K_T = F(K_e + K_s + K_{ox})\theta^{(T-20)} \qquad \text{(R-35)}$$

where $F$ = turbulence factor, from 1.00 (quiescent) to 2.05 (turbulent); $K_e$ = rate constant for evaporation = 0.013/$H$/day; $H$ = depth of flow, m; $K_s$ = rate constant for photooxidation = 0.03/day; $K_{ox}$ = rate of free radical oxidation by chlorine = 0.065/day; and $\theta$ = Arrhenius constant = 1.08. *See also* chlorine decay.

**residual concentration** The equilibrium concentration of a metal in solution, as indicated by the solubility product expression for a precipitation process.

**residual disinfectant concentration** ("C" in CT calculations). The concentration of disinfectant measured in mg/L after a given contact time (T, in minutes) in a representative sample of water.

**residual disposal** *See* desalting residual disposal.

**residual hardness** The portion of noncarbonated hardness (or permanent hardness) that is not removed from water by the lime–soda softening process, often to save on the cost of soda ash.

**residual nuclear radiation** Nuclear radiation that persists for some time after an explosion.

**residual oxygen** The equilibrium dissolved oxygen concentration in wastewater or a receiving water at the completion of oxygenation and reaeration processes.

**residual pressure** The pressure in a distribution system when water is flowing at a specified rate. *See also* pressure test and static pressure.

**residual risk** (1) The extent of health risk from air pollutants remaining after application of he maximum achievable control technology (MACT). (2) In water utility risk management and other types of emergency management, residual risk is the risk that remains after risk treatment.

**residuals** All solid, liquid, or gaseous wastes produced by water and wastewater treatment. *See* residual for detail.

**residuals blowdown** Withdrawal of residuals from a solids processing tank to maintain a desirable solids level. Also called sludge blowdown.

**residuals cake** Water or wastewater treatment solids sufficiently dewatered (by a filter press, centrifuge, or other device) to form a semidry mass, typically having a thickness of 0.25 inch and a solids concentration of 20–35%. *See* sludge cake.

**residuals characteristics** Physical characteristics affecting the design of residuals handling processes include compressibility, density, particle size, shear stress, solids content, and specific resistance. *See also* blinding sludge.

**residuals conditioning** The pretreatment of liquid residuals, mostly by chemical, but also biological or physical, means, to facilitate subsequent treatment such as thickening or dewatering and improve drainability. *See* sludge conditioning for detail.

**residuals dewatering** The process of reducing the water content of residuals by a mechanical or physical method such as centrifugation, compaction, drainage, evaporation, filtration, flotation, freezing, gravity dewatering, and squeezing. *See* sludge dewatering for detail.

**residuals disposal** The ultimate discharge of solids from water or wastewater treatment to the environment; the permanent storage of residuals at the site of a water or wastewater treatment plant, or their removal from the site for disposal in a sanitary landfill, land application, discharge to sanitary sewers, reuse, etc. *See* sludge disposal for detail. *See also* desalting residual disposal.

**residuals drying bed** A water or wastewater treatment plant unit used to separate water and solids in residuals. *See* sludge drying bed and drying bed for detail.

**residuals freezing bed** A device for dewatering sludge using the freeze–thaw process. *See* freeze–thaw bed for detail.

**residuals handling process** *See* thickening, coagulant recovery, conditioning, dewatering, drying, residuals disposal.

**residuals lagoon** A natural depression or an excavated basin or pond that is used to dewater and store stabilized wastewater sludge, with formation of a superficial crust and deposition of solids at the bottom. *See* sludge lagoon for detail.

**residuals management** The operations involved in the handling and disposal of residuals from water or wastewater treatment.

**residuals mass balance** A theoretical accounting of the liquid and solids coming into a treatment plant, added during treatment, and withdrawn from the plant in the finished water or wastewater effluent and in residuals streams.

**residual sodium carbonate (RSC)** The excess of carbonates and bicarbonates over calcium and magnesium in a water sample, all concentrations expressed in milliequivalents per liter (meq/L):

$$RSC = [CO_3^{2-}] + [HCO_3^-] - [Ca^{2+}] - [Mg^{2+}] \quad \text{(R-36)}$$

This parameter is used to determine the suitability of water for irrigation. It is desirable to have RSC < 2.5.

**residual soil** Soil formed on-site from the original rock. Also called sedentary soil.

**residuals processing** The collection and treatment of solids from water and wastewater treatment. *See* sludge processing for detail.

**residual stream** In general, the waste stream containing the pollutants removed by a treatment process. In particular, the portion of the feed stream that contains the retained, dissolved or suspended constituents in membrane filtration processes. *See* reject for more detail.

**residue** The dry solid or semisolid material remaining after the evaporation, processing, or incineration of a sample of water or sludge. *See also* total dissolved solids.

**resin** A nonvolatile organic mixture of carboxylic acids obtained from certain plants or prepared by polymerization of simple molecules. Ion-exchange water demineralizers use mostly synthetic resins in the form of polymer beads, specifically produced and characterized by functional groups of exchangeable ions that are soluble in water. Resins are subject to fouling and can be irreversibly coated by iron and manganese.

**resin beads** Spherical polymer beads used in ion-exchange demineralization.

**resin binding** The plugging of an ion-exchange bed caused by influent residual organics and high total suspended solids, and resulting in high head losses and inefficient operation. Appropriate pretreatment can alleviate this problem.

**resin degradation** *See* degradation.

**resin exchange capacity** The quantity of an exchangeable ion that can be taken up by, or the number of fixed charge sites per unit volume or weight of, an ion-exchange resin. *See also* exchange capacity, matrix, functionality, porosity.

**resin fouling** The accumulation of solids on ion-exchange resins, thus increasing their maintenance and reducing their capacity; e.g., iron and manganese can irreversibly coat the resin.

**resin matrix** The resin backbone. *See* matrix.

**resin regeneration** *See* regeneration.

**resistance** For plants and animals, the ability to withstand poor environmental conditions or attacks by chemicals or disease. May be inborn or acquired.

**resistant form** A form taken by some microorganisms to resist to adverse conditions such as heat and chemical disinfectants. *See* cyst, egg, endospore, oocyst, and spore.

**resistivity** A measure, in ohm meters, of the ability of a material to resist to the flow of electricity. It is inversely proportional to conductance and commonly used as an accurate measure of the ionic purity or degree of demineralization of water.

**resistivity meter** A battery-powered instrument that measures the resistance to the flow of electrical current in a solution or medium.

**resistivity survey** A geophysical investigation conducted to locate groundwater sources, based on the relationships between resistivity and some soil characteristics (porosity, water content, salt content). *See also* gravity survey, seismic survey.

**resolubilization** The conversion of suspended organic matter into soluble matter as a result of anaerobic or septic conditions in a sedimentation basin. *See* solubilization (3) for detail.

**resolution** (1) The smallest distance between two points visible to the aided or naked eye. (2) The capability of distinguishing between two adjacent objects or light sources. (3) The degree of sharpness of a computer-generated image.

**resonance** (1) The condition in which atoms bonded together share electrons. *See* Lewis structure. Resonance makes a substance like benzene ($C_6H_6$) stable and difficult to degrade; it shares electrons in the bonds among the carbon atoms. (2) The property of aromatic compounds that cannot be represented by one valence-bond structure but have many approximate structures.

**resorcin** Same as resorcinol.

**resorcinol ($C_6H_6O_2$)** A white, needlelike, water-soluble, solid derivative of benzene ($C_6H_6$); a phenolic compound used in dyes, resins, as a reagent, in tanning, and other applications. By reacting with hypochloric acid (HOCl) during water chlorination, resorcinol may form chloroform ($CHCl_3$), a disinfection by-product. Also called resorcin.

**resource recovery** The process of obtaining matter or energy from materials formerly discarded; the reuse of natural resources. For example, the recycling of solid wastes after collection. *See* energy recovery.

**respiration** (1) The sum of physical and chemical processes in which an organism uses oxygen from air or water for oxidation (and life processes) and gives off carbon dioxide and other products; e.g., breathing. Certain organisms, like algae, release oxygen during the day through photosynthesis, but reverse the process at night through respiration, which can lead to oxygen depletion. *See also* aerobic heterotrophic respiration, aerobic autotrophic respiration, facultative anaerobic heterotrophic respiration, facultative anaerobic autotrophic respiration, anaerobic heterotrophic respiration, metabolism, digestion. (2) A similar process occurring in the absence of oxygen, e.g., in anaerobic bacteria.

**respiration rate** A measure of the amount of oxygen used by microorganisms in the activated sludge process, expressed as mg of oxygen per gram of mixed liquor volatile suspended solids per hour, a value that correlates well with the effluent COD. It is also used as an operational indicator of the process. Also called specific oxygen uptake rate. *See also* oxygen uptake rate.

**respiratory metabolism** The mechanism used by certain organisms to generate energy through the enzyme-mediated transport of electrons to an external electron acceptor; fermentative respiration. *See also* fermentative metabolism.

**respiratory quotient** The ratio of the volume of carbon dioxide ($CO_2$) released to the volume of oxygen ($O_2$) used up by an organism during respiration. It depends on the organic matter being oxidized and on the type of respiration (aerobic or anaerobic).

**respirometer** An instrument for measuring the character and extent of respiration (e.g., oxygen consumption and carbon dioxide production), usually equipped for data collection and processing by computer. *See* manometric respirometer, volumetric respirometer, electrolytic respirometer, direct-input respirometer.

**respirometric BOD determination** The determination of the biochemical oxygen demand (BOD) of a wastewater sample by measuring the amount of oxygen consumed by bacteria while they respire. *See* respirometric method, dilution BOD.

**respirometric method** A method that directly measures oxygen consumption by microorganisms from air in a closed vessel under conditions of constant temperature and agitation. It is used, mostly as a diagnostic tool, in assessing the biodegradability of specific chemicals, the treatability of industrial wastes, the effects of toxic compounds on oxygen uptake, and similar applications. *See* respirometer for the principal types of instruments available.

**respirometry** The use of a respirometric method to measure oxygen uptake more or less continuously.

**response** (1) In water utility risk management and other types of emergency management, response is the phase that follows the disaster or the emergency; it includes immediate assistance, mitigation of secondary damages, and initiation of recovery activities. (2) The effect on an organism of exposure to a toxic substance or other risk. *See* dose–response relationship.

**response logit (RL)** A variable used in the logit dose–response mouse model; calculated as a function of the proportion ($P$) of animals infected:

$$\mathrm{RL} = \ln [P/(1 - P)] \qquad \text{(R-37)}$$

where ln is the natural logarithm.

**restabilization** One of the predominant mechanisms observed in the coagulation of particles using alum in water treatment. It happens when a high coagulant dose causes a charge reversal in the previously destabilized particles, which do not stick together anymore and do not settle well. The extent of this phenomenon depends on pH and alum dosage (C, mg/L), e.g., for $5.0 < \mathrm{pH} < 6.5$ and $2.0 < C < 25$. *See also* adsorption–destabilization, bridging, double-layer compression, and sweep-floc coagulation.

**restricted reuse** Wastewater reuse that requires a relatively low degree of treatment because of a low risk to public health, e.g., pasture irrigation or golf course irrigation. *See also* unrestricted reuse.

**retained phase** In general, the waste stream containing the pollutants removed by a treatment process. In particular, the portion of the feed stream that contains the retained, dissolved, or suspended constituents in membrane filtration processes. *See* reject for more detail.

**retardance** The phenomenon that makes a flood crest after a storm has moved out of a tributary area. Ordinarily, cresting at a point occurs when all parts of the area contribute runoff. However, the effective area may be reduced when the storm moves upstream or sweeps the tributary area before the distant runoff can reach the point of concentration.

**retardance coefficient** (1) A parameter ($n$) used to modify the growth rate of first-order reactions so that the rate ($K$), instead of being constant, decreases with time ($t$) from an initial value of $K_0$, as in the following equation:

$$K = K_0/(1 + nK_0 t) \qquad \text{(R-38)}$$

(2) A coefficient in the Kerby formula for inlet time. It is analogous to the coefficient of roughness; the higher the coefficient, the longer the inlet time (i.e., the slower the flow). It varies from 0.02 for impervious surfaces to 0.80 in timberland or dense grasses. (3) A similar coefficient in the Izzard's formula for time of concentration or time to equilibrium; it varies from 0.007 for smooth surfaces to 0.06 for dense bluegrass turf.

**retardant reaction** A chemical reaction whose rate increases or decreases with time as a result of changes in the concentration of the reactant. The overall reaction rate is such that:

$$dC/dt = \pm kC/(1 + \alpha t) \qquad \text{(R-39)}$$

where $C$ = concentration of a substance at time $t$, $t$ = time, $k$ = reaction rate constant, and $\alpha$ = rate factor due to concentration change. The quantity $[1/(1 + \alpha t)]$ is a retardation factor, acting to reduce reaction rate over time. The retardant model has been used to describe substrate removal in biological wastewater treatment. *See also* first-order retarded reaction.

**retardation** (1) The preferential retention of contaminant movement in the subsurface resulting from adsorptive processes or solubility differences. (2) The opposite of short-circuiting in flow through a basin, whereby some of the inflow takes much longer than the theoretical detention period to reach the outlet.

**retardation equations** *See* retardation factor.

**retardation factor** (1) The unitless ratio ($R_d$) of the average linear velocity ($V_w$, ft/day) of groundwater to the average linear velocity ($V_c$, ft/day) of a contaminant. The following equations have been proposed:

Unconsolidated sediments: $R_\mathrm{d} = V_\mathrm{w}/V_\mathrm{c} = 1 + BK_\mathrm{d}/n_\mathrm{e}$ (R-40)

Low-porosity fractured rocks: $R_\mathrm{d} = V_\mathrm{w}/V_\mathrm{c} = 1 + 2\,K_\mathrm{A}/b$ (R-41)

Acids: $R_\mathrm{a} = 1 + \alpha_\mathrm{a} BK_\mathrm{d}/n_\mathrm{e}$ (R-42)

Bases: $R_\mathrm{b} = 1 + \alpha_\mathrm{b} BK_\mathrm{d}/n_\mathrm{e} + CB(1 - \alpha_\mathrm{b})/(100\ \Sigma z^+ n_\mathrm{e})$ (R-43)

where $B$ = average soil bulk density (g/cm³);

$$K_\mathrm{d} = \text{distribution coefficient } (\mathrm{cm^3/g}) = f_\mathrm{oc} \cdot K_\mathrm{oc} \qquad \text{(R-44)}$$

$n_e$ = effective porosity, unitless; $f_{oc}$ = fraction of naturally occurring organic carbon in soil; $K_{oc}$ = soil sorption coefficient (soil-sedimentation coefficient); $K_A$ = distribution coefficient in fractured rocks; b = aperture of fracture, cm; $\alpha_a$ = fraction of unionized acid; $\alpha_b$ = fraction of unionized base; $C$ = cation exchange capacity of the soil, $cm^3/g$; and $\Sigma z^+$ = sum of all positively charged particles in the soil, $meq/cm^3$. (2) A coefficient that affects the rate expression of a reaction to make it decrease with time. *See* first-order retarded reaction and retardant reaction. (3) In microbial transport studies, the retardation factor is the ratio of the time for the maximum concentration of a tracer to be detected in the effluent to the corresponding time for a microorganism injected simultaneously as the tracer.

**retarded first-order reaction** *See* first-order retarded reaction.

**retarding catalyst** A substance that decreases the speed or yield of a chemical or biological reaction without being consumed or chemically changed by the reaction; e.g., dissolved metals, enzymes. Also called a negative catalyst or an inhibitor.

**Retec®** An apparatus manufactured by Wheelabrator Engineered Systems, Inc. for the recovery of heavy metals.

**retentate** In general, the waste stream containing the pollutants removed by a treatment process. In particular, the portion of the feed stream that contains the retained, dissolved or suspended constituents in membrane filtration processes. *See* reject for more detail.

**retention** The part of precipitation that does not escape as runoff, that is, the difference between precipitation and total runoff. Retention is the same as basin recharge.

**retention basin/retention pond** (1) A basin/pond designed to capture stormwater permanently, such as an infiltration basin/pond. The stormwater entering the basin or pond is not discharged directly into a receiving water. (2) Same as retarding basin. (3) A basin/pond, enclosed by artificial dikes, used for wastewater treatment and/or storage. *See also* stormwater retention.

**retention capacity** The ability of soils or basins to retain precipitation under given conditions.

**retention chamber** A structure within a flow-through test chamber that confines the test organisms, facilitating observation of test organisms and eliminating washout from test chambers (EPA-40CFR797.1950-8).

**retention coefficient (RC)** A dimensionless parameter that indicates the extent of retention of a substance by a membrane, equal to the ratio of the substance concentration in the permeate ($P$) to its concentration in the retentate ($R$):

$$\mathrm{RC} = 1 - P/R \qquad \text{(R-45)}$$

**retention factor** *See* retardation factor.

**retention period** Same as detention time, hydraulic retention time, or, simply, retention time. That is, the time ($t$) that water, wastewater, or stormwater is retained in a reactor, unit process, storage basin, or any similar facility of volume ($v$) at a given hydraulic loading or discharge rate $Q$:

$$t = v/Q \qquad \text{(R-46)}$$

**retention pond** *See* retention basin/retention pond and retention/detention ponds.

**retention time** (1) Same as detention time, hydraulic retention time or retention period. (2) The time in minutes elapsed between sample injection into the chromatograph and the maximum concentration recorded on a chromatogram. The retention time is characteristic of the substance, the liquid phase flow rate, and the stationary phase, at a given temperature (EPA-40CFR796.1570).

**retention/detention ponds** A dual system of stormwater management using first infiltration and evaporation in an offline retention pond and then temporary storage in a detention pond when the flow exceeds the capacity of the retention pond. Also called dual ponds.

**Re-Therm™** A thermal oxidation apparatus manufactured by Regenerative Environmental Equipment Co. for the treatment of volatile organic compounds.

**Retox** A regenerative oxidizer produced by Adwest Technologies, Inc. for the treatment of volatile organic compounds.

**retrofit** Addition of a pollution control device on an existing facility without making major changes to the generating plant.

**retrofit kit** *See* bathroom retrofit kit.

**Retroliner** Forms manufactured by Roberts Filter Manufacturing Co. for the rehabilitation of filter underdrains.

**retrospective analysis** *See* retrospective study.

**retrospective study** An epidemiological study that compares diseased persons with nondiseased persons and works back in time to determine exposures. It uses historical information about exposure characteristics, risk factors, events, or experiences. *See also* case-control study.

**retrovirus** Any member of the Retroviridae, a family of enveloped viruses that contain a single strand of RNA with an enzyme.

**Rettger weir** A weir with horizontal crest and sides curved such that the discharge is proportional to the head above the crest. *See also* proportional weir.

**return activated sludge (RAS)** Settled activated sludge that is returned to mix with raw or primary settled wastewater for purposes of inoculation or to promote more complete biological oxidation. Return sludge typically has a concentration between 4000 and 12,000 mg/L and a flow rate based on maintaining a specific level of mixed liquor suspended solids in the aeration basin or a given depth of sludge blanket in the secondary clarifier. Sometimes called returned floc, returned sludge, returned activated sludge, or return sludge. *See also* excess sludge.

**returned activated sludge** Same as return activated sludge.

**returned floc** Same as return activated sludge.

**returned sludge** Same as return activated sludge.

**return flow** (1) That portion of the water diverted from a stream that finds its way back to the stream channel either as surface or underground flow. (2) In wastewater studies, the term return flow is sometimes used to designate the portion of water use included in the wastewater discharge. *See also* spent water.

**return offset** A double fitting (elbow or bend) installed in plumbing to maintain the original alignment of a pipe.

**return on investment** The amount of profit, before tax and after depreciation, from an investment, usually as a percentage of the original total cost invested. Similar concepts applicable to water and wastewater utilities include return on assets and discounted cashflow return on investment. *See* financial indicators for related terms.

**return sludge** Same as return activated sludge.

**return solids** The weight or concentration of solids in the return sludge, usually the suspended volatile solids (VSS) or total suspended solids (TSS).

**reuse** (1) Using a product or component of municipal solid waste in its original form more than once, e.g., refilling a glass bottle that has been returned or using a coffee can to hold nuts and bolts. (2) The use of reclaimed water, appropriately treated wastewater, or residuals for agricultural, recreational, or other beneficial purposes.

**reuse health hazards** The risks associated with the reuse of wastewater, sludge, or excreta; the occupational hazard to reuse workers and the risk of infection from reuse products.

**reuse water** Wastewater adequately treated for reuse.

**reutilization** Same as recycling or reuse.

**revenue bond** A debt instrument used to finance a public utility project and expected to be repaid from the revenues generated by the utility (e.g., sale of water or sewer service). *See also* general obligation bond, special assessment bond.

**revenue-neutral rate** A new water or wastewater rate schedule that replaces an existing rate but produces the same revenue.

**reverse deionization** The deionization process in the reverse order, i.e., with the anion-exchange resin ahead of the cation-exchange resin.

**reversed phase chromatography** *See* high-performance liquid chromatography.

**reverse filtration** A technique for cleaning membrane equipment used in water or wastewater treatment; solids accumulated on a membrane surface are removed by applying air and/or permeate water in the opposing flow direction of filtration through the membrane fiber. *See* backpulse for detail.

**reverse osmosis (RO)** The application of hydrostatic pressure greater than the osmotic pressure to a concentrated solution that causes the passage of a liquid from the concentrated solution to a weaker solution across a semipermeable membrane. The membrane allows the passage of the solvent (water) but not the solutes (ions, salts, and other dissolved solids). The liquid produced is freshwater. The thin membrane must withstand pressures varying from 600 psi for brackish water to 1500 psi for seawater. Reverse osmosis units are available in a number of configurations, e.g., hollow-fiber, plate-and-frame, spiral-wound, and tubular. This advanced method is used to produce freshwater from brackish and saline waters, and, in wastewater treatment, by separating the water from the salts, specific ions, and organics. Reverse osmosis used to be called hyperfiltration. It is a water recovery process as compared to demineralization or desalination. *See also* osmosis, osmotic pressure, saline water conversion classification.

**reverse osmosis design equations** Basic equations used to determine the required membrane area ($A$, $m^2$), permeate total dissolved solids (TDS) concentration ($C_p$, $kg/m^3$), rejection rate ($R$, %), and concentrate stream TDS ($C_c$, $kg/m^3$):

Area: $$F_w = k_w(\Delta P_a - \Delta\Pi) = Q_p/A \quad \text{(R-47)}$$

Permeate TDS: $$F_i = k_i \Delta C_i = Q_p C_p / A \quad \text{(R-48)}$$

Concentrate TDS: $$Q_f C_f = Q_p C_p + Q_c C_c \quad \text{(R-49)}$$

Rejection rate: *See* rejection rate

where $F_w$ = water flux rate, kg/m²/sec; $k_w$ = water mass transfer coefficient, m/sec; $\Delta P_a$ = average imposed pressure gradient, kg/m/sec²; $\Delta\Pi$ = osmotic pressure gradient, kg/m/sec²; $F_i$ = flux of solute species *i*, kg/m²/sec; $k_i$ = solute mass transfer coefficient, m/sec; $\Delta C_i$ = solute concentration gradient, kg/m³; $Q_p$ = permeate stream flow, kg/sec; $Q_f$ = feed stream flow, kg/sec; $C_f$ = solute concentration in feed stream, kg/m³; and $Q_c$ = concentrate flow rate, kg/sec.

**reverse osmosis membrane** A synthetic membrane used in reverse osmosis; a thin skin supported by a porous substructure and acting as the selective layer to allow flow and retain the solute.

**reverse osmosis system** A water treatment device consisting of a semipermeable membrane supported by a structure, a pressure vessel, a pump, and, sometimes, a pretreatment unit to remove bacteria and suspended solids that can cause fouling.

**reverse osmosis water** Bottled water that has been treated by reverse osmosis to meet the definition of purified water, a labeling regulated by the U.S. Food and Drug Administration. *See* bottled purified water.

**reverse phase technique** The application of liquid chromatography to nonpolar compounds.

**reverse taintor gate** A device used to regulate flow in combined or interceptor sewers; water level is controlled by float and gate travels. Also known as mechanical regulator or automatic regulator.

**reverse transcriptase** An enzyme that can convert a retrovirus to a double-stranded form; used to clone genes from RNA strands.

**reverse transcriptase–polymerase chain reaction (RT–PCR)** A technique that combines a reverse transcriptase and polymerase chain reaction to detect RNA viruses.

**reversible effect** An effect that is not permanent, especially adverse effects that diminish when exposure to a toxic chemical ceases.

**reversible reaction** A reaction, represented by a double arrow (↔), that can proceed in the forward or reverse direction, depending on solution characteristics.

**reversion** The conversion of a higher polymeric or condensed phosphate to the uncondensed or orthophosphate form ($PO_4^{3-}$) in an aqueous solution.

**revetment** A blanket of stone, concrete, or other material laid on beaches, embankments, and shores to protect them against erosion from oncoming waves. *See* Figure R-07 for a typical revetment to protect the slope of lagoon or stabilization pond. Other coastal hydraulic structures include jetties, breakwaters, seawalls, groins, and bulkheads.

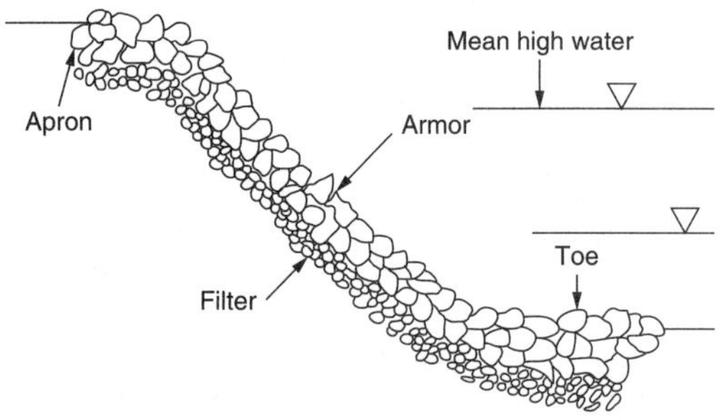


**Figure R-07.** Revetment (courtesy CRC Press).

**revolutions per minute (rpm)** A unit of angular velocity of a body that rotates through an angle of 360° so that every point in the body returns to its original position in one minute. The abbreviation rpm is commonly used to express the specific speed of pumps and turbines.

**revolution(s) per second (rps)** A unit of angular velocity of a body that rotates through an angle of 360° so that every point in the body returns to its original position in one second.

**Revolver™** A rotary unit manufactured by Vara International to remove volatile organic compounds by adsorption.

**revolving-drum filter** A widely used vacuum filter consisting of a drum with a number of cells of the same length under pressure or vacuum. The drum revolves through the sludge to be dewatered.

**revolving screen** A cylinder or a continuous belt used in screening solids out of water or wastewater, with hydraulic, mechanical, or manual removal of screenings.

**rewash connection** A valved pipe stub or similar device that branches from the filter effluent so as to waste the product water after washing the bed with raw water. *See also* filter-to-waste connection.

**Rex** Environmental control products of Envirex, Inc.

**Reynolds critical velocity** Velocity at the point where the flow condition changes from laminar (streamline) or nonturbulent and where friction becomes proportional to the square instead of the first power of the velocity. It differs from the usual definition of critical velocity or Belanger's critical velocity in open-channel flow.

**Reynolds number** A dimensionless parameter widely used in fluid mechanics to characterize the flow regime as laminar or turbulent; named in honor of Sir Osborne Reynolds who studied turbulent flows in the late nineteenth century and developed this number. As the ratio of inertia forces to viscous forces, the Reynolds number ($R_e$) is the product of the fluid density ($\rho$) by the mean velocity ($V$) by a characteristic length ($L_s$) divided by the fluid absolute viscosity ($\mu$) or simply the product of the mean velocity by the characteristic length divided by the kinematic viscosity ($\nu$), i.e.,

$$R_e = \rho \cdot V \cdot L_s/\mu = V \cdot L_s/\nu \qquad \text{(R-50)}$$

For open-channel flow the characteristic length is the hydraulic radius (R), and the flow is laminar for $R_e < 500$ and turbulent when $R_e > 2000$. Between these two numbers, the flow may be laminar or turbulent, depending on other factors. Similarly, in closed conduits flowing full (i.e., under pressure), using the diameter ($D$) as the characteristic length, the Reynolds number becomes

$$R_e = \rho \cdot V \cdot D/\mu = V \cdot D/\nu \qquad \text{(R-51)}$$

Then the flow is laminar for a Reynolds number less than 2100 and turbulent for $R_e > 4000$. For groundwater flow, Darcy's law is valid for Reynolds numbers smaller than 1, i.e., practically for all natural porous media. The Reynolds number is a consideration in several formulas or phenomena, e.g., hindered settling (Richardson and Zaki equation), flow through porous media (Kozeny equation), impeller mixer power requirement, and interstitial velocity (modified Reynolds number). *See also* hydraulics and open channel flow.

**R factor** A plasmid indicative of antibiotic resistance. Also called R plasmid.

**RF–AS process** Abbreviation of roughing filter/activated sludge process.

**RfC** Abbreviation of reference concentration.

**RfD** Abbreviation of reference dose.

**RF–PA** Abbreviation of roughing filter–plain aeration process.

**RgD** Abbreviation of regulatory dose.

**rheology** The study of the deformation and flow characteristics of matter. *See* sludge rheology method.

***Rhizobium*** A group of heterotrophic bacteria active in biological denitrification and in nitrogen fixation; commonly found in symbionic association with leguminous plants.

**rhizoplane** The external surface of roots, including the associated debris and soils; the part of the rhizosphere in direct contact with the plant roots.

**rhizosphere** The zone of intermingled roots and soil, subject to enhanced microbial activity and particular plant–microorganism interactions; it extends 5 mm or more from the root surface.

**rhizosphere effect** A factor indicative of the influence of the rhizosphere on microbial populations; numbers of microorganisms decrease from the rhizoplane outward toward bulk soil. *See* R/S ratio.

**rhodamine B** A common dye used as a tracer to assess dispersion and mixing regime in water and wastewater treatment reactors.

**Rhodesian sleeping sickness** Common name of one variety of African trypanosomiasis, a disease caused by the protozoa *Trypanosoma rhodesiense* and transmitted by the game tsetse fly (*Glossina morsitans*). It is found mainly in East Africa. *See also* Gambian sleeping sickness and Chagas' disease.

**rhodochrosite** A common mineral containing the metallic element manganese; essentially manganese carbonate, $MnCO_3$, with some iron and calcium. Also called dialogite.

***Rhodococcus*** A genus of bacteria that, in combination with *Arthrobacter* and surfactants, can biodegrade petroleum products.

***Rhodophyta*** *See* red algae.

***Rhodopseudomonas* genus** A group of heterotrophic, anaerobic, phototrophic, sulfur-oxidizing bacteria, also active in biological denitrification.

**R horizon** The massive bedrock below the C horizon on a soil profile. Also called D horizon.

**RI** Acronym of (a) Ridick index; (b) Ryzner index.

**ribonucleic acid (RNA)** A long-chain molecule that carries the genetic message from DNA to a cellular protein-producing mechanism; it contains hereditary material vital to reproduction. It is based on sequences of nucleotides of adenine ($C_5H_5N_5$), cytosine ($C_4H_5N_3O$), guanine ($C_5H_5N_5O$), and uracil ($C_4H_4N_2O_2$).

**ribonucleic acid–polymerase chain reaction (RNA–PCR)** See polymerase chain reaction.

**ribosome** A particle in the cytoplasm composed of ribonucleic acid and protein; a site of production of protein.

**Rich Tech** Products of Aerators, Inc.

**Richardson–Zaki equation** An empirical equation proposed in 1954 to represent data on two-phase flow systems. It has been applied to filter bed expansion, backwash velocity, and hindered settling:

$$e_e = (V_b/V_i)^n \qquad \text{(R-52)}$$

$$e_e = (V_s/V_t)^n \qquad \text{(R-53)}$$

where $e_e$ = porosity of the expanded medium, $V_b$ = superficial backwash velocity (the intercept at $e_e$ =

1 of a log-log plot of $e_e$ versus $V_b$), $V_i$ = intercept velocity, $n$ = coefficient related to the medium (slope of the line representing the log-log plot of $e_e$ versus $V_b$) or a power coefficient that depends on the Reynolds number of the particles, $V_s$ = settling velocity of a suspension, and $V_t$ = terminal settling velocity of unhindered particles. *See also* Carman–Kozeny equation, incipient fluidization, minimum fluidization condition, interstitial velocity.

**rickettsia** Single-cell parasitic organisms, intermediate between bacteria and viruses, that grow in the tissue of an appropriate host; e.g., *Rickettsia prowazeki.*

***Rickettsia prowazeki*** The parasite that causes louse-borne typhus.

**rickettsial disease** Louse-borne typhus, a water-washed infection carried by body lice, most commonly in poor mountainous areas. It is an acute and often fatal fever.

**Riddick index** or **Riddick's corrosion index (RI)** An empirical index proposed in 1944 for corrosion control, based on field experience and incorporating the concentrations of the following water quality parameters: alkalinity (Alk, mg/L as $CaCO_3$), carbon dioxide ($CO_2$, mg/L), hardness (Hrd, mg/L as $CaCO_3$), chloride ($Cl^-$, mg/L), nitrate ($NO_3^-$, mg/L as N), silica ($SiO_2$, mg/L), actual dissolved oxygen (DO, mg/L), and the saturation concentration of dissolved oxygen ($DO_{sat}$, mg/L):

$$RI = (75/Alk)\,[CO_2 + 0.5\,(Hrd - Alk) + Cl^- + 20\,NO_3^-\,(DO + 2)/(SiO_2 \cdot DO_{sat})] \qquad (R\text{-}54)$$

This index is applicable to water in the northeastern United States; water is noncorrosive for RI < 25, moderately corrosive for 26 < RI < 50, corrosive for 51 < RI < 75, and very corrosive for RI > 75. *See* corrosion index for a list of other corrosion control parameters.

**Riddick's corrosion index** Same as Riddick index.

**ridge-and-furrow air diffusion** An aeration method of the activated sludge process using porous tile diffusers in saw-tooth depressions of the aeration basin and in rows perpendicular to the direction of flow.

**ridge-and-furrow irrigation** One of two main types of surface application of wastewater, using short runs of piping and surface standpipes to provide the necessary head on moderately sloped land. *See also* border-strip flooding.

**Riga-Sorb™** Activated carbon adsorbers manufactured by Farr Co.

**right of entry** The authority that the Safe Drinking Water Act grants representatives of the USEPA to enter any water supply facility subject to the National Primary Drinking Water Regulations.

**right of free capture** The principle that a person owns the water under his or her land and is free to use as much of it as he or she wants. Also called law of the biggest pump.

**right-of-way** Right of passage over a piece of land, or the actual land used in the construction of roads, water and sewer facilities, railroads, etc. *See also* easement and leasehold.

**rigidity coefficient** *See* coefficient of rigidity.

**rigid layer** A rigid layer attached to a negatively charged colloidal particle, in the electrical double layer theory. *See* Stern layer for detail.

**rigid porous diffuser** A porous diffuser made of ceramic or plastic. *See also* membrane diffuser, high-density polyethylene, styrene acrylonitrile.

**rill** A small, elongated, and deep channel created by the eroding action of running water, usually dry except after a rainstorm, icemelt, or snowmelt. Rills or rivulets are similar to but smaller and less deep than gullies. *See* stream.

**rim canal** A canal at the bottom of a dike, dam, or levee, on the water side to collect overflows from the impoundment. *See also* toe ditch.

**rime** A coating of very small, white, granular ice particles formed by the rapid freezing of water droplets. *See also* graupel.

**rim-feed settling tank** A type of circular tank used in primary and secondary sedimentation; including (a) an annular space formed by a circular baffle into which the influent wastewater discharges and (b) a revolving mechanism for sludge transport and removal. The bottom of the tank slopes toward the central hopper. Also called peripheral-feed settling tank. *See also* center-feed settling tank.

**Rim-Flo** A circular clarifier manufactured by Envirex, Inc. using peripheral feed.

**Ringlace®** A proprietary integrated fixed-film activated sludge method of wastewater treatment developed by Ringlace, Inc. It uses film growth supported by PVC modules installed in about 30% of the aeration tank volume. *See also* Bio-2-Sludge®, BioMatrix®, Captor®, Kaldnes®, moving-bed biofilm reactor, submerged rotating biological contactor.

**ring lace process** A proprietary process developed in Japan for the secondary treatment of wastewater. It consists of woven plastic filaments, suspended in an aeration tank and supporting attached growth, and of a diffused-air apparatus.

**ring main** A water supply network that forms a closed loop as opposed to a branched or dendritic system.

**RingSparjer** An air diffuser manufactured by Walker Process Equipment Co.

**rinse** A step in the regeneration of softening and ion-exchange resins consisting in the removal of excess regenerant by passing freshwater through the bed.

**rinse water** (1) The water used to wash excess chemicals off metal surfaces in industrial processes or any industrial water or wastewater used for cleaning purposes. (2) The freshwater passed through ion-exchange resins during regeneration.

**riparian doctrine/riparian rights** The doctrine or common law governing the use of water resources in the Eastern United States, under which land owners are entitled to use the water on or bordering their property, including the right to prevent diversion or misuse of upstream waters. Riparian land is land that borders on surface water.

**riparian ecosystem** A transition between an aquatic ecosystem and the adjacent terrestrial ecosystem, based on soil and vegetation characteristics.

**riparian habitat** An area adjacent to rivers and streams with a high density, diversity, and productivity of plant and animal species relative to nearby uplands.

**riparian land** *See* riparian doctrine.

**riparian pollution** Water used to wash excess chemicals off metals or metal products, or wastewater from commercial or industrial cleaning operations.

**riparian right** *See* riparian doctrine.

**riparian water right** *See* riparian doctrine.

**rip current** A strong, localized, outgoing current that returns water to the sea from the incoming surf.

**ripening** *See* filter ripening.

**ripening period** The initial period of water quality degradation and improvement of a run in slow sand filters. This period may last from 6 hr to 15 days. Water quality parameters of concern include turbidity and cysts. *See also* filter-to-waste, slow start.

**rise rate** A design and operation parameter for settling tanks and clarifiers; expressed in gallons per day per square foot, it is equal to the ratio of the average flowrate (in gpd) to the surface area of tank (in $ft^2$). Operators use it to determine the proper hydraulic loading of the tanks. Also called overflow rate or surface loading rate. *See also* weir overflow rate (or weir loading).

**rising-film evaporator** An evaporator that functions like a coffee percolator. It consists of a vertical heating element made of heated tubes, a vapor body at the top (including an entrainment separator and a deflector plate), and a liquor box at the bottom. The liquor to be treated enters at the bottom of the tubes, generates vapor and moves upward; thus, the rising film inside the tubes. This device is used in the treatment of some hazardous wastes. *See also* batch pan, evaporation, falling-film evaporator, forced-circulation evaporator, and natural-circulation evaporator.

**rising rate** The rate of operation of a solids contact unit in gpm/square foot.

**rising sludge** A common operational problem of activated sludge plants whereby sludge rises or floats to the surface of a secondary clarifier after a short settling period. It is usually caused by nitrogen gas bubbles resulting from denitrification and rendering the sludge mass buoyant. A possible remedial action is to increase the rate of return activated sludge, thereby reducing the sludge blanket depth in the clarifier.

**rising time** The time it takes particles to rise to the surface and be removed in the flotation process.

**risk** (1) The potential for realization of unwanted adverse consequences or events; the product of the probability of an event and its consequences. (2) The probability of injury, disease, or death under specific circumstances. In quantitative terms, risk is expressed in values ranging from zero (representing the certainty that harm will not occur) to one (representing the certainty that harm will occur). The following are examples showing the manner in which risk is expressed in the Integrated Risk Information System (IRIS): E-4 = a risk of 1/10,000, E-5 = a risk of 1/100,000, E-6 = a risk of 1/1,000,000. Similarly, 1.3E-3 = a risk of 1.3/1000 = 1/770, 8E-3 = a risk of 1/125, and 1.2E-5 = a risk of 1/83,000. *See also* risk ratio.

**risk analysis** The use of available information to identify possible events and estimate their probabilities and consequences. *See* risk assessment.

**risk assessment** A qualitative or quantitative evaluation of the environmental and/or health risk resulting from exposure to a chemical or physical agent; it combines exposure assessment results with toxicity assessment results to estimate risk. It includes four components: hazard identification, dose–response assessment, exposure assessment, and risk characterization. Also called environmental risk analysis.

**risk characterization** The final component of risk assessment that involves integration of the data and analysis in hazard evaluation, dose–response evaluation, and human exposure evaluation to determine the likelihood that humans will experience any of the various forms of toxicity associated with a substance. *See also* toxicity terms.

**risk estimate** A description of the probability that organisms exposed to a specified dose of chemical will develop an adverse response (e.g., cancer).

**risk factor** A characteristic or variable associated with increased probability of a toxic effect, e.g., race, sex, age, obesity, smoking, or occupational exposure level.

**risk identification** The process of listing, defining, and characterizing risk elements, as part of risk analysis.

**risk indicator** Same as risk factor.

**risk management** A decision-making process that entails considerations of political, social, economic, and engineering information with risk-related information to develop, analyze, and compare regulatory options and to select the appropriate regulatory response to a potential health hazard. Risk management involves decisions whether an assessed risk is sufficiently high to present a public health concern and about appropriate means of control. *See also* toxicity terms.

**risk ratio** The ratio of incidence or risk among exposed individuals to incidence or risk among nonexposed individuals; also called relative risk. The risk ratio ($R$) may be calculated as:

$$R = d \cdot L/P \qquad \text{(R-55)}$$

where $d$ is the number of deaths per year, $L$ is the average lifetime (years), and $P$ is the population.

**risk reduction** A series of actions designed to reduce the probability and or the consequences of an identified risk.

**risk-specific dose** The dose associated with a specified risk level.

**riverbank filtration (RBF)** A technique that subjects surface water to passage through the ground, thereby improving the raw water quality, before its collection and use as a drinking water source; a practice used extensively in Europe and now being considered in the United States because of its ability to reduce the formation of disinfection byproducts, among other benefits. Systems using riverbank filtration locate water supply wells close to a surface water body; the surface water goes through the streambed and the unconsolidated materials of the aquifer before reaching the wells. RBF may be considered acceptable treatment for the reduction of *Cryptosporidium parvum,* with a treatment credit of 1.0 log under the surface water treatment rules. Also called bank filtration, natural filtration, riverbank infiltration, in-situ natural filtration. *See also* induced infiltration.

**riverbank infiltration** An alternative source of drinking water, instead of direct surface water withdrawal. Same as riverbank filtration.

**river basin** The land area drained by a river and its tributaries. *See also* catchment, drainage area, drainage basin, and watershed.

**river blindness** The common name of onchocerciasis, a chronic disease caused by the mosquito-borne helminth *Onchocerca volvulus,* which is transmitted by blackflies. These members of the *Simulium* species breed in fast-flowing water. The disease is endemic in the Volta River watershed and other areas of West and Central Africa; it is also found in Latin America and Yemen.

**riverine habitat** Any wetland and deep-water habitat within a channel, except the palustrine habitats.

**Rivers and Harbors Appropriations** A federal water pollution control law of 1899 requiring a permit from the U.S. Army Corps of Engineers before dumping solids into navigational waters.

**RKL®** A pinch valve manufactured by Robbins & Myers, Inc.

**RMCL** Acronym of recommended maximum contaminant level.

**RMP** Acronym of Roberts' Manhattan Process.

**RMS** (or **rms**) **velocity gradient** *See* root-mean-square velocity gradient.

**$R_4N^+$** Chemical formula of quaternary amine.

**RNA** Acronym of ribonucleic acid.

**$R_4N^+X^-$** Chemical formula of quaternary ammonium salts.

**RO** Acronym of reverse osmosis.

**Roberts–Haberer filter** *See* Roberts–Haberer process.

**Roberts–Haberer process** A method of application of powdered activated carbon to the treatment of water or wastewater, e.g., for the removal of total organic carbon. It uses buoyant polystyrene spheres about 2 mm in diameter, coated with activated carbon, and held by means of a screen in an upflow reactor. This unit is sometimes used as a roughing filter to reduce suspended solids ahead of a rapid filter.

**Robo™** A bar screen manufactured by Vulcan Industries, Inc.

**Robo Rover™** A bar screen by Vulcan Industries, Inc.

**Robo Stat™** A stationary bar screen by Vulcan Industries, Inc.

**robustness** "The ability of a filtration system to provide excellent particle/pathogen removal under normal operating conditions and to deviate minimally from this performance during moderate to severe process upsets" (Coffey et al., 1998, quoted in AWWA, 1999).

**rock filter** A submerged bed of rocks or other coarse medium near the outlet of a stabilization pond or lagoon to remove algae and other solids. A slow sand filter may be used for the same purpose.

**rock-reed filter** A type of constructed wetlands consisting of channels or trenches, the bottom of which supports emergent vegetation for the secondary or advanced treatment of wastewater. *See* subsurface flow system for detail.

**rock salt** Another name for halite, a soft white, red, yellow, blue, or colorless evaporite mineral that is at least 95% sodium chloride (NaCl) in cubic crystals; it occurs together with other minerals such as gypsum and sylvite. Also called native salt, fossil salt.

**rods** Bacteria shaped like cylinders. *See also* cocci, helix, pleomorphic bacteria.

**Rodac agar plate** *See* Rodac dish.

**Rodac dish** A Petri dish completely filled with agar so that its convex surface is pressed against the surface being sampled; commonly used to detect bateria on fomites. *See also* swab rinse method.

**rodenticide** Any substance or chemical used to kill or control rodents.

**Roebelt** A belt filter press manufactured by Roediger Pittsburgh, Inc.

**ROEC** Acronym of Reed odorless earth closet.

**Roedos** An apparatus manufactured by Roediger Pittsburgh, Inc. for mixing polymers.

**Roefilt** A sieve drum concentrator manufactured by Roediger Pittsburgh, Inc.

**Roeflex** A fine-bubble diffuser manufactured by Roediger Pittsburgh, Inc.

**Roemix** An apparatus manufactured by Roediger Pittsburgh, Inc. for mixing lime in the treatment of dewatered sludge.

**roentgen (R)** A unit of exposure dose of ionizing radiation; gamma or X-radiation absorbed in air, equal to 0.000258 coulomb per kg of pure, dry air (the SI unit), or to 83.6 ergs per gram of air. Named after W. K. Roentgen.

**roentgen equivalent man (rem)** *See* rem.

**Roentgen, Wilhem Konrad** German physicist (1845–1923), discoverer of X-rays in 1895, winner of Nobel prize in 1901.

**Roepress** A belt filter press manufactured by Roediger Pittsburgh, Inc.

**Ro-Flo®** A sliding vane compressor manufactured by A-C Compressor Corp.

**Rogun** A chemical solution produced by Argo Scientific for cleaning reverse osmosis membranes.

**roiliness** A milky appearance imparted to water by particles of clay or silt; turbidity, muddiness.

**RollAer** Aeration apparatus manufactured by Walker Process Equipment Co. for use in aerobic digestion units.

**Roll-Dry** A rotary fine screen manufactured by Schlueter Co. with internal feed.

**Rolling Grit** An apparatus manufactured by Walker Process Equipment Co. to wash and remove grit from aerated chambers.

**rolling-up curtain weir** A movable weir made of horizontal lathes and watertight hinges. *See also* needle weir.

**Romembra** Reverse osmosis membranes manufactured by Toray Industries, Inc.

**RO membrane** Abbreviation of reverse osmosis membrane.

**Romicon®** Products manufactured by Koch Membrane Systems, Inc. for hollow fiber membranes.

**Romi-Kon™** A device manufactured by Koch Membrane Systems, Inc. to separate oil from water.

**röntgen** Alternative spelling for roentgen.

**root crop** A plant whose edible parts are grown below the surface of the soil.

**root-mean-square (RMS) velocity gradient** Same as velocity gradient. "Root mean square" indicates that power is dissipated in all three directions.

**Roots** Centrifugal compressors and blowers manufactured by Dresser Industries/Roots Division.

**root-zone filter** A type of constructed wetlands consisting of channels or trenches, the bottom of which supports emergent vegetation for the secondary or advanced treatment of wastewater. *See* subsurface flow system for detail.

**Rose equation** An equation proposed by H. E. Rose in 1945 to estimate the head loss in a clean, stratified filter bed:

$$H/L = (1.067\, C_d/\Psi)(V^2/F^4 g) \sum_{i=1}^{n} (P_i/D_i) \quad \text{(R-56)}$$

where $H$ = head loss, $L$ = bed depth, $C_d$ = drag coefficient, $\Psi$ = particle shape factor, $V$ = approach velocity, $F$ = bed porosity, $g$ = gravitational acceleration, $n$ = number of layers of the stratified bed, $P_i$ = the fractional weight of layer $i$, and $D_i$ = the sieve size of layer $i$. *See also* Carman–Kozeny, Fair–Hatch, and Hazen equations.

**Rosep™** A reverse osmosis unit manufactured by Graver Co.

**Rossum-Merrill formula** *See* calcium carbonate precipitation potential.

**Rotadisc®** A sludge dryer manufactured by Stord, Inc.

**Rotafine** A rotary fine screen manufactured by Jones & Attwood, Inc.

**Rotamat®** Screening products of Lakeside Equipment Corp.

**rotameter** An instrument for measuring the flow rate of a fluid, consisting of a tapered vertical tube in which the fluid flows upward, lifting a weight or float until the fluid force just balances its weight.

**Rotapak** A screenings compactor manufactured by Longwood Engineering Co.

**Rota-Rake** A circular sludge collector manufactured by Graver Co.

**Rotarc** An arc-type bar screen manufactured by John Meunier, Inc.

**rotary collector** The rotating mechanism that collects and removes sludge in circular clarifiers.

**rotary disc press** A device used to dewater sludge; it consists of two parallel chrome-plated stainless steel screens, rotating on a single shaft at about 2 rpm. Under frictional force and increasing pressure, the filtrate flows through the screens as flocculated sludge moves through the channel. *See also* sludge pressing.

**rotary distributor** A rotating apparatus consisting of horizontal arms with orifices, used to distribute wastewater flow evenly and intermittently on the surface of a trickling filter. *See also* distributor, fixed-nozzle distributor, revolving distributor.

**rotary-drum screen** A cylindrical screen that rotates in a wastewater flow channel to remove coarse, medium, or fine floatable and suspended solids, with internal feed at one end (and solids collection) or external feed from the top (and solids collection)

**rotary-drum thickener (RDT)** A rotating cylindrical screen used to thicken sludge. It consists of a drum with wedge wires, perforations and a porous medium of stainless steel or polyester fabric. When flocculated sludge is processed through the unit, water drains through while solids are retained on the medium. RDTs are also used as prethickeners before belt press dewatering in small-to-medium size plants for waste activated sludge. Also called rotary screen thickener. *See also* gravity belt thickening.

**rotary-drum thickening (RDT)** A sludge thickening method that uses a rotary-drum thickener, including a polymer feeding device. *See also* centrifugal thickening, cosettling thickening, flotation thickening, gravity belt thickening, gravity thickening.

**rotary dryer** A cylindrical steel drum rotating at 5–8 rpm and equipped with air pollution control devices; it is used to dry dewatered sludge by exposing it for approximately 40 min to air heated to 650°C. *See also* direct dryer, flash drying, fluidized-bed dryer, indirect dryer, solvent extraction.

**rotary drying** The use of rotary dryers to reduce the moisture content of sludge to about 10%.

**rotary kiln** A furnace or incinerator consisting of a rotating, horizontal cylinder lined with refractory materials and an afterburner. Rotary kilns are used for the destruction of solid and hazardous wastes, the incineration of wastewater sludge, and the regeneration of spent activated carbon.

**rotary kiln furnace** A furnace used in the reactivation of spent adsorbents. It has low equipment and energy costs, but poor mass transfer characteristics.

**rotary kiln incinerator** A rotary kiln; an incinerator with a rotating combustion chamber that keeps waste moving, thereby allowing it to vaporize for easier burning.

**rotary-lobe positive-displacement blower** A commonly used blower for the aeration of wastewater. *See* positive-displacement blower.

**rotary-lobe pump** A positive-displacement pump with two rotating lobes that push fluid (wastewater or sludge) through and are easier to replace than a stator and rotor. Same as rotary pump.

**rotary-motion valve** Rotary-motion valves, as defined by the type of throttling element movement, include ball, butterfly, and plug valves. *See also* linear-motion valve.

**rotary process** A method of well sinking using a rotating hollow drill rod; rotary drilling.

**rotary pump** A self-priming displacement pump that moves fluids by the action of two rotary elements, alternatively drawing and discharging. It is usually installed for low pressures and small discharges.

**rotary-screen thickener** *See* rotary-drum thickener.

**rotary sludge drying** *See* rotary dryer.

**rotary surface washer** A device with arms attached to a rod, used to clean the surface of filter media during backwash. *See also* stationary surface washer.

**rotary vacuum filtration** A common method of mechanical sludge dewatering. The equipment consists of a continuous belt of fabric that winds around a horizontal rotating drum partially immersed in a sludge tank. *See also* cake filtration equation.

**rotary vacuum precoat filter** A rotary vacuum filter that has the filter medium covering the entire drum face. Before initiating filtration, the slurry tank is filled with a precoat material.

**rotary valve** A spherical valve with a rotating gate.

**Rotasieve** A rotary fine screen manufactured by Jones & Attwood, Inc. with external feed.

**rotating biological contactor (RBC)** A fixed-film device developed in Europe and used for biological treatment of wastewater. It consists of large, closely spaced plastic discs that provide a high specific area and rotate slowly, through wastewater and through air, about a horizontal shaft in a basin. A biological slime or biofilm—mainly bacteria, protozoa, and fungi—develops on the inert surfaces of the discs to decompose the organic matter in the waste. A typical RBC plant also includes primary and final clarifiers, no recirculation through the RBCs but return of excess biological sludge to the primary clarifiers for removal. The RBC process is similar to trickling filtration and may be designed on the same models or using the following equation:

$$Q(I - E) = PAE/(K + E) \qquad \text{(R-57)}$$

where $Q$ = influent flow, $I$ = influent BOD, $E$ = effluent BOD, $A$ = total disc area, and $K$ and $P$ = experimental constants. Also called biodisk. *See also* trickling filter, biological tower, loping.

**rotating distributor** *See* rotary distributor.

**rotating-drum thickener** Same as rotary-drum thickener.

**rotating reactor** The reactor containing rotating biological contactors.

**rotating screen** A device used in sludge composting plants to recover bulking agents while improving the appearance of the compost.

**rotational speed formula** A formula developed to determine the rotational speed ($N$, rpm) of a trickling filter's rotary distributor:

$$N = 16.67\,q(1 + R)/(AD_r) \qquad \text{(R-58)}$$

where $q$ = influent hydraulic loading, $m^3/m^2$/hr; $R$ = recycle ratio; $A$ = number of arms in the distributor assembly; and $D_r$ = dosing rate, mm/pass of distributor arm.

**rotavirus** A wheel-shape, excreted pathogen of the family Reoviridae that causes viral diarrhea and vomiting. The most important agent of infantile gastroenteritis, with an incidence of 10–30%, rotaviruses are shed in large numbers, up to $10^{10}$ particles per gram of feces and transmitted via the fecal–oral route.

**rotavirus gastroenteritis** Infection by rotaviruses, a major childhood disease, characterized by vomiting and, sometimes, diarrhea, fever, and dehydration. It is sometimes associated with pathogenic *E. coli.*

**rotavirus infection** Rotavirus gastroenteritis.

**Rotex** A rotating apparatus manufactured by Simon-Hartley, Inc. for grit removal from wastewater.

**rotifer** Any minute, multicellular aquatic animal having rings of cilia on the anterior end. Rotifers are aerobic heterotrophic animal eukaryotes that feed on bacteria and small protozoa, thus contributing to the decomposition of organic matter in wastewater. They can be found in activated sludge and in biofilms. As they require more dissolved oxygen than lower organisms, their presence indicates a certain level of waste stabilization.

**Rotifera** The phylum or class of rotifers.

**Rotobelt** An in-channel fine screen manufactured by Dontech, Inc.

**Roto-Brush** A device manufactured by Dontech, Inc. to clean rotary screen brushes.

**Roto-Channel** An apparatus manufactured by Dontech, Inc., combining a bar screen with a compacting conveyor.

**RotoClean** A device manufactured by Parkson Corp. for washing screenings.

**RotoClear** A microscreen manufactured by Walker Process Equipment Co.

**Rotoco®** A continuous granular filter manufactured by Eimco Process Equipment Co.

**Rotodip®** A volumetric feeder manufactured by Leeds & Northrup.

**RotoDip** A pipe skimmer manufactured by Walker Process Equipment Co., including a manual control.

**Rotodisintegrator** A device manufactured by Zimpro Environmental, Inc. for grinding debris.

**Roto-Drum** A rotary fine screen-thickener manufactured by Dontech, Inc., with internal feed.

**rotodynamic pumps** Continuous-flow pumps with a rotating element called an impeller, enclosed in a casing that connects to the pipeline; radial-flow or centrifugal, axial-flow or propeller, and mixed-flow pumps. Rotodynamic pumps, also called kinetic pumps, used to handle sludge include grinder pumps, nonclog mixed-flow pumps, and recessed-impeller pumps. The other broad category of pump is positive displacement.

**Rotofilter** A rotary fine screen manufactured by Sepra Tech, with external feed.

**Roto-Guard** An apparatus manufactured by Parkson Corp., combining screening and thickening in a horizontal drum.

**Rotoline** A rotary distributor manufactured by FMC Corp. for trickling filters.

**Rotomite** A sludge handling dredger manufactured by Crisafulli Pump Co.

**Rotopac®** A screw screenings compactor manufactured by John Meunier, Inc.

**Rotopass** A rotary fine screen manufactured by Passavant-Werke AG, with external feed.

**RotoPress** A screenings compactor manufactured by Parkson Corp.

**Rotopress®** A screw compactor manufactured by Andritz-Ruthner, Inc.

**Roto-Press-1** An apparatus manufactured by Dontech, Inc., combining fine screening and press dewatering.

**Roto-Press-2** A screenings compactor manufactured by Roto-Sieve AB.

**rotor** (1) Generally speaking, it is a rotating part of a machine. *See also* stator and impeller. It has also been defined as the rotating member of a turbine, blower, fan, axial or centrifugal pump, mixing apparatus, alternating electrical motor, propeller meter, electric generator, or motor. (2) A mechanical aeration device used in oxidation ditches. Also called a brush aerator.

**Rotordisk™** A wastewater treatment system manufactured by CMS Group, Inc.; uses rotating biological contactors.

**Rotorobic** A package treatment plant of Hycor Corp., using rotating biological contactors.

**Rotoscoop®** A self-cleaning volumetric feeder of Wyssmont Co., Inc.

**Roto-Scour** A sand filter underdrain manufactured by Graver Co.

**RotoSeal** A rotary distributor manufactured by Walker Process Equipment Co. for trickling filters.

**Roto-Sep™** An apparatus of Dontech, Inc. that provides primary wastewater treatment.

**Rotoshear** A rotary fine screen manufactured by Hycor Corp., including internal feed.

**Roto-Sieve®** A rotary fine screen manufactured by Roto-Sieve AB, including internal feed.

**Roto-Skim** A rotary pipe skimmer manufactured by Envirex, Inc.

**Rotosludge®** A drum rotary sludge thickener manufactured by Hycor Corp.

**Rotospir®** A shaftless screw conveyor of Andritz-Ruthner, Inc.

**Rotostep** In-channel bar screen manufactured by Hycor Corp.

**Rotostrainer** A rotary fine screen manufactured by Hycor Corp.

**Rotosweep** A surface agitating device manufactured by Roberts Filter Manufactuuring Co. for use in filters.

**Rototherm®** A thin film evaporator manufactured by Artisan Industries, Inc.

**Roto-Thickener™** A sludge thickener manufactured by Dontech, Inc. as a rotary drum.

**rototiller** *See* rototilling.

**rototilling** Tilling the soil with a rototiller (a motorized device having spoke-like spinning blades perpendicular to the ground), as it is done for sludge that is crumbled and combined with solid waste shreddings for windrow composting.

**Roto-Trak** Components manufactured by Komline-Sanderson for use in gravity sludge dewatering units.

**Roto-Trols** A pressure-operated device manufactured by Healy-Ruff Co. to control pump operation.

**Rotox** A submersible apparatus manufactured by Davis Water & Waste Industries, Inc. to provide aeration and mixing.

**rough fish** Fish not prized for eating, such as gar and suckers. Most are more tolerant of changing environmental conditions than game fish. *See also* tolerant fish.

**roughing filter** (1) A high-rate filter used as a first or intermediate treatment step for the partial removal of turbidity from drinking water. Also called contact filter, contact roughing filter, or preliminary filter. (2) A trickling filter, usually with plastic packing, used as pretreatment or a first-stage treatment to remove approximately half of the organic load of concentrated wastewaters; it operates at higher loadings than high-rate trickling filters. *See also* high-rate trickling filter, intermediate-rate trickling filter, low-rate trickling filter, roughing filter/activated-sludge process, super-rate filter, biofilter, biotower, oxidation tower, dosing tank, filter fly.

**roughing filter–activated sludge (RF–AS) process** Same as the roughing filter–plain aeration process, except that in the RF–AS settled sludge from the final clarifier is recycled to the aeration basin. *See* other dual processes under combined filtration–aeration.

**roughing filter–plain aeration (RF–PA) process** A method of biological wastewater treatment that uses a trickling filter with direct recirculation ahead of a plain aeration basin, which is followed by a final clarifier. All of these three units are usually designed with reduced dimensions compared to conventional units. *See* other dual processes under combined filtration–aeration.

**roughing tank** A settling tank of short detention time used to remove coarse suspended solids or grease from wastewater.

**roughness** *See* roughness coefficient.

**roughness coefficient** or **roughness factor** Roughness is a characteristic of channels and conduits related to their resistance to fluid flow. Roughness retards the flow, increases the potential for infiltration, and decreases erosion. The roughness factor indicates the roughness of the channel or conduit (as a result of fabrication, use, biological growth, etc.), indicates the effects of roughness on energy losses in the flowing fluid, and is used as an empirical coefficient in several hydraulic formulas (Bazin's coefficient "$\beta$," Chézy's "$C_z$," Darcy–Weisbach's friction "*f*," Hazen–Williams' friction "*C*," Kutter's roughness "*n*," Manning's roughness "*n*"). The roughness factor in the Colebrook–White equation is also called equivalent sand grain, roughness element magnitude, roughness size, or sand grain size.

**roughness factor** Same as roughness coefficient.

**rounded-crest weir** A weir that has a crest curved or rounded upward in the direction of flow, entirely or only at both ends.

**round-nosed weir** A broad-crested weir similar to the rectangular weir or the rounded-crest weir, except that it has only the downstream end rounded.

**round-robin test** The analysis of a water or wastewater sample of known composition by various laboratories to assess their competence.

**roundworm** (1) A nematode of the family Ascaridae. Distributed worldwide, it is transmitted from soils contaminated by the feces of an infected person. *See Ascaris lumbricoides* life cycle. *See also* flatworm. (2) The common name for ascariasis, a latent and persistent infestation with *Ascaris lumbricoides* or other ascarids.

**route of exposure** The avenue by which a chemical comes into contact with an organism (e.g., inhalation, ingestion, dermal contact, injection). *See also* fecal–oral route.

**routine sample** A sample collected to comply with the maximum contaminant levels of the National Drinking Water Regulations.

**Rover** Products of Hach Co. for rust removal.

**Roxidizer®** An apparatus manufactured by Tellkamp Systems, Inc. for the control of volatile organic compounds.

**R plasmid** A plasmid indicative of antibiotic resistance. Also called R factor.

**rpm or RPM** Acronym of revolution(s) per minute.

**rps** Acronym of revolution(s) per second.

**R & R** Renew and replace

**RSDS** Acronym of rapid sludge dewatering system, a vacuum-assisted apparatus manufactured by U.S. Environmental Products, Inc.

**RSI** Acronym of Ryznar saturation index.

**$R_3Sn^+$** Symbol of tryalkyltin compounds.

**R/S ratio** A parameter expressing the rhizosphere effect (numbers of microorganisms decrease from the rhizoplane outward toward bulk soil); equal to the ratio of the number of a given microorganism in the rhizosphere (*R*) to its number in the bulk soil (*S*). This ratio varies from 2 for protozoa, to 12 for fungi, to > 100 for ammonifiers, and > 1200 for denitrifiers.

**RSSCT** Acronym of rapid small-scale column test.

**r-strategist** Same as copiotroph or zymogenous organism.

**RTD curve** *See* residence time distribution curve.

**RTO** Acronym of regenerative thermal oxidation or regenerative thermal oxidizer.

**RT-PCR** Acronym of reverse transcriptase–polymerase chain reaction.

**RTU** Acronym of remote terminal unit.

**rubber barges in tow** A surface vessel system used to transport sludge from a treatment plant to an ocean disposal site; it consists of a tug towing an inflatable, partly submerged rubber cylinder and an apparatus for discharging the sludge. *See also* articulated tug barge, barges-in-tow, barge transport, and self-propelled sludge vessel.

**rubber industry waste** Wastewater generated during the production of the various types of rubber: (a) natural rubber from latex or rubber-plant sap, (b) synthetic rubber from copolymerization, (c) scrap rubber, and (d) rubber-like plastics. The wastewater originates from latex washing, rubber coagulation, and discarded impurities; it contains high concentrations or levels of BOD, odor, suspended solids, and chlorides. It is usually treated by the appropriate combination of aeration, biological methods, coagulation, sulfonation, and activated carbon adsorption.

**rubbish** Solid waste, excluding food waste and ashes, from homes, institutions, and workplaces.

**rule of thumb** A rule followed by engineers based on average conditions in common practice instead of specific computations; for example, (a) at a manhole where a sewer changes size, match inverts, keep crowns continuous, or keep the 0.8 D line continuous (D representing sewer diameter); (b) a ratio of COD:N:P = 100:5:1 on a mass basis for the activated sludge process—if the waste is deficient, nutrient supplements should be provided; (c) size air supply in sludge stabilization to provide 20–30 $m^3$/min per 1000 $m^3$ of mixing tank volume; in the operation of an activated sludge plant, a period of three sludge ages is recommended to see the results of a change in sludge age. *See also* Haber's law.

**run** (1) The continuous time period of a unit operation, a process, or a test. A model run is a single execution of a model, i.e., a single performance of all the computational and administrative procedures of the model. Model runtime or model execution time is the time required to complete one model run on the computer, i.e., the time required to perform all these procedures. Run time is a function of both the model complication and the speed of the computer used. *See also* run time (of pumps). (2) A succession of similar events preceded and followed by events of a different kind, e.g., a time series of stream flows with periods of deficit and surplus with respect to a reference value. (3) A shallow, natural stream that usually flows continuously, but is turbulent and swift. It is not as large as a river or a creek, but not as small and intermittent as a streamlet. *See also* stream.

**rundown screen** Same as static screen.

**running trap** A double-bend trap with a water seal at the bottom to prevent the reflux of gases. Also called siphon trap.

**running-water habitat** A stream, river, brook, creek, rill, etc., as compared to a standing-water habitat.

**run-of-bank sand** Natural granular material from a river bank, used in the early slow sand filters, without screening. Modern sand filters meet size and uniformity criteria by the elimination of coarse grains and fines. *See* coefficient of uniformity and effective size.

**runoff** That part of precipitation, snowmelt, or irrigation water that runs off the land into streams or other surface water. It can carry pollutants from the air and land into receiving waters. Any rainwater, leachate, or other liquid that drains over land from any part of a facility (EPA-40CFR241.101-q and 40CFR260.10). Basically, runoff is the part of precipitation that eventually reaches the surface streams, including direct runoff (overland flow plus interflow) as well as groundwater runoff, but excluding basin recharge (interception, depression storage, soil moisture) and other losses. *See also* rainfall–runoff relationship for the definition of these terms. Other runoff terms are: base runoff, channel-phase runoff, cumulative runoff, delayed runoff, direct runoff, groundwater flow, groundwater runoff, mass runoff, mean annual runoff, overland runoff, runoff depth, runoff intensity, runoff rate, subsurface runoff, surface runoff, urban runoff.

**run-of-the-river reservoir** A reservoir with a large flow rate compared to its volume. *See also* storage reservoir.

**run time** The period of time that a device, piece of machinery, unit, process, etc. is under operation. Pump runtime is an important output in sewer system modeling applications. *See also* model runtime, NAPOT.

**rupture disk** A diaphragm designed to burst at a predetermined pressure differential.

**rural waste disposal** *See* on-site wastewater management.

**rural water survey** A survey conducted by the USEPA under the Safe Drinking Water Act to determine the quantity and quality of rural water supplies.

**rush** Any emergent grasslike plant of the genus *Juncus,* having pithy or hollow stems, found in natural wetlands.

**Russell's ionic strength approximation** A relation between the specific conductance or electrical conductivity ($EC$, micromhos/cm) and the ionic strength ($I$) of a solution:

$$I = EC/62{,}500 \qquad \text{(R-59)}$$

*See also* specific conductance and Langelier's ionic strength approximation.

**Russian degree** A unit of hardness, expressed in mg/L of CaO, equivalent to 2.5 mg/L as $CaCO_3$. *See also* American degree, British degree, French degree, German degree.

**rust** (1) The red or orange coating formed on the surface of iron as a result of electrochemical action of atmospheric oxygen and moisture; it consists of ferric hydroxide and ferric oxide. Rust accumulations in pipelines reduce their hydraulic capacity; a variety of tools (e.g., pigs with wire brushes) are used to clean the lines and restore their capacity. (2) Particles of ferric iron, suspended in water as ferric oxide ($Fe_2O_3$) or ferric hydroxide [$Fe(OH)_3$].

**rusty water** Water that is colored red as a result of the corrosion of a bronze faucet attached to an iron pipe. *See* red water.

**ruthenium-103** A radionuclide of the steel-gray, rare metallic element ruthenium, of the platinum group, with a half-life of 40 days. Ruthenium's atomic weight = 101.07, atomic number = 44, and specific gravity = 12.2 at 20°C. It is sometimes found in water sources.

**Ryznar index** Same as Ryznar stability index.

**Ryznar saturation index** Same as Ryznar stability index.

**Ryznar stability index (RSI)** An empirical index used to evaluate the corrosion or scaling potential of water, i.e., whether a water tends to dissolve precipitated calcium carbonate ($CaCO_3$) or to precipitate calcium carbonate:

$$RSI = 2\ pH_s - pH \qquad (R\text{-}60)$$

where $pH_s$ and pH represent, respectively, the saturation pH for calcium carbonate and the actual pH of the water. Scale formation is expected to be heavy when RSI is less than 5.5 and average for $5.5 < RSI < 6.2$; no scale or corrosion is expected for $6.2 < RSI < 6.8$; but water is moderately aggressive for $6.8 < RSI < 8.5$ or very aggressive for $RSI > 8.5$ (Metcalf & Eddy, Inc., 2003). Also called Ryznar index, Ryznar saturation index. *See* corrosion index for a list of other corrosion control parameters.

# S

**SAB Reactor** A package wastewater treatment plant manufactured by Biosab, Inc.

**SAC** Acronym of starved air combustion.

**S.A.C.™** Acronym of sludge age control, a system designed by United Industries, Inc.

**SAC resin** *See* strong acid cation resin.

***Saccharomyces cerevisiae*** A yeast that effectively combines with *Arthrobacter* to biodegrade petroleum hydrocarbons. *See* brewer's yeast.

**saccharophylic** Characteristic of a microorganism that tolerates a high concentration of sugar.

**sacrificial anode** An easily corroded material such as magnesium, aluminum or zinc, deliberately installed in a tank or pipe. The intent of such installation is to give up this anode to corrosion while the water supply facilities remain relatively free of corrosion. The duration ($t$) of a sacrificial anode of mass ($m$) depends on the current ($i$) in coulombs/sec and on the equivalent weight ($w$) of the metal:

$$t = EFm/(iw) \qquad \text{(S-01)}$$

where $E$ is an efficiency factor (0.90 for zinc and 0.50 for magnesium) and $F$ is the Faraday constant (96,500 coulombs/equivalent weight). *See also* galvanic protection and electrolytic cathodic protection.

**sacrificial anode system** A cathodic protection apparatus using a sacrificial anode.

**sacrificial galvanic anode** *See* sacrificial anode.

**saddle** (1) A structure that supports a pipe or penstock above ground or an assembly of metal straps around a pipe at a point of connection. (2) A depression in a ridge that connects two higher elevations.

**safe** Condition of exposure under which there is a "practical certainty" that no harm will result in exposed individuals.

**safe drinking water** Water that does not pose a risk to human health, i.e., water that does not contain pathogens or toxic materials. *See* potable water.

**Safe Drinking Water Act (SDWA)** A 1974 act that requires the USEPA to (a) establish a drinking water priority list of contaminants that may adversely affect human health and to promulgate regulations for 25 new contaminants every three years, (b) set standards for maximum contaminant levels and maximum contaminant level goals of certain chemical and bacteriological pollutants in public drinking water systems, and (c) regulate underground injection systems.

**Safe Drinking Water Act (SDWA) Amendments** The SDWA was amended in 1977, 1979, 1980, 1986, 1988, and 1996.

**Safe Drinking Water Act (SDWA) Amendments of 1986** Amendments of the 1974 Act, in particular requiring the USEPA to monitor specific contaminants, set surface water filtration criteria, require disinfection of public water supplies, and establish programs for wellhead protection.

**Safe Drinking Water Act (SDWA) Amendments of 1996** The 1996 amendments require states to establish Source Water Assessment Programs, delineate source water protection areas, inventory significant contaminants in these areas, and determine the susceptibility of public water systems to contamination. These amendments also include other measures such as monitoring of backwash recycle.

**Safe Drinking Water Act Rules** Regulations issued by the USEPA pursuant to the Safe Drinking Water Act.

**Safe Drinking Water Hotline** A USEPA telephone hotline for information about the Safe Drinking Water Act requirements.

**Safe Drinking Water Information System** A computerized information system of the USEPA containing data on public water systems.

**safety** Freedom from unacceptable risk; security.

**safety factor (SF)** A factor used in the design of biological system to take into account a number of uncertainties such as variability of wastewater flow, waste characteristics, and weather conditions. Applied to a design parameter such as solids retention time (Θ, days), the safety factor is taken as the ratio of the value actually used in design ($\Theta_d$) to the lowest value at which operation is possible ($\Theta_m$):

$$SF = (\Theta_d)/(\Theta_m) \qquad \text{(S-02)}$$

SF is at lest 2.5 for nitrification; it varies from 4 to more than 20 for suspended growth processes and from 2 to 4 for sludge digestion. *See also* uncertainty factor.

**safety relief valve** Same as safety valve.

**safety valve** A valve normally closed that automatically opens when prescribed conditions are exceeded in a pipeline or other closed container of fluids. For example, an automatic valve that opens when pressure becomes excessive in a steam boiler or a pressure cooker. Also called safety relief valve. *See also* pressure-relief device and relief valve.

**safety zone** On a graph of pathogen survival as a function of temperature and time, the safety zone is defined by all the time–temperature combinations that indicate complete destruction of some common pathogens: enteric viruses, *Shigella, Taenia, Vibrio cholerae, Salmonella, Ascaris,* and *Entamoeba histolytica,* with the possible exception of spore-forming bacteria (e.g., *Clostridium perfringens*) and hepatitis-A virus. *See* Figure S-01.

**safe velocity** The fluid velocity that will not scour a conduit while maintaining solids in movement.

**safe water** Water that does not contain harmful bacteria, toxic materials, or chemicals, and is considered safe for drinking even though it may have taste, odor, color, and certain mineral problems. Drinking water should be not only safe, but also palatable, i.e., free of these problems.

**safe yield** The annual amount of water that can be taken from a source or supply over a period of years without depleting that source beyond its ability to be replenished naturally in "wet years." Safe yield is sometimes taken as the minimum yield recorded in the past or defined as the maximum dependable draft that can be made continuously upon a source of surface or ground water supply over a given period of time, during which the probable driest period and, therefore, period of greatest deficiency in water supply, is likely to occur. Dependability is relative and is a function of storage provided and drought probability. The safe yield of a surface stream is close to its average flow with adequate storage; otherwise it is its lowest dry-weather flow. The concept of safe yield does not imply 100% reliability or zero risk. Also called firm yield. *See also* groundwater yield, sustainable yield.

**Safgard** Products of Schlueter Co. for use in rotary fine screens.

**sag curve** *See* dissolved-oxygen sag analysis.

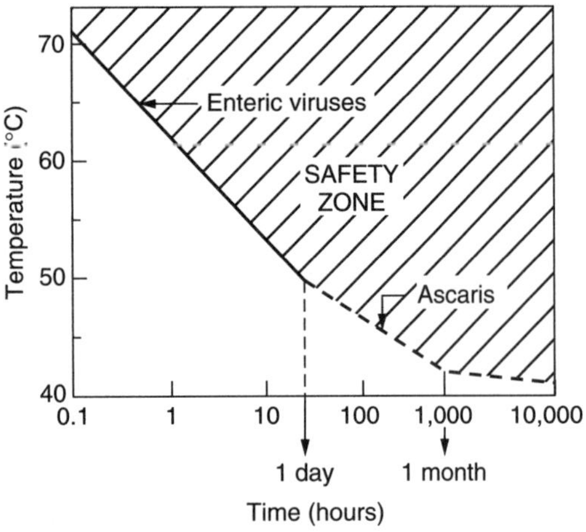


**Figure S-01.** Saftey zone (adapted from Feachem et al., 1983).

**sal ammoniac** Another name for ammonium chloride ($NH_2Cl$).

**salimeter** An instrument used to measure the density or salt content in a solution as sodium chloride (NaCl), a 100% salinity corresponding to 26.4% salt by weight at 60°F = 15.56°C. Also called salometer. *See also* hydrometer, salinometer.

**salination** *See* salinization.

**saline** (1) Pertaining to or containing salt. *See also* brackish. (2) A salty solution; salty water. (3) A sterile solution of sodium chloride (NaCl) used to dilute medication or for intravenous therapy.

**saline contamination** Contamination of a water supply by saline water.

**saline estuarine water** Those semienclosed coastal waters that have a free connection to the territorial sea, undergo net seaward exchange with ocean waters, and have salinities comparable to those of the ocean. Generally, these waters are near estuaries and have cross-sectional annual mean salinities greater than 25 parts per thousand (EPA-40CFR125.58-q). *See* salinity and Figure S-02.

**saline intrusion** The penetration of saltwater into a freshwater body; a salt wedge. *See* saltwater intrusion for more detail. *See also* Figure S-02.

**saline–sodic soil** Soil formed from the combination of salinization and alkalinization, characterized by a saturated extract conductivity higher than 4.0 dS/m and an exchangeable sodium percentage greater than 15.

**saline solution** A solution that contains a concentration of dissolved solids between 10,000 and 30,000 mg/L. *See* saline water.

**saline spring** A spring whose water has a significant concentration of sodium chloride (NaCl) and other salts.

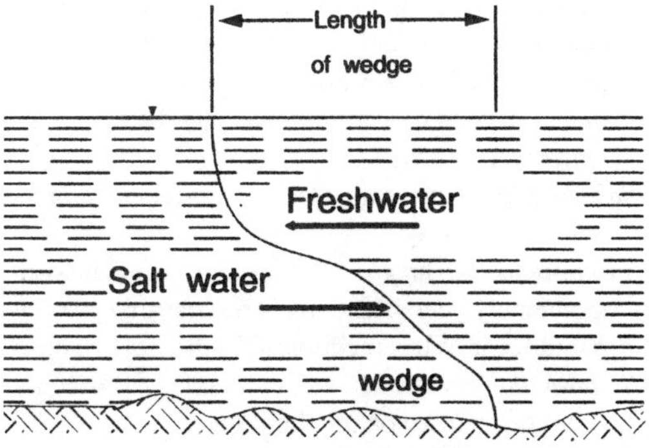


**Figure S-02.** Saline intrusion (courtesy CRC Press).

**saline water** Water with a fairly high mineral content as expressed by a concentration of dissolved solids between 10,000 and 30,000 mg/l as compared to brackish water, brine, desalination, freshwater, saline estuarine waters, and sea or ocean water. Also called saltwater. *See* salinity.

**saline water conversion** The production of freshwater from saline or brackish waters.

**saline water conversion classification** Saline water conversion processes are commonly referred to synonymously as demineralization (demineralizing) and desalination (desalinization, desalting). However, there is a basic difference between two groups of processes: demineralizing or desalting techniques remove minerals or salts from the saline water, whereas water-extracting or water-recovery techniques separate water from the saline water. *See* Table S-01 for a summary of common processes.

**saline water intrusion** Same as saltwater intrusion.

**salinity** (1) The presence or relative concentration of dissolved salts (mostly sodium chloride, NaCl) or dissolved minerals in water. Expressed as mg/L of chlorine or mg/L of total dissolved solids. More generally, a measure of the concentration of dissolved mineral substances in water, as determined by measuring the density of the solution using a salinometer. Electrical conductivity is used as a surrogate for salinity in determining the suitability of treated wastewater for irrigation. Salinity ($S$) may be computed from chlorinity ($C$) through one of these two relationships:

$$S = 0.03 + 1.805\ C \qquad \text{(S-03)}$$

$$S = 0.07 + 1.811\ C \qquad \text{(S-04)}$$

*See also* seawater composition, specific conductance, total dissolved solids, and $EC_{dw}$. Table S-02 shows one classification of water sources according to salinity. (2) More formally, the dissolved solids content of salt water in parts per thousand (‰) by weight after the oxidation of all carbonates and organics and the substitution of chlorides for all bromides and iodides.

**salinity intrusion** Same as saline intrusion.

**salinity of drainage water** *See* the symbol $EC_{dw}$.

**salinity of irrigation water** *See* the symbols $EC_w$ and $EC_{dw}$.

**salinity of saturation extract** *See* the symbols $EC_e$ and $EC_{dw}$.

**salinity–surface tension correction factor** A factor used to correct the clean-water oxygen transfer rate for solubility differences due to salts, particu-

**Table S-01.** Saline-water conversion classification

| Group | Constituent removed | Phase to which transported | Process |
|---|---|---|---|
| Desalting (demineralizing) | Salt | Vapor | None |
| | | Liquid | Electrodialysis |
| | | | Osmionic separation |
| | | | Thermal diffusion |
| | | Solid | Ion exchange |
| | | | Adsorption on carbon electrode |
| Water extraction (water recovery) | Water | Vapor | Distillation |
| | | Liquid | Solvent extraction |
| | | | Reverse osmosis |
| | | Solid | Freezing |
| | | | Contact freezing |
| | | | Adsorption |

Adapted from Fair et al. (1971).

lates, and surfactants in the water or wastewater sample. Also called the $\beta$ (beta) factor.

**salinity wedge** Same as salt wedge.

**salinization** The process of salt accumulation in a soil as a result of irrigation with brackish water, excessive irrigation of an arid area, the capillary flow of saline groundwater, or other causes.

**salinometer** An instrument used to determine the salinity of a solution through its electrical conductivity. Also called salt gage. *See also* salimeter, salometer, salinity.

***Salmonella*** Any rod-shaped, facultatively anaerobic bacteria of the genus *Salmonella* that may enter the digestive tract of humans through contaminated food and cause gastrointestinal illness, including severe diarrhea, abdominal pains, typhoid, and paratyphoid; one of the most common pathogens found in domestic wastewater. More than 2200 known serotypes. (Named after the American bacteriologist Daniel E. Salmon, 1850–1914.)

**salmonella bacteria** *See* Salmonella.

***Salmonella cholerae-suis*** A species of bacteria found in the intestines of warm-blooded animals and causing the same type of infection as *Salmonella enteritidis,* gastroenteritis, and *Salmonella* septicemia.

***Salmonella dublin*** A species of bacteria causing the same type of infection as *Salmonella enteritidis.*

**salmonellae** Plural of salmonella.

***Salmonella enteritidis*** A species of bacteria causing salmonellosis or enteric fever, sometimes accompanied by pyrogenic lesions of internal organs, particularly in patients with impaired resistance to infections.

***Salmonella hirshfeldii*** A species of bacteria, also called *S. paratyphi* C, causing the same type of infection as *Salmonella enteritidis.*

***Salmonella* microsome assay** A test used to infer potential carcinogenicity on the basis of mutagenic activity (specifically reverse mutations) induced by an agent or to determine whether a carcinogenic chemical acts through a genotoxic mechanism. The test uses microsomal or S-9 fractions of *Salmonella typhimurium* bacteria. *See* Ames test for detail.

***Salmonella*** microsome test Same as *Salmonella* microsome assay.

***Salmonella paratyphi*** A species of nonlatent, moderately persistent bacteria causing paratyphoid fever; it has medium-to-high infective dose.

***Salmonella paratyphi A*** A waterborne bacterial agent that causes paratyphoid fever.

***Salmonella paratyphi B*** *See Salmonella schottmuleri.*

**Table S-02.** Salinity classification of water sources

| Classification | Also called | Dissolved solids concentration, mg/L |
|---|---|---|
| 1. Fresh water | | < 1,000 |
| 2. Mildly brackish water | Mildly saline | 1,000–2,000 |
| 3. Brackish water | Moderately saline | 2,000–10,000 |
| 4. Saline water (saltwater) | Severely saline | 10,000–30,000 |
| 5. Saline estuarine water | | > 25,000 |
| 6. Seawater (ocean water) | | 30,000–36,000 |
| 7. Brine | | > 36,000 |

***Salmonella paratyphi C*** *See Salmonella hirshfeldii.*

***Salmonella schottmulleri*** One of three main groups of bacteria causing paratyphoid fever. Also called *Salmonella paratyphi B.*

***Salmonella sendai*** A species of bacteria causing the same type of infection as *Salmonella enteritidis.*

***Salmonella* septicemia** The invasion and persistence of *Salmonella* bacteria in the bloodstream of patients infected by *S. typhi* and *S. paratyphi;* enteric fever.

***Salmonella typhi*** The waterborne bacterial agent that causes typhoid fever, the most severe and serious infection of the *Salmonella* genus. Excreted load = $10^8$/gram of feces. Persistence = 2 months. Median infective dose > $10^6$.

***Salmonella typhimurium*** A strain of bacteria used in the Ames test and characterized by the inability to produce the essential amino acid histidine and by susceptibility to mutagens. It is the most common species of bacteria causing salmonellosis, the same type of infection as *Salmonella enteritidis.*

***Salmonella typhosa*** A waterborne bacterial agent that causes a fever of the typhoid type. It is also transmitted by food, insects, fomites, and shellfish.

**salmonellosis** Food poisoning caused by salmonella bacteria and characterized by a sudden onset of fever, vomiting, diarrhea, and abdominal pains; the most common infection associated with the *Salmonella* genus. Also called paratyphoid fever. *See also* Widal reaction.

**salometer** Same as salimeter.

**salomonid fish** A species of fish found in cold, freshwater, requiring a minimum dissolved oxygen concentration varying from 5.0 mg/L at 25°C to 8.0 mg/L at 0°C. A related effluent limitation is 0.02 mg/L of un-ionized ammonia.

**SAL-PROC™** A patented desalination process that uses sequential or selective extraction to recover beneficial salts from brackish water or seawater. *See* salt solidification and sequestration for detail.

**sal soda** *See* sodium carbonate.

**salt** (1) A mineral that water picks up as it passes through the air, over and under the ground, or from households and industry. Some natural waters are unpalatable because of an excess of chlorides (e.g., 3000 mg/L) or sulfates (e.g., 1500 mg/L) and may have laxative effects on new users. *See also* total dissolved solids. (2) An ionic compound resulting from an acid–base mixture; e.g., the reaction between hydrochloric acid (HCl) and sodium hydroxide (NaOH) produces the salt sodium chloride (NaCl) and water ($H_2O$). Also defined as (a) the compound formed by replacing the hydrogen ion of an acid by a metal; named according to the metal and the acid, or (b) a compound of a positively charged cation other than $H^+$ and a negatively charged anion other than $OH^-$:

$$HCl + NaOH \rightarrow NaCl + H_2O \qquad \text{(S-05)}$$

(3) The common salt or sodium chloride (NaCl).

**saltation** (1) The transportation of clastic (broken or fractured) sediments in air or water by intermittent leaps or bounds. (2) A sudden change or discontinuity in a line of descent; a mutation.

**salt bridging** *See* bridging.

**salt cake** A substance that contains 90–99% sodium sulfate ($NA_2SO_4$), produced, along with muriatic or hydrochloric acid (HCl), from rock salt heated with sulfuric acid ($H_2SO_4$); used in the manufacture of glass, ceramic glazes, soaps, and sodium salts.

**salt dosage** The relative quantity of regenerant used in ion exchange, expressed, e.g., in pounds of regenerant per cubic foot of ion exchange bed. Also called regeneration level.

**salt efficiency** A measure of the performance of a water softener in terms of its capacity to remove hardness, expressed, e.g., in milligrams or grains of hardness removed per kilogram or pound of salt used. *See also* operational salt efficiency.

**salt flux** The mass of salt (amount of dissolved substances) that passes through a membrane per unit area per unit time in a membrane filtration process such as reverse osmosis; expressed in pounds per square foot per second or grams per square meter per second. Also called solute flux.

**salt gage** An instrument used to determine the salinity of a solution through its electrical conductivity. Also called salinometer. *See also* salimeter.

**salt gradient pond** An evaporation pond designed to capture and retain solar energy by keeping warm water at the bottom, thus preventing or reducing the loss of energy due to turnover; used, for example, in zero liquid discharge schemes.

**salt of phosphorus** A common name of sodium ammonium phosphate ($NaNH_4HPO_4 \cdot 4\ H_2O$).

**salting out** The phenomenon observed in an aqueous solution in which the addition of a salt decreases the solubility of a solute. It is used in water analysis to increase the extraction efficiency of some compounds. *See* salting-out effect.

**salting-out constant ($K_s$)** *See* salting-out effect. Also called Setschenow constant. For example, $K_s$ = 0.195 L/mol for benzene at 20°C.

**salting-out effect** The decrease in the solubility of a volatile component A of an aqueous solution due to a high concentration of dissolved solids. This causes an increase in the activity coefficient ($\lambda_A > 1$) of this component with increasing ionic strength and an apparent Henry's law constant ($H'$) greater than the thermodynamic constant ($H$):

$$H' = H\lambda_A \quad \text{(S-06)}$$

$$\log_{10} \lambda_A = K_s I \quad \text{(S-07)}$$

where $K_s$ is the salting-out constant (in L/mol) and $I$ is the ionic strength of the aqueous solution. As an example of the salting-out effect, at 20°C in dilute water the solubility of carbon dioxide ($CO_2$) is 0.0391 mol/atm; it decreases to 0.0373 and 0.0332 mol/atm, respectively, in seawater of 1% and 3.5% salinity. *See also* Lewis and Randall correlation.

**salt marsh** A coastal marsh that is wet with saltwater or periodically flooded by the sea, and then drained as the tides rise and fall. It supports salt-tolerant grasses and plants.

**salt mass transfer coefficient** Same as salt permeability coefficient.

**Saltmaster** An apparatus manufactured by Bruner Corp. for reclaiming water softener brine.

**salt method** The injection of a salt solution of known strength and rate into water flowing under pressure to measure discharge. After dilution, the salt concentration is measured at a point downstream, and the discharge is deducted using a mass balance. *See also* dilution method, salt velocity method.

**salt mushing** *See* mushing.

**salt pan** (1) A crater, tectonic basin, or other undrained natural depression that has a salt deposit due to evaporation. (2) An area in which salt has accumulated in the soil and may be harmful to crops.

**salt passage (SP)** The fraction of salt that passes with the feedwater through a membrane and becomes part of the permeate or product water:

$$\text{SP}\ (\%) = 100\ C_p/C_f \quad \text{(S-08)}$$

where $C_p$ and $C_f$ represent, respectively, the permeate and feedwater concentrations (e.g., in mg/L). Specific solute passage is defined similarly. *See also* salt rejection.

**salt passage correction factor** A dimensionless coefficient used to calculate salt passage through a membrane by reference to the salt passage at a standard condition, depending on driving pressure and concentration gradient.

**salt permeability coefficient or constant** A parameter indicative of the flow of solute through the membrane of reverse osmosis or other pressure-driven processes. *See* solute permeability coefficient.

**salt rejection** In reverse osmosis and other pressure-driven membrane processes, salt rejection, or simply rejection is a measure of the membrane's ability to resist the passage of solutes through the membrane. It is expressed, in percent, as the ratio of salts removed to the initial salt concentration. *See* rejection, rejection rate. Salt rejection capability is a common classification parameter for pressure-driven membranes; it is high for reverse osmosis (90.0–99.9%), somewhat lower for nanofiltration, and higher for divalent than monovalent ion rejection. *See also* molecular-weight cutoff, operating pressure, osmotic pressure, and salt density index.

**salt rejection capability** *See* salt rejection.

**salt seeding** A process used in fluidized bed crystallizers and similar devices to remove the salts of the concentrate of reverse osmosis units by providing a substrate of similar characteristics. *See also* zero liquid discharge.

**salts in groundwater** Chlorides, sulfates, and other salts make water unpalatable and may cause laxative effects in new consumers.

**salt solidification and sequestration** An emerging concentrate disposal and management method (as of 2007) that includes the recovery and resale of salts as well as near-zero liquid discharge as part of a brackish water or seawater desalination project. One example is the absorption of ammonia ($NH_3$) in the brine, which is then contacted with carbon dioxide ($CO_2$) for the conversion of concentrated sodium chloride (NaCl) to such valuable products as sodium bicarbonate ($NaHCO_3$), sodium carbonate ($Na_2CO_3$), ammonium chloride ($NH_4Cl$), and magnesium chloride ($MgCl_2$). *See also* SAL-PROC™.

**salt solution** The aqueous portion of soil that contains dissolved matter leached from weathered minerals. It provides water and nutrients for plant growth. Also called soil solution.

**salt splitting** An ion-exchange process that converts salts to their corresponding acids and bases; e.g., conversion of sodium chloride (NaCl) to caustic soda (NaOH) by an anion-exchange resin or to hydrochloric acid (HCl) by a cation-exchange resin. *See* basic salt splitting, neutral salt splitting.

**salt splitting capacity** A test for determining the capacity of a used ion-exchange resin.

**salt spring** Same as saline spring.

**salt transport coefficient** Same as salt permeability coefficient.

**saltwater** Water with a fairly high mineral content as expressed by a concentration of dissolved solids between 10,000 and 30,000 mg/L as compared to brackish water, brine, desalination, freshwater, saline estuarine waters, and sea or ocean water. *See* salinity. Also called saline water.

**saltwater bulge** The body of saltwater that rises above the saltwater–freshwater interface as a result of pumping. *See* Ghyben–Herzberg principle.

**saltwater conversion** *See* saline-water conversion

**saltwater creep** The upstream movement of a saltwater wedge while freshwater continues to flow to the ocean, the two failing to mix vertically.

**saltwater encroachment** Same as saltwater intrusion.

**saltwater intrusion** The invasion of fresh, surface, or groundwater by salt water, the balance between the fresh and saline water bodies under static conditions occurring according to the U-tube principle. It typically occurs in coastal areas as seawater intrusion, as a result of excessive pumping of an aquifer. Intrusion is active when it is rapid and causes the lowering of the local freshwater potentiometric surface below the local mean sea level; it is passive otherwise. Recharge wells are sometimes used to halt saltwater intrusion. Also called saline intrusion, saline water intrusion, saltwater encroachment.

**saltwater system** A firefighting system consisting of saltwater mains.

**saltwater underrun** A rapid movement of ocean water at the bottom of a tidal estuary.

**saltwater wedge** Same as salt wedge.

**salt wedge** The volume of saltwater that intrudes into a body of freshwater as a result of density differences, as occurs in a tidal waterway where saltwater flows under freshwater. Also called saltwater wedge. *See also* saltwater intrusion.

**salt wedge estuary** Hydrodynamically speaking, an estuary that has a well-defined saltwater wedge extending upstream under the incoming freshwater, usually with large freshwater flows into a narrow estuary. *See also* partially mixed estuary, well-mixed estuary.

**salt well** A well bored or driven to obtain brine.

**salvage** The utilization of waste materials.

**salvaged water** The part of a stream or other water source that is artificially retained in the supply and made available for use.

**salvage water right** The right to use waters that would otherwise be wasted.

***Salvinia auriculata*** Scientific name of the water fern, one aquatic plant that colonizes man-made lakes and serves as a habitat for *Mansonia* spp., which transmits Malayan filariasis.

**SAM™** Acronym of status alert modem, a device manufactured by Strantrol, Inc. to monitor disinfectant dosing.

**SAMMS** Acronym of self-assembled monolayers on mesoporous supports.

**sample** A fraction of a population, material, flow, etc., collected for testing or analysis to determine the characteristics of the whole. *See also* composite, continuous, grab, proportional, representative samples.

**sample bottle** A glass or plastic bottle used to collect water or wastewater samples for examination.

**sample point** The location in a stream, water distribution network, or wastewater collection system where samples are collected.

**sample preservation** Treatment of a sample to maintain its original characteristics. Wastewater samples are better preserved by storage at 4°C and not by freezing, which affects the characteristics of solids.

**sampler** (1) A person that collects samples. (2) A device used to collect samples. *See also* scoop sampler, tube sampler.

**sample split** A portion of a sample.

**sampling** The process of collecting water or wastewater samples for analysis.

**sampling iron** A metal frame that holds bottles to collect samples from surface waters.

**sampling well** A well used to measure groundwater levels or to obtain samples for analysis, for example, at a hazardous waste management facility or Superfund site. *See also* monitoring well.

**SAN** Acronym of styrene acrilonitrile.

**SANCHO model** A computer model that simulates the behavior of hypochloric acid (HOCl) and biodegradable organic carbon in water distribution systems.

**sand** (1) A loose, gritty material of easily distinguishable grains of worn or disintegrated rock (mostly quartz or silicon dioxide), varying from about 0.05 mm to 2 mm in equivalent diameter (according to the classification of the U.S. Department of Agriculture). It is sometimes subdivided into very fine sand, fine sand, medium sand, coarse sand, and grit or fine gravel. Coarse sand particles are larger than ½ mm, whereas very fine sand particles have diameters between 0.10 mm and 0.25 mm. *See* soil classification. Sand commonly used as medium for water and wastewater filtration is characterized by its effective size of 0.45–0.55 mm and a uniformity coefficient of 1.2–1.7 for rapid filtration. *See* Table S-03. (2) A type of sediment carried by water or ice, and deposited along shores, in river beds, or in deserts.

**Table S-03.** Sand characteristics for water filtration

| | Effective size, mm | Uniformity coefficient |
|---|---|---|
| Rapid sand filtration | 0.45–0.55 | 1.2–1.7 |
| Slow sand filtration | 0.10–0.30 | 2.0–3.0 |

**sand-ballasted flocculation–sedimentation** *See* ballast and ballasted flocculation.

**sand bed** *See* sand drying bed, fluidized bed, fluidized sand bed, sludge drying bed.

**sand bed dewatering** Sludge dewatering on sand drying beds.

**sand bed** measure of the performance of a sand drying bed in terms of the weight of dry solids per unit surface area per year:

$$Y = 0.624\ ES_fS_d/(S_f - S_d) \qquad \text{(S-09)}$$

where 0.624 = a conversion factor for English customary units (1.2 for metric units); $Y$ = sand bed yield, lb/ft²/yr; $S_f$ = final percent solids concentration; $S_d$ = drained solids percent concentration (drained solids are those remaining after drainage and decanting); and $E$ = net pan evaporation, inches/month. Sand bed yield may be used to determine the bed area required.

**sand boil** (1) A difficulty sometimes encountered in the operation of a filter, when porosity and permeability differences between sand and gravel cause sand to fluidize, boil up like quicksand, and rise to the surface along with some gravel. Slow opening of backwash valves and surface washing before fluidization may help reduce the extent of this phenomenon. *See also* filter crack, mud ball, jetting, and sand leakage. (2) The phenomenon of stratification by size of the grains of a single-medium filter during backwash with fluidization, the finer grains resting on top and the coarser grains at the bottom. The tendency to stratify depends on the backwash rate and the differences in bulk densities of the various grain sizes. Also called gulf streaming, jet action. (3) A spring or geyser carrying accumulated sand or silt and caused by the pressure of floodwater against a levee. (4) A spring or geyser caused by unbalanced hydrostatic pressure in an excavation.

**sand catcher** A chamber at the upper end of a depressed sewer or at other points on combined or storm sewers; the chamber is sized and shaped to reduce the velocity of flow through it and permit the settling out of grit. Also called grit catcher. *See also* grit chamber and sand interceptor.

**Sand Dollar** A machine manufactured by Cherrington Corp. for harvesting sludge from drying beds.

**sand drying** Sludge dewatering on sand drying beds.

**sand drying bed** A water or wastewater treatment plant unit used to separate water and solids in residuals. A typical sand drying bed consists of a layer of coarse sand (about 20 cm) on which wet sludge is distributed for draining and evaporation. The sand bed is supported by a 30 cm graded gravel underdrain system. Drying beds may be open to the atmosphere or covered as in a greenhouse. They are simple to operate but require large land areas; a typical loading rate for digested sludge is 60–200 kg/m²/year. After several weeks (or months, depending on climate), the sludge cake (about 30% solids) formed on the surface is removed manually or mechanically for landfill disposal or use as a soil conditioner. The term sand drying bed is also applied to underdrained, shallow, dyked, earthen structures used for drying sludge. *See* Figure S-03. *See also* other types of drying bed.

**sand ejector** A portable device that uses water under pressure to transport sand.

**sand-filled pressure-dosed disposal field** *See* shallow sand-filled disposal field.

**sand filter** A unit consisting of beds of sand, installed underground in trenches or precast concrete boxes, or aboveground, to remove fine suspended materials and colloids from water, wastewater, or stormwater. In wastewater treatment, air and bacteria decompose additional wastes filtering through the sand so that cleaner water drains from the bed. Sand filters are commonly used to treat runoff from large buildings, access roads, and parking lots. They are also used in tertiary wastewater treatment and in sludge drying beds. A sand filter can provide some degree of preliminary treatment as a first-stage or roughing filter. *See also* granular filter media, rapid sand filter, schmutzdecke, slow sand filter.

**Sandfloat** An apparatus manufactured by Krofta Engineering Corp. for water treatment using dissolved-air flotation and sand filtration.

**sandfly** An insect that transmits the pathogens of tropical infections. *See* leishmaniasis, Oroya fever, papatasi fever, phlebotomine sandfly.

**sand incrustation** The coating and cementing of grains of the medium of a sand filter that treats inadequately stabilized water in a softening plant. Caused by precipitated calcium and magnesium, it

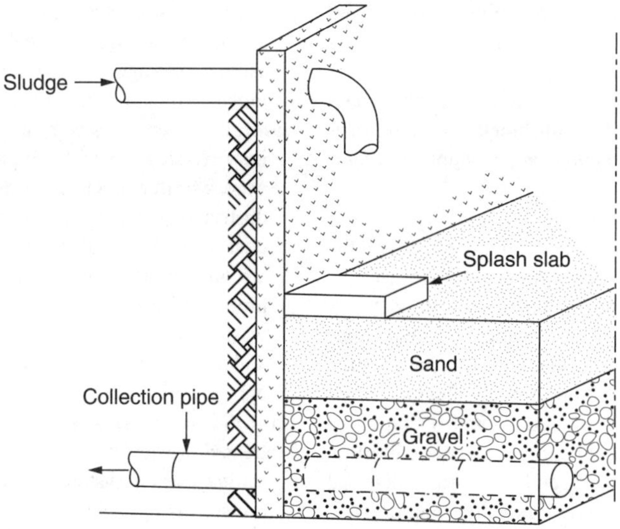


**Figure S-03.** Sand drying bed (adapted from Water Environment Federation, 1991).

can be prevented by pH adjustment to keep the metals in solution.

**sand interceptor** A device (e.g., a detention chamber) used to keep sand and other solids from entering a storm or sanitary sewer system.

**sand leakage** A difficulty sometimes encountered in the operation of a filter, when sand passes through into the underdrainage system. Proper sizing of the sand and gravel layers and careful backwashing will prevent this problem. *See also* filter crack, mud ball, jetting, and sand boil.

**sand-lined bed and fill system** A soil absorption system consisting of a bed or trenches lined with a depth of sand below and around the distribution pipes. It is used for subsurface disposal of septic tank effluent where the soil is highly permeable or the bedrock is shallow and fractured.

**SandPIPER®** A diaphragm pump manufactured by Warren Rupp, Inc.

**sand pump** A simple centrifugal-type device for pumping mud and liquids laden with sand or gravel out of boreholes, without excessive clogging or damage. It is a hollow cylinder, open at the top, with a ball or clack valve at the bottom. Also called sludger or shell pump.

**sand-pump dredge** A scow that carries a centrifugal pump for pumping sand. Also called hydraulic dredge or suction dredge.

**Sandsep®** A screw-type device manufactured by JDV Equipment Corp. for handling grit from wastewater.

**sandstone** A common sedimentary rock, usually cemented quartz sand, an aquifer material that usually yields relatively soft water of low total dissolved solids content.

**sand trap** A device installed in a conduit to trap and remove sand and soil particles from water. *See also* sediment trap (1).

**Sandwash** A hydrocyclone manufactured by Serck Baker, Inc.

**sand washer** Same as sand-washing machine.

**sand-washing machine** A device used for washing the sand that will be placed in the bed of slow sand filter or a rapid sand filter. Also called a sand washer.

**sandy** Pertaining to soil or other material that contains a large proportion of sand.

**Sanilec** An apparatus manufactured by Eltech International Corp. for the production of sodium hypochlorite.

**Sanilo™** Products of U.S. Filter Corp. for use in water treatment.

**Sani-Sieve** A static screen manufactured by Dontech, Inc., with gravity feed.

**sanitarian** A specialist in sanitation and public health; e.g., a college graduate with training in environmental health, food technology, or other physical and biological sciences. Some states require licensing of sanitarians.

**sanitary code** A code established by a government agency delineating methods of liquid and solid waste disposal, as well as measures to control air and water pollution.

**sanitary collection system** The sewer network for the collection and conveyance of municipal wastewater, including sewers, pumping stations, and their appurtenances. *See also* sanitary sewer system, storm sewer, and combined sewer.

**sanitary connection** The connection of a residence, a commercial establishment, or an industrial establishment to a public water supply or a sanitary sewer system.

**sanitary drinking fountain** A drinking fountain from which a user drinks without contact with the equipment and the unused water goes to waste.

**sanitary engineer** *See* sanitary engineering.

**sanitary engineering** The branch of civil engineering dealing with works and activities for the protection and promotion of public health, particularly the design, construction, and operation of water supply and treatment, wastewater collection and treatment, drainage, and solid waste disposal facilities. *See also* environmental engineering, public health engineering, sanitarian.

**sanitary landfill** Sanitary landfills are disposal sites for nonhazardous solid wastes, operated in accordance with environmental protection standards: wastes are spread in layers, compacted to the smallest practical volume, and covered by material (e.g., a 14-in layer of clean soil) applied at the end of each operating day to minimize blowing, fire hazards, odors, and rodent problems. The landfill also includes measures to minimize and monitor leaching to the groundwater. After reaching the design height, the landfill is decommissioned and covered with soil or grass. In addition to other municipal solid wastes, sanitary landfills are sometimes used for the disposal of biosolids, grit, screenings, and other residuals from water and wastewater treatment plants. Also called controlled tipping as a method of waste disposal. *See also* landfill, secure landfill, secure chemical.

**sanitary landfill disposal** A common method of disposal of sludge from wastewater treatment plants. Although sludge processed by drying, composting, or incineration can be deposited directly, wet sludge must be mixed with other refuse before spreading.

**sanitary revolution** *See* Great Sanitary Awakening.

**sanitary seal** A protective device against contamination. A sanitary seal is placed at the top of a well or borehole casing to prevent the entry of contaminated water or other material. The seal extends a certain distance below surface in the annular space between the sidewall and the casing.

**sanitary sewage** Wastewater containing human wastes from residences and commercial, institutional and industrial establishments, but excluding any process or hazardous wastes. It constitutes the major portion of municipal wastewater. The term sanitary sewage is becoming obsolete and being replaced by domestic or sanitary wastewater.

**sanitary sewer** A sewer intended to carry only sanitary wastewater, i.e., liquid and water-carried wastes from residences, commercial buildings, industrial plants, and institutions together with minor quantities of ground, storm, and surface waters that are not admitted intentionally. Or an underground pipe that carries off only domestic or industrial waste, not stormwater (EPA-40CFR35.2005-37 and 40CFR35.905). *See* storm and combined sewers.

**sanitary sewer overflow (SSO)** Discharge of untreated wastewater when the flow capacity of a sanitary sewer system is exceeded, usually during rainstorms or as a result of infiltration/inflow. *See also* overflow.

**sanitary sewer system** A network of facilities for the collection, transmission, treatment and disposal of sanitary wastewater. *See also* sanitary collection system, sewerage, sewer system, storm sewer, and combined sewer.

**sanitary survey** (1) An on-site review of the water sources, facilities, equipment, operation, and maintenance of a public water system to evaluate the adequacy of those elements for producing and distributing safe drinking water. A periodic sanitary survey is sometimes substituted for the increased monitoring of small water supply systems. For an untreated water supply, the sanitary survey can also reveal leaks and the proximity of sources of pollution. *See also* microbiological analysis. (2) A field investigation conducted to identify watershed conditions that may affect the quality of a potential water supply source. Other field surveys include industrial waste and pollutional surveys.

**sanitary technician** A sanitarian's assistant.

**sanitary waste** Same as sanitary wastewater.

**sanitary wastewater** Wastewater containing human wastes from residences and commercial, institutional, and industrial establishments, together with minor quantities of ground, storm, and surface waters that are not admitted intentionally, but excluding any process or hazardous wastes. It constitutes the major portion of municipal wastewater. Also called domestic wastewater or sanitary sewage.

**sanitary water** Water discharged from sinks, showers, kitchens, or other nonindustrial operations, but not from commodes. *See also* graywater or sullage.

**sanitary well seal** *See* sanitary seal.

**sanitation** Control of physical factors in the human environment that could harm development, health, or survival. More generally, sanitation is the science and practice of healthy environmental measures such as potable water supply, drainage, wastewater and solid waste disposal, and ventilation. In a more restricting sense, the word applies to the hygienic collection and disposal of community excreta and wastewater, or to the hygienic collection and disposal of refuse and solid wastes.

**San-I-Tech™** A grease interceptor manufactured by Scienco/FAST Systems.

**sanitize** To free from dirt, germs, etc., or to make less offensive.

**sanitizer** An agent that reduces the number of microbial contaminants to a safe level, e.g., a chemical agent that destroys 99.999% of specific bacteria in 30 seconds, especially on food-processing equipment. *See also* disinfectant, bacteriostat.

**Sanitron™** A unit manufactured by Atlantic Ultraviolet Corp. for water treatment using ultraviolet light.

**San Joaquin Valley fever** An acute or progressively chronic respiratory infection characterized by fever and reddish bumps on the skin. *See* coccidioidomycosis for detail.

**Sanuril®** An apparatus manufactured by Eltech International Corp. for water disinfection with hypochlorite tablets.

**saponification** The conversion of grease and fats to soap by reaction with an alkali. Saponification is one of the reactions that occur during lime stabilization of sludge, thus requiring additional lime to neutralize the glyceride of stearic acid ($C_{17}H_{35}COOH$). The saponification number or saponification value is the quantity in milligrams of potassium hydroxide (KOH) required to saponify one gram of a given substance (e.g., glyceride).

**saponification number** *See* saponification.

**saponification value** *See* saponification.

**saprobe** Same as saprophyte.

**saprobic classification** A classification of river organisms on the basis of their tolerance of organic pollution in slow-moving streams. The saprobic system, used mostly in Europe, uses saprophytes, which depend on decomposing organic matter. Also called saprobien system. *See also* alpha-mesaprobic zone, beta-mesaprobic zone, oligosaprobic zone, polysaprobic zone.

**saprobic organisms** Organisms that feed on dead organic matter; saprophytes and saprozoa contribute to the stabilization of wastewater by converting organic substances into stable matter.

**saprobien system** *See* saprobic classification.

**saprobiont** Same as saprobe.

**sapropel** A zone of mud, slimy sediment, or decomposed organic matter at the bottom of a stagnant body of water.

**saprophyte** An organism that lives on dead or decaying organic matter, e.g., the bodies of other organisms. Saprophytes, mostly certain fungi and bacteria but also some protozoa, help the natural decomposition of organic matter in water. Also called saprobe or saprobiont.

**saprophytic** Living on dead or decaying organic matter.

**saprophytic bacteria** Bacteria that feed on dead or nonliving organic matter; the group of organisms most responsible for waste decomposition and stabilization. In aerobic decomposition, they use oxygen ($O_2$) to mineralize organic matter, i.e., to produce carbon dioxide ($CO_2$), water ($H_2O$), nitrates ($NO_3^-$), phosphates ($PO_4^{3-}$), and sulfates ($SO_4^{2-}$), respectively, with the elements carbon (C), hydrogen (H), nitrogen (N), phosphorus (P), and sulfur (S). *See* saprophyte.

**saprozoa** Animallike organisms that feed on decaying organic matter; *see* saprobic organisms.

**SAR** Acronym of sodium adsorption ratio.

***Sarcina ventriculi*** A species of acidophilic, obligate anaerobic bacteria that ferment sugars.

**sarcocystiasis** A rare and little-known protozoal infection of the human intestinal tract, caused by pathogens of the genus *Sarcocystis*.

***Sarcocystis*** The protozoal genus that causes sarcocystiasis.

***Sarotheroden mossambicus*** A hardy, freshwater, food fish species of the *Tilapia* genus that survives well in stabilization ponds used for aquaculture.

***Sarotheroden niloticus*** A hardy, freshwater, food fish species of the *Tilapia* genus that survives well in stabilization ponds used for aquaculture.

**SAT** Acronym of soil-aquifer treatment.

**Satellite** A rotary distributor manufactured by Simon-Hartley, Ltd. for fixed-film reactors.

**satellite reuse plant** *See* satellite treatment.

**satellite treatment** A small-scale, membrane-based treatment system built on or near existing sewer lines in areas with high potential for reclaimed water. The wastewater collected is given adequate treatment and delivered to the reuse customers. Also called point-of-sale treatment and sewer mining. *See also* end-of-line treatment.

**Sation®** Water treatment products of the U.S. Filter Corp.

**satisfied outer shell** *See* filled electron shell.

**saturated** (1) Pertaining to a soil sample that has its pores filled with a liquid such as water or to any material that cannot absorb any more of another material. Saturated air contains the maximum amount of water vapor for a given temperature and pressure. *See also* relative humidity. (2) Pertaining to a solution that contains the maximum amount of solute that can be dissolved under given conditions. *See also* supersaturated. (3) Pertaining to an organic compound that contains no single or double bonds and in which all valence bonds of carbon (or another element) are attached to another atom or group. (4) Pertaining to an inorganic compound that has no free valence electrons.

**saturated air** *See* saturated (1).

**saturated fat** A single-bond fat, as in butter, meat, palm oil.

**saturated hydrocarbon** A chemical compound that contains only carbon (C) and hydrogen (H), with single bonds between the carbon atoms. *See* alkane, paraffin.

**saturated liquid** *See* saturated (2).

**saturated rock** A rock that has all its interstices and pores filled with water.

**saturated soil** *See* saturated (1)

**saturated solution** A solution in which the dissolved solute is in equilibrium with an excess of undissolved solute; or a solution in equilibrium such that, at a fixed temperature and pressure, the concentration of the solute in the solution is at its maximum value and will not change even in the presence of an excess of solute (EPA-40CFR796.1860-iii and 796.1840-vi). *See* Langelier index and other corrosion control indicators for solution saturation with calcium carbonate. *See also* crystal growth, nucleation, supersaturated solution.

**saturated steam** Vapor that is in equilibrium with water at the boiling temperature.

**saturated zone** That part of the Earth's crust in which all voids are filled with water; or the area below the water table where all open spaces are filled with water. Actually, a portion of this water is subsurface runoff in the temporary saturation zone; it will become overland runoff before reaching the permanent zone of saturation, which is actually the groundwater. *See* subsurface water.

**saturation** (1) The condition of a liquid (e.g., water) when it has taken into solution the maximum possible quantity of a given substance (e.g., dissolved oxygen) at a given temperature and pressure. *See also* solubility. (2) The saturation capacity or concentration.

**saturation capacity** The maximum quantity of water that an ion-exchange column or similar treatment device can process or the maximum quantity of material the unit can remove from water, expressed, e.g., in pounds of material removed per cubic foot of ion-exchange resin.

**saturation concentration** The concentration of a gas dissolved in a liquid at equilibrium, based on Henry's law; e.g., the maximum concentration of dissolved oxygen in a stream at a given temperature. *See also* supersaturated solution.

**saturation condition** A relationship between various components of a stable water:

$$\mathrm{pH_s} = \mathrm{p}K_2 - \mathrm{p}K_{sp} - \log\,[\mathrm{Ca^{2+}}] + S - \log\,[\mathrm{Alk}] \quad \text{(S-10)}$$

where $pH_s$ = pH of saturation; $K_2$ = equilibrium constant of the reaction

$$\mathrm{HCO_3^-} \leftrightarrow \mathrm{H^+} + \mathrm{CO_3^{2-}} \quad \text{(S-11)}$$

$K_{sp}$ = equilibrium constant of the reaction

$$\mathrm{CaCO_3} \leftrightarrow \mathrm{Ca^{2+}} + \mathrm{CO_3^{2-}} \quad \text{(S-12)}$$

$[Ca^{2+}]$ = molar concentration of calcium, alkalinity, mole/L; [Alk] = molar concentration of alkalinity, mole/L; and $S$ = a salinity correction factor.

**saturation constant** A parameter of the Monod equation, representing the limiting substrate concentration at one-half the maximum growth rate. Also called the half-saturation constant or half-velocity constant.

**saturation deficit** The difference between the quantity of dissolved oxygen (DO) in a body of water and the quantity of DO required for saturation under given conditions of temperature and pressure, all quantities being expressed in percent of saturation.

**saturation dissolved oxygen** *See* dissolved-oxygen saturation.

**saturation extract** The extract from a soil sample saturated with water.

**saturation index (SI)** An index reflecting the equilibrium pH of a water with respect to calcium and alkalinity. This index is used in stabilizing water to control both corrosion and the deposition of scale and is generally defined for any solid solubility reaction as:

$$SI_x = \log_{10}\,(P_x/K_x) \quad \text{(S-13)}$$

where the subscript $x$ relates to mineral $x$, $P$ = ion activity product, and $K$ = solubility product constant. Also called disequilibrium index. *See* corrosion index for a list of corrosion control parameters..

**saturation line** A horizontal line through the cross section of an earth structure that marks the limit of saturation with water and the uppermost limit of flow through the structure. *See also* phreatic line.

**saturation oxygen concentration** *See* dissolved-oxygen saturation.

**saturation pH** The pH at which a solution (e.g., water) is saturated with a compound (e.g., calcium carbonate or $CaCO_3$); commonly represented by $pH_s$:

$$pH_s = -\log (K\gamma_c[Ca^{2+}]\gamma_h[HCO_3^-]/K_{sp}) \quad (S\text{-}14)$$

where $K$ = equilibrium constant for the dissociation of bicarbonate; $\gamma_c$ = activity coefficient for calcium; $[Ca^{2+}]$ = calcium concentration, mole; $\gamma_h$ = activity coefficient for bicarbonate; $[HCO_3^-]$ = concentration of bicarbonate, mole; and $K_{sp}$ = solubility product constant for calcium carbonate dissociation. *See also* the various indices of calcium carbonate saturation such as Langelier and Ryznar.

**saturation point** The point at which a substance stops dissolving in a solution and starts precipitating.

**saturation-type equation** The equation of a reaction whose order changes depending on the relative magnitude of the concentrations of the reactants. *See* mixed-order reaction for detail.

**saturation values** *See* dissolved-oxygen saturation.

**saturation vapor pressure** The vapor pressure of a system under saturation conditions, but not supersaturation conditions. *See also* equilibrium vapor pressure.

**saturation zone** Same as saturated zone.

**saturator** (1) A device that produces a fluoride solution for the fluoridation process. It is usually a cylindrical container with granular sodium fluoride (NaF) on the bottom. Water flows either upward or downward through the sodium fluoride to produce the fluoride solution. (2) A similar device used to feed hydrated lime $[Ca(OH)_2]$.

**saturometry** The process of measuring calcium carbonate saturation in water. *See* carbonate saturometer and corrosion index.

***S. aureus*** *See Staphylococcus aureus.*

**SAV 715** A stainless steel chain manufactured by Hitach Maxco, Ltd. for use in sludge collection.

**save-all** In general, a means to prevent loss or waste. In a paper mill, a separation device used to reclaim fibers and fillers from white water while providing some treatment through sedimentation, filtration, or flotation..

**Save-All** A clarifier designed by Walker Process Equipment Co. for recovery of fibers from paper mill wastewater.

**Sb** Chemical symbol of the metallic element antimony.

**SBA resin** *See* strong-base anion resin.

**SBOD** Acronym of soluble biochemical oxygen demand.

***S. bovis*** *See Streptococcus bovis.*

***S. boydii*** *See Shigella boydii.*

**SBR** Acronym of sequencing batch reactor.

**SBR–BNR process** The sequencing batch reactor process operated for biological nitrogen removal (BNR) using the influent wastewater BOD in pre-anoxic denitrification. It includes the following steps: fill, anoxic/anaerobic mixing and fill, react/aeration, settle, decant, and idle.

**SBR clarification** The sedimentation provided in sequencing batch reactors during the settling and decanting modes.

**SBR process** *See* sequencing batch reactor process.

**SBR with biological phosphorus removal** The sequencing batch reactor process operated as follows for both BOD and biological phosphorus removal: anaerobic conditions during the fill and react steps for the development of phosphorus-accumulating bacteria, followed by a sufficient aerobic period for nitrification, and then the anoxic, settle, and decant steps. Some nitrogen removal is also possible.

**SBS** Acronym of sodium bisulfite.

**$Sb_2S_3$** Chemical symbol of antimony sulfide or stibnite.

**SC™** A deaerating heater of the spray type manufactured by Graver Co.

**scabies** A water-washed disease that can be controlled by providing sufficient water for personal hygiene.

**SCADA** Acronym of supervisory control and data acquisition.

**SCADA-Flo™** An open-channel transmitter manufactured by Marsh-McBirney, Inc. of Frederick, MD.

**SCADA system** Same as SCADA.

**scale** (1) An inorganic coating or incrustation, as on the inside of pipes, boilers, or water heaters, formed by the precipitation of salts from water: e.g., oxide flakes, carbonate precipitates, sulfate scales, and silica scales. Also called hard water scale or the scale deposited on the inside of pipes and appliances by water containing a high concentration of carbonates and bicarbonates of calcium and magnesium. *See also* protecting scale, nonprotective scale, erosion corrosion. (2) As scales, an iron or other oxide occurring in a scaly form on the surface of a metal at high temperature. (3) Mill scale is formed on iron or steel during hot-rolling.

**scale inhibitor** A chemical substance used to control the formation and buildup of scales in boilers, reverse osmosis membranes, or other industrial applications. *See* antiscalant for more detail.

**scale prevention compound** Same as scale inhibitor.

**scales** *See* scale (2).

**scaling** (1) In pressure-driven membrane processes, the precipitation of solids in a membrane element due to solute concentration in the feed stream. (2) The formation of hard deposits (e.g., calcium and magnesium carbonate, phosphate, or sulfate) on the hot surface of a cooling tower, thereby reducing the efficiency of heat exchange.

**scaling control** The control of the formation of calcium carbonate ($CaCO_3$) scales by reducing pH, alkalinity, or $CaCO_3$ concentration via ion exchange or lime softening, or increasing $CaCO_3$ solubility through the use of antiscalants.

**scaling potential** The tendency to develop calcium carbonate and sulfate scale during water or wastewater treatment; it can be approximated by the Langelier saturation index or the Ryznar stability index.

**scanning electron microscope (SEM)** A device that uses an electron probe to scan the surface of the specimen and form a magnified three-dimensional image thereof on a television screen. The image results from the electrons reflected or given off. SEM's magnification ranges from 20 to 200,000× at a resolution of 100 Å. *See also* transmission electron microscopy.

**SCAT™** An aboveground tank manufactured by Industrial Environmental Supply, Inc.

**scatter diagram, scattergraph, scatter plot** A graphical representation of the differences between observed values and model predictions. These graphs are used with statistical analysis to evaluate the degree of agreement between observations and predictions and, in a sense, judge model performance. Scattergraphs can also be used to verify the repeatability and accuracy of flow monitoring data, e.g., by plotting velocity versus depth of flow in a conduit or open channel, with a trend line through the points, and comparing the field results with the theoretical relationship. For example, in one sewer modeling application (Adrien, 2004), the consultant monitored approximately 300 pump stations, created scatter plots of water level vs. flow velocity for each meter, and used these plots to filter the data by eliminating any obviously erroneous or inconsistent points. The result was a set of "clean" dry-weather hydrographs for use as input to the model.

**scattergraph** Same as scatter diagram.

**scatter plot** Same as scatter diagram.

**scavenged-type metal** A reactive trace metal, such as aluminum or lead, that is removed from surface waters by adsorption onto particles at various depths in the water column (Stumm & Morgan, 1996). *See also* nutrient-type metal.

**scavenger** (1) A substance that consumes or inactivates another in a mixture; *see* scavenging, radical scavenger. (2) A polymer matrix or ion exchanger that removes specific organic substances from the feed stream as pretreatment to deionization. (3) *See* scavenging (1). (4) A nightsoil collector who regularly empties the contents of bucket latrines into a wheelbarrow or cart for transportation and disposal. Also called a sweeper.

**Scavenger** An apparatus manufactured by Aqua Products, Inc. for sludge scrubbing and removal.

**scavenger well** A well that is used to control saline water intrusion by pumping brackish water from beneath a freshwater lens.

**scavenging** (1) A process in which a substance such as a free radical is removed, consumed, or inactivated by converting it to another form or adsorbing it onto another compound. *See* radical scavenger. (2) The removal, sometimes unauthorized or uncontrolled, of materials from solid wastes. For an industrial firm that does not want to handle its own solid wastes, a scavenger will haul, treat, reclaim, and dispose of them under contract.

**SCD** Acronym of (a) Soil Conservation District; *see also* SWCD and (b) streaming current detector.

***Scenedesmus*** A common algal genus found in stabilization ponds; it reacts with chlorine to form trihalomethanes where biomass is present.

**scenic waterway** A river or river segment chosen for its scenic and recreational qualities to be preserved in its natural state.

**scentometer** A hand-held instrument used to measure the intensity of odors, e.g., around the site of a wastewater treatment plant; it consists of graduated orifices through which the malodorous air passes. *See also* butanol wheel and triangle olfactometer.

**Scentoscreen** A portable gas chromatograph manufactured by Sentex Systems, Inc. for the analysis of volatile organic compounds.

**SCFM** Acronym of standard cubic feet per minute or cubic feet of air per minute at standard conditions of pressure, temperature, and humidity (0°C, 14.7 psia, 50% relative humidity).

**schedule of compliance** A schedule of remedial measures included in a permit, including an enforceable sequence of interim requirements (for example, actions, operations, or milestone events) designed to achieve or maintain compliance, or correct noncompliance, with an appropriate act or regulation, e.g., an effluent limitation, other limitation, prohibition, or standard (EPA-40CFR124.2, 72.2).

**schedule (pipe)** A sizing system of arbitrary numbers that specifies the inside and outside diameters of each pipe size. This term is used for steel, wrought iron, and some types of plastic pipe, and to describe the strength of some types of plastic pipe. For example, a plastic pipe schedule 40, common in water and wastewater work, with a nominal diameter of 16 inches, actually has an inside diameter of 15 inches. *See also* pipe class.

**Scheele's green** *See* copper arsenite.

***Schistosoma*** A genus of trematodes or flukes responsible for numerous infections (schistosomiasis or bilharzia) and deaths worldwide. The larvae, developed in various freshwater snails (*Biomphalaria*), enter the body through the skin. The three species of importance to human health are *Schistosoma haematobium, Schistosoma japonicum,* and *Schistosoma mansoni.* Also called blood fluke. *See also* schistosome.

***Schistosoma haematobium*** A species of helminthic pathogen that causes schistosomiasis in Africa, India, and the Middle East. It is transmitted from the feces of an infected host through aquatic snails.

***Schistosoma intercalatum*** A species of helminthic pathogen that causes schistosomiasis in Central Africa (Cameroon, the two Congos, Gabon). It is transmitted from the feces of an infected host through aquatic snails.

***Schistosoma japonicum*** A species of helminthic pathogen that causes schistosomiasis in East Asia and the Philippines. It is transmitted from the feces of an infected host through aquatic snails.

***Schistosoma mansoni*** A species of helminthic pathogen that causes schistosomiasis in Africa, the Caribbean, the Middle East, and South America. It is transmitted from the feces of an infected host through aquatic snails.

**schistosome** An elongated trematode of the genus *Schistosoma,* parasitic in the blood vessels of humans and other mammals after infecting an intermediate snail host; a blood fluke. Schistosome cercariae can also cause a skin irritation (swimmer's itch). Also called bilharzia. *See also* cercaria, miracidia.

**schistosome cercariae** The larvae of schistosome; their snail hosts are found in natural bathing waters and transported by infected waterfowl. They can infect humans and cause swimmer's itch (a skin irritation).

**schistosome life cycle** The route followed by schistosomes from an infected host to another host: feces from an infected host (humans, and, in some cases, domestic or wild animals) reach an aquatic environment produce eggs in a few minutes, which become miracidia in 16 hours, but take 4 to 8 weeks to develop within an intermediate snail host. The snail sheds cercariae or swimming schistosomes, which can infect a human or animal host in 2 to 3 days.

**schistosomiasis** A waterborne disease of tropical and subtropical regions transmitted indirectly to humans by the schistosomes *Schistosoma haematobium, S. intercalatum, S. japonicum,* and *S. mansoni;* commonly called bilharzia and bilharziasis. Freshwater snails act as intermediate hosts, releasing schistosome larvae that then penetrate the skin of bathers and waders from shrinking water droplets. The cases found in the United States (about 0.2% of the worldwide infections) are assumed to be imported as the intermediate host, the snail *Biomphalaria sp.* is not present in the country. Various symptoms accompany the infection: itching, skin inflammation, fever, respiratory symptoms, dysentery, blood loss, and and even cancer and death. *See also* cercaria, miracidium, schistosome life cycle, *Bulinus, Oncomelania, Tricula.*

**schizogony** (1) The multiple fission of a trophozoite or schizont into merozoites; asexual reproduction of sporozoans. (2) *See* budding.

**schizont** A cell developed from the multiple fission of a trophozoite.

***Schizothrix*** A genus of blue-green algae or cyanobacteria that can cause off-tastes and odors in drinking water sources.

***Schizothrix calciola*** A species of algae that produce toxin and may cause gastroenteritis.

**Schloss green** *See* copper arsenite.

**schmutzdecke** (From the German words schmutz, meaning dirt or impurity, and decke, meaning cover or layer) The layer of solids and biological growth that forms on the surface of slow sand filters, trickling filters, infiltrating surface of spreading basins, etc., facilitating the removal of suspended solids and pathogens. In slow sand filters, this layer is periodically scraped off and removed or washed in place. Also called dirty skin. *See also* biofilm, biofilm mat, mud blanket, slime, slough.

***S. cholerae-suis*** *See Salmonella cholerae-suis.*

**Schrodinger ratio** The ratio of entropy increase to the entropy of ordered structure, equal to the ratio of respiration to biomass multiplied by calorie equivalents and divided by absolute temperature; a parameter used in studying the ecological effects of pollution on life.

**Schultze–Hardy rule** The coagulating power of an ion increases by more than one order of magnitude when its charge increases by one (Droste, 1997). For example, the relative coagulating power of sodium for positive colloids is 1 in sodium chloride (NaCl), but 30 in sodium sulfate ($Na_2SO_4$) and 1000 in sodium phosphate ($Na_3PO_4$). Also spelled Schulze–Hardy rule.

**Schultz–Germain formula** *See* Schulze formula and Germain formula, two different equations related to the design of trickling filters. *See also* RBC performance.

**Schulze equation** Same as Schulze formula.

**Schulze formula** A 1960 modification of the empirical formula proposed by Velz for the design of trickling filters, assuming that the liquid contact time is proportional to filter depth and inversely proportional to the hydraulic loading rate:

$$T = CD/q^n \quad \text{(S-15)}$$

$$L_e/L_0 = \exp(-kD/Q^n) \quad \text{(S-16)}$$

where $T$ = liquid contact time, min; $C$ = constant; $D$ = filter depth, feet; $q$ = hydraulic application rate, gpm/ft$^2$; $n$ = constant characteristic of filter medium; $L_0$ = influent $BOD_5$, mg/L; $L_e$ = $BOD_5$ of settled filter effluent, mg/L; $k$ = rate constant determined experimentally, $0.51/\text{day} < k < 0.76/\text{day}$ (also called wastewater treatability and packing coefficient); and $Q$ = hydraulic application rate, mgd/acre. Other trickling filter design formulas include British Manual, Eckenfelder, Galler–Gotaas, Germain, Howland, Kincannon and Stover, Logan, modified Velz, NRC, Rankin, and Velz.

**Schulze–Hardy rule** *See* Schultze–Hardy rule.

**Schumberger array** A dipole array.

**scintillation cocktail** A mixture of organic solvents used in the analysis of radionuclides in water.

**Scion®** Abbreviation of short-cycle ion, an ion-exchange apparatus manufactured by U.S. Filter Corp.

***Scirpus*** A genus of bulrushes, emergent vegetation commonly found in natural wetlands.

**SCONOx™** A device manufactured by Goal Line Environmental Technologies for catalytic absorption of nitrogen oxides and carbon monoxide (NOx/CO).

**Scoop-A-Fish** A trough manufactured by Norair Engineering Corp. as a traveling screen for fish collection.

**scoop sampler** A device used to obtain a sample across an entire section of a water or wastewater stream; such a sample is more representative than that from a tube sampler.

**scoop wheel** A drainage pump with flat vanes revolving in a curved channel.

**ScorGuard®** Organic water treatment products of Western Water Management, Inc. for use in cooling towers.

**Scott–Darcy process** A chemical precipitation process that uses ferric chloride ($FeCl_3$) prepared by passing an aqueous solution of chlorine gas ($Cl_2$) over iron scrap.

**scour** (1) The transport of particles in traction along the bottom or the sides of a sedimentation basin, conduit, waterway, or pipeline. Scouring of bottom deposits in a water body may sometimes exert sudden oxygen demands at the surface. *See also* auxiliary scour, bottom scour, scouring velocity. (2) The separation and transport of particles removed on the granular bed of a rapid or slow sand filter during backwashing. *See also* air scour, high-velocity wash, mechanical scour, surface scour. (3) The enlargement of the section of a waterway caused by the scouring action of a flowing liquid.

**scouring** The resuspension of settled particles. *See* scour (1).

**scouring sluice** A gated opening in a dam or other hydraulic structure for the release of accumulated material.

**scouring velocity** The minimum velocity required for a fluid in motion to dislodge and carry away material particle accumulations in a conduit, pipeline, or waterway. In the water and wastewater applications of sedimentation, the horizontal velocity ($V$) just sufficient to cause scour may be computed using the Camp–Shields equation:

$$V = [8\,\beta(\gamma - 1)g \cdot d/f]^{0.5} \quad \text{(S-17)}$$

where $\beta$ is a dimensionless constant ranging from 0.04 to 0.06, $f$ the Darcy–Weisbach friction factor (usually between 0.02 and 0.03), $\gamma$ the specific gravity of the particles, $d$ the equivalent diameter of the particles, and $g$ the gravitational acceleration. Also called scour velocity. *See also* critical velocity, critical scouring velocity, flushing velocity, horizontal velocity, minimum velocity, self-cleansing velocity, settling velocity.

**Scour-Pak®** A gravity depth filter with granular media, manufactured by Graver Co.

**scour valve** A small, gated takeoff valve installed at a low point in a pressure conduit or at a depression in a pipeline to allow drainage or flushing of the line. Also called a blow-off valve or washout valve.

**scour velocity** Same as scouring velocity.

**scraper** (1) A device that is used to remove accumulated deposits in a small pipeline or to move settled sludge from the bottom of a sedimentation basin. *See also* polypig, squeegee. (2) A blade used to separate sediment from the surface of a filter or screen.

**scraping** Filter scraping is the cleaning of a slow sand filter by removing the schmutzdecke along with a small amount of sand, which is washed and stockpiled for later reuse. *See also* filter resanding.

**screen** (1) A device with openings, usually of uniform size, installed at surface water intakes and wastewater treatment plants to retain coarse particles or materials larger than 0.25 in (6 mm), which can damage pumps and other equipment. It may be operated and cleaned manually or automatically. Some screens consist of bars, rods, wires, grating, wire mesh, or perforated plates. *See also* the following types of screen: bar rack, bar screen, coarse screen, fine screen, microscreen, rack. (2) A well screen. (3) A sieving device that segregates sand, soil, or other granular materials into different sizes.

**screen approach velocity** The flow velocity before going through a screen, different from the actual settling velocity of the solid particles and typically selected as 1.0 to 2.0 fps (0.3–0.6 m/s). It affects the head loss through the screen.

**screen blinding** The accumulation of debris on a wastewater screening device, obstructing the openings and contributing to head loss.

**screen chamber** The chamber that contains the screens in a water or wastewater treatment plant.

**screening** (1) The use of screens to remove coarse floating and suspended solids from raw wastewater or water to protect other equipment in surface water intakes, ahead of water or wastewater pumps. Screening also improves treatment and is sometimes used with outfalls for untreated wastewater. *See also* microstraining, comminution and grinding. (2) *See* sludge screening.

**screenings** The material removed by screens from water, wastewater, or sludge. Screenings from wastewater treatment are ground and returned to the influent, combined with primary sludge, or processed otherwise; eventually they are disposed in a sanitary landfill. They contain 10–20% dry solids and have a bulk density of 40–70 pounds per cubic foot. *See also* coarse screenings.

**screenings dewatering** The partial removal of water from screening materials by gravity or mechanical equipment.

**screenings grinder** A device used to grind, shred, or macerate materials removed by screens from water or wastewater treatment plants.

**screenings press** A device used to compact and dewater the material removed by screens.

**screenings shredder** *See* screenings grinder.

**screenings triturator** *See* screenings grinder.

**screen size** The diameter of the largest particle that will pass through a screen, as determined from the U.S. Sieve Series. *See* mesh size.

**Screezer** An apparatus manufactured by Jones & Attwood, Inc. combining screening and dewatering.

**screw auger** A device or mechanism used to remove grit collected in the hopper at the inlet end of a grit chamber. *See also* chain-and-flight mechanism, chain-and-bucket elevator.

**screw conveyor (or conveyer)** A device that uses a rotating helical screw in a trough to transport grit, sludge, or similar material, most commonly horizontally over short distances (e.g., 30 to 40 ft). Inclined and vertical screw conveyors can also be used for handling dewatered sludge and in truck loading. *See also* conveyor, belt conveyor, inclined screw, horizontal screw conveyor.

**screwed pipe** A pipe that has threaded ends and sections connected by threaded couplings.

**screw-feed pump** A pump that moves fluids by a runner with radial blades in a cylindrical casing. *See also* horizontal screw pump, vertical screw pump, and screw pump.

**screw impeller** The helical impeller of a screw pump.

**Screwpeller™** A centrifugal screw impeller manufactured by Aeration Industries, Inc. for use in surface aerators.

**screw press** A device used to dewater sludge; it consists of a screw-shaped conveyor that moves sludge solids through a cone-shaped screen. Its diameter shrinks to create pressure and force the filtrate out. *See* inclined screw press.

**screw pump** A low-lift, high-capacity, positive-displacement pump that raises water or wastewater by rotating a helical impeller in an inclined trough or cylinder. It can handle a wide flow range at a constant speed and prevents breakup of biological floc of return activated sludge, but requires considerable space. Also called Archimedean screw, Archimedean screw pump, Archimedes screw, Archimedes screw pump, and water snail. *See also* enclosed screw pump, open screw pump, screw-feed pump.

**scroll centrifuge** A device consisting of a rotating helical screw conveyor inside a rotating bowl, used for sludge dewatering. *See* solid-bowl centrifuge for detail.

**scrubber** An air pollution device that uses a spray of water (or reactant) or a dry process to trap pollutants in emissions (e.g., from combustion or chemical process exhaust streams). *See also* chemical scrubber.

**scrubbing** The removal of suspended solids and undesirable gases from air or gaseous emissions by entrainment in a water spray. Scrubbing is used for odor control in such units as biological towers, soil mounds, and packed towers.

**scrubbing fluid** A liquid used in a chemical scrubber to oxidize such odorous compounds as hydrogen sulfide ($H_2S$). Common scrubbing fluids include hydrogen peroxide ($H_2O_2$), potassium permanganate ($KMnO_4$), and sodium hypochlorite (NaOCl). Typical reactions are:

$$2\ KMnO_4 + 3\ H_2S \rightarrow 3\ S + 2\ KOH + 2\ MnO_2 + 2\ H_2O \quad \text{(acidic pH)} \quad \text{(S-18)}$$

$$NaOCl + H_2S \rightarrow S^0 \downarrow + NaCl + H_2O \quad \text{(S-19)}$$

**scrubbing tower** A device used to remove odors from wastewater management facilities; a scrubber.

**ScruPac™** A screw-type screenings compactor manufactured by Vulcan Industries, Inc.

**Scru-Peller®** A sludge pump manufactured by Yeomans Chicago Corp.

**scrutinyite** **($\alpha$- $PbO_2$)** One of two forms of lead dioxide ($PbO_2$), resulting from the corrosion of lead piping. *See also* cerrusite and plattnerite.

**SCS** Acronym of Soil Conservation Service.

**SCUBA™** A self-contained servomechanism manufactured by Rodney Hunt Co. to operate gates and valves.

**scum** In general, a film or layer of foul or extraneous solid matter that forms on or rises to the surface of a liquid. In particular, scum is the floatable material (specific gravity between 0.95 and 1.0) on the surface of digestion tanks, settling tanks, grit chambers, chlorine contact tanks, and ponds. For example, primary wastewater treatment produces an average of 0.000008 gallon of scum per gallon of flow; it contributes to plug pipelines. It consists mostly of foam, fats, oils, soaps, and grease. Measures for controlling scum accumulation in tanks include reducing the flowline with respect to the depth, mechanical destruction, and wetting. *See also* sloping beach, tilting trough.

**scum baffle** A baffle installed in a wastewater treatment tank to retain scum and other floating matter. It dips below the surface of the liquid but not all the way to the bottom. Also called a scum board.

**scum board** Same as scum baffle.

**scum breaker** A device designed to break up scum in sludge digestion tanks.

**Scumbuster™** A pump manufactured by Vaughn Co., Inc. to grind digester sum solids.

**scum chamber** A compartment or special space provided in a sludge digestion tank or in an Imhoff tank for the accumulation of rising scum.

**scum collector** A mechanical device used for skimming and removing scum from the surface of settling and digestion tanks.

**scum mat** A large floating mass of scum floating on the surface of wastewater treatment units.

**scum removal** A preliminary or primary treatment step that consists in the separation of scum (floating oil and grease) from the wastewater. Scum may be collected at the end of rectangular primary clarifiers and disposed of with the sludge or separately. Scraping devices, rotating pipe-through skimmers, and slotted pipes are used to remove scum from secondary clarifiers; the scum is discharged to sludge thickeners or digesters.

**scum space** *See* scum chamber.

**Scum Sucker™** A telescopic pipe manufactured by United Industries, Inc. for scum removal.

**scum trough** A trough used for scum removal in primary sedimentation units.

**scum well** A box used to store scum from the surface of a wastewater clarifier. The scum is usually drawn off by a horizontal, slotted pipe that can be rotated by a lever or by a screw.

**SCWO** Acronym of supercritical water oxidation or supercritical wet oxidation.

***S. damnosum*** Abbreviation of *Simulium damnosum.*

**SDI** Acronym of silt density index.

**SDNR** Acronym of specific denitrification rate.

**$SDNR_b$** Acronym of specific denitrification rate relative to heterotrophic biomass concentration.

**SDSI** Acronym of Stiff and Davis Scaling Index

**S&DSI** Acronym of Stiff and Davis Stability Index.

***S. dublin*** *See Salmonella dublin.*

***S. durans*** *See Streptococcus durans.*

**SDVB resin** *See* styrene divinylbenzene.

**SDVS sewer system** *See* small-diameter, variable-slope sewer system.

**SDWA** Acronym of Safe Drinking Water Act.

**SDWS** Acronym of Secondary Drinking Water Standards.

***S. dysenteriae*** *See Shigella dysenteriae.*

**Se** Chemical symbol of the metallic element selenium.

**Sea Devil** A floating skimmer manufactured by Vikoma International for oil removal. *See also* Seaskimmer.

**sea disposal** *See* ocean disposal.

**seal** (1) A device or packing material placed around a pump shaft or similar equipment to prevent air intake or water leakage. (2) An amount of water held in a siphon or a trap to prevent the passage of foul gases, e.g., from a sewer line. (3) An impermeable material placed in the annular space between the wall and casing of a borehole to prevent the downhole movement of surface water or the vertical mixing of artesian waters.

**sealant** (1) A polyelectrolyte, chemical grouting, or other material used to plug leaks, reduce seepage or infiltration, or fill soil strata. (2) Any substance applied to a surface or circulated through pipes to form a dry watertight coating.

**Sealed Register®** A series of positive displacement, turbine, compound, and fire-service meters manufactured by Invensys Metering Systems.

**sea lettuce** A seaweed that has large leaflike blades and grows to become a nuisance when nutrients are available.

**sea-level datum** The average surface level of the sea or mean sea level, uninfluenced by tidal movement or waves, adopted as a reference for heights and elevations. *See also* NGVD.

**Sealtrode** A sealed pump controller manufactured by Yeomans Chicago Corp.

**Seaskimmer** A floating skimmer manufactured by Vikoma International for oil removal. *See also* Sea Devil.

**seatworm** Another name of *Enterobius vermicularis.*

**sea urchin** A marine organism, *Champia parvula,* used in subchronic bioassays.

**seawater** General term for sea or ocean water, with a high mineral content as expressed by a concentration of dissolved solids between 30,000 and 36,000 mg/L as compared to also brackish water, brine, freshwater, saline estuarine waters, and saline water. Also called saline water.

**seawater composition** On average, seawater is mostly sodium chloride (NaCl), the prevalent ions being sodium and chloride as shown in Table S-04.

**seawater constituents** *See* seawater composition.

**Seawater Conversion Vessel** A self-contained desalination plant housed in an ocean-going ship, developed by the Water Standard Company™. It includes the following units: a pumping station to draw seawater, microfiltration, reverse osmosis, and brine disposal.

**seawater density** The mass of seawater per unit volume; it depends on salinity and temperature and is often expressed in $\sigma_t$ units, which is the density of the liquid in grams per liter minus 1000. For example, at a temperature of 20°C and a salinity of 4%, the seawater density is approximately 3.2 $\sigma_t$ units, i.e., 1003.2 g/L.

**seawater intrusion** *See* saltwater intrusion.

***sec*-butylbenzene** *See* butylbenzene.

**2-*sec*-butyl-4,6-dinitrophenol** *See* dinoseb.

**Secchi depth** The depth at which a Secchi disk becomes invisible or barely visible when it is lowered in a water body. In a lake or other impoundment, the Secchi depth corresponds approximately to the depth of the euphotic zone. Also called Secchi disk depth or Secchi disk transparency.

**Secchi disc** Same as Secchi disk.

**Secchi disk** An inexpensive tool for determining some characteristics of water bodies: water clarity, depth of the euphotic zone, and extinction and attenuation coefficients of visible light. A Secchi disk is a flat, circular metal plate of about 25 cm in diameter, divided in black and white quadrants. The disk is lowered into the water by a rope until it is just barely visible. At this point, the depth of the disk from the water surface is the recorded Secchi disk transparency.

**Secchi disk depth** Same as Secchi depth.

**Table S-04.** Seawater composition (average)

| Element | Concentration (mg/L) | Element | Concentration (mg/L) |
|---|---|---|---|
| Chloride (Cl) | 19,000 | Sodium (Na) | 10,600 |
| Magnesium (Mg) | 1,270 | Sulfur (S) | 880 |
| Calcium (Ca) | 400 | Potassium (K) | 380 |
| Bromide (Br) | 65 | Carbon (C) | 28 |
| Strontium (Sr) | 13 | Boron (B) | 5 |
| Other ions | 2,359 | Total | 35,000 |

**Secchi disk transparency** Same as Secchi depth.

**secondary benefit** A benefit that accrues to people who do not use directly the product or service resulting from an action. For example, industrial wastewater treatment contributes to the quality of the receiving water, which may be used by a downstream community for water supply or recreation. Also called indirect benefit. *See also* primary benefit, intangible benefit.

**secondary biological treatment** *See* secondary treatment and biological treatment.

**secondary carbonation** An operation in which carbon dioxide is bubbled into water being treated to lower the pH. *See* recarbonation for more detail.

**secondary clarification** Sedimentation following secondary treatment.

**secondary clarifier** A sedimentation tank that follows biological or other type of secondary treatment and is designed to remove by gravity the suspended solids in the biomass formed or present in the influent. Some sludge thickening also occurs in the clarifier. Design parameters vary from those used for primary clarifiers and depend on the type of solids to remove. A typical overflow rate for activated sludge is 800 gpd/sq ft. Also called final clarifier.

**secondary contact recreation** A recreational activity in which a person's water contact would be limited to the extent that bacterial infections of eyes, ears, respiratory or digestive systems, or urogenital areas would normally be avoided (such as wading or fishing) (EPA-40CFR131.35.d-13).

**secondary contaminant** A contaminant that does not affect the safety of drinking water but such parameters as taste, odor, and color. *See also* primary contaminant.

**secondary digester** The second unit in a two-stage digestion process; it is an unheated, unmixed, low-rate reactor that follows a high-rate digester and provides sludge thickening.

**secondary disinfectant** A chemical agent that can be used in secondary disinfection, e.g., free chlorine and chloramines, selected for its capability to maintain a reliable residual throughout transport and storage of water or wastewater without adverse health effects, as a protection against microbial regrowth. Chlorine dioxide is used in Europe, but in the United States there are health concerns about this agent as well as its inorganic by-products (chlorate and chlorite).

**secondary disinfection** A second or final disinfection step provided to maintain a disinfectant residual throughout water distribution and storage, and protect the liquid against microbial regrowth. Ammonia may be added to convert a free residual to a combined residual. *See also* primary disinfection, regrowth, baseline treatment.

**secondary drinking water regulations** Nonenforceable regulations applying to public water systems and specifying the maximum contamination levels that, in the judgment of the USEPA, are required to protect the public welfare. These regulations apply to any contaminants that may adversely affect the odor or appearance of such water and, consequently, may cause people served by the system to discontinue its use.

**Secondary Drinking Water Standards** A set of regulations issued by the USEPA to classify physical parameters. Table S-05 shows current limits for drinking water.

**secondary effluent** (1) The effluent of a secondary treatment plant, i.e., the liquid portion of the wastewater that has undergone primary and biological treatment. Solids removed by treatment are not part of the effluent. (2) The effluent of a wastewater treatment plant that meets the requirements of 30 mg/L or less in $BOD_5$ and 30 mg/L or less in suspended solids.

**secondary feeder** (1) Same as secondary main. (2) An organism (e.g., protozoa) that feeds on bacteria and other primary feeders. *See also* predator–prey relationship.

**secondary footprint** *See* carbon footprint.

**secondary main** In a pipe network, primary mains constitute its basic structure, carrying flow to and from storage tanks and usually laid in loops. Secondary mains form smaller loops within primary loops, running from one primary line to another, at an average spacing of three blocks.

**Table S-05.** Secondary Drinking Water Standards (in mg/L except as noted)

| | | | | | |
|---|---|---|---|---|---|
| Aluminum | 0.05–0.20 | Manganese | 0.05 | **Special units** | |
| Chloride | 250 | Silver | 0.10 | Color | 15 |
| Fluoride | 2.0 | Sulfate | 250 | Corrosivity | noncorrosive |
| Foaming agents | 0.5 | TDS | 500 | Odor | 3 TON |
| Iron | 0.3 | Zinc | 5 | pH | 6.5–8.5 |

**secondary maximum contaminant level (SMCL)** Nonenforceable concentration limits for secondary contaminants (taste, odor, color) established by the USEPA to protect public welfare. The SMCL means the maximum permissible level of a contaminant in water that is delivered to the free-flowing outlet of the ultimate user of a public water system. Contaminants added to the water under circumstances controlled by the user, except those resulting from corrosion of piping and plumbing caused by water quality, are excluded from this definition (EPA-40CFR143.2-f). *See* Secondary Drinking Water Standards for detail. *See also* primary maximum contaminant level.

**secondary MCL** *See* secondary maximum contaminant level.

**secondary metabolite** In a microbial culture, a substance, such as an antibiotic, that is produced at a low rate to maintain high cell numbers but is not associated with growth. *See also* primary metabolite.

**secondary minimum aggregation** The condition that prevails in a suspension when particles are held in proximity by van der Waals forces but remain apart under the influence of repulsive forces of the electrical double layers. If the kinetic energy of the particles increases, they tend to aggregate.

**secondary phosphorus release** The release of phosphorus (in the form of orthophosphate, O-$PO_4$) in an enhanced biological removal system without the concomitant uptake of volatile fatty acids by the organisms, a phenomenon that may lead to high phosphorus concentrations in the effluent. It is caused by long anaerobic retention times and poor sludge characteristics.

**secondary pollutant** A pollutant that is formed in the environment by the combination of other pollutants or of naturally occurring substances; e.g., ozone and nitrogen oxides.

**secondary production** *See* microbial loop.

**secondary propyl alcohol ($C_3H_8O$ or $CH_3$—$CH_3$—CHOH)** Same as isopropyl alcohol.

**secondary radioactive nuclide** A nuclide formed in radioactive transformations of uranium-238, uranium-235, and thorium-232. *See also* primary and induced radioactive nuclides.

**secondary release** *See* secondary phosphorus release.

**secondary salinization** The accumulation of salts at or near the surface, following an upward movement of saline groundwater and evaporation from the soil surface.

**secondary settling tank** Same as secondary clarifier.

**secondary sludge** The sludge from secondary treatment, as opposed to primary sludge or chemical sludge. *See also* biosolids and waste-activated sludge.

**secondary standards** A pollution limit based on environmental effects and not related to health. National ambient air quality standards designed to protect welfare, including effects on soils, water, crops, vegetation, man-made materials, animals, wildlife, weather, visibility, climate, damage to property, transportation hazards, effects on economic values, and personal comfort and well-being. For drinking water, *see* Secondary Maximum Contaminant Level. *See also* primary standard.

**secondary substrate** *See* cometabolism.

**secondary treatment** (1) The second step in most publicly owned wastewater treatment systems, intended to remove soluble and colloidal organic matter, following primary treatment. It is accomplished by physicochemical processes, but more commonly by biological treatment (e.g., trickling filter and activated sludge) bringing together waste, bacteria, and oxygen. This treatment removes floating and settleable solids and about 90% of the oxygen-demanding substances and suspended solids. Disinfection is the final stage of secondary treatment. Secondary treatment is sometimes used synonymously with such biological processes as activated sludge and trickling filtration. *See also* primary treatment, tertiary treatment. (2) Wastewater treatment that meets (a) removal efficiencies of 85% for BOD and suspended solids (SS) or (b) the effluent requirements of 30 mg/L or less in $BOD_5$ and 30 mg/L or less in SS. Additional requirements may include $6.0 < pH < 9.0$ and carbonaceous BOD of $CBOD_5 < 25$ mg/L as a 30-day average or < 40 mg/L as a 7-day average.

**secondary treatment equivalency** A minimum national performance standard established for publicly owned treatment works (municipal wastewater dischargers). The standard specifies limits for the BOD, suspended solids, pH, whole effluent toxicity, and fecal coliform bacteria of effluents from conventional secondary treatment, trickling filters, and stabilization ponds. For example, for conventional secondary treatment, at least 85% of the suspended solids must be removed and the effluent must contain less than 30 mg/L as a 30-day average and less than 45 mg/L as a 7-day average.

**secondary treatment standard** A regulatory requirement that defines the quality of the effluent of a secondary wastewater treatment plant, e.g., 30 mg/L or less of suspended solids and 30 mg/L or less of BOD on a monthly average basis.

**secondary wastewater treatment** *See* secondary treatment.

**second-class water quality** The classification for water that is not directly ingested but is in contact with skin or tools used in cooking. It represents approximately 25% of the daily domestic water consumption.

**Second Contaminant Candidate List (CCL2)** or **Contaminant Candidate List 2** A list of 51 drinking water contaminants proposed in 2004 by the USEPA for possible future regulations. It includes nine microbiological contaminants and 42 chemicals. *See* Contaminant Candidate List.

**second-foot (second-feet)** An old term for cubic foot (feet) per second (cfs).

**second-foot day (second-feet day)** A unit of daily volume, equivalent to the discharge during a 24-hour period when the flow rate is one cubic foot per second (1.0 cfs), i.e., a volume of 86,400 cubic feet or 1.98 acre-feet.

**second law of thermodynamics** Between two bodies in contact, heat flows from the warmer to the cooler; energy tends to be distributed evenly in a closed system, i.e., its entropy increases with time. The second law applies also to biological and chemical processes. For example, a sugar lump will dissolve in water, i.e., concentrations tend to decrease. *See also* zeroth law, first law, third law of thermodynamics.

**second-order consumer** In an aquatic food chain, sunfishes and other carnivores that consume the herbivores (first-order consumers) and are eaten by third-order consumers.

**second-order reaction** A reaction in which a variable (e.g., $r$ = rate of change) is proportional to the square of another (e.g., a concentration, C) or the product of two other variables whose exponents add up to 2. Examples of second-order equations are:

$$r = \pm k \; C^2 \qquad \text{(S-20)}$$

$$r = \pm k \cdot C_1^a \cdot C_2^{2-a} \qquad \text{(S-21)}$$

where $C_1$ and $C_2$ are concentrations, $k$ = reaction rate constant and $a$ = an exponent (reaction order).

**second-stage biochemical oxygen demand** The quantity of oxygen consumed by the biochemical oxidation of ammonia, organic, and other forms of nitrogen in wastewater, which starts after satisfaction of a good deal of the carbonaceous oxygen demand. *See* nitrogenous oxygen demand for more detail.

**second-stage BOD** *See* second-stage biochemical oxygen demand.

**section 304(a) criteria** Criteria developed by the USEPA under authority of section 304(a) of the Clean Water Act based on the latest scientific information on the relationship that the effect of a constituent concentration has on particular aquatic species and/or human health. The information is issued periodically to the states as guidance for use in developing criteria (EPA-40CFR131.3-c).

**section 404 program** A USEPA-approved state program to regulate the discharge of dredged or fill material under section 404 of the Clean Water Act in state regulated waters (EPA-40CFR124.2 and 233.2). *See also* state program.

**sectionalizing valve** A device installed in a pipeline to shut off flow in a section for inspection or repair. Also called stop valve.

**secure chemical landfill** Same as secure landfill.

**secure landfill** A disposal site for hazardous waste, selected and designed to minimize the chance of release of hazardous substances into the environment, for example, by segregating and isolating these materials and preventing them from reaching the groundwater or the atmosphere. Also called secure chemical landfill. *See also* landfill, sanitary landfill.

**secure maximum contaminant level** Maximum permissible level of a contaminant in water delivered to the free-flowing outlet of the ultimate user, or of contamination resulting from corrosion of piping and plumbing caused by water quality. *See also* maximum contaminant level and maximum contaminant level goal.

**security** *See* safety.

**sedentary soil** Soil formed on-site from the original rock. Also called residual soil.

**sedge** Any rushlike or grasslike plant growing in wet places, like the emergent plants used along the shoreline and in shallow waters of constructed wetlands.

**Sedgewick, William T.** A sanitary engineer of the Massachusetts Lawrence Experiment Station who established the link between typhoid fever and drinking water toward the end of the 19th century.

**Sedgwick–Rafter cell** A microscope cell measuring 50 mm × 20 mm × 1 mm, i.e., 1000 mm$^3$ or 1 milliliter, used for counting microorganisms larger than bacteria in water. The use of a standard unit of 8000 cubic microns has also been reported for algae and other plankton.

**Sedgwick–Rafter filter** A funnel used as a filter to concentrate and count microorganisms larger than bacteria in water.

**Sedgwick–Rafter method** The quantitative determination of microorganisms larger than bacteria in water, using the Sedgwick–Rafter cell and a low-magnification microscope.

**Sedifloat** Water and wastewater treatment units manufactured by Krofta Engineering Corp.

**Sediflotor®** An apparatus manufactured by Infilco Degremont, Inc. for water treatment by dissolved-air flotation.

**sediment** Mineral or organic material that is in suspension, is being transported, or has been moved from its site of origin by water, wind, ice, or mass wasting and has come to rest on the Earth's surface either above or below sea level (Hawley and Parsons in Dodson, 1998). Soil, sand, and minerals washed from land into water, usually after rain. They pile up in reservoirs, rivers, and harbors, destroying fish and wildlife habitat and clouding the water so that sunlight cannot reach aquatic plants. Careless farming, mining, and building activities will expose sediment materials, allowing them to wash off the land after rainfall. Streams carry sediment in the suspended form and as bedload.

**sedimentary cycle** A biogeochemical cycle in which materials move back and forth between land and sea.

**sedimentary record** A record of the accumulation of trace elements in sediments and a profile of the concentrations at various depths.

**sedimentary rock** Rock formed by the accumulation and cementation of loose mud or sand materials in water. *See also* igneous rock, metamorphic rock. These materials may include fragments of preexisting rocks, organic remains, evaporates, and fragments blown out of volcanoes.

**sedimentation** (1) The process of settling out by gravity of suspended solids (sediment) and floating material from water, wastewater, or other liquids. In water and wastewater treatment, the process is carried out in a sedimentation tank, sedimentation basin, or clarifier, and is sometimes enhanced by coagulation and flocculation. It is hindered by turbulence. Sedimentation also applies to the settling of grit in primary clarifiers and the biological floc removal in secondary clarifiers. Design criteria for sedimentation include detention time, overflow rate, (effluent) weir loading, horizontal velocity, surface loading, and surface settling rate. Also called clarification, settling. *See also* plain sedimentation. (2) The deposit of solid particles at the bottom of a stream due to their greater density than the carrying water; the solids may be resuspended if flow velocity is sufficient to produce bottom scour. *See also* sedimentation terms. (3) A major filtration mechanism: see inertial impaction–sedimentation.

**sedimentation basin/sedimentation tank** A quiescent, circular or rectangular, basin/tank for the settling of suspended (settleable) solids by gravity and their subsequent removal as sludge, usually by a motor-driven rake mechanism. This mechanism used to be considered the difference between clarifier and settling tank but this distinction no longer applies and the two terms are used synonymously. Common sedimentation design criteria include: detention time, overflow rate, weir loading, and horizontal velocity. Other design considerations apply to tank inlets (to reduce density currents), outlets (weirs), and optimum proportions. Also called clarifier, settling basin/tank, subsidence tank. *See also* drag coefficient, horizontal velocity, Lamella clarifier, primary sedimentation tank, secondary sedimentation tank, sediment tank, upflow clarifier, short-circuiting, surface overflow rate, terminal settling velocity, tube settler, weir overflow rate.

**sedimentation/clarification phase** The third step in the sequencing batch reactors process, during which solids separate from the liquid under quiescent conditions, producing a clarified supernatant. Also called settle phase. *See also* fill phase, decant phase, react phase, and idle phase.

**sedimentation compartment** The portion of a water or wastewater treatment basin used for the settling of particles, as in an Imhoff tank or a flocculation–sedimentation unit.

**sedimentation number ($N_s$)** A dimensionless number ($N_s$) relating fluid density ($\rho$), particle diameter ($d$), acceleration of gravity ($g$), particle density ($\rho'$), and absolute viscosity ($\mu$). It is the reciprocal of the Galileo number and is used in the study of sedimentation. The Galileo number is used in determining the minimum fluidization velocity in filter backwashing. *See* minimum fluidization condition.

$$N_s = \mu^2/[d^3\rho(\rho' - \rho)g] \quad \text{(S-22)}$$

**sedimentation pond** An earthen lagoon that receives the effluent of a stabilization pond for solids separation and/or effluent polishing, with adequate provision for sludge withdrawal and storage. *See also* maturation pond.

**sedimentation potential** The potential arising from an imposed movement of charged particles through a solution; the potential gradient created when charged particles settle relative to a stationary liquid; the opposite of electrophoresis.

**sedimentation tank** *See* sedimentation basin and sediment tank.

**sedimentation terms** *See* the following terms related to the process of sedimentation (3) in streams: anabranch, bed configuration, bed form, bed load, bed material, bed-material discharge, braided stream, contact load, contact-sediment discharge, critical shear stress, discharge, fine-sediment discharge, flow-duration curve, fluvial, geomorphology, hydraulic radius, meandering stream, median diameter, sediment yield, shear stress, shear velocity, slope, suspended sediment, suspended-sediment discharge, total bed-material discharge, total sediment discharge, tractive force, unmeasured sediment discharge, wash-load discharge.

**sedimentation test** A test using a soil sample in a container of water to determine the soil characteristics for excavation work.

**sedimentation type I** The gravity settling of discrete, nonflocculent particles, as separate, unhindered units; characterized by very low solids concentrations. It is the primary mechanism in grit removal. The design of ideal settling basins is based on this type. Such a basin has four zones: inlet, settling, sludge, outlet; *see* ideal sedimentation basin and Figure I-01. Underlying assumptions include: (a) no settling occurs in the inlet and outlet zones; (b) uniform dispersion and concentration of particles; (c) continuous, constant flow; (d) no resuspension of settled particles; (e) particles move forward at the same velocity as the liquid. *See also* the Bhargava–Rajagopal equation, Stokes' law, terminal settling velocity. Also called discrete settling, free settling, nonflocculent settling, type I sedimentation, type I settling.

**sedimentation type II** The settling of flocculent particles whose size and velocity increase as time passes, as a result of either their natural tendency to agglomerate or the addition of chemicals. The primary mechanism of primary sedimentation, it is characterized by low solids concentration. Also called flocculent or type II sedimentation, type II settling. *See also* settling column test, settling tube.

**sedimentation type III** The sedimentation of closely spaced particles in a suspension of intermediate concentration (e.g., sludge or wastewater) so that settling is hindered by interparticle forces and is about the same for an entire zone. It is characterized by high solids concentrations and is the primary mechanism in final or secondary sedimentation. *See* Figure S-04, which shows the two main zones: sludge blanket and concentrated sludge. The particles tend to remain in a fixed positions with respect to one another and to settle in mass as a unit, with a solids–liquid interface at the top. Terminal settling velocity depends on the frictional forces and the extent of flocculation. Type III sedimentation is normally used to design secondary clarifiers. Also called hindered settling, type III settling, or zone settling. *See also* column settling test, gravity flux, limiting flux, sludge blanket, solids flux, Stokes' law, underflow flux, wall effect, and other types of settling: I or free, II or flocculent, IV or compression.

**sedimentation type IV** The settling of particles in a concentrated suspension where further settling can occur only by compression of the existing structure of settled particles. Also called compression settling, type IV sedimentation, type IV settling. *See also* Stokes' law, and other types of settling: I or free, II or flocculent, III or hindered or zone.

**sedimentation zones** The four zones of an ideal sedimentation basin, as proposed by Thomas R. Camp in 1936 and 1953. *See* inlet zone, outlet zone, settling zone, sludge zone. *See also* sedimentation type I.

**sediment basin** A settling pond with a controlled water-release structure, designed to collect and store sediment from construction and other activities.

**sediment composition testing** *See* sludge composition testing.

**sediment concentration** The dimensionless ratio of the weight (or volume) of the sediment in a suspension to the weight (or volume) of the suspension; expressed as a percentage or in ppm.

**sediment concentration testing** *See* sludge concentration testing.

**sediment condition testing** *See* sludge condition testing.

**sediment discharge** Dry weight of sediment per unit of time. Same as sediment load.

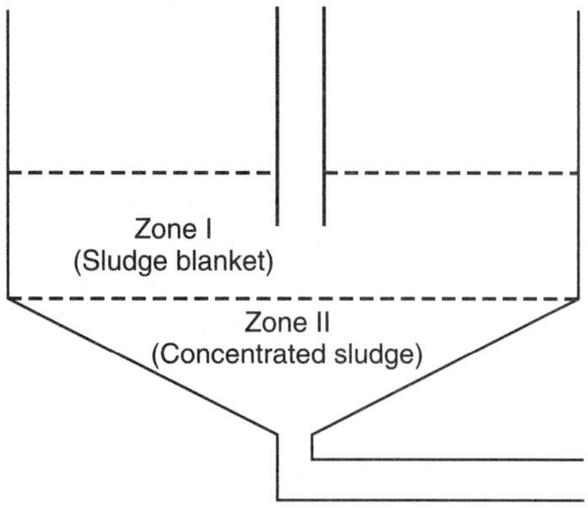


**Figure S-04.** Sedimentation Type III.

**sediment discharge curve** A graphical representation of sediment discharge versus the stage of a stream.

**sediment erosion** The process of detachment of sediment particles by water, wind, ice, glaciers, etc.

**sediment load** Dry weight of sediment per unit of time. Same as sediment discharge.

**sediment oxygen demand** The oxygen demand exerted by decomposing sediments that are partly organic, e.g., the residual solids discharged with treated wastewater effluents, which eventually settle to the bottom of the receiving water.

**sediment removal efficiency (*E*)** A formula proposed by Thomas R. Camp in 1946 and used in stormwater modeling to estimate the performance of extended detention wet ponds:

$$E = 1 - (1 + A \cdot V/nQ)^n \qquad \text{(S-23)}$$

where $A$ is the active storage surface area of the facility, $V$ the settling velocity, $n$ a turbulence or short-circuiting parameter, and $Q$ the average constant release rate from the facility.

**sediment tank** A portable metal tank providing at least ten minutes of storage for sediment to settle out discharges from a construction site. *See also* sediment basin, sedimentation tank.

**sediment tests** *See* sludge tests commonly conducted for wastewater solids deposited in collection works or accumulated in receiving waters, as well as those generated or removed in treatment processes.

**sediment trap** (1) A device installed in a conduit to trap and remove sediment from water. *See also* sand trap. (2) A temporary device to retain runoff from a small drainage area and allow silt to settle out. It consists of an excavated pond or a basin formed by an embankment across a low area or drainage swale, with an outlet or spillway of large stones or aggregates.

**sediment trapping efficiency** The percentage of sediment trapped in a reservoir or a detention basin. It may be estimated using Brune's trap efficiency curves, as a function of the ratio of reservoir capacity to annual flow.

**sediment yield** The quantity of sediment from a unit of land (field, watershed, or drainage area) arriving at a specific location per unit of time. *See also* sedimentation terms.

**seed** A well-digested or ripened sludge used to seed a digester. *See also* sludge seeding.

**seeded slurry precipitation and recycle** A variation of the reverse osmosis process that uses crystals to precipitate scaling compounds in membrane filtration. *See* slurry precipitation and recycle reverse osmosis for detail.

**seeding** The use of crystals to facilitate the precipitation of solids and to control scaling during crystallization and evaporation of water treatment brines.

**seeding material** Same as seed.

**seed slurry** A quantity of crystals maintained in suspension in crystallizers and evaporators to facilitate the precipitation of solids and control the scaling of heat transfer surfaces. Calcium sulfate ($CaSO_4$) is a common seed material.

**seep** A spot where groundwater oozes slowly to form a pool on the surface; a small spring.

**seepage** Percolation of water through the porous soils from the bottom or sides of unlined canals, ditches, laterals, watercourses, or water storage facilities. Also, the slow movement of water through small cracks, pores, or interstices of a material into or out of a body of surface or ground water.

**seepage bed** A disposal field with a bottom width greater than 3–4 ft. Also called disposal bed. *See* soil absorption system for detail.

**seepage pit** A covered excavation that is used to introduce, by seepage through the bottom and sides, effluents of septic tanks or other partial treatment devices from small installations. It consists of one or more pits dug into the porous layer and lined with open-jointed materials, and a means of access such as a manhole cover. It is sometimes assimilated to a giant cesspool that receives effluents from neighboring individual cesspools and septic tanks. Also called disposal pit. *See also* disposal bed, seepage bed, soil absorption system.

**seepage pond** A pond constructed to allow wastewater, storm or other surface water to percolate and recharge underground formations. Other means of groundwater recharge include check dams, recharge wells, spreading grounds, and underground leaching systems.

**seepage spring** A spring that occurs where the water table meets the ground surface: water flows by gravity from numerous small openings in a permeable materials or from openings in a rock formation. Also called a filtration spring.

**seepage trenches** A network of 18–36 in. deep and 12–36 in. wide trenches, containing 6–12 inches of gravel or crushed stone and open-jointed tiles for the distribution and seepage of partially treated wastewater. The trenches are backfilled to the level of the natural ground. The system provides additional treatment as well as water loss by evapotranspiration.

**seepage velocity** The ratio of the volume of groundwater through a formation in unit time to the product of the cross-sectional area by the effective porosity. *See* actual groundwater velocity for more detail.

**Seghers Pelletech®** An apparatus manufactured by Wheelabrator Clean Water Systems, Inc. for sludge drying and pelletizing.

**segregated flow model (SFM)** An equation that uses normalized microorganism inactivation dose–response data and normalized detention-time data from batch disinfection and tracer or dispersion studies to estimate the performance of a chlorine contact basin, assuming no interaction between successive blocks of water in the basin (Metcalf & Eddy, 2003):

$$dN = N(\theta) \cdot E(\theta) \cdot dt \qquad \text{(S-24)}$$

where $dN$ = reduction in the number of organisms between times $t$ and $t + dt$, $N(\theta)$ = number of organisms remaining after time $t$, $\theta$ = normalized residence time, and $E(\theta)$ = fraction of flow remaining in the basin between times $t$ and $t + dt$.

**segregated stormwater sewer system** A drain or collection system designed and operated for the sole purpose of collecting runoff at a facility, and which is segregated from all other individual drain systems (EPA-40CFR61.341).

**segregation** *See* industrial waste segregation.

**seismic survey** A geophysical investigation conducted to locate groundwater sources, based on the time it takes a shock wave to arrive at a number of detectors. *See also* resistivity survey, gravity survey.

**seize up** The phenomenon that occurs when an engine overheats and a part expands to the point where the engine will not run. Also called freezing.

**selective adsorption** A gas chromatography technique used to separate and analyze organic compounds.

**selective calcium carbonate removal** A method used to soften water that is low in magnesium hardness, e.g., less than 40 mg/L as $CaCO_3$. It consists of lime clarification, single-stage recarbonation, and filtration. Addition of soda ash may or may not be required.

**selective draft** A type of water supply that diverts and stores enough clean floodwaters to meet the demand during low-flow seasons. On the other hand, continuous draft uses a surface source without an impounding reservoir, i.e., water is pumped directly from the source to the treatment works. Continuous draft is used when available flow is high at all seasons compared to the demand.

**selective enrichment medium** A medium that contains substances or conditions that allow a specific organism to grow normally but inhibit the growth of other organisms so that the latter can be more easily detected or enumerated. *See also* differential medium, elective enrichment medium, selective plating medium, specialized isolation medium.

**selective ion electrode** An electrode used in the electrochemical analysis of water or wastewater, e.g., to determine the concentration of specific ions such as hydrogen or fluoride, based on the measured voltage in the solution. Also called selective ion probe.

**selective ion exchange** A treatment process that uses a solid phase containing free ions that can be displaced by ions of the opposite charge, e.g., chelating ion exchange resins for certain cations. The process is used, e.g., (a) to remove the ammonium ion before recarbonation, as a substitute for air stripping; and (b) to recover metals from plating solutions.

**selective ion exchanger** An ion-exchange medium that removes specific ions from solution.

**selective ion probe** Same as selective ion electrode.

**selective leaching** A form of corrosion that consists in the preferential removal of one metal from an alloy; e.g., dezincification (the removal of zinc from brass) or the removal of disseminated lead from brass and soldered joints. Also called dealloying. (2) The dissolution of asbestos–cement pipe or deterioration of cement mortar linings of iron pipe by aggressive waters.

**selective medium** A medium that allows the growth of certain microorganisms in preference to or to the exclusion of others.

**selective membrane** A membrane that selectively passes specific anions or cations.

**selective plating medium** An agar medium modified to suppress the growth of one group of organisms and promote the growth of another. Selective agents include antibiotics, heavy metals, and toxic chemicals. Environmental conditions (pH, dissolved oxygen, temperature) may also be used. *See also* selective enrichment medium.

**selective precipitation** The precipitation of target ions by control of pH or counterions, e.g., from plating solutions.

**selective toxicity** Toxicity of a substance to specific organisms.

**selective withdrawal** A phenomenon associated with the stratification of lakes and reservoirs, whereby water is withdrawn from the layer of fluid that is at the same level as the outlet, with consequences for the temperature and other characteristics of the withdrawal.

**selectivity** (1) Characteristic of an ion-exchange resin or a pressure-driven membrane to remove or retain some ions or some molecules in preference to others. (2) An experimental parameter used to evaluate the performance of pervaporation membranes, expressed as a separation factor or the selectivity coefficient defined for ion-exchange resins. *See also* permeation flux.

**selectivity band** The zone in an ion-exchange or adsorption medium where specific ions or substances accumulate or are removed from solution, based on their preference for that zone.

**selectivity coefficient** An equilibrium relationship derived from an ion-exchange reaction and characterizing the selectivity of an exchanger for a specific ion. The selectivity coefficient ($K_{ij}$) of the exchange of ion $j$ for ion $i$ is defined as:

$$K_{ij} = [\text{ion } i^*][\text{ion } j]/([\text{ion } j^*][\text{ion } i]) = (Q_i)(C_j)/(Q_j)(C_i) \quad \text{(S-25)}$$

where the asterisk (*) denotes resin phase (with omission of the matrix designation), the quantities between brackets [ ] denote concentrations in mol/L; $Q_i$ = concentration of ion $j$ on resin, eq/L; and $C_i$ = concentration of ion $i$ in water, eq/L. *See* selectivity quotient, selectivity sequence, separation factor, homovalent exchange, and ion-exchange equilibria for related terms.

**selectivity quotient** A parameter ($Q_s$), similar to an equilibrium constant, that characterizes the equilibrium relation of an ion exchange reaction such as (Droste, 1997):

$$nR^-B^+ + A^{n+} \rightarrow R_{n-}A^{n+} + nB^+ \quad \text{(S-26)}$$

$$Q_s = [B^+]^n[R_n^- A^{n+}]/[R^-B^+]^n[A^{n+}] \quad \text{(S-27)}$$

where $[R_n^- A^{n+}]$ and $[R^-B^+]$ are in moles per gram of resin or moles per volume of resin.

**selectivity reversal** The phenomenon of an ion-exchange isotherm that changes from favorable (convex) to unfavorable (concave) when the ionic solution increases and the selectivity for higher-valence ions decreases.

**selectivity sequence** The order in which a particular ion-exchange resin or porous oxide surface (e.g., activated alumina) prefers ions. Preferred ions have high separation factors and selectivity coefficients or relative affinities. Note that ions exit the ion-exchange column in the reverse order of selectivity. Examples of selectivity sequence for (a) activated alumina in the pH range of 5.5 to 8.5, (b) synthetic cationic resins, (c) synthetic anionic resins, (d) polyacrylic resins, and (e) polystyrene resins, respectively, follow.

(a) $I^- < Br^- < NO_3^- < Cl^- < HCO_3^- \ll CrO_4^{2-} \ldots$
$\ldots < SO_4^{2-} < HSeO_3^- < F^- < Si(OH)_3O^-$,
$H_2AsO_4^- < OH^-$ (S-28)

(b) $Li^+ < H^+ < Na^+ < NH_4^+ < K^+ < Rb^+ < Ag^+ \ldots$
$\ldots < Mg^{2+} < Zn^{2+} < Co^{2+} < Cu^{2+} < Ca^{2+}$
$< Sr^{2+} < Ba^{2+}$ (S-29)

(c) $OH^- < F^- < HCO_3^- < Br^- < NO_3^- < ClO_4^-$ (S-30)

(d) $HCO_3^- < Cl^- < NO_3^- < H_2AsO_4^- < ClO_4^-$
$< SO_4^{2-}$ (S-31)

(e) $HCO_3^- < Cl^- < NO_3^- < H_2AsO_4^- < SO_4^{2-}$
$< ClO_4^-$ (S-32)

**selectivity series** Same as selectivity sequence.

**Selectofilter** A revolving-drum strainer manufactured by FMC Corp.

**selector** One or more chambers or small contact tanks in series installed in a biological wastewater treatment plant to limit the growth of undesirable organisms or to promote the growth of selected organisms. For example, selectors are placed ahead of a completely mixed reactor for rapid mixing of the influent wastewater and return sludge to promote the development of (i.e., select) nonfilamentous organisms and avoid bulking by increasing the F/M ratio. Selector design is based on kinetics (high F/M ratios) or metabolism (anoxic or anaerobic conditions). Also called biological selector, selector reactor, selector tank.

**selector concept** The use of a specific reactor design to promote the growth of desirable organisms in biological wastewater treatment, e.g., floc-forming instead of filamentous bacteria.

**selector process** A wastewater treatment process using a series of small reactors (with different conditions, e.g., aerobic, anoxic, anaerobic) instead of, or ahead of, a large completely mixed or plug-flow reactor.

**selector reactor** Same as selector.

**selector tank** Same as selector.

**selector zone** In biological nitrogen removal systems, anaerobic or anoxic selector zones promote the growth of floc formers by restricting access to food by filamentous organisms.

**Selectostrainer** An in-line strainer manufactured by FMC Corp. for use in water or wastewater screening.

**Selemion** A membrane manufactured by Ashai Glass America, Inc. for use in ion-exchange treatment.

**selenate ($SeO_4^{2-}$)** A salt or ester of selenic acid ($H_2SeO_4$).

**selenide** A compound of divalent selenium with a positive element, e.g., potassium selenide ($K_2Se$).

**selenite ($SeO_3^{2-}$)** (1) A salt or ester of selenious acid ($H_2SeO_3$). (2) A variety of gypsum in colorless, transparent crystals and foliated masses.

**selenium (Se)** A gray or red, nonmetallic element of the sulfur group, occurring in certain soils and used in electronics, photography, and other industrial applications. Atomic weight =78.96. Atomic number = 34. Specific gravity = 4.79 (gray), 4.50 (red), or 4.28 (vitreous). Its compounds (selenate, selenite) are soluble in water. Selenium is an essential nutritional element at low levels, but, at sufficiently high concentrations, it may interfere with enzyme production and cause physiological effects in humans. Over the long term it may cause red staining of fingers, teeth, and hair. Regulated by the USEPA as a drinking water contaminant: MCL = MCLG = 0.05 mg/L. Effective removal processes: anion exchange, reverse osmosis, electrodialysis, activated alumina adsorption. *See also* alkali disease, blind staggers.

**Selex®** A cartridge filter manufactured by Osmonics, Inc.

**self-assembled monolayers on mesoporous supports (SAMMS)** A highly porous material, consisting of nanoporous silica coated with thiol, that reportedly removes mercury from wastewater, while ignoring other ions like calcium, magnesium, and sodium. The material is reusable and the process seems to cost only 20% as much as traditional mercury removal methods.

**self-backwashing filter** A filter that is designed hydraulically for backwashing without pumping, but using the gravity flow of filtered water.

**self-cleansing velocity** A flow velocity that is expected to prevent deposition of solids in sewers. A minimum velocity of 2.0 fps is often recommended but, in general, the self-cleansing velocity ($V$) is:

$$V = C[kd(\gamma_s - \gamma)/\gamma]^{0.5} \quad \text{(S-33)}$$

where $C$ = Chézy coefficient of roughness; $k$ = a coefficient depending on the porosity ratio of the sediment and the slope of the sewer, and varying from 0.04 to more than 0.80; $d$ = sediment diameter; $\gamma$ = specific weight of the wastewater; and $\gamma_s$ = specific weight of the sediment. *See also* scouring velocity.

**self-contained recycle system** *See* recycle treatment system.

**self-ionization** A chemical reaction in which a substance is both a proton donor and a proton acceptor, i.e., the substance acts like an acid and a base. For water ($H_2O$), ammonia ($NH_3$), and sulfuric acid ($H_2SO_4$):

$$H_2O + H_2O \rightarrow H_3O^+ + OH^- \quad \text{(S-34)}$$

$$NH_3 + NH_3 \rightarrow NH_4^+ + NH_2^- \quad \text{(S-35)}$$

$$H_2SO_4 + H_2SO_4 \rightarrow H_3SO_4^+ + HSO_4^- \quad \text{(S-36)}$$

*See also* water ionization.

**self-propelled sludge vessel** A surface vessel used to transport sludge to an ocean disposal site; it consists of a vessel divided into tank compartments, each quipped with a gate at the bottom to discharge the sludge by gravity. *See also* articulated tug barge, barges-in-tow, barge transport, and rubber barges-in-tow.

**self-purification** The slow reduction or elimination of contaminants from natural waters by such physical, chemical, and biological phenomena or factors as atmospheric reaeration, bottom deposit, currents, dilution, non-point-source photosynthesis, runoff, sedimentation, sunlight, and temperature. These natural processes result in the die-away of pathogens, the stabilization of organic matter, the replenishment of dissolved oxygen, and the restoration of normal biota. Filtration is the principal process in the self-purification of groundwater. Also called natural purification, as compared to the purification provided by treatment. *See also* bacterial self-purification.

**self-purification coefficient** In dissolved-oxygen analyses of streams, the self-purification coefficient ($f$) is the ratio of the reaeration rate ($r$) to the deoxygenation rate ($k$) or the ratio of the receiving stream's first-stage BOD at the critical point ($L_c$) to the critical dissolved-oxygen deficit ($D_c$):

$$f = r/k = L_c/D_c \quad \text{(S-37)}$$

At 20°C, the self-purification coefficient varies from less than 1.0 for small ponds and backwaters to about 2.5 for large streams of moderate velocity to more than 5.0 for rapids and waterfalls. *See also* dissolved-oxygen sag.

**self-purification constant** Same as self-purification coefficient.

**self-purification ratio** Same as self-purification coefficient.

**self-supplied water** Water withdrawn from a surface or ground water source by a user, as compared to water from a water supply agency.

**SelRO®** An apparatus manufactured by LCI Corp. for water treatment using membrane filtration.

**seltzer** or **Seltzer** (1) A naturally effervescent water containing some sodium chloride (NaCl), calcium carbonate ($CaCO_3$), and magnesium carbonate ($MgCO_3$). From the German village Selters, near Wiesbaden. (2) Tap water made effervescent by the addition of carbon dioxide ($CO_2$), without the addition of mineral salts. Also called soda water or sparkling water. *See also* bottled sparkling water.

**seltzer water** Same as seltzer (2).

**SEM** Acronym of (1) scanning electron microscope, (2) standard error of the mean.

**semibatch ozone reactor** A reactor containing a static volume of liquid to which ozone ($O_3$) is added continuously.

**semiconductor** *See* semimetal or metalloid.

**semiconfined aquifer** An aquifer partially confined by soil layers of low permeability through which recharge and discharge can still occur. Also called leaky aquifer. *See also* aquitard.

**semicontinuous filter** A filter that is periodically taken off-line for backwashing. *See also* continuous filter.

**semimetal** A metalloid.

**semipermeable** Permeable only to certain small molecules.

**semipermeable membrane** A membrane without measurable pores but permeable to small molecules or to molecules of a special nature. It is used in such processes as dialysis and reverse osmosis for the separation of constituents (e.g., colloids) in a fluid, based on differences in one or more characteristics, e.g., diffusion rate, solubility, electrical charge, or size and shape. Also called permselective membrane.

**senescence** The aging process or condition; sometimes used to describe the condition of lakes or other bodies of water in advanced stages of eutrophication, with a lot of sediment and rooted aquatic plants, and on their way to becoming marshes.

**senescent lake** *See* senescence.

**SenSafe™ manganese check** A test developed by Industrial Test Systems, Inc. of Rock Hill, SC to make manganese testing safer and without a meter. It involves dipping manganese strips in a water or wastewater sample and comparing their color to a standard color block. One PAN indicator strip uses ascorbic acid to reduce oxidized manganese ions to $Mn^{2+}$. A second strip uses an alkaline cyanide reagent to mask interferences and a third strip develops an orange color in the presence of manganese.

**sensible heat** Heat that is measurable by temperature; a critical parameter in the design of evaporators. The sensible heat ($Q_s$, in kJ or Btu) is equal to the heat capacity or specific heat ($h_s$, kJ/kg/°C or Btu/lb/°F) of a substance multiplied by its weight ($W$, in kg or lb) and by its temperature ($T$, °C or °F) above a reference point ($T_0$, °C or °F):

$$Q_s = h_s(W)(T - T_0) \qquad \text{(S-38)}$$

The reference temperature is usually set at 0°C (32°F) or 15.56°C (60°F). The sensible heat of sludge in thermal processing includes the heat content of the ash and the requirement of the flue gases. *See also* latent heat and flame temperature.

**sensitive population** A group of persons who are more easily affected than others by pathogens; e.g., individuals with compromised immune systems, infants, chronically ill, elderly, and AIDS and cancer patients. Recent water quality standards pay attention to the conditions of the sensitive population regarding new pathogenic protozoa (*Cryptosporidium*, *Giardia*), fluorides, and other parameters. *See also* opportunistic pathogens.

**sensitivity analysis** In a modeling exercise, it is an analysis to determine which factors, parameters, initial conditions, and boundary conditions affect significantly the results of the simulation. These parameters and conditions are varied individually or in groups by constant percentages. Sometimes a precalibration sensitivity analysis is conducted to determine important model factors. The postcalibration analysis will then examine the impact of errors in model parameters, input variables, or initial values of state variables on predicted values. The final result of the analysis is a comparison of percent changes in model output versus percent changes in factors or parameters. A sensitivity analysis is not as rigorous as an error analysis, which can determine model validity by assigning uncertainties to important parameters and conditions. In recent sewer modeling studies (Adrien, 2003), the consultant tested model stability and model efficiency for sensitivity to analytical and numerical parameters and found the following. (a) Model continuity and runtime improved with increasing values of the underrelaxation parameter ($\omega$) from 0.60 to 0.85. Continuity improved by 4% and model runtime decreased by 2 minutes. (b) Variations of the time-weighting factor ($\theta$) from 0.55 to 0.85 did not produce any significant changes, but the combinations ($\theta = 0.60$, $\omega = 0.85$) and ($\theta = 0.75$, $\omega = 0.85$) yielded the best results in terms of continuity, convergence, and efficiency for the gravity systems, and manifold systems, respectively. (c) Between the flow and head tolerance combinations of (0.001, 0.001) and (0.005, 0.005) the continuity error increased by 5.8%. (d)

Between time steps ($\Delta t$) of 5 and 45 seconds, the continuity error increased by 12% for a gravity system and junction convergence decreased for a manifold system. (e) Between –10% and +30% of the default value of the Manning's roughness coefficient "*n*" total pump runtime increased by an average of 1.6 hours per day and the average of all junction HGL elevations increased by 5 feet. (f) Between –50% and +20% of the default value of the minor loss coefficient or "*k*-factor" (depending on the pump station layout) the pump run time increased by 0.6 hour per pump per day, but there was no significant change in pressure.

**sensitivity level** The lowest concentration of a substance that a test method can detect and quantify. Related terms include detection limit, method detection limit, practical quantification level, quantification limit, and reliable quantification level.

**sensitization** Development of an allergic reaction, especially involving the skin or lungs, when exposed to a chemical.

**sensor** An instrument that measures a physical condition or variable of interest. Floats and thermocouples are examples of sensors. Also called a primary element.

**sensor factor** A factor that affects the required UV doses in bench-scale experiments. It depends on the sensitivity of the sensor used to measure UV light intensity. *See also* average UV irradiance.

**sensory method** The use of the human olfactory system to measure odors from wastewater treatment units: subjects are exposed to odors diluted in air to determine the number of dilutions required for a minimum detectable threshold odor concentration, also called dilutions to threshold. Also called organoleptic measurement.

***S. enteritidis*** *See Salmonella enteriditis.*

***S. enteritidis* var. *chaco*** *See Salmonella enteriditis.*

**Sentinel** An apparatus manufactured by Roberts Filter Manufacturing Co. to control filter backwashing operations.

**Sentinel™** A line of ultraviolet disinfection products of Calgon Carbon Corp. of Pittsburgh, PA for the destruction of *Cryptosporidium* and *Giardia* cysts.

**sentinel well** A groundwater monitoring well situated between a sensitive receptor downgradient and the source of a contaminant plume upgradient. Contamination is first detected in the sentinel well, which serves as a warning that contamination may be moving closer to the receptor. The sentinel well is located far enough upgradient of the receptor to allow enough time before the contamination arrives at the receptor to initiate other measures to prevent contamination from reaching the receptor, or in the case of a supply well, provide for an alternative water source.

**Sentre-Fier** A rotary fine screen manufactured by Dontech, Inc.

**$SeO_3^{2-}$** Chemical formula of selenite.

**$SeO_4^{2-}$** Chemical formula of selenate.

**SEPA®** An apparatus manufactured by Douglas Engineering to recover gasoline from groundwater.

**separate nitrification** Nitrification of wastewater in a reactor separate from the aeration basin used for the satisfaction of carbonaceous oxygen demand. *See also* separate-stage nitrification, single-stage nitrification, combined nitrification.

**separate sewer** A sewer that is designed to carry only wastewater or stormwater, as compared to a combined sewer. *See also* sanitary sewer, storm sewer.

**separate sewer system** A sewer system consisting of separate sewers to carry wastewater (sanitary sewers) and stormwater/surface water (storm sewers) as opposed to a combined sewer system.

**separate sludge digestion** (1) The digestion of sludge from wastewater treatment in a tank different than the one in which it settled. *See* Imhoff tank. (2) The digestion of primary and biological sludges in separate tanks to optimize performance by taking advantage of their different characteristics.

**separate sludge digestion tank (unit)** A tank (unit) that provides for the digestion of sludge solids separately from the sedimentation tank, as compared to a single-storied septic tank that provides both functions. A two-storied septic tank includes a separate digestion compartment. Also called a digester.

**separate-stage nitrification** An arbitrary classification for a biological wastewater treatment method that accomplishes oxidation of carbonaceous matter and nitrification in a single basin with a $BOD_5$/total Kjehldal nitrogen ratio equal to or lower than 4. Then the nitrogenous oxygen demand accounts for more than 60% of the total oxygen demand. Separate-stage nitrification differs from separate nitrification and actually corresponds to one mode of operating a single-stage or combined process. *See also* single-stage nitrification.

**separate system** *See* separate sewer system.

**separating weir** A device such as a gap or an opening in the invert of a combined sewer to let the dry-weather flow fall to a separate sanitary sewer. *See* leaping weir for detail.

**separation factor** One of two parameters that indicate quantitatively the preference of an ion-exchange resin for specific ions, the other parameter being selectivity coefficient. The separation factor ($\alpha_{ij}$) of the exchange of ion $j$ for ion $i$ is defined as:

$$\alpha_{ij} = \text{ratio of the distribution of ion } i \text{ between phases to the distribution of ion } j \text{ between phases} = (y_i/x_i)/(y_j/x_j) \quad \text{(S-39)}$$

where $y_i$ = equivalent fraction of ion $i$ in resin = $Q_i/Q$; $x_i$ = equivalent fraction of ion $i$ in water = $C_i/C$; $Q_i$ = concentration of ion $i$ on resin, eq/L; $Q$ = total exchange capacity of resin, eq/L; $C_i$ = concentration of ion $i$ in water, eq/L; and $C$ = total ionic concentration in water, eq/L. Also called binary separation factor. *See* selectivity coefficient and ion-exchange equilibria for other related terms.

**separation line** The interface between the clarified liquid and the sludge blanket of a solids-contact unit.

**separation process, separation technique, separation technology** A water or wastewater treatment method (technique or technology) used to (a) separate the liquid, solid, and gaseous components of a mixture or (b) distinguish between solutes and solvent, e.g., adsorption, chromatography, distillation, extraction, filtration, and sedimentation. In the field of pollution prevention, separation processes are used to remove or isolate constituents from streams for in-process recycling or recovery and reuse.

**separation technique** *See* separation process.

**separation technology** *See* separation process.

**separator** Generally, a treatment unit that separates a feed into two products, as compared to a reactor, in which reactions occur (oxidation, reduction, solubilization, immobilization, conditioning). (1) A device, e.g., a small settling, skimming, or holding tank, installed in a drainage system to keep sand, grease, and oil out of the system. Also called a trap. (2) A device that is used to eliminate or reduce entrainment, i.e., to remove entrained droplets of water from a vapor stream produced during evaporation (e.g., during desalting). Also called demister. demister entrainment device, entrainment separator, mist eliminator. (3) A component placed between the membranes of a stack and designed to allow controlled fluid flow. Also called a spacer.

**separator tank** A device used for the separation of two immiscible liquids.

**sepralator** A membrane element used in crossflow filtration units.

**sepsis** Presence of pathogens or toxins in the blood or tissues.

**septa** Plural of septum.

**septa fitting** A special fitting used to seal vials (a liner for a threaded cap) or gas chromatographs to provide closure. Septa fittings can be manufactured in single, double, or triple layers of silicone rubber and other plastic materials. A syringe with a measured quantity of contaminant can be injected through a septa closure and into a gas chromatograph column for separation analysis.

**septage** The liquid and solid material pumped from a septic tank, cesspool, or similar domestic wastewater treatment system, or holding tank when the system is cleaned or maintained (EPA-40CFR122.2); a combination of scum, sludge, and liquid. The application of septage to land is limited by its nitrogen content:

$$A_{\mathrm{NV}} = KU_{\mathrm{N}} \quad \text{(S-40)}$$

where $A_{\mathrm{NV}}$ = application rate of septage, Mgal/acre/yr; $K$ = 385; and $U_{\mathrm{N}}$ = crop nitrogen requirement, lb/acre/year.

**septage characteristics** *See* septage composition.

**septage composition** A typical septage has a total solids content of 34%, which is about 63% volatile, a pH of 6.9, and an alkalinity of 970 mg/L as $CaCO_3$. Additional typical septage characteristics in mg/L include:

| | | | |
|---|---|---|---|
| $BOD_5$ | = 6000 | Ammonia | = 400 |
| Chemical oxygen demand | = 30,000 | Total Kjeldahl nitrogen | = 700 |
| Suspended solids | = 15,000 | Total phosphorus | = 250 |
| Volatile suspended solids | = 7000 | Heavy metals | = 300 |
| Total solids | = 40,000 | Grease | = 8000 |

**septage treatment and disposal** Methods commonly used in the treatment and disposal of septage include surface or subsurface land application, biological or chemical cotreatment with wastewater, codisposal with solid wastes by landfilling or composting, and separate processing (biological, lime stabilization, chemical oxidation, or composting).

**septic** Characteristic of a condition produced by microbial decomposition when all oxygen supplies are depleted; synonymous with anaerobic, putrid, rotten, foul smelling. If the condition is severe, bottom deposits produce hydrogen sulfide ($H_2S$), the deposits and the water turn black, give off foul odors, and the water has a greatly increased chlorine demand.

**septicaemia** Septicemia.

**septicemia** Blood poisoning by pathogenic organisms; sepsis.

**septicity** The condition of a substance (e.g. wastewater or organic matter) undergoing anaerobic microbial decomposition, with production of hydrogen sulfide and foul odors.

**septic privy** An on-site disposal system that is not served by a water supply and in which fecal matter is placed in a septic tank that contains water. The effluent from the tank flows to a drain field.

**septic sludge** The partially digested sludge from a septic tank, Imhoff tank, or digester.

**septic system** An on-site system designed to treat and dispose of wastewater from individual residences or commercial establishments. A typical septic system consists of a septic tank that receives and treats the wastewater and a system of tile lines or a pit for disposal of the liquid effluent through the soil. *See also* soil absorption system. Malfunctioning septic systems cause surface runoff and groundwater pollution problems.

**septic tank** A watertight and structurally sound chamber, commonly in concrete or fiberglass, usually installed underground, that receives and partially treats wastewater from a single residence or building. The single-storied tank combines sedimentation, anaerobic sludge digestion, and sludge storage in one or two compartments. Settleable solids form a sludge layer at the bottom, while grease and other light materials float at the surface where a scum layer is formed. *See* Figure S-05. Bacteria in the tank decompose the solids, which must be pumped out periodically and hauled to a treatment facility. Facultative and anaerobic decomposition converts the organic material to carbon dioxide ($CO_2$), methane ($CH_4$), and hydrogen sulfide ($H_2S$). The effluent is usually disposed of by leaching, in a soakage pit, or in a drainfield; it poses a risk of groundwater contamination by bacteria and viruses. A two-storied structure includes a separate digestion compartment; see Imhoff tank. *See also* percolation test, septage, septic system.

**septic tank–absorption field system** A wastewater treatment and disposal system consisting of a septic tank followed by a soil absorption field.

**septic tank effluent pumping (STEP) system** A pressure sewer system that uses a septic tank for solids removal before the wastewater is pumped.

**septic tank evapotranspiration system** A wastewater treatment and disposal system consisting of a septic tank followed by an evapotranspiration bed.

**septic tank–mound system** A wastewater treatment and disposal method consisting of a septic tank followed by a mound system.

**septic tank–sand filter system** A wastewater treatment and disposal system consisting of a septic tank followed by a sand filter.

**septic tank system** Same as septic tank–absorption field system.

**septic wastewater** Wastewater undergoing anaerobic decomposition; it has the odor of hydrogen sulfide ($H_2S$) and a black color resulting from the formation of metallic sulfides. *See also* stale wastewater.

**septum (plural: septa)** In general, a dividing membrane or wall in a structure. (1) In a precoat filter, the septum is a 3- to 5-mm thick, permeable material that supports the filter media and is supported by a rigid structure called the filter element. Its

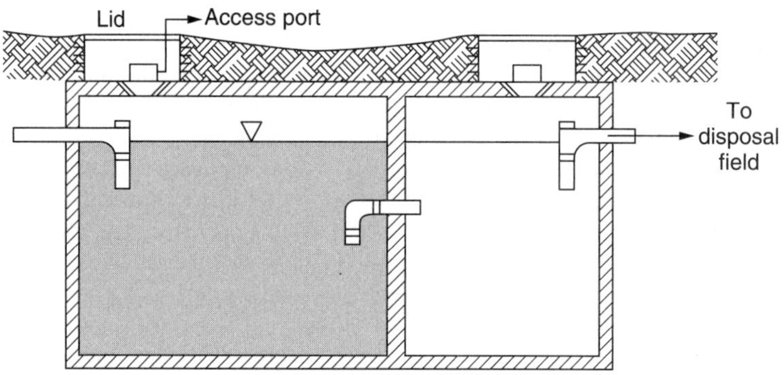


**Figure S-05.** Septic tank.

openings are larger than the precoat media particles. A typical septum consists of a flat frame covered on both sides by tightly woven stainless steel wire mesh or tightly fitted synthetic fabric and attached to a pipe manifold for collection of filtered water. (2) The lining of a bottle cap (or covering of an inlet port to an instrument) that can be pierced to draw (or inject) a sample. (3) A partition dividing a parent cell into two daughter cells during binary fission or separating two adjacent cells in the hyphae.

**sequencing batch reactor (SBR)** A fill-and-draw method of biological wastewater treatment involving the same processes (aeration and sedimentation) as conventional activated sludge but sequentially and in the same complete-mix reactor. The SBR process operates in five successive steps, the number in parentheses representing typical percentages of the cycle time that each step takes: (a) filling the reactor with raw wastewater or primary effluent (fill, 25%); (b) aeration, mixing, and completion of biological reactions (react, 35%); (c) quiescent settling and solids separation (settle, 20%); (d) removal of the clarified effluent (draw, 15%); and (e) emptying one reactor while another is being filled (idle, 5%). Sludge wasting occurs in the settle or idle step. The SBR process can be operated for the removal of carbonaceous matter as well as nitrogenous and phosphorus compounds. In the latter case, the SBR may include three additional steps: mixed fill, anaerobic react, and anoxic react. It was developed for use in remote areas with little operator attendance and has been designated by the USEPA as an innovative/alternative technology. *See also* SBR–BNR process, SBR with biological phosphorus removal, and intermittent cycle extended aeration system (ICEAS) as a variation of SBR.

**sequent depths** The depths before ($y_1$) and after ($y_2$), a hydraulic jump. Also called conjugate depths. *See also* Froude number. (It is more appropriate to call $y_1$ the initial depth, and $y_2$ the sequent depth.)

**sequential composite sample** A sample consisting of equal volumes collected continuously or at regular time intervals.

**sequential disinfection** The application of two or more disinfectants (e.g., chlorine and ultrasound) one after the other (not simultaneously).

**sequential pathogen removal** The use of a series of steps or processes (e.g., filtration, chlorination, ozonation) to achieve the desired level of disinfection. The total percent inactivation achieved by any step $n$ ($G_{t,n}$) depends on the percent inactivation of that step ($G_n$) and the total percent inactivation of the previous step ($G_{t,n-1}$):

$$G_{t,n} = G_{t,n-1} + G_n(100 - G_{t,n-1})/100 \quad \text{(S-41)}$$

**Sequest-All** A sequestering agent produced by Sper Chemical Corp. to control iron scaling and corrosion.

**sequester** (1) To undergo sequestration, i.e., to form a stable, soluble complex in water. (2) To keep a substance in solution by sequestration.

**sequestering** Same as sequestration.

**sequestering agent** A chemical that sequesters other substances or ions, e.g., a chemical that forms soluble compounds with metallic ions to prevent their precipitation. Sodium hexametaphosphate is a sequestering agent used to treat hard water to prevent calcium soap precipitates. Other common sequestering agents are ethylenediaminetetraacetic acid (EDTA) and sodium salts used for sequestering iron and manganese (silicates, phosphates, polyphosphates). Sequestering agents can also be used against concentration polarization in membrane processes.

**sequestration** In general, the inhibition or stoppage of normal ion behavior by combination with added materials. In particular, the formation of a stable, soluble substance with an ion in water to prevent its precipitation; especially the prevention of scaling and metallic ion precipitation from solution by formation of a coordination complex with a phosphate or another compound. Polyphosphates and sodium metaphosphate are sometimes used to sequester divalent ions in iron and manganese removal or in water stabilization (complexation of calcium). *See also* chelation, complexation.

***S. equinus*** *See Streptococcus equinus.*

**Seral®** Water treatment products of U.S. Filter Corp. for use in laboratories.

**Serfilco®** Equipment manufactured by Serfilco, Inc. for wastewater treatment.

**serial dilution** A series of successive dilutions that amplifies the dilution factor, beginning with a small initial quantity of material (e.g., a bacterial culture). The diluted material in each step serves as source of dilution for each subsequent step. In a serial dilution, the total dilution factor at any point is the product of the individual dilution factors in each step up to it. *See also* simple dilution.

**serial dilution endpoint method** A procedure for quantifying viruses using the cytopathogenic effect (CPE) method. It consists of adding serial dilutions of a virus suspension to host cells and observing the production of CPE over time, the endpoint being the dilution that produces a CPE in 50% of the tissue culture. This dilution is called the median (or medium) tissue culture infective dose or $TCID_{50}$. Also called $TCID_{50}$ method.

**serial filtration** Filtration using two or more units in series, as compared to an arrangement in parallel.

**series trickling filter–activated sludge (STF–AS) process** A method of wastewater treatment that is used mostly to upgrade existing activated sludge plants by adding a trickling filter and intermediate clarifier with solids recycle ahead of the aeration basin. It differs from the simple trickling filter–activated sludge process in that it includes the intermediate clarifier. *See* other dual processes under combined filtration–aeration process.

**serogroup** A group or category of organisms or cells that have some common antigens or against which common antibodies are produced. *See also* serotype.

**serologic analysis** *See* serotyping.

**serology** The study of relations between antigens and antibodies and their use to classify microorganisms into serotypes.

**seronegative** Pertaining to individuals that do not show significant levels of antibodies or similar signs in the serum of previous exposure to an infectious agent.

**seropositive** Pertaining to individuals that show significant levels of antibodies or similar signs in the serum of previous exposure to an infectious agent.

**serotype** (1) Same as serogroup, i.e., the type of an organism determined by its constituent antigens; e.g., the cholera vibrio exists in three serotypes: Inaba, Ogawa, and Hikojima. Human feces contain more than 100 virus serotypes, most commonly: adenoviruses, enteroviruses, hepatitis-A virus, reoviruses, and other viruses causing gastroenteritis and diarrhea. *See also* biotype and genotype. (2) The combination of antigens or antibodies that defines a serogroup. (3) To categorize by serotype.

**serotype analysis** *See* serotyping.

**serotype *pomona*** A virulent strain of the bacteria *Leptospira.*

**serotyping** The process of differentiating between organisms within the same species on the basis of antigenic recognition of antibodies. Also called serologic or serotyping analysis.

**serovar** Same as serotype.

**serovariety** Same as serotype.

**serpentine weir** A type of outlet weir sometimes installed in sedimentation tanks to increase the effective length of a peripheral weir by extending the weir pans inward from the effluent trough in a winding way.

**Serpentix®** A self-cleaning belt conveyor manufactured by Serpentix Conveyor Corp.

***Serratia*** A genus of opportunistic enterobacteriaceae associated with waterborne gastroenteritis, particularly in individuals with weakened immune systems.

***Serratia marcescens*** A species of rod-shaped, aerobic saprophytes; sometimes used to investigate aerosol production in flushing toilets.

**serum hepatitis** The most severe of the five forms of hepatitis; also called hepatitis-B. Its virus is transmitted through contaminated needles, blood products, or syringes.

**service** *See* water service connection.

**service age** The period of time that a structure, piece of equipment, etc., has been operating or in service. *See also* service life.

**service box** A valve box containing a corporation or curb cock to provide water service to a customer.

**service charge** The charge levied on customers for water supply or wastewater disposal services. It covers totally or partially the capital and operating costs of the facilities. Service charges for water supply are usually based on consumption, i.e., on measured flow. Sewer service charges may be based on flat fees, on water consumption, or on such fixtures as toilets, showers, laundry facilities, etc. *See also* user fee.

**service connection** *See* water service connection.

**service connector** *See* water service connection.

**service flow rate (SFR)** The rate at which the capacity of an ion exchange bed is being exhausted; the reciprocal of the empty bed contact time. It is commonly designed in the range of 1.0–5.0 gpm/cubic foot:

$$\text{SFR} = 1/\text{EBCT} = Q/V \qquad \text{(S-42)}$$

where $Q$ is the volumetric flow rate in gallons per minute and $V$ is the resin bed volume, including voids in cubic feet. Also called exhaustion rate..

**service life** The period of time between the date a structure, piece of equipment, etc., was put into service and the date it is retired. *See also* service age.

**service line** *See* water service connection.

**service line sample** A one-liter sample of water collected in accordance with CFR Section 141.86(b)(6) of the Code of Federal Regulations, that has been standing for at least 6 hours in a service line.

**service meter** *See* water service connection.

**service pipe** The pipeline that extends from the water main to the building served or to the consumer's system. Also called water service pipe. *See also* water service connection.

**service reservoir** A reservoir in a water distribution system providing local storage in case of an emergency and to respond to daily fluctuations in demand. Also called distribution or distributing reservoir. *See also* service storage.

**SERVQUAL** A computer model developed to assess water service quality as perceived by customers. It uses the following ten criteria: access, communication, competence, courtesy, credibility, reliability, responsiveness, security, tangibles, and understanding.

**SESOIL** A one-dimensional model for estimating pollutant distribution in an unsaturated soil column. SESOIL results are commonly used to estimate the source term for groundwater transport modeling of the saturated zone.

**Sessil®** A polyethylene strip medium produced by NSW Corp. for use in trickling filters.

**S-ethyl di-N,N-propylthiocarbamate** Defined under ethyl di-N,N-propylthiocarbamate.

**SETLdek** A tube settler manufactured by Munters.

**set point** The position at which the control or controller is set. This is the same as the desired value of the process variable, or an input value to be maintained by a control device.

**Setschenow coefficient** Same as salting-out constant.

**settleability** The tendency or ability of suspended solids to settle.

**settleability test** A test to determine settleability of solids in a suspension. Usually reported in mL/L, it measures the volume of solids (mL) that settle out of a given volume of sample (L) in a specified time period (e.g., 60 min). *See* Imhoff cone for settleable solids in wastewater.

**settleable solids** The portion of suspended solids that are heavy enough to sink to the bottom of a water or wastewater treatment tank within a defined period of time, as determined by a settleability test. Settleable solids can be removed by conventional sedimentation.

**settled wastewater** Wastewater that has undergone primary sedimentation. Also called clarified wastewater.

**settle phase** The third of the five steps in the operation of sequencing batch reactors, whereby solids separate from the liquid under quiescent conditions, producing a supernatant to be discharged as effluent. Also called sedimentation/clarification phase. *See also* fill phase, decant phase, react phase, and idle phase.

**settler** Same as sedimentation basin.

**settling** *See* sedimentation, discrete settling, flocculent settling, hindered settling.

**settling basin** A quiescent basin for the removal of suspended (settleable) solids by gravity. Also called clarifier and sedimentation basin. In general, settling refers to the gravity separation of heavy materials from light materials. Settling basins are used not only in stormwater and wastewater treatment, but also to trap stream sediment ahead of a reservoir or to treat factory effluents. A settling chamber is a vessel, whereas a settling reservoir consists of a series of settling basins. *See also* sedimentation basin.

**settling chamber** (1) *See* settling basin. (2) A series of screens placed in the way of flue gases to slow the stream of air, thus helping gravity to pull particles into a collection device. This air pollution control device is used to remove particulates in the thermal processing of sludge at wastewater treatment plants. *See also* mechanical collector, cyclone separator, impingement separator.

**settling column test** A test conducted to determine the settling characteristics of a suspension of flocculent particles for the design of a settling tank. It uses a column, the height of which equals the depth of the proposed tank. The column is filled with a uniform suspension representing the flocculent particles. Uniform temperature is maintained during the test, which lasts the proposed settling time. The settled solids are then drawn off. The expected performance of the tank is computed from a comparison of the total suspended solids (TSS) remaining in the liquid to the TSS of the original sample. *See also* settling tube (1).

**settling efficiency** The proportion of particles removed by sedimentation, theoretically a function of settling velocity and surface area relative to the flowrate (i.e., the surface overflow rate). Actually, it is affected by dead spaces and short-circuiting caused by eddy, surface, vertical convection, density, and outlet currents. *See also* Hazen's settling equation and trap efficiency.

**settling pond** Defined similarly to settling basin.

**settling regime** The settling of particles is classified as sedimentation type I, sedimentation type II, sedimentation type III, or sedimentation type IV.

**settling reservoir** *See* settling basin.

**settling solids** Same as settleable solids.

**settling tank** A holding tank for wastewater, where heavier particles sink to the bottom for removal and disposal. *See* sedimentation basin for more detail.

**settling time** The time necessary for suspended solids to settle by gravity or for colloidal solids to aggregate and precipitate.

**settling tube** (1) A tall, vertical, transparent tube used to study the settleability of an aqueous solution for the design of water or wastewater treatment units. Samples, drawn at various time intervals from ports inserted in the tube at 1.5 ft intervals, are analyzed for suspended solids, and the percent removal for each sample is plotted on a graph of depth vs. time to draw isopercent removal lines (or isoconcentration lines). *See also* settling column test. (2) An inclined tube installed in a settling tank to improve solids removal. *See* tube settler.

**settling velocity** The velocity at which particles settle in air, water, or wastewater. *See also* Stokes' law, standard fall velocity, terminal velocity.

**settling zone** A zone of the ideal sedimentation basin in which a uniform concentration of particles settles at the terminal velocity. *See also* Figure I-01 under ideal sedimentation basin, Camp sedimentation theory, inlet zone, outlet zone, sludge zone.

**seven-day average** The arithmetic mean of pollutant parameter values for samples collected in a period of seven consecutive days (EPA-40CFR133.101pa).

**seven-day, consecutive low flow with a ten-year return frequency (7Q10)** The lowest streamflow for seven consecutive days expected to occur once in ten years. *See* design flow.

**7Q10** Abbreviation of seven-day, consecutive low flow with a ten-year return frequency.

**severely saline water** Inland or coastal water with a high salt concentration, e.g., up to 30,000 mg/L. *See also* salinity.

**sewage** Human body wastes and the wastes from toilets and other receptacles intended to receive or retain body wastes. The waste and wastewater produced by residential and commercial sources and discharged into sewers. *See also* wastewater.

**Sewage and Industrial Wastes Federation** An association of engineers, scientists, and public works officials founded in the 1920s, later renamed Water Pollution Control Federation and then, in 1991, The Water Environment Federation.

**sewage charge** A charge levied by an agency to provide wastewater management services, including collection, treatment, and disposal. It is sometimes based on water consumption, wastewater flow and characteristics, or plumbing fixtures. Also called sewer service charge.

**sewage collection system** Each, and all, of the common lateral sewers, within a publicly owned treatment system, which are primarily installed to receive wastewater directly from facilities that convey wastewater from individual structures or from private property, and which include service connection Y fittings designed for connection with those facilities. The facilities that convey wastewater from individual structures, from private property to the public lateral sewer, or its equivalent, are specifically excluded from the definition, with the exception of pumping units, and pressurized lines, for individual structures or groups of structures when such units are cost effective and are owned and maintained by the grantee (EPA-40CFR35.905). *See also* sewer system.

**sewage composition** *See* wastewater composition.

**sewage farm** A precursor of the wastewater filter and an early development of water reuse, it is an operation in which wastewaters and/or sludges with little or no treatment are spread on agricultural land for final disposal as well as reuse of water and nutrients. With the advent of sewerage systems, there were a number of sewage farms in operation in Europe and in the United States in the late 19th and early 20th centuries. Also called sewer farm or wastewater farm. *See also* water reuse applications, sludge drying bed.

**sewage farming** An operation that applies wastewater and/or sludge to grow crops on agricultural lands. This operation serves a dual purpose: wastewater disposal as well as land irrigation and fertilization. In a similar operation, called broad irrigation, the primary purpose is wastewater disposal, whereas crop production is incidental.

**sewage fungus** *See Sphaerotilus.*

**sewage gas** The gas mixture that results from the decomposition of organic matter in wastewater or that is produced during sludge digestion. The components of the mixture vary with the type of decomposition (aerobic or anaerobic). *See also* sewer gas.

**sewage lagoon** A large, relatively shallow, earthen basin where sunlight, microbial action (by algae and bacteria), and oxygen work to purify wastewater or sludge. *See* stabilization pond for more detail.

**sewage rate** A charge or schedule of charges for wastewater management services. *See* sewage charge.

**sewage-sick** Pertaining to tight soils and overloaded farmlands that are irrigated with wastewater too intensely; they become wet, septic, and sour.

**sewage sludge** Sludge produced at publicly owned treatment works, the disposal of which is regulated under the Clean Water Act.

**sewage sludge incinerator** An enclosed device in which only sewage sludge and auxiliary fuel are fired (EPA-40CFR503.41-k).

**sewage treatment** *See* wastewater treatment.

**sewage treatment works** Municipal or domestic waste treatment facilities of any type that are publicly owned or regulated to the extent that feasible compliance schedules are determined by the availability of funding provided by federal, state, or local governments (EPA-40CFR220.2-f).

**sewage water** Wastewater, sewage, treated effluent.

**sewage works** An old term for wastewater collection, treatment, and disposal facilities.

**sewer** A channel or conduit that carries wastewater or stormwater runoff from the source to a treatment plant or receiving stream. *See also* combined sewers, sanitary sewers, and storm sewers. *See* Figures S-06 and S-07 for a sewer profile and some common sewer sections. Sewerage is the collection, conveyance, treatment, and disposal of liquid wastes; same as sewer system or wastewater facilities.

**sewerage** *See* sewer.

**sewerage system** Sewerage or sewer system, including the facilities and the administrative activities.

**sewer appurtenances** *See* appurtenances.

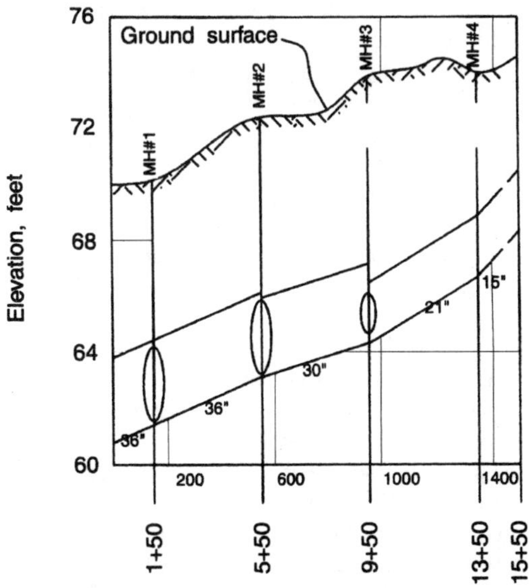


**Figure S-06.** Sewer profile (example).

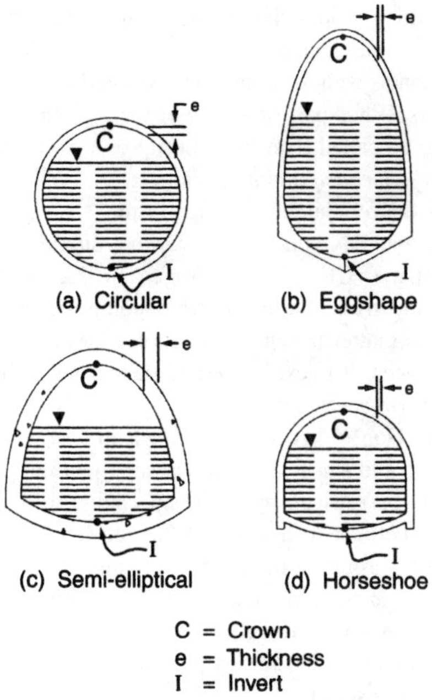


**Figure S-07.** Sewer sections (courtesy CRC Press).

**sewer atlas** *See* atlas.

**sewer authority** *See* sewer district.

**sewerbot** A contraction of sewer robot, an instrument used in the inspection of sewer lines.

**sewer charge** *See* sewer service charge.

**Sewer Chewer™** An apparatus manufactured by Yeomans Chicago Corp., combining comminution and solids grinding.

**sewer collection system** Piping, pumps, conduits, and other equipment necessary to collect and transport the flow of surface water runoff resulting from precipitation, or domestic, commercial or industrial wastewater to and from retention areas or any areas where treatment is designated to occur. The collection of stormwater and wastewater does not include treatment except where incidental to conveyance. A stormwater sewer system is a drain and collection system designed and operated for the sole purpose of collecting stormwater; it is segregated from the wastewater collection system (EPA-40CFR280.12 and 40CFR60.691).

**sewer corrosion** Corrosion of sewers that are made of concrete, iron, steel, or other acid-soluble materials. *See also* crown corrosion.

**sewer crown** The inside top of the arch of a sewer.

**sewer department** *See* sewer district.

**sewer district** A legally established agency or organization for the design, construction, financing, maintenance and operation of a sewer system. The territory within the area served by this system, which may include more than one political subdivision as well as industrial parks and other private users. When part of a state or local government, it may be called a sewer authority or a sewer department. *See also* drainage district.

**sewer farm** Same as sewage farm.

**sewer gas** A mixture of carbon dioxide ($CO_2$), methane ($CH_4$), and hydrogen sulfide ($H_2S$) resulting from the anaerobic decomposition of organic matter in sewers, manholes, or other appurtenances; it is explosive and dangerous to inhale. *See also* sewage gas.

**Sewer Hog™** A device, consisting of a slurry pump and a hose, that uses a water jet (350 gpm at 2000 psi) driven by a 600-hp engine to dislodge obstructions in a sewer. When lowered into the sewer through a manhole, the pump chews up all debris in the pipe and shoots the resulting slurry through a hose to the surface.

**sewer invert** The lowest point of the internal surface of a sewer.

**sewer line** A lateral, branch line, trunk line, or other enclosed conduit, including grates, trenches, etc., used to convey wastewater streams or residuals to a downstream waste management unit (EPA-40CFR63.111). For sewer manhole, see manhole. A sewer outfall or sewer outlet is an outlet, structure, or point of final discharge of stormwater, wastewater, or treatment plant effluent.

**sewer manhole** *See* manhole.

**sewer mining** A technique in which a small-scale, membrane-based treatment system is built on or near existing sewer lines in areas with high potential for reclaimed water. *See* satellite treatment for more detail.

**sewer ordinance** A municipal enactment that regulates the construction, operation, and use of a sewer system; specifying, for example, construction materials, requirements for grease traps, pretreatment, materials prohibited from discharge, etc.

**sewer outfall** *See* sewer.

**sewer outlet** *See* sewer.

**sewer overflow rate** A key indicator that wastewater utilities can use to assess their performance compared to established standards. It measures the condition of the sewer collection system as the number of overflows per length of collection piping, e.g., per mile or per 100 miles.

**sewer profile** A longitudinal profile along the axis of flow in a sewer, showing such elements as sewer size, distances, ground elevations, manhole locations, energy and hydraulic gradelines, and tributary lines. Typically, the horizontal scale is between 1:500 and 1:1000, and the vertical scale 10 times greater. *See* sewer and hydraulic profile. *See* Figure S-06.

**sewer rate** *See* service charge.

**sewer regulator** A structure or device that diverts stormwater or wastewater from one line to another. *See* regulator or diversion structure for detail.

**sewer rental** *See* sewer use charge.

**sewer separation** The conversion of a combined sewer system into separate sanitary and stormwater collection systems, a positive but expensive means of reducing or eliminating combined sewer overflow pollution.

**sewer service charge** A charge levied by an agency to provide wastewater management services, including collection, treatment, and disposal. It is sometimes based on water consumption, wastewater flow and characteristics, or plumbing fixtures. Also called sewage charge.

**sewershed** The land area served by a sewer or sewer system. *See* sewer territory.

**sewer surcharge** A flow condition that exists when the capacity of a sewer is exceeded.

**sewer system** All the components used in providing sewer service to customers within a given territory: land, buildings, gravity lines, force mains, manholes and other appurtenances, pumping stations, treatment facilities, etc. The sewer system may correspond to a given sewer district and may include one or more treatment plants. A regional system may include several local subsystems. For example, within Miami-Dade County (Florida), some municipalities (or volume sewer customers) collect wastewater within their boundaries and deliver it to the county facilities for transmission, treatment, and disposal.

**sewer territory** The land area within the boundaries of a sewer district.

**sewer use charge** A fee or charge levied on customers for the use of the sewers; it usually includes a charge for wastewater treatment and disposal, and may be based on the volume of water used.

**sewer utility** A utility that provides sewer service.

**Sewpadisc** A wastewater treatment apparatus manufactured by Biwater Treatment Ltd., using rotating biological contactors.

**sex pili** Hairlike protein fibers protruding from the surface of Gram-negative bacteria.

**sexual reproduction** A mode of reproduction of two free-living protozoa of compatible strains that exchange genetic materials, separate, and then divide by binary fission. Also protozoal reproduction when male sperm makes contact with female eggs to initiate fusion.

***S. faecalis*** Abbreviation of *Streptococcus faecalis.*

***S. faecalis* var. *liquefaciens*** Abbreviation of *Streptococcus faecalis* var. *liquefaciens.*

***S. faecium*** Abbreviation of *Streptococcus faecium.*

***S. flexneri*** Abbreviation of *Shigella flexneri.*

**SFM** Acronym of segregated flow model.

**SFR** Acronym of specific filtration resistance.

**SFS** Acronym of subsurface flow system.

**SFT™** Acronym of sediment flushing tank, a device manufactured by John Meunier, Inc.

***S. fuelleborni*** *See Strongyloides fuelleborni.*

***S. haematobium*** *See Schistosoma haematobium.*

**shaft horsepower** Same as brake horsepower, i.e., the horsepower measured at a pump input shaft.

**shallow lake** For seasonal density or thermal stratification considerations, a lake that is less than 20 feet deep.

**shallow reservoir** For seasonal density or thermal stratification considerations, a reservoir that is less than 20 feet deep.

**shallow sand-filled disposal field** An on-site treatment and disposal system used for septic tank effluents in areas of high water table or other unsuitable conditions. It functions as a combination intermittent sand filter and disposal field. *See* Figure S-08.

**shallow sand-filled pressure-dosed disposal field** *See* shallow sand-filled disposal field.

**ShallowTray** An aeration device manufactured by North East Environmental Products, Inc. for the removal of volatile organics (Pankratz, 1996).

**shallow-trench pump-in test** A test used to determine the assimilative capacity and the coefficient of permeability of a site relative to on-site disposal of wastewater. It uses (a) a trench approximately 8 ft. deep dug in the site, with a wooden box in the middle and washed drain rocks packed around the box; and (b) a number of observation wells drilled up and down gradient to monitor water levels. Water applied to the trench is metered and maintained at a constant level in the box. *See* Figure S-09. The capacity and permeability of the soil are estimated from the extent of the spread of water, the volumes of water applied and remaining in the trench. *See also* percolation test and soil profile examination.

**Shann-No-Corr** Zinc metaphosphate produced by Shannon Chemical Corp. for use as a corrosion inhibitor and sequestering agent.

**shape factor** (1) A factor ($S$) that reflects how closely a grain of sand or other particle approximates a true sphere; defined as the ratio of the surface area of an equivalent sphere to the surface area of the particle, for particles of the same volume. The minimum value is $S = 6.0$, for a spherical particle. For crushed materials, $S = 8.5$. *See* sphericity for more detail. *See also* the Fair–Hatch equation. (2) A factor ($SF$) that affects the coefficient of drag (and the terminal velocity) of a particle:

$$SF = c/(ab)^{0.5} \quad \text{(S-43)}$$

where $a$, $b$, and $c$ are, respectively, the longest, intermediate, and shortest dimensions of the particle. The effect of the shape factor is to increase the coefficient of drag, for example, in the laminar flow region, by a factor of 2.0 for sand, 4.0 for gypsum, and more than 20 for graphite flakes and fractal floc.

**shared system** *See* cluster sewer system.

**Sharon™ process** Sharon = acronym of single-reactor high-activity ammonia removal over nitrite, a process developed in the Netherlands to remove nitrogen biologically from digester recycle in a complete-mix reactor of short detention time, aerated intermittently to produce nitrification and

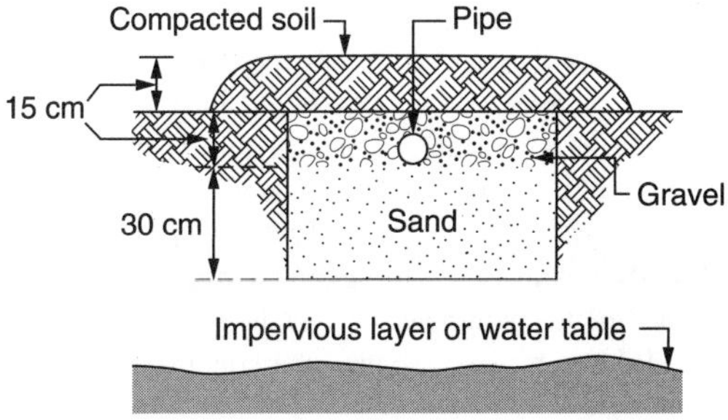


**Figure S-08.** Shallow sand-filled disposal field.

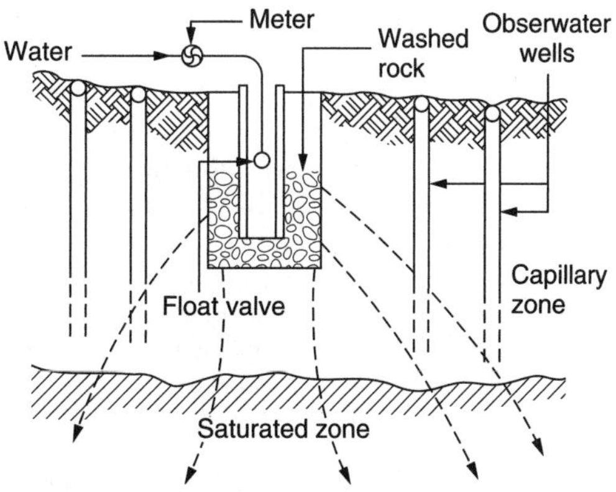


**Figure S-09.** Shallow trench pump-in test.

denitrification, with methanol addition during the anoxic period.

**sharp-crested weir** A device or structure, usually made of a plastic or metal plate and used for flow distribution or flow measurement. Materials used in wastewater treatment include also steel, stainless steel, fiberglass, and aluminum. Its crest is so thin that water flowing over it touches only a line. Also called thin-plate weir. *See also* broad-crested weir.

**Sharpshooter** A device manufactured by Norchem Industries to feed polymer.

**Shattuck, Lemuel** (1793–1859) A schoolteacher and bookseller in Boston who wrote the *Report of the Massachusetts Sanitary Commission* (1850), which indirectly led to the Metropolitan Health Law of 1866. *See also* the Great Sanitary Awakening.

***Sh. boydii*** Abbreviation of *Shigella boydii.*

***Sh. dysenteriae*** Abbreviation of *Shigella dysenteriae.*

**SHE** Acronym of standard hydrogen electrode.

**Shearfuser** A diffuser manufactured by FMC Corp. for use in sludge digestion.

**shear gradient** A measure of the mixing intensity in a water or wastewater treatment process such as flocculation; the change in velocity per unit distance along the vertical velocity curve. *See* velocity gradient for detail.

**shear-lift pump** Same as recessed-impeller pump.

**shear strength** A parameter relating to the overall ability of sludge to support itself and external loadings such as vehicular traffic. A function of solids concentration, it is an important characteristic for landfill or monofill disposal as well as for sludge reuse in backfill or as a soil substitute. *See also* Atterberg test, Büchner funnel test, capillary suction time (CST) test, compaction density, filterability constant, filterability index, specific resistance, standard jar test, time-to-filter (TTF) test.

**shear stress** (1) The force or drag per unit area exerted by flowing water on the boundaries of a stream channel. Under uniform flow conditions, shear stress is the product of the unit weight of water, the hydraulic radius, and the slope. Also called tractive force. *See also* sedimentation terms. (2) *See* plastic fluid, Bingham plastic.

**shear velocity** The square root of the shear stress at the bottom of a channel ($\tau_0$) divided by the water density ($\rho$); a parameter that has the same units as velocity. *See also* sedimentation terms. For steady uniform flow in a wide channel of bottom slope $S_0$ and mean depth $y$,

$$V^* = (g \cdot y \cdot S_0)^{0.5} \qquad \text{(S-44)}$$

$g$ being the gravitational acceleration.

**sheep liver fluke** *See Fasciola hepatica.*

**sheet erosion** The gradual erosion of the ground surface by sheets of water running down a slope, rather than by streams. Also called sheet flood erosion. *See also* surface wash (2).

**sheet flood erosion** Same as sheet erosion.

**sheet flow** Flow in a relatively thin sheet of generally uniform thickness, e.g., overland stormwater flow.

**shell** The outer wall of a hollow-fiber membrane. *See also* lumen.

**shellfish** Clams, mussels, and oysters grown in polluted waters may cause waterborne diseases such as typhoid and hepatitis, particularly when they are eaten raw. They process large volumes of water and accumulate the organisms it contains.

**shell pump** A simple centrifugal-type device for pumping mud and liquids laden with sand or gravel out of boreholes, without excessive clogging or damage. It is a hollow cylinder, open at the top, with a ball or clack valve at the bottom. Also called sludger or sand pump.

**sherardizing** The process of protecting steel with a corrosion-resistant layer of zinc by heating the steel in a mixture of sand and zinc.

**Sherwood number** A dimensionless parameter ($S_h$) characteristic of the phenomenon of concentration polarization in membrane filtration processes, equal to the product of the mass transfer coefficient ($k$, length/time) by the hydraulic diameter of the membrane element ($d$, length) divided by the coefficient of diffusivity ($D$, length squared per time). It can also be expressed in terms of the Reynolds number ($R_e$), kinematic viscosity ($\nu$), and some adjustable parameters,

$$S_h = kd/D = AR_e^B S^C (d/L)^E \qquad \text{(S-45)}$$

where $A$, $B$, $C$, and $E$ are adjustable parameters,

$$R_e = Vd/\nu \qquad \text{(S-46)}$$

$$S = \nu/D \qquad \text{(S-47)}$$

$L$ = length of membrane element, and $V$ = average cross-flow velocity. The Graetz–Leveque correlation is used under laminar flow conditions with $A$ = 1.86 and $B = C = E = 0.33$, whereas the Linton–Sherwood correlation is applicable to turbulent flow with $A = 0.023$, $B = 0.83$, $C = 0.33$, and $E = 0$.

***Sh. flexneri*** Abbreviation of *Shigella flexneri.*

**Shields parameter** In sediment transport theory, the Shields parameter ($K_s$) is the ratio of the critical shear stress ($\tau_c$) to the submerged weight of the particle:

$$K_s = \tau_c/[(S_s - 1)\gamma D] \qquad \text{(S-48)}$$

where $S_s$ = specific gravity of water, $\gamma$ = unit weight of water, and $D$ = particle diameter.

**shiga bacillus** *Shigella dysenteriae* Type 1.

**Shiga's bacillus** *Shigella dysenteriae* Type 1.

***Shigella*** A genus of Gram-negative, rod-shaped, facultatively anaerobic bacteria; some species, transmitted through fecal contamination, are pathogenic to humans and warm-blooded animals. In the water environment, this genus behaves similarly to the coliform bacteria. Excreted load = $10^7$ per gram of feces. Latency = 0. Persistence = 1 month. Median infective dose ≈ $10^4$.

***Shigella boydii*** One of the four species of *Shigella* bacteria that cause bacillary dysentery or shigellosis.

***Shigella dysenteriae*** A waterborne bacterial agent that causes bacillary dysentery, the severe form of shigellosis.

***Shigella dysenteriae* Type 1** The *Shigella* bacteria isolated by Kiyoshi Shiga during a severe epidemic of shigellosis in Japan in 1897, with a mortality rate of 25%. Also called Shiga's bacillus or shiga bacillus.

**shigellae** Plural of shigella.

***Shigella flexneri*** One of the four main species of *Shigella* bacteria that cause bacillary dysentery; along with *Shigella sonnei,* the most prevalent forms in the United States.

***Shigella sonnei*** The most common species of *Shigella* bacteria involved in waterborne bacillary dysentery in the United States; characterized by slow lactose fermentation.

***Shigella* spp.** All species of *Shigella.*

**shigellosis** A serious waterborne bacterial infection caused by *Shigella dysenteriae* and transmitted through the fecal–oral route; named after the Japanese bacteriologist Shiga. Its symptoms include diarrhea, fever, nausea, and, sometimes, cramps, vomiting, and tenesmus. Also called bacillary dysentery, it is distributed worldwide, nonlatent, moderately persistent, with an infective dose between medium and high.

**ship-based desalination** Seawater desalination conducted onboard cruise lines and military and other vessels. *See* Seawater Conversion Vessel.

**Shipek sampler** An instrument used to collect samples of undisturbed bottom sediment.

***S. hirschfeldii*** Abbreviation of *Salmonella hirschfeldii.*

**SHMP** Abbreviation of sodium hexametaphosphate.

**shoal** A shallow area in a body of water; a sandbank or sand bar exposed at low tide. *See also* bar, braided stream.

**shock load** The arrival at a water treatment plant of raw water containing unusual amounts of algae, colloidal matter, color, suspended solids, turbidity, or other pollutants. More generally, a sudden increase in the hydraulic, organic or other loading to a facility. Also called a slug load.

**Shone ejector** A type of pneumatic ejector that uses compressed air to lift water or wastewater.

**short-circuiting** (1) Uneven flow through a tank, vessel, or other unit resulting from such factors as density currents or inadequate mixing and causing the time of travel for parts of the flow to be less than the hydraulic residence time. This is usually undesirable since it may result in shorter contact, reaction, or setting times in comparison with the theoretical or presumed detention times. *See also* retardation, completely mixed flow, dead volume, plug flow. (2) The entry of ambient air into an extraction well (used for SVE and bioventing) without first passing through the contaminated zone. Short-circuiting may occur through utility trenches, incoherent well or surface seals, or layers of high permeability geologic materials.

**short-circuiting index (SCI)** One of the terms used to characterize tracer response curves and describe the hydraulic performance of reactors. Two expressions are reported for this parameter. (1) The ratio of the time interval for the first appearance of the tracer in the effluent ($T_i$) to the theoretical hydraulic residence time ($\tau$):

$$\mathrm{SCI} = T_i/\tau \qquad \text{(S-49)}$$

SCI = 1.0 for an ideal plug-flow reactor and decreases toward 0 as mixing increases. For typical sedimentation basins, $0.2 < \mathrm{SCI} < 0.3$. (2) The difference between 1.0 and the ratio of the time of the peak tracer concentration ($T_p$) to the time of the mean ($T_g$):

$$\mathrm{SCI}' = 1.0 - T_p/T_g \qquad \text{(S-50)}$$

SCI′ = 0 for an ideal plug flow reactor and SCI′ = 1 for complete mixing. *See also* residence time distribution (RTD) curve, time–concentration curve.

**short-incubation hepatitis** Same as hepatitis-A.

**short-launder approach** One of two common methods of designing weirs and launders for primary clarifiers. It assumes that the length of the weir is not important; hence, e.g., a tank-width weir at the end of the sedimentation basin for outlet control. *See also* long-launder approach.

**short-term aeration** A pretreatment process similar to a roughing filter. Also called high-rate activated sludge. *See* high-rate aeration process for detail.

**short-term exposure** Multiple or continuous exposures occurring over a week or so.

**short-term exposure limit (STEL)** The maximum allowable concentration of a substance to which a worker can be exposed in a continuous 15-minute period and a maximum of four exposures per day, according to the Occupational Safety and Health Administration.

**shortwave ultraviolet** Ultraviolet light of such a wavelength that it has germicidal properties. The germicidal wavelength range is between 200 and 300 nanometers with a peak at about 254 nanometers. Also called germicidal ultraviolet.

**shortwave ultraviolet band** The range of ultraviolet light wavelength (200–300 nanometers) that destroys germs. Also called germicidal ultraviolet band.

**shoulder effect** The phenomenon observed in wastewater chlorination according to which initially there is little reduction in the number of organisms, before a phase of straight-line inactivation (log inactivation vs. chlorine dose times contact time). Also called lag effect. *See also* tailing effect, Gard's disinfection model, Collins' disinfection model.

**shoulder** A deviation from the Chick–Watson law consisting of a time lag until the onset of disinfection. *See* Chick–Watson law deviations.

**shredder** (1) *See* screenings shredder. (2) A freshwater organism that feeds on larger particles of organic matter that it shreds into finer particles. *See also* gouger, collector, grazer, piercer.

**shrinkage limit** One of the three Atterberg limits: the water content (as a percentage of the dry weight) of a soil such that no further change in volume occurs on drying. *See also* plastic limit, liquid limit, plasticity index.

**shrink sleeve** A corrosion protective coating in the form of a wraparound or tubular sleeve, consisting of (a) a polyethylene or polypropylene backing to which an adhesive is applied and (b) an epoxy primer. Sleeves are attached onto the cutbacks between sections of pipelines during welding and construction. Also called heat-shrinkable sleeve.

**Shriver®** A plate-and-frame filter press manufactured by Eimco Process Equipment Co.

**shroud** A plate installed at the top or bottom of a radial flow turbine.

**shrouded turbine** A straight-blade turbine with a shroud mounted at the top.

***Sh. sonnei*** Abbreviation of *Shigella sonnei.*

**Shultz–Hardy rule** A series of ratios used to determine the effectiveness of potential-determining ions or counterions (Metcalf & Eddy, 2003):

$$1/1^6:1/2^6:1/3^6 \ldots \quad \text{or } 1:0.016:0.0014 \ldots$$
$$\text{or } 100:1.56:0.14 \ldots$$

**shunt wound dc motor** A type of direct current motor manufactured for pump drives used in wastewater treatment.

**shute wire** The horizontal wire in woven wire mesh; also called weft wire.

**shutoff** A shutoff valve.

**shutoff head** The head at which a pump ceases to operate, i.e., the flow rate $Q = 0$; it is a fixed upper limit corresponding to the minimum power consumption.

**shutoff valve** A device that can close off the flow of water partially or totally to a distribution system. Shutoff valves allow the sectionalization of a water system so that an area affected by a break can be isolated for repair without affecting service in other areas.

**shutter weir** A movable weir made of shutters or panels.

**Si** Chemical symbol of silicon.

**SI** Acronym of (a) saturation index; (b) surface impoundment; (c) Système International.

**SIC** Acronym of Standard Industrial Classification.

**SIC codes** Two-digit code numbers used for classification of economic activity in the Standard Industrial Classification manual.

**sick building syndrome** An illness caused by exposure to pollutants or germs in an inadequately ventilated building. Symptoms experienced by occupants include lethargy, headache, and irritations of the eyes, skin, and nose.

**side-channel clarifier** An intrachannel clarifier manufactured by Envirex (Wuakesha, WI) for oxidation ditch wastewater treatment systems. *See also* boat clarifier, Burns and McDonnell treatment system, Carrousel Intraclarifier™, pumpless integral clarifier, sidewall separator.

**side-flow weir** A diverting weir on the side of a channel or conduit; often used to control combined sewer overflows, in which case it is installed high enough to prevent the discharge of dry-weather flows. It may also be used in an outfall structure without a tide gate. In the weir equation, the exponent for a side-flow weir is taken as 5/3, as opposed to 3/2 for a transverse weir. Also called side weir. *See* Figure S-10.

**side-hill screen** A fine-to-medium screen that uses a stationary, inclined deck as a sieve to remove solids from wastewater, following coarse bar screens in small or industrial waste treatment plants. *See* static screen for more detail.

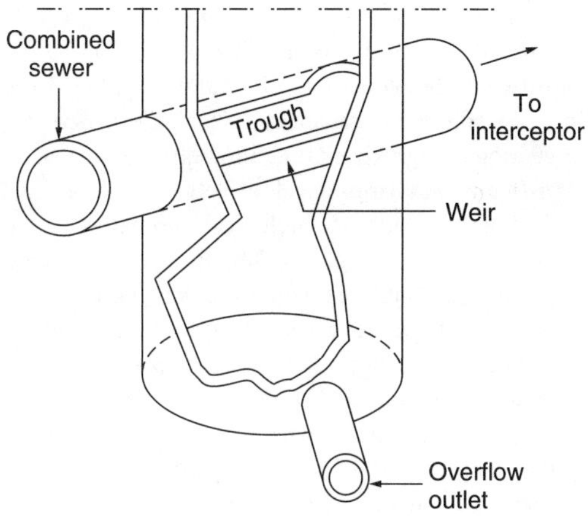


**Figure S-10.** Sideflow weir (adapted from Metcalf & Eddy, 1991).

**sideline equalization** A configuration in which only the portion of influent water or wastewater above the daily average flow or another predetermined limit passes through an equalization basin, e.g., to capture the first flush from combined sewer systems. Also called off-line equalization. *See also* in-line equalization.

**siderite ($Fe_2CO_3$)** (1) A common mineral, essentially iron carbonate, found in yellowish to deep-brown masses. (2) A compound formed during the corrosion of iron pipes; also called an iron phase or a corrosion scale. (3) A meteorite consisting almost entirely of metal.

**sidestream fermentation process** The production of volatile fatty acids for enhanced phosphorus uptake in a unit separate from the anaerobic reactor that provides time for phosphorus release; e.g., a primary clarifier, a gravity thickener of primary sludge, an activated primary, or a sidestream fermenter. *See* OWASA Nutrification™ process for an application.

**sidestream fermenter** A tank used in sidestream fermentation processes for the formation of volatile fatty acids, such as acetic and propionic acids. Also called an acid fermenter or a sludge fermenter, it may include a compartment for storage of methane gas.

**Sidewall Separator** An intrachannel clarifier manufactured by Lakeside Equipment Co. (Bartlett, IL) for oxidation ditch wastewater treatment systems. *See also* boat clarifier, Burns and McDonnell treatment system, Carrousel Intraclarifier™, pumpless integral clarifier, side-channel clarifier.

**sidewater depth (SWD)** The depth of water measured along a vertical interior wall of a reservoir, tank, basin, etc.

**side weir** Same as side-flow weir.

**Side Winder** Screens manufactured by Cook Screen Technologies, Inc. for water and wastewater treatment applications.

**sieve** A device, usually in a circular frame, with fine meshes or a perforated bottom, used to separate fine from coarse parts of sand, aggregates, or other loose matter or for straining liquids.

**sieve analysis** An analysis conducted to determine the size distribution of a granular filter medium, using a series of standard, decreasing sieve sizes. Size distribution is usually represented graphically by plotting the cumulative percent passing a sieve size on arithmetic-log or probability-log paper. *See* Figure S-11. Two important size–frequency parameters are derived from this plot: effective size and uniformity coefficient. *See also* particle size distribution.

**sieve designation number** Same as sieve size.

**sieve manufacturer's rating** A U.S. standard sieve size or sieve number as determined by direct measurement of the clear dimensions of a representative number of screen openings; usually simply called manufacturer's rating. *See also* Hazen's sieve rating.

**sieve number** Same as sieve size.

**Sievers® 800 TOC Analyzer** A device manufactured by Ionics Instrument Business Group for the monitoring of total organic carbon.

**sievert (Sv)** The Système International's unit of dose of ionizing radiation to the body or an internal organ, equivalent to 1 joule of energy per kilogram of absorbing tissue; equal to 100 rem or about 8.38 roentgens.

**sieve series** A list of sieve openings that stand successively in the ratio of $2^{1/4}$ (i.e., 1.1892) to one another. *See* standard sieve series.

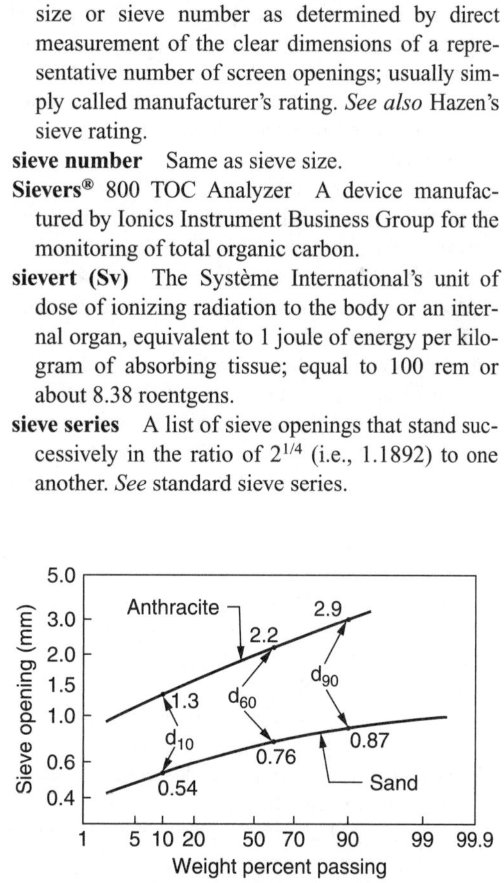


**Figure S-11.** Sieve analysis (example).

**sieve size** A number that corresponds to the opening of a sieve and denotes the approximate number of meshes per inch. The sieve size, sieve number, or sieve designation number is inversely related to the size of the opening. For example, sieve sizes 4, 10, and 50 correspond, respectively, to opening sizes of 0.187 in (4.76 mm), 0.0787 in (2.00 mm), and 0.0117 in (0.297 mm). Common U.S. sieve sizes go from 4 to 200. In the complete list of sieve sizes, openings stand successively in the ratio of $2^{1/4}$ (i.e., 1.1892) to one another; some intermediate sizes may be omitted from the list. *See* standard sieve series.

**sieve tray column** A type of packed tower stripper used to remove volatile organic chemicals (VOC) from contaminated waters. It consists of a countercurrent, 10-ft-high column with a series of perforated trays along the column, each tray ending with inlet and outlet channels. The liquid flows from the top and horizontally across the trays. Also called a low-profile air stripper. *See also* cascade air stripping, cocurrent packed tower, cascade packed tower, end effect, crossflow packed tower, minimum air-to-water ratio, packed tower design equation, wall effect.

**sieving** In membrane filtration of water, the diffusion-controlled removal of contaminants that are larger than the pore size of the membranes.

**$SiF_4$** Chemical formula of silicotetrafluoride.

**SightWell** A circular clarifier manufactured by Walker Process Equipment Co., including a device for solids removal by hydraulic suction.

**Sigma** A low-speed surface aerator manufactured by Purestream, Inc.

**sigma concept** An approach based on Stokes' law (gravity settling of discrete particles) applied to the design of full-scale centrifuges, using the results of pilot tests and considering only hydraulic loading rates. The allowable flow through the centrifuge ($Q$) is:

$$Q = [V\omega^2/\ln(R_2/R_1)][(P_p - P)D^2/18\,\mu] \quad \text{(S-51)}$$

where $V$ = volume of sludge and water in the pool; $\omega$ = radial velocity of centrifuge, rad/s; $R_2$ = radius from centerline of the centrifuge to the bowl; $R_1$ = radius from the centerline of the centrifuge to the pool level; $P_p$ = particle density; $P$ = fluid density; $D$ = particle diameter; and $\mu$ = viscosity.

**Sigma Flight** A fiberglass sludge collector manufactured by Envirex, Inc.

**$\sigma_t$ unit** A unit of seawater density, equal to the density in grams per liter minus 1000.

**significant biological treatment** Aerobic or anaerobic biological treatment process that consistently achieves a 30-day average of at least 65% removal of five-day BOD (EPA-40CFR133.101).

**significant deterioration** Pollution resulting from a new source in previously "clean" areas. *See* prevention of significant deterioration.

**significant environmental effects** (a) Any irreversible damage to biological, commercial, or agricultural resources of importance to society; (b) any reversible damage to biological, commercial, or agricultural resources of importance to society if the damage persists beyond a single generation of the damaged resource or beyond a single year; or (c) any unknown or reasonably anticipated loss of members of an endangered or threatened species, as identified by the Secretary of the Interior (EPA-40CFR723.50-11).

**significant hazard to public health** Any level of contaminant that causes or may cause an aquifer to exceed any maximum contaminant level set forth in any promulgated National Primary Drinking Water standard at any point where the water may be used for drinking purposes or that may otherwise adversely affect the health of persons, or that may require a public water system to install additional treatment to prevent such adverse effect (EPA-40CFR149.101-j).

**significant industrial user (SIU)** (a) Any industrial user subject to categorical pretreatment standards. (b) Any industrial user that discharges an average of 25,000 gallons per day or more of process wastewater to a publicly owned treatment works (POTW) (excluding sanitary, noncontact, and cooling and boiler blowdown wastewater); contributes a process wastestream that makes up 5% or more of the average dry-weather hydraulic or organic capacity of the POTW treatment plant; or is designated as such on the basis that the industrial user has a reasonable potential for adversely affecting the POTW's operation or for violating any pretreatment standard or requirement (EPA-40CFR403.3-t). *See also* categorical industrial user

**significantly** As used in the National Environmental Protection Act (NEPA), this term requires consideration of both context and intensity. (a) Context: This means the significance of an action must be analyzed in several contexts such as a whole (human, national), the affected region, the affected interests, and the locality. Significance varies with the setting of the proposed action. For instance, in the case of a site-specific action, significance would usually depend upon the effects in the locale rather than in the world as a whole. Both short- and long-term effects are relevant. (b)

Intensity: This refers to the severity of impact. Responsible officials must bear in mind that more than one agency may make decisions about partial aspects of a major action (EPA-40CFR1508.27).

**significantly greater effluent reduction than BAT** Means that the effluent reduction over BAT produced by an innovative technology is significant when compared to the effluent reduction over best practicable control technology currently available produced by BAT (EPA-40CFR125.22-c).

**significantly more stringent limitation** A $BOD_5$ and suspended solids (SS) concentration necessary to meet the percent removal requirements of at least 5 mg/L more stringent than the otherwise applicable concentration-based limitations (e.g., less than 25 mg/L in the case of the secondary treatment limits for $BOD_5$ and SS), or other percent removal limitations that would by themselves force significant construction or other significant capital expenditure (EPA-40CFR133.101-m).

**significant municipal facilities** Publicly owned wastewater treatment plants that discharge one million gallons per day or more and are therefore considered by states to have the potential to substantially affect the quality of receiving waters. *See also* majors, minors.

**significant noncompliance** The condition of a discharger that has committed a significant violation.

**significant noncomplier** A water supply or wastewater management system that has committed a significant violation, as defined by the USEPA.

**significant source of groundwater** An aquifer that (a) is saturated with water having less than 10,000 mg/L of total dissolved solids; (b) is within 2500 ft of the land surface; (c) has a transmissivity greater than 200 gal/day/ft, provided that any formation included within the source of groundwater has a hydraulic conductivity greater than 2 gal/day/ft$^2$; and (d) is capable of continuously yielding at least 10,000 gal/day to a pumped or flowing well for a period of at least a year. An aquifer that provides the primary source of water for a community water system (EPA-40CFR191.12).

**significant violation** A violation by a point source discharger of sufficient magnitude or duration to be a regulatory priority.

**SIHI-Halberg** A sludge mixer manufactured by SIHI Pumps, Inc. for use in digesters.

**silane** (1) A water-soluble gas with an unpleasant odor, used in the semiconductor industry. Also called silicon tetrahydride ($SiH_4$). (2) Any silicon hydride, analogous to the alkanes, e.g., trichloromethylsilane ($CH_3Cl_3Si$), which hydrolyzes to methylsilanol [$CH_3Si(OH)_3$] and hydrochloric acid (HCl):

$$CH_3Cl_3Si + 3\ H_2O \rightarrow CH_3Si(OH)_3 + 3\ HCl \quad (S\text{-}52)$$

**silcrete** In semiarid climates, a strongly compacted soil layer underlying soft materials and cemented with silica. Also called duricrust or duripan. *See* hardpan for other similar layers.

**Silent Check** A line of check valves manufactured by Val-Matic®, designed with threaded seats and seat O-rings for silent operation. *See also* Swing-Flex®, Dual Disc® Cushion Swing, and Titled Disc®.

**silica ($SiO_2^-$)** The common name for silicon dioxide, a natural hard, glassy mineral composed of silicon and oxygen and found as sand, quartz, flint, opal, and diatomite. It is used to produce glass and as filter medium in water and wastewater treatment. Silica results in natural waters from rock disintegration and appears as suspended solids or colloidal particles. *See also* silicon, activated silica.

**silica gel** A dehumidifying and dehydrating agent consisting of highly adsorbent gelatinous silica.

**silica sand** A granular material containing quartz sand (silica) and commonly used as a medium in granular bed filters, alone or in combination with other media.

**silica sol** Another name of activated silica, used in water and wastewater treatment; usually produced on-site from sodium silicate ($Na_2SiO_3$), alum, ammonium sulfate, and other chemicals.

**silicate** (1) Any substance containing silicon (Si), oxygen (O), and a metallic compound. Silicate minerals include amethyst, agate, asbestos, clay, feldspar, flint, granite, hornblende, jasper, and mica. Silicates are used as anodic corrosion inhibitors and as sequestering agents (in combination with chlorine) for iron and manganese. Sodium metasilicate ($Na_2SiO_3$) is used as silicon supplement in animal studies. (2) Any salt of silicic acid or derived from silica.

**silicate mineral** The major group of compounds in the Earth's crust, consisting of $SiO_2$ or $SiO_4$ groupings, e.g., quartz, beryl, garnet, feldspar, mica, and clays.

**siliceous** Containing or consisting of silica or a silicate compound.

**siliceous gel zeolite** A synthetic sodium aluminum silicate used as a cation exchange product in residential water softeners. Also called simply gel zeolite. *See also* aluminosilicate.

**silicic acid** An amorphous gelatinous mass formed by combining an alkaline silicate with an acid; it dissociates into silica ($SiO_2$) and water ($H_2O$). It is used as an adsorbing material in the laboratory measurement of organic acids.

**silicium** Former name of silicon.

**silicofluoric acid** Another name of fluorosilicic acid.

**silicon (Si)** A nonmetallic element occurring as a brown powder or dark grey crystals found in silicate minerals or in mica; the second most abundant element of the Earth's crust (more than 25% by weight), used in construction materials, glass, and other industrial applications; an essential nutritional element. Atomic weight = 28.086. Atomic number = 14. Specific gravity = 2.42.

**silicon-32** A radionuclide sometimes found in water sources from cosmic reactions; half-life = 300 years.

**silicon metal** Any silicon alloy containing more than 96% silicon by weight (EPA-40CFR60.261-x).

**silicosis** Fibrosis or scarring of lung tissues caused by prolonged inhalation of siliceous particles, as by stonecutters.

**silicotetrafluoride ($SiF_4$)** A nontoxic gas heavier than air; an impurity or intermediate product of the dissociation of fluorosilicic acid.

**Sil-Kleer®** A filter aid produced by Silbrico Corp.

**sill** (1) A horizontal piece of timber, concrete, or other material serving as a foundation, base, or other support for a hydraulic structure such as a lift gate, dam, or spillway. (2) A low dam installed in a small mountain stream to protect its soft or unconsolidated banks. (3) A structure with a level upper edge, such as a plate or a wall, installed in a channel, a ditch, or in a river pool to spread the flow evenly. Also called spreader. (4) A sheet of intrusive igneous rock between beds of sedimentary or volcanic materials. (5) A low wall constructed in a stilling basin for energy dissipation.

**silo** A tall, usually cylindrical structure for storing grain, fodder, forage, or other dry solids.

**Silo Pac** A chemical feed device manufactured by Wallace & Tiernan, Inc.

**silt** Sedimentary material composed of fine or intermediate-size mineral particles, varying from fine sand to clay, i.e., from 0.005 mm to 0.05 mm in equivalent diameter. Silts make up poor aquifers; they are highly porous but have low permeability. Suspended in still water or carried by moving water, they often accumulate on the bottom of rivers, deltas, bays, tanks, basins, and reservoirs. *See* soil classification.

**siltation** or **silting** Deposition of silt and, in general, finely divided soil and rock particles on the bed of a body of water, due to erosion and resulting in the filling up or raising of the bed. Reservoir silting results in the loss of storage capacity. *See also* trap efficiency.

**silt basin** A basin in a storm sewer designed to receive solids and reduce velocity.

**silt density index (SDI)** An empirical measure of the fouling tendency or plugging characteristics of water based on the times it takes 500 mL of filtrate to pass through a 0.45 micron pore membrane filter at a constant pressure of 30 psi at the beginning and at the end of the test period. Also called plugging factor and calculated as follows:

$$SDI = 100\,[1 - (t_i/t_f)]/t \qquad \text{(S-53)}$$

with $t_i$ = time to collect initial sample of 500 mL, sec; $t_f$ = time to collect final sample of 500 mL, sec; and $t$ = filtration test period, min; it depends on the fouling nature of the liquid and varies from 15 min to 2 hours. The SDI is the most widely used index to assess the treatability of a given water or wastewater by nanofiltration (NF) and reverse osmosis (RO). Pretreatment is recommended if SDI exceeds 2.0 for NF and hollow-fiber RO, or 3.0 for spiral-wound RO. Two other indexes are the modified fouling index and the mini plugging factor index

**silting** Same as siltation.

**silt trap** A simple structure, similar to a spring box, designed to remove silt from collected spring water before it is distributed. The silt trap receives the water from two or more adjacent spring boxes. *See* Figure S-12.

**silver** (Ag, from its Latin name argentums.) A brilliant, white, ductile, malleable, corrosion-resistant, precious metal, used in jewelry, alloys, photography, and many other industrial applications and as a disinfectant. Atomic weight = 107.87.

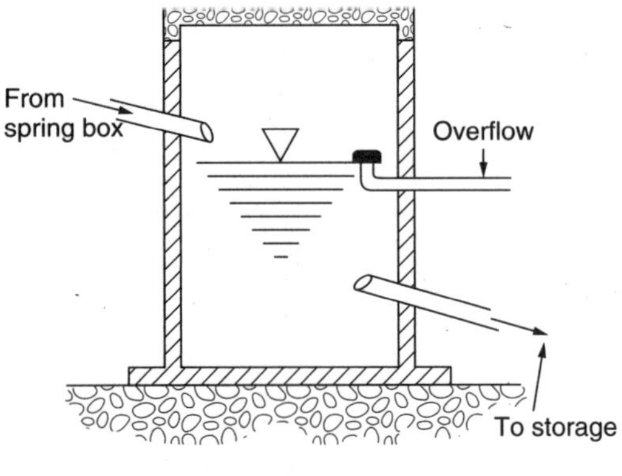


**Figure S-12.** Silt trap.

Atomic number = 47. Specific gravity = 10.5. Melting point = 962°C. Boiling point = 2212°C. Silver nitrate and sulfate are somewhat soluble in water but most of the other silver salts are insoluble. High levels of colloidal silver can be fatal. *See also* argyria. The USEPA recommends a Secondary Drinking Water Standard of 0.1 mg/L of silver. The metal is recovered industrially by ion exchange and removed from wastewater by chemical coagulation. Silver has disinfecting properties even at such a low dosage as 0.05 mg/L but is relatively expensive. It may also be used to retard microorganism growth in point-of-use filtering devices. Conventional treatment is rated as good to excellent in removing silver from drinking water.

**Silverback™** A ceramic membrane filter apparatus manufactured by U.S. Filter Corp. to recover cleaning agents.

**Silver Band** A granular pressure filter manufactured by Envirotech Co.

**silver disinfection** The use of the bactericidal power of silver to disinfect water or to impregnate point-of-use filters. Doses of 30–90 micrograms per liter of $Ag^+$ can reduce *E. coli* organisms by 3 log units.

**silver iodide (AgI)** A pale-yellow, insoluble solid used in medicine, photography, and rainmaking.

**Silverseries™** A compact desalination apparatus manufactured by Matrix Desalination, Inc.

**silvex [$Cl_3C_6H_2OCH(CH_3)COOH$]** The generic name of 2-(2,4,5-trichlorophenoxy) propionic acid, or 2,4,5-TP, a synthetic organic chemical, used as pesticide and regulated by the USEPA as a drinking water contaminant: MCL = MCLG = 0.05 mg/L.

**silvicultural point source** Any discernible, confined, and discrete conveyance related to rock crushing, gravel washing, log sorting or log storage facilities that are operated in connection with silvicultural activities and from which pollutants are discharged into waters of the United States, excluding some nonpoint silvicultural activities (EPA-40CFR122.27-1).

**simazine [$C_7H_{12}ClN_5$ or $ClC_3N_3(NHC_2H_5)_2$]** The generic name of 2-chloro-4,6-bis (ethylamino)-s-triazine, a common, colorless, crystalline, nonvolatile, potentially carcinogenic organic chemical with low solubility, used to control broadleaf and grassy weeds. It may be found in surface waters receiving agricultural runoff and is not removed effectively by conventional water treatment processes. WHO-recommended guideline value for drinking water: 0.002 mg/L. It is a derivative of cyanuric chloride; *see also* atrazine, cyanazine, propazine, and s-triazine.

**Simcar®** A turbine aerator of United Industries.

**simple continuous distillation** A continuous distillation process that uses a single stage, with the liquid and vapor leaving in equilibrium. *See* flash distillation for detail.

**simple dilution** The combination of a unit volume of a material with an appropriate volume of a solvent liquid to achieve a desired concentration. The dilution factor is the total number of unit volumes in which the material is dissolved. For example, a 1:10 dilution entails combining 1 unit volume of material and 9 unit volumes of the solvent (hence, 1 + 9 = 10 = dilution factor). *See also* serial dilution.

**simple distillation** *See* simple continuous distillation.

**simple electrolyte** A compound that forms charged ions in solution. *See also* polyelectrolyte, polymer, monomer.

**simple lognormal model** A distribution used to describe the probability of human infection by enteric microorganisms. *See also* beta-Poisson model, exponential model, annual risk, lifetime risk.

**simple pit latrine** A simple and inexpensive means of excreta disposal; it consists of a defecation hole, a slab made of concrete or other material, a hole cover, a superstructure, and a drainage channel. *See also* ventilated improved pit (VIP) latrine.

**simple stain** A positive stain consisting of a single dye such as methylene blue and resulting in all cell components appearing in one color.

**simple sugar** *See* monosaccharide.

**simple surge tank** A surge tank that does not have a restricted orifice.

**Simplex** A low-speed surface aerator manufactured by Asdor.

**simplex pump** A plunger pump having only one chamber.

**Simplified Particulate Transport Model (SIMPTM)** A continuous stormwater quality model that simulates the process of pollutant accumulation and washoff and the effectiveness of street sweeping as well as other practices in improving the quality of urban runoff.

**Simpson, James** Builder of the first sizable slow sand filter for treating Thames River water in London in 1829; one of the significant achievements of the Great Sanitary Awakening. He showed that such filters are effective in removing pathogens even without pretreatment.

**SIMPTM** Acronym of Simplified Particulate Transport Model.

**Simrake** A rotary bar screen of Simon-Hartley, Ltd..

**Simspray** A rectangular distributor manufactured by Simon-Hartley, Ltd. for fixed-film reactors.

**Simtafier process** A water treatment process developed to increase the rinse rates of clarifiers. It uses and recycles fine sand as a ballast. The same process is also known as Cyclofloc. *See* ballasted floc.

**simulate** To reproduce the action of some process, usually on a smaller scale.

**simulation** The process of mimicking some of the behavior of a system with a different system, e.g., with computers, models, and other equipment. The representation of physical systems and phenomena by mathematical models. The conduct of experiments with a model to better understand present or future conditions of the actual system, predict outcomes, or evaluate scenarios. All models are imperfect tools, with inherent errors in the representation of reality and limitations on input data adequacy. The model user must always interpret simulation results. A computer does not validate a model; it simply facilitates calculations and record keeping. *See also* modeling.

**simulation efficiency** Same as model efficiency.

***Simulium spp.*** Species of mosquitoes that carry the pathogen of river blindness.

***Simulium damnosum*** A species of mosquitoes that transmit river blindness in West Africa. Also called blackfly.

***Simulium neavei*** A species of mosquitoes that transmit river blindness in East Africa. Also called blackfly.

**simultaneous disinfection** The simultaneous application of two disinfecting agents, e.g., chlorine and ultrasound, which may produce a better result than their application in series (or sequential disinfection).

**simultaneous distillation extraction** A laboratory method using steam to extract and concentrate trace organics. Also called steam distillation extraction.

**simultaneous nitrification–denitrification (SNdN)** (1) A phenomenon that occurs in the aerobic zone of a biological nitrogen removal plant under conditions of low dissolved oxygen and long solids retention times, particularly in oxidation ditch and other complete mix systems. Nitrogen removals of as much as 90% have been observed in some activated sludge plants treating municipal wastewater. (2) A biological nitrogen removal process in which both nitrification and denitrification occur in the same reactor, which maintains separate aerobic and anoxic zones with appropriate dissolved oxygen control. *See also* Orbal™ process and Figure O-06, low-DO oxidation ditch, and Sym-Bio™ process.

**Simultech™** An apparatus manufactured by Schreiber Corp. for biological nutrient removal.

**Sinclair** A rotary fine screen manufactured by Bielomatik London, Ltd., including internal feed.

**sine-wave method** One of five methods proposed for estimating the volume required for a wastewater equalization tank.

**single-action pump** A reciprocating pump with suction action only on one side of the piston and with intermittent discharge. *See also* single-stage pump.

**single centrifugal pump** A centrifugal pump that admits water only on one side of the impeller.

**single-effect evaporator** An evaporator that includes only one step. *See also* multiple-effect evaporator.

**single-hit model** *See* one-hit model.

**single-loop controller (SLC)** A device connected to the control system of a wastewater treatment installation, as part of a distributed control system, to regulate a single variable. *See also* multiloop controller.

**single-main grid** A water distribution system consisting of only one line on one side of the street to serve all customers on both sides, as compared to a dual-main grid.

**single-medium filter** A water or wastewater filtration device with only one material as medium, e.g., a sand or activated carbon filter.

**single-pit latrine** A latrine built over one pit, as opposed to a twin-pit latrine.

**single-pit VIP latrine** A VIP latrine built over one pit, as opposed to a twin-pit VIP latrine.

**single-reactor high-activity ammonia removal over nitrite process** *See* Sharon™.

**single-sludge biological nitrogen-removal process** *See* single-sludge process.

**single-sludge process** A suspended growth biological nitrogen-removal process that includes only one solids separation device, e.g., a secondary clarifier. *See also* two-sludge system. *See* the following variations: postanoxic nitrification–denitrification, preanoxic nitrification–denitrification, simultaneous nitrification–denitrification, Wuhrman process, Ludzack–Ettinger process, modified Ludzack–Ettinger process, Bardenpho™ process, Orbal™, Bio-denitro™.

**single-stage digestion** Sludge digestion in only one step, as compared to digestion in two tanks or more in series.

**single-stage high-rate digestion** An anaerobic sludge digestion process that operates usually in the mesophilic range, but occasionally in the thermophilic range, and includes equipment for heating, auxiliary mixing, uniform feeding, and thickening of the sludge. *See also* two-stage digestion, separate sludge digestion.

**single-stage lime softening** A water treatment process that uses lime to precipitate hardness. Lime and other chemicals are added to the raw water, which then passes through a clarifier, a recarbonation unit, and a filter. In the recarbonation step, carbon dioxide ($CO_2$) gas is added to the water to reduce the pH of the solution. *See also* lime softening, two-stage lime softening, and split softening.

**single-stage nitrification** A biological wastewater treatment method that accomplishes oxidation of carbonaceous matter and nitrification in a single basin with a $BOD_5$/total Kjehldal nitrogen ratio greater than 4. Also called combined nitrification. *See also* separate-stage nitrification, separate nitrification.

**single-stage pump** A centrifugal pump with only one impeller. *See also* single-action pump.

**single-stage recarbonation** Same as recarbonation.

**single-station controller** An apparatus in a water or wastewater system that allows the operator to control a given function (e.g., water level or chemical addition) by entering a setpoint value. *See also* programmable logic controller, SCADA.

**single-stroke deep-well pump** A reciprocating pump for deep wells, with a single rod connecting the power head with the plunger.

**single-suction impeller** A pump impeller having one suction inlet.

**singlet oxygen** The single oxygen atom, represented by the symbol $O(^1D)$, as opposed to molecular oxygen ($O_2$) or ozone ($O_3$). It is formed in the intermediate reaction of ozone photolysis. *See* excited oxygen atom.

**sink** A place in the environment where a compound or material collects; e.g., the ocean, sometimes wrongly considered a bottomless sink. *See* reservoir and sources and sinks.

**sinkhole** (1) A funnel-shape hole formed in soluble rock (e.g., limestone) by water flowing from the surface to the underground. Also called swallow hole. (2) A depressed area (e.g., in marshy flat land) where waste or drainage collects.

**sinking fund** A fund established in installments for the retirement of a debt.

***S. intercalatum*** *See Schistosoma intercalatum.*

**$SiO_2$** Chemical formula of silica.

**SIP** Acronym of State Implementation Plan.

**siphon** A closed conduit in the approximate form of an inverted U or V with a shorter leg above the hydraulic gradeline where flow is forced up by atmospheric pressure and a longer leg where flow is downward (by gravity). The pressure in the shorter leg is less than atmospheric and a vacuum is necessary in the conduit to start the flow. Priming the siphon creates the vacuum at the crown. After priming, the siphon's discharge ($Q$) is a function of its cross-section area ($A$), its head ($H$), the gravitational acceleration ($g$), and a coefficient of discharge ($K$, usually about 0.9):

$$Q = K \cdot A \cdot (2\,gH)^{0.5} \qquad \text{(S-54)}$$

An inverted siphon or depressed sewer is not a siphon. It is a U- or V-shaped section of gravity sewer dropped below the hydraulic gradeline beneath an obstacle (railway, highway cut, stream, gully, subway, etc.). *See also* siphon spillway, dosing siphon.

**siphon breaker** A small valve that closes under raw-water pump pressure when the engine runs and opens when the engine stops, thus preventing the siphoning of water. Also called antisiphon valve, vented loop, vacuum valve.

**siphon spillway** (1) An automatic spillway designed to function as a siphon for the control of such vectors as *Simulium, Anopheles,* and snails in reservoirs. It causes a continuous fluctuation of water levels in the reservoir and intermittent discharges downstream. *See* Figure S-13. (2) A device used to separate excess stormwater flows from dry-weather flows in a combined sewer;

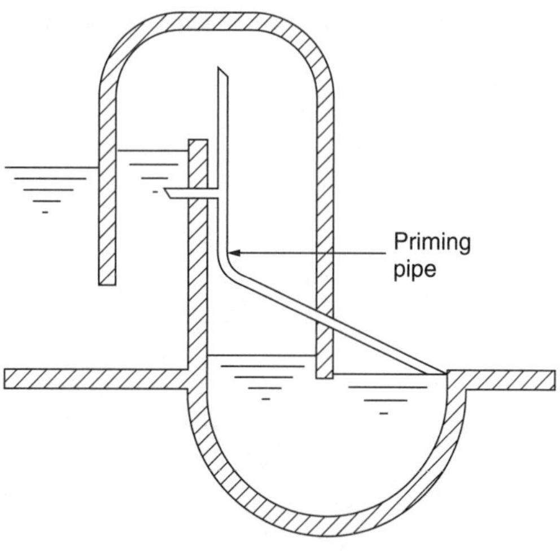


**Figure S-13.** Siphon spillway (dam). (Adapted from Cairncross & Feachem, 1992.)

it carries the flow that exceeds the capacity of the interceptor into an overflow channel. *See* Figure S-14. Other overflow diversion devices include diverting weirs, leaping weirs, and control valves.

**siphon trap** A double-bend trap with a water seal at the bottom to prevent the reflux of gases. Also called running trap.

**Sirofloc process** A proprietary water treatment process developed in Australia, using and recycling powdered magnetite as a ballasting agent, without coagulation. The magnetite is conditioned with sodium hydroxide (NaOH). *See* ballasted floc.

**Sitepro™** An apparatus manufactured by ORS Environmental Systems to control operations at remediation sites.

**siting** The process of selecting a location for a facility.

**SIU** Acronym of significant industrial user.

**SI unit** Abbreviation of Système International unit. *See* Système International.

**size exclusion** In pressure-driven membrane processes, the removal of particles by sieving (Metcalf & Eddy, 2003).

***S. japonicum*** Abbreviation of *Schistosoma japonicum.*

**skatole ($C_9H_9N$)** A white, crystalline, soluble solid having a strong fecal and nauseating odor (odor threshold = 0.019 ppm by volume); used in the preparation of perfumes; it may also be formed during anaerobic decomposition.

**SK concept** An approach used in the design of trickling filters, based on the instantaneous dosing intensity or flushing intensity of the filter, represented by the parameter *SK* (mm/pass of an arm of distributor):

$$SK = 1000\,(Q + R)/(AN) \qquad \text{(S-55)}$$

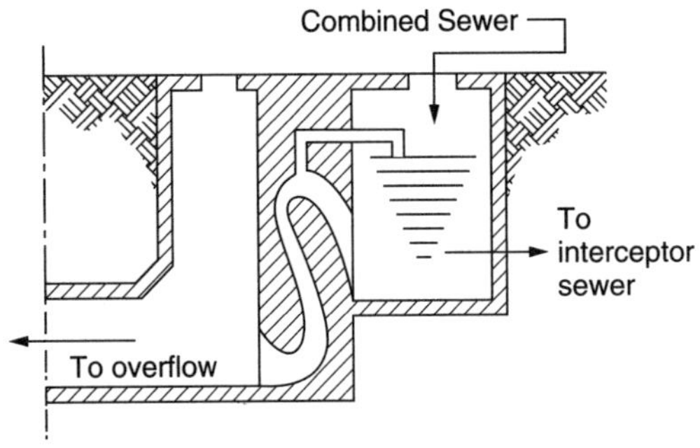


**Figure S-14.** Siphon spillway (combined sewer). (Adapted from Fair et al., 1971.)

where $Q + R$ = average hydraulic rate, $m^3/m^2/hr$; $A$ = number of arms; and $N$ = rpm. *Note:* SK stands for the German word spülkraft, or dosing rate. Recommended design SK values depend on $BOD_5$ loadings, e.g., 30–200 mm/pass for $BOD_5$ = 1.0 $kg/m^3/day$, whereas the flushing intensity would be at least 300 mm/pass. The SK concept may also be used in operation: high SK values may improve performance, control fly larvae, and enhance reduction of nitrogenous oxygen demand.

**skeletal fluorosis** A condition of brittle bones caused by excessive fluoride. *See* fluorosis.

**skewness** The tendency of a bell-shaped curve to be asymmetrical; measured by the third moment of the distribution, it represents the distance to the right or left of the mean and median from the mode. For many water and wastewater data that are skewed, their log is normally distributed. Other moments measure the mean, variance, and kurtosis. *See also* log-normal distribution.

**skewness coefficient** A statistical parameter used in the analysis of water and wastewater management data. *See also* coefficient of kurtosis.

**skid mounting** A type of installation in which equipment is placed on a horizontal, flat structure or platform for easy handling and operation.

**Skim-Kleen®** A belt skimmer manufactured by Tenco Hydro, Inc. for oil removal.

**skimming** (1) Using a machine to remove floating grease, oil, or scum from the surface of water or wastewater. (2) Diverting water from a stream, a conduit or a reservoir to avoid undesirable materials or characteristics (debris), using a shallow overflow or an outlet at an appropriate elevation. (3) Same as centrifuge skimming, i.e., the removal of soft sludge from the inner wall of sludge of an imperforate basket centrifuge.

**skimming detritus tank** A long, trough-shaped wastewater treatment tank used for the simultaneous removal of grease, floating matter, and other light materials, as well as settleable solids; it is equipped with booms, weirs, and thickening devices. *See also* skimming tank.

**skimmings** Materials such as grease, liquids, scum, and solids removed in skimming tanks or otherwise skimmed from clarifiers.

**skimming tank** A long, trough-shaped wastewater treatment tank used for the removal of grease and floating matter, which are retained in baffled subsurface compartments at the entrance and exit. *See also* aerated skimming tank, skimming detritus tank.

**skimming weir** An adjustable-crest weir that allows the control of the depth of overflow from a tank or a reservoir. *See also* diverting weir, effluent launder, effluent weir, and overflow weir.

**Skim-Pak™** A floating weir skimmer manufactured by Douglas Engineering.

**skin of water** A thin layer at the free surface of a water body that is denser than the other layers and resists tension. *See* surface tension.

**skip** A rake for cleaning bar screens.

**SKRAM** An acoustic fish control device of FMC Corp.

**SK value** Same as dosing rate for a trickling filter; from the German word spülkraft.

**slag** Waste residues resulting from metal smelting and coal gasification. Ore extraction slag contains silicon, iron, magnesium, zinc, and smaller amounts of manganese, copper, and sulfur.

**slagging furnace** A slowly rotating, cylindrical steel vessel used for heating wastewater treatment sludge beyond the point of fusion and producing a glassy slag. *See also* basicity adjustment agent, multiple-hearth fluid-bed furnace.

**slake** To cause to heat and crumble by treatment with water; to mix with water. A true chemical combination (hydrolysis) takes place, such as in the slaking of lime.

**slaked hydrated lime** Same as slaked lime.

**slaked lime** Limestone that has been burned and treated with water under controlled conditions until the calcium oxide (CaO) portion has been converted to calcium hydroxide [$Ca(OH)_2$]. Hydrated lime is quicklime combined with water. *See* hydrated lime for more detail.

**slaker** A device used to mix quicklime (CaO) with water ($H_2O$) to form slaked or hydrated lime [calcium hydroxide, $Ca(OH)_2$] or to prepare a lime slurry. *See also* slaking chamber.

**slaking** The formation of slaked or hydrated lime [$Ca(OH)_2$] or preparation of a lime slurry by heating quicklime (CaO) with water ($H_2O$):

$$CaO + H_2O \rightarrow Ca(OH)_2 \qquad \text{(S-56)}$$

**slaking chamber** An open channel used to slake quicklime (add water to lime in a given ratio) before the suspension is fed to the water to be treated. *See also* slaker.

**slaking process** *See* slaking.

**slam** *See* check valve slam.

**slaughterhouse waste** The waste resulting from slaughterhouses or abattoirs (where the killing, dressing, and some by-product processing take place). Slaughterhouse wastewater has high concentrations of BOD (about 2000 mg/L) and nitrogen (about 500 mg/L) as well as a reddish-brown color and a considerable amount of hair, dirt, etc. Waste blood may be recovered for use as a binder in wood laminating and in the production of glue. *See also* meatpacking waste, paunch manure.

**SLC** Acronym of single-loop controller.

**sleek** A thin oily film on the surface of a water body. *See* sleek field.

**sleek field** The perceptible, oily area created on the surface of a receiving water body by the submerged discharge of raw wastewater or treated effluent. Also called wastewater field. For raw wastewater, the sleek area may be estimated as follows (Fair et al., 1971):

$$A = P(11.5 - 3.5 \log P) \qquad \text{(S-57)}$$

where $A$ is the area in acres and $P$ is the population equivalent of the discharge in thousands.

**sleeping sickness** A severe, often fatal, vector-borne disease caused by a species of the pathogen *Trypanosoma* and transmitted by tsetse flies (*Glossina spp.*). There are one American and two African varieties of sleeping sickness. *See* the following common names for more detail: Chagas disease, Gambian sleeping sickness, and Rhodesian sleeping sickness.

**sleet** Precipitation as a mixture of rain and ice.

**sleeve** A tubular piece of metal, plastic or the like, designed to fit over another part such as a pipe or a rod; for example, a pipe fitting for uniting two sections of equal size.

**Slichter method** A method of measuring the velocity of groundwater using perforated pipes driven into the aquifer and a strong electrolyte to determine the electrical resistance of the water at different time intervals.

**Slide Gate** A screenings press manufactured by Andritz-Ruthner, Inc.

**sliding-panel weir** A movable weir made of wooden panels that slide in the grooves of fixed frames.

**slime** (1) A viscous organic substance formed from microbial growth as a coating on inorganic surfaces. (2) The gelatinous film of microbial growth covering the surface of a medium or spanning the interstices of a granular bed; e.g., the biofilm that is produced when wastewater is sprayed over fixed media, as in a trickling filter, or when water flows through a slow sand filter. It consists mainly of bacteria, protozoa, and fungi that use oxygen, organic matter, and other waste constituents for food and energy, releasing carbon dioxide. It also includes sludge worms, fly larvae, rotifers, and other biota. Also called bacterial slime, biofilm, biological slime or microbial film. *See* fixed-growth process, schmutzdecke, sloughing.

**slime layer** (1) A loose aggregation of carbohydrates that form a coating of macromolecules around the wall of a bacterial cell; also called glycocalyx. *See also* capsule. (2) Same as slime (2). The biological film on the rock or plastic packing of a trickling filter may be as thick as 1.0 cm. *See* sloughing.

**slimicide** A chemical substance used to control microbial slimes.

**slip** The percentage of water taken into the suction end of a pump that is not actually discharged.

**slip joint** A joint inserted between two pipe sections, with the end of one section slipped into the flared end of the other; a similar joint between a new masonry wall and an older wall, part of the new wall fitting into a groove of the other.

**sliplining** A trenchless method of rehabilitation, usually applied to large-diameter water mains. It consists of removing a portion of the pipe for access and placing a new piece of HDPE or other thermoplastic pipe into the damaged section. Sliplining is also used to repair in place a section of deteriorated sanitary sewer with plastic pipe. *See* trenchless technology, fold-and-form liner, tight-fit liner.

**Slo-Mixer** A paddle flocculator manufactured by Envirex, Inc.

**slope** (1) The ratio of the vertical distance to the horizontal distance or "rise over run," as the slope or inclination of a trench bottom or a trench side wall. *See also* grade. (2) The slope of a stream is the fall per unit length of channel bottom, water surface, or energy gradeline. *See also* sedimentation terms.

**slope factor** The slope of the dose–response curve in the low-dose region. When low-dose linearity cannot be assumed, the slope factor is the slope of the straight line from 0 dose (and 0 excess risk) to the dose at 1% excess risk. An upper bound on this slope is usually used instead of the slope itself. The units of the slope factor are usually expressed as 1/(mg/kg-day). Also called human carcinogenic potency factor or, simply, potency factor. *See also* multistage model, toxicity terms.

**sloping beach** A stationary device with a collector trough, used for removing scum from primary clarifiers. Jet sprays, scrapers, or sludge collection mechanisms direct the floating materials to the sloping beach and another scraper moves the materials into a trough and then a wet well. *See also* tilting trough.

**slop oil** The floating oil and solids that accumulate on the surface of an oil–water separator or oil separator during refinery, start-up, shut-down, or abnormal operation.

**slot** A narrow, elongated opening, like the gas vent or the opening in an Imhoff tank for the passage of solids into the digestion chamber.

**slot opening** The size of the holes in a well screen.

**slotted-tube** A nonporous, fixed-orifice diffuser made of stainless steel tubing, perforated and slotted, used to provide diffused air for the aeration of wastewater. *See also* orifice diffuser, static tube, porous diffusers.

**slough** In general, a riparian marsh along a slow-moving stream. (1) A small marsh, especially a marshy area lying in a local, shallow, closed depression on a piece of dry land, as on the prairie of the Midwest. (2) In the Mississippi Valley, a creek or sluggish body of water in a tidal flat, floodplain, or coastal marshland. (3) A sluggish channel of water, such as a side channel of a river, in which water flows slowly through low, swampy ground, or a section of an abandoned river channel that may contain stagnant water and occurs in a floodplain or delta. (4) An area of soft, miry, muddy, or waterlogged ground; a place of deep mud. (5) The slime of schmutzdecke of slow sand filters and trickling filters.

**sloughing** The process of detachment of accumulated slime and biological solids from trickling filter media and similar contact areas. The process is due to physical (e.g., shearing) or chemical action and depends on hydraulic loading and organic loading. The sloughed off material is usually removed in subsequent treatment (e.g., sedimentation). *See also* schmutzdecke, slime.

**slow infiltration system** A slow-rate wastewater treatment system designed for reuse and handling flexibility, thus requiring secondary or even advanced levels of preapplication treatment. Also called type-2 slow-rate system.

**slow-rate land application** Treatment and disposal of wastewater in a manner similar to agricultural irrigation; the liquid moves downward through the soil without surface runoff and reaches the groundwater by percolation or the atmosphere by evapotranspiration. Besides wastewater treatment and disposal (including nutrient removal), this technique produces cash-value crops and conserves irrigation water. However, it poses a risk of disease transmission through airborne droplets. Also called spray irrigation. *See* long-term acceptance rate, hydraulic loading rate.

**slow-rate land disposal** *See* slow-rate land application.

**slow-rate land treatment** *See* slow-rate land application.

**slow-rate system** A system of natural wastewater treatment. *See* slow-rate treatment.

**slow-rate system type 1** A natural, slow-rate treatment system designed for wastewater treatment without regard to the water requirement of the vegetation; soil permeability and constituent loading are the controlling design parameters. Wastewater receiving a minimum of primary treatment is applied by sprinkling or surface methods at a rate of 6–20 feet per year. *See also* slow-rate system type 2.

**slow-rate system type 2** A natural, slow-rate treatment system designed primarily for water reuse through crop production or landscape irrigation. Wastewater receiving a minimum of primary treatment is applied by sprinkling or surface methods at a rate of 2–7 feet per year. Also called wastewater irrigation system or crop irrigation system. *See also* slow-rate system type 1.

**slow-rate treatment** The application of wastewater to vegetated land for wastewater treatment and vegetation growth. The liquid percolates through the soil or is consumed by evapotranspiration, with runoff collection and recycle if necessary. *See* slow-rate system type 1, slow-rate system type 2, exchangeable sodium percentage, sodic soil, vertical permeability. Other natural wastewater treatment methods include rapid infiltration, overland flow, wetlands, floating aquatic plants, and aquaculture.

**slow-release fertilizer** A fertilizer that allows the controlled release of ammonia and minimizes nitrate formation.

**slow rinse** The portion of the rinsing stage of an ion exchanger following the introduction of the regenerant, the rinse water, and the regenerant flowing at the same rate. Its purpose is to gradually push the remaining chemical in the vessel through the resin bed, extending the contact time. *See also* fast rinse.

**slow sand filter** A device developed in Great Britain in the early 19th century for the treatment of raw water; it consists of a layer of 24 to 40 inches of sand, the top few millimeters of which constitute the schmutzdecke, a biological mat that accomplishes most of the filtration and retains or kills most of the pathogens. Effective size of the sand is about 0.2 mm and its uniformity coefficient about 2.5. The device includes an underdrainage system for the collection and removal of the filtrate. Slow sand filters, designed to operate at low rates of 3–6 mgad without coagulation pretreatment, require large land areas and have practically disappeared in the United States; a few of them are still used to treat surface water of low turbidity. Slow sand filters are effective in removing *Giardia* cysts but are not successful in treating river waters containing clay, which are common in the United States. Slow sand filters are also used in the treatment of wastewater. *See also* filter ripening.

**slow sand filter biomass** Same as schmutzdecke.

**slow sand filtration** A process involving passage of raw water, without chemical coagulation, through a bed of sand at low velocity (generally less than 0.4 m/h), resulting in substantial particulate removal by physical and biological mechanisms (EPA-40CFR141.2).

**slow sand water filter** *See* slow sand filter.

**slow start** The process of starting the operation of a filter at a rate lower than normal and gradually increasing its effluent flow to allow more time for particle removal. A slow start may also be practiced immediately after backwashing, e.g., for three empty bed volumes. *See also* delayed start, filter-to-waste.

**slow-start filtration** *See* slow start.

**SLRT** Acronym of sewer level remote telemetry.

**sludge** (1) A solid, semisolid, or liquid residue from any of a number of air, water or wastewater treatment processes; can be a hazardous waste. Sludge does not include the treated effluent from a wastewater treatment plant, solids or dissolved materials in wastewater or other significant pollutants in water resources, such as silt, dissolved or suspended solids in industrial wastewater effluents, dissolved materials in irrigation return flows, or other common water pollutants (EPA-40CFR240.101-x). The term sludge is being replaced by residuals, but sludge is still in use, particularly in reference to untreated solid materials and chemical residuals from water or wastewater treatment, or in conjunction with a specific process, e.g., primary sludge, secondary sludge, waste-activated sludge. *See also* biosolids, float, grit, screenings, scum, solids, spent filter backwash water. (2) An aggregate of oil or oil and other matter of any kind in any form other than dredged spoil having a combined specific gravity equivalent to or greater than water (EPA-40CFR110.1). (3) Organic deposits on the bottom of water bodies.

**sludge age** The average time, in days, that microorganisms spend in a biological waste treatment system, a leading design and operation parameter. Recently, the term sludge age has been replaced by mean cell residence time (MCRT) and then solids retention time (SRT). *See* solids retention time for more detail.

**Sludge Age Controller™** A device manufactured

by United Industries, Inc. to control sludge age (solids retention time) in biological wastewater treatment.

**sludge agricultural value** *See* agricultural value of sludge.

**sludge air drying** A sludge dewatering operation that removes moisture from sludges on sandbeds or other materials by letting it evaporate or drain to the drying bed. *See* air drying for more detail.

**sludge bank** An accumulation of wastewater solids on the banks, edges, or shores of a water body.

**sludge barging** The transport of sludge in barges for disposal into the ocean.

**sludge bed** A water or wastewater treatment plant unit used to separate water and solids in residuals. It consists of confined, underdrained shallow layers of sand, gravel, or other natural or artificial porous materials on which wet sludge is distributed for draining and evaporation. *See* more detail under drying bed.

**sludge blanket** (1) The zone at the bottom of a settling basin or thickening tank where solids accumulate and concentrate, with an overlying clearwater zone. In gravity thickening, the thickness of the sludge blanket controls the underflow concentration. Integrity of the sludge blanket is a consideration in the design and operation of these tanks (e.g., use of nonscouring velocities, timing of sludge withdrawal to control the blanket depth). The top of the sludge blanket determines the depth of clarification and a minimum blanket level of 2.0 ft at peak flow is recommended. *See also* sludge blanket zone, sludge volume ratio, solids blanket. (2) In a solids-contact clarifier, the sludge blanket is the zone of the tank where particle settling velocity equals the upward velocity of the fluid; it acts as a filter for falling and rising particles and serves as an area for periodic sludge wasting.

**sludge-blanket filtration** An upflow clarification process used to treat water, combining filtration with sedimentation. *See* contact filtration for more detail.

**sludge-blanket clarifier** An upflow clarifier often used in water softening plants, combining mixing (for flocculation), sedimentation, and sludge recirculation, which promotes the formation of a sludge blanket. *See* solids-contact basin for detail.

**sludge-blanket level** A method of control of the performance of a secondary clarifier by maintaining an optimum level of the sludge blanket, basically a balance between settling depth and sludge storage based on experience.

**sludge-blanket zone** In type III sedimentation or zone settling, the sludge blanket zone (or zone I) is the portion of the sedimentation vessel where solids accumulate with a relatively constant flow and solids flux through the blanket into the concentrated sludge zone (or zone II).

**sludge-blanket process** An anaerobic wastewater treatment process designed for carbonaceous BOD removal from high-strength wastes. *See* upflow anaerobic sludge blanket.

**sludge blending** The mixing of sludges from primary, secondary, and advanced wastewater treatment to produce a uniform material with consistent characteristics that is easy to process in dewatering heat treatment and incineration units. Blending can occur in the primary clarifiers, piping, sludge digesters, or separate blending tanks. *See also* sludge grinding, screening, and degritting.

**sludge blinding** A phenomenon that occurs in sludges with a bimodal particle distribution, characterized by the migration of fines through the sludge cake and a decreased cake permeability.

**sludge blowdown** Withdrawal of sludge from a solids processing tank to maintain a desirable solids level. Also called residuals blowdown.

**sludge boil** The upwelling of sludge deposits caused by the release of decomposition gases.

**sludge bulking** A condition of activated sludge treatment plants in which sludge occupies a larger volume than normal, does not settle or concentrate well, and may carry over excessively in the effluent with a concomitant increase in BOD and suspended solids. Bulking sludge, characterized by a sludge volume index (SVI) of 200 or more (whereas well-settling sludge has an SVI between 50 and 100), is often associated with filamentous microorganisms (filamentous bulking) such as *Sphaerotilus,* whose growth is promoted by high carbohydrate concentrations and whose large surface-area-to-volume ratio decreases their settling velocity. Excessive water retention in microbial cells may also cause viscous (or hydrous) bulking. Factors considered in sludge bulking investigations include low dissolved oxygen concentration, low F:M ratio, low pH, mixed liquor concentration, nitrification–denitrification, nutrient deficiency, organic and volumetric loadings, poor mixing, septic wastewater, solids retention time, temperature, and toxins. *See also* bulking microorganisms.

**sludge burial** Disposal of water or wastewater treatment solids in a landfill.

**Sludgebuster™** A shredder of International Shredder, Inc.

**sludge cake** (1) Water or wastewater treatment solids sufficiently dewatered (by a filter press, centrifuge, or other device) to form a semidry mass, typically having a thickness of 0.25 inch, a solids concentration of 15–40%, and a consistency ranging from that of pudding to that of damp cardboard. Sludge cakes do not flow by gravity from dewatering equipment; they can be pumped, but with high head losses. Also called residuals cake. *See also* trickling filter humus, hydraulically driven reciprocating piston pump, belt conveyor. (2) A similar material formed on the surface of a sand drying bed after repeated sludge applications.

**sludge calcination** A method of recovering lime form water softening sludge; precipitated calcium carbonate ($CaCO_3$) is converted into lime or calcium oxide (CaO) by heating the dry solids to drive off carbon dioxide:

$$CaCO_3 \rightarrow CaO + CO_2 \qquad \text{(S-58)}$$

See Hoover process and Lykken–Estabrook process.

**sludge cement** A material having the approximate composition of Portland cement, made from a mixture of gypsum and activated sludge flocculated with carbon dioxide and calcium dioxide.

**sludge centrifugation** The process of separating sludge solids using a centrifuge, the suspending liquid becoming a by-product. *See* centrifugation.

**sludge centrifuging** *See* sludge centrifugation.

**sludge characteristics** The specific gravity of sludge is the ratio of the weight of a sample of sludge to the weight of an equal volume of distilled water; see also sludge volume index, sludge density index, moisture–weight–volume relationship, sludge viscosity, plastic fluid.

**sludge chemical conditioning** *See* sludge conditioning.

**sludge chemical treatment** *See* sludge conditioning.

**sludge circulation** The overturning of solids in a digester using mechanical or hydraulic means to disperse scum.

**SludgeCleaner™** A device manufactured by Parkson Corp. for screening and compacting sludge and scum from wastewater treatment.

**sludge coincineration** *See* coincineration.

**sludge collector** A mechanical device used to scrape solids accumulated on the bottom of a sedimentation tank and convey them to a sump where they can be removed. *See* chain-and-flight sludge collector, sludge hopper, traveling bridge, plow-type sludge collector.

**sludge combustion** *See* sludge incineration.

**sludge compactibility** A sludge characteristic related to its dewaterability and measured by the column settling test. It represents the ratio ($\gamma$) of volatile suspended solids in waste sludge or recycled underflow ($X_{Vr}$) to the volatile suspended solids in the mixed liquor ($X_V$), both in mg/L. It may be used to determine the recycle ratio ($r$):

$$\gamma = (X_{Vr})/(X_V) \qquad \text{(S-59)}$$

$$r = (1 - \theta_d/\theta_x)/(\gamma - 1) \qquad \text{(S-60)}$$

where $\theta_d$ and $\theta_x$ represent, respectively, the hydraulic detention time and the solids retention time, both in days (Droste, 1997).

**sludge compaction** A physical characteristic of sludge relating its density and moisture content; it affects the stability, permeability, and resistance to erosion of landfills used for sludge disposal. The compaction curve is a plot of density or dry unit weight versus solids concentration. *See also* standard Proctor compaction, sludge compressibility, zero-air-void curve.

**sludge compartment** A separate compartment for sludge digestion included in a sedimentation. *See also* Imhoff tank, Travis tank.

**sludge composition** Table S-06 lists typical values for some important sludge constituents with regard to processing and disposal.

**sludge composition testing** Tests conducted to determine the characteristics of the sludge solids, particularly their pH/alkalinity/acidity, organic content, BOD, fatty acids, grease, ammonia, radioactivity. Other sludge tests relate to its concentration, condition, functional characteristics, and processing.

**Table S-06.** Sludge composition

| Constituent | Raw primary sludge | Digested primary sludge | Raw activated sludge |
|---|---|---|---|
| Total dry solids (TS*), % | 6 | 4 | 1 |
| Volatile solids (% of TS) | 65 | 40 | 70 |
| Protein (% of TS) | 25 | 18 | 36 |
| Nitrogen (N, % of TS) | 2.5 | 3 | 4 |
| Phosphorus ($P_2O_5$, % of TS) | 1.6 | 2.5 | 7 |
| Potash ($K_2O$, % of TS) | 0.4 | 1 | 0.6 |
| Cellulose (% of TS) | 10 | 10 | |
| Iron (% of TS) | 2.5 | 4 | |
| Silica ($SiO_2$, % of TS) | 17 | 15 | |
| pH | 6 | 7 | 6.5–8.0 |
| Alkalinity (mg/L as $CaCO_3$) | 600 | 3000 | 600–1100 |
| Organic acids (mg/L) | 500 | 200 | |
| 1100–1700 | | | |
| Energy content, kJ TS/kg | 25,000 | 12,000 | 21,000 |

*TS = total solids.

**sludge composting** The aerobic thermophilic decomposition of the organic matter in wastewater treatment sludge that has a moisture content between 40 and 60%. Optimum conditions: pasteurization temperature range of 50–70°C and pH between 6.0 and 7.5. The process results in a stable end product usable as soil conditioner and in substantial reduction of pathogens. It qualifies as a process to further reduce pathogens (temperature = 55°C for 15 days) or process to significantly reduce pathogens (40°C for 5 days). Composted sludge may also be disposed in sanitary landfills. *See* the following terms for more detail: aerated static pile composting, aerobic composting, agitated bed composting, amendment, *Aspergillus fumigatus,* biofilter, bulking agent, cocomposting, compost bioaerosol, composting methods, composting microbiology, composting operations, composting performance, composting stages, composting toilet, cooling stage, endotoxin, forced-aeration composting, in-vessel composting, mechanical composting, negative aeration, plug-flow in-vessel composting, positive aeration, process to further reduce pathogens, process to significantly reduce patogens, rotating screen, static bed composting, static pile composting, two-step windrow composting, vibrating deck screen, windrow composting.

**sludge compressibility** A physical characteristic of a sludge whose dewatering rate decreases as a result of floc deformation under the pressure applied. This characteristic is usually expressed by the relationship between the specific resistances corresponding to two pressures:

$$R_2 = R_1(\Delta P)^S \qquad \text{(S-61)}$$

where $R_1$ = specific resistance at pressure $P_1$, $R_2$ = specific resistance at pressure $P_2$, $\Delta P$ = pressure difference = $P_2 - P_1$, and $S$ = coefficient of compressibility, usually determined graphically by plotting values of specific resistance versus pressure on log-log paper. *See also* compaction, compression index, plasticity, shear strength, swelling index.

**sludge compression** The reduction of the volume of sludge disposed in a landfill resulting from the extrusion of water and causing the settlement of the landfill. *See also* compression curve, compression index.

**sludge concentration** A treatment process to reduce the volume of sludge before its further treatment or ultimate disposal; the resulting product is still in a fluid state. *See* centrifuge, filter press, vacuum filter.

**sludge concentration testing** Tests conducted to determine the total residue, water content, and specific gravity of sludge. Other sludge tests relate to its composition, condition, functional characteristics, and processing.

**sludge conditioner** A chemical such as lime, ferric chloride, or polymer used in sludge conditioning. *See* sludge conditioning chemicals.

**sludge conditioning** The pretreatment of liquid sludge, mostly by chemical, but also biological or physical, means, to facilitate subsequent treatment such as thickening or dewatering and improve drainability. Conditioning methods used prior to sludge dewatering include chemical treatment, aerobic digestion, anaerobic digestion, heat treatment, sludge preheating, and freeze–thaw conditioning. These methods produce a sludge with a solids content of at least 4%. Some factors affect the results of sludge conditioning: sludge characteristics, sludge handling methods, sludge coagulation and flocculation. For example, it is easier to condition raw primary sludge than biological or digested sludge. Also called residuals conditioning.

**sludge conditioning chemicals** Inorganic or organic chemicals used to cause the aggregation of fine sludge solids by coagulation and flocculation prior to dewatering. The most commonly used in the United States are ferric chloride ($FeCl_3$), lime (CaO), filter aids, chlorinated copperas ($FeSO_4Cl$), ferric sulfate [$Fe_2(SO_4)_3$], aluminum sulfate [$Al_2(SO_4)_3 \cdot 18\ H_2O$], and organic polyelectrolytes (polyacrylamide, and dry, liquid, and emulsion polymers). Ferrous sulfate and lime are used in Great Britain. While polymers do not significantly add to sludge production, inorganic chemicals increase the dry solids by 20–30 percent. *See also* Büchner funnel test, capillary suction test, conditioner requirement, standard jar test, double-layer theory.

**sludge condition testing** Tests conducted to determine the color, odor, BOD, and flora and fauna of sludge. Other sludge tests relate to its composition, concentration, functional characteristics, and processing.

**sludge degritting** A preliminary sludge processing operation consisting of the application of centrifugal force or another method to remove grit from sludge in treatment plants that do not adequately remove the grit from the wastewater, e.g., plants that handle wet-weather flows or have undersized grit chambers. *See* cyclone degritter. *See also* sludge grinding, screening, and blending.

**sludge density index (SDI)** A measure of the degree of compaction of sludge, in grams per milliliter. It is equal to the percentage weight of suspended solids in the sludge after settling the aerated liquor for 30 minutes in a graduated container. It is related to the sludge volume index (SVI):

$$SDI = 100/SVI \qquad (S-62)$$

Also called inverse sludge index.

**sludge deposit** An accumulation of wastewater solids.

**Sludge Detention Optimizer™** A process developed by Dontech, Inc. for sludge conditioning and thickening.

**sludge dewatering** The process of reducing the water content of sludge by a mechanical or physical method such as centrifugation, compaction, drainage, evaporation, filtration, flotation, freezing, gravity dewatering, and squeezing. Common dewatering units currently used are belt filter presses, centrifuges, drying beds, lagoons, and recessed-plate filter presses. Some imperforate basket centrifuges and vacuum filters are still in operation, but new units are no longer being installed. Dewatered sludge is spadable, less costly to truck away, and easier to handle than liquid sludge. Dewatering is also required for further processing, e.g., composting, incineration, landfilling, land application, and odor control. The solids content of the dewatered sludge cake varies with sludge characteristics and the technique used, e.g., from 12 to 50%. Volume reduction is in the order of 80–85%. Also called residuals dewatering.

**sludge dewatering lagoon** *See* sludge lagoon.

**sludge digester** A tank in which complex organic substances like wastewater sludges are biologically degraded. During these reactions, energy is released and much of the wastewater is converted to methane, carbon dioxide, and water; also called a sludge digestion tank. *See* digester and digester capacity for more detail.

**sludge digestion** The biochemical decomposition of the organic matter in sludge, resulting in stabilization, i.e., partial gasification, liquefaction, mineralization of pollutants, and volume reduction. Some pathogens are also destroyed during digestion. Digestion is one metabolic mechanism that organisms use to process energy and materials: they break down complex molecules to simpler substances that they can use. Digestion is a suspended-growth process operated without solids recycle, with a solids retention time dependent on temperature, e.g., 6 days at 35°C and 30 days at 20°C. Digested sludge is a relatively inert material that can be dewatered easily. Smaller installations tend to use aerobic sludge digestion, while larger installations favor anaerobic digestion. *See also* the following types of digestion: high-rate, mesophilic, single-stage, thermophilic, two-stage.

**sludge digestion chamber** A sludge compartment, as in the lower part of an Imhoff tank or Travis tank.

**sludge digestion gas** The gaseous mixture produced by the anaerobic decomposition of the organic matter in sludge from wastewater treatment. *See also* sewer gas.

**sludge digestion tank** Same as sludge digester.

**sludge disinfection** *See* processes to further reduce pathogens and processes to significantly reduce pathogens.

**sludge disposal** The ultimate discharge of solids from water or wastewater treatment to the environment. The residue from sludge processing is disposed of mostly in sanitary landfills. Other methods of disposal depend on whether the sludge is wet (raw or digested), partially dewatered, or dried. Some residue, depending on the sludge treatment methods, may be disposed of in the ocean, spread on-site or on agricultural land, or distributed as compost. Chemical precipitates from water treatment plants are sometimes discharged in sewers or streams, preferably after lagooning. Also called residuals disposal. *See also* land disposal, ocean disposal.

**sludge drainability** The relative rate of moisture release by sludge solids, as determined by a sludge filtration test. Also called sludge filterability. *See* Büchner funnel.

**sludge dryer** A device used to reduce the moisture content of sludge or screenings by heating to temperatures above 65°C directly with combustion gases (EPA-40CFCR61.51-m).

**sludge drying** The process of reducing the moisture content of dewatered sludge by methods using gravity or evaporation, such as flash drying, rotary drying, and solvent extraction. *See also* the Carver–Greenfield process.

**sludge drying bed** A water or wastewater treatment plant unit used to separate water and solids in residuals. It consists of confined, underdrained shallow layers of sand, gravel, or other natural or artificial porous materials on which wet sludge is distributed for draining and evaporation. Also called residuals drying bed. *See* more detail under drying bed.

**sludge drying equation** The theoretical expression of the constant evaporation rate ($E$, lb/hr) of sludge drying under equilibrium conditions:

$$E = KA(H_s - H_a) \qquad (S-63)$$

where $K$ = mass transfer coefficient of gas phase, lb mass/ft$^2$; $A$ = area of wetted surface exposed to drying medium, ft$^2$; $H_s$ = saturation humidity of air at sludge–air interface, lb water vapor/lb dry air; and $H_a$ = humidity of drying air, lb water vapor/lb dry air.

**sludge elutriation** A sludge conditioning process. *See* elutriation.

**sludge energy content** Same as sludge fuel value.

**sludge excess** In the activated sludge or any other biological treatment process, it is the sludge that is produced in excess of the required solids recirculation and that does not escape in the effluent. It is withdrawn from the system and wasted. Also called waste sludge, excess sludge, or waste activated sludge.

**Sludge Expert** An apparatus manufactured by Solids Technology International, Ltd. to provide automatic control of the operation of belt presses.

**sludge fermenter** A tank used in sidestream fermentation processes for the formation of volatile fatty acids, such as acetic and propionic acids. Also called a sidestream fermenter or an acid fermenter, it may include a compartment for storage of methane gas.

**sludge filter** A device that uses a vacuum or pressure to partially dewater sludge from water or wastewater treatment. Before filtration, the sludge usually undergoes chemical conditioning.

**sludge filterability** The relative rate of moisture release by sludge solids, as determined by a sludge filtration test. Also called sludge drainability. *See* Büchner funnel.

**sludge filtration** The use of a porous medium to dewater sludge in such devices as centrifuges, drying beds, filter presses, and vacuum filters. *See also* Blake–Kozeny equation.

**sludge flow** Sludge flow depends mainly on its solids concentration. Poiseuille's equation for viscous fluids is sometimes used to determine the head loss of the flow of sludge in a pipe:

$$H/L = (32/g)(V/D^2)[(\eta/\rho) + 1/6\,(\tau/\rho)(D/V) \quad \text{(S-64)}$$

where $H$ = head loss, ft; $L$ = pipe length, ft; $V$ = flow velocity (fps); $D$ = pipe diameter, ft; $g$ = gravitational acceleration = 32.2; $\eta$ = coefficient of rigidity of sludge; $\rho$ = mass density of the sludge solids; and $\tau$ = shearing stress of sludge. *See also* thixotropy.

**sludge foaming** A condition of froth and scum rising in and overflowing from an Imhoff tank or a sludge digester; it is caused by an increase in gas production.

**sludge formula** A chemical formula commonly used to represent sludge. *See* Hoover–Porges formula.

**sludge freezing** A sludge conditioning and dewatering process. *See* freezing for details.

**sludge fuel value** The quantity of heat that can be obtained from a unit mass of a sludge. It is used in the design of sludge processing units. The gross fuel value of sludge is generally comparable to that of fossil fuels and petroleum products, i.e., between 10,000 and 20,000 BTU per dry-weight pound. Based on correlation with results of bomb calorimeter tests, sludge fuel values may be estimated as follows:

$$Q = A\{[(100\,P_v)/(100 - P_c)] - B\}[(100 - P_c)/100] \quad \text{(S-65)}$$

where $Q$ = fuel value of sludge, BTU/dry-weight pound; $A$ and $B$ = coefficients depending on the type of sludge; e.g., $(A, B) = (131, 10)$ for primary municipal sludge and (107, 5) for fresh activated sludge; $P_v$ = percentage of volatile matter in sludge; and $P_c$ = percentage of precipitating or conditioning chemical in sludge. This is the total sludge fuel value, also called gross heating value, sludge energy content, sludge heating value. *See also* bomb calorimeter, Dulong formula, effective (or characteristic, net) fuel value, sludge-to-oil reactor, and Tubingen process.

**sludge gas** The gas mixture released during the decomposition of organic matter, more specifically during sludge digestion. *See* digester gas for more detail.

**sludge-gas holder** A tank used to stabilize, store, and redistribute the gas collected from sludge decomposition

**sludge-gas utilization** The use of sludge digestion gas for such purposes as heating buildings, incineration, or fueling engines.

**sludge grinding** A preliminary sludge processing operation in which large and stringy materials are cut and sheared to prevent clogging of progressive cavity pumps, heat exchangers, solid-bowl centrifuges, and belt-filter presses. Sludge grinding also enhances chlorine contact with the solid particles. *See also* sludge screening, degritting, and blending, and sludge macerator-grinder.

**Sludge Gun®** A device manufactured by Markland Specialty Engineering, Ltd. to detect the level of sludge in tanks.

**Sludge Guzzler** A sludge pump manufactured by Guzzler Mfg., Inc. with a hydraulic drive.

**sludge handling** A series of operations or processes designed to reduce the moisture content (thus the volume) of sludge from treatment plants before its final disposal. Some reduction of dry solids may also occur. Sludge handling processes include combustion, composting, conditioning, dewatering, digestion, drying, elutriation, and thickening.

**sludge hazardous constituents** Toxic or hazardous originally present in wastewater and concentrated in sludge by treatment. Contaminants of concern include arsenic, heavy metals, explosive, flammable, or corrosive materials, and organic chemicals.

**sludge heating value** *See also* sludge energy content.

**sludge hopper** A container with steep sides, up to 10 ft deep, into which settled sludge is scraped from a clarifier and from which it is removed by gravity. It is usually located at the center of a circular tank or at the inlet end of a rectangular tank.

**sludge incineration** A treatment technology involving destruction or volume reduction of sludge by controlled burning at high temperatures. *See* incineration for more detail.

**sludge index** *See* sludge density index and sludge volume index.

**sludge lagoon** A natural depression or an excavated basin or pond, equipped with overflow and underdrainage, that is used to dewater and store stabilized wastewater sludge, typically for 30–60 days, with formation of a superficial crust and deposition of solids at the bottom. The crust is periodically removed to allow drying by evaporation and the supernatant is discharged or returned ahead of the plant for treatment. The dewatered sludge has a 25–30% solids content; it is periodically removed or allowed to accumulate until the lagoon is filled and taken out of service. Also called residuals lagoon.

**sludge lagooning** The use of lagoons for sludge dewatering. *See* dewatering lagoon.

**sludge land disposal** *See* land disposal.

**sludge landfilling** A relatively simple, effective, and inexpensive method for disposing of mechanically dewatered or sand-bed-dried sludge, exclusively or along with municipal solid wastes.

**sludge macerator–grinder** A device with a rotating multiple-blade cutter or counterrotating, intermeshing cutters, similar to a meat grinder, that cuts or shears large sludge solids into smaller particles to prevent operational problems in subsequent equipment. *See* sludge grinding.

**SludgeMaster** A submersible pump manufactured by Warren-Rupp, Inc.

**sludge moisture content** The percentage by weight of moisture (water) in wet sludge.

**sludge nutritional requirements** The approximate concentrations of nutrients (mainly nitrogen, phosphorus, sulfur, sodium, potassium, calcium, magnesium, and iron) that activated sludge microorganisms require for adequate growth. Otherwise, such undesirable phenomena as bulking may occur. *See* more detail under activated sludge nutrients.

**sludge ocean disposal** The final disposal of sludge in the ocean. *See* ocean disposal.

**Sludgepactor** A sludge screen and compactor manufactured by Jones & Attwood, Inc.

**sludge pasteurization** The use of direct steam injection, heat exchange, or a combination of thermophilic aerobic digestion and anaerobic digestion to disinfect liquid sludge. Sludge pasteurization is not common in the United States, but is practiced in Germany and Switzerland.

**sludge physical characteristics** *See* compaction, compressibility, compression index, plasticity, shear strength, swelling index.

**sludge pond** A pond used for the treatment and disposal of sludge or industrial wastewater, or for aquaculture.

**sludge preheating** The use of 60°C heat to condition sludge prior to heat drying, which increases the sludge cake solids concentration.

**sludge press** A device that uses pressure to dewater sludge. *See* sludge pressing for names commonly used.

**SludgePress™** A belt filter press of Enviroquip, Inc.

**sludge pressing** Sludge dewatering under pressure through a cloth fabric. The corresponding device has various names: belt filter, belt filter press, diaphragm filter press, diaphragm press, filter press, fixed-volume recessed filter press, inclined screw press, plate-and-frame filter press, plate-and-frame press, plate press, pressure filter, pressure filter press, rotary disc press, screw press, sludge press, variable-volume recessed-plate filter press.

**sludge pretreatment** *See* sludge degritting and sludge grinding.

**sludge processing** The collection and treatment of solids from water and wastewater treatment. Sludge processing includes the following steps: pumping, preliminary operations (blending, storage, etc.), thickening, stabilization, conditioning, dewatering, drying, incineration, and conveyance and storage. Final disposal—through land application, dedicated land disposal, or landfilling—is sometimes included in processing. Also called residuals processing.

**sludge processing tests** Tests conducted to determine the treatment characteristics of sludge, namely its suspended solids content, settleability, sludge density index, and sludge volume index. Other sludge tests relate to its composition, concentration, condition, and functional characteristics.

**sludge production** Sludge production, i.e., the net waste activated sludge produced each day ($P_x$, kg VSS/day) for the activated sludge process, can be estimated from an estimate of observed sludge yield ($Y_{obs}$, g VSS/g substrate removal); with VSS = volatile suspended solids. Sludge production can also be calculated from wastewater characteristics as the sum of heterotrophic biomass ($A$), cell debris ($B$), nitrifying bacteria biomass ($C$), and nonbiodegradable VSS ($D$):

$$P_x = 0.001\ Y_{obs}Q(S_0 - S_e) \quad \text{(S-66)}$$

$$P_x = A + B + C + D \quad \text{(S-67)}$$

$$A = 0.001\ YQ(S_0 - S_e)/(1 + \theta k_d) \quad \text{(S-68)}$$

$$B = 0.001\ YQf_dk_d\theta(S_0 - S_e)/(1 + \theta k_d) \quad \text{(S-69)}$$

$$C = 0.001\ Y_nQN_x/(1 + \theta k_{dn}) \quad \text{(S-70)}$$

$$D = 0.001\ Q(\text{nb VSS}) \quad \text{(S-71)}$$

where $Q$ = influent flow rate, m$^3$/day; $S_0$ = influent substrate concentration, mg/L; $S_e$ = effluent substrate concentration, mg/L; $Y$ = synthesis yield coefficient, g VSS/g biodegradable soluble COD; $\theta$ = solids retention time, days; $k_d$ = endogenous decay coefficient, g VSS/g VSS/day; $f_d$ = fraction of biomass remaining as cell debris; $Y_n$ = synthesis yield coefficient for nitrifying bacteria, g VSS/g biodegradable soluble COD; $N_x$ = ammonium nitrogen (NH-N) concentration in influent that is nitrified, mg/L; $k_{dn}$ = endogenous decay coefficient for nitrifying organisms, g VSS/g VSS/day; and nb VSS = nonbiodegradable volatile suspended solids. *See also* sludge quantities, solids production.

**sludge pump** Devices used for pumping sludge include the following types of pumps: centrifugal pump (nonclog, chopper, recessed impeller), diaphragm pump, high-pressure piston pump, high-viscosity sludge pump, peristaltic hose pump, plunger pump, positive-displacement pump, progressing cavity pump, reciprocating plunger pump, and rotary lobe pump.

**sludge pumping headloss** The head loss resulting from pumping sludge over short distances is determined by multiplying the corresponding head loss for water by a factor varying from 1.5 to 13, depending on the percent solids by weight. *See also* sludge rheology method.

**sludge pyrolysis** A technique that uses distillation for the combustion of sludge from wastewater treatment; it produces a solid carbon residue and approximately 0.25 g of an oil and gas residue per gram of volatile solids.

**sludge quantities** The quantity of sludge produced in wastewater treatment depends on process performance and the solids content of the waste. For example, primary clarification removes about 55% of suspended solids and produces sludge that has a moisture content of about 95%. Coagulation sludge contains more solids but a moisture content of about 90%. Suspended-growth processes produce 0.2–0.5 kg of volatile suspended solids per kg of BOD removed, depending on solids retention time, with a moisture content approximating 94%; attached-growth processes produce less sludge. For water treatment, solids concentration varies from 0.1% to 2.5% of the underflow from coagulation and filtration tanks, depending on their degree of thickening. *See also* sludge production, solids production, scum.

**sludger** A simple centrifugal-type device for pumping mud and liquids laden with sand or gravel out of boreholes, without excessive clogging or damage. It is a hollow cylinder, open at the top, with a ball or clack valve at the bottom. Also called sand pump or shell pump.

**sludge reaeration** The continuous aeration of sludge to maintain its condition or sludge aeration in the contact-stabilization process.

**sludge recycle softening** A practice used in some water softening plants to improve the precipitation and settling of hardness-causing materials.

**sludge reduction** The reduction in sludge quantity due to digestion.

**sludge requirements** Statutory provisions and regulations of federal legislation (or more stringent state or local regulations) dealing with sludge processing and disposal, or permits issued thereunder: Clean Water Act, Solid Waste Disposal Act, Clean Air Act, Toxic Substances Control Act, and Marine Protection, Research, and Sanctuaries Act (EPA-40CFR403.7(a)(1)-ii).

**sludge residence time** Same as sludge age.

**sludge reuse** The use of sludge from wastewater treatment to enrich soils, fish production, and biogas production. *See also* agricultural reuse, aquaculture.

**sludge rheology method** A technique that uses the two dimensionless Reynolds ($R_e$) and Hedstrom ($H_e$) numbers to determine the pressure drop due to friction for sludge. The appropriate friction factor ($f$) is determined from curves corresponding to various values of $H_e$ and plotting $f$ vs. $R_e$, assuming that the sludge behaves like a Bingham plastic. *See* Metcalf & Eddy (2003) for details. The pressure drop ($\Delta P$, $N/m^2$) is then:

$$\Delta P = 2 f \rho L V^2 / D \qquad \text{(S-72)}$$

where $\rho$ = sludge density, $kg/m^3$; $L$ = length of pipeline carrying the sludge, m; $V$ = average velocity, m/sec; and $D$ = diameter of pipe, m.

**sludge ripening** Completion of sludge digestion.

**sludge screening** A preliminary processing operation that uses 2–5 mm step screens or in-line screens to remove fine solids from sludge, as an alternative to grinding. *See also* sludge grinding, degritting, and blending.

**sludge seeding** The inoculation of a wastewater treatment or sludge processing unit with biologically active solids to enhance initial performance.

**sludge settleability** The degree to which solids settle in a clarifier, as measured by the unstirred or traditional sludge volume index (SVI), the stirred SVI, the dilute SVI, and the stirred SVI at a mixed liquor suspended solids of 3500 mg/L.

**sludge slurry** A mixture of effluent and wastewater solids; sludge; slurry.

**sludge solids** Dissolved and suspended solid matter in sludge.

**sludge spreading** Disposal of wastewater treatment sludge by spreading it on farmland.

**sludge stability** The property of a sludge that has been converted into a product ready for use or ultimate disposal, as determined by the two criteria of volatile solids reduction and pathogen indicator reduction.

**sludge stabilization** The conversion of sludge into a stable product for use or ultimate disposal by reducing or eliminating pathogens, offensive odors, and the potential for putrefaction. Sludge stabilization uses four main mechanisms: biological reduction or chemical oxidation of volatile organic matter and chemical or thermal inactivation of pathogens. Current technologies applied in the United States for stabilization are aerobic digestion, anaerobic digestion, composting, and lime stabilization. *See also* alkaline stabilization, autothermal thermophilic digestion, chlorine oxidation, heat treatment. Anaerobic digestion is the most common, but aerobic digestion and lime stabilization are used in smaller plants, whereas composting is appropriate where the product will serve as a soil amendment.

**sludge stabilization performance** The performance of sludge digestion processes as measured by the percent reduction in volatile solids, related to the solids retention time ($T$ = 15–20 days) or to the detention time of the untreated sludge feed. The following empirical equation is often used for the percent destruction ($P$, %) of volatile solids in a high-rate, complete-mix digester:

$$P = 13.7 \ln T + 18.9 \qquad \text{(S-73)}$$

**sludge storage** The retention of solids to smooth out fluctuations in their production and processing, on a short-term basis in wastewater settling tanks or in sludge thickeners, or on a long-term basis in separate tanks or in digesters. Anaerobically digested biosolids are also stored in basins, lagoons, or storage pads before their final disposal or their reuse.

**sludge suspension** A mixture of wastewater solids and water. *See* bulk water, bound water, interstitial water, water of hydration, vicinal water.

**sludge tests** Sludge tests commonly conducted for wastewater solids deposited in collection works or accumulated in receiving waters, as well as those generated or removed in treatment processes, include sludge composition test, sludge concentration test, sludge condition test, functional sludge test, sludge processing test.

**sludge thickener** A tank or similar unit designed to concentrate the solids in sludge from water or wastewater treatment. *See* gravity thickener and flotation thickener.

**sludge thickening** The process of concentrating the solids in water or wastewater treatment sludge from 2–3% to 5–6%, by such mechanisms as gravity settling or flotation, ahead of dewatering or digestion. *See* thickening for more detail.

**sludge-to-oil reactor** A process investigated by Battelle Pacific Northwest Laboratories to recover almost three quarters of the calorific value of sludge by heating under pressure a mixture of dewatered sludge and alkali; the mixture is converted to oil, char, gas, and soluble organics. *See also* Tubingen process.

**sludge treatment** The processing of sludge from water or wastewater treatment to render it innocuous and reduce its volume to facilitate final disposal. *See* sludge handling for the various steps involved in sludge processing. Various treatment methods are available to accomplish these steps: aerobic digestion, anaerobic digestion, belt filter dewatering, bulking agents, Carver–Geenfield process, chemical treatment, chlorination, composting, elutriation, centrifugation, flash drying, freeze-assisted drying, freezing, gravity thickening, flotation thickening, heat treatment, lagooning, multiple-hearth incineration, paved-bed drying, pressure filtration, pyrolysis, rotary drying, sand bed drying, solvent extraction, starved-air incineration, thermal conditioning, vacuum filtration, vacuum-assisted drying, and wedgewire bed drying.

**sludge volume index (SVI)** The volume in milliliters occupied by one gram of suspended solids after settling a 1000 mL sample for 30 minutes in a graduated container. It is equal to 1000 times the ratio of the volume settled (in mL) to the concentration of the sludge suspended solids concentration (in mg/L). *See* sludge density index. The SVI is a measure of the settleability and compactibility of sludge. A sludge with a low SVI (between 50 and 100 mL/g) will settle well. Very low or very high solids retention times (or F:M ratios) correspond to high SVIs (> 200 mL/g), associated with filamentous organisms, pinpoint floc, and bulking sludge. The SVI can be used (e.g., in the Kraus process) as an operation control parameter in an activated sludge plant with a recirculation ratio ($r$), because

$$r = 1/\{[100/(P_w)(\mathrm{SVI})] - 1\} \qquad \text{(S-74)}$$

where $P_w$ is the concentration of suspended solids by weight of the mixed liquor. In flotation thickening, better performances correspond to SVI < 200 mL/g. *See also* aggregate volume index (AVI), diluted SVI test, stirred SVI test, zone settling velocity.

**sludge volume–mass relationships** The following equations are used to determine volumes and masses of sludge:

$$W/S = W_f/S_f + W_v/S_v \qquad \text{(S-75)}$$

$$V = M/(\rho_w S_s P) \qquad \text{(S-76)}$$

where $W$ = weight of solids; $S$ = specific gravity of solids; $W_f$ = weight of fixed solids; $S_f$ = specific gravity of fixed solids; $W_v$ = weight of volatile solids; $S_v$ = specific gravity of volatile solids; $V$ = volume, m$^3$; $M$ = mass of dry solids, kg; $\rho_w$ = specific weight of water, 1000 kg/m$^3$; $S_s$ = specific gravity of sludge; and $P$ = percent solids as a decimal.

**sludge volume ratio** In sludge thickening operations, the sludge volume ratio (SVR, days) is the ratio of the sludge blanket to the volume of the thickened sludge removed daily.

**sludge wasting** The removal of excess sludge from a water or wastewater treatment plant. In the activated sludge process, sludge can be wasted from the return sludge line ($Q_{wr}$, m$^3$/day) or from the mixed liquor in the aeration tank ($Q_{wm}$, m$^3$/day):

$$Q_{wr} \approx VX/(\theta X_r) \qquad \text{(S-77)}$$

$$Q_{wm} = V/\theta \qquad \text{(S-78)}$$

where $V$ = volume of the aeration basin, m$^3$; $X$ = aeration tank mass concentration, mg/L; $\theta$ = solids retention time, days; and $X_r$ = solids concentration in the return sludge line, mg/L.

**sludge yield** The production of sludge or solids during a biological wastewater treatment process. *See* solids yield, sludge production, activated sludge production.

**sludge zone** The bottom portion of a sedimentation unit where settled solids accumulate until they are removed. It is one of the four zones of the Camp sedimentation theory. *See also* settling, inlet, and outlet zones.

**Sludgifier** A lagoon dredge manufactured by VMI, Inc.

**Sludglite** A device manufactured by Ecolotech Corp. to detect the level of a sludge blanket.

**slug** (1) A temporary and abnormally high quantity or concentration of a substance. (2) A unit of mass = 32.2 lb = 15 kg.

**slug discharge** The discharge of a large volume of wastewater in a short period, usually from a wet manufacturing process, with concentrated contaminants and flow surges. To reduce the effects of a slug discharge, it can be retained in a holding pond and then allowed to flow uniformly over an extended period. Also called batch discharge.

**slug dose** A single application of a material such as a tracer in hydraulic measurements. *See also* step dose.

**slug-dose input** Same as slug input.

**slug dosing** *See* slug dose.

**slug input** The introduction and mixing of a quantity of a tracer into a reactor over a very short period of time compared to the detention time. Most commonly, a dye is injected into the reactor to characterize tracer response curves and describe the hydraulic performance of reactors. Also called pulse input. *See also* step input, residence time distribution (RTD) curve, time–concentration curve.

**slug load** The arrival at a water treatment plant of raw water containing unusual amounts of algae, colloidal matter, color, suspended solids, turbidity, or other pollutants. More generally, a sudden increase in the hydraulic, organic or other loading to a facility. Also called a shock load.

**slug method** A method of disinfection of a new large water main before it is placed in service. A chlorine solution of at least 300 mg/L is fed for at least 3 hours and at a rate sufficient to contact the entire interior surface of the main and disinfect appurtenant valves and hydrants. Then the main is flushed to waste using potable water. *See also* continuous-feed method, tablet method.

**slug test** A test using a slug dose of a tracer in a well to determine hydraulic properties of an aquifer. *See also* auger hole test, bail-down test.

**slurry** (1) A watery mixture or suspension of insoluble matter resulting from some pollution control techniques; a thin watery mud or any substance resembling it, such as a grit slurry or a lime slurry (milk of lime). A slurry has a suspended solids concentration larger than 5000 mg/L but can be pumped by liquid-handling machinery. *See also* sludge, sludge slurry, sludge suspension. (2) The brine concentrate from a desalination process that is returned to the sea or recycled.

**slurry precipitation and recycle reverse osmosis** A variation of the reverse osmosis process that uses crystals to precipitate scaling compounds in membrane filtration. A circulating slurry of seed crystals allows calcium salts and silicates to grow on the crystals and not within the membrane tubes. The salts precipitate when their solubility is exceeded. This technology is classified as an emerging concentrate disposal and management method. Also called seeded slurry precipitation and recycle.

**slurry precipitation and recycle RO (SPARRO)** *See* slurry precipitation and recycle reverse osmosis.

**Slurrystore** An apparatus manufactured by A. O. Smith Harvestore products, Inc. for storing slurries.

**small-bore piping** A metal pipe having a diameter of 0.5 inch.

**small-bore sewer** A sewer of small diameter (e.g., 3–4 inches) and nominal gradient (e.g. 0.5%) that carries the effluent from septic tanks. Also called an effluent drain.

**small-bore sewerage system** A system of small sewers (90–120 mm outside diameter, PVC), with house connections, collecting wastewater from interceptor tanks. *See* pressure sewerage, small-diameter variable-slope sewer.

**small-diameter, variable-grade gravity sewer** *See* small-diameter, variable-slope sewer.

**small-diameter, variable-slope (SDVS) sewer** A gravity sewer line or sewer system installed with a net positive slope and at about the same depth below the ground, regardless of grade, from inlet to outlet, with downhill and uphill sections. Air release valves are located appropriately, based on the hydraulic gradeline and the pipeline profile. The outlet must be lower than the inlet and any house connection. Also called small-diameter, variable-grade gravity sewer.

**small-footprint technology** A water or wastewater treatment method that uses space efficiently, or that uses less space than conventional technologies to achieve comparable or better results. Developed mostly in Europe and Japan, small-footprint technologies are also being implemented in the United States for treatment plant expansions and other circumstances of limited space. *See,* for example, ballasted flocculation, biological aerated filters, membrane biological reactors, and integrated fixed-film activated sludge. *See also* compact technology.

**small round virus (SRV)** A loose term that lumps together many viruses found in fecal samples from gastroenteritis patients. They are smaller than the adenoviruses and rotaviruses, e.g., on the order of 20–40 nanometers. They include the Norwalk, Montgomery County, Hawaii, and other agents.

**small-scale column test** A test conducted in a small-scale column to simulate the performance of full-scale reactors. *See* high-pressure minicolumn and rapid small-scale column test.

**Small Systems Technology Initiative** A program of the USEPA that provides information about treatment technologies applicable to small water supply systems.

**small water system** A public water supply system serving fewer than 3300 people.

***S. mansoni*** Abbreviation of *Schistosoma mansoni,* one of the pathogens causing schistosomiasis.

**SmartFilter™** A traveling bridge filter manufactured by Agency Environmental, Inc.

**SmartRO™** A reverse osmosis unit manufactured by Water & Power Technologies.

**Smart Water Application Technology (SWAT)** A technique that uses automatic devices based on evapotranspiration to control irrigation water.

**SMBS** Abbreviation of sodium metabisulfite.

**SMCL** Acronym of secondary maximum contaminant level.

**smear** A thin film of a liquid suspension of cells spread on a slide and air-dried.

**Smith–Heiftje background correction** A procedure used to compensate for interferences in atomic absorption spectrophotometry.

**smog** A mixture of pollutants, principally ground-level ozone, produced by chemical reactions in the air involving smog-forming chemicals, e.g., volatile organic compounds found in paints and solvents. Smog can harm health, damage the environment and cause poor visibility.

**Smogless™** A line of equipment manufactured by U.S. Filter Corp. for wastewater treatment, sludge drying, and incineration.

**smoke** Particles suspended in air or gaseous emissions after incomplete combustion or incomplete scrubbing. *See also* dust, fume.

**Smooth-tex** A woven mesh for wastewater screens manufactured by Envirex, Inc.

**SMR** Acronym of standardized mortality ratio.

**SMX** A belt filter press manufactured by Andritz-Ruthner, Inc.

**Sn** Chemical symbol of the metallic element tin; from its Latin name stannum.

**snail** An intermediate host of the pathogens of the water- and excreta-related disease schistosomiasis. Schistosome worms excreted by infected individuals develop in aquatic snails, which then shed them as cercariae that can infect man through the skin. Particular genera of snails infected by the schistosomes are *Biomphalaria, Bulinus, Oncomelania,* and *Tricula.* Snails sometimes create problems in the operation of rock trickling filters by feeding on the slime growth and filling the void spaces of the bed when they die.

**Snail** Grit dewatering equipment manufactured by Eutek Systems.

**snail fever** Fever caused by parasites carried by snails; schistosomiasis.

**SNARL** Acronym of suggested no adverse response level.

***S. natans*** *See Sphaerotilus natans.*

**$SnCl_2$** Chemical formula of stannous chloride.

**SND** Acronym of simultaneous nitrification/denitrification.

**SNdN** Acronym of simultaneous nitrification denitrification.

**SNdN process** *See* simultaneous nitrification denitrification process.

***S. neavei*** Abbreviation of *Simulium neavei.*

**$SnO_2$** The chemical formula of tin oxide or cassiterite, the principal source of the metal.

**snow cover** A layer of snow on the ground. *See* water equivalent.

**snow density** The specific gravity of a sample of snow, i.e., the ratio of the meltwater derived from that sample to the original volume of the sample. Snow density varies from very low values for freshly fallen snow to 0.91 for glacial ice. *See also* water content, water equivalent.

**Snowflake Packing** Plastic medium produced by Norton Co. for use in air stripping units.

**Snow, John** British physician (1813–1858) who established in 1855 the foundations of epidemiology with his observations of the occurrence of cholera in London related to the source of drinking water contaminated by wastewater. *See* Broad Street pump, Koch (Robert), and Pasteur (Louis).

**Snow Mountain agent** A waterborne calicivirus that causes acute gastrointestinal illness; named after the location of its first observed outbreak.

**snowpack** The accumulation of winter snowfall. *See* water equivalent.

**snow pellets** Precipitation in the form of crisp, white, opaque ice particles of 2–5 mm in diameter. Also called graupel, soft hail, tapioca snow.

**snow pillow** An instrument that measures the pressure of a mass of snow to provide an indication of the water equivalent of a snowpack.

**snowslide** *See* avalanche.

**SNTEMP** The Instream Water Temperature Model of the U.S. Fish and Wildlife Service. A model that predicts streamwater temperatures from hydrological conditions, meteorological conditions, and stream geometry.

**$SO_2$** Chemical formula of sulfur dioxide.

**$SO_2$ process** *See* sulfur dioxide process.

**$SO_3$** Chemical formula of sulfur trioxide.

**$SO_4^{2-}$** Chemical formula of sulfate.

**$SO_x$** Chemical formula of sulfur oxides.

**$S_2O_3^{2-}$** Chemical formula of thiosulfate.

**$S_4O_6^{2-}$** Chemical formula of tetrathionate.

**soakage pit** Same as soak pit or soakaway pit.

**soakaway** An arrangement such as a pit, trenches, or a mound, used to promote the seepage of effluents (from pour-flush toilets, aquaprivies, septic tanks, etc.) into the ground. *See also* cesspool.

**soakaway mound** A mound constructed on natural ground for the subsoil dispersion of effluents. It typically consists of a 6″ layer of grassed topsoil over the mound of sandy loam soil and sand fill, with an 8″ core of clean rock in the center. Soakaway mounds are used to avoid groundwater contamination where the natural soils are inappropriate for ordinary soakaway pits or trenches. Also called evapotranspiration mound. *See* Figure S-15 and mound system.

**soakaway pit** Same as soakpit.

**soakaway trench** A trench dug in the ground and containing open-jointed pipes on a bed of gravel or similar materials, for the subsoil dispersion of liquid wastes. Soakaway trenches are used in lieu of soakpits that would require too large an area. Soakaway trenches are designed on the same basis as soil absorption fields or drainfields.

**soakpit** A hole dug in the ground for the subsoil dispersion of liquid wastes from toilets that use small quantities of water (e.g., pour-flush latrines and cistern-flush toilets); a soakaway. A typical soakpit has a diameter of 3 ft and a concrete cover slab; the top 1-ft portion of the walls is impermeable but the bottom portion is made of open bricks for lateral dispersion. Also called soakage pit or soakaway pit. *See also* soakaway trench.

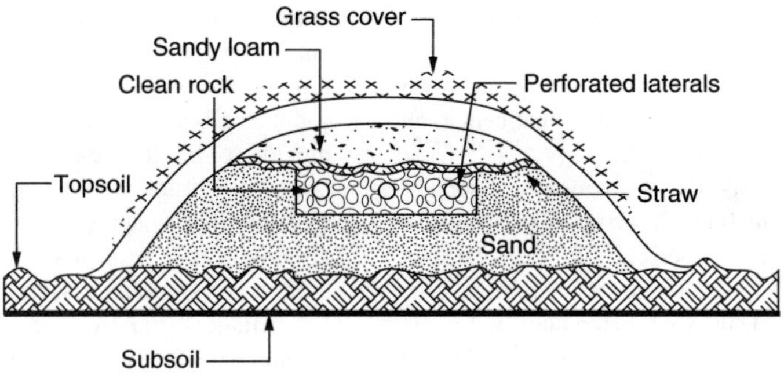


**Figure S-15.** Soakaway mound.

**soap** The alkali (sodium or potassium) salt of an animal fatty acid. A soap molecule has one end ionic and the other end polar. It dissociates in water as a micelle—a negatively charged particle surrounded by positively charged ions. Dissolved minerals in hard waters cause excessive consumption of soaps derived from animal fats. *See also* detergent.

**soap curd** *See* lime soap.

**SOC** Acronym of synthetic organic chemical.

**SOD** Acronym of sediment oxygen demand.

**soda** Another name for caustic soda, sodium carbonate, or sodium hydroxide.

**soda ash ($Na_2CO_3$)** A common name for commercial sodium carbonate, the second most important compound of the metal. It is used extensively in water treatment as a coagulant aid and in softening (to supply alkalinity and carbonate ions); see lime–soda ash process.

**soda lake** A lake that is rich in sodium bicarbonate ($NaHCO_3$), an example of an extreme environment. Compared to an average lake, a soda lake has a high pH (9.6 vs. 7.7) and high concentrations of sodium (pNa of 0 vs. 3.4), bicarbonate ($pHCO_3$ of 0.4 vs. 2.7), and chloride (pCl of 0.3 vs. 3.6). *See also* bitter lake, borax lake, freshwater.

**soda niter** *See* sodium nitrate.

**soda water** Tap water made effervescent by the addition of carbon dioxide ($CO_2$), without the addition of mineral salts. Also called seltzer water or sparkling water.

**sodic** Pertaining to or containing sodium.

**sodicity** The relative amount of sodium in a solution, an important parameter of irrigation water quality. *See* sodium adsorption ratio.

**sodic soil** A soil having a high sodium adsorption ratio or a high exchangeable sodium ratio, which reduce permeability.

**sodium** (Na, from its former Latin name: natrium.) A soft, silver-white, alkaline, metallic element, abundant in the Earth's crust but found mostly in its compounds: chloride, carbonate, bicarbonate, hydroxide, phosphate, thiosulfate, and borax. Atomic weight = 22.9898. Atomic number = 11. Specific gravity = 0.971 at 20°C. Sodium compounds are used in manufacturing (paper, glass, soap, etc.); sodium carbonate (soda ash, $Na_2CO_3$), hydroxide (caustic soda, NaOH), sodium chloride (NaCl), and sodium hypochlorite (NaOCl) are used in water treatment. Sodium is also used in the synthesis of some products and is necessary to maintain normal fluid balance and other physiological functions. High intake of sodium, generally in the form of sodium chloride, has been linked to high blood pressure, but drinking water contributes less than 1% of total salt intake. The USEPA and the European Community recommend a guide value of 20 mg/L of sodium in drinking water for health reasons, whereas the WHO recommends a limit of 200 mg/L to avoid taste complaints.

**sodium acetate ($C_2H_3NaO_2$)** A chemical used in laboratory experiments as a source of nutrients.

**sodium acid carbonate** *See* sodium bicarbonate.

**sodium adsorption ratio (SAR or $R_{Na}$)** The ratio of the concentration of sodium ($Na^+$) ion to the concentrations of calcium ($Ca^{2+}$) and magnesium ($Mg^{2+}$) ions, a major parameter of irrigation water quality:

$$RNa = [Na^+]/\{([Ca^{2+}] + [Mg^{2+}])/2\}^{0.5} \quad \text{(S-79)}$$

with all cation concentrations expressed in meq/L. A ratio of 15 is considered generally acceptable, but an excess of sodium reduces the permeability of the soil and encourages waterlogging and the formation of crusts. A high SAR is a problem for irrigation with reclaimed water. *See also* salinity, sodicity, exchangeable sodium percentage, adjusted sodium adsorption ratio.

**sodium alginate** An extract from kelp (a brown seaweed) used in food and in water treatment as an aid to alum coagulation.

**sodium alkalies** or **alkalis** Compounds of sodium (Na), some of which neutralize acids to form salts and turn red litmus paper blue. Products used in water and wastewater treatment for various purposes include caustic soda or sodium hydroxide (NaOH) and soda ash or sodium carbonate ($Na_2CO_3$).

**sodium aluminate ($NaAlO_2$ or $Na_2Al_2O_4$)** A softening agent and secondary coagulant used (a) alone or with alum in the treatment of water of low pH and fine turbidity and (b) in the removal of suspended solids and precipitation of dissolved phosphorus from wastewater. It consists of aluminum oxide ($Al_2O_3$) stabilized with caustic soda (NaOH):

$$Al_2O_3 + 2\ NaOH \rightarrow Na_2Al_2O_4 + H_2O \quad \text{(S-80)}$$

The phosphate precipitation reaction requires the presence of carbon dioxide ($CO_2$):

$$Na_2Al_2O_4 + 2\ H_2PO_4^- + 4\ CO_2 \rightarrow 2\ AlPO_4 + 2\ Na^+ + 4\ HCO_3^- \quad \text{(S-81)}$$

It is also commonly used in papermaking.

**sodium aluminosilicate** Another name for gel zeolite.

**sodium aluminum fluoride ($Na_3AlF_6$)** One of the minerals that contain fluorine; it is essentially a compound of this halogen with sodium and aluminum, occurring in white masses and used in the electrolytic production of aluminum. Also called sodium cryolite and Greenland spar.

**sodium amide ($NaNH_2$)** A white, crystalline, insoluble powder used in the manufacture of sodium cyanide (NaCN) and organic products.

**sodium ammonium phosphate ($NaNH_4HPO_4 \cdot 4\ H_2O$)** A colorless, odorless, crystalline, soluble solid that can be produced from human urine and used in testing metallic oxides. Also called microcosmic salt or salt of phosphorus.

**sodium arsenite ($NaAsO_2$ or $Na_3AsO_3$)** A white or grayish-white, soluble powder; a toxic compound of sodium and arsenic formerly used in pesticides, now used in arsenical soaps for hides and in testing for residual chlorine. Also called sodium meta-arsenite.

**sodium benzoate ($C_7H_5NaO_2$)** A white, crystalline or granular, soluble powder used in food processing, diagnostic testing, and the manufacture of fungicides. Also called benzoate of soda.

**sodium bicarbonate ($NaHCO_3$)** A white, crystalline, soluble powder or granule, prepared through the Solvay process or by the reaction of sodium carbonate (soda ash, $Na_2CO_3$) with carbon dioxide ($CO_2$). It is used in water treatment to increase alkalinity in the coagulation process, increase buffer capacity, and control corrosion. It is also used as a reagent in the laboratory, in the preparation of sodium salts, beverages, and as a fire extinguisher. Also called bicarbonate of soda, sodium acid carbonate, sodium hydrogen carbonate, and baking soda.

**sodium bichromate** Same as sodium dichromate.

**sodium bisulfate ($NaHSO_4$)** A colorless, crystalline, soluble compound used in dyes, cement, paper, soap, and as a cleaner. Also called sodium hydrogen sulfate.

**sodium bisulfite ($NaHSO_3$)** A liquid chemical (usually a 38-44% solution) used as a dechlorinating agent (antichlor), a disinfectant, a preservative, or a reductant; it is also available as a white powder or as a granular material. The stoichiometric dechlorinating reactions are:

$$NaHSO_3 + Cl_2 + H_2O \rightarrow NaHSO_4 + 2\ HCl \quad \text{(S-82)}$$

$$NaHSO_3 + NH_2Cl + H_2O \rightarrow NaHSO_4 + Cl^- + NH_4^+ \quad \text{(S-83)}$$

*See also* sodium metabisulfite, sodium hydrosulfite, sodium thiosulfate.

**sodium bisulfite treatment** A shock treatment of chlorine-sensitive membranes for the control of biofilms.

**sodium borate** *See* borax.

**sodium borohydride ($NaBH_4$)** A powerful reductant, commercially available as a white crystalline powder or a 12% solution in 40% sodium hydroxide (NaOH). At high pH, borohydride ($BH_4^-$) hydrolyzes to borate [$B(OH)_3$], hydrogen gas ($H_2$), and hydroxyl ion ($OH^-$):

$$BH_4^- + 4\ H_2O \rightarrow B(OH)_3 + 4\ H_2 + OH^- \quad \text{(S-84)}$$

It is used in the recovery of precious metals, but not to remove toxic metals from wastewater because it is expensive and the borate ion and hydrogen gas produced are undesirable.

**sodium brine** A solution of sodium chloride (NaCl) used in the regeneration of ion-exchange resins.

**sodium bromide (NaBr)** A white, crystalline, soluble solid used in photography and as a sedative.

**sodium carbonate ($Na_2CO_3$)** An anhydrous, grayish-white, odorless, soluble powder, produced through the Solvay process; a salt compound used in water treatment to increase the alkalinity or pH or to neutralize acidity. It is also used to manufacture glass, ceramics, soaps, sodium salts, and other products. Also called soda, soda ash, sal soda, washing soda. Monohydrated ($Na_2CO_3 \cdot H_2O$), heptahydrated ($Na_2CO_3 \cdot 7\ H_2O$), and decahydrated ($Na_2CO_3 \cdot 10\ H_2O$) forms of this salt are used for the same purposes.

**sodium carbonate decahydrate ($Na_2CO_3 \cdot 10\ H_2O$)** One of the four forms of sodium carbonate ($Na_2CO_3$) commercially available, commonly called washing soda, containing approximately 37% $Na_2CO_3$ and 63% water of crystallization ($H_2O$).

**sodium carbonate heptahydrate ($Na_2CO_3 \cdot 7\ H_2O$)** One of the four forms of sodium carbonate ($Na_2CO_3$) commercially available, containing approximately 45.7% $Na_2CO_3$ and 54.3% water of crystallization ($H_2O$).

**sodium carbonate monohydrate ($Na_2CO_3 \cdot H_2O$)** One of the four forms of sodium carbonate ($Na_2CO_3$) commercially available, containing approximately 85.5% $Na_2CO_3$ and 14.5% water of crystallization ($H_2O$).

**sodium cellulose glycolate** A white, water-soluble polymer produced from cellulose and used in papermaking, in foods, in textile-finishing mills, and in general-purpose synthetic detergents. Also called carboxymethylcellulose, cellulose gum, or CMC.

**sodium chlorate ($NaClO_3$)** A colorless, soluble, solid sodium compound used in explosives, matches, herbicides, and in the processing of leather, paper, and textiles. It is formed during water disinfection with chlorine dioxide ($ClO_2$). *See also* chlorate.

**sodium chloride (NaCl)** The chemical name of common salt; a crystalline compound, a constituent of seawater, used as a preservative, in food processing, and in water treatment (e.g., to regenerate ion-exchange softeners). *See also* fossil salt, halite, native salt, rock salt.

**sodium chlorite ($NaClO_2$)** A chlorine compound sometimes used in water disinfection as a dry powder or a liquid solution; unstable at high temperatures. *See* acid-chlorite process.

**sodium citrate ($Na_3C_6H_5O_7 \cdot 2\ H_2O$)** A white, crystalline or granular, soluble salt of citric acid, used in photography, in drinks, and in medicine. In water supply, it is added to copper sulfate in algae control to prevent the precipitation of copper carbonate in very alkaline waters. The citrate complexes the copper in a soluble form. Sodium citrate can also sequester hardness minerals; it is used as a builder in nonphosphate laundry detergents.

**sodium cyanide (NaCN)** A white, crystalline, soluble, poisonous powder used in alloys and in electroplating.

**sodium cycle** An ion-exchange process used for water softening by exchanging sodium ions of the resin for hardness ions of the water. *See also* hydrogen cycle.

**sodium cycle exchange** An ion-exchange treatment process based on a zeolite resin.

**sodium cycle resin** A sodium-based resin used for the removal of hardness by exchanging the divalent (hardness) ions for the sodium ion; general formula: $Na_2X$. For example, with calcium sulfate:

$$CaSO_4 + Na_2X \rightarrow Na_2SO_4 + CaX \quad \text{(S-85)}$$

Such a resin is commonly regenerated by sodium chloride (NaCl):

$$CaX + 2\ NaCl \rightarrow Na_2X + CaCl_2 \quad \text{(S-86)}$$

**sodium dichromate ($Na_2Cr_2O_7 \cdot 2\ H_2O$)** A red or orange, crystalline, soluble solid used as an oxidizing agent in dyes and inks, in electroplating, etc., and as a laboratory reagent. Also called sodium bichromate.

**sodium dithionite** Same as sodium hydrosulfite.

**sodium fluoride (NaF)** A colorless, crystalline, soluble, poisonous solid used in saturators to add approximately 1.0 mg/L of fluoride to drinking water after conventional treatment; also used as a rodenticide or pesticide. It dissociates in water into sodium ($Na^+$) and fluoride ($F^-$) ions:

$$NaF \rightarrow Na^+ + F^- \quad \text{(S-87)}$$

*See also* ammonium silicofluoride, hydrofluosilicic acid, sodium silicofluoride.

**sodium fluoroacetate ($NaC_2H_2FO_2$)** A white, amorphous, soluble poisonous powder used as a rodenticide.

**sodium fluorosilicate ($Na_2SiF_6$)** A dry, white, odorless, crystalline substance used in water fluoridation. It dissociates in water into the sodium ($Na^+$) and fluorosilicate ($SiF_6^-$) ions:

$$Na_2SiF_6 \rightarrow 2\ Na^+ + SiF_6^{2-} \quad \text{(S-88)}$$

Also called sodium silicofluoride. *See also* ammonium silicofluoride, hydrofluosilicic acid, sodium fluoride.

**sodium fluosilicate** *See* sodium fluorosilicate ($Na_2SiF_6$).

**sodium hex** Abbreviation of sodium hexametaphosphate.

**sodium hexametaphosphate [$(NaPO_3)_n$]** A water-soluble mixture of sodium monoxide ($Na_2O$)

and at least 76% phosphorus pentoxide ($P_2O_5$), used as a sequestering, dispersing, or deflocculating agent. Specific applications include the control of iron and manganese precipitates, the stabilization of water to prevent incrustation, and the inhibition of scaling in reverse osmosis units. It is also used as a bleach and in the lime–soda process as an alternative to recarbonation to lower pH and dissolve calcium carbonate. Also called sodium polyphosphate, glassy sodium phosphate. *See also* Calgon, polyphosphonate.

**sodium hydrate** Another name for caustic soda or sodium hydroxide.

**sodium hydrogen carbonate** *See* sodium bicarbonate.

**sodium hydrogen sulfate** Same as sodium bisulfate.

**sodium hydrosulfite ($Na_2S_2O_4$)** A white, crystalline, soluble powder; a compound of sodium and sulfur and a strong reducing agent used as a cleaning product for ion-exchange resins fouled by iron compounds. It hydrolyzes under acidic conditions but is relatively stable under alkaline conditions. Also called hydrosulfite, sodium dithionite, sodium hyposulfite. *See also* the following sodium salts: bisulfite, metabisulfite, thiosulfate.

**sodium hydroxide (NaOH)** A white, deliquescent, strong alkaline compound, highly soluble in water, commonly found in lumps, pellets, etc., used as a concentrated solution or a flake in treatment processes to raise the pH of water, neutralize acidity, and increase alkalinity. It is also used as a reagent, in chemicals, soaps, and in medicine. Sodium hydroxide is found in wastes from soap manufacturing, textile dyeing, and other industries. It is toxic to fish in concentrations as low as 25 mg/L and can, in boiler-feed water, cause embrittlement of pipes. Also called caustic soda, lye, soda, sodium hydrate, white caustic.

**sodium hypochlorite (NaOCl)** A pale-green, crystalline, soluble compound containing 12–17% available chlorine, used as a bleaching agent in textiles and paper, and as a fungicide or a household product. As a liquid chlorine solution, NaOCl is used for water or wastewater disinfection, particularly where safety with chlorine gas is an issue and in small plants, and for odor control. The active agent hypochlorous acid (HOCl) results from the dissociation and hydrolysis of hypochlorite:

$$NaOCl \rightarrow Na^+ + OCl^- \quad \text{(S-89)}$$

$$OCl^- + H_2O \rightarrow HOCl + OH^- \quad \text{(S-90)}$$

$$NaOCl + H_2O \rightarrow HOCl + NaOH \quad \text{(S-91)}$$

It can be generated from sodium chloride (NaCl) or seawater. Also called liquid bleach or soda bleach solution, it is an aqueous solution of chlorine and sodium hydroxide. *See also* scrubbing fluid.

**sodium hyposulfite ($Na_2S_2O_4$)** A compound of sodium and sulfur and a strong reducing agent. *See* sodium hydrosulfite, sodium thiosulfate.

**sodium iodide (NaI)** A colorless or white, deliquescent, soluble solid used in photography, organic synthesis, and as a disinfectant.

**sodium ion-exchange softening** A water treatment process that uses strong acid cation resins to exchange sodium ($Na^+$) for calcium ($Ca^{+2}$) and magnesium ($Mg^{+2}$) ions.

**sodium lactate ($NaC_3H_5O_3$)** A soluble, hygroscopic salt used in medicine.

**sodium meta-arsenite** Same as sodium arsenite.

**sodium metabisulfite ($Na_2S_2O_5$)** A dry, crystalline form of sodium bisulfite, used as a dechlorinating agent and as a biocide:

$$Na_2S_2O_5 + H_2O \rightarrow 2\ NaHSO_3 \quad \text{(S-92)}$$

$$NaHSO_3 + HOCl \rightarrow HCl + NaHSO_4 \quad \text{(S-93)}$$

*See also* sodium bisulfite, sodium hydrosulfite, sodium hyposulfite, sodium thiosulfate.

**sodium metasilicate ($Na_2SiO_3$)** A white, granular, water-soluble substance used in detergents, bleaches, cleaning products and in animal studies to establish silicon as an essential nutritional trace element.

**sodium metavanadate** A soluble compound of vanadium; solubility = approximately 211 g/L.

**sodium methoxide** *See* sodium methylate.

**sodium methylate ($NaOCH_3$)** A white, flammable powder that hydrolyzes to sodium hydroxide (NaOH) and methyl alcohol or methanol ($CH_3OH$):

$$NaOCH_3 + H_2O \rightarrow CH_3OH + NaOH \quad \text{(S-94)}$$

It is used in organic synthesis. Also called sodium methoxide.

**sodium monoxide ($Na_2O$)** A white powder that produces sodium hydroxide (NaOH) with water. Also called soda or sodium oxide.

**sodium nitrate ($NaNO_3$)** A crystalline, soluble chemical used in laboratory experiments as a source of nutrients; it is also used in explosives, fertilizers, glass, and food processing. It occurs naturally as soda niter, a white or transparent mineral used in fertilizers and in chemical synthesis (e.g., nitric acid, potassium nitrate, sulfuric acid).

**sodium nitrite ($NaNO_2$)** A yellowish or white, crystalline, soluble compound used in dyes and in food processing. Also called nitrite.

**sodium perborate ($NaBO_2 \cdot 3\ H_2O$ or $NaBO_3 \cdot 4\ H_2O$)** A white, crystalline, soluble solid used as a bleaching agent and antiseptic. Also called perborax.

**sodium peroxide ($Na_2O_2$)** A yellowish-white, soluble powder used as a bleaching or oxidizing agent.

**sodium phosphate ($NaH_2PO_4$)** A white, crystalline, soluble powder used in dyes and electroplating. Also called monobasic sodium phosphate. *See also* disodium phosphate and trisodium phosphate.

**sodium propionate ($NaC_3H_5O_2$)** A transparent crystalline, soluble powder used in foodstuffs, in medicine, and as a fungicide.

**sodium pyroborate** *See* borax.

**sodium silicate** A clear, white, or greenish, soluble combination of sodium oxide ($Na_2O$) and silica ($SiO_2$), a chemical used in water treatment for corrosion prevention, for pH control (in small systems), in the manufacture of synthetic gel zeolite, or as a coagulant aid, with various formulas reported in the literature: ($Na_2SiO_3$), ($Na_2O \cdot SiO_2$), [$Na_2O(SiO_2)_{3.25}$], [$Na_2O \cdot (SiO_2)_x$], ($Na_2Si_4O_9$), ($Na_4SiO_4$). As a corrosion inhibitor, it combines with existing pipe deposits to form a protective film. Also called liquid glass and water glass. One liquid commercial product contains about 9% $Na_2O$ and 29% $SiO_2$. *See also* activated silica.

**sodium silicate test** A test to determine the resistance to frost action of broken stone as a trickling filter medium by submerging the stone in sodium silicate and drying it.

**sodium silicofluoride ($Na_2SiF_6$)** A dry chemical used to add approximately 1.0 mg/L of fluoride to drinking water after conventional treatment. *See* sodium fluorosilicate for detail.

**sodium stearate [$NaC_{18}H_{35}O_2$ or $CH_3(CH_2)_{16}COONa$]** A salt of stearic acid; a typical soap. *See* micelle.

**sodium sulfate ($Na_2SO_4$)** A white, crystalline, soluble solid used in dyes, detergents, soaps, glass, etc.

**sodium sulfide ($Na_2S$)** A yellow or red, crystalline, soluble solid used in dyes, soaps, and in processing copper and lead ores.

**sodium sulfite ($Na_2SO_3$)** A white, crystalline, soluble solid, used as a food preservative, a bleaching agent, etc.; also a reductant used as a dechlorinating agent:

$$Na_2SO_3 + Cl_2 + H_2O \rightarrow Na_2SO_4 + 2\ HCl \qquad \text{(S-95)}$$

Sodium bisulfite is also used as an oxygen scavenger in boiler and cooling water systems.

**sodium tetraborate** *See* borax.

**sodium thiosulfate ($Na_2S_2O_3 \cdot 5\ H_2O$)** A white, crystalline, soluble powder used as a bleach and in photography. It is a strong reducing agent used as antichlor to remove excess or residual chlorine from paper pulp, textiles, water, wastewater, etc. Also called hypo, hyposulfite. *See also* sodium bisulfite, sodium hydrosulfite, sodium hyposulfite, sodium metabisulfite, sodium sulfite.

**sodium thiosulfate dechlorination** The reduction or elimination of chlorine residuals using sodium thiosulfate ($Na_2S_2O_3$). Full-scale dechlorination of wastewater effluents is not common because of mixing and dosage problems.

**sodium triphosphate** Same as sodium tripolyphosphate.

**sodium tripolyphosphate ($Na_5P_3O_{10}$)** Acronym: STP or STPP. A white powder used as a water softener, a sequestering agent, and a food additive. Also called sodium triphosphate or pentasodium tripolyphosphate.

**sodium zeolite reactions** Reactions that occur in zeolite softening (a cation-exchange process) that forms insoluble calcium and magnesium zeolites in exchange for salts of sodium:

$$Na_2Z + Ca(HCO_3)_2 \rightarrow CaZ + 2\ NaHCO_3 \qquad \text{(S-96)}$$

$$Na_2Z + Mg(HCO_3)_2 \rightarrow MgZ + 2\ NaHCO_3 \qquad \text{(S-97)}$$

$$Na_2Z + CaSO_4 \rightarrow CaZ + Na_2SO_4 \qquad \text{(S-98)}$$

$$Na_2Z + MgSO_4 \rightarrow MgZ + Na_2SO_4 \qquad \text{(S-99)}$$

$$Na_2Z + CaCl_2 \rightarrow CaZ + 2\ NaCl \qquad \text{(S-100)}$$

$$Na_2Z + MgCl_2 \rightarrow MgZ + 2\ NaCl \qquad \text{(S-101)}$$

where Z represents the zeolite radical.

**soffit** The inside top of the arch of a pipe, sewer, covered channel, or conduit; or the lower surface of a slab. Also called crown.

**soft** (1) Pertaining to wastewater that contains natural organic matter or other organic matter that is easily degraded in biological treatment. (2) Pertaining to soft water.

**soft detergent** A cleaning agent that breaks down in nature; a synthetic detergent that responds to biological action.

**soft drink bottling waste** Wastewater generated during the production and bottling of carbonated and noncarbonated, nonalcoholic beverages; it originates from bottle washing, floor and equipment cleaning, and syrup-storage-tank drains. It is characterized by high pH, BOD, and suspended solids. It may be discharged to municipal sewers after adequate screening.

**softened water** Water that has been treated to reduce its total hardness content below one grain per gallon, i.e., about 17.0 mg/L as calcium carbonate ($CaCO_3$). *See also* soft water, hard water.

**softener** A detergent additive, polyphosphate, or other chemical compound that, when added to water, counteracts the effects of hardness ions. Also called water softener.

**softening** Chemical treatment of water to remove partially or totally calcium, magnesium, and other undesirable cations that cause hardness. Lime softening and the lime–soda ash process are used in municipal water treatment plants but cation exchange is more prevalent in residential and commercial applications. Factors affecting required chemical dosages include the ionic balance and the carbonate–bicarbonate equilibria of the water. As an alternative to lime–soda, caustic soda (NaOH) may be used. Cation exchange uses mostly sodium-cycle resins, and also hydrogen-cycle resins. *See also* hardness, carbonate hardness, noncarbonated hardness, permanent hardness, natural softening, split treatment.

**softening chemistry** The chemical reactions that result in the precipitation of the main polyvalent cations responsible for water hardness (calcium as calcium carbonate and magnesium as magnesium hydroxide) are:

$$CaCO_3\ (s) = Ca^{2+} + CO_3^{2-} \qquad (S\text{-}102)$$

$$Mg(OH)_2\ (s) = Mg^{2+} + 2\ OH^- \qquad (S\text{-}103)$$

$$H_2CO_3 = HCO_3^- + H^+ \qquad (S\text{-}104)$$

$$HCO_3^- = CO_3^{2-} + H^+ \qquad (S\text{-}105)$$

$$H_2O = H^+ + OH^- \qquad (S\text{-}106)$$

$$\text{Alkalinity (eq/L)} = 2\,[CO_3^{2-}] + [HCO_3^-] + [OH^-] - [H^+] \qquad (S\text{-}107)$$

**soft hail** *See* graupel, snow pellets.

**soft Lewis acid** Same as B-type metal.

**software** The programs, routines or symbolic languages of a computer, such as the operating system or applications like word processing, games, spreadsheets, database management, network solution, mapping, and Cybernet, XP-SWMM. Also, the list of instructions for the computer to perform a given task or tasks.

**software program** Same as software.

**soft water** Water that has a low concentration of dissolved minerals such as salts of calcium and magnesium. According to U.S. Geological Survey guidelines, soft water has a hardness of 60 milligrams per liter or less. *See also* hard water, softened water.

**soil** The unconsolidated mineral and organic material on the immediate surface of the Earth that serves as a natural medium for the growth of land plants; including, but not limited to, organic matter, silts, clays, sands, gravel, and small rocks. Its formation and properties are determined by various factors such as parent material, climate, macroorganisms and microorganisms, topography, and time (EPA-40CFR192.11-d and 796.2700-iv).

**soil absorption capacity** The ability of a soil to absorb water or wastewater, a factor considered in the design of a soil absorption system. *See also* soil absorption test.

**soil absorption field** A subsurface area containing perforated pipes laid in gravel-lined, looped or lateral trenches or in a bed of clean stones, through which treated wastewater may seep into the surrounding soil for further treatment and disposal. Often combined with a septic tank to form a septic system. *See* Figure S-16. The required length of trenches ($L$, m) may be determined as follows:

$$L = Nq/2\,dI \qquad (S\text{-}108)$$

where $N$ is the number of users, $q$ the per-capita daily wastewater flow (liters per capita per day), $d$ the effective depth (m), and $I$ the infiltration rate (liters per square meter per day). Groundwater contamination sometimes results from a concentration of such fields and some pollutants (detergents, nitrates, chlorides, and pathogens) may affect nearby shallow water-supply wells. Also called absorption field, disposal bed, disposal field, disposal pit, drainfield, leachfield, percolation field, seepage bed, seepage pit, soil absorption system, soil adsorption field, subsurface soil absorption system. *See also* soakaway, soakaway trench, soakpit.

**soil absorption system** *See* soil absorption field.

**soil absorption test** A test conducted to determine whether a site is suitable for subsurface effluent disposal. It measures the rate of water absorption by the soil in flow rate per unit of surface area, e.g., gpd/acre. *See also* percolation test.

**soil adsorption field** The more common term is soil absorption field or system.

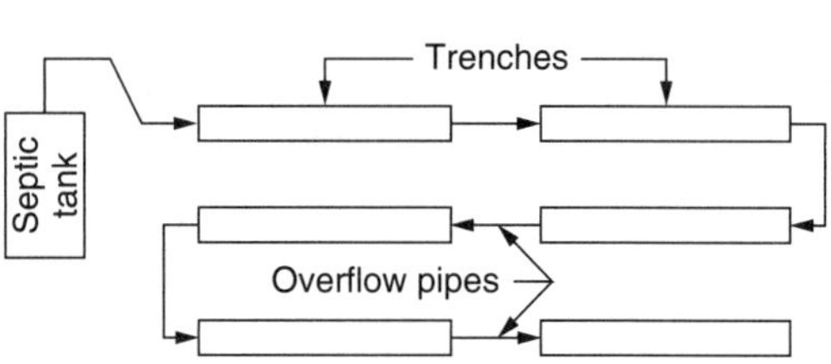


**Figure S-16.** Soil absorption field.

**soil aggregate** The combination of soil particles (sand, silt, clay) bound by microbial cells and their extracellular metabolites into secondary units. These units may be arranged in the profile in a distinctive characteristic pattern that can be classified on the basis of size, shape, and degree of distinctness into classes, types, and grades (EPA-40CFR796.2700-v and 2750-iv). Also called ped.

**soil air** The gases in the interstices of the aeration zone that are connected to the atmosphere. Also called subsurface air or ground air.

**soil alkalinity** The pH of a soil extract.

**soil amendment** (1) The use of an amendment or soil conditioner to improve soil qualities (e.g., moisture retention, pH balance, porosity) and promote plant growth. (2) Same as soil conditioner; also simply called an amendment. *See also* bulking agent.

**soil and water conservation practices** Control measures consisting of managerial, vegetative, and structural practices to reduce the loss of soil and water.

**soil-aquifer treatment (SAT)** A method of water reclamation use in hydrogeological conditions favorable for groundwater recharge with spreading basins. It includes a recharge zone and recovery of renovated water by subsurface drains, extraction wells around the spreading basin, or midway wells. Filtration by percolation through the soil and the vadose zone provides additional treatment to the pretreated effluent. Also called geopurification. *See also* high-rate infiltration, high-rate sprinkling, low-rate irrigation, overland flow, rapid infiltration extraction, rapid infiltration, and surface spreading.

**soil association** A group of related soil types forming a typical pattern in a geographical area.

**soil classification** (1) The classification of soil according to the size of its particles. The classification of the U.S. Department of Agriculture (which differs from that of the International Society of Soil Science) is shown in Table S-07. (2) The definition of various types of soil according to the proportions of sand, silt, and clay that they contain. *See* Figure S-17. The various types are: sand, silt loam, silt, light clay, loamy sand, sandy clay loam, silty clay loam, medium clay, sandy loam, sandy clay, silty clay, heavy clay, loam, clay loam.

**Table S-07.** USDA soil classification

| Soil class | Equivalent diameter, mm |
|---|---|
| Grit or fine gravel | 1.00–2.00 |
| Coarse sand | 0.50–1.00 |
| Medium sand | 0.25–0.50 |
| Fine sand | 0.10–0.25 |
| Very fine sand | 0.05–0.10 |
| Silt | 0.005–0.05 |
| Clay | $\leq$ 0.005 |

**soil conditioner** (1) A stable organic material like humus or compost that helps soil absorb and retain water, build a bacterial community, and take up mineral nutrients. Wastewater, sludge, and night soil have been used as soil conditioners after varying degrees of treatment. Also called amendment or soil amendment. (2) An organic or inorganic material added to soil to improve its structure.

**soil conditioning** The addition of solid agents (soil conditioners) to improve air or water movement in the soil.

**Soil Conservation District (SCD)** or **Soil and Water Conservation District (SWCD)** A local government entity with a defined water and soil protection area that provides assistance to farmers and other local residents in conserving natural resources, especially soil and water.

**Soil Conservation Service (SCS)** Former name of an agency of the United States Department of Agriculture that provides technical assistance for resource conservation to farmers, other federal, state, and local agencies, and to local soil conservation districts. It also publishes bulletins on hydrological methods. Currently called Natural Resources Conservation Service.

**soil corrosivity** A characteristic of soils that are aggressive toward water pipes and other materials; it can be measured by special electrodes for soil potential. Soil corrosivity decreases as soil resistivity increases from 0 to 100,000+ ohm-cm. A mildly corrosive soil has a resistivity between 10,000 and 25,000 ohm-cm..

**soil erodibility** A measure of the soil's susceptibility to raindrop impact, runoff, and other erosional processes.

**soil erosion** The removal of soil from one place and its deposition at another place by wind, water, waves, glaciers, and construction activities.

**soil filter** A filtering device used for the biological treatment of odorous air and consisting of a bed of soil. *See* bulk media filter for detail.

**soil horizon** A layer of soil or sediment, approximately parallel to the soil surface, with distinct characteristics produced by soil-forming processes. *See* the following horizons: A, B, C, D, E, O, R; and processes: eluviation, illuviation.

**soil mantle** *See* solum.

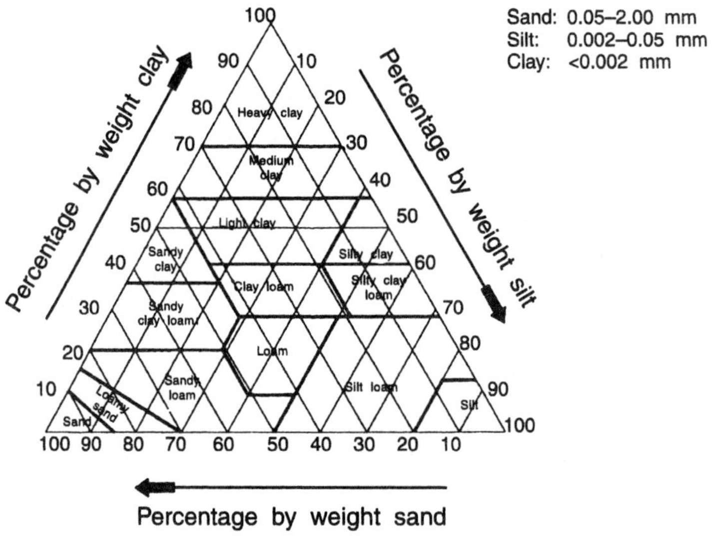


**Figure S-17.** Soil classification.

**soil moisture** Subsurface water in the form of moisture in the vadose zone, including soil water (available to root plants by abstraction) and pellicular water or the water held by molecular attraction on the surface of soil particles. The latter is not readily available to plants. *See also* subsurface water. Soil moisture contributes to evapotranspiration and to basin recharge, but not to total runoff. *See also* rainfall–runoff relationship.

**soil moisture tension** A measure of the energy required to remove water from a soil sample at a given soil dryness, resulting from gravity as well as hydrostatic, osmotic, and electrochemical adhesive forces.

**soil order** The broadest category of soil classification, based on general similarities of physical/chemical properties.

**soil percolation test** *See* percolation test.

**soil–plant barrier** The limits on crop yields exerted by excessive concentrations of metallic elements (except cadmium) in soils, which prevents them from accumulating to undesirable levels in food crops. Federal and state regulations consider the soil–plant barrier to establish the amounts of trace metals in sludge applications.

**soil profile** The vertical cross section of a soil showing its three horizons: the rich, dark topsoil of 50–300 mm and a paler and poorer layer of 400 mm forming the A horizon; the denser, brighter, and poorer subsoil of 100 mm to 2.0 m or horizon B; and, finally, a horizon C of variable thickness and composition.

**soil profile examination** A method used to determine the assimilative capacity of a site relative to onsite disposal of wastewater. It is based on a detailed analysis of soils of the site and corresponding loading rates recommended by the USEPA. *See also* percolation test and shallow trench pump-in test.

**soil salinity threshold** The maximum permissible electrical conductivity of the soil saturation extract for a given crop. Also called vegetation salinity threshold or crop salinity threshold. *See also* leaching requirement.

**soil sealant** A product used to plug porous soils and prevent leaching or percolation.

**soil–sediment partition coefficient** The ratio of adsorbed chemical per unit weight of organic carbon to the aqueous concentration of solute; an indication of the tendency of a chemical compound to partition between particles containing organic carbon and water. Also called soil sorption coefficient. *See also* retardation factor.

**soil series** The basic unit of soil classification; a subdivision of a family. A series consists of soils that were developed under comparable climactic and vegetational conditions. The soils comprising a series are essentially alike in all major profile characteristics except for the texture of the "A" horizon (i.e., the surface layer of soil) (EPA-40CFR796.2700-xi).

**soil solution** The aqueous portion of soil that contains dissolved matter leached from weathered minerals. It provides water and nutrients for plant growth. Also called salt solution.

**soil sorption coefficient** The ratio of adsorbed chemical per unit weight of organic carbon to the aqueous concentration of solute. *See* soil–sediment partition coefficient for detail.

**soil stripping** A site preparation procedure that consists in removing all topsoil that contains more than 2% of organic matter before beginning excavation for a reservoir.

**soil structure** The arrangement of soil particles into aggregates.

**soil texture** A soils classification based on the proportions of soil particles (sand, silt, clay) in a soil profile. The soil classes are: clay, sandy clay, silty clay, clay loam, silty clay loam, sandy clay loam, loam, silt loam, silt, sandy loam, loamy sand, and sand (EPA-40CFR796.2700-xii and 796.2750-xiv).

**soil-transmitted helminths** Parasitic worms whose eggs develop in moist soil infected by fecal matter. The parasites then infect man by ingestion of vegetables or through the skin of the feet. These helminths may cause such infections as hookworm, roundworm, strongylidiasis, and whipworm.

**soil type** A subdivision of a soil series based on texture.

**soil-vapor extraction (SVE)** A technique used for removing volatile organic compounds in unsaturated soils. The first phase in a pump-and-treat approach, SVE consists of volatilizing subsurface contaminants into mobile gases and extracting these gases aboveground. A typical SVE installation includes extraction wells, piping, pumps, meters, gauges, and sampling ports. Also called aeration, enhanced volatilization, in situ volatilization, soil venting, or vacuum extraction. *See also* air sparging, biofiltration, bioremediation, bioventing.

**soil venting** Same as soil-vapor extraction.

**soil water** *See* soil moisture.

**soil water potential** The work required to transfer a very small quantity of water from a given elevation and pressure to another elevation and pressure in a porous medium. Also defined as the potential energy per unit mass of water, pure water having zero potential. It is the sum of five components due to gravity, pressure, density, osmosis, and attraction of the soil matrix for water.

**Sokalan®** An antiscalant produced by BASF for use in desalination plants.

**sol** Short for hydrosol, a fluid solution, a colloidal suspension in water. *See also* aerosol, gel.

**solar bed** *See* solar drying bed.

**solar constant** The average rate at which the earth receives the sun's radiant energy, equal to 1.94 small calories per minute per square centimeter of a horizontal area at the top of the Earth's atmosphere.

**solar desalination** The use of solar energy to desalinate water in a closed system.

**solar disinfection** The use of sunlight to kill pathogens in drinking water or in wastewater.

**solar distillation** (1) Distillation using solar energy. (2) A water desalination process using a solar still. Also called solar humidification.

**solar drying bed** A device that uses evaporation to dewater water or wastewater treatment sludge that is spread 12-in deep and periodically mixed mechanically. It typically consists of a covered, asphalt, or concrete pavement overlying a porous gravel layer, with sand drains on the perimeter and in the center of the bed to collect the drainage as well as sensors, ventilation fans, air louvers, etc. Heavy equipment is used for solids agitation and removal. A microprocessor controls the operating conditions. Also called paved drying bed. *See also* mole, air drying, sand drying bed.

**solar evaporation** The use of sunlight in solar evaporation ponds or other units to further reduce the volume of concentrated hazardous wastes for ultimate disposal or for permanent burial at the same site. *See also* forced-circulation evaporator and natural-circulation evaporator.

**solar humidification** Same as solar distillation (2).

**solar pond** (1) A pond that uses direct solar heating for evaporation. (2) A pond with salt water at the bottom and freshwater at the top, used to capture solar radiation as a source of energy.

**solar salt** The common salt produced by solar evaporation and used to regenerate water softeners.

**solar still** A distillation device consisting of glass- or plastic-covered shallow basins used for the condensation of water vaporized by solar radiation; used in the treatment of brackish and saline waters.

**solder** A metallic compound used to seal joints between pipes. Until recently, most solder contained 50% lead. Use of lead solder containing more than 0.2% lead is now prohibited for pipes carrying drinking water.

**solenoid-driven metering pump** A simple, relatively accurate, and inexpensive device developed in the 1970s to deliver a controlled volume of liquid in water and wastewater treatment processes as well as other industrial applications. It uses a diaphragm actuated by a solenoid. *See also* metering pump.

**sole-source aquifer** (SSA) An aquifer that supplies 50% or more of the drinking water of an area or an aquifer that has been designated an SSA by the Administrator of the USEPA.

**solid-bowl centrifugal device** *See* solid-bowl centrifuge and high-solids centrifuge.

**solid-bowl centrifuge** A countercurrent or continuous cocurrent device that uses centrifugal force to separate water or wastewater treatment sludge into a solid cake and a liquid centrate; it is used in sludge thickening or sludge dewatering operations. It consists of a rotating helical screw conveyor inside a long, rotating bowl that is partly cylindrical and partly conical. Also called a conveyor centrifuge, decanter centrifuge, or scroll centrifuge. *See also* basket centrifuge, differential speed.

**solid-bowl conveyor centrifuge** *See* solid-bowl centrifuge, cocurrent centrifuge, countercurrent centrifuge.

**solid-bowl scroll centrifuge** Same as solid-bowl centrifuge; the internal conveyor is a helical scroll.

**Solidex** A screw press manufactured by Bepex Corp.

**solidification** The addition of materials to wastes to obtain a solid that is more suitable for long-term disposal or to encapsulate the waste constituents. Solidification may involve a combination of chemical and physical processes such as evaporation, sorption, reactions with solidification agents, vitrification, and encapsulation. *See also* solidification/stabilization.

**solidification agents** Substances used to solidify wastes and facilitate their disposal. These substances react with the wastes or provide a solid matrix to isolate them, e.g., cement, silicates, activated carbon, fly ash, kiln dust, clays, vermiculite, and some proprietary materials.

**solidification heat** *See* heat of solidification.

**solidification/stabilization** *See* the following processes that are designed to treat waste with additives and binders or to remediate contaminated sites: encapsulation, fixation, solidification, sorption, stabilization, vitrification.

**solid–liquid separation** A treatment process that removes suspended particles from water or wastewater.

**solid–phase cytometry** An effective method of microsphere enumeration, using fluorescent dyes to examine microbial cells. It detects microspheres captured on a slide or membrane. Also called laser scanning cytometry.

**solid-phase extraction (SPE)** A method used to separate constituents from solution by adsorption onto a solid matrix, e.g., the extraction of emulsified oil and grease from water by adsorption on a disk and subsequent elution with *n*-hexane. *See also* SPE/CE method.

**solid-phase extraction disk** A hydrophobic disk, similar in shape to a standard filter paper, used to adsorb organic compounds from water samples. It is then desorbed with an elution solvent for analysis.

**solid-phase immunosorbent assay (SPIA)** An ELISA test using a solid support to capture antigens or antibodies.

**solid-phase microextraction (SPME)** A separation technique used in the analysis of organic materials, e.g., the adsorption of target taste- and odor-causing compounds on a fiber coated with divinylbenzene-polydimethylsiloxane cross-link.

**solids** In water, wastewater, or stormwater engineering, the suspended, colloidal, or dissolved, volatile, or nonvolatile substances contained in the liquid or removed from it by such processes as sedimentation and filtration. Dissolved solids are molecules or ions held by the molecular structure of the water; they are generally smaller than 0.001 micrometer, pass through the 0.45-micrometer pore-diameter filter, and are sometimes called filterable residues. Suspended solids, larger than molecules, are supported by buoyant and viscous forces in the water; larger than 1 micron, they are retained on the filter and sometimes called nonfilterable residues. In the United States, the average per-capita contribution of suspended solids to domestic wastewater is approximately 90 grams per day. Colloids are intermediate between suspended and dissolved particles, ranging in size from 0.001 to 1 micron. Total solids are less than 1% of wastewater. *See also* biosolids, wastewater solids, filterable residues, nonfilterable residues, sludge.

**solids balance** An inventory of all identified solids entering, leaving, or accumulating in a system (e.g., basin, tank, reservoir), or a quantitative analysis of the changes occurring in the system. *See also* mass balance and materials balance.

**solids blanket** A layer of solid particles in a sedimentation tank or other basin that settle together while maintaining the same position with respect to each other. *See* hindered settling, sludge blanket.

**solids breakthrough** The increase of suspended solids in the effluent of a filter beyond an acceptable concentration.

**solids capture** The amount or fraction of solids removed by a sludge thickening or dewatering unit.

**solids-contact basin** A water or wastewater treatment basin where the solids-contact process occurs. *See also* solids-contact clarifier.

**solids-contact clarifier** A circular or rectangular water or wastewater treatment unit where the solids-contact process occurs. It combines rapid mix, flocculation, and sedimentation in one basin. It is often used in precipitative-lime softening plants. Common design criteria include: mixing and flocculation time ≥ 30 minutes; detention time = 2–4 hours for removal of turbidity from surface water, or 1–2 hours for softening of groundwater; weir loading of 10–20 gpm/ft; and upflow rate of 1.0–1.75 gpm/sq ft. Also called contact clarifier, flocculator–clarifier, upflow blanket clarifier, upflow solids contact clarifier, upflow tank, sludge-blanket clarifier. *See also* floc blanket clarifier, Reactor-Clarifier™, and the two types of solids-contact clarifiers: premix and premix–recirculation.

**solids-contact process** A water or wastewater treatment process that combines flash mixing, coagulation, flocculation, and sedimentation in a single basin/tank/unit, often called a solids-contact clarifier. Coagulation and softening chemicals may be added in the influent pipe, with the flow directed upward through a sludge (solids) blanket at the bottom of the tank. Some circular units incorporate an inner cylindrical flocculation compartment with a paddle- or low-speed mixer. Settling occurs in a peripheral zone, from which the overflow rises and discharges at or near the surface. Solids are continually or periodically withdrawn from the sludge blanket. This process is often used in lime softening of groundwater and the chemical treatment of surface water as well as industrial wastewater. The solids-contact process requires less space and is less costly than more conventional methods, but it offers less operational flexibility and requires an influent with fairly uniform flow and characteristics.

**solids-contact tank** A water or wastewater treatment tank where the solids-contact process occurs. *See also* solids-contact clarifier.

**solids-contact unit** A water or wastewater treatment unit where the solids-contact process occurs. *See also* solids-contact clarifier.

**solids disposal** *See* sludge disposal.

**solids flux (*G*)** A design parameter used in sedimentation and sludge thickening; it is the product of the mixed liquor suspended solids concentration ($X$) by the velocity of the solids–water interface or zone settling velocity ($V$):

$$G = VX = AXe^{-KX} \qquad \text{(S-109)}$$

with $A$ and $K$ being two constants from settling tests. The flux curve (Figure S-18) plots the variation of solids flux or mass flux vs. solids concentration. In activated sludge settling or thickening, total solids flux (in kg/m²/hr) at a given point is the sum of gravity flux and underflow flux:

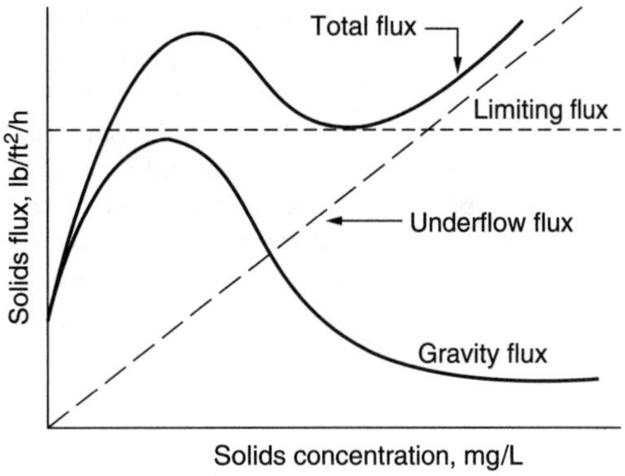


**Figure S-18.** Solids flux.

$$G = 0.001\,(C_iV_i + C_iU_b) \qquad \text{(S-110)}$$

where $C_i$ = solids concentration at the given point, mg/L; $V_i$ = settling velocity at concentration $C_i$, m/hr; and $U_b$ = bulk downward velocity or underflow velocity, m/hr. *See also* limiting flux and Vesilind equation.

**solids flux analysis** *See* solids flux method.

**solids flux method** The use of the limiting solids flux concept to determine the area required for solids thickening in clarifiers or thickeners.

**solids loading** The weight of solids per unit area imposed on a wastewater treatment or sludge processing unit, e.g., in the design of sludge thickener it may be 30 pounds per day per square foot for primary sludge or 5 pounds per day per square foot for activated sludge.

**solids loading rate (SLR)** A clarifier design and operating parameter equal to the mass of solids applied per unit time per unit clarifier area, expressed, e.g., in pounds per square foot per day. In an activated sludge system with solids recycle:

$$\text{SLR} = 8.34\,(Q + Q_r)X/A \qquad \text{(S-111)}$$

where $Q$ = influent flow, mgd; $Q_r$ = flow of return activated sludge, mgd; $X$ = concentration of mixed liquor suspended solids, mg/L; and $A$ = surface area of the clarifier; ft². *See also* overflow rate, weir loading rate, limiting flux, rising sludge blanket.

**solid-solution formation** A form of coprecipitation of secondary elements (e.g., heavy metals, radionuclides, and viruses) along with the main targets of chemical precipitation (e.g., calcium, magnesium, and organic contaminants) during water treatment. It involves the entrainment of a particle by another particle having ions of the same size. *See also* adsorption, inclusion, occlusion.

**solids production** The amount of volatile suspended solids produced during a biological treatment process, including the synthesis yield as well as other organic solids. *See* solids yield for detail. *See also* sludge production sludge quantities.

**solids residence time** Same as solids retention time.

**solids retention time (SRT)** The average time, in days, that microorganisms spend in a biological waste treatment system, a leading design and operation parameter; e.g., nitrifiers' SRTs are longer than SRTs for organisms that stabilize carbonaceous matter. If the SRT is too short, microorganisms do not have the time to reproduce and are washed out; too high SRTs promote endogenous decay and poor settleability. It is usually calculated as the total weight of suspended solids in the system (e.g., kg or lb of mixed liquor) divided by the total weight of suspended solids leaving the system (e.g., kg or lb of waste sludge) per day. Common design values are 15–25 days for aerobic digestion, 10–20 days for high-rate anaerobic digestion, 30–60 days for low-rate anaerobic digestion, 4–14 days for conventional activated sludge, and 0.8–4 days for short-term aeration (high-rate activated sludge). *See* Table S-08. This parameter was first called sludge age and then mean cell residence time (MCRT). It is also called cell residence time, sludge residence time. *See also* food-to-microorganism ratio, specific biomass growth rate, specific substrate utilization rate, volatile solids reduction.

**Table S-08.** Solids retention time formula

Generally, solids retention time (SRT) is simply defined as the mass of organisms in a biological reactor divided by the mass of organisms removed daily from the treatment system. However, a generalized formula can be established for this parameter, using a converging geometric series to derive the total residence time (TRT) in a system as a function of the hydraulic residence time ($\theta$) and the recycled fraction ($k$) of the mass of microorganisms:

$$\text{TRT} = \theta/(1 - k) \qquad \text{(S-111-a)}$$

It was shown that this generalized formula is equivalent to all preceding SRT or sludge age formulas under appropriate assumptions. In particular, if one considers only the biological reactor, neglecting the channels and sedimentation unit of an activated sludge system, the TRT expression becomes the formula commonly used for SRT:

$$\text{TRT} = \text{SRT} = (V_a X_a)/(Q_w X_w + Q_e X_e) \qquad \text{(S-111-b)}$$

where $V_a$ = volume of the biological reactor, $X_a$ = concentration of mixed liquor volatile suspended solids, $Q_w$ = flow rate of waste activated sludge, $X_w$ = concentration of waste sludge volatile suspended solids, $Q_e$ = flow rate of effluent, and $X_e$ = concentration of volatile suspended solids in the effluent.

Adapted from Adrien (1998).

**solids sandwich** The firm cake of sludge solids between the porous belts of a filter press when the belts come together in the low-pressure zone.

**solids separation** In water and wastewater treatment, a unit operation that separates liquid from solids. Clarification or sedimentation produces an unthickened sludge that is 2–6% solids. Thickening increases the solids content to 9–15%, whereas dewatering of a thickened sludge produces a cake of 18–40% solids.

**solids yield** The amount of volatile suspended solids (VSS) produced during a biological treatment process, including the synthesis yield as well as other organic solids. Also called solids production. In suspended growth processes, solids production includes heterotrophic biomass, cell debris, and nonbiodegradable influent VSS. *See also* net biomass yield, observed yield, synthesis yield, or true yield.

**Solidur®** High-molecular-weight polyethylene parts manufactured by Solidur Plastics Co. for chain-and-flight sludge collectors.

**solid waste** (1) Nonliquid, nonsoluble materials ranging from municipal garbage to industrial wastes that contain complex and sometimes hazardous substances. (2) As defined under the Resource Conservation and Recovery Act (RCRA), solid wastes include any solid, semisolid, liquid, or contained gaseous materials discarded from industrial, commercial, mining, or agricultural operations, and from community activities; for example, garbage, construction debris, commercial refuse, sludge from water supply or wastewater treatment plants or air pollution control facilities, and other discarded materials. Solid waste does not include solid or dissolved materials in irrigation return flows or industrial discharges that are point sources subject to permits under section 402 of the Clean Water Act, or source, special nuclear, or by-product material as defined in the Atomic Energy Act.

**Solo™** Well cleanup pumps manufactured by QED Environmental Systems.

**solubility** (1) The ability or tendency of a substance (solute) to dissolve in or blend uniformly with a liquid (solvent), partially or completely. Liquids and gases are said to be miscible with (and not soluble in), respectively, other liquids and gases. (2) The amount or mass of a compound that will dissolve in a unit volume of solution under specified conditions. *See also* concentration and saturation.

**solubility coefficient** The quantity of a gas absorbed by a unit volume of water or wastewater at a given temperature and under a barometric pressure of 1 atmosphere, usually expressed in mg/L; also called gas solubility coefficient.

**solubility diagram** A plot versus pH of the solubility of the solid species in an aqueous solution. It indicates which solid species predominates as a stable phase for selected conditions. For example, the solubility diagram for Fe(II) in a $10^{-3}$ M carbonate system shows that below pH ≈ 10, the carbonate [$FeCO_3$(s)] predominates and controls the solubility, whereas above that pH it is the hydroxide [$Fe(OH)_2$(s)] that is more stable. In corrosion studies, solubility diagrams provide information about the potential for precipitating passivating films on pipe surfaces or the deposition of ferric and other hydroxides.

**solubility equilibria** Relationships or conditions that indicate that a chemical reaction has reached equilibrium. *See* chemical equilibrium and related terms: activity, activity product, common-ion effect, concentration, crystal growth, dissolution reaction, equilibrium, equilibrium condition, equilibrium constant, ion product, LeChatelier principle, molar concentration, nucleation, saturated solution, solubility product, solubility product constant, stoichiometric coefficient, supersaturated solution.

**solubility in water** The saturated concentration of a compound in water at a given temperature and pressure, expressed as molarity (mol/L), molality (mol/kg), mole fraction, weight percent, or mass per unit volume (e.g., mg/L). It is an indication of the amount of a chemical that can be dissolved in water, shown as a percentage or as a description. A low percent of solubility (or a description of slight solubility or low solubility) means that only a small amount will dissolve in water. Knowing this may help firefighters or personnel cleaning a spill. If a substance is water soluble, it can very readily disperse through the environment. Water solubility varies with temperature, pH, and other dissolved constituents. It relates to other parameters: bioconcentration factor, boiling point, Henry's law constant, *n*-octanol/water partition coefficient, and soil/sediment partition coefficient.

**solubility product** (1) The product of the ionic molar concentrations in a dissolution reaction. For example, a solid substance $A_aB_b(s)$ may reach equilibrium with the A and B ions as follows:

$$A_aB_b(s) \leftrightarrow a \cdot A^{b+} + b \cdot B^{a-} \qquad \text{(S-112)}$$

The solubility product for this reaction is $[A^{b+}]^a \cdot [B^{a-}]^b$, where the brackets define molar concentrations. When the solubility product is equal to the solubility product constant, the reaction is in equilibrium. The solution is unsaturated if the product of ionic concentrations is less than the constant and supersaturated if the product is greater than the constant. (2) Solubility product is very often used to mean solubility product constant.

**solubility product constant** The equilibrium constant ($K_{sp}$) of the dissolution reaction of a solid in water. It is the same as the solubility product of a saturated solution, i.e., for equation (S-112):

$$K_{sp} = [A^{b+}]^a \cdot [B^{a-}]^b \qquad \text{(S-113)}$$

For a given reaction, the constant varies with the temperature of the solution. For example, at 25°C, the solubility product constants of calcium carbonate ($CaCO_3$), aluminum hydroxide [$Al(OH)_3$], and calcium fluoride ($CaF_2$) are, respectively, $5 \times 10^{-9}$, $1 \times 10^{-32}$, and $3 \times 10^{-11}$.

**solubilization** (1) The process of making a substance soluble or more readily dissolved, or of increasing its solubility. (2) The conversion of sludge solids to a soluble form during thermal conditioning, thereby reducing the mass of solids for further processing but producing a high-strength sidestream of decant and filtrate. (3) The conversion of suspended organic matter into soluble matter as a result of anaerobic or septic conditions in a sedimentation basin. Some of the suspended solids result from the flocculation of smaller particles, hence this phenomenon is sometimes called resolubilization. This may occur in a basin having a long detention time without continuous sludge withdrawal, particularly in hot climates. BOD removal from a wastewater with a high soluble fraction is lower than with a low soluble fraction.

**soluble** Capable of being dissolved or liquefied. In this book, when no solvent is mentioned for a constituent or substance, soluble means water-soluble.

**soluble BOD removal** The soluble BOD effluent ($S_e$, mg/L) from a trickling filter may be estimated by the following first-order equation, as a function of the influent soluble BOD ($S_0$, mg/L):

$$S_e = S_0 \exp(-K_{20}\theta^{T-20}A_sD/Q^n) \qquad \text{(S-114)}$$

where $K_{20}$ = reaction rate coefficient at 20°C; $\theta$ = temperature coefficient, usually 1.035; $T$ = wastewater temperature, °C; $A_s$ = specific surface area of trickling filter media, $ft^2/ft^3$; $D$ = depth of trickling filter media, ft; $Q$ = hydraulic loading, $gpm/ft^2$; and $n$ = empirical flow constant, e.g., 0.5 for vertical flow and cross-flow media

**soluble chemical oxygen demand (sCOD)** The soluble portion of the chemical oxygen demand of wastewater, as compared to the particulate fraction (which includes colloidal and suspended solids). Also called dissolved COD. *See* COD fractionation.

**Solu Comp®** A measuring device manufactured by Rosemount Analytical, Inc.

**solum** The upper layers of the soil profile, including the two horizons (A and B) that are influenced by plant roots and that correspond to organic, leached, and deposited materials. Also called soil mantle. *See also* substratum and soil horizon.

**solute** A substance dissolved in a solvent, the two making up the solution; in pressure-driven membrane processes, the dissolved solids in the raw, feed, permeate, and concentrate streams.

**solute distribution parameter** In the granular activated carbon (GAC) adsorption process, the solute distribution parameter ($D_g$) is the dimensionless ratio of the mass of adsorbing compound on the solid to that in water at equilibrium:

$$D_g = Bq/(EC_0) \qquad \text{(S-115)}$$

where $B$ = bulk density of the activated carbon bed, g of GAC/L of bed volume; $q$ = equilibrium solid phase loading on GAC, mg adsorbed/g of GAC; $E$ = pore volume in the bed; and $C_0$ = feed concentration of adsorbate, mg/L. The parameter $D_g$, a large number in the range 15,000–10,000,000, is used to calculate the actual time to breakthrough. *See* breakthrough time.

**solute flux** The mass of solute (amount of salt or dissolved substances) that passes through a membrane per unit area per unit time in a membrane filtration process such as reverse osmosis; expressed in pounds per square foot per second or grams per square meter per second. Also called salt flux.

**solute mass transfer coefficient** Same as solute permeability coefficient.

**solute permeability coefficient** A parameter (SuPC) indicative of the flow of solute through the membrane of reverse osmosis or other pressure-driven process. It is the ratio of solute flux (SuF) to the concentration gradient (CG) through the membrane:

$$\text{SuPC} = \text{SuF/CG} \qquad \text{(S-116)}$$

The coefficient SuPC is expressed in fps; the solute flux SuF, i.e., the flow rate of solute through a unit area of membrane, in psf/sec.; and the concentration gradient, in pounds per cubic foot. Also called solute mass transfer coefficient or solute transport coefficient. *See also* solvent permeability coefficient.

**solute rejection coefficient** A parameter that describes the performance of a water treatment membrane; a measure of the ability of a membrane to reject a constituent. *See* rejection rate for detail.

**solute stabilization** The conversion of an objectionable substance into an unobjectionable form, without actually removing it; e.g., the conversion of hydrogen sulfide into sulfate or the conversion of excess lime into bicarbonate.

**solute transport** The movement of a dissolved constituent in a solution or from one phase to another.

**solute transport coefficient** Same as solute permeability coefficient.

**solution** A liquid mixture of dissolved substances; solution = solvent + solutes. In a solution, it is impossible to see all the separate parts; solutes are homogeneously distributed and are in the form of molecules or ions.

**solution channel** An opening created in a rock by water; also called solution opening. Solution channels and other soil–groundwater connections are generally undesirable features in a biosolids application site.

**solution diffusion model** A mathematical representation of the solvent and solute transport through a semipermeable membrane: under certain conditions, solvent flux is proportional to the net driving force and solute flux is proportional to the concentration gradient. *See also* homogeneous surface diffusion model.

**solution feed** The addition of chemicals to water or wastewater as liquids or solutions instead of solids or gases. *See* liquid chemical-feed system for detail. *See also* dry feed.

**solution feeder** A small positive-displacement pump that dispenses a specific volume of a concentrated solution for each stroke of a piston or rotation of an impeller. Diaphragm and plunger metering pumps are used as solution feeders in water and wastewater treatment. *See also* metering pumps, chemical feeders, liquid feed, liquid chemical-feed system.

**solution force** The characteristic of polar compounds that makes them more soluble and less readily adsorbed in water than nonpolar compounds. Because of their positive and negative charge centers, water molecules are efficient in solubilizing other polar and nonionic materials. *See also* chemical adsorption or chemisorption, desorption, electrostatic or exchange adsorption, exchange adsorption, hydrophobic bonding, ideal adsorption, physical adsorption, polarity, solvophobic force, van der Waals force.

**solution heat** Same as heat of solution.

**solution mining** Removal of a soluble mineral by dissolving it and leaching it out. For example, metals removed from low-grade ores can then be recovered from the leachate.

**solution opening** An opening created in a rock by water. *See* solution channel.

**solution transport** The mechanism that carries the bulk of the adsorbate into and through the pores of the adsorbent during adsorption powdered activated carbon. *See also* intraparticle diffusion.

**Solvay process** A process used to generate chlorine dioxide ($ClO_2$) on the site of water treatment plants, with sodium chlorate ($NaClO_3$), methanol ($CH_3OH$), and sulfuric acid ($H_2SO_4$):

$$2\ NaClO_3 + CH_3OH + H_2SO_4 \rightarrow 2\ ClO_2 + HCHO + Na_2SO_4 + 2\ H_2O \quad \text{(S-117)}$$

Also called the methanol process. *See also* the chloride reduction, Jazka–CIP, Mathieson (or sulfur dioxide, $SO_2$) processes.

**solvency** *See* metal solvency.

**solvent** A liquid substance that is used to dissolve, dilute, suspend, or extract other materials (the solute) without chemical change to the material or solvent; solvent and solutes form the solution. Water is an inorganic solvent. Organic solvents include hydrocarbons, oxygenated solvents, and chlorinated solvents. In pressure-driven membrane processes, the liquid, usually water, contains dissolved solids.

**solvent extraction** (1) The transfer of a substance from solution in one solvent (e.g., water) to another solvent without any chemical change. Examples: (a) the removal of phenols from wastewater effluents from petroleum refineries using the solvent methyl isobutyl ketone and (b) the recovery of acetic acid from industrial wastewaters using the solvent ethyl acetate. *See* Figure S-19. *See also* leaching. (2) A selective extraction process using an organic solvent to separate a liquid constituent from wastewater or other solutions. It is a commonly used technique to prepare samples for the analysis of organic compounds. It may also be used to produce freshwater from brackish and saline waters, or to extract valuable organic compounds from water. Also called liquid–liquid extraction. *See also* continuous liquid–liquid extraction, saline water conversion classification.

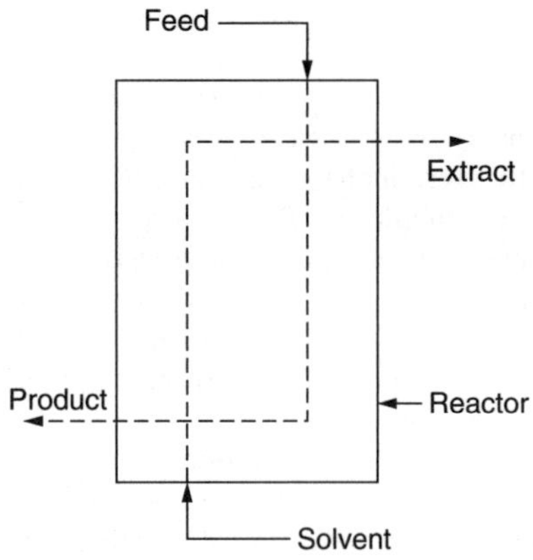


**Figure S-19.** Solvent extraction.

**solvent flux** *See* flux.

**solvent mass transfer coefficient** Same as solvent permeability coefficient.

**solvent permeability coefficient** A parameter (SvPC) indicative of the flow of solvent through the membrane of reverse osmosis or other pressure-driven process. It is equal to the ratio of the solvent flux (SvF) to the net driving pressure (NDP) across the membrane:

$$\text{SvPC} = \text{SvF/NDP} \quad \text{(S-118)}$$

The coefficient SvPC is expressed in gpd per square foot (of membrane) per psi (of driving pressure). The solvent flux (SvF), i.e., the flow rate of solvent through a unit area of membrane, is expressed in gpd/ft$^2$. The net driving pressure (NDP) is the difference between the hydraulic pressure differential and the osmotic pressure differential across the membrane, in pounds per square inch. *See also* solute permeability coefficient.

**solvent transport coefficient** Same as solvent permeability coefficient.

**solvophobic force** The result of substantial incompatibility between the solute and solvent of an aqueous solution, resulting in a thermodynamic gradient that drives the solute out of solution and in adsorption rates higher than those from surface interactions alone. *See also* chemical adsorption or chemisorption, desorption, electrostatic or exchange adsorption, exchange adsorption, hydrophobic bonding, ideal adsorption, physical adsorption, polarity, solution force, van der Waals force.

**Solvo Salveger** A vacuum-assisted distillation process developed by Hoyt Corp.

**Som-A-System** An apparatus manufactured by Somat Corp. for sludge dewatering.

**somatic bacteriophage** *See* somatic phage.

**somatic coliphage** Bacteriophage that infect strains of *E. coli* and related bacteria through cell wall receptors. It is one of the groups of bacteriophage considered as possible indicators of fecal contamination. *See also* F-specific (or male-specific) coliphage.

**somatic damage** A result of radioactivity or other phenomena affecting an organism during its life. Radiation can cause damages ranging from burns to cancer. *See also* carcinogenic and genetic damages.

**somatic phage** Bacteriophage that infect hosts through the cell wall. *See also* capsule phage, appendage phage.

**sonic sensor method** A method that uses an in-line sonic sensor to detect defects in ultrafiltration membranes; an increase in hydraulic noise in a module indicates a loss of membrane integrity.

**sonification** The inactivation of pathogens using ultrasound. *See* acoustic cavitation for detail.

**Sonix 100** A chlorinator on tank manufactured by Wallace & Tiernan, Inc.

**Sonozaire™** An electronic device manufactured by Howe-Baker Engineers, Inc. for odor control.

**—$SO_2OH$** Chemical formula of sulfonic acid.

**soot** Carbon dust formed by incomplete combustion.

**SOR** Acronym of surface overflow rate.

**sorbate** A substance that is sorbed, i.e., concentrated on the surface of another substance (sorbent) by sorption. *See also* adsorbate.

**sorbent** A solid material (e.g., activated carbon, ion-exchange resin) used to concentrate dissolved solids (sorbate) by sorption. *See also* adsorbent.

**Sorb-N-C** A sorbent produced by Church & Dwight Co., Inc. for stack gases.

**Sorbond** Products of Colloid Environmental Technologies Co. for sludge stabilization.

**sorption** (1) The transfer of a substance from a solution to a solid phase; a surface phenomenon that results in the concentration of dissolved solids by absorption, adsorption, chemisorption, ion exchange, or a combination thereof; term often used when the specific mechanism is not known. (2) The third step of the adsorption process, following macrotransport and microtransport. (3) The addition of a solid (e.g., activated carbon, anhydrous sodium silicate, clay) to take up any free liquid in a waste. *See also* solidification/stabilization.

**sorption coefficient** A distribution coefficient ($K_p$, mol/kg solid phase per mole/L liquid phase) equal to the product of the organic carbon content ($F_{oc}$, fraction by weight) and the octanol–water partition coefficient ($K_{ow}$, dimensionless):

$$K_p = b(F_{oc})(K_{ow})^a \quad \text{(S-119)}$$

**SOS** Acronym of stormwater overflow screen, a device manufactured by John Meunier, Inc.

**SOTE** Acronym of standard oxygen transfer efficiency.

**SOTR** Acronym of standard oxygen transfer rate.

**sounding line** A rope or cable attached to a weight and used for measuring depths of water during hydrographic surveying. Also called lead line.

**SOUR** Acronym of specific oxygen uptake rate.

**sour brine** Brine that contains excessive concentrations of substances (e.g., calcium and magnesium) that render it unfit for the regeneration of ion-exchange resins.

**source control** In general, the practice of reducing pollutants at their source. *See* on-site source control for stormwater reduction measures. *See also* downstream control, and end-of-pipe alternative.

**source material** (1) Uranium, thorium, or any combination thereof in any physical or chemical form. (2) Ores that contain by weight one-twentieth of one percent (0.05%) or more of uranium, thorium, or any combination of uranium and thorium, excluding special nuclear material (EPA-10 CFR 20.1003).

**source reduction** Reducing the amount of materials entering the waste stream by redesigning products or patterns of production or consumption (e.g., using returnable beverage containers). *See also* waste reduction.

**sources and sinks** The causes of change in hydraulic, hydrodynamic or water quality modeling. Also called forcing functions. The change may be related to water mass or to momentum. Sources and sinks include (a) processes such as evaporation, precipitation, inflows, outflows, nonpoint runoff of stormwater, point discharges, withdrawals, injection or extraction wells in groundwater, seepage, infiltration; (b) forces such as buoyancy, Coriolis force, friction, gravity; and (c) changes in the masses of water constituents such as hydrophobic chemicals. *See also* dendritic network for dummy sources and sinks.

**source separation** Segregating various wastes at the point of generation (e.g., separation of paper, metal, and glass from other wastes to make recycling simpler and more efficient).

**source water** Intake water prior to any treatment or use. Also, a water supply source from a spring, stream, aquifer, or lake. Also called raw water.

**Source Water Assessment Program (SWAP)** A program of the USEPA under the Safe Drinking Water Act; it requires the states to inventory public water supply systems, identify potential contaminants, and inform the public accordingly. *See,* e.g., wellhead protection planning, wellhead protection area.

**source water ozonation** The application of ozone ahead of coagulation, sedimentation, or filtration. Also called preozonation.

**sour environment** An environment that contains a high level of acid or hydrogen sulfide.

**sour gas** Natural gas that contains a significant amount of hydrogen sulfide ($H_2S$), which can be separated as a source of sulfur. *See also* sweet gas.

**sour smell** The smell of hydrogen sulfide, as found in the vicinity of petroleum refineries and characteristic of mercaptans.

**sour water stream** A stream that (a) contains ammonia ($NH_3$) or sulfur compounds (usually hydrogen sulfide, $H_2S$) at concentrations of 10 ppm by weight or more; (b) is generated from separation of water from a feed stock, intermediate, or product that contained ammonia or sulfur compounds; and (c) requires treatment to remove the ammonia or sulfur compounds (EPA-40CFR61.341).

**sour water stripper** A unit that (a) is designed and operated to remove ammonia ($NH_3$) or sulfur compounds (usually hydrogen sulfide, $H_2S$) from sour water streams; (b) has the sour water stream transferred to the stripper through hard piping or other enclosed system; and (c) is operated in such a manner that the off-gases are sent to a sulfur recovery unit, processing unit, incinerator, flare, or other combustion device (EPA-40CFR61.341).

**$SO_x$** Abbreviation or formula of sulfur oxides.

**Soxhlet extension tube** An apparatus used in the Soxhlet grease extraction method. It consists of a long glass or plastic tube fitted with a flask at the end and the accessories for heating and weighing the samples.

**Soxhlet extraction method** A procedure, using the Soxhlet extension tube, for the determination of the grease content of a wastewater sample: a one-liter sample is passed through a special filter, which is then dried in a thimble at 103°C and placed in the tube. The flask is filled with the nonflammable solvent trichlorotrifluoroethane and heated. Repeatedly vaporizing and flushing the solvent through the sample leaves the grease in the flask. Grease content is estimated based on the weight of the flask before and after the procedure.

**Soxhlet grease extraction apparatus** Same as Soxhlet extension tube.

**Soxhlet grease extraction method** Same as Soxhlet extraction method.

**sp.** Abbreviation used to designate a single, unidentified species; several species of a genus may be designated by the abbreviation spp.

**Spaans** Screw pumps and conveyors manufactured by Asdor.

**spacer** A component placed between the membranes of a stack and designed to allow controlled fluid flow. Also called a separator.

**spadable sludge** Sludge solids than can be easily removed from a drying bed using a fork or a shovel.

**span** The scale or range of values an instrument is designed to measure. *See also* range.

***S. paratyphi*** *See Salmonella paratyphi.*

***S. paratyphi* A** *See Salmonella paratyphi.*

***S. paratyphi* B** *See Salmonella paratyphi.*

***S. paratyphi* C** *See Salmonella paratyphi.*

**sparge** To inject air below the water table to strip dissolved volatile organic compounds and/or oxygenate the groundwater to facilitate aerobic biodegradation of organic compounds.

**sparger** A rapidly rotating device that breaks up coarse air bubbles for diffusion in a liquid, as used in water or wastewater treatment; unlike ordinary fine-bubble diffusers, the sparger is not prone to plugging. Spargers are used alone or in combination with mechanical aeration devices. *See also* flexible diffuser, jet aeration.

**sparingly soluble compound** A compound that has a low solubility in water, e.g., from 0.0 to a few thousand mg/L. A sparingly soluble salt is relatively insoluble in water and may precipitate even at low concentrations. Salts such as the sulfates or carbonates of barium (Ba), calcium (Ca), and strontium (Sr) are of concern in membrane filtration processes.

**sparingly soluble salt** *See* sparingly soluble compound.

**Sparjair** A package wastewater treatment manufactured by Walker Process Equipment Co.

**Sparjer** Aeration equipment manufactured by Walker Process Equipment Co.

**SparjLift** An air injection pump manufactured by Walker Process Equipment Co.

**sparkling bottled water** See bottled water and sparkling water. *See also* natural sparkling water.

**sparkling water** Effervescent water (or other beverage) artificially charged with carbon dioxide

($CO_2$). Also called seltzer water, soda water. *See also* bottled sparkling water, natural sparkling water.

**SPARRO** Acronym of slurry precipitation and recycle reverse osmosis.

**Spaulding precipitator** A premix solids-contact clarifier developed in 1935 as a modification of the Candy tank. It has a flat bottom, milder side slopes than the Candy tank, a mixing zone in the middle, and a sludge filter zone around it.

**SPC** Acronym of standard plate count.

**SPCC** Acronym of Spill Prevention Control and Countermeasures.

**SPE** Acronym of solid phase extraction.

**SPE/CE method** A method used to analyze ambient and spiked water samples from a distribution system. *See* solid phase extraction (SPE) and capillary electrophoresis (CE).

**special** A pipe section constructed for a special purpose but not of standard length or shape.

**special assessment** A direct tax levied on private property to pay the cost of water, wastewater, and other local public improvements.

**special assessment bond** A bond, usually short term, issued as a result of a special assessment, backed by the value of the property served. *See also* general obligation bond, revenue bond.

**specialized isolation medium** An enrichment medium that contains a substance that meets the nutritional requirements of a specific organism or group of organisms so that they can be more easily detected and enumerated. *See also* selective enrichment medium.

**special nuclear material (SNM)** (1) Plutonium, uranium-233, uranium enriched in the isotope 233 or in isotope 235, and any other material that the Nuclear Regulatory Commission (NRC) determines to be SNM, but excluding source material. (2) Any material artificially enriched by any of the foregoing, but excluding source material. (10 CFR 20. 1003). SNM is important in the fabrication of weapons grade materials and as such has strict licensing and handling controls.

**special source of groundwater** A class I groundwater identified in accordance with USEPA's groundwater protection strategy and that is (a) within the controlled area encompassing a disposal system or less than five kilometers beyond the controlled area; (b) supplying drinking water for thousands of persons; and (c) irreplaceable in that no reasonable alternative source of drinking water is available to that population (EPA-40CFR191.12).

**special wastes** Items such as household hazardous waste, bulky wastes, and used oil.

**speciation** (1) The formation of a new species as a result of conditions that prevent populations from continuing their historical interbreeding. (2) The chemical composition of a group of compounds; the chemical or physical form in which an element is present, e.g., as a hydrated ion, a molecule, a complex. The form of an element determines its ecological effects. For example, mercury as HgS is insoluble and deposits into the sediments of a lake, whereas mercury ion ($Hg^{2+}$) is soluble and methyl mercury ($CH_3Hg^+$) is toxic. (3) Determination of particular forms of an element, not just the total element. Metals may be speciated by oxidation state, e.g., Cr (III) vs. Cr (VI).

**species** (1) A reproductively isolated aggregate of interbreeding organisms, with common characteristics or qualities related to heredity, physiology, way of life, etc.; the basic taxonomic unit of identification of bacteria and other organisms, which are given a genus and a species name; e.g., *Bacillus subtilis* indicates the bacteria of the genus *Bacillus* and species *subtilis.* (2) The major subdivision of a genus or subgenus. *See also* biological classification. (3) The actual form in which an element, molecule, or ion is present in solution. For example, in natural waters, metals may be present as free metal ions, inorganic complexes, organic complexes, colloids, etc.

**species diversity** The state of a species as determined by the number and relative abundance of its members in an area.

**species population** A group of similar organisms in a given space and time.

**species sanitation** An approach to control vectors of such diseases as filariasis, malaria, and schistosomiasis, using methods based on the behavior, characteristics, and breeding habits of the vectors. Mosquito control methods, for example, may include vegetation clearing or engineering works.

**specifications** Documents that, along with plans, define a project, e.g., describing the materials and methods of construction or prescribing the performance to be achieved by equipment.

**specific biomass growth rate** Same as net specific growth rate.

**specific capacity** A common measure of the capacity, efficiency or productivity of a well, usually expressed in gallons per minute per foot (gpm/ft) and relating its yield (gpm) to its depth (ft). It is also defined as the ratio of discharge ($Q$) to well drawdown ($D$). The ratio $Q/D$ is the theoretical specific capacity, a function only of formation characteristics. The actual specific capacity depends on productivity and takes losses into account.

**specific conductance** The conductivity of a substance, i.e., a measure of its ability to conduct electric current. It is measured using a standard one-centimeter-wide cell. Specific conductance ($\kappa$, micromhos/cm or microsiemens) is the reciprocal of specific resistance ($\rho$) and is expressed in micromhos or in microsiemens at 25°C. It is related to the concentration of ionized substances in a water supply and is used as a rapid measure of the total dissolved solids (TDS, mg/L) content of the water:

$$TDS = k\kappa = k/\rho \quad \text{(S-120)}$$

where k is an adjustment factor varying from 0.55 to 0.90. For raw and potable waters, $50 < \kappa < 500$ microsiemens. Specific conductance, sometimes simply called conductance, is also related to the normality of a solution and to ionic strength. *See* equivalent conductance, Langelier's ionic strength approximation, and Russell's ionic strength approximation.

**specific denitrification rate (SDNR)** The nitrate reduction rate in an anoxic tank divided by the mixed liquor volatile suspended solids (MLVSS) concentration (mg/L), expressed in g $NO_3$-N/g MLVSS/day. It may also be calculated as a function of the food-to-microorganism ratio (F/M):

$$\begin{aligned} SDNR &= (NO_r)/[(V_{nox})(MLVSS)] \\ &= 0.03\ (F/M) + 0.029 \end{aligned} \quad \text{(S-121)}$$

where the F/M is expressed in g BOD applied per gram MLVSS/day in the anoxic tank; $NO_r$ = nitrate removal, g/d; and $V_{nox}$ = volume of anoxic tank, $m^3$. In the intermittent aeration process, a specific denitrification rate relative to heterotrophic biomass concentration (SDNRb) is similarly defined, but also takes into account the active heterotrophic biomass:

$$SDNR_b = 0.175\ A_n/(\theta Y_n) \quad \text{(S-122)}$$

where $SDNR_b$ is expressed in g $NO_3$-N/g biomass/day; $A_n$ = net oxygen utilization coefficient, g $O_2$/g biodegradable COD (bCOD) removed; $\theta$ = solids retention time, days; and $Y_n$ = net heterotrophic biomass yield, g VSS/g bCOD.

**specific density** Same as relative density, i.e., the ratio of the density of a substance to the density of distilled water.

**specific deposit** (1) The principal variable ($\sigma$) in the expression of the filtration rate coefficient; it is the volume of deposited matter per unit volume, as the filter run progresses. *See* Ives equation. (2) In sludge dewatering by vacuum filter, specific deposit is the volume of sludge cake formed per volume of filtrate obtained:

$$\sigma = A\Delta L/V \quad \text{(S-123)}$$

where $A$ is the area of flow, $\Delta L$ is the depth of the medium, and $V$ is the volume of water.

**specific discharge** (1) The discharge per unit area; e.g., a stream discharge in cfs divided by the corresponding drainage area in square miles; often used to define flood magnitudes. (2) Flowrate over a cross-sectional area of soil, i.e., the Darcy flux or flux per unit area per unit time in Darcy's law; also called Darcy velocity or discharge velocity.

**specific energy** Introduced by Bakhmeteff in 1912, it is the energy head ($E$) above the low point in a channel, or the sum of the depth of flow ($y$) and the velocity head, i.e.,

$$E = y + V^2/2\,g \quad \text{(S-124)}$$

where $V$ = the average velocity and $g$ = the gravitational acceleration. *See* critical flow equation.

**specific filtration resistance** *See* specific resistance.

**specific flux** In a membrane filtration process, the specific flux is the flux produced per unit net operating pressure. It is used in evaluating the performance of a membrane. For a pressure-driven membrane system, specific flux is the ratio of permeate flux to net driving pressure, expressed, e.g., in gallons per day per square foot per pound per square inch (gpd/$ft^2$/psi). *See also* membrane autopsy, normalized specific flux, temperature correction factor.

**specific gravity** or **relative density** The dimensionless ratio of the density of a substance to a standard density. For liquids, the standard density is that of water at 4°C, i.e., 1000 kg/$m^3$ or 1 kg/liter; in U.S. customary units it is 1.941 slugs/$ft^3$. The specific gravity of water is equal to 1.0 by definition. Most petroleum products have a specific gravity less than 1.0, generally between 0.6 and 0.9. As such, they will float on water; these are also referred to as LNAPL, or light nonaqueous-phase liquids. Substances with a specific gravity greater than 1.0 will sink through water; these are referred to as DNAPL, or dense nonaqueous-phase liquids. Most flammable liquids are lighter than water. *See also* specific weight.

**specific growth rate** Same as net specific growth rate.

**specific heat** (1) The quantity of energy that must be supplied to raise the temperature of a unit mass of substance one degree, or the amount of heat released by a decrease of temperature of one degree of a unit mass, expressed in calories/gram/°C or BTUs/pound/°F. Also called specific heat capacity. *See also* heat capacity, sensible heat. (2) Originally, the ratio of the heat capacity (or thermal capacity) of a substance to the heat capacity of a reference material such as water; the heat capacity of water may be taken for a temperature change from 62 to 63°F.

**specific heat capacity** Same as specific heat.

**specific humidity** The dimensionless ratio ($H_s$) of the mass of water vapor in moist air to the total mass of the mixture of air and vapor. It is related to the mixing ratio ($H_s$) as follows:

$$H_s = R_m/(1 + R_m) \qquad \text{(S-125)}$$

*See also* absolute humidity, dewpoint, relative humidity.

**specific injectivity (gpm/ft)** In aquifer storage and recovery systems, it is the ratio of the rate of injection in gallons per minute to the water-level buildup in feet; often used as a measure of the performance of an ASR well. *See also* plugging (2).

**specific interfacial area** A parameter used in the design of aeration and air stripping equipment, defined as the interfacial area available for mass transfer divided by the system unit volume. It is applied to convert a flux to a volume in the volumetric mass transfer coefficient ($K_La$).

**specific-ion electrode** An electrode that contains a membrane sensitive to a specific ion or group of ions and transmits their potential to an external circuit; e.g., the pH electrode.

**specific-ion meter** A voltmeter that uses specific electrodes to measure the concentration of ions in water or wastewater.

**specific-ion toxicity** The condition in which a specific anion or cation, instead of just osmotic effects, causes the decline of growth of sensitive crops. Ions of concern in land application of wastewaters are boron, chloride, sodium, and some trace elements.

**specific level** The level of the water surface at a specific site and discharge of a river. It varies with the cross section.

**specific oxygen uptake rate (SOUR)** (1) A parameter proposed by the USEPA to assess the odor causing and vector attraction potential of a sludge. It is determined by aerating a closed sample of the sludge, measuring the rate of oxygen depletion, and dividing this rate by the solids concentration of the sample. A SOUR $\leq$ 1.0 mg/L per hour per gram of total sludge solids is acceptable. SOUR is a standard of performance of aerobic digestion. (2) A measure of the amount of oxygen used by microorganisms in the activated sludge process, expressed as mg of oxygen per gram of mixed liquor volatile suspended solids per hour, a value that correlates well with the effluent COD. It is also used as an operational indicator of the process. Also called respiration rate. *See also* oxygen uptake rate.

**specific productivity** For a pressure-driven membrane, specific productivity is the ratio of permeate production rate to net driving pressure, in gallons per day per pound per square inch (gpd/psi). It varies with the temperature. *See also* temperature correction factor.

**specific resistance (SR)** (1) The capacity of a material to resist the flow of electricity, measured as the resistance between opposite faces of a one-centimeter cube of the material. Also defined as the ratio of electric intensity to cross-sectional area or the reciprocal of specific conductance. Also called resistivity. In water and wastewater analysis, specific resistance, expressed in ohm-cm, is used to estimate total ionized solids concentrations. Specific resistance varies inversely with concentration, e.g., from $\rho$ = 4700 ohm-cm for a solution of 100 mg/L of NaCl to $\rho$ = 500 ohm-cm for a solution of 1000 mg/L of the same salt. (2) A measure of the resistance of sludge to dewatering or filtration. The specific resistance to filtration (SRF), an indicator of sludge filterability, is measured by the time required to filter a sludge sample under specified conditions, using, e.g., a Büchner funnel. Specific resistance ($R_{wc}$) can be expressed in terms of intrinsic resistance ($R_c$), (the reciprocal of intrinsic permeability), specific deposit ($\sigma$), and the mass (w) of dry cake solids produced per unit volume of filtrate:

$$R_{wc} = \sigma R_c/w \qquad \text{(S-126)}$$

Activated sludge has a specific resistance of about $8 \times 10^{13}$ m/kg. (3) An indicator of the dewatering rate of sludge, measured as the resistance to vacuum filtration, reflecting the size of particles in the filter cake, and calculated as follows:

$$R = 2\,PA^2B/\mu C \qquad \text{(S-127)}$$

where $R$ = specific resistance; $P$ = pressure drop across the sludge cake; $A$ = surface area of the filter; $B$ = slope of a plot of $T/V$ versus $V$; $T$ = time of filtration; $V$ = volume of filtrate; $\mu$ = filtrate viscosity; and $C$ = weight of dry solids deposited per volume of filtration. *See also* Atterberg test, Büchner funnel, capillary suction time, CST test, filterability constant, filterability index, sludge compressibility, shear stress, TTF test.

**specific-resistance (SR) test** A test conducted on a 100-mL sample of sludge in a Buchner funnel to determine the specific resistance of the sludge, i.e., the resistance to fluid flow exerted by a sludge cake of unit weight of dry solids per unit area. The density of compacted sludge or the achievable dry density of a given sludge is an important characteristic for landfill or monofill disposal as well as for sludge reuse in backfill or as a soil substitute. *See also* Atterberg test, Büchner funnel test, capillary suction time (CST) test, filterability constant, filterability index, filter leaf test, shear strength, standard jar test, standard shear test, time-to-filter (TTF) test.

**specific resistance to filtration** Same as specific resistance (2).

**specific retention** The volume of groundwater stored in interconnected spores, but prevented from flowing by molecular forces and surface tension and not available for withdrawal. *See also* specific yield.

**specific solute passage** The fraction of solute that passes with the feedwater through a membrane and becomes part of the permeate or product water. *See* salt passage for detail.

**specific speed ($N_s$)** - A performance parameter used in the rating or selection of pumps and turbines. Expressed in revolutions per minute (rpm), it relates rotational speed ($N$, in rpm), total dynamic head ($H$, in ft), and discharge ($Q$, in gallons per minute) or power output ($P$, in horsepower or kilowatts) at optimum performance:

For pumps: $$N_s = NQ^{0.5}/H^{0.75} \quad \text{(S-128)}$$

For turbines: $$N_s = NP^{0.5}/H^{1.25} \quad \text{(S-129)}$$

To obtain dimensionless specific speeds ($K_N$), divide $N_s$ by an appropriate power of the gravitational acceleration ($g$) and the density ($\rho$) of the fluid:

For pumps: $$K_N = N_s/g^{0.75} \quad \text{(S-130)}$$

For turbines: $$K_N = N_s/\rho^{0.5}g^{1.25} \quad \text{(S-1310)}$$

*See also* characteristic speed, characteristic type, pump characteristic curves.

**specific substrate utilization rate** The ratio of the rate of substrate utilization to the biomass concentration in suspended growth wastewater treatment processes:

$$U = r_{su}/X = (S_0 - S_e)/TX \quad \text{(S-132)}$$

where $U$ = specific substrate utilization rate, g BOD or COD/g VSS/day; $r_{su}$ = substrate utilization, grams of biodegradable soluble COD per cubic meter per day; $X$ = biomass concentration, grams per cubic meter; VSS = volatile suspended solids; $S_0$ = influent soluble substrate concentration, g BOD or bsCOD/m$^3$; bsCOD = biodegradable soluble COD; $S_e$ = effluent soluble substrate concentration, g BOD or bsCOD/m$^3$; $T$ = hydraulic detention time = basin volume/flowrate, days; and $X$ = biomass concentration, mg/L or g/m$^3$. *See also* specific biomass growth rate.

**specific surface** The surface area per unit volume of trickling filter media; an important characteristic. It varies from 46 m$^2$/m$^3$ for slag to an average of 100 m$^2$/m$^3$ for plastic sheet. *See also* void space.

**specific surface area** Same as specific surface.

**specific thermal conductivity** A measure of the ability of a substance or material to conduct heat; the rate of heat transfer by conduction through a material or substance of unit thickness per unit area and per unit temperature gradient. *See* thermal conductivity for detail.

**specific throughput** The volume of water or wastewater treated per unit weight of adsorbent in the adsorption process, expressed in cubic meters per gram; the reciprocal of the carbon usage rate. *See also* empty-bed contact time, bed life.

**specific ultraviolet absorbance** *See* specific UV absorbance.

**specific ultraviolet adsorption (SUVA)** *See* specific UV adsorption.

**specific UV absorbance (SUVA)** The ratio of ultraviolet light absorbance to the dissolved organic carbon concentration (in mg/L of C). The absorbance is expressed in the reciprocal of light path length in meters at a wavelength of 254 nanometers, and SUVA in liters per meter per milligram (L/m/mg). This parameter is used to measure how well coagulants remove total organic carbon and precursors of disinfection by-products from drinking water or the potential for trihalomethane formation.

**specific volume** In the adsorption process, the volume of liquid that can be treated with a given amount of activated adsorbent, expressed as volume per unit mass ($V/M$); the reciprocal of the adsorbent dose:

$$V/M = q_e/(C_0 - C_e) \quad \text{(S-133)}$$

where $M$ is the mass of adsorbent (grams), $V$ the volume of liquid in the reactor (liters), $C_0$ the initial concentration of adsorbate (mg/L), $C_e$ the final equilibrium concentration of adsorbate after adsorption (mg/L), and $q_e$ the adsorbent phase concentration after equilibrium (mg adsorbate per gram adsorbent).

**specific weight or unit weight ($\gamma$)** The weight of a unit volume of fluid, equal to the product of the fluid density ($\rho$) by the gravitational acceleration ($g$):

$$\gamma = \rho g \qquad \text{(S-134)}$$

For water at 10°C or 50°F, $\gamma$ = 62.4 pounds/ft$^3$ or 9.81 kN/m$^3$. *See also* specific gravity.

**specific well yield** *See* specific yield.

**specific yield** The amount of water (as a ratio or percentage by volume) that a unit volume of saturated permeable rock will yield when drained by gravity. Specific yield applies to an unconfined aquifer; it is similar to, but less than, porosity. Specific yield is also called effective porosity or useful storage. The corresponding term for an artesian aquifer is storage coefficient. Also called effective porosity. The specific yield varies from near 0 for some clays to more than 30% for some sands and gravels, with an average around 15%. For a well, the specific well yield is the maximum flow rate at which the well can yield water under specified conditions of drawdown or pump size. *See also* specific retention.

**spectral absorption coefficient (SAC)** The ratio of ultraviolet light absorbance of a water or wastewater sample to the light path length. *See* specific ultraviolet absorbance.

**spectral UV absorbance (UVscan)** The ultraviolet absorbance of a water sample measured by a spectrophotometer at a wavelength between 190 and 400 nanometers.

**spectrometer** An optical device for measuring wavelengths and other parameters, e.g., a chemical analysis instrument used in spectroscopy.

**spectrometry** The use of a spectrometer in chemical analysis; spectroscopy.

**spectrophotometer** An accurate laboratory instrument that uses a prism or grating system to develop a monochromatic light beam for colorimetric analyses.

**spectrophotometric method** A method in which a spectrophotometer is used in laboratory analyses for such parameters as color, iron, and manganese.

**spectrophotometry** The use of a spectrophotometer for laboratory analyses. *See* spectroscopy.

**spectroscope** An optical instrument for the production and observation of a spectrum of light or radiation.

**spectroscopy** The science that deals with spectrum analysis and the use of the spectroscope. Chemical analyses via spectroscopy are based on the principle that some substances allow a unique fraction of radiation or light to pass when returning from an atomic vapor state to their fundamental state. The quantity of the light absorbed or emitted is proportional to the concentration of the substance.

**Spectrum™** Aeration equipment manufactured by Environmental Dynamics, Inc.

**Speece cone** A cone-shaped device that provides prolonged oxygen bubble contact time and high rates of oxygen transfer to aerate water or wastewater. *See* downflow bubble contactor for detail.

**spent filter backwash water** The water treatment residuals produced from filter backwashing; it is characterized by high instantaneous flow rates and low solids concentrations.

**spent regenerant** Waste materials from the regeneration of exhausted ion exchange beds or catalyst media.

**spent water** The portion of the community water supply that is discharged in sewers or individual disposal units; it is about 60 to 70% of water use in residential areas. With infiltration and inflow, it constitutes wastewater. It is the difference between the water supplied and various other uses such as fire fighting and lawn watering. *See also* return flow (2).

***Sphaerotilus*** A genus of filamentous bacteria associated with activated sludge bulking and often found around contact aerators; also called sewage fungus. They also use iron compounds to obtain energy for their metabolic needs.

***Sphaerotilus* bulking** Sludge bulking caused by an excessive number of organisms of the genus *Sphaerotilus*.

***Sphaerotilus natans*** A species of filamentous organism found in activated sludge under conditions of low dissolved oxygen concentrations. *S. natans* is also found in the slime layer of trickling filters.

**Spher-Flo** A single-stage pump manufactured by Aurora Pumps, Inc.

**sphericity** (1) A factor ($\psi$) that reflects how closely a grain of sand or other particle approximates a true sphere; defined as the ratio of the surface area of the equivalent volume sphere to the actual surface area of the particle. The maximum value is $\psi$ = 1.0, for a spherical particle. For an angular particle, $\psi$ = 0.78. Sphericity is related to the shape factor S and to the settling velocity and diameter of the particle:

$$S \cdot \psi = 6/D \qquad \text{(S-135)}$$

$$\psi = D/d = (v/V)^{0.5} \qquad \text{(S-136)}$$

where $D$ is the diameter of the equivalent volume sphere, $V$ is its settling velocity, and $d$ and $v$ are the corresponding characteristics of the granular material under consideration. Sphericity and shape factor are used, sometimes interchangeably, in the theory of water filtration on granular materials. *See also* equivalent spherical diameter, Kozeny equation, Ergun equation. (2) A measure of bead roundness or count of whole beads in ion-exchange resins or filter media.

**spheroplast** A Gram-negative bacterial cell treated with a destructive enzyme and still retaining much of its outer membrane.

**spigot** (1) The end of a pipe, fitting, or valve that enters the enlarged end of another to form a joint. *See* bell and spigot. (2) A faucet or cock that serves to control the flow of liquid from a pipe.

**spigot end** The outside threaded end of a pipe section that connects with the bell end of another section. Also called outside-threaded connection or male end.

**spiked sample** A sample of water or wastewater containing a known amount of an added substance.

**spill** (1) An accidental or intentional release of a substance into the environment; a spilling, as of a liquid, e.g., oil spill. (2) Same as spillway, i.e., a passageway for the release of water from a reservoir, lake, etc.

**Spill Prevention Control and Countermeasures Plan (SPCP)** Plan covering the release of hazardous substances as defined in the Clean Water Act.

**spinel** (1) Any of a group of minerals having the general formula $XY_2O_4$, where X is iron (Fe), magnesium (Mg), manganese (Mn), nickel (Ni), or zinc (Zn) and Y is aluminum (Al), chromium (Cr), or Fe; e.g., chromite ($FeCr_2O_4$) and magnetite ($Fe_3O_4$). (2) A mineral of this group, essentially magnesium aluminate ($MgAl_2O_4$).

**Spiracone®** An upflow clarifier, with a conical tank, manufactured by General Filter Co.

**Spiraflo** A clarifier manufactured by Lakeside Equipment Corp., including peripheral feed.

**Spirafloc** A clarifier manufactured by Lakeside Equipment Corp., including peripheral feed and a flocculation zone.

**Spiragester** An apparatus manufactured by Lakeside Equipment Corp., combining clarification and sludge digestion.

**Spiragrit** A grit removal apparatus manufactured by Lakeside Equipment Corp.

**spiral air-flow diffusion** An aeration method used in the activated sludge process using baffles and diffusers to impart a helical movement to the contents of the aeration basin. Also called spiral-flow aeration.

**spiral-flow aeration** Same as spiral air-flow diffusion.

**spiral-flow tank** A circular sedimentation tank or other type of tank that combines the advantages of horizontal-flow and upward-flow tanks: water flows from one end at an angle of 45° and makes several circuits of the circumference before exiting over a submerged weir. (2) An aeration basin equipped with spiral air-flow diffusion.

**Spiralift™** A screw pump manufactured by Zimpro Environmental, Inc.

**Spiralklean** A device manufactured by Parkson Corp. for washing screenings.

**Spiral Scoop** A skimming apparatus manufactured by Krofta Engineering Corp. for use in dissolved-air flotation units.

**spiral-wound module** A membrane configuration used in reverse osmosis units and in cartridge filter units. For reverse osmosis, it consists of a number of sheets having two layers of membranes separated by a perforated backing or coarse mesh screen, the sheets wound around a product water tube. The pressurized influent flows along the backing or screen. In cartridge filters, the filtration medium is wound around the backing or screen. Other types of reverse osmosis modules include hollow-fiber, plate-and-frame, and tubular.

**Spiraltek™** Filters manufactured by Osmonics, Inc. for dry sumps.

**Spira-Pac** A compact apparatus manufactured by Lakeside Equipment Corp., combining clarification and digestion.

**Spiratex™** A point-of-use treatment unit of Osmonics, Inc.

**Spirathickener** A gravity thickener manufactured by Lakeside Equipment Corp., including peripheral feed.

**Spiratrex** A membrane manufactured by Osmonics, Inc. for use in ultrafiltration.

**Spira-twin Spiragester** An apparatus manufactured by Lakeside Equipment Corp., combining clarification and digestion for primary and trickling filter sludges.

**Spiravac®** A peripheral feed clarifier manufactured by Lakeside Equipment Corp.

***Spirillum*** A genus of heterotrophic, rigid, spirally twisted, Gram-negative bacteria active in biological denitrification and occurring in stagnant freshwaters. Some species are pathogenic to humans. Some other similar organisms are also called spirillum.

***Spirillum NOX*** A bacterial strain used in bioassays to estimate oxalate-carbon equivalents.

***Spirillum sp.*** Any species of the *Spirillum* genus of bacteria.

**spirochaetal disease** Same as spirochetosis.

**spirochaetosis** Same as spirochetosis.

**spirochaetes** Same as spirochetes.

**spirochetal disease** Same as spirochetosis.

**spirochetes** A group of slender, spiral-shaped bacteria, such as *Borrelia, Leptospira,* and *Treponema,* which are pathogenic to humans and carried by dogs and rats. *See* spirochetosis.

**spirochetosis** An excreta-related disease caused by a spirochete, such as Weil's disease (or leptospirosis) and relapsing fever.

***Spirodela spp.*** A group of small, green floating aquatic plants (among the duckweeds) with short roots, sometimes used in natural wastewater treatment systems.

**Spirolift®** A screw vertical conveyor manufactured by JDV Equipment Corp.

***Spirometra*** A genus of tapeworms that use humans and other intermediate hosts, but infect mostly nonhuman mammals.

**Spiropac®** A screw compactor manufactured by JDV Equipment Corp.

**Spiropress®** A screw dewatering press manufactured by JDV Equipment Corp.

**Spirosand®** A shaftless grit classifier manufactured by Andritz-Ruthner, Inc.

**Spirovortex** A wastewater treatment apparatus manufactured by Dorr-Oliver, Inc. using activated sludge and tertiary filtration.

***Spirulina*** A genus of corkscrew-shaped cyanobacteria sometimes added to human food for its nutrient value.

**spit** (1) A light precipitation (rainfall or snowfall). (2) A narrow point of land projecting into a body of water or a long narrow shoal extending from the shores.

**split addition** The application of water treatment chemicals at several points along a flow path.

**split-case pump** A centrifugal pump, split horizontally along the shaft, with parallel ports for suction and discharge on opposite sides.

**Split-ClarAtor™** Equipment manufactured by Aero-Mod, Inc. for secondary clarifiers.

**split-flow pressure flotation** One of three basic modes of operation of pressure (dissolved-air) flotation. *See also* recycle-flow pressure flotation and full-flow pressure flotation.

**split neutral salts** To separate neutral salts into their constituents during strong acid cation exchange. *See* neutral salt splitting.

**split recarbonation** A treatment process used to remove magnesium from waters with high noncarbonated hardness. Carbon dioxide is added in two stages; first to produce carbonate from excess lime and then for final pH adjustment.

**split sample** A water or wastewater sample divided into two or more, equally representative, subsamples.

**split softening** A water treatment process that uses lime to precipitate hardness. Lime and other chemicals are added to the raw water, which then passes through two clarifiers in series and a filter. No recarbonation is necessary as a portion of the raw water bypasses chemical addition and the first clarifier to provide adequate pH and alkalinity control. *See also* lime softening, two-stage lime softening, and single-stage softening.

**split-stream treatment** Same as split treatment.

**splitter box** (1) A device, equipped with weirs or control valves, that divides incoming flow into two or more streams, e.g., to distribute flow from primary clarifiers equally among aeration tanks. Also called division box, splitting box. (2) A device used in wastewater sludge handling to minimize fluctuations in digester feed rates.

**splitting box** Same as splitter box.

**splitting of salt** *See* salt splitting, basic salt splitting, neutral salt splitting.

**split treatment** Also called split-stream treatment. (1) A variation of the excess-lime treatment method of water softening where chemicals are applied only to a portion of the influent; both portions are mixed in a secondary basin for recarbonation and hardness precipitation. *See* Figure S-20. Split treatment may be used when there is a limit on the final concentration of one of the hardness constituents; the fraction to be treated ($F$) is determined as:

$$F = (E - I)/(P - I) \qquad \text{(S-137)}$$

where $E$ = effluent concentration, meq/L; $I$ = influent concentration, meq/L; and $P$ = concentration of the split fraction, meq/L. (2) A similar variation of the ion-exchange softening process in which one train provides near complete hardness removal and its product is blended with a bypass stream. (3) Any other process treating two or more streams separately and then combining them, e.g., (a) parallel softening and coagulation, and (b) blended lime-softened stream with another stream.

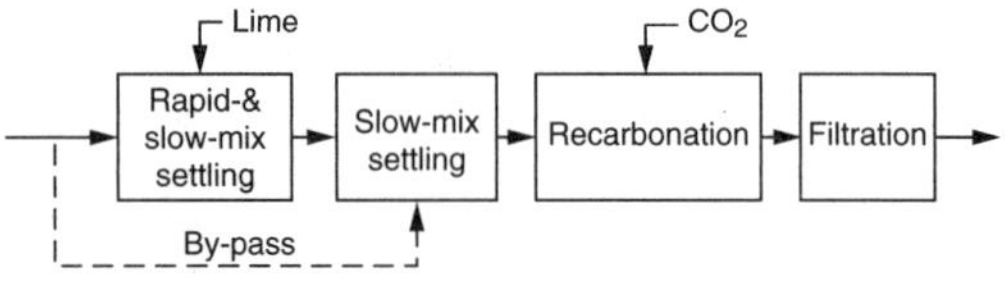


**Figure S-20.** Split treatment (excess lime).

**split treatment softening** *See* split treatment.

**spodumene** An important lithium ore.

**spoil** Material such as soil excavated from the trench of a water main.

**sponge structure** A typical structural profile of membranes used in water filtration showing pores arranged as a dense sponge. *See also* finger structure.

***Sporatichum*** A genus of fungi identified in trickling filters, active in waste stabilization but susceptible of causing problems of clogging and restriction of ventilation.

**spore** The reproductive body of an organism that is capable of giving rise to a new organism either directly or indirectly; a viable body regarded as the resting stage of an organism. A spore is usually more resistant to disinfectants, heat, and desiccation than most organisms. It is a nonvegetative form that reverts to a vegetative state under favorable conditions, a common characteristic of the *Bacillus* and *Clostridium* genera of bacteria. *See also* cyst, egg, gamete, oocyst, sporocyst, sporozoite.

**sporeformer** A microorganism that produces spores.

**spore-forming bacteria** Bacteria that form spores in response to adverse conditions, e.g., those in the genera *Bacillus* and *Clostridium.* Bacterial spores are asexual, formed within the vegetative cell.

**spore-forming protozoa** *See* Sporozoa.

**spore removal surrogate technique** The use of *Bacillus* spores as surrogates to evaluate the effectiveness of the coagulation–flocculation–filtration process in removing cysts and oocysts from drinking water.

**sporocyst** (1) A cyst resulting from the division of a sporozoan; a sporozoite. (2) A stage in the development of trematodes producing cercaria asexually.

**sporogony** A stage in the sexual reproduction of a malarial parasite in the mosquito.

**Sporozoa** A class of parasitic protozoa, all pathogenic to humans. Under adverse conditions they take a nonvegetative form similar to a bacterial spore, which reverts to a vegetative form when favorable conditions return. An example is the *Plasmodium* genus that causes malaria. *Toxoplasma* and *Isospora* are other examples of spore-forming protozoa.

**sporozoan** Any parasitic spore-forming protozoan of the class Sporozoa.

**sporozoite** A cell that results from the sexual union of spores; more specifically, an elongated, nucleated cell from the fission of zygotes of female oocysts. *See also* trophozoite, merozoite, schizont.

**spot sample** A single air, water or wastewater sample, representative of the composition of the flow at a particular time and place. *See* grab sample for detail.

**spouting velocity** The theoretical velocity ($V$) at which water flows from an orifice under a head ($H$), neglecting the effect of friction:

$$V = (2\,gH)^{0.5} \qquad \text{(S-138)}$$

where $g$ is the gravitational acceleration.

**spp.** Abbreviation used to designate several species of a genus; a single, unidentified species may be designated by the abbreviation sp.

**Sprague–Dawley rat** A laboratory rat used in toxicological studies.

**spray** A liquid droplet greater than 10 micrometers formed by mechanical action.

**spray aeration** The process of using spray aerators in water or wastewater treatment.

**spray aerator** A fixed or movable device that sprays droplets of water into the air from stationary orifices or nozzles or wastewater droplets from moving orifices or nozzles. Spray aerators are used, for example, to remove iron and manganese from groundwater, and for the aeration of wastewater in trickling filters. Also called nozzle aerator, pressure aerator. *See also* spray tower, fountain spray aerator, full-cone nozzle, hollow-cone nozzle.

**spray-applied cement mortar lining** A trenchless method of water main rehabilitation, using a spray nozzle to apply a coating of cement mortar to the walls of a damaged pipe. Water passing through the mortar becomes alkaline and produces a chemical inhibitor against oxidation. The mortar cures in 12 to 24 hours. *See also* trenchless technology, epoxy lining.

**spray-applied lining** A trenchless method of water main rehabilitation, using a rotating spray nozzle to apply a coating of liner material on the pipe walls. *See* trenchless technology, epoxy lining, spray-applied cement mortar lining.

**spray dryer** A device, such as an evaporator or a high-speed centrifugal bowl, used to concentrate and dry solids from sludge, salt water, or other suspensions. The feed stream is atomized into fine particles and sprayed into a drying chamber for transfer of moisture to hot gases.

**spray electrification** *See* Lenard effect.

**Spray-Film®** A distillation apparatus manufactured by Aqua-Chem, Inc., using vapor compression.

**spray head** A sprinkler irrigation nozzle on a riser. *See* spray nozzle.

**spray irrigation** A land treatment and disposal method for spreading treated wastewaters on agricultural land by spraying them through fixed or moving nozzles. *See* slow-rate land disposal for more detail.

**Spraymaster®** A compact deaerator manufactured by Cleaver-Brooks.

**spray method** The use of tree spraying equipment to apply copper sulfate to reservoirs. The equipment is calibrated to spray enough chemical over the water surface from a boat provide a dose of 0.3 mg/L. The application of copper sulfate controls the growth of plankton (algae) in reservoirs. *See also* blower, burlap bag, and continuous-dose methods.

**spray nozzle** (1) A sprinkler irrigation nozzle placed on a riser to apply a medium-fine spray at a relatively high rate. (2) A nozzle used in wastewater treatment units to eliminate foam or direct floatables to a convenient collection point.

**spray pond** A basin over which water is sprayed to reduce its temperature.

**spray tower** (1) A tower around a spray aerator to minimize wind action, prevent freezing, or hold the pipe grid that supports the nozzles. (2) An air pollution control device that uses water or another liquid to remove less than 50% of particulates, fumes, dusts, mists, and vapors from gas streams; for example, in sludge incinerators. It consists of a column, at the top of which the liquid is sprayed against a rising gas stream. *See also* cyclone scrubber, ejector-venturi scrubber, and venturi scrubber.

**spray tower scrubber** A device that sprays alkaline water into a chamber where acid gases are present to aid in the neutralizing of the gas.

**spreader** A structure with a level upper edge, such as a plate or a wall, installed in a channel, a ditch, or in a river pool to spread the flow evenly. Also called sill.

**spreading** The dispersal of wastes on the ground to facilitate infiltration, a method of wastewater distribution used in rapid infiltration systems. *See* spreading basin.

**spreading basin** A basin constructed to receive diverted stormwater, reclaimed water, or other surface water, which is allowed to percolate to the zone of saturation. Some decomposition of easily biodegradable matter may occur in the first 2 ft of infiltration. Also called an infiltration basin. Other means of groundwater recharge include check dams, recharge basins, recharge wells, seepage ponds, and underground leaching systems. *See also* water reuse applications.

**spread plate count method** A method of enumerating individual bacteria in wastewater. *See* spread plate method for detail.

**spread plate method** One of three laboratory procedures used in the heterotrophic plate count method of bacterial enumeration in a water or other sample. It involves the serial dilution of a sample, mixing each dilution with a warm culture medium, placing and spreading the sample on the surface of a culture dish, and counting the bacterial colonies formed. It indicates the number of colonies of heterotrophic (pathogenic and innocuous) bacteria grown on selected solid media at a given temperature and incubation period, e.g., tryptone glucose extract agar or tryptone glucose yeast for 48 hours at 35°C. The other two procedures are membrane filter and pour plate. *See also* direct count, pour plate count, membrane-filter technique, multiple-tube fermentation, presence–absence test, heterotrophic plate count.

**spring** Groundwater seeping out of the earth where the water table intersects the ground surface. Natural spring water is usually free of dissolved chemicals.

**spring box** A small structure, generally constructed of concrete or masonry to collect the water from a spring, protect it from contamination, and allow settling of suspended sediment so as to prevent blockage of the eye of the spring. It includes a screened outflow pipe, an overflow pipe below the eye, and a removable cover to allow occasional cleaning. *See also* silt trap, spring capping.

**spring cap** *See* spring capping.

**spring capping** A simple installation used to develop a spring into a drinking water supply. It includes a spring box and other protective measures such as a cover of selected sand or gravel over the eye of the spring, an impermeable layer (e.g., puddled clay) over the water-bearing material, a drainage ditch to divert surface water, and a protective fence or hedge. *See* Figure S-21.

**spring circulation** Same as spring turnover.

**spring line** Theoretical center of a pipeline. Also, the guideline for laying a course of bricks

**spring melt** The process by which warm temperatures melt winter snow and ice. Because various forms of acid deposition may have been stored in the frozen water, the melt can result in abnormally large amounts of acidity entering streams and rivers, sometimes causing fishkills. Also called spring thaw.

**spring overturn** Same as spring turnover.

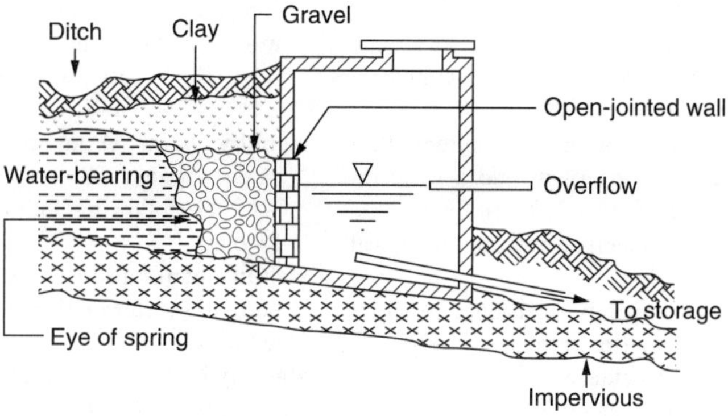


**Figure S-21.** Spring capping (adapted from Cairncross & Feachem, 1992).

**spring overturning** Same as spring turnover.

**spring thaw** Same as spring melt.

**spring turnover** The phenomenon occurring in lakes in the spring when temperatures rise above the freezing point, surface ice melts into higher-density water and tends to sink, causing the turnover. It lasts a few weeks, with circulation more pronounced when temperature is uniform at all depths and approximates 39.2°F or 4°C (at maximum water density). Also called spring overturning, spring overturn, or spring circulation. *See also* thermal stratification.

**spring water** Water, bottled or not, that comes from an underground formation, collected as it flows naturally to the surface or through a borehole that taps water that would otherwise flow to the surface.

**sprinkle** A light rain of scattered drops.

**sprinkler** (1) Any device for sprinkling water, e.g., a watering pot or a small stand with a revolving nozzle in irrigation. (2) A device for sprinkling or spraying wastewater on a filter bed.

**sprinkler distribution** A common method of wastewater distribution used in slow-rate treatment systems. *See also* sprinkler distribution, center-pivot system.

**sprinkler infiltrometer** A sprinkling device used to measure percolation rates for the design of slow-rate infiltration systems.

**sprinkler irrigation** Irrigation using a pipe with sprinklers, fed by gravity or by pumping.

**sprinkler nozzle** A nozzle used to spray wastewater to a trickling filter.

**sprinkler system** (1) A network of pipes with sprinkler heads and heat-sensitive devices used for automatic fire extinguishing in a building when the temperature reaches a certain level. (2) An irrigation system consisting of pipes with sprinklers or spray nozzles for water distribution or wastewater application.

**sprinkling filter** A trickling filter with sprinkler nozzles for the application of wastewater.

**sprinkling infiltrometer** A device used in a common test to measure hydraulic conductivity and the limiting application rate of wastewater to land through sprinklers. It consists of a revolving sprinkler and a shield that has an opening through which water is sprayed. *See also* flooding basin technique, cylinder infiltrometer.

**Sprint™** A submersible pump manufactured by Crane Pumps & Systems.

**spülkraft** The German term for dosing rate of a trickling filter. *See* SK concept.

**square brackets [...]** A notation used to designate the concentration of a species or substance.

**square horizontal-flow grit chamber** Same as a detritus tank.

**square pitch** The pitch of an impeller is its advance per revolution, assimilating the device to a screw; it is a square pitch when the pitch equals the diameter of the device.

**square-tank degritter** Same as detritus tank.

**Squarex** A circular mechanism manufactured by Dorr-Oliver, Inc. for sludge collection in square basins.

**squat hole** A hole in the floor (squatting plate) of a latrine through which excreta fall.

**squeegee** An implement with a soft rubber edge used to remove water from windows after window washing, to sweep water from surfaces, to dislodge wastewater solids or floating grease and scum from walls and bottoms of sedimentation tanks, etc.

**squirrel cage induction motor** A common type of direct current motor manufactured for pump drives used in wastewater treatment. It has a low initial cost and is easy to maintain.

**SR** Acronym of specific resistance.

**Sr** Chemical symbol of strontium.

**SR-7™** An ion-exchange resin manufactured by Sybron Chemicals, Inc.

**Sr-90** Symbol or abbreviation of strontium-90, a radioactive isotope of strontium and a toxic radionuclide; it has a half-life of 28 years.

**SRBC** Acronym of submerged rotating biological contactor.

**SRF** Acronym of (1) specific resistance to filtration, (b) state revolving (loan) fund.

**SR test** *See* specific resistance test.

**SRT** Acronym of solids retention time.

**ss** Acronym of stainless steel.

**SS** Acronym of suspended solids.

***S. schottmuelleri*** *See Salmonella schottmuelleri.*

***S. sendai*** *See Salmonella sendai.*

**SSO** Acronym of sanitary sewer overflow.

***S. sonnei*** *See Schigella sonnei.*

***S. stercoralis*** *See Strongyloides stercoralis.*

**Stabilaire** A package wastewater treatment plant manufactured by the FMC Corp., using the contact stabilization process.

**stability** (1) Behavior of a stable numerical solution, i.e., the characteristic of a numerical procedure in which errors decrease with succeeding steps, or in which round-off errors of the algorithm are negligible. *See* convergence. (2) Characteristic of a watercourse or channel that has not suffered, or is protected from, erosion or deposition. *See also* channel stability, regime theory, extremal hypothesis (or variational principle), and tractive force theory. A stable channel is a channel whose cross section, slope, and alignment do not vary significantly, and in which sediment deposition and scour are in equilibrium. (3) The ability of water, wastewater, sludge, or any substance to resist change. *See also* putrescibility. (4) The condition of a filter medium or ion-exchange resin that resists chemical and physical degradation. (5) The condition of a stratified water body that resists overturning.

**stability constants** The overall formation constants for the reactions of the formation of complexes. For example, for the $Cu^{2+}$ reactions with ammonia ($NH_3$):

$$Cu^{2+} + NH_3 \leftrightarrow CuNH_3^{2+} \quad \text{(S-139)}$$

$$CuNH_3^{2+} + NH_3 \leftrightarrow Cu(NH_3)_2^{2+} \quad \text{(S-140)}$$

$$Cu(NH_3)_2^{2+} + NH_3 \leftrightarrow Cu(NH_3)_3^{2+} \quad \text{(S-141)}$$

$$Cu(NH_3)_3^{2+} + NH_3 \leftrightarrow Cu(NH_3)_4^{2+} \quad \text{(S-142)}$$

the stability constants are

$$\beta_i = [Cu(NH_3)_i^{2+}]/[Cu(NH_3)_{i-1}^{2+}][NH_3] \quad \text{(S-143)}$$

compared to the equilibrium constants (or stepwise formation constants) for the same reactions:

$$K_i = [Cu(NH_3)_i^{2+}]/[Cu^{2+}][NH_3]^i \quad \text{(S-144)}$$

**stability index** A number that indicates to what extent a water will cause scales or corrosion. *See also* the following indexes: Langelier, Riddick, Ryznar.

**stability of flotation** The resistance of a floating body to overturning under the influence of external forces.

**stability ratio** Same as colloid stability ratio.

**stabilization** (1) Adjustment of the ionic balance of water to prevent corrosion or the deposition of calcium carbonate and other scales. [Note, however, that a thin deposit of calcium carbonate ($CaCO_3$) in distribution pipes may help to retard corrosion]. Water may be corrosive naturally or as a result of treatment. Stabilization may be achieved through the adjustment of pH (e.g., recarbonation), alkalinity, or calcium content, or by the addition of phosphate compounds (*see* sodium hexametaphosphate). *See also* Langelier index, marble test. (2) Conversion of a waste to a physically and chemically more stable form, e.g., conversion of the active organic matter in sludge into inert, harmless material. Lime, with or without ferric chloride, is often used in sludge conditioning prior to further processing. *See also* aerobic digestion, anaerobic digestion, chemical stabilization. *See also* solidification/stabilization. (3) Maintenance of a relatively nonfluctuating condition in a hydraulic structure; e.g., equalization of effluent flow prior to discharge or equalization of wastewater flow prior to treatment. (4) Protection of dikes and shorelines against erosion with riprap, sod, etc.

**stabilization basin** A natural or artificial structure used for the biological treatment of wastewater. *See* stabilization pond.

**stabilization lagoon** A relatively shallow structure used in the treatment, storage, or equalization of wastewater. *See* stabilization pond. *See also* lagoons and ponds.

**stabilization of water** *See* stabilization (1).

**stabilization pond** A large, relatively shallow, earthen basin where sunlight, microbial action (by algae and bacteria), and oxygen work to purify wastewater or sludge. The growth of algae (e.g., *Chlorella* and other green or blue-green algae) and photosynthesis are predominant processes in stabilization ponds, which also may involve digestion, evaporation, gas exchange, oxidation, sedimentation, and seepage. Stabilization ponds, which are simple, inexpensive and easy-to-operate structures, use suspended growth, attached growth, or a combination of these processes. However, the performance of the pond is often affected by an excess of algal cells in the effluent, resulting in high concentrations of suspended solids and BOD. In addition to decomposition of organic matter, the long detention times in stabilization ponds provide an opportunity for pathogen destruction, e.g., an effluent with 100 fecal coliforms/100 mL, well beyond the performance of more conventional treatment methods. Also called oxidation ponds or sewage lagoons, they are used mostly in small communities (because of low costs) and for the treatment of industrial or mixed domestic–industrial wastes. *See also* aquaculture, *Culex tarsalis,* lagoons and ponds, and the following types of ponds: aerobic, aerobic–anaerobic, anaerobic, anaerobic lagoon, anaerobic pretreatment, facultative, high-rate, low-rate, maturation, pond system, tertiary.

**stabilization pond detention time** *See* the Hermann–Gloyna formula.

**stabilization pond microorganisms** In addition to bacteria, microorganisms active in stabilization ponds include diatoms, green and blue-green algae, mostly chlorella, but also the *Anaboena, Chlamydomonas, Euglena, Oscillatoria* and *Scedenesmus* genera.

**stabilization tank** A water treatment basin used for pH adjustment.

**stabilized channel** An earth channel that does not suffer appreciable erosion or deposition over time.

**stable** *See* stability.

**stable channel** *See* stability.

**stable effluent** Treated wastewater that can resist change, e.g., an effluent with sufficient dissolved oxygen to satisfy its BOD.

**stable water** Water that does not dissolve or deposit calcium carbonate ($CaCO_3$). *See also* aggressive water, chemical stabilization, saturation condition, saturation index.

**stack** (1) A chimney, a smokestack, or vertical pipe that discharges used air or gaseous emissions. (2) A vertical line that receives wastewater flow from horizontal branches or drains in a plumbing system. *See also* vent stack. (3) A basic electrodialysis unit consisting of membrane cells and the accessories necessary for operation.

**stacked clarifiers** Two or more primary or secondary settling tanks, one on top of the other, fed independently but operating on a common water surface and in series or in parallel. This arrangement saves space and facilitates odor control. Also called multilevel clarifiers, or tray clarifiers. *See also* multiple-tray clarifier.

**stacked-pan aerator** A device used to transfer oxygen to water or wastewater, consisting of stacks of perforated pans or troughs that are sometimes filled with coke, stone, or other media. The liquid flows by gravity from an inlet pipe through the pans or troughs, over the media, and into a collecting pan ending in an outlet pipe.

**stacked sedimentation tanks** *See* stacked clarifiers.

**stack gas** *See* flue gas.

**stage** (1) In membrane filtration processes, a stage or bank is a series of pressure vessels arranged in parallel. Interconnected stages constitute an array or train. (2) Any of several units that operate at successively lower pressure in a flash evaporator.

**Stage 1 D/DBP rule** A USEPA rule issued in 1998 for all public water supply systems to reduce the risks to consumers from disinfectants and disinfection by-products. Specifically, it establishes a maximum residual disinfectant level goal and maximum residual disinfectant level of 4.0 mg/L for residual free chlorine in drinking water.

**Stage 2D/DBPR** Abbreviation of Stage 2 Disinfectants/Disinfection By-Products Rule.

**Stage 2 Disinfectants/Disinfection By-Products Rule** or **Stage 2D/DBPR** A draft proposal of the USEPA to establish a maximum contaminant level goal (MCLG) of 0.070 mg/L of chloroform in drinking water. It also requires community water systems to conduct an initial year-long identification of monitoring sites and eventually to comply with standards of 80 micrograms per liter of total trihalomethanes (TTHM) and 60 micrograms per liter of five haloacetic acids (HAA5).

**stage aeration** The activated sludge process with two or more aeration basins in series and sludge recycle at each stage.

**staged activated sludge process** A variation of the conventional plug-flow activated-sludge system with the tank divided into a number of complete-mix zones in series by the installation of baffle walls.

**stage digestion** The biochemical decomposition of wastewater sludge in two or more tanks in series, with each tank operated under different mixing conditions. Also called multiple-stage sludge digestion. *See also* stage treatment (4).

**staged mesophilic digestion** One of a number of variations of the two-phased anaerobic digestion process, designed to optimize conditions for the separate groups of microorganisms involved by using two reactors in series. Acidogenesis and methanogenesis take place in separate reactors: the first reactor at a solids retention time (SRT) of 7–10 days and the second reactor at a variable SRT, both under mesophilic conditions. The process yields a more stable, digested sludge. *See also* acid/gas phased digestion, staged thermophilic digestion, temperature-phased digestion.

**staged thermophilic digestion** One of a number of variations of the two-phased anaerobic digestion process, designed to optimize conditions for the separate groups of microorganisms involved by using two reactors in series. Acidogenesis and methanogenesis take place in separate reactors: a first reactor with a solids retention time (SRT) of about 20 days followed by one or more reactors with an SRT of 2 days each, all under thermophilic conditions. This arrangement reduces short-circuiting and achieves a Class A sludge. *See also* acid/gas phased digestion, staged mesophilic digestion, temperature-phased digestion.

**stage treatment** (1) Water or wastewater treatment carried out in series or stages of similar unit processes. (2) A variation of the activated sludge process including two or more aeration basin–sedimentation tank combinations in series. (3) Wastewater treatment in two or more trickling filters in series, with or without intermediate clarification. (4) The anaerobic digestion of wastewater treatment sludge conducted in a complete mixing tank followed by a tank for the separation of supernatant liquor from the solids. *See also* stage digestion.

**stage trickling filter** *See* stage treatment (3).

**staggers** An acute livestock disease characterized by impaired vision, wandering in circles, staggering gait, respiratory failure, and death; caused by an excess of selenium (Se). Also called blind staggers. *See also* alkali disease.

**stagnant** Characteristic of fluids that are not flowing or running and that often become stale or foul.

**stagnant layer** The stagnant liquid layer that separates the biofilm from the bulk liquid in a biological attached growth treatment process. Through this layer, substrate, oxygen, and nutrients diffuse to the biofilm, while biodegradation products diffuse to the bulk liquid. Also called bound water layer, diffusion layer, stagnant liquid layer, or fixed water layer. *See* fixed-growth process.

**stagnant liquid layer** Same as stagnant layer.

**stagnation** Lack of motion in a mass of air or water. Stagnation may occur in finished water storage reservoirs or in distribution system piping. In wastewater facilities, stagnation holds pollutants in place.

**stagnation point** The upstream or upwind point of zero flow velocity on the surface of an object that divides a fluid flow.

**stagnation zone** The lowest layer in a thermally stratified body of water. It consists of colder, more dense water, has a constant temperature, and no mixing occurs; water is almost stagnant. The hypolimnion of a eutrophic lake is usually low or lacking in oxygen. *See* hypolimnion for more detail.

**Stahlermatic®** Equipment manufactured by IBERO Anlagentechnik, GmbH to provide aeration in rotating biological contactors.

**staining** (1) Discoloration with spots or streaks caused on laundry and household fixtures by water that contains iron, manganese, or copper in solution. Depending on the dissolved metal, the stain may be red, red-brown, dark brown, black, or blue. (2) *See* staining technique.

**staining technique** A method that uses a dye to identify cell components. *See* acid-fast stain, Gram stain.

**stainless** Same as stainless steel.

**stainless steel (ss)** A 12–25% chromium alloy of steel that is resistant to corrosion and rust. Also called stainless.

**stakeholder** Someone who has an interest in or responsibility for the outcome of a process. For water supply or wastewater management services, stakeholders include consumers or direct users, financing agencies, indirect users, owners, policy-making bodies, "pressure groups" (environmental, consumer protection, etc.), regulatory agencies, and utilities.

**STAKfilter** A pressure screen filter manufactured by Everfilt Corp.

**Stak-Tracker™** A continuous monitoring apparatus manufactured by Reuter-Stokes.

**stale wastewater** Wastewater that is not fresh, that contains little or no oxygen, but has not yet undergone putrefaction. *See also* septic wastewater.

**stale water** Stagnant water that may have a problem of taste and odor from standing in distribution or storage facilities.

**standard** (1) A recommended practice in manufacturing, engineering, science, laboratory analysis, etc. *See* standard method. (2) A physical or chemical quantity whose value is known exactly, and is used to calibrate or standardize instruments. *See also* reference. (3) Standards are norms that impose limits on the amount of pollutants or emissions produced. The USEPA establishes minimum standards, but states are allowed to be stricter. Along with characteristics and intended uses of receiving waters, standards determine the degree of wastewater treatment required. (4) Same as national pretreatment standards or, simply, pretreatment standards. *See also* aesthetic quality standards, drinking water standards.

**standard addition method** An atomic absorption technique that uses calibration curves to compensate for interferences in the analysis of certain constituents.

**standard alum** A chemical commonly used in water and wastewater treatment and produced by the direct reaction of sulfuric acid with an aluminum ore. It may contain higher concentrations of other metals (sodium and zinc) than two other water treatment chemicals (low-iron alum and polyaluminum chloride). *See also* aluminum-based coagulants.

**standard amino acid** *See* amino acid.

**standard atmosphere** (1) An arbitrary vertical distribution of atmospheric conditions (temperature, pressure, density). (2) A unit of atmospheric pressure = 1013.2 millibars = 29.9213 in or 760 mm of mercury.

**standard atmospheric pressure** *See* standard pressure.

**standard biochemical oxygen demand** The biochemical oxygen demand of a sample determined for 5 days at 20°C.

**standard boiling point** The temperature at which the pressure of the saturated vapor of a liquid is the same as the standard pressure. The measured boiling point is dependent on the atmospheric pressure (EPA-40CR796.1220).

**standard cell potential** The sum ($E^0_{net}$) of the half-cell potentials of a redox reaction, i.e., the potentials of the oxidation ($E^0_{ox}$) and reduction ($E^0_{red}$) half-reactions. It is related to the equilibrium constant (K) of the reaction:

$$(E^0_{net}) = (E^0_{ox}) + (E^0_{red}) = (RT/nF) \ln K \quad \text{(S-145)}$$

where $R$ is the universal gas constant, $T$ the absolute temperature, $n$ the number of moles, and $F$ the Faraday constant.

**standard color solution** A solution of chloroplatinate ($K_2PtCl_6$) tinted with a small amount of cobalt chloride ($CoCl_2$).

**standard color unit** The color produced by 1 mg/L of platinum combined with 0.5 mg/L of metallic cobalt. *See* color unit for detail.

**standard conditions** Standard temperature and pressure.

**standard cubic feet per minute** A unit of air flow at standard conditions of pressure (1 atm), temperature (0°C), and relative humidity (50%).

**standard electrode potential** The standard cell potential ($E^0$) for an electrode reaction that involves the oxidation of molecular hydrogen ($H$) to solvated protons, e.g.:

$$Zn^{2+}(aq) + H_2(g) \rightarrow 2\,H^+(aq) + Zn(s) \quad \text{(S-146)}$$

It is a measure of the tendency of a half-reaction to proceed as written. *See also* standard potential.

**standard electromotive force** *See* electromotive force.

**standard fall diameter** The diameter of a sphere of specific gravity 2.65 and having the standard fall velocity of a given particle.

**standard fall velocity** The average velocity of a particle falling alone in quiescent distilled water at a temperature of 24°C.

**standard filtration rate** A filtration rate equal to 2 gpm/ft$^2$ or 5 m/h recommended in 1898 for chemically pretreated surface waters. Also called traditional filtration rate.

**standard free-energy ($G^0$)** *See* Gibbs' free energy.

**standard free-energy change ($\Delta G^0$)** *See* Gibbs' free energy.

**standard half-cell potential** The standard potential of an oxidation or reduction half-reaction. *See* standard cell potential.

**standard hydrogen electrode** An electrode consisting of a platinum-black tube covered with hydrogen bubbles or through which hydrogen gas is bubbled continuously into a solution for a pressure of 1.0 atm and a unit activity of [H$^+$] ions. *See* Figure S-22. It is used as a standard reference, with zero potential to measure pH or the potential for oxidation–reduction reactions. The platinum serves as a catalyst for the reaction between the hydrogen gas [$H_2(g)$] and hydrogen ions ($H^+$):

$$H_2 \rightarrow 2\,H^+ + 2\,e^- \quad \text{(S-147)}$$

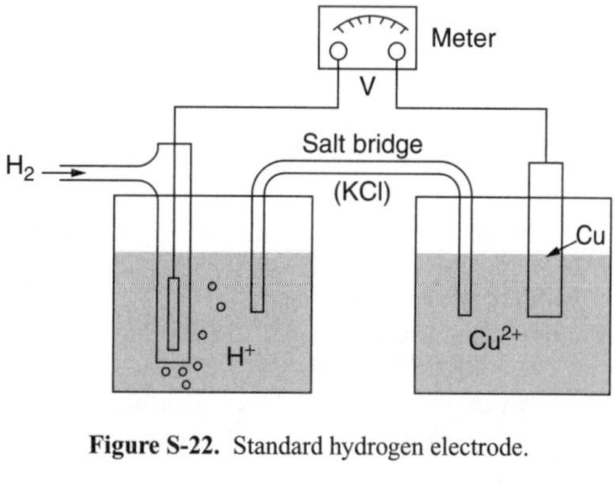


**Figure S-22.** Standard hydrogen electrode.

*See also* electrochemical cell, standard electrode potential.

**Standard Industrial Classification (SIC)** A system used by the U.S. government to classify business activities or industrial facilities for analytical and reporting purposes. *See* SIC codes, which are sometimes used in wastewater studies to estimate industrial wastewater volumes and characteristics.

**standardization** A procedure to produce a substance or a device to satisfy a standard; e.g., the preparation of standard solutions for laboratory analyses or the comparison of an instrument to a standard.

**standardize** To compare a solution or an instrument with a standard. (1) In wet chemistry, to find out the exact strength of a solution by comparing it to a standard of known strength. (2) To set up an instrument or device to read a standard. This allows one to adjust the instrument so that it reads accurately or enables one to apply a correction factor to the readings.

**standardized monitoring framework** A compliance-monitoring framework established by the USEPA. *See* Phase II Rule.

**standardized mortality ratio (SMR)** The ratio of observed deaths to expected deaths.

**standard jar test** A test of standard volumes of sludge samples, using the jar test technique, followed by rapid mixing, flocculation, and sedimentation to determine the required concentrations of conditioning chemicals. Often simply called jar test. *See also* Atterberg test, Büchner funnel test, capillary suction time (CST) test, filterability constant, filterability index, filter leaf test, shear strength, specific resistance test, standard shear test, time-to-filter (TTF) test.

**standard method** A technique generally accepted for the examination or treatment of water and wastewater.

**Standard Methods** *See* Standard Methods for the Examination of Water and Wastewater.

**Standard Methods for the Examination of Water and Wastewater** A joint publication of the American Public Health Association, American Water Works Association, and the Water Environment Federation that outlines the procedures used to analyze the impurities in water and wastewater. It is a standard reference for the characterization of water and wastewater, compiled in consultation with experts around the world.

**standard monitoring** Initial monitoring required by the USEPA under the Safe Drinking Water Act.

**standard of performance** A standard for the control of the discharge of pollutants that reflects the greatest degree of effluent reduction that the USEPA Administrator determines to be achievable through best available demonstrated control technology.

**standard or limitation** Any prohibition, effluent limitation, emission standard, or toxic, pretreatment, or new source performance standard established or publicly proposed pursuant to a federal legislation, including limitations or prohibitions in a permit issued or proposed by the USEPA (EPA-40CFR2.301-3 and 2.302-3).

**standard oxygen transfer efficiency (SOTE)** The clean water oxygen transfer efficiency at a diffuser depth of submergence of 4.5 feet. Transfer efficiency varies with depth and with diffuser type.

**standard oxygen transfer rate (SOTR)** The oxygen transfer rate of a given diffuser in tap water, at 20°C and zero dissolved oxygen. *See also* actual oxygen transfer rate.

**standard plate count (SPC)** An expression of the density of bacteria in a water sample.

**standard plate count (SPC) technique** A laboratory method of bacterial enumeration in a water or other sample. It indicates the number of bacteria that will develop into colonies at a given temperature and during a given incubation period (35°C for 24 hours or 20°C for 48 hours) in a nutrient (e.g., tryptone glucose extract agar), usually expressed in number of bacteria per milliliter of sample. It is also called aerobic plate count, bacterial count, plate count, total bacterial count, total count, total plate count, water plate count. *See also* heterotrophic plate count.

**standard potential ($E^0$)** A characteristic of species in redox reactions such that if the concentration of each species is unity, the species with the higher standard potential is the oxidant. It is related to the standard free energy change ($\Delta G^0$):

$$E^0 = \Delta G^0/nF \qquad \text{(S-148)}$$

where $n$ is the number of electrons transferred and $F$ is the Faraday constant. *See also* standard electrode potential.

**standard pressure** Atmospheric pressure at sea level at a temperature of 0°C.

**standard Proctor compaction** The compaction of a material corresponding to a certain standard Proctor density. *See* Proctor density.

**standard Proctor density** *See* Proctor density.

**standard-rate digestion** A single-stage anaerobic sludge stabilization process operated in the mesophilic range (30–38°C), including digestion, thickening, and supernatant formation in one tank;

used mostly in small installations. *See* low-rate anaerobic digestion for detail.

**standard-rate filter** Same as standard-rate trickling filter. Also called low-rate filter.

**standard-rate trickling filter** A secondary wastewater treatment unit designed to operate without recirculation at loadings of 0.3–1.5 kg BOD/day/$m^3$ and 45–90 gpd/$ft^2$. Also called low-rate trickling filter or conventional trickling filter. *See also* high-rate trickling filter, intermediate-rate trickling filter, roughing filter, super-rate filter, biofilter, biotower, oxidation tower, dosing tank, filter fly.

**standard redox potential** The potential of a redox equation involving substances that are all in their standard states of unit activity. *See* Nernst equation.

**standard reference material** Same as certified reference material.

**standard rock trickling filter** A trickling filter characterized by the type of medium used: crushed stone, gravel, or slag, free of sand and clay, and ranging from 60 to 90 mm.

**standards** *See* standard.

**standard sample** The part of finished drinking water that is examined for the presence of coliform bacteria.

**standard seawater** (1) A typical seawater with an approximate concentration of total dissolved solids of 36 grams per liter or 36,000 mg/L. (2) A specially standardized water, prepared by the Hydrographical Laboratories of Copenhagen (Denmark), widely accepted as a control sample for the analysis of seawater salinity. *See* Copenhagen water for detail.

**standards for sewage sludge use or disposal** Regulations promulgated pursuant to section 405(d) of the Clean Water Act that govern minimum requirements for sludge quality, management practices, and monitoring and reporting applicable to the generation or treatment of sewage sludge from a treatment works treating domestic sewage or use or disposal of that sewage sludge by any person (EPA-40CFR122.2 and 501.2).

**standard shear test** A test to determine the physical strength of conditioned sludge floc by subjecting the material to vigorous stirring and measuring the change in capillary suction time (CST); sludge strength varies inversely with CST change. *See also* Atterberg test, Büchner funnel test, capillary suction time (CST) test, filterability constant, filterability index, filter leaf test, shear strength, specific resistance test, standard jar test, time-to-filter (TTF) test.

**standard short tube** A tube with a ratio of $D/L = 1/3$, where $D$ is the diameter and $L$ is the length. Also called a standard tube.

**standard sieve series** The American standard sieve series is a list of openings that stand successively in the ratio of $2^{1/4}$ (i.e., 1.1892) to one another; a sieve opening of 1 mm corresponds to 18 meshes to the inch. *See* Table S-09.

**standard solution** A solution in which the exact strength, concentration, or reacting value of a chemical or compound is known; it is used in laboratories to determine the properties of other substances.

**standard state** The actual or hypothetical condition of a solution in which the concentration and the activity coefficient are both unity. For a pure solvent, the standard state and reference state are the same.

**standard temperature** A temperature commonly used in the water and wastewater field as a reference: 20°C or 68°F. *See* standard biochemical oxygen demand, temperature effect, van't Hoff equation.

**standard temperature and pressure (STP)** A condition commonly used to state the properties of gases or to develop chemical relationships: a temperature of 0°C and a pressure of 1.0 atmosphere. Also called standard conditions.

**standard tests** Laboratory tests commonly conducted for water, wastewater, or sludge and falling into several categories, e.g.: composition, concentration, condition, economic usefulness, functional tests, palatability or esthetic acceptability, safety and wholesomeness, treatability. *See* Table S-10.

**Table S-09.** Standard sieve series

| Sieve designation number (meshes/inch) | Opening size, mm |
|---|---|
| 400 | 0.037 |
| 200 | 0.074 |
| 140 | 0.105 |
| 100 | 0.149 |
| 70 | 0.210 |
| 50 | 0.297 |
| 40 | 0.42 |
| 30 | 0.59 |
| 20 | 0.84 |
| 18 | 1.00 |
| 16 | 1.19 |
| 12 | 1.68 |
| 8 | 2.38 |
| 6 | 3.36 |
| 4 | 4.76 |
| 3.5 | 5.66 |

**Table S-10.** Standard tests (typical)

| Category | Water | Wastewater | Sludge |
|---|---|---|---|
| Composition, concentration, condition | | N, $NH_3$, DO, Cl, $PO_4$, Solids, BOD, COD, TOC, sulfide, acidity, alkalinity, bioassays, pH, coliforms, pollution indicators | Solids content, moisture, specific gravity, organic matter, volatile residue, BOD, pH, $NH_3$, color, odor, flora and fauna |
| Economic usefulness and functional tests | Hardness, pH, DO, metals, $H_2S$, jar test, Cl residual | BOD, SS, chlorine demand | Settleability, gas production, filterability, bioassays |
| Palatability and safety | Temperature, turbidity, color, odor, taste, coliform, toxicity, organics, pollution indicators, radioactivity | | |
| Treatability | Alkalinity, pH, taste and odor, metals | Suspended solids, BOD, COD, DO, N, pH | Sludge volume index, sludge density index |

**standard test conditions** Conditions specified for testing the performance of equipment or instruments.

**standard tube** Same as standard short tube.

**standby** Same as backup.

**standby pump** Same as backup pump.

**standby system** Same as backup system.

**standing crop** The biota (amount of living matter) present in an area at a given time, or the total number of individuals of a species alive in an area at a given time, usually expressed as biomass.

**standing level** The water level in a well that is not pumping. When it is not influenced by pumping of other wells, it is the same as the static level; otherwise it is an elevation on the cone of depression of the pumping well. Also called standing water level.

**standing sample** A one-liter sample of tap water collected in accordance with CFR Section 141.86(b)(2), that has been standing in plumbing pipes at least 6 hours and is collected without flushing the tap. *See* first-draw sample for detail.

**standing water level** *See* standing level.

**standing-water habitat** A lake, pond, reservoir, or other type of impoundment, as compared to a running-water habitat.

**standing wave** (1) A wave or distortion in fluid flow with a stationary shape with respect to the surface. Also called stationary wave. (2) A wave created by two progressive waves traveling through each other in opposite direction with vertical oscillations of water between fixed nodes. (3) The wave formed by a stream on the surface of a water body upon entering the body at high velocity. (4) A hydraulic jump or similar sudden and stationary rise in water surface. **standing-wave flume** An outlet control device, such as a Parshall flume, that measures discharge and does not require special calibration.

**standpipe** (1) A vertical, cylindrical pipe, tower, or tank with a height greater than its diameter, used to store water in a distribution system, maintain a required head in a water supply system, or to provide relief from surges of pressure in pipelines (then, also called a surge tank). Its useful capacity is the volume above the required pressure in the distribution system. *See* Figure S-23. Other types of distribution service storage include elevated tanks and ground-level reservoirs. (2) A fixed vertical pipe in a building or structure for supplying fire hoses, connected usually with a siamese outside the building. (3) A public water point from which users collect water; it usually consists of a masonry structure supporting a water pipe with one or more taps and accessories such as a platform for buckets and a drainage apparatus for spilt water. *See* Figure S-24.

**stannous chloride ($SnCl_2$)** A chemical used as antioxidant and preservative in food products. It is also used in combination with ammonium molybdate in the colorimetric determination of orthophosphate and is under consideration as a corrosion inhibitor in drinking water systems.

**stannum** Latin name of tin (Sn).

**Stanton number** A parameter ($\Phi$) used in the design of diffused or bubble aeration systems and defined as follows:

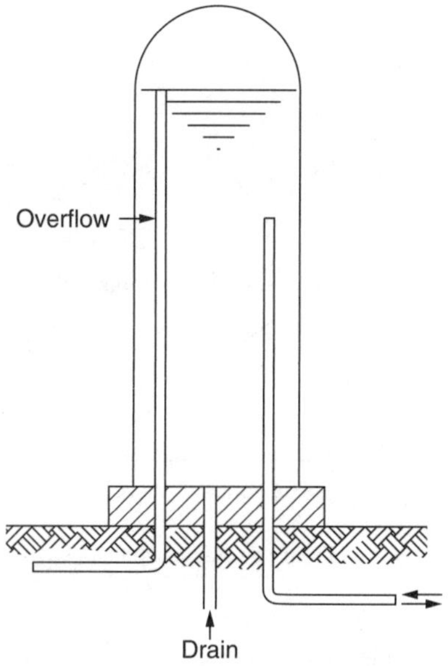


**Figure S-23.** Standpipe (reservoir).

$$\Phi = B(K_L a)/(HV) \quad \text{(S-149)}$$

where $B$ = volume of the bubble aeration tank, $(K_L a)$ = overall mass transfer coefficient for bubble aeration, $H$ = dimensionless Henry's law constant, and $V$ = air flow rate.

**staphylococci** Plural of staphylococcus.

***Staphylococcus*** A genus of spherical bacteria, some of which are pathogenic to humans; a group of nonmethanogenic bacteria responsible for hydrolysis and fermentation in anaerobic digestion. They occur in pairs, tetrads, or clusters. When these bacteria form clumps, the inner cells are protected from chemical disinfection.

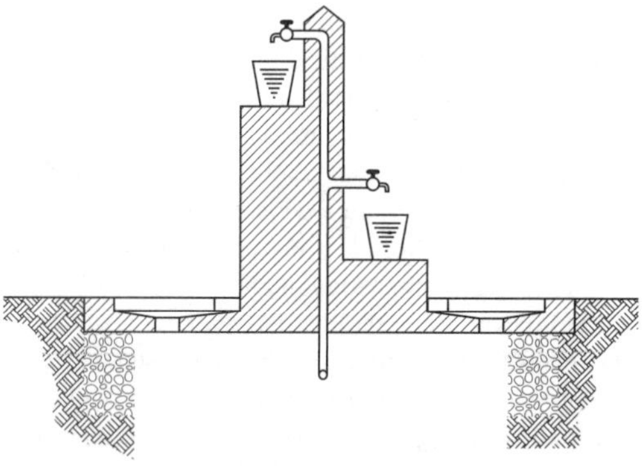

**Figure S-24.** Standpipe (water point).

***Staphylococcus aureus*** A species of Gram-positive bacteria responsible for water-related infections transmitted through the consumption of contaminated shellfish. It has been proposed, instead of coliforms, as an indicator of swimming water pollution associated with infections of the respiratory tract, skin, and eyes. *See also Candida albicans.*

**stapling** The entanglement of stringy or fibrous debris on a mesh or bar rack (Pankratz, 1996).

**Star™** A package plant manufactured by EnviroSystems Supply, Inc. for anaerobic wastewater treatment.

**starch** $[(C_6H_{10}O_5)_n]$ An easily degradable type of polysaccharide carbohydrate that microorganisms use as a source of energy; it is abundant in potatoes, rice, corn, and other edible plants.

**Star Filter** A filter press of Star Systems, Inc.

**StarScreen** An in-line fine screen manufactured by OVRC Environmental.

**start-action notice** A notice of the USEPA that it will initiate work on a rule or a regulation under the Safe Drinking Water Act.

**start date** The end of the four-month monitoring period within a compliance period when a water supply system must collect samples from target taps for implementation of the Lead and Copper Rule.

**starter** A device used to start up large motors gradually to avoid severe mechanical shock to a driven machine and to prevent disturbance to the electrical lines, (which can cause dimming and flickering of lights.

**starved-air combustion** Incomplete combustion because of insufficient oxygen, a thermal reduction process used in sludge handling with such products as combustible gases, tars, oils, and a solid char. Also called partial pyrolysis. *See also* pyrolysis, wet oxidation.

**starved-air incineration** Incomplete sludge combustion in a multiple-hearth furnace that is operated with less than the stoichiometrically required amount of air, which allows for higher loading rates and reduced fuel consumption.

**Sta-Sieve** A static fine screen manufactured by SWECO Engineering Co.

**state** (1) One of the United States of America, the District of Columbia, the Commonwealth of Puerto Rico, the Virgin Islands, Guam, American Samoa, the Trust Territory of the Pacific Islands, the Commonwealth of the Northern Mariana Islands, and an Indian Tribe eligible for treatment as a state pursuant to regulations promulgated under the authority of section 5189e) of the Clean Water

Act (EPA-40CFR503.9-x). (2) The condition of matter with respect to structure, form, constitution, etc. For example, solid particles in water or wastewater may be in the suspended, dissolved, or colloidal states. *See also* phase.

**state 404 program** A USEPA-approved state program to regulate the discharge of dredged or fill material under section 404 of the Clean Water Act in state-regulated waters (EPA-40CFR124.2 and 233.2). Same as section 404 program. *See also* state program.

**state agency** The state water pollution control agency designated by the Governor as having responsibility for enforcing state laws relating to the abatement of pollution (EPA-40CFR35.905).

**state certifying authority** (a) The state water pollution control agency defined in section 502 of the Clean Water Act; or (b) the state air pollution control agency designated pursuant to the Clean Air Act; or any interstate agency authorized to act in place of the certifying agency of a state (EPA-40CFR20.2-b).

**state equation** *See* equation of state.

**state facility plan** A requirement of some states of the United States for hazardous waste generators and toxic substance users to prepare a plan to reduce the use of toxic substances and the generation of hazardous wastes. Such a plan includes a review of industrial processes, an identification and ranking of pollution prevention opportunities, and an implementation schedule. *See also* facilities plans.

**state implementation plan (SIP)** A USEPA-approved state plan for the establishment, regulation, and enforcement of air pollution standards. It is a detailed description of the programs a state will use to carry out its responsibilities under the Clean Air Act.

**state parameter** *See* state variable.

**state point** In a solids flux diagram (the plot of solids flux versus mixed liquor suspended solids concentration), the state point is the intersection of the clarifier overflow solids flux rate and underflow solids flux rate lines, i.e., the actual operating point of a clarifier. Operation is satisfactory when the state point is within the flux curve and the underflow rate line is below the descending limb of the curve. *See* Figure S-25.

**state point analysis** A procedure used in the design and operation of final clarifiers. The analysis uses site-specific data to examine the behavior and expected performance of a clarifier under various mixed-liquor concentrations and clarifier operating conditions. *See* state point, overflow solids flux, underflow operating line, solids flux theory, zone settling velocity.

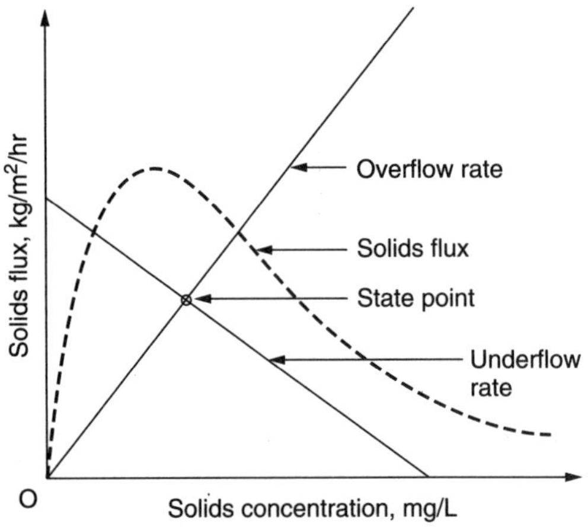


**Figure S-25.** State point.

**state-regulated waters** Those waters of the United States in which the Corps of Engineers suspends the issuance of section 404 permits upon approval of a state's section 404 permit program by the EPA (EPA-40CFR232.2).

**state revolving loan fund program** A mechanism for providing federal grants and low-interest loans through the states to finance water supply and wastewater management improvements. It was established by the 1987 Water Quality Act to replace the construction grants program.

**state variable** Any of the variables in the equation of state (2) or ideal gas law: pressure, temperature, and specific volume. Also called state parameter, thermodynamic variable. *See also* thermodynamic function of state.

**state water quality certification** Certification by a state agency under the Clean Water Act.

**static-bed composting** A method of sludge stabilization using an enclosed mechanical (in-vessel) system of top-fed vertical silos. The composting mixture, aerated by forced air, consists of dewatered sludge, sawdust, and recycled compost. The total detention time is about 32 days, including compost curing. Static-bed is not the same as static-pile composting. *See* in-vessel composting for other sludge composting methods.

**static burster** *See* static bursting.

**static bursting** A trenchless process used in the replacement of water or sewer pipes: a bursting head is pulled through the excavation to break the existing pipe, push the pieces into the surrounding soil, and install the new pipe that is attached at the rear of the head. The static burster is a cone-shaped or bullet-shaped tool that is pulled with a winch and cable to burst or split the existing pipe. *See* trenchless technology, open cut, hydraulic bursting, pneumatic bursting.

**static composting** A method of composting in which air is blown into the material to be composted, which remains static. *See* aerated static pile and agitated-bed composting.

**static discharge head** The vertical distance between the centerline of a pump and the water level in the discharge tank. *See also* pump head terms.

**static feed-forward control** In wastewater process control applications, a control method in which a variable control parameter is based on a linear relationship with a controlled parameter; a special case of feed-forward control.

**static head** Head resulting from elevation differences (e.g., the difference in elevations of the headwater and tailwater in a pipeline or a power plant), or the vertical distance between a fluid's supply surface level and its free discharge level. Also called fixed system head because it does not vary with flow. *See also* dynamic head.

**static in-line mixer** An in-line mixer that does not have a moving part; a device used in water or wastewater treatment for mixing and blending chemicals instantaneously, in one second or less (e.g., alum, ferric chloride, chlorine, cationic polymers). *See also* mixing and flocculation devices.

**static lift** Static head.

**Static Mixaerator™** A static mixing aerator manufactured by Ralph B. Carter Co.

**static mixer** A device that creates turbulence to mix chemicals in water using fixed sloping baffles or vanes without the application of power. *See also* mixing and flocculation devices.

**static pile** *See* aerated static pile.

**static pile composting** A sludge composting method in which piles of a sludge–bulking agent mixture are aerated. Also called forced aeration composting. *See* aerated static pile sludge composting for more detail.

**static pressure** (1) The pressure exerted by a body at rest, for example, hydrostatic pressure. In still water, it is the vertical distance (in feet) from a specific point to the water surface. Static pressure in psi = static head in ft times 0.433 psi/ft. (2) The pressure measured in a water distribution system when there is no flow. *See also* pressure test.

**static screen** A fine-to-medium screen that uses a stationary, inclined deck as a sieve to remove solids from wastewater, following coarse bar screens in small or industrial waste treatment plants. The medium consists of stainless-steel wedge-shaped bars, with the flat part facing the flow. Also called fixed screen, sidehill screen, static wedgewire screen. Sometimes confused with, but actually different from, tangential flow screen. Other fine screens used in wastewater treatment include drum or rotary screens, step screens, and horizontal reciprocating screens.

**static suction head** *See* pump head terms.

**static suction lift** *See* pump head terms.

**static toxicity test** The determination of acute toxicity of a wastewater effluent by placing it in a container with the test organisms and holding the mixture in a controlled environment for 24 hours.

**static tube** The main element of a static tube diffuser or a static tube mixer.

**static tube aerator** Same as static tube diffuser.

**static tube diffuser** A nonporous, coarse bubble diffuser that consists of a stationary vertical tube, about 3 ft long, with internal baffles to promote the mixing of air water and increase contact time. The tubes, which function like air lift pumps, are mounted on the floor of the aeration basin.

**static tube mixer** A wastewater aeration device, used mainly in aerated lagoons and activated sludge processes, and consisting of short tubes. Internal baffles maintain contact between the liquid and the air injected at the bottom of the tubes. *See* wastewater aeration systems for other types of aeration device.

**static water depth** The vertical distance from the centerline of the pump discharge down to the surface level of the free pool while no water is being drawn from the pool or water table.

**static water level** Elevation or level of the water table in a well when the pump is not operating. Also, the level or elevation to which water would rise in an open tube connected to an artesian aquifer or a conduit under pressure.

**static wedgewire screen** A fine-to-medium screen that uses a stationary, inclined deck as a sieve to remove solids from wastewater, following coarse bar screens in small or in industrial waste treatment plants. *See* static screen for more detail.

**stationary phase** (1) A constituent or substance that remains in place during chromatographic analysis, bonded to particles or to the walls of capillary tubes. (2) The third phase of batch culture, the phase that follows the declining growth phase and in which the number of microorganisms as well as the biomass remains constant.

**stationary state** The condition of a constituent in an aqueous solution, the concentration ($C$) of which does not change with time ($t$), i.e., for which $dC/dt = 0$. *See also* time-invariant condition for detail.

**stationary surface washer** A stationary apparatus used to clean the surface the media of a filter. *See* surface wash, rotary surface washer.

**stationary tidal wave A tidal wave that oscillates about a vertical axis.** ***See*** standing wave.

**stationary wave** *See* standing wave.

**statistical approach to design** A method of selecting design values for parameters or variables based on coefficients of reliability (COR). *See* COR method.

**statistical hydrology** The application of statistical methods (e.g., frequency analysis, least-squares fitting) to reduce large amounts of hydrological data to a few representative parameters (e.g., annual precipitation, storm runoff, flood flow) that can be used for engineering design and decision making.

**stator** (1) A stationary machine part in or about which a rotor turns, especially the stationary parts in the magnetic circuits. It applies to an electric generator or motor. *See* rotor and impeller. (2) A baffle installed along a wall of a paddle flocculator perpendicularly to the direction of flow.

**Stato-Screen** A static screen manufactured by Vulcan Industries, Inc.

**steady inflow** A steady flow of stormwater to the sanitary sewer system that cannot be identified or measured separately from groundwater infiltration; e.g., discharges from cellar and foundation drains, cooling water, and drains from springs and swamps. *See also* delayed inflow, direct inflow, total inflow, infiltration/inflow.

**steady state** *See* steady-state condition.

**steady-state condition** or simply **steady state** A condition of equilibrium, i.e., no net accumulation, no change over time. It applies to the laws of conservation of properties such as energy, mass, and momentum. Steady state is reached when the effects of transport and all processes of a given property, including sources and sinks, are in equilibrium. Steady-state flow is fluid flow without any change in composition or phase equilibrium relationships.

**steady-state flow** *See* steady-state condition.

**steady-state removal** The removal of contaminants (e.g., TOC and THM precursors) by biological action in a granular activated carbon unit during adsorption and after the adsorptive capacity is reached, at which time performance of the unit levels off at a steady rate. *See also* biological activated carbon.

**steady-state simplification** An assumption that simplifies the solution to the mass balance equation in most water or wastewater applications; specifically, the rate of accumulation is zero, and

$$r_c = Q(C_e - C_0)/V \qquad \text{(S-150)}$$

where $r_c$ = first-order reaction rate, $Q$ = flow rate in and out of a control volume, $C_e$ = concentration of a constituent leaving the control volume, $C_0$ = concentration of a constituent entering the control volume, and $V$ = control volume, e.g., volume of a reactor.

**Steady Stream** A turbidimeter manufactured by Great Lakes Instruments, Inc.

**steam box** Same as steam chest.

**steam chest** The steam chamber adjacent to a heat exchanger tube sheet or from which steam enters the cylinder of an engine. Also called steam box.

**steam distillation extraction** A laboratory method using steam to extract and concentrate trace organics. Also called simultaneous distillation extraction.

**steam-generating heavy-water reactor** A nuclear reactor using enriched uranium as fuel, heavy water as moderator, and light water as coolant.

**steam power plant waste** Wastewater generated during the operation of a steam power plant, including cooling water, boiler blowdown, and coal drainage; a high-volume and high-temperature waste that contains elevated concentrations of inorganic and dissolved solids. It may be treated by aeration (for cooling), storage of ashes, and neutralization of acid fractions.

**steam pump** A pump that has a steam engine, with the steam and water cylinders in the same unit.

**steam stripper** An apparatus that uses steam to remove volatile contaminants from liquids. It consists mainly of a packed or tray tower through which liquid and steam flow in a countercurrent, a decanter and pumps for the treatment and recovery of overhead vapor, and a feed-bottoms exchanger (Freeman et al., 1995, 1998).

**steam stripping** A continuous, fractional distillation process, similar to and more versatile than air stripping, that uses steam to remove ammonia, carbon dioxide, oxygen, hydrogen, and a variety of volatile organic compounds from liquid streams. It produces a treated bottoms stream and an overhead vapor that is further treated after condensation or recycled. The process is also used in the regeneration of activated carbon.

**steam vacuum pump** A displacement pump in which condensing steam creates a vacuum that draws water into the cylinder.

**stearic acid [$CH_3(CH_2)_{16}COOH$]** A typical fatty acid, colorless, waxlike, sparingly soluble in water, derived from animal fats and oils, and used in the production of soap, candles, cosmetics, etc.

**stearic acid glyceride** A common form of grease and fat found in wastewater treatment sludge.

**steel mill waste** Wastewater generated during the production of steel and originating from the coking of coal, washing of blast-furnace flue gases, and pickling of steel. It may contain high concentrations of acids, cyanogens, phenol, coke, alcalies, oils, mill scales, and fine suspended solids. Appropriate treatment methods include neutralization, chemical coagulation, recovery, and reuse.

**steel pickle liquor** Iron and steel products used in the production of iron-based coagulants for water and wastewater treatment. *See also* ferric chloride and ferric sulfate.

**Stefan–Boltzmann constant** The coefficient used to express the Stefan–Boltzmann law stating that the energy radiated from a body is proportional to the fourth power of its absolute temperature; named after two Austrian physicists who independently formulated the law: Josef Stefan (1835–1893) and Ludwig E. Boltzmann (1844–1906). *See* radiation drying.

**STEL** Acronym of short-term exposure limit.

**St. Elmo's fire** A luminous discharge of electricity at the surface of a conductor or between two conductors, causing the ionization of oxygen and the formation of ozone. Also called corona, corona discharge, electric glow. *See also* brush discharge.

**Stengel Baffle** An inlet baffle manufactured by Zimpro Environmental, Inc. for rectangular sludge collectors.

**stenothermophiles** Stenothermophilic organisms, which grow in a narrow or close temperature range. Also defined as bacteria that grow best above 60°C.

**stenothermophilic bacteria** *See* stenothermophiles.

**step 1 facilities planning** The preparation of a facilities plan under the Clean Water Act. Step 2 and step 3 are, respectively, the preparation of design plans and specifications and the construction of publicly owned treatment works (EPA-40CFR6.501-a).

**step aeration** A variation of the activated sludge process in which settled wastewater is added to the aeration tank in increments at several points. *See* step feed for detail.

**Stepaire** A package wastewater treatment plant manufactured by the FMC Corp. using the step aeration procedure.

**step dose** (1) The continuous application of water or wastewater treatment chemicals at progressively larger or smaller dosages. Step doses of dyes are also used in tracer studies. (2) The continuous introduction of a tracer until the outlet concentration approaches the feed concentration. *See also* slug dose.

**step dosing** *See* step dose.

**step feed** A variation of the activated sludge process in which wastewater is introduced in increments at several points into the aeration tank for a more uniform load distribution and oxygen demand; used in treating domestic waste to reduce the size of the tank while maintaining the same treatment capacity and solids retention time. Return sludge is usually added to the first portion of the aeration tank. Compared to conventional or tapered aeration systems, step aeration reduces the peak hydraulic and organic loads as well as the high initial oxygen demand. *See* Figure S-26 and also step aeration, step-feed BNR process.

**step-feed activated sludge process** *See* step feed and step-feed process.

**step-feed aeration process** *See* step aeration and step feed.

**step-feed BNR process** A step-feed plant designed and/or operated with an anoxic zone in each pass to achieve biological nitrogen removal.

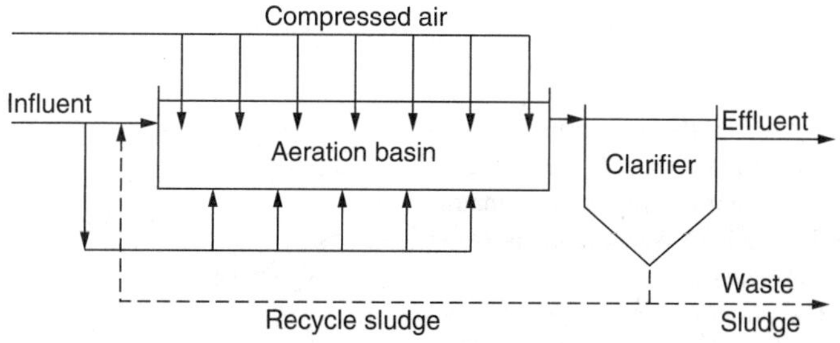


**Figure S-26.** Step-feed process.

**step-feed nitrogen removal** *See* step-feed BNR process.

**step-feed process** *See* step feed.

**step input** The introduction and mixing of a quantity of a tracer into a reactor until the effluent concentration equals the influent concentration. Most commonly, a dye is injected into the reactor to characterize tracer response curves and describe the hydraulic performance of reactors. Also called continuous injection or continuous-step input. *See also* pulse input, residence-time distribution (RTD) curve, time–concentration curve.

**step-lag logistic function** An equation used to determine the time to breakthrough ($t$) in measuring the performance of granular activated carbon units, as required by the Information Collection Rule. It represents effluent concentration as a function of time [$C(t)$]:

$$C(t) = A_0 + A_1/(1 + Be - Dt) \qquad \text{(S-151)}$$

where $A_0$, $A_1$, $B$, and $D$ are curve-fitting coefficients.

**step screen** A fine-screening device used in wastewater treatment (mostly in Europe); it consists of two step-shaped sets of fixed and movable vertical plates alternating across an open channel. Step screens are also used in the treatment of septage, primary sludge, and biosolids. *See* filter mat.

**Step Screen** An in-channel fine screen manufactured by Hycor Corp.

**STEP system** *See* septic tank effluent pumping system.

**step testing** The measurement of water flow in successive parts of a distribution system to detect possible leaks as indicated by large increases in successive readings.

**step valve** A check valve inserted in the head of a packed-plunger metering pump.

**stepwise formation constant** The equilibrium constants of the various steps in the formation of complexes. *See* stability constants.

**stepwise regeneration** Regeneration of a cation exchange resin with successively higher concentrations of sulfuric acid ($H_2SO_4$) to prevent the precipitation of calcium sulfate ($CaSO_4$).

**steric effect** *See* reaction site.

**steric interaction** The repulsion between particles in a suspension, caused by the adsorption of polymers at the solid–liquid interface. *See also* DLVO theory, electrostatic stabilization, hydrodynamic retardation, London–van der Waals force, polymer bridging, steric stabilization.

**steric rejection** The principal mechanism of rejection of particles by filtration membranes. *See* mechanical sieving for detail.

**steric stabilization** A condition caused by the adsorption of stabilizing polymers at the solid–liquid interface in a suspension, resulting in a repulsive interaction between particles. *See* Figure S-27, steric interaction.

**sterile water** Water that is free from bacteria and other microorganisms, as compared to potable water, which is free from pathogens. Sterile water is used in research, medicine, and for the preparation of pharmaceutical and specialized chemical products. *See also* distilled water.

**sterile water for inhalation, sterile water for injection, and sterile water for irrigation** Water that contains no added substance and meets the quality requirements of U.S. Pharmacopeia after purification by an appropriate water treatment process such as distillation, ion exchange or other USEPA-approved drinking water treatment. *See* pharmaceutical-grade water.

**sterile water for injection** *See* sterile water for inhalation.

**sterile water for irrigation** *See* sterile water for inhalation.

**sterilization** The removal or destruction of all living organisms, including pathogenic and other bacteria, vegetative forms, and spores. *See also* disinfection, pasteurization, sanitizer, bacteriostat.

**sterilized wastewater** Effluent or raw wastewater in which all microorganisms are destroyed or removed by sterilization.

**sterilized water** Bottled water that meets the sterility requirements of the *U.S. Pharmacopeia,* 23rd edition, of January 1, 1995; a labeling regulated by the U.S. Food and Drug Administration. Also called bottled sterile water.

**Stern layer** A compact, rigid layer attached to a negatively charged colloidal particle and in which the potential drops from the Nernst potential to the Stern potential, in the electrical double layer theory; also called bound-water layer. *See* Guoy–Stern colloidal model.

**SternPAC™** A polyaluminum coagulant of Sternson, Ltd.

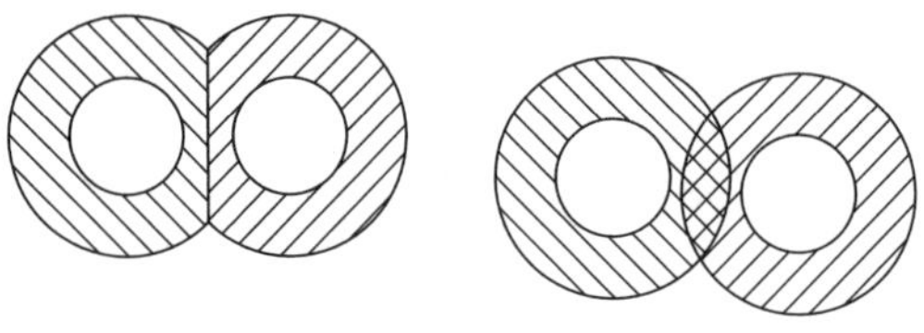

**Figure S-27.** Steric stabilization.

**Stern potential** The potential at the top of the diffuse layer in the electrical double layer theory, before the rapid drop in potential that occurs within the bound-water layer around a (negatively charged) colloidal particle, caused by the action of positively charged ions from the bulk solution. *See* Guoy–Stern colloidal model, zeta potential.

**stewardship** Management of resources so as to preserve their basic integrity for future generations. *See also* sustainability.

**STF-AS** Acronym of series trickling filter-activated sludge.

**stibium** Latin name of antimony (Sb).

**stibnite ($Sb_2S_3$)** A natural compound, sulfide of antimony.

**stick** In rendering and other meatpacking operations, it is the waste wash water in which the bones of cattle and hogs are cooked; one of two major sources of water pollution from that industry. *See also* packers pollution, rendering pollution, meatpacking waste.

**sticking coefficient** A dimensionless coefficient that indicates the fraction of successful collisions or the probability of attachment of particles flowing in a porous medium; it is based on chemical characteristics and is usually determined experimentally. *See* attachment probability and collision efficiency.

**sticky limit** The limit at which a soil loses its ability to adhere to a metal blade.

**Stiff and Davis index** *See* Stiff and Davis scaling index and Stiff and Davis stability index.

**Stiff and Davis scaling index (SDSI)** An index used to determine the solubility of calcium carbonate as a function of the pH of the solution, the negative logarithm of the molar concentration of calcium (pCa), the negative logarithm of alkalinity in equivalents per liter (pAlk), and an empirical constant ($K$) that depends on temperature and ionic strength:

$$\mathrm{SDSI} = \mathrm{pH} - \mathrm{pCa} - \mathrm{pAlk} - K \qquad \text{(S-152)}$$

*See also* Stiff and Davis stability index.

**Stiff and Davis stability index (S&DSI)** An index used to determine whether a saline water (with a TDS higher than 10,000 mg/L) is in equilibrium with calcium carbonate ($CaCO_3$). It is equal to the difference between the actual pH of the water (pHa) and the pH of saturation ($pH_s$) and at equilibrium with $CaCO_3$, i.e.,

$$\mathrm{S\&DSI} = (\mathrm{pH_a}) - (\mathrm{pH_s}) \qquad \text{(S-153)}$$

This equation is the same as the one for the Langelier saturation index, but there is a different method to calculate the saturation pH. At equilibrium, the index is equal to 0, whereas the water will tend to deposit $CaCO_3$ if S&DSI > 0 or dissolve it if S&DI < 0. *See* other similar indexes: Riddick and Ryznar. *See also* Stiff and Davis scaling index.

**still** An apparatus used in distillation (alcoholic beverages, distilled water, etc.), consisting of a vessel for vaporizing a liquid and a device for condensing the vapor.

**stillage** The residual grain mash from a distillation column, the principal source of pollutional load from breweries, usually recovered to produce animal feed or chemical products.

**still bottoms** By-products of distillation, often hazardous pollutants, consisting of unevaporated solids, semisolid tars, and sludges. Also called distillation bottoms.

**still water** The portion of a water body in which there is no apparent or viable current because it is level or the level of inclination is very slight.

**stimulation** *See* well stimulation.

**Sti-P3®** A standard established by the Steel Tank Institute for cathodic protection of underground tanks.

**stirred sludge volume index** The sludge volume index of a suspension determined for a sample that is slowly stirred. *See* stirred SVI test.

**stirred SVI test** A sludge volume index test conducted in a small-diameter apparatus that is equipped with a slow-speed stirring device to eliminate wall effects on the settling solids. *See also* SVI test and diluted SVI test.

**stirred-tank reactor** *See* continuous stirred tank. *See also* Wetox® system.

**STM Aerotor™** A wastewater treatment system designed or manufactured by WesTech Engineering, Inc. of Salt Lake City, UT for BOD and biological nitrogen removal applications.

**STM-PA** A medium used in the presence–absence test for multiple indicators; it contains lactose, lauryl tryptose, and tryptone. *See also* MacConley-PA.

**stochastic** Based on the assumption that the actions of a chemical substance result from probabilistic events. A stochastic process is governed by chance or involves random variables; the opposite of a deterministic process.

**stock pond** An impoundment consisting of an excavation and an earthen dam to provide water for livestock. Also called an earth tank.

**stock solution** A concentrated chemical solution, often used as a reagent after appropriate dilution.

**stockyard waste** The waste resulting from stockyards, where the animals are kept before they are killed; a major source of pollution of the meatpacking industry. Stockyard wastewater is lower in BOD (< 100 mg/L) and suspended solids (< 200 mg/L) than domestic waste because of its high hydraulic population equivalent.

**stoichiometric** Of or pertaining to stoichiometry; related to substances that are in the exact proportions or quantities required for a given reaction. A stoichiometric compound (or pure compound) contains elements in the precise proportions represented in its chemical formula. A stoichiometric mixture corresponds to a stoichiometric compound. In a stoichiometric combustion (or ideal combustion), all consumable elements are burned or oxidized in accordance with the law of energy conservation. The stoichiometric ratio denotes the exact or fixed proportions in which elements or compounds combine to form new compounds; e.g., the stoichiometric ratio of carbon to oxygen is 1:2 in carbon dioxide ($CO_2$) and that of carbon to hydrogen is 1:4 in methane ($CH_4$).

**stoichiometric coefficients** (1) The numbers of moles of reactants and products required to balance a chemical reaction. For example, in the COD reaction, dichromate ($Cr_2O_7^{2-}$) oxidizes organic matter (simplified formula $CH_2O$), with chromium, carbon dioxide ($CO_2$), and water ($H_2O$) as end products:

$$CH_2O + a \cdot (Cr_2O_7^{2-}) + b \cdot H^+ \rightarrow c \cdot Cr^{3+} + d \cdot CO_2 + e \cdot H_2O \quad \text{(S-154)}$$

the stoichiometric coefficients are a, b, c, d, and e, which can be determined from mass balance relations and electron balance considerations between the left hand and right hand sides (Droste, 1997). *See also* equation (S-156) under stoichiometric equation, rate law, chemical equilibrium. Also called stoichiometry coefficient. (2) Two parameters used in the design of suspended-growth biological treatment: the true cell yield ($Y$, mass/mass) and the endogenous decay coefficient ($b$, inverse of time), which relate the observed cell yield ($Y'$, mass/mass) to the solids retention time ($\Theta_c$) as follows (Water Environment Federation et al., 1998):

$$Y' = Y/(1 - b \cdot \Theta_c) \quad \text{(S-155)}$$

**stoichiometric combustion** *See* stoichiometric.

**stoichiometric compound** *See* stoichiometric.

**stoichiometric equation** An equation that represents a balanced chemical reaction, e.g., the COD reaction:

$$3\ CH_2O + 2\ (Cr_2O_7^{2-}) + 16\ H^+ \rightarrow 4\ Cr^{3+} + 3\ CO_2 + 11\ H_2O \quad \text{(S-156)}$$

Compare to equation (S-154).

**stoichiometric mixture** *See* stoichiometric.

**stoichiometric numbers** The coefficients that multiply the formulas of substances on both sides of a chemical reaction so that the reaction is balanced

**stoichiometric ratio** *See* stoichiometric.

**stoichiometry** (1) The definition or calculation of the quantities of chemical elements or compounds involved in a balanced reaction. (2) The branch of chemistry that studies the quantitative relationships between combining elements.

**stoichiometry coefficient** Same as stoichiometric coefficient (1).

**stoker** A mechanical device for supplying solid fuel to a furnace.

**Stokes' conversion factor** A factor ($k$) that combines the effects of density and viscosity on the settling or rising velocity of a discrete particle. *See* Stokes law:

$$k = g \cdot (\rho - \rho')/18 \cdot \mu \quad \text{(S-157)}$$

**Stokes–Einstein's law of diffusion** An equation that relates the coefficient of molecular diffusion ($D$, m²/s) of spherical particles to terms of their frictional coefficients (viscosity and radius), as defined by Stokes' law:

$$D = kT/6\pi\mu r = RT/6\pi\eta rN \quad \text{(S-158)}$$

where $k$ = Boltzmann constant; $T$ = temperature = 273.15 + °C; $\mu$ = dynamic viscosity, N · s/m²; $r$ = radius of particle, m; $R$ = universal gas law constant; and $N$ = Avogadro's number.

**Stokes' law** A law of physics as stated by Sir George G. Stokes in 1845: the force that retards the movement of a spherical particle through a fluid is directly proportional to the velocity and radius of the particle and the viscosity of the fluid. In the water and wastewater treatment field (sedimentation and flotation processes), Stokes' law is used to derive the formula that expresses the terminal settling or rising velocity ($V$) of a discrete particle in a fluid as a function of particle diameter ($d$) and density ($\rho$), gravitational acceleration ($g$), and the fluid density ($\rho'$) and absolute viscosity ($\mu$):

$$V = g \cdot d^2(\rho - \rho')/18 \cdot \mu \quad \text{(S-159)}$$

*See also* drag force, drag coefficient.

**Stokes' range** The range of Reynolds numbers ($R_e$) within which flow is laminar and the drag coefficient ($C_D$) is equal to 24 divided by the Reynolds number:

$$R_e < 1 \qquad C_D = 24/R_e \quad \text{(S-160)}$$

The coefficient of drag affects the terminal settling velocity of particles in water and wastewater treatment. Also called laminar range. *See also* transition range and turbulent range.

**Stokes' settling velocity** The terminal settling velocity as determined from Stokes' law.

**Stokes, Sir George Gabriel** A British physicist and mathematician (1819–1903). *See* Stokes' law.

**stomach worm** Another name for *Ascaris lumbricoides,* a species of nematodes or roundworms that cause ascariasis worldwide.

**stomatitis** Inflammation of the mucous membrane of the mouth.

**stone filter berm** A temporary ridge constructed of loose stone or crushed rock to capture sediment from runoff by slowing and filtering flow, and diverting it from an exposed traffic area.

**stop box** A metallic or concrete box, housing, or vault installed near the curb, containing and providing access to a shutoff valve in a customer's water service line, with a cover to keep out dirt and debris. It is usually located on the property line, between the main water line and the meter or between the curb and the sidewalk. The valve may be operated to start or stop water service to the customer. This receptacle is also called curb cock or curb stop, curb stop and box, valve box. *See also* valve vault.

**stopcock** A cock; a hand-operated device, such as a valve or faucet, for regulating flow in a pipe.

**stop log** A removable beam or bulkhead used to stop water flow in a dam, channel, or conduit.

**stop plank** A removable wooden or steel plank used to stop water flow from one channel to another.

**stop valve** A device installed in a pipeline to shut off flow in a section for inspection or repair. Also called sectionalizing valve.

**storage** The impounding of water, stormwater, or wastewater for future use or release. Storage implies a longer retention time than pondage, regulation, or detention. *See* elevated storage, ground storage, prism storage, wedge storage, and reservoir storage.

**Storage and Retrieval of U.S. Waterways Parametric Data (STORET)** A program initiated by the U.S. Public Health Service in 1964 and currently managed by the U.S. Environmental Protection Agency to collect, store, and disseminate data dealing with the quality and quantity of surface and ground waters within and contiguous to the United States. Such data include records of precipitation and other climatological information, storm runoff and flood flow data, effluents, and biological, chemical, and physical water quality; they may be provided by the U.S. Geological Survey, the U.S. Environmental Data Service, the U.S. Army Corps of Engineers, and individual states.

**storage basin** *See* sludge storage.

**storage coefficient** (1) A coefficient that accounts for storage and other runoff conditions in the Snyder method for the establishment of synthetic hydrographs. (2) The volume of water taken from or into storage by a vertical column of an artesian aquifer having a base of one square foot and a drop of one foot in the piezometric surface. For an artesian aquifer, the change in storage results from the lowering of the water table and from pressure changes in the aquifer. Jacob's equation expresses the storage coefficient ($S$) of a confined aquifer:

$$S = \theta\gamma b[\beta + (\alpha/\theta)] \qquad \text{(S-161)}$$

where $\theta$ = average porosity of the aquifer, $\gamma$ = specific weight of water, $b$ = saturated thickness of the aquifer, $\beta$ = compressibility of water, and $\alpha$ = vertical compressibility of the aquifer material. Also called storativity, the storage coefficient varies from 0.00005 to 0.0005 for most aquifers. The corresponding term for a water-table aquifer is specific yield.

**storage lagoon** A lagoon whose bottom is lined or sealed for the collection or storage of wastewater solids.

**storage of sewage sludge** The placement of sewage sludge on land on which the sewage sludge remains for two years or less. This does not include the placement of sewage sludge on land for treatment (EPA-40CFR503.9-y).

**storage pad** *See* sludge storage.

**storage reservoir** A reservoir with a relatively large volume, compared to the inflow and outflow rates. *See also* run-of-the-river reservoir.

**storage SI** Storage surface impoundment.

**storativity** Same as storage coefficient.

**STORET** Abbreviation of Storage and Retrieval of U.S. Waterways Parametric Data.

**Stormceptor®** A stormwater treatment apparatus manufactured by Rinker Materials using a patented internal bypass to prevent previously accumulated pollutants from being resuspended during peak flows.

**storm drain** A drain designed to carry stormwater, surface water, groundwater, subsurface water, building drainage, condensate, cooling water, and other similar discharges to a storm sewer or a combined sewer. Also called storm sewer.

**Storm King™** A vortex-type separator used in stormwater management and manufactured by H. I. L. Technology, Inc. *See* vortex separator.

**storm sewage** Refuse liquids and wastes carried by sewers during or following rainfall; it is heavily polluted, particularly during the first flush. Also called storm wastewater. For storm sewer, see storm drain. Storm sewer discharge is the discharge of water from a storm sewer into a receiving water. *See also* storm water, combined sewer.

**storm sewer** *See* storm drain.

**StormTreat™** A system designed by StormTreat Systems for the collection and treatment of stormwater.

**storm wastewater** Same as storm sewage.

**stormwater** Stormwater runoff, snowmelt runoff, and surface runoff and drainage, all resulting from precipitation, which either runs off from the surface into a stream, are captured by a storm or combined sewer, or percolate into the soil. Sometimes called storm sewage.

**stormwater characteristics** Table S-11 shows typical characteristics of stormwater runoff.

**Storm Water Management Model (SWWM)** A public-domain computer model that assesses combined sewer systems by simulating the processes of precipitation, rainfall conversion to runoff, and collection and transport of stormwater runoff and wastewater through the system. SWMM performs both hydraulic and contaminant routing.

**stormwater pollution prevention plan (SWPPP)** An important part of the permit application for industrial stormwater discharges, designed to eliminate, minimize, or reduce pollution due to stormwater discharges from a site. Dodson (1999) recommends the consideration of the following categories of BMPs in preparing a SWPPP: good housekeeping, preventive maintenance, spill prevention and response procedures, inspections, employee training, and record keeping and internal reporting procedures.

**Table S-11.** Stormwater characteristics (typical)

| Constituent | Concentration |
|---|---|
| Total suspended solids, mg/L TSS | 85 |
| Biochemical oxygen demand, mg/L BOD | 9 |
| Chemical oxygen demand, mg/L COD | 55 |
| Fecal coliform bacteria, MPN/100 mL | 6000 |
| Total Kjeldahl nitrogen, mg/L N | 0.7 |
| Nitrate nitrogen, mg/L N | 0.7 |
| Total phosphorus, mg/L P | 1.2 |
| Copper, μg/L Cu | 30 |
| Lead, μg/L Pb | 85 |
| Zinc, μg/L Zn | 180 |

**stormwater quality control facilities** Facilities designed to reduce the pollutant loads from stormwater into receiving waters, e.g., retention basins, sedimentation ponds, wet settling ponds. *See also* best management practices.

**stormwater quality ponds** Wet or dry ponds designed to treat stormwater.

**stormwater retention** Retention/detention basins, ponds, reservoirs, and tanks are used in stormwater management to reduce the flooding and water quality impacts of storm runoff. Some of these structures can achieve a high degree of removal of such pollutants as BOD, sediment, metals, and nutrients. In most of them, the most important design and control parameter is the retention period, also called detention time or retention time, i.e., the time ($t$) required to displace the volume ($v$) of the structure at a given flow rate ($Q$):

$$t = v/Q \qquad \text{(S-162)}$$

Stormwater retention facilities include: dry detention basin, extended dry detention pond, infiltration basin/infiltration pond, sedimentation pond with displacement, and wet retention basin/wet retention pond/wet settling basin/wet settling pond.

**stormwater runoff** *See* stormwater and runoff.

**STP** Acronym of (a) sewage treatment plant, (b) sodium trypolyphosphate, or (c) standard temperature and pressure.

**straight-edged weir** A weir having a straight crest, as compared to a V-notch weir.

**straight-flow pump** A device that has the pump as well as the suction and discharge pipes all in line.

**Straightline®** A line of products of FMC Corp. for use in wastewater treatment.

**straightway valve** A valve through which fluid passes without deviation.

**strain** A particular type of a microorganism; e.g., there are several strains of *Vibrio cholerae,* the pathogen that causes cholera.

**strainer** A filtration device that separates and retains solid particles but allows a fluid stream through.

**strainer head** A perforated device installed in a rapid sand filter for the flow of filtered water and backwash water. Also called filter strainer or rapid sand filter strainer.

**strainer system** An apparatus with strainer heads to collect filtered water in underdrains.

**strainer well** A tube well with a strainer to keep debris from entering the well.

**straining** (1) A water or wastewater treatment operation using screens and racks to separate floating and suspended solids from the liquid; sometimes associated with shredding devices to reduce the size of the solids. (2) A major filtration mechanism contributing to the removal of particulate matter within a granular filter. *See* chance contact straining, mechanical straining.

**Strain-O-Matic** A self-cleaning strainer manufactured by Hayward Industrial Products, Inc.

**StrainPress** A compactor manufactured by Parkson Corp.

**Strantrol** A device manufactured by Stranco, Inc. for the control of the operation of disinfection systems.

**Strata Clear** A device manufactured by Smith & Loveless, Inc. for solids separation by flotation.

**StrataMix** A fine-bubble diffuser manufactured by Wilfley Weber, Inc.

**Strata-Sand™** A downflow (gravity) sand filter manufactured by F. B. Leopold Co., Inc., including continuous backwashing.

**Stratavap** An evaporator manufactured by Licon, Inc.

**stratification** (1) The formation of layers of different densities or different temperatures in a body of water. Each layer has similar characteristics and plant or animal life. Most lakes and reservoirs deeper than 15 feet stratify during part of the year. *See* density stratification, density current, thermal stratification, lake stratification, critical stratification temperature, mixing time, thermocline, epilimnion, hypolimnion, fall overturn, selective withdrawal. (2) The formation of layers in a sludge digestion tank that is not adequately stirred: a scum layer, a supernatant layer, a digesting layer, and an inactive layer. (3) The arrangement of materials in layers or strata, one above the other, as in the formation of sedimentary rocks. (4) *See also* stratified bed. (5) The division of the sea into feeding or trophic levels of marine life by virtue of changes in water pressure.

**stratified bed** A column of two ion-exchange resins of different densities and classes, one on top of the other.

**stratosphere** The portion of the atmosphere that is 10 to 25 miles above the Earth's surface. *See also* troposphere. It contains most of the Earth's ozone, which filters out harmful sun rays, including a type of sunlight called ultraviolet B that has been linked to health and environmental damage.

**stratum** A horizontal layer of geologic material of similar composition, especially one of several parallel layers arranged one on top of another.

**stray current** A direct electrical current that passes through the earth or other conductor around an underground structure such as a water pipe, which may cause pitting and/or corrosion.

**stray current corrosion** Localized corrosion caused in a device by a stray current from another source, e.g., when home appliances or electrical circuits are grounded through water pipes, which affects not only the pipes but also faucets and valves. *See also* electrolysis (2).

**stream aeration** The injection of air or pure oxygen into a stream using vertical pumps or other means. Aeration of a receiving stream may be used as an alternative to aeration of the organic waste being discharged, e.g., where draft tubes of the turbines of a power dam across the stream are available to draw oxygen into the water.

**stream augmentation** *See* flow augmentation, low-flow augmentation.

**streambank erosion** The wearing away of streambanks by flowing water.

**stream biota** The plants and animals living in a stream.

**stream classification** The classification of a stream according to its current or intended uses, with appropriate water quality criteria and effluent standards to protect such uses. *See* water body classification.

**streamflow source zone** The upstream headwaters area that drains into the recharge zone (EPA-40CFR149.101-i).

**streaming current** The current gradient that occurs on passage of a solution or suspension of electrolytes and polyelectrolytes, or charged particles through a capillary space, as influenced by adsorption and electrical double layers; it is used to monitor and control coagulation and flocculation processes. Also defined as the net ionic and colloidal surface charges of suspended solid particles in solution. *See also* electrokinetic potential, electroosmosis, electrophoresis, electrophoretic mobility, sedimentation potential, streaming potential, surface potential, zeta potential.

**streaming current detector (SCD)** An instrument that can detect and measure online the streaming current (or net electrical charge) generated by particles in a solution. It is used to adjust coagulant dosage based on the degree of particle destabilization detected.

**streaming potential** The tendency of a solvent, under the influence of an electrical field, to move through capillaries or through the walls of a porous membrane. *See also* electroosmosis and sedimentation potential.

**Streamline-1** A device manufactured by American Sigma, Inc. for the proportional sampling of flows.

**Streamline-2** A chain and flight collector manufactured by Purestream, Inc. for rectangular tanks.

**stream quality** *See* stream standard, water quality, effluent standard.

**stream sanitation** The branch of environmental science and engineering that deals with the water quality of streams and the impact of wastewater discharges into the stream. *See,* e.g., the Streeter–Phelps formulation and the Churchill method.

**Stream Saver™** An apparatus manufactured by ILC Dover for the automatic control of spills.

**stream specialization** A procedure for improving stream water quality by using one stream as an open sewer for all waste discharges while preserving another stream for other uses.

**stream standards** Criteria set by a government for the desired conditions of a stream, with appropriate regulations to maintain the stream classification or quality.

**Streeter, H. W.** Coauthor of the classic oxygen-sag analysis.

**Streeter–Phelps equation** *See* Streeter–Phelps formulation.

**Streeter–Phelps formula** *See* Streeter–Phelps formulation.

**Streeter–Phelps formulation** An equation proposed by H. B. Streeter and E. B. Phelps in the 1920s to describe oxygen depletion in a stream with a point discharge. For a continuous discharge of wastewater into a stream, neglecting the effects of photosynthesis, respiration, and sediment oxygen demand, the dissolved-oxygen deficit as a function of travel time is:

$$D = [KL_0/(K' - K)](e^{-Kt} - e^{-K't}) + D_0e^{-K't} \quad \text{(S-163)}$$

The coordinates of the point of maximum (or critical) dissolved-oxygen deficit are:

$$t^* = ([1/(K' - K)] \ln \{(K'/K)[1 - D_0(K' - K)/KL_0]\} \quad \text{(S-164)}$$

$$D^* = (K/K')(L_0e^{-Kt^*}) \quad \text{(S-165)}$$

where $D$ = dissolved-oxygen deficit = the difference between the saturation concentration and the actual concentration, mg/L; $D_0$ = initial dissolved-oxygen deficit, i.e., deficit at time $t = 0$, mg/L; $D^*$ = maximum dissolved-oxygen deficit, mg/L; $K$ = decay coefficient or deoxygenation constant, per day; $K'$ = surface reaeration rate, per day; $L_0$ = ultimate BOD of the mixture of stream and wastewater, mg/L; $t$ = travel time, days; and $t^*$ = time of maximum deficit, days. Also called classic oxygen-sag formulation. *See also* dissolved-oxygen sag curve.

**Streeter–Phelps oxygen-sag analysis** *See* Streeter–Phelps formulation.

**streptobacilli** Rod-shaped bacteria that remain attached after cell division.

**streptococci** Plural of streptococcus.

**streptococcus** (plural: **streptococci)** A genus of spherical and oval, pair or chain bacteria that includes some common human pathogens, e.g., those that cause scarlet fever and tonsillitis.

***Streptococcus bovis*** A species of true fecal streptococci predominately found in animals.

***Streptococcus equinus*** A species of true fecal streptococci predominately found in animals.

***Streptococcus faecalis*** A species of (enterococci total fecal streptococci).

***Streptococcus* genus** *See* streptococcus.

***Streptomyces*** A group of moldlike bacteria usually associated with algae and responsible for taste and odor problems in surface waters. *See also* actinomycetes.

**stress corrosion** Corrosion that affects metal at points of internal or external tensile stress, caused, for example, by welding. Internal stresses can be reduced by heat treatment. Also called fatigue corrosion.

**stress corrosion cracking** Formation of cracks in water pipes due to stress corrosion.

**stressed water** A receiving environment in which an applicant can demonstrate to the satisfaction of the USEPA that the absence of a balanced, indigenous population is caused solely by human perturbations other than the applicant's modified discharge (EPA-40CFR125.58-t).

**Stress-Key** A package wastewater treatment plant manufactured by Marolf, Inc. in precast concrete.

**stressor** A substance, circumstance, energy factor, etc., that can cause an adverse effect on a biological system. Stressors and their impacts are identified during an ecological risk assessment. *See also* toxicity terms.

**stress testing** An experiment conducted by operating a process and its equipment under extreme conditions to improve the calibration of a treatment plant performance model.

***Str. faecalis*** *See Streptococcus faecalis.*

**s-triazines** A group of organic chemicals (atrazine, cyanazine, propazine, and simazine) used to control broadleaf and grassy weeds. They may be found in surface waters receiving agricultural runoff and are not removed effectively by conventional water treatment processes.

**strict aerobe** *See* strict organisms.

**strict anaerobe** *See* strict organisms.

**strict organisms** Organisms that cannot grow in either the presence or absence of a specific environmental factor; for example, strict or obligate aerobes require molecular (dissolved) oxygen for their metabolism, whereas strict or obligate anaerobes cannot survive in the presence of free oxygen. *See also* facultative organisms. Most of the bacteria that carry out the decomposition of organic matter in wastewater treatment are facultative as opposed to strict or obligate organisms. Some specific biological reactions such as nitrification and methanogenesis are mediated by strict organisms.

**string filter** A vacuum filter in which the sludge cake passes through or is removed by moving strings.

**string-wound element** A type of cartridge filter.

**strip-chart recorder** A recording device used to display instantaneously and store monitoring and laboratory data. *See also* circular chart recorder, flow totalizer.

**strip cropping** A crop production system that involves planting alternating strips of row crops and close-growing forage crops; the forage strips intercept and slow runoff from the less-protected row-crop strips.

**strip mining** An open-pit mining method that uses machines to scrape the overburden (vegetation, soil, or rock) away from mineral deposits just under the Earth's surface. It exposes metal-sulfide ores to the atmosphere and results in the formation of acid mine drainage. Also called open-cut mining, surface mining.

**stripper** (1) A chemical substance used to remove varnish, paint, wax, etc. from surfaces. (2) An apparatus used in stripping processes to remove volatile and semivolatile contaminants from water. *See also* aerator, air stripper, steam stripper. (3) *See* stripper tank or anaerobic stripper.

**Stripper®** An aeration apparatus manufactured by Lowry Aeration Systems, Inc., using multistage diffuse bubbles.

**Stripperator** A treatment device manufactured by Ejector Systems, Inc. to remove hydrocarbons from contaminated water.

**stripper tank** An anaerobic reactor in which phosphorus is released to solution in enhanced phosphorus-uptake processes; *see,* e.g., PhoStrip™. The stripped biomass returns to the aeration basin while the supernatant is further processed for phosphorus precipitation. Also called anaerobic stripper.

**stripping** (1) A process that uses an air stream to remove ammonia, volatile organics, and other compounds from water, wastewater, waste regenerant, etc. It transfers target compounds from the liquid phase to the gas phase. *See* air stripping. (2) The removal of topsoil and vegetation before excavating for construction or mining activities.

**stripping factor** A parameter ($S$) used in the design of stripping towers, defined as a function of the moles of incoming gas per unit time ($G$), moles of incoming liquid (water or wastewater) per unit time ($L$), Henry's law constant ($H$), and total pressure ($P_T$):

$$S = (GH)/(LP_T) \qquad \text{(S-166)}$$

**stripping section** The cascade or set of plates below the feed introduction point of a continuous distillation column that involves only one feed stream. *See also* enriching or rectifying section.

**stripping tower** A tall, usually cylindrical device used to transfer gas from a liquid phase (e.g., water, wastewater) to the gas phase, including a support plate for packing materials, a distribution system for the liquid above the packing, and an air supply at the bottom of the tower. *See* packed tower air stripper and Figure P-01.

**stripping tower design** Table S-12 shows the customary variables and parameters used in the following design equations of stripping towers:

$$Z = NH_t \qquad \text{(S-167)}$$

$$N = [S/(S-1)] \ln \{[(C_0/C_e)(S-1)+1]/S\} \qquad \text{(S-168)}$$

$$H_t = L/(AK_La) \qquad \text{(S-169)}$$

where $Z$ = packing depth, m; $N$ = number of transfer units; $H_t$ = height of a transfer unit, m; $S$ = stripping factor, unitless; $C_0$ = concentration of solute in liquid entering the tower, moles/mole; $C_e$ = concentration of solute in liquid leaving the tower, moles/mole; $L$ = liquid volumetric flow rate,

**Table S-12.** Stripping tower design (variables and parameters)

| | |
|---|---|
| Packing material (pall rings, rashig rings, etc.) | Packing factor ($C_f$, unitless, based on material type and size) |
| Stripping factor ($S$, unitless) | Allowable air pressure drop ($\Delta P$, N/m$^2$/m) |
| Diameter of tower ($D$, m) | Height-to-diameter ratio ($Z/D$, m/m) |
| Packing depth ($Z$, m) | Safety factor ($SF$, %$D$, %$Z$) |
| Liquid loading rate ($L$/m$^2$/min) | Water or wastewater pH (unitless) |
| Gas loading rate ($G$) | Air-to-liquid ratio ($G/L$, m$^3$/m$^3$) |

$m^3$/sec; $A$ = cross-sectional area of tower, $m^2$; and $K_La$ = volumetric mass transfer coefficient,/sec.

**stripping tower mass balance** A materials balance for a continuous tower stripping dissolved gases from wastewater, indicating an equilibrium between the mass of solute entering in the gas and liquid streams and the mass of solute leaving in both streams:

$$y_0 - y_e = L(C_e - C_0)/G \qquad \text{(S-170)}$$

where $y_0$ = entering solute concentration in gas, moles of solute per mole of solute-free gas; $y_e$ = leaving solute concentration in gas, moles of solute per mole of air; $L$ = flow of influent wastewater, moles of liquid per unit time; $C_e$ = leaving solute concentration, moles of solute per mole of liquid; $C_0$ = entering solute concentration in wastewater, moles of solute per mole of liquid; and $G$ = flow of incoming gas, moles per unit time. The equilibrium line is based on Henry's law, while the operating line represents actual operating conditions between the points ($C_0$, $y_e$) and ($C_e$, $y_0$). *See also* theoretical air-to-liquid ratio.

**strong acid** An acid that has a high degree (approaching 100%) of ionization in water or other dilute solutions. Also defined as an acid that tends to transfer a proton (a hydrogen ion, $H^+$) to another molecule. The strength of the acid is its degree of dissociation or acid dissociation constant. Examples of common strong acids include hydrochloric (HCl), nitric ($HNO_3$), and sulfuric ($H_2SO_4$) acids. They dissociate in water as follows:

$$HA + H_2O = H_3O^+ + A^- \qquad \text{(S-171)}$$

where A represents an anion (Cl) or a radical ($NO_3$, $SO_4$). Strong acids have high dissociation constants (K values) corresponding to inverse logarithms (pK values) less than 3.

**strong-acid cation (SAC) exchanger** Same as strong-acid cation-exchange resin.

**strong-acid cation (SAC) exchange resin** A cation-exchange resin that contains exchangeable functional groups derived from sulfuric acid ($H_2SO_4$) or another strong acid. It operates over a wide pH range (1–14) and is regenerated using a concentrated solution of hydrochloric acid (HCl) or sodium chloride (NaCl). *See also* salt splitting, weak-acid cation (WAC) exchanger.

**strong-acid cation (SAC) resin** Same as strong-acid cation-exchange resin.

**strong-acid ion exchange** *See* strong-acid cation-exchange resin.

**strong-acid resin** *See* strong-acid cation-exchange resin.

**strong base** A base that has a high degree (approaching 100%) of ionization in water or other dilute solutions. Also defined as a base that tends to accept a proton (a hydrogen ion, $H^+$) from another molecule. The strength of the base is its degree of dissociation or basicity constant. Examples of common strong bases include potassium hydroxide (KOH) and sodium hydroxide (NaOH). They dissociate in water as follows:

$$B + H_2O = OH^- + HB^+ \qquad \text{(S-172)}$$

where B represents a cation (K or Na).

**strong-base anion (SBA) exchanger** *See* strong-base anion-exchange resin.

**strong-base anion (SBA) exchange resin** An anion-exchange resin that contains exchangeable functional groups derived from a quaternary ammonium compound. It is used, within a wide pH range (0–13), in water demineralization and for nitrate removal. *See also* salt splitting, weak-base anion (WBA) exchanger.

**strong-base ion exchange** *See* strong-base anion-exchange resin.

**strong-base ion resin** *See* strong-base anion-exchange resin.

**strong chelating agents** All compounds that, by virtue of their chemical structure and amount present, form soluble metal complexes that are not removed by subsequent metal control techniques such as pH adjustment followed by clarification or filtration (EPA-40CFR413.02-f).

**strongyl** Strongyle.

**strongyle** A nematode of the family Strongylidae, intestinal parasites of horses and other mammals. Also spelled strongyl. *See Strongyloides stercoralis.*

**strongylodiasis** *See* strongyloidiasis.

***Strongyloides*** A genus of helminths that multiply within human hosts.

***Strongyloides stercoralis*** The helminth species that causes strongyloidiasis, transmitted between human hosts through soils mainly in warm, wet climates; also called threadworm. Excreted load: 10 per gram of feces. Latency: 3 days. Persistence: 3 weeks, longer in free-living stage. Median infective dose < 100.

**strongyloidiasis** An excreted helminth infection caused by the pathogen *Strongyloides stercoralis; see* its common name, threadworm, for more detail.

**strongylosis** Weakness and anemia caused by strongyle infestation of horses or other mammals. *See also* strongyloidiasis and threadworm.

**strontium (Sr)** A silver-white, reactive metallic element with characteristics similar to calcium, used in glass, metal refining, and other applications. Atomic weight = 87.62. Atomic number = 38. Specific gravity = 2.54. Melting point = 769°C. Boiling point = 1384°C. Valence = 2.

**strontium-89 (Sr-89)** A radioactive isotope of strontium; half-life = 51 days.

**strontium-90 (Sr-90)** A radioactive isotope of strontium present in the fallout from nuclear explosions; a toxic radionuclide, it has a half-life of 28 years. The WHO recommends a maximum acceptable limit of 30 pCi/L in drinking water.

**structural formula** The chemical formula of a compound showing the order in which the atoms in a molecule are arranged; a graphical representation of atoms or groups of atoms relative to each other. *See also* molecular formula.

**structural isomerism** Same as isomerism, the property of compounds that have the same molecular formula but different structural formulas.

**structure** *See* soil structure.

**struvite ($MgNH_4PO_4 \cdot 6\ H_2O$)** A natural compound that forms in solutions saturated with ammonium ($NH_4$), magnesium (Mg), and phosphate ($PO_4$); also found as kidney stones or as hard deposits or scales in wastewater facilities, such as clogged pipes, damaged pumps, filtration membranes, centrifuges, and sludge digesters. It can also cause problems in biological nutrient removal plants that include anaerobic digesters.

**Stuart-Carter** A walking beam flocculator manufactured by Ralph B. Carter Co.

**stuffing box** A box in a pump assembly, located between the impeller and the radial bearing, and designed to prevent the flow of liquid along the shaft. It contains packing rings and the water seal ring.

**Stumm, Werner** Well-known author and professor (1924–1999) in the United States (Harvard University) and in his native Switzerland; considered the father of aquatic chemistry, a term he first coined and a field he helped develop.

**Stuttgart disease** Same as canine leptospirosis.

**Stuttgart's disease** Same as canine leptospirosis.

***S. typhi*** *See Salmonella typhi.*

***S. typhimurium*** *See Salmonella typhimurium.*

**styrene ($C_8H_8$ or $C_6H_5CH{=}CH_2$)** A colorless, insoluble liquid with a penetrating aromatic odor prepared from ethylene and benzene or ethylbenzene that polymerizes to other compounds; used in the production of plastics, resins, and rubber. It is an organic contaminant that affects growth, the kidney, liver, and lungs. Regulated by the USEPA in drinking water: MCLG = MCL = 0.1 mg/L. Also called cinnamene, phenylethylene, styrene monomer, vinyl benzene. *See also* polystyrene.

**styrene acrylonitrile** One of two common thermosetting polymers used to manufacture porous diffusers.

**styrene divinylbenzene resin (SDVB)** An adsorbent resin without functional groups in addition to those in the matrix; it can be used for the remove pesticides and other organic compounds from water. Regenerants used for reactivation include acetone and isopropanol.

**styrene monomer** *See* styrene.

**Styrofoam® pellet** A proprietary bulking material used to maintain the porosity of compost and peat biofilters.

**subaqueous pipe** A pipe that is submerged, laid under water, e.g., on or under the bed of a body of water.

**subchronic** Of intermediate duration, usually used to describe studies or levels of exposure between 5 and 90 days. Health effects of chemicals in drinking water are sometimes considered subchronic when they last 2 to 13 weeks. *See also* chronic.

**subchronic exposure** Multiple or continuous exposures occurring usually over 3 months.

**subchronic study** A toxicity study designed to measure effects from subchronic exposure to a chemical.

**subchronic health effect** A health effect of intermediate duration, as defined under subchronic.

**subchronic toxicity test** A toxicity study conducted in rodents for a period of 90 days.

**subclinical** Pertaining to the early stage of an infection or other disease before symptoms and signs become clinically apparent. *See also* asymptomatic, inapparent infection.

**subcritical fluidization backwash** A backwash procedure for cleaning a filter at a rate that does not fluidize the bed completely, thus leaving the media particles in contact with one another and providing some scouring action. *See also* collapse pulsing backwash, subfluidization.

**subdrainage** A piping system designed to control or remove excess groundwater or seepage, or to lower the water table. *See also* underdrainage.

**subfloor** A reinforced concrete slab that supports the underdrainage system, media, and water load of a trickling filter; it slopes downward to a drainage channel.

**subfluidization** In the extended terminal subfluidization wash procedure, subfluidization refers to the reduction or elimination of filter bed expansion. *See also* subcritical fluidization backwash.

**sublethal toxicity** The characteristic of a constituent that causes adverse effects other than death. *See also* toxicity terms.

**sublimation** (1) The direct conversion of a solid substance to vapor, which then condenses back to a solid, without going through the liquid phase. (2) Also defined as the direct conversion of a gas or vapor to the solid state. (3) The direct passage of water from snow or ice to atmospheric vapor.

**sublimation heat** *See* heat of sublimation.

**submain** A sewer line that collects wastewater from laterals and house sewers and discharges it into a main or trunk sewer. *See also* force main.

**submain sewer** A submain.

**submarine outfall** A pipeline that discharges wastewater or treated effluent into the sea. *See* ocean outfall for detail.

**submerged aerator** A porous or nonporous diffuser installed underwater in biological wastewater treatment as opposed to a surface aerator.

**submerged aquatic vegetation (SAV)** Aquatic vegetation, such as sea grasses, that cannot withstand excessive drying and therefore live with their leaves at or below the water surface. SAVs provide an important habitat for young fish and other aquatic organisms.

**submerged attached growth process** Any of a number of upflow or downflow wastewater treatment processes using a packed-bed or fluidized-bed reactor without secondary clarification. *See* Biocarbone®, Biofor®, Biostyr®, fluidized-bed bioreactor. *See also* nonsubmerged attached growth process, suspended growth with fixed-film packing.

**submerged combustion** A process used to produce carbon dioxide ($CO_2$) by burning gas underwater. The carbon dioxide produced can be applied to the neutralization of nylon and other alkaline wastes. *See also* alkaline waste fermentation.

**submerged-contact aerator** A biological treatment unit that consists of an aeration tank containing stones, plastic sheets, or asbestos-cement sheets on the surface of which microorganisms grow for wastewater stabilization. *See* aerated-contact bed for more detail.

**submerged crib outlet** A waterworks intake built on the bed of a water body and below normal water level.

**submerged horizontal aerator** A mechanical aerator consisting of disks or paddles attached to a rotating shaft and designed to entrain oxygen into a liquid from the atmosphere or from air, or pure oxygen introduced at the bottom of the basin.

**submerged jet flocculation** A flocculation technique in which the coagulated water flows upward into a chamber through a central nozzle to reduce head loss and energy use.

**submerged mechanical aerator** An aeration device that introduces air or oxygen by diffusion into wastewater beneath the impeller or downflow of radial aerators. *See* Figure S-28, draft tube.

**submerged membrane system** A pressure-driven system using hollow-fiber membranes mounted on racks and submerged in a feedwater tank. Vacuum is used to draw water from the membranes and residuals are removed by intermittent backwashing or continuous bleeding of concentrate from the tank. Also called immersed membrane system.

**submerged paddle** A simple mechanical aerator that circulates water and renews its air–water interface. *See also* surface paddle or brush, propeller blade, turbine blade.

**submerged pipe** A pipe that is laid underwater, e.g., on or under the bed of a body of water. Also called subaqueous pipe.

**submerged rotating biological contactor** (1) An integrated fixed-film activated sludge method of wastewater treatment It uses rotating biological contactors submerged about 85% in the aeration tank. *See also* Bio-2-Sludge®, BioMatrix®, Captor®, Kaldnes®, moving-bed biofilm reactor, Ringlace®. (2) A similar unit included in an advanced wastewater treatment facility and operated with methanol addition for denitrification. The 24-mgd plant also includes preliminary treatment, primary sedimentation, BOD removal and nitrification by other RBCs, alum addition for phosphorus removal, secondary clarification, filtration, chlorination, and postaeration.

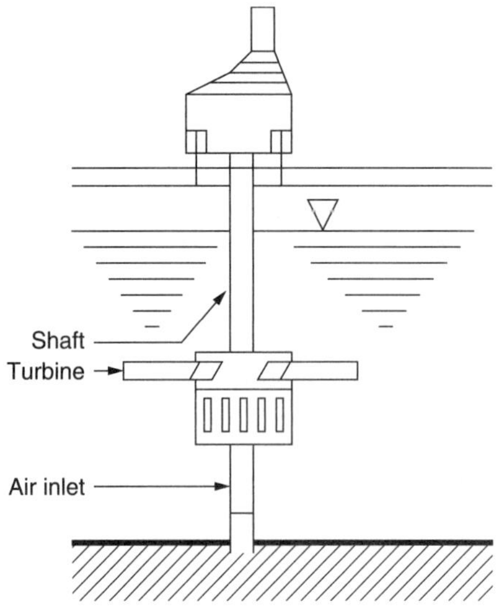


**Figure S-28.** Submerged mechanical aerator (turbine).

**submerged tube evaporator** An apparatus in which steam enters tubes submerged in the fluid to be evaporated.

**submerged turbine aerator** An aeration device consisting of a motor and gear box drive and submerged impeller with piping for air from a blower.

**submerged vertical aerator** A mechanical aerator consisting of an impeller and a turbine or a draft tube; it is designed to entrain oxygen into a liquid from the atmosphere or from air or pure oxygen introduced at the bottom of the basin. *See also* submerged turbine aerator, submerged horizontal aerator, and surface aerators.

**submerged weir** A weir (or dam) whose crest is lower than or at the same elevation as the downstream water level. Submerged weirs are usually not used for flow measurement because of the absence of free-flow conditions. Also called drowned weir. For a submerged weir, the flow equation is modified by the addition of a submergence coefficient. Alternatively, the Villemonte equation may be used; it relates the submerged discharge to the free discharge and to the upstream and downstream heads. The depth of submergence ($h_s$) is the elevation difference between the downstream water surface and the weir crest. *See* weir equation. *See also* Figure W-12.

**submergence** The distance between the water surface and the media surface in a filter.

**submersible pump** A pump, including its motor, installed in a protective housing and operating underwater, e.g., the motor and pump unit installed in a well. A submersible pump station houses submersible pumps. *See also* flooded suction, suction lift, and booster stations.

**submersible pump station** *See* submersible pump.

**submicrometer filter** A cartridge membrane filter that removes particles smaller than 1 micrometer in diameter.

**subnatant** In flotation and other water and wastewater treatment processes, the liquid that remains beneath the surface of floating solids. In the biological flotation of primary sludge, the subnatant is removed as a by-product. *See also* supernatant.

**subnatant liquid** Same as subnatant.

**Suboscreen®** A rotary fine screen manufactured by Andritz-Ruthner, Inc.

**Sub Part H utility** A water supply system that uses surface water or groundwater under the influence of surface water. It is subject to the USEPA's Surface Water Treatment Rule, under EPA-40CFR141, Sub-Part H.

**Subrotor** A progressing cavity pump manufactured by Ingersoll-Dresser Pump.

**subsidence** (1) The lowering of the natural land surface in response to: earth movement; lowering of fluid pressure; removal of underlying supporting material by mining or solution of solids, either artificially or from natural causes; compaction due to wetting (hydrocompaction); oxidation of organic matter in soils; or added load on the land (EPA-40CFR146.3). (2) Same as compression settling or type-4 sedimentation.

**subsidence tank** A sedimentation tank.

**subsidence velocity** *See* settling velocity.

**subsiding basin** A sedimentation basin.

**subsistence agriculture** Production of subsistence crops.

**subsistence crops** Crops intended for the basic needs of a farmer, with little surplus for marketing, or crops that provide only for a marginal livelihood; e.g., the vegetables produced by villagers in developing countries, which are sometimes fertilized by nightsoil and wastewater treatment sludge obtained free of charge. *See also* fodder crops.

**subsistence farming** Production of subsistence crops.

**subspecies** A group of organisms genetically different from the rest of their species.

**substitutive dehalogenation** A nucleophilic reaction in which the hydroxyl group ($OH^-$) replaces the halogen atoms in a mono- or dihalogenated compound, e.g.:

$$CH_3 - CH_2Cl + H_2O \rightarrow CH_3CH_2OH + H^+ + Cl^- \quad \text{(S-173)}$$

*See also* oxidative dehalogenation, reductive dehalogenation.

**substrate** (1) The surface, base or material on which a nonmotile organism lives or grows; e.g., a culture medium, a host organism, and, particularly, the carbonaceous organic matter, energy sources, and the nutrients that microorganisms use for food during biological wastewater treatment. In growth kinetics, the term substrate usually refers to electron donors, although electron acceptors or nutrients can be limiting factors. (2) The surface to which an organism is attached or upon which it moves; also called substratum. (3) The substance or group of substances activated by an enzyme. (4) The surface onto which a coating is applied or into which a coating is impregnated (EPA-40CFR52.741).

**substrate denitrification** Same as preanoxic denitrification, a biological nitrogen removal process in which the wastewater provides organic matter as the electron donor in the redox reactions.

**substrate flux** In an attached-growth process, the mass of substrate that diffuses across the stagnant layer to the biofilm per unit surface area per day ($R_{sf}$, g/m²/day):

$$R_{sf} = -D(S_b - S_s)/L \quad \text{(S-174)}$$

where $D$ = diffusion coefficient of substrate in water; $S_b$ = substrate concentration in the bulk liquid, mg/L; $S_s$ = substrate concentration at the outer layer of the biofilm, mg/L; and $L$ = effective thickness of the stagnant layer.

**substrate limitation** A limiting factor in a biological process due to the insufficiency of an electron donor or an electron acceptor concentrations in the bulk liquid.

**substrate surface flux** Same as substrate flux.

**substrate utilization** The use of wastewater nutrients by organisms for growth in their number and weight of biomass.

**substrate utilization model** The Michaelis–Menten equation applied to the rate of soluble substrate utilization or soluble substrate change ($R_{su}$, g/m²/day) in biological wastewater treatment:

$$R_{su} = kXS/(K_s + S) = kXS_f/(K_s + S_f) \quad \text{(S-175)}$$

where $k$ = maximum specific substrate utilization rate, g substrate/g microorganisms/day; $X$ = concentration of biomass (microorganisms), mg/L; $S$ = growth limiting substrate concentration, mg/L; $K_s$ = half-velocity constant; substrate concentration at one-half the maximum specific substrate utilization rate, mg/L; and $S_f$ = substrate concentration at any point in the biofilm of an attached-growth process, mg/L.

**substrate utilization rate** *See* substrate utilization model.

**substratum** (1) An underlying stratum; a basis or foundation. The soil layer beneath the solum, corresponding to horizons C and D; the subsoil. *See also* soil horizon. (2) Same as substrate (2).

**subsurface air** Same as soil air.

**subsurface disposal** (1) Disposal of wastewater, particularly effluents of septic tanks, in a drainfield, soil absorption system, or subsurface filter. *See* drainfield, soil absorption system. (2) Disposal of wastes through deep-well injection, spreading basins, or in underground cavities. (3) Disposal of salt water into subsurface formations from which oil has been extracted.

**subsurface filter** (1) A more or less horizontal gallery into a water-bearing formation with side and bottom openings. (2) A sand filter installed in an excavation for the treatment of wastewater or septic tank effluent at a loading of about 1.0 gallon per square foot per day. An underdrain system collects the treated effluent for surface discharge.

**subsurface filtration** The use of a subsurface filter to collect source water or to treat wastewater.

**subsurface flow system** One of two common types of constructed wetlands used in the secondary or advanced treatment of wastewater. It consists of channels or trenches on a slope of about 1%, the bottom of which is relatively impermeable and filled with sand or rocks to support emergent vegetation. *See* Figure S-29. The required surface area ($A$, ha) depends on the wastewater flow ($Q$, m³/day), the BOD concentrations in the influent ($C_0$, mg/L) and effluent ($C_e$, mg/L), the first-order rate coefficient ($K$,/day), the depth of media ($D$, m), and the fraction of drainable voids ($F$):

$$A = 0.0001\, Q(\ln C_0 - \ln C_e)/(KDF) \quad \text{(S-176)}$$

Also called rock-reed filter, root-zone filter, or vegetated submerged bed. *See also* free water surface system, reed bed.

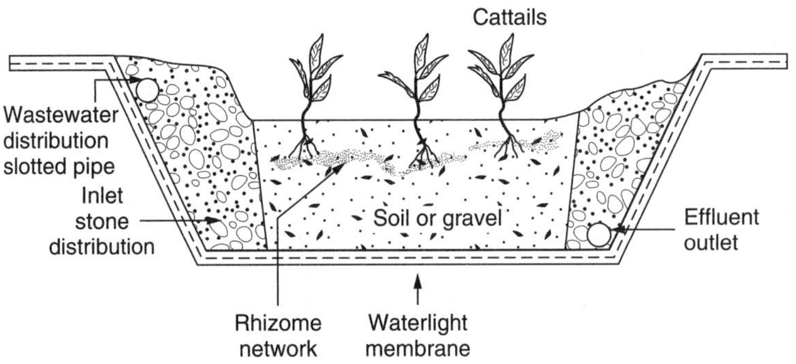


**Figure S-29.** Subsurface flow system.

**subsurface-flow wetland** A constructed wetland in which water flows through a medium that supports emergent vegetation. Same as subsurface-flow system.

**subsurface injection** The disposal of digested sludge approximately 6 inches below the ground surface. *See also* deep-well disposal and underground injection.

**subsurface irrigation** (1) Irrigation using underground porous tiles. (2) A method of treating and disposing of wastewater or effluent by distributing it below the ground surface through open-jointed pipes. *See also* tile field.

**subsurface soil absorption** A method of final treatment and disposal of effluent from septic tanks or other individual treatment units. *See* soil absorption field.

**subsurface soil absorption system** *See* soil absorption field.

**subsurface soil liquefaction** A change in the condition of subsurface soil and other materials below the water table, whereby they become fluidlike during seismic activities, which may cause underground structures to float and above-ground structures to tilt and settle.

**subsurface wastewater disposal** *See* subsurface irrigation (2).

**subsurface water** All water below the earth surface. *See* Figure S-30. According to Meinzer's and other classifications, it includes two main components: interstitial and internal. Related terms listed and defined alphabetically include: aquifer, capillary water, connate water, fixed groundwater, free water, gravitational water, internal water, interstitial water, pellicular water, phreatic water, soil water, unproductive formations, vadose water. Other subsurface water terms are: aeration zone, confining bed, intermediate groundwater, intermediate vadose water, saturation zone, unsaturated zone, water of adhesion.

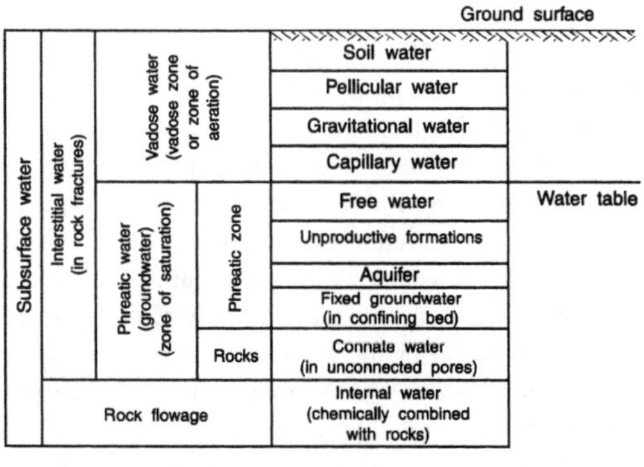


**Figure S-30.** Subsurface water (courtesy CRC Press).

**sucrose ($C_{12}H_{22}O_{11}$)** A crystalline disaccharide from sugarcane, sugar beets, etc., as in common table sugar, which is glucose plus fructose.

**suction dredge** A scow that carries a centrifugal pump for pumping sand; often compared to a powerful underwater vacuum cleaner that sucks up streambed materials and passes them up through a suction hose; commonly used in gold mining. Also called hydraulic dredge, sand-pump dredge, vacuum dredge.

**suction head** *See* dynamic head.

**suction lift** The negative pressure (in feet or meters of water, or in inches or centimeters of mercury vacuum) on the suction side of a pump. The pressure can be measured from the centerline of the pump down to the elevation of the hydraulic gradeline on the suction side of the pump. *See also* dynamic head.

**suction lift pump** A pump set above the surface of the water to be pumped. *See* suction pump for detail.

**suction pipe** The inlet pipeline of a pump.

**suction piping** The inlet piping or suction side of a pump.

**suction pit** A sump, wet well, or a pit that contains the inlet opening(s) of a pump.

**suction pump** A device that consists of a vertical cylinder in which a piston moves up and down to raise water or other fluids, with valves controlling the flow. The pump is set above the surface of the fluid and uses atmospheric pressure to push it into a partial vacuum. Also called suction lift pump.

**suction valve** A valve installed on the suction side of a pump to open or close the suction line.

**sugar** (1) A sweet, crystalline, water-soluble substance obtained from sugar cane, sugar beets, etc. and having the formula $C_{12}H_{22}O_{11}$; a carbohydrate of the same class as fructose, glucose, and lactose. (2) More generally, sugars include monosaccharides, disaccharides, and trisaccharides.

**sugar of iron** A common name of ferrous sulfate ($FeSO_4$). *See* sugar sulfate.

**sugar of lead** Lead acetate.

**sugar of milk** Lactose.

**sugar pollution** The discharge of industrial effluents from such factories as breweries, candy making, dairies, jam and jelly, and glucose processes. Sugar may be harmful to aquatic biota.

**sugar sand** Sand of small grains, which can pass through grit removal units and cause problems in downstream equipment.

**sugar sulfate** A common name for ferrous sulfate ($FeSO_4$), a bluish-green, saline-tasting, coagulant commonly used in water treatment and for other industrial purposes (fertilizer, ink, medicine, etc.); it reacts with the natural alkalinity of water or added lime to form a ferrous hydroxide floc.

**suggested no adverse response level (SNARL)** The concentration of a chemical in water that is expected not to cause an adverse health effect.

**Sul-biSul process** A proprietary process using an anionic-exchange resin to treat brackish and saline waters by converting sulfates to bisulfates; the resin is regenerated with raw water.

**sulfamic acid ($HSO_3NH_2$)** A chemical related to sulfanilamide and often used as a cleaning product.

**sulfanilamide ($C_6H_8N_2O_2S$)** A white, crystalline amide of sulfanilic acid having bactericidal properties. It is also used as a reagent in the laboratory to determine nitrite produced by the reduction of nitrate.

**sulfanilic acid ($C_6H_7NO_3S$)** A grayish-white, crystalline solid used in the production of dyes.

**sulfarsenide** Any compound of arsenide and a sulfide; an arsenic ore.

**sulfate ($SO_4^{2-}$)** A salt or ester of sulfuric acid ($H_2SO_4$) derived from sulfur; an inorganic anion widely distributed in nature. Sulfates are highly soluble in water, particularly those of ammonium, potassium, and sodium. Water pollution by sulfates is caused by detergents and industrial effluents (e.g., from tanneries, steel mills, sulfate pulp mills, and textile plants). Some natural waters are unpalatable because of an excess of sulfates (e.g., 1500 mg/L) and may have laxative effects on new users. The USEPA has set a secondary maximum contaminant level of 250 mg/L for sulfates in drinking water. Sulfates may cause metals in solution to remain soluble or interfere with the formation of a protective film, thereby increasing metallic corrosion. Sulfate is analyzed in wastewater to assess the potential for the formation of odors and the impact on waste sludge treatability; it is reduced to sulfide in sludge digesters and may upset the digestion process. *See* transitory diarrhea.

**sulfate ion** *See* sulfate.

**sulfate of aluminum** *See* aluminum sulfate or alum.

**sulfate process** A chemical process for making wood pulp, using an alkaline liquor [a mixture of sodium sulfate ($Na_2SO_4$) and caustic soda (NaOH)] to digest wood chips. *See* kraft process for detail.

**sulfate-reducing bacteria** Bacteria, mostly of the genera *Desulfovibrio* and *Thiobacillus,* that can reduce sulfates or other forms of oxidized sulfur compounds to sulfides under anaerobic conditions. In sewers and treatment tanks, other bacteria can oxidize hydrogen sulfide and produce a much stronger acid ($H_2SO_4$) that causes microbial corrosion ("microbially induced corrosion") of concrete and metal surfaces with concomitant taste and odor problems:

$$H_2S + 2\ O_2 + \text{bacteria} \rightarrow H_2SO_4 \quad \text{(S-177)}$$

Sulfate-reducing bacteria can also replace oxygen, resulting in a localized increase in pH and a reduction of the effectiveness of disinfectants:

$$SO_4^{2-} + 8\ H^+ + 8\ e^- + \text{bacteria} \rightarrow S^{2-} + 4\ H_2O \quad \text{(S-178)}$$

*See also* sulfur bacteria.

**sulfate reduction** The reduction of the sulfate ion ($SO_4^{2-}$) by anaerobic bacteria, with formation of hydrogen sulfide ($H_2S$) and sulfide ion ($S^{2-}$).

**sulfation** The conversion of sulfuric acid ($H_2SO_4$) and calcium carbonate ($CaCO_3$) into calcium sulfate ($CaSO_4$) in the limestone treatment of acid wastes.

**SulfaTreat®** A process developed by SulfaTreat Co., for the removal of hydrogen sulfide.

**Sulfaver** A reagent produced by Hach Co. for the determination of phosphates in water.

**Sulf Control®** A product of NuTech Environmental Corp. for the inhibition of hydrogen sulfide.

**Sulfex®** A process developed by U.S. Filter Corp. for the precipitation of sulfide and the the removal of heavy metals.

**sulfhydryl group** The univalent group—SH. A mechanism of chlorine disinfection is believed to be its interference with certain enzymes containing the sulfhydryl group. Also called sulfhydryl radical, mercapto group, mercapto radical. *See also* thiol.

**sulfhydryl radical** *See* sulfhydryl group.

**sulfide ($S^{2-}$)** A compound of sulfur with a more electropositive element or a group; an anion present in groundwater and in hot springs; found in wastewater from the decomposition of organic matter and the reduction of sulfate ($SO_4^{2-}$). Total sulfide as measured by the potassium ferricyanide titration method. Sulfide is usually removed from drinking water by air stripping. Undissociated hydrogen sulfide is toxic to aquatic life Sulfide compounds have high odor potentials and, in excess of 200 mg/L, are inhibitory to methanogens and to anaerobic metabolism. *See also* metallic sulfide.

**sulfide pollution** The discharge of industrial effluents from textile dyeing operations, tanneries, and rayon manufacture.

**sulfite ($SO_3^{2-}$)** (1) A salt or ester of sulfurous acid ($H_2SO_3$). (2) A compound containing sulfites, used as a preservative in foods or drug products. Sulfites are found in some industrial wastewaters. *See also* sodium sulfite.

**sulfite dechlorination** The use of sulfite compounds to reduce or eliminate chlorine or combined chlorine residuals from water or wastewater. Compounds used include sodium sulfite ($Na_2SO_3$), sodium bisulfite ($NaHSO_3$), and sodium metabisulfite ($Na_2S_2O_5$):

$$Na_2SO_3 + Cl_2 + H_2O \rightarrow Na_2SO_4 + 2\ HCl \quad (S\text{-}179)$$

$$Na_2SO_3 + NH_2Cl + H_2O \rightarrow Na_2SO_4 + Cl^- + NH_4^+ \quad (S\text{-}180)$$

$$NaHSO_3 + Cl_2 + H_2O \rightarrow NaHSO_4 + 2\ HCl \quad (S\text{-}181)$$

$$NaHSO_3 + NH_2Cl + H_2O \rightarrow NaHSO_4 + Cl^- + NH_4^+ \quad (S\text{-}182)$$

$$Na_2S_2O_5 + Cl_2 + 3\ H_2O \rightarrow 2\ NaHSO_4 + 4\ HCl \quad (S\text{-}183)$$

$$Na_2S_2O_5 + 2\ NH_2Cl + 3\ H_2O \rightarrow Na_2SO_4 + 2\ Cl^- + H_2SO_4 + 2\ NH_4^+ \quad (S\text{-}184)$$

**sulfite ion** *See* sulfite. The sulfite ion may exist as $SO_3^{2-}$ or $HSO_3^-$.

**sulfite pollution** The discharge of industrial effluents from the manufacture of wood and paper products and viscose.

**sulfite process** The production of wood pulp by digesting wood chips in an acidic mixture of sulfurous acid ($H_2SO_3$) and the bisulfite ion ($HSO_3^-$) or a salt such as calcium bisulfite. *See also* kraft process.

**sulfite-reducing *Clostridium*** Any member of a group of anaerobic organisms that form endospores that are very resistant to disinfection and other adverse conditions. This group includes the *Clostridium* genus, particularly the *C. perfringens* species, and the genus *Desulfotomaculum.* Also called anaerobic sporeformers.

***Sulfolobus* genus** A genus of thermophilic archaebacteria associated with acid mine drainage.

**sulfonate** (1) A salt or ester derived from a sulfonic acid. *See* alkyl sulfonate. (2) To convert an organic compound into a sulfonic acid by treatment with sulfuric acid ($H_2SO_4$) or to introduce the sulfonic group (—$SO_3H$) into an organic compound.

**sulfonation** (1) *See* sulfonate (2). (2) The addition of sulfur dioxide to water to remove excess chlorine:

$$Cl_2 + SO_2 + 2\ H_2O \rightarrow H_2SO_4 + 2\ HCl \quad (S\text{-}185)$$

*See also* dechlorination. Also spelled sulphonation.

**sulfonator** A device (e.g., an orifice flowmeter or a dosimeter) used to inject and monitor sulfur dioxide ($SO_2$) in dechlorination processes. *See also* chemical feeder.

**sulfonic acid (—$SO_2OH$)** Any of a group of organic compounds, strong acids that form neutral sodium salts used in the production of phenols, ion-exchange resins, dyes, etc.

**sulfonic group** *See* sulfonic acid.

**sulfonium ($H_2S^+$)** A positively charged group containing the sulfur and hydrogen elements, its salts or substitute products.

**sulfotransferase** A class of enzymes that transfer sulfate ($SO_4^{2-}$) to form esters.

**sulfur (S)** A pale, yellow, odorless, brittle, nonmetallic element, insoluble but found in drinking water as sulfate ($SO_4^{2-}$) and sulfide ($S^{2-}$), causing taste, odor, and other problems in distribution systems; it burns with a blue flame and a suffocating odor. Sulfur is required for the synthesis of proteins and is released in their decomposition. It is used in gunpowder, matches, rubber, and in medicine. Air stripping removes sulfide but not sulfate. Atomic weight = 32.064. Atomic number = 16. Specific gravity = 2.07 at 20°C. Valence = 2, 4, and 6. Also spelled sulphur. Also called brimstone.

**sulfurated hydrogen** Same as hydrogen sulfide ($H_2S$).

**sulfur bacteria** Bacteria, mostly of the genera *Beggiatoa* and *Thiobacillus,* that can derive energy from dissolved sulfur or inorganic sulfur compounds. Some of them can oxidize hydrogen sulfide to the sulfate ion, which is eventually converted to sulfuric acid, a strong acid that causes corrosion:

$$H_2S + 2\ O_2 + \text{bacteria} \rightarrow 2\ H^+ + SO_4^{2-} \rightarrow H_2SO_4 \quad (S\text{-}186)$$

*See also* crown corrosion, sulfate-reducing bacteria.

**sulfur bacterium** *See* sulfur bacteria.

**sulfur compound joint** A bell-and-spigot pipe joint sealed with a melted sulfur compound instead of lead. The jointing compound is sometimes called mineral lead.

**sulfur cycle** The chemical transformation of sulfur in nature through various stages of decomposition and assimilation, including the different forms of the element found in wastewater treatment and sludge processing. The main species involved in the cycle are hydrogen sulfide ($H_2S$), mineral sulfides (e.g., FeS), sulfuric acid ($H_2SO_4$), and the organic sulfur in some proteins.

**sulfur dioxide ($SO_2$)** A pungent, colorless, nonflammable, soluble, suffocating, gaseous pollutant formed primarily by the combustion of fossil fuels (when sulfur burns), but also by some industrial processes, such as paper production and metal smelting. Sulfur dioxide is closely related to sulfuric acid, a strong acid, and plays an important role in acid rain. It is used as an antichlor (to remove excess chlorine residual from water or wastewater), as a food additive, and in chemical synthesis (e.g., production of sulfuric acid, $H_2SO_4$).

**sulfur dioxide dechlorination** The removal or reduction of chlorine (in the form of hypochloric acid, HOCl), chloramines, or chlorine dioxide ($ClO_2$) residuals by reaction with sulfur dioxide ($SO_2$):

$$SO_2 + H_2O \rightarrow HSO_3^- + H^+ \quad (S\text{-}187)$$

$$HOCl + HSO_3^- \rightarrow Cl^- + SO_4^{2-} + 2\,H^+ \quad (S\text{-}188)$$

$$HOCl + SO_2 + H_2O \rightarrow Cl^- + SO_4^{2-} + 3\,H^+ \quad (S\text{-}189)$$

$$NH_2Cl + SO_2 + 2\,H_2O \rightarrow Cl^- + SO_4^{2-} + NH_4^+ + 2\,H^+ \quad (S\text{-}190)$$

$$2\,ClO_2 + 5\,SO_2 + 6\,H_2O \rightarrow 2\,HCl + 5\,H_2SO_4 \quad (S\text{-}191)$$

**sulfur dioxide process** A process used to generate chlorine dioxide ($ClO_2$) on the site of water treatment plants, with sodium chlorate ($NaClO_3$), sulfur dioxide ($SO_2$), and sulfuric acid. *See* Mathieson process for more detail.

**sulfuric acid ($H_2SO_4$)** A clear, colorless to brownish, dense, oily liquid that is miscible in water; used extensively in industrial as well as agricultural chemicals, in petroleum refining, as a regenerant of ion exchange systems, in laboratory tests (e.g., COD test), and to lower the pH of alkaline solutions (e.g., in alum coagulation of colored water). A strong, highly corrosive acid, usually produced from sulfur dioxide ($SO_2$), which is cooled and converted to sulfite ($SO_3$ gas) and hydrolized to $H_2SO_4$. Sulfuric acid droplets result in the atmosphere from the combination of sulfur trioxide and water. Special types of stainless or lined steel are used as materials resistant to $H_2SO_4$. It can precipitate calcium sulfate ($CaSO_4$) in filtration units and cause fouling of the media. Also called oil of vitriol. *See also* acid rain, crown corrosion, sulfate-reducing bacteria, sulfur bacteria.

**sulfuric acid treatment** A method of neutralization of alkaline wastes by addition of sulfuric acid ($H_2SO_4$). The neutralization reaction, in which alkalinity is represented by sodium hydroxide (NaOH), is:

$$2\,NaOH + H_2SO_4 \rightarrow Na_2SO_4 + 2\,H_2O \quad (S\text{-}192)$$

**sulfuric anhydride** Same as sulfur trioxide ($SO_3$).

**sulfuric ether** A colorless, highly volatile, flammable liquid, with an aromatic odor and sweet, burning taste, resulting from the action of sulfuric acid ($H_2SO_4$) on ethyl alcohol ($C_2H_5OH$); used as a solvent in analytical chemistry to extract polar organics. *See* diethyl ether for detail.

**sulfuritic material** A compound of iron and sulfur found in coal mines and in acid mine water.

**sulfur mineralization** The aerobic or anaerobic release of sulfur from organic matter in such volatile compounds as hydrogen sulfide or dimethylsulfide, which can then be photooxidized to sulfate.

**sulfurous acid ($H_2SO_3$)** A colorless liquid with a suffocating odor produced from the dissolution of sulfur dioxide ($SO_3$) in water or by the combination of sulfur oxides with atmospheric moisture. It can further combine with water in air to form sulfuric acid ($H_2SO_4$).

**sulfur oxides (SOx)** The air contaminants sulfur dioxide ($SO_2$) and trioxide ($SO_3$) from natural sources (e.g., volcanic eruptions, sulfur springs, phytoplankton activity in decaying organic matter) and man-made sources (principally combustion of sulfur-containing fuels in the presence of oxygen). A pungent gas, $SO_2$ is used as a disinfectant and preservative in the food industry; it is the predominant form but it may convert to $SO_3$. Both affect the respiratory tract and can cause bronchial spasms above certain concentrations. These oxides are a cause of acid rain.

**sulfur oxidizer** A microorganism, such as *Thiobacillus thiooxidans,* that oxidizes sulfide and forms sulfuric acid ($H_2SO_4$). *See* sulfur-oxidizing bacteria, crown corrosion.

**sulfur-oxidizing bacteria** Bacteria that use reduced sulfur compounds for growth, including (a) a group of strictly aerobic chemoautotrophs that oxidize sulfide ($H_2S$) to elemental sulfur ($S^0$) or elemental sulfur to sulfuric acid ($H_2SO_4$), including

*Achromatium, Beggiatoa, Thermothrix, Thiobacillus,* and *Thiomicrospira:*

$$H_2S + \frac{1}{2}\, O_2 \rightarrow S^0 + H_2O \qquad \text{(S-193)}$$

$$S^0 + 1.5\, O_2 + H_2O \rightarrow H_2SO_4 \qquad \text{(S-194)}$$

(b) strictly anaerobic photoautotrophic, green and purple bacteria that fix carbon using light energy and oxidize sulfide to sulfur ($S^0$), including *Chlorobium, Chromatium, Ectothiorhodospira, Rhodopseudomonas,* and *Thiopedia:*

$$CO_2 + H_2S \rightarrow S^0 + \text{fixed carbon} \qquad \text{(S-195)}$$

and (c) some aerobic heterotrophic bacteria that can oxidize sulfur to thiosulfate or sulfate.

**sulfur-reducing agent** A reducing agent that contains sulfur. Common agents include sodium bisulfite, sodium sulfite, sodium thiosulfate, and sulfur dioxide.

**sulfur-reducing bacteria** Bacteria that use sulfate as terminal electron acceptor and hydrogen as electron donor:

$$4\, H_2 + SO_4^{2-} \rightarrow S^{2-} + 4\, H_2O \qquad \text{(S-196)}$$

Sulfur-reducing bacteria include the following genera: *Desulfobacter, Desulfobulbus, Desulfococcus, Desulfonema, Desulfosarcina, Desulfotomaculum,* and *Desulfovibrio. See also* sulfate-reducing bacteria.

**sulfur reservoir** Sulfur is out-gassed from the Earth's core through volcanic activity and found in the Earth's crust as inert elemental sulfur deposits, sulfur–metal precipitates, and buried fossil fuels.

**sulfur respiration** One of two forms of anaerobic sulfur reduction whereby bacteria like *Desulfuromonas* use elemental sulfur ($S^0$) as terminal electron acceptor and a low-molecular-weight compound such as acetate ($CH_3COOH$), ethanol, or propanol as a carbon source:

$$CH_3COOH + 2\, H_2O + 4\, S^0 \rightarrow 2\, CO_2 + 4\, S^{2-} + 8\, H^+ \qquad \text{(S-197)}$$

*See also* assimilatory sulfate reduction and dissimilatory sulfate reduction.

**sulfur spring** A spring, the water of which contains sulfur compounds naturally, often with an emanating odor of hydrogen sulfide.

**sulfur trioxide ($SO_3$)** An irritant, corrosive solid resulting from the oxidation of sulfur dioxide ($SO_2$):

$$2\, SO_2 + O_2 \rightarrow 2\, SO_3 \qquad \text{(S-198)}$$

This reaction may readily occur in the atmosphere on the surface of water droplets and lead to the formation of sulfuric acid ($H_2SO_4$). Sulfur trioxide is used in the production of sulfuric acid. Also called sulfuric anhydride.

**sulfur water** Water with a rotten-egg odor because of excessive concentration of hydrogen sulfide ($H_2S$).

**sullage** All nontoilet household wastewater from sinks, basins, baths, and showers; that is, domestic wastewater that does not contain excreta and is expected to contain considerably fewer pathogenic microorganisms than sewage. Sullage ponding on the ground is a health risk because it can promote the breeding of such vectors as *Culex pipiens* and *Anopheles* mosquitoes. Also called graywater. *See also* sanitary water.

**sullage soakaway** An arrangement such as a pit, trench, or mound, used to promote the seepage of sullage into the ground.

**sulphate** British spelling of sulfate.

**sulphide** British spelling of sulfide.

**sulphite** British spelling of sulfite.

**sulphonate** British spelling of sulfonate.

**sulphonation** *See* sulfonation (2).

**sulphur** British spelling of sulfur.

**sulphur dioxide** British spelling of sulfur dioxide. *See* sulfur oxides.

**sulphuric** British spelling of sulfuric.

**sulphuric acid** British spelling of sulfuric acid.

**sulphur trioxide** British spelling of sulfur trioxide. *See* sulfur oxides.

**Sulzer** Equipment manufactured by Dorr Oliver, Inc. for grinding screenings from wastewater treatment.

**Sumigate®** A rubber dam manufactured by Rodney Hunt Co. for use in the control of combined sewer overflows.

**summer stagnation** The phenomenon occurring in lakes in the summer when a stable equilibrium is established between a warmer, lighter layer above a colder, denser layer. It extends a few months between April and November, with the bottom layer almost stagnant near 39.2°F or 4°C (at maximum water density). Mechanical or pneumatic aeration may be used to eliminate or reduce this phenomenon. *See also* thermal stratification.

**sum of five haloacetic acids (HAA5)** The sum of the concentrations of five haloacetic acids (monochloroacetic, dichloroacetic, trichloroacetic, monobromoacetic, and dibromoacetic acids), regulated under the Disinfectant/Disinfection By-Products Rule.

**sum of six haloacetic acids (HAA6)** Symbol representing the sum of the six haloacetic acids considered in the formation of disinfection by-products: trichloroacetic, dichloroacetic, chloroacetic, bromochloroacetic, bromoacetic, and dibromoacetic acids.

**sump** (1) A pit, tank, or reservoir that receives and temporarily stores stormwater, wastewater, or water for subsequent removal (usually by pumping). (2) Any pit or reservoir that meets the definition of tank and those troughs/trenches connected to it that serve to collect hazardous waste for transport to hazardous waste storage, treatment, or disposal facilities (EPA-40CFR260.10).

**Sumpaire™** An automatic jet aerator manufactured by Framco Environmental Technologies.

**Sump-Gard®** A vertical centrifugal pump manufactured by Vanton Pump & Equipment Corp.

**sump pit** Same as sump.

**sump pump** A small, single-stage vertical pump, a submerged centrifugal pump, or an ejector used to drain sumps, shallow pits, or wet wells.

**sunlight disinfection** *See* solar disinfection, ultraviolet disinfection.

**sunlight effect** Sunlight affects water quality as a disinfectant; it also promotes algal growth with a corresponding production of oxygen during the day (possibly leading to oversaturation of dissolved oxygen) and deoxygenation at night.

**Suparator®** A device manufactured by Lemacon Techniek B. V. for the separation of oil from water.

**Super Blend™** A membrane made of cellulose acetate by TriSep Corp. for use in reverse osmosis processes.

**Superblock II®** An underdrain system manufactured by F. B. Leopold Co. with a device for filter backwashing using either water or air and water.

**Super Blue®** A steel storage tank manufactured by U.S. Filter Corp.

**superbug** A genetically engineered organism capable of degrading contaminants at a very high rate or specifically adapted for a particular contaminant. *See also* bioaugmentation.

**Super-Cel®** Diatomaceous earth manufactured by Celite Corp. for use as filter media.

**Supercell®** An apparatus manufactured by Krofta Engineering Corp. for water treatment using dissolved-air flotation.

**superchlorination** Chlorination with high doses that are deliberately selected to destroy tastes and odors or to produce free or combined residuals so large as to require dechlorination.

**superchlorination/dechlorination** A treatment method used for waters of poor quality (e.g., high ammonia content and/or taste and odor problems) in which chlorine is added beyond the breakpoint and the residual is reduced by the application of a dechlorinating agent such as a sulfur compound or activated carbon.

**supercool** To cool a liquid below its freezing point without solidifying or crystallizing it.

**supercritical fluid** (1) A fluid that possesses the characteristics of both a liquid and a gas above a certain temperature and pressure, e.g., carbon dioxide ($CO_2$) above 31.1°C and 73.8 atm. Supercritical fluids are used to transfer hazardous substances by solvent extraction or leaching. (2) More generally, supercritical fluids exist in a critical temperature and pressure region:

$$0.9 < T_r < 1.2 \quad \text{and} \quad 1.0 < PT_r < 3.0 \qquad \text{(S-199)}$$

where

$$T_r = T/T_c = \text{reduced temperature} \qquad \text{(S-200)}$$

$T$ = actual temperature, $T_c$ = critical temperature,

$$P_r = P/P_c \text{ reduced pressure} \qquad \text{(S-201)}$$

$P$ = actual pressure, and $P_c$ = critical pressure. They are gases at ambient temperatures.

**supercritical fluid chromatography (SFC)** A technique using a supercritical fluid like carbon dioxide ($CO_2$) to analyze nonvolatile compounds.

**supercritical water** Water that is near its critical temperature, $T_c$ = 374°C or 704°F and critical pressure $P_c$ = 22.1 MPa or 3208 psia. Its characteristics make it an excellent solvent for organic compounds: low density (< 0.6 g/cc), weak hydrogen bonding, and low polarity. It is used in the destruction of organics.

**supercritical water oxidation (SCWO)** (1) A type of thermal treatment using moderate temperatures and high pressures to enhance the ability of water to break down large organic molecules into smaller, less toxic ones. Oxygen injected during this process combines with simple organic compounds to form carbon dioxide and water. (2) A treatment process that oxidizes organic materials in wastewater sludge and hazardous wastes at high temperature and pressure above the fluid's critical point. The process converts the materials to carbon dioxide, water, nitrogen, and organic acids, with efficiencies ranging from 99.99 to 99.9999%. It consists of six distinct unit operations: feed pressurization, heating, reaction, effluent cooling, depressurization, and separation. Also called supercritical wet oxidation and hydrothermal oxidation.

**supercritical wet oxidation** Same as supercritical water oxidation.

**Super Detox™** A process developed by Conversion Systems, Inc. using encapsulation for the chemical stabilization of furnace dust and heavy metal residues.

**Super Dome™** A ceramic dome diffuser manufactured by Ferro Corp.

**Superdraw** A device manufactured by Walker Process Equipment Co. for the withdrawal of supernatant.

**superficial velocity** The velocity ($V_s$) of flow through a filter related to the surface area ($A_s$) of the filter; also called the surface loading rate:

$$V_s = Q/A_s = eV \qquad \text{(S-202)}$$

where $Q$ is the volumetric flow rate, $e$ is the porosity of the bed, and $V$ is the velocity related to the depth.

**Superfloc®** A line of flocculants in liquid, dry, and emulsion forms produced by Cytec Industries, Inc. for use in water treatment, including the high-charge cationic flocculant Superfloc Excel.

**Superfloc Excel** *See* Superfloc®.

**superheat** (1) To heat a liquid above the boiling point without forming bubbles or vapor; or to heat a gas to a point where one can reduce its temperature or increase its pressure without liquefying it. (2) The state of being superheated or the amount of superheating.

**supernatant** Liquid removed from settled sludge. Supernatant commonly refers to the liquid between the settled sludge on the bottom and the water surface of a sedimentation basin or a gravity thickener. In a sludge digester, supernatant refers to the most liquid portion, i.e., the water of separation between a top scum layer and a bottom layer of actively digesting sludge and digested concentrate. Depending on its origin, supernatant may be recycled to a preceding unit of the plant, discharged to a water course, or submitted to further treatment before discharge. In all cases, supernatant quality is judged in terms of its clarity or its total suspended solids (TSS) concentration and its organic loading: e.g., a typical supernatant of aerobic sludge digestion has a median TSS of 3400 mg/L and a COD of 2600 mg/L. Also called supernatant liquid or supernatant water. *See also* subnatant.

**supernatant liquid** Same as supernatant.

**supernatant liquor** The liquid overlying deposited solids or situated between the sludge at the bottom and the floating scum at the top of digestion tank.

**supernatant water** (1) Same as supernatant. (2) The liquid portion of a body of water that has mud and sludge deposits at the bottom.

**superoxide** A compound that contains the univalent ion ($O_2$); peroxide. Also called hyperoxide.

**superoxide anion** A highly reactive form of reduced oxygen, a product of biological oxidations. *See* superoxide ion.

**superoxide ion ($O_2^-$)** An intermediate product in the chain of reactions of ozone autodecomposition:

$$HO_2 \rightarrow H^+ + O_2^- \qquad \text{(S-203)}$$

*See also* hydroperoxyl radical.

**superparasite** A parasite that feeds on other parasites. Also called a hyperparasite.

**superphosphate** A product of the treatment of phosphate rock with sulfuric acid ($H_2SO_4$); it contains soluble phosphorus pentoxide ($P_2O_5$). Further treatment of superphosphate with phosphoric acid increases the $P_2O_5$ content and produces triple superphosphate.

**Superpulsator®** A solids contact clarifier manufactured by Infilco Degremont, Inc., with inclined plates and intermittent pulsing.

**super-rate filter** A trickling filter packed with plastic media and used as a roughing filter for strong wastewaters or as a secondary treatment unit. *See also* high-rate trickling filter, intermediate-rate trickling filter, low-rate trickling filter, roughing filter, biofilter, biotower, oxidation tower, dosing tank, filter fly.

**supersaturated** Pertaining to an unstable solution (water) that contains a dissolved substance (solute) at a concentration greater than the saturation concentration for the substance.

**supersaturated solution** A solution that contains a greater concentration of a solute than is possible at equilibrium under fixed conditions of temperature and pressure (EPA-40CFR796.1840-v). Also called an oversaturated solution. *See also* saturated solution, crystal growth, nucleation.

**supersaturation** (1) The unstable condition of a solution that contains a solute at a concentration exceeding saturation at a given temperature and pressure. (2) The unstable condition of a vapor with a density greater than its equilibrium density at a given temperature and pressure. (3) The unstable condition of a space that contains more water vapor than at saturation under a given temperature and pressure.

**Super Shredder®** An in-line macerator of Franklin Miller.

**Super Sieve Screen** A sieve screen manufactured by Sizetec, Inc.

**Superslant™** An inclined-plate clarifier manufactured by Infilco Degremont, Inc.

**superstructure** The structure built above the floor of a latrine to provide privacy and protection.

**Superthickener** A large circular gravity thickener manufactured by Dorr Oliver, Inc.

**supervisory control and data acquisition (SCADA)** An information system used at wastewater treatment plants to control and continuously monitor the operation of sewer collection networks by transmitting data over telephone lines, by radio, or over cable lines. SCADA allows the remote control of such devices as pumps and valves and the remote acquisition of such data as flows, pressures, and water levels. Current telemetry systems allow direct input of field data into collection system models. *See also* sewer-level remote telemetry. SCADA has advanced to a level where the computer systems can handle many, if not most, of the routine issues facing a utility.

**supplemental hypochlorination** The installation of a chlorinator in an office or institutional building to reinforce the chlorine residual in the water system as a preventive measure against the regrowth of *Legionella* bacteria.

**supplemental *Legionella* control** In addition to primary disinfection to limit entry into the distribution system and secondary disinfection to inhibit regrowth of *Legionella,* disinfection of susceptible home plumbing fittings is a consideration. Localized control measures against *Legionella* colonization include supplemental hypochlorination, thermal inactivation through higher temperatures in water heaters, and supplemental disinfection using ozone or ultraviolet light.

**supplier of water** Any person who owns or operates a public water system.

**supply line** A conduit between the water source and a treatment plant or distribution system.

**support gravel** A layer of graded gravel installed underneath the filter media to separate the media from the underdrainage system.

**support media bed** A secondary medium used in a granular filter or an ion exchange unit to support the primary medium and improve backwashing.

**suppression** An ion chromatography technique used to reduce the conductivity of the eluant and increase the conductivity of the constituent being analyzed.

**Supracell** A treatment device manufactured by Krofta Engineering Corp.; uses dissolved-air flotation.

**Suprex™** A device manufactured by Graver Co. for the treatment of condensate using a mixed bed.

**surcharge** (1) The hydraulic or organic load on a system beyond what is anticipated or designed for. (2) *See* surcharge pricing.

**surcharged condition** Same as surcharging.

**surcharge pricing** A method of charging customers for water or wastewater service when they exceed set quantity or quality limits, or during times of peak usage.

**surcharging** A flow condition that exists when the capacity of a sewer is exceeded.

**SURF®** A two-stage treatment apparatus manufactured by General Filter Co. and combining contact clarification and filtration.

**surface-active agent** *See* surfactant.

**surface aeration** The absorption of air through the surface of a liquid; the use of a mechanical device (e.g., a brush aerator or turbine aerator) to provide air through the surface of water, wastewater, or stormwater. Surface aeration is also used to strip gases and volatile contaminants; see air stripping.

**surface aerator** A mechanical device that is designed to transfer oxygen from the atmosphere into a liquid by surface renewal and exchange, and to provide mixing to the contents of the basin or container. Surface aerators may have radial flow at low speed or axial flow at high speed. There are also aspirating devices and aerators with horizontal rotors. Another type of mechanical aerator is submerged. *See also* horizontal surface aerator, submerged aerator, vertical surface aerator.

**surface biological adsorption** A minor mechanism that contributes to the removal of particulate matter within a granular filter; see biological growth.

**surface brush** A simple mechanical aeration device that dips lightly into the aeration chamber to circulate water, release air bubbles, and spray droplets on the surface. Surface paddles and surface brushes are similar devices. *See also* submerged paddle, propeller blade, turbine blade.

**surface charge** The electrical charge on the surface of most particles in water. The charge depends on the concentration of protons in the solution, but most particles are negatively charged. In water filtration, surface charges may aid or inhibit attachment; they are affected by pH changes and coagulant addition. Factors or mechanisms causing the development of surface charges include chemical reactions at the surface, isomorphous replacement, structural imperfections, preferential adsorption, and ionization. *See also* electrical double layer, isoelectric point, net proton charge, point of zero net proton charge, point of zero salt effect, potential surface, surface potential, zeta potential.

**surface charge neutralization** The reduction of the net surface charge of particles in a suspension to decrease the energy required for them to come into contact with one another. It is one of two common methods of mixing chemicals for coagulation; coagulants are dispersed rapidly (e.g., in less than one second) and mixed at high intensity, forming hydrolysis products that will contact the colloidal particles for precipitating hydroxides. *See also* double-layer compression, heterocoagulation, interparticle bridging.

**surface charge production** The charge on the surface of a particle results from (a) the exchange of protons between the particle and the surrounding water, (b) the exchange of protons between the particle and the solutes, or (c) the structural imperfections of the particle (isomorphic replacement).

**surface complexation model** A quantitative model that describes the formation of charge, potential, and ion adsorption at the particle–water interface. *See* constant capacitance model.

**surface condenser** A shell-and-tube device for condensing steam or vapor by passing it over a cool surface.

**surface deposit** Unconsolidated material that covers the bedrock after in situ weathering or transport from another site by wind, water, ice, or gravity. Also called surficial deposit.

**surface disposal** The disposal of sludge from water or wastewater treatment in dedicated land disposal sites, in monofills, in piles or mounds, and in impoundments or lagoons, excluding the placement of solids for storage purposes.

**surface disposal site** An area of land that contains one or more active sewage sludge units (EPA-40CFR503.21-p).

**surface film** The stationary layer of liquid immediately surrounding an adsorbing particle. *See* film diffusion transport. Also called hydrodynamic boundary layer.

**surface filtration** Filtration occurring at the surface of a granular media filter, used for the removal of residual suspended solids from wastewater. Surface filtration occurs by mechanical sieving when the liquid passes through a thin septum such as a metal or cloth fabric, as in an automobile air filter or a kitchen colander. *See also* filter septum, membrane filter, Cloth-Media Disk Filter®, Discfilter®, depth filtration.

**surface flow wetland** A man-made wetland that has a free water surface, functions like a natural wetland, and supports emergent vegetation. A first-order expression is commonly used to model the removal performance of this type of wetland:

$$C = C_0 e^{-KAD/Q} \quad \text{(S-204)}$$

where $C$ = effluent BOD concentration, $C_0$ = influent BOD concentration, $K$ = rate constant, $A$ = surface area of the bed, $D$ = depth of submergence of the bed, and $Q$ = influent flow rate. *See also* subsurface wetland.

**surface flux limitation** *See* substrate flux and surface limitation.

**surface flux substrate limitation** *See* substrate flux and surface limitation.

**surface impoundment** (1) Treatment, storage, or disposal of liquid hazardous wastes in ponds. (2) A natural topographic depression, man-made excavation, or diked area formed primarily of earthen materials that is designed to hold, treat or dispose of liquid wastes or wastes containing free liquids and that is not an injection well. A surface impoundment may also be lined with man-made materials. Examples of surface impoundments are holding, storage, settling, and aeration pits, ponds, and lagoons (EPA-40CFR 280.12, 61.341, 63.111, and 260.10). Also simply called an impoundment.

**Surface Impoundment Assessment** A survey conducted by the USEPA in 1983 to account for, locate, and characterize more than 180,000 surface impoundments in the United States.

**surface impoundment closure** A set of activities conducted to close a surface impoundment that is no longer being used, including a clean closure or an in-place (landfill) closure, the installation of a final cover, and postclosure care.

**surface infiltration** A method of groundwater recharge. *See* surface spreading for detail. *See also* direct injection.

**surface irrigation** The application of water other than by spraying and so that the irrigation water does not come in contact with the edible portion of any food crop.

**surface loading** One of the guidelines for the design of settling tanks and clarifiers in treatment plants. Used by operators to determine if tanks and clarifiers are hydraulically over- or underloaded. *See* overflow rate, surface loading rate, surface overflow rate.

**surface loading rate** In water and wastewater treatment, the surface loading rate (also called the overflow rate) is one of the design criteria for settling tanks; expressed in gallons per day per square foot, it is equal to the ratio of the average flow rate (in gpd) to the surface area of tank (in $ft^2$). For a rectangular tank of length $L$ (ft) and width $W$ (ft), the surface loading rate or overflow rate ($OR$, gpd/$ft^2$) is:

$$OR = Q/(WL) \quad \text{(S-205)}$$

where $Q$ (gpd) is the flow rate. The surface loading rate or surface overflow rate of a secondary clarifier can also be determined knowing the zone or interface settling velocity ($V_i$, m/hr or $m^3/m^2/hr$) and applying a safety factor ($F$, e.g. 1.75 – 2.5):

$$SLR = 24\ V_i/F \qquad \text{(S-206)}$$

The surface loading rate also applies to filters; *see* superficial velocity. *See also* overflow rate, surface overflow rate, zone settling velocity, solids loading rate, weir loading rate.

**surface mechanical aerator** Same as surface aerator.

**surface mining** Same as strip mining or open-cut mining.

**Surface Mining Control and Reclamation Act (SMCRA)** A 1977 U.S. Public Law (No. 95-87) that establishes performance standards and requires permits to protect the environment against adverse effects of surface coal-mining operations.

**surface overflow rate** Same as surface loading rate.

**surface paddle** A simple mechanical aeration device that dips lightly into the aeration chamber to circulate water, release air bubbles, and spray droplets on the surface. Surface paddles and surface brushes are similar devices. *See also* submerged paddle, propeller blade, turbine blade.

**surface potential** *See* zeta potential.

**surface precipitation** A mechanism of ion transfer between two substances that is different from ion exchange; e.g., the precipitation of iron and manganese on manganese zeolite and the regeneration of the latter with potassium permanganate. Also called contact precipitation.

**surface protonation** Same as net proton charge.

**surface pump** A device for removing water or wastewater from a sump or wet well.

**surface reaeration** The net flux of dissolved oxygen (DO) from the atmosphere through the free surface of a water body. It is a function of the surface reaeration rate ($K_2$/day) and the DO deficit in the water, which is the difference between the DO saturation concentration ($C_s$, mg/L) and the actual DO concentration ($C$, mg/L):

$$R = K_2\,(C_s - C) \qquad \text{(S-207)}$$

where $R$ is the rate of oxygen gain due to reaeration over unit time per unit volume, e.g., in mg/L/day.

**surface reaeration rate ($K_2$)** *See* reaeration constant.

**surface resistance** Resistance to fluid flow causing head losses by friction over surfaces; such losses in conduits are significant for long distances but negligible compared to form losses over short distances. Also called frictional resistance. *See* the Darcy–Weisbach and Hazen–Williams equations.

**surface runoff** Precipitation, snowmelt, or irrigation in excess of what can infiltrate the soil surface and be stored in small surface depressions; a major transporter of nonpoint source pollutants. It includes overland runoff and the precipitation that falls directly into the channels. Also called overland flow, excess rainfall, or storm flow. *See* rainfall–runoff relationship, quick-response runoff, surface wash (2).

**Surface Scatter** An on-line turbidimeter manufactured by Hach Co.

**surface screen** A device used to keep debris out of a surface water intake.

**surface settling rate** A sedimentation tank design parameter defined as the average daily overflow divided by the surface area of the tank, expressed in $gpd/ft^2$. *See* surface loading rate, surface overflow rate.

**surface spreading** A common method of groundwater recharge, whereby water percolates from a spreading basin through the vadose zone. Also called surface infiltration. *See also* geopurification, high-rate infiltration, high-rate sprinkling, low-rate irrigation, overland flow, rapid infiltration extraction, rapid infiltration, soil aquifer treatment, and direct subsurface recharge.

**surface storage** The method of control of combined sewer overflows in open basins to collect stormwater before it enters the collection system, e.g., the basins ("lakes") in new subdivisions to retard peak runoff rates to predevelopment levels. *See also* in-system storage, offline storage, in-receiving water flow balance method.

**surface tension** The tensile force per unit length at the surface of a liquid, resulting from the interaction of the molecules, which tends to produce a meniscus at the surface. Surface tension explains why a glass can be filled with liquid slightly above the brim without spilling or why a needle can float on the surface of a liquid. It is expressed in pounds force per foot, newtons per meter, or kilograms per second squared. Surface tension is usually negligible, except in low flows. In water in contact with air, interfacial tension varies from 75.6 dynes/cm at 0°C to 71.2 dynes/cm at 30°C. *See* Weber number.

**surface vessel disposal** The transport of liquid sludge to an ocean disposal site; one of two basic methods of sludge discharge into the oceans. *See* articulated tug barges, barges in tow, rubber barges in tow, and self-propelled sludge vessel. *See also* ocean outfall.

**surface wash** (1) A high-pressure water device that is applied at or near the surface of a rapid granular filter to agitate and wash the medium; it is used as a supplementary wash system to improve filter performance, particularly in removing mudballs. *See also* surface washing, air scour, air wash, Baylis (John R.). (2) The surface runoff into a drain or ditch. *See also* sheet erosion.

**surface washer** A granular filter appurtenance, fixed or mounted on a rotary sweep, used to provide surface washing. A common design features nozzles attached to a rotating pipe, sometimes with a check-valve flap.

**surface washing** An effective method for cleaning granular filters by removing attached materials from the medium, using water or a mixture of air and water at high pressure to agitate the media and loosen the materials. Surface washing, which takes place at the beginning of backwash and during bed expansion, helps prevent media caking and the formation of mudballs. Also called auxiliary scour, filter agitation. *See also* air wash.

**surface washing system** Same as surface washer.

**surface water** (1) All water open to the atmosphere and subject to surface runoff. All water naturally open to the atmosphere (rivers, lakes, reservoirs, ponds, streams, impoundments, seas, estuaries, etc.) and all springs, wells, or other collectors directly influenced by surface water (EPA-40CFR141.2). Surface waters come from precipitation or street washing. (2) Water that is bound to or held on the surface of solid particles during sludge dewatering; such water cannot be removed by mechanical means. Also called vicinal water. *See also* bulk water, interstitial water, and water of hydration.

**surface water disinfection** *See* Surface Water Treatment Rule.

**surface water quality impacts** Factors affecting the quality of surface water include climate, watershed conditions, geology, nutrient distribution, seasonal or thermal stratification, deforestation, and saltwater intrusion.

**Surface Water Treatment Rule (SWTR)** A regulation issued by the USEPA in 1989 establishing (a) maximum contaminant levels goals for public systems using surface water or groundwater under the direct influence of surface water, and (b) requirements for the application of filtration and disinfection. Specifically, the rule requires filtration and disinfection treatment that achieves a 3-log (99.9%) removal/inactivation of *Giardia lamblia* cysts and a 4-log (99.99%) removal/inactivation of enteric viruses. *See also* CT concept, CT credit.

**Surface Water Treatment Rule credit** The performance assumed for some filtration units concerning the removal/inactivation of microbial contaminants from drinking water. *See* Table S-13.

**Surfact** A process developed by Envirex, Inc. to upgrade activated sludge units using air-driven rotating biological contactors.

**surfactant** Abbreviation for surface-active agent. (1) A large organic molecule that is slightly soluble in water; the active agent in detergents that possesses a high cleaning ability, promotes lathering, reduces surface tension, and increases the "wetting" ability of water; also called detergent, syndet, wetting agent. This synthetic chemical compound is made from aromatic sulfonates, alkyl sulfates, etc. It causes foaming in wastewater treatment units and in surface waters; it may interfere with coagulation. *See also* alkyl benzene sulfonate, linear alkyl sulfonate, methylene blue active substance test, cobalt thiocyanate active substance test. (2) A methylene blue active substance amenable to measurement by USEPA-recommended methods (EPA-40CFR417.91-c).

**Surfaer** A low-speed surface aerator manufactured by Aerators, Inc.

**surficial deposit** Unconsolidated material that covers the bedrock after in situ weathering or transport from another site by wind, water, ice, or gravity. Also called surface deposit.

**Surfpac™** A PVC medium manufactured by American Surfpac Corp. for use in trickling filters. It allows a void space of about 97% and an exposed surface area of 27–37 square feet per cubic foot, or more than twice the capacity of rocks and fieldstones. *See also* Koroseal® and FloCor®.

**surge anticipation valve** A pressure relief valve that is designed to open on a downsurge that is expected to be followed by an upsurge. *See* other types of pressure surge control devices.

**Table S-13.** Surface Water Treatment Rule credit

| | Log removal credit | |
|---|---|---|
| Treatment | *Giardia* cysts | Enteric viruses |
| Conventional | 2.5 | 2.0 |
| Slow sand filtration | 2.0 | 1.0 |
| Direct filtration | 2.0 | 2.0 |
| Diatomaceous earth filtration | 2.0 | 1.0 |

**SurgeBuster®** A check valve manufactured by Val-Matic® in sizes 2″–36″ and designed for high head applications and multiple pump and surge tank installations. With a valve seat on a 45° angle while maintaining a full flow area equal to the mating pipe, it achieves rapid closure through a short disc stroke of 35° and adjustable Disc Accelerator™.

**surge chamber** A chamber or tank connected to a pipe and located at or near a valve that may quickly open or close or a pump that may suddenly start or stop. When the flow of water in a pipe starts or stops quickly, the surge chamber allows water to flow into or out of the pipe and minimize any sudden positive or negative pressure waves or surges in the pipe. *See also* Figure S-31.

**surge control tank** (1) A large pipe or storage reservoir sufficient to contain the surging liquid discharge of the process tank to which it is connected (EPA-40CFR264.1031). (2) A standpipe or storage reservoir designed to absorb the sudden pressure increases in a closed conduit and to provide water when the pressure drops. *See also* surge chamber. Also called surge tank.

**surge control vessel** (1) A device used to protect a piping system from pressure surges, e.g., a compressor vessel, bladder tank, or hybrid tank. *See* pressure surge control for other similar devices. (2) A vessel used within a chemical manufacturing process unit when in-process storage, mixing, or management of flow rates or volumes is needed to assist in the production of a product; e.g., feed drums, recycle drums, and intermediate vessels (EPA-40CFR63.101, 63.111, 63.161, and 63.191).

**surge-flow irrigation** Intermittent application of water to a field for part of an irrigation cycle and then diversion to another field for the remainder of the cycle.

**surge protection** The reduction of the effects of unsteady flows by mechanical or hydraulic means such as reducing valve closures, air chambers, surge tanks, or surge towers.

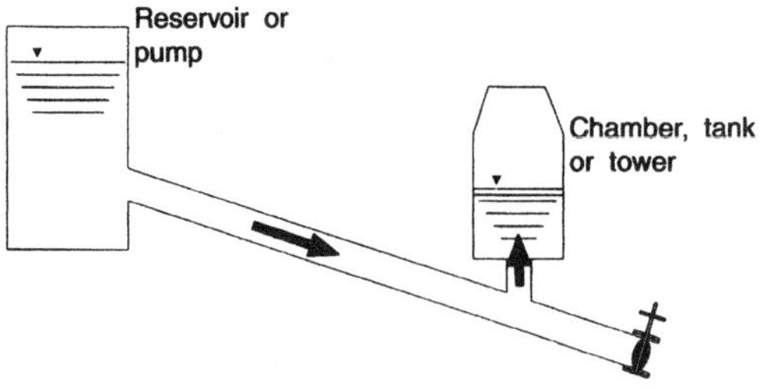


**Figure S-31.** Surge chamber (courtesy CRC Press).

**surger** A tightly fitting device used to develop a well by successively pushing it down and pulling it up like a piston.

**surge relief valve** *See* relief valve.

**surge suppressor** A valve that opens or closes slowly to reduce surges when a pump starts or stops.

**surge tank** Same as surge control tank. *See also* standpipe (1).

**surge vessel** Same as surge control vessel.

**surgical sterilizer** A fixture installed in older buildings for the sterilization of instruments, sometimes with inlet below its water level and thus a potential source of contamination by cross-connection.

**surging** *See* well surging.

**surrogate chemical indicator** A chemical substance used as a substitute for another. *See* surrogate indicator.

**surrogate compound** A substance that is chemically similar to a contaminant but more easily measured for evaluation purposes. *See* surrogate indicator.

**surrogate indicator** An organism or a chemical substance that is similar to a contaminant but easier to measure for evaluation purposes. Examples of surrogate indicators include: (a) total organic halide (TOX) and nonspecific bioassays for disinfection by-products, (b) total coliforms for pathogenic organisms in drinking water, and (c) microspheres for *Cryptosporidium* oocysts.

**surrogate measurement** The measurement of a nonspecific or indicator parameter in lieu of a specific contaminant. *See* surrogate indicator.

**surrogate organism** *See* surrogate indicator.

**surveillance** "Surveillance is not merely finding out what is wrong and putting matters right, it includes understanding remedial action to reduce or eliminate health hazards and advising on, assisting with, or stimulating improvements whenever possible. Surveillance also includes more general activities to promote the safety of water supplies—operator training and health education of the public in the prevention of waterborne enteric disease, for example" (WHO as cited in McJunkin, 1981).

**surveillance sample** A water or wastewater sample collected and analyzed by a regulatory agency to determine compliance with regulations. *See also* monitoring sample.

**surveillance system** A series of monitoring devices designed to check on environmental conditions.

**survey meter** A portable instrument used to detect and measure radiation. *See* Geiger counter, ionization chamber.

**suspended** Pertaining to elements that are retained by a 0.45 micrometer (μm) membrane filter.

**suspended-bed reactor** A three-phase reactor in which the solid catalyst is suspended and moving.

**suspended-frame weir** A timber weir whose steel frame can be raised during floods.

**suspended growth** One of two types of microbial development used in biological wastewater treatment; microorganisms grow in suspension in the liquid medium and use the contaminants as substrate. The other type is attached or fixed growth.

**suspended-growth aerated lagoon** A shallow earthen basin, 6–15 ft deep, equipped with mechanical aerators for the biological treatment of wastewater. *See* the following types: facultative partially mixed lagoon, aerobic partially mixed lagoon, and aerobic lagoon with solids recycle. *See also* lagoons and ponds

**suspended-growth process** A biological wastewater treatment process that uses suspended growth by maintaining the biological mass in suspension through appropriate natural or mechanical mixing. The solids retention time is a basic design and operating parameter of suspended-growth processes. Examples include aerobic processes (activated sludge, aerated lagoon, aerobic digestion), an anoxic denitrification process, and anaerobic processes (anaerobic contact and anaerobic digestion). Some single or multistage treatment techniques combine these three types. *See also* fixed-growth processes, lagoon processes.

**suspended-growth system** A wastewater treatment system that uses a suspended-growth process such as activated sludge. It usually includes some form of pretreatment or primary treatment, an aeration basin, and a solids separation unit.

**suspended growth with fixed-film packing process** A wastewater treatment process that uses packing materials suspended in the mixed liquor of an aeration basin, thereby combining attached growth and suspended growth properties. *See also* nonsubmerged attached growth process, submerged attached growth process.

**suspended load** The load of suspended sediment maintained in the water column by turbulence and carried with the flow of water; generally expressed as mass per unit time (e.g. kilograms or pounds per day). It is fine sediment (particles small enough to be carried by the moving water) as compared to coarse material, which is called bedload. Also called suspended sediment load. *See also* washload.

**suspended matter** Same as suspended solids.

**suspended metal** As compared to dissolved metals, a suspended metal is retained on a 0.45-micron membrane filter. *See also* suspended solids.

**suspended particles** Same as suspended solids.

**suspended sediment** (1) Fine soil particles that remain in suspension for some time in a body of water as a result of turbulence. *See also* sedimentation terms. (2) Same as suspended solids.

**suspended-sediment discharge** The quantity of suspended sediment passing through a stream cross section per unit of time. *See also* sedimentation terms.

**suspended sediment load** *See* suspended load.

**suspended solids (SS)** (1) Small insoluble particles of solid pollutants that float on the surface of, or are suspended in, water, wastewater, or other liquids. They are retained on standard glass wool mats or a 0.45-micron membrane filter, but sometimes suspended particles are defined as those larger than 1.0 micron and removable by gravity settling. Also called suspended matter or suspended sediment. Specific gravity varies from 1.0–1.5 for waterlogged vegetable matter, to 1.20–1.35 for chemical precipitates from coagulation, to 2.65 for sand. In the United States, domestic wastewater contributes approximately 90 grams of suspended solids per capita per day. Clarification or sedimentation units remove suspended solids from wastewater. *See also* colloids, dissolved solids, filter paper, nonfilterable residues, and settleable solids. (2) Fine dust particles distributed as an aerosol in the air.

**suspended solids contact clarifier** *See* solids contact clarifier.

**suspension** (1) The state of particles that are mixed, but not dissolved, in a liquid or gaseous medium; a substance that is in such a state. (2) A system of small particles dispersed by agitation or by molecular motion in a surrounding medium. *See* flocculent suspension.

**suspension culture** Cells growing in a liquid nutrient medium.

**suspensoid** A sol that has a solid disperse phase, i.e., a dispersion of colloidal particles in a liquid. *See also* emulsoid.

**sustainability** Characteristic of a sustainable project or action. Sustainability implies a process that can continue indefinitely, e.g., the "use of resources by our generation in a manner that will not diminish the ability of future generations to meet their needs" (quotation in Bloetscher and Muniz, 2006). *See also* water resources sustainability.

**sustainable capacity** *See* sustainable yield (1).

**sustainable development** Development that meets the needs of the present without compromising the ability of future generations to meet their own needs (WHO, 1997). Theoretically, sustainable development can continue indefinitely because it is based on the use of renewable resources and does not cause permanent environmental damage. *See also* appropriate technology, environmental equity, pollution prevention.

**Sustainable Infrastructure Initiative** A collaborative initiative of the USEPA, the AWWA, and other industry groups that emphasizes better utility management, full-cost pricing, efficiency, source water protection, and watershed-based planning.

**sustainable management** A method of resource exploitation that can be carried on indefinitely.

**sustainable water** *See* safe yield, sustainable yield.

**sustainable yield** (1) The maximum extent to which a renewable resource may be exploited without depletion. Also called sustainable capacity. (2) A hydrologic concept relating to the amount of water that can be withdrawn from a source at rates lower than their recharge potential and without impairing the quality of the source. *See also* safe yield.

**sustained flow rate** A water or wastewater flow rate that is equaled or exceeded for a certain number of consecutive days. *See also* sustained peaking factor, sustained loadings.

**sustained loading** A sustained peak loading is a hydraulic or mass loading that is equaled or exceeded at a treatment plant for a specified number of consecutive days. Long-term operating data may be analyzed to determine the ratios of averaged sustained peak flow rates to average flow rates (or loadings of BOD, TSS, nitrogen, or phosphorus) and plot these ratios versus the corresponding consecutive days. Sustained averaged low-flow flows can be similarly defined and plotted.

**sustained mass loading** A mass loading (in terms of BOD, total suspended solids, nitrogen, phosphorus, etc.) that is equaled or exceeded at a treatment plant for a certain number of consecutive days. *See also* sustained loading.

**sustained overdraft** Long-term withdrawal from an aquifer that exceeds its recharge.

**sustained peaking factor** The ratio of peak (hourly, daily, etc.) flow rates to long-term average flows. Also simply called peaking factor. *See also* peak rate factor, sustained loading.

**Sutorbilt®** An air blower manufactured by Gardner-Denver/Cooper Industries.

**Sutro weir** A weir or dam having a horizontal crest and at least one curved side, and such that the head above the crest is directly proportional to the discharge. The following equations are used in the design of Sutro weirs in grit chambers:

$$Q_r = (2/3)\, bC_d(2\, g)^{0.5}[H^{1.5} - (H - a)^{1.5}] \quad \text{(S-208)}$$

$$x = b[1 - (2/\pi)\tan^{-1}(y/a - 1)^{-0.5}] \quad \text{(S-209)}$$

$$Q_t = bC_d(2\, ag)^{0.5}(h + 2\, a/3) \quad \text{(S-210)}$$

(the Pratt formula)

where $Q_r$ = discharge through the rectangular section; $b$ = width of the weir at its crest; $C_d$ = weir discharge coefficient, approximately 0.62; $g$ = gravitational acceleration; $H$ = depth of flow; $a$ = height of the rectangular section; $x$ = width of the weir at any point, corresponding to distance $y$; $y$ = distance above the weir crest, corresponding to width $x$; $\tan^{-1}$ = inverse tangent, or arc that has a tangent equal to the indicated quantity (arc tan); $Q_t$ = total discharge through the weir; and $h$ = depth of water above the elevation $y$. *See also* proportional weir.

**SUVA** Acronym of (a) specific ultraviolet light absorbance or (b) specific ultraviolet adsorption.

**$SUVA_{254}$** The specific ultraviolet light absorbance of a filtered water sample at the wavelength of 254 nm, a parameter used in the identification of the aromatic fraction of dissolved organic matter. It is calculated as the ratio of the $UVA_{254}$ absorbance (1/cm) to the concentration of dissolved organic carbon (DOC, mg/L). *See also* DOC and $UVA_{254}$:

$$SUVA_{254} = 100 \times UVA_{254}/DOC \quad \text{(S-211)}$$

**SV** Abbreviation of sievert, a unit of ionizing radiation.

**SVE** Acronym of soil-vapor extraction.

**SVI** Acronym of sludge volume index.

**SVI test** Sludge volume index test.

**SVR** Acronym of sludge volume ratio.

**SVVS™** A groundwater bioremediation apparatus designed by Brown & Root Environmental.

**swab–rinse method** A method using a sterile cotton swab moistened with a buffer and rubbed over a surface to be sampled for bacterial or viral contamination. The tip of the swab is placed in a container of rinse fluid, which is then assayed.

**swallow hole** *See* sinkhole (1).

**swamp** A type of wetland dominated by woody vegetation but without appreciable peat deposits. Swamp water contains decaying vegetation with resulting high levels of taste, odor, and color. Swamps may be fresh or saltwater, and tidal or nontidal. They are created on flat lands by rainwater accumulation or river overflows, in sluggish streams by backwater, or by seepage outcrops over an impervious formation. The word is often used to include areas with nonwoody vegetation, e.g., marshes, bogs, etc.

**swamp gas** The gaseous product of the anaerobic degradation of organic matter, as in a swamp or other wetland area, consisting primarily of methane ($CH_4$), but also containing ammonia ($NH_3$) and sulfur compounds. Also called marsh gas, used as a common name for methane. *See also* biogas.

**SWAP** Acronym of source water assessment program.

**SWAT** Acronym of Smart Water Application Technology.

**sweat joint** Melted solder used to join copper pipe or tubing and fittings.

**Swebebet™** An ion-exchange resin manufactured by Bayer for use in countercurrent packed-bed systems; first intended for water treatment but also applicable to heavy metals recovery from wastewater.

**sweep coagulation** *See* sweep flocculation.

**sweeper** A night-soil collector who regularly empties the contents of bucket latrines into a wheelbarrow or cart for transportation and disposal. Also called a scavenger.

**sweep floc** The gelatinous floc that results from an iron or aluminum coagulant dose sufficient to cause the formation of metal hydroxides that enmesh colloidal particles while precipitating. The sweep-floc region or zone corresponds to a combination of pH values and coagulant dosages (e.g., $5.5 < pH < 9.0$ and $12.0$ mg/L $<$ alum $< 120$ mg/L) such that this phenomenon predominates over adsorption, destabilization, and restabilization. The sweep-floc mechanism plays a role in the removal of not only turbidity but also color and trihalomethane precursors. *See also* adsorption–destabilization, bridging, double-layer compression, and restabilization.

**sweep-floc coagulation** Same as sweep flocculation.

**sweep flocculation** A variation of the coagulation and flocculation processes often used in conventional water treatment, by which a large dose of coagulant (e.g., four orders of magnitude of the solubility of the metal) promotes the formation of large floc particles that trap (sweep) smaller particles and settle out of the suspension in less than 10 seconds. Sweep flocculation enhances coagulation beyond the adsorption–destabilization mechanism. Also called sweep-floc coagulation. *See also* enmeshment, charge neutralization.

**sweep-floc region** *See* sweep floc.

**sweep-floc zone** *See* sweep floc.

**sweep-gas membrane distillation** Membrane distillation is a desalination method using a temperature-driven, hydrophobic membrane to separate water, in the form of condensed vapor, from contaminants. This configuration uses a sweep gas to pull water vapor out of the membrane gap for outside condensation.

**sweet brine** Brine that is fit for use in ion-exchange resin regeneration because it contains enough sodium (Na) or potassium (K) and is low in interfering substances. *See also* sour brine.

**sweeten** To make soil less acidic or to remove sulfur and its compounds from oil and gas products.

**sweet (environment)** The opposite of sour (environment), i.e., containing little or no acid or hydrogen sulfide ($H_2S$).

**sweet gas** Natural gas that contains little hydrogen sulfide. *See also* sour gas.

**sweet water** (1) The solution of 8–10% crude glycerine and 90–92% water that is a byproduct of saponification or fat splitting (EPA-40CFR417.41-c). (2) Freshwater as compared to saltwater. (3) Water whose salt concentration does not meet potable standards but that may nevertheless be used for drinking.

**swelling** (1) The reversible expansion of some ion-exchange resins during exhaustion or regeneration. (2) The condition of a particle as a result of water intrusion.

**swelling index** A parameter used in studying the compressibility of sludge disposed in a landfill and determined from the slope of the swelling or rebound portion of the compression curve. It is used to indicate the landfill rebound when the consolidation pressure is reduced. *See also* sludge compressibility and compression index.

**swimmer's itch** A skin disease, cercarial dermatitis, caused by snail hosts and schistosome larvae, found in some waters of the United States and transported by infected water fowl. Copper sulfate is effective against the snail host.

**swimming-pool bacteriology** Tests conducted for swimming-pool water include the 24-hour, 37°C agar plate count and the coliform test.

**Swing-Flex®** A line of check valves manufactured by Val-Matic®, with only one moving part and a short, 35° stroke. *See also* Tilted Disc®, Dual Disc® Cushion Swing, and Silent Check.

**Swingfuser®** A header–drop pipe assembly manufactured by FMC Corp. for installation in aeration devices.

**Swingtherm®** A regenerative catalytic oxidizer manufactured by MoDoChementics.

**SwingUp** A header–drop pipe assembly manufactured by Aerators, Inc. for installation in aeration devices.

**Swingwirl** A vortex flowmeter manufactured by Endress+Hauser.

**swirl concentrator** A cylindrical device used, with or without chemical additions, to treat combined sewer overflows and peak wet-weather wastewater flows. *See* vortex separator for more detail.

**Swirl-Flo™** A solids separation process developed by H. I. L. Technology, Inc.

**SwirlMix** A compact wastewater treatment plant manufactured by Walker Process Equipment Co. using a complete mix reactor.

**Swiss Combi System** An apparatus manufactured by Wheelabrator Clean Water Systems, Inc. for sludge drying and pelletization.

**SW membrane** Spiral-wound membrane.

**SWMM** *See* Stormwater Management Model.

**SWRO** Acronym of seawater reverse osmosis.

**SWS wetland** *See* subsurface-flow system wetland.

**SWTR** Acronym of Surface Water Treatment Rule.

**sylvin** *See* sylvite.

**sylvine** *See* sylvite.

**sylvite (KCl)** A common mineral, essentially potassium chloride, colorless to milky-white or red, found in crystals or masses; a major source of potassium and an essential fertilizer. Also called sylvin, sylvine. *See also* carnallite.

**symba process** A wastewater treatment process that uses a fungus to hydrolyze starch wastes from the food industry and then a yeast to produce proteins.

**symbiont** An organism that lives in close association (symbiosis) with another organism, e.g., intestinal bacteria. Also called symbiote. *See also* parasite.

**Sym-Bio™ process** An oxidation ditch operated under conditions that promote simultaneous nitrification/denitrification. It uses a DO probe to measure dissolved oxygen concentrations and a nicotinamide adenine dinucleotide hydrogen (NADH) probe to monitor the content of this substance in sludge and to vary the DO level accordingly. *See* low-DO oxidation ditch.

**symbiosis** The condition of two dissimilar organisms living together in an association that is advantageous or even necessary to one or both and not harmful to either. This condition is present in biological wastewater treatment, e.g., between bacteria and algae in stabilization ponds. Another example is the occurrence of cellulose-digesting protozoans in the guts of wood-eating cockroaches and termites. *See also* commensalism, parasite, symbiotic cycle.

**symbiote** Same as symbiont.

**symbiotic** Characterized by symbiosis; pertaining particularly to organisms of different species whose activities benefit each other.

**symbiotic cycle** The mutually beneficial association between bacteria and algae in a stabilization pond.

**symmetrical membrane** A membrane with a uniform structure. *See also* asymmetrical membrane.

**synchronous motor** A type of direct current motor manufactured for pump drives used in wastewater treatment, particularly in large sizes (e.g., more than 500 hp) and with variable-speed control.

**syndet** Abbreviation of synthetic detergent.

**synecology** The study of the relations between natural communities and their environments.

***Synedra*** A diatom species that lives in eutrophic fresh surface waters.

**synergism** An interaction of two or more entities (e.g., organisms, chemicals) that results in an effect that is greater than the sum of their effects taken independently. *See also* antagonism.

**synergistic inactivation** The combined effect of two disinfectants applied sequentially in water or wastewater treatment. Often, this combined effect is greater than the sum of the individual effects. *See also* downstream disinfectant and upstream disinfectant.

**synergy** Cooperative or combined action or functioning of two or more agents, members, drugs, contaminants, etc., such action having a result greater than the sum of the individual results; synergism.

**synfuel** Abbreviation of synthetic fuel.

**synthesis** The fabrication of complex molecules from simpler ones; also called anabolism or assimilation. In biological wastewater treatment, synthesis occurs as the conversion of a portion of the organic matter into biomass, along with energy release by oxidation:

$$\mathrm{COHNS} + O_2 + \text{bacteria} + \text{energy} \rightarrow C_5H_7NO_2 \qquad \text{(S-212)}$$

where COHNS represents the principal elements of organic waste (carbon, oxygen, hydrogen, nitrogen, sulfur) and the formula $C_5H_7NO_2$ represents cell tissue. *See also* oxidation, autooxidation, metabolism, catabolism.

**synthesis biomass yield** Same as synthesis yield.

**synthesis yield** The amount of biomass produced immediately by microorganisms upon consuming a substrate or upon oxidation of an electron donor. For example, the synthesis yield of the aerobic decomposition of an organic compound is about 0.40 gram of volatile suspended solids per gram of COD. Synthesis yield exceeds observed yield by the amount of concurrent cell loss. Also called true yield. *See also* net biomass yield, net specific growth rate, observed yield, solids production or solids yield.

**synthetic** (1) Pertaining to synthesis, as opposed to analysis, or to substances that are formed artificially through a chemical process, as opposed to natural substances (e.g., synthetic fiber, synthetic resins). (2) A substance or product resulting from chemical synthesis (e.g., plastics).

**synthetic contaminants** Man-made products or elements thereof, e.g., pharmaceutical residuals and synthetic organics, that end up in the environment, as opposed to natural contaminants.

**synthetic detergent (abbreviation: syndet)** Any manufactured cleaning substance, other than soap. Actually, all detergents are synthetic cleaning agents. Syndets contain surface active agents (surfactants) and may be anionic, cationic, or nonionic. *See* more detail under detergent and surface active agent.

**synthetic fuel** A liquid or gaseous fuel produced from coal, lignite, shale, tar sands, or other solid carbon sources. Abbreviation: synfuel.

**synthetic ion-exchange resin** An insoluble, tough ion-exchange material manufactured by copolymerization of two substances, e.g., styrene as basic matrix and divinylbenzene as cross-linking compound. *See* the following types of resin: heavy-metal selective chelating, strong-acid cation, strong-base anion, weak-acid cation, and weak-base anion.

**synthetic-medium filter** A wastewater treatment filter packed with a highly porous medium, manufactured from polyvinalydene. *See* Fuzzy filter.

**synthetic organic chemicals (SOC)** A large group of man-made organic chemicals that may be highly toxic or may cause cancer; more than 700 have been identified in drinking water, with pesticide contamination causing the greatest concern (e.g., alachlor, aldicarb, atrazine, carbofuran). They are used in industry, agriculture, and household applications. Some SOCs are volatile, whereas others tend to stay dissolved in water instead of evaporating. These substances may be found in drinking water sources in very low concentrations but some may have significant health effects. Conventional water treatment processes are usually not effective, but membrane filtration and advanced oxidation have been used successfully. Also called synthetic organic compounds.

**synthetic organic compound (SOC)** Same as synthetic organic chemical.

**synthetic polyelectrolyte** A high-molecular-weight material formed by the polymerization of simple monomers.

**synthetic polymer** A long-chain, high-molecular-weight, organic chemical. *See* anionic polymer, cationic polymer, nonionic polymer.

**synthetic sludge** Experimental sludge produced by the combination of organic materials (e.g., milk solids), nutrients, and seed microorganisms.

**syntrophic relationship** A mutually beneficial relationship, e.g., between methanogens and acidogens in the fermentation process. *See* interspecies hydrogen transfer.

***Synura*** A group of algae that cause a bitter taste and odor problem in drinking water, even at low concentrations, a problem that is difficult to eliminate and is intensified when the water undergoes insufficient chlorination. *See* marginal chlorination. Breakpoint chlorination is effective against *Synura* and similar organisms; it destroys not only the organisms but also their oils and cells, which are responsible for the taste and odor problem.

**syphonic bowl** A type of flush toilet that includes a bowl with a siphon.

**System-3** A unit manufactured by Megator for the treatment of oily water.

**system development charge** A one-time charge assessed by a water, sewer, or other utility against new users, developers, or applicants for new service to recover the direct and indirect connection costs as well as part or all of the costs of existing or additional facilities. *See also* capacity charge, capital recovery charge, impact fee, facility expansion charge, connection charge, and development impact fee.

**systemic toxicant** A chemical that affects entire organ systems, often operating far from the original site of entry.

**Système International** or **SI** A decimal system of units used in scientific work and commercially throughout the world, based on the meter as unit of length, the second as unit of time, and the kilogram as unit of mass. One meter (m) = 100 cm = 1000 mm = 0.001 km = 3.281 feet. One kilogram (kg) = 1000 g = 0.001 metric ton = 2.2046 pounds. Other basic SI units include the ampere (A) for electric current, the kelvin (K) for temperature, the mole (mol) for amount of matter, and the candela (cd) for luminous intensity. SI also has supplementary, additional, and derived units for acceleration, angles, force, pressure, surface area, velocity, volume, etc. (The English or United States customary units are the foot, second, and pound). *See* metric system.

**system head** The total head (or total dynamic head) against which a pump or system of pumps works. It is the sum of the static and dynamic heads. *See also* system head curve.

**system head-capacity curve** or **system head curve** A graphical representation of the system head in a pumping installation. It is the graph of system total dynamic head ($H$) as a function of static head ($H_{ts}$) and the discharge ($Q$) according to the equation

$$H = H_{ts} + 0.0252[fLQ^2/d^5) + (\Sigma kQ^2/d^4)] \quad \text{(S-213)}$$

where $f$ = Darcy–Weisbach friction factor, $L$ = pipe length, $d$ = pipe diameter, and $k$ = minor loss coefficient.

**system head curve** Same as system head-capacity curve.

**systemic effects** Systemic effects are those that require absorption and distribution of the toxicant to a site distant from its entry point, at which point effects are produced. Most chemicals that produce systemic toxicity do not cause a similar degree of toxicity in all organs, but usually demonstrate major toxicity to one or two organs. These are referred to as the target organs of toxicity for that chemical.

**systemic toxicity** *See* systemic effects.

**system with a single service connection** A system that supplies drinking water to consumers via a single service line.

# T

**303d-listing** A list published by a state showing waters that are not fishable and swimmable and the applicable maximum daily loads, pursuant to Section 303 (d) of the Clean Water Act.

**3DP™** A dewatering filter having three belts, manufactured by Eimco Process Equipment Co.

**3-log removal** *See* three-log removal.

**2,4,5-T** *See* trichlorophenoxyacetic acid.

$T_{10}$ The time it takes 10% of the liquid in a reactor to leave the reactor or the time it takes 10% of a tracer to pass through the reactor. It is used as an empirical safety factor for reactor contact time in the design of contactors. It is also used as tank and reservoir contact time ($T$) in computing the disinfection $C \times T$ factor. *See also* Morrill dispersion index.

**$T_{10}$ time** or **$t_{10}$ time** *See* $T_{10}$.

$T_{50}$ or $t_{50}$ (1) The time it takes 50% of the liquid in a reactor to leave the reactor or the time it takes 50% of a tracer to pass through the reactor. *See also* index of mean retention time. (2) The time required for a 50% reduction, or half-life.

$T_{90}$ or $t_{90}$ (1) The time it takes 90% of the liquid in a reactor to leave the reactor or the time it takes 90% of a tracer to pass through the reactor. *See also* Morrill dispersion index. (2) The time required for a 90% reduction, or one log reduction, in pathogen concentration:

$$T_{90} = (\ln 10)/k = 2.3/k \qquad \text{(T-01)}$$

where $k$ is the die-off rate constant.

**$T_{90}$ value** The time required for a 90% reduction.

**$T_{90}/T_{10}$** The Morrill dispersion index.

***Tabellaria*** A genus of filter-clogging algae (diatoms) sometimes found in drinking water sources; they may cause short direct filtration runs.

**tableau** (Plural: tableaux.) A table representing the equilibrium composition of a solution, listing all the components, species, stoichiometric coefficients, and formation equilibrium constants. *See* Table T-01 for an example, corresponding to the equilibrium concentration of a $5 \times 10^{-4}$ M solution of boric acid [$B(OH)_3$ or $H_3BO_3$].

**tablet chlorinator** A simple device used in the disinfection of drinking water by the erosion of tablets of calcium hypochlorite [$Ca(OCl)_2$]. It may consist of a self-venting hopper containing the tablets and placed in the path of flowing water in a dissolving chamber of the storage tank of a gravity supply or pumped supply, or it may

**Table T-01.** Tableau (Equilibrium of a $5 \times 10^{-4}$ M solution of boric acid ($H_3BO_3$)

| Component species | HB | $H^+$ | $H_2O$ | log *K* (25°) |
|---|---|---|---|---|
| $B^-$ | 1 | –1 | 0 | –9.2 |
| HB | 1 | 0 | 0 | 0 |
| $H_2O$ | 0 | 0 | 1 | 0 |
| $OH^-$ | 0 | –1 | 1 | –14.0 |
| $H^+$ | 0 | 1 | 0 | 0 |
| Total | $5 \times 10^{-4}$ M | 0 | 55.4 M | |

Reprinted with permission from Stumm and Morgan (1996). Copyright © John Wiley & Sons, Inc.

be adapted for use with a hand pump. *See* Figure T-01.

**tablet method** A method of disinfection of a new small water main before it is placed in service; calcium hypochlorite tablets are placed in the pipe sections, valves, and hydrants during construction; the main is then filled slowly to dissolve the tablets and leave a residual of at least 50 mg/L for a minimum of 24 hours. Finally, the main is flushed to waste using potable water. *See also* slug method, continuous-feed method.

**TAB vaccine** A combination of *Salmonella typhi, S. paratyphi A,* and *S. paratyphi B* organisms to protect against all three forms of enteric fever; it is discouraged by WHO.

**TAC** Acronym of toxic air contaminant.

***Taenia*** A genus of tapeworms that infect humans who consume infected beef and pork that are insufficiently cooked.

***Taenia saginata*** One of the pathogens that cause taeniasis; common name: beef tapeworm.

***Taenia saginata* life cycle** The route of transmission of beef tapeworm from one host to another: an infected person's feces contain worm segments that release eggs when ruptured; a cow eats vegetation that contains these eggs and the parasites are transmitted to consumers of insufficiently cooked beef

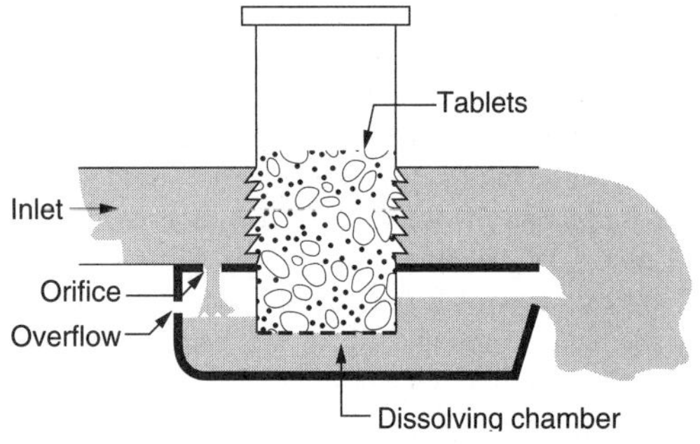


**Figure T-01.** Tablet chlorinator.

**taeniasis** The excreted helminth infection caused by taenia; *see* its common names, beef tapeworm and pork tapeworm, for more detail.

***Taenia solium*** One of the pathogens that cause taeniasis; common name, pork tapeworm. Excreted load = 10,000. Latency = 2 months. Persistence = 9 months. Median infective dose < 100.

**tailing effect** The result of a phenomenon observed in some water and wastewater treatment processes: the curves representing such processes are not symmetrical but have a long tail. Examples: (a) the disinfection curve plotting log inactivation vs. chlorine dose; (b) the ion-exchange breakthrough curve plotting relative contaminant concentration vs. volume of water treated. *See also* shoulder effect.

**tailing off** A deviation from the Chick–Watson law in which the rate of inactivation of microorganisms decreases progressively. *See* Chick–Watson law deviations. Same as inactivation rate tailing.

**tailing region** The area of a curve where the tailing effect is present. For example, on the disinfection curve, beyond a chlorine dose of 60 mg.min/L, log inactivation flattens around 5.0 instead of following a log-linear function; this is probably due to the shielding of microorganisms by solid particles.

**tailings** (1) Residue of raw material or waste separated out during the processing of crops or mineral ores. (2) Aggregates, gravel, etc. that are retained on a given screen.

**tailpipe** The suction pipe of a pump.

**tail water** The runoff of irrigation water from the lower end of an irrigated field.

**tamper** (1) To introduce a contaminant into a public water system with the intention of harming persons. (2) To otherwise interfere with the operation of a public water system with the intention of harming persons (Safe Drinking Water Act).

**tandem mass spectrometry** The use of two mass spectrometers in series. Also called mass spectrometry–mass spectrometry.

**tangential-flow filtration** Same as crossflow filtration.

**tangential-flow screen** A fine-mesh cylindrical screen designed for the treatment of combined sewer overflows. It uses a circular motion to trap floatables and push solids toward a central sump for removal. *See also* horizontal reciprocating screen.

**tangential screen** *See* static screen and tangential-flow screen.

**tangible asset** Permanent physical property.

**tank** In general, a tank is a large stationary container, receptacle, or structure constructed primarily of nonearthen materials (e.g., wood, concrete, steel, plastic) for holding, storing, or transporting a liquid or gas. In water and wastewater conveyance and treatment, tanks are used for such operations or processes as aeration, disinfection, equalization, sedimentation, holding, mixing, and chemical feeding.

**tankage** (1) The capacity or contents of a storage tank. (2) Dried animal by-product residues used in foodstuffs.

**tank mixing time** *See* mixing time.

**tank treatment** The detention of wastewater in a tank.

**tank water** In rendering or other meatpacking operations, tank water is the waste stream remaining after grease and residue are removed from the process water. It is one of the most significant sources of water pollution from this industry. To reduce pollution load, tank water is usually evaporated, in ponds, for example.

**tannery waste** Wastewater generated during tanning operations (the conversion of animal skins to leather), including unhairing, soaking, deliming, and bating of hides. Tannery wastes have high BOD, chromium, hardness, pH, precipitated lime, sulfides, and total solids. Common treatment methods include equalization, sedimentation, and trickling filtration or activated sludge.

**tannin** A substance that is responsible for color in water and that results from the degradation of plant materials. Tannins include the reddish compound that is used in tanning.

**tanning** The process of converting animals hides or skins into leather. It includes hide preparation for the absorption of tannin or chromium, and hide dehydration. *See* tannery waste for contaminants resulting from tanning.

**tap** (1) A faucet or a cock. (2) A corporation cock used to connect a service line to a water main.

**tapered aeration** A variation of the conventional, plug flow activated sludge process; more precisely, a method of supplying air to the aeration tank—air is supplied to match the oxygen demand of the mixed liquor from inlet to outlet. Tapered aeration, usually achieved by varying the spacing of the diffusers, results in reduced blower capacity, greater operational control, and inhibition of nitrification. *See* Figure T-02. *See also* step aeration.

**tapered flocculation** A variation of the flocculation process using multiple chambers and increasing velocity gradients.

**tapeworm** A flat or tapelike worm of the class Cestoidea or Cestoda and phylum Platyhelminthes, lacking an alimentary canal, parasitic when adult in the digestive tract of humans and other vertebrates. Also called cestode. Tapeworms include three parasites of the human intestine: the genera *Diphyllobothrium, Hymenolepis,* and *Taenia. See also* fluke (trematode).

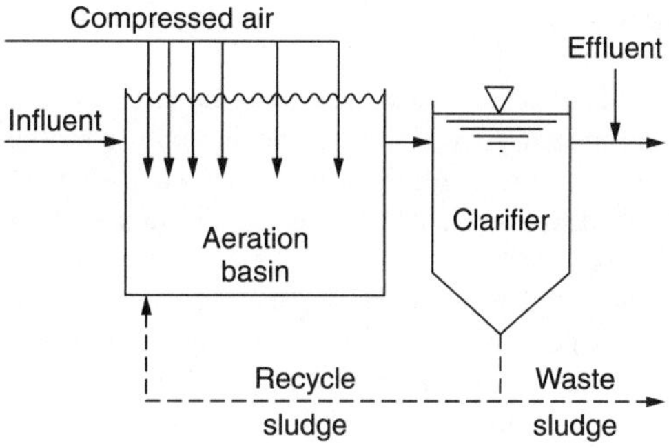


**Figure T-02.** Tapered aeration.

**tap fee** A connection fee charge to help finance capital improvements to a water supply system.

**tapioca snow** *See* snow pellets.

**tapping** Drilling and threading a hole in a water line to install a corporation cock.

**tapping machine** A tool attached to a water main for tapping; a drilling machine. Also called pressure-tapping machine if there is flow in the main, or dry-tapping machine otherwise.

**tapping sleeve** A split sleeve used in making a wet connection to a water line under pressure.

**tapping valve** A valve connected to a drilling machine for tapping into a water main.

**tap saddle** A fitting around a water main for the connection of a corporation cock.

**Targa** A water desalination apparatus manufactured by Mechanical Equipment Co., Inc. using vapor compression.

**target compound** The compound selected for analysis, monitoring, or the establishment of an analytical protocol.

**target concentration** The selected concentration of a constituent that indicates when to stop a treatment process, e.g., when to initiate backwashing of a filter or regeneration of a resin bed.

**target value** In European practice, the expected natural or background concentration of a given contaminant in groundwater, which normally does not require remediation. Also called reference value. *See also* intervention value.

**Taskmaster®** A screenings grinder manufactured by Franklin Miller.

**taste** *See* taste and odor.

**taste and odor** An objectionable characteristic resulting mostly from natural biological processes and affecting surface waters more often than groundwater. The words taste and odor are often used interchangeably or in combination when applied to water quality, both relating to sensations perceived by the mouth and nose. Depending on the responsible organisms and substances, a water's taste and odor may be swampy, grassy, medicinal, septic, phenolic, musty, fishy, or sweet. Actually, there are four basic tastes and at least four fundamental odors; see Table T-02. Algae, actinomycetes, other microorganisms, natural and synthetic organic substances (e.g., chloroform, chlorophenols, geosmin, methylisoborneol), hydrogen sulfide, landfill leachate, dissolved iron and manganese, chlorine, etc. may all contribute to impart taste and odor to water. Many water treatment processes contribute to the reduction of taste and odor, e.g., adsorption, aeration, chemical oxidation, clarification, prechlorination. *See* flavor profile analysis, odor control, threshold odor number, Weber–Fechner law.

**taste and odor control** The use of measures such as watershed supervision and reservoir supervision (through the application of algicides or aeration) to prevent the formation of taste and odor-causing substances, as well as pretreatment and such effective processes as adsorption on powdered activated carbon. Breakpoint chlorination is effective in destroying most taste and odor substances.

**taste and odor removal** *See* taste and odor control.

**taste rating test** A test conducted to determine the acceptability of drinking water, based on a sensory evaluation.

**taste threshold** The minimum concentration of a taste-causing substance that can be tasted or the level at which one or more individuals in a panel can detect the presence of the substance.

**taste threshold test** A test conducted to determine the minimum detectable taste in a drinking water.

***Tatlockia*** A genus of bacteria of the family *Legionellaceae,* sometimes implicated in pneumonia-like diseases. *See* legionellosis.

**Table T-02.** Tastes and odors (basic)

| Basic tastes | Fundamental odors |
|---|---|
| 1. Bitter | 1. Acid or sour |
| 2. Salty | 2. Burnt or empyreumatic |
| 3. Sour | 3. Caprylic or goaty |
| 4. Sweet | 4. Fragrant or sweet |

**Taulman/Weiss®** An in-vessel composting unit manufactured by Taulman Co.

**tax-exempt lease** An installment sale transaction whereby a manufacturer or other provider leases equipment to a municipality or utility, does not pay income tax on the interest portion of the lease, and passes on the savings to the lessee in the form of a lower interest rate.

**tax-exempt leasing** Financing a water or wastewater project through a tax-exempt lease.

**taxonomic classification** The classification of organisms based on physical properties and metabolic characteristics, with the species as the basic taxonomic unit, and progressing to the genus, family, order, class, phylum, and kingdom. Bacteria are identified by their species and genus, e.g., *Vibrio cholerae. See also* biological classification, environmental classification, phylogenetic classification.

**taxonomy** The science or technique of classification, particularly, the description, identification, naming and classification of organisms; *see* biological taxonomy for detail.

**TBC** Acronym of total bacteria count.

***T. brockianus*** Abbreviation of *Thermus brockianus.*

**TBS** Acronym of tetrapropylene benzene sulfonate.

**TBT** Acronym of top brine temperature.

**TC** Acronym of total carbon.

**TCA** Acronym of trichloroacetic acid.

**TCAN** Acronym of trichloroacetonitrile.

**TCDD** Acronym of tetrachlorodibenzo-*p*-dioxin (dioxin).

**2,3,7,8-TCDD** *See* dioxin.

**TCE** Acronym of (a) tetrachloroethylene; (b) trichloroethane; (c) trichloroethene; (d) trichloroethylene.

**TCF®** A horizontal tube cartridge filter manufactured by Ropur AG.

**TCID** Acronym of tissue culture infective dose.

**$TCID_{50}$** Symbol of tissue culture infectious dose 50 or median tissue culture infective dose.

**$TCID_{50}$ method** *See* serial dilution endpoint method.

**TCLP** Acronym of toxicity characteristic leaching procedure.

**TCU** Acronym of total color unit.

**TDH** Acronym of total dynamic head.

**T-Ditch process** A proprietary wastewater treatment technique that combines the oxidation ditch and sequencing batch reactor processes for the removal of biochemical oxygen demand, suspended solids, and ammonia nitrogen ($NH_3$-N). As shown in Figure T-03, it includes two oxidation ditches in parallel, both linked to a third ditch, from which sludge is drawn for processing and disposal. Each ditch is equipped with a rotor aerator. There is no recirculation and no separate clarification. *See* phased isolation ditch for other related processes. *See also* biological nitrogen removal.

**TDS** Acronym of total dissolved solids.

**TDT** Acronym of thermal death time.

**TEA** Acronym of terminal electron acceptor.

**Teacup™** A rotating cylindrical device manufactured by Eutek Systems for grit removal. Flow entering tangentially at the top generates a vortex. Centrifugal and gravitational forces cause the separation of organics, which exit with the effluent, from grit, which settles to the bottom. *See* accelerated gravity separator, Pista® Grit.

**tea water** Water of better quality than is commonly available, used for making tea, in reference to a situation that existed in the early settlements in the New York City area.

**Technasand™** Backwash equipment manufactured by WesTech Engineering of Salt Lake City, UT for use in municipal and industrial water filtration.

**technological obsolescence** The condition of a process or equipment that works even though there are newer and better performing versions thereof.

**technology-based limitations** Industry-specific effluent limitations applied to a discharge when it will not cause a violation of water quality standards at low stream flows. Usually applied to discharges into large rivers.

**technology-based standards** Effluent limitations applied to direct and indirect sources that are developed on a category-by-category basis using statutory factors, not including water-quality effects.

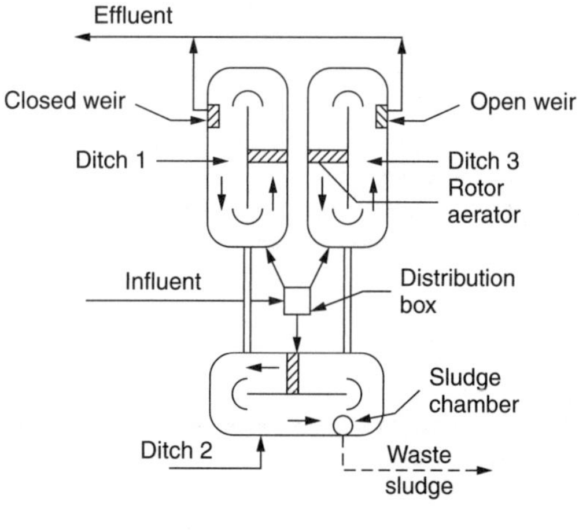


**Figure T-03.** T-Ditch process.

**technology-based treatment requirements** Discharge permit requirements based on the application of a specified technology or category of technology, e.g., best available technology economically achievable, best conventional technology, or best practicable technology.

**TechXtract™** A process developed by EET, Inc. for chemical decontamination.

**TecTank** A storage tank manufactured by Peabody TecTank in bolted steel.

**Tecweigh®** A volumetric feeder manufactured by Tecnetics Industries, Inc.

**Tekleen™** A filter screen manufactured by Automatic Filters, Inc.

**telemeter** or **telemetering system** (1) The equipment for measuring, transmitting, and receiving data (such as flow, water level, temperature, pressure, radiation, etc.) at a distance, by wires, radio waves, or other means. (2) An instrument for measuring the distance to a remote object.

**telemetering** The application of a telemetering system.

**telemetering system** *See* telemeter.

**telemetry** The science or process of telemetering or the use of a telemetering system. *See also* SCADA and sewer level remote telemetry.

**telemetry system** Same as telemetering system.

**TeleTote™** An open-channel electromagnetic flowmeter manufactured by Marsh-McBirney, Inc. of Frederick, MD.

**TEM** Acronym of transmission electron microscope or microscopy.

**Temik** The trade name of aldicarb.

**temperate phage** A lysogenic phage; a phage in which the nucleic acid is integrated with the chromosome of the host, which transmits it to its descendants and daughter cells.

**temperate virus** *See* temperate phage.

**temperature** A measure of the thermal condition (warmth or coldness) of an object or a substance with respect to a standard. An important physical characteristic that affects the acceptability of water as well as water chemistry and water treatment. Temperature affects biological and chemical reaction rates, dissolved oxygen concentrations, taste and odor, solubility, corrosion, etc. *See* temperature effect.

**temperature, absolute** *See* absolute temperature.

**temperature approach** In a heat exchanger or a cooling tower, the lowest difference between the two fluids.

**temperature coefficient** (1) A dimensionless coefficient (constant over small temperature ranges) that expresses the effect of temperature on biological and chemical reaction rates. Also called Arrhenius constant, Arrhenius parameter, theta factor, or $\theta$ factor. *See* Arrhenius equation, Arrhenius constant. (2) A temperature coefficient ($Q_{10}$) that indicates the increase in the rate of a process that is due to a temperature increase of 10°C.

**temperature correction** *See* temperature correction coefficient and temperature correction factor.

**temperature correction coefficient** A coefficient commonly used in water and wastewater engineering to adjust rate constants for the effects of temperature. *See* temperature effect.

**temperature correction factor (TCF)** A dimensionless coefficient used to convert a membrane flux ($J$) at temperature $T$, °C to a normalized flux ($J_{25C}$) at the standard temperature of 25°C:

$$J_{25C} = J \exp\{-\mathrm{TCF}\,[1/298 + 1/(T + 273)]\} \quad \text{(T-02)}$$

**temperature-driven membrane** A hydrophobic membrane that uses a thermal gradient as driving force.

**temperature effect** Temperature affects all water and wastewater treatment processes to some degree. For example:

- Adsorption potential.
- Arrhenius equation.
- Biological waste treatment: cold weather reduces the efficiency of high-rate trickling filters by approximately 30%; low temperatures inhibit nitrification (by 75% from 30°C to 10°C) more than BOD removal.
- BOD exertion. The following expression is commonly used for the rate ($k_T$) of oxidation of organic matter as a function of temperature ($T$, °C), $\theta$ being a constant and $k_{20}$ the rate at 20°C:

$$k_T = k_{20}\,\theta^{T-20} \quad \text{(T-03)}$$

- Buffering capacity.
- Coagulation—lower reduction of turbidity and suspended solids at low temperatures.
- Corrosion.
- Digestion—the minimum solids retention time varies from 2 days at 35°C to 10 days at 20°C; the heat requirement of the digester depends on outside temperature.
- Disinfection—the rate and degree of completion of oxidation reactions, and consequently disinfectant dosages, increase with temperature.
- Disinfection by-product formation.
- Enzyme activity.
- Equilibrium constants.
- Henry's law constant.
- Membrane flux (Hagen–Poiseuille equation).
- Microbial growth.
- Oxygen transfer rates.
- Sedimentation—temperature differences between incoming wastewater and the contents of a sedimentation basin can cause density currents.
- Solubility product constants.
- Specific denitrification rate (SDNR).
- Water quality—oxygen solubility and dissolved oxygen saturation decrease with increasing temperatures.

**temperature effect on reaction rate** *See* van't Hoff–Arrhenius equation.

**temperature monitoring device** A unit of equipment used to monitor temperature and having an accuracy of ±1% of the temperature being monitored expressed in degrees Celsius or ±0.5°C, whichever is greater (EPA-40CFR63.111).

**temperature-phased anaerobic digestion (TPAD)** A variation of the two-phased anaerobic digestion process, developed in Germany and designed to optimize conditions for the separate groups of microorganisms involved by using two reactors in series. As shown in Figure T-04, acidogenesis takes place in the first reactor at a solids retention time (SRT) of about 4 days and a temperature of 55°C, whereas the second reactor provides an SRT of 10 days and a temperature of 35°C for methanogenesis. *See also* acid/gas phased digestion, staged mesophilic digestion, staged thermophilic digestion.

**temperature-phased digestion** *See* temperature-phased anaerobic digestion.

**temperature quotient ($Q_{10}$)** The ratio of two reaction rate constants corresponding to a temperature difference of 10°C. For rates of chemically or bio-

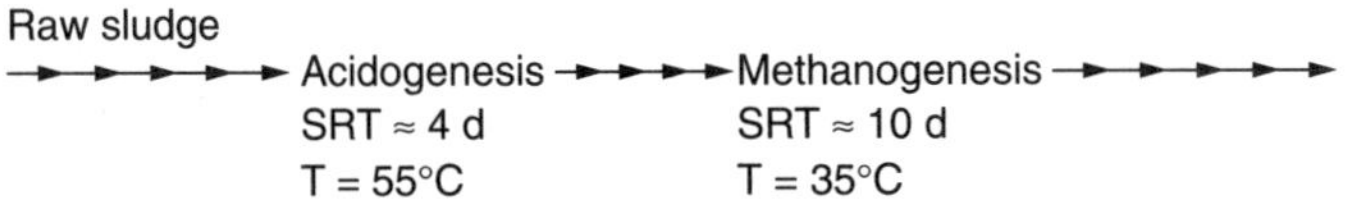


**Figure T-04.** Temperature-phased digestion.

logically activated reactions, see the vant't Hoff–Arrhenius equation. *See also* activation energy, adsorption rate.

**temperature sensor** A device that opens and closes a switch in response to changes in the temperature. This device might be a metal contact, a thermocouple that generates a minute electrical current proportional to the difference in heat, or a variable resistor whose value changes in response to changes in temperature. Also called a heat sensor.

**temporary hardness** The hardness caused by the carbonate and bicarbonate ions (mainly those of calcium and magnesium), which may be eliminated by boiling the water, resulting in a scale deposit, e.g.:

$$Ca^{2+} + 2\ HCO_3^- \rightarrow CaCO_3\ (s) + CO_2\ (g) + H_2O \quad \text{(T-04)}$$

$$Mg^{2+} + 2\ HCO_3^- \rightarrow MgCO_3\ (s) + CO_2\ (g) + H_2O \quad \text{(T-05)}$$

Temporary hardness is sometimes called carbonate hardness because they are approximately the same. *See also* permanent or noncarbonate hardness.

**tenorite (CuO)** The mineral cupric oxide.

**Ten States' formula** *See* the Rankin formulas developed for the design of trickling filters, based on the Ten-State Standards.

**Ten States' Standards** Same as Ten-State Standards.

**Ten-State Standards** Short title for the publications (a) *Recommended Standards for Sewage Works* and (b) *Recommended Standards for Water Works of the Great Lakes,* issued by the Upper Mississippi River Board (GLUMRB) of State Sanitary Engineers. They contain guidelines for the review and approval of plans and specifications for water and sewer works. The states are Illinois, Indiana, Iowa, Michigan, Minnesota, Missouri, New York, Ohio, Pennsylvania, and Wisconsin, plus the Canadian province of Ontario.

**Tenten** A gravity sand filter manufactured by F. B. Leopold Co., Inc. with continuous backwash.

**teratogen** (From the Greek prefix terat, meaning monster.) A chemical substance that triggers a mechanism that causes fetus or embryo abnormalities observed at birth; e.g., xenobiotics can inhibit enzymes. Dioxin (in herbicides) is a strong teratogen. Water disinfection by-products (DBPs) are suspected teratogens. *See also* carcinogen, mutagen.

**teratogenesis** The production or induction of nonhereditary congenital malformations (birth defects) or monstrosities in a developing fetus by exogenous factors acting in the womb; interference with normal embryonic development. Also called teratogeny. Teratogenetic mechanisms include enzyme inhibition, deprivation of the fetus of essential nutrients, and interference with energy supply.

**teratogenetic effect** The result of a teratogenetic mechanism, i.e., a specific birth defect caused by a chemical substance. For example, shortened limbs of a newborn may be a teratogenic effect of the drug thalidomide when taken during pregnancy.

**teratogenic** Of or pertaining to teratogens, drugs or other chemicals that can interfere with the development of a fetus and cause nonhereditary congenital malformations (birth defects) in offspring. *See also* carcinogenic, genotoxic, mutagenic.

**teratogenicity** The capacity of a physical or chemical agent to cause teratogenesis in offspring.

**teratogeny** Same as teratogenesis.

**teratology** The study of chemical substances that cause birth defects. *See* teratogenesis.

**terminal backwash velocity** *See* terminal velocity (2).

**terminal disinfection** Same as postchlorination.

**terminal disinfection by-product concentration** *See* instantaneous disinfection by-product concentration.

**terminal electron acceptor (TEA)** A compound or molecule that accepts an electron (is reduced) during metabolism (oxidation) of a carbon source. Under aerobic conditions, molecular oxygen is the terminal electron acceptor. Under anaerobic conditions, a variety of terminal electron acceptors may be used. In order of decreasing redox potential, these TEAs include nitrate, manganic manganese, ferric iron, sulfate, and carbon dioxide. Microorganisms preferentially utilize electron acceptors that provide the maximum free energy during respiration. Of the common terminal electron acceptors listed above, oxygen has the highest redox potential and provides the most free energy during electron transfer. The disappearance or reduction of TEA reflects the level of microbial activity. It may be measured by oxygen uptake, oxygen probes, Winkler titration, manometry, and other methods.

**terminal head loss** The head loss recorded at the end of a filter run, i.e., when the buildup of solids in the bed increases resistance to flow and prevents filtration from proceeding at the desired rate and efficiency.

**terminal manhole** The most upstream manhole in a gravity sewer line. Terminal manholes are used to determine the boundaries of service areas.

**terminal particle velocity** *See* terminal rise velocity, terminal settling velocity.

**terminal rise velocity** The velocity ($V_\infty$) of a rising air bubble in water or wastewater, similar to the settling velocity of a particle and achieved almost immediately upon release. It depends on the Reynolds number ($R_e$) of the bubble and other factors; e.g., for $R_e < 2.0$:

$$V_\infty = 2\,R^2 g(\rho - \rho')/(9\,\mu) \qquad \text{(T-06)}$$

where $R$ = nominal radius of the bubble, $g$ = gravitational acceleration, $\rho$ = density of the liquid, $\rho'$ = density of the gas bubble, and $\mu$ = dynamic viscosity of the liquid.

**terminal settling velocity** The vertical velocity ($V$) of a suspended particle or its maximum rate of unhindered sedimentation. It may be determined from Stokes law or from Newton's law:

$$V = 2[g(\rho - \rho')d/(3\,\rho C_D)]^{0.5} \qquad \text{(T-07)}$$

where $g$ is the acceleration of gravity, $d$ and $\rho$ the particle diameter and mass density, $\rho'$ the fluid density, and $C_D$ the drag coefficient. *See also* fall velocity.

**terminal temperature** The maximum temperature of the feed water during distillation.

**terminal total trihalomethane concentration** *See* instantaneous total trihalomethane concentration.

**terminal velocity** (1) Same as terminal settling velocity. (2) The velocity ($V_t$, in m/min) that will cause the filter medium to wash out and thus damage the filter during backwash. *See* backwash velocity for more detail. Also called washout velocity.

**Terra-Gator** An apparatus manufactured by Ag-Chem Equipment Co. for waste sludge injection under pressure.

**terra rossa** Red earth formed on limestone by deposits of insoluble iron oxides.

**TERRA Series** A preintegrated, flexible SCADA platform manufactured by Federal Signal Controls of University Park, IL.

***tert*-butylbenzene** *See* butylbenzene.

**tertiary effluent** The effluent of a tertiary wastewater treatment plant.

**tertiary filtration** An advanced wastewater treatment method used to remove residual solids from a secondary effluent by filtration on granular media. Process variations range from plain filtration (no chemicals, no further sedimentation), to direct filtration (with coagulation and optional flocculation), to traditional filtration. Tertiary filters are used as polishing filters (final treatment step) or as intermediate units before additional treatment (e.g., activated carbon adsorption or ammonia removal).

**tertiary lagoon** Same as tertiary pond.

**tertiary nitrification** The oxidation of ammonia nitrogen ($NH_4$-N) to nitrates ($NO_3^-$) in a tall trickling filter filled with plastic packing. It requires an influent with a $BOD_5$/TKN ratio equal to or less than 1.0 and a soluble $BOD_5$ equal to or less than 12.0 mg/L. *See also* the Gujer–Boller equation and the Parker et al. formula.

**tertiary pond** A shallow, aerobic waste stabilization pond used for polishing wastewater effluents from an activated sludge plant, a trickling filter plant, or a facultative pond. *See* maturation pond for more detail.

**tertiary treatment** The application of physical, chemical, or biological processes to the effluent of a secondary treatment plant. This treatment improves the quality of a secondary effluent by removing residual suspended solids and such nutrients as nitrogen and phosphorus. *See also* advanced waste treatment.

**test** A laboratory procedure to determine the concentration or presence of a constituent in water or wastewater, e.g., absorption, iodometric, and orthotholidine tests.

**test organism** An organism such as the fathead minnow or water flea used in testing the toxicity of wastewater effluents.

**test tube** A hollow glass or plastic cylinder with one end closed; used in chemical and biological analyses.

**tetanus bacillus** The bacterium, *Clostridium tetani,* that causes tetanus, an often fatal infection contracted through wounds and characterized by respiratory paralysis, tonic spasms, and rigidity of muscles. The pathogen is commonly found in human feces.

**1,1,3,3-tetrachloroacetone** A haloketone that may cause mutagenic effects.

**tetrachlorocarbon** Same as carbon tetrachloride.

**tetrachlorodibenzo-*p*-dioxin** *See* dioxin.

**2,3,7,8-tetrachlorodibenzo-*p*-dioxin ($C_{12}H_4O_2Cl_4$)** Acronym: TCDD. *See* dioxin.

**tetrachlorodiphenylethane [$(ClC_6H_4)_2CHCHCl_2$]** *See* dichlorodiphenyl dichloroethane.

**tetrachloroethane ($CHC_{12}CHCl_2$)** A chlorinated hydrocarbon and volatile organic compound used as an industrial cleaner or a solvent in pesticides, soil fumigants, bleaches, and paints; a potential water disinfection by-product. Also listed as 1,1,1,2-tetrachloroethane and 1,1,2,2-tetrachloroethane.

**tetrachloroethene** Same as tetrachloroethylene. Also listed as 1,1,2,2-tetrachloroethene.

**tetrachloroethylene ($Cl_2C{=}CCl_2$)** A volatile organic solvent insoluble in water; used as a dry-cleaning or vapor-degreasing solvent and in fluorocarbons; it is a water chlorination by-product. Also listed as 1,1,2,2-tetrachloroethylene. Same as tetrachloroethene or perchloroethylene (PCE).

**tetrachloromethane** Same as carbon tetrachloride.

**tetraethyl lead [$Pb(C_2H_5)_4$]** An antiknock compound added to leaded gas to improve combustion efficiency.

**Tetra® filter** A proprietary denitrification filter of Tetra Technologies, Inc. of Houston, TX, which requires occasional backwash bumping to release accumulated nitrogen gas.

**tetraglycine hydroperiodide** A substance containing iodine, used in the field in tablet form for water disinfection. Proprietary formulas include Coghlan, Globaline, and Potable-Agua.

**tetrahedrite ($Cu_{12}Sb_4S_{13}$)** A steel-gray or blackish mineral that is essentially the sulfide of copper and antinomy, found in hydrothermal vents.

**TetraPace** A patented device of Tetra Process Div. of Severn Trent Water Purification used to dose methanol in denitrification filters.

**tetrathionate ($S_4O_6^{2-}$)** A compound of thiosulfuric acid ($H_2S_2O_3$). *See Thiobacillus thiooxidans* and thiosulfate ($S_2O_3^=$).

**Texas Star** A circular membrane diffuser manufactured by Aeration Research Co.

**textile waste** Wastewater generated during textile mill operations from cooking of fibers, weaving, dyeing, printing, and finishing. Such a waste has high alkalinity, BOD, color, temperature, and suspended solids. It may be treated using neutralization, chemical precipitation, and trickling filtration or activated sludge.

**texture** A property of rocks, soils, and surface deposits defined by the size, shape, arrangement, and distribution of their sand, clay, and silt particles; an important factor that governs the movement of water and contaminants in a porous medium. *See also* structure.

**TF/AS process** Abbreviation of trickling filter/activated sludge process.

**TFC** Acronym of thin-film composite.

**TFC®** A thin-film composite membrane manufactured by Fluid Systems Corp. for use in reverse osmosis units.

**TFCL®** A thin-film composite membrane manufactured by Fluid Systems Corp. for use in reverse osmosis units.

**TFC membrane** Abbreviation of thin-film composite membrane.

**TFCS™** Thin-film composite membrane components manufactured by Fluid Systems Corp. for use in reverse osmosis units.

***T. ferrooxidans*** Abbreviation of *Thiobacillus ferrooxidans.*

**TFM®** Membranes manufactured by Desalination Systems, Inc. for use in reverse osmosis units.

**TFS** Acronym of total fixed solids.

**TF/SC process** Abbreviation of trickling filter–solids contact process.

***T. gondii*** Abbreviation of *Toxoplasma gondii.*

**Th** Chemical symbol of the metallic element thorium.

**thallium (Tl)** A rare, soft, malleable, bluish-white metal that resembles lead; it is found mostly in the minerals crooksite, hutchinsonite, lorandite, and pyrites; used as a rodenticide, in glass manufacturing, pharmaceutical products, mercury alloys, etc. All its compounds are very poisonous. Atomic weight = 204.38. Atomic number = 81. Specific gravity = 11.85. Valence = 1 or 3. Thallium may be carcinogenic.

**thalweg** The line connecting the lowest points in each cross section of a valley. (2) The line of maximum depth in a stream.

**theoretical air-to-liquid ratio** In gas stripping operations, the theoretical air-to-liquid ratio is the minimum amount of air that can be used for stripping and corresponds to the condition in which there is no solute in the influent (both the gas and liquid streams). This theoretical ratio also assumes a 100% efficient operation in an infinitely high stripping tower. It corresponds to

$$G/L = P_T/H \qquad \text{(T-08)}$$

where $G$ = flow of incoming gas, moles per unit time; $L$ = flow of influent wastewater, moles of liquid per unit time; $P_T$ = total pressure, atm; and $H$ = Henry's law constant, atm (mole gas/mole air)/(mole gas/mole liquid). *See also* air/liquid ratio, air-to-water ratio, stripping tower mass balance.

**theoretical flame temperature** The temperature ($T$, °C or °F) of a substance as calculated from the sensible heat equation, assuming that no heat is lost during combustion:

$$T = T_0 + Q_s/(h_s W) \qquad \text{(T-09)}$$

where $T_0$ (°C or °F) is a reference point, $Q_s$ (kJ or Btu) is the sensible heat, $h_s$ (kJ/kg/°C or Btu/lb/°F) is heat capacity or specific heat, and $W$ (kg or lb) is the weight of the substance.

**theoretical oxygen demand (ThOD)** The oxygen demand of a compound based on its reaction with oxygen ($O_2$) to produce carbon dioxide ($CO_2$) and water ($H_2O$), or the oxygen demand of a water or wastewater sample based on the chemical formula of its organic matter. For example, the ThOD of glucose ($C_6H_{12}O_6$) and glycine ($CH_2 \cdot NH_2 \cdot COOH$) is based on their respective total oxidation reactions:

$$\underset{180}{C_6H_{12}O_6} + \underset{192}{6\,O_2} \rightarrow \underset{264}{6\,CO_2} + \underset{108}{6\,H_2O} \quad \text{(T-10)}$$

$$CH_2 \cdot NH_2 \cdot COOH + 7/2\,O_2 \rightarrow 2\,CO_2 + 2\,H_2O + HNO_3 \quad \text{(T-11)}$$

Thus, the ThOD is 192/180 = 1.07 mg $O_2$ per mg of glucose and 7/2 moles $O_2$ per mole of glycine. Other measures of organic content are BOD, COD, TOC, and TOD.

**theoretical pump displacement** The capacity of a pump, as rated by its manufacturer, in gpm/100 RPM (gallons per minute per 100 revolutions per minute).

**theoretical specific capacity** *See* specific capacity.

**theoretical velocity** The velocity that a fluid would attain without friction and other losses.

**therapeutic index** The ratio of the dose required to produce toxic or lethal effects to dose required to produce nonadverse or therapeutic response.

**ThermaGrid™** A regenerative thermal oxidizer manufactured by United McGill Corp.

**thermal capacity** The quantity of energy that must be supplied to raise the temperature of a substance one degree, or the amount of heat released by a decrease of one degree in the temperature of the substance, expressed in calories/°C or BTUs/°F. *See* more detail under heat capacity.

**thermal conditioning** The processing of sludge through heat treatment (about 200°C for 25 minutes at a pressure of 200 psi) or low-pressure oxidation to destroy biological cells and release moisture so that the sludge can be more easily thickened or dewatered. *See,* for example, the Zimpro process.

**thermal conductivity** (1) A measure of the ability of a substance or material to conduct heat; the rate of heat transfer by conduction through a material or substance of unit thickness per unit area and per unit temperature gradient. Thermal conductivity is one of the parameters used in the design of evaporators. Also called heat conductivity, specific thermal conductivity, coefficient of thermal conduction. (2) A method used in gas chromatography to detect trace organic chemicals in water or wastewater samples. *See also* electron capture, flame ionization, mass spectrometry.

**thermal conductivity detector** A nonselective instrument used in gas chromatography.

**thermal crystallizer** An apparatus used to distill the concentrate of a desalination plant into dry salts. *See also* brine concentrator, fluidized-bed crystallizer, zero liquid discharge.

**thermal death point** A point on the curve relating sludge digestion time to temperature, at approximately 37–40°C, where the range of mesophilic organisms ends and the range of thermophilic bacteria begins. *See* Figure T-05.

**thermal death time (TDT)** The time necessary for the thermal destruction of a number of organisms at a given temperature. *See also* decimal reduction time or $D$ value, and $z$ value.

**thermal desorber** The primary treatment unit that heats petroleum-contaminated materials and desorbs the organic materials into a purge gas or off-gas.

**thermal desorption system** A thermal desorber and associated systems for handling materials and treated soils and treating off-gases and residuals.

**thermal destruction** The complete or partial conversion of sludge organic solids to carbon dioxide ($CO_2$), water ($H_2O$), and other oxidized products, using such methods as incineration and starved-air combustion (the partial oxidation and volatilization of organics). Thermal destruction reduces the quantity of sludge for final disposal. Concepts associated with thermal destruction include available heat, bomb calorimeter, characteristic heating value, Dulong formula, effective heating value, gross heating value, latent heat of vaporization, la-

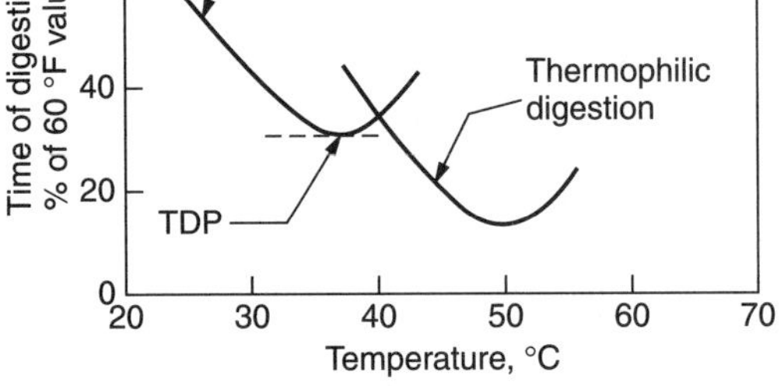


**Figure T-05.** Thermal death point (TDP). (Adapted from Fair et al., 1971.)

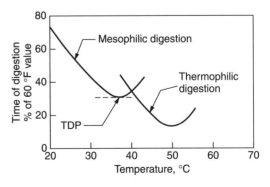

tent heat, sensible heat, sludge fuel value, sludge heating value, theoretical flame temperature, total heating value. *See also* multiple-hearth furnace, fluid-bed furnace, infrared or electric furnace, multiple-hearth fluid-bed furnace, slagging furnace.

**thermal diffusion** (1) A process that uses a temperature gradient to produce freshwater by removing dissolved salts from saline or brackish waters. Also called thermodiffusion. *See also* saline-water conversion classification. (2) Heat conduction in water.

**thermal disinfection** The use of heat to disinfect water, usually on a small scale. *See* thermal inactivation.

**thermal dryers** *See* conduction (or indirect) dryer, convection (or direct) dryer, radiation (or infrared) dryer.

**thermal drying** The application of heat to evaporate moisture from sludge, which is then used as a fertilizer or soil conditioner, disposed in a landfill, or incinerated. Thermal sludge drying occurs in three stages: warm up, constant rate, and falling rate. *See also* convection drying, conduction drying, and radiation drying.

**thermal efficiency** (1) The condition of an incinerator that operates at the design temperature and produces optimum burnout. (2) The ratio of total useful energy produced by a power plant as a percentage of the calorific value of the fuel consumed.

**thermal gradient** Temperature difference between two points or two areas.

**thermal inactivation** The use of heat to inactivate microorganisms by the denaturation of their enzymes, capsids, and nucleic acids.

**thermal odor processing** Common thermal methods used to process odors at wastewater treatment plants include catalytic oxidation, recuperative oxidation, regenerative oxidation, and thermal oxidation.

**thermal oxidation** The complete oxidation of odorous gases in a combustion chamber at temperatures of 800–1400°F and after adequate preheating. The oxidation reaction for methane ($CH_4$), for example, is:

$$CH_4 \text{ (gas)} + 2\,O_2 \text{ (gas)} \rightarrow CO_2 \text{ (gas)} + 2\,H_2O \text{ (vapor)} + \text{heat} \qquad \text{(T-12)}$$

**thermal oxidizer** An emissions control device using heat to oxidize volatile organic compounds in odor management. In thermal oxidation, the odorous gases are preheated before going into the combustion chamber. *See* recuperative oxidizer and regenerative oxidizer.

**thermal ozone destructor** A device used to destroy excess ozone at high temperature.

**thermal pollution** Discharge of heated substances into the environment from industrial processes, causing an increase in ambient temperature that can kill or injure organisms. For example, cooling water discharges from a coal-fired steam plant often have temperatures 11–17°F higher than the temperature of the receiving stream.

**thermal reactivation** *See* thermal regeneration.

**thermal reduction** A sludge processing technique that uses (a) incineration or wet-air oxidation to totally or partially convert organic matter to carbon dioxide (CO) and water ($H_2O$) or (b) pyrolysis or starved-air combustion to convert organic solids to end products with energy content.

**thermal regeneration** The use of combustion in the absence of oxygen to regenerate or reactivate granular carbon. It is more widely used than chemical reactivation. Thermal regeneration takes place in multiple-hearth furnaces, fluidized-bed furnaces, electric infrared furnaces, rotary kilns, or other devices.

**thermal sludge conditioning** *See* thermal conditioning.

**thermal spring** A spring that has a temperature significantly above the mean annual ambient temperature, originating from volcanic areas or great depths. Also defined as a spring that has mineralized water with a temperature above that of local groundwater. *See also* cold spring, hot spring, warm spring.

**thermal stratification** The formation of layers of different temperatures in a lake, reservoir, or other body of water, as in Figure T-06. With the effects of density, wind, and surface currents, thermal density produces a sequence of seasonal patterns

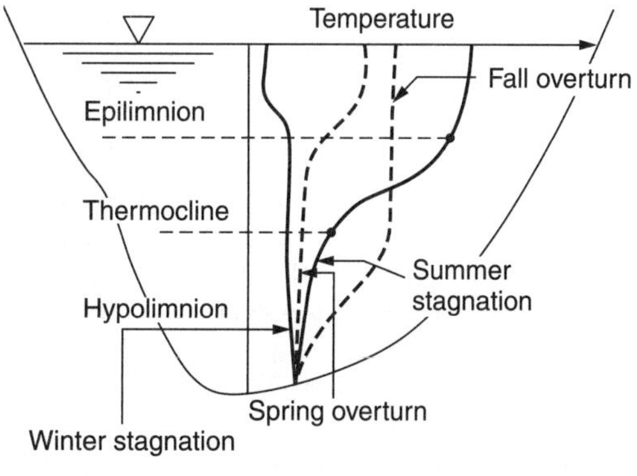


**Figure T-06.** Thermal stratification (adapted from Fair et al., 1971, and Metcalf & Eddy, 1991).

and seasonal gradients in water quality, particularly the dissolved oxygen content of the water body. The quality gradient is more pronounced during the summer than during the winter. *See also* algal bloom, density current, density stratification, epilimnion, fall turnover, great turnover, hypolimnion, lake stratification, selective withdrawal, spring circulation, spring turnover, stratification, summer stagnation, thermocline, transition zone, winter stagnation, zonal differentiation.

**thermal treatment** The treatment of hazardous waste in a device that uses elevated temperatures as the primary means to change the chemical, physical, or biological character or composition of the hazardous waste. Examples of thermal treatment processes are incineration, molten salt, pyrolysis, calcination, wet air oxidation, and microwave discharge (EPA-40CFR260.10).

**thermal value** The quantity of heat that can be obtained from a material per unit mass, i.e., the number of units of heat obtained by the complete combustion of a unit mass of the material. *See* calorific value for detail.

**thermoacidophilic group** A group of organisms that live in severe conditions of high temperature and high acidity, e.g., in sulfurous thermal vents.

**ThermoBlender™** An apparatus manufactured by RDP Co. for mixing quicklime and sludge.

**thermocline** The middle layer in a thermally stratified lake or reservoir. In this layer, there is a rapid decrease in temperature with depth, e.g., 0.5°F or more per foot. Also called meta-limnion, mesolimnion, or transition zone. *See also* chemocline, thermal stratification.

**thermocline zone** The thermocline.

**thermocouple** A heat-sensing device made of two conductors of different metals joined at their ends. An electric current is produced when there is a difference in temperature between the ends. Thermocouples are used to monitor temperatures in multiple-hearth furnaces and fluid-bed furnaces.

**thermodiffusion** A process that uses a temperature gradient to produce freshwater by removing dissolved salts from saline or brackish waters. Also called thermal diffusion. *See also* saline-water conversion classification.

**thermodynamic function of state** Any of the thermodynamic quantities that depend on state variables.

**thermodynamic potential** Same as Gibbs' free energy.

**thermodynamics** The science concerned with the relations between heat and energy or work. Thermodynamics indicates whether a chemical reaction will proceed as written. Thermodynamic considerations applied in water chemistry include standard cell potential, electrochemical potential, Gibbs free energy of reaction, Faraday's constant, equilibrium constant, equilibrium quotient, kinetic law of mass action, reaction rate constant, rate law.

**thermodynamics first law** Energy is conserved; it cannot be created or lost. It can be transformed from chemical to heat energy, from potential to kinetic energy, etc.

**thermodynamics laws** *See* zeroth law, first law, second law, third law of thermodynamics.

**thermodynamics second law** The useful work obtained from an energy transformation is less than the energy transformed.

**thermodynamics third law** *See* third law of thermodynamics.

**thermodynamics zeroth law** *See* zeroth law of thermodynamics.

**thermodynamic variable** *See* state variable.

**thermogenesis** *See* organic thermogenesis.

**thermograph** An instrument that records temperatures automatically.

**thermophile** (1) Any of a group of organisms that grow best in a range of warm temperatures (from the prefix thermo, meaning heat or hot). The literature reports different temperatures for this range; they generally fall between 35°C (95°F) and 85°C (185°F); the optimum range for growth is between 50 and 60°C (AWWA, 2000; Pankratz, 1996; Manahan, 1997; Metcalf & Eddy, 1991; Metcalf & Eddy, 2003; Hammer & Hammer, 1996; Droste, 1997; Maier, 2000; WEF, 1991). *See also* mesophile, psychrophile, thermal death point. (2) Same as thermophilic.

**thermophilic** Pertaining to a thermophile. Thermophilic bacteria active in waste digestion have an optimum temperature of 50–60°C. *See also* psychrophilic and mesophilic.

**thermophilic aerobic digestion (TAD)** A solids handling process in which liquid sludge is agitated with air or oxygen to maintain aerobic conditions, with a mean cell residence time of 10 days at a temperature within the thermophilic range. TAD is one of the processes to further reduce pathogens (PFRP) as defined by the USEPA. *See also* autothermal thermophilic aerobic digestion.

**thermophilic anaerobic digestion** Anaerobic sludge digestion carried out within the thermophilic range, e.g., between 50 and 57°C. Because of process instability, odors, high energy requirements, and high supernatant dissolved solids, thermophilic anaerobic digestion is not widely used. The USEPA classifies thermophilic digestion as a process to further reduce pathogens but not as a process to significantly reduce pathogens. *See also* two-phased anaerobic digestion.

**thermophilic bacteria** Bacteria that grow at relatively high temperatures. *See* thermophile and thermophilic.

**thermophilic composting range** Same as thermophilic stage.

**thermophilic digestion** Sludge digestion within the thermophilic range, e.g., 50–57°C. *See* thermophilic aerobic digestion, thermophilic anaerobic digestion, process to further reduce pathogens, process to significantly reduce pathogens.

**thermophilic (micro)organism** A (micro)organism that grows at warm temperatures. Thermophilic microorganisms active in composting include bacteria, actinomycetes, and fungi. *See* thermophile, thermophilic, and thermophilic range.

**thermophilic organism** *See* thermophilic microorganism.

**thermophilic range** (1) The range of warm temperatures over which some organisms can exist. *See* thermophile. (2) The optimum temperature range to operate waste digestion or sludge composting units so as to enhance the growth of thermophilic organisms; reported as variously anywhere between 40 and 80°C (between 122 and 176°F).

**thermophilic stage** The second of three stages of the sludge composting process, during which temperature of the composting material increases from 40 to 60°C (104 to 140°F), with a preponderance of bacteria, actinomycetes, and fungi. This is the most productive composting range in terms of degradation and stabilization of organic material. Also called thermophilic composting range. *See also* cooling stage and mesophilic stage.

**Thermopipe®** A lining system made of polyethylene reinforced by polyester, manufactured by Insituform for the rehabilitation of water mains and other pressurized installations.

**thermoplastic** Soft and pliable when heated, without a change in properties; pertaining to some plastics and resins that are used in pipes and other materials used in water and wastewater treatment.

**thermoset** Thermoplastic material that will not resoften once solidified.

**Thermo-Sludge Dewatering™** A sludge thickening process developed by Dontech, Inc.

***Thermothrix*** A genus of strictly aerobic, chemoautotrophic, sulfur-oxidizing bacteria.

**thermotolerant** Pertaining to microorganisms (bacteria and archaebacteria) that can survive at temperatures higher than 70°C, including the genera *Methanobacterium, Pyrococcus, Pyrodictium, Sulfolobus,* and *Thermus.*

**Thermox®** An instrument manufactured by Ametek, Inc. to analyze flue gases.

***Thermus aquaticus*** A species of thermotolerant bacteria.

***Thermus brockianus*** A microbe that produces a catalase enzyme that can chemically convert hydrogen peroxide ($H_2O_2$) into water and oxygen. It can potentially be used to solve wastewater treatment problems in the bleaching industry.

***Thermus* genus** A genus of thermotolerant bacteria

**theta ($\theta$) factor** A dimensionless coefficient (constant over small temperature ranges) that expresses the effect of temperature on biological and chemical reaction rates. Also called Arrhenius constant, Arrhenius parameter, temperature coefficient, or $\theta$ factor. *See* Arrhenius equation, Arrhenius constant.

**thickener** A water or wastewater treatment unit (tank, vessel, apparatus) that increases the solids concentration of sludge from gravity sedimentation. *See* flotation thickener, gravity thickener, gravity-belt thickener, rotary-drum (or rotating-drum) thickener, solid-bowl centrifuge, thickening clarifier.

**thickening** The process of concentrating the solids in water or wastewater treatment sludge from 2–3% to 5–6%, by such mechanisms as gravity settling, flotation, or centrifugation, ahead of dewatering or digestion. Stirring promotes the formation of larger solids and facilitates thickening. The thickener supernatant is usually returned to the primary clarifier and the settled solids are further processed. *See also* centrifugal thickening, cosettling thickening, flotation thickening, gravity belt dewatering, gravity thickening, dissolved-air flotation, drying.

**thickening centrifuge** A centrifuge used for sludge thickening as compared to a dewatering centrifuge.

**thickening clarifier** A primary clarifier designed to achieve both sedimentation and thickening of sludge solids in wastewater treatment by incorporating a sludge blanket and providing retention times of 12 to 24 hours. *See* cosettling thickening for detail.

**thickening of sludge** *See* thickening.

**thief** An instrument consisting of two stainless steel or brass, slotted concentric tubes, used to sample dry granule or powdered hazardous wastes with a diameter that is less than one-third the size of the slots. *See also* auger, bailer, composite liquid-waste sampler, dipper, trier, weighted bottle.

**thief sampler** An instrument consisting of two slotted concentric tubes of stainless steel or brass used to sample dry granule or powdered hazardous wastes.

**thin-film composite (TFC) membrane** A semipermeable membrane used in water treatment by reverse osmosis and consisting of more than one polymer, as opposed to an anisotropic or asymmetric membrane. It has a rejecting layer of one polymer (e.g., polyamide, a thin cellulose acetate, about 0.20 micron) supported by layers of thicker, different porous materials. Also called composite membrane or thin-film membrane.

**thin-film evaporation** A process that removes volatile constituents from a liquid waste by heating a thin layer of the liquid spread on a heated surface.

**thin-film evaporator** An apparatus for evaporation of liquids in thin films over heat transfer tubes.

**thin-film membrane** *See* thin-film composite membrane.

**thin-plate weir** A device or structure, usually made of a plastic or metal plate, used for flow distribution or flow measurement. Materials used in wastewater treatment include also steel, stainless steel, fiberglass, and aluminum. Its crest is so thin that water flowing over it touches only a line. Also called sharp-crested weir. *See* broad-crested weir.

**thio** (1) A prefix indicating that some or all of the oxygen atoms of a compound have been replaced by sulfur. (2) A combining form meaning sulfur.

**thioalcohol** Any of a number of aliphatic organic compounds that contain sulfur, have a disagreeable garlicky odor, and are found in certain wastewaters. *See* more detail under mercaptan. Also called thiol, which may be used for the removal of heavy metals from wastewater; see, e.g., self-assembled monolayers on mesoporous supports.

***Thiobacillus*** A genus of sulfur oxidizers; bacteria that need little oxygen to convert hydrogen sulfide ($H_2S$) to sulfuric acid ($H_2SO_4$) and cause corrosion when moisture accumulates at the crown of a sewer. It is very common in acid mine drainage.

***Thiobacillus denitrificans*** A species of facultative anaerobic, sulfur-oxidizing bacteria that can use nitrate ($NO_3^-$) as a terminal electron acceptor (instead of oxygen); it is active in denitrification processes and is said to achieve facultative anaerobic autotrophic respiration:

$$S^0 + 2\,KNO_3 \rightarrow K_2SO_4 + N_2 + O_2 \quad \text{(T-13)}$$

***Thiobacillus ferrooxidans*** A species of iron- and sulfur-oxidizing, acid-tolerant (optimum pH = 2.0) chemoautotroph bacteria, active in the formation of acid mine drainage and in the microbial process used to recover metals in copper and uranium mining.

***Thiobacillus thiooxidans*** A species of colorless, aerobic, autotrophic, sulfur-oxidizing, acid-tolerant (optimum pH = 2.0) bacteria, most responsible for the formation of acid mine drainage; it produces sulfate and sulfuric using sulfur, thiosulfate, or tetrathionate for food. It is active in the microbial process used to recover metals in copper and uranium mining. *See also* concrete corrosion, crown corrosion.

**thiol** Same as thioalcohol.

***Thiomicrospira*** A genus of chemoautotrophic, strictly aerobic, sulfur-oxidizing bacteria.

***Thiopedia*** A genus of photoautotrophic, strictly anaerobic, green or purple sulfur-oxidizing bacteria.

**thiosulfate ($S_2O_3^{2-}$)** A salt or ester of thiosulfuric acid; it may result from the oxidation of sulfur by aerobic heterotrophic bacteria and fungi. Thiosulfate is used in the azide modification of the iodometric method to measure dissolved oxygen; it reacts with free iodine ($I_2^0$) to form tetrathionate ($S_4O_6^{2-}$) and iodide ion ($I^-$):

$$2\,S_2O_3^{2-} + I_2^0 \rightarrow S_4O_6^{2-} + 2\,I^- \quad \text{(T-14)}$$

*See also Thiobacillus thiooxidans.*

**thiosulfuric acid ($H_2S_2O_3$)** An acid corresponding to sulfuric acid ($H_2SO_4$) in which sulfur replaces one oxygen atom.

***Thiothrix*** A genus of filamentous sulfur bacteria that grow well on volatile fatty acids and reduced sulfur compounds. They cause bulking in activated sludge systems as well as operating problems in trickling filters and rotating biological contactors.

***Thiothrix spp.* type 021N** A type of filamentous sulfur bacteria that grow in septic wastewater, deficient in nutrients but with sulfide available.

***Thiothrix spp.* type 0914** A type of filamentous sulfur bacteria that grow in septic wastewater, deficient in nutrients but with sulfide available.

**thiram ($C_6H_{12}N_2S_4$)** A white, crystalline compound, insoluble in water; used as an insecticide and in industry.

**third-class water quality** The quality of a water that is not intended for drinking or cooking (i.e., first class) or for contact with skin or cooking tools (second class). Such water may be used for air conditioning, lawn watering, or floor cleaning; its applicable uses represent approximately 73% of domestic water consumption.

**Third Contaminant Candidate List (CCL3)** A list of chemical and microbiological drinking water contaminants that the USEPA intends to issue in 2008 for possible future regulations. *See* Contaminant Candidate Lists 1, 2.

**third law of thermodynamics** Absolute zero (°K) can never be attained.

**third-order consumer** An organism (e.g., a bass) that feeds on second-order consumers (e.g., sunfishes) in a food chain.

**thirty-day average** The arithmetic mean of pollutant parameter values of samples collected in a period of 30 consecutive days (EPA-40CFR133.101-b). The thirty-day limitation is the value that should not be exceeded by the average of daily measurements taken during any 30-day period (EPA-40CFR429.11-j). The thirty-day rolling average is any value arithmetically averaged over any consecutive thirty days (EPA-40CFR52.74-1).

**thirty-day limitation** *See* thirty-day average.

**thirty-day rolling average** *See* thirty-day average.

**33/50 Program** A successful pollution prevention program of the USEPA. Begun in 1991 to encourage industries to reduce emission of 17 toxic chemicals by 33% in 1992 and 50% in 1995, in exchange for assistance, recognition, and awards. *See also* Toxics Release Inventory and 33/50 Program chemicals.

**33/50 Program chemicals** *See* Table T-03.

**thixotrophy** The reversible property of certain materials (gels, clay mixed with water, sludges) of becoming viscous liquids when stirred or shaken. *See also* rheology. Most sludges are thixothropic because their plastic properties change during processing, which makes it difficult to determine their coefficient of rigidity and shearing stress; these parameters are necessary to determine head losses by the sludge flow equation.

**THM** Acronym of trihalomethane.

**THM4** Symbol representing the sum of the four trihalomethanes considered in the formation of disinfection by-products: chloroform, bromodichloromethane, chlorodibromomethane, and bromoform.

**Table T-03.** 33/50 Program chemicals

| | |
|---|---|
| benzene | methyl ethyl ketone |
| cadmium and its compounds | methyl isobutyl ketone |
| carbon tetrachloride | nickel and its compounds |
| chloroform (trichloromethane) | tetrachloroethylene (perchloroethylene) |
| chromium and its compounds | toluene |
| cyanide and its compounds | 1,1,1-trichloroethane (methyl chloroform) |
| dichloromethane (methylene chloride) | trichloroethylene |
| lead and its compounds | xylene |
| mercury and its compounds | |

**THMFP** Acronym of trihalomethane formation potential.

**THM precursor** *See* precursor.

**THMt** Abbreviation of terminal trihalomethane concentration.

**ThOD** Abbreviation of theoretical oxygen demand.

**Thomas BOD method** A graphical method proposed by H. A. Thomas, Jr. in 1950 to determine the parameters $L$ and $K$ of the BOD exertion equation:

$$\mathrm{BOD}_t = L(1 - e^{-Kt}) \qquad \text{(T-15)}$$

It replaces this equation by a function whose series expansion has first terms similar to those in the expansion of equation (T-15) and that can be linearized:

$$\mathrm{BOD} = LKt(1 + Kt/6)^{-3} \qquad \text{(T-16)}$$

or

$$(t/\mathrm{BOD})^{1/3} = (KL)^{-1/3} + tK^{2/3}/6\,L^{1/3} \qquad \text{(T-17)}$$

with $\mathrm{BOD}_t$ = BOD exerted at time $t$; $L$ = BOD at time $t = 0$; $K$ = a rate constant that varies with temperature; and $t$ = time, days. Given a set of pairs of values of time ($t$) and BOD, $(t/\mathrm{BOD})^{1/3}$ plots as a straight line against time, which allows the graphical determination of the intercept $(KL)^{-1/3}$ (at $t$ = 0) and slope $K^{2/3}/6\,L^{1/3}$; thus, there are two equations that can be solved for $K$ and $L$. *See also* Fujimoto BOD analysis method, least-squares BOD analysis.

**Thomas' formulation** *See* Thomas BOD method.

**Thomas, Harold A., Jr.** American author, professor and researcher (Harvard University), with particular contributions in sanitary engineering, hydrology, and resource economics (1913–2002).

**Thomas' method** *See* Thomas BOD method and Thomas' nomogram..

**Thomas' MPN equation** An approximation proposed by H. A. Thomas, Jr. in 1942 (*Journal AWWA,* vol. 34, No. 4, p. 572) to estimate the most probable number (MPN) in the absence of the Poisson parameters or MPN tables:

$$\mathrm{MPN}/100\ \mathrm{mL} = 100\,P/(NT)^{1/2} \qquad \text{(T-18)}$$

where $P$ = number of positive tubes, $N$ = mL of sample in negative tubes, and $T$ = mL of samples in all tubes. Also called most probable number formula.

**Thomas' nomogram** A nomogram proposed in 1948 for the graphical solution of the Streeter–

Phelps equation. It plots $(D/L_a)$ versus $(Rt)$ for various ratios of $(R/K)$. The dissolved oxygen (DO) deficit can be determined at any time $t$ downstream of a pollution source or, conversely, the pollution load can be determined, given a critical DO deficit: $D$ = dissolved oxygen deficit at a distance $x$ from the point of pollution or at a time $t$ corresponding to that distance; $L_a$ = initial first-stage BOD of the mixture of stream and discharge; $K$ = the decay constant or deoxygenation rate; $R$ = reaeration rate; and $T$ = time of stream flow. *See also* Thomas BOD method.

**thorium (Th)** A silvery-white, radioactive, metallic element, used as a nuclear fuel and in industry. Atomic weight = 232.04. Atomic number = 90. Specific gravity = 11.72. Valence = 2, 3, 4. Long-term exposure to thorium salts may cause neoplasms.

**thorium-230** A radioisotope of thorium whose daily contribution from drinking water is less than 0.08 pCi based on a consumption of 2.0 liters. Half-life = 75,200 years. It results from the decomposition of uranium-238 in three steps.

**thorium-232** A natural radionuclide whose daily intake through drinking water is less than 0.02 pCi based on a consumption of 2.0 liters.

**thorium-234** A radioisotope of thorium, resulting from the decomposition of uranium-238; half-life = 24 days.

**threadworm** (1) The helminth species, *Strongyloides stercoralis,* that causes strongyloidiasis, transmitted between human hosts through soils, mainly in warm, wet climates. (2) Common name of strongyloidiasis, the disease caused by this pathogen. Symptoms are often absent but sometimes include diarrhea, abdominal pain, rash, and respiratory problems.

**threatened species** A species that is likely to become an endangered species in the foreseeable future.

**threatened water** Water that may not support its designated uses in the future unless pollution control measures are taken.

**three-log removal** or **3-log removal** 99.9% removal or inactivation of a pathogen, more specifically Giardia cysts, as specified in USEPA's Surface Water Treatment Rule.

**three-stage A²/O™ process** Same as A²O™ process.

**threshold** (1) The lowest dose of a chemical at which a specified measurable effect is observed and below which it is not observed. (2) The dose or exposure below which a significant adverse effect is not expected. Carcinogens are thought to be nonthreshold chemicals, to which no exposure can be presumed to be without some risk of adverse effect.

**threshold CT** The threshold exposure, i.e., the threshold product of disinfectant concentration and exposure time $CT_{th}$, at which a given microbial cell is inactivated. *See* disinfection kinetics.

**threshold dose** The minimum dose that does not produce an observable adverse effect. It is assumed that there is no threshold dose for carcinogens.

**threshold level** Time-weighted average pollutant concentration value, exposure beyond which is likely to adversely affect human health. *See also* environmental exposure.

**threshold limit value (TLV)** (1) The concentration of an airborne substance that an average person can be repeatedly exposed to without adverse effects. TLVs may be expressed in three ways. (a) TLV-TWA = time-weighted average, based on an allowable exposure averaged over a normal 8-hour work day or 40-hour workweek; (b) TLV-STEL = short-term exposure limit or maximum concentration for a brief specified period of time, depending on a specific chemical (TWA must still be met); and (c) TLV-C = ceiling exposure limit or maximum exposure concentration not to be exceeded under any circumstances (TWA must still be met). (2) Recommended guidelines for occupational exposure to airborne contaminants published by the American Conference of Governmental Industrial Hygienists (ACGIH). The TLVs represent the average concentration (in mg/cu.m) for an 8-hour workday and a 40-hour workweek to which nearly all workers may be repeatedly exposed, day after day, without adverse effect. *See also* permissible exposure limit and recommended exposure limit.

**threshold odor** The minimum odor of a water sample that can just be detected after successive dilutions with odorless water. Odor threshold may also be defined as the lowest concentration of the responsible substance that the olfactory sense can detect; it is commonly reported in μg/L or μg/kg. Odor threshold in water varies widely; e.g., 1 μg/L for geosmin and 20 mg/L for chloroform; chlorine taste complaints are often reported by consumers at a concentration as low as 0.2–0.4 mg/L. Also called odor threshold. *See also* dilutions to threshold, minimum detectable threshold odor concentration, odor threshold concentration, and threshold odor number.

**threshold odor concentration** Same as odor threshold concentration.

**threshold odor number (TON)** A parameter used to express the concentration of odor-causing materials in water. It is the greatest dilution of a sample with odor-free water that still yields a just-detectable odor. The TON is sometimes erroneously defined as the dilution that is required to eliminate the odor. As an example, if odor is perceptible in a 5:1 dilution but not in a 6:1 dilution, the TON is 5, not 6. The TON may be calculated as

$$\text{TON} = 1 + V_d/V_t \quad \text{(T-19)}$$

where $V_d$ and $V_t$ are, respectively, the volume tested and the volume of dilution with odor-free distilled water. The USEPA sets a secondary maximum contaminant level of TON = 3 for drinking water. The TON of sludge is much higher, on the order of 8000 for raw primary or secondary sludges and 1000 for lime-stabilized sludge. *See also* flavor profile analysis (FPA).

**threshold pollutant** A substance that is harmful to a certain organism only above a certain concentration or threshold level.

**threshold substance** Same as trace substance.

**threshold treatment** The addition of small quantities of sodium hexametaphosphate [$(NaPO_3)_n$] to softened water to prevent the deposition of calcium carbonate ($CaCO_3$).

**throttle valve** A valve that reduces flow through a pipeline.

**throughput volume** The volume of water processed by a treatment unit during a cycle, e.g., a filter run or before an exchanger or bed reaches exhaustion.

**Thru Clean** A back-cleaned mechanical bar screen manufactured by FMC Corp.

**thrust block** A mass of concrete or similar material appropriately placed around a pipe to prevent movement when the pipe is carrying water. Usually placed at bends and valve structures.

**thymine ($C_5H_6O_2N_2$)** A pyrimidine base that pairs with adenine as important constituents of DNA.

**thymine dimer** A compound of two thymine molecules formed under the action of ultraviolet light.

**thymine dimerization** The formation of a thymine dimer, a mechanism of ultraviolet disinfection that damages microbial DNA or RNA by blocking nucleic acid replication and inactivates microorganisms. *See also* photoreactivation.

**thymol blue** A reagent used in water analyses as an acid–base indicator, with pH transition ranges of 1.2 (red)–2.8 (yellow) and 8.0 (yellow)–9.6 (blue).

**thymol blue indicator** *See* thymol blue.

**TIC** Total inorganic carbon.

**tick** An arthropod vector of such diseases as relapsing fever, affecting mostly people in East and Southern Africa. *See Ornithodorus moubata.*

**tick-borne relapsing fever** The disease caused by the pathogen *Borrelia duttoni,* which is spread by the tick *Ornithodorus moubata.*

**tidal marsh** Low, flat marshlands traversed by channels and tidal hollows, subject to tidal inundation; normally, the only vegetation present is salt-tolerant bushes and grasses. *See* wetlands.

**tidal prism** In an estuary, the body of fresh and salt water that lies between high tide and low tide.

**Tideflex™** A diffuser check valve manufactured by Red Valve Co., Inc.

**tide gate** A gate used to prevent the backflow into a combined sewer whose overflow outlet discharges below the high-water level of the receiving water. Also called backwater gate.

**Tier 1 violation** A serious violation of the National Primary Drinking Water Regulations, e.g., exceeding a maximum contaminant level, which requires an extensive public notification. A tier 2 violation, such as violating a reporting requirement, is less serious.

**Tier 2 violation** *See* Tier 1 violation.

**tiered rate** Same as inclining block rate.

**tight-fit lining** A modified version of sliplining, a trenchless method of gas and water main rehabilitation; the HDPE liner pipe is pulled into the damaged pipe through a series of reducing rollers and then allowed to expand so as to fit tightly within the existing main. *See* trenchless technology, modified sliplining, fold-and-form lining.

***Tilapia*** A genus of hardy, freshwater food fish that survive well in stabilization ponds used for aquaculture.

**tile** A ceramic tube or pipe used for drainage.

**Tile** A filter underdrain in tiles, manufactured by Roberts Filter Manufacturing Co.

**tile drainage** The collection and removal of excess groundwater through unsealed joints and/or perforations in drains. *See also* drain tile.

**tile field** A system of open-jointed drain tiles laid on a rock or gravel fill to distribute septic tank or other wastewater effluents over an absorption area or for providing drainage in wet areas. *See also* subsurface irrigation.

**tile filter bottom** The underdrainage system that collects filtered water below the granular bed.

**tile underdrainage** A network of tile drains in underground trenches to collect and remove excess groundwater.

**tile wastewater disposal lines** Lines of open-jointed tiles in underground trenches for the dispersal of wastewater or septic tank effluent. Subsurface disposal, subsurface irrigation.

**Tilted Disc®** A check valve manufactured by Val-Matic® for low head loss operation of centrifugal and turbine pumps.

**tilted plate separator** A device consisting of inclined plates used to separate oil from water based on their density differences.

**tilting trough** A mechanism, activated by a tripping device on sludge collectors, used for removing scum from primary clarifiers and directing it to a wet well for pumping to another treatment unit. *See also* sloping beach.

**time composite** A sample made of equal-volume subsamples collected over time. *See* composite sample.

**time–concentration curve** A tracer response curve, i.e., a graphical representation of the distribution of a tracer versus time of flow; also called a C curve, flow-through curve, fluid residence time distribution curve, residence time distribution curve, or RTD curve. It is used in studying short-circuiting, retardation, and longitudinal mixing in sedimentation basins and other reactors; e.g., (a) in the absence of short-circuiting, the mean, median, and mode of the curve coincide; (b) percentile ratios correspond to variability of exposure to sedimentation; (c) flow through the basin is unstable and performance is erratic if the curve does not reproduce well in repeated tests (Fair et al., 1971). *See* Figure T-07. *See also* average retention time index, cumulative residence time distribution curve, dispersion number, dispersion symmetry index, exit age curve, mean retention time index, modal retention time index, Morrill dispersion index, residence time distribution curve, short-circuiting index, volumetric efficiency.

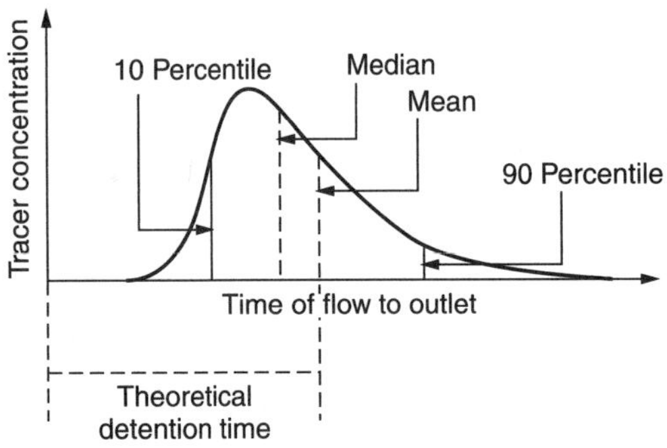


**Figure T-07.** Time–concentration curve (adapted from Fair et al., 1971).

**time constant** In the control of wastewater treatment processes using on-line instrumentation, time constant is "the time required for a parameter to achieve 63.2% of the difference between the initial and final conditions after a disturbance is introduced" (Metcalf & Eddy, 2003).

**time-invariant condition** The condition of a constituent in an aqueous solution, the concentration ($C$) of which does not change with time ($t$), i.e., for which $dC/dt = 0$. The time-invariant condition is equilibrium for a closed system and steady state for a continuous, open system. Also called stationary state, time-invariant state.

**time-invariant state** Same as time-invariant condition.

**time of breakthrough** *See* time to breakthrough.

**time of concentration ($t_c$)** (1) The time required for water to move from the hydraulically most distant point of the drainage area to the outlet or design point under consideration. It is the sum of inlet time and time of flow. Used in the rational formula to determine the peak runoff flow, under the assumption that the flow reaches a peak when the entire drainage area is contributing runoff. *See,* e.g., the Kerby formula. (2) Sometimes called time of equilibrium or time to equilibrium, which is the time when the rate of runoff equals the rate of rainfall. (3) Based on a rainfall hydrograph, the resulting runoff hydrograph, the rainfall excess, and the direct runoff; the time of concentration is, on the time scale, the distance between the center of mass of rainfall excess and the inflection point on the recession of the direct runoff hydrograph or $t_c$ = the difference between the end of rainfall excess and the inflection point. *See also* hydraulic residence time and hydrograph times.

**time of exposure** An important parameter in water and wastewater treatment processes, sometimes measured by dyes or other tracers. Examples of time of exposure include the hydraulic detention time (ratio of volume to flow), sludge age, and filtration time (ratio of filter depth to the velocity through the bed).

**time-of-flight mass spectrometer** A mass spectrometer that separates ions on the basis of their velocity through a draft tube.

**time of flow** (1) The time for water to flow in a stream, sewer, canal, etc. from one point to another. *See also* travel time, hydraulic residence time. (2) The time of flow in a sewer, canal, etc. from the point of concentration to the outfall; a component of the time of concentration.

**time of passage** Same as time of exposure.

**time of travel** Same as travel time; a parameter in the Streeter–Phelps formulation.

**timer** A device for automatically starting or stopping a machine or other device at a given time.

**time-rate factor** A factor, with dimension of inverse of time (e.g., per sec or per day), affecting treatment and reaction rates and showing a dependency on time of exposure. *See* the general purification equation and the reaeration rate.

**time–temperature safety zone** *See* safety zone.

**time-to-bed exhaustion** A parameter used in the design of fixed-bed adsorbers ($t_e$), equal to the ratio of the volume passing through the bed at exhaustion ($V_e$) to the flow rate ($Q$):

$$t_e = V_e/Q \qquad \text{(T-20)}$$

**time to breakthrough** (1) The time to reach breakthrough in the carbon adsorption process, i.e., the duration of an adsorption run. It can be estimated from the breakthrough curve and other adsorption data:

$$T_b = (KM)/[Q(C_0 - C_b/2)] \qquad \text{(T-21)}$$

where $T_b$ = time to breakthrough, days; $K$ = breakthrough adsorption capacity, grams/gram; $M$ = mass of carbon in the column, grams; $Q$ = flow rate, $m^3$/day; $C_0$ = influent concentration, mg/L; and $C_b$ = breakthrough concentration, mg/L. (2) The time of breakthrough may also be determined from the maximum number of bed volumes that can be fed prior to breakthrough ($B_v$) and the empty bed contact time ($T_e$):

$$T_b = B_v \cdot T_e \qquad \text{(T-22)}$$

**time to filter** The time to filter a sludge sample through a porous membrane.

**time-to-filter (TTF) test** A simplified version of the specific resistance test to determine sludge dewaterability and filterability, the TTF test uses also the Büchner funnel but measures only the time to filter one-half of the sludge sample. *See also* Atterberg test, Büchner funnel test, capillary suction time (CST) test, filterability constant, filterability index, filter leaf test, shear strength, specific resistance test, standard jar test, standard shear test.

**time-weighted average (TWA)** The average value of a parameter (e.g., concentration of BOD in wastewater) that varies over time.

**time-weighted composite** A composite sample consisting of a mixture of equal-volume aliquots collected at a constant time interval.

**TIN** Acronym of total inorganic nitrogen.

**tin** (Sn from its Latin name, stannum.) A scarce, silvery-white, malleable metal, found mostly in its oxide (cassiterite); used as an alloy and in fungicides, glass, etc. Atomic weight = 118.69. Atomic number = 50. Specific gravity = 7.31. Valence = 2 or 4. Most tin salts are insoluble. At sufficiently high concentrations (e.g., from liquids in cans), tin may cause food poisoning, interfere with enzyme production and cause physiological human effects.

**tines** The sharp projecting teeth or prongs of a fork or a cleaning rake. Also spelled tynes.

**$TiO_2$** Chemical formula of titanium dioxide.

**tip** British for (1) to dispose (e.g., refuse) by dumping or (2) a dump for refuse.

**tipping** The dumping of the contents of a waste truck at a waste disposal facility. A tipping fee is charged to dispose of solid waste at a sanitary landfill or a controlled tipping site.

**tipping bucket rain gauge** A rain gage that continuously records rainfall in increments of 0.01 inch on a clock-driven chart.

**tipping (controlled)** Same as sanitary landfill.

**tipping fee** *See* tipping.

**tipping-plate regulator** A flow-regulating device used for the control of combined sewer overflows and consisting of a plate pivoted off-center, its motion being controlled by the difference of water levels above and below the gate.

**Tipping Scum Weir** A pivoting weir manufactured by F. B. Leopold Co., Inc. for scum removal.

**tissue** A group of similar cells.

**tissue culture** (1) The technique or process of cultivating living tissue in a prepared medium outside the body for experimental research. (2) The tissue cultivated. *See* buffalo green monkey kidney cell.

**tissue culture infectious dose 50 ($TCID_{50}$)** The concentration of virus that kills or infects 50% of the cells in a tissue culture. *See* serial dilution endpoint method.

**titanium dioxide ($TiO_2$)** A white, water-insoluble powder used in pigments, plastics, ceramics, and in photocalytic oxidation.

**titanium dioxide ($TiO_2$) manufacturing** A process whose by-products include iron-based coagulants used in water and wastewater treatment. *See also* ferric chloride and ferric sulfate.

**titer** Also spelled titre. (1) The strength of a solution or the concentration of a substance in a sample, as determined by titration. (2) The concentration of a substance based on a series of dilutions.

**Titeseal** A gate manufactured by Plasti-Fab, Inc. for the control of liquid polymer.

**Titled Disc®** A line of check valves manufactured by Val-Matic. *See also* Swing-Flex®, Dual Disc® Cushion Swing, and Silent Check.

**titrant** The reagent added to a sample during a titration to reach an end point. *See also* acid–base indicator.

**titrate** To titrate a sample, a chemical solution of known strength is added drop by drop until a certain color change, precipitate, or pH change is observed in the sample (end point). Titration is the process of adding the chemical reagent in increments until completion of the reaction, as signaled by the end point.

**titration** A laboratory method used to determine the concentration of a dissolved substance; for a given substance, known volumes of an appropriate reagent are added to the solution until it reaches an endpoint. Titration is commonly used in water analysis to determine acidity, alkalinity, hardness, chlorine residual, etc. *See* Figure T-08. *See also* reverse titration and composite titration curve.

**titration endpoint** *See* endpoint (1). *See also* carbonic acid endpoint, bicarbonate endpoint.

**Titraver** A line of chemicals produced by Hach Co. for the analysis of hardness in water.

**titre** Same as titer.

**titrimetric method** A laboratory method using titration to determine the concentration or strength of a substance.

**TKN** Acronym of total Kjeldahl nitrogen.

**Tl** Chemical symbol of the metallic element thallium.

**TLC™** Acronym of thin-layer composite, a reverse osmosis membrane manufactured by Osmonics.

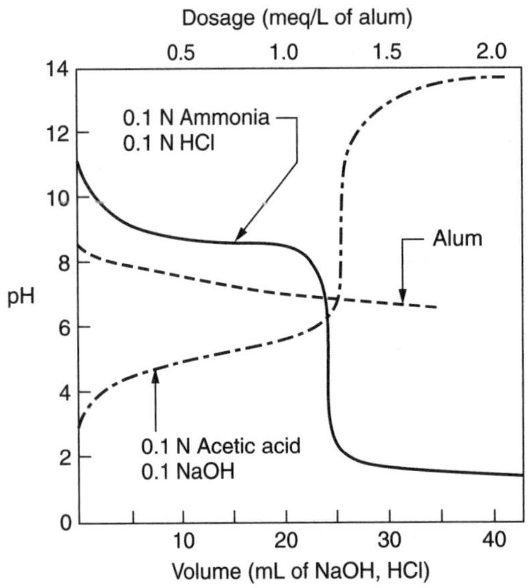


**Figure T-08.** Titration curves.

**$TL_m$** Symbol of median tolerance limit.

**TLV** Acronym of threshold limit value.

**TMDL** Acronym of total maximum daily load.

**TMDL rule** A USEPA regulation promulgated in 2000 to protect ambient water quality as a shift from technology-based controls. *See* total maximum daily load.

**TMP** Acronym of transmembrane pressure.

**TNCWS** Acronym of transient noncommunity water system.

**TNT** Acronym of trinitrotoluene.

**TNTC** Too numerous to count.

**T&O** Acronym of taste and odor.

**TOC** Acronym of total organic carbon.

**TOD** Acronym of total oxygen demand.

**toe ditch** A ditch along the foot of a dike, dam, or levee, on the land side, to collect seepage. *See also* rim canal.

**toilet** A place for defecation and urination, e.g., the superstructure of a latrine; sometimes used synonymously with latrine. *See* the following types of toilets: chemical, cistern-flush, communal, composting, pour-flush, twin-pit pour-flush, and vault.

**toilet (chemical)** *See* chemical toilet.

**toilet dam** A partition or a flexible insert installed in a toilet tank to reduce the amount of water used per flush. *See also* flow-reducing devices and appliances.

**toilet leak detector** A tablet that releases a dye while dissolving in a toilet tank and indicates whether the flush valve leaks. *See also* flow-reducing devices and appliances.

**tolerance** (1) A permissible residue level for pesticides in raw agricultural produce and processed foods. Whenever a pesticide is registered for use on a food or a feed crop, a tolerance (or exemption from the tolerance requirement) must be established. The USEPA establishes the tolerance levels, which are enforced by the Food and Drug Administration and the Department of Agriculture. (2) The ability of an organism to endure or resist the action of a drug, poison, etc.; tolerance level may be defined in terms of a period of exposure.

**tolerance level** *See* tolerance (2).

**tolerant fish species** Fish species that can withstand wide fluctuations in environmental conditions, warm water, low dissolved oxygen concentration, high carbon dioxide, and high turbidity, e.g., carp, black bullhead, bowfin. *See also* intolerant species, game fish.

**Tolhurst** A basket centrifuge manufactured by Ketema, Inc. for laboratories.

**tolidine $\{[C_6H_3(CH_3)NH_2]_2\}$** A derivative of biphenyl containing two methyl and two amino groups, e.g., the ortho isomer used as a reagent and in dyes. *See* orthotolidine.

**tolidine arsenite test** *See* orthotolidine arsenite test.

**toluene ($C_6H_5CH_3$)** A colorless, flammable liquid; an aromatic solvent with a benzenelike odor that is only slightly soluble in water; used in the production of benzene derivatives, perfumes, explosives (e.g., trinitrotoluene or TNT), and in other industrial applications. Acute exposure to this product may cause damages to the nervous system and certain organs. A volatile organic compound regulated by the USEPA as a drinking water contaminant: MCL = MCLG = 1.0 mg/L. Also called phenylmethane and methylbenzene.

**toluidine ($C_7H_9N$)** Any of three isomeric amines derived from toluene: metatoluidine, orthotoluidine, paratoluidine.

**TON** Acronym of (a) threshold odor number, (b) total organic nitrogen.

**ton** A unit of mass or weight equal to 2000 pounds; also called short ton. *See also* tonne.

**ton container** A storage unit that contains one ton of chlorine.

**Tonkaflo** A series of pumps manufactured by Osmonics for use in reverse osmosis, ultrafiltration, and deionization units.

**tonnage oxygen** A gas that is 95% oxygen; used in industry and in biological wastewater treatment; also called pure oxygen. It is produced by the liquefaction and fractional distillation of air.

**tonne** Same as metric ton, a unit of mass or weight equal to 1000 kg or approximately 2205 pounds. *See also* ton.

**too numerous to count (TNTC)** A phrase used to express coliform density in the membrane filtration technique when the total number of bacterial colonies exceeds 200 per 100 ml on a 47-micrometer diameter membrane filter (EPA-40CFR141.2).

**top brine temperature** The maximum temperature that a fluid under evaporation can reach.

**topsoil** (1) The fertile, cultivated, upper layer of the soil, enriched by humus, varying in depth from 8 to 45 cm. It also contains earthworms and other soil organisms. (2) Horizon A in a soil profile.

**Tordon acid™** The synthetic herbicide 4-amino-3,5,6-trichloropicolinic acid, picloram. The category includes chlorination processes utilized in Tordon™ acid production (EPA-40CFR6363.191).

**Tornado®** An aspirating surface aerator manufactured by Aeromix Systems, Inc.

**Torpedo™** An apparatus manufactured by United Industries, Inc. for the removal of foam and grease from clarifiers.

**Torpedo Filter** A floating microfilter manufactured by BTG, Inc.

**Torque-Flow** A vortex pump manufactured by Envirotech Co. for grit removal.

**torque-flow pump** A centrifugal pump with a fully recessed impeller that develops a vortex for the liquid to propel itself; used in sludge handling. Also called recessed-impeller pump, shear-lift pump, or vortex pump.

**TorusDisc** A device manufactured by Bepex Corp. for sludge drying and cooling.

**Torvex®** A catalytic oxidation apparatus manufactured by CSM Environmental Systems, Inc.

**total acidity** The total concentration of acid salts, carbon dioxide ($CO_2$), and mineral acidity in a solution. *See also* methyl orange acidity and phenolphthalein acidity.

**total alkalinity** The sum of bicarbonate, carbonate, and hydroxide alkalinity. *See also* phenolphthalein alkalinity.

**total ammonia nitrogen (TAN)** The sum of the concentrations of ammonia ($NH_3$) and ammonium ion ($NH_4^+$).

**total available chlorine** The sum of chlorine compounds available for disinfection and/or oxidation, including chloramines and other chlorine species, but excluding organic chloramines, which are not available as disinfectants. *See also* free available chlorine, total chlorine.

**total bacteria count (TBC)** A method used to determine the total number of viable microorganisms in a water sample: a known volume of water is filtered on a membrane and the retained microorganism colonies are counted after being cultured for several hours on an appropriate medium. *See also* total bacterial plate count.

**total bacterial plate count** An estimate of the number of bacteria in a sample. *See* standard plate count.

**total bed-material discharge** The sum of the suspended sediment bed-material discharge and the contact sediment (bed-load) discharge. *See also* sedimentation terms.

**total body clearance** The rate at which the body eliminates a chemical in volume or weight per unit time.

**total bromine-substituted trihalomethanes** Same as total trihalomethane bromine.

**total carbon** The sum of the concentrations of total inorganic carbon and total organic carbon in a water or wastewater sample. It may be measured by chemical oxidation to carbon dioxide ($CO_2$) and detection in a carbon analyzer.

**total carbon dioxide** The sum of free carbon dioxide ($CO_2$) and the carbon dioxide as bicarbonate ($HCO_3^-$) and carbonate ($CO_3^{2-}$).

**total carbonic species** The sum ($C_T$) of the concentrations of carbon dioxide ($CO_2$), carbonic acid ($H_2CO_3$), carbonates ($CO_3^{2-}$), and bicarbonates ($HCO_3^-$) in an aqueous solution:

$$C_T = [H_2CO_3^*] + [HCO_3^-] + [CO_3^{2-}] \quad \text{(T-23)}$$

where $[H_2CO_3^*]$ represents the combined concentrations of carbonic acid and carbon dioxide.

**total chlorine** The total concentration of chlorine in water or wastewater, i.e., the sum of free available chlorine and combined chlorine (inorganic and organic chloramines).

**total chlorine residual** Same as total residual chlorine.

**total chlorine-substituted trihalomethanes** Same as total trihalomethane chlorine.

**total chromium** The sum of hexavalent and trivalent chromium.

**total coliform bacteria** *See* total coliforms.

**total coliform count** The number of total coliform bacteria in a sample.

**total coliform group** *See* total coliforms.

**Total Coliform Rule (TCR)** A regulation of the USEPA, promulgated in June 1989 and amended in January 1991, setting standards for coliforms in drinking water. *See* National Drinking Water Regulations.

**total coliforms (TC)** A group of aerobic and facultatively anaerobic, gram-negative, nonspore-forming, rod-shaped bacteria (of the *Enterobacteriaceae* family) that ferment lactose to produce acid and gas within 48 hours at 35°C. They are used as an indicator of fecal pollution of drinking water from warm-blooded animals and regulated by the USEPA. They serve also as a measure of water treatment effectiveness and as a measure of public health risk. *See also* fecal coliforms, Total Coliform Rule.

**total coliform technique** The total coliform test.

**total coliform test** A test to detect the presence of coliform bacteria in water. *See* multiple-tube fermentation test, membrane filter test, presence–absence test.

**total color unit (TCU)** A common unit used to express color in water, including both true and apparent color, such that color is not noticeable at 10–15 TCU and 100 TCU corresponds to the appearance of tea. One color unit (CU) corresponds to the color produced by 1.0 mg/L of platinum. *See also* the platinum–cobalt method.

**total-containment pond** A pond used for wastewater disposal in areas where, annually, evaporation exceeds precipitation; the liquid is discharged in the pond and allowed to dry under the influence of solar heat. Regarding effluent discharge, there are also continuous-discharge ponds, controlled-discharge ponds, and hydrograph-controlled-release ponds. Also called an evaporation pond. *See also* lagoons and ponds.

**total count** An estimate of the number of bacteria in a sample. *See* standard plate count.

**total dissolved phosphorus** Total phosphorus content of material that will pass through a filter of a specific size, which is determined as orthophosphate without prior digestion or hydrolysis. Also called soluble P or ortho P.

**total dissolved solids (TDS)** The total filterable residue that passes through a standard glass fiber filter disk and remains after evaporation and drying to a constant weight at 180°C; it is considered to be a measure of the dissolved salt content of the water (EPA-40CFR131.35.d-16). TDS is one of the quality parameters used for drinking water as well as irrigation water; it may interfere with agricultural and industrial processes. It consists mainly of inorganic salts (carbonate, bicarbonate, chloride, sulfate, nitrate of sodium, potassium, calcium, magnesium), and small amounts of organic matter. *See also* electrical conductivity, salinity.

**total dissolved solids buildup** The increase in total dissolved solids (TDS) concentration created by the breakpoint chlorination of wastewater, which may affect effluent suitability for agricultural or industrial reuse. The buildup is measured in terms of units of ammonium ion ($NH_4^+$) consumed, varying from 6.2 mg/L of TDS for each mg/L of $NH_4^+$ for chlorine gas alone to almost 15:1 for chlorine gas plus addition of caustic soda (NaOH) to neutralize acidity.

**total dissolved solids test** A test to measure the total dissolved solids (TDS) of a water or wastewater sample by filtering out the suspended solids and evaporating the filtered water; the residue represents TDS.

**total dose** Total dose is usually taken as the product of the substance concentration ($C$, mg/L), the intake rate ($R$, L/day), the exposure duration ($T$, days), and an absorption factor ($f$, dimensionless). *See* chronic daily intake for detail.

**total dynamic discharge head** or **total dynamic head (TDH)** Same as dynamic head.

**total dynamic head (TDH)** Same as dynamic head.

**total energy** The sum of the three forms of hydraulic energy: potential, kinetic, and pressure. Same as total head.

**total filterable residue** *See* total dissolved solids.

**total fixed solids (TFS)** The residue that remains after ignition of total solids to 500 ± 50°C. *See also* total volatile solids.

**total fluorides** Elemental fluorine and all fluoride compounds as measured by appropriate methods.

**total flux** The sum of gravity flux and underflow flux in the theory of sedimentation of solids suspensions. *See* solids flux.

**total hardness (TH)** The total of calcium, magnesium, and other divalent metallic cations (strontium $Sr^{2+}$, iron $Fe^{2+}$, and manganese $Mn^{2+}$), usually expressed as calcium carbonate ($CaCO_3$) equivalent. The major anions associated with hardness include bicarbonate ($HCO_3^-$), sulfate ($SO_4^{2-}$), chloride ($Cl^-$), nitrate ($NO_3^-$), and silicate ($SiO_3^{2-}$).

**total head** Same as dynamic head. In open-channel flow, the flow depth plus the velocity head.

**total heating value** Same as sludge heating value. *See also* available heat.

**total heterotrophic bacteria** Bacteria detected or enumerated using the heterotrophic plate count technique.

**total human exposure concept** The concept that all relevant exposures to a hazard must be identified in order to develop effective preventive actions against such hazard. For example, lead can be found in drinking water, food, soil, or air; potential routes for a cholera epidemic include seafood, drinking water, and foodstuffs.

**total hydrocarbons** The organic compounds in the exit gas from a sewage sludge incinerator stack measured using a flame ionization detection instrument referenced to propane (EPA-40CFR503.41-m).

**total inactivation** *See* CT.

**total inflow** The total inflow of stormwater and surface water to a sanitary sewer system, including direct inflow, overflows, delayed inflow, and pumping station bypasses.

**total inorganic carbon (TIC)** Inorganic carbon species in aqueous solutions include bicarbonates ($HCO_3^-$), carbonates ($CO_3^{2-}$), and dissolved carbon dioxide ($CO_2$).

**total inorganic nitrogen** The sum of the concentrations of inorganic nitrogen in a sample, including ammonia ($NH_3$), ammonium ($NH_4^+$), nitrite ($NO_2^-$), and nitrate ($NO_3^-$).

**total irrigation water requirement** The amount of water, expressed in depth ($D$, in), applied for irrigation to satisfy not only the net requirement ($R$, in) but also the losses and inefficiencies of distribution. It is usually calculated as

$$D = 100\ R/E_u \qquad \text{(T-24)}$$

where $E_u$ is the percent unit application efficiency of the distribution system; it varies from 65–85% for a surface system to 75–90% for a level-border system.

**totalizer** A device or meter that continuously measures and calculates total flows in gallons, million gallons, cubic feet, or other unit of volume measurement. *See* flow totalizer for detail.

**total Kjeldahl nitrogen (TKN)** The sum of organic nitrogen and ammonia nitrogen, which is determined by digesting a sample and distilling the resulting ammonia nitrogen. (Digestion converts the original organic N to ammonia N.) The combination of ammonia and organic nitrogen is sometimes simply called Kjeldahl nitrogen. TKN in wastewater averages 20 to 30 mg/L.

**totally enclosed fan cooled (TEFC)** The designation of a motor enclosure that is not airtight and does not allow air exchange with the outside, an attached fan providing external cooling (Pankratz, 1996).

**totally enclosed nonventilated (TENV)** The designation of a motor enclosure that is not airtight and that allows air exchange between the inside and outside (Pankratz, 1996).

**total matter** The sum of all suspended, colloidal, and dissolved solids in a water or wastewater sample.

**total maximum daily load (TMDL)** The sum of the individual nonpoint sources and natural background. It is a tool for implementing state water quality standards based on the relationship between pollution sources and in-stream water quality conditions. If a receiving water has only one point source discharger, the TMDL is the sum of that point source waste load allocation plus the load allocations for any nonpoint sources of pollution and natural background sources, tributaries, or adjacent segments. TMDL can be expressed in terms of mass per time, toxicity, or other appropriate measure. If best management practices or other nonpoint source pollution controls make more stringent load allocations practicable, then waste load allocations can be made less stringent. Thus, the TMDL process provides for nonpoint source control tradeoffs (EPA-40CFR130.2-i). *See also* TMDL rule.

**total metals** (1) The sum of the concentration or mass of copper (Cu), nickel (Ni), chromium (Cr) (total), and zinc (Zn) (EPA-40CFR413.02-e). (2) The total concentration of dissolved and suspended metals or the concentration of metals determined on an unfiltered wastewater sample after digestion. *See also* acid extractable metals, dissolved metals, suspended metals.

**total nitrate plus nitrite** The total inorganic nitrogen, i.e., the sum of nitrate nitrogen ($NO_3^-$-N) and nitrite nitrogen ($NO_2^-$-N), which is a drinking water contaminant regulated by the USEPA.

**total nitrogen (TN or $N_{tot}$)** The total nitrogen content of water or wastewater: ammonia gas ($NH_3$), ammonium ion ($NH_4^+$), nitrate ($NO_3^-$), nitrite ($NO_2^-$), and organic nitrogen. *See also* total ammonia nitrogen, total inorganic nitrogen, total Kjeldahl nitrogen.

**total organic carbon (TOC)** A measure of dissolved and particulate organic matter in a water or wastewater sample, as represented by the covalently bonded organic carbon. It is determined by eliminating the inorganic carbon from a sample, and then oxidizing the organic carbon to carbon dioxide ($CO_2$) by a combination of heat, oxygen, ultraviolet radiation, and chemical oxidants. The TOC test is often used as a substitute for the BOD test. TOC can also be measured as the difference between total carbon and total inorganic carbon. In natural drinking water sources, TOC consists mainly of humic materials and is used as a surrogate for disinfection by-product precursors. Most groundwaters have a TOC < 2.0 mg of carbon (C) per liter, while lakes and rivers have a TOC between 2.0 and 10.0 mg C/L. Coagulation and, less effectively, softening can remove total organic carbon from drinking water. *See also* purgeable organics, total organic halogen. Other measures of organic content are BOD, COD, TOD, and ThOD.

**total organic halides (TOX)** Same as total organic halogen.

**total organic halide analysis** Abbreviation: TOX analysis, a widely used analysis for the determination of water disinfection byproducts. It measures the total organic halogen content of a water sample, usually through a surrogate such as carbon-adsorbable organic halogen (CAOX) or total organic carbon (TOC).

**total organic halogen (TOX)** A surrogate measurement for the total amount of organic compounds that contain one or more halogen atoms. Such compounds are found in treated or raw water as disinfection by-products, halogenated humic substances, or synthetic organic chemicals. Also called adsorbable organic halogen (AOX), carbon-adsorbable organic halogen (CAOX), dissolved organic halogen (DOX), or total organic halide. *See also* halogen-substituted organic material, total organic carbon.

**total organic halogen formation potential (TOHFP)** The maximum amount of total organic halogen that can be formed in a water sample as determined by laboratory testing when the sample is treated with a large dose of chlorine or other disinfectant, taking into account any prior organic halogen concentration. It is used to indicate the amount of organic halogen precursors present in the sample. *See also* disinfection by-product formation potential.

**total organic nitrogen (TON)** A measure of the organic nitrogen in water or wastewater, commonly determined with ammonia using the Kjeldahl method. *See also* organic nitrogen and total Kjeldahl nitrogen.

**total-oxidation process** A method of treatment for organic wastes, particularly in small installations, consisting of comminution, long-period aeration (1–3 days), final settling, and high sludge recycle rate (100–300% of flow). Also called high-rate aerobic treatment, but different from high-rate aeration process.

**total oxygen demand (TOD)** A quantitative measure of oxidizable organic matter in a water or wastewater sample. It is obtained by determining the depletion of oxygen in a combustion chamber at high temperature. Other measures of organic content are BOD, COD, ThOD, and TOC. *See also* carbonaceous, nitrogenous, and ultimate oxygen demands.

**total particulate phosphorus** Total phosphorus content of material retained on a filter of a specific size.

**total petroleum hydrocarbons (TPH)** A measure of the concentration or mass of petroleum hydrocarbon constituents present in a given amount of air, soil, or water. The term total is a misnomer, in that few, if any, of the procedures for quantifying hydrocarbons are capable of measuring all fractions of petroleum hydrocarbons present in the sample. Volatile hydrocarbons are usually lost in the process and not quantified. Additionally, some nonpetroleum hydrocarbons may be included in the analysis.

**total phenols** Total phenolic compounds as measured by an appropriate method such as distillation followed by colorimetric (EPA-40CFR464.02-g).

**total phosphorus** The sum of all phosphorus forms.

**total pumping head** A measure of the total energy imparted by a pump to a fluid, equal to the algebraic difference between the total discharge head and the total suction head.

**total recoverable petroleum hydrocarbons (TRPH)** A USEPA method (#418.1) for measuring total petroleum hydrocarbons in samples of soil or water. Hydrocarbons are extracted from the sample using a chlorofluorocarbon solvent (typically Freon-113) and quantified by infrared spectrophotometry. The method specifies that the extract be passed through silica gel to remove the nonpetroleum fraction of the hydrocarbons.

**total residual** The sum of combined and free chlorine residuals. In water and wastewater chlorination, initially the combined residuals predominate, but as the chlorine dose increases, free residuals predominate. *See* breakpoint chlorination.

**total residual chlorine** The amount of available chlorine remaining after a given contact time. The sum of the combined available residual chlorine and the free available residual chlorine. *See also* residual chlorine.

**total residuals content** The total solids content of a sample, i.e., the sum of floatable, dissolved, colloidal, and suspended solids.

**total residue on evaporation** Same as total solids.

**total retention system** An oxidation pond in which treated wastewater is allowed to evaporate without an outlet; however, most such systems lose or purposely discharge some flow by percolation and total retention is not feasible where annual precipitation is excessive compared to pan evaporation. The required surface area ($A$) is:

$$A = Q/(E + I - P) \qquad \text{(T-25)}$$

where $Q$ = annual wastewater flow, $E$ = annual evaporation, $I$ = annual precipitation, and $P$ = allowable annual percolation. *See also* wastewater evaporation.

**total sediment discharge** The sum of the suspended sediment discharge and the contact sediment (bed-load) discharge, the sum of the bed-material discharge and the wash-load discharge, or the sum of the measured sediment discharge and the unmeasured sediment discharge. *See also* sedimentation terms.

**total solids (TS)** The materials in wastewater or sludge that remain as residue when a sample is evaporated and dried at 103–105°C (EPA-40CFR503.31-I). Also, the sum of dissolved, colloidal, and suspended solids in a water or wastewater sample. Also called total residue on evaporation. *See also* wastewater solids.

**total solids content** Same as total residuals content, i.e., the sum of floatable, dissolved, colloidal, and suspended solids. It is the most important physical characteristic of wastewater. *See also* wastewater solids.

**total static head** *See* definition under pump head terms.

**total suspended particles (TSP)** A method of monitoring particulate matter by total weight.

**total suspended solids (TSS)** A measure of the suspended solids in wastewater, effluent or water bodies, determined by tests for "total suspended nonfilterable solids." *See* suspended solids. *See also* wastewater solids.

**total suspended solids recovery** A measure of the performance of a centrifuge (used in sludge thickening and dewatering), defined as the percentage of thickened dry solids with respect to feed solids. *See* recovery for detail.

**total toxic organics (TTO)** A regulated parameter that is the sum of the masses or concentrations of specified toxic organic compounds found at a concentration greater than ten (10) micrograms per liter (EPA-40CFR468.02-r, 469.12-a, and 469.31-b).

**total trihalomethane bromine (TTHM-Br)** The molar sum of bromine in a trihalomethane, i.e., the number of bromine atoms present in a molecule of the trihalomethane. For example, in bromoform (CHBr), TTHM-Br = 3.

**total trihalomethane chlorine (TTHM-Cl)** The molar sum of chlorine in a trihalomethane, i.e., the number of chlorine atoms present in a molecule of the trihalomethane. For example, in bromodichloromethane ($CHBrCl_2$), TTHM-Cl = 2.

**total trihalomethanes (TTHM)** The sum of the concentrations, in milligrams per liter, of the several trihalommethane compounds [trichloromethane (chloroform), dibromochloromethane, bromodichloromethane, and tribromomethane (bromoform)], rounded to two significant figures (EPA-40CFR141.2).

**total volatile dissolved solids** Solid matter that can be volatilized and burned off by igniting total dissolved solids at 500 ± 50°C. *See also* wastewater solids.

**total volatile solids (TVS)** An approximate measure of the organic content of a water or wastewater sample; determined by igniting the residue from the total solids test to constant weight at 500 ± 50°C and measuring the weight remaining (fixed solids):

$$\text{Volatile solids} = \text{total solids} - \text{fixed solids} \qquad \text{(T-26)}$$

*See also* wastewater solids.

**total water mixing ratio** The mixing ratio all the water components: liquid, solid, and vapor.

**TouchRead®** A flowmeter reading system manufactured by Invensys Metering Systems; it uses the Intelligent Communications Encoder. *See also* RadioRead® and PhonRead®.

**Toveko®** A continuous sand filter manufactured by Kruger, Inc.

**Tow-Bro®** An apparatus manufactured by Envirex, Inc. for sludge removal by suction from circular clarifiers.

**tower aeration** *See* packed-bed aerator.

**tower aerator** *See* gravity aerator, packed-bed aerator

**Tower Biology®** A biological wastewater treatment plant manufactured by Biwater Treatment, Ltd.

**Towerbrom®** A bromine biocide produced by Calgon Corp.

**tower crib** A river or lake intake structure that extends from the bottom to the high-water level.

**Towermaster** A pressurized sand filter manufactured by Serck Baker, Inc. for use with cooling towers.

**Tower Press** A belt filter press manufactured by Roediger Pittsburgh, Inc.

**tower resistance** A parameter used in determining the airflow requirement for the operation of a trickling filter. Tower resistance ($N_p$) is the sum of all individual headlosses related to airflow through the inlet, underdrain, and packing material; it is usually expressed in terms of velocity heads. *See* Dow Chemical formula.

**TOX** Acronym of total organic halide and total organic halogen.

**ToxAlarm™** A device manufactured by Anatel Corp. for monitoring toxicity online.

**TOX analysis** Abbreviation of total organic halide analysis.

**toxaphene ($C_{10}H_{10}Cl_8$)** An amber, waxy solid; a mixture of chlorinated camphene derivatives, slightly soluble in water, used as an insecticide and rodenticide. This chemical causes adverse health effects (mainly liver damage) in domestic water supplies and is also toxic to freshwater and marine aquatic life. Regulated by the USEPA as a drinking water contaminant: MCL = 0.005 mg/L and MCLG = 0.

**toxic** (1) Pertaining to a substance that can cause adverse effects on absorption or contact. (2) A substance that is poisonous to an organism.

**toxic air contaminant (TAC)** Any volatile organic compounds and other contaminants released to the atmosphere from wastewater collection and treatment facilities, a serious health concern addressed in the Clean Air Act.

**toxicant** A substance or agent that may kill or injure an exposed organism through contact, inhalation, or ingestion, particularly a substance that is hazardous to aquatic life or human health. Examples of toxicants that may injure by biological, chemical, or physical action include heavy metals (e.g., cadmium, copper, lead), synthetic organic compounds (e.g., cyanides, pesticides, volatile organics), oxidants (residual chlorine), and poisonous gases (e.g., ammonia, hydrogen sulfide). Also called poison. *See also* toxin, priority toxic water pollutant.

**toxic chemical** Any chemical listed in USEPA rules as "Toxic Chemicals Subject to Section 313 of the Emergency Planning and Community Right-to-Know Act of 1986."

**toxic effect** An adverse change in the structure or function of an experimental animal as a result of exposure to a chemical substance (EPA-40CFR798. 6050-2).

**toxicity** (1) The property, quality or degree of being poisonous or harmful to plant, animal, or human life. (2) The potential for a constituent to cause adverse effects on living organisms. *See* acute toxicity, chronic toxicity, toxicity terms.

**toxicity assessment** Characterization of the toxicological properties and effects of a chemical, including all aspects of its absorption, metabolism, excretion, and mechanism of action, with special emphasis on establishment of dose–response characteristics. In water quality, toxicity may be assessed by testing raw and treated samples for specific substances or by bioassays to expose selected organisms to the samples. *See also* bioassay, flow-through test, test organism, toxicity terms.

**toxicity characteristic leach procedure** Same as toxicity characteristic leaching procedure.

**toxicity characteristic leaching procedure (TCLP)** An extraction method specified by the USEPA under the Resource Recovery and Conservation Act for the determination of hazardous substances that will leach from a solid waste. It uses a vacuum-sealed extraction vessel to determine the toxicity of a sample due to metals, herbicides, pesticides, and volatile organic chemicals. *See also* extraction procedure toxicity test and extraction procedure toxicity threshold.

**toxicity curve** The curve produced from toxicity tests when $LC_{50}$ values are plotted against duration of exposure. (This term is also used in aquatic toxicology, but in a less precise sense, to describe the curve produced when the median period of survival is plotted against test concentrations) (EPA-40CFR797.1350-9).

**toxicity rating** An indication of the estimated toxicity of a substance to humans. *See* Table T-04.

**Table T-04.** Toxicity rating

| Rating | Dose (mg of substance per kg of body weight) | Example |
|---|---|---|
| Practically nontoxic | > 15,000 | |
| Slightly toxic | 5,000–15,000 | Ethanol, sodium chloride |
| Moderately toxic | 500–5000 | Malathion |
| Very toxic | 50–500 | Heptachlor, chlordane |
| Extremely toxic | 5–50 | Parathion |
| Supertoxic | < 5 | Tetraethylpyrophosphate, tetrodotoxin, dioxin |

**toxicity reduction evaluation (TRE)** An investigation conducted by a publicly owned treatment works to determine the sources of effluent toxicity, pollutants responsible for a toxicity violation, and applicable pollution control alternatives.

**toxicity terms** Terms related to toxicity testing and toxicity assessment, as often defined and updated by the USEPA or in such authoritative publications as *Standard Methods:* acceptable risk, acute-to-chronic ratio, acute toxicity, chronic daily intake, chronic toxicity, chronic value, control, criterion continuous concentration, criterion maximum concentration, critical initial dilution, cumulative toxicity, dose, dose–response assessment, dose–response models, ecological risk assessment, effective concentration, exposure time, inhibiting concentration, in vitro test, in vivo test, lethal concentration, lethal dose, lifetime risk, lowest-observed-effect concentration, maximum-allowable-toxicant concentration, median effective concentration, median lethal concentration, median tolerance limit, no-observed-effect concentration, potency factor, reference dose, risk characterization, risk management, slope factor, stressor, sublethal effect, sublethal toxicity, total dose, toxicity, toxicity assessment, toxic unit acute, toxic unit chronic, toxic units, water reuse reliability.

**toxicity test** A test conducted to determine the risk that exposure to a substance or constituent poses to human health, aquatic biota, and other environmental components. Such tests are also conducted for more specific reasons, e.g., to establish water quality criteria, assess the degree of wastewater treatment required to meet water quality standards, or determine compliance with federal and state standards. *See also* toxicity terms.

**toxicity testing** Biological testing (usually with an invertebrate, fish, or small mammal) to determine the adverse effects of a compound or effluent.

**toxic metal** One of three classes of metals, based on their biological functions and effects, including silver (Ag), cadmium (Cd), tin (Sn), gold (Au), mercury (Hg), thallium (Tl), lead (Pb), aluminum (Al), and the metalloids arsenic (As), selenium (Se), antimony (Sb), and germanium (Ge). They have no known biological functions and are harmful through the displacement of essential metals, inhibition of enzymatic functions, and disruption of nucleic acid activity. *See also* essential metal.

**toxic metal ions** Metallic elements such as chromium ($Cr^{6+}$), copper ($Cu^{2+}$), and zinc ($Zn^{2+}$) that interfere with biological oxidation of organic matter by tying up the required enzymes.

**toxicological chemistry** The branch of chemistry that studies the origins, uses, exposure aspects, fates, and disposal of toxic substances.

**toxicological profile** An examination, summary, and interpretation of a hazardous substance to determine levels of exposure and associated health effects.

**toxicology** The science and study of adverse effects and control of poisons or toxic agents on living organisms.

**toxic pollutants** Materials contaminating the environment that can cause death, disease, and birth defects in organisms that ingest or absorb them. The quantities and length of exposure necessary to cause these effects can vary widely.

**Toxics Release Inventory (TRI)** A database of toxic releases in the United States compiled from SARA Title III section 313 reports. It contains information on the release of 312 chemicals from the manufacturing sector.

**toxic substance** A chemical or mixture that may represent an unreasonable risk of injury to health or the environment; e.g., benzene, dioxin, lead, radon. *See* harmful organics and harmful inorganics.

**Toxic Substances Control Act (TSCA)** A 1976 U.S. law (Public Law 94-469) that authorizes the USEPA to identify potentially harmful substances, products, or uses for federal control.

**toxic unit acute ($TU_a$)** *See* toxic units, toxicity terms.

**toxic unit chronic ($TU_c$)** *See* toxic units, toxicity terms.

**toxic units** A series of parameters used to express toxicity test results and formulate standards for the protection of aquatic life. Toxic unit acute ($TU_a$) is the reciprocal of the effluent dilution that causes an acute effect by the end of the acute exposure period and is calculated as 100 divided by the median lethal concentration ($LC_{50}$), which is the concentration that causes mortality to 50% of the test population:

$$TU_a = 100/LC_{50} \quad \text{(T-27)}$$

Toxic unit chronic ($TU_c$) is the reciprocal of the effluent dilution that causes no unacceptable effect (i.e., no effect different from the control) on the test organisms by the end of the chronic exposure period and is calculated as 100 divided by the no-observed-effect concentration (NOEC):

$$TU_c = 100/NOEC \quad \text{(T-28)}$$

The acute-to-chronic ratio (ACR) is

$$ACR = LC_{50}/NOEC \quad \text{(T-29)}$$

The criterion maximum concentration (CMC) is designed to protect aquatic life against short-term/acute effects, whereas the criterion continuous concentration (CCC) protects against long-term/chronic effects. They are determined as follows:

$$CMC = TU_a/CID \leq 0.3\, TU_a \quad \text{(T-30)}$$

$$CCC = TU_c/CID \leq TU_c \quad \text{(T-31)}$$

where CID is the critical initial dilution, i.e., the dilution achieved under worst-case ambient conditions.

**toxic waste** A waste that can produce injury to a living organism if inhaled, swallowed or absorbed through the skin.

**toxigenicity** The degree to which an organism can produce toxic symptoms.

**Toxilog** A portable instrument manufactured by Biosystems, Inc. for the individual detection of gases.

**toxin** (1) A chemical substance, having a protein structure, that is secreted by certain organisms and is capable of causing toxicosis when introduced into the body tissues of another organism, but is also capable of introducing a counteragent or an antitoxin. (2) Sometimes used synonymously with toxicant or toxic agent.

***Toxoplasma*** A genus of nonspecific, invasive protozoa that infect humans, other mammals, and birds, particularly individuals with weak or impaired immune systems. *See* toxoplasmosis, opportunistic pathogen.

***Toxoplasma gondii*** An intracellular coccidian protozoan of felines that causes toxoplasmosis. Its life cycle includes felines as definitive hosts and sheep, cattle, goats, chickens, etc. as intermediate hosts. The latter shed fecal oocysts in water or moist soil. Humans acquire the parasite through raw or undercooked meat, contaminated milk, or contaminated water. It is estimated that 13% of the world population and 40% of cats in the United States are infected with *T. gondii.*

**toxoplasmosis** An infection, sometimes asymptomatic, transmitted by the protozoan parasite *Toxoplasma gondii* and affecting birds and mammals. When symptoms are present, the disease resembles infectious mononucleosis; it may cause blindness or death in unborn fetuses, brain damage in children, and encephalitis. Humans contract the disease by consuming insufficiently cooked contaminated food, or by contact with infected cats or contaminated soils. Some waterborne outbreaks have also been documented.

**2,4,5-TP** *See* silvex.

**TPAD** Acronym of temperature-phased anaerobic digestion.

**TPC®** A potassium permanganate produced by Technical Products Corp.

**TPH** Acronym of total petroleum hydrocarbons.

**trace** (1) A very small amount of a constituent; a trace element. (2) An amount of precipitation that cannot be measured by an instrument such as a rain gage, e.g., less than 0.005 inch or 0.127 mm. Also called a trace of precipitation.

**trace concentration** The concentration of a trace element. Current analytical methods allow the measurement of trace concentrations in the range of $10^{-12}$ to 1.0 mg/L. *See also* gross concentration.

**trace element** (1) Any element that is present in water or wastewater samples in very low concentrations, considering such factors as natural distribution, solubility, or industrial use. Some organics and inorganics are harmful even in trace concentrations; *see* harmful inorganics, harmful organics. Trace elements commonly found in natural waters include arsenic (As), beryllium (Be), boron (B), cadmium (Cd), chromium (Cr), copper (Cu), fluorine (F) or fluoride ion, iodine (I) or iodide ion, iron (Fe), lead (Pb), manganese (Mg), mercury (Hg), molybdenum (Mo), selenium (Se), silver (Ag), and zinc (Zn). (2) Any element that is essential at a concentration of 1% or less to the growth of plants and animals. Examples include boron, cobalt, copper, iron, manganese, molybdenum, potassium, sodium, and zinc; most trace elements are constituents of enzymes. The absence or deficiency of trace elements in diet and drinking water may be responsible for some diseases; see, e.g., an association is postulated between chromium deficiency and juvenile diabetes. *See also* essential element, micronutrient, trace nutrient, macronutrient.

**trace metal** A metal that is present at a very low concentration in water or wastewater and that originates from residences, groundwater infiltration, and commercial and industrial discharges. Even such low concentrations may be toxic. Potentially hazardous trace metals include copper (Cu), zinc (Zn), silver (Ag), antimony (Sb), tin (Sn), mercury (Hg), and lead (Pb). Surface waters may also contain traces of iron, manganese, molybdenum, and cobalt. On the other hand, traces of cobalt (Co), nickel (Ni), and zinc may stimulate methanogenic activity in anaerobic processes. Atomic absorption spectrophotometry is used to detect toxic trace metals in water. *See also* essential trace metal.

**trace metal analyzer** A single-cell instrument that automatically determines the amounts of some potentially hazardous trace metals (e.g., lead, mercury, zinc) in air, water, soil, or food.

**trace nutrient** Such nutrients as boron, calcium, cobalt, and magnesium that organisms require at very low concentrations. *See also* trace element, macronutrient.

**trace of precipitation** An amount of precipitation that cannot be measured by an instrument such as a rain gage, e.g., less than 0.005 inch or 0.127 mm. Also called trace (2).

**trace organic chemicals** Same as trace organics.

**trace organics** Natural and synthetic organic materials present in water or wastewater at low concentrations, in the range of $10^{-12}$ to $10^{-3}$ mg/L, determined using such laboratory methods as gas chromatography and mass spectroscopy; e.g., humic acid, fulvic acid, pesticides. Natural trace organics are precursors to disinfection by-products. *See also* trace substance.

**tracer** A material such as a dye, an electrolyte, or an isotope used to study microbial transport, residence time distribution, short-circuiting, retardation, and longitudinal mixing in sedimentation basins and other reactors. *See* conservative tracer, retardation factor, time–concentration curve.

**tracer response curve** A graphical representation of the distribution of a tracer versus time of flow. *See* time–concentration curve for more detail.

**tracer study** The introduction of a tracer into the influent of a reactor and the determination of its arrival time at the effluent. The results are usually plotted on a curve of tracer concentration versus time. Salt solutions, dyes, radioactive and other tracers are used to study short-circuiting in clarifiers. Such studies indicate the importance of some clarifier design details, particularly the inlet and outlet structures. *See also* continuous step input, pulse input, slug input.

**tracer test method** A method involving (a) the injection of a measured amount of a conservative tracer to a digester to evaluate the mixing efficiency of the digester, (b) the collection of digested sludge samples, and (c) their analysis for tracer content. The data are used to draw a tracer washout curve and estimate the percent active volume of the digester.

**trace substance** A substance that is found in water or wastewater in such a concentration that it can be detected but not quantified accurately by standard testing methods. Also called threshold substance. *See also* trace concentration, trace element.

**trachoma** An eye infection, often transmitted by flies, that may lead to blindness. It is prevalent in tropical areas where poor hygiene is associated with a lack of water. *See also* conjunctivitis, water-washed disease.

**TrackOne™** A device manufactured by Hydro Flow Products, Inc. for use in fire flow testing and flushing.

**Trac Pump** A sludge dredging device manufactured by H & H Pump and Dredge Co.

**tractive force** *See* shear stress (1).

**Tractor Drive** A drive apparatus manufactured by Walker Process Equipment Co. for use in circular clarifiers.

**TracVac™** A sludge removal mechanism manufactured by Eimco Process Equipment Co.

**TracWare™** A procedure developed by Chemtrac Systems, Inc. for counting particles.

**trade effluent** Wastewater from an industrial process; trade waste. *See* industrial wastewater for detail.

**trade waste** Wastewater generated in an industrial process or practice (chemical industry, food processing, manufacturing, materials industry), as opposed to domestic or sanitary wastewater. *See* industrial wastewater for more detail.

**traditional electrodialysis** The electrodialysis process carried out through alternating anion- and cation-permeable membranes, as compared to a variation that uses ion-selective membranes in an electrolytic cell.

**traditional filtration** Tertiary wastewater treatment that includes chemical coagulation, flocculation, sedimentation, filtration, and disinfection. *See also* direct filtration and plain filtration.

**traditional filtration rate** A filtration rate equal to 2 gpm/ft$^2$ or 5 m/h recommended in 1898 for chemically pretreated surface waters. Also called standard filtration rate.

**train** (1) In pressure-driven membrane processes, a train consists of multiple interconnected stages in series and sometimes controlled as a single unit. (A stage consists of parallel pressure vessels.) Also called array. *See also* high-recovery array. (2) An assemblage of membrane modules installed in parallel or in series and operated as a single unit. Also called rack, unit.

**tramp element** A noble or less reactive element (e.g., copper, nickel, tin) that tends to remain in an alloy during steel making or recycling operations, while the more reactive elements (e.g., silicon, aluminum) tend to be removed in the slag.

***trans-*** A prefix in chemical nomenclature denoting a geometric isomer having two identical atoms or groups attached to opposite sides of a double bond. Examples: *trans*-1,2-dichloroethylene, *trans*-2-heptene, *trans*-2-pentene. *See also* geometric isomer, *cis*-. In this text, entries are listed alphabetically as though these prefixes did not exist.

**trans-1,2-dichloroethylene** *See* under dichloroethylene

**transboundary pollutant** A pollutant, more specifically an air pollutant, that travels from one jurisdiction to another, often crossing state or international boundaries.

**transducer** A device or substance that converts energy from one form to another; e.g., from mechanical energy to electrical energy. Commonly used in flow monitoring.

**transduction** The multiplication of a bacteriophage in a host cell and subsequent infection of another cell by transfer of genetic material.

**transfer pipette** A pipette calibrated for a single volume.

**transfer pump** A device that pumps water, wastewater, or chemicals from one tank to another within a treatment plant.

**transient** A phenomenon that lasts only a short time. Hydraulic transients like water hammer or surge are caused when valves are opened or closed suddenly or when pumps start or stop. *See also* pressure transient.

**transient chlorine species** Chlorine compounds that are formed during chlorination, under certain conditions and in addition to the major species of free chlorine. Transient species include $H_2OCl^+$, $Cl^+$, and $Cl_3^+$.

**transient noncommunity water system** A noncommunity water system that does not regularly serve at least 25 of the same persons over six months per year (EPA-40CFR141.2).

**transient pressure** *See* pressure transient.

**transient water system** A noncommunity water system that does not serve 25 of the same nonresident persons per day for more than six months per year. Also called a transient noncommunity water system.

**transition metal cation** Any metal cation with characteristics between those of type A or hard and type B or soft ions, as to polarizability, toxicity, and preferences for forming complexes. Transition metal ions include: vanadium ($V^{2+}$), chromium ($Cr^{2+}$), manganese ($Mn^{2+}$), iron ($Fe^{2+}$), cobalt ($Co^{2+}$), nickel ($Ni^{2+}$), copper ($Cu^{2+}$), titanium ($Ti^{3+}$), vanadium ($V^{3+}$), chromium ($Cr^{3+}$), manganese ($Mn^{3+}$), iron ($Fe^{3+}$), and cobalt ($Co^{3+}$). Some of them are also called borderline metal ions, which include all micronutrient metals. *See also* class A metal ion, class B metal ion, Irving–Williams order.

**transition range** The range of Reynolds numbers ($R_e$) within which flow is between laminar and turbulent; the drag coefficient ($C_D$), a function of the Reynolds number, affects the terminal settling velocity of particles in water and wastewater treatment:

$$1 < R_e < 1000 \qquad \text{(T-32)}$$

$$C_D = 24/R_e + 3/R_e^{0.5} + 0.34 \qquad \text{(T-33)}$$

or

$$C_D = 18.5/R_e^{0.6} \qquad \text{(T-34)}$$

*See also* Stokes range and turbulent range.

**transition region** The region intermediate between the nearfield (or initial mixing zone) and the farfield of a wastewater effluent discharge into the ocean or a large lake.

**transition zone** The middle layer in a thermally stratified lake or reservoir. In this layer there is a rapid decrease in temperature with depth, e.g., 0.5°F or more per foot. *See* thermocline for more detail.

**transitory diarrhea** The laxative effect caused by high concentrations of sulfate in drinking water, e.g., above 1000 mg/L for adults and 600 mg/L for bottle-fed infants. However, adults living in a high-sulfate area eventually adjust to this condition.

**Transmax®** An apparatus manufactured by Enviroquip, Inc. for air diffusion using medium-to-coarse bubbles.

**transmembrane flux** Flux through a membrane.

**transmembrane pressure (TMP)** The net driving force across a pressure-driven membrane or the hydraulic pressure differential from feed to permeate in a microfiltration or ultrafiltration membrane. TMP is a parameter used to measure the degree of fouling of a membrane; the higher the TMP, the more severe the fouling. TMP is also defined by the following equations:

crossflow filtration: $P_{tm} = 0.5\,(P_f + P_c) - P_p$ (T-35)

direct-feed filtration: $P_{tm} = P_f - P_p$ (T-36)

where $P_{tm}$ = transmembrane pressure gradient, kPa; $P_f$ = inlet pressure of feed stream, kPa; $P_c$ = pressure of concentrate stream, kPa; and $P_p$ = pressure of permeate stream; kPa. *See also* critical permeate flux.

**transmission** Any mechanism by which a susceptible human host is exposed to an infectious agent.

**transmission characteristics** Pathogen transmission characteristics include latency, infective dose, and persistence.

**transmission electron microscope (TEM)** A microscope that has a magnification range between 220 and 1,000,000 times at a resolution of 2 Å, in which an electron beam through the specimen forms the magnified image to be observed. *See also* scanning electron microscope.

**transmission electron microscopy (TEM)** The use of a transmission electron microscope to examine specimens that are too fine for ordinary instruments.

**transmission line** A pipeline that transports raw water from its source to a water treatment plant, then to the distribution grid system.

**transmission of infection** A water-related infectious agent can be transmitted (a) directly, (b) indirectly through an intermediate host, (c) indirectly through a vector, and indirectly through the air.

**transmission works** (1) Transmission lines and pumps that convey water from its source. (2) Sewers, force mains, and pump stations that transmit wastewater from local service areas to larger or regional facilities. *See also* interceptors.

**transmissivity** (1) The hydraulic conductivity integrated over the saturated thickness of an underground formation. The transmissivity of a series of formations is the sum of the individual transmissivities of each formation comprising the series (EPA-40CFR191.12). (2) The ability of an aquifer to transmit water through its entire thickness. Transmissivity ($T$) is the product of thickness ($b$) by hydraulic conductivity ($K$) for an aquifer of uniform thickness:

$$T = Kb \quad \text{(T-37)}$$

**transmissivity coefficient** Same as transmissivity.

**transmittance (*T*)** The capacity of a substance to transmit light or radiation, or the amount of light or radiation actually transmitted; defined as the ratio of light intensity at a distance from the surface ($I$) to light intensity at the surface ($I_0$):

$$T = I/I_0 \quad \text{(T-38)}$$

It is sometimes expressed as a percentage:

$$T, \% = 100\,(I/I_0) \quad \text{(T-39)}$$

All substances in water and wastewater that absorb or scatter light affect transmittance, including iron, copper, humic compounds, and organic dyes. *See also* absorbance, Beer–Lambert law.

**transmutation** The conversion of one chemical element into another by changing the number of protons in the nucleus.

**Trans-Pak®** An apparatus manufactured by Harris Waste Management Group, Inc. for compacting and baling solid waste.

**transpiration** The process by which living plants lose water vapor to the atmosphere through their leaves. The term can also be applied to the quantity of water thus dissipated. Plant transpiration is similar to animal perspiration. Transpiration is generally larger than evaporation and depends on vegetation types, soil moisture, and meteorological conditions. In most engineering applications, transpiration and evaporation are lumped into evapotranspiration because it is difficult to measure them separately.

**transpiration ratio** The ratio of the weight of water transpired by a plant to the weight of dry matter produced. Also called water-use ratio.

**transport and attachment** The two-step process of particulate matter removal within a granular filter: the transport of the suspended particles to or near the surface of the filtering medium and their actual removal by straining, sedimentation, adsorption, etc. *See also* filtration mechanisms.

**transport pore** A pore that is larger than the largest adsorption pore of a granular activated carbon bed, serving as a diffusion path but not for actual adsorption. *See also* adsorption pore, macropore, mesopore, micropore.

**transuranic element** All chemical elements having an atomic number greater than that of uranium, which is 92; they are all radioactive. Neptunium and plutonium are naturally found; the others are synthesized. Also called transuranium element.

**transuranic radioactive waste (TRW)** Waste that contains more than 100 nanocuries of alpha-emitting transuranic isotopes, with half-lives greater than twenty years, per gram of waste, except for (a) high-level radioactive waste, (b) wastes that do not require isolation, or (c) waste that has been approved for land disposal.

**transuranic waste** Wastes that contain transuranium elements (those that are above uranium in the periodic table), by-products of weapons fabrication and other industrial operations.

**transuranium element** Same as transuranic element.

**Transvap** A mobile apparatus manufactured by Licon, Inc. for the reduction or recovery of wastewaters.

**transverse flow** The condition of a pressure-driven membrane unit in which the feedwater flows tangentially across the membrane and the permeate passes perpendicularly through the membrane. *See also* crossflow.

**transverse weir** A weir or small dam installed across a combined sewer, perpendicularly to the line of flow, to control overflows and direct dry-weather flows to the interceptor sewer. *See* Figure T-09.

**trap** (1) A device installed in a conduit to trap and remove sediment or a temporary device to retain runoff and allow silt to settle out. *See* sediment trap for more detail. *See also* separator. 2) Same as water seal.

**trap efficiency** The ratio or percentage of sediment retained to the sediment load. Trap efficiency of settling basins varies from close to 100% when capacity equals one full year's tributary flow to 30–60% for 5 weeks' inflow. *See also* Brune's trap efficiency curves.

**trapezoidal weir** A weir with a trapezoidal notch. *See also* the Cipolletti and compound weirs.

**Trasar®** Monitoring and control procedures developed by Nalco Chemical Co.

**trash** (1) Debris that may be removed by coarse racks in water treatment plants. (2) Combustible waste containing up to 10% by weight of plastic and rubber materials.

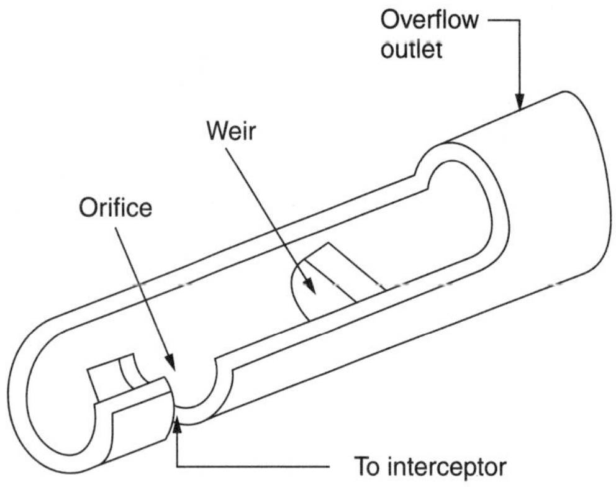


**Figure T-09.** Transverse weir.

**trash gate** A gate that allows the removal or discharge of floating debris.

**Trash Hog®** A self-priming pump manufactured by ITT Marlow for use in solids handling units.

**trash rack** A bar screen having large openings, 1.5–6 inches; installed in combined sewer systems and usually followed by screens with smaller openings.

**trash rake** A mechanical device that removes coarse debris from a trash rack.

**Travalift™** A traversing apparatus manufactured by FMC Corp. for the collection and pumping of wastewater sludge from package treatment plants.

**Traveco®** A traveling screen manufactured by Eco Equipment for use in water treatment plants.

**travelers' diarrhea** An infection caused by heat-labile or heat-stable enterotoxins of foodborne or waterborne enterotoxigenic *E. coli.* Other pathogens (*Salmonella, Shigella, Cryptosporidium, Giardia, Cyclospora,* etc.) may also cause travelers' diarrhea. It is a troublesome, acute gastrointestinal upset contracted away from home, commonly called Bali belly, Mexican quickstep, Montezuma's revenge, and Thai trots. Persistent and often severe, the illness, experienced by a traveler unaccustomed to the causative bacteria, has cholera-like syndromes (cramping, vomiting, diarrhea, prostration, dehydration), an incubation period of 10–72 hours, and lasts 3–5 days.

**traveling bridge** A mechanism, similar to an overhead crane, that travels along the sides of a rectangular sedimentation tank or on a support structure to collect sludge from the bottom of the tank. The sludge is removed by a scraper or a suction manifold and discharged to a collection trough. *See* Figure T-10. *See also* traveling flight, chain-and-flight sludge collector, plow-type sludge collection.

**traveling-bridge clarifier** A rectangular clarifier in which a traversing bridge supports the sludge removal mechanism. *See* Figure T-11.

**traveling-bridge filter** A proprietary continuous downflow, low-head, granular media filter with multiple compartments that can be individually cleaned by a movable, bridge-mounted backwashing device without taking the entire filter out of service.

**traveling bridge sludge collector** *See* traveling bridge.

**traveling flight** A sludge collection device or mechanism used in rectangular clarifiers, similar to the chain-and-flight sludge collector.

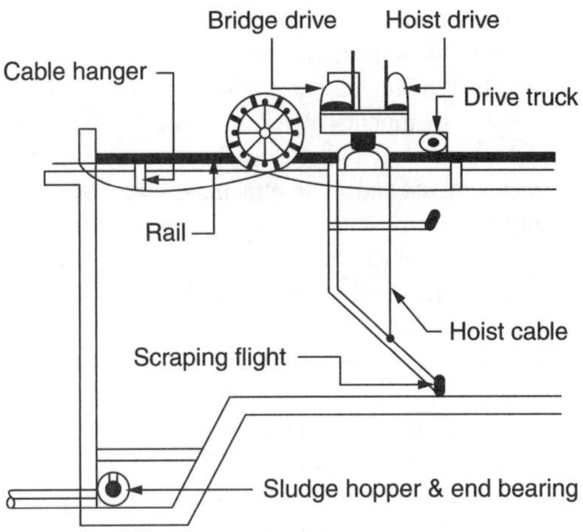


**Figure T-10.** Traveling bridge.

**traveling screen** A revolving trash screen that is made of chain-mounted wire mesh panels that are automatically cleaned; also called traveling water screen, moving screen.

**traveling water screen (TWS)** Same as traveling screen.

**travellers' diarrhea** *See* travelers' diarrhea

**travel time** (1) The time it takes a contaminant to travel from the source to a particular point downgradient. (2) The time it takes a water particle to travel from one point to another along a flow path. *See also* time of flow, time of travel, hydraulic residence time.

**Travis tank** A two-story septic tank with contact elements called colloiders in the settling compartment, a predetermined fraction of the influent wastewater flowing to the smaller lower chamber. This device was later improved in Germany; see Imhoff tank. Also called two-story septic tank.

**Figure T-11.** Traveling bridge clarifier (top, view; bottom, festoon electrical system). (Courtesy Enviroquip, Inc.)

**Travis, W. O.** An engineer in Hampton, England who conceived a two-story septic tank, the precursor of the Imhoff tank.

**tray aerator** *See* gravity aerator for detail.

**tray clarifiers** Two or more primary or secondary settling tanks, one on top of the other to save space. *See* stacked clarifiers for detail.

**Traypak™** Deaerator trays manufactured by Graver Co.

**tray-type aerator** A tray aerator.

**TRC** Acronym of total residual chlorine.

**TRE** Acronym of toxicity reduction evaluation.

**treatability (wastewater)** *See* wastewater treatability.

**treatability and packing coefficient** *See* the Germain equation used in the design of trickling filters with plastic packing.

**treatability study** A test of potential water or wastewater treatment processes, or hazardous waste cleanup technologies conducted in a laboratory. Treatability studies determine (a) whether water, wastewater, or solid waste is amenable to a treatment process; (b) what pretreatment is required; (c) the optimal process conditions needed to achieve the desired treatment; (d) the efficiency of a treatment process for a specific water, wastewater, or solid waste; or (e) the characteristics and volumes of residuals from a particular treatment process (adapted from EPA-40CFR260.10). *See also* pilot plant.

**treated drinking water standards** Rigorous requirements set for the quality of treated water used for drinking. One example is the guidelines of the World Health Organization for microbiological quality: water treated for distribution should not contain any coliform organism and 95% of tap water samples should be free of coliforms in any one year .

**treated sewage** Same as treated wastewater.

**treated wastewater** Wastewater that has been subjected to one or more physical, chemical, and biological processes to reduce its pollution or health hazard.

**treated water** Water that has undergone a water treatment process.

**treatment** (1) Any method, technique, or process designed to remove solids and/or pollutants from water, solid waste, wastestreams, effluents, and air emissions. (2) A method used to change the biological character or composition of any regulated medical waste so as to substantially reduce or eliminate its potential for causing disease. *See also* the following types of treatment: advanced treatment, aerobic treatment, anaerobic treatment, anticorrosion treatment, biological treatment, chemical treatment, chlorine–ammonia treatment, complete treatment, conventional treatment, intermediate treatment, physical-chemical treatment, preliminary treatment, pretreatment, primary treatment, secondary treatment, sludge treatment, split treatment, stage treatment, tertiary treatment, wastewater water treatment.

**treatment-chemical contamination** Contamination of drinking water caused by impurities in chemicals used for treatment, e.g., carbon tetrachloride ($CCl_4$) in chlorine and bromate ($BrO_3^-$) in sodium hypochlorite (NaOCl). *See also* aluminum- and iron- based coagulants.

**treatment chemicals** *See* important chemicals.

**treatment facility and treatment system** All structures that contain, convey, and, as necessary, physically or chemically treat coal mine drainage, coal preparation plant process wastewater, or drainage from coal preparation plant associated areas, which remove regulated pollutants from such waters. This includes all pipes, channels, ponds, basins, tanks and all other equipment serving such structures (EPA-40CFR434.11-o).

**treatment plant** An important component of water supply or wastewater disposal systems, containing a number of units and processes to produce a finished water or a wastewater effluent of a specified quality. *See* Figure T-12.

**treatment plant reliability** The probability that a treatment plant or a treatment process can consistently meet performance objectives over the design or planning periods under specified conditions; more specifically, the probability or percent of time that effluent concentrations will meet permit requirements. For example, a 95% reliability suggests that the plant will not meet daily permit limits about 18 times a year (5% × 365 days). *See* mechanical reliability.

**treatment plant residuals** *See* residuals.

**treatment plant residue** Solid material removed from water or wastewater during treatment.

**treatment process** A specific technique that removes or destroys specific constituents in water, wastewater, or a residual stream (such as steam stripping unit, waste incinerator, biological treatment unit). Most treatment processes are conducted in tanks or reactors; they are a subset of waste management units (EPA-40CFR63.111).

**treatment SI** Treatment surface impoundment. Most surface impoundments provide some level of treatment. *See also* disposal SI, storage SI.

**treatment system** *See* treatment facility.

**treatment technique requirement** A requirement

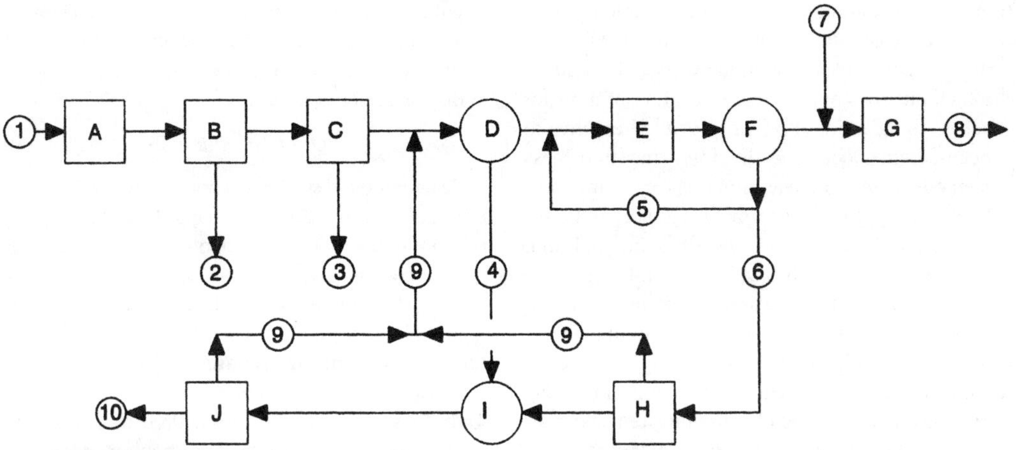

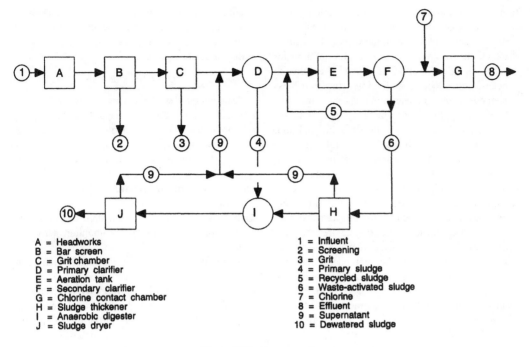


**Figure T-12.** Treatment plant.

of the national primary drinking water regulations that specifies for a contaminant a specific treatment technique known to the USEPA to lead to a reduction in the level of such contaminant sufficient to comply with relevant requirements (EPA-40CFR142.2).

**treatment technology** Any unit operation or series of unit operations that alters the composition of a hazardous substance, pollutant, or contaminant through chemical, biological, or physical means so as to reduce toxicity, mobility, or volume of the contaminated materials being treated. Treatment technologies are an alternative to land disposal of hazardous wastes without treatment (EPA-40CFR300.5).

**treatment works** Any devices and systems for the storage, treatment, recycling, and reclamation of municipal sewage, domestic sewage, or liquid industrial wastes. These include intercepting sewers, outfall sewers, sewage collection systems, individual systems, pumping, power or other equipment and their appurtenances; extensions, improvement, remodeling, additions, and alterations thereof; and land that is part of the treatment process or is used for ultimate disposal of treatment residues (EPA-40CFR35.2005-49 and 35.905). Also called purification works.

**treatment works treating domestic sewage** A publicly owned treatment works or any other sludge or wastewater treatment device or system, regardless of ownership, used in the storage, treatment, recycling, and reclamation of municipal or domestic wastewater, including land dedicated for sludge disposal; excluding septic tanks or similar devices (EPA-40CFR122.2).

**Trebler** An apparatus manufactured by Lakeside Equipment Co. for automatic sampling.

**Trematoda** The class of parasitic flatworms or trematodes.

**trematode** A fluke or other parasitic flatworm of the class Trematoda having leaflike bodies, hooks, and/or suckers. Schistosomes and other trematodes of sanitary significance use snails as intermediate breeding sites. *See* blood fluke, cercaria, liver fluke, miracidium, schistosomiasis.

**trematode worm** A trematode.

**tremie** A device used to place concrete or grout under water; a small-diameter pipe placed in the annular space at the bottom of a well hole for pumping cement.

**trench filling** The disposal of water or wastewater treatment residuals in narrow or wide trenches of a sanitary landfill (monofill). *See also* area filling.

**trench improvement** A corrosion control method using an electrolytic backfill material to protect buried pipes. Also called environmental alteration.

**trenchless technology** A method commonly used for the rehabilitation or replacement of water mains. A trenchless technique does not require a trench but does require the excavation of a pit for access to the point of repair. It is an alternative to open-cut methods. *See* the following trenchless methods: cured-in-place lining, epoxy lining, felt-based lining, fold-and-form lining, horizontal directional drilling, hydraulic bursting, internal-joint sealing, membrane lining, microtunneling, pipe bursting, pneumatic bursting, sliplining, spray-applied lining, static bursting, tight-fit lining, woven-hose lining. *See also* no-dig technology, open cut.

**trench method** One of three common landfilling methods: the flat or gently sloping site is excavated and filled with successive parallel trenches separated by 3–4 ft soil embankments. Solid wastes are dumped into the trenches, compacted, and covered with earth. *See also* area method and ramp method.

**trench pump-in method** A method used to determine the assimilative capacity of a site with regard to on-site disposal of wastewater. *See* shallow trench pump-in test for detail.

**trench pump-in test** Same as shallow trench pump-in test.

**Trey Deaerator™** A three-stage deaerator manufactured by U.S. Filter Corp.

***T. rhodesiense*** Abbreviation of *Trypanosoma rhodesiense.*

**TRI** Acronym of Toxics Release Inventory.

**trial burn** An incinerator test in which emissions are monitored for the presence of specific organic compounds, particulates, and hydrogen chloride and show compliance with the requirements of RCRA.

**trialkyltin compound** A compound of tin (Sn) containing three atoms of an alkyl group (e.g., Me for methyl or Bu for butyl), highly toxic to aquatic organisms. Common tryalkyltin compounds include $Bu_3SnCl$, $Bu_3SnOH$, $Me_3SnCl$, and $Me_3SnOH$. Some trialkyltins are used as antifouling agents.

**Triangle Brand®** Copper sulfate produced by Phelps Dodge Refining Co.

**triangle olfactometer** A device used to measure the intensity of odors; it contains six cups, each with a different dilution ratio and three ports—two ports with purified air and the third with a diluted sample. An odor panel sniffs and selects the port containing the sample. *See also* butanol wheel and scentometer.

**triangular-notch weir** A weir with a V-shaped notch for measuring small flows. Discharge varies with the 2.5 power of weir head. Also called triangular weir, V-notch weir, Vee weir, V-notched weir. The discharge $Q$ of a free-flow triangular weir is a function of the head $H$, the notch angle $\theta$, and the discharge coefficient $K$:

$$Q = KH^{2.5} (\tan \theta/2) \qquad \text{(T-40)}$$

See compound weir and Thomson weirs.

**triangular weir** Same as triangular-notch weir.

**triatomid bug** The arthropod vector of American trypanosomiasis or Chagas' disease; an insect that emerges nightly from cracks in walls and furniture to suck blood.

**tribasic sodium phosphate** Same as trisodium phosphate.

**Triboflow** A device manufactured by Auburn International to continuously monitor particulate emissions.

**tribromamine ($NBr_3$)** A compound formed during the chloramination of water containing bromide.

**tribromoacetic acid ($CBr_3COOH$)** Acronym: TBAA. A haloacetic acid containing three bromine atoms; sometimes formed during the chlorination or ozonation of bromide-containing water.

**tribromomethane ($CHBr_3$)** A trihalomethane and a disinfection by-product. *See* bromoform for detail.

**tribromonitromethane ($CBr_3NO_2$)** A by-product of chlorination and ozonation formed during the disinfection of waters containing bromide. Common name: bromopicrin.

**tributary** A stream that discharges into another stream or body of water.

**tributary population load** A convenient expression of the load on wastewater treatment works, equal to the sum of population equivalents of the sources (domestic, industrial, institutional, etc.).

**tricalcium silicate [$(CaO)_3 \cdot SiO_2$ or $Ca_3SiO_5$]** A compound of lime and silica, used in cement and food products to prevent caking.

**Tricanter®** A device manufactured by Flottweg for sludge centrifugation.

**tricarboxylic acid cycle** A series of enzyme-catalyzed reactions occurring in the aerobic metabolism of carbohydrates, proteins, and fatty acids, with the conversion of pyruvic acid into carbon dioxide and water, the buildup of ATP, and the reduction of oxygen. Also called Krebs cycle, citric acid cycle.

**Tricellorator** A dissolved-air flotation device manufactured in three compartments by Pollution Control Engineering, Inc.

**Tricep™** A process developed by Grave Co. for the removal of iron and copper oxides with oil, using granular resin filtration.

**trichloramine ($NCl_3$)** A compound of nitrogen and chlorine formed during the chlorination of water that contains ammonia ($NH_3$); it can cause objectionable taste and odor if not removed (e.g., by aeration, dechlorination, or exposure to sunlight). Also called chlorine nitride, nitrogen chloride, nitrogen trichloride, trichlorine nitride. *See* breakpoint chlorination.

**trichlorine nitride** *See* trichloramine.

**trichloroacetaldehyde** Same as chloral hydrate.

**trichloroacetate** The salt and main species of trichloroacetic acid present in drinking water at common pHs.

**trichloroacetic acid ($CCl_3COOH$ or $HC_2Cl_3O_2$)** A toxic, colorless, crystalline, soluble organic compound used in pharmaceuticals and herbicides, and as a reagent for the detection of albumin. It is one of two principal haloacetic acids formed during water chlorination and a contaminant regulated by the USEPA under the Disinfectants and Disinfection By-products Rule: MCLG = 0.3. Abbreviation TCA, TCAA, or $Cl_3AA$. *See also* dichloroacetic acid.

**1,1,1-trichloroacetone** Same as 1,1,1-trichloropropanone.

**trichlorobenzene ($C_6H_3Cl_3$)** A volatile organic compound, with three chlorine atoms substituting for hydrogen atoms in benzene. 1,2,4-trichlorobenzene is regulated by the USEPA as a drinking water contaminant under Phase I: MCL = MCLG = 0.07 mg/L.

**trichloroethane (TCA)** (1) An organic chemical used as a cleaning solvent; it causes adverse health effects in domestic water supplies. *See* 1,1,1-trichloroethane and 1,1,2-trichloroethane.

**1,1,1-trichloroethane ($CH_3CCl_3$)** A nonflammable, volatile organic liquid chemical, insoluble in water, used as an industrial solvent, cleaner, and degreaser. Its inhalation may cause depression as well as liver and cardiovascular effects. Regulated by the USEPA: MCLG = MCL = 0.2 mg/L. Also called chlorotene or methyl chloroform.

**1,1,2-trichloroethane ($CHCl_2CH_2Cl$)** A volatile organic liquid chemical, insoluble in water, used as an industrial solvent. Regulated by the USEPA: MCLG = 0.003 mg/L and MCL = 0.005 mg/L. Also called vinyl trichloride.

**trichloroethene (TCE)** Same as trichloroethylene.

**trichloroethylene (TCE) ($CHCl{=}CCl_2$)** A volatile organic compound/chlorinated hydrocarbon, with a low boiling-point, frequently detected in contaminated groundwater. It is a colorless liquid, toxic if inhaled, that is used as a solvent or metal degreaser, and in other industrial applications. It may cause organ damage and tumors; MCL in drinking water as established by the USEPA: 0.005 mg/L. Also called ethinyl trichloride, trichloroethene, tri-clene.

**1,1(2,2,2-trichloroethylidene) bis (4-methoxy-benzene)** Same as methoxychlor.

**trichlorofluoromethane ($CCl_3F$)** A volatile organic compound, commonly found in drinking water sources; used as a refrigerant, aerosol propellant, solvent, etc. Also called fluorotrichloromethane.

**trichloromethane ($CHCl_3$)** Acronym: TCM. Same as chloroform.

**trichloromethylsilane** *See* silane (2).

**trichloronitromethane** Same as chloropicrin.

**trichlorophenol (TCP)** A chlorophenol containing three chlorine atoms, e.g., 2,4,6-trichlorophenol.

**2,4,6-trichlorophenol ($C_6H_2Cl_3OH$)** An insecticide and a disinfection by-product, cause of objectionable taste and odor in water.

**2,4,5-trichlorophenoxyacetic acid ($C_8H_5Cl_3O_3$)** Abbreviation: 2,4,5-T. A light-tan, insoluble solid; an active herbicide ingredient, with a WHO-recommended guideline value of 0.009 mg/L for drinking water. Effective treatment process: adsorption on granular activated carbon. *See also* dioxin.

**1,2,3-trichloropropane** A volatile organic compound regulated by the USEPA as a drinking water contaminant under Phase I: and Phase VIB: MCL = 0.0008 mg/L and MCLG = 0.

**1,1,1-trichloropropanone ($CCl_3COCH_3$)** Abbreviation: 1,1,1-TCP; commonly called 1,1,1-trichloroacetone. The principal haloketone disinfection by-product of chlorination; it causes mutations in strains of *Salmonella*. It may be converted to chloroform by hydrolysis.

**trichlorotrifluoroethane** A nonflammable solvent used in the Soxhlet method of grease extraction.

**trichocephaliasis** Another name of trichuriasis or human whipworm.

***Trichocephalus dispar*** Former name of *Trichuris trichiura.*

***Tricho. hominis*** *Trichocephalus hominis,* former name of *Trichuris trichiura.*

***Tricho. trichiura*** *Trichocephalus trichiura,* former name of *Trichuris trichiura.*

**trichuriasis** An excreted helminth infection caused by *Trichuris trichiura. See* its common name, whipworm, for more detail.

***Trichuris*** The genus of whipworms.

***Trichuris trichiura*** The 30–50 mm, waterborne or foodborne, pathogen that causes whipworm disease or trichuriasis. It is distributed worldwide and is transmitted through the soil. Excreted load = 1000. Latency = 20 days. Persistence = 9 months. Median infective dose < 100. Former names: *Trichocephalus dispar, Trichocephalus hominis, Trichocephalus trichiura.*

**trickle-bed reactor** A common type of all gas–liquid fixed-bed reactor, operated in the downflow cocurrent mode.

**trickle irrigation** An irrigation method using perforated plastic pipes at the base of the plants and realizing water savings (due to considerable reduction of evaporation and percolation) and sometimes economy of nutrients as well as salinity reduction. Also called drip irrigation. *See* other basic irrigation methods: flooding, furrow, sprinkler, and subirrigation.

**trickling contact** Wastewater treatment by trickling filters.

**trickling filter** A coarse aerobic, nonsubmerged fixed-film treatment system in which wastewater is trickled over a bed of stones or other material covered with bacteria that break down the organic waste, with oxygen provided by diffusion through the void spaces. The biofilm sloughs off periodically and is separated in a sedimentation basin. The hydraulic and organic loadings vary from 45–90 gpd/ft$^2$ and 0.3–1.5 kg BOD/day/m$^3$ for conventional filters to 230–690 gpd/ft$^2$ and 1.5–19 kg BOD/day/m$^3$ for high-rate units. The system is kept aerobic by spray aeration and natural ventilation. Also called biological bed. *See also* aerated contact bed, bacteria bed, biofilter, contact aerator, contact filter, conventional trickling filter, crossflow, dosing chamber, fixed-nozzle distributor, high-rate trickling filter, intermittent sand filter, low-rate trickling filter, nidus rack, percolating filter, rotary distributor, roughing filter, standard rock filter, submerged-contact aerator.

**trickling filter–activated sludge process (TF–AS)** Same as trickling filtration–activated sludge process. *See also* series trickling filtration–activated sludge process.

**trickling filter BOD removal** *See* soluble BOD removal.

**trickling filter contact time** The time it takes wastewater to flow through a trickling filter. The nominal time ($T_n$, sec) is the ratio of filter depth (D, ft) to the surface velocity ($V$, fps):

$$T_n = D/V = D/(Q/A) \qquad \text{(T-41)}$$

where $Q$ (cfs) is the influent flow, including any recirculation, and $A$ (ft$^2$) is the surface area of the filter. Because of flow tortuosity and the effect of media geometry and packing characteristics, the actual contact time ($T_a$, sec) is greater:

$$T_a = CD/(Q/A)^n \qquad \text{(T-42)}$$

where $C$ and $n$ are constants characteristic of the medium. *See also* Howland's equation.

**trickling filter design formulation** Trickling filter design formulas include British Manual, Eckenfelder, Galler–Gotaas, Germain, Howland, Kincannon–Stover, Logan, modified Velz, Rankin, Schulze, Velz.

**trickling filter distributor** *See* distributor.

**trickling filter dosing** *See* dosing rate.

**trickling filter formulas** *See* trickling filter design formulation.

**trickling-filter humus** The surface film of settleable solids sloughed off intermittently by trickling filters and intercepted in secondary clarifiers or added to primary sludge. *See also* interfacial sloughing.

**trickling filter medium** Common trickling filter media include rocks and plastics. *See also* standard rock filter and crossflow.

**trickling filter nitrification rate** The nitrification rate ($R_n$) achievable in a trickling filter may be estimated in function of the influent BOD to total Kjehldal nitrogen (TKN) ratio:

$$R_n = 0.82\ (\text{BOD/TKN})^{-0.44} \qquad \text{(T-43)}$$

**trickling filter operation** Problems associated with the operation of a trickling filter include low efficiency at low temperatures, production of odors, nuisance of filter flies near the plant, and media clogging by snail shells.

**trickling filter organisms** Predominating organisms in trickling filters are facultative bacteria, along with some aerobic and anaerobic species, fungi, algae, and protozoa, e.g.: *Achromobacter, Alcaligenes, Beggiatoa, Flavobacterium, Pseudomonas, Sphaerotilus natans, Fusazium, Geotrichum, Mucor, Penicillium, Sporatichum, Chlorella, Phormidium, Ulothrix, Epistylis, Opercularia, Vorticella.*

**trickling filter performance** The BOD removal efficiency of a trickling filter. *See* trickling filter design formulation.

**trickling filter performance equation** *See* Velz equation.

**trickling filter plant** A trickling filter plant includes not only the filter but also, usually, at least a primary clarifier and a secondary clarifier with sludge return. Flowchart variations include (a) one-stage and two-stage plants, (b) intermediate clarifiers, and (c) flow or sludge recirculation ahead of the primary clarifier or around the filter.

**trickling filter rotational speed** *See* rotational speed formula.

**trickling filter sloughing** *See* sloughing.

**trickling filter sludge** The brownish, flocculent, humus-like material produced during wastewater treatment in a trickling filter. It decomposes relatively slowly, but digests readily.

**trickling filter–solids contact process** *See* trickling filtration–solids contact process.

**trickling filter solids production** A trickling filter plant produces 0.2–0.5 kg of volatile suspended solids per kg of BOD removed, with a moisture content of 94–98%; both factors (weight and moisture) increase with loading.

**trickling filter underdrain** A structure made of perforated pipes, blocks, and similar materials to support the filter medium, collect the effluent and sloughed solids, and distribute air through the bed.

**trickling filtration** The wastewater treatment process that uses a trickling filter. *See* fixed-growth process. Microorganisms, mainly bacteria, grow on the surface of the filter media and form a slime layer. Wastewater flows in a thin film over the bacterial growth. Oxygen is provided by diffusion through media voids and is carried, along with nutrients, to the aerobic zone of the slime, while waste products are transferred to the adjacent water layer. Eventually, the slime layer sloughs off and settles in the clarifier. For most wastewaters, oxygen is the rate-limiting factor. *See* controlled filtration, trickling filter plant.

**trickling filtration–activated sludge (TF–AS) process** A wastewater treatment method that includes a trickling filter, an aeration basin, and a final clarifier. A portion of the settled sludge is returned to the aeration basin, while a portion of the filter underflow is recycled. This process differs from the series trickling filtration–activated sludge process because it does not include an intermediate clarifier. It does not include sludge reaeration, either. *See* Figure T-13. *See also* combined filtration–aeration process.

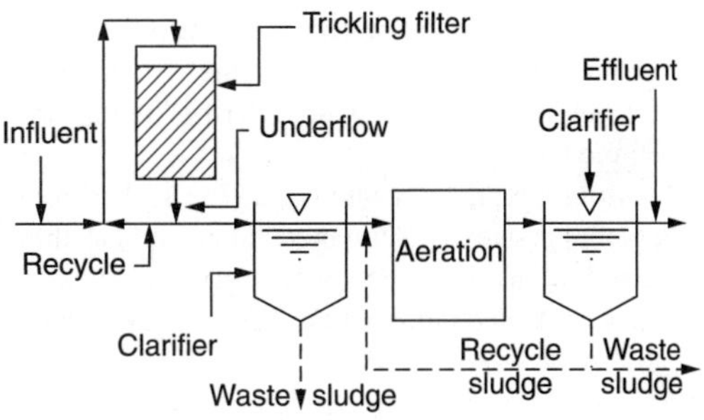


**Figure T-13.** Trickling filtration—activated sludge.

**trickling filtration–solids contact (TF–SC) process** One of a few dual wastewater treatment processes that combine the trickling filtration and activated sludge methods. A trickling filter or biological tower of reduced size receives the primary effluent and is followed by an aeration basin and a final clarifier. Some settled sludge and a portion of the filter underflow are recycled ahead of the filter. This process differs from the biofiltration–activated sludge process only because it includes a small unit for reaeration of recycle solids and, sometimes, a flocculating well in the clarifier. *See* Figure T-14. *See also* combined filtration–aeration process.

**tri-clene** Same as trichloroethylene.

**triclosan ($C_{12}H_7Cl_3O_2$)** A white powder with a slight phenolic odor; a chlorinated aromatic compound slightly soluble in water, used in soaps, deodorants, and toothpastes; listed among the xenobiotics.

**Tricon** A water treatment plant manufactured by Wheelabrator Engineered Systems, Inc., including a flocculator–clarifier with buoyant media.

**tricresol** A mixture of the three isomeric cresols ($C_7H_8O$), synthetic organic chemicals derived from coal or wood tar. *See also* orthocresol.

***Tricula*** A genus of aquatic snails that can be infected by the trematode worm *Schistosoma japonicum,* which causes schistosomiasis in East Asia and the Philippines. *See also Oncomelania.*

**Tridair** A treatment apparatus manufactured by Engineered Specialties, Inc., using dissolved-air flotation.

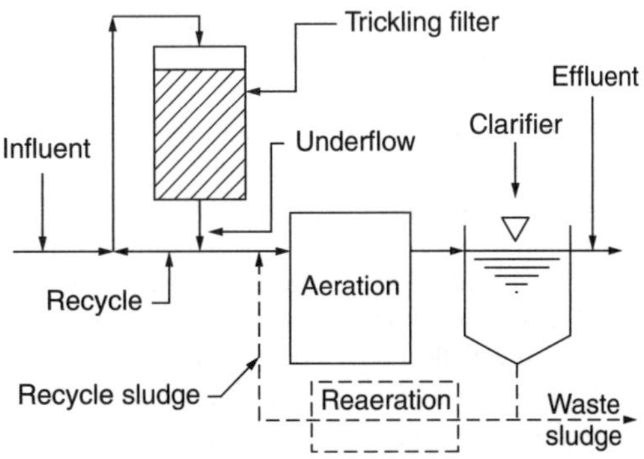


**Figure T-14.** Trickling filtration—solids contact.

**Trident®** A modular water treatment plant manufactured by Wheelabrator Engineered Systems, Inc., including a flocculator–clarifier with buoyant media.

**tridentate ligand** An anion or molecule that forms a complex with a central metal atom and has three ligand atoms, i.e., three atoms responsible for its basic or nucleophilic nature. *See* multidentate ligand.

**trier** An instrument consisting of a tube cut in half lengthwise with a sharpened tip, used to sample hazardous wastes present as moist or sticky solids with a diameter that is less than one-half the size of the instrument. *See also* auger, bailer, composite liquid–waste sampler, dipper, thief, weighted bottle.

**trifluorochloromethane** Same as chlorotrifluoromethane.

**trifluralin** A pesticide with a WHO-recommended guideline value of 0.02 mg/L for drinking water. The USEPA has lower limits for this chemical: MCL = MCLG = 0.005 mg/L.

**trigger point** A condition that causes the automatic regeneration of a water softener.

**triglyceride** A simple fat obtained from glycerol by the esterification of three hydroxyl groups with fatty acids.

**trihalomethane (THM)** One of a family of organic compounds named as derivatives of methane, wherein three of the four hydrogen atoms of methane are each replaced by a halogen (bromine, chlorine, iodine) atom in the molecular structure. THMs are generally the by-products from chlorination of drinking water that contains organic material. The resulting compounds (THMs) are suspected of causing cancer and are regulated by the USEPA, which sets an MCL of 0.08 for total THM. They are of concern more in a developed country like the United States than in developing countries where waterborne pathogens are a more serious risk. Effective water treatment processes: aeration and stripping, adsorption on granular activated carbon. *See* disinfection by-products, humic substances, and total trihalomethanes. Table T-05 shows 10 THMs of concern in drinking water.

**Table T-05.** Trihalomethanes

| |
|---|
| *THM4* |
| Bromodichloromethane ($CHBrCl_2$)—FC = 241* |
| Bromoform (tribromomethane, $CHBr_3$)—FC = 929 |
| Chloroform (trichloromethane, $CHCl_3$)—FC = 92.5 |
| Dibromochloromethane (chlorodibromomethane, $CHBr_2Cl$)—FC = 585 |
| *Other THMs* |
| Bromochloroiodomethane (CHBrClI) |
| Bromodiiodomethane ($CHBrI_2$) |
| Chlorodiiodomethane ($CHClI_2$) |
| Dibromoiodomethane ($CHBr_2I$) |
| Dichloroiodomethane ($CHCl_2I$) |
| Iodoform (triiodomethane, $CHI_3$) |

*FC = Freundlich constant.

**trihalomethane formation potential (THMFP)** The amount of trihalomethanes that can be formed during disinfection of source or treated water. THMFP is determined by a test in which the water is dosed with the disinfectant to produce a residual of 3.0 mg/L and incubated under conditions to promote maximum THM formation. The result is indicative of the amount of THM precursors in the sample tested. Total organic carbon (TOC, in mg of C/L) has been proposed as a surrogate measurement for THMFP (in μg/L):

$$THMFP = 43.78\,(TOC)^{1.248} \qquad (T\text{-}44)$$

*See also* DBP formation potential, SUVA test.

**trihalomethane precursors** Natural organic compounds, mainly humic and fulvic acids, found in all surface and ground waters; during disinfection, they may be converted to disinfection by-products. They react with halogens to form trihalomethanes. *See* precursor (1) for more detail.

**triiodomethane ($CHI_3$)** A trihalomethane and a disinfection by-product. Also called iodoform.

**1,2,4-trimethylbenzene [$C_6H_3(CH_3)_3$]** A volatile organic compound used in dyes and pharmaceuticals, regulated by the USEPA as a drinking water contaminant under Phase I list 3 (monitoring at the discretion of states).

**1,3,5-trimethylbenzene** A volatile organic compound regulated by the USEPA as a drinking water contaminant under Phase I list 3 (monitoring at the discretion of states).

**Trimite™** A package water treatment plant manufactured by Wheelabrator Engineered Systems, Inc., including a flocculator-clarifier with buoyant media.

**trinitrotoluene [$CH_3C_6H_2(NO_2)_3$]** Acronym TNT. *See* toluene.

**Tri-NOx®** A process developed by Tri-Mer Corp. for the removal of nitrogen oxides.

**Trio-Denipho™** A phased isolation ditch treatment process designed for the removal of phosphorus and nitrogen from wastewater. It provides anaerobic, anoxic, and aerobic conditions in three oxidation ditches and other reactors.

**Tri-Packs®** Plastic media manufactured by Jaeger Products.

**triple-expansion steam pump** A pump with three cylinders for steam exhaustion and expansion.

**triple-media filter** A dual-media filter with the addition of a third, finer layer of high-density material (e.g., garnet or ilmenite) at the bottom to improve filtrate quality at high filtration rates. Sometimes called, erroneously, multimedia filter.

**triple point** The particular condition of temperature and pressure of a given substance whose three states (gaseous, liquid, solid) coexist in equilibrium.

**triple-sludge process** A wastewater treatment system consisting of three biological reactor–clarifier sets in series and designed separately for carbonaceous BOD removal, nitrification, and denitrification. Methanol is added for denitrification.

**triple superphosphate** *See* superphosphate.

**Triplex™** A multistage air scrubber manufactured by Davis Water and Waste Industries, Inc.

**triplex pump** A reciprocating pump with three cylinders in line, connected to the same suction and discharge line, providing continuous intake and discharge.

**tripolyphosphate** Any soluble salt with the $P_3O_{10}^{5-}$ anion; most common is sodium tripolyphosphate ($Na_5P_3O_{10}$). Tripolyphosphates are used as chelating agents for certain metals in solution.

**Trisep™** A filter manufactured by Graver Co. for the separation of oil, solids, and water.

**trisodium phosphate ($Na_3PO_4 \cdot 12\ H_2O$)** Acronym: TSP. A colorless, soluble, crystalline compound used in the production of water-softening agents, detergents, paper, and textiles. Also used as an alkaline additive in membrane-cleaning solutions and called tribasic sodium phosphate. *See also* sodium phosphate and disodium phosphate.

**tristimulus filter method** An acceptable laboratory method for the analysis of color in raw water and industrial and domestic wastewaters. *See also* sphectrophotometric method, visual comparison method.

**tritiated water** *See* tritium.

**tritium (T, $H^3$, $^3H$, $H^c$, or H-3)** A radioisotope of hydrogen, having two neutrons and one proton as nucleus and emitting beta particles; used in hydrogen bombs and as a radioactive tracer. Atomic weight = 3. Half-life = 12.5 years. Regulated by the USEPA as a drinking water contaminant: MCL = 20,000 pCi/L. Tritium, in the form of tritiated water, may be used as a tracer in water quality modeling; *see* reaeration constant.

**Triton** A circular membrane diffuser manufactured by Aeration Research Co.

**Triton®** A lateral underdrain apparatus manufactured by Wheelabrator Engineered Systems, Inc. for use in wedgewire screens.

**Tritor** A device manufactured by FMC Corp. for frontal cleaning of bar screens and grit chambers.

**triturator** A type of screenings grinder.

**Triturator** A device manufactured by Envirex, Inc. to grind screenings from treatment plants.

**trivalent** (1) Having a valence of 3. (2) Having three binding sites; pertaining to certain antigens in immunology.

**trivalent chromium ($Cr^{3+}$)** One of the two forms of chromium in natural water and an essential nutrient, safe at 0.20 mg/day, unstable in chlorinated water; it slowly oxidizes to hexavalent chromium, which is toxic to humans. Also called chromium III. *See also* hexavalent chromium.

**TriZone™** A compact water treatment plant manufactured by Wheelabrator Engineered Systems, Inc.

**TrojanUVSwift™ ECT** A process that uses a combination of ultraviolet (UV) light and hydrogen peroxide ($H_2O_2$) for the simultaneous disinfection of drinking water as well as the removal of taste, odor, and other contaminants.

**Troll** A submersible instrument manufactured by In-Situ, Inc. for monitoring water level and temperature.

**Tromax™** A trammel screen manufactured by Norkot Mfg. Co.

**trommel** A cylindrical or conical, rotating screen used to separate solid wastes or other materials (ore, coal, gravel, etc.) according to size and density.

**trommel screen** A trommel.

**troph** A trophozoite.

**trophic condition** A relative description of a lake's biological productivity based on the availability of plant nutrients. The range of trophic conditions is characterized by the terms oligotrophic for the least biologically productive to eutrophic for the most biologically productive (EPA-40CFR35.1605-6). *See also* trophic state, trophic state index.

**trophic level** Any of various stages in a food chain; the position occupied by a species or group of species in a food chain. *See* producer, primary consumer, secondary consumer, tertiary consumer. (2) The nutrient status of a water body: dystrophic, oligotrophic, mesotrophic, or eutrophic.

**trophic state** The state of nourishment of a water body for plant growth. *See* dystrophic, eutrophic, oligotrophic, trophic conditions.

**trophic state index (TSI)** A numerical index on a scale of 0 to 100 developed by R. E. Carlson in

1977 to represent the trophic state of a lake, based on its content of chlorophyll a, total phosphorus, and Secchi disk transparency. The TSI is widely used in studies assessing the status lakes with respect to eutrophication. Also called Carlson index.

**trophic status** Same as trophic condition.

**trophozoite** A protozoan in the vegetative, active growth stage; one of the forms of amebiasis, giardiasis, or other intestinal protozoal infections. Also called troph. *See also* cyst, schizont, sporozoite, merozoite.

**tropoparasite** An organism that is an obligate parasite for only part of its life cycle, and a nonparasite for the remainder.

**tropopause** The boundary between the troposphere and the stratosphere.

**troposphere** The layer of the atmosphere closest to the earth surface and extending about 9 to 16 km, containing most of the clouds and moisture. *See also* stratosphere.

**TroubleShooter™** A portable container manufactured by ThermaFab, Inc. for use in hazardous waste spills.

**trough** A long, narrow, open structure (channel or conduit) for holding or conveying water or other liquids; e.g., the steep open channel of a chute spillway (also called chute).

**TRPH** Acronym of total recoverable petroleum hydrocarbons.

**true aqueous solubility** The solubility of a chemical constituent in pure water. *See also* apparent solubility.

**true color** In water, the portion of total color that is attributable to substances in solution (humic and fulvic acids) after the removal of turbidity (suspended materials) by centrifugation or filtration. *See also* apparent color.

**true declining-rate filtration** *See* constant-pressure filtration.

**true facultative anaerobes** Microorganisms, particularly bacteria, that can grow in the absence of molecular oxygen but shift their metabolism from fermentation to respiration when free oxygen is available. *See also* aerotolerant, facultative, and obligate anaerobes.

**true groundwater velocity** The ratio of the volume of groundwater through a formation in unit time to the product of the cross-sectional area by the effective porosity. *See* actual groundwater velocity for more detail.

**true residence time** A parameter ($T_a$, hr) used in the design and analysis of bulk media filters or biofilters. *See* actual residence time for detail.

**true solubility** *See* true aqueous solubility.

**true vapor pressure** The equilibrium partial pressure exerted by a volatile organic liquid as determined in accordance with methods described by the American Petroleum Institute (EPA-40CFR52.741, 60.111-I and 60.111a-1).

**true yield** The amount of biomass produced immediately by microorganisms upon consuming a substrate or upon oxidation of an electron donor. *See* synthesis yield for detail.

**true yield coefficient** The ratio ($Y$, g/g) of maximum specific growth rate of an organism ($\mu_m$, g of new cells/g of cells/day) to its maximum specific substrate utilization rate ($k$, g/g/day):

$$Y = \mu_m/k \qquad \text{(T-45)}$$

**Tru-Grit** An apparatus manufactured by Hycor Corp. for the separation and washing of grit.

**Tru-Gritter** An apparatus manufactured by Vulcan Industries, Inc. for grit removal.

**Tru-Test** An automatic sampler manufactured by FMC Corp.

***Trypanosoma brucei gambiense*** A hemoflagellate that causes trypanosomiasis (Gambian sleeping sickness) in Africa.

***Trypanosoma brucei rhodesiense*** A hemoflagellate that causes trypanosomiasis (Rhodesian sleeping sickness) in Africa.

***Trypanosoma cruzi*** A protozoan species that causes the American trypanosomiasis or Chagas disease.

***Trypanosoma gambiense*** A protozoan species that causes the Gambian sleeping sickness.

***Trypanosoma rhodesiense*** A protozoan species that causes the Rhodesian sleeping sickness.

**trypanosomiasis** A severe, sometimes fatal, vector-borne disease caused by a species of the pathogen *Trypanosoma.* There are one American and two African varieties of trypanosomiasis. *See* the following common names for more detail: Chagas disease, Gambian sleeping sickness, and Rhodesian sleeping sickness.

**tryptophan** **[$(C_8H_6N)CH_2CH(NH_2)COOH$]** A colorless, crystalline, and aromatic amino acid released from protein. *See* indole.

**tryptose** An enzymatic digest of protein used in preparing microbiological culture media. *See* lauryl tryptose broth.

**TS** Acronym of total solids.

***T. saginata*** *See Taenia saginata.*

**TSCA** Acronym of Toxic Substances Control Act.

**tsetse fly** A fly that transmits the protozoan pathogen that causes trypanosomiasis or sleeping sickness. *See Glossina cruzi, G. morsitans, G. palpalis.*

**TSI** Acronym of trophic state index.

***T. solium*** *See Taenia solium.*

**TSS** Acronym of total suspended solids.

**TSSc** Symbol of total suspended solids concentration in the centrate of a thickener, % by weight.

**TSSf** (1) Symbol of total suspended solids concentration in the feed of a thickener, % by weight. (2) A factor used to convert turbidity readings to total suspended solids, expressed in mg/L TSS/NTU. *See* turbidity (2).

**$TSS_p$** Symbol of total suspended solids concentration in the thickened solids product, % by weight.

**TSS recovery** A measure of the performance of a centrifuge (used in sludge thickening and dewatering), defined as the the percentage of thickened dry solids with respect to feed solids. *See* recovery for detail.

***T. suis*** *See Trichuris suis.*

**TTF test** *See* time-to-filter test.

**TTHM** Acronym of total trihalomethanes.

**TTHM-Br** Total trihalomethane bromine.

**TTHM-Cl** Total trihalomethane chlorine.

**TTO** Acronym of total toxic organics.

***T. trichiura*** *See Trichuris trichiura.*

**$TU_a$** Symbol of toxic unit acute.

**tube-and-plate settler** *See* inclined plate settler, tube settler.

**tube clarifier** A clarifier whose capacity has been increased by the installation of inclined, parallel bundles of metal or plastic tubes. *See also* lamella clarifier.

**tubercle** (1) A protective crust or knob-like mound of products (rust) that builds up over a pit caused by the loss of metal (corrosion). In iron or steel pipes, the tubercles are iron oxides and oxyhydroxides. Tubercles in water pipes can offer a medium for pathogen regrowth, but free chlorine has been observed to penetrate tubercles under slightly alkaline conditions. (2) A deposit of metallic salts, like a bunch of grapes, in pipes.

**tubercle bacilli** The bacteria, *Mycobacterium tuberculosis,* that causes tuberculosis; commonly found in wastewater, effluent, and sludge from areas with tuberculosis sanitaria.

**tubercular corrosion** *See* tuberculation.

**tuberculation** The development or formation of small mounds of pitting corrosion products (rust) on the inside of iron pipes. These mounds (tubercles) increase the roughness of the inside of the pipe, thus increasing resistance to water flow and reducing the pipe's capacity. Tuberculation is often associated with high pH, hence the recommendation of a balance of minimum hardness and alkalinity to prevent it. The presence of dissolved oxygen in water is also believed to promote tuberculation. *See also* pitting.

**tuberculosis** A disease caused by the bacterial agent *Mycobacterium* carried by humans and animals.

**tube sampler** A tube used to obtain a sample from one point of a flow of water or wastewater; because of differences in solids concentrations across a section, such a sample is not representative of the entire flow unless the stream velocity is uniform. *See also* scoop sampler.

**tube settler** A shallow device that uses bundles of small bore (50 to 75 mm) plastic tubes installed on an incline as an aid to sedimentation. The tubes may come in a variety of shapes including circular and rectangular. As water rises within the tubes, settling solids fall to the tube surface. As the sludge (from the settled solids) in the tube gains weight, it moves down the tubes and settles to the bottom of the basin for removal by conventional sludge collection means. Tube settlers are sometimes installed in sedimentation basins and clarifiers to improve particle removal, on the theory that settling depends on area rather than detention time. In essence, the tubes improve performance by increasing the surface area of the basin and reducing the surface overflow rate, while suppressing wind currents and reducing turbulence. Tube settlers are efficient sedimentation devices, but they are subject to clogging in biological treatment. *See also* settling tube, Lamella clarifier.

**tubesheet** A flat plate that holds in place the tubes of an evaporator or heat exchanger.

**tubing** Material in the form of a tube, e.g., glass tubing, copper tubing; particularly, a small and short, flexible pipe, or a high-test pipe with couplings and fittings.

**Tubingen process** A batch reactor process developed at the University of Tubingen (West Germany) to convert wastewater sludge to oil and char by heating dried sludge to temperatures of 300–350°C for 30 minutes in an oxygen-free container. *See also* sludge-to-oil reactor.

**Tub Scrubber** An air scrubber manufactured by Purafil, Inc.

**tubular conveyor** A sloped mechanism or device used in large wastewater treatment plants to lift grit out of a grit chamber, and wash and dewater the materials removed. Also called inclined screw. *See also* horizontal screw conveyor, chain-and-flight mechanism, clamshell bucket, chain-and-bucket elevator.

**tubular membrane** *See* tubular module.

**tubular module** An arrangement of reverse osmosis membranes in perforated or porous tubes that provide support and allow feedwater to pass through on one side; product water is collected on the other side of the tube. Tubular membranes are commonly used in the treatment of wastewater or water of high suspended solids content. They are easy to clean but produce water at a low rate. Other membrane configurations include hollow-fiber, plate-and-frame, and spiral-wound.

**TUc** Symbol of toxic unit chronic.

**Tuff-Span** Tank cover manufactured by Composite Technology, Inc.

**tularemia** An intestinal disease caused by the bacterial agents *Francisella tularemia* and *Pasturella tularensis,* often through the urine of infected rabbits or other wild animals. Symptoms may include fever, skin ulcer, pharyngitis, intestinal pain, vomiting, diarrhea, and even pneumonia and typhoid fever.

**Tulsion®** A resin manufactured by Thermax, Ltd. for use at high temperatures in ion exchange units.

**tungsten** (W, from its German name, wolfram.) A bright-gray element with a metallic luster and high melting point, used in alloys and in lamp filaments. Atomic weight = 183.85. Atomic number = 74. Specific gravity = 19.3. Also called wolfram.

**Tunnel Reactor®** An in-vessel composting apparatus manufactured by Waste Solutions.

**turbid** Having a cloudy or muddy appearance; thick or opaque with matter in suspension, as pertaining to a river or lake after a rainfall.

**turbidimeter** An instrument that is used for the determination of turbidity. It measures the amount or intensity of light scattered at a fixed angle by the particles in suspension in a water sample and uses a standard suspension as a reference. *See* Jackson candle turbidimeter and nephelometer.

**turbidimetry** (1) The determination of water turbidity using a turbidimeter. (2) The limiting depth for the visibility of an object in a turbid medium.

**turbidity** (1) A cloudy condition in water or wastewater due to suspended silt or colloidal matter that interferes with the passage of light (EPA-40CFR131.35,d-19). Turbidity cannot be directly equated to suspended solids because white particles reflect more light than dark-colored particles and many small particles will reflect more light than an equivalent large particle. Turbidity blocks sunlight in flowing waters, thus impeding photosynthesis and eutrophication, and can shelter microorganisms from disinfection. Turbidity is closely related to the presence of bacteria and *Giardia* cysts. (2) A measure of the suspended solids that cause the cloudy condition. The clarity of water is expressed as nephelometric turbidity units (NTU) and measured with a calibrated turbidimeter. Drinking water adequately treated by coagulation and filtration has a low turbidity, e.g., 0.1 NTU. The USEPA set a drinking water standard of 1–5 NTU. The total suspended solids (TSS, mg/L) of a settled and filtered secondary effluent is related to its turbidity (NTU) as follows:

$$\mathrm{TSS} = (\mathrm{TSS}_f)(T) \qquad \text{(T-46)}$$

where $T$ is turbidity in NTU and $\mathrm{TSS}_f$ is a conversion factor in mg/L TSS per NTU. This factor varies from 1.3 to 1.5 for a filtered secondary effluent and from 2.0 to 2.4 for a settled effluent.

**turbidity breakthrough** A problem encountered in the operation of depth filters, whereby the effluent shows unacceptable levels of turbidity even before reaching the terminal head loss. Other depth filtration operation problems include emulsified grease buildup, filter medium loss, gravel mounding, mudball formation, cracks and contraction of filter bed.

**turbidity measurement** Turbidity is measured to monitor the aesthetic quality of treated water, to assess the performance of the filtration units, and as an indicator of the removal of disinfectant-resistant pathogens.

**turbidity removal** Filtration, preceded or not by coagulation, is effective in removing turbidity (suspended and colloidal solids). Coarser materials can be reduced by sedimentation.

**turbidity sensors** Devices installed in water filters to trigger an alarm when turbidity of the treated water exceeds 1.0 NTU.

**turbidity units** *See* formazin turbidity unit (FTU), Jackson turbidity unit (JTU), and nephelometric turbidity unit (NTU).

**turbine** A type of impeller used in water treatment for mixing or stirring, consisting of flat or curved blades connected to vertical or horizontal shaft. Other types of impellers include paddles and propellers.

**turbine aerator** A device used in wastewater aeration. It consists of a submerged propeller system installed in the center of an aeration basin and surrounded by draft tubes. The propeller rotates and draws water through the inner section and into the air. *See also* brush aerator, submerged mechanical aerator.

**turbine blade** A type of mechanical aerator with blades capping a central tube in a chamber and spraying droplets over the water. *See also* propeller blade, submerged paddle, surface brush or surface paddle.

**turbine flocculator** A turbine mixer with three or four blades attached to a vertical shaft, used for the flocculation of coagulated colloidal particles. *See also* mixing and flocculation devices.

**turbine mixer** A device with a turbine impeller and short blades, used for blending, rapid mixing, and aeration in water or wastewater treatment and functioning like a centrifugal pump without a casing. It generates radial and tangential currents, including vortexing and swirling, the effects of which can be minimized by baffles. Mixing time varies from 2–20 seconds for mixing chemicals to 10–30 minutes for the flocculation of coagulated particles. *See also* mixing and flocculation devices.

**turbine pump** (1) A centrifugal pump with fixed vanes that converts the velocity energy into pressure head. (2) A device consisting of small centrifugal pumps, called bowls, connected through one shaft and driven by a single motor. *See also* turbo pump, regenerative pump.

**turbine wheel** A rotor designed to convert fluid energy into rotational energy. Hydraulic turbines are used to extract energy from water as the water velocity increases due to a change in head or kinetic energy at the expense of the potential energy as the water flows from a higher elevation to a lower elevation. The fluid velocity tangential component contributes to the rotation of the rotor in a turbomachine.

**Turbo™** A booster pump manufactured by Pump Engineering, Inc. for reverse osmosis units.

**TURBO™** A floating surface aerator manufactured by Aeration Industries, Inc.

**TurboBlade™** A low-speed mixer manufactured by Eimco Process Equipment Co.

**TURBO-Dryer®** A sludge dryer manufactured by Wyssmont Co., Inc.

**Turbofill™** Media manufactured by Diversified Remediation Controls, Inc. for packing air strippers.

**TurboFlow™** Equipment manufactured by Invincible AirFlow Systems, Inc. for use in soil remediation.

**turbopump** A pumping device powered by a turbine. *See also* turbine pump.

**Turbo-Scour** A package sand filter manufactured by Smith & Loveless, Inc., including a mechanism for continuous cleaning.

**Turboshredder** A cutting device manufactured for Homa Pumps, Inc. for grinder pumps.

**Turbostripper™** An air stripper of Diversified Remediation Controls, Inc.

**Turbotron®** A blower with rotary lobes manufactured by Lamson Corp.

**Turbulator** A device manufactured by Walker Process Equipment Co. for rapid mixing.

**turbulence** Irregular fluctuations in fluid motion.

**turbulent diffusion** Mass transfer is called eddy diffusion or turbulent diffusion when it is caused by turbulence without flow. *See also* diffusion and dispersion.

**turbulent flow** The opposite of laminar flow, i.e., the unsteady, random motion of fluid particles, with mixing between adjacent layers and predominance of higher velocities and energy head losses. The Reynolds number ($R_e$) determines whether flow is laminar or turbulent. In most cases water or wastewater flow in conduits is turbulent. For turbulent flow of incompressible fluids like water, the friction factor ($f$) can be determined from the Moody diagram as a function of $R_e$ and the relative pipe roughness. Also called eddy flow or sinuous flow.

**turbulent range** The range of Reynolds numbers ($R_e$) within which flow is turbulent; the drag coefficient ($C_D$), a function of the Reynolds number, affects the terminal settling velocity of particles in water and wastewater treatment (Droste, 1997):

$$R_e > 1000 \quad \text{and} \quad 0.34 < C_D < 0.40 \qquad \text{(T-47)}$$

*See also* Stokes range and transition range.

**turbulent transport** One of the mechanisms that cause relative motion and collisions between particles in a suspension, thereby promoting flocculation: turbulent flow gives rise to eddies and velocity gradients that cause relative motion of entrained particles and flocculation. *See also* Kolmogoroff microscale, flocculation rate constant, Brownian diffusion, orthokinetic flocculation, flocculation rate correction factor, differential settling, G value concept.

**turndown** The ratio of the maximum to the minimum capacity of a unit or a process.

**turndown ratio** *See* burner turndown ratio.

**turnkey construction** Construction that uses the turnkey procurement approach.

**turnkey procurement approach** A procurement process in which a single entity, such as a consultant or a contractor, is responsible for the design, construction, start-up, and, sometimes, financing of a project and then turns over to the owner the "keys," i.e., the responsibility for operation and maintenance. Other procurement approaches include architect/engineer, design–build–operate, privatization, and a combination thereof.

**turnkey project** or **turnkey system** (1) A project that is implemented, or a system that is installed, on a turnkey procurement approach; a general contractor assembles components from various vendors and assumes responsibility for their performance, but the owner takes over for operation and maintenance. Performance specifications usually define a turnkey project, with a reduced role for the consulting engineer. (2) A project that is operational immediately after installation.

**turnover** The phenomenon occurring in lakes in the spring (spring turnover) when temperatures rise above the freezing point and surface ice melts into higher-density water and tends to sink, causing the turnover and the recirculation of materials. In the fall, density also rises at the surface, and the water becomes colder, causing a similar turnover. Turnover may stir up nutrients and anoxic water, resulting in an increase in some water characteristics (temperature, color, sediment, turbidity, nutrients). Also called overturn or overturning. *See also* destratification.

**turnover ratio** The ratio of respiration to biomass. *See also* Schrodinger ratio.

**turnover time** The time required to recirculate the entire volume of a reactor, a sludge digester, or any other vessel or basin. It is the total volume of the vessel divided by the flow through it. For example, a common turnover time for digesters is 30–45 minutes.

**tuyer** Same as tuyère.

**tuyère** An orifice through which air is injected into a fluidized-bed incinerator to fluidize the bed; also used in a blast furnace, forge, cupola, or similar devices to facilitate combustion. Also spelled tuyer.

**TVS** Acronym of total volatile solids.

**twin dosing tanks** Two dosing tanks of equal capacity, equipped for alternate use.

**twin-pit pour-flush toilet** A pour-flush latrine has a U-pipe filled with water serving as a seal below the seat or squatting plate to block flies and odors. A twin-pit pour-flush latrine has two pits connected to the U-pipe and used alternately.

**twin-pit VIP latrine** A VIP latrine that is built over two adjacent pits, that are used alternately. One pit is used until it is full and then it is closed. The second pit is used the same way and then the first pit is emptied and put to use.

**two-bed** A water treatment device consisting of a cation-exchange and anion-exchange tanks operating in series.

**two-film theory** A theory of gas transfer between a liquid phase and a gas phase. Gas transfer is assumed to occur by molecular diffusion within a gas–liquid interface consisting of a gas film and a liquid film. *See* Figure T-15. The rate of mass transfer by diffusion ($dm/dt$) is:

$$dm/dt = KA(C_i - C_l) \qquad \text{(T-48)}$$

where $K$ = mass transfer coefficient, $A$ = interfacial area, $C_i$ = gas concentration in the liquid at the interface, and $C_l$ = gas concentration in the liquid bulk phase.

**two-main system** *See* dual distribution system.

**two-pass nanofiltration** A brackish water or seawater desalination technique that uses two nanofiltration units in series, at respective pressures of 525 and 250 psi. It is classified as an emerging concentrate disposal and minimization method.

**two-phase anaerobic digestion** An improvement over single-stage anaerobic digestion by carrying out the process in two reactors in series. It takes advantage of the two-stage nature of anaerobic metabolism (hydrolysis–acidogenesis and methanogenesis) to optimize conditions (e.g., pH) for separate groups of microorganisms. *See* the following two-phase variations: acid/gas phased digestion, staged mesophilic digestion, staged thermophilic digestion, temperature-phased digestion. *See also* two-stage digestion.

**two-phase reverse osmosis—biological** A brackish water or seawater desalination technique that uses a biological process and a second reverse osmosis unit to enhance recovery. A biological reactor reduces the sulfate to sulfide, which is subsequently air-stripped. The system includes a gravity thickener and a filter for solids removal. It is classified as an emerging concentrate disposal and minimization method.

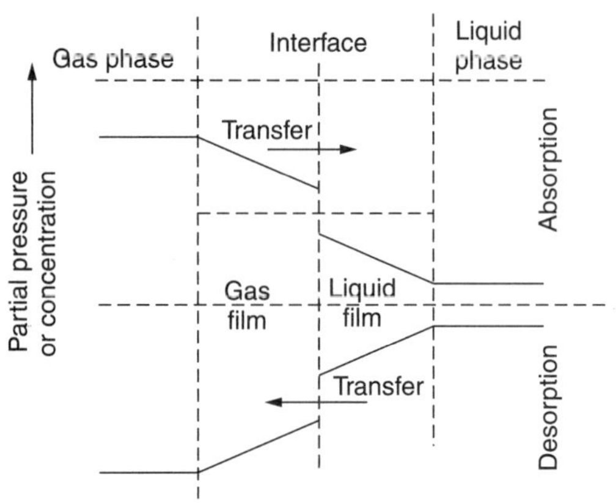


**Figure T-15.** Two-film theory.

**two-phase reverse osmosis—physical–chemical** A brackish water or seawater desalination technique that uses an intermediate physical–chemical step and a second reverse osmosis unit to enhance recovery. The intermediate step includes chemical treatment, precipitation, and filtration. It is classified as an emerging concentrate disposal and minimization method.

**two-port nonclog radial pump** A centrifugal pump having two vanes in its impeller and used for raw wastewater, sludge, and effluent.

**two-port radial clean water pump** A centrifugal pump having two vanes in its impeller and used for flushing cleaning water, spraying water to disperse foam, and providing water for solutions.

**two-sludge BNR process** *See* two-stage BNR process.

**two-sludge process** *See* two-stage activated sludge process and two-stage BNR process.

**two-sludge system** *See* two-stage activated sludge process and two-stage BNR process.

**two-stage A/O™ process** A proprietary biological treatment process developed by Kruger, Inc. (patented and marketed by Air Products and Chemicals, Inc.) for BOD removal and enhanced phosphorus uptake. *See* A/O™ process for detail.

**two-stage activated sludge process** A variation of the activated sludge process designed with two stages, each stage consisting of an aeration tank and a clarifier, a high-rate first stage for BOD removal and a second stage for nitrification. Also called two-sludge process or two-sludge system.

**two-stage BNR process** A variation of the activated sludge process designed with two stages for biological nitrogen removal. The first stage is a nitrification tank and a clarifier with solids recycle; the second stage consists of a denitrification tank, also with a clarifier and sludge recycle. Methanol is added ahead of denitrification as a carbon source. Also called two-sludge BNR process. *See also* two-sludge activated sludge process, simultaneous nitrification–denitrification process.

**two-stage digestion** A sludge digestion process using a high-rate, heated digester to accomplish the biological decomposition of organic matter followed by another tank for solids–liquid separation and storage. *See also* two-stage anaerobic digestion.

**two-stage filtration** (1) A water treatment process that includes coagulation, a first-stage coarse-medium bed serving as a contact flocculator and a roughing filter, followed by a conventional multimedia downflow filter with optional addition of chemicals. This process is usually applied to source water of poor quality, particularly in package plants. (2) An advanced wastewater treatment process that includes two deep-bed filters in series: a large-size sand filter followed by a smaller-sized sand filter. *See* DSS Environmental two-stage filter.

**two-stage recarbonation** The addition of carbon dioxide ($CO_2$) at two points in the two-stage lime softening process: ahead of the second rapid mix unit and then ahead of the filters; often used with excess lime to soften high-magnesium waters.

**two-stage process** A wastewater treatment method that incorporates two processes, sometimes different, in series; also called a dual process. Most plants commonly using dual processes combine a fixed-film unit with a suspended-growth reactor. *See also* combined filtration–aeration process and dual biological treatment system.

**two-stage softening** A water treatment process that uses lime to precipitate hardness. Lime and other chemicals are added to the raw water, which then passes through two clarifiers in series, a recarbonation unit, and a filter. In the recarbonation step, carbon dioxide ($CO_2$) gas is added to the water to reduce the pH of the solution. *See also* lime softening, single-stage lime softening, and split softening.

**two-stage trickling filter system** A system of two trickling filters in series, with an intermediate clarifier and a final clarifier; commonly used to treat high-strength wastes or for nitrification. The intermediate clarifier is sometimes omitted.

**two-step windrow process** A variation of the windrow composting method. To facilitate initial aeration and disinfection, the composting operation proceeds in small windrows for 2–3 weeks, reaching a temperature of 50–55°C. Then two or more small adjacent windrows are combined into one large windrow and mixed for continued composting for another 3 weeks or longer.

**two-story septic tank** *See* Imhoff tank, Travis tank.

**two-tray clarifier** A wastewater sedimentation unit consisting of two clarifiers in series, one on top of the other. This is usually an arrangement to save space, as compared to a one-story basin.

**TWS** Acronym of traveling water screen.

**TxPro™** A solids transmission device manufactured by BTG, Inc.

**Tyndall effect** The characteristic property of colloidal particles scattering white light as a light blue hue observed at a right angle, resulting from their being about the same size as the wavelength of light and from the fact that the dispersed phase has a different refractive index from that of the dispersion medium.

**tynes** British spelling of tines.

**type 021N bacteria** A type of filamentous organism that causes activated sludge bulking when the food-to-microorganism ratio (F:M) is low.

**type 1 resin** A strong base polystyrene/divinylbenzene anion exchange resin with a trimethylamine [$—N(CH_3)_3$] as exchange site.

**type 1701 bacteria** A type of filamentous organism that causes activated sludge bulking when dissolved oxygen concentration is low.

**type I sedimentation** The settling of discrete, nonflocculent particles, as separate, unhindered units. *See* sedimentation type I for more detail.

**type I settling** The settling of discrete, nonflocculent particles, as separate, unhindered units. *See* sedimentation type I for more detail.

**type 1 slow-rate system** A slow-rate wastewater treatment system that requires only a minimum of preapplication treatment to protect public health and avoid nuisance conditions; e.g., screening and primary sedimentation. Soil permeability or allowable loading rate is the limiting design factor. Also called crop irrigation system.

**type 2 resin** A strong base polystyrene/divinylbenzene anion exchange resin with a dimethylethanol [$—N(CH_3)_2(CH_2CH_2OH)$] as exchange site.

**type 2 slow-rate system** A slow-rate wastewater treatment system designed for reuse and handling flexibility, thus requiring secondary or even advanced levels of preapplication treatment. Also called slow infiltration system.

**type II sedimentation** *See* sedimentation type II.

**type II settling** *See* sedimentation type II.

**type III sedimentation** *See* sedimentation type III.

**type III settling** *See* sedimentation type III.

**type IV settling** *See* sedimentation type IV.

***Typha*** Cattail, an emergent plant commonly found in natural wetlands.

**typhoid** Same as typhoid fever.

**typhoid fever** A highly infectious disease of the gastrointestinal tract, caused by the waterborne bacterium *Salmonella typhi*. Its symptoms may include headache, diarrhea, muscular weakness, nosebleed, etc. It was widespread until the advent of adequate drinking water treatment and wastewater disposal practices (e.g., water disinfection by chlorine and chlorine compounds in the United States starting early in the 19th century), but it still persists in areas where such facilities are not available. Also called enteric fever, or simply typhoid, it is transmitted via the fecal–oral route and has worldwide distribution. *See also* paratyphoid fever.

**Typhoid Mary** Ms. Mary Mallon, a cook in residences and hospitals in New York State, and a typhoid carrier around 1930. She reportedly exposed thousands of victims to the disease without showing the symptoms herself.

**typhus** A disease associated with the lack of water for personal hygiene or with poor refuse disposal practices. *See* louse-borne typhus and endemic typhus.

**typing** Any epidemiological method of grouping organisms, e.g., antibiogram, bacteriophage typing, biotyping, serotyping.

**Tysul** Hydrogen peroxide produced by E. I. du Pont de Nemours, Inc.

# U

**u** *See* atomic mass unit.

**U-235** Abbreviation or symbol of uranium-235.

**U-237** Abbreviation or symbol of uranium-237.

**U-238** Abbreviation or symbol of uranium-238.

**UASB reactor** *See* upflow anaerobic sludge blanket reactor.

**UBOD** Acronym of ultimate biochemical oxygen demand.

**UC** Acronym of uniformity coefficient.

**UCBF** Acronym of upflow continuous backwash filter.

**UCT process** *See* University of Cape Town process.

**UDF** *See* unidirectional flushing.

**UF** Acronym of ultrafiltration.

**UHR Filter** A dual-media sand filter manufactured by Idreco USA, Ltd.

**UIC** Acronym of Underground Injection Control.

***Ulothrix*** A genus of green algae abundant in waste stabilization ponds and in trickling filters; resistant to anaerobiosis and to temperature extremes. They do not participate directly in waste degradation, but they contribute oxygen during daylight hours. They may also cause clogging of the filter surface and odors.

**Ultima** A nonmetallic device manufactured by Budd Co. for sludge collection.

**ultimate adsorptive capacity** The ultimate amount of a solute that can be adsorbed by a unit weight of adsorbent; sometimes simply called adsorptive capacity. For example, granular activated carbon can adsorb an approximate maximum (or saturation value) of 100 mg of phenol per gram. This can be determined from an adsorption isotherm such as Langmuir's equation. *See also* adsorption rate.

**ultimate analysis** A technique used in the laboratory to determine approximate percentages of the predominant elements of carbohydrates, fats, and proteins that make up the volatile matter of sludge (carbon, oxygen, hydrogen, and nitrogen).

**ultimate biochemical oxygen demand (ultimate BOD or BODu)** The quantity of oxygen needed to meet the total (carbonaceous and nitrogenous) demand. It used to denote the complete first-stage BOD.

**ultimate biodegradability** The breakdown of an organic compound to $CO_2$, water, the oxides or mineral salts of other elements, and/or to products associated with normal metabolic processes of microorganisms (EPA-40CFR796.3100-ii).

**ultimate BOD (UBOD, $BOD_L$, $BOD_u$)** *See* ultimate biochemical oxygen demand.

**ultimate carbonaceous biochemical oxygen demand** A measure of the amount of oxygen required to stabilize a waste.

**ultimate carbonaceous BOD** *See* ultimate carbonaceous biochemical oxygen demand.

**ultimate disposal** The final release of residuals onto the environment after processing and treatment. *See also* codisposal, land application, landfilling, ocean discharge.

**ultracentrifugation** The centrifugation of sols and solutions at very high speeds that create forces many times the force of gravity to facilitate the separation of proteins, viruses, and other molecules from a fluid.

**Ultracept®** A device manufactured by Jay R. Smith Mfg. Co. for separating oil from water.

**UltraChem™** A membrane manufactured by Membrex, Inc. for the treatment of oily wastes.

**ultrafiltration (UF)** A low-pressure membrane filtration process, similar in operation to reverse osmosis, used to separate from water finely divided and colloidal particles of less than 0.01 up to 0.1 micron and dissolved solutes down to a molecular weight cutoff of 1000 daltons from a liquid stream. In wastewater treatment, at 8 gallons per square foot per day, it removes practically 100% of a treated effluent's suspended solids, BOD, COD, and TOC. *See also* microfiltration, nanofiltration.

**UltraFlex®** A containment liner manufactured in LPDE by SLT North America, Inc.

**Ultraflow** A programmable instrument manufactured by Monitek Technologies, Inc. for flow measurements in open channels.

**ultra-low-flush toilet** A toilet that uses 1.6 gallons per flush (gpf) or less, as allowed by the 1992 U.S. Energy Policy Act. *See* low-flow toilet for detail.

**ultramicrobacteria** Marine or soil bacteria that divide and shrink to a fraction of their size (less than 0.3 μm) in reaction to starvation or lack of nutrients.

**Ultramix™** An efficient motor manufactured by Lightnin for use in mixing devices.

**ultrapure water** A high-quality water with low solids content and a high specific resistance (> 1 megohm-cm) to meet the requirements of some industrial applications (semiconductor, medical, etc.). *See also* reagent-grade water, 18-megohm water, and pharmaceutical-grade water.

**Ultra-Pyrolysis System®** A closed-loop, mobile batch pyrolysis apparatus developed by Wangtec, Inc. and using a plasma reactor to produce temperatures as high as 1700°C for the destruction of liquid wastes from storage drums. Also called UPS-2000.

**UltraScrub** A wet scrubber manufactured by Tri-Mer Corp.

**Ultrasep** A membrane manufactured by Ionics, Inc. for use in pressure-driven processes.

**ultrasonation** The use of ultrasonic waves at frequencies of 20 to 400 kHz and a detention time of 60 minutes to sterilize water; it provides no residual for the distribution system. Disinfection is feasible at lower retention times.

**ultrasonic** Pertaining to sound or acoustic waves with frequencies above 20 kHz, the approximate upper limit of human hearing.

**ultrasonic cleaning** A technique that uses high-frequency sound waves to improve the performance of aqueous cleaners. The sound waves generate zones of high and low pressures, create microscopic vacuum bubbles in the liquid, and loosen contaminants. Ultrasonic cleaning is used in industrial cleaning and degreasing processes. Also called cavitation process.

**ultrasonic disinfection** The inactivation of pathogens using ultrasound. *See* acoustic cavitation for detail.

**ultrasonic Doppler velocity meter** *See* Doppler current meter.

**ultrasonic meters** *See* Doppler current meter and ultrasonic transit time velocity meter.

**ultrasonic nebulizer** A device used to efficiently introduce samples to atomic spectrometers as aerosol particles.

**ultrasonic transit time velocity meter** A device installed in pipelines to measure the flow of liquids. It consists of two ultrasonic, diagonally opposed transducers mounted to a precalibrated flow tube. The device measures the average fluid velocity as a function of the time difference of the pressure pulses transmitted against and with the direction of flow. *See also* Doppler current meter.

**ultrasound inactivation** The inactivation of pathogens using ultrasound. *See* acoustic cavitation for detail.

**UltraStrip** An air stripper manufactured by GeoPure Continental.

**Ultratest** A polyelectrolyte produced by Ashland Chemical, Drew Division to improve solids separation.

**ultraviolet** Beyond the violet in the spectrum; pertaining to light of a wavelength shorter than 4000 angstroms, or to any device that produces or uses such light.

**ultraviolet A (UVA, UV-A)** Ultraviolet radiation in the waveband of 0.32–0.40 μm, which can tan or cause redness to the human skin.

**ultraviolet absorbance at 254 nanometers (UV-254 or $UV_{254}$)** (1) A measure of water's ultraviolet absorption at a wavelength of 254 nanometers; an excellent surrogate for total organic carbon (TOC). (2) A method used to determine the presence in water of organic and aromatic compounds that absorb ultraviolet radiation.

**ultraviolet B (UVB)** Ultraviolet radiation in the wavelength band of 0.29–0.32 μm. A type of sunlight associated with sunburn, skin cancer, eye cataracts, and damage to the environment. The ozone in the stratosphere, high above the Earth, filters out UVB rays and keeps them from reaching the Earth; thinning of the ozone layer in the stratosphere results in increased amounts of UVB reaching the Earth.

**ultraviolet C (UVC, UV-C)** Ultraviolet radiation in the wavelength band of 0.20–0.29 μm, which can cause chromosome mutations; a source of low-pressure ultraviolet (UV) light at 254 nm or short-wave or germicidal UV light at less than 280 nm.

**ultraviolet demand** *See* ultraviolet light demand equivalent and ultraviolet dosage.

**ultraviolet disinfection** *See* ultraviolet light disinfection.

**ultraviolet dosage** The amount of ultraviolet irradiation delivered to disinfect water or wastewater; the product of UV intensity and contact time, in watt-seconds per square centimeter.

**ultraviolet dose** *See* ultraviolet light dose.

**ultraviolet inactivation** *See* ultraviolet light inactivation.

**ultraviolet (UV) irradiation** Radiation by ultraviolet light. *See* ultraviolet light disinfection, ultraviolet light irradiation, ultraviolet radiation.

**ultraviolet (UV) light** Light rays with an approximate wavelength range of 10–400 nanometers, beyond the violet region of the visible spectrum. UV light is an effective disinfectant of bacteria and viruses. It is also used to produce hydroxyl radicals (OH) in advanced oxidation processes. *See* actinometer and pyrheliometer for the measurement of UV light intensity.

**ultraviolet light absorbance** *See* transmittance, UV-254 and SUVA-254.

**ultraviolet-light-absorbing organics** Organic constituents of wastewater that strongly absorb ultraviolet radiation, e.g., humic substances, lignin, tanning, and aromatic compounds. Hence, ultraviolet light absorption is sometimes used as a surrogate measure for organics.

**ultraviolet-light-demand equivalent** The amount of ultraviolet radiation required to react with dissolved and suspended materials (particularly proteins, humic materials, and iron compounds) that may shield microorganisms. This is similar to chlorine demand. Also called ultraviolet demand. *See also* ultraviolet dosage.

**ultraviolet (UV) light disinfection** The use of ultraviolet light to inactivate microorganisms in water or wastewater, including oocysts of *Cryptosporidium parvum* and cysts of *Giardia lamblia.* In kinetic models, incident light intensity is applied the same way as the concentration of chlorine or other chemical disinfectant. UV light promotes the formation of pyrimidine dimers within the nucleic acids of irradiated cells and disrupts their metabolism. The peak absorption wavelength for DNA is between 250 and 265 nanometers; low-pressure mercury arc lamps emit 85% of their output at about 254 nm and are effective in disinfection. UV light disinfection needs very low retention times, does not use chemicals, but does not leave a residual. Effective irradiation requires a liquid that has a low solids content. The process is used in industrial applications, small water systems, and to treat secondary treatment effluents.

**ultraviolet-light-disinfection model** An equation proposed to describe the inactivation of coliform bacteria by ultraviolet light:

$$N(t) = N_D(0)\, e^{-kD} + N_P(0)\,(1 - e^{-kD})/(kD) \quad \text{(U-01)}$$

where $N(t)$ = number of surviving bacteria at time $t$, $N_D(0)$ = number of dispersed bacteria before the application of UV light at time $t = 0$, $N_P(0)$ = number of particles containing at least one bacterium at time $t = 0$, $k$ = inactivation rate coefficient, $cm^2/mJ$; and $D$ = ultraviolet light dose, $mJ/cm^2$. *See also* ultraviolet light inactivation.

**ultraviolet light disinfection system** A water disinfection apparatus whose principal components are ultraviolet lamps in quartz sleeves on a supporting structure, ballasts providing power to the lamps, a reactor, a cleaning device, and a control system. A validation protocol is required to obtain regulatory approval; it relates to monitoring data and dose delivery.

**ultraviolet (UV) light dose** or UV dose The product ($D$) of the incident irradiance or UV intensity ($I$) expressed as $mW/cm^2$ and the time of exposure in seconds ($t$):

$$D = It \quad \text{(U-02)}$$

The unit $mW \cdot s/cm^2$ is equivalent to $mJ/cm^2$; e.g., 3–10 $mJ/cm^2$ for vegetative bacteria and 20–50 $mJ/cm^2$ for enteric viruses. The UV dose, a measure of the UV energy delivered, is analogous to the

CT product and is a basis to measure the effectiveness of ultraviolet light disinfection.

**ultraviolet light dose–response** or **UV dose–response** A measure of a pathogen's log inactivation as a function of the UV dose, as determined in a laboratory using a collimated-beam apparatus.

**ultraviolet light–hydrogen peroxide** A combination used in advanced oxidation processes for the treatment of synthetic organics, xenobiotics, and endocrine disruptors.

**ultraviolet light inactivation** The rate of inactivation of microorganisms by ultraviolet light, commonly assumed to follow a first-order reaction:

$$N = N_0 e^{-kD} \quad \text{(U-03)}$$

where $N$ = concentration of microorganisms at time $t$, $N_0$ = initial concentration of microorganisms (at time $t$ = 0), $k$ = inactivation coefficient, and $D$ = radiation dose = product of UV intensity ($E$) and detention time ($t$) in the reactor. *See also* ultraviolet light model.

**ultraviolet light irradiation** The use of ultraviolet light radiation for water or wastewater disinfection, or for other purposes; sometimes used synonymously with ultraviolet radiation.

**ultraviolet light lamp** or **UV lamp** A mercury arc lamp housed in a quartz sleeve; it may operate under low or medium pressures.

**ultraviolet light radiation disinfection mechanism** Ultraviolet light radiation destroys microorganisms by penetrating their cell walls and causing their death or preventing their replication.

**ultraviolet light radiation production** Ultraviolet light is produced when mercury vapor lamps are excited by an electric arc. *See* low-pressure, low-intensity; low-pressure, high-intensity; and medium-pressure, high-intensity lamps.

**ultraviolet light spectroscopy** An analytical technique used to determine the electronic structure of a molecule based on its absorption of ultraviolet radiation.

**ultraviolet oxidation (UVOX)** An advanced oxidation process that is effective in the removal of MTBE, xenobiotics, and drug residuals. It uses UV light, alone or in combination with other oxidants, to irradiate the target compounds.

**ultraviolet photooxidation** Same as ultraviolet oxidation.

**ultraviolet radiation** (1) Electromagnetic radiation in the wavelength range of 10 to 400 nm, generated by a current flowing between electrodes in ionized mercury vapor. It is a component of the sunlight spectrum and is not as penetrating as ionizing radiation. (2) A proven method of water and wastewater disinfection. It inactivates common waterborne bacteria as well as some viruses and chlorine-resistant pathogens (*Cryptosporidium parvum* and *Giardia lamblia*), without contributing significantly to the formation of disinfection by-products. Sometimes called ultraviolet irradiation. *See* ultraviolet light disinfection, germicidal unit, coefficient of attenuation or extinction, sunlight disinfection. *See also* Beer–Lambert law, photoreactivation, ultraviolet irradiation.

**ultraviolet radiation disinfection** *See* ultraviolet light disinfection.

**ultraviolet radiation treatment** *See* ultraviolet light disinfection.

**ultraviolet (UV) ray** A light ray beyond the violet of the spectrum; radiation from the sun that can be useful or potentially harmful. UV rays from one part of the spectrum (UV-A) enhance plant life and are useful in some medical and dental procedures; UV rays from other parts of the spectrum (UV-B) can cause skin cancer or other tissue damage. The ozone layer in the atmosphere partly shields us from ultraviolet rays reaching the Earth's surface. *See* ultraviolet radiation.

**ultraviolet ray microscope** *See* fluorescence microscope.

**ultraviolet spectroscopy** *See* ultraviolet light spectroscopy.

**Ultrex®** A tubular membrane diffuser manufactured by Aeration Research Co.

**Ultrion®** Cationic coagulants produced by Nalco Chemical Co.

**Ultrox®** A process developed by Ultrox International for the destruction of toxic organic materials using ultraviolet light and oxidizing agents.

**UMB** Acronym of ultramicrobacteria.

**UMERGE** A simple mixing model developed for the USEPA and used to establish effluent limits.

**unacceptable adverse effect** Impact on an aquatic or wetland ecosystem that is likely to result in significant degradation of municipal surface or ground water supplies or significant loss of or damage to fisheries, shellfishing, wildlife habitat, or recreational areas. In evaluating the unacceptability of such impacts, consideration should be given to the relevant portions of section 404(b)(1) of the Clean Water Act (EPA-40CFR231.2-e). *See also* unreasonable adverse effect on the environment.

**unaccounted flow** or **unaccounted-for water** The portion of water withdrawn from a supply source that is not assigned to a specific user. This includes all losses from the point of withdrawal to the customer meters, from leaks in the distribution lines, to unauthorized connections, and to malfunctioning meters or errors in meter readings.

**unaccounted-for water** *See* unaccounted flow.

**unburned lime** Calcium carbonate.

**uncertainty factor (UF)** (1) A number (equal to or greater than one) used to divide NOAEL or LOAEL values derived from measurements in animals or small groups of humans, in order to estimate a NOAEL value for the whole human population. (2) One of several, generally 10-fold, factors used in operationally deriving the reference dose (RfD) from experimental data. UFs are intended to account for (a) the variation in sensitivity among the members of the human population; (b) the uncertainty in extrapolating animal data to the case of humans; (c) the uncertainty in extrapolating from data obtained in a study that is of less-than-lifetime exposure; and (d) the uncertainty in using LOAEL data rather than NOAEL data. (3) *See* safety factor.

**uncharged hydroxyl radical (OH')** A very reactive species, a radical with an unpaired electron, formed by ozone ($O_3$). *See also* free radical.

**unconfined aquifer** An aquifer containing water that is not under pressure; the water level in a well through an unconfined aquifer is the same as the water table outside the well. Also called phreatic aquifer or water-table aquifer.

**unconsolidated formation** An earth formation that is not lithified, consisting of alluvium, soil, gravel, clay, and overburden.

**underbed** A layer of gravel at the base of a filter or softener tank.

**undercurrent** A current below the surface of a fluid body or below other currents.

**underdrain** (1) A drain to remove groundwater, stormwater or the drainage from water or wastewater structures; for example, the perforated plastic pipes installed at the bottom of an infiltration basin to collect and remove excess stormwater. *See also* groundwater drain, land drain. (2) In a granular media filter, a piping system that supports the bed, collects filtered water, and distributes backwash water. It consists of perforated pipes or split tiles laid in coarse stones or gravel, precast tile or concrete blocks, plastic nozzles, porous plates, etc., Also called filter bottom, filter underdrain, underdrainage, or underdrainage system. *See* fabricated self-supporting underdrain, false-floor underdrain with nozzles, manifold-lateral underdrain. (3) Similarly, the underdrainage system that supports a sand filter or the graded gravel of a biological bed. (4) The underdrainage system of a trickling filter, comprising a subfloor, filter blocks, and a drainage channel.

**underdrainage** Same as underdrain (2), (3), or (4).

**underdrainage system** Same as underdrain (2), (3), or (4).

**underflow** (1) Movement of water through subsurface materials. (2) The concentrated solids removed from the bottom of a tank or basin, or the washwater distributed throughout a filter for backwash.

**underflow flow** Flow of water at the bottom of an impoundment or tank, the moving water being denser than the rest.

**underflow flux** The product ($N_u$) of the underflow velocity ($U_b$) and the solids concentration ($C$) of a suspension. *See also* solids flux.

**underflow operating line** In state-point analysis, the underflow operating line represents the negative slope of the clarifier underflow velocity.

**underground drinking water source** *See* underground sources of drinking water.

**underground injection** The subsurface emplacement of fluids (e.g., brines and other wastes) through a bored, drilled, or driven well, or a dug well, where the depth of the dug well is greater than the largest surface dimension (EPA-40CFR260.10). *See also* injection well, deep-well disposal.

**underground injection control (UIC)** (1) The program under the Safe Drinking Water Act that regulates the use of wells to pump fluids into the ground. (2) A means, procedure, or program of regulating waste disposal by underground injection.

**Underground Injection Control Program** A program of the USEPA, under the Safe Drinking Water Act, to protect groundwater sources against contamination by waste injection.

**underground injection well** A well used for the subsurface emplacement of fluids. *See* injection well for detail.

**underground source of drinking water** An aquifer currently used as a source of drinking water or capable of supplying a public water system. It has a total dissolved solids content of 10,000 mg/L or less, and is not an "exempted aquifer."

**underground storage tank (UST)** A tank located at least partially (10%) underground and designed to hold gasoline, other petroleum products, chemicals, or regulated substances.

**underground utilities** Buried pipes and conduits that convey water, wastewater, stormwater, gas, electricity, communications, and similar services to customers.

**underground water** Water in the lithosphere (the solid portion of the Earth, as distinguished from the hydrosphere and the atmosphere). *See* subsurface water, groundwater.

**underground water source protection program** A program for the adoption and enforcement of regulations under section 1421 of the Safe Drinking Water Act and for keeping records and making required reports.

**undertow** A current below the surface, seaward or along the beach when waves are breaking on the shore.

**undigested sludge** Settled sludge that has been removed recently from a sedimentation unit before any processing and before significant decomposition. Also called raw sludge.

**unidirectional flushing (UDF)** Flushing of water distribution piping in one direction and at high velocities to scour and remove deposits and debris.

**UNEP** The United Nations Environment Programme, a UN agency with water quality programs. Its mission is "To provide leadership and encourage partnership in caring for the environment by inspiring, informing, and enabling nations and peoples to improve their quality of life without compromising that of future generations."

**unfavorable isotherm** *See* ion-exchange isotherm.

**UNICEF** The United Nations Children Fund, a UN agency that carries out water supply and sanitation projects.

**unidentate ligand** An anion or molecule that has only one site that bonds to a metal ion to form a complex, e.g., the cyanide ion ($CN^-$). Also called monodentate ligand. *See also* multidentate ligand.

**Uni-Dose™** An electronic metering pump manufactured by Liquid Metronics, Inc.

**Uniflap** A urethane flap valve manufactured by Ashbrook Corp.

**Uniflow** A settling tank manufactured by FMC Corp., including a chain and flight mechanism to collect sludge.

**uniform block rate** Same as uniform rate.

**uniform corrosion** A condition in which material loss through corrosion is about the same over an entire surface, e.g., when anodic and cathodic areas are microscopic and close to each other. *See* cuprosolvency, plumbosolvency.

**uniform-flow filtration** The operation of filtration units to maintain constant flow rates despite variable head losses and variable heads. *See also* declining-rate filtration.

**uniformity coefficient (UC)** A number that represents the size composition of a granular material or how well the material is graded. It is computed as the ratio of two grain sizes:

$$UC = d_{60}/d_{10} \qquad \text{(U-04)}$$

where $d_{60}$ is the diameter of a grain or grain size for which 60% of the material by weight has a finer grain, i.e., 60% of the material will pass through a sieve of that size; similarly, 10% of the material will pass through a sieve of size $d_{10}$. If media sizes follow a log-normal distribution,

$$d_{90} = d_{10}\,(UC)^{1.67} \qquad \text{(U-05)}$$

The uniformity coefficient represents also the degree of variation of the material; the closer it is to unity, the more uniform the material. Fair et al. (1971) suggest calling this parameter the nonuniformity coefficient because it increases with nonuniformity instead of uniformity. UC is one of two parameters used to specify granular media for filtration; the other is effective size, which is the 10th percentile diameter or $d_{10}$ (Allen Hazen recommended the two parameters in 1892). A low UC reduces backwash flowrates. UCs less than 1.7 are recommended, but 1.5 is a practical lowest limit of commercially available materials. On the other hand, large uniformity coefficients (about 1.9) in granular activated carbon promote stratification during backwashing. *See also* manufacturer's rating, mechanical amalysis.

**uniform particle size** The particle size of ion exchange and filtration media based on U.S. Mesh Standards.

**uniform rate** A rate schedule that charges a constant price per unit consumption. Also called flat fee or uniform block rate.

**UniMix** A flocculation unit manufactured by Walker Process Equipment Co.

**union** A threaded device used to connect pipe sections.

**un-ionized ammonia** One of the two forms of ammonia in aqueous solutions, $[NH_3(aq)]$, the other being the ammonium ion ($NH_4^+$). The un-ionized form is toxic, and the ionized form nontoxic, to fish and other aquatic organisms.

**Uni-Pac** Water treatment equipment manufactured by Cochrane Environmental Systems.

**UniPro™** A pump manufactured by Union Pump Co. for use in reverse osmosis units.

**Unipure** A package waste treatment plant manufactured by Unocal Corp. for the removal of heavy metals.

**Uni-Scour** A dual-media filter manufactured by Smith & Loveless, Inc, including a backwash storage unit.

**Unisweep™** An apparatus manufactured by Roberts Filter for washing filter surfaces.

**Unisystem®** A wastewater treatment apparatus manufactured by Aeration Industries, Inc.; uses the activated sludge process.

**unit** An assemblage of membrane modules installed in parallel or in series and operated as a single unit. Also called rack, train.

**Unitank®** A wastewater treatment apparatus manufactured by Seghers-Dinamec, Inc.

**unit application efficiency** A coefficient ($E_u$) used to estimate the total irrigation water requirement ($D$) by adjusting the net requirement ($R$) for distribution losses and inefficiencies.

**Unitech** A series of products manufactured by Graver Co. for use in vaporators and crystallizers.

**United States Environmental Protection Agency (USEPA)** An agency of the United States government created in 1970, primarily responsible for coordinating environmental protection activities and enforcing federal environmental laws. It also conducts research in the field, issues helpful guidelines, and develops tools such as the SWMM and other models.

**United States Pharmacopeia** (or **U.S. Pharmacopoeia)** An official government publication that lists drugs, their formulas, methods of preparation, etc., and related products. *See also* pharmaceutical-grade water.

**United States Public Health Service (USPHS)** An agency within the U.S. Department of Health and Human Services that is responsible for the protection of public health; e.g., it sponsors and administers programs for the prevention and control of waterborne and other diseases.

**unit filter backwash volume** The product of backwash rate and time, a parameter used to determine the design volume of a filter backwash system.

**unit filter run volume (URFV)** The volume of water produced during a filter run, an expression of filtration efficiency in volume per unit surface area.

**Unitherm™** An apparatus manufactured by Regenerative Environmental Equipment Co. for the thermal oxidation of volatile organic compounds.

**unit isotherm** *See* ion-exchange isotherm.

**unitless Henry's law** A form of Henry's law written with a unitless constant ($H_u$) as the ratio of Henry's law constant ($H$) expressed in atm · m³/mole to the product of the universal gas constant ($R$ = 0.000082057 atm · m³/mole · K) by the temperature ($T$, in degree K = 273.15 + °C):

$$H_u = H/(RT) \qquad \text{(U-06)}$$

**unit loading factors** Per capita values used for various wastewater parameters in the planning and design of treatment facilities. *See* Table U-01 for some typical values.

**Table U-01.** Unit loading factors (typical, with ground-up kitchen waste)

| Constituent | Grams/ capita/day | Constituent | Grams/ capita/day |
|---|---|---|---|
| $BOD_5$ | 100 | TKN | 14.3 |
| COD | 220 | Organic P | 1.3 |
| TSS | 110 | Inorganic P | 2.2 |
| $NH_3$-N | 8.4 | Total P | 3.5 |
| Organic N | 5.9 | FOG | 34 |

**unit operation** (1) A chemical engineering concept, first introduced by Arthur D. Little in 1915, such that any chemical process may be resolved into a limited number of basic unit operations. This approach has been applied to water and wastewater treatment, in which most processes involve unit operations that are essentially phase transfers. (2) A method of water or wastewater treatment in which physical forces predominate, such as sedimentation or filtration, as compared to a unit process, such as activated sludge or disinfection, in which contaminant removal results from chemical or biological reactions. A treatment plant combines a number of operations and processes to achieve its objective. *See* Table U-02 for a list of unit operations.

**unit process** A method of water or wastewater treatment, such as activated sludge or disinfection, in which contaminant removal results from chemical or biological reactions, as compared to a unit operation, such as sedimentation or filtration, in which physical forces predominate. A treatment

**Table U-02.** Unit operations of water and wastewater treatment

| Operation | Equipment/Device |
|---|---|
| Adsorption | Adsorption column |
| Aeration | Aeration basin, aerator |
| Air stripping | Packed tower |
| Comminution | Comminutor |
| Filtration | Filter |
| Filtration (membrane) | Membrane filter |
| Flocculation | Flocculator |
| Flotation | Flotation tank, flotation thickener |
| Flow equalization | Equalization tank |
| Grinding | Screenings grinder |
| Maceration | Macerator |
| Mixing | Rapid mixer |
| Screening (coarse) | Bar rack |
| Screening (fine) | Fine screen |
| Screening (micro) | Microscreen |
| Sedimentation | Clarifier |
| Sedimentation (accelerated) | Vortex separator |
| Thickening | Thickener |

plant combines a number of operations and processes to achieve its objective. *See* Table U-03 for a list of biological unit processes and Table U-04 for chemical unit processes.

**unit process redundancy** In a wastewater treatment plant, unit process redundancy means that the plant effluent will not be adversely affected with one or more reactors out of service under average flow conditions. *See* redundancy for detail.

**Unitube** An apparatus manufactured by Envirex, Inc. for the removal of sludge from circular clarifiers.

**unit waste-loading factor** *See* unit loading factors.

**unit water use** The quantity of water used per unit of product.

**univalent** (1) Having a valence of one; monovalent. (2) Pertaining to a single or unpaired chromosome.

**Univer** A series of chemicals produced by Hach Co. for the determination of water hardness.

**UniversaLevel** A liquid level transmitter manufactured by Drexelbrook Engineering Co.

**universal gas constant** *See* universal gas law constant.

**universal gas law** The principle that the product of the pressure ($P$) and volume ($V$) of one gram molecule of an ideal gas equals the product of its absolute temperature ($T$, °K) and the universal gas constant ($R$). *See* ideal gas law for more detail.

**Table U-03.** Unit processes (biological) of wastewater treatment

| Biological process | Application |
|---|---|
| Activated sludge | CBOD removal, nitrification |
| Aerated lagoon | CBOD removal, nitrification |
| Aerobic digestion | CBOD removal, solids stabilization |
| Anaerobic bed | CBOD removal, denitrification, solids stabilization |
| Anaerobic contact | CBOD removal |
| Anaerobic digestion | Solids stabilization, pathogen destruction |
| Anoxic attached growth | Denitrification |
| Anoxic suspended growth | Denitrification |
| Lagoon (aerobic) | CBOD removal |
| Lagoon (anaerobic) | CBOD removal |
| Lagoon (facultative) | CBOD removal |
| Lagoon (maturation) | CBOD removal, nitrification |
| Packed-bed reactor | CBOD removal, nitrification |
| Rotating biological contact | CBOD removal, nitrification |
| Sludge blanket | CBOD removal |
| Trickling filter/activated sludge | CBOD removal, nitrification |
| Trickling filtration | CBOD removal, nitrification |

**Table U-04.** Unit processes (chemical) of water and wastewater treatment

| Chemical process | Application |
|---|---|
| Coagulation | Colloids |
| Dechlorination | Removal of excess chlorine |
| Disinfection | Destruction of pathogens |
| Fluoridation | Addition of fluoride |
| Ion exchange | $NH_3$, metals, softening, solids, organics |
| Neutralization | pH control |
| Oxidation (advanced) | Refractory organics |
| Oxidation (chemical) | BOD, grease, $NH_3$, odor, microorganisms |
| Precipitation | BOD, solids, metals, softening, $PO_4$ |
| Scale control | Control of scaling |
| Stabilization | pH control, effluent stabilization |

**universal gas law constant** A proportionality constant used in the universal (or ideal) gas equation: $R = 0.082057$ atm · L/mole/°K = 8.3145 J/mole/°K. Also called gas constant or universal gas constant.

**universal gravitational constant** *See* law of gravitation.

**Universal RAI** A series of rotary blowers manufactured by Dresser Industries/Roots Division.

**Universal Underdrain®** An underdrain manufactured by F. B. Leopold Co., Inc. for use in sand filters.

**University of Cape Town (UCT) process** A biological wastewater treatment process designed to remove nitrogen and phosphorus beyond the metabolic requirements of the organisms. It consists of three biological reactors (in the sequence anaerobic → anoxic → aerobic) followed by a final clarifier, with multiple sludge recycles: (a) internally from anoxic to anaerobic, (b) internally from aerobic to anoxic, and (c) from the clarifier underflow to anoxic. *See also* combined nutrient removal, VIP process, $A^2O$ process.

**Unizone** An ozone generator manufactured by Emery-Trailigaz Ozone Co.

**unloading** Sloughing of biological film from the medium of a trickling filter.

**unmeasured sediment discharge** The sediment discharge that is not measured by suspended-sediment samplers: sediment in the unsampled zone and contact sediment. Sediment samplers do not typically measure to the bed of the stream. *See also* sedimentation terms.

**unmetered rate** A charge assessed for water or wastewater service based on a factor other metered consumption.

**unmixed flow** Same as plug flow, or the opposite of mixed flow.

**UNOX process** A pure oxygen activated sludge treatment process developed by the Union Carbide Corporation.

**unpaired electron** *See* free radical.

**unplanned reuse** Unintentional water reuse, e.g., use of a wastewater receiving stream by a downstream community for water supply.

**unreasonable adverse effect on the environment** Any unreasonable risk to humans or the environment, taking into account the economic, social, and environmental costs and benefits of the proposed action (EPA-40CFR166.3-j). *See also* unacceptable adverse effect.

**unreasonable degradation of the marine environment** (a) Significant adverse changes in ecosystem diversity, productivity and stability of the biological community within the area of discharge and surrounding biological communities; (b) threat to human health through direct exposure to pollutants or through consumption of exposed aquatic organisms; or (c) loss of esthetic, recreational, scientific, or economic values that is unreasonable in relation to the benefit derived from the discharge (EPA-40CFR125.121-e).

**unregulated contaminants** Contaminants for which the USEPA has not issued removal requirements under the Safe Drinking Water Act (i.e., no MCL or MCLG). However, unregulated monitoring requirements have been set for bromacil, methyl-*t*-butyl-ether, and prometon under Phase VIb.

**Unregulated Contaminant Monitoring Rule** A requirement under the Safe Drinking Water Act for the USEPA to issue every five years a list of not more than 30 unregulated contaminants.

**unrestricted irrigation** The use of reclaimed water to irrigate parks, playgrounds, schoolyards, residential lawns, and commercial landscaping. It requires a high degree of treatment and disinfection. *See also* restricted irrigation.

**unrestricted reuse** The use of reclaimed water where public contact is expected, e.g., unrestricted irrigation, toilet flushing, fire protection, and construction. It requires a high degree of treatment and disinfection. *See also* restricted reuse.

**unsanitary** Unhealthy, unsafe, tending to spread or harbor disease, e.g., contaminated water.

**unsaturated** (1) Pertaining to a solution that can dissolve more of a substance; undersaturated. *See* unsaturated solution. (2) The characteristic of a carbon atom in a hydrocarbon molecule that shares a double bond with another carbon atom. For example, ethylene ($CH_2{=}CH_2$, or $C_2H_4$) is unsaturated, while ethane ($CH_3{=}CH_3$, or $C_2H_6$) is saturated.

**unsaturated compound** An aliphatic compound that contains a double or a triple carbon—carbon bond; e.g., alkenes and alkynes. Many synthetic organic substances are formed by breaking these multiple bonds to add a halogen.

**unsaturated hydrocarbon** A compound of carbon (C) and hydrogen (H), with multiple bonds between the carbon atoms, e.g., olefins like ethene or ethylene ($CH_2{=}CH_2$) and acetylene ($CH{=}CH$).

**unsaturated solution** A solution that can still dissolve more solute, i.e., for which the product of the ionic molar concentrations is less than the solubility product constant.

**unsaturated zone** The area between the water table and the ground surface where soil pores are not fully saturated but contain vadose water. The moisture content is less than saturation and pressure is less than atmospheric. Soil pore spaces typically contain air or other gases. Also called aeration zone (or zone of aeration), vadose zone, or undersaturated zone. *See* subsurface water. (Vadose water includes all suspended water in the form of soil, pellicular, gravitational, and capillary water.)

**unshared electron** *See* free radical.

**unslaked lime** Another name for lime, a white or grayish-white material, slightly soluble in water; mostly calcium oxide (CaO) or calcium oxide in natural association with a lesser amount of magnesium oxide. Quicklime is capable of combining with water to form hydrated lime. It may be obtained from ground limestone (calcite) or calcium carbonate ($CaCO_3$) and has many applications, e.g., in water treatment, mortars, cements, etc. *See* quicklime for detail.

**unstabilized solids** Organic materials in sewage sludge that have not been treated in either an aerobic or an anaerobic process (EPA-40CFR503.31-j).

**unstable** Characteristic of a substance that reacts spontaneously or decomposes or changes biologically or chemically to form other substances; also said of corrosive or scale-forming elements or compounds.

**unwater** To remove or drain the water from a tank, a trench, a riverbed, a structure, a cofferdam, an excavation, etc. *See* dewater.

**$(UO_2)^{2+}$** Formula of the uranyl ion.

**upcomer** The outer annular space between the concentric vertical tubes of a deep-well WAO process.

**upconing** The local rise of the interface between freshwater and saltwater that occurs when an aquifer containing an underlying layer of saline water is pumped by a well that penetrates only the upper freshwater portion. *See also* saltwater intrusion, saltwater wedge.

**Upcore™** Acronym of upflow countercurrent regeneration, a process and resin developed by Dow Chemical; first intended for water treatment but also applicable to heavy metals recovery from wastewater.

**upflow** A flow pattern in water and wastewater treatment units (e.g., filters, carbon adsorption columns) in which the liquid, chemical solution, or air enters at the bottom and leaves at the top. Also called countercurrent operation.

**upflow anaerobic filter** A cylindrical or rectangular tank, fully or partially (50–70%) filled with enclosed fixed media, used in biological wastewater treatment.

**upflow anaerobic sludge blanket (UASB) process** A high-rate anaerobic treatment process, developed in the Netherlands; it uses a dense, active sludge mass (dense granulated sludge) of microorganism granules at the bottom of a cylindrical or rectangular reactor where volatile suspended solids concentrations are high (up to 100,000 mg/L). Waste is fed at the bottom. Rising bubbles of the carbon dioxide ($CO_2$) and methane ($CH_4$) formed provide sufficient mixing. An upper compartment provides separation of solids, some of which are recycled to the bottom. *See also* anaerobic baffled reactor and anaerobic migrating blanket reactor.

**upflow attached-growth anaerobic treatment** *See* upflow packed-bed anaerobic filter, anaerobic expanded-bed reactor, anaerobic fluidized-bed reactor, downflow attached-growth process.

**upflow blanket clarifier** A circular or rectangular water or wastewater treatment unit in which the solids-contact process occurs. *See* solids-contact clarifier and sludge blanket for more detail.

**upflow brining** A mode of operation of a cation exchanger whereby the brine solution moves upward for resin regeneration.

**upflow clarification** A water or wastewater unit operation for the separation of solids from the liquid. *See* upflow clarifier, contact filtration.

**upflow clarifier** (1) The settling basin used in most water treatment plants, in which flocculated water rises for discharge through effluent channels, floc settles to the bottom, and settled solids are removed mechanically. (2) Sometimes used synonymously with other units that combine mixing, flocculation, and clarification in one basin, and that may or may not involve a sludge or solids blanket: flocculator–clarifier, upflow contact clarifier, sludge blanket clarifier, solids-contact clarifier. For more clarity, it is preferable to reserve the term upflow clarifier for a unit that receives water that is already flocculated.

**upflow coagulation** Coagulation in a basin with upward flow of the liquid and chemicals through a sludge blanket.

**upflow contact clarifier** A water treatment unit that provides flocculation and clarification in an upward-flow mode through a sludge blanket.

**upflow continuous backwash filter (UCBF)** A 4–6.5 ft deep filter, consisting of a sand and gravel bed supported by underdrains; used to remove nitrogen and suspended solids from a nitrified wastewater effluent. It consists of modular steel tanks or multiple cells in concrete basins, with the influent flowing upward, countercurrent to the movement of the bed. *See also* bump, downflow denitrification filter, Astrasand filter, DynaSand Filter.

**upflow expanded-bed mode** A mode of operation of fixed-bed and pulsed-bed adsorbers; it provides sufficient fluid velocity to expand the bed by approximately 10% and allow suspended solids to pass through. Thus there is no need for prefiltration, but carbon losses are higher than in the downflow mode.

**upflow filter** A gravity or pressure, granular media filter in which the liquid flows upward through the bed before discharge; used commonly for waters with high suspended solids concentrations, for industrial water, or for wastewater. Upflow filters allow longer filter runs than standard filters. *See also* biflow filter.

**upflow fixed-film reactor** *See* packed biological reactor.

**upflow moving-bed filter** A moving-bed filter in which influent enters at the bottom and flows upward through riser tubes. The clean filtrate exits the bed over a weir, whereas the bed is drawn downward into the suction of an airlift pipe for separation of the accumulated impurities by scouring.

**upflow packed-bed anaerobic filter** A cylindrical or rectangular tank, fully or partially packed with plastic cross-flow or tubular modules, plastic pall rings, or similar materials for wastewater treatment, with or without effluent recycle. The liquid flows upward and exits at the top. The active biomass is partly attached to the packing material and partly held loosely in the packing void spaces.

**upflow reactor** *See* packed biological reactor.

**upflow softening** Water softening using an ion-exchange bed with upward flow.

**upflow solids-contact clarifier** A circular or rectangular water or wastewater treatment unit in which the solids-contact process occurs. *See* solids-contact clarifier and sludge blanket for more detail.

**upflow tank** (1) Same as vertical-flow tank, i.e., any sedimentation tank in which the influent enters near the bottom, flows vertically upward, and exits at the top; also called upward-flow tank. (2) Sometimes used synonymously with upflow solids-contact clarifier, a circular or rectangular water or wastewater treatment unit in which the solids-contact process occurs. *See* upflow clarifier (2).

**UPLUME** A simple mixing model developed for the USEPA and used to establish effluent limits.

**upper alternate stage** *See* alternate stages or alternate depths.

**upper distributor** The piping arrangement at the top of a water filtration or softening device.

**upper explosion limit (UEL)** Same as upper explosive limit.

**upper explosive limit (UEL)** The maximum concentration (percent volume in air) of a flammable gas or vapor required for ignition or explosion to occur in the presence of an ignition source. Also called upper explosive limit. *See also* flash point, lower explosive limit.

**upper filter remnant stage** The third stage in the filter-ripening sequence whereby dislodged backwash remnant particles remain on top of the media.

**uppermost aquifer** The geologic formation nearest the natural ground surface that is an aquifer, as well as lower aquifers that are hydraulically interconnected with this aquifer within the facility's property boundary (EPA-40CFR258.2).

**UPS-2000®** *See* Ultra-Pyrolysis System.

**upset** (1) Generally, an unexpected malfunction of a process or operation in a water or wastewater treatment plant. (2) An exceptional incident in which there is unintentional and temporary noncompliance with technology-based permit effluent limitations because of factors beyond the reasonable control of a permittee, or noncompliance with categorical pretreatment standards because of factors beyond the control of an industrial user. An upset does not include noncompliance to the extent caused by operational error, improperly designed treatment facilities, inadequate treatment facilities, lack of preventive maintenance, or careless improper operation (EPA-40CFR122.41-n and 403.16-a).

**upstream disinfectant** The disinfectant applied first in the sequential use of two or more disinfecting agents. *See also* synergistic inactivation and downstream disinfectant.

**uptake** (1) The sorption of a test substance into and onto aquatic organisms during exposure. The uptake phase is the initial portion of a bioconcentration test when test organisms are being exposed to the test solution. (2) The rate of buildup of a substance in an organism through a combination of absorption, inhalation, and ingestion. The uptake rate constant ($k_1$) is the mathematically determined value that is used to define the uptake of test material by exposed test organisms, usually reported in units of liters/gram/hour (EPA-40CFR797.1520-17, 797.160-7 and 797.1560-8).

**uptake phase** *See* uptake (1).

**uptake rate** Same as uptake (2).

**uptake rate constant** *See* uptake (2).

**U pump** A reciprocating pump with valves in the piston and unidirectional flow through the cylinder.

**upward-flow tank** A sedimentation tank in which water flows from the bottom to the surface; often, a circular tank used in the sludge blanket or solids contact units; same as upflow tank. *See also* horizontal-flow and spiral-flow tanks.

**upwelling region** The area adjacent to a continent where ocean bottom waters rich in nutrients are brought to the surface.

**UQT** *See* utility quick test.

**uranium (U)** A hard, heavy, ductile, silvery-white, moderately malleable metal with 14 radioactive isotopes, including three occurring naturally: U-234, U-235, and U-238. It is used to produce atomic energy: one pound of fissioned uranium has the fuel equivalent of 1500 tons of coal. It is also used in nuclear weapons and in peaceful applications. Atomic weight = 238.03. Atomic number = 92. Specific gravity = 19. Valence = 2, 3, 4, 5, or 6. Melting point = 1132°C. Boiling point = 3818°C. At sufficiently high concentrations, uranium may interfere with enzyme production and cause physiological effects in humans, including cancer. The USEPA regulates uranium as a drinking water contaminant: MCL = 20 ug/L; MCLG = 0. Effective water treatment processes: ion exchange, reverse osmosis, electrodialysis, and adsorption on activated alumina. *See also* natural uranium.

**uranium-235** An important radioisotope of uranium (U-235) that can undergo continuous fission; used to produce atomic energy.

**uranium-237** An important and plentiful radioisotope of uranium (U-237), used to produce plutonium by nuclear conversion.

**uranium-238** An important radioisotope of uranium (U-238), used to produce atomic energy; it has a half-life of $4.51 \times 10^9$ years. It is one of four radionuclides of most concern, based on health effects and occurrence. U-238 is a weak carcinogen; it accumulates in the bones and has toxic effects on the kidneys. Drinking water contributes 0.03–1 pCi/day of U-238 radioactivity.

**uranium mill-tailings waste piles** Licensed active mills with tailings piles and evaporation ponds created by acid or alkaline leaching processes.

**uranyl ion ($UO_2^{2+}$)** The most common oxygen-containing compound of uranium and a stable form of the metal in aqueous solutions. The group $UO_2$ forms salts with acids.

**uranyl sulfate** A substance used in a chemical actinometer to react with oxalic acid ($H_2C_2O_4$) for the measurement of the intensity of ultraviolet light and other forms of radiation.

**urban runoff** Stormwater from city streets and adjacent domestic or commercial properties; it carries pollutants of various kinds into the sewer systems and receiving waters.

**urea [$CO(NH_2)_2$ or $CH_4ON_2$]** A soluble nitrogen compound found in urine (about 24,000 mg/L) or produced by some plants and an important constituent of fresh wastewater; an amide derivative of carboxylic acid. Urea readily hydrolizes to ammonia ($NH_3$) or converts to ammonium carbonate. Also called carbamide.

**uric acid** (1) A breakdown product of proteins and nucleic acids ($C_5H_4N_4O_3$); excreted in small amounts by mammals and in large amounts by some other animals. (2) A white, crystalline, tasteless, odorless powder form of this product.

**urinary pathogen** A pathogen that can be passed through urine, which is normally a sterile substance; e.g., those that cause leptospirosis, typhoid, and urinary schistosomiasis. *See also* cystitis.

**urinary schistosomiasis** The form of schistosomiasis caused by *Schistosoma haematobium,* the eggs of which pass chiefly through urine.

**urine** The fluid excreted by the kidneys and stored in the bladder. Human urine is yellowish and slightly acid; it contains about 550 mg/L of ammonia and 24,000 mg/L of urea and few organisms because of the high ammonia content. An adult produces about 1.15 kg of urine per day. On a dry-weight basis, human urine is approximately 5.25% calcium as CaO, 14% carbon, 6.17% nitrogen, 75% organic matter, 3.75% phosphorus as $P_2O_5$, and 3.75% potassium as $K_2O$. *See also* fecal composition.

**usable alkalinity** A parameter recommended to estimate the alkalinity in an anaerobic reactor, based on the observation that titrating to pH = 5.75 converts approximately 80% of the bicarbonate ion to carbon dioxide. Usable alkalinity is taken as the total bicarbonate alkalinity determined by titration to pH 5.75 ($TBA_{5.75}$):

$$TBA_{5.75} = 1.25\ Alk_{5.75} \quad \text{(U-07)}$$

where $Alk_{5.75}$ is total alkalinity determined by titration to pH 5.75. The error in this estimate due to the concentration of volatile acids is less than 20%. *See also* anaerobic alkalinity correction.

**USDW** Acronym of underground source of drinking water.

**useful life** The estimated period during which a treatment works or other facility will be operated, as opposed to the design life, which is the period for which the facility is planned and designed to be operated (EPA-40CFR35.2005-51). At start-up of an installation, component, or equipment, its useful life is equivalent to its design life, that is, its expected number of years of service, as estimated by the design engineer. At any other time, the useful life is the remaining number of years the installation can provide the intended level of service.

**useful power input** The portion of the total power input to a biological treatment unit that is actually used to speed up the transfer of nutrients to flocs and films. The useful proportion of power input is low in treatment units because of frictional resistance, vortex formation, and other factors. It is about the same for trickling filtration and activated sludge, but lower for intermittent sand filtration. For an impeller (Fair et al., 1971),

$$P = 0.5\ C_D \rho A V^3 \quad \text{(U-08)}$$

where $P$ = useful power input, $C_D$ = coefficient of drag, $\rho$ = fluid density, $A$ = area of impeller blade, and $V$ = relative velocity of impeller and fluid. *See also* power number and impeller mixer power requirement.

**U.S. Environmental Protection Agency** Abbreviation of United States Environmental Protection Agency.

**USEPA** Acronym of United States Environmental Protection Agency or U.S. Environmental Protection Agency.

**USEPA WTP model** An empirical computer model used by the USEPA to develop the Disinfection/Disinfectant By-products Rule under the Safe Drinking Water Act. The model simulates by-product formation, removal of natural organic matter, disinfectant decay, and other parameters.

**user** The consumer, as far as water supply and wastewater management services are concerned, is the entity that is billed for the services. Thus, one user or consumer may represent more than one person.

**user charge** A charge levied on users of a treatment works, or that portion of the ad valorem taxes paid by a user, for users' proportionate share of the cost of operation and maintenance of such works under sections 2(b)(1)(A) and 201(h)(2) of the Clean Water Act (EPA-40CFR35.905).

**user fee** A fee collected from only those persons who use a particular service (e.g., water supply and sewerage), as opposed to one collected from the public in general. User fees generally vary in proportion to the degree of use of the service.

**USGS** Acronym of United States Geological Survey.

**USP** Acronym of United States Pharmacopeia.

**USP grade water** Water that contains no added substance and meets the quality requirements of the U.S. Pharmacopeia after purification by an appropriate water treatment process such as distillation, ion exchange, or other USEPA approved drinking water treatment. *See* pharmaceutical-grade water. Also called USP purified water.

**U.S. Pharmacopeia** Abbreviation of United States Pharmacopeia.

**USPHS** Acronym of United States Public Health Service or U.S. Public Health Service.

**USP-purified water** Same as USP grade water.

**U.S. Public Health Service** Abbreviation of United States Public Health Service.

**UST** Acronym of underground storage tank.

**utilidor** A heated conduit used in cold climates to install water and sewer pipes and protect them from freezing.

**utility** (1) A business organization, often a monopoly in its service area, subject to governmental regulation and providing an essential commodity or service to the public, such as drinking water, gas, electricity, drainage, or sewerage. (2) A program that improves the performance of other programs, such as an antivirus program in a computer, a program to facilitate file organization or retrieval, or a graphical user interface. Examples of utilities are several programs used to manipulate or display data stored in the Hydraulic Engineering Center Data Storage System (HEC-DSS): DSSUTL, DSPLAY, REPGEN, and DSSMATH.

**utility quick test (UQT)** A leaching/migration test recommended for utilities to evaluate taste and odor characteristics of materials before installation in a water distribution system

**U-tube aeration** The use of a U-tube aerator in wastewater treatment.

**U-tube aerator** A 30–500 ft deep shaft used to introduce air into wastewater under high pressure, forcing the mixture to the bottom and back up, and resulting in high oxygen transfer rates.

**U-tube contactor** (1) An ozone contactor in the form of a vertical U-tube that takes advantage of the pressure of the column of water. (2) A deep device (e.g., 100 ft) used to transfer oxygen to wastewater at a high throughput velocity and short contact time.

**U-tube manometer** A U-shaped transparent tube partly filled with a liquid, used for the measurement of gage or differential pressures.

**UV** Acronym of ultraviolet.

**UV-254** or **$UV_{254}$** Abbreviation of ultraviolet absorbance measured at a wavelength of 254 nanometers; the ultraviolet light absorbance of a filtered water sample at the wavelength of 254 nm, a parameter used in the identification of the aromatic fraction of dissolved organic matter. *See also* DOC and $SUVA_{254}$.

**UV-254 absorbance** Same as UV-254.

**UV4000™** An apparatus manufactured by Trojan Technologies, Inc. for water disinfection using ultraviolet light.

**UVA, UV-A** Acronym of ultraviolet A.

**UV absorbance** *See* ultraviolet light absorbance.

**UV-absorbing organic constituents** *See* ultraviolet light-absorbing organics.

**UVB, UV-B** Acronym of ultraviolet B.

**UVC, UV-C** Acronym of ultraviolet C.

**UV disinfection** *See* ultraviolet light disinfection.

**UV disinfection model** *See* ultraviolet light disinfection model.

**UV disinfection system** *See* ultraviolet light disinfection system.

**UV dose** *See* ultraviolet light dose.

**UV dose–response** *See* ultraviolet light dose–response.

**$UV–H_2O_2$** Ultraviolet light–hydrogen peroxide.

**UV inactivation** *See* ultraviolet light inactivation.

**UV lamp** *See* ultraviolet light lamp.

**UV radiation** *See* ultraviolet radiation.

**UV radiation disinfection mechanism** *See* ultraviolet light radiation disinfection mechanism.

**UV radiation production** *See* ultraviolet light radiation production.

**$UV_{scan}$** Abbreviation of spectral ultraviolet (UV) absorbance.

**UV spectrophotometer** *See* ultraviolet light spectrophotometer.

# V

**V** Chemical symbol of vanadium.

**vaccination** Artificial introduction of a killed or attenuated pathogen to induce specific protective immunity.

**Vacflush®** An apparatus manufactured by Jet Tech, Inc. to clean aeration equipment.

**Vacuator** A device operated by vacuum, manufactured by Dorr-Oliver, Inc. to remove floating solids and scum.

**vacuole** A cavity in the cytoplasm of a cell that may contain ingested bacteria, yeast cells, or debris.

**Vacu-Treat** A rotary vacuum filter of Envirex, Inc.

**vacuum** A space entirely devoid of matter or an enclosed space from which air has been partially removed and the remaining gas exerts less than atmospheric pressure. In a closed water system, discharge of water at a low point without a point for air to enter creates a vacuum that may result in backflow into the system.

**vacuum-assisted dewatering** *See* vacuum-assisted drying bed.

**vacuum-assisted drying bed** A device for dewatering chemically conditioned sludge or for accelerating sludge drying. It consists of a granular bed installed on a rigid, porous underdrain through which water is extracted by an apparatus that creates a vacuum. The application rate is about 10 $kg/m^2$ per one-day cycle, resulting in a sludge of about 10–25% solids.

**vacuum-assisted sludge drying bed** *See* vacuum-assisted drying bed.

**vacuum breaker** A special valve or other mechanical device installed in a pipeline to prevent backflow or backsiphonage by relieving a complete or partial vacuum. A vacuum breaker is insufficient to prevent backpressure and it may be subject to corrosion. *See also* backflow preventer and air gap.

**vacuum-breaking valve** *See* air-release valve.

**vacuum collection system** *See* vacuum sewer system.

**vacuum deaeration** The use of a vacuum deaerator.

**vacuum deaerator** An apparatus operating under a vacuum and used to remove dissolved gases from a liquid.

**vacuum dewatering** The use of vacuum filters to reduce the volume of wastewater sludge before further processing or final disposal.

**vacuum distillation** A process using a vacuum to lower the pressure on a liquid and allow volatilization at a temperature lower than normal.

**vacuum draft tube** A narrow tube lowered into an extraction well through which a strong vacuum is pulled via a suction pump at the ground surface. Fluids (gas, water, and/or free product) are drawn into the draft tube and conveyed to the surface for treatment or disposal. Depending upon the configuration of the extraction system, the inlet of the draft tube may be either above or below the static level of the liquid in the well.

**vacuum ejector** A special vacuum valve used in vacuum sewer systems to seal the line and maintain the necessary vacuum level.

**vacuum evaporator** A device that uses an eductor or a vacuum pump to boil and evaporate water at relatively low temperatures (110–130°F). *See also* atmospheric evaporator, multiple-effect evaporator, single-effect evaporator.

**vacuum extraction** A technique used for removing volatile organic compounds in unsaturated soils; *see* soil-vapor extraction.

**vacuum filter** (1) A device used mostly in wastewater treatment to separate water from chemically conditioned sludge by applying suction through a filtering material. It consists of a cylindrical drum, covered with a fine filter fabric, mounted on a horizontal axis and rotating slowly and continuously in a tank of sludge. It includes a mechanism for scraping off and removing the dewatered sludge cake and for the removal or recycling of the filtrate. Belt filters are more common in new installations. The solids content of the product varies from 10–15% for activated sludge to 25–38% for primary sludge. Other sludge dewatering units include pressure filters, centrifuges, drying beds or lagoons, and composting units. *See also* Coilfilter. (2) An open diatomaceous earth filter installed on the suction side of a pump.

**vacuum filter performance** The yield of solids produced by a vacuum filter on a dry weight basis, expressed in pounds per square foot per hour, or the solids content of the cake expressed as a percentage on a dry weight basis. A common design rate is 3.5 $lb/ft^2/hr$.

**vacuum filter system** A vacuum filter (filter drum) and appurtenant devices such as sludge-feed pump, chemical feeders, sludge conditioning tank, sludge cake conveyor or hopper, vacuum system, and filtrate removal mechanism.

**vacuum filter yield** *See* filter yield.

**vacuum filtration** A treatment operation using a vacuum filter to dewater wastewater sludge or to separate water from lime sludge in water softening. The vacuum filter rotates partially submerged in a tank of chemically treated sludge while the partial vacuum applied to the filter medium draws solids to the surface and allows water to pass. The filtrate is discharged and the solids are removed from the nonsubmerged part. In new installations, vacuum filtration has been replaced by more competitive processes such as pressure filtration because of lower energy consumption, lower chemical requirements, a dryer sludge cake, and better performance with dilute sludges.

**vacuum flotation** A process designed to concentrate low-density particles in water and wastewater treatment residuals. It generates through a vacuum small gas bubbles (50–100 micron diameter) in a supersaturated solution to absorb particles and cause them to float to the surface for separation. Pressure flotation is more commonly used. *See also* dissolved air flotation and dispersed air flotation.

**vacuum freezing** A desalination process that uses a vacuum to cool and freeze saline water; solids separation occurs as they concentrate in the portion of the liquid that freezes last.

**vacuum hold test** A test used to determine the integrity of a wetted membrane.

**vacuum membrane distillation** Membrane distillation is a desalination method using a temperature-driven, hydrophobic membrane to separate water, in the form of condensed vapor, from contaminants. This configuration pulls water vapor out of the system by applying a vacuum to the membrane.

**vacuum pan** An airtight container equipped with a vacuum pump for rapid evaporation by boiling a substance at low temperature and reduced pressure. Such a vessel is used to produce water softening salt from brine.

**vacuum pump** A pump that creates a partial vacuum in a closed container, which forces up liquid by pressure differential; often used in sludge dewatering, groundwater degasification, and the maintenance of suction lifts.

**vacuum relief valve** *See* vacuum valve.

**vacuum sewer system** A system of plastic pipes, 2 to 6 inches in diameter, installed just below the freezing line and operating under a vacuum that is maintained at a central collection tank. Gravity service lines from households or other customers are connected to the vacuum mains via an automatic valve or vacuum ejector that seals the line to maintain the required vacuum level. Vacuum systems are used in sparsely populated areas to avoid the installation of deep gravity lines. *See* Figure V-01. *See also* pressurized sewer system.

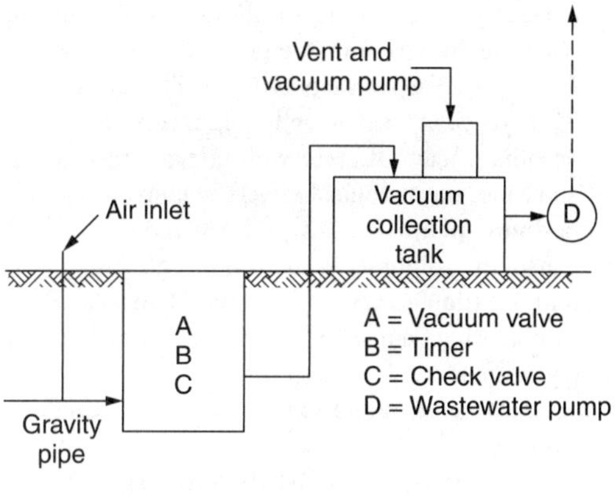


**Figure V-01.** Vacuum sewer system.

**vacuum tanker** A tank or other container mounted on a truck that is equipped with a vacuum pump to draw the contents of aqua-privies, cesspits, septic tanks, vaults, etc. into the container for transporting them to a treatment or disposal site. *See also* vacuum truck.

**vacuum toilet** A toilet that uses a vacuum and a small amount of water to flush solids. *See also* flow-reducing devices and appliances.

**vacuum truck** A tank truck that pumps out the inside air and replaces it with wastewater to be removed for disposal elsewhere (Pankratz, 1996). *See also* vacuum tanker.

**vacuum valve** A valve used in water and wastewater networks to let and maintain air into an empty pipe to counteract a vacuum. Also called an air-and-vacuum valve or a vacuum relief valve. *See also* siphon breaker.

**vadose water** (From the Latin vadosus, meaning shallow.) In the undersaturated zone (or aeration zone), vadose water includes all suspended water in the form of soil, pellicular, gravitational, and capillary water. Also called suspended water. *See* subsurface water.

**vadose zone** The area between the water table and the ground surface where soil pores are not fully saturated but contain some water. Also called aeration zone (or zone of aeration), unsaturated zone, or undersaturated zone. *See* subsurface water.

**valence** The number of electrons that an atom or group of atoms has lost or gained. A negative valence also represents the number of hydrogen atoms with which one atom of an element or radical can combine; e.g., sulfate ($SO_4^{2-}$), which is divalent, combines with two hydrogen atoms to form sulfuric acid ($H_2SO_4$). A positive valence represents the number of hydrogen atoms that one atom of an element or radical can displace; e.g., sodium (Na), which is monovalent, can displace one hydrogen atom from hydrochloric acid (HCl) to form sodium chloride (NaCl). *See* elements.

**valence number** Same as valence or oxidation number.

**valeric acid ($C_4H_9COOH$)** The common name of one of the simplest carboxylic acids (pentanoic acid).

**validation** The verification that a sample of water or wastewater treated by a given process or equipment model meets the applicable standards.

**validation factor (VF)** A factor applied to the ultraviolet (UV) light dose predicted by a dose-monitoring equation to account for uncertainties in the validation and dose-monitoring steps. The USEPA's UV Disinfection Guidance Manual recommends the following formula:

$$VF = B_{RED}(1 + U/100) \qquad \text{(V-01)}$$

where $B_{RED}$ is the RED bias (bias of the reduction equivalent dose) and $U$ is the percent uncertainty of the predicted RED. For example, the measured value for the test microbe MS2 might be 32 $mJ/cm^2$ with a RED bias factor of 2.0 and uncertainty of 8.0 $mJ/cm^2$ or 8.0/32.0 = 25%. Then VF = 2.0 (1 + 0.25) = 2.5 and the adjusted dose = 32/2.5 = 12.8 $mJ/cm^2$.

**validation monitoring** Water quality monitoring to determine if model coefficients are adequate.

**valley fever** An acute or progressively chronic respiratory infection characterized by fever and reddish bumps on the skin. *See* coccidioidomycosis for detail.

**value engineering** A specialized cost-control technique that uses a systematic and creative approach to identify and to focus on unnecessarily high-cost items in a project to arrive at a cost saving without sacrificing the reliability or efficiency of the project. Also, a similar analysis of each contract term or task to ensure that its essential function is provided at the overall lowest cost (EPA-40CFR35.2005-53 and 40CFR35.6015-53).

**valve** (1) A device used to regulate the magnitude and direction of fluid flow in machinery and piping systems. It consists essentially of a shell and a movable control element fitted to the shell such that it can open, close, or obstruct ports and passageways. The following types of valve are common in water and wastewater engineering: actuation, air, air and vacuum, air-release, air-relief, altitude, angle, automatic, back-pressure, ball, blowoff, butterfly, bypass, check, flow-control, foot, four-way, gate, globe, hydraulic, negative-pressure, plug, pressure-reducing, pressure-regulating, pressure-relief, reducing, relief, rotary, safety, stop, throttle, vacuum. (2) The shell of a mollusk or each complete part of the shell of a bivalve mollusk.

**valve box** A metal or plastic box that houses and provides access to a small, buried valve; larger valves are usually installed in manholes. *See* stop box, valve vault.

**valve capacity coefficient** A parameter (Kv, in metric units) used to determine the required size of a valve:

$$K_v = Q/(\Delta P/S)^{0.5} = 0.856\ C_v \qquad \text{(V-02)}$$

where $Q$ = maximum design flow through the valve, $m^3/h$; $\Delta P$ = head loss through the valve at maximum design flow, $kg/cm^2$; $S$ = specific gravity relative to water; and $C_v$ = valve capacity coefficient in U.S. customary units. Also called $K_v$ (or $C_v$) value.

**valved orifice diffuser** An air diffuser that has a check valve to prevent backflow when the air is shut off. It may also be used to adjust air flow rates.

**valve-exercising program** A program of scheduled inspection and operation of valves in a water or wastewater system to make sure that they are in proper operating condition.

**valve key** A metal wrench used to operate gate valves from a distance.

**Valve PAC™** A device manufactured by F. B. Leopold Co., Inc. to control valve positioning.

**valve selection** *See* valve capacity coefficient. Valves are selected on the basis of the $K_v$ or $C_v$ value; if necessary, two parallel valves of different sizes may be installed to take into account peak flows as well as low flows.

**valve stem** The rod or spindle by which a valve is operated from outside.

**valve tower** A hollow tower in a reservoir with pipes for drawing water.

**valve vault** A concrete structure that houses (a) a valve on a water main or (b) two or more valves at the intersection of large water mains. It allows access to the valves for inspection, operation, and maintenance. *See also* valve box.

**valving** The arrangement and system of control of valves in a treatment plant, irrigation scheme, etc.

**vanadate ($VO_4$)** A salt or ester of vanadic acid, or an anion formed from soluble vanadium salts, a potential source of water contamination.

**vanadic acid** An acid containing vanadium, e.g., $H_3VO_4$.

**vanadium (V)** A white, abundant, bright, soft, ductile, element found in several minerals; used in nuclear production, as an alloy, in glass, and in photography. Atomic weight = 50.94. Atomic number = 23. Specific gravity = 6.11. Valence = 2, 3, 4, or 5. At sufficiently high concentrations, vanadium may interfere with enzyme production and cause physiological effects in humans.

**vanadium pentoxide** A soluble compound of vanadium; solubility = about 700 mg/L.

**vanadyl cation (VO)** A compound formed from soluble vanadium salts, a potential source of water contamination.

**Van Bemmelan equation** Same as Freundlich isotherm.

**Vandermeyden–Cornwell–Schenkelberg equation** A formula proposed in 1997; *see* sand bed yield.

**van der Waals attraction** *See* van der Waals force. Also called physical attraction, ideal adsorption, or physical adsorption.

**van der Waals force** A weak, natural force of attraction that acts between any two molecules or masses. It creates a random motion or Brownian movement of colloidal particles and enhances chemical coagulation when the force of attraction is greater than the force of repulsion. It is also one of the forces holding adsorbates on the surface of adsorbents. This force depends on the kind and number of atoms in the colloidal particles and in the surrounding liquid, is independent of the composition of the solution, and its magnitude varies inversely with the distance between particles. Also called London–van der Waals force. *See also* chemical adsorption or chemisorption, desorption, DLVO theory, electrical double layer, electrostatic or exchange adsorption, electrostatic stabilization, hydrodynamic retardation, hydrophobic bonding, ideal adsorption, physical adsorption, polarity, polymer bridging, solution force, solvophobic force, steric interaction.

**Vanguard®** A line of airstrippers manufactured by Delta Cooling Towers, Inc. of Rockaway, NJ for the removal of volatile organic chemicals from groundwater and wastewater.

**van't Hoff–Arrhenius equation** A basic relationship of chemical reaction rate constants to temperature, formulated by Arrhenius in 1889, based on empirical observations. *See* Arrhenius equation for detail.

**van't Hoff–Arrhenius factor** The temperature correction coefficient ($\theta$) in the Arrhenius equation.

**van't Hoff–Arrhenius relationship** *See* Arrhenius equation.

**van't Hoff equation** An equation derived by van't Hoff from thermodynamic considerations, expressing the effects of temperature ($T$, in °K) on the equilibrium constant ($K$) of a chemical reaction:

$$d(\ln K)/dT = \Delta H^0/RT^2 \quad \text{(V-03)}$$

or

$$\ln(K_1/K_2) = (\Delta H^0/R)(1/T_2 - 1/T_1) \quad \text{(V-04)}$$

where $\Delta H^0$ = the standard enthalpy change for the reaction (enthalpy = heat released or taken up by the reaction), in cal/mole; $\Delta H^0$ is sometimes replaced by the symbol $E$, for activation energy; $R$ = the universal gas constant = 1.987 cal/°K/mole; $K_1$ = the equilibrium constant at temperature $T_1$; $K_2$ = the equilibrium constant at temperature $T_2$; and $T_1$ and $T_2$ = temperatures in °K. *See also* temperature quotient.

**van't Hoff, Jacobus Hendricus** Dutch chemist (1852–1911), winner of Nobel prize in 1901.

**vapor** (1) The gaseous state of any substance that is a liquid or a solid at standard conditions of temperature and pressure; the result of evaporation. (2) Visible condensation or moisture particles suspended in air, e.g., fog, mist, steam.

**Vapor Combustor™** An apparatus manufactured by QED Environmental Systems to extract vapors.

**vapor compression** The compression of vapor using a mechanical or a steam jet device.

**vapor-compression distillation** A saline water conversion process that uses the vapor-compression technique to extract freshwater from saline or brackish waters; it is also used in advanced wastewater treatment. Heated tubes in a chamber at an appropriate pressure vaporize part of the feedwater. Compression increases the temperature of the vapor sufficiently to supply heat through condensation to evaporate the feed water. A typical unit includes a boiler for steam production using condensate return, a heat exchanger, and the distillation unit from which the product water and the concentrated waste exit. *See also* multiple-effect and multistage flash evaporation.

**vapor-compression process** *See* vapor-compression distillation.

**vapor-compression still** An evaporation apparatus in which the vapor produced is compressed for reuse in heating.

**vapor compressor** Same as mechanical vapor-recompression evaporator.

**vapor density** The amount or mass of a vapor per unit volume of the vapor; a measure of the heaviness of a substance compared to the weight of a similar amount of air. A vapor density of 1.0 corresponds to air. Vapors that are heavier than air may build up in low-lying areas, such as along floors, in sewers, or in elevator shafts. Vapors that are lighter than air rise and may collect near the ceiling.

**vapor dispersion** The movement of vapor clouds in air due to wind, thermal action, gravity spreading, and mixing.

**Vapor Guard** A fabric membrane manufactured by ILC Dover, Inc. as a tank cover for odor control.

**vaporimeter** An instrument for measuring vapor pressure or volume.

**vapor incinerator** An enclosed combustion device that is used for destroying organic compounds and does not extract energy in the form of steam or process heat (EPA-40CFR 264.1031).

**vaporization** The process of a substance changing from a liquid or solid to a gas, e.g., the vaporization of sludge moisture into air. *See also* evaporation, thermal drying, latent heat of vaporization.

**vaporization heat** *See* heat of vaporization.

**vaporization latent heat** Same as heat of vaporization.

**VaporMate™** A carbon adsorption apparatus manufactured by Northeast Environmental Products, Inc. for the control of volatile organics.

**Vapor Pacs** Replacement activated carbon canisters manufactured by Calgon Carbon Corp. for the control of volatile organic compounds.

**vapor-phase biological treatment** A process used to treat odorous gases in the vapor phase. *See* biofilter and biotrickling filter.

**vapor plume** A trail of flue gases visible because they contain water droplets.

**vapor pressure** (1) The force per unit area exerted by a vapor in an equilibrium state with its pure solid, liquid, or solution at a given temperature. Vapor pressure is a measure of a substance's propensity to evaporate; it increases exponentially with temperature. Chemicals with low boiling points have high vapor pressures. If a chemical with a high vapor pressure spills, there is an increased risk of explosion and a greater risk that workers will inhale toxic fumes. (2) The partial pressure of water in the atmosphere or the partial pressure of any liquid. The vapor pressure of water ($P_w$, in mm of Hg) varies from 4.58 at 0°C to 9.21 at 10°C, 17.50 at 20°C, and 31.80 at 30°C.

**vapor-pressure deficit** The difference between the actual vapor pressure and the vapor pressure in a saturated atmosphere, under the same conditions of temperature and pressure.

**vapor recovery system** An apparatus designed to collect and treat vapors and gases from a storage tank.

**vapour** *See* vapor.

**VaPure®** A vapor compression still manufactured by Mueller International Sales Corp.

**variable costs** Input costs that change as the nature of the production activity or its circumstances change; for example, as production levels vary.

**variable declining-rate filtration** A method of filter operation whereby the flow rate declines and the liquid level rises in the unit from the beginning to the end of the filter run. Often simply called declining-rate filtration. Other common methods of filter operation include constant pressure and constant rate.

**variable-displacement pump** A centrifugal or other type of pump whose discharge is inversely proportional to total head.

**variable-frequency controller** A device used in induction pumps that causes a motor to vary its output speed whether the load is connected or not; a variable-speed or variable-frequency drive.

**variable-frequency drive** Same as variable-speed drive.

**variable membrane operation** One of three operating modes of a membrane process, which allows both the flux rate and the transmembrane pressure to vary with time. *See also* constant-flux operation, constant-TMP operation.

**variable motor speed control system** A variable-speed drive that adjusts the motor speed without mechanical couplings.

**variable-speed drive** A device used in induction pumps that adjusts the applied power frequency to control motor speed; frequently used in wastewater applications where water and air volumes vary widely. It is the general type of device driving wastewater pumps at varying speeds. Also called a variable-frequency drive. *See also* variable-torque transmission system, variable motor speed control system.

**variable-speed pump** A pump that is designed to operate at a variable speed and discharge at a rate that varies with the electrical current input. The opposite is a constant-speed pump.

**variable-torque transmission system** A device that consists of separate fixed-speed input and variable-speed output shafts and is used to drive wastewater pumps at varying speeds. *See* eddy current clutch, liquid clutch.

**variable-volume filter press** A filter press used for sludge dewatering and having a flexible membrane across the filter plate face. *See also* diaphragm filter press, fixed-volume filter press.

**variable-volume, recessed-plate filter press** Same as diaphragm filter press.

**Variair** A diffuser manufactured by Enviroquip, Inc.

**variance** A state with primacy may relieve a public water system from a requirement respecting an MCL by granting a variance if certain conditions exist. These are: (a) the system cannot meet the MCL in spite of the application of best available treatment technology, treatment techniques or other means (taking costs into consideration), due to the characteristics of the raw water sources that are reasonably available to the system, and (b) the variance will not result in an unreasonable public health risk. A system may also be granted a variance if it can show that, due to the nature of the system's raw water source, such treatment is not necessary to protect public health. *See also* exemption.

**Vari-Ator** A device manufactured by Aerators, Inc. to control the energy to floating aerators.

**Vari-Cant™** A jet aeration apparatus manufactured by Jet Tech, Inc.

**Variflo®** A wastewater distribution nozzle manufactured by Wheelabrator Engineered Systems, Inc. for use in high-rate trickling filters.

**VariSieve™** A device manufactured by Krebs Engineers for solids classification.

**Varivoid™** A unit manufactured by Graver Co. for high-rate filtration on granular media.

**vault** (1) An arched structure that frames a ceiling or roof over an enclosed construction. (2) A concrete chamber, underground or at street level, used in a utility distribution system to house pumps, meters, valves, and other appurtenances. (3) A reinforced concrete structure, installed above or below ground, for storing radioactive and other hazardous wastes. (4) A watertight tank for the storage of excreta; *see* vault toilet.

**vault and vacuum-truck cartage system** A system that uses vault toilets, under or beside the house, for collection and storage of excreta generated in a community, and vacuum trucks for carrying the excreta away for treatment and disposal. *See* Figure V-02.

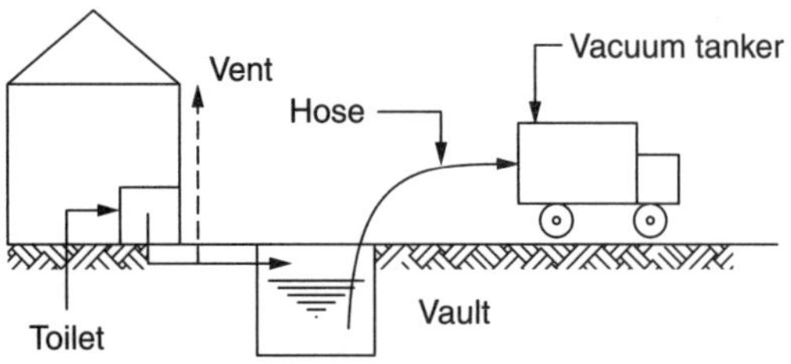


**Figure V-02.** Vault and vacuum-truck cartage system.

**vault latrine** Vault toilet.

**VaultSafe™** A line of products designed by Val-Matic® to protect air valve vaults from flooding, tampering, etc. *See* FloodSafe™ Inflow Preventer, FrostSafe™, VentSafe™.

**vault toilet** A toilet that discharges into a watertight container, located under or beside the house, large enough for storing excreta for a period of about four weeks, after which a vacuum truck takes the waste away for treatment and disposal. It is similar to a pour-flush toilet but does not require any water. This cartage system is common in urban areas of the Far East.

**V-Auto™** An in-line strainer manufactured by Andritz-Ruthner, Inc.

***V. cholerae*** *See Vibrio cholerae.*

***V. cholerae* O1** *See Vibrio cholerae* 01.

***V. cholerae* O139** *See Vibrio cholerae* 0139.

**VCM** Acronym of vinyl chloride monomer.

**VDS** (1) Acronym of vertical drum screen, a screening device manufactured by Jones and Attwood, Inc. (2) Acronym of volatile dissolved solids.

**vector** (1) An organism, often an insect, arthropod, or rodent, that carries disease from one source to another, e.g., from one person to another or from an infected animal to a person. (2) A plasmid, virus, or bacterium used to transport genes into a host cell. A gene is placed in the vector and the vector then "infects" the bacterium.

**vector attraction** The characteristic of sewage sludge that attracts rodents, flies, mosquitoes, or other organisms capable of transporting infectious agents (EPA-40CFR503.31-k).

**vector-attraction reduction** A measure required by the USEPA, along with Class A criteria, before biosolids can be applied to lawns and home gardens or given away in bags. Options for vector-attraction reduction include volatile solids reduction, required specific oxygen uptake rate, aerobic treatment, pH adjustment by alkali addition, and injection or incorporation into soils.

**vectorborne** Pertaining to diseases transmitted to humans directly or indirectly by nonhuman animate carriers.

**vector-borne protozoa** Protozoal pathogen that transmits infections via mosquitoes, flies, bugs, etc. Vector-borne, water related protozoans include *Plasmodium, Trypanosoma cruzi, Trypanosoma gambiense,* and *Trypanosoma rhodesiense. See also* protozoal disease, excreted protozoa.

**vector-borne transmission** Transmission of an infection by vectors that carry the infecting agent. *See also* vehicle-borne transmission.

**vector-transmitted excreted infection** A disease that is spread by excreta-related insects, e.g., Bancroftian filariasis.

**Vee-wire®** A wedge-shaped wire manufactured by Wheelabrator Engineered Systems, Inc.

**vegetated submerged bed** A constructed wetland used for the treatment of municipal and industrial wastewater. *See* subsurface flow system for detail.

**vegetation controls** (1) Nonpoint source pollution control practices that involve vegetative cover to reduce erosion and minimize loss of pollutants. (2) Herbicide sprays or other means of controlling the growth of weeds, grasses, and other vegetation on the surface of sand drying beds.

**vegetation salinity threshold** The maximum permissible electrical conductivity of the soil-saturation extract for a given crop. Also called soil salinity threshold or crop salinity threshold. *See also* leaching requirement.

**vegetative controls** Same as vegetation controls.

**vehicle** A nonliving source of pathogens that can infect a large number of individuals, e.g., food, water. *See also* fomite.

**vehicle-borne transmission** Transmission of a disease agent through a contaminated material, e.g., airborne, foodborne, waterborne, diapers, etc. *See also* vector borne.

**Vekton** Nylon sprockets manufactured by Norton Performance Plastics for use in sludge collection mechanisms.

**Vektor™** A mixer manufactured by Lighnin.

**veliger** The larval stage of a mollusk characterized by the presence of a velum (a veillike or curtain-like membranous partition).

**velocity** The speed of flow or the ratio of the distance traveled to the time of travel, usually expressed in feet per second (fps) or meters per second (m/s). Velocity in any channel varies with channel characteristics. For convenience, an average flow velocity ($V$) at any cross section is defined as the ratio of the discharge ($Q$) to the cross-sectional area ($A$). *See* celerity.

**velocity cap** A cap on a vertical water intake to reduce fish entrainment. *See also* fish screen.

**velocity coefficient** A factor, less than 1.0, equal to the ratio of the actual velocity to the theoretical velocity of flow from an orifice, weir, hydraulic structure, etc. *See also* discharge coefficient.

**velocity factor** The ratio of the flow rate ($Q$) to the applicable surface area ($A$), used (a) as million gallons per day per acre for trickling filters, land treatment, intermittent sand filters, and stabilization ponds, and (b) as gallons per day per square foot (surface overflow rate) for settling tanks. *See also* hydraulic loading intensity.

**velocity gradient (*G* or *G* value)** (1) The change in velocity per unit distance along the vertical velocity curve. (2) A measure of the mixing intensity or mechanical agitation in a water or wastewater treatment process such as flocculation. It is used in the design of mixing (e.g., $G = 300$/sec.), flocculation (e.g., $G = 25$–75/sec.), or horizontal sedimentation (e.g., $G = 10$–50/sec.) units and to describe their performance characteristics. However, $G$ values between 30 and 70/sec do not affect the performance of flotation units. $G$ is a function of power input, fluid viscosity, and basin volume:

$$G = (P/\mu V)^{0.5} \qquad \text{(V-05)}$$

where $P$ = power input or power dissipated, $\mu$ = absolute viscosity, and $V$ = volume of mixing or flocculating basin. Also called shear gradient, mean velocity gradient, or mean temporal velocity gradient. *See also* GT, power dissipation function, mixing opportunity parameter.

**velocity gradient concept** The use of the velocity gradient or $G$ value as a flocculation design parameter applicable to turbulent fluid motion.

**velocity head** Variously defined term as (a) head of a moving fluid, (b) kinetic energy in a hydraulic system, (c) the vertical distance through which a body would have to fall under the force of gravity to acquire a given velocity, and (d) the theoretical height that a fluid may be raised by its kinetic energy. The velocity head is calculated as the ratio of the square of the mean velocity ($V$) to twice the gravitational acceleration ($g$), or $V^2/2\,g$. It is a factor in several formulas: critical flow or specific energy, Darcy–Weisbach, dynamic head, energy gradeline elevation, equivalent pipes, and weir. Velocity head is one of two convenient ways to express head losses, the other being equivalent pipe lengths. Velocity head converts to static head when flow velocity decreases or stops; conversely, some static head converts to velocity head when flow velocity increases.

**velocity of approach** The velocity of a liquid as it is entering a conduit, dam, venturi tube, weir, or other hydraulic structure. For example, the approach velocity at water intake screens affects head losses through the screens and is typically 1–2 fps.

**velocity pressure** The kinetic pressure in the direction of flow necessary to cause a fluid at rest to flow at a given velocity. It is usually expressed in inches of water gauge (EPA-29CFR1910.94b).

**Velz equation** Same as Velz formula.

**Velz formula** An empirical formula proposed by C. J. Velz in 1948 for the design of trickling filters:

$$L_D/L_0 = 10^{-KD} \qquad \text{(V-06)}$$

$$K_T = K_{20}[1.047^{(T-20)}] \qquad \text{(V-07)}$$

where $L_0$ = influent $BOD_5$, mg/L; $L_D$ = $BOD_5$ removed at depth $D$, mg/L; $D$ = filter depth, feet; $K$ = first-order rate constant,/day; $K_T$ = first-order rate constant at temperature $T$,/day; and $K_{20}$ = first-order rate constant at temperature $T$ = 20°C,/day. Also called trickling filter performance equation. Other trickling filter design formulas include British Manual, Eckenfelder, Galler–Gotaas, Germain, Howland, Kincannon–Stover, Logan, modified Velz, Rankin, Schulze.

**vena contracta** (Latin for contracted vein) The minimum cross section of the jet stream emerging from an orifice, the reduction of the jet below the orifice diameter being caused by the convergence of the streamlines.

**vent** An opening serving as an outlet for air, smoke, etc. Also, a pipeline for air flow in a plumbing system. Vents in drainage systems help equalize air pressures and prevent damage to the seals and traps.

**vented loop** *See* siphon breaker.

**ventilated double-pit latrine** *See* double-pit VIP latrine.

**ventilated improved pit (VIP) latrine** A pit latrine improved by the addition of a screened vent pipe and a partially dark superstructure to reduce mosquitoes and odors. *See also* single-pit VIP latrine, double-pit VIP latrine.

**ventox** Acrylonitrile.

**VentOXAL®** An apparatus manufactured by Liquid Air for oxygenation.

**vent pipe** A vertical pipe installed in a latrine, septic tank, waste or soil pipe, chamber, or similar unit to facilitate the escape of gases and odors, or to prevent negative pressures due to siphoning.

**VentSafe™** A vent pipe security cage designed by Val-Matic® to help prevent bugs, birds, and small animals from nesting in the vent pipe, while minimizing the potential for malicious tampering of liquids and other matter into the vent pipe.

**Vent-Scrub™** A water treatment unit manufactured by Wheelabrator Clean Water Systems, Inc. using adsorption on powdered activated carbon.

**VentSorb** Disposable filters manufactured by Calgon Carbon Corp. using granular activated carbon.

**vent stack** A vertical waste pipe or vent pipe serving a number of floors of a building, to provide air circulation and release of odors. A vent stack may also provide rapid draining by gravity. *See also* stack.

**venturi or Venturi** A flow-measuring device, such as a short tube with a constricted throat for determining fluid pressures and velocities by measuring differential pressures generated at the throat as a fluid traverses the tube. A venturi flume is an open flume, used to measure flow, with a constriction that causes a drop in the hydraulic gradeline. *See also* Parshall flume. A venturi meter is an instrument for measuring fluid flow in closed conduits, consisting of a venturi tube (a narrowed section tapering out to the conduit diameter at each end), and a flow- or pressure-registering device. Velocity increases and pressure decreases in the throat section. Discharge ($Q$) is estimated from the pressure head differential ($\Delta P$) indicated by a manometer (*see* Figure V-03):

$$Q = 1.111\,KD^2\{g(\gamma' - \gamma)/[(r^4 - 1)]\}^{0.5} \quad \text{(V-08)}$$

where $K$ is a discharge coefficient, $D$ the conduit diameter, $g$ the gravitational acceleration, $\gamma'$ the density of the manometer fluid, $\gamma$ the density of the fluid being measured, and $r = D/d$ = the ratio of the conduit diameter ($D$) to the throat diameter ($d$). Flow through a venturi meter is similar to flow through an orifice plate but with a higher discharge coefficient.

**venturi effect** An increase in the velocity of a fluid as it passes through a constriction in a pipe or channel. *See* venturi principle.

**venturi flume** *See* venturi.

**Venturi, G. B.** Italian physicist (1746–1822) whose work led to the invention of the venturi tube.

**venturi meter** *See* venturi.

**venturi principle** The increase in flow velocity and decrease in pressure, caused by a constricted throat in a pipe or channel; used to measure flow in a venturi meter and similar devices.

**venturi scrubber** An air pollution control device that uses water or another liquid to remove up to 90% of particulates, fumes, dusts, mists, and vapors from gas streams; for example, in sludge incinerators. The device includes a venturi or orifice across which the gas stream accelerates and mixes turbulently with the sprayed liquid. *See also* ejector–venturi scrubber, spray tower, and cyclone scrubber.

**venturi tube** *See* venturi.

**vermiculture** An emerging technology of sludge stabilization and disposal that uses earthworms to consume the organic matter in municipal wastewater treatment sludge. After screening, the worms are used as animal feed and the worm castings as soil conditioner. Also called earthworm stabilization or vermistabilization.

**vermistabilization** *See* vermiculture.

**VerTech™** A wet oxidation process developed by Air Products and Chemicals, Inc.

**vertical-column reactor** A cocurrent bubble column that uses wet-air oxidation for the treatment of hazardous wastes.

**vertical-flow tank** Any sedimentation tank in which the influent enters near the bottom, flows vertically upward, and exits at the top. Also called upflow tank or upward-flow tank. *See also* horizontal-flow tank, spiral-flow tank, sludge blanket, solids contact.

**vertical permeability** A coefficient of proportionality describing the rate at which water can move through a permeable medium. It is an important

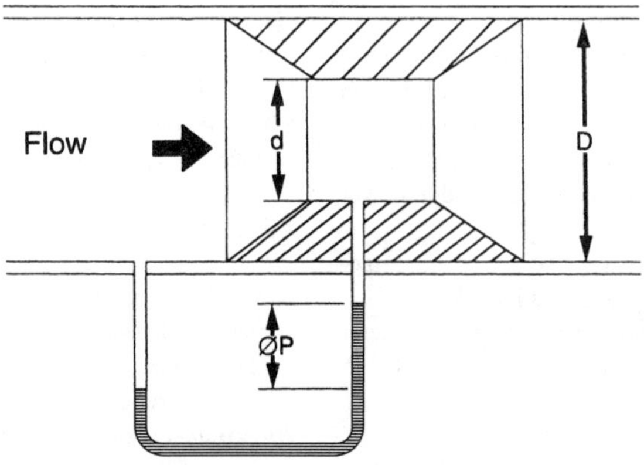


**Figure V-03.** Venturi meter (courtesy CRC Press).

factor in the evaluation and selection of sites for the slow-rate application of wastewater to land. *See* hydraulic conductivity for detail.

**vertical pump** A reciprocating pump whose piston or plunger moves vertically, or a centrifugal pump with a vertical shaft. Its motor is mounted on a pedestal or above the top of the wet well.

**vertical screw pump** A pump that moves fluids by a vertical runner with radial blades in a cylindrical casing. *See also* screw pump, horizontal screw pump, and wood-screw pump.

**vertical shaft pump** A dry- or wet-pit, vertical pump with submerged suction.

**vertical surface aerator** A mechanical device that is designed to transfer oxygen from the atmosphere into a liquid by surface renewal and exchange, and to provide mixing to the contents of the basin or container. It consists of a vertical axis on which are mounted submerged or partially submerged impellers that are attached to fixed or floating motors. It induces an updraft or a downdraft through a pumping action. *See also* horizontal surface aerator, submerged aerator.

**vertical throat** A flow-control device installed at the end of a grit chamber.

**vertical-tube distillation** Distillation in vertical evaporation tubes.

**vertical turbine pump** A centrifugal pump with a vertical shaft.

**vertical velocity curve** A plot of velocity versus depth of water flowing in an open channel, at a given point and along a vertical line. Also called mean velocity curve.

**Verti-Flo** A rectangular clarifier with vertical flow manufactured by Envirex, Inc.

**Verti-Jet™** A vertical pressure leaf filter manufactured by U.S. Filter Corp.

**Vertimatic** An upflow sand filter of U.S. Filter Corp.

**Vertimill™** An apparatus manufactured by Svedala Industries, Inc. for slaking lime.

**Verti-Press (1)** An inclined screw press manufactured by Dontech, Inc. for handling solids and screenings removed in wastewater treatment.

**Verti-Press (2)** A high-pressure filtration device manufactured by Filtra-Systems for sludge dewatering.

**Vertiscreen** A vertical stationary screen manufactured by Black Clawson-Shartles Division.

**very fine sand** Sediment particles of a diameter between 0.0025 inch and 0.005 inch (0.062–0.125 mm).

**very hard water** *See* water hardness.

**very small system** A public water supply system serving fewer than 500 people.

**very, very small system** A public water supply system serving 25 to 100 people.

**Vesilind equation** A relationship that defines the zone settling velocity (ZSV) as a function of mixed liquor suspended solids concentration ($X$) and two constants ($A$, $K$) from settling tests:

$$\mathrm{ZSV} = Ae^{-KX} \qquad \text{(V-09)}$$

**vested water right** The right granted by a state agency to use surface or ground water; a right not subject to defeat or cancellation by the act of any other person.

**veterite** A soluble, crystalline form of calcium carbonate, less stable than calcite.

**VF** Acronym of validation factor.

**VFA** Acronym of volatile fatty acid

**VFAR** Acronym of virulence factor activity relationship.

**viability** (1) The ability of a water supply or wastewater management system to provide services in accordance with federal and state regulations. (2) The ability of an organism to survive, grow, and develop. *See* viable.

**viability count** *See* direct viability count.

**viability index** In toxicology studies related to exposure to a substance, the viability index is the percentage of rodent offspring that survive four days past birth.

**viable** Pertaining, in general, to an organism that can survive, grow, and develop. However, some microbes are viable but nonculturable, having been dormant for some time and injured by adverse conditions such as nutrient deficiency. Viable but nonculturable pathogens can still infect and cause disease. That is the case of more than 99% of all soil microorganisms. *See also* culturable count, direct count, direct viability count.

**viable but nonculturable** *See* viable.

**viable population** A population having an adequate number and distribution of reproductive individuals to ensure its continued existence.

**Vibra-Matic™** A vibrating apparatus manufactured by U.S. Filter Corp. for cleaning pressure leaf filters.

**Vibrasieve®** A vibrating fine screen manufactured by Andritz-Ruthner, Inc.

**vibrating deck screen** A device equipped with cleaning brushes and used in sludge composting plants to recover bulking agents while improving the appearance of the compost.

**vibrating screen** A screening device with a vibrating surface to improve solids separation; sometimes used as a solids classifier.

**vibrio** Any of several Gram-negative, facultatively anaerobic, primarily aquatic bacteria having a curved shape (in the form of an S or a comma) of the genus *Vibrio;* some species are pathogenic for humans and other animals.

***Vibrio cholerae*** The genus of bacteria that cause epidemic cholera; they produce a substance that upsets the fluid balance in the digestive system leading to excessive fluid loss and death within hours. They are found in many habitats and are very resistant. Excreted load = 10,000,000 per g of feces. Persistence = 1 month. Median infective dose > 1,000,000. *See also* classical cholera, El Tor cholera.

***Vibrio cholerae* 01** *See Vibrio cholerae* serogroup 01.

***Vibrio cholerae* 0139** A strain of *Vibrio cholerae* that caused an epidemic in southern Asia beginning in 1992; practically indisguishable from *Vibrio cholerae* 01.

***Vibrio cholerae El Tor*** A particular strain of the Gram-negative bacteria *Vibrio cholerae* that causes Asiatic cholera in humans; Koch's bacillus. Also known as O1, it has been the dominant strain in the recent pandemic. It is distinguished from the classic strain at a genetic level, although both are in the O1 serogroup. It was first identified in 1905 at a camp in El-Tor, Egypt.

***Vibrio cholerae* serogroup O1** The vibrio strain that includes the variations classical and El Tor.

***Vibrio comma*** The species of comma-shaped bacteria that causes endemic cholera in Asia.

***Vibrio* genus** A group of heterotrophic organisms active in biological denitrification. *See* vibrio.

***Vibrio parahaemolyticus*** A species of bacteria, usually foodborne, that causes watery diarrhea, abdominal cramps, and sometimes nausea, vomiting, fever, and headache.

**vibrionaceae** A family of human enteric bacteria that includes the genus Vibrio.

***Vibrio vulnificus*** A species of bacteria responsible for water-related infections transmitted through the consumption of contaminated shellfish; its fatality rate is very high.

**vicinal water** Water that is bound to or held on the surface of solid particles during sludge dewatering; such water cannot be removed by mechanical means. Also called surface water. *See also* bulk water, interstitial water, and water of hydration.

**Viggers Valve** An automatic sludge blow-off valve manufactured by Walker Process Equipment Co.

**village-level operation and maintenance (VLOM)** Operation and maintenance activities that can be carried out by a village or local mechanic. The VLOM concept, developed at the World Bank in the 1980s, is indicative of simple and inexpensive technology or equipment that does not require a high level of knowledge, which is normally not available in a village in a developing country. An example in the rural water supply field is a hand-dug well or a borehole equipped with a hand pump that can be repaired by a bicycle or moped mechanic. A capped spring or a spring-fed gravity water supply system would also qualify as VLOM.

**Vincent Press** A horizontal dewatering screw press manufactured by Vincent Corp.

**vinegar acid** Same as acetic acid.

**vinyl benzene** Another name of styrene.

**vinyl chloride ($CH_2{=}CHCl$)** A colorless, flammable, easily liquefied, volatile organic, gaseous compound resulting from the halogenation of ethylene and used in producing some plastics, e.g., polyvinyl chloride or PVC for pipes, flooring, PVC resins, etc. It is believed to be oncogenic. Regulated by the USEPA: MCLG = 0 and MCL = 0.002 mg/L. Also called chloroethene, chloroethylene, monochloroethene, monochloroethylene, vinyl chloride monomer.

**vinyl chloride monomer (VCM)** *See* vinyl chloride.

**Vinyl Core** A PVC medium produced by B. F. Goodrich Co. for use in biological filters.

**vinylcyanide** Acrylonitrile.

**vinylethylene** Another name of butadiene.

**vinylidene chloride** Same as 1,1-dichloroethylene.

**vinyl trichloride** Another name of 1,1,2-trichloroethane.

**VIP latrine** Abbreviation of ventilated improved pit latrine.

**VIP process** *See* Virginia Initiative Plant process.

**VIRADEL** Acronym of virus adsorption elution.

**viradel method** An analytical technique using electropositive cartridge filters and an organic flocculation procedure to concentrate viruses in water.

**viral** Pertaining to or caused by viruses.

**viral aerosol** *See* aerosolized excreted virus, airborne droplet, airborne virus.

**viral diarrhea** A water-related and excreta-related disease caused by viruses, particularly rotavirus and Norwalk agent. *See also* gastroenteritis.

**viral disease** A disease caused by a virus; see excreted virus and mosquito-borne virus.

**viral gastroenteritis** Gastroenteric disease caused by viruses (astroviruses, rotaviruses, caliciviruses, coronaviruses, Norwalk agent).

**viral hepatitis** All types of hepatitis are caused by viruses. Hepatitis-A is sometimes called infectious hepatitis or viral hepatitis.

**viral infection** Infection caused by a virus.

**viral pathogen** A virus pathogenic to humans. More than 100 virus strains from a dozen groups are excreted in feces. Known pathogenic viruses include norovirus, hepatitis-A, rotavirus, adenovirus, astrovirus, Echo group, Reo group, poliovirus, Adeno group, Coxsackie A, and Coxsackie B.

**VIRALT** Abbreviation of viral transport, a semianalytical model developed for the USEPA to predict the concentrations of viruses in groundwater after subsurface transport. It may be used to define wellhead protection areas and contaminant breakthrough curves for wells.

**Virchem®** A chemical produced by Technical Products Corp. for corrosion control.

**viremia** The presence of a virus (or viruses) in the blood.

**virgin flow** The stream flow corresponding to natural conditions of a stream or its drainage basin. *See* natural flow.

**virgin granular activated carbon** Granular activated carbon (GAC) that has not been used; it has a higher adsorptive capacity than regenerated or reactivated GAC.

**virgin materials** Resources extracted from nature in their raw form, such as timber or metal ore.

**Virginia Initiative Plant (VIP) process** A variation of the University of Cape Town process, a wastewater treatment process designed to remove nitrogen and phosphorus beyond the metabolic requirements of the organisms. It consists of three biological reactors (in the sequence anaerobic → anoxic → aerobic) followed by a final clarifier. Each reactor is subdivided into two or more completely mixed cells in series. Sludge is recycled: (a) from the anoxic zone to the influent line and (b) from the mixed liquor and from the clarifier underflow to the inlet of the anoxic zone. *See also* combined nutrient removal, VIP process, UCT process, $A^2O$ process.

**virion** A mature virus; the infectious form of a virus outside a host cell; it consists of a protein-coated, nucleic acid core, sometimes in an envelope. Enteric virions find their way into wastewater via the feces of an infected host.

**viroid** An infectious agent of plants, similar to a virus, consisting of a short, single strand of RNA, without the protein coat. It is able to reproduce in several genetically different hosts and exists in vivo as free nucleic acid. *See also* prion.

**virology** The study of viruses and the diseases they cause.

**virulence** The ability of a microorganism to cause disease; the degree of pathogenicity of a disease-causing organism.

**virulence factor activity relationship (VFAR)** A genomic-based concept that can be used to assess risks posed by microbial contaminants such as unregulated waterborne pathogens, e.g., in establishing contaminant candidate lists.

**virus** (1) The smallest form of microorganism capable of causing disease (by reproduction), especially a virus of fecal origin that is infectious to humans by waterborne transmission. A virus is an obligate parasite that depends on nutrients inside host cells (plants, animals, bacteria); it consists of a strand of either DNA or RNA with a capsid (protein covering); it cannot reproduce outside the host cell, but it can commandeer the host cell's machinery for its reproduction. Viruses of interest in water and wastewater treatment are enteric viruses; they have a diameter between 20 and 100 nanometers (millimicrons, mμ) and are about 50 times smaller than bacteria. The largest virus particle (smallpox) is 200 nm long. The most common are adenovirus, coxsackievirus, echovirus, hepatitis-A virus, Norwalk-like viruses, poliovirus, and rotavirus; but human excreta contain more than 100 known virus serotypes. Removal of hepatitis-A virus and rotavirus by water coagulation and sedimentation exceeds 90%. The USEPA has set an MCLG = 0 for viruses in drinking water. *See also* acute virus infection, asymptomatic infection, bacterial density threshold, bacteriophage, capsid, chronic infection, cytopathogenic virus, emerging viral pathogen, human virus, isoelectric point, lysogeny, oncogenic virus, phage, prion, prophage, temperate phage, transduction, virion, viroid. (2) A computer program that reproduces itself by infecting other programs.

**virus adsorption elution (VIRADEL)** A method used to collect viruses from large volumes of water by passing the water through a microporous filter with pore size larger than the viruses, adsorption occurring through electrostatic and hydrophobic interactions. *See* electronegative filter, electropositive filter.

**virus concentration method** Any of a number of techniques used to concentrate viruses for enumeration and identification. Because of low numbers of viruses in water and wastewater, large volumes are used (e.g., 1000 gallons of water and 1+ gal of wastewater) are passed through a special filter that absorbs the viruses, which are then eluted and assayed.

**virus detection** Viruses are detected in water in two stages: concentration and then assay for identification and enumeration. *See* cytopathogenic effect method, plaque assay, plaque-forming unit method, serial dilution endpoint method.

**virus particle** Essentially, an inner nucleic acid genome in an outer protein capsid and, sometimes, an additional membrane envelope.

**viscera** Internal organs of animals.

**Viscomatic** An apparatus manufactured by Infilco Degremont, Inc. for slaking lime.

**viscometer** An instrument for measuring fluid viscosity.

**viscosity** A measure of the internal friction of a fluid that provides resistance to shear within the fluid. The greater the forces of internal friction (i.e., the greater the viscosity), the less easily the fluid will flow. Molecular viscosity or simply viscosity is the property of a fluid that makes it resist the tendency to deform or flow under external forces. For water and other Newtonian fluids, Newton's law of viscosity states that the viscous shearing stress ($\tau$) is the product of the coefficient of viscosity ($\mu$) by the velocity gradient ($\partial V/\partial s$) between the fluid layers:

$$\tau = \mu \cdot \partial V/\partial s \quad \text{(V-10)}$$

where $V$ is the mean velocity and $s$ the vertical distance. Absolute viscosity $\mu$ (or dynamic viscosity, also called coefficient of viscosity) is a measure of internal resistance to flow. Kinematic viscosity $\nu$ is the ratio of absolute viscosity to the fluid density $\rho$, i.e.,

$$\nu = \mu/\rho \quad \text{(V-11)}$$

Viscosity is an element of the Reynolds number, which indicates whether flow is laminar or turbulent, and of the hydraulic conductivity formula. Absolute viscosity varies with temperature; e.g., $\mu$ = 0.0008 kg/m/sec at 30°C and $\mu$ = 0.0018 kg/m/sec at 0°C for water.

**viscosity coefficient** *See* coefficient of viscosity.

**viscosity law** (1) Same as Newton's law. (2) *See* plastic fluid.

**viscous bulking** An operational problem of activated sludge plants in which an excess of extracellular hydrophilic biopolymer produces a slimy, jellylike sludge that has a low density and does not settle or compact well. The secretions constitute viscous cushions between flocs that prevent compaction. Viscous bulking is associated with nutrient deficiencies, very high organic loading, foaming, scum formation, and a rising sludge blanket in the clarifier. Also called hydrous bulking (because of the amount of water retained by the sludge), nonfilamentous bulking, and zoogleal bulking. *See also* dispersed growth, filamentous bulking, pin floc.

**viscous flow** Same as laminar flow.

**Viser** A computer program developed by OVRC Environmental, Inc. for sludge management.

**Vistex** An apparatus manufactured by Vulcan Industries, Inc. for grit removal.

**visual comparison method** An acceptable procedure for the analysis of color in water samples from the distribution system or from raw sources; the samples are compared to known concentrations of colored solutions or to calibrated colored glass disks.

**visual (turbidity) method** A standard measure of turbidity based on the observation of the outline of a candle. *See* Jackson turbidity unit.

**vitalistic theory** An interpretation proposed in the early 1900s for observed disinfection results; specifically, the gradual decline in the number of viable microorganisms with increased disinfectant concentration and/or exposure time (CT) is due to a distribution in the degree of resistance of the microorganisms to the disinfectant. Also called microbial diversity theory. *See also* disinfection kinetics.

**vitamin** An organic substance essential in small quantities to normal metabolism of heterotrophic organisms, found naturally in small quantities or produced synthetically.

**Vitox®** A wastewater treatment device manufactured by BOC Gases for oxygen injection.

**vitrification** The process of converting materials into a glass-like substance; typically a thermal process: the application of high temperatures (1000°C) pyrolizes, combusts, and stabilizes the materials into a moltenlike state. Radionuclides and other inorganics are chemically bonded in the glass matrix. Consequently, vitrified materials generally perform very well in leach tests. The USEPA has specified, under the land disposal restrictions, vitrification to be the treatment technology for high-level waste. Also called glassification. *See also* solidification/stabilization.

**vitrified clay** A heat-treated, brittle clay product, with an inert, impervious, and corrosion-resistant surface; used to make pipes, filter blocks, and bricks.

**vitrified tile block underdrain** Filter underdrainage system that includes a shallow gravel layer and no air scour.

**vitriol** A metallic sulfate of glassy appearance. *See* blue vitriol, green vitriol, white vitriol, oil of vitriol.

**vitriol oil** Same as oil of vitriol or sulfuric acid.

**vivax malaria** The type of severe malaria that is caused by the parasite *Plasmodium vivax* in humid tropical areas.

**vivianite [$Fe_3(PO_4)_2 \cdot 8\ H_2O$]** A secondary mineral, hydrous ferrous phosphate, in pale blue crystals or powder. It is also a compound formed during the corrosion of iron pipes (called an iron phase or a corrosion scale) or upon the addition of iron salts to heated sludge piping.

**VLOM** Acronym of village-level operation and maintenance.

**V-notched weir** or **V-notch weir** A weir with a V-shaped notch for measuring small flows. It is also used in water treatment units, e.g., mounted on the edges of clarifier effluent channels. The discharge equation for a V-notch weir is:

$$Q = 2.362\ C\ (\tan \alpha/2)H^{2.5} \qquad \text{(V-12)}$$

$$Q = 1.382\ H^{2.5} \qquad \text{for } \alpha = 90° \qquad \text{(V-13)}$$

where $Q$ = discharge, m/sec; $C$ = contraction coefficient, depending on the angle $\alpha$; $C = 0.585$ for $\alpha = 90°$; $\alpha$ = central angle of the weir; and $H$ = height of water surface above the bottom of the notch, m. Also called triangular weir, triangular-notch weir or Vee weir.

**V-notch weir** Same as V-notched weir.

**VO** Chemical formula of the vanadyl cation.

**$VO_4$** Chemical formula of the vanadate anion.

**VOC** Acronym of volatile organic compound.

**VOCarb** A unit manufactured by Wheelabrator Clean Water Systems for the removal of volatile organic compounds by activated carbon.

**VOC control** The removal of volatile or semi-volatile organic compounds from water or wastewater by such methods as gas stripping and diffused-air or mechanical aeration.

**VOC polarizability** A parameter used in the design of air stripping equipment. *See* Lorentz–Lorentz equation.

**VOC refractive index** A parameter used in the design of air stripping equipment. *See* Lorentz–Lorentz equation.

**VOC Wagon™** A self-contained thermal oxidizer manufactured by NAO, Inc. for the removal of volatile organic compounds.

**void** The pore space or other openings in a rock or granular bed.

**void ratio** The ratio of void volume in a filter medium to the total volume occupied. *See also* porosity, specific surface.

**void space** The volume of void in a filter medium, ion exchanger, or other granular bed; often expressed as a percent of total volume of the bed.

**void volume** *See* void space.

**volatile** Pertaining to a substance such as benzene or chloroform that readily evaporates or vaporizes at a relatively low temperature and a given pressure (having a low boiling point or subliming pressure).

**volatile acid** An acid produced during digestion; a fatty acid that is soluble in water and can be steam-distilled at atmospheric pressure. Also called organic acid. Volatile acids are commonly reported as equivalent to acetic acid; in the anaerobic digestion process they further break down into methane and carbon dioxide.

**volatile constituent** A component of a substance that is readily lost by evaporation, e.g., dissolved gases and substances with low boiling points.

**volatile fatty acid (VFA)** An end product of fermentation processes; e.g., acetic acid, formic acid, propionic acid. VFAs are present in wastewater and play an important role in such processes as denitrification and enhanced biological phosphorus removal. *See also* sidestream fermentation.

**volatile liquid** A liquid that easily vaporizes or evaporates at room temperature.

**volatile matter** The matter that is apparently lost from a residue ignited at 550 ± 25°C for about 10–15 minutes.

**volatile organic carbon** The portion of total organic carbon removed by gas stripping; purgeable organic carbon.

**volatile organic chemical** *See* volatile organic compound.

**volatile organic compound (VOC)** A group of lightweight gases and volatile liquids that produce vapors readily. At room temperature and normal atmospheric pressure, vapors escape easily from volatile liquid chemicals. Organic compounds with a boiling point (BP) lower than 100°C and a vapor pressure (VP) greater than 1.0 mm Hg at 25°C are generally classified as VOCs; e.g., vinyl chloride: BP = –13.9°C and VP = 2548 mm Hg at 20°C. Volatile organic chemicals include gasoline, industrial chemicals such as benzene, solvents such as toluene and xylene, mercaptans, butyric acid, and tetrachloroethylene. Industrial wastewater may contain VOCs that are flammable, toxic, and odorous. Many volatile organic chemicals are also hazardous air pollutants; e.g., benzene causes cancer. The USEPA designates as VOC any organic compound that participates in atmospheric photochemical reactions, except those having negligible photochemical reactivity. VOCs can cause cancer or chronic damage to the liver, kidneys, and nervous system. Maximum contaminant levels established for VOCs in drinking water are very low, from 0.002 mg/L for vinyl chloride to 0.7 mg/L for ethylbenzene. Drinking water treatment by coagulation–sedimentation–filtration does not remove or alter volatile organic chemicals, which include a number of trihalomethanes.

VOCs are all synthetic organic chemicals (SOCs) but the two groups are treated separately in drinking water requlations.

**volatile organic contaminant** A volatile organic compound.

**volatile organic liquid (VOL)** Any organic liquid that can emit volatile organic compounds into the atmosphere that contribute appreciably to the formation of ozone (EPA-40CFR60.111b-k).

**volatile organic material (VOL)** Same as volatile organic compound.

**volatile organics** *See* volatile organic compound.

**volatile organic substance** Same as volatile organic compound.

**volatile residue** Volatile solids.

**volatile solids (VS)** Those solids (generally organic) suspended or dissolved in water or other liquid that are lost on ignition of a dry sample in the presence of excess air, at a specified temperature (e.g., 500 ± 50°C) and for a specified time (e.g., 15 to 20 minutes). The residue is ash, fixed solids, or nonvolatile inorganic solids. In wastewater treatment, VS provides a measure of biodegradable organic matter but the two are not always equivalent; ignition may volatilize some inorganics and convert some pure organics to ashes.

**volatile solids content** The weight or concentration of volatile solids in a substance; commonly used as an indicator of the amount of organic matter, e.g., in wastewater treatment sludge.

**volatile solids destruction** *See* sludge stabilization performance, anaerobic digestion.

**volatile solids loading** The mass of volatile solids (VS) applied per day per unit volume of a digester, expressed in kg VS/m$^3$/day or in lb/ft$^3$/day. It is one of the digester design parameters; *see also* solids retention time (SRT) and hydraulic retention time (HRT).

**volatile solids reduction (VSR)** The change in biodegradable volatile solids as a result of aerobic or anaerobic digestion, a function of the mass of solids in the digester and of a reaction rate constant. It also varies with temperature and solids retention time. Another criterion of sludge stabilization is pathogen-indicator-organism reduction.

**volatile suspended solids (VSS)** The volatile fraction of suspended solids in water or wastewater. It includes organic matter and volatile inorganic salts that ignite and burn when a dry sample is heated at 500 ± 50°C for 60 minutes in the presence of oxygen. In the activated sludge process, volatile suspended solids accumulate at a rate $\Delta y$ (pounds per day) that depends on the BOD of the influent ($X$, pounds per day) and the weight of mixed liquor suspended solids ($Z$, pounds):

$$\Delta y = 0.5\,X - 0.055\,Z \qquad \text{(V-14)}$$

**volatile synthetic organic chemical** A chemical that tends to volatilize or evaporate.

**volatilization** (1) Loss of a substance to the air from a surface or from solution through evaporation. (2) The process of transfer of a chemical from the aqueous or solid phase to the gas phase. Solubility, molecular weight, and vapor pressure of the liquid and the nature of the gas–liquid interface affect the rate of volatilization. The rate of volatilization per unit time per unit volume ($R_v$) is

$$R_v = -K_v(C - C_s) \qquad \text{(V-15)}$$

where $K_v$ = volatilization constant,/time; $C$ = constituent concentration in the liquid phase, e.g., mg/L; and $C_s$ = saturation concentration of the constituent in the liquid, e.g., mg/L. The release of volatile organic compounds from wastewater surfaces is an example of volatilization.

**volatilization factor** A parameter that indicates the fraction of ammonia nitrogen that volatilizes during land application of wastewater treatment sludge. *See* plant available nitrogen.

**volcanic spring** A spring with water brought to the surface from considerable depths by volcanic forces.

**volcanic water** *See* volcanic spring.

**Volcano™** A downflow sand filter manufactured by Lighthouse Separation Systems, Inc.

**Volclay®** Liners in sodium bentonite manufactured by Colloid Environmental Technologies, Inc. for use as soil sealants.

**voltage-driven membrane** A membrane used in water treatment by electrodialysis.

**volume cap** A limit imposed by federal laws on the total amount of tax-exempt bonds that private activities (including water and sewer activities) can issue annually within a state. Utilities often lose to schools and housing in the competition for such bonds.

**volume charge** A water or wastewater service rate, or the portion thereof, that is based on volume.

**volume ratio** A measure of concentration, e.g., milliliter of a constituent per liter of a solvent (mL/L).

**volume reduction** A method of processing waste materials to decrease the amount of space they occupy, usually by compacting or shredding, incineration, or composting. Wastewater volume may also be reduced through reuse, recycling, and reduction of water uses.

**Volumeter™** A flow monitoring device manufactured by Marsh-McBirney, Inc.

**volumetric** Pertaining to a measurement based on the volume of some factor. *See also* gravimetric.

**volumetric analysis** A quantitative analysis, such as volumetric titration, based on the addition of known volumes of standardized solutions.

**volumetric concentration** A parameter used in the theory of flocculation, representing the floc volume fraction or the volume of floc divided by the volume suspension. If $N$ is the number of flocs and $D$ their diameter, the volumetric concentration ($\Phi$) is:

$$\Phi = N\pi D^3/6 \qquad \text{(V-16)}$$

Volumetric concentration contributes to the probability of particle collision and, thus, to floc formation and flocculation performance.

**volumetric dry feeder** Same as volumetric feeder.

**volumetric efficiency** (1) One of the terms used to characterize tracer response curves and describe the hydraulic performance of reactors, it is equal to 100 times the reciprocal of the Morrill dispersion index, proposed by A. B. Morrill in 1932. *See* Morrill dispersion index for more detail. (2) The volumetric efficiency of a hydraulically driven reciprocating piston pump is the ratio of volume of sludge solids pumped per piston stroke to the total volume displaced per piston stroke.

**volumetric feeder** A mechanical device used in water and wastewater treatment to measure and supply specific volumes of powdered chemicals (e.g., lime in sludge stabilization), which may be preset or determined proportionally to the flow. Volumetric feeders are simpler and less expensive, but also less accurate than gravimetric feeders (the other type of dry feeder). Also called volumetric dry feeder. *See also* solution feeder, gravimetric feeder, and slaker.

**volumetric flask** A bottle with a long and narrow neck used in the laboratory to prepare single-volume solutions.

**volumetric loading** (1) The amount of organic matter, expressed as BOD or COD, applied to the aeration basin of a suspended-growth process per unit volume per day:

$$L_0 = 0.001\ QS_0/V \qquad \text{(V-17)}$$

where $L_0$ is the volumetric organic loading, kg of BOD or COD/m$^3$/day; $Q$ is the influent wastewater flowrate, m$^3$/day; $S_0$ is the influent BOD or COD concentration, mg/L; and V is the volume of aeration basin, m$^3$. *See also* organic loading, hydraulic loading. (2) The amount of organic matter, expressed as BOD or COD, applied per unit weight of mixed liquor suspended solids per day to the aeration basin of a suspended growth process or to the aerobic biomass of a trickling filter.

**volumetric mass transfer coefficient** A parameter used in mass transfer formulas; it depends on water quality and type of equipment. *See* $K_La$.

**volumetric organic loading** *See* volumetric loading.

**volumetric oxidation rate (VOR)** A parameter that characterizes the oxidation of biochemical oxygen demand (BOD) and ammonia nitrogen ($NH_3$-N) in plastic-packing trickling filters; expressed in kilograms per cubic meter per day (or in pounds per cubic foot per day):

$$\text{VOR} = 0.001\ Q\,(S_0 + 4.6\,N)/V \qquad \text{(V-18)}$$

where $Q$ = influent flowrate, m$^3$/day; $S_0$ = influent BOD concentration, mg/L; $N$ = amount of ammonia nitrogen oxidized, g/m$^3$; and $V$ = volume of trickling filter packing, m$^3$. *See also* trickling filter nitrification rate.

**volumetric performance** The rate of BOD removal by a trickling filter with respect to the hydraulic application rate. Deeper towers tend to promote better hydraulic distribution, higher hydraulic rates, and better performance.

**volumetric pipette** A pipette calibrated for a single volume; also called a transfer pipette.

**volumetric respirometer** An instrument for measuring the character and extent of respiration (e.g., oxygen consumption and carbon dioxide production), usually equipped for data collection and processing by computer. It measures oxygen uptake in incremental changes in gas volume while maintaining pressure constant. *See* respirometric method, Gilson respirometer, Warburg respirometer.

**volumetric tank test** One of several tests to determine the physical integrity of a storage tank; the volume of fluid in the tank is measured directly or calculated from product-level changes. A marked drop in volume indicates a leak.

**volumetric titration** A means of measuring unknown concentrations of water quality indicators in a sample by determining the volume of titrant or liquid reagent needed to complete particular reactions.

**volute** The spiral casing surrounding the impeller of a centrifugal pump, designed for a smooth conversion of pump rotation to pressure.

**volute pump** A centrifugal pump in a spiral casing; *see* volute.

**volutin** Protein granules in the cytoplasm that certain bacteria use to store inorganic phosphates.

**VO Nozzle®** A variable-orifice spray nozzle manufactured by FMC Corp.

**von Siemens, Werner** German electrical engineer (1816–1892) who introduced the production of ozone by corona discharge.

**VOR** Acronym of volumetric oxidation rate.

**Vortair®** A low-speed surface aerator manufactured by Infilco Degremont, Inc.

**vortex** A revolving mass of water that forms a whirlpool, caused by water flowing out of a small opening in the bottom of a basin or reservoir. A funnel-shaped opening is created downward from the water surface. In a vortex, the streamlines are concentric circles and solid particles are drawn toward the center of the cavity.

**Vortex™** A circular grit-removal device manufactured by Infilco Degremont, Inc., including a turbine-type rotor.

**vortex flow pump** A centrifugal pump, with a multivaned, semiopen impeller recessed from the volute area that generates a whirling motion; used for grit pumping and sludge recirculation. Also called torque flow pump.

**vortex grit chamber** *See* vortex grit removal.

**vortex grit removal** A mechanism that uses a mechanically induced vortex to capture grit solids in the center hopper of a cylindrical chamber with flat or sloping bottom. The mechanism combines flow through an inlet flume, an inlet baffle, and rotating paddles in the center to create a spiraling, doughnut-shaped flow pattern. The resulting centrifugal and gravitational forces cause the grit to separate from the liquid. *See* Pista® grit chamber, Teacup™.

**vortexing** The phenomenon of mass swirling of water or wastewater, observed in propeller or turbine mixers when the liquid rotates with the impeller and causes a decrease in mixing effectiveness. The installation of vertical baffles helps to reduce vortexing.

**vortex pump** A pump with a recessed impeller, standard concentric casing, axial suction, and tangential discharge opening, used in handling sludges and slurries; it draws liquid and solid particles into the pump cavity in a swirling motion. Also called recessed impeller pump.

**vortex separator** A cylindrical device equipped with a tangential inlet, used, with or without chemical additions, to separate suspended solids from water, combined sewer overflows, and peak wet-weather wastewater flows. Sludge accumulates at the center of the bottom by gravity and under the action of secondary currents and centrifugal forces. Vortex separators are manufactured under several designs, e.g., the USEPA Swirl concentrator, the FluidSep™, and the Storm King™. Also called swirl concentrator or hydrodynamic vortex separator. *See also* continuous deflection separator.

***Vorticella*** One of the groups of ciliate protozoa active in trickling filters, feeding on the biological film and maintaining its high growth rate.

**Vorti-Mix®** A submerged turbine aerator manufactured by Infilco Degremont, Inc.

**Vostrip™** An air stripping tower manufactured by EnviroSystems Supply, Inc.

**Voxsan®** A proprietary bleach that contains about 4% of available chlorine; it can be used to make a 1% stock solution for water disinfection. Such a solution is unstable under warm conditions.

***V. parahaemolyticus*** *See Vibrio parahaemolyticus.*

**VPS™** Acronym of vapor phase system, an apparatus manufactured by NuTech Environmental Corp. for odor control.

**VR-ditch process** A proprietary wastewater treatment technique that combines the oxidation ditch and sequencing batch reactor processes for the removal of biochemical oxygen demand, suspended solids, and ammonia nitrogen ($NH_3$-N). It includes two concentric oxidation ditches equipped with a rotor aerator. There is no recirculation and no separate clarification. *See* phased isolation ditch for other related processes. *See also* biological nitrogen removal.

**VS** Acronym of volatile solids.

**VSB** Acronym of vegetated submerged bed.

**V*Sep®** A membrane filtration process manufactured by New Logic International; uses vibration to control fouling.

**VSR** Acronym of volatile solids reduction.

**VSS** Acronym of volatile suspended solids.

**VTC™** Acronym of vertical tube coalescing, a device manufactured by AFL Industries, Inc. for oil/water separation.

**VTSH®** Acronym of vertical turbine solids handling, a pump manufactured by Fairbanks Morse Pump Corp.

**VTX®** A vertical vortex pump manufactured by Yeomans Chicago Corp.

**vulnerability** The likelihood that an action or a condition will adversely affect a public water supply system. As defined in the USEPA's rules for standards and monitoring requirements related to contaminants in drinking water, the states determine vulnerability based on (a) previously recorded monitoring results, (b) proximity of the water supply infrastructure to sources of contamination, and (c) level of protection of water sources.

**vulnerability analysis** Assessment of elements in the community that are susceptible to damage should a release of hazardous materials occur.

**vulnerability assessment** Assessment of the susceptibility of a system to damage. (1) A requirement of the Public Health Security and Bioterrorism Preparedness and Response Act of 2002, according to which drinking water suppliers are to identify threats that would "substantially disrupt the ability of a system to provide a safe and reliable supply of drinking water." The assessment concerns facilities as well as information systems. (2) A process that allows a state, a water supply system, or a third party to request a waiver from some of the monitoring requirements of USEPA's Phase II Rule.

**V. vulnificus** *See Vibrio vulnificus.*

**Vydate®** An insecticide and nematicide produced by the DuPont Company. *See* oxamyl.

# W

W W W W W W W W W W W W W W W W W W W W W

**W** Chemical symbol of tungsten (from its German name wolfram).

***W. bancrofti*** *See Wuchereria bancrofti.*

**WAC** Acronym of weak-acid cation exchanger.

**WAC®** Acronym of weak-acid cation exchanger, a polyaluminum chloride flocculant produced by Elf Atochem North America.

**WAC resin** Weak-acid cation resin.

**WACX** Water and wastewater treatment products of Dean Wacx.

**wadi** In Northern African and Arabian countries, a river bed or intermittent stream that carries flash flood water; the equivalent of arroyo or wash in the western or southwest United States.

**waffle-bottom anaerobic digester** An anaerobic digestion with a bottom in the shape of a waffle to reduce the accumulation of grit and the frequency of cleaning. It is an alternative to the cylindrical sludge digestion tank, which has a conical bottom sloping toward the center.

**Wagner** An underdrainage system manufactured by Infilco Degremont, Inc. for sand filters.

**WAIE** Acronym of wind-aided intensified evaporation.

**waiver** A document issued by a primacy agency allowing a water supply system to reduce or eliminate the required monitoring for a specific contaminant.

**waiver for filtration** A document issued by a primacy agency allowing a water supply system to distribute water from a given source without filtration.

**walking beam flocculator** (1) A flocculation apparatus consisting of mixing paddles attached to an oscillating horizontal beam. (2) A conical device mounted on rods to produce slow mixing of water by the alternating raising and lowering of the rods. Also called a reciprocating flocculator.

**wall demand** A source of disinfectant loss due to reaction with scales on the surface of pipes.

**wall effect** (1) The interference of the walls of narrow containers in the settling of particles. *See* hindered settling. (2) In a packed tower used for stripping volatile organic chemicals (VOC) from water or wastewater, wall effect refers to the increase in the volumetric mass transfer coefficient ($K_L a$), as the tower diameter increases, due to water channeling down the inside of the column walls. *See also* cascade air stripping, cocurrent packed tower, crossflow packed tower, cascade packed tower, low-profile air stripper, minimum air-to-water ratio, packed tower design equation, sieve tray column, end effect.

**Wand Inductor** An apparatus manufactured by Biological Systems USA, Inc. for the remediation of contaminated soils.

**WAO** Acronym of wet-air oxidation.

**WAR process** *See* wet-air regeneration process.

**Warburg apparatus** Same as Warburg respirometer.

**Warburg method** A manometric technique that measures, from a change in pressure, the oxygen consumed by bacteria as they respire. *See* Warburg respirometer.

**Warburg respirometer** A device that measures oxygen consumption while the gas volume is held constant, e.g., the biochemical oxygen demand (BOD) of a wastewater sample. It consists of a manometer to measure gas consumption, connected to a flask that contains the sample and seed culture. In a separate container, a potassium hydroxide (KOH) solution absorbs the carbon dioxide ($CO_2$) produced. Currently, the electrolysis cell respirometer is preferred for the determination of BOD. *See also* respirometric method.

**warm spring** A thermal spring that discharges water with a temperature below that of the human body (37°C or 98.6°F).

**warm-up stage** The first of three stages of thermal sludge drying, in which sludge temperature and drying rate increase steadily until they reach a constant rate. *See also* constant-rate stage, falling-rate stage.

**warp** The vertical wire in woven wire mesh.

**WAS** Acronym of waste activated sludge.

**wash** Any of a number of water courses or water bodies: bog, depression or channel formed by flowing water, fen, land washed by the sea or a river, marsh, shallow arm of the sea, shallow part of a river, shallow pool, or small stream. (2) The dry bed of an intermittent stream in the western United States. Also called dry wash. *See also* arroyo, wadi. (3) Alluvial material carried and deposited by flowing water.

**washbox** A baffled compartment incorporated in a moving-bed filter to allow cleaned sand to separate by gravity from the concentrated waste solids. The sand returns to the top of the bed and the solids are removed for disposal.

**washer** *See* bedpan washer.

**washing soda [$Na_2CO_3 \cdot 10\ H_2O$]** *See* sodium carbonate decahydrate.

**washload** The portion of the suspended solid load composed of smaller particles than those generally found in the streambed, particles coming from the banks and the watershed, transported without deposition. Also known as fine sediment load. *See also* bed load.

**washload discharge** *See* fine-sediment discharge.

**washoff** The transport of sediment or other pollutants out of ponds, watersheds, etc. during a storm event.

**washout** *See* rainout.

**washout valve** A valve used to drain or flush a pipeline. Also called blowoff valve, scour valve.

**washout velocity** The velocity ($V_t$, in m/min) that will cause the filter medium to wash out and thus damage the filter during backwash. *See* backwash velocity for more detail. Also called terminal velocity.

**Washpactor** An apparatus manufactured by Jones & Attwood, Inc. for washing and compacting screenings from wastewater treatment.

**Wash Press** An apparatus manufactured by Lakeside Equipment Corp. for washing and compacting screenings from wastewater treatment.

**wash trough** Same as washwater trough.

**washwater** Same as backwash water. Also, water used to clean any water treatment process equipment.

**washwater gullet** An overflow structure placed on the side of a filter box to carry rising water and accumulated contaminants from the top of the bed during backwash; more common in Europe than in the United States. *See* washwater trough.

**washwater gutter** An overflow structure placed above the fluidized bed of a filter to carry away rising water and accumulated contaminants from the top of the bed during backwash. *See* washwater trough for more detail.

**washwater rate** *See* backwash rate.

**washwater tank** An elevated tank used in a water treatment plant to store water for filter backwash.

**washwater trough** An overflow structure placed above the fluidized bed of a filter to carry away rising water (backwash water) and accumulated contaminants from the top of the bed during backwash; more common in the United States than in Europe. The trough runs the length or width of the filter and discharges into a gullet. Washwater troughs are often open conduits constructed of fiberglass, plastic, or sheet metal, or of concrete with adjustable weir plates. Also called washwater gutter, wash trough. *See* washwater gullet.

**waste** Unwanted materials left over from a manufacturing process; refuse from places of human or animal habitation; wastewater, household refuse, industrial waste.

**waste activated sludge (WAS)** The solids removed from an activated sludge treatment plant for further processing and disposal, thus preventing an excessive build-up. Its solids content is about 1.0%. Sludge can be wasted either from the aeration tank mixed liquor or from the return activated sludge line. Also called secondary sludge. *See also* biosolids.

**waste assimilation** The process of reducing the concentration of wastewater or the ability of a stream to absorb a waste discharge.

**waste-boiler-flue gas treatment** A method used to neutralize alkaline wastes by blowing boiler-flue gas through them. The gas contains carbon dioxide ($CO_2$), which forms carbonic acid ($H_2CO_3$) in the wastewater and reacts with the caustic waste, alkalinity being represented by sodium hydroxide (NaOH) or soda ash (sodium carbonate, $Na_2CO_3$):

$$CO_2 + H_2O \rightarrow H_2CO_3 \quad \text{(W-01)}$$

$$H_2CO_3 + 2\,NaOH \rightarrow Na_2CO_3 + 2\,H_2O \quad \text{(W-02)}$$

$$H_2CO_3 + Na_2CO_3 + H_2O \rightarrow 2\,NaHCO_3 + H_2O \quad \text{(W-03)}$$

When the wastewater contains a sufficient concentration of sulfur, the flue gas will cause the formation of hydrogen sulfide, which must be burned or otherwise disposed of. Other methods of neutralization of alkaline wastes include carbon dioxide treatment, submerged combustion, alkaline waste fermentation, and sulfuric acid treatment.

**waste brine** The waste stream resulting from the regeneration of ion-exchange resins. A strong sodium solution or another regenerant with a high salt content is typically used. The volume of waste brine is about 5% of the feedwater. *See also* reject brine.

**waste characterization** Identification of chemical and microbiological constituents of a waste material.

**waste disposal plant** A plant equipped to treat and dispose of wastes. *See* wastewater treatment plant.

**waste effluent** The mixture of liquid and solid matter discharged from industrial processes. *See also* emission.

**waste equalization** *See* equalization, flow equalization.

**waste exchange** An arrangement used in waste minimization or pollution prevention programs; a public or private organization matches the waste or by-product of one manufacturer with the need of another manufacturer, thus reducing disposal costs and promoting reuse. Also called materials exchange.

**waste feed** The continuous or intermittent flow of wastes into an incinerator.

**waste-gas burner** A device used to burn gas from a sludge digestion tank.

**waste gate** A gate used to discharge surplus water from a canal.

**waste heat** Heat released to the environment, e.g., in the form of cooling water discharge with a temperature 10–15°C above ambient temperature. Waste heat can be reused in desalination, aquaculture, or for heating.

**waste-heat evaporator** An evaporator that uses the heat from another device (e.g., a gas turbine, diesel engine).

**waste industry** A relatively new activity that involves the recycling of materials, e.g., the rebuilding and resale of carburetors and generators from junked cars.

**waste load allocation (WLA)** The maximum load of pollutants each discharger of waste is allowed to release into a particular waterway; the portion of a stream's assimilative capacity assigned to an individual discharge Discharge limits are usually required for each specific water-quality criterion being, or expected to be, violated. WLAs constitute a type of water-quality-based effluent limitation.

**waste management** The task of identifying the sources of air, water, and solid wastes in a community and implementing appropriate methods of pollution control and abatement.

**waste metering** The process of identifying the leaks and other water losses in a distribution system, e.g., by measuring the water flowing to an isolated portion of the network and the actual water billed to consumers.

**waste minimization** An activity that involves measures or techniques to reduce the amount of wastes generated during industrial production processes. The term is also applied to recycling and other efforts to reduce the amount of material going into the waste stream.

**waste mixing** The neutralization of acid and alkaline wastes by mixing them in appropriate proportion before industrial waste treatment or discharge into a municipal sewer.

**waste neutralization** The conversion of organic matter in liquid or solid wastes to inert, harmless material. *See* neutralization, stabilization.

**waste oil** (1) Used lubricating or automotive crankcase oil that is usually treated for reuse or disposal. (2) Oily water mixture originating from ship discharges.

**waste oil emulsion** A thick and viscous mixture that is an emulsion of water in waste oil.

**waste-oxidation basin** An earlier name given to lagoons and ponds used for the storage of industrial wastes when it was observed that decomposition takes place in these basins; same as oxidation pond.

**waste pipe** A pipe conveying wastewater from building fixtures, except from water closets, to the house drain.

**waste proportioning** *See* industrial waste proportioning.

**waste ratio** The ratio of waste to total output of an industrial activity, a measure of manufacturing efficiency.

**waste recovery** *See* recycling.

**waste recycling** *See* recycling.

**waste reduction** The use of such measures as source reduction, recycling, or composting to prevent or reduce waste generation. *See also* volume reduction.

**waste reduction techniques** Common methods used by industry to reduce manufacturing wastes include (a) inventory management (inventory control and material control), (b) production process modification (operation and maintenance procedures, material change, process equipment modification), (c) volume reduction through source segregation and concentration, and (d) on-site or off-site recovery. *See* 33/50 Program.

**waste segregation** *See* industrial waste segregation.

**waste sludge** In the activated sludge or any other biological treatment process, it is the sludge that is produced in excess of the required solids recirculation and that does not escape in the effluent. It is withdrawn from the system and wasted. Also called sludge excess or waste activated sludge. *See also* return sludge.

**waste solids** The by-product slurries or sludges separated from water or wastewater during treatment, for example: detritus, floating debris, grit, mineral solids from river water, precipitates from chemical treatment, rakings, screenings, skimmings, slurries from chemical softening, and underflows from settling tanks. Residues from treatment plants range from incinerator ash to raw solids having a moisture content above 95%. Solids handling and disposal accounts for approximately 35% of the capital costs, 50% of the operation and maintenance costs, and 90% of the operational problems of a wastewater treatment plant.

**waste stabilization** The treatment of waste to make it fit for disposal, e.g., by converting its active organic matter to inert, harmless materials.

**waste stabilization pond** A large, shallow pond that receives raw or partially treated wastewater. *See* stabilization pond, anaerobic pretreatment pond, facultative pond, maturation pond for more detail.

**waste stream** (1) The total flow of solid waste (from homes, businesses, institutions, and manufacturing plants) that is recycled, burned, or disposed of in landfills, or segments thereof, such as the residential waste stream or the recyclable waste stream. (2) In general, the waste stream containing the pollutants removed by a treatment process. In particular, the portion of the feed stream that contains the retained, dissolved or suspended constituents in membrane filtration processes. *See* reject for more detail.

**waste treatability** *See* wastewater treatability.

**waste treatment** *See* wastewater treatment.

**waste treatment lagoon** An impoundment made by excavation or earthfill for biological treatment of wastewater.

**waste treatment plant** A facility containing a series of tanks, screens, filters and other processes by which pollutants are removed from water or wastewater.

**waste treatment stream** The continuous movement of waste from generator to treater and disposer.

**waste utilization** The use of disposed-of solid waste for some beneficial purpose, e.g., the conversion of processed solid waste into compost or soil amendment, or the use of a completed sanitary landfill as a golf course. *See also* recycling.

**Wastewarrior** A wastewater treatment apparatus manufactured by Hyde Products, Inc. using ultrafiltration.

**wastewater** (1) The spent or used water from residences, community, farms, institutions, commercial establishments, or industry that contains dissolved or suspended matter as well as a certain amount of groundwater, surface water, and stormwater. *See also* black water, graywater, industrial wastewater, infiltration/inflow, municipal wastewater, sewage, sullage. (2) Organic hazardous air-pollutant-containing water, raw material, intermediate product, byproduct, coproduct, or waste material that exits equipment in a chemical manufacturing process; and either (a) contains a total volatile organic hazardous air-pollutant concentration of at least 5 parts per million by weight and has a flow rate of 0.02 liter per minute or greater; or (b) contains a total volatile organic hazardous air-pollutant concentration of at least 10,000 parts per million by weight at any flow rate. Wastewater includes process wastewater and maintenance wastewater (EPA-40CFR63.101).

**wastewater aeration systems** *See* diffused-air aeration, mechanical aeration, high-purity oxygen, submerged aeration, surface aeration, cascade aeration.

**wastewater age** The relative amount of ammonia present in wastewater, based on the observation that bacterial decomposition readily converts organic nitrogen to ammonia. The age or condition of a wastewater may also be determined qualitatively by its color and odor. Fresh wastewater is light brownish, whereas septic wastewater is black.

**wastewater analysis** The examination of a sample of wastewater to determine its biological condition, chemical composition, and physical composition. Tests generally conducted include concentration or strength parameters (e.g. $BOD_5$, $NH_3$, DO), treatment efficiency, and functional tests.

**wastewater biomonitoring** The use of bioassays to determine the toxicity of wastewater effluents.

**wastewater blending** *See* blending.

**wastewater characteristics** *See* constituents, wastewater constituents.

**wastewater characterization** The determination of the constituents and composition of wastewater, an important step in the planning, design, and performance evaluation of a treatment plant.

**wastewater charge** A specific charge made for providing wastewater collection, treatment, and disposal services.

**wastewater clarification (primary)** *See* primary treatment.

**wastewater clarification (secondary)** *See* secondary clarifier.

**wastewater collection system** *See* sewer collection system.

**wastewater collection system integrity** A key indicator that wastewater utilities can use to assess their performance compared to established standards. It measures the frequency of collection system failures per mile or per 100 miles.

**wastewater composition** The various liquid, solid, gaseous, chemical, physical, and biological constituents of wastewater, including their concentrations. Common tests measuring the composition of wastewater include solids, $BOD_5$, COD, TOC, N, $NH_3$, and $PO_4$. *See* Table W-01 for the composition of an average domestic wastewater, the COD of which is about 18% protein, 12% carbohydrates, and 21% lipids.

**wastewater condition** A biological and physical characterization of the state of a wastewater sample, as to its stage of decomposition, odor, pH, whether it is fresh or stale, aerobic or anaerobic, etc. *See also* wastewater age.

**Table W-01.** Wastewater composition (average domestic)*

| Constituent | Concentration |
|---|---|
| Bacteria | $10^8$ |
| Ammonia ($NH_3$), as N | 35 |
| $BOD_5$ | 200 |
| Nitrate ($NO_3$) and nitrite ($NO_2$), as N | 0.6 |
| Organic nitrogen, as N | 55 |
| Phosphorus, soluble, as P | 5 |
| Phosphorus, total, as P | 9 |
| Solids, total | 450 |
| Solids, suspended | 250 |
| Solids, total dissolved | 200 |
| Solids, volatile suspended | 200 |
| Solids, total volatile | 300 |

*Concentrations in mg/L but bacteria as number per 100 mL.

**wastewater constituents** Individual components, elements, or entities of wastewater. Table W-02 lists the principal biological, chemical, and physical characteristic of wastewater.

**wastewater constituents of concern** Constituents of concern for wastewater treatability include: biodegradable organics, dissolved inorganics, heavy metals, nutrients, pathogens, priority pollutants, refractory organics, and suspended solids.

**wastewater control** One of the branches of public health engineering and environmental engineering, dealing with (a) domestic and industrial wastewater discharges and their impact on the quality of receiving waters, and (b) the design, operation, and monitoring of wastewater collection and wastewater treatment facilities.

**wastewater decomposition** The biological or chemical transformation of organic and inorganic matter in wastewater.

**wastewater dilution** *See* dilution, a method of reducing the concentration of pollutants in an effluent by discharge into a receiving stream.

**wastewater disposal** The final discharge of wastewater by such methods as dilution, dispersion, evaporation, broad irrigation, privy, cesspool.

**wastewater ejector** *See* wastewater ejector.

**wastewater engineering** The branch of environmental engineering that deals with the collection, conveyance, treatment, disposal, or reuse of wastewater.

**wastewater evaporation** A method of final wastewater disposal, using for example an oxidation pond without an outlet, where annual precipitation does not exceed a certain level (e.g., 70% of the pan evaporation). *See also* total retention system.

**Table W-02.** Wastewater constituents (common analyses)

| | | |
|---|---|---|
| *Biological* | *Chemical (inorganic)* | *Physical* |
| Coliforms | Alkalinity | Colloidal solids |
| Bacteria, helminths, protozoa, viruses, and other specific organisms | Ammonia (free) | Color |
| Toxic substances | Chloride | Conductivity |
| | Gases (ammonia, carbon dioxide, hydrogen sulfide, methane, oxygen) | Density |
| *Chemical (organic)* | Metals* | Fixed suspended solids |
| Antibodies (human and veterinary)† | Nitrogen (Kjeldahl) | Odor |
| Carbonaceous BOD | Nitrogen (organic) | Particle size distribution |
| COD | Nitrogen (nitrate) | Settleable solids |
| Drugs (prescription and other)† | Nitrogen (nitrite) | Temperature |
| Endocrine disruptors (other)† | Nitrogen (organic) | Total dissolved solids* |
| Home care products† | Nitrogen (total) | Total fixed dissolved solids |
| Industrial and household products† | pH | Total fixed solids |
| Nitrogenous oxygen demand | Phosphorus (inorganic) | Total solids |
| Organic compounds (MBAS, CTAS) | Phosphorus (organic) | Total suspended solids |
| Refractory organics* | Phosphorus (total) | Total volatile solids |
| Surfactants* | Sulfate | Transmittance |
| Total organic carbon | | Turbidity |
| Ultimate BOD | | Volatile dissolved solids |
| Volatile organic compounds* | | Volatile suspended solids |

*Denotes a nonconventional constituent; † Denotes an emerging constituent

**wastewater facilities** Sewers, pumping stations, treatment plants. More generally, the structures, equipment, and processes required to collect, convey, treat, and dispose of wastewater, including sludge handling and disposal.

**wastewater facilities plans** Plans and studies related to the construction of wastewater treatment works necessary to comply with the Clean Water Act or RCRA. A facilities plan investigates needs and provides information on the cost-effectiveness of alternatives, a recommended plan, an environmental assessment of the recommendations, descriptions of the treatment works and costs, and a completion schedule.

**wastewater farm** A precursor of the wastewater filter, it is an operation in which wastewaters and/or sludges with little or no treatment are spread on agricultural land for final disposal as well as reuse of water and nutrients. *See* sewage farm for more detail.

**wastewater farming** An operation that applies wastewater and/or sludge to grow crops on agricultural lands. This operation serves a dual purpose: wastewater disposal as well as land irrigation and fertilization. In a similar operation, called broad irrigation, the primary purpose is wastewater disposal; crop production is incidental.

**wastewater field** The perceptible, oily area created on the surface of a receiving water body by the submerged discharge of raw wastewater or treated effluent. *See* sleek field for more detail.

**wastewater filtration** Any of the following unit operations or processes commonly used in wastewater treatment to remove suspended solids: (a) depth filtration—intermittent filtration, rapid filtration, recirculating filtration, slow sand filtration; (b) surface filtration—cloth filtration, diatomaceous earth filtration; (c) membrane filtration—microfiltration, ultrafiltration, reverse osmosis, nanofiltration. *See also* bioflocculation, filtration efficiency, granular media filtration.

**wastewater filtration problems** For problems that affect the operation and performance of depth, surface, or membrane filtration of wastewater, *see* cracks and contraction, emulsified grease buildup, gravel mounding, membrane fouling, mudball formation, residuals disposal, scale formation, and turbidity breakthrough.

**wastewater flocculation** (1) The agitation of wastewater by air or mechanically to promote the destabilization and aggregation of finely divided solid particles. *See also* macroflocculation, microflocculation. (2) A wastewater treatment process that uses flocculation and digestion under anaerobic conditions. It includes the creation and maintenance of a biological floc, sludge return, disposal of excess sludge, and production of stable end products. *See* anaerobic contact for more detail.

**wastewater flow components** The three main contributions to wastewater flow are base wastewater flow or BWWF (sometimes called return flow), groundwater infiltration or GWI, and rainfall-dependent infiltration/inflow or RDI/I. The sum of BWWF and GWI makes up dry-weather flow. *See* hydrograph analysis.

**wastewater influent** The wastewater that is flowing into a pumping station or a wastewater treatment plant.

**wastewater infrastructure** The network for the collection, treatment, and disposal of wastewater in a community. The level of treatment will depend on the size of the community, the type of discharge, and/or the designated use of the receiving water.

**wastewater inorganics** Inorganic elements or compounds found in wastewater, originally present in the water supply or added through water use. Inorganics of importance include nitrogen, phosphorus, heavy metals, and monovalent cations.

**wastewater irrigation** The use of wastewater or effluent to irrigate crops. *See also* land disposal, surface irrigation, spray irrigation.

**wastewater irrigation system** A natural, slow-rate treatment system designed primarily for water reuse through crop production or landscape irrigation. *See* slow-rate system type 2 for detail.

**wastewater lagoon** An impoundment used for the oxidation of organic matter in wastewater. *See* oxidation pond.

**wastewater land disposal** *See* land application, land disposal, land treatment.

**wastewater management** All the activities involved in the collection, treatment, reuse and disposal of wastewater, including administration, financing, planning and engineering, operation and maintenance, and monitoring and evaluation.

**wastewater mud** A sludge deposit. *See* sludge bank.

**wastewater operations and maintenance** Actions taken after construction to assure that facilities constructed to treat wastewater will be operated, maintained, and managed to reach prescribed effluent levels in optimum manner.

**wastewater organics** Organic compounds found in wastewater, originally present in the water supply or added through water use. Organics of importance include carbohydrates, proteins, fats, and refractory organics. *See also* biochemical oxygen demand, chemical oxygen demand, total organic carbon

**wastewater outfall** The outlet or structure of final wastewater disposal.

**wastewater outlet** The point of final discharge of wastewater or effluent.

**wastewater oxidation** The conversion of the organic matter in wastewater to a stable form.

**wastewater ozonation by-product** Principal ozonation by-products include aldehydes, aldoacids, ketoacids, brominated by-products, and hydrogen peroxide.

**wastewater phosphorus** Phosphorus is present in wastewater mainly as phosphates ($PO_4$), salts of phosphoric acid ($H_3PO_4$).

**wastewater plume** *See* plume.

**wastewater processing** *See* wastewater treatment.

**wastewater purification** Obsolete term for wastewater treatment.

**wastewater rate** A schedule of charges for wastewater management services. *See* wastewater charge.

**wastewater reclamation** The recovery of wastewater and its improvement to meet the requirements of the intended reuse. *See* water reclamation for more detail.

**wastewater recycle** *See* wastewater reuse.

**wastewater recycling** *See* wastewater reuse.

**wastewater renovation** Treatment of wastewater for reuse.

**wastewater return rate** The ratio of base wastewater flow (BWWF or return flow) to water consumption; usually less than 1.0 because some water used, for lawn watering and car washing, for example, may not reach the sewers.

**wastewater reuse** The use of reclaimed wastewater in such applications as agricultural irrigation, landscape irrigation, industrial cooling, boiler feed, process water, heavy construction, groundwater recharge, stream-flow augmentation, fire protection, toilet flushing, etc. *See also* reclamation, water reuse applications.

**wastewater sample preservation** *See* sample preservation.

**wastewater sludge** Sludge produced from wastewater treatment. Its composition and condition depend on the treatment process. Fresh primary sludge contains settleable solids from the raw waste. Chemical coagulation sludge contains about 80% of the suspended solids of the raw waste. Sludge from a secondary clarifier contains 55–85% of the solids from trickling filters or 85% of the solids from activated sludge. Solids content varies from about 1% for fresh activated sludge to 8% for thickened trickling filter sludge and 15% for digested primary sludge.

**wastewater solids** *See* total solids, total volatile solids, total fixed solids, total suspended solids, volatile suspended solids, fixed suspended solids, total dissolved solids, total volatile dissolved solids, fixed dissolved solids, settleable solids.

**wastewater stream** A stream that contains only wastewater.

**wastewater survey** A field investigation conducted to determine (a) the quality of each waste stream of an industry or a municipality, and (b) the effects of wastewater discharges on the quality of receiving waters; also called a pollutional survey. Other field investigations include industrial waste and sanitary surveys.

**wastewater system** A combination of components, equipment, or installations that receive, convey, treat, discharge, or process wastewater and its solids. *See also* publicly owned treatment works.

**wastewater tank** A stationary waste management unit that is designed to contain an accumulation of wastewater or residuals and is constructed primarily of nonearthen materials (e.g., wood, concrete, steel, plastic) that provide structural support; including flow equalization tanks (EPA-40CFR63.111).

**wastewater treatability** The degree to which a given wastewater may be treated by a particular process. For trickling filtration, for example, treatability of a wastewater depends on the ratio of particulate to soluble organics and may be expressed as a reaction rate coefficient, simple organics being removed at a higher rate than complex organics. *See also* treatability study.

**wastewater treatment** The use of unit operations and processes to alter the constituents and characteristics of wastewater, e.g., to remove BOD, suspended solids and phosphates, or to increase dissolved oxygen. Wastewater treatment may be preliminary, primary, secondary, intermediate, tertiary, or advanced; biological, chemical, physical, or physicochemical.

**wastewater treatment chemicals** *See* important chemicals.

**wastewater treatment plant (WWTP)** A facility that receives wastewater (and sometimes runoff) from domestic and/or industrial sources, and by a combination of physical, chemical, and biological processes reduces the wastewater to less harmful by-products. Also known as sewage treatment plant (STP), water pollution control plant, water pollution control facility, and publicly owned treatment works (POTW).

**wastewater treatment plant residuals** *See* residuals.

**wastewater treatment process** Any process that modifies characteristics such as BOD, COD, TSS, and pH, usually to meet effluent guidelines and standards. Wastewater treatment uses a combination of the following unit operations and processes to achieve the desired objective: activated carbon adsorption, air flotation, aerobic treatment, air stripping, ammonia stripping, anaerobic treatment, chemical feeding, coagulant recovery, coagulation, comminution, disinfection, equalization, filtration, flocculation, grit removal, incineration, lagooning, membrane filtration, neutralization, oxidation ditch, recarbonation, screening, sedimentation, sludge dewatering, sludge digestion, sludge thickening, stabilization pond.

**wastewater treatment reactors** Common reactor types are batch, complete-mix, complete-mix in series, complete-mix with recycle, fluidized-bed, packed-bed, plug flow, and plug flow with recycle.

**wastewater treatment tank** A tank that is designed to receive and treat an influent wastewater through physical, chemical or biological methods (EPA-40CFR280.12).

**wastewater treatment unit** A device that (a) is part of a wastewater treatment facility that is subject to Clean Water Act regulations; and (b) receives and treats or stores an influent wastewater that is a hazardous waste or that generates and accumulates a wastewater treatment sludge that is a hazardous waste; and (c) meets the definition of tank or tank system (EPA-40CFR260.10).

**waste weir** A short section of a canal or open conduit designed with a level crest to allow spilling of excess water. *See also* spillway and diverting weir.

**WATEQ** Acronym of water equilibrium, a series of computer programs developed by the U.S. Geological Survey to determine equilibrium aqueous speciation, saturation states of solids, and other parameters, based on water quality data

**WATEQX** A version of the WATEQ programs that can be run on mainframe as well as personal computers.

**water ($H_2O$)** An odorless, colorless, tasteless, virtually incompressible liquid compound of oxygen and hydrogen. Freezing point = 0°C (or 32°F). Boiling point = 100°C (or 212°F). Maximum density: 1 kilogram per liter at 4°C (or 39°F). Used as a reference to express the specific gravity of solids and liquids, i.e., its specific gravity is 1. Besides its basic chemical formula ($H_2O$), water has other molecular combinations, e.g., $H_8O_4$. Principal biological, chemical, and physical properties: (a) biotic index, coliform count; (b) dissolved oxygen, hardness, pH (including acidity and alkalinity); (c) color, taste and odor, temperature, turbidity. Water combines with certain metals or metal oxides to form bases (e.g., sodium hydroxide), with nonmetal oxides to form acids (e.g., sulfuric acid), and with salts to form hydrates (e.g., washing soda):

$$H_2O + Na_2O \rightarrow 2\ NaOH \quad (W\text{-}04)$$

$$H_2O + SO_3 \rightarrow H_2SO_4 \quad (W\text{-}05)$$

$$10\ H_2O + Na_2CO_3 \rightarrow Na_2CO_3 \cdot 10\ H_2O \quad (W\text{-}06)$$

Water, particularly when hot, is an efficient solvent for salts and polar molecules. Natural waters contain a variety of solid, liquid, and gaseous materials in solution or in suspension. Water makes up 60–70% of the human body, with about 25 liters within the cells, 12 liters in tissue fluid, and 3 liters in blood plasma. A loss of 4 liters is enough to cause hallucinations; a loss of 8–10 liters may cause death.

**water activity** The relative availability of water in a substance; equal to 1.0 for pure water.

**water age** The period of time that water spends in a storage facility or water distribution system. It depends on the inflow and outflow rates. It affects bacterial regrowth and the formation of disinfection by-products. Regular flushing is sometimes used to reduce water age. *See also* groundwater age.

**water analysis** The examination of a sample of wastewater to determine its biological, chemical, and physical characteristics. Tests generally conducted include coliforms, toxicity, organics, pollution indicators, palatability, economic usefulness, and functional tests.

**water- and excreta-related diseases** For a summary of common water- and excreta- related diseases, see Table W-03.

**water audit** An examination of the records or accounts of a water supply agency to determine the efficiency of water distribution.

**water balance** (1) A mass balance for water in a hydrologic system, or in a water supply/wastewater system; i.e., an inventory of all identified water or wastewater quantities entering, leaving, or accumulating in the system, or a quantitative analysis of the changes occurring in the system. It may be used to estimate evaporation ($E$) from a water surface, as inflow ($I$) minus outflow ($O$), minus storage ($S$):

$$E = I - O - S \quad (W\text{-}07)$$

*See also* water budget, groundwater mining. (2) The method used to estimate leachate production by a landfill site or the capacity of the site for additional liquid wastes. It is basically an application of the above equation; leachate is estimated to come from abstraction from storage, surface springs and seepages, and groundwater recharge.

**water banking** Saving current unused water allocations for future use.

**water-based disease** A disease that is caused by a pathogen that originates in water or spends part of its life cycle in an aquatic animal and comes in direct contact with humans in water or by inhalation; e.g., (a) a parasitic worm or helminth that spends a part of its life in a cyclopoid or a snail, and (b) bacteria of the genus *Legionella.* Some water-based pathogens infect through the skin, whereas others are ingested or inhaled. *See* schistosomiasis, clonorchiasis, diphyllobothriasis, fasciolopsiasis, Guinea worm, legionnaires' disease, paragonimiasis, primary meningoencephalitis. *See also* Table W-03, excreta-related disease, and other categories of water-related diseases: fecal–oral, insect-borne, waterborne, and water-washed.

**water-based helminth** A parasitic worm that spends a part of its life in an aquatic animal. *See* water-based disease.

**water-based pathogens** *See* water-based disease.

**water-based route** The route followed by water-based pathogens to infect their hosts; they spend a part of their life cycle in an aquatic animal.

**Water Blaze** A wastewater evaporator manufactured by Landa, Inc.

**water blending** The use of water from different sources, with different characteristics, in the same water supply system. For example, the Pinellas County Utilities (Clearwater, FL) is building (2006) a 100-mgd plant to blend, store, and treat groundwater, surface water, and desalinated water. One consequence of water blending is the risk of instability of the mixture with respect to calcium carbonate precipitation, even though the components may be stable individually. *See also* interface zone.

**water bloom** A bloom, i.e., a sudden development of conspicuous masses of blue-green algae and other plankton, often with a greenish, yellowish, or brownish color on the surface of a lake, pond, or reservoir. *See* algal bloom.

**water body classification** The classification of a water body according to its current or intended uses, with appropriate water quality criteria and effluent standards to protect such uses. For example, inland waters may be classified for drinking, bathing and other recreational purposes, fish and wildlife habitat, industrial use, etc.; coastal and marine waters may be classified for bathing, shell-

**Table W-03.** Water- and excreta-related diseases

| Infection/disease (common name) | Pathogen | Main transmission mechanism | Epidemiology* | | | |
|---|---|---|---|---|---|---|
| | | | A | B | C | D |
| *A. Fecal–oral (waterborne, water-washed, or excreta-related)* | | | | | | |
| Amebiasis (amoebic dysentery) | *Entammoeba histolytica* | Fecal–oral, humans to humans | 5 | 0 | 0.8 | L |
| Bacterial enteritis (diarrhea, gastroenteritis) | *Campylobacter jejuni, Salmonella spp.* | Fecal–oral, humans or animal to humans | 7–8 | 0 | 0.2–2 | H |
| Balantidiasis (diarrhea) | *Balantidium coli* | Fecal–oral, humans or pig to humans | | 0 | | L |
| Cholera (cholera) | *Vibrio cholerae* | Fecal–oral, humans to humans | 7 | 0 | 1 | H |
| Cryptosporidiosis (diarrhea) | *Cryptosporidium spp.* | Fecal–oral, humans or animal to humans | | | | |
| *E. coli* diarrhea (diarrhea, gastroenteritis) | *Escherichia coli* | Fecal–oral, humans or animal to humans | 8 | 0 | 3 | H |
| Enterobiasis (pinworm) | *Enterobius vermicularis* | Humans to humans | | 0 | 0.2 | L |
| Giardiasis (diarrhea) | *Giardia lamblia* | Fecal–oral, humans to humans | 5 | 0 | 0.8 | L |
| Hepatitis-A (infectious hepatitis, jaundice) | Hepatitis-A virus | Fecal–oral, humans to humans | 6 | 0 | | L |
| Hymenolepiasis (dwarf tapeworm) | *Hymenolepis nana* | Humans or rodent to humans | | 0 | 1 | L |
| Leptospirosis (Weil's disease) | *Leptospira interrogans* | Animal to humans | | 0 | 0.3 | L |
| Paratyphoid fever (paratyphoid) | *Salmonella paratyphi* | Fecal–oral, humans to humans | 8 | 0 | 2–3 | H |
| Poliomyelitis (polio) | Poliovirus | Fecal–oral, humans to humans | 7 | 0 | 3 | L |
| Rotavirus diarrhea, viral diarrhea (diarrhea) | Rotavirus, Norwalk agent, other viruses | Fecal–oral, humans to humans | 6 | 0 | | L |
| Salmonellosis (diarrhea, gastroenteritis) | *Salmonella spp.* | Fecal–oral, humans or animal to humans | 8 | 0 | 3 | H |
| Shigellosis (bacillary dysentery) | *Shigella spp.* | Fecal–oral, humans to humans | 7 | 0 | 1 | M |
| Typhoid fever, enteric fever (typhoid, enteric fever) | *Salmonella typhi* | Fecal–oral, humans to humans | 8 | 0 | 2 | H |
| Yersiniosis (diarrhea, gastroenteritis) | *Yersinia enterocolitica* | Fecal–oral, humans or animal to humans | 5 | 0 | 3 | H |
| *B. Water-washed (skin, eye, and other infections)* | | | | | | |
| Infectious skin disease | Miscellaneous | | | | | |
| Infectious eye disease | Miscellaneous | | | | | |
| Louseborne typhus (epidemic typhus, classical typhus) | *Rickettsia prowazeki* | Louseborne, humans to louse to humans | | | | |
| Louseborne relapsing fever (relapsing fever) | *Borrelia recurrentis* | Louseborne, humans to louse to humans | | | | |
| *C. Water-based (through skin, or ingestion)* | | | | | | |
| Clonorchiasis (Chinese liver fluke) | *Clonorchis sinensis* | Animal or humans to aquatic snail to fish to humans | 2 | 1.5 | | L |
| Diphyllobothriasis (fish tapeworm) | *Diphyllobothrium latum* | Animal or humans to copepod to fish to humans | 4 | 2 | | L |
| Dracunculiasis, dracontiasis (Guinea worm, Guinea worm disease) | *Dracunculus medinensis* | Humans to through *Cyclops* to humans | | | | |
| Fascioliasis (sheep liver fluke, liver rot) | *Fasciola gigantica, F. hepatica* | Animal or humans to aquatic snail to aquatic plant to humans | | 2 | 4 | L |
| Fasciolopsiasis (giant intestinal fluke) | *Fasciolopsis buski* | Pig or humans to aquatic snail to aquatic plant to humans | 3 | 2 | | L |
| Gastrodiscoidiasis | *Gastrodiscoides hominis* | Animal or humans to aquatic snail to aquatic plant to humans | | 2 | | L |
| Heterophyasis | *Heterophyes heterophyes* | Animal or humans to aquatic snail to aquatic plant to humans | | 1.5 | | L |
| Metagonimiasis | *Metagonimus yokogawai* | Animal or humans to aquatic snail to aquatic plant to humans | | 1.5 | | L |

**Table W-03.** *Continued*

| Infection/disease (common name) | Pathogen | Main transmission mechanism | Epidemiology* | | | |
|---|---|---|---|---|---|---|
| | | | A | B | C | D |
| Paragonimiasis (lung fluke) | *Paragonimus westermani* | Animal to aquatic snail to crab or crayfish to humans | | 4 | | L |
| Schistosomiasis (bilharziasis) | *Schistosoma haematobium* | Humans or animal to snail to humans | < 1 | 1.2 | < 0.1 | L |
| Schistosomiasis (bilharziasis) | *Schistosoma japonicum* | Humans or animal to snail to humans | < 1 | 1.8 | < 0.1 | L |
| Schistosomiasis (bilharziasis) | *Schistosoma mansoni* | Humans or animal to snail to humans | < 1 | 1 | < 0.1 | L |
| *D. Water-related insect vector (biting near or breeding in water)* | | | | | | |
| Dengue (breakbone fever) | Dengue virus | *Aedes aegypti* and other mosquitoes to humans to mosquito to humans | | | | |
| Bancroftian filariasis | *Wucheria bancrofti* | Culex pipiens, Anopheles, Aedes mosquitoes to humans to mosquito to humans | | | | |
| Malayan filariasis | Brugia malayi | Humanssonia, Anopheles, Aedes mosquitoes to humans to mosquito to humans | | | | |
| Loiasis | *Loa loa* | *Chrysops spp.* (mangrove fly) to humans to fly to humans | | | | |
| Malaria | *Plasmodium spp.* | Anopheles mosquitoes to humans to mosquito to humans | | | | |
| Onchocerciasis (river blindness) | *Onchocerca volvulus* | *Simulium spp.* (black fly) to humans to fly to humans | | | | |
| African trypanosomiasis-1 | *Trypanosome gambiense* | *Glossina spp.* (riverine tse tse) to humans to fly to humans | | | | |
| African trypanosomiasis-2 (Rhodesian sleeping sickness) | *Trypanosome rhodesiense* | *Glossina spp.* (riverine tse tse) to animal to fly to humans | | | | |
| American trypanosomiasis (Chagas' disease) | *Trypanosome cruzi* | Reduviidae bugs, humans or animal to bug to humans | | | | |
| Yellow fever | Yellow fever virus | *Aedes aegypti, Haemagogus spp.,* and other mosquitoes to humans or monkey to mosquito to humans | | | | |
| Other arboviral diseases (encephalitic and haemorrhagic infections) | Miscellaneous viruses | Animal to arthropods to humans | | | | |
| *E. Soil-transmitted helminths* | | | | | | |
| Ascariasis (roundworm) | *Ascaris lumbricoides* | Humans to soil to humans | 4 | 0.3 | 12 | L |
| Hookworm | *Ancylostoma duodenale, Necator americanus* | Humans to soil to humans | 2 | 0.2 | 3 | L |
| Strongyloidiasis (threadworm) | *Strongyloides stercoralis* | Humans to soil to humans | 1 | 0.1 | 0.75 | L |
| Trichuriasis (whipworm) | *Trichuris trichiura* | Humans to soil to humans | 3 | 0.7 | 9 | L |
| *F. Beef and porc tapeworms* | | | | | | |
| Taeniasis-1 (beef tapeworm) | *Taenia saginata* | Humans to cow to humans | 4 | 2 | 9 | L |
| Taeniasis-2 (pork tapeworm) | *Taenia solium* | Humans to pig to humans | 4 | 2 | 9 | L |

*Epidemiology:

A = excreted load, average number of organisms excreted per gram of feces (or gram of urine for *S. haematobium* and *Leptospira*). The reported number represents a power of 10, i.e., the number 7 represents $10^7$ = 10,000,000 organisms per gram.

B = latency or minimum time from excretion to infectivity, in months.

C = persistence or estimated maximum life of infective eggs at 20–30°C, in months.

D = median infective dose ($ID_{50}$): L = low, < 100; M = medium, ≈ 10,000; and H = high, > 1,000,000.

Adapted from (1) Cairncross and Feachem (1993), Table 1.2, Table 1.3, and Appendix C, and (2) Feachem et al. (1981), Table 1-9 and Table 2-3.

fish harvesting, cooling, and fish and wildlife habitat, etc. *See also* water quality parameter, water quality standards, effluent standards, chemical-specific standard, whole-effluent standard.

**waterborne** Carried by water, e.g., a waterborne disease or a waterborne contaminant.

**waterborne bacterial agent** The most common bacteria that cause gastrointestinal diseases in humans when ingested through drinking water include *Salmonella* (typhoid, paratyphoid, enteric fever), *Shigella* (bacillary dysentery), *Vibrio cholerae* (cholera), *Escherischia coli* and *Yersinia enterocolitica* (gastroenteritis), and *Leptospira.*

**waterborne-chemical disease** A disease contracted through the consumption of water that contains natural or man-made toxic substances.

**waterborne disease** A disease that is contracted by drinking contaminated water; the infecting pathogen is in the water consumed. The best known waterborne diseases, cholera and typhoid, are also excreta-related and may be spread via contaminated food. The term waterborne has been, incorrectly, used synonymously with water-related disease. Beside cholera and typhoid, waterborne diseases include amebic dysentery, bacillary dysentery (shigellosis), balantidiasis, *Campylobacter* enteritis, cholera, cryptosporidiosis, *E. coli* diarrhea, giardiasis, infectious hepatitis (hepatitis A), leptospirosis (Weil's disease), paratyphoid, poliomyelitis, rotavirus diarrhea, salmonellosis, and yersiniosis. *See also* Table W-03, excreta-related disease, and other categories of water-related diseases: fecal–oral, insect-borne, water-based, and water-washed.

**waterborne disease outbreak** The significant occurrence of acute infectious illness, epidemiologically associated with the ingestion of water from a public water system that is deficient in treatment, as determined by the appropriate local or state agency. Examples of documented waterborne outbreaks are giardiasis in Rome, NY (1975), cryptosporidiosis that affected 400,000 individuals in Milwaukee, and 7 deaths in a *Salmonella* outbreak in 1993 in Giddeon, GA.

**waterborne illness** A waterborne disease.

**waterborne microbiological disease** A disease caused by pathogens in drinking water, e.g., amebic dysentery, cholera, and hepatitis-A.

**waterborne poison** Waterborne poisons include toxic substances of mineral origin (e.g., fluoroapatite), phytotoxins, heavy metals, industrial toxics, and pesticides.

**waterborne route** *See* waterborne disease.

**waterbox** The inlet chamber of a condenser.

**Water Boy®** A compact water treatment plant manufactured by Wheelabrator Engineered Systems, Inc.

**water budget** An average annual water balance, including such components as precipitation, snowmelt and rainfall runoff, evapotranspiration, infiltration, and surficial and deeper groundwater recharge. This accounting of inflows and outflows applies to such hydrologic units as an aquifer, a drainage basin, or an impoundment. Determination of the water budget requires the quantification of the average volumes of storage and rates of movement among the various components. It uses the storage form of the continuity equation. For example, the surface water budget may be stated as the change in surface water storage is equal to rainfall, plus snowmelt, minus upper soil infiltration, minus surface runoff, all the quantities being expressed in mm or in inches. Through its RUNOFF and EXTRAN blocks, the SWMM program can be used to study surface water interactions, whereas MODFLOW allows the evaluation of steady-state groundwater conditions. Water budgets are sometimes represented by the equation

$$R_e = P + R_a - ET + Q_{ro} + Q_{gw} \qquad \text{(W-08)}$$

where $R_e$ = effective or net recharge, in length or depth ($L$) per unit time ($T$), i.e., $L/T$; $P$ = precipitation, $L/T$; $R_a$ = artificial recharge, $L/T$; $ET$ = evapotranspirration, $L/T$; $Q_{ro}$ = flow out or into an area due to surface water runoff, $L/T$; and $Q_{gw}$ = groundwater flow out or in, $L/T$. Also called hydrologic budget. *See also* water balance.

**Water Buffalo™** A treatment unit manufactured by Mechanical Equipment Co.; uses reverse osmosis.

**Water Champ** A chemical induction apparatus manufactured by GARDINER Equipment Co., Inc.

**water chemistry** The science that deals with the chemical aspects of water and wastewater, e.g., their composition, reactions with various elements and compounds, etc. It draws heavily on other fields such as physical chemistry, biology, geology, thermodynamics, and hydrology. Important water chemistry considerations for health include the absence or deficiency of necessary chemical elements (e.g., fluoride or iodine) and the presence or excess of harmful substances (e.g., toxics, THMs, or arsenic). Chemical reactions in natural waters or in treatment processes are assumed to occur in dilute solution; molar concentrations are used instead of chemical activities. An important convention is the use of calcium carbonate ($CaCO_3$) equivalents to express alkalinity, hardness, and other parameters. *See also* the following concepts: acid–base relationship, bicarbonate–carbonate equilibria, chlorination, coagulation, common-ion effect, equilibrium constant, equivalent weight, ion exchange, iron and manganese removal, oxidation–reduction, pH, phosphorus removal, softening, solubility product. Also called aquatic chemistry or aqueous chemistry.

**Water-Chex®** A chlorine indicator of PyMah Corp.

**water ciculation coefficient** The ratio of total precipitation in a region to the precipitation due to evaporation from the oceans.

**water clarification** A series of physicochemical processes (coagulation, sedimentation, filtration) used to remove microorganisms, turbidity, humic substances, and color.

**water closet (WC)** A pan from which excreta are flushed by water into a drain; a flushable toilet.

**water cloud** A cloud containing only water, no ice crystals.

**water conditioning** Water treatment or processing to meet a specific quality.

**water conditioning diagram** Same as Caldwell–Lawrence diagram.

**water conservation** The efficient use of water resources, including such measures as the reduction of water consumption and the reduction of distribution system losses. *See also* flow-reduction devices.

**water consumption** The quantity of water supplied to a community for all uses.

**water-contact disease** A disease transmitted by skin contact with pathogen-contaminated or toxin-infested water, e.g., schistosomiasis, leptospirosis, tularemia.

**water content** (1) The quantity of water present in soil, sludge, screenings, etc., expressed as a percentage of wet weight. *See* moisture content for detail. (2) The percent by weight of liquid water present in a snow sample. Also called free-water content, liquid water content. *See also* water equivalent, snow density.

**water correction** Treatment of water specifically to stabilize it and reduce its tendency to corrode pipes and form incrustations.

**WATER CO$T model** A computer program developed by the USEPA to estimate capital and operating costs of a treatment plant having a capacity larger than 1.0 mgd. *See also* WATER model.

**water cycle** Movement or exchange of water between the atmosphere and the earth. Also called hydrologic cycle. The hydrogeologic cycle is similarly defined: the natural process recycling water from the atmosphere down to (and through) the earth and back to the atmosphere again. The water cycle comprises the unending processes that control the distribution and movement of water on the Earth's surface, in the soil, and in the atmosphere: evaporation from the oceans and the earth, transport over the land masses, condensation of the water vapor, fog or cloud formation, precipitation as rainfall or snowfall, evapotranspiration, depression storage, infiltration, percolation, ultimate runoff to the oceans, etc. Evaporation and transpiration account for two-thirds of the precipitation over land, and runoff to oceans for the remaining third. *See* rainfall–runoff relationship, hydrologic cycle, water reuse applications, water reuse pathways.

**water deficit** The cumulative difference between reference (or potential) evapotranspiration and precipitation in a given period when the latter is smaller.

**water demand** A schedule of requirements for all purposes: municipal, agricultural, power, etc. Water demand often increases as a result of improvements in availability and/or quality. *See also* water use.

**water-demand curve** A graphical representation of cumulative water demand versus time.

**water demand management** An approach to produce economies of water use and eliminate, reduce, or delay capacity expansion. *See* demand management for more detail.

**water dew point** The temperature at which moisture in an emission (from a stack or a flue) condenses. *See also* acid dew point.

**water dissociation** The splitting of water molecules into negatively and positively charged ions. *See* water ionization for detail.

**water distribution system integrity** A key indicator that water utilities can use to assess their performance compared to established standards. It quantifies the condition of the distribution network, i.e., the number of breaks and leaks per length of piping.

**water district** An agency created to provide water supply services to a specific area; the land area within the jurisdiction of such an agency.

**water dumping** The disposal of pesticides in or on lakes, ponds, rivers, sewers, or other water systems as defined in Public Law 92-500 (EPA-40CFR165.1-z).

**Water Eater** A wastewater evaporator manufactured by Equipment Manufacturing Co.

**water-effects ratio (WER)** A test to determine

the ability of a site to dilute metal toxicity by comparing its toxic endpoint to that of laboratory water.

**Water Environment Federation** Formerly the Sewage and Industrial Wastes Federation (1920s–1960) and Water Pollution Control Federation (WPCF, 1960–1991), a technical association of professionals involved in activities to enhance and preserve water quality. It disseminates technical information and regulations, alone or with others, e.g., Standard Methods, a manual of municipal wastewater treatment, and various design and/or operation manuals.

**Water Equilibrium programs** *See* WATEQ and WATEQX.

**water equivalent** The amount or depth of water obtained from the complete melting of a snow sample, snow cover, or snowpack. *See also* water content, snow density.

**water-extracting techniques** *See* saline-water conversion classification.

**water factor** A correction factor applied to incident irradiance to obtain the average UV irradiance used in bench-scale experiments. It accounts for the effects of water absorbance and sample depth.

**Water Factory 21 (WF21)** An advanced wastewater treatment plant of Orange County Water District, CA, designed for a high-quality effluent that is injected into an aquifer to prevent saltwater intrusion. After adequate flow equalization, effluents from activated sludge plants go through the following processes: lime clarification and recalcination, recarbonation, granular media filtration, activated carbon adsorption and regeneration, chlorination, and desalting by reverse osmosis. The effluent, which reportedly meets state and federal drinking water standards, is blended with deep well water before underground injection.

**waterfall** A sudden and nearly vertical drop in a stream.

**waterfall effect** *See* Lenard effect.

**water farming** The practice of purchasing prior agricultural rights to groundwater reserves for urban water supply.

**water fern** Common name of *Salvinia auriculata,* a species of aquatic plant that colonizes man-made lakes and serves as a habitat for *Mansonia* spp., which transmits Malayan filariasis. It also provides a habitat for aquatic organisms that metabolize wastewater organics.

**water filtration plant** A water treatment plant that includes filtration and other processes. Although the plant includes other processes, to underscore the importance of filtration, it is often called by the type of filtration used, such as conventional, direct, slow sand, rapid sand, pressure, or diatomaceous earth. Also called a filter plant, a filtration plant, or the filters.

**water flea** A small crustacean that moves in water like a flea, including the genus *Daphnia.*

**waterflood** *See* water flooding.

**water flooding** An underground mining method used to recover secondary minerals by injecting water into a formation to force additional materials to the surface or into a producing well. Also called water flood. *See also* oil well flooding.

**water-flow formula** Any one of a number of formulas used to determine discharge or velocity of water. *See* the Chezy, Darcy–Weisbach, Hazen–Williams, Kutter, and Manning formulas.

**water for injection** Water that contains no added substance and meets the quality requirements of U.S. Pharmacopeia after purification by an appropriate water treatment process such as distillation, ion exchange or other USEPA-approved drinking water treatment. Such water is intended for the preparation of solutions that will enter the body, but not through the digestive canal or within the intestine. *See* pharmaceutical-grade water.

**Water Free™** A gate manufactured by TETRA Technologies, Inc. for grit removal and screening units.

**water gain** The addition of water from natural or man-made flows (e.g., precipitation, transfers) net of such withdrawals as evapotranspiration, to existing sources in a given area. There is a water loss when withdrawals exceed additions.

**water glass [$Na_2O \cdot (SiO_2)_x$]** Basically, a combination of sodium oxide ($Na_2O$) and silica ($SiO_2$), a chemical used in water treatment for corrosion prevention, for pH control (in small systems), or as a coagulant aid. *See* sodium silicate for more detail.

**water habitat vector disease** A disease caused by a pathogen that spends part of its life cycle n an animal vector that lives all or part of its life in or near a water habitat, e.g., schistosomiasis, filariasis, malaria, onchocerciasis, trypanosomiasis.

**water hammer** The fluctuation in pressure that accompanies a sudden change in velocity. Also, the sound, like someone hammering on a pipe, that occurs when a valve is opened or closed very rapidly. When a valve position is changed quickly, the water pressure in a pipe will increase and decrease back and forth very quickly. This rise and fall in pressures can do serious damage to the system. The maximum pressure ($P^*$) produced by water hammer is:

$$P^* = \rho V[\rho(1/K + D/ET)]^{-0.5} \qquad \text{(W-09)}$$

where $\rho$ = water density, $V$ = velocity of flow, $K$ = bulk modulus of elasticity of water, $D$ = pipe diameter, $E$ = modulus of elasticity of pipe, and $T$ = pipe wall thickness. *See also* hydraulic transient, pressure surge, pressure transient.

**water hardness** A characteristic of water caused mainly by the salts of calcium and magnesium, such as bicarbonate ($HCO_3$), carbonate ($CO_3$), sulfate ($SO_4$), chloride (Cl), and nitrate ($NO_3$); usually expressed as calcium carbonate ($CaCO_3$) equivalent. *See* hardness for detail.

**water hole** (1) A depression, pool, pond, or cavity containing water in the surface of the ground or in the bed of a river. (2) A hole in the frozen surface of a water body. (3) A spring, well, or other source of drinking water in the desert. *See also* watering hole, watering place.

**water horsepower** The theoretical power required of a motor to drive a pump, assuming 100% efficiency. Also called hydraulic horsepower. *See* horsepower.

**water hyacinth** A perennial, floating aquatic plant with flowering top that provides a habitat for aquatic organisms that metabolize wastewater organics; water hyacinth also removes nutrients, metals, and phenols from wastewater. It is the common name of *Eichornia crassipes,* a species that colonizes ponds, natural and constructed wetlands, and man-made lakes with an extensive root system and rapid growth rate. It can be harvested for use as animal feed or to produce fertilizers and methane. It serves as a habitat for *Mansonia* spp., which transmits Malayan filariasis. Also simply called hyacinth.

**water hygiene disease** A disease that prevails in areas where water is in short supply for personal and domestic hygiene. It is caused by pathogens that can be reduced or eliminated by the use of water, regardless of its quality, for hygienic purposes. *See* water-washed disease for detail.

**Water Information Sharing and Analysis Center (WaterISAC)** An organization intended to promote the security of water and wastewater utilities through information gathering and dissemination; it maintains a public Web site: www.waterisac.org. *See also* Water Security Channel.

**watering hole** A gathering place where alcoholic drinks are sold. Also called watering place, watering spot. *See also* water hole.

**watering place** (1) A place for humans or animals to obtain drinking water, e.g., a spring or a water hole. (2) Same as watering hole.

**watering spot** Same as watering hole.

**water intake** The quantity of water pumped from the source for water supply in a given period of time, e.g., in gallons or million gallons per day, month, or year. *See also* intake structures.

**water ionization** The splitting or dissociation of water molecules into negatively and positively charged ions (hydroxyl $OH^-$, hydrogen $H^+$, or hydronium $H_3O^+$):

$$H_2O \rightarrow H^+ + OH^- \qquad \text{(W-10)}$$

This is a simplified representation of one molecule of water dissolving in another molecule (solvent):

$$H_2O + H_2O \rightarrow H_3O^+ + OH^- \qquad \text{(W-11)}$$

Also called water dissociation. *See also* self-ionization.

**WaterISAC** Abbreviation of Water Information Sharing and Analysis Center.

**water jacket** A casing that surrounds an apparatus to hold water or to circulate water for cooling purposes.

**water law** The field of law that deals with water as a resource, both surface and groundwater, its ownership, control, and use for various purposes. *See also* water rights.

**water lettuce** Common name of *Pistia stratiotes,* a species of floating aquatic plant, with thick spongy leaves, that colonizes man-made lakes and serves as a habitat for *Mansonia spp.,* which transmits Malayan filariasis. It also provides a habitat for aquatic organisms that metabolize wastewater organics.

**water-level recorder** An instrument that records water levels in wells and water bodies.

**waterlogging** A condition of soil saturation, with the water table rising close to the land surface and reaching into the root zone.

**water loss** (1) *See* water gain. (2) Same as evapotranspiration.

**water main** The water pipe beneath a street from which water is delivered to customers through service lines.

**watermaker** A package water treatment plant using vapor compression for thermal desalination. *See also* rainmaking.

**water mass transfer coefficient** Same as water permeability coefficient.

**watermaster** An agent who administers the distribution of available water supply and collects hydrographic data.

**Water Maze®** A wastewater clarifier manufactured by Landa, Inc.

**water meter** A mechanical or electronic device that measures and records the quantity of water passing through a pipe or an outlet. A meter may also play other roles in the operation of a water or wastewater treatment unit.

**water meter address matching** A procedure used in the estimation and projection of water uses and wastewater flows. With a GIS program, water meters are referenced to their true location in a street network. Within that program, such operations as the computation of flows by pumping station area can be easily carried out.

**water-meter load factor** The ratio of the average water use recorded by a meter to the discharge capacity of the meter over a period of one year, expressed as a percentage.

**WATER model** A computer program developed by the USEPA to estimate capital and operating costs of a treatment plant having a capacity less than 1.0 mgd. *See also* WATER CO$T model.

**water molecule** The electrostatic dipole formed by a central oxygen atom and two hydrogen atoms along an angle of 104°.

**water of adhesion** Subsurface water that adheres to soil particles after drainage by gravity. Also called pellicular water or adhesive water, it is found between the soil and gravity subzones. It can be absorbed by roots and is subject to evapotranspiration. *See* subsurface water.

**water of crystallization** Water that combines with salts when they crystallize, i.e., a part of a crystalline compound. Also called water of hydration.

**water of dehydration** Water that has been released from its chemical combination with minerals.

**water of hydration** (1) Water that is chemically bound to solid particles in a sludge suspension. *See also* bound water, bulk water, interstitial water, and vicinal water. (2) Water that combines with salts when they crystallize, i.e., a part of the crystalline compound. Also called water of crystallization.

**water of separation** In a sludge digester, the most liquid portion, between a top scum layer and a bottom layer of actively digesting sludge and digested concentrate. *See* supernatant for more detail.

**water permeability coefficient** A parameter ($P$) indicative of the flow of water through the membrane of reverse osmosis or other pressure-driven process. Also called water mass transfer coefficient or water transport coefficient. *See* solvent permeability coefficient.

**water physical properties** *See* Table W-04 for the following physical properties of water at 0, 25, and 100°C: specific weight ($\gamma$), density ($\rho$), modulus of elasticity ($E$), dynamic viscosity ($\mu$), kinematic viscosity ($\nu$), surface tension ($\sigma$), and vapor pressure ($P_v$).

**water plant** (1) A water treatment plant. (2) An aquatic plant.

**water plate count** An estimate of the number of bacteria in a sample. *See* standard plate count.

**water point** A point of water supply, private or public, e.g., a well, standpipe, capped spring, or household connection.

**WaterPOINT™ 870 Optical** An instrument manufactured by the Sensicore Company, combining electrochemistry, colorimetry, and turbidity/color measurement to analyze several physical and chemical water quality parameters.

**water pollutant** *See* pollutant.

**water pollution** The presence in water of enough harmful or objectionable material to damage the water's quality.

**Water Pollution Control Federation** Former name of the Water Environment Federation.

**water pollution control plant** A wastewater treatment plant.

**water potential** *See* soil water potential.

**water power** The energy of water descending by gravity (e.g., from a dammed stream) or water rising or falling from tides.

**water processing** Water treatment.

**water production waste** Residuals generated by municipal or industrial water treatment plants. *See* water treatment plant residuals.

**water provider** A private or public agency that supplies water to customers; a water purveyor.

**water purification** Water treatment.

**water purveyor** An agency or person that supplies water (usually potable water).

**Table W-04.** Water physical properties

| Property | 0°C | 25°C | 100°C |
|---|---|---|---|
| Specific weight ($\gamma$), kN/m$^3$ | 9.805 | 9.777 | 9.399 |
| Density ($\rho$), kg/m$^3$ | 999.8 | 997.0 | 958.4 |
| Modulus of elasticity ($E$), $10^6$ kN/m$^2$ | 1.98 | 2.22 | 2.07 |
| Dynamic viscosity ($\mu$) × $10^3$, N.sec/m$^2$ | 1.781 | 0.890 | 0.282 |
| Kinematic viscosity ($\nu$) × $10^6$, m$^2$/sec | 1.785 | 0.893 | 0.294 |
| Surface tension ($\sigma$), N/m | 0.076 | 0.072 | 0.059 |
| Vapor pressure (Pv), kN/m$^2$ | 0.61 | 3.17 | 101.33 |

**water quality** The biological, chemical, and physical characteristics of water as related to its intended use. Water quality standards and tests depend on intended uses such as drinking and bathing. Water quality parameters are usually classified as chemical, microbiological, and physical.

**water-quality-based effluent standard** *See* water-quality-based limitations.

**water-quality-based limitations** Effluent limitations applied to dischargers when mere technology-based limitations would cause violations of water quality standards. Usually applied to discharges into small streams.

**water-quality-based permit** A permit with an effluent limit more stringent than one based on technology performance. Such limits may be necessary to protect the designated use of receiving waters (i.e., recreation, irrigation, industry or water supply).

**water-quality-based standard** *See* water-quality-based limitations.

**water-quality-based toxics control** An integrated strategy to regulate the discharge of toxic pollutants into surface waters. *See* whole-effluent approach, chemical-specific approach.

**water quality criteria** Levels of water quality expected to render a body of water suitable for its designated uses. Criteria are based on specific levels of pollutants that would make the water harmful if used for drinking, swimming, farming, fish production, or industrial processes.

**water quality determinant** Water quality parameter.

**water quality guideline** A concentration limit or statement corresponding to a designated water use. *See also* water quality objective.

**water quality index** A single number, understandable and usable by the public, that expresses overall water quality at a certain site and time, based on several water quality parameters. It indicates whether a water body poses a potential threat to various uses (e.g., aquatic life habitat, drinking water, recreation). An example is the National Sanitation Foundation water quality index, a 100 point scale based on nine parameters: temperature, pH, dissolved oxygen, turbidity, fecal coliform, biochemical oxygen demand, total phosphates, nitrates, and total suspended solids.

**water quality indicator** (1) A microorganism or constituent selected as a surrogate for pathogens in routine monitoring of water quality. Current bacteriological water quality indicators include total coliforms, fecal coliforms, *Escherichia coli*, heterotrophic plate count, *Clostridium perfringens,* coliphages, Bacteroides, particle count, turbidity, aerobic sporeformers, and microscopic particulates. (2) Water quality parameter.

**water-quality-limited impact** The impact exerted on a stream by the discharge of an effluent that has received the equivalent of secondary wastewater treatment and still violates the receiving stream's water quality criteria. *See also* effluent-limited pollutant, whole-effluent toxicity.

**water-quality-limited segment** Any segment in which it is known that water quality does not meet applicable water quality standards, and/or is not expected to meet applicable water quality standards, even after the application of the technology-based effluent limitations required by sections 301(b) and 306 of the Clean Water Act (EPA-40CFR130-2.j).

**water quality management plan** A state or areawide waste management plan developed and updated in accordance with the provisions of sections 205(j), 208 and 303 of the Clean Water Act (EPA-40CFR130.2-k).

**water quality model** A mathematical equation or a set of equations that describe the physical, chemical, and biological processes that take place in a receiving water body. Water quality models may be empirical or deterministic, one-dimensional or multidimensional. *See,* for example, the Churchill–Buckingham and Streeter–Phelps formulations.

**water quality monitoring** The collection of information at specific sites and regular intervals over a period of time to provide the data necessary to define current water quality conditions, trends, etc. Data evaluation and reporting may also be included in monitoring or in water quality assessment. In addition to preliminary surveys, a monitoring program includes field operations (sample collection and in situ measurements), laboratory operations (pretreatment and analysis of samples), and hydrologic measurements (e.g., discharge, water level). *See also* surveillance.

**water quality objective** A concentration limit or statement of criteria corresponding to a designated water use. *See also* water quality guideline.

**water quality parameter** A measurable biological, chemical, or physical characteristic of water. *See* Table W-05 for a list of typical water quality parameters reported by water treatment plants for raw and treated water. In addition to the constituents or substances listed in this table, parameters of concern include nutrients, natural organic matter, synthetic organics, coliform bacteria, and oil and grease. Also called water quality determi-

**Table W-05.** Water quality parameters (typical)

| Parameter | Unit | Frequency | Long-term average | Parameter | Unit | Frequency | Long-term average |
|---|---|---|---|---|---|---|---|
| Alkalinity (total) | mg/L | Weekly | 60 | Phosphate | mg/L | Yearly | < 0.1 |
| Aluminum | mg/L | Monthly | 0.33 | Potassium | mg/L | Yearly | 1.2 |
| Arsenic | μg/L | Yearly | < 5 | Radiation (gross $\alpha$) | pCi/L | Yearly | < 4 |
| Asbestos | mg/L | Yearly | 3 | Radiation (gross $\beta$) | pCi/L | Yearly | < 4 |
| Barium | mg/L | Yearly | < 0.05 | Radium-228 | pCi/L | Yearly | < 3 |
| Beryllium | mg/L | Yearly | < 0.005 | Selenium | μg/L | Yearly | < 5 |
| Boron | mg/L | Yearly | < 0.05 | Silica ($SiO_2$) | mg/L | Yearly | < 0.5 |
| Cadmium | mg/L | Yearly | < 0.005 | Silver | mg/L | Yearly | < 0.01 |
| Calcium | mg/L | 2 × Weekly | 27 | Sodium | mg/L | Yearly | 13.8 |
| Chemical oxygen demand (COD) | mg/L | Quarterly | 11 | Specific conductance | umhos/cm | 2 × Weekly | 200 |
| Chloride | mg/L | 2 × Weekly | 15 | Strontium | mg/L | Yearly | 0.15 |
| Chromium (total) | mg/L | Yearly | < 0.01 | Sulfate | mg/L | Yearly | 20 |
| Color | CU | Weekly | 10 | Temperature | °C | Daily | 12 |
| Copper | mg/L | Yearly | 0.02 | Total dissolved solids (TDS) | mg/L | Quarterly | 130 |
| Cyanide | mg/L | Quarterly | < 0.01 | | | | |
| Detergent | mg/L | Yearly | < 0.1 | Turbidity | NTU | Daily | 29 |
| Dissolved oxygen | mg/L | Yearly | 6 | Aldrin | μg/L | Yearly | < 0.05 |
| Hardness (total) | mg/L | 2 × Weekly | 90 | Chloroform | μg/L | Yearly | < 0.5 |
| Iron | mg/L | Yearly | 0.3 | Chlordane | μg/L | Yearly | < 0.5 |
| Lead | mg/L | Monthly | < 0.001 | 2,4-D | μg/L | Yearly | < 0.05 |
| Lithium | mg/L | Yearly | < 0.1 | DDD | μg/L | Yearly | < 0.1 |
| Magnesium | mg/L | Yearly | 4.2 | DDT | μg/L | Yearly | < 0.1 |
| Manganese | mg/L | Yearly | 0.07 | Endrin | μg/L | Yearly | < 0.1 |
| Mercury | μg/L | Quarterly | < 0.4 | Heptachlor | μg/L | Yearly | < 0.05 |
| Nickel | mg/L | Yearly | < 0.04 | Methylene chloride | μg/L | Yearly | < 0.5 |
| Nitrogen (ammonia) | mg/L | Quarterly | 0.4 | Toluene | μg/L | Yearly | < 0.5 |
| Nitrogen (nitrate) | mg/L | Quarterly | 0.7 | Toxaphene | μg/L | Yearly | < 1 |
| Nitrogen (nitrite) | mg/L | Quarterly | 0.02 | 2,4,5-TP Silvex | μg/L | Yearly | < 0.05 |
| Odor | Odor units | Weekly | 2 | Trichloroethylene | μg/L | Yearly | < 0.5 |
| pH | pH unit | Daily | 7.7 | Standard plate count | #/ml | Daily | 1500 |
| Phenol | mg/L | Yearly | < 0.01 | Total coliform | #/100 ml | Daily | 1400 |

nant, water quality indicator, or water quality variable.

**water quality standards** State-adopted and USEPA-approved ambient standards for water bodies. The standards prescribe the use of the water body and establish the water quality criteria that must be met to protect designated uses. Water quality standards are meant to protect the public health or welfare, enhance the quality of water, and serve the purposes of the Clean Water Act. *See also* effluent standards, chemical-specific approach, whole-effluent approach, stream classification.

**water quality surveillance** *See* surveillance, water quality monitoring.

**water quality trading system** A policy of the USEPA to facilitate implementation of the Clean Water Act and federal regulations. It is designed to improve water quality while lowering costs. To meet its pollution control requirements, one facility can use credits earned by other facilities that lower their pollution levels beyond their own requirements.

**water quality variable** Water quality parameter.

**water rate** (1) The charge to customers for water supply per unit of water use. *See also* user fee, lifeline rate. (2) The amount of water vapor required by a steam plant per unit of energy produced.

**water ratio** The ratio of the amount of water in a sample of soil, sediment, sludge, etc. to the volume or weight of the entire sample.

**water reclamation** The recovery of wastewater and stormwater and their renovation to make them reusable and meet the requirements of the intended use (e.g., cooling water, irrigation water, groundwater recharge). Water reclamation involves a combination of conventional and advanced waste treatment processes, with pretreatment of industrial wastes and an added consideration for the removal of heavy metals, organics, inorganic salts, and pathogens. Water reclamation also includes delivery to the place of use and actual use. *See,* e.g., Water Factory 21 and Fred Harvey Water Reclamation Plant. *See also* wastewater reuse and water reuse applications.

**water reclamation plant reliability** The ability of a plant to produce acceptable reclaimed water consistently. Plant performance is affected by the variability of influent characteristics and equipment failures.

**water recovery** One of two parameters used to express the performance of a pressure-driven membrane process used in water treatment. *See* recovery rate (3) for more detail.

**water-recovery techniques** *See* saline-water conversion classification.

**water recycle** The reuse of reclaimed water in the same application as originally used, usually in such industries as manufacturing. *See also* water reuse applications.

**water recycling** *See* water recycle.

**water reducible** Characteristic of a coating that contains more than 5% of water by weight in its volatile fraction (EPA-40CFR60.391).

**water regain** The amount of water retained within the resin bead and on the surface of the media of an ion exchanger, expressed as a percent of the weight of the latter. Also called water retention.

**water-related disease** A disease that is related to the quality (bacteriological, chemical, or physical), quantity, or presence of water. *See* the following categories of water-related diseases: water-based, waterborne, water-washed, and insect-borne. Waterborne is often, incorrectly, used synonymously with water-related. *See also* water- and excreta-related diseases and excreta-related disease.

**water-related infection** *See* water-related disease.

**water-related insect vector route** The transmission route followed by a pathogen in causing a water-related, insect-borne infection, i.e., from an infected host through an insect vector, to the new host.

**water-related route** The transmission route followed by a pathogen in causing a water-related infection, i.e., the fecal–oral, water-based, water-washed, and insect-vector routs.

**water resources** Potentially useful forms or sources of water as found in all states (liquid, solid, vapor) and in the various points of the hydrologic cycle, e.g., clouds, rain, snow, ice, surface water, ground water, reclaimed or reused water, and seawater. Water resources systems include natural elements (atmosphere, watersheds, channels, wetlands, floodplains, lakes, estuaries, etc.) and also constructed facilities (dams, canals, etc.)

**water resources sustainability** "The planning, development, and management of water resources to provide an adequate and reliable supply of water with a quality suitable to meet their economic, environmental and social needs for current and future generations" (American Water Works Association quoted in Bloetscher and Muniz, 2006). *See also* safe yield, sustainable yield.

**water resources systems** *See* water resources.

**water reuse** The use of reclaimed water/treated wastewater in such beneficial applications as agricultural irrigation or industrial cooling. *See* water reuse applications,

**water reuse applications** *See* the following terms: agricultural irrigation, beneficial use, closed-loop recycle, direct potable reuse, direct reuse, dual distribution system, environmental water reuse, groundwater recharge, indirect potable reuse, indirect reuse, industrial recycling/reuse, landscape irrigation, municipal wastewater reuse, nonpotable urban reuse, nonpotable water reuse, pipe-to-pipe potable reuse, planned reuse, potable water reuse, reclaimed water, recreational water reuse, recycled water, sewage farm, spreading basin, streamflow augmentation, unplanned reuse, wastewater reuse, water reclamation, water recycle or recycling, water reuse.

**water reuse pathway** The ways reclaimed water is used, e.g., irrigation, industry, groundwater recharge.

**water reuse reliability** The probability that the risk of infection through water reuse does not exceed an acceptable level.

**water right** A legal right to use the water from a particular stream, lake, canal, or source.

**water rights** A body of legislation regarding water ownership. *See* appropriation doctrine, English rule, prior appropriation, riparian right, water law.

**water-right value** The monetary value of a water right.

**water saver** A toilet that uses 3.5 gallons per flush, as defined by the American Society of Mechanical Engineers and the American National Standards Institute. It differs from so-called low-flow or high-efficiency toilets.

**water-saving device** Any of a number of devices that can be installed on connections or taps to reduce water flow or restrict it to a fixed quantity, e.g., shower nozzles, spray taps. Water-saving devices also include low-flush toilets.

**water-saving fixture** A plumbing fixture designed to reduce water use.

**water-saving kit** A set of devices that can help reduce water use.

**water-saving taps and fittings** *See* water-saving device.

**WaterSC** Abbreviation of Water Security Channel.

**water-scarce route** The route followed by water-washed pathogens to infect their hosts, i.e., the fecal–oral route, through the skin, or through the eyes. Also water-washed route.

**water seal** A water barrier between two media, e.g., between a sewer line and the atmosphere or between the atmosphere and the pit of a latrine. The water may be contained in a U-shaped pipe or hemispherical bowl to prevent the passage of odors and insects. Also called a trap. *See* Figure W-01, aqua-privy, pour-flush latrine, water seal control.

**water seal control** A seal pot, p-leg trap, or other type of trap filled with water (e.g. flooded sewers that maintain water levels adequate to prevent airflow through the system) that creates a water barrier between the sewer line and the atmosphere. The water level of the seal must be maintained in the vertical leg of a drain in order to be considered a water seal (EPA-40CFR63.111).

**WaterSense** A program of the USEPA that promotes water-efficient technologies, e.g., the specification of toilets that use less than 1.3 gpf, including single- and dual-flush tanks and flushometer and electrohydraulic types.

**water separator** Any device used to recover perchloroethylene from a water–perchloroethylene mixture (EPA-40CFR63.321).

**Water Security Channel (WaterSC)** A free e-mail notification service designed to disseminate water security alerts and other information from federal agencies. *See also* Water Information Sharing and Analysis.

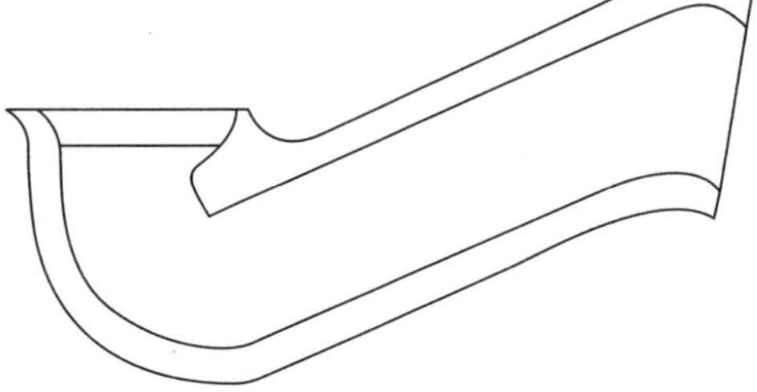

**Figure W-01.** Water seal

**water service connection** All the piping and appurtenances required to provide water to a consumer, from the public distribution main to the consumer's system; also called service line or sometimes simply service, including: service connection or service connector, the pipe that carries water from the main to a curb stop; service pipe, the pipeline from the curb stop to the meter; meter and appurtenances, such as corporation stop, curb box, and shutoff valve.

**water service pipe** The pipe extending from the water main to the customer's premises. Also called service pipe.

**water service schedule** A schedule setting forth the rates charged customers for water supply service.

**watershed** (1) The land area that drains into a stream; an area of land that contributes runoff to a specific delivery point. Large watersheds may be composed of several smaller "subsheds," each of which contributes runoff to different locations that ultimately combine at a common delivery point (EPA Glossaries). A "small" watershed is sometimes defined as having less than 30 square miles or 80 square kilometers. A divide or ridge topographically defines a watershed. Also called drainage area, drainage basin, or river basin. *See also* catchment. Watershed divide is the line that separates one drainage basin from another; it follows the ridges or summits forming the exterior of the drainage basins. Also called drainage divide or topographic divide. (In British usage, the drainage area, drainage basin, or river basin is called a catchment, whereas the word watershed refers to the divide between two catchments.) (2) A reserved area where natural or artificial lakes are protected for water supply and used for storage, natural sedimentation, and seasonal pretreatment.

**watershed-based approach** An integrated approach promoted by the USEPA for the management of stormwater and other discharges in a watershed. It focuses on the coordination of private and public actions to address the highest-priority water quality problems within hydrologically defined areas.

**watershed-based trading** The trading of credits among point source, nonpoint source, and indirect dischargers; buyers purchase pollutant reductions for less than it would cost them to achieve the reductions themselves, whereas sellers receive compensation for reducing pollution beyond the required level. The total pollution reduction, however, is the same as or greater than without trading.

**watershed divide** *See* watershed.

**watershed protection** *See* watershed sanitation.

**watershed sanitation** The practice of healthy environmental measures for the protection of the quality of drinking water sources, e.g., the regulation of recreational uses and residential development within the watershed.

**waterside cleaning wastes** Wastes generated by the removal of deposits and corrosion products collected on boiler tubes. *See* boiler chemical cleaning wastes for detail.

**water skin** A thin layer at the free surface of a water body that is denser than the other layers and resists tension. *See* surface tension.

**water snail** (1) A low-lift, high-capacity, positive-displacement pump that raises water or wastewater by rotating a helical impeller in an inclined trough or cylinder. *See* Archimedes screw for more detail. (2) A mollusk, the species *Limnaea truncatula,* that serves as intermediate host to the parasitic flatworm *Fasciola hepatica,* or sheep liver fluke.

**water snow** Snow that yields a greater amount of water than average upon melting. Also called cooking snow.

**water softener** (1) A pressurized device that passes water through a bed of cation-exchange media to reduce its hardness by exchanging calcium and magnesium ions for sodium or potassium ions. (2) A detergent additive—polyphosphate, or other chemical compound—that, when added to water, counteracts the effects of hardness ions. Also called softener.

**water softener salt** A salt for regenerating cation exchange water softeners. Common softener salts include sodium chloride (NaCl) and potassium chloride (KCl).

**water softening** The complete or partial removal of the ions (mainly calcium and magnesium) that cause water hardness. *See* softening for more detail.

**waters of the United States** As defined in the Clean Water Act, "waters of the United States" includes (a) all waters used currently or in the past, or susceptible to use by the United States in interstate or foreign commerce; (b) all interstate waters; (c) all other waters for which use, degradation, or destruction could affect interstate or foreign commerce; (d) all impoundments and tributaries of waters defined as "waters of the United States"; (e) the territorial seas; (f) wetlands adjacent to these waters.

**water softness** Pertaining to water that contains relatively low concentrations of calcium and magnesium salts; the opposite of hardness. It may be associated with the high incidence of cardiovascular disease in developed countries.

**water solubility** The maximum concentration of a chemical compound that can result when it is dissolved in water. *See* solubility in water for detail.

**water-soluble carbon test** A test that uses potassium persulfate ($K_2S_2O_5$ or $K_2S_2O_8$) to analyze soluble carbonaceous matter in wastewater, excluding cellulose, long-chain fatty acids, certain proteins, and ring compounds.

**water solution** A homogeneous liquid comprising water as the solvent and salts and other substances as solutes.

**water source** A surface water (stream, lake, reservoir, seawater) or groundwater (spring, well, aquifer) that can provide a supply of water. Also called raw water, source water, or water supply source.

**water-splitting membrane** A membrane comprising two attached layers, one with a cation exchanger (e.g., a strong or a weak acid) and the other with an ion exchanger (e.g., a strong or a weak base); used industrially for the production of acid and base chemicals from salts and in the removal of ions from aqueous solutions. *See,* e.g., LINX®.

**water spotting** A film, spot or other mark left on a surface after water has dried from it.

**waterspout** (1) A pipe that runs down the side of a building to carry away water from the roof; a spout or duct. (2) A funnel-shaped cloud that resembles a solid column over a body of water.

**water spreading** (1) The application of water to land for storage and subsequent withdrawal to grow crops. (2) A method of artificial recharge. *See* spreading basin.

**water stabilization** The adjustment of the pH, alkalinity, or calcium content of water to reduce its corrosiveness. *See* stabilization (1).

**water-stage recorder** An instrument that continuously records the stage of a stream at a given point; a river gage.

**water standards** *See* water quality standard.

**water still** A device that produces distilled water by evaporation and condensation.

**water stop** Same as curb stop and box.

**water storage** Water storage, even in small domestic containers, may contribute to mosquito breeding (e.g., *Aedes aegypti*) and the spread of water-related diseases.

**water storage pond** An impound for liquid wastes, so designed as to accomplish some degree of biochemical treatment of the wastes.

**water stripping** A metal products industry technique that removes paint by spraying a high-pressure water stream at the surface of a metal part.

**water supplier** One who owns or operates a public water system. *See also* water purveyor.

**water supply** (1) The potable water supplied to a community. (2) The sources of water for public or private use. (3) The facilities required for providing water of good quality, under satisfactory pressure, and in adequate quantity for domestic, commercial, industrial, firefighting, and other municipal purposes.

**water supply augmentation** The increase of the quantity or the improvement of the utility of fresh water available, by such methods as weather modification (to increase precipitation or reduce evapotranspiration), seawater desalting, urban and agricultural water conservation, wastewater reclamation and reuse, and groundwater storage.

**water supply development** The activities required to plan, design, and construct water supply sources.

**water supply engineering** The branch of civil engineering dealing with the design, construction, and operation of water supply and treatment facilities. *See also* sanitary engineering.

**water supply facility** Any of the works, structures, and equipment required to supply water to a community. Water supply facilities include intakes, wells, pump stations, storage reservoirs, treatment plants, etc.

**water supply source** Same as water source.

**water supply system** The water sources, structures, land, equipment, and organization required to collect, treat, store, and distribute potable water from source to consumer.

**water system** Same as water supply system.

**water system appurtenance** Auxiliary components, other than pipes and conduits, used in a water distribution system, e.g., valves and fire hydrants.

**water table** The level of groundwater; the upper surface of the saturation zone of groundwater above an impermeable layer of soil or rock (through which water cannot move) as in an unconfined aquifer. This level can be near the surface of the ground or far below it. In an unconfined aquifer, the fluid at the water surface is at atmospheric pressure. Also called groundwater table. Same as phreatic surface.

**water-tank indicator** A device used to indicate the depth of water in a tank, e.g., a wood or metal strip with appropriate markings and combined with a float.

**water tower** A standpipe or tower with a tank for storing water at an adequate pressure for distribution. It provides storage for local distribution where a ground-level reservoir would be inadequate. Also called an elevated tank.

**water transport coefficient** Same as water permeability coefficient.

**water treatment** The use of unit operations and processes to alter the constituents and characteristics of water, e.g., to remove turbidity, hardness, and pathogens, or to add a disinfectant residual and fluoride. Water treatment may be conventional or advanced.

**water treatment chemicals** Any combination of chemical substances used to treat water, including coagulants, disinfectants, corrosion inhibitors, antiscalants, dispersants, and any other chemical substances (EPA-40CFR749.68-16). *See* important chemicals.

**water treatment facilities** All the installations, equipment, structures, etc. that are necessary to treat and distribute potable water to consumers.

**water treatment lagoon** An impound for liquid wastes designed to accomplish some degree of biological treatment.

**water treatment plant** All the installations, equipment, structures, etc. that are necessary to treat drinking water for distribution to consumers.

**water treatment plant residuals** Organic and inorganic, liquid, solid, and gaseous waste products generated by water treatment processes, e.g., colloidal and suspended particles, precipitates, spent activated carbon, resins, and filter media. *See also* residuals.

**water treatment plant waste** Water treatment plant residuals.

**water treatment process** Any combination of the following unit operations and processes to achieve the desired objective: activated carbon adsorption, aeration, air flotation, biological treatment, chemical feeding, coagulant recovery, coagulation, dechlorination, disinfection, filtration, flocculation, fluoridation, ion exchange, membrane filtration, neutralization, prechlorination, preoxidation, recarbonation, screening, sedimentation, softening, stabilization, super-chlorination.

**water treatment wastes** *See* water treatment plant residuals.

**water treatment works** Water treatment plant.

**water type** A combination of characteristics used to describe water masses, e.g., temperature, salinity, dissolved oxygen, nutrients.

**water usage fee** A fee for water service based on the quantity consumed.

**water use** (1) An activity that is conducted in or on the water; but does not mean or include the establishment of any water quality standard or criteria or the regulation of the discharge or runoff of water pollutants except the standards, criteria, or regulations that are incorporated in any program as required by the Coastal Zone Management Act. (2) A classification of the utilization of water resources, e.g., potable water supply, recreation, fishery, industrial, transportation, waste disposal, and power production.

**water-use ratio** The ratio of the weight of water transpired by a plant to the weight of dry matter produced. Also called transpiration ratio.

**water utility** Same as waterworks, i.e., the system of facilities used in the supply, transmission, treatment, storage, and distribution of water to consumers, plus the administrative structure.

**water vapor** The gaseous form of water, subject to vapor pressure laws. Also called aqueous vapor, moisture.

**water-vectored disease** A disease transmitted by a vector related to water, e.g., insects that breed or bite in or near water. *See* dengue fever, filariasis, malaria, onchocerciasis, yellow fever. *See also* water habitat vector disease.

**water-vector habitat disease** *See* water habitat vector disease.

**waterwall incineration** Waste incineration in a furnace whose walls are lined with steel tubes containing circulating water. The device includes an overhead crane that distributes the waste evenly into a hopper and pistons that ram the waste from the hopper onto a grate, which moves it across the combustion chamber.

**water-washed disease** A disease that prevails in areas where water is in short supply for personal and domestic hygiene. It is caused by pathogens that can be reduced or eliminated by the use of water, regardless of its quality, for hygienic purposes. Water-washed diseases include the fecal–oral infections of the intestinal tract, skin and eye infections, and louse-borne infections (typhus, relapsing fever). Also called water hygiene disease. *See also* water- and excreta-related diseases, excreta-related disease, and other categories of water-related diseases: fecal–oral, insect-borne, water-based, and waterborne.

**water-washed route** The route followed by water-washed pathogens to infect their hosts, i.e., the fecal–oral route, through the skin, or through the eyes. Also water-scarce route.

**water waste survey** An investigation conducted to determine the extent and location of leaks and other sources of unaccounted-for water in a distribution system. It is undertaken usually when the difference between pumping records and reported consumer uses is excessive. *See also* leak survey.

**Waterweb®** A mesh manufactured by Misonix, Inc. for use in air scrubbers.

**water weed** Aquatic weeds can be controlled by chemicals (e.g., copper sulfate and chlorine), draining the water body, dredging, or cutting.

**water well** An excavation where the intended use is for location, acquisition, development, or artificial recharge of groundwater (excluding sandpoint wells).

**water wheel distributor** A mechanically driven device that distributes wastewater over a trickling filter; e.g., a rotary wheel energized by the influent flow and installed in the center or at the periphery of the filter.

**waterwork** *See* waterworks (3).

**waterworks** (1) With a singular or plural verb, the system of facilities used in the supply, transmission, treatment, storage, and distribution of water to consumers, e.g., pipes, reservoirs, pumps, and appurtenances. *See also* water utility, water supply system. (2) With a singular verb, a pumping station or a treatment. (3) With a plural verb, a spectacular display of water. Also called waterwork.

**waterworks aluminum** Another name for aluminum sulfate $[Al_2\ (SO_4)_3 \cdot 14H_2O]$ in reference to alum, which is used in water treatment.

**water year** A continuous period of 12 months for recording hydrological and climatic data, e.g., October 1 to September 30 for the Northern Hemisphere (and as selected by the U.S. Geological Survey) and July 1 to June 30 for the Southern Hemisphere. Also called climatic year, hydrologic year.

**WATMAN** A computer program developed to estimate the costs of water treatment alternatives, based on average and peak flows, and on treatment methods. *See also* CAPDET.

**Watson's law** In disinfection operations, the inactivation rate constant ($k$) is related to the concentration ($C$) of disinfectant, the die-off constant ($k'$), and the coefficient of dilution ($n$):

$$k = k'C^n \qquad \text{(W-12)}$$

*See also* Chick–Watson equation.

**Wave Oxidation®** A wastewater treatment apparatus manufactured by Parkson Corp.; uses alternating aerobic and anaerobic processes.

**WBA** Acronym of weak-base anion.

**WBA resin** Abbreviation of weak-base anion resin.

**WC** Acronym of water closet.

**weak acid** An acid that has a low degree of ionization and results in relatively few hydrogen ions in water or other dilute solutions. Also defined as an acid that has a weak proton-donating tendency. The strength of the acid is its degree of dissociation or acid dissociation constant. Examples of common weak acids include acetic ($CH_3COOH$), hypochlorous ($HClO$), and carbonic ($H_2CO_3$) acids. Weak acids have low dissociation constants (K values) corresponding to high inverse logarithms (pK values).

**weak-acid cation (WAC) exchanger** Same as weak-acid cation-exchange resin.

**weak-acid cation (WAC) exchange resin** A resin that has a weak-acid functional group (—COOH), e.g., a carboxylic group with high affinity for $H^+$ ions; it can exchange only cations associated with alkalinity (e.g., in carbonate hardness removal) but not those associated with strong acids (e.g., in noncarbonated hardness removal). *See also* strong-acid cation (SAC) exchanger.

**weak-acid cation (WAC) resin** Same as weak-acid cation-exchange resin.

**weak-acid resin** Same as weak-acid cation-exchange resin.

**weak base** A base that has a low degree of ionization in water or other dilute solutions. Also defined as a base that has a low tendency to accept a proton (a hydrogen ion, $H^+$) from another molecule. Ammonia ($NH_3$) is considered a weak base.

**weak-base anion (WBA) exchanger** Same as weak-base anion-exchange resin.

**weak-base anion (WBA) exchange resin** A resin that has a weak-base functional group; it can exchange only anions associated with strong mineral acids (e.g., sulfates, chlorides, nitrates) but not those associated with weak acids (carbonates, bicarbonates, silicates) or organic acids.

**weak-base anion (WBA) resin** Same as weak-base anion-exchange resin.

**weanling diarrhea** Pediatric diarrhea caused by contaminated food of poor nutritional quality prepared in unsanitary conditions.

**wearing ring** A ring placed between the casing and the impeller of a centrifugal pump to improve efficiency.

**weathering** The process during which a complex compound is reduced to its simpler component parts, transported via physical processes, or biodegraded over time; the disintegration of rocks and unconsolidated mineral particles by wind, water, ice, chemical action, or biological activity. *See also* chemical weathering.

**Weber–Fechner law** A law stated by two German philosophers in the nineteenth century: a stimulus increases geometrically when the sensation increases arithmetically. It is used to conclude that in taste and odor studies, it is the threshold ratios, not the threshold differences, that matter.

**Wedge-Flow™** Water filtration products of LEEM Filtration Products, Inc. made of wedgewire.

**WedgePress** A belt filter press manufactured by Gray Engineering Co.

**wedgewater bed** Same as wedgewire bed.

**Wedgewater Filter bed™** A filter bed manufactured by Gravity Flow Systems, Inc. for gravity sludge dewatering.

**Wedgewater Sieve** A static bar screen manufactured by Gravity Flow Systems, Inc.

**wedgewire** A special wire that has a triangular or trapezoidal cross section. Also called profile wire.

**wedgewire bed** A device used to dewater water or wastewater treatment sludge, using as medium a stainless steel septum with wedge-shaped slots that supports the sludge cake and allows drainage. Also called wedgewater bed. *See also* air drying and Figure W-02.

**wedgewire drying bed** Same as wedgewire bed.

**wedge zone** The third area of a belt filter press in which the two porous belts come together, with the sludge solids in between as in a sandwich. Also called low-pressure zone.

**Wedging Roll** A belt filter press manufactured by OVRC Environmental.

**weep hole, weephole** A drainage opening, i.e., a hole left through a sill, retaining wall, apron, lining, foundation, etc. to permit drainage (e.g., of accumulated condensation or seepage) and/or reduce pressure. Also called weeping hole.

**weeping hole** Same as weep hole.

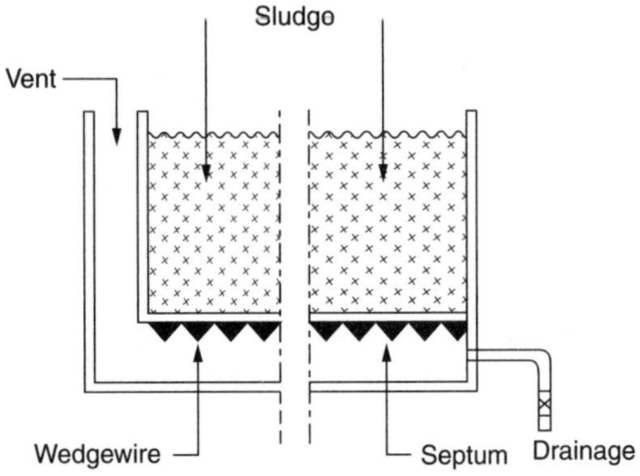


**Figure W-02.** Wedgewire bed.

**WEF** Acronym of Water Environment Federation.

**weft wire** The horizontal wire in woven wire mesh; also called shute wire.

**Wehner–Wilhelm method** *See* Wehner–Wilhem's equation.

**Wehner–Wilhelm's equation** An equation proposed in 1958 for the design of aerobic ponds, based on first-order removal kinetics:

$$S/S_0 = 4\,Ae^{1/2d}/[(1+A)^2 e^{A/2d} - (1-A)^2\, e^{-A/2d}] \qquad \text{(W-13)}$$

where $S$ = effluent substrate concentration, mg/L; $S_0$ = influent substrate concentration, mg/L; $A = (1 + 4\,kdt)^{1/2}$; $k$ = first-order reaction rate constant,/hr; $t$ = detention time, hr; $d$ = dispersion factor = $D/uL$, dimensionless; $D$ = axial dispersion coefficient, ft$^2$/hr; $u$ = fluid velocity; ft/hr; and $L$ = characteristic length, ft. *See also* aerobic–anaerobic lagoon.

**Weibull model** A dose–response model of the form

$$P(d) = 1 - \exp\,[-b(d^m)] \qquad \text{(W-14)}$$

where $P(d)$ is the probability of cancer due to continuous dose rate $d$, and $b$ and $m$ are constants. *See* dose–response model for related distributions.

**weight** The gravitational force with which the Earth (or another celestial body) attracts an object, equal to the product of the object's mass by the acceleration of gravity.

**weight concentration ratio** The ratio of the weight of feedwater to the weight of reject water at any time during an ultrafiltration run.

**weighted average** An average computed with a weighting factor (smaller or larger than 1.0) assigned to one or more of the elements of the sample.

**weighted bottle** A glass or plastic bottle equipped with a sinker and a stopper, used to sample hazardous wastes occurring as liquids or free-flowing slurries. A line is used to lower, raise, and open the bottle. *See also* auger, bailer, composite liquid-waste sampler, dipper, thief, trier.

**weighted composite sample** The flow-weighted average of several samples taken over a given period of time, as compared to a grab sample, which is a single sample. Often simply called composite sample or proportional sample.

**weighting agent** A coagulant aid used in the treatment of waters that yield light floc that is difficult to settle. Such waters may be high in color but low in mineral content. Weighting agents include activated carbon, bentonite clay, limestone, and powdered silica; they increase the density of the suspension and the surface available for adsorption. Also called adsorbent-weighting agent.

**weight-loss method** A common method used to measure the rate of metal corrosion over a period of time. The rate of corrosion is expressed in mils (0.001 inch) penetration per year. *See* coupon weight-loss method and loop system weight-loss method.

**Weil's disease** A water-related disease, the human form of leptospirosis, caused by the spirochaete *Leptospira icterohaemorrhagiae;* named after the German physician Adolph Weil (1848–1916). Its symptoms include fever, jaundice, and muscle pain; it may also affect the liver and kidneys. Poor refuse disposal practices may promote the proliferation of rodents and the transmission of the disease. Also called hemorrhagic jaundice, it is transmitted from the urine of rodents and other animals through a person's skin, eyes, or mouth. Related diseases include leptospirosis, canicola fever, and canine leptospirosis.

**Weil's syndrome** *See* Weil's disease.

**weir** (1) A wall or plate placed in an open channel to measure the flow of water. (2) A wall or obstruction used to control flow from settling tanks and clarifiers to assure a uniform flow rate and avoid short-circuiting. (3) Any natural or man-made, regular obstruction over which flow occurs. Typically, while flowing over the weir, water passes from subcritical to critical conditions, and the weir equation describes the relationship between the subcritical depth and the critical flow. The crest of the weir is the upper edge of the weir plate or the bottom edge of the opening over which water flows. The crest height ($h$) is its distance to the bottom of the body of water, and the crest width or length ($L$) is the width or length of the weir. The nappe is the overfalling stream of water (from the French term for sheet). The weir head or depth ($H$) is the height of the pool above the crest, measured upstream where unaffected by water surface curvature. *See* Figures W-03 through W-13. Weirs are inexpensive, easy to install, and easy to use, as compared to other means of flow measurement. However, they cause high head losses, require periodical cleaning, and are not always accurate, particularly when affected by excessive approach velocities and debris. *See* the following types of weirs: broad-crested, Cipolletti, clarifier, compound, contracted, crib, Crump, diverting, division, drowned, effluent, flow equalization, free, friction, influent, leaping, log, long-based, movable, needle, outlet, overfall, overflow, parabolic, parallel finger, peripheral, proportional, proportional flow, rectangular, Rettger, rolling-up curtain, rounded-crest, round-nosed, separating, serpentine, sharp-crested, shutter, side-flow, slid-

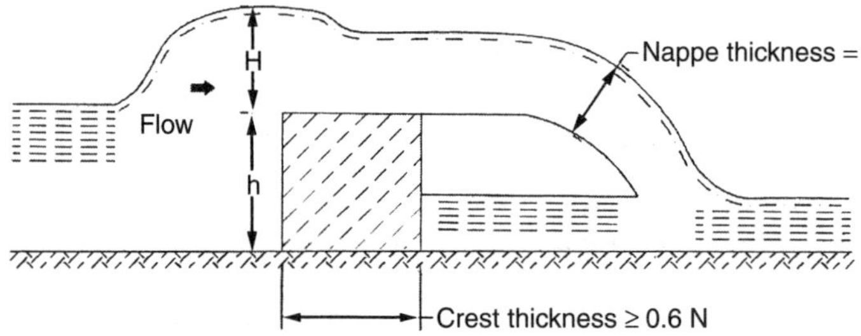


**Figure W-03.** Weir (broad crested) (courtesy CRC Press).

ing-panel, straight-edged, submerged, suppressed, suspended-frame, Sutro, thin plate, Thomson, transverse, trapezoidal, triangular, triangular-notch, vee, V-notch, V-notched, waste, wide-crested.

**weir box** A box used to enlarge a narrow waterway upstream of a weir.

**weir crest** *See* weir.

**weir crest height** *See* weir.

**weir depth** *See* weir.

**weir diameter** Many circular clarifiers have a circular weir within their outside edge. All the water leaving the clarifier flows over this weir. The diameter of the weir is the length of a line from one edge to the opposite edge and passing through the center of the circle formed by the weir.

**weir end contraction** The condition of a rectangular weir when its ends project inward from the sides of the channel. The equation of rectangular weirs is usually modified to account for end contractions by subtracting one-tenth of the height ($H$, m or ft) from the length ($L$, m or ft) for each contraction:

$$H = [Q/C_w(L - 0.2\,H)]^{0.67} \qquad \text{(W-15)}$$

where $Q$ is the flow (in m$^3$/sec or in cfs) and $C_w$ is the weir coefficient, which accounts for the approach velocity and ranges from 1.8 to 2.3 in metric units and from 3.2 to 4.2 in English units.

**weir equation** An equation used to determine the discharge ($Q$) over a weir, as a function of the weir geometry and the head ($H$) above the weir. For example,

$$Q = K(2\,g)^{1/2} L H^{3/2} \qquad \text{(W-16)}$$

(for a rectangular weir)

$$Q = 0.179\,(2\,g)^{1/2} H^{5/2} \qquad \text{(W-17)}$$

(for a 60° triangular weir)

where $K$ is a flow coefficient, $L$ the length of the weir or the width of the channel, g the acceleration of gravity, and $H$ the head on the weir. The weir formulas used in the SWMM model are:

$$Q = KL[(H + V^2/2\,g)^a - (V^2/2\,g)^a] \qquad \text{(W-18)}$$

in general

$$Q = K'KL(y_1 - y_0)^{3/2} \qquad \text{(W-19)}$$

for a submerged weir

$$Q = K''L(y_3 - y_0)[2\,g\forall(y_2, y_0)]^{1/2} \qquad \text{(W-20)}$$

for a surcharged weir

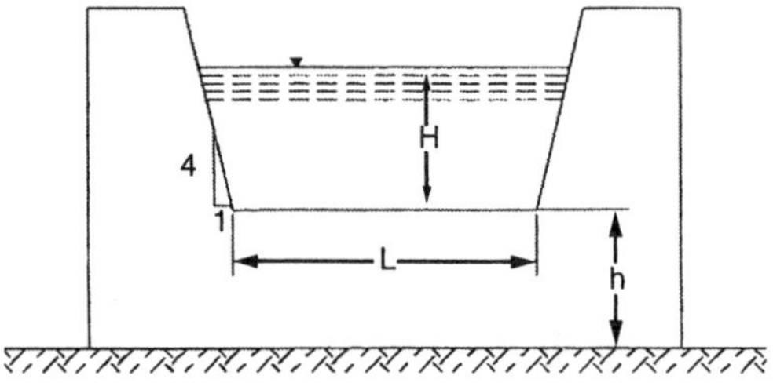


**Figure W-04.** Weir (Cipoletti) (courtesy CRC Press).

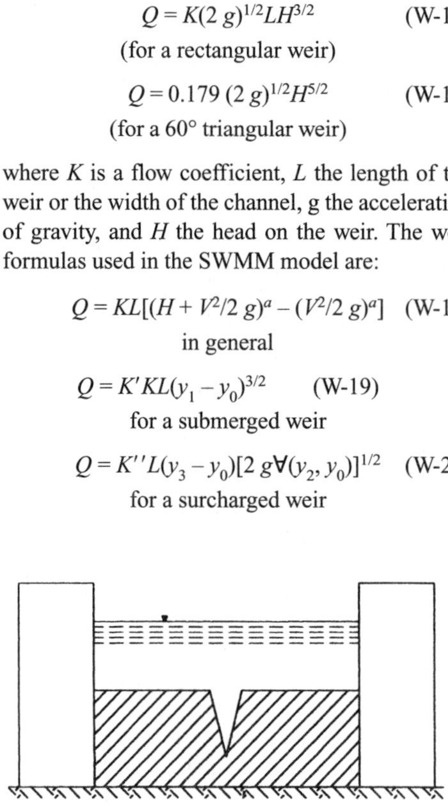

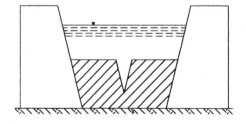

**Figure W-05.** Weir (compound) (courtesy CRC Press).

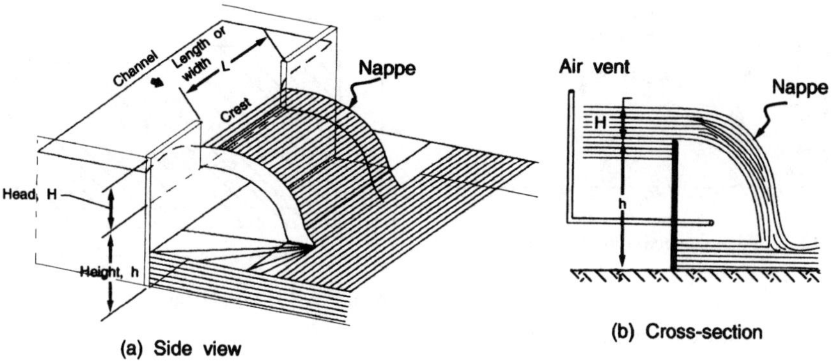


**Figure W-06.** Weir (contracted, side view and cross section) (courtesy CRC Press).

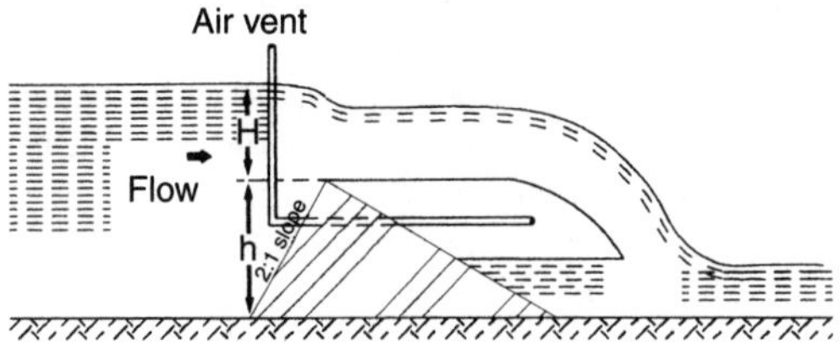


**Figure W-07.** Weir (Crump) (courtesy CRC Press).

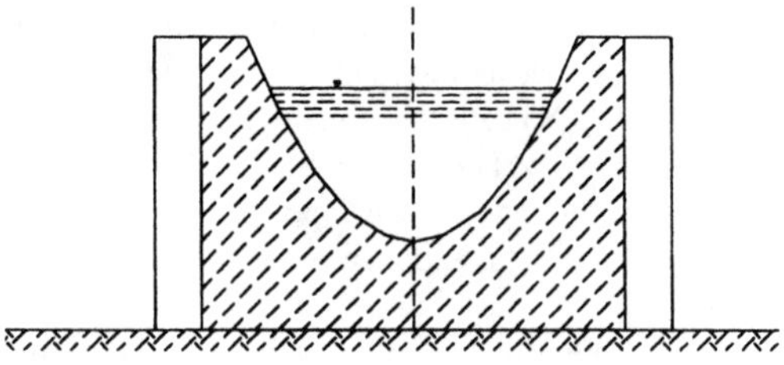

**Figure W-08.** Weir (parabolic) (courtesy CRC Press).

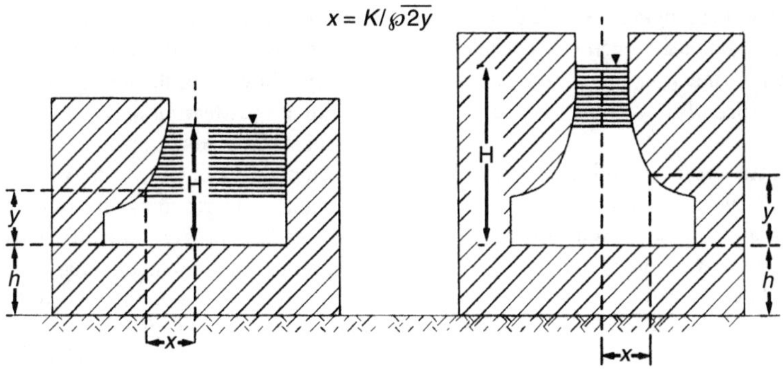


**Figure W-09.** Weir (proportional) (courtesy CRC Press).

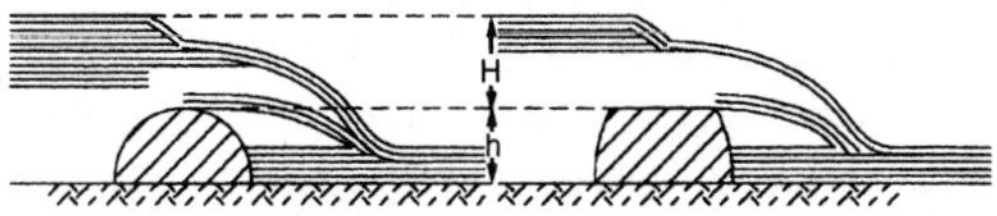


**Figure W-10.** Weir (rounded crest) (courtesy CRC Press).

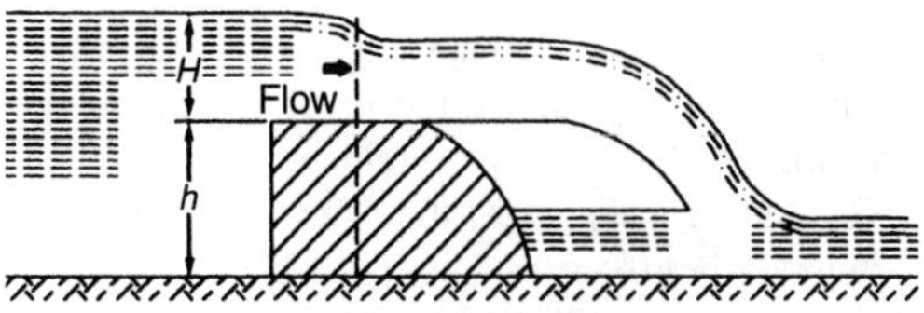


**Figure W-11.** Weir (rounded nose) (courtesy CRC Press).

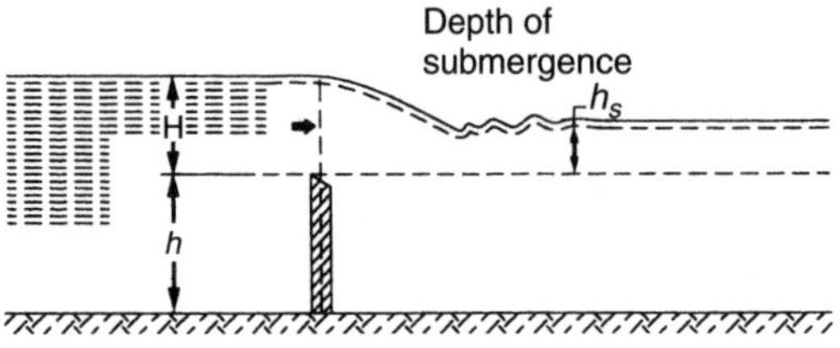


**Figure W-12.** Weir (submerged) (courtesy CRC Press).

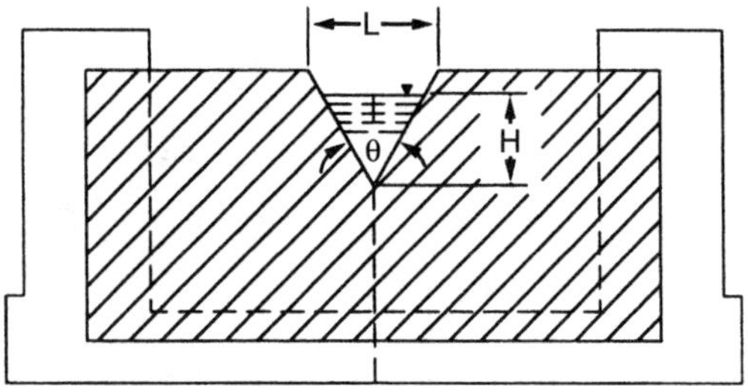


**Figure W-13.** Weir (triangular or V-notched) (courtesy CRC Press).

where a is the weir exponent ($a$ = 3/2 for a transverse weir and $a$ = 5/3 for a side-flow weir), $K'$ is the submergence coefficient, $K''$ is the weir surcharge coefficient, $V$ is the approach velocity; $y_0$), $y_1$, $y_2$, and $y_3$ are, respectively, the height of the weir crest above the node invert, the water depth on the upstream side of the weir, the downstream depth, and the distance to the top of the weir opening. The term $\forall(y_2, y_0)$ means the larger of the two depths $y_2$ and $y_0$). *See also* Francis formula, discharge coefficient, Thomson weirs, and Cipoletti weir.

**weir exponent** An exponent used in a weir equation.

**weir flow** Flow over a bridge or culvert and the immediate sections of the approach roadway, computed with a weir equation; a flow condition in flow routing problems.

**weir head** *See* weir.

**weir loading** The ratio of the flow rate out of a basin or tank to the length of weir, e.g., expressed in gpd/ft, a parameter sometimes considered in clarifier design and operation; also called weir overflow rate. *See also* overflow rate, solids loading rate.

**weir loading rate** Same as weir loading.

**weir nappe** *See* weir.

**weir overflow rate** Same as weir loading.

**weir rate** Same as weir loading.

**weir splitter box** A device used in wastewater sludge handling to minimize fluctuations in digester feed rates.

**weir surcharge coefficient** A coefficient that modifies the orifice equation for application to a surcharged weir.

**weir teeth** The notches of a V-notch weir plate. *See also* alligator teeth.

**Weisbach, Julius** German engineer (1806–1871) who taught engineering mechanics at the School of Mines of Erzgebirge, Freiberg, near Dresden, Germany. He suggested the Darcy–Weisbach formula for head losses in pipes in 1845.

**Weiss** An in-vessel composting apparatus manufactured by Taulman Co.

**WEKOSEAL®** A pipe joint rehabilitation product fabricated by Miller Pipeline Corp. of Indianapolis, IN; it is used in trenchless pipe renewal.

**well** A bored, drilled, or driven shaft, or a dug hole (generally of a cylindrical form), often walled with bricks or tubing to prevent the earth from caving in, whose depth is greater than the largest surface dimension, and whose purpose is to reach underground water supplies or oil, or to store or bury fluids below ground (EPA-40CFR260.10).

**well acidization** A well stimulation procedure that uses chemicals (e.g., a solution of hydrochloric acid, HCl) to break down matter that clogs the rock pores in an injection zone, thereby improving the flow of injection fluids. Also called acid treatment. *See also* well surging.

**wellfield** Area containing one or more wells that produce usable amounts of water or oil.

**wellhead** The area immediately surrounding the top of a well, or the top of the well casing.

**wellhead protection area** A protected surface and subsurface zone surrounding a well or wellfield supplying a public water system to keep contaminants from reaching the well water. A wellhead protection program identifies potential sources of groundwater contamination; it delineates a safe infiltration area on the basis of hydrologic and geologic data that allow the establishment of time-of-travel zones or catchment areas around pumping wells. *See also* well protection.

**Wellhead Protection Planning (WHPP)** A process designed to protect groundwater sources from contamination. It was established under the Safe Drinking Water Act (SDWA) amendments of 1987 and became a part of the USEPA's Source Water Protection Program (SDWA's 1996 amendments).

**wellhead protection program (WHPP)** *See* wellhead protection area, Wellhead Protection Planning.

**well injection** The subsurface emplacement of fluids or disposal of liquid wastes through a bored, drilled, or driven well; or through a dug well whose depth is greater than the largest surface dimension (EPA-40CFR144.3 and 40CFR165.1-aa). *See also* underground injection.

**well-mixed estuary** Hydrodynamically speaking, an estuary in which salinity is almost uniform vertically but increases toward the sea. One-dimensional models are often used to represent well-mixed estuaries; see, e.g., continuous discharge and dissolved-oxygen sag curve. *See also* partially mixed estuary.

**well monitoring** The measurement, by on-site instruments or laboratory methods, of the quality of water in a well.

**well plug** A watertight and gastight seal installed in a borehole or well to prevent the movement of fluids.

**wellpoint** (1) A reusable, perforated or slotted tube that is driven into the ground to construct a well. (2) A small driven well. A line of wellpoints on either side or both sides of a trench, attached in series to a common pump, can be used to dewater excavations and prevent quicksand.

**well profiling** An investigation conducted to draw the profile of a well from samples at various depths. The flow and chemistry data, collected under normal pumping conditions, are used to identify the zones that contribute water of undesirable characteristics (e.g., arsenic, iron, nitrates) to the well and to block off these zones. Well profiling is often used as an alternative to other options such as treatment, drilling new wells, or alternate sources.

**well protection** The safeguarding of a well from surface and subsurface contamination sources by such measures as location at a safe distance from waste disposal installations, adequate casing and sealing, and thorough disinfection of the well and accessories. *See* wellhead protection.

**well screen** A perforated casing installed in a well to prevent the passage of solids while allowing water to flow.

**well stimulation** (1) A procedure used to restore the capacity of an injection well that has become clogged or plugged and less receptive to injection fluids. It consists of a series of processes to clean the well bore, enlarge channels, and increase pore space in the interval to be injected, thus making it possible for wastewater to move more readily into the formation, including: surging, jetting, blasting, acidizing, and hydraulic fracturing (EPA-40CFR146.3). *See* well acidization and well surging. (2) Any method used to rehabilitate a water supply well and restore its performance, including the use of hydrochloric or other mineral acids, mechanical means, or carbon dioxide.

**well surging** A well-stimulation procedure that pumps water into or out of an injection well at a fluctuating rate to loosen matter that clogs rock openings and impedes the flow of injection fluids. *See also* well acidization.

**well treatment fluid** Any fluid used to restore or improve productivity by chemically or physically altering hydrocarbon-bearing strata after a well has been drilled (EPA-40CFR435.11-y).

**Well Wizard®** An instrument manufactured by QED Environmental Systems to sample and monitor groundwater.

**Wemco** Water and wastewater treatment products of Envirotech Co.

**Westates Carbon** Carbon products and services of Wheellabrator Clean Air Systems, Inc.

**Westchar™** Activated carbon produced by Western Filter Co.

**Westchlor®** A coagulant containing polyaluminum manufactured by Westwood Chemical Corp.

**Westfalia** A centrifuge manufactured by Centrico, Inc. for sludge dewatering.

**WestRO™** Membranes and other products manufactured by Western Filter Co. for use in reverse osmosis units.

**WET** Acronym of whole effluent toxicity.

**wet-air oxidation (WAO)** The incomplete oxidation of volatile solids under high temperature (540°C) and pressure (1200–1800 psig) without vaporizing the liquid portion, using pumps, compressors, heat exchangers, and vapor separators. WAO was developed in Norway for pulp mill wastes and has been adapted to the thermal reduction of sludge from primary clarifiers or thickeners. It reduces COD by 80% and volatile solids by 90%. By-products are an effluent suspension and exhaust gases. WAO is the same as wet combustion but it can operate at higher pressures, up to 3000 psig. Also called wet oxidation process. *See,* e.g., Zimmerman Process. *See also* high-pressure oxidation, intermediate-pressure oxidation, deep-well WAO process, incomplete combustion, starved-air combustion, wet combustion.

**wet air regeneration (WAR)** The application of the wet-air oxidation technology to oxidize excess biological solids and regenerate spent powdered activated carbon. *See also* PACT process.

**wet-bulb temperature** The temperature indicated by a thermometer that has a wetted wick around its bulb.

**wet chemical feed** The introduction of dissolved or suspended chemicals in feed lines of water or wastewater treatment plants. The chemicals are fed in solutions of known strength or slurries of known concentration. Also called liquid chemical feed. *See also* dry chemical feed, gas feed.

**wet chemistry analysis** Laboratory procedures used to analyze a sample of water using liquid chemical solutions instead of, or in addition to, laboratory instruments.

**wet combustion** The incomplete oxidation of volatile solids under high temperature (540°C) and pressure (1200–1800 psig) without vaporizing the liquid portion, using pumps, compressors, heat exchangers, and vapor separators. It is similar to wet-air oxidation but the latter can operate at higher pressures.

**wet connection** A connection to a water main under pressure.

**wet day** Any 24-hour period with some level of precipitation, e.g., at least one inch or 25 mm of rainfall. Consequently, a dry day has less than the specified minimum.

**wet deposition** The flux of dissolved and particulate substances carried from the atmosphere to the Earth's surface by rain or snow; rainout, washout. *See also* dry deposition, acid deposition.

**Wetec™** A wet-aeration apparatus manufactured by Aeration Technologies, Inc.

**wet excreta disposal system** The use of water to carry wastes to treatment or disposal facilities; a sewerage system. *See also* dry or night soil disposal system.

**wet intake** A towerlike structure built to draw water from a lake or a river, consisting of an outer well, a central well, and pipes that bridge them and open into the outer well to transfer water to the intake conduit. *See also* dry intake.

**wetlands** (1) Areas that are saturated or nearly saturated by surface or ground water, at least intermittently, with vegetation adapted for life under those soil conditions. They include swamps, bogs, fens, freshwater marshes around ponds and channels, brackish and saltwater marshes, estuaries, beaches, wet meadows, sloughs, potholes, mudflats, and river overflows. Water depth is usually less than 2.0 ft. (2) Areas that are inundated or saturated by surface or groundwater with a frequency sufficient to support, or that under normal hydrologic conditions does or would support, a prevalence of vegetation or aquatic life typically adapted for life in saturated or seasonally saturated soil conditions. This definition includes those wetlands areas separated from their natural supply of water as a result of activities such as the construction of structural flood protection methods or solid-fill roadbeds and activities such as mineral extraction and navigation improvements. (Definition from the code of federal regulations on floodplain management and the protection of wetlands.)

**wetland treatment** A method of wastewater treatment that uses aquatic plants to absorb organic matter, heavy metals, nutrients, and other contaminants in artificial or natural wetlands. The plants are periodically harvested and removed. Performance is comparable to that of secondary treatment.

**wetland value** A characteristic of a wetland such as area or volume, aquatic productivity, pollutant assimilation, flood storage capacity, and wildlife habitat.

**wetland water budget** A relationship between the various flow elements into and out of a wetland, which determines the rate of change of the volume of water with respect to time:

$$dV/dV = P - A + Q_i + G_i - Q_0 - G_0 - ET \quad \text{(W-21)}$$

where $V$ = volume of water storage in the wetland, $t$ = time, $P$ = precipitation, $A$ = abstractions from precipitation, $Q_i$ = surface inflow, $G_i$ = groundwater inflow, $Q_0$ = surface outflow, $G_0$ = groundwater outflow; and $ET$ = evapotranspiration.

**wet meadow** Grassland with soil that is waterlogged after a rainfall.

**wet oxidation** The oxidation of sludge volatile solids under high temperature (180 to 270°C) and pressure (1000 to 3000 psi) without vaporizing the liquid portion. *See also* thermal conditioning, wet-air oxidation.

**Wetox® system** A patented device of U.S. Filter/Zimpro Inc., consisting of a series of stirred-tank reactors, used for the wet oxidation of wastes.

**wet packed-bed scrubber system** *See* packed-bed scrubber.

**wet pit** *See* wet well.

**wet pressure loss** The loss of air pressure due to the resistance of wet diffuser plates in a wastewater aeration basin.

**wet rendering** A method used in rendering plants to process meat; it produces a waste liquor that contains a high concentration of nitrogenous and other organic matter. This liquor may be disposed in evaporation ponds and the residue mixed with the plant products. *See* rendering pollution, meatpacking waste.

**wet-salt saturator tank** A tank containing saturated brine above the level of unsaturated salt at the bottom.

**wet scrubber** An air pollution control device that uses water or another liquid to remove particulates, fumes, dusts, mists, and vapors from gas streams; for example, in sludge incinerators. *See* cyclone scrubber, ejector–venturi scrubber, electrostatic precipitator, fabric filter, mechanical collector, spray tower, and venturi scrubber. (2) A device or method used to reduce the emission of sludge combustion acid gases (hydrogen bromide, hydrogen chloride, hydrogen fluoride, sulfur oxides) by diffusion of the gases into the liquid phase. *See also* fluid bed combustion, dry scrubber.

**wet scrubber absorption** The treatment of odorous air with a scrubbing liquid. *See* packed-bed scrubber and mist scrubber system.

**wet scrubbing** A process used in air pollution control to remove particulates and gaseous contaminants by passing the air stream through an aqueous spray. The removal mechanism may be absorption, dissolution, entrainment, or chemical reaction.

**wet season** The period of the year when atmospheric precipitation is more intense than average. For example, the wet season in southeastern Florida is May through October, when two-thirds of the annual precipitation of 60 inches fall.

**wet sludge** The slurry removed from water or wastewater treatment, including moisture and solids content. The volume of wet sludge ($V$, gallons) may be calculated from the estimated weight of dry solids ($W$, pounds) and the percent solids content (S) or percent water content ($P$):

$$V = 12\ W/S = 12\ W/(100 - P) \qquad \text{(W-22)}$$

**wet steam** Steam that contains water droplets.

**wet system** *See* wet excreta disposal system.

**wetted-in-water density** In granular activated carbon adsorption systems, the particle density wetted in water is the mass of activated carbon including the water required to fill the internal pores per unit volume of particle, typically 1300–1500 $kg/m^3$.

**wetted perimeter** The length of the wetted contact between a flowing fluid and its channel, measured perpendicularly to the direction of flow; used in the computation of hydraulic radius. *See* open channel flow for the wetted perimeters of some particular cross sections.

**wetting** The absorption of a liquid by a solid surface and the formation of a liquid film.

**wetting agent** Same as surfactant.

**wetting effectiveness** Same as wetting efficiency.

**wetting efficiency** The ratio of the wetted surface area of the medium of a trickling filter to the clean surface area; a parameter indicative of the performance of media. Some limited studies indicate that, depending on the media and on the hydraulic application rate, the wetting efficiency may vary within the range 0.2–0.6. *See also* SK concept.

**wet-volume capacity** The capacity of an ion-exchange resin, expressed in milliequivalents per milliliter (meq/mL) of wet resin; 1 meq/mL corresponds to $6.023 \times 10^{20}$ exchange sites per milliliter. *See also* dry-weight capacity, operating capacity.

**wet-weather flow** The average or peak flow during the wet season, or as a result of rainfall. In a sanitary sewer system, wet-weather flow includes dry-weather flow, infiltration, and inflow. In combined sewers, wet-weather flow is predominantly stormwater runoff.

**wet well** A compartment or chamber where water or wastewater is collected and to which the suction pipe of a pump is connected. Also called inlet well.

**wet-well pumping station** A wastewater lift station with the submersible or turbine pumps mounted on a wet well, without a dry well, for pumping small-to-moderate flows.

**WF21** Acronym of Water Factory 21.

**WFI** Acronym of water-for-injection.

**Whatman glass fiber filter** A filter having a nominal pore size of 1.58 micron, used in the determination of total suspended solids.

**Wheeler** An underdrainage system manufactured by Roberts Filter Manufacturing Co. for rapid sand filters. It consists of a concrete bottom with inverted conical or pyramidal openings that are filled with heavy balls. Also called Wheeler bottom.

**Wheeler bottom** *See* Wheeler.

**Whipple disk** An ocular disk with various grid sizes, used with a microscope to count plankton.

**Whipple, G. C.** American sanitary engineer (1866–1924) in practice and teaching (Harvard University), with a particular interest in waterborne typhoid fever; he proposed a standard areal unit to count algae and other plankton.

**whipworm** The common name of *Trichuris trichiura,* the excreted helminth that causes trichuriasis. It is distributed worldwide and is transmitted from the feces of infected hosts through the soil; it is occasionally waterborne. Also the common name for the disease.

**whipworm infection** Same as the disease trichiuriasis, or simply whipworm. Also called trichocephaliasis.

**Whirl-Flo®** A vortex pump manufactured by ITT Marlow for handling solids.

**whirl vortex** The circular motion of a liquid passing through an orifice at the bottom of a basin, creating a downward opening from the liquid surface.

**Whispair** A series of blowers and pumps manufactured by Dresser Industries/Roots Division.

**white caustic** Another name for caustic soda or sodium hydroxide (NaOH).

**white clay** Same as kaolin, or china clay. *See* hydrolysis (5).

**whitefish** Any of a group of intolerant fishes, similar to the trout but with a smaller mouth and larger scales; they require cold water, high dissolved oxygen, and low turbidity.

**white liquor** A mixture of chemicals, sodium hydroxide (NaOH), and sodium sulfide ($Na_2S$), used to dissolve the lignin from wood fibers in making wood pulp for paper. When cooked, the mixture produces black liquor and brown pulp. *See also* brownstock washing.

**white smoker** A buildup of nutrients and chemical precipitates surrounding a hot hydrothermal vent (water temperature up to 400°C).

**white vitriol** Zinc sulfate.

**whitewash** Another name for slaked lime or calcium hydroxide [$Ca(OH)_2$]. The product is used in water or wastewater treatment as a wet suspension or slurry in a water-to-lime ratio between 2:1 and 6:1 by weight. Also called milk of lime. *See* lime for detail.

**white water** (1) Filtrate from a wire screen on which paper or board is formed, usually recycled for density control and conservation, e.g., as spray and wash water. (2) In the dissolved-air flotation process, the mixture of liquid and rising microbubbles in the contact zone. (3) Frothy water in whitecaps and rapids, or light-colored seawater over a sandy bottom.

**WhiteWater™** An air stripper manufactured by QED Treatment Systems.

**white water collector model** A set of equations used to describe the mass transport of particles to bubble surfaces in the dissolved-air flotation process, based on the concept of single collector efficiency:

$$dN_{p,i}/dt = -1.5\ \alpha_{pb}\eta_T N_p \nu_b \Phi_b / d_b \qquad \text{(W-23)}$$

$$N_{p,i}/N_{p,e} = \exp\left(-1.5\ \alpha_{pb}\eta_T \nu_b \Phi_b t_{cz}/d_b\right) \qquad \text{(W-24)}$$

where $N_{p,i}$ = influent particle concentration to the contact zone, $t$ = time, $\alpha_{pb}$ = particle-bubble attachment efficiency factor, $\eta_T$ = total single collector collision efficiency, $N_p$ = particle number concentration, $\nu_b$ = bubble rise velocity, $\Phi_b$ = bubble volume concentration, $d_b$ = bubble diameter, $N_{p,e}$ = effluent particle concentration, and $t_{cz}$ = detention time of the contact zone assuming plug flow. *See also* heterogeneous flocculation model.

**whiting** A white solid, essentially powdered calcium carbonate ($CaCO_3$), used instead of clay as a coagulant aid (with alum) in the treatment of cold, soft waters of low turbidity. *See also* calcite.

**WHO** Acronym of World Health Organization.

**whole-effluent approach** The use of whole-effluent toxicity to establish water quality standards and criteria.

**whole-effluent toxicity (WET)** The aggregate toxic effect of an effluent measured directly by a toxicity test; a bioassay conducted in accordance with USEPA guidelines to determine the toxicity of an effluent to selected organisms of a receiving water by exposing these organisms to various dilutions of the effluent and measuring such effects as lower reproduction rates, reduced growth, or death. *See also* toxicity reduction evaluation.

**wholesale contract** An agreement to receive or provide water or sewer services on a wholesale basis. *See* wholesale customer.

**wholesale customer** A customer that pays for water or sewer service on a wholesale basis, i.e., usually a large user that pays at a rate lower than the retail rate. For example, in Miami–Dade County, FL, the Homestead Air Force Base (a Federal establishment), which collects its own wastewater and uses the county facilities for treatment and disposal, is a wholesale customer, whereas the Miami International Airport (a county establishment) does the same thing but is a retail customer.

**wholesome water** Potable water; water that is safe and palatable for human consumption. *See also* drinking water.

**WHPA** Acronym of wellhead protection area.

**WHPP** Acronym of (a) wellhead protection planning, (b) Wellhead Protection Program.

**Widal reaction** A procedure used for the diagnosis of gastroenteritis in the absence of bacteriological facilities. It is based on the observation that the agglutination of antibodies commonly accompanies enteric fever.

**wide-crested weir** *See* broad-crested weir.

**Wiese-Flo®** A self-cleaning filter screen manufactured by Wheelabrator Engineered Systems, Inc.

**Wild and Scenic Rivers Act** A U.S. public law that protects rivers that have "outstandingly remarkable scenic, recreational, geologic, fish and wildlife, historic, cultural, or other similar values".

**Williams, Gardner S.** American engineer (1866–1931) who developed jointly with Allen Hazen a widely used formula for flow in pipes; professor of hydraulics at Cornell University and the University of Michigan. *See* Hazen–Williams formula.

**Willowtech®** A sludge mixer manufactured by Ashbrook Corp.

**Wilson–Lee equation** An equation proposed in 1982 to determine the maximum allowable hydraulic loading rate ($Q$, m$^3$/day) of a suspended-growth wastewater treatment unit (WEF & ASCE, 1991):

$$Q = 24\ V_i A/F \qquad \text{(W-25)}$$

where $V_i$ = initial settling velocity (m/hr) at the design mixed liquor suspended solids; $A$ = clarifier area, m$^2$; and $F$ = clarifier safety factor, typically at least 1.5–2.0.

**Wilson's disease** A disorder that affects copper metabolism in some individuals.

**wilt** A plant disease caused by bacteria that invade its vessels, interfere with water and nutrient movement, and produce toxins; eventually, the plant leaves dry out, droop, and wither.

**wilting coefficient** The percentage soil moisture content at which plants cannot obtain moisture for growth.

**windage** Water lost from a cooling tower or spray pond under the influence of the wind.

**windage emission** Emission of saturated vapors from a tank due to the action of wind.

**wind-aided intensified evaporation (WAIE)** A technique that uses wind energy to enhance the evaporation on wetted surfaces. It is considered an emerging concentrate-minimization and disposal method; the concentrate from a desalination plant is sprayed over vertical surfaces in an evaporation pond.

**windbox** *See* cold windbox, hot windbox.

**Windjammer™** A brush aerator manufactured by United Industries, Inc.

**windmill** A machine driven by wind energy acting on a number of vanes, and used for such operations as pumping water or grinding. A wind pump is driven by a windmill, which rotates the pump's multiple-blade propeller.

**wind pump** *See* windmill.

**windrow** A low, elongated row of material (e.g., leaves, dust, grain) swept together by the wind or left uncovered to dry. Windrows are typically arranged in parallel.

**windrow composting** A composting method in which a mixture of dewatered sludge and wood chips or other bulking agent is arranged in long triangular-shaped rows, called windrows, that are mechanically turned and remixed periodically to introduce oxygen. As aerobic conditions cannot always be maintained, the biomass may be a combination of aerobic, facultative, and anaerobic microorganisms. Each windrow is about 3–6 ft high by a base of 6–16 ft. The operation lasts several weeks at a temperature of 55°C. *See also* two-step windrow process, sludge composting, briquette composting.

**Windrow-Mate** An apparatus manufactured by Nu-Tech Environmental Corp. for the control of odors from composting operations.

**windrow process** *See* windrow composting.

**windrow sludge composting** *See* windrow composting.

**windrow system** *See* windrow composting.

**wind water** (1) Water drawn from ambient air by condensation of atmospheric moisture. The apparatus uses a condenser along the shore with piped-in seawater. Condensation occurs when the warm, humid wind hits the cold pipe. (2) Freshwater extracted from brackish water using the energy of the wind.

**winery waste** *See* brewery/distillery/winery wastes.

**wing screen** A screening device consisting of curved vanes rotating on a horizontal axis.

**Winklepress®** A belt filter press manufactured by Ashbrook Corp.

**Winkler method** *See* Winkler titration.

**Winkler titration** A widely used volumetric titration method to determine the dissolved oxygen (DO) content of a water or wastewater sample and to calibrate other dissolved oxygen methods. It uses manganous sulfate ($MnSO_4$) and a mixture of potassium hydroxide (KOH) and potassium iodide (KI) to precipitate the DO as manganic basic oxide, which reacts with sulfuric acid ($H_2SO_4$) to form manganic sulfate. The latter reacts with KI to liberate iodine. The number of moles of liberated iodine is equal to the number of moles of DO originally present in the sample.

**winter chapping** A skin irritation (cracks and redness) caused by the causticity of highly mineralized or hard waters in sensitive persons when it is cold. *See also* dishpan hands.

**winter stagnation** The phenomenon occurring in ice-covered lakes in the winter when a stable equilibrium is established between the layer just below the ice at 32°F and the bottom layer almost stagnant near 39.2°F or 4°C (at maximum water density). Mechanical or pneumatic aeration may be used to eliminate or reduce this phenomenon. *See also* thermal stratification.

**wire-mesh screen** A screening device consisting of a wire fabric attached to a metal frame; it is used to remove fine debris. *See also* bar screen.

**wire-to-water efficiency (*E*)** The efficiency of a pumping plant (pump, motor, and associated fittings); the ratio of mechanical output (work done) of the pump to the electrical (energy) input, from the water source through the pump to the discharge point; expressed as a percentage:

$$E(\%) = 100\, QH/(3960\, I) = QH/(39.6\, I) \quad \text{(W-26)}$$

where $Q$, $H$, and $I$ designate, respectively, the flow rate in gallons per minute, the total dynamic head in feet, and the input horsepower delivered to drive the motor. The constant 3960 converts the units to horsepower. Also called overall efficiency, total efficiency, wire-to-wire efficiency. *See also* brake horsepower, water horsepower, power terms, pump efficiency, motor efficiency.

**wire-to-wire efficiency** Same as wire-to-water efficiency; the efficiency of a pump and motor together.

**withdrawal** The process of taking water from a source and conveying it to a place for a particular type of use.

**witherite ($BaCO_3$)** The mineral barium carbonate, a secondary ore of barium; it is found in white to grayish crystals or masses.

***Wolffia spp.*** A group of duckweeds commonly found in wetlands.

**wolfram** German name of tungsten (W).

**Wolman, Abel** Widely known and respected American sanitary engineer and educator (1892–1989), associated with the Johns Hopkins University; prolific writer with major contributions in water supply, wastewater disposal, and public health. Developed in 1918, with Linn H. Enslow, a formula for treating water with chlorine at filtration plants.

**wood alcohol** Methanol or methyl alcohol.

**wood-based granular activated carbon** Granular activated carbon made from wood, offering a relatively large pore structure.

**wood-screw pump** A pump that has a radial-blade runner in a horizontal casing, often used in drainage. Also called horizontal screw pump.

**Wood's lamp** A test performed in a dark room, shining ultraviolet light on the area of interest. Also called ultraviolet light test.

**workable sludge** Sludge from water or wastewater treatment that can easily be forked or shoveled from a drying bed because of its relatively low moisture content (e.g., less than 75%).

**working pressure** The pressure range within which a water or wastewater processing device is designed to function, as indicated by its manufacturer, in psi or in any other unit of pressure. Also called operating pressure.

**working pressure head** The actual head of water in a conduit; the difference in elevation between the hydraulic gradeline and the centerline of the conduit at any point.

**working volume** Same as active storage. *See* reservoir storage.

**works** (With a singular or plural verb) An assemblage of structures and devices used for the conveyance or treatment of water, wastewater, or stormwater. *See* waterworks, headworks, treatment works, water treatment works.

**World Health Organization (WHO)** A specialized agency of the United Nations, headquartered in Geneva, Switzerland, with several regional branches, dedicated to public health. Its objective is "the attainment by all peoples of the highest possible levels of health." It does not issue enforceable standards but has published several documents on water quality (e.g., *Guidelines for Drinking-Water Quality,* 1984), water treatment, wastewater treatment, and excreta disposal.

**worm** Any of numerous elongated, soft-bodied invertebrates that move in an undulating manner. Worms are one of the five categories of waterborne parasitic organisms pathogenic to humans. *See Ascaris,* guinea worm, helminth, nematode worm, tapeworm, trematode worm. Granular filtration and stabilization ponds are effective against the eggs and adults of common intestinal parasitic worms, but these pathogens survive ordinary wastewater treatment and are found in the sludge. *See also* worms.

**worm burden** The weight of worms or the intensity of a helminthic infection, usually determined from the output of eggs in the excreta, which is also a guide to pathology and transmissibility. Heavy worm burdens can build up over the early years and affect mental and physical development.

**worm casting** The coil of sand and waste passed by an earthworm; more generally, something that is shed, ejected, or cast off or out by an animal. *See* earthworm stabilization or vermiculture.

**worm infection** Intestinal infection caused by a worm whose eggs and larvae reach water bodies via improper disposal of wastewater or excreta. The infection is transmitted by the ingestion of contaminated water, the consumption of crops irrigated with such water, or bathing in it. *See,* e.g., schistosomiasis, ascariasis, guinea worm, swimmer's itch.

**worm load** The number or weight of parasitic worms that a person carries. *See* worm burden.

**worms** Any disease caused by parasitic worms in the intestines or other tissues; used as a singular noun. *See* helminthic infection for more detail.

**worst-case discharge** (a) In the case of a vessel, a discharge in adverse weather conditions of its entire cargo; and (b) in the case of an offshore facility, the largest foreseeable discharge in adverse weather conditions (Clean Water Act).

**wound-rotor induction motor** A type of direct current motor manufactured for pump drives used in wastewater treatment, particularly in large sizes (e.g., more than 500 hp) and with variable-speed control.

**woven-hose lining** A trenchless water main rehabilitation method using a woven polyester fiber hose, impregnated with a resin that is cured in place by hot water or steam. It is used to restore the integrity of a main that is damaged by breaks, corrosion, pinholes, or joint failure. *See* cured-in-place lining, felt-based lining, membrane lining. *See also* trenchless technology.

**WPCF** Acronym of Water Pollution Control Federation.

**WQA** (1) Acronym of Water Quality Act (of 1987). (2) Acronym of Water Quality Association.

**WQM** Acronym of water quality management.

**Wring-Dry** A rotary fine screen manufactured by Schlueter Co., including internal feed.

**wrought-iron pipe** Pipe made with iron that is almost entirely free of carbon, has a fibrous structure, and is readily forged or welded.

**wrought pipe** Pipe made of welded steel plates.

**WTP** Acronym of water treatment plant or drinking water treatment plant.

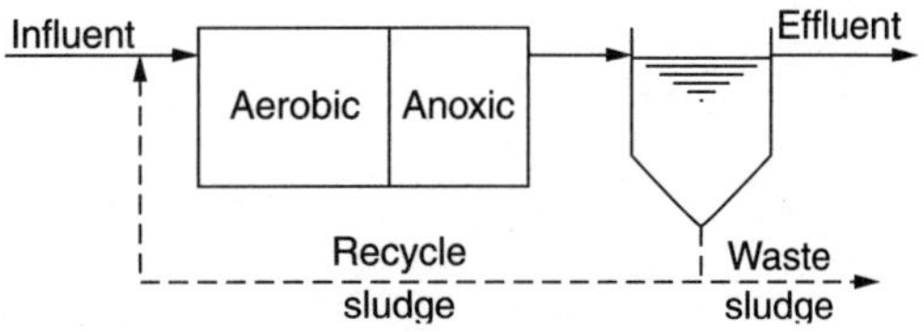


**Figure W-14.** Wuhrmann process.

**WTP residuals** *See* water treatment plant residuals.

***Wuchereria bancrofti*** The mosquito-borne helminth species of pathogens that cause Bancroftian filariasis.

**Wuhrmann process** A single-sludge process designed to remove nitrogen beyond the metabolic requirements of the organisms. It relies on the organic matter in the influent or on endogenous respiration for the energy to drive denitrification. It consists of an aerobic reactor followed by an anoxic reactor and a final clarifier, with recycling of the clarifier underflow to the influent line. Also called postdenitrification.

**wulfenite** A minor molybdenum mineral that also contains lead, essentially $PbMoO_4$.

**WWEMA** Acronym of Water and Wastewater Equipment Manufacturers Association.

**WWTP** Acronym of wastewater treatment plant.

**wye** A pipe branching off a straight main run at an angle of 45 degrees.

**wye connection** *See* Y connection.

**Wyss®** A fine-bubble diffuser manufactured by Parkson Corp.

# X

**XAD** An adsorbent resin of macroporous polystyrene and acrylic manufactured by Rohm and Haas and used in the treatment of organic wastes.

***Xanthobacter autotrophicus*** A species of bacteria that can degrade some haloacetic acids.

**Xanthophyta, Xanthophytaceae** A group of yellow-green, mainly freshwater and terrestrial, algae.

***X. autotrophicus*** Abbreviation of *Xanthobacter autotrophicus.*

**xenobiote** Any biotum displaced from its normal habitat; a chemical foreign to a biological system.

**xenobiotic** A chemical substance that is foreign to the human body, to an organism, or to a biological system. (The prefix xeno is from the Greek xenos meaning alien, foreign, stranger, etc.) Xenobiotics include endocrine disruptors, MTBE, NDMA, perchlorate, pharmaceutically active compounds, personal care products, and many others. These substances are usually difficult to remove from drinking water or wastewater by conventional treatment processes, but membrane filtration and advanced oxidation may be effective. A 1993 survey found more than 70,000 synthetic organic chemicals classified as xenobiotics in general use, including the endocrinologically active compounds and the pharmaceutical and personal care products. *See also* refractory.

**xenobiotic compound** *See* xenobiotic.

**xenobiotic substance** *See* xenobiotic.

**xenon (Xe)** A heavy, colorless, monatomic gaseous element used in electronics and in narrow-band excimer lamps with possible applications to wastewater disinfection. Atomic weight = 131.3. Atomic number = 54.

**xenon chloride (XeCl)** A compound of xenon and chloride, used in narrow-band excimer lamps, with possible applications to wastewater disinfection.

**xenoparasite** A parasite that infests uncommon hosts or only injured organisms.

***Xenopsylla cheopis*** The tropical rat flea, an arthropod vector.

**xeric** (From the Greek word xeros, meaning dry or arid.) Pertaining to or adapted to a dry environment; e.g., an organism that requires very little water or moisture to survive.

**xeric landscape** A landscape designed to minimize water use (e.g., 25% of a turf grass landscape), typically including a high proportion of desert-adapted shrubs, trees, ornamental grasses, and crushed rock mulch.

**xeriscape** A xeric landscape; xeriscaping.

**Xeriscape™** A program of the Denver water supply agency designed to minimize the use of water in landscaping, based on seven key principles: sound planning and design, limitation of turf, use of water-efficient plants, efficient irrigation, use of soil amendment, use of mulches, and appropriate landscape maintenance. *See also* cash for grass.

**xeriscaping** Designing land uses (residential, recreational, etc.) to minimize water use.

**xerophile** An organism that tolerates dry habitats.

**X-Flo™** A plastic medium manufactured by American Surfpac Corp. for use in crossflow trickling filters.

**X-linking** *See* cross-linking.

**$(x/m)_b$** Symbol of breakthrough adsorption capacity.

**XP** Abbreviation of explosion proof.

**X-Pando Joint Compounds** A series of products of the X-Pando Products Company (Trenton, NJ) designed for threaded or flanged joints of iron, steel, brass, copper, stainless steel, cement-lined, and PVC pipes. The tasteless, odorless, and nontoxic materials expand as they set, from 0.002″ to 0.004″ per cubic inch to insure leak-proof joints.

**X-Pruf™** A portable, explosion-proof submersible pump manufactured by Crane Pumps and Systems.

**X-ray** Electromagnetic radiation of shorter wavelength than light and ultraviolet radiation, capable of penetrating solids and producing ionizing gases; discovered by W. Roentgen. (2) A radiograph made by X-rays, as used in spectrometry and in medical therapy.

**X-ray diffraction (XRD)** The diffraction of X-rays by the regularly spaced atoms of a crystal; a technique used in the identification of crystalline substances and to study the structure of crystals. When an X-ray beam of a given wavelength strikes a sample of material, the X-rays are diffracted in a pattern characteristic of the crystalline structure of the material. Mixtures of compounds can also be analyzed.

**X-ray fluorescence** A method used to identify and analyze particles and solutions, e.g., heavy metals and corrosion products in water. It is based on the property of atoms to emit characteristic fluorescence when a sample absorbs X-rays.

**X-ray photoelectron spectroscopy** A precise technique using an X-ray beam to determine the type and amount of residue on a surface by exciting the contaminant atoms in a sample and causing the emission of electrons. Also called electron spectroscopy chemical analysis.

**XRD** Acronym of X-ray diffraction.

**Xtractor™** A process developed by Davis Water & Waste Industries, Inc. for sludge removal.

**xylene** An organic contaminant made of three isomers of dimethylbenzene. Chemical formula $C_6H_4(CH_3)_2$. Used as a solvent, in gasoline, and in pharmaceutical products. Potentially harmful to the liver and kidney, it is regulated by the USEPA as a drinking water contaminant: MCLG = MCL = 10 mg/L (total xylenes).

***m*-xylene** *Meta*-xylene or 1,3-dimethylbenzene.

***o*-xylene** *Ortho*-xylene or 1,2-dimethylbenzene.

***p*-xylene** *Para*-xylene or 1,4-dimethylbenzene.

# Y

**yard waste** Grass clippings, leaves, miscellaneous vegetative matter, and other wastes from lawns and backyard gardens, some of which may be composted.

**Y branch** Same as wye.

**Y connection** A pipe fitting with three branches, like the letter Y, or a similar junction of three conduits.

**yeast** A type of nonfilamentous, unicellular fungus that reproduces asexually by budding and causes fermentation; some yeasts are used in brewing, winemaking, and bread-making. Yeasts metabolize sugar anaerobically to produce alcohol. Some yeasts have been considered as alternative indicators to the coliform group of organisms.

**yellowboy** Iron oxide flocculent [$Fe(OH)_3$], usually observed as unsightly orange-yellow, amorphous, semigelatinous deposits in surface streams with excessive iron content.

**yellow fever** An acute, infectious, often fatal, water-related disease that is caused by a mosquito-borne virus, which is present in humans and such animals as monkeys. It is characterized by liver damage and jaundice. Also called yellow jack.

**yellow fever virus** The virus that causes yellow fever. It is transmitted by the mosquitoes *Aedes aegyptii* as well as other *Aedes* and *Haemagogus* species.

**yellow-green algae** *See* Xanthophyta.

**yellow jack** Yellow fever.

**yellow jaundice** Another name of infectious hepatitis, from the yellow color in the white of the eyes and skin of hepatitis patients. Also simply called jaundice.

**yellow water** Water of a yellowish color imparted by dissolved iron compounds from the source water or from piping corrosion. *See* red water for detail.

***Y. enterocolitica*** *See Yersinia enterocolitica.*

***Yersinia*** A genus of gram-negative, fermentative, facultatively anaerobic, rod-shaped Enterobacteriaceae, widely distributed in the aquatic environment; essentially an animal parasite with few species pathogenic for humans.

***Yersinia enterocolitica*** A species of Gram-negative, fermentative, facultatively anaerobic, rod-shaped Enterobacteriaceae, widely distributed in the aquatic environment that causes waterborne gastroenteritis. Incubation period = 3 – 7 days. Excreted load = 100,000 per gram of feces. Persistence = 3 months. Median infective dose > 1,000,000. It can grow at low temperatures and survive in the refrigerator. Pigs are the principal reservoirs of this pathogen. *See* yersiniosis.

***Yersinia* genus** *See Yersinia.*

***Yersinia pestis*** The species of bacteria that cause human plague, but is primarily a parasite of rodents. *See* bubonic plague.

***Yersinia pseudotuberculosis*** A species of rod-shaped, Gram-negative bacteria that mainly cause zoonosis (e.g., in pigs, birds, beavers, cats, dogs) but incidentally affects humans. *See* yersiniosis.

**yersiniosis** An acute enteric infection caused by the species *Yersinia enterocolitica* (affecting mainly children and infants) and *Yersinia pseudotuberculosis* (affecting mainly animals and, incidentally, humans); characterized by abdominal pain, joint pain, fever, diarrhea, and septicemia.

**yield** (1) The quantity of water (expressed as a flow rate or total volume per year) that can be collected for a given use from surface or groundwater sources. The yield may vary with the proposed use, with the plan of development, and also with economic considerations. *See also* safe yield and watershed yield. (2) The amount of biomass produced per unit weight of substrate consumed, usually represented by the symbol Y and expressed in g biomass per g substrate. *See* biomass yield, net biomass yield, observed solids yield, observed yield, sludge yield, solids yield, synthesis yield, true yield, and yield factor. Also called yield coefficient.

**yield coefficient** Same as yield (2).

**yield factor (*Y*)** In biological wastewater treatment, the mass of microorganisms formed by unit mass of substrate removed:

$$dS/dt = -(1/Y)\, dN/dt \qquad \text{(Y-01)}$$

where $S$ is the substrate concentration and $N$ the number of microorganisms at time $t$. The yield factor is used to estimate the production of biomass ($R_{xp}$, mg/L/day) or volatile suspended solids (VSS) as a function of substrate removal ($R_s$, mg/L/day):

$$R_{xp} = -YR_s \qquad \text{(Y-02)}$$

For a given substrate, the yield factor is larger for aerobic organisms than for anaerobic organisms, typically 0.15 and 0.50 gram VSS formed by gram of COD removed.

**yield stress** *See* plastic fluid, Bingham plastic, shear stress, Hedstrom number.

**Y junction** A chamber in which liquid can be directed to either of two pipes or channels. *See also* distribution box.

**$Y_{obs}$** Symbol of observed yield.

***Y. pestis*** *See Yersinia pestis.*

***Y. pseudotuberculosis*** *See Yersinia pseudotuberculosis.*

# Z

**ZAVC** Acronym of zero-air-void curve.

**Z Chlor™** A device manufactured by Fischer & Porter for the monitoring chlorine and sulfite.

**ZD** Acronym of zero discharge.

**zebra mussel** A freshwater bivalve, catalogued as *Dreissena polymorpha,* that multiplies rapidly in freshwater and can clog intake screens and pipes. It has been found in the Great Lakes, in the Mississippi and Ohio Rivers, and may have been imported from Asian/European waterways; it looks like the quagga mussel. Hypochlorite and permanganate are effective means of control. (A bivalve is a mollusk that has two hinged shells, a soft body, and lamellate gills.)

**Zeeman background correction** Compensation for interferences in graphite furnace atomic absorption spectrophotometry, a technique based on the Zeeman effect.

**Zeeman effect** The division of a spectral line, by a radiation source in a magnetic field, into three equally spaced lines or into three or more unequally spaced lines.

**ZeeWeed®** A series of products (ZeeWeed® 500 Reinforced Membranes™ and ZeeWeed® 1000 Membrane Media™) manufactured by Zenon for use in water or wastewater treatment. They are hollow-fiber, ultrafiltration membranes that are immersed in treatment process tanks. Membranes in individual chambers are taken out of service for cleaning in a bleach solution.

**Zenobox** An apparatus manufactured by Zenon Environmental, Inc. for water treatment by membrane filtration.

**Zenofloc** An apparatus manufactured by Zenon Environmental, Inc. for polymer control in belt filter presses.

**Zenogem® process** A proprietary membrane bioreactor system developed by Zenon Environmental, Inc. It consists of a microfiltration cassette of hollow-fiber membranes with appurtenant pumping, storage, feed, and air-scour facilities.

**Zeobest™ and Zeobest™ Ultra** Respectively, a natural zeolite and a natural antimicrobial zeolite produced by Northern Filter Media International/Northern Gravel Co. of Muscatine, IA.

**Zeobest™ Ultra** *See* Zeobest™

**Zeo-Karb®** A cation exchange process developed by U.S. Filter Corp., using sulfonated coal.

**Zeol™** A concentrator manufactured by Zeol Division, Munters Corp. for the control of volatile organic compounds.

**zeolite** A natural or synthetic, hydrated sodium alumina silicate (crystalline aluminosilicate), used in water treatment as a cation-exchange resin that exchanges sodium for divalent ions (e.g., calcium and magnesium). Natural (greensand) and synthetic manganese zeolites are also used in filters for the removal of iron, manganese, and hydrogen sulfide. Synthetic zeolites are produced from a mixture of sodium silicate and sodium aluminate, which is then dried and crushed. Activated alumina is granular natural zeolite. *See also* clinoptilolite.

**zeolite filter** A water softening device containing zeolite as exchange medium for the removal of hardness ions.

**zeolite process** A cation-exchange process that uses a natural or synthetic zeolite to remove hardness from water. The sodium ions of the zeolite are exchanged for the hardness-causing calcium and magnesium ions. It used to be called base exchange or cation exchange. Cation-exchange softening has largely replaced zeolite softening in residential and commercial applications. Same as zeolite process, zeolite softening. *See also* softening.

**zeolite softener** A treatment unit that uses natural greensand or synthetic zeolites to soften water. In this process, calcium and magnesium ions are exchanged for the sodium in the zeolites. Lime pretreatment may be used, if necessary, to reduce bicarbonate hardness. Backwashing or regeneration is necessary to restore the capacity of the unit. A split treatment is usually implemented to maintain some hardness in the final product and reduce corrosiveness. *See also* sodium zeolite reactions.

**zeolite softening** A treatment method for the removal of calcium and magnesium hardness from water through ion exchange with zeolite resins. *See* zeolite process for detail.

**Zeo-Rex®** A filter manufactured by U.S. Filter Corp. for the removal of iron and manganese.

**Zephyr** An apparatus manufactured by Aeromix Systems, Inc. for solids separation by induced-air flotation.

**zero-air-void curve (ZAVC)** In studying the compaction behavior of sludge, ZAVC is a curve that outlines the upper limit of the line that plots the dry unit weight ($W$, lb/ft$^3$) versus the percent solids concentration of the sludge, regardless of the compaction effort or consolidation pressure. The curve is based on the following relationship:

$$W = GU/[(1 + G)(C/100)] \qquad \text{(Z-01)}$$

where $G$ = specific gravity of sludge, $U$ = unit weight of water, and $C$ = percent water content of sludge. *See also* consolidation curve, standard Proctor compaction.

**zero discharge (ZD)** (1) The condition of a facility that, through the practice of complete recycling, does not discharge any material to the environment, or that discharges effluents that either are essentially pure or contain no substance with a concentration higher than that found normally in the local environment. (2) Same as zero discharge water.

**zero discharge plant** A desalination or other facility that does not discharge any liquid from its processes; the plant recycles or evaporates all its liquid waste, typically using a brine concentrator and a crystallizer.

**zero discharge water** The condition of an industrial or commercial facility that does not discharge any process liquid effluent to a municipal wastewater system. The facility may have a permit to discharge sanitary wastewaters. *See also* zero liquid discharge.

**Zerofuel** An apparatus manufactured by Seghers Dinamec, Inc. for sludge incineration using fluidized beds.

**zero liquid discharge (ZLD)** The condition of a desalination or other plant that does not discharge any liquid effluent to the environment; the facility recycles or evaporates all its liquid waste. It is also an approach used by inland communities to eliminate discharges of concentrate from desalination plants; for example, combining reverse osmosis with a fluidized bed crystallizer and enhanced evaporation. *See also* brine concentrator, salt gradient pond, salt seeding, thermal crystallizer, zero discharge plant, zero discharge water.

**zero nitrate recycle** A condition maintained in the operation of the UCT process, whereby the internal recycle is controlled so that there is no nitrate in the effluent from the anoxic reactor and no nitrate is returned to the anaerobic reactor.

**zero ODP** Abbreviation of zero ozone depletion potential.

**zero-order kinetics** The condition of a zero-order reaction, i.e., a reaction that proceeds at a rate independent of the concentration of the reactants and products. *See also* reaction order.

**zero-order reaction** A reaction that proceeds at a rate that is independent of the concentration of any reactant or product. For a single reactant or product, the concentration ($C$, mg/L) at any time ($t$, days) may be expressed by:

$$C = C_0 - k \cdot t \qquad \text{(Z-02)}$$

$$r = \pm k \qquad \text{(Z-03)}$$

where $C_0$ is the initial concentration (mg/L), $k$ is a reaction rate constant (per day), and $r$ is the reaction rate.

**zero ozone depletion potential** The condition of a commercial or industrial facility that does not discharge any material that can damage the ozone layer.

**zero point of charge (ZPC)** The pH of a solution that corresponds to a surface charge of zero on a particle. For example, for activated alumina a typical ZPC = 8.2, below which the alumina surface has a net positive charge and excess protons are available; above a pH of 8.2, the alumina surface has a net negative charge. *See also* point of zero charge, point of zero net proton charge.

**zero proton condition** The condition corresponding to the point of zero charge.

**zero soft water** Water treated by the cation exchange process and having a very low hardness, e.g., less than 1.0 grain per gallon as calcium carbonate ($CaCO_3$).

**zeroth law of thermodynamics** Thermal equilibrium between two objects in contact with each other is reached when no heat passes from one to the other. *See also* first law, second law, and third law of thermodynamics.

**zeta meter** A zeta potential meter.

**ζp** Symbol for electrokinetic potential or zeta potential.

**Zeta-Pak** A cartridge filter manufactured by Alsop Engineering Co.

**zeta plus filter** A filter used to concentrate viruses from large volumes of water or wastewater; it consists of a bed of cellulose–diatomaceous earth. Also called electropositive filter.

**zeta potential (ζp)** In coagulation and flocculation procedures, the difference in the electrical charge between the dense layer of ions surrounding a particle and the charge of the bulk of the suspended fluid surrounding this particle. Another definition of zeta potential is the measure of the potential at the surface of the cloud of ions dragged by a particle, when the particle is placed in an electrolyte through which an electric current passes and one of the two electrodes attracts the particle. The zeta potential, usually measured in millivolts, can be determined from the electrophoretic mobility or from the following equation:

$$\zeta p = 4\,\pi \delta q/D \tag{Z-04}$$

where $\pi = 3.1416$, $\delta$ = thickness of the zone of influence of the charge, $q$ = charge of the particle, and $D$ = dielectric constant of the liquid. The zeta potential from electrophoretic measurements is typically lower than the surface potential from the double-layer theory. The magnitude of this parameter often controls the stability of colloids, but it is not the only factor that explains how coagulation proceeds. Also called electrokinetic potential. *See also* electrical double layer, electrokinetic charge, electroosmosis, electrophoresis, electrophoretic mobility, Guoy–Stern colloidal model, Nernst potential, sedimentation potential, Stern layer, Stern potential, streamimg current, streaming potential, surface of shear.

**zeta potential meter** An instrument that measures the zeta potential (or electrophoretic mobility) of a water sample. It is used to control the dosage of water treatment coagulants. Also called zeta meter. *See also* streaming current detector.

**ZID** Acronym of zone of initial dilution.

**Ziehl–Neelsen stain** A special stain or technique used to identify mycobacteria, which do not stain with common dyes. This procedure uses a solution of ethanol (95%) and hydrochloric acid (3%) to wash the organisms, which eliminates the stain carbol fuchsin from all organisms except the acid-fast bacteria. Also called acid-fast stain. *See also* Gram stain.

**Zimmerman process** A wet-air oxidation process developed by Zimpro Environmental, Inc.

**Zimpress** A plate-and-frame sludge press manufactured by Zimpro Environmental, Inc.

**ZIMPRO** The Zimpro process.

**Zimpro process** A proprietary wet combustion process developed by Zimpro Environmental, Inc. for the thermal conditioning or thermal reduction of sludge. For thermal conditioning (or heat treatment), it uses a low-pressure system that includes a sludge-to-sludge heat exchanger, air injection, and live steam injection into the reactor. *See* Figure Z-01. For thermal reduction, the process is similar but uses wet-air oxidation under higher pressures and temperatures. Application of this process has been limited because of high capital cost, skills required for operation and maintenance, production of a high-strength recycle liquor, scale formation on the exchangers, pipes, etc., and odor problems.

**Zimpro system** *See* Zimpro process.

**zinc (Zn)** A bluish-white metal that is brittle at ordinary temperatures but malleable around 125°C; often associated with mine drainage. Atomic weight = 65.38. Atomic number = 30. Specific gravity = 7.14 at 20°C. Valence = 2. Melting point = 420°C. Boiling point = 907°C. It is used in alloys, batteries, electroplating, and fungicides. Its salts are only slightly soluble in water. The metal is an

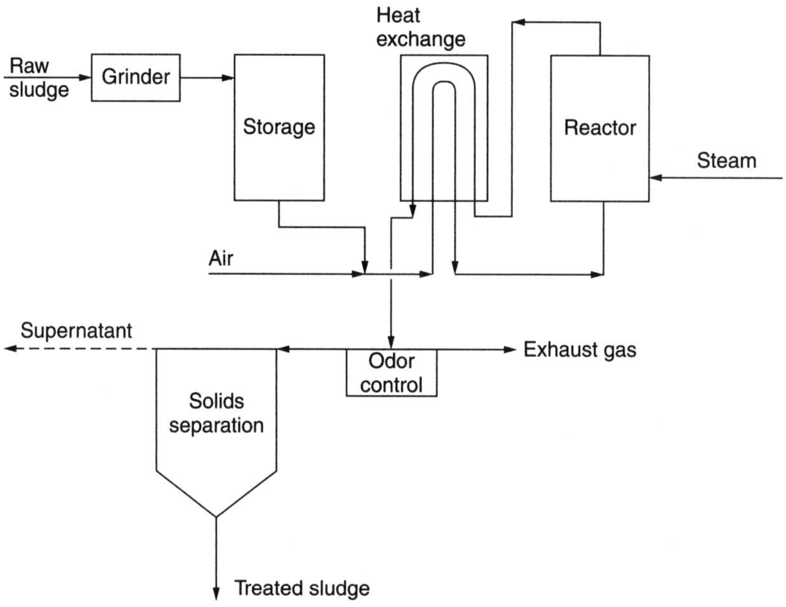


**Figure Z-01.** Zimpro process.

essential nutritional element and there do not seem to be health effects of the metal in drinking water; however, at concentrations higher than 25 mg/L it may impart an objectionable taste. It is toxic to plants at high concentrations and toxic to aquatic organisms. USEPA issued a secondary drinking water standard of 5.0 mg/L. Zinc in drinking water is likely caused by corrosion of galvanized metal pipes. On the other hand, like iron and manganese, zinc may combine with the cement matrix of asbestos–cement pipe to form a protective coating against corrosion. Effective removal processes: lime softening, ion exchange, reverse osmosis, electrodialysis, adsorption on activated alumina.

**zinc chloride ($ZnCl_2$)** A white, crystalline, soluble, poisonous solid used as a wood preservative, a disinfectant, and an antiseptic.

**zinc green** Cobalt green.

**zinc equivalent** A parameter ($Z_e$) used to assess the cumulative effect of metals in soils on the growth of crops. It is calculated as follows:

$$Z_e = 2\,(\mathrm{Cu}) + 8\,(\mathrm{Ni}) + (\mathrm{Zn}) \qquad \text{(Z-05)}$$

where (Cu), (Ni), and (Zn) represent, respectively, the concentrations of copper, nickel, and zinc in sludge, expressed in mg/kg. A limit of 250 mg/kg in topsoil is recommended.

**zinc orthophosphate [$Zn_3(PO_4)_2$]** A phosphate compound of zinc used to form a protective coating of water pipes against corrosion.

**zinc oxide (ZnO)** A white or yellowish-white, amorphous, odorless, insoluble powder used as a paint pigment, in cosmetics, etc. Also called flowers of zinc, zinc white.

**zinc phosphide ($Zn_3P_2$)** A dark gray, gritty, insoluble, poisonous powder used as a rodenticide.

**zinc pollution** The discharge of industrial effluents containing zinc, e.g., from galvanizing and plating plants, rubber and polymer processing, and rayon manufacture.

**zinc stearate [$Zn(C_{18}H_{35}O_2)_2$]** A white, fine, soft, insoluble powder used in cosmetics, ointments, etc.

**zinc sulfate ($ZnSO_4$)** A colorless, crystalline, soluble powder used in the preservation of skins and wood, in paper making, in medicine, etc. Also called white vitriol, zinc vitriol.

**zinc sulfide (ZnS)** A white to yellow, crystalline, soluble powder used as a pigment.

**zinc vitriol** Zinc sulfate.

**zinc white** Zinc oxide.

**zinkgeriesel** The zinc precipitate from the corrosion of galvanized pipe.

**zirconium-95** A radionuclide sometimes found in natural waters.

**ZLD** Acronym of zero liquid discharge.

**Z-metal** A pearlitic malleable iron chain material manufactured by Envirex, Inc.

**$ZnCl_2$** Chemical formula of zinc chloride.

**ZnO** Chemical formula of zinc oxide.

**$Zn_3P_2$** Chemical formula of zinc phosphide.

**$Zn(PO_4)_2$** Chemical formula of zinc orthophosphate.

**ZnS** Chemical formula of zinc sulfide.

**$ZnSO_4$** Chemical formula of zinc sulfate.

**zonal differentiation** The formation of layers of different temperatures and densities in a lake, reservoir, or other body of water, particularly during the summer. *See* thermal stratification.

**zone of aeration** The comparatively dry soil or rock located between the ground surface and the top of the water table. Also called vadose zone. *See* subsurface water.

**zone of influence** The land area within the horizontal projection of the cones of depression of the wells in a wellfield; corresponding to the area of influence for a single well.

**zone of initial dilution (ZID)** The region of initial mixing surrounding or adjacent to the end of the outfall pipe or diffuser ports, provided that the ZID may not be larger than allowed by mixing zone restrictions in applicable water quality standards (EPA-40CFR125.58-w). This zone is used to collect samples to determine compliance with water quality standards.

**zone of saturation** *See* saturation zone.

**zone sedimentation** Same as zone settling.

**zone settling** The sedimentation of closely spaced particles in a suspension of intermediate concentration (e.g., sludge or wastewater) so that settling is hindered by interparticle forces and is about the same for an entire zone; flocculated particles settle together as a layer or zone. *See* sedimentation type III for more detail.

**zone settling velocity (ZSV)** The settling velocity ($V_i$, m/hr) of the sludge/water interface at the beginning of a sludge settleability test; it is used in calculating the surface overflow rate to design secondary clarifiers. It can be estimated from the following equation:

$$V_i = V_{max} \exp(-10^{-6}K)X \qquad \text{(Z-06)}$$

where $V_{max}$ = maximum interface settling velocity ≈ 7.0 m/hr; $K$ = a constant ≈ 600 L/mg; and $X$ = average mixed liquor suspended solids concentration, mg/L. Also called interface settling velocity. *See also* overflow rate, solids loading rate.

**Zonite®** A proprietary disinfectant that contains about 1% of available chlorine by weight; it can be used directly as a stock solution for water disinfection. Such a solution is unstable under warm conditions.

**zooglea** A gelatinous matrix developed by some of the floc-forming bacteria responsible for waste assimilation in trickling filter slime and activated sludge floc. *See also* biomass.

**zoogleal bulking** An operational problem of activated sludge plants in which an excess of extracellular hydrophilic biopolymer produces a slimy, jellylike sludge that has a low density and does not settle well. *See* viscous bulking for detail.

**zoogleal form** Slime growth active in trickling filtration. *See* zooglea.

**zooglea-like microorganism** *See* zooglea.

**zoogleal matrix** The floc of slime-producing bacteria in the activated sludge process or in trickling filters.

***Zooglea ramigera*** A species of biofilm-producing bacteria that form a gelatinous mass organized in a branching shape for protection against predators or better food processing and oxygen transfer.

**zoogloea** *See* zooglea.

**zoonosis (plural: zoonoses)** An infectious disease (enzootic or epizootic) transmissible to humans by animals, e.g., balantidiasis, brucellosis, salmonellosis, yersiniosis.

**zoonotic disease** Same as zoonosis.

**zoophilic** Characteristic of fungi that grow preferentially on animals.

**zooplankton** Small, usually microscopic animals, such as protozoa, with little or no means of propulsion, found in lakes and reservoirs and eaten by fish. They include herbivores like the planktonic arthropods that feed on phytoplankton and carnivores that feed on the latter.

**ZP** Acronym of zeta potential.

**ZPC** Acronym of zero point of charge.

**ZSV** Acronym of zone settling velocity.

**z value** When the thermal destruction of microorganisms is expressed graphically by D value (decimal reduction time) vs. temperature, the z value is the number of degrees Fahrenheit required for the thermal destruction curve to traverse one log cycle.

**zygote** The stage after fertilization in the development of a sporozoan; the ovum surrounds itself with an enclosing cyst wall and constitutes an oocyst.

**zymogenous** Pertaining to organisms that can survive in periods of dormant or rapid growth depending on the availability of organic matter in the soil. *See also* autochthonous and allochthonous. A zymogenous organism is also called a copiotroph or an r-strategist.

# References


Adrien, Nicolas G. Derivation of Mean Cell Residence Time Formula. *ASCE Journal of Environmental Engineering Division, 124:*5, May 1998.

Adrien, Nicolas G. *Computational Hydraulics and Hydrology—An Illustrated Dictionary.* CRC Press LLC, Boca Raton, FL, 2004.

Alspatch, Brent and Ian Watson. Sea Change. *Civil Engineering,* February, 70–75, 2004.

Amburgey, James E., Appiah Amirtharajah, Barbara M. Brouckaert, and Neal C. Spivey. An Enhanced Backwashing Technique for Improved Filter Ripening. *Journal of AWWA, 95:*12, 81–94, 2003.

APHA (American Public Health Association, American Society of Civil Engineers, American Water Works Association, and Water Pollution Control Federation). *Glossary of Water and Wastewater Control Engineering.* Third Edition, Denver, New York, and Washington, DC, 1981.

AWWA (American Water Works Association). *Water Quality and Treatment—A Handbook of Community Water Supplies.* Fifth Edition. McGraw-Hill, New York, 1999.

American Water Works Association. *Benchmarking Performance Indicators for Water and Wastewater Utilities: Survey Data and Analyses Report.* Denver, 2005.

Batch, Lawrence F., Christopher R. Schulz, and Karl G. Linden. Evaluating Water Quality Effects on UV Disinfection of MS2 Coliphage. *Journal of AWWA, 96:*7, 75–87, 2004.

Bloetscher, Frederick, and Albert Muniz. Defining Sustainability. *Florida Water Resources Journal,* 13–19, October 2006.

Bond, Rick J. and Francis A. Digiano. Evaluating GAC Performance Using the ICR Database. *Journal of AWWA, 96:*6, 96–104, 2004.

Brown, Jess C. and Andrew Salveson. Emerging Disinfection Technologies. *Florida Water Resources Journal,* 4–9, July 2006.

Bryant, Edward A., George P. Fulton, and George C. Budd. *Disinfection Alternatives for Safe Drinking Water.* Prepared for Hazen and Sawyer Environmental Engineering and Scientists. Van Nostrand Reinhold, New York, 1992.

Cairncross, Sandy, and Richard Feachem. *Environmental Health Engineering in the Tropics—An Introductory Text.* Second Edition. Wiley, Chichester, England, 1993.

Chadwick, Andrew, and John Morfett. *Hydraulics in Civil and Environmental Engineering.* Third Edition. E & FN Spon, an imprint of Routledge, London, 1998.

Contech Stormwater Solutions. Equipment brochures or picture files. (9025 Centre Pointe Dr., Suite 400, West Chester, OH 45069.)

Crittenden, John C., R. Rhodes Trussel, David W. Hand, Kerry J. Howe, and George Tchobanoglous, for Montgomery Watson Harza. *Water Treatment: Principles and Design.* Second Edition. Wiley, Hoboken, 2005.

CUES, Inc. Equipment brochures or picture files. (3600 Rio Vista Ave., Orlando, FL 32805.)

DeZuane, John. *Handbook of Drinking Water Quality.* Second Edition. Van Nostrand Reinhold–International Thomson Publishing, New York, 1997.

Dodson, Roy D. *Storm Water Pollution Control.* Second Edition. McGraw-Hill, New York, 1999.

DorrOliver Eimco. Equipment brochures or picture files. (1.801.526.2000 or www.glv.com.)

Droste, Ronald L. *Theory and Practice of Water and Wastewater Treatment.* Wiley, New York, 1997.

Eimco Water Technologies. Equipment brochures or picture files. (1.801.526.2000 or www.glv.com.)

Enviroquip, Inc. Equipment brochures or picture files. (2404 Rutland Dr., Austin, TX 78758, www.enviroquip.com)

EPA-40CFRxxx 40 Code of Federal (environmental) Regulations. Federal Register.

Fair, Gordon M., John C. Geyer, and Daniel A. Okun. *Water and Wastewater Engineering. Volume 1. Water Supply and Wastewater Removal.* Wiley, New York, 1966.

Fair, Gordon M., John C. Geyer, and Daniel A. Okun. *Elements of Water Supply and Wastewater Disposal.* Second Edition. Wiley, New York, 1971.

Farahbakhsh, K., Adham, S. S., and Smith, D. W. Monitoring the Integrity of Low-Pressure Membranes. *Journal of AWWA, 95:*6, 2003.

Feachem, R. G., D. J. Bradley, H. Garelick, and D. D. Mara. *Sanitation and Disease—Health Aspects of Excreta and Wastewater Management.* World Bank Studies in Water Supply and Sanitation 3. International Bank for Reconstruction and Development/The World Bank, Washington, DC, 1983.

Freeman, Harry M. (Editor). *Industrial Pollution Prevention Handbook.* McGraw-Hill, New York, 1995.

Freeman, Harry M. (Editor). *Standard Handbook of Hazardous Waste Treatment and Disposal.* Second Edition. McGraw-Hill, New York, 1998.

Grayman, Walter M., Lewis A. Rosman, Rolf A. Deininger, Charlotte D. Smith, Clifford N. Arnold, and James F. Smith. Mixing and Aging of Water in Distribution System Facilities. *Journal of AWWA, 96:*9, 70–80, 2004.

Hammer, M. J. and M. J. Hammer, Jr. *Water and Wastewater Technology.* Third Edition. Prentice-Hall, Englewood Cliffs, NJ, 1996.

Hazen and Sawyer Environmental Engineers and Scientists. *Disinfection Alternatives for Safe Drinking Water.* Prepared by Edward A. Bryant, George P. Fulton, and George C. Budd. Van Nostrand-Reinhold, New York, 1992.

Howard, Alan G. *Aquatic Environmental Chemistry.* Oxford University Press, New York, 1998.

Jones + Atwood. Equipment brochures or picture files. (Titan Works, Sourbridge, West Midlands, DY8 4LR England.)

Linsley, Ray K., Joseph B. Franzini, David L. Freyberg, and George Tchobanoglous. *Water Resources Engineering.* Fourth Edition. McGraw-Hill, New York, 1992.

Long, Bruce W., Robert A. Hulsey, and Jeff J. Neemann. Mixing It Up: Integrated Disinfection Scenarios in Drinking Water. *Journal of AWWA, 97:*10, 30–33, 2005.

Maier, Raina M., Ian L. Pepper, and Charles P. Gerba. *Environmental Microbiology.* Academic Press, San Diego, CA, 2000.

Manahan, Stanley E. *Environmental Science and Technology.* CRC Press, Boca Raton, FL, 1997.

Martin, J. L., and S. McCutcheon. *Hydrodynamics and Transport for Water Quality Modeling.* CRC Press/Lewis Publishers, Boca Raton, FL, 1999.

McGhee, Terence J. *Water Supply and Sewerage.* Sixth Edition. McGraw-Hill, New York, 1991.

Metcalf & Eddy, Inc. *Wastewater Engineering—Treatment and Reuse.* Fourth Edition. Revised by George Tchobanoglous, Franklin L. Burton, and H. David Stensel. McGraw-Hill, New York. 2003.

Metcalf & Eddy, Inc. *Wastewater Engineering—Treatment, Disposal, and Reuse.* Third Edition. Revised by George Tchobanoglous and Franklin L. Burton. McGraw-Hill, New York, 1991.

Montgomery, John H. *Groundwater Chemicals—Desk Reference.* Second Edition. CRC Press, Boca Raton, FL, 1996.

Najm, Issam. An Alternative Interpretation of Disinfection Kinetics. *Journal of AWWA, 98:*10, 93–101, 2006.

Nemerow, Nelson L., and Avijit Dasgupta. *Industrial and Hazardous Waste Treatment.* Van Nostrand-Reinhold, New York, 1991.

Nix, S. J. *Urban Stormwater Modeling and Simulation.* CRC Press/Lewis Publishers. Boca Raton, FL, 1994.

Pankratz, Thomas M. Concise Dictionary of Environmental Engineering. CRC Press/Lewis Publishers, Boca Raton, FL 1996.

Porteous, Andrew. *Dictionary of Environmental Science and Technology.* Revised Edition. Wiley, West Sussex, England. 1992.

Reardon, Roderick, Brian Karmasin, and Lisa Prieto. A Look at Compact Liquid Treatment Technologies and Their Impanct on Minimizing Plant Site Requirements. *Florida Water Resources Journal,* 45–48, January 2004.

Ruiz, Alejandro, Theodore W. Sammis, Geno A. Pichioni, John G. Mexal, and Wayne A. Mackay. An Irrigation Scheduling Protocol for Treated Industrial Effluent in the Chihuahuan Desert. *Journal of AWWA, 98:*2, 122–133, 2006.

Sarnoff, Paul. *The New York Times Encyclopedic Dictionary of the Environment.* Quadrangle Books Inc., New York, 1971.

Sethi, Sandeep, Steve Walker, Jorg Drewes, and Pei Xu. Existing and Emerging Concentrate Minimization & Disposal Practices for Membrane Systems. *Florida Water Resources Journal,* 38–48, June 2006.

Simon, Andrew L. and Scott F. Korom. *Hydraulics.* Fourth Edition. Prentice-Hall, Englewood Cliffs, NJ, 1997.

Stumm, Werner and James J. Morgan. *Aquatic Chemistry—Chemical Equilibria and Rates in Natural Waters.* Third Edition. Wiley, New York, 1996.

Symons, J. M., Bradley, L. C. Jr., and Cleveland, T. C. *The Drinking Water Dictionary.* American Water Works Association, Denver, 2000.

Val-Matic, Inc. Equipment brochures or picture files. (905 Riverside Dr., Elmhurst, IL 60126.)

Walski, Thomas, William Bezts, Emanuel T. Posluszny, Mark Weir, and Brian E. Whitman. Modeling Leakage Reduction through Pressure Control. *Journal of AWWA, 98:*4, 147–155, 2006.

*Webster's New World Dictionary of American English,* 3rd ed., V. Neufeldt, Ed., Webster's New World, Cleveland, 1991.

WEF & ASCE (Water Environment Federation and American Society of Civil Engineers). *Design of Municipal Wastewater Treatment Plants.* Volume II, Chapters 13–20, 1991.